Comparison of *Bacteria, Archaea,* and *Eucarya*

Property	*Bacteria*	*Archaea*	*Eucarya*
Membrane-Enclosed Nucleus with Nucleolus	Absent	Absent	Present
Complex Internal Membranous Organelles	Absent	Absent	Present
Cell Wall	Almost always have peptidoglycan containing muramic acid	Variety of types, no muramic acid	No muramic acid
Membrane Lipid	Have ester-linked, straight-chained fatty acids	Have ether-linked, branched aliphatic chains	Have ester-linked, straight-chained fatty acids
Gas Vesicles	Present	Present	Absent
Transfer RNA	Thymine present in most tRNAs	No thymine in T or TψC arm of tRNA	Thymine present
	N-formylmethionine carried by initiator tRNA	Methionine carried by initiator tRNA	Methionine carried by initiator tRNA
Polycistronic mRNA	Present	Present	Absent
mRNA Introns	Absent	Absent	Present
mRNA Splicing, Capping, and Poly A Tailing	Absent	Absent	Present
Ribosomes			
Size	70S	70S	80S (cytoplasmic ribosomes)
Elongation factor 2	Does not react with diphtheria toxin	Reacts	Reacts
Sensitivity to chloramphenicol and kanamycin	Sensitive	Insensitive	Insensitive
Sensitivity to anisomycin	Insensitive	Sensitive	Sensitive
DNA-Dependent RNA Polymerase			
Number of enzymes	One	Several	Three
Structure	Simple subunit pattern (4 subunits)	Complex subunit pattern similar to eucaryotic enzymes (8–12 subunits)	Complex subunit pattern (12–14 subunits)
Rifampicin sensitivity	Sensitive	Insensitive	Insensitive
Polymerase II Type Promoters	Absent	Present	Present
Metabolism			
Similar ATPase	No	Yes	Yes
Methanogenesis	Absent	Present	Absent
Nitrogen fixation	Present	Present	Absent
Chlorophyll-based photosynthesis	Present	Absent	Present[a]
Chemolithotrophy	Present	Present	Absent

[a]Present in chloroplasts (of bacterial origin).

(repeated as Table 19.1)

Prescott, Harley, and Klein's

Microbiology

Seventh Edition

Joanne M. Willey
Hofstra University

Linda M. Sherwood
Montana State University

Christopher J. Woolverton
Kent State University

Boston Burr Ridge, IL Dubuque, IA New York San Francisco St. Louis
Bangkok Bogotá Caracas Kuala Lumpur Lisbon London Madrid Mexico City
Milan Montreal New Delhi Santiago Seoul Singapore Sydney Taipei Toronto

The McGraw-Hill Companies

 Higher Education

PRESCOTT, HARLEY, AND KLEIN'S MICROBIOLOGY, SEVENTH EDITION

Published by McGraw-Hill, a business unit of The McGraw-Hill Companies, Inc., 1221 Avenue of the Americas, New York, NY 10020.

 This book is printed on recycled, acid-free paper containing 10% postconsumer waste.

1 2 3 4 5 6 7 8 9 0 DOW/DOW 0 9 8 7 6

ISBN 978-0-07-299291-5
MHID 0-07-299291-3

Publisher: *Colin Wheatley/Janice Roerig-Blong*
Senior Developmental Editor: *Lisa A. Bruflodt*
Senior Marketing Manager: *Tami Petsche*
Senior Project Manager: *Jayne Klein*
Lead Production Supervisor: *Sandy Ludovissy*
Senior Media Project Manager: *Jodi K. Banowetz*
Senior Media Producer: *Eric A. Weber*
Designer: *John Joran*
(USE) Cover Image: *Dennis Kunkel Microscopy, Inc.*
Lead Photo Research Coordinator: *Carrie K. Burger*
Photo Research: *Mary Reeg*
Supplement Producer: *Mary Jane Lampe*
Compositor: *Carlisle Publishing Services*
Typeface: *10/12 Times Roman*
Printer: *R. R. Donnelley Willard, OH*

Library of Congress Cataloging-in-Publication Data

Willey, Joanne M.
Prescott, Harley, and Klein's microbiology / Joanne M. Willey, Linda M. Sherwood, Christopher J. Woolverton. — 7th ed.
p. cm.
Includes index.
ISBN 978-0-07-299291-5 — ISBN 0-07-299291-3 (hard copy : alk. paper)
1. Microbiology. I. Sherwood, Linda M. II. Woolverton, Christopher J. III. Prescott, Lansing M. Microbiology. IV. Title.

QR41.2.P74 2008
616.9'041—dc22 2006027152
CIP

www.mhhe.com

This text is dedicated to our mentors—John Waterbury, Richard Losick, Thomas Bott, Hank Heath, Pete Magee, Lou Rigley, Irv Snyder, and R. Balfour Sartor. And to our students.

—Joanne M. Willey
—Linda M. Sherwood
—Christopher J. Woolverton

Brief Contents

Contents

Part III Microbial Metabolism

Part IV Microbial Molecular Biology and Genetics

Part V DNA Technology and Genomics

Part VI The Viruses

Part VII The Diversity of the Microbial World

About the Authors

Joanne M. Willey has been a professor at Hofstra University on Long Island, New York since 1993, and was recently promoted to full professor. Dr. Willey received her B.A. in Biology from the University of Pennsylvania, where her interest in microbiology began with work on cyanbacterial growth in eutrophic streams. She earned her Ph.D. in biological oceanography (specializing in marine microbiology) from the Massachusetts Institute of Technology-Woods Hole Oceanographic Institution Joint Program in 1987. She then went to Harvard University where she spent four years as a postdoctoral fellow studying the filamentous soil bacterium *Streptomyces coelicolor.* Dr. Willey continues to actively investigate this fascinating microbe and has co-authored a number of publications that focus on its complex developmental cycle. She is an active member of the American Society for Microbiology and has served on the editorial board of the journal *Applied and Environmental Microbiology* since 2000. Dr. Willey regularly teaches microbiology to biology majors as well as allied health students. She also teaches courses in cell biology, marine microbiology, and laboratory techniques in molecular genetics. Dr. Willey lives on the north shore of Long Island with her husband and two sons. She is an avid runner and enjoys skiing, hiking, sailing, and reading. She can be reached at biojmw@hofstra.edu.

Linda M. Sherwood is a member of the Department of Microbiology at Montana State University. Her interest in microbiology was sparked by the last course she took to complete a B.S. degree in Psychology at Western Illinois University. She went on to complete an M.S. degree in Microbiology at the University of Alabama, where she studied histidine utilization by *Pseudomonas acidovorans*. She subsequently earned a Ph.D. in Genetics at Michigan State University where she studied sporulation in *Saccharomyces cerevisiae*. She briefly left the microbial world to study the molecular biology of *dunce* fruit flies at Michigan State University before her move to Montana State University. Dr. Sherwood has always had a keen interest in teaching, and her psychology training has helped her to understand current models of cognition and learning and their implications for teaching. Over the years, she has taught courses in general microbiology, genetics, biology, microbial genetics, and microbial physiology. She has served as the editor for ASM's *Focus on Microbiology Education*, and has participated in and contributed to numerous ASM Conferences for Undergraduate Educators (ASMCUE). She also has worked with K-12 teachers to develop a kit-based unit to introduce microbiology into the elementary school curriculum, and has coauthored with Barbara Hudson a general microbiology laboratory manual, *Explorations in Microbiology: A Discovery Approach*, published by Prentice-Hall. Her association with McGraw-Hill began when she prepared the study guides for the fifth and sixth editions of *Microbiology*. Her non-academic interests focus primarily on her family. She also enjoys reading, hiking, gardening, and traveling. She can be reached at lsherwood@montana.edu.

Christopher J. Woolverton is Professor of Biological Sciences and a member of the graduate faculty in Biological Sciences and The School of Biomedical Sciences at Kent State University in Kent, Ohio. Dr. Woolverton also serves as the Director of the KSU Center for Public Health Preparedness, overseeing its BSL-3 Training Facility. He earned his B.S. from Wilkes College, Wilkes-Barre, Pennsylvania and a M.S. and a Ph.D. in Medical Microbiology from West Virginia University, College of Medicine. He spent two years as a postdoctoral fellow at the University of North Carolina at Chapel Hill studying cellular immunology. Dr. Woolverton's research interests are focused on the detection and control of bacterial pathogens. Dr. Woolverton and his colleagues have developed the first liquid crystal biosensor for the immediate detection and identification of microorganisms, and a natural polymer system for controlled antibiotic delivery. He publishes and frequently lectures on these two technologies. Dr. Woolverton has taught microbiology to science majors and allied health students, as well as graduate courses in Immunology and Microbial Physiology. He is an active member of ASM, serving on the editorial boards of ASM's *Microbiology Education* and *Focus on Microbiology Education*. He has participated in and contributed to numerous ASM conferences for Undergraduate Educators, serving as co-chair of the 2001 conference. Dr. Woolverton resides in Kent, Ohio with his wife and three daughters. When not in the lab or classroom, he enjoys hiking, biking, tinkering with technology, and just spending time with his family. His email address is cwoolver@kent.edu.

Preface

Prescott, Harley, and Klein's Microbiology has acquired the reputation of covering the broad discipline of microbiology at a depth not found in any other textbook. The seventh edition introduces a new author team. As new authors, we were faced with the daunting task of making a superior textbook even better. We bring over 40 years of combined research and teaching experience. Our keen interest in teaching has been fostered by our involvement in workshops and conferences designed to explore, implement, and assess various pedagogical approaches. Thus one of our goals for this edition was to make the book more accessible to students. To accomplish this we focused on three specific areas: readability, artwork, and the integration of several key themes throughout the text.

Our Strengths

Readability

We have retained the relatively simple and direct writing style used in previous editions of *Prescott, Harley, and Klein's Microbiology*. However, for the seventh edition, we have added style elements designed to further engage students. For example, we have introduced the use of the first person to describe the flow of information (e.g., see chapter openers) and we pose questions within the text, prompting students to reflect on the matter at hand. Each chapter is divided into numbered section headings and organized in an outline format. Some chapters have been significantly reorganized to present the material in a more logical format (e.g., chapters 12, 28, and 39). As in previous editions, key terminology is boldfaced and clearly defined. In addition, some words are now highlighted in red font: these include names of scientists with whom the students should be acquainted, as well as names of techniques and microbes. Every term in the extensive glossary, which includes over 200 new and revised entries, includes a page reference.

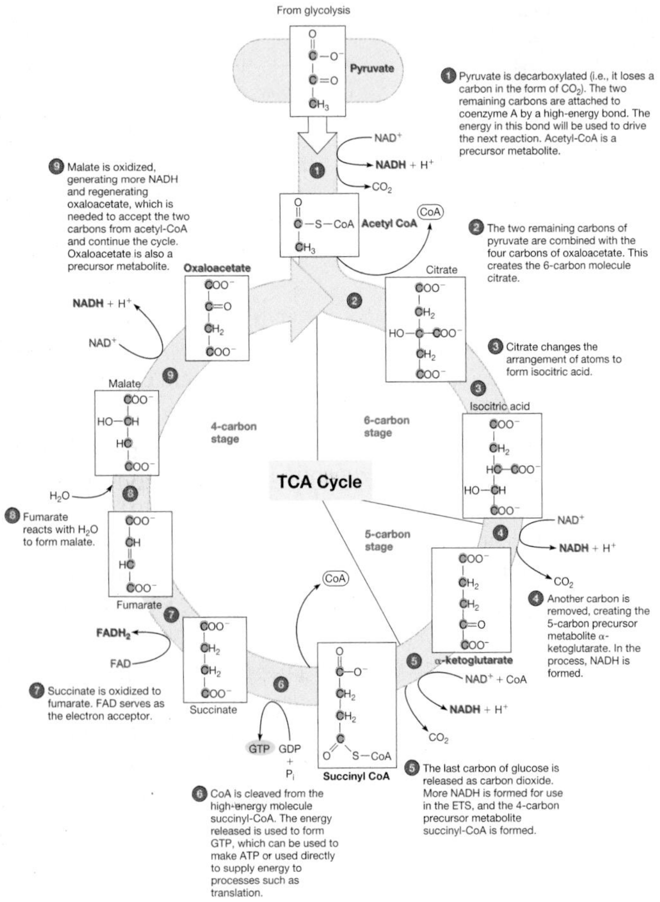

From glycolysis
Pyruvate
1 Pyruvate is decarboxylated (i.e., it loses a carbon in the form of CO_2). The two remaining carbons are attached to coenzyme A by a high-energy bond. The energy in this bond will be used to drive the next reaction. Acetyl-CoA is a precursor metabolite.
NAD+
NADH + H+
CO_2
Acetyl CoA
CoA
2 The two remaining carbons of pyruvate are combined with the four carbons of oxaloacetate. This creates the 6-carbon molecule citrate.
Citrate
3 Citrate changes the arrangement of atoms to form isocitric acid.
Isocitric acid
9 Malate is oxidized, generating more NADH and regenerating oxaloacetate, which is needed to accept the two carbons from acetyl-CoA and continue the cycle. Oxaloacetate is also a precursor metabolite.
Oxaloacetate
Malate
4-carbon stage
6-carbon stage
5-carbon stage
TCA Cycle
8 Fumarate reacts with H_2O to form malate.
Fumarate
$FADH_2$
FAD
7 Succinate is oxidized to fumarate. FAD serves as the electron acceptor.
Succinate
GTP GDP + P_i
Succinyl CoA
α-ketoglutarate
NAD+ + CoA
4 Another carbon is removed, creating the 5-carbon precursor metabolite α-ketoglutarate. In the process, NADH is formed.
5 The last carbon of glucose is released as carbon dioxide. More NADH is formed for use in the ETS, and the 4-carbon precursor metabolite succinyl-CoA is formed.
6 CoA is cleaved from the high-energy molecule succinyl-CoA. The energy released is used to form GTP, which can be used to make ATP or used directly to supply energy to processes such as translation.

Artwork

To engage today's students, a textbook must do more than offer text and images that just adequately describe the topic at hand. Our goal is to make the students *want* to read the text because they find the material interesting and appealing. The seventh edition brings a new art program that features three-dimensional renditions and bright, attractive colors. However, not only have existing figures been updated, over 200 new figures have been added. The updated art program also includes new pedagogical features such as concept maps (*see figures 8.1, 12.1, and 31.1*) and annotation of key pathways and processes (*see figures 9.9 and 11.17*).

Leading strand
β clamp
Core
β clamp waiting to be loaded
γ complex
Parental DNA strands
DnaB helicase
DNA primase
Lagging strand
Previously synthesized Okazaki fragment
RNA primer
1 The leading-strand core polymerase synthesizes DNA as the parental DNA strands are unwound by DnaB helicase. The lagging strand core polymerase is nearing completion of an Okazaki fragment. DNA primase begins synthesis of the RNA primer for the next Okazaki fragment to be synthesized.
β clamp being loaded onto template primer
2 Upon completion of the new RNA primer, DNA primase dissociates, and the γ complex (clamp loader) loads a β clamp onto the template primer.
3 The lagging-strand core polymerase reaches the previously synthesized Okazaki fragment and dissociates from the DNA.
Discarded β clamp
DNA polymerase I (not shown) eventually removes primer and fills gap
4 The lagging-strand core polymerase associates with the newly loaded β clamp and synthesis of a new Okazaki fragment begins.

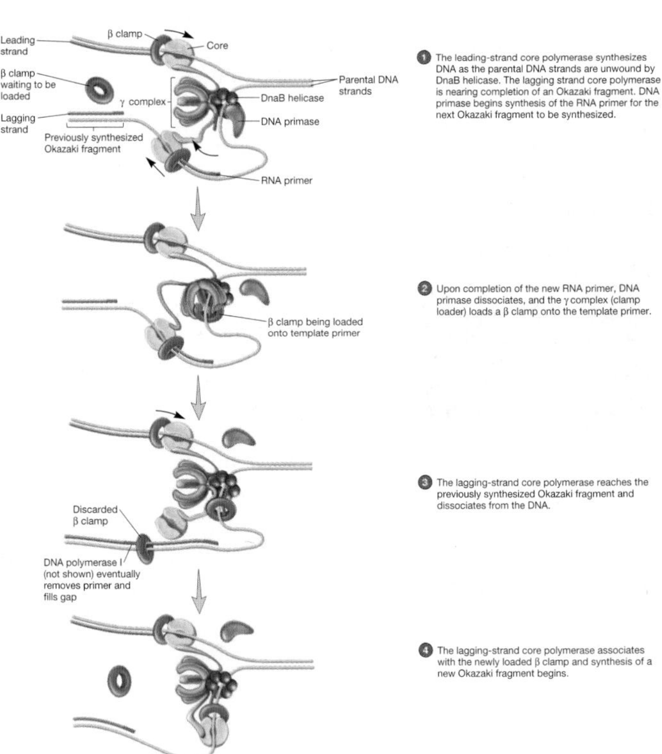

About the Authors

Joanne M. Willey has been a professor at Hofstra University on Long Island, New York since 1993, and was recently promoted to full professor. Dr. Willey received her B.A. in Biology from the University of Pennsylvania, where her interest in microbiology began with work on cyanbacterial growth in eutrophic streams. She earned her Ph.D. in biological oceanography (specializing in marine microbiology) from the Massachusetts Institute of Technology-Woods Hole Oceanographic Institution Joint Program in 1987. She then went to Harvard University where she spent four years as a postdoctoral fellow studying the filamentous soil bacterium *Streptomyces coelicolor.* Dr. Willey continues to actively investigate this fascinating microbe and has co-authored a number of publications that focus on its complex developmental cycle. She is an active member of the American Society for Microbiology and has served on the editorial board of the journal *Applied and Environmental Microbiology* since 2000. Dr. Willey regularly teaches microbiology to biology majors as well as allied health students. She also teaches courses in cell biology, marine microbiology, and laboratory techniques in molecular genetics. Dr. Willey lives on the north shore of Long Island with her husband and two sons. She is an avid runner and enjoys skiing, hiking, sailing, and reading. She can be reached at biojmw@hofstra.edu.

Linda M. Sherwood is a member of the Department of Microbiology at Montana State University. Her interest in microbiology was sparked by the last course she took to complete a B.S. degree in Psychology at Western Illinois University. She went on to complete an M.S. degree in Microbiology at the University of Alabama, where she studied histidine utilization by *Pseudomonas acidovorans*. She subsequently earned a Ph.D. in Genetics at Michigan State University where she studied sporulation in *Saccharomyces cerevisiae*. She briefly left the microbial world to study the molecular biology of *dunce* fruit flies at Michigan State University before her move to Montana State University. Dr. Sherwood has always had a keen interest in teaching, and her psychology training has helped her to understand current models of cognition and learning and their implications for teaching. Over the years, she has taught courses in general microbiology, genetics, biology, microbial genetics, and microbial physiology. She has served as the editor for ASM's *Focus on Microbiology Education*, and has participated in and contributed to numerous ASM Conferences for Undergraduate Educators (ASMCUE). She also has worked with K-12 teachers to develop a kit-based unit to introduce microbiology into the elementary school curriculum, and has coauthored with Barbara Hudson a general microbiology laboratory manual, *Explorations in Microbiology: A Discovery Approach*, published by Prentice-Hall. Her association with McGraw-Hill began when she prepared the study guides for the fifth and sixth editions of *Microbiology*. Her non-academic interests focus primarily on her family. She also enjoys reading, hiking, gardening, and traveling. She can be reached at lsherwood@montana.edu.

Christopher J. Woolverton is Professor of Biological Sciences and a member of the graduate faculty in Biological Sciences and The School of Biomedical Sciences at Kent State University in Kent, Ohio. Dr. Woolverton also serves as the Director of the KSU Center for Public Health Preparedness, overseeing its BSL-3 Training Facility. He earned his B.S. from Wilkes College, Wilkes-Barre, Pennsylvania and a M.S. and a Ph.D. in Medical Microbiology from West Virginia University, College of Medicine. He spent two years as a postdoctoral fellow at the University of North Carolina at Chapel Hill studying cellular immunology. Dr. Woolverton's research interests are focused on the detection and control of bacterial pathogens. Dr. Woolverton and his colleagues have developed the first liquid crystal biosensor for the immediate detection and identification of microorganisms, and a natural polymer system for controlled antibiotic delivery. He publishes and frequently lectures on these two technologies. Dr. Woolverton has taught microbiology to science majors and allied health students, as well as graduate courses in Immunology and Microbial Physiology. He is an active member of ASM, serving on the editorial boards of ASM's *Microbiology Education* and *Focus on Microbiology Education*. He has participated in and contributed to numerous ASM conferences for Undergraduate Educators, serving as co-chair of the 2001 conference. Dr. Woolverton resides in Kent, Ohio with his wife and three daughters. When not in the lab or classroom, he enjoys hiking, biking, tinkering with technology, and just spending time with his family. His email address is cwoolver@kent.edu.

Preface

Prescott, Harley, and Klein's Microbiology has acquired the reputation of covering the broad discipline of microbiology at a depth not found in any other textbook. The seventh edition introduces a new author team. As new authors, we were faced with the daunting task of making a superior textbook even better. We bring over 40 years of combined research and teaching experience. Our keen interest in teaching has been fostered by our involvement in workshops and conferences designed to explore, implement, and assess various pedagogical approaches. Thus one of our goals for this edition was to make the book more accessible to students. To accomplish this we focused on three specific areas: readability, artwork, and the integration of several key themes throughout the text.

Our Strengths

Readability

We have retained the relatively simple and direct writing style used in previous editions of *Prescott, Harley, and Klein's Microbiology*. However, for the seventh edition, we have added style elements designed to further engage students. For example, we have introduced the use of the first person to describe the flow of information (e.g., see chapter openers) and we pose questions within the text, prompting students to reflect on the matter at hand. Each chapter is divided into numbered section headings and organized in an outline format. Some chapters have been significantly reorganized to present the material in a more logical format (e.g., chapters 12, 28, and 39). As in previous editions, key terminology is boldfaced and clearly defined. In addition, some words are now highlighted in red font: these include names of scientists with whom the students should be acquainted, as well as names of techniques and microbes. Every term in the extensive glossary, which includes over 200 new and revised entries, includes a page reference.

Artwork

To engage today's students, a textbook must do more than offer text and images that just adequately describe the topic at hand. Our goal is to make the students *want* to read the text because they find the material interesting and appealing. The seventh edition brings a new art program that features three-dimensional renditions and bright, attractive colors. However, not only have existing figures been updated, over 200 new figures have been added. The updated art program also includes new pedagogical features such as concept maps (*see figures 8.1, 12.1, and 31.1*) and annotation of key pathways and processes (*see figures 9.9 and 11.17*).

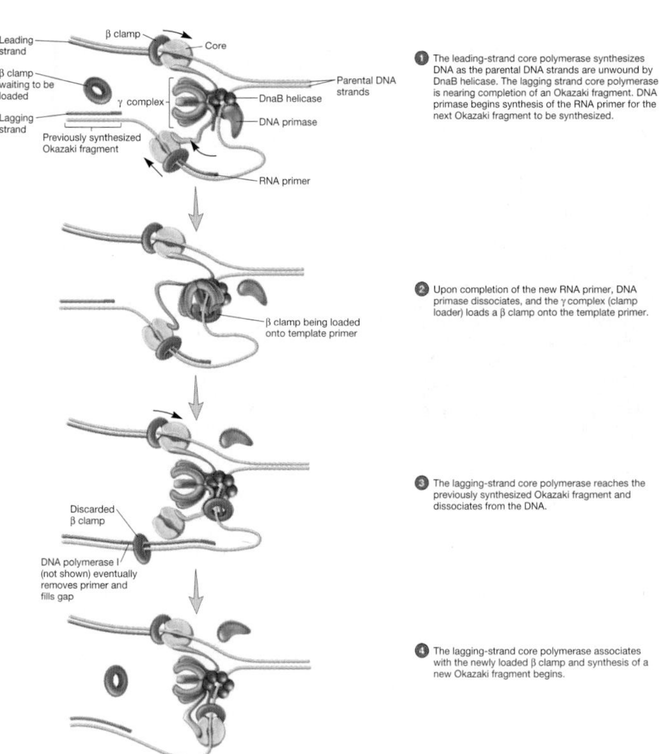

Thematic Integration

With the advent of genomics and the increased reach of cell biology, the divisions among microbiology subdisciplines have become blurred; for instance, the microbial ecologist must also be well-versed in microbial physiology, evolution, and the principles and practices of molecular biology. In addition, the microbiologist must be acquainted with all major groups of microorganisms: viruses, bacteria, archaea, protists, and fungi. Students new to microbiology are asked to assimilate vocabulary, facts, and most importantly, concepts, from a seemingly vast array of subjects. The challenge to the professor of microbiology is to integrate essential concepts throughout the presentation of material while conveying the beauty of microbes and excitement of this dynamic field.

While previous editions of *Microbiology* excelled in incorporating genetics and metabolism throughout the text, in this edition we have attempted to bring the diversity of the microbial world into each chapter. Of course this was most easily done in those chapters devoted to microbial evolution, diversity, and ecology (chapters 19 to 30), but we challenged ourselves to bring microbial diversity into chapters that are traditionally *E. coli*-based. So, although the chapters on genetics (chapters 11 to 13) principally review processes as they are revealed in *E. coli,* we also explore other systems as well, such as the regulation of sporulation in *Bacillus subtilis* and quorum sensing in *V. fischerii* (*see figures 12.19 through 12.21*).

We also thought it was important to weave the thread of evolution throughout the text. We start in the first chapter with a discussion of the universal tree of life (*see figure 1.1*), with various renditions of "the big tree" appearing in later chapters. Importantly, we remind students that structures and processes *evolved* to their current state; that natural selection is always at work (e.g., the title and the tone of chapter 13—now called *Microbial Genetics: Mechanisms of Genetic Variations*—have been changed). Finally, the seventh edition of *Microbiology* explores theories regarding the origin of life at a depth not seen in other microbiology texts (chapter 19).

Indeed, depth of coverage has been one of the mainstays of *Prescott, Harley, and Klein's Microbiology.* The text was founded on two fundamental principles: (1) students need an introduction to the whole of microbiology before concentrating on specialized areas, and (2) this introduction should provide the level of understanding required for students to grasp the conceptual underpinning of facts. We remain committed to this approach. Thus the seventh edition continues to provide a balanced and thorough introduction to all major areas of microbiology. This book is suitable for courses with orientations ranging from basic microbiology to medical and applied microbiology. Students preparing for careers in medicine, dentistry, nursing, and allied health professions will find the text as useful as will those aiming for careers in research, teaching, and industry. While two courses each of biology and chemistry are assumed, we provide a strong overview of the relevant chemistry in appendix I.

Changes to the Seventh Edition

The seventh edition of *Prescott, Harley, and Klein's Microbiology* is the result of extensive review and analysis of previous editions, the input from reviewers, and casual discussions with our colleagues. As a new author team, we were committed to keeping the in-depth coverage that *Microbiology* is known for, while at the same time bringing a fresh perspective not only to specific topics but to the overall presentation as well.

Up-to-Date Coverage

Each year exciting advances are made in microbiology. While we understand that not all of these are appropriate for discussion in an introductory textbook, we have incorporated the most up-to-date information and exciting, recent discoveries to maintain accurate descriptions of structures and processes and to illustrate essential points. A few specific examples include a current description of the structure and function of DNA polymerase III, the role of viruses in marine ecosystems, the ubiquitous nature of type III secretion systems, an updated coverage of the inflammatory response, and the current understanding of HIV origins and avian influenza epidemiology.

Increased Emphasis on Microbial Evolution and Diversity

Microbial evolution, diversity, and ecology are no longer subdisciplines to be ignored by those interested in microbial genetics, physiology, or pathogenesis. For example, within the last 10 years, polymicrobial diseases, intercellular communication, and biofilms have been recognized as important microbial processes that closely tie evolution to genetics, ecology to physiology, and ecology to pathogenesis. The seventh edition strives to integrate these themes throughout the text. We begin chapter 1 with a discussion of the universal tree of life and whenever possible, bring diverse microbial species into discussions so that students can begin to appreciate the tremendous variation in the microbial world. Chapter 19 now covers microbial evolution in greater depth than other texts. It has been retitled *Microbial Evolution, Taxonomy, and Diversity* and the content significantly revised so that microbial evolution is presented as a key component of microbiology. We also introduce and frequently remind students of the enormity of microbial diversity. Like previous editions, the seventh edition features specific chapters that review the members of the microbial world. The chapters that are specifically devoted to ecology (chapters 27 through 29) have undergone significant revisions. We continue to use the classification scheme set forth in the second edition of *Bergey's Manual of Systematic Bacteriology;* in addition, we have introduced the Baltimore System of virus classification and the International Society of Protistologists' new classification scheme for eucaryotes in chapters 18 and 25, respectively.

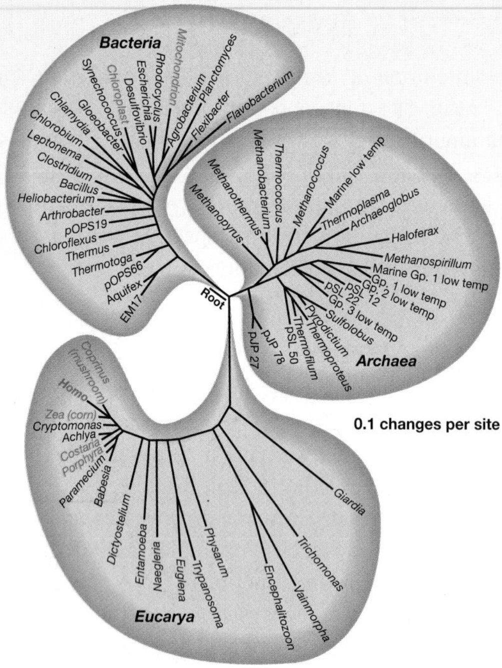

Writing for Student Understanding

Our goal as a new author team was to retain the straightforward writing style of previous editions while at the same time making the text more readable for the average college student. We have thus added style elements designed to help the reader understand the larger context of the topic at hand. For example, the opening text in several chapters is accompanied by a concept map, enabling the student to visualize the relationships among component topics found within a chapter. Parts of the text are now written in first person; we want students to appreciate that we, as authors, understand that learning is a process that needs to be guided.

Significantly Enhanced Art Program

Today's student must be visually engaged. The artwork in each chapter of the seventh edition has been revised and updated to include realistic, three-dimensional images designed to spark student interest and curiosity. This new program uses bright and appealing colors that give the text an attractive look. We have taken the opportunity to both update and annotate a number of images so that students can picture a complex process step-by-step. New pedagogical features such as concept maps and annotation of key pathways have been added. The three-dimensional renderings help the student appreciate the beauty and elegance of the cell, while at the same time making the material more comprehendible.

Questions for Review and Reflection

Our belief that concepts are just as important as facts, if not more, is also reflected in the questions for review and reflection that appear throughout each chapter. Those who have used previous editions of *Microbiology* may notice that in addition to questions that quiz the retention of key facts, new questions designed to be more thought provoking have been added.

Content Changes By Part

Each chapter has been thoroughly reviewed and almost all have undergone significant revision. In some chapters, there are changes in both organization and content (e.g., chapters 11 to 13), while many other chapters retain the same order of presentation but the content has been updated. A summary of important new material by parts includes:

Part I

Chapter 1—Expanded introduction to the three domains of life and the microbes found in each domain.

Chapter 3—Increased coverage of the difference between archaeal and bacterial cellular structure.

Chapter 4—Reorganized and updated discussion of the biosynthetic-secretory pathway and endocytosis.

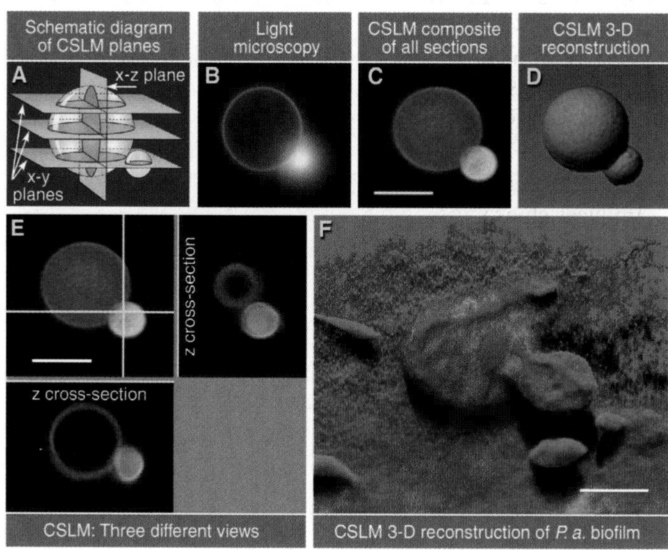

Part II

Chapter 6—Updated discussion of the procaryotic cell cycle, including current models of chromosome partitioning and septation; updated and expanded coverage of biofilms and quorum sensing.

Part III

Chapter 8—A new section providing an overview of metabolism and a framework for the more detailed discussions of metabolism that follow; chemotaxis is introduced as an example of regulation of a behavioral response by covalent modification of enzymes.

Chapter 9—Reorganized discussion of chemoorganotrophic metabolism to illustrate the connections among the pathways used and how these pathways supply the materials needed for anabolism; addition of a discussion of rhodopsin-based phototrophy.

Chapter 10—Reorganized to more clearly correlate N-, P-, and S-assimilation mechanisms with the synthesis of amino acids and nucleotides; discussion of peptidoglycan synthesis is included in the discussion of polysaccharide biosynthesis.

Part IV

Chapter 11—Reorganized to focus solely on genome structure and replication, gene structure, and gene expression.

Chapter 12—Focuses exclusively on the regulation of gene expression; reorganized according to level at which regulation occurs; updated and expanded discussion of riboswitches and regulation by small RNA molecules.

Chapter 13—Covers mutation, repair, and recombination in the context of processes that introduce genetic variation into populations.

Part V

Chapter 14—Begins with, and then builds upon, a concept map describing the principal steps involved in the construction of recombinant DNA molecules with emphasis that recombinant DNA technology is not confined to a few model and industrial microorganisms.

Chapter 15—Rewritten to explore the many ways in which genomics has changed microbiology. Expanded sections on bioinformatics and functional genomics, and a new section introduces environmental genomics (metagenomics).

Part VI

Chapter 16—A new section describing virus reproduction in general terms, so that this chapter can now stand alone as an introduction to viruses.

Part VII

Chapter 19—Rewritten and re-titled *Microbial Evolution, Taxonomy, and Diversity;* the chapter now opens with an in-depth discussion of the origin of life. Discussion of molecular techniques and their importance in microbial taxonomy has also been expanded.

Chapter 20—In keeping with recent discoveries describing the ubiquity of archaea, the seventh edition presents the differences between microbes in the bacterial and archaeal domains in chapter 3. Thus chapter 20 now presents a more in-depth look at some of the specifics of archaeal physiology, genetics, taxonomy, and diversity.

Chapter 25—The protist chapter has been completely rewritten in accordance with the 2005 reclassification of the *Eucarya* by the International Society of Protistologists. Emphasis is placed on medically and environmentally important protists. Thus the chapter entitled *The Algae* found in previous editions has been eliminated and photosynthetic protists are now covered in chapter 25.

Part VIII

Chapter 27—Rewritten and re-titled *Biogeochemical Cycling and Introductory to Microbial Ecology*. Expanded coverage of biogeochemical cycling now includes the phosphorus cycle. Discussion on microbial ecology emphasizes the importance and application of culture-independent approaches. Discussion of water purification and wastewater treatment has been moved to chapter 41, *Applied and Industrial Microbiology.*

Chapter 28—Expanded and reorganized to cover the microbial communities found in the major biomes within marine and freshwater environments. The role of the oceans in regulating global warming is introduced.

Chapter 29—Reorganized to first introduce soils as an environment, is followed by more in-depth and updated treatment of mycorrhizae, the rhizobia, and plant pathogens. Approaches to studying the subsurface environment and new discoveries in this growing field are now included.

Chapter 30—Microbial interactions previously introduced in chapter 27 have been moved to this chapter, where they are presented along with human-microbe interactions (previously presented with innate immunity), helping to convey the concept that the human body is an ecosystem.

Part IX

Chapter 31—Reorganized and updated "nonspecific host resistance" as its own chapter (normal microflora is now in chapter 30); enhanced sections on natural antimicrobial substances.

Chapter 32—Reorganized and updated to enhance linkages between innate and acquired immune activities; integrated medical immunology concepts.

Chapter 33—Most virulence mechanisms have been either updated and/or expanded; added section on host defenses to microbial invasion to link infectious disease processes with host immunity.

Part X

Chapter 34—Content focuses on mechanism of action of each antimicrobial agent; added section on anti-protozoan drugs.

Chapter 35—Now includes both clinical microbiology and immunology; reorganized and updated to reflect current clinical laboratory practices.

Chapter 36—New focus on the important role of epidemiology in preventative medicine, thus vaccines are now covered in this chapter (formerly found in chapter 32); new section on bioterorrism preparedness added.

Chapter 37—Reorganized and updated to reflect viral pathogenesis; select (potential bioterrorism) agents highlighted; influenza section augmented to include the most current information regarding avian influenza; HIV etiology, pathogenesis and treatment sections updated; new section on viral zoonoses.

Chapter 38—Expanded coverage of bacterial pathogenesis; select (potential bioterrorism) agents highlighted; new sections on group B streptococcal disease and bacterial zoonoses.

Chapter 39—Reorganized and updated to reflect disease transmission routes (similar to chapters 37 and 38); new sections on cyclospora and microsporidia.

Part XI

Chapter 40—Expanded discussion of lactic acid bacteria, probiotics: chocolate fermentation now featured in a Techniques & Applications box.

Chapter 41—Revised to include water purification and wastewater treatment. New section on nanotechnology; expanded section on the biochemistry of bioremediation.

Tools For Learning

Chapter Preview

- Each chapter begins with a preview—a list of important concepts discussed in the chapter.

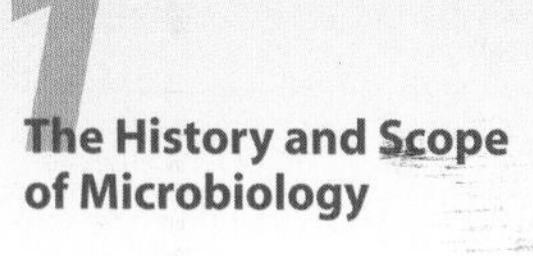

Louis Pasteur, one of the greatest scientists of the nineteenth century, maintained that "Science knows no country, because knowledge belongs to humanity, and is a torch which illuminates the world."

Preview

- Microbiology is defined not only by the size of its subjects but the techniques it uses to study them.
- Microorganisms include acellular entities (e.g., viruses), procarytic cells, and eucaryotic cells. Cellular microorganisms are found in all three domains of life: *Bacteria, Archaea, Eucarya.*
- The development of microbiology as a scientific discipline has depended on the availability of the microscope and the ability to isolate and grow pure cultures of microorganisms. The development of these techniques in large part grew out of studies disproving the Theory of Spontaneous Generation and others establishing that microorganisms can cause disease.
- Microbiology is a large discipline; it has had and will continue to have a great impact on other areas of biology and general human welfare.

The importance of microorganisms can't be overemphasized. In terms of sheer number and mass—it is estimated that microbes contain 50% of the biological carbon and 90% of the biological nitrogen on Earth—they greatly exceed every other group of organisms on the planet. Furthermore, they are found everywhere: from geothermal vents in the ocean depths to the coldest arctic ice, to every person's skin. They are major contributors to the functioning of the biosphere, being indispensable for the cycling of the elements essential for life. They also are a source of nutrients at the base of all ecological food chains and webs. Most importantly, certain microorganisms carry out photosynthesis, rivaling plants in their role of capturing carbon dioxide and releasing oxygen into the atmosphere. Those microbes that inhabit humans also play important roles, including helping the body digest food and producing vitamins B and K. In addition, society in general benefits from microorganisms, as they are necessary for the production of bread, cheese, beer, antibiotics, vaccines, vitamins, enzymes, and many other important products. Indeed, modern biotechnology rests upon a microbiological foundation.

Although the majority of microorganisms play beneficial or benign roles, some harm humans and have disrupted society over the millennia. Microbial diseases undoubtedly played a major role in historical events such as the decline of the Roman Empire and the conquest of the New World. In 1347, plague or black death, an arthropod-borne disease, struck Europe with brutal force, killing 1/3 of the population (about 25 million people) within four years. Over the next 80 years, the disease struck again and again, eventually wiping out 75% of the European population. The plague's effect was so great that some historians believe it changed European culture and prepared the way for the Renaissance. Today the struggle by microbiologists and others against killers like AIDS and malaria continues.

In this introductory chapter, we introduce the microbial world to provide a general idea of the organisms and agents that microbiologists study. Then we describe the historical development of the science of microbiology and its relationship to medicine and other areas of biology. Finally, we discuss the scope, relevance, and future of modern microbiology.

1.1 Members of the Microbial Wo

Microbiology often has been defined as the study and agents too small to be seen clearly by the unaid is, the study of **microorganisms.** Because objects le one millimeter in diameter cannot be seen clearly

Dans les champs de l'observation, le hasard ne favorise que les esprits préparés.
(In the field of observation, chance favors only prepared minds.)

—Lou

Cross-Referenced Notes

- In-text notes in blue type refer students to other parts of the book to review.

Review and Reflection Questions within Narrative

- Review questions throughout each chapter assist students in mastering section concepts before moving on to other topics.

the color they fluoresce after treatment with a special mixture of stains (**figure 2.13*a***). Thus the microorganisms can be viewed and directly counted in a relatively undisturbed ecological niche. Identification of microorganisms from specimens: Immunologic techniques (section 35.2)

1. List the parts of a light microscope and describe their functions.
2. Define resolution, numerical aperture, working distance, and fluorochrome.
3. If a specimen is viewed using a 5X objective in a microscope with a 15X eyepiece, how many times has the image been magnified?
4. How does resolution depend on the wavelength of light, refractive index, and numerical aperture? How are resolution and magnification related?
5. What is the function of immersion oil?
6. Why don't most light microscopes use 30X ocular lenses for greater magnification?
7. Briefly describe how dark-field, phase-contrast, differential interference contrast, and epifluorescence microscopes work and the kind of image provided by each. Give a specific use for each type.

2.3 Preparation and Staining of Specimens

Although living microorganisms can be directly examined with the light microscope, they often must be fixed and stained to increase visibility, accentuate specific morphological features, and preserve them for future study.

Fixation

The stained cells seen in a microscope should resemble living cells as closely as possible. **Fixation** is the process by which the internal and external structures of cells and microorganisms are preserved and fixed in position. It inactivates enzymes that might disrupt cell morphology and toughens cell structures so that they do not change during staining and observation. A microorganism usually is killed and attached firmly to the microscope slide during fixation.

There are two fundamentally different types of fixation. **Heat fixation** is routinely used to observe procaryotes. Typically, a film of cells (a smear) is gently heated as a slide is passed

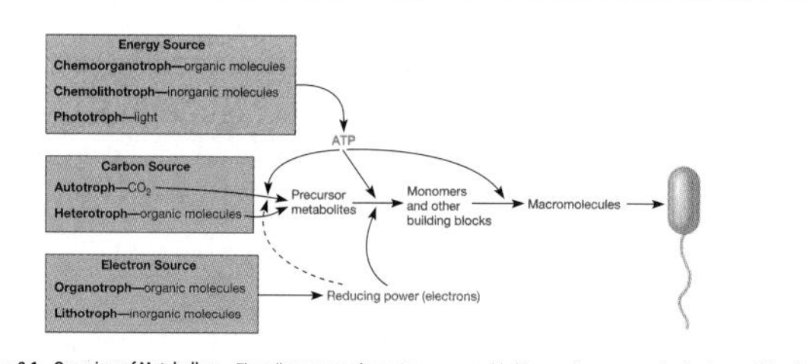

Figure 8.1 Overview of Metabolism. The cell structures of organisms are assembled from various macromolecules (e.g., nucleic acids and proteins). Macromolecules are synthesized from monomers and other building blocks (e.g., nucleotides and amino acids), which are the products of biochemical pathways that begin with precursor metabolites (e.g., pyruvate and α-ketoglutarate). In autotrophs, the precursor metabolites arise from CO_2-fixation pathways and related pathways; in heterotrophs, they arise from reactions of the central metabolic pathways. Reducing power and ATP are consumed in many metabolic pathways. All organisms can be defined metabolically in terms of their energy source, carbon source, and electron source. In the case of chemoorganotrophs, the energy source is an organic molecule that is also the source of carbon and electrons. For chemolithotrophs, the energy sourc the carbon source can be either CO_2 (autotrophs) or an organic molecule (heter bon source can be CO_2 or organic molecules, and the electron source can be wa such as hydrogen sulfide (anoxygenic phototrophs).

BACTERIA

Transcription — Gene — Genetic regulatory proteins can bind to the DNA and control whether or not transcription begins. Attenuation: Transcription can terminate very early after it has begun due to the formation of a transcriptional terminator. Binding of a metabolite to a riboswitch in mRNA can cause premature termination of transcription.

Translation — mRNA — Translational repressor proteins can bind to the mRNA and prevent translation from starting. Antisense RNA can bind to mRNA and control whether or not translation begins. Binding of a metabolite to a riboswitch in mRNA can block translation.

Posttranslation — Protein — Small molecules can bind (noncovalently) to a protein and affect its function. An example is feedback inhibition, in which the product of a metabolic pathway inhibits the first enzyme in the pathway. The structure and function of a protein can be altered by covalent changes to the protein. These can be reversible (e.g., phosphorylation/dephosphorylation) or irreversible (e.g., removal of amino acid residues). These are called posttranslational modifications. — Functional protein

(a)

ARCHAEA

Transcription — Gene — Genetic regulatory proteins can bind to the DNA and control whether or not transcription begins The compaction level of chromatin may influence transcription.

Translation — mRNA — Antisense RNA can bind to mRNA and control whether or not translat begins.

Posttranslation — Protein — Feedback inhibition and covalent modifications (reversible and irreversible) may regulate protein function. — Functional protein

(b)

Figure 12.1 Gene Expression and Common Regulatory Mechanisms in the Three Domains of Life.

1. Isolate DNA to be cloned.
2. Use a restriction enzyme or PCR to generate fragments of DNA.
3. Generate a recombinant molecule by inserting DNA fragments into a cloning vector.
4. Introduce recombinant molecule into new host.

Vector — Linear vector — New host

Figure 14.1 Steps in Cloning a Gene. Each step shown in this overview is discussed in more detail in Chapter 14.

Concept Maps

- Many chapters include a concept map, new to this edition, that outlines critical themes.

Special Interest Essays

- Interesting essays on relevant topics are included in each chapter. Readings are organized into these topics: Historical Highlights, Techniques & Applications, Microbial Diversity & Ecology, Disease, and Microbial Tidbits.

Microbial Tidbits

35.2 Biosensors: The Future Is Now

The 120-plus-year-old pathogen detection systems based on culture and biochemical phenotyping are being challenged. Fueled by the release of anthrax spores in the U.S. postal system, government agencies have been calling for newer technologies for the near-immediate detection and identification of microbes. In the past, detection technologies have traded speed for cost and complexity. The agar plate technique, refined by Robert Koch and his contemporaries in the 1880s, is a trusted and highly efficient method for the isolation of bacteria into pure cultures. Subsequent phenotyping biochemical methods, often using differential media in a manner similar to that used in the isolation step, then identifies common bacterial pathogens. Unfortunately, reliable results from this process often take several days. More rapid versions of the phenotyping systems can be very efficient, yet still require pure culture inoculations. The rapid immunological tests offer faster detection responses but may sacrifice sensitivity. Even DNA sequence comparisons, which are extremely accurate, may require significant time for DNA amplification and significant cost for reagents and sensitive readers. As usual, necessity has begat invention.

The more recent microbial detection systems, many of which are still untested in the clinical arena, sound like science fiction gizmos, yet promise a new age for near-immediate detection and identification of pathogens. These technologies are collectively referred to as "biosensors," and if the biosensor is integrated with a computer microchip for information management, it is then called a "biochip." Biosensors should ideally be capable of highly specific recognition so as to discriminate between nearest relatives, and "communicate" detection through some type of transducing system. Biosensors that detect specific DNA sequences, expressed proteins, and metabolic products have been developed that use optical (mostly fluorescence), electrochemical, or even mass displacement, to report detection. The high degree of recognition required to reduce false-positive results has demanded the uniquely specific, receptor-like capture that is associated with nucleic acid hybridization and antibody binding. Several microbial biosensors employ single-stranded DNA or RNA sequences, or antibody, for the detection component. The transducing or sensing component of biosensors may be markedly different, however. For example, microcantilever systems detect the increased mass of the receptor-bound ligand; the surface acoustic wave device detects change in specific gravity; the bulk quartz resonator monitors fluid density and viscosity; the quartz crystal microbalance measures frequency change in proportion to the mass of material deposited on the crystal; the micromirror sensor uses an optical fiber waveguide that changes reflectivity; and the liquid crystal-based system reports the reorientation of polarized light. Thus the specific capture of a ligand is reflected in the net change measured by each system and results in a signal that announces the initial capture event. Microchip control of the primary and subsequent secondary signals has resulted in automation of the detection process. The reliable detection of pathogens in complex specimens will be the real test as each of these technologies continues to compete for a place in the clinical laboratory.

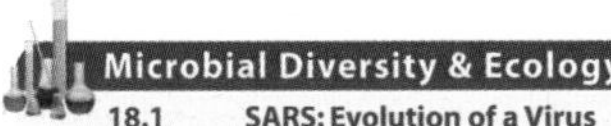

Microbial Diversity & Ecology

18.1 SARS: Evolution of a Virus

In November 2002, a mysterious pneumonia was seen in the Guangdong Province of China, but the first case of this new type of pneumonia was not reported until February 2003. Thanks to the ease of global travel, it took only a couple of months for the pneumonia to spread to more than 25 countries in Asia, Europe, and North and South America. This newly emergent pneumonia was labeled Severe Acute Respiratory Syndrome (SARS) and its causative agent was identified as a previously unrecognized member of the coronavirus family, the SARS-CoV. Almost 10% of the roughly 8,000 people with SARS died. However, once the epidemic was contained, the virus appeared to "die out," and with the exception of a few mild, sporadic cases in 2004, no additional cases have been identified. From where does a newly emergent virus come? What does it mean when a virus "dies out"?

We can answer these questions thanks to the availability of the complete SARS-CoV genome sequence and the power of molecular modeling. Coronaviruses are large, enveloped viruses with positive-strand RNA genomes. They are known to infect a variety of mammals and birds. Researchers suspected that SARS-CoV had "jumped" from its animal host to humans, so samples of animals at open markets in Guangdong were taken for nucleotide sequencing. These studies revealed that cat-like animals called masked palm civits (*Paguma larvata*) harbored variants of the SARS-CoV. Although thousands of civits were then slaughtered, further studies failed to find widespread infection of domestic or wild civits. In addition, experimental infection of civits with human SARS-CoV strains made these animals ill, making the civit an unlikely candidate for the reservoir species. Such a species would be expected to harbor SARS-CoV without symptoms so that it could efficiently spread the virus.

Bats are reservoir hosts of several zoonotic viruses (viruses spread from animals to people) including the emerging Hendra and Nipah viruses that have been found in Australia and East Asia, respectively. Thus it was perhaps not too surprising when in 2005, two groups of international scientists independently demonstrated that Chinese horseshoe bats (genus *Rhinolophus*) are the natural reservoir of a SARS-like coronavirus. When the genomes of the human and bat SARS-CoV are aligned, 92% of the nucleotides are identical. More revealing is alignment of the translated amino acid sequences of the proteins encoded by each virus. The amino acid sequences are 96 to 100% identical for all proteins except the receptor-binding spike proteins, which are only 64% identical. The SARS-CoV spike protein mediates both host cell surface attachment and membrane fusion. Thus a mutation of the spike protein allowed the virus to "jump" from bat host cells to those of another species. It is not clear if the SARS-

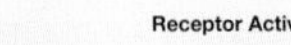

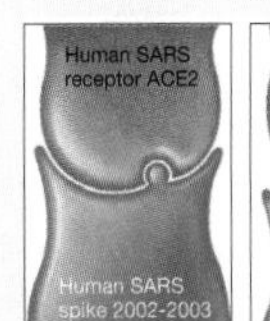

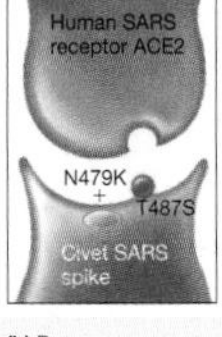

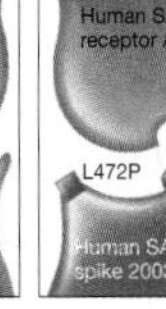

(a) Good (b) Poor (c) Poor

Host Range of SARS-CoV Is Determined by Several Amino Acid Residues in the Spike Protien. (a) The spike protien of the SARS-CoV that caused the SARS epidemic in 2002–2003 fits tightly to the human host cell receptor ACE2. (b) The civit SARS-CoV has two different amino acids at positions 4 and 487. This spike protein binds very poorly to human ACE2, thus the receptor is only weakly activated. (c) The spike protein on the human SARS-CoV that was isolated from patients in 200 and 2004 also differs from that seen in the epidemic-causing SARS-CoV by two amino acids. This SARS-CoV variant caused on mild, sporadic cases.

within the RBD, only four differ between civit and human. Two these amino acids appear to be critical. As shown in the **Box figu** compared to the spike RBD in the SARS-CoV that caused 2002–2003 epidemic, the civit spike has a serine (S) substituted a threonine (T) at position 487 (T487S) and a lysine (K) at positi 479 instead of asparagine (N), N479K. This causes a 1,000-fold crease in the capacity of the virus to bind to human ACE2. Furth more, the spike found in SARS-CoV isolated from patients in 20 and 2004 also has a serine at position 487 as well as a proline (P) leucine (L) substitution at position 472 (L472P). These amino a substitutions could be responsible for the reduced virulence of virus found in these more recent infections. In other words, th mutations could be the reason the SARS virus "died out."

Meanwhile a SARS vaccine based on the virulent 2002–20 strain is being tested. This raises additional questions. Does

Techniques & Applications

40.3 Chocolate: The Sweet Side of Fermentation

Chocolate could be characterized as the "world's favorite food," and yet few people realize that fermentation is an essential part of chocolate production. The Aztecs were the first to develop chocolate fermentation, serving a chocolate drink made from the seeds of the chocolate tree, *Theobroma cocao* [Greek *theos*, god and *broma*, food, or "food of the gods"]. Chocolate trees now grow in West Africa as well as South America.

The process of chocolate fermentation has changed very little over the past 500 years. Each tree produces a large pod that contains 30 to 40 seeds in a sticky pulp (see **Box Figure**). Ripe pods are harvested and slashed open to release the pulp and seeds. The sooner the fermentation begins, the better the product, so fermentation occurs on the farm where the trees are grown. The seeds and pulp are placed in "sweat boxes" or in heaps in the ground and covered, usually with banana leaves.

Like most fermentations, this process involves a succession of microbes. First, a community of yeasts, including *Candida rugosa* and *Kluyveromyces marxianus*, hydrolyze the pectin that covers the seeds and ferment the sugars to release ethyl alcohol and Co_2. As the temperature and the alcohol concentration increase, the yeasts are inhibited and lactic acid bacteria increase in number. The mixture is stirred to aerate the microbes and ensure an even temperature distribution. Lactic acid production drives the pH down; this encourages the growth of bacteria that produce acetic acid as a fermentation end product. Acetic acid is critical to the production of fine chocolate because it kills the sprout inside the seed and releases enzymes that cause further degradation of proteins and carbohydrates, contributing to the overall taste of the chocolate. In addition, acetate esters, derived from acetic acid, are important for the development of good flavor. Fermentation takes five to seven days. An experienced chocolate grower will know when the fermentation is complete—if it is stopped too soon the chocolate will be bitter and astringent. On the other hand, if fermentation lasts too long, microbes start growing on the seeds instead of in the pulp. "Off-tastes" arise when the gram-positive bacterium *Bacillus* and the filamentous fungi *Aspergillis*, Penicillium, and *Mucor* hydrolyze lipids in the seeds to release short-chain fatty acids. As the pH begins to rise, the bacteria of the genera *Pseudomonas*, *Enterobacter*, and *Escherichia* also contribute to bad tastes and odor.

After fermentation, the seeds, now called beans, are spread out to dry. Ideally this is done in the sun, although drying ovens are also used. The oven-drying method is considered inferior because the beans can acquire a smoky taste. The dried beans are brown and lack the pulp. They are bagged and sold to chocolate manufacturers, who first roast the beans to further reduce the bitter taste and kill most of the microbes (some *Bacillus* spores may remain). The beans are then ground and the nibs—the inner part of each bean—are removed. The nibs are crushed into a thick paste called a chocolate liquor, which contains cocoa solids and cocoa butter, but no alcohol. Cocoa solids

Historical Highlights

5.1 The Discovery of Agar as a Solidifying Agent and the Isolation of Pure Cultures

The earliest culture media were liquid, which made the isolation of bacteria to prepare pure cultures extremely difficult. In practice, a mixture of bacteria was diluted successively until only one organism, as an average, was present in a culture vessel. If everything went well, the individual bacterium thus isolated would reproduce to give a pure culture. This approach was tedious, gave variable results, and was plagued by contamination problems. Progress in isolating pathogenic bacteria understandably was slow.

The development of techniques for growing microorganisms on solid media and efficiently obtaining pure cultures was due to the efforts of the German bacteriologist Robert Koch and his associates. In 1881 Koch published an article describing the use of boiled potatoes, sliced with a flame-sterilized knife, in culturing bacteria. The surface of a sterile slice of potato was inoculated with bacteria from a needle tip, and then the bacteria were streaked out over the surface so that a few individual cells would be separated from the remainder. The slices were incubated beneath bell jars to prevent airborne contamination, and the isolated cells developed into pure colonies. Unfortunately many bacteria would not grow well on potato slices.

At about the same time, Frederick Loeffler, an associate of Koch, developed a meat extract peptone medium for cultivating pathogenic bacteria. Koch decided to try solidifying this medium. Koch was an amateur photographer—he was the first to take photomicrographs of bacteria—and was experienced in preparing his own photographic plates from silver salts and gelatin. Precisely the same approach was employed for preparing solid media. He spread a mixture of Loeffler's medium and gelatin over a glass plate, allowed it to harden, and inoculated the surface in the same way he had inoculated his sliced potatoes. The new solid medium worked well, but it could not be incubated at 37°C (the best temperature for most human bacterial pathogens) because the gelatin would Furthermore, some bacteria digested the gelatin.

About a year later, in 1882, agar was first used as a solidif agent. It had been discovered by a Japanese innkeeper, Mi Tarazaemon. The story goes that he threw out extra seaweed and discovered the next day that it had jelled during the cold w night. Agar had been used by the East Indies Dutch to make je and jams. Fannie Eilshemius Hesse (*see figure 1.7*), the New Je born wife of Walther Hesse, one of Koch's assistants, had lea of agar from a Dutch acquaintance and suggested its use whe heard of the difficulties with gelatin. Agar-solidified medium an instant success and continues to be essential in all area microbiology.

Disease

1.2 Koch's Molecular Postulates

Although the criteria that Koch developed for proving a causal relationship between a microorganism and a specific disease have been of great importance in medical microbiology, it is not always possible to apply them in studying human diseases. For example, some pathogens cannot be grown in pure culture outside the host; because other pathogens grow only in humans, their study would require experimentation on people. The identification, isolation, and cloning of genes responsible for pathogen virulence have made possible a new molecular form of Koch's postulates that resolves some of these difficulties. The emphasis is on the virulence genes present in the infectious agent rather than on the agent itself. The molecular postulates can be briefly summarized as follows:

1. The virulence trait under study should be associated much more with pathogenic strains of the species than with nonpathogenic strains.
2. Inactivation of the gene or genes associated with the suspected virulence trait should substantially decrease pathogenicity.
3. Replacement of the mutated gene with the normal wild-type gene should fully restore pathogenicity.
4. The gene should be expressed at some point during the infection and disease process.
5. Antibodies or immune system cells directed against the gene products should protect the host.

The molecular approach cannot always be applied because of problems such as the lack of an appropriate animal system. It also is difficult to employ the molecular postulates when the pathogen is not well characterized genetically.

has been used to study the interactions between the *E. coli* GroES and GroEL chaperone proteins, to map plasmids by locating restriction enzymes bound to specific sites, to follow the behavior of living bacteria and other cells, and to visualize membrane proteins (**figure 2.29**).

1. How does a confocal microscope operate? Why does it provide better images of thick specimens than does the standard compound microscope?
2. Briefly describe the scanning probe microscope and compare and contrast its most popular versions—the scanning tunneling microscope and the atomic force microscope. What are these microscopes used for?

Summary

2.1 Lenses and the Bending of Light

a. A light ray moving from air to glass, or vice versa, is bent in a process known as refraction.

b. Lenses focus light rays at a focal point and magnify images (**figure 2.2**).

2.2 The Light Microscope

a. In a compound microscope like the bright-field microscope, the primary image is formed by an objective lens and enlarged by the eyepiece or ocular lens to yield the final image (**figure 2.3**).

b. A substage condenser focuses a cone of light on the specimen.

c. Microscope resolution increases as the wavelength of radiation used to illuminate the specimen decreases. The maximum resolution of a light microscope is about 0.2 μm.

d. The dark-field microscope uses only refracted light to form an image (**figure 2.7**), and objects glow against a black background.

e. The phase-contrast microscope converts variations in the refractive index and density of cells into changes in light intensity and thus makes colorless, unstained cells visible (**figure 2.9**).

f. The differential interference contrast microscope uses two beams of light to create high-contrast, three-dimensional images of live specimens.

g. The fluorescence microscope illuminates a fluorochrome-labeled specimen and forms an image from its fluorescence (**figure 2.12**).

2.3 Preparation and Staining of Specimens

a. Specimens usually must be fixed and stained before viewing them in the bright-field microscope.

b. Most dyes are either positively charged basic dyes or negative acidic dyes and bind to ionized parts of cells.

c. In simple staining a single dye is used to stain microorganisms.

d. Differential staining procedures like the Gram stain and acid-fast stain distinguish between microbial groups by staining them differently (**figure 2.15**).

e. Some staining techniques are specific for particular structures like bacterial capsules, flagella, and endospores (**figure 2.14**).

2.4 Electron Microscopy

a. The transmission electron microscope uses magnetic lenses to form an image from electrons that have passed through a very thin section of a specimen (**figure 2.19**). Resolution is high because the wavelength of electrons is very short.

b. Thin section contrast can be increased by treatment with solutions of heavy metals like osmium tetroxide, uranium, and lead.

c. Specimens are also prepared for the TEM by negative staining, shadowing with metal, or freeze-etching.

d. The scanning electron microscope (**figure 2.23**) is used to study external surface features of microorganisms.

2.5 Newer Techniques in Microscopy

a. The confocal scanning laser microscope (**figure 2.25**) is used to study thick, complex specimens.

b. Scanning probe microscopes reach very high magnifications that allow scientists to observe biological molecules (**figures 2.27** and **2.29**).

Key Terms

acidic dyes 26
acid-fast staining 26
atomic force microscope 36
basic dyes 26
bright-field microscope 18
capsule staining 26
chemical fixation 26
chromophore groups 26
confocal scanning laser microscope (CSLM) 34
dark-field microscope 21
differential interference contrast (DIC) microscope 23
differential staining 26
endospore staining 26
eyepieces 18
fixation 25
flagella staining 28
fluorescence microscope 23
fluorescent light 23
fluorochromes 24
focal length 18
focal point 18
freeze-etching 30
Gram stain 26
heat fixation 25
mordant 26
negative staining 26
numerical aperture 19
objective lenses 18
ocular lenses 18
parfocal 18
phase-contrast microscope 21
refraction 17
refractive index 17
resolution 18
scanning electron microscope (SEM) 30
scanning probe microscope 35
scanning tunneling microscope 35
shadowing 29
simple staining 26
substage condenser 18
transmission electron microscope (TEM) 29
working distance 20

Chapter Summaries

- End-of-chapter summaries are organized by numbered headings and provide a snapshot of important chapter concepts.

End-of-Chapter Material

- *Key Terms* highlight chapter terminology and list term location in the chapter.
- *Critical Thinking Questions* supplement the questions for review and reflection found throughout each chapter; they are designed to stimulate analytical problem solving skills.
- *Learn More* includes a short list of recent and relevant papers for the interested student and professor. Additional references can be found at the Prescott website at *www.mhhe.com/prescott7*.

Critical Thinking Questions

1. If you prepared a sample of a specimen for light microscopy, stained with the Gram stain, and failed to see anything when you looked through your light microscope, list the things that you may have done incorrectly.
2. In a journal article, find an example of a light micrograph, a scanning or transmission electron micrograph, or a confocal image. Discuss why the figure was included in the article and why that particular type of microscopy was the method of choice for the research. What other figures would you like to see used in this study? Outline the steps that the investigators would take in order to obtain such photographs or figures.

Learn More

Binning, G., and Rohrer, H. 1985. The scanning tunneling microscope. *Sci. Am.* 253(2):50–56.

Dufrêne, Y. F. 2003. Atomic force microscopy provides a new means for looking at microbial cells. *ASM News* 69(9):438–42.

Hörber, J.K.H., and Miles, M. J. 2003. Scanning probe evolution in biology. *Science* 302:1002–5.

Lillie, R. D. 1969. *H. J. Conn's biological stains*, 8th ed. Baltimore: Williams & Wilkins.

Rochow, T. G. 1994. *Introduction to microscopy by means of light, electrons, X-rays, or acoustics*. New York: Plenum.

Scherrer, Rene. 1984. Gram's staining reaction, Gram types and cell walls of bacteria. *Trends Biochem. Sci.* 9:242–45.

Stephens, D. J., and Allan, V. J. 2003. Light microscopy techniques for live cell imaging. *Science* 300:82–6.

Please visit the Prescott website at www.mhhe.com/prescott7 for additional references.

Student Resources

Student Study Guide

The **Student Study Guide** is a valuable resource that provides learning objectives, study outlines, learning activities, and self-testing material to help students master course content.

Laboratory Exercises in Microbiology

The seventh edition of ***Laboratory Exercises in Microbiology*** by John P. Harley has been prepared to accompany the text. Like the text, the laboratory manual provides a balanced introduction in each area of microbiology. The class-tested exercises are modular and short so that an instructor can easily choose those exercises that fit his or her course.

ARIS

McGraw-Hill's ARIS—Assessment, Review, and Instruction System for *Prescott, Harley, and Klein's Microbiology, www.mhhe.com/prescott7*. This online resource provides helpful study materials that support each chapter in the book. Features include:

Self-quizzes
Animations (with quizzing)
Flashcards
Clinical case studies
Additional course content and more!

Instructor Resources

ARIS (www.mhhe.com/prescott7)

McGraw-Hill's ARIS—Assessment, Review, and Instruction System for *Prescott, Harley, and Klein's Microbiology* is a complete, online tutorial, electronic homework, and course management system. Instructors can create and share course materials and assignments with colleagues with a few clicks of the mouse. All PowerPoint lectures, assignments, quizzes, and tutorials are directly tied to text-specific materials. Instructors can also edit questions, import their own content, and create announcements and due dates for assignments. ARIS has automatic grading and reporting of easy-to-assign homework, quizzing, and testing. All student activity within McGraw-Hill's ARIS is automatically recorded and available to the instructor through a fully integrated grade book that can be downloaded to Excel. Contact your local McGraw-Hill Publisher's representative for more information on getting started with ARIS.

ARIS Presentation Center

Build instructional materials wherever, whenever, and however you want!

ARIS Presentation Center is an online digital library containing assets such as photos, artwork, animations, PowerPoints, and other media types that can be used to create customized lectures, visually enhanced tests and quizzes, compelling course websites, or attractive printed support materials.

Access to your book, access to all books!

The Presentation Center library includes thousands of assets from many McGraw-Hill titles. This ever-growing resource gives instructors the power to utilize assets specific to an adopted textbook as well as content from all other books in the library.

Nothing could be easier!

Accessed from the instructor side of your textbook's ARIS website, Presentation Center's dynamic search engine allows you to explore by discipline, course, textbook chapter, asset type, or keyword. Simply browse, select, and download the files you need to build engaging course materials. All assets are copyright McGraw-Hill Higher Education but can be used by instructors for classroom purposes.

- **Art Library**—Color-enhanced, digital files of all illustrations in the book can be readily incorporated into lecture presentations, exams, or custom-made classroom materials. The large, bolded labels make the images appropriate for use in large lecture halls.
- **TextEdit Art Library**—Every line art piece is placed into a PowerPoint presentation that allows the user to revise, move, or delete labels as desired for creation of customized presentations or for testing purposes.
- **Photo Library**—Like the Art Library, digital files of all the photographs from the book are available.
- **Table Library**—Every table that appears in the book is provided in electronic form.
- **Animations Library**—Full-color presentations involving key process figures in the book have been brought to life via animations. These animations offer flexibility for instructors and were designed to be used in lecture or for self-study. Instructors can pause, rewind, fast forward, and turn audio off/on to create dynamic lecture presentations.
- **PowerPoint Lecture Outlines**—These ready-made presentations combine art and lecture notes for each of the 41 chapters of the book. The presentations can be used as they are, or they can be customized to reflect your preferred lecture topics and organization.
- **PowerPoint Outlines**—The art, photos, and tables for each chapter are inserted into blank PowerPoint presentations to which you can add your own notes.

Instructor Testing and Resource CD-ROM

This cross-platform CD contains the *Instructor's Manual and Test Bank,* both available in Word and PDF formats. The *Instructor's Manual* contains chapter overviews, objectives, and answer guidelines for Critical Thinking Questions. The Test Bank provides questions that can be used for homework assignments or the preparation of exams. The computerized test bank allows the user to quickly create customized exams. This user-friendly program allows instructors to search for questions by topic, format, or difficulty level; edit existing questions or add new ones; and scramble questions and answer keys for multiple versions of the same test.

Transparencies

A set of 250 full-color acetate transparencies is available to supplement classroom lectures. These have been enhanced for projection and are available to adopters of the seventh edition.

Acknowlegments

Focus Group Participants

Jeffrey Isaacson, *Nebraska Wesleyan University*
Janice Knepper, *Villanova University*
Donald Lehman, *University of Delaware*
Susan Lovett, *Brandeis University*
Anne Morris Hooke, *Miami University of Ohio*
Mark Schneegurt, *Wichita State University*
Daniel Smith, *Seattle University*
Michael Troyan, *Pennsylvania State University*
Russell Vreeland, *West Chester University*
Stephen Wagner, *Stephen F. Austin State University*
Darla Wise, *Concord University*

Reviewers

Phillip M. Achey, *University of Florida*
Shivanti Anandan, *Drexel University*
Cynthia Anderson, *Mt. San Antonio College*
Michelle L. Badon, *University of Texas*
Larry L. Barton, *University of New Mexico*
Mary Burke, *Oregon State University*
Frank B. Dazzo, *Michigan State University*
Johnny El-Rady, *University of South Florida*
Paul G. Engelkirk, *Central Texas College*
Robert H. Findlay, *University of Alabama*
Steven Foley, *University of Central Arkansas*
Bernard Lee Frye, *University of Texas at Arlington*
Amy M. Grunden, *North Carolina State University*
Janet L. Haynes, *Long Island University*
Diane Herson, *University of Delaware*
D. Mack Ivey, *University of Arkansas*
Colin R. Jackson, *Southeastern Louisiana University*
Shubha Kale Ireland, *Xavier University of Louisiana*
Judith Kandel, *California State University–Fullerton*
James Kettering, *Loma Linda University*
Madhukar B. Khetmalas, *Texas Tech University*
Peter J. King, *Stephen F. Austin State University*
Tina M. Knox, *University of Illinois–Urbana*
Duncan Krause, *University of Georgia*
Don Lehman, *University of Delaware*
Rita B. Moyes, *Texas A&M University*
Ronald D. Porter, *Pennsylvania State University*
Sabine Rech, *San Jose State University*
Pratibha Saxena, *University of Texas at Austin*
Geoffrey B. Smith, *New Mexico State University*
Fred Stutzenberger, *Clemson University*
Karen Sullivan, *Louisiana State University*
Jennifer R. Walker, *University of Georgia*
William Whalen, *Xavier University of Louisiana*
John M. Zamora, *Middle Tennessee State University*

International Reviewers

Judy Gnarpe, *University of Alberta*
Shaun Heaphy, *University of Leicester*
Edward E. Ishiguro, *University of Victoria*
Kuo-Kau Lee, *National Taiwan Ocean University*
Jong-Kang Liu, *National Sun Yat-sen University*
Cheryl L. Patten, *University of New Brunswick*
Clive Sweet, *University of Birmingham*
Chris Upton, *University of Victoria*
Fanus Venter, *University of Pretonia*
Shang-Shyng Yang, *National Taiwan University*
Guang-yu Zheng, *Beijing Normal University*

As a new group of authors, we encountered a very steep learning curve that we could not have overcome without the help of our editors, Lisa Bruflodt, Jayne Klein, Janice Roerig-Blong, and Colin Wheatley. We would also like to thank our photo editor Mary Reeg and the tremendous talent and patience displayed by the artists. We are also very grateful to Professor Norman Pace for his helpful discussions, and the many reviewers who provided helpful criticism and analysis. Finally, we thank our spouses and children who provided support and tolerated our absences (mental if not physical) while we completed this demanding project.

1 The History and Scope of Microbiology

Louis Pasteur, one of the greatest scientists of the nineteenth century, maintained that "Science knows no country, because knowledge belongs to humanity, and is a torch which illuminates the world."

Preview

- Microbiology is defined not only by the size of its subjects but the techniques it uses to study them.
- Microorganisms include acellular entities (e.g., viruses), procarytic cells, and eucaryotic cells. Cellular microorganisms are found in all three domains of life: *Bacteria, Archaea, Eucarya.*
- The development of microbiology as a scientific discipline has depended on the availability of the microscope and the ability to isolate and grow pure cultures of microorganisms. The development of these techniques in large part grew out of studies disproving the Theory of Spontaneous Generation and others establishing that microorganisms can cause disease.
- Microbiology is a large discipline; it has had and will continue to have a great impact on other areas of biology and general human welfare.

The importance of microorganisms can't be overemphasized. In terms of sheer number and mass—it is estimated that microbes contain 50% of the biological carbon and 90% of the biological nitrogen on Earth—they greatly exceed every other group of organisms on the planet. Furthermore, they are found everywhere: from geothermal vents in the ocean depths to the coldest arctic ice, to every person's skin. They are major contributors to the functioning of the biosphere, being indispensable for the cycling of the elements essential for life. They also are a source of nutrients at the base of all ecological food chains and webs. Most importantly, certain microorganisms carry out photosynthesis, rivaling plants in their role of capturing carbon dioxide and releasing oxygen into the atmosphere. Those microbes that inhabit humans also play important roles, including helping the body digest food and producing vitamins B and K. In addition, society in general benefits from microorganisms, as they are necessary for the production of bread, cheese, beer, antibiotics, vaccines, vitamins, enzymes, and many other important products. Indeed, modern biotechnology rests upon a microbiological foundation.

Although the majority of microorganisms play beneficial or benign roles, some harm humans and have disrupted society over the millennia. Microbial diseases undoubtedly played a major role in historical events such as the decline of the Roman Empire and the conquest of the New World. In 1347, plague or black death, an arthropod-borne disease, struck Europe with brutal force, killing 1/3 of the population (about 25 million people) within four years. Over the next 80 years, the disease struck again and again, eventually wiping out 75% of the European population. The plague's effect was so great that some historians believe it changed European culture and prepared the way for the Renaissance. Today the struggle by microbiologists and others against killers like AIDS and malaria continues.

In this introductory chapter, we introduce the microbial world to provide a general idea of the organisms and agents that microbiologists study. Then we describe the historical development of the science of microbiology and its relationship to medicine and other areas of biology. Finally, we discuss the scope, relevance, and future of modern microbiology.

1.1 Members of the Microbial World

Microbiology often has been defined as the study of organisms and agents too small to be seen clearly by the unaided eye—that is, the study of **microorganisms.** Because objects less than about one millimeter in diameter cannot be seen clearly and must be

Dans les champs de l'observation, le hasard ne favorise que les esprits préparés.
(In the field of observation, chance favors only prepared minds.)

—*Louis Pasteur*

examined with a microscope, microbiology is concerned primarily with organisms and agents this small and smaller. However, some microorganisms, particularly some eucaryotic microbes, are visible without microscopes. For example, bread molds and filamentous algae are studied by microbiologists, yet are visible to the naked eye, as are the two bacteria *Thiomargarita* and *Epulopiscium*. Microbial Diversity & Ecology 3.1: Monstrous Microbes

The difficulty in setting the boundaries of microbiology has led to the suggestion of other criteria for defining the field. For instance, an important characteristic of microorganisms, even those that are large and multicellular, is that they are relatively simple in their construction, lacking highly differentiated cells and distinct tissues. Another suggestion, made by Roger Stanier, is that the field also be defined in terms of its techniques. Microbiologists usually first isolate a specific microorganism from a population and then culture it. Thus microbiology employs techniques—such as sterilization and the use of culture media—that are necessary for successful isolation and growth of microorganisms.

Microorganisms are diverse, and their classification has always been a challenge for microbial taxonomists. Their early descriptions as either plants or animals were too simple. For instance, some microbes are motile like animals, but also have cell walls and are photosynthetic like plants. Such microbes cannot be placed easily into one kingdom or another. Another important factor in classifying microorganisms is that some are composed of procaryotic cells and others of eucaryotic cells. **Procaryotic cells** [Greek *pro,* before, and *karyon,* nut or kernel; organisms with a primordial nucleus] have a much simpler morphology than eucaryotic cells and lack a true membrane-delimited nucleus. In contrast, **eucaryotic cells** [Greek, *eu,* true, and *karyon,* nut or kernel] have a membrane-enclosed nucleus; they are more complex morphologically and are usually larger than procaryotes. These observations eventually led to the development of a classification scheme that divided organisms into five kingdoms: the *Monera, Protista, Fungi, Animalia,* and *Plantae*. Microorganisms (except for viruses, which are acellular and have their own classification system) were placed in the first three kingdoms.

In the last few decades, great progress has been made in three areas that profoundly affect microbial classification. First, much has been learned about the detailed structure of microbial cells from the use of electron microscopy. Second, microbiologists have determined the biochemical and physiological characteristics of many different microorganisms. Third, the sequences of nucleic acids and proteins from a wide variety of organisms have been compared. The comparison of ribosomal RNA (rRNA), begun by Carl Woese in the 1970s, was instrumental in demonstrating that there are two very different groups of procaryotic organisms: *Bacteria* and *Archaea,* which had been classified together as *Monera* in the five-kingdom system. Later, studies based on rRNA comparisons suggested that *Protista* was not a cohesive taxonomic unit and that it should be divided into three or more kingdoms. These studies and others have led many taxonomists to conclude that the five-kingdom system is too simple. A number of alternatives have been suggested, but currently, most

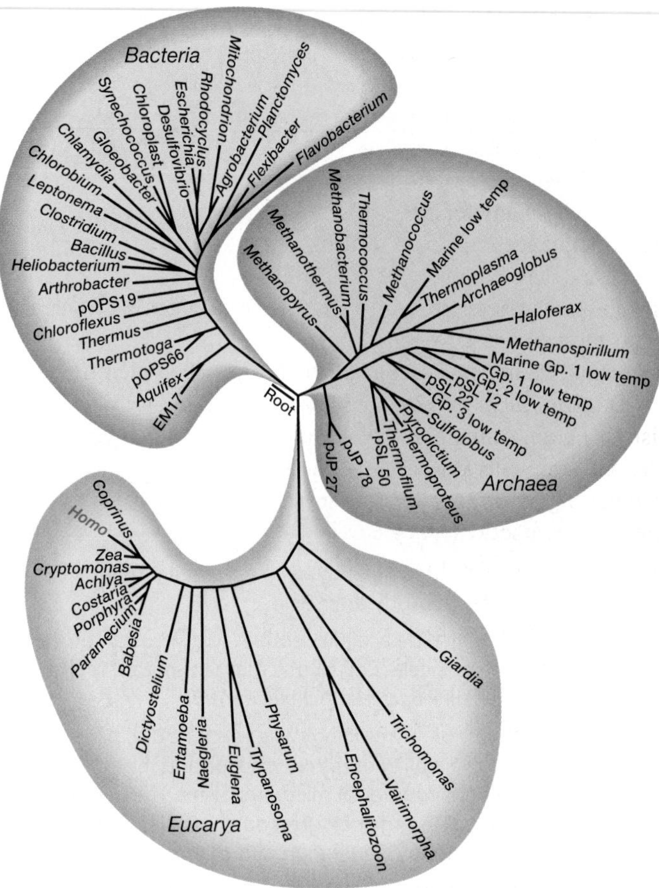

Figure 1.1 Universal Phylogenetic Tree. These evolutionary relationships are based on rRNA sequence comparisons. Man (Homo) is highlighted in red.

microbiologists believe that organisms should be divided among three domains: *Bacteria* (the true bacteria or eubacteria), *Archaea,*[1] and *Eucarya* (all eucaryotic organisms) (**figure 1.1**). This system, which we shall use here, and the results leading to it are discussed in chapter 19. A brief description of the three domains and of the microorganisms placed in them follows.

Bacteria[2] are procaryotes that are usually single-celled organisms. Most have cell walls that contain the structural molecule peptidoglycan. They are abundant in soil, water, and air and are also major inhabitants of our skin, mouth, and intestines. Some bacteria live in environments that have extreme temperatures,

[1] Although this will be discussed further in chapter 19, it should be noted here that several names have been used for the *Archaea*. The two most important are archaeobacteria and archaebacteria. In this text, we shall use only the name *Archaea*.

[2] In this text, the term bacteria (s., bacterium) will be used to refer to procaryotes that belong to domain *Bacteria,* and the term archaea (s., archaeon) will be used to refer to procaryotes that belong to domain *Archaea*. It should be noted that in some publications, the term bacteria is used to refer to all procaryotes. That is not the case in this text.

pH, or salinity. Although some bacteria cause disease, many play more beneficial roles such as cycling elements in the biosphere, breaking down dead plant and animal material, and producing vitamins. Cyanobacteria produce significant amounts of oxygen through the process of photosynthesis.

Archaea are procaryotes that are distinguished from *Bacteria* by many features, most notably their unique ribosomal RNA sequences. They also lack peptidoglycan in their cell walls and have unique membrane lipids. Some have unusual metabolic characteristics, such as the methanogens, which generate methane gas. Many archaea are found in extreme environments. Pathogenic archaea have not yet been identified.

Domain *Eucarya* includes microorganisms classified as protists or *Fungi*. Animals and plants are also placed in this domain. **Protists** are generally larger than procaryotes and include unicellular algae, protozoa, slime molds, and water molds. **Algae** are photosynthetic protists that together with the cyanobacteria produce about 75% of the planet's oxygen. They are also the foundation of aquatic food chains. **Protozoa** are unicellular, animal-like protists that are usually motile. Many free-living protozoa function as the principal hunters and grazers of the microbial world. They obtain nutrients by ingesting organic matter and other microbes. They can be found in many different environments and some are normal inhabitants of the intestinal tracts of animals, where they aid in digestion of complex materials such as cellulose. A few cause disease in humans and other animals. **Slime molds** are protists that are like protozoa in one stage of their life cycle, but are like fungi in another. In the protozoan phase, they hunt for and engulf food particles, consuming decaying vegetation and other microbes. **Water molds,** as their name implies, are found in the surface water of freshwater sources and moist soil. They feed on decaying vegetation such as logs and mulch. Some water molds have produced devastating plant infections, including the Great Potato Famine of 1846–1847. ***Fungi*** are a diverse group of microorganisms that range from unicellular forms (yeasts) to molds and mushrooms. Molds and mushrooms are multicellular fungi that form thin, threadlike structures called hyphae. They absorb nutrients from their environment, including the organic molecules that they use as a source of carbon and energy. Because of their metabolic capabilities, many fungi play beneficial roles, including making bread rise, producing antibiotics, and decomposing dead organisms. Other fungi cause plant diseases and diseases in humans and other animals.

Viruses are acellular entities that must invade a host cell in order to replicate. They are the smallest of all microbes (the smallest is 10,000 times smaller than a typical bacterium), but their small size belies their power—they cause many animal and plant diseases and have caused epidemics that have shaped human history. The diseases they cause include smallpox, rabies, influenza, AIDS, the common cold, and some cancers.

The development of microbiology as a science is described in sections 1.2 to 1.5. **Figure 1.2** presents a summary of some of the major events in this process and their relationship to other historical landmarks.

1. Describe the field of microbiology in terms of the size of its subject material and the nature of its techniques.
2. Describe and contrast procaryotic and eucaryotic cells.
3. Describe and contrast the five-kingdom classification system with the three-domain system. Why do you think viruses are not included in either system?

1.2 The Discovery of Microorganisms

Even before microorganisms were seen, some investigators suspected their existence and responsibility for disease. Among others, the Roman philosopher Lucretius (about 98–55 B.C.) and the physician Girolamo Fracastoro (1478–1553) suggested that disease was caused by invisible living creatures. The earliest microscopic observations appear to have been made between 1625 and 1630 on bees and weevils by the Italian Francesco Stelluti, using a microscope probably supplied by Galileo. In 1665, the first drawing of a microorganism was published in Robert Hooke's *Micrographia*. However, the first person to publish extensive, accurate observations of microorganisms was the amateur microscopist Antony van Leeuwenhoek (1632–1723) of Delft, The Netherlands (**figure 1.3*a***). Leeuwenhoek earned his living as a draper and haberdasher (a dealer in men's clothing and accessories), but spent much of his spare time constructing simple microscopes composed of double convex glass lenses held between two silver plates (figure 1.3*b*). His microscopes could magnify around 50 to 300 times, and he may have illuminated his liquid specimens by placing them between two pieces of glass and shining light on them at a 45° angle to the specimen plane. This would have provided a form of dark-field illumination in which the organisms appeared as bright objects against a dark background and made bacteria clearly visible (figure 1.3*c*). Beginning in 1673, Leeuwenhoek sent detailed letters describing his discoveries to the Royal Society of London. It is clear from his descriptions that he saw both bacteria and protozoa.

As important as Leeuwenhoek's observations were, the development of microbiology essentially languished for the next 200 years. Little progress was made primarily because microscopic observations of microorganisms do not provide sufficient information to understand their biology. For the discipline to develop, techniques for isolating and culturing microbes in the laboratory were needed. Many of these techniques began to be developed as scientists grappled with the conflict over the Theory of Spontaneous Generation. This conflict and the subsequent studies on the role played by microorganisms in causing disease ultimately led to what is now called the Golden Age of Microbiology.

1. Give some examples of the kind of information you think can be provided by microscopic observations of microorganisms.
2. Give some examples of the kind of information you think can be provided by isolating microorganisms from their natural environment and culturing them in the laboratory.

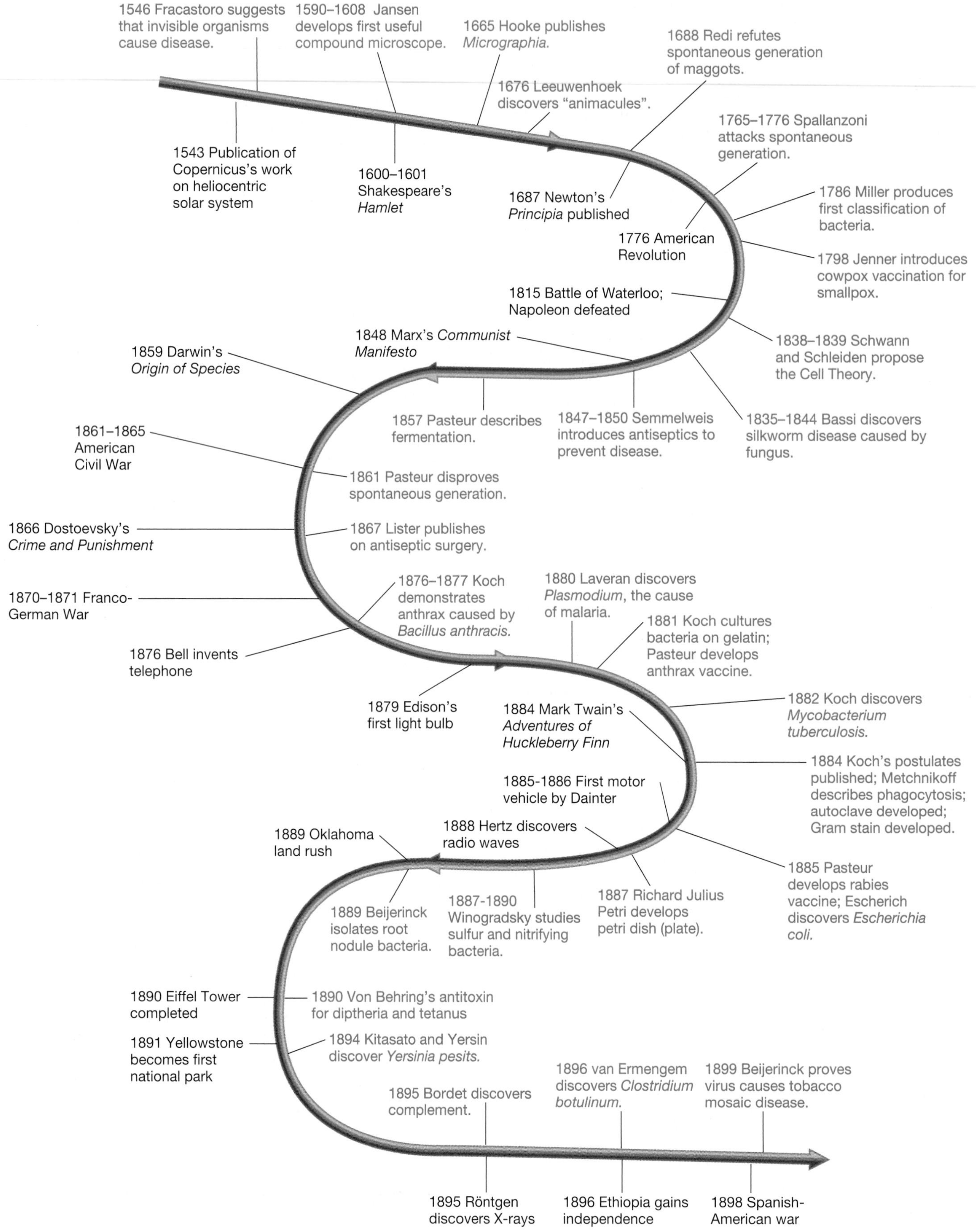

Figure 1.2(a) Some Important Events in the Development of Microbiology (1546–1899). Milestones in microbiology are marked in red; other historical events are in black.

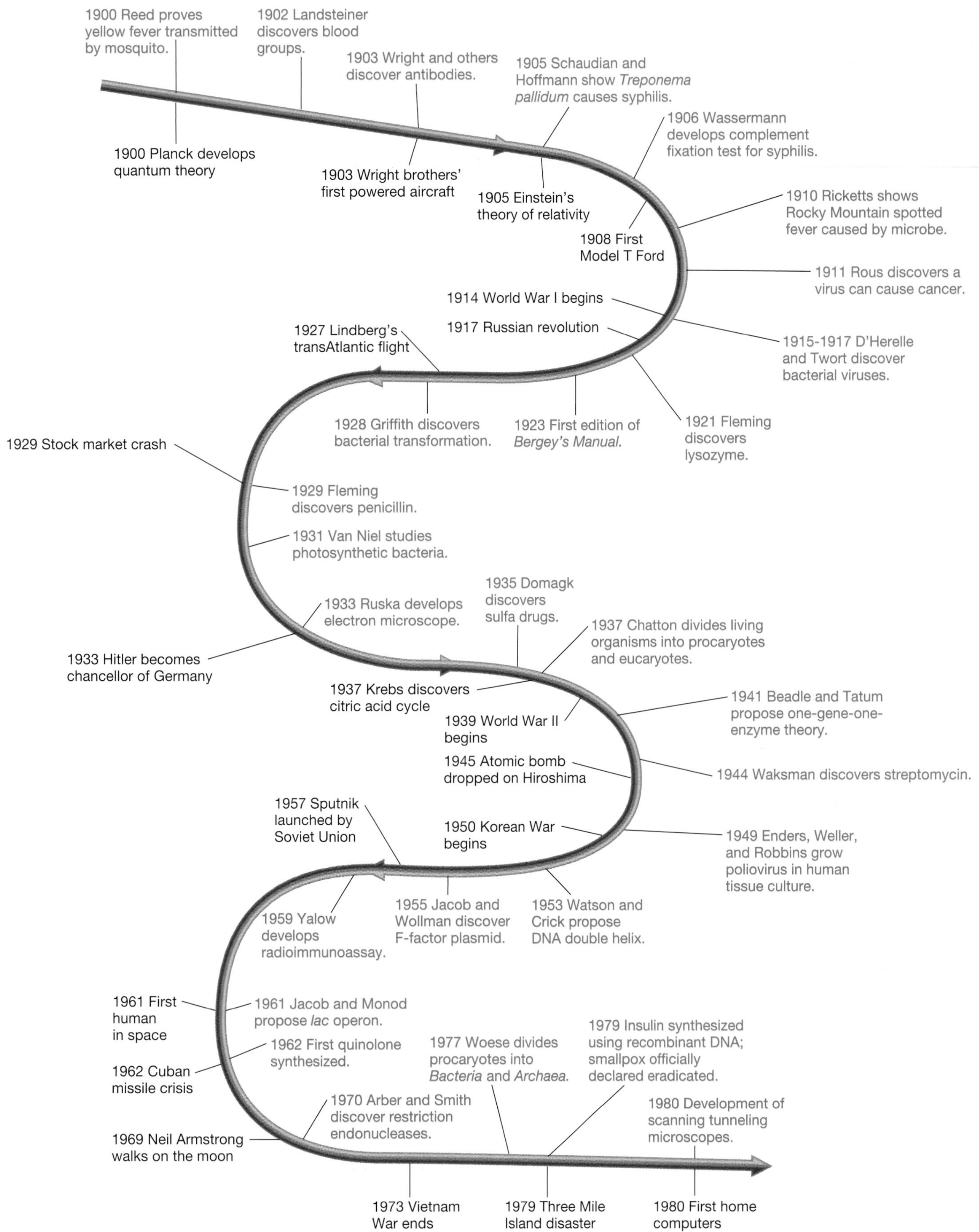

Figure 1.2(b) Some Important Events in the Development of Microbiology (1900–1980). Milestones in microbiology are marked in red; other historical events are in black.

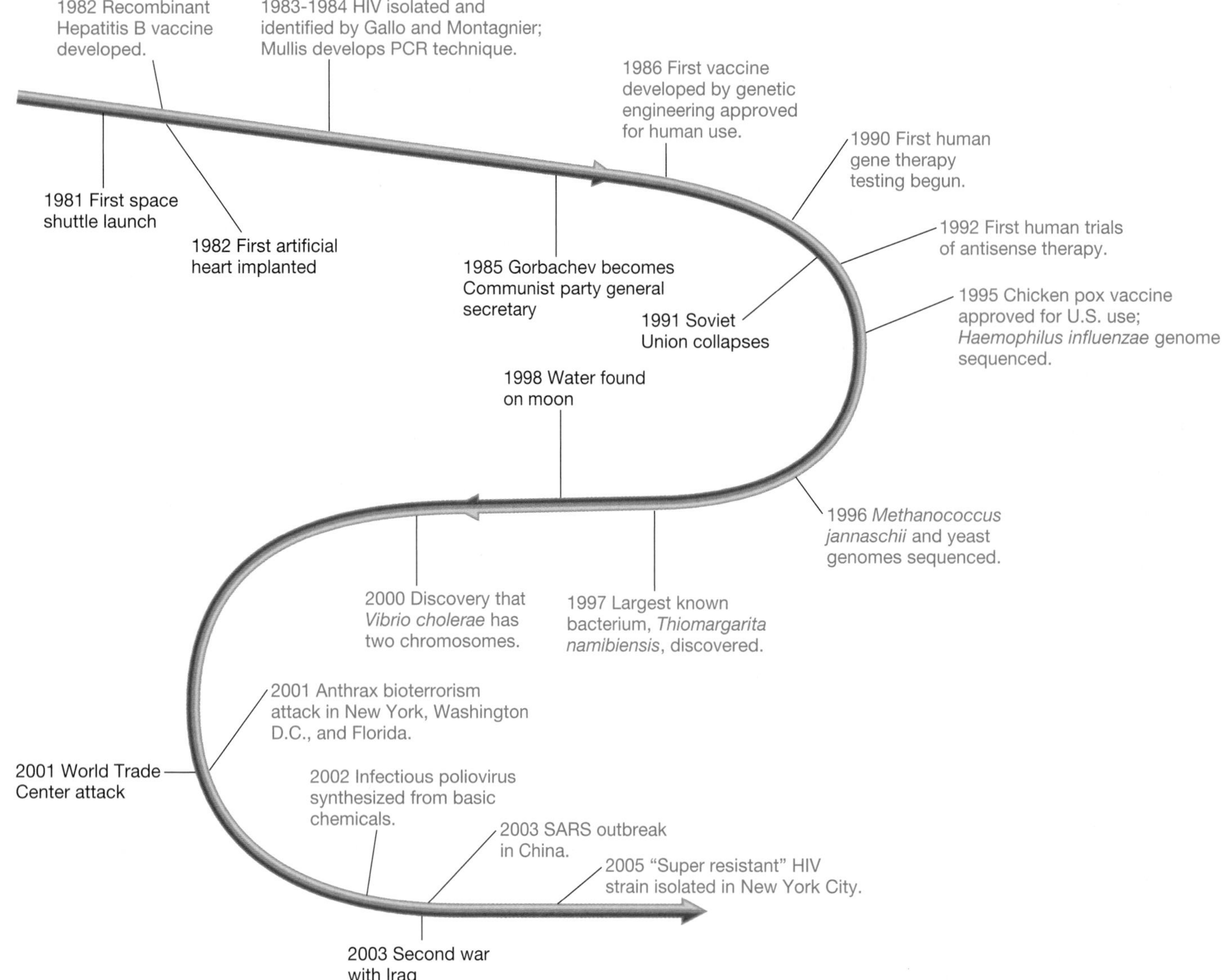

Figure 1.2(c) Some Important Events in the Development of Microbiology (1981–2005). Milestones in microbiology are marked in red; other historical events are in black.

1.3 The Conflict Over Spontaneous Generation

From earliest times, people had believed in **spontaneous generation**—that living organisms could develop from nonliving matter. Even Aristotle (384–322 B.C.) thought some of the simpler invertebrates could arise by spontaneous generation. This view finally was challenged by the Italian physician Francesco Redi (1626–1697), who carried out a series of experiments on decaying meat and its ability to produce maggots spontaneously. Redi placed meat in three containers. One was uncovered, a second was covered with paper, and the third was covered with a fine gauze that would exclude flies. Flies laid their eggs on the uncovered meat and maggots developed. The other two pieces of meat did not produce maggots spontaneously. However, flies were attracted to the gauze-covered container and laid their eggs on the gauze; these eggs produced maggots. Thus the generation of maggots by decaying meat resulted from the presence of fly eggs, and meat did not spontaneously generate maggots as previously believed. Similar experiments by others helped discredit the theory for larger organisms.

Leeuwenhoek's discovery of microorganisms renewed the controversy. Some proposed that microorganisms arose by spontaneous generation even though larger organisms did not. They pointed out that boiled extracts of hay or meat would give rise to microorganisms after sitting for a while. In 1748, the English priest John Needham (1713–1781) reported the results of his experiments on spontaneous generation. Needham boiled mutton broth and then tightly stoppered the flasks. Eventually many of the flasks became cloudy and contained microorganisms. He thought organic matter contained a vital force that could confer the properties of life on nonliving matter. A few years later, the Italian priest and naturalist Lazzaro Spallanzani (1729–1799) improved on Needham's experimental design by first sealing glass flasks that contained water and seeds. If the sealed flasks were placed in boiling water for 3/4 of an hour, no growth took place as long as the flasks remained sealed. He proposed that air carried germs to the culture medium, but also commented that the external air might be required for growth of animals already in the medium. The supporters of spontaneous generation maintained that heating the air in sealed flasks destroyed its ability to support life.

Several investigators attempted to counter such arguments. Theodore Schwann (1810–1882) allowed air to enter a flask containing a sterile nutrient solution after the air had passed through a red-hot tube. The flask remained sterile. Subsequently Georg Friedrich Schroder and Theodor von Dusch allowed air to enter a flask of heat-sterilized medium after it had passed through sterile cotton wool. No growth occurred in the medium even though the air had not been heated. Despite these experiments the French naturalist Felix Pouchet claimed in 1859 to have carried out experiments conclusively proving that microbial growth could occur without air contamination. This claim provoked Louis Pasteur (1822–1895) to settle the matter once and for all. Pasteur (**figure 1.4**) first filtered air through cotton and found that objects resembling plant spores had been trapped. If a piece of the cotton was placed in sterile medium after air had been filtered through it, microbial growth occurred. Next he placed nutrient solutions in flasks, heated their necks in a flame, and drew them out into a variety of curves, while keeping the ends of the necks open to the atmosphere (**figure 1.5**). Pasteur then boiled the solutions for a few minutes and allowed them to cool. No growth took place even though the contents of the flasks were exposed to the air. Pasteur pointed out that no growth occurred because dust and germs had been trapped on the walls of the curved necks. If the necks were broken, growth commenced immediately. Pasteur had not only resolved the controversy by 1861 but also had shown how to keep solutions sterile.

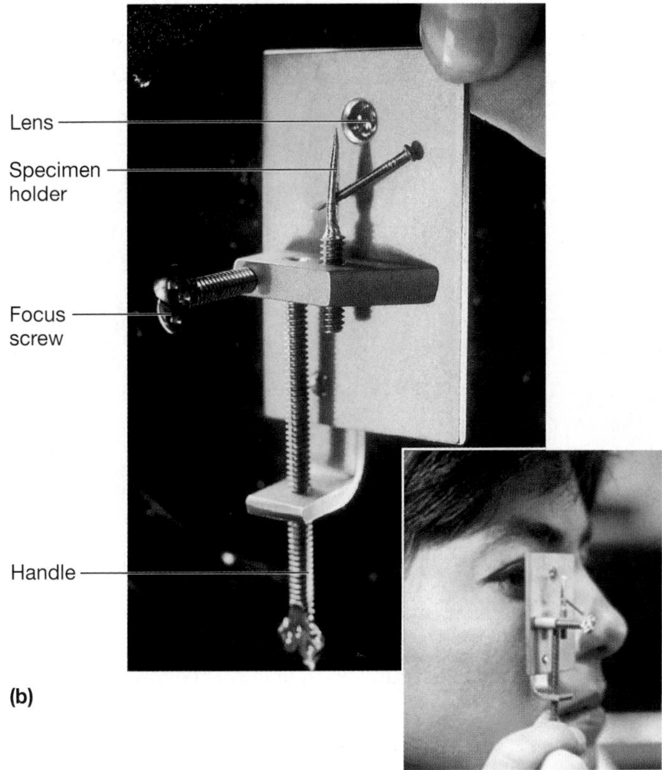

(b)

(a)

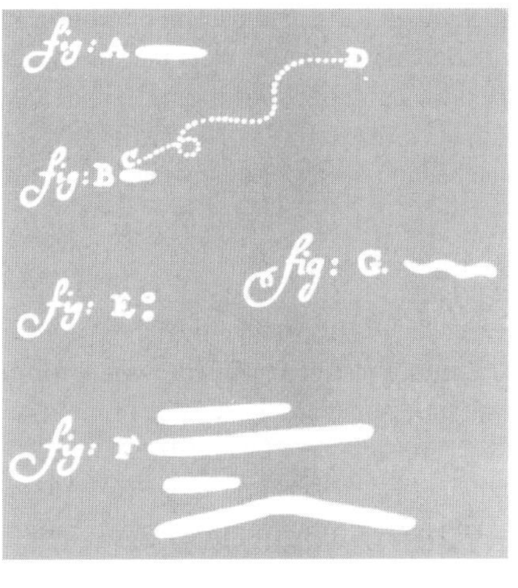

(c)

Figure 1.3 Antony van Leeuwenhoek. **(a)** An oil painting of Leeuwenhoek (1632–1723). **(b)** A brass replica of the Leeuwenhoek microscope. Inset photo shows how it is held. **(c)** Leeuwenhoek's drawings of bacteria from the human mouth.

Figure 1.4 Louis Pasteur. Pasteur (1822–1895) working in his laboratory.

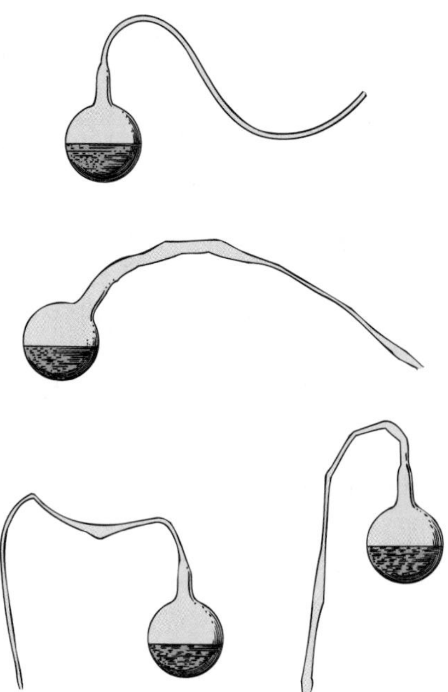

Figure 1.5 The Spontaneous Generation Experiment. Pasteur's swan neck flasks used in his experiments on the spontaneous generation of microorganisms. *Source: Annales Sciences Naturelle, 4th Series, Vol. 16, pp. 1–98, Pasteur, L., 1861, "Mémoire sur les Corpuscules Organisés Qui Existent Dans L'Atmosphère: Examen de la Doctrine des Générations Spontanées."*

The English physicist John Tyndall (1820–1893) dealt a final blow to spontaneous generation in 1877 by demonstrating that dust did indeed carry germs and that if dust was absent, broth remained sterile even if directly exposed to air. During the course of his studies, Tyndall provided evidence for the existence of exceptionally heat-resistant forms of bacteria. Working independently, the German botanist Ferdinand Cohn (1828–1898) discovered the existence of heat-resistant bacterial endospores. The bacterial endospore (section 3.11)

1. How did Pasteur and Tyndall finally settle the spontaneous generation controversy?
2. Why was the belief in spontaneous generation an obstacle to the development of microbiology as a scientific discipline?

1.4 The Golden Age of Microbiology

Pasteur's work with swan neck flasks ushered in the Golden Age of Microbiology. Within 60 years (1857–1914), a number of disease-causing microbes were discovered, great strides in understanding microbial metabolism were made, and techniques for isolating and characterizing microbes were improved. Scientists also identified the role of immunity in preventing disease and controlling microbes, developed vaccines, and introduced techniques used to prevent infection during surgery.

Recognition of the Relationship between Microorganisms and Disease

Although Fracastoro and a few others had suggested that invisible organisms produced disease, most believed that disease was due to causes such as supernatural forces, poisonous vapors called miasmas, and imbalances among the four humors thought to be present in the body. The role of the four humors (blood, phlegm, yellow bile [choler], and black bile [melancholy]) in disease had been widely accepted since the time of the Greek physician Galen (129–199). Support for the idea that microorganisms cause disease—that is, the germ theory of disease—began to accumulate in the early nineteenth century. Agostino Bassi (1773–1856) first showed a microorganism could cause disease when he demonstrated in 1835 that a silkworm disease was due to a fungal infection. He also suggested that many diseases were due to microbial infections. In 1845, M. J. Berkeley proved that the great Potato Blight of Ireland was caused by a water mold, and in 1853, Heinrich de Bary showed that smut and rust fungi caused cereal crop diseases. Following his successes with the study of fermentation, Pasteur was asked by the French government to investigate the pèbrine disease of silkworms that was disrupting the silk industry. After several years of work, he showed that the disease was due to a protozoan parasite. The disease was controlled by raising caterpillars from eggs produced by healthy moths.

Indirect evidence for the germ theory of disease came from the work of the English surgeon Joseph Lister (1827–1912) on the prevention of wound infections. Lister, impressed with Pasteur's studies on the involvement of microorganisms in fermentation and putrefaction, developed a system of antiseptic surgery designed to prevent microorganisms from entering wounds. Instruments were heat sterilized, and phenol was used on surgical dressings and at times sprayed over the surgical area. The approach was remarkably successful and transformed surgery after Lister published his findings in 1867. It also provided strong indirect evidence for the role of microorganisms in disease because phenol, which kills bacteria, also prevented wound infections.

Koch's Postulates

The first direct demonstration of the role of bacteria in causing disease came from the study of anthrax by the German physician Robert Koch (1843–1910). Koch (**figure 1.6**) used the criteria proposed by his former teacher, Jacob Henle (1809–1885), to establish the relationship between *Bacillus anthracis* and anthrax, and published his findings in 1876 (**Techniques & Applications 1.1** briefly discusses the scientific method). Koch injected healthy mice with material from diseased animals, and the mice became ill. After transferring anthrax by inoculation through a series of 20 mice, he incubated a piece of spleen containing the anthrax bacillus in beef serum. The bacilli grew, reproduced, and produced endospores. When the isolated bacilli or their spores were injected into mice, anthrax developed. His criteria for proving the causal relationship between a microorganism and a specific disease are known as **Koch's postulates** (**table 1.1**). Koch's proof that *B. anthracis* caused anthrax was independently confirmed by Pasteur and his coworkers. They discovered that after burial of dead animals, anthrax spores survived and were brought to the surface by earthworms. Healthy animals then ingested the spores and became ill.

Although Koch used the general approach described in the postulates during his anthrax studies, he did not outline them fully until his work on the cause of tuberculosis (table 1.1). In 1884, he reported that this disease was caused by a rod-shaped bacterium, *Mycobacterium tuberculosis;* he was awarded the Nobel Prize in Physiology or Medicine in 1905 for his work. Koch's postulates quickly became the cornerstone of connecting many diseases to their causative agent. However, their use is at times not feasible (**Disease 1.2**). For instance, some organisms, like *Mycobacterium leprae,* the causative agent of leprosy, cannot be isolated in pure culture.

Figure 1.6 Robert Koch. Koch (1843–1910) examining a specimen in his laboratory.

The Development of Techniques for Studying Microbial Pathogens

During Koch's studies on bacterial diseases, it became necessary to isolate suspected bacterial pathogens in pure culture—a culture containing only one type of microorganism. At first Koch cultured bacteria on the sterile surfaces of cut, boiled potatoes, but this was unsatisfactory because the bacteria would not always grow well. Eventually he developed culture media using meat extracts and protein digests because of their similarity to body fluids. He first tried to solidify the media by adding gelatin. Separate bacterial colonies developed after the surface of the solidified medium had been streaked with a bacterial sample. The sample could also be mixed with liquefied gelatin medium.

Table 1.1 Koch's Application of His Postulates to Demonstrate that *Mycobacterium tuberculosis* is the Causative Agent of Tuberculosis.

Postulate	Experimentation
1. The microorganism must be present in every case of the disease but absent from healthy organisms.	Koch developed a staining technique to examine human tissue. *M. tuberculosis* cells could be identified in diseased tissue.
2. The suspected microorganisms must be isolated and grown in a pure culture.	Koch grew *M. tuberculosis* in pure culture on coagulated blood serum.
3. The same disease must result when the isolated microorganism is inoculated into a healthy host.	Koch injected cells from the pure culture of *M.tuberculosis* into guinea pigs. The guinea pigs subsequently died of tuberculosis.
4. The same microorganism must be isolated again from the diseased host.	Koch isolated *M. tuberculosis* from the dead guinea pigs and was able to again culture the microbe in pure culture on coagulated blood serum.

Techniques & Applications

1.1 The Scientific Method

Although biologists employ a variety of approaches in conducting research, microbiologists and other experimentally oriented biologists often use the general approach known as the scientific method. They first gather observations of the process to be studied and then develop a tentative hypothesis—an educated guess—to explain the observations (see **Box figure**). This step often is inductive and creative because there is no detailed, automatic technique for generating hypotheses. Next they decide what information is required to test the hypothesis and collect this information through observation or carefully designed experiments. After the information has been collected, they decide whether the hypothesis has been supported or falsified. If it has failed to pass the test, the hypothesis is rejected, and a new explanation or hypothesis is constructed. If the hypothesis passes the test, it is subjected to more severe testing. The procedure often is made more efficient by constructing and testing alternative hypotheses and then refining the hypothesis that survives testing. This general approach is often called the hypothetico-deductive method. One deduces predictions from the currently accepted hypothesis and tests them. In deduction the conclusion about specific cases follows logically from a general premise ("if . . ., then . . ." reasoning). Induction is the opposite. A general conclusion is reached after considering many specific examples. Both types of reasoning are used by scientists.

When carrying out an experiment, it is essential to use a control group as well as an experimental group. The control group is treated precisely the same as the experimental group except that the experimental manipulation is not performed on it. In this way one can be sure that any changes in the experimental group are due to the experimental manipulation rather than to some other factor not taken into account.

If a hypothesis continues to survive testing, it may be accepted as a valid theory. A theory is a set of propositions and concepts that provides a reliable, systematic, and rigorous account of an aspect of nature. It is important to note that hypotheses and theories are never absolutely proven. Scientists simply gain more and more confidence in their accuracy as they continue to survive testing, fit with new observations and experiments, and satisfactorily explain the observed phenomena. Ultimately, if the support for a hypothesis or theory becomes very strong, it is considered to be a scientific law. Examples include the laws of thermodynamics discussed in section 8.3.

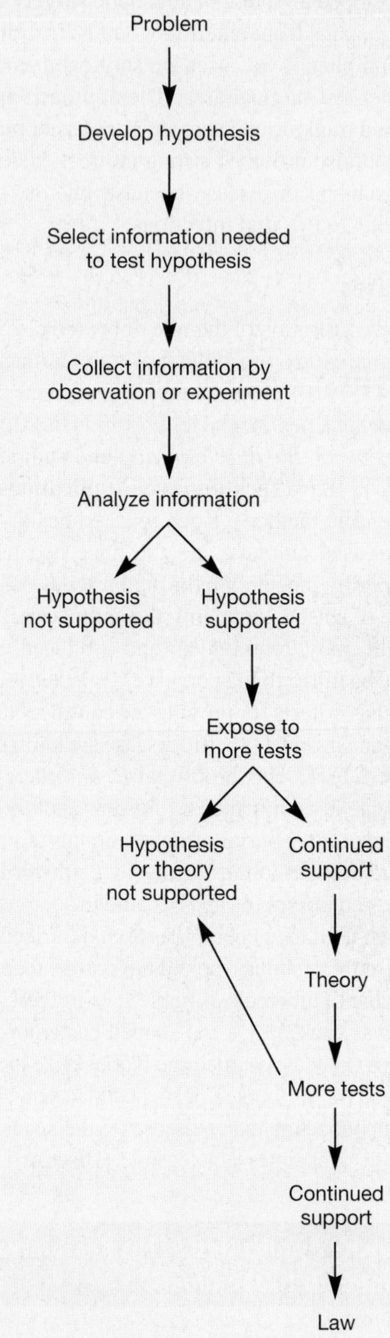

When the gelatin medium hardened, individual bacteria produced separate colonies. Despite its advantages, gelatin was not an ideal solidifying agent because it can be digested by many bacteria and melts at temperatures above 28°C. A better alternative was provided by Fannie Eilshemius Hesse, the wife of Walther Hesse, one of Koch's assistants (**figure 1.7**). She suggested the use of agar as a solidifying agent—she had been using it successfully to make jellies for some time. Agar was not attacked by most bacteria and did not melt until reaching a temperature of 100°C. Furthermore, once melted, it did not solidify until it reached a temperature of 50°C, eliminating the need to handle boiling liquid and providing time for manipulation of the medium. Some of the media developed by Koch and his associates, such as nutrient broth and nutrient agar, are still widely

Disease

1.2 Koch's Molecular Postulates

Although the criteria that Koch developed for proving a causal relationship between a microorganism and a specific disease have been of great importance in medical microbiology, it is not always possible to apply them in studying human diseases. For example, some pathogens cannot be grown in pure culture outside the host; because other pathogens grow only in humans, their study would require experimentation on people. The identification, isolation, and cloning of genes responsible for pathogen virulence have made possible a new molecular form of Koch's postulates that resolves some of these difficulties. The emphasis is on the virulence genes present in the infectious agent rather than on the agent itself. The molecular postulates can be briefly summarized as follows:

1. The virulence trait under study should be associated much more with pathogenic strains of the species than with nonpathogenic strains.
2. Inactivation of the gene or genes associated with the suspected virulence trait should substantially decrease pathogenicity.
3. Replacement of the mutated gene with the normal wild-type gene should fully restore pathogenicity.
4. The gene should be expressed at some point during the infection and disease process.
5. Antibodies or immune system cells directed against the gene products should protect the host.

The molecular approach cannot always be applied because of problems such as the lack of an appropriate animal system. It also is difficult to employ the molecular postulates when the pathogen is not well characterized genetically.

Figure 1.7 Fannie Eilshemius (1850–1934) and Walther Hesse (1846–1911). Fannie Hesse suggested to her husband Walther (a physican and bacteriologist) that he should try using agar in his culture medium when more typical media failed to meet his needs.

used. Another important tool developed in Koch's laboratory was a container for holding solidified media—the petri dish (plate), named after Richard Petri, who devised it. These developments directly stimulated progress in all areas of bacteriology. Culture media (section 5.7); Isolation of pure cultures (section 5.8)

Viral pathogens were also studied during this time. The discovery of viruses and their role in disease was made possible when Charles Chamberland (1851–1908), one of Pasteur's associates, constructed a porcelain bacterial filter in 1884. Dimitri Ivanowski and Martinus Beijerinck (pronounced "by-a-rink") used the filter to study tobacco mosaic disease. They found that plant extracts and sap from diseased plants were infectious, even after being filtered with Chamberland's filter. Because the infectious agent passed through a filter that was designed to trap bacterial cells, the agent must be something smaller than a bacterium. Beijerinck proposed that the agent was a "filterable virus." Eventually viruses were shown to be tiny, acellular infectious agents. Early development of virology (section 16.1)

Immunological Studies

In this period progress also was made in determining how animals resisted disease and in developing techniques for protecting humans and livestock against pathogens. During studies on chicken cholera, Pasteur and Roux discovered that incubating their cultures for long intervals between transfers would attenuate the bacteria, which meant they had lost their ability to cause the disease. If the chickens were injected with these attenuated cultures, they remained healthy but developed the ability to resist the disease. He called the attenuated culture a *vaccine* [Latin *vacca,* cow] in honor of Edward Jenner because, many years earlier, Jenner had used material from cowpox lesions to protect people against smallpox. Shortly after this, Pasteur and Chamberland developed an attenuated anthrax vaccine in two ways: by treating cultures with potassium bichromate and by incubating the bacteria at 42 to 43°C. Control of epidemics: Vaccines and immunizations (section 36.8)

Pasteur next prepared rabies vaccine by a different approach. The pathogen was attenuated by growing it in an abnormal host, the rabbit. After infected rabbits had died, their brains and spinal cords were removed and dried. During the course of these studies, Joseph Meister, a nine-year-old boy who had been bitten by a rabid dog, was brought to Pasteur. Since the boy's death was certain in the absence of treatment, Pasteur agreed to try vaccination. Joseph was injected 13 times over the next 10 days with increasingly virulent preparations of the attenuated virus. He survived.

In gratitude for Pasteur's development of vaccines, people from around the world contributed to the construction of the

Pasteur Institute in Paris, France. One of the initial tasks of the Institute was vaccine production.

After the discovery that the diphtheria bacillus produced a toxin, Emil von Behring (1854–1917) and Shibasaburo Kitasato (1852–1931) injected inactivated toxin into rabbits, inducing them to produce an antitoxin, a substance in the blood that would inactivate the toxin and protect against the disease. A tetanus antitoxin was then prepared and both antitoxins were used in the treatment of people.

The antitoxin work provided evidence that immunity could result from soluble substances in the blood, now known to be antibodies (humoral immunity). It became clear that blood cells were also important in immunity (cellular immunity) when Elie Metchnikoff (1845–1916) discovered that some blood leukocytes could engulf disease-causing bacteria (**figure 1.8**). He called these cells phagocytes and the process phagocytosis [Greek *phagein,* eating].

Figure 1.8 Elie Metchnikoff. Metchnikoff (1845–1916) shown here at work in his laboratory.

1. Discuss the contributions of Lister, Pasteur, and Koch to the germ theory of disease and to the treatment or prevention of diseases.
2. What other contributions did Koch make to microbiology?
3. Describe Koch's postulates. What is a pure culture? Why are pure cultures important to Koch's postulates?
4. Would microbiology have developed more slowly if Fannie Hesse had not suggested the use of agar? Give your reasoning.
5. What are Koch's molecular postulates? Why are they important?
6. Some individuals can be infected by a pathogen yet not develop disease. In fact, some become chronic carriers of the pathogen. How does this observation impact Koch's postulates? How might the postulates be modified to account for the existence of chronic carriers?
7. Describe the scientific method in your own words. How does a theory differ from a hypothesis? Why is it important to have a control group?
8. How did von Behring and Metchnikoff contribute to the development of immunology?

1.5 The Development of Industrial Microbiology and Microbial Ecology

Although humans had unknowingly exploited microbes for thousands of years, industrial microbiology developed in large part from the work of Louis Pasteur and others on the alcoholic fermentations that yielded wine and other alcoholic beverages. In 1837, when Theodore Schwann and others proposed that yeast cells were responsible for the conversion of sugars to alcohol, the leading chemists of the time believed microorganisms were not involved. They were convinced that fermentation was due to a chemical instability that degraded the sugars to alcohol. Pasteur did not agree; he believed that fermentations were carried out by living organisms. In 1856 M. Bigo, an industrialist in Lille, France, where Pasteur worked, requested Pasteur's assistance. His business produced ethanol from the fermentation of beet sugars, and the alcohol yields had recently declined and the product had become sour. Pasteur discovered that the fermentation was failing because the yeast normally responsible for alcohol formation had been replaced by microorganisms that produced lactic acid rather than ethanol. In solving this practical problem, Pasteur demonstrated that all fermentations were due to the activities of specific yeasts and bacteria, and he published several papers on fermentation between 1857 and 1860. His success led to a study of wine diseases and the development of pasteurization to preserve wine during storage. Pasteur's studies on fermentation continued for almost 20 years. One of his most important discoveries was that some fermentative microorganisms were anaerobic and could live only in the absence of oxygen, whereas others were able to live either aerobically or anaerobically. Controlling food spoilage (section 40.3)

Microbial ecology developed when a few of the early microbiologists chose to investigate the ecological role of microorganisms. In particular they studied microbial involvement in the carbon, nitrogen, and sulfur cycles taking place in soil and aquatic habitats. The Russian microbiologist Sergei Winogradsky (1856–1953) made many contributions to soil microbiology. He discovered that soil bacteria could oxidize iron, sulfur, and ammonia to obtain energy, and that many bacteria could incorporate CO_2 into organic matter much like photosynthetic organisms do. Winogradsky also isolated anaerobic nitrogen-fixing soil bacteria and studied the decomposition of cellulose. Martinus Beijerinck (1851–1931) was one of the great general microbiologists who made fundamental contributions to microbial ecology and many other fields. He isolated the aerobic nitrogen-fixing bacterium *Azotobacter,* a root nodule bacterium also capable of fixing nitrogen (later named *Rhizobium*), and sulfate-reducing bacteria. Beijerinck and Winogradsky also developed the enrichment-culture technique and the use of selective media, which have been of such great importance in microbiology. Biogeochemical cycling (section 27.2); Culture media (section 5.7)

1. Briefly describe Pasteur's work on microbial fermentations.
2. How did Winogradsky and Beijerinck contribute to the study of microbial ecology?

3. Leeuwenhoek is often referred to as the Father of Microbiology. However, many historians feel that Louis Pasteur, Robert Koch, or perhaps both, deserve that honor. Who do you think is the Father of Microbiology? Why?
4. Consider the discoveries described in sections 1.2 to 1.5. Which do you think were the most important to the development of microbiology? Why?

1.6 The Scope and Relevance of Microbiology

As the late scientist-writer Steven Jay Gould emphasized, we live in the Age of *Bacteria*. They were the first living organisms on our planet and live virtually everywhere life is possible. Furthermore, the whole biosphere depends on their activities, and they influence human society in countless ways. Because microorganisms play such diverse roles, modern microbiology is a large discipline with many different specialties; it has a great impact on fields such as medicine, agricultural and food sciences, ecology, genetics, biochemistry, and molecular biology. One indication of the importance of microbiology is the Nobel Prize given for work in physiology or medicine. About one-third of these have been awarded to scientists working on microbiological problems (*see inside front cover*).

Microbiology has both basic and applied aspects (**figure 1.9**). The basic aspects are concerned with the biology of microorganisms themselves and include such fields as bacteriology, virology, mycology (study of fungi), phycology or algology (study of algae), protozoology, microbial cytology and physiology, microbial genetics and molecular biology, microbial ecology, and microbial taxonomy. The applied aspects are concerned with practical problems such as disease, water and wastewater treatment, food spoilage and food production, and industrial uses of microbes. It is important to note that the basic and applied aspects of microbiology are intertwined. Basic research is often conducted in applied fields and applications often arise out of basic research. A discussion of some of the major fields of microbiology and the occupations they provide follows.

One of the most active and important fields in microbiology is medical microbiology, which deals with diseases of humans and animals. Medical microbiologists identify the agents causing infectious diseases and plan measures for their control and elimination. Frequently they are involved in tracking down new, unidentified pathogens such as the agent that causes variant Creutzfeldt-Jakob disease, (the human version of "mad cow disease") the hantavirus, the West Nile virus, and the virus responsible for SARS. These microbiologists also study the ways in which microorganisms cause disease. Arthropod-borne viral diseases (section 37.2); Microbial Diversity & Ecology 18.1: SARS: Evolution of a virus

Public health microbiology is closely related to medical microbiology. Public health microbiologists try to identify and control the spread of communicable diseases. They often monitor community food establishments and water supplies in an attempt to keep them safe and free from infectious disease agents.

Immunology is concerned with how the immune system protects the body from pathogens and the response of infectious agents. It is one of the fastest growing areas in science; for example, techniques for the production and use of monoclonal antibodies have developed extremely rapidly. Immunology also deals with practical health problems such as the nature and treatment of allergies and autoimmune diseases like rheumatoid arthritis. Techniques & Applications 32.2: Monoclonal Antibody Technology

Agricultural microbiology is concerned with the impact of microorganisms on agriculture. Agricultural microbiologists try to combat plant diseases that attack important food crops, work on methods to increase soil fertility and crop yields, and study the role of microorganisms living in the digestive tracts of ruminants such as cattle. Currently there is great interest in using bacterial and viral insect pathogens as substitutes for chemical pesticides.

Microbial ecology is concerned with the relationships between microorganisms and the components of their living and nonliving habitats. Microbial ecologists study the global and local contributions of microorganisms to the carbon, nitrogen, and sulfur cycles. The study of pollution effects on microorganisms also is important because of the impact these organisms have on the environment. Microbial ecologists are employing microorganisms in bioremediation to reduce pollution.

Scientists working in food and dairy microbiology try to prevent microbial spoilage of food and the transmission of food-borne diseases such as botulism and salmonellosis. They also use microorganisms to make foods such as cheeses, yogurts, pickles, and beer. In the future, microorganisms themselves may become a more important nutrient source for livestock and humans. Microbiology of food (chapter 40)

In 1929, Alexander Fleming discovered that the fungus *Penicillium* produced what he called penicillin, the first antibiotic that could successfully control bacterial infections. Although it took World War II for scientists to learn how to mass produce it, scientists soon found other microorganisms capable of producing additional antibiotics as well as compounds such as citric acid, vitamin B_{12}, and monosodium glutamate. Today, industrial microbiologists use microorganisms to make products such as antibiotics, vaccines, steroids, alcohols and other solvents, vitamins, amino acids, and enzymes. Industrial microbiologists identify microbes of use to industry. They also engineer microbes with desirable traits and devise systems for culturing them and isolating the products they make.

Microbiologists working in microbial physiology and biochemistry study many aspects of the biology of microorganisms. They may study the synthesis of antibiotics and toxins, microbial energy production, the ways in which microorganisms survive harsh environmental conditions, microbial nitrogen fixation, and the effects of chemical and physical agents on microbial growth and survival.

Microbial genetics and molecular biology focus on the nature of genetic information and how it regulates the development and function of cells and organisms. The use of microorganisms has been very helpful in understanding gene structure and function. Microbial geneticists play an important role in applied microbiology because they develop techniques that are useful in agricultural microbiology, industrial microbiology, food and dairy microbiology, and medicine.

1. Briefly describe the major subdisciplines in microbiology.
2. Why do you think microorganisms are so useful to biologists as experimental models?
3. List all the activities or businesses you can think of in your community that are directly dependent on microbiology.

(a) Rita Colwell
(b) R. G. E. Murray
(c) Stanley Falkow
(d) Martha Howe
(e) Frederick Neidhardt
(f) Jean Brenchley

Figure 1.9 Important Contributors to Microbiology. **(a)** Rita Colwell has studied the genetics and ecology of marine bacteria such as *Vibrio cholerae* and helped establish the field of marine biotechnology. **(b)** R. G. E. Murray has contributed greatly to the understanding of bacterial cell envelopes and bacterial taxonomy. **(c)** Stanley Falkow has advanced our understanding of how bacterial pathogens cause disease. **(d)** Martha Howe has made fundamental contributions to our knowledge of the bacteriophage Mu. **(e)** Frederick Neidhardt has contributed to microbiology through his work on the regulation of *E. coli* physiology and metabolism, and by coauthoring advanced textbooks. **(f)** Jean Brenchley has studied the regulation of glutamate and glutamine metabolism, helped found the Pennsylvania State University Biotechnology Institute, and is now finding biotechnological uses for psychrophilic (cold-loving) microorganisms.

1.7 The Future of Microbiology

As the preceding sections have shown, microbiology has had a profound influence on society. What of the future? Science writer Bernard Dixon is very optimistic about microbiology's future for two reasons. First, microbiology has a clearer mission than do many other scientific disciplines. Second, microbiology has great practical significance. Dixon notes that microbiology is required both to face the threat of new and reemerging human infectious diseases and to develop industrial technologies that are more efficient and environmentally friendly.

What are some of the most promising areas for future microbiological research and their potential practical impacts? What kinds of challenges do microbiologists face? A discussion of some aspects of the future of microbiology follows.

Medical microbiology, public health microbiology, and immunology will continue to be areas of intense research. New infectious diseases are continually arising and old diseases are once again becoming widespread and destructive. AIDS, SARS, hemorrhagic fevers, and tuberculosis are excellent examples of new and reemerging infectious diseases. Microbiologists will have to respond to these threats, many of them presently unknown. They will also need to find ways to stop the spread of established infectious diseases, as well as the spread of multiple antibiotic resistance, which can render a pathogen resistant to current medical treatment. Microbiologists will also be called upon to create new drugs and vaccines, to study the association between infectious agents and chronic disease (e.g., autoimmune and cardiovascular diseases), and to further our understanding of host defenses and how pathogens interact with host cells. It will be necessary to use techniques in molecular biology and recombinant DNA technology to solve many of these problems.

Industrial microbiology and environmental microbiology also face many challenges and opportunities. Microorganisms are increasingly important in industry and environmental control, and we must learn how to use them in a variety of new ways. For example, microorganisms can serve as sources of high-quality food and other practical products such as enzymes for industrial applications. They may also be used to degrade pollutants and toxic wastes and as vectors to treat diseases and enhance agricultural productivity. There also is a continuing need to protect food and crops from microbial damage.

The development of techniques, especially DNA-based techniques, that allow the study of microorganisms in their natural environment has greatly stimulated research in microbial ecology. Several areas of research will continue to be important. Understanding microbial diversity is one area that requires further research. It is estimated that less than 1% of Earth's microbes have been cultured. Greater efforts to grow previously uncultivated microbes will be required. Much work also needs to be done on microorganisms living in extreme environments. The discovery of new and unusual microorganisms may well lead to further advances in the development of new antimicrobial agents, industrial processes, and bioremediation. Another area of increasing interest to microbial ecologists is biofilms. Microbes often form biofilms on surfaces, and in doing so exhibit a physiology that differs from that observed when they live freely or planktonically. For instance, microbes in a biofilm are often more resistant to killing agents than they are when not in a biofilm. Biofilms are not only of interest to microbial ecologists; they can form on human tissues, on indwelling catheters, and on other man-made medical devices. In fact, microbial ecologists and medical microbiologists now understand that microorganisms are essential partners with higher organisms. Greater knowledge of the nature of these symbiotic relationships can help improve our appreciation of the living world. It also will lead to new approaches in treating infectious diseases in livestock and in humans.

The fields of genomics and proteomics have and will continue to have a tremendous impact on microbiology. The genomes of many microorganisms have already have been sequenced and many more will be determined in the coming years. These sequences are ideal for learning how the genome is related to cell structure and function and for providing insights into fundamental questions in biology, such as how complex cellular structures develop and how cells communicate with one another and respond to the environment. Analysis of the genome and its activity will require continuing advances in the field of bioinformatics and the use of computers to investigate biological problems.

Perhaps the biggest challenge facing microbiologists will be to assess the implications of new discoveries and technological developments. The pace of these discoveries and developments is very rapid, and sometimes it is difficult for nonscientists to follow and assess them. Microbiologists will need to communicate a balanced view of both the positive and the negative long-term impacts of these developments on society.

Clearly, the future of microbiology is bright. The microbiologist René Dubos has summarized well the excitement and promise of microbiology:

> How extraordinary that, all over the world, microbiologists are now involved in activities as different as the study of gene structure, the control of disease, and the industrial processes based on the phenomenal ability of microorganisms to decompose and synthesize complex organic molecules. Microbiology is one of the most rewarding of professions because it gives its practitioners the opportunity to be in contact with all the other natural sciences and thus to contribute in many different ways to the betterment of human life.

1. **What do you think are the five most important research areas to pursue in microbiology? Give reasons for your choices.**

Summary

1.1 Members of the Microbial World

a. Microbiology studies microscopic organisms that are often unicellular, or if multicellular, do not have highly differentiated tissues. The discipline is also defined by the techniques it uses—in particular, those used to isolate and culture microorganisms.

b. Procaryotic cells differ from eucaryotic cells in lacking a membrane-delimited nucleus, and in other ways as well.

c. Microbiologists divide organisms into three domains: *Bacteria, Archaea,* and *Eucarya*.

d. Domains *Bacteria* and *Archaea* consist of procaryotic microorganisms. The eucaryotic microbes (protists and fungi) are placed in *Eucarya*. Viruses are acellular entities that are not placed in any of the domains but are classified by a separate system.

1.2 The Discovery of Microorganisms

a. Antony van Leeuwenhoek was the first person to extensively describe microorganisms.

1.3 The Conflict Over Spontaneous Generation

a. Experiments by Redi and others disproved the theory of spontaneous generation in regard to larger organisms.

b. The spontaneous generation of microorganisms was disproved by Spallanzani, Pasteur, Tyndall, and others.

1.4 The Golden Age of Microbiology

a. Support for the germ theory of disease came from the work of Bassi, Pasteur, Koch, and others. Lister provided indirect evidence with his development of antiseptic surgery.

b. Koch's postulates and molecular Koch's postulates are used to prove a direct relationship between a suspected pathogen and a disease.

c. Koch developed the techniques required to grow bacteria on solid media and to isolate pure cultures of pathogens.

d. Vaccines against anthrax and rabies were made by Pasteur; von Behring and Kitasato prepared antitoxins for diphtheria and tetanus.

e. Metchnikoff discovered some blood leukocytes could phagocytize and destroy bacterial pathogens.

1.5 The Development of Industrial Microbiology and Microbial Ecology

a. Pasteur showed that fermentations were caused by microorganisms and that some microorganisms could live in the absence of oxygen.

b. The role of microorganisms in carbon, nitrogen, and sulfur cycles was first studied by Winogradsky and Beijerinck.

1.6 The Scope and Relevance of Microbiology

a. In the twentieth century, microbiology contributed greatly to the fields of medicine, genetics, agriculture, food science, biochemistry, and molecular biology.

b. There is a wide variety of fields in microbiology, and many have a great impact on society. These include the more applied disciplines such as medical, public health, industrial, food, and dairy microbiology. Microbial ecology, physiology, biochemistry, and genetics are examples of basic microbiological research fields.

1.7 The Future of Microbiology

a. Microbiologists will be faced with many exciting and important future challenges such as finding new ways to combat disease, reduce pollution, and feed the world's population.

Key Terms

Critical Thinking Questions

1. Consider the impact of microbes on the course of world history. History is full of examples of instances or circumstances under which one group of people lost a struggle against another. In fact, when examined more closely, the "losers" often had the misfortune of being exposed to, more susceptible to, or unable to cope with an infectious agent. Thus weakened in physical strength or demoralized by the course of a devastating disease, they were easily overcome by human "conquerors."
   a. Choose an example of a battle or other human activity such as exploration of new territory and determine the impact of microorganisms, either indigenous or transported to the region, on that activity.
   b. Discuss the effect that the microbe(s) had on the outcome in your example.
   c. Suggest whether the advent of antibiotics, food storage and preparation technology, or sterilization technology would have made a difference in the outcome.

2. Vaccinations against various childhood diseases have contributed to the entry of women, particularly mothers, into the full-time workplace.
   a. Is this statement supported by data—comparing availability and extent of vaccination with employment statistics in different places or at different times?
   b. Before vaccinations for measles, mumps, and chickenpox, what was the incubation time and duration of these childhood diseases? What impact would such diseases have on mothers with several elementary schoolchildren at home if they had fulltime jobs and lacked substantial child care support?
   c. What would be the consequence if an entire generation of children (or a group of children in one country) were not vaccinated against any diseases? What do you predict would happen if these children went to college and lived in a dormitory in close proximity with others who had received all of the recommended childhood vaccines?

Learn More

Brock, T. D. 1961. *Milestones in microbiology*. Englewood Cliffs, N.J.: Prentice-Hall.

Brock, T. D. 1988. *Robert Koch: A life in medicine and bacteriology*. Madison, Wisc.: Science Tech Publishers.

Chung, K. T., and Ferris, D. H. 1996. Martinus Willem Beijerinck (1851–1931): Pioneer of general microbiology. *ASM News* 62(10):539–43.

de Kruif, P. 1937. *Microbe hunters*. New York: Harcourt, Brace.

Dixon, B. 1997. Microbiology present and future. *ASM News* 63(3):124–25.

Ford, B. J. 1998. The earliest views. *Sci. Am.* 278(4):50–53.

Fredricks, D. N., and Relman, D. A. 1996. Sequence-based identification of microbial pathogens: A reconsideration of Koch's postulates. *Clin. Microbiol. Rev.* 9(1):18–33.

Geison, G. L. 1995. *The private science of Louis Pasteur*. Princeton, N.J.: Princeton University Press.

Stanier, R. Y. 1978. What is microbiology? In *Essays in microbiology*, J. R. Norris and M. H. Richmond, editors, 1/1–1/32. New York: John Wiley and Sons.

Woese, C. R. 2000. Interpreting the universal phylogenetic tree. *Proc. Natl. Acad. Sci.* 97(15):8392–96.

Please visit the Prescott website at www.mhhe.com/prescott7 for additional references.

2 The Study of Microbial Structure: Microscopy and Specimen Preparation

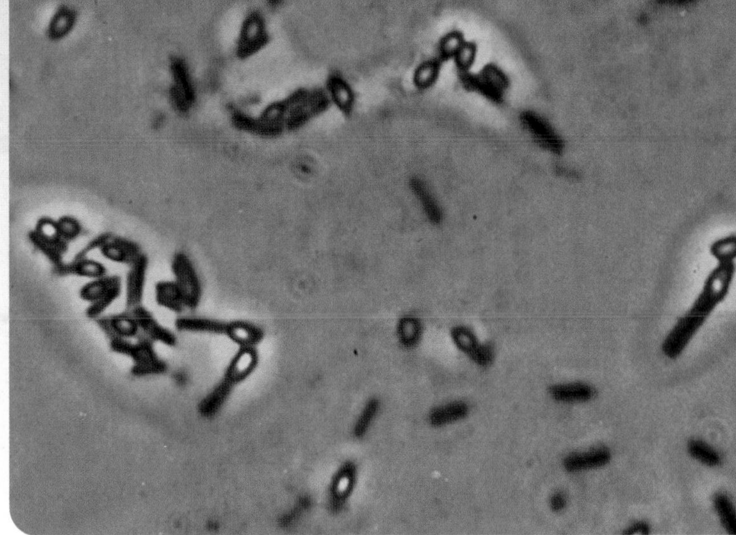

Clostridium botulinum is a rod-shaped bacterium that forms endospores and releases botulinum toxin, the cause of botulism food poisoning. In this phase-contrast micrograph, the endospores are the bright, oval objects located at the ends of the rods; some endospores have been released from the cells that formed them.

Preview

- Light microscopes use glass lenses to bend and focus light rays to produce enlarged images of small objects. The maximum resolution of a light microscope is about 0.2 μm.
- Many types of light microscopes have been developed, including bright-field, dark-field, phase-contrast, and fluorescence microscopes. Each yields a distinctive image.
- Bright-field microscopy requires the application of stains to microorganisms for easy viewing. Stains are also used to determine the nature of bacterial cell walls or to visualize specific procaryotic structures such as flagella and capsules.
- The useful magnification of a light microscope is limited by its resolving power. The resolving power is limited by the wavelength of the illuminating beam.
- Electron microscopes use beams of electrons rather than light to achieve very high resolution (up to 0.5 nm) and magnification.
- New forms of microscopy are improving our ability to observe microorganisms and molecules. Two examples are the confocal scanning laser microscope and the scanning probe microscope.

Microbiology usually is concerned with organisms so small they cannot be seen distinctly with the unaided eye. Because of the nature of this discipline, the microscope is of crucial importance. Thus it is important to understand how the microscope works and the way in which specimens are prepared for examination.

In this chapter we begin with a detailed treatment of the standard bright-field microscope and then describe other common types of light microscopes. Next we discuss preparation and staining of specimens for examination with the light microscope. This is followed by a description of transmission and scanning electron microscopes, both of which are used extensively in current microbiological research. We close the chapter with a brief introduction to two newer forms of microscopy: confocal microscopy and scanning probe microscopy.

2.1 Lenses and the Bending of Light

To understand how a light microscope operates, one must know something about the way in which lenses bend and focus light to form images. When a ray of light passes from one medium to another, **refraction** occurs—that is, the ray is bent at the interface. The **refractive index** is a measure of how greatly a substance slows the velocity of light; the direction and magnitude of bending is determined by the refractive indices of the two media forming the interface. When light passes from air into glass, a medium with a greater refractive index, it is slowed and bent toward the normal, a line perpendicular to the surface (**figure 2.1**). As light leaves glass and returns to air, a medium with a lower refractive

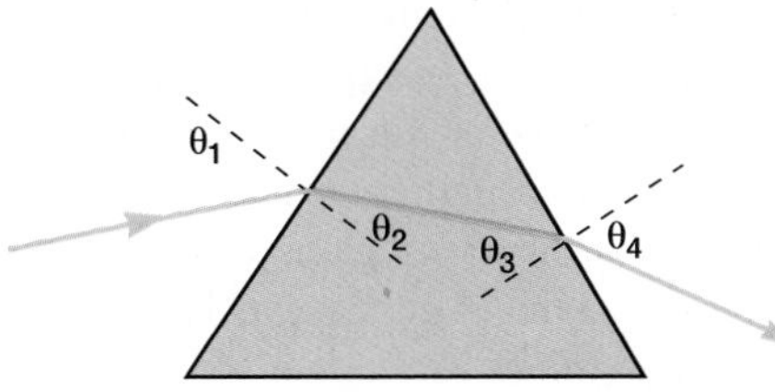

Figure 2.1 The Bending of Light by a Prism. Normals (lines perpendicular to the surface of the prism) are indicated by dashed lines. As light enters the glass, it is bent toward the first normal (angle θ_2 is less than θ_1). When light leaves the glass and returns to air, it is bent away from the second normal (θ_4 is greater than θ_3). As a result the prism bends light passing through it.

There are more animals living in the scum on the teeth in a man's mouth than there are men in a whole kingdom.

—*Antony van Leeuwenhoek*

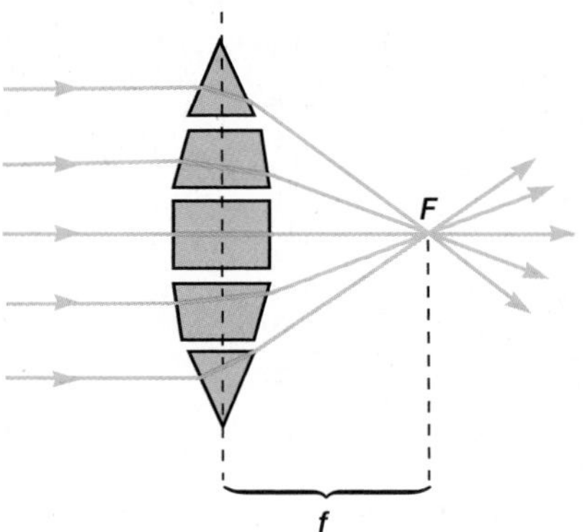

Figure 2.2 Lens Function. A lens functions somewhat like a collection of prisms. Light rays from a distant source are focused at the focal point *F*. The focal point lies a distance *f*, the focal length, from the lens center.

Table 2.1 Common Units of Measurement

Unit	Abbreviation	Value
1 centimeter	cm	10^{-2} meter or 0.394 inches
1 millimeter	mm	10^{-3} meter
1 micrometer	μm	10^{-6} meter
1 nanometer	nm	10^{-9} meter
1 Angstrom	Å	10^{-10} meter

index, it accelerates and is bent away from the normal. Thus a prism bends light because glass has a different refractive index from air, and the light strikes its surface at an angle.

Lenses act like a collection of prisms operating as a unit. When the light source is distant so that parallel rays of light strike the lens, a convex lens will focus these rays at a specific point, the **focal point** (*F* in **figure 2.2**). The distance between the center of the lens and the focal point is called the **focal length** (*f* in figure 2.2).

Our eyes cannot focus on objects nearer than about 25 cm or 10 inches (**table 2.1**). This limitation may be overcome by using a convex lens as a simple magnifier (or microscope) and holding it close to an object. A magnifying glass provides a clear image at much closer range, and the object appears larger. Lens strength is related to focal length; a lens with a short focal length will magnify an object more than a weaker lens having a longer focal length.

1. Define refraction, refractive index, focal point, and focal length.
2. Describe the path of a light ray through a prism or lens.
3. How is lens strength related to focal length?

2.2 The Light Microscope

Microbiologists currently employ a variety of light microscopes in their work; bright-field, dark-field, phase-contrast, and fluorescence microscopes are most commonly used. Modern microscopes are all compound microscopes. That is, the magnified image formed by the objective lens is further enlarged by one or more additional lenses.

The Bright-Field Microscope

The ordinary microscope is called a **bright-field microscope** because it forms a dark image against a brighter background. The microscope consists of a sturdy metal body or stand composed of a base and an arm to which the remaining parts are attached (**figure 2.3**). A light source, either a mirror or an electric illuminator, is located in the base. Two focusing knobs, the fine and coarse adjustment knobs, are located on the arm and can move either the stage or the nosepiece to focus the image.

The stage is positioned about halfway up the arm and holds microscope slides by either simple slide clips or a mechanical stage clip. A mechanical stage allows the operator to move a slide around smoothly during viewing by use of stage control knobs. The **substage condenser** is mounted within or beneath the stage and focuses a cone of light on the slide. Its position often is fixed in simpler microscopes but can be adjusted vertically in more advanced models.

The curved upper part of the arm holds the body assembly, to which a nosepiece and one or more **eyepieces** or **ocular lenses** are attached. More advanced microscopes have eyepieces for both eyes and are called binocular microscopes. The body assembly itself contains a series of mirrors and prisms so that the barrel holding the eyepiece may be tilted for ease in viewing (**figure 2.4**). The nosepiece holds three to five **objective lenses** of differing magnifying power and can be rotated to position any objective beneath the body assembly. Ideally a microscope should be **parfocal**—that is, the image should remain in focus when objectives are changed.

The image one sees when viewing a specimen with a compound microscope is created by the objective and ocular lenses working together. Light from the illuminated specimen is focused by the objective lens, creating an enlarged image within the microscope (figure 2.4). The ocular lens further magnifies this primary image. The total magnification is calculated by multiplying the objective and eyepiece magnifications together. For example, if a 45× objective is used with a 10× eyepiece, the overall magnification of the specimen will be 450×.

Microscope Resolution

The most important part of the microscope is the objective, which must produce a clear image, not just a magnified one. Thus resolution is extremely important. **Resolution** is the ability of a lens to separate or distinguish between small objects that are close together.

Resolution is described mathematically by an equation developed in the 1870s by Ernst Abbé, a German physicist responsible for much of the optical theory underlying microscope design. The Abbé equation states that the minimal distance *(d)* between two

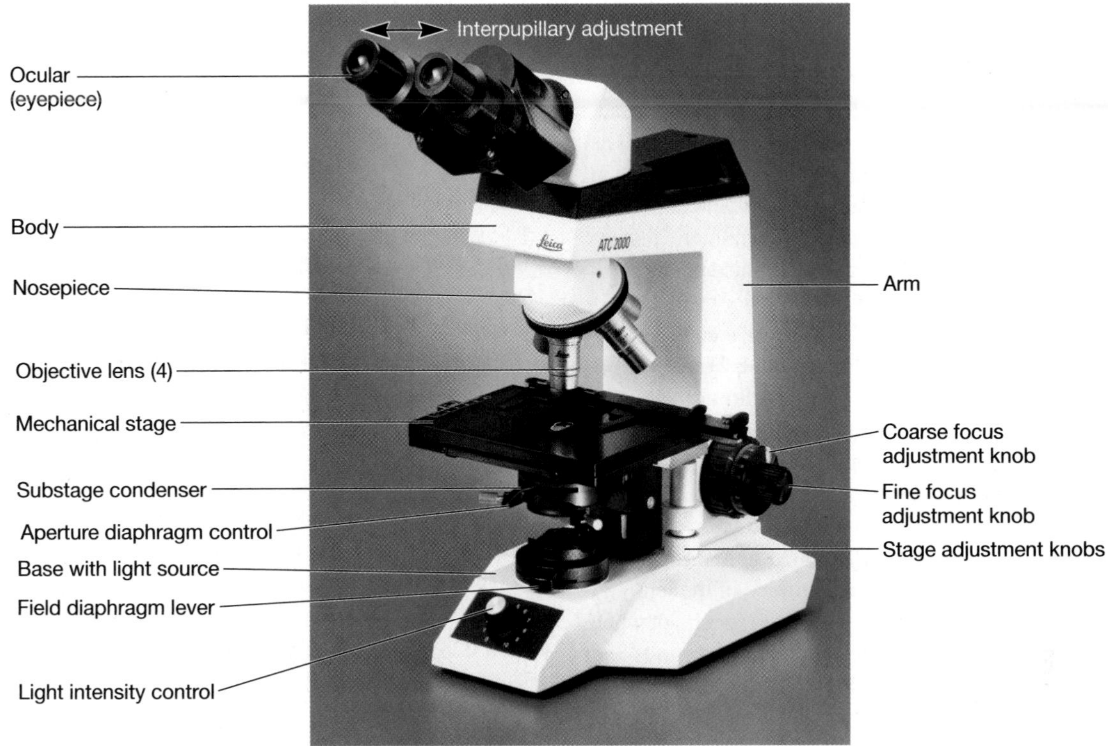

Figure 2.3 A Bright-Field Microscope. The parts of a modern bright-field microscope. The microscope pictured is somewhat more sophisticated than those found in many student laboratories. For example, it is binocular (has two eyepieces) and has a mechanical stage, an adjustable substage condenser, and a built-in illuminator.

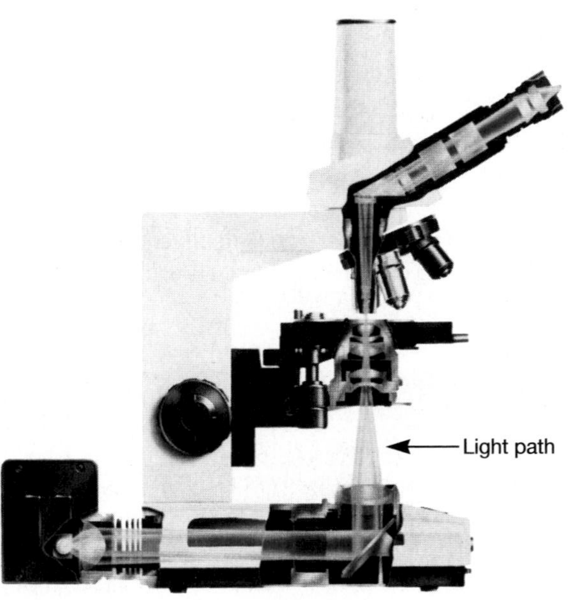

Figure 2.4 A Microscope's Light Path. The light path in an advanced bright-field microscope (see also figure 2.19).

objects that reveals them as separate entities depends on the wavelength of light (λ) used to illuminate the specimen and on the **numerical aperture** of the lens ($n \sin \theta$), which is the ability of the lens to gather light.

$$d = \frac{0.5\lambda}{n \sin \theta}$$

As d becomes smaller, the resolution increases, and finer detail can be discerned in a specimen; d becomes smaller as the wavelength of light used decreases and as the numerical aperture (NA) increases. Thus the greatest resolution is obtained using a lens with the largest possible NA and light of the shortest wavelength, light at the blue end of the visible spectrum (in the range of 450 to 500 nm; *see figure 6.25*).

The numerical aperture ($n \sin \theta$) of a lens is a complex concept that can be difficult to understand. It is defined by two components: n is the refractive index of the medium in which the lens works (e.g., air) and θ is 1/2 the angle of the cone of light entering an objective (**figure 2.5**). When this cone has a narrow angle and tapers to a sharp point, it does not spread out much after leaving the slide and therefore does not adequately separate images of closely packed objects. If the cone of light has a very wide angle and spreads out rapidly after passing through a specimen, closely packed objects appear widely separated and are resolved. The angle of the cone of light that can enter a lens depends on the refractive index (n) of the medium in which the lens works, as well as upon the objective itself. The refractive index for air is 1.00 and sin θ cannot be greater than 1 (the maximum θ is 90° and sin 90° is 1.00). Therefore no lens working in air can have a numerical aperture greater than 1.00. The only practical way to raise the numerical aperture above 1.00, and therefore achieve higher resolution, is to increase the refractive index with immersion oil, a colorless liquid with the same refractive index as glass (**table 2.2**). If air is replaced with immersion oil, many light rays that did not enter the objective due to reflection and refraction at the surfaces of the objective lens and slide will now do so (**figure 2.6**). An increase in numerical aperture and resolution results.

Numerical aperture is related to another characteristic of an objective lens, the working distance. The **working distance** of an objective is the distance between the front surface of the lens and the surface of the cover glass (if one is used) or the specimen when it is in sharp focus. Objectives with large numerical apertures and great resolving power have short working distances (table 2.2).

The preceding discussion has focused on the resolving power of the objective lens. The resolution of an entire microscope must take into account the numerical aperture of its condenser as is evident from the equation below.

$$d_{\text{microscope}} = \frac{\lambda}{(\text{NA}_{\text{objective}} + \text{NA}_{\text{condenser}})}$$

The condenser is a large, light-gathering lens used to project a wide cone of light through the slide and into the objective lens. Most microscopes have a condenser with a numerical aperture between 1.2 and 1.4. However, the condenser numerical aperture will not be much above about 0.9 unless the top of the condenser is oiled to the bottom of the slide. During routine microscope operation, the condenser usually is not oiled and this limits the overall resolution, even with an oil immersion objective.

Although the resolution of the microscope must consider both the condenser and the objective lens, in most cases the limit of resolution of a light microscope is calculated using the Abbé equation, which considers the objective lens only. The maximum theoretical resolving power of a microscope with an oil immersion objective (numerical aperture of 1.25) and blue-green light is approximately 0.2 μm.

$$d = \frac{(0.5)(530 \text{ nm})}{1.25} = 212 \text{ nm or } 0.2\mu\text{m}$$

At best, a bright-field microscope can distinguish between two dots about 0.2 μm apart (the same size as a very small bacterium).

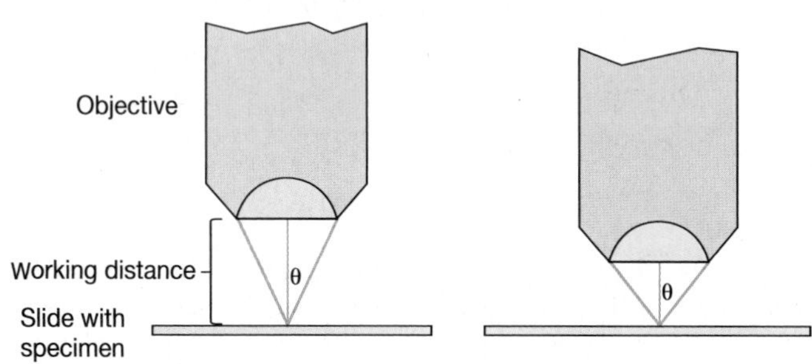

Figure 2.5 Numerical Aperture in Microscopy. The angular aperture θ is 1/2 the angle of the cone of light that enters a lens from a specimen, and the numerical aperture is $n \sin \theta$. In the right-hand illustration the lens has larger angular and numerical apertures; its resolution is greater and its working distance smaller.

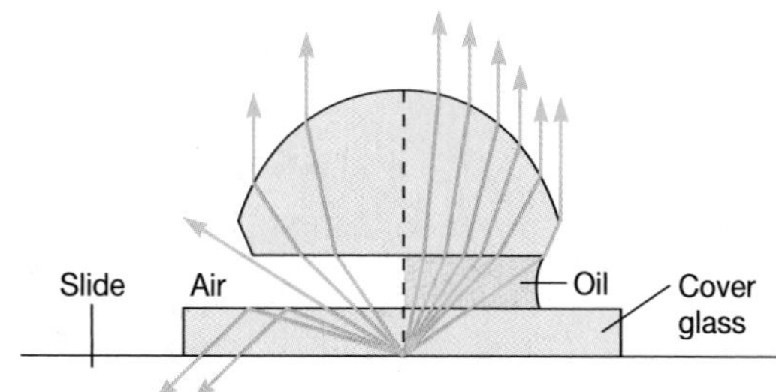

Figure 2.6 The Oil Immersion Objective. An oil immersion objective operating in air and with immersion oil.

Table 2.2 The Properties of Microscope Objectives

	Objective			
Property	**Scanning**	**Low Power**	**High Power**	**Oil Immersion**
Magnification	4×	10×	40–45×	90–100×
Numerical aperture	0.10	0.25	0.55–0.65	1.25–1.4
Approximate focal length (f)	40 mm	16 mm	4 mm	1.8–2.0 mm
Working distance	17-20 mm	4-8 mm	0.5–0.7 mm	0.1 mm
Approximate resolving power with light of 450 nm (blue light)	2.3 μm	0.9 μm	0.35 μm	0.18 μm

Given the limit of resolution of a light microscope, the largest useful magnification—the level of magnification needed to increase the size of the smallest resolvable object to be visible with the light microscope—can be determined. Our eye can just detect a speck 0.2 mm in diameter, and consequently the useful limit of magnification is about 1,000 times the numerical aperture of the objective lens. Most standard microscopes come with 10× eyepieces and have an upper limit of about 1,000× with oil immersion. A 15× eyepiece may be used with good objectives to achieve a useful magnification of 1,500×. Any further magnification does not enable a person to see more detail. Indeed, a light microscope can be built to yield a final magnification of 10,000×, but it would simply be magnifying a blur. Only the electron microscope provides sufficient resolution to make higher magnifications useful.

The Dark-Field Microscope

The **dark-field microscope** allows a viewer to observe living, unstained cells and organisms by simply changing the way in which they are illuminated. A hollow cone of light is focused on the specimen in such a way that unreflected and unrefracted rays do not enter the objective. Only light that has been reflected or refracted by the specimen forms an image (**figure 2.7**). The field surrounding a specimen appears black, while the object itself is brightly illuminated (**figure 2.8*a,b***).The dark-field microscope can reveal considerable internal structure in larger eucaryotic microorganisms (figure 2.8*b*). It also is used to identify certain bacteria like the thin and distinctively shaped *Treponema pallidum* (figure 2.8*a*), the causative agent of syphilis.

The Phase-Contrast Microscope

Unpigmented living cells are not clearly visible in the bright-field microscope because there is little difference in contrast between the cells and water. As will be discussed in section 2.3, one solution to this problem is to kill and stain cells before observation to increase contrast and create variations in color between cell structures. But what if an investigator must view living cells in order to observe a dynamic process such as movement or phagocytosis? Phase-contrast microscopy can be used in this situation. A **phase-contrast microscope** converts slight differences in refractive index and cell density into easily detected variations in light intensity and is an excellent way to observe living cells (figure 2.8*c–e*).

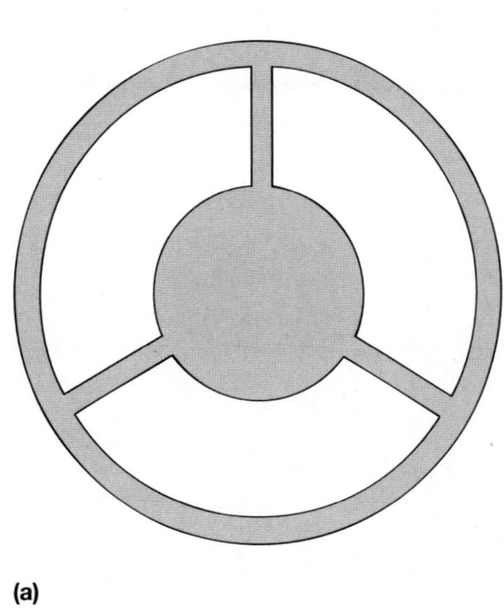

(a)

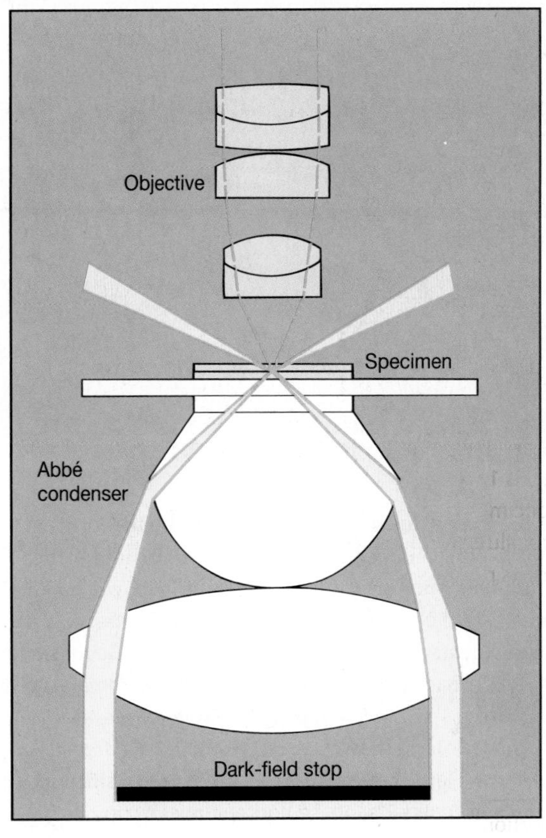

(b)

Figure 2.7 Dark-Field Microscopy. The simplest way to convert a microscope to dark-field microscopy is to place **(a)** a dark-field stop underneath **(b)** the condenser lens system. The condenser then produces a hollow cone of light so that the only light entering the objective comes from the specimen.

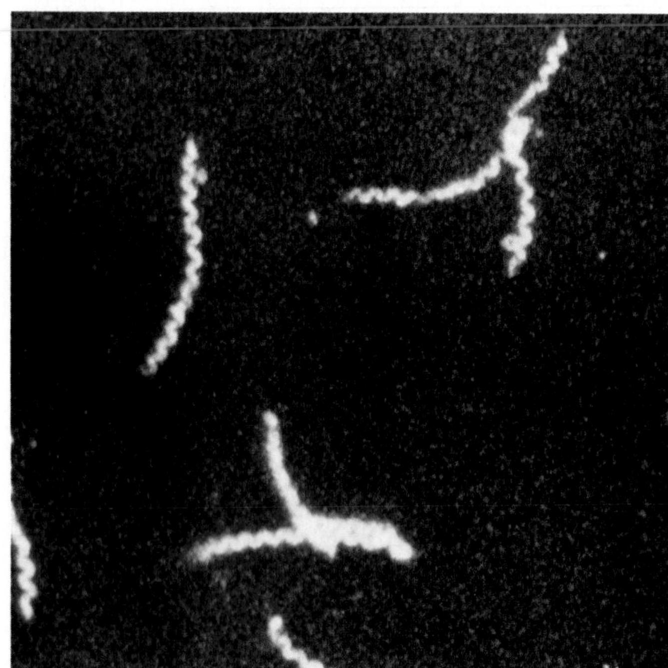

(a) *T. pallidum:* dark-field microscopy

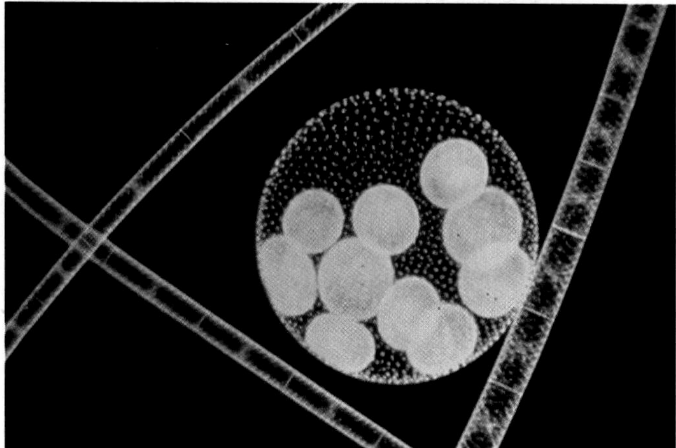

(b) *Volvox* and *Spirogyra:* dark-field microscopy

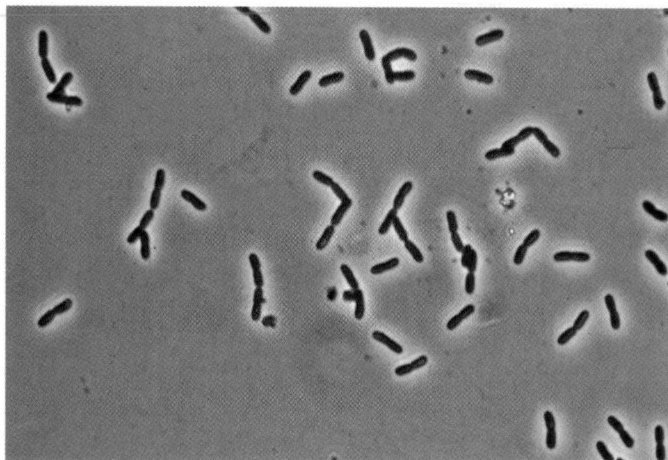

(c) *Pseudomonas:* phase-contrast microscopy

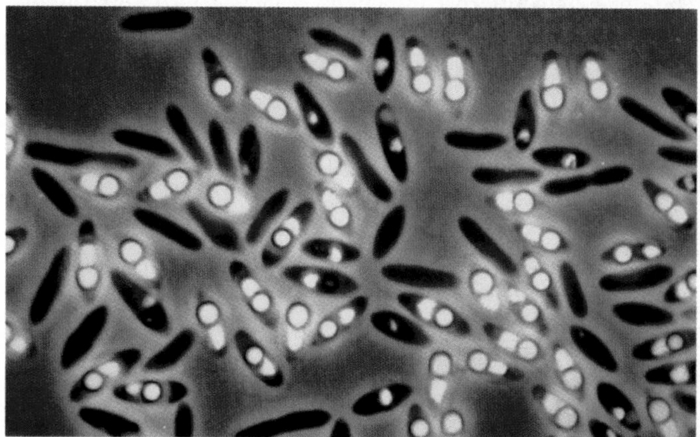

(d) *Desulfotomaculum:* phase-contrast microscopy

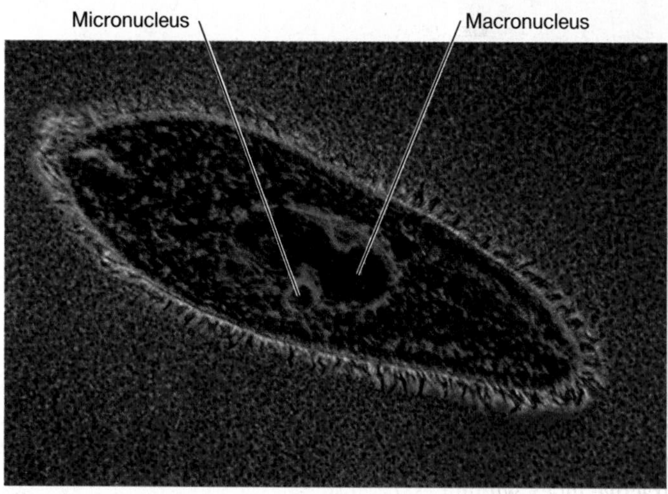

(e) *Paramecium:* phase-contrast microscopy

Figure 2.8 Examples of Dark-Field and Phase-Contrast Microscopy. **(a)** *Treponema pallidum,* the spirochete that causes syphilis; dark-field microscopy. **(b)** *Volvox and Spirogyra;* dark-field microscopy (×175). Note daughter colonies within the mature *Volvox* colony (center) and the spiral chloroplasts of *Spirogyra* (left and right). **(c)** A phase-contrast micrograph of *Pseudomonas* cells, which range from 1–3 μm in length. **(d)** *Desulfotomaculum acetoxidans* with endospores; phase contrast (×2,000). **(e)** *Paramecium* stained to show a large central macronucleus with a small spherical micronucleus at its side; phase-contrast microscopy (×100).

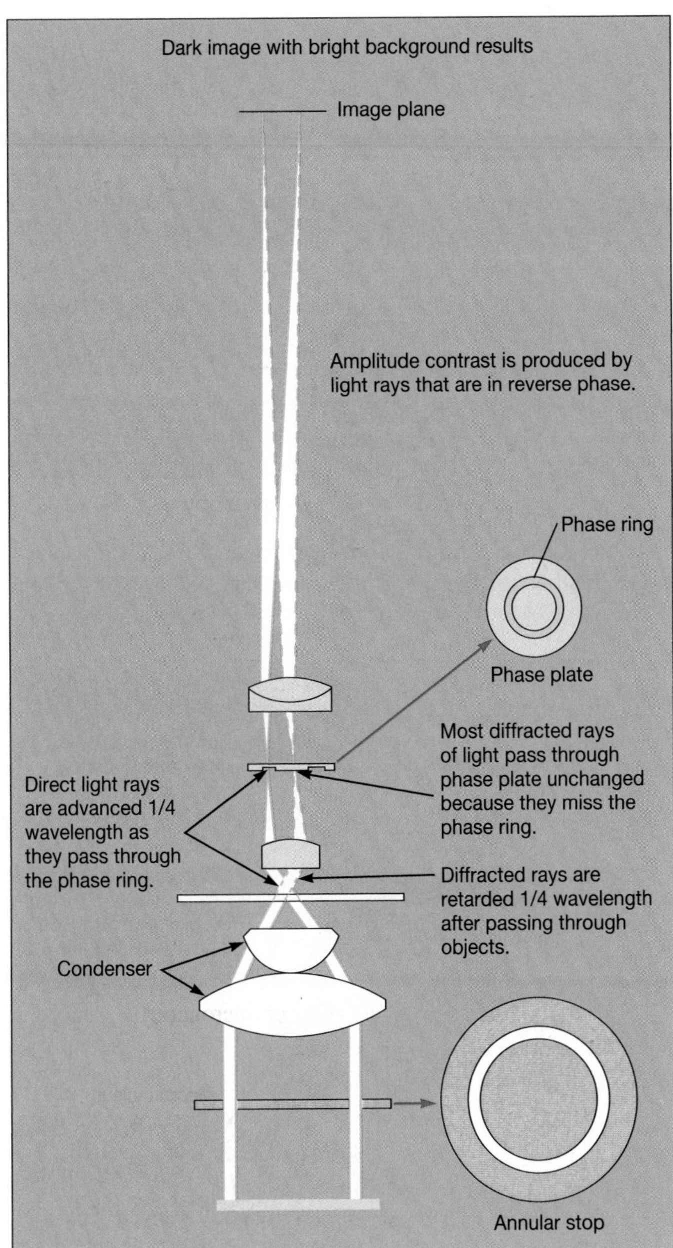

Figure 2.9 Phase-Contrast Microscopy. The optics of a dark-phase-contrast microscope.

The condenser of a phase-contrast microscope has an annular stop, an opaque disk with a thin transparent ring, which produces a hollow cone of light (**figure 2.9**). As this cone passes through a cell, some light rays are bent due to variations in density and refractive index within the specimen and are retarded by about 1/4 wavelength. The deviated light is focused to form an image of the object. Undeviated light rays strike a phase ring in the phase plate, a special optical disk located in the objective, while the deviated rays miss the ring and pass through the rest of the plate. If the phase ring is constructed in such a way that the undeviated light passing through it is advanced by 1/4 wavelength, the deviated and undeviated waves will be about 1/2 wavelength out of phase and will cancel each other when they come together to form an image (**figure 2.10**). The background, formed by undeviated light, is bright, while the unstained object appears dark and well-defined. This type of microscopy is called dark-phase-contrast microscopy. Color filters often are used to improve the image (figure 2.8*d*).

Phase-contrast microscopy is especially useful for studying microbial motility, determining the shape of living cells, and detecting bacterial components such as endospores and inclusion bodies that contain poly-β-hydroxyalkanoates (e.g., poly-β-hydroxybutyrate), polymetaphosphate, sulfur, or other substances. These are clearly visible (figure 2.8*d*) because they have refractive indices markedly different from that of water. Phase-contrast microscopes also are widely used in studying eucaryotic cells. The cytoplasmic matrix: Inclusion bodies (section 3.3)

The Differential Interference Contrast Microscope

The **differential interference contrast (DIC) microscope** is similar to the phase-contrast microscope in that it creates an image by detecting differences in refractive indices and thickness. Two beams of plane-polarized light at right angles to each other are generated by prisms. In one design, the object beam passes through the specimen, while the reference beam passes through a clear area of the slide. After passing through the specimen, the two beams are combined and interfere with each other to form an image. A live, unstained specimen appears brightly colored and three-dimensional (**figure 2.11**). Structures such as cell walls, endospores, granules, vacuoles, and eucaryotic nuclei are clearly visible.

The Fluorescence Microscope

The microscopes thus far considered produce an image from light that passes through a specimen. An object also can be seen because it actually emits light, and this is the basis of fluorescence microscopy. When some molecules absorb radiant energy, they become excited and later release much of their trapped energy as light. Any light emitted by an excited molecule will have a longer wavelength (or be of lower energy) than the radiation originally absorbed. **Fluorescent light** is emitted very quickly by the excited molecule as it gives up its trapped energy and returns to a more stable state.

The **fluorescence microscope** exposes a specimen to ultraviolet, violet, or blue light and forms an image of the object with the resulting fluorescent light. The most commonly used fluorescence microscopy is epifluorescence microscopy, also called incident light or reflected light fluorescence microscopy. Epifluorescence microscopes employ an objective lens that also acts as a condenser (**figure 2.12**). A mercury vapor arc lamp or other source produces an intense beam of light that passes through an exciter filter. The exciter filter transmits only the desired wavelength of excitation light. The excitation light is directed down the microscope by a special mirror called the dichromatic mirror. This mirror reflects light of shorter wavelengths (i.e., the excitation light),

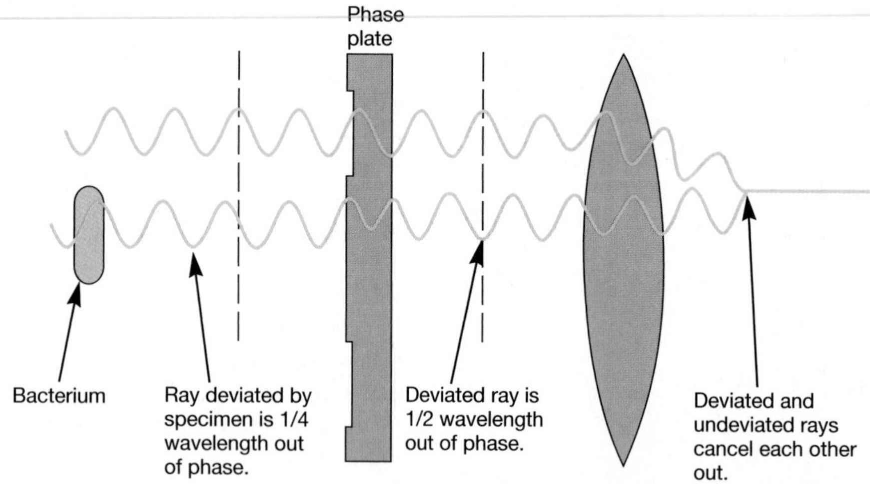

Figure 2.10 The Production of Contrast in Phase Microscopy. The behavior of deviated and undeviated or undiffracted light rays in the dark-phase-contrast microscope. Because the light rays tend to cancel each other out, the image of the specimen will be dark against a brighter background.

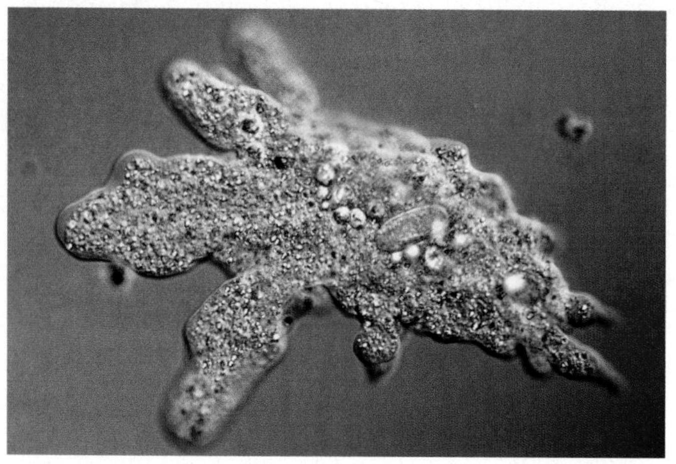

Figure 2.11 Differential Interference Contrast Microscopy. A micrograph of the protozoan *Amoeba proteus*. The three-dimensional image contains considerable detail and is artificially colored (×160).

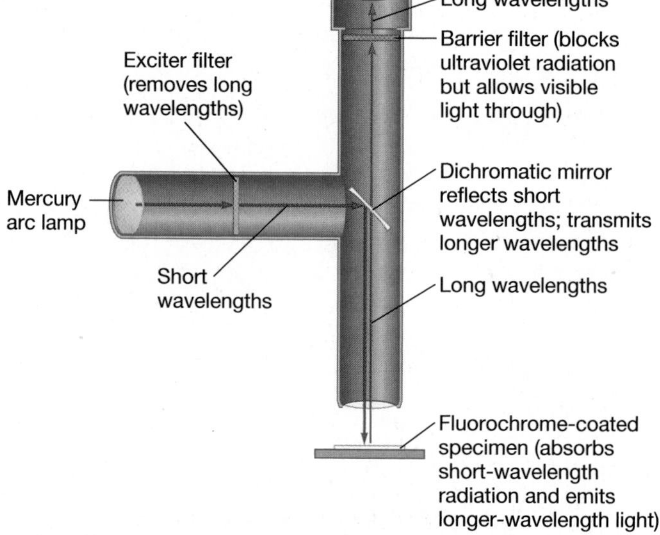

Figure 2.12 Epifluorescence Microscopy. The principles of operation of an epifluorescence microscope.

but allows light of longer wavelengths to pass through. The excitation light continues down, passing through the objective lens to the specimen, which is usually stained with special dye molecules called **fluorochromes (table 2.3).** The fluorochrome absorbs light energy from the excitation light and fluoresces brightly. The emitted fluorescent light travels up through the objective lens into the microscope. Because the emitted fluorescent light has a longer wavelength, it passes through the dichromatic mirror to a barrier filter, which blocks out any residual excitation light. Finally, the emitted light passes through the barrier filter to the eyepieces.

The fluorescence microscope has become an essential tool in medical microbiology and microbial ecology. Bacterial pathogens (e.g., *Mycobacterium tuberculosis,* the cause of tuberculosis) can be identified after staining them with fluorochromes or specifically labeling them with fluorescent antibodies using immunofluorescence procedures. In ecological studies the fluorescence microscope is used to observe microorganisms stained with fluorochrome-labeled probes or fluorochromes that bind specific cell constituents (table 2.3). In addition, microbial ecologists use epifluorescence microscopy to visualize photosynthetic microbes, as their pigments naturally fluoresce when excited by light of specific wavelengths. It is even possible to distinguish live bacteria from dead bacteria by

Table 2.3 Commonly Used Fluorochromes

Fluorochrome	Uses
Acridine orange	Stains DNA; fluoresces orange
Diamidino-2-phenyl indole (DAPI)	Stains DNA; fluoresces green
Fluorescein isothiocyanate (FITC)	Often attached to antibodies that bind specific cellular components or to DNA probes; fluoresces green
Tetramethyl rhodamine isothiocyanate (TRITC or rhodamine)	Often attached to antibodies that bind specific cellular components; fluoresces red

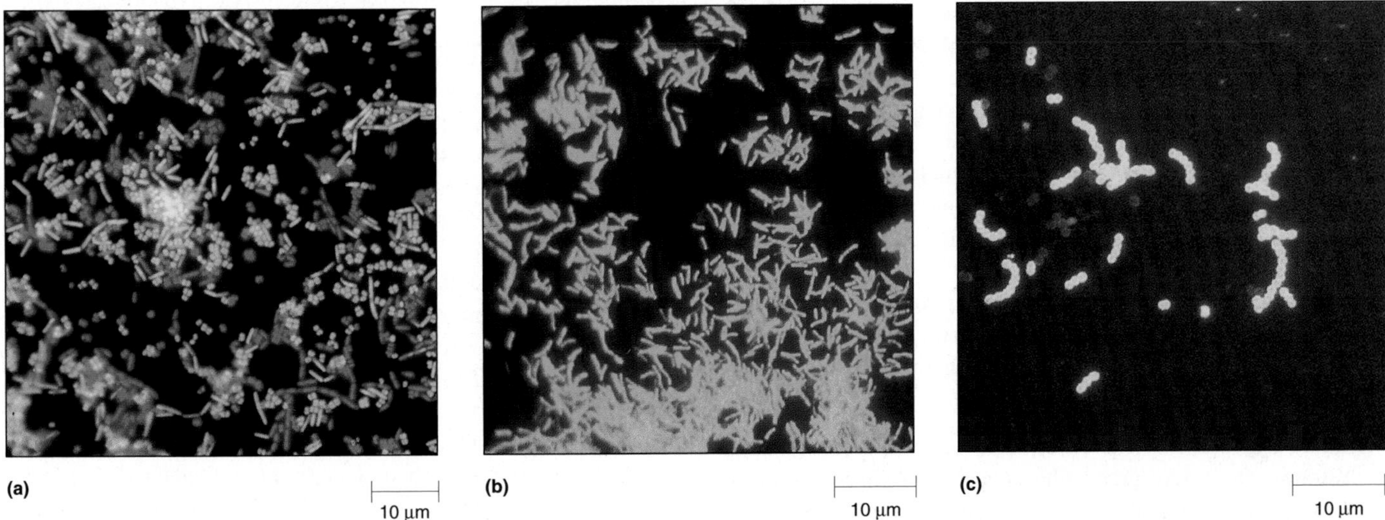

Figure 2.13 Fluorescent Dyes and Tags. **(a)** Dyes that cause live cells to fluoresce green and dead ones red; **(b)** Auramine is used to stain *Mycobacterium* species in a modification of the acid-fast technique; **(c)** Fluorescent antibodies tag specific molecules. In this case, the antibody binds to a molecule that is unique to *Streptococcus pyogenes*.

the color they fluoresce after treatment with a special mixture of stains (**figure 2.13*a***). Thus the microorganisms can be viewed and directly counted in a relatively undisturbed ecological niche. Identification of microorganisms from specimens: Immunologic techniques (section 35.2)

1. List the parts of a light microscope and describe their functions.
2. Define resolution, numerical aperture, working distance, and fluorochrome.
3. If a specimen is viewed using a 5X objective in a microscope with a 15X eyepiece, how many times has the image been magnified?
4. How does resolution depend on the wavelength of light, refractive index, and numerical aperture? How are resolution and magnification related?
5. What is the function of immersion oil?
6. Why don't most light microscopes use 30X ocular lenses for greater magnification?
7. Briefly describe how dark-field, phase-contrast, differential interference contrast, and epifluorescence microscopes work and the kind of image provided by each. Give a specific use for each type.

2.3 Preparation and Staining of Specimens

Although living microorganisms can be directly examined with the light microscope, they often must be fixed and stained to increase visibility, accentuate specific morphological features, and preserve them for future study.

Fixation

The stained cells seen in a microscope should resemble living cells as closely as possible. **Fixation** is the process by which the internal and external structures of cells and microorganisms are preserved and fixed in position. It inactivates enzymes that might disrupt cell morphology and toughens cell structures so that they do not change during staining and observation. A microorganism usually is killed and attached firmly to the microscope slide during fixation.

There are two fundamentally different types of fixation. **Heat fixation** is routinely used to observe procaryotes. Typically, a film of cells (a smear) is gently heated as a slide is passed through a flame. Heat fixation preserves overall morphology but not structures within cells. **Chemical fixation** is used to protect fine cellular substructure and the morphology of larger, more delicate microorganisms. Chemical fixatives penetrate cells and react with cellular components, usually proteins and lipids, to render them inactive, insoluble, and immobile. Common fixative mixtures contain such components as ethanol, acetic acid, mercuric chloride, formaldehyde, and glutaraldehyde.

Dyes and Simple Staining

The many types of dyes used to stain microorganisms have two features in common: they have **chromophore groups,** groups with conjugated double bonds that give the dye its color, and they can bind with cells by ionic, covalent, or hydrophobic bonding. Most dyes are used to directly stain the cell or object of interest, but some dyes (e.g., India ink and nigrosin) are used in **negative staining,** where the background but not the cell is stained; the unstained cells appear as bright objects against a dark background.

Dyes that bind cells by ionic interactions are probably the most commonly used dyes. These ionizable dyes may be divided into two general classes based on the nature of their charged group.

1. **Basic dyes**—methylene blue, basic fuchsin, crystal violet, safranin, malachite green—have positively charged groups (usually some form of pentavalent nitrogen) and are generally sold as chloride salts. Basic dyes bind to negatively charged molecules like nucleic acids, many proteins, and the surfaces of procaryotic cells.
2. **Acidic dyes**—eosin, rose bengal, and acid fuchsin—possess negatively charged groups such as carboxyls (—COOH) and phenolic hydroxyls (—OH). Acidic dyes, because of their negative charge, bind to positively charged cell structures.

The staining effectiveness of ionizable dyes may be altered by pH, since the nature and degree of the charge on cell components change with pH. Thus acidic dyes stain best under acidic conditions when proteins and many other molecules carry a positive charge; basic dyes are most effective at higher pHs.

Dyes that bind through covalent bonds or because of their solubility characteristics are also useful. For instance, DNA can be stained by the Feulgen procedure in which the staining compound (Schiff's reagent) is covalently attached to its deoxyribose sugars. Sudan III (Sudan Black) selectively stains lipids because it is lipid soluble but will not dissolve in aqueous portions of the cell.

Microorganisms often can be stained very satisfactorily by **simple staining,** in which a single dye is used (**figure 2.14*a,b***). Simple staining's value lies in its simplicity and ease of use. One covers the fixed smear with stain for a short period of time, washes the excess stain off with water, and blots the slide dry. Basic dyes like crystal violet, methylene blue, and carbolfuchsin are frequently used in simple staining to determine the size, shape, and arrangement of procaryotic cells.

Differential Staining

The **Gram stain,** developed in 1884 by the Danish physician Christian Gram, is the most widely employed staining method in bacteriology. It is an example of **differential staining**—procedures that are used to distinguish organisms based on their staining properties. Use of the Gram stain divides *Bacteria* into two classes—gram negative and gram positive.

The Gram-staining procedure is illustrated in **figure 2.15.** In the first step, the smear is stained with the basic dye crystal violet, the primary stain. This is followed by treatment with an iodine solution functioning as a **mordant.** The iodine increases the interaction between the cell and the dye so that the cell is stained more strongly. The smear is next decolorized by washing with ethanol or acetone. This step generates the differential aspect of the Gram stain; gram-positive bacteria retain the crystal violet, whereas gram-negative bacteria lose their crystal violet and become colorless. Finally, the smear is counterstained with a simple, basic dye different in color from crystal violet. Safranin, the most common counterstain, colors gram-negative bacteria pink to red and leaves gram-positive bacteria dark purple (figures 2.14*c* and 2.15*b*). The bacterial cell wall (section 3.6)

Acid-fast staining is another important differential staining procedure. It is most commonly used to identify *Mycobacterium tuberculosis* and *M. leprae* (figure 2.14*d*), the pathogens responsible for tuberculosis and leprosy, respectively. These bacteria have cell walls with high lipid content; in particular, mycolic acids—a group of branched-chain hydroxy lipids, which prevent dyes from readily binding to the cells. However, *M. tuberculosis* and *M. leprae* can be stained by harsh procedures such as the Ziehl-Neelsen method, which uses heat and phenol to drive basic fuchsin into the cells. Once basic fuchsin has penetrated, *M. tuberculosis* and *M. leprae* are not easily decolorized by acidified alcohol (acid-alcohol), and thus are said to be acid-fast. Non-acid-fast bacteria are decolorized by acid-alcohol and thus are stained blue by methylene blue counterstain.

Staining Specific Structures

Many special staining procedures have been developed to study specific structures with the light microscope. One of the simplest is **capsule staining** (figure 2.14*f*), a technique that reveals the presence of capsules, a network usually made of polysaccharides that surrounds many bacteria and some fungi. Cells are mixed with India ink or nigrosin dye and spread out in a thin film on a slide. After air-drying, the cells appear as lighter bodies in the midst of a blue-black background because ink and dye particles cannot penetrate either the cell or its capsule. Thus capsule staining is an example of **negative staining.** The extent of the light region is determined by the size of the capsule and of the cell itself. There is little distortion of cell shape, and the cell can be counterstained for even greater visibility. Components external to the cell wall: Capsules, slime layers, and S-layers (section 3.9)

Endospore staining, like acid-fast staining, also requires harsh treatment to drive dye into a target, in this case an endospore. An endospore is an exceptionally resistant structure produced by some bacterial genera (e.g., *Bacillus* and *Clostrid-*

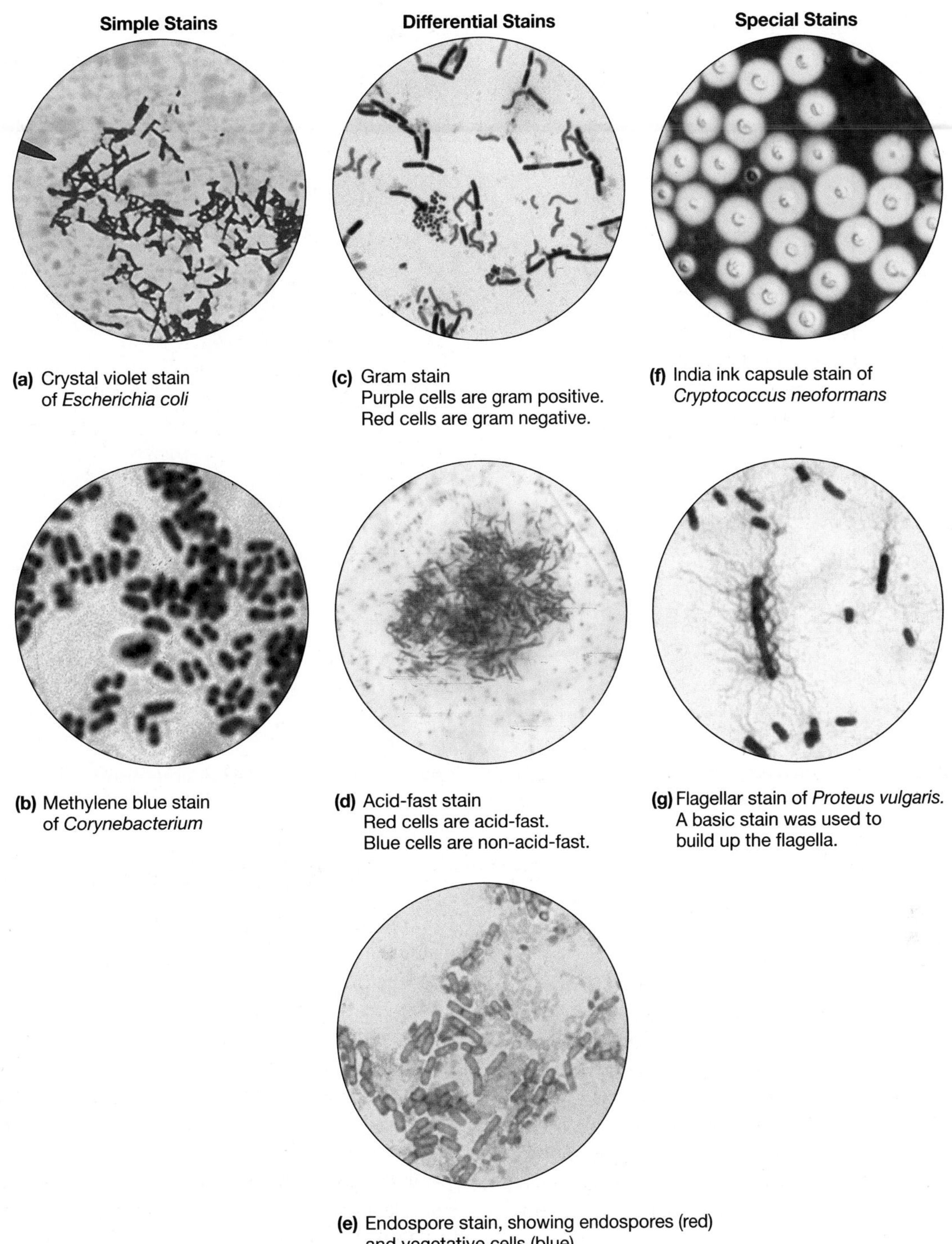

(a) Crystal violet stain of *Escherichia coli*

(c) Gram stain
Purple cells are gram positive.
Red cells are gram negative.

(f) India ink capsule stain of *Cryptococcus neoformans*

(b) Methylene blue stain of *Corynebacterium*

(d) Acid-fast stain
Red cells are acid-fast.
Blue cells are non-acid-fast.

(g) Flagellar stain of *Proteus vulgaris*. A basic stain was used to build up the flagella.

(e) Endospore stain, showing endospores (red) and vegetative cells (blue)

Figure 2.14 Types of Microbiological Stains.

ium). It is capable of surviving for long periods in an unfavorable environment and is called an endospore because it develops within the parent bacterial cell. Endospore morphology and location vary with species and often are valuable in identification; endospores may be spherical to elliptical and either smaller or larger than the diameter of the parent bacterium. Endospores are not stained well by most dyes, but once stained, they strongly resist decolorization. This property is the basis of most endospore staining methods (figure 2.14*e*). In the Schaeffer-Fulton procedure, endospores are first stained by heating bacteria with malachite green, which is a very strong stain that can penetrate endospores. After malachite green treatment, the rest of the cell

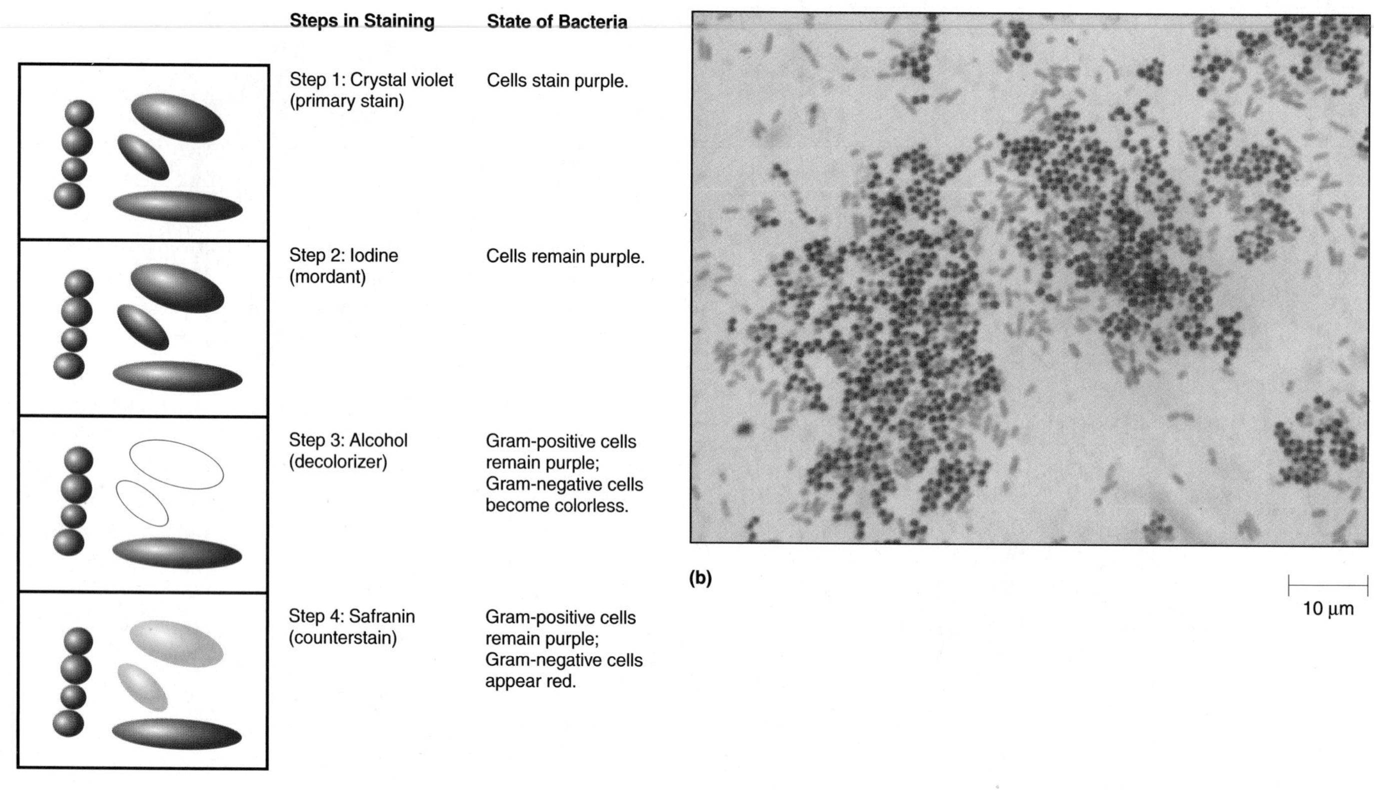

Steps in Staining
State of Bacteria
Step 1: Crystal violet (primary stain)
Cells stain purple.
Step 2: Iodine (mordant)
Cells remain purple.
Step 3: Alcohol (decolorizer)
Gram-positive cells remain purple; Gram-negative cells become colorless.
Step 4: Safranin (counterstain)
Gram-positive cells remain purple; Gram-negative cells appear red.
(a)
(b)
10 μm

Figure 2.15 Gram Stain. **(a)** Steps in the Gram stain procedure. **(b)** Results of a Gram stain. The Gram-positive cells (purple) are *Staphylococcus aureus;* the Gram-negative cells (reddish-pink) are *Escherichia coli.*

is washed free of dye with water and is counterstained with safranin. This technique yields a green endospore resting in a pink to red cell. The bacterial endospore (section 3.11); Class *Clostridia* (section 23.4); and Class *Bacilli* (section 23.5)

Flagella staining provides taxonomically valuable information about the presence and distribution pattern of flagella on procaryotic cells (figure 2.14*g: see also figure 3.39*). Procaryotic flagella are fine, threadlike organelles of locomotion that are so slender (about 10 to 30 nm in diameter) they can only be seen directly using the electron microscope. To observe them with the light microscope, the thickness of flagella is increased by coating them with mordants like tannic acid and potassium alum, and then staining with pararosaniline (Leifson method) or basic fuchsin (Gray method). Components external to the cell wall: Flagella and motility (section 3.9)

1. Define fixation, dye, chromophore, basic dye, acidic dye, simple staining, differential staining, mordant, negative staining, and acid-fast staining.
2. Describe the two general types of fixation. Which would you normally use for procaryotes? For protozoa?
3. Why would one expect basic dyes to be more effective under alkaline conditions?
4. Describe the Gram stain procedure and explain how it works. What step in the procedure could be omitted without losing the ability to distinguish between gram-positive and gram-negative bacteria? Why?
5. How would you visualize capsules, endospores, and flagella?

2.4 Electron Microscopy

For centuries the light microscope has been the most important instrument for studying microorganisms. However, even the very best light microscopes have a resolution limit of about 0.2 μm, which greatly compromises their usefulness for detailed studies of many microorganisms. Viruses, for example, are too small to be seen with light microscopes. Procaryotes can be observed, but because they are usually only 1 μm to 2 μm in diameter, just their general shape and major morphological features are visible. The detailed internal structure of larger microorganisms also cannot be effectively studied by light microscopy. These limitations arise from the nature of visible light waves, not from any inadequacy of the light microscope itself. Electron microscopes have much greater resolution. They have transformed microbiology and added immeasurably to our knowledge. The nature of the electron microscope and the ways in which specimens are prepared for observation are reviewed briefly in this section.

The Transmission Electron Microscope

Electron microscopes use a beam of electrons to illuminate and create magnified images of specimens. Recall that the resolution of a light microscope increases with a decrease in the wavelength of the light it uses for illumination. Electrons replace light as the

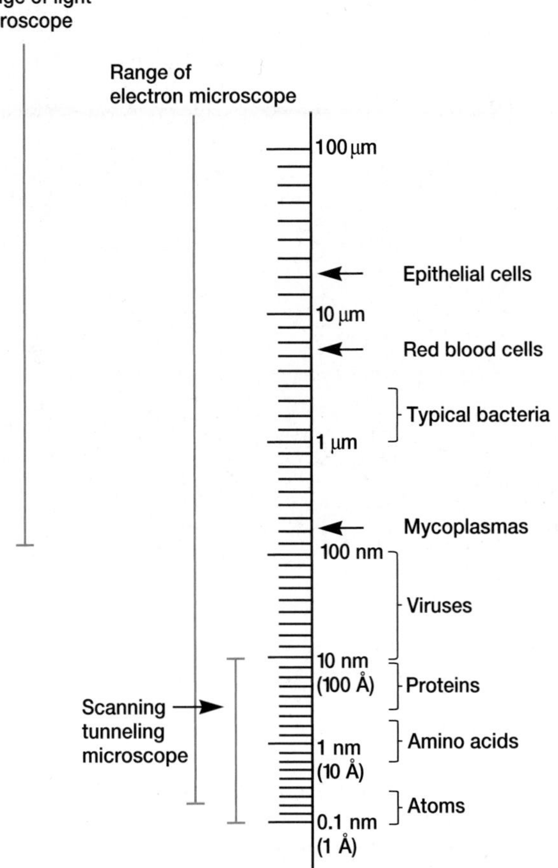

Figure 2.16 The Limits of Microscopic Resolution. Dimensions are indicated with a logarithmic scale (each major division represents a tenfold change in size). To the right side of the scale are the approximate sizes of cells, bacteria, viruses, molecules, and atoms.

illuminating beam. They can be focused, much as light is in a light microscope, but their wavelength is around 0.005 nm, approximately 100,000 times shorter than that of visible light. Therefore, electron microscopes have a practical resolution roughly 1,000 times better than the light microscope; with many electron microscopes, points closer than 0.5 nm can be distinguished, and the useful magnification is well over 100,000× (**figure 2.16**). The value of the electron microscope is evident on comparison of the photographs in **figure 2.17,** microbial morphology can now be studied in great detail.

A modern **transmission electron microscope (TEM)** is complex and sophisticated (**figure 2.18**), but the basic principles behind its operation can be readily understood. A heated tungsten filament in the electron gun generates a beam of electrons that is then focused on the specimen by the condenser (**figure 2.19**). Since electrons cannot pass through a glass lens, doughnut-shaped electromagnets called magnetic lenses are used to focus the beam. The column containing the lenses and specimen must be under high vacuum to obtain a clear image because electrons are deflected by collisions with air molecules. The specimen scatters some electrons, but those that pass through are used to form an enlarged image of the specimen on a fluorescent screen. A denser region in the specimen scatters more electrons and therefore appears darker in the image since fewer electrons strike that area of the screen; these regions are said to be "electron dense." In contrast, electron-transparent regions are brighter. The image can also be captured on photographic film as a permanent record.

Table 2.4 compares some of the important features of light and transmission electron microscopes. The TEM has distinctive features that place harsh restrictions on the nature of samples that can be viewed and the means by which those samples must be prepared. Since electrons are deflected by air molecules and are easily absorbed and scattered by solid matter, only extremely thin slices (20 to 100 nm) of a microbial specimen can be viewed in the average TEM. Such a thin slice cannot be cut unless the specimen has support of some kind; the necessary support is provided by plastic. After fixation with chemicals like glutaraldehyde or osmium tetroxide to stabilize cell structure, the specimen is dehydrated with organic solvents (e.g., acetone or ethanol). Complete dehydration is essential because most plastics used for embedding are not water soluble. Next the specimen is soaked in unpolymerized, liquid epoxy plastic until it is completely permeated, and then the plastic is hardened to form a solid block. Thin sections are cut from this block with a glass or diamond knife using a special instrument called an ultramicrotome.

As with bright-field microscopy, cells usually must be stained before they can be seen clearly. The probability of electron scattering is determined by the density (atomic number) of the specimen atoms. Biological molecules are composed primarily of atoms with low atomic numbers (H, C, N, and O), and electron scattering is fairly constant throughout the unstained cell. Therefore specimens are prepared for observation by soaking thin sections with solutions of heavy metal salts like lead citrate and uranyl acetate. The lead and uranium ions bind to cell structures and make them more electron opaque, thus increasing contrast in the material. Heavy osmium atoms from the osmium tetroxide fixative also "stain" cells and increase their contrast. The stained thin sections are then mounted on tiny copper grids and viewed.

Two other important techniques for preparing specimens are negative staining and shadowing. In negative staining, the specimen is spread out in a thin film with either phosphotungstic acid or uranyl acetate. Just as in negative staining for light microscopy, heavy metals do not penetrate the specimen but render the background dark, whereas the specimen appears bright in photographs. Negative staining is an excellent way to study the structure of viruses, bacterial gas vacuoles, and other similar objects (figure 2.17*c*). In **shadowing,** a specimen is coated with a thin film of platinum or other heavy metal by evaporation at an angle of about 45° from horizontal so that the metal strikes the microorganism on only one side. In one commonly used imaging method, the area coated with metal appears dark in photographs, whereas the uncoated side and the shadow region created by the object is light (**figure 2.20**). This technique is particularly useful in studying virus morphology, procaryotic flagella, and DNA.

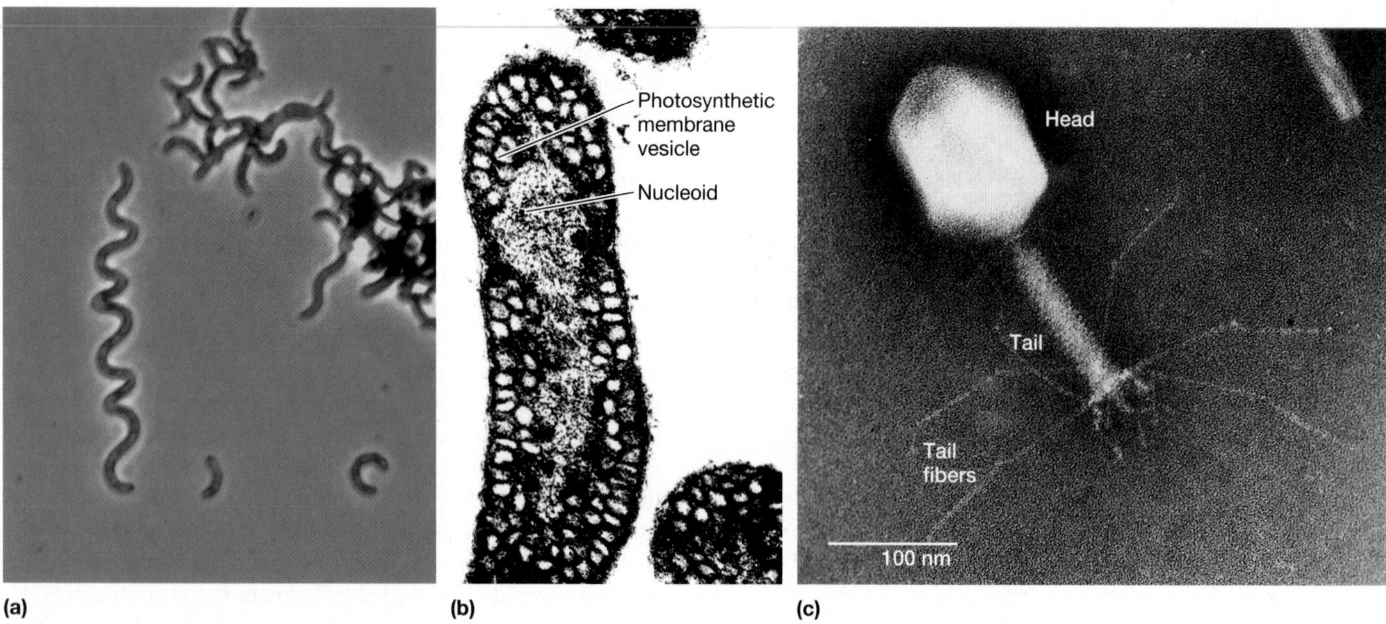

Figure 2.17 Light and Electron Microscopy. A comparison of light and electron microscopic resolution. **(a)** *Rhodospirillum rubrum* in phase-contrast light microscope (×600). **(b)** A thin section of *R. rubrum* in transmission electron microscope (×100,000). **(c)** A transmission electron micrograph of a negatively stained T4 bacteriophage.

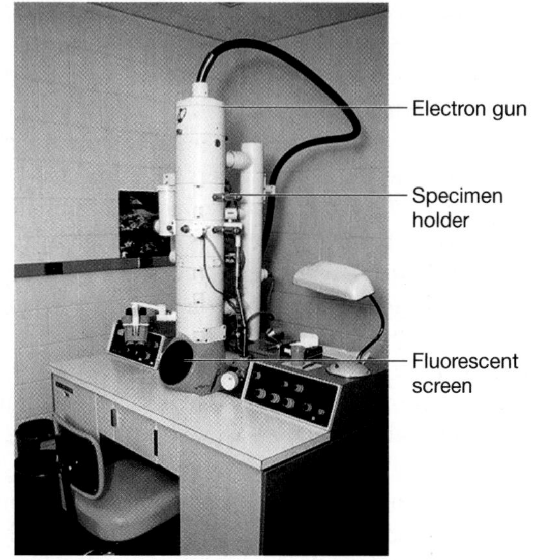

Figure 2.18 A Transmission Electron Microscope. The electron gun is at the top of the central column, and the magnetic lenses are within the column. The image on the fluorescent screen may be viewed through a magnifier positioned over the viewing window. The camera is in a compartment below the screen.

The TEM will also disclose the shape of organelles within microorganisms if specimens are prepared by the **freeze-etching** procedure. First, cells are rapidly frozen in liquid nitrogen and then warmed to −100°C in a vacuum chamber. Next a knife that has been precooled with liquid nitrogen (−196°C) fractures the frozen cells, which are very brittle and break along lines of greatest weakness, usually down the middle of internal membranes (**figure 2.21**). The specimen is left in the high vacuum for a minute or more so that some of the ice can sublimate away and uncover more structural detail. Finally, the exposed surfaces are shadowed and coated with layers of platinum and carbon to form a replica of the surface. After the specimen has been removed chemically, this replica is studied in the TEM and provides a detailed, three-dimensional view of intracellular structure (**figure 2.22**). An advantage of freeze-etching is that it minimizes the danger of artifacts because the cells are frozen quickly rather than being subjected to chemical fixation, dehydration, and plastic embedding.

The Scanning Electron Microscope

Transmission electron microscopes form an image from radiation that has passed through a specimen. The **scanning electron microscope (SEM)** works in a different manner. It produces an image from electrons released from atoms on an object's surface. The SEM has been used to examine the surfaces of microorganisms in great detail; many SEMs have a resolution of 7 nm or less.

Specimen preparation for SEM is relatively easy, and in some cases air-dried material can be examined directly. Most often, however, microorganisms must first be fixed, dehydrated, and dried to preserve surface structure and prevent collapse of the cells when they are exposed to the SEM's high vacuum. Before viewing, dried samples are mounted and coated with a thin layer of metal to prevent the buildup of an electrical charge on the surface and to give a better image.

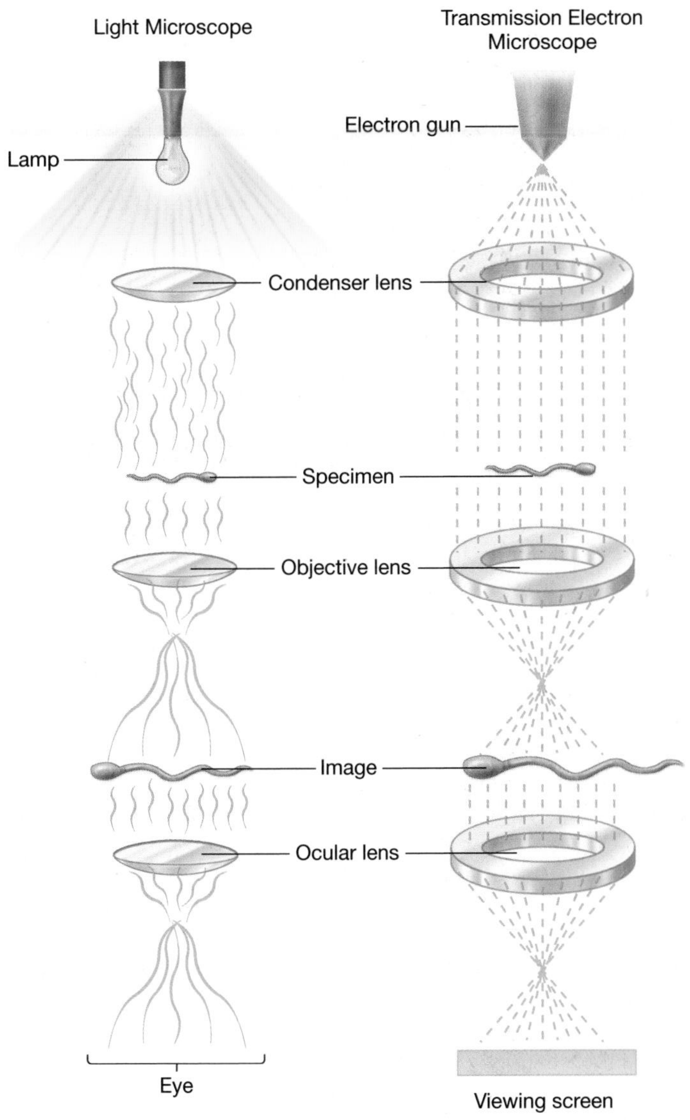

Figure 2.19 Transmission Electron Microscope Operation. An overview of TEM operation and a comparison of the operation of light and transmission electron microscopes.

To create an image, the SEM scans a narrow, tapered electron beam back and forth over the specimen (**figure 2.23**). When the beam strikes a particular area, surface atoms discharge a tiny shower of electrons called secondary electrons, and these are trapped by a special detector. Secondary electrons entering the detector strike a scintillator causing it to emit light flashes that a photomultiplier converts to an electrical current and amplifies. The signal is sent to a cathode-ray tube and produces an image like a television picture, which can be viewed or photographed.

The number of secondary electrons reaching the detector depends on the nature of the specimen's surface. When the electron beam strikes a raised area, a large number of secondary electrons enter the detector; in contrast, fewer electrons escape a depression in the surface and reach the detector. Thus raised areas appear lighter on the screen and depressions are darker. A realistic three-dimensional image of the microorganism's surface results (**figure 2.24**). The actual in situ location of microorganisms in ecological niches such as the human skin and the lining of the gut also can be examined.

1. Why does the transmission electron microscope have much greater resolution than the light microscope?
2. Describe in general terms how the TEM functions. Why must the TEM use a high vacuum and very thin sections?
3. Material is often embedded in paraffin before sectioning for light microscopy. Why can't this approach be used when preparing a specimen for the TEM?
4. Under what circumstances would it be desirable to prepare specimens for the TEM by use of negative staining? Shadowing? Freeze-etching?
5. How does the scanning electron microscope operate and in what way does its function differ from that of the TEM? The SEM is used to study which aspects of morphology?

2.5 Newer Techniques in Microscopy

Confocal Microscopy

Like the large and small beads illustrated in **figure 2.25*a***, biological specimens are three-dimensional. When three-dimensional objects are viewed with traditional light microscopes, light from

Table 2.4 Characteristics of Light and Transmission Electron Microscopes

Feature	Light Microscope	Transmission Electron Microscope
Highest practical magnification	About 1,000–1,500	Over 100,000
Best resolution[a]	0.2 μm	0.5 nm
Radiation source	Visible light	Electron beam
Medium of travel	Air	High vacuum
Type of lens	Glass	Electromagnet
Source of contrast	Differential light absorption	Scattering of electrons
Focusing mechanism	Adjust lens position mechanically	Adjust current to the magnetic lens
Method of changing magnification	Switch the objective lens or eyepiece	Adjust current to the magnetic lens
Specimen mount	Glass slide	Metal grid (usually copper)

[a]The resolution limit of a human eye is about 0.2 mm.

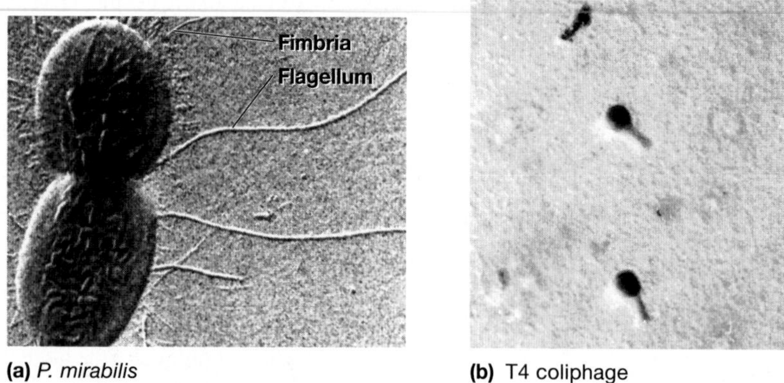

Figure 2.20 Specimen Shadowing for the TEM. Examples of specimens viewed in the TEM after shadowing with uranium metal. **(a)** *Proteus mirabilis* (×42,750); note flagella and fimbriae. **(b)** T4 coliphage (×72,000).

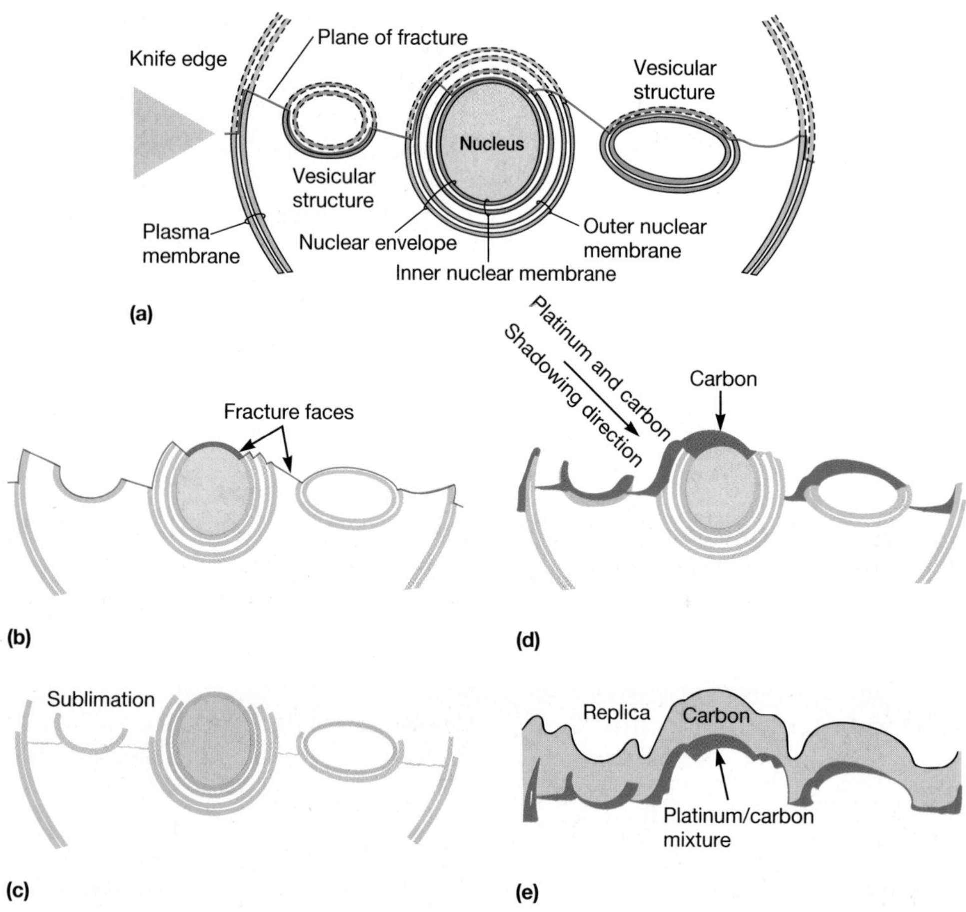

Figure 2.21 The Freeze-Etching Technique. In steps **(a)** and **(b)**, a frozen eucaryotic cell is fractured with a cold knife. Etching by sublimation is depicted in **(c)**. Shadowing with platinum plus carbon and replica formation are shown in **(d)** and **(e)**. See text for details.

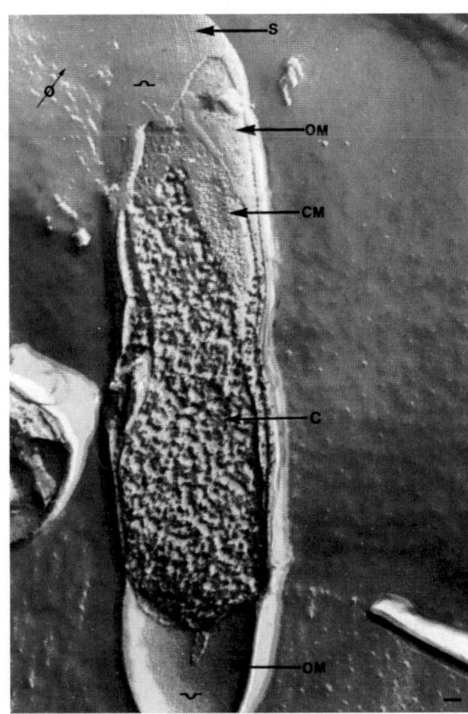

Figure 2.22 Example of Freeze-Etching. A freeze-etched preparation of the bacterium *Thiobacillus kabobis*. Note the differences in structure between the outer surface, S; the outer membrane of the cell wall, OM; the cytoplasmic membrane, CM; and the cytoplasm, C. Bar = 0.1 μm.

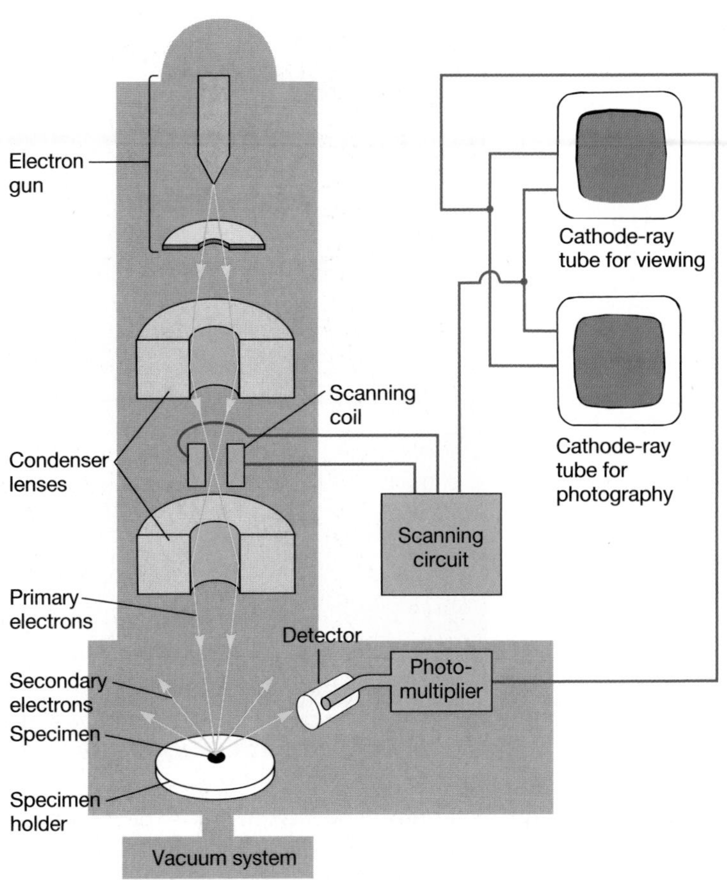

Figure 2.23 The Scanning Electron Microscope. See text for explanation.

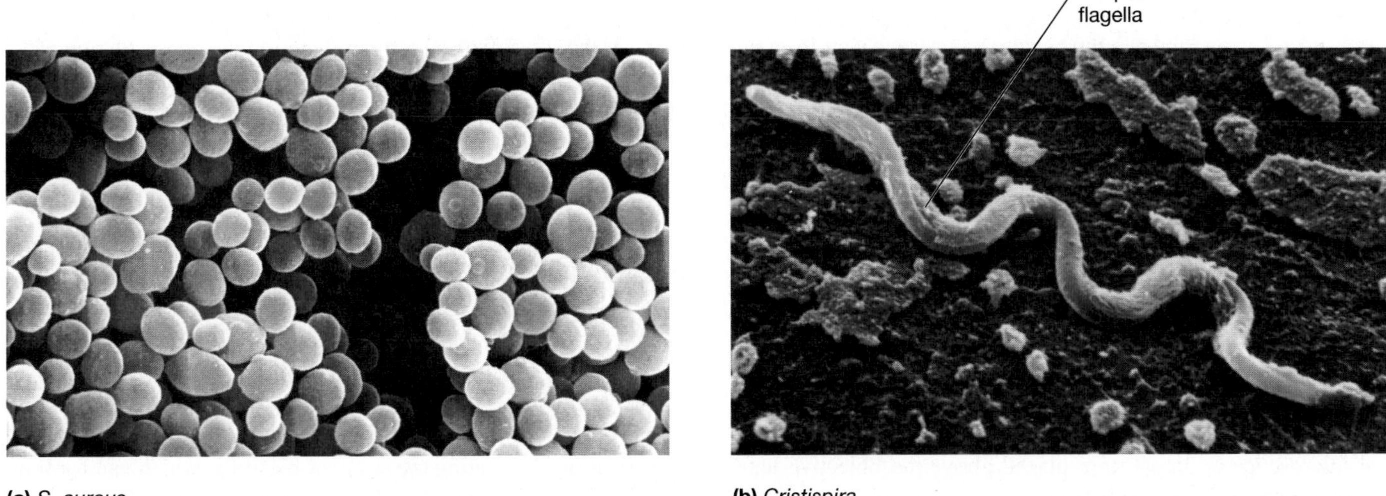

Figure 2.24 Scanning Electron Micrographs of Bacteria. **(a)** *Staphylococcus aureus* (×32,000). **(b)** *Cristispira,* a spirochete from the crystalline style of the oyster, *Ostrea virginica.* The axial fibrils or periplasmic flagella are visible around the protoplasmic cylinder (×6,000).

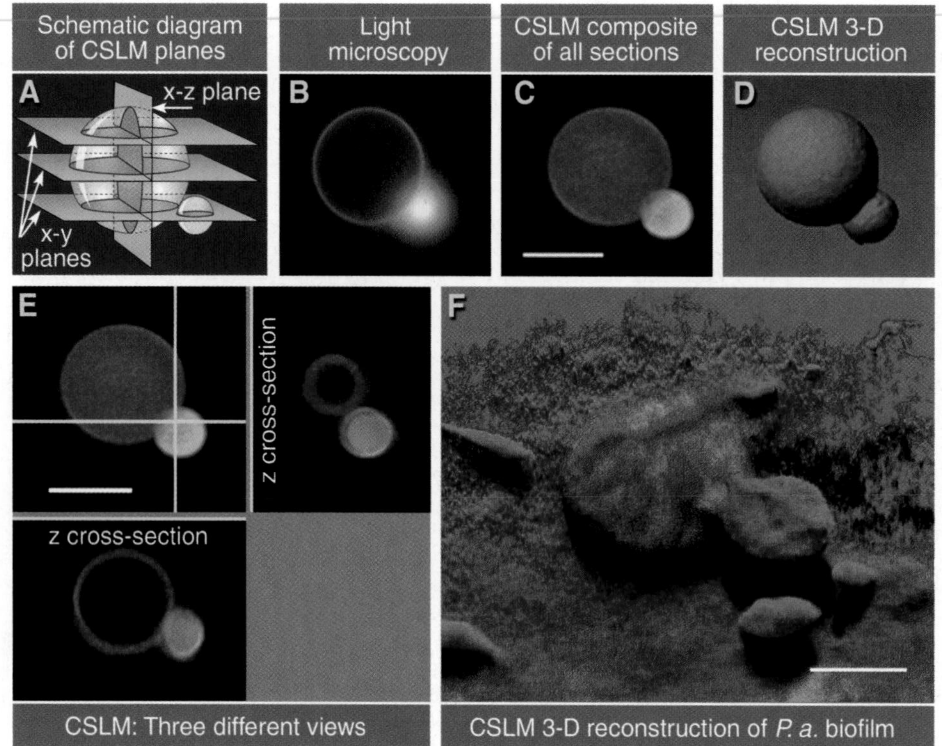

Figure 2.25 Light and Confocal Microscopy. Two beads examined by light and confocal microscopy. Light microscope images are generated from light emanating from many areas of a three-dimensional object. Confocal images are created from light emanating from only a single plane of focus. Multiple planes within the object can be examined and used to construct clear, finely detailed images. **(a)** The planes observable by confocal microscopy. **(b)** The light microscope image of the two beads shown in (a). Note that neither bead is clear and that the smaller bead is difficult to recognize as a bead. **(c)** A computer connected to a confocal microscope can make a composite image of the two beads using digitized information collected from multiple planes within the beads. The result is a much clearer and more detailed image. **(d)** The computer can also use digitized information collected from multiple planes within the beads to generate a three-dimensional reconstruction of the beads. **(e)** Computer generated views of a specimen: the top left panel is the image of a single x-y plane (i.e., looking down from the top of the specimen). The two lines represent the two x-z planes imaged in the other two panels. The vertical line indicates the x-z plane shown in the top right panel (i.e., a view from the right side of the specimen) and the horizontal line indicates the x-z plane shown in the bottom panel (i.e., a view from the front face of the specimen). **(f)** A three-dimensional reconstruction of a *Pseudomonas aeruginosa* biofilm. The biofilm was exposed to an antibacterial agent and then stained with dyes that distinguish living (green) from dead (red) cells. The cells on the surface of the biofilm have been killed, but those in the lower layers of the biofilm are still alive.

all areas of the object, not just the plane of focus, enter the microscope and are used to create an image. The resulting image is murky and fuzzy (figure 2.25*b*). This problem has been solved by the development of the **confocal scanning laser microscope (CSLM),** or simply, confocal microscope. The confocal microscope uses a laser beam to illuminate a specimen, usually one that has been fluorescently stained. A major component of the confocal microscope is an aperture placed above the objective lens, which eliminates stray light from parts of the specimen that lie above and below the plane of focus (**figure 2.26**). Thus the only light used to create the image is from the plane of focus, and a much clearer sharp image is formed.

Computers are integral to the process of creating confocal images. A computer attached to the confocal microscope receives digitized information from each plane in the specimen that is examined. This information can be used to create a composite image that is very clear and detailed (figure 2.25*c*) or to create a three-dimensional reconstruction of the specimen (figure 2.25*d*). Images of x-z plane cross-sections of the specimen can also be generated, giving the observer views of the specimen from three perspectives (figure 2.25*e*). Confocal microscopy has numerous applications, including the study of biofilms, which can form on many different types of surfaces including in-dwelling medical devices such as hip joint replacements. As shown in figure 2.25*f*, it is difficult to kill all cells in a biofilm. This makes them a particular concern to the medical field because formation of biofilms on medical devices can result in infections that are difficult to treat. Microbial growth in natural environments: Biofilms (section 6.6)

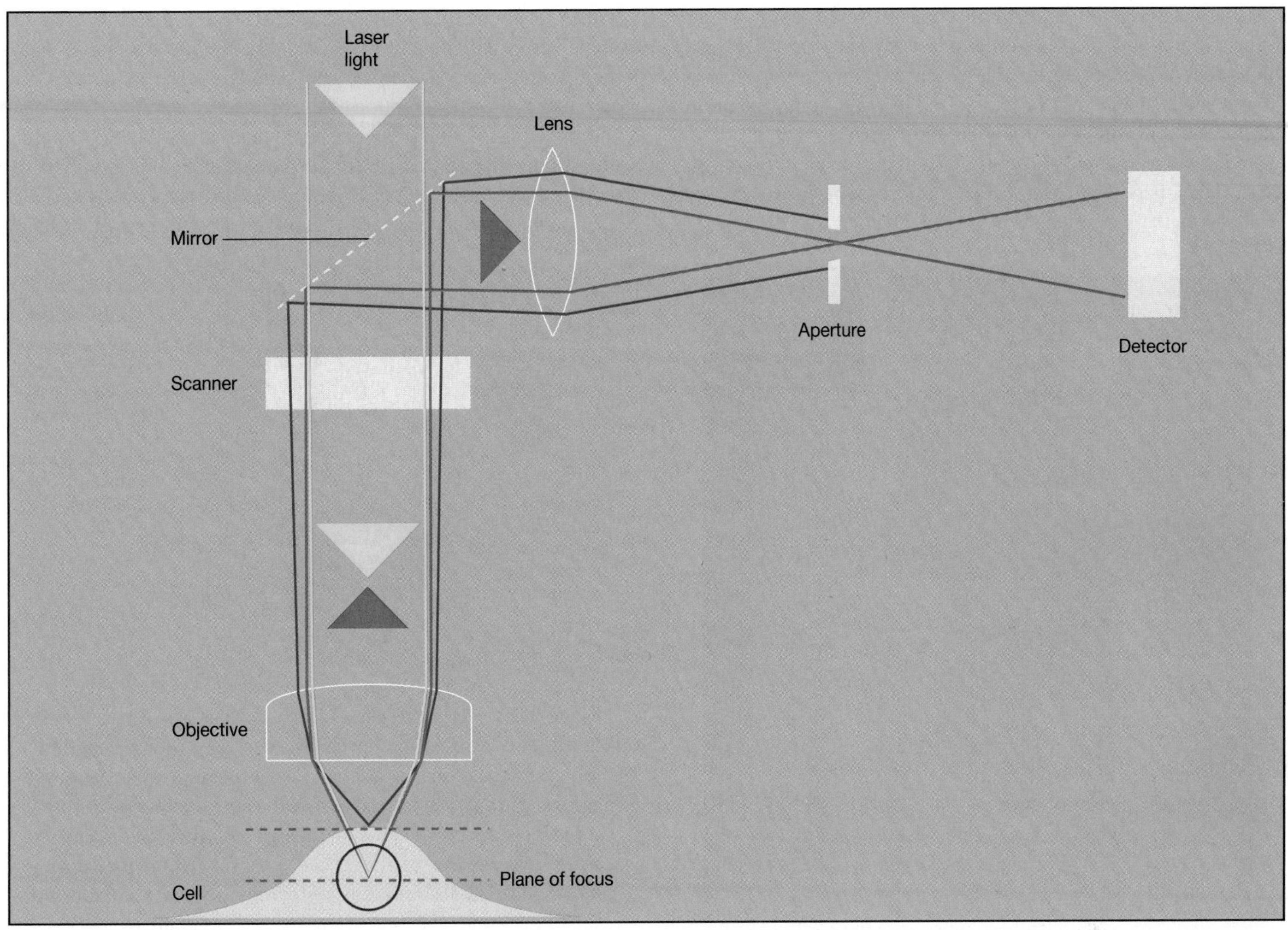

Figure 2.26 A Ray Diagram of a Confocal Laser Scanning Microscope. The yellow lines represent laser light used for illumination. Red lines symbolize the light arising from the plane of focus, and the blue lines stand for light from parts of the specimen above and below the focal plane. See text for explanation.

Scanning Probe Microscopy

Although light and electron microscopes have become quite sophisticated and have reached an advanced state of development, powerful new microscopes are still being created. A new class of microscopes, called **scanning probe microscopes,** measure surface features by moving a sharp probe over the object's surface. The **scanning tunneling microscope,** invented in 1980, is an excellent example of a scanning probe microscope. It can achieve magnifications of 100 million and allow scientists to view atoms on the surface of a solid. The electrons surrounding surface atoms tunnel or project out from the surface boundary a very short distance. The scanning tunneling microscope has a needlelike probe with a point so sharp that often there is only one atom at its tip. The probe is lowered toward the specimen surface until its electron cloud just touches that of the surface atoms. If a small voltage is applied between the tip and specimen, electrons flow through a narrow channel in the electron clouds. This tunneling current, as it is called, is extraordinarily sensitive to distance and will decrease about a thousandfold if the probe is moved away from the surface by a distance equivalent to the diameter of an atom.

The arrangement of atoms on the specimen surface is determined by moving the probe tip back and forth over the surface while keeping it at a constant height above the specimen by adjusting the probe distance to maintain a steady tunneling current. As the tip moves up and down while following the surface contours, its motion is recorded and analyzed by a computer to create an accurate three-dimensional image of the surface atoms. The surface map can be displayed on a computer screen or plotted on paper. The resolution is so great that individual atoms are observed easily. The microscope's inventors, Gerd Binnig and Heinrich Rohrer, shared the 1986 Nobel Prize in Physics for their work, together with Ernst Ruska, the designer of the first transmission electron microscope.

The scanning tunneling microscope is already having a major impact in biology. It can be used to directly view DNA and other biological molecules (**figure 2.27**). Since the microscope can examine objects when they are immersed in water, it may be particularly useful in studying biological molecules.

More recently a second type of scanning probe microscope has been developed. The **atomic force microscope** moves a sharp probe over the specimen surface while keeping the distance between the probe tip and the surface constant. It does this by exerting a very small amount of force on the tip, just enough to maintain a constant distance but not enough force to damage the surface. The vertical motion of the tip usually is followed by measuring the deflection of a laser beam that strikes the lever holding the probe (**figure 2.28**). Unlike the scanning tunneling microscope, the atomic force microscope can be used to study surfaces that do not conduct electricity well. The atomic force microscope

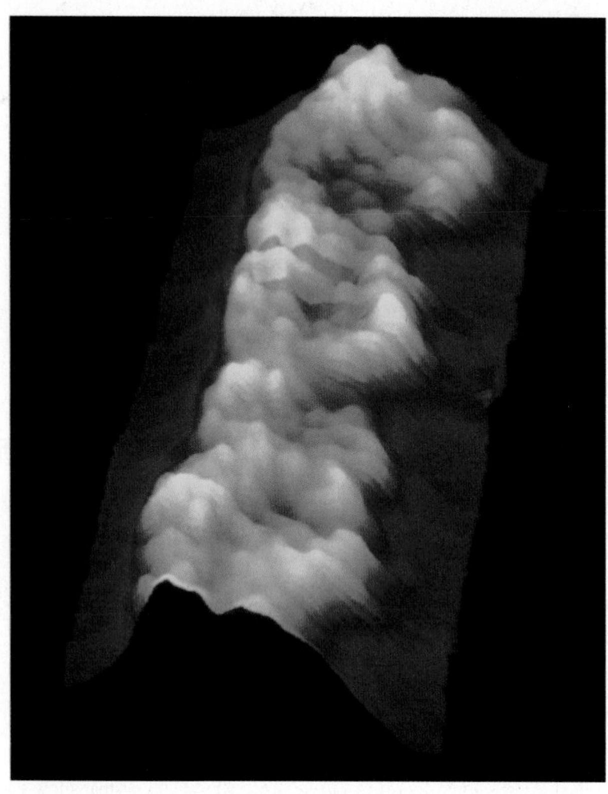

Figure 2.27 Scanning Tunneling Microscopy of DNA. The DNA double helix with approximately three turns shown (false color; ×2,000,000).

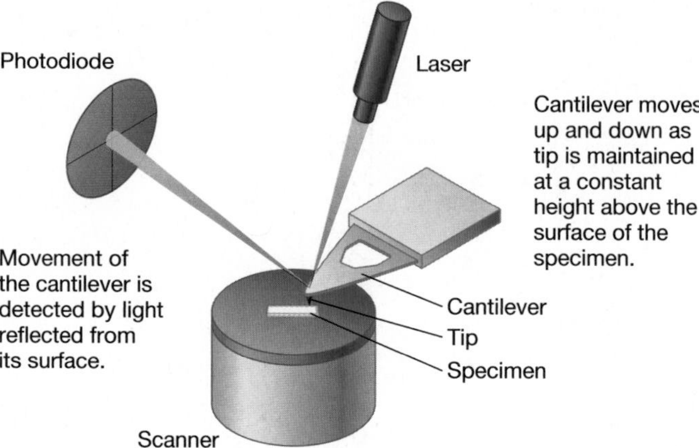

Figure 2.28 Atomic Force Microscopy—The Basic Elements of an Atomic Force Microscope. The tip used to probe the specimen is attached to a cantilever. As the probe passes over the "hills and valleys" of the specimen's surface, the cantilever is deflected vertically. A laser beam directed at the cantilever is used to monitor these vertical movements. Light reflected from the cantilever is detected by the photodiode and used to generate an image of the specimen.

(a)

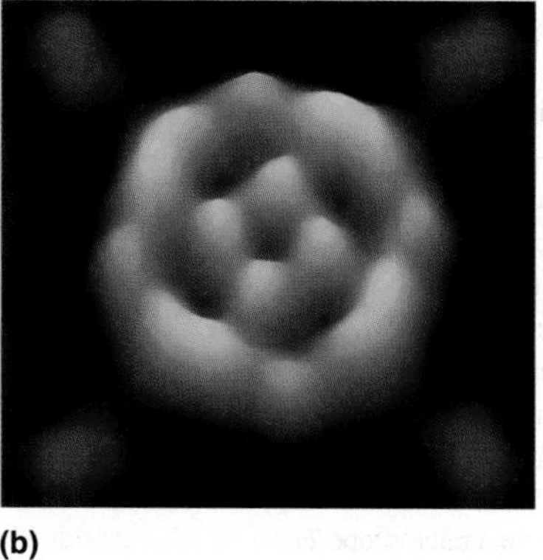

(b)

Figure 2.29 The Membrane Protein Aquaporin Visualized by Atomic Force Micrscopy. Aquaporin is a membrane-spanning protein that allows water to move across the membrane. **(a)** Each circular structure represents the surface view of a single aquaporin protein. **(b)** A single aquaporin molecule observed in more detail and at higher magnification.

has been used to study the interactions between the *E. coli* GroES and GroEL chaperone proteins, to map plasmids by locating restriction enzymes bound to specific sites, to follow the behavior of living bacteria and other cells, and to visualize membrane proteins (**figure 2.29**).

1. How does a confocal microscope operate? Why does it provide better images of thick specimens than does the standard compound microscope?
2. Briefly describe the scanning probe microscope and compare and contrast its most popular versions—the scanning tunneling microscope and the atomic force microscope. What are these microscopes used for?

Summary

2.1 Lenses and the Bending of Light

a. A light ray moving from air to glass, or vice versa, is bent in a process known as refraction.

b. Lenses focus light rays at a focal point and magnify images (**figure 2.2**).

2.2 The Light Microscope

a. In a compound microscope like the bright-field microscope, the primary image is formed by an objective lens and enlarged by the eyepiece or ocular lens to yield the final image (**figure 2.3**).

b. A substage condenser focuses a cone of light on the specimen.

c. Microscope resolution increases as the wavelength of radiation used to illuminate the specimen decreases. The maximum resolution of a light microscope is about 0.2 μm.

d. The dark-field microscope uses only refracted light to form an image (**figure 2.7**), and objects glow against a black background.

e. The phase-contrast microscope converts variations in the refractive index and density of cells into changes in light intensity and thus makes colorless, unstained cells visible (**figure 2.9**).

f. The differential interference contrast microscope uses two beams of light to create high-contrast, three-dimensional images of live specimens.

g. The fluorescence microscope illuminates a fluorochrome-labeled specimen and forms an image from its fluorescence (**figure 2.12**).

2.3 Preparation and Staining of Specimens

a. Specimens usually must be fixed and stained before viewing them in the bright-field microscope.

b. Most dyes are either positively charged basic dyes or negative acidic dyes and bind to ionized parts of cells.

c. In simple staining a single dye is used to stain microorganisms.

d. Differential staining procedures like the Gram stain and acid-fast stain distinguish between microbial groups by staining them differently (**figure 2.15**).

e. Some staining techniques are specific for particular structures like bacterial capsules, flagella, and endospores (**figure 2.14**).

2.4 Electron Microscopy

a. The transmission electron microscope uses magnetic lenses to form an image from electrons that have passed through a very thin section of a specimen (**figure 2.19**). Resolution is high because the wavelength of electrons is very short.

b. Thin section contrast can be increased by treatment with solutions of heavy metals like osmium tetroxide, uranium, and lead.

c. Specimens are also prepared for the TEM by negative staining, shadowing with metal, or freeze-etching.

d. The scanning electron microscope (**figure 2.23**) is used to study external surface features of microorganisms.

2.5 Newer Techniques in Microscopy

a. The confocal scanning laser microscope (**figure 2.25**) is used to study thick, complex specimens.

b. Scanning probe microscopes reach very high magnifications that allow scientists to observe biological molecules (**figures 2.27** and **2.29**).

Key Terms

acidic dyes 26
acid-fast staining 26
atomic force microscope 36
basic dyes 26
bright-field microscope 18
capsule staining 26
chemical fixation 26
chromophore groups 26
confocal scanning laser microscope (CSLM) 34
dark-field microscope 21
differential interference contrast (DIC) microscope 23
differential staining 26
endospore staining 26
eyepieces 18
fixation 25
flagella staining 28
fluorescence microscope 23
fluorescent light 23
fluorochromes 24
focal length 18
focal point 18
freeze-etching 30
Gram stain 26
heat fixation 25
mordant 26
negative staining 26
numerical aperture 19
objective lenses 18
ocular lenses 18
parfocal 18
phase-contrast microscope 21
refraction 17
refractive index 17
resolution 18
scanning electron microscope (SEM) 30
scanning probe microscope 35
scanning tunneling microscope 35
shadowing 29
simple staining 26
substage condenser 18
transmission electron microscope (TEM) 29
working distance 20

Critical Thinking Questions

1. If you prepared a sample of a specimen for light microscopy, stained with the Gram stain, and failed to see anything when you looked through your light microscope, list the things that you may have done incorrectly.
2. In a journal article, find an example of a light micrograph, a scanning or transmission electron micrograph, or a confocal image. Discuss why the figure was included in the article and why that particular type of microscopy was the method of choice for the research. What other figures would you like to see used in this study? Outline the steps that the investigators would take in order to obtain such photographs or figures.

Learn More

Binning, G., and Rohrer, H. 1985. The scanning tunneling microscope. *Sci. Am.* 253(2):50–56.

Dufrêne, Y. F. 2003. Atomic force microscopy provides a new means for looking at microbial cells. *ASM News* 69(9):438–42.

Hörber, J.K.H., and Miles, M. J. 2003. Scanning probe evolution in biology. *Science* 302:1002–5.

Lillie, R. D. 1969. *H. J. Conn's biological stains,* 8th ed. Baltimore: Williams & Wilkins.

Rochow, T. G. 1994. *Introduction to microscopy by means of light, electrons, X-rays, or acoustics*. New York: Plenum.

Scherrer, Rene. 1984. Gram's staining reaction, Gram types and cell walls of bacteria. *Trends Biochem. Sci.* 9:242–45.

Stephens, D. J., and Allan, V. J. 2003. Light microscopy techniques for live cell imaging. *Science* 300:82–6.

Please visit the Prescott website at www.mhhe.com/prescott7 for additional references.

3 Procaryotic Cell Structure and Function

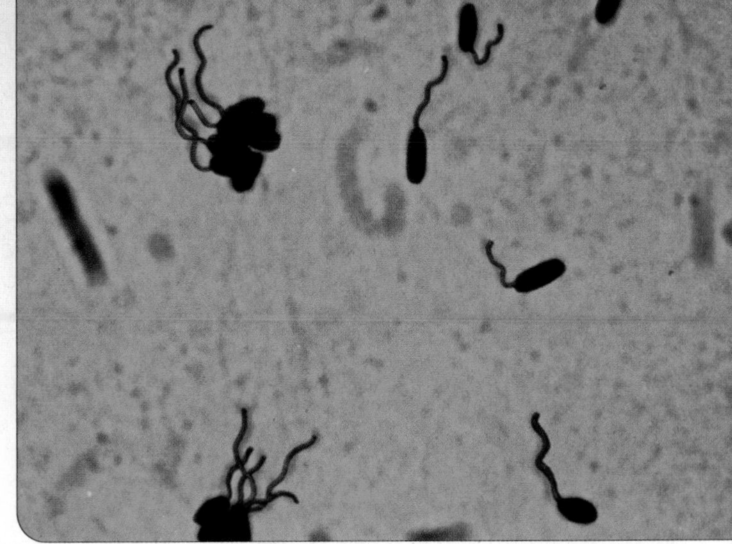

Bacterial species may differ in their patterns of flagella distribution. These *Pseudomonas* cells have a single polar flagellum used for locomotion.

Preview

- Procaryotes can be distinguished from eucaryotes in terms of their size, cell structure, and molecular make-up. Most procaryotes lack extensive, complex, internal membrane systems.
- Although some cell structures are observed in both eucaryotic and procaryotic cells, some structures are unique to procaryotes.
- Procaryotes can be divided into two major groups: *Bacteria* and *Archaea*. Although similar in overall structure, *Bacteria* and *Archaea* exhibit important differences in their cell walls and membranes.
- Most bacteria can be divided into two broad groups based on cell wall structure; the differences in cell wall structure correlate with the reaction to the Gram staining procedure.
- Many procaryotes are motile; several mechanisms for motility have been identified.
- Some bacteria form resistant endospores to survive harsh environmental conditions in a dormant state.

Even a superficial examination of the microbial world shows that procaryotes are one of the most important groups by any criterion: numbers of organisms, general ecological importance, or practical importance for humans. Indeed, much of our understanding of phenomena in biochemistry and molecular biology comes from research on procaryotes. Although considerable space in this text is devoted to eucaryotic microorganisms, the major focus is on procaryotes. Therefore the unit on microbial morphology begins with the structure of procaryotes. As mentioned in chapter 1, there are two quite different groups of procaryotes: *Bacteria* and *Archaea*. Although considerably less is known about archaeal cell structure and biochemistry, certain features distinguish the two domains. Whenever possible, these distinctions will be noted. A more detailed discussion of the *Archaea* is provided in chapter 20. A comment about nomenclature is necessary to avoid confusion. The word procaryote will be used in a general sense to include both the *Bacteria* and *Archaea;* the term bacterium will refer specifically to a member of the *Bacteria* and archaeon to a member of the *Archaea.* Members of the microbial world (section 1.1)

3.1 An Overview of Procaryotic Cell Structure

Because much of this chapter is devoted to a discussion of individual cell components, a preliminary overview of the procaryotic cell as a whole is in order.

Shape, Arrangement, and Size

One might expect that small, relatively simple organisms like procaryotes would be uniform in shape and size. This is not the case, as the microbial world offers almost endless variety in terms of morphology (**figures 3.1** and **3.2**). However, most commonly encountered procaryotes have one of two shapes. **Cocci** (s., **coccus**) are roughly spherical cells. They can exist as individual cells, but also are associated in characteristic arrangements that are frequently useful in their identification. **Diplococci** (s., **diplococcus**) arise when cocci divide and remain together to form pairs. Long chains of cocci result when cells adhere after repeated divisions in one plane; this pattern is seen in the genera *Streptococcus, Enterococcus,* and *Lactococcus* (figure 3.1*b*). *Staphylococcus* divides in random planes to generate irregular grapelike clumps (figure 3.1*a*). Divisions in two or three planes

The era in which workers tended to look at bacteria as very small bags of enzymes has long passed.

—*Howard J. Rogers*

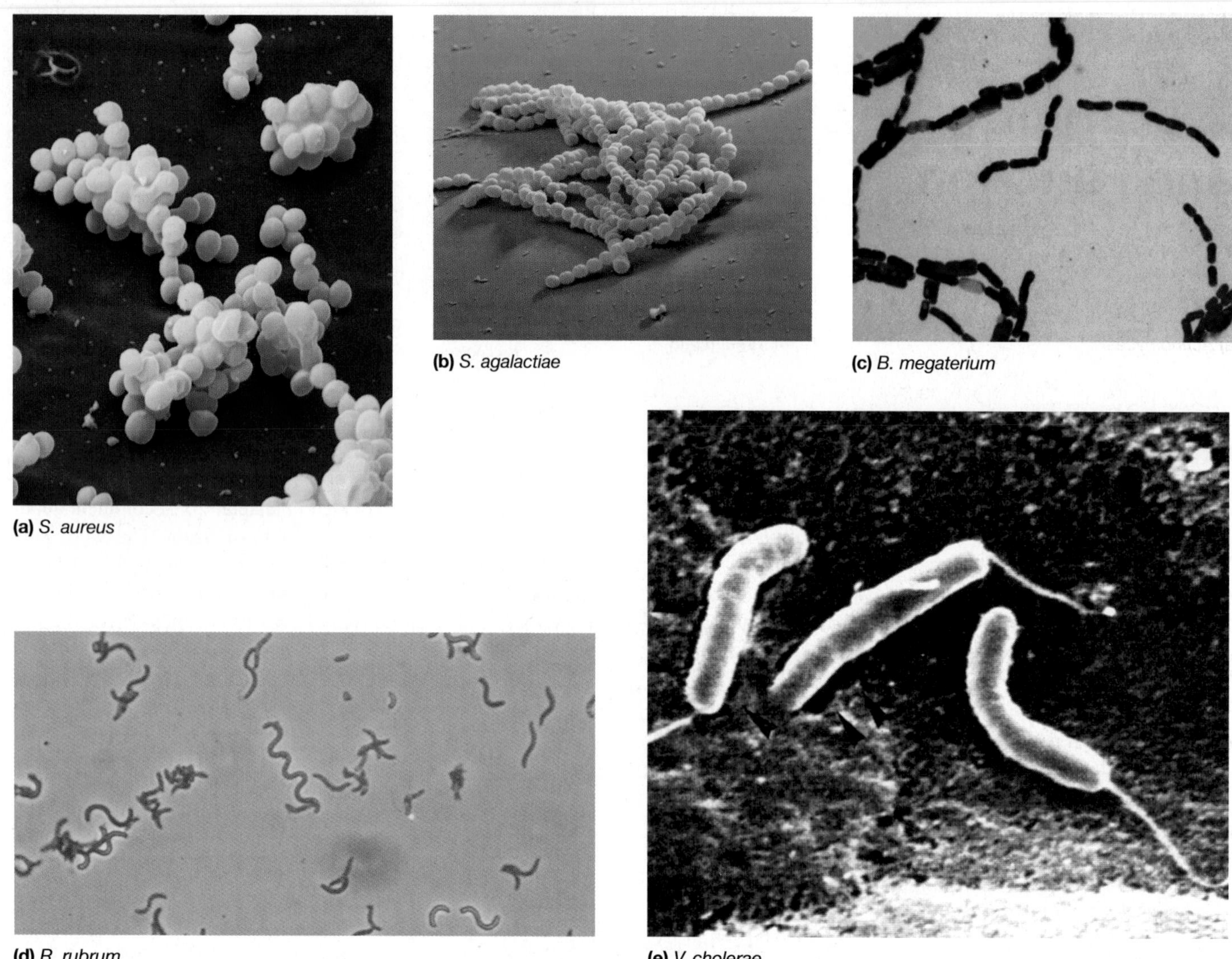

(a) *S. aureus*

(b) *S. agalactiae*

(c) *B. megaterium*

(d) *R. rubrum*

(e) *V. cholerae*

Figure 3.1 Common Procaryotic Cell Shapes. **(a)** *Staphylococcus aureus* cocci arranged in clusters; color-enhanced scanning electron micrograph; average cell diameter is about 1 μm. **(b)** *Streptococcus agalactiae,* the cause of Group B streptococcal infections; cocci arranged in chains; color-enhanced scanning electron micrograph (×4,800). **(c)** *Bacillus megaterium,* a rod-shaped bacterium arranged in chains, Gram stain (×600). **(d)** *Rhodospirillum rubrum,* phase contrast (×500). **(e)** *Vibrio cholera,* curved rods with polar flagella; scanning electron micrograph.

can produce symmetrical clusters of cocci. Members of the genus *Micrococcus* often divide in two planes to form square groups of four cells called tetrads. In the genus *Sarcina,* cocci divide in three planes producing cubical packets of eight cells.

The other common shape is that of a **rod,** sometimes called a **bacillus** (pl., **bacilli**). *Bacillus megaterium* is a typical example of a bacterium with a rod shape (figure 3.1*c*). Bacilli differ considerably in their length-to-width ratio, the coccobacilli being so short and wide that they resemble cocci. The shape of the rod's end often varies between species and may be flat, rounded, cigar-shaped, or bifurcated. Although many rods occur singly, some remain together after division to form pairs or chains (e.g., *Bacillus megaterium* is found in long chains).

Although procaryotes are often simple spheres or rods, other cell shapes and arrangements are not uncommon. **Vibrios** most closely resemble rods, as they are comma-shaped (figure 3.1*e*). Spiral-shaped procaryotes can be either classified as **spirilla,** which usually have tufts of flagella at one or both ends of the cell (figure 3.1*d* and 3.2*c*), or spirochetes. **Spirochetes** are more flexible and have a unique, internal flagellar arrangement. Actinomycetes typically form long filaments called hyphae that may branch to produce a network called a **mycelium** (figure 3.2a). In this sense, they are

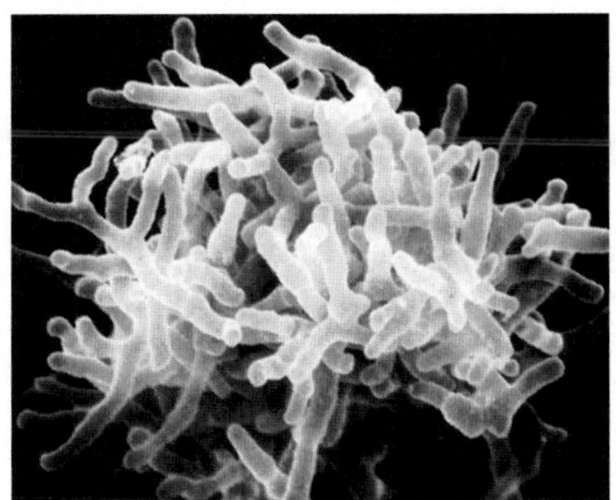
(a) *Actinomyces*

(b) *M. pneumoniae*

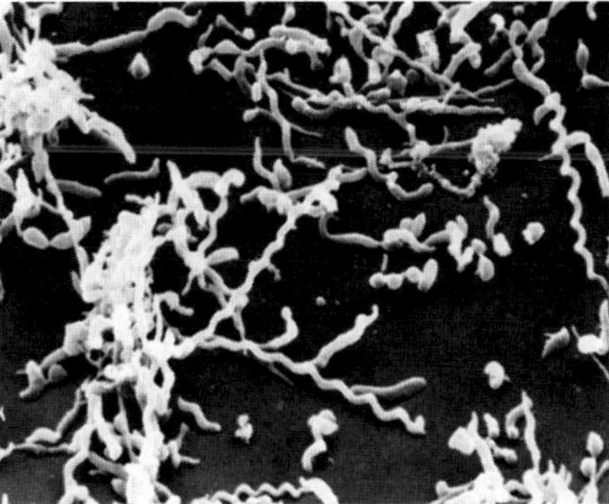
(c) *Spiroplasma*

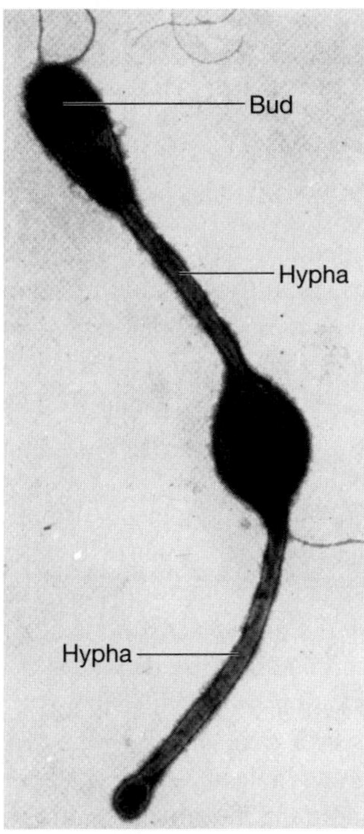

(d) *Hyphomicrobium*

(e) Walsby's square archaeon

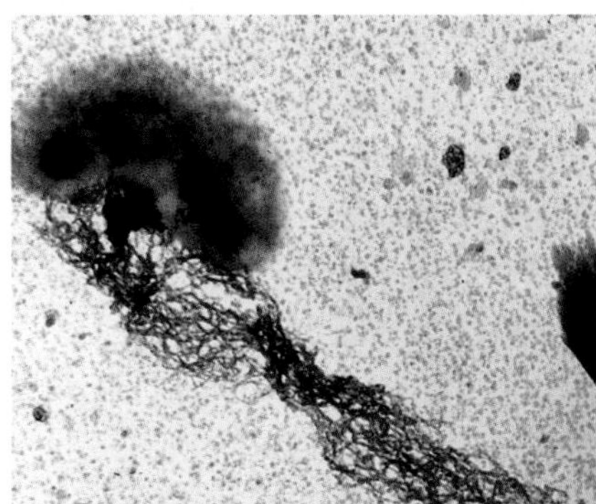
(f) *G. ferruginea*

Figure 3.2 Unusually Shaped Procaryotes. Examples of procaryotes with shapes quite different from bacillus and coccus types. **(a)** *Actinomyces,* SEM (×21,000). **(b)** *Mycoplasma pneumoniae,* SEM (×62,000). **(c)** *Spiroplasma,* SEM (×13,000). **(d)** *Hyphomicrobium* with hyphae and bud, electron micrograph with negative staining. **(e)** Walsby's square archaeon. **(f)** *Gallionella ferruginea* with stalk.

similar to filamentous fungi, a group of eucaryotic microbes. The oval- to pear-shaped *Hyphomicrobium* (figure 3.2*d*) produces a bud at the end of a long hypha. Other bacteria such as *Gallionella* produce nonliving stalks (figure 3.2*f*). A few procaryotes actually are flat. For example, Anthony Walsby has discovered square archaea living in salt ponds (figure 3.2*e*). They are shaped like flat, square-to-rectangular boxes about 2 μm by 2 to 4 μm, and only 0.25 μm thick. Finally, some procaryotes are variable in shape and lack a single, characteristic form (figure 3.2*b*). These are called **pleomorphic** even though they may, like *Corynebacterium,* have a generally rod-like form. Phylum *Spirochaetes* (section 21.6)

Bacteria vary in size as much as in shape (**figure 3.3**). *Escherichia coli* is a rod of about average size, 1.1 to 1.5 μm wide by 2.0 to 6.0 μm long. Near the small end of the size continuum are members of the genus *Mycoplasma,* an interesting group of bacteria that lack cell walls. For many years, it was thought that they were the smallest procaryotes at about 0.3 μm in diameter, approximately the size of the poxviruses. However, even smaller procaryotes have been discovered. Nanobacteria range from around 0.2 μm to less than 0.05 μm in diameter. Only a few strains have been cultured, and these appear to be very small, bacteria-like organisms. The discovery of nanobacteria was quite surprising because theoretical calculations predicted that the smallest cells were about 0.14 to 0.2 μm in diameter. At the other end of the continuum are bacteria such as the spirochaetes, which can reach 500 μm in length, and the photosynthetic bacterium *Oscillatoria,* which is about 7 μm in diameter (the same diameter as a red blood cell). A huge bacterium lives in the intestine of

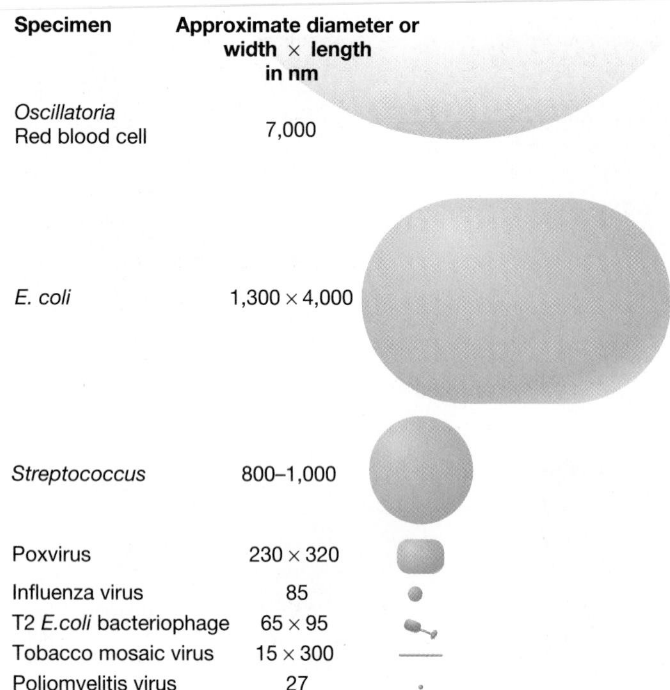

Specimen
Approximate diameter or width × length in nm
Oscillatoria
Red blood cell
7,000
E. coli
1,300 × 4,000
Streptococcus
800–1,000
Poxvirus
230 × 320
Influenza virus
85
T2 E.coli bacteriophage
65 × 95
Tobacco mosaic virus
15 × 300
Poliomyelitis virus
27

Figure 3.3 Sizes of Procaryotes and Viruses. The sizes of selected bacteria relative to a red blood cell and viruses.

the brown surgeonfish, *Acanthurus nigrofuscus. Epulopiscium fishelsoni* grows as large as 600 by 80 μm, a little smaller than a printed hyphen. More recently an even larger bacterium, *Thiomargarita namibiensis,* has been discovered in ocean sediment (**Microbial Diversity & Ecology 3.1**). Thus a few bacteria are much larger than the average eucaryotic cell (typical plant and animal cells are around 10 to 50 μm in diameter).

Procaryotic Cell Organization

Procaryotic cells are morphologically simpler than eucaryotic cells, but they are not just simpler versions of eucaryotes. Although many structures are common to both cell types, some are unique to procaryotes. The major procaryotic structures and their functions are summarized and illustrated in **table 3.1** and **figure 3.4,** respectively. Note that no single procaryote possesses all of these structures at all times. Some are found only in certain cells in certain conditions or in certain phases of the life cycle. However, despite these variations procaryotes are consistent in their fundamental structure and most important components.

Procaryotic cells almost always are bounded by a chemically complex cell wall. Interior to this wall lies the plasma membrane. This membrane can be invaginated to form simple internal membranous structures such as the light-harvesting membrane of some photosynthetic bacteria. Since the procaryotic cell does not contain internal membrane-bound organelles, its interior appears morphologically simple. The genetic material is localized in a discrete region, the nucleoid, and usually is not separated from the surrounding cytoplasm by membranes. Ribosomes and larger masses called inclusion bodies are scattered about the cytoplasmic matrix. Many procaryotes use flagella for locomotion. In addition, many are surrounded by a capsule or slime layer external to the cell wall.

In the remaining sections of this chapter we describe the major procaryotic structures in more detail. We begin with the plasma membrane, a structure that defines all cells. We then proceed inward to consider structures located within the cytoplasm. Then the discussion moves outward, first to the cell wall and then to structures outside the cell wall. Finally, we consider a structure unique to bacteria, the bacterial endospore.

1. What characteristic shapes can bacteria assume? Describe the ways in which bacterial cells cluster together.
2. Draw a bacterial cell and label all important structures.

3.2 Procaryotic Cell Membranes

Membranes are an absolute requirement for all living organisms. Cells must interact in a selective fashion with their environment, whether it is the internal environment of a multicellular organism or a less protected and more variable external environment. Cells must not only be able to acquire nutrients and eliminate wastes, but they also have to maintain their interior in a constant, highly organized state in the face of external changes.

The **plasma membrane** encompasses the cytoplasm of both procaryotic and eucaryotic cells. It is the chief point of contact with the cell's environment and thus is responsible for much of its relationship with the outside world. The plasma membranes of procaryotic cells are particularly important because they must fill an incredible variety of roles. In addition to retaining the cytoplasm, the plasma membrane also serves as a selectively permeable barrier: it allows particular ions and molecules to pass, either into or out of the cell, while preventing the movement of others. Thus the membrane prevents the loss of essential components through leakage while allowing the movement of other molecules. Because many substances cannot cross the plasma membrane without assistance, it must aid such movement when necessary. Transport systems are used for such tasks as nutrient uptake, waste excretion, and protein secretion. The procaryotic plasma membrane also is the location of a variety of crucial metabolic processes: respiration, photosynthesis, and the synthesis of lipids and cell wall constituents. Finally, the membrane contains special receptor molecules that help procaryotes detect and respond to chemicals in their surroundings. Clearly the plasma membrane is essential to the survival of microorganisms. Uptake of nutrients by the cell (section 5.6)

As will be evident in the following discussion, all membranes apparently have a common, basic design. However, procaryotic membranes can differ dramatically in terms of the lipids they contain. Indeed, membrane chemistry can be used to identify particular bacterial species. To understand these chemical differences and to understand the many functions of the

Microbial Diversity & Ecology

3.1 Monstrous Microbes

Biologists often have distinguished between procaryotes and eucaryotes based in part on cell size. Generally, procaryotic cells are supposed to be smaller than eucaryotic cells. Procaryotes grow extremely rapidly compared to most eucaryotes and lack the complex vesicular transport systems of eucaryotic cells described in chapter 4. It has been assumed that they must be small because of the slowness of nutrient diffusion and the need for a large surface-to-volume ratio. Thus when Fishelson, Montgomery, and Myrberg discovered a large, cigar-shaped microorganism in the intestinal tract of the Red Sea brown surgeonfish, *Acanthurus nigrofuscus,* they suggested in their 1985 publication that it was a protist. It seemed too large to be anything else. In 1993 Esther Angert, Kendall Clemens, and Norman Pace used rRNA sequence comparisons to identify the microorganism, now called *Epulopiscium fishelsoni,* as a procaryote related to the gram-positive bacterial genus *Clostridium.*

E. fishelsoni [Latin, *epulum,* a feast or banquet, and *piscium,* fish] can reach a size of 80 μm by 600 μm, and normally ranges from 200 to 500 μm in length (see **Box figure**). It is about a million times larger in volume than *Escherichia coli.* Despite its huge size the organism has a procaryotic cell structure. It is motile and swims at about two body lengths a second (approximately 2.4 cm/min) using the flagella that cover its surface. The cytoplasm contains large nucleoids and many ribosomes, as would be required for such a large cell. *Epulopiscium* appears to overcome the size limits set by diffusion by having a highly convoluted plasma membrane. This increases the cell's surface area and aids in nutrient transport.

It appears that *Epulopiscium* is transmitted between hosts through fecal contamination of the fish's food. The bacterium can be eliminated by starving the surgeonfish for a few days. If juvenile fish that lack the bacterium are placed with infected hosts, they are reinoculated. Interestingly this does not work with uninfected adult surgeonfish.

In 1997, Heidi Schulz discovered an even larger procaryote in the ocean sediment off the coast of Namibia. *Thiomargarita namibiensis* is a spherical bacterium, between 100 and 750 μm in diameter, that often forms chains of cells enclosed in slime sheaths. It is over 100 times larger in volume than *E. fishelsoni.* A vacuole occupies about 98% of the cell and contains fluid rich in nitrate; it is surrounded by a 0.5 to 2.0 μm layer of cytoplasm filled with sulfur granules. The cytoplasmic layer is the same thickness as most bacteria and sufficiently thin for adequate diffusion rates. It uses sulfur as an energy source and nitrate as the electron acceptor for the electrons released when sulfur is oxidized in energy-conserving processes.

The discovery of these procaryotes greatly weakens the distinction between procaryotes and eucaryotes based on cell size. They are certainly larger than a normal eucaryotic cell. The size distinction between procaryotes and eucaryotes has been further weakened by the discovery of eucaryotic cells that are smaller than previously thought possible. The best example is *Nanochlorum eukaryotum. Nanochlorum* is only about 1 to 2 μm in diameter, yet is truly eucaryotic and has a nucleus, a chloroplast, and a mitochondrion. Our understanding of the factors limiting procaryotic cell size must be reevaluated. It is no longer safe to assume that large cells are eucaryotic and small cells are procaryotic.

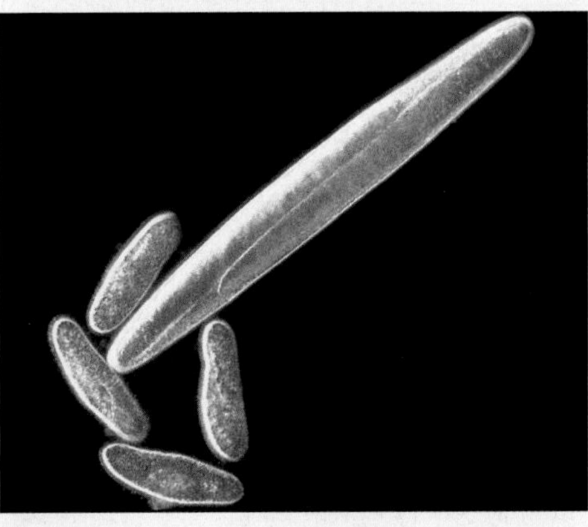

(a)

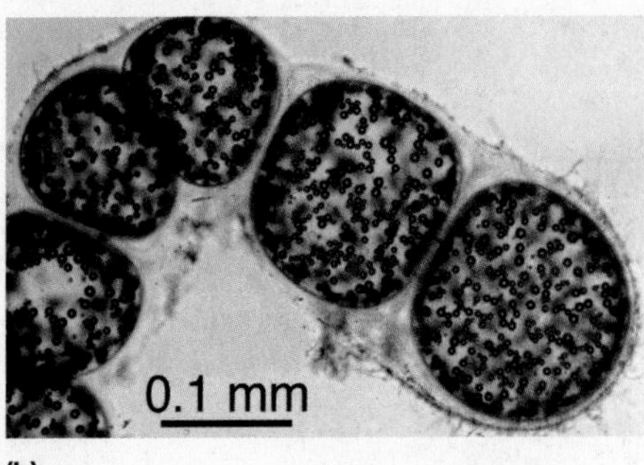

(b)

Giant Bacteria. (a) This photograph, taken with pseudo dark-field illumination, shows *Epulopiscium fishelsoni* at the top of the figure dwarfing the paramecia at the bottom (×200). **(b)** A chain of *Thiomargarita namibiensis* cells as viewed with the light microscope. Note the external mucous sheath and the internal sulfur globules.

Sources: Angert, E. R., Clements, K. D., and Pace, N. R. 1993 The largest bacterium. *Nature* 362:239–41; and Shulz, H. N., Brinkhoff, T., Ferdelman, T. G., Mariné, M. H., Teske, A., and Jorgensen, B.B. 1999. Dense populations of a giant sulfur bacterium in Namibian shelf sediments. *Science* 284:493–95.

Table 3.1 Functions of Procaryotic Structures

Plasma membrane	Selectively permeable barrier, mechanical boundary of cell, nutrient and waste transport, location of many metabolic processes (respiration, photosynthesis), detection of environmental cues for chemotaxis
Gas vacuole	Buoyancy for floating in aquatic environments
Ribosomes	Protein synthesis
Inclusion bodies	Storage of carbon, phosphate, and other substances
Nucleoid	Localization of genetic material (DNA)
Periplasmic space	Contains hydrolytic enzymes and binding proteins for nutrient processing and uptake
Cell wall	Gives procaryotes shape and protection from osmotic stress
Capsules and slime layers	Resistance to phagocytosis, adherence to surfaces
Fimbriae and pili	Attachment to surfaces, bacterial mating
Flagella	Movement
Endospore	Survival under harsh environmental conditions

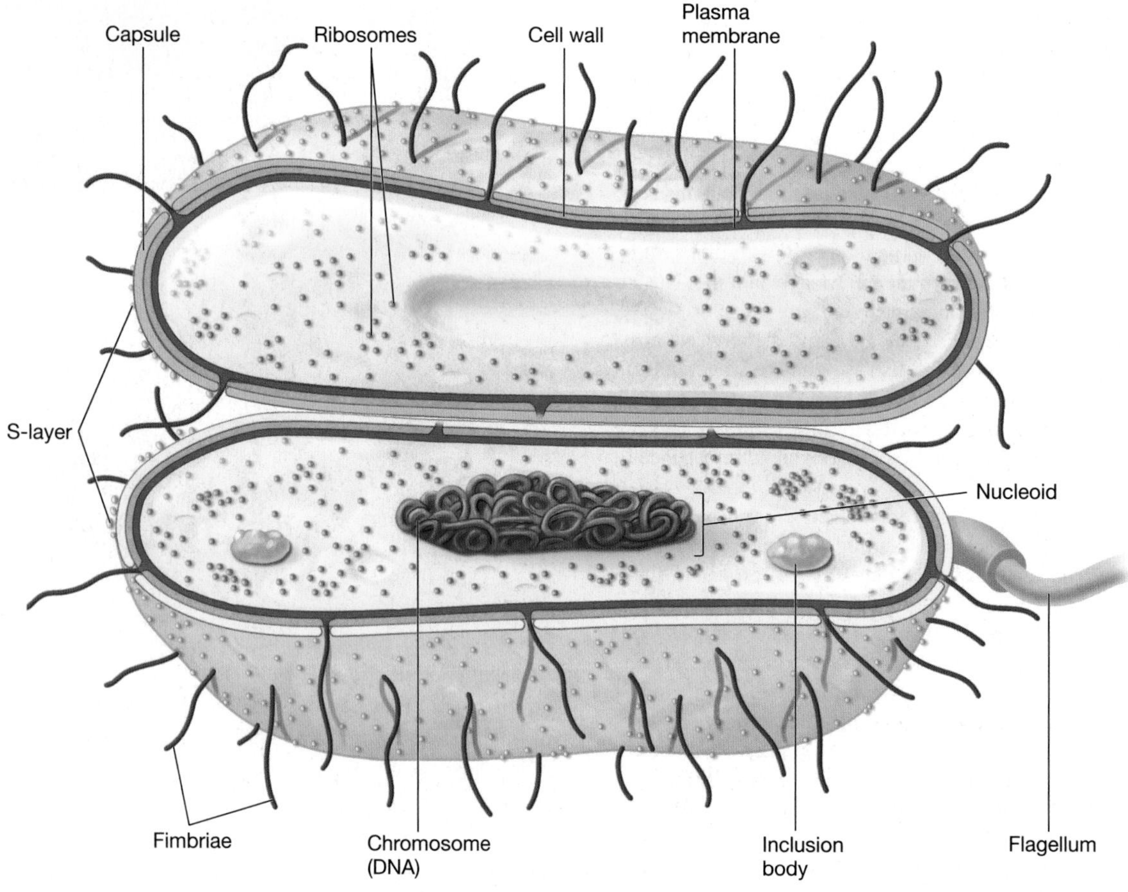

Figure 3.4 Morphology of a Procaryotic Cell.

plasma membrane, it is necessary to become familiar with membrane structure. In this section, the common basic design of all membranes is discussed. This is followed by a consideration of the significant differences between bacterial and archaeal membranes.

The Fluid Mosaic Model of Membrane Structure

The most widely accepted model for membrane structure is the **fluid mosaic model** of Singer and Nicholson (**figure 3.5**), which proposes that membranes are lipid bilayers within which proteins float. The model is based on studies of eucaryotic and bac-

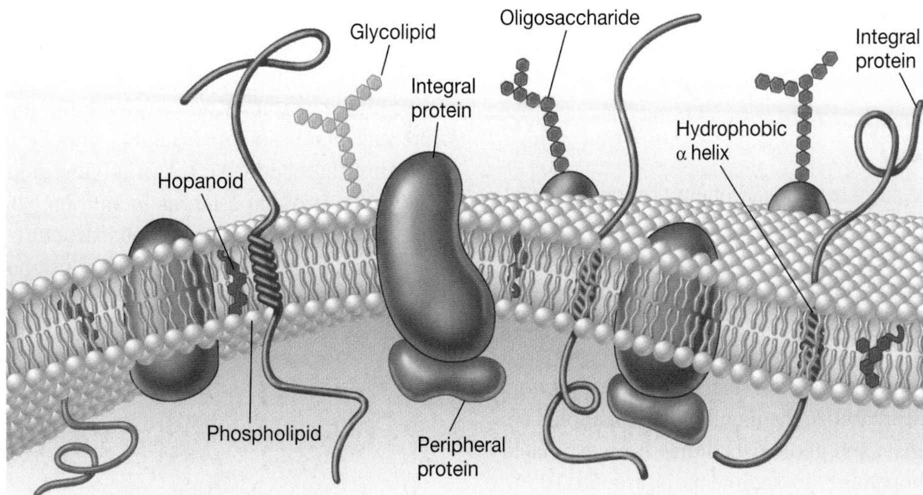

Figure 3.5 Bacterial Plasma Membrane Structure. This diagram of the fluid mosaic model of bacterial membrane structure shows the integral proteins (blue) floating in a lipid bilayer. Peripheral proteins (purple) are associated loosely with the inner membrane surface. Small spheres represent the hydrophilic ends of membrane phospholipids and wiggly tails, the hydrophobic fatty acid chains. Other membrane lipids such as hopanoids (red) may be present. For the sake of clarity, phospholipids are shown in proportionately much larger size than in real membranes.

terial membranes, and a variety of experimental approaches were used to establish it. Transmission electron microscopy (TEM) studies were particularly important. When membranes are stained and examined by TEM, it can be seen that cell membranes are very thin structures, about 5 to 10 nm thick, and that they appear as two dark lines on either side of a nonstained interior. This characteristic appearance has been interpreted to mean that the membrane lipid is organized in two sheets of molecules arranged end-to-end (figure 3.5). When membranes are cleaved by the freeze-etching technique, they can be split down the center of the lipid bilayer, exposing the complex internal structure. Within the lipid bilayer, small globular particles are visible; these have been suggested to be membrane proteins lying within the membrane lipid bilayer. The use of atomic force microscopy has provided powerful images to support this interpretation. Electron microscopy (section 2.4); Newer techniques in microscopy: Scanning probe microscopy (section 2.5)

The chemical nature of membrane lipids is critical to their ability to form bilayers. Most membrane-associated lipids are structurally asymmetric, with polar and nonpolar ends (**figure 3.6**) and are called **amphipathic.** The polar ends interact with water and are **hydrophilic;** the nonpolar **hydrophobic** ends are insoluble in water and tend to associate with one another. In aqueous environments, amphipathic lipids can interact to form a bilayer. The outer surfaces of the bilayer membrane are hydrophilic, whereas hydrophobic ends are buried in the interior away from the surrounding water (figure 3.5). Lipids (appendix I)

Two types of membrane proteins have been identified based on their ability to be separated from the membrane. **Peripheral proteins** are loosely connected to the membrane and can be easily removed (figure 3.5). They are soluble in aqueous solutions

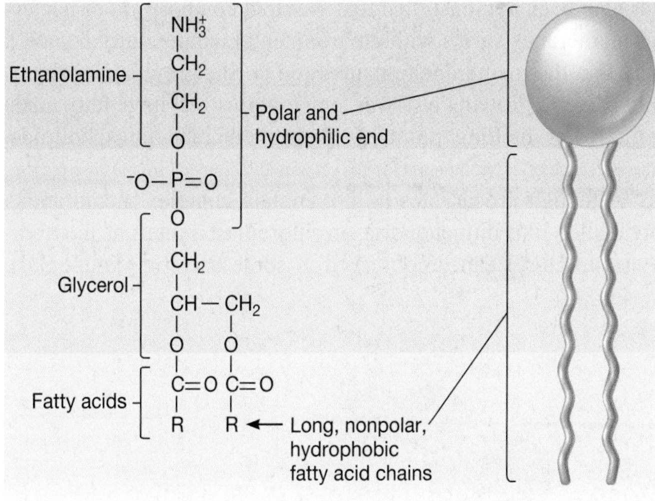

Figure 3.6 The Structure of a Polar Membrane Lipid. Phosphatidylethanolamine, an amphipathic phospholipid often found in bacterial membranes. The R groups are long, nonpolar fatty acid chains.

and make up about 20 to 30% of total membrane protein. About 70 to 80% of membrane proteins are **integral proteins.** These are not easily extracted from membranes and are insoluble in aqueous solutions when freed of lipids. Integral proteins, like membrane lipids, are amphipathic; their hydrophobic regions are buried in the lipid while the hydrophilic portions project from the membrane surface (figure 3.5). Integral proteins can diffuse laterally in the membrane to new locations, but do not flip-flop or rotate through the lipid layer. Carbohydrates often are attached to the outer surface of plasma membrane proteins, where they have important functions. Proteins (appendix I)

Bacterial Membranes

Bacterial membranes are similar to eucaryotic membranes in that many of their amphipathic lipids are phospholipids (figure 3.6), but they usually differ from eucaryotic membranes in lacking sterols (steroid-containing lipids) such as cholesterol (**figure 3.7*a***). However, many bacterial membranes contain sterol-like molecules called **hopanoids** (figure 3.7*b*). Hopanoids are synthesized from the same precursors as steroids, and like the sterols in eucaryotic membranes, they probably stabilize the membrane. Hopanoids are also of interest to ecologists and geologists: it has been estimated that the total mass of hopanoids in sediments is around 10^{11-12} tons—about as much as the total mass of organic carbon in all living organisms (10^{12} tons)—and there is evidence that hopanoids have contributed significantly to the formation of petroleum.

The emerging picture of bacterial plasma membranes is one of a highly organized and asymmetric system that also is flexible and dynamic. Numerous studies have demonstrated that lipids are not homogeneously distributed in the plasma membrane. Rather, there are domains in which particular lipids are concentrated. It has also been demonstrated that the lipid composition of bacterial membranes varies with environmental temperature in such a way that the membrane remains fluid during growth. For example, bacteria growing at lower temperatures will have fatty acids with lower melting points in their membrane phospholipids. The influence of environmental factors on growth: Temperature (section 6.5)

Although procaryotes do not contain complex membranous organelles like mitochondria or chloroplasts, internal membranous structures can be observed in some bacteria (**figure 3.8**). Plasma membrane infoldings are common in many bacteria and can become extensive and complex in photosynthetic bacteria such as the cyanobacteria and purple bacteria or in bacteria with very high respiratory activity, like the nitrifying bacteria. These internal membranous structures may be aggregates of spherical vesicles, flattened vesicles, or tubular membranes. Their function may be to provide a larger membrane surface for greater metabolic activity. One membranous structure sometimes reported in bacteria is the mesosome. Mesosomes appear to be invaginations of the plasma membrane in the shape of vesicles, tubules, or lamellae. Although a variety of functions have been ascribed to

HO

(a) Cholesterol (a steroid) is found in eucaryotes

OH OH

OH OH

(b) A bacteriohopanetetrol (a hopanoid), as found in bacteria

Figure 3.7 Membrane Steroids and Hopanoids. Common examples.

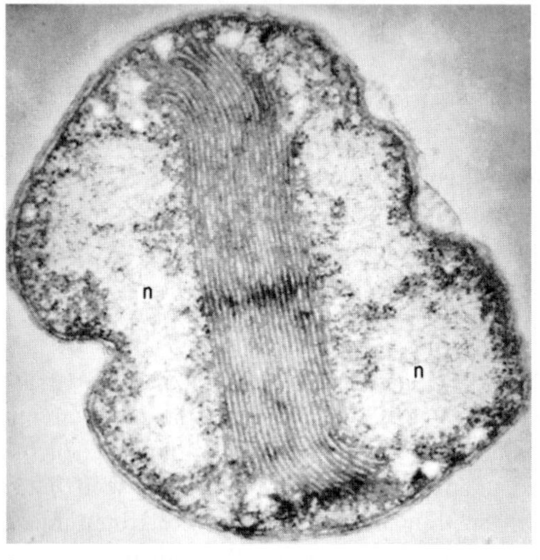

(a)

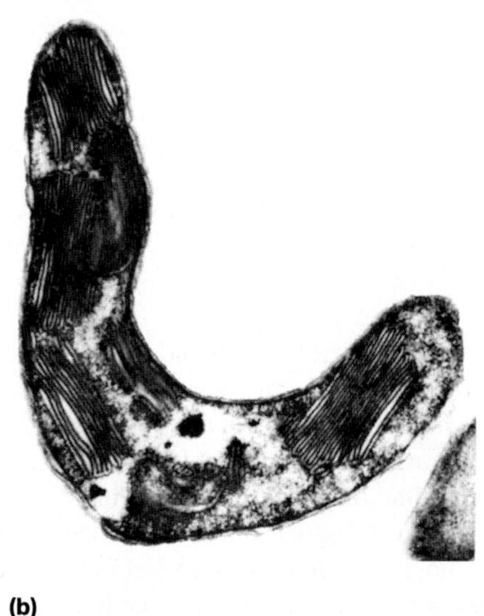

(b)

Figure 3.8 Internal Bacterial Membranes. Membranes of nitrifying and photosynthetic bacteria. **(a)** *Nitrocystis oceanus* with parallel membranes traversing the whole cell. Note nucleoplasm (n) with fibrillar structure. **(b)** *Ectothiorhodospira mobilis* with an extensive intracytoplasmic membrane system (×60,000).

mesosomes, many bacteriologists believe that they are artifacts generated during the chemical fixation of bacteria for electron microscopy.

Archaeal Membranes

One of the most distinctive features of the *Archaea* is the nature of their membrane lipids. They differ from both *Bacteria* and *Eucarya* in having branched chain hydrocarbons attached to glycerol by ether links rather than fatty acids connected by ester links (**figure 3.9**). Sometimes two glycerol groups are linked to form an extremely long tetraether. Usually the diether hydrocarbon chains are 20 carbons in length, and the tetraether chains are 40 carbons. Cells can adjust the overall length of the tetraethers by cyclizing the chains to form pentacyclic rings (figure 3.9).

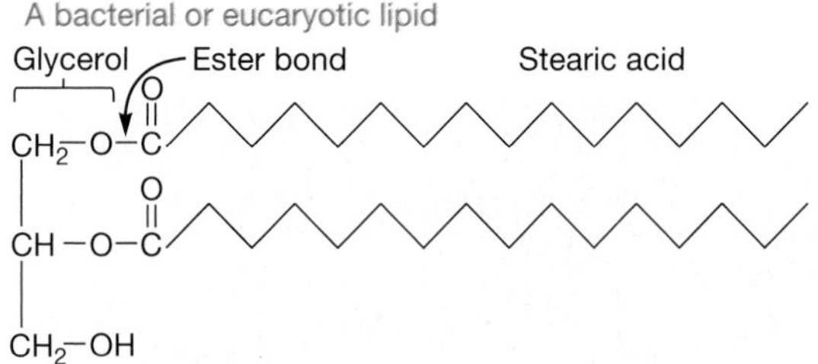

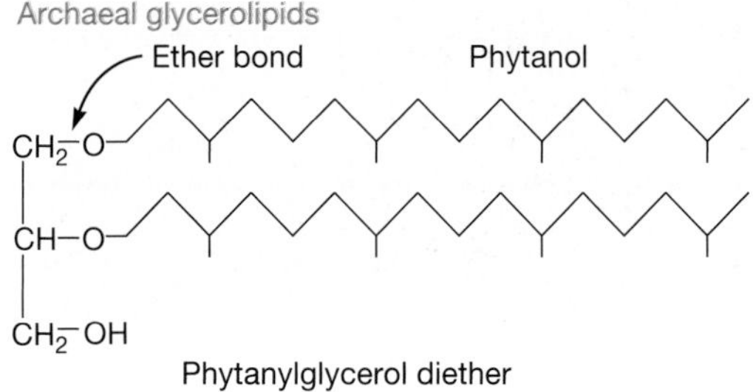

HO–CH_2
–O–CH
CH_2–O
CH–O
–O–CH_2
CH_2–OH
Dibiphytanyldiglycerol tetraether

CH_2–O
HO–CH_2
CH–O
–O–CH
CH_2–OH
–O–CH_2
Tetraether with bipentacyclic C_{40} biphytanyl chains

Figure 3.9 Archaeal Membrane Lipids. An illustration of the difference between archaeal lipids and those of *Bacteria*. Archaeal lipids are derivatives of isopranyl glycerol ethers rather than the glycerol fatty acid esters in *Bacteria*. Three examples of common archaeal glycerolipids are given.

Phosphate-, sulfur- and sugar-containing groups can be attached to the third carbons of the diethers and tetraethers, making them polar lipids. These predominate in the membrane, and 70 to 93% of the membrane lipids are polar. The remaining lipids are nonpolar and are usually derivatives of squalene (**figure 3.10**).

Despite these significant differences in membrane lipids, the basic design of archaeal membranes is similar to that of *Bacteria* and eucaryotes—there are two hydrophilic surfaces and a hydrophobic core. When C_{20} diethers are used, a regular bilayer membrane is formed (**figure 3.11*a***). When the membrane is constructed of C_{40} tetraethers, a monolayer membrane with much more rigidity is formed (figure 3.11*b*). As might be expected from

Squalene

Tetrahydrosqualene

Figure 3.10 Nonpolar Lipids of *Archaea*. Two examples of the most predominant nonpolar lipids are the C_{30} isoprenoid squalene and one of its hydroisoprenoid derivatives, tetrahydrosqualene.

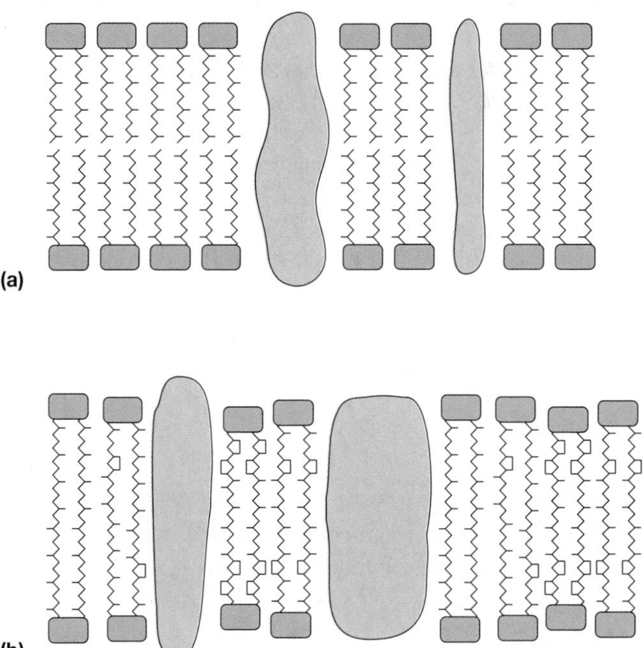

Figure 3.11 Examples of Archaeal Membranes. **(a)** A membrane composed of integral proteins and a bilayer of C_{20} diethers. **(b)** A rigid monolayer composed of integral proteins and C_{40} tetraethers.

their need for stability, the membranes of extreme thermophiles such as *Thermoplasma* and *Sulfolobus,* which grow best at temperatures over 85°C, are almost completely tetraether monolayers. Archaea that live in moderately hot environments have a mixed membrane containing some regions with monolayers and some with bilayers. Phylum *Euryarchaeota:* Thermoplasms (section 20.3); Phylum *Crenarchaeota: Sulfolobus* (section 20.2)

1. List the functions of the procaryotic plasma membrane.
2. Describe in words and with a labeled diagram the fluid mosaic model for cell membranes.
3. Compare and contrast bacterial and archaeal membranes.
4. Discuss the ways bacteria and archaea adjust the lipid content of their membranes in response to environmental conditions.

3.3 The Cytoplasmic Matrix

The **cytoplasmic matrix** is the substance in which the nucleoid, ribosomes, and inclusion bodies are suspended. It lacks organelles bound by lipid bilayers (often called unit membranes), and is largely water (about 70% of bacterial mass is water). Until recently, it was thought to lack a cytoskeleton. The plasma membrane and everything within is called the **protoplast;** thus the cytoplasmic matrix is a major part of the protoplast.

The Procaryotic Cytoskeleton

When examined with the electron microscope, the cytoplasmic matrix of procaryotes is packed with ribosomes. For many years it was thought that procaryotes lacked the high level of cytoplasmic organization present in eucaryotic cells because they lacked a cytoskeleton. Recently homologs of all three eucaryotic cytoskeletal elements (microfilaments, intermediate filaments, and microtubules) have been identified in bacteria, and one has been identified in archaea (**table 3.2**). The cytoskeletal filaments of procaryotes are structurally similar to their eucaryotic counterparts and carry out similar functions: they participate in cell division, localize proteins to certain sites in the cell, and determine cell shape (table 3.2 and **figure 3.12**). The procaryotic cell cycle: Cytokinesis (section 6.1)

Inclusion Bodies

Inclusion bodies, granules of organic or inorganic material that often are clearly visible in a light microscope, are present in the cytoplasmic matrix. These bodies usually are used for storage (e.g., carbon compounds, inorganic substances, and energy), and also reduce osmotic pressure by tying up molecules in particulate form. Some inclusion bodies lie free in the cytoplasm—for example, polyphosphate granules, cyanophycin granules, and some glycogen granules. Other inclusion bodies are enclosed by a shell about 2.0 to 4.0 nm thick, which is single-layered and may consist of proteins or a membranous structure composed of proteins and phospholipids. Examples of enclosed inclusion bodies are poly-β-hydroxybutyrate granules, some glycogen and sulfur granules, carboxysomes, and gas vacuoles. Many inclusion bodies are used for storage; their quantity will vary with the nutritional status of the cell. For example, polyphosphate granules will

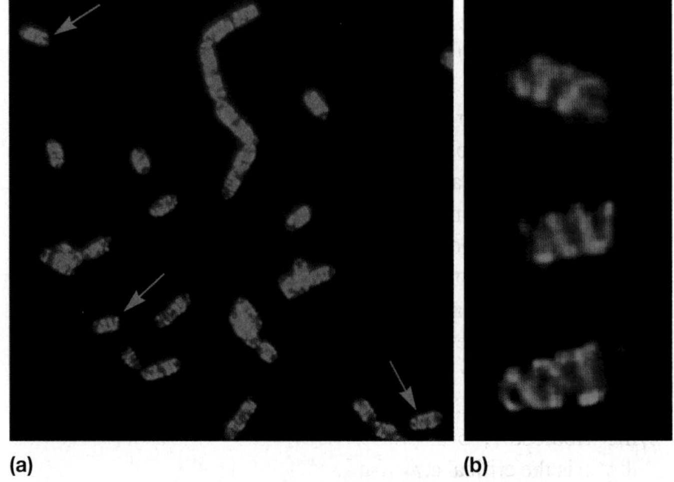

Figure 3.12 The Procaryotic Cytoskeleton. Visualization of the MreB-like cytoskeletal protein (Mbl) of *Bacillus subtilis.* The Mbl protein has been fused with green fluorescent protein and live cells have been examined by fluorescence microscopy. **(a)** Arrows point to the helical cytoskeletal cables that extend the length of the cells. **(b)** Three of the cells from (a) are shown at a higher magnification.

Table 3.2 Procaryotic Cytoskeletal Proteins

Procaryotic Protein (Eucaryotic Counterpart)	Function	Comments
FtsZ (tubulin)	Cell division	Widely observed in *Bacteria* and *Archaea*
MreB (actin)	Cell shape	Observed in many rod-shaped bacteria; in *Bacillus subtilis* is called Mbl
Crescentin (intermediate filament proteins)	Cell shape	Discovered in *Caulobacter crescentus*

be depleted in freshwater habitats that are phosphate limited. A brief description of several important inclusion bodies follows.

Organic inclusion bodies usually contain either glycogen or poly-β-hydroxyalkanoates (e.g., poly-β-hydroxybutyrate). **Glycogen** is a polymer of glucose units composed of long chains formed by α(1→4) glycosidic bonds and branching chains connected to them by α(1→6) glycosidic bonds. **Poly-β-hydroxybutyrate (PHB)** contains β-hydroxybutyrate molecules joined by ester bonds between the carboxyl and hydroxyl groups of adjacent molecules. Usually only one of these polymers is found in a species, but some photosynthetic bacteria have both glycogen and PHB. Poly-β-hydroxybutyrate accumulates in distinct bodies, around 0.2 to 0.7 μm in diameter, that are readily stained with Sudan black for light microscopy and are seen as empty "holes" in the electron microscope (**figure 3.13*a***). This is because the solvents used to prepare specimens for electron microscopy dissolve these hydrophobic inclusion bodies. Glycogen is dispersed more evenly throughout the matrix as small granules (about 20 to 100 nm in diameter) and often can be seen only with the electron microscope. If cells contain a large amount of glycogen, staining with an iodine solution will turn them reddish-brown. Glycogen and PHB inclusion bodies are carbon storage reservoirs providing material for energy and biosynthesis. Many bacteria also store carbon as lipid droplets. Carbohydrates (appendix I)

Cyanobacteria, a group of photosynthetic bacteria, have two distinctive organic inclusion bodies. **Cyanophycin granules** (figure 3.13*b*) are composed of large polypeptides containing approximately equal amounts of the amino acids arginine and aspartic acid. The granules often are large enough to be visible in the light microscope and store extra nitrogen for the bacteria. **Carboxysomes** are present in many cyanobacteria and other CO_2-fixing bacteria. They are polyhedral, about 100 nm in diameter, and contain the enzyme ribulose-1, 5-bisphosphate carboxylase, called Rubisco. Rubisco is the critical enzyme for CO_2 fixation, the process of converting CO_2 from the atmosphere into sugar. The enzyme assumes a paracrystalline arrangement in the carboxysome, which serves as a reserve of the enzyme. Carboxysomes also may be a site of CO_2 fixation. The fixation of CO_2 by autotrophs (section 10.3)

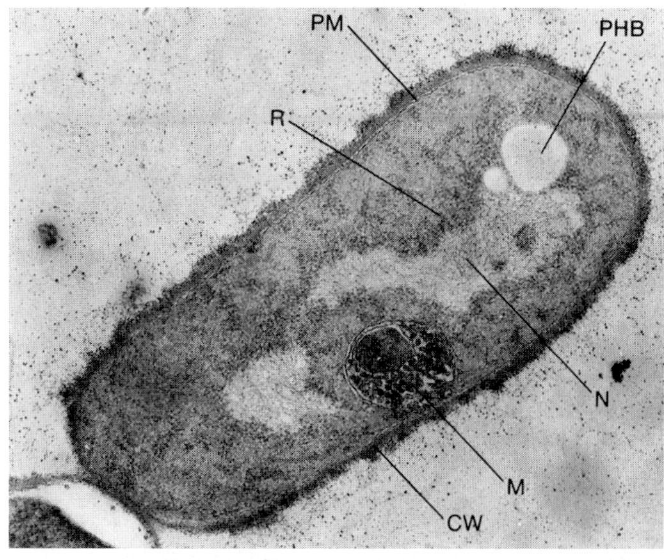

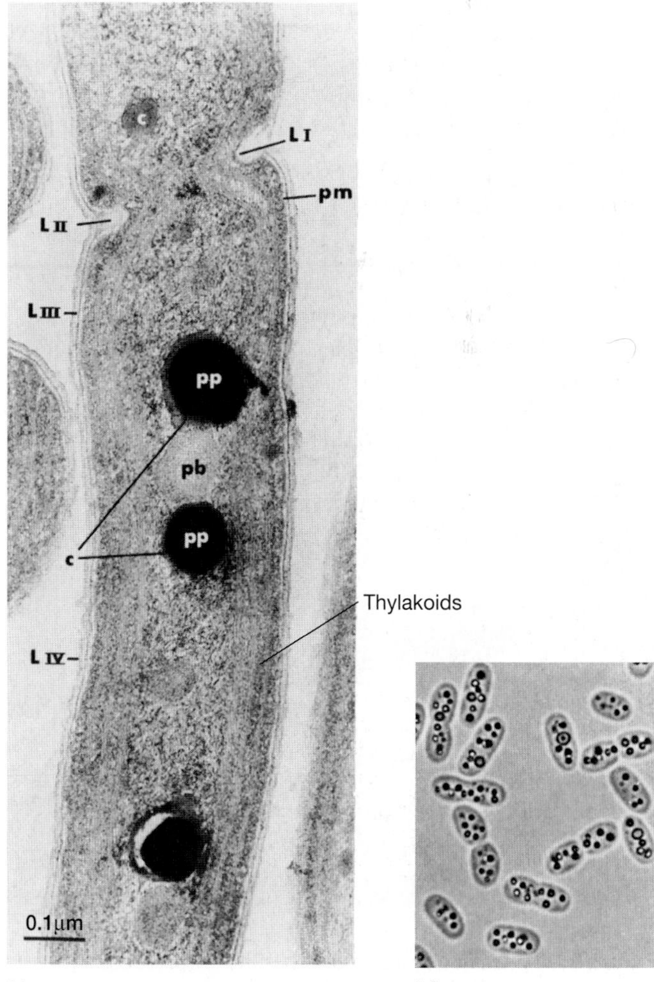

Figure 3.13 Inclusion Bodies in Bacteria. **(a)** Electron micrograph of *Bacillus megaterium* (×30,500). Poly-β-hydroxybutyrate inclusion body, PHB; cell wall, CW; nucleoid, N; plasma membrane, PM; "mesosome," M; and ribosomes, R. **(b)** Ultrastructure of the cyanobacterium *Anacystis nidulans*. The bacterium is dividing and a septum is partially formed, LI and LII. Several structural features can be seen, including cell wall layers, LIII and LIV; polyphosphate granules, pp; a polyhedral body, pb; cyanophycin material, c; and plasma membrane, pm. Thylakoids run along the length of the cell. **(c)** *Chromatium vinosum*, a purple sulfur bacterium, with intracellular sulfur granules, bright-field microscopy(×2,000).

A most remarkable organic inclusion body is the **gas vacuole,** a structure that provides buoyancy to some aquatic procaryotes. Gas vacuoles are present in many photosynthetic bacteria and a few other aquatic procaryotes such as *Halobacterium* (a salt-loving archaeon) and *Thiothrix* (a filamentous bacterium). Gas vacuoles are aggregates of enormous numbers of small, hollow, cylindrical structures called **gas vesicles** (**figure 3.14**). Gas vesicle walls are composed entirely of a single small protein. These protein subunits assemble to form a rigid enclosed cylinder that is hollow and impermeable to water but freely permeable to atmospheric gases. Procaryotes with gas vacuoles can regulate their buoyancy to float at the depth necessary for proper light intensity, oxygen concentration, and nutrient levels. They descend by simply collapsing vesicles and float upward when new ones are constructed.

Two major types of inorganic inclusion bodies are seen in procaryotes: polyphosphate granules and sulfur granules. Many bacteria store phosphate as **polyphosphate granules** or **volutin granules** (figure 3.13*b*). Polyphosphate is a linear polymer of orthophosphates joined by ester bonds. Thus volutin granules function as storage reservoirs for phosphate, an important component of cell constituents such as nucleic acids. In some cells they act as an energy reserve, and polyphosphate can serve as an energy source in reactions. These granules are sometimes called **metachromatic granules** because they show the metachromatic effect; that is, they appear red or a different shade of blue when stained with the blue dyes methylene blue or toluidine blue. Sulfur granules are used by some procaryotes to store sulfur temporarily (figure 3.13*c*). For example, photosynthetic bacteria can use hydrogen sulfide as a photosynthetic electron donor and accumulate the resulting sulfur in either the periplasmic space or in special cytoplasmic globules. Phototrophy: The light reaction in anoxygenic photosynthesis (section 9.12)

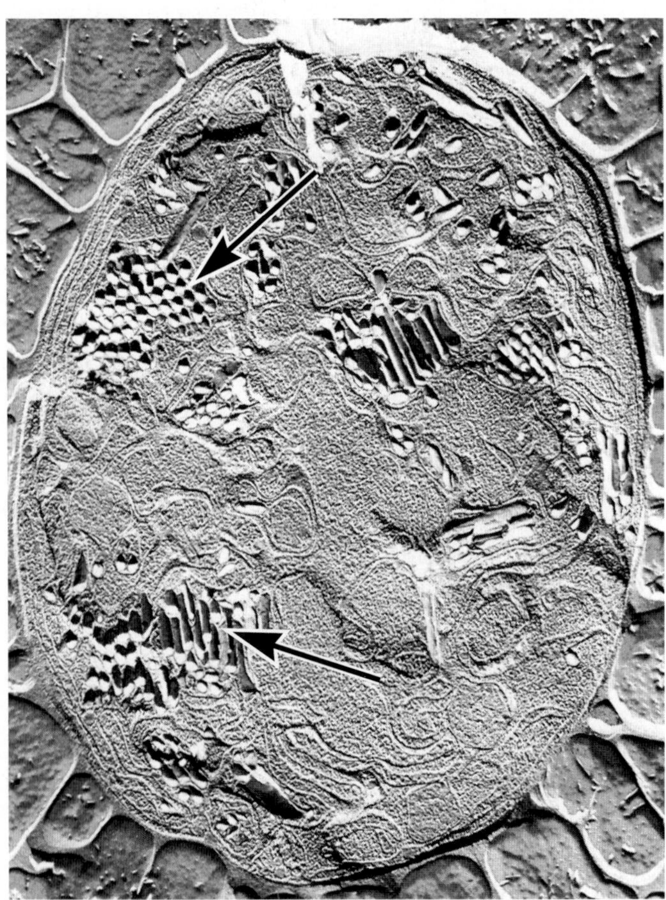

Figure 3.14 Gas Vesicles and Vacuoles. A freeze-fracture preparation of *Anabaena flosaquae* (×89,000). Clusters of the cigar-shaped vesicles form gas vacuoles. Both longitudinal and cross-sectional views of gas vesicles can be seen (arrows).

Inorganic inclusion bodies can be used for purposes other than storage. An excellent example is the **magnetosome,** which is used by some bacteria to orient in the Earth's magnetic field. Many of these inclusion bodies contain iron in the form of magnetite (**Microbial Diversity & Ecology 3.2**).

Ribosomes

As mentioned earlier, the cytoplasmic matrix often is packed with **ribosomes;** they also may be loosely attached to the plasma membrane. Ribosomes are very complex structures made of both protein and ribonucleic acid (RNA). They are the site of protein synthesis; cytoplasmic ribosomes synthesize proteins destined to remain within the cell, whereas plasma membrane ribosomes make proteins for transport to the outside. The newly formed polypeptide folds into its final shape either as it is synthesized by the ribosome or shortly after completion of protein synthesis. The shape of each protein is determined by its amino acid sequence. Special proteins called molecular chaperones, or simply chaperones, aid the polypeptide in folding to its proper shape. Protein synthesis, including a detailed treatment of ribosomes and chaperones, is discussed at considerable length in chapter 11.

Procaryotic ribosomes are smaller than the cytoplasmic or endoplasmic reticulum-associated ribosomes of eucaryotic cells. Procaryotic ribosomes are called 70S ribosomes (as opposed to 80S in eucaryotes), have dimensions of about 14 to 15 nm by 20 nm, a molecular weight of approximately 2.7 million, and are constructed of a 50S and a 30S subunit (**figure 3.15**). The S in 70S and similar values stands for **Svedberg unit.** This is the unit of the sedimentation coefficient, a measure of the sedimentation velocity in a centrifuge; the faster a particle travels when centrifuged, the greater its Svedberg value or sedimentation coefficient. The sedimentation coefficient is a function of a particle's molecular weight, volume, and shape (*see figure 16.19*). Heavier and more compact particles normally have larger Svedberg numbers or sediment faster.

1. Briefly describe the nature and function of the cytoplasmic matrix.
2. List and describe the functions of cytoskeletal proteins, inclusion bodies, and ribosomes.
3. List the most common kinds of inclusion bodies.
4. Relate the structure of a gas vacuole to its function.

Microbial Diversity & Ecology

3.2 Living Magnets

Bacteria can respond to environmental factors other than chemicals. A fascinating example is that of the aquatic magnetotactic bacteria that orient themselves in the Earth's magnetic field. Most of these bacteria have intracellular chains of magnetite (Fe_3O_4) particles that are called magnetosomes. Magnetosomes are around 35 to 125 nm in diameter and are bounded by a lipid bilayer (see **Box figure**). Some species from sulfidic habitats have magnetosomes containing greigite (Fe_3S_4) and pyrite (FeS_2). Since each iron particle is a tiny magnet, the Northern Hemisphere bacteria use their magnetosome chain to determine northward and downward directions, and swim down to nutrient-rich sediments or locate the optimum depth in freshwater and marine habitats. Magnetotactic bacteria in the Southern Hemisphere generally orient southward and downward, with the same result. Magnetosomes also are present in the heads of birds, tuna, dolphins, green turtles, and other animals, presumably to aid navigation. Animals and bacteria share more in common behaviorally than previously imagined.

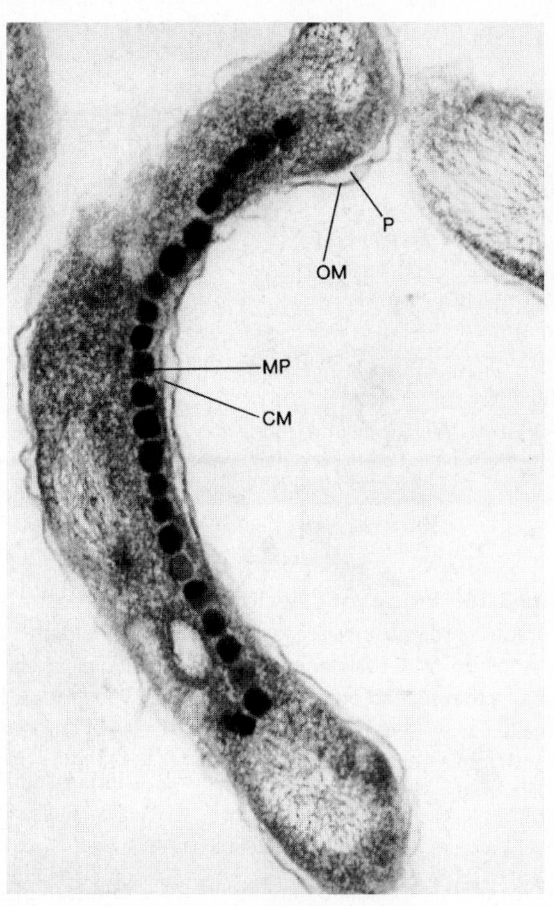

(a)

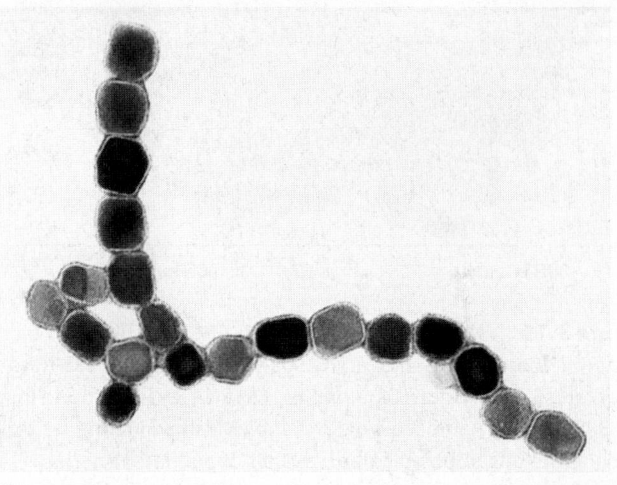

(b)

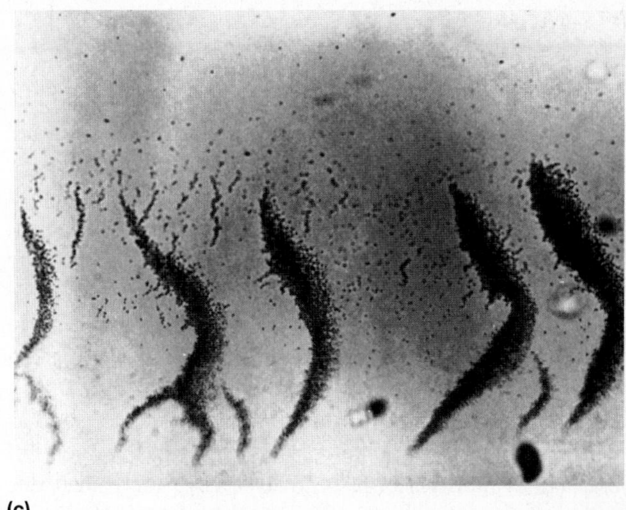

(c)

Magnetotactic Bacteria. (**a**) Transmission electron micrograph of the magnetotactic bacterium *Aquaspirillum magnetotacticum* (×123,000). Note the long chain of electron-dense magnetite particles, MP. Other structures: OM, outer membrane; P, periplasmic space; CM, cytoplasmic membrane. **(b)** Isolated magnetosomes (×140,000). (**c**) Bacteria migrating in waves when exposed to a magnetic field.

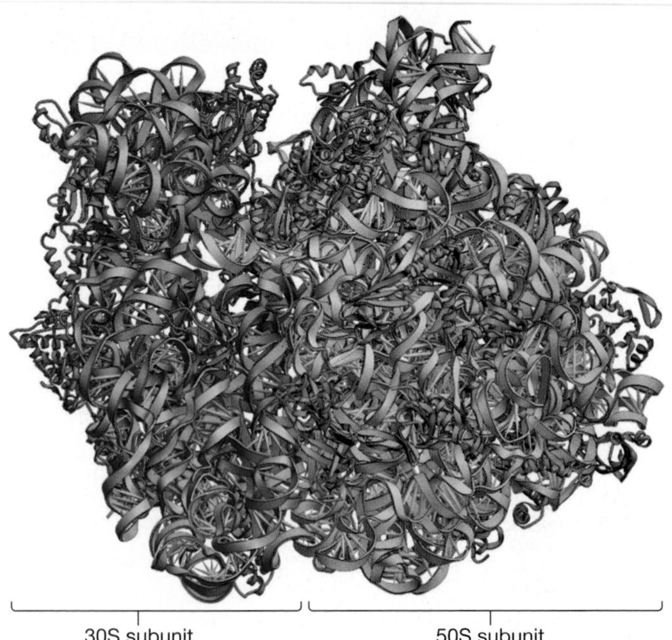

Figure 3.15 Procaryotic Ribosome. The two subunits of a bacterial ribosome are shown. The 50S subunit includes 23S rRNA (gray) and 5S rRNA (light blue), while 16S rRNA (cyan) is found in the 30S subunit. A molecule of tRNA (gold) is shown in the A site. To generate this ribbon diagram, crystals of purified bacterial ribosomes were grown, exposed to X rays, and the resulting diffraction pattern analyzed.

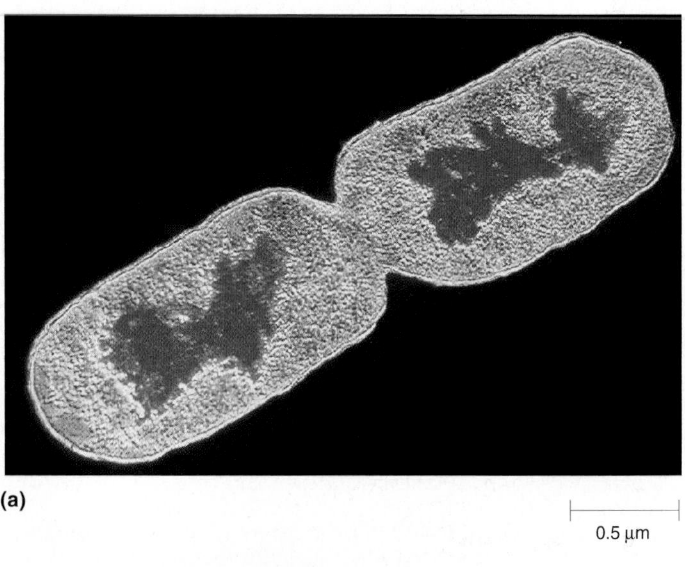

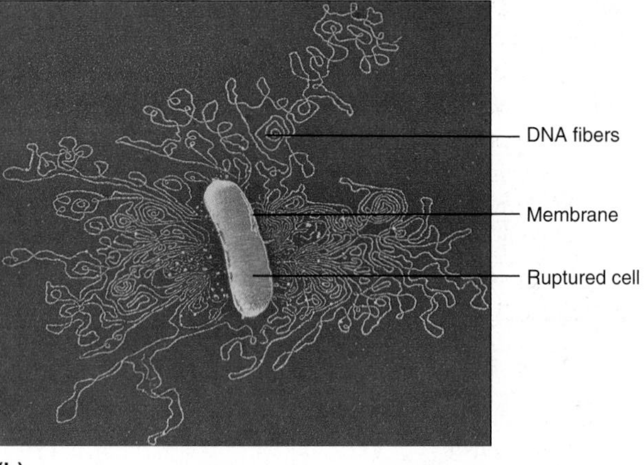

Figure 3.16 Procaryotic Nucleoids and Chromosomes. Procaryotic chromosomes are located in the nucleoid, an area in the cytoplasm. **(a)** A color-enhanced transmission electron micrograph of a thin section of a dividing *E. coli* cell. The red areas are the nucleoids present in the two daughter cells. **(b)** Chromosome released from a gently lysed *E. coli* cell. Note how tightly packaged the DNA must be inside the cell.

3.4 The Nucleoid

Probably the most striking difference between procaryotes and eucaryotes is the way in which their genetic material is packaged. Eucaryotic cells have two or more chromosomes contained within a membrane-delimited organelle, the nucleus. In contrast, procaryotes lack a membrane-delimited nucleus. The procaryotic chromosome is located in an irregularly shaped region called the **nucleoid** (other names are also used: the nuclear body, chromatin body, nuclear region) (**figure 3.16**). Usually procaryotes contain a single circle of double-stranded **deoxyribonucleic acid (DNA),** but some have a linear DNA chromosome and some, such as *Vibrio cholerae* and *Borrelia burgdorferi* (the causative agents of cholera and Lyme disease, respectively), have more than one chromosome.

Both electron and light microscopic studies have been important for understanding nucleoid structure and function, especially during active cell growth and division. The nucleoid has a fibrous appearance in electron micrographs; the fibers are probably DNA. In actively growing cells, the nucleoid has projections that extend into the cytoplasmic matrix. Presumably these projections contain DNA that is being actively transcribed to produce mRNA. Other studies have shown that more than one nucleoid can be observed within a single cell when genetic material has been duplicated but cell division has not yet occurred (figure 3.16*a*).

It is possible to isolate pure nucleoids. Chemical analysis of purified nucleoids reveals that they are composed of about 60% DNA, 30% RNA, and 10% protein by weight. In *Escherichia coli,* the closed DNA circle measures approximately 1,400 μm or about 230–700 times longer than the cell (figure 3.16*b*). Obviously it must be very efficiently packaged to fit within the nucleoid. The DNA is looped and coiled extensively (*see figure 11.8*), probably with the aid of RNA and a variety of nucleoid proteins. These include condensing proteins, which are conserved in both *Bacteria* and *Archaea*. Unlike the eucaryotes and some archaea, *Bacteria* do not use histone proteins to package their DNA.

There are a few exceptions to the preceding picture. Membrane-bound DNA-containing regions are present in two genera of the unusual bacterial phylum *Planctomycetes* (*see figure 21.12*). *Pirellula* has a single membrane that surrounds a region, the pirellulosome, which contains a fibrillar nucleoid and ribosome-like particles. The nuclear body of *Gemmata obscuriglobus* is bounded by two membranes. More work will be required to determine the functions of these membranes and how widespread this phenomenon is. Phylum *Planctomycetes* (section 21.4)

1. Describe the structure and function of the nucleoid and the DNA it contains.
2. List three genera that are exceptional in terms of their chromosome or nucleoid structure. Suggest how the differences observed in these genera might impact how they function.

3.5 Plasmids

In addition to the genetic material present in the nucleoid, many procaryotes (and some yeasts and other fungi) contain extrachromosomal DNA molecules called plasmids. Indeed, most of the bacterial and archaeal genomes sequenced thus far include plasmids. In some cases, numerous different plasmids within a single species have been identified. For instance, *B. burgdorferi,* carries 12 linear and 9 circular plasmids. Plasmids play many important roles in the lives of the organisms that have them. They also have proved invaluable to microbiologists and molecular geneticists in constructing and transferring new genetic combinations and in cloning genes, as described in chapter 14. This section discusses the different types of procaryotic plasmids.

Plasmids are small, double-stranded DNA molecules that can exist independently of the chromosome. Both circular and linear plasmids have been documented, but most known plasmids are circular. Linear plasmids possess special structures or sequences at their ends to prevent their degradation and to permit their replication. Plasmids have relatively few genes, generally less than 30. Their genetic information is not essential to the host, and cells that lack them usually function normally. However, many plasmids carry genes that confer a selective advantage to their hosts in certain environments.

Plasmids are able to replicate autonomously. Single-copy plasmids produce only one copy per host cell. Multicopy plasmids may be present at concentrations of 40 or more per cell. Some plasmids are able to integrate into the chromosome and are thus replicated with the chromosome. Such plasmids are called **episomes.** Plasmids are inherited stably during cell division, but they are not always equally apportioned into daughter cells and sometimes are lost. The loss of a plasmid is called **curing.** It can occur spontaneously or be induced by treatments that inhibit plasmid replication but not host cell reproduction. Some commonly used curing treatments are acridine mutagens, UV and ionizing radiation, thymine starvation, antibiotics, and growth above optimal temperatures.

Plasmids may be classified in terms of their mode of existence, spread, and function. A brief summary of the types of bacterial plasmids and their properties is given in **table 3.3. Conjugative plasmids** are of particular note. They have genes for the construction of hairlike structures called pili and can transfer copies of themselves to other bacteria during conjugation. Perhaps the best-studied conjugative plasmid is the **F factor** (fertility factor or F plasmid) of *E. coli,* which was the first conjugative factor to be described. The F factor contains genes that direct the formation of sex pili that attach an F^+ cell (a cell containing an F plasmid) to an F^- cell (a cell lacking an F plasmid). Other plasmid-encoded gene products aid DNA transfer from the F^+ cell to the F^- cell. The F factor also has several segments called insertion sequences that enable it to integrate into the host cell chromosome. Thus the F factor is an episome. Transposable elements (section 13.5); Bacterial conjugation (section 13.7)

Resistance factors (R factors, R plasmids) are another group of important plasmids. They confer antibiotic resistance on the cells that contain them. R factors typically have genes that code for enzymes capable of destroying or modifying antibiotics. Some R plasmids have only a single resistance gene, whereas others have as many as eight. Often the resistance genes are within mobile genetic elements called transposons, and thus it is possible for multiple-resistance plasmids to evolve. R factors usually are not integrated into the host chromosome.

R factors are of major concern to public health officials because they can spread rapidly throughout a population of cells. This is possible for several reasons. One is that many R factors also are conjugative plasmids. However, a nonconjugative R factor can be spread to other cells if it is present in a cell that also contains a conjugative plasmid. In such a cell, the R factor can sometimes be transferred when the conjugative plasmid is transferred—that is, it is "mobilized." Even more troubling is the fact that some R factors are readily transferred *between* species. When humans and other animals consume antibiotics, the growth of host bacteria with R factors is promoted. The R factors then can be transferred to more pathogenic genera such as *Salmonella* or *Shigella,* causing even greater public health problems. Drug resistance (section 34.6)

Several other important types of plasmids have been discovered. These include bacteriocin-encoding plasmids, virulence plasmids, and metabolic plasmids. Bacteriocin-encoding plasmids may give the bacteria that harbor them a competitive advantage in the microbial world. **Bacteriocins** are bacterial proteins that destroy other bacteria. They usually act only against closely related strains. Some bacteriocins kill cells by forming channels in the plasma membrane, thus breaching the critical selective permeability required for cell viability. They also may degrade DNA and RNA or attack peptidoglycan and weaken the cell wall. **Col plasmids** contain genes for the synthesis of bacteriocins known as colicins, which are directed against *E. coli.* Other plasmids carry genes for bacteriocins against other species. For example, cloacins kill *Enterobacter* species. Some Col plasmids are conjugative and carry resistance genes. It should be noted that not all bacteriocin genes are on plasmids. For example, the bacteriocin genes of *Pseudomonas aeruginosa,* which code for proteins called pyocins, are located on the chromosome. Bacteriocins produced by the normal flora of humans (and other animals) also are

Table 3.3 Major Types of Bacterial Plasmids

Type	Representatives	Approximate Size (kbp)	Copy Number (Copies/ Chromosome)	Hosts	Phenotypic Features[a]
Fertility Factor[b]	F factor	95–100	1–3	*E. coli, Salmonella, Citrobacter*	Sex pilus, conjugation
R Plasmids	RP4	54	1–3	*Pseudomonas* and many other gram-negative bacteria	Sex pilus, conjugation, resistance to Amp, Km, Nm, Tet
	R1	80	1–3	Gram-negative bacteria	Resistance to Amp, Km, Su, Cm, Sm
	R6	98	1–3	*E. coli, Proteus mirabilis*	Su, Sm, Cm, Tet, Km, Nm
	R100	90	1–3	*E. coli, Shigella, Salmonella, Proteus*	Cm, Sm, Su, Tet, Hg
	pSH6	21		*Staphylococcus aureus*	Gm, Tet, Km
	pSJ23a	36		*S. aureus*	Pn, Asa, Hg, Gm, Km, Nm, Em, etc.
	pAD2	25		*Enterococcus faecalis*	Em, Km, Sm
Col Plasmids	ColE1	9	10–30	*E. coli*	Colicin E1 production
	ColE2		10–15	*Shigella*	Colicin E2
	CloDF13			*Enterobacter cloacae*	Cloacin DF13
Virulence Plasmids	Ent (P307)	83		*E. coli*	Enterotoxin production
	K88 plasmid			*E. coli*	Adherence antigens
	ColV-K30	2		*E. coli*	Siderophore for iron uptake; resistance to immune mechanisms
	pZA10	56		*S. aureus*	Enterotoxin B
	Ti	200		*Agrobacterium tumefaciens*	Tumor induction
Metabolic Plasmids	CAM	230		*Pseudomonas*	Camphor degradation
	SAL	56		*Pseudomonas*	Salicylate degradation
	TOL	75		*Pseudomonas putida*	Toluene degradation
	pJP4			*Pseudomonas*	2,4-dichlorophenoxyacetic acid degradation
				E. coli, Klebsiella, Salmonella	Lactose degradation
				Providencia	Urease
	sym			*Rhizobium*	Nitrogen fixation and symbiosis

[a] Abbreviations used for resistance to antibiotics and metals: Amp, ampicillin; Asa, arsenate; Cm, chloramphenicol; Em, erythromycin; Gm, gentamycin; Hg, mercury; Km, kanamycin; Nm, neomycin; Pn, pencillin; Sm, streptomycin; Su, sulfonamides; Tet, tetracycline.

[b] Many R plasmids, metabolic plasmids and others are also conjugative.

components of our defenses against invading pathogens. **Virulence plasmids** encode factors that make their hosts more pathogenic. For example, enterotoxigenic strains of *E. coli* cause traveler's diarrhea because they contain a plasmid that codes for an enterotoxin. **Metabolic plasmids** carry genes for enzymes that degrade substances such as aromatic compounds (toluene), pesticides (2,4-dichlorophenoxyacetic acid), and sugars (lactose). Metabolic plasmids even carry the genes required for some strains of *Rhizobium* to induce legume nodulation and carry out nitrogen fixation.

1. Give the major features of plasmids. How do they differ from chromosomes?
2. What is an episome? A conjugative plasmid?
3. Describe each of the following plasmids and explain their importance: F factor, R factor, Col plasmid, virulence plasmid, and metabolic plasmid.

3.6 The Bacterial Cell Wall

The cell wall is the layer, usually fairly rigid, that lies just outside the plasma membrane. It is one of the most important procaryotic structures for several reasons: it helps determine the shape of the cell; it helps protect the cell from osmotic lysis; it can protect the cell from toxic substances; and in pathogens, it can contribute to pathogenicity. The importance of the cell wall is reflected in the fact that relatively few procaryotes lack cell walls. Those that do have other features that fulfill cell wall function. The procaryotic cell wall also is the site of action of several antibiotics. Therefore, it is important to understand its structure.

The cell walls of *Bacteria* and *Archaea* are distinctive and are another example of the important features distinguishing these organisms. In this section, we focus on bacterial cell walls. An overview of bacterial cell wall structure is provided first. This is followed by more detailed discussions of particular aspects of cell wall structure and function. Archaeal cell walls are discussed in section 3.7.

Overview of Bacterial Cell Wall Structure

After Christian Gram developed the Gram stain in 1884, it soon became evident that most bacteria could be divided into two major groups based on their response to the Gram-stain procedure (*see table 19.9*). Gram-positive bacteria stained purple, whereas gram-negative bacteria were colored pink or red by the technique. The true structural difference between these two groups did not become clear until the advent of the transmission electron microscope. The gram-positive cell wall consists of a single 20 to 80 nm thick homogeneous layer of **peptidoglycan (murein)** lying outside the plasma membrane **(figure 3.17).** In contrast, the gram-negative cell wall is quite complex. It has a 2 to 7 nm peptidoglycan layer covered by a 7 to 8 nm thick **outer membrane.** Because of the thicker peptidoglycan layer, the walls of gram-positive cells are more resistant to osmotic pressure than those of gram-negative bacteria. Microbiologists often call all the structures from the plasma membrane outward the **cell envelope.** Therefore this includes the plasma membrane, cell wall, and structures like capsules (p. 65) when present. Preparation and staining of specimens: Differential staining (section 2.3)

One important feature of the cell envelope is a space that is frequently seen between the plasma membrane and the outer membrane in electron micrographs of gram-negative bacteria, and is sometimes observed between the plasma membrane and the wall in gram-positive bacteria. This space is called the **periplasmic space.** The substance that occupies the periplasmic space is the **periplasm.** The nature of the periplasmic space and periplasm differs in gram-positive and gram-negative bacteria. These differences are pointed out in the more detailed discussions that follow.

Peptidoglycan Structure

Peptidoglycan, or murein, is an enormous meshlike polymer composed of many identical subunits. The polymer contains two sugar derivatives, *N*-acetylglucosamine and *N*-acetylmuramic acid (the lactyl ether of *N*-acetylglucosamine), and several different amino acids. Three of these amino acids are not found in proteins: D-glutamic acid, D-alanine, and *meso*-diaminopimelic acid. The presence of D-amino acids protects against degradation by most peptidases, which recognize only the L-isomers of amino acid residues. The peptidoglycan subunit present in most gram-negative bacteria and many gram-positive ones is shown in **figure 3.18.** The backbone of this

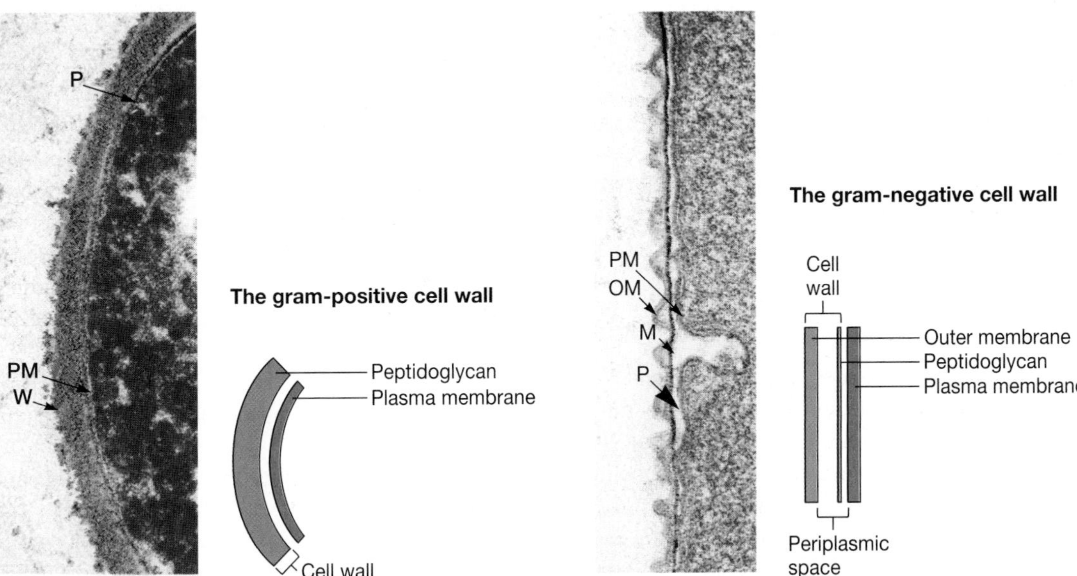

Figure 3.17 Gram-Positive and Gram-Negative Cell Walls. The gram-positive envelope is from *Bacillus licheniformis* (left), and the gram-negative micrograph is of *Aquaspirillum serpens* (right). M; peptidoglycan or murein layer; OM, outer membrane; PM, plasma membrane; P, periplasmic space; W, gram-positive peptidoglycan wall.

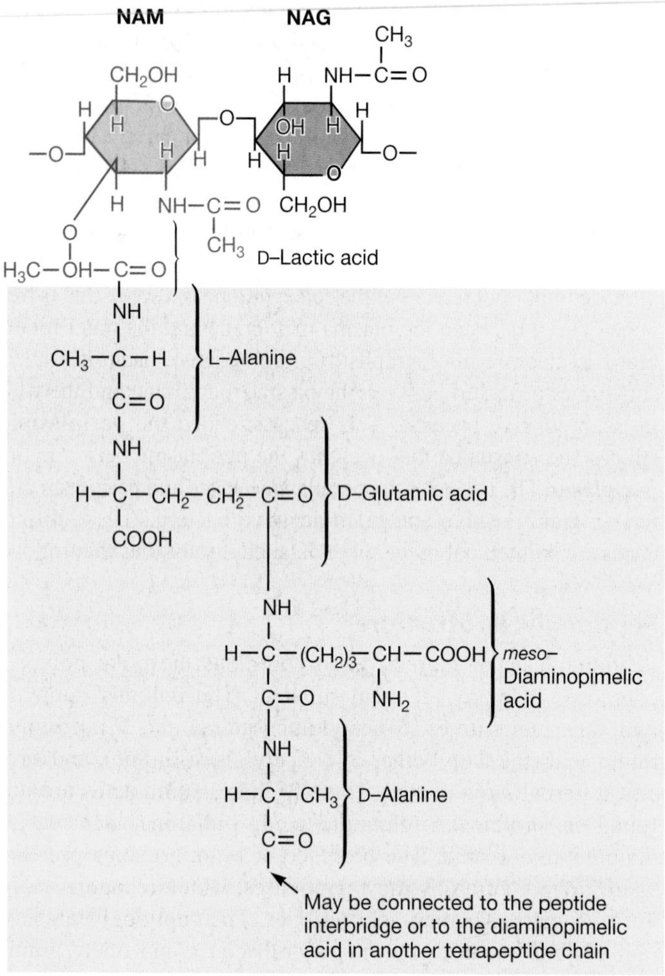

Figure 3.18 Peptidoglycan Subunit Composition. The peptidoglycan subunit of *E. coli,* most other gram-negative bacteria, and many gram-positive bacteria. NAG is *N*-acetylglucosamine. NAM is *N*-acetylmuramic acid (NAG with lactic acid attached by an ether linkage). The tetrapeptide side chain is composed of alternating D- and L-amino acids since *meso*-diaminopimelic acid is connected through its L-carbon. NAM and the tetrapeptide chain attached to it are shown in different shades of color for clarity.

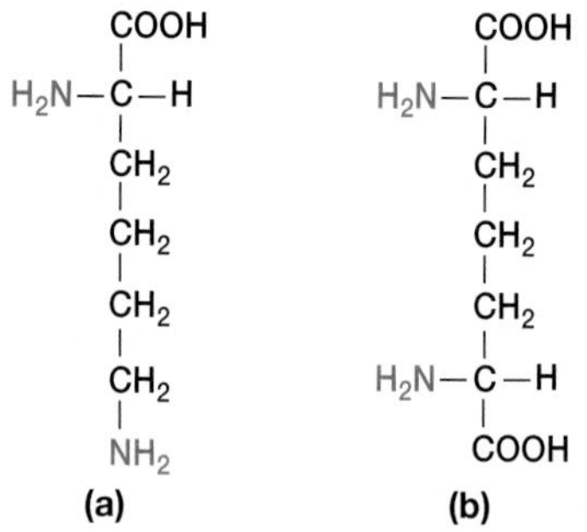

Figure 3.19 Diaminoacids Present in Peptidoglycan. **(a)** L-Lysine. **(b)** *meso*-Diaminopimelic acid.

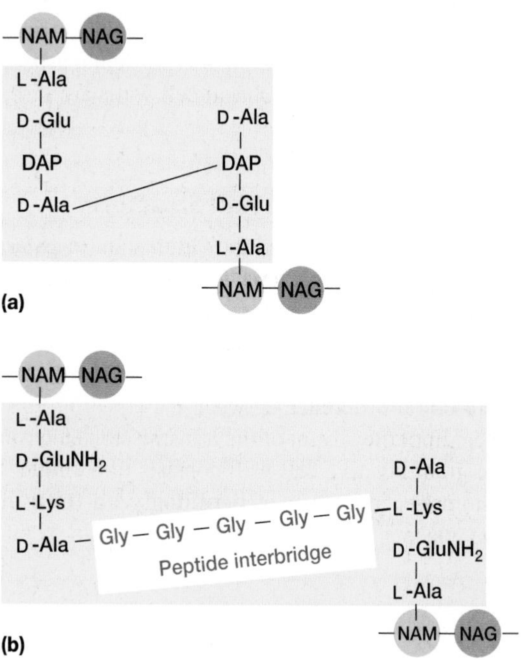

Figure 3.20 Peptidoglycan Cross-Links. (**a**) *E. coli* peptidoglycan with direct cross-linking, typical of many gram-negative bacteria. (**b**) *Staphylococcus aureus* peptidoglycan. *S. aureus* is a gram-positive bacterium. NAM is *N*-acetylmuramic acid. NAG is *N*-acetylglucosamine. Gly is glycine. Although the polysaccharide chains are drawn opposite each other for the sake of clarity, two chains lying side-by-side may be linked together (see figure 3.21).

polymer is composed of alternating *N*-acetylglucosamine and *N*-acetylmuramic acid residues. A peptide chain of four alternating D- and L-amino acids is connected to the carboxyl group of *N*-acetylmuramic acid. Many bacteria replace *meso*-diaminopimelic acid with another diaminoacid, usually L-lysine (**figure 3.19**). Carbohydrates (appendix I); Peptidoglycan and endospore structure (section 23.3); Proteins (appendix I)

In order to make a strong, meshlike polymer, chains of linked peptidoglycan subunits must be joined by cross-links between the peptides. Often the carboxyl group of the terminal D-alanine is connected directly to the amino group of diaminopimelic acid, but a **peptide interbridge** may be used instead (**figure 3.20**). Most gram-negative cell wall peptidoglycan lacks the peptide interbridge. This cross-linking results in an enormous peptidoglycan sac that is actually one dense, interconnected network (**figure 3.21**). These sacs have been isolated from gram-positive bacteria and are strong enough to retain their shape and integrity (**figure 3.22**), yet they are relatively porous, elastic, and somewhat stretchable.

Gram-Positive Cell Walls

Gram-positive bacteria normally have cell walls that are thick and composed primarily of peptidoglycan. Peptidoglycan in gram-positive bacteria often contains a peptide interbridge (figure 3.21 and **figure 3.23**). In addition, gram-positive cell walls usually contain large amounts of **teichoic acids,** polymers of glycerol or ribitol joined by phosphate groups (figure 3.23 and **figure 3.24**). Amino acids such as D-alanine or sugars like glucose are attached to the glycerol and ribitol groups. The teichoic acids are covalently connected to either the peptidoglycan itself or to plasma membrane lipids; in the latter case they are called lipoteichoic acids. Teichoic acids appear to extend to the surface of the peptidoglycan, and, because they are negatively charged, help give the gram-positive cell

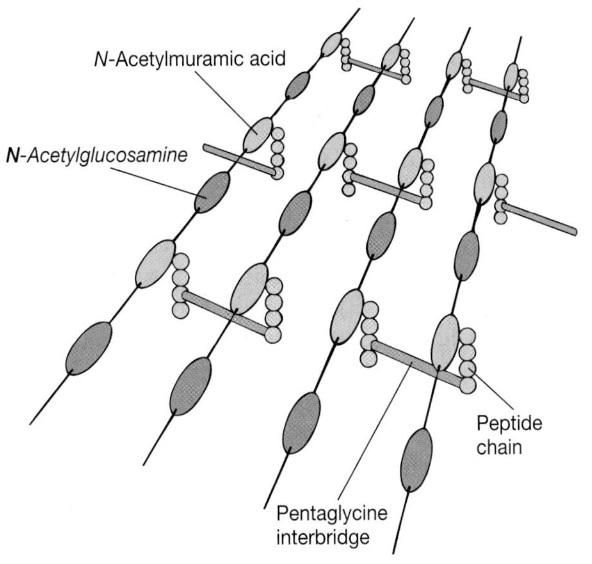

Figure 3.21 Peptidoglycan Structure. A schematic diagram of one model of peptidoglycan. Shown are the polysaccharide chains, tetrapeptide side chains, and peptide interbridges.

Figure 3.22 Isolated Gram-Positive Cell Wall. The peptidoglycan wall from *Bacillus megaterium,* a gram-positive bacterium. The latex spheres have a diameter of 0.25 μm.

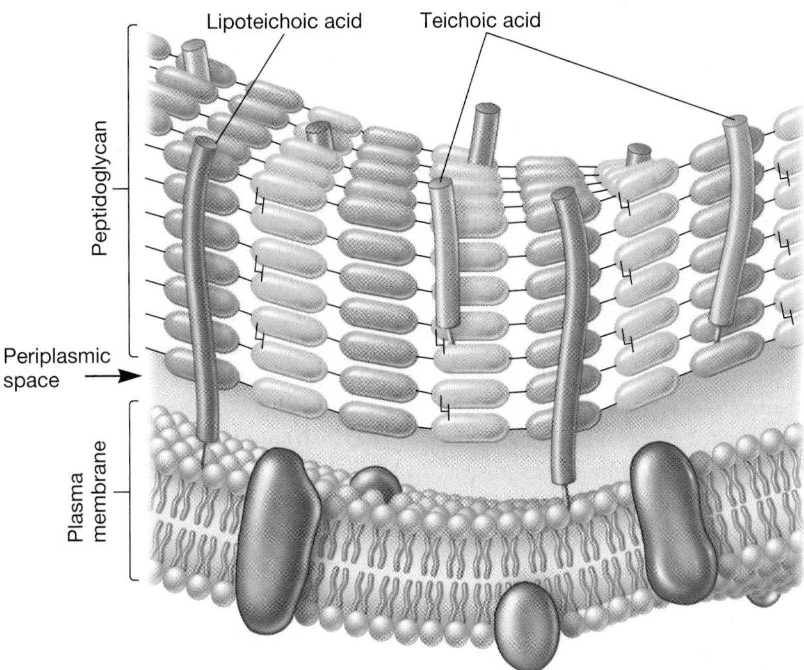

Figure 3.23 The Gram-Positive Envelope.

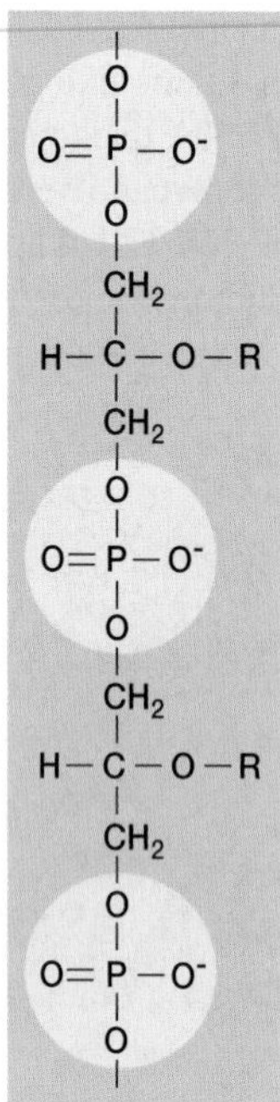

Figure 3.24 Teichoic Acid Structure. The segment of a teichoic acid made of phosphate, glycerol, and a side chain, R. R may represent D-alanine, glucose, or other molecules.

wall its negative charge. The functions of these molecules are still unclear, but they may be important in maintaining the structure of the wall. Teichoic acids are not present in gram-negative bacteria.

The periplasmic space of gram-positive bacteria, when observed, lies between the plasma membrane and the cell wall and is smaller than that of gram-negative bacteria. Even if gram-positive bacteria lack a discrete, obvious periplasmic space, they may have periplasm. The periplasm has relatively few proteins; this is probably because the peptidoglycan sac is porous and any proteins secreted by the cell usually pass through it. Enzymes secreted by gram-positive bacteria are called **exoenzymes.** They often serve to degrade polymeric nutrients that would otherwise be too large for transport across the plasma membrane. Those proteins that remain in the periplasmic space are usually attached to the plasma membrane.

Staphylococci and most other gram-positive bacteria have a layer of proteins on the surface of their cell wall peptidoglycan. These proteins are involved in the interactions of the cell with its environment. Some are noncovalently attached by binding to the peptidoglycan, teichoic acids, or other receptors. For example, the S-layer proteins (see p. 66) bind noncovalently to polymers scattered throughout the wall. Enzymes involved in peptidoglycan synthesis and turnover also seem to interact noncovalently with the cell wall. Other surface proteins are covalently attached to the peptidoglycan. Many covalently attached proteins, such as the M protein of pathogenic streptococci, have roles in virulence, such as aiding in adhesion to host tissues and interfering with host defenses. In staphylococci, these surface proteins are covalently joined to the pentaglycine bridge of the cell wall peptidoglycan. An enzyme called sortase catalyzes the attachment of these surface proteins to the gram-positive peptidoglycan. Sortases are attached to the plasma membrane of the bacterial cell.

Gram-Negative Cell Walls

Even a brief inspection of figure 3.17 shows that gram-negative cell walls are much more complex than gram-positive walls. The thin peptidoglycan layer next to the plasma membrane and bounded on either side by the periplasmic space may constitute not more than 5 to 10% of the wall weight. In *E. coli* it is about 2 nm thick and contains only one or two sheets of peptidoglycan.

The periplasmic space of gram-negative bacteria is also strikingly different than that of gram-positive bacteria. It ranges in size from 1 nm to as great as 71 nm. Some recent studies indicate that it may constitute about 20 to 40% of the total cell volume, and it is usually 30 to 70 nm wide. When cell walls are disrupted carefully or removed without disturbing the underlying plasma membrane, periplasmic enzymes and other proteins are released and may be easily studied. Some periplasmic proteins participate in nutrient acquisition—for example, hydrolytic enzymes and transport proteins. Some periplasmic proteins are involved in energy conservation. For example, the denitrifying bacteria, which convert nitrate to nitrogen gas, and bacteria that use inorganic molecules as energy sources (chemolithotrophs) have electron transport proteins in their periplasm. Other periplasmic proteins are involved in peptidoglycan synthesis and the modification of toxic compounds that could harm the cell. Chemolithotrophy (section 9.10); Biogeochemical cycling: The nitrogen cycle (section 27.2)

The outer membrane lies outside the thin peptidoglycan layer (**figures 3.25** and **3.26**) and is linked to the cell in two ways. The first is by Braun's lipoprotein, the most abundant protein in the outer membrane. This small lipoprotein is covalently joined to the underlying peptidoglycan, and is embedded in the outer membrane by its hydrophobic end. The outer membrane and peptidoglycan are so firmly linked by this lipoprotein that they can be isolated as one unit. The second linking mechanism involves the many adhesion sites joining the outer membrane and the plasma membrane. The two membranes appear to be in direct contact at these sites. In *E. coli,* 20 to 100 nm areas of contact between the two membranes can be seen. Adhesion sites may be regions of direct contact or possibly true membrane fusions. It has been proposed that substances can move directly into the cell through these adhesion sites, rather than traveling through the periplasm.

Possibly the most unusual constituents of the outer membrane are its **lipopolysaccharides (LPSs).** These large, complex molecules contain both lipid and carbohydrate, and consist of three parts: (1) lipid A, (2) the core polysaccharide, and (3) the O side chain. The LPS from *Salmonella* has been studied most, and its general structure is described here (**figure 3.27**). The **lipid A** region contains two glucosamine sugar derivatives, each with three fatty acids and phosphate or pyrophosphate attached. The fatty acids attach the lipid A to the outer membrane, while the remainder of the LPS molecule projects from the surface. The **core polysaccharide** is joined to lipid A. In *Salmonella* it is constructed of

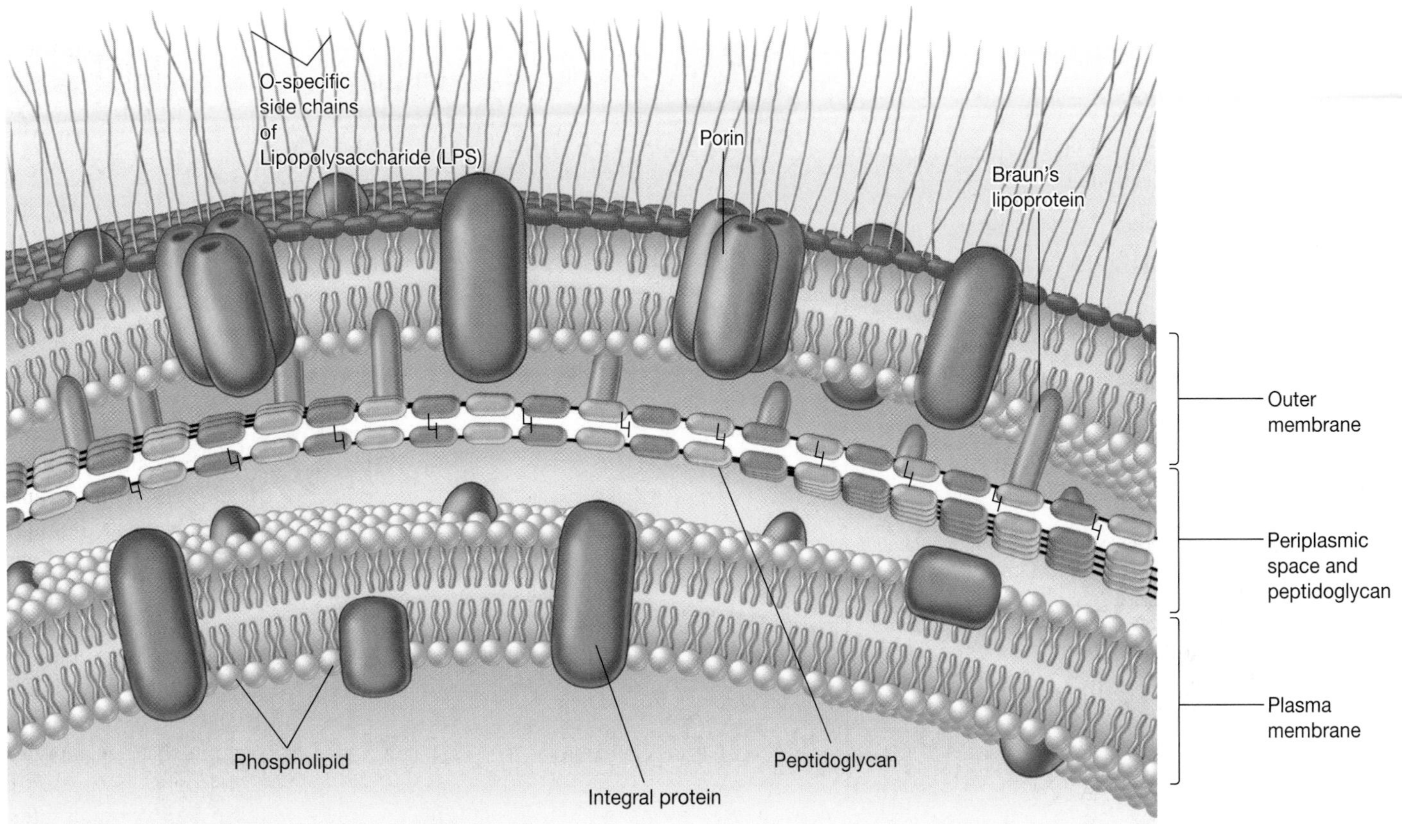

Figure 3.25 The Gram-Negative Envelope.

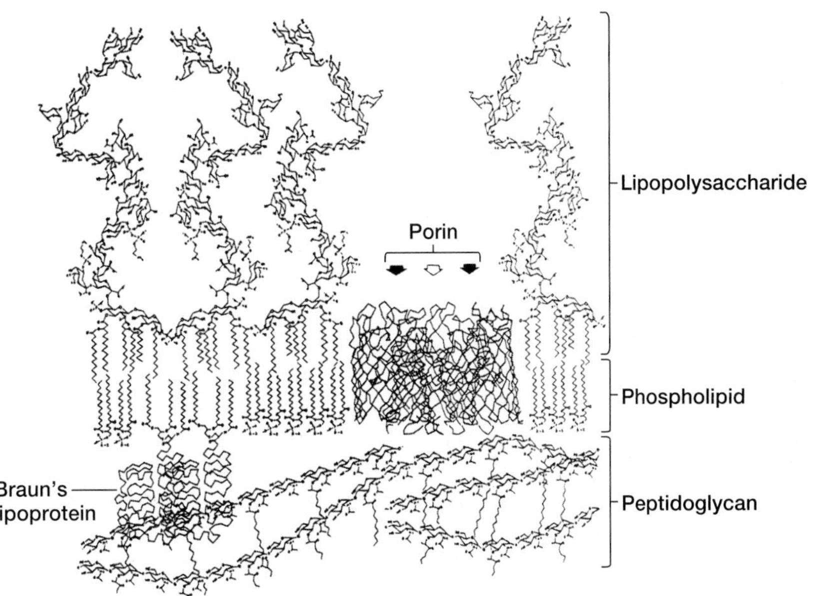

Figure 3.26 A Chemical Model of the *E. coli* Outer Membrane and Associated Structures. This cross-section is to scale. The porin OmpF has two channels in the front (solid arrows) and one channel in the back (open arrow) of the trimeric protein complex. LPS molecules can be longer than the ones shown here.

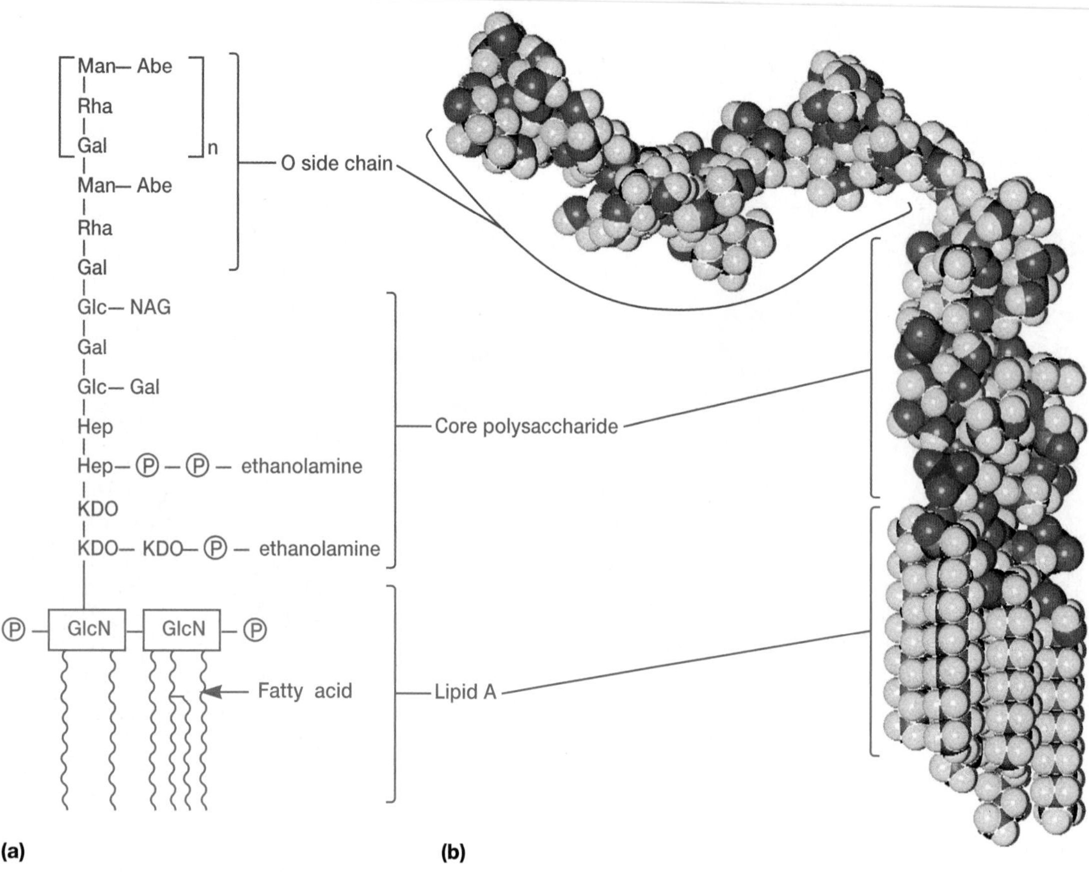

Figure 3.27 Lipopolysaccharide Structure. **(a)** The lipopolysaccharide from *Salmonella*. This slightly simplified diagram illustrates one form of the LPS. Abbreviations: Abe, abequose; Gal, galactose; Glc, glucose; GlcN, glucosamine; Hep, heptulose; KDO, 2-keto-3-deoxyoctonate; Man, mannose; NAG, *N*-acetylglucosamine; P, phosphate; Rha, L-rhamnose. Lipid A is buried in the outer membrane. **(b)** Molecular model of an *Escherichia coli* lipopolysaccharide. The lipid A and core polysaccharide are straight; the O side chain is bent at an angle in this model.

10 sugars, many of them unusual in structure. The **O side chain** or **O antigen** is a polysaccharide chain extending outward from the core. It has several peculiar sugars and varies in composition between bacterial strains.

LPS has many important functions. Because the core polysaccharide usually contains charged sugars and phosphate (figure 3.27), LPS contributes to the negative charge on the bacterial surface. As a major constituent of the exterior leaflet of the outer membrane, lipid A also helps stabilize outer membrane structure. LPS may contribute to bacterial attachment to surfaces and biofilm formation. A major function of LPS is that it aids in creating a permeability barrier. The geometry of LPS (figure 3.27*b*) and interactions between neighboring LPS molecules are thought to restrict the entry of bile salts, antibiotics, and other toxic substances that might kill or injure the bacterium. LPS also plays a role in protecting pathogenic gram-negative bacteria from host defenses. The O side chain of LPS is also called the O antigen because it elicits an immune response. This response involves the production of antibodies that bind the strain-specific form of LPS that elicited the response. However, many gram-negative bacteria are able to rapidly change the antigenic nature of their O side chains, thus thwarting host defenses. Importantly, the lipid A portion of LPS often is toxic; as a result, the LPS can act as an endotoxin and cause some of the symptoms that arise in gram-negative bacterial infections. If the bacterium enters the bloodstream, LPS endotoxin can cause a form of septic shock for which there is no direct treatment. Overview of bacterial pathogenesis (section 33.3)

Despite the role of LPS in creating a permeability barrier, the outer membrane is more permeable than the plasma membrane and permits the passage of small molecules like glucose and other monosaccharides. This is due to the presence of **porin proteins** (figures 3.25 and 3.26). Most porin proteins cluster together to form a trimer in the outer membrane (figure 3.25 and **figure 3.28**). Each porin protein spans the outer membrane and is more or less tube-shaped; its narrow channel allows passage of molecules smaller than about 600 to 700 daltons. However, larger molecules such as vitamin B_{12} also cross the outer membrane. Such large molecules do not pass through porins; instead, specific carriers transport them across the outer membrane.

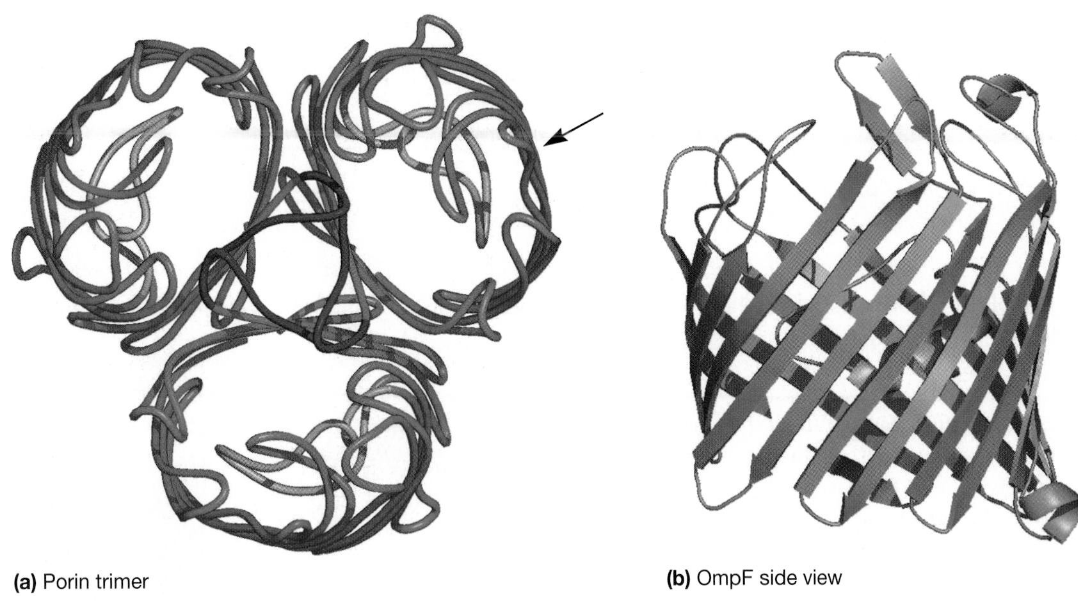

Figure 3.28 Porin Proteins. Two views of the OmpF porin of *E. coli.* **(a)** Porin structure observed when looking down at the outer surface of the outer membrane (i.e., top view). The three porin proteins forming the protein each form a channel. Each porin can be divided into three loops: the green loop forms the channel, the blue loop interacts with other porin proteins to help form the trimer, and the orange loop narrows the channel. The arrow indicates the area of a porin molecule viewed from the side in panel **(b).** Side view of a porin monomer showing the β-barrel structure characteristic of porin proteins.

The Mechanism of Gram Staining

Although several explanations have been given for the Gram-stain reaction results, it seems likely that the difference between gram-positive and gram-negative bacteria is due to the physical nature of their cell walls. If the cell wall is removed from gram-positive bacteria, they stain gram negative. Furthermore, genetically wall-less bacteria such as the mycoplasmas also stain gram negative. The peptidoglycan itself is not stained; instead it seems to act as a permeability barrier preventing loss of crystal violet. During the procedure the bacteria are first stained with crystal violet and next treated with iodine to promote dye retention. When gram-positive bacteria then are treated with ethanol, the alcohol is thought to shrink the pores of the thick peptidoglycan. Thus the dye-iodine complex is retained during this short decolorization step and the bacteria remain purple. In contrast, recall that gram-negative peptidoglycan is very thin, not as highly cross-linked, and has larger pores. Alcohol treatment also may extract enough lipid from the gram-negative outer membrane to increase its porosity further. For these reasons, alcohol more readily removes the purple crystal violet-iodine complex from gram-negative bacteria. Thus gram-negative bacteria are then easily stained red or pink by the counterstain safranin.

The Cell Wall and Osmotic Protection

Microbes have several mechanisms for responding to changes in osmotic pressure. This pressure arises when the concentration of solutes inside the cell differs from that outside, and the adaptive responses work to equalize the solute concentrations. However, in certain situations, the osmotic pressure can exceed the cell's ability to adapt. In these cases, additional protection is provided by the cell wall. When cells are in hypotonic solutions—ones in which the solute concentration is less than that in the cytoplasm—water moves into the cell, causing it to swell. Without the cell wall, the pressure on the plasma membrane would become so great that it would be disrupted and the cell would burst—a process called **lysis.** Conversely, in hypertonic solutions, water flows out and the cytoplasm shrivels up—a process called **plasmolysis.**

The protective nature of the cell wall is most clearly demonstrated when bacterial cells are treated with lysozyme or penicillin. The enzyme **lysozyme** attacks peptidoglycan by hydrolyzing the bond that connects *N*-acetylmuramic acid with *N*-acetylglucosamine (figure 3.18). **Penicillin** works by a different mechanism. It inhibits peptidoglycan synthesis. If bacteria are treated with either of these substances while in a hypotonic solution, they will lyse. However, if they are in an isotonic solution, they can survive and grow normally. If they are gram positive, treatment with lysozyme or penicillin results in the complete loss of the cell wall, and the cell becomes a protoplast. When gram-negative bacteria are exposed to lysozyme or penicillin, the peptidoglycan layer is lost, but the outer membrane remains. These cells are called **spheroplasts.** Because they lack a complete cell wall, both protoplasts and spheroplasts are osmotically sensitive. If they are transferred to a dilute solution, they will lyse due to uncontrolled water influx (**figure 3.29**). Antibacterial drugs (section 34.4)

Although most bacteria require an intact cell wall for survival, some have none at all. For example, the mycoplasmas lack a cell wall and are osmotically sensitive, yet often can grow in dilute

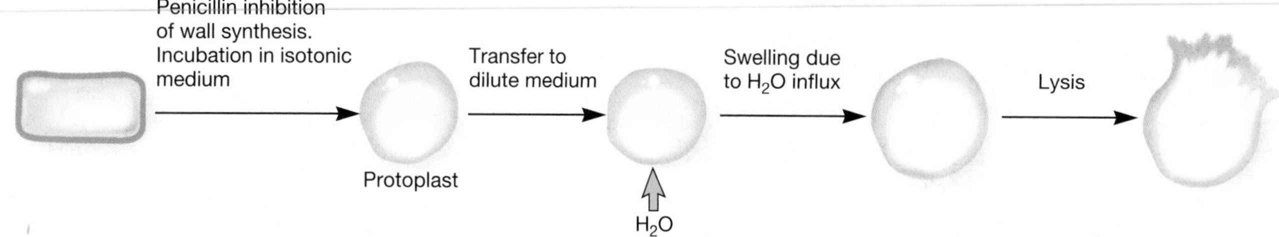

Figure 3.29 Protoplast Formation and Lysis. Protoplast formation induced by incubation with penicillin in an isotonic medium. Transfer to dilute medium will result in lysis.

media or terrestrial environments because their plasma membranes are more resistant to osmotic pressure than those of bacteria having walls. The precise reason for this is not clear, although the presence of sterols in the membranes of many species may provide added strength. Without a rigid cell wall, mycoplasmas tend to be pleomorphic or variable in shape.

1. List the functions of the cell wall.
2. Describe in detail the composition and structure of peptidoglycan. Why does peptidoglycan contain the unusual D isomers of alanine and glutamic acid rather than the L isomers observed in proteins?
3. Compare and contrast the cell walls of gram-positive bacteria and gram-negative bacteria. Include labeled drawings in your discussion.
4. Define or describe the following: outer membrane, periplasmic space, periplasm, envelope, teichoic acid, adhesion site, lipopolysaccharide, porin protein, protoplasts, and spheroplasts.
5. Design an experiment that illustrates the cell wall's role in protecting against lysis.
6. With a few exceptions, the cell walls of gram-positive bacteria lack porins. Why is this the case?

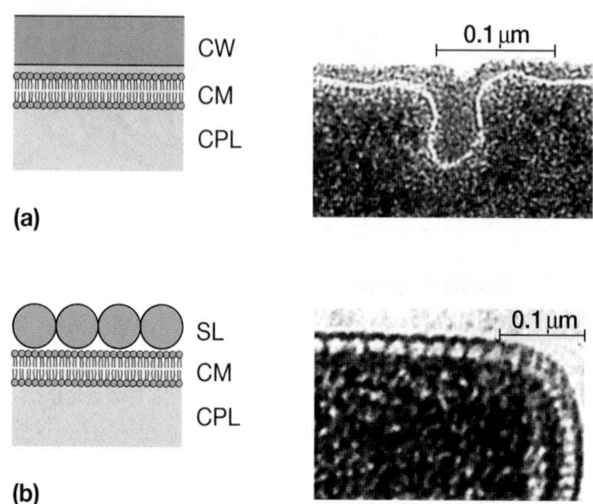

Figure 3.30 Cell Envelopes of *Archaea*. Schematic representations and electron micrographs of **(a)** *Methanobacterium formicicum*, and **(b)** *Thermoproteus tenax*. CW, cell wall; SL, surface layer; CM, cell membrane or plasma membrane; CPL, cytoplasm.

3.7 Archaeal Cell Walls

Before they were distinguished as a unique domain of life, the *Archaea* were characterized as being either gram positive or gram negative. However, their staining reaction does not correlate as reliably with a particular cell wall structure as does the Gram reaction of *Bacteria*. Archaeal wall structure and chemistry differ from those of the *Bacteria*. Archaeal cell walls lack peptidoglycan and also exhibit considerable variety in terms of their chemical make-up. Some of the major features of archaeal cell walls are described in this section.

Many archaea have a wall with a single, thick homogeneous layer resembling that in gram-positive bacteria (**figure 3.30*a***). These archaea often stain gram positive. Their wall chemistry varies from species to species but usually consists of complex heteropolysaccharides. For example, *Methanobacterium* and some other methane-generating archaea (methanogens) have walls containing **pseudomurein** (**figure 3.31**), a peptidoglycan-like polymer that has L-amino acids instead of D-amino acids in its cross-links, *N*-acetyltalosaminuronic acid instead of *N*-acetylmuramic acid, and β (1→3) glycosidic bonds instead of β (1→4) glycosidic bonds. Other archaea, such as *Methanosarcina* and the salt-loving *Halococcus*, contain complex polysaccharides similar to the chondroitin sulfate of animal connective tissue. Phylum *Euryarchaeota:* The methanogens; The halobacteria (section 20.3)

Many archaea that stain gram negative have a layer of glycoprotein or protein outside their plasma membrane (figure 3.30*b*). The layer may be as thick as 20 to 40 nm. Sometimes there are two layers—an electron-dense layer and a sheath surrounding it. Some methanogens (*Methanolobus*), salt-loving archaea (*Halobacterium*), and extreme thermophiles (*Sulfolobus, Thermoproteus,* and *Pyrodictium*) have glycoproteins in their walls. In contrast, other methanogens (*Methanococcus, Methanomicrobium,* and *Methanogenium*) and the extreme thermophile *Desulfurococcus* have protein walls. Phylum *Crenarchaeota* (section 20.2); Phylum *Euryarchaeota:* Extremely thermophilic S^0-metabolizers (section 20.3)

1. How do the cells walls of *Archaea* differ from those of *Bacteria?*
2. What is pseudomurein? How is it similar to peptidoglycan? How is it different?
3. Archaea with cell walls consisting of a thick, homogeneous layer of complex polysaccharides often retain the crystal violet dye when stained using the Gram-staining procedure. Why do you think this is so?

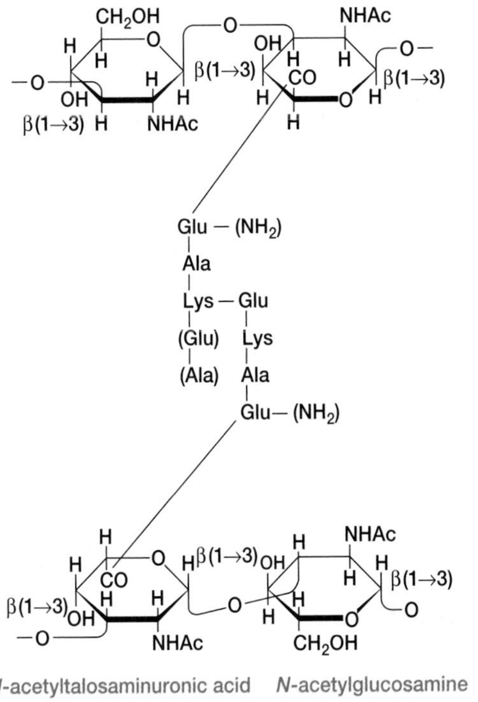

N-acetyltalosaminuronic acid N-acetylglucosamine

Figure 3.31 The Structure of Pseudomurein. The amino acids and amino groups in parentheses are not always present. Ac represents the acetyl group.

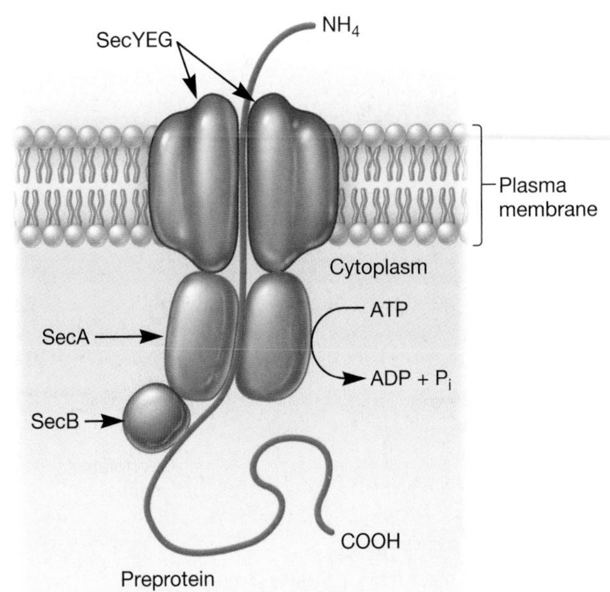

Figure 3.32 The Sec-Dependent Pathway. The Sec-dependent pathway of *E. coli*.

3.8 Protein Secretion in Procaryotes

The membranes of the procaryotic cell envelope (i.e., the plasma membrane of *Archaea* and gram-positive bacteria and the plasma and outer membranes of gram-negative bacteria) present a considerable barrier to the movement of large molecules into or out of the cell. Yet, as will be discussed in section 3.9, many important structures are located outside the wall. How are the large molecules from which some of these structures are made transported out of the cell for assembly? Furthermore, exoenzymes and other proteins are released from procaryotes into their surroundings. How do these proteins get through the membrane(s) of the cell envelope? Clearly procaryotes must be able to secrete proteins. The research on protein secretion pathways has mushroomed in the last few decades in part because of the fundamental importance of protein secretion, but also because certain protein secretion mechanisms are common to pathogenic bacteria. Furthermore, an understanding of protein secretion can be exploited for vaccine development and a variety of industrial processes. Because relatively little is known about archaeal protein secretion systems, this section provides an overview of bacterial protein secretion pathways.

Overview of Bacterial Protein Secretion

Protein secretion poses different difficulties depending on whether the bacterium is gram-positive or gram-negative. In order for gram-positive bacteria to secrete proteins, the proteins must be transported across the plasma membrane. Once across the plasma membrane, the protein either passes through the relatively porous peptidoglycan into the external environment or it becomes embedded in or attached to the peptidoglycan. Gram-negative bacteria have more hurdles to jump when they secrete proteins. They, too, must transport the proteins across the plasma membrane, but in order to complete the secretion process, the proteins must be able to escape attack from protein-degrading enzymes in the periplasmic space, and they must be transported across the outer membrane. In both gram-positive and gram-negative bacteria, the major pathway for transporting proteins across the plasma membrane is the Sec-dependant (*sec*retion-dependent) pathway (**figure 3.32**). In gram-negative bacteria, proteins can be transported across the outer membrane by several different mechanisms, some of which bypass the Sec-dependent pathway altogether, moving proteins directly from the cytoplasm to the outside of the cell (**figure 3.33**). All protein secretion pathways described here require the expenditure of energy at some step in the process. The energy is usually supplied by the hydrolysis of high-energy molecules such as ATP and GTP, but another form of energy, the proton motive force, also sometimes plays a role. The role of ATP in metabolism (section 8.5); Electron transport and oxidative phosphorylation (section 9.5)

The Sec-Dependent Pathway

The **Sec-dependent pathway,** sometimes called the general secretion pathway, is highly conserved and has been identified in all three domains of life (figure 3.32). It translocates proteins across the plasma membrane or integrates them into the membrane itself. Proteins to be transported across the plasma membrane by this pathway are synthesized as presecretory proteins called preproteins. The preprotein has a **signal peptide** at its

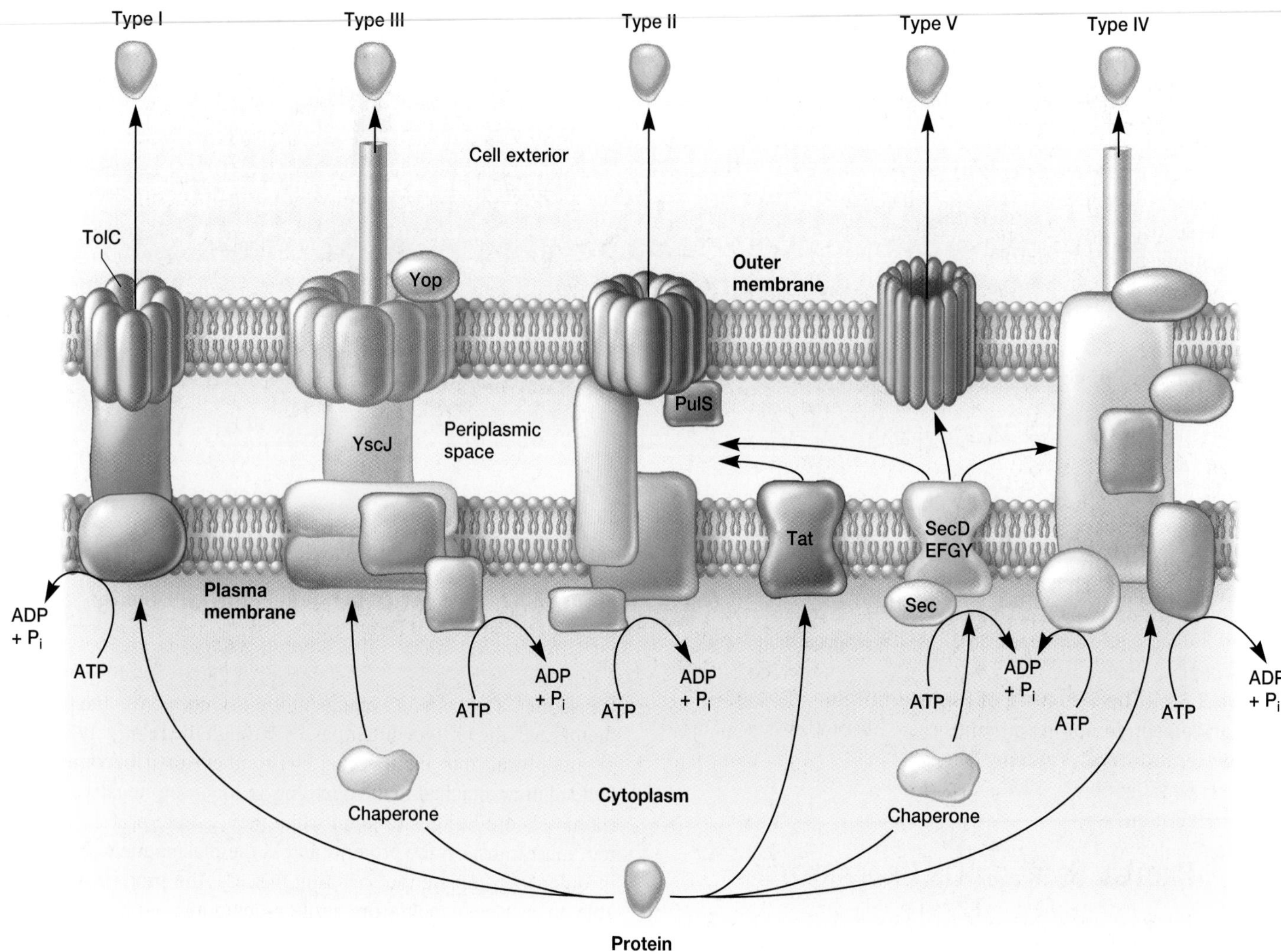

Figure 3.33 The Protein Secretion Systems of Gram-Negative Bacteria. The five secretion systems of gram-negative bacteria are shown. The Sec-dependent and Tat pathways deliver proteins from the cytoplasm to the periplasmic space. The type II, type V, and sometimes type IV systems complete the secretion process begun by the Sec-dependent pathway. The Tat system appears to deliver proteins only to the type II pathway. The type I and type III systems bypass the Sec-dependent and Tat pathways, moving proteins directly from the cytoplasm, through the outer membrane, to the extracellular space. The type IV system can work either with the Sec-dependent pathway or can work alone to transport proteins to the extracellular space. Proteins translocated by the Sec-dependent pathway and the type III pathway are delivered to those systems by chaperone proteins.

amino-terminus, which is recognized by the Sec machinery. Soon after the signal peptide is synthesized, special proteins called chaperone proteins (e.g., SecB) bind it. This helps delay protein folding, thereby helping the preprotein reach the Sec transport machinery in the conformation needed for transport. There is evidence that translocation of proteins can begin before the completion of their synthesis by ribosomes. Certain Sec proteins (SecY, SecE and SecG) are thought to form a channel in the membrane through which the preprotein passes. Another protein (SecA) binds to the SecYEG proteins and the SecB-preprotein complex. SecA acts as a motor to translocate the preprotein (but not the chaperone protein) through the plasma membrane using ATP hydrolysis. When the preprotein emerges from the plasma membrane, free from chaperones, an enzyme called signal peptidase removes the signal peptide. The protein then folds into the proper shape, and disulfide bonds are formed when necessary.

Protein Secretion in Gram-Negative Bacteria

Currently five protein secretion pathways have been identified in gram-negative bacteria (figure 3.33). Recall that gram-negative bacteria have a second, outer membrane that proteins must cross.

Gram-negative bacteria use the type II and type V pathways to transport proteins across the outer membrane after the protein has first been translocated across the plasma membrane by the Sec-dependant pathway. The type I and type III pathways do not interact with proteins that are first translocated by the Sec system, so they are said to be Sec-independent. The type IV pathway sometimes is linked to the Sec-dependent pathway but usually functions on its own.

The **type II protein secretion pathway** is present in a number of plant and animal pathogens, including *Erwinia carotovora, Klebsiella pneumoniae, Pseudomonas aeruginosa,* and *Vibrio cholerae.* It is responsible for the secretion of proteins such as the degradative enzymes pullulanases, cellulases, pectinases, proteases, and lipases, as well as other proteins like cholera toxin and pili proteins. Type II systems are quite complex and can contain as many as 12 to 14 proteins, most of which appear to be integral membrane proteins (figure 3.33). Even though some components of type II systems span the plasma membrane, they apparently translocate proteins only across the outer membrane. In most cases, the Sec-dependent pathway first translocates the protein across the plasma membrane and then the type II system completes the secretion process. In gram-negative and gram-positive bacteria, another plasma membrane translocation system called the Tat pathway can move proteins across the plasma membrane. In gram-negatives, these proteins are then delivered to the type II system. The Tat pathway is distinct from the Sec system in that it translocates already folded proteins.

The **type V protein secretion pathways** are the most recently discovered protein secretion systems. They, too, rely on the Sec-dependent pathway to move proteins across the plasma membrane. However, once in the periplasmic space, many of these proteins are able to form a channel in the outer membrane through which they transport themselves; these proteins are referred to as autotransporters. Other proteins are secreted by the type V pathway with the aid of a separate helper protein.

The **ABC protein secretion pathway,** which derives its name from *A*TP *b*inding *c*assette, is ubiquitous in procaryotes—that is, it is present in gram-positive and gram-negative bacteria as well as *Archaea.* It is sometimes called the **type I protein secretion pathway** (figure 3.33). In pathogenic gram-negative bacteria, it is involved in the secretion of toxins (α-hemolysin), as well as proteases, lipases, and specific peptides. Secreted proteins usually contain C-terminal secretion signals that help direct the newly synthesized protein to the type I machinery, which spans the plasma membrane, the periplasmic space, and the outer membrane. These systems translocate proteins in one step across both membranes, bypassing the Sec-dependent pathway. Gram-positive bacteria use a modified version of the type I system to translocate proteins across the plasma membrane. Analysis of the *Bacillus subtilis* genome has identified 77 ABC transporters. This may reflect the fact that ABC transporters transport a wide variety of solutes in addition to proteins, including sugars and amino acids, as well as exporting drugs from the cell interior.

Several gram-negative pathogens have the **type III protein secretion pathway,** another secretion system that bypasses the Sec-dependent pathway. Most type III systems inject virulence factors directly into the plant and animal host cells these pathogens attack. These virulence factors include toxins, phagocytosis inhibitors, stimulators of cytoskeleton reorganization in the host cell, and promoters of host cell suicide (apoptosis). However, in some cases the virulence factor is simply secreted into the extracellular milieu. Type III systems also transport other proteins, including (1) some of the proteins from which the system is built, (2) proteins that regulate the secretion process, and (3) proteins that aid in the insertion of secreted proteins into target cells. Type III systems are structurally complex and often are shaped like a syringe (figure 3.33). The slender, needlelike portion extends from the cell surface; a cylindrical base is connected to both the outer membrane and the plasma membrane and looks somewhat like the flagellar basal body (see p. 67). It is thought that proteins may move through a translocation channel. Important examples of bacteria with type III systems are *Salmonella, Yersinia, Shigella, E. coli, Bordetella, Pseudomonas aeruginosa,* and *Erwinia.* The participation of type III systems in bacterial virulence is further discussed in chapter 33.

Type IV protein secretion pathways are unique in that they are used to secrete proteins as well as to transfer DNA from a donor bacterium to a recipient during bacterial conjugation. Type IV systems are composed of many different proteins, and like the type III systems, these proteins form a syringelike structure. Type IV systems and conjugation are described in more detail in chapter 13.

1. Give the major characteristics and functions of the protein secretion pathways described in this section.
2. Which secretion pathway is most widespread?
3. What is a signal peptide? Why do you think a protein's signal peptide is not removed until after the protein is translocated across the plasma membrane?

3.9 Components External to the Cell Wall

Procaryotes have a variety of structures outside the cell wall that can function in protection, attachment to objects, and cell movement. Several of these are discussed.

Capsules, Slime Layers, and S-Layers

Some procaryotes have a layer of material lying outside the cell wall. This layer has different names depending on its characteristics. When the layer is well organized and not easily washed off, it is called a **capsule** (**figure 3.34*a***). It is called a **slime layer** when it is a zone of diffuse, unorganized material that is removed easily. When the layer consists of a network of polysaccharides extending from the surface of the cell, it is referred to as the **glycocalyx** (figure 3.34*b*), a term that can encompass both capsules and slime layers because they usually are composed of polysaccharides. However, some slime layers and capsules are constructed of other materials. For example, *Bacillus anthracis* has a

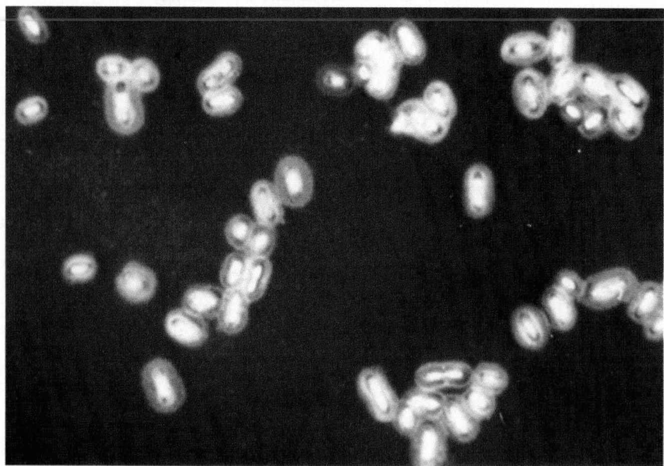

(a) *K. pneumoniae*

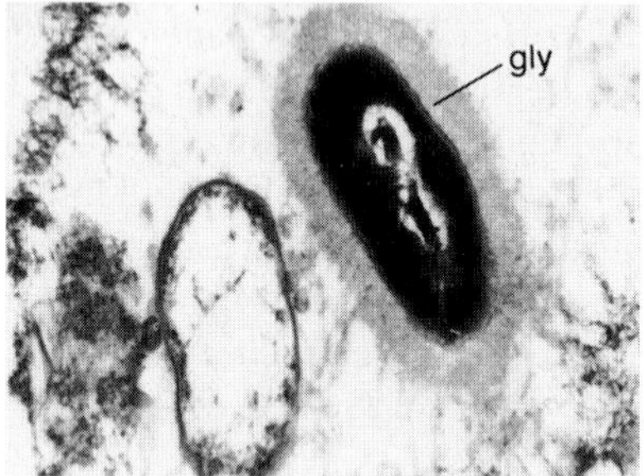

(b) *Bacteroides*

Figure 3.34 Bacterial Capsules. **(a)** *Klebsiella pneumoniae* with its capsule stained for observation in the light microscope (×1,500). **(b)** *Bacteroides* glycocalyx (gly), TEM (×71,250).

proteinaceous capsule composed of poly-D-glutamic acid. Capsules are clearly visible in the light microscope when negative stains or special capsule stains are employed (figure 3.34*a*); they also can be studied with the electron microscope (figure 3.34*b*).

Although capsules are not required for growth and reproduction in laboratory cultures, they do confer several advantages when procaryotes grow in their normal habitats. They help pathogenic bacteria resist phagocytosis by host phagocytes. *Streptococcus pneumoniae* provides a dramatic example. When it lacks a capsule, it is destroyed easily and does not cause disease, whereas the capsulated variant quickly kills mice. Capsules contain a great deal of water and can protect against desiccation. They exclude viruses and most hydrophobic toxic materials such as detergents. The glycocalyx also aids in attachment to solid surfaces, including tissue surfaces in plant and animal hosts (**figure 3.35**). Gliding bacteria often produce slime, which in some cases, has been shown to facilitate motility. Microbial Diversity & Ecology 21.1: The mechanism of gliding motility; Phagocytosis (section 31.3); Overview of bacterial pathogenesis (section 33.3)

Many procaryotes have a regularly structured layer called an **S-layer** on their surface. In bacteria, the S-layer is external to the cell wall. In archaea, the S-layer may be the only wall structure outside the plasma membrane. The S-layer has a pattern something like

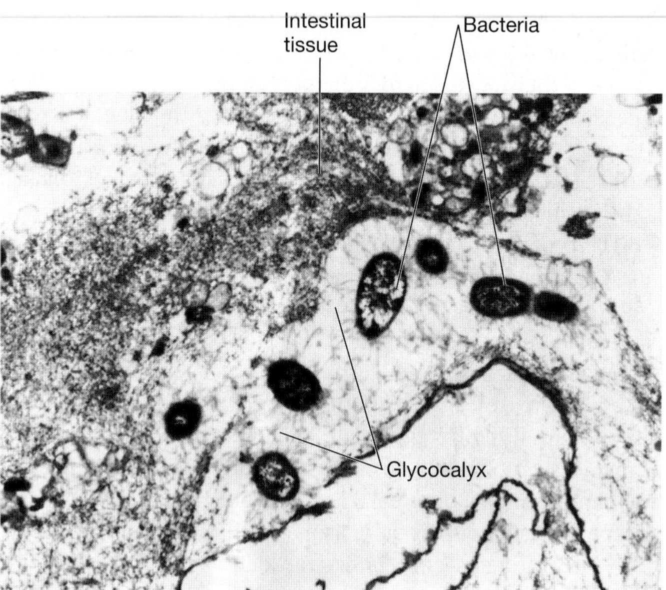

Figure 3.35 Bacterial Glycocalyx. Bacteria connected to each other and to the intestinal wall, by their glycocalyxes, the extensive networks of fibers extending from the cells (×17,500).

floor tiles and is composed of protein or glycoprotein (**figure 3.36**). In gram-negative bacteria the S-layer adheres directly to the outer membrane; it is associated with the peptidoglycan surface in gram-positive bacteria. It may protect the cell against ion and pH fluctuations, osmotic stress, enzymes, or the predacious bacterium *Bdellovibrio*. The S-layer also helps maintain the shape and envelope rigidity of some cells. It can promote cell adhesion to surfaces. Finally, the S-layer seems to protect some bacterial pathogens against host defenses, thus contributing to their virulence. Class *Deltaproteobacteria*: Order *Bdellovibrionales* (section 22.4)

Pili and Fimbriae

Many procaryotes have short, fine, hairlike appendages that are thinner than flagella. These are usually called **fimbriae** (s., **fimbria**). Although many people use the terms fimbriae and pili interchangeably, we shall distinguish between fimbriae and sex pili. A cell may be covered with up to 1,000 fimbriae, but they are only visible in an electron microscope due to their small size (**figure 3.37**). They are slender tubes composed of helically arranged protein subunits and are about 3 to 10 nm in diameter and up to several micrometers long. At least some types of fimbriae attach bacteria to solid surfaces such as rocks in streams and host tissues.

Fimbriae are responsible for more than attachment. Type IV fimbriae are present at one or both poles of bacterial cells. They can aid in attachment to objects, and also are required for the twitching motility that occurs in some bacteria such as *P. aeruginosa, Neisseria gonorrhoeae,* and some strains of *E. coli.* Movement is by short, intermittent jerky motions of up to several micrometers in length and normally is seen on very moist surfaces. There is evidence that the fimbriae actively retract to move these bacteria. Type

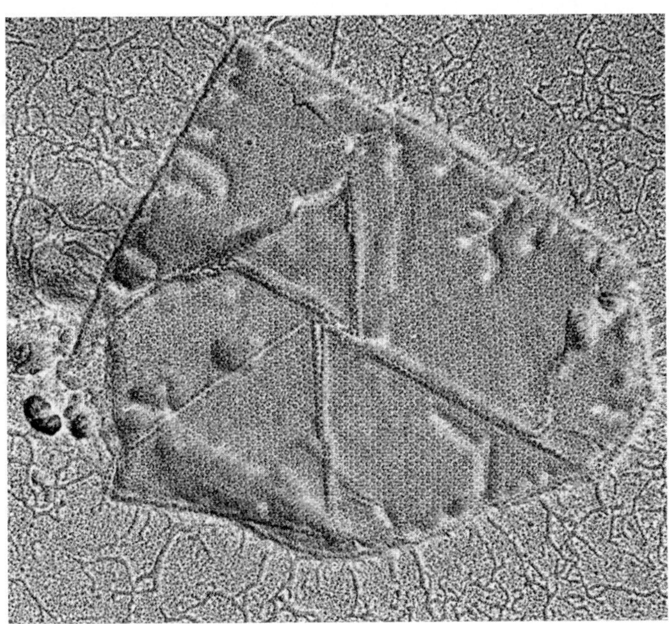

Figure 3.36 The S-Layer. An electron micrograph of the S-layer of the bacterium *Deinococcus radiodurans* after shadowing.

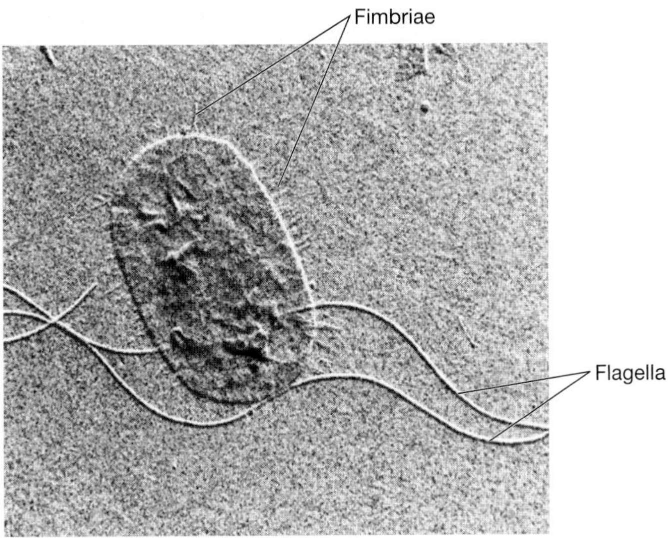

Figure 3.37 Flagella and Fimbriae. The long flagella and the numerous shorter fimbriae are very evident in this electron micrograph of the bacterium *Proteus vulgaris* (×39,000).

IV fimbriae are also involved in gliding motility by myxobacteria. These bacteria are also of interest because they have complex life cycles that include the formation of a fruiting body. Class *Deltaproteobacteria:* Order *Myxococcales* (section 22.4).

Many bacteria have about 1-10 **sex pili** (s., **pilus**) per cell. These are hairlike structures that differ from fimbriae in the following ways. Pili often are larger than fimbriae (around 9 to 10 nm in diameter). They are genetically determined by conjugative plasmids and are required for conjugation. Some bacterial viruses attach specifically to receptors on sex pili at the start of their reproductive cycle. Bacterial conjugation (section 13.7)

Flagella and Motility

Most motile procaryotes move by use of **flagella** (s., **flagellum**), threadlike locomotor appendages extending outward from the plasma membrane and cell wall. Bacterial flagella are the best studied and they are the focus of this discussion.

Bacterial flagella are slender, rigid structures, about 20 nm across and up to 15 or 20 μm long. Flagella are so thin they cannot be observed directly with a bright-field microscope, but must be stained with special techniques designed to increase their thickness. The detailed structure of a flagellum can only be seen in the electron microscope (figure 3.37).

Bacterial species often differ distinctively in their patterns of flagella distribution and these patterns are useful in identifying bacteria. **Monotrichous** bacteria (*trichous* means hair) have one flagellum; if it is located at an end, it is said to be a **polar flagellum** (**figure 3.38*a***). **Amphitrichous** bacteria (*amphi* means on both sides) have a single flagellum at each pole. In contrast, **lophotrichous** bacteria (*lopho* means tuft) have a cluster of flagella at one or both ends (figure 3.38*b*). Flagella are spread fairly evenly over the whole surface of **peritrichous** (*peri* means around) bacteria (figure 3.38*c*).

Flagellar Ultrastructure

Transmission electron microscope studies have shown that the bacterial flagellum is composed of three parts. (1) The longest and most obvious portion is the **flagellar filament,** which extends from the cell surface to the tip. (2) A **basal body** is embedded in the cell; and (3) a short, curved segment, the **flagellar hook,** links the filament to its basal body and acts as a flexible coupling. The filament is a hollow, rigid cylinder constructed of subunits of the protein **flagellin,** which ranges in molecular weight from 30,000 to 60,000 daltons, depending on the bacterial species. The filament ends with a capping protein. Some bacteria have sheaths surrounding their flagella. For example, *Bdellovibrio* has a membranous structure surrounding the filament. *Vibrio cholerae* has a lipopolysaccharide sheath.

The hook and basal body are quite different from the filament (**figure 3.39**). Slightly wider than the filament, the hook is made of different protein subunits. The basal body is the most complex part of a flagellum. In *E. coli* and most gram-negative bacteria, the basal body has four rings connected to a central rod (figure 3.39*a,d*). The outer L and P rings associate with the lipopolysaccharide and peptidoglycan layers, respectively. The inner M ring contacts the plasma membrane. Gram-positive bacteria have only two basal body rings—an inner ring connected to the plasma membrane and an outer one probably attached to the peptidoglycan (figure 3.39*b*).

Flagellar Synthesis

The synthesis of bacterial flagella is a complex process involving at least 20 to 30 genes. Besides the gene for flagellin, 10 or more genes code for hook and basal body proteins; other genes

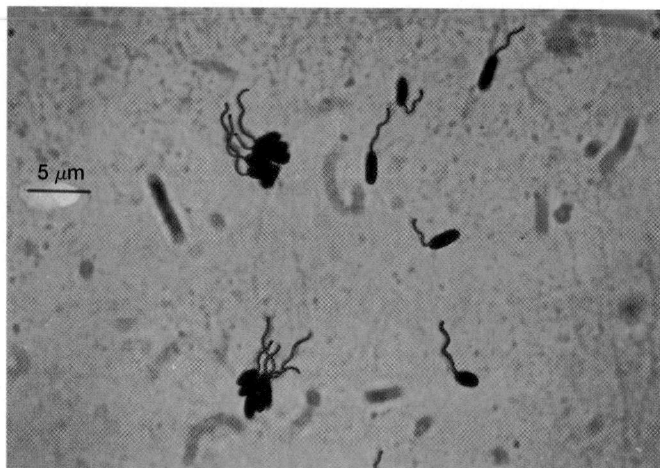

(a) *Pseudomonas*—monotrichous polar flagellation

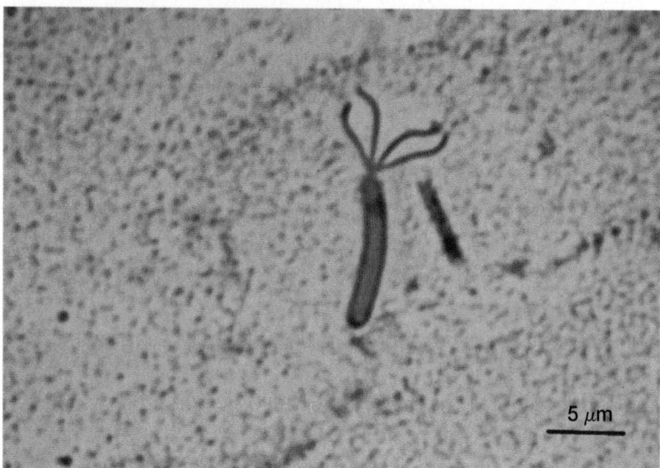

(b) *Spirillum*—lophotrichous flagellation

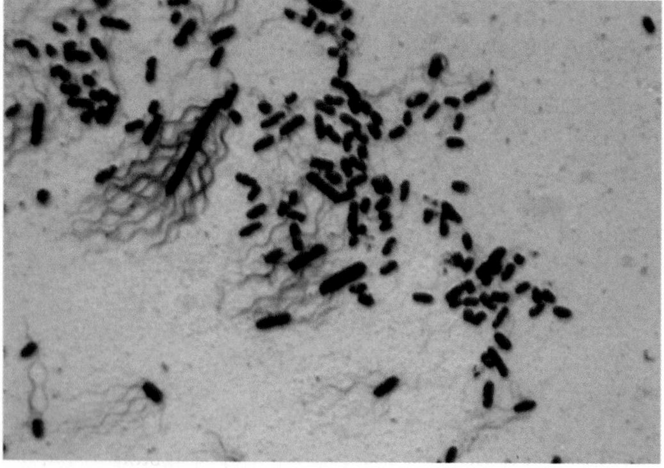

(c) *P. vulgaris*—peritrichous flagellation

Figure 3.38 Flagellar Distribution. Examples of various patterns of flagellation as seen in the light microscope. **(a)** Monotrichous polar (*Pseudomonas*). **(b)** Lophotrichous (*Spirillum*). **(c)** Peritrichous (*Proteus vulgaris*, ×600).

are concerned with the control of flagellar construction or function. How the cell regulates or determines the exact location of flagella is not known.

When flagella are removed, the regeneration of the flagellar filament can then be studied. Transport of many flagellar components is carried out by an apparatus in the basal body that is a specialized type III protein secretion system. It is thought that flagellin subunits are transported through the filament's hollow internal core. When they reach the tip, the subunits spontaneously aggregate under the direction of a special filament cap so that the filament grows at its tip rather than at the base (**figure 3.40**). Filament synthesis is an excellent example of **self-assembly.** Many structures form spontaneously through the association of their component parts without the aid of any special enzymes or other factors. The information required for filament construction is present in the structure of the flagellin subunit itself.

The Mechanism of Flagellar Movement

Procaryotic flagella operate differently from eucaryotic flagella. The filament is in the shape of a rigid helix, and the cell moves when this helix rotates. Considerable evidence shows that flagella act just like propellers on a boat. Bacterial mutants with straight flagella or abnormally long hook regions cannot swim. When bacteria are tethered to a glass slide using antibodies to filament or hook proteins, the cell body rotates rapidly about the stationary flagellum. If polystyrene-latex beads are attached to flagella, the beads spin about the flagellar axis due to flagellar rotation. The flagellar motor can rotate very rapidly. The *E. coli* motor rotates 270 revolutions per second; *Vibrio alginolyticus* averages 1,100 rps. Cilia and flagella (section 4.10)

The direction of flagellar rotation determines the nature of bacterial movement. Monotrichous, polar flagella rotate counterclockwise (when viewed from outside the cell) during normal forward movement, whereas the cell itself rotates slowly clockwise. The rotating helical flagellar filament thrusts the cell forward in a run with the flagellum trailing behind (**figure 3.41**). Monotrichous bacteria stop and tumble randomly by reversing the direction of flagellar rotation. Peritrichously flagellated bacteria operate in a somewhat similar way. To move forward, the flagella rotate counterclockwise. As they do so, they bend at their hooks to form a rotating bundle that propels the cell forward. Clockwise rotation of the flagella disrupts the bundle and the cell tumbles.

Because bacteria swim through rotation of their rigid flagella, there must be some sort of motor at the base. A rod extends from the hook and ends in the M ring, which can rotate freely in the plasma membrane (**figure 3.42**). It is thought that the S ring is attached to the cell wall in gram-positive cells and does not rotate. The P and L rings of gram-negative bacteria would act as bearings for the rotating rod. There is some evidence that the basal body is a passive structure and rotates within a membrane-embedded protein complex much like the rotor of an electrical motor turns in the center of a ring of electromagnets (the stator).

The exact mechanism that drives basal body rotation is not entirely clear. Figure 3.42 provides a more detailed depiction of

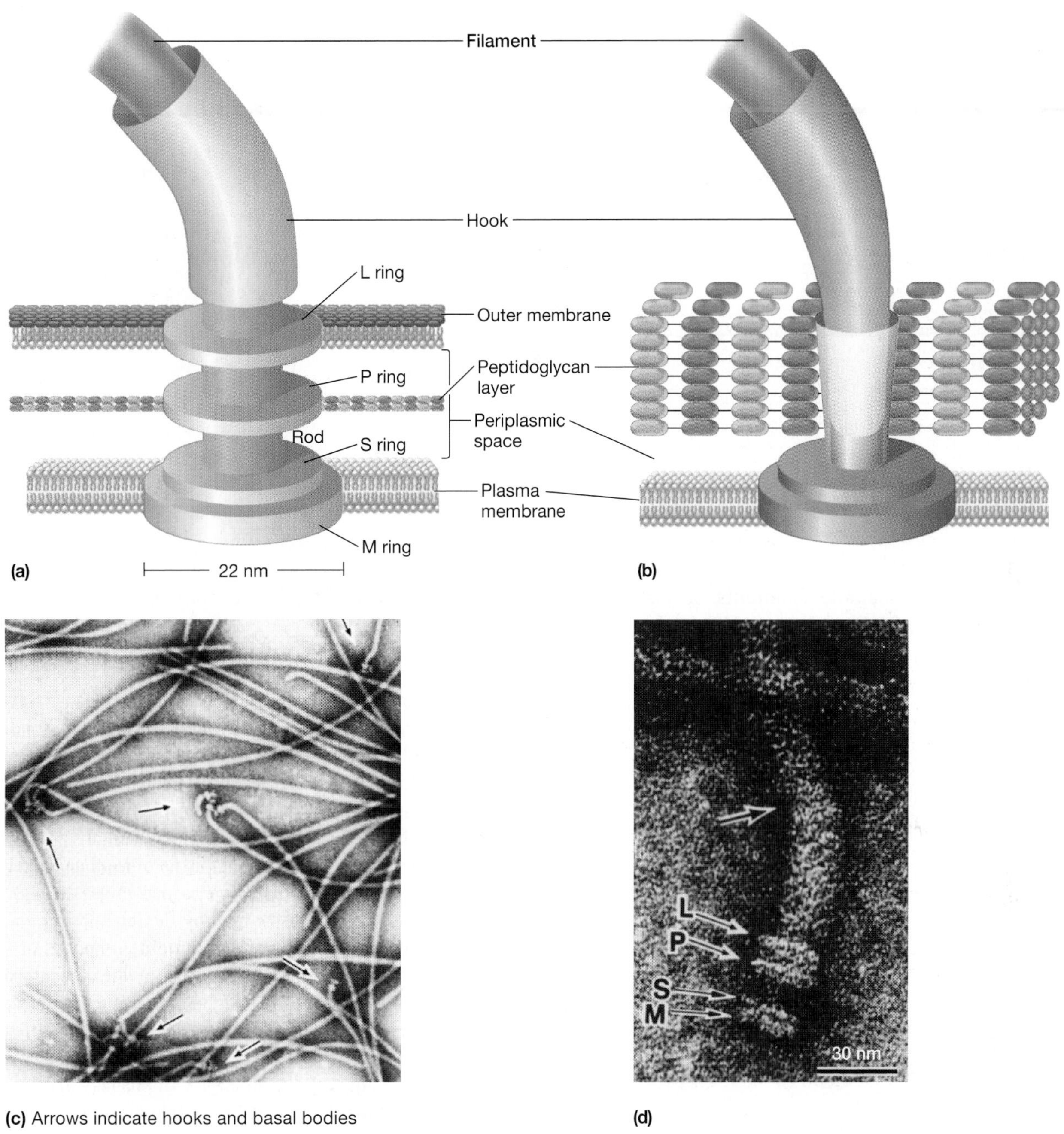

(c) Arrows indicate hooks and basal bodies

(d)

Figure 3.39 The Ultrastructure of Bacterial Flagella. Flagellar basal bodies and hooks in **(a)** gram-negative and **(b)** gram-positive bacteria. **(c)** Negatively stained flagella from *Escherichia coli* (×66,000). **(d)** An enlarged view of the basal body of an *E. coli* flagellum (×485,000). All four rings (L, P, S, and M) can be clearly seen. The uppermost arrow is at the junction of the hook and filament.

the basal body in gram-negative bacteria. The rotor portion of the motor seems to be made primarily of a rod, the M ring, and a C ring joined to it on the cytoplasmic side of the basal body. These two rings are made of several proteins; FliG is particularly important in generating flagellar rotation. The two most important proteins in the stator part of the motor are MotA and MotB. These form a proton channel through the plasma membrane, and MotB also anchors the Mot complex to cell wall peptidoglycan. There is some evidence that MotA and FliG directly interact during flagellar rotation. This rotation is driven by proton or sodium gradients in procaryotes, not directly by ATP as is the case with eucaryotic flagella. The electron transport chain and oxidative phosphorylation (section 9.5)

The flagellum is a very effective swimming device. From the bacterium's point of view, swimming is quite a task because the surrounding water seems as thick and viscous as molasses. The cell

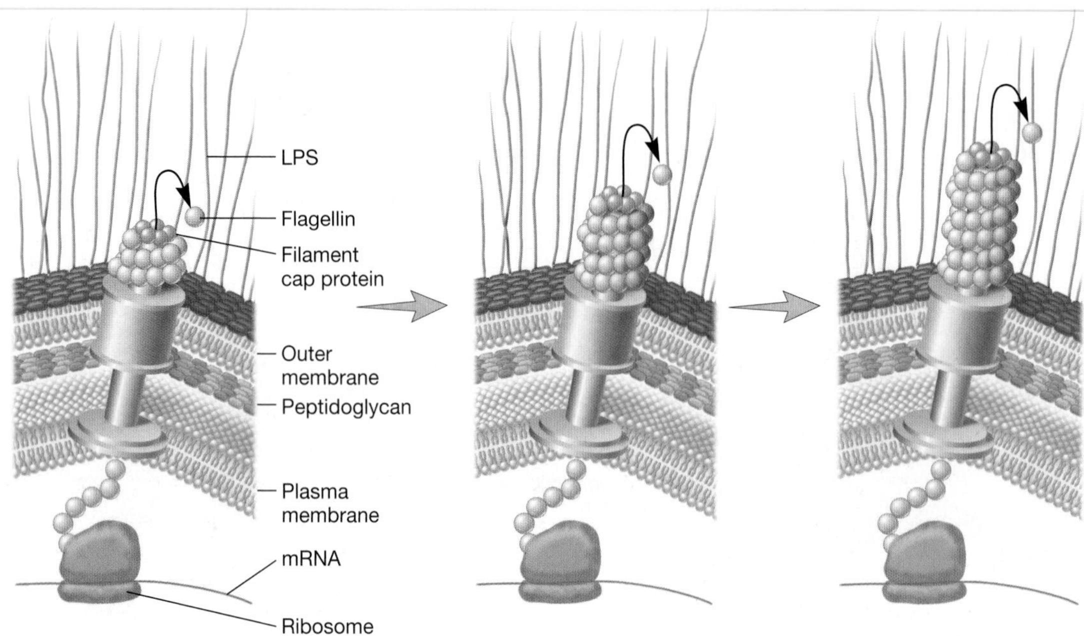

Figure 3.40 Growth of Flagellar Filaments. Flagellin subunits travel through the flagellar core and attach to the growing tip. Their attachment is directed by the filament cap protein.

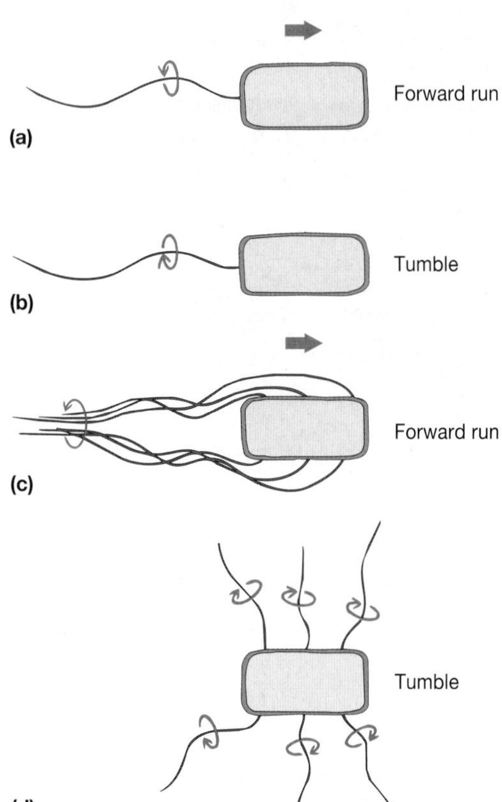

Figure 3.41 Flagellar Motility. The relationship of flagellar rotation to bacterial movement. Parts **(a)** and **(b)** describe the motion of monotrichous, polar bacteria. Parts **(c)** and **(d)** illustrate the movements of peritrichous organisms.

must bore through the water with its corkscrew-shaped flagella, and if flagellar activity ceases, it stops almost instantly. Despite such environmental resistance to movement, bacteria can swim from 20 to almost 90 μm/second. This is equivalent to traveling from 2 to over 100 cell lengths per second. In contrast, an exceptionally fast 6-ft human might be able to run around 5 body lengths per second.

Bacteria can move by mechanisms other than flagellar rotation. Spirochetes are helical bacteria that travel through viscous substances such as mucus or mud by flexing and spinning movements caused by a special **axial filament** composed of periplasmic flagella. The swimming motility of the helical bacterium *Spiroplasma* is accomplished by the formation of kinks in the cell body that travel the length of the bacterium. A very different type of motility, **gliding motility,** is employed by many bacteria: cyanobacteria, myxobacteria and cytophagas, and some mycoplasmas. Although there are no visible external structures associated with gliding motility, it enables movement along solid surfaces at rates up to 3 μm/second. Microbial Diversity & Ecology 21.1: The mechanism of gliding motility; Phylum *Spirochaetes* (section 21.6); Photosynthetic bacteria (section 21.3); Class *Deltaproteobacteria: Order Myxococcales* (section 22.4); Class *Mollicutes* (the Mycoplasmas) (section 23.2)

1. Briefly describe capsules, slime layers, glycocalyxes, and S-layers. What are their functions?
2. Distinguish between fimbriae and sex pili, and give the function of each.
3. Be able to discuss the following: flagella distribution patterns, flagella structure and synthesis, and the way in which flagella operate to move a bacterium.
4. What is self-assembly? Why does it make sense that the flagellar filament is assembled in this way?

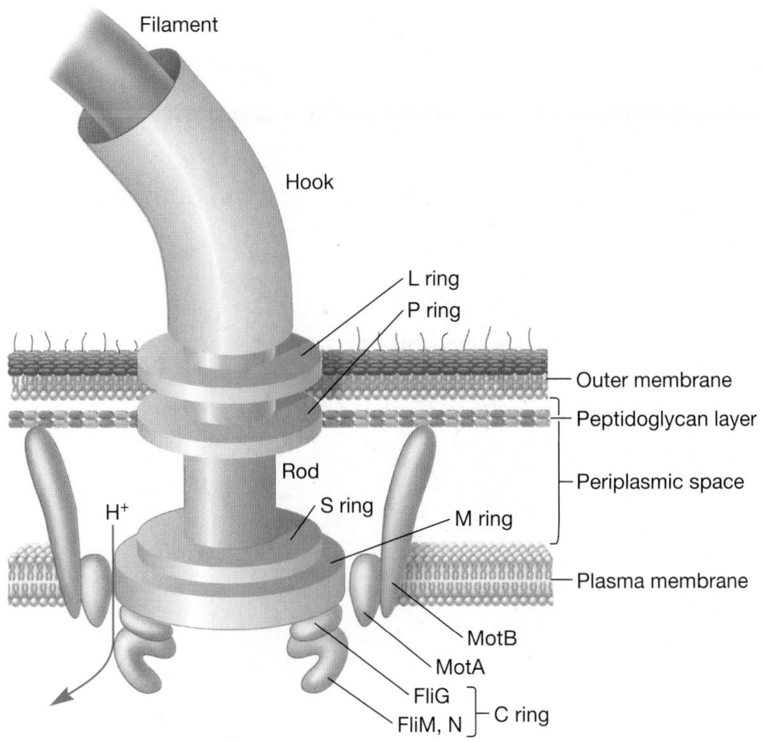

Figure 3.42 Mechanism of Flagellar Movement. This diagram of a gram-negative flagellum shows some of the more important components and the flow of protons that drives rotation. Five of the many flagellar proteins are labeled (MotA, MotB, FliG, FliM, FliN).

3.10 Chemotaxis

Bacteria do not always move aimlessly but are attracted by such nutrients as sugars and amino acids, and are repelled by many harmful substances and bacterial waste products. Bacteria also can respond to other environmental cues such as temperature (thermotaxis), light (phototaxis), oxygen (aerotaxis), osmotic pressure (osmotaxis), and gravity; (Microbial Diversity & Ecology 3.2.) Movement toward chemical attractants and away from repellents is known as **chemotaxis.** Such behavior is of obvious advantage to bacteria.

Chemotaxis may be demonstrated by observing bacteria in the chemical gradient produced when a thin capillary tube is filled with an attractant and lowered into a bacterial suspension. As the attractant diffuses from the end of the capillary, bacteria collect and swim up the tube. The number of bacteria within the capillary after a short length of time reflects the strength of attraction and rate of chemotaxis. Positive and negative chemotaxis also can be studied with petri dish cultures (**figure 3.43**). If bacteria are placed in the center of a dish of semisolid agar containing an attractant, the bacteria will exhaust the local supply and then swim outward following the attractant gradient they have created. The result is an expanding ring of bacteria. When a disk of repellent is placed in a petri dish of semisolid agar and bacteria, the bacteria will swim away from the repellent, creating a clear zone around the disk (**figure 3.44**).

Bacteria can respond to very low levels of attractants (about 10^{-8} M for some sugars), the magnitude of their response increasing with attractant concentration. Usually they sense repellents only at higher concentrations. If an attractant and a repellent are present together, the bacterium will compare both signals and respond to the chemical with the most effective concentration.

Attractants and repellents are detected by **chemoreceptors,** special proteins that bind chemicals and transmit signals to the other components of the chemosensing system. About 20 attractant chemoreceptors and 10 chemoreceptors for repellents have been discovered thus far. These chemoreceptor proteins may be located in the periplasmic space or the plasma membrane. Some receptors participate in the initial stages of sugar transport into the cell.

The chemotactic behavior of bacteria has been studied using the tracking microscope, a microscope with a moving stage that automatically keeps an individual bacterium in view. In the absence of a chemical gradient, *E. coli* and other bacteria move randomly. For a few seconds, the bacterium will travel in a straight or slightly curved line called a **run.** When a bacterium is running, its flagella are organized into a coordinated, corkscrew-shaped bundle (figure 3.41*c*). Then the flagella "fly apart" and the bacterium will stop and **tumble.** The tumble results in the random reorientation of the bacterium so that it often is facing in a different direction. Therefore when it begins the next run, it usually goes in a different direction (**figure 3.45*a***). In contrast, when the bacterium is exposed to an attractant, it tumbles less frequently (or has longer runs) when traveling towards the attractant. Although the tumbles can still orient the bacterium away from the attractant, over time, the bacterium gets closer and closer to the attractant (figure 3.45*b*). The opposite response occurs with a repellent. Tumbling frequency decreases (the run time lengthens) when the bacterium moves away from the repellent.

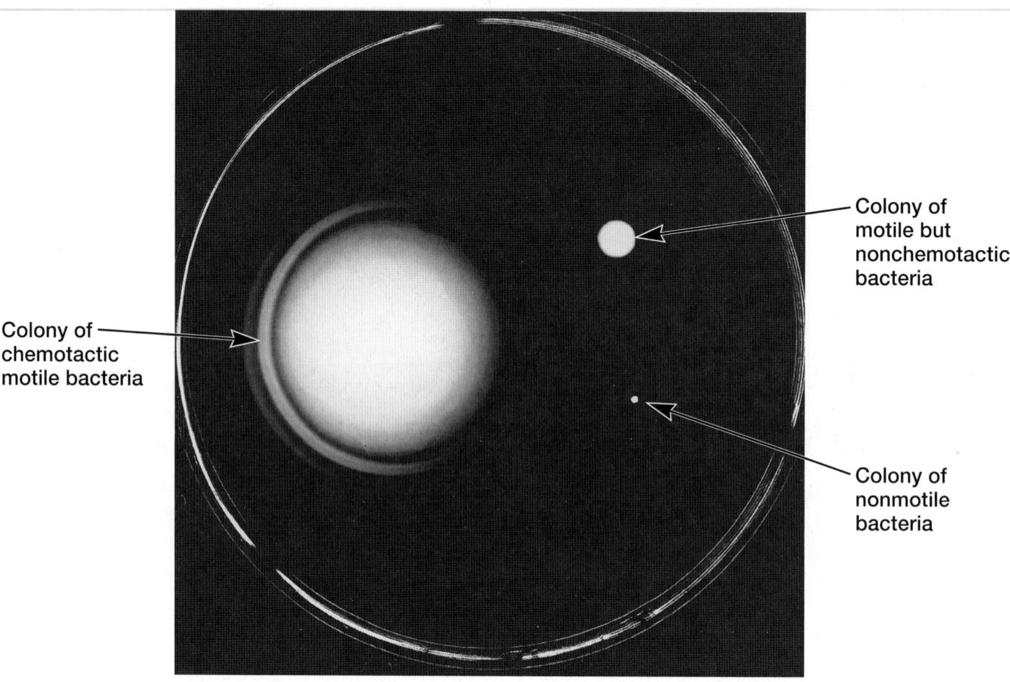

Figure 3.43 Positive Bacterial Chemotaxis. Chemotaxis can be demonstrated on an agar plate that contains various nutrients. Positive chemotaxis by *E. coli* on the left. The outer ring is composed of bacteria consuming serine. The second ring was formed by *E. coli* consuming aspartate, a less powerful attractant. The upper right colony is composed of motile, but nonchemotactic mutants. The bottom right colony is formed by nonmotile bacteria.

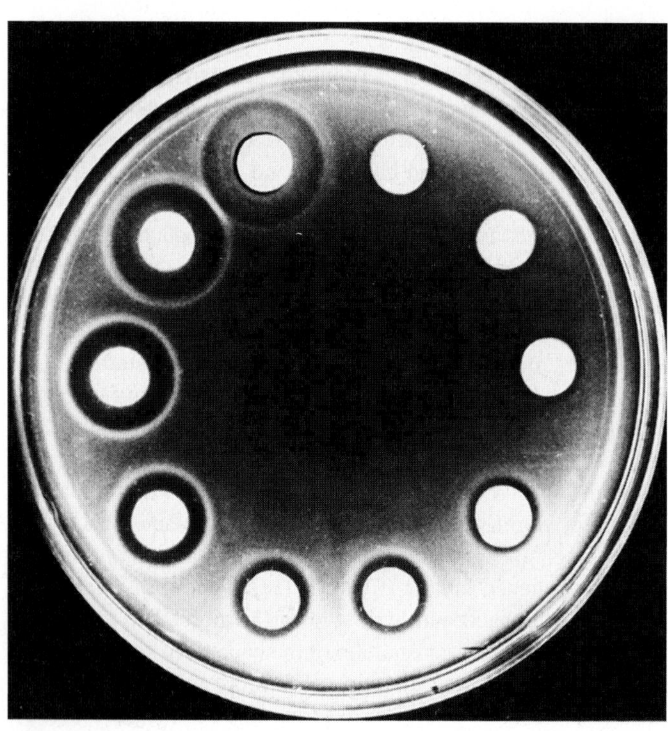

Figure 3.44 Negative Bacterial Chemotaxis. Negative chemotaxis by *E. coli* in response to the repellent acetate. The bright disks are plugs of concentrated agar containing acetate that have been placed in dilute agar inoculated with *E. coli.* Acetate concentration increases from zero at the top right to 3 M at top left. Note the increasing size of bacteria-free zones with increasing acetate. The bacteria have migrated for 30 minutes.

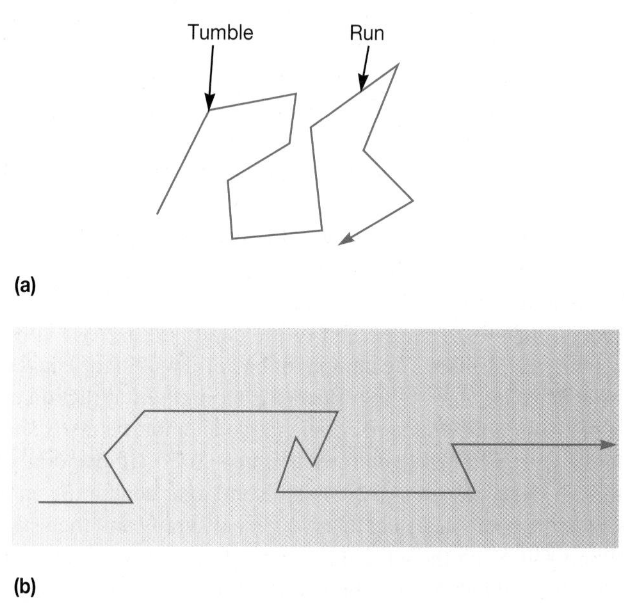

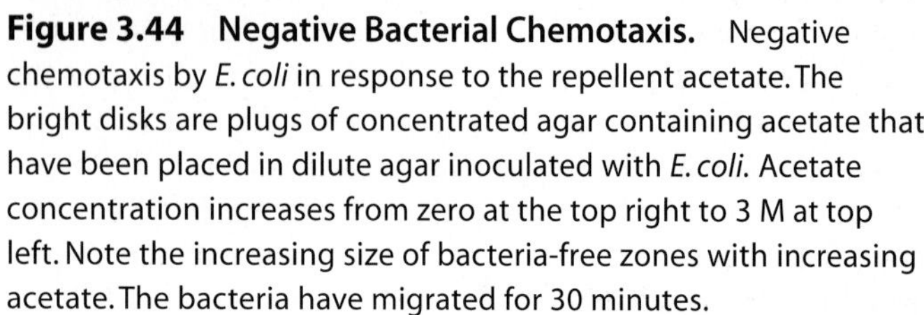

Figure 3.45 Directed Movement in Bacteria. **(a)** Random movement of a bacterium in the absence of a concentration gradient. Tumbling frequency is fairly constant. **(b)** Movement in an attractant gradient. Tumbling frequency is reduced when the bacterium is moving up the gradient. Therefore, runs in the direction of increasing attractant are longer.

Clearly, the bacterium must have some mechanism for sensing that it is getting closer to the attractant (or is moving away from the repellent). The behavior of the bacterium is shaped by temporal changes in chemical concentration. The bacterium moves toward the attractant because it senses that the concentration of the attractant is increasing. Likewise, it moves away from a repellent because it senses that the concentration of the repellent is decreasing. The bacterium's chemoreceptors play a critical role in this process. The molecular events that enable bacterial cells to sense a chemical gradient and respond appropriately are presented in chapter 8.

1. Define chemotaxis, run, and tumble.
2. Explain in a general way how bacteria move toward substances like nutrients and away from toxic materials.

3.11 The Bacterial Endospore

A number of gram-positive bacteria can form a special resistant, dormant structure called an **endospore.** Endospores develop within vegetative bacterial cells of several genera: *Bacillus* and *Clostridium* (rods), *Sporosarcina* (cocci), and others. These structures are extraordinarily resistant to environmental stresses such as heat, ultraviolet radiation, gamma radiation, chemical disinfectants, and desiccation. In fact, some endospores have remained viable for around 100,000 years. Because of their resistance and the fact that several species of endospore-forming bacteria are dangerous pathogens, endospores are of great practical importance in food, industrial, and medical microbiology. This is because it is essential to be able to sterilize solutions and solid objects. Endospores often survive boiling for an hour or more; therefore autoclaves must be used to sterilize many materials. Endospores are also of considerable theoretical interest. Because bacteria manufacture these intricate structures in a very organized fashion over a period of a few hours, spore formation is well suited for research on the construction of complex biological structures. In the environment, endospores aid in survival when moisture or nutrients are scarce. The use of physical methods in control: Heat (section 7.4)

Endospores can be examined with both light and electron microscopes. Because endospores are impermeable to most stains, they often are seen as colorless areas in bacteria treated with methylene blue and other simple stains; special endospore stains are used to make them clearly visible. Endospore position in the mother cell (**sporangium**) frequently differs among species, making it of considerable value in identification. Endospores may be centrally located, close to one end (subterminal), or definitely terminal (**figure 3.46**). Sometimes an endospore is so large that it swells the sporangium. Preparation and staining of specimens (section 2.3)

Electron micrographs show that endospore structure is complex (**figure 3.47**). The spore often is surrounded by a thin, delicate covering called the **exosporium.** A **spore coat** lies beneath the exosporium, is composed of several protein layers, and may be fairly thick. It is impermeable to many toxic molecules and is responsible for the spore's resistance to chemicals. The coat also is thought to contain enzymes that are involved in germination. The **cortex,** which may occupy as much as half the spore volume, rests beneath the spore coat. It is made of a peptidoglycan that is less cross-linked than that in vegetative cells. The **spore cell wall** (or core wall) is inside the cortex and surrounds the protoplast or **spore core.** The core has normal cell structures such as ribosomes and a nucleoid, but is metabolically inactive.

It is still not known precisely why the endospore is so resistant to heat and other lethal agents. As much as 15% of the spore's dry weight consists of dipicolinic acid complexed with calcium ions (**figure 3.48**), which is located in the core. It has long been thought that dipicolinic acid was directly involved in

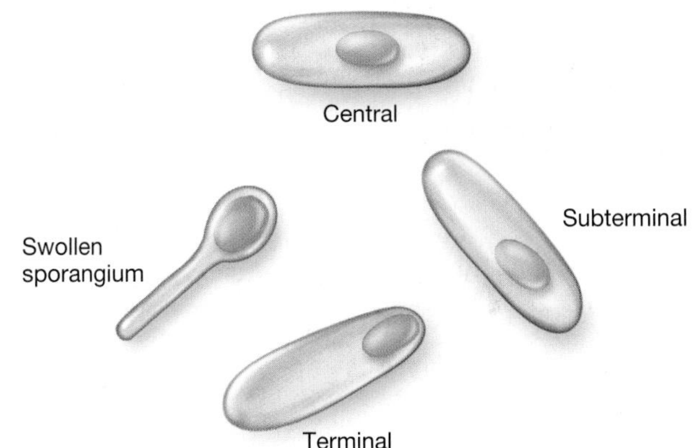

Figure 3.46 Examples of Endospore Location and Size.

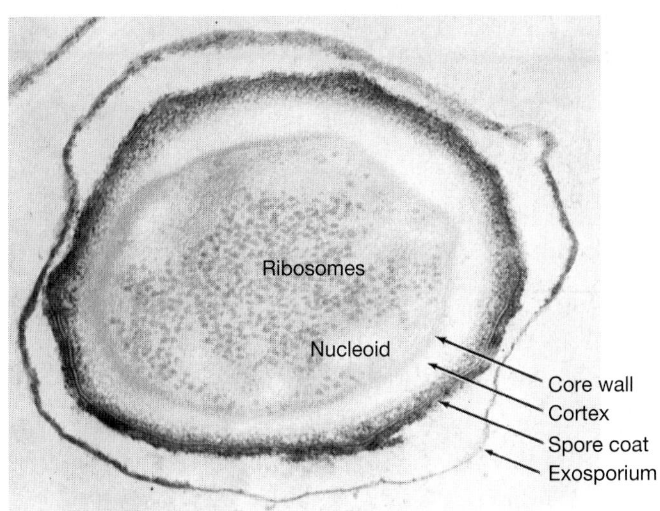

Figure 3.47 Endospore Structure. *Bacillus anthracis* endospore (×151,000).

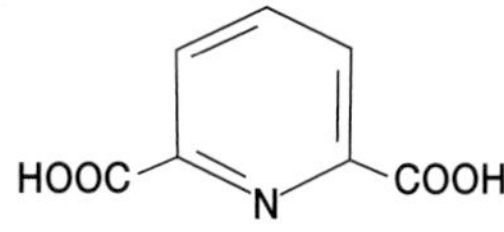

Figure 3.48 Dipicolinic Acid.

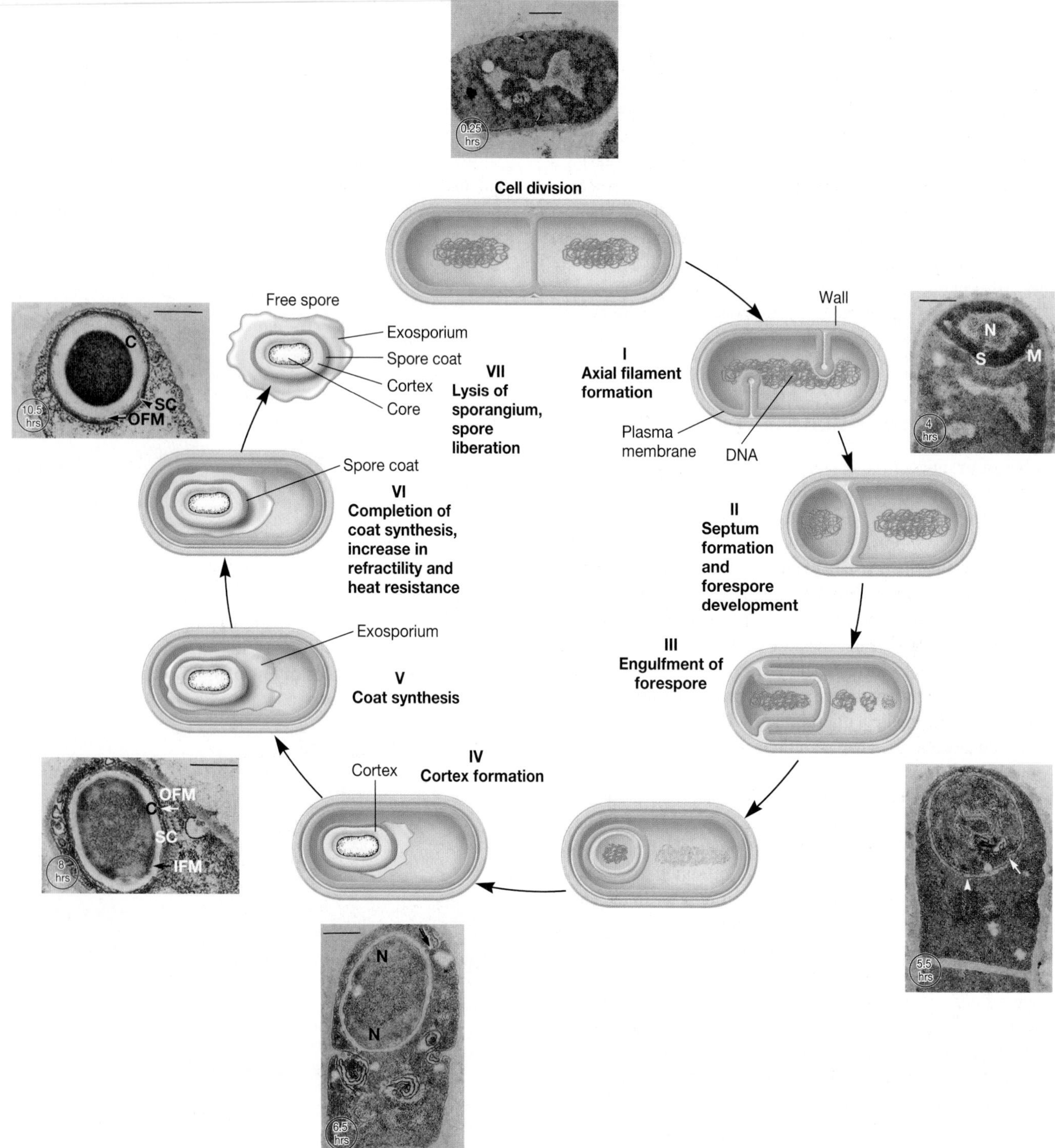

Figure 3.49 Endospore Formation: Life Cycle of *Bacillus megaterium*. The stages are indicated by Roman numerals. The circled numbers in the photographs refer to the hours from the end of the logarithmic phase of growth: 0.25 h—a typical vegetative cell; 4 h–stage II cell, septation; 5.5 h–stage III cell, engulfment; 6.5 h–stage IV cell, cortex formation; 8 h–stage V cell, coat formation; 10.5 h–stage VI cell, mature spore in sporangium. Abbreviations used: C, cortex; IFM and OFM, inner and outer forespore membranes; M, mesosome; N, nucleoid; S, septum; SC, spore coats. Bars = 0.5 μm.

heat resistance, but heat-resistant mutants lacking dipicolinic acid have been isolated. Calcium does aid in resistance to wet heat, oxidizing agents, and sometimes dry heat. It may be that calcium-dipicolinate stabilizes the spore's nucleic acids. In addition, specialized *s*mall, *a*cid-*s*oluble DNA-binding *p*roteins (SASPs), are found in the endospore. They saturate spore DNA and protect it from heat, radiation, dessication, and chemicals. Dehydration of the protoplast appears to be very important in heat resistance. The cortex may osmotically remove water from the protoplast, thereby protecting it from both heat and radiation damage. The spore coat also seems to protect against enzymes and chemicals such as hydrogen peroxide. Finally, spores contain some DNA repair enzymes. DNA is repaired once the spore germinates and the cell becomes active again. In summary, endospore heat resistance probably is due to several factors: calcium-dipicolinate and acid-soluble protein stabilization of DNA, protoplast dehydration, the spore coat, DNA repair, the greater stability of cell proteins in bacteria adapted to growth at high temperatures, and others.

Endospore formation, also called **sporogenesis** or **sporulation,** normally commences when growth ceases due to lack of nutrients. It is a complex process and may be divided into seven stages (**figure 3.49**). An axial filament of nuclear material forms (stage I), followed by an inward folding of the cell membrane to enclose part of the DNA and produce the forespore septum (stage II). The membrane continues to grow and engulfs the immature endospore in a second membrane (stage III). Next, cortex is laid down in the space between the two membranes, and both calcium and dipicolinic acid are accumulated (stage IV). Protein coats then are formed around the cortex (stage V), and maturation of the endospore occurs (stage VI). Finally, lytic enzymes destroy the sporangium releasing the spore (stage VII). Sporulation requires about 10 hours in *Bacillus megaterium.* Global regulatory systems: Sporulation in *Bacillus subtilus* (section 12.5)

The transformation of dormant spores into active vegetative cells seems almost as complex a process as sporogenesis. It occurs in three stages: (1) activation, (2) germination, and (3) outgrowth (**figure 3.50**). Often a spore will not germinate successfully, even in a nutrient-rich medium, unless it has been activated. **Activation** is a process that prepares spores for germination and usually results from treatments like heating. It is followed by **germination,** the breaking of the spore's dormant state. This process is characterized by spore swelling, rupture or absorption of the spore coat, loss of resistance to heat and other stresses, loss of refractility, release of spore components, and increase in metabolic activity. Many normal metabolites or nutrients (e.g., amino acids and sugars) can trigger germination after activation. Germination is followed by the third stage, **outgrowth.** The spore protoplast makes new components, emerges from the remains of the spore coat, and develops again into an active bacterium.

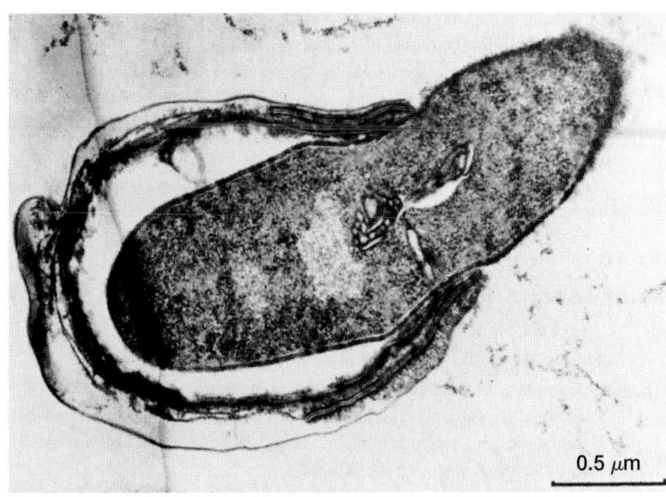

Figure 3.50 Endospore Germination. *Clostridium pectinovorum* emerging from the spore during germination.

1. Describe the structure of the bacterial endospore using a labeled diagram.
2. Briefly describe endospore formation and germination. What is the importance of the endospore? What might account for its heat resistance?
3. How might one go about showing that a bacterium forms true endospores?
4. Why do you think dehydration of the protoplast is an important factor in the ability of endospores to resist environmental stress?

Summary

3.1 An Overview of Procaryotic Cell Structure

a. Procaryotes may be spherical (cocci), rod-shaped (bacilli), spiral, or filamentous; they may form buds and stalks; or they may even have no characteristic shape at all (pleomorphic) (**figure 3.1** and **3.2**).

b. Procaryotic cells can remain together after division to form pairs, chains, and clusters of various sizes and shapes.

c. Procaryotes are much simpler structurally than eucaryotes, but they do have unique structures. **Table 3.1** summarizes the major functions of procaryotic cell structures.

3.2 Procaryotic Cell Membranes

a. The plasma membrane fulfills many roles, including acting as a semipermeable barrier, carrying out respiration and photosynthesis, and detecting and responding to chemicals in the environment.

b. The fluid mosaic model proposes that cell membranes are lipid bilayers in which integral proteins are buried (**figure 3.5**). Peripheral proteins are loosely associated with the membrane.

c. Bacterial membranes are composed of phospholipids constructed of fatty acids connected to glycerol by ester linkages (**figure 3.6**). Bacterial membranes usually lack sterols, but often contain hopanoids (**figure 3.7**).

d. The plasma membrane of some bacteria invaginates to form simple membrane systems containing photosynthetic and respiratory assemblies. Other bacteria, like the cyanobacteria, have internal membranes (**figure 3.8**).

e. Archaeal membranes are composed of glycerol diether and diglycerol tetraether lipids (**figure 3.9**). Membranes composed of glycerol diether are lipid bilayers. Membranes composed of diglycerol tetraethers are lipid monolayers (**figure 3.11**). The overall structure of a monolayer membrane is similar to that of the bilayer membrane in that the membrane has a hydrophobic core and its surfaces are hydrophilic.

3.3 The Cytoplasmic Matrix

a. The cytoplasm of procaryotes contains proteins that are similar in structure and function to the cytoskeletal proteins observed in eucaryotes.

b. The cytoplasmic matrix of procaryotes contains inclusion bodies. Most are used for storage (glycogen inclusions, PHB inclusions, cyanophycin granules, carboxysomes, and polyphosphate granules), but others are used for other purposes (magnetosomes and gas vacuoles).

c. The cytoplasm of procaryotes is packed with 70S ribosomes (**figure 3.15**).

3.4 The Nucleoid

a. Procaryotic genetic material is located in an area called the nucleoid and is not usually enclosed by a membrane (**figure 3.16**).

b. In most procaryotes, the nucleoid contains a single chromosome. The chromsosome consists of a double-stranded, covalently closed, circular DNA molecule.

3.5 Plasmids

a. Plasmids are extrachromosomal DNA molecules. They are found in many procaryotes.

b. Although plasmids are not required for survival in most conditions, they can encode traits that confer selective advantage in some environments.

c. Episomes are plasmids that are able to exist freely in the cytoplasm or can be integrated into the chromosome.

d. Conjugative plasmids encode genes that promote their transfer from one cell to another.

e. Resistance factors are plasmids that have genes conferring resistance to antibiotics.

f. Col plasmids contain genes for the synthesis of colicins, proteins that kill *E. coli*. Other plasmids encode virulence factors or metabolic capabilities.

3.6 The Bacterial Cell Wall

a. The vast majority of procaryotes have a cell wall outside the plasma membrane to give them shape and protect them from osmotic stress.

b. Bacterial walls are chemically complex and usually contain peptidoglycan (**figures 3.17–3.21**).

c. Bacteria often are classified as either gram-positive or gram-negative based on differences in cell wall structure and their response to Gram staining.

d. Gram-positive walls have thick, homogeneous layers of peptidoglycan and teichoic acids (**figure 3.23**). Gram-negative bacteria have a thin peptidoglycan layer surrounded by a complex outer membrane containing lipopolysaccharides (LPSs) and other components (**figure 3.25**).

e. The mechanism of the Gram stain is thought to depend on the peptidoglycan, which binds crystal violet tightly, preventing the loss of crystal violet during the ethanol wash.

3.7 Archaeal Cell Walls

a. Archaeal cell walls do not contain peptidoglycan (**figure 3.30**).

b. Archaea exhibit great diversity in their cell wall make-up. Some archaeal cell walls are composed of heteropolysaccharides, some are composed of glycoprotein, and some are composed of protein.

3.8 Protein Secretion in Procaryotes

a. The Sec-dependent protein secretion pathway (**figure 3.32**) has been observed in all domains of life. It transports proteins across or into the cytoplasmic membrane.

b. Gram-negative bacteria have additional protein secretion systems that allow them to move proteins from the cytoplasm, across both the cytoplasmic and outer membranes, to the outside of the cell (**figure 3.33**). Some of these systems work with the Sec-dependent pathway to accomplish this (Type II, Type V, and usually Type IV). Some pathways function alone to move proteins across both membranes (Types I and III).

c. ABC transporters (Type I protein secretion system) are used by all procaryotes for protein translocation.

3.9 Components External to the Cell Wall

a. Capsules, slime layers, and glycocalyxes are layers of material lying outside the cell wall. They can protect procaryotes from certain environmental conditions, allow procaryotes to attach to surfaces, and protect pathogenic bacteria from host defenses (**figures 3.34** and **3.35**).

b. S-layers are observed in some bacteria and many archaea. They are composed of proteins or glycoprotein and have a characteristic geometric shape. In many archaea the S-layer serves as the cell wall (**figure 3.36**).

c. Pili and fimbriae are hairlike appendages. Fimbriae function primarily in attachment to surfaces, but some types of bacterial fimbriae are involved in a twitching motility. Sex pili participate in the transfer of DNA from one bacterium to another (**figure 3.37**).

d. Many procaryotes are motile, usually by means of threadlike, locomotory organelles called flagella (**figure 3.38**).

e. Bacterial species differ in the number and distribution of their flagella.

f. In bacteria, the flagellar filament is a rigid helix that rotates like a propeller to push the bacterium through water (**figure 3.41**).

3.10 Chemotaxis

a. Motile procaryotes can respond to gradients of attractants and repellents, a phenomenon known as chemotaxis.

b. A bacterium accomplishes movement toward an attractant by increasing the length of time it spends moving toward the attractant, shortening the time it spends tumbling. Conversely, a bacterium increases its run time when it moves away from a repellent.

3.11 The Bacterial Endospore

a. Some bacteria survive adverse environmental conditions by forming endospores, dormant structures resistant to heat, desiccation, and many chemicals (**figure 3.47**).

b. Both endospore formation and germination are complex processes that begin in response to certain environmental signals and involve numerous stages (**figures 3.49** and **3.50**).

Key Terms

ABC protein secretion pathway 65
activation 75
amphipathic 45
amphitrichous 67
axial filament 70
bacillus 40
bacteriocin 53
basal body 67
capsule 65
carboxysomes 49
cell envelope 55
chemoreceptors 71
chemotaxis 71
coccus 39
Col plasmid 53
conjugative plasmid 53
core polysaccharide 58
cortex 73
curing 53
cyanophycin granules 49
cytoplasmic matrix 48
deoxyribonucleic acid (DNA) 52
diplococcus 39
endospore 73
episome 53
exoenzyme 58
exosporium 73
F factor 53
fimbriae 66
flagellar filament 67
flagellar hook 67
flagellin 67
flagellum 67
fluid mosaic model 44
gas vacuole 50
gas vesicles 50
germination 75
gliding motility 70
glycocalyx 65
glycogen 49
hopanoids 46
hydrophilic 45
hydrophobic 45
inclusion body 48
integral proteins 45
lipid A 58
lipopolysaccharides (LPSs) 58
lophotrichous 67
lysis 61
lysozyme 61
magnetosomes 50
metabolic plasmid 54
metachromatic granules 50
monotrichous 67
murein 55
mycelium 40
nucleoid 52
O antigen 60
O side chain 60
outer membrane 55
outgrowth 75
penicillin 61
peptide interbridge 56
peptidoglycan 55
peripheral proteins 45
periplasm 55
periplasmic space 55
peritrichous 67
plasma membrane 42
plasmid 53
plasmolysis 61
pleomorphic 41
polar flagellum 67
poly-β-hydroxybutyrate (PHB) 49
polyphosphate granules 50
porin proteins 60
protoplast 48
pseudomurein 62
resistance factor (R factor, R plasmid) 53
ribosome 50
rod 40
run 71
Sec-dependent pathway 63
self-assembly 68
sex pili 67
signal peptide 63
S-layer 66
slime layer 65
spheroplast 61
spirilla 40
spirochete 40
sporangium 73
spore cell wall 73
spore coat 73
spore core 73
sporogenesis 75
sporulation 75
Svedberg unit 50
teichoic acid 57
tumble 71
type I protein secretion pathway 65
type II protein secretion pathway 65
type III protein secretion pathway 65
type IV protein secretion pathway 65
type V protein secretion pathway 65
vibrio 40
virulence plasmid 54
volutin granules 50

Critical Thinking Questions

1. Propose a model for the assembly of a flagellum in a gram-positive cell envelope. How would that model need to be modified for the assembly of a flagellum in a gram-negative cell envelope?
2. If you could not use a microscope, how would you determine whether a cell is procaryotic or eucaryotic? Assume the organism can be cultured easily in the laboratory.
3. The peptidoglycan of bacteria has been compared with the chain mail worn beneath a medieval knight's suit of armor. It provides both protection and flexibility. Can you describe other structures in biology that have an analogous function? How are they replaced or modified to accommodate the growth of the inhabitant?

Learn More

Cannon, G. C.; Bradburne, C. E.; Aldrich, H. C.; Baker, S. H.; Heinhorst, S.; and Shively, J. M. 2001. Microcompartments in prokaryotes: Carboxysomes and related polyhedra. *Appl. Env. Microbiol.* 67(12):5351–61.

Drews, G. 1992. Intracytoplasmic membranes in bacterial cells: Organization, function and biosynthesis. In *Prokaryotic structure and function,* S. Mohan, C. Dow, and J. A. Coles, editors, 249–74. New York: Cambridge University Press.

Frankel, R. B., and Bazylinski, D. A. 2004. Magnetosome mysteries. *ASM News* 70(4):176–83.

Ghuysen, J.-M., and Hekenbeck, R., editors. 1994. *Bacterial cell wall.* New York: Elsevier.

Gital, Z. 2005. The new bacterial cell biology: Moving parts and subcellular architecture. *Cell* 120:577–86.

Harshey, R. M. 2003. Bacterial motility on a surface: Many ways to a common goal. *Annu. Rev. Microbiol.* 57:249–73.

Henderson, I. R.; Navarro-Garcia, F.; Desvaux, M.; Fernandez, R. C.; and Ala'Aldeen, D. 2004. Type V protein secretion pathway: The autotransporter story. *Microbiol. Mol. Biol. Rev.* 68(4):692–744.

Hoppert, M., and Mayer, F. 1999. Prokaryotes. *American Scientist* 87:518–25.

Kerfeld, C. A.; Sawaya, M. R.; Tanaka, A.; Nguyen, C. V.; Phillips, M.; Beeby, M.; and Yeates, T. O. 2005. Protein structures forming the shell of primitive bacterial organelles. *Science* 309:936–38.

Kostakioti, M.; Newman, C. L.; Thanassi, D. G.; and Stathopoulos, C. 2005. Mechanisms of protein export across the bacterial outer membrane. *J. Bacteriol.* 187(13):4306–14.

Macnab, R. M. 2003. How bacteria assemble flagella. *Annu. Rev. Microbiol.* 57:77–100.

Mattick, J. S. 2002. Type IV pili and twitching motility. *Annu. Rev. Microbiol.* 56:289–314.

Nicholson, W. L.; Munakata, N.; Horneck, G.; Melosh, H. J.; and Setlow, P. 2000. Resistance of *Bacillus* endospores to extreme terrestrial and extraterrestrial environments. *Microbiol. Mol. Biol. Rev.* 64(3):548–72.

Nikaido, H. 2003. Molecular basis of bacterial outer membrane permeability revisited. *Microbiol. Mol. Biol. Rev.* 67(4):593–656.

Parkinson, J. S. 2004. Signal amplification in bacterial chemotaxis through receptor teamwork. *ASM News* 70(12):575–82.

Robinow, C., and Kellenberger, E. 1994. The bacterial nucleoid revisited. *Microbiol. Rev.* 58(2):211–32.

Sára, M., and Sleytr, U. B. 2000. S-layer proteins. *J. Bacteriol.* 182(4):859–68.

Scherrer, R. 1984. Gram's staining reaction, Gram types and cell walls of bacteria. *Trends Biochem. Sci.* 9:242–45.

Schulz, H. N., and Jorgensen, B. B. 2001. Big bacteria. *Annu. Rev. Microbiol.* 55:105–37.

Trun, N. J., and Marko, J. F. 1998. Architecture of a bacterial chromosome. *ASM News* 64(5):276–83.

Walsby, A. E. 1994. Gas vesicles. *Microbiol. Rev.* 58(1):94–144.

Walsby, A. E. 2005. Archaea with square cells. *Trends Microbiol.* 13(5):193–95.

Wätermann, M., and Steinbüchel, A. 2005. Neutral lipid bodies in prokaryotes: Recent insights into structure, formation, and relationship to eukaryotic lipid depots. *J. Bacteriol.* 187(11):3607–19.

Please visit the Prescott website at www.mhhe.com/prescott7 for additional references.

4 Eucaryotic Cell Structure and Function

Often we emphasize procaryotes and viruses, but eucaryotic microorganisms also have major impacts on human welfare. For example, the protozoan parasite *Trypanosoma brucei gambiense* is a cause of African sleeping sickness. The organism invades the nervous system and the victim frequently dies after suffering several years from symptoms such as weakness, headache, apathy, emaciation, sleepiness, and coma.

Preview

- Eucaryotic cells differ most obviously from procaryotic cells in having a variety of complex membranous organelles in the cytoplasmic matrix and the majority of their genetic material within membrane-delimited nuclei. Each organelle has a distinctive structure directly related to specific functions.
- A cytoskeleton composed of microtubules, microfilaments, and intermediate filaments helps give eucaryotic cells shape; the cytoskeleton is also involved in cell movements, intracellular transport, and reproduction.
- When eucaryotes reproduce, genetic material is distributed between cells by the highly organized, complex processes called mitosis and meiosis.
- Despite great differences between eucaryotes and procaryotes with respect to such things as morphology, they are similar on the biochemical level.

In chapter 3 considerable attention is devoted to procaryotic cell structure and function because procaryotes are immensely important in microbiology and have occupied a large portion of microbiologists' attention in the past. Nevertheless, protists and fungi also are microorganisms and have been extensively studied. These eucaryotes often are extraordinarily complex, interesting in their own right, and prominent members of ecosystems (**figure 4.1**). In addition, many protists and fungi are important model organisms, as well as being exceptionally useful in industrial microbiology. A number of protists and fungi are also major human pathogens; one only need think of candidiasis, malaria, or African sleeping sickness to appreciate the significance of eucaryotes in medical microbiology. So although this text emphasizes procaryotes, eucaryotic microorganisms also demand attention and are briefly discussed in this chapter.

Chapter 4 focuses on eucaryotic cell structure and its relationship to cell function. Because many valuable studies on eucaryotic cell ultrastructure have used organisms other than microorganisms, some work on nonmicrobial cells is presented. At the end of the chapter, procaryotic and eucaryotic cells are compared in some depth.

4.1 An Overview of Eucaryotic Cell Structure

The most obvious difference between eucaryotic and procaryotic cells is in their use of membranes. Eucaryotic cells have membrane-delimited nuclei, and membranes also play a prominent part in the structure of many other organelles (**figures 4.2** and **4.3**). **Organelles** are intracellular structures that perform specific functions in cells analogous to the functions of organs in the body. The name organelle (little organ) was coined because biologists saw a parallel between the relationship of organelles to a cell and that of organs to the whole body. It is not satisfactory to define organelles as membrane-bound structures because this would exclude such components as ribosomes and bacterial flagella. A comparison of figures 4.2 and 4.3 with figures 3.4 and 3.13*a* shows how structurally complex the eucaryotic cell is. This complexity is due chiefly to the use of internal membranes for several purposes. The partitioning of the eucaryotic cell interior by membranes makes possible the placement of different biochemical and physiological functions in separate compartments so that they can more easily take place simultaneously under independent control and proper coordination. Large membrane surfaces

The key to every biological problem must finally be sought in the cell.

—E. B. Wilson

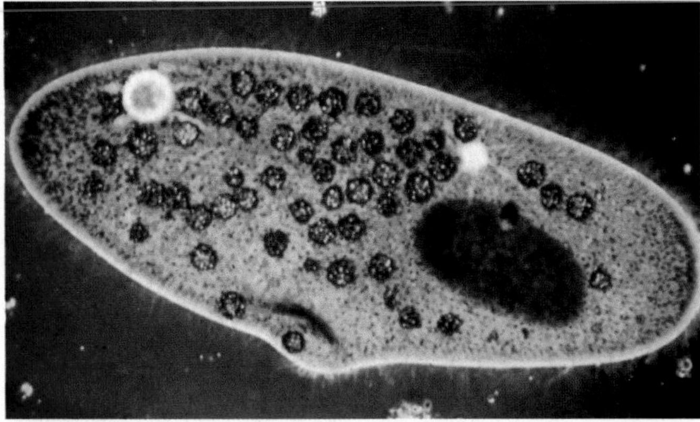

(a) *Paramecium*

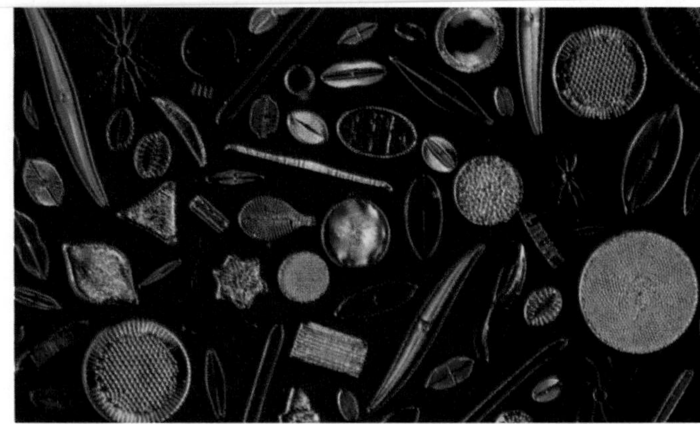

(b) Diatom frustules

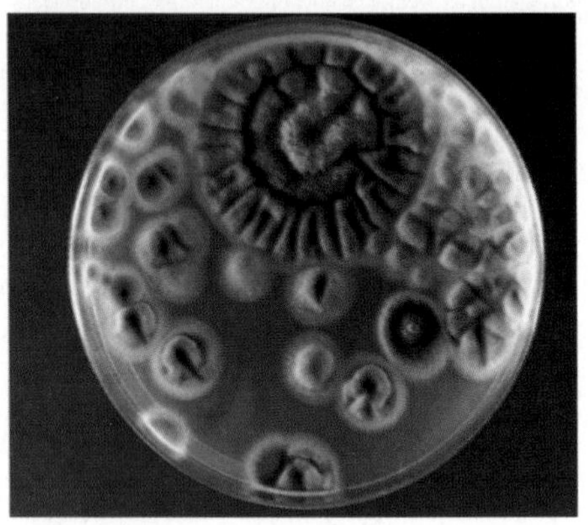

(c) *Penicillium*

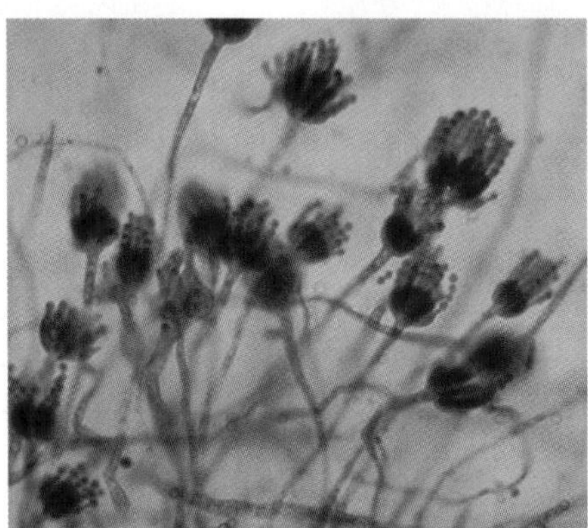

(d) *Penicillium*

(e) *Stentor*

(f) *Amanita muscaria*

Figure 4.1 Representative Examples of Eucaryotic Microorganisms.
(a) *Paramecium* as seen with interference-contrast microscopy (×115). **(b)** Mixed diatom frustules (×100). **(c)** *Penicillium* colonies, and **(d)** a microscopic view of the mold's hyphae and conidia (×220). **(e)** *Stentor.* The ciliated protozoa are extended and actively feeding, dark-field microscopy (×100). **(f)** *Amanita muscaria,* a large poisonous mushroom (×5).

make possible greater respiratory and photosynthetic activity because these processes are located exclusively in membranes. The intracytoplasmic membrane complex also serves as a transport system to move materials between different cell locations. Thus abundant membrane systems probably are necessary in eucaryotic cells because of their large volume and the need for adequate regulation, metabolic activity, and transport.

Figures 4.2 and 4.3 illustrate most of the organelles to be discussed here. **Table 4.1** briefly summarizes the functions of the major eucaryotic organelles. Our detailed discussion of eucaryotic cell structure begins with the eucaryotic membrane. We then proceed to organelles within the cytoplasm, and finally to components outside the membrane.

1. What is an organelle?
2. Why is the compartmentalization of the cell interior advantageous to eucaryotic cells?

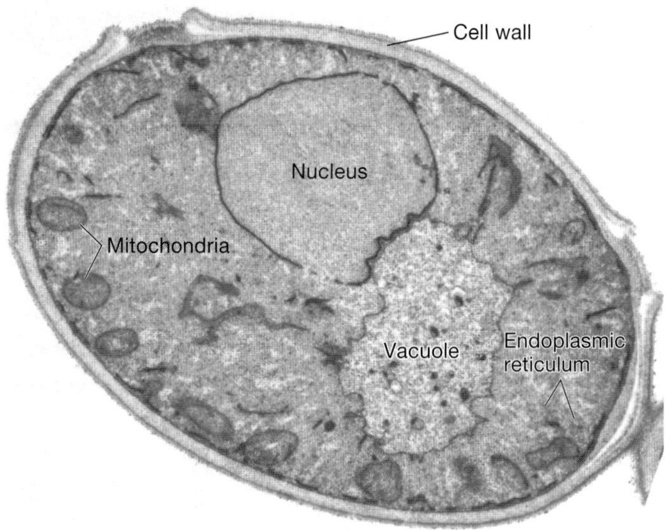

Figure 4.2 Eucaryotic Cell Ultrastructure. The yeast *Saccharomyces* (×7,200). Note the nucleus, mitochondrion, vacuole, endoplasmic reticulum, and cell wall.

4.2 The Plasma Membrane and Membrane Structure

As discussed in chapter 3, the fluid mosaic model of membrane structure is based largely on studies of eucaryotic membranes. In eucaryotes, the major membrane lipids are phosphoglycerides, sphingolipids, and cholesterol (**figure 4.4**). The distribution of these lipids is asymmetric. Lipids in the outer monolayer differ from those of the inner monolayer. Although most lipids in individual monolayers mix freely with each other, there are microdomains that differ in lipid and protein composition. One such microdomain is the **lipid raft,** which is enriched in cholesterol and lipids with many saturated fatty acids including some sphingolipids. The lipid raft spans the membrane bilayer, and lipids in the adjacent monolayers interact. These lipid rafts appear to participate in a variety of cellular processes (e.g., cell movement and signal transduction). They also may be involved in the entrance of some viruses into their host cells and the assembly of some viruses before they are released from their host cells.

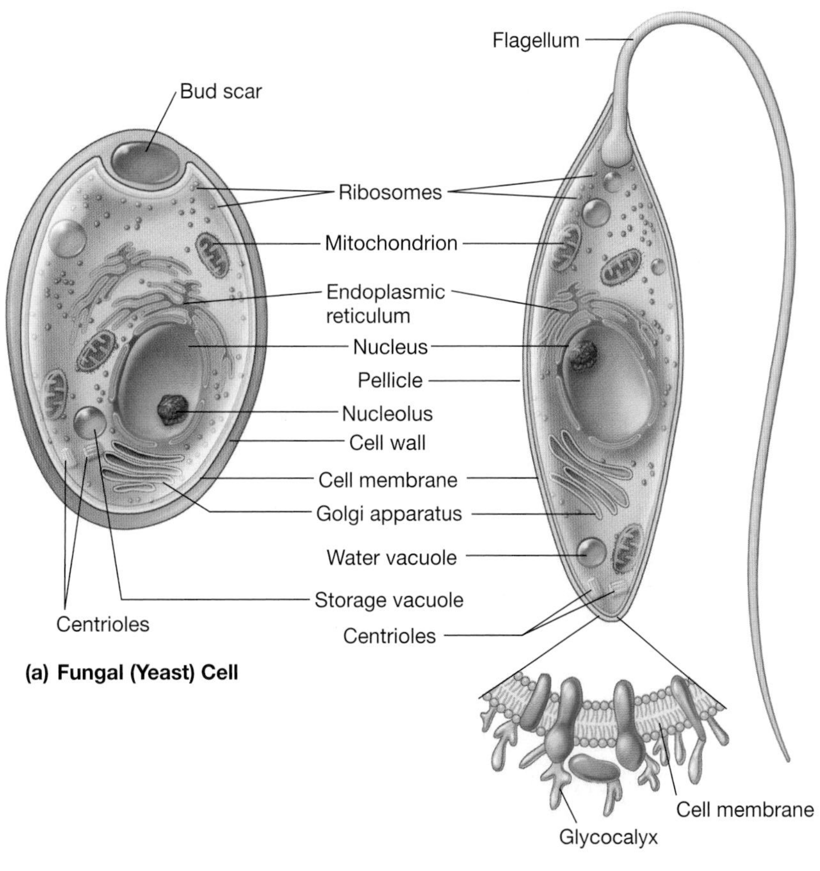

Figure 4.3 The Structure of Two Representative Eucaryotic Cells. Illustrations of a yeast cell (fungus) **(a)** and the flagellated protozoan *Peranema* **(b)**.

Table 4.1 Functions of Eucaryotic Organelles

Plasma membrane	Mechanical cell boundary, selectively permeable barrier with transport systems, mediates cell-cell interactions and adhesion to surfaces, secretion
Cytoplasmic matrix	Environment for other organelles, location of many metabolic processes
Microfilaments, intermediate filaments, and microtubules	Cell structure and movements, form the cytoskeleton
Endoplasmic reticulum	Transport of materials, protein and lipid synthesis
Ribosomes	Protein synthesis
Golgi apparatus	Packaging and secretion of materials for various purposes, lysosome formation
Lysosomes	Intracellular digestion
Mitochondria	Energy production through use of the tricarboxylic acid cycle, electron transport, oxidative phosphorylation, and other pathways
Chloroplasts	Photosynthesis—trapping light energy and formation of carbohydrate from CO_2 and water
Nucleus	Repository for genetic information, control center for cell
Nucleolus	Ribosomal RNA synthesis, ribosome construction
Cell wall and pellicle	Strengthen and give shape to the cell
Cilia and flagella	Cell movement
Vacuole	Temporary storage and transport, digestion (food vacuoles), water balance (contractile vacuole)

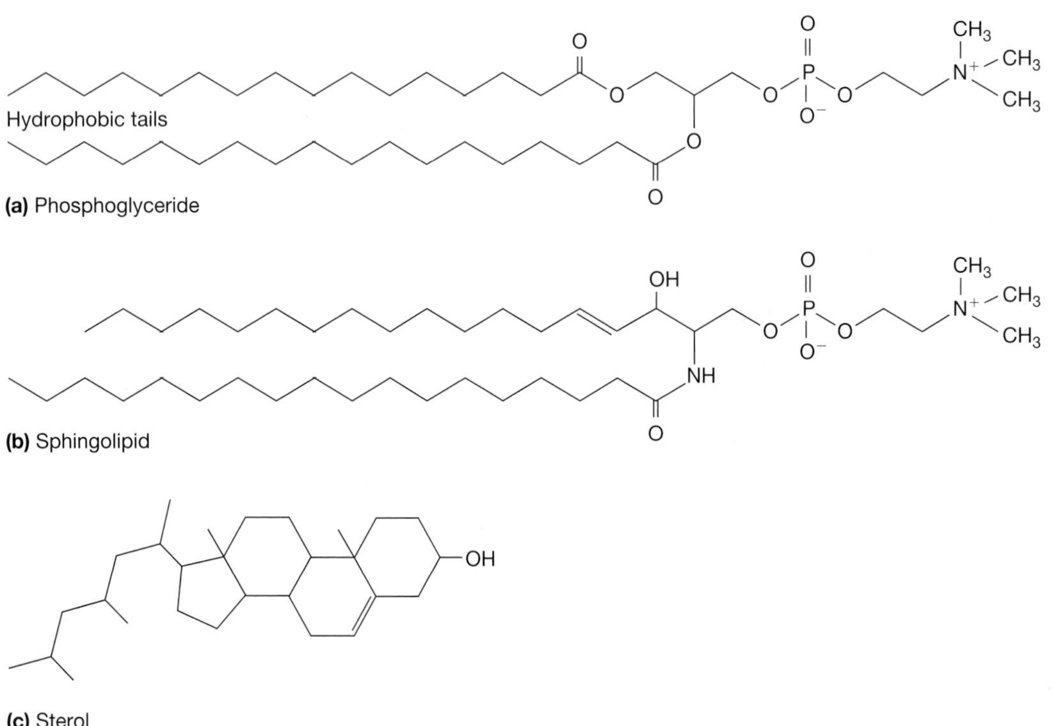

Figure 4.4 Examples of Eucaryotic Membrane Lipids. **(a)** Phosphatidylcholine, a phosphoglyceride. **(b)** Sphingomyelin, a sphingolipid. **(c)** Cholesterol, a sterol.

4.3 The Cytoplasmic Matrix, Microfilaments, Intermediate Filaments, and Microtubules

The many organelles of eucaryotic cells lie in the **cytoplasmic matrix.** The matrix is one of the most important and complex parts of the cell. It is the "environment" of the organelles and the location of many important biochemical processes. Several physical changes seen in cells—viscosity changes, cytoplasmic streaming, and others—also are due to matrix activity.

A major component of the cytoplasmic matrix is a vast network of interconnected filaments called the **cytoskeleton.** The cytoskeleton plays a role in both cell shape and movement. Three types of filaments form the cytoskeleton: microfilaments, microtubules, and intermediate filaments (**figure 4.5**). **Microfilaments** are minute protein filaments, 4 to 7 nm in diameter, that may be either scattered within the cytoplasmic matrix or organized into networks and parallel arrays. Microfilaments are composed of an actin protein that is similar to the actin contractile protein of muscle tissue. Microfilaments are involved in cell motion and shape changes such as the motion of pigment granules, amoeboid movement, and protoplasmic streaming in slime molds. Interestingly, some pathogens use the actin proteins of their eucaryotic hosts to move rapidly through the host cell and to propel themselves into new host cells (**Disease 4.1: Getting Around**). Protist classification: *Eumycetozoa* and *Stramenopiles* (section 25.6)

Microtubules are shaped like thin cylinders about 25 nm in diameter. They are complex structures constructed of two spherical protein subunits—α-tubulin and β-tubulin. The two proteins are the same molecular weight and differ only slightly in terms of their amino acid sequence and tertiary structure. Each tubulin is approximately 4 to 5 nm in diameter. These subunits are assembled in a helical arrangement to form a cylinder with an average of 13 subunits in one turn or circumference (figure 4.5).

Microtubules serve at least three purposes: (1) they help maintain cell shape, (2) are involved with microfilaments in cell movements, and (3) participate in intracellular transport processes. Microtubules are found in long, thin cell structures requiring support such as the axopodia (long, slender, rigid pseudopodia) of protists (**figure 4.6**). Microtubules also are present in structures that participate in cell or organelle movements—the mitotic spindle, cilia, and flagella.

Intermediate filaments are heterogeneous elements of the cytoskeleton. They are about 10 nm in diameter and are assembled from a group of proteins that can be divided into several classes. Intermediate filaments having different functions are assembled from one or more of these classes of proteins. The role

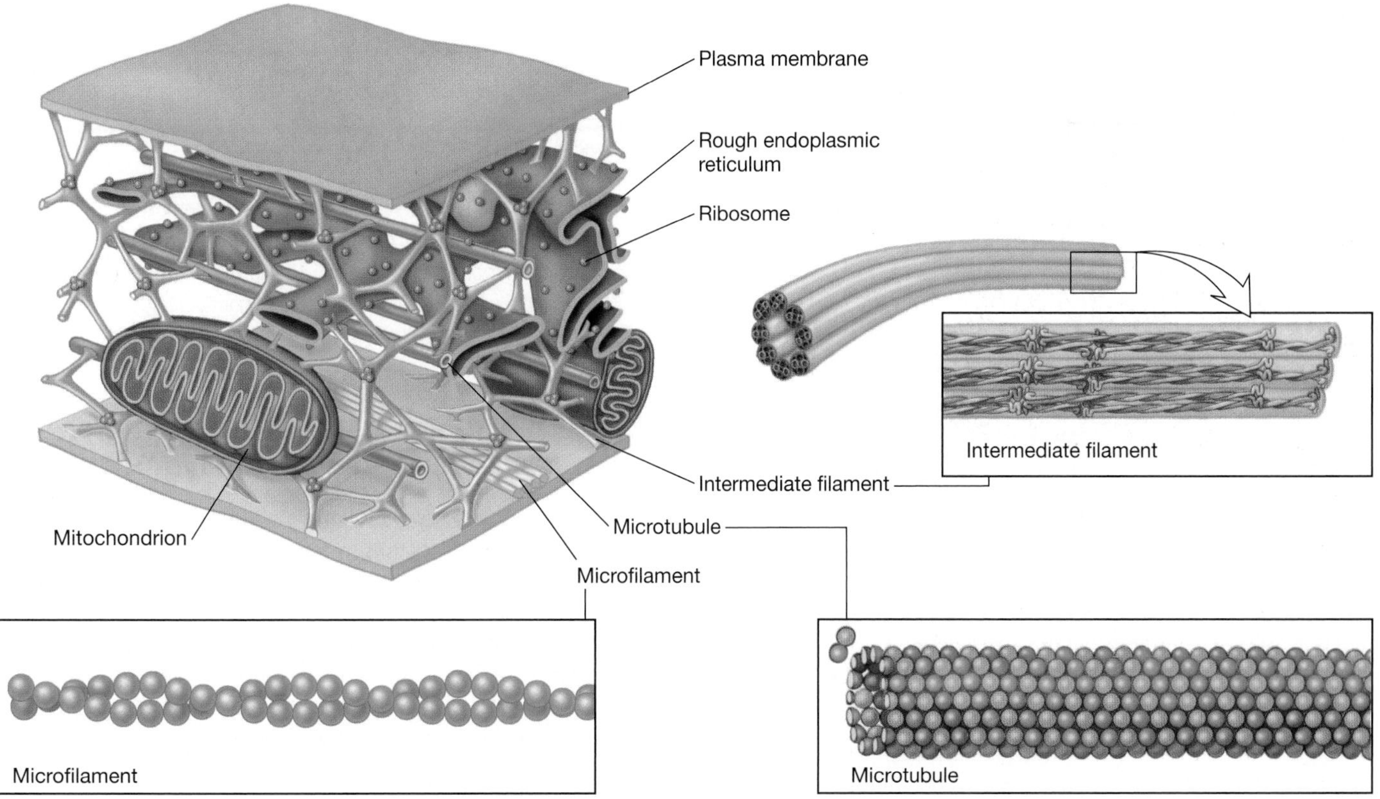

Figure 4.5 The Eucaryotic Cytoplasmic Matrix and Cytoskeleton. The cytoplasmic matrix of eucaryotic cells contains many important organelles. The cytoskeleton helps form a framework within which the organelles lie. The cytoskeleton is composed of three elements: microfilaments, microtubules, and intermediate filaments.

Disease

4.1 Getting Around

Listeria monocytogenes is a gram-positive, rod-shaped bacterium responsible for the disease listeriosis. Listeriosis is a food-borne infection that is usually mild but can cause serious disease (meningitis, sepsis, and stillbirth) in immunocompromised individuals and pregnant women. *L. monocytogenes* is an intracellular pathogen that has a number of important virulence factors. One virulence factor is the protein ActA, which the bacterium releases after entering a host cell. ActA causes actin proteins to polymerize into filaments that form a tail at one end of the bacterium (see **Box figure**). As more and more of the actin proteins are polymerized, the growing tail pushes the bacterium through the host cell at rates up to 11 μm/minute. The bacterium can even be propelled through the cell surface and into neighboring cells.

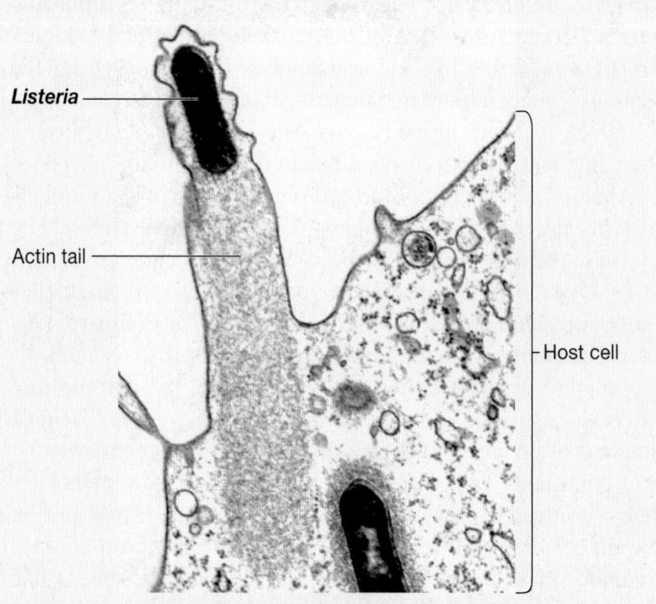

***Listeria* Motility and Actin Filaments.** A *Listeria* cell is propelled through the cell surface by a bundle of actin filaments.

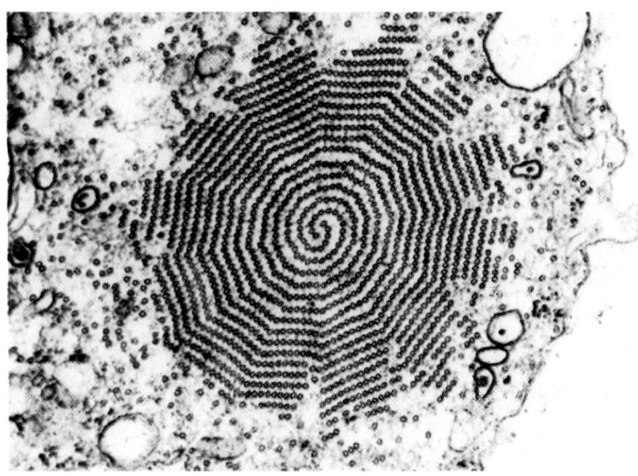

Figure 4.6 Cytoplasmic Microtubules. Electron micrograph of a transverse section through the axopodium of a protist known as a heliozoan (×48,000). Note the parallel array of microtubules organized in a spiral pattern.

of intermediate filaments, if any, in eucaryotic microorganisms is unclear. Thus far, they have been identified and studied only in animals: some intermediate filaments have been shown to form the nuclear lamina, a structure that provides support for the nuclear envelope (see p. 91); and other intermediate filaments help link cells together to form tissues.

1. Compare the membranes of *Eucarya*, *Bacteria*, and *Archaea*. How are they similar? How do they differ?
2. What are lipid rafts? What roles do they play in eucaryotic cells?
3. Define cytoplasmic matrix, microfilament, microtubule, and tubulin. Discuss the roles of microfilaments, intermediate filaments, and microtubules.
4. Describe the cytoskeleton. What are its functions?

4.4 Organelles of the Biosynthetic-Secretory and Endocytic Pathways

In addition to the cytoskeleton, the cytoplasmic matrix is permeated with an intricate complex of membranous organelles and vesicles that move materials into the cell from the outside (endocytic pathway), and from the inside of the cell out, as well as from location to location within the cell (biosynthetic-secretory pathway). In this section, some of these organelles are described. This is followed by a summary of how the organelles function in the biosynthetic-secretory and endocytic pathways.

The Endoplasmic Reticulum

The **endoplasmic reticulum (ER)** (figure 4.3 and **figure 4.7**) is an irregular network of branching and fusing membranous tubules, around 40 to 70 nm in diameter, and many flattened sacs called **cisternae** (s., **cisterna**). The nature of the ER varies with the functional and physiological status of the cell. In cells synthesizing a great deal of protein for purposes such as secretion, a large part of the ER is studded on its outer surface with ribosomes and is called **rough endoplasmic reticulum** (**RER**). Other cells, such as those producing large quantities of lipids, have ER that lacks ribosomes. This is **smooth ER** (**SER**).

The endoplasmic reticulum has many important functions. Not only does it transport proteins, lipids, and other materials through the cell, it is also involved in the synthesis of many of the materials it transports. Lipids and proteins are synthesized by ER-associated enzymes and ribosomes. Polypeptide chains synthesized on RER-bound ribosomes may be inserted either into the ER membrane or into its lumen for transport elsewhere. The ER is also a major site of cell membrane synthesis.

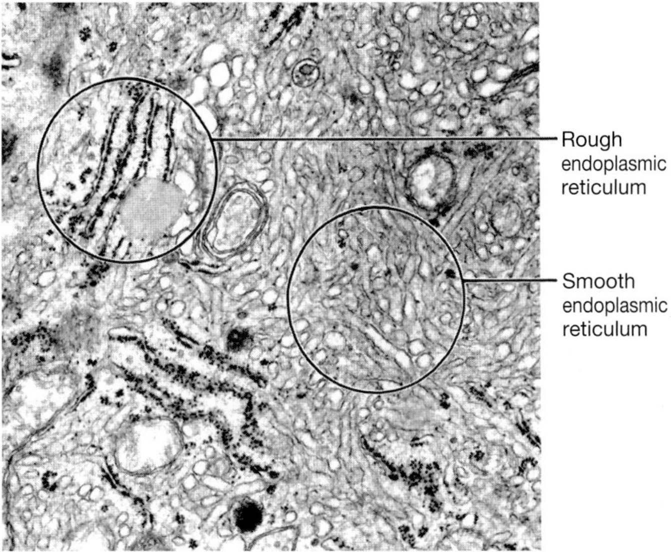

Figure 4.7 The Endoplasmic Reticulum. A transmission electron micrograph of the corpus luteum in a human ovary showing structural variations in eucaryotic endoplasmic reticulum. Note the presence of both rough endoplasmic reticulum lined with ribosomes and smooth endoplasmic reticulum without ribosomes (×26,500).

The Golgi Apparatus

The **Golgi apparatus** is composed of flattened, saclike cisternae stacked on each other (**figure 4.8**). These membranes, like the smooth ER, lack bound ribosomes. There are usually around 4 to 8 cisternae in a stack, although there may be many more. Each is 15 to 20 nm thick and separated from other cisternae by 20 to 30 nm. A complex network of tubules and vesicles (20 to 100 nm in diameter) is located at the edges of the cisternae. The stack of cisternae has a definite polarity because there are two faces that are quite different from one another. The sacs on the cis or forming face often are associated with the ER and differ from the sacs on the trans or maturing face in thickness, enzyme content, and degree of vesicle formation.

The Golgi apparatus is present in most eucaryotic cells, but many fungi and ciliate protozoa lack a well-formed structure. Sometimes the Golgi consists of a single stack of cisternae; however, many cells may contain up to 20, and sometimes more, separate stacks. These stacks of cisternae, often called **dictyosomes,** can be clustered in one region or scattered about the cell.

The Golgi apparatus packages materials and prepares them for secretion, the exact nature of its role varying with the organism. For instance, the surface scales of some flagellated photosynthetic and radiolarian protists appear to be constructed within the Golgi apparatus and then transported to the surface in vesicles. The Golgi often participates in the development of cell membranes

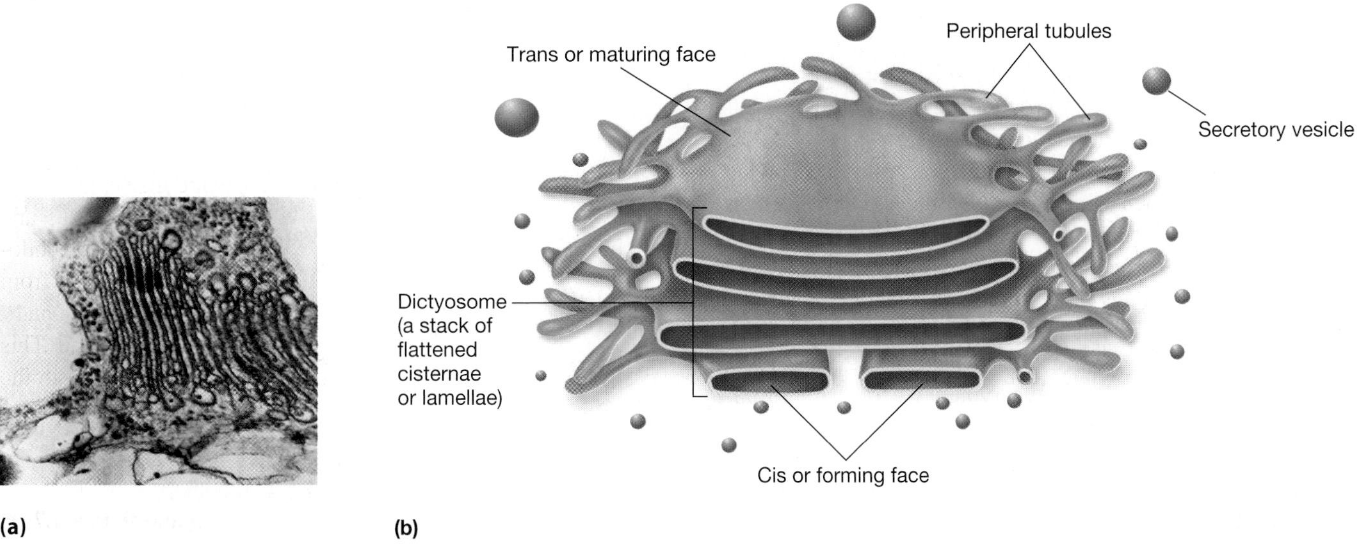

Figure 4.8 Golgi Apparatus Structure. Golgi apparatus of *Euglena gracilis.* Cisternal stacks are shown in the electron micrograph (×165,000) in **(a)** and diagrammatically in **(b).**

and in the packaging of cell products. The growth of some fungal hyphae occurs when Golgi vesicles contribute their contents to the wall at the hyphal tip. The protists (chapter 25)

Lysosomes

Lysosomes, or structures very much like them, are found in most eucaryotic organisms, including protists, fungi, plants, and animals. Lysosomes are roughly spherical and enclosed in a single membrane; they average about 500 nm in diameter, but range from 50 nm to several μm in size. They are involved in intracellular digestion and contain the enzymes needed to digest all types of macromolecules. These enzymes, called hydrolases, catalyze the hydrolysis of molecules and function best under slightly acidic conditions (usually around pH 3.5 to 5.0). Lysosomes maintain an acidic environment by pumping protons into their interior.

The Biosynthetic-Secretory Pathway

The **biosynthetic-secretory pathway** is used to move materials to lysosomes as well as from the inside of the cell to either the cell membrane or cell exterior. The process is complex and not fully understood. The movement of proteins is of particular importance and is the focus of this discussion.

Proteins destined for the cell membrane, lysosomes, or secretion are synthesized by ribosomes attached to the rough endoplasmic reticulum (RER) (figure 4.7). These proteins have sequences of amino acids that target them to the lumen of the RER through which they move until released in small vesicles that bud from the ER. As the proteins pass through the ER, they are often modified by the addition of sugars—a process known as glycosylation.

The vesicles released from the ER travel to the cis face of the Golgi apparatus (figure 4.8). A popular model of the biosynthetic-secretory pathway posits that these vesicles fuse to form the cis face of the Golgi. The proteins then proceed to the trans face of the Golgi by a process called cisternal maturation. As the proteins proceed from the cis to trans side, they are further modified. Some of these modifications target the proteins for their final location. For instance, lysosomal proteins are modified by the addition of phosphates to their mannose sugars.

Transport vesicles are released from the trans face of the Golgi. Some deliver their contents to lysosomes. Others deliver proteins and other materials to the cell membrane. Two types of vesicles transport materials to the cell membrane. One type constitutively delivers proteins in an unregulated manner, releasing them to the outside of the cell as the transport vesicle fuses with the plasma membrane. Other vesicles, called secretory vesicles, are found only in multicellular eucaryotes, where they are observed in secretory cells such as mast cells and other cells of the immune system. Secretory vesicles store the proteins to be released until the cell receives an appropriate signal. Once received, the secretory vesicles move to the plasma membrane, fuse with it, and release their contents to the cell exterior. Cells, tissues, and organs of the immune system (section 31.2)

One interesting and important feature of the biosynthetic-secretory pathway is its quality-assurance mechanism. Proteins that fail to fold or have misfolded are not transported to their intended destination. Instead they are secreted into the cytosol, where they are targeted for destruction by the attachment of several small ubiquitin polypeptides as detailed in **figure 4.9.** Ubiquitin marks the protein for degradation, which is accomplished by a huge, cylindrical complex called a **26S proteasome.** The protein is broken down to smaller peptides in an ATP-dependent process as the ubiquitins are released. The proteasome also is involved in producing peptides for antigen presentation during many immunological responses described in chapter 31.

The Endocytic Pathway

Endocytosis is used to bring materials into the cell from the outside. During endocytosis a cell takes up solutes or particles by enclosing them in vesicles pinched off from the plasma membrane. In most cases, these materials are delivered to a lysosome where they are digested. Endocytosis occurs regularly in all cells as a mechanism for recycling molecules in the membrane. In addition, some cells have specialized **endocytic pathways** that allow them to concentrate materials outside the cell before bringing them in. Others use endocytic pathways as a feeding mechanism. Many viruses and other intracellular pathogens use endocytic pathways to enter host cells.

Numerous types of endocytosis have been described. **Phagocytosis** involves the use of protrusions from the cell surface to surround and engulf particulates. It is carried out by certain immune system cells and many eucaryotic microbes. The endocytic vesicles formed by phagocytosis are called **phagosomes** (**figure 4.10**). Other types of endocytosis also involve invagination of the plasma membrane. As the membrane invaginates, it encloses liquid, soluble matter, and, in some cases, particulates in the resulting endocytic vesicle. One example of endocytosis by invagination is **clathrin-dependent endocytosis.** Clathrin-dependent endocytosis begins with coated pits, which are specialized membrane regions coated with the protein **clathrin** on the cytoplasmic side. The endocytic vesicles formed when these regions invaginate are called coated vesicles. Coated pits have receptors on their extracellular side that specifically bind macromolecules, concentrating them before they are endocytosed. Therefore this endocytic mechanism is referred to as **receptor-mediated endocytosis.** Clathrin-dependent endocytosis is used to ingest such things as hormones, growth factors, iron, and cholesterol. Another example of endocytosis by invagination is **caveolae-dependent endocytosis. Caveolae** ("little caves") are tiny, flask-shaped invaginations of the plasma membrane (about 50 to 80 nm in diameter) that are enriched in cholesterol and the membrane protein caveolin. The vesicles formed when caveolae pinch off are called caveolar vesicles. Caveolae-dependent endocytosis has been implicated in signal transduction, transport of small molecules such as folic acid, as well as transport of macromolecules. There is evidence that toxins such as cholera toxin enter their target cells via caveolae. Caveolae also appear to be used by many viruses, bacteria, and protozoa to enter host cells.

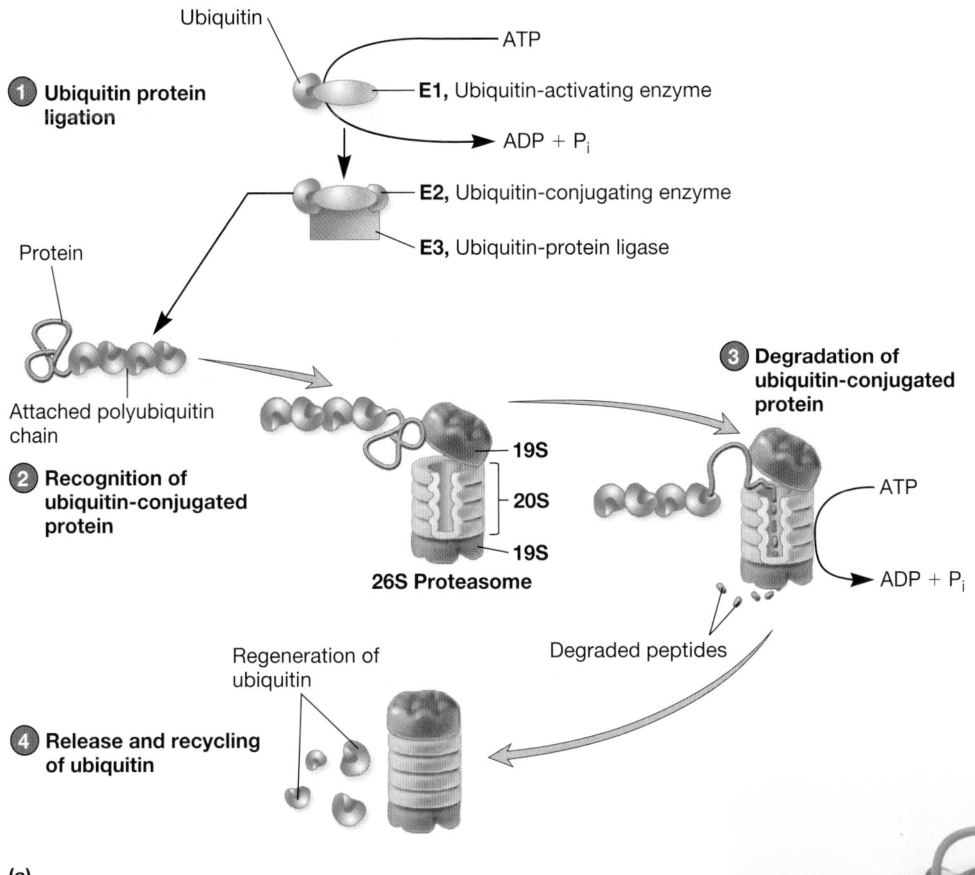

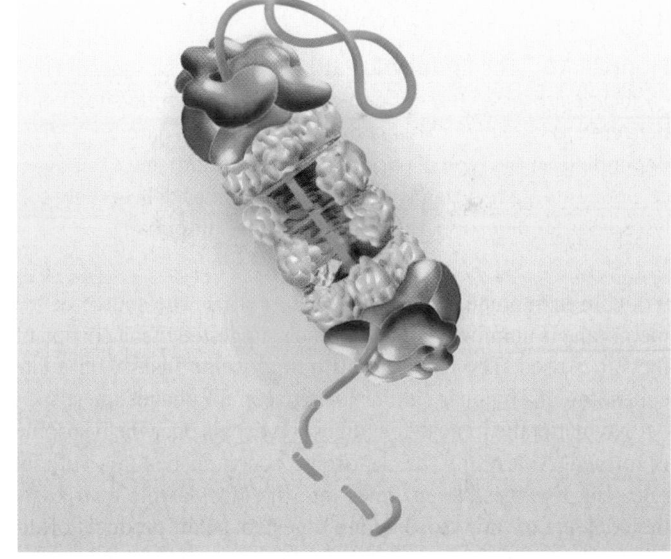

Figure 4.9 Proteasome Degradation of Proteins. **(a)** The first step in this protein degradation pathway is to tag the target protein with a small polypeptide called ubiquiton. This requires the action of two enzymes and energy is consumed. Once tagged, the protein is recognized by the 26S proteasome. It passes into the large, cylindrical proteasome and is cleaved into smaller peptides, which are released into the cytoplasm. The amino acids in the small peptides can be recycled and used in the synthesis of new proteins. The ubiquitin polypeptides are regenerated and can participate in the degradation of other proteins. **(b)** A model of the 26S proteasome, showing its cylindrical structure and the location of the tagged protein within the cylinder.

With the exception of caveolar vesicles, all other endocytic vesicles eventually deliver their contents to lysosomes. However, the route used varies. **Coated vesicles** fuse with small organelles containing lysosomal enzymes. These organelles are called **early endosomes** (figure 4.10). Early endosomes mature into late endosomes, which fuse with transport vesicles from the Golgi delivering additional lysosomal enzymes. **Late endosomes** eventually become lysosomes. The development of endosomes into lysosomes is not well understood. It appears that maturation involves the movement of the organelles to a more central location in the cell and the selective retrieval of membrane proteins. Phagosomes take a slightly different route to lysosomes; they fuse with late endosomes rather than early endosomes (figure 4.10).

Materials for digestion can also be delivered to lysosomes by another route that does not involve endocytosis. Cells selectively digest and recycle cytoplasmic components (including organelles such as mitochondria) by a process called **autophagy.** It is believed that the cell components to be digested are surrounded by

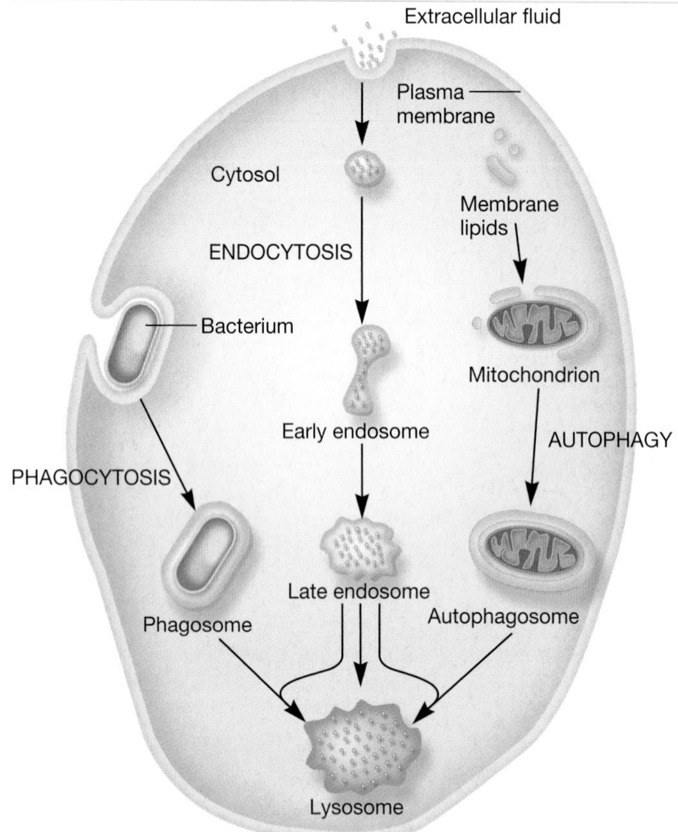

Figure 4.10 The Endocytic Pathway. Materials ingested by endocytic processes (except caveolae-dependent endocytosis) are delivered to lysosomes. The pathway to lysosomes differs, depending on the type of endocytosis. In addition, cell components are recycled when autophagosomes deliver them to lysosomes for digestion. This process is called autophagy.

a double membrane, as shown in figure 4.10. The source of the membrane is unknown, but it has been suggested that a portion of the ER is used. The resulting **autophagosome** fuses with a late endosome in a manner similar to that seen for phagosomes.

No matter the route taken, digestion occurs once the lysosome is formed. Amazingly, the lysosome accomplishes this without releasing its digestive enzymes into the cytoplasmic matrix. As the contents of the lysosome are digested, small products of digestion leave the lysosome, where they are used as nutrients or for other purposes. The resulting lysosome containing undigested material is often called a **residual body.** In some cases, the residual body can release its contents to the cell exterior by a process called lysosome secretion.

1. How do the rough and smooth endoplasmic reticulum differ from one another in terms of structure and function? List the processes in which the ER is involved.
2. Describe the structure of a Golgi apparatus in words and with a diagram. How do the cis and trans faces of the Golgi apparatus differ? List the major Golgi apparatus functions discussed in the text.
3. What is a proteasome? Why is it important to the proper functioning of the endoplasmic reticulum?
4. What are lysosomes? How do they participate in intracellular digestion?
5. Describe the biosynthetic-secretory pathway. To what destinations does this pathway deliver proteins and other materials?
6. Define endocytosis. Describe the endocytic pathway and the three routes that deliver materials to lysosomes for digestion. Which type of endocytosis does not deliver ingested material to lysosomes?
7. Define autophagy, autophagosome, phagosome, phagocytosis, and residual body.
8. Caveolae-mediated endocytosis is used by a number of pathogens to enter their host cells. Why might this route of entry be advantageous to the pathogens that use it?

4.5 Eucaryotic Ribosomes

The eucaryotic ribosome (i.e., one not found in mitochondria and chloroplasts) is larger than the procaryotic 70S ribosome. It is a dimer of a 60S and a 40S subunit, about 22 nm in diameter, and has a sedimentation coefficient of 80S and a molecular weight of 4 million. Eucaryotic ribosomes can be either associated with the endoplasmic reticulum or free in the cytoplasmic matrix. When bound to the endoplasmic reticulum to form rough ER, they are attached through their 60S subunits.

Both free and ER-bound ribosomes synthesize proteins. Proteins made on the ribosomes of the RER are often secreted or are inserted into the ER membrane as integral membrane proteins. Free ribosomes are the sites of synthesis for nonsecretory and nonmembrane proteins. Some proteins synthesized by free ribosomes are inserted into organelles such as the nucleus, mitochondrion, and chloroplast. As discussed in chapters 3 and 11, molecular chaperones aid the proper folding of proteins after synthesis. They also assist the transport of proteins into eucaryotic organelles such as mitochondria.

1. Describe the structure of the eucaryotic 80S ribosome and contrast it with the procaryotic ribosome.
2. How do free ribosomes and those bound to the ER differ in function?

4.6 Mitochondria

Found in most eucaryotic cells, **mitochondria** (s., **mitochondrion**) frequently are called the "powerhouses" of the cell (**figure 4.11**). Tricarboxylic acid cycle activity and the generation of ATP by electron transport and oxidative phosphorylation take place here. In the transmission electron microscope, mitochondria usually are cylindrical structures and measure approximately 0.3 to 1.0 μm by 5 to 10 μm. (In other words, they are about the same size as procaryotic cells.) Although some cells possess 1,000 or more mitochondria, others, including some yeasts, unicellular algae, and trypanosome protozoa, have a single, giant, tubular mitochondrion twisted into a continuous net-

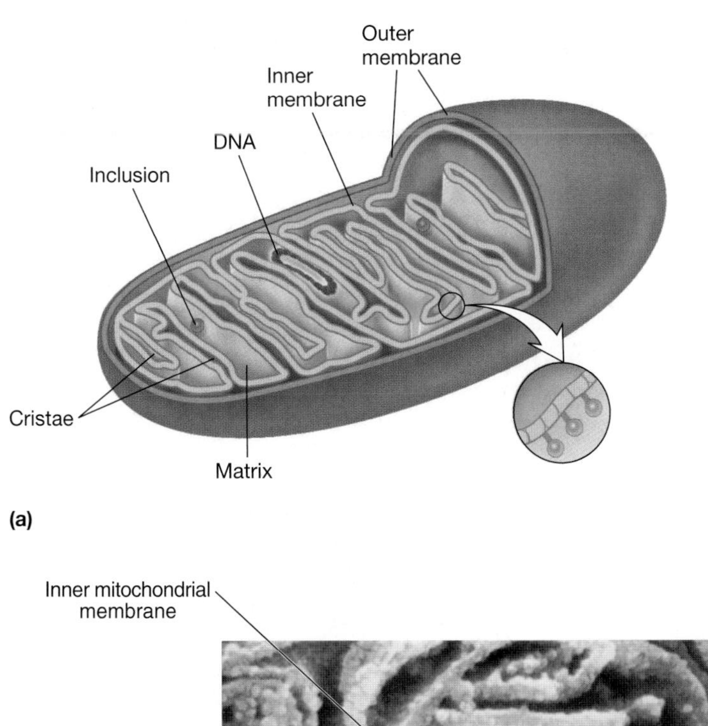

(a)

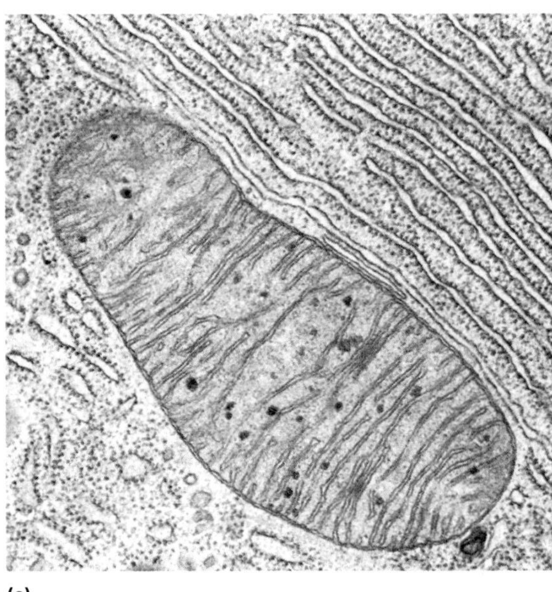

(c)

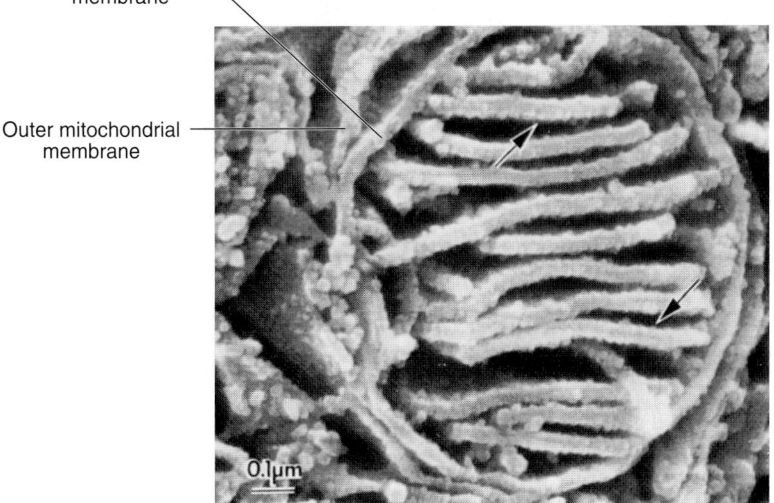

(b)

Figure 4.11 Mitochondrial Structure. **(a)** A diagram of mitochondrial structure. The insert shows the ATP-synthesizing enzyme ATP synthase lining the inner surface of the cristae. **(b)** Scanning electron micrograph (×70,000) of a freeze-fractured mitochondrion showing the cristae (arrows). The outer and inner mitochondrial membranes also are evident. **(c)** Transmission electron micrograph of a mitochondrion from a bat pancreas (×85,000). Note outer and inner mitochondrial membranes, cristae, and inclusions in the matrix. The mitochondrion is surrounded by rough endoplasmic reticulum.

work permeating the cytoplasm (**figure 4.12**). The tricarboxylic acid cycle (section 9.4); Electron transport and oxidative phosphorylation (section 9.5)

The mitochondrion is bounded by two membranes, an outer mitochondrial membrane separated from an inner mitochondrial membrane by a 6 to 8 nm intermembrane space (figure 4.11). The outer mitochondrial membrane contains porins and thus is similar to the outer membrane of gram-negative bacteria. The inner membrane has special infoldings called **cristae** (s., **crista**), which greatly increase its surface area. The shape of cristae differs in mitochondria from various species. Fungi have platelike (laminar) cristae, whereas euglenoid flagellates may have cristae shaped like disks. Tubular cristae are found in a variety of eucaryotes; however, amoebae can possess mitochondria with cristae in the shape of vesicles (**figure 4.13**). The inner membrane encloses the mitochondrial matrix, a dense matrix containing ribosomes, DNA, and often large calcium phosphate granules. Mitochondrial ribosomes are smaller than cytoplasmic ribosomes and resemble those of bacteria in several ways, including their size and subunit composition. In many organisms, mitochondrial DNA is a closed circle, like bacterial DNA. However, in some protists, mitochondrial DNA is linear.

Each mitochondrial compartment is different from the others in chemical and enzymatic composition. The outer and inner mitochondrial membranes, for example, possess different lipids. Enzymes and electron carriers involved in electron transport and oxidative phosphorylation (the formation of ATP as a consequence of electron transport) are located only in the inner membrane. The enzymes of the tricarboxylic acid cycle and catabolism of fatty acids are located in the matrix. Lipid catabolism (section 9.9)

The mitochondrion uses its DNA and ribosomes to synthesize some of its own proteins. In fact, mutations in mitochondrial DNA often lead to serious diseases in humans. Most mitochondrial proteins, however, are manufactured under the direction of

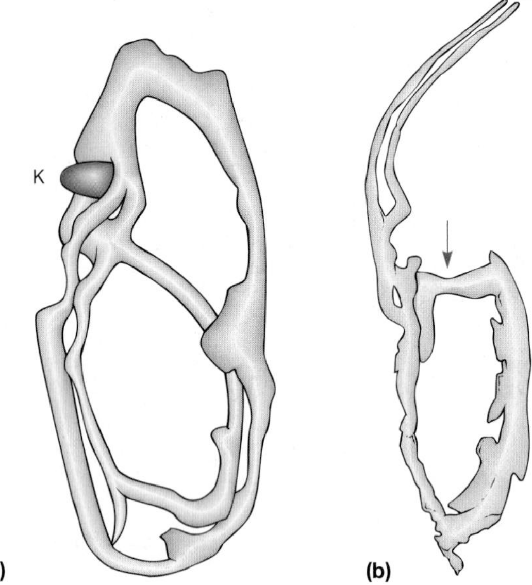

Figure 4.12 Trypanosome Mitochondria. The giant mitochondria from trypanosomes. **(a)** *Crithidia fasciculata* mitochondrion with kinetoplast, K. The kinetoplast contains DNA that codes for mitochondrial RNA and protein. **(b)** *Trypanosoma cruzi* mitochondrion with arrow indicating position of kinetoplast.

the nucleus. Mitochondria reproduce by binary fission. Because mitochondria resemble bacteria to some extent, it is thought that they arose from symbiotic associations between bacteria and larger cells (**Microbial Diversity & Ecology 4.2**). Microbial evolution: Endosymbiotic origin of mitochondria and chloroplasts (section 19.1)

4.7 Chloroplasts

Plastids are cytoplasmic organelles of photosynthetic protists and plants that often possess pigments such as chlorophylls and carotenoids, and are the sites of synthesis and storage of food reserves. The most important type of plastid is the chloroplast. **Chloroplasts** contain chlorophyll and use light energy to convert CO_2 and water to carbohydrates and O_2. That is, they are the site of photosynthesis.

Although chloroplasts are quite variable in size and shape, they share many structural features. Most often they are oval with dimensions of 2 to 4 μm by 5 to 10 μm, but some algae possess one huge chloroplast that fills much of the cell. Like mitochondria, chloroplasts are encompassed by two membranes (**figure 4.14**). A matrix, the **stroma,** lies within the inner membrane. It contains DNA, ribosomes, lipid droplets, starch granules, and a complex internal membrane system whose most prominent components are flattened, membrane-delimited sacs, the **thylakoids.** Clusters of two or more thylakoids are dispersed within the stroma of most algal chloroplasts (figures 4.14 and 4.24*b*). In some photosynthetic protists, several disklike thylakoids are stacked on each other like coins to form **grana** (s., **granum**).

(a)

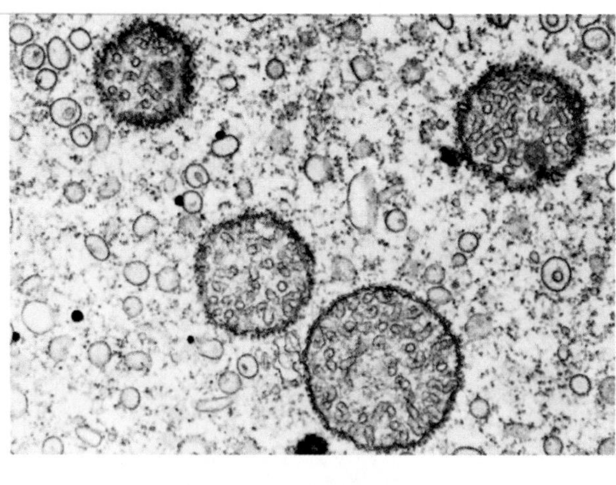

(b)

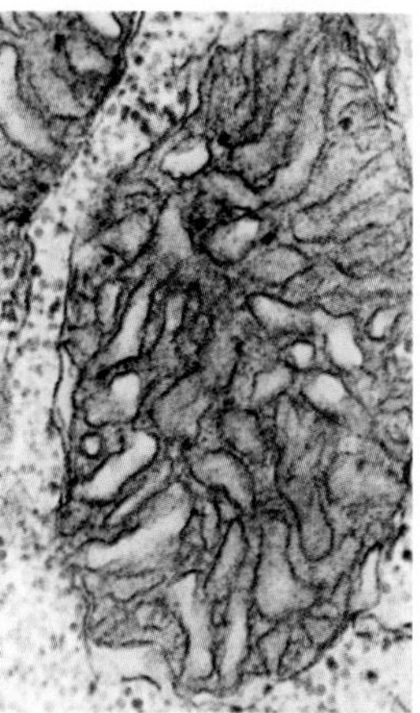

Figure 4.13 Mitochondrial Cristae. Mitochondria with a variety of cristae shapes. **(a)** Mitochondria from the slime mold *Schizoplasmodiopsis micropunctata.* Note the tubular cristae (×49,500). **(b)** The protist *Actinosphaerium* with vesicular cristae (×75,000).

Photosynthetic reactions are separated structurally in the chloroplast just as electron transport and the tricarboxylic acid cycle are in the mitochondrion. The trapping of light energy to generate ATP, NADPH, and O_2 is referred to as the light reactions. These reactions are located in the thylakoid membranes, where chlorophyll and electron transport components are also found. The ATP and NADPH formed by the light reactions are used to form carbohydrates from CO_2 and water in the dark reactions. The dark reactions take place in the stroma. Phototrophy (section 9.12)

The chloroplasts of many algae contain a **pyrenoid** (figure 4.24*b*), a dense region of protein surrounded by starch or another polysaccharide. Pyrenoids participate in polysaccharide synthesis.

1. Describe in detail the structure of mitochondria and chloroplasts. Where are the different components of these organelles' energy-trapping systems located?
2. Define plastid, dark reactions, light reactions, and pyrenoid.

Microbial Diversity & Ecology

4.2 The Origin of the Eucaryotic Cell

The profound differences between eucaryotic and procaryotic cells have stimulated much discussion about how the more complex eucaryotic cell arose. Some biologists believe the original "protoeucaryote" was a large, aerobic archaeon or bacterium that formed mitochondria, chloroplasts, and nuclei when its plasma membrane invaginated and enclosed genetic material in a double membrane. The organelles could then evolve independently. It also is possible that a large cyanobacterium lost its cell wall and became phagocytic. Subsequently, primitive chloroplasts, mitochondria, and nuclei would be formed by the fusion of thylakoids and endoplasmic reticulum cisternae to enclose specific areas of cytoplasm.

By far the most popular theory for the origin of eucaryotic cells is the endosymbiotic theory. In brief, it is supposed that the ancestral procaryotic cell, which may have been an archaeon, lost its cell wall and gained the ability to obtain nutrients by phagocytosing other procaryotes. When photosynthetic cyanobacteria arose, the environment slowly became oxic. If an anaerobic, amoeboid, phagocytic procaryote—possibly already possessing a developed nucleus—engulfed an aerobic bacterial cell and established a permanent symbiotic relationship with it, the host would be better adapted to its increasingly oxic environment. The endosymbiotic aerobic bacterium eventually would develop into the mitochondrion. Similarly, symbiotic associations with cyanobacteria could lead to the formation of chloroplasts and photosynthetic eucaryotes. Some have speculated that cilia and flagella might have arisen from the attachment of spirochete bacteria (*see chapter 21*) to the surface of eucaryotic cells, much as spirochetes attach themselves to the surface of the motile protozoan *Myxotricha paradoxa* that grows in the digestive tract of termites.

There is evidence to support the endosymbiotic theory. Both mitochondria and chloroplasts resemble bacteria in size and appearance, contain DNA in the form of a closed circle like that of bacteria, and reproduce semiautonomously. Mitochondrial and chloroplast ribosomes resemble procaryotic ribosomes more closely than those in the eucaryotic cytoplasmic matrix. The sequences of the chloroplast and mitochondrial genes for ribosomal RNA and transfer RNA are more similar to bacterial gene sequences than to those of eucaryotic rRNA and tRNA nuclear genes. Finally, there are symbiotic associations that appear to be bacterial endosymbioses in which distinctive procaryotic characteristics are being lost. For example, the protozoan flagellate *Cyanophora paradoxa* has photosynthetic organelles called cyanellae with a structure similar to that of cyanobacteria and the remains of peptidoglycan in their walls. Their DNA is much smaller than that of cyanobacteria and resembles chloroplast DNA. The endosymbiotic theory is discussed in more detail in chapter 19.

3. What is the role of mitochondrial DNA?
4. What features of chloroplasts and mitochondria support the endosymbiotic theory of their evolution?

4.8 The Nucleus and Cell Division

The **nucleus** is by far the most visually prominent organelle in eucaryotic cells. It was discovered early in the study of cell structure and was shown by Robert Brown in 1831 to be a constant feature of eucaryotic cells. The nucleus is the repository for the cell's genetic information and is its control center.

Nuclear Structure

Nuclei are membrane-delimited spherical bodies about 5 to 7 μm in diameter (figures 4.2 and 4.24*b*). Dense fibrous material called **chromatin** can be seen within the nucleoplasm of the nucleus of a stained cell. This is the DNA-containing part of the nucleus. In nondividing cells, chromatin is dispersed, but it condenses during cell division to become visible as **chromosomes.** Some chromatin, the euchromatin, is loosely organized and contains those genes that are actively expressed. In contrast, heterochromatin is coiled more tightly, appears darker in the electron microscope, and is not genetically active most of the time.

The nucleus is bounded by the **nuclear envelope** (figures 4.2 and 4.24*b*), a complex structure consisting of inner and outer membranes separated by a 15 to 75 nm perinuclear space. The envelope is continuous with the ER at several points and its outer membrane is covered with ribosomes. A network of intermediate filaments, called the nuclear lamina, is observed in animal cells. It lies against the inner surface of the envelope and supports it. Chromatin usually is associated with the inner membrane.

Many **nuclear pores** penetrate the envelope (**figure 4.15**), and each pore is formed by a fusion of the outer and inner membranes. Pores are about 70 nm in diameter and collectively occupy about 10 to 25% of the nuclear surface. A complex ringlike arrangement of granular and fibrous material called the annulus is located at the edge of each pore.

The nuclear pores serve as a transport route between the nucleus and surrounding cytoplasm. Particles have been observed moving into the nucleus through the pores. Although the function of the annulus is not understood, it may either regulate or aid the movement of material through the pores. Substances also move directly through the nuclear envelope by unknown mechanisms.

Often the most noticeable structure within the nucleus is the **nucleolus** (**figure 4.16**). A nucleus may contain from one to many nucleoli. Although the nucleolus is not membrane-enclosed, it is a complex organelle with separate granular and fibrillar regions.

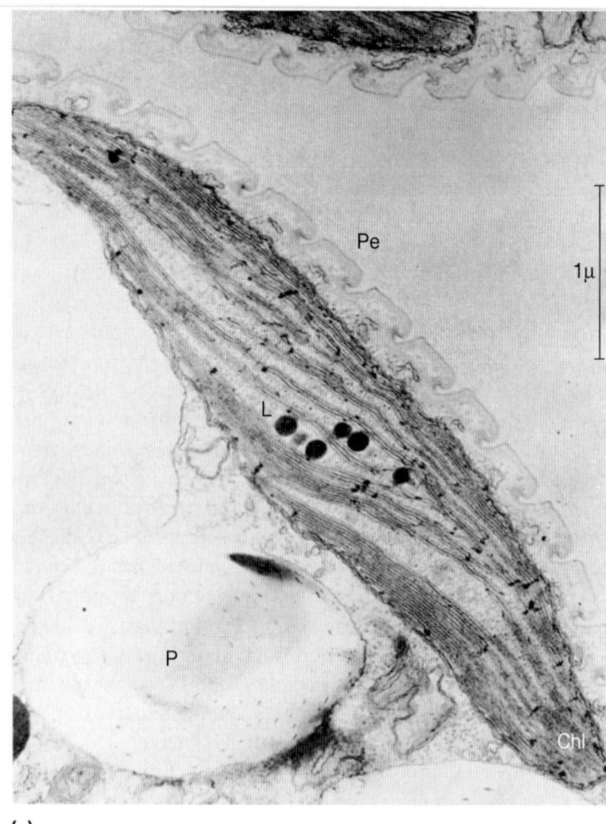

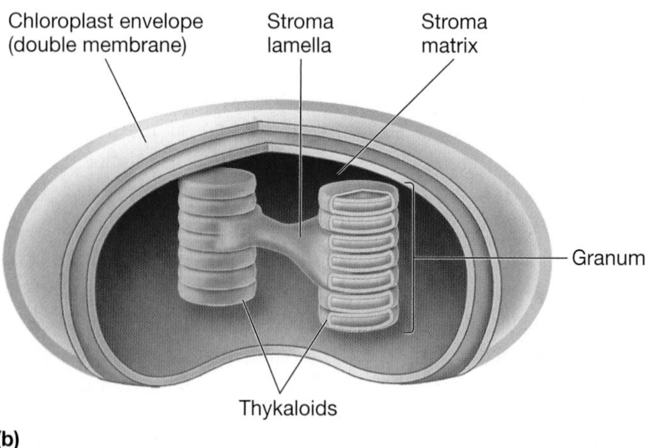

Figure 4.14 Chloroplast Structure. **(a)** The chloroplast (Chl), of the euglenoid flagellate *Colacium cyclopicolum.* The chloroplast is bounded by a double membrane and has its thylakoids in groups of three or more. A paramylon granule (P), lipid droplets (L), and the pellicular strips (Pe) can be seen (×40,000). **(b)** A diagram of chloroplast structure.

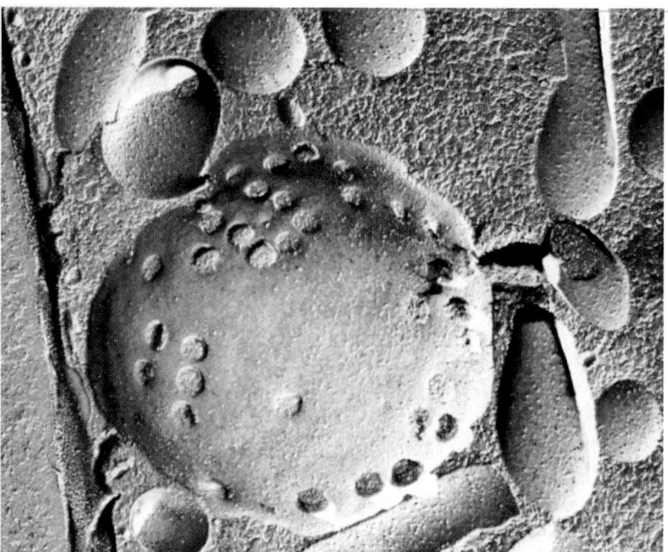

Figure 4.15 The Nucleus. A freeze-etch preparation of the conidium of the fungus *Geotrichum candidum* (×44,600). Note the large, convex nuclear surface with nuclear pores scattered over it.

It is present in nondividing cells, but frequently disappears during mitosis. After mitosis the nucleolus reforms around the nucleolar organizer, a particular part of a specific chromosome.

The nucleolus plays a major role in ribosome synthesis. The nucleolar organizer DNA directs the production of ribosomal RNA (rRNA). This RNA is synthesized in a single long piece that is cut to form the final rRNA molecules. The processed rRNAs next combine with ribosomal proteins (which have been synthesized in the cytoplasmic matrix) to form partially completed ribosomal subunits. The granules seen in the nucleolus are probably these subunits. Immature ribosomal subunits then leave the nucleus, presumably by way of the nuclear envelope pores, and mature in the cytoplasm.

Mitosis and Meiosis

When a eucaryotic microorganism reproduces asexually, its genetic material must be duplicated and then separated so that each new nucleus possesses a complete set of chromosomes. This process of nuclear division and chromosome distribution in eucaryotic cells is called **mitosis.** Mitosis actually occupies only a small portion of a microorganism's life as can be seen by examining the **cell cycle** (**figure 4.17**). The cell cycle is the total sequence of events in the growth-division cycle between the end of one division and the end of the next. Cell growth takes place in the **interphase,** that portion of the cycle between periods of mitosis. Interphase is composed of three parts. The G_1 period (gap 1 period) is a time of active synthesis of RNA, ribosomes, and other cytoplasmic constituents accompanied by considerable cell growth. This is followed by the S period (synthesis period) in which DNA is replicated and doubles in quantity. Finally, there is a second gap, the G_2 period, when the cell prepares for mitosis, the M period, by activities such as the synthesis of special division proteins. The total length of the cycle differs considerably between microorganisms, usually due to variations in the length of G_1.

Mitotic events are summarized in figure 4.17. During mitosis, the genetic material duplicated during the S period is distributed

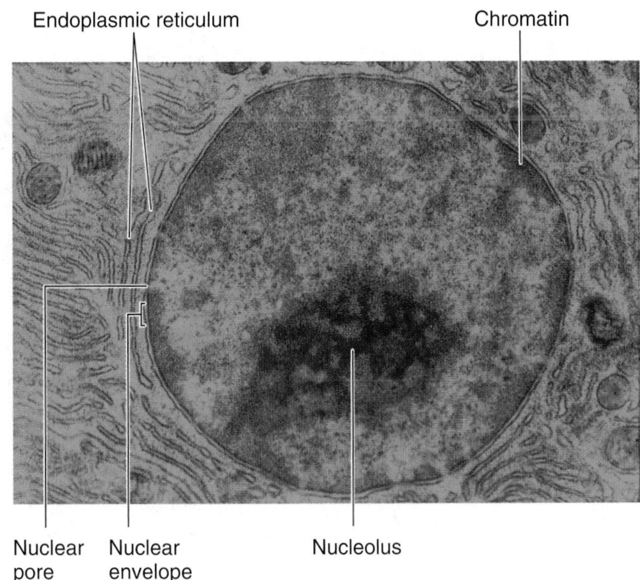

Figure 4.16 The Nucleolus. The nucleolus is a prominent feature of the nucleus. It functions in rRNA synthesis and the assembly of ribosomal subunits. Chromatin, nuclear pores, and the nuclear envelope are also visible in this electron micrograph of an interphase nucleus.

equally to the two new nuclei by cytoskeletal elements so that each has a full set of genes. There are four phases in mitosis. In prophase, the chromosomes—each with two chromatids—become visible and move toward the equator of the cell. The mitotic spindle forms, the nucleolus disappears, and the nuclear envelope begins to dissolve. The chromosomes are arranged in the center of the spindle during metaphase and the nuclear envelope disappears. During anaphase the chromatids in each chromosome separate and move toward the opposite poles of the spindle (**figure 4.18**). Finally during telophase, the chromatids become less visible, the nucleolus reappears, and a nuclear envelope reassembles around each set of chromatids to form two new nuclei. The resulting progeny cells have the same number of chromosomes as the parent. Thus after mitosis, a diploid organism will remain diploid.

Mitosis in some eucaryotic microorganisms can differ from that pictured in figure 4.17. For example, the nuclear envelope does not disappear in many fungi and some protists. Frequently cytokinesis, the division of the parental cell's cytoplasm to form new cells, begins during anaphase and finishes by the end of telophase. However, mitosis can take place without cytokinesis to generate multinucleate or coenocytic cells.

Many microorganisms have a sexual phase in their life cycles (**figure 4.19**). In this phase, they must reduce their chromosome number by half, from the diploid state to the haploid or 1N (a single copy of each chromosome). Haploid cells may immediately act as gametes and fuse to reform diploid organisms or may form gametes only after a considerable delay (figure 4.19). The process

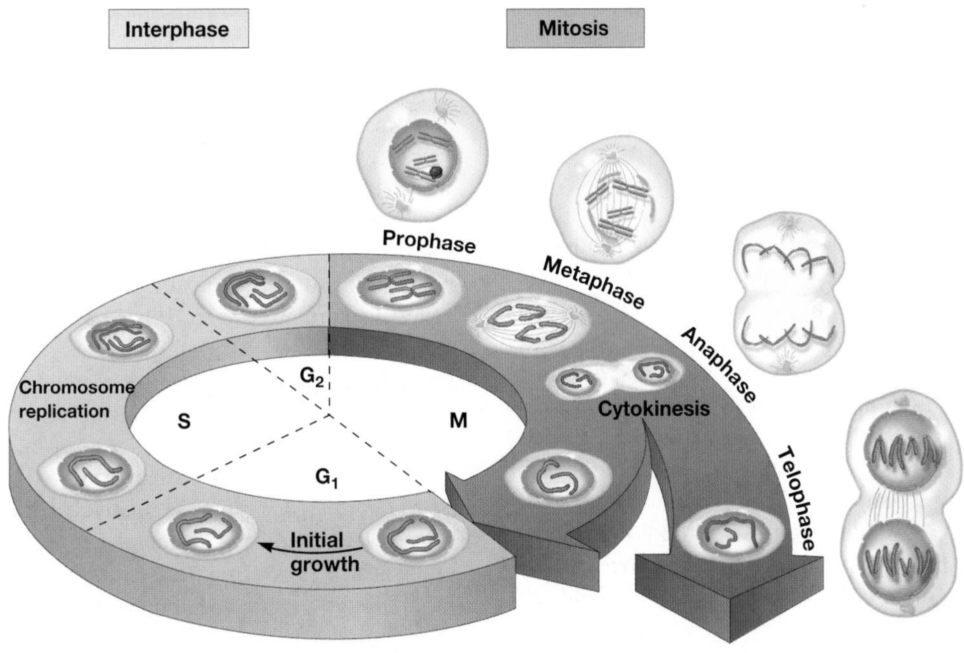

Figure 4.17 The Eucaryotic Cell Cycle. The length of the M period has been increased disproportionately in order to show the phases of mitosis. G_1 period: synthesis of mRNA, tRNA, ribosomes, and cytoplasmic constituents. Nucleolus grows rapidly. S period: rapid synthesis and doubling of nuclear DNA and histones. G_2 period: preparation for mitosis and cell division. M period: mitosis (prophase, metaphase, anaphase, telophase) and cytokinesis.

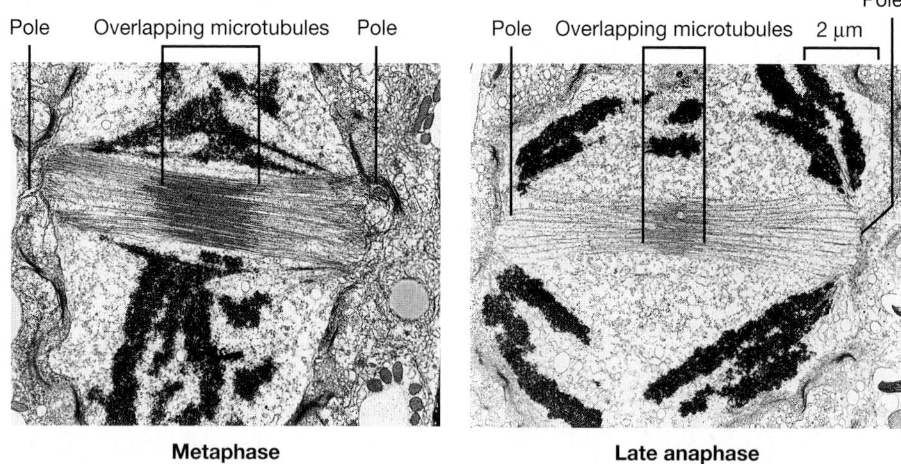

Figure 4.18 Mitosis. In these electron micrographs of dividing diatoms, the overlap of the microtubules lessens markedly during spindle elongation as the cell passes from metaphase to anaphase.

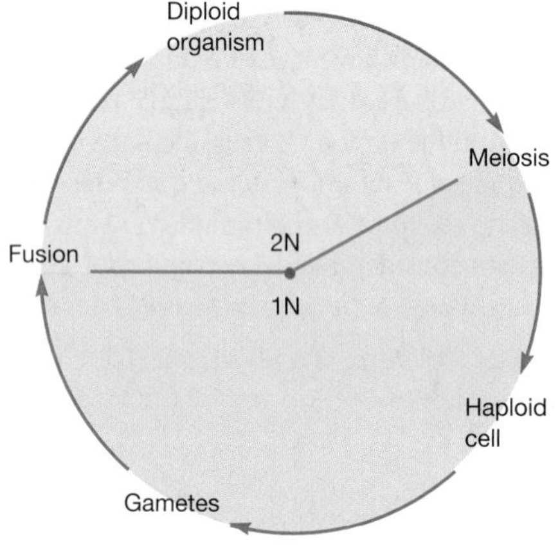

Figure 4.19 Generalized Eucaryotic Life Cycle.

by which the number of chromosomes is reduced in half with each daughter cell receiving one complete set of chromosomes is called **meiosis.** Life cycles can be quite complex in eucaryotic microorganisms and are discussed in more detail in chapters 25 and 26.

Meiosis is quite complex and involves two stages. The first stage differs markedly from mitosis. During prophase, homologous chromosomes come together and lie side-by-side, a process known as synapsis. The homologues move to opposite poles in anaphase, thus reducing the number of chromosomes by half. The second stage of meiosis is similar to mitosis in terms of mechanics, and chromatids of each chromosome are separated. After completion of meiosis I and meiosis II, the original diploid cell has been transformed into four haploid cells.

1. Describe the structure of the nucleus. What are euchromatin and heterochromatin? What is the role of the pores in the nuclear envelope?
2. Briefly discuss the structure and function of the nucleolus. What is the nucleolar organizer?
3. Describe the eucaryotic cell cycle, its periods, and the process of mitosis.
4. What is meiosis, how does it take place, and what is its role in the microbial life cycle?

4.9 External Cell Coverings

Eucaryotic microorganisms differ greatly from procaryotes in the supporting or protective structures they have external to the plasma membrane. In contrast with most bacteria, many eucaryotes lack an external cell wall. The amoeba is an excellent example. Eucaryotic cell membranes, unlike most procaryotic membranes, contain sterols such as cholesterol in their lipid bilayers, and this may make them mechanically stronger, thus reducing the need for external support. Of course many eucaryotes do have a rigid external **cell wall.** The cell walls of photosynthetic protists usually have a layered appearance and contain large quantities of polysaccharides such as cellulose and pectin. In addition, inorganic substances like silica (in diatoms) or calcium carbonate may be present. Fungal cell walls normally are rigid. Their exact composition varies with the organism; usually cellulose, chitin, or glucan (a glucose polymer different from cellulose) are present. Despite their nature the rigid materials in eucaryotic cell walls are chemically simpler than procaryotic peptidoglycan. The bacterial cell wall (section 3.6)

Many protists have a different supportive mechanism, the **pellicle** (figure 4.14*a*). This is a relatively rigid layer of components just beneath the plasma membrane (sometimes the plasma membrane is also considered part of the pellicle). The pellicle may be fairly simple in structure. For example, *Euglena* has a series of overlapping strips with a ridge at the edge of each strip fitting into a groove on the adjacent one. In contrast, the pellicles of ciliate protozoa are exceptionally complex with two membranes and a variety of associated structures. Although pellicles are not as strong and rigid as cell walls, they give their possessors a characteristic shape.

4.10 Cilia and Flagella

Cilia (s., **cilium**) and **flagella** (s., **flagellum**) are the most prominent organelles associated with motility. Although both are whiplike and beat to move the microorganism along, they differ from one another in two ways. First, cilia are typically only 5 to 20 μm in length, whereas flagella are 100 to 200 μm long. Second, their patterns of movement are usually distinctive (**figure 4.20**). Flagella move in an undulating fashion and generate planar or helical waves originating at either the base or the tip. If the wave moves from base to tip, the cell is pushed along; a beat traveling from the tip toward the base pulls the cell through the water. Sometimes the flagellum will have lateral hairs called flimmer filaments (thicker, stiffer hairs are called mastigonemes). These filaments change flagellar action so that a wave moving down the filament toward the tip pulls the cell along instead of pushing it. Such a flagellum often is called a tinsel flagellum, whereas the naked flagellum is referred to as a whiplash flagellum (**figure 4.21**). Cilia, on the other hand, normally have a beat with two distinctive phases. In the effective stroke, the cilium strokes through the surrounding fluid like an oar, thereby propelling the organism along in the water. The cilium next bends along its length while it is pulled forward during the recovery stroke in preparation for another effective stroke. A ciliated microorganism actually coordinates the beats so that some of its cilia are in the recovery phase while others are carrying out their effective stroke (**figure 4.22**). This coordination allows the organism to move smoothly through the water.

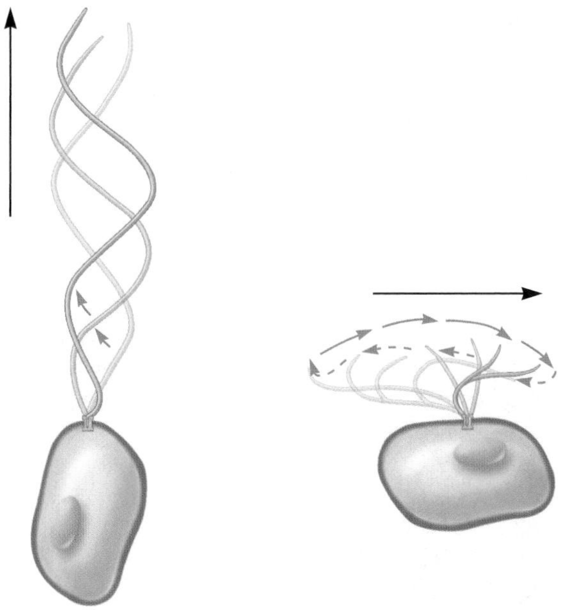

Figure 4.20 Patterns of Flagellar and Ciliary Movement. Flagellar and ciliary movement often takes the form of waves. Flagella (left illustration) move either from the base of the flagellum to its tip or in the opposite direction. The motion of these waves propels the organism along. The beat of a cilium (right illustration) may be divided into two phases. In the effective stroke, the cilium remains fairly stiff as it swings through the water. This is followed by a recovery stroke in which the cilium bends and returns to its initial position. The black arrows indicate the direction of water movement in these examples.

Despite their differences, cilia and flagella are very similar in ultrastructure. They are membrane-bound cylinders about 0.2 μm

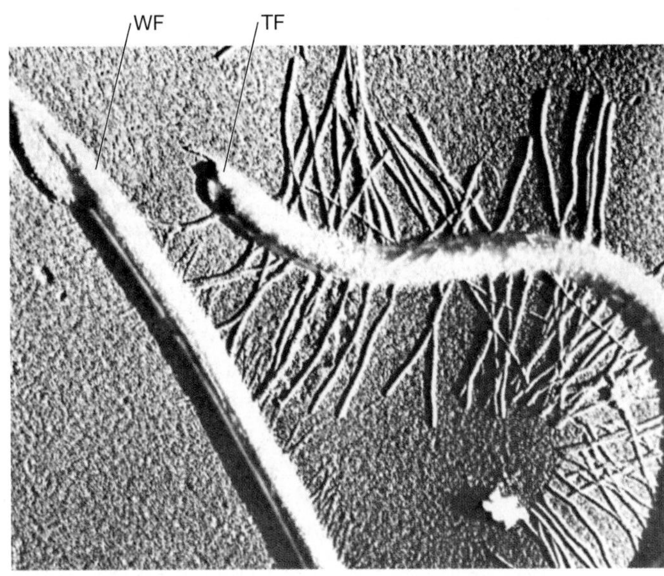

Figure 4.21 Whiplash and Tinsel Flagella. Transmission electron micrograph of a shadowed whiplash flagellum, WF, and a tinsel flagellum, TF, with mastigonemes.

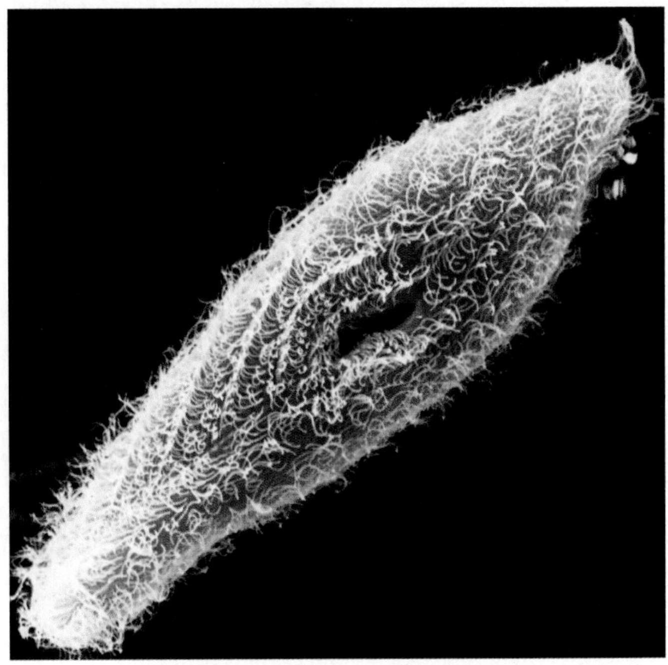

Figure 4.22 Coordination of Ciliary Activity. A scanning electron micrograph of *Paramecium* showing cilia (×1,500). The ciliary beat is coordinated and moves in waves across the protozoan's surface, as can be seen in the photograph.

in diameter. Located in the matrix of the organelle is a complex, the **axoneme,** consisting of nine pairs of microtubule doublets arranged in a circle around two central tubules (**figure 4.23**). This is called the 9 + 2 pattern of microtubules. Each doublet also has pairs of arms projecting from subtubule A (the complete microtubule) toward a neighboring doublet. A radial spoke extends from subtubule A toward the internal pair of microtubules with their central sheath. These microtubules are similar to those found in the cytoplasm. Each is constructed of two types of tubulin subunits, α- and β-tubulins, that resemble the contractile protein actin in their composition. Components external to the cell wall: Flagella and motility (section 3.9)

A **basal body** lies in the cytoplasm at the base of each cilium or flagellum. It is a short cylinder with nine microtubule triplets around its periphery (a 9 + 0 pattern) and is separated from the rest of the organelle by a basal plate. The basal body directs the construction of these organelles. Cilia and flagella appear to grow through the addition of preformed microtubule subunits at their tips.

Cilia and flagella bend because adjacent microtubule doublets slide along one another while maintaining their individual lengths. The doublet arms (figure 4.23), about 15 nm long, are made of the protein **dynein.** ATP powers the movement of cilia and flagella, and isolated dynein hydrolyzes ATP. It appears that dynein arms interact with the B subtubules of adjacent doublets to cause the sliding. The radial spokes also participate in this sliding motion.

Cilia and flagella beat at a rate of about 10 to 40 strokes or waves per second and propel microorganisms rapidly. The record holder is the flagellate *Monas stigmatica,* which swims at a rate of 260 μm/second (approximately 40 cell lengths per second); the common euglenoid flagellate, *Euglena gracilis,* travels at around 170 μm or 3 cell lengths per second. The ciliate protozoan *Paramecium caudatum* swims at about 2,700 μm/second (12 lengths per second). Such speeds are equivalent to or much faster than those seen in higher animals, but not as fast as those in procaryotes.

1. How do eucaryotic microorganisms differ from procaryotes with respect to supporting or protective structures external to the plasma membrane? Describe the pellicle and indicate which microorganisms have one.
2. Prepare and label a diagram showing the detailed structure of a cilium or flagellum. How do cilia and flagella move, and what is dynein's role in the process? Contrast the ways in which flagella and cilia propel microorganisms through water.
3. Compare the structure and mechanism of action of procaryotic and eucaryotic flagella.

4.11 Comparison of Procaryotic and Eucaryotic Cells

A comparison of the cells in **figure 4.24** demonstrates that there are many fundamental differences between eucaryotic and procaryotic cells. **Eucaryotic cells** have a membrane-enclosed nucleus. In

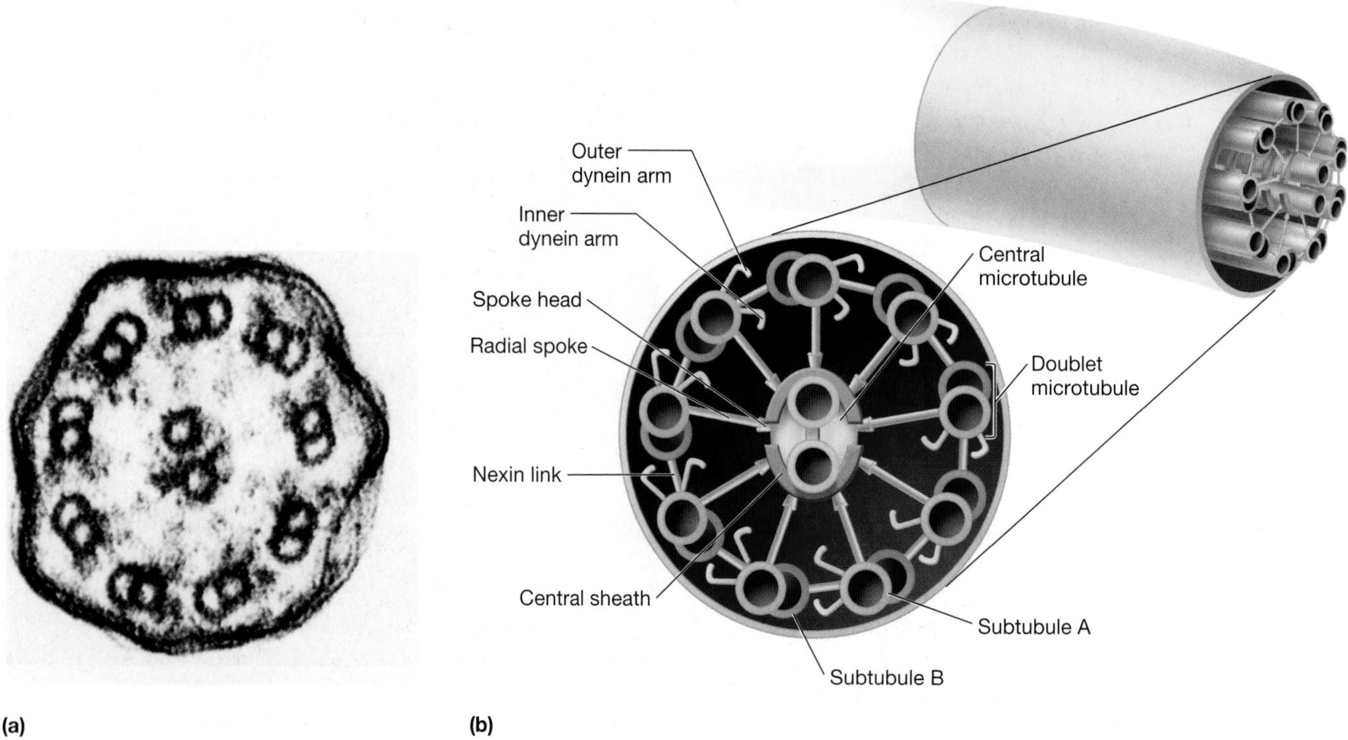

Figure 4.23 Cilia and Flagella Structure. **(a)** An electron micrograph of a cilium cross section. Note the two central microtubules surrounded by nine microtubule doublets (×160,000). **(b)** A diagram of cilia and flagella structure.

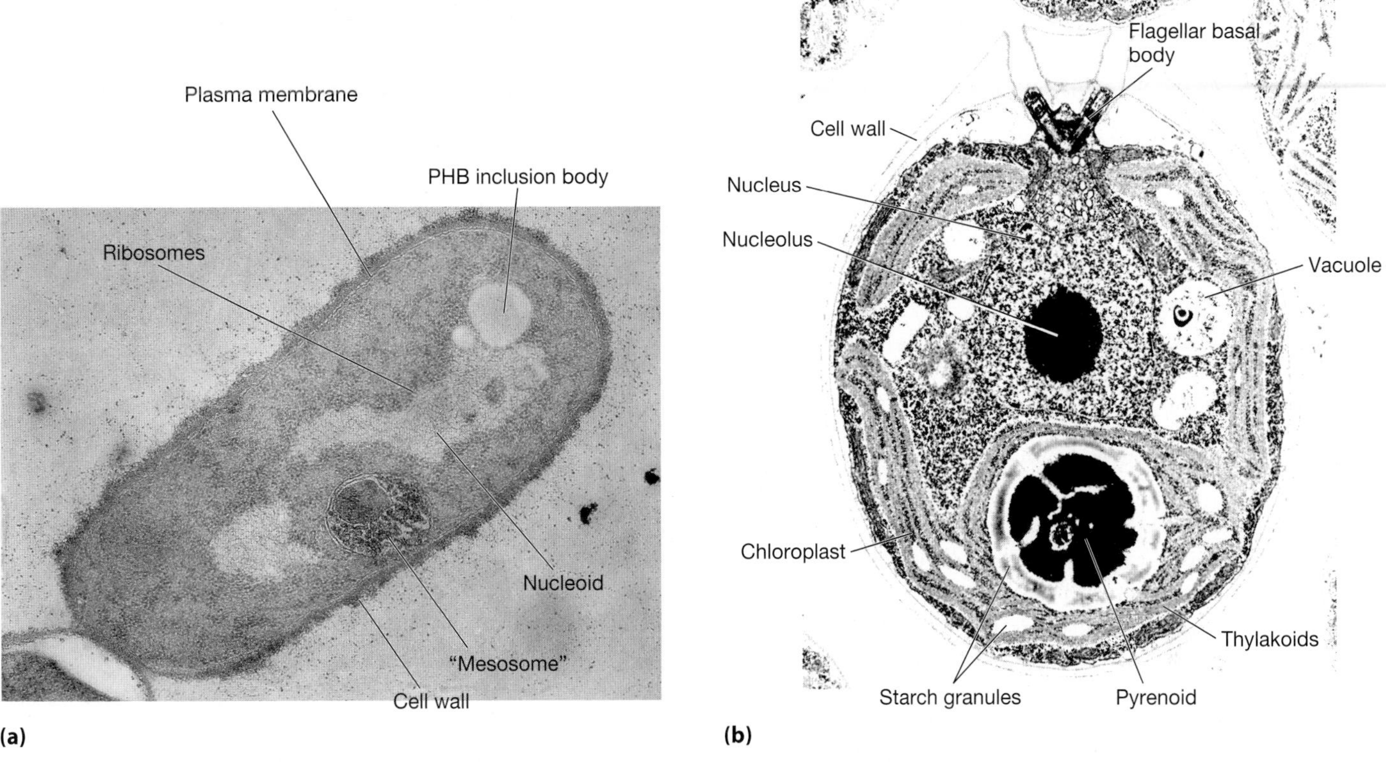

Figure 4.24 Comparison of Procaryotic and Eucaryotic Cell Structure. **(a)** The procaryote *Bacillus megaterium* (×30,500). **(b)** The eucaryotic alga *Chlamydomonas reinhardtii,* a deflagellated cell. Note the large chloroplast with its pyrenoid body (×30,000).

contrast, **procaryotic cells** lack a true, membrane-delimited nucleus. *Bacteria* and *Archaea* are procaryotes; all other organisms—fungi, protists, plants, and animals—are eucaryotic. Most procaryotes are smaller than eucaryotic cells, often about the size of eucaryotic mitochondria and chloroplasts.

The presence of the eucaryotic nucleus is the most obvious difference between these two cell types, but many other major distinctions exist. It is clear from **table 4.2** that procaryotic cells are much simpler structurally. In particular, an extensive and diverse collection of membrane-delimited organelles is missing. Furthermore, procaryotes are simpler functionally in several ways. They lack mitosis and meiosis, and have a simpler genetic organization. Many complex eucaryotic processes are absent in procaryotes: endocytosis, intracellular digestion, directed cytoplasmic streaming, and ameboid movement, are just a few.

Despite the many significant differences between these two basic cell forms, they are remarkably similar on the biochemical level as we discuss in succeeding chapters. Procaryotes and eucaryotes are composed of similar chemical constituents. With a few exceptions, the genetic code is the same in both, as is the way in which the genetic information in DNA is expressed. The principles underlying metabolic processes and many important metabolic pathways are identical. Thus beneath the profound structural and functional differences between procaryotes and eucaryotes, there is an even more fundamental unity: a molecular unity that is basic to all known life processes.

1. Outline the major differences between procaryotes and eucaryotes. How are they similar?
2. What characteristics make *Archaea* more like eucaryotes? What features make them more like *Bacteria?*

Table 4.2 Comparison of Procaryotic and Eucaryotic Cells

	Procaryotes		Eucaryotes
Property	***Bacteria***	***Archaea***	***Eukarya***
Organization of Genetic Material			
True membrane-bound nucleus	No	No	Yes
DNA complexed with histones	No	Some	Yes
Chromosomes	Usually one circular chromosome	Usually one circular chromosome	More than one; chromosomes are linear
Plasmids	Very common	Very common	Rare
Introns in genes	No	No	Yes
Nucleolus	No	No	Yes
Mitochondria	No	No	Yes
Chloroplasts	No	No	Yes
Plasma Membrane Lipids	Ester-linked phospholipids and hopanoids; some have sterols	Glycerol diethers and diglycerol tetraethers; some have sterols	Ester-linked phospholipids and sterols
Flagella	Submicroscopic in size; composed of one protein fiber	Submicroscopic in size; composed of one protein fiber	Microscopic in size; membrane bound; usually 20 microtubules in 9 + 2 pattern
Endoplasmic Reticulum	No	No	Yes
Golgi Apparatus	No	No	Yes
Peptidoglycan in Cell Walls	Yes	No	No
Ribosome Size	70S	70S	80S
Lysosomes	No	No	Yes
Cytoskeleton	Rudimentary	Rudimentary	Yes
Gas Vesicles	Yes	Yes	No

Summary

4.1 An Overview of Eucaryotic Cell Structure

a. The eucaryotic cell has a true, membrane-delimited nucleus and many membranous organelles (**table 4.1; figure 4.2**).

b. The membranous organelles compartmentalize the cytoplasm of the cell. This allows the cell to carry out a variety of biochemical reactions simultaneously. It also provides more surface area for membrane-associated activities such as respiration.

4.2 The Plasma Membrane and Membrane Structure

a. Eucaryotic membranes are similar in structure and function to those of bacteria. The two differ in terms of their lipid composition.

b. Eucaryotic membranes contain microdomains called lipid rafts. They are enriched for certain lipids and proteins and participate in a variety of cellular processes.

4.3 The Cytoplasmic Matrix, Microfilaments, Intermediate Filaments, and Microtubules

a. The cytoplasmic matrix contains microfilaments, intermediate filaments, and microtubules, small organelles partly responsible for cell structure and movement. These and other types of filaments are organized into a cytoskeleton (**figure 4.5**).

b. Microfilaments and microtubules have been observed in eucaryotic microbes. Microfilaments are composed of actin proteins; microtubules are composed of α-tubulin and β-tubulin.

c. Intermediate filaments are assembled from a heterogeneous family of proteins. They have not been identified or studied in eucaryotic microbes.

4.4 Organelles of the Biosynthetic-Secretory and Endocytic Pathways

a. The cytoplasmic matrix is permeated with a complex of membranous organelles and vesicles. Some are involved in the synthesis and secretion of materials (biosynthetic-secretory pathway). Some are involved in the uptake of materials from the extracellular millieux (endocytic pathway).

b. The endoplasmic reticulum (ER) is an irregular network of tubules and flattened sacs (cisternae). The ER may have attached ribosomes and may be active in protein synthesis (rough endoplasmic reticulum), or it may lack ribosomes (smooth ER) (**figure 4.7**).

c. The ER can donate materials to the Golgi apparatus, an organelle composed of one or more stacks of cisternae (**figure 4.8**). This organelle prepares and packages cell products for secretion.

d. The Golgi apparatus also buds off vesicles that deliver hydrolytic enzymes and other proteins to lysosomes. Lysosomes are organelles that contain digestive enzymes and aid in intracellular digestion of extracellular materials delivered to them by endocytosis (**figure 4.10**).

e. Eucaryotes ingest materials using several kinds of endocytosis. These include phagocytosis, clathrin-dependent endocytosis, and caveolae-dependent endocytosis. Some macromolecules are bound to receptors prior to endocytosis in a process called receptor-mediated endocytosis.

4.5 Eucaryotic Ribosomes

a. Eucaryotic ribosomes are either found free in the cytoplasmic matrix or bound to the ER.

b. Eucaryotic ribsomes are 80S in size.

4.6 Mitochondria

a. Mitochondria are organelles bounded by two membranes, with the inner membrane folded into cristae (**figure 4.11**).

b. Mitochondria are responsible for energy generation by the tricarboxylic acid cycle, electron transport, and oxidative phosphorylation.

4.7 Chloroplasts

a. Chloroplasts are pigment-containing organelles that serve as the site of photosynthesis (**figure 4.14**).

b. The trapping of light energy takes place in the thylakoid membranes of the chloroplast, whereas CO_2 fixation occurs in the stroma.

4.8 The Nucleus and Cell Division

a. The nucleus is a large organelle containing the cell's chromosomes. It is bounded by a complex, double-membrane envelope perforated by pores through which materials can move (**figure 4.15**).

b. The nucleolus lies within the nucleus and participates in the synthesis of ribosomal RNA and ribosomal subunits (**figure 4.16**).

c. Eucaryotic chromosomes are distributed to daughter cells during cell division by mitosis (**figure 4.18**). Meiosis is used to halve the chromosome number during sexual reproduction.

4.9 External Cell Coverings

a. When a cell wall is present, it is constructed from polysaccharides, like cellulose, that are chemically simpler than peptidoglycan, the molecule found in bacterial cell walls.

b. Many protozoa have a pellicle rather than a cell wall.

4.10 Cilia and Flagella

a. Many eucaryotic cells are motile because of cilia and flagella, membrane-delimited organelles with nine microtubule doublets surrounding two central microtubules (**figure 4.23**).

b. The cell moves when the microtubule doublets slide along each other, causing the cilium or flagellum to bend.

4.11 Comparison of Procaryotic and Eucaryotic Cells

a. Despite the fact that eucaryotes and procaryotes differ structurally in many ways (**table 4.2**), they are quite similar biochemically.

Key Terms

autophagosome 88
autophagy 87
axoneme 96
basal body 96
biosynthetic-secretory pathway 86
caveolae 86
caveolae-dependent endocytosis 86
cell cycle 92
cell wall 94
chloroplast 90
chromatin 91
chromosome 91
cilia 95
cisternae 84
clathrin 86
clathrin-dependent endocytosis 86
coated vesicles 87
cristae 89
cytoplasmic matrix 83
cytoskeleton 83
dictyosome 85
dynein 96
early endosome 87
endocytic pathway 86
endocytosis 86
endoplasmic reticulum (ER) 84
eucaryotic cells 96
flagella 95
Golgi apparatus 85
grana 90
intermediate filament 83
interphase 92
late endosomes 87
lipid raft 81
lysosome 86
meiosis 94
microfilament 83
microtubule 83
mitochondrion 88
mitosis 92
nuclear envelope 91
nuclear pores 91
nucleolus 91
nucleus 91
organelle 79
pellicle 94
phagocytosis 86
phagosomes 86
plastid 90
procaryotic cells 97
pyrenoid 90
receptor-mediated endocytosis 86
residual body 88
rough endoplasmic reticulum (RER) 85
smooth endoplasmic reticulum (SER) 85
stroma 90
thylakoid 90
26S proteasome 86

Critical Thinking Question

1. Discuss the statement: "The most obvious difference between eucaryotic and procaryotic cells is in their use of membranes." What general roles do membranes play in eucaryotic cells?

Learn more:

Alberts, B.; Johnson, A.; Lewis, J.; Raff, M.; Roberts, K.; and Walter, P. 2002. *Molecular biology of the cell,* 4th ed. New York: Garland Science.

Lee, M. C. S.; Miller, E. A.; Goldberg, J.; Orci, L.; and Schekman, R. 2004. Bidirectional protein transport between the ER and Golgi. *Annu. Rev. Cell Dev. Biol.* 20:87–123.

Lodish, H.; Berk, A.; Matsudaira, P.; Kaiser, C. A.; Krieger, M.; Scott, M. P.; Zipursky, S. L.; and Darnell, J. 2004. *Molecular cell biology,* 5th ed. New York: W. H. Freeman.

McCollum, D. 2002. Coordinating cytokinesis and nuclear division in *S. pombe. ASM News* 68(7):325–29.

McConville, M. J.; Mullin, K. A.; Ilgoutz, S. C.; and Teasdale, R. D. 2002. Secretory pathway of trypanosomatid parasites. *Microbiol. Mol. Biol. Rev.* 66(1):122–54.

Shin, J.-S., and Abraham, S. N. 2001. Caveolae—Not just craters in the cellular landscape. *Science* 293:1447–48.

Steigmeier, F., and Amon. A. 2004. Closing mitosis: The functions of the CDC14 phosphatase and its regulation. *Annu. Rev. Genet.* 38:203–32.

Please visit the Prescott website at www.mhhe.com/prescott7 for additional references.

5 Microbial Nutrition

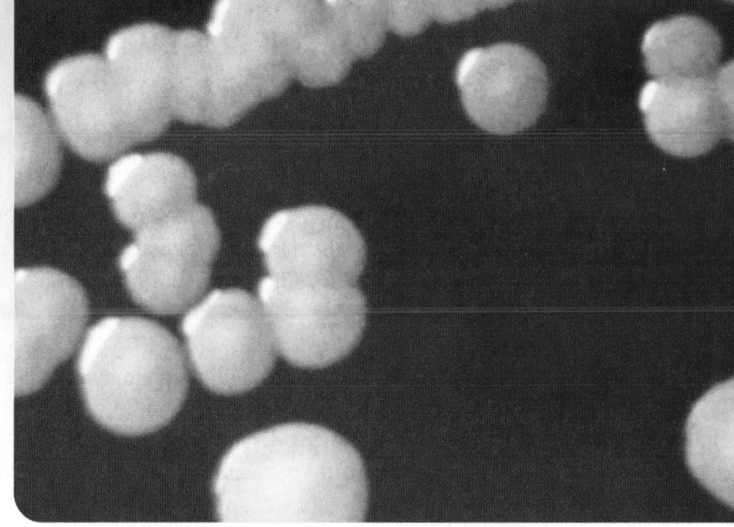

Staphylococcus aureus forms large, golden colonies when growing on blood agar. This human pathogen causes diseases such as boils, abscesses, bacteremia, endocarditis, food poisoning, pharyngitis, and pneumonia.

Preview

- Microorganisms require about 10 elements in large quantities for the synthesis of macromolecules. Several other elements are needed in very small amounts and are parts of enzymes and cofactors.
- All microorganisms can be placed in one of a few nutritional categories on the basis of their requirements for carbon, energy, and electrons.
- Most nutrient molecules must be transported through the plasma membrane by one of three major mechanisms involving the use of membrane carrier proteins. Eucaryotic microorganisms also employ endocytosis for nutrient uptake.
- Culture media are needed to grow microorganisms in the laboratory and to carry out specialized procedures like microbial identification, water and food analysis, and the isolation of specific microorganisms. Many different media are available for these and other purposes.
- Pure cultures can be obtained through the use of spread plates, streak plates, or pour plates and are required for the careful study of an individual microbial species.

As discussed in chapters 3 and 4, microbial cells are structurally complex and carry out numerous functions. In order to construct new cellular components and do cellular work, organisms must have a supply of raw materials or nutrients and a source of energy. **Nutrients** are substances used in biosynthesis and energy release and therefore are required for microbial growth. In this chapter we describe the nutritional requirements of microorganisms, how nutrients are acquired, and the cultivation of microorganisms.

5.1 The Common Nutrient Requirements

Analysis of microbial cell composition shows that over 95% of cell dry weight is made up of a few major elements: carbon, oxygen, hydrogen, nitrogen, sulfur, phosphorus, potassium, calcium, magnesium, and iron. These are called **macroelements** or macronutrients because they are required by microorganisms in relatively large amounts. The first six (C, O, H, N, S, and P) are components of carbohydrates, lipids, proteins, and nucleic acids. The remaining four macroelements exist in the cell as cations and play a variety of roles. For example, potassium (K^+) is required for activity by a number of enzymes, including some of those involved in protein synthesis. Calcium (Ca^{2+}), among other functions, contributes to the heat resistance of bacterial endospores. Magnesium (Mg^{2+}) serves as a cofactor for many enzymes, complexes with ATP, and stabilizes ribosomes and cell membranes. Iron (Fe^{2+} and Fe^{3+}) is a part of cytochromes and a cofactor for enzymes and electron-carrying proteins.

In addition to macroelements, all microorganisms require several nutrients in small amounts. These are called **micronutrients** or **trace elements.** The micronutrients—manganese, zinc, cobalt, molybdenum, nickel, and copper—are needed by most cells. However, cells require such small amounts that contaminants from water, glassware, and regular media components often are adequate for growth. In nature, micronutrients are ubiquitous and probably do not usually limit growth. Micronutrients are normally a part of enzymes and cofactors, and they aid in the catalysis of reactions and maintenance of protein structure. For example, zinc (Zn^{2+}) is present at the active site of some enzymes but can also be involved in the association of regulatory and catalytic subunits

The whole of nature, as has been said, is a conjugation of the verb to eat, in the active and passive.

—*William Ralph Inge*

(e.g., *E. coli* aspartate carbamoyltransferase). Manganese (Mn^{2+}) aids many enzymes that catalyze the transfer of phosphate groups. Molybdenum (Mo^{2+}) is required for nitrogen fixation, and cobalt (Co^{2+}) is a component of vitamin B_{12}. Enzymes (section 8.7); Control of protein activity (section 8.10)

Besides the common macroelements and trace elements, microorganisms may have particular requirements that reflect their specific morphology or environment. Diatoms need silicic acid (H_4SiO_4) to construct their beautiful cell walls of silica [$(SiO_2)_n$]. Although most procaryotes do not require large amounts of sodium, many archaea growing in saline lakes and oceans depend on the presence of high concentrations of sodium ion (Na^+). Protist classification: *Stramenopiles* (section 25.6); Phylum *Euryarchaeota:* The Halobacteria (section 20.3)

Finally, it must be emphasized that microorganisms require a balanced mixture of nutrients. If an essential nutrient is in short supply, microbial growth will be limited regardless of the concentrations of other nutrients.

5.2 Requirements for Carbon, Hydrogen, Oxygen, and Electrons

All organisms need carbon, hydrogen, oxygen, and a source of electrons. Carbon is needed for the skeletons or backbones of all the organic molecules from which organisms are built. Hydrogen and oxygen are also important elements found in organic molecules. Electrons are needed for two reasons. As will be described more completely in chapter 9, the movement of electrons through electron transport chains and during other oxidation-reduction reactions can provide energy for use in cellular work. Electrons also are needed to reduce molecules during biosynthesis (e.g., the reduction of CO_2 to form organic molecules).

The requirements for carbon, hydrogen, and oxygen often are satisfied together because molecules serving as carbon sources often contribute hydrogen and oxygen as well. For instance, many **heterotrophs**—organisms that use reduced, preformed organic molecules as their carbon source—can also obtain hydrogen, oxygen, and electrons from the same molecules. Because the electrons provided by these organic carbon sources can be used in electron transport as well as in other oxidation-reduction reactions, many heterotrophs also use their carbon source as an energy source. Indeed, the more reduced the organic carbon source (i.e., the more electrons it carries), the higher its energy content. Thus lipids have a higher energy content than carbohydrates. However, one carbon source, carbon dioxide (CO_2), supplies only carbon and oxygen, so it cannot be used as a source of hydrogen, electrons, or energy. This is because CO_2 is the most oxidized form of carbon, lacks hydrogen, and is unable to donate electrons during oxidation-reduction reactions. Organisms that use CO_2 as their sole or principal source of carbon are called **autotrophs.** Because CO_2 cannot supply their energy needs, they must obtain energy from other sources, such as light or reduced inorganic molecules.

A most remarkable nutritional characteristic of heterotrophic microorganisms is their extraordinary flexibility with respect to carbon sources. Laboratory experiments indicate that there is no naturally occurring organic molecule that cannot be used by some microorganism. Actinomycetes, common soil bacteria, will degrade amyl alcohol, paraffin, and even rubber. Some bacteria seem able to employ almost anything as a carbon source; for example, *Burkholderia cepacia* can use over 100 different carbon compounds. Microbes can degrade even relatively indigestible human-made substances such as pesticides. This is usually accomplished in complex microbial communities. These molecules sometimes are degraded in the presence of a growth-promoting nutrient that is metabolized at the same time—a process called cometabolism. Other microorganisms can use the products of this breakdown process as nutrients. In contrast to these bacterial omnivores, some microbes are exceedingly fastidious and catabolize only a few carbon compounds. Cultures of methylotrophic bacteria metabolize methane, methanol, carbon monoxide, formic acid, and related one-carbon molecules. Parasitic members of the genus *Leptospira* use only long-chain fatty acids as their major source of carbon and energy. Biodegradation and bioremediation by natural communities (section 41.6)

1. What are nutrients? On what basis are they divided into macroelements and trace elements?
2. What are the six most important macroelements? How do cells use them?
3. List two trace elements. How do cells use them?
4. Define heterotroph and autotroph.

5.3 Nutritional Types of Microorganisms

Because the need for carbon, energy, and electrons is so important, biologists use specific terms to define how these requirements are fulfilled. We have already seen that microorganisms can be classified as either heterotrophs or autotrophs with respect to their preferred source of carbon (**table 5.1**). There are only two sources of energy available to organisms: (1) light energy, and (2) the energy derived from oxidizing organic or inorganic molecules.

Table 5.1 Sources of Carbon, Energy, and Electrons

Carbon Sources	
Autotrophs	CO_2 sole or principal biosynthetic carbon source (section 10.3)
Heterotrophs	Reduced, preformed, organic molecules from other organisms (*chapters 9 and 10*)
Energy Sources	
Phototrophs	Light (section 9.12)
Chemotrophs	Oxidation of organic or inorganic compounds (*chapter 9*)
Electron Sources	
Lithotrophs	Reduced inorganic molecules (section 9.11)
Organotrophs	Organic molecules (chapter 9)

Phototrophs use light as their energy source; **chemotrophs** obtain energy from the oxidation of chemical compounds (either organic or inorganic). Microorganisms also have only two sources for electrons. **Lithotrophs** (i.e., "rock-eaters") use reduced inorganic substances as their electron source, whereas **organotrophs** extract electrons from reduced organic compounds.

Despite the great metabolic diversity seen in microorganisms, most may be placed in one of five nutritional classes based on their primary sources of carbon, energy, and electrons (**table 5.2**). The majority of microorganisms thus far studied are either photolithotrophic autotrophs or chemoorganotrophic heterotrophs.

Photolithotrophic autotrophs (often called **photoautotrophs** or photolithoautotrophs) use light energy and have CO_2 as their carbon source. Photosynthetic protists and cyanobacteria employ water as the electron donor and release oxygen (**figure 5.1*a***). Other photolithoautotrophs, such as the purple and green sulfur bacteria (figure 5.1*b,c*), cannot oxidize water but extract electrons from inorganic donors like hydrogen, hydrogen sulfide, and elemental sulfur. **Chemoorganotrophic heterotrophs** (often called **chemoheterotrophs,** chemoorganoheterotrophs, or just heterotrophs) use organic compounds as sources of energy, hydrogen, electrons, and carbon. Frequently the same organic nutrient will satisfy all these requirements. Essentially all pathogenic microorganisms are chemoheterotrophs.

The other nutritional classes have fewer known microorganisms but often are very important ecologically. Some photosynthetic bacteria (purple and green bacteria) use organic matter as their electron donor and carbon source. These **photoorganotrophic heterotrophs** (photoorganoheterotrophs) are common inhabitants of polluted lakes and streams. Some of these bacteria also can grow as photoautotrophs with molecular hydrogen as an electron donor. **Chemolithotrophic autotrophs** (chemolithoautotrophs), oxidize reduced inorganic compounds such as iron, nitrogen, or sulfur molecules to derive both energy and electrons for biosynthesis (**figure 5.2*a***). Carbon dioxide is the carbon source. **Chemolithoheterotrophs,** also known as **mixotrophs** (figure 5.2*b*), use reduced inorganic molecules as their energy and electron source, but derive their carbon from organic sources. Chemolithotrophs contribute greatly to the chemical transformations of elements (e.g., the conversion of ammonia to nitrate or sulfur to sulfate) that continually occur in ecosystems. Photosynthetic bacteria (section 21.3); Class *Alphaproteobacteria:* Nitrifying bacteria (section 22.1)

Although a particular species usually belongs in only one of the nutritional classes, some show great metabolic flexibility and alter their metabolic patterns in response to environmental changes. For example, many purple nonsulfur bacteria act as photoorganotrophic heterotrophs in the absence of oxygen but oxidize organic molecules and function chemoorganotrophically at normal oxygen levels. When oxygen is low, photosynthesis and chemoorganotrophic metabolism may function simultaneously. This sort of flexibility seems complex and confusing, yet it gives these microbes a definite advantage if environmental conditions frequently change.

1. Discuss the ways in which microorganisms are classified based on their requirements for energy, carbon, and electrons.
2. Describe the nutritional requirements of the major nutritional groups and give some microbial examples of each.

Table 5.2 Major Nutritional Types of Microorganisms

Nutritional Type	Carbon Source	Energy Source	Electron Source	Representative Microorganisms
Photolithoautotrophy (photolithotrophic autotrophy)	CO_2	Light	Inorganic e^- donor	Purple and green sulfur bacteria, cyanobacteria
Photoorganoheterotrophy (photoorganotrophic heterotrophy)	Organic carbon, but CO_2 may also be used	Light	Organic e^- donor	Purple nonsulfur bacteria, green nonsulfur bacteria
Chemolithoautotrophy (chemolithotrophic autotrophy)	CO_2	Inorganic chemicals	Inorganic e^- donor	Sulfur-oxidizing bacteria, hydrogen-oxidizing bacteria, methanogens, nitrifying bacteria, iron-oxidizing bacteria
Chemolithoheterotrophy or mixotrophy (chemolithotrophic heterotrophy)	Organic carbon, but CO_2 may also be used	Inorganic chemicals	Inorganic e^- donor	Some sulfur-oxidizing bacteria (e.g., *Beggiatoa*)
Chemoorganoheterotrophy (chemoorganotrophic heterotrophy)	Organic carbon	Organic chemicals often same as C source	Organic e^- donor, often same as C source	Most nonphotosynthetic microbes, including most pathogens, fungi, many protists, and many archaea

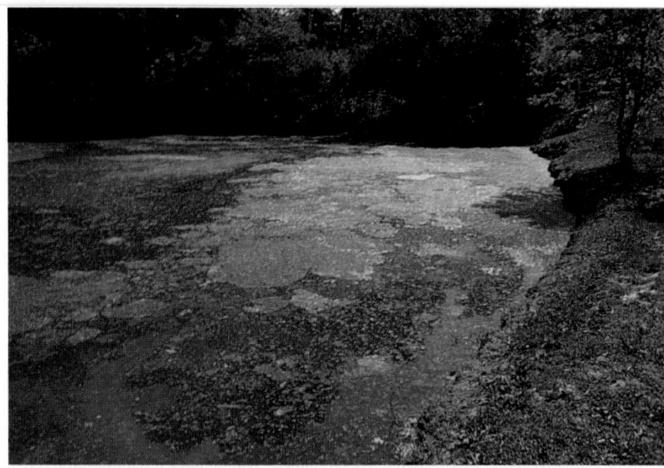

(a) Bloom of cyanobacteria (photolithoautotrophic bacteria)

(b) Purple sulfur bacteria (photoheterotrophs)

(c) Purple sulfur bacteria

Figure 5.1 Phototrophic Bacteria. Phototrophic bacteria play important roles in aquatic ecosystems, where they can cause blooms. **(a)** A cyanobacterial and an algal bloom in a eutrophic pond. **(b)** Purple sulfur bacteria growing in a bog. **(c)** A bloom of purple sulfur bacteria in a sewage lagoon.

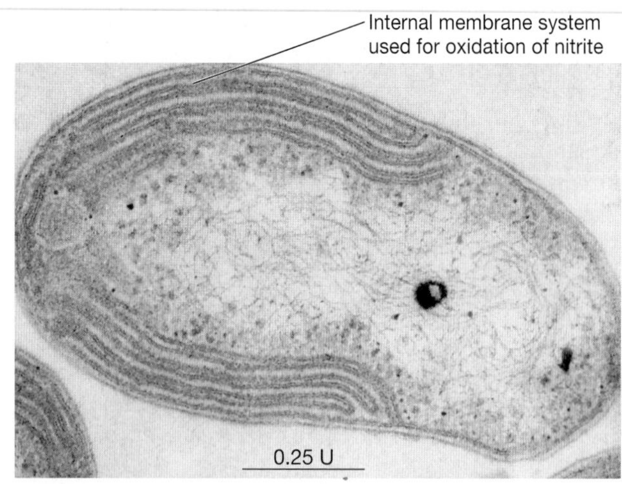

(a) *Nitrobacter winogradskyi*, a chemolithoautotroph

(b) *Beggiatoa alba*, a chemolithoheterotroph (mixotroph)

Figure 5.2 Chemolithotrophic Bacteria. **(a)** Transmission electron micrograph of *Nitrobacter winogradskyi*, an organism that uses nitrite as its source of energy (×213,000). **(b)** Light micrograph of *Beggiatoa alba*, an organism that uses hydrogen sulfide as its energy source and organic molecules as carbon sources. The dark spots within the filaments are granules of elemental sulfur produced when hydrogen sulfide is oxidized.

5.4 Requirements for Nitrogen, Phosphorus, and Sulfur

To grow, a microorganism must be able to incorporate large quantities of nitrogen, phosphorus, and sulfur. Although these elements may be acquired from the same nutrients that supply carbon, microorganisms usually employ inorganic sources as well.

Nitrogen is needed for the synthesis of amino acids, purines, pyrimidines, some carbohydrates and lipids, enzyme cofactors, and other substances. Many microorganisms can use the nitrogen in amino acids. Others can incorporate ammonia directly through the action of enzymes such as glutamate dehydrogenase or glutamine synthetase and glutamate synthase (*see figures 10.11 and 10.12*). Most phototrophs and many chemotrophic microorganisms reduce nitrate to ammonia and incorporate the ammonia in a process known as assimilatory nitrate reduction (*see p. 235*). A variety of bacteria (e.g., many cyanobacteria and the symbiotic bacterium *Rhizobium*) can assimilate atmospheric nitrogen (N_2) by reducing it to ammonium (NH_4^+). This is called nitrogen fixation. Synthesis of amino acids (section 10.5)

Phosphorus is present in nucleic acids, phospholipids, nucleotides like ATP, several cofactors, some proteins, and other cell components. Almost all microorganisms use inorganic phosphate as their phosphorus source and incorporate it directly. Low phosphate levels actually limit microbial growth in many aquatic environments. Some microbes, such as *Escherichia coli,* can use both organic and inorganic phosphate. Some organophosphates such as hexose 6-phosphates can be taken up directly by the cell. Other organophosphates are hydrolyzed in the periplasm by the enzyme alkaline phosphatase to produce inorganic phosphate, which then is transported across the plasma membrane. Synthesis of purines, pyrimidines, and nucleotides (section 10.6)

Sulfur is needed for the synthesis of substances like the amino acids cysteine and methionine, some carbohydrates, biotin, and thiamine. Most microorganisms use sulfate as a source of sulfur and reduce it by assimilatory sulfate reduction; a few microorganisms require a reduced form of sulfur such as cysteine.

1. Briefly describe how microorganisms use the various forms of nitrogen, phosphorus, and sulfur.
2. Why do you think ammonia (NH_3) can be directly incorporated into amino acids while other forms of combined nitrogen (e.g., NO_2^- and NO_3^-) are not?

5.5 Growth Factors

Some microorganisms have the enzymes and biochemical pathways needed to synthesize all cell components using minerals and sources of energy, carbon, nitrogen, phosphorus, and sulfur. Other microorganisms lack one or more of the enzymes needed to manufacture indispensable constituents. Therefore they must obtain these constituents or their precursors from the environment. Organic compounds that are essential cell components or precursors of such components but cannot be synthesized by the organism are called **growth factors.** There are three major classes of growth factors: (1) amino acids, (2) purines and pyrimidines, and (3) vitamins. Amino acids are needed for protein synthesis; purines and pyrimidines for nucleic acid synthesis. **Vitamins** are small organic molecules that usually make up all or part of enzyme cofactors and are needed in only very small amounts to sustain growth. The functions of selected vitamins, and examples of microorganisms requiring them, are given in **table 5.3.** Some microorganisms require many vitamins; for example, *Enterococcus faecalis* needs eight different vitamins for growth. Other growth factors are also seen; heme (from hemoglobin or cytochromes) is required by *Haemophilus influenzae,* and some mycoplasmas need cholesterol. Enzymes (section 8.7)

Understanding the growth factor requirements of microbes has important practical applications. Both microbes with known, specific requirements and those that produce large quantities of a substance (e.g., vitamins) are useful. Microbes with a specific growth factor requirement can be used in bioassays for the factor they need. A typical assay is a growth-response assay, which allows the amount of growth factor in a solution to be determined. These assays are based on the observation that the amount of growth in a culture is related to the amount of growth factor present. Ideally, the amount of growth is directly proportional to the amount of growth factor; if the growth factor concentration doubles the amount of microbial growth doubles. For example, species from the bacterial genera *Lactobacillus* and *Streptococcus* can be used in microbiological assays of most vitamins and amino acids. The appropriate bacterium is grown in a series of culture vessels, each containing medium with an excess amount of all required components except the growth factor to be assayed. A different amount of growth factor is added to each vessel. The standard curve is prepared by plotting the growth factor quantity or concentration against the total extent of bacterial growth. The quantity of the growth factor in a test sample is determined by comparing the extent of growth caused by the unknown sample with that resulting from the standards. Microbiological assays are specific, sensitive, and simple. They still are used in the assay of substances like vitamin B_{12} and biotin, despite advances in chemical assay techniques.

On the other hand, those microorganisms able to synthesize large quantities of vitamins can be used to manufacture these compounds for human use. Several water-soluble and fat-soluble vitamins are produced partly or completely using industrial fermentations. Good examples of such vitamins and the microorganisms that synthesize them are riboflavin (*Clostridium, Candida, Ashbya, Eremothecium*), coenzyme A (*Brevibacterium*), vitamin B_{12} (*Streptomyces, Propionibacterium, Pseudomonas*), vitamin C (*Gluconobacter, Erwinia, Corynebacterium*), β-carotene (*Dunaliella*), and vitamin D (*Saccharomyces*). Current research focuses on improving yields and finding microorganisms that can produce large quantities of other vitamins.

1. What are growth factors? What are vitamins?
2. How can humans put to use a microbe with a specific growth factor requirement?
3. List the growth factors that microorganisms produce industrially.
4. Why do you think amino acids, purines, and pyrimidines are often growth factors, whereas glucose is not?

5.6 Uptake of Nutrients by the Cell

The first step in nutrient use is uptake of the required nutrients by the microbial cell. Uptake mechanisms must be specific—that is, the necessary substances, and not others, must be acquired. It

Table 5.3 Functions of Some Common Vitamins in Microorganisms

Vitamin	Functions	Examples of Microorganisms Requiring Vitamin[a]
Biotin	Carboxylation (CO_2 fixation) One-carbon metabolism	*Leuconostoc mesenteroides* (B) *Saccharomyces cerevisiae* (F) *Ochromonas malhamensis* (P) *Acanthamoeba castellanii* (P)
Cyanocobalamin (B_{12})	Molecular rearrangements One-carbon metabolism—carries methyl groups	*Lactobacillus* spp. (B) *Euglena gracilis* (P) Diatoms (P) *Acanthamoeba castellanii* (P)
Folic acid	One-carbon metabolism	*Enterococcus faecalis* (B) *Tetrahymena pyriformis* (P)
Lipoic acid	Transfer of acyl groups	*Lactobacillus casei* (B) *Tetrahymena* spp. (P)
Pantothenic acid	Precursor of coenzyme A—carries acyl groups (pyruvate oxidation, fatty acid metabolism)	*Proteus morganii* (B) *Hanseniaspora* spp. (F) *Paramecium* spp. (P)
Pyridoxine (B_6)	Amino acid metabolism (e.g., transamination)	*Lactobacillus* spp. (B) *Tetrahymena pyriformis* (P)
Niacin (nicotinic acid)	Precursor of NAD and NADP—carry electrons and hydrogen atoms	*Brucella abortus, Haemophilus influenzae* (B) *Blastocladia pringsheimii* (F) *Crithidia fasciculata* (P)
Riboflavin (B_2)	Precursor of FAD and FMN—carry electrons or hydrogen atoms	*Caulobacter vibrioides* (B) *Dictyostelium* spp. (P) *Tetrahymena pyriformis* (P)
Thiamine (B_1)	Aldehyde group transfer (pyruvate decarboxylation, α-keto acid oxidation)	*Bacillus anthracis* (B) *Phycomyces blakesleeanus* (F) *Ochromonas malhamensis* (P) *Colpidium campylum* (P)

[a] The representative microorganisms are members of the following groups: *Bacteria* (B), *Fungi* (F), and protists (P).

does a cell no good to take in a substance that it cannot use. Because microorganisms often live in nutrient-poor habitats, they must be able to transport nutrients from dilute solutions into the cell against a concentration gradient. Finally, nutrient molecules must pass through a selectively permeable plasma membrane that prevents the free passage of most substances. In view of the enormous variety of nutrients and the complexity of the task, it is not surprising that microorganisms make use of several different transport mechanisms. The most important of these are facilitated diffusion, active transport, and group translocation. Eucaryotic microorganisms do not appear to employ group translocation but take up nutrients by the process of endocytosis. Organelles of the biosynthetic-secretory and endocytic pathways (section 4.4)

Passive Diffusion

A few substances, such as glycerol, can cross the plasma membrane by **passive diffusion.** Passive diffusion, often called diffusion or simple diffusion, is the process in which molecules move from a region of higher concentration to one of lower concentration. The rate of passive diffusion is dependent on the size of the concentration gradient between a cell's exterior and its interior (**figure 5.3**). A fairly large concentration gradient is required for adequate nutrient uptake by passive diffusion (i.e., the external nutrient concentration must be high while the internal concentration is low), and the rate of uptake decreases as more nutrient is acquired unless it is used immediately. Very small molecules such as H_2O, O_2, and CO_2 often move across membranes by passive diffusion. Larger molecules, ions, and polar substances must enter the cell by other mechanisms.

Facilitated Diffusion

The rate of diffusion across selectively permeable membranes is greatly increased by using carrier proteins, sometimes called **permeases,** which are embedded in the plasma membrane. Diffusion involving carrier proteins is called **facilitated diffusion.** The rate of facilitated diffusion increases with the concentration gradient

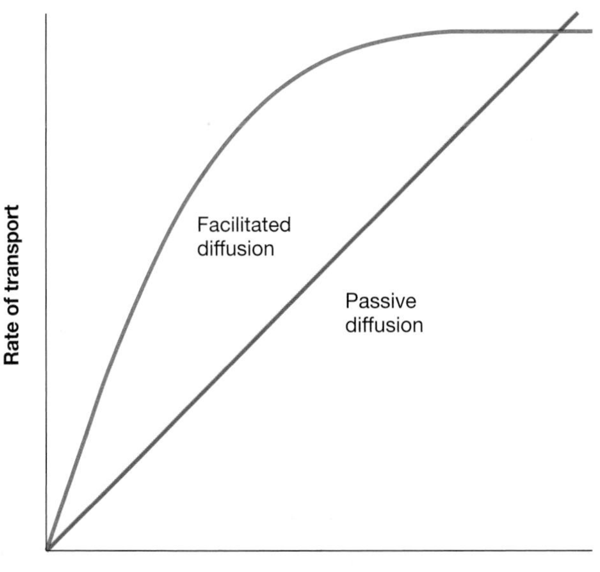

Figure 5.3 Passive and Facilitated Diffusion. The dependence of diffusion rate on the size of the solute's concentration gradient (the ratio of the extracellular concentration to the intracellular concentration). Note the saturation effect or plateau above a specific gradient value when a facilitated diffusion carrier is operating. This saturation effect is seen whenever a carrier protein is involved in transport.

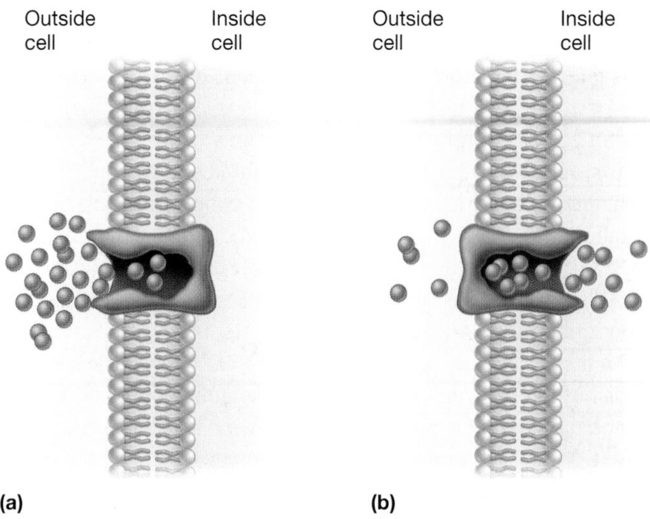

Figure 5.4 A Model of Facilitated Diffusion. The membrane carrier **(a)** can change conformation after binding an external molecule and subsequently release the molecule on the cell interior. **(b)** It then returns to the outward oriented position and is ready to bind another solute molecule. Because there is no energy input, molecules will continue to enter only as long as their concentration is greater on the outside.

much more rapidly and at lower concentrations of the diffusing molecule than that of passive diffusion (figure 5.3). Note that the diffusion rate levels off or reaches a plateau above a specific gradient value because the carrier is saturated—that is, the carrier protein is binding and transporting as many solute molecules as possible. The resulting curve resembles an enzyme-substrate curve (*see figure 8.18*) and is different from the linear response seen with passive diffusion. Carrier proteins also resemble enzymes in their specificity for the substance to be transported; each carrier is selective and will transport only closely related solutes. Although a carrier protein is involved, facilitated diffusion is truly diffusion. A concentration gradient spanning the membrane drives the movement of molecules, and no metabolic energy input is required. If the concentration gradient disappears, net inward movement ceases. The gradient can be maintained by transforming the transported nutrient to another compound. Once the nutrient is inside a eucaryotic cell, the gradient can be maintained by moving the nutrient to another membranous compartment. Some permeases are related to the major intrinsic protein (MIP) family of proteins. MIPs facilitate diffusion of small polar molecules. They are observed in virtually all organisms. The two most widespread MIP channels in bacteria are aquaporins (*see figure 2.29*), which transport water. Other important MIPs are the glycerol facilitators, which aid glycerol diffusion.

Although much work has been done on the mechanism of facilitated diffusion, the process is not yet understood completely. It appears that the carrier protein complex spans the membrane (**figure 5.4**). After the solute molecule binds to the outside, the carrier may change conformation and release the molecule on the cell interior. The carrier subsequently changes back to its original shape and is ready to pick up another molecule. The net effect is that a hydrophilic molecule can enter the cell in response to its concentration gradient. Remember that the mechanism is driven by concentration gradients and therefore is reversible. If the solute's concentration is greater inside the cell, it will move outward. Because the cell metabolizes nutrients upon entry, influx is favored.

Although glycerol is transported by facilitated diffusion in many bacteria, facilitated diffusion does not seem to be the major uptake mechanism. This is because nutrient concentrations often are lower outside the cell. Facilitated diffusion is much more prominent in eucaryotic cells where it is used to transport a variety of sugars and amino acids.

Active Transport

Because facilitated diffusion can efficiently move molecules to the interior only when the solute concentration is higher on the outside of the cell, microbes must have transport mechanisms that can move solutes against a concentration gradient. This is important because microorganisms often live in habitats characterized by very dilute nutrient sources. Microbes use two important transport processes in such situations: active transport and group translocation. Both are energy-dependent processes.

Active transport is the transport of solute molecules to higher concentrations, or against a concentration gradient, with the input of metabolic energy. Because active transport involves permeases, it resembles facilitated diffusion in some ways. The permeases bind particular solutes with great specificity for the molecules transported. Similar solute molecules can compete for the same carrier protein in both facilitated diffusion and active transport. Active transport is also characterized by the carrier saturation effect at high solute concentrations (figure 5.3). Nevertheless, active transport differs from facilitated diffusion in its use of metabolic energy and in its ability to concentrate substances. Metabolic inhibitors that block energy production will inhibit active transport but will not immediately affect facilitated diffusion.

A**TP-*****b*****inding *****c*****assette transporters (ABC transporters)** are important examples of active transport systems. They are observed in *Bacteria, Archaea,* and eucaryotes. Usually these transporters consist of two hydrophobic membrane-spanning domains associated on their cytoplasmic surfaces with two ATP-binding domains (**figure 5.5**). The membrane-spanning domains form a pore in the membrane and the ATP-binding domains bind and hydrolyze ATP to drive uptake. ABC transporters employ special substrate binding proteins, which are located in the periplasmic space of gram-negative bacteria (*see figure 3.25*) or are attached to membrane lipids on the external face of the gram-positive plasma membrane. These binding proteins bind the molecule to be transported and then interact with the membrane transport proteins to move the solute molecule inside the cell. *E. coli* transports a variety of sugars (arabinose, maltose, galactose, ribose) and amino acids (glutamate, histidine, leucine) by this mechanism. They can also pump antibiotics out using a multidrug-resistance ABC transporter.

Substances entering gram-negative bacteria must pass through the outer membrane before ABC transporters and other active transport systems can take action. There are several ways in which this is accomplished. When the substance is small, a generalized porin protein such as OmpF (*o*uter *m*embrane *p*rotein) can be used. An example of the movement of small molecules across the outer membrane is provided by the phosphate uptake systems of

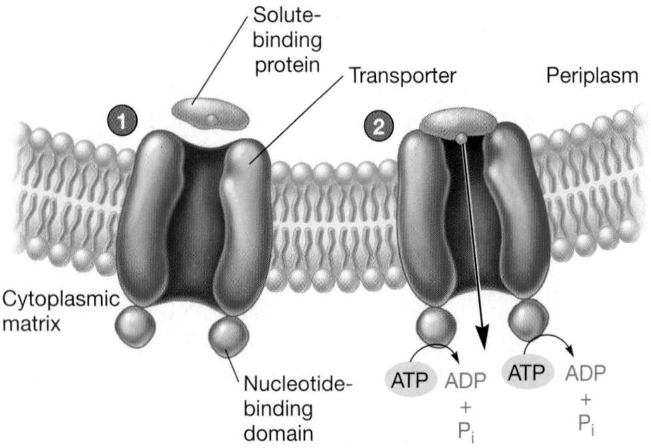

Figure 5.5 ABC Transporter Function. (*1*) The solute binding protein binds the substrate to be transported and approaches the ABC transporter complex. (*2*) The solute binding protein attaches to the transporter and releases the substrate, which is moved across the membrane with the aid of ATP hydrolysis. See text for details.

E. coli. Inorganic phosphate crosses the outer membrane by the use of a porin protein channel. Then, one of two transport systems moves the phosphate across the plasma membrane. Which system is used depends on the concentration of phosphate. The PIT system functions at high phosphate concentrations. When phosphate concentrations are low, an ABC transporter system called PST (*p*hosphate-*s*pecific *t*ransport) brings phosphate into the cell, using a periplasmic binding protein. In contrast to small molecules like phosphate, the transport of larger molecules, such as vitamin B_{12}, requires the use of specialized, high-affinity outer-membrane receptors that function in association with specific transporters in the plasma membrane.

As will be discussed in chapter 9, electron transport during energy-conserving processes generates a proton gradient (in procaryotes, the protons are at a higher concentration outside the cell than inside). The proton gradient can be used to do cellular work including active transport. The uptake of lactose by the lactose permease of *E. coli* is a well-studied example. The permease is a single protein that transports a lactose molecule inward as a proton simultaneously enters the cell. Such linked transport of two substances in the same direction is called **symport.** Here, energy in the form of a proton gradient drives solute transport. Although the mechanism of transport is not completely understood, X-ray diffraction studies show that the transport protein exists in outward- and inward-facing conformations. When lactose and a proton bind to separate sites on the outward-facing conformation, the protein changes to its inward-facing conformation. Then the sugar and proton are released into the cytoplasm. *E. coli* also uses proton symport to take up amino acids and organic acids like succinate and malate. Electron transport and oxidative phosphorylation (section 9.5)

A proton gradient also can power active transport indirectly, often through the formation of a sodium ion gradient. For example, an *E. coli* sodium transport system pumps sodium outward in response to the inward movement of protons (**figure 5.6**). Such linked transport in which the transported substances move in opposite directions is termed **antiport.** The sodium gradient generated by this proton antiport system then drives the uptake of sugars and amino acids. Although not well understood, it is thought that a sodium ion attaches to a carrier protein, causing it to change shape. The carrier then binds the sugar or amino acid tightly and orients its binding sites toward the cell interior. Because of the low intracellular sodium concentration, the sodium ion dissociates from the carrier, and the other molecule follows. *E. coli* transport proteins carry the sugar melibiose and the amino acid glutamate when sodium simultaneously moves inward. Sodium symport or cotransport also is an important process in eucaryotic cells where it is used in sugar and amino acid uptake. However, ATP, rather than proton motive force, usually drives sodium transport in eucaryotic cells.

Often a microorganism has more than one transport system for each nutrient, as can be seen with *E. coli.* This bacterium has at least five transport systems for the sugar galactose, three systems each for the amino acids glutamate and leucine, and two potassium transport complexes. When there are several transport systems for the same substance, the systems differ in such properties

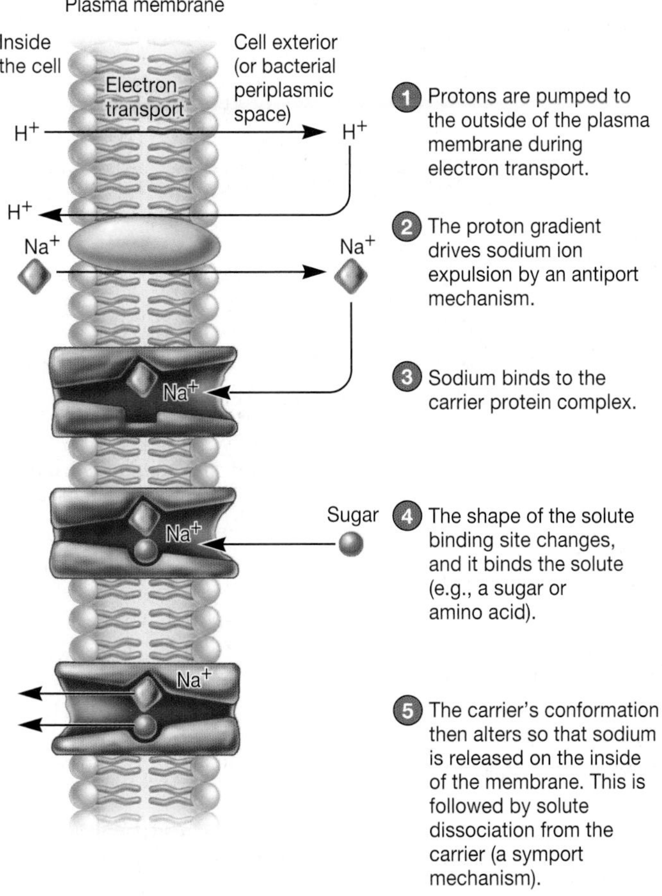

Plasma membrane
Inside the cell
Cell exterior (or bacterial periplasmic space)
Electron transport
H+
H+
H+
Na+
Na+
Na+
Na+
Sugar
Na+
1 Protons are pumped to the outside of the plasma membrane during electron transport.
2 The proton gradient drives sodium ion expulsion by an antiport mechanism.
3 Sodium binds to the carrier protein complex.
4 The shape of the solute binding site changes, and it binds the solute (e.g., a sugar or amino acid).
5 The carrier's conformation then alters so that sodium is released on the inside of the membrane. This is followed by solute dissociation from the carrier (a symport mechanism).

Figure 5.6 Active Transport Using Proton and Sodium Gradients.

as their energy source, their affinity for the solute transported, and the nature of their regulation. This diversity gives the microbe an added competitive advantage in a variable environment.

Group Translocation

In active transport, solute molecules move across a membrane without modification. Another type of transport, called **group translocation,** chemically modifies the molecule as it is brought into the cell. Group translocation is a type of active transport because metabolic energy is used during uptake of the molecule. This is clearly demonstrated by the best-known group translocation system, the **phosphoenolpyruvate: sugar *p*hospho*t*ransferase *s*ystem (PTS),** which is observed in many bacteria. The PTS transports a variety of sugars while phosphorylating them, using phosphoenolpyruvate (PEP) as the phosphate donor.

$$\text{PEP} + \text{sugar (outside)} \rightarrow \text{pyruvate} + \text{sugar-phosphate (inside)}$$

PEP is an important intermediate of a biochemical pathway used by many chemoorganoheterotrophs to extract energy from organic energy sources. PEP is a high-energy molecule that can be used to synthesize ATP, the cell's energy currency. However, when it is used in PTS reactions, the energy present in PEP is used to energize uptake rather than ATP synthesis. The role of ATP in metabolism (section 8.5); The breakdown of glucose to pyruvate (section 9.3)

The transfer of phosphate from PEP to the incoming molecule involves several proteins and is an example of a **phosphorelay system.** In *E. coli* and *Salmonella,* the PTS consists of two enzymes and a low molecular weight heat-stable protein (HPr). HPr and enzyme I (EI) are cytoplasmic. Enzyme II (EII) is more variable in structure and often composed of three subunits or domains. EIIA is cytoplasmic and soluble. EIIB also is hydrophilic and frequently is attached to EIIC, a hydrophobic protein that is embedded in the membrane. A phosphate is transferred from PEP to enzyme II with the aid of enzyme I and HPr (**figure 5.7**). Then, a sugar molecule is phosphorylated as it is carried across the membrane by enzyme II. Enzyme II transports only specific sugars and varies with the PTS, whereas enzyme I and HPr are common to all PTSs. Control of enzyme activity (section 8.10)

PTSs are widely distributed in bacteria. Most members of the genera *Escherichia, Salmonella, Staphylococcus,* as well as many other facultatively anaerobic bacteria (bacteria that grow either in the presence or absence of O_2) have phosphotransferase systems; some obligately anaerobic bacteria (e.g., *Clostridium*) also have PTSs. However, most aerobic bacteria, with the exception of some species of *Bacillus,* seem to lack PTSs. Many carbohydrates are transported by PTSs. *E. coli* takes up glucose, fructose, mannitol, sucrose, *N*-acetylglucosamine, cellobiose, and other carbohydrates by group translocation. Besides their role in transport, PTS proteins can bind chemical attractants, toward which bacteria move by the process of chemotaxis. The influence of environmental factors on growth: Oxygen concentration (section 6.5); Chemotaxis (section 3.10)

Iron Uptake

Almost all microorganisms require iron for use in cytochromes and many enzymes. Iron uptake is made difficult by the extreme insolubility of ferric iron (Fe^{3+}) and its derivatives, which leaves little free iron available for transport. Many bacteria and fungi have overcome this difficulty by secreting siderophores [Greek for iron bearers]. **Siderophores** are low molecular weight organic molecules that are able to complex with ferric iron and supply it to the cell. These iron-transport molecules are normally either hydroxamates or phenolates-catecholates. Ferrichrome is a hydroxamate produced by many fungi; enterobactin is the catecholate formed by *E. coli* (**figure 5.8*a,b***). It appears that three siderophore groups complex with iron to form a six-coordinate, octahedral complex (figure 5.8*c*).

Microorganisms secrete siderophores when iron is scarce in the medium. Once the iron-siderophore complex has reached the cell surface, it binds to a siderophore-receptor protein. Then the iron is either released to enter the cell directly or the whole iron-siderophore complex is transported inside by an ABC transporter. In *E. coli* the siderophore receptor is in the outer membrane of the cell envelope; when the iron reaches the periplasmic space, it moves through the plasma membrane with the aid of the transporter. After

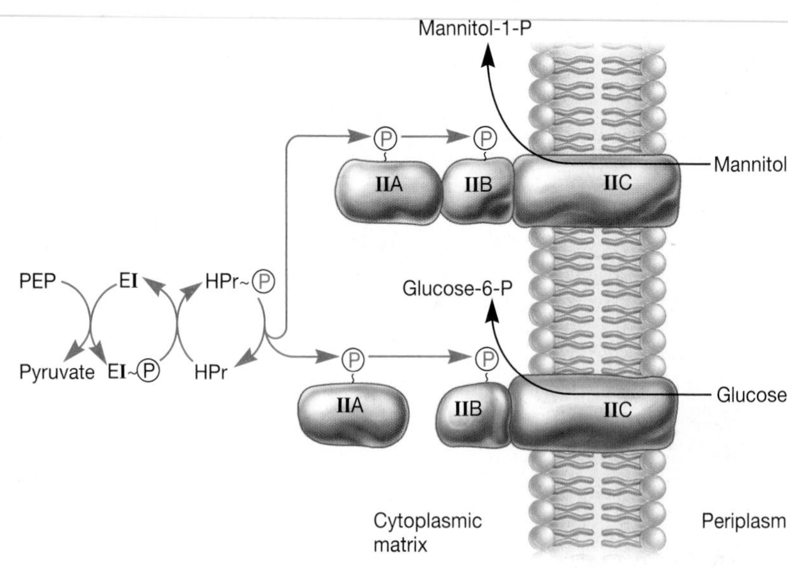

Figure 5.7 Group Translocation: Bacterial PTS Transport. Two examples of the phosphoenolpyruvate: sugar phosphotransferase system (PTS) are illustrated. The following components are involved in the system: phosphoenolpyruvate (PEP), enzyme I (EI), the low molecular weight heat-stable protein (HPr), and enzyme II (EII). The high-energy phosphate is transferred from HPr to the soluble EIIA. EIIA is attached to EIIB in the mannitol transport system and is separate from EIIB in the glucose system. In either case the phosphate moves from EIIA to EIIB, and then is transferred to the sugar during transport through the membrane. Other relationships between the EII components are possible. For example, IIA and IIB may form a soluble protein separate from the membrane complex; the phosphate still moves from IIA to IIB and then to the membrane domain(s).

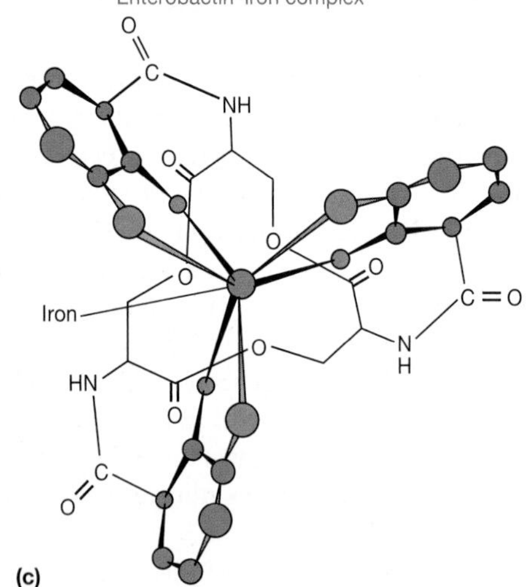

Figure 5.8 Siderophore Ferric Iron Complexes. **(a)** Ferrichrome is a cyclic hydroxamate [—CO—N(O^-)—] molecule formed by many fungi. **(b)** *E. coli* produces the cyclic catecholate derivative, enterobactin. **(c)** Ferric iron probably complexes with three siderophore groups to form a six-coordinate, octahedral complex as shown in this illustration of the enterobactin-iron complex.

the iron has entered the cell, it is reduced to the ferrous form (Fe^{2+}). Iron is so crucial to microorganisms that they may use more than one route of iron uptake to ensure an adequate supply.

1. Describe facilitated diffusion, active transport, and group translocation in terms of their distinctive characteristics and mechanisms. What advantage does a microbe gain by using active transport rather than facilitated diffusion?
2. What are symport and antiport processes?
3. What two mechanisms allow the passage of nutrients across the outer membrane of gram-negative bacteria before they are actively transported across the plasma membrane?
4. What is the difference between an ABC transporter and a porin in terms of function and cellular location?
5. What are siderophores? Why are they important?

5.7 Culture Media

Much of the study of microbiology depends on the ability to grow and maintain microorganisms in the laboratory, and this is possible only if suitable culture media are available. A culture medium is a solid or liquid preparation used to grow, transport, and store microorganisms. To be effective, the medium must contain all the nutrients the microorganism requires for growth. Specialized media are essential in the isolation and identification of microorganisms, the testing of antibiotic sensitivities, water and food analysis, industrial microbiology, and other activities. Although all microorganisms need sources of energy, carbon, nitrogen, phosphorus, sulfur, and various minerals, the precise composition of a satisfactory medium will depend on the species one is trying to cultivate because nutritional requirements vary so greatly. Knowledge of a microorganism's normal habitat often is

Table 5.4 Types of Media

Physical Nature	Chemical Composition	Functional Type
Liquid	Defined (synthetic)	Supportive (general purpose)
Semisolid	Complex	Enriched
Solid		Selective
		Differential

Table 5.5 Examples of Defined Media

BG–11 Medium for Cyanobacteria	Amount (g/liter)
$NaNO_3$	1.5
$K_2HPO_4 \cdot 3H_2O$	0.04
$MgSO_4 \cdot 7H_2O$	0.075
$CaCl_2 \cdot 2H_2O$	0.036
Citric acid	0.006
Ferric ammonium citrate	0.006
EDTA (Na_2Mg salt)	0.001
Na_2CO_3	0.02
Trace metal solution[a]	1.0 ml/liter
Final pH 7.4	
Medium for *Escherichia coli*	**Amount (g/liter)**
Glucose	1.0
Na_2HPO_4	16.4
KH_2PO_4	1.5
$(NH_4)_2SO_4$	2.0
$MgSO_4 \cdot 7H_2O$	200.0 mg
$CaCl_2$	10.0 mg
$FeSO_4 \cdot 7H_2O$	0.5 mg
Final pH 6.8–7.0	

Sources: Data from Rippka, et al. *Journal of General Microbiology,* 111:1–61, 1979; and S. S. Cohen, and R. Arbogast, *Journal of Experimental Medicine,* 91:619, 1950.

[a]The trace metal solution contains H_3BO_3, $MnCl_2 \cdot 4H_2O$, $ZnSO_4 \cdot 7H_2O$, $Na_2Mo_4 \cdot 2H_2O$, $CuSO_4 \cdot 5H_2O$, and $Co(NO_3)_2 \cdot 6H_2O$.

useful in selecting an appropriate culture medium because its nutrient requirements reflect its natural surroundings. Frequently a medium is used to select and grow specific microorganisms or to help identify a particular species. In such cases the function of the medium also will determine its composition.

Culture media can be classified on the basis of several parameters: the chemical constituents from which they are made, their physical nature, and their function (**table 5.4**). The types of media defined by these parameters are described here.

Chemical and Physical Types of Culture Media

A medium in which all chemical components are known is a **defined** or **synthetic medium.** It can be in a liquid form (broth) or solidified by an agent such as agar, as described in the following sections. Defined media are often used to culture photolithotrophic autotrophs such as cyanobacteria and photosynthetic protists. They can be grown on relatively simple media containing CO_2 as a carbon source (often added as sodium carbonate or bicarbonate), nitrate or ammonia as a nitrogen source, sulfate, phosphate, and a variety of minerals (**table 5.5**). Many chemoorganotrophic heterotrophs also can be grown in defined media with glucose as a carbon source and an ammonium salt as a nitrogen source. Not all defined media are as simple as the examples in table 5.5 but may be constructed from dozens of components. Defined media are used widely in research, as it is often desirable to know what the experimental microorganism is metabolizing.

Media that contain some ingredients of unknown chemical composition are **complex media.** Such media are very useful, as a single complex medium may be sufficiently rich to completely meet the nutritional requirements of many different microorganisms. In addition, complex media often are needed because the nutritional requirements of a particular microorganism are unknown, and thus a defined medium cannot be constructed. This is the situation with many fastidious bacteria that have complex nutritional or cultural requirements; they may even require a medium containing blood or serum.

Complex media contain undefined components like peptones, meat extract, and yeast extract. **Peptones** are protein hydrolysates prepared by partial proteolytic digestion of meat, casein, soya meal, gelatin, and other protein sources. They serve as sources of carbon, energy, and nitrogen. Beef extract and yeast extract are aqueous extracts of lean beef and brewer's yeast, respectively. Beef extract contains amino acids, peptides, nucleotides, organic acids, vitamins, and minerals. Yeast extract is an excellent source of B vitamins as well as nitrogen and carbon compounds. Three commonly used complex media are (1) nutrient broth, (2) tryptic soy broth, and (3) MacConkey agar (**table 5.6**).

Although both liquid and solidified media are routinely used in microbiology labs, solidified media are particularly important. Solidified media can be used to isolate different microbes from each other in order to establish pure cultures. As discussed in chapter 1, this is a critical step in demonstrating the relationship between a microbe and a disease using Koch's postulates. Both defined and complex media can be solidified with the addition of 1.0 to 2.0% agar; most commonly 1.5% is used. **Agar** is a sulfated polymer composed mainly of D-galactose, 3,6-anhydro-L-galactose, and D-glucuronic acid (**Historical Highlights 5.1**). It usually is extracted from red algae. Agar is well suited as a solidifying agent for several reasons. One is that it melts at about 90°C but once melted does not harden until it reaches about 45°C. Thus after being melted in boiling water, it can be cooled to a temperature that is tolerated by human hands as well as microbes. Furthermore, microbes growing on agar medium can be incubated at a wide range of temperatures. Finally, agar is an excellent hardening agent because most microorganisms cannot degrade it.

Other solidifying agents are sometimes employed. For example, silica gel is used to grow autotrophic bacteria on solid media

Table 5.6 Some Common Complex Media

Nutrient Broth	Amount (g/liter)
Peptone (gelatin hydrolysate)	5
Beef extract	3
Tryptic Soy Broth	
Tryptone (pancreatic digest of casein)	17
Peptone (soybean digest)	3
Glucose	2.5
Sodium chloride	5
Dipotassium phosphate	2.5
MacConkey Agar	
Pancreatic digest of gelatin	17.0
Pancreatic digest of casein	1.5
Peptic digest of animal tissue	1.5
Lactose	10.0
Bile salts	1.5
Sodium chloride	5.0
Neutral red	0.03
Crystal violet	0.001
Agar	13.5

in the absence of organic substances and to determine carbon sources for heterotrophic bacteria by supplementing the medium with various organic compounds.

Functional Types of Media

Media such as tryptic soy broth and tryptic soy agar are called general purpose media or **supportive media** because they sustain the growth of many microorganisms. Blood and other special nutrients may be added to general purpose media to encourage the growth of fastidious microbes. These specially fortified media (e.g., blood agar) are called **enriched media** (**figure 5.9**).

Selective media favor the growth of particular microorganisms (**table 5.7**). Bile salts or dyes like basic fuchsin and crystal violet favor the growth of gram-negative bacteria by inhibiting the growth of gram-positive bacteria; the dyes have no effect on gram-negative organisms. Endo agar, eosin methylene blue agar, and MacConkey agar (tables 5.6 and 5.7) are three media widely used for the detection of *E. coli* and related bacteria in water supplies and elsewhere. These media contain dyes that suppress gram-positive bacterial growth. MacConkey agar also contains bile salts. Bacteria also may be selected by incubation with nutrients that they specifically can use. A medium containing only cellulose as a carbon and energy source is quite effective in the isolation of cellulose-digesting bacteria. The possibilities for selection are endless, and there are dozens of special selective media in use.

Historical Highlights

5.1 The Discovery of Agar as a Solidifying Agent and the Isolation of Pure Cultures

The earliest culture media were liquid, which made the isolation of bacteria to prepare pure cultures extremely difficult. In practice, a mixture of bacteria was diluted successively until only one organism, as an average, was present in a culture vessel. If everything went well, the individual bacterium thus isolated would reproduce to give a pure culture. This approach was tedious, gave variable results, and was plagued by contamination problems. Progress in isolating pathogenic bacteria understandably was slow.

The development of techniques for growing microorganisms on solid media and efficiently obtaining pure cultures was due to the efforts of the German bacteriologist Robert Koch and his associates. In 1881 Koch published an article describing the use of boiled potatoes, sliced with a flame-sterilized knife, in culturing bacteria. The surface of a sterile slice of potato was inoculated with bacteria from a needle tip, and then the bacteria were streaked out over the surface so that a few individual cells would be separated from the remainder. The slices were incubated beneath bell jars to prevent airborne contamination, and the isolated cells developed into pure colonies. Unfortunately many bacteria would not grow well on potato slices.

At about the same time, Frederick Loeffler, an associate of Koch, developed a meat extract peptone medium for cultivating pathogenic bacteria. Koch decided to try solidifying this medium. Koch was an amateur photographer—he was the first to take photomicrographs of bacteria—and was experienced in preparing his own photographic plates from silver salts and gelatin. Precisely the same approach was employed for preparing solid media. He spread a mixture of Loeffler's medium and gelatin over a glass plate, allowed it to harden, and inoculated the surface in the same way he had inoculated his sliced potatoes. The new solid medium worked well, but it could not be incubated at 37°C (the best temperature for most human bacterial pathogens) because the gelatin would melt. Furthermore, some bacteria digested the gelatin.

About a year later, in 1882, agar was first used as a solidifying agent. It had been discovered by a Japanese innkeeper, Minora Tarazaemon. The story goes that he threw out extra seaweed soup and discovered the next day that it had jelled during the cold winter night. Agar had been used by the East Indies Dutch to make jellies and jams. Fannie Eilshemius Hesse (*see figure 1.7*), the New Jersey-born wife of Walther Hesse, one of Koch's assistants, had learned of agar from a Dutch acquaintance and suggested its use when she heard of the difficulties with gelatin. Agar-solidified medium was an instant success and continues to be essential in all areas of microbiology.

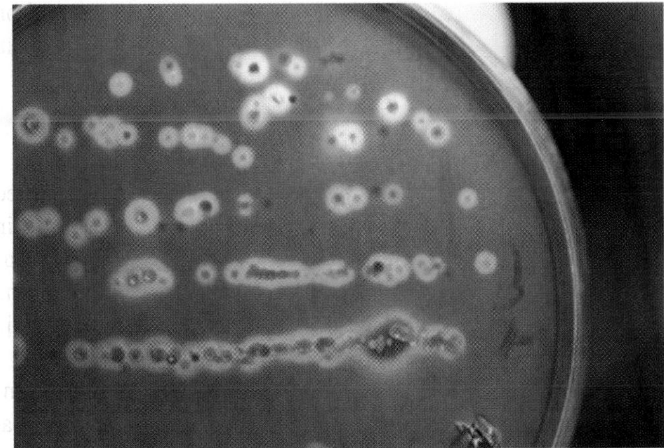

(a)

(b)

Figure 5.9 Enriched Media. **(a)** Blood agar culture of bacteria from the human throat. **(b)** Chocolate agar, an enriched medium used to grow fastidious organisms such as *Neisseria gonorrhoeae*. The brown color is the result of heating red blood cells and lysing them before adding them to the medium. It is called chocolate agar because of its chocolate brown color.

Differential media are media that distinguish among different groups of microbes and even permit tentative identification of microorganisms based on their biological characteristics. Blood agar is both a differential medium and an enriched one. It distinguishes between hemolytic and non-hemolytic bacteria. Hemolytic bacteria (e.g., many streptococci and staphylococci isolated from throats) produce clear zones around their colonies because of red blood cell destruction (figure 5.9*a*). MacConkey agar is both differential and selective. Since it contains lactose and neutral red dye, lactose-fermenting colonies appear pink to red in color and are easily distinguished from colonies of nonfermenters.

1. Describe the following kinds of media and their uses: defined media, complex media, supportive media, enriched media, selective media, and differential media. Give an example of each kind.
2. What are peptones, yeast extract, beef extract, and agar? Why are they used in media?

5.8 Isolation of Pure Cultures

In natural habitats microorganisms usually grow in complex, mixed populations with many species. This presents a problem for microbiologists because a single type of microorganism cannot be studied adequately in a mixed culture. One needs a **pure culture,** a population of cells arising from a single cell, to characterize an individual species. Pure cultures are so important that the development of pure culture techniques by the German bacteriologist Robert Koch transformed microbiology. Within about 20 years after the development of pure culture techniques most pathogens responsible for the major human bacterial diseases had been isolated (*see figure 1.2*). There are several ways to prepare pure cultures; a few of the more common approaches are reviewed here.

The Spread Plate and Streak Plate

If a mixture of cells is spread out on an agar surface at a relatively low density, every cell grows into a completely separate **colony,** a macroscopically visible growth or cluster of microorganisms on a solid medium. Because each colony arises from a single cell, each colony represents a pure culture. The **spread plate** is an easy, direct way of achieving this result. A small volume of dilute microbial mixture containing around 30 to 300 cells is transferred to the center of an agar plate and spread evenly over the surface with a sterile bent-glass rod (**figure 5.10**). The dispersed cells develop into isolated colonies. Because the number of colonies should equal the number of viable organisms in the sample, spread plates can be used to count the microbial population.

Pure colonies also can be obtained from **streak plates.** The microbial mixture is transferred to the edge of an agar plate with an inoculating loop or swab and then streaked out over the surface in one of several patterns (**figure 5.11**). After the first sector is streaked, the inoculating loop is sterilized and an inoculum for the second sector is obtained from the first sector. A similar process is followed for streaking the third sector, except that the inoculum is from the second sector. Thus this is essentially a dilution process. Eventually, very few cells will be on the loop, and single cells will drop from it as it is rubbed along the agar surface. These develop into separate colonies. In both spread-plate and streak-plate techniques, successful isolation depends on spatial separation of single cells.

Table 5.7 Mechanisms of Action of Selective and Differential Media

Medium	Functional Type	Mechanism of Action
Blood agar	Enriched and differential	Blood agar supports the growth of many fastidious bacteria. These can be differentiated based on their ability to produce hemolysins—proteins that lyse red blood cells. Hemolysis appears as a clear zone around the colony (β-hemolysis) or as a greenish halo around the colony (α-hemolysis) (e.g., *Streptococcus pyogenes,* a β-hemolytic streptococcus).
Eosin methylene blue (EMB) agar	Selective and differential	Two dyes, eosin Y and methylene blue, inhibit the growth of gram-positive bacteria. They also react with acidic products released by certain gram-negative bacteria when they use lactose or sucrose as carbon and energy sources. Colonies of gram-negative bacteria that produce large amounts of acidic products have a green, metallic sheen (e.g., fecal bacteria such as *E. coli*).
MacConkey (MAC) agar	Selective and differential	The selective components in MAC are bile salts and crystal violet, which inhibit the growth of gram-positive bacteria. The presence of lactose and neutral red, a pH indicator, allows the differentiation of gram-negative bacteria based on the products released when they use lactose as a carbon and energy source. The colonies of those that release acidic products are red (e.g., *E. coli*).
Mannitol salt agar	Selective and differential	A concentration of 7.5% NaCl selects for the growth of staphylococci. Pathogenic staphylococci can be differentiated based on the release of acidic products when they use mannitol as a carbon and energy source. The acidic products cause a pH indicator (phenol red) to turn yellow (e.g., *Staphylococcus aureus*).

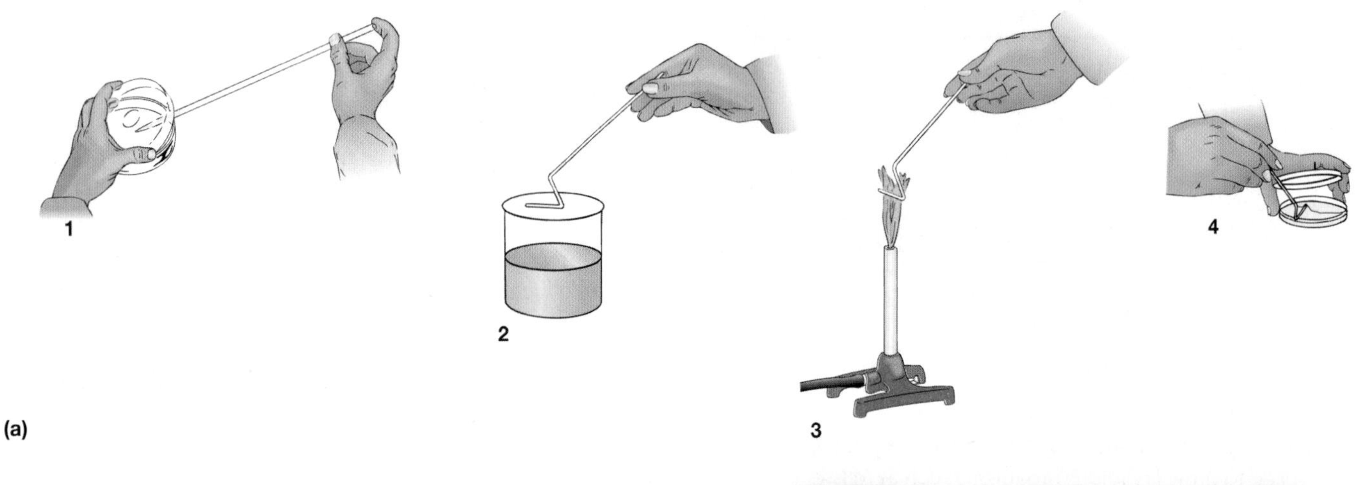

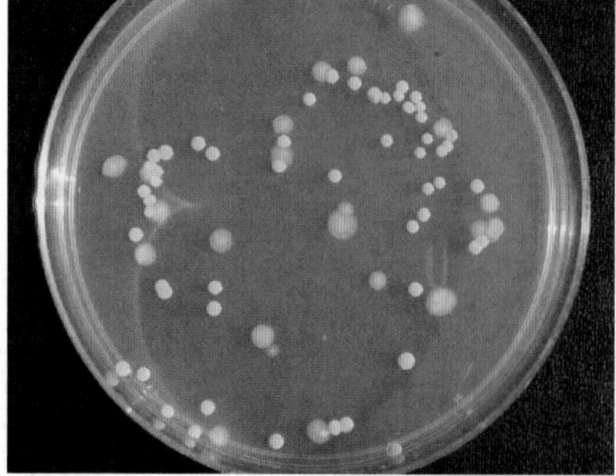

Figure 5.10 Spread-Plate Technique. **(a)** The preparation of a spread plate. (*1*) Pipette a small sample onto the center of an agar medium plate. (*2*) Dip a glass spreader into a beaker of ethanol. (*3*) Briefly flame the ethanol-soaked spreader and allow it to cool. (*4*) Spread the sample evenly over the agar surface with the sterilized spreader. Incubate. **(b)** Typical result of spread-plate technique.

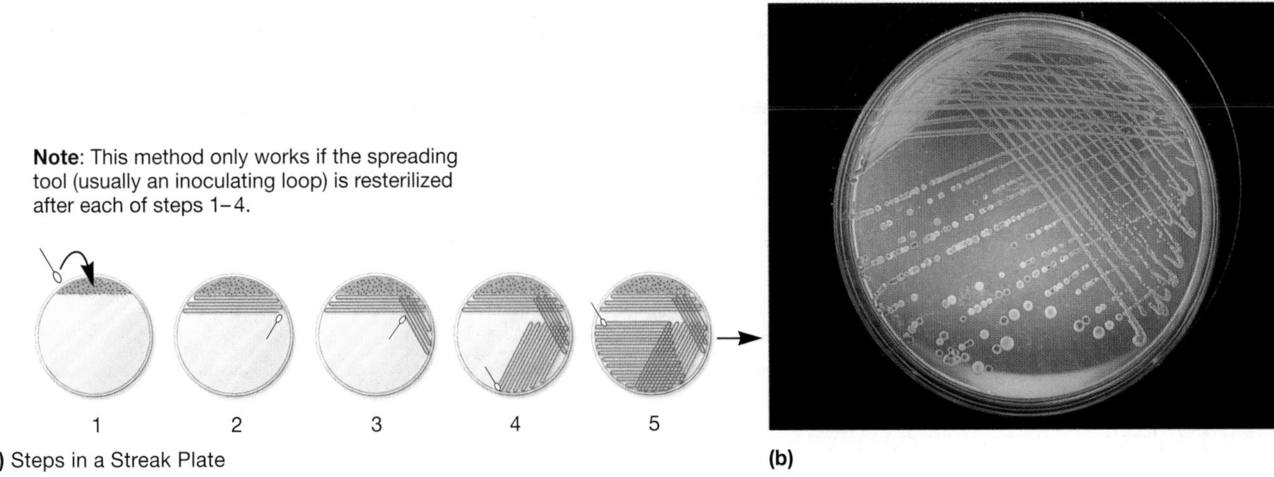

Figure 5.11 Streak-Plate Technique. A typical streaking pattern is shown **(a)** as well as an example of a streak plate **(b).**

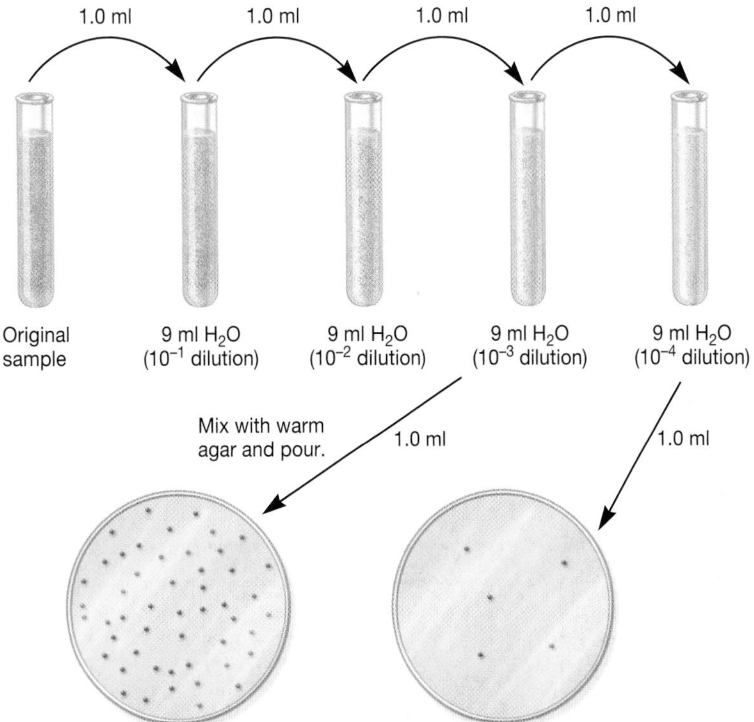

Figure 5.12 The Pour-Plate Technique. The original sample is diluted several times to thin out the population sufficiently. The most diluted samples are then mixed with warm agar and poured into petri dishes. Isolated cells grow into colonies and can be used to establish pure cultures. The surface colonies are circular; subsurface colonies are lenticular (lens shaped).

The Pour Plate

Extensively used with procaryotes and fungi, a **pour plate** also can yield isolated colonies. The original sample is diluted several times to reduce the microbial population sufficiently to obtain separate colonies when plating (**figure 5.12**). Then small volumes of several diluted samples are mixed with liquid agar that has been cooled to about 45°C, and the mixtures are poured immediately into sterile culture dishes. Most bacteria and fungi are not killed by a brief exposure to the warm agar. After the agar has hardened, each cell is fixed in place and forms an individual colony. Like the spread plate, the pour plate can be used to determine the number of cells in a population. Plates containing between 30 and 300 colonies are counted. The total number of colonies equals the number of viable microorganisms in the sample that are capable of growing in the medium used. Colonies growing on the surface also can be used to inoculate fresh medium and prepare pure cultures (**Techniques & Applications 5.2**).

Techniques & Applications

5.2 The Enrichment and Isolation of Pure Cultures

A major practical problem is the preparation of pure cultures when microorganisms are present in very low numbers in a sample. Plating methods can be combined with the use of selective or differential media to enrich and isolate rare microorganisms. A good example is the isolation of bacteria that degrade the herbicide 2,4-dichlorophenoxyacetic acid (2,4-D). Bacteria able to metabolize 2,4-D can be obtained with a liquid medium containing 2,4-D as its sole carbon source and the required nitrogen, phosphorus, sulfur, and mineral components. When this medium is inoculated with soil, only bacteria able to use 2,4-D will grow. After incubation, a sample of the original culture is transferred to a fresh flask of selective medium for further enrichment of 2,4-D metabolizing bacteria. A mixed population of 2,4-D degrading bacteria will arise after several such transfers. Pure cultures can be obtained by plating this mixture on agar containing 2,4-D as the sole carbon source. Only bacteria able to grow on 2,4-D form visible colonies and can be subcultured. This same general approach is used to isolate and purify a variety of bacteria by selecting for specific physiological characteristics.

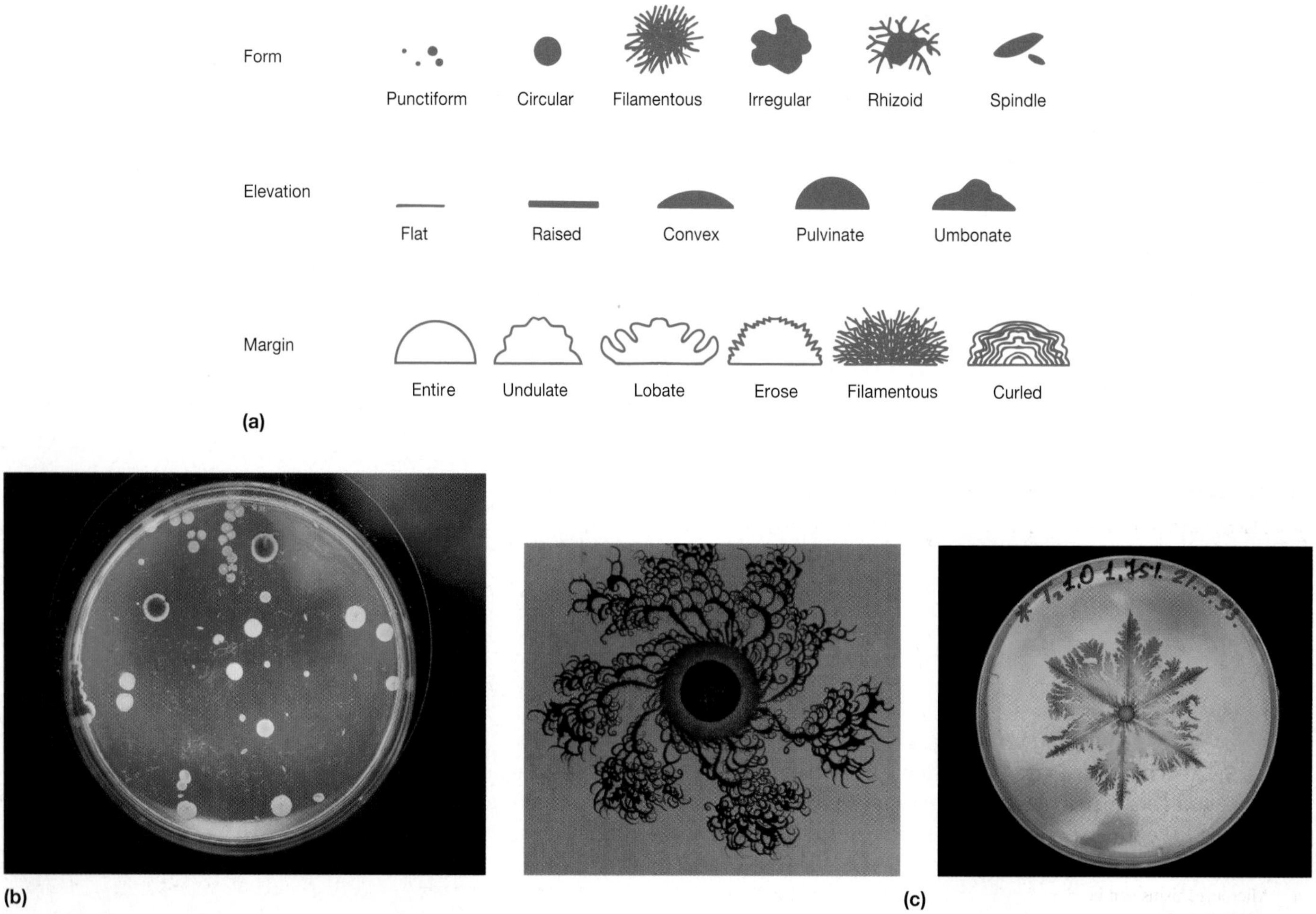

Form
Punctiform Circular Filamentous Irregular Rhizoid Spindle

Elevation
Flat Raised Convex Pulvinate Umbonate

Margin
Entire Undulate Lobate Erose Filamentous Curled

(a)

(b)

(c)

Figure 5.13 Bacterial Colony Morphology. **(a)** Variations in bacterial colony morphology seen with the naked eye. The general form of the colony and the shape of the edge or margin can be determined by looking down at the top of the colony. The nature of colony elevation is apparent when viewed from the side as the plate is held at eye level. **(b)** Examples of commonly observed colony morphologies. **(c)** Colony morphology can vary dramatically with the medium on which the bacteria are growing. These beautiful snowflakelike colonies were formed by *Bacillus subtilis* growing on nutrient-poor agar. The bacteria apparently behave cooperatively when confronted with poor growth conditions, and often the result is an intricate structure that resembles the fractal patterns seen in nonliving systems.

The preceding techniques require the use of special culture dishes named **petri dishes** or plates after their inventor Julius Richard Petri, a member of Robert Koch's laboratory; Petri developed these dishes around 1887 and they immediately replaced agar-coated glass plates. They consist of two round halves, the top half overlapping the bottom. Petri dishes are very easy to use, may be stacked on each other to save space, and are one of the most common items in microbiology laboratories.

Microbial Growth on Agar Surfaces

Colony development on agar surfaces aids microbiologists in identifying microorganisms because individual species often form colonies of characteristic size and appearance (**figure 5.13**). When a mixed population has been plated properly, it sometimes is possible to identify the desired colony based on its overall appearance and use it to obtain a pure culture. The structure of bacterial colonies also has been examined with the scanning electron microscope. The microscopic structure of colonies is often as variable as their visible appearance.

In nature, microorganisms often grow on surfaces in biofilms—slime-encased aggregations of microbes. However, sometimes they form discrete colonies. Therefore an understanding of colony growth is important, and the growth of colonies on agar has been frequently studied. Generally the most rapid cell growth occurs at the colony edge. Growth is much slower in the center, and cell autolysis takes place in the older central portions of some colonies. These differences in growth are due to gradients of oxygen, nutrients, and toxic products within the colony. At the colony edge, oxygen and nutrients are plentiful. The colony center is much thicker than the edge. Consequently oxygen and nutrients do not diffuse readily into the center, toxic metabolic products cannot be quickly eliminated, and growth in the colony center is slowed or stopped. Because of these environmental variations within a colony, cells on the periphery can be growing at maximum rates while cells in the center are dying. Microbial growth in natural environments: Biofilms (section 6.6)

It is obvious from the colonies pictured in figure 5.13 that bacteria growing on solid surfaces such as agar can form quite complex and intricate colony shapes. These patterns vary with nutrient availability and the hardness of the agar surface. It is not yet clear how characteristic colony patterns develop. Nutrient diffusion and availability, bacterial chemotaxis, and the presence of liquid on the surface all appear to play a role in pattern formation. Cell-cell communication is important as well. Much work will be required to understand the formation of bacterial colonies and biofilms.

1. What are pure cultures, and why are they important? How are spread plates, streak plates, and pour plates prepared?
2. In what way does microbial growth vary within a colony? What factors might cause these variations in growth?
3. How might an enrichment culture be used to isolate bacteria capable of degrading pesticides and other hazardous wastes?

Summary

Microorganisms require nutrients, materials that are used in biosynthesis and to make energy available.

5.1 The Common Nutrient Requirements

a. Macronutrients or macroelements (C, O, H, N, S, P, K, Ca, Mg, and Fe) are needed in relatively large quantities.

b. Micronutrients or trace elements (e.g., Mn, Zn, Co, Mo, Ni, and Cu) are used in very small amounts.

5.2 Requirements for Carbon, Hydrogen, Oxygen, and Electrons

a. All organisms require a source of carbon, hydrogen, oxygen, and electrons.

b. Heterotrophs use organic molecules as their source of carbon. These molecules often supply hydrogen, oxygen, and electrons as well. Some heterotrophs also derive energy from their organic carbon source.

c. Autotrophs use CO_2 as their primary or sole carbon source; they must obtain hydrogen and electrons from other sources.

5.3 Nutritional Types of Microorganisms

a. Microorganisms can be classified based on their energy and electron sources (**table 5.1**). Phototrophs use light energy, and chemotrophs obtain energy from the oxidation of chemical compounds.

b. Electrons are extracted from reduced inorganic substances by lithotrophs and from organic compounds by organotrophs (**table 5.2**).

5.4 Requirements for Nitrogen, Phosphorus, and Sulfur

a. Nitrogen, phosphorus, and sulfur may be obtained from the same organic molecules that supply carbon, from the direct incorporation of ammonia and phosphate, and by the reduction and assimilation of oxidized inorganic molecules.

5.5 Growth Factors

a. Many microorganisms need growth factors.

b. The three major classes of growth factors are amino acids, purines and pyrimidines, and vitamins. Vitamins are small organic molecules that usually are components of enzyme cofactors.

c. Knowing whether a microbe requires a particular growth factor has practical applications: those needing a growth factor can be used in bioassays that detect and quantify the growth factor; those that do not need a particular growth factor can sometimes be used to produce the growth factor in industrial settings.

5.6 Uptake of Nutrients by the Cell

a. Although some nutrients can enter cells by passive diffusion, a membrane carrier protein is usually required.

b. In facilitated diffusion the transport protein simply carries a molecule across the membrane in the direction of decreasing concentration, and no metabolic energy is required (**figure 5.4**).

c. Active transport systems use metabolic energy and membrane carrier proteins to concentrate substances actively by transporting them against a gradient. ATP is used as an energy source by ABC transporters (**figure 5.5**). Gradients of protons and sodium ions also drive solute uptake across membranes (**figure 5.6**).

d. Bacteria also transport organic molecules while modifying them, a process known as group translocation. For example, many sugars are transported and phosphorylated simultaneously (**figure 5.7**).

e. Iron is accumulated by the secretion of siderophores, small molecules able to complex with ferric iron (**figure 5.8**). When the iron-siderophore complex reaches the cell surface, it is taken inside and the iron is reduced to the ferrous form.

5.7 Culture Media

a. Culture media can be constructed completely from chemically defined components (defined media or synthetic media) or constituents like peptones and yeast extract whose precise composition is unknown (complex media).

b. Culture media can be solidified by the addition of agar, a complex polysaccharide from red algae.

c. Culture media are classified based on function and composition as supportive media, enriched media, selective media, and differential media. Supportive media are used to culture a wide variety of microbes. Enriched media are supportive media that contain additional nutrients needed by fastidious microbes. Selective media contain components that select for the growth of some microbes. Differential media contain components that allow microbes to be differentiated from each other, usually based on some metabolic capability.

5.8 Isolation of Pure Cultures

a. Pure cultures usually are obtained by isolating individual cells with any of three plating techniques: the spread-plate, streak-plate, and pour-plate methods. The spread-plate (**figure 5.10**) and pour-plate (**figure 5.12**) methods usually involve diluting a culture or sample and then plating the dilutions. In the spread-plate technique, a specially shaped rod is used to spread the cells on the agar surface; in the pour-plate technique, the cells are first mixed with cooled agar-containing media before being poured into a petri dish. The streak-plate technique (**figure 5.11**) uses an inoculating loop to spread cells across an agar surface.

b. Microorganisms growing on solid surfaces tend to form colonies with distinctive morphology (**figure 5.13**). Colonies usually grow most rapidly at the edge where larger amounts of required resources are available.

Key Terms

active transport 107
agar 111
antiport 108
ATP-binding cassette transporters (ABC transporters) 108
autotrophs 102
chemoheterotrophs 103
chemolithoheterotrophs 103
chemolithotrophic autotrophs 103
chemoorganotrophic heterotrophs 103
chemotrophs 103
colony 113
complex medium 111
defined medium 111
differential media 113
enriched media 112
facilitated diffusion 106
group translocation 109
growth factors 105
heterotrophs 102
lithotrophs 103
macroelements 101
micronutrients 101
mixotrophs 103
nutrient 101
organotrophs 103
passive diffusion 106
peptones 111
permease 106
petri dish 117
phosphoenolpyruvate: sugar phosphotransferase system (PTS) 109
phosphorelay system 109
photoautotrophs 103
photolithotrophic autotrophs 103
photoorganotrophic heterotrophs 103
phototrophs 103
pour plate 115
pure culture 113
selective media 112
siderophores 109
spread plate 113
streak plate 113
supportive media 112
symport 108
synthetic medium 111
trace elements 101
vitamins 105

Critical Thinking Questions

1. Discuss the advantages and disadvantages of group translocation versus endocytosis.
2. If you wished to obtain a pure culture of bacteria that could degrade benzene and use it as a carbon and energy source, how would you proceed?

Learn More

Abramson, J.; Smirnova, I.; Kasho, V.; Verner, G.; Kaback, H. R.; and Iwata, S. 2003. Structure and mechanism of the lactose permease of *Escherichia coli. Science* 301:610–15.

Becton, Dickinson and Co. 2005. *Difco and BBL manual: Manual of microbiological culture media,* 1st ed., Franklin Lakes, NJ: BD.

Davidson, A. L., and Chen, J. 2004. ATP-binding cassette transporters in bacteria. *Annu. Rev. Biochem.* 73:241–68.

Gottschall, J. C.; Harder, W.; and Prins, R. A. 1992. Principles of enrichment, isolation, cultivation, and preservation of bacteria. In *The Prokaryotes,* 2d ed., A. Balows et al., editors, 149–96. New York: Springer-Verlag.

Gutnick, D. L., and Ben-Jacob, E. 1999. Complex pattern formation and cooperative organization of bacterial colonies. In *Microbial ecology and infectious disease,* E. Rosenberg, editor, 284–99. Washington, D.C.: ASM Press.

Hancock, R. E. W., and Brinkman, F. S. L. 2002. Function of *Pseudomonas* porins in uptake and efflux. *Annu. Rev. Microbio.* 56:17–38.

Hohman, S.; Bill, R. M.; Kayingo, G.; and Prior, B. A. 2000. Microbial MIP channels. *Trends in Microbiol.* 8(1):33–38.

Holt, J. G., and Krieg, N. R. 1994. Enrichment and isolation. In *Methods for general and molecular bacteriology,* 2d ed. P. Gerhardt, editor, 179–215. Washington, D.C.: American Society for Microbiology.

Kelly, D. P. 1992. The chemolithotrophic prokaryotes. In *The Prokaryotes,* 2d ed., A. Balows et al., editors, 331–43. New York: Springer-Verlag.

Postma, P. W.; Lengeler, J. W.; and Jacobson, G. R. 1996. Phosphoenolpyruvate: Carbohydrate phosphotransferase systems. In *Escherichia coli and Salmonella: Cellular and molecular biology,* 2d ed., F. C. Neidhardt, et al., editors, 1149–74. Washington, D.C.: ASM Press.

Wandersman, C., and Delepelaire, P. 2004. Bacterial iron sources: From siderophores to hemophores. *Annu. Rev. Microbiol.* 58:611–47.

Please visit the Prescott website at www.mhhe.com/prescott7 for additional references.

6 Microbial Growth

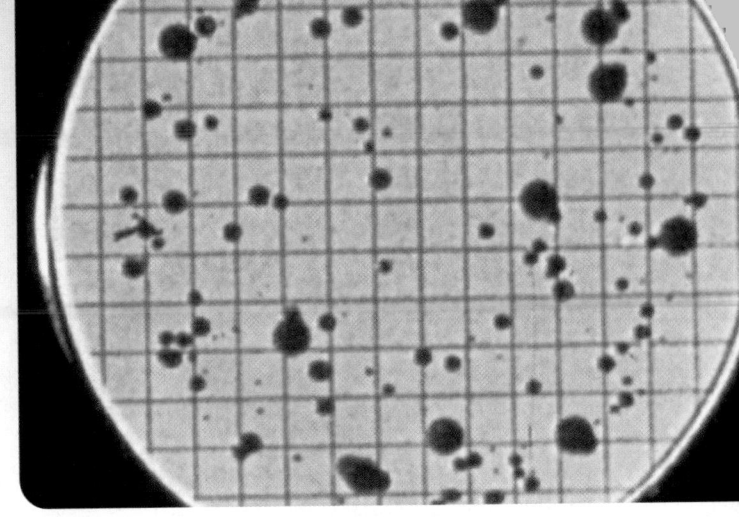

Membrane filters are used in counting microorganisms. This membrane has been used to obtain a total bacterial count using an indicator to color colonies for easy counting.

Preview

- Most procaryotes reproduce by binary fission. Although simpler than mitosis and meiosis, binary fission and the procaryotic cell cycle are still poorly understood.
- Growth is defined as an increase in cellular constituents and may result in an increase in a microorganism's size, population number, or both.
- When microorganisms are grown in a closed system, population growth remains exponential for only a few generations and then enters a stationary phase due to factors such as nutrient limitation and waste accumulation. In an open system with continual nutrient addition and waste removal, the exponential phase can be maintained for long periods.
- A wide variety of techniques can be used to study microbial growth by following changes in the total cell number, the population of viable microorganisms, or the cell mass.
- Water availability, pH, temperature, oxygen concentration, pressure, radiation, and a number of other environmental factors influence microbial growth. Yet many microorganisms, and particularly procaryotes, have managed to adapt and flourish under environmental extremes that would destroy most higher organisms.
- In the natural environment, growth is often severely limited by available nutrient supplies and many other environmental factors.
- Many microorganisms form biofilms in natural environments. This is an important survival strategy.
- Microbes can communicate with each other and behave cooperatively using population density-dependent signals.

In chapter 5 we emphasize that microorganisms need access to a source of energy and the raw materials essential for the construction of cellular components. All organisms must have carbon, hydrogen, oxygen, nitrogen, sulfur, phosphorus, and a variety of minerals; many also require one or more special growth factors. The cell takes up these substances by membrane transport processes, the most important of which are facilitated diffusion, active transport, and group translocation. Eucaryotic cells also employ endocytosis.

Chapter 6 concentrates more directly on procaryotic reproduction and growth. First we describe binary fission, the type of cell division most frequently observed among procaryotes, and the procaryotic cell cycle. Cell reproduction leads to an increase in population size, so we consider growth and the ways in which it can be measured next. Then we discuss continuous culture techniques. An account of the influence of environmental factors on microbial growth and microbial growth in natural environments completes the chapter.

Growth may be defined as an increase in cellular constituents. It leads to a rise in cell number when microorganisms reproduce by processes like budding or binary fission. Growth also results when cells simply become longer or larger. If the microorganism is **coenocytic**—that is, a multinucleate organism in which nuclear divisions are not accompanied by cell divisions—growth results in an increase in cell size but not cell number. It is usually not convenient to investigate the growth and reproduction of individual microorganisms because of their small size. Therefore, when studying growth, microbiologists normally follow changes in the total population number.

6.1 The Procaryotic Cell Cycle

The **cell cycle** is the complete sequence of events extending from the formation of a new cell through the next division. Most procaryotes reproduce by **binary fission,** although some procaryotes

The paramount evolutionary accomplishment of bacteria as a group is rapid, efficient cell growth in many environments.

—*J. L. Ingraham, O. Maaløe, and F. C. Neidhardt*

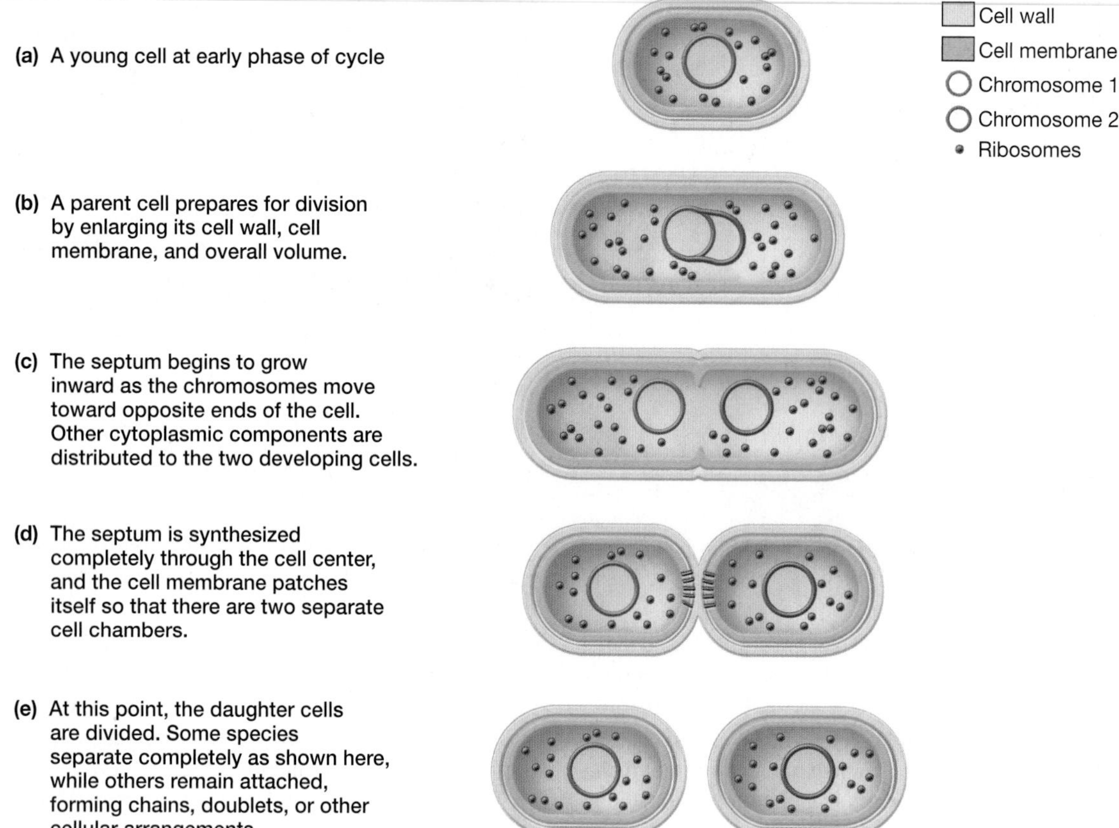

Cell wall
Cell membrane
Chromosome 1
Chromosome 2
Ribosomes
(a) A young cell at early phase of cycle
(b) A parent cell prepares for division by enlarging its cell wall, cell membrane, and overall volume.
(c) The septum begins to grow inward as the chromosomes move toward opposite ends of the cell. Other cytoplasmic components are distributed to the two developing cells.
(d) The septum is synthesized completely through the cell center, and the cell membrane patches itself so that there are two separate cell chambers.
(e) At this point, the daughter cells are divided. Some species separate completely as shown here, while others remain attached, forming chains, doublets, or other cellular arrangements.

Figure 6.1 Binary Fission.

reproduce by budding, fragmentation, and other means (**figure 6.1**). Binary fission is a relatively simple type of cell division: the cell elongates, replicates its chromosome, and separates the newly formed DNA molecules so there is one chromosome in each half of the cell. Finally, a septum (or cross wall) is formed at midcell, dividing the parent cell into two progeny cells, each having its own chromosome and a complement of other cellular constituents.

Despite the apparent simplicity of the procaryotic cell cycle, it is poorly understood. The cell cycles of *Escherichia coli, Bacillus subtilis,* and the aquatic microbe *Caulobacter crescentus* have been examined extensively, and our understanding of the cell cycle is based largely on these studies. Two pathways function during the cell cycle (**figure 6.2**): one pathway replicates and partitions the DNA into the progeny cells, the other carries out cytokinesis (septum formation and formation of progeny cells). Although these pathways overlap, it is easiest to consider them separately.

Chromosome Replication and Partitioning

Recall that most procaryotic chromosomes are circular. Each circular chromosome has a single site at which replication starts called the **origin of replication,** or simply the origin (**figure 6.3**). Replication is completed at the **terminus,** which is located directly opposite the origin. In a newly formed *E. coli* cell, the chromosome is compacted and organized so that the origin and terminus are in opposite halves of the cell. Early in the cell cycle, the origin and terminus move to midcell and a group of proteins needed for DNA synthesis assemble to form the **replisome** at the origin. DNA replication proceeds in both directions from the origin and the parent DNA is thought to spool through the replisome, which remains relatively stationary. As progeny chromosomes are synthesized, the two newly formed origins move toward opposite ends of the cell, and the rest of the chromosome follows in an orderly fashion.

Although the process of DNA synthesis and movement seems rather straightforward, the mechanism by which chromosomes are partitioned to each daughter cell is not well understood. Surprisingly, a picture is emerging in which components of the cytoskeleton are involved. For many years, it was assumed that procaryotes were too small for eucaryotic-like cytoskeletal structures. However, a protein called **MreB,** which is similar to eucaryotic actin, seems to be involved in several processes, including determining cell shape and chromosome movement. MreB polymerizes (that is to say, MreB units are linked together) to form a spiral around the inside periphery of the cell (**figure 6.4*a***). One model suggests that the origin of each newly replicated chromosome associates with

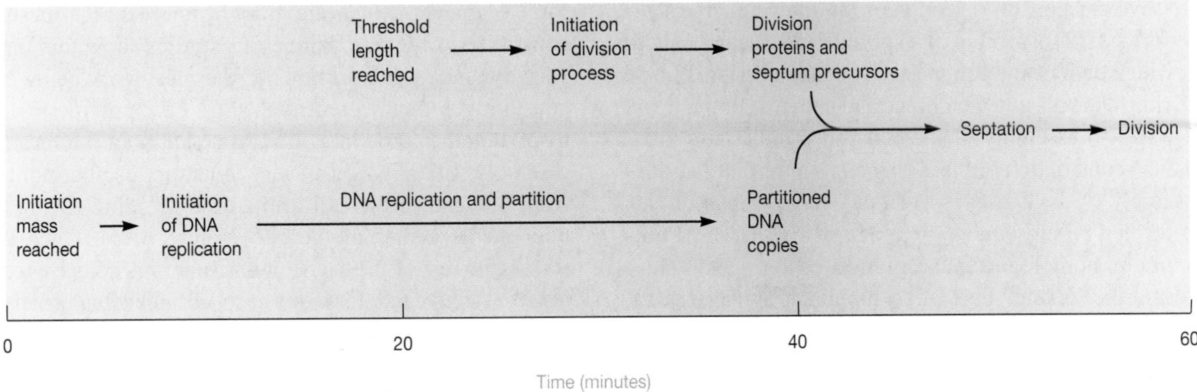

Figure 6.2 The Cell Cycle in *E. coli.* A 60-minute interval between divisions has been assumed for purposes of simplicity (the actual time between cell divisions may be shorter). *E. coli* requires about 40 minutes to replicate its DNA and 20 minutes after termination of replication to prepare for division. The position of events on the time line is approximate and meant to show the general pattern of occurrences.

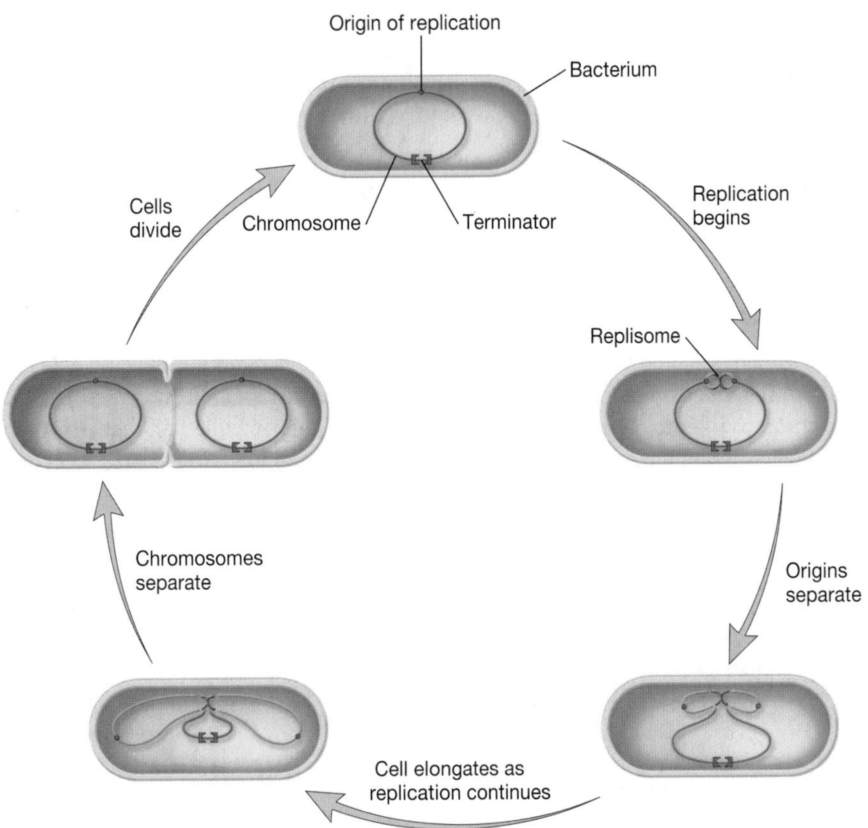

Figure 6.3 Cell Cycle of Slow-Growing *E. coli.* As the cell readies for replication, the origin migrates to the center of the cell and proteins that make up the replisome assemble. As replication proceeds, newly synthesized chromosomes move toward poles so that upon cytokinesis, each daughter cell inherits only one chromosome.

MreB, which then moves them to opposite poles of the cell. The notion that procaryotic chromosomes may be actively moved to the poles is further suggested by the fact that if MreB is mutated so that it can no longer hydrolyze ATP, its source of energy, chromosomes fail to segregate properly.

Cytokinesis

Septation is the process of forming a cross wall between two daughter cells. **Cytokinesis,** a term that has traditionally been used to describe the formation of two eucaryotic daughter cells, is now used to describe this process in procaryotes as well. Septation is divided into several steps: (1) selection of the site where the septum will be formed; (2) assembly of a specialized structure called the **Z ring,** which divides the cell in two by constriction; (3) linkage of the Z ring to the plasma membrane and perhaps components of the cell wall; (4) assembly of the cell wall-synthesizing machinery; and (5) constriction of the Z ring and septum formation.

The assembly of the Z ring is a critical step in septation, as it must be formed if subsequent steps are to occur. The **FtsZ protein,** a tubulin homologue found in most bacteria and many archaea, forms the Z ring. FtsZ, like tubulin, polymerizes to form filaments, which are thought to create the meshwork that constitutes the Z ring. Numerous studies show that the Z ring is very dynamic, with portions of the meshwork being exchanged constantly with newly formed, short FtsZ polymers from the cytosol. Another protein, called MinCD, is an inhibitor of Z-ring assembly. Like FtsZ, it is very dynamic, oscillating its position from one end of the cell to the other, forcing Z-ring formation only at the center of the cell (figure 6.4*b*). Once the Z-ring forms, the rest of the division machinery is constructed, as illustrated in **figure 6.5**. First one or more anchoring proteins link the Z ring to the cell membrane. Then the cell wall-synthesizing machinery is assembled. The cytoplasmic matrix: The procaryotic cytoskeleton (section 3.3)

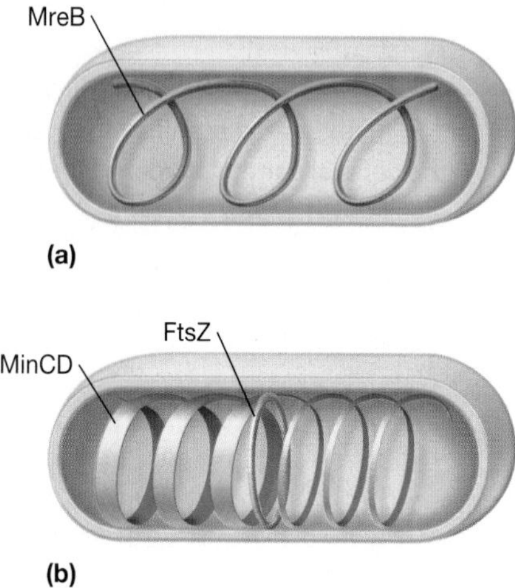

Figure 6.4 Cytoskeletal Proteins Involved in Cytokinesis in Rod-Shaped Bacteria. (a) The actin homolog MreB forms spiral filaments around the inside of the cell that help determine cell shape and may serve to move chromosomes to opposite cell poles. **(b)** The tubulin-like protein FtsZ assembles in the center of the cell to form a Z ring, which is essential for septation. MinCD, together with other Min proteins, oscillates from pole to pole, thereby preventing the formation of an off-center Z ring.

The final steps in division involve constriction of the Z ring, accompanied by invagination of the cell membrane and synthesis of the septal wall. Several models for Z-ring constriction have been proposed. One model holds that the FtsZ filaments are shortened by losing FtsZ subunits (i.e., depolymerization) at sites where the Z ring is anchored to the plasma membrane. This model is supported by the observation that Z rings of cells producing an excessive amount of FtsZ subunits fail to constrict.

DNA Replication in Rapidly Growing Cells

The preceding discussion of the cell cycle describes what occurs in slowly growing *E. coli* cells. In these cells, the cell cycle takes approximately 60 minutes to complete: 40 minutes for DNA replication and partitioning and about 20 minutes for septum formation and cytokinesis. However, *E. coli* can reproduce at a much more rapid rate, completing the entire cell cycle in about 20 minutes, despite the fact that DNA replication always requires at least 40 minutes. How can *E. coli* complete an entire cell cycle in 20 minutes when it takes 40 minutes to replicate its chromosome? *E. coli* accomplishes this by beginning a second round of DNA replication (and sometimes even a third or fourth round) before

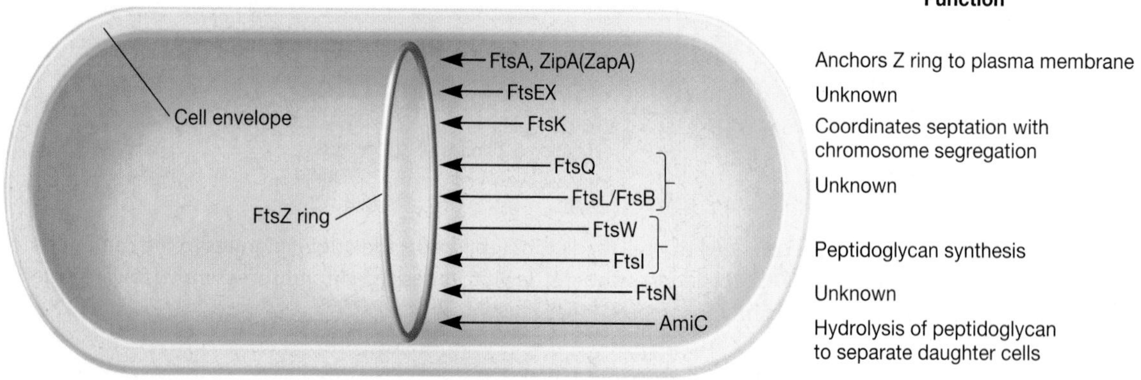

Figure 6.5 Formation of the Cell Division Apparatus in *E. coli.* The cell division apparatus is composed of numerous proteins that are thought to assemble in the order shown. The process begins with the polymerization of FtsZ to form the Z ring. Then FtsA and ZipA (possibly ZapA in *Bacillus subtilis*) proteins anchor the Z ring to the plasma membrane. Although numerous proteins are known to be part of the cell division apparatus, the functions of relatively few are known.

the first round of replication is completed. Thus the progeny cells receive two or more replication forks, and replication is continuous because the cells are always copying their DNA.

1. What two pathways function during the procaryotic cell cycle?
2. How does the procaryotic cell cycle compare with the eucaryotic cell cycle? List two ways they are similar; list two ways they differ.

6.2 The Growth Curve

Binary fission and other cell division processes bring about an increase in the number of cells in a population. Population growth is studied by analyzing the growth curve of a microbial culture. When microorganisms are cultivated in liquid medium, they usually are grown in a **batch culture** or closed system—that is, they are incubated in a closed culture vessel with a single batch of medium. Because no fresh medium is provided during incubation, nutrient concentrations decline and concentrations of wastes increase. The growth of microorganisms reproducing by binary fission can be plotted as the logarithm of the number of viable cells versus the incubation time. The resulting curve has four distinct phases (**figure 6.6**).

Lag Phase

When microorganisms are introduced into fresh culture medium, usually no immediate increase in cell number occurs, so this period is called the **lag phase.** Although cell division does not take place right away and there is no net increase in mass, the cell is synthesizing new components. A lag phase prior to the start of cell division can be necessary for a variety of reasons. The cells may be old and depleted of ATP, essential cofactors, and ribosomes; these must be synthesized before growth can begin. The medium may be different from the one the microorganism was growing in previously. Here new enzymes would be needed to use different nutrients. Possibly the microorganisms have been injured and require time to recover. Whatever the causes, eventually the cells retool, replicate their DNA, begin to increase in mass, and finally divide.

The lag phase varies considerably in length with the condition of the microorganisms and the nature of the medium. This phase may be quite long if the inoculum is from an old culture or one that has been refrigerated. Inoculation of a culture into a chemically different medium also results in a longer lag phase. On the other hand, when a young, vigorously growing exponential phase culture is transferred to fresh medium of the same composition, the lag phase will be short or absent.

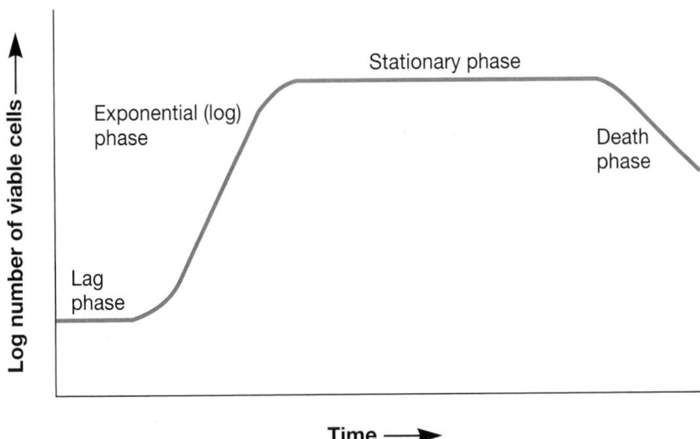

Figure 6.6 Microbial Growth Curve in a Closed System. The four phases of the growth curve are identified on the curve and discussed in the text.

Exponential Phase

During the **exponential** or **log phase,** microorganisms are growing and dividing at the maximal rate possible given their genetic potential, the nature of the medium, and the conditions under which they are growing. Their rate of growth is constant during the exponential phase; that is, the microorganisms are dividing and doubling in number at regular intervals. Because each individual divides at a slightly different moment, the growth curve rises smoothly rather than in discrete jumps (figure 6.6). The population is most uniform in terms of chemical and physiological properties during this phase; therefore exponential phase cultures are usually used in biochemical and physiological studies.

Exponential growth is **balanced growth.** That is, all cellular constituents are manufactured at constant rates relative to each other. If nutrient levels or other environmental conditions change, **unbalanced growth** results. This is growth during which the rates of synthesis of cell components vary relative to one another until a new balanced state is reached. Unbalanced growth is readily observed in two types of experiments: shift-up, where a culture is transferred from a nutritionally poor medium to a richer one; and shift-down, where a culture is transferred from a rich medium to a poor one. In a shift-up experiment, there is a lag while the cells first construct new ribosomes to enhance their capacity for protein synthesis. This is followed by increases in protein and DNA synthesis. Finally, the expected rise in reproductive rate takes place. In a shift-down experiment, there is a lag in growth because cells need time to make the enzymes required for the biosynthesis of unavailable nutrients. Consequently cell division and DNA replication continue after the shift-down, but net protein and RNA synthesis slow. The cells become smaller and reorganize themselves metabolically until they are able to grow again. Then balanced growth is resumed and the culture enters the exponential phase. These shift-up and shift-down experiments demonstrate that microbial growth is under precise, coordinated control and responds quickly to changes in environmental conditions.

When microbial growth is limited by the low concentration of a required nutrient, the final net growth or yield of cells increases with the initial amount of the limiting nutrient present (**figure 6.7*a***). This is the basis of microbiological assays for vitamins and other growth

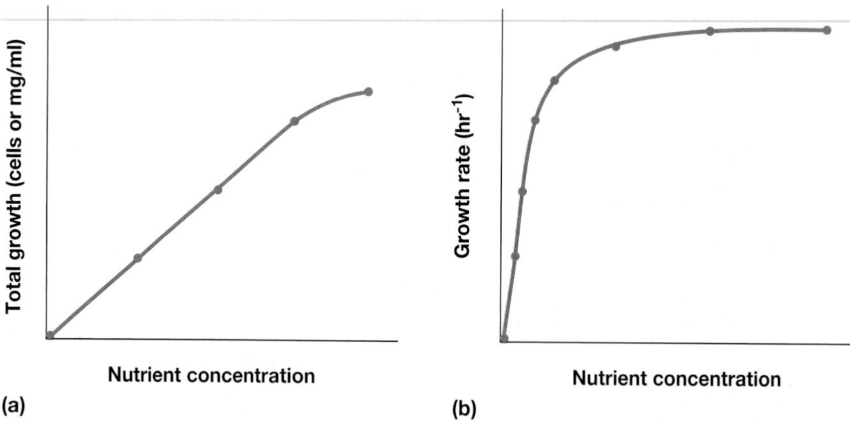

Figure 6.7 Nutrient Concentration and Growth. **(a)** The effect of changes in limiting nutrient concentration on total microbial yield. At sufficiently high concentrations, total growth will plateau. **(b)** The effect on growth rate.

factors. The rate of growth also increases with nutrient concentration (figure 6.7*b*), but in a hyperbolic manner much like that seen with many enzymes (*see figure 8.18*). The shape of the curve seems to reflect the rate of nutrient uptake by microbial transport proteins. At sufficiently high nutrient levels the transport systems are saturated, and the growth rate does not rise further with increasing nutrient concentration. Uptake of nutrients by the cell (section 5.6)

Stationary Phase

Because this is a closed system, eventually population growth ceases and the growth curve becomes horizontal (figure 6.6). This **stationary phase** usually is attained by bacteria at a population level of around 10^9 cells per ml. Other microorganisms normally do not reach such high population densities; protist cultures often have maximum concentrations of about 10^6 cells per ml. Of course final population size depends on nutrient availability and other factors, as well as the type of microorganism being cultured. In the stationary phase the total number of viable microorganisms remains constant. This may result from a balance between cell division and cell death, or the population may simply cease to divide but remain metabolically active.

Microbial populations enter the stationary phase for several reasons. One obvious factor is nutrient limitation; if an essential nutrient is severely depleted, population growth will slow. Aerobic organisms often are limited by O_2 availability. Oxygen is not very soluble and may be depleted so quickly that only the surface of a culture will have an O_2 concentration adequate for growth. The cells beneath the surface will not be able to grow unless the culture is shaken or aerated in another way. Population growth also may cease due to the accumulation of toxic waste products. This factor seems to limit the growth of many anaerobic cultures (cultures growing in the absence of O_2). For example, streptococci can produce so much lactic acid and other organic acids from sugar fermentation that their medium becomes acidic and growth is inhibited. Streptococcal cultures also can enter the stationary phase due to depletion of their sugar supply. Finally, there is some evidence that growth may cease when a critical population level is reached. Thus entrance into the stationary phase may result from several factors operating in concert.

As we have seen, bacteria in a batch culture may enter stationary phase in response to starvation. This probably often occurs in nature because many environments have low nutrient levels. Procaryotes have evolved a number of strategies to survive starvation. Many do not respond with obvious morphological changes such as endospore formation, but only decrease somewhat in overall size, often accompanied by protoplast shrinkage and nucleoid condensation. The more important changes are in gene expression and physiology. Starving bacteria frequently produce a variety of **starvation proteins,** which make the cell much more resistant to damage in a variety of ways. They increase peptidoglycan crosslinking and cell wall strength. The Dps (*DNA*-binding *p*rotein from *s*tarved cells) protein protects DNA. Chaperone proteins prevent protein denaturation and renature damaged proteins. As a result of these and many other mechanisms, the starved cells become harder to kill and more resistant to starvation itself, damaging temperature changes, oxidative and osmotic damage, and toxic chemicals such as chlorine. These changes are so effective that some bacteria can survive starvation for years. There is even evidence that *Salmonella enterica* serovar Typhimurium (*S. typhimurium*), and some other bacterial pathogens become more virulent when starved. Clearly, these considerations are of great practical importance in medical and industrial microbiology.

Senescence and Death

For many years, the decline in viable cells following stationary cells was described simply as the "death phase." It was assumed that detrimental environmental changes like nutrient deprivation and the buildup of toxic wastes caused irreparable harm resulting in loss of viability. That is, even when bacterial cells were transferred to fresh medium, no cellular growth was observed. Because loss of viability was often not accompanied by a loss in total cell number, it was assumed that cells died but did not lyse.

This view is currently under debate. There are two alternative hypotheses (**figure 6.8**). Some microbiologists believe starving cells that show an exponential decline in density have not irreversibly lost their ability to reproduce. Rather, they suggest that microbes are temporarily unable to grow, at least under the labora-

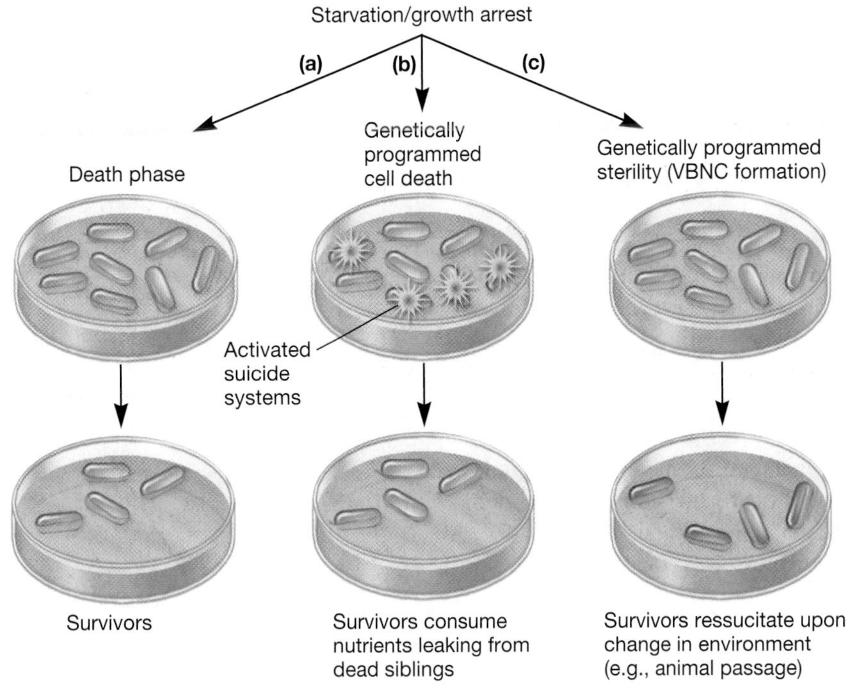

Figure 6.8 Loss of Viability. **(a)** It has long been assumed that as cells leave stationary phase due to starvation or toxic waste accumulation, the exponential decline in culturability is due to cellular death. **(b)** Some believe that a fraction of a microbial population dies due to activation of programmed cell death genes. The nutrients that are released by dying cells supports the growth of other cells. **(c)** The viable but nonculturable (VBNC) hypothesis posits that when cells are starved, they become temporarily nonculturable under laboratory conditions. When exposed to appropriate conditions, some cells will regain the capacity to reproduce.

tory conditions used. This phenomenon, in which the cells are called **viable but nonculturable (VBNC),** is thought to be the result of a genetic response triggered in starving, stationary phase cells. Just as some bacteria form spores as a survival mechanism, it is argued that others are able to become dormant without changes in morphology (figure 6.8*c*). Once the appropriate conditions are available (for instance, a change in temperature or passage through an animal), VBNC microbes resume growth. VBNC microorganisms could pose a public health threat, as many assays that test for food and drinking water safety are culture-based.

The second alternative to a simple death phase is **programmed cell death** (figure 6.8*b*). In contrast to the VBNC hypothesis whereby cells are genetically programmed to survive, programmed cell death predicts that a fraction of the microbial population is genetically programmed to commit suicide. In this case, nonculturable cells are dead (as opposed to nonculturable) and the nutrients that they leak enable the eventual growth of those cells in the population that did not initiate suicide. The dying cells are thus altruistic—that is to say, they sacrifice themselves for the benefit of the larger population.

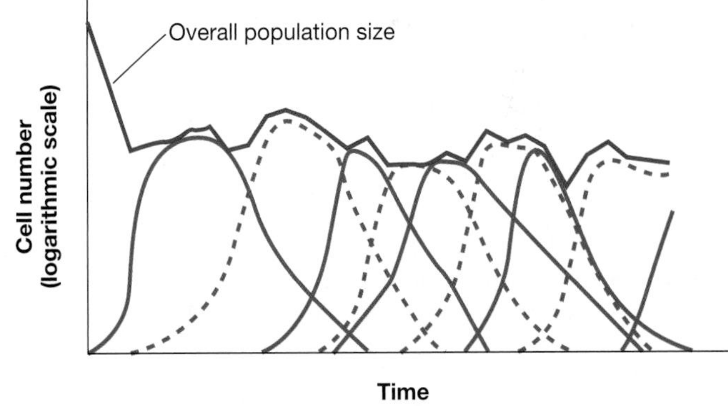

Figure 6.9 Prolonged Decline in Growth. Instead of a distinct death phase, successive waves of genetically distinct subpopulations of microbes better able to use the released nutrients and accumulated toxins survive. Each successive solid or dashed curve represents the growth of a new subpopulation.

Phase of Prolonged Decline

Long-term growth experiments reveal that an exponential decline in viability is sometimes replaced by a gradual decline in the number of culturable cells. This decline can last months to years (**figure 6.9**). During this time the bacterial population continually evolves so that actively reproducing cells are those best able to use the nutrients released by their dying brethren and best able to tolerate the accumulated toxins. This dynamic process is marked by successive waves of genetically distinct variants. Thus natural selection can be witnessed within a single culture vessel.

The Mathematics of Growth

Knowledge of microbial growth rates during the exponential phase is indispensable to microbiologists. Growth rate studies contribute to basic physiological and ecological research and are applied in industry. The quantitative aspects of exponential phase growth discussed here apply to microorganisms that divide by binary fission.

During the exponential phase each microorganism is dividing at constant intervals. Thus the population will double in number during a specific length of time called the **generation time** or **doubling time.** This situation can be illustrated with a simple example. Suppose that a culture tube is inoculated with one cell that divides every 20 minutes (**table 6.1**). The population will be 2 cells after 20 minutes, 4 cells after 40 minutes, and so forth. Because the population is doubling every generation, the increase in population is always 2^n where n is the number of generations. The resulting population increase is exponential or logarithmic (**figure 6.10**).

These observations can be expressed as equations for the generation time.

Let N_0 = the initial population number

N_t = the population at time t

n = the number of generations in time t

Then inspection of the results in table 6.1 will show that

$$N_t = N_0 \times 2^n.$$

Solving for n, the number of generations, where all logarithms are to the base 10,

$$\log N_t = \log N_0 + n \cdot \log 2, \text{ and}$$

$$n = \frac{\log N_t - \log N_0}{\log 2} = \frac{\log N_t - \log N_0}{0.301}$$

The rate of growth during the exponential phase in a batch culture can be expressed in terms of the **mean growth rate constant *(k)*.**

Table 6.1 An Example of Exponential Growth

Time[a]	Division Number	2^n	Population ($N_o \times 2^n$)	$\log_{10}N_t$
0	0	$2^0 = 1$	1	0.000
20	1	$2^1 = 2$	2	0.301
40	2	$2^2 = 4$	4	0.602
60	3	$2^3 = 8$	8	0.903
80	4	$2^4 = 16$	16	1.204
100	5	$2^5 = 32$	32	1.505
120	6	$2^6 = 64$	64	1.806

[a]The hypothetical culture begins with one cell having a 20-minute generation time.

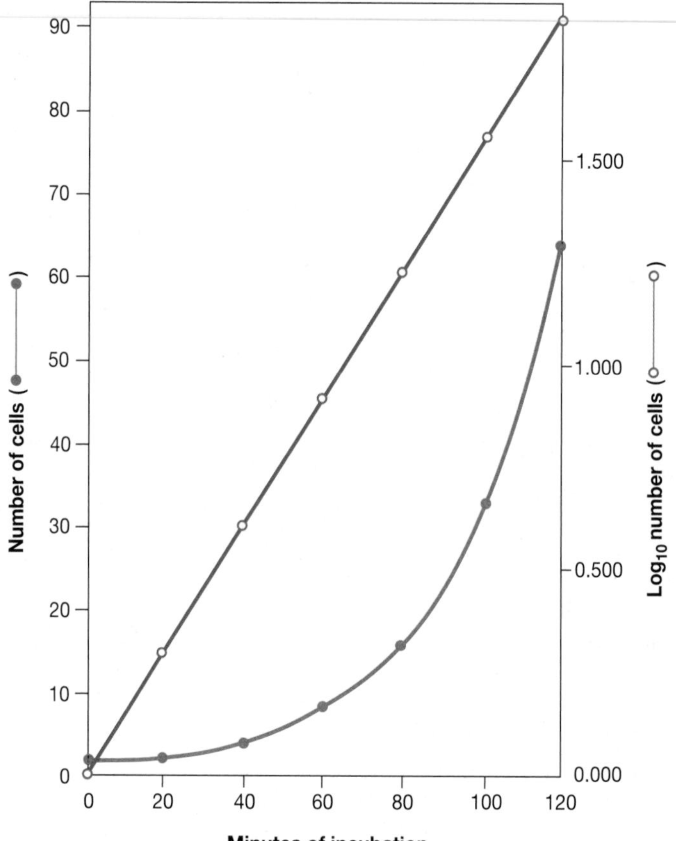

Figure 6.10 Exponential Microbial Growth. The data from table 6.1 for six generations of growth are plotted directly (●—●) and in the logarithmic form (○—○). The growth curve is exponential as shown by the linearity of the log plot.

This is the number of generations per unit time, often expressed as the generations per hour.

$$k = \frac{n}{t} = \frac{\log N_t - \log N_0}{0.301t}$$

The time it takes a population to double in size—that is, the **mean generation time** or mean doubling time (g)—can now be calculated. If the population doubles ($t = g$), then

$$N_t = 2N_0.$$

Substitute $2N_0$ into the mean growth rate equation and solve for k.

$$k = \frac{\log(2N_0) - \log N_0}{0.301\,g} = \frac{\log 2 + \log N_0 - \log N_0}{0.301\,g}$$

$$k = \frac{1}{g}$$

The mean generation time is the reciprocal of the mean growth rate constant.

$$g = \frac{1}{k}$$

The mean generation time (g) can be determined directly from a semilogarithmic plot of the growth data (**figure 6.11**) and the growth rate constant calculated from the g value. The generation time also may be calculated directly from the previous equations. For example, suppose that a bacterial population increases from 10^3 cells to 10^9 cells in 10 hours.

$$k = \frac{\log 10^9 - \log 10^3}{(0.301)(10 \text{ hr})} = \frac{9 - 3}{3.01 \text{ hr}} = 2.0 \text{ generations/hr}$$

$$g = \frac{1}{2.0 \text{ gen./hr}} = 0.5 \text{ hr/gen. or } 30 \text{ min/gen.}$$

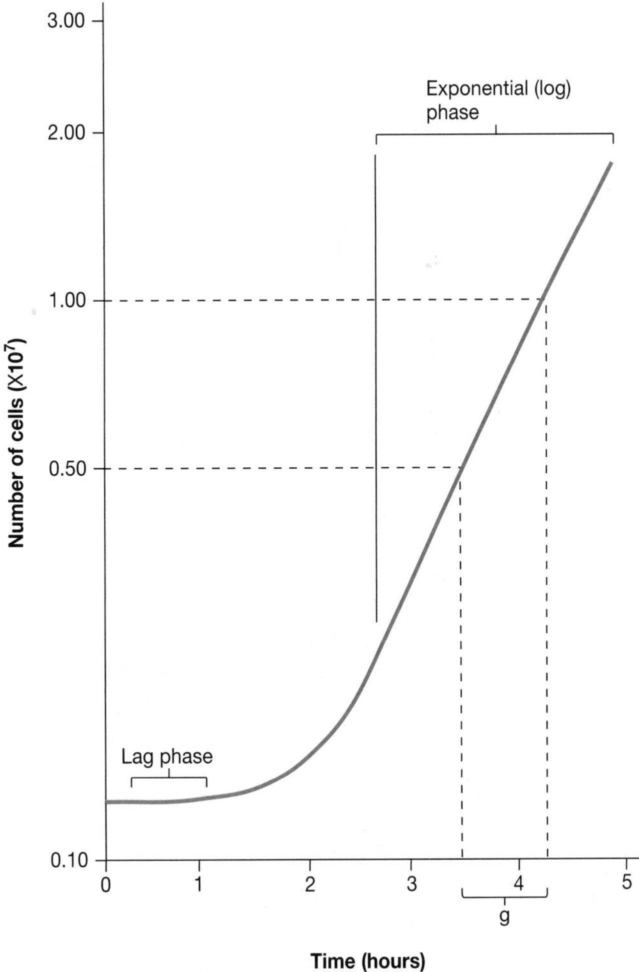

Figure 6.11 Generation Time Determination. The generation time can be determined from a microbial growth curve. The population data are plotted with the logarithmic axis used for the number of cells. The time to double the population number is then read directly from the plot. The log of the population number can also be plotted against time on regular axes.

Table 6.2 Examples of Generation Times[a]

Microorganism	Incubation Temperature (°C)	Generation Time (Hours)
Bacteria		
Beneckea natriegens	37	0.16
Escherichia coli	40	0.35
Bacillus subtilis	40	0.43
Staphylococcus aureus	37	0.47
Pseudomonas aeruginosa	37	0.58
Clostridium botulinum	37	0.58
Rhodospirillum rubrum	25	4.6–5.3
Anabaena cylindrica	25	10.6
Mycobacterium tuberculosis	37	≈12
Treponema pallidum	37	33
Protists		
Tetrahymena geleii	24	2.2–4.2
Scenedesmus quadricauda	25	5.9
Chlorella pyrenoidosa	25	7.75
Asterionella formosa	20	9.6
Leishmania donovani	26	10–12
Paramecium caudatum	26	10.4
Euglena gracilis	25	10.9
Acanthamoeba castellanii	30	11–12
Giardia lamblia	37	18
Ceratium tripos	20	82.8
Fungi		
Saccharomyces cerevisiae	30	2
Monilinia fructicola	25	30

[a] Generation times differ depending on the growth medium and environmental conditions used.

Generation times vary markedly with the species of microorganism and environmental conditions. They range from less than 10 minutes (0.17 hours) for a few bacteria to several days with some eucaryotic microorganisms (**table 6.2**). Generation times in nature are usually much longer than in culture.

1. Define growth. Describe the four phases of the growth curve in a closed system and discuss the causes of each.
2. Why might a culture have a long lag phase after inoculation? Why would cells that are vigorously growing when inoculated into fresh culture medium have a shorter lag phase than those that have been stored in a refrigerator?
3. List two physiological changes that are observed in stationary cells. How do these changes impact the organism's ability to survive?
4. Define balanced growth and unbalanced growth. Why do shift-up and shift-down experiments cause cells to enter unbalanced growth?
5. Define the generation or doubling time and the mean growth rate constant. Calculate the mean growth rate and generation time of a culture that increases in the exponential phase from 5×10^2 to 1×10^8 in 12 hours.

6. Suppose the generation time of a bacterium is 90 minutes and the initial number of cells in a culture is 10^3 cells at the start of the log phase. How many bacteria will there be after 8 hours of exponential growth?
7. What effect does increasing a limiting nutrient have on the yield of cells and the growth rate?
8. Contrast and compare the viable but nonculturable status of microbes with that of programmed cell death as a means of responding to starvation.

6.3 Measurement of Microbial Growth

There are many ways to measure microbial growth to determine growth rates and generation times. Either population number or mass may be followed because growth leads to increases in both. Here the most commonly employed techniques for growth measurement are examined briefly and the advantages and disadvantages of each noted. No single technique is always best; the most appropriate approach will depend on the experimental situation.

Measurement of Cell Numbers

The most obvious way to determine microbial numbers is through direct counting. Using a counting chamber is easy, inexpensive, and relatively quick; it also gives information about the size and morphology of microorganisms. Petroff-Hausser counting chambers can be used for counting procaryotes; hemocytometers can be used for both procaryotes and eucaryotes. These specially designed slides have chambers of known depth with an etched grid on the chamber bottom (**figure 6.12**). Procaryotes are more easily counted in these chambers if they are stained, or when a phase-contrast or a fluorescence microscope is employed. The number of microorganisms in a sample can be calculated by taking into account the chamber's volume and any sample dilutions required. One disadvantage to the technique is that the microbial population must be fairly large for accuracy because such a small volume is sampled.

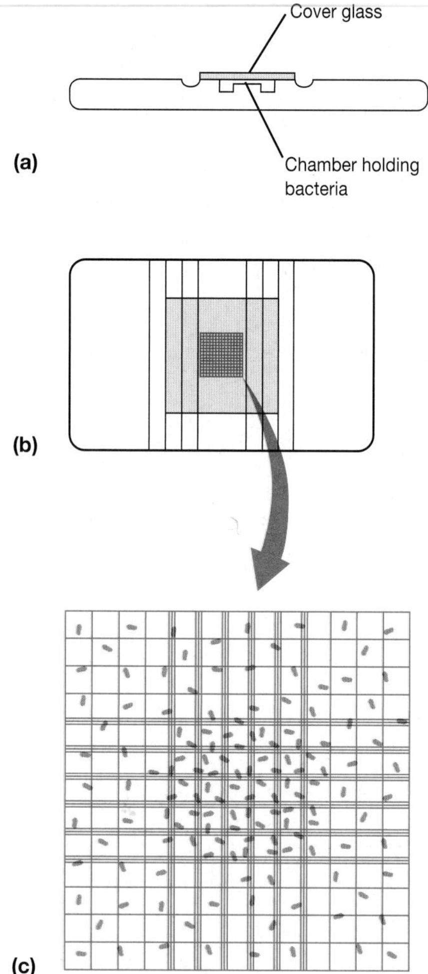

Figure 6.12 The Petroff-Hausser Counting Chamber. **(a)** Side view of the chamber showing the cover glass and the space beneath it that holds a bacterial suspension. **(b)** A top view of the chamber. The grid is located in the center of the slide. **(c)** An enlarged view of the grid. The bacteria in several of the central squares are counted, usually at ×400 to ×500 magnification. The average number of bacteria in these squares is used to calculate the concentration of cells in the original sample. Since there are 25 squares covering an area of 1 mm^2, the total number of bacteria in 1 mm^2 of the chamber is (number/square)(25 squares). The chamber is 0.02 mm deep and therefore,

$$\text{bacteria/mm}^3 = (\text{bacteria/square})(25 \text{ squares})(50).$$

The number of bacteria per cm^3 is 10^3 times this value. For example, suppose the average count per square is 28 bacteria:

$$\text{bacteria/cm}^3 = (28 \text{ bacteria})\ (25 \text{ squares})(50)(10^3) = 3.5 \times 10^7.$$

Larger microorganisms such as protists and yeasts can be directly counted with electronic counters such as the Coulter Counter, although more recently the flow cytometer is increasingly used. The microbial suspension is forced through a small hole or orifice in the Coulter Counter. An electrical current flows through the hole, and electrodes placed on both sides of the orifice measure its electrical resistance. Every time a microbial cell passes through the orifice, electrical resistance increases (or the conductivity drops) and the cell is counted. The Coulter Counter gives accurate results with larger cells and is extensively used in hospital laboratories to count red and white blood cells. It is not as useful in counting bacteria because of interference by small debris particles, the formation of filaments, and other problems. Identification of microorganisms from specimens (section 35.2)

The number of bacteria in aquatic samples is frequently determined from direct counts after the bacteria have been trapped on special membrane filters. In the membrane filter technique, the sample is first filtered through a black polycarbonate membrane filter. Then the bacteria are stained with a fluorescent dye such as acridine orange or the DNA stain DAPI, and observed microscopically. The stained cells are easily observed against the black background of the membrane filter and can be counted when viewed with an epifluorescence microscope. The light microscope: The fluorescence microscope (section 2.2)

Traditional methods for directly counting microbes in a sample usually yield cell densities that are much higher than the plating methods described next because direct counting procedures do not distinguish dead cells from live cells. Newer methods for direct counts avoid this problem. Commercial kits that use fluorescent reagents to stain live and dead cells differently are now available, making it possible to directly count the number of live and dead microorganisms in a sample (*see figures 2.13a and 27.16*).

Several plating methods can be used to determine the number of viable microbes in a sample. These are referred to as viable counting methods because they count only those cells that are alive and able to reproduce. Two commonly used procedures are the spread-plate technique and the pour-plate technique. In both of these methods, a diluted sample of bacteria or other microorganisms is dispersed over a solid agar surface. Each microorganism or group of microorganisms develops into a distinct colony. The original number of viable microorganisms in the sample can be calculated from the number of colonies formed and the sample dilution. For example, if 1.0 ml of a 1×10^{-6} dilution yielded 150 colonies, the original sample contained around 1.5×10^{8} cells per ml. Usually the count is made more accurate by use of a special colony counter. In this way the spread-plate and pour-plate techniques may be used to find the number of microorganisms in a sample. Isolation of pure cultures: The spread plate and streak plate; The pour plate (section 5.8)

Another commonly used plating method first traps bacteria in aquatic samples on a membrane filter. The filter is then placed on an agar medium or on a pad soaked with liquid media (**figure 6.13**) and incubated until each cell forms a separate colony. A colony count gives the number of microorganisms in the filtered sample, and special media can be used to select for specific microorganisms (**figure 6.14**). This technique is especially useful in analyzing water purity. Water purification and sanitary analysis (section 41.1)

Plating techniques are simple, sensitive, and widely used for viable counts of bacteria and other microorganisms in samples of food, water, and soil. Several problems, however, can lead to inaccurate counts. Low counts will result if clumps of cells are not

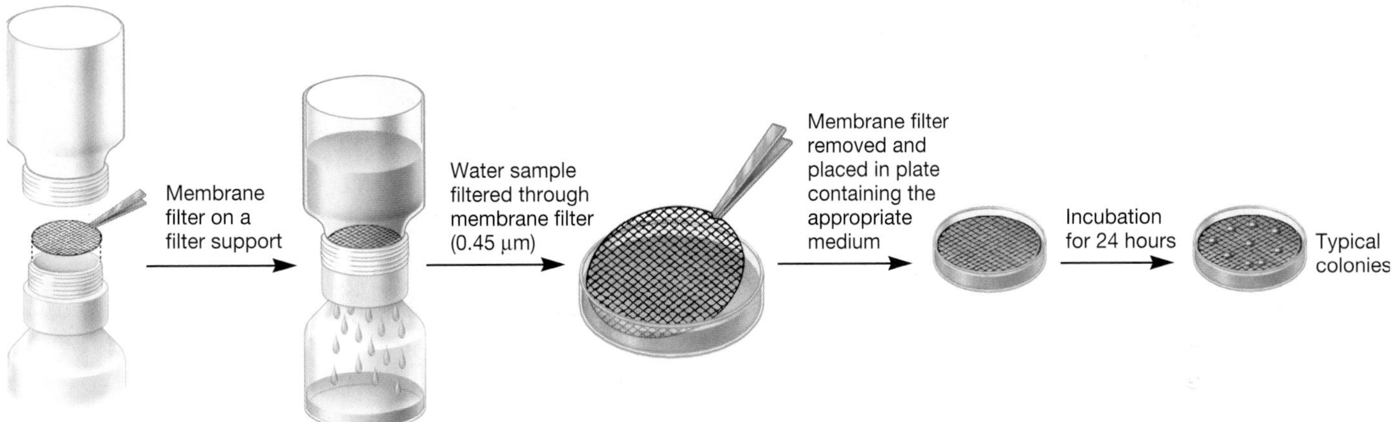

Figure 6.13 The Membrane Filtration Procedure. Membranes with different pore sizes are used to trap different microorganisms. Incubation times for membranes also vary with the medium and microorganism.

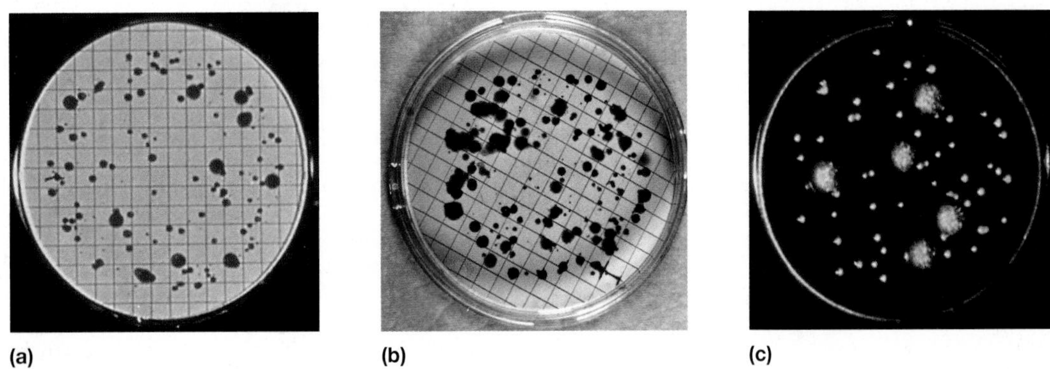

Figure 6.14 Colonies on Membrane Filters. Membrane-filtered samples grown on a variety of media. **(a)** Standard nutrient media for a total bacterial count. An indicator colors colonies red for easy counting. **(b)** Fecal coliform medium for detecting fecal coliforms that form blue colonies. **(c)** Wort agar for the culture of yeasts and molds.

broken up and the microorganisms well dispersed. Because it is not possible to be absolutely certain that each colony arose from an individual cell, the results are often expressed in terms of **colony forming units (CFU)** rather than the number of microorganisms. The samples should yield between 30 and 300 colonies for most accurate counting. Of course the counts will also be low if the agar medium employed cannot support growth of all the viable microorganisms present. The hot agar used in the pour-plate technique may injure or kill sensitive cells; thus spread plates sometimes give higher counts than pour plates.

Measurement of Cell Mass

Increases in the total cell mass, as well as in cell numbers, accompany population growth. Therefore techniques for measuring changes in cell mass can be used in following growth. The most direct approach is the determination of microbial dry weight. Cells growing in liquid medium are collected by centrifugation, washed, dried in an oven, and weighed. This is an especially useful technique for measuring the growth of filamentous fungi. It is time-consuming, however, and not very sensitive. Because bacteria weigh so little, it may be necessary to centrifuge several hundred milliliters of culture to collect a sufficient quantity.

Spectrophotometry can also be used to measure cell mass. These methods are more rapid and sensitive. They depend on the fact that microbial cells scatter light that strikes them. Because microbial cells in a population are of roughly constant size, the amount of scattering is directly proportional to the biomass of cells present and indirectly related to cell number. When the concentration of bacteria reaches about 10 million cells (10^7) per ml, the medium appears slightly cloudy or turbid. Further increases in concentration result in greater turbidity and less light is transmitted through the medium. The extent of light scattering can be measured by a spectrophotometer and is almost linearly related to cell concentration at low absorbance levels (**figure 6.15**). Thus population growth can be easily measured as long as the population is high enough to give detectable turbidity.

If the amount of a substance in each cell is constant, the total quantity of that cell constituent is directly related to the total microbial cell mass. For example, a sample of washed cells collected from a known volume of medium can be analyzed for total protein or nitrogen. An increase in the microbial population will be reflected in higher total protein levels. Similarly, chlorophyll determinations can be used to measure algal and cyanobacterial populations, and the quantity of ATP can be used to estimate the amount of living microbial mass.

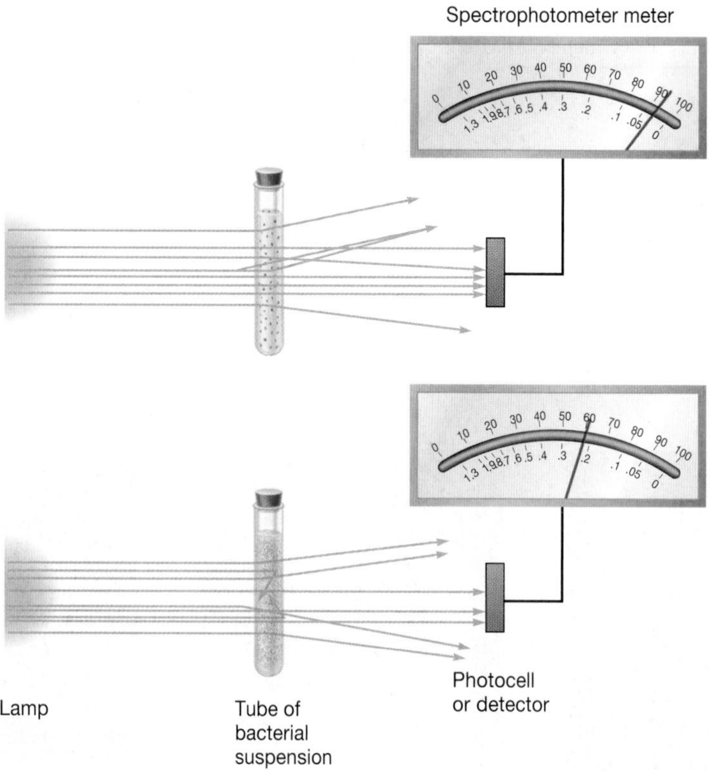

Figure 6.15 Turbidity and Microbial Mass Measurement. Determination of microbial mass by measurement of light absorption. As the population and turbidity increase, more light is scattered and the absorbance reading given by the spectrophotometer increases. The spectrophotometer meter has two scales. The bottom scale displays absorbance and the top scale, percent transmittance. Absorbance increases as percent transmittance decreases.

1. Briefly describe each technique by which microbial population numbers may be determined and give its advantages and disadvantages.
2. When using direct cell counts to follow the growth of a culture, it may be difficult to tell when the culture enters the phase of senescence and death. Why?
3. Why are plate count results expressed as colony forming units?

6.4 The Continuous Culture of Microorganisms

Up to this point the focus has been on closed systems called batch cultures in which nutrient supplies are not renewed nor wastes removed. Exponential growth lasts for only a few generations and soon the stationary phase is reached. However, it is possible to grow microorganisms in an open system, a system with constant environmental conditions maintained through continual provision of nutrients and removal of wastes. These conditions are met in the laboratory by a **continuous culture system.** A microbial population can be maintained in the exponential growth phase and at a constant biomass concentration for extended periods in a continuous culture system.

The Chemostat

Two major types of continuous culture systems commonly are used: (1) chemostats and (2) turbidostats. A **chemostat** is constructed so that sterile medium is fed into the culture vessel at the same rate as the media containing microorganisms is removed (**figure 6.16**). The culture medium for a chemostat possesses an essential nutrient (e.g., an amino acid) in limiting quantities. Because one nutrient is limiting, the growth rate is determined by the rate at which new medium is fed into the growth chamber, and the final cell density depends on the concentration of the limiting nutrient. The rate of nutrient exchange is expressed as the dilution rate (D), the rate at which medium flows through the culture vessel relative to the vessel volume, where f is the flow rate (ml/hr) and V is the vessel volume (ml).

$$D = f/V$$

For example, if f is 30 ml/hr and V is 100 ml, the dilution rate is 0.30 hr^{-1}.

Both the microbial population level and the generation time are related to the dilution rate (**figure 6.17**). The microbial population density remains unchanged over a wide range of dilution rates. The generation time decreases (i.e., the rate of growth increases) as the dilution rate increases. The limiting nutrient will be almost completely depleted under these balanced conditions. If the dilution rate rises too high, the microorganisms can actually be washed out of the culture vessel before reproducing because the dilution rate is greater than the maximum growth rate. This occurs because fewer microorganisms are present to consume the limiting nutrient.

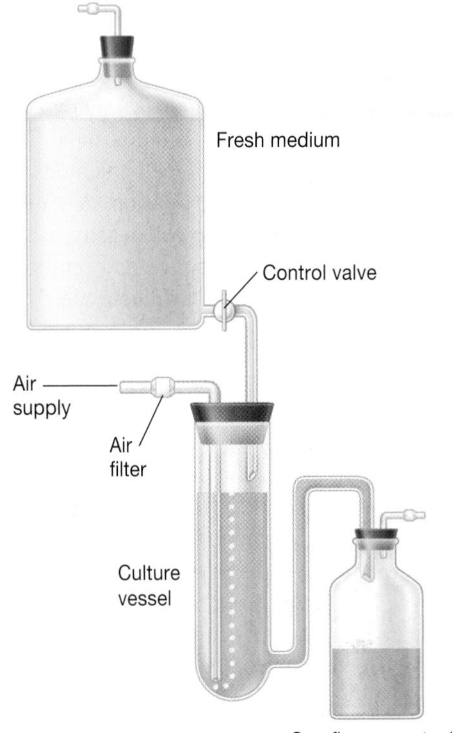

Figure 6.16 A Continuous Culture System: The Chemostat. Schematic diagram of the system. The fresh medium contains a limiting amount of an essential nutrient. Growth rate is determined by the rate of flow of medium through the culture vessel.

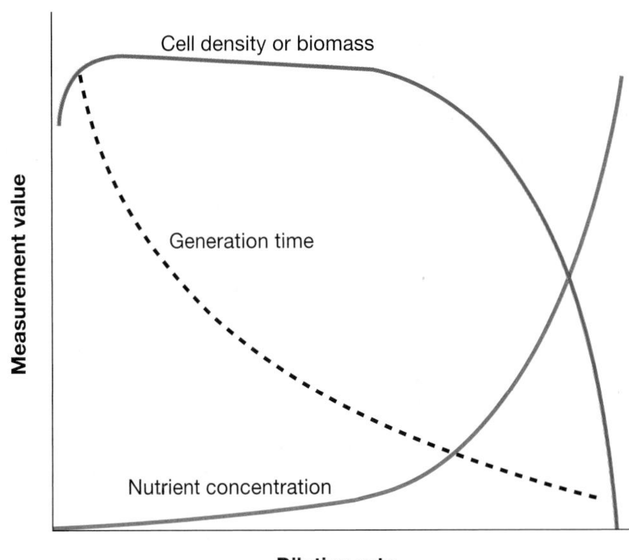

Figure 6.17 Chemostat Dilution Rate and Microbial Growth. The effects of changing the dilution rate in a chemostat.

At very low dilution rates, an increase in D causes a rise in both cell density and the growth rate. This is because of the effect of nutrient concentration on the growth rate, sometimes called the Monod relationship (figure 6.7*b*). Only a limited supply of nutrient is available at low dilution rates. Much of the available energy must be used for cell maintenance, not for growth and reproduction. As the dilution rate increases, the amount of nutrients and the resulting cell density rise because energy is available for both maintenance and reproduction. The growth rate increases when the total available energy exceeds the maintenance energy.

The Turbidostat

The second type of continuous culture system, the **turbidostat,** has a photocell that measures the absorbance or turbidity of the culture in the growth vessel. The flow rate of media through the vessel is automatically regulated to maintain a predetermined turbidity or cell density. The turbidostat differs from the chemostat in several ways. The dilution rate in a turbidostat varies rather than remaining constant, and its culture medium contains all nutrients in excess. That is, none of the nutrients is limiting. The turbidostat operates best at high dilution rates; the chemostat is most stable and effective at lower dilution rates.

Continuous culture systems are very useful because they provide a constant supply of cells in exponential phase and growing at a known rate. They make possible the study of microbial growth at very low nutrient levels, concentrations close to those present in natural environments. These systems are essential for research in many areas—for example, in studies on interactions between microbial species under environmental conditions resembling those in a freshwater lake or pond. Continuous systems also are used in food and industrial microbiology (chapters 40 and 41, respectively).

1. How does an open system differ from a closed culture system or batch culture?
2. Describe how the two different kinds of continuous culture systems, the chemostat and turbidostat, operate.
3. What is the dilution rate? What is maintenance energy?
4. How is the rate of growth of a microbial population controlled in a chemostat? In a turbidostat?

6.5 The Influence of Environmental Factors on Growth

As we have seen, microorganisms must be able to respond to variations in nutrient levels, and particularly to nutrient limitation. The growth of microorganisms also is greatly affected by the chemical and physical nature of their surroundings. An understanding of environmental influences aids in the control of microbial growth and the study of the ecological distribution of microorganisms.

The ability of some microorganisms to adapt to extreme and inhospitable environments is truly remarkable. Procaryotes are present anywhere life can exist. Many habitats in which procaryotes thrive would kill most other organisms. Procaryotes such as *Bacillus infernus* are even able to live over 1.5 miles below the Earth's surface, without oxygen and at temperatures above 60°C. Microorganisms that grow in such harsh conditions are often called **extremophiles.**

In this section we shall briefly review how some of the most important environmental factors affect microbial growth. Major emphasis will be given to solutes and water activity, pH, temperature, oxygen level, pressure, and radiation. **Table 6.3** summarizes the way in which microorganisms are categorized in terms of their response to these factors. It is important to note that for most environmental factors, a range of levels supports growth of a microbe. For example, a microbe might exhibit optimum growth at pH 7, but will grow, though not optimally, at pH values down to pH 6 (its pH minimum) and up to pH 8 (its pH maximum). Furthermore, outside this range, the microbe might cease reproducing but will remain viable for some time. Clearly, each microbe must possess adaptations that allow it to adjust its physiology within its preferred range, and it may also have adaptations that protect it in environments outside this range. These adaptations will also be discussed in this section.

Solutes and Water Activity

Because a selectively permeable plasma membrane separates microorganisms from their environment, they can be affected by changes in the osmotic concentration of their surroundings. If a microorganism is placed in a hypotonic solution (one with a lower osmotic concentration), water will enter the cell and cause it to burst unless something is done to prevent the influx. Conversely if it is placed in a hypertonic solution (one with a higher osmotic concentration), water will flow out of the cell. In microbes that have cell walls (i.e., most procaryotes, fungi, and algae), the membrane shrinks away from the cell wall—a process called plasmolysis. Dehydration of the cell in hypertonic environments may damage the cell membrane and cause the cell to become metabolically inactive.

It is important, then, that microbes be able to respond to changes in the osmotic concentrations of their environment. For instance, microbes in hypotonic environments can reduce the osmotic concentration of their cytoplasm. This can be achieved by the use of inclusion bodies. Some bacteria and archaea also have mechanosensitive (MS) channels in their plasma membrane. In a hypotonic environment, the membrane stretches due to an increase in hydrostatic pressure and cellular swelling. MS channels then open and allow solutes to leave. Thus they can act as escape valves to protect cells from bursting. Since many protists do not have a cell wall, they must use contractile vacuoles (*see figures 25.5 and 25.17b*) to expel excess water. Many microorganisms, whether in hypotonic or hypertonic environments, keep the osmotic concentration of their protoplasm somewhat above that of the habitat by the use of compatible solutes, so that the plasma membrane is always pressed firmly against their cell wall. **Compatible solutes** are solutes that do not interfere with metabolism and growth when at high intracellular concentrations. Most procaryotes increase their

Table 6.3 Microbial Responses to Environmental Factors

Descriptive Term	Definition	Representative Microorganisms
Solute and Water Activity		
Osmotolerant	Able to grow over wide ranges of water activity or osmotic concentration	*Staphylococcus aureus, Saccharomyces rouxii*
Halophile	Requires high levels of sodium chloride, usually above about 0.2 M, to grow	*Halobacterium, Dunaliella, Ectothiorhodospira*
pH		
Acidophile	Growth optimum between pH 0 and 5.5	*Sulfolobus, Picrophilus, Ferroplasma, Acontium, Cyanidium caldarium*
Neutrophile	Growth optimum between pH 5.5 and 8.0	*Escherichia, Euglena, Paramecium*
Alkalophile	Growth optimum between pH 8.0 and 11.5	*Bacillus alcalophilus, Natronobacterium*
Temperature		
Psychrophile	Grows well at 0°C and has an optimum growth temperature of 15°C or lower	*Bacillus psychrophilus, Chlamydomonas nivalis*
Psychrotroph	Can grow at 0–7°C; has an optimum between 20 and 30°C and a maximum around 35°C	*Listeria monocytogenes, Pseudomonas fluorescens*
Mesophile	Has growth optimum around 20–45°C	*Escherichia coli, Neisseria gonorrhoeae, Trichomonas vaginalis*
Thermophile	Can grow at 55°C or higher; optimum often between 55 and 65°C	*Geobacillus stearothermophilus, Thermus aquaticus, Cyanidium caldarium, Chaetomium thermophile*
Hyperthermophile	Has an optimum between 80 and about 113°C	*Sulfolobus, Pyrococcus, Pyrodictium*
Oxygen Concentration		
Obligate aerobe	Completely dependent on atmospheric O_2 for growth	*Micrococcus luteus, Pseudomonas, Mycobacterium;* most protists and fungi
Facultative anaerobe	Does not require O_2 for growth, but grows better in its presence	*Escherichia, Enterococcus, Saccharomyces cerevisiae*
Aerotolerant anaerobe	Grows equally well in presence or absence of O_2	*Streptococcus pyogenes*
Obligate anaerobe	Does not tolerate O_2 and dies in its presence	*Clostridium, Bacteroides, Methanobacterium, Trepomonas agilis*
Microaerophile	Requires O_2 levels below 2–10% for growth and is damaged by atmospheric O_2 levels (20%)	*Campylobacter, Spirillum volutans, Treponema pallidum*
Pressure		
Barophilic	Growth more rapid at high hydrostatic pressures	*Photobacterium profundum, Shewanella benthica, Methanocaldococcus jannaschii*

internal osmotic concentration in a hypertonic environment through the synthesis or uptake of choline, betaine, proline, glutamic acid, and other amino acids; elevated levels of potassium ions are also involved to some extent. Photosynthetic protists and fungi employ sucrose and polyols—for example, arabitol, glycerol, and mannitol—for the same purpose. Polyols and amino acids are ideal solutes for this function because they normally do not disrupt enzyme structure and function. The cytoplasmic matrix: Inclusion bodies (section 3.3)

Some microbes are adapted to extreme hypertonic environments. **Halophiles** grow optimally in the presence of NaCl or other salts at a concentration above about 0.2 M (**figure 6.18**). Extreme halophiles have adapted so completely to hypertonic, saline conditions that they require high levels of sodium chloride to grow—concentrations between about 2 M and saturation (about 6.2 M). The archaeon *Halobacterium* can be isolated from the Dead Sea (a salt lake between Israel and Jordan and the lowest lake in the world), the Great Salt Lake in Utah, and other aquatic habitats with salt concentrations approaching saturation. *Halobacterium* and other extremely halophilic procaryotes accumulate enormous quantities of potassium in order to remain hypertonic to their environment; the internal potassium concentration may reach 4 to 7 M. Furthermore, their enzymes, ribosomes, and transport proteins require high potassium levels for stability and activity. In addition, the plasma membrane and cell wall of *Halobacterium* are stabilized by high concentrations of sodium ion. If the sodium concentration decreases too much, the wall and plasma membrane disintegrate. Extreme halophiles have successfully adapted to environmental conditions that would destroy most organisms. In the process they have become so specialized that they have lost ecological flexibility and can prosper only in a few extreme habitats. Phylum *Euryarchaeota*: The Halobacteria (section 20.3)

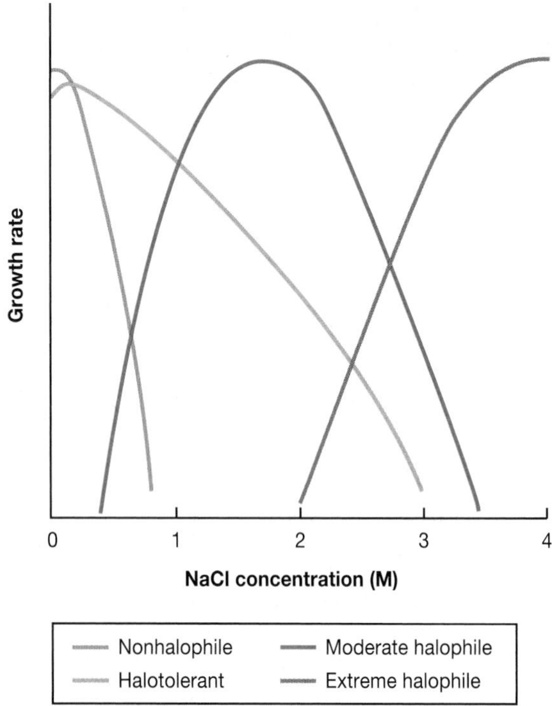

Figure 6.18 The Effects of Sodium Chloride on Microbial Growth. Four different patterns of microbial dependence on NaCl concentration are depicted. The curves are only illustrative and are not meant to provide precise shapes or salt concentrations required for growth.

Because the osmotic concentration of a habitat has such profound effects on microorganisms, it is useful to be able to express quantitatively the degree of water availability. Microbiologists generally use **water activity (a_w)** for this purpose (water availability also may be expressed as water potential, which is related to a_w). The water activity of a solution is 1/100 the relative humidity of the solution (when expressed as a percent). It is also equivalent to the ratio of the solution's vapor pressure (P_{soln}) to that of pure water (P_{water}).

$$a_w = \frac{P_{soln}}{P_{water}}$$

The water activity of a solution or solid can be determined by sealing it in a chamber and measuring the relative humidity after the system has come to equilibrium. Suppose after a sample is treated in this way, the air above it is 95% saturated—that is, the air contains 95% of the moisture it would have when equilibrated at the same temperature with a sample of pure water. The relative humidity would be 95% and the sample's water activity, 0.95. Water activity is inversely related to osmotic pressure; if a solution has high osmotic pressure, it's a_w is low.

Microorganisms differ greatly in their ability to adapt to habitats with low water activity (**table 6.4**). A microorganism must expend extra effort to grow in a habitat with a low a_w value because it must maintain a high internal solute concentration to retain water. Some microorganisms can do this and are **osmotolerant;** they will grow over wide ranges of water activity or osmotic concentration. For example, *Staphylococcus aureus* is halotolerant (figure 6.18) and can be cultured in media containing sodium chloride concentration up to about 3 M. It is well adapted for growth on the skin. The yeast *Saccharomyces rouxii* will grow in sugar solutions with a_w values as low as 0.6. The photosynthetic protist *Dunaliella viridis* tolerates sodium chloride concentrations from 1.7 M to a saturated solution.

Although a few microorganisms are truly osmotolerant, most only grow well at water activities around 0.98 (the approximate a_w for seawater) or higher. This is why drying food or adding large quantities of salt and sugar is so effective in preventing food spoilage. As table 6.4 shows, many fungi are osmotolerant and thus particularly important in the spoilage of salted or dried foods. Controlling food spoilage (section 40.3)

1. How do microorganisms adapt to hypotonic and hypertonic environments? What is plasmolysis?
2. Define water activity and briefly describe how it can be determined.
3. Why is it difficult for microorganisms to grow at low a_w values?
4. What are halophiles and why does *Halobacterium* require sodium and potassium ions?

pH

pH is a measure of the hydrogen ion activity of a solution and is defined as the negative logarithm of the hydrogen ion concentration (expressed in terms of molarity).

$$pH = -\log [H^+] = \log(1/[H^+])$$

The pH scale extends from pH 0.0 (1.0 M H^+) to pH 14.0 (1.0×10^{-14} M H^+), and each pH unit represents a tenfold change in hydrogen ion concentration. **Figure 6.19** shows that the habitats in which microorganisms grow vary widely—from pH 0 to 2 at the acidic end to alkaline lakes and soil that may have pH values between 9 and 10.

It is not surprising that pH dramatically affects microbial growth. Each species has a definite pH growth range and pH growth optimum. **Acidophiles** have their growth optimum between pH 0 and 5.5; **neutrophiles,** between pH 5.5 and 8.0; and **alkalophiles** prefer the pH range of 8.0 to 11.5. Extreme alkalophiles have growth optima at pH 10 or higher. In general, different microbial groups have characteristic pH preferences. Most bacteria and protists are neutrophiles. Most fungi prefer more acidic surroundings, about pH 4 to 6; photosynthetic protists also seem to favor slight acidity. Many archaea are acidophiles. For example, the archaeon *Sulfolobus acidocaldarius* is a common inhabitant of acidic hot springs; it grows well around pH 1 to 3 and at high temperatures. The archaea *Ferroplasma acidarmanus* and *Picrophilus oshimae* can actually grow at pH 0, or very close to it.

Although microorganisms will often grow over wide ranges of pH and far from their optima, there are limits to their tolerance.

Table 6.4 Approximate Lower a_w Limits for Microbial Growth

Water Activity	Environment	Procaryotes	Fungi	Photosynthetic protists
1.00—Pure water	Blood, Plant wilt { Vegetables, meat, fruit; Seawater	Most gram-negative bacteria and other nonhalophiles		
0.95	Bread	Most gram-positive rods	*Basidiomycetes*	Most genera
0.90	Ham	Most cocci, *Bacillus*	*Fusarium* *Mucor, Rhizopus* Ascomycetous yeasts	
0.85	Salami	*Staphylococcus*	*Saccharomyces rouxii* (in salt)	
0.80	Preserves		*Penicillium*	
0.75	Salt lakes Salted fish	*Halobacterium* *Actinospora*	*Aspergillus*	*Dunaliella*
0.70	Cereals, candy, dried fruit		*Aspergillus*	
0.60	Chocolate Honey Dried milk		*Saccharomyces rouxii* (in sugars) *Xeromyces bisporus*	
0.55—DNA disordered				

Drastic variations in cytoplasmic pH can harm microorganisms by disrupting the plasma membrane or inhibiting the activity of enzymes and membrane transport proteins. Most procaryotes die if the internal pH drops much below 5.0 to 5.5. Changes in the external pH also might alter the ionization of nutrient molecules and thus reduce their availability to the organism.

Microorganisms respond to external pH changes using mechanisms that maintain a neutral cytoplasmic pH. Several mechanisms for adjusting to small changes in external pH have been proposed. The plasma membrane is impermeable to protons. Neutrophiles appear to exchange potassium for protons using an antiport transport system. Extreme alkalophiles like *Bacillus alcalophilus* maintain their internal pH closer to neutrality by exchanging internal sodium ions for external protons. Internal buffering also may contribute to pH homeostasis. However, if the external pH becomes too acidic, other mechanisms come into play. When the pH drops below about 5.5 to 6.0, *Salmonella enterica* serovar Typhimurium and *E. coli* synthesize an array of new proteins as part of what has been called their acidic tolerance response. A proton-translocating ATPase contributes to this protective response, either by making more ATP or by pumping protons out of the cell. If the external pH decreases to 4.5 or lower, chaperone proteins such as acid shock proteins and heat shock proteins are synthesized. These prevent the acid denaturation of proteins and aid in the refolding of denatured proteins. Uptake of nutrients by the cell (section 5.6); Translation: Protein folding and molecular chaperones (section 11.8)

Microorganisms frequently change the pH of their own habitat by producing acidic or basic metabolic waste products. Fermentative microorganisms form organic acids from carbohydrates, whereas chemolithotrophs like *Thiobacillus* oxidize reduced sulfur components to sulfuric acid. Other microorganisms make their environment more alkaline by generating ammonia through amino acid degradation. Fermentation (section 9.7); Chemolithotrophy (section 9.11)

Because microorganisms change the pH of their surroundings, buffers often are included in media to prevent growth inhibition by large pH changes. Phosphate is a commonly used buffer and a good example of buffering by a weak acid ($H_2PO_4^-$) and its conjugate base (HPO_4^{2-}).

$$H^+ + HPO_4^{2-} \longrightarrow H_2PO_4^-$$

$$OH^- + H_2PO_4^- \longrightarrow HPO_4^{2-} + HOH$$

If protons are added to the mixture, they combine with the salt form to yield a weak acid. An increase in alkalinity is resisted because the weak acid will neutralize hydroxyl ions through proton donation to give water. Peptides and amino acids in complex media also have a strong buffering effect.

1. Define pH, acidophile, neutrophile, and alkalophile.
2. Classify each of the following organisms as an alkalophile, a neutrophile, or an acidophile: *Staphylococcus aureus, Microcycstis aeruginosa, Sulfolobus acidocaldarius,* and *Pseudomonas aeruginosa.* Which might be pathogens? Explain your choices.

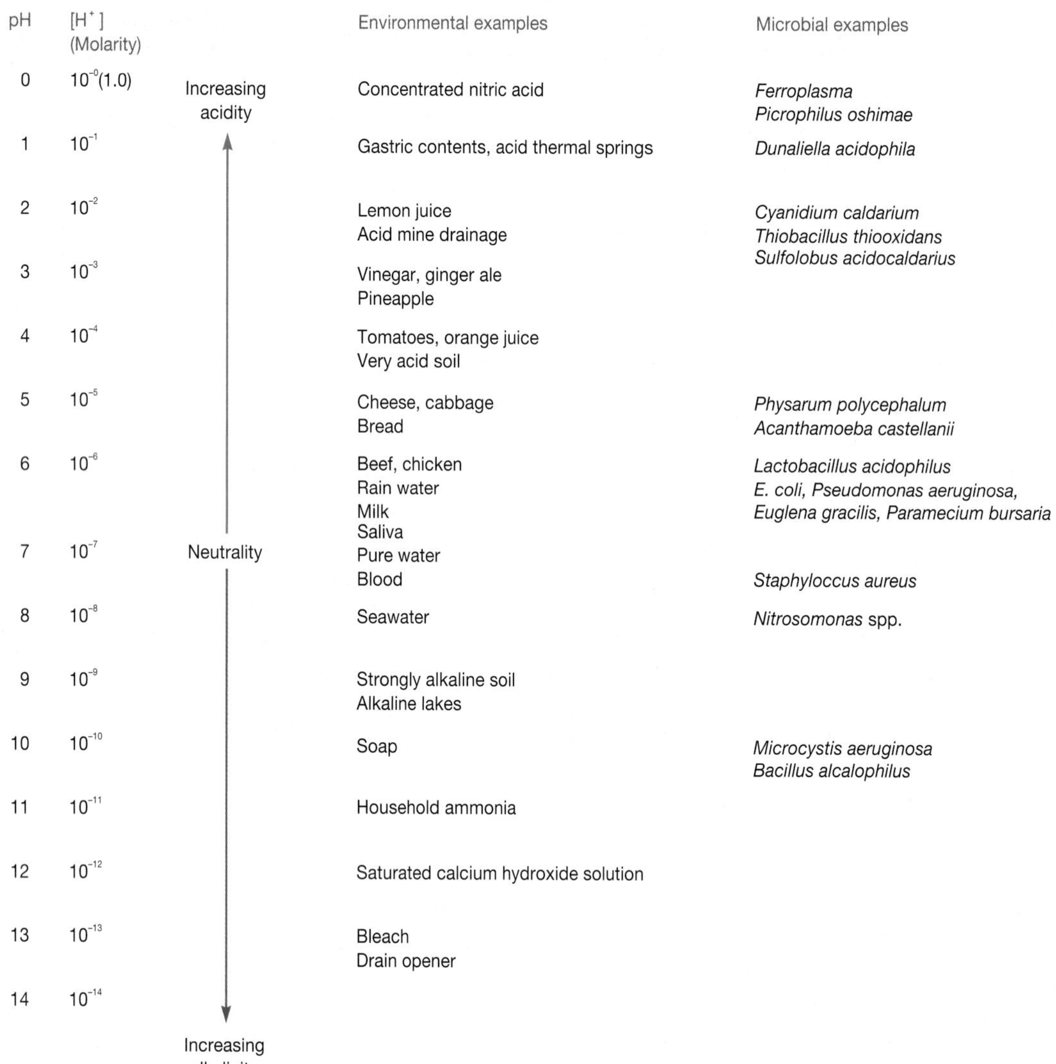

pH
[H⁺] (Molarity)
Environmental examples
Microbial examples
0
10⁻⁰(1.0)
Increasing acidity
Concentrated nitric acid
Ferroplasma
Picrophilus oshimae
1
10⁻¹
Gastric contents, acid thermal springs
Dunaliella acidophila
2
10⁻²
Lemon juice
Acid mine drainage
Cyanidium caldarium
Thiobacillus thiooxidans
Sulfolobus acidocaldarius
3
10⁻³
Vinegar, ginger ale
Pineapple
4
10⁻⁴
Tomatoes, orange juice
Very acid soil
5
10⁻⁵
Cheese, cabbage
Bread
Physarum polycephalum
Acanthamoeba castellanii
6
10⁻⁶
Beef, chicken
Rain water
Milk
Saliva
Lactobacillus acidophilus
E. coli, Pseudomonas aeruginosa,
Euglena gracilis, Paramecium bursaria
7
10⁻⁷
Neutrality
Pure water
Blood
Staphyloccus aureus
8
10⁻⁸
Seawater
Nitrosomonas spp.
9
10⁻⁹
Strongly alkaline soil
Alkaline lakes
10
10⁻¹⁰
Soap
Microcystis aeruginosa
Bacillus alcalophilus
11
10⁻¹¹
Household ammonia
12
10⁻¹²
Saturated calcium hydroxide solution
13
10⁻¹³
Bleach
Drain opener
14
10⁻¹⁴
Increasing alkalinity

Figure 6.19 The pH Scale. The pH scale and examples of substances with different pH values. The microorganisms are placed at their growth optima.

3. Describe the mechanisms microbes use to maintain a neutral pH. Explain how extreme pH values might harm microbes.
4. How do microorganisms change the pH of their environment? How does the microbiologist minimize this effect when culturing microbes in the lab?

Temperature

Environmental temperature profoundly affects microorganisms, like all other organisms. Indeed, microorganisms are particularly susceptible because their temperature varies with that of the external environment. A most important factor influencing the effect of temperature on growth is the temperature sensitivity of enzyme-catalyzed reactions. Each enzyme has a temperature at which it functions optimally (*see figure 8.19b*). At some temperature below the optimum, it ceases to be catalytic. As the temperature rises from this low temperature, the rate of catalysis increases to that observed for the optimal temperature. The velocity of the reaction will roughly double for every 10°C rise in temperature. When all enzymes in a microbe are considered together, as the rate of each reaction increases, metabolism as a whole becomes more active, and the microorganism grows faster. However, beyond a certain point, further increases actually slow growth, and sufficiently high temperatures are lethal. High temperatures damage microorganisms by denaturing enzymes, transport carriers, and other proteins. Temperature also has a significant effect on microbial membranes. At very low temperatures, membranes solidify. At high temperatures, the lipid bilayer simply melts and disintegrates. In summary, when organisms are above their optimum temperature, both function and cell structure are affected. If temperatures are very low, function is affected but not necessarily cell chemical composition and structure.

Because of these opposing temperature influences, microbial growth has a fairly characteristic temperature dependence with distinct **cardinal temperatures**—minimum, optimum, and max-

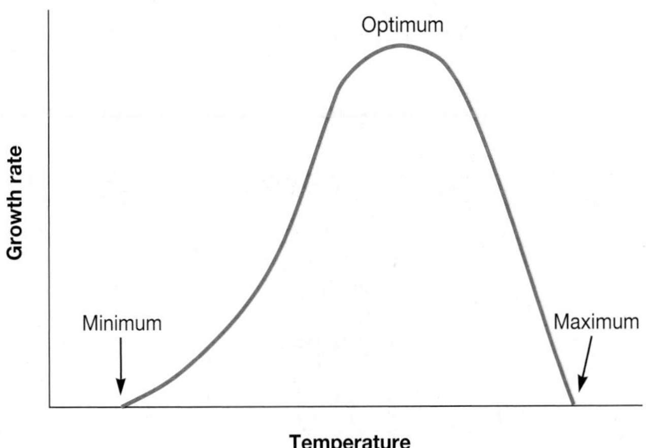

Figure 6.20 Temperature and Growth. The effect of temperature on growth rate.

Table 6.5 Temperature Ranges for Microbial Growth			
	Cardinal Temperatures (°C)		
Microorganism	**Minimum**	**Optimum**	**Maximum**
Nonphotosynthetic Procaryotes			
Bacillus psychrophilus	−10	23–24	28–30
Micrococcus cryophilus	−4	10	24
Pseudomonas fluorescens	4	25–30	40
Staphylococcus aureus	6.5	30–37	46
Enterococcus caecalis	0	37	44
Escherichia coli	10	37	45
Neisseria gonorrhoeae	30	35–36	38
Thermoplasma acidophilum	45	59	62
Bacillus stearothermophilus	30	60–65	75
Thermus aquaticus	40	70–72	79
Sulfolobus acidocaldarius	60	80	85
Pyrococcus abyssi	67	96	102
Pyrodictium occultum	82	105	110
Pyrolobus fumarii	0	106	113
Photosynthetic Bacteria			
Rhodospirillum rubrum	ND[a]	30–35	ND
Anabaena variabilis	ND	35	ND
Oscillatoria tenuis	ND	ND	45–47
Synechococcus eximius	70	79	84
Protists			
Chlamydomonas nivalis	−36	0	4
Fragilaria sublinearis	−2	5–6	8–9
Amoeba proteus	4–6	22	35
Euglena gracilis	ND	23	ND
Skeletonema costatum	6	16–26	>28
Naegleria fowleri	20–25	35	40
Trichomonas vaginalis	25	32–39	42
Paramecium caudatum		25	28–30
Tetrahymena pyriformis	6–7	20–25	33
Cyclidium citrullus	18	43	47
Cyanidium caldarium	30–34	45–50	56
Fungi			
Candida scotti	0	4–15	15
Saccharomyces cerevisiae	1–3	28	40
Mucor pusillus	21–23	45–50	50–58

[a]ND, no data.

imum growth temperatures (**figure 6.20**). Although the shape of the temperature dependence curve can vary, the temperature optimum is always closer to the maximum than to the minimum. The cardinal temperatures for a particular species are not rigidly fixed but often depend to some extent on other environmental factors such as pH and the available nutrients. For example, *Crithidia fasciculate,* a flagellated protist living in the gut of mosquitoes, will grow in a simple medium at 22 to 27°C. However, it cannot be cultured at 33 to 34°C without the addition of extra metals, amino acids, vitamins, and lipids.

The cardinal temperatures vary greatly between microorganisms (**table 6.5**). Optima usually range from 0°C to 75°C, whereas microbial growth occurs at temperatures extending from less than −20°C to over 120°C. Some archaea can even grow at 121°C (250°F), the temperature normally used in autoclaves (**Microbial Diversity and Ecology 6.1**). The major factor determining this growth range seems to be water. Even at the most extreme temperatures, microorganisms need liquid water to grow. The growth temperature range for a particular microorganism usually spans about 30 degrees. Some species (e.g., *Neisseria gonorrhoeae*) have a small range; others, like *Enterococcus faecalis,* will grow over a wide range of temperatures. The major microbial groups differ from one another regarding their maximum growth temperatures. The upper limit for protists is around 50°C. Some fungi can grow at temperatures as high as 55 to 60°C. Procaryotes can grow at much higher temperatures than eucaryotes. It has been suggested that eucaryotes are not able to manufacture organellar membranes that are stable and functional at temperatures above 60°C. The photosynthetic apparatus also appears to be relatively unstable because photosynthetic organisms are not found growing at very high temperatures.

Microorganisms such as those listed in table 6.5 can be placed in one of five classes based on their temperature ranges for growth (**figure 6.21**).

1. **Psychrophiles** grow well at 0°C and have an optimum growth temperature of 15°C or lower; the maximum is around 10°C. They are readily isolated from Arctic and Antarctic habitats; because 90% of the ocean is 5°C or colder, it constitutes an enormous habitat for psychrophiles. The psychrophilic protist *Chlamydomonas nivalis* can actually turn a snowfield or glacier pink with its bright red spores. Psychrophiles are widespread among bacterial taxa and are found in such genera as *Pseudomonas, Vibrio, Alcaligenes, Bacillus, Arthrobacter,*

Microbial Diversity & Ecology

6.1 Life above 100°C

Until recently the highest reported temperature for procaryotic growth was 105°C. It seemed that the upper temperature limit for life was about 100°C, the boiling point of water. Now thermophilic procaryotes have been reported growing in sulfide chimneys or "black smokers," located along rifts and ridges on the ocean floor, that spew sulfide-rich super-heated vent water with temperatures above 350°C (see **Box figure**). Evidence has been presented that these microbes can grow and reproduce at 121°C and can survive temperatures to 130°C for up to 2 hours. The pressure present in their habitat is sufficient to keep water liquid (at 265 atm; seawater doesn't boil until 460°C).

The implications of this discovery are many. The proteins, membranes, and nucleic acids of these procaryotes are remarkably temperature stable and provide ideal subjects for studying the ways in which macromolecules and membranes are stabilized. In the future it may be possible to design enzymes that can operate at very high temperatures. Some thermostable enzymes from these organisms have important industrial and scientific uses. For example, the Taq polymerase from the thermophile *Thermus aquaticus* is used extensively in the polymerase chain reaction. The polymerase chain reaction (section 14.3)

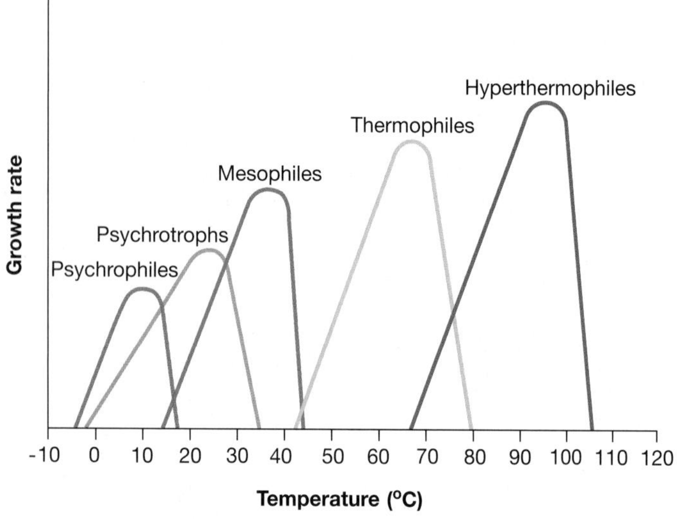

Figure 6.21 Temperature Ranges for Microbial Growth. Microorganisms can be placed in different classes based on their temperature ranges for growth. They are ranked in order of increasing growth temperature range as psychrophiles, psychrotrophs, mesophiles, thermophiles, and hyperthermophiles. Representative ranges and optima for these five types are illustrated here.

Moritella, Photobacterium, and *Shewanella*. A psychrophilic archaeon, *Methanogenium,* has been isolated from Ace Lake in Antarctica. Psychrophilic microorganisms have adapted to their environment in several ways. Their enzymes, transport systems, and protein synthetic mechanisms function well at low temperatures. The cell membranes of psychrophilic microorganisms have high levels of unsaturated fatty acids and remain semifluid when cold. Indeed, many psychrophiles begin to leak cellular constituents at temperatures higher than 20°C because of cell membrane disruption.

2. Many species can grow at 0 to 7°C even though they have optima between 20 and 30°C, and maxima at about 35°C. These are called **psychrotrophs** or **facultative psychrophiles.** Psychrotrophic bacteria and fungi are major factors in the spoilage of refrigerated foods as described in chapter 40.
3. **Mesophiles** are microorganisms with growth optima around 20 to 45°C; they often have a temperature minimum of 15 to 20°C. Their maximum is about 45°C or lower. Most microorganisms probably fall within this category. Almost all human pathogens are mesophiles, as might be expected because their environment is a fairly constant 37°C.
4. Some microorganisms are **thermophiles;** they can grow at temperatures of 55°C or higher. Their growth minimum is usually around 45°C and they often have optima between 55 and 65°C. The vast majority are procaryotes although a few photosynthetic protists and fungi are thermophilic (table 6.5). These organisms flourish in many habitats including composts, self-heating hay stacks, hot water lines, and hot springs.

Thermophiles differ from mesophiles in many ways. They have more heat-stable enzymes and protein synthesis systems, which function properly at high temperatures. These proteins are stable for a variety of reasons. Heat-stable proteins have highly organized, hydrophobic interiors; more hydrogen bonds and other noncovalent bonds strengthen the structure. Larger quantities of amino acids such as proline also make the polypeptide chain less flexible. In addition, the proteins are stabilized and aided in folding by special chaperone proteins. There is evidence that in thermophilic bacteria, DNA is stabilized by special histonelike proteins. Their membrane lipids are also quite temperature stable. They tend to be more saturated, more branched, and of higher molecular weight. This increases the melting points of membrane lipids. Archaeal thermophiles have membrane lipids with ether linkages, which protect the lipids from hydrolysis at high temperatures. Sometimes archaeal lipids actually span the membrane to form a rigid, stable monolayer. Proteins (appendix I); Procaryotic cell membranes (section 3.2)

5. As mentioned previously, a few thermophiles can grow at 90°C or above and some have maxima above 100°C. Procaryotes that have growth optima between 80°C and about 113°C are called **hyperthermophiles.** They usually do not grow well below 55°C. *Pyrococcus abyssi* and *Pyrodictium occultum* are examples of marine hyperthermophiles found in hot areas of the seafloor.

1. What are cardinal temperatures?
2. Why does the growth rate rise with increasing temperature and then fall again at higher temperatures?
3. Define psychrophile, psychrotroph, mesophile, thermophile, and hyperthermophile.
4. What metabolic and structural adaptations for extreme temperatures do psychrophiles and thermophiles have?

Oxygen Concentration

The importance of oxygen to the growth of an organism correlates with its metabolism—in particular, with the processes it uses to conserve the energy supplied by its energy source. Almost all energy-conserving metabolic processes involve the movement of electrons through an electron transport system. For chemotrophs, an externally supplied terminal electron acceptor is critical to the functioning of the electron transport system. The nature of the terminal electron acceptor is related to an organism's oxygen requirement.

An organism able to grow in the presence of atmospheric O_2 is an **aerobe,** whereas one that can grow in its absence is an **anaerobe.** Almost all multicellular organisms are completely dependent on atmospheric O_2 for growth—that is, they are **obligate aerobes** (table 6.3). Oxygen serves as the terminal electron acceptor for the electron-transport chain in aerobic respiration. In addition, aerobic eucaryotes employ O_2 in the synthesis of sterols and unsaturated fatty acids. **Facultative anaerobes** do not require O_2 for growth but grow better in its presence. In the presence of oxygen they use aerobic respiration. **Aerotolerant anaerobes** such as *Enterococcus faecalis* simply ignore O_2 and grow equally well whether it is present or not. In contrast, **strict** or **obligate anaerobes** (e.g., *Bacteroides, Fusobacterium, Clostridium pasteurianum, Methanococcus, Neocallimastix*) do not tolerate O_2 at all and die in its presence. Aerotolerant and strict anaerobes cannot generate energy through aerobic respiration and must employ fermentation or anaerobic respiration for this purpose. Finally, there are aerobes such as *Campylobacter,* called **microaerophiles,** that are damaged by the normal atmospheric level of O_2 (20%) and require O_2 levels below the range of 2 to 10% for growth. The nature of bacterial O_2 responses can be readily determined by growing the bacteria in culture tubes filled with a solid culture medium or a special medium like thioglycollate broth, which contains a reducing agent to lower O_2 levels (**figure 6.22**). Oxidation-reduction reactions, electron carriers, and electron transport systems (section 8.6); Aerobic respiration (section 9.2); Anaerobic respiration (section 9.6); Fermentation (section 9.7)

A microbial group may show more than one type of relationship to O_2. All five types are found among the procaryotes and protozoa. Fungi are normally aerobic, but a number of species—particularly among the yeasts—are facultative anaerobes. Photosynthetic protists are almost always obligate aerobes. It should be noted that the ability to grow in both oxic and anoxic environments provides considerable flexibility and is an ecological advantage.

Although obligate anaerobes are killed by O_2, they may be recovered from habitats that appear to be oxic. In such cases they associate with facultative anaerobes that use up the available O_2 and thus make the growth of strict anaerobes possible. For example, the strict anaerobe *Bacteroides gingivalis* lives in the mouth where it grows in the anoxic crevices around the teeth.

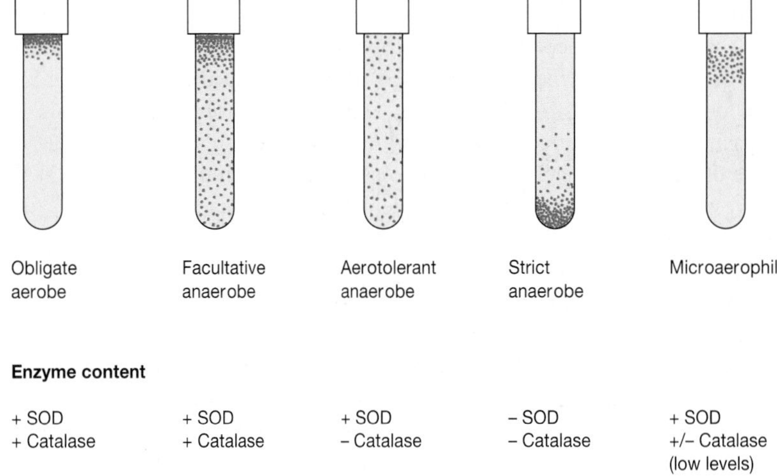

Figure 6.22 Oxygen and Bacterial Growth. Each dot represents an individual bacterial colony within the agar or on its surface. The surface, which is directly exposed to atmospheric oxygen, will be oxic. The oxygen content of the medium decreases with depth until the medium becomes anoxic toward the bottom of the tube. The presence and absence of the enzymes superoxide dismutase (SOD) and catalase for each type are shown.

These different relationships with O_2 are due to several factors, including the inactivation of proteins and the effect of toxic O_2 derivatives. Enzymes can be inactivated when sensitive groups like sulfhydryls are oxidized. A notable example is the nitrogen-fixation enzyme nitrogenase, which is very oxygen sensitive. Synthesis of amino acids: Nitrogen assimilation (section 10.5)

Oxygen accepts electrons and is readily reduced because its two outer orbital electrons are unpaired. Flavoproteins, which function in electron transport, several other cell constituents, and radiation promote oxygen reduction. The result is usually some combination of the reduction products **superoxide radical, hydrogen peroxide,** and **hydroxyl radical.**

$$O_2 = e^- \rightarrow O_2\bar{\bullet} \text{ (superoxide radical)}$$

$$O_2\bar{\bullet} + e^- + 2H^+ \rightarrow H_2O_2 \text{ (hydrogen peroxide)}$$

$$H_2O_2 + e^- + H^+ \rightarrow H_2O + OH\bullet \text{ (hydroxyl radical)}$$

These products of oxygen reduction are extremely toxic because they oxidize and rapidly destroy cellular constituents. A microorganism must be able to protect itself against such oxygen products or it will be killed. Indeed, neutrophils and macrophages, two important immune system cells, use these toxic oxygen products to destroy invading pathogens. Phagocytosis (section 31.3)

Many microorganisms possess enzymes that afford protection against toxic O_2 products (figure 6.22). Obligate aerobes and facultative anaerobes usually contain the enzymes **superoxide dismutase (SOD)** and **catalase,** which catalyze the destruction of superoxide radical and hydrogen peroxide, respectively. Peroxidase also can be used to destroy hydrogen peroxide.

$$2O_2\bar{\bullet} + 2H^+ \xrightarrow{\text{superoxide dismutase}} O_2 + H_2O_2$$

$$2H_2O_2 \xrightarrow{\text{catalase}} 2H_2O + O_2$$

$$H_2O_2 + NADH + H^+ \xrightarrow{\text{peroxidase}} 2H_2O + NAD^+$$

Aerotolerant microorganisms may lack catalase but almost always have superoxide dismutase. The aerotolerant *Lactobacillus plantarum* uses manganous ions instead of superoxide dismutase to destroy the superoxide radical. All strict anaerobes lack both enzymes or have them in very low concentrations and therefore cannot tolerate O_2.

Because aerobes need O_2 and anaerobes are killed by it, radically different approaches must be used when growing the two types of microorganisms. When large volumes of aerobic microorganisms are cultured, either the culture vessel is shaken to aerate the medium or sterile air must be pumped through the culture vessel. Precisely the opposite problem arises with anaerobes; all O_2 must be excluded. This can be accomplished in several ways. (1) Special anaerobic media containing reducing agents such as thioglycollate or cysteine may be used. The medium is boiled during preparation to dissolve its components; boiling also drives off oxygen very effectively. The reducing agents will eliminate any dissolved O_2 remaining within the medium so that anaerobes can grow beneath its surface. (2) Oxygen also may be eliminated from an anaerobic system by removing air with a vacuum pump and flushing out residual O_2 with nitrogen gas (**figure 6.23**). Often CO_2 as well as nitrogen is added to the chamber since many anaerobes require a small amount of CO_2 for best growth. (3) One of the most popular ways of culturing small numbers of anaerobes is by use of a GasPak jar (**figure 6.24**). In this procedure the environment is made anoxic by using hydrogen and a palladium catalyst to remove O_2 through the formation of water. The reducing agents in anaerobic agar also remove oxygen, as mentioned previously. (4) Plastic bags or pouches make convenient containers when only a few samples are to be incubated anaerobically. These have a catalyst and calcium carbonate to produce an anoxic, carbon-dioxide-rich atmosphere. A special solution is added to the pouch's reagent compartment; petri dishes or other containers are placed in the pouch; it then is clamped shut and placed in an incubator. A laboratory may make use of all these techniques since each is best suited for different purposes.

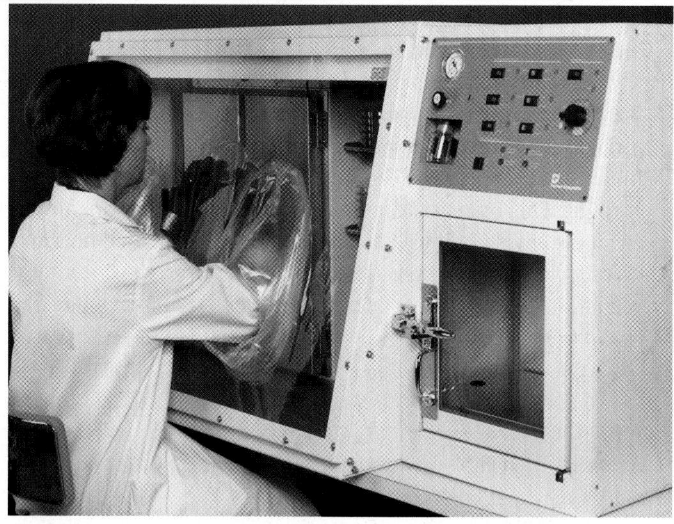

Figure 6.23 An Anaerobic Work Chamber and Incubator. This anaerobic system contains an oxygen-free work area and an incubator. The interchange compartment on the right of the work area allows materials to be transferred inside without exposing the interior to oxygen. The anaerobic atmosphere is maintained largely with a vacuum pump and nitrogen purges. The remaining oxygen is removed by a palladium catalyst and hydrogen. The oxygen reacts with hydrogen to form water, which is absorbed by desiccant.

1. Describe the five types of O_2 relationships seen in microorganisms.
2. How do chemotrophic aerobes use O_2?
3. What are the toxic effects of O_2? How do aerobes and other oxygen-tolerant microbes protect themselves from these effects?
4. Describe four ways in which anaerobes may be cultured.

Pressure

Organisms that spend their lives on land or on the surface of water are always subjected to a pressure of 1 atmosphere (atm), and are never affected significantly by pressure. Yet many procaryotes live in the deep sea (ocean of 1,000 m or more in depth)

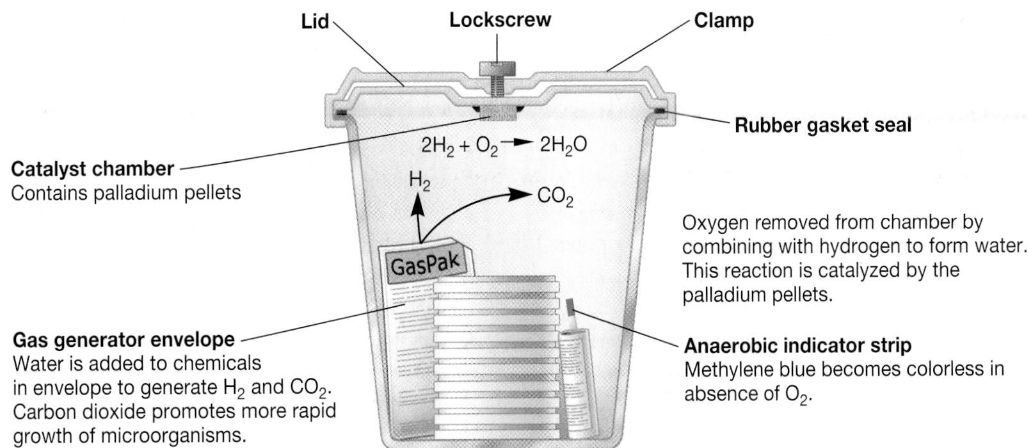

Figure 6.24 The GasPak Anaerobic System. Hydrogen and carbon dioxide are generated by a GasPak envelope. The palladium catalyst in the chamber lid catalyzes the formation of water from hydrogen and oxygen, thereby removing oxygen from the sealed chamber.

where the hydrostatic pressure can reach 600 to 1,100 atm and the temperature is about 2 to 3°C. Many of these procaryotes are **barotolerant:** increased pressure adversely affects them but not as much as it does nontolerant microbes. Some procaryotes in the gut of deep-sea invertebrates such as amphipods (shrimplike crustaceans) and holothurians (sea cucumbers) are truly **barophilic**—they grow more rapidly at high pressures. These microbes may play an important role in nutrient recycling in the deep sea. A barophile recovered from the Mariana trench near the Philippines (depth about 10,500 m) is actually unable to grow at pressures below about 400 to 500 atm when incubated at 2°C. Thus far, barophiles have been found among several bacterial genera (e.g., *Photobacterium, Shewanella, Colwellia*). Some archaea are thermobarophiles (e.g., *Pyrococcus* spp., *Methanocaldococcus jannaschii*). Microorganisms in marine environments (section 28.3)

Radiation

Our world is bombarded with electromagnetic radiation of various types (**figure 6.25**). This radiation often behaves as if it were composed of waves moving through space like waves traveling on the surface of water. The distance between two wave crests or troughs is the wavelength. As the wavelength of electromagnetic radiation decreases, the energy of the radiation increases—gamma rays and X rays are much more energetic than visible light or infrared waves. Electromagnetic radiation also acts like a stream of energy packets called photons, each photon having a quantum of energy whose value will depend on the wavelength of the radiation.

Sunlight is the major source of radiation on the Earth. It includes visible light, ultraviolet (UV) radiation, infrared rays, and radio waves. Visible light is a most conspicuous and important aspect of our environment: most life is dependent on the ability of photosynthetic organisms to trap the light energy of the sun. Almost 60% of the sun's radiation is in the infrared region rather than the visible portion of the spectrum. Infrared is the major source of the Earth's heat. At sea level, one finds very little ultraviolet radiation below about 290 to 300 nm. UV radiation of

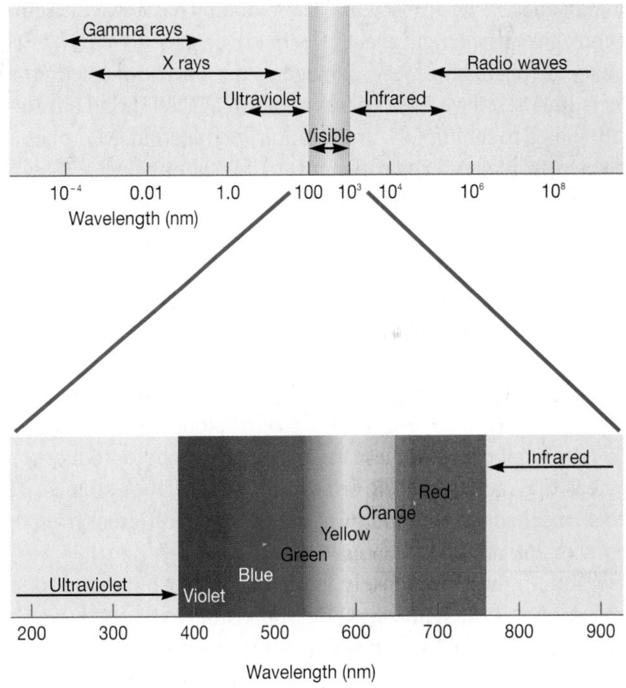

Figure 6.25 The Electromagnetic Spectrum. A portion of the spectrum is expanded at the bottom of the figure.

wavelengths shorter than 287 nm is absorbed by O_2 in the Earth's atmosphere; this process forms a layer of ozone between 25 and 30 miles above the Earth's surface. The ozone layer then absorbs somewhat longer UV rays and reforms O_2. The fairly even distribution of sunlight throughout the visible spectrum accounts for the fact that sunlight is generally "white." Phototrophy (section 9.12)

Many forms of electromagnetic radiation are very harmful to microorganisms. This is particularly true of **ionizing radiation,** radiation of very short wavelength and high energy, which can cause atoms to lose electrons (ionize). Two major forms of ionizing radiation are (1) X rays, which are artificially produced, and (2) gamma rays, which are emitted during radioisotope decay. Low levels of ionizing radiation will produce mutations and may indirectly result in death, whereas higher levels are directly lethal. Although microorganisms are more resistant to ionizing radiation than larger organisms, they will still be destroyed by a sufficiently large dose. Ionizing radiation can be used to sterilize items. Some procaryotes (e.g., *Deinococcus radiodurans*) and bacterial endospores can survive large doses of ionizing radiation. The use of physical methods in control: Radiation (section 7.4); *Deinococcus-Thermus* (section 21.2)

A variety of changes in cells are due to ionizing radiation; it breaks hydrogen bonds, oxidizes double bonds, destroys ring structures, and polymerizes some molecules. Oxygen enhances these destructive effects, probably through the generation of hydroxyl radicals (OH·). Although many types of constituents can be affected, it is reasonable to suppose that destruction of DNA is the most important cause of death.

Ultraviolet (UV) radiation can kill all kinds of microorganisms due to its short wavelength (approximately from 10 to 400 nm) and high energy. The most lethal UV radiation has a wavelength of 260 nm, the wavelength most effectively absorbed by DNA. The primary mechanism of UV damage is the formation of thymine dimers in DNA. Two adjacent thymines in a DNA strand are covalently joined to inhibit DNA replication and function. Microbes are protected from shorter wavelengths of UV light because they are absorbed by oxygen, as described previously. The damage caused by UV light that reaches Earth's surface can be repaired by several DNA repair mechanisms, which are discussed in chapter 13. Excessive exposure to UV light outstrips the organism's ability to repair the damage and death results. Longer wavelengths of UV light (near-UV radiation; 325 to 400 nm) are not absorbed by oxygen and so reach the Earth's surface. They can also harm microorganisms because they induce the breakdown of tryptophan to toxic photoproducts. It appears that these toxic tryptophan photoproducts plus the near-UV radiation itself produce breaks in DNA strands. The precise mechanism is not known, although it is different from that seen with 260 nm UV. Mutations and their chemical basis (section 13.1)

Visible light is immensely beneficial because it is the source of energy for photosynthesis. Yet even visible light, when present in sufficient intensity, can damage or kill microbial cells. Usually pigments called photosensitizers and O_2 are required. All microorganisms possess pigments like chlorophyll, bacteriochlorophyll, cytochromes, and flavins, which can absorb light energy, become excited or activated, and act as photosensitizers. The excited photosensitizer (P) transfers its energy to O_2 generating **singlet oxygen** ($^1O_2$).

$$P \xrightarrow{\text{light}} P\,(\text{activated})$$

$$P\,(\text{activated}) + O_2 \longrightarrow P + {}^1O_2$$

Singlet oxygen is a very reactive, powerful oxidizing agent that will quickly destroy a cell. It is probably the major agent employed by phagocytes to destroy engulfed bacteria. Phagocytosis (section 31.3)

Many microorganisms that are airborne or live on exposed surfaces use carotenoid pigments for protection against photooxidation. Carotenoids effectively quench singlet oxygen—that is, they absorb energy from singlet oxygen and convert it back into the unexcited ground state. Both photosynthetic and nonphotosynthetic microorganisms employ pigments in this way.

1. What are barotolerant and barophilic bacteria? Where would you expect to find them?
2. List the types of electromagnetic radiation in the order of decreasing energy or increasing wavelength.
3. Why is it so important that the Earth receives an adequate supply of sunlight? What is the importance of ozone formation?
4. How do ionizing radiation, ultraviolet radiation, and visible light harm microorganisms? How do microorganisms protect themselves against damage from UV and visible light?

6.6 Microbial Growth in Natural Environments

Section 6.5 surveys the effects on microbial growth of individual environmental factors such as water availability, pH, and temperature. Although microbial ecology is introduced in more detail in chapters 27 to 30, we now briefly consider the effect of the environment as a whole on microbial growth.

Growth Limitation by Environmental Factors

The microbial environment is complex and constantly changing. It often contains low nutrient concentrations **(oligotrophic environment)** and exposes microbes to many overlapping gradients of nutrients and other environmental factors. The growth of microorganisms depends on both the nutrient supply and their tolerance of the environmental conditions present in their habitat at any particular time. Two laws clarify this dependence. Liebig's law of the minimum states that the total biomass of an organism will be determined by the nutrient present in the lowest concentration relative to the organism's requirements. This law applies in both the laboratory (figure 6.7) and in terrestrial and aquatic environments. An increase in a limiting essential nutrient such as phosphate will result in an increase in the microbial population until some other nutrient becomes limiting. If a specific nutrient is limiting, changes in other nutrients will have no effect. Shelford's law of tolerance states that there are limits to environmental factors below and above which a microorganism cannot survive and grow, regardless of the nutrient supply. This can readily be seen for temperature as shown in figure 6.21. Each microorganism has a specific temperature range in which it can grow. The same rule applies to other factors such as pH, oxygen level, and hydrostatic pressure in the marine environment. Inhibitory substances in the environment can also limit microbial growth. For instance, rapid, unlimited growth ensues if a microorganism is exposed to excess nutrients. Such growth quickly depletes nutrients and often results in the release of toxic products. Both nutrient depletion and the toxic products limit further growth. Another example is seen with microbes growing in nutrient-poor or oligotrophic environments, where the growth of microbes can be directly inhibited by a variety of natural substances including phenolics, tannins, ammonia, ethylene, and volatile sulfur compounds.

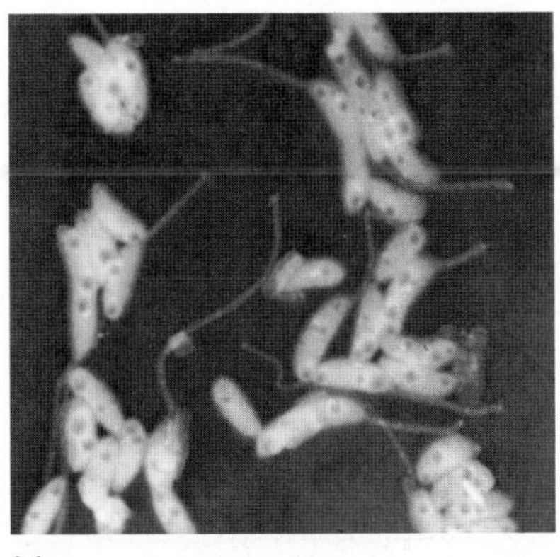
(a)

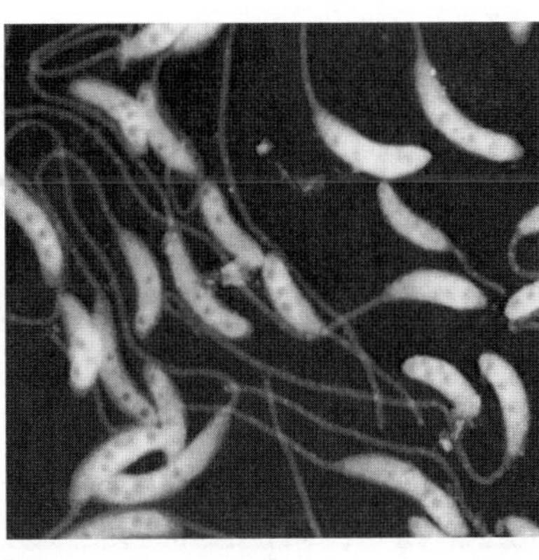
(b)

Figure 6.26 Morphology and Nutrient Absorption. Microorganisms can change their morphology in response to starvation and different limiting factors to improve their ability to survive. **(a)** *Caulobacter* has relatively short stalks when phosphorous is plentiful. **(b)** The stalks are extremely long under phosphorus-limited conditions.

In response to oligotrophic environments and intense competition, many microorganisms become more competitive in nutrient capture and exploitation of available resources. Often the organism's morphology will change in order to increase its surface area and ability to absorb nutrients. This can involve conversion of rod-shaped procaryotes to "mini" and "ultramicro" cells or changes in the morphology of prosthecate or stalked bacteria, in response to starvation. Nutrient deprivation induces many other changes as discussed previously (**figure 6.26**). For example, microorganisms can undergo a step-by-step shutdown of metabolism except for housekeeping maintenance genes.

Many factors can alter nutrient levels in oligotrophic environments. Microorganisms may sequester critical limiting nutrients, such as iron, making them less available to competitors. The atmosphere can contribute essential nutrients and support microbial growth. This is seen in the laboratory as well as natural environments. Airborne organic substances have been found to stimulate microbial growth in dilute media, and enrichment of growth media by airborne organic matter can allow significant populations of microorganisms to develop. Even distilled water, which contains traces of organic matter, can absorb one-carbon compounds from the atmosphere and grow microorganisms. The presence of such airborne nutrients and microbial growth, if not detected, can affect experiments in biochemistry and molecular biology, as well as studies of microorganisms growing in oligotrophic environments.

Counting and Identifying Microorganisms in Natural Environments

Microbial ecologists ask two important questions: What microbes are in a microbial habitat, and how many there are? Although microbiologists have developed numerous techniques for identifying and counting microbes, these questions are not easily answered. There are two reasons for this. First, many identification and counting methods rely on the ability of a microbe to form colonies. This presupposes that the microbiologist knows how to construct a growth medium and create environmental conditions that will support all the microbes in a habitat. Unfortunately, this knowledge eludes microbiologists, and it is estimated that only about 1% of the microbes in natural environments have been cultured. Increasingly, molecular methods are being used to analyze the diversity of microbial populations. The second reason is related to the "stress" microbes experience in natural environments. John Postgate of the University of Sussex in England was one of the first to note that microorganisms stressed by survival in natural habitats—or in many selective laboratory media—were particularly sensitive to secondary stresses. Such stresses can produce viable microorganisms that have lost the ability to grow on media normally used for their cultivation. To determine the growth potential of such microorganisms, Postgate developed what is now called the Postgate microviability assay, in which microorganisms are cultured in a thin agar film under a coverslip. The ability of a cell to change its morphology, even if it does not grow beyond the single-cell stage, indicates that the microorganism does show "life signs."

Since that time many workers have developed additional sensitive microscopic, isotopic, and molecular genetic procedures to evaluate the presence and significance of these viable but nonculturable procaryotes (VBNC) in both lab and field. The new field of environmental genomics, or metagenomics, is discussed in chapter 15. In a more routine approach, levels of fluorescent antibody and acridine orange-stained cells often are compared with population counts obtained by the most probable number (MPN) method and plate counts using selective and nonselective media. The release of radioactive-labeled cell materials also is used to monitor stress effects on microorganisms. Despite these advances, the estimation of substrate-responsive viable cells by Postgate's method is still important. These studies show that even when pathogenic bacteria such as *Escherichia coli, Vibrio cholerae, Klebsiella pneumoniae, Enterobacter aerogenes,* and *Enterococcus faecalis* have lost their ability to grow on conventional laboratory media using standard cultural techniques, they still might be able to play a role in infectious disease. Microbial ecology and its methods: An overview (section 27.4); Water purification and sanitary analysis (section 41.1)

Biofilms

Although scientists observed as early as the 1940s that more microbes in aquatic environments were found attached to surfaces (sessile) rather than were free-floating (planktonic), only relatively recently has this fact gained the attention of microbiologists. These attached microbes are members of complex, slime-encased communities called **biofilms.** Biofilms are ubiquitous in nature. There they are most often seen as layers of slime on rocks or other objects in water (**figure 6.27*a***). When they form on the hulls of

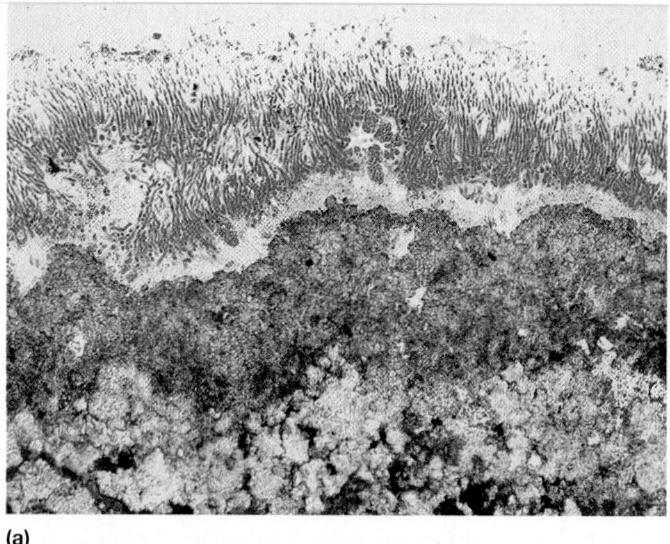
(a)

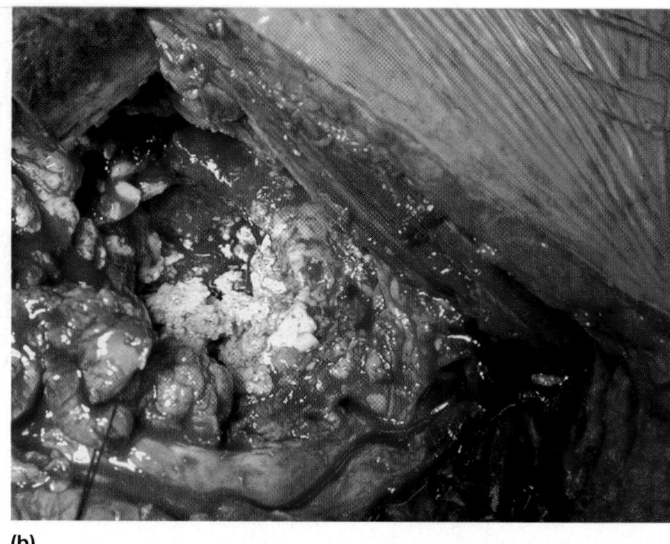
(b)

Figure 6.27 Examples of Biofilms. Biofilms form on almost any surface exposed to microorganisms. **(a)** Biofilm on the surface of a stromatolite in Walker Lake (Nevada, USA), an alkaline lake. The biofilm consists primarily of the cyanobacterium *Calothrix.* **(b)** Photograph taken during surgery to remove a biofilm-coated artificial joint. The white material is composed of pus, bacterial and fungal cells, and the patient's white blood cells.

boats and ships, they cause corrosion, which limits the life of the ships and results in economic losses. Of major concern is the formation of biofilms on medical devices such as hip and knee implants (figure 6.27*b*). These biofilms often cause serious illness and failure of the medical device. Biofilm formation is apparently an ancient ability among the procaryotes, as evidence for biofilms can be found in the fossil record from about 3.4 billion years ago.

Biofilms can form on virtually any surface, once it has been conditioned by proteins and other molecules present in the environment (**figure 6.28**). Microbes reversibly attach to the conditioned surface and eventually begin releasing polysaccharides, proteins, and DNA. These polymers allow the microbes to stick more stably to the surface. As the biofilm thickens and matures, the microbes reproduce and secrete additional polymers. The end result is a complex, dynamic community of microorganisms. The microbes interact in a variety of ways. For instance, the waste products of one microbe may be the energy source for another microbe. The cells also communicate with each other as described next. Finally, the presence of DNA in the extracellular slime can be taken up by members of the biofilm community. Thus genes can be transferred from one cell (or species) to another.

While in the biofilm, microbes are protected from numerous harmful agents such as UV light, antibiotics, and other antimicrobial agents. This is due in part to the extracellular matrix in which they are embedded, but it also is due to physiological changes. Indeed, numerous proteins synthesized or activated in biofilm cells are not observed in planktonic cells and vice versa. The resistance of biofilm cells to antimicrobial agents has serious consequences. When biofilms form on a medical device such as a hip implant (figure 6.27*b*), they are difficult to kill and can cause serious illness. Often the only way to treat patients in this situation is by removing the implant. Another problem with biofilms is that cells are regularly sloughed off (figure 6.28). This can have many consequences. For instance, biofilms in a city's water distribution pipes can serve as a source of contamination after the water leaves a water treatment facility.

Cell-Cell Communication Within Microbial Populations

For decades, microbiologists tended to think of bacterial populations as collections of individual cells growing and behaving independently. But about 30 years ago, it was discovered that the marine luminescent bacterium *Vibrio fischeri* controls its ability to glow by producing a small, diffusible substance called autoinducer. The autoinducer molecule was later identified as an **acylhomoserine lactone (AHL).** It is now known that many gram-negative bacteria make AHL molecular signals that vary in length and substitution at the third position of the acyl side chain (**figure 6.29**). In many of these species, the AHL is freely diffusible across the plasma membrane. Thus at a low cell density it diffuses out of the cell. However, when the cell population increases and AHL accumulates outside the cell, the diffusion gradient is reversed so that the AHL enters the cell. Because the influx of AHL is cell-density-dependent, it enables individual cells to assess population density. This is referred to as **quorum sensing;** a quorum usually refers to the minimum number of members in an organization, such as a legislative body, needed to conduct business. When AHL reaches a threshold level inside the cell, it serves to induce the expression of target genes that regulate a number of functions, depending on the microbe. These functions are most effective only if a large number of microbes are present. For instance, the light produced by one cell

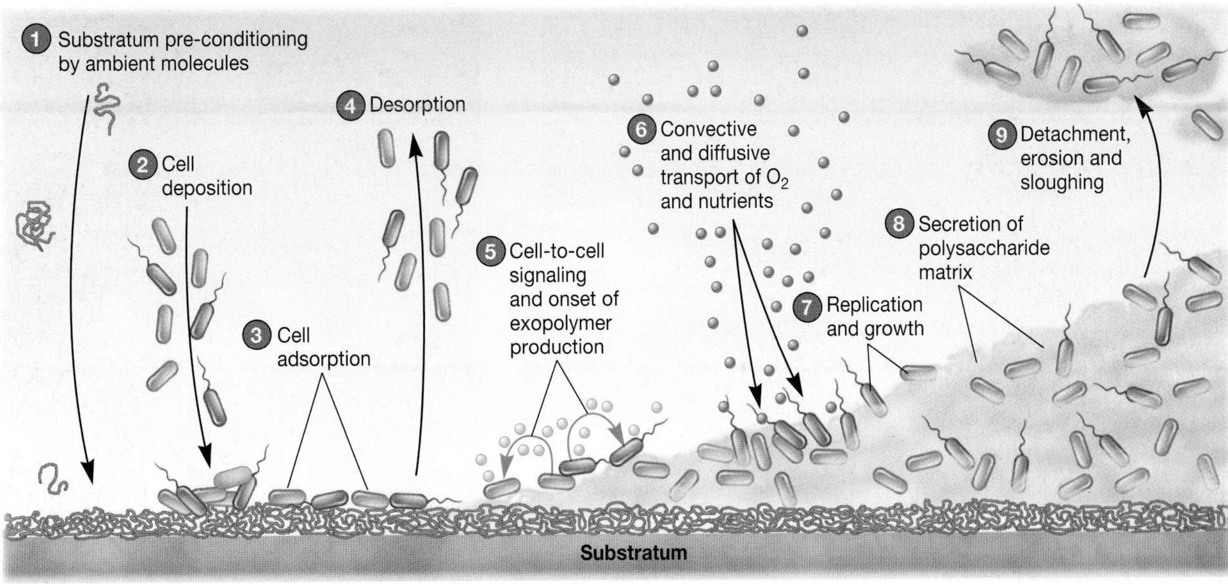

Figure 6.28 Biofilm Formation.

Signal and Structure	Representative Organism	Function Regulated
N-acyl homoserine lactone (AHL)	*Vibrio fischeri*	Bioluminescence
	Agrobacterium tumefaciens	Plasmid transfer
	Erwinia carotovora	Virulence and antibiotic production
	Pseudomonas aeruginosa	Virulence and biofilm formation
	Burkholderia cepacia	Virulence
Furanosylborate (AI-2) 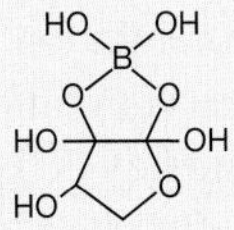	*Vibrio harveyi*	Virulence
Cyclic thiolactone (AIP-II) 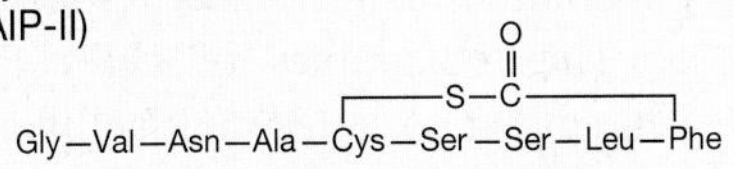 Gly—Val—Asn—Ala—Cys—Ser—Ser—Leu—Phe	*Staphylococcus aureus*	Virulence
Hydroxy-palmitic acid methyl ester (PAME) 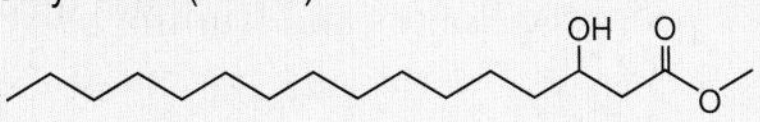	*Ralstonia solanacearum*	Virulence
Methyl dodecenoic acid (DSF) 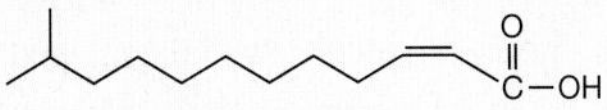	*Xanthomonas campestris*	Virulence
Farnesoic acid 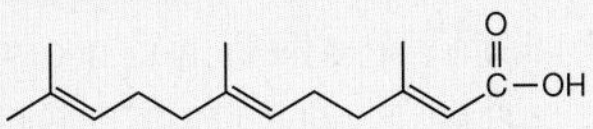	*Candida albicans*	Dimorphic transition and virulence

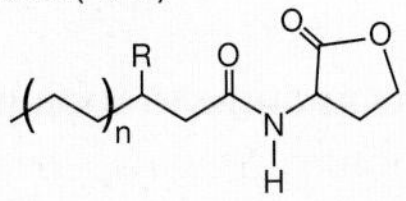

Figure 6.29 Representative Cell-Cell Communication Molecules.

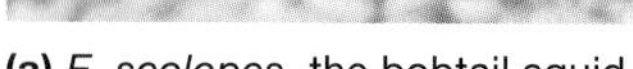

(a) *E. scolopes,* the bobtail squid

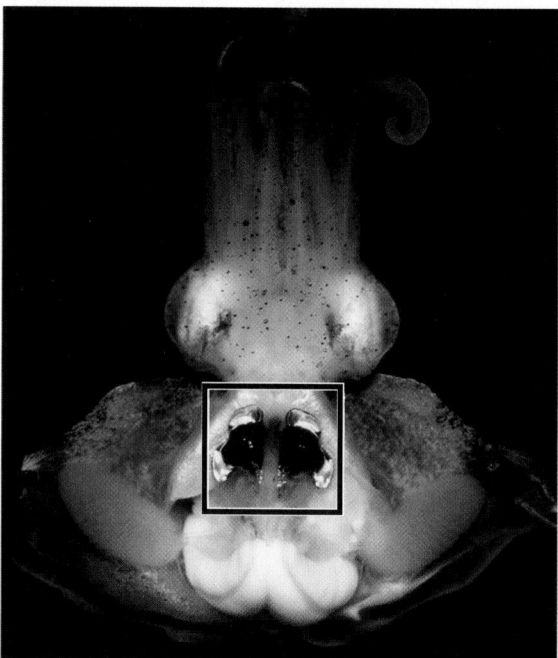

(b) Light organ

Figure 6.30 ***Euprymna scolopes.*** **(a)** *E. scolopes* is a warm-water squid that remains buried in sand during the day and feeds at night. **(b)** When feeding it uses its light organ (boxed, located on its ventral surface) to provide camouflage by projecting light downward. Thus the outline of the squid appears as bright as the water's surface to potential predators looking up through the water column. The light organ is colonized by a large number of *Vibrio fischeri* so autoinducer accumulates to a threshold concentration, triggering light production.

is not visible, but cell densities within the light organ of marine fish and squid reach 10^{10} cells per milliliter. This provides the animal with a flashlight effect while the microbes have a safe and nutrient-enriched habitat (**figure 6.30**). In fact, many of the processes regulated by quorum sensing involve host-microbe interactions such as symbioses and pathogenicity. Global regulatory systems: Quorum sensing (section 12.5)

Many different bacteria use AHL signals. In addition to *V. fischeri* bioluminescence, the opportunistic pathogens *Burkholderia cepacia* and *Pseudomonas aeruginosa* use AHLs to regulate the expression of virulence factors (figure 6.29). These gram-negative bacteria cause debilitating pneumonia in people who are immunocompromised, and are important pathogens in cystic fibrosis patients. The plant pathogens *Agrobacterium tumefaciens* will not infect a host plant and *Erwinia carotorvora* will not produce antibiotics without AHL signaling. Finally, *B. cepacia, P. aeruginosa,* as well as *Vibrio cholerae* use AHL intercellular communication to control biofilm formation, an important strategy to evade the host's immune system.

The discovery of additional molecular signals made by a variety of microbes underscores the importance of cell-cell communication in regulating procaryotic processes. For instance, while only gram-negative bacteria are known to make AHLs, both gram-negative and gram-positive bacteria make autoinducer-2 (AI-2). Gram-positive bacteria usually exchange short peptides called oligopeptides instead of autoinducer-like molecules. Examples include *Enterococcus faecalis,* whose oligopeptide signal is used to determine the best time to conjugate (transfer genes). Oligopeptide communication by *Staphylococcus aureus* and *Bacillus subtilis* is used to trigger the uptake of DNA from the environment. The soil microbe *Streptomyces griseus* produces a gamma-butyrolactone known as A-factor. This small molecule regulates both morphological differentiation and the production of the antibiotic streptomycin. Eucaryotic microbes also rely on cell-cell communication to coordinate key activities within a population. For example, the pathogenic fungus *Candida albicans* secretes farnesoic acid to govern morphology and virulence.

These examples of cell-cell communication demonstrate what might be called multicellular behavior in that many individual cells communicate and coordinate their activities to act as a unit. Other examples of such complex behavior is pattern formation in colonies and fruiting body formation in the myxobacteria. Isolation of pure cultures: Microbial growth on agar surfaces (section 5.8); Class *Deltaproteobacteria:* Order *Myxococcales* (section 22.4)

1. How are Liebig's law of the minimum and Shelford's law of tolerance related? Why are generation times in nature usually much longer than in culture?
2. Describe how microorganisms respond to oligotrophic environments.
3. Briefly discuss the Postgate microviability assay and other ways in which viable but nonculturable microorganisms can be counted or studied.
4. What is a biofilm? Why might life in a biofilm be advantageous for microbes?
5. What is quorum sensing? Describe how it occurs and briefly discuss its importance to microorganisms.

Summary

Growth is an increase in cellular constituents and results in an increase in cell size, cell number, or both.

6.1 The Procaryotic Cell Cycle

a. Most procaryotes reproduce by binary fission, a process in which the cell elongates and the chromosome is replicated and segregates to opposite poles of the cell prior to the formation of a septum, which divides the cell into two progeny cells (**figures 6.1** and **6.3**).

b. Two overlapping pathways function during the procaryotic cell cycle: the pathway for chromosome replication and segregation and the pathway for septum formation (**figure 6.2**). Both are complex and poorly understood. The partitioning of the progeny chromosomes may involve homologues of eucaryotic cytoskeletal proteins.

c. In rapidly dividing cells, initiation of DNA synthesis may occur before the previous round of synthesis is completed. This allows the cells to shorten the time needed for completing the cell cycle.

6.2 The Growth Curve

a. When microorganisms are grown in a closed system or batch culture, the resulting growth curve usually has four phases: the lag, exponential or log, stationary, and death phases (**figure 6.6**).

b. In the exponential phase, the population number of cells undergoing binary fission doubles at a constant interval called the doubling or generation time (**figure 6.10**). The mean growth rate constant (k) is the reciprocal of the generation time.

c. Exponential growth is balanced growth, cell components are synthesized at constant rates relative to one another. Changes in culture conditions (e.g., in shift-up and shift-down experiments) lead to unbalanced growth. A portion of the available nutrients is used to supply maintenance energy.

6.3 Measurement of Microbial Growth

a. Microbial populations can be counted directly with counting chambers, electronic counters, or fluorescence microscopy. Viable counting techniques such as the spread plate, the pour plate, or the membrane filter can be employed (**figures 6.12** and **6.14**).

b. Population changes also can be followed by determining variations in microbial mass through the measurement of dry weight, turbidity, or the amount of a cell component (**figure 6.15**).

6.4 The Continuous Culture of Microorganisms

a. Microorganisms can be grown in an open system in which nutrients are constantly provided and wastes removed.

b. A continuous culture system is an open system that can maintain a microbial population in the log phase. There are two types of these systems: chemostats and turbidostats (**figure 6.16**).

6.5 The Influence of Environmental Factors on Growth

a. Most bacteria, photosynthetic protists, and fungi have rigid cell walls and are hypertonic to the habitat because of solutes such as amino acids, polyols, and potassium ions. The amount of water actually available to microorganisms is expressed in terms of the water activity (a_w).

b. Although most microorganisms will not grow well at water activities below 0.98 due to plasmolysis and associated effects, osmotolerant organisms survive and even flourish at low a_w values. Halophiles actually require high sodium chloride concentrations for growth (**figure 6.18** and **table 6.3**).

c. Each species of microorganism has an optimum pH for growth and can be classified as an acidophile, neutrophile, or alkalophile (**figure 6.19**).

d. Microorganisms can alter the pH of their surroundings, and most culture media must be buffered to stabilize the pH.

e. Microorganisms have distinct temperature ranges for growth with minima, maxima, and optima—the cardinal temperatures. These ranges are determined by the effects of temperature on the rates of catalysis, protein denaturation, and membrane disruption (**figure 6.20**).

f. There are five major classes of microorganisms with respect to temperature preferences: (1) psychrophiles, (2) facultative psychrophiles or psychrotrophs, (3) mesophiles, (4) thermophiles, and (5) hyperthermophiles (**figure 6.21** and **table 6.3**).

g. Microorganisms can be placed into at least five different categories based on their response to the presence of O_2: obligate aerobes, facultative anaerobes, aerotolerant anaerobes, strict or obligate anaerobes, and microaerophiles (**figure 6.22** and **table 6.3**).

h. Oxygen can become toxic because of the production of hydrogen peroxide, superoxide radical, and hydroxyl radical. These are destroyed by the enzymes superoxide dismutase, catalase, and peroxidase.

i. Most deep-sea microorganisms are barotolerant, but some are barophilic and require high pressure for optimal growth.

j. High-energy or short-wavelength radiation harms organisms in several ways. Ionizing radiation—X rays and gamma rays—ionizes molecules and destroys DNA and other cell components. Ultraviolet (UV) radiation induces the formation of thymine dimers and strand breaks in DNA.

k. Visible light can provide energy for the formation of reactive singlet oxygen, which will destroy cells.

6.6 Microbial Growth in Natural Environments

a. Microbial growth in natural environments is profoundly affected by nutrient limitations and other adverse factors. Some microorganisms can be viable but unculturable and must be studied with special techniques.

b. Many microbes form biofilms, aggregations of microbes growing on surfaces and held together by extracellular polysaccharides (**figure 6.28**). Life in a biofilm has several advantages, including protection from harmful agents.

c. Often, bacteria will communicate with one another in a density-dependent way and carry out a particular activity only when a certain population density is reached. This phenomenon is called quorum sensing (**figures 6.29** and **6.30**).

Key Terms

acidophile 134
acylhomoserine lactone (AHL) 144
aerobe 139
aerotolerant anaerobe 139
alkalophile 134
anaerobe 139
balanced growth 123
barophilic 141
barotolerant 141
batch culture 123
binary fission 119
biofilm 143
cardinal temperatures 136
catalase 140
cell cycle 119
chemostat 131
coenocytic 119
colony forming units (CFU) 130
compatible solutes 132
continuous culture system 131
cytokinesis 121
doubling time 126
exponential phase 123
extremophiles 132
facultative anaerobe 139
facultative psychrophiles 138
FtsZ protein 122
generation time 126
growth 119
halophile 133
hydrogen peroxide 140
hydroxyl radical 140
hyperthermophile 139
ionizing radiation 141
lag phase 123
log phase 123
mean generation time 126
mean growth rate constant (k) 126
mesophile 138
microaerophile 139
MreB protein 120
neutrophile 134
obligate aerobe 139
obligate anaerobe 139
oligotrophic environment 142
origin of replication 120
osmotolerant 134
programmed cell death 125
psychrophile 137
psychrotroph 138
quorum sensing 144
replisome 120
septation 121
singlet oxygen 142
starvation proteins 124
stationary phase 124
strict anaerobe 139
superoxide dismutase (SOD) 140
superoxide radical 140
terminus 120
thermophile 138
turbidostat 132
ultraviolet (UV) radiation 142
unbalanced growth 123
viable but nonculturable (VBNC) 125
water activity (a_w) 134
Z ring 122

Critical Thinking Questions

1. As an alternative to diffusable signals, suggest another mechanism by which bacteria can quorum sense.
2. Design an "enrichment" culture medium and a protocol for the isolation and purification of a soil bacterium (e.g., *Bacillus subtilis*) from a sample of soil. Note possible contaminants and competitors. How will you adjust conditions of growth and what conditions will be adjusted to differentially enhance the growth of the *Bacillus?*
3. Design an experiment to determine if a slow-growing microbial culture is just exiting lag phase or is in exponential phase.
4. Why do you think the cardinal temperatures of some microbes change depending on other environmental conditions (e.g., pH)? Suggest one specific mechanism underlying such change.

Learn More

Atlas, R. M., and Bartha, R. 1997. *Microbial ecology: Fundamentals and applications,* 4th ed. Menlo Park, CA: Benjamin/Cummings.

Bartlett, D. H., and Roberts, M. F. 2000. Osmotic stress. In *Encyclopedia of microbiology,* 2d ed., vol. 3, J. Lederberg, editor-in-chief, 502–15. San Diego: Academic Press.

Cavicchioli, R., and Thomas, T. 2000. Extremophiles. In *Encyclopedia of microbiology,* 2d ed., vol. 2, J. Lederburg, editor-in-chief, 317–37. San Diego: Academic Press.

Cotter, P. D., and Hill, C. 2003. Surviving the acid test: Responses of gram-positive bacteria to low pH. *Microbiol. Mol. Biol. Rev.* 67(3):429–53.

Gitai, Z.; Thanbichler, M.; and Shapiro, L. 2005. The choreographed dynamics of bacterial chromosomes. *Trends Microbiol.* 13(5):221–28.

Hall-Stoodley, L.; Costerton, J. W.; and Stoodley, P. 2004. Bacterial biofilms: From the natural environment to infectious diseases. *Nature Rev. Microbiol.* 2:95–108.

Hoskisson, P. A., and Hobbs, G. 2005. Continuous culture—making a comeback? *Microbiology* 151:3153–59.

Kashefi, K., and Lovley, D. R. 2003. Extending the upper temperature limit for life. *Science* 301:934.

Krieg, N. R., and Hoffman, P. S. 1986. Microaerophily and oxygen toxicity. *Annu. Rev. Microbiol.* 40:107–30.

Krulwich, T. A., and Guffanti, A. A. 1989. Alkalophilic bacteria. *Annu. Rev. Microbiol.* 43:435–63.

Marr, A. G. 2000. Growth kinetics, bacterial. In *Encyclopedia of microbiology,* 2d ed., vol. 2, J. Lederberg, editor-in-chief, 584–89. San Diego: Academic Press.

Morita, R. Y. 2000. Low-temperature environments. In *Encyclopedia of microbiology,* 2d ed., vol. 3, J. Lederberg, editor-in-chief, 93–98. San Diego: Academic Press.

Nyström, T. 2004. Stationary-phase physiology. *Annu. Rev. Microbiol.* 58:161–81.

Visick, K. L., and Fuqua, C. 2005. Decoding microbial chatter: Cell-cell communication in bacteria. *J. Bact.* 187(16):5507–19.

Weiss, D. S. 2004. Bacterial cell division and the septal ring. *Molec. Microbiol.* 54(3):588–97.

Please visit the Prescott website at www.mhhe.com/prescott7 for additional references.

7 Control of Microorganisms by Physical and Chemical Agents

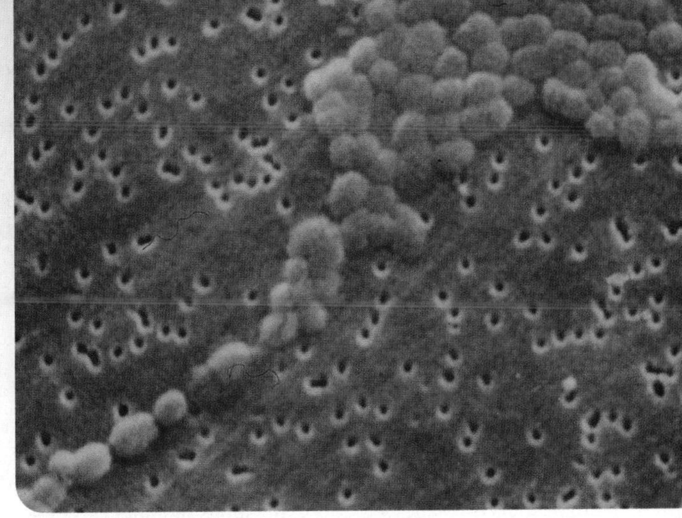

Bacteria are trapped on the surface of a membrane filter used to remove microorganisms from fluids.

Preview

- Microbial population death is exponential, and the effectiveness of an agent is not fixed but influenced by many environmental factors.
- Solid objects can be sterilized by physical agents such as heat and radiation; liquids and gases are sterilized by heat, radiation, and filtration.
- Most chemical agents do not readily destroy bacterial endospores and therefore cannot sterilize objects; they are used as disinfectants, sanitizers, and antiseptics. Objects can be sterilized by gases like ethylene oxide and vaporized hydrogen peroxide that destroy endospores.
- Chemotherapeutic agents are chemicals used to kill or inhibit the growth of microorganisms within host tissues.

Chapters 5 and 6 are concerned with microbial nutrition and growth. In this chapter we address the subject of the control and destruction of microorganisms, a topic of immense practical importance. Although most microorganisms are beneficial and necessary for human well-being, microbial activities may have undesirable consequences, such as food spoilage and disease. Therefore it is essential to be able to kill a wide variety of microorganisms or inhibit their growth to minimize their destructive effects. The goal is twofold: (1) to destroy pathogens and prevent their transmission, and (2) to reduce or eliminate microorganisms responsible for the contamination of water, food, and other substances.

From the beginning of recorded history, people have practiced disinfection and sterilization, even though the existence of microorganisms was unknown. The Egyptians used fire to sterilize infectious material and disinfectants to embalm bodies, and the Greeks burned sulfur to fumigate buildings. Mosaic law commanded the Hebrews to burn any clothing suspected of being contaminated with leprosy. Today the ability to destroy microorganisms is no less important: it makes possible the aseptic techniques used in microbiological research, the preservation of food, and the treatment and prevention of disease. The techniques described in this chapter are also essential to personal safety in both the laboratory and hospital (**Techniques & Applications 7.1**).

This chapter focuses on the control of microorganisms by physical and chemical agents, including chemotherapeutic agents, which are discussed in more detail in chapter 35. However, microbes can be controlled by many mechanisms that will not be considered in this chapter. For instance, the manipulation of environmental parameters is used extensively in the food industry to preserve foods. Increased solutes, such as salt and sugar, preserve meats, jams, and jellies. Microbial fermentations of milk and vegetables decrease the pH of these foods, creating new foods such as yogurt, cheese, and pickles—all of which have a longer shelf life than the milk and vegetables from which they are made. Heat and the generation of anoxic conditions are important in the preservation of canned foods, and ionizing radiation is used to extend the shelf life of seafood, fruits, and vegetables. The use of these control measures is described in more detail in chapter 40.

7.1 Definitions of Frequently Used Terms

Terminology is especially important when the control of microorganisms is discussed because words like disinfectant and antiseptic often are used loosely. The situation is even more confusing because a particular treatment can either inhibit growth or kill depending on the conditions. The types of control agents and their uses are outlined in **figure 7.1.**

We all labour against our own cure, for death is the cure of all diseases.

—*Sir Thomas Browne*

Techniques & Applications

7.1 Safety in the Microbiology Laboratory

Personnel safety should be of major concern in all microbiology laboratories. It has been estimated that thousands of infections have been acquired in the laboratory, and many persons have died because of such infections. The two most common laboratory-acquired bacterial diseases are typhoid fever and brucellosis. Most deaths have come from typhoid fever (20 deaths) and Rocky Mountain spotted fever (13 deaths). Infections by fungi (histoplasmosis) and viruses (Venezuelan equine encephalitis and hepatitis B virus from monkeys) are also not uncommon. Hepatitis is the most frequently reported laboratory-acquired viral infection, especially in people working in clinical laboratories and with blood. In a survey of 426 U.S. hospital workers, 40% of those in clinical chemistry and 21% in microbiology had antibodies to the hepatitis B virus, indicating their previous exposure (though only about 19% of these had disease symptoms).

Efforts have been made to determine the causes of these infections in order to enhance the development of better preventive measures. Although often it is not possible to determine the direct cause of infection, some major potential hazards are clear. One of the most frequent causes of disease is the inhalation of an infectious aerosol. An aerosol is a gaseous suspension of liquid or solid particles that may be generated by accidents and laboratory operations such as spills, centrifuge accidents, removal of closures from shaken culture tubes, and plunging of contaminated loops into a flame. Accidents with hypodermic syringes and needles, such as self-inoculation and spraying solutions from the needle, also are common. Hypodermics should be employed only when necessary and then with care. Pipette accidents involving the mouth are another major source of infection; pipettes should be filled with the use of pipette aids and operated in such a way as to avoid creating aerosols.

People must exercise care and common sense when working with microorganisms. Operations that might generate infectious aerosols should be carried out in a biological safety cabinet. Bench tops and incubators should be disinfected regularly. Autoclaves must be maintained and operated properly to ensure adequate sterilization. Laboratory personnel should wash their hands thoroughly before and after finishing work.

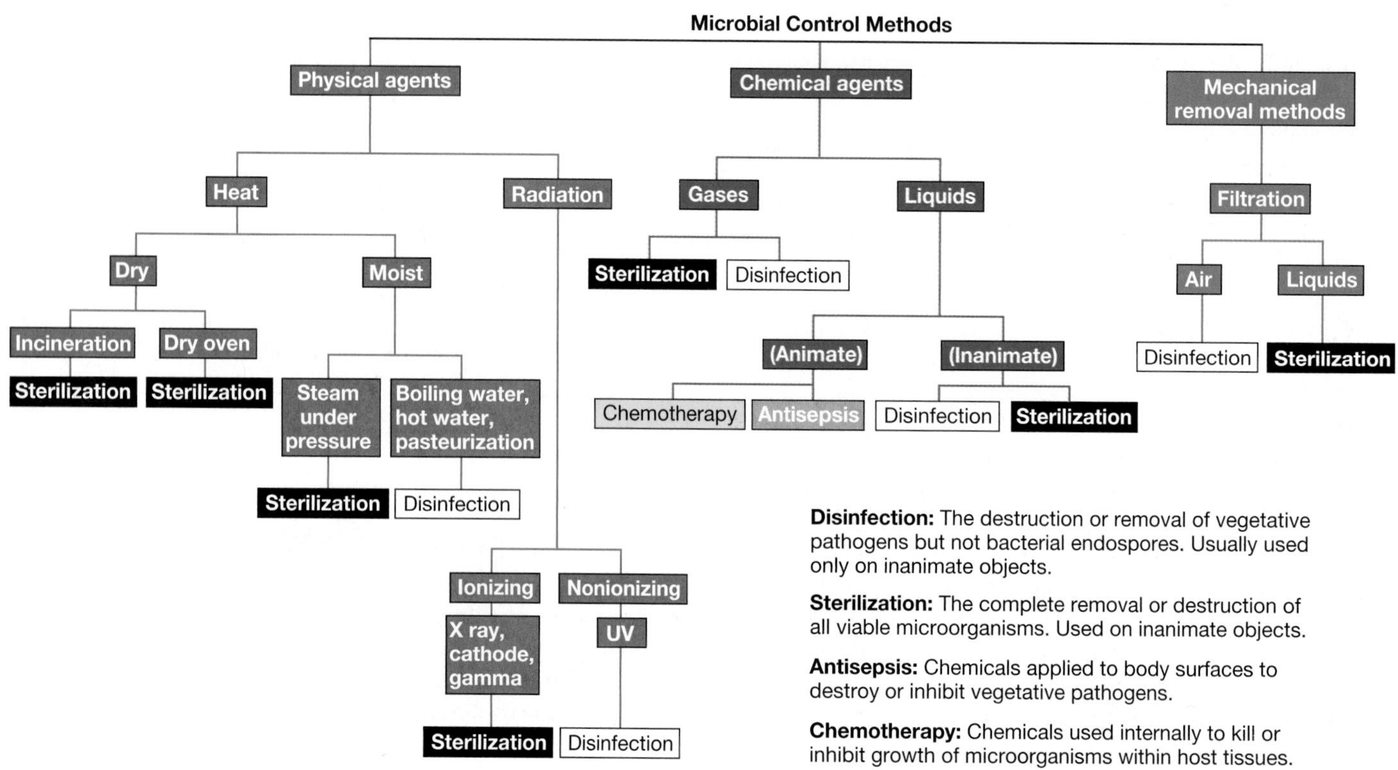

Figure 7.1 Microbial Control Methods.

The ability to control microbial populations on inanimate objects, like eating utensils and surgical instruments, is of considerable practical importance. Sometimes it is necessary to eliminate all microorganisms from an object, whereas only partial destruction of the microbial population may be required in other situations. **Sterilization** [Latin *sterilis,* unable to produce offspring or barren] is the process by which all living cells, spores, and acellular entities (e.g., viruses, viroids, and prions) are either destroyed or removed from an object or habitat. A sterile object is totally free of viable microorganisms, spores, and other infectious agents. When sterilization is achieved by a chemical agent, the chemical is called a sterilant. In contrast, **disinfection** is the killing, inhibition, or removal of microorganisms that may cause disease. The primary goal is to destroy potential pathogens, but disinfection also substantially reduces the total microbial population. **Disinfectants** are agents, usually chemical, used to carry out disinfection and are normally used only on inanimate objects. A disinfectant does not necessarily sterilize an object because viable spores and a few microorganisms may remain. **Sanitization** is closely related to disinfection. In sanitization, the microbial population is reduced to levels that are considered safe by public health standards. The inanimate object is usually cleaned as well as partially disinfected. For example, sanitizers are used to clean eating utensils in restaurants. Prions (section 18.10); Viroids and virusoids (section 18.9)

It also is frequently necessary to control microorganisms on or in living tissue with chemical agents. **Antisepsis** [Greek *anti,* against, and *sepsis,* putrefaction] is the prevention of infection or sepsis and is accomplished with **antiseptics.** These are chemical agents applied to tissue to prevent infection by killing or inhibiting pathogen growth; they also reduce the total microbial population. Because they must not destroy too much host tissue, antiseptics are generally not as toxic as disinfectants. **Chemotherapy** is the use of chemical agents to kill or inhibit the growth of microorganisms within host tissue.

A suffix can be employed to denote the type of antimicrobial agent. Substances that kill organisms often have the suffix -cide [Latin *cida,* to kill]; a **germicide** kills pathogens (and many nonpathogens) but not necessarily endospores. A disinfectant or antiseptic can be particularly effective against a specific group, in which case it may be called a **bactericide, fungicide, algicide,** or **viricide.** Other chemicals do not kill, but they do prevent growth. If these agents are removed, growth will resume. Their names end in -static [Greek *statikos,* causing to stand or stopping]—for example, **bacteriostatic** and **fungistatic.**

Although these agents have been described in terms of their effects on pathogens, it should be noted that they also kill or inhibit the growth of nonpathogens as well. Their ability to reduce the total microbial population, not just to affect pathogen levels, is quite important in many situations.

1. Define the following terms: sterilization, sterilant, disinfection, disinfectant, sanitization, antisepsis, antiseptic, chemotherapy, germicide, bactericide, bacteriostatic.

7.2 The Pattern of Microbial Death

A microbial population is not killed instantly when exposed to a lethal agent. Population death, like population growth, is generally exponential or logarithmic—that is, the population will be reduced by the same fraction at constant intervals (**table 7.1**). If the logarithm of the population number remaining is plotted against the time of exposure of the microorganism to the agent, a straight-line plot will result (**figure 7.2**). When the population has been greatly reduced, the rate of killing may slow due to the survival of a more resistant strain of the microorganism.

To study the effectiveness of a lethal agent, one must be able to decide when microorganisms are dead, a task by no means as easy as with macroorganisms. It is hardly possible to take a bacterium's pulse. A bacterium is often defined as dead if it does not grow and reproduce when inoculated into culture medium that would normally support its growth. In like manner, an inactive virus cannot infect a suitable host. This definition has flaws, however. It has been demonstrated that when bacteria are exposed to certain conditions, they can remain alive but are temporarily unable to reproduce. When in this state, they are referred to as viable but nonculturable (VBNC) *(see figure 6.8).* In conventional tests to demonstrate killing by an antimicrobial agent, VBNC bacteria would be thought to be dead. This is a serious problem because

Table 7.1 A Theoretical Microbial Heat-Killing Experiment

Minute	Microbial Number at Start of Minute[a]	Microorganisms Killed in 1 Minute (90% of total)[a]	Microorganisms at End of 1 Minute	Log_{10} of Survivors
1	10^6	9×10^5	10^5	5
2	10^5	9×10^4	10^4	4
3	10^4	9×10^3	10^3	3
4	10^3	9×10^2	10^2	2
5	10^2	9×10^1	10	1
6	10^1	9	1	0
7	1	0.9	0.1	−1

[a]Assume that the initial sample contains 10^6 vegetative microorganisms per ml and that 90% of the organisms are killed during each minute of exposure. The temperature is 121°C.

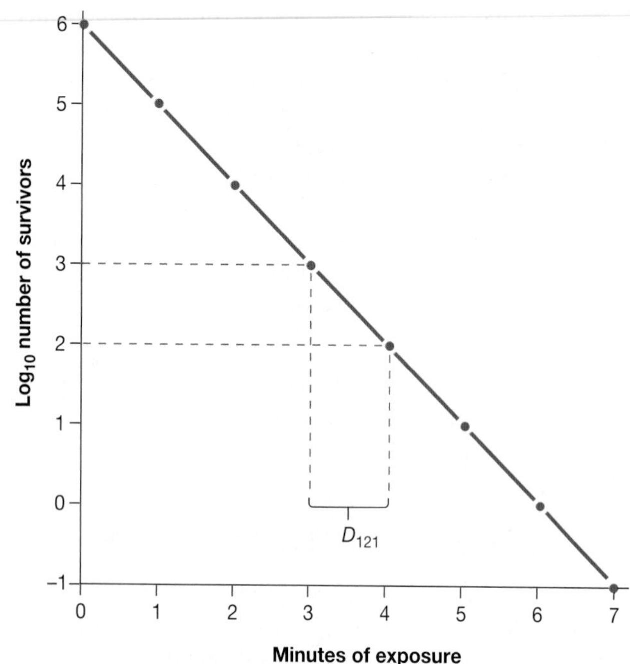

Figure 7.2 The Pattern of Microbial Death. An exponential plot of the survivors versus the minutes of exposure to heating at 121°C. In this example the D_{121} value is 1 minute. The data are from table 7.1.

after a period of recovery, the bacteria may regain their ability to reproduce and cause infection. The growth curve: Senescence and death (section 6.2)

1. Describe the pattern of microbial death and how one decides whether microorganisms are actually dead.

7.3 Conditions Influencing the Effectiveness of Antimicrobial Agents

Destruction of microorganisms and inhibition of microbial growth are not simple matters because the efficiency of an **antimicrobial agent** (an agent that kills microorganisms or inhibits their growth) is affected by at least six factors.

1. Population size. Because an equal fraction of a microbial population is killed during each interval, a larger population requires a longer time to die than a smaller one. This can be seen in the theoretical heat-killing experiment shown in table 7.1 and figure 7.2. The same principle applies to chemical antimicrobial agents.
2. Population composition. The effectiveness of an agent varies greatly with the nature of the organisms being treated because microorganisms differ markedly in susceptibility. Bacterial endospores are much more resistant to most antimicrobial agents than are vegetative forms, and younger cells are usually more readily destroyed than mature organisms. Some species are able to withstand adverse conditions better than others. For instance, *Mycobacterium tuberculosis,* which causes tuberculosis, is much more resistant to antimicrobial agents than most other bacteria.
3. Concentration or intensity of an antimicrobial agent. Often, but not always, the more concentrated a chemical agent or intense a physical agent, the more rapidly microorganisms are destroyed. However, agent effectiveness usually is not directly related to concentration or intensity. Over a short range a small increase in concentration leads to an exponential rise in effectiveness; beyond a certain point, increases may not raise the killing rate much at all. Sometimes an agent is more effective at lower concentrations. For example, 70% ethanol is more effective than 95% ethanol because its activity is enhanced by the presence of water.
4. Duration of exposure. The longer a population is exposed to a microbicidal agent, the more organisms are killed (figure 7.2). To achieve sterilization, an exposure duration sufficient to reduce the probability of survival to 10^{-6} or less should be used.
5. Temperature. An increase in the temperature at which a chemical acts often enhances its activity. Frequently a lower concentration of disinfectant or sterilizing agent can be used at a higher temperature.
6. Local environment. The population to be controlled is not isolated but surrounded by environmental factors that may either offer protection or aid in its destruction. For example, because heat kills more readily at an acidic pH, acidic foods and beverages such as fruits and tomatoes are easier to pasteurize than foods with higher pHs like milk. A second important environmental factor is organic matter, which can protect microorganisms against heating and chemical disinfectants. Biofilms are a good example. The organic matter in a biofilm protects the biofilm's microorganisms, and the biofilm and its microbes often are hard to remove. Furthermore, it has been clearly documented that bacteria in biofilms are altered physiologically, and this makes them less susceptible to many antimicrobial agents. Because of the impact of organic matter, it may be necessary to clean objects, especially syringes and medical or dental equipment, before they are disinfected or sterilized. The same care must be taken when pathogens are destroyed during the preparation of drinking water. When a city's water supply has a high content of organic material, steps are taken to decrease the organic matter or to add more chlorine. Microbial growth in natural environments: Biofilms (section 6.6)

1. Briefly explain how the effectiveness of antimicrobial agents varies with population size, population composition, concentration or intensity of the agent, treatment duration, temperature, and local environmental conditions.
2. How does being in a biofilm affect an organism's susceptibility to antimicrobial agents?
3. Suppose hospital custodians have been assigned the task of cleaning all showerheads in patient rooms in order to prevent the spread of infectious disease. What two factors would have the greatest impact on the effectiveness of the disinfectant the custodians use? Explain what that impact would be.

7.4 The Use of Physical Methods in Control

Heat and other physical agents are normally used to control microbial growth and sterilize objects, as can be seen from the continual operation of the autoclave in every microbiology laboratory. The four most frequently employed physical agents are heat, low temperatures, filtration, and radiation.

Heat

Fire and boiling water have been used for sterilization and disinfection since the time of the Greeks, and heating is still one of the most popular ways to destroy microorganisms. Either moist or dry heat may be applied.

Moist heat readily kills viruses, bacteria, and fungi (**table 7.2**). Moist heat is thought to kill by degrading nucleic acids and by denaturing enzymes and other essential proteins. It may also disrupt cell membranes. Exposure to boiling water for 10 minutes is sufficient to destroy vegetative cells and eucaryotic spores. Unfortunately the temperature of boiling water (100°C or 212°F at sea level) is not high enough to destroy bacterial endospores, which may survive hours of boiling. Therefore boiling can be used for disinfection of drinking water and objects not harmed by water, but boiling does not sterilize.

In order to destroy bacterial endospores, moist heat sterilization must be carried out at temperatures above 100°C, and this requires the use of saturated steam under pressure. Steam sterilization is carried out with an **autoclave** (**figure 7.3**), a device somewhat like a fancy pressure cooker. The development of the autoclave by Chamberland in 1884 tremendously stimulated the growth of microbiology. Water is boiled to produce steam, which is released through the jacket and into the autoclave's chamber (figure 7.3*b*). The air initially present in the chamber is forced out until the chamber is filled with saturated steam and the outlets are closed. Hot, saturated steam continues to enter until the chamber reaches the desired temperature and pressure, usually 121°C and 15 pounds of pressure. At this temperature saturated steam destroys all vegetative cells and endospores in a small volume of liquid within 10 to 12 minutes. Treatment is continued for at least 15 minutes to provide a margin of safety. Of course, larger containers of liquid such as flasks and carboys require much longer treatment times.

Table 7.2 Approximate Conditions for Moist Heat Killing

Organism	Vegetative Cells	Spores
Yeasts	5 minutes at 50–60°C	5 minutes at 70–80°C
Molds	30 minutes at 62°C	30 minutes at 80°C
Bacteria[a]	10 minutes at 60–70°C	2 to over 800 minutes at 100°C 0.5–12 minutes at 121°C
Viruses	30 minutes at 60°C	

[a]Conditions for mesophilic bacteria.

Autoclaving must be carried out properly or the processed materials will not be sterile. If all air has not been flushed out of the chamber, it will not reach 121°C even though it may reach a pressure of 15 pounds. The chamber should not be packed too tightly because the steam needs to circulate freely and contact everything in the autoclave. Bacterial endospores will be killed only if they are kept at 121°C for 10 to 12 minutes. When a large volume of liquid must be sterilized, an extended sterilization time is needed because it takes longer for the center of the liquid to reach 121°C; 5 liters of liquid may require about 70 minutes. In view of these potential difficulties, a biological indicator is often autoclaved along with other material. This indicator commonly consists of a culture tube containing a sterile ampule of medium and a paper strip covered with spores of *Geobacillus stearothermophilus*. After autoclaving, the ampule is aseptically broken and the culture incubated for several days. If the test bacterium does not grow in the medium, the sterilization run has been successful. Sometimes either special tape that spells out the word *sterile* or a paper indicator strip that changes color upon sufficient heating is autoclaved with a load of material. If the word appears on the tape or if the color changes after autoclaving, the material is supposed to be sterile. These approaches are convenient and save time but are not as reliable as the use of bacterial endospores.

Many substances, such as milk, are treated with controlled heating at temperatures well below boiling, a process known as **pasteurization** in honor of its developer Louis Pasteur. In the 1860s the French wine industry was plagued by the problem of wine spoilage, which made wine storage and shipping difficult. Pasteur examined spoiled wine under the microscope and detected microorganisms that looked like the bacteria responsible for lactic acid and acetic acid fermentations. He then discovered that a brief heating at 55 to 60°C would destroy these microorganisms and preserve wine for long periods. In 1886 the German chemists V. H. and F. Soxhlet adapted the technique for preserving milk and reducing milk-transmissible diseases. Milk pasteurization was introduced into the United States in 1889. Milk, beer, and many other beverages are now pasteurized. Pasteurization does not sterilize a beverage, but it does kill any pathogens present and drastically slows spoilage by reducing the level of nonpathogenic spoilage microorganisms.

Many objects are best sterilized in the absence of water by **dry heat sterilization.** Some items are sterilized by incineration. For instance, inoculating loops, which are used routinely in the laboratory, can be sterilized in a small, bench-top incinerator (**figure 7.4**). Other items are sterilized in an oven at 160 to 170°C for 2 to 3 hours. Microbial death apparently results from the oxidation of cell constituents and denaturation of proteins. Dry air heat is less effective than moist heat. The spores of *Clostridium botulinum*, the cause of botulism, are killed in 5 minutes at 121°C by moist heat but only after 2 hours at 160°C with dry heat. However, dry heat has some definite advantages. It does not corrode glassware and metal instruments as moist heat does, and it can be used to sterilize powders, oils, and similar items. Most laboratories sterilize glassware and pipettes with dry heat. Despite these advantages, dry heat sterilization is slow and not suitable for heat-sensitive materials like many plastic and rubber items.

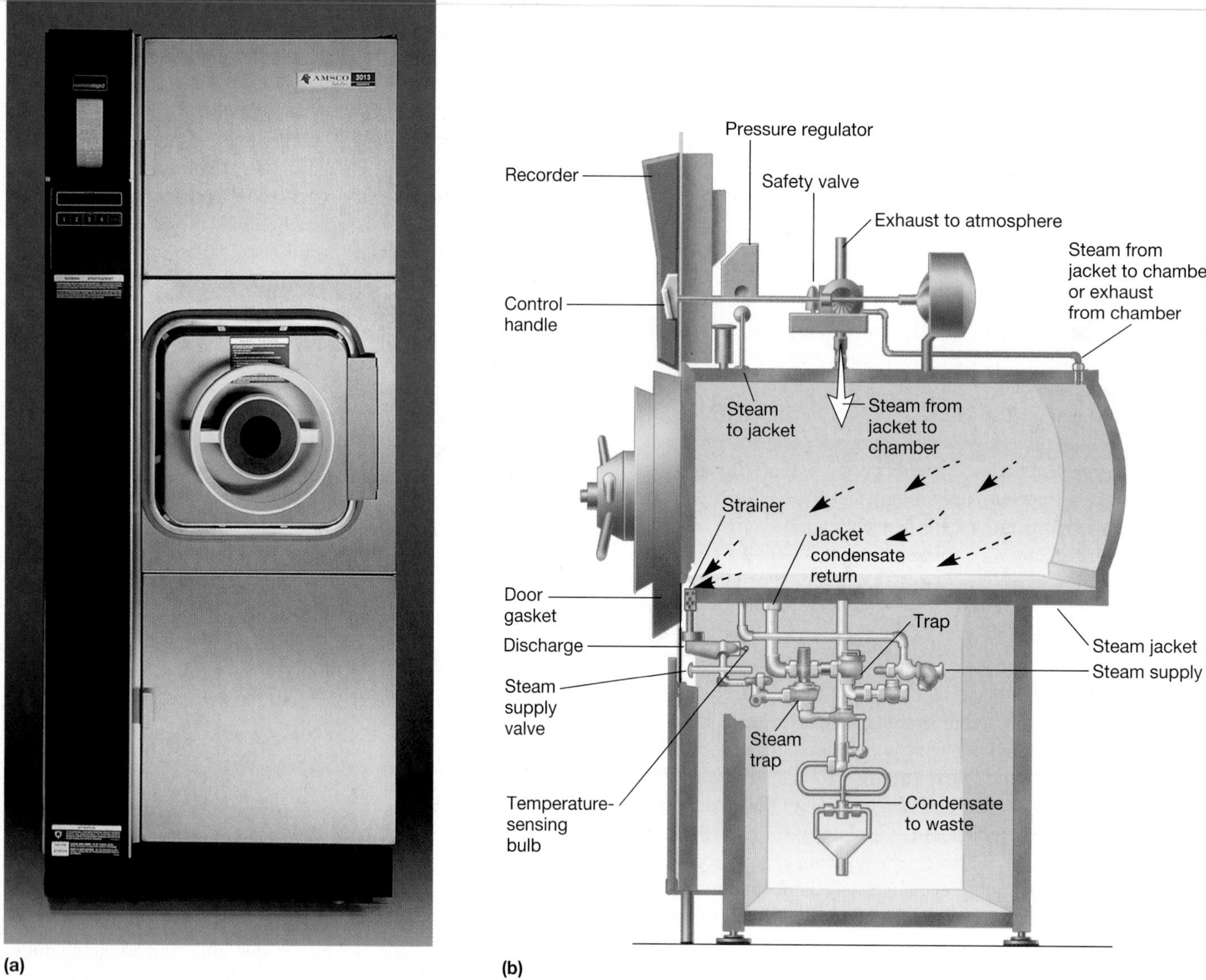

Figure 7.3 The Autoclave or Steam Sterilizer. **(a)** A modern, automatically controlled autoclave or sterilizer. **(b)** Longitudinal cross section of a typical autoclave showing some of its parts and the pathway of steam. From John J. Perkins, *Principles and Methods of Sterilization in Health Science,* 2nd edition, 1969. Courtesy of Charles C. Thomas, Publisher, Springfield, Illinois.

Because heat is so useful in controlling microorganisms, it is essential to have a precise measure of the heat-killing efficiency. Initially effectiveness was expressed in terms of thermal death point (TDP), the lowest temperature at which a microbial suspension is killed in 10 minutes. Because TDP implies that a certain temperature is immediately lethal despite the conditions, **thermal death time (TDT)** is now more commonly used. This is the shortest time needed to kill all organisms in a microbial suspension at a specific temperature and under defined conditions. However, such destruction is logarithmic, and it is theoretically not possible to completely destroy microorganisms in a sample, even with extended heating. Therefore an even more precise figure, the **decimal reduction time (*D*)** or ***D* value** has gained wide acceptance. The decimal reduction time is the time required to kill 90% of the microorganisms or spores in a sample at a specified temperature. In a semilogarithmic plot of the population remaining versus the time of heating, the *D* value is the time required for the line to drop by one log cycle or tenfold (figure 7.2). The *D* value is usually written with a subscript, indicating the temperature for which it applies. *D* values are used to estimate the relative resistance of a microorganism to different temperatures through calculation of the ***z* value.** The *z* value is the increase in temperature required to reduce *D* to 1/10 its value or to reduce it by one log cycle when log *D* is plotted against temperature (**figure 7.5**). Another way to describe heating effectiveness is with the *F* value. The ***F* value** is the time in minutes at a specific tem-

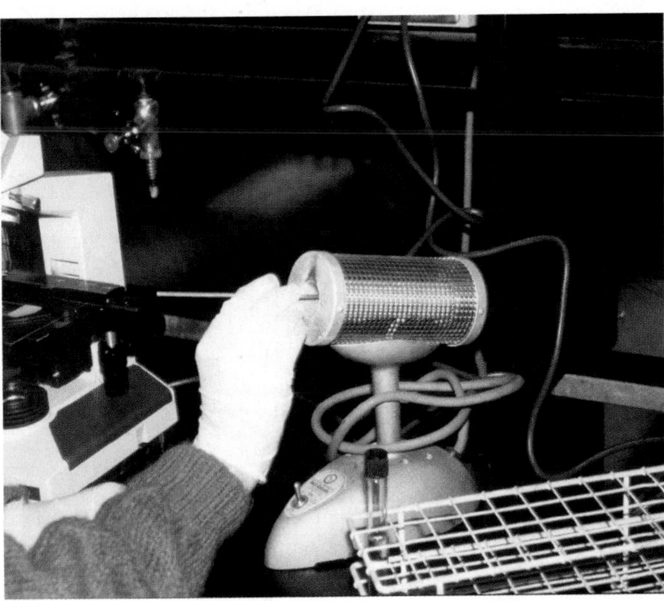

Figure 7.4 Dry Heat Incineration. Bench-top incinerators are routinely used to sterilize inoculating loops used in microbiology laboratories.

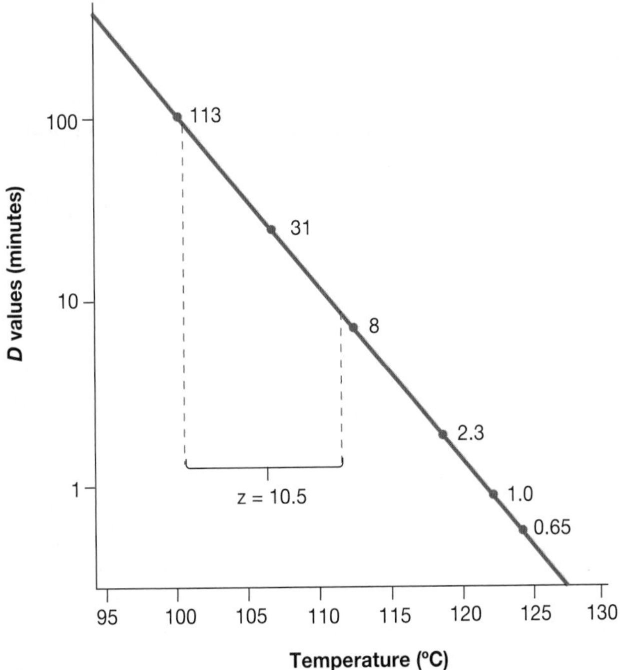

Figure 7.5 *z* Value Calculation. The *z* value used in calculation of time-temperature relationships for survival of a test microorganism, based on *D* value responses at various temperatures. The *z* value is the increase in temperature needed to reduce the decimal reduction time (*D*) to 10% of the original value. For this homogeneous sample of a test microorganism the *z* value is 10.5°. The *D* values are plotted on a logarithmic scale.

perature (usually 250°F or 121.1°C) needed to kill a population of cells or spores.

The food processing industry makes extensive use of *D* and *z* values. After a food has been canned, it must be heated to eliminate the risk of botulism arising from *Clostridium botulinum* spores. Heat treatment is carried out long enough to reduce a population of 10^{12} *C. botulinum* spores to 10^{0} (one spore); thus there is a very small chance of any can having a viable spore. The *D* value for these spores at 121°C is 0.204 minute. Therefore it would take 12*D* or 2.5 minutes to reduce 10^{12} spores to one spore by heating at 121°C. The *z* value for *C. botulinum* spores is 10°C—that is, it takes a 10°C change in temperature to alter the *D* value tenfold. If the cans were to be processed at 111°C rather than at 121°C, the *D* value would increase by tenfold to 2.04 minutes and the 12*D* value to 24.5 minutes. *D* values and *z* values for some common food-borne pathogens are given in **table 7.3.** Three *D* values are included for *Staphylococcus aureus* to illustrate the variation of killing rate with environment and the protective effect of organic material. Controlling food spoilage (section 40.3)

Low Temperatures

Although our emphasis is on the destruction of microorganisms, often the most convenient control technique is to inhibit their growth and reproduction by the use of either freezing or refrigeration. This approach is particularly important in food microbiology. Freezing items at −20°C or lower stops microbial growth because of the low temperature and the absence of liquid water. Some microorganisms will be killed by ice crystal disruption of cell membranes, but freezing does not destroy all contaminating microbes. In fact, freezing is a very good method for long-term storage of microbial samples when carried out properly, and many laboratories have a low-temperature freezer for culture storage at −30 or −70°C. Because frozen food can contain many microorganisms, it should be thawed in a refrigerator and consumed promptly in order to avoid spoilage and pathogen growth. The influence of environmental factors on growth: Temperature (section 6.5)

Refrigeration greatly slows microbial growth and reproduction, but does not halt it completely. Fortunately most pathogens are mesophilic and do not grow well at temperatures around 4°C. Refrigerated items may be ruined by growth of psychrophilic and psychrotrophic microorganisms, particularly if water is present. Thus refrigeration is a good technique only for shorter-term storage of food and other items.

1. Describe how an autoclave works. What conditions are required for sterilization by moist heat? What three things must one do when operating an autoclave to help ensure success?
2. In the past, spoiled milk was responsible for a significant proportion of infant deaths. Why is untreated milk easily spoiled? Why is boiling milk over prolonged periods not a desirable method for controlling spoilage and spread of milk-borne pathogens?
3. Define thermal death point (TDP), thermal death time (TDT), decimal reduction time (*D*) or *D* value, *z* value, and the *F* value.

Table 7.3 ***D* Values and *z* Values for Some Food-Borne Pathogens**

Organism	Substrate	*D* Value (°C) in Minutes	*z* Value (°C)
Clostridium botulinum	Phosphate buffer	$D_{121} = 0.204$	10
Clostridium perfringens (heat-resistant strain)	Culture media	$D_{90} = 3–5$	6–8
Salmonella spp.	Chicken à la king	$D_{60} = 0.39–0.40$	4.9–5.1
Staphylococcus aureus	Chicken à la king	$D_{60} = 5.17–5.37$	5.2–5.8
	Turkey stuffing	$D_{60} = 15.4$	6.8
	0.5% NaCl	$D_{60} = 2.0–2.5$	5.6

Values taken from F. L. Bryan, 1979, "Processes That Affect Survival and Growth of Microorganisms," *Time-Temperature Control of Foodborne Pathogens*, Atlanta: Centers for Disease Control and Prevention, Atlanta, GA.

4. How can the *D* value be used to estimate the time required for sterilization? Suppose that you wanted to eliminate the risk of salmonellosis by heating your food ($D_{60} = 0.4$ minute, *z* value $= 5.0$). Calculate the 12*D* value at 60°C. How long would it take to achieve the same results by heating at 50, 55, and 65°C?
5. In table 7.3, why is the *D* value so different for the three conditions in which *S. aureus* might be found?
6. How can low temperatures be used to control microorganisms? Compare the control goal for using heat with that for using low temperatures.

Filtration

Filtration is an excellent way to reduce the microbial population in solutions of heat-sensitive material, and sometimes it can be used to sterilize solutions. Rather than directly destroying contaminating microorganisms, the filter simply removes them. There are two types of filters. **Depth filters** consist of fibrous or granular materials that have been bonded into a thick layer filled with twisting channels of small diameter. The solution containing microorganisms is sucked through this layer under vacuum, and microbial cells are removed by physical screening or entrapment and also by adsorption to the surface of the filter material. Depth filters are made of diatomaceous earth (Berkefield filters), unglazed porcelain (Chamberlain filters), asbestos, or other similar materials.

Membrane filters have replaced depth filters for many purposes. These circular filters are porous membranes, a little over 0.1 mm thick, made of cellulose acetate, cellulose nitrate, polycarbonate, polyvinylidene fluoride, or other synthetic materials. Although a wide variety of pore sizes are available, membranes with pores about 0.2 μm in diameter are used to remove most vegetative cells, but not viruses, from solutions ranging in volume from 1 ml to many liters. The membranes are held in special holders (**figure 7.6**) and are often preceded by depth filters made of glass fibers to remove larger particles that might clog the membrane filter. The solution is pulled or forced through the filter with a vacuum or with pressure from a syringe, peristaltic pump, or nitrogen gas bottle, and collected in previously sterilized containers. Membrane filters remove microorganisms by screening them out much as a sieve separates large sand particles from small ones (**figure 7.7**). These filters are used to sterilize pharmaceuticals, ophthalmic solutions, culture media, oils, antibiotics, and other heat-sensitive solutions.

Air also can be sterilized by filtration. Two common examples are surgical masks and cotton plugs on culture vessels that let air in but keep microorganisms out. Other important examples are **laminar flow biological safety cabinets,** which employ **high-efficiency particulate air (HEPA) filters** (a type of depth filter) to remove 99.97% of 0.3 μm particles. Laminar flow biological safety cabinets or hoods force air through HEPA filters, then project a vertical curtain of sterile air across the cabinet opening. This protects a worker from microorganisms being handled within the cabinet and prevents contamination of the room (**figure 7.8**). A person uses these cabinets when working with dangerous agents such as *Mycobacterium tuberculosis* and tumor viruses. They are also employed in research labs and industries, such as the pharmaceutical industry, when a sterile working surface is needed for conducting assays, preparing media, examining tissue cultures, and the like.

Radiation

In chapter 6, the types of radiation and the ways in which radiation damages or destroys microorganisms were discussed. Microbiologists take advantage of the effects of ultraviolet and ionizing radiation to sterilize or disinfect objects.

Ultraviolet (UV) radiation around 260 nm (*see figure 6.25*) is quite lethal but does not penetrate glass, dirt films, water, and other substances very effectively. Because of this disadvantage, UV radiation is used as a sterilizing agent only in a few specific situations. UV lamps are sometimes placed on the ceilings of rooms or in biological safety cabinets to sterilize the air and any exposed surfaces. Because UV radiation burns the skin and damages eyes, people working in such areas must be certain the UV lamps are off when the areas are in use. Commercial UV units are available for water treatment (**figure 7.9**). Pathogens and other microorganisms are destroyed when a thin layer of water is passed under the lamps.

Ionizing radiation is an excellent sterilizing agent and penetrates deep into objects. It will destroy bacterial endospores and vegetative cells, both procaryotic and eucaryotic; however, ion-

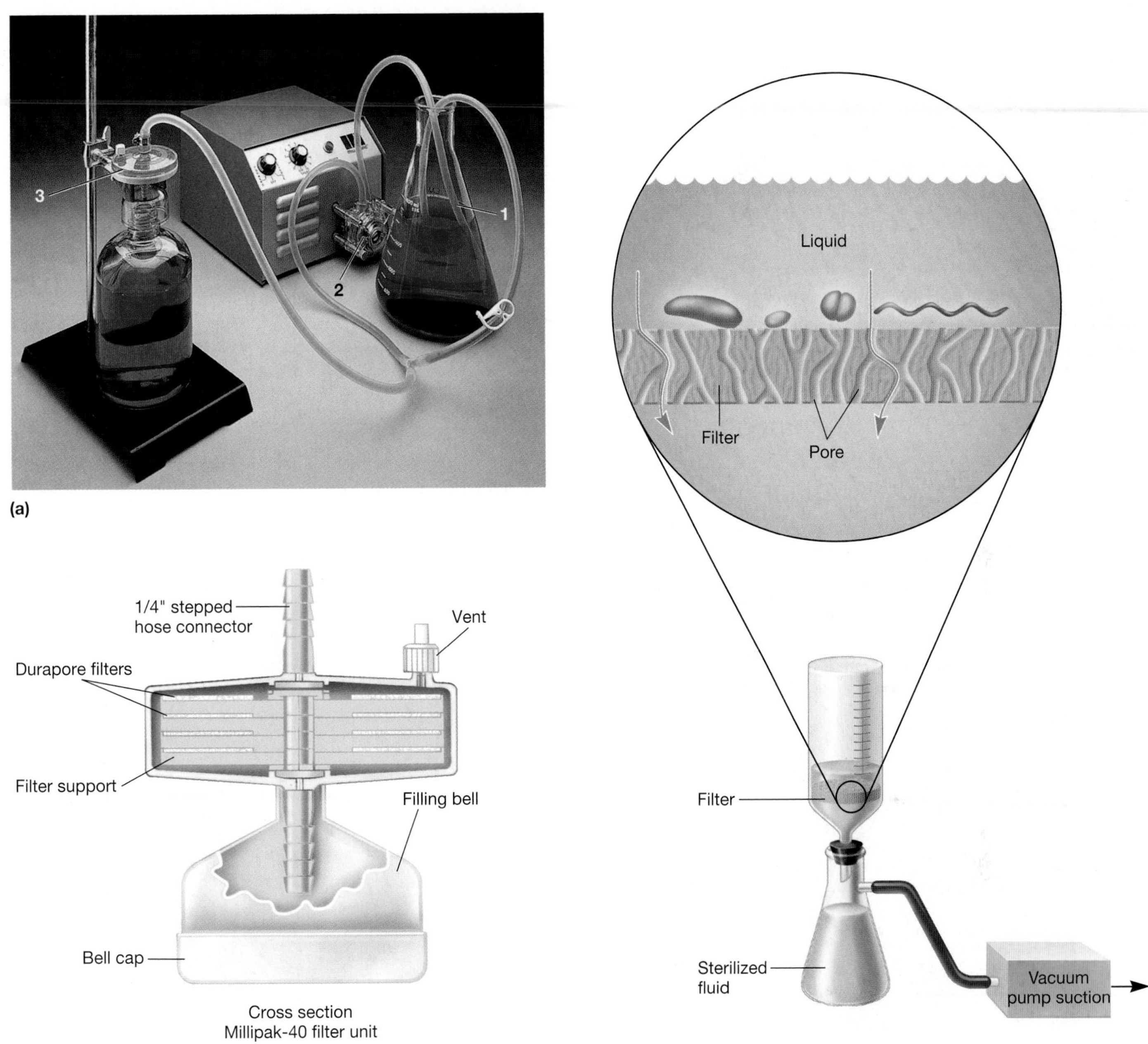

Figure 7.6 Membrane Filter Sterilization. The liquid to be sterilized is pumped through a membrane filter and into a sterile container. **(a)** A complete filtering setup. The nonsterile solution is in the Erlenmeyer flask, *1*. A peristaltic pump, *2*, forces the solution through the membrane filter unit, *3*. **(b)** Schematic representation of a membrane filtration setup that uses a vacuum pump to force liquid through the filter. The inset shows a cross section of the filter and its pores, which are too small for microbes to pass through. **(c)** Cross section of a membrane filtration unit. Several membranes are used to increase its capacity.

izing radiation is not always effective against viruses. Gamma radiation from a cobalt 60 source is used in the cold sterilization of antibiotics, hormones, sutures, and plastic disposable supplies such as syringes. Gamma radiation has also been used to sterilize and "pasteurize" meat and other food (**figure 7.10**). Irradiation can eliminate the threat of such pathogens as *Escherichia coli* O157:H7, *Staphylococcus aureus,* and *Campylobacter jejuni.* Based on the results of numerous studies, both the Food and Drug Administration and the World Health Organization have approved food irradiation and declared it safe. Currently irradiation is being used to treat poultry, beef, pork, veal, lamb, fruits, vegetables, and spices.

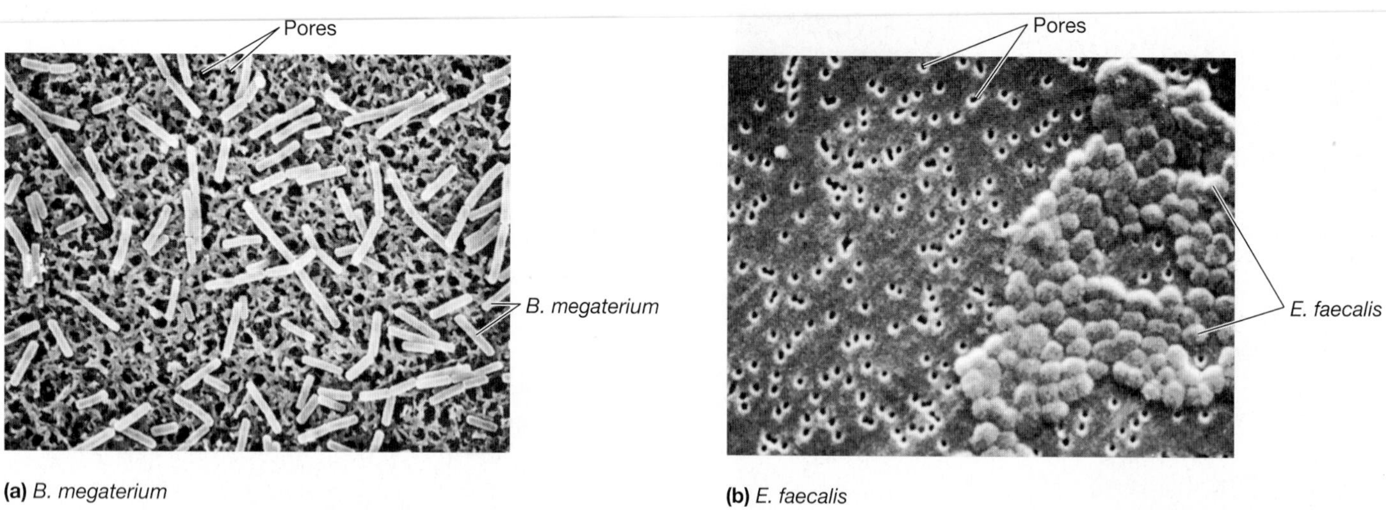

Figure 7.7 Membrane Filter Types. **(a)** *Bacillus megaterium* on an Ultipor nylon membrane with a bacterial removal rating of 0.2 μm (×2,000). **(b)** *Enterococcus faecalis* resting on a polycarbonate membrane filter with 0.4 μm pores (×5,900).

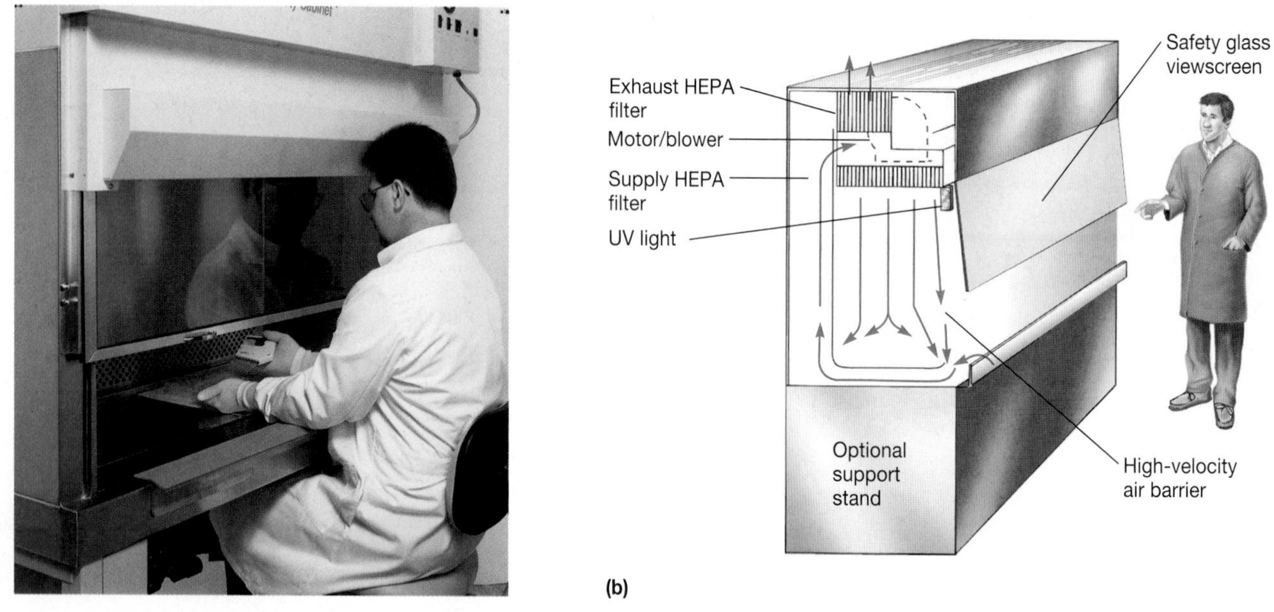

Figure 7.8 A Laminar Flow Biological Safety Cabinet. **(a)** A technician pipetting potentially hazardous material in a safety cabinet. **(b)** A schematic diagram showing the airflow pattern.

1. What are depth filters and membrane filters, and how are they used to sterilize liquids? Describe the operation of a biological safety cabinet.
2. Give the advantages and disadvantages of ultraviolet light and ionizing radiation as sterilizing agents. Provide a few examples of how each is used for this purpose.

7.5 The Use of Chemical Agents in Control

Physical agents are generally used to sterilize objects. Chemicals, on the other hand, are more often employed in disinfection and antisepsis. The proper use of chemical agents is essential to laboratory and hospital safety (**Techniques & Applications 7.2**). Chemicals also are employed to prevent microbial growth in food, and certain chemicals are used to treat infectious disease.
Techniques & Applications 35.1: Standard microbiological practices

Many different chemicals are available for use as disinfectants, and each has its own advantages and disadvantages. In selecting an agent, it is important to keep in mind the characteristics of a desirable disinfectant. Ideally the disinfectant must be effective against a wide variety of infectious agents (gram-positive and gram-negative bacteria, acid-fast bacteria, bacterial endospores, fungi, and viruses) at low concentrations and in the presence of organic matter. Although the chemical must be toxic for infectious agents, it should not be toxic to people or corrosive for common materials. In practice, this balance between effectiveness and low toxicity for animals is hard to achieve. Some chemicals are used despite their low effectiveness because they are relatively nontoxic. The ideal disinfectant should be stable upon storage, odorless or with a pleasant odor, soluble in water and lipids for penetration into microorganisms, have a low surface tension so that it can enter cracks in surfaces, and be relatively inexpensive.

One potentially serious problem is the overuse of antiseptics. For instance, the antibacterial agent triclosan is found in products such as deodorants, mouthwashes, soaps, cutting boards, and baby toys. Unfortunately, the emergence of triclosan-resistant bacteria has become a problem. For example, *Pseudomonas aeruginosa* actively pumps the antiseptic out of the cell. There is now evidence that extensive use of triclosan also increases the frequency of bacterial resistance to antibiotics. Thus overuse of antiseptics can have unintended harmful consequences. Drug resistance (section 34.6)

Figure 7.9 Ultraviolet (UV) Treatment System for Disinfection of Water. Water flows through racks of UV lamps and is exposed to 254 nm UV radiation. This system has a capacity of several million gallons per day and can be used as an alternative to chlorination.

The properties and uses of several groups of common disinfectants and antiseptics are surveyed next. Chemotherapeutic agents are briefly introduced at the end of this section. Many of the characteristics of disinfectants and antiseptics are summarized in **tables 7.4** and **7.5.** Structures of some common agents are given in **figure 7.11.**

Phenolics

Phenol was the first widely used antiseptic and disinfectant. In 1867 Joseph Lister employed it to reduce the risk of infection during surgery. Today phenol and phenolics (phenol derivatives) such as cresols, xylenols, and orthophenylphenol are used as disinfectants in laboratories and hospitals. The commercial disinfectant Lysol is made of a mixture of phenolics. Phenolics act by denaturing proteins and disrupting cell membranes. They have some real advantages as disinfectants: phenolics are tuberculocidal, effective in the presence of organic material, and remain active on surfaces long after application. However, they have a disagreeable odor and can cause skin irritation.

Hexachlorophene (figure 7.11) has been one of the most popular antiseptics because once applied it persists on the skin and reduces skin bacteria for long periods. However, it can cause brain damage and is now used in hospital nurseries only in response to a staphylococcal outbreak.

Alcohols

Alcohols are among the most widely used disinfectants and antiseptics. They are bactericidal and fungicidal but not sporicidal; some lipid-containing viruses are also destroyed. The two most popular alcohol germicides are ethanol and isopropanol, usually used in

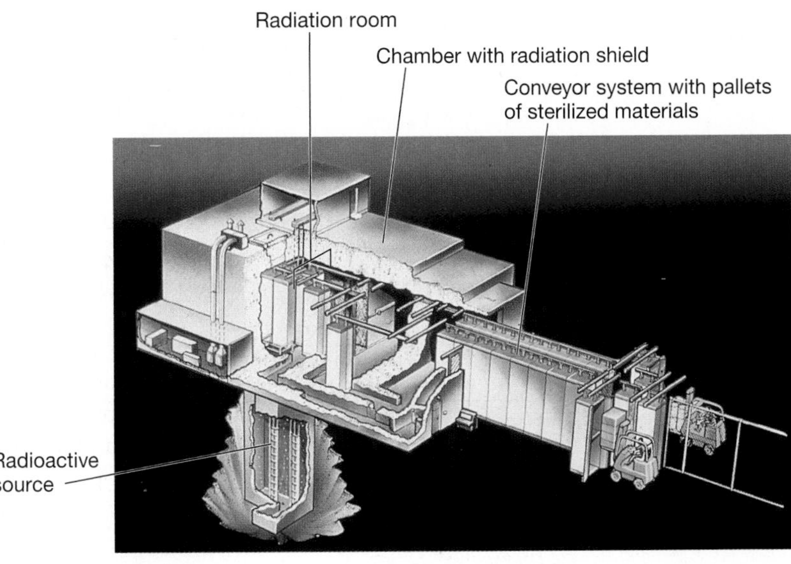

Figure 7.10 Sterilization with Ionizing Radiation. **(a)** An irradiation machine that uses radioactive cobalt 60 as a gamma radiation source to sterilize fruits, vegetables, meats, fish, and spices. **(b)** The universal symbol for irradiation that must be affixed to all irradiated materials.

Techniques & Applications

7.2 Universal Precautions for Microbiology Laboratories

Blood and other body fluids from all patients should be considered infective.

1. All specimens of blood and body fluids should be put in a well-constructed container with a secure lid to prevent leaking during transport. Care should be taken when collecting each specimen to avoid contaminating the outside of the container and of the laboratory form accompanying the specimen.
2. All persons processing blood and body-fluid specimens should wear gloves. Masks and protective eyewear should be worn if mucous membrane contact with blood or body fluids is anticipated. Gloves should be changed and hands washed after completion of specimen processing.
3. For routine procedures, such as histologic and pathological studies or microbiologic culturing, a biological safety cabinet is not necessary. However, biological safety cabinets should be used whenever procedures are conducted that have a high potential for generating droplets. These include activities such as blending, sonicating, and vigorous mixing.
4. Mechanical pipetting devices should be used for manipulating all liquids in the laboratory. Mouth pipetting must not be done.
5. Use of needles and syringes should be limited to situations in which there is no alternative, and the recommendations for preventing injuries with needles outlined under universal precautions should be followed. Techniques & Applications 35.1: Standard microbiological practices
6. Laboratory work surfaces should be decontaminated with an appropriate chemical germicide after a spill of blood or other body fluids and when work activities are completed.
7. Contaminated materials used in laboratory tests should be decontaminated before reprocessing or be placed in bags and disposed of in accordance with institutional policies for disposal of infective waste.
8. Scientific equipment that has been contaminated with blood or other body fluids should be decontaminated and cleaned before being repaired in the laboratory or transported to the manufacturer.
9. All persons should wash their hands after completing laboratory activities and should remove protective clothing before leaving the laboratory.
10. There should be no eating, drinking, or smoking in the work area.

Source: Adapted from *Morbidity and Mortality Weekly Report,* 36 (Suppl. 2S) 5S–10S, 1987, the Centers for Disease Control and Prevention Guidelines.

Table 7.4 Activity Levels of Selected Germicides

Class	Use Concentration of Active Ingredient	Activity Level[a]
Gas		
Ethylene oxide	450–500 mg/liter[b]	High
Liquid		
Glutaraldehyde, aqueous	2%	High to intermediate
Formaldehyde + alcohol	8 + 70%	High
Stabilized hydrogen peroxide	6–30%	High to intermediate
Formaldehyde, aqueous	6–8%	High to intermediate
Iodophors	750–5,000 mg/liter[c]	High to intermediate
Iodophors	75–150 mg/liter[c]	Intermediate to low
Iodine + alcohol	0.5 + 70%	Intermediate
Chlorine compounds	0.1–0.5%[d]	Intermediate
Phenolic compounds, aqueous	0.5–3%	Intermediate to low
Iodine, aqueous	1%	Intermediate
Alcohols (ethyl, isopropyl)	70%	Intermediate
Quaternary ammonium compounds	0.1–0.2% aqueous	Low
Chlorhexidine	0.75–4%	Low
Hexachlorophene	1–3%	Low
Mercurial compounds	0.1–0.2%	Low

Source: From Seymour S. Block, *Disinfection, Sterilization and Preservation.* Copyright © 1983 Lea & Febiger, Malvern, Pa. Reprinted by permission.

[a]High-level disinfectants destroy vegetative bacterial cells including *M. tuberculosis,* bacterial endospores, fungi, and viruses. Intermediate-level disinfectants destroy all of the above except endospores. Low-level agents kill bacterial vegetative cells except for *M. tuberculosis,* fungi, and medium-sized lipid-containing viruses (but not bacterial endospores or small, nonlipid viruses).

[b]In autoclave-type equipment at 55 to 60°C.

[c]Available iodine.

[d]Free chlorine.

Table 7.5 Relative Efficacy of Commonly Used Disinfectants and Antiseptics

Class	Disinfectant	Antiseptic	Comment
Gas			
Ethylene oxide	3–4[a]	0[a]	Sporicidal; toxic; good penetration; requires relative humidity of 30% or more; microbicidal activity varies with apparatus used; absorbed by porous material; dry spores highly resistant; moisture must be present, and presoaking is most desirable
Liquid			
Glutaraldehyde, aqueous	3	0	Sporicidal; active solution unstable; toxic
Stabilized hydrogen peroxide	3	0	Sporicidal; solution stable up to 6 weeks; toxic orally and to eyes; mildly skin toxic; little inactivation by organic matter
Formaldehyde + alcohol	3	0	Sporicidal; noxious fumes; toxic; volatile
Formaldehyde, aqueous	1–2	0	Sporicidal; noxious fumes; toxic
Phenolic compounds	3	0	Stable; corrosive; little inactivation by organic matter; irritates skin
Chlorine compounds	1–2	0	Fast action; inactivation by organic matter; corrosive; irritates skin
Alcohol	1	3	Rapidly microbicidal except for bacterial spores and some viruses; volatile; flammable; dries and irritates skin
Iodine + alcohol	0	4	Corrosive; very rapidly microbicidal; causes staining; irritates skin; flammable
Iodophors	1–2	3	Somewhat unstable; relatively bland; staining temporary; corrosive
Iodine, aqueous	0	2	Rapidly microbicidal; corrosive; stains fabrics; stains and irritates skin
Quaternary ammonium compounds	1	0	Bland; inactivated by soap and anionics; compounds absorbed by fabrics; old or dilute solution can support growth of gram-negative bacteria
Hexachlorophene	0	2	Bland; insoluble in water, soluble in alcohol; not inactivated by soap; weakly bactericidal
Chlorhexidine	0	3	Bland; soluble in water and alcohol; weakly bactericidal
Mercurial compounds	0	±	Bland; greatly inactivated by organic matter; weakly bactericidal

Source: From Seymour S. Block, *Disinfection, Sterilization and Preservation.* Copyright © 1983 Lea & Febiger, Malvern, Pa. Reprinted by permission.

[a]Subjective ratings of practical usefulness in a hospital environment—4 is maximal usefulness; 0 is little or no usefulness; ± signifies that the substance is sometimes useful but not always.

about 70 to 80% concentration. They act by denaturing proteins and possibly by dissolving membrane lipids. A 10 to 15 minute soaking is sufficient to disinfect thermometers and small instruments.

Halogens

A halogen is any of the five elements (fluorine, chlorine, bromine, iodine, and astatine) in group VIIA of the periodic table. They exist as diatomic molecules in the free state and form saltlike compounds with sodium and most other metals. The halogens iodine and chlorine are important antimicrobial agents. Iodine is used as a skin antiseptic and kills by oxidizing cell constituents and iodinating cell proteins. At higher concentrations, it may even kill some spores. Iodine often has been applied as tincture of iodine, 2% or more iodine in a water-ethanol solution of potassium iodide. Although it is an effective antiseptic, the skin may be damaged, a stain is left, and iodine allergies can result. More recently iodine has been complexed with an organic carrier to form an **iodophor.** Iodophors are water soluble, stable, and nonstaining, and release iodine slowly to minimize skin burns and irritation. They are used in hospitals for preoperative skin degerming and in hospitals and laboratories for disinfecting. Some popular brands are Wescodyne for skin and laboratory disinfection and Betadine for wounds.

Chlorine is the usual disinfectant for municipal water supplies and swimming pools and is also employed in the dairy and food industries. It may be applied as chlorine gas, sodium hypochlorite (bleach), or calcium hypochlorite, all of which yield hypochlorous acid (HClO) and then atomic oxygen. The result is oxidation of cellular materials and destruction of vegetative bacteria and fungi, although not spores.

$$Cl_2 + H_2O \longrightarrow HCl + HClO$$
$$Ca(OCl)_2 + 2H_2O \longrightarrow Ca(OH)_2 + 2HClO$$
$$HClO \longrightarrow HCl + O$$

Death of almost all microorganisms usually occurs within 30 minutes. Since organic material interferes with chlorine action by reacting with chlorine and its products, an excess of chlorine is added to ensure microbial destruction. One potential problem is that chlorine reacts with organic compounds to form carcinogenic trihalomethanes, which must be monitored in drinking water. Ozone

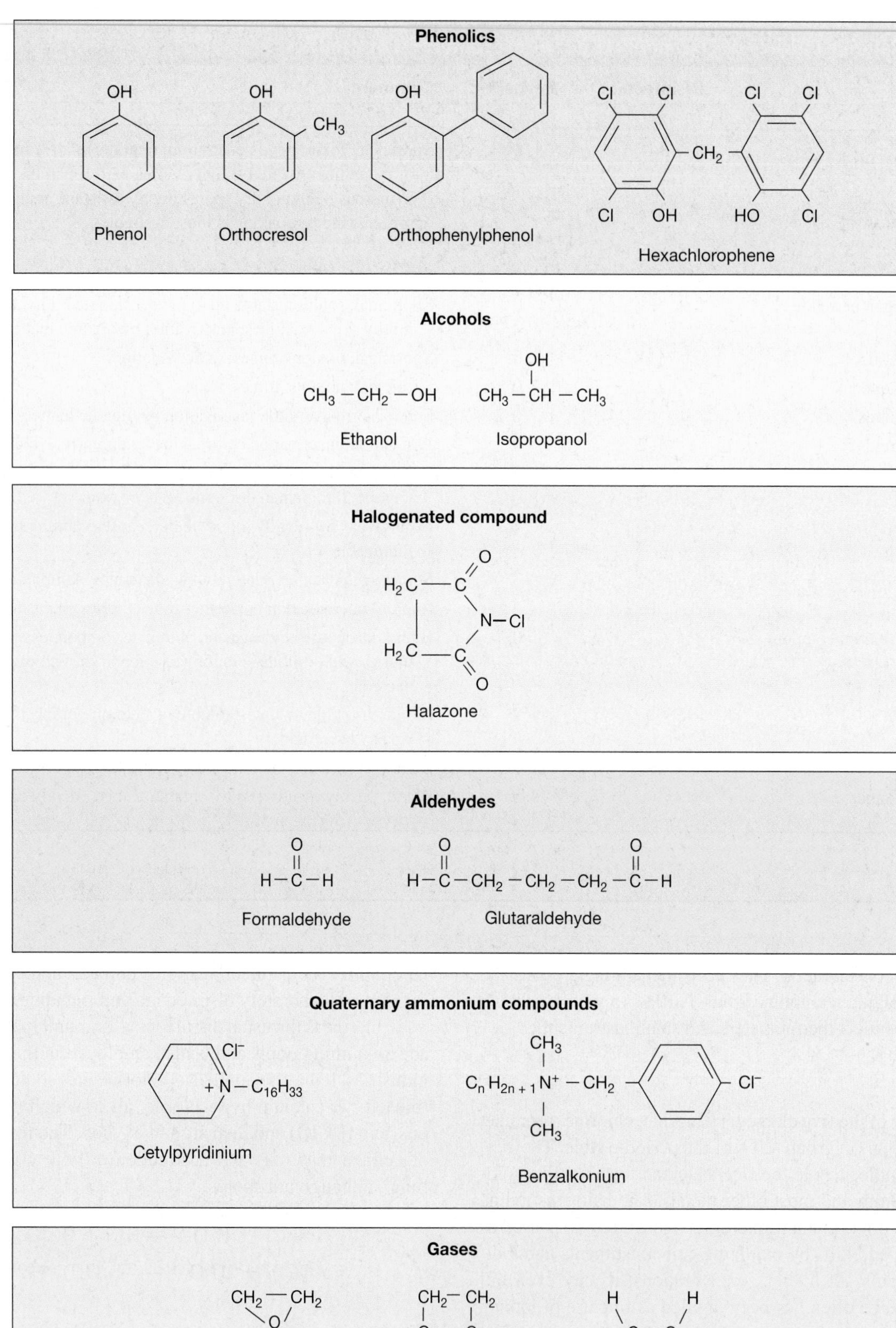

Figure 7.11 Disinfectants and Antiseptics. The structures of some frequently used disinfectants and antiseptics.

sometimes has been used successfully as an alternative to chlorination in Europe and Canada.

Chlorine is also an excellent disinfectant for individual use because it is effective, inexpensive, and easy to employ. Small quantities of drinking water can be disinfected with halazone tablets. Halazone (parasulfone dichloramidobenzoic acid) slowly releases chloride when added to water and disinfects it in about a half hour. It is frequently used by campers lacking access to uncontaminated drinking water.

Chlorine solutions make very effective laboratory and household disinfectants. An excellent disinfectant-detergent combination can be prepared if a 1/40 dilution of household bleach is combined with a nonionic detergent, such as a dishwashing detergent, to give a 0.8% detergent concentration. This mixture will remove both dirt and bacteria.

Heavy Metals

For many years the ions of heavy metals such as mercury, silver, arsenic, zinc, and copper were used as germicides. These have now been superseded by other less toxic and more effective germicides (many heavy metals are more bacteriostatic than bactericidal). There are a few exceptions. In some hospitals, a 1% solution of silver nitrate is added to the eyes of infants to prevent ophthalmic gonorrhea. Silver sulfadiazine is used on burns. Copper sulfate is an effective algicide in lakes and swimming pools.

Heavy metals combine with proteins, often with their sulfhydryl groups, and inactivate them. They may also precipitate cell proteins.

Quaternary Ammonium Compounds

Quaternary ammonium compounds are detergents that have antimicrobial activity and are effective disinfectants. **Detergents** [Latin *detergere,* to wipe away] are organic cleansing agents that are amphipathic, having both polar hydrophilic and nonpolar hydrophobic components. The hydrophilic portion of a quaternary ammonium compound is a positively charged quaternary nitrogen; thus quaternary ammonium compounds are cationic detergents. Their antimicrobial activity is the result of their ability to disrupt microbial membranes; they may also denature proteins.

Cationic detergents like benzalkonium chloride and cetylpyridinium chloride kill most bacteria but not *M. tuberculosis* or endospores. They have the advantages of being stable and nontoxic but they are inactivated by hard water and soap. Cationic detergents are often used as disinfectants for food utensils and small instruments and as skin antiseptics. Several brands are on the market. Zephiran contains benzalkonium chloride and Ceepryn, cetylpyridinium chloride.

Aldehydes

Both of the commonly used aldehydes, formaldehyde and glutaraldehyde (figure 7.11), are highly reactive molecules that combine with nucleic acids and proteins and inactivate them, probably by cross-linking and alkylating molecules (**figure 7.12**). They are sporicidal and can be used as chemical sterilants. Formaldehyde is usually dissolved in water or alcohol before use. A 2% buffered solution of glutaraldehyde is an effective disinfectant. It is less irritating than formaldehyde and is used to disinfect hospital and laboratory equipment. Glutaraldehyde usually disinfects objects within about 10 minutes but may require as long as 12 hours to destroy all spores.

Sterilizing Gases

Many heat-sensitive items such as disposable plastic petri dishes and syringes, heart-lung machine components, sutures, and catheters are sterilized with ethylene oxide gas (figure 7.11). Ethylene oxide (EtO) is both microbicidal and sporicidal and kills by combining with cell proteins. It is a particularly effective sterilizing agent because it rapidly penetrates packing materials, even plastic wraps.

Sterilization is carried out in a special ethylene oxide sterilizer, very much resembling an autoclave in appearance, that controls the EtO concentration, temperature, and humidity (**figure 7.13**). Because pure EtO is explosive, it is usually supplied in a 10 to 20% concentration mixed with either CO_2 or dichlorodifluoromethane. The ethylene oxide concentration, humidity, and temperature influence the rate of sterilization. A clean object can be sterilized if treated for 5 to 8 hours at 38°C or 3 to 4 hours at 54°C when the relative humidity is maintained

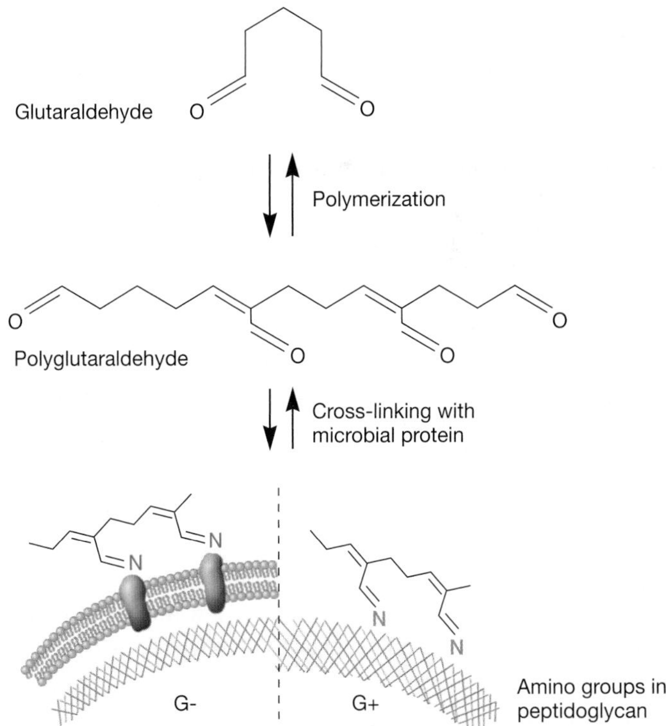

Figure 7.12 Effects of Glutaraldehyde. Glutaraldehyde polymerizes and then interacts with amino acids in proteins (left) or in peptidoglycan (right). As a result, the proteins are alkylated and cross-linked to other proteins, which inactivates them. The amino groups in peptidoglycan are also alkylated and cross-linked, which prevents them from participating in other chemical reactions such as those involved in peptidoglycan synthesis.

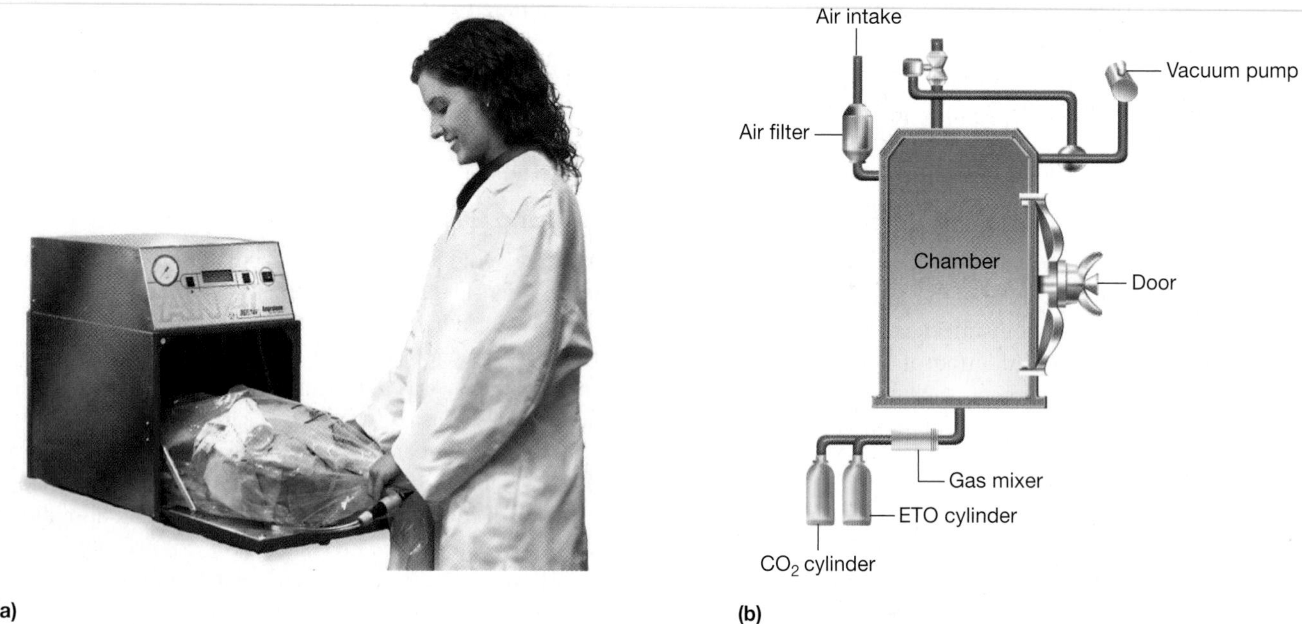

Figure 7.13 An Ethylene Oxide Sterilizer. **(a)** An automatic ethylene oxide (EtO) sterilizer. **(b)** Schematic of an EtO sterilizer. Items to be sterilized are placed in the chamber and EtO and carbon dioxide are introduced. After the sterilization procedure is completed, the EtO and carbon dioxide are pumped out of the chamber and air enters.

at 40 to 50% and the EtO concentration at 700 mg/liter. Extensive aeration of the sterilized materials is necessary to remove residual EtO because it is so toxic.

Betapropiolactone (BPL) is occasionally employed as a sterilizing gas. In the liquid form it has been used to sterilize vaccines and sera. BPL decomposes to an inactive form after several hours and is therefore not as difficult to eliminate as EtO. It also destroys microorganisms more readily than ethylene oxide but does not penetrate materials well and may be carcinogenic. For these reasons, BPL has not been used as extensively as EtO.

Vaporized hydrogen peroxide can be used to decontaminate biological safety cabinets, operating rooms, and other large facilities. These systems introduce vaporized hydrogen peroxide into the enclosure for some time, depending on the size of the enclosure and the materials within. Hydrogen peroxide is toxic and kills a wide variety of microorganisms. However, during the course of the decontamination process, it breaks down to water and oxygen, both of which are harmless. Other advantages of these systems are that they can be used at a wide range of temperatures (4 to 80°C) and they do not damage most materials.

Chemotherapeutic Agents

The chemicals discussed thus far are appropriate for use either on inanimate objects or external host tissues. **Chemotherapeutic agents** are chemicals that can be used internally to kill or inhibit the growth of microbes within host tissues. They can be used internally because they have **selective toxicity;** that is, they target the microbe and do relatively little if any harm to the host. Most chemotherapeutic agents are **antibiotics**—chemicals synthesized by microbes that are effective in controlling the growth of bacteria. Since the discovery of the first antibiotics, pharmaceutical companies have developed numerous derivatives and many synthetic antibiotics. Chemotherapeutic agents for treating diseases caused by fungi, protists, and viruses have also been developed. Chemotherapeutic agents are described in more detail in chapter 34.

1. Why are most antimicrobial chemical agents disinfectants rather than sterilants? What general characteristics should one look for in a disinfectant?
2. Describe each of the following agents in terms of its chemical nature, mechanism of action, mode of application, common uses and effectiveness, and advantages and disadvantages: phenolics, alcohols, halogens, heavy metals, quaternary ammonium compounds, aldehydes, and ethylene oxide.
3. Which disinfectants or antiseptics would be used to treat the following: oral thermometer, laboratory bench top, drinking water, patch of skin before surgery, small medical instruments (probes, forceps, etc.)? Explain your choices.
4. How do chemotherapeutic agents differ from the other chemical control agents described in this chapter?
5. Which physical or chemical agent would be the best choice for sterilizing the following items: glass pipettes, tryptic soy broth tubes, nutrient agar, antibiotic solution, interior of a biological safety cabinet, wrapped package of plastic petri plates? Explain your choices.

7.6 Evaluation of Antimicrobial Agent Effectiveness

Testing of antimicrobial agents is a complex process regulated by two different federal agencies. The U.S. Environmental Protection Agency regulates disinfectants, whereas agents used on humans and animals are under the control of the Food and Drug

Administration. Testing of antimicrobial agents often begins with an initial screening test to see if they are effective and at what concentrations. This may be followed by more realistic in-use testing.

The best-known disinfectant screening test is the **phenol coefficient test** in which the potency of a disinfectant is compared with that of phenol. A series of dilutions of phenol and the disinfectant being tested are prepared. A standard amount of *Salmonella typhi* and *Staphylococcus aureus* are added to each dilution; the dilutions are then placed in a 20 or 37°C water bath. At 5-minute intervals, samples are withdrawn from each dilution and used to inoculate a growth medium, which is incubated for two or more days and then examined for growth. If there is no growth in the growth medium, the dilution at that particular time of sampling killed the bacteria. The highest dilution (i.e., the lowest concentration) that kills the bacteria after a 10-minute exposure, but not after 5 minutes, is used to calculate the phenol coefficient. This is done by dividing the reciprocal of the appropriate dilution for the disinfectant being tested by the reciprocal of the appropriate phenol dilution. For instance, if the phenol dilution was 1/90 and maximum effective dilution for disinfectant X was 1/450, then the phenol coefficient of X would be 5. The higher the phenol coefficient value, the more effective the disinfectant under these test conditions. A value greater than 1 means that the disinfectant is more effective than phenol. A few representative phenol coefficient values are given in **table 7.6.**

The phenol coefficient test is a useful initial screening procedure, but the phenol coefficient can be misleading if taken as a direct indication of disinfectant potency during normal use. This is because the phenol coefficient is determined under carefully controlled conditions with pure bacterial strains, whereas disinfectants are normally used on complex populations in the presence of organic matter and with significant variations in environmental factors like pH, temperature, and presence of salts.

To more realistically estimate disinfectant effectiveness, other tests are often used. The rates at which selected bacteria are destroyed with various chemical agents may be experimentally determined and compared. A **use dilution test** can also be carried out. Stainless steel cylinders are contaminated with specific bacterial species under carefully controlled conditions. The cylinders are dried briefly, immersed in the test disinfectants for 10 minutes, transferred to culture media, and incubated for two days. The disinfectant concentration that kills the organisms in the sample with a 95% level of confidence under these conditions is determined. Disinfectants also can be tested under conditions designed to simulate normal in-use situations. In-use testing techniques allow a more accurate determination of the proper disinfectant concentration for a particular situation.

Table 7.6 Phenol Coefficients for Some Disinfectants

	Phenol Coefficients[a]	
Disinfectant	***Salmonella typhi***	***Staphylococcus aureus***
Phenol	1	1
Cetylpyridinium chloride	228	337
O-phenylphenol	5.6 (20°C)	4.0
p-cresol	2.0–2.3	2.3
Hexachlorophene	5–15	15–40
Merthiolate	600	62.5
Mercurochrome	2.7	5.3
Lysol	1.9	3.5
Isopropyl alcohol	0.6	0.5
Ethanol	0.04	0.04
2% I_2 solution in EtOH	4.1–5.2 (20°C)	4.1–5.2 (20°C)

[a]All values were determined at 37°C except where indicated.

1. Briefly describe the phenol coefficient test.
2. Why might it be necessary to employ procedures like the use dilution and in-use tests?

Summary

7.1 Definitions of Frequently Used Terms

a. Sterilization is the process by which all living cells, viable spores, viruses, and viroids are either destroyed or removed from an object or habitat. Disinfection is the killing, inhibition, or removal of microorganisms (but not necessarily endospores) that can cause disease.

b. The main goal of disinfection and antisepsis is the removal, inhibition, or killing of pathogenic microbes. Both processes also reduce the total number of microbes. Disinfectants are chemicals used to disinfect inanimate objects; antiseptics are used on living tissue.

c. Antimicrobial agents that kill organisms often have the suffix -cide, whereas agents that prevent growth and reproduction have the suffix -static.

7.2 The Pattern of Microbial Death

a. Microbial death is usually exponential or logarithmic (**figure 7.2**).

7.3 Conditions Influencing the Effectiveness of Antimicrobial Agents

a. The effectiveness of a disinfectant or sterilizing agent is influenced by population size, population composition, concentration or intensity of the agent, exposure duration, temperature, and nature of the local environment.

7.4 The Use of Physical Methods in Control

a. Moist heat kills by degrading nucleic acids, denaturing enzymes and other proteins, and disrupting cell membranes.

b. Although treatment with boiling water for 10 minutes kills vegetative forms, an autoclave must be used to destroy endospores by heating at 121°C and 15 pounds of pressure (**figure 7.3**).

c. Glassware and other heat-stable items may be sterilized by dry heat at 160 to 170°C for 2 to 3 hours.

d. The efficiency of heat killing is often indicated by the thermal death time or the decimal reduction time.

e. Refrigeration and freezing can be used to control microbial growth and reproduction.

f. Microorganisms can be efficiently removed by filtration with either depth filters or membrane filters (**figure 7.6**).

g. Biological safety cabinets with high-efficiency particulate filters sterilize air by filtration (**figure 7.8**).

h. Radiation of short wavelength or high-energy ultraviolet and ionizing radiation can be used to sterilize objects (**figures 7.9** and **7.10**).

7.5 The Use of Chemical Agents in Control

a. Chemical agents usually act as disinfectants because they cannot readily destroy bacterial endospores. Disinfectant effectiveness depends on concentration, treatment duration, temperature, and presence of organic material (**tables 7.4** and **7.5**).

b. Phenolics and alcohols are popular disinfectants that act by denaturing proteins and disrupting cell membranes (**figure 7.11**).

c. Halogens (iodine and chlorine) kill by oxidizing cellular constituents; cell proteins may also be iodinated. Iodine is applied as a tincture or iodophor. Chlorine may be added to water as a gas, hypochlorite, or an organic chlorine derivative.

d. Heavy metals tend to be bacteriostatic agents. They are employed in specialized situations such as the use of silver nitrate in the eyes of newborn infants and copper sulfate in lakes and pools.

e. Cationic detergents are often used as disinfectants and antiseptics; they disrupt membranes and denature proteins.

f. Aldehydes such as formaldehyde and glutaraldehyde can sterilize as well as disinfect because they kill spores.

g. Ethylene oxide gas penetrates plastic wrapping material and destroys all life forms by reacting with proteins. It is used to sterilize packaged, heat-sensitive materials.

h. Chemotherapeutic agents are chemicals such as antibiotics that can be ingested by or injected into a host. They kill or inhibit the growth of microbes within host tissues.

7.6 Evaluation of Antimicrobial Agent Effectiveness

a. A variety of procedures can be used to determine the effectiveness of disinfectants, among them the following: phenol coefficient test, measurement of killing rates with germicides, use dilution testing, and in-use testing.

Key Terms

algicide 151
antibiotics 164
antimicrobial agent 152
antisepsis 151
antiseptics 151
autoclave 153
bactericide 151
bacteriostatic 151
chemotherapeutic agents 164
chemotherapy 151
decimal reduction time (*D*) 154
depth filters 156
detergent 163
disinfectant 151
disinfection 151
dry heat sterilization 153
D value 154
fungicide 151
fungistatic 151
F value 154
germicide 151
high-efficiency particulate air (HEPA) filters 156
iodophor 161
ionizing radiation 156
laminar flow biological safety cabinets 156
membrane filters 156
pasteurization 153
phenol coefficient test 165
sanitization 151
selective toxicity 164
sterilization 151
thermal death time (TDT) 154
ultraviolet (UV) radiation 156
use dilution test 165
viricide 151
z value 154

Critical Thinking Questions

1. Throughout history, spices have been used as preservatives and to cover up the smell/taste of food that is slightly spoiled. The success of some spices led to a magical, ritualized use of many of them and possession of spices was often limited to priests or other powerful members of the community.
   a. Choose a spice and trace its use geographically and historically. What is its common-day use today?
   b. Spices grow and tend to be used predominantly in warmer climates. Explain.
2. Design an experiment to determine whether an antimicrobial agent is acting as a cidal or static agent. How would you determine whether an agent is suitable for use as an antiseptic rather than as a disinfectant?
3. Suppose that you are testing the effectiveness of disinfectants with the phenol coefficient test and obtained the following results. What disinfectant can you safely say is the most effective? Can you determine its phenol coefficient from these results?

	Bacterial Growth after Treatment		
Dilution	**Disinfectant A**	**Disinfectant B**	**Disinfectant C**
1/20	–	–	–
1/40	+	–	–
1/80	+	–	+
1/160	+	+	+
1/320	+	–	+

Learn More

Barkley, W. E., and Richardson, J. H. 1994. Laboratory safety. In *Methods for general and molecular bacteriology,* P. Gerhardt, et al., editors, 715–34. Washington, D.C.:American Society for Microbiology.

Gilbert, P., and McBain, A. J. 2003. Potential impact of increased use of biocides in consumer products on prevalence of antibiotic resistance. *Clin. Microbiol. Rev.* 16(2):189–208.

Sewell, D. L. 1995. Laboratory-associated infections and biosafety. *Clin. Microbiol. Rev.* 8(3):389–405.

Sondossi, M. 2000. Biocides. In *Encyclopedia of microbiology,* 2d ed., vol. I, J. Lederberg, editor-in-chief, 445–60. San Diego: Academic Press.

Widmer, A. F., and Frei, R. 1999. Decontamination, disinfection, and sterilization. In *Manual of clinical microbiology,* 7th ed., P. R. Murray, et al., editors, 138–64. Washington, D.C.: ASM Press.

Please visit the Prescott website at www.mhhe.com/prescott7 for additional references.

8 Metabolism: Energy, Enzymes, and Regulation

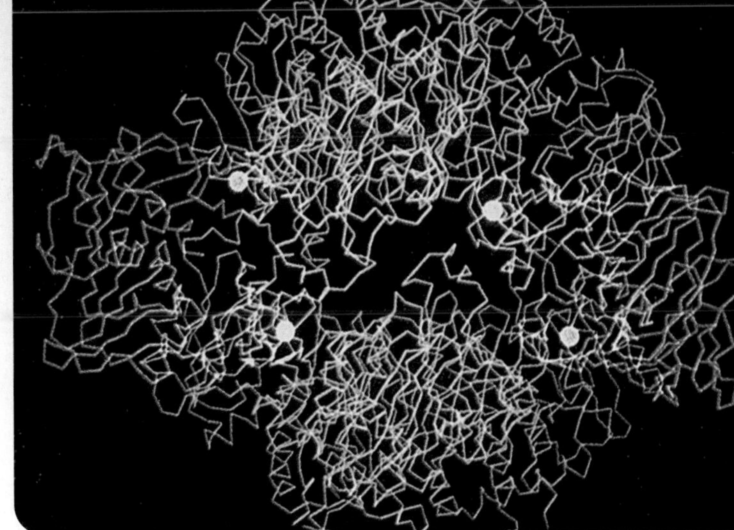

This model shows *Escherichia coli* aspartate carbamoyltransferase in the less active T state. The catalytic polypeptide chains are in blue and the regulatory chains are colored red.

Preview

- Metabolism is the total of all chemical reactions that occur in cells. It is divided into two major parts: energy-conserving reactions that release and conserve the energy provided by an organism's energy source; and anabolism, the reactions that consume energy in order to build large, complex molecules from smaller, simpler molecules.
- Cells use energy to do cellular work. Living organisms do three major types of work: chemical work, transport work, and mechanical work.
- All living organisms obey the laws of thermodynamics. These laws can be used to predict the spontaneity of chemical reactions that occur in cells, and the amount of energy released or energy consumed during a reaction.
- ATP is a high-energy molecule that serves as the cell's energy currency. It links energy-yielding exergonic reactions to energy-consuming endergonic reactions.
- Oxidation-reduction reactions are important in the energy-conserving processes that cells carry out. When electrons are transferred from an electron donor with a more negative reduction potential to an electron acceptor with a more positive potential, energy is made available for work.
- Enzymes are protein catalysts that make life possible by increasing the rate of reactions. They do this by lowering the activation energy of the reactions they catalyze.
- Metabolic pathways are regulated to maintain cell components in proper balance, even in the face of a changing environment, and to conserve energy and raw materials. Metabolic pathways are regulated by one of three methods: metabolic channeling, regulating the activity of certain enzymes, and regulating the amount of an enzyme that is synthesized.
- The activity of enzymes and proteins involved in complex behaviors such as chemotaxis can be regulated by the same mechanisms used to control metabolic pathways.

In the early chapters of this text, we focus on a series of "what" questions about microorganisms: what are they; what do they look like; what are they made of? In chapters 8 through 13 we begin to consider a number of "how" questions: how do microbes extract energy from their energy source; how do they use the nutrients obtained from their environment; how do they build themselves? To begin to answer these "how" questions, we must turn our attention more fully to the chemistry of cells; that is, their metabolism. Chapters 8 through 10 introduce metabolism, focusing on those processes that conserve the energy supplied by an organism's energy source and on how that energy is used to synthesize the building blocks from which an organism is constructed. Chapters 11 through 13 consider the synthesis of three important macromolecules: DNA, RNA, and proteins.

This chapter begins with a brief overview of metabolism. In order to understand metabolism, the nature of energy and the laws of thermodynamics must be considered, so a discussion of these topics follows. As will be seen, microorganisms display an amazing array of metabolic diversity, especially in terms of the energy sources and energy-conserving processes they employ. Yet, despite this diversity, there are several basic principles and processes common to the metabolism of all microbes. These will be the focus of most of the remaining sections of the chapter. The chapter ends with a discussion of metabolic regulation.

8.1 An Overview of Metabolism

Metabolism is the total of all chemical reactions occurring in the cell. These chemical reactions are summarized in **figure 8.1.** Metabolism may be divided into two major parts: energy-conserving

Fresh oxygen flows
From the open stomata
The whole world inhales

—*Crystal Cunningham*

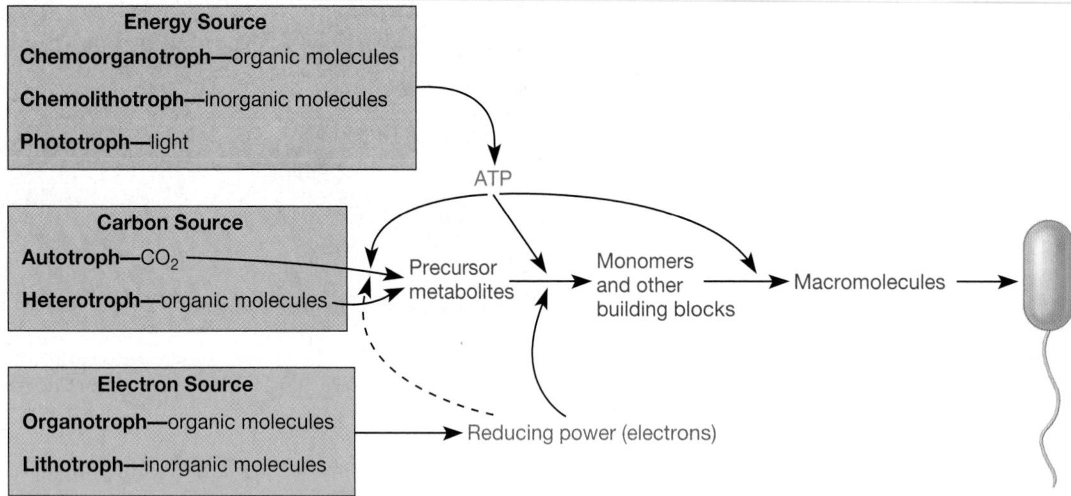

Figure 8.1 Overview of Metabolism. The cell structures of organisms are assembled from various macromolecules (e.g., nucleic acids and proteins). Macromolecules are synthesized from monomers and other building blocks (e.g., nucleotides and amino acids), which are the products of biochemical pathways that begin with precursor metabolites (e.g., pyruvate and α-ketoglutarate). In autotrophs, the precursor metabolites arise from CO_2-fixation pathways and related pathways; in heterotrophs, they arise from reactions of the central metabolic pathways. Reducing power and ATP are consumed in many metabolic pathways. All organisms can be defined metabolically in terms of their energy source, carbon source, and electron source. In the case of chemoorganotrophs, the energy source is an organic molecule that is also the source of carbon and electrons. For chemolithotrophs, the energy source is an inorganic molecule that is also the electron source; the carbon source can be either CO_2 (autotrophs) or an organic molecule (heterotrophs). For phototrophs, the energy source is light, the carbon source can be CO_2 or organic molecules, and the electron source can be water (oxygenic phototrophs) or another reduced molecule such as hydrogen sulfide (anoxygenic phototrophs).

reactions and anabolism. In the **energy-conserving reactions** or fueling reactions, the energy provided to the cell by its energy source is released and conserved as ATP. These reactions are sometimes referred to as **catabolism** [Greek *cata,* down, and *ballein,* to throw], since they can involve the breakdown of relatively large, complex organic molecules into smaller, simpler molecules. **Anabolism** [Greek, *ana,* up] is the synthesis of complex organic molecules from simpler ones. It involves a series of steps: (1) conversion of the organism's carbon source into a set of small molecules called precursor metabolites; (2) synthesis of monomers and other building blocks (i.e., amino acids, nucleotides, simple carbohydrates, and simple lipids) from the precursor metabolites; (3) synthesis of macromolecules (i.e., proteins, nucleic acids, complex carbohydrates, and complex lipids); and (4) assembly of macromolecules into cellular structures. Anabolism requires energy, which is transferred from the energy source to the synthetic systems of the cell by ATP. Anabolism also requires a source of electrons stored in the form of **reducing power.** Reducing power is needed because anabolism is a reductive process; that is, electrons are added to small molecules as they are used to build macromolecules (figure 8.1). Energy conservation and the provision of reducing power are the focus of chapter 9; the initial steps in anabolism are the focus of chapter 10.

As discussed in chapter 5, there are five major nutritional types of microorganisms based on their sources of energy, carbon, and electrons (figure 8.1). Animals and many microbes are chemoorganoheterotrophs. These organisms use organic molecules as their source of energy, carbon, and electrons. In other words, the same molecule that supplies them with energy also supplies them with carbon and electrons. Chemoorganoheterotrophs (often simply referred to as chemoorganotrophs or chemoheterotrophs) can use one or more of the following catabolic processes: fermentation, aerobic respiration, or anaerobic respiration. Chemolithoautotrophs use CO_2 as a carbon source and reduced inorganic molecules as sources of both energy and electrons. Their energy-conserving processes are sometimes referred to as respiration because they are similar to the respiratory processes carried out by chemoorganoheterotrophs. Photolithotrophic microbes use light as their source of energy and inorganic molecules as a source of electrons. When they use water as their electron source, as do plants, they release oxygen into the atmosphere by a process called oxygenic photosynthesis. Certain photosynthetic bacteria do not use water as an electron source; they do not release oxygen into the atmosphere and are called anoxygenic phototrophs. Photolithotrophs are usually autotrophic, using CO_2 as a carbon source. However, some phototrophic microbes are heterotrophic. Aerobic respiration (section 9.2); Anaerobic respiration (section 9.6); Fermentation (section 9.7); Chemolithotrophy (section 9.11); Phototrophy (section 9.12)

The interactions of the nutritional types of microorganisms are critical to the functioning of the biosphere. The ultimate source of most biological energy is visible sunlight. Light energy is trapped and reducing power is generated by photoautotrophs and used to transform CO_2 into organic molecules such as glucose. The or-

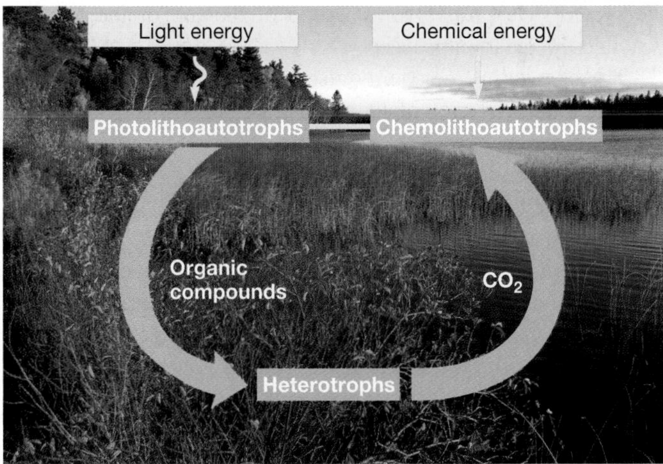

Figure 8.2 The Flow of Carbon and Energy in an Ecosystem. This diagram depicts the flow of energy and carbon in general terms. See text for discussion.

ganic molecules then serve as energy, carbon, and electron sources for chemoorganoheterotrophs. The breakdown of the organic molecules by chemoorganotrophs releases CO_2 back into the atmosphere (**figure 8.2**). In a similar cycle, chemolithoautotrophs use the energy and reducing power derived from inorganic energy sources to synthesize organic molecules, which "feed" chemoorganoheterotrophs (figure 8.2). Thus the flows of carbon and energy in ecosystems are intimately related.

8.2 Energy and Work

Energy may be most simply defined as the capacity to do work. This is because all physical and chemical processes are the result of the application or movement of energy. Living cells carry out three major types of work, and all are essential to life processes. **Chemical work** involves the synthesis of complex biological molecules from much simpler precursors (i.e., anabolism); energy is needed to increase the molecular complexity of a cell. **Transport work** requires energy in order to take up nutrients, eliminate wastes, and maintain ion balances. Energy input is needed because molecules and ions often must be transported across cell membranes against an electro chemical gradient. For example, molecules move into a cell even though their concentration is higher internally. Similarly a solute may be expelled from the cell against a concentration gradient. The third type of work is **mechanical work,** perhaps the most familiar of the three. Energy is required for cell motility and to move structures within cells.

1. Define metabolism, energy, energy-conserving reactions, catabolism, anabolism, and reducing power.
2. Describe in general terms how energy from sunlight is spread throughout the biosphere. What sources of energy, other than sunlight, do microorganisms use?
3. What kinds of work are carried out in a cell? Suppose a bacterium was doing the following: synthesizing peptidoglycan, rotating its flagellum and swimming, and secreting siderophores. What type of work is the bacterium doing in each case?

8.3 The Laws of Thermodynamics

To understand how energy is trapped as ATP and how ATP is used to do cellular work, some knowledge of the basic principles of thermodynamics is required. The science of **thermodynamics** analyzes energy changes in a collection of matter (e.g., a cell or a plant) called a system. All other matter in the universe is called the surroundings. Thermodynamics focuses on the energy differences between the initial state and the final state of a system. It is not concerned with the rate of the process. For instance, if a pan of water is heated to boiling, only the condition of the water at the start and at boiling is important in thermodynamics, not how fast it is heated or on what kind of stove.

Two important laws of thermodynamics must be understood. The **first law of thermodynamics** says that energy can be neither created nor destroyed. The total energy in the universe remains constant although it can be redistributed, as it is during the many energy exchanges that occur during chemical reactions. For example, heat is given off by exothermic reactions and absorbed during endothermic reactions. However, the first law alone cannot explain why heat is released by one chemical reaction and absorbed by another. Nor does it explain why gas will flow from a full cylinder to an empty cylinder until the gas pressure is equal in both (**figure 8.3**). Explanations for these phenomena require the **second law of thermodynamics** and a condition of matter called entropy. **Entropy** may be considered a measure of the randomness or disorder of a system. The greater the disorder of a system, the greater is its entropy. The second law states that physical and chemical processes proceed in such a way that the randomness or disorder of the universe (the system and its surroundings)

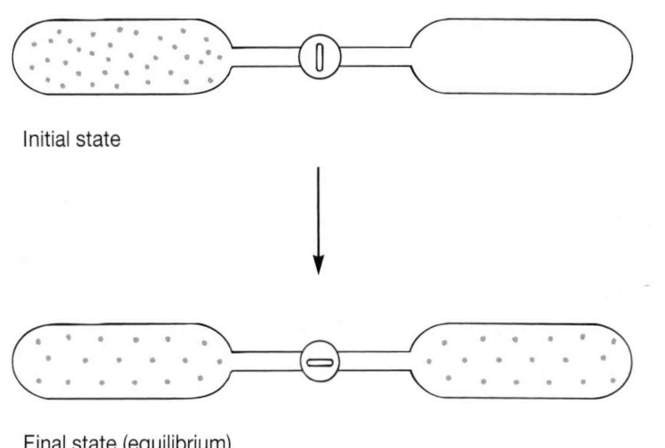

Figure 8.3 A Second Law Process. The expansion of gas into an empty cylinder simply redistributes the gas molecules until equilibrium is reached. The total number of molecules remains unchanged.

increases to the maximum possible. Gas will always expand into an empty cylinder.

It is necessary to specify quantitatively the amount of energy used in or evolving from a particular process, and two types of energy units are employed. A **calorie** (cal) is the amount of heat energy needed to raise one gram of water from 14.5 to 15.5°C. The amount of energy also may be expressed in terms of **joules** (J), the units of work capable of being done. One cal of heat is equivalent to 4.1840 J of work. One thousand calories or a kilocalorie (kcal) is enough energy to boil 1.9 ml of water. A kilojoule is enough energy to boil about 0.44 ml of water, or enable a person weighing 70 kg to climb 35 steps. The joule is normally used by chemists and physicists. Because biologists most often speak of energy in terms of calories, this text will employ calories when discussing energy changes.

8.4 Free Energy and Reactions

The first and second laws can be combined in a useful equation, relating the changes in energy that can occur in chemical reactions and other processes.

$$\Delta G = \Delta H - T \cdot \Delta S$$

ΔG is the change in free energy, ΔH is the change in enthalpy, T is the temperature in Kelvin (°C + 273), and ΔS is the change in entropy occurring during the reaction. The change in **enthalpy** is the change in heat content. Cellular reactions occur under conditions of constant pressure and volume. Thus the change in enthalpy is about the same as the change in total energy during the reaction. The **free energy change** is the amount of energy in a system (or cell) available to do useful work at constant temperature and pressure. Therefore the change in entropy (ΔS) is a measure of the proportion of the total energy change that the system cannot use in performing work. Free energy and entropy changes do not depend on how the system gets from start to finish. A reaction will occur spontaneously if the free energy of the system decreases during the reaction or, in other words, if ΔG is negative. It follows from the equation that a reaction with a large positive change in entropy will normally tend to have a negative ΔG value and therefore occur spontaneously. A decrease in entropy will tend to make ΔG more positive and the reaction less favorable.

It can be helpful to think of the relationship between entropy (ΔS) and change in free energy (ΔG) in terms that are more concrete. Consider the Greek myth of Sisyphus, king of Corinth. For his assorted crimes against the gods, he was condemned to roll a large boulder to the top of a steep hill for all eternity. This represents a very negative change in entropy—a boulder poised at the top of a hill is neither random nor disordered—and this activity (reaction) has a very positive ΔG. That is to say, Sisyphus had to put a lot of energy into the system. Unfortunately for Sisyphus, as soon as the boulder was at the top of the hill, it *spontaneously* rolled back down the hill. This represents a positive change in entropy and a negative ΔG. Sisyphus did not need to put energy into the system. He probably just stood at the top of the hill and watched the reaction proceed.

The change in free energy also has a definite, concrete relationship to the direction of chemical reactions. Consider this simple reaction.

$$A + B \rightleftharpoons C + D$$

If the molecules A and B are mixed, they will combine to form the products C and D. Eventually C and D will become concentrated enough to combine and produce A and B at the same rate as C and D are formed from A and B. The reaction is now at **equilibrium:** the rates in both directions are equal and no further net change occurs in the concentrations of reactants and products. This situation is described by the **equilibrium constant (K_{eq})**, relating the equilibrium concentrations of products and substrates to one another.

$$K_{eq} = \frac{[C][D]}{[A][B]}$$

If the equilibrium constant is greater than one, the products are in greater concentration than the reactants at equilibrium—that is, the reaction tends to go to completion as written.

The equilibrium constant of a reaction is directly related to its change in free energy. When the free energy change for a process is determined at carefully defined standard conditions of concentration, pressure, pH, and temperature, it is called the **standard free energy change** (ΔG°). If the pH is set at 7.0 (which is close to the pH of living cells), the standard free energy change is indicated by the symbol $\Delta G^{\circ\prime}$. The change in standard free energy may be thought of as the maximum amount of energy available from the system for useful work under standard conditions. Using $\Delta G^{\circ\prime}$ values allows one to compare reactions without worrying about variations in the ΔG due to differences in environmental conditions. The relationship between $\Delta G^{\circ\prime}$ and K_{eq} is given by this equation.

$$\Delta G^{\circ\prime} = -2.303RT \cdot \log K_{eq}$$

R is the gas constant (1.9872 cal/mole-degree or 8.3145 J/mole-degree), and T is the absolute temperature. Inspection of this equation shows that when $\Delta G^{\circ\prime}$ is negative, the equilibrium constant is greater than one and the reaction goes to completion as written. It is said to be an **exergonic reaction** (**figure 8.4**). In an **endergonic**

Exergonic reactions

$$A + B \rightleftharpoons C + D$$

$$K_{eq} = \frac{[C]\,[D]}{[A]\,[B]} > 1.0$$

$\Delta G^{\circ\prime}$ is negative.

Endergonic reactions

$$A + B \rightleftharpoons C + D$$

$$K_{eq} = \frac{[C]\,[D]}{[A]\,[B]} < 1.0$$

$\Delta G^{\circ\prime}$ is positive.

Figure 8.4 $\Delta G^{\circ\prime}$ and Equilibrium. The relationship of $\Delta G^{\circ\prime}$ to the equilibrium of reactions. Note the differences between exergonic and endergonic reactions.

reaction $\Delta G^{\circ\prime}$ is positive and the equilibrium constant is less than one. That is, the reaction is not favorable, and little product will be formed at equilibrium under standard conditions. Keep in mind that the $\Delta G^{\circ\prime}$ value shows only where the reaction lies at equilibrium, not how fast the reaction reaches equilibrium.

8.5 The Role of ATP in Metabolism

As already noted, considerable metabolic diversity exists in the microbial world. However, there are several biochemical principles common to all types of metabolism. These are (1) the use of ATP to store energy captured during exergonic reactions so it can be used to drive endergonic reactions; (2) the organization of metabolic reactions into pathways and cycles; (3) the catalysis of metabolic reactions by enzymes; and (4) the importance of oxidation-reduction reactions in energy conservation. This section considers the role of ATP in metabolism.

Energy is released from a cell's energy source in exergonic reactions (i.e., those reactions with a negative ΔG). Rather than wasting this energy, much of it is trapped in a practical form that allows its transfer to the cellular systems doing work. These systems carry out endergonic reactions (i.e., anabolism), and the energy captured by the cell is used to drive these reactions to completion. In living organisms, this practical form of energy is **adenosine 5′-triphosphate (ATP; figure 8.5)**. In a sense, cells carry out certain processes so that they can "earn" ATP and carry out other processes in which they "spend" their ATP. Thus ATP is often referred to as the cell's energy currency. In the cell's economy, ATP serves as the link between exergonic reactions and endergonic reactions (**figure 8.6**).

What makes ATP suited for this role as energy currency? ATP is a **high-energy molecule.** That is, it breaks down or hydrolyzes almost completely to the products **adenosine diphosphate (ADP)** and orthophosphate (P_i) with a $\Delta G^{\circ\prime}$ of −7.3 kcal/mole.

$$ATP + H_2O \rightleftharpoons ADP + P_i$$

The reference to ATP as a high-energy molecule does not mean that there is a great deal of energy stored in a particular bond of ATP. It simply indicates that the removal of the terminal phosphate goes to completion with a large negative standard free energy change; that is, the reaction is strongly exergonic. Because ATP readily transfers its phosphate to water, it is said to have a high **phosphate group transfer potential,** defined as the negative of $\Delta G^{\circ\prime}$ for the hydrolytic removal of phosphate. A molecule with a higher group transfer potential will donate phosphate to one with a lower potential.

Although the free energy change for the hydrolysis of ATP is quite large, there are numerous reactions that release even greater amounts of free energy. This energy is used to resynthesize ATP from ADP and P_i during catabolism and other energy-conserving processes. Likewise, catabolism can generate molecules with a phosphate group transfer potential that is even higher than that of ATP. Cells use these molecules to regenerate ATP from ADP by a mechanism called substrate-level phosphorylation. Thus ATP, ADP, and P_i form an energy cycle (**figure 8.7**). The fueling reactions conserve energy released from an energy source by using it to synthesize ATP from ADP and P_i. When ATP is hydrolyzed, the energy released drives endergonic processes such as anabolism, transport, and mechanical work. The mechanisms for synthesizing ATP will be described in more detail in chapter 9.

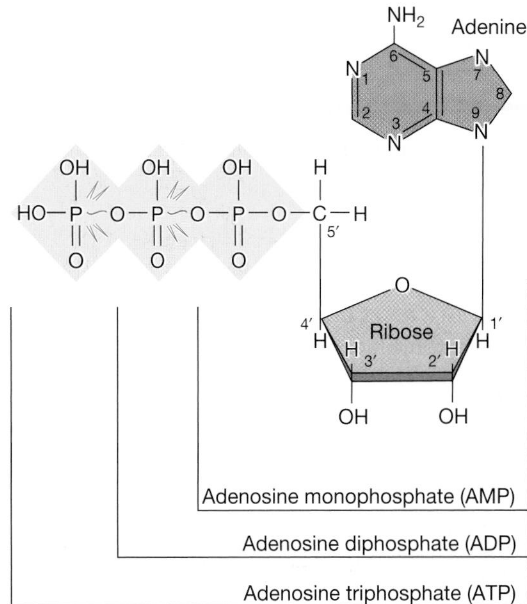

(a) ~ Bond that releases energy when broken

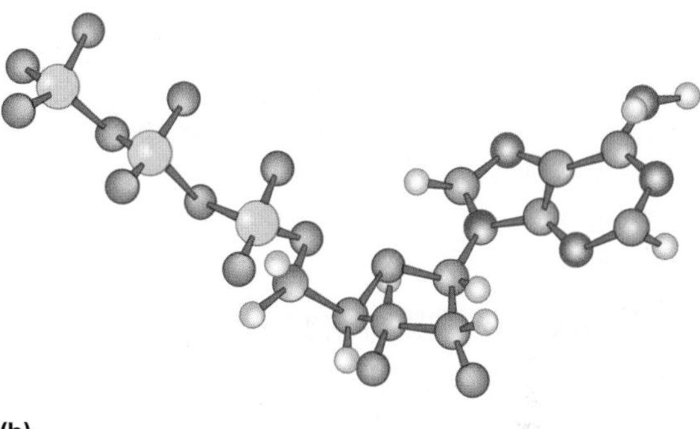

(b)

Figure 8.5 Adenosine Triphosphate and Adenosine Diphosphate. **(a)** Structure of ATP, ADP, and AMP. The two red bonds (~) are more easily broken or have a high phosphate group transfer potential (see text). The pyrimidine ring atoms have been numbered as have the carbon atoms in ribose. **(b)** A model of ATP. Carbon is in green; hydrogen in light blue; nitrogen in dark blue; oxygen in red; and phosphorus in yellow.

1. What is thermodynamics? Summarize the first and second laws of thermodynamics.
2. Define entropy and enthalpy. Do living cells increase entropy within themselves? Do they increase entropy in the environment?

Endergonic reaction alone

A + B ⇌ C + D

Endergonic reaction coupled to ATP breakdown

ATP → ADP + P_i

A + B ⇌ C + D

Figure 8.6 ATP as a Coupling Agent. The use of ATP to make endergonic reactions more favorable. It is formed by exergonic reactions and then used to drive endergonic reactions.

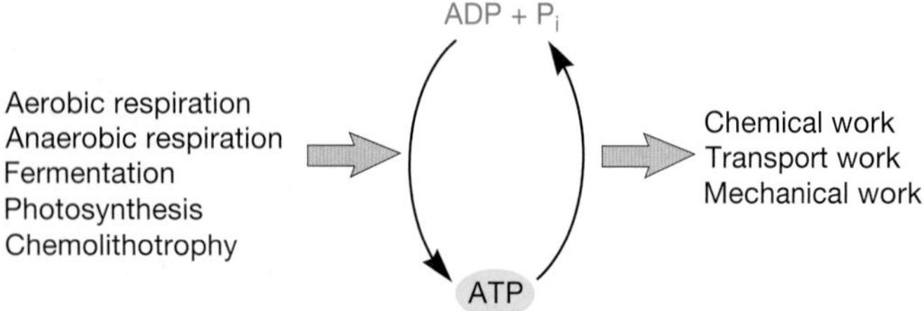

Figure 8.7 The Cell's Energy Cycle. ATP is formed from energy made available during aerobic respiration, anaerobic respiration, fermentation, chemolithotrophy, and photosynthesis. Its breakdown to ADP and phosphate (P_i) makes chemical, transport, and mechanical work possible.

3. Define free energy. What are exergonic and endergonic reactions?
4. Suppose that a chemical reaction had a large negative $\Delta G^{\circ\prime}$ value. Is the reaction endergonic or exergonic? What would this indicate about its equilibrium constant?
5. Describe the energy cycle and ATP's role in it. What characteristics of ATP make it suitable for this role? Why is ATP called a high-energy molecule?

8.6 Oxidation-Reduction Reactions, Electron Carriers, and Electron Transport Systems

Free energy changes are related to the equilibria of all chemical reactions including the equilibria of oxidation-reduction reactions. The release of energy from an energy source normally involves oxidation-reduction reactions. **Oxidation-reduction (redox) reactions** are those in which electrons move from an **electron donor** to an **electron acceptor.**[1] By convention such a reaction is written with the donor to the right of the acceptor and the number (n) of electrons (e^-) transferred.

$$\text{Acceptor} + ne^- \rightleftharpoons \text{donor}$$

The acceptor and donor pair is referred to as a redox couple (**table 8.1**). When an acceptor accepts electrons, it then becomes the donor of the couple. The equilibrium constant for the reaction is called the **standard reduction potential** (E_0) and is a measure of the tendency of the donor to lose electrons. The reference standard for reduction potentials is the hydrogen system with an E'_0 (the reduction potential at pH 7.0) of −0.42 volts or −420 millivolts.

$$2H^+ + 2e^- \rightleftharpoons H_2$$

In this reaction each hydrogen atom provides one proton (H^+) and one electron (e^-). As just noted, the standard reduction potential is measured in volts or millivolts. The volt is a unit of electrical potential or electromotive force. Therefore redox couples like the hydrogen system are a potential source of energy.

The reduction potential has a concrete meaning. Redox couples with more negative reduction potentials will donate electrons to couples with more positive potentials and greater affinity for electrons. Thus electrons tend to move from donors at the top of the list in table 8.1 to acceptors at the bottom because the latter have more positive potentials. This may be expressed visually in the form of an electron tower in which the most negative reduc-

Table 8.1 Selected Biologically Important Redox Couples

Redox Couple	E'_0 (Volts)[a]
$2H^+ + 2e^- \rightarrow H_2$	−0.42
Ferredoxin (Fe^{3+}) + e^- → ferredoxin (Fe^{2+})	−0.42
$NAD(P)^+ + H^+ + 2e^- \rightarrow NAD(P)H$	−0.32
$S + 2H^+ + 2e^- \rightarrow H_2S$	−0.274
Acetaldehyde + $2H^+ + 2e^-$ → ethanol	−0.197
$\text{Pyruvate}^- + 2H^+ + 2e^- \rightarrow \text{lactate}^{2-}$	−0.185
$FAD + 2H^+ + 2e^- \rightarrow FADH_2$	−0.18[b]
$\text{Oxaloacetate}^{2-} + 2H^+ + 2e^- \rightarrow \text{malate}^{2-}$	−0.166
$\text{Fumarate}^{2-} + 2H^+ + 2e^- \rightarrow \text{succinate}^{2-}$	0.031
Cytochrome *b* (Fe^{3+}) + e^- → cytochrome *b* (Fe^{2+})	0.075
Ubiquinone + $2H^+ + 2e^-$ → ubiquinone H_2	0.10
Cytochrome *c* (Fe^{3+}) + e^- → cytochrome *c* (Fe^{2+})	0.254
Cytochrome *a* (Fe^{3+}) + e^- → cytochrome *a* (Fe^{2+})	0.29
Cytochrome a_3 (Fe^{3+}) + e^- → cytochrome a_3 (Fe^{2+})	0.35
$NO_3^- + 2H^+ + 2e^- \rightarrow NO_2^- + H_2O$	0.421
$NO_2^- + 8H^+ + 6e^- \rightarrow NH_4^+ + 2H_2O$	0.44
$Fe^{3+} + e^- \rightarrow Fe^{2+}$	0.771[c]
$O_2 + 4H^+ + 4e^- \rightarrow 2H_2O$	0.815

[a] E'_0 is the standard reduction potential at pH 7.0.

[b] The value for FAD/$FADH_2$ applies to the free cofactor because it can vary considerably when bound to an apoenzyme.

[c] The value for free Fe, not Fe complexed with proteins (e.g., cytochromes).

[1] In an oxidation-reduction reaction, the electron donor is often called the reducing agent or reductant because it is donating electrons to the acceptor and thus reducing it. The electron acceptor is called the oxidizing agent or oxidant because it is removing electrons from the donor and oxidizing it.

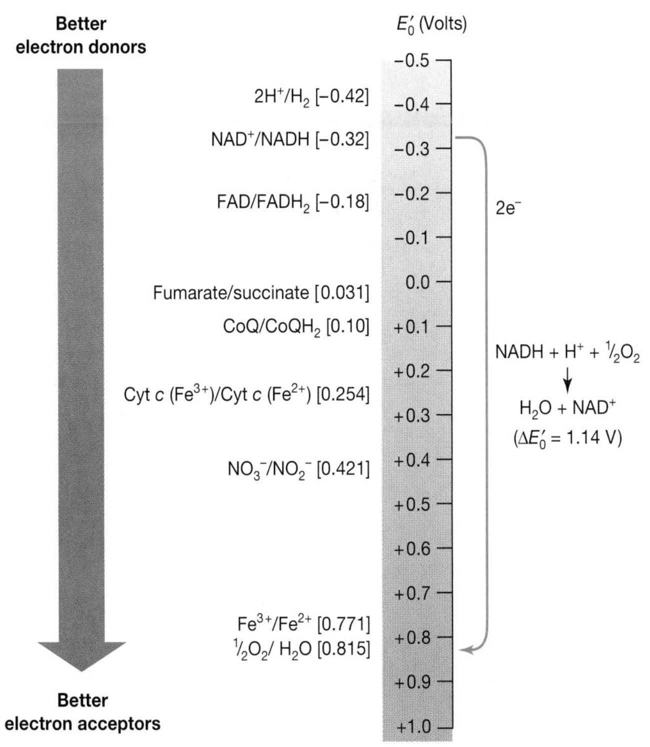

Figure 8.8 Electron Movement and Reduction Potentials. The vertical electron tower in this illustration has the most negative reduction potentials at the top. Electrons will spontaneously move from donors higher on the tower (more negative potentials) to acceptors lower on the tower (more positive potentials). That is, the donor is always higher on the tower than the acceptor. For example, NADH will donate electrons to oxygen and form water in the process. Some typical donors and acceptors are shown on the left, and their redox potentials are given in brackets.

tion potentials are at the top (**figure 8.8**). Electrons move from donors to acceptors down the potential gradient or fall down the tower to more positive potentials. Consider the case of the electron carrier **nicotinamide adenine dinucleotide (NAD^+)**. The $NAD^+/NADH$ couple has a very negative E'_0 and can therefore give electrons to many acceptors, including O_2.

$$NAD^+ + 2H^+ + 2e^- \rightleftharpoons NADH + H^+ \qquad E'_0 = -0.32 \text{ volts}$$

$$1/2O_2 + 2H^+ + 2e^- \rightleftharpoons H_2O \qquad E'_0 = +0.82 \text{ volts}$$

Because the reduction potential of $NAD^+/NADH$ is more negative than that of $1/2\ O_2/H_2O$, electrons will flow from NADH (the donor) to O_2 (the acceptor) as shown in figure 8.8.

$$NADH + H^+ + 1/2O_2 \rightarrow H_2O + NAD^+$$

Because the $NAD^+/NADH$ couple has a relatively negative E'_0, it stores more potential energy than redox couples with less negative (or more positive) E'_0 values. It follows that when electrons move from a donor to an acceptor with a more positive redox potential, free energy is released. The $\Delta G^{\circ\prime}$ of the reaction is directly related to the magnitude of the difference between the reduction potentials of the two couples ($\Delta E'_0$). The larger the $\Delta E'_0$, the greater the amount of free energy made available, as is evident from the equation

$$\Delta G^{\circ\prime} = -nF \cdot \Delta E'_0$$

in which n is the number of electrons transferred and F is the Faraday constant (23,062 cal/mole-volt or 96,494 J/mole-volt). For every 0.1 volt change in $\Delta E'_0$, there is a corresponding 4.6 kcal change in $\Delta G^{\circ\prime}$ when a two-electron transfer takes place. This is similar to the relationship of $\Delta G^{\circ\prime}$ and K_{eq} in other chemical reactions—the larger the equilibrium constant, the greater the $\Delta G^{\circ\prime}$. The difference in reduction potentials between $NAD^+/NADH$ and $1/2O_2/H_2O$ is 1.14 volts, a large $\Delta E'_0$ value. When electrons move from NADH to O_2, a large amount of free energy is made available to synthesize ATP.

We have focused our attention on the reduction of O_2 by NADH because NADH plays a central role in the metabolism of many organisms, especially chemoorganotrophs. Many chemoorganotrophs use glucose as a source of energy. As glucose is catabolized, it is oxidized. Many of the electrons released from glucose are accepted by NAD^+, which is then reduced to NADH. NADH next transfers the electrons to O_2. However, it does not do so directly. Instead, the electrons are transferred to O_2 via a series of electron carriers. The electron carriers are organized into a system called an **electron transport system (ETS)** or **electron transport chain (ETC)**. The carriers are organized such that the first electron carrier has the most negative E'_0, and each successive carrier is slightly less negative (**figure 8.9**). In this way, the potential energy stored in the redox couple whose electrons initiate electron flow is released and used to form ATP.

The ETSs of chemoorganotrophs are located in the plasma membrane in procaryotes and the internal mitochondrial membranes in eucaryotes. Electron transport systems also play a pivotal role in the metabolism of chemolithotrophs and phototrophs, where they are used to conserve energy from inorganic energy sources and light, respectively. The ETSs are located in the plasma membrane or internal membrane systems of chemolithotrophs, which are all procaryotes. They are located in the plasma membrane and internal membrane systems of procaryotic phototrophs and in the thylakoid membranes of chloroplasts in eucaryotic phototrophs (figure 8.9).

The carriers that make up ETSs differ in terms of their chemical nature and the way they carry electrons. NAD^+, and its chemical relative **nicotinamide adenine dinucleotide phosphate ($NADP^+$)**, contain a nicotinamide ring (**figure 8.10**). This ring accepts two electrons and one proton from a donor (e.g., an intermediate formed during the catabolism of glucose), and a second proton is released. **Flavin adenine dinucleotide (FAD)** and **flavin mononucleotide (FMN)** bear two electrons and two protons on the complex ring system shown in **figure 8.11.** Proteins bearing FAD and FMN are often called flavoproteins. **Coenzyme Q (CoQ)** or **ubiquinone** is a quinone that transports two electrons and two protons in many electron transport chains

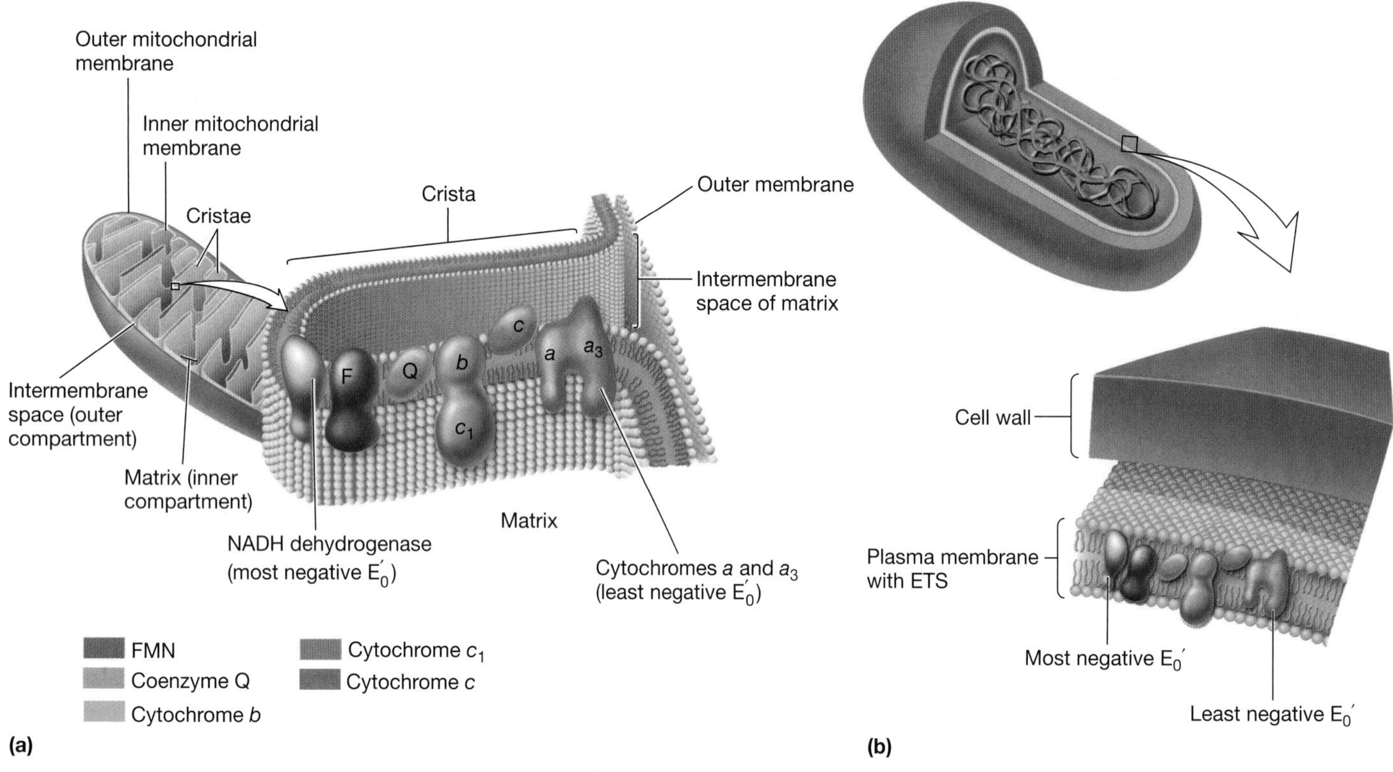

Figure 8.9 Electron Transport Systems. Electron transport systems (ETSs) are located in membranes. Electrons flow from the electron carrier having the most negative reduction potential to the carrier having the most positive reduction potential. During respiratory processes (aerobic respiration, anaerobic respiration, and chemolithotrophy), an exogenous molecule such as oxygen serves as the terminal electron acceptor. **(a)** The mitochondrial ETS. **(b)** A typical bacterial ETS.

(**figure 8.12**). **Cytochromes** and several other carriers use iron atoms to transport electrons one electron at a time by reversible oxidation and reduction reactions.

$$Fe^{3+}(\text{ferric iron}) + e^{+} \rightleftharpoons Fe^{2+}(\text{ferrous iron})$$

In the cytochromes these iron atoms are part of a heme group (**figure 8.13**) or other similar iron-porphyrin rings. Several different cytochromes, each of which consists of a protein and an ironporphyrin ring, are a prominent part of electron transport chains. Some iron containing electron-carrying proteins lack a heme group and are called **nonheme iron proteins. Ferredoxin** is a nonheme iron protein active in photosynthetic electron transport and several other electron transport processes. Even though its iron atoms are not bound to a heme group, they still undergo reversible oxidation and reduction reactions. Like cytochromes, they carry only one electron at a time. This difference in the number of electrons and protons carried is of great importance in the operation of electron transport chains and is discussed further in chapter 9.

1. Write a generalized equation for a redox reaction. Define standard reduction potential.
2. How is the direction of electron flow between redox couples related to the standard reduction potential and the release of free energy?
3. When electrons flow from the $NAD^+/NADH$ redox couple to the O_2/H_2O redox couple, does the reaction begin with NAD^+ or with NADH? What is produced—O_2 or H_2O?
4. Which among the following would be the best electron donor? Which would be the worst? ubiquinone/ubiquinoneH_2, $NAD^+/NADH$, FAD/$FADH_2$, NO_3^-/NO_2^-. Explain your answers.
5. In general terms, how is $\Delta G^{\circ\prime}$ related to $\Delta E'_0$? What is the $\Delta E'_0$ when electrons flow from the $NAD^+/NADH$ redox couple to the Fe^{3+}/Fe^{2+} redox couple? How does this compare to the $\Delta E'_0$ when electrons flow from the Fe^{3+}/Fe^{2+} redox couple to the O_2/H_2O couple? Which will yield the largest amount of free energy to the cell?
6. Name and briefly describe the major electron carriers found in cells. Why is NAD^+ a good electron carrier? Why is ferredoxin an even better electron carrier?

8.7 Enzymes

Recall that an exergonic reaction is one with a negative $\Delta G^{\circ\prime}$ and an equilibrium constant greater than one. An exergonic reaction will proceed to completion in the direction written (that is, toward the right of the equation). Nevertheless, one often can combine the reactants for an exergonic reaction with no obvious result. For instance, the hydrolysis of polysaccharides into

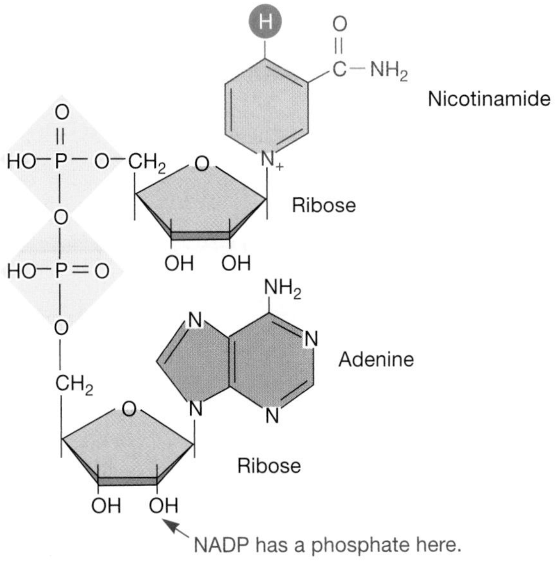

Figure 8.10 The Structure and Function of NAD. **(a)** The structure of NAD and NADP. NADP differs from NAD in having an extra phosphate on one of its ribose sugar units. **(b)** NAD can accept electrons and a hydrogen from a reduced substrate (SH_2). These are carried on the nicotinamide ring. **(c)** Model of NAD^+ when bound to the enzyme lactate dehydrogenase.

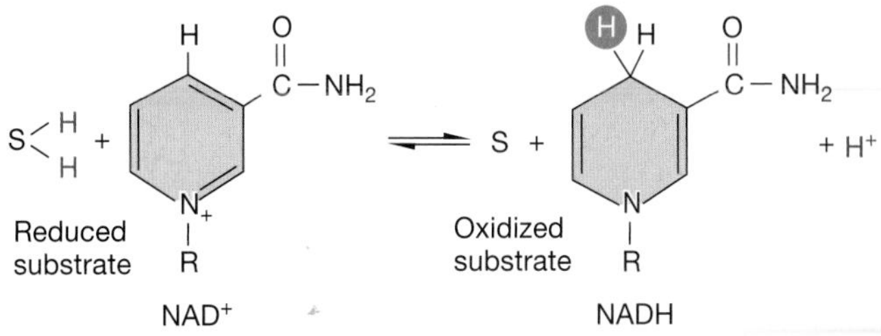

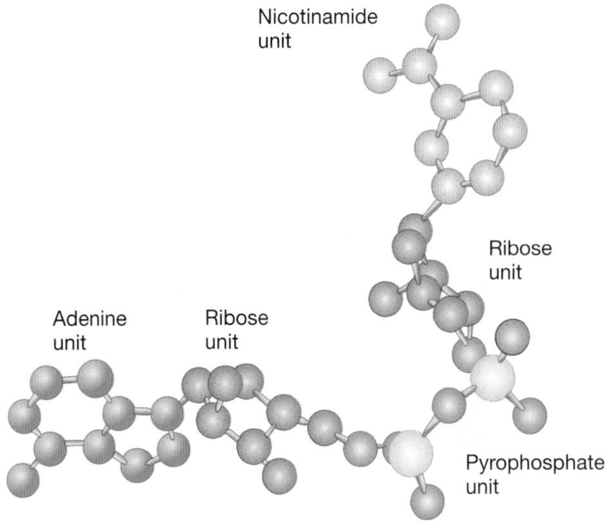

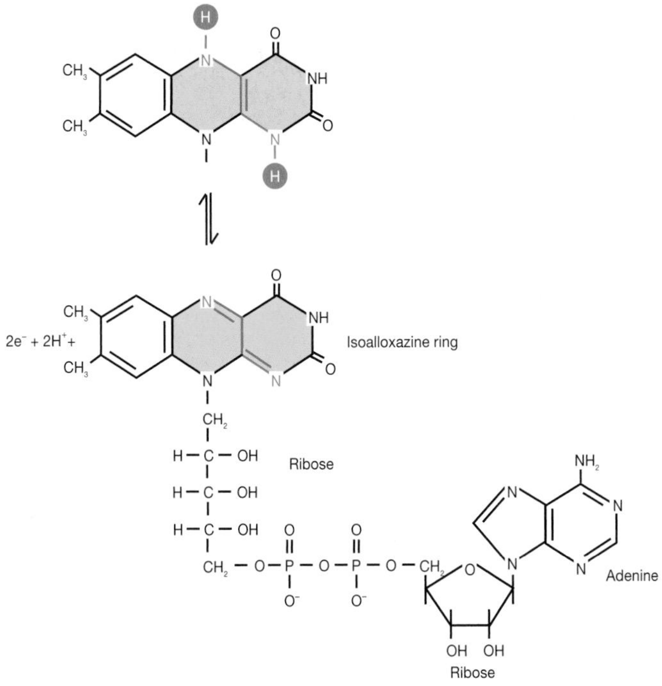

Figure 8.11 The Structure and Function of FAD. The vitamin riboflavin is composed of the isoalloxazine ring and its attached ribose sugar. FMN is riboflavin phosphate. The portion of the ring directly involved in oxidation-reduction reactions is in color.

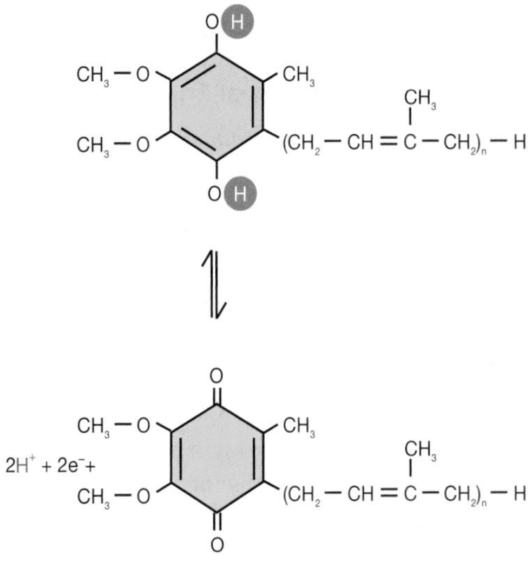

Figure 8.12 The Structure and Function of Coenzyme Q or Ubiquinone. The length of the side chain varies among organisms from n = 6 to n = 10.

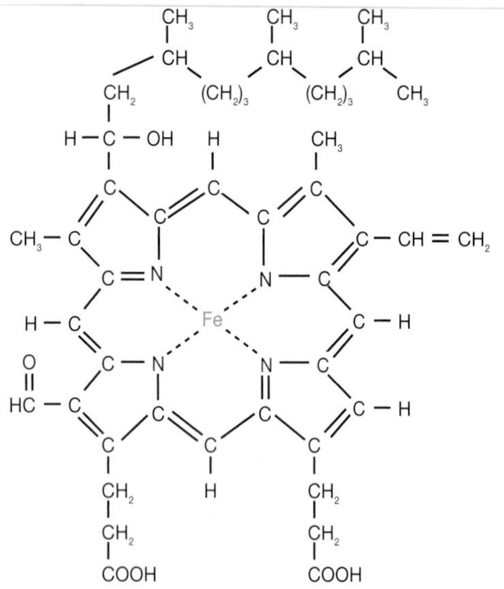

Figure 8.13 The Structure of Heme. Heme is composed of a porphyrin ring and an attached iron atom. It is the nonprotein component of many cytochromes. The iron atom alternatively accepts and releases an electron.

their component monosaccharides is exergonic and will occur spontaneously. However, an organic chemist would have to carry out this reaction in 6 M HCl and at 100°C for several hours to get it to go to completion. A cell, on the other hand, can accomplish the same reaction at neutral pH, at a much lower temperature, and in just fractions of a second. How are cells able to do this? They can do so because they manufacture proteins called enzymes that speed up chemical reactions. Enzymes are critically important to cells, since most biological reactions occur very slowly without them. Indeed, enzymes make life possible.

Structure and Classification of Enzymes

Enzymes may be defined as protein catalysts that have great specificity for the reaction catalyzed and the molecules acted on. A **catalyst** is a substance that increases the rate of a chemical reaction without being permanently altered itself. Thus enzymes speed up cellular reactions. The reacting molecules are called **substrates,** and the substances formed are the **products.** Proteins (appendix I)

Many enzymes are composed only of proteins. However, some enzymes consist of a protein, the **apoenzyme,** and a nonprotein component, a **cofactor,** required for catalytic activity. The complete enzyme consisting of the apoenzyme and its cofactor is called the **holoenzyme.** If the cofactor is firmly attached to the apoenzyme it is a **prosthetic group.** If the cofactor is loosely attached to the apoenzyme and can dissociate from the protein after products have been formed, it is called a **coenzyme.** Many coenzymes can carry one of the products to another enzyme (**figure 8.14**). For example, NAD^+ is a coenzyme that carries electrons within the cell. Many vitamins that humans require serve as

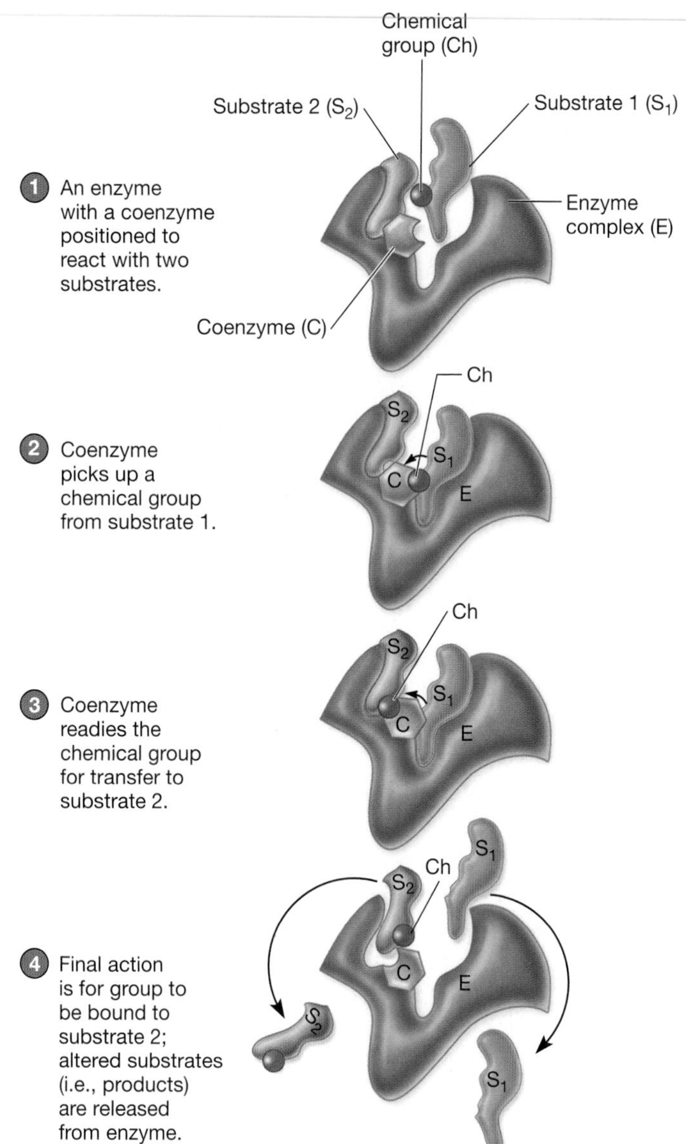
Chemical group (Ch)
Substrate 2 (S2)
Substrate 1 (S1)
1 An enzyme with a coenzyme positioned to react with two substrates.
Enzyme complex (E)
Coenzyme (C)
2 Coenzyme picks up a chemical group from substrate 1.
3 Coenzyme readies the chemical group for transfer to substrate 2.
4 Final action is for group to be bound to substrate 2; altered substrates (i.e., products) are released from enzyme.
Ch
S2
S1
C
E

Figure 8.14 Coenzymes as Carriers.

coenzymes or as their precursors. Niacin is incorporated into NAD^+ and riboflavin into FAD. Metal ions may also be bound to apoenzymes and act as cofactors.

Despite the large number and bewildering diversity of enzymes present in cells, they may be placed in one of six general classes (**table 8.2**). Enzymes usually are named in terms of the substrates they act on and the type of reaction catalyzed. For example, lactate dehydrogenase (LDH) removes hydrogens from lactate.

$$\text{Lactate} + \text{NAD}^+ \xrightleftharpoons{\text{LDH}} \text{pyruvate} + \text{NADH} + \text{H}^+$$

Lactate dehydrogenase can also be given a more complete and detailed name, L-lactate:NAD oxidoreductase. This name describes the substrates and reaction type with even more precision.

Table 8.2 Enzyme Classification

Type of Enzyme	Reaction Catalyzed by Enzyme	Example of Reaction
Oxidoreductase	Oxidation-reduction reactions	Lactate dehydrogenase: Pyruvate + NADH + H $\rightleftharpoons$ lactate + NAD^+
Transferase	Reactions involving the transfer of groups between molecules	Aspartate carbamoyltransferase: Aspartate + carbamoylphosphate $\rightleftharpoons$ carbamoylaspartate + phosphate
Hydrolase	Hydrolysis of molecules	Glucose-6-phosphatase: Glucose-6-phosphate + H_2O → glucose + P_i
Lyase	Removal of groups to form double bonds or addition of groups to double bonds $>C{=}C< + X{-}Y \rightleftharpoons -\overset{X}{\underset{\mid}{\overset{\mid}{C}}}-\overset{Y}{\underset{\mid}{\overset{\mid}{C}}}-$	Fumarate hydratase: L-malate $\rightleftharpoons$ fumarate + H_2O
Isomerase	Reactions involving isomerizations	Alanine racemase: L-alanine $\rightleftharpoons$ D-alanine
Ligase	Joining of two molecules using ATP energy (or that of other nucleoside triphosphates)	Glutamine synthetase: Glutamate + NH_3 + ATP → glutamine + ADP + P_i

The Mechanism of Enzyme Reactions

It is important to keep in mind that enzymes increase the rates of reactions but do not alter their equilibrium constants. If a reaction is endergonic, the presence of an enzyme will not shift its equilibrium so that more products can be formed. Enzymes simply speed up the rate at which a reaction proceeds toward its final equilibrium.

How do enzymes catalyze reactions? Although a complete answer would be long and complex, some understanding of the mechanism can be gained by considering the course of a simple exergonic chemical reaction.

$$A + B \rightleftharpoons C + D$$

When molecules A and B approach each other to react, they form a **transition-state complex,** which resembles both the substrates and the products (**figure 8.15**). **Activation energy** is required to bring the reacting molecules together in the correct way to reach the transition state. The transition-state complex can then resolve to yield the products C and D. The difference in free energy level between reactants and products is $\Delta G^{\circ\prime}$. Thus the equilibrium in our example will lie toward the products because $\Delta G^{\circ\prime}$ is negative (i.e., the products are at a lower energy level than the substrates).

As seen in figure 8.15, A and B will not be converted to C and D if they are not supplied with an amount of energy equivalent to the activation energy. Enzymes accelerate reactions by lowering the activation energy; therefore more substrate molecules will have sufficient energy to come together and form products. Even though the equilibrium constant (or $\Delta G^{\circ\prime}$) is unchanged, equilibrium will be reached more rapidly in the presence of an enzyme because of this decrease in the activation energy.

Researchers have worked hard to discover how enzymes lower the activation energy of reactions, and the process is becoming clearer. Enzymes bring substrates together at a specific place on their surface called the **active site** or **catalytic site** to form an **enzyme-substrate complex** (**figures 8.16, 8.17;** *see also appendix figure AI.19*). An enzyme can interact with its substrate in two general ways. It may be rigid and shaped to precisely fit the substrate so that the correct substrate binds specifically and is positioned properly for reaction. This mechanism is referred to as the lock-and-key model (figure 8.16). An enzyme also may change shape when it binds the substrate so that the active site surrounds and precisely fits the substrate. This has been called the

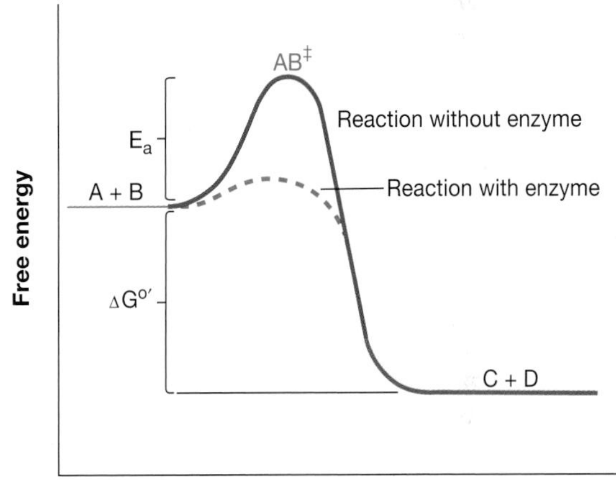

Figure 8.15 Enzymes Lower the Energy of Activation. This figure traces the course of a chemical reaction in which A and B are converted to C and D. The transition-state complex is represented by $AB^{\ddagger}$, and the activation energy required to reach it, by E_a. The red line represents the course of the reaction in the presence of an enzyme. Note that the activation energy is much lower in the enzyme-catalyzed reaction.

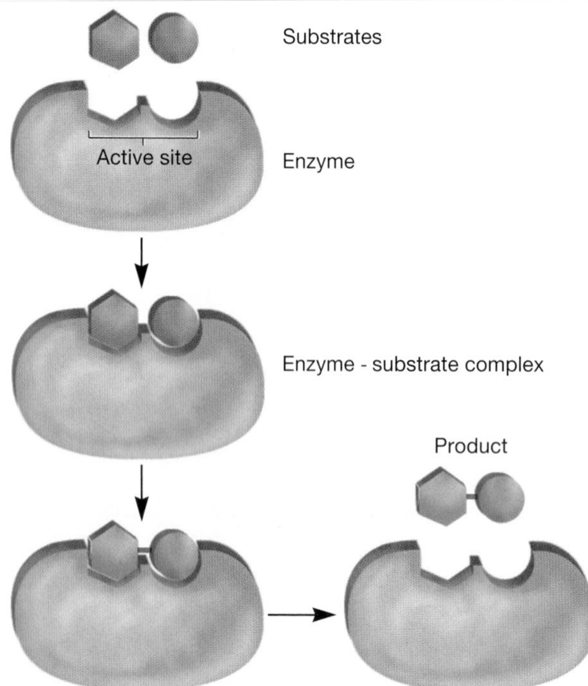

Figure 8.16 Lock-and-Key Model of Enzyme Function. In this model, the active site is a relatively rigid structure that accommodates only those molecules with the correct corresponding shape. The formation of the enzyme-substrate complex and its conversion to product is shown.

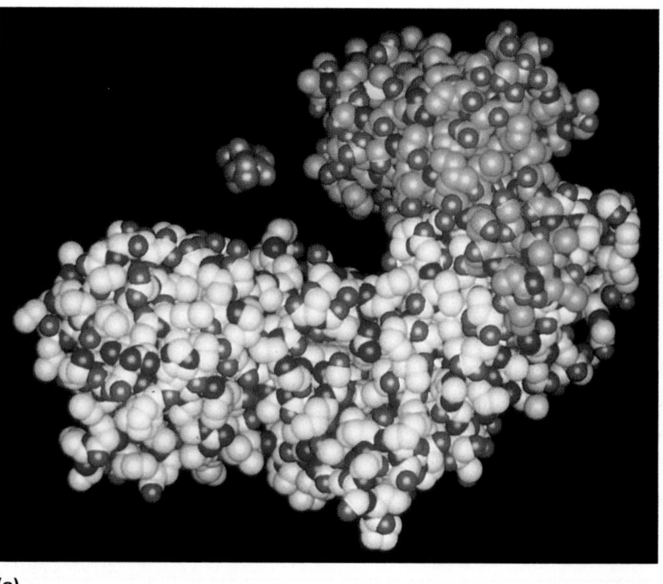
(a)

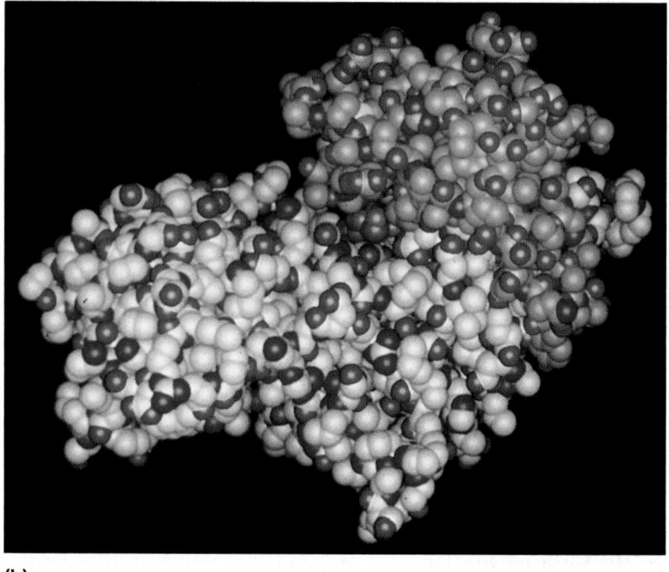
(b)

Figure 8.17 The Induced Fit Model of Enzyme Function. **(a)** A space-filling model of yeast hexokinase and its substrate glucose (purple). The active site is in the cleft formed by the enzyme's small lobe (green) and large lobe (blue). **(b)** When glucose binds to form the enzyme-substrate complex, hexokinase changes shape and surrounds the substrate.

induced fit model and is used by hexokinase and many other enzymes (figure 8.17). The formation of an enzyme-substrate complex can lower the activation energy in many ways. For example, by bringing the substrates together at the active site, the enzyme is, in effect, concentrating them and speeding up the reaction. An enzyme does not simply concentrate its substrates, however. It also binds them so that they are correctly oriented with respect to each other in order to form a transition-state complex. Such an orientation lowers the amount of energy that the substrates require to reach the transition state. These and other catalytic site activities speed up a reaction hundreds of thousands of times.

The Effect of Environment on Enzyme Activity

Enzyme activity varies greatly with changes in environmental factors, one of the most important being the substrate concentration. As will be emphasized later, substrate concentrations are usually low within cells. At very low substrate concentrations, an enzyme makes product slowly because it seldom contacts a substrate molecule. If more substrate molecules are present, an enzyme binds substrate more often, and the reaction velocity (usually expressed in terms of the rate of product formation) is greater than at a lower substrate concentration. Thus the rate of an enzyme-catalyzed reaction increases with substrate concentration (**figure 8.18**). Eventually further increases in substrate concentration do not result in a greater reaction velocity because the available enzyme molecules are binding substrate and converting it to product as rapidly as possible. That is, the enzyme is saturated with substrate and operating at maximal velocity (V_{max}). The resulting substrate concentration curve is a hyperbola (figure 8.18). It is useful to know the substrate concentration an enzyme needs to function adequately. Usually the **Michaelis constant (K_m)**, the substrate concentration required for the enzyme to achieve half maximal velocity, is used as a measure of the apparent affinity of an enzyme for its substrate. The lower the K_m value, the lower the substrate concentration at which an enzyme catalyzes its reaction. Enzymes with a low K_m value are said to have a high affinity for their substrates.

Enzymes also change activity with alterations in pH and temperature (**figure 8.19**). Each enzyme functions most rapidly at a specific pH optimum. When the pH deviates too greatly from an enzyme's optimum, activity slows and the enzyme may be damaged. Enzymes likewise have temperature optima for maximum activity. If the temperature rises too much above the optimum, an enzyme's structure will be disrupted and its activity lost. This phenomenon, known as **denaturation,** may be caused by extremes of pH and temperature or by other factors. The pH and temperature optima of a microorganism's enzymes often reflect the pH and temperature of its habitat. Not surprisingly bacteria growing best at high temperatures often have enzymes with high temperature optima and great heat stability. The influence of environmental factors on growth (section 6.5)

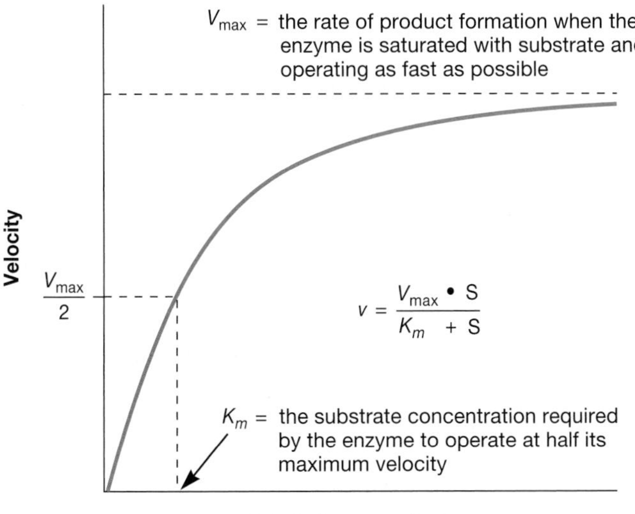

Figure 8.18 Michaelis-Menten Kinetics. The dependence of enzyme activity upon substrate concentration. This substrate curve fits the Michaelis-Menten equation given in the figure, which relates reaction velocity (v) to the substrate concentration (S) using the maximum velocity and the Michaelis constant (K_m).

Enzyme Inhibition

Microorganisms can be poisoned by a variety of chemicals, and many of the most potent poisons are enzyme inhibitors. A **competitive inhibitor** directly competes with the substrate at an enzyme's catalytic site and prevents the enzyme from forming product (**figure 8.20**). Competitive inhibitors usually resemble normal substrates, but they cannot be converted to products.

Competitive inhibitors are important in the treatment of many microbial diseases. Sulfa drugs like sulfanilamide (figure 8.20*b*) resemble *p*-aminobenzoate, a molecule used in the formation of the coenzyme folic acid. The drugs compete with *p*-aminobenzoate for the catalytic site of an enzyme involved in folic acid synthesis. This blocks the production of folic acid and inhibits bacterial growth. Humans are not harmed because they do not synthesize folic acid but rather obtain it in their diet. Antimicrobial drugs: Metablic antagonists (section 34.4)

Noncompetitive inhibitors also can affect enzyme activity by binding to the enzyme at some location other than the active site. This alters the enzyme's shape, rendering it inactive or less active. These inhibitors are called noncompetitive because they do not directly compete with the substrate. Heavy metal poisons like mercury frequently are noncompetitive inhibitors of enzymes.

1. What is an enzyme? How does it speed up reactions? How are enzymes named? Define apoenzyme, holoenzyme, cofactor, coenzyme, prosthetic group, active or catalytic site, and activation energy.
2. Draw a diagram showing how enzymes catalyze reactions by altering the activation energy. What is a transition state complex? Use the diagram to explain why enzymes do not change the equilibria of the reactions they catalyze.
3. What is the difference between the lock-and-key and the induced-fit models of enzyme-substrate complex formation?
4. Define the terms Michaelis constant and maximum velocity. How does enzyme activity change with substrate concentration, pH, and temperature?
5. What special properties might an enzyme isolated from a psychrophilic bacterium have? Will enzymes need to lower the activation energy more or less in thermophiles than in psychrophiles?
6. What are competitive and noncompetitive inhibitors and how do they inhibit enzymes?

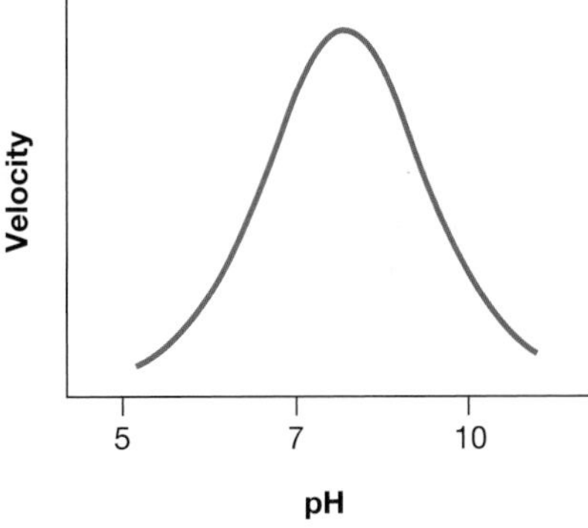

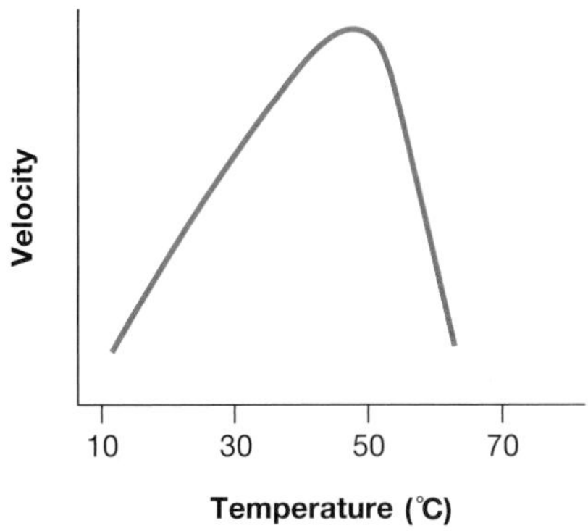

Figure 8.19 pH, Temperature, and Enzyme Activity. The variation of enzyme activity with changes in pH and temperature. The ranges in pH and temperature are only representative. Enzymes differ from one another with respect to the location of their optima and the shape of their pH and temperature curves.

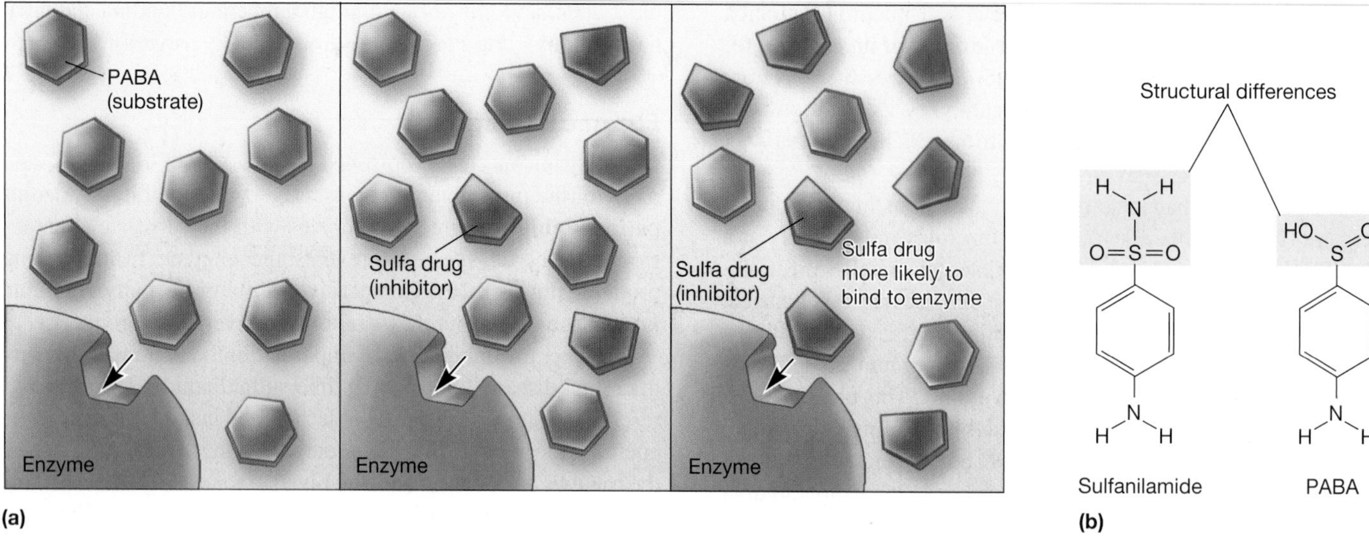

Figure 8.20 Competitive Inhibition of Enzyme Activity. **(a)** A competitive inhibitor is usually similar in shape to the normal substrate of the enzyme, and therefore can bind the active site of the enzyme. This prevents the substrate from binding, and the reaction is blocked. **(b)** Structure of sulfanilamide, a structural analog of PABA. PABA is the substrate of an enzyme involved in folic acid biosynthesis. When sulfanilamide binds the enzyme, activity of the enzyme is inhibited and synthesis of folic acid is stopped.

8.8 The Nature and Significance of Metabolic Regulation

Microorganisms must regulate their metabolism to conserve raw materials and energy and to maintain a balance among various cell components. Because they live in environments where the nutrients, energy sources, and physical conditions often change rapidly, they must continuously monitor internal and external conditions and respond accordingly. This involves activating or inactivating pathways as needed. For instance, if a particular energy source is unavailable, the enzymes required for its use are not needed and their further synthesis is a waste of carbon, nitrogen, and energy. Similarly it would be extremely wasteful for a microorganism to synthesize the enzymes required to manufacture a certain end product if that end product were already present in adequate amounts.

The drive to maintain balance and conserve energy and material is evident in the regulatory responses of a bacterium like *E. coli.* If the bacterium is grown in a very simple medium containing only glucose as a carbon and energy source, it will synthesize all needed cell components in balanced amounts. However, if the amino acid tryptophan is added to the medium, the pathway synthesizing tryptophan will be immediately inhibited and synthesis of the pathway's enzymes also will slow or cease. Likewise, if *E. coli* is transferred to a medium containing only the sugar lactose, it will synthesize the enzymes required for catabolism of this nutrient. In contrast, when *E. coli* grows in a medium possessing both glucose and lactose, glucose (the sugar supporting most rapid growth) is catabolized first. The culture will use lactose only after the glucose supply has been exhausted.

Metabolic pathways can be regulated in three major ways:

1. **Metabolic channeling**—this phenomenon influences pathway activity by localizing metabolites and enzymes into different parts of a cell.
2. Regulation of the amount of synthesis of a particular enzyme—in other words, transcription and translation can be regulated. These two processes function in synthesizing enzymes. Regulation at this level is relatively slow, but it saves the cell considerable energy and raw material.
3. Direct stimulation or inhibition of the activity of critical enzymes—this type of regulation rapidly alters pathway activity. It is often called **posttranslational regulation** because it occurs after the enzyme has been synthesized.

In this chapter we introduce metabolic channeling and direct control of enzyme activity. Discussion of the regulation enzyme synthesis follows the descriptions of DNA, RNA, and protein synthesis and can be found in chapter 12.

8.9 Metabolic Channeling

One of the most common metabolic channeling mechanisms is that of **compartmentation,** the differential distribution of enzymes and metabolites among separate cell structures or organelles. Compartmentation is particularly important in eucaryotic microorganisms with their many membrane-bound organelles. For example, fatty acid catabolism is located within the mitochondrion, whereas fatty acid synthesis occurs in the cytoplasmic matrix. The periplasm in procaryotes can also be considered an example of compartmentation. Compartmentation makes possible the simultaneous, but separate, operation and regulation of similar pathways. Furthermore, pathway activities can be coordinated through regulation of the transport of metabolities and coenzymes between cell compartments. Suppose two pathways in different cell compartments require NAD for activity. The distribution of NAD between the two compartments will then determine the relative activity of these competing pathways, and the pathway with access to the most NAD will be favored. The bacterial cell wall (section 3.6); Archaeal cell wall (section 3.7)

Channeling also occurs within compartments such as the cytoplasmic matrix. The matrix is a structured dense material with many subcompartments. In eucaryotes it also is subdivided by the endoplasmic reticulum and cytoskeleton. Metabolites and coenzymes do not diffuse rapidly in such an environment, and metabolite gradients will build up near localized enzymes or enzyme systems. This occurs because enzymes at a specific site convert their substrates to products, resulting in decreases in the concentration of one or more metabolites and increases in others. For example, product concentrations will be high near an enzyme and decrease with increasing distance from it. The cytoplasmic matrix, microfilaments, intermediate filaments, and microtubules (section 4.3)

Channeling can generate marked variations in metabolite concentrations and therefore directly affect enzyme activity. Substrate levels are generally around 10^{-3} moles/liter (M) to 10^{-6} M or even lower. Thus they may be in the same range as enzyme concentrations and equal to or less than the Michaelis constants (K_m) of many enzymes (figure 8.18). Under these conditions the concentration of an enzyme's substrate may control its activity because the substrate concentration is in the rising portion of the hyperbolic substrate saturation curve (**figure 8.21**). As the substrate level increases, it is converted to product more rapidly; a decline in substrate concentration automatically leads to lower enzyme activity. If two enzymes in different pathways use the same metabolite, they may directly compete for it. The pathway winning this competition—the one with the enzyme having the lowest K_m value for the metabolite—will operate closer to full capacity. Thus channeling within a cell compartment can regulate and coordinate metabolism through variations in metabolite and coenzyme levels.

1. Give three ways in which a metabolic pathway may be regulated.
2. Define the terms metabolic channeling and compartmentation. How are they involved in the regulation of metabolism?

8.10 Control of Enzyme Activity

Adjustment of the activity of regulatory enzymes and other proteins controls the functioning of many metabolic pathways and cellular processes. This type of regulation is an example of posttranslational regulation because it occurs after the protein is synthesized. There are a number of posttranslational regulatory mechanisms. Some are irreversible—for instance, cleavage of a protein can either activate or inhibit its activity. Other types of posttranslational control are reversible. In this section, we consider examples of two important, reversible control measures: allosteric regulation and covalent modification. Our focus will be on the regulation of metabolic pathways, but it is important to remember that not all proteins or enzymes function in metabolic pathways. Instead, they are involved in cellular behaviors. At the end of this section, we will consider the regulation of one of these behaviors—chemotaxis.

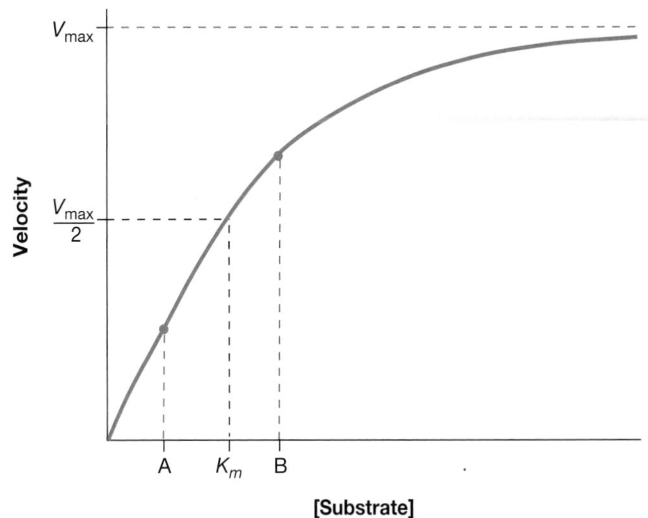

Figure 8.21 Control of Enzyme Activity by Substrate Concentration. An enzyme-substrate saturation curve with the Michaelis constant (K_m) and the velocity equivalent to half the maximum velocity (V_{max}) indicated. The initial velocity of the reaction (v) is plotted against the substrate concentration [Substrate]. The maximum velocity is the greatest velocity attainable with a fixed amount of enzyme under defined conditions. When the substrate concentration is equal to or less than the K_m, the enzyme's activity will vary almost linearly with the substrate concentration. Suppose the substrate increases in concentration from level A to B. Because these concentrations are in the range of the K_m, a significant increase in enzyme activity results. A drop in concentration from B to A will lower the rate of product formation.

Allosteric Regulation

Most regulatory enzymes are **allosteric enzymes.** The activity of an allosteric enzyme is altered by a small molecule known as an **effector** or **modulator.** The effector binds reversibly by noncovalent forces to a **regulatory site** separate from the catalytic site and causes a change in the shape or conformation of the enzyme (**figure 8.22**). The activity of the catalytic site is altered as a result. A positive effector increases enzyme activity, whereas a negative effector decreases activity or inhibits the enzyme. These changes in activity often result from alterations in the apparent affinity of the enzyme for its substrate, but changes in maximum velocity also can occur.

The substrate saturation curve for an allosteric enzyme is often sigmoidal rather than hyperbolic like that of a nonregulatory enzyme. Therefore the substrate concentration required for a regulatory enzyme to function at half its maximal velocity is given its own name: $[S]_{0.5}$ or $K_{0.5}$. The impact of positive effectors and negative effectors on the $K_{0.5}$ of an allosteric enzyme can be readily

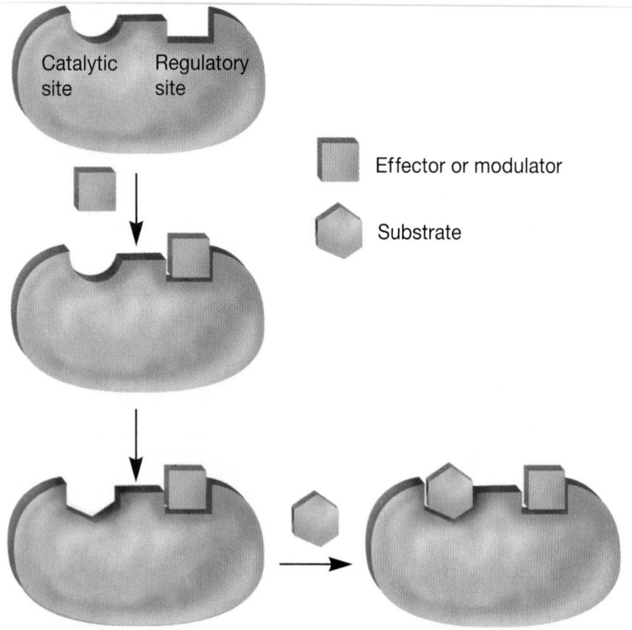

Figure 8.22 Allosteric Regulation. The structure and function of an allosteric enzyme. In this example the effector or modulator first binds to a separate regulatory site and causes a change in enzyme conformation that results in an alteration in the shape of the active site. The active site can now more effectively bind the substrate. This effector is a positive effector because it stimulates substrate binding and catalytic activity.

seen with one of the best-studied allosteric regulatory enzymes—aspartate carbamoyltransferase (ACTase) from *E. coli.* The enzyme catalyzes the condensation of carbamoyl phosphate with aspartate to form carbamoylaspartate (**figure 8.23**). This is the rate-determining reaction of the pyrimidine nucleotide biosynthetic pathway in *E. coli.* The substrate saturation curve is sigmoidal when the concentration of either substrate is varied (**figure 8.24**). This is because the enzyme has more than one active site, and the binding of a substrate molecule to an active site increases the binding of substrate at the other sites. In addition, cytidine triphosphate (CTP), an end product of pyrimidine biosynthesis, inhibits the enzyme, while the purine ATP activates it. Both effectors alter the $K_{0.5}$ value of the enzyme but not its maximum velocity. CTP inhibits by increasing $K_{0.5}$ (i.e., by shifting the substrate saturation curve to higher values). This causes the enzyme to operate more slowly at a particular substrate concentration when CTP is present. ATP activates the enzyme by moving the curve to lower substrate concentration values so that the enzyme is maximally active over a wider substrate concentration range. Thus when the pathway is so active that the CTP concentration rises too high, CTP acts as a brake to decrease ACTase activity. In contrast, when the purine end product ATP increases relative to CTP, it stimulates CTP synthesis through its effects on ACTase. Synthesis of purine, pyrimidines and nucleotides (section 10.6)

E. coli aspartate carbamoyltransferase provides a clear example of separate regulatory and catalytic sites in allosteric enzymes. The enzyme is a large protein composed of two catalytic subunits and three regulatory subunits (**figure 8.25**). The catalytic

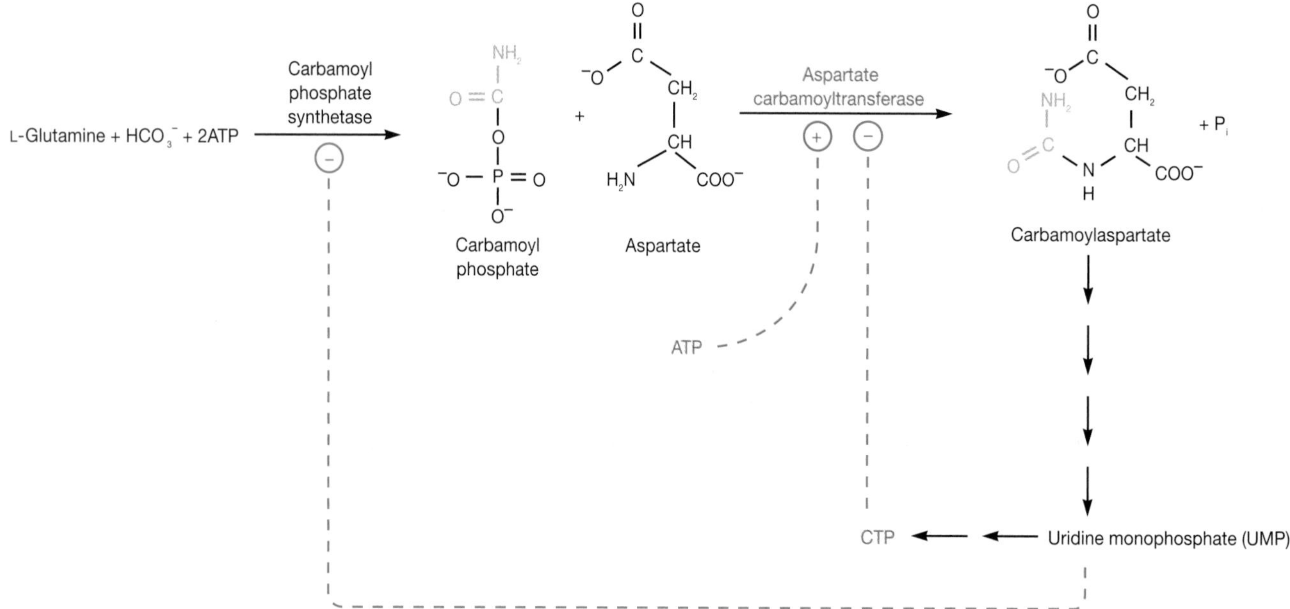

Figure 8.23 ACTase Regulation. The aspartate carbamoyltransferase reaction and its role in the regulation of pyrimidine biosynthesis. The end product CTP inhibits its activity (−) while ATP activates the enzyme (+). Carbamoyl phosphate synthetase is also inhibited by pathway end products such as UMP.

subunits contain only catalytic sites and are unaffected by CTP and ATP. Regulatory subunits do not catalyze the reaction but possess regulatory sites to which CTP and ATP bind. When these effectors bind to the regulatory subunits, they cause conformational changes in both the regulatory and catalytic subunits. The enzyme can change reversibly between a less active T form and a more active R form (figure 8.25*b, c*). Thus the regulatory site influences a catalytic site that is about 6.0 nm away.

Covalent Modification of Enzymes

Regulatory enzymes also can be switched on and off by **reversible covalent modification.** Usually this occurs through the addition and removal of a particular group, typically a phosphoryl, methyl or adenyl group. The enzyme with an attached group can be either activated or inhibited.

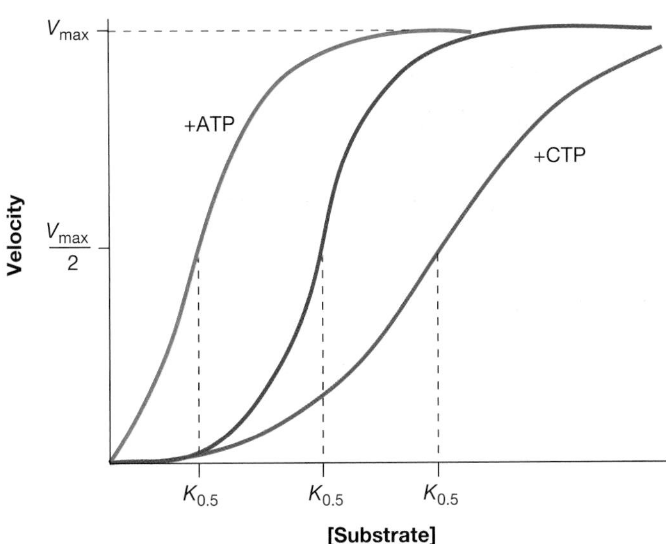

Figure 8.24 The Kinetics of *E. coli* Aspartate Carbamoyltransferase. CTP, a negative effector, increases the $K_{0.5}$ value while ATP, a positive effector, lowers the $K_{0.5}$. The V_{max} remains constant.

One of the most intensively studied regulatory enzymes is *E. coli* glutamine synthetase, an enzyme involved in nitrogen assimilation. It is a large, complex enzyme consisting of 12 subunits, each of which can be covalently modified by an adenylic acid residue (**figure 8.26**). When an adenylic acid residue is attached to all of its 12 subunits, glutamine synthetase is not very active. Removal of AMP groups produces more active deadenylylated glutamine synthetase, and glutamine is formed. Synthesis of amino acids: Nitrogen assimilation (section 10.5)

There are some advantages to using covalent modification for the regulation of enzyme activity. These interconvertible enzymes often are also allosteric. For instance, glutamine synthetase also is regulated allosterically. Because each form can respond differently to allosteric effectors, systems of covalently modified enzymes are able to respond to more stimuli in varied and sophisticated ways. Regulation can also be exerted on the enzymes that catalyze the covalent modifications, which adds a second level of regulation to the system.

Feedback Inhibition

The rate of many metabolic pathways is adjusted through control of the activity of the regulatory enzymes described in the preceding section. Every pathway has at least one **pacemaker enzyme** that catalyzes the slowest or rate-limiting reaction in the pathway. Because other reactions proceed more rapidly than the pacemaker reaction, changes in the activity of this enzyme directly alter the speed with which a pathway operates. Usually the first step in a pathway is a pacemaker reaction catalyzed by a regulatory enzyme. The end product of the pathway often inhibits this regulatory enzyme, a process known as **feedback inhibition** or **end product inhibition.** Feedback inhibition ensures balanced production of a pathway end product. If the end product becomes too concentrated, it inhibits the regulatory enzyme and slows its own synthesis. As the end product concentration decreases, pathway

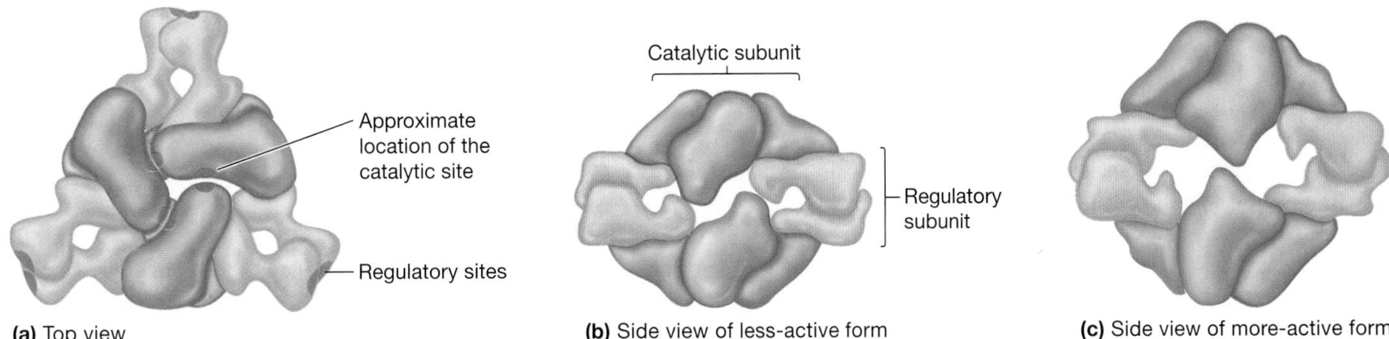

Figure 8.25 The Structure and Regulation of *E. coli* Aspartate Carbamoyltransferase. **(a)** A schematic diagram of the enzyme showing the six catalytic polypeptide chains (blue), the six regulatory chains (tan), and the catalytic and regulatory sites. The enzyme is viewed from the top. Each catalytic subunit contains three catalytic chains, and each regulatory subunit has two chains. **(b)** The less active T state of ACTase viewed from the side. **(c)** The more active R state of ACTase. The regulatory subunits have rotated and pushed the catalytic subunits apart.

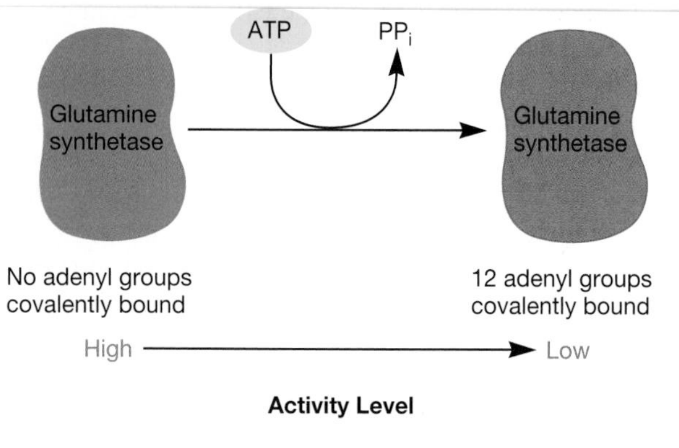

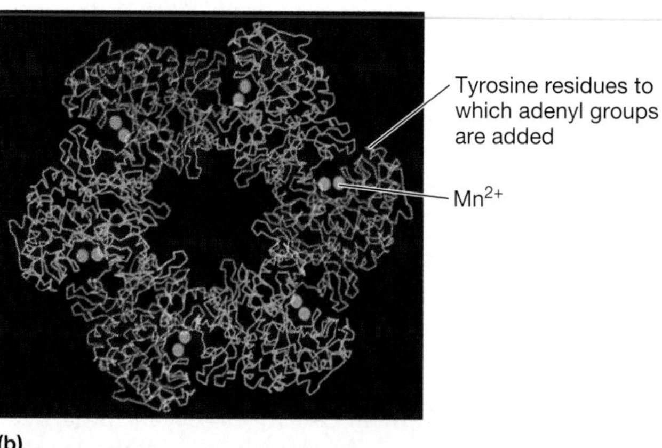

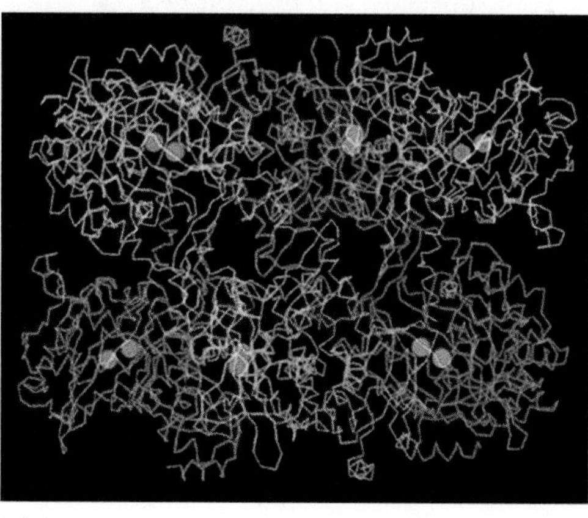

Figure 8.26 Regulation of Glutamine Synthetase Activity. Glutamine synthetase consists of 12 subunits, each of which can be adenylylated. **(a)** As the number of adenyl groups increases, the activity of the enzyme decreases. **(b)** Top view of glutamine synthetase showing 6 of the 12 subunits. The other six subunits lie directly below the six shown here. For clarity, the subunits are colored alternating green and blue. Each of the six catalytic sites shown has a pair of Mn^{2+} ions (red circles). Adenyl groups can be attached to certain tyrosine residues (small red structures) in each subunit. **(c)** Side view.

activity again increases and more product is formed. In this way feedback inhibition automatically matches end product supply with the demand. The previously discussed *E. coli* aspartate carbamoyltransferase is an excellent example of end product or feedback inhibition.

Frequently a biosynthetic pathway branches to form more than one end product. In such a situation the synthesis of pathway end products must be coordinated precisely. It would not do to have one end product present in excess while another is lacking. Branching biosynthetic pathways usually achieve a balance between end products through the use of regulatory enzymes at branch points (**figure 8.27**). If an end product is present in excess, it often inhibits the branch-point enzyme on the sequence leading to its formation, in this way regulating its own formation without affecting the synthesis of other products. In figure 8.27 notice that both products also inhibit the initial enzyme in the pathway. An excess of one product slows the flow of carbon into the whole pathway while inhibiting the appropriate branch-point enzyme. Because less carbon is required when a branch is not functioning, feedback inhibition of the initial pacemaker enzyme helps match the supply with the demand in branching pathways. The regulation of multiple branched pathways is often made even more sophisticated by the presence of **isoenzymes,** different forms of an enzyme that catalyze the same reaction. The initial pacemaker step may be catalyzed by several isoenzymes, each under separate and independent control. In such a situation an excess of a single end product reduces pathway activity but does not completely block pathway function because some isoenzymes are still active.

1. Define the following: allosteric enzyme, effector or modulator, and $[S]_{0.5}$ or $K_{0.5}$.
2. How can regulatory enzymes be influenced by reversible covalent modification? What group is used for this purpose with glutamine synthetase, and which form of this enzyme is active?
3. What is a pacemaker enzyme? Feedback inhibition? How does feedback inhibition automatically adjust the concentration of a pathway end product?
4. What is the significance of the fact that regulatory enzymes often are located at pathway branch points? What are isoenzymes and why are they important in pathway regulation?

Chemotaxis

Thus far in our discussion of the regulation of enzyme activity, we have focused on the control of metabolic pathways brought about by modulating the activity of certain regulatory enzymes that function in the pathway. However, not all enzymes function in metabolic pathways. Rather, some enzymes are involved in processes that are more complex. These include behavioral changes made by microbes in response to their environment. Chemotaxis is an example of the roles enzymes play in micro-

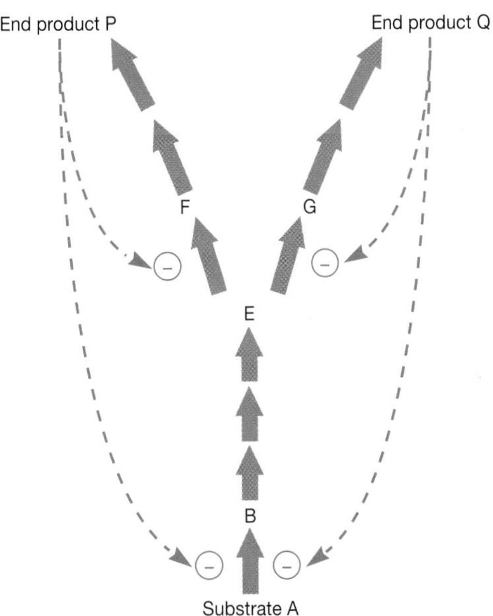

Figure 8.27 Feedback Inhibition. Feedback inhibition in a branching pathway with two end products. The branch-point enzymes, those catalyzing the conversion of intermediate E to F and G, are regulated by feedback inhibition. Products P and Q also inhibit the initial reaction in the pathway. A colored line with a minus sign at one end indicates that an end product, P or Q, is inhibiting the enzyme catalyzing the step next to the minus. See text for further explanation.

bial behavior, and how controlling enzyme activity changes that behavior.

In chapter 3, chemotaxis was briefly introduced. Recall that microorganisms are able to sense chemicals in their environment and either move toward them or away from them, depending on whether the chemical is an attractant or a repellant. For simplicity of discussion, we will only concern ourselves with movement toward an attractant. The best-studied chemotactic system is that of *E. coli,* which, like many other bacteria, exhibits two movement modalities: a forward-swimming motion called a run and a tumbling motion called a tumble. A run occurs when the flagellum rotates in a counterclockwise direction (CCW), and a tumble occurs when the flagellum rotates clockwise (CW) (*see figures 3.41 and 3.45*). The cell alternates between these two types of movements, with the tumble establishing the direction of movement in the run that follows. When *E. coli* is in an environment that is homogenous—that is, the concentration of all chemicals in the environment is the same throughout its habitat—the cell moves about randomly, with no apparent direction or purpose; this is called a random walk. However, if a chemical gradient exists in its environment, the frequency of tumbles decreases as long as the cell is moving toward the attractant. In other words, the length of time spent moving toward the attractant is increased and eventually the cell gets closer to the attractant. The process is not perfect, however. Because bacteria are small, they often can be knocked off course by the movement of molecules in their environment. Therefore they must continually readjust their direction through a trial and error process that is mediated by tumbling. When one examines the path taken by the cell, it is similar to a random walk, but is biased toward the attractant. Thus the movement of the bacterium toward the attractant often is referred to as a biased random walk.

For over three decades, scientists have been dissecting this complex behavior in order to understand how *E. coli* senses the presence of an attractant, how it switches from a run to a tumble and back again, and how it knows it is heading in the correct direction. Many aspects of chemotaxis are now understood, at least superficially, but many questions remain. However, one thing is clear: the chemotactic response of *E. coli* involves a number of enzymes and other proteins that are regulated by covalent modification. One important component is a phosphorelay system. **Phosphorelay systems** consist of at least two proteins: a **sensor kinase** and a **response regulator.** As described here, a phosphorelay system is used to regulate enzyme activity. Other phosphorelay systems are used to regulate protein synthesis and generally use only these two components. These systems are described in chapter 12.

In order for chemotaxis to occur, *E. coli* must determine if an attractant is present and then modulate the activity of the phosphorelay system that dictates the rotational direction of the flagellum (i.e., either run or tumble). *E. coli* senses chemicals in its environment when they bind to chemoreceptors (**figure 8.28**). Numerous chemoreceptors have been identified. We will focus on one class of receptors called **methyl-accepting chemotaxis proteins (MCPs)**. The phosphorelay system that controls direction of flagellar rotation consists of the sensor kinase CheA and the response regulator CheY. When activated, CheA phosphorylates itself using ATP (figure 8.28*c*). The phosphoryl group is then quickly transferred to CheY. Phosphorylated CheY diffuses through the cytoplasm to the flagellar motor. Upon interacting with the motor, the direction of rotation is switched from CCW to CW, and a tumble ensues. When CheA is inactive, the flagellum rotates in its default mode (CCW), and the cell moves forward in a smooth run.

As implied by the preceding discussion, the state of the MCPs must be communicated to the CheA/CheY phosphorelay system. How is this accomplished? The MCPs are buried in the plasma membrane with different parts exposed on each side of the membrane (figure 8.28*c*). The periplasmic side of each MCP has a binding site for one or more attractant molecules. The cytoplasmic side of an MCP interacts with two proteins, CheW and CheA. The CheW protein binds to the MCP and helps attach the CheA protein. Together with CheW and CheA, the MCP receptors form large clusters at one or both poles of the cell (figure 8.28*b*). It is thought that smaller aggregations of the MCPs, CheA and CheW, which function as signaling teams, are the building blocks of the receptor clusters. The number of each of these molecules in the signaling team is not clear, but it has been suggested that each team includes three receptors, often of different types (**figure 8.29**), two CheW molecules, and one CheA dimer. It has also been suggested that the signaling teams aggregate with each other to form "signaling leagues." Finally, the signaling teams (and perhaps signaling

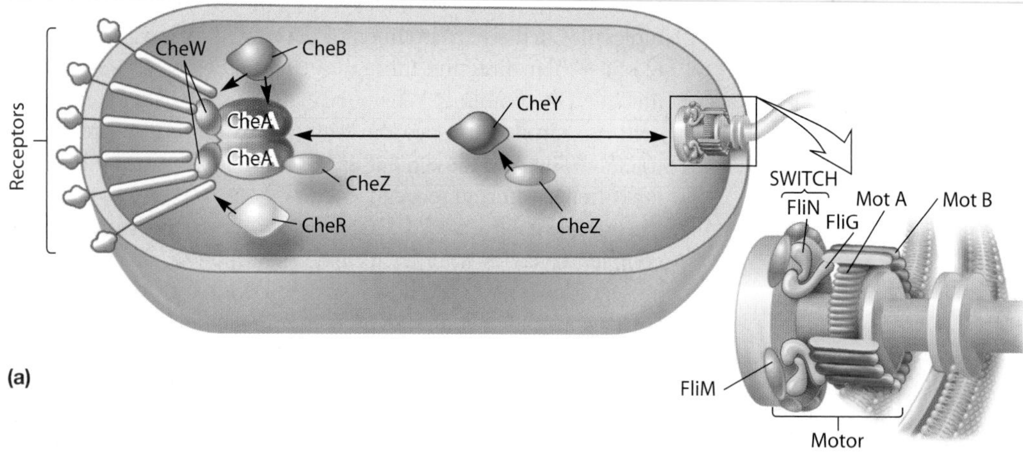

(a)

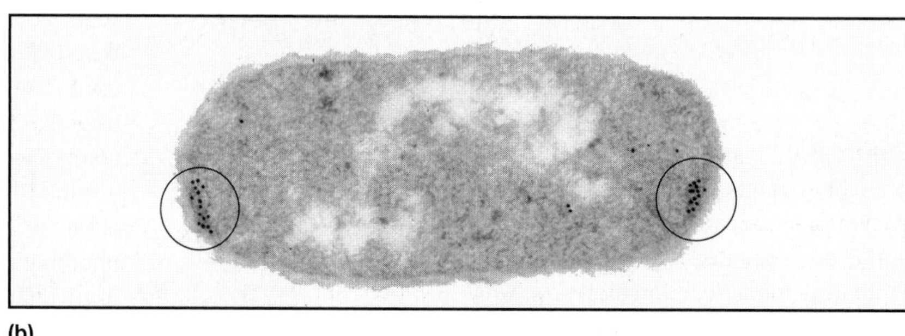

(b)

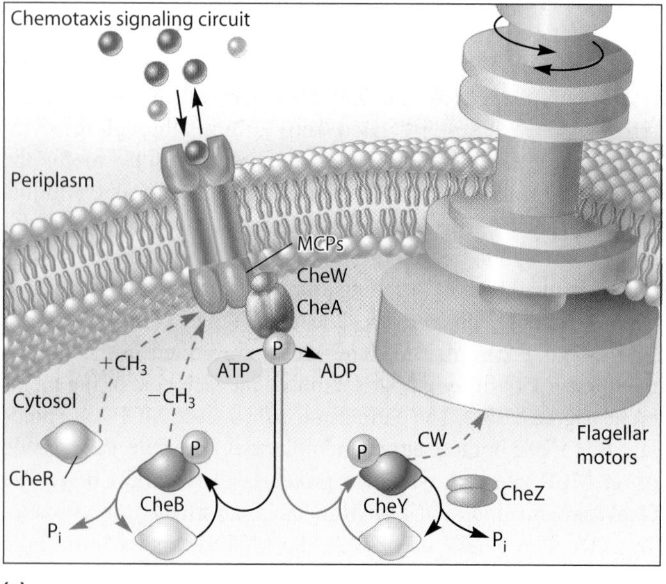

(c)

Figure 8.28 Proteins and Signaling Pathways of the Chemotaxis Response in *E. coli*. **(a)** The methyl-accepting chemotaxis proteins (MCPs) form clusters associated with the CheA and CheW proteins. CheA is a sensor kinase that when activated phosphorylates CheB, a methylesterase, or CheY. Phosphorylated CheY interacts with the FliM protein of the flagellar motor, causing rotation of the flagellum to switch from counterclockwise (CCW) to clockwise (CW). This results in a switch from a run (CCW rotation) to a tumble (CW rotation). **(b)** MCPs, CheW, CheA complexes form large clusters of receptors at either end of the cell, as shown in this electron micrograph of *E. coli*. Gold-tagged antibodies were used to label the receptor clusters, which appear as black dots (encircled). **(c)** The chemotactic signaling pathways of *E. coli*. The pathways that increase the probability of CCW rotation are shown in red. CCW rotation is the default rotation. It is periodically interrupted by CW rotation, which causes tumbling. The pathways that lead to CW rotation are shown in green. Molecules shown in gray are unphosphorylated and inactive. Note that MCP, CheA, and CheZ are homodimers. CheW, CheB, CheY, and CheR are monomers.

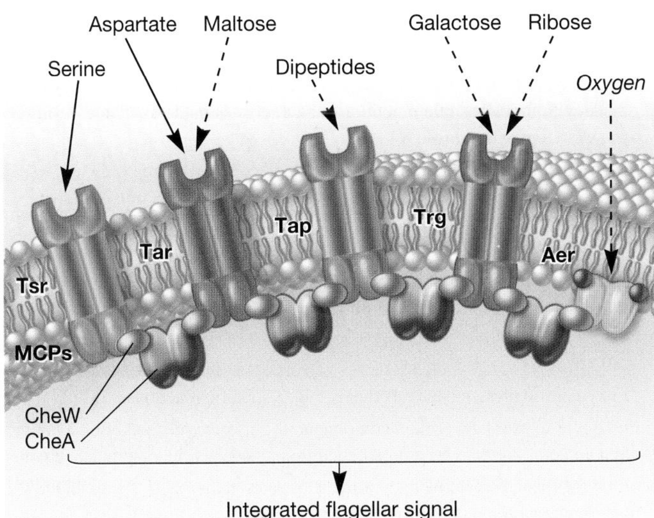

Figure 8.29 The Methyl-Accepting Chemotaxis Proteins of *E. coli*. The attractants sensed by each methyl-accepting chemotaxis protein (MCP) are shown. Some are sensed directly, when the attractant binds the MCP (solid lines). Others are sensed indirectly (dashed lines). The attractants maltose, dipeptides, galactose, and ribose are detected by their interaction with periplasmic binding proteins. Oxygen is detected indirectly by the Aer chemoreceptor, which differs from other MCPs in that it lacks a periplasmic sensing domain. Instead, the cytoplasmic domain has a binding site for FAD. FAD is an important electron carrier found in many electron transport systems. The redox state of the MCP-bound FAD molecule is used to monitor the functioning of the electron transport system. This in turns mediates a tactic response to oxygen.

leagues) become interconnected by an unknown mechanism to form the receptor clusters visible at the poles of the cell.

No matter what the precise stoichiometry or architecture of the receptor clusters, there is evidence that the MCPs in each signaling team work cooperatively to modulate CheA activity. When any one of the MCPs in the signaling team is bound to an attractant, CheA autophosphorylation is inhibited, the flagellum continues rotating CCW, and the cell continues in its run. Because of this cooperation, the cell can respond to very low concentrations of attractant. Furthermore, it can integrate signals from all receptors in the team (figure 8.29). On the other hand, if attractant levels decrease, so that the level of attractant bound to the MCPs in a signaling team decreases, CheA is stimulated to autophosphorylate, the phosphorelay is set into motion, and the cell begins to tumble. However, tumbling does not continue indefinitely. About 10 seconds after the switch to CW rotation occurs, the phosphoryl group is removed from CheY by the CheZ protein, and CCW rotation is resumed.

But how does *E. coli* measure the concentration of attractant in its environment, and how does it know when it is moving toward the attractant? *E. coli* measures the concentration of an attractant every few seconds and determines if the concentration is increasing or decreasing over time. As long as the concentration increases, the cell continues a run. If the concentration decreases, a tumble is triggered. In order to compare concentrations of the attractant over time, *E. coli* must have a mechanism for remembering the previous concentration. *E. coli* accomplishes this by comparing the overall methylation level of the MCPs (on the cytoplasmic side) with the overall amount of attractant bound (on the periplasmic face). The cytoplasmic portion of each MCP has four to six glutamic acid residues that can be methylated. Addition and removal of methyl groups is catalyzed by two different enzymes. Methylation is catalyzed by the MCP-specific methyltransferase CheR. Demethylation is catalyzed by the MCP-specific methylesterase CheB. Methylation occurs at a fairly steady rate regardless of the attractant level. However, an MCP-attractant complex is a better substrate for CheR than is an MCP that is not bound to attractant. Thus when attractant is bound, methylation of the MCP is favored. The methylesterase activity of CheB is also modified by the CheA protein. As long as the concentration of the attractant keeps increasing, the number of MCPs bound to attractant remains high, and the MCP methylation level remains high. However, if the attractant concentration decreases, the level of methylation will exceed the level of attractant bound. This disparity in methylation level and MCP-bound attractant stimulates CheA to autophosphorylate. As a result, the phosphorelay signal for CW flagellar rotation is initiated and the cell tumbles in an attempt to reorient itself in the gradient so that it is moving up the gradient (toward the attractant) rather than down the gradient (away from the attractant). At the same time, some of the phosphoryl groups on CheA are transferred to CheB. This activates CheB, and it removes methyl groups from the MCPs. This lowers the methylation level so that it is commensurate with the number of MCPs bound to attractant. A few seconds later, the number of MCPs bound to attractant will be compared to this new methylation level. Based on the correspondence of the two, the cell will determine if it is again moving up the gradient. If it is, tumbling will be suppressed (as will methylesterase activity) and the run will continue.

1. What is a phosphorelay system?
2. Describe the MCP-CheW-CheA receptor complex. What two proteins are phosphorylated by CheA? What is the role of each?
3. How does the MCP regulate the rate of CheA autophosphorylation? How does this mediate chemotaxis?

Summary

8.1 An Overview of Metabolism

a. Metabolism is the total of all chemical reactions that occur in cells. It can be divided into two parts: energy-conserving reactions (sometimes called catabolism) and anabolism.

b. Organisms are defined nutritionally based on how they fulfill their carbon, energy, and electron needs. The interactions of organisms belonging to different nutritional types is the basis for the flow of energy, carbon, and electrons in the biosphere. The ultimate source of energy for most microbes is sunlight trapped by photoautotrophs and used to form organic material from CO_2. Photoautotrophs and the organic molecules they have synthesized are consumed by chemoorganoheterotrophs (**figures 8.1** and **8.2**).

8.2 Energy and Work

a. Energy is the capacity to do work. Living cells carry out three major kinds of work: chemical work of biosynthesis, transport work, and mechanical work.

8.3 The Laws of Thermodynamics

a. The first law of thermodynamics states that energy is neither created nor destroyed.

b. The second law of thermodynamics states that changes occur in such a way that the randomness or disorder of the universe increases to the maximum possible. That is, entropy always increases during spontaneous processes.

8.4 Free Energy and Reactions

a. The first and second laws can be combined to determine the amount of energy made available for useful work.

$$\Delta G = \Delta H - T \cdot \Delta S$$

In this equation the change in free energy (ΔG) is the energy made available for useful work, the change in enthalpy (ΔH) is the change in heat content, and the change in entropy is ΔS.

b. The standard free energy change ($\Delta G^{\circ\prime}$) for a chemical reaction is directly related to the equilibrium constant.

c. In exergonic reactions $\Delta G^{\circ\prime}$ is negative and the equilibrium constant is greater than one; the reaction goes to completion as written. Endergonic reactions have a positive $\Delta G^{\circ\prime}$ and an equilibrium constant less than one (**figure 8.4**).

8.5 The Role of ATP in Metabolism

a. ATP is a high-energy molecule that serves as an energy currency; it transports energy in a useful form from one reaction or location in a cell to another. (**figure 8.5**)

b. ATP is readily synthesized from ADP and P_i using energy released from exergonic reactions; when hydrolyzed back to ADP and P_i, it releases the energy, which is used to drive endergonic reactions. This cycling of ATP with ADP and P_i is called the cell's energy cycle (**figure 8.7**).

8.6 Oxidation-Reduction Reactions, Electron Carriers, and Electron Transport Systems

a. In oxidation-reduction (redox) reactions, electrons move from an electron donor to an electron acceptor. The standard reduction potential measures the tendency of the donor to give up electrons.

b. Redox couples with more negative reduction potentials donate electrons to those with more positive potentials, and energy is made available during the transfer (**figure 8.8, table 8.1**).

c. Some of the most important electron carriers in cells are NAD^+, $NADP^+$, FAD, FMN, coenzyme Q, cytochromes, and the nonheme iron proteins.

d. Electron carriers are often organized into electron transport systems that are located in membranes. These systems are critical to the energy-conserving processes observed during aerobic respiration, anaerobic respiration, chemolithotrophy, and photosynthesis (**figure 8.9**).

8.7 Enzymes

a. Enzymes are protein catalysts that catalyze specific reactions.

b. Enzymes consist of a protein component, the apoenzyme, and often a nonprotein cofactor that may be a prosthetic group, a coenzyme, or a metal activator.

c. Enzymes speed reactions by binding substrates at their active sites and lowering the activation energy (**figure 8.15**).

d. The rate of an enzyme-catalyzed reaction increases with substrate concentration at low substrate levels and reaches a plateau (the maximum velocity) at saturating substrate concentrations. The Michaelis constant is the substrate concentration that the enzyme requires to achieve half maximal velocity (**figure 8.18**).

e. Enzymes have pH and temperature optima for activity (**figure 8.19**).

f. Enzyme activity can be slowed by competitive and noncompetitive inhibitors (**figure 8.20**).

8.8 The Nature and Significance of Metabolic Regulation

a. The regulation of metabolism keeps cell components in proper balance and conserves metabolic energy and material.

8.9 Metabolic Channeling

a. The localization of metabolites and enzymes in different parts of the cell, called metabolic channeling, influences pathway activity. A common channeling mechanism is compartmentation.

8.10 Control of Enzyme Activity

a. Many regulatory enzymes are allosteric enzymes, enzymes in which an effector or modulator binds noncovalently and reversibly to a regulatory site separate from the catalytic site and causes a conformational change in the enzyme to alter its activity (**figure 8.22**).

b. Enzyme activity also can be regulated by reversible covalent modification. Usually a phosphoryl, methyl, or adenyl group is attached to the enzyme.

c. The first enzyme in a pathway and enzymes at branch points often are subject to feedback inhibition by one or more end products. Excess end product slows its own synthesis (**figure 8.27**).

d. Complex behaviors such as chemotaxis can also be regulated by altering enzyme activity (**figure 8.28** and **8.29**).

Key Terms

activation energy 177
active site 177
adenosine diphosphate (ADP) 171
adenosine 5′-triphosphate (ATP) 171
allosteric enzymes 181
anabolism 168
apoenzyme 176
calorie 170
catabolism 168
catalyst 176
catalytic site 177
chemical work 169
coenzyme 176
coenzyme Q or CoQ (ubiquinone) 173
cofactor 176
compartmentation 180
competitive inhibitor 179
cytochrome 174
denaturation 179
effector or modulator 181
electron acceptor 172
electron donor 172
electron transport chain (ETC) 173
electron transport system (ETS) 173
endergonic reaction 170
end product inhibition 183
energy 169
energy-conserving reactions 168
enthalpy 170
entropy 169
enzyme 176
enzyme-substrate complex 177
equilibrium 170
equilibrium constant (K_{eq}) 170
exergonic reaction 170
feedback inhibition 183
ferredoxin 174
first law of thermodynamics 169
flavin adenine dinucleotide (FAD) 173
flavin mononucleotide (FMN) 173
free energy change 170
high-energy molecule 171
holoenzyme 176
isoenzymes 184
joule 170
mechanical work 169
metabolic channeling 180
metabolism 167
methyl-accepting chemotaxis proteins (MCPs) 185
Michaelis constant (K_m) 178
nicotinamide adenine dinucleotide (NAD^+) 173
nicotinamide adenine dinucleotide phosphate ($NADP^+$) 173
noncompetitive inhibitor 179
nonheme iron protein 174
oxidation-reduction (redox) reaction 172
pacemaker enzyme 183
phosphate group transfer potential 171
phosphorelay system 185
posttranslational regulation 181
product 176
prosthetic group 176
reducing power 168
regulatory site 181
response regulator 185
reversible covalent modification 183
second law of thermodynamics 169
sensor kinase 185
standard free energy change 170
standard reduction potential 172
substrate 176
thermodynamics 169
transition-state complex 177
transport work 169

Critical Thinking Questions

1. How could electron transport be driven in the opposite direction? Why would it be desirable to do this?
2. Suppose that a chemical reaction had a large negative $\Delta G^{\circ\prime}$ value. What would this indicate about its equilibrium constant? If displaced from equilibrium, would it proceed rapidly to completion? Would much or little free energy be made available?
3. Take a look at the structures of macromolecules (appendix I). Which type has the most electrons to donate? Why are carbohydrates usually the primary source of electrons for chemooganotrophic bacteria?
4. Most enzymes do not operate at their biochemical optima inside cells. Why not?
5. Examine the branched pathway shown here for the synthesis of the amino acids aspartate, methionine, lysine, threonine, and isoleucine. For each of these two scenarios answer the following questions:
   a. Which portion(s) of the pathway would need to be shut down in this situation?
   b. How might allosteric control be used to accomplish this?

   Scenario 1: The microbe is cultured in a medium containing aspartate and lysine, but lacking methionine, threonine, and isoleucine.

   Scenario 2: The microbe is cultured in a medium containing a rich supply of all five amino acids.

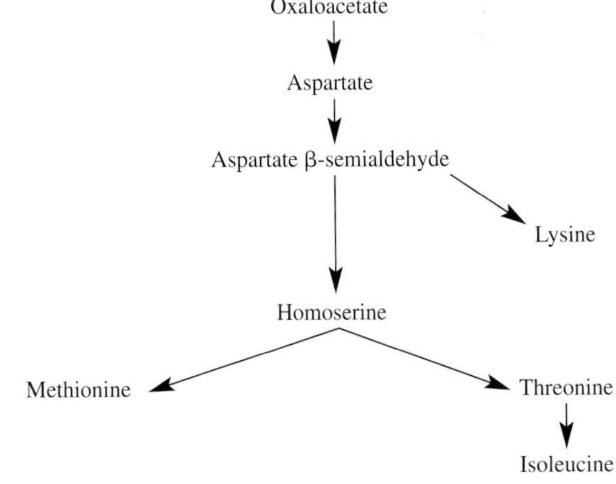

Learn More

Bren, A., and Eisenbach, M. 2000. How signals are heard during bacterial chemotaxis: Protein-protein interactions in sensory signal propagation. *J. Bact.* 182(24):6865–73.

International Union of Biochemistry and Molecular Biology. 1992. *Enzyme nomenclature*. San Diego: Academic Press.

Kantrowitz, E. R., and Lipscomb, W. N. 1988. *Escherichia coli* aspartate transcarbamylase: The relation between structure and function. *Science* 241:669–74.

Lodish, H.; Berk, A.; Matsudaira, P.; Kaiser, C. S.; Drieger, M.; Scott, M. P.; Zipursky, S. L.; and Darnell, J. 2004. *Molecular cell biology,* 5th ed. New York: W.H. Freeman.

McKee, T., and McKee, J. R. 2003. *Biochemistry: The molecular basis of life,* 3d ed. New York: McGraw-Hill.

Nelson, D. L., and Cox, M. M. 2005. *Lehninger principles of biochemistry,* 4th ed. New York: W.H. Freeman.

Parkinson, J. S. 2004. Signal amplification in bacterial chemotaxis through receptor teamwork. *ASM News* 70(12):575–82.

Please visit the Prescott website at www.mhhe.com/prescott7 for additional references.

9 Metabolism: Energy Release and Conservation

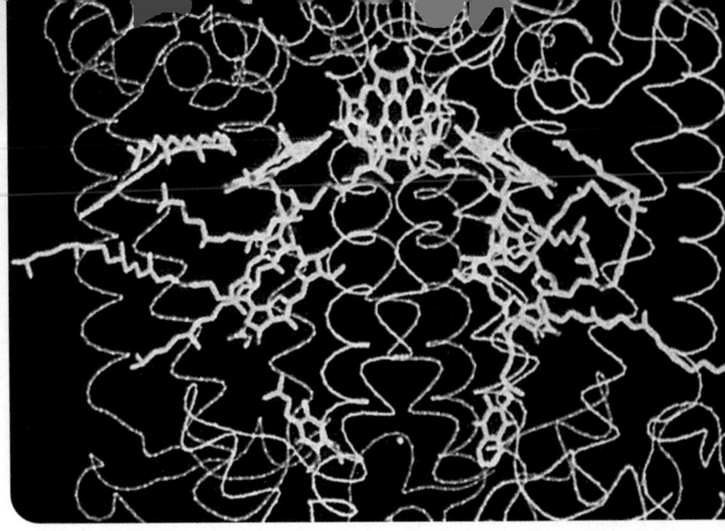

The reaction center of the purple nonsulfur bacterium, *Rhodopseudomonas viridis,* with the bacteriochlorophylls and other prosthetic groups in yellow. These pigments trap light during photosynthesis.

Preview

- Chemoorganotrophs have three fueling-process options. They are differentiated by the electron acceptor used. Respiration uses exogenous electron acceptors—O_2 for aerobic respiration, and other molecules for anaerobic respiration. Fermentation uses endogenous electron acceptors.
- During catabolism, nutrients are funneled into a few common pathways for more efficient use of enzymes (a few pathways process a wide variety of nutrients).
- The most widely used pathways are the Embden-Meyerhof pathway, the pentose phosphate pathway, and the tricarboxylic acid cycle. These pathways are amphibolic, functioning both catabolically and anabolically. The tricarboxylic acid cycle is the final pathway for the aerobic oxidation of nutrients to CO_2.
- The majority of energy released during aerobic and anaerobic respiration is generated by the movement of electrons from electron transport carriers with more negative reduction potentials to ones with more positive reduction potentials. Because the O_2/H_2O redox couple has a very positive standard reduction potential, aerobic respiration is much more efficient than anaerobic catabolism.
- Chemolithotrophs use reduced inorganic molecules as electron donors for electron transport and ATP synthesis.
- In chlorophyll-based photosynthesis, trapped light energy boosts electrons to more negative reduction potentials (i.e., higher energy levels). These energized electrons are then used to make ATP. Some procaryotes carry out rhodopsin-based phototrophy. Electron flow is not involved in this process.
- Proton motive force is a type of potential energy generated by: (1) oxidation of chemical energy sources coupled to electron transport; (2) light-driven electron transport; and (3) light-driven pumping of protons during rhodopsin-based phototrophy; it is used to power the production of ATP and other processes such as transport and bacterial motility.

From the open seas to a eutrophic lake, from a log rotting in a forest to a microbrewery, and from a hydrothermal vent to the tailings near a coal mine, the impact of the fueling reactions of microorganisms can be seen. Phototrophs convert the energy of the sun into chemical energy (**figure 9.1**), which feeds chemoorganotrophs. Chemoorganotrophs recycle the wastes of other organisms and play important roles in industry; chemolithotrophs oxidize inorganic molecules and in the process contribute to biogeochemical cycles such as the iron and sulfur cycles. All can contribute to pollution and all can help in the maintenance of pristine environments.

This chapter examines the fueling reactions of these diverse nutritional types. We begin with an overview of the metabolism of chemoorganotrophs. This is followed by an introduction to the oxidation of carbohydrates, especially glucose, and a discussion of the generation of ATP by aerobic and anaerobic respiration. Fermentation is then described, followed by a survey of the breakdown of other carbohydrates and organic substances such as lipids, proteins, and amino acids. We end the chapter with sections on chemolithotrophy (the oxidation of inorganic energy sources) and phototrophy (the conversion of light energy into chemical energy).

9.1 Chemoorganotrophic Fueling Processes

Recall from chapter 8 that chemoorganotrophs oxidize an organic energy source and conserve the energy released in the form of ATP. The electrons released are accepted by a variety of electron

It is in the fueling reactions that bacteria display their extraordinary metabolic diversity and versatility. Bacteria have evolved to thrive in almost all natural environments, regardless of the nature of available sources of carbon, energy, and reducing power. . . . The collective metabolic capacities of bacteria allow them to metabolize virtually every organic compound on this planet. . . .

—*F. C. Neidhardt, J. L. Ingraham, and M. Schaechter*

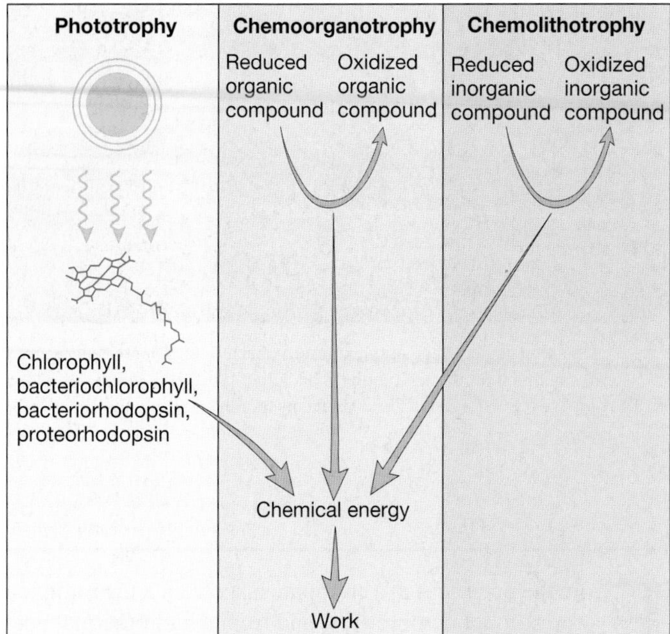

Figure 9.1 Sources of Energy for Microorganisms. Most microorganisms employ one of three energy sources. Phototrophs trap radiant energy from the sun using pigments such as bacteriochlorophyll and chlorophyll. Chemotrophs oxidize reduced organic and inorganic nutrients to liberate and trap energy. The chemical energy derived from these three sources can then be used in work as discussed in chapter 8.

acceptors, and whether the acceptor is exogenous (that is, externally supplied) or endogenous (internally supplied) defines the energy-conserving process used by the organism. When the electron acceptor is exogenous, the metabolic process is called **respiration** and may be divided into two different types (**figure 9.2**). In aerobic respiration, the final electron acceptor is oxygen, whereas the terminal acceptor in anaerobic respiration is a different exogenous acceptor such as NO_3^-, SO_4^{2-}, CO_2, Fe^{3+}, and SeO_4^{2-}. Organic acceptors such as fumarate and humic acids also may be used. Respiration involves the activity of an electron transport chain. As electrons pass through the chain to the final electron acceptor, a type of potential energy called the proton motive force (PMF) is generated and used to synthesize ATP from ADP and P_i. In contrast, fermentation [Latin *fermentare,* to cause to rise or ferment] uses an endogenous electron acceptor and does not involve an electron transport chain or the generation of PMF. The endogenous electron acceptor is usually an intermediate (e.g., pyruvate) of the catabolic pathway used to degrade and oxidize the organic energy source. During fermentation, ATP is synthesized only by substrate-level phosphorylation, a process in which a phosphate group is transferred to ADP from a high-energy molecule (e.g., phosphoenolpyruvate) generated by catabolism of the energy source.

In the following sections, we explore the metabolism of chemoorganotrophs in more detail. The discussion begins with aerobic respiration and introduces a number of metabolic pathways and other processes that also occur during anaerobic respi-

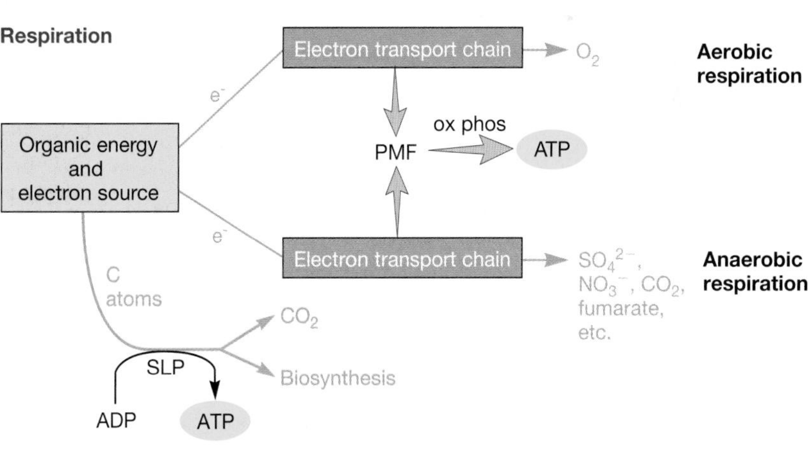

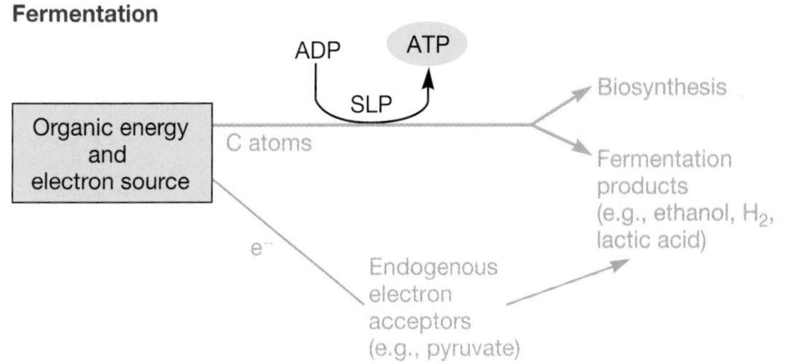

Figure 9.2 Chemoorganotrophic Fueling Processes. Organic molecules serve as energy and electron sources for all three fueling processes used by chemoorganotrophs. In aerobic respiration and anaerobic respiration, the electrons pass through an electron transport system. This generates a proton motive force (PMF), which is used to synthesize most of the cellular ATP by a mechanism called oxidative phosphorylation (ox phos); a small amount of ATP is made by a process called substrate-level phosphorylation (SLP). In aerobic respiration, O_2 is the terminal electron acceptor, whereas in anaerobic respiration exogenous molecules other than O_2 serve as electron acceptors. During fermentation, endogenous organic molecules act as electron acceptors, the electron flow is not coupled with ATP synthesis, and ATP is synthesized only by substrate-level phosphorylation.

ration. Fermentation involves only a subset of the pathways that function during respiration, but is widely used by microbes and has important practical applications.

9.2 Aerobic Respiration

Aerobic respiration involves several important catabolic pathways. Before learning about some of the more important ones, it is best to look at the "lay of the land" and get our bearings. Albert Lehninger, a biochemist who worked at Johns Hopkins Medical School, helped considerably by pointing out that aerobic respiration may be divided into three stages (**figure 9.3**). In the first stage, larger nutrient molecules (proteins, polysaccharides, and lipids) are hydrolyzed or otherwise broken down into their constituent parts. The chemical reactions occurring during this stage do not release much energy. Amino acids, monosaccharides, fatty acids, glycerol, and other products of the first stage are degraded to a few simpler molecules in the second stage. Usually metabolites like acetyl coenzyme A and pyruvate are formed. In addition, the second stage produces some ATP as well as NADH and/or $FADH_2$. Finally during the third stage of catabolism, partially oxidized carbon is fed into the tricarboxylic acid cycle and oxidized completely to CO_2 with the production of ATP, NADH, and $FADH_2$. Most of the ATP derived from aerobic respiration comes from the oxidation of NADH and $FADH_2$ by the electron transport chain, which uses oxygen as the terminal electron acceptor.

Although this picture is somewhat oversimplified, it is useful in discerning the general pattern of respiration. Notice that the microorganism begins with a wide variety of molecules and reduces their number and diversity at each stage. That is, nutrient molecules are funneled into ever fewer metabolic intermediates until they are finally fed into the tricarboxylic acid cycle. A common pathway often degrades many similar molecules (e.g., several different sugars). These metabolic pathways consist of enzyme-catalyzed reactions arranged so that the product of one reaction serves as a substrate for the next. The existence of a few common catabolic pathways, each degrading many nutrients,

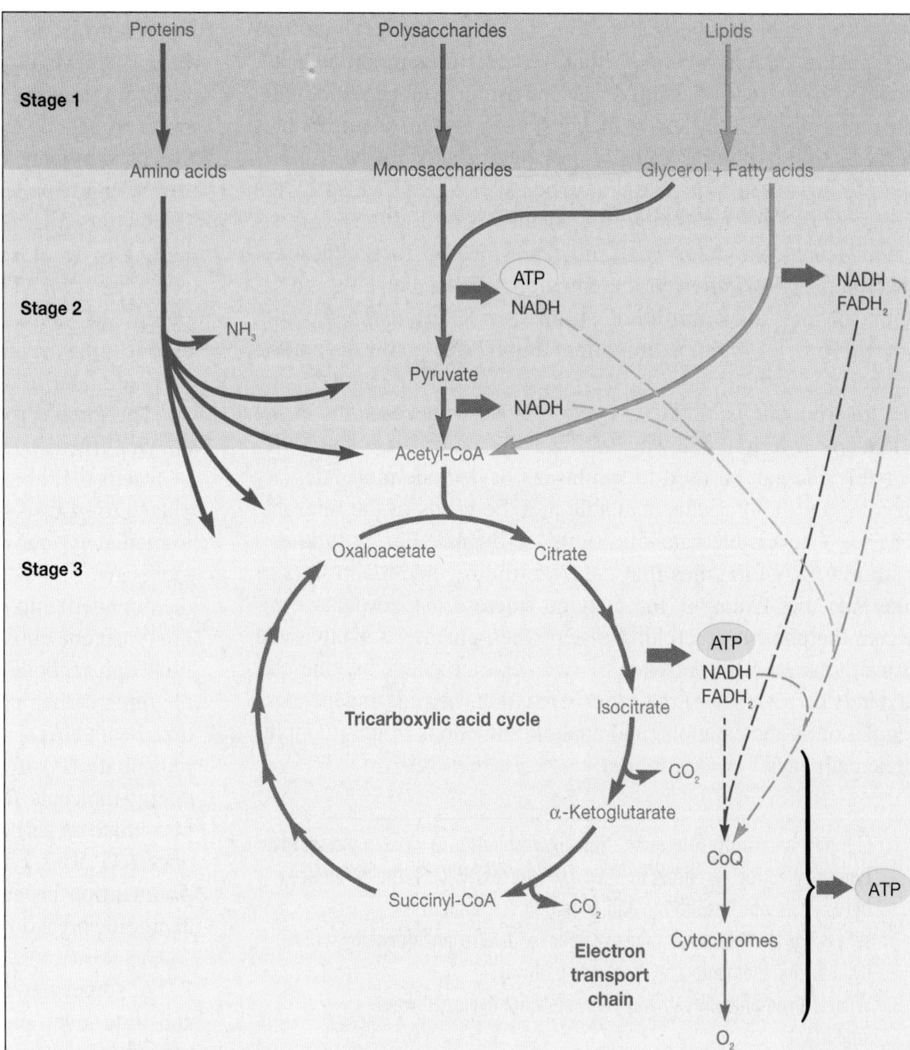

Figure 9.3 The Three Stages of Aerobic Respiration. A general diagram of aerobic respiration in a chemoorganoheterotroph showing the three stages in this process and the central position of the tricarboxylic acid cycle. Although there are many different proteins, polysaccharides, and lipids, they are degraded through the activity of a few common metabolic pathways. The dashed lines show the flow of electrons, carried by NADH and $FADH_2$, to the electron transport chain.

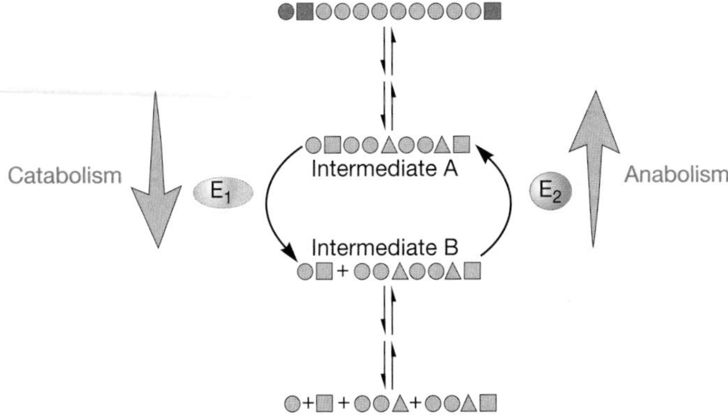

Figure 9.4 Amphibolic Pathway. A simplified diagram of an amphibolic pathway, such as glycolysis. Note that the interconversion of intermediates A and B is catalyzed by two separate enzymes, E_1 operating in the catabolic direction, and E_2 in the anabolic.

greatly increases metabolic efficiency by avoiding the need for a large number of less metabolically flexible pathways.

Although biosynthesis is the topic of chapter 10, it is important to point out that many catabolic pathways also are important in anabolism. They supply materials needed for biosynthesis, including precursor metabolites and reducing power. Precursor metabolites serve as the starting molecules for biosynthetic pathways. Reducing power is used to reduce the carbon skeletons provided by the precursor metabolites as they are transformed into amino acids, nucleotides, and the other small molecules needed for synthesis of macromolecules. Pathways that function both catabolically and anabolically are called **amphibolic pathways** [Greek *amphi,* on both sides]. Three of the most important amphibolic pathways are the Embden-Meyerhof pathway, the pentose phosphate pathway, and the tricarboxylic acid (TCA) cycle. Many of the reactions of the Embden-Meyerhof pathway and the TCA cycle are freely reversible and can be used to synthesize or degrade molecules depending on the nutrients available and the needs of the microbe. The few irreversible catabolic steps are bypassed in biosynthesis with alternate enzymes that catalyze the reverse reaction (**figure 9.4**). For example, the enzyme fructose bisphosphatase reverses the phosphofructokinase step when glucose is synthesized from pyruvate. The presence of two separate enzymes, one catalyzing the reversal of the other's reaction, permits independent regulation of the catabolic and anabolic functions of these amphibolic pathways. The precursor metabolites (section 10.2)

1. Compare and contrast fermentation and respiration. Give examples of the types of electron acceptors used by each process. What is the difference between aerobic respiration and anaerobic respiration?
2. Why is it to the cell's advantage to catabolize diverse organic energy sources by funneling them into a few common pathways?
3. What is an amphibolic pathway? Why are amphibolic pathways important?

9.3 The Breakdown of Glucose to Pyruvate

Microorganisms employ several metabolic pathways to catabolize glucose and other sugars. Because of this metabolic diversity, their metabolism is often confusing. To avoid confusion as much as possible, the ways in which microorganisms degrade sugars to pyruvate and similar intermediates are introduced by focusing on only three routes: (1) the Embden-Meyerhof pathway, (2) the pentose phosphate pathway, and (3) the Entner-Doudoroff pathway. In this text, these three pathways will be referred to collectively as **glycolytic pathways** or as **glycolysis** [Greek *glyco,* sweet, and *lysis,* a loosening]. However, in other texts, the term glycolysis is often reserved for use in reference only to the Embden-Meyerhof pathway. For the sake of simplicity, the detailed structures of metabolic intermediates are not used in pathway diagrams. Common metabolic pathways (appendix II)

The Embden-Meyerhof Pathway

The **Embden-Meyerhof pathway** is undoubtedly the most common pathway for glucose degradation to pyruvate in stage two of aerobic respiration. It is found in all major groups of microorganisms and functions in the presence or absence of O_2. As noted earlier, it is also an important amphibolic pathway and provides several precursor metabolites. The Embden-Meyerhof pathway occurs in the cytoplasmic matrix of procaryotes and eucaryotes.

The pathway as a whole may be divided into two parts (**figure 9.5** and *appendix II*). In the initial six-carbon phase, energy is consumed as glucose is phosphorylated twice, and is converted to fructose 1,6-bisphosphate. This preliminary phase consumes two ATP molecules for each glucose and "primes the pump" by adding phosphates to each end of the sugar. In essence, the organism invests some of its ATP so that more can be made later in the pathway.

The three-carbon, energy-conserving phase begins when the enzyme fructose 1,6-bisphosphate aldolase catalyzes the cleavage of fructose 1,6-bisphosphate into two halves, each with a phosphate group. One of the products, dihydroxyacetone phosphate, is immediately converted to glyceraldehyde 3-phosphate. This yields two molecules of glyceraldhyde 3-phosphate, which are then converted to pyruvate in a five-step process. Because dihydroxyacetone phosphate can be easily changed to glyceraldehyde 3-phosphate, both halves of fructose 1,6-bisphosphate are used in the three-carbon phase. First, glyceraldehyde 3-phosphate is oxidized with NAD^+ as the electron acceptor (to form NADH), and a phosphate (P_i) is simultaneously incorporated to give a high-energy molecule called 1,3-bisphosphoglycerate. The high-energy phosphate on carbon one is subsequently donated to ADP to produce ATP. This synthesis of ATP is called **substrate-level phosphorylation** because ADP phosphorylation is coupled with the exergonic breakdown of a high-energy bond. The role of ATP in metabolism (section 8.5)

A somewhat similar process generates a second ATP by substrate-level phosphorylation. The phosphate group on 3-phosphoglycerate shifts to carbon two, and 2-phosphoglycerate is

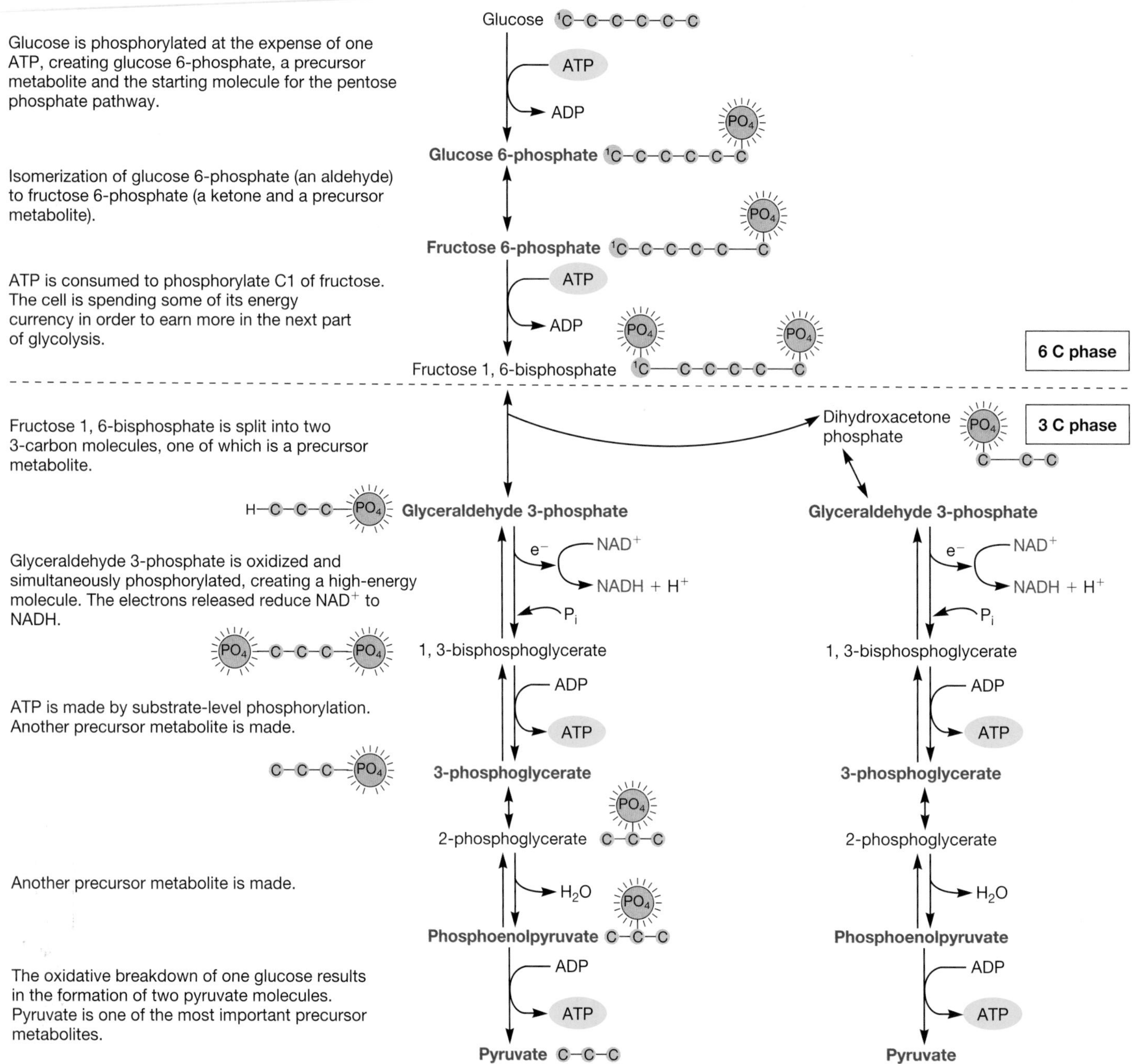
Glucose is phosphorylated at the expense of one ATP, creating glucose 6-phosphate, a precursor metabolite and the starting molecule for the pentose phosphate pathway.
Isomerization of glucose 6-phosphate (an aldehyde) to fructose 6-phosphate (a ketone and a precursor metabolite).
ATP is consumed to phosphorylate C1 of fructose. The cell is spending some of its energy currency in order to earn more in the next part of glycolysis.
Fructose 1, 6-bisphosphate is split into two 3-carbon molecules, one of which is a precursor metabolite.
Glyceraldehyde 3-phosphate is oxidized and simultaneously phosphorylated, creating a high-energy molecule. The electrons released reduce NAD+ to NADH.
ATP is made by substrate-level phosphorylation. Another precursor metabolite is made.
Another precursor metabolite is made.
The oxidative breakdown of one glucose results in the formation of two pyruvate molecules. Pyruvate is one of the most important precursor metabolites.
Glucose
ATP
ADP
Glucose 6-phosphate
Fructose 6-phosphate
Fructose 1, 6-bisphosphate
6 C phase
3 C phase
Dihydroxacetone phosphate
Glyceraldehyde 3-phosphate
NAD+
NADH + H+
Pi
1, 3-bisphosphoglycerate
3-phosphoglycerate
2-phosphoglycerate
H2O
Phosphoenolpyruvate
Pyruvate

Figure 9.5 Embden-Meyerhof Pathway. This is one of three glycolytic pathways used to catabolize glucose to pyruvate, and it can function during aerobic respiration, anaerobic respiration, and fermentation. When used during a respiratory process, the electrons accepted by NAD^+ are transferred to an electron transport chain and are ultimately accepted by an exogenous electron acceptor. When used during fermentation, the electrons accepted by NAD^+ are donated to an endogenous electron acceptor (e.g., pyruvate). The Embden-Meyerhof pathway is also an important amphibolic pathway, as it generates several precursor metabolites (shown in blue).

dehydrated to form a second high-energy molecule, phosphoenolpyruvate. This molecule donates its phosphate to ADP forming a second ATP and pyruvate, the final product of the pathway.

The Embden-Meyerhof pathway degrades one glucose to two pyruvates by the sequence of reactions just outlined and shown in figure 9.5. ATP and NADH are also produced. The yields of ATP and NADH may be calculated by considering the two phases separately. In the six-carbon phase, two ATPs are used to form fructose 1,6-bisphosphate. For each glyceraldehyde 3-phosphate transformed into pyruvate, one NADH and two ATPs are formed.

Because two glyceraldehyde 3-phosphates arise from a single glucose (one by way of dihydroxyacetone phosphate), the three-carbon phase generates four ATPs and two NADHs per glucose. Subtraction of the ATP used in the six-carbon phase from that produced by substrate-level phosphorylation in the three-carbon phase gives a net yield of two ATPs per glucose. Thus the catabolism of glucose to pyruvate can be represented by this simple equation.

$$\text{Glucose} + 2\text{ADP} + 2\text{P}_i + 2\text{NAD}^+ \longrightarrow$$
$$2\text{ pyruvate} + 2\text{ATP} + 2\text{NADH} + 2\text{H}^+$$

The Pentose Phosphate Pathway

A second pathway, the **pentose phosphate** or **hexose monophosphate pathway,** may be used at the same time as either the Embden-Meyerhof or the Entner-Doudoroff pathways. It can operate either aerobically or anaerobically and is important in both biosynthesis and catabolism.

The pentose phosphate pathway begins with the oxidation of glucose 6-phosphate to 6-phosphogluconate followed by the oxidation of 6-phosphogluconate to the pentose sugar ribulose 5-phosphate and CO_2 (**figure 9.6** and *appendix II*). NADPH is produced during these oxidations. Ribulose 5-phosphate is then

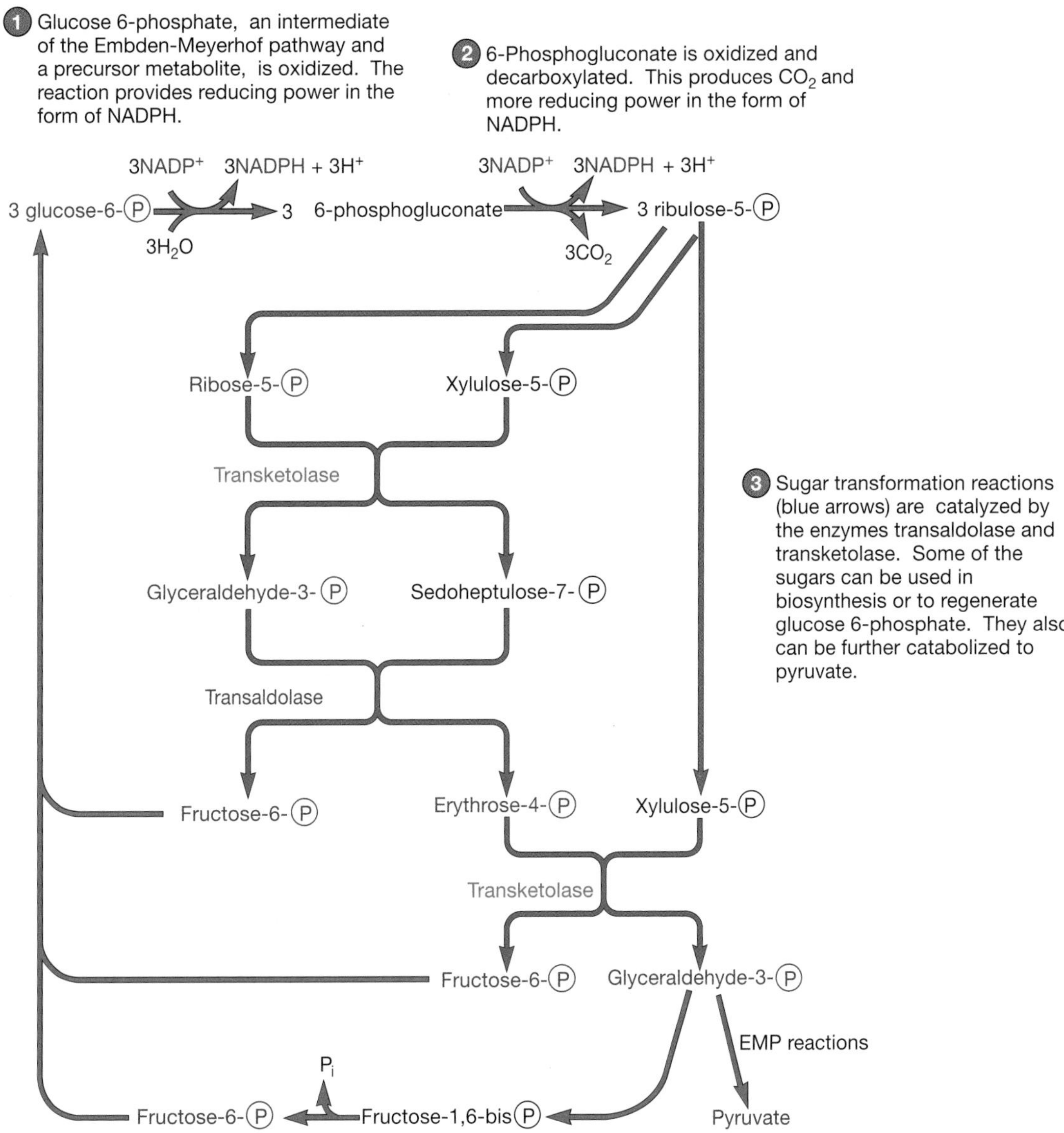

Figure 9.6 The Pentose Phosphate Pathway. The catabolism of three glucose 6-phosphate molecules to two fructose 6-phosphates, a glyceraldehyde 3-phosphate, and three CO_2 molecules is traced. Note that the pentose phosphate pathway generates several intermediates that are also intermediates of the Embden-Meyerhof pathway (EMP). These intermediates can be fed into the EMP with two results: (1) continued degradation to pyruvate or (2) regeneration of glucose 6-phosphate. The pentose phosphate pathway also plays a major role in producing reducing power (NADPH) and several precursor metabolites (shown in blue). The sugar transformations are indicated with blue arrows. These reactions are catalyzed by the enzymes transketolase and transaldolase and are shown in more detail in figure 9.7.

converted to a mixture of three- through seven-carbon sugar phosphates. Two enzymes play a central role in these transformations: (1) transketolase catalyzes the transfer of two-carbon ketol groups, and (2) transaldolase transfers a three-carbon group from sedoheptulose 7-phosphate to glyceraldehyde 3-phosphate (**figure 9.7**). The overall result is that three glucose 6-phosphates are converted to two fructose 6-phosphates, glyceraldehyde 3-phosphate, and three CO_2 molecules, as shown in this equation.

$$3 \text{ glucose 6-phosphate} + 6NADP^+ + 3H_2O \longrightarrow$$
$$2 \text{ fructose 6-phosphate} + \text{glyceraldehyde 3-phosphate} + 3CO_2 + 6NADPH + 6H^+$$

These intermediates are used in two ways. The fructose 6-phosphate can be changed back to glucose 6-phosphate while glyceraldehyde 3-phosphate is converted to pyruvate by enzymes of the

The transketolase reactions

Xylulose 5-phosphate + Ribose 5-phosphate → Sedoheptulose 7-phosphate + Glyceraldehyde 3-phosphate

1 Two 5-carbon molecules react (10 carbons total), producing a 7-carbon molecule and a 3-carbon molecule (10 carbons total).

Xylulose 5-phosphate + Erythrose 4-phosphate → Fructose 6-phosphate + Glyceraldehyde 3-phosphate

2 A 5-carbon molecule and a 4-carbon molecule react (9 carbons total), producing a 6-carbon molecule and a 3-carbon molecule (9 carbons total).

The transaldolase reaction

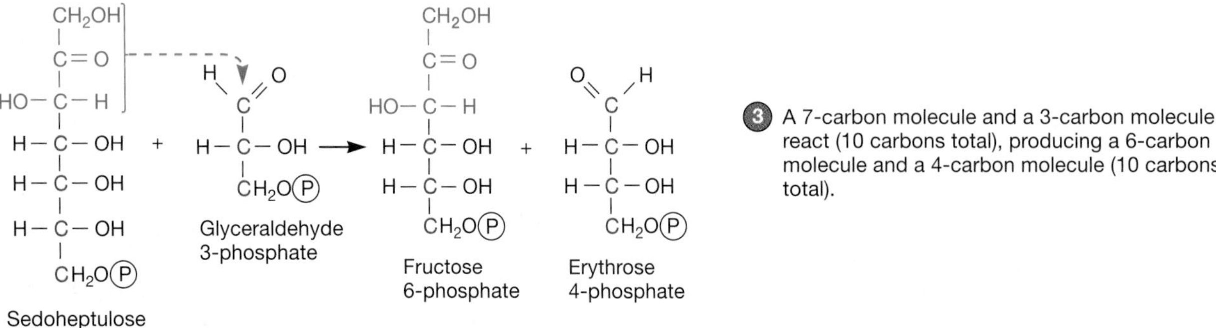

3 A 7-carbon molecule and a 3-carbon molecule react (10 carbons total), producing a 6-carbon molecule and a 4-carbon molecule (10 carbons total).

Figure 9.7 Transketolase and Transaldolase Reactions. Examples of the transketolase and transaldolase reactions of the pentose phosphate pathway. The groups transferred in these reactions are in green.

Embden-Meyerhof pathway. Alternatively two glyceraldehyde 3-phosphates may combine to form fructose 1,6-bisphosphate, which is eventually converted back into glucose 6-phosphate. This results in the complete degradation of glucose 6-phosphate to CO_2 and the production of a great deal of NADPH.

$$\text{Glucose 6-phosphate} + 12\text{NADP}^+ + 7\text{H}_2\text{O} \longrightarrow 6\text{CO}_2 + 12\text{NADPH} + 12\text{H}^+ + \text{P}_i$$

The pentose phosphate pathway is a good example of an amphibolic pathway as it has several catabolic and anabolic functions that are summarized as follows:

1. NADPH from the pentose phosphate pathway serves as a source of electrons for the reduction of molecules during biosynthesis.
2. The pathway produces two important precursor metabolites: erythrose 4-phosphate, which is used to synthesize aromatic amino acids and vitamin B_6 (pyridoxal) and ribose 5-phosphate, which is a major component of nucleic acids. Note that when a microorganism is growing on a pentose carbon source, the pathway can function biosynthetically to supply hexose sugars (e.g., glucose needed for peptidoglycan synthesis).
3. Intermediates in the pentose phosphate pathway may be used to produce ATP. Glyceraldehyde 3-phosphate from the pathway can enter the three-carbon phase of the Embden-Meyerhof pathway and be converted to pyruvate, as ATP is produced by substrate-level phosphorylation. Pyruvate may be oxidized in the tricarboxylic acid cycle to provide more energy.

Although the pentose phosphate pathway may be a source of energy in many microorganisms, it is more often of greater importance in biosynthesis. Several functions of the pentose phosphate pathway are mentioned again in chapter 10 when biosynthesis is considered more directly.

The Entner-Doudoroff Pathway

Although the Embden-Meyerhof pathway is the most common route for the conversion of hexoses to pyruvate, the **Entner-Doudoroff pathway** is used by soil microbes, such as *Pseudomonas, Rhizobium, Azotobacter,* and *Agrobacterium*, and a few other gram-negative bacteria. Very few gram-positive bacteria have this pathway, with the intestinal bacterium *Enterococcus faecalis* being a rare exception.

The Entner-Doudoroff pathway begins with the same reactions as the pentose phosphate pathway: the formation of glucose 6-phosphate, which is then converted to 6-phosphogluconate (**figure 9.8** and *appendix II*). Instead of being further oxidized, 6-phosphogluconate is dehydrated to form 2-keto-3-deoxy-6-phosphogluconate or KDPG, the key intermediate in this pathway. KDPG is then cleaved by KDPG aldolase to pyruvate and glyceraldehyde 3-phosphate. The glyceraldehyde 3-phosphate is converted to pyruvate in the Embden-Meyerhof pathway. If the Entner-Doudoroff pathway degrades glucose to pyruvate in this way, it yields one ATP, one NADPH, and one NADH per glucose metabolized.

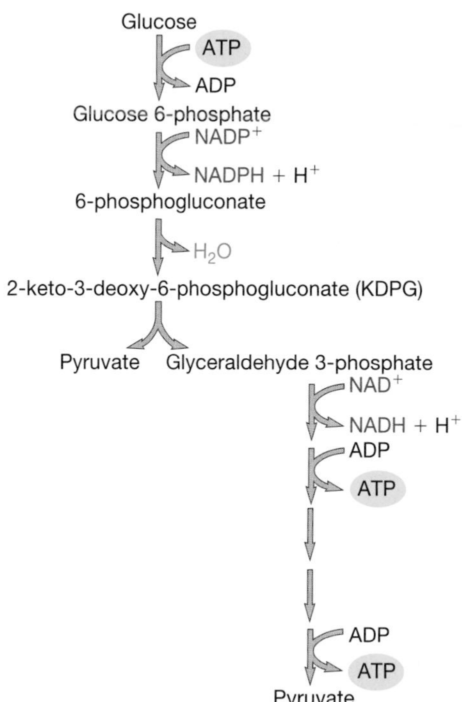

Figure 9.8 The Entner-Doudoroff Pathway. The sequence leading from glyceraldehyde 3-phosphate to pyruvate is catalyzed by enzymes common to the Embden-Meyerhof pathway.

1. Summarize the major features of the Embden-Meyerhof, pentose phosphate, and Entner-Doudoroff pathways. Include the starting points, the products of the pathways, any critical or unique enzymes, the ATP yields, and the metabolic roles each pathway has.
2. What is substrate-level phosphorylation?

9.4 The Tricarboxylic Acid Cycle

In the glycolytic pathways, the energy captured by the oxidation of glucose to pyruvate is limited to no more than two ATP generated by substrate-level phosphorylation. During aerobic respiration, the catabolic process continues by oxidizing pyruvate to three CO_2. The first step of this process employs a multienzyme system called the pyruvate dehydrogenase complex. It oxidizes and cleaves pyruvate to form one CO_2 and the two-carbon molecule **acetyl-coenzyme A (acetyl-CoA)** (**figure 9.9**). Acetyl-CoA is energy-rich because a high-energy thiol links acetic acid to coenzyme A. Note that during stage three of aerobic respiration (figure 9.3), carbohydrates as well as fatty acids and amino acids can be converted to acetyl-CoA.

Acetyl-CoA then enters the **tricarboxylic acid (TCA) cycle,** which is also called the **citric acid cycle** or the **Krebs cycle** (figure 9.9 and *appendix II*). In the first reaction acetyl-CoA is condensed with (i.e., added to) a four-carbon intermediate, oxaloacetate, to form citrate, a molecule with six carbons. Cit-

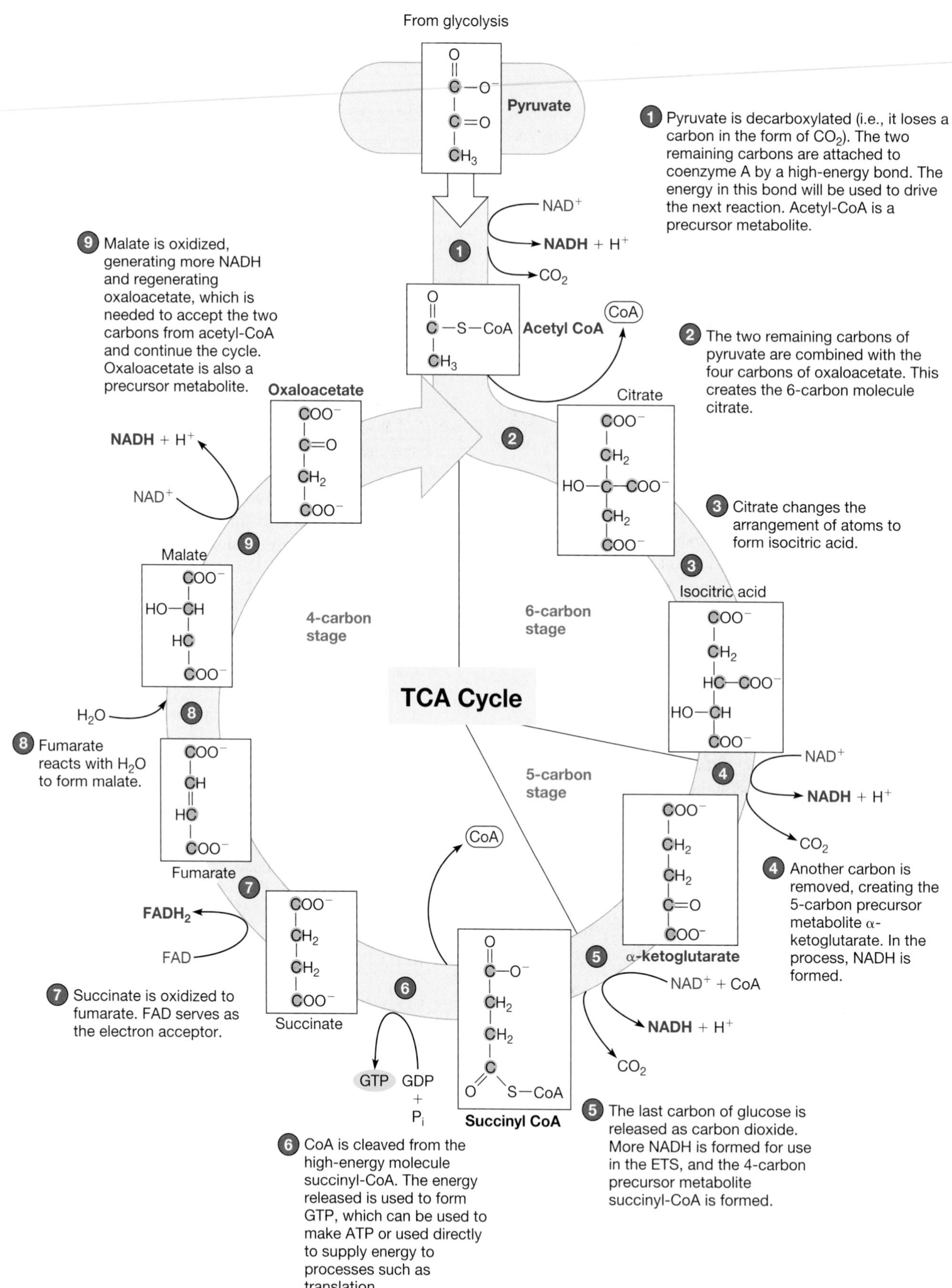

Figure 9.9 The Tricarboxylic Acid Cycle. The TCA cycle is linked to glycolysis by a connecting reaction catalyzed by the pyruvate dehydrogenase complex. The reaction decarboxylates pyruvate (removes a carboxyl group as CO_2) and generates acetyl-CoA. The cycle may be divided into three stages based on the size of its intermediates. The three stages are separated from one another by two decarboxylation reactions. Precursor metabolites, carbon skeletons used in biosynthesis, are shown in blue. NADH and $FADH_2$ are shown in purple; all can transfer electrons to the electron transport chain (ETC).

rate (a tertiary alcohol) is rearranged to give isocitrate, a more readily oxidized secondary alcohol. Isocitrate is subsequently oxidized and decarboxylated twice to yield α-ketoglutarate (five carbons), and then succinyl-CoA (four carbons), a molecule with a high-energy bond. At this point two NADH molecules have been formed and two carbons lost from the cycle as CO_2. The cycle continues when succinyl-CoA is converted to succinate. This involves breaking the high-energy bond in succinyl-CoA and using the energy released to form one GTP by substrate-level phosphorylation. GTP is also a high-energy molecule, and it is functionally equivalent to ATP. Two oxidation steps follow, yielding one $FADH_2$ and one NADH. The last oxidation step regenerates oxaloacetate, and as long as there is a supply of acetyl-CoA the cycle can repeat itself. Inspection of figure 9.9 shows that the TCA cycle generates two CO_2 molecules, three NADH molecules, one $FADH_2$, and one GTP for each acetyl-CoA molecule oxidized.

TCA cycle enzymes are widely distributed among microorganisms. In procaryotes, they are located in the cytoplasmic matrix. In eucaryotes, they are found in the mitochondrial matrix. The complete cycle appears to be functional in many aerobic bacteria, free-living protists, and fungi. This is not surprising because the cycle is such an important source of energy. Even those microorganisms that lack the complete TCA cycle usually have most of the cycle enzymes, because the TCA cycle is also a key source of carbon skeletons for use in biosynthesis. Synthesis of amino acids: Anaplerotic reactions and amino acid biosynthesis (section 10.5)

1. Give the substrate and products of the tricarboxylic acid cycle. Describe its organization in general terms. What are its two major functions?
2. What chemical intermediate links pyruvate to the TCA cycle?
3. How many times must the TCA cycle be performed to completely oxidize one molecule of glucose to six molecules of CO_2? Why?
4. In what eucaryotic organelle is the TCA cycle found? Where is the cycle located in procaryotes?
5. Why might it be desirable for a microbe with the Embden-Meyerhof pathway and the TCA cycle also to have the pentose phosphate pathway?
6. Why is GTP functionally equivalent to ATP?

9.5 Electron Transport and Oxidative Phosphorylation

During the oxidation of glucose to six CO_2 molecules by glycolysis and the TCA cycle, four ATP molecules are generated by substrate-level phosphorylation. Thus at this point, the work done by the cell has yielded relatively little ATP. However, in oxidizing glucose, the cell has also generated numerous molecules of NADH and $FADH_2$. Both of these molecules have a relatively negative E'_0 and can be used to conserve energy (*see table 8.1*). In fact, most of the ATP generated during aerobic respiration comes from the oxidation of these electron carriers in the electron transport chain. The mitochondrial electron transport chain will be examined first because it has been so well studied. Then we will turn to bacterial chains, and finish with a discussion of ATP synthesis.

The Electron Transport Chain

The mitochondrial **electron transport chain** is composed of a series of electron carriers that operate together to transfer electrons from donors, like NADH and $FADH_2$, to acceptors, such as O_2 (**figure 9.10**). The electrons flow from carriers with more negative reduction potentials to those with more positive potentials and eventually combine with O_2 and H^+ to form water. This pattern of electron flow is exactly the same as seen in the electron tower that is described in chapter 8 (*see figure 8.8*). The electrons move down this potential gradient much like water flowing down a series of rapids. The difference in reduction potentials between O_2 and NADH is large, about 1.14 volts, which makes possible the release of a great deal of energy. The differences in reduction potential at several points in the chain are large enough to provide sufficient energy for ATP production, much like the energy from waterfalls can be harnessed by waterwheels and used to generate electricity. Thus the electron transport chain breaks up the large overall energy release into small steps. As will be seen shortly, electron transport at these points generates proton and electrical

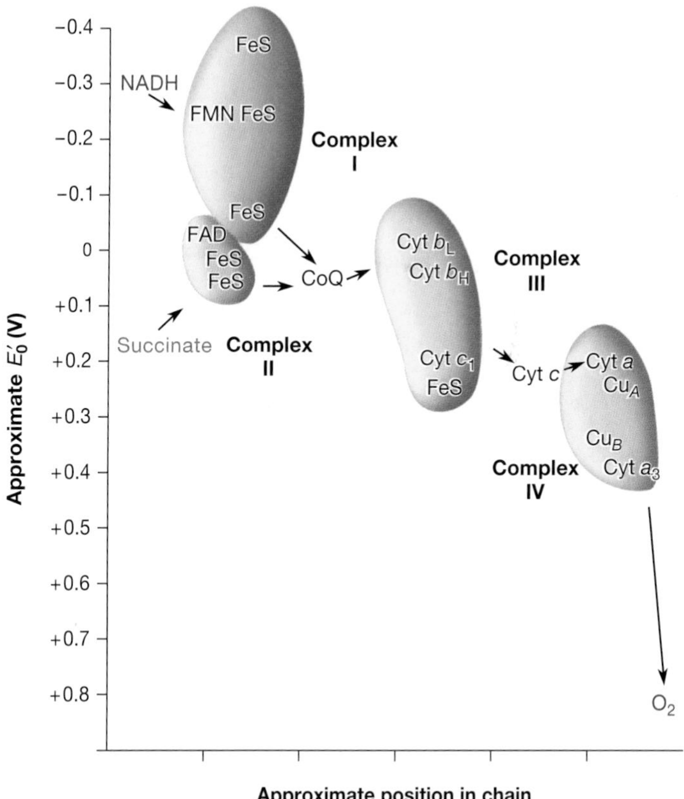

Figure 9.10 The Mitochondrial Electron Transport Chain. Many of the more important carriers are arranged at approximately the correct reduction potential and sequence. In the eucaryotic mitochondrion, they are organized into four complexes that are linked by coenzyme Q (CoQ) and cytochrome *c* (Cyt *c*). Electrons flow from NADH and succinate down the reduction potential gradient to oxygen. See text for details.

gradients. These gradients can drive ATP synthesis and perform other work.

In eucaryotes, the electron transport chain carriers reside within the inner membrane of the mitochondrion. In procaryotes, they are located within the plasma membrane. The mitochondrial system is arranged into four complexes of carriers, each capable of transporting electrons part of the way to O_2 (**figure 9.11**). Coenzyme Q and cytochrome *c* connect the complexes with each other. Although some bacterial chains resemble the mitochondrial chain, they are frequently very different. As already noted, bacterial chains are located within the plasma membrane. They also can be composed of different electron carriers (e.g., their cytochromes) and may be extensively branched. Electrons often can enter the chain at several points and leave through several terminal oxidases. Bacterial chains also may be shorter, resulting in the release of less energy. Although procaryotic and eucaryotic electron transport chains differ in details of construction, they operate using the same fundamental principles.

The electron transport chain of *E. coli* will serve as an example of these differences. A simplified view of the *E. coli* transport chain is shown in **figure 9.12.** The NADH generated by the oxidation of organic substrates (during glycolysis and the TCA cycle) is donated to the electron transport chain, where it is oxidized to NAD^+ by the membrane-bound NADH dehydrogenase. The electrons are then transferred to carriers with progressively more positive reduction potentials. As electrons move through the carriers, protons are moved across the plasma membrane to the periplasmic space (i.e., outside the cell) rather than to an intermembrane space as seen in the mitochondria (compare figures 9.11 and 9.12). Another significant difference between the *E. coli* chain and the mitochondrial chain is that the bacterial electron transport chain contains a different array of cytochromes. Furthermore, *E. coli* has evolved two branches of the electron transport chain that operate under different aeration conditions. When oxygen is readily available, the cytochrome *bo* branch is used. When oxygen levels are reduced, the cytochrome *bd* branch is used because it has a higher affinity for oxygen. However, it is less efficient than the *bo* branch because the *bd* branch moves fewer protons into the periplasmic space (figure 9.12).

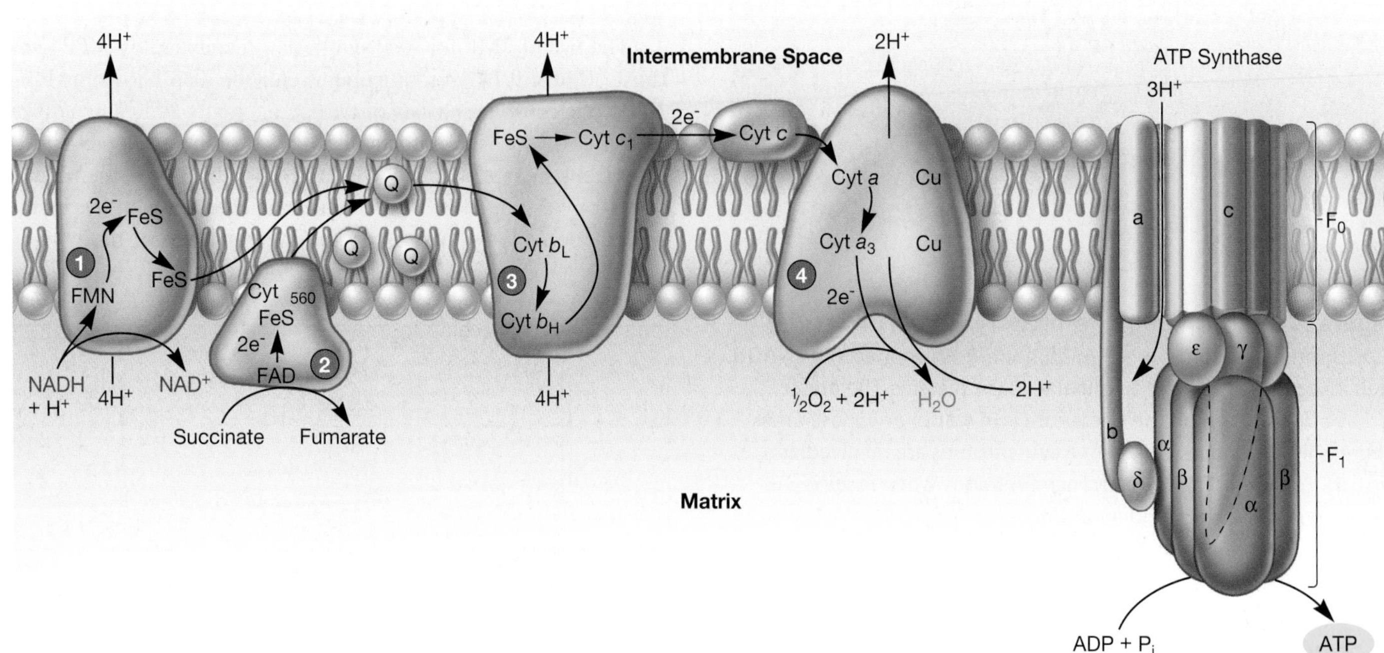

Figure 9.11 The Chemiosmotic Hypothesis Applied to Mitochondria. In this scheme the carriers are organized asymmetrically within the inner membrane so that protons are transported across as electrons move along the chain. Proton release into the intermembrane space occurs when electrons are transferred from carriers, such as FMN and coenzyme Q (Q), that carry both electrons and protons to components like nonheme iron proteins (FeS proteins) and cytochromes (Cyt) that transport only electrons. Complex IV pumps protons across the membrane as electrons pass from cytochrome *a* to oxygen. Coenzyme Q transports electrons from complexes I and II to complex III. Cytochrome *c* moves electrons between complexes III and IV. The number of protons moved across the membrane at each site per pair of electrons transported is still somewhat uncertain; the current consensus is that at least 10 protons must move outward during NADH oxidation. One molecule of ATP is synthesized and released from the enzyme ATP synthase for every three protons that cross the membrane by passing through it.

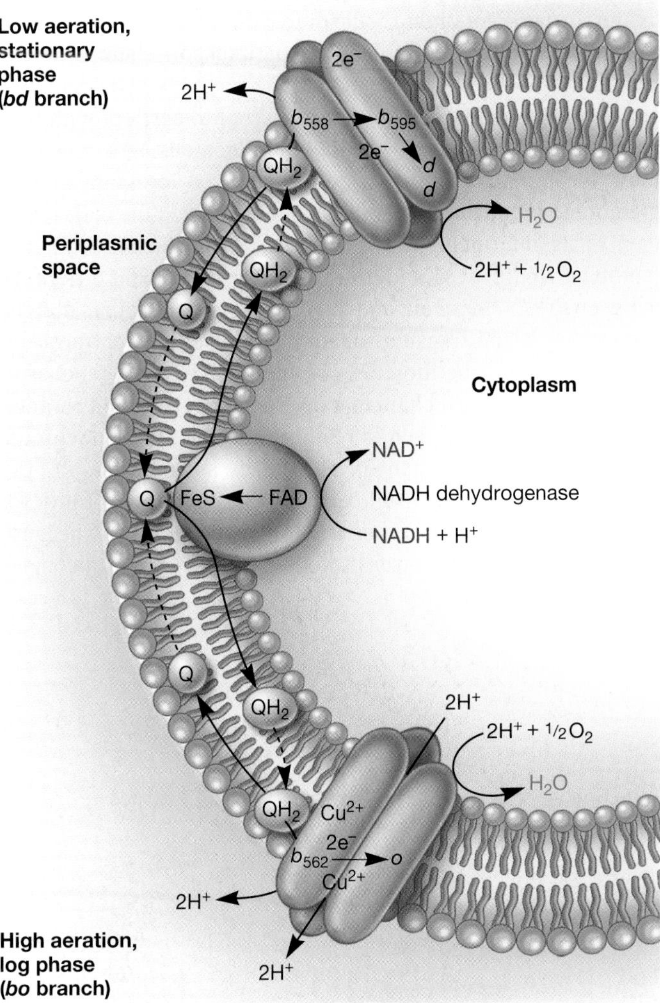

Figure 9.12 The Aerobic Respiratory System of *E. coli.* NADH is the electron source. Ubiquinone-8 (Q) connects the NADH dehydrogenase with two terminal oxidase systems. The upper branch operates when the bacterium is in stationary phase and there is little oxygen. At least five cytochromes are involved: b_{558}, b_{595}, b_{562}, *d*, and *o*. The lower branch functions when *E. coli* is growing rapidly with good aeration.

Oxidative Phosphorylation

Oxidative phosphorylation is the process by which ATP is synthesized as the result of electron transport driven by the oxidation of a chemical energy source. The mechanism by which oxidative phosphorylation takes place has been studied intensively for years. The most widely accepted hypothesis is the chemiosmotic hypothesis, which was formulated by British biochemist Peter Mitchell. According to the **chemiosmotic hypothesis,** the electron transport chain is organized so that protons move outward from the mitochondrial matrix as electrons are transported down the chain (figure 9.11).

The movement of protons across the membrane is not completely understood. However, in some cases, the protons are actively pumped across the membrane (e.g., by complex IV of the mitochondrial chain; figure 9.11). In other cases, translocation of protons results from the juxtaposition of carriers that accept both electrons and protons with carriers that accept only electrons. For instance, coenzyme Q carries two electrons and two protons to cytochrome *b* in complex III of the mitochondrial chain. Cytochrome *b* accepts only one electron at a time, and will not accept protons. Thus for each electron transferred by coenzyme Q, one proton is released to the intermembrane space.

The result of proton expulsion during electron transport is the formation of a concentration gradient of protons (ΔpH; chemical potential energy) and a charge gradient (Δψ; electrical potential energy). Thus the mitochondrial matrix is more alkaline and more negative than the intermembrane space. Likewise with procaryotes, the cytoplasm is more alkaline and more negative than the periplasmic space. The combined chemical and electrical potential differences make up the **proton motive force (PMF).** The PMF is used to perform work when protons flow back across the membrane, down the concentration and charge gradients, and into the mitochondrial matrix (or procaryotic cytoplasm). This flow is exergonic and is often used to phosphorylate ADP to ATP. The PMF is also used to transport molecules into the cell directly (i.e., without the hydrolysis of ATP) and to rotate the flagellar motor. Thus the PMF plays a central role in procaryotic physiology (**figure 9.13**). Uptake of nutrients by the cell (section 5.6); Components external to the cell wall: Flagella and motility (section 3.9)

The use of PMF for ATP synthesis is catalyzed by **ATP synthase** (**figure 9.14**), a multisubunit enzyme also known as F_1F_0 ATPase because it consists of two components and can catalyze ATP hydrolysis. The mitochondrial F_1 component appears as a spherical structure attached to the mitochondrial inner membrane surface by a stalk. The F_0 component is embedded in the membrane. The ATP synthase is on the inner surface of the plasma

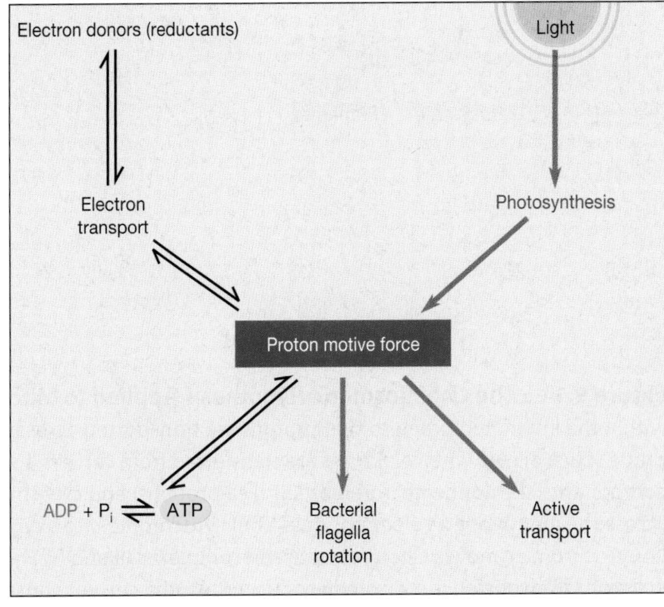

Figure 9.13 The Central Role of Proton Motive Force. It should be noted that active transport is not always driven by PMF.

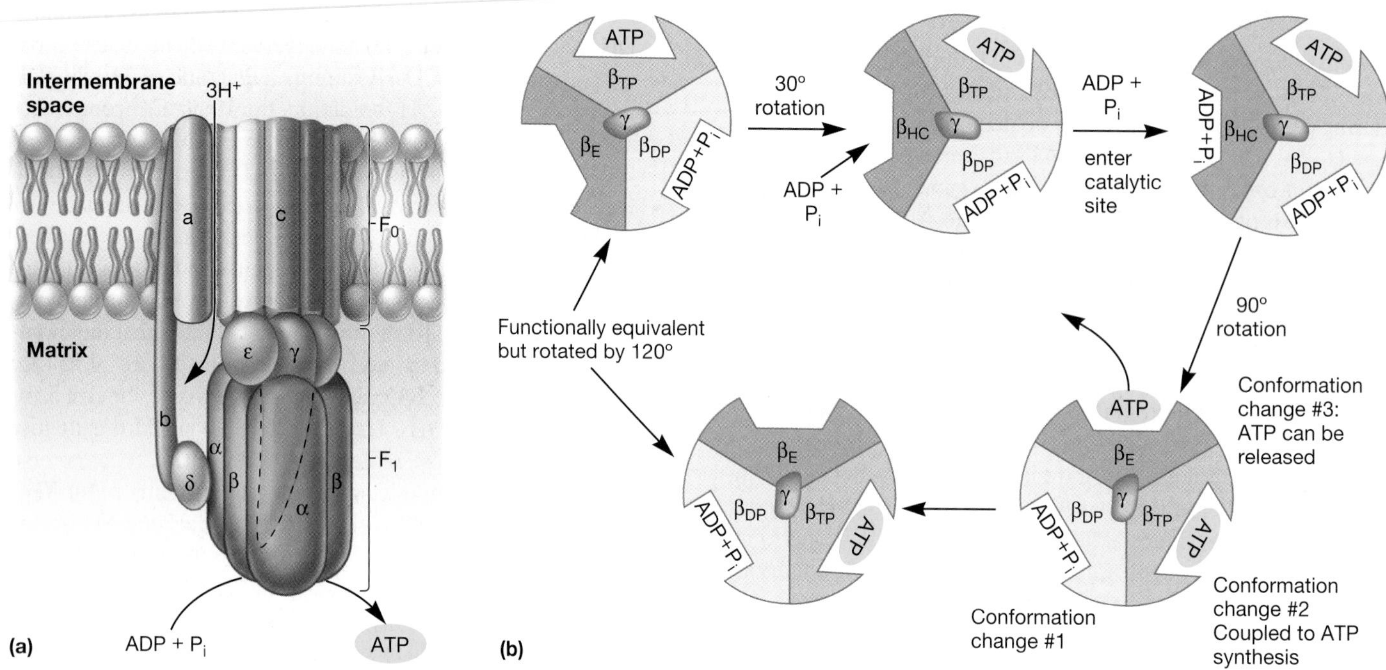

Figure 9.14 ATP Synthase Structure and Function. **(a)** The major structural features of ATP synthase deduced from X-ray crystallography and other studies. F_1 is a spherical structure composed largely of alternating α and β subunits; the three active sites are on the β subunits. The γ subunit extends upward through the center of the sphere and can rotate. The stalk (γ and ϵ subunits) connects the sphere to F_0, the membrane embedded complex that serves as a proton channel. F_0 contains one a subunit, two b subunits, and 9–12 c subunits. The stator arm is composed of subunit a, two b subunits, and the δ subunit; it is embedded in the membrane and attached to F_1. A ring of c subunits in F_0 is connected to the stalk and may act as a rotor and move past the a subunit of the stator. As the c subunit ring turns, it rotates the shaft ($\gamma\epsilon$ subunits). **(b)** The binding change mechanism is a widely accepted model of ATP synthesis. This simplified drawing of the model shows the three catalytic β subunits and the γ subunit, which is located at the center of the F_1 complex. As the γ subunit rotates, it causes conformational changes in each subunit. The β_E (*e*mpty) conformation is an open conformation, which does not bind nucleotides. When the γ subunit rotates 30°, β_E is converted to the β_{HC} (*h*alf *c*losed) conformation. P_i and ADP can enter the catalytic site when it is in this conformation. The subsequent 90° rotation by the γ subunit is critical because it brings about three significant conformational changes: (1) β_{HC} to β_{DP} (A*D*P bound), (2) β_{DP} to β_{TP} (A*T*P bound), and (3) β_{TP} to β_E. Change from β_{DP} to β_{TP} is accompanied by the formation of ATP; change from β_{TP} to β_E allows for release of ATP from ATP synthase.

membrane in procaryotes. F_0 participates in proton movement across the membrane. F_1 is a large complex in which three α subunits alternate with three β subunits. The catalytic sites for ATP synthesis are located on the β subunits. At the center of F_1 is the γ subunit. The γ subunit extends through F_1 and interacts with F_0.

It is now known that ATP synthase functions like a rotary engine. It is thought the flow of protons down the proton gradient through the F_0 subunit causes F_0 and the γ subunit to rotate. As the γ subunit rotates rapidly within the F_1 (much like a car's crankshaft), the rotation causes conformation changes in the β subunits (figure 9.14*b*). One conformation change (β_E to β_{HC}) allows entry of ADP and P_i into the catalytic site. Another conformation change (β_{HC} to β_{DP}) loosely binds ADP and P_i in the catalytic site. ATP is synthesized when the β_{DP} conformation is changed to the β_{TP} conformation, and ATP is released when β_{TP} changes to the β_E conformation, to start the synthesis cycle anew.

Much of the evidence supporting the chemiosmotic hypothesis comes from studies using chemicals that inhibit the aerobic synthesis of ATP. These chemicals can even kill cells at sufficiently high concentrations. The inhibitors generally fall into two categories. Some directly block the transport of electrons. The antibiotic piericidin competes with coenzyme Q; the antibiotic antimycin A blocks electron transport between cytochromes *b* and *c;* and both cyanide and azide stop the transfer of electrons between cytochrome *a* and O_2 because they are structural analogs of O_2. Another group of inhibitors known as **uncouplers** stops ATP synthesis without inhibiting electron transport itself. Indeed, they may even enhance the rate of electron flow. Normally electron transport is tightly coupled with oxidative phosphorylation so that the rate of ATP synthesis controls the rate of electron transport. The more rapidly ATP is synthesized during oxidative phosphorylation, the faster the electron transport chain operates to supply the required energy. Uncouplers disconnect oxidative phosphorylation from electron transport; therefore the energy released by the chain is given off as heat rather than as ATP. Many uncouplers like dinitrophenol and valinomycin allow hydrogen ions, potassium ions, and other ions to cross the membrane without activating ATP synthase. In this way they destroy the pH and ion gradients. Valinomycin also may bind directly to ATP synthase and inhibit its activity.

ATP Yield During Aerobic Respiration

It is possible to estimate the number of ATP molecules synthesized per NADH or $FADH_2$ oxidized by the electron transport chain. During aerobic respiration, a pair of electrons from NADH is donated to the electron transport chain and ultimately used to reduce an atom of oxygen to H_2O. This releases enough energy to drive the synthesis of three ATP. This is referred to as the phosphorus to oxygen (P/O) ratio because it measures the number of ATP (phosphorus) generated per oxygen (O) reduced. Because $FADH_2$ has a more positive reduction potential than NADH (*see figure 8.8*), electrons arising from its oxidation flow down a shorter chain, releasing less energy. Thus while the P/O ratio for NADH is 3, only two ATP can be made from the oxidation of a single $FADH_2$.

We can thus calculate the maximum ATP yield (**figure 9.15**) of aerobic respiration. Substrate-level phosphorylation during glycolysis yields at most two ATP molecules per glucose converted to pyruvate (figures 9.5, 9.6, and 9.8). Two additional GTP (ATP equivalents) are generated by substrate-level phosphorylation during the two turns of the TCA cycle needed to oxidize two acetyl-CoA molecules (figure 9.9). However, most of the ATP made during aerobic respiration is generated by oxidative phosphorylation. Up to 10 NADH (2 from glycolysis, 2 from pyruvate conversion to acetyl-CoA, and 6 from the TCA cycle) and 2 $FADH_2$ (from the TCA cycle) are generated when glucose is oxidized completely to 6 CO_2. Assuming a P/O ratio of 3 for NADH oxidation and 2 for $FADH_2$ oxidation, the 10 NADH could theoretically drive the synthesis of 30 ATP, while oxidation of the 2 $FADH_2$ molecules would add another 4 ATP for a maximum of 34 ATP generated via oxidative phosphorylation. Because substrate-level phosphorylation contributes only four ATP per glucose molecule oxidized, oxidative phosphorylation accounts for at least eight times more. The maximum total yield of ATP during aerobic respiration is 38 ATPs. It must be remembered that the calculations just summarized and presented in figure 9.15 are theoretical. In fact, the P/O ratios are more likely about 2.5 for NADH and 1.5 for $FADH_2$. Thus the total ATP yield from glucose may be closer to 30 ATPs rather than 38.

Because procaryotic electron transport systems often have lower P/O ratios than eucaryotic systems, procaryotic ATP yields can be less. For example, *E. coli,* with its truncated electron transport chains, has a P/O ratio around 1.3 when using the cytochrome *bo* path at high oxygen levels and only a ratio of about 0.67 when employing the cytochrome *bd* branch (figure 9.12) at low oxygen concentrations. In this case ATP production varies

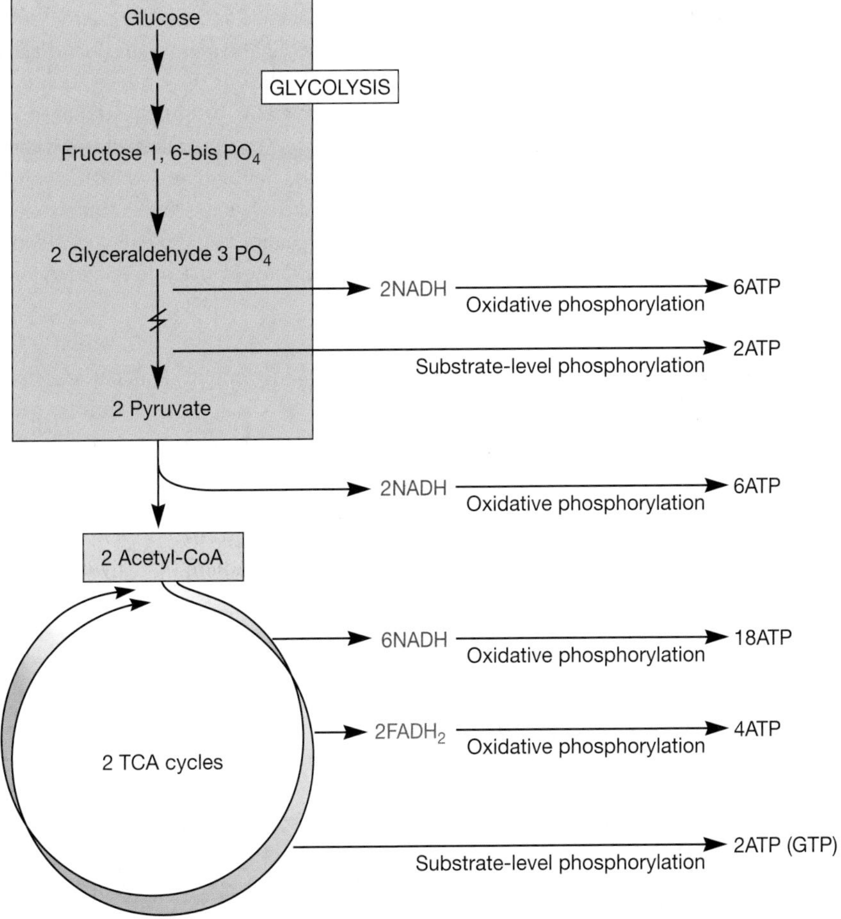

Figure 9.15 Maximum Theoretic ATP yield from Aerobic Respiration. To attain the theoretic maximum yield of ATP, one must assume a P/O ratio of 3 for the oxidation of NADH and 2 for $FADH_2$. The actual yield is probably significantly less and varies between eucaryotes and procaryotes and among procaryotic species.

with environmental conditions. Perhaps because *E. coli* normally grows in habitats such as the intestinal tract that are very rich in nutrients, it does not have to be particularly efficient in ATP synthesis. Presumably the electron transport chain functions when *E. coli* is in an oxic freshwater environment between hosts.

1. Briefly describe the structure of the electron transport chain and its role in ATP formation. How do mitochondrial and bacterial chains differ?
2. Describe the current model of oxidative phosphorylation. Briefly describe the structure of ATP synthase and explain how it is thought to function. What is an uncoupler?
3. How do substrate-level phosphorylation and oxidative phosphorylation differ?
4. Calculate the ATP yield when glucose is catabolized completely to six CO_2 by a eucaryotic microbe. How does this value compare to the ATP yield observed for a bacterium? Suppose a bacterium used the Entner-Doudoroff pathway to degrade glucose to pyruvate. How would this impact the total ATP yield? Explain your reasoning.

9.6 Anaerobic Respiration

As we have seen, during aerobic respiration sugars and other organic molecules are oxidized and their electrons transferred to NAD^+ and FAD to generate NADH and $FADH_2$, respectively. These electron carriers then donate the electrons to an electron transport chain that uses O_2 as the terminal electron acceptor. However, it is also possible for other terminal electron acceptors to be used for electron transport. **Anaerobic respiration,** a process whereby an exogenous terminal electron acceptor other than O_2 is used for electron transport, is carried out by many bacteria and archaea. The most common terminal electron acceptors used during anaerobic respiration are nitrate, sulfate, and CO_2, but metals and a few organic molecules can also be reduced (**table 9.1**).

Although some bacteria and archaea grow using only anaerobic respiration, many can perform both aerobic and anaerobic respiration, depending on the availability of oxygen. One example is *Paracoccus denitrificans,* a gram-negative, facultative anaerobic soil bacterium that is extremely versatile metabolically. It can degrade a wide variety of organic compounds and can even grow chemolithotrophically. Under anoxic conditions, *P. denitrificans* uses NO_3^- as its electron acceptor. As shown in **figure 9.16,** two different electron transport chains are used by this bacterium, one for aerobic respiration and the second for anaerobic respiration. Notice that during chemoorgantrophic growth, the source of electrons in both chains is NADH. The aerobic chain has four complexes that correspond to the mitochondrial chain (figure 9.16*a*). When *P. denitrificans* grows without oxygen, using NO_3^- as the terminal electron acceptor, the electron transport chain is more complex (figure 9.16*b*). The chain is highly branched and the cytochrome *aa* complex is replaced. Electrons are passed from coenzyme Q to cytochrome *b* for the reduction of nitrate to nitrite (catalyzed by nitrate reductase). Electrons then flow through cytochrome *c* for the sequential oxidation of nitrite to gaseous dinitrogen (N_2). Not as many protons are pumped across the membrane during anaerobic growth, but nonetheless a PMF is established.

Table 9.1 Some Electron Acceptors Used in Respiration

	Electron Acceptor	Reduced Products	Examples of Microorganisms
Aerobic	O_2	H_2O	All aerobic bacteria, fungi, and protists
Anaerobic	NO_3^-	NO_2^-	Enteric bacteria
	NO_3^-	NO_2^-, N_2O, N_2	*Pseudomonas, Bacillus,* and *Paracoccus*
	SO_4^{2-}	H_2S	*Desulfovibrio* and *Desulfotomaculum*
	CO_2	CH_4	All methanogens and acetogens
	S^0	H_2S	*Desulfuromonas* and *Thermoproteus*
	Fe^{3+}	Fe^{2+}	*Pseudomonas, Bacillus,* and *Geobacter*
	$HAsO_4^{2-}$	$HAsO_2$	*Bacillus, Desulfotomaculum, Sulfurospirillum*
	SeO_4^{2-}	Se, $HSeO_3^-$	*Aeromonas, Bacillus, Thauera*
	Fumarate	Succinate	*Wolinella*

The anaerobic reduction of nitrate makes it unavailable to the cell for assimilation or uptake. Therefore this process is called **dissimilatory nitrate reduction.** Nitrate reductase replaces cytochrome oxidase to catalyze the reaction:

$$NO_3^- + 2e^- + 2\,H^+ \rightarrow NO_2^- + H_2O$$

However, reduction of nitrate to nitrite is not a particularly efficient way of making ATP because a large amount of nitrate is required for growth (a nitrate molecule will accept only two electrons). Furthermore, nitrite is quite toxic. Bacteria such as *P. denitrificans* avoid the toxic effects of nitrite by reducing it to nitrogen gas, a process known as **denitrification.** By donating five electrons to a nitrate molecule, NO_3^- is converted into a nontoxic product.

$$2NO_3^- + 10e^- + 12H^+ \rightarrow N_2 + 6H_2O$$

As illustrated in figure 9.16, denitrification is a multistep process with four enzymes participating: nitrate reductase, nitrite reductase, nitric oxide reductase, and nitrous oxide reductase.

$$NO_3^- \rightarrow NO_2^- \rightarrow NO \rightarrow N_2O \rightarrow N_2$$

Two types of bacterial nitrite reductases catalyze the formation of NO in bacteria. One contains cytochromes *c* and d_1 (e.g.,

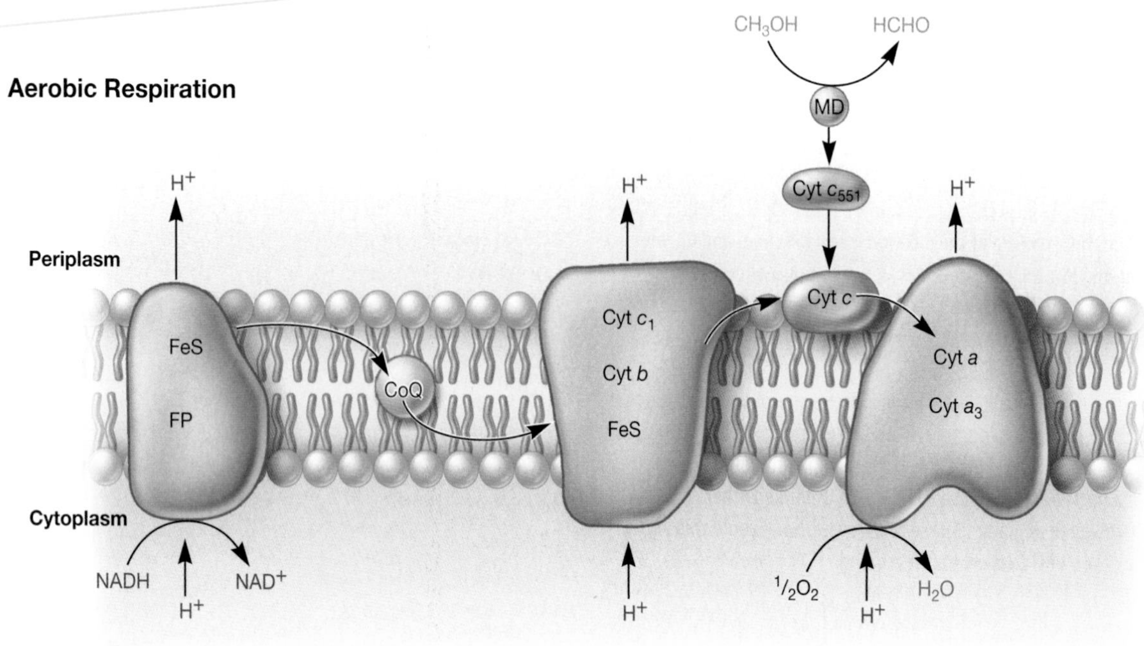

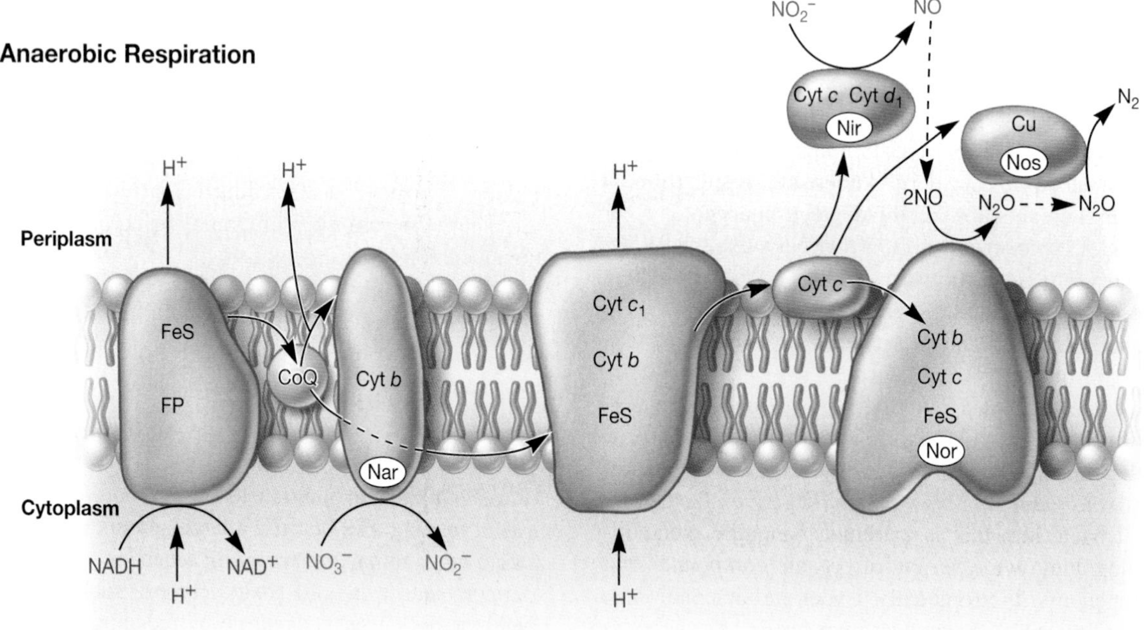

Figure 9.16 ***Paracoccus denitrificans*** **Electron Transport Chains.** **(a)** The aerobic transport chain resembles a mitochondrial electron transport chain and uses oxygen as its acceptor. Methanol and methylamine can contribute electrons at the cytochrome *c* level. **(b)** The highly branched anaerobic chain is made of both membrane and periplasmic proteins. Nitrate is reduced to diatomic nitrogen by the collective action of four different reductases that receive electrons from CoQ and cytochrome *c*. Locations of proton movement are shown, but the number of protons involved has not been indicated. Abbreviations used: flavoprotein (FP), methanol dehydrogenase (MD), nitrate reductase (Nar), nitrite reductase (Nir), nitric oxide reductase (Nor), and nitrous oxide reductase (Nos).

Paracoccus and *Pseudomonas aeruginosa*), and the other is a copper-containing protein (e.g., *Alcaligenes*). Nitrite reductase seems to be periplasmic in gram-negative bacteria. Nitric oxide reductase catalyzes the formation of nitrous oxide from NO and is a membrane-bound cytochrome *bc* complex. In *P. denitrificans,* the nitrate reductase and nitric oxide reductase are membrane-bound, whereas nitrite reductase and nitrous oxide reductase are periplasmic (figure 9.16*b*).

In addition to *P. denitrificans,* some members of the genera *Pseudomonas* and *Bacillus* carry out denitrification. All three genera use denitrification as an alternative to aerobic respiration and may be considered facultative anaerobes. Indeed, if O_2 is present, these bacteria use aerobic respiration, which is far more efficient in capturing energy. In fact, the synthesis of nitrate reductase is repressed by O_2. Denitrification in anoxic soil results in the loss of soil nitrogen and adversely affects soil fertility. Biogeochemical cycling: Nitrogen cycle (section 27.2)

Not all microbes employ anaerobic respiration facultatively. Some are obligate anaerobes that can carry out only anaerobic respiration. The methanogens are an example. These archaea use CO_2 or carbonate as a terminal electron acceptor. They are called methanogens because the electron acceptor is reduced to methane. Bacteria such as *Desulfovibrio* are another example. They donate eight electrons to sulfate, reducing it to sulfide (S_2^- or H_2S).

$$SO_4^{2-} + 8e^- + 8H^+ \rightarrow S^{2-} + 4H_2O$$

It should be noted that both methanogens and *Desulfovibrio* are able to function as chemolithotrophs, using H_2 as an energy source (section 9.11).

As we saw for denitrification, anaerobic respiration using sulfate or CO_2 as the terminal electron acceptors is not as efficient in ATP synthesis as is aerobic respiration. Reduction in ATP yield arises from the fact that these alternate electron acceptors have less positive reduction potentials than O_2 (*see table 8.1*). The difference in standard reduction potential between a donor like NADH and nitrate is smaller than the difference between NADH and O_2. Because energy yield is directly related to the magnitude of the reduction potential difference, less energy is available to make ATP in anaerobic respiration. Nevertheless, anaerobic respiration is useful because it allows ATP synthesis by electron transport and oxidative phosphorylation in the absence of O_2. Anaerobic respiration is prevalent in oxygen-depleted soils and sediments.

The ability of microbes to use a variety of electron acceptors has ecological consequences. Often one sees a succession of microorganisms in an environment when several electron acceptors are present. For example, if O_2, nitrate, manganese ion, ferric ion, sulfate, and CO_2 are available in a particular environment, a predictable sequence of electron acceptor use takes place when an oxidizable substrate is available to the microbial population. Oxygen is employed as an electron acceptor first because it inhibits nitrate use by microorganisms capable of respiration with either O_2 or nitrate. While O_2 is available, sulfate reducers and methanogens are inhibited because these groups are obligate anaerobes.

Once the O_2 and nitrate are exhausted and fermentation products (section 9.7), including hydrogen, have accumulated, competition for use of other electron acceptors begins. Manganese and iron are used first, followed by competition between sulfate reducers and methanogens. This competition is influenced by the greater energy yield obtained with sulfate as an electron acceptor. Differences in enzymatic affinity for hydrogen, an important energy and electron source used by both groups, also are important. The sulfate reducer *Desulfovibrio* grows rapidly and uses the available hydrogen at a faster rate than *Methanobacterium.* When the sulfate is exhausted, *Desulfovibrio* no longer oxidizes hydrogen, and the hydrogen concentration rises. The methanogens finally dominate the habitat and reduce CO_2 to methane. The subsurface biosphere (section 29.7)

1. Describe the process of anaerobic respiration. Is as much ATP produced in anaerobic respiration as in aerobic respiration? Why or why not?
2. What is denitrification? Why do farmers dislike this process?
3. *E. coli* can use O_2, fumarate^{2-}, or nitrate as a terminal electron acceptor under different conditions. What is the order of energy yield from highest to lowest for these electron acceptors? Explain your answer in thermodynamic terms.

9.7 Fermentations

Despite the tremendous ATP yield obtained by oxidative phosphorylation, some chemoorganotrophic microbes do not respire because either they lack electron transport chains or they repress the synthesis of electron transport chain components under anoxic conditions, making anaerobic respiration impossible. Yet NADH produced by the Embden-Meyerhof pathway reactions during glycolysis (figure 9.5) must still be oxidized back to NAD^+. If NAD^+ is not regenerated, the oxidation of glyceraldehyde 3-phosphate will cease and glycolysis will stop. Many microorganisms solve this problem by slowing or stopping pyruvate dehydrogenase activity and using pyruvate or one of its derivatives as an electron acceptor for the reoxidation of NADH in a **fermentation** process (**figure 9.17**). There are many kinds of fermentations, and they often are characteristic of particular microbial groups (**figure 9.18**). A few of the more common fermentations are introduced here, and several others are discussed at later points. Three unifying themes should be kept in mind when microbial fermentations are examined: (1) NADH is oxidized to NAD^+, (2) the electron acceptor is often either pyruvate or a pyruvate derivative, and (3) oxidative phosphorylation cannot operate, reducing the ATP yield per glucose significantly. In fermentation, the substrate is only partially oxidized, ATP is formed exclusively by substrate-level phosphorylation, and oxygen is not needed.

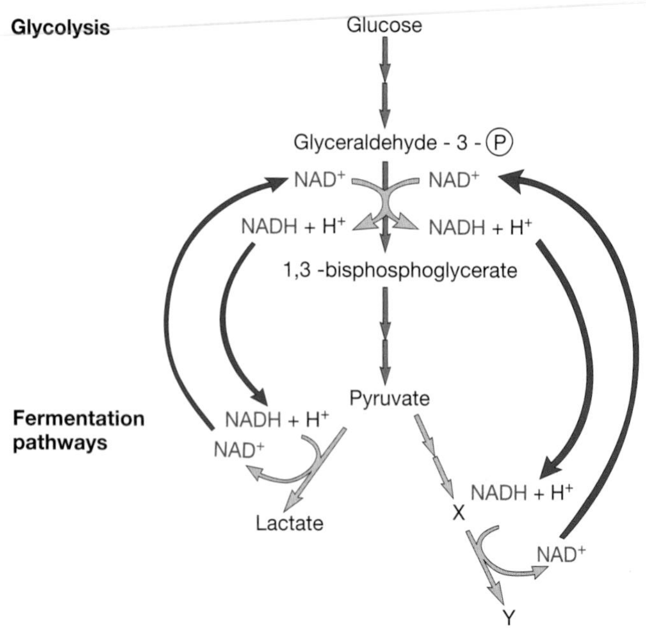

Figure 9.17 Reoxidation of NADH During Fermentation. NADH from glycolysis is reoxidized by being used to reduce pyruvate or a pyruvate derivative (X). Either lactate or reduced product Y result.

Many fungi, protists, and some bacteria ferment sugars to ethanol and CO_2 in a process called **alcoholic fermentation.** Pyruvate is decarboxylated to acetaldehyde, which is then reduced to ethanol by alcohol dehydrogenase with NADH as the electron donor (figure 9.18, number 2). **Lactic acid fermentation,** the reduction of pyruvate to lactate (figure 9.18, number 1), is even more common. It is present in bacteria (lactic acid bacteria, *Bacillus*), protists (*Chlorella* and some water molds), and even in animal skeletal muscle. Lactic acid fermenters can be separated into two groups. **Homolactic fermenters** use the Embden-Meyerhof pathway and directly reduce almost all their pyruvate to lactate with the enzyme lactate dehydrogenase. **Heterolactic fermenters** form substantial amounts of products other than lactate; many produce lactate, ethanol, and CO_2. Class *Gammaproteobacteria:* Order *Enterobacteriales* (section 22.3)

Alcoholic and lactic acid fermentations are quite useful. Alcoholic fermentation by yeasts produces alcoholic beverages; CO_2 from this fermentation causes bread to rise. Lactic acid fermentation can spoil foods, but also is used to make yogurt, sauer-

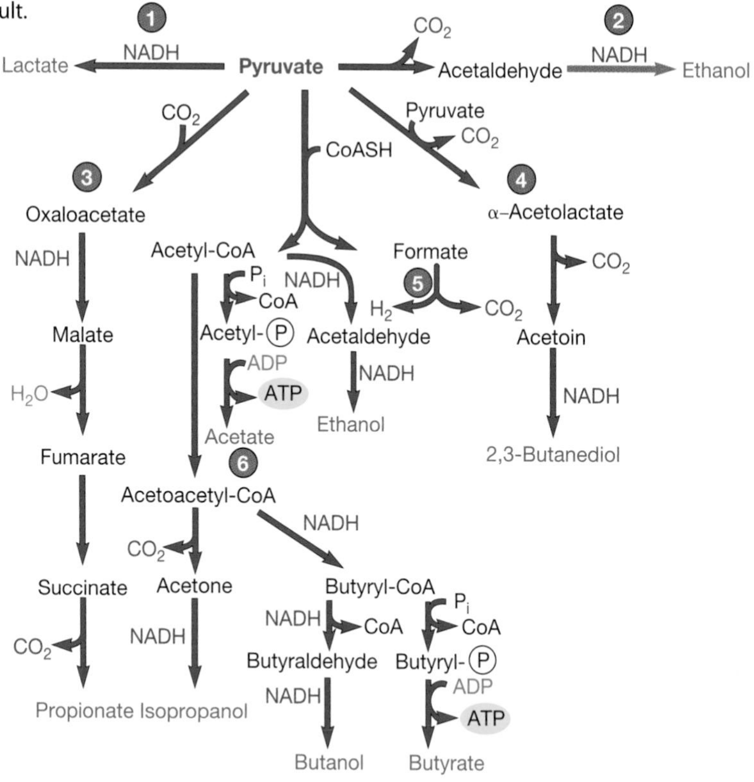

Figure 9.18 Some Common Microbial Fermentations. Only pyruvate fermentations are shown for the sake of simplicity; many other organic molecules can be fermented. Most of these pathways have been simplified by deletion of one or more steps and intermediates. Pyruvate and major end products are shown in color.

kraut, cheese, and pickles. The role of fermentations in food production is discussed in chapter 40.

Many bacteria, especially members of the family *Enterobacteriaceae,* can metabolize pyruvate to formic acid and other products in a process sometimes called the formic acid fermentation (figure 9.18, number 5). Formic acid may be converted to H_2 and CO_2 by formic hydrogenlyase (a combination of at least two enzymes).

$$HCOOH \longrightarrow CO_2 + H_2$$

There are two types of formic acid fermentation. **Mixed acid fermentation** results in the excretion of ethanol and a complex mixture of acids, particularly acetic, lactic, succinic, and formic acids (**table 9.2**). If formic hydrogenlyase is present, the formic acid will be degraded to H_2 and CO_2. This pattern is seen in *Escherichia, Salmonella, Proteus,* and other genera. The second type, **butanediol fermentation,** is characteristic of *Enterobacter, Serratia, Erwinia,* and some species of *Bacillus* (figure 9.18, number 4). Pyruvate is converted to acetoin, which is then reduced to 2,3-butanediol with NADH. A large amount of ethanol is also produced, together with smaller amounts of the acids found in a mixed acid fermentation. Class *Bacilli* (section 23.5)

Microorganisms carry out a vast array of fermentations using numerous sugars and other organic substrates as their energy source (**Historical Highlights 9.1**). Protozoa and fungi often ferment sugars to lactate, ethanol, glycerol, succinate, formate, acetate, butanediol, and additional products. Some members of the genus *Clostridium* ferment mixtures of amino acids. Proteolytic clostridia such as the pathogens *C. sporogenes* and *C. botulinum* carry out the **Stickland reaction** in which one amino acid is oxidized and a second amino acid acts as the electron acceptor. **Figure 9.19** shows the way in which alanine is oxidized and glycine reduced to produce acetate, CO_2, and NH_3. Some ATP is formed from acetyl phosphate by substrate-level phosphorylation, and the fermentation is quite useful for growing in anoxic, protein-rich environments. The Stickland reaction is used to oxidize several amino acids: alanine, leucine, isoleucine, valine, phenylalanine, tryptophan, and histidine. Bacteria also ferment amino acids (e.g., alanine, glycine, glutamate, threonine, and arginine) by other mechanisms. In addition to sugars and amino acids, organic acids such as acetate, lactate, propionate, and citrate are fermented. Some of these fermentations are of great practical importance. For example, citrate can be converted to diacetyl and give flavor to fermented milk. Class *Clostridia* (section 23.4); Microbiology of fermented foods (section 40.6)

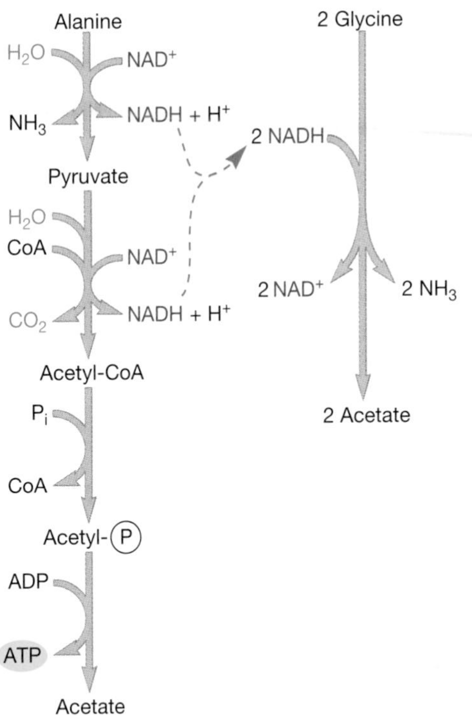

Figure 9.19 The Stickland Reaction. Alanine is oxidized to acetate and glycine is used to reoxidize the NADH generated during alanine degradation. The fermentation also produces some ATP.

Table 9.2 Mixed Acid Fermentation Products of *Escherichia coli*

	Fermentation Balance (μM Product/100 μM Glucose)	
	Acid Growth (pH 6.0)	**Alkaline Growth (pH 8.0)**
Ethanol	50	50
Formic acid	2	86
Acetic acid	36	39
Lactic acid	80	70
Succinic acid	11	15
Carbon dioxide	88	2
Hydrogen gas	75	0.5
Butanediol	0	0

1. What are fermentations and why are they so useful to many microorganisms?
2. How do the electron acceptors used in fermentation differ from the terminal electron acceptors used during either aerobic respiration or anaerobic respiration?
3. Briefly describe alcoholic, lactic acid, and formic acid fermentations. How do homolactic fermenters and heterolactic fermenters differ? How do mixed acid fermenters and butanediol fermenters differ?
4. What is the net yield of ATP during homolactic, acetate, and butyrate fermentations? How do these yields compare to aerobic respiration in terms of both quantity and mechanism of phosphorylation?
5. Some bacteria carry out fermentation only because they lack electron transport chains. Yet these bacteria still have a membrane-bound ATPase. Why do you think this is the case? How do you think these bacteria use the ATPase?
6. When bacteria carry out fermentation, only a few reactions of the TCA cycle operate. What purpose do you think these reactions might serve? Why do you think some parts of the cycle are shut down?

Historical Highlights

9.1 Microbiology and World War I

The unique economic pressures of wartime sometimes provide incentive for scientific discovery. Two examples from the First World War involve the production of organic solvents by the microbial fermentation of readily available carbohydrates, such as starch or molasses.

The German side needed glycerol to make nitroglycerin. At one time the Germans had imported their glycerol, but such imports were prevented by the British naval blockade. The German scientist Carl Neuberg knew that trace levels of glycerol were usually produced during the alcoholic fermentation of sugar by *Saccharomyces cerevisiae.* He sought to develop a modified fermentation in which the yeasts would produce glycerol instead of ethanol. Normally acetaldehyde is reduced to ethanol by NADH and alcohol dehydrogenase (figure 9.18, pathway 2). Neuberg found that this reaction could be prevented by the addition of 3.5% sodium sulfite at pH 7.0. The bisulfite ions reacted with acetaldehyde and made it unavailable for reduction to ethanol. Because the yeast cells still had to regenerate their NAD^+ even though acetaldehyde was no longer available, Neuberg suspected that they would simply increase the rate of glycerol synthesis. Glycerol is normally produced by the reduction of dihydroxyacetone phosphate (a glycolytic intermediate) to glycerol phosphate with NADH, followed by the hydrolysis of glycerol phosphate to glycerol. Neuberg's hunch was correct, and German breweries were converted to glycerol manufacture by his procedure, eventually producing 1,000 tons of glycerol per month. Glycerol production by *S. cerevisiae* was not economically competitive under peacetime conditions and was ended. Today glycerol is produced microbially by the halophilic protist *Dunaliella salina,* in which high concentrations of intracellular glycerol accumulate to counterbalance the osmotic pressure from the high level of extracellular salt. *Dunaliella* grows in habitats such as the Great Salt Lake of Utah and seaside rock pools.

The British side needed the organic solvents acetone and butanol. Butanol was required for the production of artificial rubber, whereas acetone was used as a solvent from nitrocellulose in the manufacture of the smokeless explosive powder cordite. Prior to 1914 acetone was made by the dry heating (pyrolysis) of wood. Between 80 and 100 tons of birch, beech, or maple wood were required to make 1 ton of acetone. When war broke out, the demand for acetone quickly exceeded the existing world supply. However, by 1915 Chaim Weizmann, a young Jewish scientist working in Manchester, England, had developed a fermentation process by which the anaerobic bacterium *Clostridium acetobutylicum* converted 100 tons of molasses or grain into 12 tons of acetone and 24 tons of butanol (most clostridial fermentations stop at butyric acid).

$$\text{2 pyruvate} \longrightarrow \text{acetoacetate} \longrightarrow \text{acetone} + CO_2$$

$$\text{Acetoacetate} \xrightarrow{\text{NADH}} \text{butyrate} \xrightarrow{\text{NADH}} \text{butanol}$$

This time the British and Canadian breweries were converted until new fermentation facilities could be constructed. Weizmann improved the process by finding a convenient way to select high-solvent producing strains of *C. acetobutylicum.* Because the strains most efficient in these fermentations also made the most heat-resistant spores, Weizmann merely isolated the survivors from repeated 100°C heat shocks. Acetone and butanol were made commercially by this fermentation process until it was replaced by much cheaper petrochemicals in the late 1940s and 1950s. In 1948 Chaim Weizmann became the first president of the State of Israel.

9.8 Catabolism of Carbohydrates and Intracellular Reserve Polymers

Thus far our main focus has been on the catabolism of glucose. However, microorganisms can catabolize many other carbohydrates. These carbohydrates may come either from outside the cell or from internal sources generated during normal metabolism. Often the initial steps in the degradation of external carbohydrate polymers differ from those employed with internal reserves.

Carbohydrates

Figure 9.20 outlines some catabolic pathways for the monosaccharides (single sugars) glucose, fructose, mannose, and galactose. The first three are phosphorylated using ATP and easily enter the Embden-Meyerhof pathway. In contrast, galactose must be converted to uridine diphosphate galactose (*see figure 10.11*) after initial phosphorylation, then changed into glucose 6-phosphate in a three-step process (figure 9.20).

The common disaccharides are cleaved to monosaccharides by at least two mechanisms (figure 9.20). Maltose, sucrose, and lactose can be directly hydrolyzed to their constituent sugars. Many disaccharides (e.g., maltose, cellobiose, and sucrose) are also split by a phosphate attack on the bond joining the two sugars, a process called phosphorolysis.

Polysaccharides, like disaccharides, are cleaved by both hydrolysis and phosphorolysis. Procaryotes and fungi degrade external polysaccharides by secreting hydrolytic enzymes. These exoenzymes cleave polysaccharides that are too large to cross the plasma membrane into smaller molecules that can then be assimilated. Starch and glycogen are hydrolyzed by amylases to glucose, maltose, and other products. Cellulose is more difficult to digest; many fungi and a few bacteria (some gliding bacteria, clostridia, and actinomycetes) produce extracellular cellulases that hydrolyze cellulose to cellobiose and glucose. Some actinomycetes and members of the bacterial genus *Cytophaga,* isolated from marine habitats, excrete an agarase that degrades agar. Many soil bacteria and bacterial plant pathogens degrade pectin, a polymer of galacturonic acid (a galactose derivative) that is an important constituent of plant cell walls and tissues. Lignin, another important component of plant cell walls, is usually degraded only by certain fungi that release peroxide-generating enzymes. Microorganisms in the soil environment (section 29.3)

In the context of compounds that are recalcitrant or difficult to digest, it should be noted that microorganisms also can degrade xenobiotic compounds (foreign substances not formed by natural biosynthetic processes) such as pesticides and various aromatic compounds. They transform these molecules to normal metabolic intermediates by use of special enzymes and pathways, then continue catabolism in the usual way. Biodegradation and bioremediation are discussed in chapter 41. The fungus *Phanerochaete chrysosporium* is an extraordinary example of the ability to degrade xenobiotics. Microbial Diversity & Ecology 41.4: A fungus with a voracious appetite

Reserve Polymers

Microorganisms often survive for long periods in the absence of exogenous nutrients. Under such circumstances they catabolize intracellular stores of glycogen, starch, poly-β-hydroxybutyrate, and other carbon and energy reserves. Glycogen and starch are degraded by phosphorylases. Phosphorylases catalyze a phosphorolysis reaction that shortens the polysaccharide chain by one glucose and yields glucose 1-phosphate.

$$(\text{Glucose})_n + P_i \longrightarrow (\text{glucose})_{n-1} + \text{glucose-1-P}$$

Glucose 1-phosphate can enter glycolytic pathways by way of glucose 6-phosphate (figure 9.20).

Poly-β-hydroxybutyrate (PHB) is an important, wide-spread reserve material. Its catabolism has been studied most thoroughly in the soil bacterium *Azotobacter.* This bacterium hydrolyzes PHB to 3-hydroxybutyrate, then oxidizes the hydroxybutyrate to acetoacetate. Acetoacetate is converted to acetyl-CoA, which can be oxidized in the TCA cycle.

9.9 Lipid Catabolism

Chemoorganotrophic microorganisms frequently use lipids as energy sources. Triglycerides or triacylglycerols, esters of glycerol and fatty acids (**figure 9.21**), are common energy sources and serve as our examples. They can be hydrolyzed to glycerol and fatty acids by microbial lipases. The glycerol is then phosphorylated, oxidized to dihydroxyacetone phosphate, and catabolized in the Embden-Meyerhof pathway (figure 9.5).

Fatty acids from triacylglycerols and other lipids are often oxidized in the **β-oxidation pathway** after conversion to coenzyme A esters (**figure 9.22**). In this pathway fatty acids are shortened by two carbons with each turn of the cycle. The two carbon units are released as acetyl-CoA, which can be fed into the TCA cycle or used in biosynthesis. One turn of the cycle produces acetyl-CoA, NADH, and $FADH_2$; NADH and $FADH_2$ can be oxidized by the electron transport chain to provide more ATP. The fatty acyl-CoA, shortened by two carbons, is ready for another turn of the cycle. Fatty acids are a rich source of energy for microbial growth. In a similar fashion some microorganisms grow well on petroleum hydrocarbons under oxic conditions.

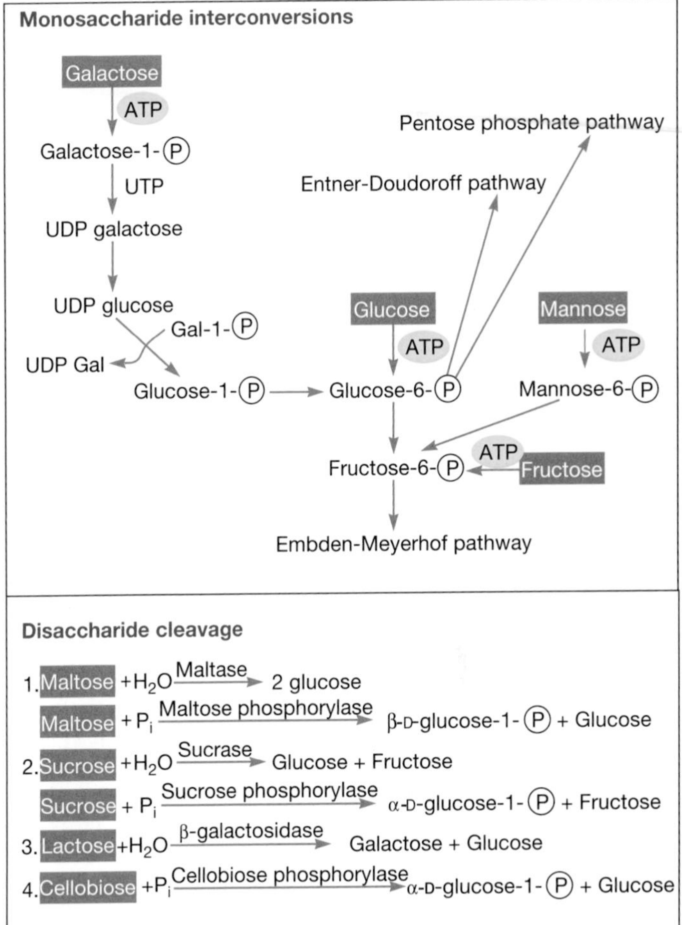

Figure 9.20 Carbohydrate Catabolism. Examples of enzymes and pathways used in disaccharide and monosaccharide catabolism. UDP is an abbreviation for uridine diphosphate.

```
              O
              ||
CH2 — O —  C — R1
 |            O
 |            ||
CH  — O —  C — R2
 |            O
 |            ||
CH2 — O —  C — R3
```

Figure 9.21 A Triacylglycerol or Triglyceride. The R groups represent the fatty acid side chains.

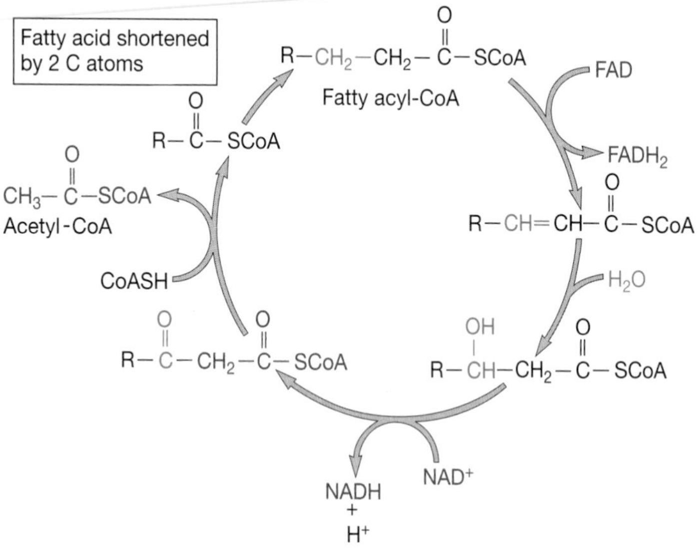

Figure 9.22 Fatty Acid β-Oxidation. The portions of the fatty acid being modified are shown in red.

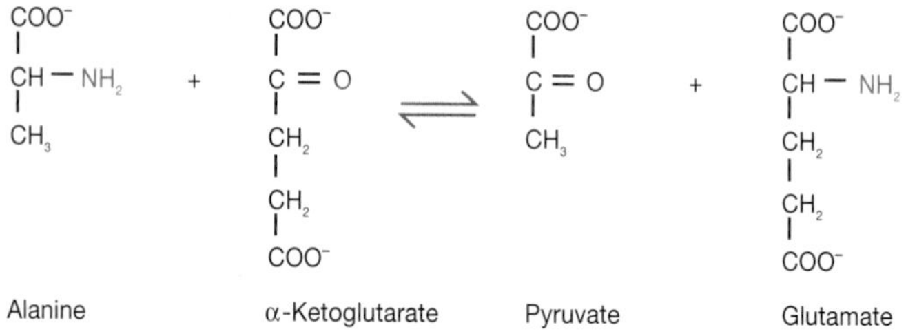

9.10 Protein and Amino Acid Catabolism

Some bacteria and fungi—particularly pathogenic, food spoilage, and soil microorganisms—can use proteins as their source of carbon and energy. They secrete **protease** enzymes that hydrolyze proteins and polypeptides to amino acids, which are transported into the cell and catabolized.

The first step in amino acid use is **deamination,** the removal of the amino group from an amino acid. This is often accomplished by **transamination.** The amino group is transferred from an amino acid to an α-keto acid acceptor (**figure 9.23**). The organic acid resulting from deamination can be converted to pyruvate, acetyl-CoA, or a TCA cycle intermediate and eventually oxidized in the TCA cycle to release energy. It also can be used as a source of carbon for the synthesis of cell constituents. Excess nitrogen from deamination may be excreted as ammonium ion, thus making the medium alkaline.

1. Briefly discuss the ways in which microorganisms degrade and use common monosaccharides, disaccharides, and polysaccharides from both external and internal sources.
2. Describe how a microorganism might derive carbon and energy from the lipids and proteins in its diet. What is β-oxidation? Deamination? Transamination?

9.11 Chemolithotrophy

So far, we have considered microbes that synthesize ATP with the energy liberated when they oxidize organic substrates such as carbohydrates, lipids, and proteins. The electron acceptor is: (1) O_2 in aerobic respiration, (2) an oxidized exogenous molecule other than O_2 in anaerobic respiration, or (3) another more oxidized endogenous organic molecule (usually pyruvate) in fermentation (figure 9.2). In fermentation, ATP is synthesized only by substrate-level phosphorylation; in both aerobic and anaerobic respiration, most of the ATP is formed using the PMF derived from electron transport chain activity. Additional metabolic diversity among bacteria and archaea is reflected in the form of energy metabolism performed by **chemolithotrophs.** These microbes obtain electrons for the electron transport chain from the oxidation of inorganic molecules rather than NADH generated by the oxidation of organic nutrients (**figure 9.24**). Each species is rather specific in its preferences for electron donors and acceptors (**table 9.3**). The acceptor is usually O_2, but sulfate and nitrate are also used. The most common electron donors are hydrogen, reduced nitrogen compounds, reduced sulfur compounds, and ferrous iron (Fe^{2+}).

Figure 9.23 Transamination. A common example of this process. The α-amino group (blue) of alanine is transferred to the acceptor α-ketoglutarate forming pyruvate and glutamate. The pyruvate can be catabolized in the tricarboxylic acid cycle or used in biosynthesis.

Much less energy is available from the oxidation of inorganic molecules than from the complete oxidation of glucose to CO_2, which is accompanied with a standard free energy change of −686 kcal/mole (**table 9.4**). This is because the NADH that donates electrons to the chain following the oxidation of an organic substrate like glucose has a more negative reduction potential than most of the inorganic substrates that chemolithotrophs use as direct electron donors to their electron transport chains. Thus the P/O ratios for oxidative phosphorylation in chemolithotrophs are probably around 1.0 (although in the oxidation of hydrogen it is considerably higher). Because the yield of ATP is so low, chemolithotrophs must oxidize a large quantity of inorganic material to grow and reproduce. This is particularly true of autotrophic chemolithotrophs, which fix CO_2 into carbohydrates. For each molecule of CO_2 fixed, these microbes expend three ATP and two NADPH molecules. Because they must consume a large amount of inorganic material, chemolithotrophs have significant ecological impact.

Several bacterial genera can oxidize hydrogen gas to produce energy because they possess a hydrogenase enzyme that catalyzes the oxidation of hydrogen (table 9.3).

$$H_2 \longrightarrow 2H^+ + 2e^-$$

Because the $H_2/2H^+$, $2e^-$ redox couple has a very negative standard reduction potential, the electrons are donated either to an electron transport chain or to NAD^+, depending on the hydrogenase. If

NADH is produced, it can be used in ATP synthesis by electron transport and oxidative phosphorylation, with O_2, Fe^{3+}, S^0, and even carbon monoxide (CO) as the terminal electron acceptors. Often these hydrogen-oxidizing microorganisms will use organic compounds as energy sources when such nutrients are available.

Some bacteria use the oxidation of nitrogenous compounds as a source of electrons. Among these chemolithotrophs, the **nitrifying bacteria,** which carry out nitrification are best understood. These are soil and aquatic bacteria of considerable ecological significance. **Nitrification** is the oxidation of ammonia to nitrate. It is a two-step process that depends on the activity of at least two different genera. In the first step, ammonia is oxidized to nitrite by a number of genera including *Nitrosomonas:*

$$NH_4^+ + 1\ 1/2\ O_2 \rightarrow NO_2^- + H_2O + 2H^+$$

In the second step, the nitrite is oxidized to nitrate by genera such as *Nitrobacter:*

$$NO_2^- + 1/2\ O_2 \rightarrow NO_3^-$$

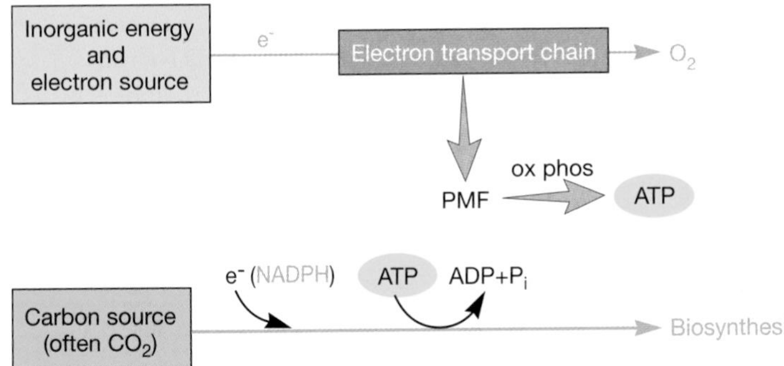

Figure 9.24 Chemolithotrophic Fueling Processes. Chemolithotrophic bacteria and archaea oxidize inorganic molecules (e.g., H_2S and NH_3), which serve as energy and electron sources. The electrons released pass through an electron transport system, generating a proton motive force (PMF). ATP is synthesized by oxidative phosphorylation (ox phos). Most chemolithotrophs use O_2 as the terminal electron acceptor. However, some can use other exogenous molecules as terminal electron acceptors. Note that a molecule other than the energy source provides carbon for biosynthesis. Many chemolithotrophs are autotrophs.

Nitrification differs from denitrification in that nitrification involves the oxidation of inorganic nitrogen compounds to yield nitrate. On the other hand, denitrification is the reduction of oxidized nitrogenous compounds to nitrogen gas (see p. 205). In nitrification, electrons are donated to the electron transport chain, while in denitrification, nitrogen species are used as electron acceptors and nitrogen is lost to the atmosphere. Biogeochemical cycling: Nitrogen cycle (section 27.2)

Energy released upon the oxidation of both ammonia and nitrite is used to make ATP by oxidative phosphorylation. However, autotrophic microorganisms also need NAD(P)H (reducing power) as well as ATP in order to reduce CO_2 and other molecules (figure 9.24). Since molecules like ammonia and nitrite have more positive reduction potentials than NAD^+, they cannot directly donate their electrons to form the required NADH and NADPH. Recall that electrons spontaneously move only from donors with more negative reduction potentials to acceptors with more positive potentials (*see figure 8.8*). Sulfur-oxidizing bacteria face the same difficulty. Both types of chemolithotrophs solve this problem by moving the electrons derived from the oxidation of their inorganic substrate (reduced nitrogen or sulfur compounds) up the electron transport chain to reduce $NAD(P)^+$ to NAD(P)H (**figure 9.25**). This is called **reverse electron flow.** Of course, this is not thermodynamically favorable, so energy in the form of the proton motive force must be diverted from performing other cellular work (e.g., ATP synthesis, transport, motility)

Table 9.4 Energy Yields from Oxidations Used by Chemolithotrophs

Reaction	$\Delta G^{\circ\prime}$ (kcal/mole)[a]
$H_2 + 1/2\ O_2 \longrightarrow H_2O$	−56.6
$NO_2^- + 1/2\ O_2 \longrightarrow NO_3^-$	−17.4
$NH_4^+ + 1\ 1/2\ O_2 \longrightarrow NO_2^- + H_2O + 2H^+$	−65.0
$S^0 + 1\ 1/2\ O_2 + H_2O \longrightarrow H_2SO_4$	−118.5
$S_2O_3^{2-} + 2O_2 + H_2O \longrightarrow 2SO_4^{2-} + 2H^+$	−223.7
$2Fe^{2+} + 2H^+ + 1/2\ O_2 \longrightarrow 2Fe^{3+} + H_2O$	−11.2

[a]The $\Delta G^{\circ\prime}$ for complete oxidation of glucose to CO_2 is −686 kcal/mole. A kcal is equivalent to 4.184kJ.

Table 9.3 Representative Chemolithotrophs and Their Energy Sources

Bacteria	Electron Donor	Electron Acceptor	Products
Alcaligenes, Hydrogenophaga, and *Pseudomonas* spp.	H_2	O_2	H_2O
Nitrobacter	NO_2^-	O_2	NO_3^-, H_2O
Nitrosomonas	NH_4^+	O_2	NO_2^-, H_2O
Thiobacillus denitrificans	S^0, H_2S	NO_3^-	SO_4^{2-}, N_2
Thiobacillus ferrooxidans	Fe^{2+}, S^0, H_2S	O_2	Fe^{3+}, H_2O, H_2SO_4

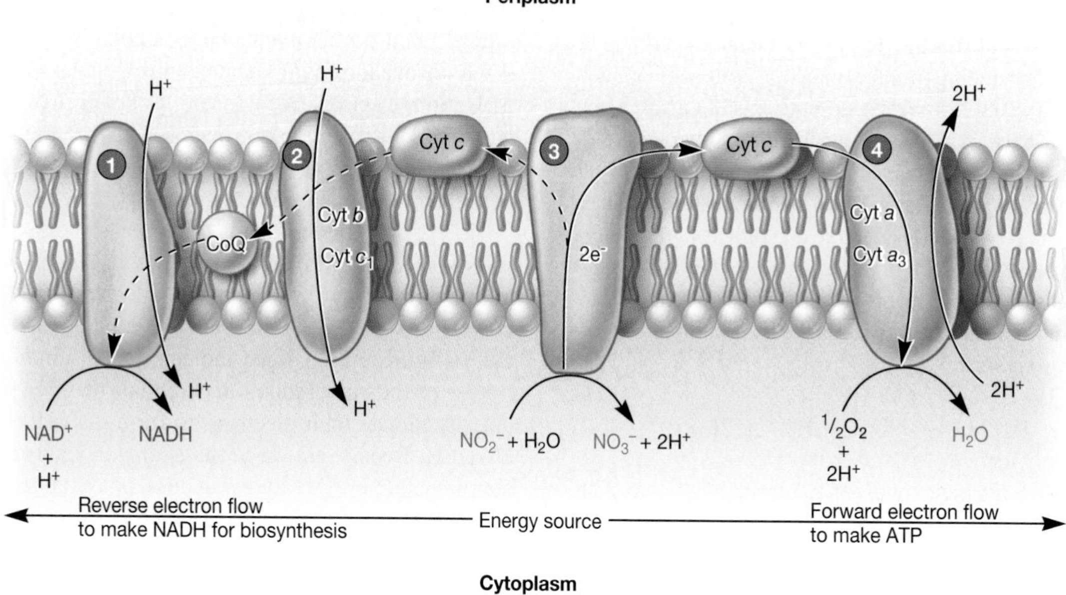

Figure 9.25 Electron Flow in *Nitrobacter* Electron Transport Chain. *Nitrobacter* oxidizes nitrite and carries out normal electron transport to generate proton motive force for ATP synthesis. This is the right-hand branch of the diagram. Some of the proton motive force also is used to force electrons to flow up the reduction potential gradient from nitrite to NAD^+ (left-hand branch). Cytochrome *c* and four complexes are involved: NADH-ubiquinone oxidoreductase (1), ubiquinol-cytochrome *c* oxidoreductase (2), nitrite oxidase (3), and cytochrome aa_3 oxidase (4).

to "push" the electrons from molecules of relatively positive reduction potentials to those that are more negative. Because this energy is used to generate NADH as well as ATP, the net yield of ATP is fairly low. Chemolithotrophs can afford this inefficiency as they have no serious competitors for their unique energy sources. Oxidation-reduction reactions, electron carriers, and electron transport systems (section 8.6)

Sulfur-oxidizing microbes are the third major group of chemolithotrophs. The metabolism of *Thiobacillus* has been best studied. These bacteria oxidize sulfur (S^0), hydrogen sulfide (H_2S), thiosulfate ($S_2O_3^{2-}$), and other reduced sulfur compounds to sulfuric acid; therefore they have a significant ecological impact (**Microbial Diversity & Ecology 9.2**). Interestingly they generate ATP by both oxidative phosphorylation and substrate-level phosphorylation involving **adenosine 5′-phosphosulfate (APS).** APS is a high-energy molecule formed from sulfite and adenosine monophosphate (**figure 9.26**).

Some sulfur-oxidizing procaryotes are extraordinarily flexible metabolically. For example, *Sulfolobus brierleyi,* an archaeon, and some bacteria can grow aerobically by oxidizing sulfur with oxygen as the electron acceptor; in the absence of O_2, they carry out anaerobic respiration and oxidize organic material with sulfur as the electron acceptor.

Sulfur-oxidizing bacteria and archaea, like other chemolithotrophs, can use CO_2 as their carbon source. Many will grow heterotrophically if they are supplied with reduced organic carbon sources like glucose or amino acids.

1. How do chemolithotrophs obtain their ATP and NADH? What is their most common source of carbon?
2. Describe energy production by hydrogen-oxidizing bacteria, nitrifying bacteria, and sulfur-oxidizing bacteria.
3. Why can hydrogen-oxidizing bacteria and archaea donate electrons to NAD^+ while sulfur- and ammonia-oxidizing bacteria and archaea cannot?
4. What is reverse electron flow and why do most chemolithotrophs perform it?
5. Arsenate is a compound that inhibits substrate-level phosphorylation. Compare the effect of this compound on a H_2-oxidizing chemolithotroph, on a sulfite-oxidizing chemolithotroph, and on a chemoorganotroph carrying out fermentation.

9.12 Phototrophy

Microorganisms derive energy not only from the oxidation of inorganic and organic compounds, but also from light energy, which they capture and use to synthesize ATP and reduce power (e.g., NADPH) (figure 9.1; *see also figure 8.10*). The process by which light energy is trapped and converted to chemical energy is called **photosynthesis.** Usually a phototrophic organism reduces and incorporates CO_2. Photosynthesis is one of the most significant metabolic processes on Earth because almost all our energy is ultimately derived from solar energy. It provides photosynthetic organisms with the ATP and reducing power necessary to synthesize the organic material required for growth. In turn

Microbial Diversity & Ecology

9.2 Acid Mine Drainage

Each year millions of tons of sulfuric acid flow to the Ohio River from the Appalachian Mountains. This sulfuric acid is of microbial origin and leaches enough metals from the mines to make the river reddish and acidic. The primary culprit is *Thiobacillus ferrooxidans*, a chemolithotrophic bacterium that derives its energy from oxidizing ferrous ion to ferric ion and sulfide ion to sulfate ion. The combination of these two energy sources is important because of the solubility properties of iron. Ferrous ion is somewhat soluble and can be formed at pH values of 3.0 or less in moderately reducing environments. However, when the pH is greater than 4.0 to 5.0, ferrous ion is spontaneously oxidized to ferric ion by O_2 in the water and precipitates as a hydroxide. If the pH drops below about 2.0 to 3.0 because of sulfuric acid production by spontaneous oxidation of sulfur or sulfur oxidation by thiobacilli and other bacteria, the ferrous ion remains reduced, soluble, and available as an energy source. Remarkably, *T. ferrooxidans* grows well at such acidic pHs and actively oxidizes ferrous ion to an insoluble ferric precipitate. The water is rendered toxic for most aquatic life and unfit for human consumption.

The ecological consequences of this metabolic life-style arise from the common presence of pyrite (FeS_2) in coal mines. The bacteria oxidize both elemental components of pyrite for their growth and in the process form sulfuric acid, which leaches the remaining minerals.

$$2FeS_2 + 7O_2 + 2H_2O \xrightarrow{\text{Autoxidation or bacterial action}} 2Fe^{2+} + 4SO_4^{2-} + 4H^+$$

$$2Fe^{2+} + 1/2\,O_2 + 2H^+ \xrightarrow{\textit{T. ferrooxidans}} 2Fe^{3+} + H_2O$$

Pyrite oxidation is further accelerated because the ferric ion generated by bacterial activity readily oxidizes more pyrite to sulfuric acid and ferrous ion. In turn the ferrous ion supports further bacterial growth. It is difficult to prevent *T. ferrooxidans* growth as it requires only pyrite and common inorganic salts. Because *T. ferrooxidans* gets its O_2 and CO_2 from the air, the only feasible method of preventing its damaging growth is to seal the mines to render the habitat anoxic.

(a) Direct oxidation of sulfite

$$SO_3^{2-} \xrightarrow{\text{sulfite oxidase}} SO_4^{2-} + 2e^-$$

(b) Formation of adenosine 5′-phosphosulfate

$$2SO_3^{2-} + 2AMP \longrightarrow 2APS + 4e^-$$

$$2APS + 2P_i \longrightarrow 2ADP + 2SO_4^{2-}$$

$$2ADP \longrightarrow AMP + ATP$$

$$2SO_3^{2-} + AMP + 2P_i \longrightarrow 2SO_4^{2-} + ATP + 4e^-$$

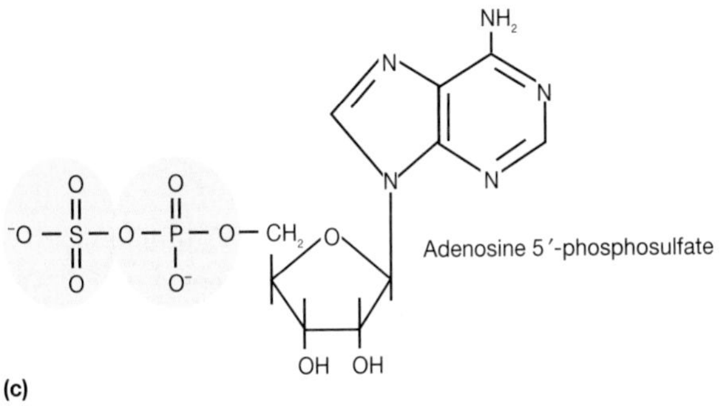

(c) Adenosine 5′-phosphosulfate

Figure 9.26 Energy Generation by Sulfur Oxidation. **(a)** Sulfite can be directly oxidized to provide electrons for electron transport and oxidative phosphorylation. **(b)** Sulfite can also be oxidized and converted to adenosine 5′-phosphosulfate (APS). This route produces electrons for use in electron transport and ATP by substrate-level phosphorylation with APS. **(c)** The structure of APS.

Table 9.5 Diversity of Phototrophic Organisms

Eucaryotic Organisms	Procaryotic Organisms
Plants	Cyanobacteria
Multicellular green, brown, and red algae	Green sulfur bacteria
	Green nonsulfur bacteria
Unicellular protists (e.g., euglenoids, dinoflagellates, diatoms)	*Halobacterium* (archaeon)
	Purple sulfur bacteria
	Purple nonsulfur bacteria
	Prochloron

these organisms serve as the base of most food chains in the biosphere. One type of photosynthesis is also responsible for replenishing our supply of O_2, a remarkable process carried out by a variety of organisms, both eucaryotic and bacterial (**table 9.5**). Although most people associate photosynthesis with the larger, more obvious plants, over half the photosynthesis on Earth is carried out by microorganisms.

Photosynthesis as a whole is divided into two parts. In the **light reactions** light energy is trapped and converted to chemical energy. This energy is then used to reduce or fix CO_2 and synthesize cell constituents in the **dark reactions.** In this section three types of phototrophy are discussed: oxygenic photosynthesis, anoxygenic photosynthesis, and rhodopsin-based phototrophy (**figure 9.27**). The dark reactions of photosynthesis are reviewed in chapter 10. The fixation of CO_2 by autotrophs (section 10.3)

The Light Reaction in Oxygenic Photosynthesis

Phototrophic eucaryotes and the cyanobacteria carry out **oxygenic photosynthesis,** so named because oxygen is generated when light energy is converted to chemical energy. Central to this process, and to all other phototrophic processes, are light-absorbing pigments (**table 9.6**). In oxygenic phototrophs, the most important pigments are the **chlorophylls.** Chlorophylls are large planar rings composed of four substituted pyrrole rings with a magnesium atom coordinated to the four central nitrogen atoms (**figure 9.28**). Several chlorophylls are found in eucaryotes, the two most important are chlorophyll *a* and chlorophyll *b*. These two molecules differ slightly in their structure and spectral properties. When dissolved in acetone, chlorophyll *a* has a light absorption peak at 665 nm; the corresponding peak for chlorophyll *b* is at 645 nm. In addition to absorbing red light, chlorophylls also absorb

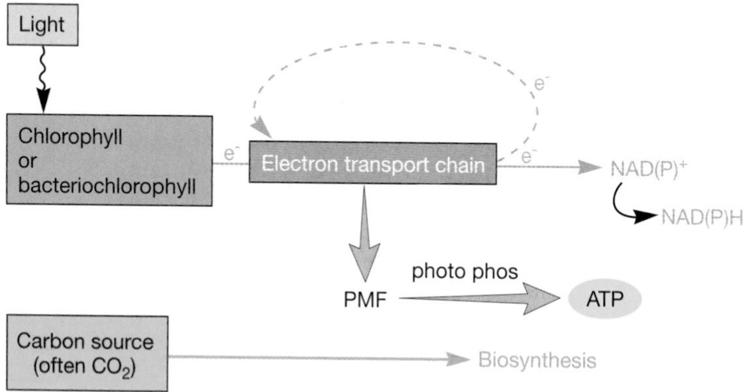

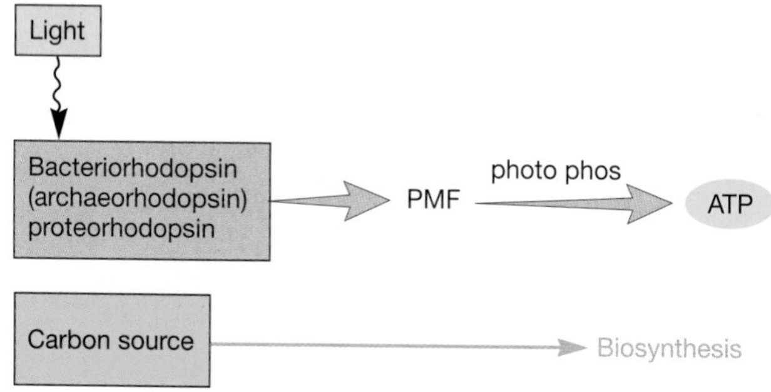

Figure 9.27 Phototrophic Fueling Reactions. Phototrophs use light to generate a proton motive force (PMF), which is then used to synthesize ATP by a process called photophosphorylation (photo phos). The process requires light-absorbing pigments. When the pigments are chlorophyll or bacteriochlorophyll, the absorption of light triggers electron flow through an electron transport chain, accompanied by the pumping of protons across a membrane. The electron flow can be either cyclic (dashed line) or noncyclic (solid line), depending on the organism and its needs. Rhodopsin-based phototrophy differs in that the PMF is formed directly by the light-absorbing pigment, which is a light-driven proton pump. Many phototrophs are autotrophs and must use much of the ATP and reducing power they make to fix CO_2.

Table 9.6 Properties of Chlorophyll-Based Photosynthetic Systems

Property	Eucaryotes	Cyanobacteria	Green Bacteria, Purple Bacteria, and Heliobacteria
Photosynthetic pigment	Chlorophyll *a*	Chlorophyll *a*	Bacteriochlorophyll
Photosystem II	Present	Present	Absent
Photosynthetic electron donors	H_2O	H_2O	H_2, H_2S, S, organic matter
O_2 production pattern	Oxygenic	Oxygenic[a]	Anoxygenic
Primary products of energy conversion	ATP + NADPH	ATP + NADPH	ATP
Carbon source	CO_2	CO_2	Organic and/or CO_2

[a]Some cyanobacteria can function anoxygenically under certain conditions. For example, *Oscillatoria* can use H_2S as an electron donor instead of H_2O.

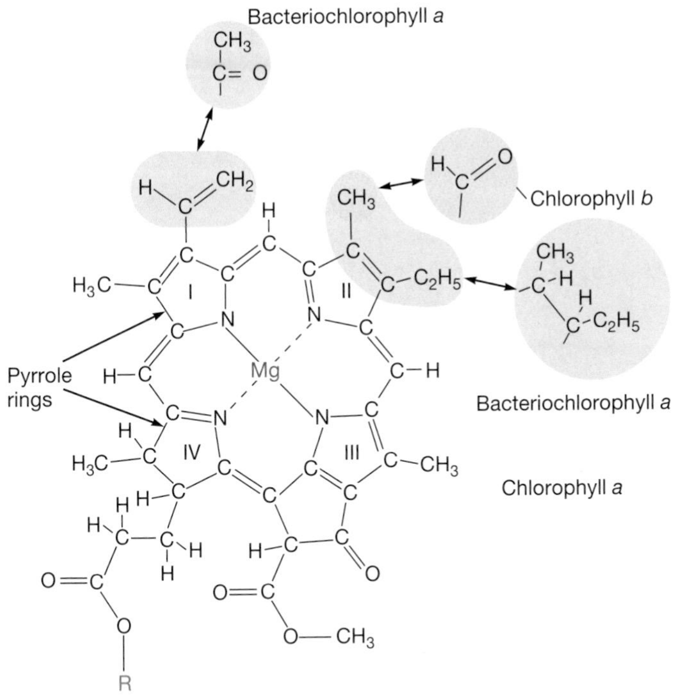

Figure 9.28 Chlorophyll Structure. The structures of chlorophyll *a*, chlorophyll *b*, and bacteriochlorophyll *a*. The complete structure of chlorophyll *a* is given. Only one group is altered to produce chlorophyll *b*, and two modifications in the ring system are required to change chlorophyll *a* to bacteriochlorophyll *a*. The side chain (R) of bacteriochlorophyll *a* may be either phytyl (a 20-carbon chain also found in chlorophylls *a* and *b*) or geranylgeranyl (a 20-carbon side chain similar to phytyl, but with three more double bonds).

blue light strongly (the second absorption peak for chlorophyll *a* is at 430 nm). Because chlorophylls absorb primarily in the red and blue ranges, green light is transmitted. Consequently many oxygenic phototrophs are green in color. The long hydrophobic tail attached to the chlorophyll ring aids in its attachment to membranes, the site of the light reactions.

Other photosynthetic pigments also trap light energy. The most widespread of these are the **carotenoids,** long molecules, usually yellowish in color, that possess an extensive conjugated double bond system (**figure 9.29**). β-Carotene is present in cyanobacteria belonging to the genus *Prochloron* and most photosynthetic protists; fucoxanthin is found in protists such as diatoms and dinoflagellates. Red algae and cyanobacteria have photosynthetic pigments called **phycobiliproteins,** consisting of a protein with a linear tetrapyrrole attached (figure 9.29). **Phycoerythrin** is a red pigment with a maximum absorption around 550 nm, and **phycocyanin** is blue (maximum absorption at 620 to 640 nm).

Carotenoids and phycobiliproteins are often called **accessory pigments** because of their role in photosynthesis. Accessory pigments are important because they absorb light in the range not absorbed by chlorophylls (the blue-green through yellow range; about 470–630 nm) (*see figure 21.4*). This light is very efficiently transferred to chlorophyll. In this way accessory pigments make photosynthesis more efficient over a broader range of wavelengths. In addition, this allows organisms to use light not used by other phototrophs in their habitat. For instance, the microbes below a canopy of plants can use light that passes through the canopy. Accessory pigments also protect microorganisms from intense sunlight, which could oxidize and damage the photosynthetic apparatus.

Chlorophylls and accessory pigments are assembled in highly organized arrays called **antennas,** whose purpose is to create a large surface area to trap as many photons as possible. An antenna has about 300 chlorophyll molecules. Light energy is captured in an antenna and transferred from chlorophyll to chlorophyll until it reaches a special **reaction-center chlorophyll pair** directly involved in photosynthetic electron transport. In oxygenic phototrophs, there are two kinds of antennas associated with two different photosystems (**figure 9.30**). **Photosystem I** absorbs longer wavelength light (≥680 nm) and funnels the energy to a special chlorophyll *a* pair called P700. The term P700 signifies that this molecule most effectively absorbs light at a wavelength of 700 nm. **Photosystem II** traps light at shorter wavelengths (≤680 nm) and transfers its energy to the special chlorophyll pair P680.

When the photosystem I antenna transfers light energy to the reaction-center P700 chlorophyll pair, P700 absorbs the energy and is excited; its reduction potential becomes very negative. This allows it to donate its excited, high-energy electron to a specific acceptor, probably a special chlorophyll *a* molecule or an iron-sulfur protein. The electron is eventually transferred to ferredoxin and can then travel in either of two directions. In the cyclic pathway (the dashed lines in figure 9.30), the electron moves in a cyclic route through a series of electron carriers and back to the oxidized P700. The pathway is termed cyclic because the electron from P700 returns to P700 after traveling through the photosynthetic electron transport chain. PMF is formed during cyclic electron transport in the region of cytochrome b_6 and used to synthesize ATP. This process is called **cyclic photophosphorylation** because electrons travel in a cyclic pathway and ATP is formed. Only photosystem I participates.

Electrons also can travel in a noncyclic pathway involving both photosystems. P700 is excited and donates electrons to ferredoxin as before. In the noncyclic route, however, reduced ferredoxin reduces $NADP^+$ to NADPH (figure 9.30). Because the electrons contributed to $NADP^+$ cannot be used to reduce oxidized P700, photosystem II participation is required. It donates electrons to oxidized P700 and generates ATP in the process. The photosystem II antenna absorbs light energy and excites P680, which then reduces pheophytin *a*. Pheophytin *a* is chlorophyll *a* in which two hydrogen atoms have replaced the central magnesium. Electrons subsequently travel to the plastoquinone pool and down the electron transport chain to P700. Although P700 has been reduced, P680 must also be reduced if it is to accept more light energy. Figure 9.30 indicates that the standard reduction potential of P680 is more positive than that of the O_2/H_2O redox couple. Thus H_2O can be used to donate electrons to P680 resulting in the release of oxygen. Because electrons flow from

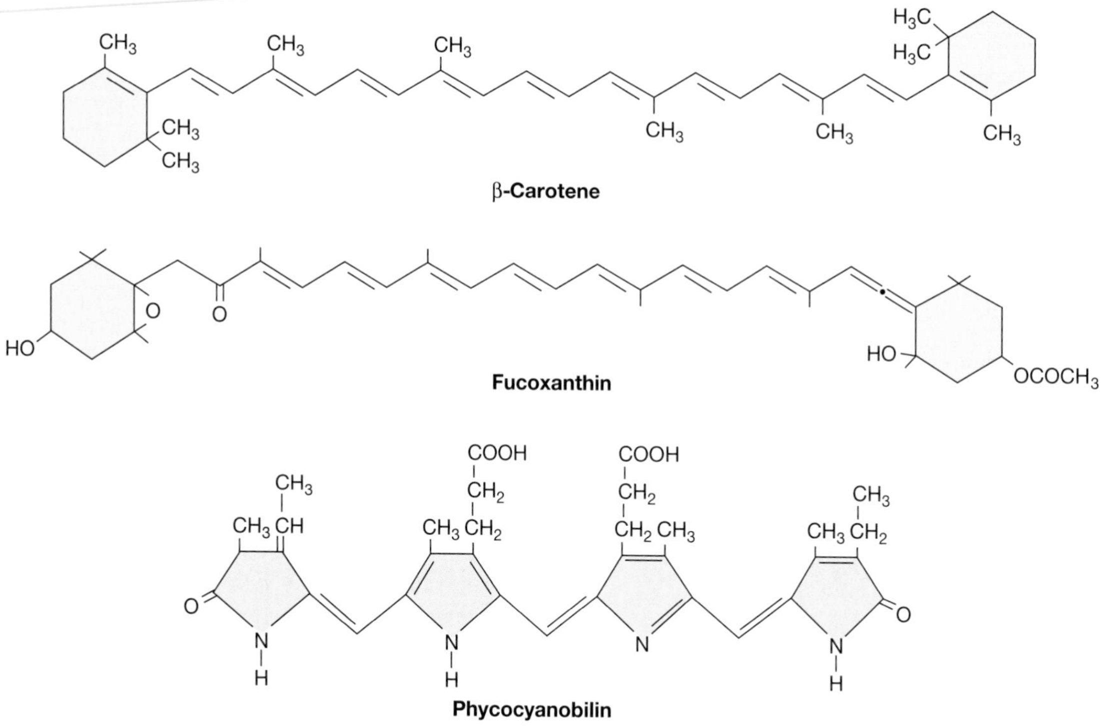

β-Carotene

Fucoxanthin

Phycocyanobilin

Figure 9.29 Representative Accessory Pigments. Beta-carotene is a carotenoid found in photosynthetic protists and plants. Note that it has a long chain of alternating double and single bonds called conjugated double bonds. Fucoxanthin is a carotenoid accessory pigment in several divisions of algae (the dot in the structure represents a carbon atom). Phycocyanobilin is an example of a linear tetrapyrrole that is attached to a protein to form a phycobiliprotein.

water to $NADP^+$ with the aid of energy from two photosystems, ATP is synthesized by **noncyclic photophosphorylation.** It appears that one ATP and one NADPH are formed when two electrons travel through the noncyclic pathway.

Just as is true of mitochondrial electron transport, photosynthetic electron transport takes place within a membrane. Chloroplast granal membranes contain both photosystems and their antennas. **Figure 9.31** shows a thylakoid membrane carrying out noncyclic photophosphorylation by the chemiosmotic mechanism. Protons move to the thylakoid interior during photosynthetic electron transport and return to the stroma when ATP is formed. It is believed that stromal lamellae possess only photosystem I and are involved in cyclic photophosphorylation alone. In cyanobacteria, photosynthetic light reactions are located in thylakoid membranes within the cell.

The dark reactions require three ATPs and two NADPHs to reduce one CO_2 and use it to synthesize carbohydrate (CH_2O).

$$CO_2 + 3ATP + 2NADPH + 2H^+ + H_2O \longrightarrow (CH_2O) + 3ADP + 3P_i + 2NADP^+$$

The noncyclic system generates one NADPH and one ATP per pair of electrons; therefore four electrons passing through the system will produce two NADPHs and two ATPs. A total of 8 quanta of light energy (4 quanta for each photosystem) is needed to propel the four electrons from water to $NADP^+$. Because the ratio of ATP to NADPH required for CO_2 fixation is 3:2, at least one more ATP must be supplied. Cyclic photophosphorylation probably operates independently to generate the extra ATP. This requires absorption of another 2 to 4 quanta. It follows that around 10 to 12 quanta of light energy are needed to reduce and incorporate one molecule of CO_2 during photosynthesis.

The Light Reaction in Anoxygenic Photosynthesis

Certain bacteria carry out a second type of photosynthesis called **anoxygenic photosynthesis.** This phototrophic process derives its name from the fact that water is not used as an electron source and therefore O_2 is not produced. The process also differs in terms of the photosynthetic pigments used, the participation of just one photosystem, and the mechanisms used to generate reducing power. Three groups of bacteria carry out anoxygenic photosynthesis: phototrophic green bacteria, phototrophic purple bacteria, and heliobacteria. The biology and ecology of these organisms is described in much more detail in chapters 21, 22, and 23.

Anoxygenic phototrophs have photosynthetic pigments called **bacteriochlorophylls** (figure 9.28). The absorption maxima of bacteriochlorophylls (Bchl) are at longer wavelengths than those of chlorophylls. Bacteriochlorophylls *a* and *b* have maxima

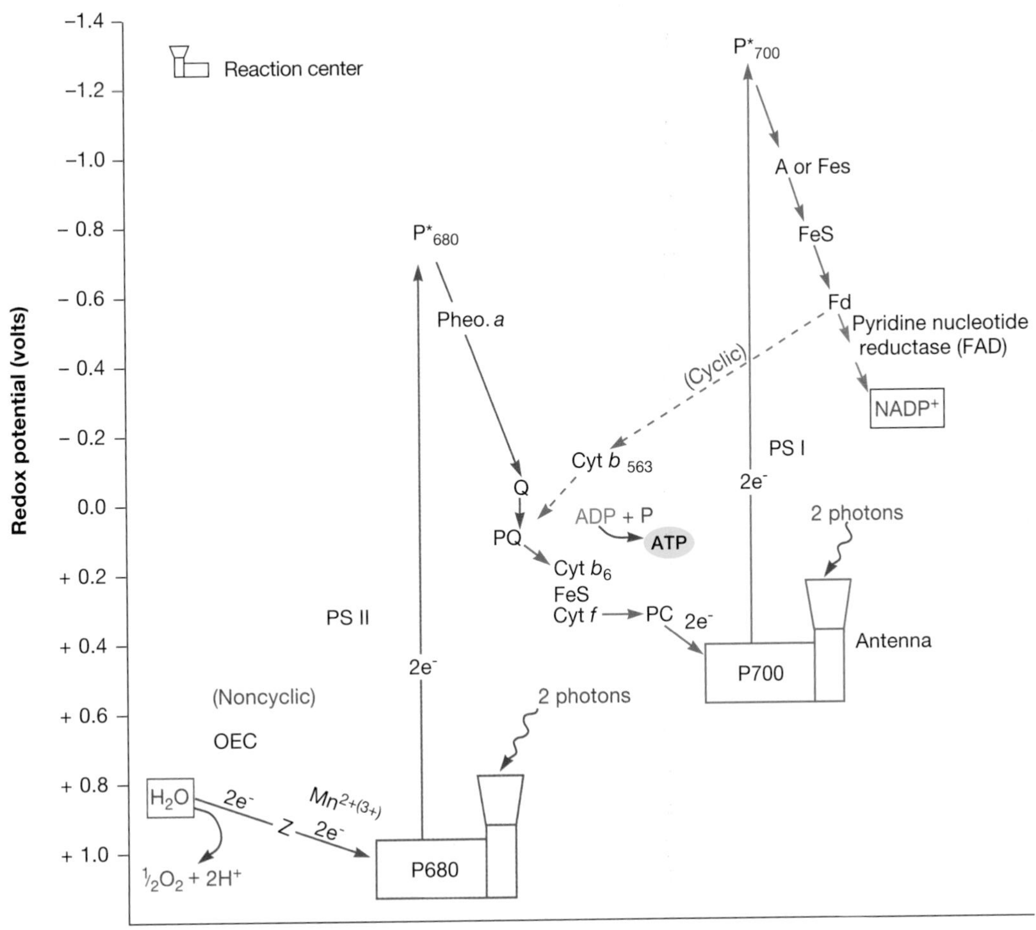

Figure 9.30 Green Plant Photosynthesis. Electron flow during photosynthesis in higher plants. Cyanobacteria and eucaryotic algae are similar in having two photosystems, although they may differ in some details. The carriers involved in electron transport are ferredoxin (Fd) and other FeS proteins; cytochromes b_6, b_{563}, and *f*; plastoquinone (PQ); copper containing plastocyanin (PC); pheophytin *a* (Pheo. *a*); possibly chlorophyll *a* (A); and the unknown quinone Q, which is probably a plastoquinone. Both photosystem I (PS I) and photosystem II (PS II) are involved in noncyclic photophosphorylation; only PS I participates in cyclic photophosphorylation. The oxygen evolving complex (OEC) that extracts electrons from water contains manganese ions and the substance Z, which transfers electrons to the PS II reaction center. See the text for further details.

in ether at 775 and 790 nm, respectively. In vivo maxima are about 830 to 890 nm (Bchl *a*) and 1,020 to 1,040 nm (Bchl *b*). This shift of absorption maxima into the infrared region better adapts these bacteria to their ecological niches.

Many differences found in anoxygenic phototrophs are due to their having a single photosystem (**figure 9.32**). Because of this, they are restricted to cyclic electron flow and are unable to produce O_2 from H_2O. Indeed, almost all anoxygenic phototrophs are strict anaerobes. A tentative scheme for the photosynthetic electron transport chain of a purple nonsulfur bacterium is given in **figure 9.33.** When the special reaction-center bacteriochlorophyll P870 is excited, it donates an electron to bacteriopheophytin. Electrons then flow to quinones and through an electron transport chain back to P870 while generating sufficient PMF to drive ATP synthesis. Note that although both green and purple bacteria lack two photosystems, the purple bacteria have a photosynthetic apparatus similar to photosystem II, whereas the green sulfur bacteria have a system similar to photosystem I.

Anoxygenic phototrophs face a further problem because they also require reducing power (NAD[P]H or reduced ferredoxin) for CO_2 fixation and other biosynthetic processes. They are able to generate reducing power in at least three ways, depending on the bacterium. Some have hydrogenases that are used to produce NAD(P)H directly from the oxidation of hydrogen gas. This is possible because hydrogen gas has a more negative reduction potential than NAD^+ (*see table 8.1*). Others, such as the photosynthetic purple bacteria, must use reverse electron flow to generate NAD(P)H (figure 9.33). In this mechanism, electrons are drawn off the photosynthetic electron transport chain and "pushed" to $NAD(P)^+$ using PMF or the hydrolysis of ATP. Electrons from electron donors such as hydrogen sulfide, elemental sulfur, and organic compounds replace the electrons removed from the electron

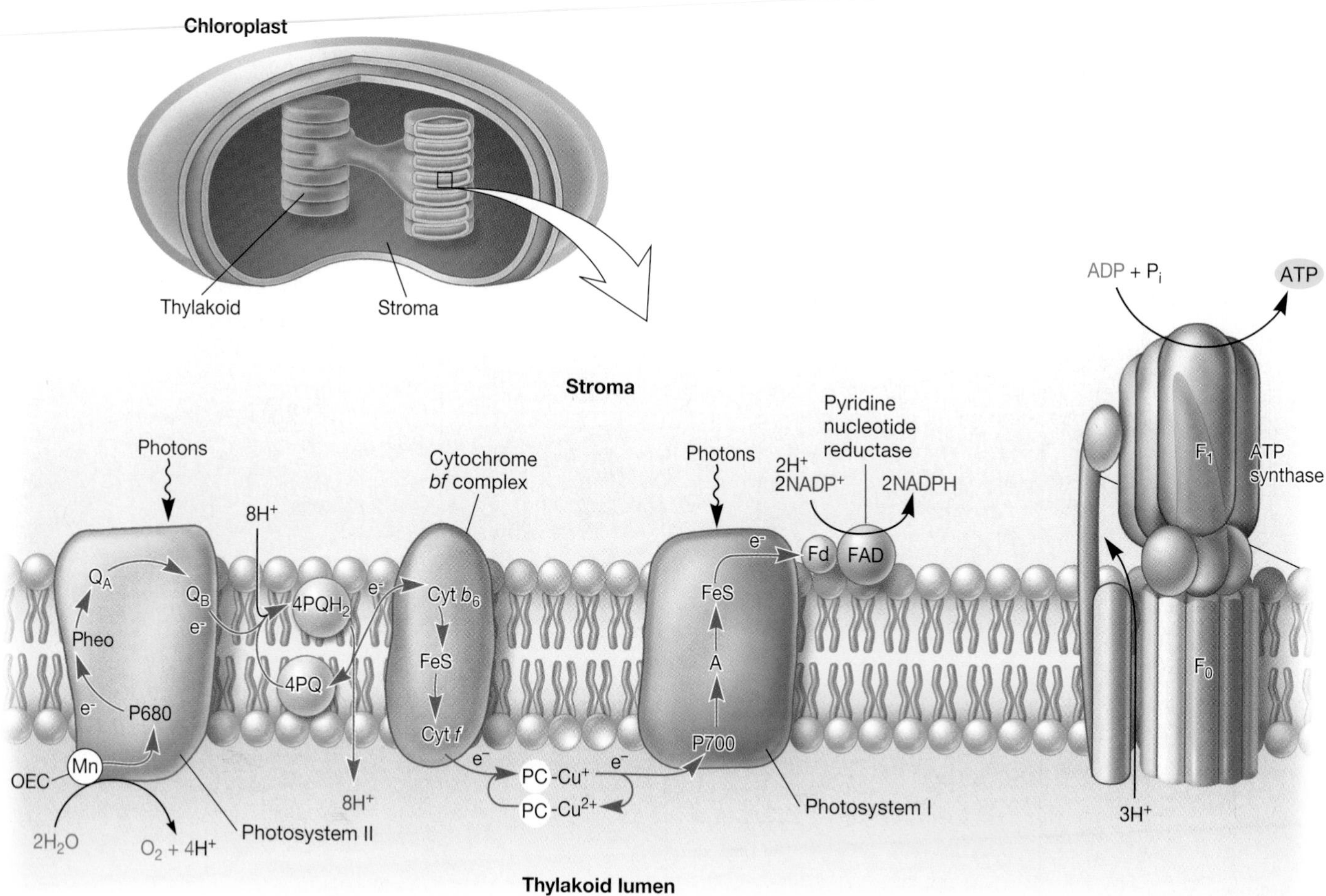

Figure 9.31 The Mechanism of Photosynthesis. An illustration of the chloroplast thylakoid membrane showing photosynthetic electron transport chain function and noncyclic photophosphorylation. The chain is composed of three complexes: PS I, the cytochrome *bf* complex, and PS II. Two diffusible electron carriers connect the three complexes. Plastoquinone (PQ) connects PS I with the cytochrome *bf* complex, and plastocyanin (PC) connects the cytochrome *bf* complex with PS II. The light-driven electron flow pumps protons across the thylakoid membrane and generates an electrochemical gradient, which can then be used to make ATP. Water is the source of electrons and the oxygen-evolving complex (OEC) produces oxygen.

transport chain in this way. Phototrophic green bacteria and heliobacteria also must draw off electrons from their electron transport chains. However, because the reduction potential of the component of the chain where this occurs is more negative than NAD^+ and oxidized ferredoxin, the electrons flow spontaneously to these electron acceptors. Thus these bacteria exhibit a simple form of noncyclic photosynthetic electron flow (**figure 9.34**).

Rhodopsin-Based Phototrophy

Oxygenic and anoxygenic photosynthesis are chlorophyll-based types of phototrophy—that is, chlorophyll or bacteriochlorophyll is the major pigment used to absorb light and initiate the conversion of light energy to chemical energy. This type of phototrophy is observed only in eucaryotes and bacteria; it has not been observed in any archaea, to date. However, some archaea are able to use light as a source of energy. Instead of using chlorophyll, these microbes use a membrane protein called **bacteriorhodopsin** (more correctly called archaeorhodopsin). One such archaeon is the halophile *Halobacterium salinarum*.

H. salinarum normally depends on aerobic respiration for the release of energy from an organic energy source. However, under conditions of low oxygen and high light intensity, it synthesizes bacteriorhodopsin, a deep-purple pigment that closely resembles the rhodopsin from the rods and cones of vertebrate eyes. Bacteriorhodopsin's chromophore is retinal, a type of carotenoid. The chromophore is covalently attached to the pigment protein, which is embedded in the plasma membrane in such a way that the retinal is in the center of the membrane.

Bacteriorhodopsin functions as a light-driven proton pump. When retinal absorbs light, a proton is released and the bacteriorhodopsin undergoes a sequence of conformation changes that translocate the proton into the periplasmic space (*see figure 20.13*). The light-driven proton pumping generates a pH gradient

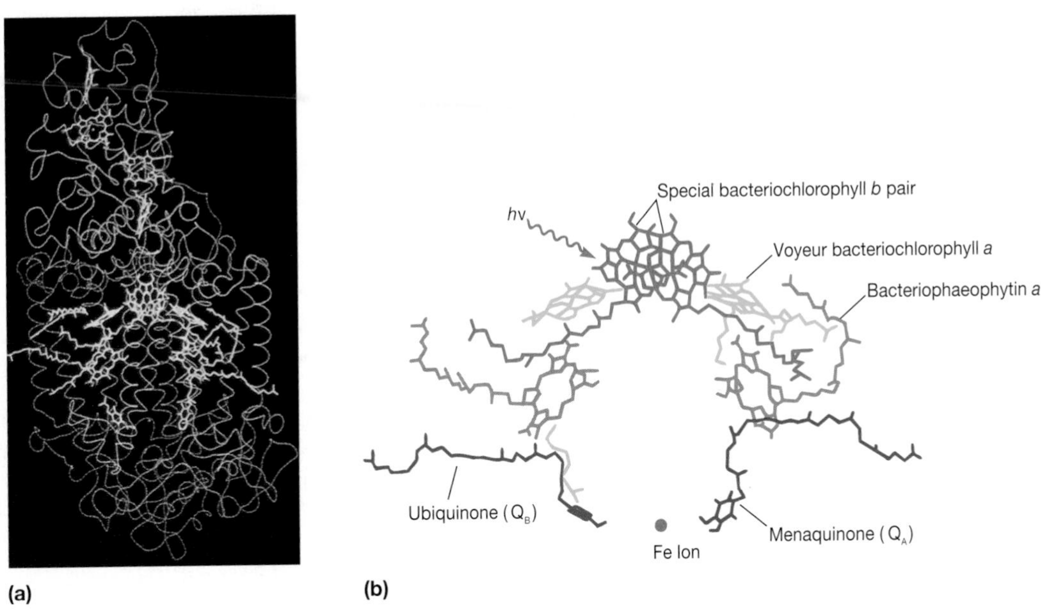

Figure 9.32 A Photosynthetic Reaction Chain. The reaction center of the purple nonsulfur bacterium, *Rhodopseudomonas viridis.* **(a)** The structure of the C_α backbone of the center's polypeptide chains with the bacteriochlorophylls and other prosthetic groups in yellow. **(b)** A close-up view of the reaction center prosthetic groups. A photon is first absorbed by the "special pair" of bacteriochlorophyll *a* molecules, thus exciting them. An excited electron then moves to the bacteriopheophytin molecule in the right arm of the system.

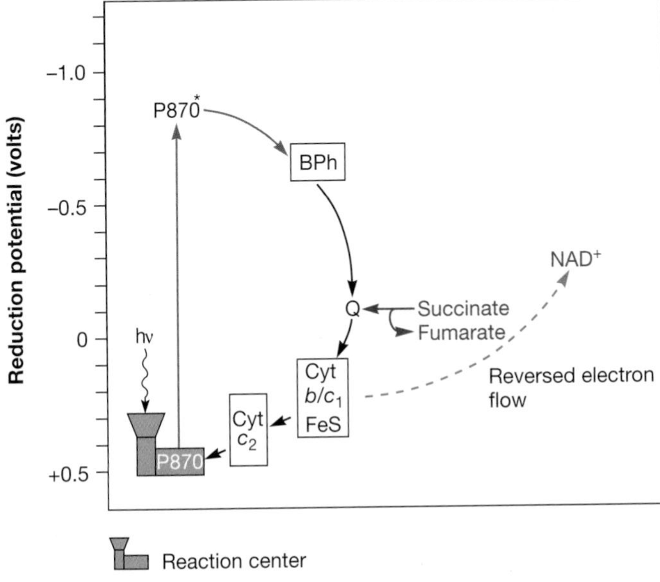

Figure 9.33 Purple Nonsulfur Bacterial Photosynthesis. The photosynthetic electron transport system in the purple nonsulfur bacterium, *Rhodobacter sphaeroides.* This scheme is incomplete and tentative. Ubiquinone (Q) is very similar to coenzyme Q. BPh stands for bacteriopheophytin. The electron source succinate is in blue.

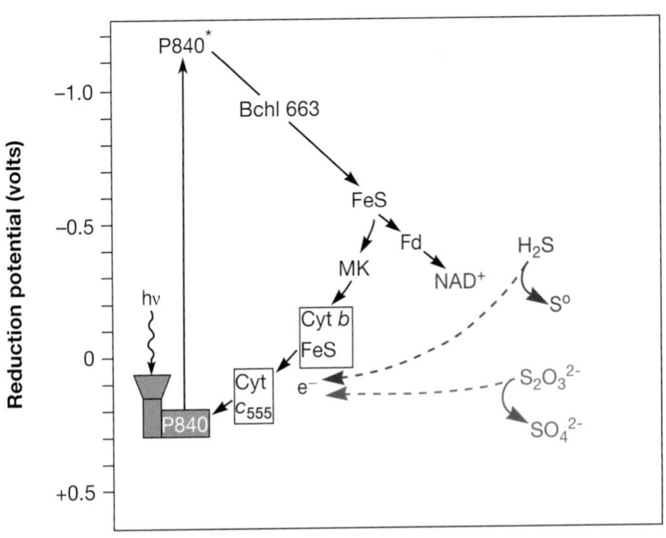

Figure 9.34 Green Sulfur Bacterial Photosynthesis. The photosynthetic electron transport system in the green sulfur bacterium, *Chlorobium limicola.* Light energy is used to make ATP by cyclic photophosphorylation and to move electrons from thiosulfate ($S_2O_3^{2-}$) and H_2S (green and blue) to NAD^+. The electron transport chain has a quinone called menaquinone (MK).

that can be used to power the synthesis of ATP by chemiosmosis. This phototrophic capacity is particularly useful to *Halobacterium* because oxygen is not very soluble in concentrated salt solutions and may decrease to an extremely low level in *Halobacterium*'s habitat. When the surroundings become temporarily anoxic, the archaeon uses light energy to synthesize sufficient ATP to survive until oxygen levels rise again. *Halobacterium* cannot grow anaerobically by anaerobic respiration or fermentation because it needs oxygen for synthesis of retinal. However, it can survive the stress of temporary oxygen limitation by means of phototrophy. Note, however, that this type of phototrophy does not involve electron transport. It had been thought that rhodopsin-based phototrophy is unique to *Archaea.* However, proton-pumping rhodopsins have recently been discovered in some proteobacteria (proteorhodopsin) and a fungus. Environmental genomics (section 15.9)

1. Define the following terms: light reaction, chlorophyll, carotenoid, phycobiliprotein, antenna, and photosystems I and II.
2. What happens to a reaction center chlorophyll pair, like P700, when it absorbs light?
3. What is the function of accessory pigments?
4. What is photophosphorylation? What is the difference between cyclic and noncyclic photophosphorylation?
5. Why is the light reaction in green bacteria, purple bacteria, and heliobacteria termed anoxygenic?
6. Compare and contrast anoxygenic photosynthesis and oxygenic photosynthesis. How do these two types of phototrophy differ from rhodopsin-based phototrophy?
7. Suppose you isolated a bacterial strain that carried out oxygenic photosynthesis. What photosystems would it possess and what group of bacteria would it most likely belong to?

Summary

9.1 Chemoorganotrophic Fueling Processes

a. Chemotrophic microorganisms can use three kinds of electron acceptors during energy metabolism (**figure 9.2**). The nutrient may be oxidized with an endogenous electron acceptor (fermentation), with oxygen as an exogenous electron acceptor (aerobic respiration), or with another external electron acceptor (anaerobic respiration).

9.2 Aerobic Respiration

a. Aerobic respiration can be divided into three stages: (1) breakdown of macromolecules into their constituent parts, (2) catabolism to pyruvate, acetyl-CoA, and other molecules by pathways that converge on glycolytic pathways and the TCA cycle, and (3) completion of catabolism by the TCA cycle. Most energy is produced at this stage and results from oxidation of NADH and $FADH_2$ by the electron transport chain and oxidative phosphorylation (**figure 9.3**).

b. The pathways used during aerobic respiration are amphibolic, having both catabolic and anabolic functions (**figure 9.4**).

9.3 The Breakdown of Glucose to Pyruvate

a. Glycolysis, used in its broadest sense, refers to all pathways used to break down glucose to pyruvate.

b. The Embden-Meyerhof pathway has a net production of two NADHs and two ATPs, the latter being produced by substrate-level phosphorylation. It also produces several precursor metabolites (**figure 9.5**).

c. In the pentose phosphate pathway, glucose 6-phosphate is oxidized twice and converted to pentoses and other sugars. It is a source of NADPH, ATP, and several precursor metabolites (**figure 9.6**).

d. In the Entner-Doudoroff pathway, glucose is oxidized to 6-phosphogluconate, which is then dehydrated and cleaved to pyruvate and glyceraldehyde 3-phosphate (**figure 9.8**). The latter product can be oxidized by glycolytic enzymes to provide ATP, NADH, and another molecule of pyruvate.

9.4 The Tricarboxylic Acid Cycle

a. The tricarboxylic acid cycle is the final stage of catabolism in most aerobic cells (**figure 9.9**). It oxidizes acetyl-CoA to CO_2 and forms one GTP, three NADHs, and one $FADH_2$ per acetyl-CoA. It also generates several precursor metabolites.

9.5 Electron Transport and Oxidative Phosphorylation

a. The NADH and $FADH_2$ produced from the oxidation of carbohydrates, fatty acids, and other nutrients can be oxidized in the electron transport chain. Electrons flow from carriers with more negative reduction potentials to those with more positive potentials (**figure 9.10** *see also figure 8.8*), and free energy is released for ATP synthesis by oxidative phosphorylation.

b. Bacterial electron transport chains are often different from eucaryotic chains with respect to such aspects as carriers and branching. In eucaryotes the P/O ratio for NADH is about 3 and that for $FADH_2$ is around 2; P/O ratios are usually much lower in bacteria.

c. ATP synthase catalyzes the synthesis of ATP. In eucaryotes, it is located on the inner surface of the inner mitochondrial membrane. Bacterial ATP synthase is on the inner surface of the plasma membrane.

d. The most widely accepted mechanism of oxidative phosphorylation is the chemiosmotic hypothesis in which proton motive force (PMF) drives ATP synthesis (**figure 9.11**).

e. Aerobic respiration in eucaryotes can yield a maximum of 38 ATPs (**figure 9.15**).

9.6 Anaerobic Respiration

a. Anaerobic respiration is the process of ATP production by electron transport in which the terminal electron acceptor is an exogenous, molecule other than O_2. The most common acceptors are nitrate, sulfate, and CO_2 (**figure 9.16**).

9.7 Fermentations

a. During fermentation, an endogenous electron acceptor is used to reoxidize any NADH generated by the catabolism of glucose to pyruvate (**figure 9.17**).

b. Flow of electrons from the electron donor to the electron acceptor does not involve an electron transport chain, and ATP is synthesized only by substrate-level phosphorylation.

9.8 Catabolism of Carbohydrates and Intracellular Reserve Polymers

a. Microorganisms catabolize many extracellular carbohydrates. Monosaccharides are taken in and phosphorylated; disaccharides may be cleaved to monosaccharides by either hydrolysis or phosphorolysis.

b. External polysaccharides are degraded by hydrolysis and the products are absorbed. Intracellular glycogen and starch are converted to glucose 1-phosphate by phosphorolysis (**figure 9.20**).

9.9 Lipid Catabolism

a. Fatty acids from lipid catabolism are usually oxidized to acetyl-CoA in the β-oxidation pathway (**figure 9.22**).

9.10 Protein and Amino Acid Catabolism

a. Proteins are hydrolyzed to amino acids that are then deaminated; their carbon skeletons feed into the TCA cycle (**figure 9.23**).

9.11 Chemolithotrophy

a. Chemolithotrophs synthesize ATP by oxidizing inorganic compounds—usually hydrogen, reduced nitrogen and sulfur compounds, or ferrous iron—with an electron transport chain and O_2 as the electron acceptor (**figure 9.24** and **table 9.3**).

b. Many of the energy sources used by chemolithotrophs have a more positive standard reduction potential than the NAD^+/NADH redox pair. These chemolithotrophs must expend energy (PMF or ATP) to drive reverse electron flow and produce the NADH they need for CO_2 fixation and other processes (**figure 9.25**).

9.12 Phototrophy

a. In oxygenic photosynthesis, eucaryotes and cyanobacteria trap light energy with chlorophyll and accessory pigments and move electrons through photosystems I and II to make ATP and NADPH (the light reactions).

b. Cyclic photophosphorylation involves the activity of photosystem I alone and generates ATP only. In noncyclic photophosphorylation photosystems I and II operate together to move electrons from water to $NADP^+$ producing ATP, NADPH, and O_2 (**figure 9.30**).

c. Anoxygenic phototrophs differ from oxygenic phototrophs in possessing bacteriochlorophyll and having only one photosystem (**figures 9.33** and **9.34**). They use cyclic photophosphorylation to make ATP. They are anoxygenic because they do not use water as an electron donor for electron flow though the photosynthetic electron transport chain.

d. Some archaea and bacteria use a type of phototrophy that involves the proton-pumping pigment bacteriorhodopsin and proteorhodopsin, respectively. This type of phototrophy generates PMF but does not involve an electron transport chain.

Key Terms

accessory pigments 217
acetyl-coenzyme A (acetyl-CoA) 198
adenosine 5′-phosphosulfate (APS) 214
aerobic respiration 193
alcoholic fermentation 208
amphibolic pathways 194
anaerobic respiration 205
anoxygenic photosynthesis 218
antenna 217
ATP synthase 202
bacteriochlorophyll 218
bacteriorhodopsin 220
β-oxidation pathway 211
butanediol fermentation 209
carotenoids 217
chemiosmotic hypothesis 202
chemolithotroph 212
chlorophylls 216
citric acid cycle 198
cyclic photophosphorylation 217
dark reactions 215
deamination 212
denitrification 205
dissimilatory nitrate reduction 205
electron transport chain 200
Embden-Meyerhof pathway 194
Entner-Doudoroff pathway 198
fermentation 207
glycolysis 194
glycolytic pathway 194
heterolactic fermenters 208
hexose monophosphate pathway 196
homolactic fermenters 208
Krebs cycle 198
lactic acid fermentation 208
light reactions 215
mixed acid fermentation 209
nitrification 213
nitrifying bacteria 213
noncyclic photophosphorylation 218
oxidative phosphorylation 202
oxygenic photosynthesis 216
pentose phosphate pathway 196
photosynthesis 214
photosystem I 217
photosystem II 217
phycobiliproteins 217
phycocyanin 217
phycoerythrin 217
protease 212
proton motive force (PMF) 202
reaction-center chlorophyll pair 217
respiration 192
reverse electron flow 213
Stickland reaction 209
substrate-level phosphorylation 194
transamination 212
tricarboxylic acid (TCA) cycle 198
uncouplers 203

Critical Thinking Questions

1. Without looking in chapter 21, predict some characteristics that would describe niches occupied by green and purple photosynthetic bacteria.
2. From an evolutionary perspective, discuss why most microorganisms use aerobic respiration to generate ATP.
3. How would you isolate a thermophilic chemolithotroph that uses sulfur compounds as a source of electrons? What changes in the incubation system would be needed to isolate bacteria using sulfur compounds in anaerobic respiration? How can one tell which process is taking place through an analysis of the sulfur molecules present in the medium?
4. Certain uncouplers block ATP synthesis by allowing protons and other ions to "leak across membranes," disrupting the charge and proton gradients established by electron flow through an electron transport chain. Does this observation support the chemiosmosis hypothesis? Explain your reasoning.
5. Two flasks of *E. coli* are grown in batch culture in the same medium (2% glucose and amino acids; no nitrate) and at the same temperature (37°C). Culture #1 is well aerated. Culture #2 is anoxic. After 16 hours the following observations are made:
   - Culture #1 has a high cell density; the cells appear to be in stationary phase, and the glucose level in the medium is reduced to 1.2%.
   - Culture #2 has a low cell density; the cells appear to be in logarithmic phase, although their doubling time is prolonged (over one hour). The glucose level is reduced to 0.2%.

   What type of glucose catabolism was used in each culture? Why does culture #2 have so little glucose remaining relative to culture #1, even though culture #2 displayed slower growth and has less biomass?

Learn More

Anraku, Y. 1988 Bacterial electron transport chains. *Annu. Rev. Biochem.* 57:101–32.

Baker, S. C.; Ferguson, S. J.; Ludwig, B.; Page, M. D.; Richter, O.-M. H.; and van Spanning, R. J. M.1998. *Microbiol. Mol. Biol. Rev.* 62(4):1046–78.

Faxén, K.; Gilderson, G.; Ädelroth, P.; and Brzezinski, P. 2005. A mechanistic principle for proton pumping by cytochrome *c* oxidase. *Nature* 437:286–89.

Gao, Y. Q.; Yang, W.; and Karplus, M. 2005. A structure-based model for the synthesis and hydrolysis of ATP by F_1-ATPase. *Cell* 123:195–205.

Gottschalk, G. 1986. *Bacterial metabolism,* 2d ed. New York: Springer-Verlag.

Gunsalus, R. P. 2000. Anaerobic respiration. In *Encyclopedia of microbiology,* 2d ed., vol. 1, J. Lederberg, editor-in-chief, 180–88. San Diego: Academic Press.

Kinosita, K., Jr.; Adachi, K.; and Itoh, H. 2004. Rotation of F_1-ATPase: How an ATP-driven molecular machine may work. *Annu. Rev. Biophys. Biomol. Struct.* 33:245–68.

McKee, T., and McKee, J. R. 2003. *Biochemistry: The molecular basis of life,* 3d ed. Dubuque, Iowa: McGraw-Hill.

Nelson, D. L., and Cox, M. M. 2005. *Lehninger: Principles of biochemistry*, 4th ed. New York: W. H. Freeman.

Nicholls, D. G., and Ferguson, S. J. 2002. *Bioenergetics,* 3d ed. San Diego: Academic Press.

Peschek, G. A.; Obinger, C.; and Paumann, M. 2004. The respiratory chain of blue-green algae (cyanobacteria). *Physiologia Plantarum* 120:358–69.

Saier, M. H., Jr. 1997. Peter Mitchell and his chemiosmotic theories. *ASM News* 63(1):13–21.

Please visit the Prescott website at www.mhhe.com/prescott7 for additional references.

10 Metabolism:
The Use of Energy in Biosynthesis

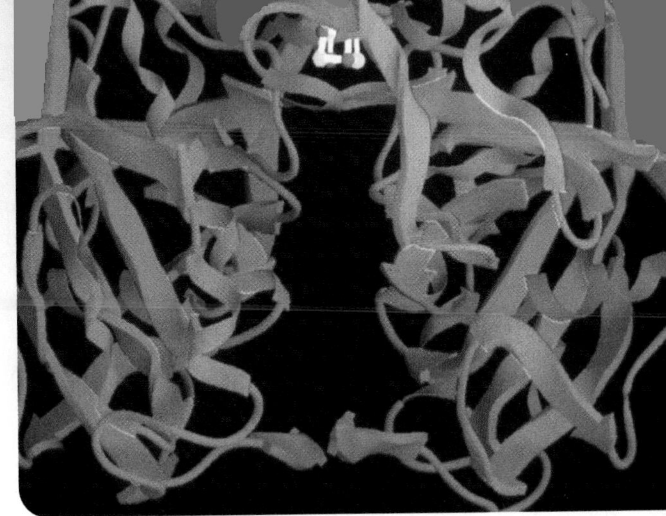

The nitrogenase Fe protein's subunits are arranged like a pair of butterfly wings. Nitrogenase consists of the Fe protein and the MoFe protein; it catalyzes the reduction of atmospheric nitrogen during nitrogen fixation.

Preview

- In anabolism, cells use free energy to construct more complex molecules and structures from smaller, simpler precursors.
- Biosynthetic pathways are organized to optimize efficiency by conserving biosynthetic raw materials and energy. This is accomplished in a number of ways, including the use of amphibolic pathways that function in both catabolic and anabolic directions. Some key reactions in amphibolic pathways require two enzymes: one for the catabolic reaction and another for the anabolic reaction.
- Precursor metabolites are carbon skeletons that serve as the starting substrates for biosynthetic pathways. They are intermediates of the central metabolic pathways.
- Four different pathways for CO_2 fixation have been identified in microorganisms. The most commonly used pathway is the Calvin cycle. All CO_2-fixation pathways consume ATP and reducing power (e.g., NADPH).
- Gluconeogenesis is used to synthesize glucose from noncarbohydrate organic molecules. Glucose and other hexoses serve as precursor metabolites for the synthesis of other sugars and polysaccharides. The synthesis of the polysaccharide peptidoglycan is particularly complex, requiring many steps and occurring at several locations in the cell.
- Many of the precursor metabolites are used in amino acid biosynthetic pathways. The carbon skeletons are remodeled and amended by the addition of nitrogen and sometimes sulfur. Many amino acid biosynthetic pathways are branched. Thus a single precursor metabolite can produce a family of related amino acids. Anaplerotic reactions ensure that an adequate supply of precursor metabolites is available for amino acid biosynthesis.
- Certain amino acids and precursor metabolites contribute to the synthesis of nucleotides. Phosphorus assimilation is also required.
- The acetyl-CoA and malonyl-CoA pathways synthesize fatty acids. These pathways are not amphibolic, as they only function in the synthesis of lipids, not in their degradation.

As chapter 9 makes clear, microorganisms can obtain energy in many ways. Much of this energy is used in anabolism. During anabolism, a microorganism begins with simple inorganic molecules and a carbon source and constructs ever more complex molecules until new organelles and cells arise (**figure 10.1**). A microbial cell must manufacture many different kinds of molecules; here, we discuss the synthesis of only the most important types of cell constituents.

In this chapter we begin with a general introduction to anabolism and the role played by the precursor metabolites in biosynthetic pathways. We then focus on CO_2 fixation and the synthesis of carbohydrates, amino acids, purines and pyrimidines, and lipids. Because protein and nucleic acid synthesis is so significant and complex, the polymerization reactions that yield these macromolecules is described separately in chapter 11.

Anabolism is the creation of order. Because a cell is highly ordered and immensely complex, a lot of energy is required for biosynthesis. This is readily apparent from estimates of the biosynthetic capacity of rapidly growing *Escherichia coli* (**table 10.1**). Although most ATP dedicated to biosynthesis is employed in protein synthesis, ATP is also used to make other cell constituents.

It is intuitively obvious why rapidly growing cells need a large supply of ATP. But even nongrowing cells need energy for the biosynthetic processes they carry out. This is because nongrowing cells continuously degrade and resynthesize cellular molecules during a process known as **turnover.** Thus cells are never the same from one instant to the next. In addition, many nongrowing cells use energy to synthesize enzymes and other substances for release into

Biological structures are almost always constructed in a hierarchical manner, with subassemblies acting as important intermediates en route from simple starting molecules to the end products of organelles, cells, and organisms.

—*W. M. Becker and D. W. Deamer*

their surroundings. Clearly, metabolism must be carefully regulated if the rate of turnover is to be balanced by the rate of biosynthesis. It must also be regulated in response to a microbe's environment. Some of the mechanisms of metabolic regulation have already been introduced in chapter 8; others are discussed in chapter 12.

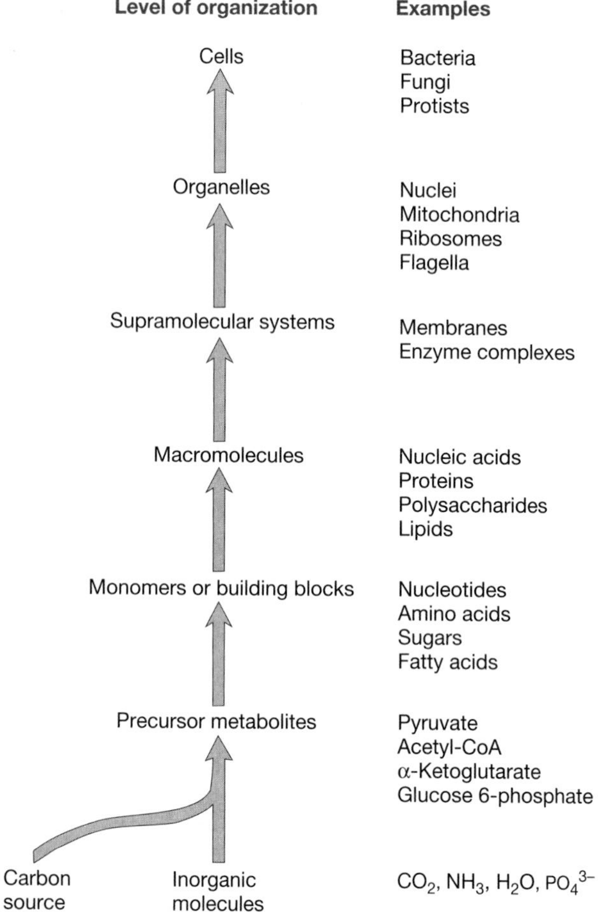

Figure 10.1 The Construction of Cells. The biosynthesis of procaryotic and eucaryotic cell constituents. Biosynthesis is organized in levels of ever greater complexity.

10.1 Principles Governing Biosynthesis

Biosynthetic metabolism generally follows certain patterns and is shaped by a few basic principles. Six of these are now briefly discussed.

1. The construction of large **macromolecules** (complex molecules) from a few simple structural units (**monomers**) saves much genetic storage capacity, biosynthetic raw material, and energy. A consideration of protein synthesis clarifies this. Proteins—whatever size, shape, or function—are made of only 20 common amino acids joined by peptide bonds. Different proteins simply have different amino acid sequences but not new and dissimilar amino acids. Suppose that proteins were composed of 40 different amino acids instead of 20. The cell would then need the enzymes to manufacture twice as many amino acids (or would have to obtain the extra amino acids in its diet). Genes would be required for the extra enzymes, and the cell would have to invest raw materials and energy in the synthesis of these additional genes, enzymes, and amino acids. Clearly the use of a few monomers linked together by a single type of covalent bond makes the synthesis of macromolecules a highly efficient process. Proteins and amino acids (appendix I)
2. The use of many of the same enzymes for both catabolic and anabolic processes saves additional materials and energy. For example, most glycolytic enzymes are involved in both the synthesis and the degradation of glucose.
3. The use of separate enzymes to catalyze the two directions of a single step in an amphibolic pathway allows independent regulation of catabolism and anabolism (**figure 10.2**). Thus catabolic and anabolic pathways are never identical although many enzymes are shared. Although this is discussed in more detail in sections 8.8 through 8.10, note that the regulation of anabolism is somewhat different from that of catabolism. Both types of pathways can be regulated by their end products as well as by the concentrations of ATP, ADP, AMP, and NAD^+. Nevertheless, end product regulation generally assumes more importance in anabolic pathways.
4. To synthesize molecules efficiently, anabolic pathways must operate irreversibly in the direction of biosynthesis. Cells can

Table 10.1 Biosynthesis in *Escherichia coli*

Cell Constituent	Number of Molecules per Cell[a]	Molecules Synthesized per Second	Molecules of ATP Required per Second for Synthesis
DNA	1[b]	0.00083	60,000
RNA	15,000	12.5	75,000
Polysaccharides	39,000	32.5	65,000
Lipids	15,000,000	12,500.0	87,000
Proteins	1,700,000	1,400.0	2,120,000

[a] Estimates for a cell with a volume of 2.25 μm^3, a total weight of 1×10^{-12}g, a dry weight of 2.5×10^{-13}g, and a 20 minute cell division cycle.

[b] It should be noted that bacteria can contain multiple copies of their genomic DNA.

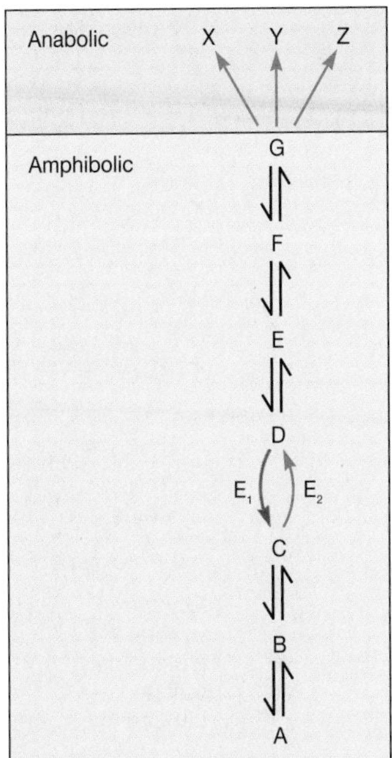

Figure 10.2 A Hypothetical Biosynthetic Pathway. The routes connecting G with X, Y, and Z are purely anabolic because they are used only for synthesis of the end products. The pathway from A to G is amphibolic—that is, it has both catabolic and anabolic functions. Most reactions are used in both roles; however, the interconversion of C and D is catalyzed by two separate enzymes, E_1 (catabolic) and E_2 (anabolic).

achieve this by connecting some biosynthetic reactions to the breakdown of ATP and other nucleoside triphosphates. When these two processes are coupled, the free energy made available during nucleoside triphosphate breakdown drives the biosynthetic reaction to completion. The role of ATP in metabolism (section 8.5)

5. Compartmentation in eucaryotic cells—that is, localization of biosynthetic pathways into certain cellular compartments and catabolic pathways into others—makes it easier for catabolic and anabolic pathways to operate simultaneously yet independently. For example, fatty acid biosynthesis occurs in the cytoplasmic matrix, whereas fatty acid oxidation takes place within the mitochondrion.
6. Finally, anabolic and catabolic pathways often use different cofactors. Usually catabolic oxidations produce NADH, a substrate for electron transport. In contrast, when an electron donor is needed during biosynthesis, NADPH rather than NADH normally serves as the donor. Fatty acid metabolism provides a second example. Fatty acyl-CoA molecules are oxidized to generate energy, whereas fatty acid synthesis involves acyl carrier protein thioesters (see p. 242).

After macromolecules have been constructed from simpler precursors, they are assembled into larger, more complex structures such as supramolecular systems and organelles (figure 10.1). Macromolecules normally contain the necessary information to form supramolecular systems spontaneously in a process known as **self-assembly.** For example, ribosomes are large assemblages of many proteins and ribonucleic acid molecules, yet they arise by the self-assembly of their components without the involvement of extra factors.

1. Define anabolism, turnover, and self-assembly.
2. Summarize the six principles by which biosynthetic pathways are organized.

10.2 The Precursor Metabolites

The generation of the **precursor metabolites** is a critical step in anabolism. Precursor metabolites are carbon skeletons (i.e., carbon chains) used as the starting substrates for the synthesis of monomers and other building blocks needed for the synthesis of macromolecules. Precursor metabolites are referred to as carbon skeletons because they are molecules that lack functional moieties such as amino and sulfhydryl groups; these are added during the biosynthetic process. The precursor metabolites and their use in biosynthesis are shown in **figure 10.3.** Several things should be noted in this figure. First, all the precursor metabolites are intermediates of the glycolytic pathways (Embden-Meyerhof pathway or the Entner-Doudoroff pathway, and the pentose phosphate pathway) and the tricarboxylic acid (TCA) cycle. Therefore these pathways play a central role in metabolism and are often referred to as the **central metabolic pathways.** Note, too, that most of the precursor metabolites are used for synthesis of amino acids and nucleotides.

From careful examination of figure 10.3, it should be clear that if an organism is a chemoorganotroph using glucose as its energy, electron, and carbon source, it generates the precursor metabolites as it generates ATP and reducing power. But what if the chemoorganotroph is using an amino acid as its sole source of carbon, electrons, and energy? And what about autotrophs? How do they generate precursor metabolites from CO_2, their carbon source? Heterotrophs growing on something other than glucose degrade that carbon and energy source into one or more intermediates of the central metabolic pathways. From there, they can generate the remaining precursor metabolites. Autotrophs must first convert CO_2 into organic carbon from which they can generate the precursor metabolites. Many of the reactions that autotrophs use to generate the precursor metabolites are reactions of the central metabolic pathways, operating either in the catabolic direction or in the anabolic direction. Thus the central metabolic pathways are important to the anabolism of both heterotrophs and autotrophs.

We begin our discussion of anabolism by first considering CO_2 fixation by autotrophs. Once CO_2 is converted to organic carbon, the synthesis of other precursor metabolites, amino acids, nucleotides, and additional building blocks is essentially the same in both autotrophs and heterotrophs. Recall that the precursor metabolites provide the carbon skeletons for the synthesis of other important organic molecules. In the process of transforming a precursor metabolite into an amino acid or a nucleotide, the carbon skeleton is modified in a number of ways, including the

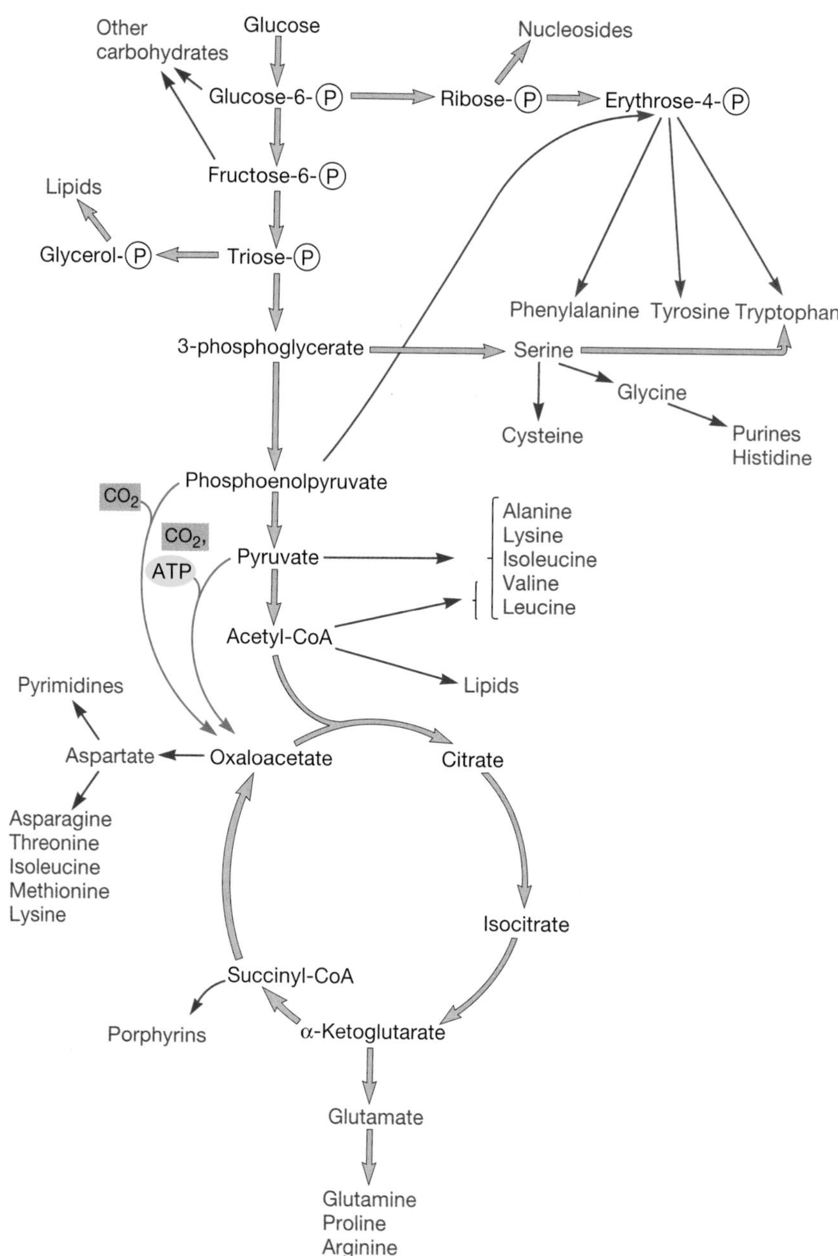

Figure 10.3 The Organization of Anabolism. Biosynthetic products (in blue) are derived from precursor metabolites, which are intermediates of amphibolic pathways. Two major anaplerotic reactions are shown in red. These reactions ensure an adequate supply of TCA cycle-derived precursor metabolites and are especially important to fermentative organisms, in which only certain TCA cycle reactions operate.

addition of nitrogen, phosphorus, and sulfur. Thus as we discuss the synthesis of monomers from precursor metabolites, we will also address the assimilation of nitrogen, sulfur, and phosphorus.

10.3 The Fixation of CO_2 by Autotrophs

Autotrophs use CO_2 as their sole or principal carbon source and the reduction and incorporation of CO_2 requires much energy. Many autotrophs obtain energy by trapping light during photosynthesis, but some derive energy from the oxidation of reduced inorganic electron donors. Autotrophic CO_2 fixation is crucial to life on Earth because it provides the organic matter on which heterotrophs depend. Chemolithotrophy (section 9.11); Phototrophy (section 9.12)

Four different CO_2-fixation pathways have been identified in microorganisms. Most autotrophs use the **Calvin cycle,** which is also called the Calvin-Benson cycle or the reductive pentose phosphate cycle. The Calvin cycle is found in photosynthetic eucaryotes and most photosynthetic bacteria. It is absent in some obligatory anaerobic and microaerophilic bacteria. Autotrophic archaea also use an alternative pathway for CO_2 fixation. We consider the Calvin cycle first, and then briefly introduce the three other CO_2-fixation pathways.

The Calvin Cycle

The Calvin cycle is also called the reductive pentose phosphate cycle because it is essentially the reverse of the pentose phosphate pathway. Thus many of the reactions are similar, in particular the sugar transformations. The reactions of the Calvin cycle occur in the chloroplast stroma of eucaryotic microbial autotrophs. In cyanobacteria, some nitrifying bacteria, and thiobacilli (sulfur-oxidizing chemolithotrophs), the Calvin cycle is associated with inclusion bodies called **carboxysomes.** These are polyhedral structures that contain the enzyme critical to the Calvin cycle and may be the site of CO_2 fixation. The breakdown of glucose to pyruvate: The pentose-phosphate pathway (section 9.3)

The Calvin cycle is divided into three phases: carboxylation phase, reduction phase, and regeneration phase (**figure 10.4** and appendix II). During the carboxylation phase, the enzyme **ribulose-1,5-bisphosphate carboxylase**, also called ribulose bisphosphate carboxylase/oxygenase (Rubisco), catalyzes the addition of CO_2 to the 5-carbon molecule ribulose-1,5-bisphosphate (RuBP), forming a six-carbon intermediate that rapidly and spontaneously splits into two molecules of 3-phosphoglycerate (PGA) (**figure 10.5**). Note that PGA is an intermediate of the Embden-Meyerhof pathway (EMP), and in the reduction phase, PGA is reduced to glyceraldehyde 3-phosphate by two reactions that are essentially the reverse of two EMP reactions. The difference is that the Calvin cycle enzyme glyceraldehyde 3-phosphate dehydrogenase uses $NADP^+$ rather than NAD^+ (compare figures 10.4 and 9.5). Finally, in the regeneration phase, RuBP is regenerated, so that the cycle can repeat. In addition, this phase produces carbohydrates such as glyceraldehyde 3-phosphate, fructose 6-phosphate, and glucose 6-phosphate, all of which are precursor metabolites (figure 10.4). This portion of the cycle is similar to the pentose phosphate pathway and involves the transketolase and transaldolase reactions.

To synthesize fructose 6-phosphate or glucose 6-phosphate from CO_2, the cycle must operate six times to yield the desired hexose and reform the six RuBP molecules.

$$6\text{RuBP} + 6\text{CO}_2 \longrightarrow 12\text{PGA} \longrightarrow 6\text{RuBP} + \text{fructose 6-P}$$

The incorporation of one CO_2 into organic material requires three ATPs and two NADPHs. The formation of glucose from CO_2 may be summarized by the following equation.

$$6\text{CO}_2 + 18\text{ATP} + 12\text{NADPH} + 12\text{H}^+ + 12\text{H}_2\text{O} \longrightarrow \text{glucose} + 18\text{ADP} + 18\text{P}_i + 12\text{NADP}^+$$

The precursor metabolites formed in the Calvin cycle can then be used to synthesize other precursor metabolites and essential molecules, as described in sections 10.4 through 10.7.

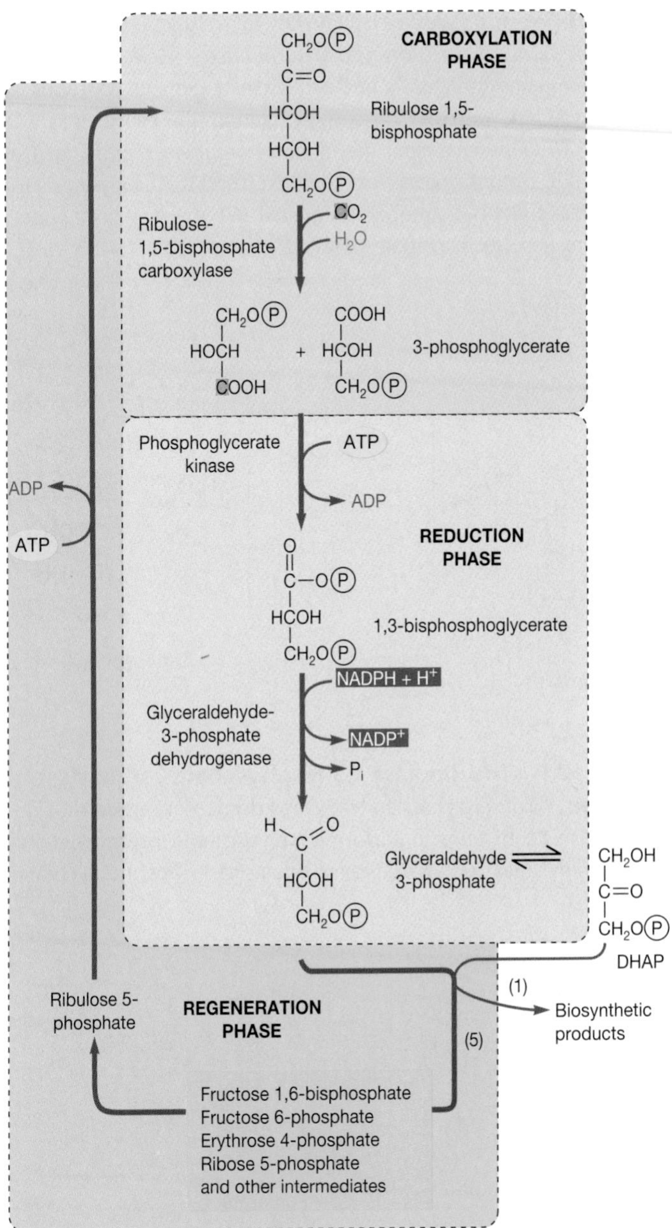

Figure 10.4 The Calvin Cycle. This is an overview of the cycle with only the carboxylation and reduction phases in detail. Three ribulose 1,5-bisphosphates are carboxylated to give six 3-phosphoglycerates in the carboxylation phase. These are converted to six glyceraldehyde 3-phosphates, which can be converted to dihydroxyacetone phosphate (DHAP). Five of the six trioses (glyceraldehyde phosphate and dihydroxyacetone phosphate) are used to reform three ribulose 1,5-bisphosphates in the regeneration phase. The remaining triose is used in biosynthesis. The numbers in parentheses at the lower right indicate this carbon flow.

Other CO_2-Fixation Pathways

Certain bacteria and archaea fix CO_2 using the reductive TCA cycle, the 3-hydroxypropionate cycle, or the acetyl-CoA pathway. The **reductive TCA cycle** (**figure 10.6**) is used by some chemolithoautotrophs (e.g., *Thermoproteus* and *Sulfolobus*, two archaeal genera, and the bacterial genus *Aquifex*) and anoxygenic phototrophs such as *Chlorobium*, a green sulfur bacterium. The reductive TCA cycle is so named because it runs in the reverse direction of the normal, oxidative TCA cycle (compare figures 10.6 and 9.9). A few archaeal genera and the green nonsulfur bacteria (another group of anoxygenic phototrophs) use the **3-hydroxypropionate cycle** to fix CO_2.

Figure 10.7 shows the cycle as it is thought to function in the green nonsulfur bacterium *Chloroflexus aurantiacus*. How its product, glyoxylate, is assimilated is unclear. Methanogens use portions of the **acetyl-CoA pathway** for carbon fixation; the pathway as it is used by *Methanobacterium thermoautotrophicum* is illustrated in **figure 10.8.** Both the acetyl-CoA pathway and methanogenesis involve the activity of a number of unusual enzymes and coenzymes. These are described in more detail in chapter 20. Phylum *Crenarchaeota* (section 20.2); *Aquificae* and *Thermotogae* (section 21.1); Photosynthetic bacteria (section 21.3)

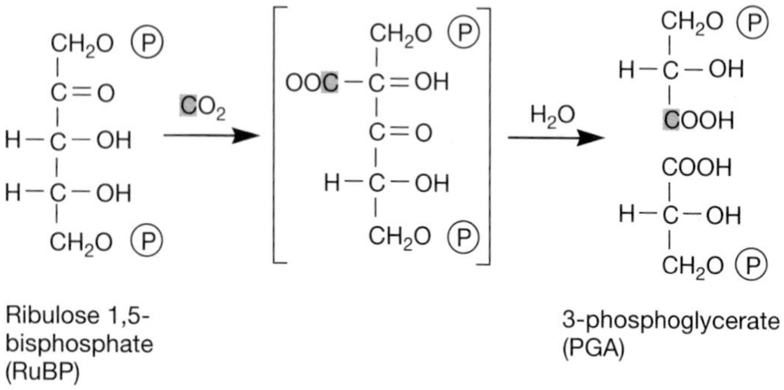

Ribulose 1,5-bisphosphate (RuBP)

3-phosphoglycerate (PGA)

Figure 10.5 The Ribulose 1,5-Bisphosphate Carboxylase Reaction. This enzyme catalyzes the addition of carbon dioxide to ribulose 1,5-bisphosphate, forming an unstable intermediate, which then breaks down to two molecules of 3-phosphoglycerate.

1. Briefly describe the three phases of the Calvin cycle. What other pathways are used to fix CO_2?
2. Which two enzymes are specific to the Calvin cycle?

10.4 Synthesis of Sugars and Polysaccharides

Autotrophs using CO_2-fixation processes other than the Calvin cycle and heterotrophs growing on carbon sources other than sugars must be able to synthesize glucose. The synthesis of glucose from noncarbohydrate precursors is called **gluconeogenesis.** The gluconeogenic pathway shares seven enzymes with the Embden-Meyerhof pathway. However, the two pathways are not identical (**figure 10.9**). Three glycolytic steps are irreversible in the cell: (1) the conversion of phosphoenolpyruvate to pyruvate, (2) the formation of fructose 1,6-bisphosphate from fructose 6-phosphate, and (3) the phosphorylation of glucose. These must be bypassed when the pathway is operating biosynthetically. For example, the formation of fructose 1,6-bisphosphate by phosphofructokinase is reversed by a different enzyme, fructose bisphosphatase, which hydrolytically removes a phosphate from fructose bisphosphate. Usually at least two enzymes are involved in the conversion of pyruvate to phosphoenolpyruvate (the reversal of the pyruvate kinase step).

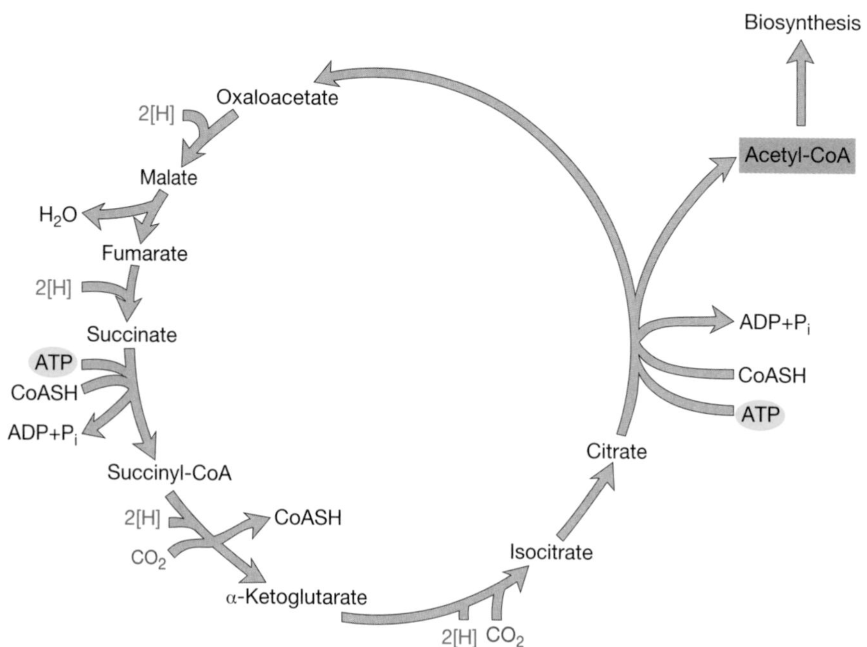

Figure 10.6 The Reductive TCA Cycle. This cycle is used by green sulfur bacteria and some chemolithotrophic archaea to fix CO_2. The cycle runs in the opposite direction as the TCA cycle. ATP and reducing equivalents [H] power the reversal. In green sulfur bacteria, the reducing equivalents are provided by reduced ferredoxin. The product of this process is acetyl-CoA, which can be used to synthesize other organic molecules and precursor metabolites.

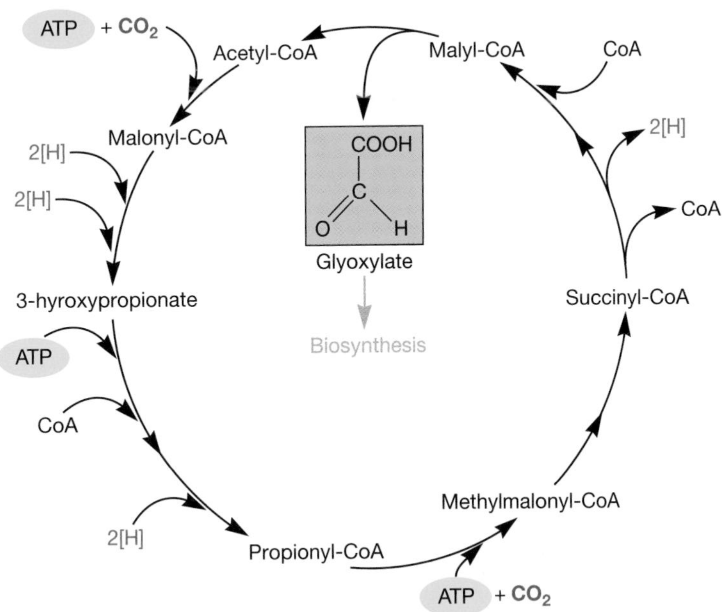

Figure 10.7 The 3-Hydroxypropionate Pathway. This pathway functions in green nonsulfur bacteria, a group of anoxygenic phototrophs. The product of the cycle is glyoxylate, which is used in biosynthesis by mechanisms that have not been definitively elucidated.

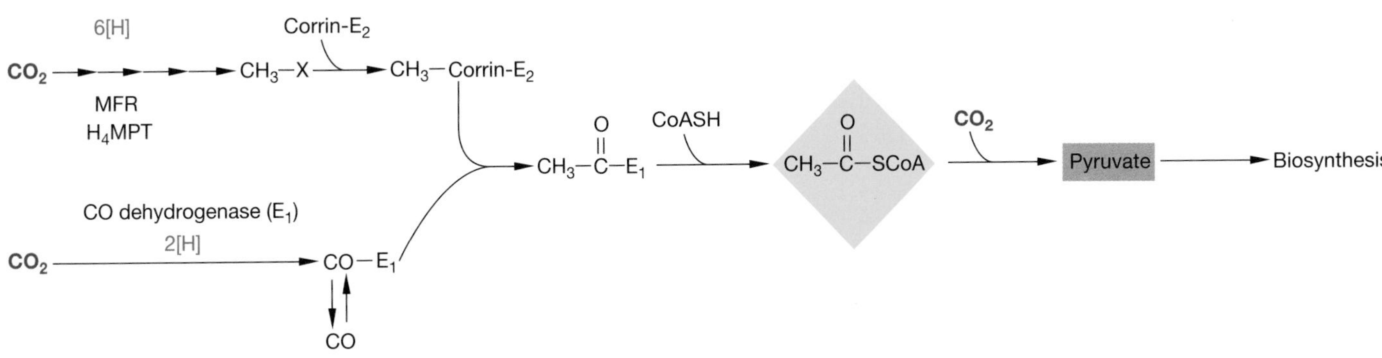

Figure 10.8 The Acetyl-CoA Pathway. Methanogens reduce two molecules of CO_2, each by a different mechanism, and combine them to form acetyl and then acetyl-CoA. Acetogenic bacteria use a slightly different version of the pathway.

Synthesis of Monosaccharides

As can be seen in figure 10.9, gluconeogenesis synthesizes fructose 6-phosphate and glucose 6-phosphate. Once these two precursor metabolites have been formed, other common sugars can be manufactured. For example, mannose comes directly from fructose 6-phosphate by a simple rearrangement.

$$\text{Fructose 6-phosphate} \rightleftharpoons \text{mannose 6-phosphate}$$

Several sugars are synthesized while attached to a nucleoside diphosphate. The most important nucleoside diphosphate sugar is **uridine diphosphate glucose (UDPG).** Glucose is activated by attachment to the pyrophosphate of uridine diphosphate through a reaction with uridine triphosphate (**figure 10.10**). The UDP portion of UDPG is recognized by enzymes and carries glucose around the cell for participation in enzyme reactions much like ADP bears phosphate in the form of ATP. UDP-galactose is synthesized from UDPG through a rearrangement of one hydroxyl group. A different enzyme catalyzes the synthesis of UDP-glucuronic acid through the oxidation of UDPG (**figure 10.11**).

Synthesis of Polysaccharides

Nucleoside diphosphate sugars also play a central role in the synthesis of polysaccharides such as starch and glycogen. Again, biosynthesis is not simply a direct reversal of catabolism. Glycogen and starch catabolism proceeds either by hydrolysis to form free sugars or by the addition of phosphate to these polymers with the production of glucose 1-phosphate. Nucleoside diphosphate sugars are not involved. In contrast, during the synthesis of

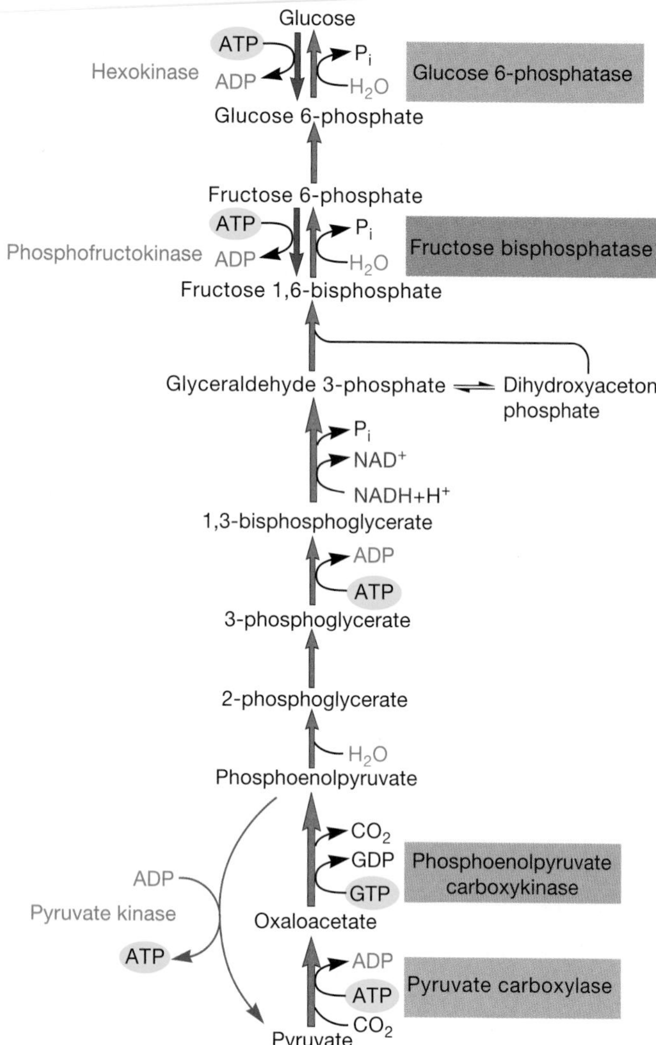

Figure 10.9 Gluconeogenesis. The gluconeogenic pathway used in many microorganisms. The names of the four enzymes catalyzing reactions different from those found in the Embden-Meyerhof pathway (EMP) are in shaded boxes. EMP steps are shown in blue for comparison.

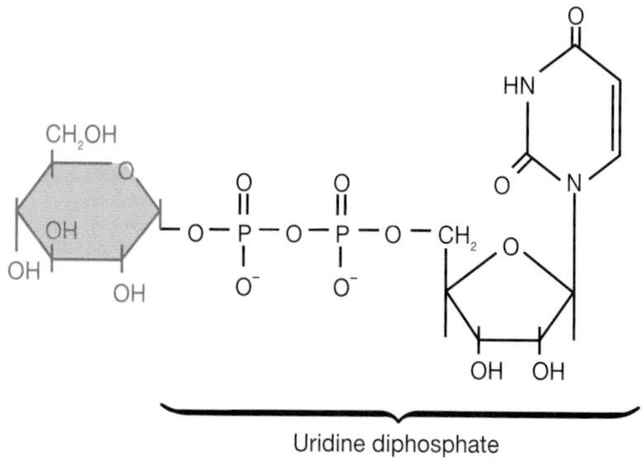

Figure 10.10 Uridine Diphosphate Glucose. Glucose is in color.

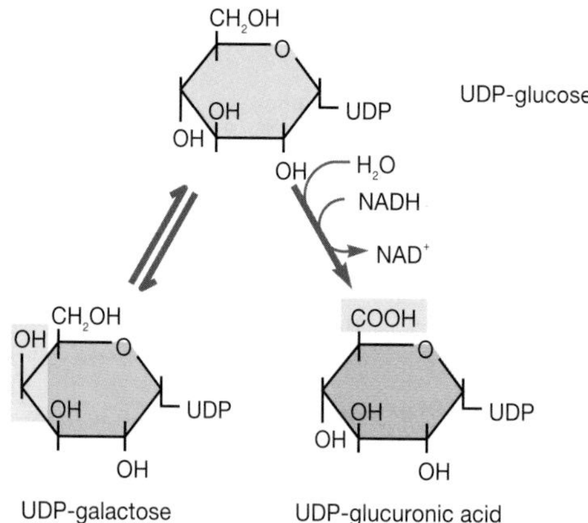

Figure 10.11 Uridine Diphosphate Galactose and Glucuronate Synthesis. The synthesis of UDP-galactose and UDP-glucuronic acid from UDP-glucose. Structural changes are indicated by blue boxes.

glycogen and starch in bacteria and algae, adenosine diphosphate glucose is formed from glucose 1-phosphate and then donates glucose to the end of growing glycogen and starch chains.

$$\text{ATP} + \text{glucose 1-phosphate} \longrightarrow \text{ADP-glucose} + \text{PP}_i$$

$$(\text{Glucose})_n + \text{ADP-glucose} \longrightarrow (\text{glucose})_{n+1} + \text{ADP}$$

Synthesis of Peptidoglycan

Nucleoside diphosphate sugars also participate in the synthesis of peptidoglycan. Recall that peptidoglycan is a large, complex molecule consisting of long polysaccharide chains made of alternating *N*-acetylmuramic acid (NAM) and *N*-acetylglucosamine (NAG) residues. Pentapeptide chains are attached to the NAM groups. The polysaccharide chains are connected through their pentapeptides or by interbridges (*see figures 3.20 and 3.21*). The bacterial cell wall (section 3.6)

Not surprisingly, such an intricate structure requires an equally intricate biosynthetic process, especially because some reactions occur in the cytoplasm, others in the membrane, and others in the periplasmic space. Peptidoglycan synthesis involves two carriers (**figure 10.12**). The first, uridine diphosphate (UDP) functions in the cytoplasmic reactions. In the first step of peptidoglycan synthesis, UDP derivatives of *N*-acetylmuramic acid and *N*-acetylglucosamine are formed. Amino acids are then added sequentially to UDP-NAM to form the pentapeptide chain. NAM-pentapeptide is then transferred to the second carrier, bactoprenol phosphate, which is located at the cytoplasmic side of the plasma membrane. The resulting intermediate is often called **Lipid I. Bactoprenol** (**figure 10.13**) is a 55-carbon alcohol and is linked to NAM by a pyrophosphate group. Next, UDP transfers NAG to the bactoprenol-NAM-pentapeptide complex (Lipid I), to generate **Lipid II.** This creates the peptidoglycan repeat unit. The

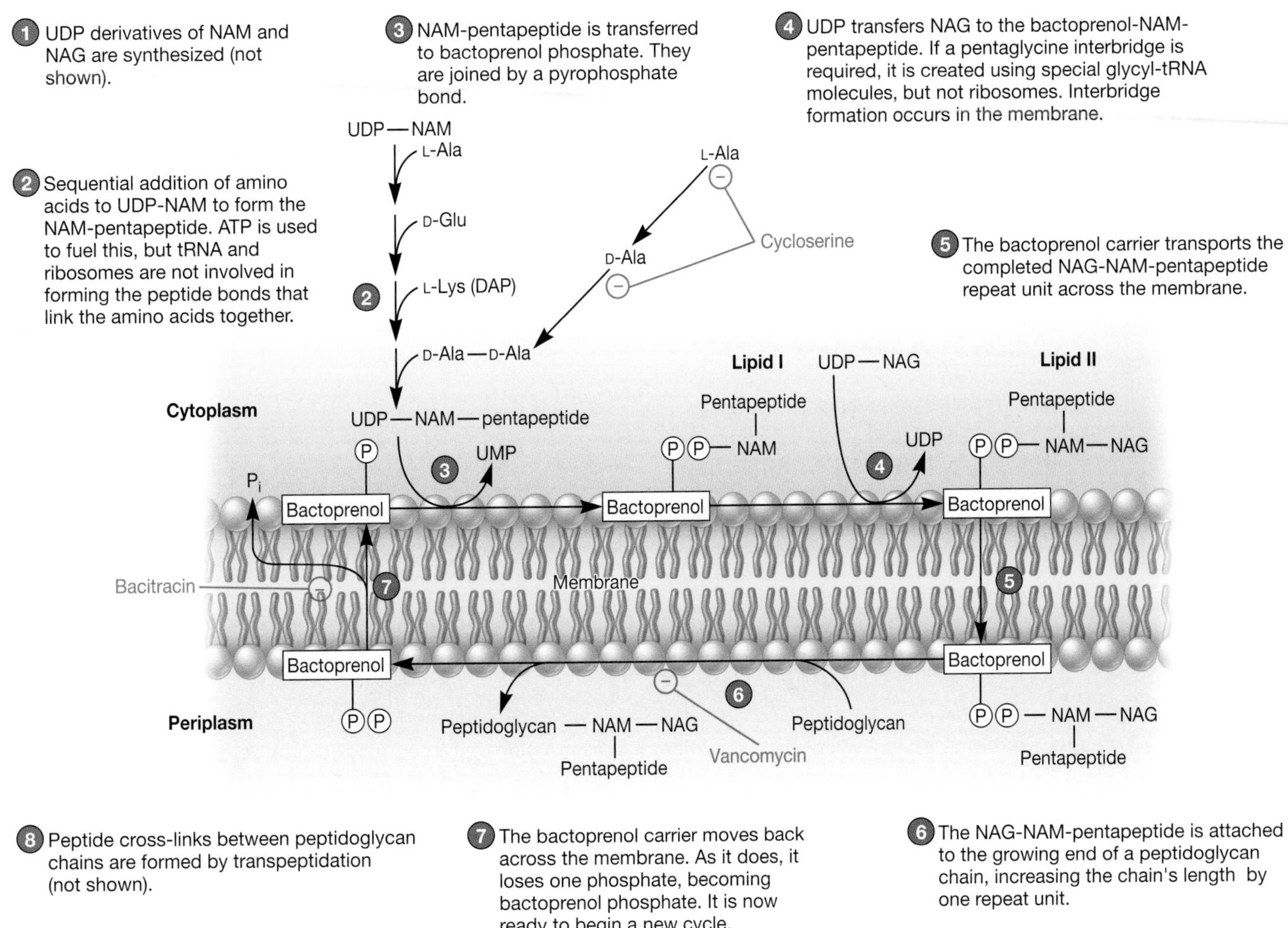

1 UDP derivatives of NAM and NAG are synthesized (not shown).
2 Sequential addition of amino acids to UDP-NAM to form the NAM-pentapeptide. ATP is used to fuel this, but tRNA and ribosomes are not involved in forming the peptide bonds that link the amino acids together.
3 NAM-pentapeptide is transferred to bactoprenol phosphate. They are joined by a pyrophosphate bond.
4 UDP transfers NAG to the bactoprenol-NAM-pentapeptide. If a pentaglycine interbridge is required, it is created using special glycyl-tRNA molecules, but not ribosomes. Interbridge formation occurs in the membrane.
5 The bactoprenol carrier transports the completed NAG-NAM-pentapeptide repeat unit across the membrane.
UDP—NAM
L-Ala
D-Glu
L-Lys (DAP)
D-Ala—D-Ala
L-Ala
D-Ala
Cycloserine
UDP—NAM—pentapeptide
UMP
Cytoplasm
Lipid I
Pentapeptide
NAM
UDP—NAG
UDP
Lipid II
Pentapeptide
NAM—NAG
P
P
i
Bactoprenol
Bactoprenol
Bactoprenol
Bacitracin
Membrane
Bactoprenol
Bactoprenol
Periplasm
Peptidoglycan — NAM—NAG
Pentapeptide
Vancomycin
Peptidoglycan
NAM—NAG
Pentapeptide
8 Peptide cross-links between peptidoglycan chains are formed by transpeptidation (not shown).
7 The bactoprenol carrier moves back across the membrane. As it does, it loses one phosphate, becoming bactoprenol phosphate. It is now ready to begin a new cycle.
6 The NAG-NAM-pentapeptide is attached to the growing end of a peptidoglycan chain, increasing the chain's length by one repeat unit.

Figure 10.12 Peptidoglycan Synthesis. NAM is *N*-acetylmuramic acid and NAG is *N*-acetylglucosamine. The pentapeptide contains L-lysine in *Staphylococcus aureus* peptidoglycan, and diaminopimelic acid (DAP) in *E. coli*. Inhibition by bacitracin, cycloserine, and vancomycin also is shown. The numbers correspond to six of the eight stages discussed in the text. Stage eight is depicted in figure 10.14.

$$CH_3-\overset{\overset{CH_3}{|}}{C}=CH-CH_2-(CH_2-\overset{\overset{CH_3}{|}}{C}=CH-CH_2)_9-CH_2-\overset{\overset{CH_3}{|}}{C}=CH-CH_2-O-\underset{\underset{O^-}{|}}{\overset{\overset{O}{\|}}{P}}-O-\underset{\underset{O^-}{|}}{\overset{\overset{O}{\|}}{P}}-O-\text{NAM}$$

Figure 10.13 Bactoprenol Pyrophosphate. Bactoprenol pyrophosphate connected to *N*-acetylmuramic acid (NAM).

repeat unit is transferred across the membrane by bactoprenol. If the peptidoglycan unit requires an interbridge, it is added while the repeat unit is within the membrane. Bactoprenol stays within the membrane and does not enter the periplasmic space. After releasing the peptidoglycan repeat unit into the periplasmic space, bactoprenol-pyrophosphate is dephosphorylated and returns to the cytoplasmic side of the plasma membrane, where it can function in the next round of synthesis. Meanwhile, the peptidoglycan repeat unit is added to the growing end of a peptidoglycan chain. The final step in peptidoglycan synthesis is **transpeptidation** (**figure 10.14**), which creates the peptide cross-links between the peptidoglycan chains. The enzyme that catalyzes the reaction removes the terminal D-alanine as the cross-link is formed.

To grow and divide efficiently, a bacterial cell must add new peptidoglycan to its cell wall in a precise and well-regulated way while maintaining wall shape and integrity in the presence of high osmotic pressure. Because the cell wall peptidoglycan is essentially a single, enormous network, the growing bacterium must be able to degrade it just enough to provide acceptor ends for the incorporation of new peptidoglycan units. It must also

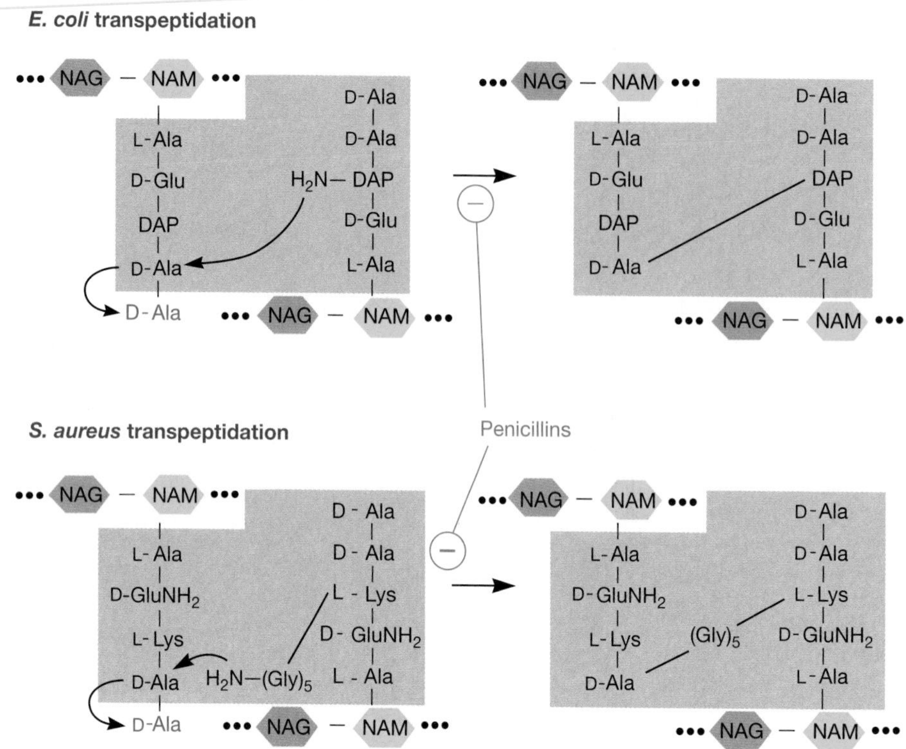

Figure 10.14 Transpeptidation. The transpeptidation reactions in the formation of the peptidoglycans of *Escherichia coli* and *Staphylococcus aureus.*

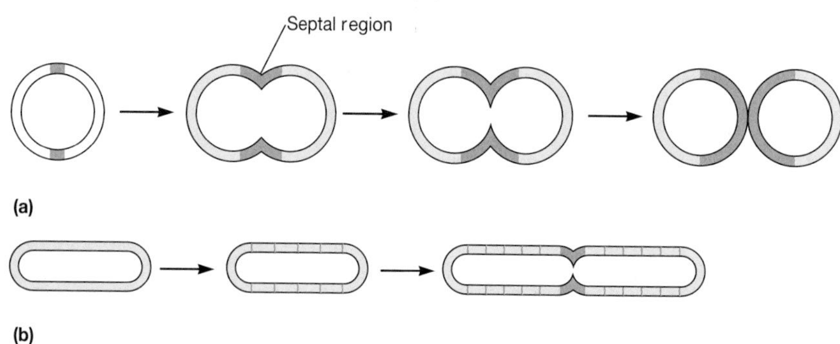

Figure 10.15 Wall Synthesis Patterns. Patterns of new cell wall synthesis in growing and dividing bacteria. **(a)** Streptococci and some other gram-positive cocci. **(b)** Synthesis in rod-shaped bacteria (*E. coli, Salmonella, Bacillus*). The zones of growth are in turquoise. The actual situation is more complex than indicated because cells can begin to divide again before the first division is completed.

reorganize peptidoglycan structure when necessary. This limited peptidoglycan digestion is accomplished by enzymes known as **autolysins,** some of which attack the polysaccharide chains, while others hydrolyze the peptide cross-links. Autolysin inhibitors are produced to keep the activity of these enzymes under tight control.

Although the location and distribution of cell wall synthetic activity varies with species, there seem to be two general patterns (**figure 10.15**). Many gram-positive cocci (e.g., *Enterococcus faecalis* and *Streptococcus pyogenes*) have only one to a few zones of growth. The principal growth zone is usually at the site of septum formation, and new cell halves are synthesized back-to-back. The second pattern of synthesis occurs in the rod-shaped bacteria *Escherichia coli, Salmonella,* and *Bacillus.* Active peptidoglycan synthesis occurs at the site of septum formation, but growth sites also are scattered along the cylindrical portion of the rod. Thus growth is distributed more diffusely in rod-shaped bacteria than in the streptococci. Synthesis must lengthen rod-shaped cells as well as divide them. Presumably this accounts for the differences in wall growth pattern. The procaryotic cell cycle (section 6.1)

Because of the importance of peptidoglycan to cell wall structure and function, its synthesis is a particularly effective target for antimicrobial agents. Inhibition of any stage of synthesis weakens the cell wall and can lead to osmotic lysis. Many commonly used antibiotics interfere with peptidoglycan synthesis. For example, penicillin inhibits the transpeptidation reaction (figure 10.14), and bacitracin blocks the dephosphorylation of bactoprenol pyrophosphate (figure 10.12). Antibacterial drugs (section 34.4)

1. What is gluconeogenesis? Why is it important?
2. Describe the formation of mannose, galactose, starch, and glycogen. What are nucleoside diphosphate sugars? How do microorganisms use them?
3. Suppose that a microorganism is growing on a medium that contains amino acids but no sugars. In general terms, how would it synthesize the pentoses and hexoses it needs? How might it generate all the precursor metabolites it needs?
4. Diagram the steps involved in the synthesis of peptidoglycan and show where they occur in the cell. What are the roles of bactoprenol and UDP? What is unusual about the synthesis of the pentapeptide chain?
5. What is the function of autolysins in cell wall synthesis? Describe the patterns of peptidoglycan synthesis seen in gram-positive cocci and in rod-shaped bacteria such as *E. coli*.

10.5 Synthesis of Amino Acids

Many of the precursor metabolites (figure 10.3) serve as starting substrates for the synthesis of amino acids. In the amino acid biosynthetic pathways, the carbon skeleton is remodeled and an amino group, and sometimes sulfur, are added. In this section, we first examine the mechanisms by which nitrogen and sulfur are assimilated and incorporated into amino acids. This is followed by a brief consideration of the organization of amino acid biosynthetic pathways.

Nitrogen Assimilation

Nitrogen is a major component not only of proteins, but of nucleic acids, coenzymes, and many other cell constituents as well. Thus the cell's ability to assimilate inorganic nitrogen is exceptionally important. Although nitrogen gas is abundant in the atmosphere, few microorganisms can reduce the gas and use it as a nitrogen source. Most must incorporate either ammonia or nitrate. We examine ammonia and nitrate assimilation first, and then briefly discuss nitrogen assimilation in microbes that fix N_2.

Ammonia Incorporation

Ammonia nitrogen can be incorporated into organic material relatively easily and directly because it is more reduced than other forms of inorganic nitrogen. Ammonia is initially incorporated into carbon skeletons by one of two mechanisms: reductive amination or by the glutamine synthetase-glutamate synthase system. Once incorporated, the nitrogen can be transferred to other carbon skeletons by enzymes called transaminases. The major reductive amination pathway involves the formation of glutamate from α-ketoglutarate, catalyzed in many bacteria and fungi by **glutamate dehydrogenase** when the ammonia concentration is high.

$$\alpha\text{-ketoglutarate} + NH_4^+ + \text{NADPH (NADH)} + H^+ \rightleftharpoons \text{glutamate} + NADP^+(NAD^+) + H_2O$$

Once glutamate has been synthesized, the newly formed α-amino group can be transferred to other carbon skeletons by transamination reactions to form different amino acids. **Transaminases** possess the coenzyme pyridoxal phosphate, which is responsible for the amino group transfer. Microorganisms have a number of transaminases, each of which catalyzes the formation of several amino acids using the same amino acid as an amino group donor. When glutamate dehydrogenase works in cooperation with transaminases, ammonia can be incorporated into a variety of amino acids (**figure 10.16**).

The **glutamine synthetase-glutamate synthase (GS-GOGAT) system** is observed in *E. coli*, *Bacillus megaterium*, and other bacteria (**figure 10.17**). It functions when ammonia levels are low. Incorporation of ammonia by this system begins when ammonia is used to synthesize glutamine from glutamate in a reaction catalyzed by **glutamine synthetase** (**figure 10.18**). Then the amide nitrogen of glutamine is transferred to α-ketoglutarate to generate a new glutamate molecule. This reaction is catalyzed by **glutamate synthase.** Because glutamate acts as an amino donor in transaminase reactions, ammonia may be used to synthesize all common amino acids when suitable transaminases are present.

Assimilatory Nitrate Reduction

The nitrogen in nitrate (NO_3^-) is much more oxidized than that in ammonia. Therefore nitrate must first be reduced to ammonia before the nitrogen can be converted to an organic form. This reduction of nitrate is called **assimilatory nitrate reduction,** which is not the same as that occurring during anaerobic respiration (dissimilatory nitrate reduction). In assimilatory nitrate reduction, nitrate is incorporated into organic material and does not participate in energy generation. The process is widespread among bacteria, fungi, and photosynthetic protists. Anaerobic respiration (section 9.6); Biogeochemical cycling: Nitrogen cycle (section 27.2)

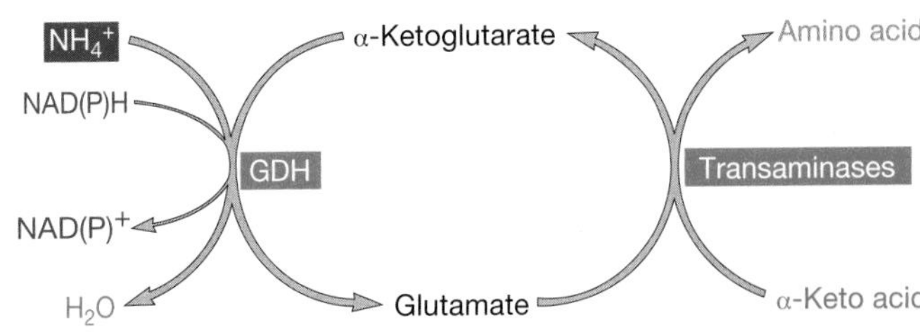

Figure 10.16 The Ammonia Assimilation Pathway. Ammonia assimilation by use of glutamate dehydrogenase (GDH) and transaminases. Either NADP- or NAD-dependent glutamate dehydrogenases may be involved. This route is most active at high ammonia concentrations.

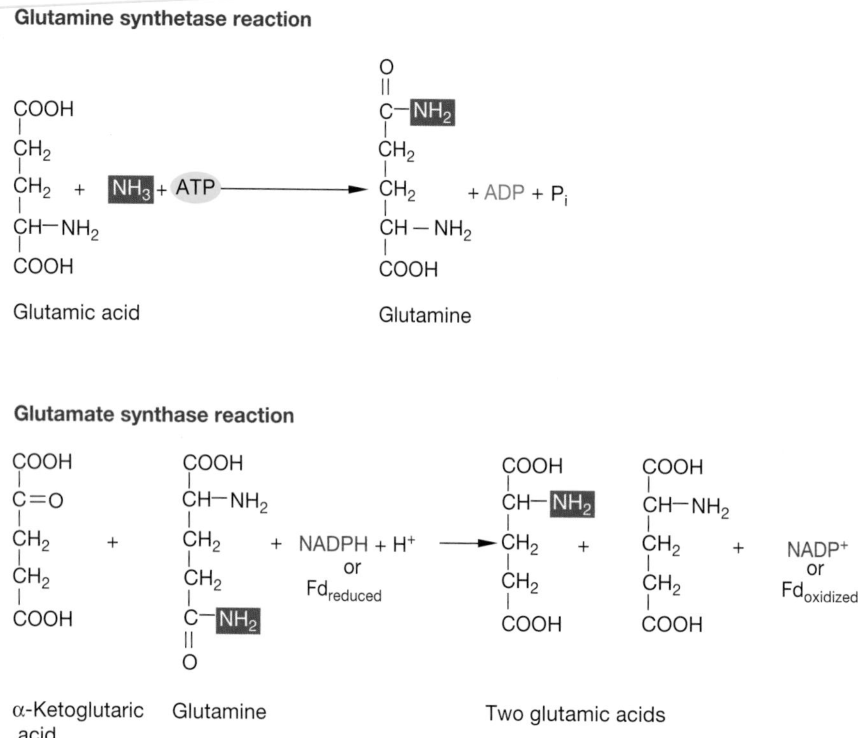

Figure 10.17 Glutamine Synthetase and Glutamate Synthase. The glutamine synthetase and glutamate synthase reactions involved in ammonia assimilation. Some glutamine synthases use NADPH as an electron source; others use reduced ferredoxin (Fd). The nitrogen being incorporated and transferred is shown in turquoise.

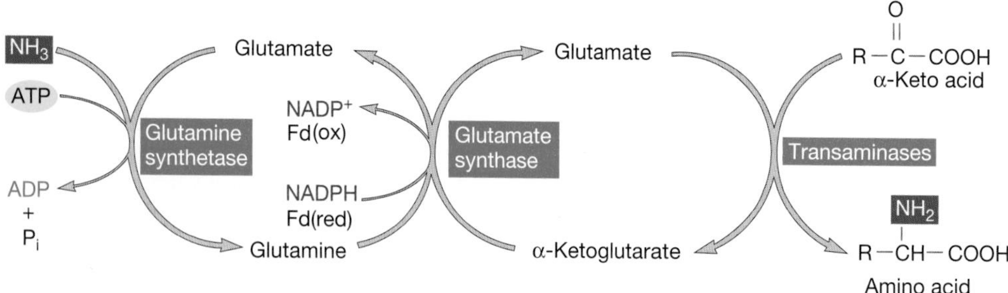

Figure 10.18 Ammonia Incorporation Using Glutamine Synthetase and Glutamate Synthase. This route is effective at low ammonia concentrations.

Assimilatory nitrate reduction takes place in the cytoplasm in bacteria. The first step in nitrate assimilation is its reduction to nitrite by **nitrate reductase,** an enzyme that contains both FAD and molybdenum (**figure 10.19**). NADPH is the electron source.

$$NO_3^- + NADPH + H^+ \longrightarrow NO_2^- + NADP^+ + H_2O$$

Nitrite is next reduced to ammonia with a series of two electron additions catalyzed by **nitrite reductase** and possibly other enzymes. The ammonia is then incorporated into amino acids by the routes already described.

Nitrogen Fixation

The reduction of atmospheric gaseous nitrogen to ammonia is called **nitrogen fixation.** Because ammonia and nitrate levels often are low and only a few bacteria and archaea can carry out nitrogen fixation (eucaryotic cells completely lack this ability), the rate of this process limits plant growth in many situations. Nitrogen fixation occurs in (1) free-living bacteria and archaea (e.g., *Azotobacter, Klebsiella, Clostridium,* and *Methanococcus*), (2) bacteria living in symbiotic association with plants such as legumes (*Rhizobium*), and (3) cyanobacteria (*Nostoc, Anabaena,* and *Trichodesmia*). The biological aspects of nitrogen fixation

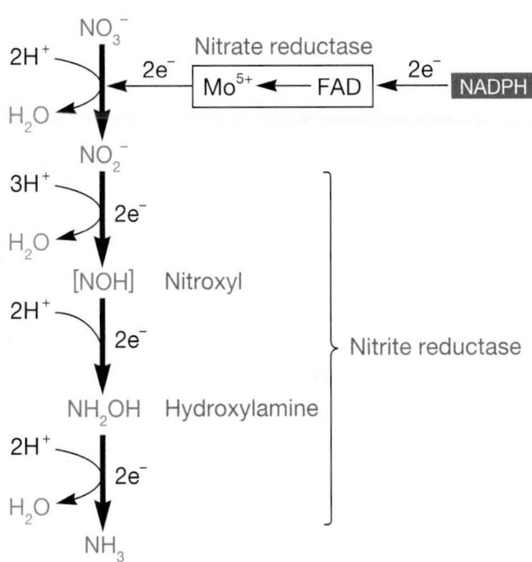

Figure 10.19 Assimilatory Nitrate Reduction. This sequence is thought to operate in bacteria that can reduce and assimilate nitrate nitrogen. See text for details.

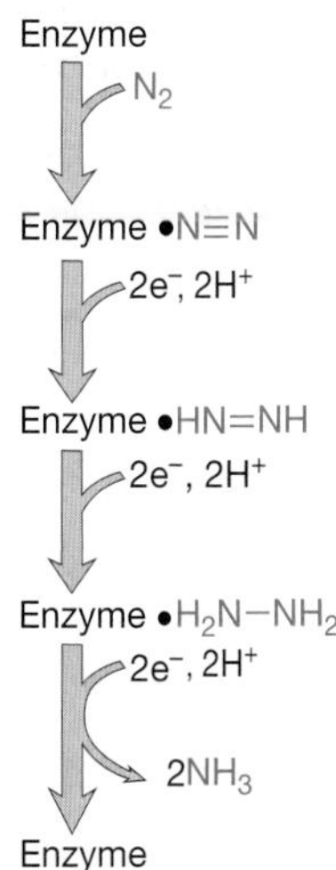

Figure 10.20 Nitrogen Reduction. A hypothetical sequence of nitrogen reduction by nitrogenase.

are discussed in chapters 28 and 29. The biochemistry of nitrogen fixation is the focus of this section.

The reduction of nitrogen to ammonia is catalyzed by the enzyme **nitrogenase.** Although the enzyme-bound intermediates in this process are still unknown, it is believed that nitrogen is reduced by two-electron additions in a way similar to that illustrated in **figure 10.20.** The reduction of molecular nitrogen to ammonia is quite exergonic, but the reaction has a high activation energy because molecular nitrogen is an unreactive gas with a triple bond between the two nitrogen atoms. Therefore nitrogen reduction is expensive and requires a large ATP expenditure. At least 8 electrons and 16 ATP molecules, 4 ATPs per pair of electrons, are required.

$$N_2 + 8H^+ + 8e^- + 16ATP \longrightarrow 2NH_3 + H_2 + 16ADP + 16P_i$$

The electrons come from ferredoxin that has been reduced in a variety of ways: by photosynthesis in cyanobacteria, respiratory processes in aerobic nitrogen fixers, or fermentations in anaerobic bacteria. For example, *Clostridium pasteurianum* (an anaerobic bacterium) reduces ferredoxin during pyruvate oxidation, whereas the aerobic *Azotobacter* uses electrons from NADPH to reduce ferredoxin.

Nitrogenase is a complex system consisting of two major protein components, a MoFe protein (MW 220,000) joined with one or two Fe proteins (MW 64,000). The MoFe protein contains 2 atoms of molybdenum and 28 to 32 atoms of iron; the Fe protein has 4 iron atoms (**figure 10.21**). Fe protein is first reduced by ferredoxin, then it binds ATP (**figure 10.22**). ATP binding changes the conformation of the Fe protein and lowers its reduction potential, enabling it to reduce the MoFe protein. ATP is hydrolyzed when this electron transfer occurs. Finally, reduced MoFe protein donates electrons to atomic nitrogen. Nitrogenase is quite sensitive to O_2 and must be protected from O_2 inactivation within the cell. In many cyanobacteria, this protection against oxygen is provided by a special structure called the heterocyst (*see figure 21.9*).

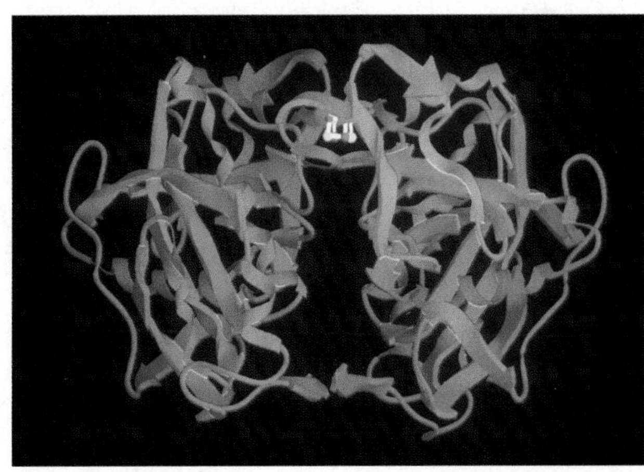

Figure 10.21 Structure of the Nitrogenase Fe Protein. The Fe protein's two subunits are arranged like a pair of butterfly wings with the iron sulfur cluster between the wings and at the "head" of the butterfly. The iron sulfur cluster is very exposed, which helps account for nitrogenase's sensitivity to oxygen. The oxygen can readily attack the exposed iron atoms.

The reduction of N_2 to NH_3 occurs in three steps, each of which requires an electron pair (figures 10.20 and 10.22). Six electron transfers take place, and this requires a total 12 ATPs per N_2 reduced. The overall process actually requires at least 8 electrons and 16 ATPs because nitrogenase also reduces protons to H_2. The H_2 reacts with diimine ($HN = NH$) to form N_2 and H_2. This futile cycle produces some N_2 even under favorable conditions and makes nitrogen fixation even more expensive. Symbiotic

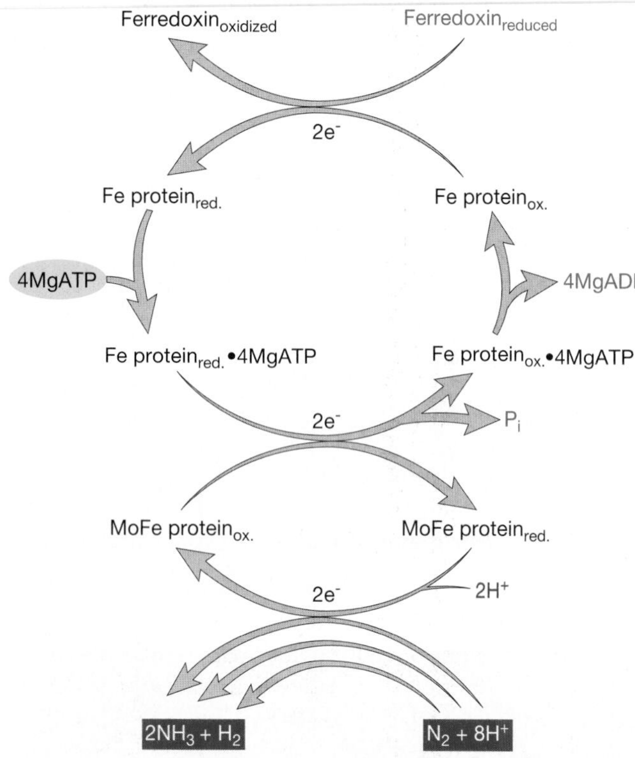

Figure 10.22 Mechanism of Nitrogenase Action. The flow of two electrons from ferredoxin to nitrogen is outlined. This process is repeated three times in order to reduce N_2 to two molecules of ammonia. The stoichiometry at the bottom includes proton reduction to H_2. See the text for a more detailed explanation.

nitrogen-fixing bacteria can consume almost 20% of the ATP produced by the host plant.

Nitrogenase can reduce a variety of molecules containing triple bonds (e.g., acetylene, cyanide, and azide).

$$HC \equiv CH + 2H^+ + 2e^- \longrightarrow H_2C = CH_2$$

The rate of reduction of acetylene to ethylene is often used to estimate nitrogenase activity.

Once molecular nitrogen has been reduced to ammonia, the ammonia can be incorporated into organic compounds. The mechanisms by which heterocystous cells exchange NH_3 with the vegetative cyanobacterial cells, as well as how symbiotic nitrogen-fixing rhizobia share ammonia with host plants, comprise an area of active research. Microorganism associations with vascular plants: Nitrogen fixation (section 29.5)

Sulfur Assimilation

Sulfur is needed for the synthesis of the amino acids cysteine and methionine. It is also needed for the synthesis of several coenzymes (e.g., coenzyme A and biotin). Sulfur is obtained from two sources. Many microorganisms use cysteine and methionine, obtained from either external sources or intracellular amino acid reserves. In addition, sulfate can provide sulfur for biosynthesis. The sulfur atom in sulfate is more oxidized than it is in cysteine and other organic molecules; thus sulfate must be reduced before it can be assimilated. This process is known as **assimilatory sulfate reduction** to distinguish it from the **dissimilatory sulfate reduction,** which takes place when sulfate acts as an electron acceptor during anaerobic respiration. Anaerobic respiration (section 9.6); Biogeochemical cycling: Sulfur cycle (section 27.2)

Assimilatory sulfate reduction involves sulfate activation through the formation of phosphoadenosine 5'-phosphosulfate (**figure 10.23**), followed by reduction of the sulfate. The process is complex (**figure 10.24**). Sulfate is first reduced to sulfite (SO_3^{2-}), then to hydrogen sulfide. Cysteine can be synthesized from hydrogen sulfide in two ways. Fungi appear to combine hydrogen sulfide with serine to form cysteine

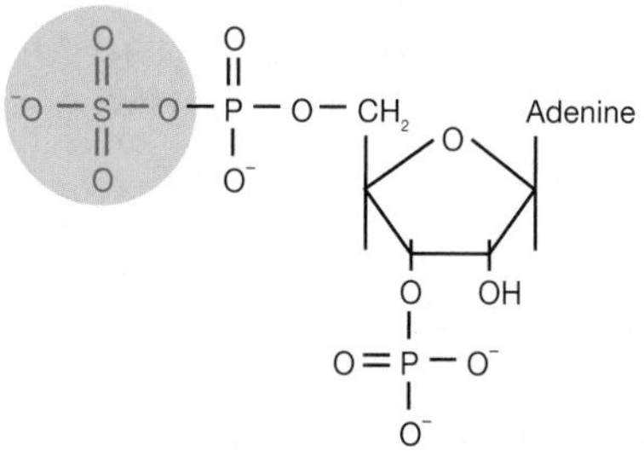

Figure 10.23 Phosphoadenosine 5'-phosphosulfate (PAPS). The sulfate group is in color.

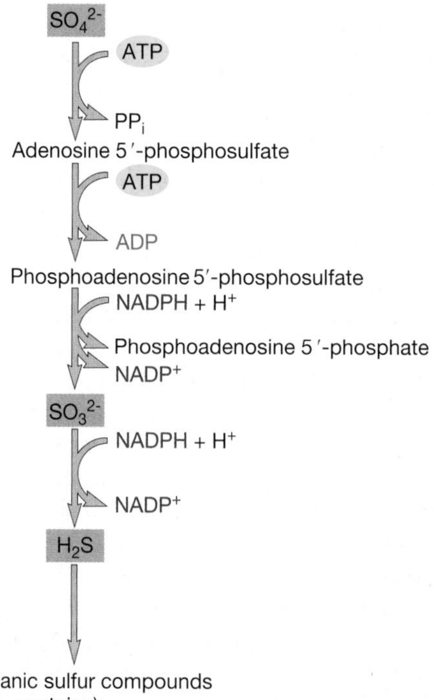

Figure 10.24 The Sulfate Reduction Pathway.

(process 1), whereas many bacteria join hydrogen sulfide with O-acetylserine instead (process 2).

(1) H_2S + serine ⟶ cysteine + H_2O

(2) Serine ⟶ (acetyl-CoA → CoA)

O-acetylserine ⟶ cysteine (H_2S → acetate)

Once formed, cysteine can be used in the synthesis of other sulfur-containing organic compounds including the amino acid methionine.

Amino Acid Biosynthetic Pathways

Some amino acids are made directly by transamination of a precursor metabolite. For example, alanine and aspartate are made directly from pyruvate and oxaloacetate, respectively, using glutamate as the amino group donor. However, for most amino acids, the precursor metabolite from which they are synthesized must be altered by more than just the addition of an amino group. In many cases, the carbon skeleton must be reconfigured, and for cysteine and methionine, the carbon skeleton must be amended by the addition of sulfur. These biosynthetic pathways are more complex. They often involve many steps and are branched. By using branched pathways, a single precursor metabolite can be used for the synthesis of a family of related amino acids. For example, the amino acids lysine, threonine, isoleucine, and methionine are synthesized from oxaloacetate by a branching anabolic route (**figure 10.25**). The biosynthetic pathways for the aromatic amino acids phenylalanine, tyrosine, and tryptophan also share many intermediates (**figure 10.26**). Because of the need to conserve nitrogen, carbon, and energy, amino acid synthetic pathways are usually tightly regulated by allosteric and feedback mechanisms. Control of enzyme activity (section 8.10)

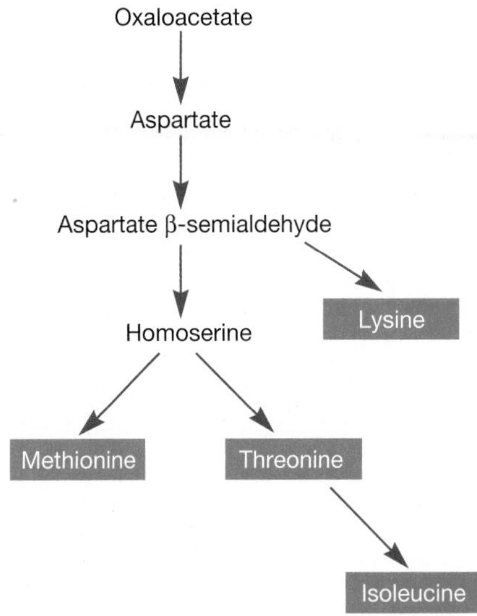

Figure 10.25 A Branching Pathway of Amino Acid Synthesis. The pathways to methionine, threonine, isoleucine, and lysine. Although some arrows represent one step, most interconversions require the participation of several enzymes. Also not shown is the consumption of reducing power and ATP. For instance, the synthesis of isoleucine consumes two ATP and three NADPH.

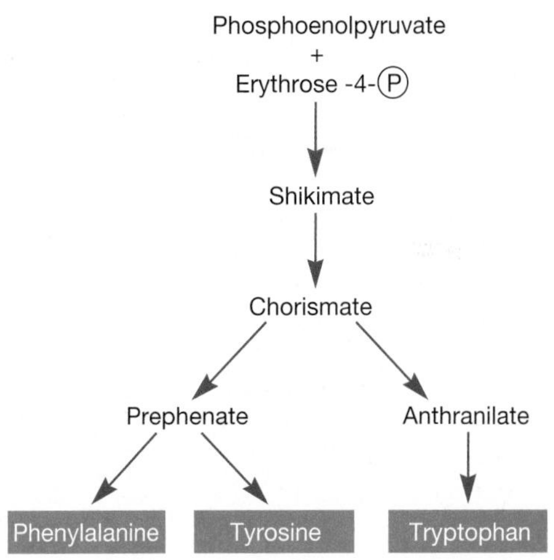

Figure 10.26 Aromatic Amino Acid Synthesis. The synthesis of the aromatic amino acids phenylalanine, tyrosine, and tryptophan. Most arrows represent more than one enzyme reaction.

Anaplerotic Reactions and Amino Acid Biosynthesis

When an organism is actively synthesizing amino acids, a heavy demand for precursor metabolites is placed on the central metabolic pathways. Because many amino acid biosynthetic pathways begin with TCA cycle intermediates, it is critical that they be readily available. This is especially true for organisms carrying out fermentation, where the TCA cycle does not function in the catabolism of glucose. To ensure an adequate supply of TCA cycle-generated precursor metabolites, microorganisms use reactions that replenish TCA cycle intermediates. Reactions that replace cycle intermediates are called **anaplerotic reactions** [Greek *anaplerotic*, filling up].

Most microorganisms can replace TCA cycle intermediates using two reactions that generate oxaloacetate from either phosphoenolpyruvate or pyruvate, both of which are intermediates of the Embden-Meyerhof pathway (figure 10.3). These 3-carbon molecules are converted to oxaloacetate by a carboxylation reaction (i.e., CO_2 is added to the molecule, forming a carboxyl group).

The conversion of pyruvate to oxaloacetate is catalyzed by the enzyme pyruvate carboxylase, which requires the cofactor biotin.

$$\text{Pyruvate} + CO_2 + \text{ATP} + H_2O \xrightarrow{\text{biotin}} \text{oxaloacetate} + \text{ADP} + P_i$$

Biotin is often the cofactor for enzymes catalyzing carboxylation reactions. Because of its importance, biotin is a required growth factor for many species. The pyruvate carboxylase reaction is observed in yeasts and some bacteria. Other microorganisms, such as the bacteria *E. coli* and *Salmonella* spp., have the enzyme phosphoenolpyruvate carboxylase, which catalyzes the carboxylation of phosphoenolpyruvate.

$$\text{Phosphoenolpyruvate} + CO_2 \longrightarrow \text{oxaloacetate} + P_i$$

Other anaplerotic reactions are part of the **glyoxylate cycle (figure 10.27)**, which functions in some bacteria, fungi, and protists. This cycle is made possible by two unique enzymes, isocitrate lyase and malate synthase, that catalyze the following reactions.

$$\text{Isocitrate} \xrightarrow{\text{isocitrate lyase}} \text{succinate} + \text{glyoxylate}$$

$$\text{Glyoxylate} + \text{acetyl-CoA} \xrightarrow{\text{malate synthase}} \text{malate} + \text{CoA}$$

The glyoxylate cycle is actually a modified TCA cycle. The two decarboxylations of the TCA cycle (the isocitrate dehydrogenase and α-ketoglutarate dehydrogenase steps) are bypassed, making possible the conversion of acetyl-CoA to form oxaloacetate without loss of acetyl-CoA carbon as CO_2. In this fashion acetate and any mol-

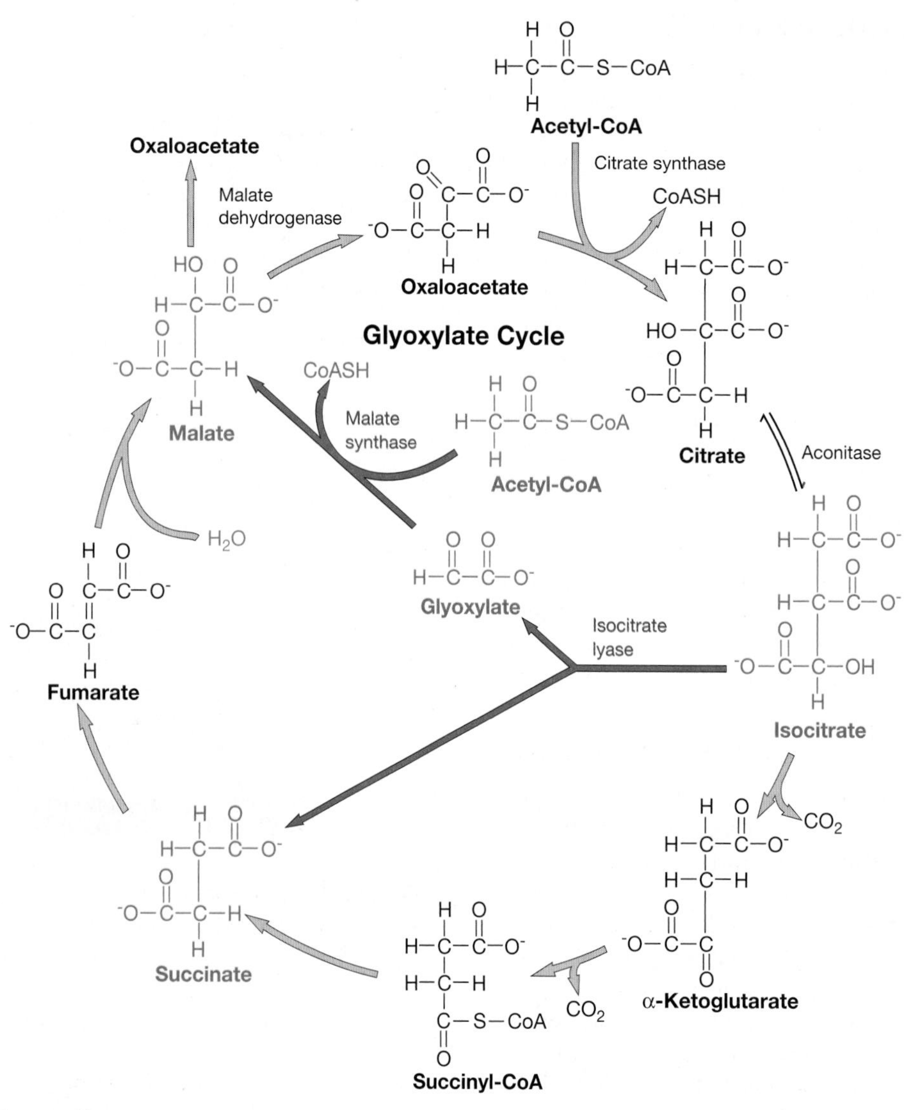

Overall equation:

$$2\ \text{Acetyl-CoA} + \text{FAD} + 2\text{NAD}^+ + 3H_2O \longrightarrow \text{Oxaloacetate} + 2\text{CoA} + \text{FADH}_2 + 2\text{NADH} + 2H^+$$

Figure 10.27 The Glyoxylate Cycle. The reactions and enzymes unique to the cycle are shown in red. The tricarboxylic acid cycle enzymes that have been bypassed are at the bottom.

ecules that give rise to it can contribute carbon to the cycle and support microbial growth. The tricarboxylic acid cycle (section 9.4)

1. Describe the roles of glutamate dehydrogenase, glutamine synthetase, glutamate synthase, and transaminases in ammonia assimilation.
2. How is nitrate assimilated? How does assimilatory nitrate reduction differ from dissimilatory nitrate reduction? What is the fate of nitrate following assimilatory nitrate reduction versus its fate following denitrification?
3. What is nitrogen fixation? Briefly describe the structure and mechanism of action of nitrogenase.
4. How do organisms assimilate sulfur? How does assimilatory sulfate reduction differ from dissimilatory sulfate reduction?
5. Why is using branched pathways an efficient mechanism for synthesizing amino acids?
6. Define an anaplerotic reaction. Give three examples of anaplerotic reactions.
7. Describe the glyoxylate cycle. How is it similar to the TCA cycle? How does it differ from the TCA cycle?

10.6 Synthesis of Purines, Pyrimidines, and Nucleotides

Purine and pyrimidine biosynthesis is critical for all cells because these molecules are used in the synthesis of ATP, several cofactors, ribonucleic acid (RNA), deoxyribonucleic acid (DNA), and other important cell components. Nearly all microorganisms can synthesize their own purines and pyrimidines as these are so crucial to cell function. DNA replication (section 11.4); Transcription (section 11.6)

Purines and **pyrimidines** are cyclic nitrogenous bases with several double bonds and pronounced aromatic properties. Purines consist of two joined rings, whereas pyrimidines have only one. The purines **adenine** and **guanine** and the pyrimidines **uracil, cytosine,** and **thymine** are commonly found in microorganisms. A purine or pyrimidine base joined with a pentose sugar, either ribose or deoxyribose, is a **nucleoside.** A **nucleotide** is a nucleoside with one or more phosphate groups attached to the sugar.

As discussed in section 10.6, amino acids participate in the synthesis of nitrogenous bases and nucleotides in a number of ways, including providing the nitrogen that is part of all purines and pyrimidines. The phosphorus present in nucleotides is provided by other mechanisms. We begin this section by examining phosphorus assimilation. We then examine the pathways for synthesis of nitrogenous bases and nucleotides.

Phosphorus Assimilation

In addition to nucleic acids, phosphorus is found in proteins, phospholipids, ATP, and coenzymes like NADP. The most common phosphorus sources are inorganic phosphate and organic phosphate esters. Inorganic phosphate is incorporated through the formation of ATP in one of three ways: by (1) photophosphorylation, (2) oxidative phosphorylation, and (3) substrate-level phosphorylation. The breakdown of glucose to pyruvate (section 9.3); Electron transport and oxidative phosphorylation (section 9.5); Phototrophy (section 9.12)

Microorganisms may obtain organic phosphates from their surroundings in dissolved or particulate form. **Phosphatases** very often hydrolyze organic phosphate esters to release inorganic phosphate. Gram-negative bacteria have phosphatases in the periplasmic space, which allows phosphate to be taken up immediately after release. On the other hand, protists can directly use organic phosphates after ingestion or hydrolyze them in lysosomes and incorporate the phosphate.

Purine Biosynthesis

The biosynthetic pathway for purines is a complex, 11-step sequence (*see appendix II*) in which seven different molecules contribute parts to the final purine skeleton (**figure 10.28**). Because the pathway begins with ribose 5-phosphate and the purine skeleton is constructed on this sugar, the first purine product of the pathway is the nucleotide inosinic acid, not a free purine base. The cofactor folic acid is very important in purine biosynthesis. Folic acid derivatives contribute carbons two and eight to the purine skeleton. In fact, the drug sulfonamide inhibits bacterial growth by blocking folic acid synthesis. This interferes with purine biosynthesis and other processes that require folic acid. Antibacterial drugs: Metabolic antagonists (section 34.4)

Once inosinic acid has been formed, relatively short pathways synthesize adenosine monophosphate and guanosine monophosphate (**figure 10.29**) and produce nucleoside diphosphates and triphosphates by phosphate transfers from ATP. DNA contains deoxyribonucleotides (the ribose lacks a hydroxyl group on carbon two) instead of the ribonucleotides found in RNA. Deoxyribonucleotides arise from the reduction of nucleoside diphosphates or nucleoside triphosphates by two different routes. Some microorganisms reduce the triphosphates with a system requiring vitamin B_{12} as a cofactor. Others, such as *E. coli,* reduce the ribose in nucleoside diphosphates. Both systems employ a small sulfur-containing protein called thioredoxin as their reducing agent.

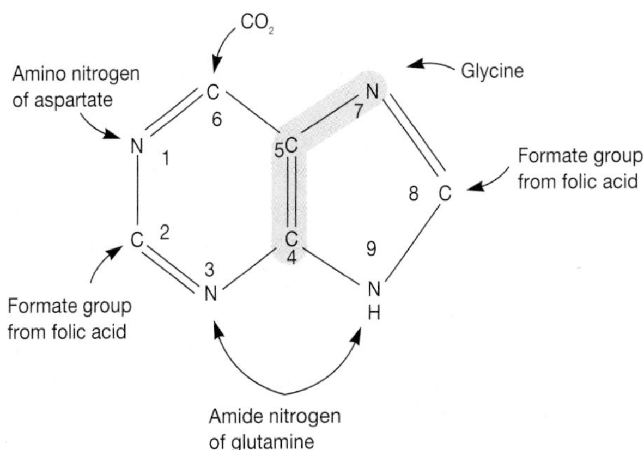

Figure 10.28 Purine Biosynthesis. The sources of purine skeleton nitrogen and carbon are indicated. The contribution of glycine is shaded in blue.

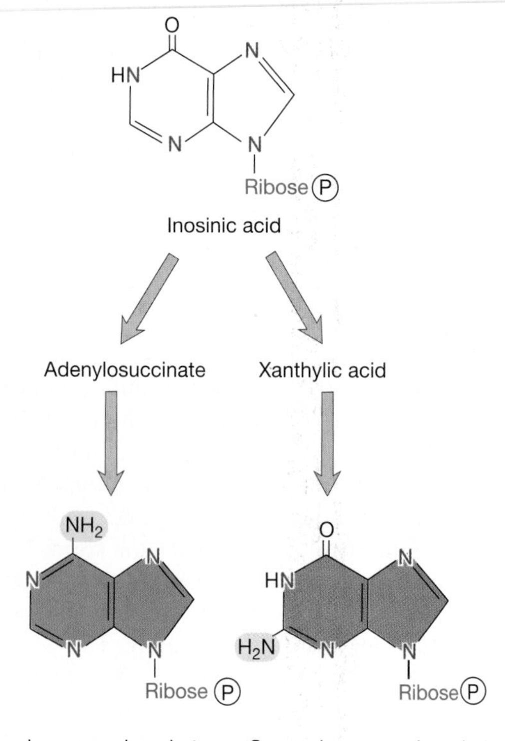

Figure 10.29 Synthesis of Adenosine Monophosphate and Guanosine Monophosphate. The highlighted groups are the ones differing from those in inosinic acid.

Pyrimidine Biosynthesis

Pyrimidine biosynthesis begins with aspartic acid and carbamoyl phosphate, a high-energy molecule synthesized from CO_2 and ammonia (**figure 10.30**). Aspartate carbamoyltransferase catalyzes the condensation of these two substrates to form carbamoylaspartate, which is then converted to the initial pyrimidine product, orotic acid.

After synthesis of the pyrimidine skeleton, a nucleotide is produced by the addition of ribose 5-phosphate, using the high-energy intermediate 5-phosphoribosyl 1-pyrophosphate. Thus construction of the pyrimidine ring is completed before ribose is added, in contrast with purine ring synthesis, which begins with ribose 5-phosphate. Decarboxylation of orotidine monophosphate yields uridine monophosphate and eventually uridine triphosphate and cytidine triphosphate.

The third common pyrimidine is thymine, a constituent of DNA. The ribose in pyrimidine nucleotides is reduced in the same way as it is in purine nucleotides. Then deoxyuridine monophosphate is methylated with a folic acid derivative to form deoxythymidine monophosphate (**figure 10.31**).

1. How is phosphorus assimilated? What roles do phosphatases play in phosphorus assimilation? Why can phosphate be directly incorporated into cell constituents, whereas nitrate, nitrogen gas, and sulfate cannot?
2. Define purine, pyrimidine, nucleoside, and nucleotide.
3. Outline the way in which purines and pyrimidines are synthesized. How is the deoxyribose component of deoxyribonucleotides made?

10.7 Lipid Synthesis

A variety of lipids are found in microorganisms, particularly in cell membranes. Most contain **fatty acids** or their derivatives. Fatty acids are monocarboxylic acids with long alkyl chains that usually have an even number of carbons (the average length is 18 carbons). Some may be unsaturated—that is, have one or more double bonds. Most microbial fatty acids are straight chained, but some are branched. Gram-negative bacteria often have cyclopropane fatty acids (fatty acids with one or more cyclopropane rings in their chains). Lipids (appendix I)

Fatty acid synthesis is catalyzed by the **fatty acid synthase** complex with acetyl-CoA and malonyl-CoA as the substrates and NADPH as the electron donor. Malonyl-CoA arises from the ATP-driven carboxylation of acetyl-CoA (**figure 10.32**). Synthesis takes place after acetate and malonate have been transferred from coenzyme A to the sulfhydryl group of the **acyl carrier protein (ACP),** a small protein that carries the growing fatty acid chain during synthesis. The synthase adds two carbons at a time to the carboxyl end of the growing fatty acid chain in a two-stage process (figure 10.32). First, malonyl-ACP reacts with the fatty acyl-ACP to yield CO_2 and a fatty acyl-ACP two carbons longer. The loss of CO_2 drives this reaction to completion. Notice that ATP is used to add CO_2 to acetyl-CoA, forming malonyl-CoA. The same CO_2 is lost when malonyl-ACP donates carbons to the chain. Thus carbon dioxide is essential to fatty acid synthesis but it is not permanently incorporated. Indeed, some microorganisms require CO_2 for good growth, but they can do without it in the presence of a fatty acid like oleic acid (an 18-carbon unsaturated fatty acid). In the second stage of synthesis, the β-keto group arising from the initial condensation reaction is removed in a three-step process involving two reductions and a dehydration. The fatty acid is then ready for the addition of two more carbon atoms.

Unsaturated fatty acids are synthesized in two ways. Eucaryotes and aerobic bacteria like *B. megaterium* employ an aerobic pathway using both NADPH and O_2.

$$\mathrm{R{-}(CH_2)_9{-}\overset{\overset{\displaystyle O}{\|}}{C}{-}SCoA + NADPH + H^+ + O_2 \longrightarrow}$$

$$\mathrm{R{-}CH{=}CH{-}(CH_2)_7{-}\overset{\overset{\displaystyle O}{\|}}{C}{-}SCoA + NADPH^+ + 2H_2O}$$

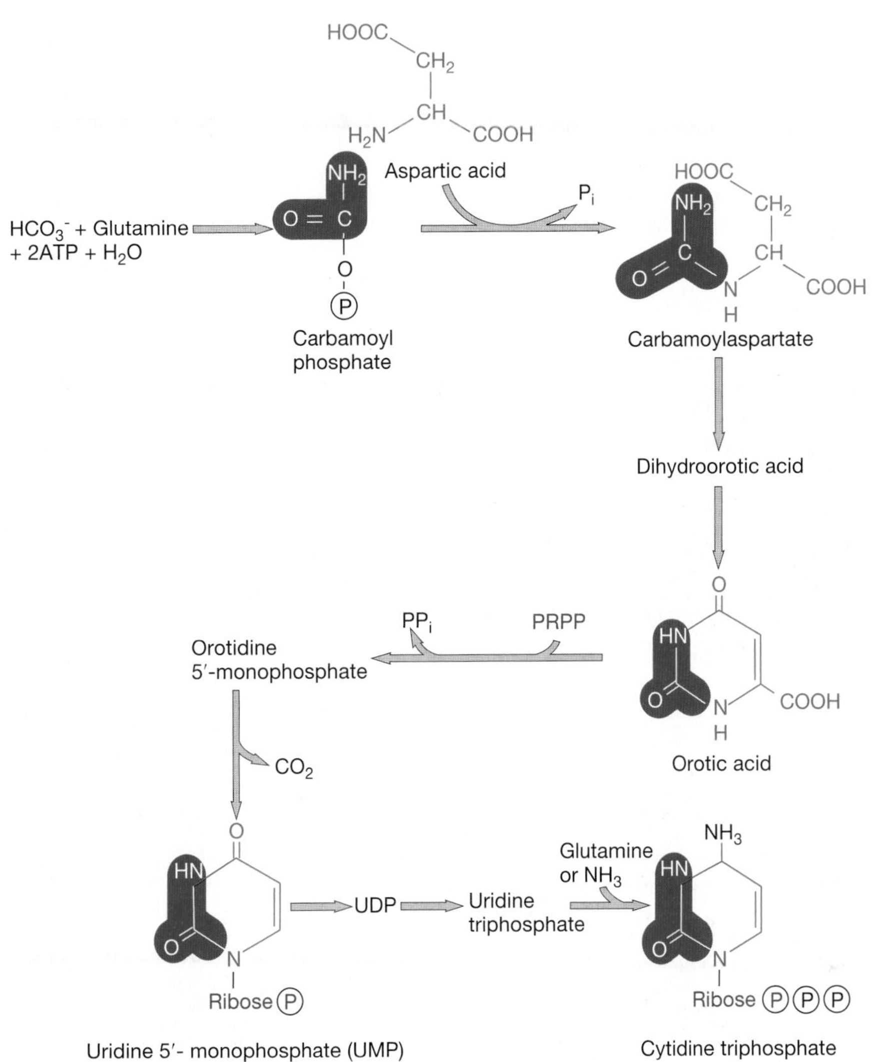

Figure 10.30 Pyrimidine Synthesis. PRPP stands for 5-phosphoribosyl 1-pyrophosphoric acid, which provides the ribose 5-phosphate chain. The part derived from carbamoyl phosphate is shaded in turquoise.

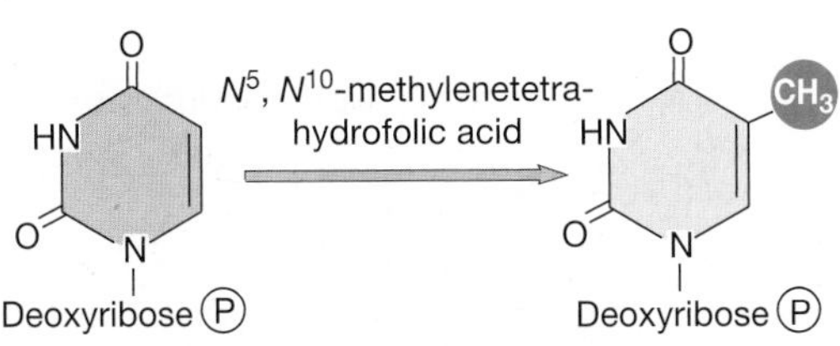

Figure 10.31 Deoxythymidine Monophosphate Synthesis. Deoxythymidine differs from deoxyuridine in having the shaded methyl group.

A double bond is formed between carbons nine and ten, and O_2 is reduced to water with electrons supplied by both the fatty acid and NADPH. Anaerobic bacteria and some aerobes create double bonds during fatty acid synthesis by dehydrating hydroxy fatty acids. Oxygen is not required for double bond synthesis by this pathway. The anaerobic pathway is present in a number of common gram-negative bacteria (e.g., *E. coli* and *Salmonella* spp.), gram-positive bacteria (e.g., *Lactobacillus plantarum* and *Clostridium pasteurianum*), and cyanobacteria.

Eucaryotic microorganisms frequently store carbon and energy as **triacylglycerol,** glycerol esterified to three fatty acids. Glycerol arises from the reduction of the precursor metabolites dihydroxyacetone phosphate to glycerol 3-phosphate, which is then esterified with two fatty acids to give **phosphatidic acid** (**figure 10.33**).

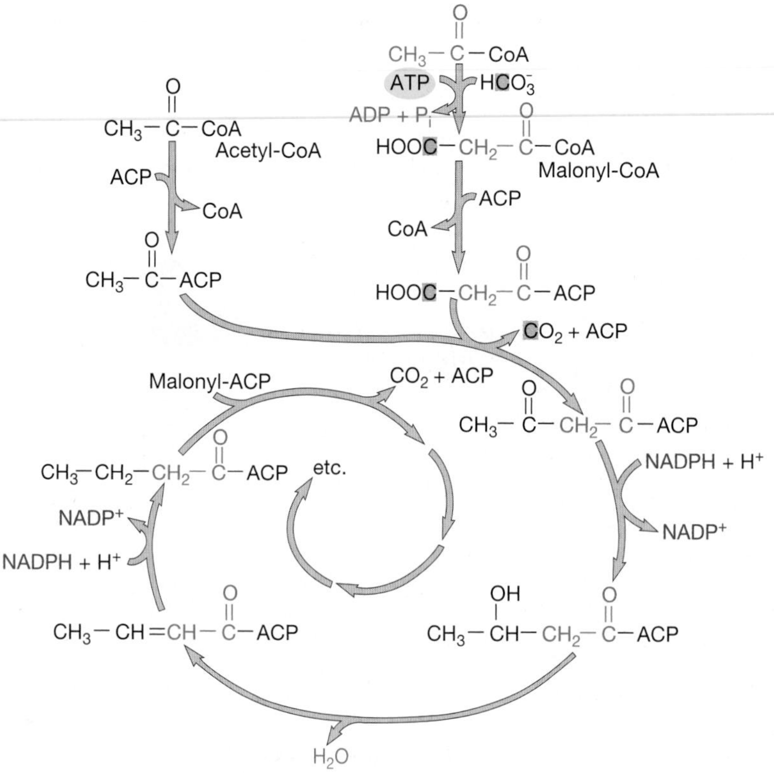

Figure 10.32 Fatty Acid Synthesis. The cycle is repeated until the proper chain length has been reached. Carbon dioxide carbon and the remainder of malonyl-CoA are shown in red. ACP stands for acyl carrier protein.

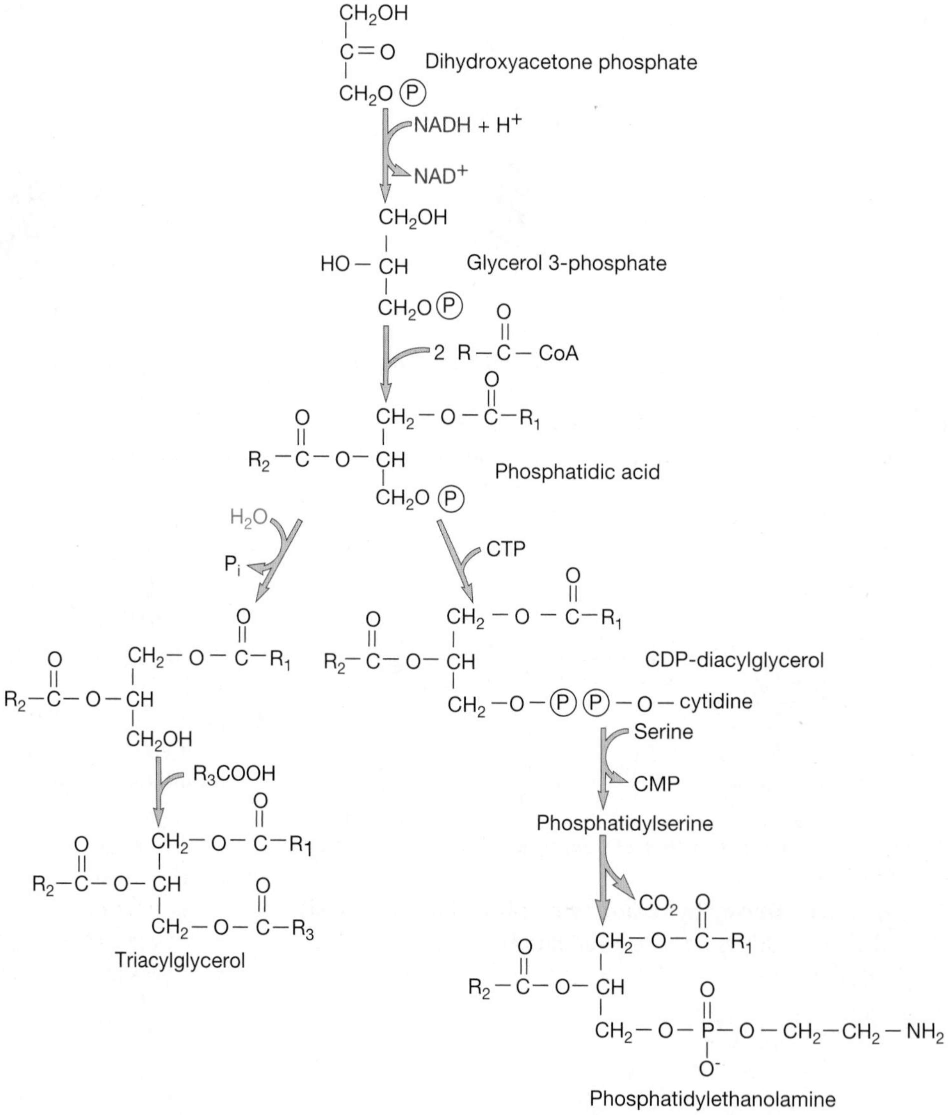

Figure 10.33 Triacylglycerol and Phospholipid Synthesis.

Phosphate is hydrolyzed from phosphatidic acid giving a diacylglycerol, and the third fatty acid is attached to yield a triacylglycerol.

Phospholipids are major components of eucaryotic and bacterial cell membranes. Their synthesis also usually proceeds by way of phosphatidic acid. A special cytidine diphosphate (CDP) carrier plays a role similar to that of uridine and adenosine diphosphate carriers in carbohydrate biosynthesis. For example, bacteria synthesize phosphatidylethanolamine, a major cell membrane component, through the initial formation of CDP-diacylglycerol (figure 10.33). This CDP derivative then reacts with serine to form the phospholipid phosphatidylserine, and decarboxylation yields phosphatidylethanolamine. In this way a complex membrane lipid is constructed from the products of glycolysis, fatty acid biosynthesis, and amino acid biosynthesis.

1. What is a fatty acid? Describe in general terms how the fatty acid synthase manufactures a fatty acid.
2. How are unsaturated fatty acids made?
3. Briefly describe the pathways for triacylglycerol and phospholipid synthesis. Of what importance are phosphatidic acid and CDP-diacylglycerol?
4. Activated carriers participate in carbohydrate, peptidoglycan, and lipid synthesis. Briefly describe these carriers and their roles. Are there any features common to all the carriers? Explain your answer.

Summary

In biosynthesis or anabolism, cells use energy to construct complex molecules from smaller, simpler precursors.

10.1 Principles Governing Biosynthesis

a. Many important cell constituents are macromolecules, large polymers constructed of simple monomers.

b. Although many catabolic and anabolic pathways share enzymes for the sake of efficiency, some of their enzymes are separate and independently regulated.

c. Macromolecular components often undergo self-assembly to form the final molecule or complex.

10.2 The Precursor Metabolites

a. Precursor metabolites are carbon skeletons used as the starting substrates for biosynthetic pathways. They are intermediates of glycolytic pathways and the TCA cycle (i.e., the central metabolic pathways) (**figure 10.3**).

b. Most precursor metabolites are used for amino acid biosynthesis.

10.3 The Fixation of CO_2 by Autotrophs

a. Four different CO_2-fixation pathways have been identified in autotrophic microorganisms: the Calvin cycle, the reductive TCA cycle, the acetyl-CoA pathway, and the hydroxypropionate cycle.

b. The Calvin cycle is used by most autotrophs to fix CO_2. It can be divided into three phases: the carboxylation phase, the reduction phase, and the regeneration phase (**figure 10.4**). Three ATPs and two NADPHs are used during the incorporation of one CO_2.

10.4 Synthesis of Sugars and Polysaccharides

a. Gluconeogenesis is the synthesis of glucose and related sugars from nonglucose precursors.

b. Glucose, fructose, and mannose are gluconeogenic intermediates or made directly from them; galactose is synthesized with nucleoside diphosphate derivatives. Bacteria and photosynthetic protists synthesize glycogen and starch from adenosine diphosphate glucose (**figure 10.9**).

c. Peptidoglycan synthesis is a complex process involving both UDP derivatives and the lipid carrier bactoprenol, which transports NAG-NAM-pentapeptide units across the cell membrane. Cross-links are formed by transpeptidation (**figures 10.12** and **10.14**).

d. Peptidoglycan synthesis occurs in discrete zones in the cell wall. Existing peptidoglycan is selectively degraded by autolysins so new material can be added (**figure 10.15**).

10.5 Synthesis of Amino Acids

a. The addition of nitrogen to the carbon chain is an important step in amino acid biosynthesis. Ammonia, nitrate, or nitrogen can serve as the source of nitrogen.

b. Ammonia nitrogen can be directly assimilated by the activity of transaminases and either glutamate dehydrogenase or the glutamine synthetase-glutamate synthase system (**figures 10.16–10.18**).

c. Nitrate is incorporated through assimilatory nitrate reduction catalyzed by the enzymes nitrate reductase and nitrite reductase (**figure 10.19**).

d. Nitrogen fixation is catalyzed by the nitrogenase complex. Atmospheric molecular nitrogen is reduced to ammonia, which is then incorporated into amino acids (**figures 10.20** and **10.22**).

e. Microorganisms can use cysteine, methionine, and inorganic sulfate as sulfur sources. Sulfate must be reduced to sulfide before it is assimilated. This occurs during assimilatory sulfate reduction (**figure 10.24**).

f. Although some amino acids are made directly by the addition of an amino group to a precursor metabolite, most amino acids are made by more complex pathways. Many amino acid biosynthetic pathways are branched. Thus a single precursor metabolite can give rise to several amino acids (**figures 10.25** and **10.26**).

g. Anaplerotic reactions replace TCA cycle intermediates to keep the cycle in balance while it supplies biosynthetic precursors. The anaplerotic reactions include the glyoxylate cycle (**figure 10.27**).

10.6 Synthesis of Purines, Pyrimidines, and Nucleotides

a. Purines and pyrimidines are nitrogenous bases found in DNA, RNA, and other molecules. The nitrogen is supplied by certain amino acids that participate in purine and pyrimidine biosynthesis. Phosphorus is provided by either inorganic phosphate or organic phosphate.

b. Phosphorus can be assimilated directly by phosphorylation reactions that form ATP from ADP and P_i. Organic phosphorus sources are the substrates of phosphatases that release phosphate from the organic molecule.

c. The purine skeleton is synthesized beginning with ribose 5-phosphate and initially produces inosinic acid. Pyrimidine biosynthesis starts with carbamoyl phosphate and aspartate, and ribose is added after the skeleton has been constructed (**figures 10.28–10.30**).

10.7 Lipid Synthesis

a. Fatty acids are synthesized from acetyl-CoA, malonyl-CoA, and NADPH by the fatty acid synthase system. During synthesis the intermediates are attached to the acyl carrier protein. Double bonds can be added in two different ways (**figure 10.32**).

b. Triacylglycerols are made from fatty acids and glycerol phosphate. Phosphatidic acid is an important intermediate in this pathway (**figure 10.33**).

c. Phospholipids like phosphatidylethanolamine can be synthesized from phosphatidic acid by forming CDP-diacylglycerol, then adding an amino acid.

Key Terms

acetyl-CoA pathway 230
acyl carrier protein (ACP) 242
adenine 241
anaplerotic reactions 239
assimilatory nitrate reduction 235
assimilatory sulfate reduction 238
autolysins 234
bactoprenol 232
Calvin cycle 228
carboxysomes 229
central metabolic pathways 227
cytosine 241
dissimilatory sulfate reduction 238
fatty acid 242
fatty acid synthase 242
gluconeogenesis 230
glutamate dehydrogenase 235
glutamate synthase 235
glutamine synthetase 235
glutamine synthetase-glutamate synthase (GS-GOGAT) system 235
glyoxylate cycle 240
guanine 241
3-hydroxypropionate cycle 229
Lipid I 232
Lipid II 232
macromolecule 226
monomers 226
nitrate reductase 236
nitrite reductase 236
nitrogenase 237
nitrogen fixation 236
nucleoside 241
nucleotide 241
phosphatase 241
phosphatidic acid 243
precursor metabolites 227
purine 241
pyrimidine 241
reductive TCA cycle 229
ribulose-1,5-bisphosphate carboxylase 229
self-assembly 227
thymine 241
transaminases 235
transpeptidation 233
triacylglycerol 243
turnover 225
uracil 241
uridine diphosphate glucose (UDPG) 231

Critical Thinking Questions

1. Discuss the relationship between catabolism and anabolism. How does anabolism depend on catabolism?
2. In metabolism, important intermediates are covalently attached to carriers, as if to mark these as important so the cell does not lose track of them. Think about a hotel placing your room key on a very large ring. List a few examples of these carriers and indicate whether they are involved primarily in anabolism or catabolism.
3. Intermediary carriers are in a limited supply—when they cannot be recycled because of a metabolic block, serious consequences ensue. Think of some examples of these consequences.

Learn More

Gottschalk, G. 1986. *Bacterial metabolism,* 2d ed. New York: Springer-Verlag.

Herter, S.; Farfsing, J.; Gad'on, N.; Rieder, C.; Eisenreich, W.; Bacher, A.; and Fuchs, G. 2001. Autotrophic CO_2 fixation by *Chloroflexus aurantiacus:* Study of glyoxylate formation and assimilation via the 3-hydroxypropionate cycle. *J. Bacteriol.* 183(14):4305–16.

Höltje, J.-V. 2000. Cell walls, bacterial. In *Encyclopedia of microbiology,* 2d ed., vol. 1, J. Lederberg, editor-in-chief, 759–71. San Diego: Academic Press.

Kuykendall, L. D.; Dadson, R. B.; Hashem, F. M.; and Elkan, G. H. 2000. Nitrogen fixation. In *Encyclopedia of microbiology,* 2d ed., vol. 3, J. Lederberg, editor-in-chief, 392–406. San Diego: Academic Press.

Lens, P., and Pol, L. H. 2000. Sulfur cycle. In *Encyclopedia of microbiology,* 2d ed., vol. 4, J. Lederberg, editor-in-chief, 495–505. San Diego: Academic Press.

McKee, T., and McKee, J. R. 2003. *Biochemistry: The molecular basis of life,* 3d ed. Dubuque, Iowa: McGraw-Hill.

Nelson, D. L., and Cox, M. M. 2005. *Lehninger: Principles of biochemistry,* 4th ed. New York: W. H. Freeman.

Schweizer, E., and Hofmann, J. 2004. Microbial type I fatty acid synthases (FAS): Major players in a network of cellular FAS systems. *Microbiol. Mol. Biol. Rev.* 68(3):501–17.

White, S. W.; Zheng, J.; Zhang, Y.-M.; and Rock, C. O. 2005. The structural biology of type II fatty acid biosynthesis. *Annu. Rev. Biochem.* 74:791–831.

Yoon, K.-S.; Hanson, T. E.; Gibson, J. L.; and Tabita, F. R. 2000. Autotrophic CO_2 metabolism. In *Encyclopedia of microbiology,* 2d ed., vol. 1, J. Lederberg, editor-in-chief, 349–58. San Diego: Academic Press.

Please visit the Prescott website at www.mhhe.com/prescott7 for additional references.

11 Microbial Genetics: Gene Structure, Replication, and Expression

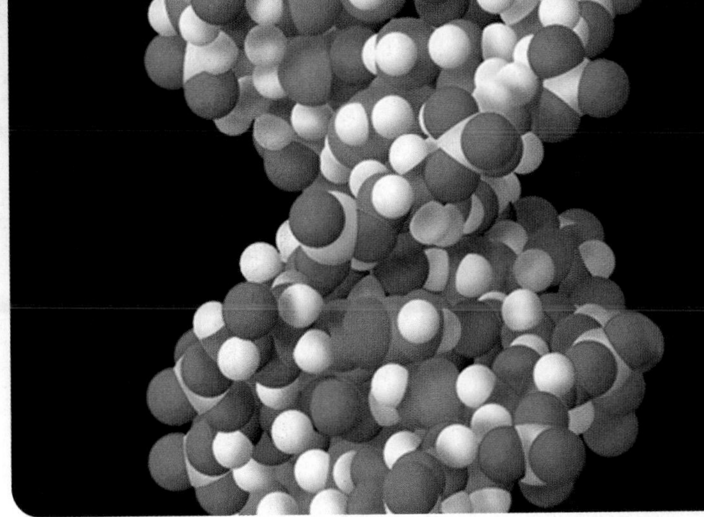

This model illustrates double-stranded DNA. DNA is the genetic material for procaryotes and eucaryotes. Genetic information is contained in the sequence of base pairs that lie in the center of the helix.

Preview

- The two kinds of nucleic acid, deoxyribonucleic acid (DNA) and ribonucleic acid (RNA), differ from one another in chemical composition and structure. In cells, DNA serves as the repository for genetic information.
- The flow of genetic information usually proceeds from DNA through RNA to protein. A protein's amino acid sequence reflects the nucleotide sequence of its mRNA, which is complementary to a portion of the DNA genome.
- DNA replication is a very complex process involving a variety of proteins and a number of steps. It is designed to operate rapidly while minimizing errors and correcting those that arise when DNA is copied.
- A gene is a nucleotide sequence that codes for a polypeptide, tRNA, or rRNA. Most procaryotic genes have at least four major parts, each with different functions: promoters, leaders, coding regions, and trailers. When a gene directs the synthesis of a polypeptide, each amino acid is specified by a triplet codon.
- In transcription the RNA polymerase copies the appropriate sequence on the DNA template strand to produce a complementary RNA copy of the gene. Transcription differs in a number of ways among *Bacteria, Archaea,* and eucaryotes, even though the basic mechanism of RNA polymerase action is essentially the same.
- Translation is the process by which the nucleotide sequence of mRNA is converted into the amino acid sequence of a polypeptide through the action of ribosomes, tRNAs, aminoacyl-tRNA synthetases, ATP and GTP energy, and a variety of protein factors. As in the case of DNA replication, this complex process is designed to minimize errors.

Chapters 8 through 10 introduce the essentials of microbial metabolism. They focus on processes that provide the energy and metabolic precursors used by cells to synthesize the macromolecules needed for construction of chromosomes, ribosomes, cell walls, and other cellular components. We now turn our attention to the synthesis of three major macromolecules—DNA, RNA, and proteins—from their constituent monomers. DNA serves as the storage molecule for the genetic instructions that allow organisms to carry out metabolism and reproduction. RNA functions in the expression of genetic information so that enzymes and other proteins can be made. These proteins are used to build certain cellular structures and to do other cellular work. The study of the synthesis of DNA, RNA, and protein falls into the realms of genetics and molecular biology.

In the mid-1800s, the discipline of genetics was born from the work of Gregor Mendel, who studied the inheritance of various traits in pea plants. In the early twentieth century, Mendel's work was rediscovered and furthered by scientists working with fruit flies and plants such as corn. The use of microorganisms as models for genetic studies soon followed. Microorganisms, especially bacteria, have significant advantages as model organisms, in part because of their unique characteristics. One important feature is the nature of their genomes. The term **genome** refers to all the DNA present in a cell or virus. Procaryotes normally have one set of genes; that is, they are haploid (1N). In addition, they often

But the most important qualification of bacteria for genetic studies is their extremely rapid rate of growth. . . . a single E. coli *cell will grow overnight into a visible colony containing millions of cells, even under relatively poor growth conditions. Thus, genetic experiments on* E. coli *usually last one day, whereas experiments on corn, for example, take months. It is no wonder that we know so much more about the genetics of* E. coli *than about the genetics of corn, even though we have been studying corn much longer.*

—R. F. Weaver and P. W. Hedrick

carry extrachromosomal genetic elements called plasmids. Eucaryotic organisms, including eucaryotic microorganisms, usually have two sets of genes, or are diploid (2N), and they rarely have plasmids. Viral genomes differ significantly from those of cellular organisms, and their genetics and molecular biology are discussed in chapters 16–18. Plasmids (section 3.5)

In this chapter we review some of the most basic concepts of molecular genetics: how genetic information is stored and organized in the DNA molecule, the way in which DNA is replicated, gene structure, and how genes function (i.e., gene expression). Based on the foundation provided in chapter 11, chapter 12 considers the regulation of gene expression. The regulation of gene expression is important because it links the **genotype** of an organism—the specific set of genes it possesses—to the **phenotype** of an organism—the collection of characteristics that are observable. All genes are not expressed at the same time or in the same place, and the environment profoundly influences which genes are expressed at any given time. Finally, chapter 13 contains information on the nature of mutation, DNA repair, and genetic recombination. These three chapters provide the background needed for understanding the material on recombinant DNA technology (chapter 14) and microbial genomics (chapter 15). Much of the information presented in chapters 11 through 13 will be familiar to those who have taken an introductory genetics course. Because of the importance of bacteria as model organisms, primary emphasis is placed on their genetics. The genetics of the *Archaea* is discussed in chapter 20.

Although modern genetic analysis began with studies of fruit flies and corn, the nature of genetic information, gene structure, the genetic code, and mutation were elucidated by elegant experiments involving bacteria and bacterial viruses. We will first review a few of these early experiments and then summarize the view of DNA, RNA, and protein relationships—sometimes called the "central dogma"—which have guided much of modern research.

Historical Highlights

11.1 The Elucidation of DNA Structure

The basic chemical composition of nucleic acids was elucidated in the 1920s through the efforts of P. A. Levene. Despite his major contributions to nucleic acid chemistry, Levene mistakenly believed that DNA was a very small molecule, probably only four nucleotides long, composed of equal amounts of the four different nucleotides arranged in a fixed sequence. Partly because of his influence, biologists believed for many years that nucleic acids were too simple in structure to carry complex genetic information. They concluded that genetic information must be encoded in proteins because proteins are large molecules with complex amino acid sequences that vary among different proteins.

As so often happens, further advances in our understanding of DNA structure awaited the development of significant new analytical techniques in chemistry. One development was the invention of paper chromatography by Archer Martin and Richard Synge between 1941 and 1944. By 1948 the chemist Erwin Chargaff had begun using paper chromatography to analyze the base composition of DNA from a number of species. He soon found that the base composition of DNA from genetic material did indeed vary among species just as he expected. Furthermore, the total amount of purines always equaled the total amount of pyrimidines; and the adenine/thymine and guanine/cytosine ratios were always 1. These findings, known as Chargaff's rules, were a key to the understanding of DNA structure.

Another turning point in research on DNA structure was reached in 1951 when Rosalind Franklin arrived at King's College, London, and joined Maurice Wilkins in his efforts to prepare highly oriented DNA fibers and study them by X-ray crystallography. By the winter of 1952–1953, Franklin had obtained an excellent X-ray diffraction photograph of DNA.

The same year that Franklin began work at King's College, the American biologist James Watson went to Cambridge University and met Francis Crick. Although Crick was a physicist, he was very interested in the structure and function of DNA, and the two soon began to work on its structure. Their attempts were unsuccessful until Franklin's data provided them with the necessary clues. Her photograph of fibrous DNA contained a crossing pattern of dark spots, which showed that the molecule was helical. The dark regions at the top and bottom of the photograph showed that the purine and pyrimidine bases were stacked on top of each other and separated by 0.34 nm. Franklin had already concluded that the phosphate groups lay to the outside of the cylinder. Finally, the X-ray data and her determination of the density of DNA indicated that the helix contained two strands, not three or more as some had proposed.

Without actually doing any experiments themselves, Watson and Crick constructed their model by combining Chargaff's rules on base composition with Franklin's X-ray data and their predictions about how genetic material should behave. By building models, they found that a smooth, two-stranded helix of constant diameter could be constructed only when an adenine hydrogen bonded with thymine and when a guanine bonded with cytosine in the center of the helix. They immediately realized that the double helical structure provided a mechanism by which genetic material might be replicated. The two parental strands could unwind and direct the synthesis of complementary strands, thus forming two new identical DNA molecules (figure 11.10). Watson, Crick, and Wilkins received the Nobel Prize in 1962 for their discoveries. Unfortunately, Franklin could not be considered for the prize because she had died of cancer in 1958 at the age of thirty-seven.

11.1 DNA as Genetic Material

Although it is now hard to imagine, it was once thought that DNA was too simple a molecule to store genetic information (**Historical Highlights 11.1**). The early work of Fred Griffith in 1928 on the transfer of virulence in the pathogen *Streptococcus pneumoniae,* commonly called pneumococcus (**figure 11.1**), set the stage for research showing that DNA was indeed the genetic material. Griffith found that if he boiled virulent bacteria and injected them into mice, the mice were not affected and no pneumococci could be recovered from the animals. When he injected a combination of killed virulent bacteria and a living nonvirulent strain, the mice died; moreover, he could recover living virulent bacteria from the dead mice. Griffith called this change of nonvirulent bacteria into virulent pathogens **transformation.**

Oswald Avery and his colleagues then set out to discover which constituent in the heat-killed virulent pneumococci was responsible for Griffith's transformation. These investigators selectively destroyed constituents in purified extracts of virulent pneumococci (S cells), using enzymes that would hydrolyze DNA, RNA, or protein. They then exposed nonvirulent pneumococcal strains (R strains) to the treated extracts. Transformation of the nonvirulent bacteria was blocked only if the DNA was destroyed, suggesting that DNA was carrying the information required for transformation (**figure 11.2**). The publication of these studies by Avery, C. M. MacLeod, and M. J. McCarty in 1944

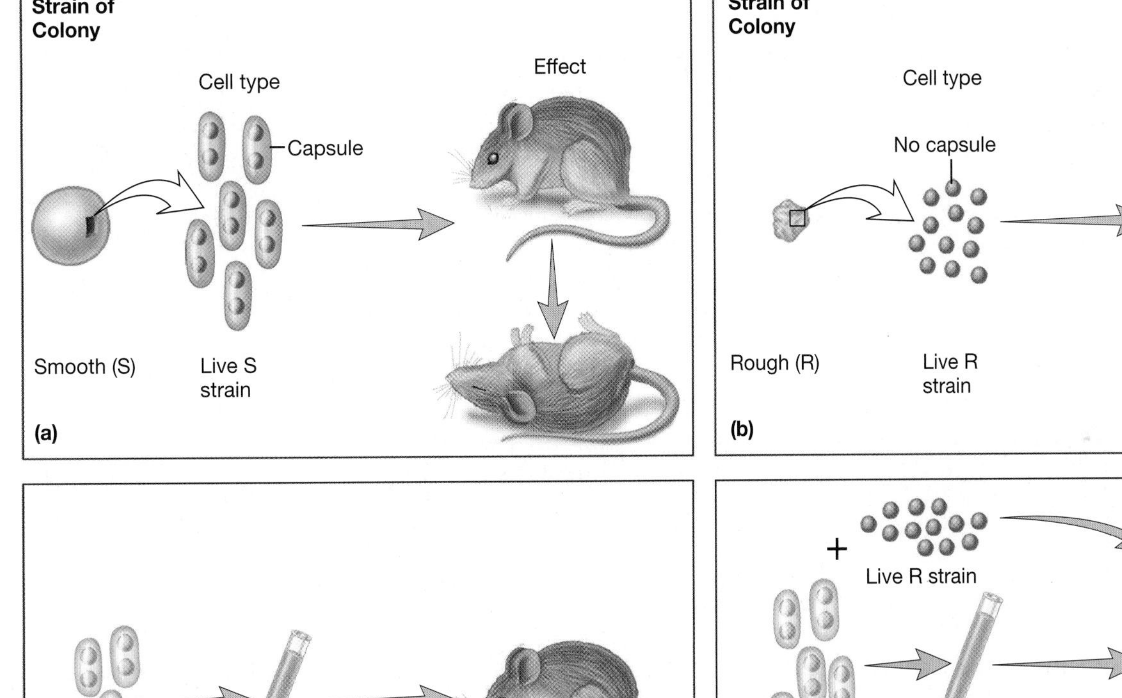

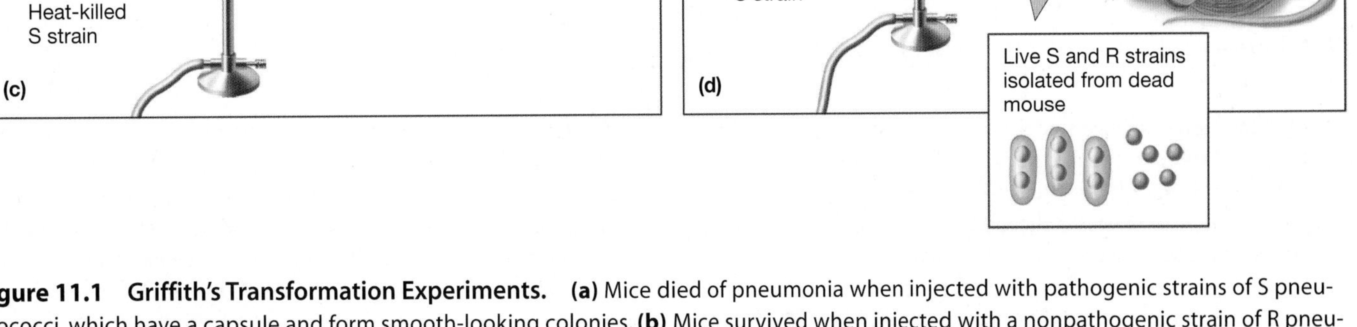

Figure 11.1 Griffith's Transformation Experiments. **(a)** Mice died of pneumonia when injected with pathogenic strains of S pneumococci, which have a capsule and form smooth-looking colonies. **(b)** Mice survived when injected with a nonpathogenic strain of R pneumococci, which lacks a capsule and forms rough colonies. **(c)** Injection with heat-killed strains of S pneumococci had no effect. **(d)** Injection with a live R strain and a heat-killed S strain gave the mice pneumonia, and live S strain pneumococci could be isolated from the dead mice.

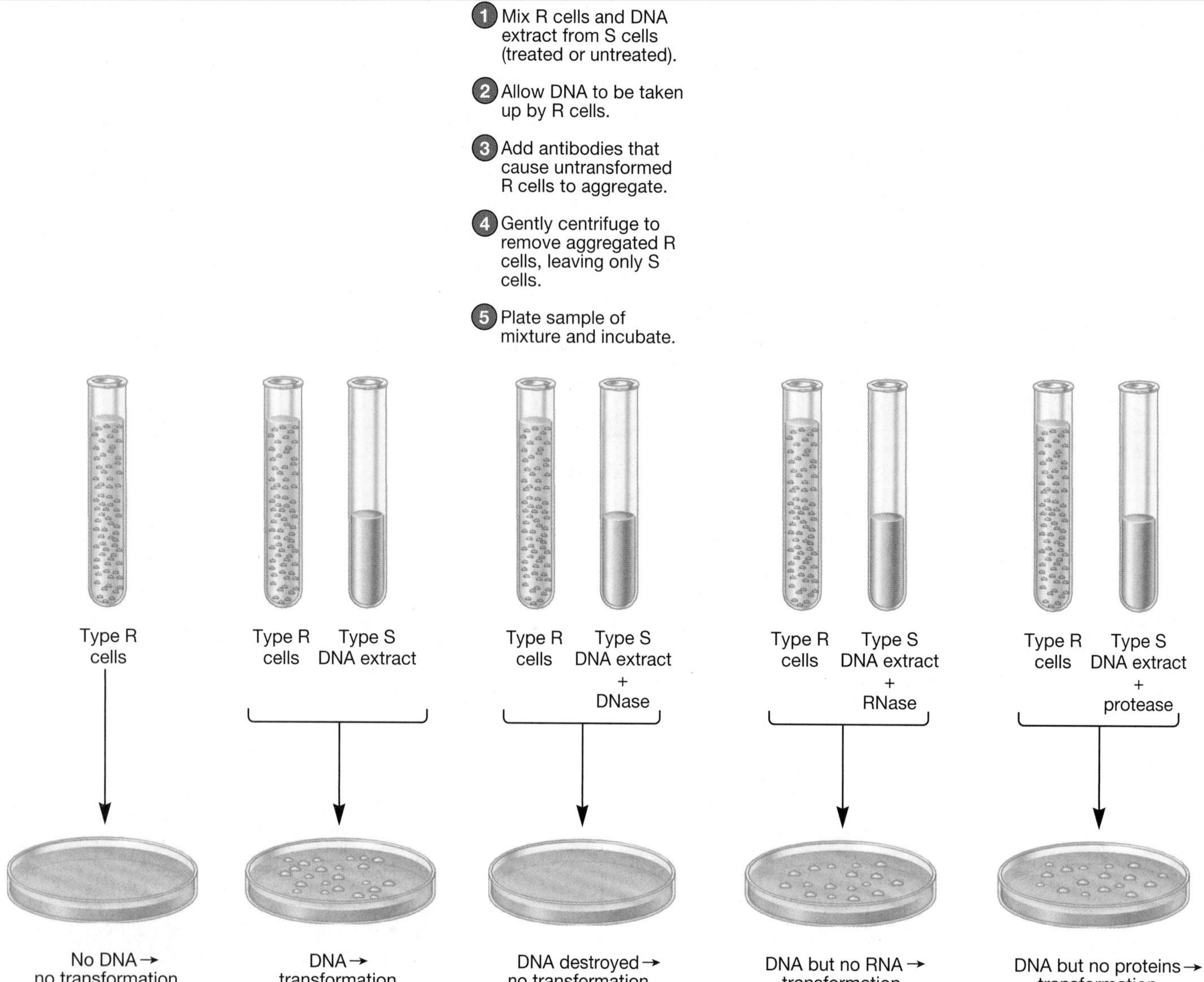
1 Mix R cells and DNA extract from S cells (treated or untreated).
2 Allow DNA to be taken up by R cells.
3 Add antibodies that cause untransformed R cells to aggregate.
4 Gently centrifuge to remove aggregated R cells, leaving only S cells.
5 Plate sample of mixture and incubate.
Type R cells
Type R cells
Type S DNA extract
Type R cells
Type S DNA extract + DNase
Type R cells
Type S DNA extract + RNase
Type R cells
Type S DNA extract + protease
No DNA→ no transformation
DNA→ transformation
DNA destroyed→ no transformation
DNA but no RNA→ transformation
DNA but no proteins→ transformation

Figure 11.2 Some Experiments on the Transforming Principle. Earlier experiments done by Avery, MacLeod, and McCarty had shown that only DNA extracts from S cells caused transformation of R cells to S cells. To demonstrate that contaminating molecules in the DNA extract were not responsible for transformation, the DNA extract from S cells was treated with RNase, DNase, and protease and then mixed with R cells. Time was allowed for the DNA from S cells to be taken up by the R cells and expressed, transforming R cells into S cells. Then, antibodies (immune system proteins that recognize specific structures) that recognized R cells, but not S cells, were added to the mixture. The addition of antibodies caused the R cells (i.e., those R cells that had not been transformed) to aggregate. These aggregated R cells were removed from the mixture by gentle centrifugation. Thus, the only cells remaining in the mixture were cells that had been transformed and were now S cells. Only treatment of the DNA extract from S cells with DNase destroyed the ability of the extract to transform the R cells.

provided the first evidence that Griffith's transforming principle was DNA and therefore that DNA carried genetic information.

Some years later (1952), Alfred Hershey and Martha Chase performed several experiments indicating that DNA was the genetic material in a bacterial virus called T2 bacteriophage. Some luck was involved in their discovery, for the genetic material of many viruses is RNA and the researchers happened to select a DNA virus for their studies. Imagine the confusion if T2 had been an RNA virus! The controversy surrounding the nature of genetic information might have lasted considerably longer than it did. Hershey and Chase made the virus's DNA radioactive with ^{32}P or they labeled the its protein coat with ^{35}S. They mixed radioactive bacteriophage with *Escherichia coli* and incubated the mixture for a few minutes. The suspension was then agitated violently in a blender to shear off any adsorbed bacteriophage particles (**figure 11.3**). After centrifugation, radioactivity in the supernatant (where the virus remained) versus the bacterial cells in the pellet was determined. They found that most radioactive protein was released into the supernatant,

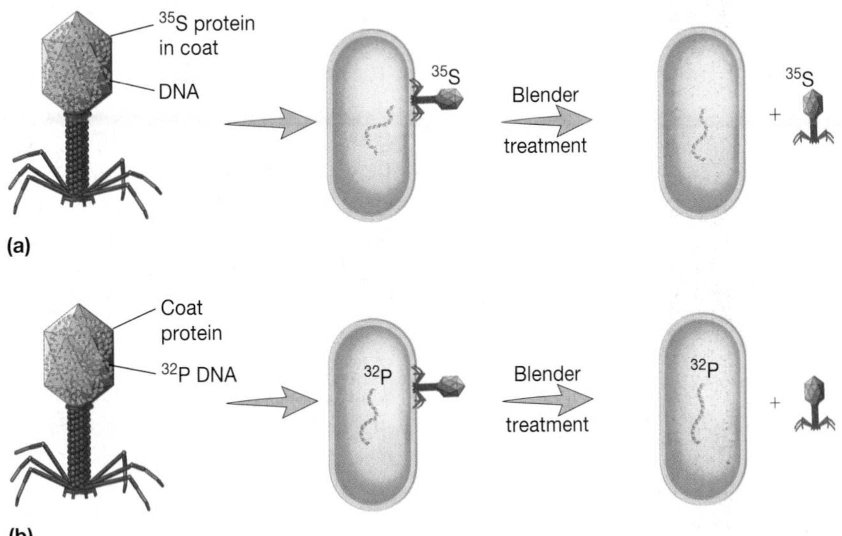

Figure 11.3 The Hershey-Chase Experiment. **(a)** When *E. coli* was infected with a T2 phage containing ^{35}S protein, most of the radioactivity remained outside the host cell. **(b)** When a T2 phage containing ^{32}P DNA was mixed with the host bacterium, the radioactive DNA was injected into the cell and phages were produced. Thus DNA was carrying the virus's genetic information.

whereas ^{32}P DNA remained within the bacteria. Since genetic material was injected and T2 progeny were produced, DNA must have been carrying the genetic information for T2. Virulent double-stranded DNA phages (section 17.2)

Subsequent studies on the genetics of viruses and bacteria were largely responsible for the rapid development of molecular genetics. Furthermore, much of the recombinant DNA technology described in chapter 14 has arisen from studies of bacterial and viral genetics. Research in microbial genetics has had a profound impact on biology as a science and on technology that affects everyday life.

1. Define genome, genotype, and phenotype.
2. Briefly summarize the experiments of Griffith; Avery, MacLeod, and McCarty; and Hershey and Chase. What did each show, and why were these experiments important to the development of microbial genetics?

11.2 The Flow of Genetic Information

Biologists have long recognized a relationship among DNA, RNA, and protein, and this recognition has guided a vast amount of research over the past decades. The pathway from DNA to RNA and RNA to protein is conserved in all forms of life and is often called the central dogma. **Figure 11.4** illustrates two essential concepts: the flow of genetic information from one generation to the next (replication); and the flow of information within a single cell, a process also called gene expression.

The transmission of genetic information from one generation to the next is shown in figure 11.4*a*. DNA functions as a storage molecule, holding genetic information for the lifetime of a cellular organism, and allowing that information to be duplicated and passed on to its progeny. Synthesis of the duplicate DNA is directed by the parental molecule and is called **replication.** The process is catalyzed by DNA polymerase enzymes.

The genetic information stored in DNA is divided into units called **genes.** In order for an organism to function properly and reproduce, its genes must be expressed at the appropriate time and place. Gene expression begins with the synthesis of an RNA copy of the gene. This process of DNA-directed RNA synthesis is called **transcription** because the DNA base sequence is being written into an RNA base sequence. RNA polymerase enzymes catalyze transcription. Although DNA has two complementary strands, only one strand, the template strand, of a particular gene is transcribed. If both strands of a single gene were transcribed, two different RNA molecules would result and cause genetic confusion. However, *different* genes may be encoded on opposite strands, thus both strands of DNA can serve as templates for RNA synthesis depending on the orientation of the gene on the DNA. Transcription yields three different types of RNA depending on the gene being transcribed. These are messenger RNA (mRNA), transfer RNA (tRNA), and ribosomal RNA (rRNA) (figure 11.4*b*).

During the last phase of gene expression, **translation,** genetic information in the form of an RNA base sequence in a **messenger RNA (mRNA)** is decoded and used to govern the synthesis of a polypeptide. Thus the amino acid sequence of a protein is a direct reflection of the base sequence in mRNA. In turn, the mRNA nucleotide sequence is complementary to a portion of the DNA genome. In addition to mRNA, translation also requires the activities of transfer RNA and ribosomal RNA. Thus all three types of RNA are involved in the production of protein, based on the code present in the DNA.

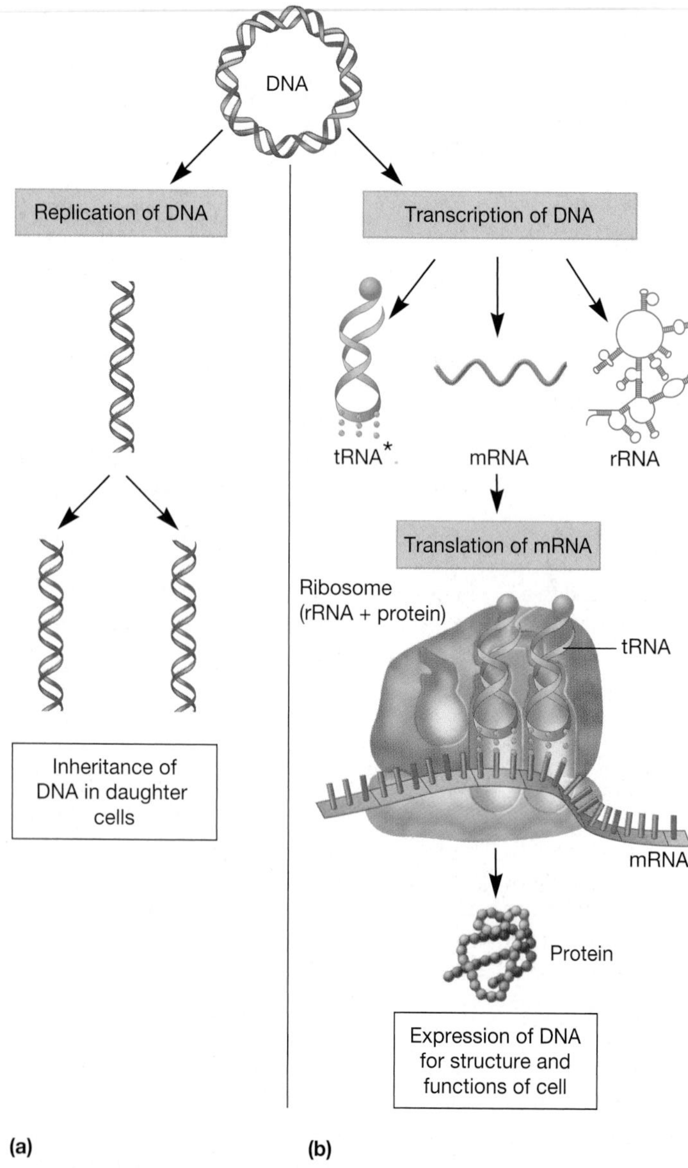

*The sizes of RNA are not to scale—tRNA and mRNA are enlarged to show details.

Figure 11.4 Summary of the Flow of Genetic Information in Cells. DNA serves as the storehouse for genetic information. **(a)** During cellular reproduction, DNA is replicated and passed to progeny cells. **(b)** In order to function properly, a cell must express the genetic information stored in DNA. This is accomplished when the genetic code is transcribed into mRNA molecules. The information in the mRNA is then translated into protein.

1. Describe the general relationship between DNA, RNA, and protein.
2. What are the products of replication, transcription, and translation?
3. Until relatively recently, the "one gene-one protein" hypothesis was used to define the role of genes in organisms. Refer to figure 11.4 and explain why this description of a gene no longer applies.

11.3 Nucleic Acid Structure

The nucleic acids, DNA and RNA, are polymers of nucleotides (**figure 11.5**) linked together by phosphodiester bonds (**figure 11.6*a***). However, DNA and RNA differ in terms of the nitrogenous bases they contain, the sugar component of their nucleotides, and whether they are double or single stranded. **Deoxyribonucleic acid (DNA)** contains the bases adenine, guanine, cytosine, and thymine. The sugar found in the nucleotides is deoxyribose, and DNA molecules are usually double stranded. **Ribonucleic acid (RNA),** on the other hand, contains the bases adenine, guanine, cytosine, and uracil (instead of thymine, although tRNA contains a modified form of thymine). Its sugar is ribose, and most RNA molecules are single stranded. The structure of DNA and RNA is described in more detail next.

DNA Structure

The discovery that DNA is the genetic material set into motion a fierce competition to determine the precise structure of DNA (Historical Highlights 11.1). DNA molecules are very large and are usually composed of two polynucleotide chains coiled together to form a double helix 2.0 nm in diameter (figure 11.6 and **figure 11.7**). Each chain contains purine and pyrimidine deoxyribonucleosides joined by phosphodiester linkages (figure 11.6*a*). That is, a phosphoric acid molecule forms a bridge between a 3′-hydroxyl of one sugar and a 5′-hydroxyl of an adjacent sugar. Purine and pyrimidine bases are attached to the 1′-carbon of the deoxyribose sugars and extend toward the middle of the cylinder formed by the two chains. They are stacked on top of each other in the center, one base pair every 0.34 nm. The purine adenine (A) of one strand is always paired with the pyrimidine thymine (T) of the opposite strand by two hydrogen bonds. The purine guanine (G) pairs with cytosine (C) by three hydrogen bonds. This AT and GC base pairing means that the two strands in a DNA double helix are **complementary.** In other words, the bases in one strand match up with those of the other according to specific base pairing rules. Because the sequences of bases in these strands encode genetic information, considerable effort has been devoted to determining the base sequences of DNA and RNA from many organisms, including a variety of microbes. Microbial genomics (chapter 15)

The two polynucleotide strands fit together much like the pieces in a jigsaw puzzle because of complementary base pairing. Inspection of figure 11.6*b,c*, depicting the B form of DNA (probably the most common form in cells), shows that the two strands are not positioned directly opposite one another in the helical cylinder. Therefore when the strands twist about one another, a wide **major groove** and narrower **minor groove** are formed by the backbone. Each base pair rotates 36° around the cylinder with respect to adjacent pairs so that there are 10 base pairs per turn of the helical spiral. Each turn of the helix has a vertical length of 3.4 nm. The helix is right-handed—that is, the chains turn counterclockwise as they approach a viewer looking down the longitudinal axis. The two backbones are antiparallel, which means they run in opposite directions with respect to the orientation of their sugars. One end of each strand has

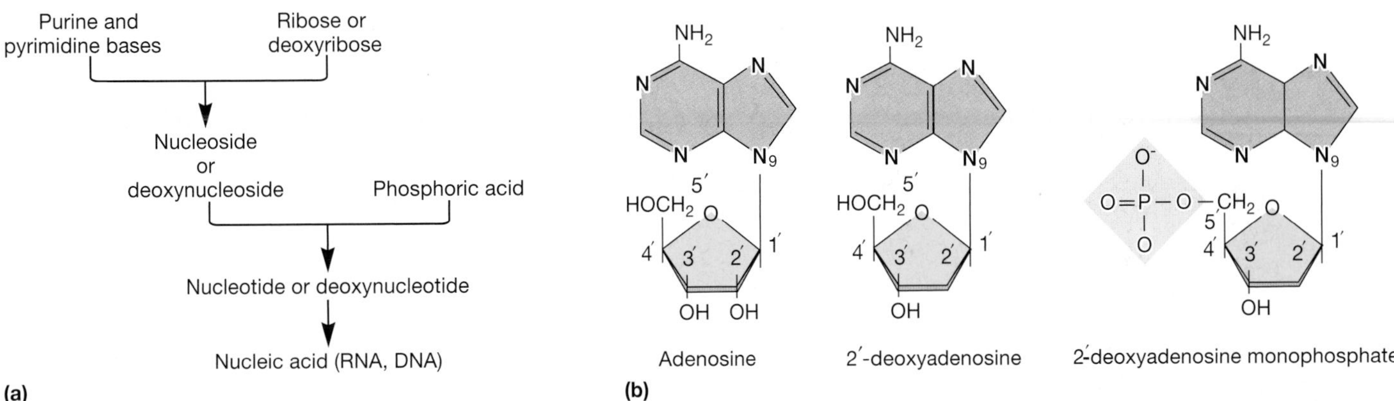

Figure 11.5 The Composition of Nucleic Acids. **(a)** A diagram showing the relationships of various nucleic acid components. Combination of a purine or pyrimidine base with ribose or deoxyribose gives a nucleoside (a ribonucleoside or deoxyribonucleoside). A nucleotide contains a nucleoside and one or more phosphoric acid molecules. Nucleic acids result when nucleotides are connected together in polynucleotide chains. **(b)** Examples of nucleosides—adenosine and 2′-deoxyadenosine—and the nucleotide 2′-deoxyadenosine monophosphate. The carbons of nucleoside and nucleotide sugars are indicated by numbers with primes.

an exposed 5′-hydroxyl group, often with phosphates attached, whereas the other end has a free 3′-hydroxyl group (figure 11.6a). If one end of a double helix is examined, the 5′ end of one strand and the 3′ end of the other are visible. In a given direction one strand is oriented 5′ to 3′ and the other, 3′ to 5′ (figure 11.6).

RNA Structure

RNA differs chemically from DNA, and is usually single stranded rather than double stranded. However, an RNA strand can coil back on itself to form a hairpin-shaped structure with complementary base pairing and helical organization. The three different types of RNA—messenger RNA, ribosomal RNA, and transfer RNA—differ from one another in function, site of synthesis in eucaryotic cells, and structure.

The Organization of DNA in Cells

Although DNA exists as a double helix in all cells, its organization differs among cells in the three domains of life. DNA is organized in the form of a closed circle in all *Archaea* and most bacteria. This circular double helix is further twisted into supercoiled DNA (**figure 11.8**). In *Bacteria,* DNA is associated with basic proteins that appear to help organize it into a coiled, chromatinlike structure.

DNA is much more highly organized in eucaryotic chromatin and is associated with a variety of proteins, the most prominent of which are **histones.** These are small, basic proteins rich in the amino acids lysine and/or arginine. There are five types of histones in almost all eucaryotic cells studied: H1, H2A, H2B, H3, and H4. Eight histone molecules (two each of H2A, H2B, H3, and H4) form an ellipsoid about 11 nm long and 6.5 to 7 nm in diameter (**figure 11.9*a***). DNA coils around the surface of the ellipsoid approximately 1 3/4 turns or 166 base pairs before proceeding on to the next. This combination of histones plus DNA, or nucleoprotein complex, is called a **nucleosome.** DNA gently isolated from chromatin looks like a string of beads. The stretch of DNA between the beads or nucleosomes, the linker region, varies in length from 14 to over 100 base pairs. Histone H1 associates with the linker regions to aid the folding of DNA into more complex chromatin structures (figure 11.9*b*). When folding reaches a maximum, the chromatin takes the shape of the visible chromosomes seen in eucaryotic cells during mitosis and meiosis.

Although the *Archaea* share the procaryotic style of cellular organization with the *Bacteria,* there are some important differences. Thus far, all archaeal genomes examined are circular. In many archaea, the DNA is complexed with histone proteins. Like eucaryotic histones, archaeal histones form nucleoprotein complexes called **archaeal nucleosomes.** In archaea, tetramers of histones (that is, four histone proteins) interact with about 60 base pairs each, protecting the DNA from DNA-digesting nucleases. The structure of archaeal histones and their interaction with DNA strongly suggests that archaeal nucleosomes are evolutionarily similar to the eucaryotic nucleosomes formed by DNA and histones H3 and H4. Microbial evolution (section 19.1)

1. What are nucleic acids? How do DNA and RNA differ in structure?
2. Describe in some detail the structure of the DNA double helix. What does it mean to say that the two strands are complementary and antiparallel?
3. What are histones and nucleosomes? Describe the way in which DNA is organized in the chromosomes of *Bacteria, Archaea,* and eucaryotes.

11.4 DNA Replication

DNA replication is an extraordinarily important and complex process upon which all life depends. At least 30 proteins are required to replicate the *E. coli* chromosome (**table 11.1**). Presumably, much of the complexity is necessary for accuracy in copying

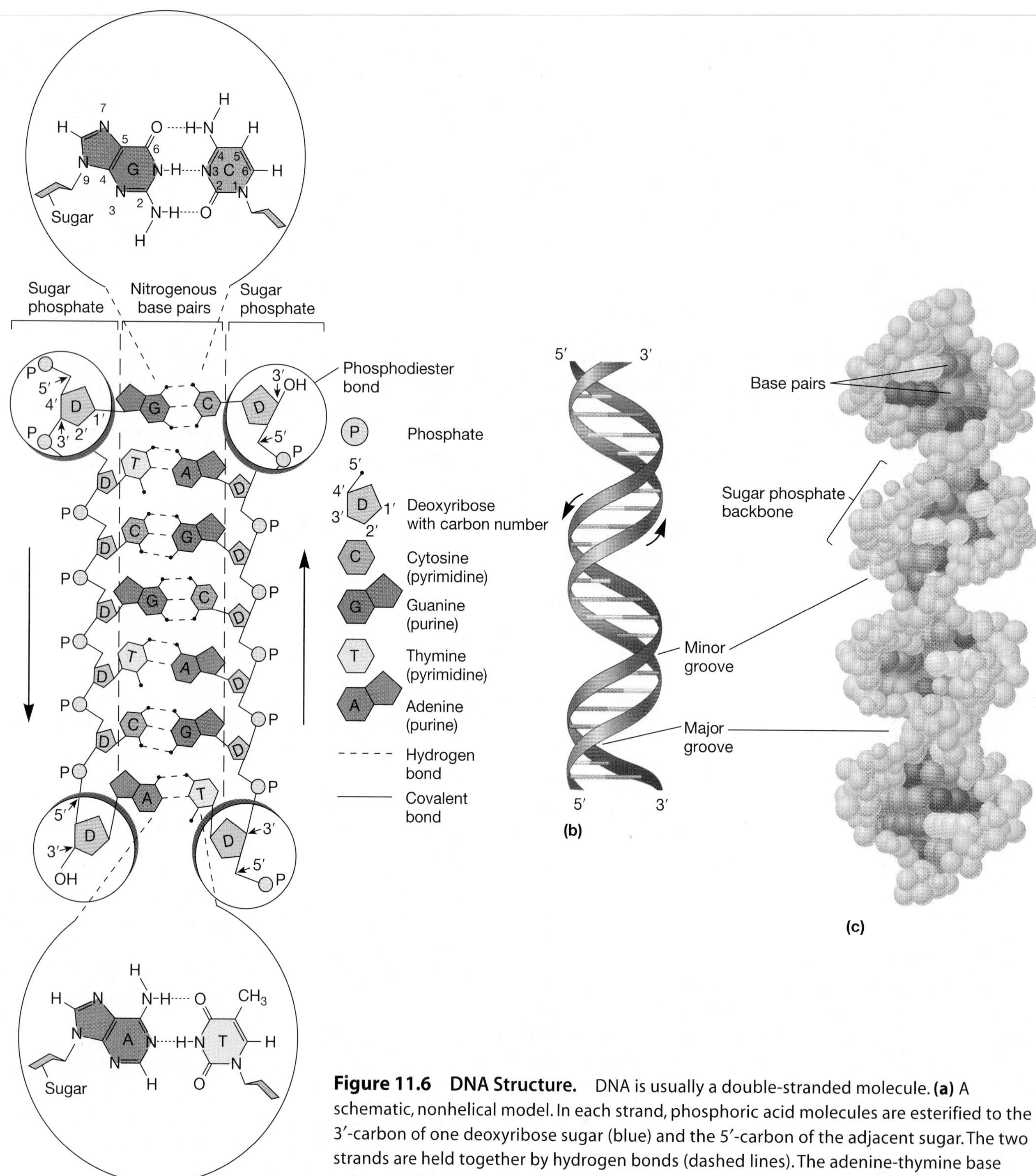

Figure 11.6 DNA Structure. DNA is usually a double-stranded molecule. **(a)** A schematic, nonhelical model. In each strand, phosphoric acid molecules are esterified to the 3′-carbon of one deoxyribose sugar (blue) and the 5′-carbon of the adjacent sugar. The two strands are held together by hydrogen bonds (dashed lines). The adenine-thymine base pairs are joined by two hydrogen bonds and guanine-cytosine base pairs have three hydrogen bonds. Because of the specific base pairing, the base sequence of one strand determines the sequence of the other. The two strands are antiparallel—that is, the backbones run in opposite directions as indicated by the two arrows, which point in the 5′ to 3′ direction. **(b)** Simplified model that highlights the antiparallel arrangement and the major and minor grooves. **(c)** Space-filling model of the B form of DNA. Note that the sugar-phosphate backbone spirals around the outside of the helix and the base pairs are embedded inside.

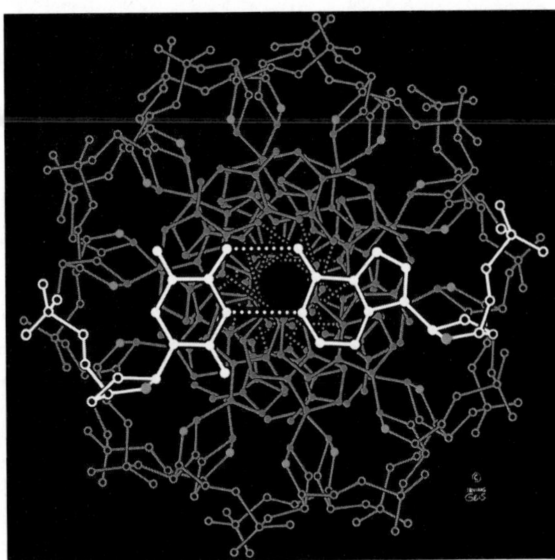

Figure 11.7 Structure of the DNA Double Helix. End view of a double helix showing the outer backbone and the bases stacked in the center of the cylinder. The ribose ring oxygens are red. The nearest base pair, an AT base pair, is highlighted in white.

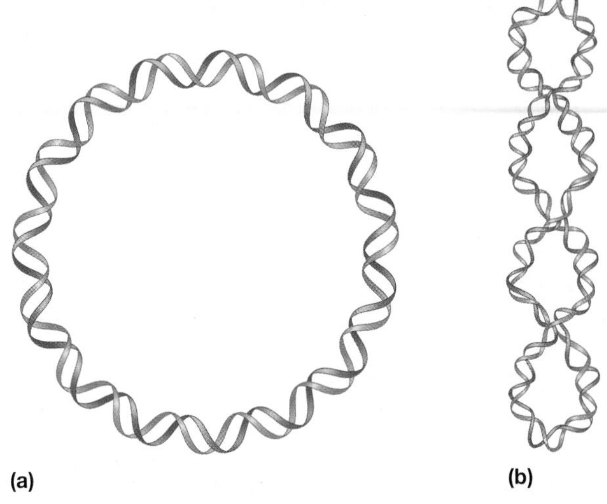

Figure 11.8 DNA Forms. **(a)** The DNA double helix of most procaryotes is in the shape of a closed circle. **(b)** The circular DNA strands, already coiled in a double helix, are twisted a second time to produce supercoils.

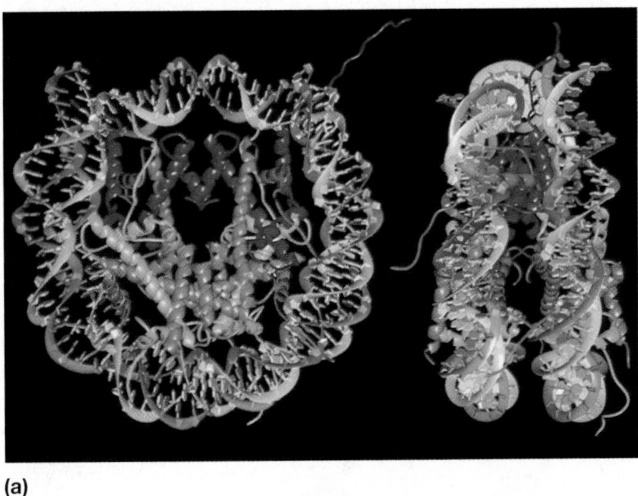

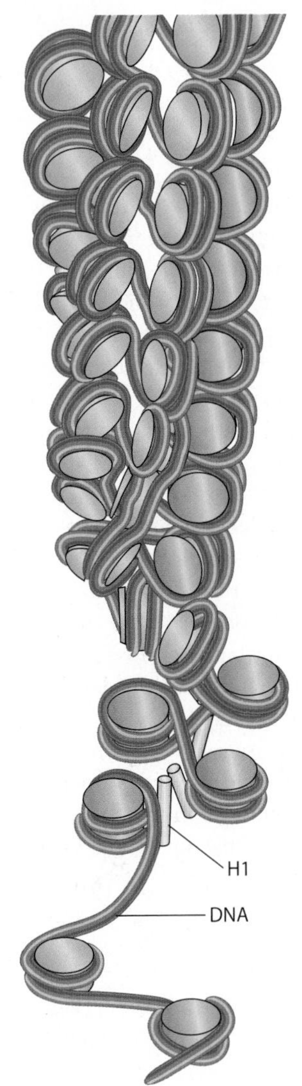

Figure 11.9 Nucleosome Internal Organization and Function. **(a)** The nucleosome core particle is a histone octamer surrounded by the 146 base pair DNA helix (brown and turquoise). The octamer is a disk-shaped structure composed of two H2A-H2B dimers and two H3-H4 dimers. The eight histone proteins are colored differently: blue, H3; green, H4; yellow, H2A; and red, H2B. Histone proteins interact with the backbone of the DNA minor groove. The DNA double helix circles the histone octamer in a left-handed helical path. **(b)** An illustration of how a string of nucleosomes, each associated with a histone H1, might be organized to form a highly supercoiled chromatin fiber. The nucleosomes are drawn as cylinders.

Table 11.1 Components of the *E. coli* Replication Machinery

Protein	Function
DnaA protein	Initiation of replication; binds origin of replication (*oriC*)
DnaB protein	Helicase (5′→3′); breaks hydrogen bonds holding two strands of double helix together; promotes DNA primase activity; involved in primosome assembly
DNA gyrase	Relieves supercoiling of DNA produced as DNA strands are separated by helicases; separates daughter molecules in final stages of replication
SSB proteins	Bind single-stranded DNA after strands are separated by helicases
DnaC protein	Helicase loader; helps direct DnaB protein (helicase) to DNA template
n′ protein	Component of primosome; helicase (3′→5′)
n protein	Primosome assembly; component of primosome
n″ protein	Primosome assembly
I protein	Primosome assembly
DNA primase	Synthesis of RNA primer; component of primosome
DNA polymerase III holoenzyme	Complex of about 20 polypeptides; catalyzes most of the DNA synthesis that occurs during DNA replication; has 3′→5′ exonuclease (proofreading) activity
DNA polymerase I	Removes RNA primers; fills gaps in DNA formed by removal of RNA primer
Ribonuclease H	Removes RNA primers
DNA ligase	Seals nicked DNA, joining DNA fragments together
DNA replication terminus site-binding protein	Termination of replication
Topoisomerase IV	Segregation of chromosomes upon completion of DNA replication

DNA. It would be dangerous for an organism to make many errors during replication because that would certainly be lethal. In fact, *E. coli* makes errors with a frequency of only 10^{-9} or 10^{-10} per base pair replicated (or about one in a million [10^{-6}] per gene per generation). Despite its complexity and accuracy, replication is very rapid. In *Bacteria,* replication rates approach 750 to 1,000 base pairs per second. Eucaryotic replication is slower, about 50 to 100 base pairs per second.

During DNA replication, the two strands of the double helix are separated; each then serves as a template for the synthesis of a complementary strand according to the base pairing rules. Each of the two progeny DNA molecules consists of one new strand and one old strand. Thus DNA replication is semiconservative (**figure 11.10**). Watson and Crick suggested semiconservative replication of DNA just one month after they published their paper on DNA structure in April 1953; subsequent research confirmed their hypothesis and elucidated the details of replication observed in procaryotes and eucaryotes.

In this section, we first discuss the various patterns of DNA replication observed in cells and viruses. We will then consider the mechanism of DNA replication in *E. coli,* beginning with an examination of the replication machinery and then events at the replication fork.

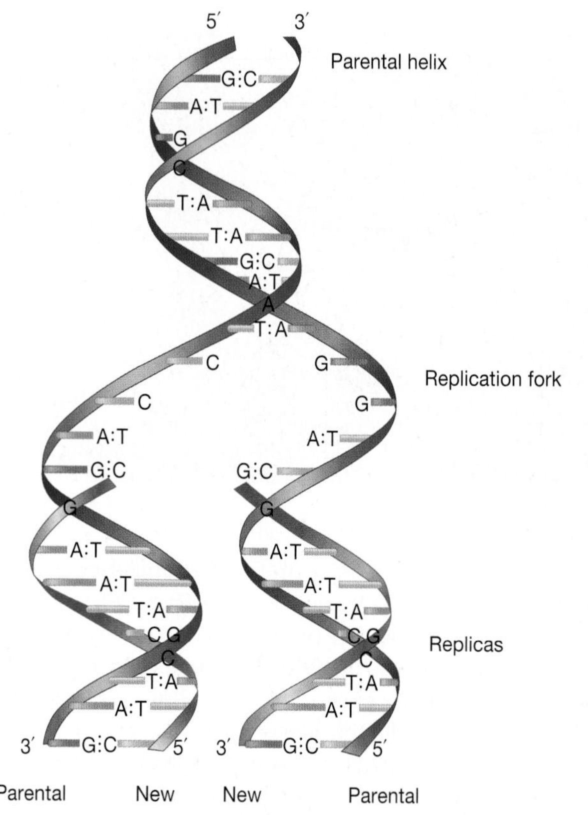

Figure 11.10 Semiconservative DNA Replication. The replication fork of DNA showing the synthesis of two progeny strands. Newly synthesized strands are purple. Each copy contains one new and one old strand. This process is called semiconservative replication.

Patterns of DNA Synthesis

Replication patterns are somewhat different in *Bacteria, Archaea,* and eucaryotes. For example, when the circular DNA chromosome of *E. coli* is copied, replication begins at a single point, the origin. Synthesis occurs at the **replication fork,** the place at

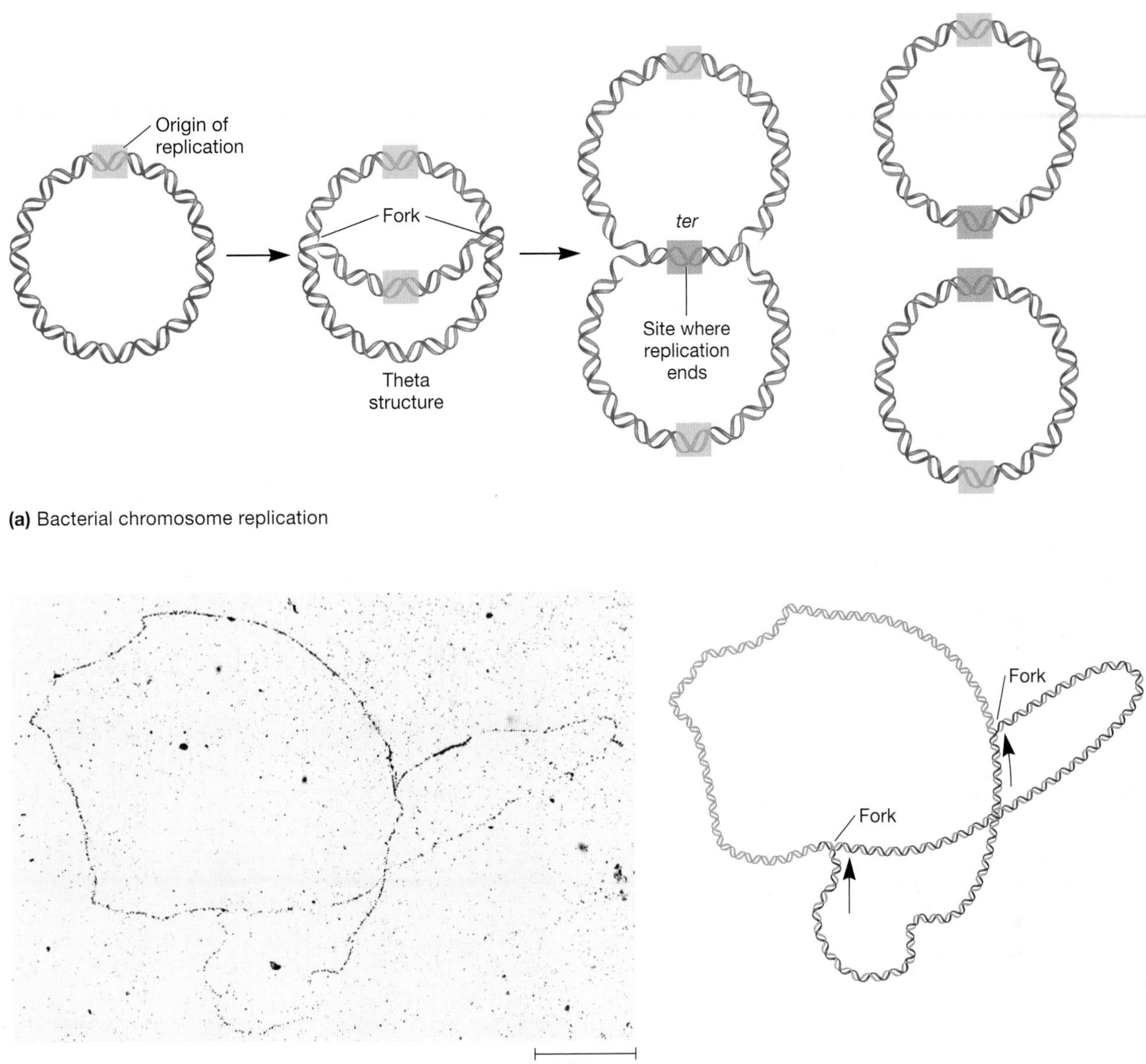

Figure 11.11 Bidirectional Replication of the *E. coli* Chromosome. **(a)** Replication begins at one site on the chromosome, called the origin of replication. Two replication forks proceed in opposite directions from the origin until they meet at a special site called the replication termination site (*ter*). The theta structure is a commonly observed intermediate of the process. **(b)** An autoradiograph of a replicating *E. coli* chromosome; about one-third of the chromosome has been replicated. To the right is a schematic representation of the chromosome. Parental DNA is blue; new DNA strands are purple, arrow represents direction of fork movement.

which the DNA helix is unwound and individual strands are replicated. Two replication forks move outward from the origin until they have copied the whole **replicon,** that portion of the genome that contains an origin and is replicated as a unit. When the replication forks move around the circle, a structure shaped like the Greek letter theta (θ) is formed (**figure 11.11**). Finally, since the bacterial chromosome is a single replicon, the forks meet on the other side and two separate chromosomes are released. Until recently, it was thought that all procaryotes have a single origin of replication. However, two members of the archaeal genus *Solfolobus* have more than one origin.

A different pattern of DNA replication occurs during *E. coli* conjugation, a type of genetic exchange mechanism observed in many bacteria. The pattern is called **rolling-circle replication,** and it is also observed during plasmid replication and the reproduction of some viruses (e.g., phage lambda). During rolling-circle replication (**figure 11.12**), one strand is nicked and the free 3′-hydroxyl end is extended by replication enzymes. As the 3′ end is lengthened while the growing point rolls around the circular template, the 5′ end of the strand is displaced and forms an ever-lengthening tail, much like the peel of an apple is displaced by a knife as an apple is pared. The single-stranded tail may be converted to the double-stranded

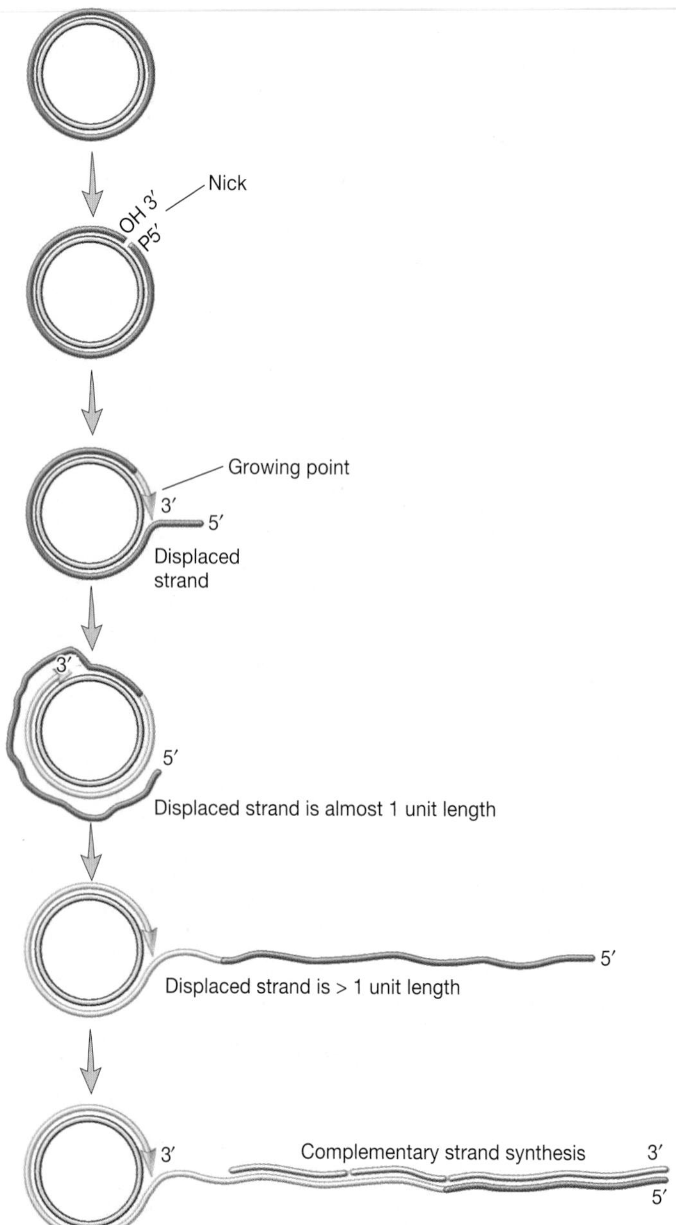

Figure 11.12 Rolling-Circle Replication. A single-stranded tail, often composed of more than one genome copy, is generated and can be converted to the double-stranded form by synthesis of a complementary strand. The "free end" of the rolling-circle strand is probably bound to the primosome. OH 3′ is the 3′-hydroxyl and P 5′ is the 5′-phosphate group created when the DNA strand is nicked.

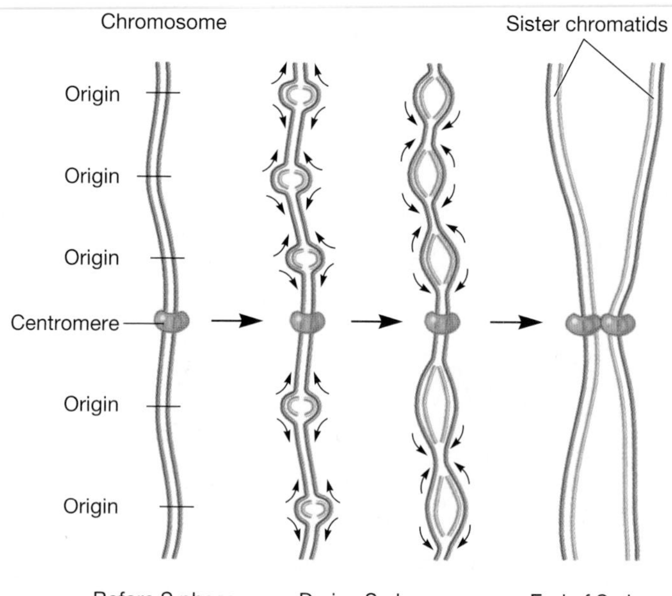

(a) DNA replication from multiple origins of replication

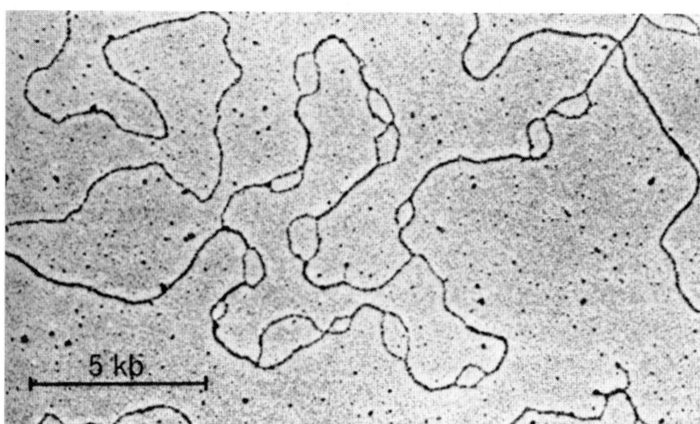

(b) A micrograph of a replicating, eucaryotic chromosome

Figure 11.13 The Replication of Eucaryotic DNA. Replication is initiated every 10 to 100 μm and the replication forks travel away from the origin. Newly copied DNA is in blue.

form by complementary strand synthesis. This mechanism is particularly useful to viruses because it allows the rapid, continuous production of many genome copies from a single initiation event.

The pattern of chromosome replication in eucaryotes differs from that in procaryotes in part because eucaryotic DNA is much longer than procaryotic DNA. For instance, *E. coli* DNA is about 1,300 μm in length, whereas the 46 chromosomes in the human nucleus have a total length of 1.8 m (almost 1,400 times longer). Clearly many replication forks must copy eucaryotic DNA simultaneously so that the molecule can be duplicated in a relatively short period. Therefore many replicons are spaced such that there is an origin about every 10 to 100 μm along the DNA. Replication forks move outward from these sites and eventually meet forks that have been copying the adjacent DNA stretch (**figure 11.13**). In this fashion a large molecule is copied quickly.

Another reason for the different pattern of replication in eucaryotes is that their chromosomes are linear. Linear chromosomes present cells with a dilemma: how to replicate the ends of the chromosomes. However, the reason for this dilemma and the mechanisms by which it is resolved can only be grasped by first understanding the mechanisms of DNA replication. Therefore we will consider that first, and then return to the problem of replicating the ends of linear chromosomes.

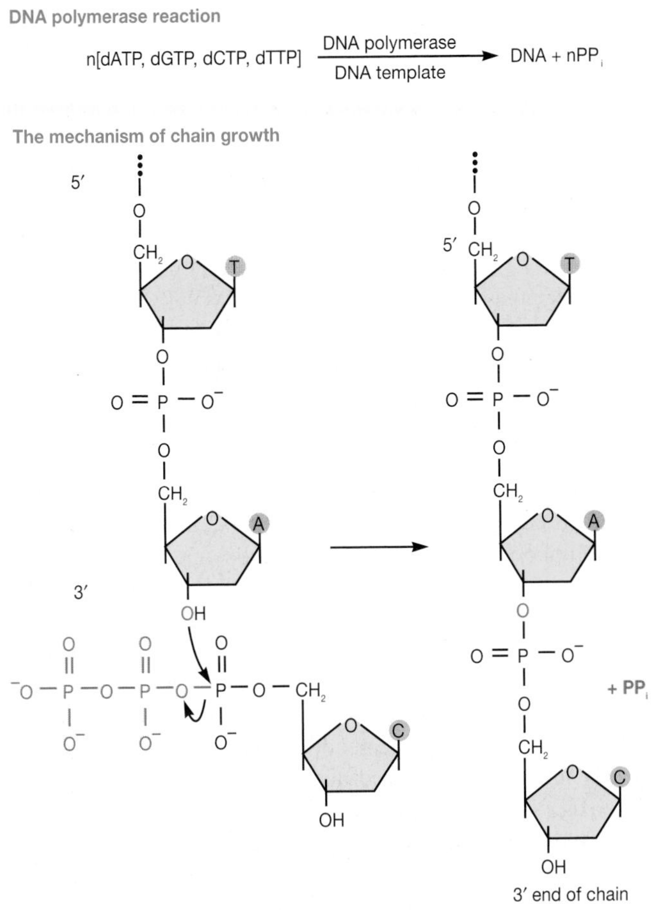

DNA polymerase reaction

$$n[\text{dATP, dGTP, dCTP, dTTP}] \xrightarrow[\text{DNA template}]{\text{DNA polymerase}} \text{DNA} + n\text{PP}_i$$

The mechanism of chain growth

Figure 11.14 The DNA Polymerase Reaction and Its Mechanism. The mechanism involves a nucleophilic attack by the hydroxyl of the 3′ terminal deoxyribose on the alpha phosphate group of the nucleotide substrate (in this example, adenosine attacks cytidine triphosphate).

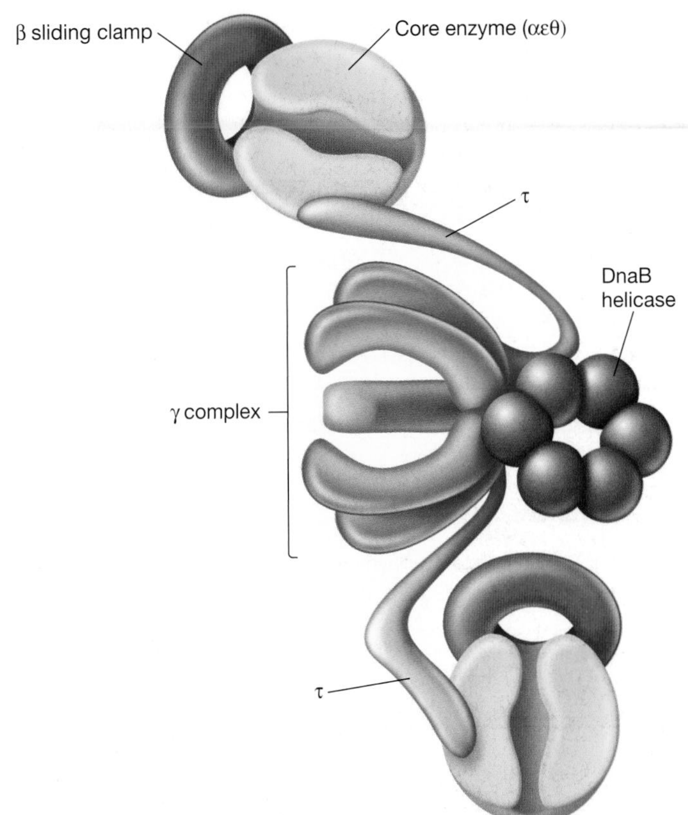

Figure 11.15 DNA Polymerase III Holoenzyme. The holoenzyme consists of two core enzymes (three subunits each; α, ϵ, θ, not shown) and several other subunits. The two tau (τ) subunits connect the two core enzymes to a large complex called the gamma (γ) complex. Each core enzyme is associated with a β sliding clamp, which tethers a DNA template to each core.

The Replication Machinery

Because DNA replication is so essential to organisms, a great deal of effort has been devoted to understanding its mechanism. The replication of *E. coli* DNA is probably best understood and is the focus of attention in this discussion. The overall process in other bacteria, *Archaea,* and eucaryotes is thought to be similar.

Enzymes called **DNA polymerase** catalyze DNA synthesis. All known DNA polymerase enzymes catalyze the synthesis of DNA in the 5′ to 3′ direction. This is because the 3′-hydroxyl group of the deoxyribose of the nucleotide at the end of the growing DNA strand attacks the alpha phosphate (the phosphate closest to the 5′ carbon) of the deoxynucleoside triphosphate to be incorporated (**figure 11.14**). This results in the formation of a phosphodiester bond; the energy needed to form this covalent bond is provided by the release of the terminal diphosphate (the beta and gamma phosphates) from the nucleotide that is added to the growing chain. Thus deoxynucleoside *tri*phosphates (dNTPs: dATP, dTTP, dCTP, dGTP) serve as DNA polymerase substrates while deoxynucleoside *mono*phosphates (dNMPs: dAMP, dTMP, dCMP, dGMP) are incorporated into the growing chain.

In order for DNA polymerases to catalyze the synthesis of a complementary strand of DNA, three things are needed: (1) a template, read in the 3′ to 5′ direction, that directs the synthesis of a complementary DNA strand; (2) a primer (e.g., an RNA strand or a DNA strand) to provide a free 3′-hydroxyl group to which nucleotides can be added (figure 11.14); and (3) dNTPs. *E. coli* has five different DNA polymerase enzymes (DNA Pol I-V). DNA polymerase III plays the major role in replication, although it is assisted by DNA polymerase I.

DNA polymerase III holoenzyme is a complex of 10 proteins including two core enzymes, each composed of three protein subunits (**figure 11.15**). As we shall see, the core enzymes are responsible for catalyzing DNA synthesis and proofreading the product to ensure fidelity of replication. A dimer of another subunit (tau) connects the two core enzymes. Associated with each core enzyme is a subunit called the β sliding clamp. This protein tethers the core enzyme to one strand of the DNA molecule. Another complex of proteins, called the γ complex, is responsible for loading the β sliding clamp onto the DNA. Because there are two core enzymes, both strands of DNA are bound by a single DNA polymerase III holoenzyme.

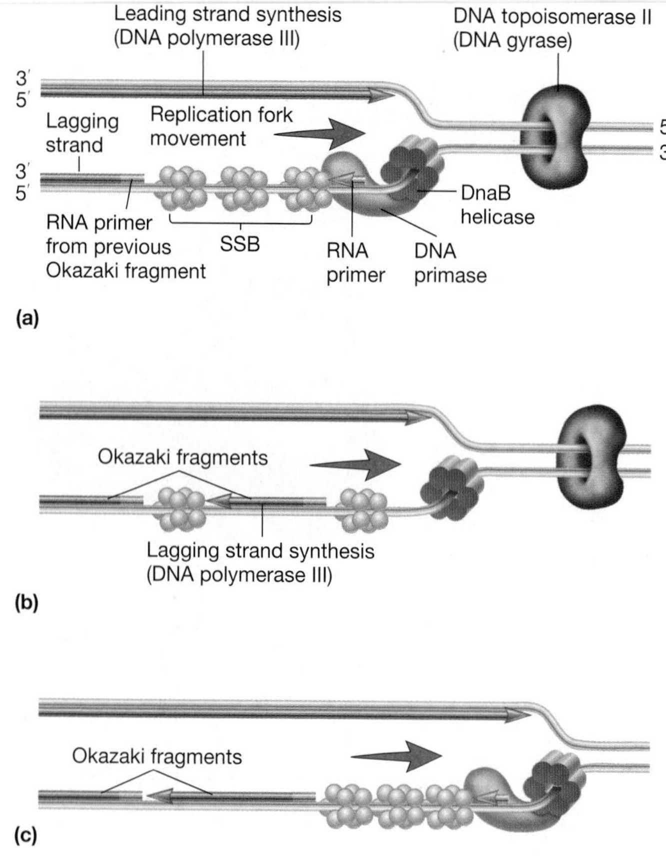

Figure 11.16 Bacterial DNA Replication. A general diagram of DNA replication in *E. coli*. A single replication fork showing both leading strand and lagging strand synthesis is illustrated. The lagging strand is synthesized in short fragments called Okazaki fragments. A new primer is required for the synthesis of each Okazaki fragment.

In *E. coli,* replication begins when a collection of **DnaA proteins** binds to specific nucleotide sequences (DnaA boxes) within the origin of replication. The DnaA proteins hydrolyze ATP to break or "melt" the hydrogen bonds between the DNA strands, thus making this localized region single stranded. Although this provides the initial template for replication, DNA polymerase III cannot by itself unwind and maintain the single-stranded DNA. These activities are provided by the action of other proteins, many of which are found in the **replisome,** a huge complex of proteins that includes DNA polymerase III holoenzyme. These other proteins include helicases, single-stranded DNA binding proteins, and topoisomerases (**figure 11.16**). **Helicases** are responsible for separating (unwinding) the DNA strands. These enzymes also use energy from ATP to unwind short stretches of helix just ahead of the replication fork. **Single-stranded DNA binding proteins (SSBs)** keep the strands apart once they have been separated, and **topoisomerases** relieve the tension generated by the rapid unwinding of the double helix (the replication fork may rotate as rapidly as 75 to 100 revolutions per second). This is important because rapid unwinding can lead to the formation of supercoils or supertwists in the helix (just as rapid separation of two strands of a rope can lead to knotting or coiling of the rope), and these can impede replication if not removed. Topoisomerases change the structure of DNA by transiently breaking one or two strands in such a way that the nucleotide sequence of the DNA remains unaltered as its shape is changed (e.g., a topoisomerase might tie or untie a knot in a DNA strand). **DNA gyrase** is an important topoisomerase in *E. coli*.

Once the template is prepared, the primer needed by DNA polymerase III can be synthesized. A special polymerase called **primase** synthesizes a short RNA strand, usually around 10 nucleotides long and complementary to the DNA; this serves as the primer (figure 11.16). RNA is used as the primer because unlike DNA polymerase, RNA polymerases (such as primase) can initiate RNA synthesis without adding a nucleotide to an existing 3′-OH. It appears that the primase requires the assistance of several other proteins, and the complex of the primase and its accessory proteins is called the **primosome** (table 11.1). The primosome is another important component of the replisome.

Because DNA polymerase enzymes must synthesize DNA in the 5′ to 3′ direction, only one of the strands, called the **leading strand,** can be synthesized continuously at its 3′ end as the DNA unwinds (figure 11.16). The other strand, called the **lagging strand,** cannot be extended in the same direction because there is no free 3′-OH to which a nucleotide can be added. As a result, the lagging strand is synthesized discontinuously in the 5′ to 3′ direction as a series of fragments, called **Okazaki fragments** after their discoverer, Reiji Okazaki. Discontinous synthesis occurs as primase adds many RNA primers along the single-stranded lagging strand. DNA polymerase III then extends these primers with DNA to form short fragments. These fragments are finally joined to form a complete strand; the steps of this process are detailed next. Thus while the leading strand requires only one RNA primer (and only one primosome) to initiate synthesis, the lagging strand has many RNA primers (and primosomes) that must eventually be removed. Okazaki fragments are about 1,000 to 2,000 nucleotides long in *Bacteria* and approximately 100 nucleotides long in eucaryotic cells.

Events at the Replication Fork

The details of DNA replication are outlined in a diagram of the replication fork (**figure 11.17**). In *E. coli,* DNA replication is initiated at specific nucleotides called the *oriC* locus (for *ori*gin of *c*hromosomal replication). Here we present replication as a series of discrete steps, but it should be remembered that synthesis is extremely rapid and occurs simultaneously on both the leading and lagging strands.

1. To initiate replication, as many as 40 DnaA proteins bind *oriC* while hydrolyzing ATP. Binding of the DnaA proteins causes the DNA to bend around the protein complex, resulting in separation of the double-stranded DNA at regions within the origin that have many A-T base pairs. Recall that adenines pair with thymines using only two hydrogen bonds, so A-T rich segments of DNA become single stranded more readily than do G-C rich regions. Once the

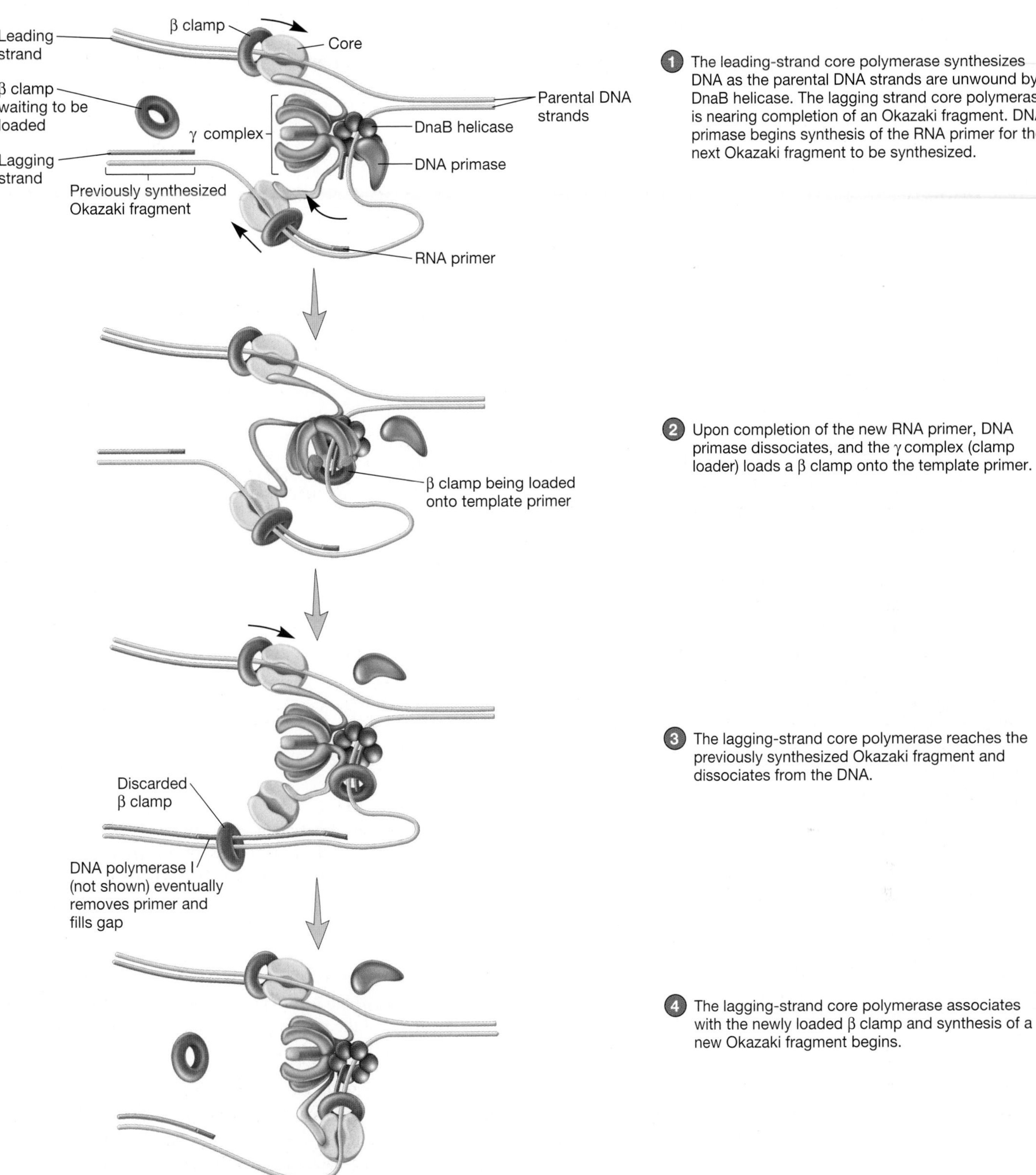

Figure 11.17 A Model for Activity at the Replication Fork. DNA polymerase III holoenzyme and other components of the replisome are responsible for the synthesis of both leading and lagging strands. The arrows show the movement of each DNA core polymerase. After completion of each new Okazaki fragment, the old β sliding clamp is discarded and a new one loaded onto the template DNA (step 3). This is achieved by the activity of the γ complex (see figure 11.15), which is also known as the clamp loader. The Okazaki fragments are eventually joined together (see figure 11.18) after removal of the RNA primer and synthesis of DNA to fill the gap, both catalyzed by DNA polymerase I; DNA ligase then seals the nick and joins the two fragments (see figure 11.19).

strands have separated, the replication process then proceeds through four stages.

2. Helicases unwind the helix with the aid of topoisomerases like DNA gyrase (figure 11.17, step 1). It appears that the DnaB protein is the helicase most actively involved in replication, but the n′ protein also may participate in unwinding. The single strands are kept separate by SSBs.
3. Primase synthesizes RNA primers as needed (figure 11.17, step 1) and a single DNA polymerase III holoenzyme catalyzes both leading strand and lagging strand synthesis from the RNA primers. Lagging strand synthesis is particularly amazing because of the "gymnastic" feats performed by the replisome. It must discard old β sliding clamps (figure 11.17, step 3), load new β sliding clamps (figure 11.17, step 2), and tether the template to the core enzyme with each new round of Okazaki fragment synthesis. All of this occurs as DNA polymerase III is synthesizing DNA. Thus DNA polymerase III is a multifunctional enzyme.
4. After most of the single-stranded region of the lagging strand has been replicated by the formation of Okazaki fragments, DNA polymerase I or (more rarely) RNaseH removes the RNA primer. DNA polymerase I can to do this because, unlike other DNA polymerases, it has 5′ to 3′ exonuclease activity—that is, it can snip off nucleotides one at a time starting at the 5′ end. Thus DNA polymerase I begins its exonuclease activity at the free end of the RNA primer. With the removal of each ribonucleotide, the adjacent 3′-OH from the deoxynucleotide is used by DNA polymerase I to fill the gap between Okazaki fragments (**figure 11.18**).

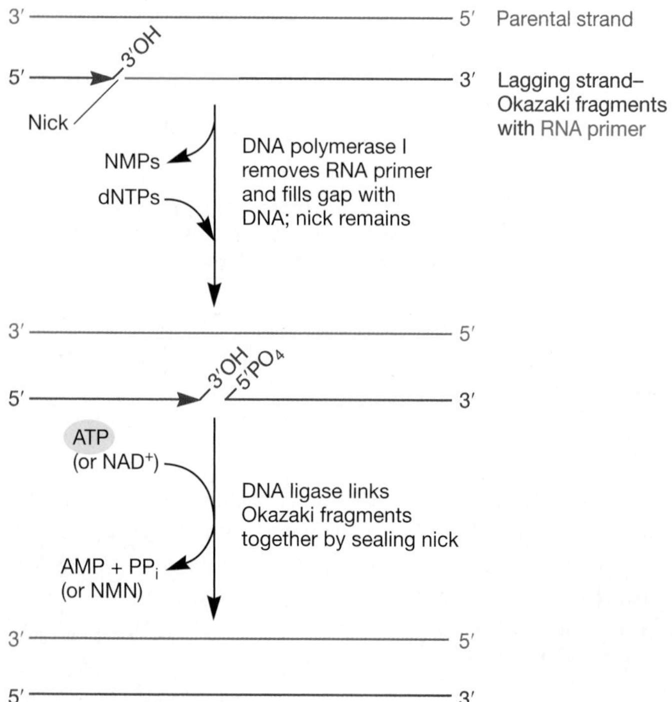

Figure 11.18 Completion of Lagging Strand Synthesis.

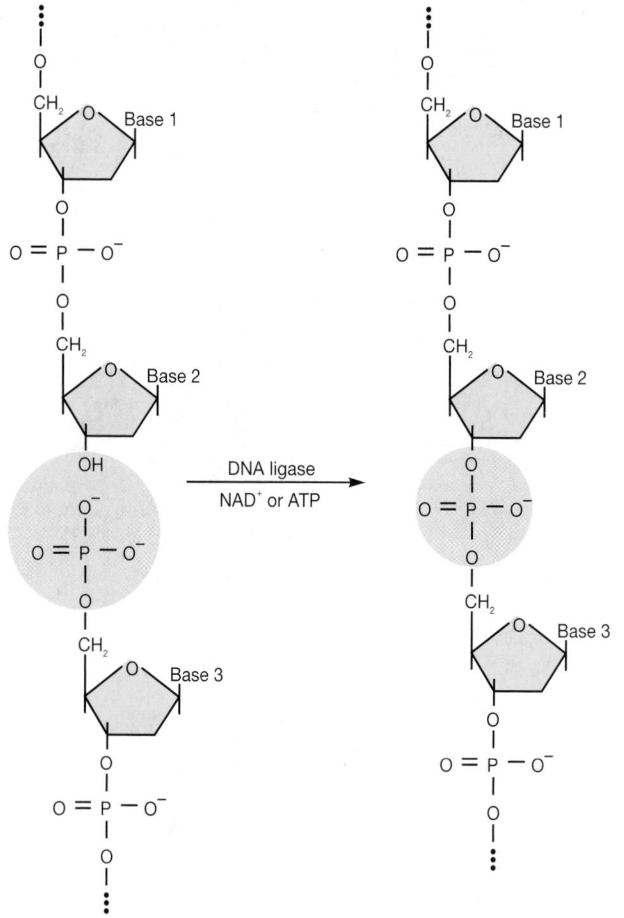

Figure 11.19 The DNA Ligase Reaction. The groups being altered are shaded in blue. Bacterial ligases use the pyrophosphate bond of NAD^+ as an energy source; many other ligases employ ATP.

5. Finally, the Okazaki fragments are joined by the enzyme **DNA ligase,** which forms a phosphodiester bond between the 3′-hydroxyl of the growing strand and the 5′-phosphate of an Okazaki fragment (**figure 11.19**).

As we have seen, DNA polymerase III is an amazing multiprotein complex, with multiple enzymatic activities. In *E. coli,* the polymerase component is encoded by the *dnaE* gene. Genome sequencing of other bacteria has revealed that some have a second *dnaE* gene. In *Bacillus subtilis,* a gram-positive bacterium that is another important experimental model, this second polymerase gene is called $dnaE_{Bs}$, and its protein product appears to be responsible for replicating the lagging strand. Thus while the overall mechanism by which DNA is replicated is highly conserved, there can be variations in replisome components.

Amazingly, DNA polymerase III, like all DNA polymerases, has an additional function that is critically important: **proofreading.** Proofreading is the removal of a mismatched base immediately after it has been added; its removal must occur before the next base is incorporated. Recall that the polymerase III core is composed of three subunits: α, ϵ, and θ. While we have discussed the α subunit polymerase activity, the ϵ subunit has 3′ to 5′ exonuclease activity. This enables it to check each newly incorporated base to see that it forms stable hydrogen bonds. In this way mismatched bases can be detected. If the wrong base has been mistakenly added, this subunit is able to remove it. Because it has *exo*nuclease activity (exo meaning outside or in this case, from the end), it can remove a mismatched base, as long as it is still at the 3′ end of the growing strand. Once removed, holoenzyme backs up and adds the proper nucleotide in its place. DNA proofreading is not 100% efficient and, as discussed in chapter 12, the mismatch repair system is the cell's second line of defense.

Termination of Replication

In *E. coli,* DNA replication stops when the replisome reaches a termination site (*ter*) on the DNA. The Tus protein binds to the *ter* sites and halts progression of the forks. In many other bacteria, replication stops randomly when the forks meet. Regardless of how fork movement is stopped, there is often a problem to be solved by the replisome: separation of daughter molecules. When replication of a circular chromosome is complete, the two circular daughter chromosomes may remain intertwined. Such interlocked chromosomes are called **catenanes.** This is obviously a problem if each daughter cell is to inherit a single chromosome. Fortunately, topoisomerases solve the problem by temporarily breaking DNA molecules, so that the strands can be separated.

Replication of Linear Chromosomes

The fact that eucaryotic chromosomes are linear poses a problem during replication because of DNA polymerase's need for a primer, providing a free 3′-OH. At the ends (telomeres) of eucaryotic chromosomes, space is not available for synthesis of a primer on the lagging strand, and therefore it should be impossible to replicate the end of that strand. Over numerous rounds of DNA replication and cell division, this would lead to a progressively shortened chromosome. Ultimately the chromosome would lose critical genetic information, which would be lethal to the cell.

Clearly, eucaryotic cells must have evolved a mechanism for replicating their telomeres. The solution to the "end replication problem" is the enzyme **telomerase.** Telomerase has two components: a protein that can synthesize DNA using an RNA template (telomerase reverse transcriptase) and an internal RNA template. The internal RNA is complementary to the single strand of DNA jutting out from the end of the chromosome (**figure 11.20**) and acts as the template for DNA synthesis to elongate that strand (i.e., the 3′-OH of the telomere DNA strand serves as the primer for DNA synthesis). After being lengthened sufficiently, the single strand of telomere DNA can serve as the template for synthesis of the complementary strand by DNA polymerase III. Thus the length of the chromosome is maintained.

Telomerase has solved the problem of end replication for eucaryotes, but recall that some bacteria also have linear chromosomes. How do they replicate the ends of their chromosomes?

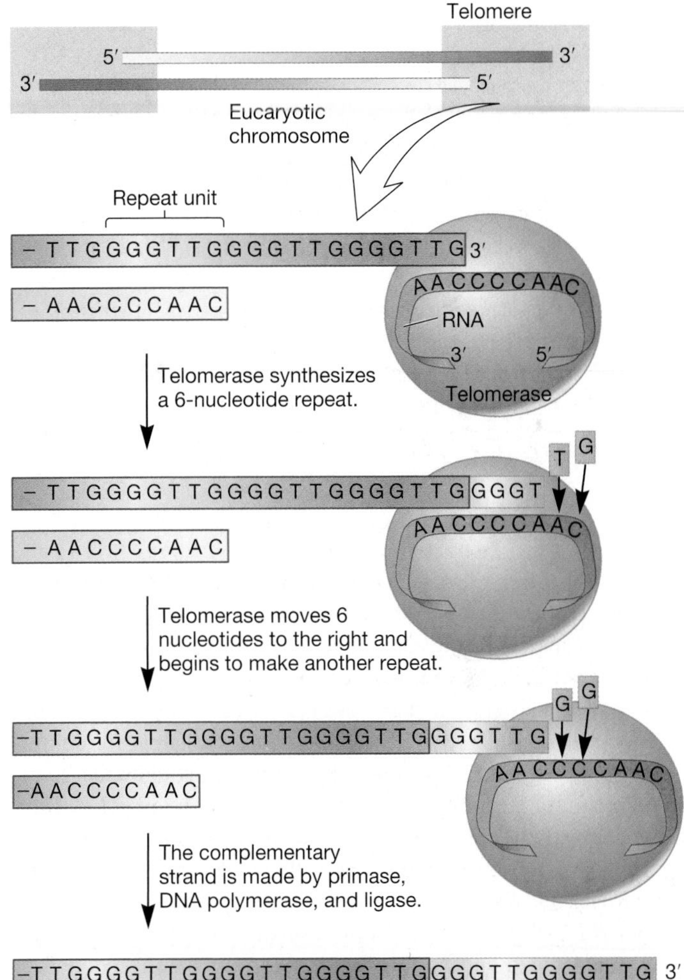

Figure 11.20 Replication of the Telomeres of Eucaryotic Chromosomes by Telomerase. Telomerase contains an RNA molecule that can base pair with a small portion of the 3′ overhang. The RNA serves as a template for DNA synthesis catalyzed by the reverse transcriptase activity of the enzyme. The 3′-OH of the telomere DNA serves as the primer and is lengthened. The process shown is repeated many times until the 3′ overhang is long enough to serve as the template for the complementary telomere DNA strand.

Unfortunately, little is known about the replication of linear bacterial chromosomes. However, a recent discovery in *Streptomyces,* an important group of soil bacteria, has led to speculation that a telomerase-like process may function in these bacteria. The ends of the linear chromosome of *Streptomyces coelicolor* are associated with a complex of proteins, including one with in vitro reverse transcriptase activity. No RNA has been found in the *Streptomyces* complex, so it is unclear if the protein functions as a reverse transcriptase in cells and what it might use as a template, if it does.

1. Define the following terms: origin of replication, replicon, replication fork, primosome, and replisome.
2. Describe the nature and functions of the following replication components and intermediates: DNA polymerases I and III, topoisomerase, DNA gyrase, helicase, single-stranded DNA binding protein, Okazaki fragment, DNA ligase, leading strand, lagging strand, primase, and telomerase.
3. How do replication patterns differ between procaryotes and eucaryotes? Describe the operation of replication forks in the generation of theta-shaped intermediates.
4. How does rolling-circle replication differ from the usual type of replication observed for cellular chromosomes?
5. Outline the steps involved in DNA synthesis at the replication fork. How do DNA polymerases correct their mistakes?

11.5 Gene Structure

DNA replication allows genetic information to be passed from one generation to the next. But how is the genetic information used? To answer that question, we must first look at how genetic information is organized. The basic unit of genetic information is the gene. The gene has been defined in several ways. Initially geneticists considered it to be the entity responsible for conferring traits on the organism and the entity that could undergo recombination. Recombination involves exchange of DNA from one source (e.g., virus, bacterium) with that from another and is responsible for generating much of the genetic variability found in viruses and living organisms. With the discovery and characterization of DNA, the gene was defined more precisely as a linear sequence of nucleotides with fixed start and end points. Creating genetic variability: Recombination at the molecular level (section 13.4)

At first, it was thought that a gene contained information for the synthesis of one enzyme, the one gene-one enzyme hypothesis. This was next modified to the one gene-one polypeptide hypothesis because of the existence of enzymes and other proteins composed of two or more different polypeptide chains coded for by separate genes. Historically, a segment of DNA that encodes a single polypeptide was termed a **cistron;** this term is still sometimes used. However, not all genes encode proteins; some code instead for rRNA and tRNA (see figure 11.4). In addition, it is now known that some eucaryotic genes encode more than one protein. Thus a gene might be defined as a polynucleotide sequence that codes for a functional product (i.e., a polypeptide, tRNA, or rRNA). The nucleotide sequences of protein-coding genes are distinct from RNA-coding genes and noncoding regions because when transcribed, the resulting mRNA can be "read" in discrete sequences of sets of three nucleotides, each set being a **codon.** Each codon codes for a single amino acid. The sequence of codons is "read" in only one way to produce a single product. That is, the code is not overlapping and there is a single starting point with one **reading frame** or way in which nucleotides are grouped into codons (**figure 11.21**). Each strand of DNA therefore usually consists of gene sequences that do not overlap one another (**figure 11.22*a***). However, there are exceptions to the rule. Some viruses such as the phage ϕX174 have overlapping genes (figure 11.22*b*), and parts of genes overlap in some bacterial genomes.

Procaryotic and viral gene structure differs greatly from that of eucaryotes. In procaryotic and viral systems, the coding information within a gene normally is continuous. However, in eucaryotic organisms, many genes contain coding information (exons) interrupted periodically by noncoding sequences (in-

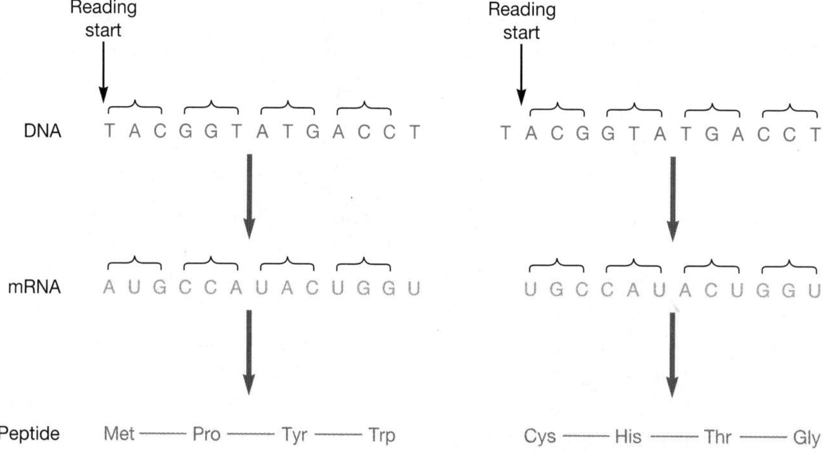

Figure 11.21 Reading Frames and Their Importance. The place at which DNA sequence reading begins determines the way nucleotides are grouped together in clusters of three (outlined with brackets), and this specifies the mRNA codons and the peptide product. In the example, a change in the reading frame by one nucleotide yields a quite different mRNA and final peptide.

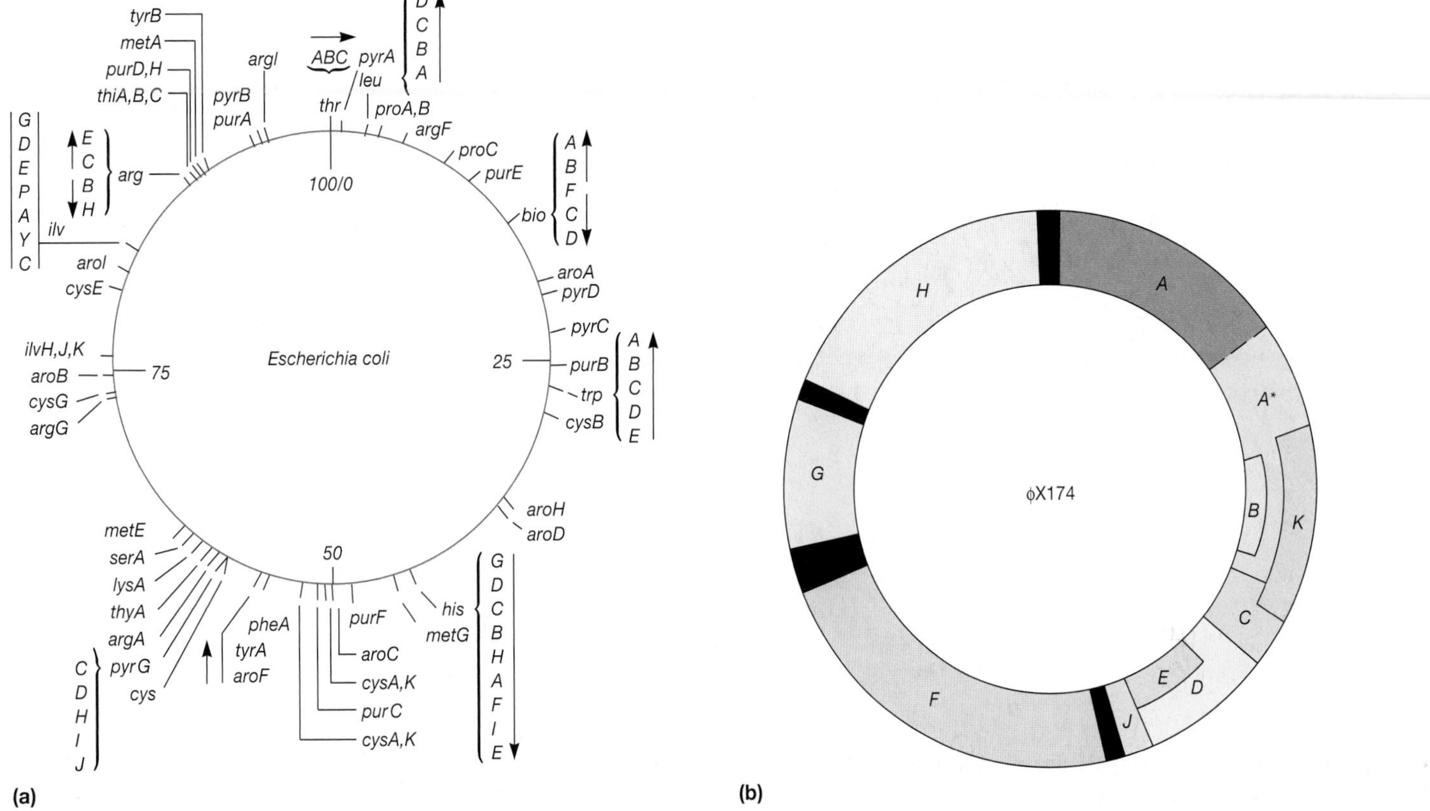

Figure 11.22 Chromosomal Organization in Bacteria and Viruses. **(a)** Simplified genetic map of *E. coli*. The *E. coli* map is divided into 100 minutes. **(b)** The map of phage φX174 shows the overlap of gene *B* with *A*, *K* with *A* and *C*, and *E* with *D*. The solid regions are spaces lying between genes. Protein A* consists of the last part of protein A and arises from reinitiation of transcription within gene *A*.

trons). The introns must be cut, or spliced, out of the mRNA before the protein is made. As we will see, this affords eucaryotes the ability to cut and paste mRNA molecules so that they can encode more than one polypeptide, a process known as **alternative splicing.** An interesting exception to this rule is eucaryotic histone genes, which lack introns. Because procaryotic and viral systems are the best characterized, the more detailed description of gene structure that follows will focus on *E. coli* genes.

Genes That Code for Proteins

In order for genetic information in the DNA to be used, it must first be transcribed to form an RNA molecule. The RNA product of a gene that codes for a protein is messenger RNA (mRNA). Recall from the discussion of information flow that although DNA is double stranded, only one strand of a gene contains coded information and directs RNA synthesis. This strand is called the **template strand,** and the complementing strand is known as the coding strand because it is the same nucleotide sequence as the mRNA, except in DNA bases (**figure 11.23**). Because the mRNA is made from the 5′ to the 3′ end, the polarity of the DNA template strand is 3′ to 5′. Therefore the beginning of the gene is at the 3′ end of the template strand. An important site, the **promoter,** is located at the start of the gene. The promoter is a recognition/binding site for RNA polymerase, the enzyme that synthesizes RNA. The promoter is neither transcribed nor translated; it functions strictly to orient RNA polymerase a specific distance from the first DNA nucleotide that will serve as a template. As we will see in chapter 12, the promoter is also very important in regulating when and where a gene will be transcribed or expressed.

The transcription start site (labeled +1 in figure 11.23) represents the first nucleotide in the mRNA synthesized from the gene. However, the initially transcribed portion of the gene does not necessarily code for amino acids. Instead it is a **leader sequence** that is transcribed into mRNA, but is not translated into amino acids. The leader sequence includes a region called the **Shine-Dalgarno sequence** that is important in the initiation of translation. The leader sometimes is also involved in regulation of transcription and translation. Regulation of transcription elongation (section 12.3); Regulation at the level of translation (section 12.4)

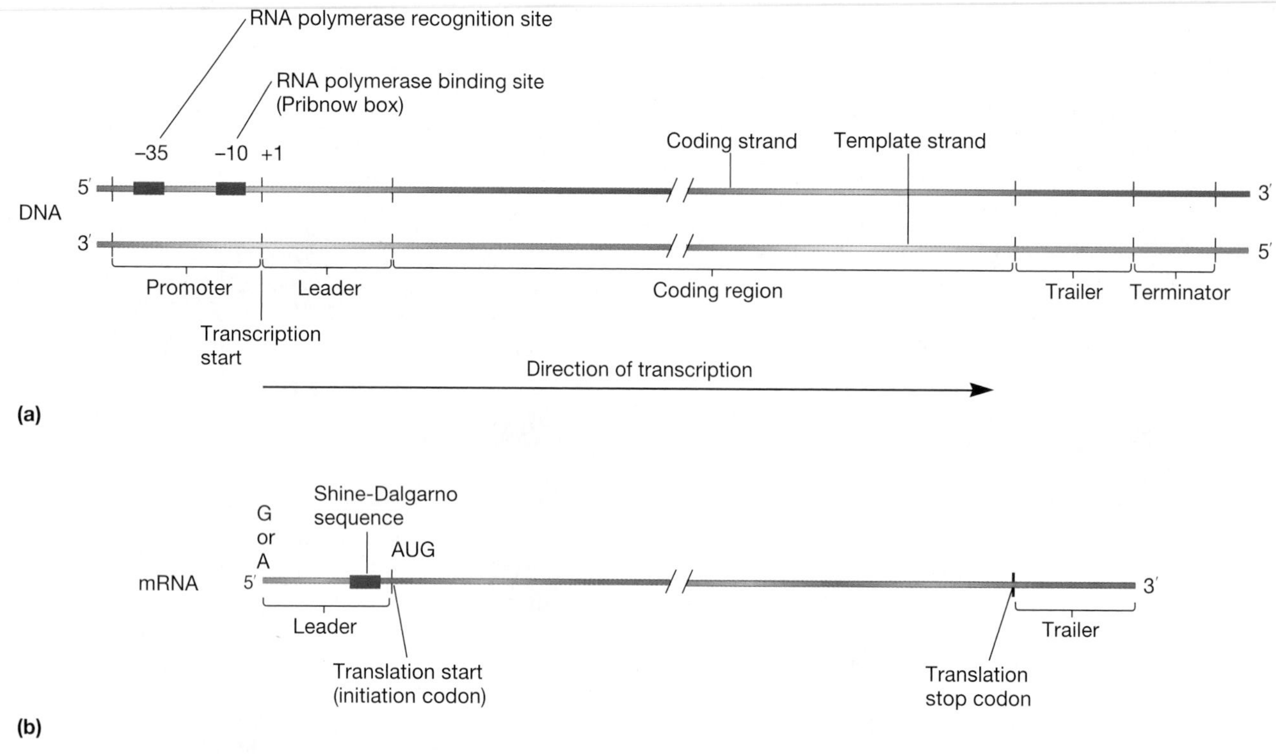

Figure 11.23 A Bacterial Structural Gene and Its mRNA Product. **(a)** The organization of a typical structural gene in bacteria. Leader and trailer sequences are included even though some genes lack one or both. Transcription begins at the +1 position in DNA and proceeds to the right as shown. The template is read in the 3′ to 5′ direction. **(b)** Messenger RNA product of the gene shown in part **a.** The first nucleotide incorporated into mRNA is usually GMP or AMP. Translation of the mRNA begins with the AUG initiation codon. Regulatory sites are not shown.

Immediately next to (and downstream of) the leader is the most important part of the gene, the **coding region** (figure 11.23). In genes that direct the synthesis of proteins, the coding region typically begins with the template DNA sequence 3′-TAC-5′. This produces the codon 5′-AUG-3′, which in bacteria codes for *N*-formylmethionine, a specially modified amino acid used to initiate protein synthesis. The remainder of the coding region consists of a sequence of codons that specifies the sequence of amino acids for that particular protein. The coding region ends with a special codon called the **stop codon,** which signals the end of the protein and stops the ribosome during translation. The stop codon is immediately followed by the **trailer sequence** (figure 11.23), which is needed for proper expression of the coding region of the gene. The stop codon is not recognized by RNA polymerase during transcription. Instead, a **terminator sequence** is used to stop transcription by dislodging the RNA polymerase from the template DNA.

Besides these basic components—the promoter, leader, coding region, trailer, and terminator—many bacterial genes have a variety of regulatory sites. These are locations where DNA-recognizing regulatory proteins bind to stimulate or prevent gene expression. Regulatory sites often are associated with promoter function, and some consider them to be parts of special promoters. Two such sites, operator and activator binding sites, are discussed in sections 12.2 and 12.5. Certainly everything is not known about genes and their structure. With the ready availability of cloned genes and DNA sequencing technology, major discoveries continue to be made in this area.

Genes That Code for tRNA and rRNA

The DNA segments that code for tRNA and rRNA also are considered genes, although they give rise to important RNA rather than protein. In *E. coli* the genes for tRNA are fairly typical, consisting of a promoter and transcribed leader and trailer sequences that are removed during the process of tRNA maturation (**figure 11.24*a***). The precise function of the leader is not clear; however, the trailer is required for transcription termination. Genes coding for tRNA may code for more than a single tRNA molecule or type of tRNA (figure 11.24*a*). The segments coding for tRNAs are separated by short spacer sequences that are removed after transcription by special ribonu-

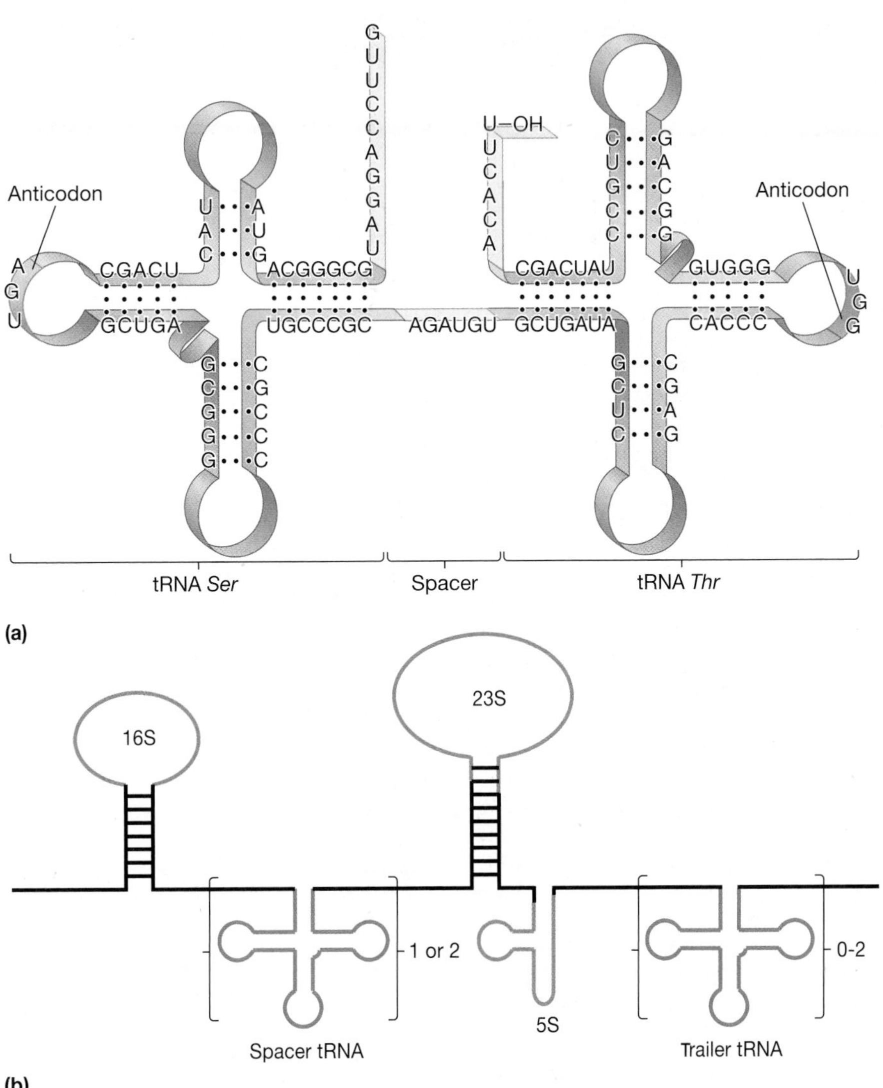

Figure 11.24 tRNA and rRNA Genes. **(a)** A tRNA precursor from *E. coli* that contains two tRNA molecules. The spacer and extra nucleotides at both ends are removed during processing. **(b)** The *E. coli* ribosomal RNA gene codes for a large transcription product that is cleaved into three rRNAs and one to three tRNAs. The 16S, 23S, and 5S rRNA segments are represented by blue lines, and tRNA sequences are placed in brackets. The seven copies of this gene vary in the number and kind of tRNA sequences.

cleases, at least one of which contains catalytic RNA. RNA molecules with catalytic activity are called **ribozymes** (**Microbial Tidbits 11.2**).

The genes for rRNA also are similar in organization to genes coding for proteins because they have promoters, trailers, and terminators (figure 11.24*b*). Interestingly all the rRNAs are transcribed as a single, large precursor molecule that is cut up by ribonucleases after transcription to yield the final rRNA products. *E. coli* pre-rRNA spacer and trailer regions even contain tRNA genes. Thus the synthesis of tRNA and rRNA involve posttranscriptional modification, a relatively rare process in procaryotes.

1. Define or describe the following: gene, template and coding strands, promoter, leader, coding region, reading frame, trailer, and terminator.
2. How do the genes of procaryotes and eucaryotes usually differ from each other?
3. Briefly discuss the general organization of tRNA and rRNA genes. How does their expression differ from that of structural genes with respect to posttranscriptional modification of the gene product?

Microbial Tidbits

11.2 Catalytic RNA (Ribozymes)

Biologists once thought that all cellular reactions were catalyzed by proteins called enzymes (*see section 8.7*). The discovery during 1981–1984 by Thomas Cech and Sidney Altman that RNA also can sometimes catalyze reactions has transformed our way of thinking about topics as diverse as catalysis and the origin of life. It is now clear that some RNA molecules, called ribozymes, catalyze reactions that alter either their own structure or that of other RNAs.

This discovery has stimulated scientists to hypothesize that the early Earth was an "RNA world" in which RNA acted as both the genetic material and a reaction catalyst. Experiments showing that introns from *Tetrahymena thermophila* can catalyze the formation of polycytidylic acid under certain circumstances have further encouraged such speculations. Some have suggested that RNA viruses are "living fossils" of the original RNA world. The first self-replicating entity: The RNA world (section 19.1)

The best-studied ribozyme activity is the self-splicing of RNA. This process is widespread and occurs in *Tetrahymena* pre-rRNA; the mitochondrial rRNA and mRNA of yeast and other fungi; chloroplast tRNA, rRNA, and mRNA; in mRNA from some bacteriophages (e.g., the T4 phage of *E. coli*); and in the hepatitis delta virusoid. The 413-nucleotide rRNA intron of *T. thermophila* provides a good example of the self-splicing reaction. The reaction occurs in three steps and requires the presence of guanosine (see **Box figure**). First, the 3′-OH group of guanosine attacks the intron's 5′-phosphate group and cleaves the phosphodiester bond. Second, the new 3′-hydroxyl on the left exon attacks the 5′-phosphate of the right exon. This joins the two exons and releases the intron. Finally, the intron's 3′-hydroxyl attacks the phosphate bond of the nucleotide 15 residues from its end. This releases a terminal fragment and cyclizes the intron. Self-splicing of this rRNA occurs about 10 billion times faster than spontaneous RNA hydrolysis. Just as with enzyme proteins, the RNA's shape is essential to catalytic efficiency. The ribozyme even has Michaelis-Menten kinetics *(see figure 8.18)*. The ribozyme from the hepatitis delta virusoid catalyzes RNA cleavage that is involved in its replication. It is unusual in that the same RNA can fold into two shapes with quite different catalytic activities: the regular RNA cleavage activity and an RNA ligation reaction.

The discovery of ribozymes has many potentially important practical consequences. Ribozymes act as "molecular scissors" and will enable researchers to manipulate RNA easily in laboratory experiments. It also might be possible to protect hosts by specifically removing RNA from pathogenic viruses, bacteria, and fungi. For example, ribozymes are being tested against the AIDS, herpes, and tobacco mosaic viruses.

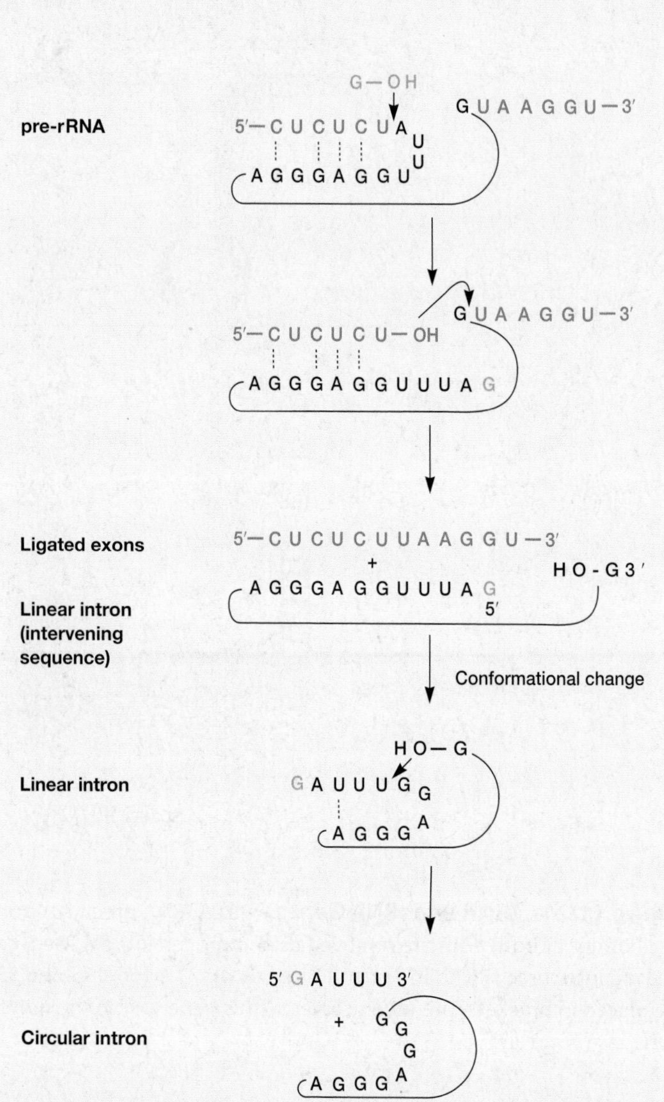

Ribozyme Action. The mechanism of *Tetrahymena thermophila* pre-rRNA self-splicing. See text for details.

11.6 Transcription

As mentioned earlier, synthesis of RNA under the direction of DNA is called transcription. The RNA product has a sequence complementary to the DNA template directing its synthesis (**table 11.2**). Thymine is not normally found in mRNA and rRNA. Although adenine directs the incorporation of thymine during DNA replication, it usually codes for uracil during RNA synthesis. Transcription generates three kinds of RNA. Messenger RNA (mRNA) bears the message for protein synthesis. In *Bacteria* and *Archaea,* the mRNA often bears coding information transcribed from adjacent genes. Therefore it is said to be polygenic or polycistronic (**figure 11.25**). Eucaryotic mRNAs, on the other hand, are usually monocistronic, containing infor-

Table 11.2 RNA Bases Coded for by DNA

DNA Base	Purine or Pyrimidine Incorporated into RNA
Adenine	Uracil
Guanine	Cytosine
Cytosine	Guanine
Thymine	Adenine

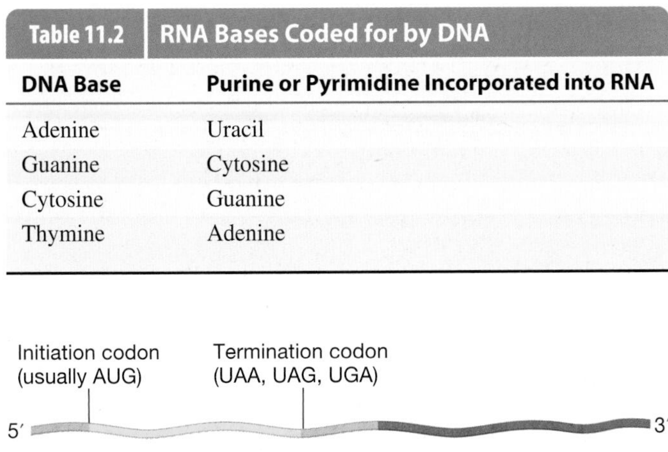

Figure 11.25 A Polycistronic Bacterial Messenger RNA.

mation for a single polypeptide. **Transfer RNA (tRNA)** carries amino acids during protein synthesis, and **ribosomal RNA (rRNA)** molecules are components of ribosomes. The synthesis of bacterial mRNA is described first.

Transcription in *Bacteria*

RNA is synthesized under the direction of DNA by the enzyme **RNA polymerase.** The reaction is quite similar to that catalyzed by DNA polymerase (figure 11.14). ATP, GTP, CTP, and UTP are used to produce an RNA complementary to the DNA template. As mentioned earlier, these nucleotides contain ribose rather than deoxyribose (figure 11.15).

$$n[\text{ATP, GTP, CTP, UTP}] \xrightarrow[\text{DNA template}]{\text{RNA polymerase}} \text{RNA} + n\text{PP}_i$$

RNA synthesis, like DNA synthesis, proceeds in a 5′ to 3′ direction with new nucleotides being added to the 3′ end of the growing chain at a rate of about 40 nucleotides per second at 37°C (**figure 11.26**). In both DNA and RNA polymerase reactions, pyrophosphate (PP_i) is produced. It is then hydrolyzed to orthophosphate in a reaction catalyzed by the pyrophosphatase enzyme. Hydrolysis of the pyrophosphate product makes DNA and RNA synthesis irreversible. If the pyrophosphate level were too high, DNA and RNA would be degraded by a reversal of the polymerase reactions.

Most bacterial RNA polymerases contain five types of polypeptide chains: α, β, β′, ω, and σ (**figure 11.27**). The **core enzyme** is composed of five chains (α_2, β, β′, and ω) and catalyzes RNA synthesis. The **sigma factor** (σ) has no catalytic activity but helps the core enzyme recognize the start of genes. When sigma is bound to core enzyme, the six-subunit complex is termed **RNA polymerase holoenzyme.** Only holoenzyme can begin transcription, but as we will see, core enzyme completes RNA synthesis once it has been initiated. The precise functions of the α, β, β′, and ω polypeptides are not yet clear. The α subunits seem to be involved in the assembly of the core enzyme, recognition of promoters, and interaction with some regulatory factors. The binding site for DNA is on β′, and the ω subunit seems to be involved in stabilizing the conformation of the β′ subunit. The β subunit binds ribonucleotide substrates. Rifampin, an RNA polymerase inhibitor, binds to the β subunit.

Recently the atomic structure of RNA polymerase from *Thermus aquaticus* has been determined (figure 11.27*a,b*). In this bacterium, the core enzyme is composed of four different subunits (α_2, β, β′, and ω) and is complexed with the sigma factor (σ). The enzyme is claw-shaped with a clamp domain that closes on an internal channel, which contains an essential magnesium and the active site. Sigma interacts extensively with the core enzyme and specifically binds to elements of the promoter. It also may widen the channel so that DNA can enter the interior of the polymerase complex.

Transcription involves three separate processes: initiation, elongation, and termination. Only a relatively short segment of DNA is transcribed (unlike replication in which the entire chromosome must be copied), and initiation begins when the RNA polymerase binds to the promoter for the gene. RNA polymerase core enzyme is not able to bind DNA tightly or specifically. This situation is drastically changed when sigma is bound to core to make the holoenzyme, which binds the promoter tightly. Recall that the promoter serves only as a target for the binding of the RNA polymerase and is not transcribed. Bacterial promoters have two characteristic features: a sequence of six bases (often TTGACA) about 35 bases pairs before the transcription starting point and a TATAAT sequence, or **Pribnow box,** usually about 10 base pairs upstream of the transcriptional start site (**figure 11.28;** also 11.23). These regions are called the −35 and −10 sites, respectively, while the first nucleotide to be transcribed is referred to as the +1 site. As noted previously, RNA polymerase holoenzyme recognizes the specific sequences at the −10 and −35 sites of promoters. Because the sites must be similar in all promoters, they are called **consensus sequences.**

Once bound to the promoter site, RNA polymerase is able to unwind the DNA without the aid of helicases (**figure 11.29**). The −10 site is rich in adenines and thymines, making it easier to break the hydrogen bonds that keep the DNA double stranded; when the DNA is unwound at this region, it is called **open complex.** A region of unwound DNA equivalent to about two turns of the helix (about 16–20 bases pairs) becomes the "transcription bubble," which moves with the RNA polymerase as it proceeds to transcribe mRNA from the template DNA strand during elongation (**figure 11.30**). Within the transcription bubble, a temporary RNA:DNA hybrid is formed. As the RNA polymerase progresses in the 3′ to 5′ direction along the DNA template, the sigma factor soon dissociates from core RNA polymerase and is available to aid another unit of core enzyme initiate transcription. The mRNA is made in the 5′ to 3′ direction so it is complementary and antiparallel to the template DNA. As elongation of the mRNA continues, single-stranded mRNA is released and the two strands of DNA behind the transcription bubble resume their double helical structure. As shown in figure 11.26, RNA polymerase is a remarkable

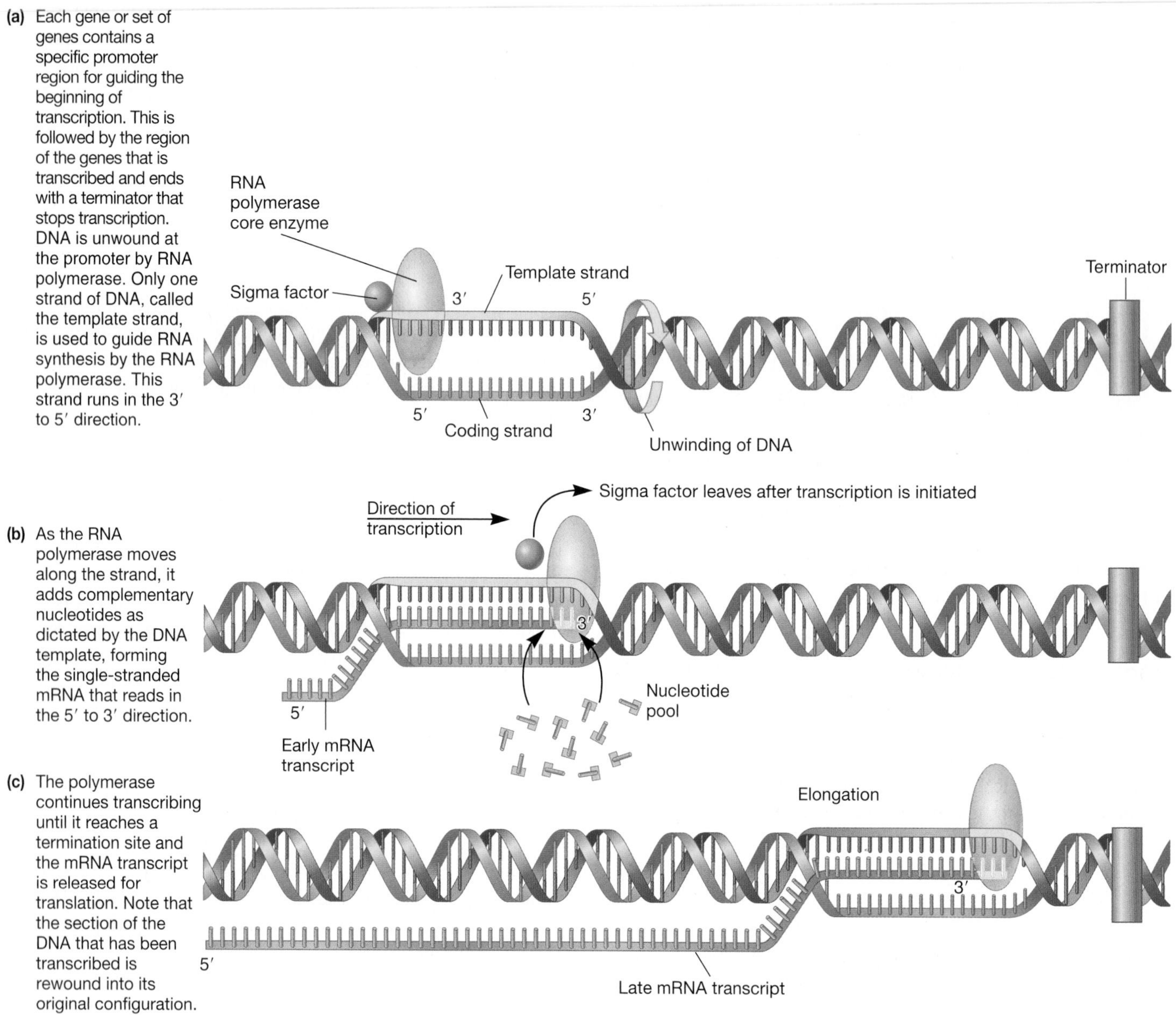

(a) Each gene or set of genes contains a specific promoter region for guiding the beginning of transcription. This is followed by the region of the genes that is transcribed and ends with a terminator that stops transcription. DNA is unwound at the promoter by RNA polymerase. Only one strand of DNA, called the template strand, is used to guide RNA synthesis by the RNA polymerase. This strand runs in the 3′ to 5′ direction.
RNA polymerase core enzyme
Sigma factor
Template strand
3′
5′
Terminator
5′
Coding strand
3′
Unwinding of DNA
(b) As the RNA polymerase moves along the strand, it adds complementary nucleotides as dictated by the DNA template, forming the single-stranded mRNA that reads in the 5′ to 3′ direction.
Direction of transcription
Sigma factor leaves after transcription is initiated
3′
5′
Early mRNA transcript
Nucleotide pool
(c) The polymerase continues transcribing until it reaches a termination site and the mRNA transcript is released for translation. Note that the section of the DNA that has been transcribed is rewound into its original configuration.
Elongation
3′
5′
Late mRNA transcript

Figure 11.26 The Major Events in Transcription.

enzyme capable of several activities, including unwinding the DNA, moving along the template, and synthesizing RNA.

Termination of transcription occurs when the core RNA polymerase dissociates from the template DNA. The end of a gene or group of genes is marked by DNA sequences in the trailer (which is transcribed but not translated) and the terminator. The sequences within procaryotic terminators often contain nucleotides that, when transcribed into RNA, form hydrogen bonds within the single-stranded RNA. This intrastrand base pairing creates a hairpin-shaped loop-and-stem structure. This structure appears to cause the RNA polymerase to pause or stop transcribing DNA. There are two kinds of terminators. The first type causes intrinsic or rho-independent termination (**figure 11.31**). It features the mRNA hairpin followed by a stretch of about six uridine residues. Once the RNA polymerase has paused at the hairpin loop, the A-U base pairs in the uracil-rich region are too weak to hold the RNA:DNA duplex together and the RNA polymerase falls off. The second kind of terminator lacks a poly-U region, and often the hairpin; it requires the aid of a special protein, the **rho factor** (ρ). This terminator causes rho-dependent termination. It is thought that rho binds to mRNA and moves along the molecule until it reaches the RNA polymerase that has halted at a terminator (**figure 11.32**). The rho factor, which has hybrid RNA:DNA helicase activity, then causes the polymerase to dissociate from the mRNA, probably by unwinding the mRNA-DNA complex.

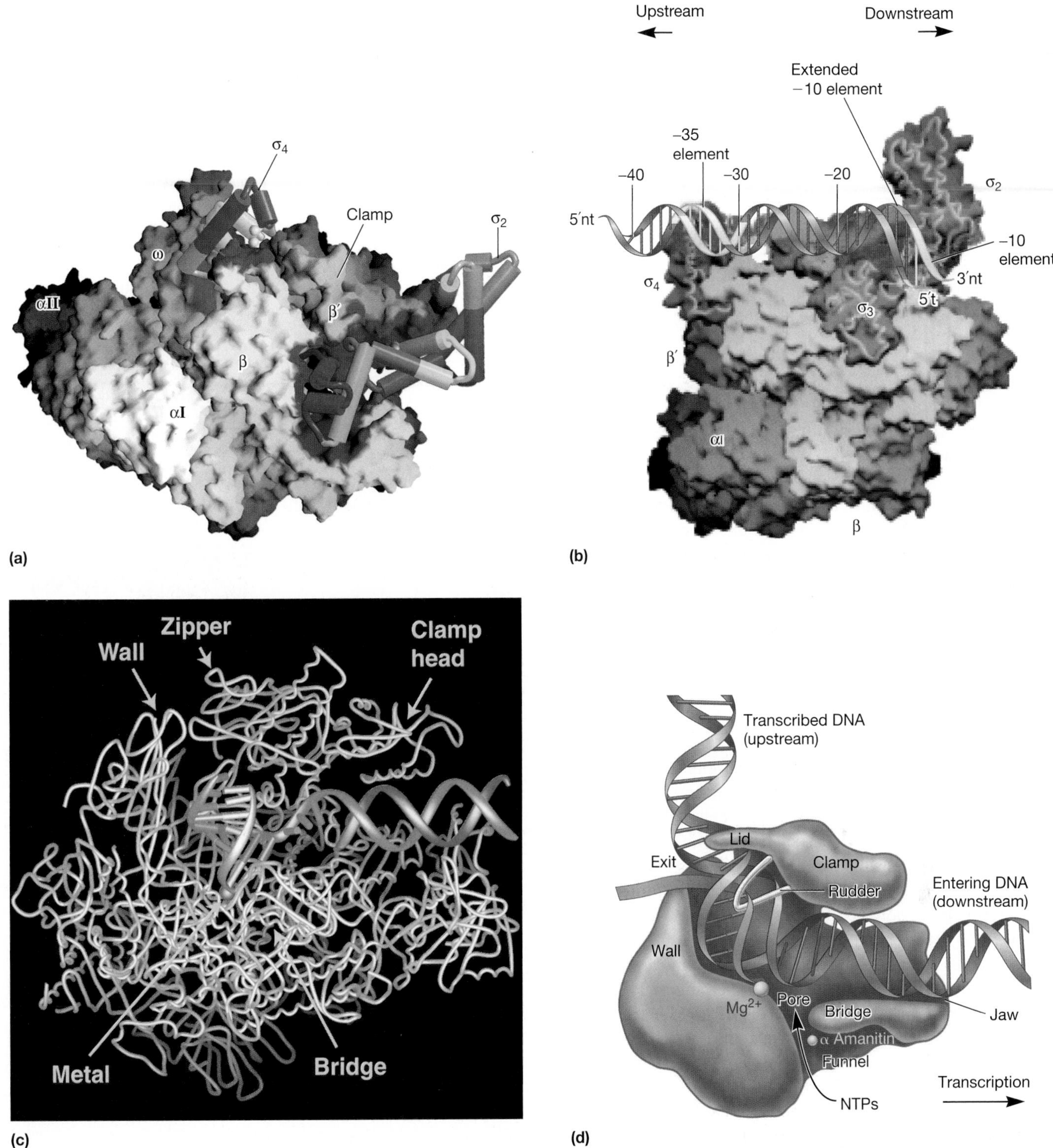

Figure 11.27 RNA Polymerase Structure. The atomic structures of RNA polymerase from the bacterium *Thermus aquaticus* **(a** and **b)** and yeast RNA polymerase II **(c** and **d)** are presented here. **(a)** The *Taq* RNA polymerase holoenzyme is shown with the σ subunit depicted as an α-carbon backbone with cylinders for α-helices. Two of the three σ factor domains are labeled. **(b)** The holoenzyme-DNA complex with the σ surface rendered slightly transparent to show the α-carbon backbone inside. Protein surfaces that contact the DNA are in green and are located on the σ factor. The −10 and −35 elements in the promoter are in yellow. The internal active site is covered by the β subunit in this view. **(c)** Yeast RNA polymerase II transcribing complex with some peptide chains removed to show the DNA. The active site metal is a red sphere. A short stretch of DNA-RNA hybrid (blue and red) lies above the metal. **(d)** A cutaway side view of the polymerase II transcribing complex with the pathway of the nucleic acids and some of the more important parts shown. The enzyme is moving from left to right and the DNA template strand is in blue. A protein "wall" forces the DNA into a right-angle turn and aids in the attachment of nucleoside triphosphates to the growing 3′ end of the RNA. The newly synthesized RNA (red) is separated from the DNA template strand and exits beneath the rudder and lid of the polymerase protein complex. The binding site of the inhibitor α-amanitin also is shown.

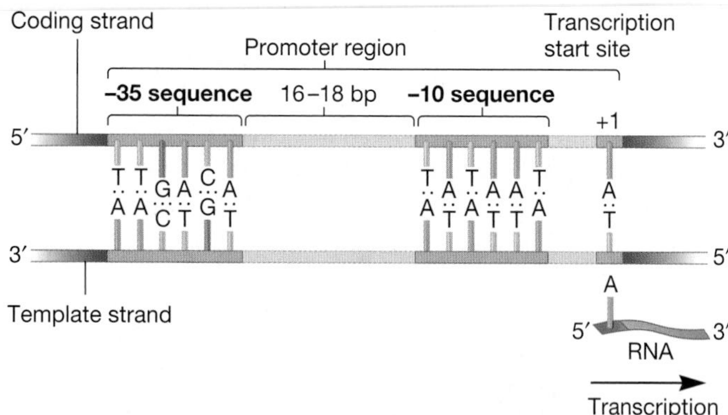

Figure 11.28 The conventional numbering system of promoters. The first nucleotide that acts as a template for transcription is designated +1. The numbering of nucleotides to the left of this spot is in a negative direction, while the numbering to the right is in a positive direction. For example, the nucleotide that is immediately to the left of the +1 nucleotide is numbered −1, and the nucleotide to the right of the +1 nucleotide is numbered +2. There is no zero nucleotide in this numbering system. In many bacterial promoters, sequence elements at the −35 and −10 regions play a key role in promoting transcription.

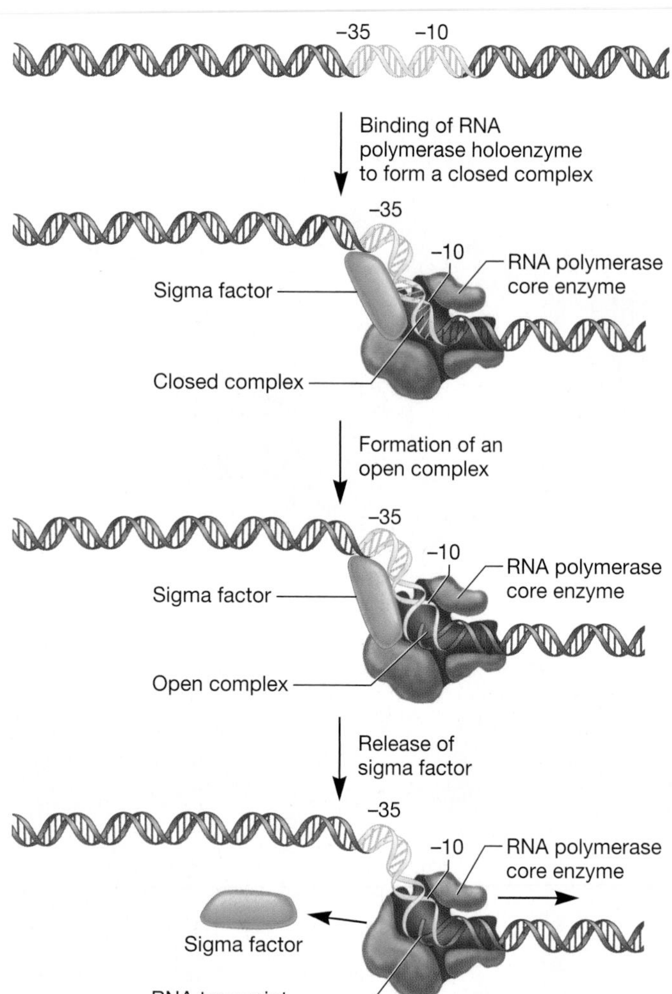

Figure 11.29 The Initiation of Transcription in *Bacteria*. The sigma factor of the RNA polymerase holoenzyme is responsible for positioning the core enzyme properly at the promoter. Sigma factor recognizes two regions in the promoter, one centered at −35 and the other centered at −10. Once positioned properly, the DNA at the −10 region unwinds to form an open complex. The sigma factor dissociates from the core enzyme as it begins transcribing the gene.

Transcription in Eucaryotes

Transcriptional processes in eucaryotic microorganisms (and in other eucaryotic cells) differ in several ways from bacterial transcription. There are three major RNA polymerases, not one as in *Bacteria.* RNA polymerase II, associated with chromatin in the nuclear matrix, is responsible for mRNA synthesis. Polymerases I and III synthesize rRNA and tRNA, respectively (**table 11.3**). The eucaryotic RNA polymerase II is a large aggregate, at least 500,000 daltons in size, with about 10 or more subunits. The atomic structure of the 10-subunit yeast RNA polymerase II associated with DNA and RNA has been determined (figure 11.27*c,d*). The entering DNA is held in a clamp that closes down on it. A magnesium ion is located at the active site and the 9 base pair DNA-RNA hybrid in the transcription bubble is bound in a cleft formed by the two large polymerase subunits. The newly synthesized RNA exits the polymerase beneath the rudder and lid regions. The substrate nucleoside triphosphates probably reach the active site through a pore in the complex. Unlike bacterial polymerase, RNA polymerase II requires extra transcription factors to recognize its promoters (**figure 11.33**). The polymerase binds near the start point; the transcription factors bind to the rest of the promoter. Eucaryotic promoters also differ from those in *Bacteria.* They have combinations of several elements. Three of the most common are the TATA box (located about 30 base pairs before or upstream of the start point), and the GC and CAAT boxes located between 50 to 100 base pairs upstream of the start site (**figure 11.34**). The TFIID transcription factor (figure 11.33) plays an important role in transcription initiation in eucaryotes. This multiprotein complex contains the TATA-binding protein (TBP). TBP has been shown to sharply bend the DNA on attachment. This makes the DNA more accessible to other initiation factors. A variety of general transcription factors, promoter specific factors, and promoter elements have been discovered in different eucaryotic cells. Each eucaryotic gene seems to be regulated differently, and more research will be required to fully understand the regulation of eucaryotic gene transcription.

Unlike bacterial mRNA, eucaryotic mRNA arises from **post-transcriptional modification** of large RNA precursors, about 5,000 to 50,000 nucleotides long, sometimes called heteroge-

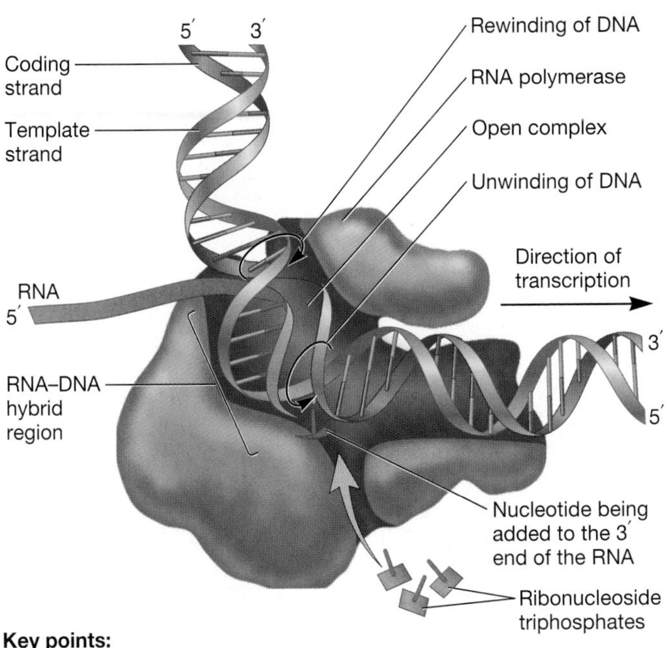
5′ 3′
Coding strand
Template strand
Rewinding of DNA
RNA polymerase
Open complex
Unwinding of DNA
Direction of transcription
RNA
5′
3′
RNA–DNA hybrid region
5′
Nucleotide being added to the 3′ end of the RNA
Ribonucleoside triphosphates
Key points:
• RNA polymerase slides along the DNA, creating an open complex as it moves.
• The DNA strand known as the template strand is used to make a complementary copy of RNA as an RNA–DNA hybrid.
• The RNA is synthesized in a 5′ to 3′ direction using ribonucleoside triphosphates as precursors. Pyrophosphate is released (not shown).
• The complementarity rule is the same as the AT/GC rule except that U is substituted for T in the RNA.

Figure 11.30 The "Transcription Bubble."

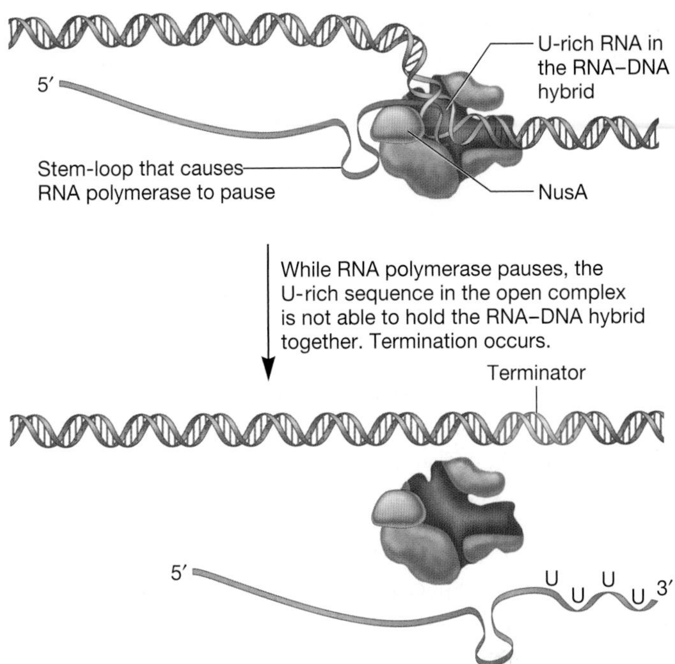
5′
U-rich RNA in the RNA–DNA hybrid
Stem-loop that causes RNA polymerase to pause
NusA
While RNA polymerase pauses, the U-rich sequence in the open complex is not able to hold the RNA–DNA hybrid together. Termination occurs.
Terminator
5′
U U U U 3′

Figure 11.31 Intrinsic Termination of Transcription. This type of terminator contains a U-rich sequence downstream from a stretch of nucleotides that can form a stem-loop and stem structure. Formation of the stem loop in the newly synthesized RNA causes RNA polymerase to pause. This pausing is stabilized by the NusA protein. The U-A bonds in the uracil-rich region are not strong enough to hold the RNA and DNA together. Therefore, the RNA, DNA, and RNA polymerase dissociate and transcription stops.

neous nuclear RNA (hnRNA) (**figure 11.35**). As hnRNA is synthesized, a 5′ cap is added. After synthesis is completed, the precursor RNA is modified by the addition of a 3′ poly-A tail. It is also processed, if necessary, to remove any introns. The 5′ cap is the unusual nucleotide 7-methylguanosine. It is attached to the 5′-hydroxyl of the hnRNA by a triphosphate linkage (**figure 11.36**). Addition of a poly-A tail is initiated by an endonuclease that shortens the hnRNA and generates a 3′-OH group. The enzyme polyadenylate polymerase then catalyzes the addition of adenylic acid to the 3′ end of hnRNA to produce a poly-A sequence about 200 nucleotides long. The functions of the 5′ cap and poly-A tail are not completely clear. The 5′ cap on eucaryotic mRNA may promote the initial binding of ribosomes to the mRNA. It also may protect the mRNA from enzymatic attack. Poly-A protects mRNA from rapid enzymatic degradation. The poly-A tail must be shortened to about 10 nucleotides before mRNA can be degraded. Poly-A also seems to aid in mRNA translation.

As noted earlier, many eucaryotic genes are split or interrupted, which leads to the final type of posttranscriptional processing. **Split** or **interrupted genes** have **exons** (*ex*pressed sequences), regions coding for RNA that end up in the mRNA. Exons are separated from one another by **introns** (*int*ervening sequences), sequences coding for RNA that is never translated into protein (**figure 11.37**). The initial RNA transcript contains both exon and intron sequences. Genes coding for rRNA and tRNA may also be interrupted. Except for cyanobacteria and *Archaea* (*see chapters 20 and 21*), interrupted genes have not been found in procaryotes.

Introns are removed from the initial RNA transcript (also called pre-mRNA or primary transcript) by a process called **RNA splicing** (figure 11.37). The intron's borders are clearly marked for accurate removal. Exon-intron junctions have a GU sequence at the intron's 5′ boundary and an AG sequence at its 3′ end. These two sequences define the splice junctions and are recognized by special RNA molecules. The nucleus contains several **small nuclear RNA (snRNA)** molecules, about 60 to 300 nucleotides long. These complex with proteins to form *s*mall *n*uclear *r*ibo*n*ucleo*p*rotein particles called snRNPs or snurps. Some of the snRNPs recognize splice junctions and ensure splicing accuracy. For example, U1-snRNP recognizes the 5′ splice junction, and U5-snRNP recognizes

Table 11.3 Eucaryotic RNA Polymerases

Enzyme	Location	Product
RNA polymerase I	Nucleolus	rRNA (5.8S, 18S, 28S)
RNA polymerase II	Chromatin, nuclear matrix	mRNA
RNA polymerase III	Chromatin, nuclear matrix	tRNA, 5S rRNA

the 3′ junction. Splicing of pre-mRNA occurs in a large complex called a **spliceosome** that contains the pre-mRNA, at least five kinds of snRNPs, and non-snRNP splicing factors. Sometimes a pre-mRNA will be spliced so that different patterns of exons remain. This alternative splicing allows a single gene to code for more than one protein. The splice pattern determines which protein will be synthesized. Splice patterns can be cell-type specific or determined by the needs of the cell. The importance of alternative splicing in multicellular eucaryotes was emphasized when it was discovered that the human genome has only about 20,000 genes rather than the anticipated 100,000. It is thought that alternative splicing is one mechanism by which human cells produce such a vast array of proteins.

As just mentioned, a few rRNA genes also have introns. Some of these pre-rRNA molecules are self-splicing, that is, the pre-rRNA is a ribozyme (Microbial Tidbits 11.2). Thomas Cech first discovered that pre-rRNA from the ciliate protozoan *Tetrahymena thermophila* is self-splicing. Sidney Altman then showed that ribonuclease P, which cleaves a fragment from one end of pre-tRNA, contains a piece of RNA that catalyzes the reaction. Several other self-splicing rRNA introns have since been discovered. Cech and Altman received the 1989 Nobel Prize in chemistry for these discoveries. Microbial evolution (section 19.1)

Transcription in the *Archaea*

Transcription in the *Archaea* is similar to and distinct from what is observed in *Bacteria* and eucaryotes. Each archaeon has a single RNA polymerase responsible for transcribing all genes in the cell (as in *Bacteria*). However, the RNA polymerase is larger and contains more subunits, many of which are similar to subunits in RNA polymerase II of eucaryotes. The promoters of archaeal genes are similar to those of eucaryotes in having a TATA box; binding of the archaeal RNA polymerase to its promoter requires a TATA-binding protein, just as in eucaryotes. Like the eucaryotic counterpart, the archaeal RNA polymerase also needs several additional transcription factors to function properly. Furthermore, some archaeal genes have introns, which must be removed by posttranscriptional processing. Finally, the mRNA molecules produced by transcription in *Archaea* are usually polycistronic, as in *Bacteria*. This intriguing mixture of bacterial and eucaryotic features in the *Archaea* has fueled a great deal of speculation about the evolution of all three domains of life. Introduction to the *Archaea:* Genetics and molecular biology (section 20.1)

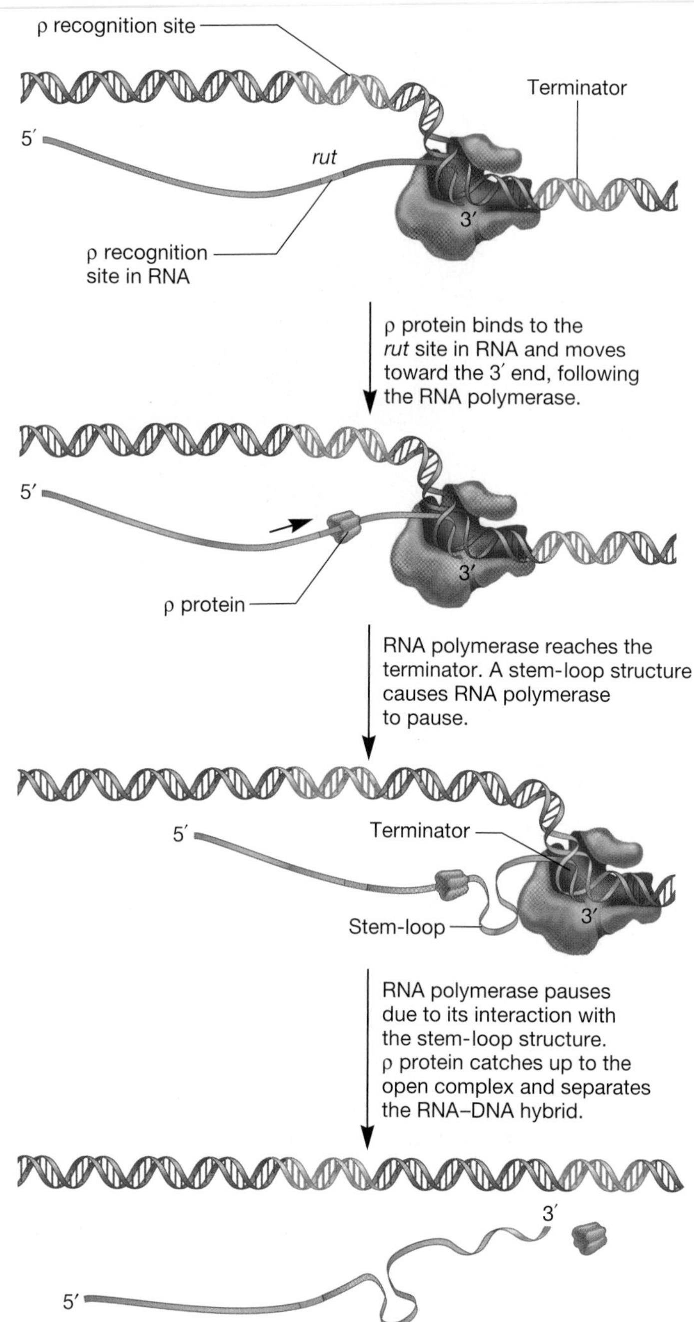

Figure 11.32 Rho-Factor (ρ)-Dependent Termination of Transcription. The *rut* site stands for *r*ho *ut*ilization site.

1. Define the following terms: polygenic mRNA, RNA polymerase core enzyme, sigma factor, RNA polymerase holoenzyme, and rho factor.
2. Define or describe posttranscriptional modification, heterogeneous nuclear RNA, 3′ poly-A sequence, 5′ capping, split or interrupted genes, exon, intron, RNA splicing, snRNA, spliceosome, and ribozyme.
3. Describe how RNA polymerase transcribes bacterial DNA. How does the polymerase know when to begin and end transcription?
4. How do bacterial RNA polymerases and promoters differ from those of *Archaea* and eucaryotes?

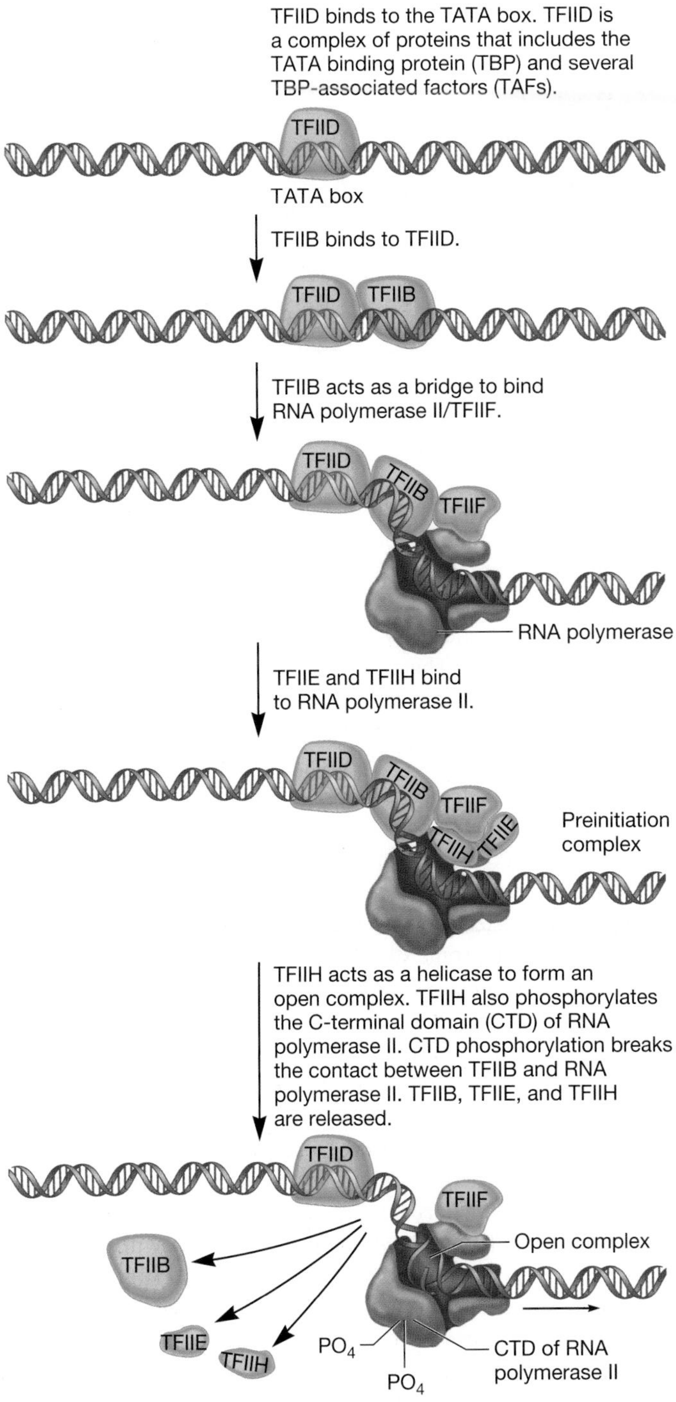

Figure 11.33 Initiation of Transcription in Eucaryotes. The TATA box is a major component of eucaryotic promoters. *TATA-binding protein* (TBP), which is a component of a complex of proteins called TFIID (*transcription factor IID*), binds the TATA box. Note that numerous other transcription factors are required for initiation of transcription, unlike *Bacteria* where only the sigma factor is needed. Initiation of transcription in *Archaea* is similar to that seen in eucaryotes.

11.7 The Genetic Code

The final step in the expression of genes that encode proteins is translation. The mRNA nucleotide sequence is translated into the amino acid sequence of a polypeptide chain. Protein synthesis is called translation because it is a decoding process. The information encoded in the language of nucleic acids must be rewritten in the language of proteins. Therefore, before we discuss protein synthesis, we will examine the nature of the genetic code.

Establishment of the Genetic Code

The realization that DNA is the genetic material triggered efforts to understand how genetic instructions are stored and organized in the DNA molecule. Early studies on the nature of the genetic code showed that the DNA base sequence corresponds to the amino acid sequence of the polypeptide specified by the gene. That is, the nucleotide and amino acid sequences are colinear. It also became evident that many mutations are the result of changes of single amino acids in a polypeptide chain. However, the exact nature of the code was still unclear.

Theoretical considerations directed much of the early work on deciphering the code. Scientists reasoned that because only 20 amino acids normally are present in proteins, there must be at least 20 different code words in DNA. Therefore the code must be contained in some sequence of the four nucleotides commonly found in DNA.

If the code words were two nucleotides in length, there would be only 16 possible combinations (4^2) of the four nucleotides and this would not be enough to code for all 20 amino acids. Therefore a code word, or codon, had to consist of at least nucleotide triplets even though this would give 64 possible combinations (4^3), many more than the minimum of 20 needed to specify the common amino acids. Research eventually confirmed this and the code was deciphered in the early 1960s by Marshall Nirenberg, Heinrich Matthaei, Philip Leder, and Har Gobind Khorana. In 1968 Nirenberg and Khorana shared the Nobel Prize with Robert Holley, the first person to sequence a nucleic acid (phenylalanyl-tRNA).

Organization of the Code

The genetic code, presented in RNA form, is summarized in **table 11.4.** Note that there is **code degeneracy.** That is, there are up to six different codons for a given amino acid. Only 61 codons, the **sense codons,** direct amino acid incorporation into protein. The remaining three codons (UGA, UAG, and UAA) are involved in the termination of translation and are called stop or **nonsense codons.** Despite the existence of 61 sense codons, there are not 61 different tRNAs, one for each codon. The 5′ nucleotide in the anticodon can vary, but generally, if the nucleotides in the second and third anticodon positions complement the first two bases of the mRNA codon, an aminoacyl-tRNA with the proper amino acid will bind to the mRNA-ribosome complex. This pattern is evident on inspection of changes in the amino acid specified

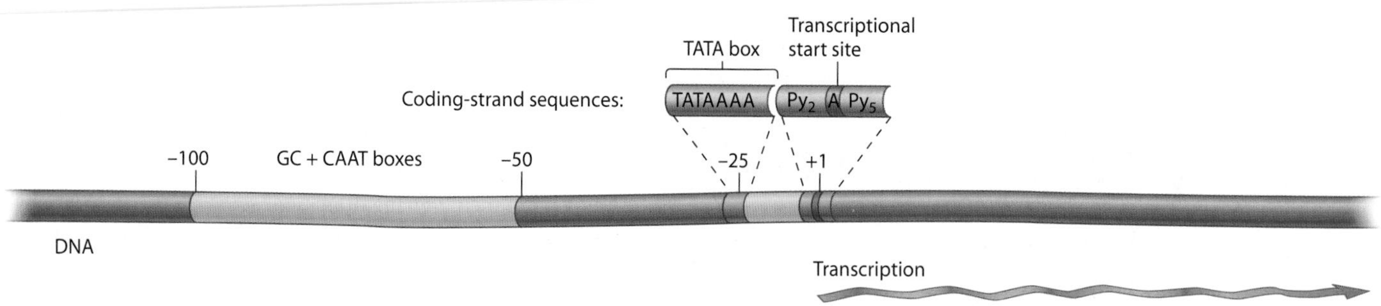

Figure 11.34 The TATA Box and Other Elements of Eucaryotic Promoters.

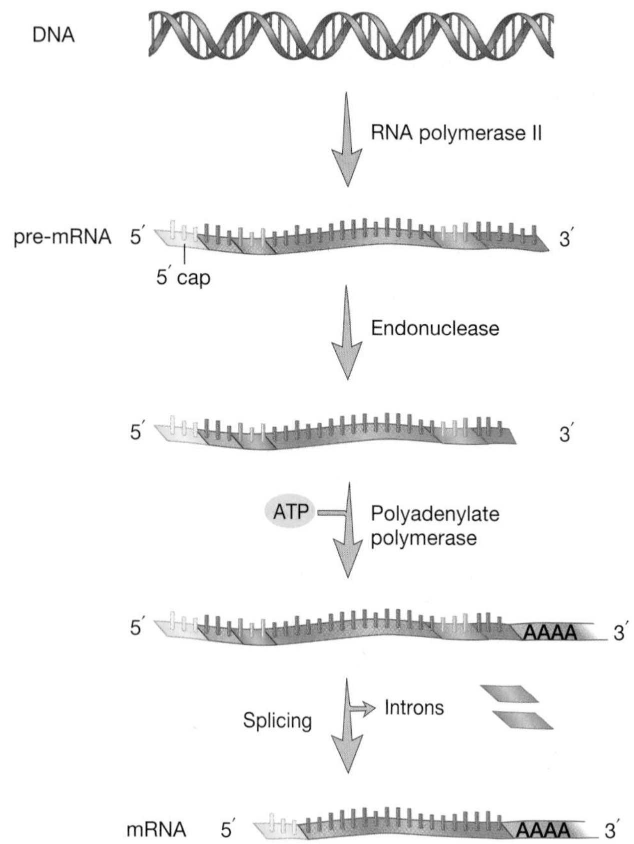

Figure 11.35 Eucaryotic mRNA Synthesis. The production of eucaryotic messenger RNA. The 5′ cap is added shortly after synthesis of the mRNA begins.

with variation in the third position (table 11.4). This somewhat loose base pairing is known as **wobble** and relieves cells of the need to synthesize so many tRNAs (**figure 11.38**). Wobble also decreases the effects of DNA mutations.

1. Why must a codon contain at least three nucleotides?
2. Define the following: code degeneracy, sense codon, stop or nonsense codon, and wobble.

11.8 Translation

Translation involves decoding mRNA and covalently linking amino acids together to form a polypeptide. Just as DNA and RNA synthesis proceeds in one direction (5′ to 3′), so too does protein synthesis. Polypeptides are synthesized by the addition of amino acids to the end of the chain with the free α-carboxyl group (the C-terminal end). That is, the synthesis of polypeptides begins with the amino acid at the end of the chain with a free amino group (the N-terminal) and moves in the C-terminal direction. The ribosome is the site of protein synthesis. Protein synthesis is not only quite accurate but also very rapid. In *E. coli* synthesis occurs at a rate of at least 900 residues per minute; eucaryotic translation is slower, about 100 residues per minute. Proteins (appendix I)

Cells that grow quickly must use each mRNA with great efficiency to synthesize proteins at a sufficiently rapid rate. The two subunits of the ribosome (the 50S subunit and the 30S subunit in *Bacteria* and *Archaea;* 60S and 40S in eucaryotes) are free in the cytoplasm if protein is not being synthesized. They come together to form the complete ribosome only when translation occurs. Frequently mRNAs are simultaneously complexed with several ribosomes, each ribosome reading the mRNA message and synthesizing a polypeptide. At maximal rates of mRNA use, there may be a ribosome every 80 nucleotides along the messenger or as many as 20 ribosomes simultaneously reading an mRNA that codes for a 50,000 dalton polypeptide. A complex of mRNA with several ribosomes is called a **polyribosome** or polysome (**figure 11.39**). Polysomes are present in both procaryotes and eucaryotes. *Bacteria* and *Archaea* can further increase the efficiency of gene expression through coupled transcription and translation (figure 11.39*b*). While RNA polymerase is synthesizing an mRNA, ribosomes can already be attached to the messenger so that transcription and translation occur simultaneously. Coupled transcription and translation is possible in procaryotes because a nuclear envelope does not separate the translation machinery from DNA as it does in eucaryotes (*see figure 3.16*).

Transfer RNA and Amino Acid Activation

The first stage of protein synthesis is **amino acid activation,** a process in which amino acids are attached to transfer RNA molecules. These RNA molecules are normally between 73 and 93 nu-

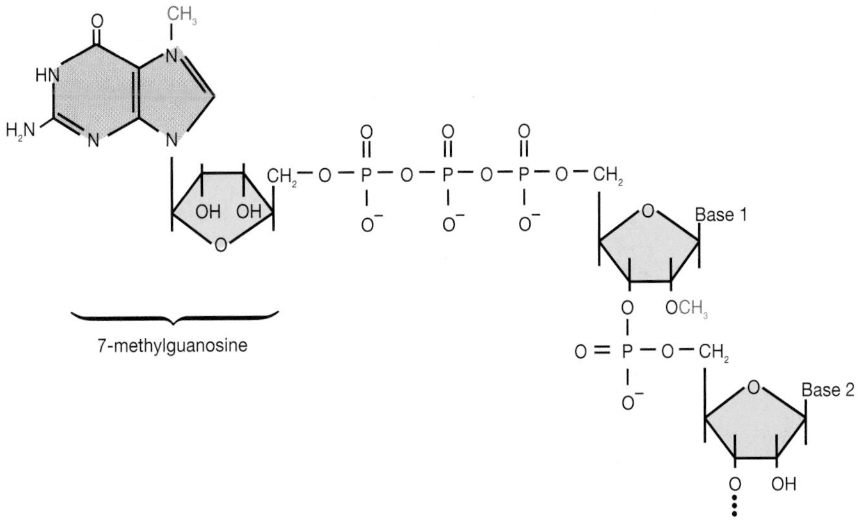

Figure 11.36 The 5′ Cap of Eucaryotic mRNA. Methyl groups are in blue.

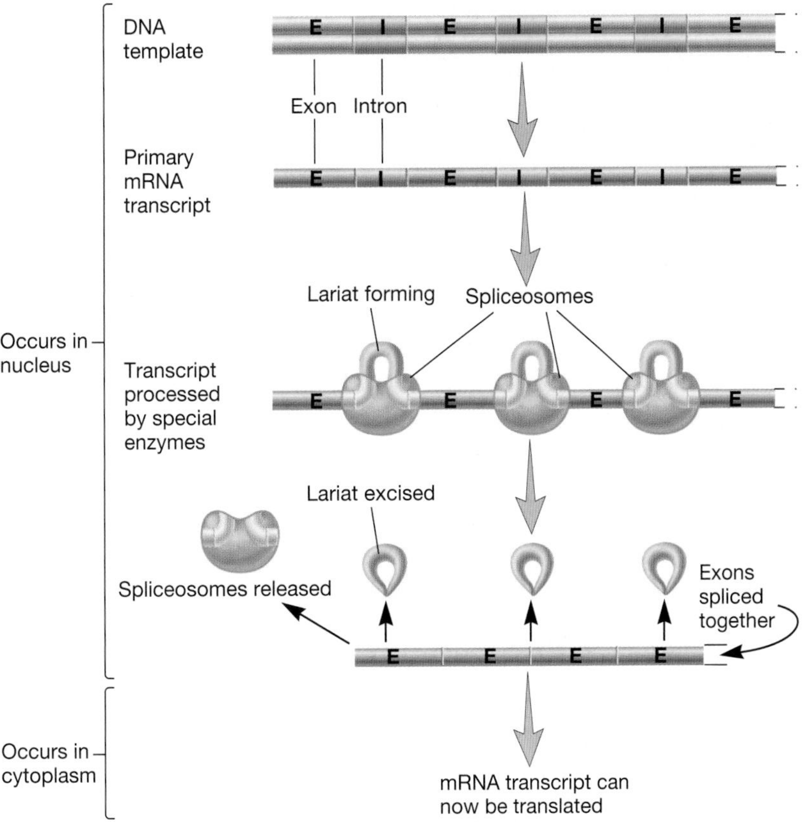

Figure 11.37 Splicing of Eucaryotic mRNA Molecules.

cleotides in length and possess several characteristic structural features. The structure of tRNA becomes clearer when its chain is folded in such a way to maximize the number of normal base pairs, which results in a cloverleaf conformation of five stems and loops (**figure 11.40**). The acceptor or amino acid stem holds the activated amino acid on the 3′ end of the tRNA. The 3′ end of all tRNAs has the same —C—C—A sequence; the amino acid is attached to the terminal adenylic acid. At the other end of the cloverleaf is the anticodon arm, which contains the **anticodon** triplet complementary to the mRNA codon triplet. There are two

Table 11.4 The Genetic Code

First Position (5′ End)[a]	Second Position U		Second Position C		Second Position A		Second Position G		Third Position (3′ End)
U	UUU	Phe	UCU	Ser	UAU	Tyr	UGU	Cys	U
	UUC		UCC		UAC		UGC		C
	UUA	Leu	UCA		UAA	STOP	UGA	STOP	A
	UUG		UCG		UAG		UGG	Trp	G
C	CUU	Leu	CCU	Pro	CAU	His	CGU	Arg	U
	CUC		CCC		CAC		CGC		C
	CUA		CCA		CAA	Gln	CGA		A
	CUG		CCG		CAG		CGG		G
A	AUU	Ile	ACU	Thr	AAU	Asn	AGU	Ser	U
	AUC		ACC		AAC		AGC		C
	AUA		ACA		AAA	Lys	AGA	Arg	A
	AUG	Met	ACG		AAG		AGG		G
G	GUU	Val	GCU	Ala	GAU	Asp	GGU	Gly	U
	GUC		GCC		GAC		GGC		C
	GUA		GCA		GAA	Glu	GGA		A
	GUG		GCG		GAG		GGG		G

[a]The code is presented in the RNA form. Codons run in the 5′ to 3′ direction. See text for details.

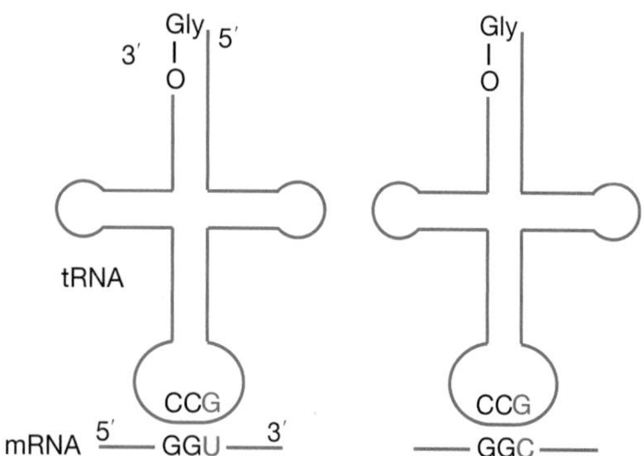

(a) Base pairing of one glycine tRNA with two codons due to wobble

Glycine mRNA codons: GGU, GGC, GGA, GGG (5′ → 3′)

Glycine tRNA anticodons: CCG, CCU, CCC (3′ → 5′)

(b) Glycine codons and anticodons

Figure 11.38 Wobble and Coding. The use of wobble in coding for the amino acid glycine. **(a)** Because of wobble, G in the 5′ position of the anticodon can pair with either C or U in the 3′ position of the codon. Thus two codons can be recognized by the same tRNA. **(b)** Because of wobble, only three tRNA anticodons are needed to translate the four glycine (Gly) codons.

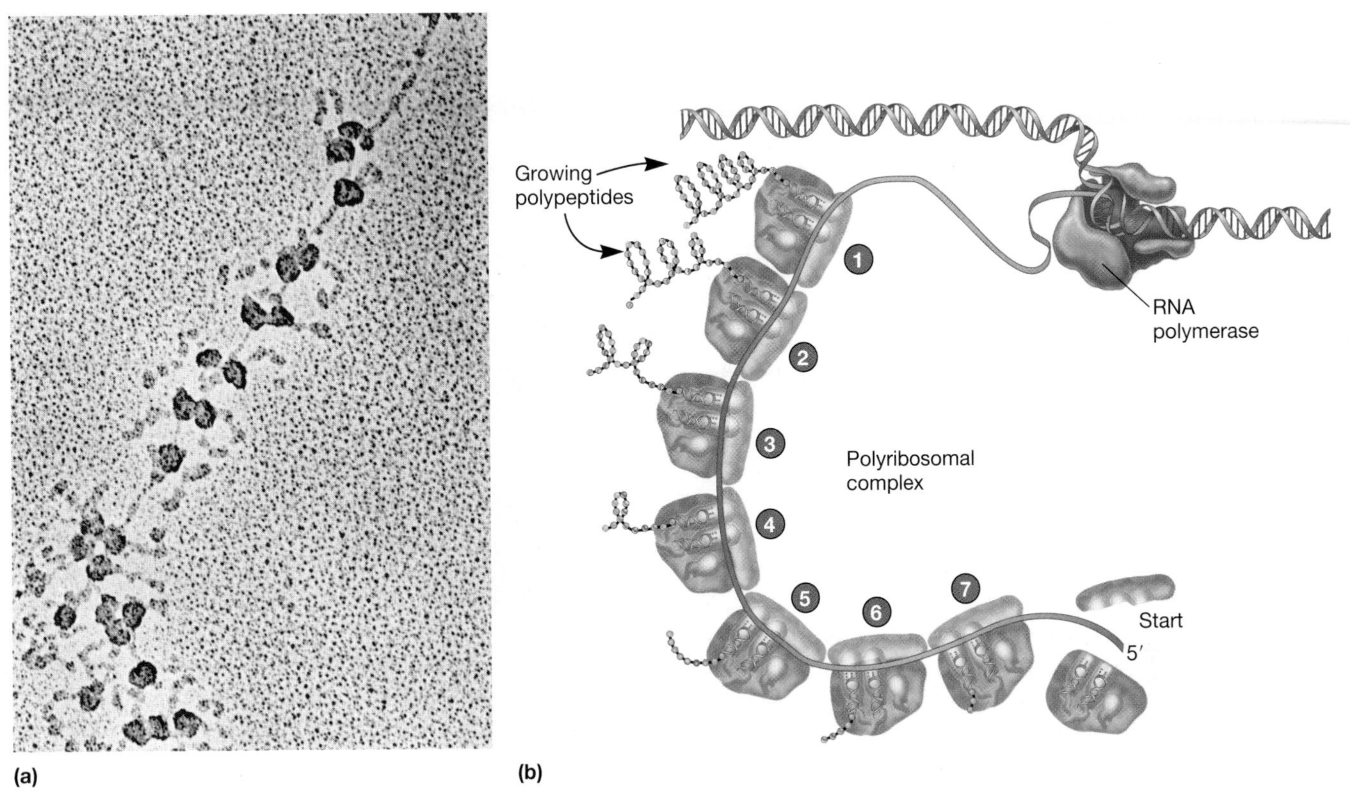

Figure 11.39 Coupled Transcription and Translation in Procaryotes. **(a)** A transmission electron micrograph showing coupled transcription and translation. **(b)** A schematic representation of coupled transcription and translation. As the DNA is transcribed, ribosomes bind the free 5′ end of the mRNA. Thus translation is started before transcription is completed. Note that there are multiple ribosomes bound to the mRNA, forming a polyribosome.

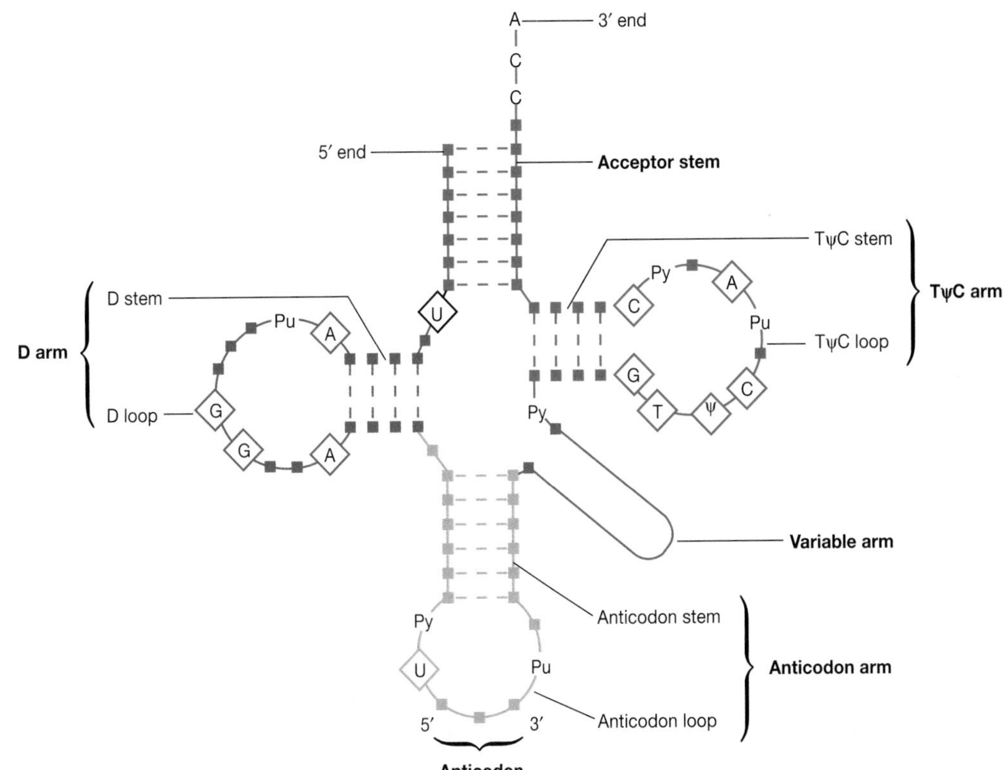

Figure 11.40 tRNA Structure. The cloverleaf structure for tRNA in procaryotes and eucaryotes. Bases found in all tRNAs are in diamonds; purine and pyrimidine positions in all tRNAs are labeled Pu and Py respectively.

other large arms: the D or DHU arm has the unusual pyrimidine nucleoside dihydrouridine; and the T or TΨC arm has ribothymidine (T) and pseudouridine (Ψ), both of which are unique to tRNA. Finally, the cloverleaf has a variable arm whose length changes with the overall length of the tRNA; the other arms are fairly constant in size.

Transfer RNA molecules are folded into an L-shaped structure (**figure 11.41**). The amino acid is held on one end of the L, the anticodon is positioned on the opposite end, and the corner of the L is formed by the D and T loops. Because there must be at least one tRNA for each of the 20 amino acids incorporated into proteins, at least 20 different tRNA molecules are needed. Actually more tRNA species exist.

Amino acids are activated for protein synthesis through a reaction catalyzed by **aminoacyl-tRNA synthetases (figure 11.42)**.

$$\text{Amino acid} + \text{tRNA} + \text{ATP} \xrightarrow{Mg^{2+}} \text{aminoacyl-tRNA} + \text{AMP} + \text{PP}_i$$

Just as is true of DNA and RNA synthesis, the reaction is driven to completion when the pyrophosphate product is hydrolyzed to two orthophosphates. The amino acid is attached to the 3′-hydroxyl of the terminal adenylic acid on the tRNA by a high-energy bond (**figure 11.43**), and is readily transferred to the end of a growing peptide chain. This is why the amino acid is said to be activated.

There are at least 20 aminoacyl-tRNA synthetases, each specific for a single amino acid and its tRNAs (cognate tRNAs). It is critical that each tRNA attach the corresponding amino acid because if an incorrect amino acid is attached to a tRNA, it will be incorporated into a polypeptide in place of the correct amino acid. The protein synthetic machinery recognizes only the anticodon of the aminoacyl-tRNA and cannot tell whether the correct amino acid is attached. Some aminoacyl-tRNA synthetases proofread just like DNA polymerases do. If the wrong amino acid is attached to tRNA, the enzyme hydrolyzes the amino acid from the tRNA rather than release the incorrect product.

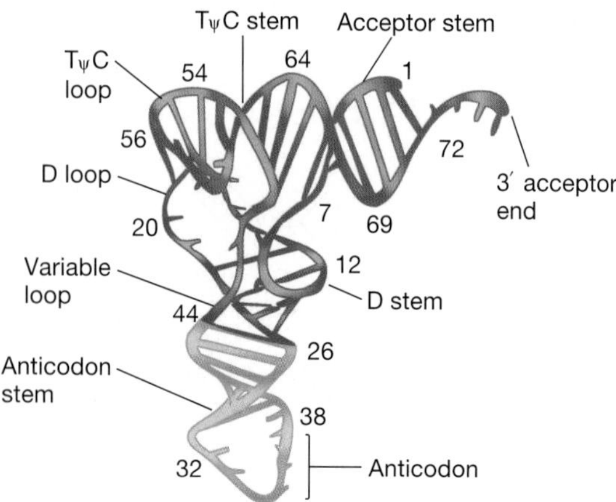

Figure 11.41 Transfer RNA Conformation. The three-dimensional structure of tRNA. The various regions are distinguished with different colors.

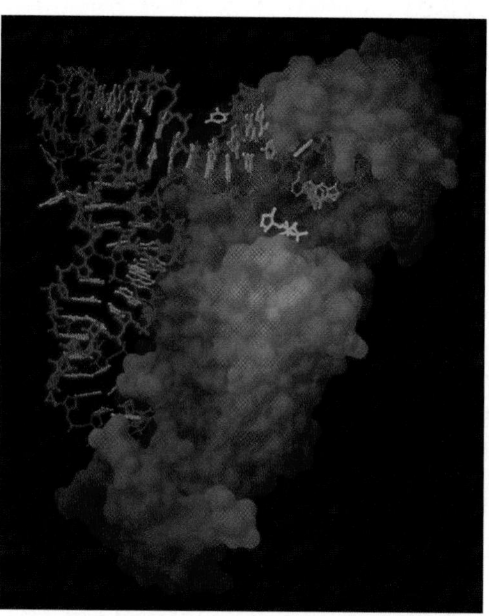

Figure 11.42 An Aminoacyl-tRNA Synthetase. A model of *E. coli* glutamyl-tRNA synthetase complexed with its tRNA and ATP. The enzyme is in blue, the tRNA in red and yellow, and ATP in green.

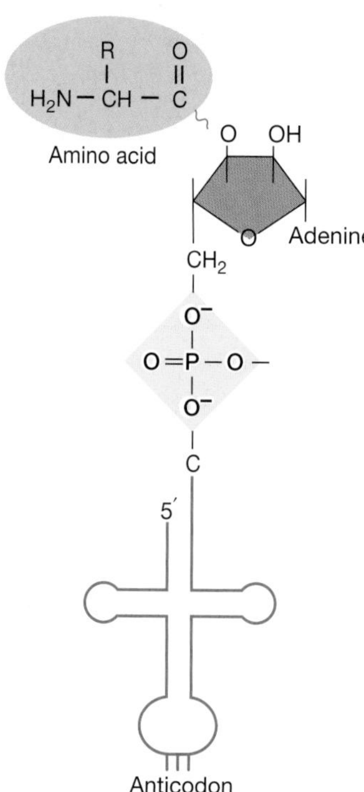

Figure 11.43 Aminoacyl-tRNA. The amino acid is attached to the 3′-hydroxyl of adenylic acid by a high-energy bond (red).

The Ribosome

The actual process of protein synthesis takes place on ribosomes that serve as workbenches, with mRNA acting as the blueprint. Procaryotic ribosomes have a sedimentation value of 70S and a mass of 2.8 million daltons. A rapidly growing *E. coli* cell may have as many as 15,000 to 20,000 ribosomes, about 15% of the cell mass.

Each of the two subunits of the procaryotic ribosome is an extraordinarily complex structure constructed from one or two rRNA molecules and many polypeptides (**figure 11.44**). The shape of the subunits and their association to form the 70S ribosome are depicted in **figure 11.45.** The region of the ribosome directly responsible for translation is called the translational domain (figure 11.45*d*). Both subunits contribute to this domain. The growing peptide chain emerges from the large subunit at the exit domain. This is located on the side of the subunit opposite the central protuberance (figure 11.45*b*). X-ray diffraction studies have now confirmed this general picture of ribosome stucture (figure 11.45*e–g*).

Ribosomal RNA is thought to have three roles. It obviously contributes to ribosome structure. The 16S rRNA of the 30S subunit is needed for the initiation of protein synthesis in *Bacteria.* There is evidence that the 3′ end of the 16S rRNA complexes with a site on the mRNA called the Shine-Dalgarno sequence, which is located in the **ribosome-binding site (RBS).** This helps position the mRNA on the ribosome. The 16S rRNA also binds a protein needed to initiate translation, initiation factor 3 and the 3′ CCA end of aminoacyl-tRNA. Finally, it appears that the 23S rRNA has a catalytic role in protein synthesis.

Initiation of Protein Synthesis

Like transcription, protein synthesis may be divided into three stages: initiation, elongation, and termination. In the initiation stage, *E. coli* and most bacteria begin protein synthesis with a specially modified aminoacyl-tRNA, *N*-formylmethionyl-tRNA^{fMet} (**figure 11.46**). Because the α-amino is blocked by a formyl group, this aminoacyl-tRNA can be used only for initiation. When methionine is to be added to a growing polypeptide chain, a normal methionyl-tRNA^{Met} is employed. Eucaryotic protein synthesis (except in the mitochondrion and chloroplast) and archaeal protein synthesis begin with a special initiator methionyl-tRNA^{Met}. Although most bacteria start protein synthesis with formylmethionine, the formyl group does not remain but is hydrolytically removed. In fact, one to three amino acids may be removed from the amino terminal end of the polypeptide after synthesis.

The initiation stage is crucial for the translation of the mRNA into the correct polypeptide (**figure 11.47**). In *Bacteria,* it begins when initiator *N*-formylmethionyl-tRNA^{fMet} (fMet-tRNA) binds to a free 30S ribosomal subunit. As noted earlier, the 30S subunit possesses a molecule of 16S rRNA with nucleotide sequences that are complementary to the Shine-Dalgarno sequence in the leader sequence of the mRNA. Recall that the leader is transcribed but not translated. This is because the role of the leader sequence is to align the mRNA with complementary bases on the 16S rRNA of the 30S ribosomal subunit such that the codon for the initiator fMet-tRNA is translated first. Messenger RNAs have a special **initiator codon** (AUG or sometimes GUG) that specifically binds with the fMet-tRNA anticodon. Finally, the 50S subunit binds to the 30S subunit-mRNA forming an active ribosome-mRNA complex. The fMet-tRNA must be positioned at the peptidyl or P site (see description of the elongation cycle). There is some uncertainty about the exact initiation sequence, and mRNA may bind before fMet-tRNA in *Bacteria.* Eucaryotic and archaeal initiation appears to begin with the binding of the initiator Met-tRNA to the small subunit, followed by attachment of the mRNA.

In *Bacteria,* three protein **initiation factors** are required (figure 11.47). Initiation factor 3 (IF-3) prevents 30S subunit binding to the 50S subunit and promotes the proper mRNA binding to the 30S subunit. IF-2, the second initiation factor, binds GTP and fMet-tRNA and directs the attachment of fMet-tRNA to the P site of the 30S subunit. GTP is hydrolyzed during association of the 50S and 30S subunits. The third initiation factor, IF-1, appears to be needed for release of IF-2 and GDP from the completed 70S ribosome. IF-1 may aid in the binding of the 50S subunit to the 30S subunit. It also blocks tRNA binding to the A site. Eucaryotes require more initiation factors; otherwise the process is quite similar to that of *Bacteria.*

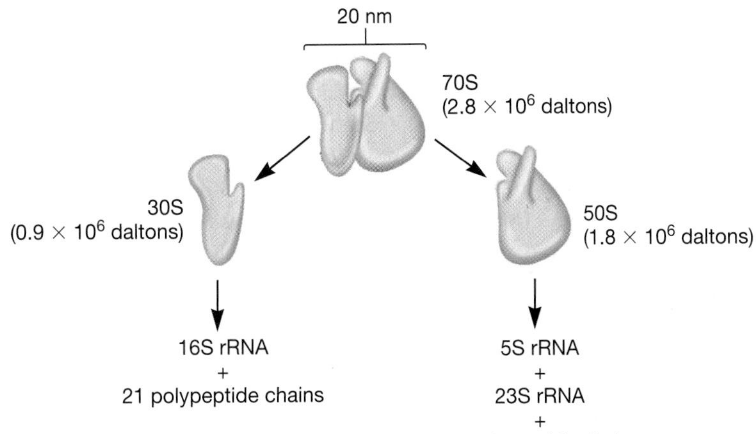

Figure 11.44 The 70S Ribosome.

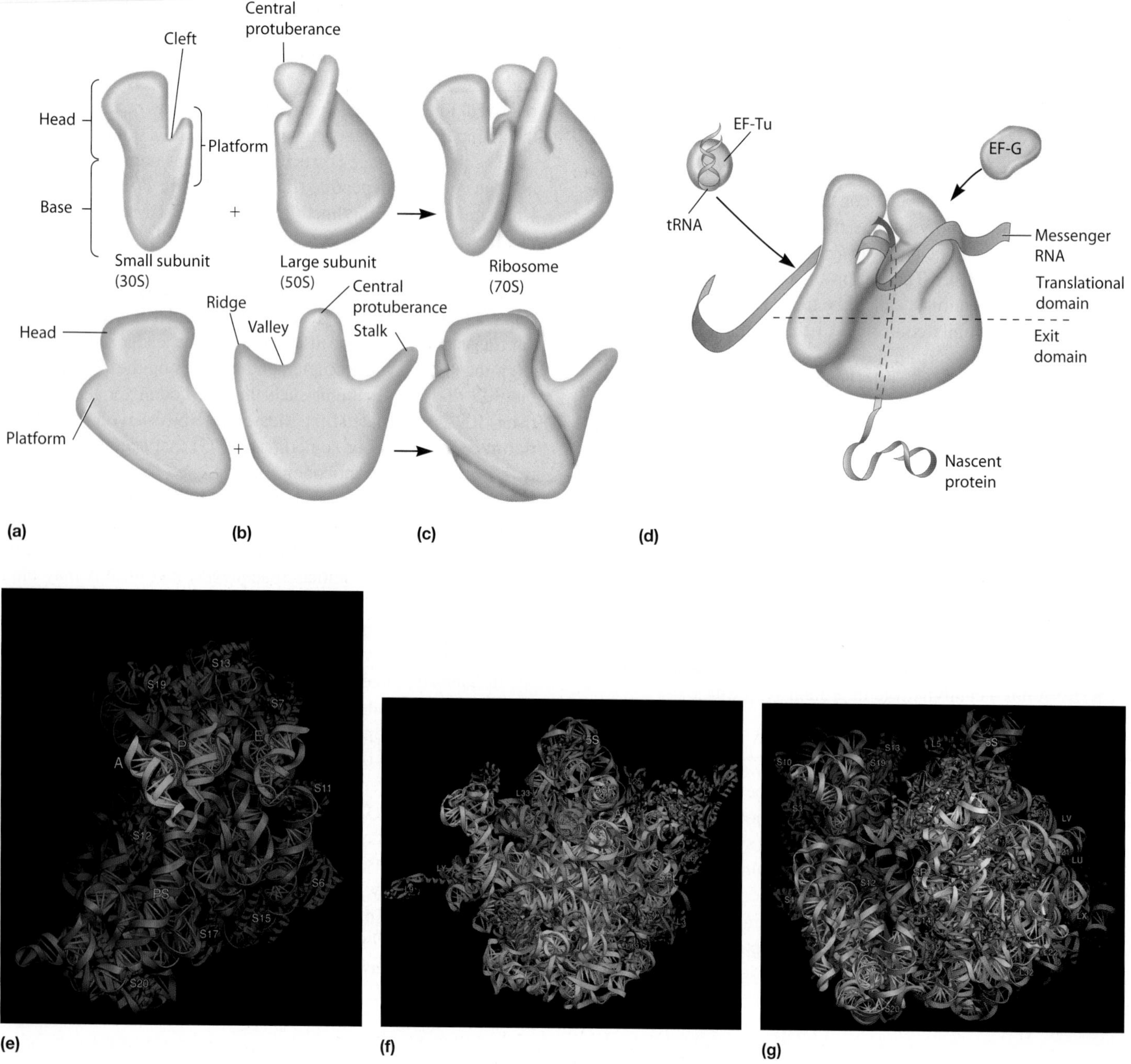

Figure 11.45 Procaryotic Ribosome Structure. Parts **(a)**–**(d)** illustrate *E. coli* ribosome organization; parts **(e)**–**(g)** show the molecular structure of the *Thermus thermophilus* ribosome. **(a)** The 30S subunit. **(b)** The 50S subunit. **(c)** The complete 70S ribosome. **(d)** A diagram of ribosomal structure showing the translational and exit domains. The locations of elongation factor and mRNA binding are indicated. The growing peptide chain probably remains unfolded and extended until it leaves the large subunit. **(e)** Interior interface view of the 30S subunit of the *T. thermophilus* 70S ribosome showing the positions of the A, P, and E site tRNAs. **(f)** Interior interface view of the *T. thermophilus* 50S subunit and portions of its three tRNAs. **(g)** The complete *T. thermophilus* 70S ribosome viewed from the right-hand side with the 30S subunit on the left and the 50S subunit on the right. The anticodon arm of the A site tRNA is visible in the interface cavity. The components in figures **(e)**–**(g)** are colored as follows: 16S rRNA, cyan; 23S rRNA, gray; 5S rRNA, light blue; 30S proteins, dark blue; 50S proteins, magenta; and A, P, and E site tRNAs (gold, orange, and red, respectively).

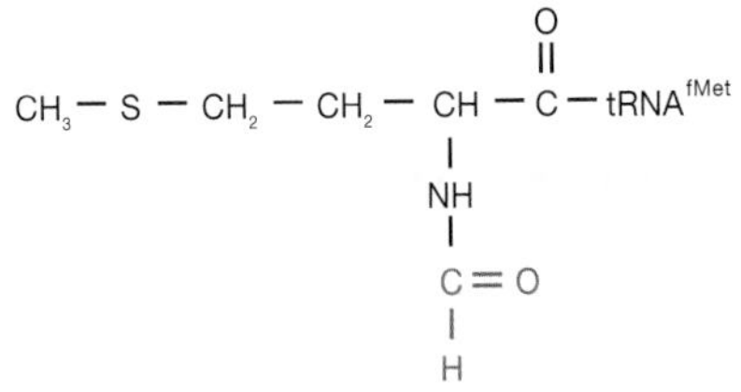

$$CH_3-S-CH_2-CH_2-\underset{\underset{\underset{\underset{H}{|}}{C=O}}{\underset{|}{NH}}}{\underset{|}{CH}}-\overset{\overset{O}{\|}}{C}-tRNA^{fMet}$$

Figure 11.46 Bacterial Initiator tRNA. The initiator aminoacyl-tRNA, *N*-formylmethionyl-tRNA ^{fMet}, is used by *Bacteria.* The formyl group is in color. *Archaea* and eucaryotes use methionyl-tRNA for initiation.

The initiation of protein synthesis is very elaborate. Apparently the complexity is necessary to ensure that the ribosome does not start synthesizing a polypeptide chain in the middle of a gene—a disastrous error.

Elongation of the Polypeptide Chain

Every amino acid addition to a growing polypeptide chain is the result of an **elongation cycle** composed of three phases: aminoacyl-tRNA binding, the transpeptidation reaction, and translocation. The process is aided by special protein **elongation factors** (just as with the initiation of protein synthesis). In each turn of the cycle, an

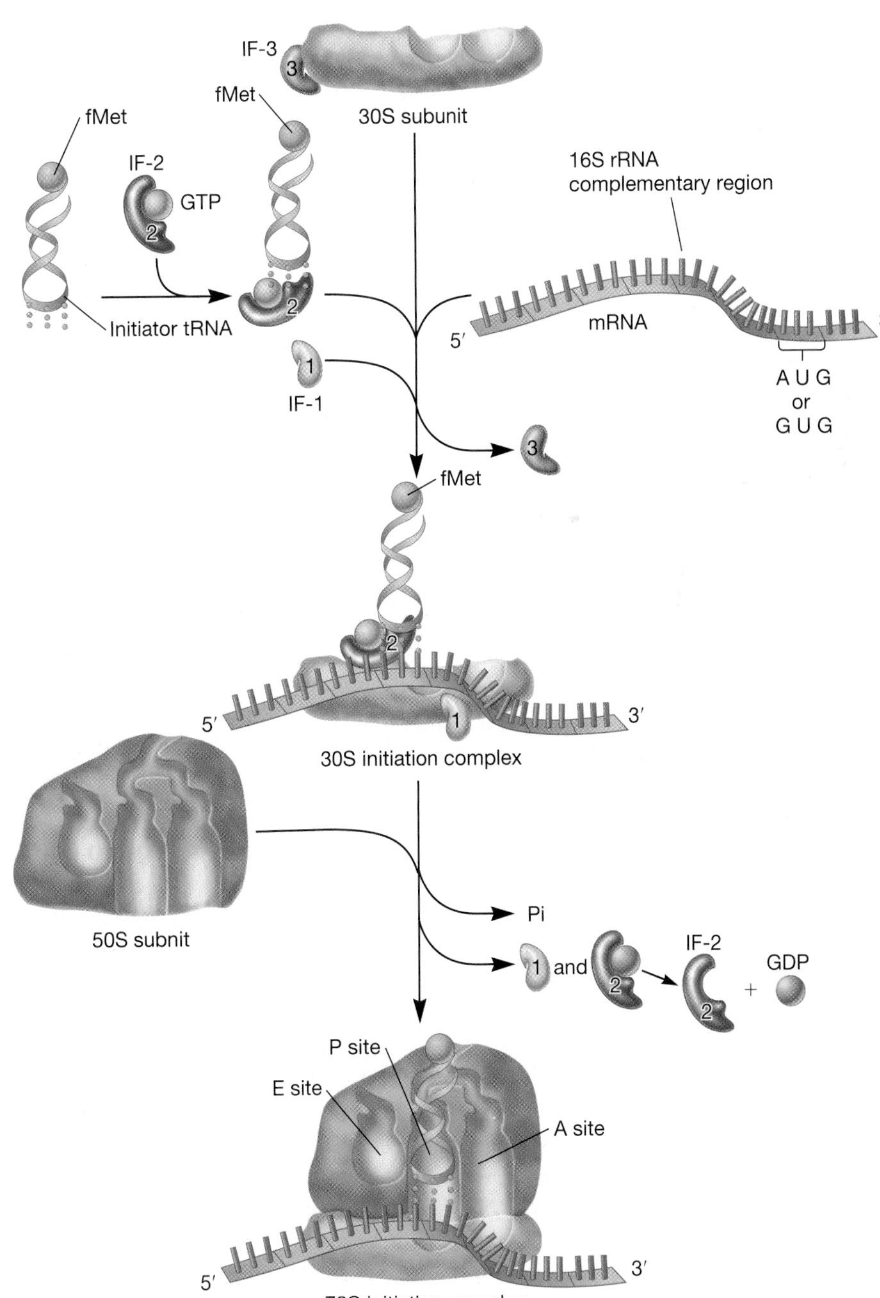

Figure 11.47 Initiation of Protein Synthesis. The initiation of protein synthesis in *Bacteria.* The following abbreviations are employed: IF-1, IF-2, and IF-3 stand for initiation factors 1, 2, and 3; initiator tRNA is *N*-formylmethionyl-tRNA^{fMet}. The ribosomal locations of initiation factors are depicted for illustration purposes only. They do not represent the actual initiation factor binding sites. See text for further discussion.

amino acid corresponding to the proper mRNA codon is added to the C-terminal end of the polypeptide chain. The bacterial elongation cycle is described next.

The ribosome has three sites for binding tRNAs: (1) the **peptidyl** or **donor site** (the **P site**), (2) the **aminoacyl** or **acceptor site** (the **A site**), and (3) the **exit site** (the **E site**). At the beginning of an elongation cycle, the peptidyl site is filled with either *N*-formylmethionyl-$tRNA^{fMet}$ or peptidyl-tRNA, and the aminoacyl and exit sites are empty (**figure 11.48**). Messenger RNA is bound to the ribosome in such a way that the proper codon interacts with the P site tRNA (e.g., an AUG codon for fMet-tRNA). The next codon is located within the A site and is ready to accept an aminoacyl-tRNA.

The first phase of the elongation cycle is the aminoacyl-tRNA binding phase. The aminoacyl-tRNA corresponding to the codon in the A site is inserted so its anticodon is aligned with the codon on the mRNA. GTP and the elongation factor EF-Tu, which carries the aminoacyl-tRNA to the ribosome, are required for this insertion. When GTP is bound to EF-Tu, the protein is in its active state and delivers aminoacyl-tRNA to the A site. This is followed by GTP hydrolysis, and the EF-Tu · GDP complex leaves the ribosome. EF-Tu · GDP is converted to EF-Tu · GTP with the aid of a second elongation factor, EF-Ts. Subsequently another aminoacyl-tRNA binds to EF-Tu · GTP (figure 11.48).

Aminoacyl-tRNA binding to the A site initiates the second phase of the elongation cycle, the transpeptidation reaction (figure 11.48 and **figure 11.49**). This is catalyzed by the 23S rRNA ribozyme activity called **peptidyl transferase,** located on the 50S subunit. The α-amino group of the A site amino acid nucleophilically attacks the α-carboxyl group of the C-terminal amino acid on the P site tRNA (figure 11.49). The peptide chain attached to the tRNA in the P site is transferred to the A site as a peptide bond is formed between the chain and the incoming amino acid. No extra energy source is required for peptide bond formation because the bond linking an amino acid to tRNA is high in energy (figure 11.43). Evidence strongly suggests that 23S rRNA contains the peptidyl transferase function, and is therefore a ribozyme. Almost all protein can be removed from the 50S subunit, leaving the 23S rRNA and protein fragments and the remaining complex still has peptidyl transferase activity. The high-resolution structure of the large subunit has been obtained by X-ray crystallography. There is no protein in the active site region. A specific adenine base seems to participate in catalyzing peptide bond formation. Thus the 23S rRNA appears to be the major component of the peptidyl transferase and contributes to both A and P site functions.

The final phase in the elongation cycle is **translocation.** Three things happen simultaneously: (1) the peptidyl-tRNA moves about 20 Å from the A site to the P site; (2) the ribosome moves one codon along mRNA so that a new codon is positioned in the A site; and (3) the empty tRNA leaves the P site. Instead of immediately being ejected from the ribosome when the ribosome moves along the mRNA, the empty tRNA is moved from the P site to the E site and then leaves the ribosome. Ribosomal proteins are involved in these tRNA movements. The intricate process also requires the participation of the EF-G or translocase protein and GTP hydrolysis. The ribosome changes shape as it moves down the mRNA in the 5′ to 3′ direction.

Termination of Protein Synthesis

Protein synthesis stops when the ribosome reaches one of three nonsense codons—UAA, UAG, and UGA (**figure 11.50**). The nonsense (stop) codon is found on the mRNA immediately before the trailer region. Three **release factors** (RF-1, RF-2, and RF-3) aid the ribosome in recognizing these codons. Because there is no cognate tRNA for a nonsense codon, the ribosome stops. The peptidyl transferase hydrolyzes the peptide free from the tRNA in the P site, and the empty tRNA is released. GTP hydrolysis seems to be required during this sequence, although it may not be needed for termination in *Bacteria.* Next the ribosome dissociates from its mRNA and separates into 30S and 50S subunits. IF-3 binds to the 30S subunit to prevent it from reassociating with the 50S subunit until the proper stage in initiation is reached. Thus ribosomal subunits associate during protein synthesis and separate afterward. The termination of eucaryotic protein synthesis is similar except that only one release factor appears to be active.

Protein synthesis is a very expensive process. Three GTP molecules probably are used during each elongation cycle, and two ATP high-energy bonds are required for amino acid activation (ATP is converted to AMP rather than to ADP). Therefore five high-energy bonds are required to add one amino acid to a growing polypeptide chain. GTP also is used in initiation and termination of protein synthesis (figures 11.47 and 11.50). Presumably this large energy expenditure is required to ensure the fidelity of protein synthesis. Very few mistakes can be tolerated.

Although the mechanism of protein synthesis is similar in *Bacteria* and eucaryotes, bacterial ribosomes differ substantially from those in eucaryotes. This explains the effectiveness of many important antibacterial agents. Either the 30S or the 50S subunit may be affected. For example, streptomycin binding to the 30S ribosomal subunit inhibits protein synthesis and causes mRNA misreading. Erythromycin binds to the 50S subunit and inhibits peptide chain elongation. Antibacterial drugs (section 34.4)

Protein Folding and Molecular Chaperones

For many years it was believed that polypeptides spontaneously folded into their final native shape, either as they were synthesized by ribosomes or shortly after completion of protein synthesis. Although the amino acid sequence of a polypeptide does determine its final conformation, it is now clear that special helper proteins aid the newly formed or nascent polypeptide in folding to its proper functional shape. These proteins, called **molecular chaperones** or chaperones, recognize only unfolded polypeptides or partly denatured proteins and do not bind to normal, functional proteins. Their role is essential because the cytoplasmic matrix is filled with nascent polypeptide chains and proteins. Under such conditions it is quite likely that new polypeptide chains often will fold improperly and aggregate to form nonfunctional complexes. Molecular chaperones suppress incorrect folding and may reverse

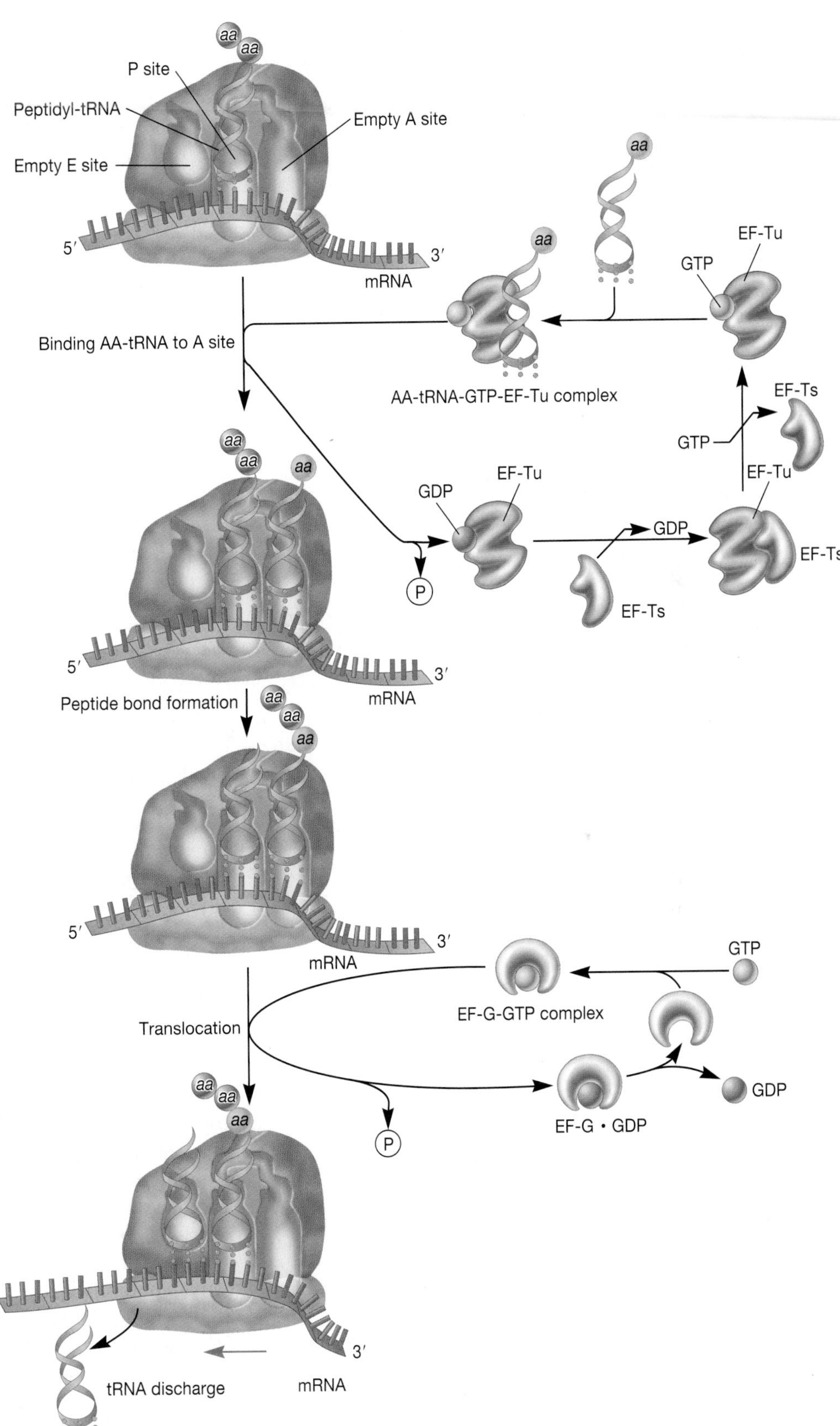

Figure 11.48 Elongation Cycle. The elongation cycle of protein synthesis. The ribosome possesses three sites, a peptidyl or donor site (P site), an aminoacyl or acceptor site (A site), and an exit site (E site). The arrow below the ribosome in the translocation step shows the direction of mRNA movement. See text for details.

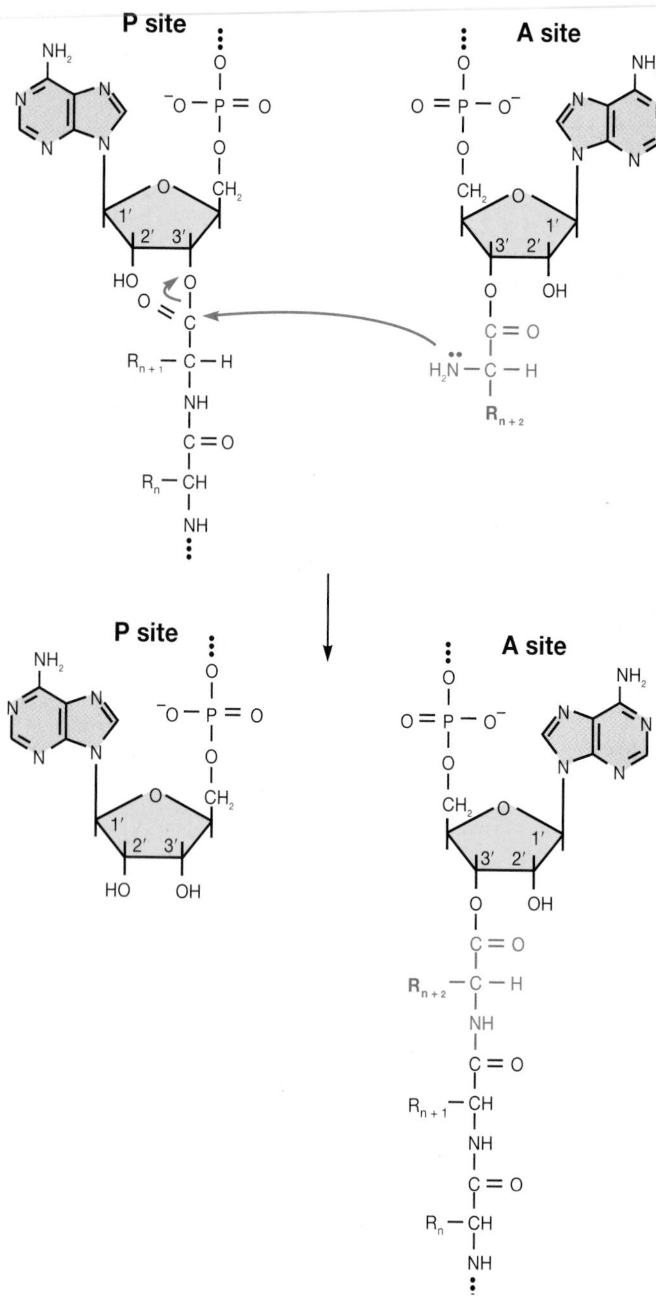

Figure 11.49 Transpeptidation. The peptidyl transferase reaction. The peptide grows by one amino acid and is transferred to the A site.

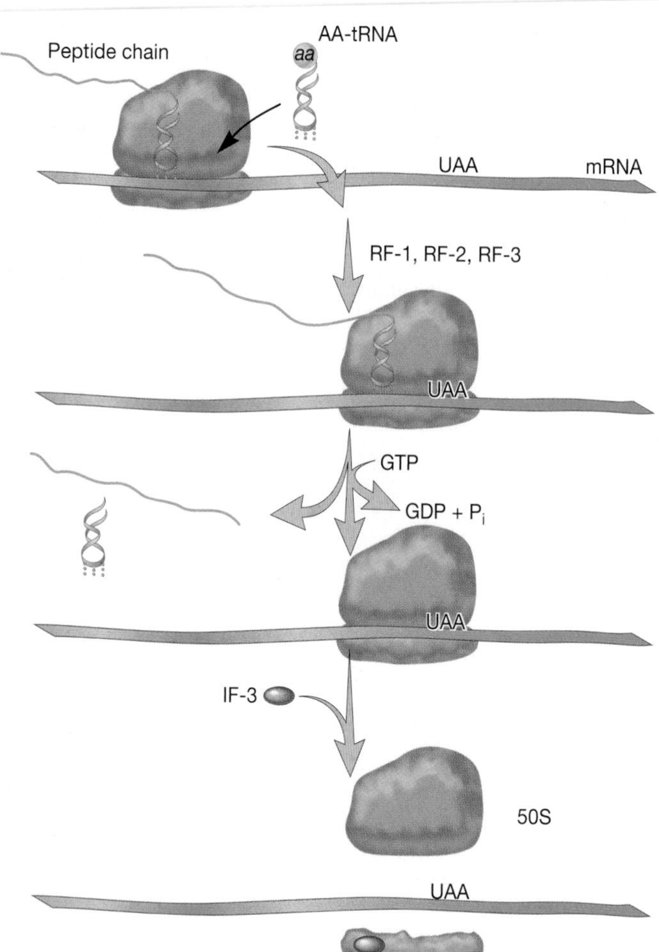

Figure 11.50 Termination of Protein Synthesis in *Bacteria*. Although three different nonsense codons can terminate chain elongation, UAA is most often used for this purpose. Three release factors (RF) assist the ribosome in recognizing nonsense codons and terminating translation. GTP hydrolysis is probably involved in termination. Transfer RNAs are in pink.

any incorrect folding that has already taken place. They are so important that chaperones are present in all cells.

Several chaperones and cooperating proteins aid proper protein folding in *Bacteria*. The process has been well studied in *E. coli* and involves at least four chaperones—DnaK, DnaJ, GroEL, and GroES—and the stress protein GrpE. After a sufficient length of nascent polypeptide extends from the ribosome, DnaJ binds to the unfolded chain (**figure 11.51**). DnaK, which is complexed with ATP, then attaches to the polypeptide. These two chaperones prevent the polypeptide from folding improperly as it is synthesized. When synthesis of the polypeptide is complete, the GrpE protein binds to the chaperone-polypeptide complex and causes DnaK to release ADP. DnaJ may also be released at this step. Then ATP binds to DnaK and DnaK dissociates from the polypeptide. The polypeptide has been folding during this sequence of events and may have reached its final native conformation. If it is still only partially folded, it can bind DnaJ and DnaK again and repeat the process, or be transferred to another set of chaperones, GroEL and GroES, where the final folding takes place. As with DnaK, ATP binding to GroEL and ATP hydrolysis change the chaperone's affinity for the folding polypeptide and regulate polypeptide binding and release (polypeptide release is ATP-dependent). GroES binds to GroEL and assists in its binding and release of the refolding polypeptide.

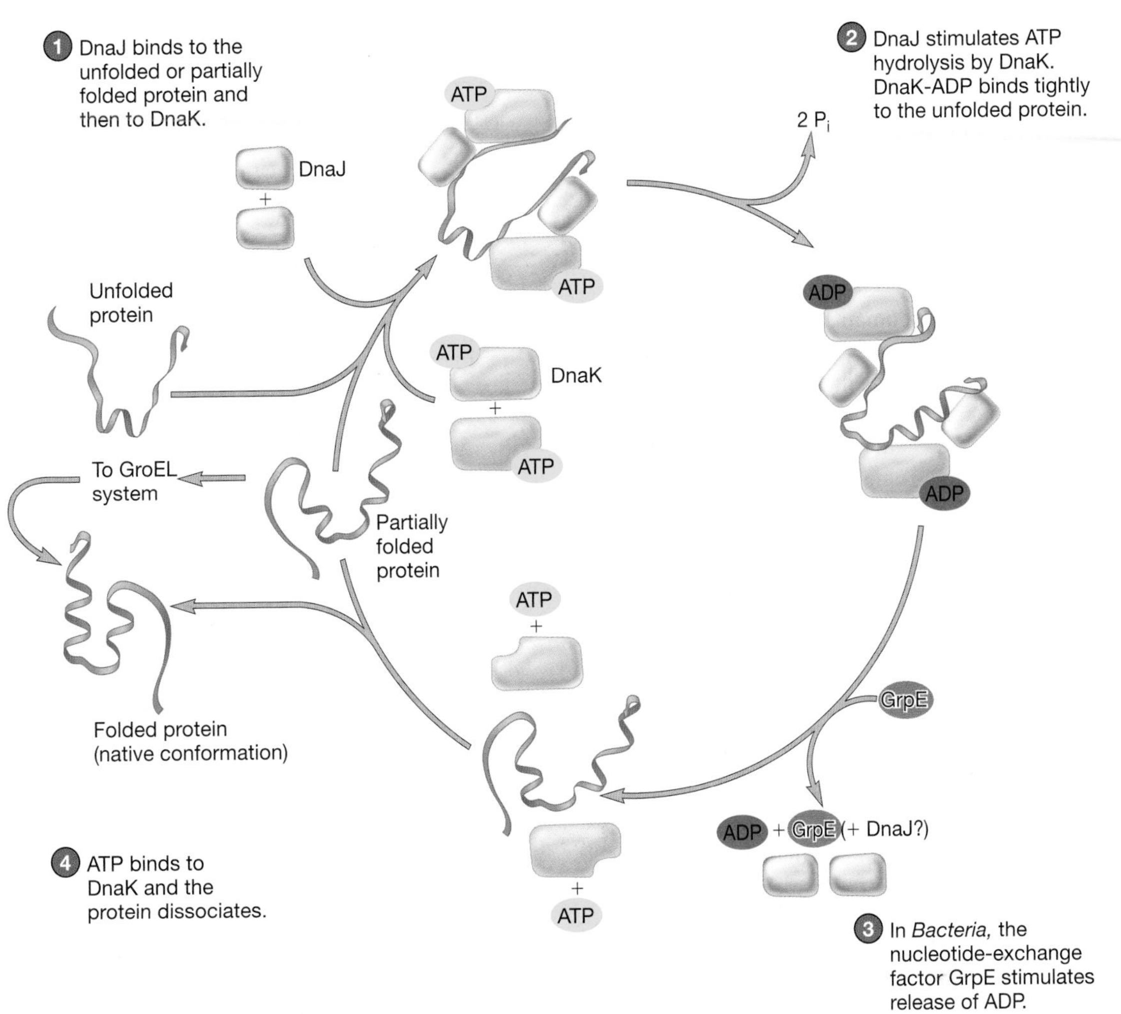
1 DnaJ binds to the unfolded or partially folded protein and then to DnaK.
DnaJ
+
ATP
ATP
2 Pi
2 DnaJ stimulates ATP hydrolysis by DnaK. DnaK-ADP binds tightly to the unfolded protein.
Unfolded protein
ATP
DnaK
+
ATP
ADP
ADP
To GroEL system
Partially folded protein
ATP
+
GrpE
Folded protein (native conformation)
ADP + GrpE (+ DnaJ?)
4 ATP binds to DnaK and the protein dissociates.
+
ATP
3 In Bacteria, the nucleotide-exchange factor GrpE stimulates release of ADP.

Figure 11.51 Chaperones and Polypeptide Folding. The involvement of bacterial chaperones in the proper folding of a newly synthesized polypeptide chain is depicted in this diagram. Three possible outcomes of a chaperone reaction cycle are shown. A native protein may result, the partially folded polypeptide may bind again to DnaK and DnaJ, or the polypeptide may be transferred to GroEL and GroES.

Chaperones were first discovered because they dramatically increase in concentration when cells are exposed to high temperatures, metabolic poisons, and other stressful conditions. Thus many chaperones often are called **heat-shock proteins** or stress proteins. When an *E. coli* culture is switched from 30 to 42°C, the concentrations of some 20 different heat-shock proteins increase greatly within about 5 minutes. If the cells are exposed to a lethal temperature, the heat-shock proteins are still synthesized but most proteins are not. Thus chaperones protect the cell from thermal damage and other stresses as well as promote the proper folding of new polypeptides. For example, DnaK protects *E. coli* RNA polymerase from thermal inactivation in vitro. In addition, DnaK reactivates thermally inactivated RNA polymerase, especially if ATP, DnaJ, and GrpE are present. GroEL and GroES also protect intracellular proteins from aggregation. As one would expect, large quantities of chaperones are present in hyperthermophiles such as *Pyrodictium occultum,* an archaeon that will grow at temperatures as high as 110°C. *P. occultum* has a chaperone similar to the GroEL of *E. coli.* The chaperone hydrolyzes ATP most rapidly at 100°C and makes up almost 3/4 of the cell's soluble protein when *P. occultum* grows at 108°C.

Chaperones have other functions as well. They are particularly important in the transport of proteins across membranes. For example, in *E. coli* the chaperone SecB binds to the partially unfolded forms of many proteins and keeps them in an export-competent state until they are translocated across the plasma membrane. Proteins translocated by the Sec-dependent system are synthesized with an amino-terminal signal sequence. The signal sequence is a short stretch of amino acids that helps direct the completed polypeptide to its final destination. Polypeptides associate with SecB and the chaperone then attaches to the membrane translocase. The polypeptides are transported through the membrane as ATP is hydrolyzed. When they enter the periplasm, a signal peptidase enzyme removes the signal sequence and the protein moves to its final location. DnaK, DnaJ, and GroEL/GroES also can aid in protein translocation across membranes. Protein secretion in procaryotes (section 3.8)

Research indicates that procaryotes and eucaryotes may differ with respect to the timing of protein folding. In terms of conformation, proteins are composed of compact, self-folding, structurally independent regions. These regions, normally around 100 to 300 amino acids in length, are called **domains.** Larger proteins such as immunoglobulins (important proteins in the immune response) may have two or more domains that are linked by less structured portions of the polypeptide chain. In eucaryotes, domains fold independently right after being synthesized by the ribosome. It appears that procaryotic polypeptides, in contrast, do not fold until after the complete chain has been synthesized. Only then do the individual domains fold. This difference in timing may account for the observation that chaperones seem to be more important in the folding of procaryotic proteins. Folding a whole polypeptide is more complex than folding one domain at a time and would require the aid of chaperones. Antibodies (section 32.7)

Protein Splicing

A further level of complexity in the formation of proteins has been discovered. Some microbial proteins are spliced after translation. In **protein splicing,** a part of the polypeptide is removed before the polypeptide folds into its final shape. Self-splicing proteins begin as larger precursor proteins composed of one or more internal intervening sequences called **inteins** flanked by external sequences or **exteins,** the N-exteins and C-exteins (**figure 11.52*a***). Inteins, which are between about 130 and 600 amino acids in length, are removed in an autocatalytic process involving a branched intermediate (figure 11.52*b*). Thus far, more than 130 inteins in 34 types of self-splicing proteins have been discovered. Over 120 inteins have been found in bacteria and archaea. Some examples are an ATPase in the yeast *Saccharomyces cerevisiae,* the RecA protein of *Mycobacterium tuberculosis,* and DNA polymerase in the archaeon *Pyrococcus.* Thus self-splicing proteins are present in all three domains of life.

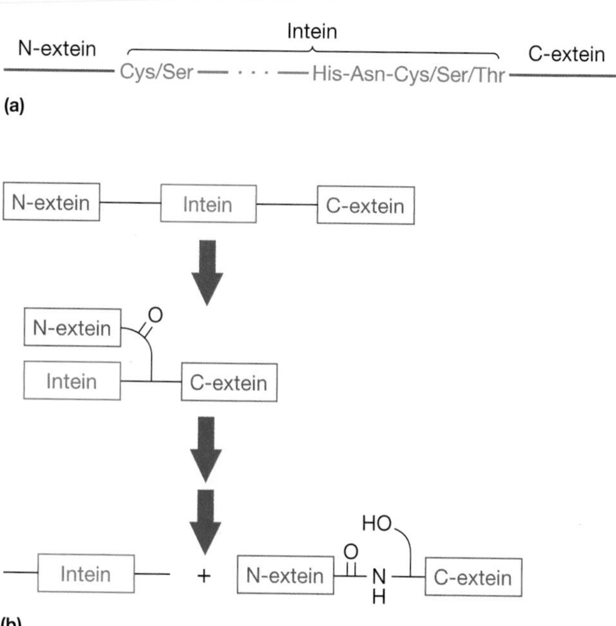

Figure 11.52 Protein Splicing. **(a)** A generalized illustration of intein structure. The amino acids that are commonly present at each end of the inteins are shown. Note that many are thiol or hydroxyl-containing amino acids. **(b)** An overview of the proposed pattern or sequence of splicing. The precise mechanism is not yet known but presumably involves the hydroxyls or thiols located at each end of the intein.

1. In which direction are polypeptides synthesized? What is a polyribosome and why is it useful?
2. Briefly describe the structure of transfer RNA and relate this to its function. How are amino acids activated for protein synthesis, and why is the specificity of the aminoacyl-tRNA synthetase reaction so important?
3. What are the translational and exit domains of the ribosome? Contrast procaryotic and eucaryotic ribosomes in terms of structure. What roles does ribosomal RNA have?
4. Describe the nature and function of the following: fMet-tRNA, initiator codon, IF-3, IF-2, IF-1, elongation cycle, peptidyl and aminoacyl sites, EF-Tu, EF-Ts, transpeptidation reaction, peptidyl transferase, translocation, EF-G or translocase, nonsense codon, and release factors.
5. What are molecular chaperones and heat-shock proteins? Describe their functions.

Summary

11.1 DNA as Genetic Material

a. The knowledge that DNA is the genetic material for cells came from studies on transformation by Griffith and Avery and from experiments on T2 phage reproduction by Hershey and Chase (**figures 11.1–11.3**).

11.2 The Flow of Genetic Information

a. DNA serves as the storage molecule for genetic information. DNA replication is the process by which DNA is duplicated so that it can be passed on to the next generation (**figure 11.4**).

b. During transcription, genetic information in DNA is rewritten as an RNA molecule. The three products of transcription are messenger RNA, ribosomal RNA, and transfer RNA.

c. Translation converts genetic information in the form of a messenger RNA molecule into a polypeptide. Ribosomal RNA and transfer RNA participate in the decoding of genetic information during translation.

11.3 Nucleic Acid Structure

a. DNA differs in composition from RNA in having deoxyribose and thymine rather than ribose and uracil.

b. DNA is double stranded, with complementary AT and GC base pairing between the strands. The strands run antiparallel and are twisted into a right-handed double helix (**figures 11.6**).
c. RNA is normally single stranded, although it can coil upon itself and base pair to form hairpin structures.
d. In almost all procaryotes DNA exists as a closed circle that is twisted into supercoils. In *Bacteria,* the DNA is associated with basic proteins but not with histones.
e. Eucaryotic DNA is associated with five types of histone proteins. Eight histones associate to form ellipsoidal octamers around which the DNA is coiled to produce the nucleosome. The DNA of many archaea is complexed with archaeal histones (**figure 11.9**).

11.4 DNA Replication

a. Most circular procaryotic DNAs are copied by two replication forks moving around the circle to form a theta-shaped (θ) figure (**figure 11.11**). Sometimes a rolling-circle mechanism is employed instead (**figure 11.12**).
b. Eucaryotic DNA has many replicons and replication origins located every 10 to 100 μm along the DNA (**figure 11.13**).
c. The replisome is a huge complex of proteins and is responsible for DNA replication.
d. DNA polymerase enzymes catalyze the synthesis of DNA in the 5′ to 3′ direction while reading the DNA template in the 3′ to 5′ direction.
e. The double helix is unwound by helicases with the aid of topoisomerases like DNA gyrase. DNA binding proteins keep the strands separate.
f. DNA polymerase III holoenzyme synthesizes a complementary DNA copy beginning with a short RNA primer made by the enzyme primase.
g. The leading strand is replicated continuously, whereas DNA synthesis on the lagging strand is discontinuous and forms Okazaki fragments (**figures 11.16** and **11.17**).
h. DNA polymerase I excises the RNA primer and fills in the resulting gap. DNA ligase then joins the fragments together (**figures 11.18** and **11.19**).
i Telomerase is responsible for repeating the ends of eucaryotic chromosomes (**figure 11.20**).

11.5 Gene Structure

a. A gene may be defined as the nucleic acid sequence that codes for a polypeptide, tRNA, or rRNA.
b. The template strand of DNA carries genetic information and directs the synthesis of the RNA transcript.
c. The gene also contains a coding region and a terminator; it may have a leader and a trailer (**figure 11.23**).
d. RNA polymerase binds to the promoter region, which contains RNA polymerase recognition and RNA polymerase binding sites (**figure 11.28**).
e. The genes for tRNA and rRNA often code for a precursor that is subsequently processed to yield several products (**figure 11.24**).

11.6 Transcription

a. RNA polymerase synthesizes RNA that is complementary to the DNA template strand (**figure 11.26**).
b. The sigma factor helps the bacterial RNA polymerase bind to the promoter region at the start of a gene (**figure 11.29**).
c. A terminator marks the end of a gene. A rho factor is needed for RNA polymerase release from some terminators (**figures 11.31** and **11.32**).
d. In eucaryotes, RNA polymerase II synthesizes pre-mRNA, which then undergoes posttranscriptional modification by RNA cleavage and addition of a 3′ poly-A sequence and a 5′ cap to generate mRNA (**figure 11.36**).
e. Many eucaryotic genes are split or interrupted genes that have exons and introns. Exons are joined by RNA splicing. Splicing involves small nuclear RNA molecules, spliceosomes, and sometimes ribozymes (**figure 11.37**).

11.7 The Genetic Code

a. Genetic information is carried in the form of 64 nucleotide triplets called codons (**table 11.4**); sense codons direct amino acid incorporation, and stop or nonsense codons terminate translation.
b. The code is degenerate—that is, there is more than one codon for most amino acids.

11.8 Translation

a. In translation, ribosomes attach to mRNA and synthesize a polypeptide beginning at the N-terminal end. A polysome or polyribosome is a complex of mRNA with several ribosomes (**figure 11.39**).
b. Amino acids are activated for protein synthesis by attachment to the 3′ end of transfer RNAs. Activation requires ATP, and the reaction is catalyzed by aminoacyl-tRNA synthetases (**figure 11.43**).
c. Ribosomes are large, complex organelles composed of rRNAs and many polypeptides. Amino acids are added to a growing peptide chain at the translational domain (**figure 11.45**).
d. Protein synthesis begins with the binding of fMet-tRNA (*Bacteria*) or an initiator methionyl-$tRNA^{Met}$ (eucaryotes and *Archaea*) to an initiator codon on mRNA and to the two ribosomal subunits. This involves the participation of protein initiation factors (**figure 11.47**).
e. In the elongation cycle the proper aminoacyl-tRNA binds to the A site with the aid of EF-Tu and GTP (**figure 11.48**). Then the transpeptidation reaction is catalyzed by peptidyl transferase. Finally, during translocation, the peptidyl-tRNA moves to the P site and the ribosome travels along the mRNA one codon. Translocation requires GTP and EF-G or translocase. The empty tRNA leaves the ribosome by way of the exit site.
f. Protein synthesis stops when a nonsense codon is reached. Bacteria require three release factors for codon recognition and ribosome dissociation from the mRNA (**figure 11.50**).
g. Molecular chaperones help proteins fold properly, protect cells against environmental stresses, and transport proteins across membranes (**figure 11.51**).
h. Procaryotic proteins may not fold until completely synthesized, whereas eucaryotic protein domains fold as they leave the ribosome.
i. Some proteins are self-splicing and excise portions of themselves before folding into their final shape.

Key Terms

heat-shock proteins 287
helicase 260
histone 253
initiation factors 281
initator codon 281
inteins 288
introns 273
lagging strand 260
leader sequence 265
leading strand 260
major groove 252
messenger RNA (mRNA) 251
minor groove 252
molecular chaperones 284
nonsense codon 275
nucleosome 253
Okazaki fragment 260
open complex 269
peptidyl (donor; P) site 284
peptidyl transferase 284
phenotype 248
polyribosome 276
posttranscriptional modification 272
Pribnow box 269
primase 260
primosome 260
promoter 265
proofreading 262
protein splicing 288
reading frame 264
release factors 284
replication 251
replication fork 256
replicon 257
replisome 260
rho factor 270
ribonucleic acid (RNA) 252
ribosomal RNA (rRNA) 269
ribosome binding site (RBS) 281
ribozyme 267
RNA polymerase 269
RNA polymerase holoenzyme 269
RNA splicing 273
rolling-circle replication 257
sense codons 275
Shine-Dalgarno sequence 265
sigma factor 269
single-stranded DNA binding proteins (SSBs) 260
small nuclear RNA (snRNA) 273
split (interrupted) genes 273
spliceosome 274
stop codon 266
telomerase 263
template strand 265
terminator sequence 266
topoisomerase 260
trailer sequence 266
transcription 251
transfer RNA (tRNA) 269
transformation 249
translation 251
translocation 284
transpeptidation reaction 284
wobble 276

Critical Thinking Questions

1. Many scientists say that RNA was the first of the information molecules (i.e., RNA, DNA, protein) to arise during evolution. Given the information in this chapter, what evidence is there to support this hypothesis?
2. *Streptomyces coelicolor* has a linear chromosome. Interestingly, there are no genes that encode essential proteins near the ends of the chromosome in this bacterium. Why do you think this is the case?
3. You have isolated several *E. coli* mutants:

   Mutant #1 has a mutation in the −10 region of the promoter of a structural gene encoding an enzyme needed for synthesis of the amino acid serine.

   Mutant #2 has a mutation in the −35 region in the promoter of the same gene.

   Mutant #3 is a double mutant with mutations in both the −10 and −35 region of the promoter of the same gene.

   Only Mutant #3 is unable to make serine. Why do you think this is so?
4. Suppose that you have isolated a microorganism from a soil sample. Describe how you would go about determining the nature of its genetic material.

Learn More

Bao, K., and Cohen, S. N. 2004. Reverse transcriptase activity innate to DNA polymerase I and DNA topoisomerse I proteins of *Streptomyces* telomere complex. *Proc. Natl. Acad. Sci.* 101(40):14361–66.

Borukhov, S., and Nudler, E. 2003. RNA polymerase holoenzyme: Structure, function and biological implications. *Current Opinion Microbiol.* 6:93–100.

Brooker, R. J. 2005. *Genetics: Analysis and principles,* 2d ed. New York: McGraw-Hill.

Gogarten, J. P.; Senejani, A. G.; Zhaxybayeva, O.; Olendzenski, L.; and Hilario, E. 2002. Inteins: Structure, function, and evolution. *Annu. Rev. Microbiol.* 56:263–87.

Grabowski, B., and Kelman, Z. 2003. Archaeal DNA replication: Eukaryal proteins in a bacterial context. *Annu. Rev. Microbiol.* 57:487–516.

Hartl, F. J., and Hayer-Hartl, M. 2002. Molecular chaperones in the cytosol: From nascent chain to folded protein. *Science* 295:1852–58.

Johnson, A., and O'Donnell, M. 2005. Cellular DNA replicases: Components and dynamics at the replication fork. *Annu. Rev. Biochem.* 74:283–315.

Kelman, L. M., and Kelman, Z. 2004. Multiple origins of replication in archaea. *Trends Microbiol.* 12(9):399–401.

Laursen, B. S.; Søjorensen, H. P.; Mortensen, K. K.; and Sperling-Petersen, H. U. 2005. Initiation of protein synthesis in bacteria. *Microbiol. Mol. Biol. Rev.* 69(1):101–23.

McKee, T., and McKee, J. R. 2003. *Biochemistry: The molecular basis of life,* 3d ed. Dubuque, Iowa: McGraw-Hill.

Narberhaus, F. 2002. Crystallin-type heatshock proteins: Socializing minichaperones in the context of a multichaperone network. *Microbiol. Mol. Biol. Rev.* 66(1):64–93.

Nelson, D. L., and Cox, M. M. 2005. *Lehninger: Principles of biochemistry,* 4th ed. New York: W. H. Freeman.

Neylon, C.; Kralicek, A. V.; Hill, T. M.; and Dixon, N. E. 2005. Replication termination in *Escherichia coli:* Structure and antihelicase activity of the Tus-*Ter* complex. *Microbiol. Mol. Biol. Rev.* 69(3):501–26.

Paul, B. J.; Ross, W.; Gaal, T.; and Gourse, R. L. 2004. rRNA transcription in *Escherichia coli. Annu. Rev. Genet.* 38:749–70.

Reeve, J. N. 2003. Archaeal chromatin and transcription. *Mol. Microbiol.* 48(3):587–98.

Reeve, J. N.; Bailey, K. A.; Li, W-T.; Marc, F.; Sandman, K.; and Soares, D. J. 2004. Archaeal histones: Structures, stability and DNA binding. *Biochem. Soc. Trans.* 32 (part 2):227–30.

Smogorzewska, A., and deLange, T. 2004. Regulation of telomerase by telomeric proteins. *Annu. Rev. Biochem.* 73:177–208.

Yarus, M.; Caporaso, J. G.; and Knight, R. 2005. Origins of the genetic code: The escaped triplet theory. *Annu. Rev. Biochem.* 74:179–98.

Please visit the Prescott website at www.mhhe.com/prescott7 for additional references.

12 Microbial Genetics: Regulation of Gene Expression

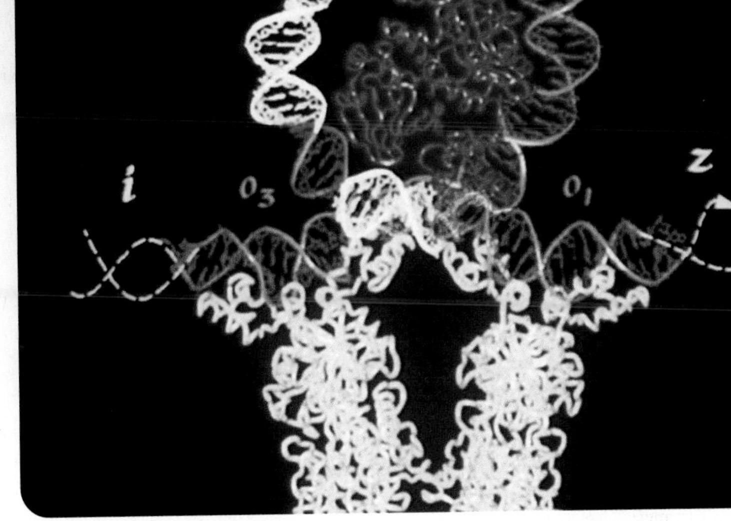

Lactose operon activity is under the control of repressor and activator proteins. The *lac* repressor (pink) and catabolite activator protein (blue) are bound to the *lac* operon. The repressor blocks transcription when bound to the operators (red).

Preview

- The long-term regulation of metabolism, behavior, and morphology is brought about by control of gene expression. This can occur at many levels including transcription initiation, transcription elongation, translation, and posttranslation.
- In *Bacteria,* control at the level of transcription initiation is often achieved by regulatory proteins. Some block transcription and others promote transcription. Furthermore, transcription can be terminated prematurely by a process called attenuation, in which ribosome behavior affects RNA polymerase activity. Translation in *Bacteria* can be regulated by small molecules that bind the leader of mRNA. The conformation changes that result block ribosome binding. Translation can also be blocked by antisense RNA molecules.
- Microorganisms must be able to respond rapidly to changing environmental conditions. Their responses often involve many genes or operons. The simultaneous control of many genes or operons is called global control. Important examples of bacterial global control systems are catabolite repression, quorum sensing, and endospore formation.
- Eucaryotic gene expression involves more steps than does bacterial gene expression; thus there are more points in the process where regulation can occur.
- Although archaeal genome organization is similar to that seen in *Bacteria*, the machinery used by the *Archaea* during information flow is more like that of the *Eucarya*. Thus archaeal regulatory mechanisms may be similar to those observed in the other domains of life.

The gram-positive soil bacterium *Bacillus subtilis* senses that the nutrient levels in its environment are decreasing, and it must determine if it should initiate sporulation. An *Escherichia coli* cell is in an environment rich in carbon and energy sources, and it must determine which to use and when to use them. A pathogen is transmitted from a stream to the intestinal tract of its animal host, and it must adjust to the warmer temperature, increased nutrient supply, and defenses of the host. These are just a few examples of situations to which microbes must respond. To make the most efficient use of the resources in the current environment and their own cellular machinery, microbes must respond to changes by altering physiological and behavioral processes. How is this accomplished?

The control of cellular processes by regulation of the activity of enzymes and other proteins is a fine-tuning mechanism: it acts rapidly to adjust metabolic activity from moment to moment. Microorganisms also are able to control the expression of their genome, although over longer intervals. For example, the *E. coli* chromosome can code for about 4500 polypeptides, yet not all proteins are produced at the same time. Regulation of gene expression serves to conserve energy and raw materials, to maintain balance between the amounts of various cell proteins, and to adapt to long-term environmental change. Thus control of gene expression complements the regulation of enzyme activity. Control of enzyme activity (section 8.10)

The particular field which excites my interest is the division between the living and the non-living, as typified by, say, proteins, viruses, bacteria and the structure of chromosomes. The eventual goal, which is somewhat remote, is the description of these activities in terms of their structure, i.e., the spatial distribution of their constituent atoms, in so far as this may prove possible. This might be called the chemical physics of biology.

—*Francis Crick*

In this chapter we explore the various mechanisms organisms used to regulate gene expression. We begin with a brief discussion of the many levels at which regulation can occur. We then introduce some important examples of the regulation of transcription initiation, transcription elongation, and translation. Finally, we examine how cells use these various regulatory mechanisms to control suites of genes in response to changes in their environments.

12.1 Levels of Regulation of Gene Expression

Figure 12.1 summarizes the expression of bacterial, archaeal, and eucaryotic genes and highlights points in the process where regulation often occurs. Although the overall processes of transcription and translation in the three domains of life are similar, there are differences that affect gene expression. For instance, chromatin structure varies. Bacterial chromosomes lack histones, whereas eucaryotic chromosomes and some archaeal chromosomes are associated with histones. DNA condensed by histones is less accessible to RNA polymerase, and expression of eucaryotic genes involves the additional step of opening up the chromatin to expose promoters. It is also important to remember that in procaryotes, genes of related function are often transcribed from a single promoter, giving rise to a polycistronic mRNA. In addition, transcription and translation are tightly coupled in procaryotes. Eucaryotic mRNA molecules, on the other hand, are monocistronic and are the product of RNA processing, which adds the 5′ cap and poly-A tail and removes introns. Fur-

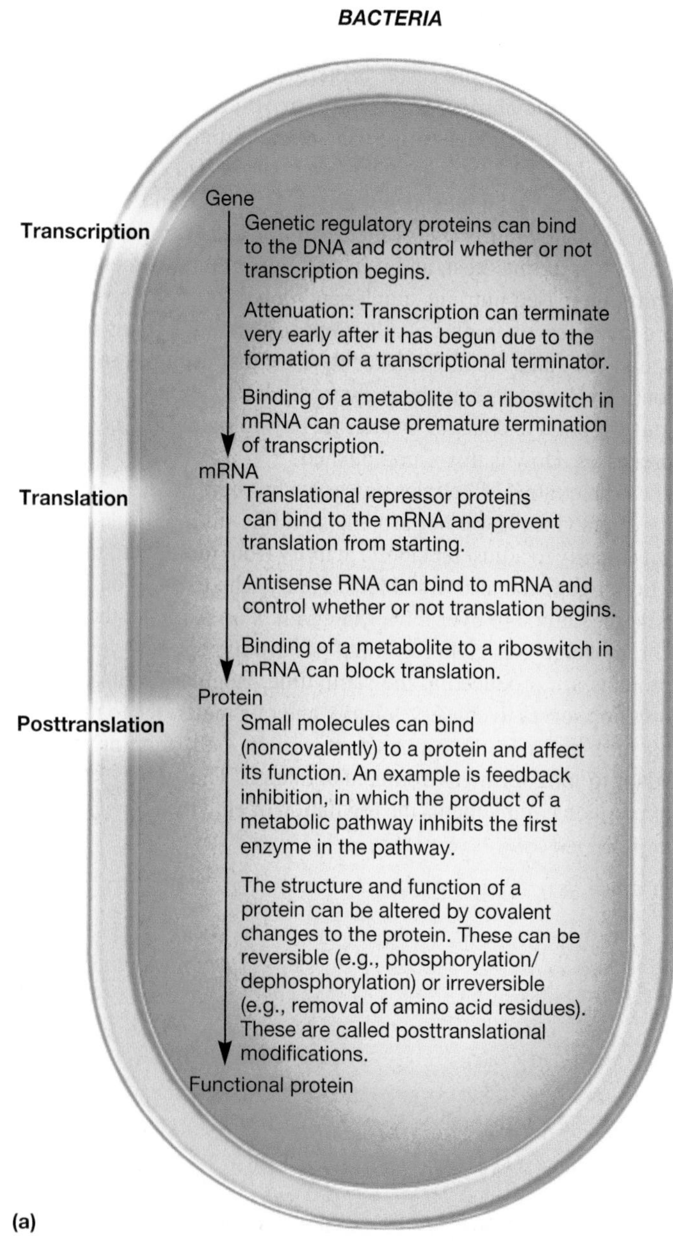

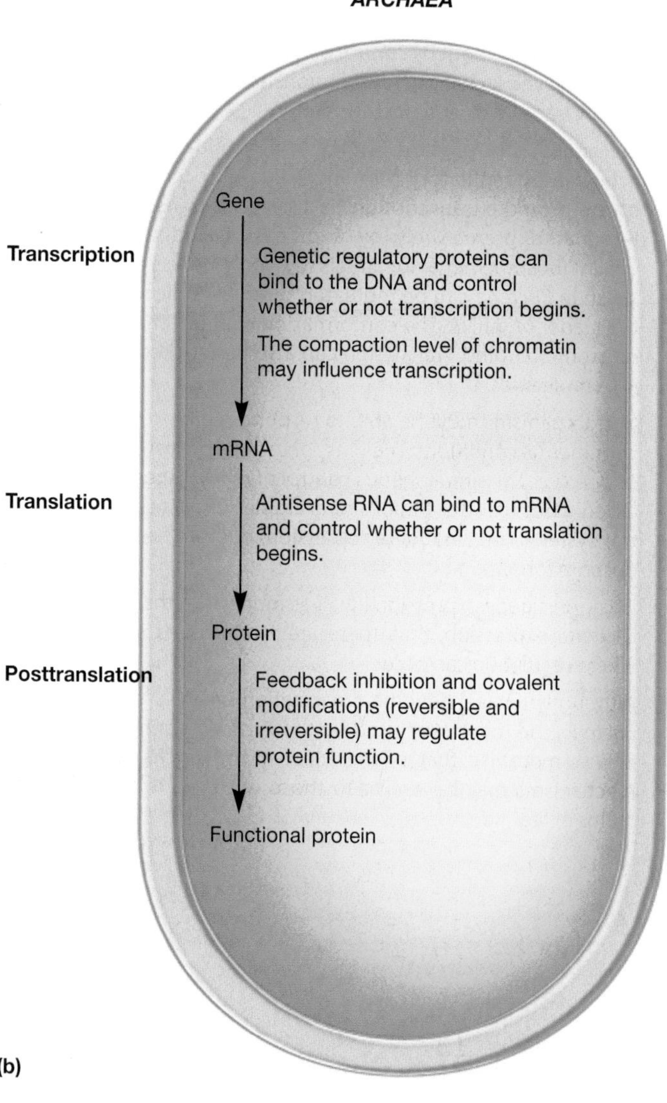

Figure 12.1 Gene Expression and Common Regulatory Mechanisms in the Three Domains of Life.

Figure 12.1 *(Continued).*

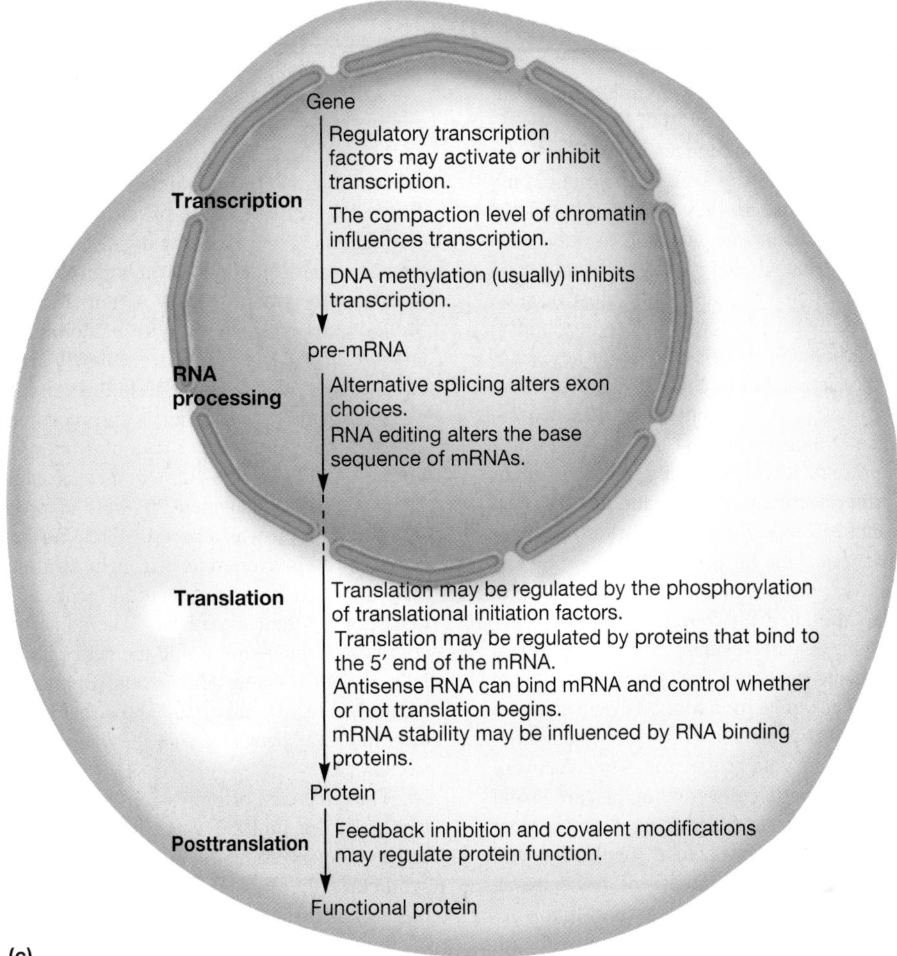
EUCARYA
Gene
Regulatory transcription factors may activate or inhibit transcription.
Transcription
The compaction level of chromatin influences transcription.
DNA methylation (usually) inhibits transcription.
pre-mRNA
RNA processing
Alternative splicing alters exon choices.
RNA editing alters the base sequence of mRNAs.
Translation
Translation may be regulated by the phosphorylation of translational initiation factors.
Translation may be regulated by proteins that bind to the 5′ end of the mRNA.
Antisense RNA can bind mRNA and control whether or not translation begins.
mRNA stability may be influenced by RNA binding proteins.
Protein
Posttranslation
Feedback inhibition and covalent modifications may regulate protein function.
Functional protein
(c)

thermore, although genes are organized in a similar fashion in the *Bacteria* and the *Archaea,* the archaeal enzymes, molecules, and signaling sequences that function in transcription and translation are more like those of the *Eucarya.*

Because of these differences, regulation of gene expression is somewhat different in each domain of life. Our focus in this chapter is on well-understood bacterial regulatory processes. We begin our discussion by introducing two phenomena: induction of enzyme synthesis and repression of enzyme synthesis. Induction and repression provided the first models for gene regulation (**Historical Highlights 12.1**). These early models involved the action of regulatory proteins, and the notion that gene expression is regulated solely by proteins persisted for many years. Eventually it was clearly demonstrated that RNA molecules also can have regulatory functions. Induction and repression also demonstrate the regulation of transcription initiation. Although many regulatory processes occur at this level, there are numerous regulatory mechanisms occurring at other levels. We describe some of these mechanisms as well.

12.2 Regulation of Transcription Initiation

Induction and repression are historically important, as they were the first regulatory processes to be understood in any detail. In this section, we first describe these phenomena and then examine the underlying regulatory events.

Induction and Repression of Enzyme Synthesis

Many enzymes are produced almost all of the time because they catalyze reactions in the cell that are needed routinely. These enzymes include those of the central metabolic pathways. Their functions are often referred to as "housekeeping functions" and the genes that encode them are often referred to as **housekeeping genes.** Those housekeeping genes that are expressed continuously are said to be **constitutive genes.** Many genes, however, are expressed only when needed. The β-galactosidase gene is an example of a regulated gene.

β-galactosidase catalyzes the hydrolysis of the disaccharide sugar lactose to glucose and galactose (**figure 12.2**). When *E. coli*

Historical Highlights

12.1 The Discovery of Gene Regulation

The ability of microorganisms to adapt to their environments by adjusting enzyme levels was first discovered by Emil Duclaux, a colleague of Louis Pasteur. He found that the fungus *Aspergillus niger* would produce the enzyme that hydrolyzes sucrose (invertase) only when grown in the presence of sucrose. In 1900 F. Dienert found that yeast contained the enzymes for galactose metabolism only when grown with lactose or galactose and would lose these enzymes upon transfer to a glucose medium. Such a response made sense because the yeast cells would not need enzymes for galactose metabolism when using glucose as its carbon and energy source. Further examples of adaptation were discovered, and by the 1930s H. Karström divided enzymes into two classes: (1) adaptive enzymes that are formed only in the presence of their substrates, and (2) constitutive enzymes that are always present. It was originally thought that enzymes might be formed from inactive precursors and that the presence of the substrate simply shifted the equilibrium between precursor and enzyme toward enzyme formation.

In 1942 Jacques Monod, working at the Pasteur Institute in Paris, began a study of adaptation in the bacterium *E. coli*. It was already known that the enzyme β-galactosidase, which hydrolyzes the sugar lactose to glucose and galactose, was present only when *E. coli* was grown in the presence of lactose. Monod discovered that nonmetabolizable analogues of β-galactosides, such as thiomethylgalactoside, also could induce enzyme production. This discovery made it possible to study induction in cells growing on carbon and energy sources other than lactose so that the growth rate and inducer concentration would not depend on the lactose supply. He next demonstrated that induction involved the synthesis of new enzyme, not just the conversion of already available precursor. Monod accomplished this by making *E. coli* proteins radioactive with ^{35}S, then transferring the labeled bacteria to nonradioactive medium and adding inducer. The newly formed β-galactosidase was nonradioactive and must have been synthesized after addition of inducer.

A study of the genetics of lactose induction in *E. coli* was begun by Joshua Lederberg a few years after Monod had started his work. Lederberg isolated not only mutants lacking β-galactosidase but also a constitutive mutant in which synthesis of the enzyme proceeded in the absence of an inducer (*lacI*$^-$). During bacterial conjugation, genes from the donor bacterium enter the recipient to temporarily form an organism with two copies of those genes provided by the donor. When Arthur Pardee, François Jacob, and Monod transferred the gene for inducibility to a constitutive recipient not sensitive to inducers, the newly acquired gene made the recipient bacterium sensitive to inducer again. This functional gene was not a part of the recipient's chromosome. Thus the special gene directed the synthesis of a cytoplasmic product that inhibited the formation of β-galactosidase in the absence of the inducer. In 1961 Jacob and Monod named this special product the repressor and suggested that it was a protein. They further proposed that the repressor protein exerted its effects by binding to the operator, a special site next to the structural genes. They provided genetic evidence for their hypothesis. The name operon was given to the complex of the operator and the genes it controlled. Several years later in 1967, Walter Gilbert and Benno Müller-Hill managed to isolate the lac repressor and show that it was indeed a protein and did bind to a specific site in the *lac* operon. Bacterial conjugation (section 13.7)

The existence of repression was discovered by Monod and G. Cohen-Bazire in 1953 when they found that the presence of the amino acid tryptophan would repress the synthesis of tryptophan synthetase, the final enzyme in the pathway for tryptophan biosynthesis. Subsequent research in many laboratories showed that induction and repression were operating by quite similar mechanisms, each involving repressor proteins that bound to operators on the genome. Jacob, Monod, and Lederberg all became Nobel laureates for their work on gene regulation.

grows with lactose as its only carbon source, each cell contains about 3,000 β-galactosidase molecules, but it has less than three molecules in the absence of lactose. The enzyme β-galactosidase is an **inducible enzyme**—that is, its level rises in the presence of a small effector molecule called an **inducer** (in this case the lactose derivative allolactose). Likewise the genes that encode inducible enzymes such as β-galactosidase are referred to as **inducible genes.**

β-galactosidase is an enzyme that functions in a catabolic pathway and many catabolic enzymes are inducible enzymes. The genes for enzymes involved in the biosynthesis of amino acids and other substances, on the other hand, are often called **repressible enzymes.** For instance, an amino acid present in the surroundings may inhibit the formation of enzymes responsible for its biosynthesis. This makes good sense because the microorganism does not need the biosynthetic enzymes for a particular substance if it is already available. Generally, repressible enzymes are necessary for synthesis and always are present unless the end product of their pathway is available. Inducible enzymes, in contrast, are required only when their substrate is available; they are missing in the absence of the inducer.

Although variations in enzyme levels could be due to changes in the rates of enzyme degradation, most enzymes are relatively stable in growing bacteria. Induction and repression result principally from changes in the rate of transcription. When *E. coli* is growing in the absence of lactose, it lacks mRNA molecules coding for the synthesis of β-galactosidase. In the presence of lactose, however, each cell has 35 to 50 β-galactosidase mRNA molecules. The synthesis of mRNA is dramatically influenced by the presence of lactose.

Control of Transcription Initiation by Regulatory Proteins

The action of regulatory proteins is most often responsible for induction and repression. Regulatory proteins can exert either negative or positive control. **Negative transcriptional control** occurs

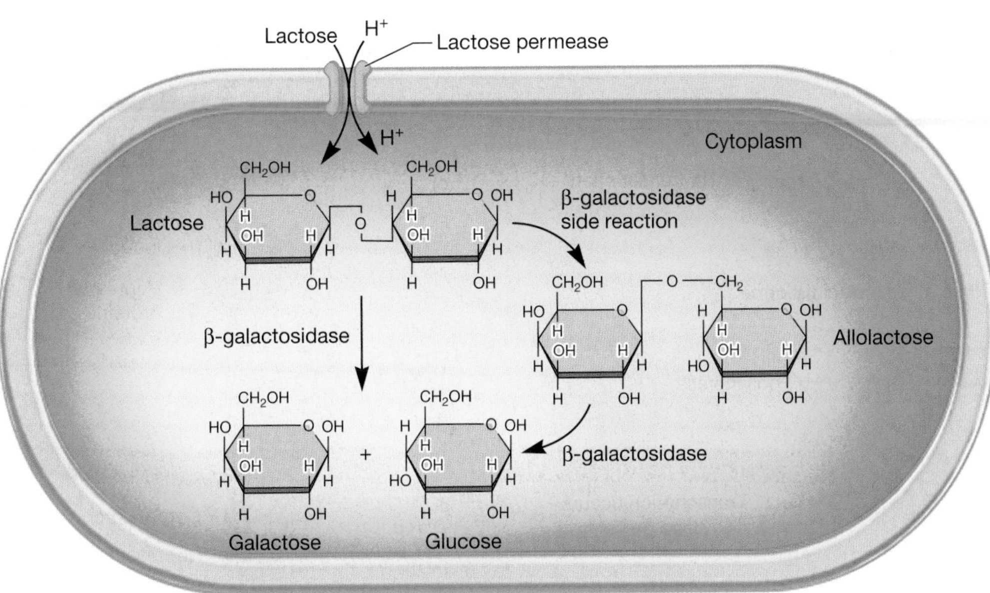

Figure 12.2 The Reactions of β-Galactosidase. The main reaction catalyzed by β-galactosidase is the hydrolysis of lactose, a disaccharide, into the monosaccharides galactose and glucose. The enzyme also catalyzes a minor reaction that converts lactose to allolactose. Allolactose acts as the inducer of β-galactosidase synthesis.

when the protein inhibits initiation of transcription. Regulatory proteins that act in this fashion are called **repressor proteins. Positive transcriptional control** occurs when the protein promotes transcription initiation. These proteins are called **activator proteins.**

Repressor and activator proteins usually act by binding DNA at specific sites. Repressor proteins bind a region called the **operator,** which usually overlaps or is downstream of the promoter (i.e., closer to the coding region) (**figure 12.3*a,b***). When bound, the repressor protein either blocks binding of RNA polymerase to the promoter or prevents its movement. Activator proteins bind **activator-binding sites** (figure 12.3*c,d*). These are often upstream of the promoter (i.e., farther away from the coding region). Binding of an activator to its regulatory site generally promotes RNA polymerase binding.

Repressor and activator proteins must exist in both active and inactive forms if transcription initiation is to be controlled appropriately. The activity of regulatory proteins is modified by small effector molecules, most of which bind the regulatory protein noncovalently. Figure 12.3 shows the four basic ways in which the interactions of an effector and a regulatory protein can affect transcription: (1) For negatively controlled inducible genes (e.g., those encoding enzymes needed for catabolism of a sugar), the repressor protein is active and prevents transcription when the substrate of the pathway is not available (figure 12.3*a*). It is inactivated by binding of the inducer (e.g., the substrate of the pathway). (2) For negatively controlled repressible genes (e.g., those encoding enzymes needed for the synthesis of an amino acid), the repressor protein is initially synthesized in an inactive form called the **aporepressor.** It is activated by binding of the **corepressor** (figure 12.3*b*). For repressible enzymes that function in a biosynthetic pathway, the corepressor is often the product of the pathway (e.g., an amino acid). (3) The activator of a positively regulated inducible gene is activated by the inducer (figure 12.3*c*); whereas (4) the activator of a positively regulated repressible gene is inactivated by an inhibitor (figure 12.3*d*). Control of enzyme activity: Allosteric regulation (section 8.10)

Recall that functionally related bacterial and archaeal genes are often transcribed from a single promoter. The **structural genes**—the genes coding for polypeptides—are simply lined up together on the DNA, and a single, polycistronic mRNA carries all the messages. The sequence of bases coding for one or more polypeptides, together with the promoter and operator or activator-binding sites, is called an **operon.** Many operons have been discovered and studied. Three well studied operons are discussed next. They demonstrate different ways that regulatory proteins can be used to control gene expression at the level of transcription initiation.

The Lactose Operon: Negative Transcriptional Control of Inducible Genes

The best-studied negative control system is the lactose (*lac*) operon of *E. coli.* The *lac* operon contains three structural genes controlled by the *lac* repressor, which is encoded by *lacI* (**figure 12.4**). One gene codes for β-galactosidase; a second gene directs the synthesis of β-galactoside permease, the protein responsible for lactose uptake. The third gene codes for the enzyme β-galactoside transacetylase, whose function still is uncertain. The presence of the first two genes in the same operon ensures that the rates of lactose uptake and breakdown will vary together.

Before we describe the regulation of the *lac* operon, we must consider two general aspects of regulation. The first is that gene expression is rarely an all-or-nothing phenomenon; it is a continuum.

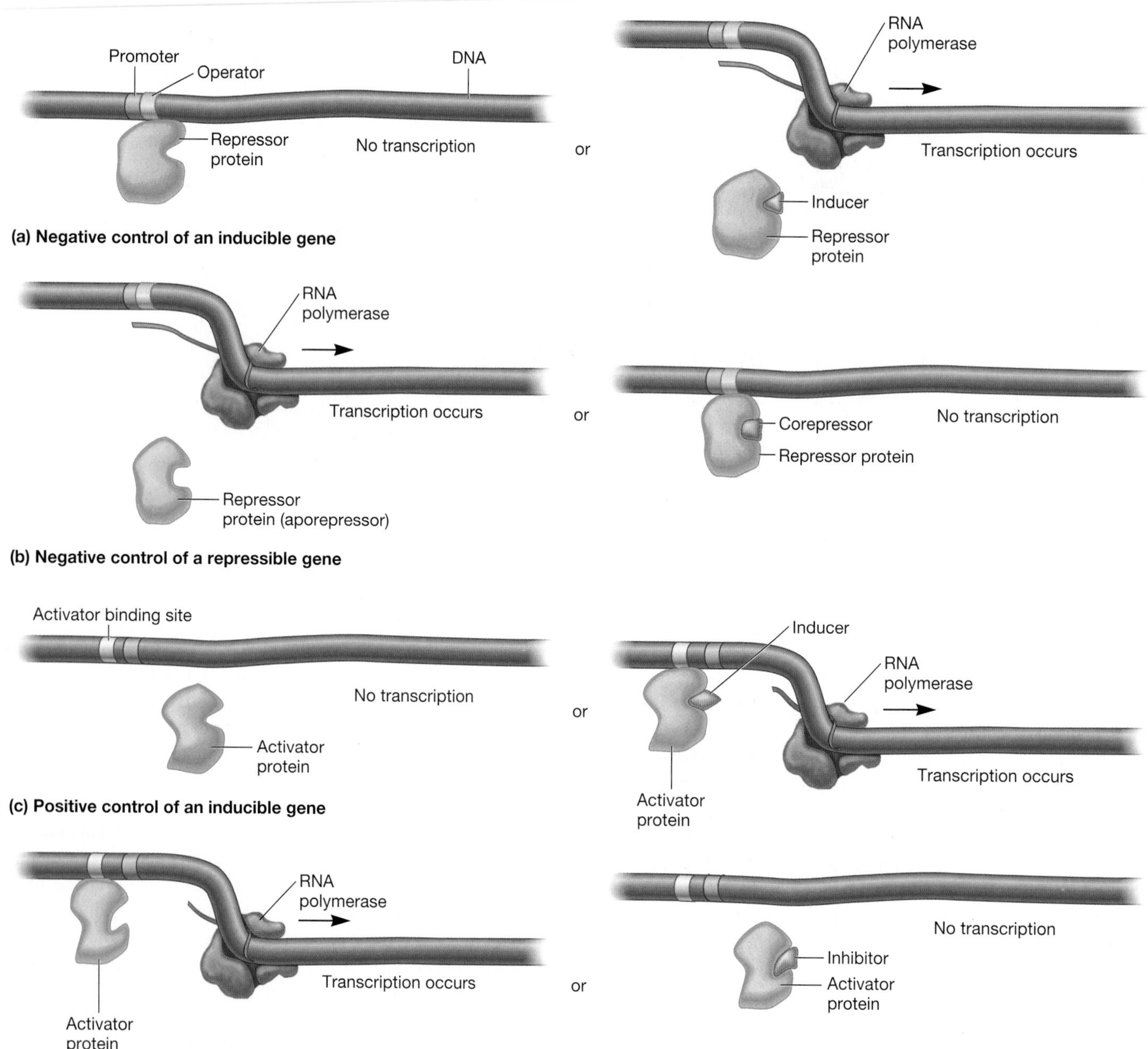

Figure 12.3 Action of Bacterial Regulatory Proteins. Bacterial regulatory proteins have two binding sites—one for a small effector molecule and one for DNA. The binding of the effector molecule changes the regulatory protein's ability to bind DNA. **(a)** In the absence of inducer, the repressor protein blocks transcription. The presence of inducer prevents the repressor from binding DNA and transcription occurs. **(b)** In the absence of a corepressor, the repressor in unable to bind DNA and transcription occurs. When the corepressor is bound to the repressor, the repressor is able to bind DNA and transcription is blocked. **(c)** The activator protein is able to bind DNA and activate transcription only when it is bound to the inducer. **(d)** The activator binds DNA and promotes transcription unless the inhibitor is present. When inhibitor is present, the activator undergoes a conformational change that prevents it from binding DNA; this inhibits transcription.

Inhibition of transcription usually does not mean that genes are "turned off" (though this terminology is frequently used). Rather it means the level of mRNA synthesis is decreased significantly, and in most cases is occurring at very low levels. In other words, many promoters of regulated genes and operons are considered "leaky," in that there is always some low, basal level of transcription. The second aspect of regulation to be considered is the "decision-making" process used by microbial cells. Consider the regulatory decisions made by an *E. coli* cell. It need only synthesize the enzymes of a specific catabolic pathway if the substrate of the pathway is

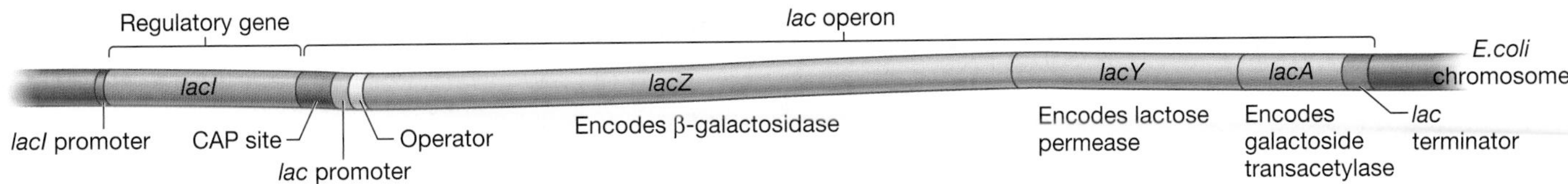

Figure 12.4 The *lac* Operon. The *lac* operon consists of three genes: *lacZ*, *lacY*, and *lacA*, which are transcribed as a single unit from the *lac* promoter. The operon is regulated both negatively and positively. Negative control is brought about by the *lac* repressor, which is the product of the *lacI* gene. The operator is the site of *lac* repressor binding. Positive control results from the action of CAP. CAP binds the CAP site located just upstream from the *lac* promoter. CAP is, in part, responsible for a phenomenon called catabolite repression, an example of a global control network, in which numerous operons are controlled by a single protein.

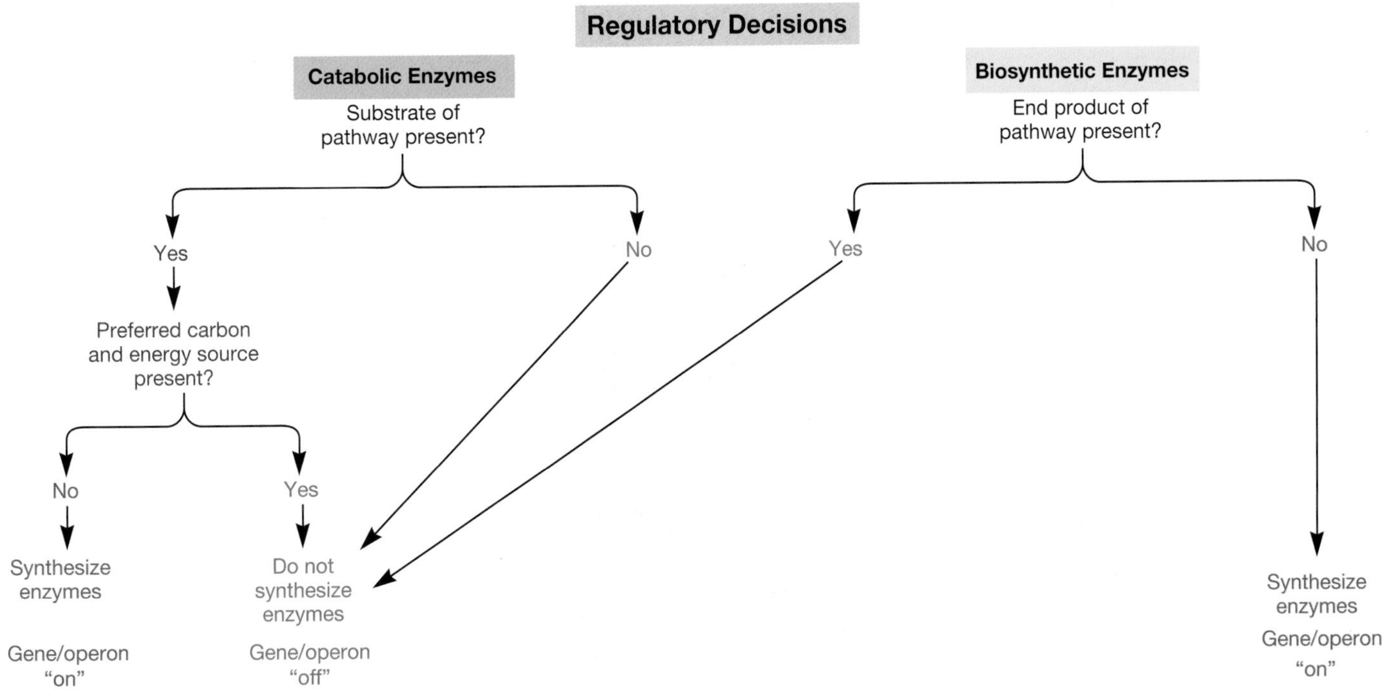

Figure 12.5 Examples of Regulatory Decisions Made by Cells.

present in the environment and a preferred carbon source (e.g., glucose) is not (**figure 12.5**). Conversely, synthesis of the enzymes involved in biosynthetic pathways is inhibited when the end product of the pathway is present.

How do these two aspects of regulation affect expression of the *lac* operon? Lactose is one of many organic molecules *E. coli* can use as a carbon and energy source. It is wasteful to synthesize enzymes of the *lac* operon when lactose is not available. Therefore the cell only expresses this operon at high levels when lactose is the only carbon and energy source present in the environment; the *lac* repressor is responsible for inhibiting transcription when there is no lactose.

The *lac* repressor is a tetramer composed of four identical subunits. The tetramer is formed when two dimers interact. When lactose catabolism is not required, each dimer recognizes and tightly binds one of three different *lac* operator sites: O_1, O_2, and O_3 (**figure 12.6*a***). O_1 is the main operator site and must be bound by the repressor if transcription is to be inhibited. When one dimer is at O_1 and another is at one of the two other operator sites, the dimers bring the two operator sites close together, with a loop of DNA forming between them. The binding of *lac* repressor is a two-step process. First, the repressor binds nonspecifically to DNA. Then it rapidly slides along the DNA until it reaches an operator site. A portion of the repressor fits into the major groove of operator-site DNA (figure 12.6*b*). Thus the shape of the repressor is ideally suited for specific binding to the DNA double helix.

How does the repressor inhibit transcription? The promoter to which RNA polymerase binds is located near the *lac* operator sites. When there is no lactose, the repressor binds O_1 and one of the other operator sites, bending the DNA in the promoter region. This prevents initiation of transcription either because RNA polymerase cannot access the promoter or because it is blocked from moving into the coding region (**figure 12.7*a***). When lactose is available, it is taken up by the lactose permease. Once inside the

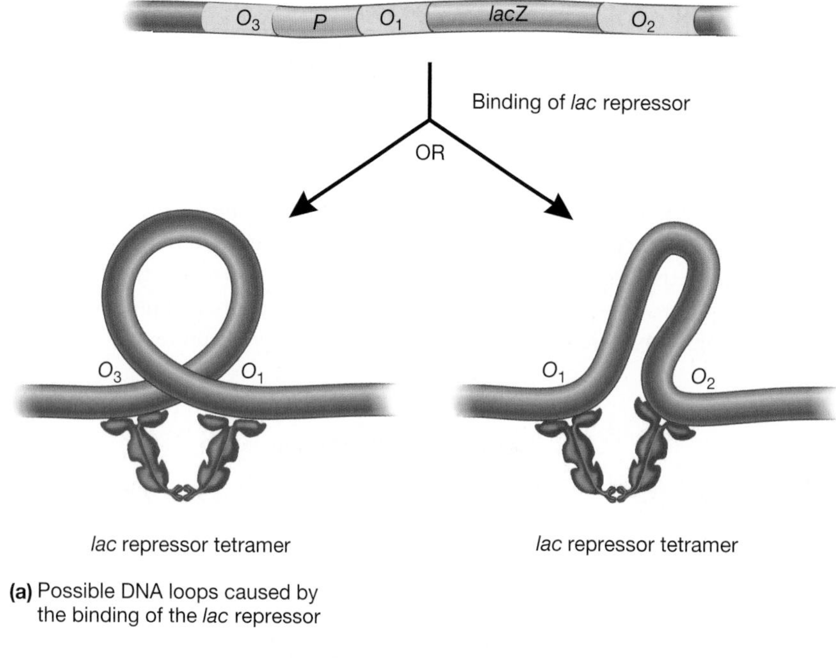

(a) Possible DNA loops caused by the binding of the *lac* repressor

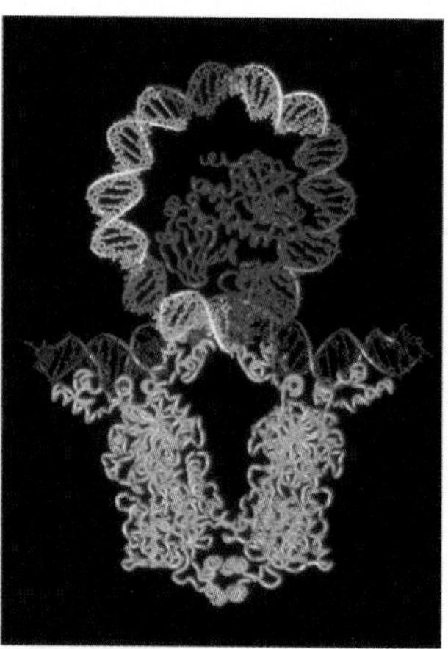

(b) Proposed model of the *lac* repressor binding to O_1 and O_3 based on crystallography studies

Figure 12.6 The *lac* Operator Sites. The *lac* operon has three operator sites: O_1, O_2, and O_3 **(a)**. O_1 is the same operator shown in figure 12.4. As shown in **(a)** and **(b)**, the *lac* repressor (violet) binds O_1 and one of the other operator sites (red) to block transcription, forming a DNA loop. The DNA loop contains the −35 and −10 binding sites (green) recognized by RNA polymerase. Thus these sites are inaccessible and transcription is blocked. The DNA loop also contains the CAP binding site and CAP (blue) is shown bound to the DNA **(b)**. When the *lac* repressor is bound to the operator, CAP is unable to activate transcription.

cell, β-galactosidase converts lactose to allolactose, the inducer of the operon (figure 12.2). This occurs because, as noted previously, there is always a low level of permease and β-galactosidase synthesis. Allolactose binds to the *lac* repressor and causes the repressor to change to an inactive shape that is unable to bind any operator sites. The inactivated repressor leaves the DNA and transcription occurs (figure 12.7*b*).

Close examination of figures 12.4 and 12.6 clearly shows that the regulation of the *lac* operon is not as simple as has just been described. That is because the *lac* operon is regulated by a second regulatory protein called CAP. CAP functions in a global regulatory network that allows *E. coli* to use glucose preferentially over all other carbon and energy sources by a mechanism called catabolite repression. The use of two different regulatory proteins to control the synthesis of an operon illustrates a point that is important in the discussion of the regulation of gene expression—there are often layers of regulation of any operon. As described in section 12.5, the use of two

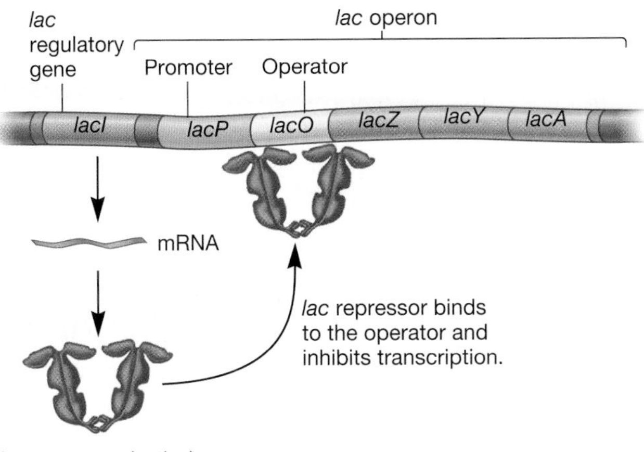

(a) No lactose in the environment

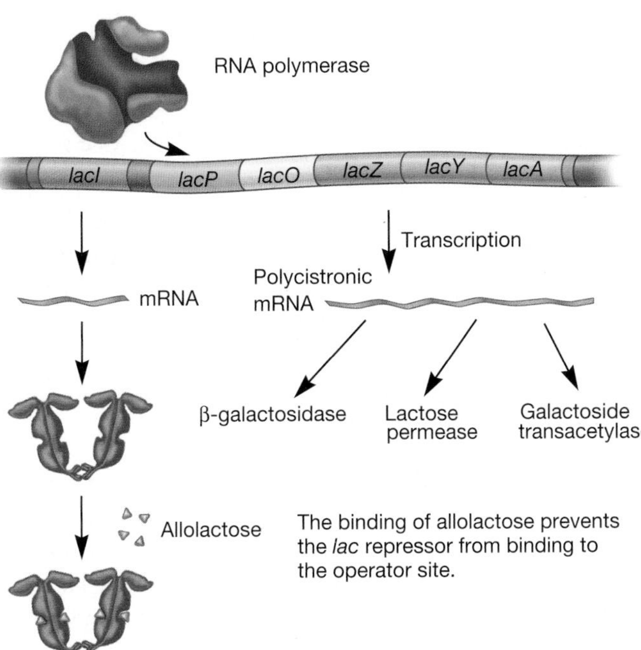

(b) Lactose present

Figure 12.7 Regulation of the *lac* Operon by the *lac* Repressor. **(a)** The *lac* repressor is active and can bind the operator as long as the inducer of the operon, allolactose, is not present. Binding of the repressor to the operator inhibits transcription of the operon by RNA polymerase. **(b)** When lactose is available, some of it is converted to allolactose by β-galactosidase. When sufficient amounts of allolactose are present, it binds and inactivates the *lac* repressor. The repressor leaves the operator and RNA polymerase is free to initiate transcription.

regulatory proteins generates a continuum of expression levels. The highest levels of transcription occur when lactose is available and glucose is not; the lowest levels occur when lactose is not available and glucose is.

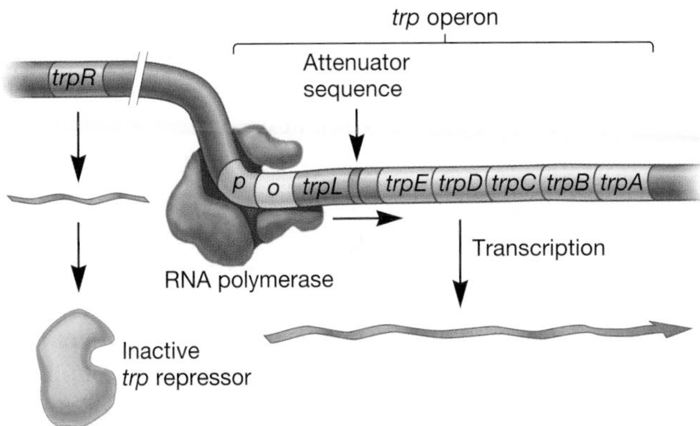

(a) Low trytophan levels, transcription of the entire *trp* operon occurs

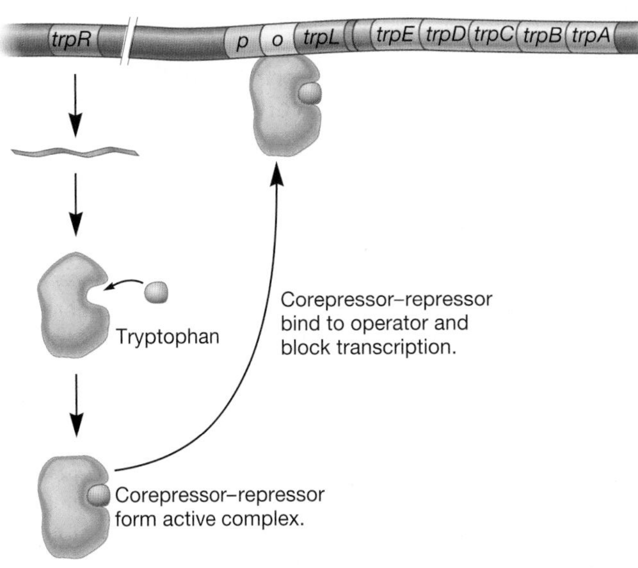

(b) High tryptophan levels, repression occurs

Figure 12.8 Regulation of the *trp* Operon by Tryptophan and the *trp* Repressor. The *trp* repressor is inactive when first synthesized and therefore is unable to bind the operator. It is activated by the binding of tryptophan, which serves as the corepressor. **(a)** When tryptophan levels are low, the repressor is inactive and transcription occurs. The enzymes encoded by the operon catalyze the reactions needed for tryptophan biosynthesis. **(b)** When tryptophan levels are sufficiently high, it binds the repressor. The repressor-corepressor complex binds the operator and transcription of the operon is inhibited.

The Tryptophan Operon: Negative Transcriptional Control of Repressible Genes

The tryptophan (*trp*) operon of *E. coli* consists of five structural genes that encode enzymes needed for synthesis of the amino acid tryptophan (**figure 12.8**). It is regulated by the *trp* repressor, which is encoded by the *trpR* gene. Because the enzymes encoded by the *trp* operon function in a biosynthetic pathway, it is wasteful to make the enzymes needed for tryptophan synthesis when tryptophan is readily available. Therefore, the operon functions only when tryptophan is not present and must be made *de novo* from precursor molecules (figure 12.5). To accomplish this regulatory goal, the *trp* repressor is synthesized in an inactive form that cannot bind the *trp* operator as long as tryptophan levels are low (figure 12.8*a*). When tryptophan levels increase, tryptophan acts as a corepressor, binding the repressor and activating it. The repressor-corepressor complex then binds the operator, blocking transcription initiation (figure 12.8*b*).

Like the *lac* operon, the *trp* operon is subject to another layer of regulation. In addition to being controlled at the level of transcription initiation by the *trp* repressor, expression of the *trp* operon is also controlled at the level of transcription elongation by a process called attenuation. This mode of regulation is discussed in section 12.3.

The Arabinose Operon: Transcriptional Control by a Protein that Acts Both Positively and Negatively

Many regulatory proteins are versatile and can function as repressors for one operon and activators for others. The regulation of the *E. coli* arabinose (*ara*) operon illustrates how the same protein can function either positively or negatively depending on the environmental conditions. The *ara* operon encodes enzymes needed for the catabolism of arabinose to xylulose 5-phosphate, an intermediate of the pentose phosphate pathway. The *ara* operon is regulated by AraC, which can bind three different regulatory sequences: $araO_2$, $araO_1$, and *araI* (**figure 12.9**). When arabinose is not present, one molecule of AraC binds *araI*, and another binds $araO_2$. The two AraC proteins interact, causing the DNA to bend. This prevents RNA polymerase from binding to the promoter of the *ara* operon, thereby blocking transcription. In these conditions, AraC acts as a repressor (figure 12.9*a*). However, when arabinose is present, it binds AraC and prevents AraC molecules from interacting. This breaks the DNA loop. Furthermore, binding of two AraC-arabinose complexes to the *araI* site promotes transcription. Thus when arabinose is present, AraC acts as an activator (figure 12.9b). The *ara* operon, like the *lac* operon, is also subject to catabolite repression (*see section 12.5*). The breakdown of glucose to pyruvate: The pentose phosphate pathway (section 9.3)

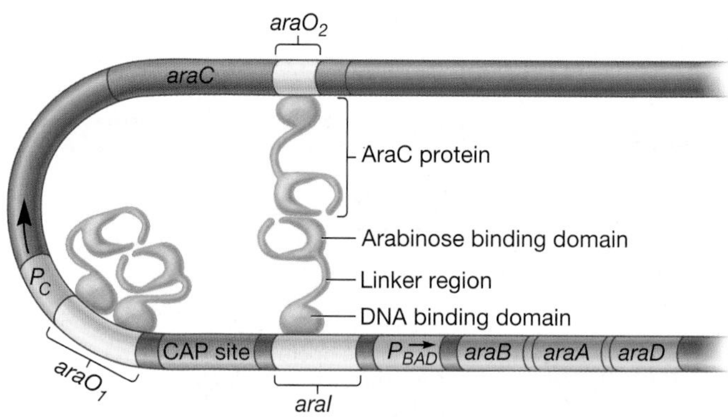

(a) Operon inhibited in the absence of arabinose

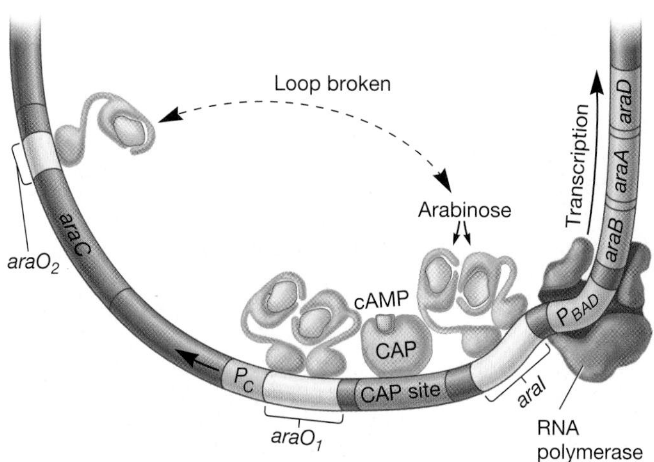

(b) Operon activated in the presence of arabinose

Figure 12.9 Regulation of the *ara* Operon by the AraC Protein. The AraC protein can act both as a repressor and as an activator, depending on the presence or absence of arabinose. **(a)** When arabinose is not available, the protein acts as a repressor. Two AraC proteins are involved. One binds the *araI* site and the other binds the $araO_2$ site. The two proteins interact in such a way that the DNA between the two operator sites is bent, making it inaccessible to RNA polymerase. **(b)** When arabinose is present, it binds AraC, disrupting the interaction between the two AraC proteins. Subsequently, two AraC proteins, each bound to arabinose, form a dimer, which binds to the *araI* site. The AraC dimer functions as an activator and transcription occurs.

Two-Component Regulatory Systems and Phosphorelay Systems

The activity levels of the *lac* repressor, *trp* repressor, and AraC protein are controlled by metabolites of those pathways. However, many environmental conditions do not produce a metabolite that can interact directly with a regulatory protein. These include temperature, osmolarity, and oxygen levels. How do organisms sense and respond to such stimuli? Many genes and operons are turned on or switched off in response to these types of signals by regulatory proteins that are part of a **two-component signal transduction system.** These systems link events occurring outside the cell to the regulation of gene expression. Some of the best-studied signal transduction systems are found in multicellular eucaryotes. However, important signal transduction systems have been identified in procaryotes. They serve as models for understanding the more complex systems of eucaryotes as well as the mechanisms by which many pathogens regulate genes encoding virulence factors. Some of these procaryotic signal transduction systems are the focus of our discussion here.

Two-component signal transduction systems are found in both the *Archaea* and the *Bacteria* and are named after the two proteins that govern the regulatory pathway. The first is a **sensor**

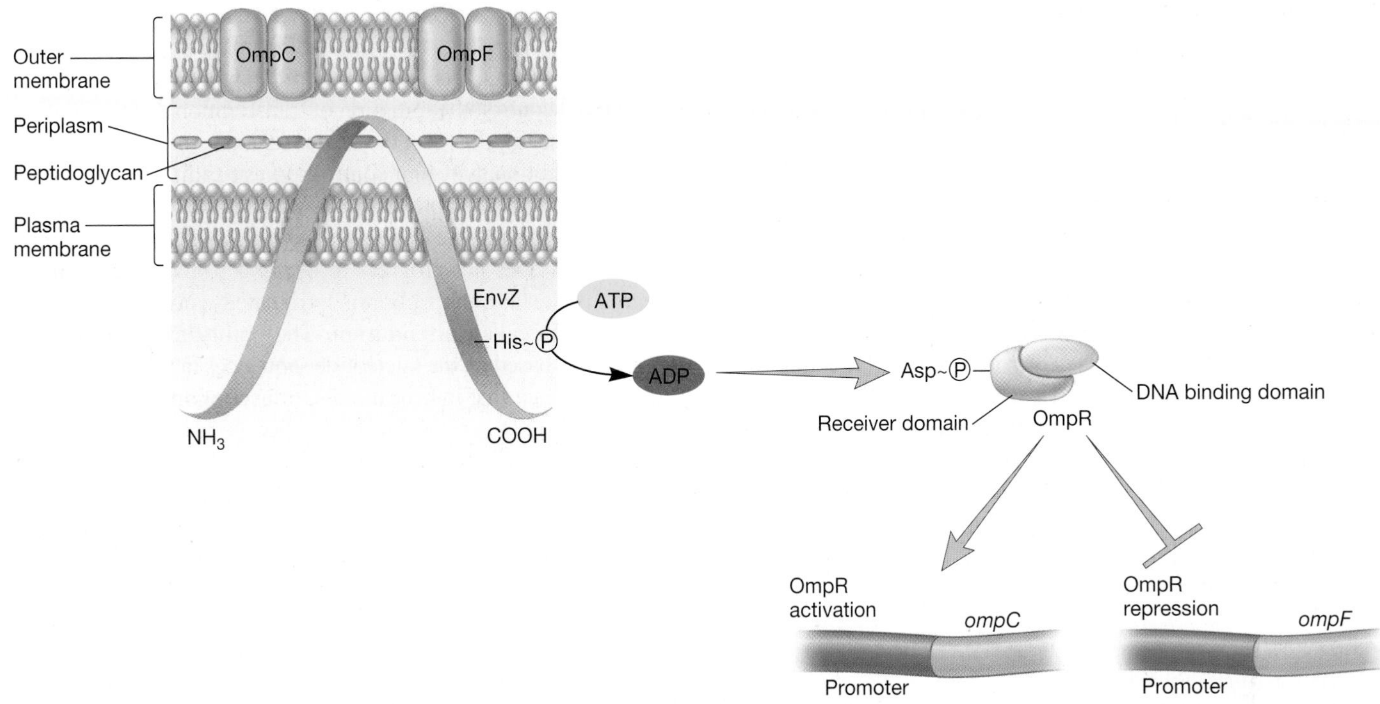

Figure 12.10 Two Component Signal Transduction System and the Regulation of Porin Proteins. In this system, the sensor kinase protein EnvZ loops through the cytoplasmic membrane so that both its C- and N-termini are in the cytoplasm. When EnvZ senses an increase in osmolarity, it autophosphorylates a histidine residue at its C-terminus. EnvZ then passes the phosphoryl group to the response regulator OmpR, which accepts it on an aspartic acid residue located in its N-terminus. This activates OmpR so that it is able to bind DNA and repress *ompF* expression and enhance that of *ompC*.

kinase protein that spans the cytoplasmic membrane so that part of it is exposed to the extracellular environment (periplasm, in gram-negative bacteria) while another part is exposed to the cytoplasm (**figure 12.10**). In this way, it can sense specific changes in the environment and communicate information to the cell's interior. The second component is the **response-regulator protein,** a DNA-binding protein that, when activated by the sensor kinase, promotes transcription of genes or operons whose expression is needed for adaptation to the detected environmental stimulus. The response-regulator protein may also inhibit transcription of genes or operons that are not needed under the current environmental conditions.

The regulation of the ratio of OmpF:OmpC porin proteins in *E. coli* is one of the best-understood two-component signal transduction systems (figure 12.10). Recall that the outer membrane of gram-negative bacteria contains channels made of porin proteins. The two most important porins in *E. coli* are OmpF and OmpC (Omp for *o*uter *m*embrane *p*rotein). OmpC pores are slightly smaller and are made when the bacterium grows at high osmotic pressures. It is the dominant porin when *E. coli* is in the intestinal tract. The larger OmpF pores are favored when *E. coli* grows in a dilute environment; OmpF allows solutes to diffuse into the cell more readily. The cell must maintain a constant level of porin protein in the membrane, but the relative levels of the two porins change to correspond with the osmolarity of the medium. Clearly, the cell must have a way of sensing increases in osmolarity so that *ompF* expression is repressed and *ompC* transcription is enhanced.

The sensor kinase in the OmpF:OmpC two-component regulatory system is the EnvZ protein (env for cell *env*elope). It is an integral membrane protein anchored to the membrane by two membrane-spanning domains. EnvZ is looped through the membrane such that a central domain protrudes into the periplasm, while the amino and carboxyl termini are exposed to the cytoplasm. The second component, OmpR, is the response-regulator protein. It is a soluble, cytoplasmic protein that regulates the transcription of the *ompF* and *ompC* structural genes. The N-terminal end of OmpR is called the receiver domain because it possesses a specific aspartic acid residue that accepts the signal (a phosphoryl group) from the sensor kinase. Upon receipt of the signal, the C-terminal end of OmpR is able to regulate transcription by binding DNA. At low osmolarity, EnvZ is inactive, but when EnvZ senses that osmolarity has increased, EnvZ phosphorylates itself (autophosphorylation) on a specific histidine residue. This phosphoryl group is quickly transferred to the N-terminus of OmpR. Once OmpR is phosphorylated, it is able to regulate transcription of the porin genes so that *ompF* transcription is repressed and *ompC* transcription is activated.

Two-component signal transduction systems are simple in design: the signal recognized by the sensor kinase is directly transduced (sent) to the response regulator that mediates the required changes in gene expression; in many cases numerous genes and operons may be regulated by the same response regulator. Thus two-component systems often function in global regulatory networks. The effectiveness of two-component systems is illustrated by their abundance: most procaryotic cells use a variety of two-component signal transduction systems to respond to an array of environmental stresses. For example, the morphologically complex genus *Streptomyces* has over 50 such systems!

Two-component signal transduction systems involve a simple phosphorelay where the sensor kinase transfers its phosphoryl group directly to the response-regulator protein. However, there are instances when more proteins participate in the transfer of phosphoryl groups. These longer pathways are called **phosphorelay systems.** An important and well-studied phosphorelay system functions during sporulation in *Bacillus subtilis* and is described in section 12.5. It should be noted that some phosphorelay systems control protein activity rather than gene transcription. An example of this type of system is chemotaxis in *E. coli,* which is described in chapter 8.

1. Many genes and operons are regulated at the level of transcription initiation. Why do you think this is the case?
2. What are induction and repression? How do bacteria use them to respond to changing nutrient supplies?
3. Define negative control and positive control of transcription initiation. Describe how regulatory proteins function in these regulatory mechanisms. Define repressor protein, activator protein, operator, activator-binding site, inducer, corepressor, structural gene, and operon.
4. Using figure 12.5 as a guide, trace the "decision-making" pathway of an *E. coli* cell that is growing in a medium containing arabinose but lacking tryptophan.
5. Describe a two-component signal transduction system. How does it differ from a phosphorelay system? How are the *lac* repressor, the *trp* repressor, and the AraC protein similar to the response regulators of two-component and phosphorelay systems? How are they different?

12.3 Regulation of Transcription Elongation

Organisms can also regulate transcription by controlling the termination of transcription. In this type of regulation, transcription is initiated but prematurely stopped depending on the environmental conditions and the needs of the organism. The first demonstration of this level of regulation, called attenuation, occurred in the 1970s in studies of the *trp* operon. More recently, riboswitches have been discovered. These regulatory sequences in the leader of an mRNA both sense and respond to environmental conditions by either prematurely terminating transcription or blocking translation. Both attenuation and riboswitches are described in this section.

Attenuation

As noted earlier, the tryptophan (*trp*) operon of *E. coli* is under the control of a repressor protein, and excess tryptophan inhibits transcription of operon genes by acting as a corepressor and activating the repressor protein. Although the operon is regulated mainly by repression, the continuation of transcription also is controlled. That is, there are two decision points involved in transcriptional control, the initiation of transcription and the continuation of transcription past the leader region.

This additional level of control serves to adjust levels of transcription in a more subtle fashion. When the repressor is not active, RNA polymerase begins transcription of the leader region but it often does not progress to the first structural gene in the operon. Instead, transcription is terminated within the leader region; this is called **attenuation.** The ability to attenuate transcription is based on the nucleotide sequences in the leader region and on the fact that in procaryotes, transcription is coupled with translation (*see figure 11.39*). The leader of the *trp* operon mRNA is unusual in that it is translated. The product, which has never been isolated, is called the leader peptide. In addition to encoding the leader peptide, the leader contains **attenuator** sequences (**figure 12.11**). When transcribed, these sequences form stem-loop secondary structures in the newly formed mRNA. We define these sequences numerically (regions 1, 2, 3, and 4). When regions 1 and 2 pair with one another (1:2; figure 12.11*a*), they form a secondary structure called the pause loop, which causes RNA polymerase to slow down. The pause loop forms just prior to the formation of a second structure called the terminator loop, which is made when regions 3 and 4 base pair (3:4; figure 12.11*a*). A poly(U) sequence follows the 3:4 terminator loop, just as it does in other rho-independent transcriptional terminators (*see figure 11.31*). However, in this case, the terminator is in the leader rather than at the end of the gene. Another stem-loop structure can be formed in the leader region by the pairing of regions 2 and 3 (2:3, figure 12.11*b*). The formation of this antiterminator loop prevents the generation of both the 1:2 pause and 3:4 terminator loops.

How do these various loops control transcription termination? Three scenarios describe the process. In the first, translation is not coupled to transcription because protein synthesis is not occuring. In other words, no ribosome is associated with the mRNA. In this scenario, the pause and terminator loops form, stopping transcription before RNA polymerase reaches the *trpE* gene (figure 12.11*a*).

In the next two scenarios, translation and transcription are coupled; that is, a ribosome associates with the leader mRNA as the rest of the mRNA is being synthesized. The interaction between RNA polymerase and the nearest ribosome determines which stem-loop structures are formed. As a ribosome translates the mRNA, it will follow the RNA polymerase. Among the first several nucleotides of region 1 are two tryptophan (trp) codons; this is unusual because normally there is only one trp per 100 amino acids in *E. coli* proteins. If tryptophan levels are low, the ribosome will stall when it encounters the two trp codons. It stalls because the paucity of charged $tRNA^{trp}$ molecules delays the filling of the A site of the ribosome (figure 12.11*b*). Meanwhile the RNA polymerase continues to transcribe mRNA, moving away from the stalled ribosome. The presence of the ribosome on region 1 will prevent it from base pairing with region 2. As RNA polymerase continues, region 3 is transcribed, enabling the formation of the 2:3 antiterminator loop. This prevents the formation of the

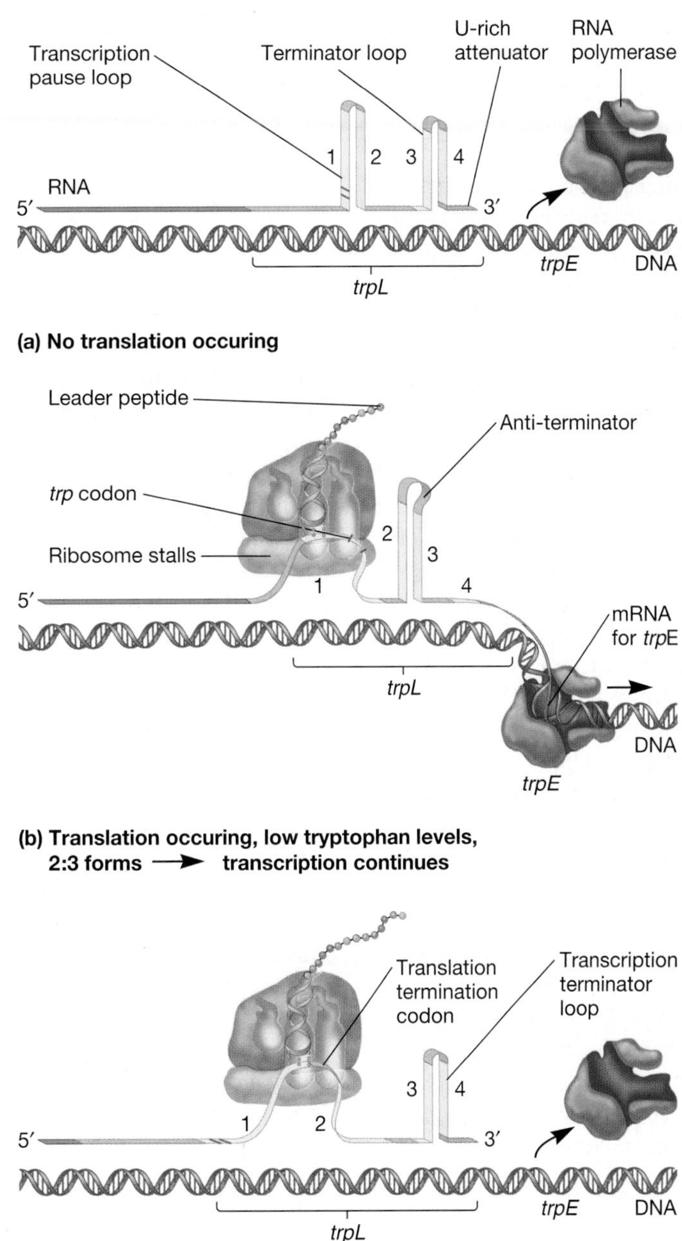

Figure 12.11 Attenuation of the *trp* Operon. **(a)** When protein synthesis has slowed, transcription and translation are not tightly coupled. Under these conditions, the most stable form of the mRNA occurs when region 1 hydrogen bonds to region 2 (RNA polymerase pause loop) and region 3 hydrogen bonds to region 4 (transcription terminator or attenuator loop). The formation of the transcription terminator causes transcription to stop just beyond *trpL* (*trp* leader). **(b)** When protein synthesis is occurring, transcription and translation are coupled, and the behavior of the ribosome on *trpL* influences transcription. If tryptophan levels are low, the ribosome pauses at the *trp* codons in *trpL* because of insufficient amounts of charged $tRNA^{trp}$. This blocks region 1 of the mRNA, so that region 2 can hydrogen bond only with region 3. Because region 3 is already hydrogen bonded to region 2, the 3:4 terminator loop cannot form. Transcription proceeds and the *trp* biosynthetic enzymes are made. **(c)** If tryptophan levels are high, translation of *trpL* progresses to the stop codon, blocking region 2. Regions 3 and 4 can hydrogen bond and transcription terminates.

3:4 terminator loop. Because the terminator loop is not formed, RNA polymerase is not ejected from the DNA and transcription continues into the *trp* biosynthetic genes. If, on the other hand, there is plenty of tryptophan in the cell, there will be an abundance of charged $tRNA^{trp}$ and the ribosome will translate these two trp codons in the leader peptide sequence without hesitation. Thus the ribosome remains close to the RNA polymerase. As RNA polymerase and the ribosome continue through the leader region, regions 1 and 2 are transcribed and readily form a pause loop. Then regions 3 and 4 are transcribed, the terminator loop forms, and RNA polymerase is ejected from the DNA template. Finally, the presence of a UGA stop codon between regions 1 and 2 will cause early termination of translation (figure 12.11*c*). Although the leader peptide will be synthesized, it appears to be rapidly degraded. The genetic code (section 11.7)

Attenuation's usefulness is apparent. If the bacterium is deficient in an amino acid other than tryptophan, protein synthesis will slow and tryptophanyl-tRNA will accumulate. Transcription of the tryptophan operon will be inhibited by attenuation. When the bacterium begins to synthesize protein rapidly, tryptophan may be scarce and the concentration of tryptophanyl-tRNA may be low. This would reduce attenuation activity and stimulate operon transcription, resulting in larger quantities of the tryptophan biosynthetic enzymes. Acting together, repression and attenuation can coordinate the rate of synthesis of amino acid biosynthetic enzymes with the availability of amino acid end products and with the overall rate of protein synthesis. When tryptophan is present at high concentrations, any RNA polymerases not blocked by the activated repressor protein probably will not get past the attenuator sequence. Repression decreases transcription about seventyfold and attenuation slows it another eight- to tenfold; when both mechanisms operate together, transcription can be slowed about 600-fold.

Attenuation is important in regulating at least five other operons that include amino acid biosynthetic enzymes. In all cases, the leader peptide sequences resemble the tryptophan system in organization. For example, the leader peptide sequence of the histidine operon codes for seven histidines in a row and is followed by an attenuator that is a terminator sequence.

Riboswitches

Regulation by riboswitches, or sensory RNAs, is a specialized form of transcription attenuation that does not involve ribosome behavior. In this case, the leader region of the mRNA can fold in different ways. If folded one way, transcription continues; if folded another, transcription is terminated. This leader region is called a **riboswitch** because it turns transcription on or off. What makes riboswitches unique and exciting is that they alter their folding pattern in direct response to the binding of an effector molecule—a capability previously thought to be associated only with proteins.

One of the first discoveries of this type of regulation was the riboflavin (*rib*) biosynthetic operon of *B. subtilis*. The synthesis of riboflavin biosynthetic enzymes is repressed by flavin mononucleotide (FMN), which is derived from riboflavin. When transcription of the *rib* operon begins, sequences in the leader region of the mRNA fold into a structure called the RFN-element. This element binds FMN, and in doing so alters the folding of the leader region, creating a terminator that stops transcription (**figure 12.12**).

It now appears that controlling transcription attenuation with sensory RNAs is an important method used by gram-positive bacteria to regulate amino acid-related genes. As in the case of the *rib* operon, the leader regions of these mRNAs contain a regulatory element. In this case, the region is called the T box. T box sequences give rise to competing terminator and antiterminator loops. The development of either a terminator or an antiterminator is determined by the binding of uncharged tRNA corresponding to the relevant amino acid (i.e., tRNA that is not carrying its cognate amino acid). For instance, expression of a tyrosyl-tRNA synthetase gene (i.e., a gene that encodes the enzyme that links tyrosine to a tRNA molecule) is governed by the presence of $tRNA^{Tyr}$. When the level

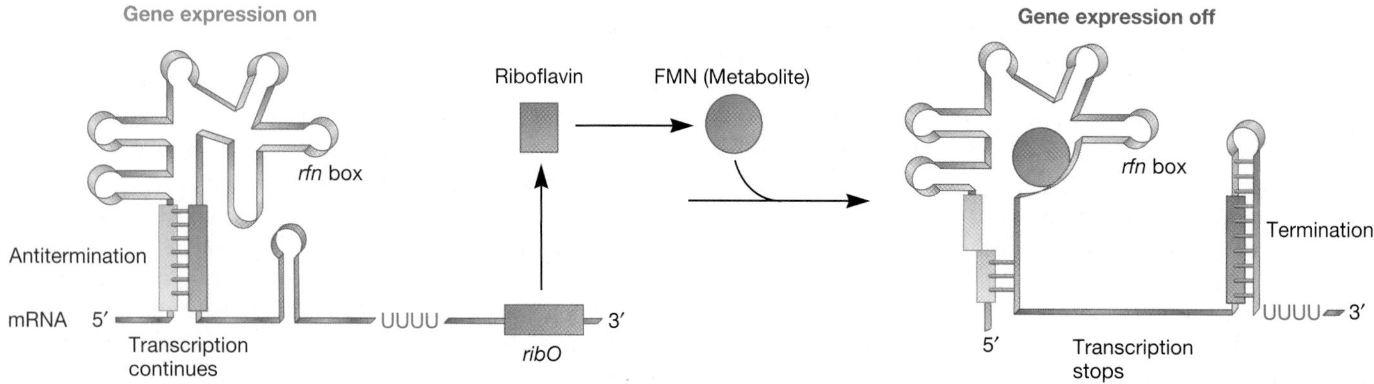

Figure 12.12 Riboswitch Control of the Riboflavin (*rib*) Operon of *Bacillus subtilis*. The *rib* operon produces enzymes needed for the synthesis of riboflavin, a component of flavin mononucleotide (FMN). Binding of FMN to the rfn (*rifampin*) box in the leader of *rib* mRNA causes a change in mRNA folding, which results in the formation of a transcription terminator and cessation of transcription.

Table 12.1 Regulation of Gene Expression by Riboswitches

System	Microbe (s)	Target genes encode:	Effector & Regulatory Response
T box	Many gram-positive bacteria	Amino acid biosynthetic enzymes	Uncharged tRNA; anticodon base pairs to 5′ end of mRNA, preventing formation of transcriptional terminator
Vitamin B_{12} element	*E. coli*	Cobalamine biosynthetic enzymes	Adenosylcobalamine (AdoCbl) binds to *btuB* mRNA and blocks translation
THI box	*Rhizobium etli* *E. coli* *B. subtilis*	Thiamin (Vitamin B_1) biosynthetic and transport proteins	Thiamin pyrophosphate (TPP) causes either premature transcriptional termination (*R. etli, B. subtilis*) or blocks ribosome binding (*E. coli*)
RFN-element	*B. subtilis*	Riboflavin biosynthetic enzymes	Flavin mononucleotide (FMN) cases premature transcriptional termination
S box	Low G + C gram-positive bacteria	Methionine biosynthetic enzymes	S-adenosylmethionine (SAM) causes premature transcriptional termination

of charged $tRNA^{Tyr}$ falls, the anticodon of an uncharged tRNA binds directly to the "specifier sequence" codon in the leader of the mRNA. At the same time, the antiterminator loop is stabilized by base pairing between sequences in the loop and the acceptor end of the tRNA, which normally binds the amino acid. This prevents formation of the terminator structure, and transcription of the tyrosyl-tRNA synthetase gene continues. Genomic analysis now suggests that the T box mechanism may be involved in regulating over 300 genes and/or operons; some other genes that bear sensory RNA in their leader regions are listed in **table 12.1.** Other riboswitches have been shown to function at the level of translation. They are described in the next section.

12.4 Regulation at the Level of Translation

It appears that in general, the riboswitches found in gram-positive bacteria function by transcriptional termination, while the riboswitches discovered in gram-negative bacteria regulate the translation of mRNA. Translation is usually regulated by blocking its initiation. As noted previously, some riboswitches work at this level. In addition, some small RNA molecules can control translation initiation. Both are described in this section.

Regulation of Translation by Riboswitches

Similar to the riboswitches described earlier, riboswitches that function at the translational level contain effector-binding elements at the 5′ end of the mRNA. Binding of the effector molecule alters the folding pattern of the leader region of the mRNA, which often results in occlusion of the Shine-Dalgarno sequence and other elements of the ribosome-binding site. This inhibits ribosome binding and initiation of translation (**figure 12.13**). An example of this type of regulation is observed for the thiamine biosynthetic operons of numerous bacteria and some archaea. The leader regions of thiamine operons contain a structure called the THI-element, which can bind thiamin pyrophosphate. Binding of thiamin pyrophosphate to the THI-element causes a conformational change in the leader region that sequesters the Shine-Dalgarno sequence and blocks translation initiation.

Regulation of Translation by Small RNA Molecules

A large number of RNA molecules have been discovered that do not function as mRNAs, tRNAs, or rRNAs. Microbiologists often refer to them as **small RNAs (sRNAs)** or as noncoding RNAs (ncRNAs). In *E. coli,* there are more than 40 sRNAs, ranging in size from around 40 to 400 nucleotides. It is thought that eucaryotes may have hundreds to thousands of sRNAs with lengths from 21 to over 10,000 nucleotides. Although some sRNAs have been implicated in the regulation of DNA replication and transcription, many function at the level of translation.

In *E. coli,* most sRNAs regulate translation by base pairing to the leader region of a target mRNA. Thus they are complementary to the mRNA and are called **antisense RNAs.** It seems intuitive that by binding to the leader, antisense RNAs would block ribosome binding and inhibit translation. Indeed, many antisense RNAs work in this manner. However, some antisense RNAs actually promote translation upon binding to the mRNA. Whether inhibitory or activating, most *E. coli* antisense RNAs work with a protein called Hfq to regulate their target RNAs. The Hfq protein is an RNA chaperone—that is, it interacts with RNA to promote changes in its structure. In addition, the Hfq protein may promote RNA-RNA interactions.

The regulation of synthesis of OmpF and OmpC porin proteins provides an example of translation control by an antisense RNA. In addition to regulation by the OmpR protein described previously (figure 12.10), expression of the *ompF* gene is regulated by an antisense RNA called MicF RNA, which is the product of the *micF* gene (mic for *m*RNA-*i*nterfering *c*omplementary RNA). The MicF RNA is complementary to *ompF* at the translation initiation site (**figure 12.14**). It base pairs with *ompF* mRNA and represses translation. MicF RNA is produced under conditions such as high

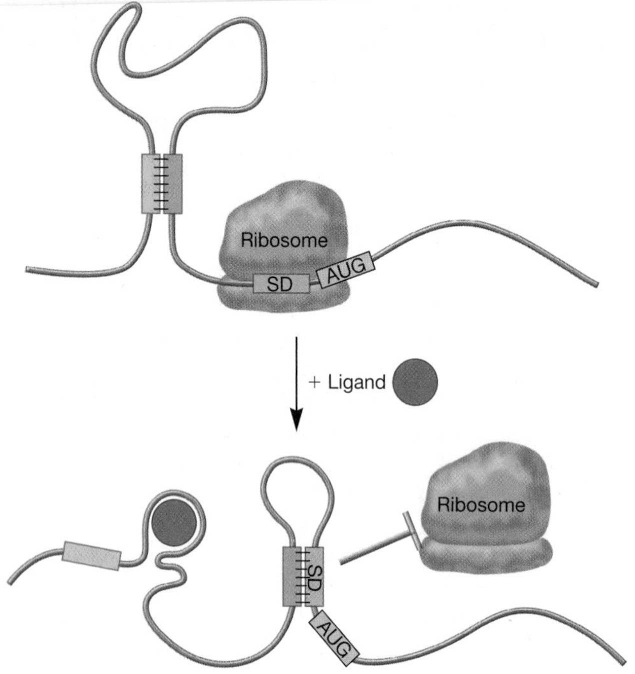

Figure 12.13 Regulation of Translation by a Riboswitch. In the absence of a relevant metabolite, an effector binding site is formed in the leader of the mRNA (red) when complementary sequences (orange box and green box) hydrogen bond. This folding pattern exposes important sequences in the ribosome-binding site (e.g., the Shine-Dalgarno sequence; blue box) and translation occurs. When the appropriate effector molecule is present, it binds the leader, disrupting the existing structure and creating a new structure with the ribosome-binding-site sequences. Thus the ribosome-binding site becomes inaccessible and translation is blocked.

osmotic pressure or the presence of some toxic material, both of which favor *ompC* expression. Production of MicF RNA helps ensure that OmpF protein is not produced at high levels at the same time as OmpC protein. Some other antisense RNAs are listed in **table 12.2.**

1. Define attenuation. What are the functions of the leader region and ribosome in attenuation?
2. Of what practical importance is attenuation in coordinating the synthesis of amino acids and proteins? Describe how attenuation activity would vary when protein synthesis is suddenly accelerated, then later rapidly decelerated.
3. What are riboswitches? How are they similar to attenuation as described for the *trp* operon? How do they differ?

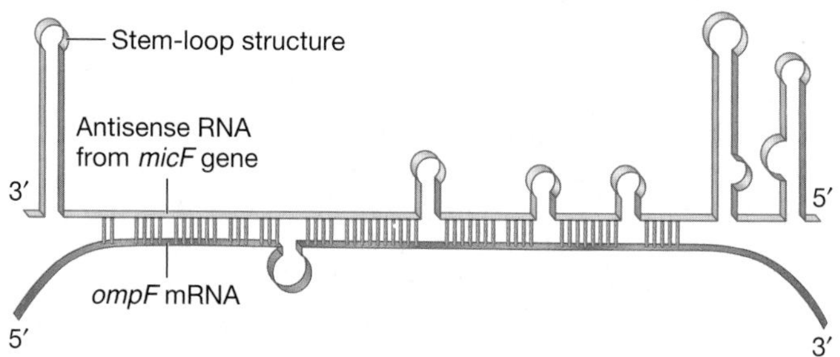

Figure 12.14 Regulation of Translation by Antisense RNA. The *ompF* mRNA encodes the porin OmpF. Translation of this message is regulated by the antisense RNA MicF, the product of the *micF* gene. MicF is complementary to the *ompF* mRNA and, when bound to it, prevents translation from occurring.

Table 12.2 Regulation of Gene Expression by Small Regulatory RNAs

Small RNA	Size	Bacterium	Function
RhyB	90 nt[1]	*E. coli*	Represses translation of iron-containing proteins (e.g., *sodB*) when iron availability is low
Spot 42	109 nt	*E. coli*	Inhibits translation of *galK* (encodes galactokinase)
RprA	105 nt	*E. coli*	Promotes translation of *rpoS* mRNA; antisense repressor of global negative regular H-NS (involved in stress responses)
MicF	109 nt	*E. coli*	Inhibits *ompF* mRNA translation
OxyS	109 nt	*E. coli*	Inhibits translation of transcriptional regulator *fhlA* mRNA and *rpoS* mRNA (encodes σ^s, a stationary phase sigma factor)
DsrA	85 nt	*E. coli*	Increases translation of *rpoS* mRNA
CsrB	366 nt	*E. coli*	Inhibits CsrA, a translational regulatory protein that positively regulates flagella synthesis, acetate metabolism, and glycolysis
RNAIII	512 nt	*Staphylococcus aureus*	Activates genes encoding secreted proteins (e.g., α hemolysin)
			Represses genes encoding surface proteins
RNA α	650 nt	*Vibrio anguillarum*	Decreased expression of Fat, an iron uptake protein
RsmB′	259 nt	*Erwinia carotovora* subsp. *carotovora*	Stabilizes mRNA of virulence proteins (e.g., cellulases, proteases, pectinolytic enzymes)

[1] nt: nucleotides

12.5 Global Regulatory Systems

Thus far, we have been considering the function of isolated operons. However, organisms must respond rapidly to a wide variety of changing environmental conditions and be able to cope with such stressors as nutrient deprivation, dessication, and major temperature fluctuations. They also have to compete successfully with other organisms for scarce nutrients and use these nutrients efficiently. These challenges require a regulatory system that can rapidly control many operons at the same time. Regulatory systems that affect many genes and pathways simultaneously are called **global regulatory systems.**

Although it is usually possible to regulate all the genes of a metabolic pathway in a single operon, there are good reasons for more complex global systems. Some processes involve too many genes to be accommodated in a single operon. For example, the machinery required for protein synthesis is composed of 150 or more gene products, and coordination requires a regulatory network that controls many separate operons. Sometimes two levels of regulation are required because individual operons must be controlled independently and also cooperate with other operons. Regulation of sugar catabolism in *E. coli* is a good example. *E. coli* uses glucose when it is available; in such a case, operons for other catabolic pathways are repressed. If glucose is unavailable and another nutrient is present, the appropriate operon is activated.

Global regulatory systems are so complex that a specialized nomenclature is used to describe the various kinds. Perhaps the most basic type is the **regulon.** A regulon is a collection of genes or operons that is controlled by a common regulatory protein. Usually the operons are associated with a single pathway or function (e.g., the production of heat-shock proteins or the catabolism of glycerol). A somewhat more complex situation is seen with a modulon. This is an operon network under the control of a common global regulatory protein, but whose constituent operons also are controlled separately by their own regulators. A good example of a modulon is catabolite repression, which is discussed on page 308. The most complex global systems are referred to as stimulons. A stimulon is a regulatory system in which all operons respond together in a coordinated way to an environmental stimulus. It may contain several regulons and modulons, and some of these may not share regulatory proteins. For instance, the genes involved in a response to phosphate limitation are scattered among several regulons and are part of one stimulon.

Mechanisms Used for Global Regulation

Global regulation is complex and often involves more than one regulatory mechanism. Most global regulatory networks are controlled by one or more regulatory proteins. Two-component regulatory systems and phosphorelay systems also play important roles in global control. In *Bacteria,* many global regulatory networks make use of **alternate sigma factors,** which can immediately change expression of many genes as they direct RNA polymerase to specific subsets of a bacterium's genome. This is possible because RNA polymerase core enzyme needs the assistance of a sigma factor to bind a promoter and initiate transcription. Each sigma factor recognizes promoters that differ in sequence, especially at the −10 and −35 positions. The specific sequences recognized by a given sigma factor are called its consensus sequences. When a complex process requires a radical change in transcription or a precisely timed sequence of transcription, it may be regulated by a series of sigma factors. Transcription (section 11.6)

E. coli synthesizes several sigma factors (**table 12.3**). Under normal conditions, a sigma factor called σ^{70} directs RNA polymerase activity. (The superscript number or letter indicates the size or function of the sigma factor; 70 stands for 70,000 Da.) When flagella and chemotactic proteins are needed, *E. coli* produces σ^F (σ^{28}). σ^F then binds its consensus sequences in promoters of genes whose products are needed for flagella biosynthesis and chemotaxis. If the temperature rises too high, σ^H (σ^{32}) is produced and stimulates the formation of around 17 heat-shock proteins that protect the cell from thermal destruction. Importantly, each sigma factor has its own set of promoters to which it binds.

In the discussions that follow, we describe three global regulatory networks. The first relatively simple example is the catabolite repression modulon, which involves regulation of transcription by both repressors and activators. The second is quorum sensing, which was introduced in chapter 6. Although the example we discuss regulates a single operon, many quorum-sensing systems regulate the transcription of suites of genes and operons. Finally, we examine a more complex process, that of sporulation in the gram-positive bacterium *B. subtilis*. Regulation of endospore formation involves numerous control mechanisms, including phosphorelay and sequential use of sigma factors.

Table 12.3 *E. coli* Sigma Factors

Sigma Factor	Genes Transcribed
σ^{70}	Genes needed during exponential growth
σ^S	Genes needed during the general stress response and during stationary phase
σ^E	Genes needed to restore membrane integrity and the proper folding of membrane proteins
σ^H (σ^{32})	Genes needed to protect against heat shock and other stresses, including chaperones that help maintain or restore proper folding of cytoplasmic proteins and proteases that degrade damaged proteins
FecI sigma factor	Genes that encode the iron citrate transport machinery in response to iron starvation and the availability of iron citrate
σ^F (σ^{28})	Genes involved in flagellum assembly
σ^{60}	Genes involved in nitrogen metabolism

Catabolite Repression

If *E. coli* grows in a medium containing both glucose and lactose, it uses glucose preferentially until the sugar is exhausted. Then after a short lag, growth resumes with lactose as the carbon source (**figure 12.15**). This biphasic growth pattern or response is called **diauxic growth.** The cause of diauxic growth or diauxie is complex and not completely understood, but **catabolite repression** plays a part. The enzymes for glucose catabolism are constitutive. However, operons that encode enzymes required for the catabolism of carbon sources that must first be modified before entering glycolysis (e.g., the *lac* operon) are regulated by catabolite repression. These include the *ara, mal* (maltose), and *gal* (galactose) operons, as well as the *lac* operon. Collectively, these can be called catabolite operons, and their expression is coordinately (or globally) repressed when glucose is plentiful and activated when it is not.

The coordinated regulation of catabolite operons is brought about by **catabolite activator protein (CAP),** which is also called **cyclic AMP receptor protein (CRP).** CAP exists in two states: it is active when the small cyclic nucleotide **3′, 5′-cyclic adenosine monophosphate (cAMP; figure 12.16)** is bound, and it is inactive when it is free of cAMP. The levels of cAMP are controlled by the enzyme adenyl cyclase, which converts ATP to cAMP and PP_i. Adenyl cyclase is active only when little or no glucose is available. Thus the level of cAMP varies inversely with that of glucose: when glucose is unavailable and the catabolism of another sugar might be needed, the amount of cAMP in the cell increases allowing cAMP to bind to and activate CAP.

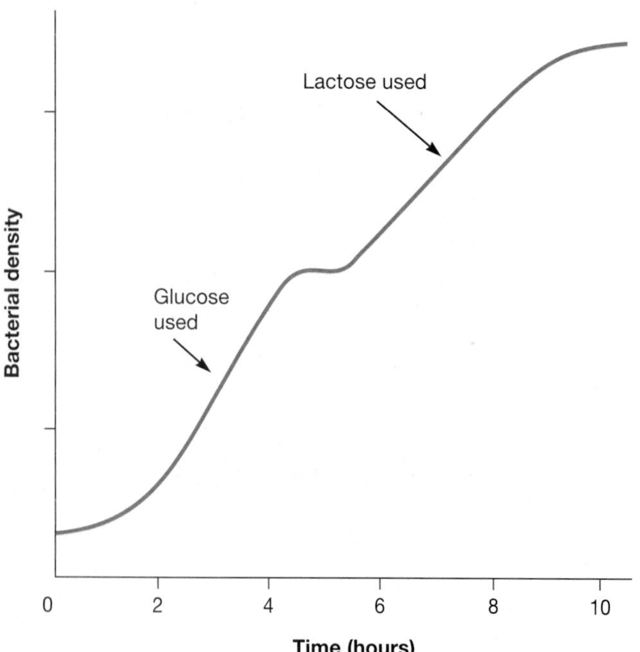

Figure 12.15 Diauxic Growth. The diauxic growth curve of *E. coli* grown with a mixture of glucose and lactose. Glucose is first used, then lactose. A short lag in growth is present while the bacteria synthesize the enzymes needed for lactose use.

All catabolite operons contain a CAP binding site, and CAP must be bound to this site before RNA polymerase can bind the promoter and begin transcription. Upon binding, CAP bends the DNA within two helical turns (figure 12. 6*b* and **figure 12.17**). Interaction of CAP with RNA polymerase then stimulates transcription. Thus all catabolite operons are controlled by two regulatory proteins: the regulatory protein specific to each operon (e.g., *lac* repressor and AraC protein) and CAP. In the case of the *lac* operon, if glucose is absent and lactose is present, the inducer allolactose will bind to and inactivate the *lac* repressor protein, CAP will be in the active form (with cAMP bound), and transcription will proceed (**figure 12.18*a***). However, if glucose and lactose are both in short supply, even though CAP binds to the *lac* promoter, transcription will be inhibited by the presence of the repressor protein, which remains bound to the operator in the absence of inducer (figure 12.18*c*). Dual control ensures that the *lac* operon is expressed only when lactose catabolic genes are needed.

We have seen how CAP controls catabolite operons; now let us turn our attention to the regulation of the levels of cAMP. The decrease in cAMP levels that occurs when glucose is present is due to the effect of the phosphoenolpyruvate: phosphotransferase system (PTS) on the activity of adenyl cyclase. Recall from chap-

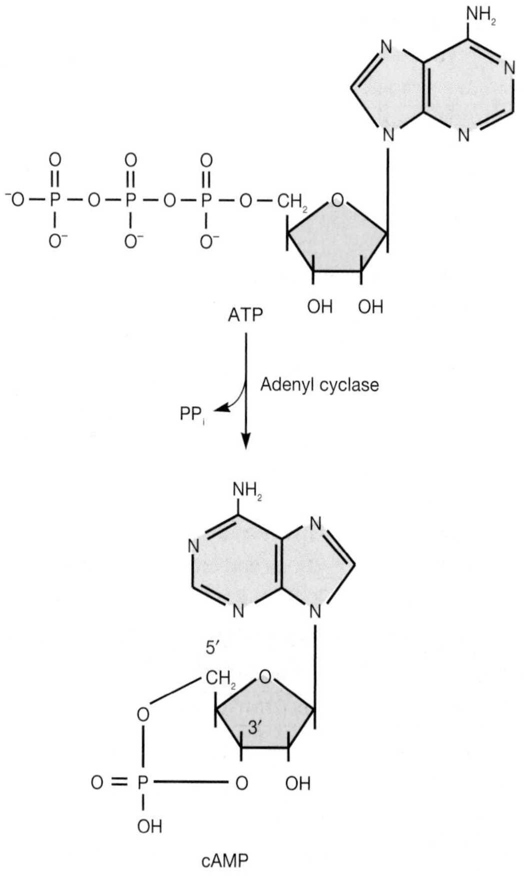

ATP
Adenyl cyclase
PP_i
cAMP

Figure 12.16 Cyclic Adenosine Monophosphate (cAMP). The phosphate group extends between the 3′ and 5′ hydroxyls of the ribose sugar. The enzyme adenyl cyclase forms cAMP from ATP.

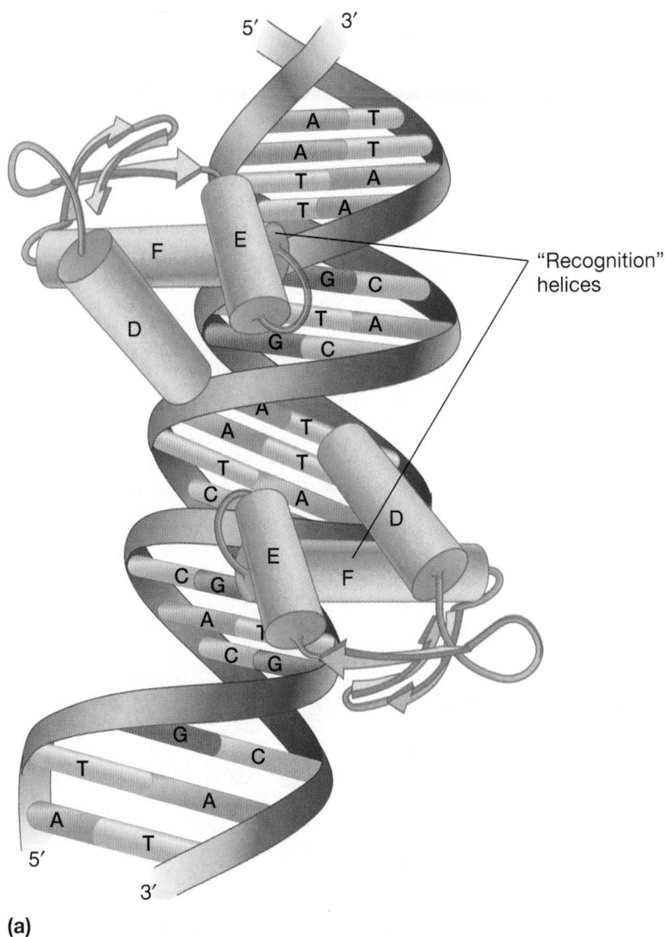

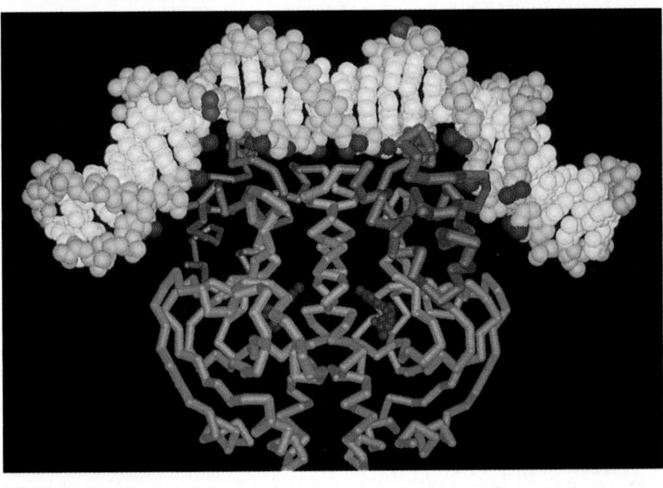

Figure 12.17 CAP Structure and DNA Binding. **(a)** The CAP dimer binding to DNA at the *lac* operon promoter. The recognition helices fit into two adjacent major grooves on the double helix. **(b)** A model of the *E. coli* CAP-DNA complex derived from crystal structure studies. The cAMP-binding domain is in blue and the DNA-binding domain, in purple. The cAMP molecules bound to CAP are in red. Note that the DNA is bent when complexed with CAP.

ter 5 that in the PTS, a phosphoryl group is transferred by a series of proteins from phosphoenolpyruvate (PEP) to glucose, which then enters the cell as glucose 6-phosphate. When glucose is present, enzyme IIA transfers the phosphoryl group to enzyme IIB, which phosphorylates glucose. Because enzyme IIA rapidly transfers phosphoryl groups, it exists largely in an unphosphorylated state. Unphosphorylated enzyme IIA inhibits the permeases for many sugars, and in doing so inhibits sugar uptake. However, when glucose is absent, the phosphoryl groups from PEP are transferred to enzyme IIA, but are not transferred to enzyme IIB. The phosphorylated form of enzyme IIA accumulates. This form of the enzyme activates adenyl cyclase, stimulating cAMP production. Uptake of nutrients by the cell: Group translocation (section 5.6)

Catabolite repression is of considerable advantage to the bacterium. It will use the most easily catabolized sugar (glucose) first rather than synthesize the enzymes necessary for catabolism of another carbon and energy source. These control mechanisms are present in a variety of bacteria and metabolic pathways.

Quorum Sensing

Cell to cell communication among procaryotes occurs by the exchange of small molecules often termed signals or signaling molecules. The exchange of signaling molecules is essential in the coordination of gene expression in microbial populations. This was first recognized in the marine bioluminescent bacterium *Vibrio fischeri,* which produces light only if cells are at high density. It has since been discovered that intercellular communication plays an essential role in the regulation of genes whose products are needed for the establishment of virulence, symbiosis, biofilm production, plasmid transfer, and morphological differentiation in a wide range of microorganisms. Here we describe how signals that are secreted by one group of cells can regulate the genetic expression of another. Our focus is on the regulation of a single operon. However, it should be kept in mind that **quorum sensing** can regulate multiple genes and operons. Microbial growth in natural environments: Cell-cell communication within microbial populations (section 6.6)

Quorum sensing in *V. fischeri* and many other gram-negative bacteria uses an **N-acyl homoserine lactone (AHL)** signal (**figure 12.19**). Synthesis of this small molecule is catalyzed by an enzyme called AHL synthase, the product of the *luxI* gene. The *luxI* gene is subject to positive autoregulation. That is to say, transcription of *luxI* increases as AHL accumulates in the cell. This is accomplished through a transcriptional activator, LuxR, which is active only when it binds AHL (figure 12.19). Thus a simple feedback loop is created. Without AHL-activated LuxR, the *luxI* gene will be transcribed only at basal levels. AHL freely diffuses out of the cell and accumulates in the environment. As cell density increases, the concentration of AHL outside the cell eventually exceeds that inside the cell, and the concentration gradient is reversed. As AHL flows back into the cell, it binds and activates LuxR. LuxR can now activate high-level transcription of *luxI* and the genes whose products are needed for bioluminescence (*luxCDABEG*). Quorum sensing is often called **autoinduction** and the AHL signal is termed **autoinducer** to reflect the autoregulatory nature of this system.

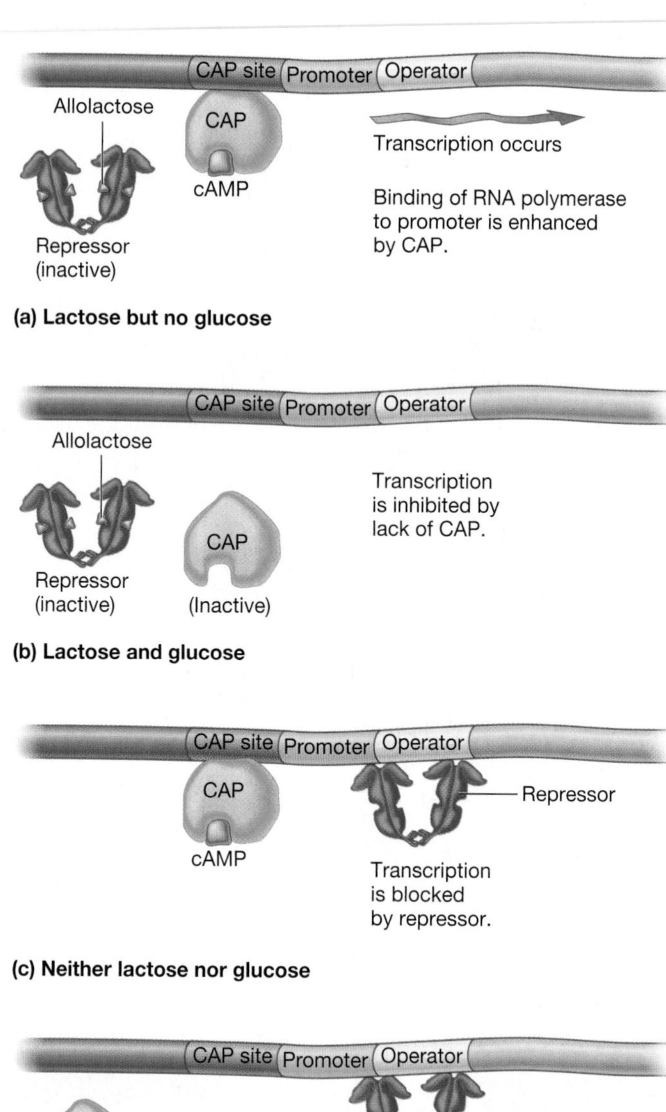

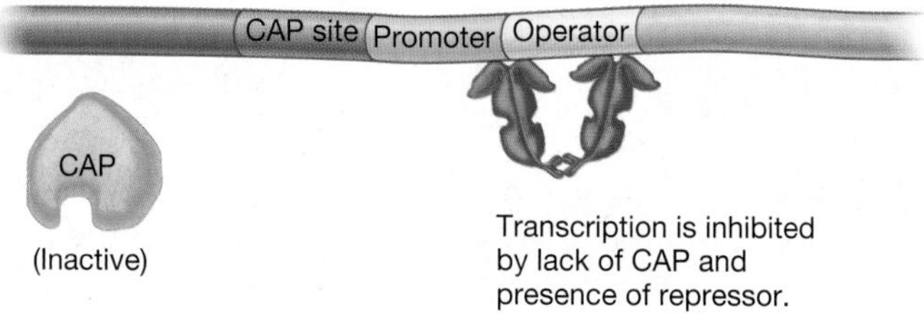

Figure 12.18 Regulation of the *lac* Operon by the *lac* Repressor and CAP. A continuum of *lac* mRNA synthesis is brought about by the action of CAP, an activator protein, and the *lac* repressor. **(a)** When lactose is available and glucose is not, the repressor is inactivated and cAMP levels increase. Cyclic AMP binds CAP, activating it. CAP binds the CAP binding site near the *lac* promoter and facilitates binding of RNA polymerase. Under these conditions, transcription occurs at maximal levels. **(b)** When both lactose and glucose are available, both CAP and the *lac* repressor are inactive. Because RNA polymerase cannot bind the promoter efficiently without the aid of CAP, transcription levels are low. **(c)** When neither glucose nor lactose is available, both CAP and the *lac* repressor are active. In this situation, both proteins are bound to their regulatory sites. CAP binding enhances the binding of RNA polymerase to the promoter. However, the repressor blocks transcription. Transcription levels are low. **(d)** When glucose is available and lactose is not, CAP is inactive and the *lac* repressor is active. Thus RNA polymerase binds inefficiently, and those polymerase molecules that do bind are blocked by the repressor. This condition results in the lowest levels of transcription observed for the *lac* operon.

Another kind of quorum sensing depends on an elaborate, two-component signal transduction system. It is found in both gram-negative and gram-positive bacteria including (but not limited to) *Staphylococcus aureus, Ralstonia solanacearum, Salmonella enterica, Vibrio cholerae*, and *E. coli*. It has been best studied in the bioluminescent bacterium *Vibrio harveyi*. Unlike *V. fischeri, V. harveyi* responds to two autoinducer molecules: AI-1 and AI-2. AI-1 is a homoserine lactone and its synthesis depends on the *luxM* gene. AI-2 is furanosylborate, a small molecule that contains a boron atom—quite an unusual component in an organic molecule. Its synthesis relies on the product of the *luxS* gene. As shown in **figure 12.20,** AI-1 and AI-2 are secreted by the cell, which then uses separate proteins called LuxN and LuxPQ to detect their presence. At low cell density in the absense of either AI-1 or AI-2, LuxN and LuxPQ autophosphorylate and converge on a single phosphotransferase protein called LuxU. LuxU accepts phosphates from each sensor kinase and then phosphorylates the response regulator LuxO. Phosphorylated LuxO in turn activates the transcription of genes encoding several small RNAs that destabilize *luxR* mRNA. Because LuxR is a transcriptional activator of the operon *luxCDABE,* which encodes proteins needed for bioluminescence, cells do not make light at low cell density. An interesting thing happens as cell and autoinducer densities increase: Lux N binds AI-1 and LuxPQ binds AI-2, and the proteins switch from functioning as kinases to phosphatases, proteins that dephosphorylate rather than phosphorylate their substrates. The flow of phosphates is now reversed; LuxO is inactivated by dephosphorylation and *luxR* mRNA is translated. LuxR now activates

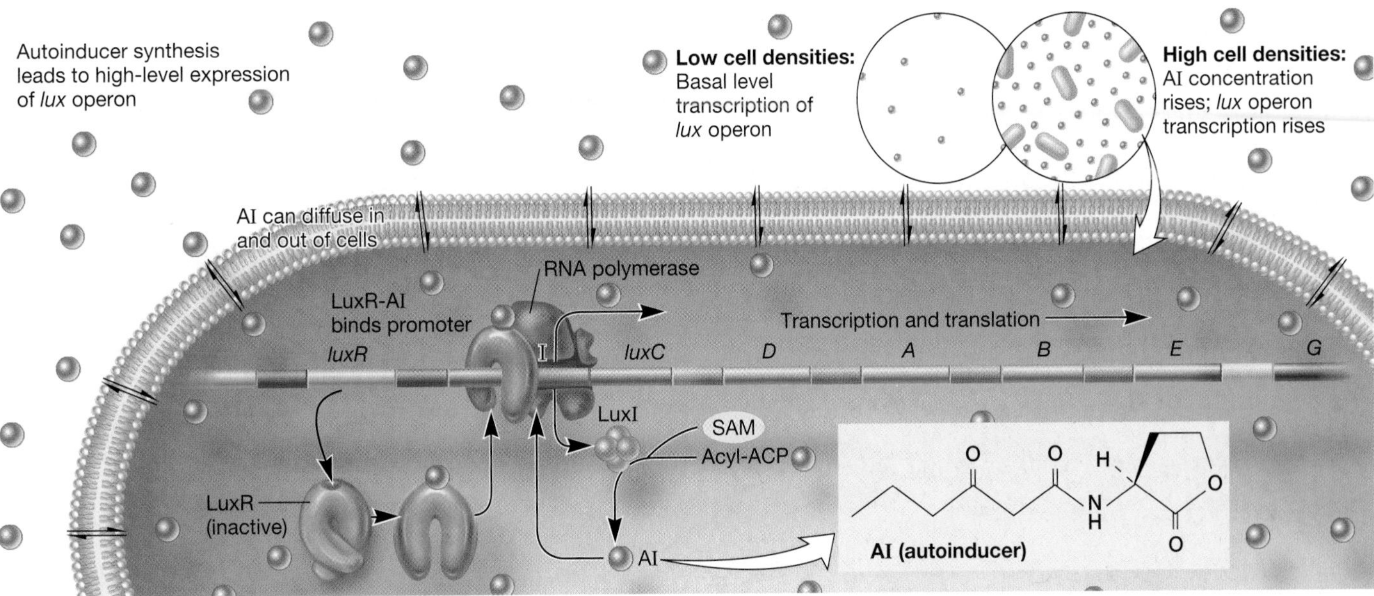

Figure 12.19 Quorum Sensing in *V. fischeri*. The AHL signaling molecule diffuses out of the cell; when cell density is high, the concentration of AHL diffuses back into the cell where it binds to and activates the transcriptional regulator LuxR. Active LuxR then stimulates transcripton of the gene coding for AHL synthase (*luxI*) as well as the genes encoding proteins needed for light production.

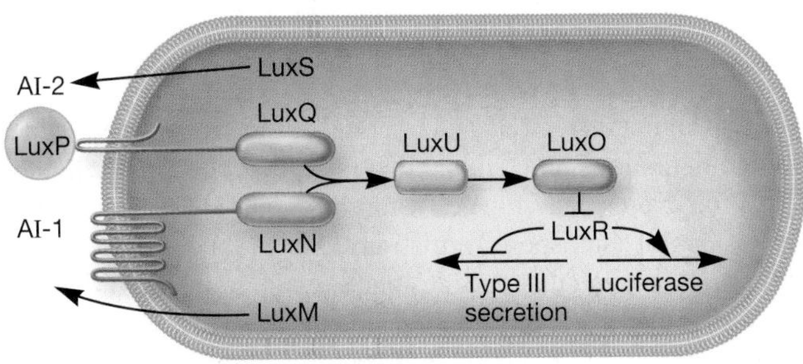

Figure 12.20 Quorum Sensing in *V. harveyi*. Two autoinducing signals, AI-1 and AI-2, are produced. At low cell density, the two component signal transduction system consisting of the signal kinases LuxPQ and LuxN initiate a phosphorelay that results in inhibition of the transcriptional regulator LuxR. Without LuxR, bioluminescence genes are repressed but genes for a type III secretion system (TTSS) are transcribed. At high cell density, LuxPQ and LuxN function as phosphatases, reversing the flow of phosphates. This results in activation of LuxR, transcriptional activation of the bioluminescence operon and repression of the TTSS genes.

transcription of *luxCDABE* and light is produced. Careful inspection of figure 12.20 reveals that another set of genes is controlled by the AI-1, AI-2 system of *V. harveyi*. In this microbe, genes for a type III protein secretion system (TTSS) are controlled in the opposite manner as those for bioluminescence.

LuxS-type autoinducers are found in a number of bacteria, but the precise AI-2 structure is specific for each species. Nonetheless, bacteria of different species can "talk" to each other because the products of the LuxS enzymes spontaneously rearrange. Thus in a bacterial community consisting of several AI-2-producing bacterial species, AI-2 molecules in the environment interconvert. This means that individual cells may be responding to their own signal and that produced by other species. LuxS-producing bacteria can also interfere with each other's communication. This has been shown experimentally. When *V. harveyi* and *E. coli* are grown together, *E. coli* consumes AI-2 produced by *V. harveyi,* thereby inhibiting light production and promoting TTSS production at high cell density. *E. coli* can also limit AI-2 regulated gene expression in *V. cholerae* by consuming the vibroid AI-2. Thus although the regulation of gene expression is generally considered in the context of a single cell, or at least a single species, it appears that in nature, microbes are most likely responding not only directly to the environment, but to each other as well.

Sporulation in *Bacillus subtilis*

As discussed in chapter 3, endospore formation is a complex process that involves asymmetric division of the cytoplasm to yield a large mother cell and a smaller forespore, engulfment of the forespore by the mother cell, and construction of additional layers of spore coverings (**figure 12.21*a*** and *figure 3.49*). Sporulation takes approximately 8 hours. It is controlled by phosphorelay, posttranslational modification of proteins, numerous transcription initiation regulatory proteins, and alternate sigma factors. The latter are particularly important. When growing vegetatively, *B. subtilis* RNA polymerase uses sigma factors σ^A and σ^H to recognize genes for normal survival. However, when cells sense a starvation signal, a cascade of events is initiated that results in the production of other sigma factors that are differentially expressed in the developing endospore and mother cell.

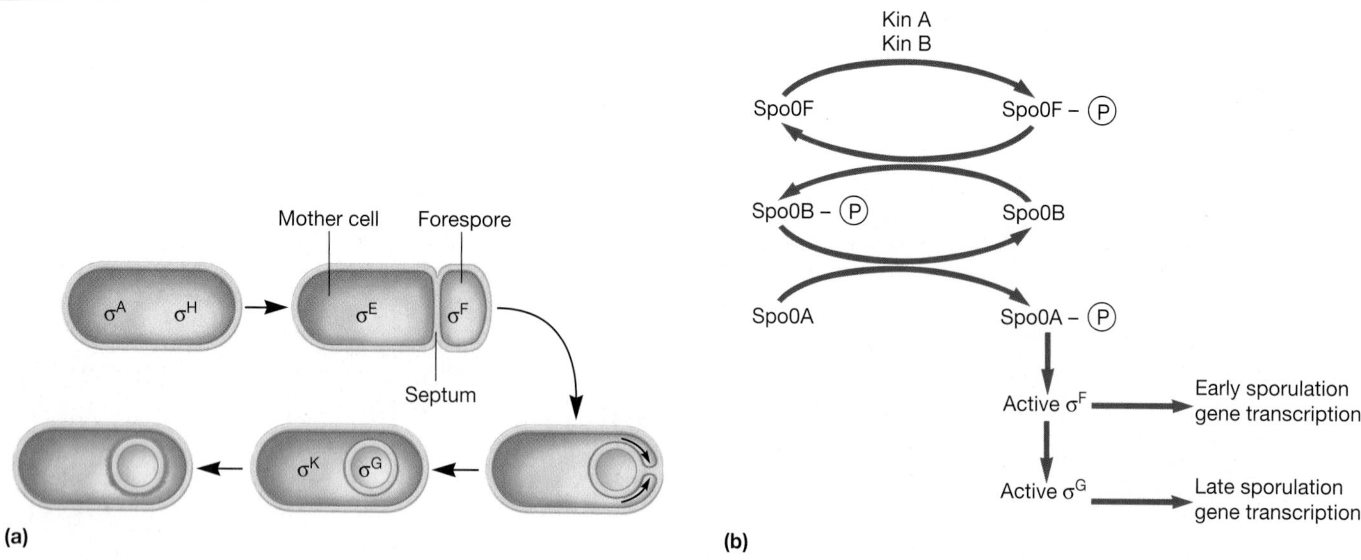

Figure 12.21 Genetic Regulation of Sporulation in *Bacillus subtilis*. **(a)** The initiation of sporulation is governed in part by the activities of two spatially separated sigma factors. σ^F is located in the forespore, while σ^E is confined to the mother cell. These sigma factors direct the initiation of transcription of genes whose products are needed for early events in sporulation. Later, σ^G and σ^K are localized to the developing endospore and mother cell, respectively. They control the expression of genes whose products are involved in the later steps of sporulation. **(b)** The activation of σ^F is accomplished through a phosphorelay system that is triggered by the activation of the sensor kinase protein KinA. When KinA senses starvation, it autophosphorylates a specific histidine residue. The phosphoryl group is then passed in relay fashion from Spo0F to Spo0B and finally to Spo0A. See text for details.

Initiation of sporulation is controlled by the protein Spo0A, a response-regulator protein that is part of a phosphorelay system. Sensor kinases associated with this system detect environmental stimuli that trigger sporulation. One of the most important sensor kinases is KinA, which senses nutrient starvation. When *B. subtilis* finds its nutrients are depleted, KinA autophosphorylates a specific histidine residue. The phosphoryl group is then transferred to an aspartic acid residue on Spo0F. However, Spo0F cannot directly regulate gene expression; instead, Spo0F donates the phosphoryl group to a histidine on Spo0B. Spo0B in turn relays the phosphoryl group to Spo0A. Phosphorylated Spo0A positively controls genes needed for sporulation and negatively controls genes that are not needed. In response to Spo0A, the expression of over 500 genes is altered. Among the genes whose expression is stimulated by Spo0A is *sigF,* the gene encoding sigma factor σ^F (figure 12.21*b*), and *spoIIGB,* the gene encoding an inactive form of σ^E (pro-σ^E).

When sporulation starts, the chromosome has replicated, with one copy remaining in the mother cell and another to be partitioned in the forespore. Shortly after the formation of the spore septum, σ^F is found in the forespore, and pro-σ^E is localized in the mother cell. Pro-σ^E is cleaved by a protease to form active σ^E. The two sigma factors, σ^F and σ^E, bind to the promoters of genes needed in the forespore and mother cell, respectively. There they direct the expression of genes whose products are needed for the early steps of endospore formation. One of the many genes σ^F regulates is a gene that encodes another sigma factor, σ^G, which will replace σ^F in the developing endospore. Likewise, σ^E directs the transcription of a mother-cell-specific sigma factor, σ^K. Like σ^E, σ^K is first produced in an inactive form, pro-σ^K. Upon activation of pro-σ^K by proteolysis, σ^K ensures that genes encoding late-stage sporulation products are transcribed. Overall, temporal regulation is achieved because σ^F and σ^E direct transcription of genes needed early in the sporulation process, while σ^G and σ^K are needed for the transcription of genes whose products are needed later. In addition, spatial control of gene expression is accomplished because σ^F and σ^G are located in the forespore and σ^E and σ^K are found only in the mother cell.

1. What are global regulatory systems and why are they necessary? Briefly describe regulons, modulons, and stimulons.
2. What is diauxic growth? Explain how catabolite repression causes diauxic growth.

3. Describe the events that occur in each of the following growth conditions: *E. coli* in a medium containing glucose, but not lactose; in a medium containing both sugars; in a medium containing lactose but no glucose; and in a medium containing neither sugar.
4. What would be the phenotype of a *V. fischeri* mutant strain that could not regulate *luxI*, so that it was constantly producing autoinducer?
5. Why do you think bacteria use quorum sensing to regulate genes needed for virulence? How might this reason be related to the rationale behind using quorum sensing to establish a symbiotic relationship?
6. Briefly describe how a phosphorelay system and sigma factors are used to control sporulation in *B. subtilis*. Give one example of posttranslational modification as a means to regulate this process.

12.6 Regulation of Gene Expression in *Eucarya* and *Archaea*

As is the case in *Bacteria*, the regulation of gene expression in *Eucarya* and *Archaea* can occur at transcriptional, translational, and posttranslational levels. However, because of chromatin structure and the additional steps needed to produce a functional protein, regulation of eucaryotic gene expression is even more complex than what has already been discussed. Much of the work on gene regulation in *Eucarya* has focused on transcription initiation. More recently, regulation by small RNA molecules has attracted considerable attention. Our understanding of the regulation of archaeal gene expression unfortunately lags considerably behind what we know for *Eucarya* and *Bacteria*. However, some intriguing discoveries are briefly introduced here.

As noted in chapter 11, transcription initiation in *Eucarya* involves numerous transcription factors (*see figure 11.33*). Many transcription factors, such as TFIID, are general transcription factors that are part of the machinery common to transcription initiation of all eucaryotic genes. On the other hand, **regulatory transcription factors** are specific to one or more genes and alter the rate of transcription. Regulatory transcription factors are in some ways analogous to bacterial regulatory proteins. As just mentioned, they too alter the rate of transcription of their target genes by binding specific DNA sequences that are usually located near the promoter. In eucaryotes, transcription factors that function as activators bind regulatory sites called **enhancers,** whereas those that function as repressors bind sites called **silencers** (**figure 12.22**). However, the manner by which regulatory transcription factors control the rate of transcription is not the same as the mechanism used by bacterial regulatory proteins. After binding an enhancer or silencer, regulatory transcription factors act indirectly to increase or decrease the rate of transcription. Many regulatory transcription factors control transcription initiation by interacting with general transcription factors, in particular TFIID (figure 12.22*a*) and a multisubunit protein complex called mediator (figure 12.22*b*). Others recruit chromatin-remodeling enzymes to a promoter. These enzymes change the degree of compaction of the DNA, making the promoter more or less accessible to RNA polymerase.

Another regulatory mechanism common to both *Bacteria* and eucaryotes is the use of sRNA molecules to control gene expression. This approach appears to be widely prevalent in eucaryotes. Many sRNAs act as antisense RNAs and function at the level of translation, as previously described. Some eucaryotic antisense RNAs are much smaller than typical bacterial antisense RNAs and are called microRNAs (miRNAs). Some sRNA molecules are important components of the spliceosome. These regulatory RNAs may contribute to the selection of splice sites used during mRNA processing to remove introns. In doing so, different proteins can be made at certain times in the life cycle of the organism by combining different protein-coding sequences.

The regulation of gene expression in the *Archaea* has garnered a great deal of interest. This is because the archaeal transcription and translation machinery is most similar to that of the *Eucarya,* yet functions in cells with typical, bacteria-like genome organization. The question being asked is whether archaeal regulation of gene expression is more like bacterial regulation or more like eucaryotic regulation. Thus far, the answer is mixed. Most of the archaeal regulatory proteins function much like bacterial activators and repressors—that is, they bind DNA sites near the promoter and enhance or block binding of RNA polymerase, respectively. However, a few seem to function more like eucaryotic regulatory transcription factors in that they bring about their effects by interacting with a general transcription factor, such as the TATA binding protein. Small RNA molecules also have been identified in some archaea; their role in regulation is still being elucidated. Introduction to the *Archaea:* Genetics and molecular biology (section 20.1)

1. List the differences among the *Bacteria, Archaea,* and *Eucarya* that affect the way each regulates gene expression. Which domain has the most levels at which gene expression can be regulated? Why?
2. How are bacterial regulatory proteins and eucaryotic regulatory transcription factors similar? How do they differ?
3. What regulatory sequences in bacterial genomes are analogous to the enhancers and silencers observed in eucaryotic genomes?

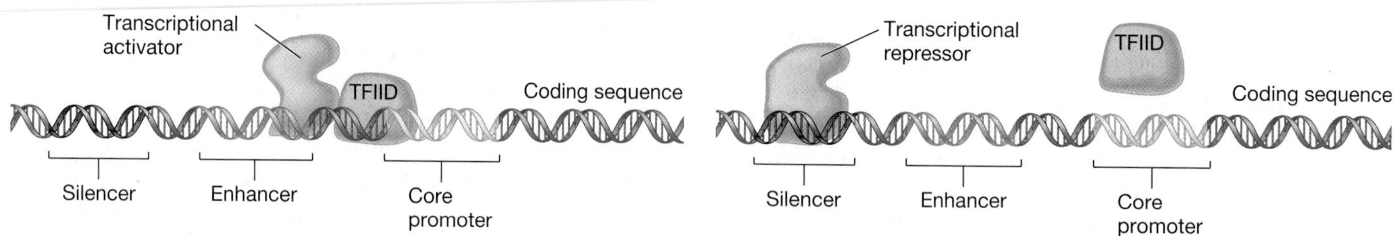

The transcriptional activator recruits TFIID to the core promoter and/or activates its function. Transcription will be activated.

The transcriptional repressor inhibits the binding of TFIID or inhibits its function. Transcription is repressed.

(a) Regulatory transcription factors and TFIID

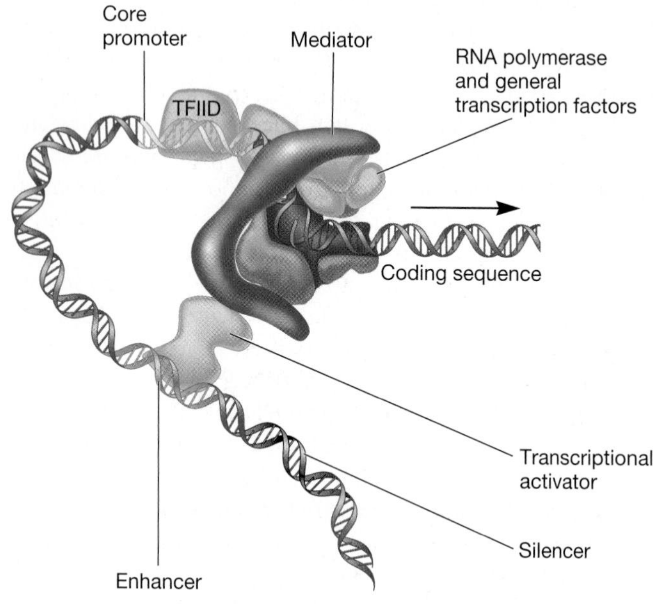

The transcriptional activator interacts with mediator. This enables RNA polymerase to form a preinitiation complex that can proceed to the elongation phase of transcription.

The transcriptional repressor interacts with mediator so that transcription is repressed.

(b) Regulatory transcription factors and mediator

Figure 12.22 The Activity of Eucaryotic Regulatory Transcription Factors. Regulatory transcription factors do not exert their effects directly on RNA polymerase. Instead they act via other proteins, most commonly the general transcription factor TFIID and a protein called mediator. Mediator's role in transcription is to aid RNA polymerase in switching from the initiation stage of transcription to the elongation stage. **(a)** A regulatory protein acting through TFIID. Activators could influence transcription-enhancing TFIID recruitment of RNA polymerase to the promoter. Repressors would inhibit this ability. **(b)** A regulatory protein acting through the mediator protein. An activator would stimulate mediator activity; a repressor would decrease mediator activity.

Summary

12.1 Levels of Regulation of Gene Expression

a. Regulation of gene expression can be controlled at many levels, including transcription initiation, transcription elongation, translation, and posttranslation (**figure 12.1**).

b. The three domains of life differ in terms of their genome structure and the steps required to complete gene expression. These differences affect the regulatory mechanisms they use.

12.2 Regulation of Transcription Initiation

a. Induction and repression of enzyme activity are two important regulatory phenomena. They usually occur because of the activity of regulatory proteins.

b. Regulatory proteins can either inhibit transcription (negative control) or promote transcription (positive control). Their activity is modulated by small effector molecules called inducers, corepressors, and inhibitors (**figure 12.3**).

c. Repressors are responsible for negative control. They block transcription by binding an operator and interfering with the binding of RNA polymerase to its promoter. They can also block transcription by blocking the movement of RNA polymerase after it binds DNA.

d. Activator proteins are responsible for positive control. They bind DNA sequences called activator-binding sites, and in doing so, promote binding of RNA polymerase to its promoter.

e. The *lac* operon of *E. coli* is an example of a negatively controlled inducible operon. When there is no lactose in the surroundings, the *lac* repressor is active and transcription is blocked. When lactose is available, it is converted to allolactose by the enzyme β-galactosidase. Allolactose acts as the inducer of the *lac* operon by binding the repressor and inactivating it. The inactive repressor cannot bind the operator and transcription occurs (**figure 12.7**).

f. The *trp* operon of *E. coli* is an example of a negatively controlled repressible operon. When there is no tryptophan available, the *trp* repressor is inactive and transcription occurs. When tryptophan levels are high, tryptophan acts as a corepressor and binds the *trp* repressor, activating it. The *trp* repressor binds the operator and blocks transcription (**figure 12.8**).

g. The *ara* operon of *E. coli* is an example of an inducible operon that is regulated by a dual-function regulatory protein, the AraC protein. AraC functions as a repressor when arabinose is not available. It functions as an activator when arabinose, the inducer, is available (**figure 12.9**).

h. Some regulatory proteins are members of two-component signal transduction systems and phosphorelay systems. These systems have a sensor kinase that detects an environmental change. The sensor kinase transduces the environmental signal to the response-regulator protein either directly (two-component system) or indirectly (phosphorelay) by transferring a phosphoryl group to it. The response regulator then activates genes needed to adapt to the new environmental conditions and inhibits expression of those genes that are not needed (**figure 12.10**).

12.3 Regulation of Transcription Elongation

a. In the tryptophan operon, a leader region lies between the operator and the first structural gene (**figure 12.11**). It codes for the synthesis of a leader peptide and contains an attenuator, a rho-independent termination site.

b. The synthesis of the leader peptide by a ribosome while RNA polymerase is transcribing the leader region regulates transcription: therefore the tryptophan operon is expressed only when there is insufficient tryptophan available. This mechanism of transcription control is called attenuation.

c. The leader regions of some mRNA molecules can bind metabolites that act as effector molecules. Binding of the metabolite to the mRNA causes a change in the leader structure, which can terminate transcription. This regulatory mechanism is called a riboswitch (**figure 12.12**).

12.4 Regulation at the Level of Translation

a. Some riboswitches regulate gene expression at the level of translation. For these riboswitches, the binding of a small molecule to specific sequences in the leader region of the mRNA alters leader structure and prevents ribosome binding (**figure 12.13**).

b. Translation can also be controlled by antisense RNAs. These small RNA molecules are noncoding. They base pair to the mRNA and usually inhibit translation (**figure 12.14**).

12.5 Global Regulatory Systems

a. Global regulatory systems can control many operons simultaneously and help microbes respond rapidly to a wide variety of environmental challenges.

b. Global regulatory systems often involve many layers of regulation. Regulatory mechanisms such as regulatory proteins, alternate sigma factors, two-component signal transduction systems, and phosphorelay systems are often used.

c. Diauxic growth is observed when *E. coli* is cultured in the presence of glucose and another sugar such as lactose (**figure 12.15**). This growth pattern is the result of catabolite repression, where glucose is used preferentially over other sugars. Operons that are part of the catabolite repression system are regulated by the activator protein CAP. CAP activity is modulated by cAMP, which is produced only when glucose is not available. Thus when there is no glucose, CAP is active and promotes transcription of operons needed for the catabolism of other sugars (**figure 12.18**).

d. Quorum sensing is a type of cell-cell communication mediated by small signaling molecules such as N-acyl-homoserine lactone (AHL). Quorum sensing couples cell density to regulation of transcription. Well-studied quorum-sensing systems include the regulation of bioluminescence in *Vibrio* spp. (**figures 12.19** and **12.20**). Other systems regulate virulence genes and biofilm formation.

e. Endospore formation in *B. subtilis* is another example of a global regulatory system. Two important regulatory mechanisms used during sporulation are a phosphorelay system that is important in initiation of sporulation and the use of alternate sigma factors (**figure 12.21**).

12.6 Regulation of Gene Expression in *Eucarya* and *Archaea*

a. Regulatory transcription factors are used by *Eucarya* to control transcription initiation. They can exert either positive or negative control (**figure 12.22**).

b. Antisense RNAs are used by *Eucarya* to regulate translation.

c. Microbiologists know relatively little about archaeal regulation of gene expression. Some archaea use regulatory proteins that are similar to bacterial regulatory systems to control initiation of transcription.

Key Terms

activator protein 295
activator-binding site 295
alternate sigma factors 307
antisense RNA 305
aporepressor 295
attenuation 302
attenuator 302
autoinducer 309
autoinduction 309
catabolite activator protein (CAP) 308
catabolite repression 308
constitutive gene 293
corepressor 295
3′, 5′-cyclic adenosine monophosphate (cAMP) 308
cyclic AMP receptor protein (CRP) 308
diauxic growth 308
enhancer 313
global regulatory systems 307
housekeeping gene 293
inducer 294
inducible enzyme 294
inducible gene 294
N-acyl homoserine lactone (AHL) 309
negative transcriptional control 294
operator 295
operon 295
phosphorelay system 302
positive transcriptional control 295
quorum sensing 309
regulatory transcription factor 313
regulon 307
repressible enzyme 294
repressor protein 295
response-regulator protein 301
riboswitch 304
sensor kinase protein 300
silencer 313
small RNAs (sRNAs) 305
structural gene 295
two-component signal transduction system 300

Critical Thinking Questions

1. Attenuation affects anabolic pathways, whereas repression can affect either anabolic or catabolic pathways. Provide an explanation for this.
2. Describe the phenotype of the following strains of *E. coli* mutants when grown in two different media: glucose only and lactose only. Explain the reasoning behind your answer.
   a. A strain with a mutation in the gene encoding the *lac* repressor such that it cannot bind allolactose.
   b. A strain with a mutation in the gene encoding CAP such that it does not release cAMP.
   c. A strain in which the Shine-Dalgarno sequence has been deleted from the gene encoding adenyl cyclase.
3. What would be the phenotype of an *E. coli* strain in which the tandem trp codons in the leader region were mutated so that they coded for serine instead?
4. What would be the phenotype of a *B. subtilis* strain whose gene for σ^G has been deleted? Consider the ability of the mutant to survive in nutrient-rich versus nutrient-depleted conditions.
5. Propose a mechanism by which a cell might sense and respond to levels of Na^+ in its environment.

Learn More

Bell, S. D. 2005. Archaeal transcriptional regulation—variation on a bacterial theme? *Trends Microbiol.* 13(6):262–65.

Brantl, S. 2004. Bacterial gene regulation: From transcription attenuation to riboswitches and ribozymes. *Trends Microbiol.* 12(11):473–75.

Brooker, R. J. 2005. *Genetics: Analysis and principles,* 2d ed. Boston: McGraw-Hill.

Feng, X.; Oropeza, R.; Walthers, D.; and Kenney, L. J. 2003. OmpR phosphorylation and its role in signaling and pathogenesis. *ASM News* 69(3):390–95.

Galperin, M. Y., and Gomelsky, M. 2005. Bacterial signal transduction modules: From genomics to biology. *ASM News* 71:326–33.

Gruber, T. M., and Gross, C. A. 2003. Multiple sigma subunits and the partitioning of bacterial transcription space. *Annu. Rev. Microbiol.* 57:441–66.

Hilbert, D. W., and Piggot, P. J. 2004. Compartmentalization of gene expression during *Bacillus subtilis* spore formation. *Microbiol. Mol. Biol. Rev.* 68(2):234–62.

Johansson, J. 2005. RNA molecules: More than mere information intermediaries. *ASM News* 71(11):515–20.

Llamas, I.; Quesada, E.; Martínez-Cánovas, M. J.; Gronquist, M.; Eberhard, A.; and González, J. W. 2005. Quorum sensing in halophilic bacteria: Detection of *N*-acyl homoserine lactones in the exopolysaccharide-producing species of *Halomonas. Extremophiles* 9:333–41.

Storz, G.; Altuvia, S.; and Wassarman, K. M. 2005. An abundance of RNA regulators. *Annu Rev. Biochem.* 74:199–217.

Warner, J. B., and Lolkema, J. S. 2003. CcpA-dependent carbon catabolite repression in bacteria. *Microbiol. Mol. Biol. Rev.* 67(4):475–90.

Xavier, K., and Bassler, B. L. 2005. Interference with AI-2-mediated bacterial cell-cell communication. *Nature* 437:750–53.

Yanofsky, C. 2004. The different roles of tryptophan transfer RNA in regulating *trp* operon expression in *E. coli* versus *B. subtilis. Trends Microbiol.* 20(8):367–74.

Please visit the Prescott website at www.mhhe.com/prescott7 for additional references.

13 Microbial Genetics: Mechanisms of Genetic Variation

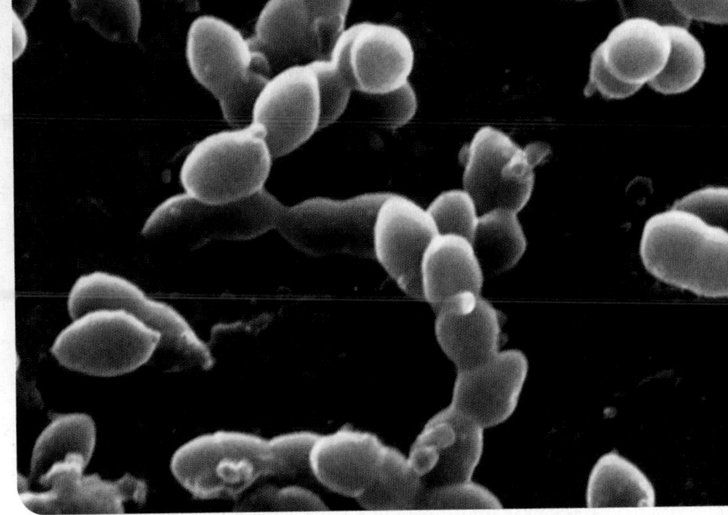

The scanning electron micrograph shows *Streptococcus pneumoniae,* the bacterium first used to study transformation and obtain evidence that DNA is the genetic material of organisms.

Preview

- Mutations are stable, heritable alterations in DNA sequence. In procaryotes, they usually produce phenotypic changes and can occur spontaneously or are induced by chemical mutagens or radiation.
- Microorganisms have several repair mechanisms designed to detect alterations in their genetic material and restore it to its original state. Despite these repair systems, some alterations remain uncorrected and provide material and opportunity for evolutionary change.
- Recombination is the process in which one or more nucleic acid molecules are rearranged or combined to produce new combinations of genes or a new nucleotide sequence.
- Gene transfer is a one-way process in procaryotes: a piece of genetic material (the exogenote) is donated to the chromosome of a recipient cell (the endogenote) and integrated into it.
- The transfer of genetic material between procaryotes is called horizontal gene transfer. It takes place in one of three ways: conjugation, transformation, or transduction.
- Transposable elements and plasmids can move genetic material between chromosomes and within chromosomes to cause rapid changes in genomes and drastically alter phenotypes.
- Bacterial chromosomes have been mapped with great precision, using Hfr conjugation in combination with transformational and transductional mapping techniques.
- Recombination of virus genomes occurs when two viruses with homologous chromosomes infect a host cell at the same time.

Chapters 11 and 12 introduce the fundamentals of molecular genetics—the way genetic information is organized, stored, replicated, and expressed. As demonstrated in these chapters, considerable information is embedded in the precise order of nucleotides in DNA. For life to exist with stability, it is essential that the nucleotide sequence of genes is not disturbed to any great extent. However, sequence changes do occur and can result in altered phenotypes. These changes may be detrimental, but those that are not are important in generating new variability in populations and in contributing to the process of evolution.

In this chapter we focus on processes that contribute to genetic variation in populations of microbes. We begin with an overview of the chemical nature of mutations and the effects of mutations at both the molecular and organismal levels. Because mutants have been put to important uses in the laboratory and in industry, the generation and isolation of mutant organisms are considered. The chapter continues with a discussion of DNA repair mechanisms. Although these repair mechanisms evolved to prevent the occurrence of mutations, as will be seen, some cellular attempts to correct DNA damage actually generate mutations. Finally, we examine microbial recombination and gene transfer in *Bacteria*. These processes have practical implications in terms of antibiotic and drug resistance. In addition, recombination and gene transfer mechanisms observed in *Bacteria* and viruses have been useful for mapping microbial genomes, and these techniques are discussed as well.

13.1 Mutations and Their Chemical Basis

Mutations [Latin *mutare,* to change] were initially characterized as altered phenotypes, but they are now understood at the molecular level. Several types of mutations exist. Some mutations arise from the alteration of single pairs of nucleotides and from the addition or deletion of one or two nucleotide pairs in the coding regions of a

Deep in the cavern of the infant's breast
The father's nature lurks, and lives anew.

—*Horace, Odes*

gene. Such small changes in DNA are sometimes called microlesions, and the smallest of these are called **point mutations** because they affect only one base pair in a given location. Larger mutations (macrolesions) are also recognized, but are less common. These include large insertions, deletions, inversions, duplications, and translocations of nucleotide sequences.

Mutations occur in one of two ways: (1) **Spontaneous mutations** arise occasionally in all cells and occur in the absence of any added agent. (2) **Induced mutations,** on the other hand, are the result of exposure to a **mutagen,** which can be either a physical or a chemical agent. Mutations can be characterized according to either the kind of genotypic change that has occurred or their phenotypic consequences. In this section, the molecular basis of mutations and mutagenesis is first considered. Then the phenotypic effects of mutations are discussed.

Spontaneous Mutations

Spontaneous mutations arise without exposure to external agents. This class of mutations may result from errors in DNA replication or from the action of mobile genetic elements such as transposons. A few of the more prevalent mechanisms are described here.

Replication errors can occur when the nitrogenous base of a template nucleotide takes on a rare tautomeric form. Tautomerism is the relationship between two structural isomers that are in chemical equilibrium and readily change into one another. Bases typically exist in the keto form. However, they can at times take on either an imino or enol form (**figure 13.1*a***). These tautomeric shifts change the hydrogen-bonding characteristics of the bases, allowing purine for purine or pyrimidine for pyrimidine substitutions that can eventually lead to a stable alteration of the nucleotide sequence (figure 13.1*b*). Such substitutions are known as

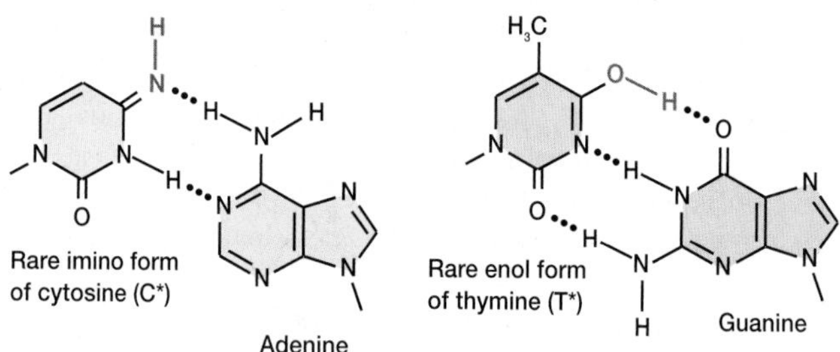

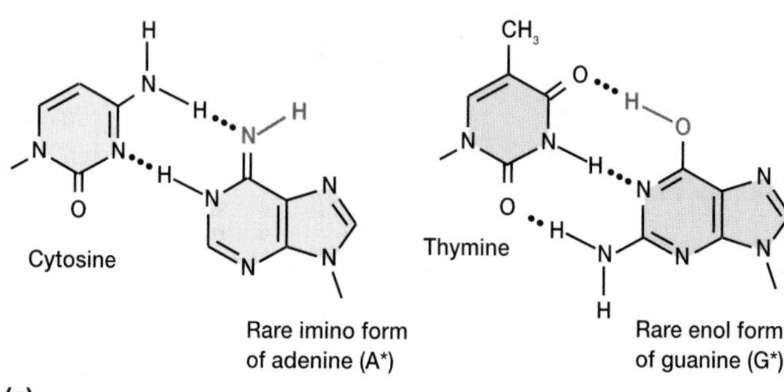

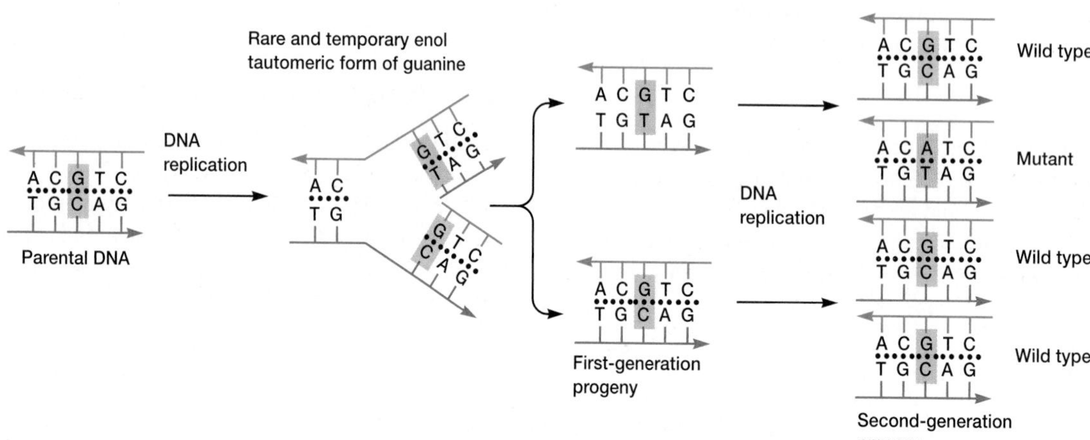

Figure 13.1 Tautomerization and Transition Mutations. Errors in replication due to base tautomerization. **(a)** Normally AT and GC pairs are formed when keto groups participate in hydrogen bonds. In contrast, enol tautomers produce AC and GT base pairs. **(b)** Mutation as a consequence of tautomerization during DNA replication. The temporary enolization of guanine leads to the formation of an AT base pair in the mutant, and a GC to AT transition mutation occurs. The process requires two replication cycles. The mutation only occurs if the abnormal first-generation GT base pair is missed by repair mechanisms.

transition mutations and are relatively common. On the other hand, **transversion mutations,** mutations where a purine is substituted for a pyrimidine, or a pyrimidine for a purine, are rarer due to the steric problems of pairing purines with purines and pyrimidines with pyrimidines.

Replication errors can also result in addition and deletion of nucleotides. These mutations generally occur where there is a short stretch of the same nucleotide. In such a location, the pairing of template and new strand can be displaced by the distance of the repeated sequence leading to additions or deletions of bases in the new strand (**figure 13.2**).

Spontaneous mutations can also originate from lesions in DNA as well as from replication errors. For example, it is possible for purine nucleotides to be depurinated—that is, to lose their base. This results in the formation of an **apurinic site,** which does not base pair normally and may cause a transition type mutation after the next round of replication. Likewise, pyrimidines can be lost, forming an **apyrimidinic site.** Other lesions are caused by reactive forms of oxygen such as oxygen free radicals and peroxides produced during aerobic metabolism. These may alter DNA bases and cause mutations. For example, guanine can be converted to 8-oxo-7,8-dihydrodeoxyguanine, which often pairs with adenine rather than cytosine during replication.

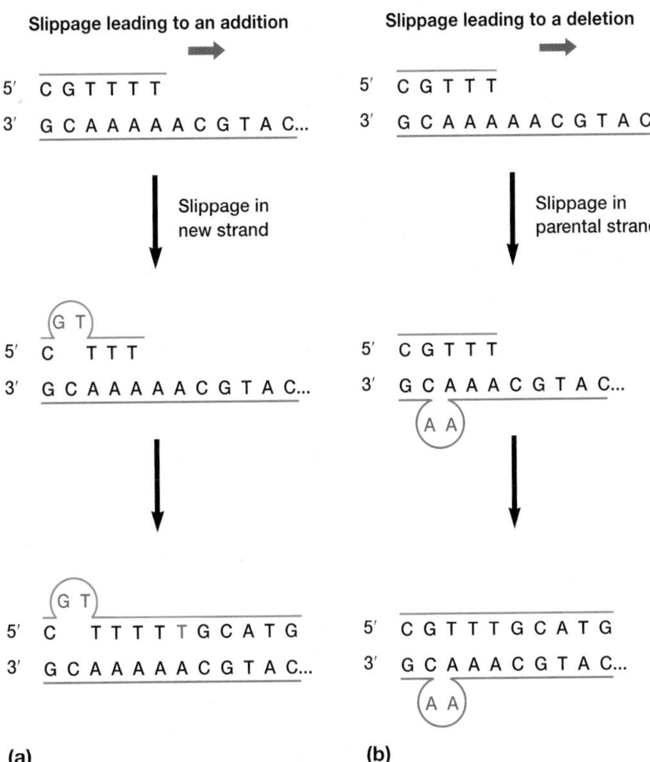

Figure 13.2 Additions and Deletions. A hypothetical mechanism for the generation of additions and deletions during replication. The direction of replication is indicated by the blue arrow. In each case there is strand slippage resulting in the formation of a small loop that is stabilized by the hydrogen bonding in the repetitive sequence, the AT stretch in this example. **(a)** If the new strand slips, an addition of one T results. **(b)** Slippage of the parental strand yields a deletion (in this case, a loss of two Ts).

Finally, spontaneous mutations can result from the insertion of DNA segments into genes. Insertions usually inactivate genes. They are caused by the movement of insertion sequences and transposons. Insertion mutations are very frequent in *Escherichia coli* and many other bacteria. These genetic elements are described in more detail in section 13.5.

Although most geneticists believe that spontaneous mutations occur randomly in the absence of an external agent and are then selected, observations by some microbiologists have led to an alternate and controversial hypothesis. The controversy began when John Cairns and his collaborators reported that a mutant *E. coli* strain, unable to use lactose as a carbon and energy source, could regain the ability to do so more rapidly when lactose was added to the culture medium as the only carbon source. Lactose appeared to induce mutations that allow *E. coli* to use the sugar again. One interpretation of these observations is that the mutations are examples of **directed** or **adaptive mutation**—that is, some bacteria seem able to select which mutations occur so that they can better adapt to their surroundings.

Many explanations have been offered to account for this phenomenon without depending on induction of particular mutations. One is the proposal that hypermutation can produce such results. Some starving bacteria might rapidly generate multiple mutations through activation of special mutator genes. This would produce many mutant bacterial cells. In such a random process, the rate of production of favorable mutants would increase, with many of these mutants surviving to be counted. There would appear to be directed or adaptive mutation because only mutants with the favorable mutations would survive. There is support for this hypothesis. Mutator genes have been discovered and have been shown to cause hypermutation under nutritional stress. Even if the directed mutation hypothesis is incorrect, it has stimulated much valuable research and led to the discovery of new phenomena.

Induced Mutations

Virtually any agent that directly damages DNA, alters its chemistry, or in some way interferes with its functioning will induce mutations. Mutagens can be conveniently classified according to their mode of action. Three common types of chemical mutagens are base analogs, DNA-modifying agents, and intercalating agents. A number of physical agents (e.g., radiation) damage DNA and also are mutagens.

Base analogs are structurally similar to normal nitrogenous bases and can be incorporated into the growing polynucleotide chain during replication (**table 13.1**). Once in place, these compounds typically exhibit base pairing properties different from the bases they replace and can eventually cause a stable mutation. A widely used base analog is 5-bromouracil (5-BU), an analog of thymine. It undergoes a tautomeric shift from the normal keto form to an enol much more frequently than does a normal base.

Table 13.1 Examples of Mutagens

Mutagen	Effect(s) on DNA Structure
Chemical	
5-Bromouracil	Base analog
2-Aminopurine	Base analog
Ethyl methanesulfonate	Alkylating agent
Hydroxylamine	Hydroxylates cytosine
Nitrogen mustard	Alkylating agent
Nitrous oxide	Deaminates bases
Proflavin	Intercalating agent
Acridine orange	Intercalating agent
Physical	
UV light	Promotes pyrimidine dimer formation
X rays	Causes base deletions, single-strand nicks, cross-linking, and chromosomal breaks

The enol tautomer forms hydrogen bonds like cytosine and directs the incorporation of guanine rather than adenine (**figure 13.3**). The mechanism of action of other base analogs is similar to that of 5-bromouracil.

DNA-modifying agents change a base's structure and therefore alter its base pairing characteristics. Some mutagens in this category are fairly selective; they preferentially react with some bases and produce a specific kind of DNA damage. An example of this type of mutagen is methyl-nitrosoguanidine, an alkylating agent that adds methyl groups to guanine, causing it to mispair with thymine (**figure 13.4**). A subsequent round of replication could then result in a GC-AT transition. Hydroxylamine is another example of a DNA-modifying agent. It hydroxylates the C-4 nitrogen of cytosine, causing it to base pair like thymine. There are many other DNA modifying agents that can cause mispairing.

Intercalating agents distort DNA to induce single nucleotide pair insertions and deletions. These mutagens are planar and insert themselves (intercalate) between the stacked bases of the helix. This results in a mutation, possibly through the formation of a loop in DNA. Intercalating agents include acridines such as proflavin and acridine orange.

Many mutagens, and indeed many carcinogens, directly damage bases so severely that hydrogen bonding between base pairs is impaired or prevented and the damaged DNA can no longer act as a template for replication. For instance, UV radiation generates cyclobutane type dimers, usually thymine dimers, between adjacent pyrimidines (**figure 13.5**). Other examples are ionizing radiation and carcinogens such as the fungal toxin aflatoxin B1 and other benzo(a)pyrene derivatives.

Retention of proper base pairing is essential in the prevention of mutations. Cells have developed extensive repair mechanisms. Often the damage can be repaired before a mutation is permanently established. If a complete DNA replication cycle takes place before the initial lesion is repaired, the mutation frequently becomes stable and inheritable. Repair mechanisms are discussed in section 13.3.

Effects of Mutations

The effects of a mutation can be described at the protein level and in terms of traits or other easily observed phenotypes. In all cases, the impact is readily noticed only if it produces a change in phenotype. In general, the more prevalent form of a gene and its associated phenotype is called the **wild type.** A mutation from wild type to a mutant form is called a **forward mutation** (**table 13.2**). A forward mutation can be reversed by a second mutation that restores the wild-type phenotype. When the second mutation is at the same site as the original mutation, it is called a **reversion mutation.** A true reversion converts the mutant nucleotide sequence back to the wild-type sequence. If the second mutation is at a different site than the original mutation, it is called a **suppressor mutation.** Suppressor mutations may be within the same gene (intragenic suppressor mutation) or in a different gene (extragenic suppressor mutation). Because point mutations are the most common types of mutations, their effects will be the focus here.

Mutations in Protein-Coding Genes

Point mutations in protein-coding genes can affect protein structure in a variety of ways (table 13.2). Point mutations are named according to if and how they change the encoded protein. The most common types of point mutations are silent mutations, missense mutations, nonsense mutations, and frameshift mutations. These are described in more detail below.

Silent mutations change the nucleotide sequence of a codon, but do not change the amino acid encoded by that codon. This is possible because of the degeneracy of the genetic code. Therefore, when there is more than one codon for a given amino acid, a single base substitution may result in the formation of a new codon for the same amino acid. For example, if the codon CGU were changed to CGC, it would still code for arginine even though a mutation had occurred. The mutation can only be detected at the level of the DNA or mRNA. When there is no change in the protein, there is no change in the phenotype of the organism. The genetic code (section 11.7)

Missense mutations involve a single base substitution that changes a codon for one amino acid into a codon for another. For example, the codon GAG, which specifies glutamic acid, could be changed to GUG, which codes for valine. The effects of missense mutations vary. They alter protein structure, but the effect of this change may range from complete loss of activity to no change at all. This is because the effect of missense mutations on protein function depends on the type and location of the amino acid substitution. For instance, replacement of a nonpolar amino acid in the protein's interior with a polar amino acid can drastically alter the protein's three-dimensional structure and therefore its function. Similarly the replacement of a critical amino acid at the active site of an enzyme often destroys its activity. However, the replacement of one polar amino

Br O···H—N H N H H N—H···N N Sugar N Sugar O H

5-bromouracil (keto form) Adenine

Br O—H···O N H H N···H—N N Sugar N Sugar O···H—N H

5-bromouracil (enol form) Guanine

(a) Base pairing of 5-BU with adenine or guanine

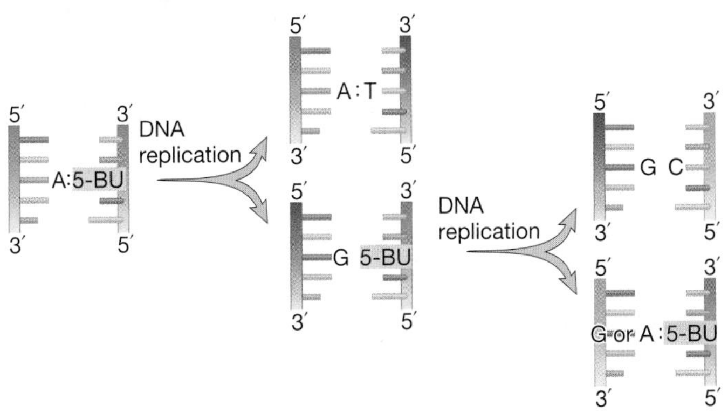

(b) How 5-BU causes a mutation in a base pair during DNA replication

Figure 13.3 Mutagenesis by the Base Analog 5-Bromouracil. **(a)** Base pairing of the normal keto form of 5-BU is shown in the top illustration. The enol form of 5-BU (bottom illustration) base pairs with guanine rather than with adenine as might be expected for a thymine analog. **(b)** If the keto form of 5-BU is incorporated in place of thymine, its occasional tautomerization to the enol form (BU_e) will produce an AT to GC transition mutation.

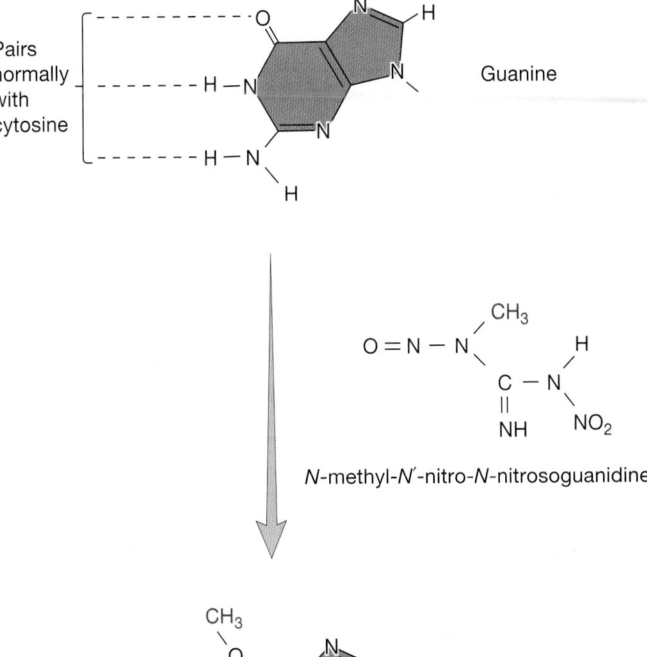

Figure 13.4 Methyl-Nitrosoguanidine Mutagenesis. Mutagnesis by methyl-nitrosoguanidine due to the methylation of guanine.

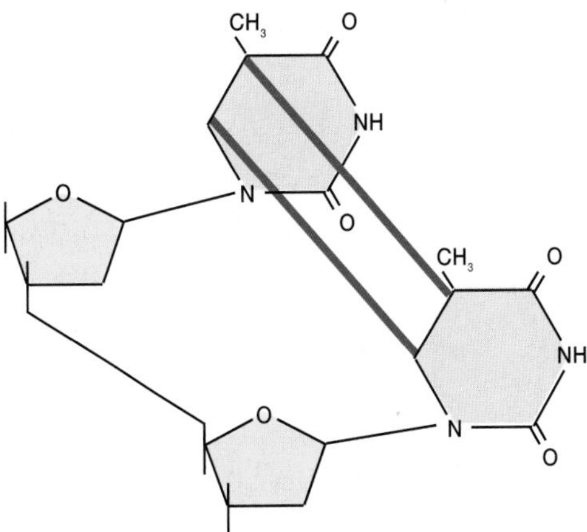

Figure 13.5 Thymine Dimer. Thymine dimers are formed by ultraviolet radiation.

acid with another at the protein surface may have little or no effect. Missense mutations actually play a very important role in providing new variability to drive evolution because they often are not lethal and therefore remain in the gene pool. Proteins (appendix I)

Nonsense mutations cause the early termination of translation and therefore result in a shortened polypeptide. They are called nonsense mutations because they convert a sense codon to a nonsense or stop codon. Depending on the relative location of the mutation, the phenotype may be more or less severely affected. Most proteins retain some function if they are shortened by only one or two amino acids; complete loss of normal function will almost certainly result if the mutation occurs closer to the beginning or middle of the gene.

Table 13.2 Types of Point Mutations

Type of Mutation	Change in DNA	Example
Forward Mutations		
None	None	5′-A-T-G-A-C-C-T-C-C-C-C-G-A-A-A-G-G-G-3′ Met - Thr - Ser - Pro - Lys - Gly
Silent	Base substitution	5′-A-T-G-A-C-A-T-C-C-C-C-G-A-A-A-G-G-G-3′ Met - Thr - Ser - Pro - Lys - Gly
Missense	Base substitution	5′-A-T-G-A-C-C-T-G-C-C-C-G-A-A-A-G-G-G-3′ Met - Thr - Cys - Pro - Lys - Gly
Nonsense		5′-A-T-G-A-C-C-T-C-C-C-C-G-T-A-A-G-G-G-3′ Met - Thr - Ser - Pro - STOP!
Frameshift	Addition/deletion	5′-A-T-G-A-C-C-T-C-C-G-C-C-G-A-A-A-G-G-G-3′ Met - Thr - Ser - Ala - Glu - Arg
Reverse Mutations		
True reversion	Base substitution	5′-A-T-G-A-C-C-T-C-C —forward→ A-T-G-C-C-C-T-C-C —reverse→ A-T-G-A-C-C-T-C-C Met - Thr - Ser → Met - Pro - Ser → Met - Thr - Ser
	Base substitution	5′-A-T-G-A-C-C-T-C-C —forward→ A-T-G-A-C-C-T-G-C —reverse→ A-T-G-A-C-C-A-G-C Met - Thr - Ser → Met - Thr - Cys → Met - Thr - Ser
Equivalent Reversion	Base substitution	5′-A-T-G-A-C-C-T-C-C —forward→ A-T-G-C-C-C-T-C-C —reverse→ A-T-G-C-T-C-T-C-C Met - Thr - Ser (polar amino acid) → Met - Pro - Ser (nonpolar amino acid) → Met - Leu - Ser (polar amino acid) pseudo-wild type
Suppressor Mutations		
		5′-A-T-G-A-C-C-T-C-C-C-C-G-A-A-A-G-G-G-3′ Met - Thr - Ser - Pro - Lys - Gly ↓ Forward mutation
Frameshift of opposite sign (intragenic suppressor)	Addition/deletion	5′-A-T-G-A-C-C-T-C-C-G-C-C-G-A-A-A-G-G-G-3′ Met - Thr - Ser - Ala - Glu - Arg ↓ Suppressor mutation (deletion) 5′-A-T-G-A-C-C-C-C-G-C-C-G-A-A-A-G-G-G-3′ Met - Thr - Pro - Pro - Lys - Gly
Extragenic suppressor		
Nonsense suppressor		Gene (e.g., for tyrosine tRNA) undergoes mutational event in its anticodon region that enables it to recognize and align with a mutant nonsense codon (e.g., UAG) to insert an amino acid (tyrosine) and permit completion of translation.
Physiological suppressor		A defect in one chemical pathway is circumvented by another mutation—for example, one that opens up another chemical pathway to the same product, or one that permits more efficient uptake of a compound produced in small quantities because of the original mutation.

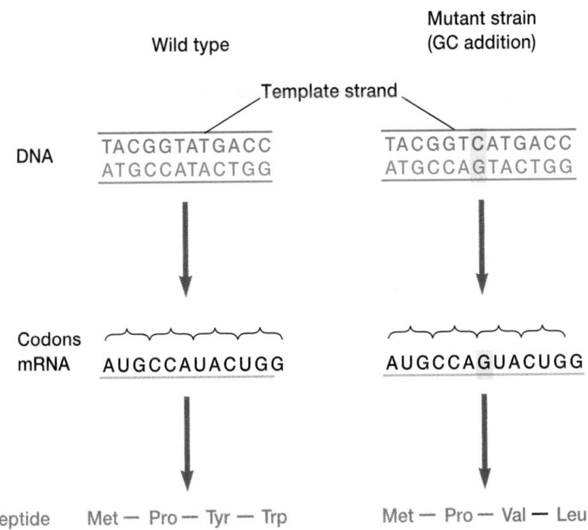

Figure 13.6 Frameshift Mutation. A frameshift mutation resulting from the insertion of a GC base pair. The reading frameshift produces a different peptide after the addition.

Frameshift mutations arise from the insertion or deletion of one or two base pairs within the coding region of the gene. Since the code consists of a precise sequence of triplet codons, the addition or deletion of fewer than three base pairs will cause the reading frame to be shifted for all codons downstream. **Figure 13.6** shows the effect of a frameshift mutation on a short section of mRNA and the amino acid sequence it codes for.

Frameshift mutations usually are very deleterious and yield mutant phenotypes resulting from the synthesis of nonfunctional proteins. In addition, frameshift mutations often produce a nonsense or stop codon so that the peptide product is shorter as well as different in sequence. Of course if the frameshift occurred near the end of the gene, or if there were a second frameshift shortly downstream from the first that restored the reading frame, the phenotypic effect might not be as drastic. A second nearby frameshift that restores the proper reading frame is a good example of an intragenic suppressor mutation (table 13.2).

As noted previously, changes in protein structure can lead to changes in protein function, which in turn can alter the phenotype of an organism. The phenotype of a microorganism can be affected in several different ways. Morphological mutations change the microorganism's colonial or cellular morphology. Lethal mutations, when expressed, result in the death of the microorganism. Because a microbe must be able to grow in order to be isolated and studied, lethal mutations are recovered only if they are recessive in diploid organisms or are conditional mutations in haploid organisms. **Conditional mutations** are those that are expressed only under certain environmental conditions. For example, a conditional lethal mutation in *E. coli* might not be expressed under permissive conditions such as low temperature but would be expressed under restrictive conditions such as high temperature. Thus the hypothetical mutant would grow normally at the permissive temperature but would die at high temperatures.

Biochemical mutations are those causing a change in the biochemistry of the cell. Since these mutations often inactivate a biosynthetic pathway, they frequently eliminate the capacity of the mutant to make an essential macromolecule such as an amino acid or nucleotide. A strain bearing such a mutation has a conditional phenotype: it is unable to grow on medium lacking that molecule, but grows when the molecule is provided. Such mutants are called **auxotrophs,** and they are said to be auxotrophic for the molecule they cannot synthesize. If the wild-type strain from which the mutant arose is a chemoorganotroph able to grow on a minimal medium containing only salts (to supply needed elements such as nitrogen and phosphorus) and a carbon source, it is called a **prototroph.** Another type of biochemical mutant is the resistance mutant. These mutants have acquired resistance to some pathogen, chemical or antibiotic. Auxotrophic and resistance mutants are quite important in microbial genetics due to the ease of their selection and their relative abundance.

Mutations in Regulatory Sequences

Some of the most interesting and informative mutations studied by microbial geneticists are those that occur in the regulatory sequences responsible for controlling gene expression. Constitutive lactose operon mutants in *E. coli* are excellent examples. Many of these mutations map in the operator site and produce altered operator sequences that are not recognized by the repressor protein. Therefore the operon is continuously transcribed, and β-galactosidase is always synthesized. Mutations in promoters also have been identified. If the mutation renders the promoter sequence nonfunctional, the mutant will be unable to synthesize the product even though the coding region of the structural gene is completely normal. Without a fully functional promoter, RNA polymerase rarely transcribes a gene as well as wild type. Regulation of transcription initiation (section 12.2)

Mutations in tRNA and rRNA Genes

Mutations in rRNA and tRNA alter the phenotype of an organism through disruption of protein synthesis. In fact, these mutants often are initially identified because of their slow growth. On the other hand, a suppressor mutation involving tRNA will restore normal (or near normal) growth rates. Here a base substitution in the anticodon region of a tRNA allows the insertion of the correct amino acid at a mutant codon (table 13.2).

1. Define or describe the following: mutation, conditional mutation, auxotroph and prototroph, spontaneous and induced mutations, mutagen, transition and transversion mutations, apurinic and apyrimidinic sites, base analog, DNA-modifying agent, intercalating agent, thymine dimer, wild type, forward and reverse mutations, suppressor mutation, point mutation, silent mutation, missense and nonsense mutations, directed or adaptive mutation, and frameshift mutation.
2. List four ways in which spontaneous mutations might arise.
3. How do the mutagens 5-bromouracil, methyl-nitrosoguanidine, proflavin, and UV radiation induce mutations?
4. Give examples of intragenic and extragenic suppressor mutations.

5. Sometimes a point mutation does not change the phenotype. List all the reasons why this is so.
6. Why might a missense mutation at a protein's surface not affect the phenotype of an organism, while the substitution of an internal amino acid does?

13.2 Detection and Isolation of Mutants

In order to study microbial mutants, one must be able to detect them readily, even when they are rare, and then efficiently isolate them from wild-type organisms and other mutants that are not of interest. Microbial geneticists typically increase the likelihood of obtaining mutants by using mutagens to increase the rate of mutation from the usual one mutant per 10^7 to 10^{11} cells to about one per 10^3 to 10^6 cells. Even at this rate, mutations are rare and carefully devised means for detecting or selecting a desired mutation must be used. This section describes some techniques used in mutant detection, selection, and isolation.

Mutant Detection

When collecting mutants of a particular organism, one must know the normal or wild-type characteristics so as to recognize an altered phenotype. A suitable detection system for the mutant phenotype also is needed. Detection systems in procaryotes and other haploid organisms are straightforward because any mutation should be seen immediately, even if it is a recessive mutation. Sometimes detection of mutants is direct. For instance, if albino mutants of a normally pigmented bacterium are being studied, detection simply requires visual observation of colony color. Other direct detection systems are more complex. For example, the **replica plating** technique is used to detect auxotrophic mutants. It distinguishes between mutants and the wild-type strain based on their ability to grow in the absence of a particular biosynthetic end product (**figure 13.7**). A lysine auxotroph, for instance, will grow on lysine-supplemented media but not on a medium lacking an adequate supply of lysine because it cannot synthesize this amino acid.

Once a detection method is established, mutants are collected. However, mutant collection can present practical problems. Consider a search for the albino mutants mentioned previously. If the mutation rate were around one in a million, on the average a million or more organisms would have to be tested to find one albino mutant. This probably would require several thousand plates. The task of isolating auxotrophic mutants in this way would be even more taxing with the added labor of replica plating. Thus if possible, it is more efficient to use a selection system employing some environmental factor to separate mutants from wild-type microorganisms. Examples of selection systems are described next.

Mutant Selection

An effective selection technique uses incubation conditions under which the mutant grows, because of properties given it by the mutation, whereas the wild type does not. Selection methods often involve reversion mutations or the development of resistance to an

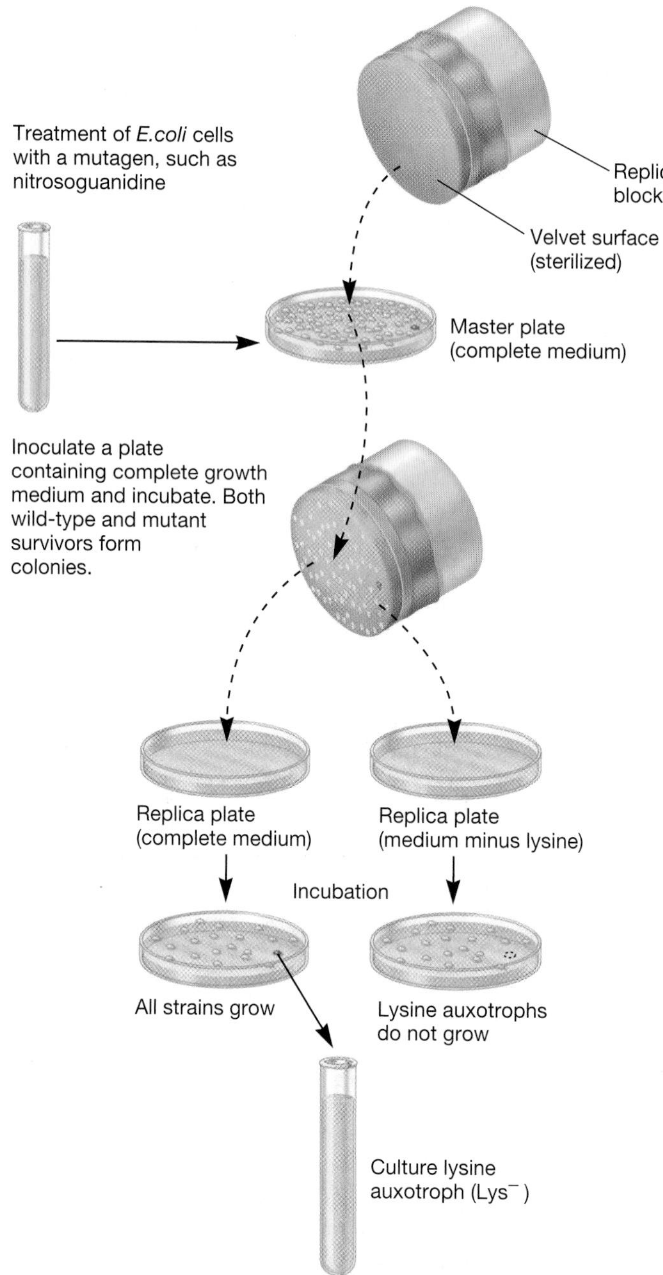
Treatment of E.coli cells with a mutagen, such as nitrosoguanidine
Replica block
Velvet surface (sterilized)
Master plate (complete medium)
Inoculate a plate containing complete growth medium and incubate. Both wild-type and mutant survivors form colonies.
Replica plate (complete medium)
Replica plate (medium minus lysine)
Incubation
All strains grow
Lysine auxotrophs do not grow
Culture lysine auxotroph (Lys$^-$)

Figure 13.7 Replica Plating. The use of replica plating in isolating a lysine auxotroph. Mutants are generated by treating a culture with a mutagen. The culture containing wild type and auxotrophs is plated on complete medium. After the colonies have developed, a piece of sterile velveteen is pressed on the plate surface to pick up bacteria from each colony. Then the velvet is pressed to the surface of other plates and organisms are transferred to the same position as on the master plate. After determining the location of Lys$^-$ colonies growing on the replica with complete medium, the auxotrophs can be isolated and cultured.

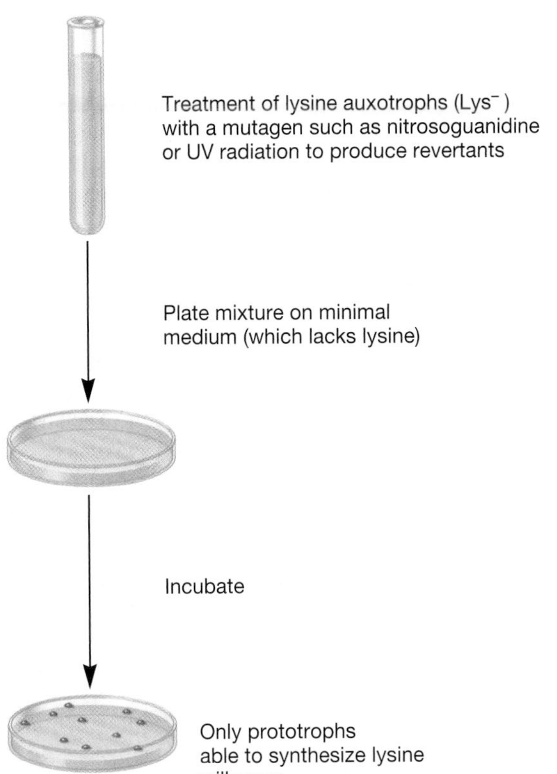
Treatment of lysine auxotrophs (Lys$^-$)
with a mutagen such as nitrosoguanidine
or UV radiation to produce revertants
Plate mixture on minimal
medium (which lacks lysine)
Incubate
Only prototrophs
able to synthesize lysine
will grow

Figure 13.8 Mutant Selection. The production and direct selection of auxotroph revertants. In this example, lysine revertants are selected after treatment of a lysine auxotroph culture because the agar contains minimal medium that does not support auxotroph growth.

environmental stress. For example, if the intent is to isolate revertants from a lysine auxotroph (Lys$^-$), the approach is quite easy. A large population of lysine auxotrophs is plated on minimal medium lacking lysine, incubated, and examined for colony formation. Only cells that have mutated to restore the ability to manufacture lysine will grow on minimal medium (**figure 13.8**). Several million cells can be plated on a single petri dish, but only the rare revertant cells will grow. Thus many cells can be tested for mutations by scanning a few petri dishes for growth. This method has proven very useful in determining the relative mutagenicity of many substances.

Methods for selecting mutants resistant to a particular environmental stress follow a similar approach. Often wild-type cells are susceptible to virus attack or antibiotic treatment, so it is possible to grow the bacterium in the presence of the agent and look for surviving organisms. Consider the example of a phage-sensitive wild-type bacterium. When the organism is cultured in medium lacking the virus and then plated on selective medium containing viruses, any colonies that form are resistant to virus attack and very likely are mutants in this regard. This type of selection can be used for virtually any environmental parameter; resistance to bacteriophages (bacterial viruses), antibiotics, or temperature are most commonly employed.

Substrate utilization mutations also are employed in bacterial selection. Many bacteria use only a few primary carbon sources. With such bacteria, it is possible to select mutants by plating a culture on medium containing an alternate carbon source. Any colonies that appear can use the substrate and are probably mutants.

Mutant detection and selection methods are used for purposes other than understanding more about the nature of genes or the biochemistry of a particular microorganism. One very important role of mutant selection and detection techniques is in the study of carcinogens. The next section briefly describes one of the first and perhaps best known of the carcinogen testing systems.

Carcinogenicity Testing

An increased understanding of the mechanisms of mutation and cancer induction has stimulated efforts to identify environmental carcinogens. The observation that many carcinogenic agents also are mutagenic is the basis for detecting potential carcinogens by testing for mutagenicity while taking advantage of bacterial selection techniques and short generation times. The **Ames test,** developed by Bruce Ames in the 1970s, has been widely used to test for carcinogens. The Ames test is a mutational reversion assay employing several special strains of *Salmonella enterica* serovar Typhimurium, each of which has a different mutation in the histidine biosynthesis operon; that is to say, they are histidine auxotrophs. The bacteria also have mutational alterations of their cell walls that make them more permeable to test substances. To further increase assay sensitivity, the strains are defective in the ability to repair DNA correctly.

In the Ames test these special tester strains of *Salmonella* are plated with the substance being tested and the appearance of visible colonies followed (**figure 13.9**). To ensure that DNA replication can take place in the presence of the potential mutagen, the bacteria and test substance are mixed in dilute molten top agar to which a trace of histidine has been added. This molten mix is then poured on top of minimal agar plates and incubated for 2 to 3 days at 37°C. All of the histidine auxotrophs grow for the first few hours in the presence of the test compound until the histidine is depleted. This is necessary because, as previously discussed, replication is required for the development of a mutation (figure 13.3). Once the histidine supply is exhausted, only revertants that have mutationally regained the ability to synthesize histidine continue to grow and produce visible colonies. These colonies need only be counted and compared to controls in order to estimate the relative mutagenicity of the compound: the more colonies, the greater the mutagenicity.

A mammalian liver extract is also often added to the molten top agar prior to plating. The extract converts potential carcinogens into electrophilic derivatives that readily react with DNA. This process occurs naturally when foreign substances are metabolized in the liver. Because bacteria do not have this activation system, addition of the liver extract promotes the same kind of enzymatic transformations that occur in mammals. Many potential carcinogens, such as aflatoxins, are not actually carcinogenic until they are modified in the liver. The addition of extract shows which

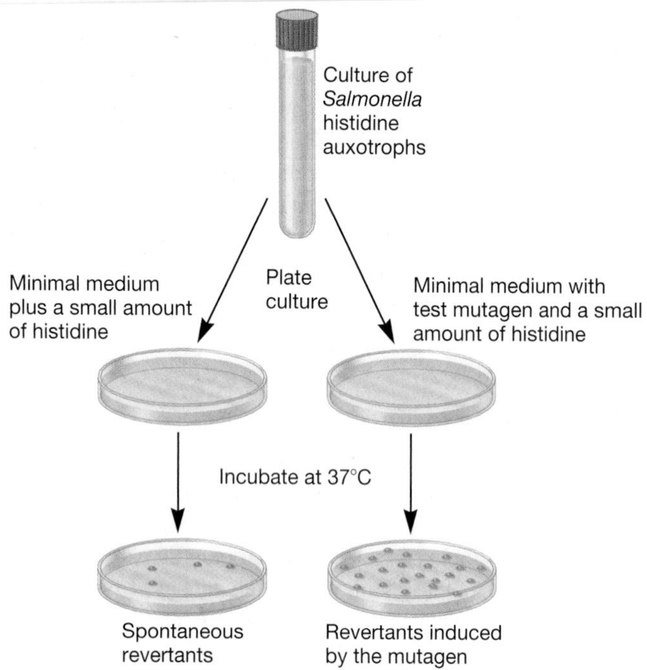

Figure 13.9 The Ames Test for Mutagenicity. See text for details.

compounds have intrinsic mutagenicity and which need activation after uptake. Despite the use of liver extracts, only about half the potential animal carcinogens are detected by the Ames test.

1. Describe how replica plating is used to detect and isolate auxotrophic mutants.
2. Why are mutant selection techniques generally preferable to the direct detection and isolation of mutants?
3. Briefly discuss how reversion mutations, resistance to an environmental factor, and the ability to use a particular nutrient can be employed in mutant selection.
4. Describe how you would isolate a mutant that required histidine for growth and was resistant to penicillin.
5. What is the Ames test and how is it carried out? What assumption concerning mutagenicity and carcinogenicity is it based upon?

13.3 DNA Repair

Because replication errors and a variety of mutagens can alter the nucleotide sequence, a microorganism must be able to repair changes in the sequence that might be lethal. DNA is repaired by several different mechanisms besides **proofreading** by replication enzymes (DNA polymerases can remove an incorrect nucleotide immediately after its addition to the growing end of the chain). Repair in *E. coli* is best understood and is briefly described in this section. DNA replication (section 11.4)

Excision Repair

Excision repair corrects damage that causes distortions in the double helix. Two types of excision repair systems have been described: nucleotide excision repair and base excision repair. They are distinguished by the enzymes used to correct DNA damage. However, they both use the same approach to repair: remove the damaged portion of a DNA strand and use the intact complementary strand as the template for synthesis of new DNA.

In **nucleotide excision repair,** a repair enzyme called UvrABC endonuclease removes damaged bases and some bases on either side of the lesion. The resulting single-stranded gap, about 12 nucleotides long, is filled by DNA polymerase I, and DNA ligase joins the fragments. **Figure 13.10** presents the process in detail. This system can remove thymine dimers (figure 13.5) and repair almost any other injury that produces a detectable distortion in DNA.

Base excision repair employs DNA glycosylases to remove damaged or unnatural bases yielding apurinic or apyrimidinic (AP) sites. Special endonucleases called AP endonucleases recognize the damaged DNA and nick the backbone at the APsite (**figure 13.11**). DNA polymerase I removes the damaged region, using its 5′ to 3′ exonuclease activity. It then fills in the gap, and DNA ligase joins the DNA fragments.

Direct Repair

Thymine dimers and alkylated bases often are corrected by **direct repair. Photoreactivation** is the repair of thymine dimers by splitting them apart into separate thymines with the help of visible light in a photochemical reaction catalyzed by the enzyme photolyase (**figure 13.12*a***). Methyls and some other alkyl groups that have been added to the O^6 position of guanine can be removed with the help of an enzyme known as alkyltransferase or methylguanine methyltransferase (figure 13.12*b*). Thus damage to guanine from mutagens such as methyl-nitrosoguanidine (figure 13.4) can be repaired directly.

Mismatch Repair

Despite the accuracy of DNA polymerase and continual proofreading, errors still are made during DNA replication. Remaining mismatched bases are usually detected and repaired by the **mismatch repair system** in *E. coli* (**figure 13.13**). The mismatch correction enzyme MutS scans the newly replicated DNA for mismatched pairs. Another enzyme, MutH, removes a stretch of newly synthesized DNA around the mismatch. A DNA polymerase then replaces the excised nucleotides, and the resulting nick is sealed with a ligase. In this regard, mismatch repair is similar to excision repair.

Successful mismatch repair depends on the ability of enzymes to distinguish between old and newly replicated DNA strands. This distinction is possible because newly replicated DNA strands lack methyl groups on their bases, whereas older DNA has methyl groups on the bases of both strands. **DNA methylation** is catalyzed by DNA methyltransferases and results in three different products: *N*6-methyladenine, 5-methylcytosine, and *N*4-methylcytosine. After strand synthesis, the *E. coli* DNA adenine methyltransferase

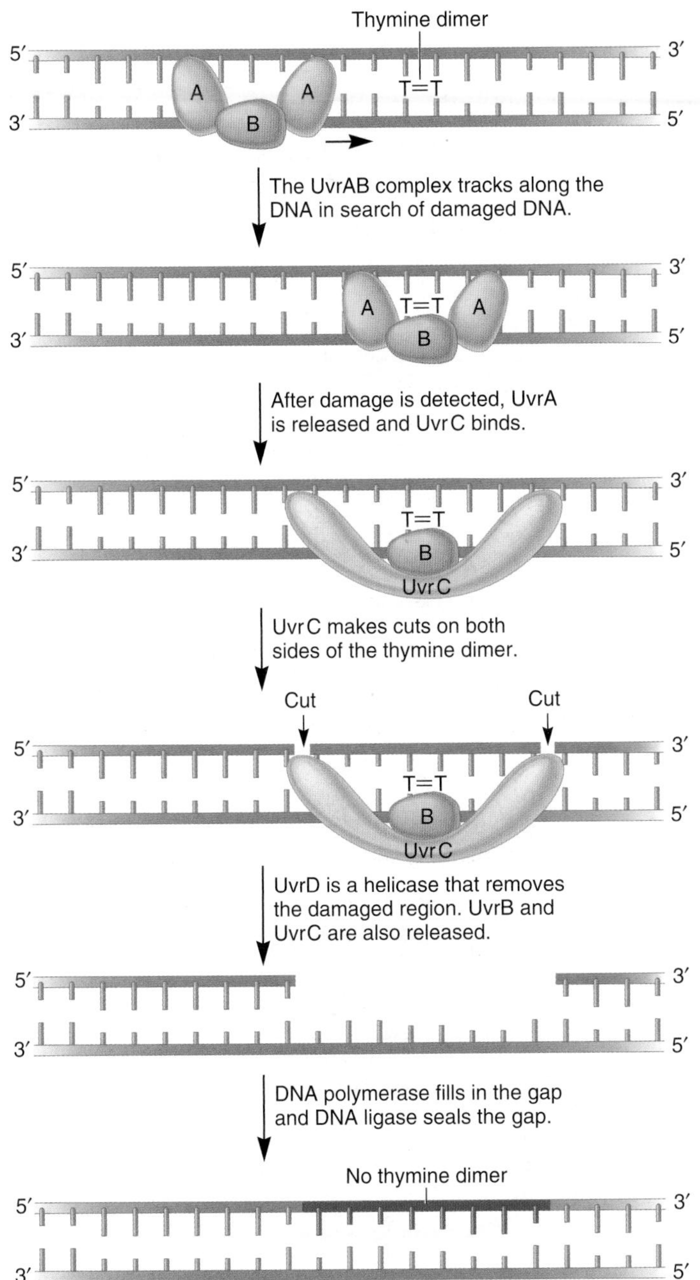

Figure 13.10 Nucleotide Excision Repair in *E. coli.*

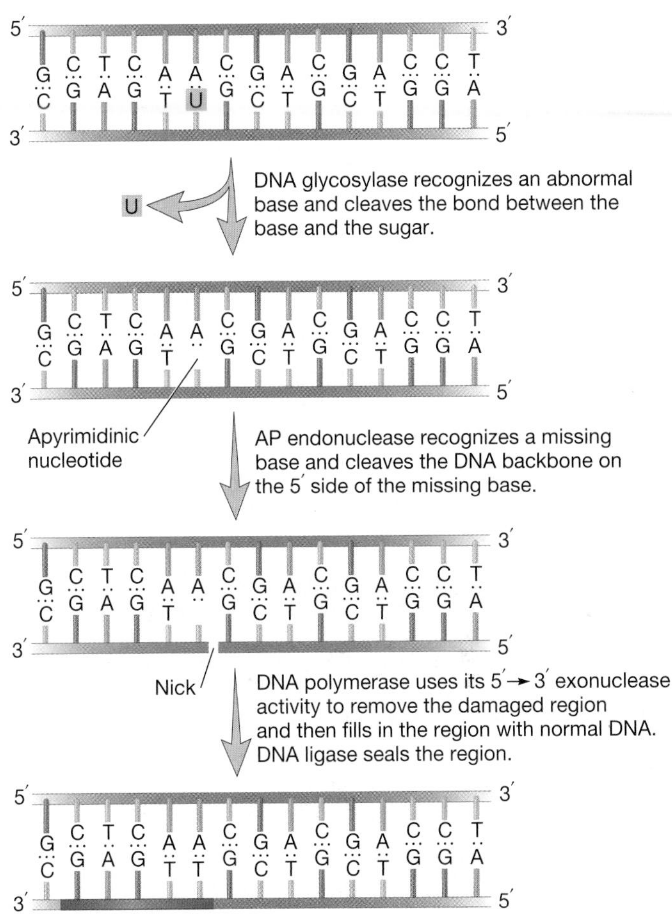

Figure 13.11 Base Excision Repair.

(DAM) methylates adenine bases in GATC sequences to form *N*6-methyladenine. For a short time after the replication fork has passed, the new strand lacks methyl groups while the template strand is methylated. In other words, the DNA is temporarily hemi-methylated. The repair system cuts out the mismatch from the unmethylated strand.

Recombinational Repair

Recombinational repair corrects damaged DNA in which both bases of a pair are missing or damaged, or where there is a gap opposite a lesion. In this type of repair the **RecA protein** cuts a piece of template DNA from a sister molecule and puts it into the gap or uses it to replace a damaged strand (**figure 13.14**). Although procaryotes are haploid, another copy of the damaged segment often is available because either it has recently been replicated or the cell is growing rapidly and has more than one copy of its chromosome. Once the template is in place, the remaining damage can be corrected by another repair system.

The SOS Response

Despite having multiple repair systems, sometimes the damage to an organism's DNA is so great that the normal repair mechanisms just described cannot repair all the damage. As a result, DNA synthesis stops completely. In such situations, a global control network called the **SOS response** is activated. The SOS response, like recombinational repair, is dependent on the activity of the RecA protein. RecA binds to single- or double-stranded DNA breaks and gaps generated by cessation of DNA synthesis. RecA binding initiates recombinational repair. Simultaneously, RecA takes on a proteolytic function that destroys a repressor protein called LexA. LexA negatively regulates the function of many genes involved in DNA repair and synthesis. Destruction of LexA increases transcription of genes for excision repair and

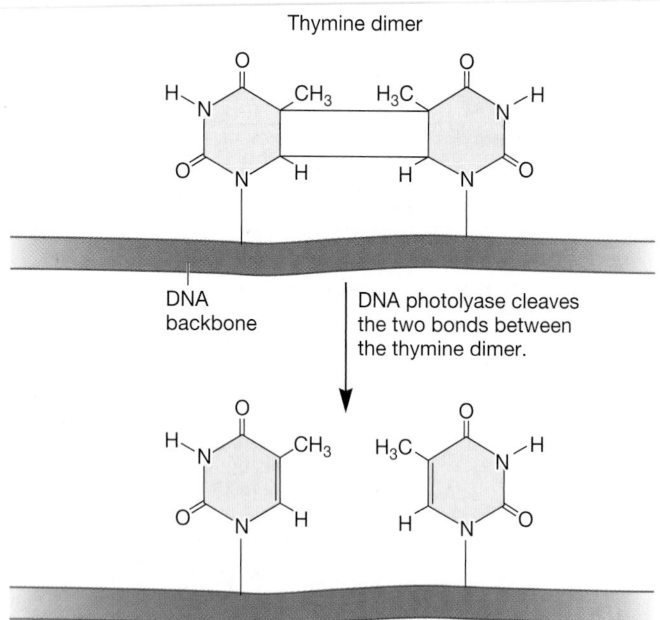

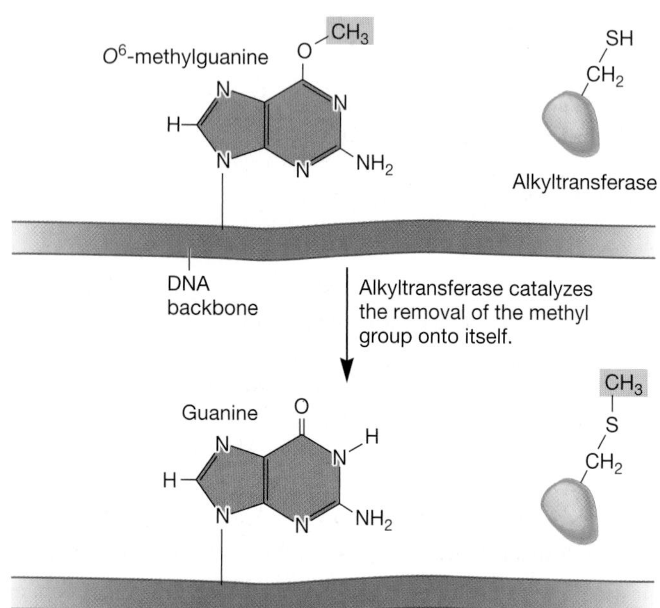

Figure 13.12 Direct Repair. **(a)** The repair of thymine dimers by photolyase. **(b)** The repair of methylguanine by the transfer of the methyl group to alkyltransferase.

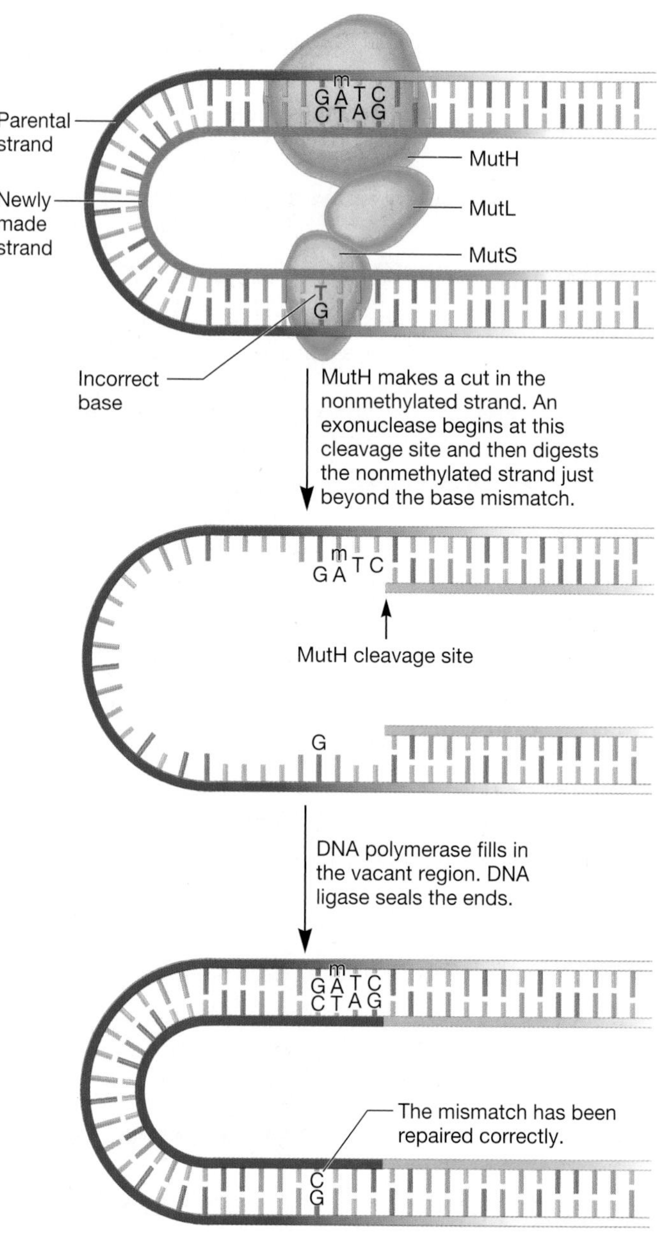

Figure 13.13 Methyl-Directed Mismatch Repair in *E. coli*. MutS slides along the DNA and recognizes base mismatches in the double helix. MutL binds to MutS and acts as a linker between MutS and MutH. The DNA must loop for this interaction to occur. The role of MutH is to identify the methylated strand of DNA, which is the nonmutated parental strand. The methylated adenine is designated with an m.

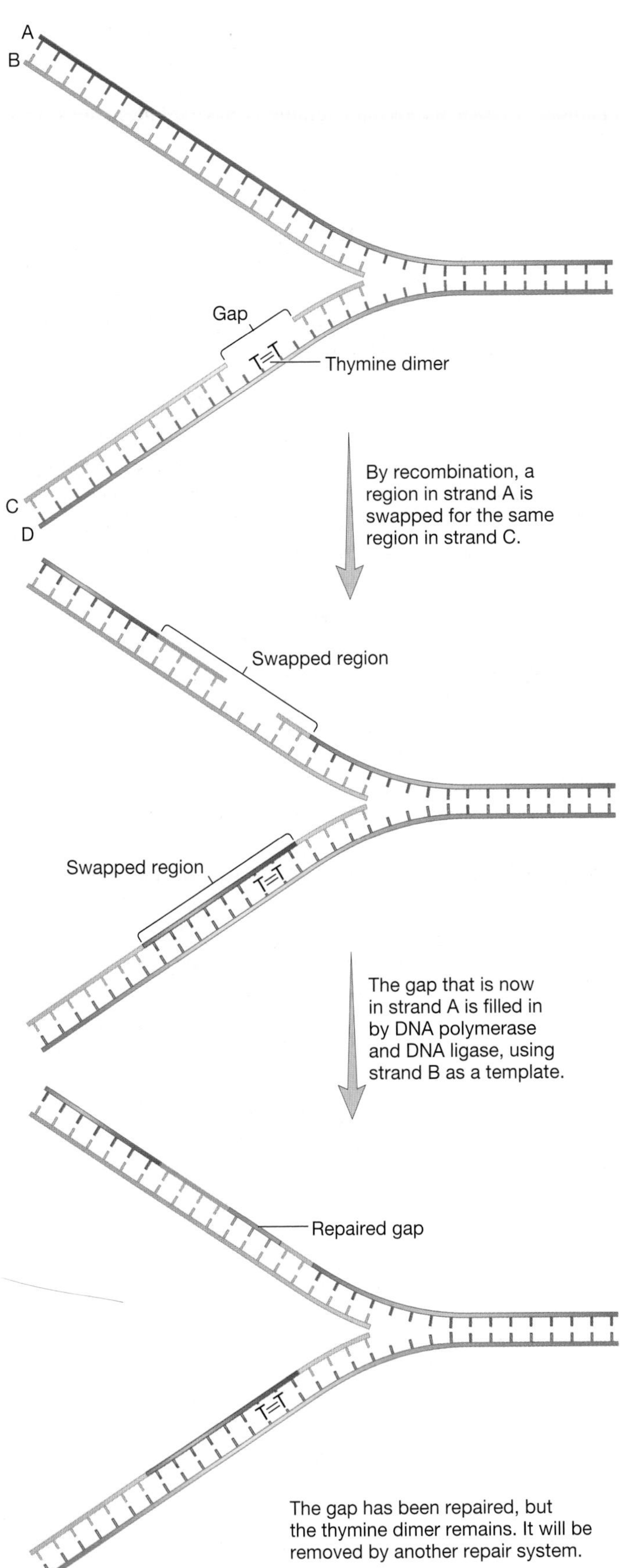
A
B
Gap
T=T
Thymine dimer
C
D
By recombination, a region in strand A is swapped for the same region in strand C.
Swapped region
Swapped region
T=T
The gap that is now in strand A is filled in by DNA polymerase and DNA ligase, using strand B as a template.
Repaired gap
T=T
The gap has been repaired, but the thymine dimer remains. It will be removed by another repair system.

Figure 13.14 Recombinational Repair.

recombinational repair, in particular. The first genes to be transcribed are those that encode the Uvr proteins needed for nucleotide excision repair (figure 13.10). Then genes involved in recombinational repair are further upregulated. To give the cell time to repair its DNA, the protein SfiA is produced; SfiA blocks cell division. Finally, if the DNA has not been fully repaired after about 40 minutes, a process called **translesion DNA synthesis** is triggered. In this process, DNA polymerases IV (also known as DinB) and V (UmuCD) synthesize DNA across gaps and other lesions (e.g., thymine dimers) that had stopped DNA polymerase III. However, because an intact template does not exist, these SOS response polymerases often insert incorrect bases. Furthermore, they lack proofreading activity. Therefore even though DNA synthesis continues, it is highly error prone and results in the generation of numerous mutations. The SOS response is so named because it is a response made in a life-or-death situation. The response increases the likelihood that some cells will survive by allowing DNA synthesis to continue. For the cell, the risk of dying because of failure to replicate DNA is greater than the risk posed by the mutations generated by this error-prone process.

1. Define the following: proofreading, excision repair, photoreactivation, methylguanine methyltransferase, mismatch repair, direct repair, DNA methylation, recombinational repair, RecA protein, SOS response, and LexA repressor.
2. Describe in general terms the mechanisms of the following repair processes: excision repair, recombinational repair, direct repair, and SOS response.
3. Explain how the following DNA alterations and replication errors would be corrected (there may be more than one way): base addition errors by DNA polymerase III during replication, thymine dimers, AP sites, methylated guanines, and gaps produced during replication.

13.4 Creating Genetic Variability

As discussed previously, the consequences of mutations can range from no effect to being lethal, depending not only on the nature of the mutation but also on the environment in which the organism lives. Thus all mutations are subject to selective pressure, and this determines if a mutation will survive in a population. Each mutant form that survives is called an **allele,** an alternate form of the gene. Mutant alleles, as well as the wild-type allele, can be combined with other genes, leading to an increase in the genetic variability within a population. Each genotype in a population can be selected for or selected against. Organisms with genotypes, and therefore phenotypes, that are best suited to the environment survive and are able to pass on their genes. Shifts in environmental pressures can lead to changes in the population and ultimately result in the evolution of new species. The mechanisms by which new combinations of genes are generated are the topic of this section. All involve **recombination,** the process in which one or more nucleic acid molecules are rearranged or combined to produce a new nucleotide sequence. This is normally accompanied by a phenotypic change. Geneticists refer to organisms produced following a recombination event as recombinant organisms or simply **recombinants.**

Recombination in Eucaryotes

The processes that create genetic variability in eucaryotes differ from those in procaryotes. Recombinant genotypes can arise from the integration of viruses into the host chromosomes and movement of mobile genetic elements. However, the most important recombination events occur during the sexual cycle, including meiosis, of those eucaryotes capable of sexual reproduction. During meiosis, **crossing-over** between homologous chromosomes—chromosomes containing identical sequences of genes (**figure 13.15**)—generates new combinations of alleles. This is followed by segregation of chromosomes into gametes and then by zygote formation, which further increases genetic variability. This transfer of genes from parents to progeny is sometimes called vertical gene transfer.

Horizontal Gene Transfer in Procaryotes

Unlike eucaryotes, procaryotes do not reproduce sexually, nor do they undergo meiosis. This would suggest that genetic variation in populations of procaryotes would be relatively limited, only occurring with the advent of a new mutation or by the integration of viruses and mobile genetic elements into the chromosome. However, this is not the case. Procaryotes have evolved three different mechanisms for creating recombinants. These mechanisms are referred to collectively as **horizontal** (or **lateral**) **gene transfer (HGT)**. HGT is distinctive from vertical gene transfer because genes from one independent, mature organism are transferred to another, often creating a stable recombinant having characteristics of both the donor and the recipient.

It was once thought that HGT occurred primarily between members of the same species. However, it is increasingly clear that HGT has been important in the evolution of many species, and that it is still commonplace in many environments. Furthermore, there are clear examples of DNA from one species being transferred to distantly related species. The importance of HGT cannot be overstated. Its recognition as an evolutionary force has caused evolutionary biologists to reconsider the universal tree of life first proposed by Carl Woese in the 1970s. It is felt by some that phylogenetic relationships are better represented by a web or network of relationships rather than a tree (*see figure 19.15*). HGT is still shaping genomes. For instance, it has been demonstrated that procaryotes sharing an ecological niche can exchange genes and this alters the nature of the microbial community in a habitat. Another important example is the evolution and spread of antibiotic-resistance genes among pathogenic bacteria. Microbial evolution (section 19.1)

During HGT, a piece of donor DNA, the **exogenote,** must enter and become a stable part of the recipient cell. This can be accomplished in two ways, depending on the nature of the exogenote. If the exogenote is a DNA fragment that is incapable of replicating itself and is susceptible to degradation by nucleases present in the recipient (e.g., a small, linear piece of the donor's chromosome), then the exogenote must integrate into the recipient cell's chromosome (**endogenote**), replacing a portion of the recipient cell's genetic material. As this occurs, the recipient becomes temporarily diploid for a portion of its genome and is called a **merozygote** (**figure 13.16**). However, if the exogenote is capable of self-replication and is resistant to attack by the recipient cell's nucleases (e.g., a plasmid), then it need not integrate into the recipient cell's chromosome. Instead, it is maintained independent of the endogenote.

Horizontal gene transfer can take place in three ways: direct transfer between two bacteria temporarily in physical contact (conjugation), transfer of a naked DNA fragment (transformation), and transport of bacterial DNA by bacterial viruses (transduction). Whatever the mode of transfer, the exogenote has only four possible fates in the recipient (figure 13.16). First, when the exogenote has a sequence homologous to that of the endogenote, integration may occur; that is, it may pair with the recipient DNA and be incorporated to yield a recombinant genome. Second, the foreign DNA sometimes persists outside the endogenote and replicates to produce a clone of partially diploid cells. Third, the exogenote may survive, but not replicate, so that only one cell is a partial diploid. Finally, host cell nucleases may degrade the exogenote, a process called **host restriction.**

Recombination at the Molecular Level

Although different processes are used in eucaryotes and procaryotes to create recombinant organisms, the mechanisms of recombination at the molecular level are remarkably similar.

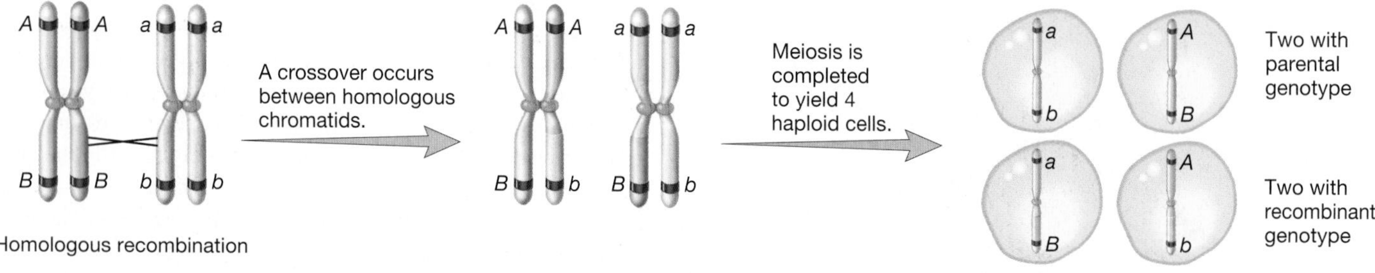

Figure 13.15 Recombination During Meiosis. During meiosis, homologous chromosomes pair and crossing-over can occur. The recombinant genotypes formed are inherited by progeny organisms, where they can result in recombinant phenotypes. Crossing-over involving similar DNA sequences is called homologous recombination.

Three types of recombination are observed: homologous recombination, site-specific recombination, and transposition. **Homologous recombination,** the most common form of recombination, usually involves a reciprocal exchange between a pair of DNA molecules with the same nucleotide sequence. It can occur anywhere on the chromosome, and it results from DNA strand breakage and reunion leading to crossing-over. Homologous recombination is carried out by the products of the *rec* genes, including the RecA protein, which is also important for DNA repair (**table 13.3**). The most widely accepted model of homologous recombination is the **double-strand break model** (**figure 13.17**). It proposes that duplex DNA with a double-stranded break is processed to create DNA with single-stranded ends. RecA promotes the insertion of one single-stranded end into an intact, homologous piece of DNA. This is called strand invasion. As can be seen in figure 13.17, strand invasion results in the formation of two gaps in the two parent DNA molecules. The gaps are filled, yielding a structure with **heteroduplex DNA;** that is, it contains strands derived from both parent molecules. The two parental DNA molecules are now linked together by two structures called Holliday junctions. These structures move along the DNA molecule during branch migration until they are finally cut and the two DNA molecules are separated. Depending on how this occurs, the resulting DNA molecules will be either recombinant or nonrecombinant. In some cases, a nonreciprocal form of homologous recombination occurs (**figure 13.18**). In **nonreciprocal homologous recombination,** a piece of genetic material is inserted into the chromosome through the incorporation of a single strand to form a stretch of heteroduplex DNA. The second type of recombination, **site-specific recombination,** is particularly important in the integration of virus genomes into host chromosomes. In site-specific recombination, the genetic material bears only a small region of homology with the chromosome it joins. The enzymes responsible for this event are often specific for sequences within the particular virus and its host. The third kind of recombination is **transposition,** which also does not depend on sequence homology. It can occur at many sites in the genome and will be discussed in more detail in section 13.5.

Until about 1945, the primary focus in genetic analysis was on the recombination of genes in plants and animals. The early work on recombination in higher eucaryotes led to the foundation of classical genetics, but it was the development of bacterial and phage genetics between about 1945 and 1965 that really stimulated a rapid advance in our understanding of molecular genetics. Therefore recombination in the *Bacteria* and viruses is the major focus of the following discussion of recombination. We begin with a consideration of transposons and plasmids—genetic elements that can be involved in recombination events—and then turn to mechanisms of horizontal gene transfer in *Bacteria.*

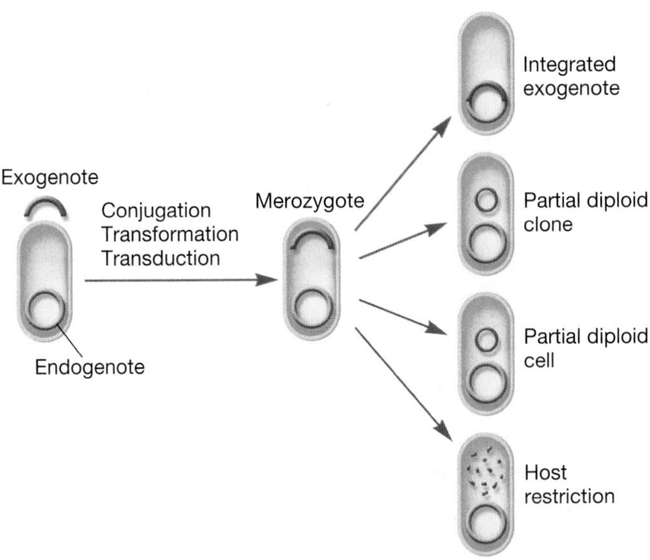

Figure 13.16 The Production and Fate of Merozygotes. See text for discussion.

1. Define the following terms: recombination, crossing-over, homologous recombination, site-specific recombination, transposition, exogenote, endogenote, horizontal (lateral) gene transfer, merozygote, and host restriction.
2. Distinguish among the three forms of recombination mentioned in this section.
3. What four fates can DNA have after entering a bacterium?

Table 13.3 *E. coli* **Homologous Recombination Proteins**

Protein	Description
Rec BCD	Recognizes double-stranded breaks and then generates single-stranded regions at the break site that are involved in strand invasion
Single-strand binding protein	Prevents excessive strand degradation by RecBCD
RecA	Promotes strand invasion and displacement of complementary strand to generate D loop
RecG	Helps form Holliday junctions and promotes branch migration
RuvABC	Endonuclease that binds Holliday junctions, promotes branch migration, and cuts strands in the Holliday junction in order to separate chromosomes

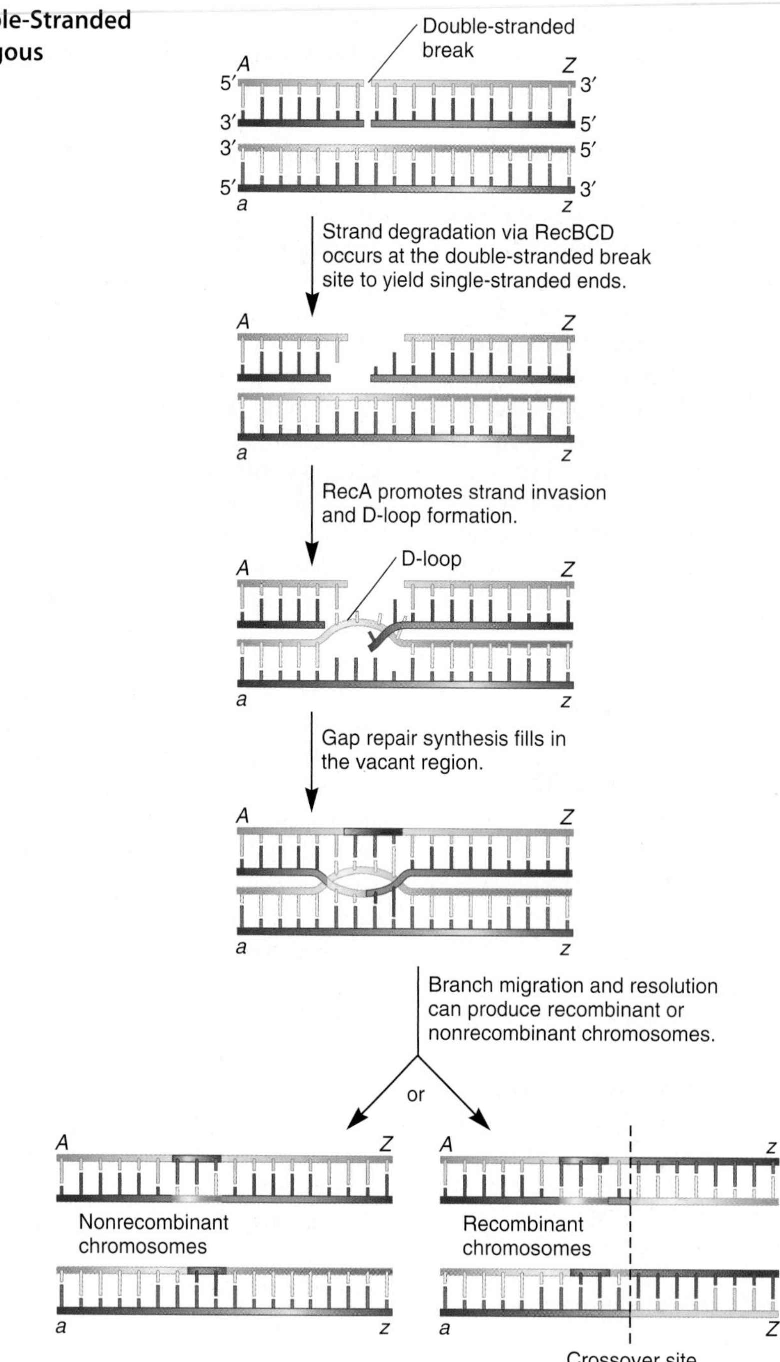

Figure 13.17 The Double-Stranded Break Model of Homologous Recombination.

13.5 Transposable Elements

The chromosomes of procaryotes, viruses, and eucaryotic cells contain pieces of DNA that can move and integrate into different sites in the chromosomes. Such movement is called transposition, and it plays important roles in the generation of new gene combinations. DNA segments that carry the genes required for transposition are **transposable elements** or **transposons,** sometimes called "jumping genes." Unlike other processes that reorganize DNA, transposition does not require extensive areas of homology between the transposon and its destination site. Transposons were first discovered in the 1940s by Barbara McClintock during her studies on maize genetics (a discovery for which she was awarded the Nobel prize in 1983). They have been most intensely studied in *Bacteria.*

The simplest transposable elements are **insertion sequences** or IS elements (**figure 13.19*a***). An IS element is a short sequence of DNA (around 750 to 1,600 base pairs [bp] in length) containing only the genes for those enzymes required for its transposition and bounded at both ends by identical or very similar sequences of nucleotides in reversed orientation known as inverted repeats. Inverted repeats are usually about 15 to 25 base pairs long and vary among IS elements so that each type of IS has its own characteristic inverted repeats. Between the inverted repeats is a gene that codes for an enzyme called **transposase.** This enzyme is required for transposition and accurately recognizes the ends of the IS. Each type of element is named by giving it the prefix IS followed by a number. In *E. coli* several copies of different IS elements have been observed; some of their properties are given in **table 13.4.**

Transposable elements also can contain genes other than those required for transposition (for example, antibiotic resistance or toxin genes). These elements often are called **composite transposons.** Composite transposons often consist of a central region containing the extra genes, flanked on both sides by IS elements that are identical or very similar in sequence (figure 13.19*b*). Many composite transposons are simpler in organization. They are bounded by short inverted repeats, and the coding region contains both transposition genes and the extra genes. It is believed that composite transposons are formed when two IS elements associate with a central segment containing one or more genes. This association could arise if an IS element replicates and moves only a gene or two down the chromosome. Composite transposon names begin with the prefix Tn. Some properties of selected composites are given in **table 13.5.**

Figure 13.18 Nonreciprocal Homologous Recombination. The Fox model for nonreciprocal homologous recombination. This mechanism has been proposed for the recombination occurring during transformation in some bacteria.

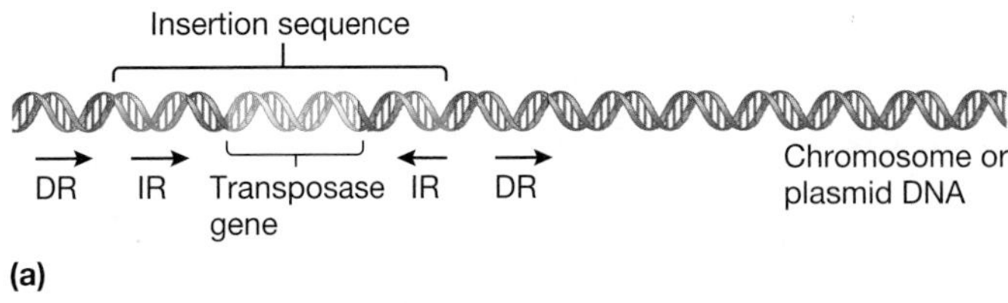

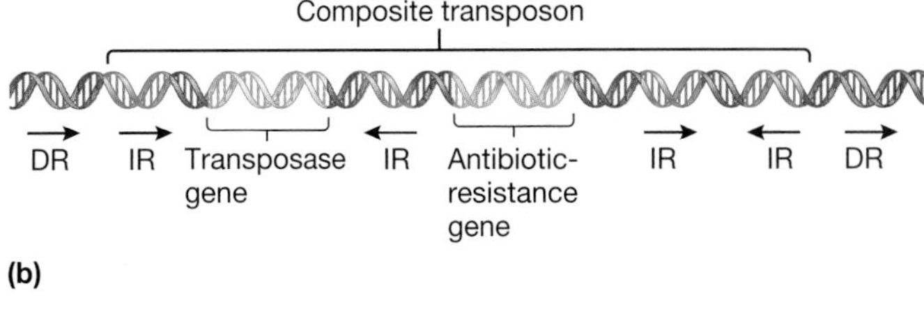

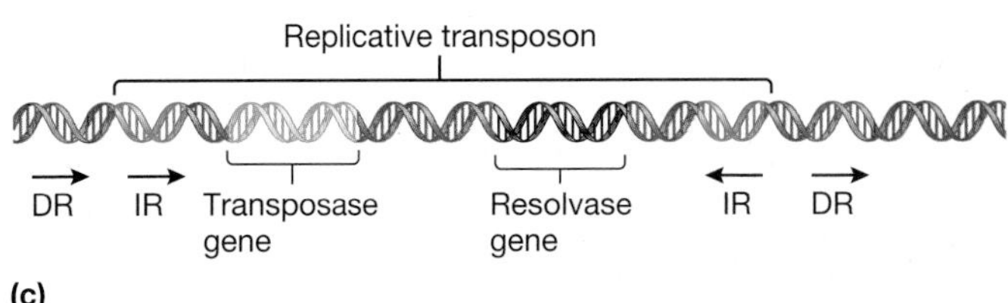

Figure 13.19 Transposable Elements. All transposable elements contain common elements. These include inverted repeats (IRs) at the ends of the element and a transposase gene. **(a)** Insertion sequences consist only of IRs on either side of the transposase gene. **(b)** Composite transposons and **(c)** replicative tranposons contain additional genes. Insertion sequences and composite transposons move by simple (cut-and-paste) transposition. Replicative transposons move by replicative transposition. DRs, direct repeats in host DNA, flank a transposable element.

Table 13.4 The Properties of Selected Insertion Sequences

Insertion Sequence	Length (bp)	Inverted Repeat (Length in bp)	Target Site (Length in bp)	Number of Copies on *E. coli* Chromosome
IS*1*	768	23	9 or 8	6–10
IS*2*	1,327	41	5	4–13(1)[a]
IS*3*	1,400	38	3–4	5–6(2)
IS*4*	1,428	18	11 or 12	1–2
IS*5*	1,195	16	4	10–11

[a]The value in parentheses indicates the number of IS elements on the F factor plasmid.

Table 13.5 The Properties of Selected Composite Transposons

Transposon	Length (bp)	Terminal Repeat Length	Terminal Module	Genetic Markers
Tn*3*	4,957	38		Ampicillin resistance
Tn*501*	8,200	38		Mercury resistance
Tn*951*	16,500	Unknown		Lactose utilization
Tn*5*	5,700		IS*50*	Kanamycin resistance
Tn*9*	2,500		IS*1*	Chloramphenicol resistance
Tn*10*	9,300		IS*10*	Tetracycline resistance
Tn*903*	3,100		IS*903*	Kanamycin resistance
Tn*1681*	2,061		IS*1*	Heat-stable enterotoxin
Tn*2901*	11,000		IS*1*	Arginine biosynthesis

The process of transposition in procaryotes can occur by two basic mechanisms. **Simple transposition,** also called **cut-and-paste transposition,** involves transposase-catalyzed excision of the transposon, followed by cleavage of a new target site and ligation of the transposon into this site (**figure 13.20**). Target sites are specific sequences about five to nine base pairs long. When a transposon inserts at a target site, the target sequence is duplicated so that short, direct-sequence repeats flank the transposon's terminal inverted repeats.

The second transposition mechanism is **replicative transposition.** In this mechanism, the original transposon remains at the parental site on the chromosome and a replicate is inserted at the target DNA site (**figure 13.21**).The transposition of the Tn*3* transposon is a well-studied example of replicative transposition. In the first stage, DNA containing Tn*3* fuses with the target DNA to form a cointegrate molecule (figure 13.21, step 1). This process requires the Tn*3* transposase enzyme coded for by the *tnpA* gene (**figure 13.22**). Note that the cointegrate has two copies of the Tn*3* transposon. In the second stage the cointegrate is resolved to yield two DNA molecules, each with a copy of the transposon (figure 13.21, step 3). Resolution involves a crossover and is catalyzed by a resolvase enzyme coded for by the *tnpR* gene (figure 13.22).

Transposable elements produce a variety of important effects. They can insert within a gene to cause a mutation or stimulate DNA rearrangement, leading to deletions of genetic material. Because some transposons carry stop codons or termination sequences, when transposed into genes they may block translation or transcription, respectively. Likewise, other transposons carry promoters and can activate genes near the point of insertion. Thus transposons can turn genes on or off. Transposons also are located in plasmids and participate in such processes as plasmid fusion, insertion of plasmids into chromosomes, and plasmid evolution.

The role of transposons in plasmid evolution is of particular note. Plasmids can contain several different transposon-target sites. Therefore, transposons frequently move between plasmids. Of concern is the fact that many transposons contain antibiotic-resistance genes. Thus as they move from one plasmid to another, resistance genes are introduced into the target plasmid, creating a resistance (R) plasmid. Multiple drug-resistance plasmids can arise from the accumulation of transposons in a plasmid (figure 13.22). Many R plasmids are able to move from one cell to another during conjugation, which spreads the resistance genes throughout a population. Finally, because transposons also move between plasmids and chromosomes, drug resistance genes can exchange between these two molecules, resulting in the further spread of antibiotic resistance.

Some transposons bear transfer genes and can move between bacteria through the process of conjugation, as discussed in section 13.7. A well-studied example of a **conjugative transposon** is Tn*916* from *Enterococcus faecalis.* Although Tn*916* cannot replicate autonomously, it can transfer itself from *E. faecalis* to a variety of recipients and integrate into their chromosomes. Because it carries a gene for tetracycline resistance, this conjugative transposon also spreads drug resistance.

13.6 Bacterial Plasmids

Conjugation, the transfer of DNA between bacteria involving direct contact, depends on the presence of an "extra" piece of DNA known as a plasmid. Plasmids play many important roles in the lives of procaryotes. They also have proved invaluable to microbiologists and molecular geneticists in constructing and transferring new genetic combinations and in cloning genes as described in chapter 14.

Recall from chapter 3 that **plasmids** are small double-stranded DNA molecules that can exist independently of host chromosomes. They have their own replication origins and autonomously replicate and are stably inherited. Some plasmids are **episomes,** plasmids that can exist either with or without being integrated into host chromosomes. Although there are a variety of plasmid types, our concern here is with **conjugative plasmids.** These plasmids can transfer copies of themselves to other bacteria during the process of conjugation, which is discussed in section 13.7.

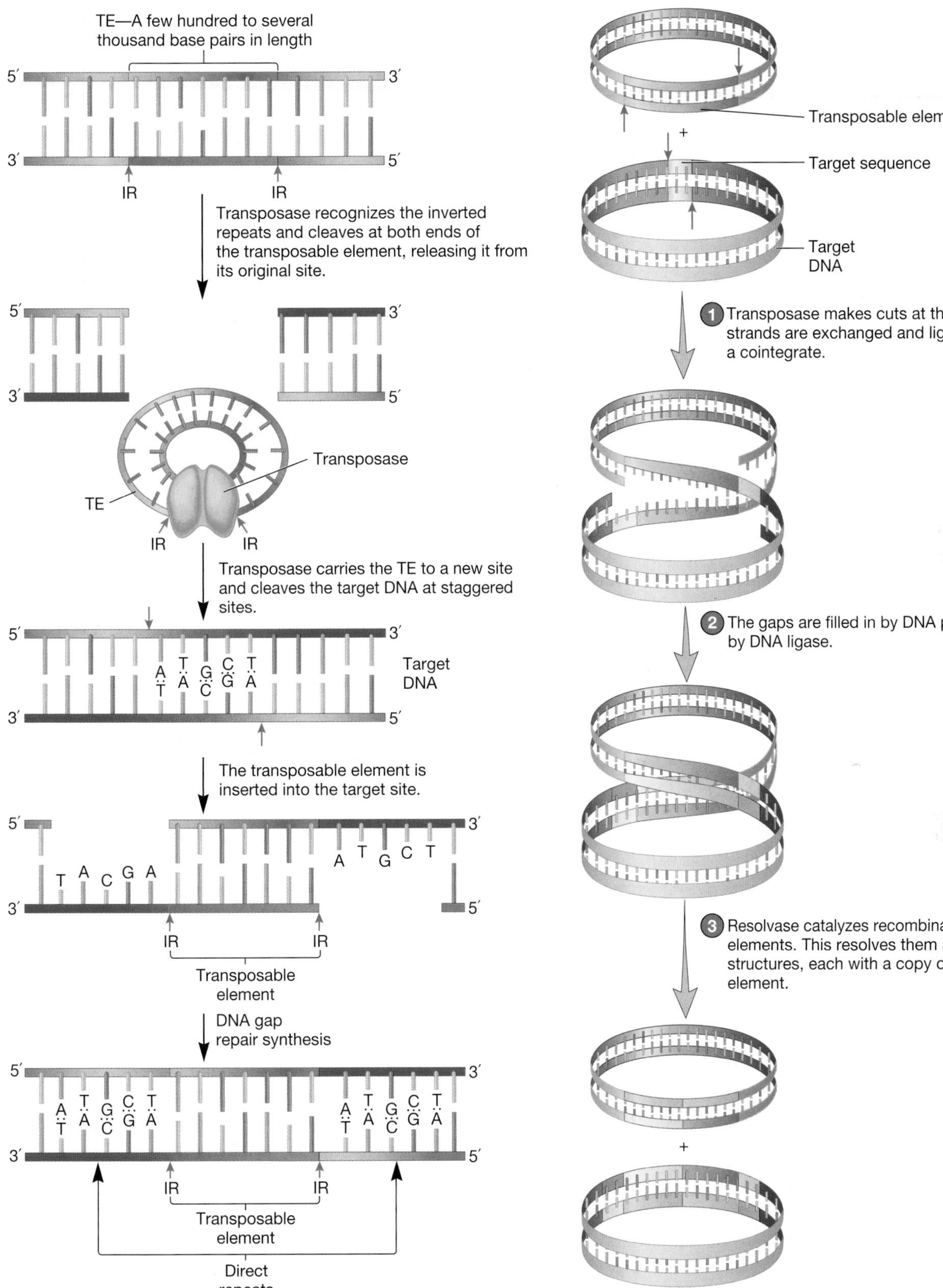

Figure 13.20 Simple Transposition. TE, transposable element; IR, inverted repeat.

Figure 13.21 Replicative Transposition.

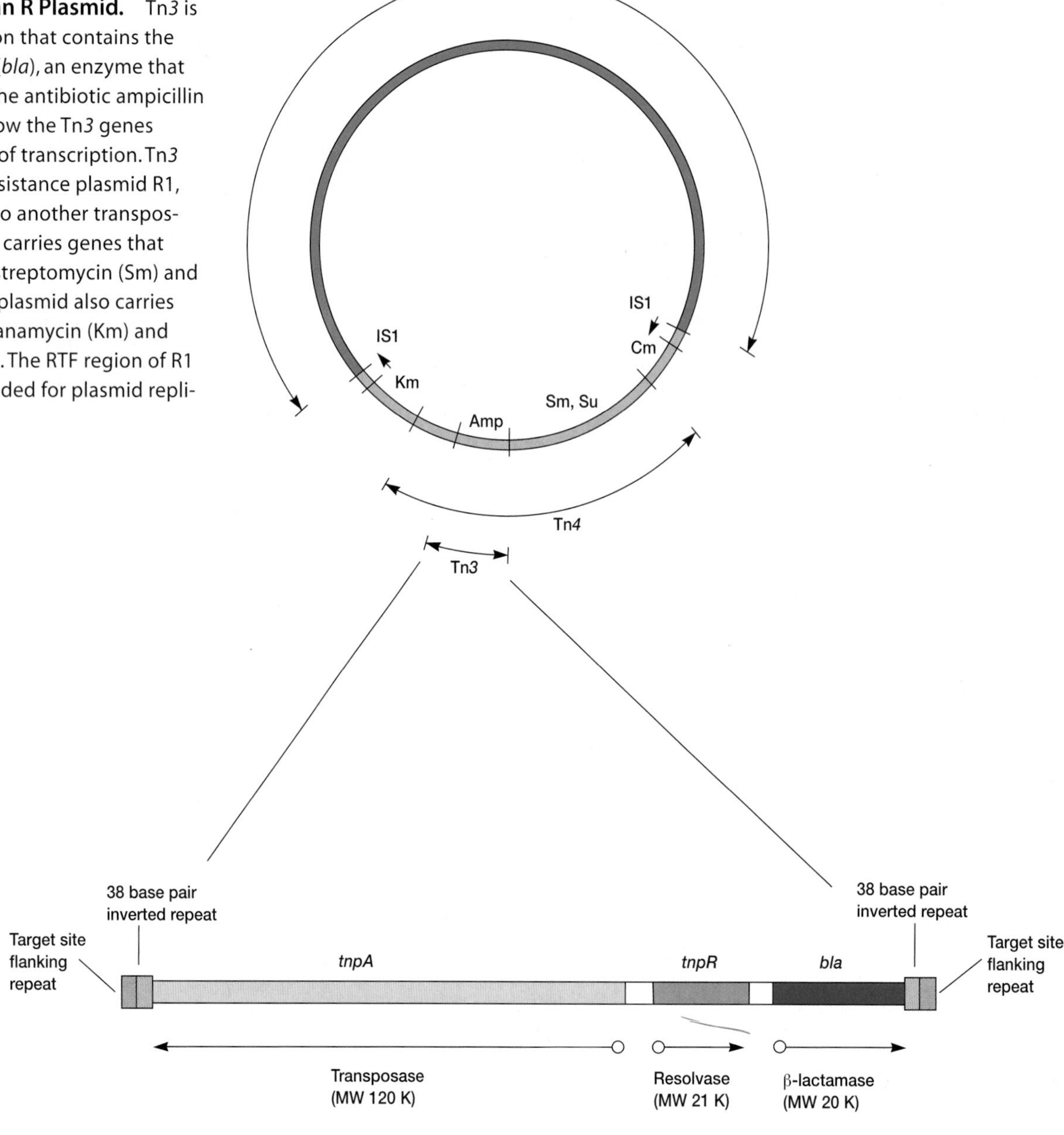

Figure 13.22 The Tn*3* Composite Transposon within an R Plasmid. Tn*3* is a replicative transposon that contains the gene for β-lactamase (*bla*), an enzyme that confers resistance to the antibiotic ampicillin (Amp). The arrows below the Tn*3* genes indicate the direction of transcription. Tn*3* can be found in the resistance plasmid R1, where it is inserted into another transposable element, Tn*4*. Tn*4* carries genes that provide resistance to streptomycin (Sm) and sulfonamide (Su). The plasmid also carries resistance genes for kanamycin (Km) and chloramphenicol (Cm). The RTF region of R1 codes for proteins needed for plasmid replication and transfer.

Perhaps the best-studied conjugative plasmid is **F factor.** It plays a major role in conjugation in *E. coli* and was the first conjugative plasmid to be described (**figure 13.23**). The F factor is about 100 kilobases long and bears genes responsible for cell attachment and plasmid transfer between specific bacterial strains during conjugation. Most of the information required for plasmid transfer is located in the *tra* operon, which contains at least 28 genes. Many of these direct the formation of sex pili that attach the F^+ cell (the donor cell containing an F plasmid) to an F^- cell (**figure 13.24**). Other gene products aid DNA transfer. In addition, the F factor has several segments called insertion sequences that assist plasmid integration into the host cell chromosome. Thus the F factor is an episome that can exist outside the bacterial chromosome or can be integrated into it (**figure 13.25**).

1. Define the following: episome, conjugative plasmid, transposition, transposase, and conjugative transposon.
2. Compare and contrast plasmids and transposable elements. Compare and contrast insertion sequences, composite transposons, and replicative transposons.
3. How might one demonstrate the presence of a plasmid in a host cell?
4. What is simple (cut-and-paste) transposition? What is replicative transposition? How do the two mechanisms of transposition differ? What happens to the target site during transposition?
5. What effect would you expect the existence of transposable elements and plasmids to have on the rate of microbial evolution? Give your reasoning.
6. How do multiple-drug-resistant plasmids often arise?

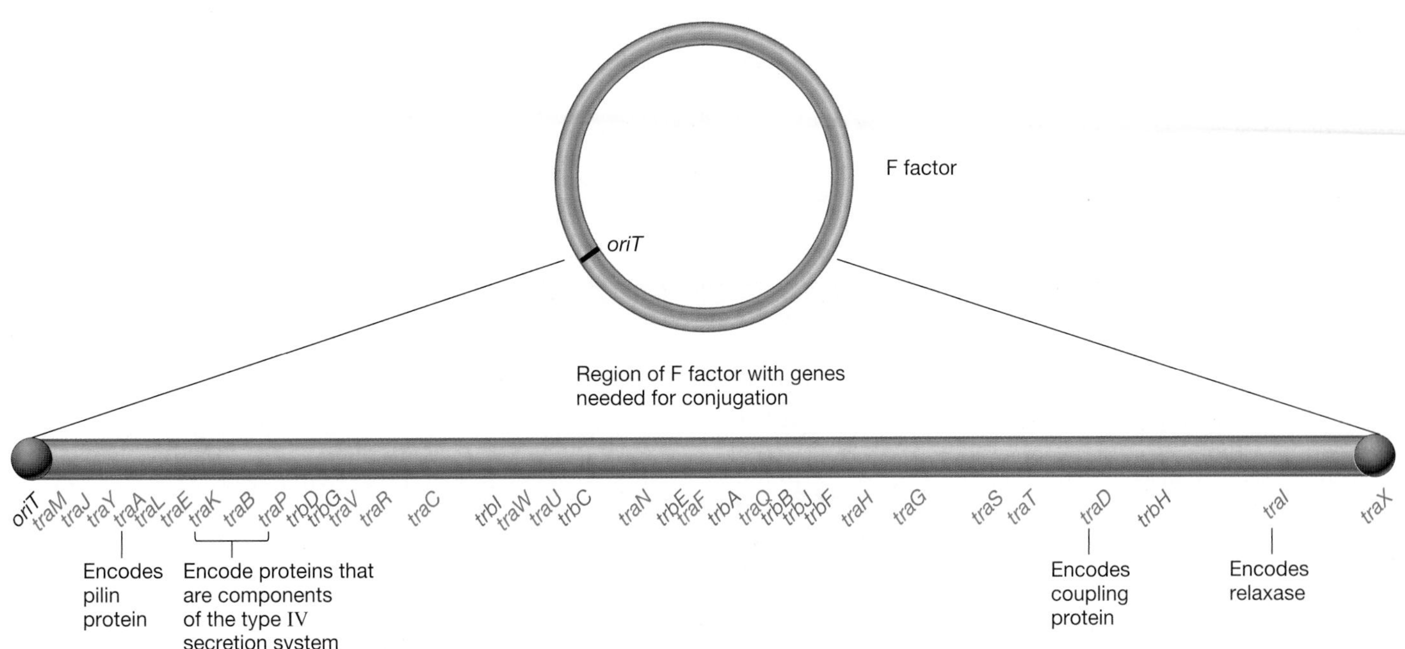

Figure 13.23 The F plasmid. Genes that play a role in conjugation are shown, and some of their functions are indicated. The plasmid also contains three insertion sequences and a transposon. The site for initiation of rolling-circle replication and gene transfer during conjugation is *oriT*.

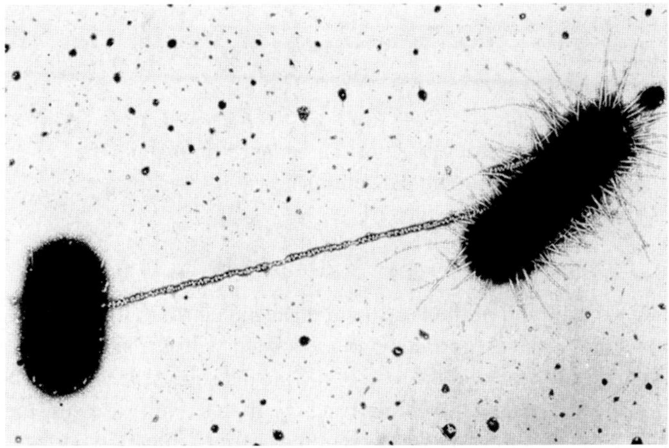

Figure 13.24 Bacterial Conjugation. An electron micrograph of two *E. coli* cells in an early stage of conjugation. The F^+ cell to the right is covered with small pili or fimbriae, and a sex pilus connects the two cells.

13.7 Bacterial Conjugation

The initial evidence for bacterial **conjugation,** the transfer of DNA by direct cell to cell contact, came from an elegant experiment performed by Joshua Lederberg and Edward Tatum in 1946. They mixed two auxotrophic strains, incubated the culture for several hours in nutrient medium, and then plated it on minimal medium. To reduce the chance that their results were due to simple reversion, they used double and triple auxotrophs on the assumption that two or three simultaneous reversions would be extremely rare. For example, one strain required biotin (Bio^-), phenylalanine (Phe^-), and cysteine (Cys^-) for growth, and another needed threonine (Thr^-), leucine (Leu^-), and thiamine (Thi^-). Recombinant prototrophic colonies appeared on the minimal medium after incubation (**figure 13.26**). Thus the chromosomes of the two auxotrophs were able to associate and undergo recombination.

Lederberg and Tatum did not directly prove that physical contact of the cells was necessary for gene transfer. This evidence was provided by Bernard Davis (1950), who constructed a U tube consisting of two pieces of curved glass tubing fused at the base to form a U shape with a fritted glass filter between the halves. The filter allowed the passage of media but not bacteria. The U tube was filled with nutrient medium and each side inoculated with a different auxotrophic strain of *E. coli* (**figure 13.27**). During incubation, the medium was pumped back and forth through the filter to ensure medium exchange between the halves. After a 4 hour incubation, the bacteria were plated on minimal medium. Davis discovered that when the two auxotrophic strains were separated from each other by the fine filter, gene transfer could not take place. Therefore direct contact was required for the recombination that Lederberg and Tatum had observed. F factor-mediated conjugation is one of the best-studied conjugation systems. It is the focus of this section.

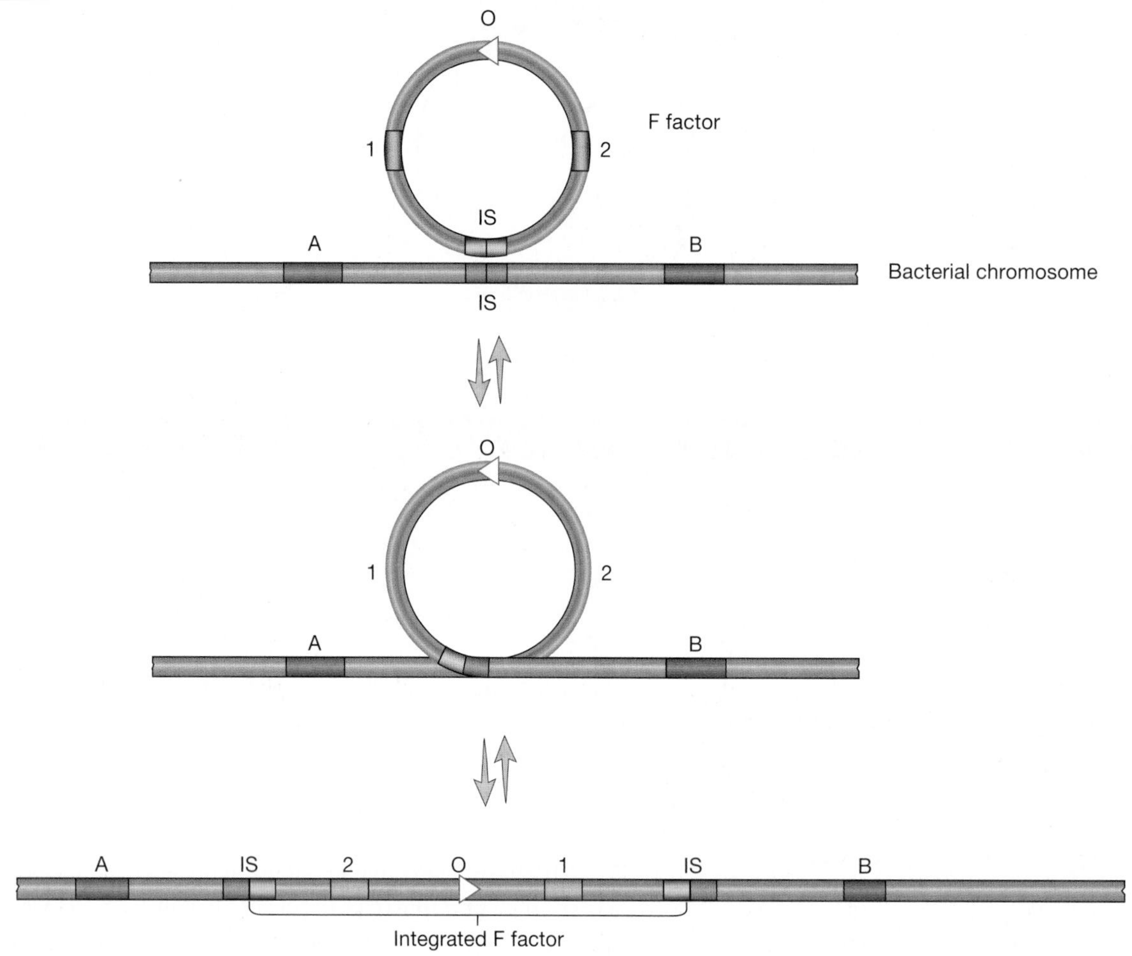

Figure 13.25 F Plasmid Integration. The reversible integration of an F plasmid or factor into a host bacterial chromosome. The process begins with association between plasmid and bacterial insertion sequences. The O arrowhead (white) indicates the site at which oriented transfer of chromosome to the recipient cell begins. A, B, 1, and 2 represent genetic markers.

$F^+ \times F^-$ Mating

In 1952 William Hayes demonstrated that the gene transfer observed by Lederberg and Tatum was polar. That is, there were definite donor (F^+, or fertile) and recipient (F^-, or nonfertile) strains, and gene transfer was nonreciprocal. He also found that in $F^+ \times F^-$ mating the progeny were only rarely changed with regard to auxotrophy (that is, chromosomal genes were not often transferred), but F^- strains frequently became F^+.

These results are readily explained in terms of the F factor previously described (figure 13.23). The F^+ strain contains an extrachromosomal F factor carrying the genes for sex pilus formation and plasmid transfer. The **sex pilus** is used to establish contact between the F^+ and F^- cells (**figure 13.28*a***). Once contact is made, the pilus retracts, bringing the cells into close physical contact. The F^+ cell then prepares for DNA transfer by assembling a type IV secretion apparatus, using many of the same genes used for sex pilus biogenesis; the sex pilus is embedded in the secretion structure (**figure 13.29**). The F factor then replicates by a rolling-circle mechanism (*see figure 11.12*). Replication is initiated by a complex of proteins called the relaxosome, which nicks one strand of the F factor at a site called *oriT* (for *ori*gin of *t*ransfer). Relaxase, an enzyme associated with the relaxosome, remains attached to the 5' end of the nicked strand. As F factor is replicated, the displaced strand and the attached relaxase enzyme move through the type IV secretion system to the recipient cell. Because the pilus is embedded in the secretion apparatus, it has been suggested that the DNA moves through a lumen in the pilus. However, studies of a related conjugation system, that of the plant pathogen *Agrobacterium tumefaciens,* provide strong evidence that the DNA does not move through the sex pilus. However, it should be noted that although the F factor system and the *Agrobacterium* system are related, there is one important difference between the two. The F factor system is used to transfer DNA from one bacterium to another, whereas the *Agrobacterium* system moves DNA from the bacterium into its plant host. Whatever the route of transfer, as the plasmid is transferred, the entering strand is copied to produce double-stranded DNA. The recombination frequency is low because chromosomal genes are rarely transferred with the independent F factor. Microorganism associations with vascular plants: *Agrobacterium* (section 29.5)

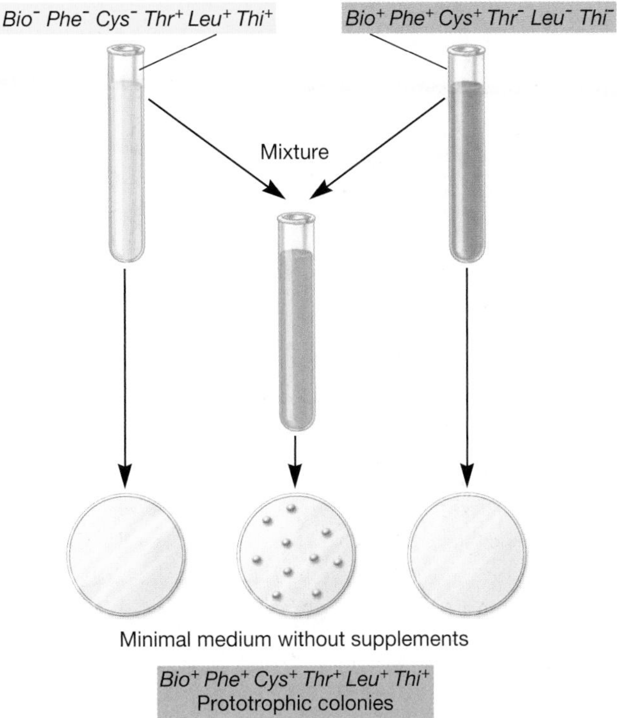

Figure 13.26 Evidence for Bacterial Conjugation. Lederberg and Tatum's demonstration of genetic recombination using triple auxotrophs. See text for details.

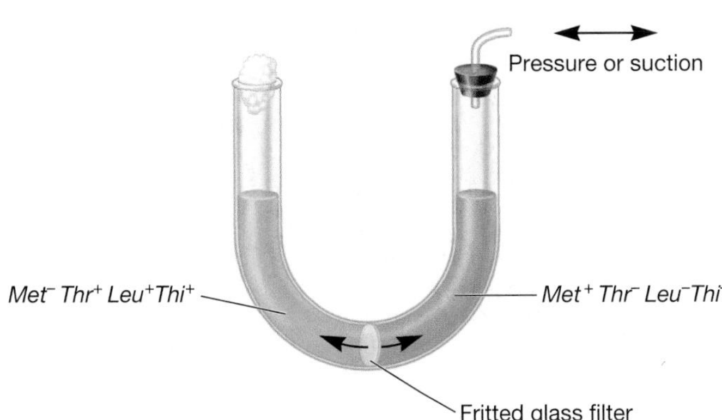

Figure 13.27 The U-tube Experiment. The U-tube experiment used to show that genetic recombination by conjugation requires direct physical contact between bacteria. See text for details.

Hfr Conjugation

Not long after the discovery of $F^+ \times F^-$ mating, a second type of F factor-mediated conjugation was discovered. In this type of conjugation, the donor transfers chromosomal genes with great efficiency, but does not change the recipient bacteria into F^+ cells. Because of the *h*igh *fr*equency of recombinants produced by this mating, it is referred to as **Hfr conjugation** and the donor is called an **Hfr strain.** Although initially the mechanism of Hfr conjugation was not known, eventually it was determined that Hfr strains contain the F factor integrated into their chromosome, rather than free in the cytoplasm (figure 13.28*b*). When integrated, the F plasmid's *tra* operon is still functional; the plasmid can direct the synthesis of pili, carry out rolling-circle replication, and transfer genetic material to an F^- recipient cell. However, rather than transferring itself, the F factor directs the transfer of host chromosome. DNA transfer begins when the integrated F factor is nicked at its site of transfer origin. As it is replicated, the chromosome moves to the recipient (figure 13.28c). Because only part of the F factor is transferred, the F^- recipient does not become F^+ unless the whole chromosome is transferred. Transfer of the entire chromosome with the integrated F factor requires about 100 minutes in *E. coli,* and the connection between the cells usually breaks before this process is finished. Thus a complete F factor usually is not transferred, and the recipient remains F^-.

As mentioned earlier, when an Hfr strain participates in conjugation, bacterial genes are frequently transferred to the recipient. Gene transfer can be in either a clockwise or counterclockwise direction around the circular chromosome, depending on the orientation of the integrated F factor. After the replicated donor chromosome enters the recipient cell, it may be degraded or incorporated into the F^- genome by recombination.

F′ Conjugation

Because the F plasmid is an episome, it can leave the bacterial chromosome and resume status as an autonomous F factor. Sometimes during this process the plasmid makes an error in excision and picks up a portion of the chromosome. Because it is now genotypically distinct from the original F factor, it is called an **F′ plasmid** (**figure 13.30*a***). It is not unusual to observe the inclusion of one or more chromosomal genes in excised F plasmids. A cell containing an F′ plasmid retains all of its genes, although some of them are on the plasmid. It mates only with an F^- recipient. $F' \times F^-$ conjugation is similar to $F^+ \times F^-$ mating. Once again, the plasmid is transferred as it is copied by rolling-circle replication. However, bacterial genes on the chromosome usually are not transferred (figure 13.30*b*). Bacterial genes acquired during excision of the F′ plasmid are transferred with it and need not be incorporated into the recipient chromosome to be expressed. The recipient becomes F′ and is a partially diploid merozygote because the same bacterial genes present on the F′ plasmid are also found on the recipient's chromosome. In this way specific bacterial genes may spread rapidly throughout a bacterial population.

F′ conjugation is very important to the microbial geneticist. A partial diploid's behavior shows whether the allele carried by an F′ plasmid is dominant or recessive to the chromosomal gene. The formation of F′ plasmids also is useful in mapping the chromosome because if two genes are picked up by an F factor they must be neighbors.

Other Examples of Bacterial Conjugation

Although most research on plasmids and conjugation has been done using *E. coli* and other gram-negative bacteria, self-transmissible

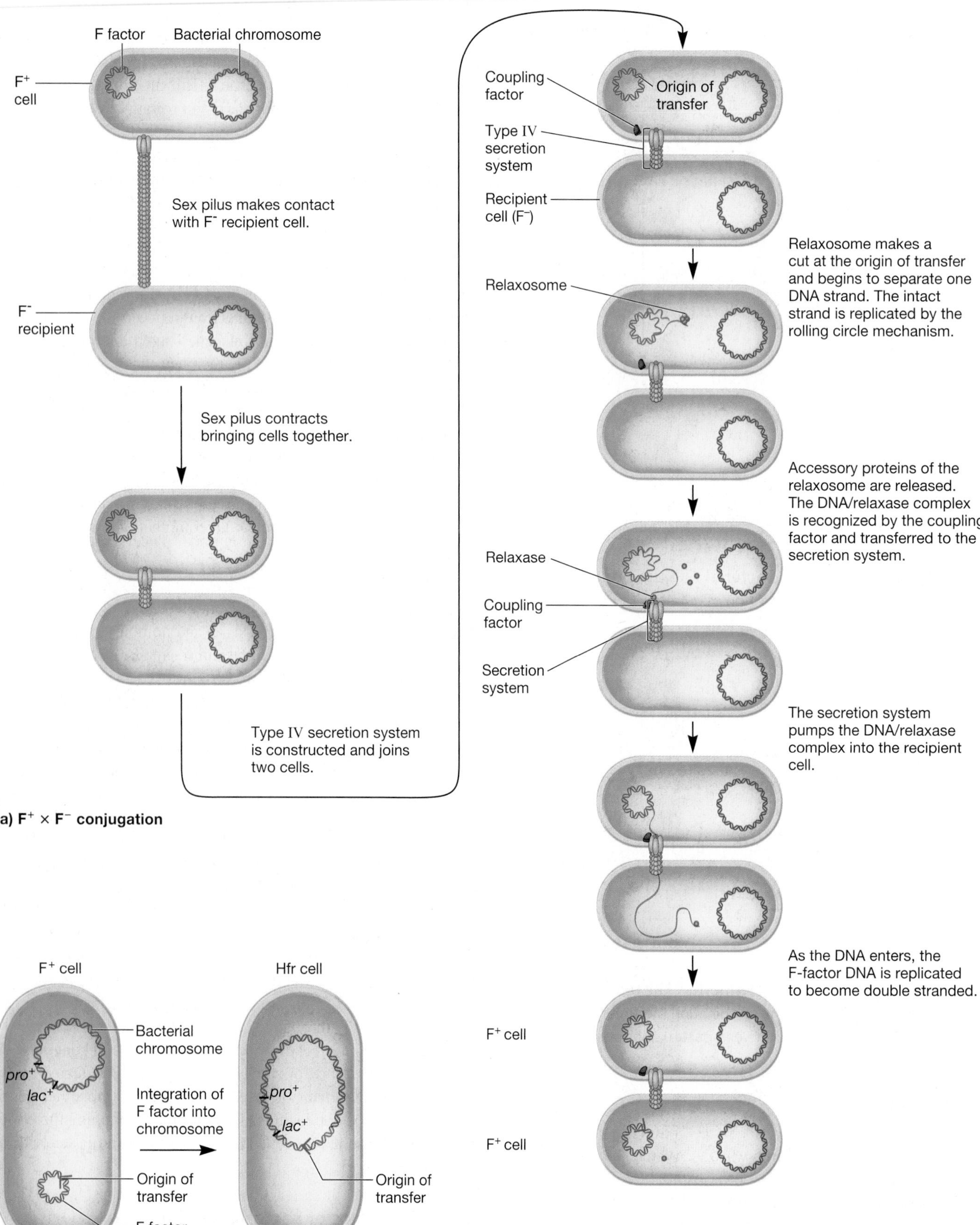

F factor
Bacterial chromosome
F⁺ cell
Sex pilus makes contact with F⁻ recipient cell.
F⁻ recipient
Sex pilus contracts bringing cells together.
Type IV secretion system is constructed and joins two cells.
Coupling factor
Origin of transfer
Type IV secretion system
Recipient cell (F⁻)
Relaxosome
Relaxosome makes a cut at the origin of transfer and begins to separate one DNA strand. The intact strand is replicated by the rolling circle mechanism.
Accessory proteins of the relaxosome are released. The DNA/relaxase complex is recognized by the coupling factor and transferred to the secretion system.
Relaxase
Coupling factor
Secretion system
The secretion system pumps the DNA/relaxase complex into the recipient cell.
As the DNA enters, the F-factor DNA is replicated to become double stranded.
F⁺ cell
F⁺ cell
F⁺ cell
Hfr cell
Bacterial chromosome
pro⁺
lac⁺
Integration of F factor into chromosome
Origin of transfer
F factor
pro⁺
lac⁺
Origin of transfer

(a) $F^+ \times F^-$ conjugation

(b) Insertion of F factor into chromosome

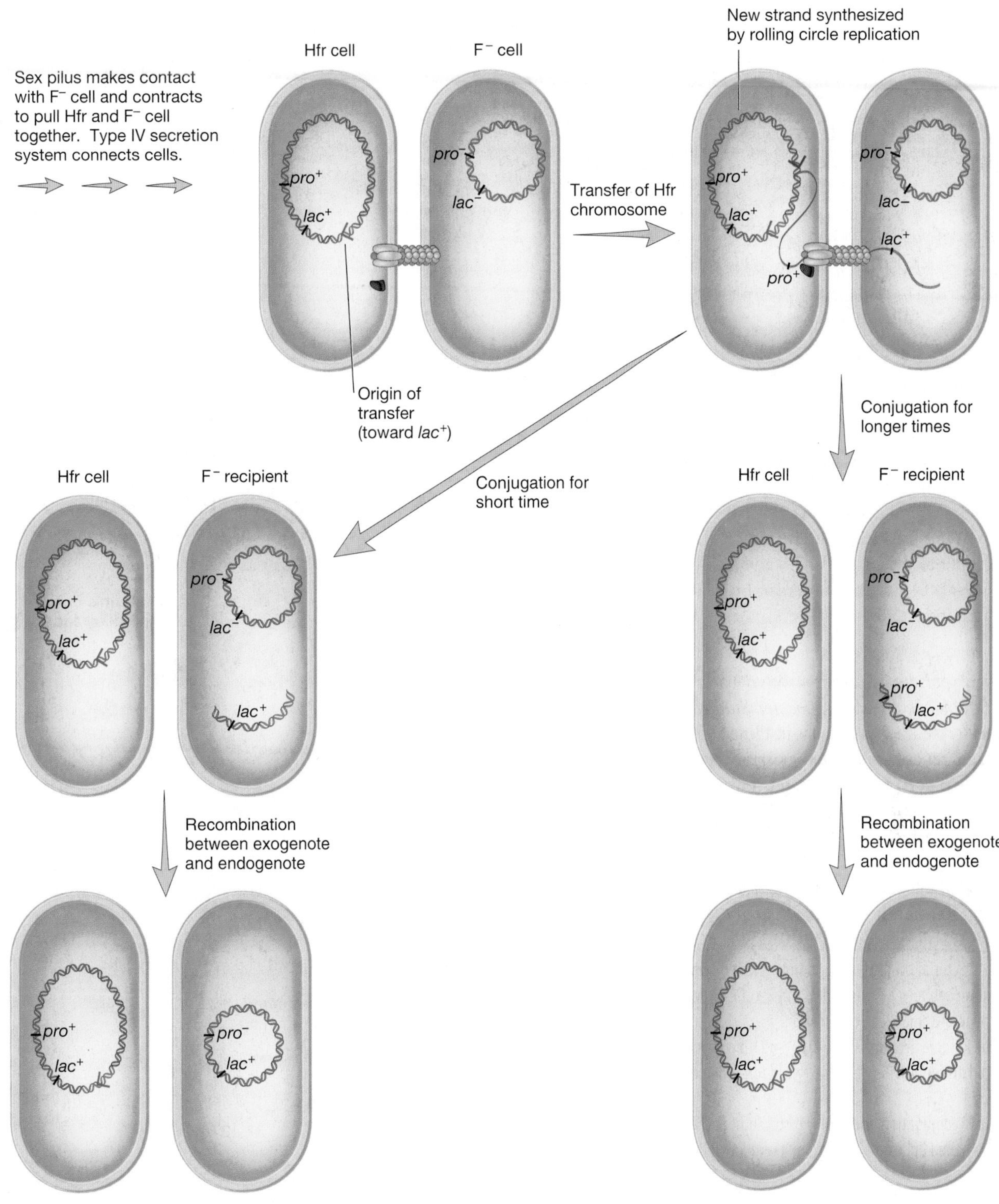

Figure 13.28 F Factor-Mediated Conjugation. The F factor encodes proteins for building the sex pilus and proteins needed to construct the type IV secretion system that will transfer DNA from the donor to the F^- recipient. One protein, the coupling factor, is thought to guide the DNA to the secretion system. **(a)** During $F^+ \times F^-$ conjugation, only the F factor is transferred because the plasmid is extrachromosomal. The recipient cell becomes F^+. **(b)** Integration of the F factor into the chromosome creates an Hfr cell. **(c)** During Hfr $\times$ F^- conjugation, some plasmid genes and some chromosomal genes are transferred to the recipient. Note that only a portion of the F factor moves into the recipient. Because the entire plasmid is not transferred, the recipient remains F^-. In addition, the incoming DNA must recombine into the recipient's chromosome if it is to be stably maintained.

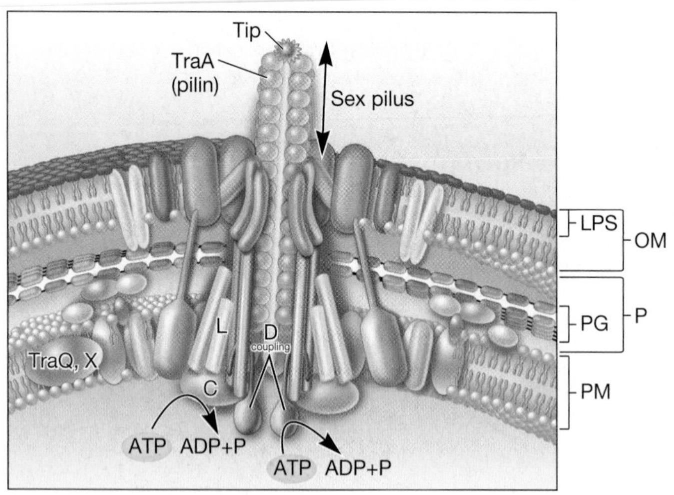

Figure 13.29 The Type IV Secretion System Encoded by F Factor. The F factor-encoded type IV secretion system is composed of numerous Tra proteins, including TraA proteins, which form the sex pilus, and TraD, which is the coupling factor. Some Tra proteins are located in the plasma membrane (PM), others extend into the periplasm (P) and pass through the peptidoglycan layer (PG) into the outer membrane (OM) and its lipopolysaccharide (LPS) layer.

plasmids are present in gram-positive bacterial genera such as *Bacillus, Streptococcus, Enterococcus, Staphylococcus,* and *Streptomyces.* Much less is known about these systems. It appears that fewer transfer genes are involved, possibly because a sex pilus may not be required for plasmid transfer. For example, *Enterococcus faecalis* recipient cells release short peptide chemical signals that activate transfer genes in donor cells containing the proper plasmid. Donor and recipient cells directly adhere to one another through special plasmid-encoded proteins released by the activated donor cell. Plasmid transfer then occurs.

1. What is bacterial conjugation and how was it discovered?
2. Distinguish between F^+, Hfr, and F^- strains of *E. coli* with respect to their physical nature and role in conjugation.
3. Describe in some detail how $F^+ \times F^-$ and Hfr conjugation processes proceed, and distinguish between the two in terms of mechanism and the final results.
4. What is F′ conjugation and why is it so useful to the microbial geneticist? How does the F′ plasmid differ from a regular F plasmid?

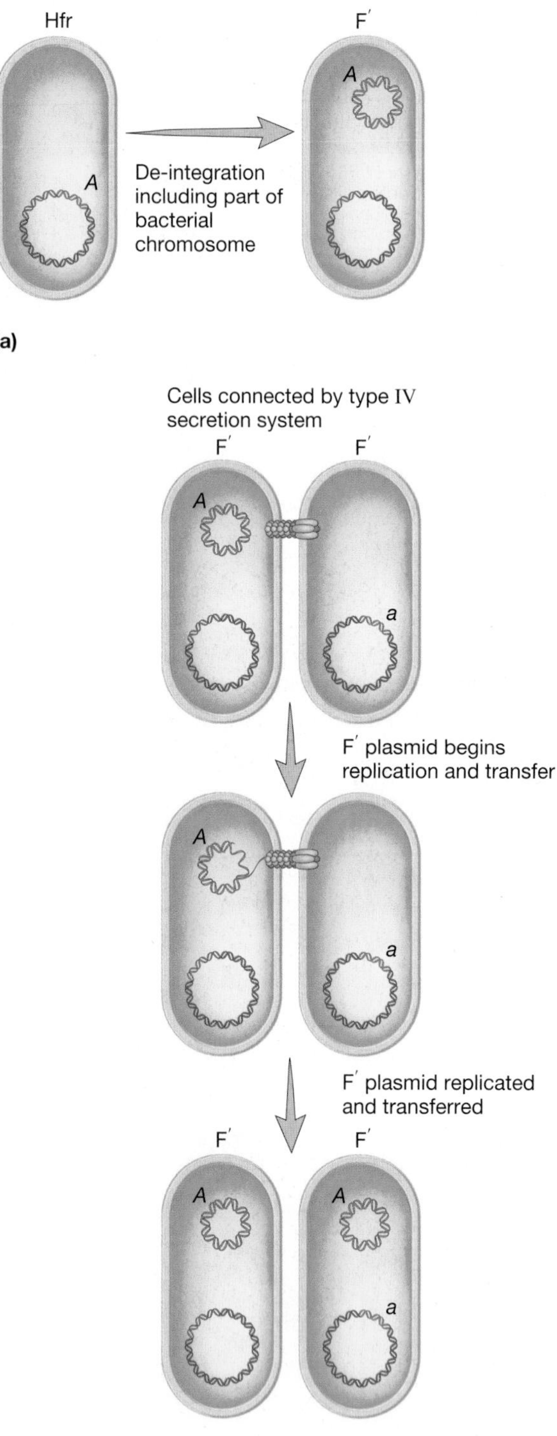

Figure 13.30 F' Conjugation. **(a)** Due to an error in excision, the *A* gene of an Hfr cell is picked up by the F factor. **(b)** The *A* gene is then transferred to a recipient during conjugation. See text for explanation.

13.8 DNA Transformation

The second way DNA can move between bacteria is through transformation, discovered by Fred Griffith in 1928. **Transformation** is the uptake by a cell of a naked DNA molecule or fragment from the medium and the incorporation of this molecule into the recipient chromosome in a heritable form. In natural transformation the DNA comes from a donor bacterium. The process is random, and any portion of a genome may be transferred between bacteria. DNA as genetic material (section 11.1)

When bacteria lyse, they release considerable amounts of DNA into the surrounding environment. These fragments may be relatively large and contain several genes. If a fragment contacts a **competent cell,** a cell that is able to take up DNA and be transformed, the DNA can be bound to the cell and taken inside (**figure 13.31*a***). The transformation frequency of very competent cells is around 10^{-3} for most genera when an excess of DNA is used. That is, about one cell in every thousand will take up and integrate the gene. Competency is a complex phenomenon and is dependent on several conditions. Bacteria need to be in a certain stage of growth; for example, *Streptococcus pneumoniae* becomes competent during the exponential phase when the population reaches about 10^7 to 10^8 cells per ml. When a population becomes competent, bacteria such as *S. pneumoniae* secrete a small protein called the competence factor that stimulates the production of 8 to 10 new proteins required for transformation. Natural transformation has been discovered so far only in certain genera including *Streptococcus, Bacillus, Thermoactinomyces, Haemophilus, Neisseria, Moraxella, Acinetobacter, Azotobacter, Helicobacter,* and *Pseudomonas.* Gene transfer by this process occurs in soil and aquatic ecosystems and may be an important route of genetic exchange in biofilm and other microbial communities.

The mechanism of transformation has been intensively studied in *S. pneumoniae* (**figure 13.32**). A competent cell binds a double-stranded DNA fragment if the fragment is moderately large; the process is random, and donor fragments compete with each other. The DNA then is cleaved by endonucleases to double-stranded fragments about 5 to 15 kilobases in size. DNA uptake requires energy expenditure. One strand is hydrolyzed by an envelope-associated exonuclease during uptake; the other strand associates with small proteins and moves through the plasma membrane. The single-stranded fragment can then align with a homologous region of the genome and be integrated, probably by a mechanism similar to that depicted in figure 13.18.

Transformation in *Haemophilus influenzae,* a gram-negative bacterium, differs from that in *S. pneumoniae* in several respects. *H. influenzae* does not produce a competence factor to stimulate the development of competence, and it takes up DNA from only closely related species (*S. pneumoniae* is less particular about the source of its DNA). Double-stranded DNA, complexed with proteins, is taken in by membrane vesicles. The specificity of *H. influenzae* transformation is due to a special 11 base pair sequence (5′AAGTGCGGTCA3′) that is repeated over 1,400 times in *H. influenzae* DNA. DNA must have this sequence to be bound by a competent cell.

The protein complexes that take up free DNA must be able to move it through gram-negative and gram-positive walls, which may be both thick and complex. As expected, the machinery is quite large and complicated and appears related to protein secretion systems. **Figure 13.33*a*** shows a schematic diagram of the complex used by the gram-negative bacterium *Neisseria gonorrhoeae.* PilQ aids in the movement across the outer membrane, and the pilin complex PilE moves the DNA through the periplasm and peptidoglycan. ComE is a DNA binding protein; N is the nuclease that degrades one strand before the DNA enters the cytoplasm through the transmembrane channel formed by

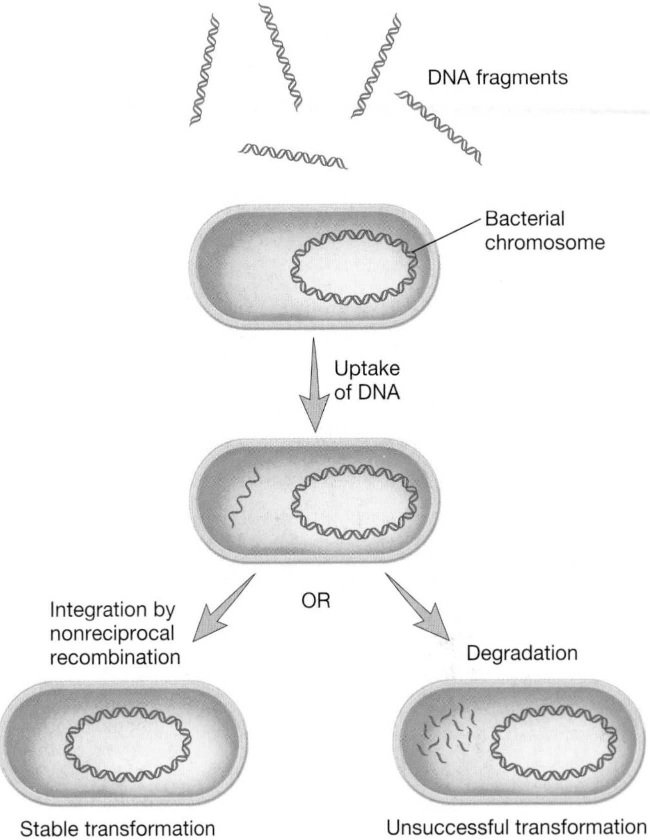

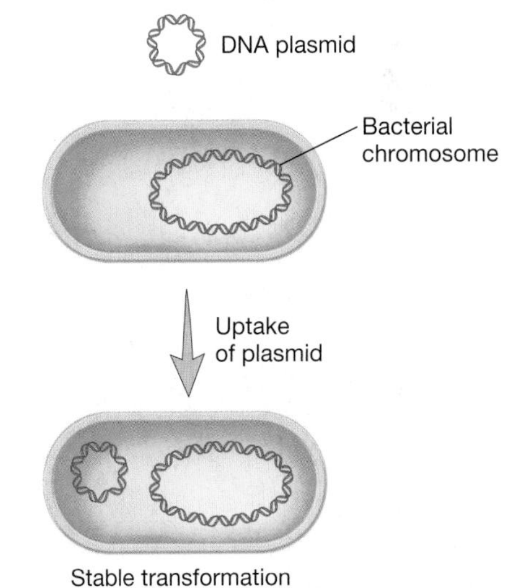

Figure 13.31 Bacterial Transformation. Transformation with **(a)** DNA fragments and **(b)** plasmids. Transformation with a plasmid often is induced artificially in the laboratory. The transforming DNA is in purple and integration is at a homologous region of the genome.

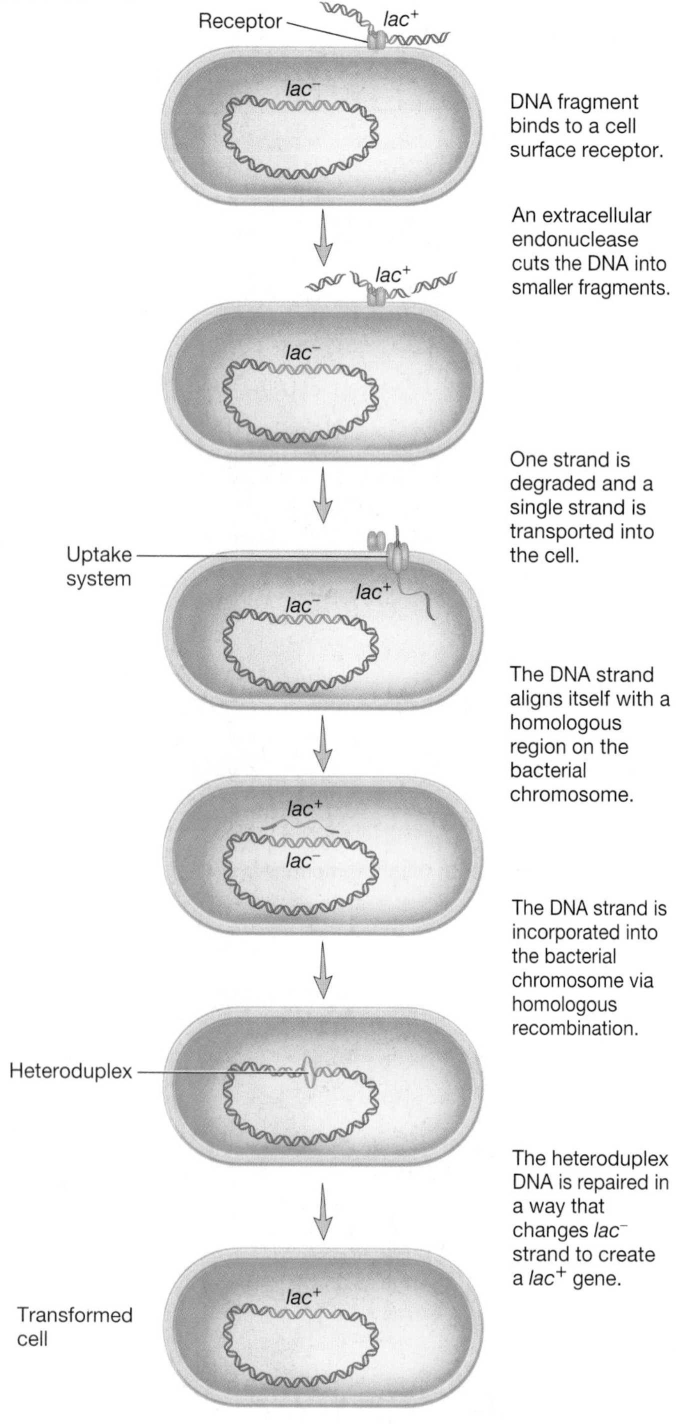
Receptor
lac+
lac−
DNA fragment binds to a cell surface receptor.
An extracellular endonuclease cuts the DNA into smaller fragments.
lac+
lac−
One strand is degraded and a single strand is transported into the cell.
Uptake system
lac−
lac+
The DNA strand aligns itself with a homologous region on the bacterial chromosome.
lac+
lac−
The DNA strand is incorporated into the bacterial chromosome via homologous recombination.
Heteroduplex
The heteroduplex DNA is repaired in a way that changes lac− strand to create a lac+ gene.
Transformed cell
lac+

Figure 13.32 Bacterial Transformation as Seen in *S. pneumoniae.*

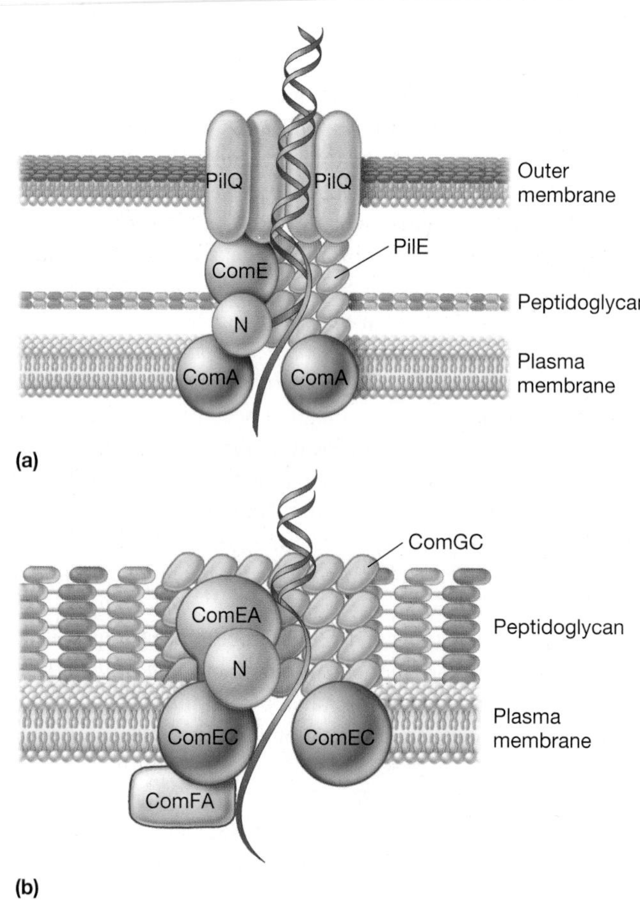
PilQ
PilQ
Outer membrane
PilE
ComE
Peptidoglycan
N
Plasma membrane
ComA
ComA
(a)
ComGC
ComEA
Peptidoglycan
N
Plasma membrane
ComEC
ComEC
ComFA
(b)

Figure 13.33 DNA Uptake Systems. (a) DNA uptake machinery in *N. gonorrhoeae.* **(b)** Uptake machinery in *B. subtilis.* See text for details.

ComA. The machinery in the gram-positive bacterium *Bacillus subtilis* is depicted in figure 13.33*b*. It is localized to the poles of the cell, and as can be seen, many of the components are similar to those of *N. gonorrhoeae:* the pilin complex (ComGC), DNA binding protein (ComEA), nuclease (N), and channel protein (ComEC). ComFA is a DNA translocase that moves the DNA into the cytoplasm. A gram-negative equivalent of ComFA has not been identified yet in *N. gonorrhoeae*.

Microbial geneticists exploit transformation to move DNA (usually recombinant DNA) into cells. However, as already noted, many species, including *E. coli,* are not naturally transformation competent. Fortunately, these bacteria can be made artificially competent by certain treatments. Two common techniques are electrical shock and exposure to calcium chloride. Both approaches render the cell membrane more permeable to DNA and both have been used to make artificially competent *E. coli* cells. To increase the transformation frequency with *E. coli,* strains that lack one or more nucleases are used. These strains are especially important when transforming the cells with linear DNA, which is vulnerable to attack by nucleases. It is easier to transform bacteria with plasmid DNA since plasmids are not as easily degraded as linear fragments and can replicate within the host (figure 13.31*b*).

1. Define transformation and competence.
2. Describe how transformation occurs in *S. pneumoniae.* How does the process differ in *H. influenzae?*
3. Discuss two ways in which artificial transformation can be used to place functional genes within bacterial cells.

13.9 TRANSDUCTION

The third mode of bacterial gene transfer is **transduction.** It is a frequent mode of horizontal gene transfer in nature and is mediated by viruses. The morphology and life cycle of bacterial viruses or bacteriophages is not discussed in detail until chapter 17. Nevertheless, it is necessary to briefly describe the life cycle here as background for a consideration of their role in gene transfer.

Viruses are structurally simple, often composed of just a nucleic acid genome protected by a protein coat called the capsid. They are unable to replicate autonomously. Instead, they infect and take control of a host cell, forcing the host to make many copies of the virus. Viruses that infect bacteria are called bacteriophages, or phages for short. Some phages are replicated by their bacterial host immediately after entry. After the number of replicated phages reaches a certain number, they cause the host to lyse, so they can be released and infect new host cells (**figure 13.34**). These phages are called **virulent bacteriophages** and the process is called the **lytic cycle.** Other bacteriophages do not immediately kill their host. Many of these viruses enter the host bacterium and, instead of replicating, insert their genomes into the bacterial chromosome. Once inserted, the viral genome is called a **prophage.** The host bacterium is unharmed by this, and the phage genome is passively replicated as the host cell's genome is replicated. These bacteriophages are called **temperate bacteriophages** and the relationship between these viruses and their host is called **lysogeny** (figure 13.34). Bacteria that have been lysogenized are called **lysogens.** Temperate phages can remain inactive in their hosts for many generations. However, they can be induced to switch to a lytic cycle of growth under certain conditions, including UV irradiation. When this occurs, the prophage is excised from the bacterial genome and the lytic cycle proceeds.

Transduction is the transfer of bacterial genes by viruses. Bacterial genes are incorporated into a phage capsid because of errors made during the virus life cycle. The virus containing these genes then injects them into another bacterium, completing the transfer. There are two different kinds of transduction: generalized and specialized.

Generalized Transduction

Generalized transduction occurs during the lytic cycle of virulent and some temperate phages and can transfer any part of the bacterial genome (**figure 13.35**). During the assembly stage, when the viral chromosomes are packaged into protein capsids, random fragments of the partially degraded bacterial chromosome also may be packaged by mistake. Because the capsid can contain only a limited quantity of DNA, the viral DNA is left behind. The quantity of bacterial DNA carried depends primarily on the size of the capsid. The P22 phage of *Salmonella enterica* serovar Typhimurium usually carries about 1% of the bacterial genome; the P1 phage of *E. coli* and a variety of gram-negative bacteria carries about 2.0 to 2.5% of the genome. The resulting virus particle often injects the DNA into another bacterial cell but cannot initiate a lytic cycle. This phage is known as a **generalized transducing particle** or phage and is simply a carrier of genetic information from the original bacterium to another cell. As in transformation, once the DNA has been injected, it must be incorporated into the recipient cell's chromosome to preserve the transferred genes. The DNA remains double stranded during

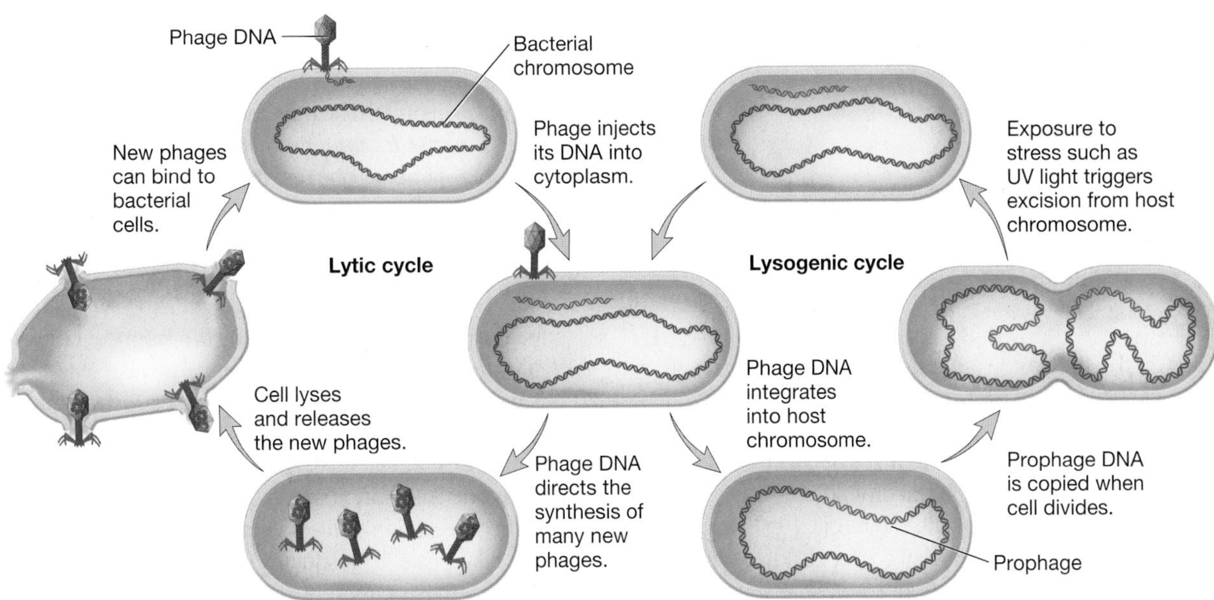

Figure 13.34 Lytic and Lysogenic Cycles of Temperate Phages. Virulent phages undergo only the lytic cycle. Temperate phages have two phases to their life cycles. The lysogenic cycle allows the genome of the virus to be replicated passively as the host cell's genome is replicated. Certain environmental factors such as UV light can cause a switch from the lysogenic cycle to the lytic cycle. In the lytic cycle, new virus particles are made and released when the host cell lyses. Virulent phages are limited to just the lytic cycle.

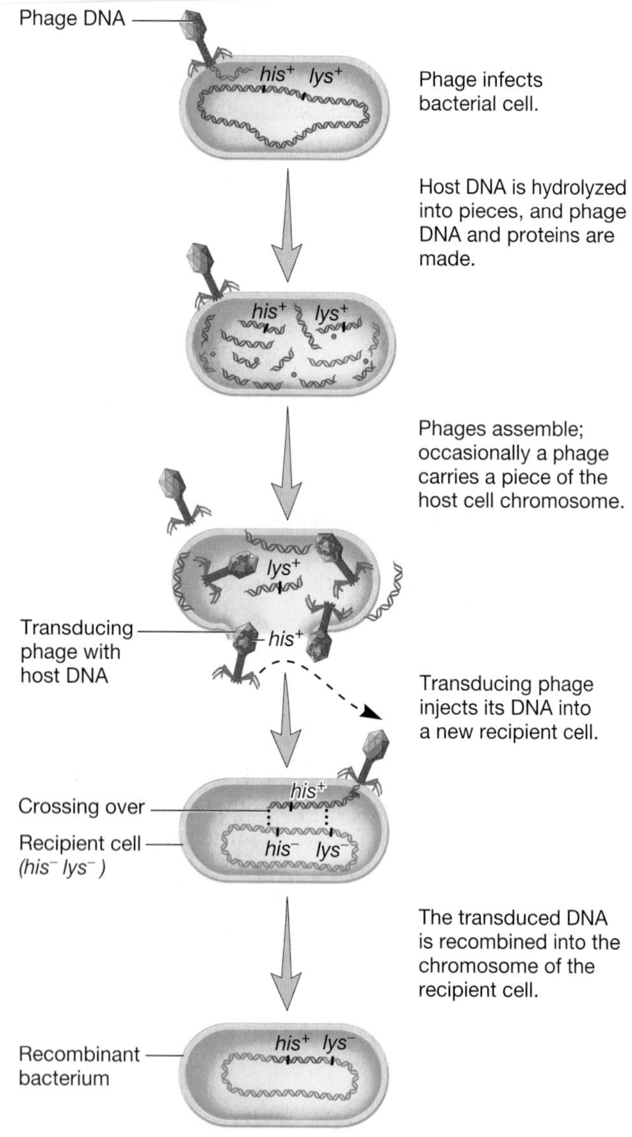

Phage DNA
his^+ lys^+
Phage infects bacterial cell.
Host DNA is hydrolyzed into pieces, and phage DNA and proteins are made.
his^+ lys^+
Phages assemble; occasionally a phage carries a piece of the host cell chromosome.
lys^+
Transducing phage with host DNA
his^+
Transducing phage injects its DNA into a new recipient cell.
Crossing over
Recipient cell (his^- lys^-)
his^+
his^- lys^-
The transduced DNA is recombined into the chromosome of the recipient cell.
Recombinant bacterium
his^+ lys^-
The recombinant bacterium has a genotype (his^+lys^-) that is different from recipient bacterial cell (his^- lys^-).

Figure 13.35 Generalized Transduction in Bacteria.

transfer, and both strands are integrated into the endogenote's genome. About 70 to 90% of the transferred DNA is not integrated but often is able to survive temporarily and be expressed. **Abortive transductants** are bacteria that contain this nonintegrated, transduced DNA and are partial diploids.

Generalized transduction was discovered in 1951 by Joshua Lederberg and Norton Zinder during an attempt to show that conjugation, discovered several years earlier in *E. coli,* could occur in other bacterial species. Lederberg and Zinder were repeating the earlier experiments with *S. enterica* serovar Typhimurium. They found that incubation of a mixture of two multiply auxotrophic strains yielded prototrophs at the level of about one in 10^5. This seemed like good evidence for bacterial recombination, and indeed it was, but their initial conclusion that the transfer resulted from conjugation was not borne out. When these investigators performed the U-tube experiment (figure 13.27) with *Salmonella,* they still recovered prototrophs. The filter in the U tube had pores that were small enough to block the movement of bacteria between the two sides but allowed phage P22 to pass. Lederberg and Zinder had intended to confirm that conjugation was present in another bacterial species but instead discovered a completely new mechanism of bacterial gene transfer. This seemingly routine piece of research led to surprising and important results. A scientist must always keep an open mind about results and be prepared for the unexpected.

Specialized Transduction

In **specialized transduction,** the transducing particle carries only specific portions of the bacterial genome. Specialized transduction is made possible by an error in the lysogenic life cycle of phages that insert their genomes into a specific site in the host chromosome. When a prophage is induced to leave the host chromosome, excision is sometimes carried out improperly. The resulting phage genome contains portions of the bacterial chromosome (about 5 to 10% of the bacterial DNA) next to the integration site, much like the situation with F′ plasmids (**figure 13.36**). A transducing phage genome usually is defective and lacks some part of its attachment site. The transducing particle will inject bacterial genes into another bacterium, even though the defective phage cannot reproduce without assistance. The bacterial genes may become stably incorporated under the proper circumstances.

The best-studied example of specialized transduction is carried out by the *E. coli* phage lambda. The lambda genome inserts into the host chromosome at specific locations known as attachment or *att* sites (**figure 13.37,** *see also figures 17.19 and 17.22*). The phage *att* sites and bacterial *att* sites are similar and can complex with each other. The *att* site for lambda is next to the *gal* and *bio* genes on the *E. coli* chromosome; consequently, specialized transducing lambda phages most often carry these bacterial genes. The lysate, or product of cell lysis, resulting from the induction of lysogenized *E. coli* contains normal phage and a few defective transducing particles. These particles are called either lambda *dgal* because they carry the galactose utilization genes or lambda *dbio* because they carry the *bio* from the other side of the *att* site (figure 13.37). Because these lysates contain only a few transducing particles, they often are called ***l*****ow-*****f*****requency *t*****ransduction lysates (LFT lysates).** Whereas the normal phage has a complete *att* site, defective transducing particles have a nonfunctional hybrid integration site that is part bacterial and part phage in origin. Integration of the defective phage chromosome does not readily take place. Transducing phages also may have lost some genes essential for reproduction. Stable transductants can arise only if there is a double cross-over event on each side of the *gal* site (figure 13.37). Temperate bacteriophages and lysogeny (section 17.5)

Defective lambda phages carrying the *gal* or *bio* genes can integrate if there is a normal lambda phage in the same cell. We will continue our discussion of this with a phage carrying the *gal* gene.

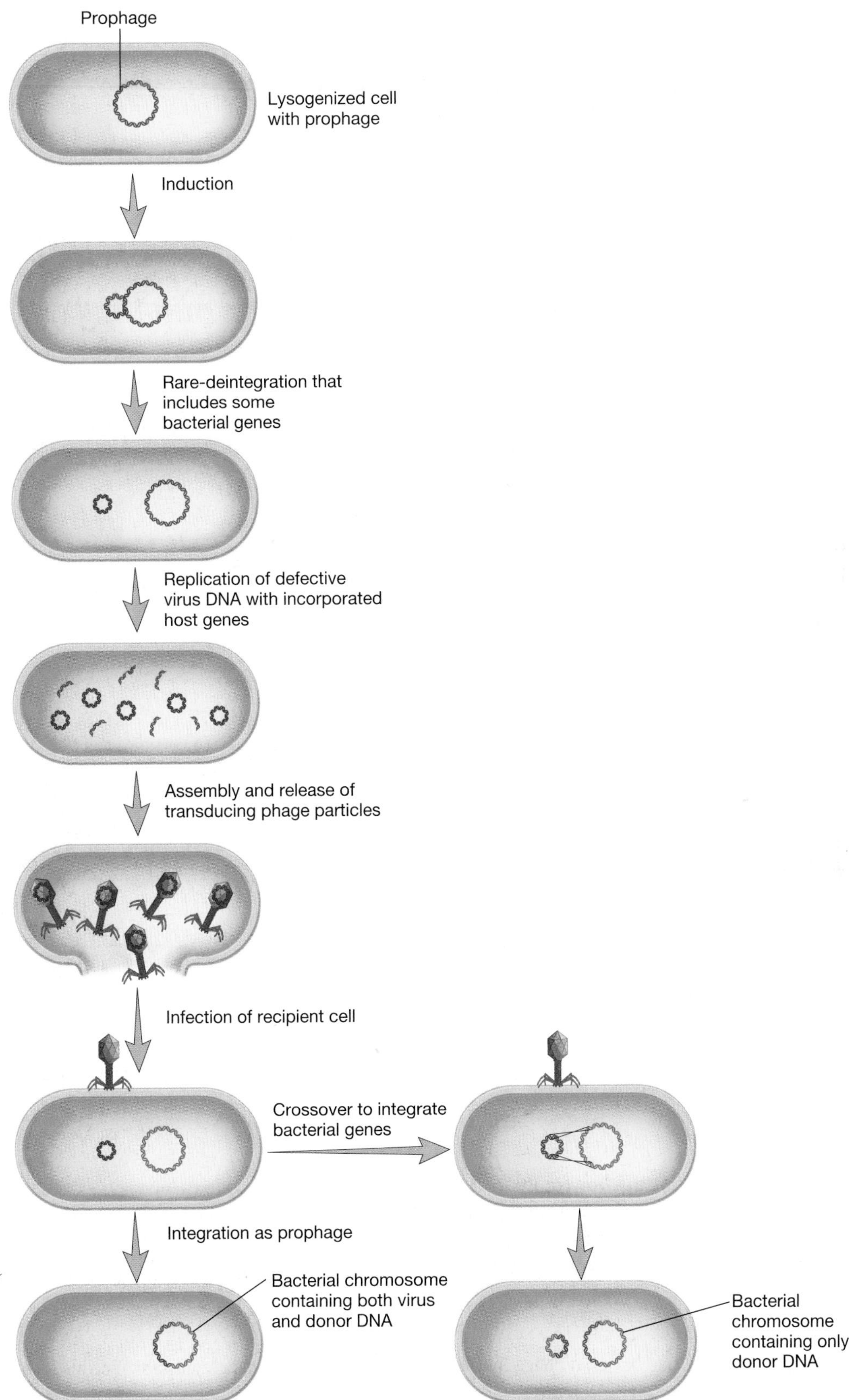

Figure 13.36 Specialized Transduction by a Temperate Bacteriophage. Recombination can produce two types of transductants.

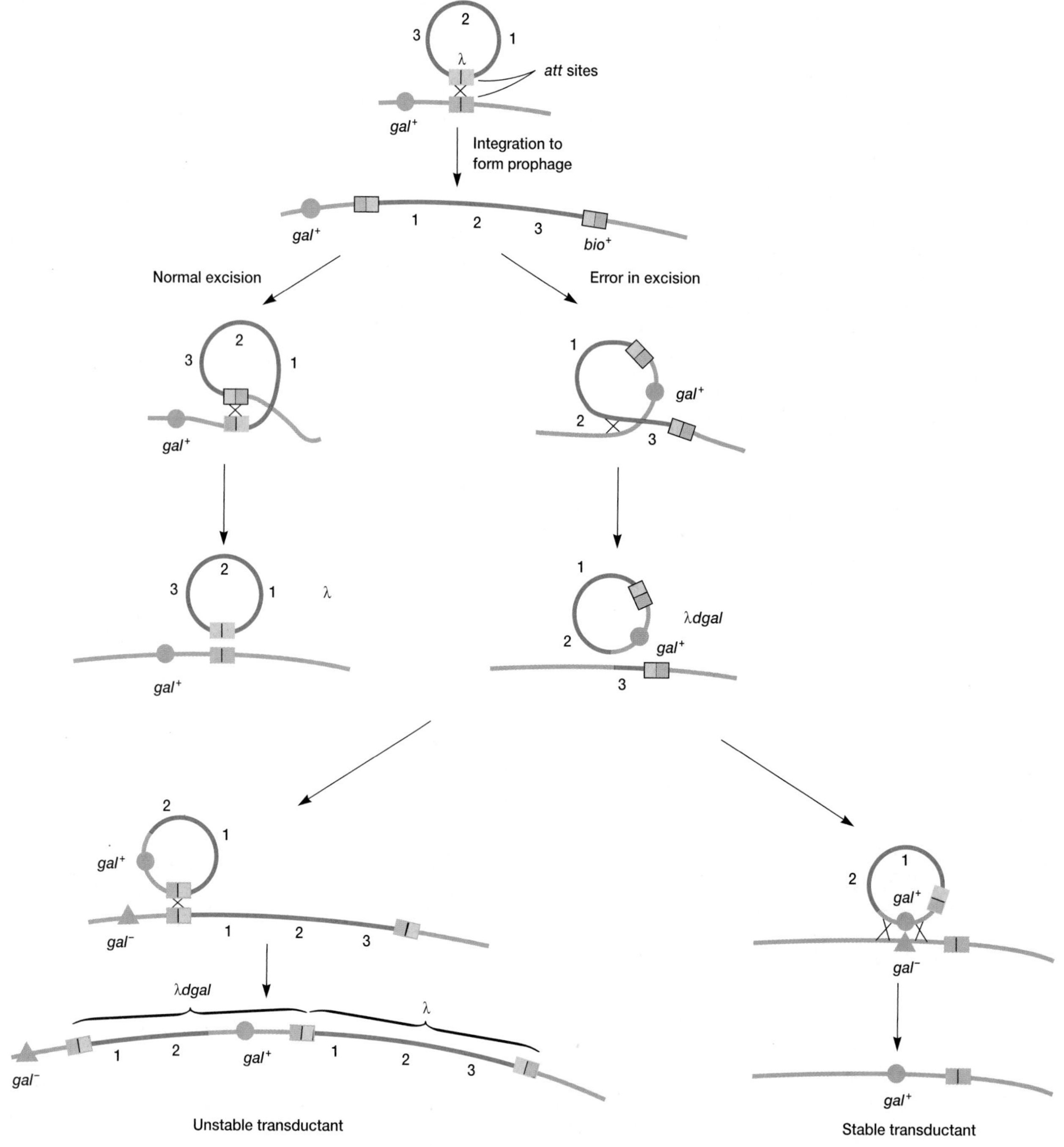

Figure 13.37 The Mechanism of Transduction for Phage Lambda and *E. coli*. Integrated lambda phage lies between the *gal* and *bio* genes. When it excises normally (top left), the new phage is complete and contains no bacterial genes. Rarely excision occurs asymmetrically (top right), and either the *gal* or *bio* genes are picked up and some phage genes are lost (only aberrant excision involving the *gal* genes is shown). The result is a defective lambda phage that carries bacterial genes and can transfer them to a new recipient.

The normal phage integrates, yielding two bacterial/phage hybrid *att* sites where the defective lambda *dgal* phage can insert (figure 13.37). It also supplies the genes missing in the defective phage. The normal phage in this instance is termed the **helper phage** because it aids integration and reproduction of the defective phage. These transductants are unstable because the prophages can be induced to excise by agents such as UV radiation. Excision, however, produces a lysate containing a fairly equal mixture of defective lambda *dgal* phage and normal helper phage. Because it is very effective in transduction, the lysate is called a ***high-frequency trans-***

duction lysate (HFT lysate). Reinfection of bacteria with this mixture will result in the generation of considerably more transductants. LFT lysates and those produced by generalized transduction have one transducing particle in 10^5 or 10^6 phages; HFT lysates contain transducing particles with a frequency of about 0.1 to 0.5.

1. Briefly describe the lytic and lysogenic viral reproductive cycles. Define lysogeny, lysogen, temperate phage, prophage, and transduction.
2. Describe generalized transduction, how it occurs, and the way in which it was discovered. What is an abortive transductant?
3. What is specialized transduction and how does it come about? Distinguish between LFT and HFT lysates and describe how they are formed.
4. How might one tell whether horizontal gene transfer was mediated by generalized or specialized transduction?
5. Why doesn't a cell lyse after successful transduction with a temperate phage?
6. Describe how conjugation, transformation, and transduction are similar. How are they different?

13.10 Mapping the Genome

Before the advent of genome sequencing, microbial geneticists only had one general approach for elucidating the organization of genes in a bacterial chromosome—to carry out linkage analysis. Such analyses yield a genetic map showing the position of genes relative to each other. Genetic mapping using linkage analysis is a very complex task. This section surveys approaches to mapping the bacterial genome, using *E. coli* as an example. All three modes of gene transfer and recombination have been used in mapping.

Hfr conjugation is frequently used to map the relative location of bacterial genes. This technique rests on the observation that during conjugation, the chromosome moves from donor to recipient at a constant rate. In an **interrupted mating experiment** the conjugation bridge is broken and Hfr $\times$ F^- mating is stopped at various intervals after the start of conjugation by mixing the culture vigorously in a blender (**figure 13.38*a***). The order and timing of gene transfer can be determined because they are a direct reflection of the order of genes on the bacterial chromosome (figure 13.38*b*). For example, extrapolation of the curves in figure 13.38*b* back to the x-axis gives the time at which each gene just began to enter the recipient. The result is a circular chromosome map with distances expressed in terms of the minutes elapsed until a gene is transferred. This technique can fairly precisely locate genes 3 minutes or more apart. The heights of the plateaus in figure 13.38*b* are lower for genes that are more distant from the F factor (the origin of transfer) because there is an ever-greater chance that the conjugation bridge will spontaneously break before these genes are transfered. Because of the relatively large size of the *E. coli* genome, it is not possible to generate a map from one Hfr strain. Therefore several Hfr strains with the F plasmid integrated at different locations must be used and their maps superimposed on one another. The overall map is adjusted to 100 minutes, although complete transfer may require somewhat more than 100 minutes. In a sense, minutes are an indication of map distance and not strictly a measure of time. Zero time is set at the threonine (*thr*) locus.

Gene linkage, or the proximity of two genes on a chromosome, can be determined from transformation by measuring the frequency with which two or more genes simultaneously transform a recipient cell. Consider the case for cotransformation by two genes. In theory, a bacterium could simultaneously receive two genes, each carried on a separate DNA fragment. However, it is much more likely that genes residing on the same fragment will be simultaneously transferred. If two genes are closely linked on the chromosome, then they should be able to cotransform. The closer the genes are together, the more often they will be carried on the same fragment and the higher will be the frequency of cotransformation. If genes are spaced a great distance apart, they will be carried on separate DNA fragments and the frequency of double transformants will equal the product of the individual transformation frequencies.

Generalized transduction can be used to obtain linkage information in much the same way as transformation. Linkages usually are expressed as cotransduction frequencies, using the argument that the closer two genes are to each other, the more likely they both will reside on the DNA fragment incorporated into a single phage capsid. The *E. coli* phage P1 is often used in such mapping because it can randomly transduce up to 1 to 2% of the genome (**figure 13.39**).

Specialized transduction is used to find which phage attachment site is close to a specific gene. The relative locations of specific phage *att* sites are known from conjugational mapping, and the genes linked to each *att* site can be determined by means of specialized transduction. These data allow precise placement of genes on the chromosome.

A simplified genetic map of *E. coli* K12 is given in **figure 13.40.** Because conjugation data are not high resolution and cannot be used to position genes that are very close together, the map was developed using several mapping techniques. Interrupted mating data were combined with those from cotransduction and cotransformation studies. Data from recombination studies also were used. New genetic markers in the *E. coli* genome were located within a relatively small region of the genome (10 to 15 minutes long) using a series of Hfr strains with F factor integration sites scattered throughout the genome. Once the genetic marker was located with respect to several genes in the same region, its position relative to nearby neighbors was more accurately determined using transformation and transduction studies. Such analyses are no longer performed in *E. coli* and other microbes for which a genome sequence has been published.

Using these techniques, researchers mapped about 2,200 genes of *E. coli* K12 and compared this with the actual nucleotide sequence of the genome (i.e., a physical map of the genome). Genome sequencing has revealed about 4,300 possible genes. Thus genetic analysis defined over half of the potential genes. The genetic map approximates the physical map, but they do not correspond perfectly. This is because the genetic map is derived from genetic linkage frequencies that do not correlate exactly with the number of nucleotides that separate two genes. Roughly speaking, one minute of the *E. coli* genetic map corresponds to 40 kilobases of DNA sequence.

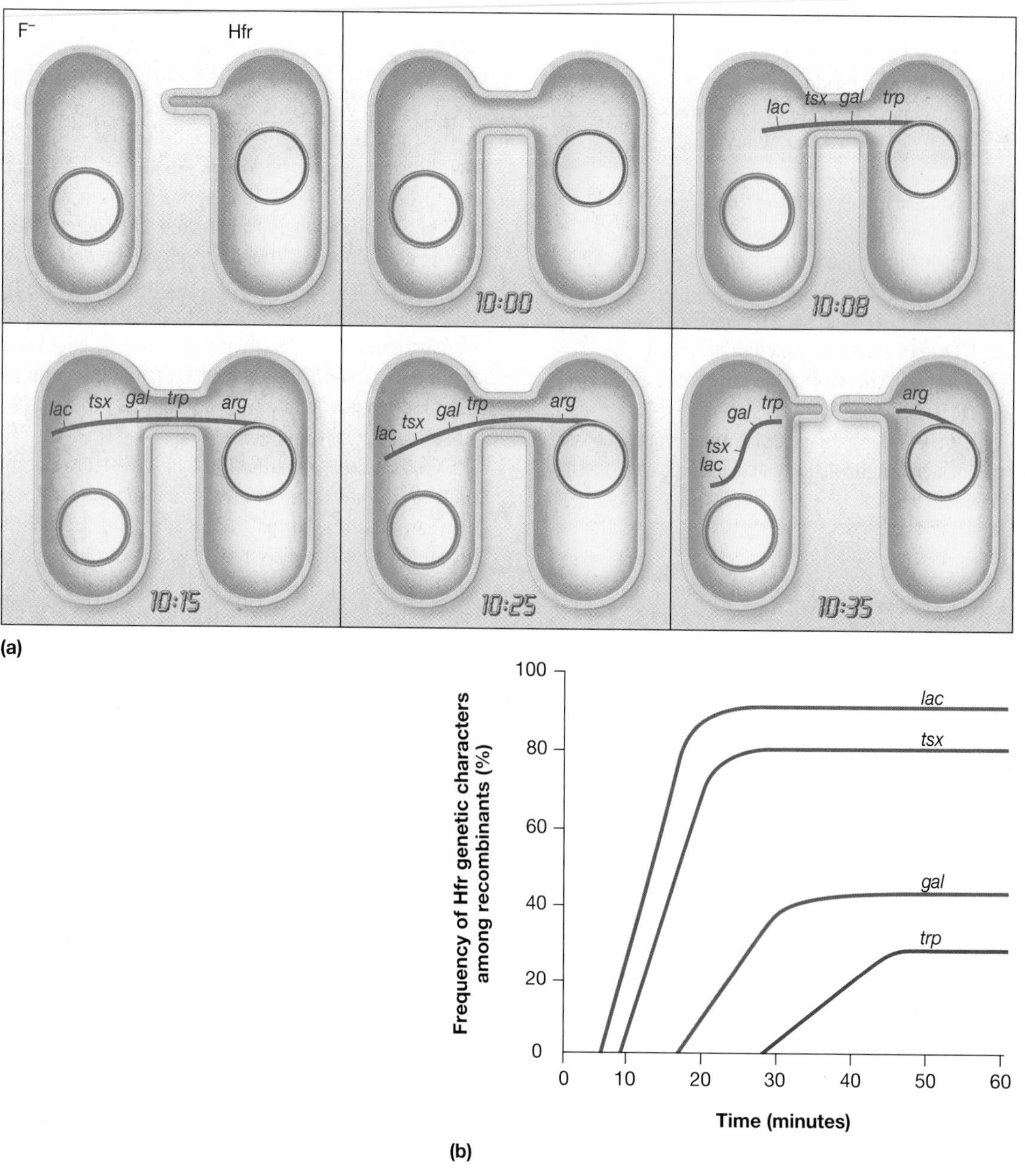

Figure 13.38 An Interrupted Mating Experiment. An interrupted mating experiment using Hfr × F⁻ conjugation. **(a)** The linear transfer of genes is stopped by breaking the conjugation bridge to study the sequence of gene entry into the recipient cell. **(b)** An example of the results obtained by an interrupted mating experiment. The gene order is *lac-tsx-gal-trp.*

13.11 Recombination and Genome Mapping in Viruses

Bacteriophage genomes also undergo recombination, although the process is different from that in bacteria. Because phages reproduce within cells and cannot recombine directly, crossing-over must occur inside a host cell. In principle, a virus recombination experiment is easy to carry out. If bacteria are mixed with enough phages so that on average at least two viruses will infect each cell, genetic recombination should be observed. Phage progeny in the resulting lysate can be checked for alternate combinations of the initial parental genotypes.

Alfred Hershey initially demonstrated recombination in the phage T2, using two strains with differing phenotypes. Two of the parental strains in Hershey's crosses were $h^+ r^+$ and *hr* (**figure 13.41**). The gene *h* influences host range; when gene *h* changes, T2 infects different strains of *E. coli.* The *r* gene of phage T2 affects plaque morphology. Plaques are visible manifestations of the phage lytic cycle, when the host is cultured on a solid growth medium (**figure 13.42*a***). Phages with the r^+ gene

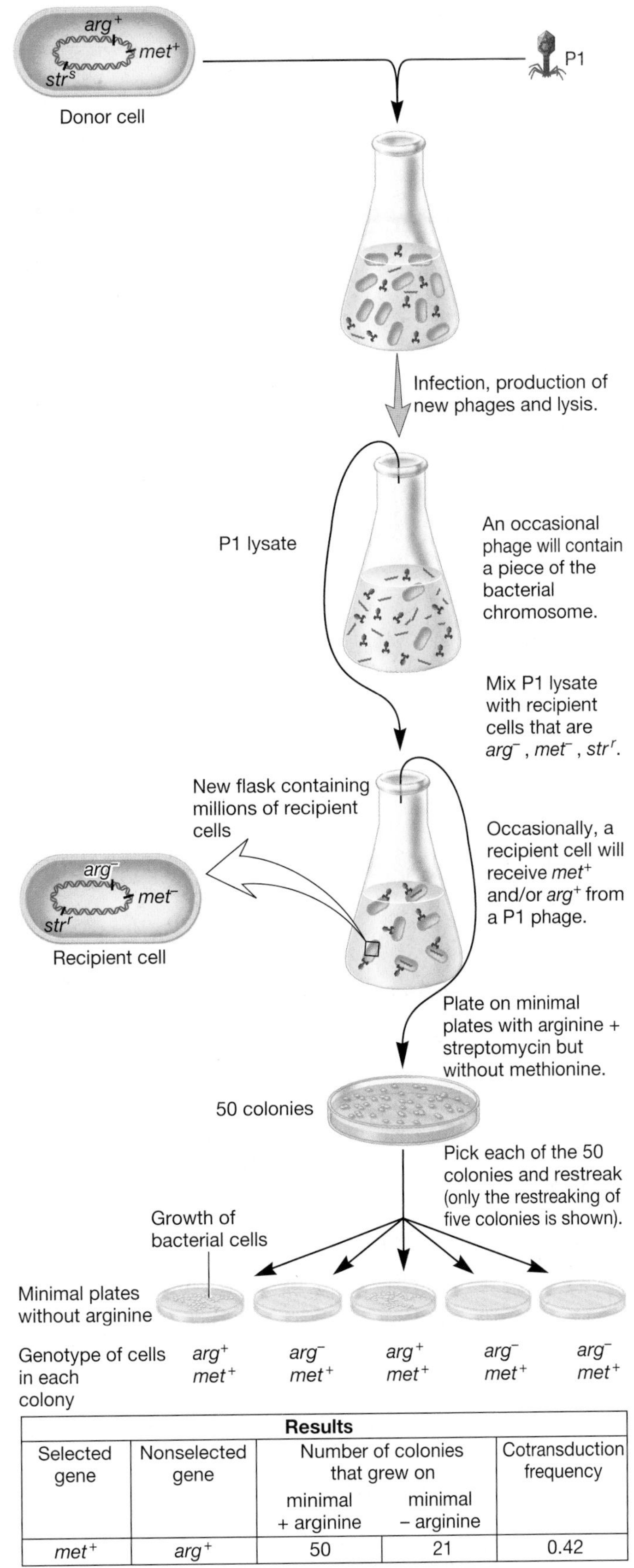
arg+
met+
strS
Donor cell
P1
Infection, production of new phages and lysis.
P1 lysate
An occasional phage will contain a piece of the bacterial chromosome.
Mix P1 lysate with recipient cells that are arg−, met−, strr.
New flask containing millions of recipient cells
arg−
met−
strr
Recipient cell
Occasionally, a recipient cell will receive met+ and/or arg+ from a P1 phage.
Plate on minimal plates with arginine + streptomycin but without methionine.
50 colonies
Pick each of the 50 colonies and restreak (only the restreaking of five colonies is shown).
Growth of bacterial cells
Minimal plates without arginine
Genotype of cells in each colony
arg+ met+
arg− met+
arg+ met+
arg− met+
arg− met+

Results				
Selected gene	Nonselected gene	Number of colonies that grew on		Cotransduction frequency
		minimal + arginine	minimal − arginine	
met$^+$	*arg*$^+$	50	21	0.42

Figure 13.39 A Cotransduction Experiment. In this experiment, the donor is able to synthesize the amino acids arginine and methionine (*arg*$^+$ and *met*$^+$) but is killed by the antibiotic streptomycin (*str*^{s}). The recipient is unable to synthesize arginine and methionine, but is resistant to streptomycin (*str*^{r}). The phage lysate made by infecting the donor bacterium is mixed with the recipient bacterium. The mixture is then plated onto a medium containing streptomycin but lacking methionine. Therefore, the only cells able to grow are those recipient cells that have received the functional methionine gene from the donor. The colonies that grow are then tested to see if they also received the gene for arginine biosynthesis from the donor. This is determined by plating the cells on a minimal medium lacking arginine. Only those cells that can synthesize arginine grow.

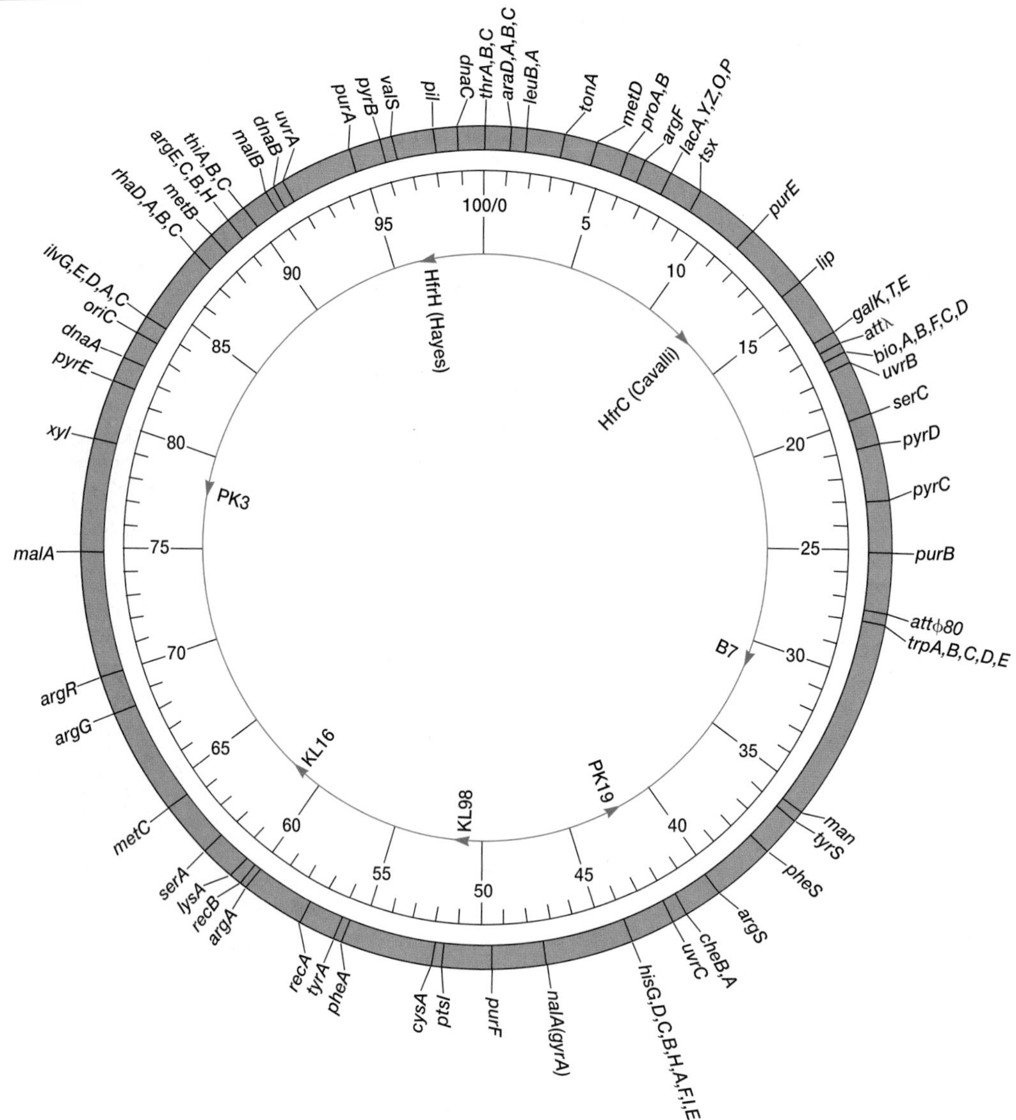

Figure 13.40 *E. coli* Genetic Map. A circular genetic map of *E. coli* K12 with the location of selected genes. The inner circle shows the origin and direction of transfer of several Hfr strains. The map is divided into 100 minutes, the time required to transfer the chromosome from an Hfr cell to F^- at 37°C.

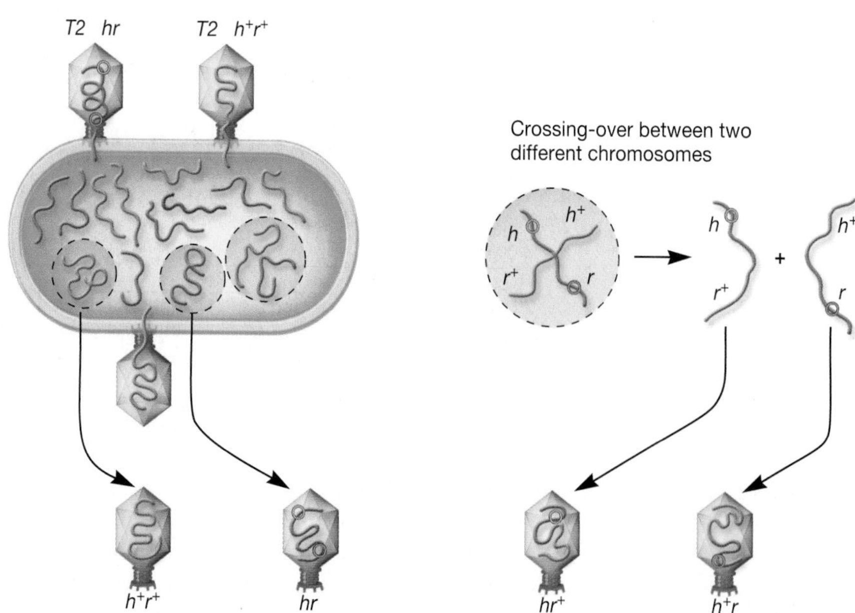

Figure 13.41 Genetic Recombination in Bacteriophages. A summary of a genetic recombination experiment with the *hr* and h^+r^+ strains of the T2 phage. The *hr* chromosome is red; the h^+r^+ chromosome is blue.

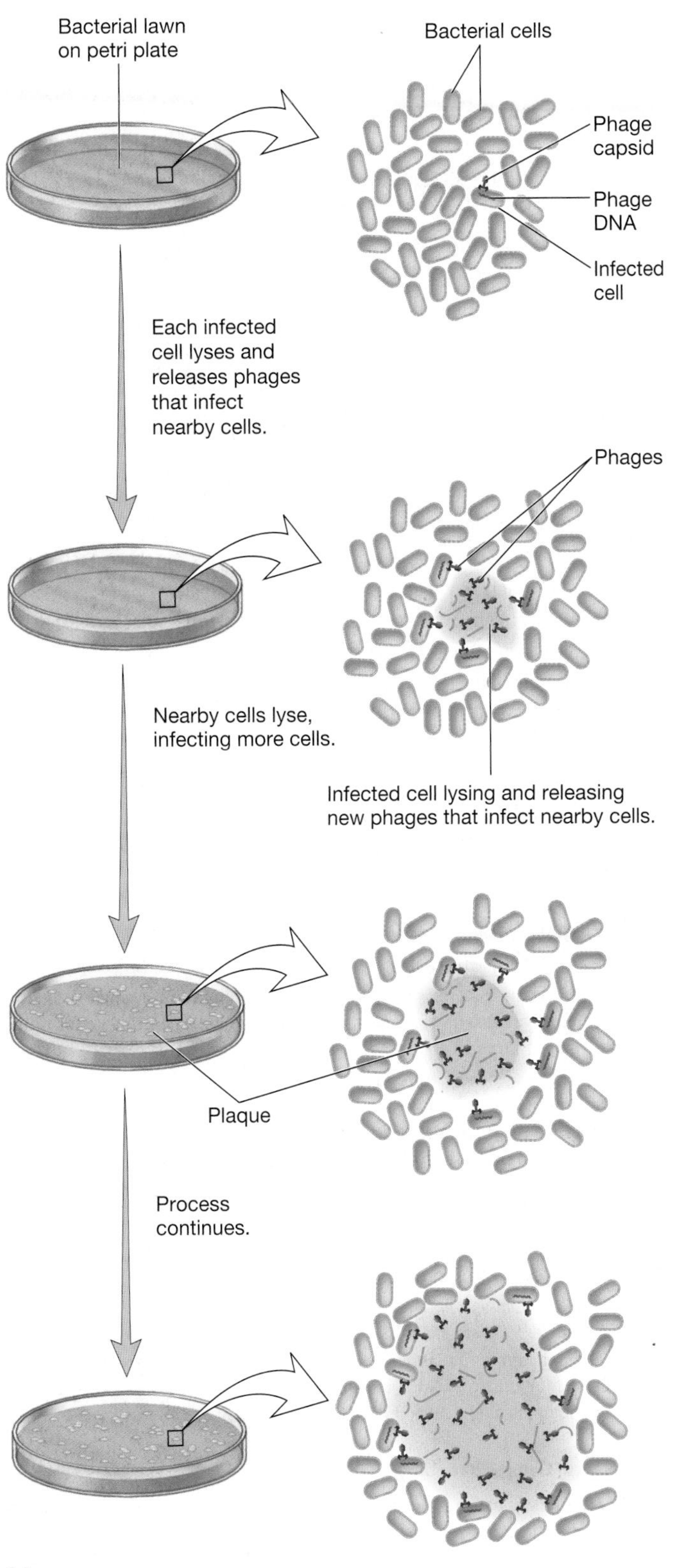

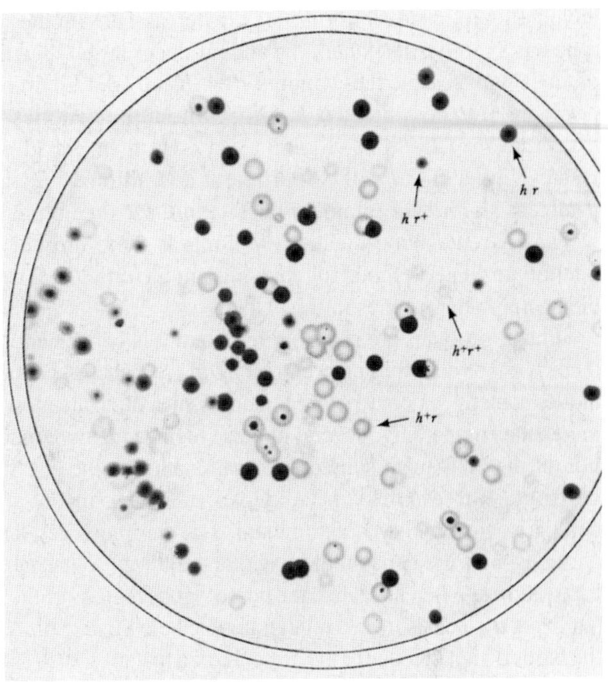

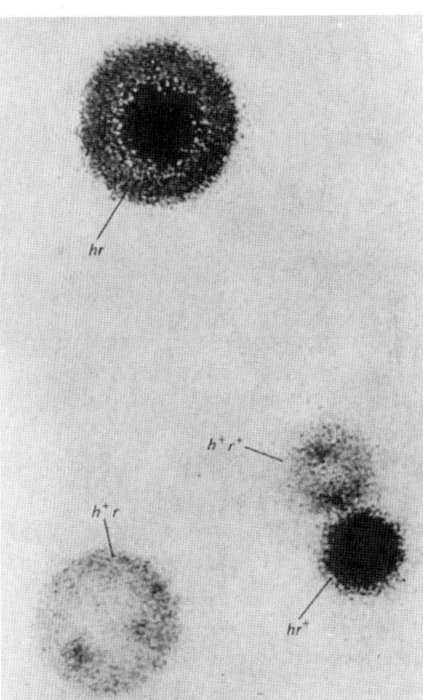

Figure 13.42 The Formation of Phage Plaques. (a) When phages and host bacterial cells are mixed at an appropriate ratio, only a portion of the cells will be initially infected. When this mixture is plated, the infected cells will be separated from each other. The infected cells eventually lyse, releasing progeny phages. They infect nearby cells, which eventually lyse, releasing more phages. This continues and ultimately gives rise to a clear area within a lawn of bacteria. The clear area is a plaque. **(b)** The types of plaques produced by a recombination experiment between T2 *hr* and T2 h^+r^+ on a lawn of *E. coli* cells. **(c)** A close-up of the four plaque types.

have wild type plaque morphology, whereas T2 with the *r* genotype has a rapid lysis phenotype and produces larger than normal plaques with sharp edges (figures 13.42*b* and 13.42*c*). In one experiment Hershey infected *E. coli* with large quantities of the h^+r^+ and *hr* T2 strains (figure 13.41). He then plated out the lysates with a mixture of two different host strains and was able to detect significant numbers of h^+r and hr^+ recombinants, as well as parental type plaques. As long as there are detectable phenotypes and methods for carrying out the crosses, it is possible to map phage genes in this way.

Phage genomes are so small that often it is convenient to map them without determining recombination frequencies. Some techniques actually generate physical maps, which often are most useful in genetic engineering. Several of these methods require manipulation of the DNA with subsequent examination in the electron microscope. For example, heteroduplex mapping involves direct comparison of wild-type and mutant viral chromosomes. The two chromosomes are denatured, mixed, and allowed to anneal due to base pairing. When annealed, the homologous regions of the different DNA molecules form a regular double helix. In locations where the bases do not pair due to the presence of a mutation such as a deletion or insertion, bubbles are visible in the electron microscope.

Several other direct techniques are used to generate physical maps of viral genomes or parts of them. Certain enzymes called restriction endonucleases can be used to cut viral DNA at specific sites. The fragments of DNA can be separated from each other based on size by gel electrophoresis—a process in which molecules move in an electrical field (*see figures 14.3 and 14.11*). By comparing genomes of different virus strains, deletions, insertions, and other mutations can be located. Phage genomes also can be directly sequenced to locate particular mutations and analyze the changes that have taken place.

1. Describe how the bacterial genome can be mapped using Hfr conjugation, transformation, generalized transduction, and specialized transduction. Include both a description of each technique and any assumptions underlying its use.
2. Why is it necessary to use several different techniques in genome mapping? How is this done in practice?
3. Describe how you would precisely locate the *recA* gene and show that it was between 58 and 58.5 minutes on the *E. coli* chromosome.
4. How does recombination in viruses differ from that in bacteria? How did Hershey first demonstrate virus recombination?
5. Describe heteroduplex mapping.

Summary

13.1 Mutations and Their Chemical Basis

a. A mutation is a stable, heritable change in the nucleotide sequence of the genetic material.

b. Spontaneous mutations can arise from replication errors (transition, transversion, and addition and deletion of nucleotides), from DNA lesions (apurinic sites, apyrimidinic sites, oxidation of DNA), and from insertions (**figures 13.1** and **13.2**).

c. Induced mutations are caused by mutagens. Mutations may result from the incorporation of base analogs, specific mispairing due to alterations of a base caused by DNA-modifying agents, the presence of intercalating agents, and severe damage to the DNA caused by exposure to radiation.

d. Mutations are usually recognized when they cause a change from the more prevalent wild-type phenotype. A mutant phenotype can be restored to wild type by either reversions or suppressor mutations (**table 13.2**).

e. There are four important types of point mutations: silent mutations, missense mutations, nonsense mutations, and frameshift mutations (**table 13.2**).

f. Mutations can affect phenotype in numerous ways. Some major types of mutations categorized based on their effects on phenotype are morphological, lethal, conditional, biochemical, and resistance mutations.

13.2 Detection and Isolation of Mutants

a. A sensitive and specific detection method is needed for detecting and isolating mutants. An example is replica plating for the detection of auxotrophs (**figure 13.7**).

b. One of the most effective mutant isolation techniques is to select for a specific mutation by adjusting environmental conditions so that the mutant will grow while the wild type does not.

c. Because many carcinogens are also mutagenic, one can test for mutagenicity with the Ames test and use the results as an indirect indication of carcinogenicity (**figure 13.9**).

13.3 DNA Repair

a. Cells have multiple mechanisms for correcting mispaired and damaged DNA.

b. Excision repair systems remove damaged portions from a single strand of DNA (e.g., thymine dimers), and use the other strand as a template for filling in the gap (**figures 13.10** and **13.11**).

c. Direct repair systems correct damaged DNA without removing damaged regions. For instance, during photoreactivation thymine dimers are repaired by splitting the two thymines apart. This is catalyzed in the presence of light by the enzyme photolyase (**figure 13.12**).

d. Mismatch repair is similar to excision repair, except that it replaces mismatched based pairs (**figure 13.13**).

e. Recombinational repair removes damaged DNA by recombination of the damaged DNA with a normal DNA strand elsewhere in the cell (**figure 13.14**).

f. When DNA damage is severe, DNA replication is halted. This triggers the SOS response. During the SOS response, genes of the repair systems are transcribed at a higher rate. In addition, special DNA polymerases are produced. These are able to replicate damaged DNA. However, they do so without a proper template and therefore create mutations.

13.4 Creating Genetic Variability

a. In recombination, genetic material from two different DNA molecules is combined to form a new hybrid molecule.

b. In eucaryotes capable of sexual reproduction, crossing-over during meiosis is important in creating genetic variation (**figure 13.15**).

c. Horizontal gene transfer is an important mechanism for creating genetic diversity in procaryotes. It is a one-way process in which the exogenote is transferred from the donor to a recipient and integrated into the endogenote (**figure 13.16**).

d. There are three types of recombination: homologous recombination, site-specific recombination, and transposition.

13.5 Transposable Elements

a. Transposons or transposable elements are DNA segments that move about the genome in a process known as transposition.

b. There are three types of transposable elements: insertion sequences, composite transposons, and replicative transposons (**figure 13.19**).

c. Simple (cut-and-paste) transposition and replicative transposition are two distinct mechanisms of transposition (**figures 13.20** and **13.21**).

d. Transposable elements can cause mutations, turn genes on and off, aid F plasmid insertion, and carry antibiotic resistance genes.

13.6 Bacterial Plasmids

a. Plasmids are small, autonomously replicating DNA molecules that can exist impendent of the host chromosome.

b. Episomes are plasmids that can be reversibly integrated with the host chromosome.

c. The F factor is one type of conjugative plasmid; that is, it is able to transfer itself from one bacterium to another (**figure 13.23**).

13.7 Bacterial Conjugation

a. Conjugation is the transfer of genes between bacteria that depends upon direct cell-cell contact. F factor conjugation is mediated by a sex pilus and a type IV secretion system.

b. In $F^+ \times F^-$ mating the F factor remains independent of the chromosome and a copy is transferred to the F^- recipient; donor genes are not usually transferred (**figure 13.28*a***).

c. Hfr strains transfer bacterial genes to recipients because the F factor is integrated into the host chromosome. A complete copy of the F factor is not often transferred (**figure 13.28*b*, *c***).

d. When the F factor leaves an Hfr chromosome, it occasionally picks up some bacterial genes to become an F′ plasmid, which readily transfers these genes to other bacteria (**figure 13.30**).

13.8 DNA Transformation

a. Transformation is the uptake of naked DNA by a competent cell and its incorporation into the genome (**figure 13.31** and **13.32**).

13.9 Transduction

a. Bacterial viruses or bacteriophages can reproduce and destroy the host cell (lytic cycle) or become a latent prophage that remains within the host (lysogenic cycle) (**figure 13.34**).

b. Transduction is the transfer of bacterial genes by viruses.

c. In generalized transduction any host DNA fragment can be packaged in a virus capsid and transferred to a recipient (**figure 13.35**).

d. Certain temperate phages carry out specialized transduction by incorporating bacterial genes during prophage induction and then donating those genes to another bacterium (**figure 13.37**).

13.10 Mapping the Genome

a. The bacterial genome can be mapped by following the order of gene transfer during Hfr conjugation (**figure 13.38**); transformational and transductional mapping techniques also may be used (**figure 13.39**).

13.11 Recombination and Genome Mapping in Viruses

a. When two viruses simultaneously enter a bacterial cell, their chromosomes can undergo recombination (**figure 13.41**).

b. Virus genomes are mapped by recombination (genetic mapping). Physical maps can be created by heteroduplex mapping and other techniques.

Key Terms

abortive transductants 346
adaptive (directed) mutation 319
allele 329
Ames test 325
apurinic site 319
apyrimidinic site 319
auxotroph 323
base analog 319
base excision repair 326
competent cell 343
composite transposon 333
conditional mutation 323
conjugation 337
conjugative plasmid 334
conjugative transposon 334
crossing-over 330
directed (adaptive) mutation 319
direct repair 326
DNA methylation 326
DNA-modifying agent 320
double-strand break model 331
endogenote 330
episome 334
excision repair 326
exogenote 330
F factor 336
F′ plasmid 339
forward mutation 320
frameshift mutation 323
generalized transducing particle 345
generalized transduction 345
helper phage 348
heteroduplex DNA 331
Hfr conjugation 339
Hfr strain 339
high-frequency transduction lysates (HFT lysates) 348
homologous recombination 331
horizontal (lateral) gene transfer (HGT) 330
host restriction 330
induced mutations 318
insertion sequence 332
intercalating agent 320
interrupted mating experiment 349
lateral (horizontal) gene transfer 330
low-frequency transduction lysate (LFT lysates) 346
lysogen 345
lysogeny 345
lytic cycle 345
merozygote 330
mismatch repair system 326
missense mutation 320
mutagen 318
mutation 317
nonreciprocal homologous recombination 331
nonsense mutation 321
nucleotide excision repair 326
photoreactivation 326
plasmid 334
point mutation 318
proofreading 326
prophage 345
prototroph 323
RecA protein 327
recombinants 329
recombination 329
recombinational repair 327
replica plating 324
replicative transposition 334
reversion mutation 320
sex pilus 338
silent mutation 320
simple (cut-and-paste) transposition 334
site-specific recombination 331
SOS response 327
specialized transduction 346
spontaneous mutations 318
suppressor mutation 320
temperate bacteriophage 345
transduction 345
transformation 342
transition mutation 319
translesion DNA synthesis 329
transposable element 332
transposase 333
transposition 331
transposon 332
transversion mutation 319
virulent bacteriophage 345
wild type 320

Critical Thinking Questions

1. Mutations are often considered harmful. Give an example of a mutation that would be beneficial to a microorganism. What gene would bear the mutation? How would the mutation alter the gene's role in the cell, and what conditions would select for this mutant allele?
2. Mistakes made during transcription affect the cell, but are not considered "mutations." Why not?
3. Given what you know about the differences between bacterial and eucaryotic cells, give two reasons why the Ames test detects only about half of potential carcinogens, even when liver extracts are used.
4. Diagram a double crossover event and a single crossover event. Which is more infrequent and why? Suggest experiments in which you would use one or the other event and what types of genetic markers you would employ. What kind of recognition features and catalytic capabilities would the recombination machinery need to possess?
5. Suppose that transduction took place when a U-tube experiment was conducted. How would you confirm that something like a virus was passed through the filter and transduced the recipient?
6. Suppose that you carried out a U-tube experiment with two auxotrophs and discovered that recombination was not blocked by the filter but was stopped by treatment with deoxyribonuclease. What gene transfer process is responsible? Why would it be best to use double or triple auxotrophs in this experiment?
7. What would be the evolutionary advantage of having a period of natural "competence" in a bacterial life cycle? What would be possible disadvantages?

Learn More

Barkay, T., and Smets, B. F. 2005. Horizontal gene flow in microbial communities. *ASM News* 71(9):412–19.

Brock, T. D. 1990. *The emergence of bacterial genetics.* Cold Spring Harbor, N.Y.: Cold Spring Harbor Laboratory Press.

Brooker, R. J. 2005. *Genetics: Analysis and principles,* 2d ed. Boston: McGraw-Hill.

Foster, P. L. 2004. Adaptive mutation in *Escherichia coli. J. Bacteriol.* 186(15): 4846–52.

Gogarten, J. P., and Townsend, J. P. 2005. Horizontal gene transfer, genome innovation and evolution. *Nature Rev. Microbiol.* 3:679–87.

Grohmann, E.; Muth, G.; and Espinosa, M. 2003. Conjugative plasmid transfer in gram-positive bacteria. *Microbiol. Molec. Biol. Rev.* 67(2):277–301.

Hahn, J.; Maier, B.; Haijema, B. J.; Sheetz, M.; and Dubnau, D. 2005. Transformation proteins and DNA uptake localize to the cell poles in *Bacillus subtilis. Cell* 122:59–71.

Lawley, T. D.; Klimke, W. A., Gubbins, M. J.; and Frost, L. S. 2003. F factor conjugation is a true type IV secretion system. *FEMS Microbiol. Lett.* 224:1–15.

Roth, J. R., and Andersson, D. I. 2004. Adaptive mutation: How growth under selection stimulates lac^+ reversion by increasing target copy number. *J. Bacteriol.* 186(15):4855–60.

Schröder, G., and Lanka, E. 2005. The mating pair formation system of conjugative plasmids—A versatile secretion machinery for transfer of proteins and DNA. *Plasmid* 54:1–25.

Sutton, M. D.; Smith, B. T.; Godoy, V. G.; and Walker, G. C. 2000. The SOS response: Recent insights into *umuDC*-dependent mutagenesis and DNA damage tolerance. *Annu. Rev. Genet.* 34:479–97.

Please visit the Prescott website at www.mhhe.com/prescott7 for additional references.

14 Recombinant DNA Technology

A scientist examines DNA following agarose gel electrophoresis. Each bright band is a fragment of DNA stained with ethidium bromide, so that upon illumination with ultraviolet light, the DNA fluoresces.

Preview

- Genetic engineering makes use of recombinant DNA technology to fuse genes with vectors and then clone them in host cells. In this way isolated genes can be replicated in high copy and large quantities of their products can be synthesized.
- The isolation of individual genes or DNA fragments depends on the ability of restriction endonucleases to cleave DNA at specific sites.
- Plasmids, bacteriophages and other viruses, cosmids, and artificial chromosomes are used as cloning vectors. They can replicate within a host cell while carrying foreign DNA. These vectors carry genes that confer phenotypic traits that allow them to be detected and maintained by the host cell.
- Genetic engineering contributes substantially to biological research, medicine, industry, and agriculture. Benefits from this technology will continue to grow.
- Genetic engineering also is accompanied by challenges in such areas as safety, the ethics of its use with human subjects, environmental impact, and biological warfare.

Chapters 11 through 13 introduce the essentials of microbial genetics. In this chapter we focus on the practical applications of microbial genetics and the technologies arising from it.

Although human beings have been altering the genetic makeup of organisms for centuries by selective breeding, only recently has the direct manipulation of DNA been possible. The deliberate modification of an organism's genetic information by directly changing the sequence of nucleic acids in its genome is called **genetic engineering** and is accomplished by a collection of methods known as **recombinant DNA technology.** The generation of a large number of genetically identical DNA molecules is called **cloning.** The most commonly used steps to clone a gene or other DNA element are outlined in **figure 14.1.** First, the DNA responsible for a particular phenotype is identified and isolated (figure 14.1, steps 1 and 2). Once purified, the gene or genes are fused with another piece of DNA called a cloning vector to form recombinant DNA molecules (step 3). These are propagated by insertion into an organism that may not even be in the same domain as the original gene donor (step 4). Recombinant DNA technology opens up totally new areas of research and applied biology. It is an essential part of **biotechnology,** which is experiencing exceptionally rapid growth and development. Although the term has several definitions, here biotechnology refers to those processes in which living organisms are manipulated, particularly at the molecular genetic level, to form useful products. The promise for medicine, agriculture, and industry is great; yet not without controversy. Applied and industrial microbiology (chapter 41)

Recombinant DNA technology is very much the result of several key discoveries in microbial genetics. Section 14.1 briefly reviews some landmarks in the development of recombinant technology (**table 14.1**).

14.1 Historical Perspectives

Recombinant DNA is DNA with a new sequence formed by joining fragments from two or more different sources. One of the first breakthroughs leading to recombinant DNA technology was the discovery in the late 1960s by Werner Arber and Hamilton Smith of bacterial enzymes that make cuts in double-stranded DNA. These enzymes, known as **restriction enzymes** or restriction

The recombinant DNA breakthrough has provided us with a new and powerful approach to the questions that have intrigued and plagued man for centuries.

—*Paul Berg*

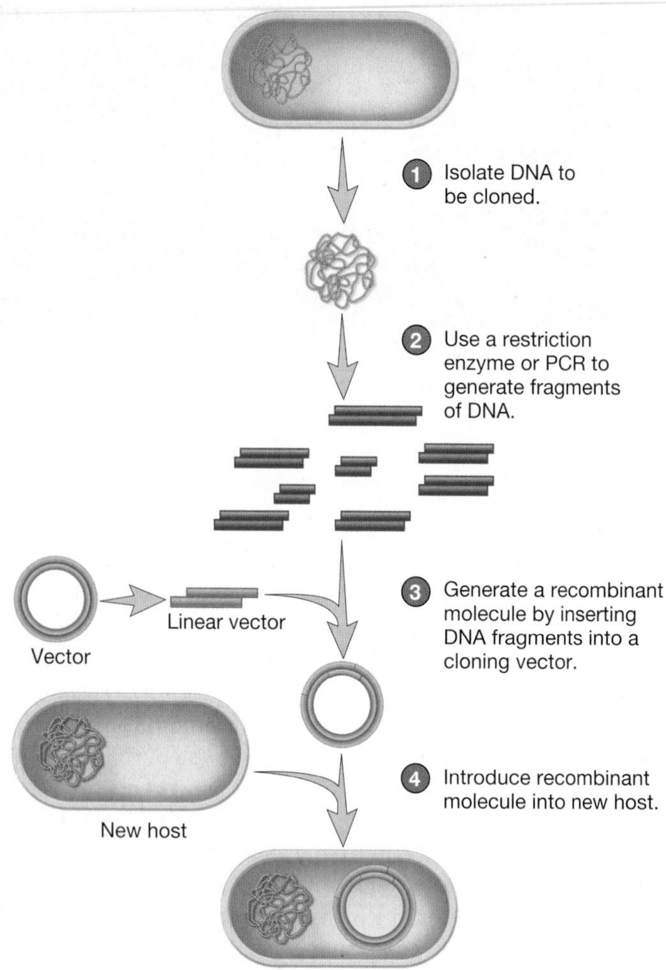
1 Isolate DNA to be cloned.
2 Use a restriction enzyme or PCR to generate fragments of DNA.
3 Generate a recombinant molecule by inserting DNA fragments into a cloning vector.
Vector
Linear vector
4 Introduce recombinant molecule into new host.
New host

Figure 14.1 Steps in Cloning a Gene. Each step shown in this overview is discussed in more detail in this chapter.

endonucleases, recognize and cleave specific sequences about 4 to 8 base pairs long (**figure 14.2**). They normally protect the host cell by destroying phage DNA after its entrance. Cells protect their own DNA from restriction enzymes by methylating specific nucleotides. Incoming foreign DNA that is not methylated in the same pattern as the host may be cleaved by host restriction enzymes. Restriction enzymes recognize specific DNA sequences called recognition sites. Each restriction enzyme has its own recognition site. Hundreds of different restriction enzymes have been purified and are commercially available (**table 14.2**). Type I and type III endonucleases identify their unique recognition sites and then cleave DNA at a defined distance from it. The more common type II endonucleases cut DNA directly at their recognition sites. These enzymes can be used to prepare DNA fragments containing specific genes or portions of genes. For example, the restriction enzyme *Eco*RI, isolated by Herbert Boyer in 1969 from *Escherichia coli,* cleaves DNA between G and A in the base sequence 5′-GAATTC-3′ (**figure 14.3**). Because DNA is antiparallel, this sequence is reversed on opposite strands of DNA. When *Eco*RI cleaves between the G and A residues, the remaining unpaired 5′-AATTC-3′ remains at the end of each strand. The complementary bases on two *Eco*RI-cut fragments can hydrogen bond, thus *Eco*RI and other endonucleases like it generate cohesive or **sticky ends.** In contrast, cleavage by restriction enzymes like *Alu*I and *Hae*III leave blunt ends. A few restriction enzymes and their recognition sites are listed in table 14.2. Note that each enzyme is named after the bacterium from which it is purified.

Very early in the development of recombinant DNA technology, it was evident that cloning eucaryotic DNA into procaryotic hosts would be desirable but problematic. This is because eucaryotic pre-mRNA must be processed (e.g., introns spliced out), and procaryotes lack the molecular machinery to perform this task. In 1970, Howard Temin and David Baltimore independently discovered the enzyme that solved this dilemma. They isolated the enzyme **reverse transcriptase (RT)** from retroviruses. These viruses have an RNA genome that is copied into DNA prior to replication. The mechanism by which reverse transcriptase accomplishes this is outlined in **figure 14.4.** By using processed mRNA as a template for **complementary DNA (cDNA)** synthesis, RNA processing is not required when cloned cDNA is expressed. Reproduction of vertebrate viruses: Genome replication, transcription, and protein synthesis in RNA viruses (section 18.2)

The next advance came in 1972, when David Jackson, Robert Symons, and Paul Berg reported that they had successfully generated recombinant DNA molecules. They allowed the sticky ends of fragments to anneal—that is, to base pair with one another—and then covalently joined the fragments with the enzyme DNA ligase. Within a year, plasmid **vectors** that carry foreign DNA fragments during gene cloning had been developed and combined with foreign DNA (**figure 14.5**). The first such recombinant plasmid capable of being replicated within a bacterial host was the pSC101 plasmid constructed by Stanley Cohen and Herbert Boyer in 1973 (SC in the plasmid name stands for *S*tanley *C*ohen).

Once genes could be recombined into cloning vectors, biologists sought to clone specific genes from various organisms. But how could one distinguish the fragment of DNA possessing the gene of interest from the numerous chromosomal fragments produced by restriction enzyme digestion? In 1975, Edwin Southern solved this problem with his **Southern blot procedure.** This technique enables the detection of specific DNA fragments from a mixture of DNA molecules (**figure 14.6**).

In the Southern blot procedure, DNA fragments are first separated by size with agarose gel electrophoresis (see section 14.4). The fragments are then denatured (rendered single stranded) and transferred to a nylon membrane and treated so that each fragment is firmly bound to the filter at the same position as on the gel. The transfer occurs when buffer flows through the gel and the membrane as shown in figure 14.6. Alternatively, the negatively charged DNA fragments can be electrophoresed from the gel onto the blotting membrane. The filter is bathed with a solution containing a radioactive **probe,** which is a fragment of labeled, single-stranded nucleic acid that is complementary to the DNA of interest. Those fragments to which the probe hydrogen bonds become radioactive and are

Table 14.1 Some Milestones in Biotechnology and Recombinant DNA Technology

1958	DNA polymerase purified
1970	A complete gene synthesized in vitro Discovery of the first sequence-specific restriction endonuclease and the enzyme reverse transcriptase
1972	First recombinant DNA molecules generated
1973	Use of plasmid vectors for gene cloning
1975	Southern blot technique for detecting specific DNA sequences
1976	First prenatal diagnosis using a gene-specific probe
1977	Methods for rapid DNA sequencing Discovery of "split genes" and somatostatin synthesized using recombinant DNA
1978	Human genomic library constructed
1979	Insulin synthesized using recombinant DNA First human viral antigen (hepatitis B) cloned
1981	Foot-and-mouth disease viral antigen cloned First monoclonal antibody-based diagnostic kit approved for use
1982	Commercial production by *E. coli* of genetically engineered human insulin Isolation, cloning, and characterization of a human cancer gene Transfer of gene for rat growth hormone into fertilized mouse eggs
1983	Engineered Ti plasmids used to transform plants
1985	Tobacco plants made resistant to the herbicide glyphosate through insertion of a cloned gene from *Salmonella* Development of the polymerase chain reaction technique
1987	Insertion of a functional gene into a fertilized mouse egg cures the shiverer mutation disease of mice, a normally fatal genetic disease
1988	The first successful production of a genetically engineered staple crop (soybeans) Development of the gene gun
1989	First field test of a genetically engineered virus (a baculovirus that kills cabbage looper caterpillars)
1990	Production of the first fertile corn transformed with a foreign gene (a gene for resistance to the herbicide bialaphos)
1991	Development of transgenic pigs and goats capable of manufacturing proteins such as human hemoglobin First test of gene therapy on human cancer patients
1994	The Flavr Savr tomato introduced, the first genetically engineered whole food approved for sale Fully human monoclonal antibodies produced in genetically engineered mice
1995	*Haemophilus influenzae* genome sequenced
1996	*Methanocaldococcus jannaschii* and *Saccharomyces cerevisiae* genomes sequenced
1997	Human clinical trials of antisense drugs and DNA vaccines begun; *E. coli* genome sequenced
1998	First cloned mammal (the sheep Dolly)
2002	*Plasmodium falciparum* genome sequenced
2003	Completion of the draft of the human genome
2005	Reconstruction of 1918 influenza virus

readily detected by **autoradiography.** In this technique a sheet of photographic film is placed over the filter. When developed, bands appear wherever a radioactive fragment is located because the energy released by the isotope causes the formation of dark-silver grains. Nonradioactive probes may also be used to detect specific DNAs. They are more rapid and safer than using radioisotopes.

By the late 1970s the techniques for cloning DNA were harnessed to produce recombinant human insulin and, by 1982, commercial production of insulin from genetically engineered *E. coli* began. This was an important development for several reasons: first, diabetic individuals no longer had to depend on insulin from pigs or other animals; second, it demonstrated the commercial feasibility of using recombinant DNA to make a better product. Other important innovations that followed are listed in table 14.1; sections 14.2 through 14.8 discuss how these and other techniques are currently used in the important field of genetic engineering.

1. Describe restriction enzymes, sticky ends, and blunt ends. Can you think of a cloning situation where blunt-ended DNA might be more useful than DNA with sticky ends?
2. What is cDNA? How does it differ from the DNA isolated from a procaryote?
3. What is the purpose of Southern blotting? How is a probe selected? Why do you think the Southern blot technique was an important breakthrough when it was first introduced?

Figure 14.2 Restriction Endonuclease Binding to DNA. The structure of *Bam*HI binding to DNA viewed down the DNA axis. The enzyme's two subunits lie on each side of the DNA double helix. The α-helices are in green, the β conformations in purple, and DNA is in orange.

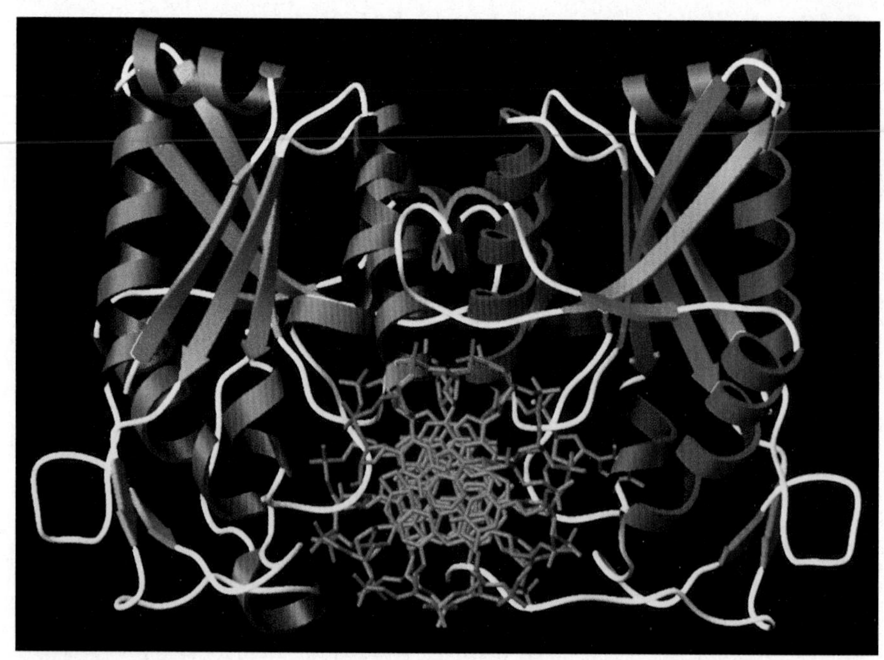

Table 14.2 Some Type II Restriction Endonucleases and Their Recognition Sequences

Enzyme	Microbial Source	Recognition Sequence[a]	End Produced
*Alu*I	*Arthrobacter luteus*	5′ AG↓CT 3′ 3′ TC↑GA 5′	5′ AG CT 3′ 3′ TC GA 5′
*Bam*HI	*Bacillus amyloliquefaciens* H	5′ G↓GATCC 3′ 3′ CCTAG↑G5′	5′ G GATCC 3′ 3′ CCTAG G 5′
*Eco*RI	*Escherichia coli*	5′ G↓AATTC 3′ 3′ CTTAA↑G 5′	5′ G AATTC 3′ 3′ CTTAA G 5′
*Hae*III	*Haemophilus aegyptius*	5′ GG↓CC 3′ 3′ CC↑GG 5′	5′ GG CC 3′ 3′ CC GG 5′
*Hind*III	*Haemophilus influenzae* d	5′ A↓AGCTT 3′ 3′ TTCGA↑A 5′	5′ A AGCTT 3′ 3′ TTCGA A 5′
*Not*I	*Nocardia otitidis-caviarum*	5′ GC↓GGCCGC 3′ 3′ CGCCGG↑CG 5′	5′ GC GGCCGC 3′ 3′ CGCCGG CG 5′
*Pst*I	*Providencia stuartii*	5′ CTGCA↓G 3′ 3′ G↑ACGTC 5′	5′ CTGCA G3′ 3′ G ACGTC 5′
*Sal*I	*Streptomyces albus*	5′ G↓TCGAC 3′ 3′ CAGCT↑G 5′	5′ G TCGAC 3′ 3′ CAGCT G 5′

[a]The arrows indicate the sites of cleavage on each strand.

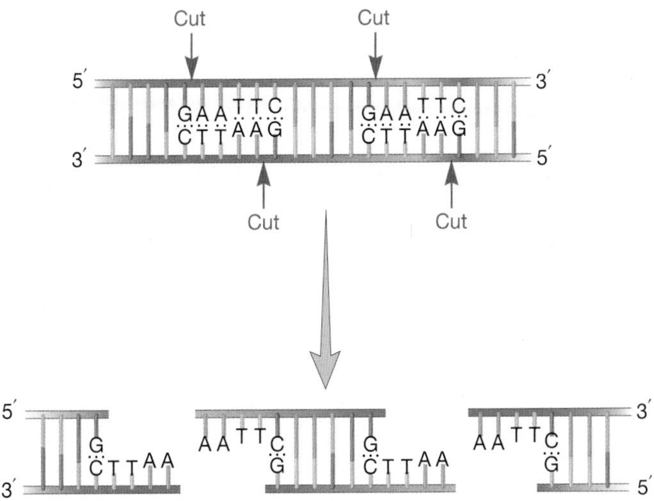

Figure 14.3 Restriction Endonuclease Action. The cleavage catalyzed by the restriction endonuclease *Eco*RI. The enzyme makes staggered cuts on the two DNA strands to form sticky ends.

14.2 Synthetic DNA

So far, the manipulation of DNA purified from living cells has been reviewed. However, the ability to synthesize short pieces of DNA called **oligonucleotides** [Greek *oligo*, few or scant] was another important advance. Oligonucleotides are generally between 15 and 30 nucleotides long, and can be either RNA or DNA. They are used in a variety of molecular techniques, such as the polymerase chain reaction (PCR) discussed in section 14.3. Nucleic acid structure (section 11.3)

DNA oligonucleotides are synthesized by a stepwise process in which single nucleotides are added to the end of the growing chain (**figure 14.7**). The 3′ end of the chain is attached to a solid support such as a silica gel particle. A DNA synthesizer or "gene machine" carries out the solid-phase synthesis. A specially activated nucleotide derivative is added to the 5′ end of the chain in a series of steps. At the end of an addition cycle, the growing chain is separated from the reaction mixture by filtration or centrifugation. The process is then repeated to attach another nucleotide. In a relatively short time, chains 50 to 100 nucleotides long can be synthesized.

Advances in DNA synthetic techniques have accelerated progress in the study of protein function. One of the most effective ways of studying the relationship of protein structure to function is by altering a specific part of the protein and observing functional changes. In the past this was accomplished either by chemically modifying individual amino acids or by inducing random mutations in the gene coding for the protein under study. There are problems with these two approaches. Chemical modification of a protein is not always specific; several amino acids may be altered, not just the one desired. It is not always possible to produce the

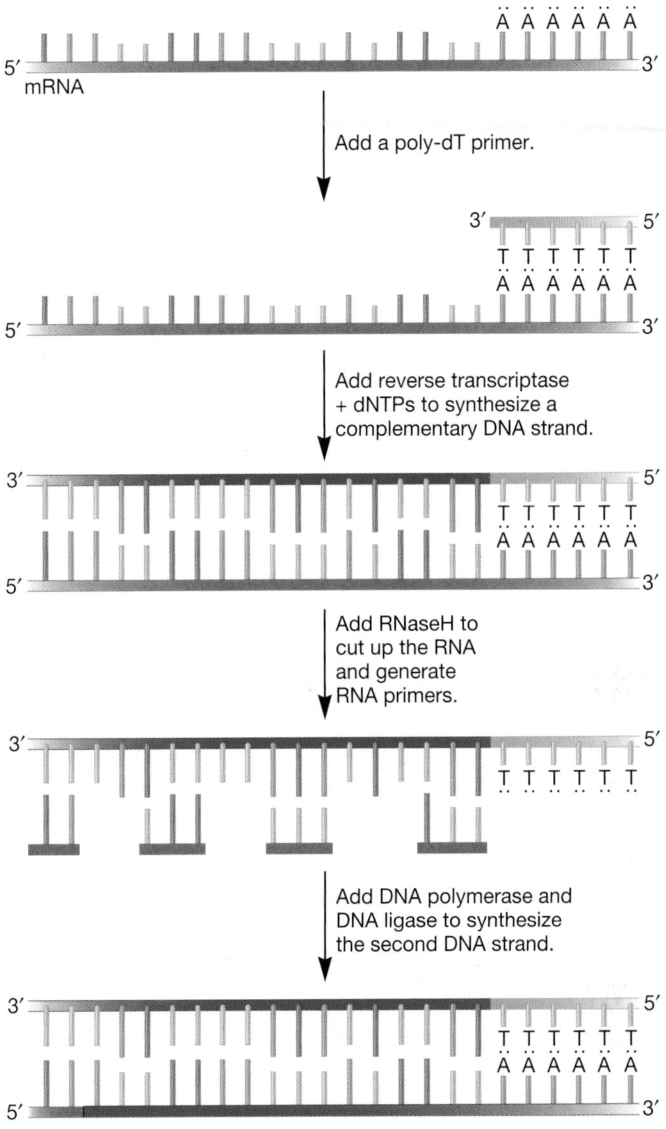

Figure 14.4 Synthesis of cDNA. A poly-dT primer anneals to the 3′ end of mRNAs. Reverse transcriptase then catalyzes the synthesis of a complementary DNA strand (cDNA). RNaseH digests the mRNA into short pieces that are used as primers by DNA polymerase to synthesize the second DNA strand. The 5′ to 3′ exonuclease function removes all of the RNA primers except the one at the 5′ end (because there is no primer upstream from this site). This RNA primer can be removed by the subsequent addition of another RNase. After the double stranded cDNA is made, it can then be inserted into vectors as described in figure 14.13.

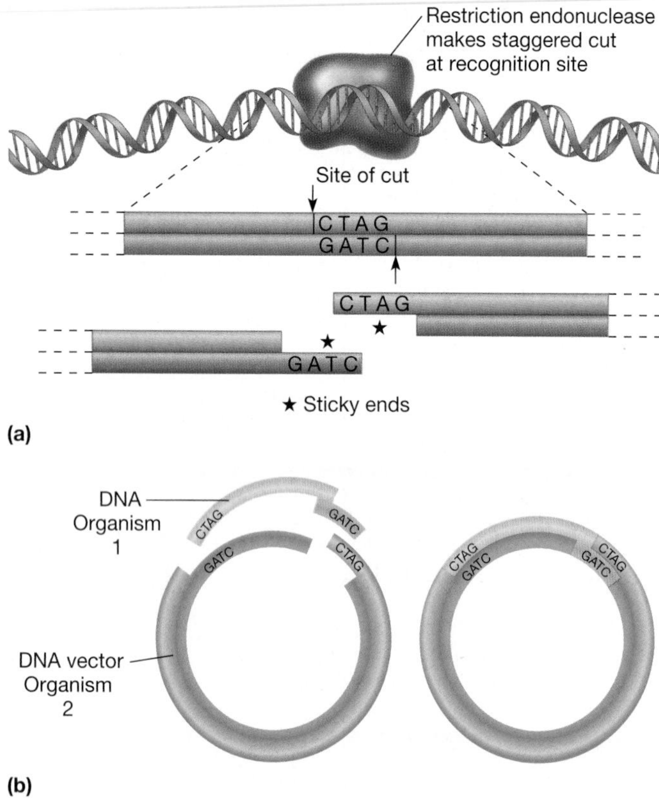

Figure 14.5 Recombinant Plasmid Construction. **(a)** A restriction endonuclease recognizes and cleaves DNA at its specific recognition site. Cleavage produces sticky ends that accept complementary tails for gene splicing. **(b)** The sticky ends can be used to join DNA from different organisms by cutting it with the same restriction enzyme, ensuring that all fragments have complementary ends.

proper mutation in the desired gene location. These difficulties can overcome with a technique called **site-directed mutagenesis.**

In site-directed mutagenesis an oligonucleotide of about 20 nucleotides that contains the desired sequence change is synthesized. The altered oligonucleotide with its artificially mutated sequence is allowed to bind to a single-stranded copy of the complete gene (**figure 14.8**). DNA polymerase is added to the gene-primer complex. The polymerase extends the primer and replicates the remainder of the target gene to produce a new gene copy with the desired mutation. The DNA is then cloned using the techniques described in sections 14.5 and 14.6. This yields large quantities of the mutant protein for study.

14.3 The Polymerase Chain Reaction

The synthesis of oligonucleotides is a process that evolved over a number of years (the first report of chemically synthesized DNA was published in 1955, just 2 years after Watson and Crick resolved the structure of DNA). In contrast, the **polymerase chain reaction (PCR)**, invented by Kary Mullis in the early 1980s, exploded onto the biotechnology landscape. It has had such a profound impact on biology, biochemistry, and medicine that Mullis and Michael Smith, who developed the technique of site-directed mutagenesis, shared a Nobel Prize in 1993. Why is PCR so important? Quite simply, it enables the rapid synthesis of many, many copies of a specific DNA fragment from a complex mixture of DNA. Researchers can thus obtain large quantities of specific pieces of DNA for experimental and diagnostic purposes.

Figure 14.9 outlines how the PCR technique works. Suppose that one wishes to make large quantities of a particular DNA sequence, a process known as DNA or gene amplification. The first step is to synthesize DNA fragments with sequences identical to those flanking the targeted sequence. This is accomplished with a DNA synthesizer. These synthetic oligonucleotides are usually about 20 nucleotides long and serve as DNA **primers** for DNA synthesis. The primers are one component of the reaction mixture, which also contains the target DNA (often copies of an entire genome), a thermostable DNA polymerase, and each of the four deoxyribonucleoside triphosphates (dNTPs). PCR requires a series of repeated reactions, called cycles. Each cycle has three steps that are precisely executed in a machine called a **thermocycler.** In the first step, the target DNA containing the sequence to be amplified is heat denatured to make it single-stranded. Next, the temperature is lowered so that the primers can hydrogen bond or anneal to the DNA on both sides of the target sequence. Because the primers are very small and are present in excess, the targeted DNA strands anneal to the primers rather than to each other. Finally, DNA polymerase extends the primers and synthesizes copies of the target DNA sequence using dNTPs. Only polymerases able to function at the high temperatures employed in the PCR technique can be used. Two popular enzymes are the *Taq* polymerase from the thermophilic bacterium *Thermus aquaticus* and the Vent polymerase from *Thermococcus litoralis.* At the end of one cycle, the targeted sequences on both strands have been copied. When the three-step cycle is repeated (figure 14.9), the two strands from the first cycle are copied to produce four fragments. These are amplified in the third cycle to yield eight double-stranded products. Thus, each cycle increases the number of target DNA molecules exponentially. Depending on the initial concentration of the template DNA and other parameters such as the G+C composition of the DNA to be amplified, it is theoretically possible to produce about *one million* copies of targeted DNA sequence after 20 cycles, and as many as *one billion* after 30 cycles. Pieces ranging in size from less than 100 base pairs to several thousand base pairs in length can be amplified, and only 10 to 100 picomoles of primer are required. The concentration of target DNA can be as low as 10^{-20} to 10^{-15} M (or 1 to 10^5 DNA copies per 100 μl). The whole reaction mixture is often 50 μl or less in volume. DNA replication (section 11.4)

PCR is most frequently used in one of two ways. If one wants to generate large quantities of a specific sequence, the reaction products are collected and purified at the end of a designated number of cycles. The final number of DNA fragments amplified

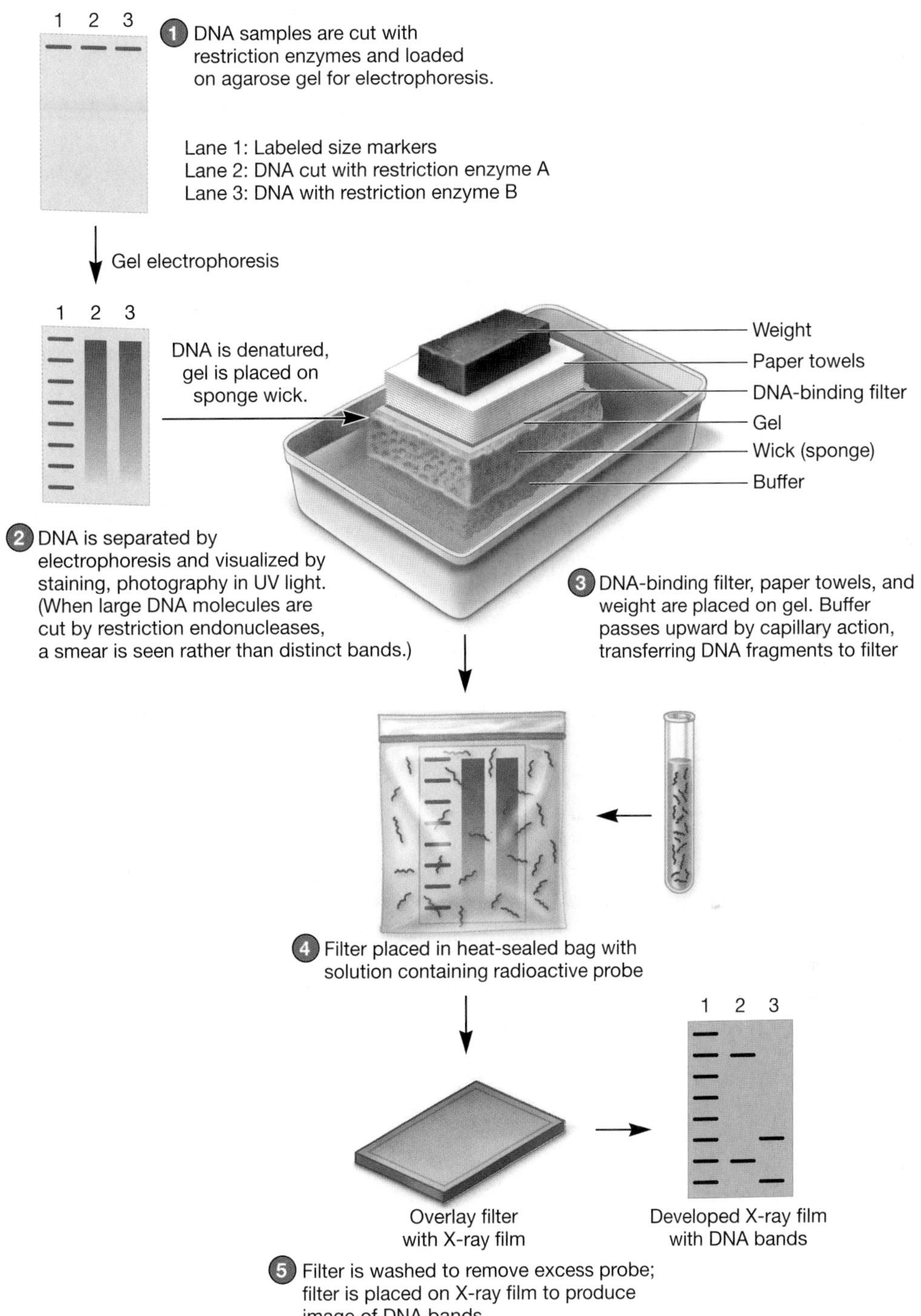
1 2 3
1 DNA samples are cut with restriction enzymes and loaded on agarose gel for electrophoresis.
Lane 1: Labeled size markers
Lane 2: DNA cut with restriction enzyme A
Lane 3: DNA with restriction enzyme B
Gel electrophoresis
1 2 3
DNA is denatured, gel is placed on sponge wick.
Weight
Paper towels
DNA-binding filter
Gel
Wick (sponge)
Buffer
2 DNA is separated by electrophoresis and visualized by staining, photography in UV light. (When large DNA molecules are cut by restriction endonucleases, a smear is seen rather than distinct bands.)
3 DNA-binding filter, paper towels, and weight are placed on gel. Buffer passes upward by capillary action, transferring DNA fragments to filter
4 Filter placed in heat-sealed bag with solution containing radioactive probe
1 2 3
Overlay filter with X-ray film
Developed X-ray film with DNA bands
5 Filter is washed to remove excess probe; filter is placed on X-ray film to produce image of DNA bands.

Figure 14.6 The Southern Blotting Technique.

is not quantitative, meaning that the amount of final product does not always reflect the amount of template DNA present before amplification. In contrast, **real-time PCR (RT-PCR)** is quantitative. That is, it allows one to ask how much DNA or RNA template (which is converted to DNA with reverse transcriptase) is present in a given sample. This is accomplished by adding a fluorescently labeled probe to the reaction mixture and measuring its signal quantitatively during the exponential phase of the reaction. The fluorescence increases as PCR products accumulate during the initial cycles. This is when the rate of DNA amplification is logarithmic. However, as the PCR cycles continue, substrates are consumed and polymerase efficiency declines. So although the amount of product increases, its rate of synthesis is no longer exponential (this is why end-point collection of PCR products is not

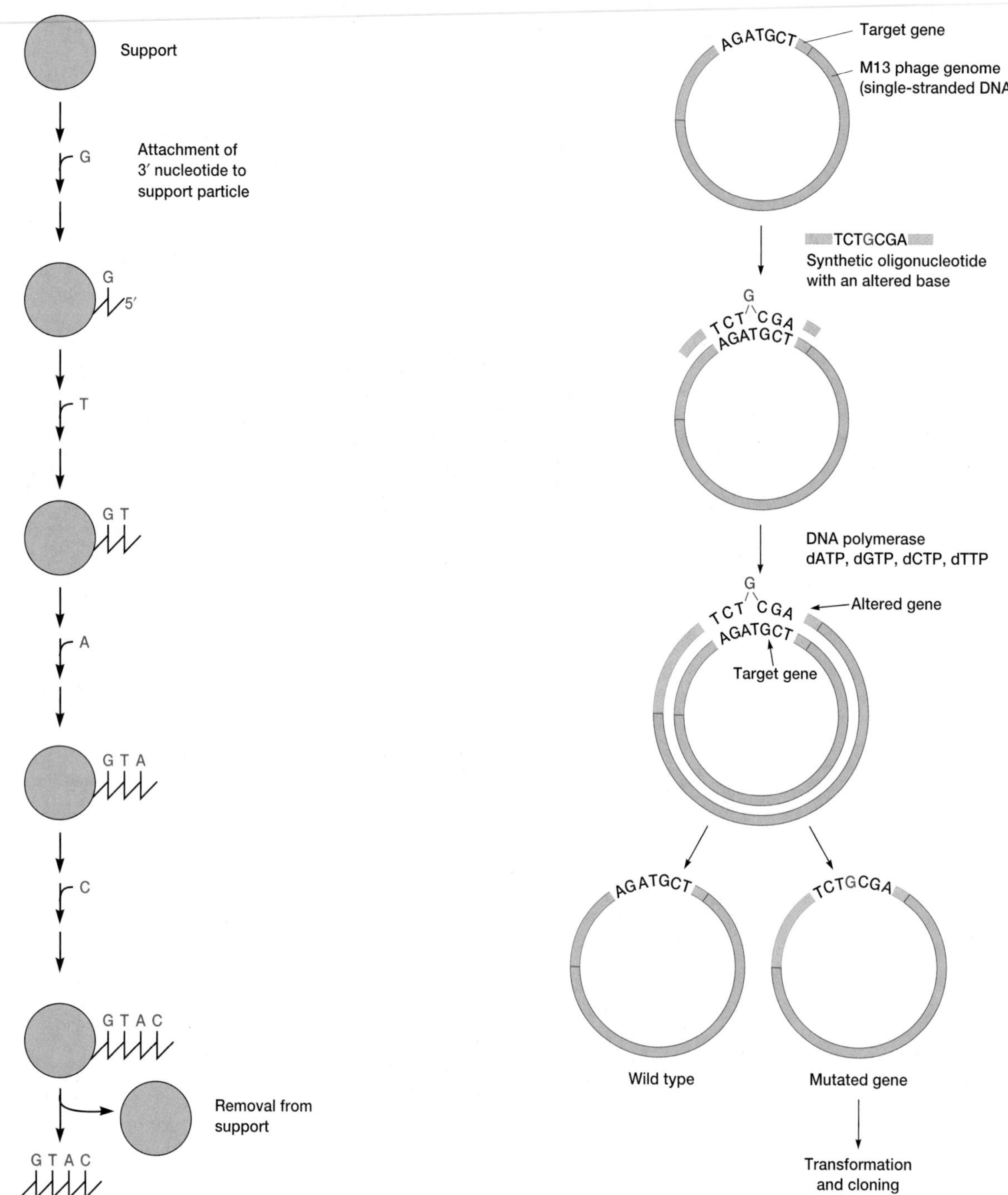

Figure 14.7 The Synthesis of a DNA Oligonucleotide. During each cycle, the DNA synthesizer adds an activated nucleotide (A, T, G, or C) to the growing end of the chain. At the end of the process, the oligonucleotide is removed from its support.

Figure 14.8 Site-Directed Mutagenesis. A synthetic oligonucleotide is used to add a specific mutation to a gene. See text for details.

quantitative). Thermocyclers specifically designed for real-time PCR are used that record the amount of PCR product generated as it happens; thus the term real-time PCR. Gene expression studies often rely on real time-PCR, because mRNA transcripts can be copied and amplified by reverse transcriptase (RT). Therefore the procedure monitors the level of gene transcription of the gene targeted by the primers. This is sometimes called RT-RT-PCR.

The PCR technique is an essential tool in many areas of molecular biology, medicine, and biotechnology. As shown in **figure 14.10,** when PCR is used to obtain DNA for cloning, a

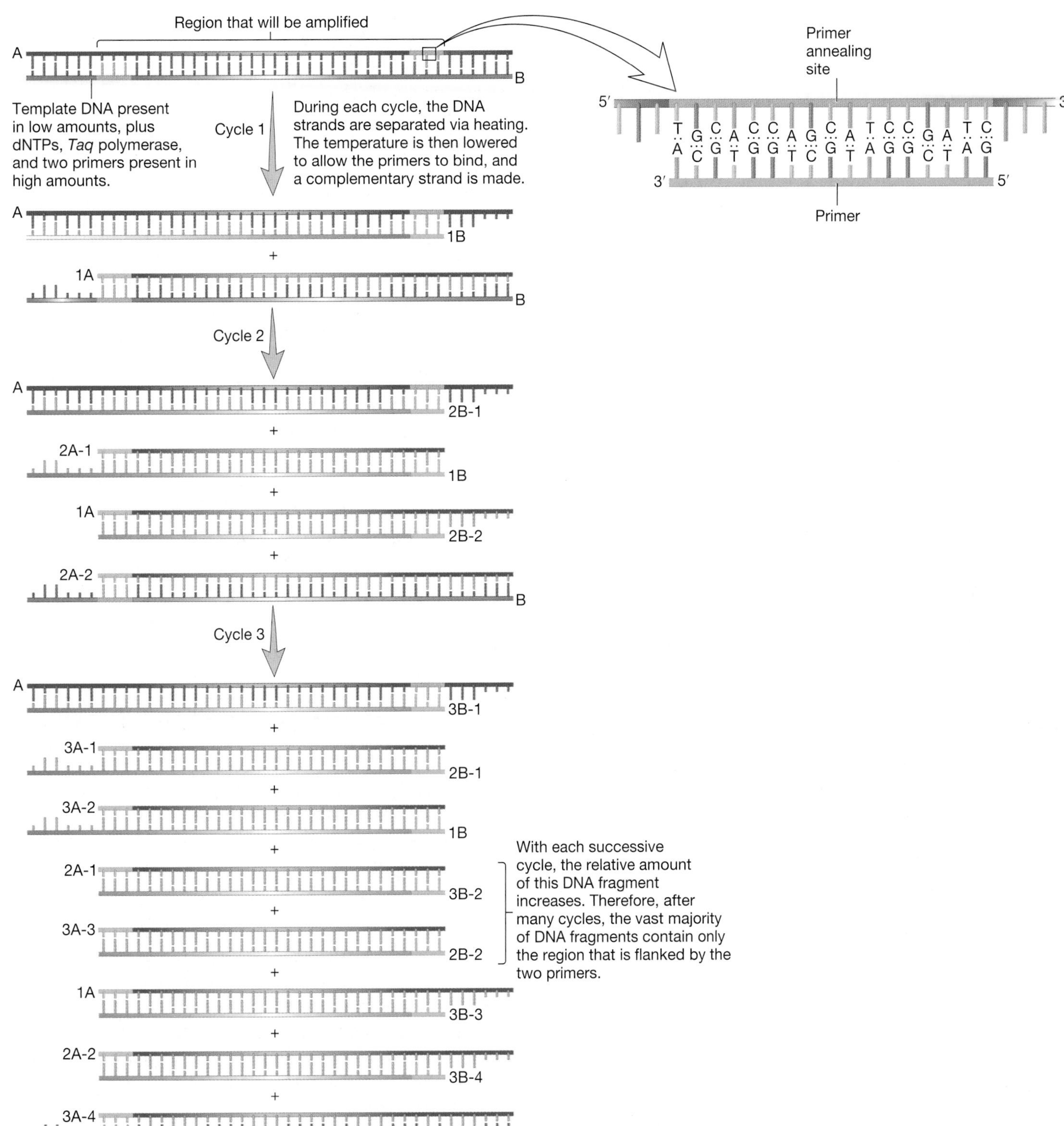

Figure 14.9 The Technique of Polymerase Chain Reaction (PCR). During each cycle, oligonucleotides that are complementary to the ends of the targeted DNA sequence bind to the DNA and act as primers for the synthesis of this DNA region. The primers used in actual PCR experiments are usually 15 to 20 nucleotides in length. The region between the two primers is typically hundreds of nucleotides in length, not just several nucleotides as shown here. The net result of PCR is the synthesis of many copies of DNA in the region that is flanked by the two primers.

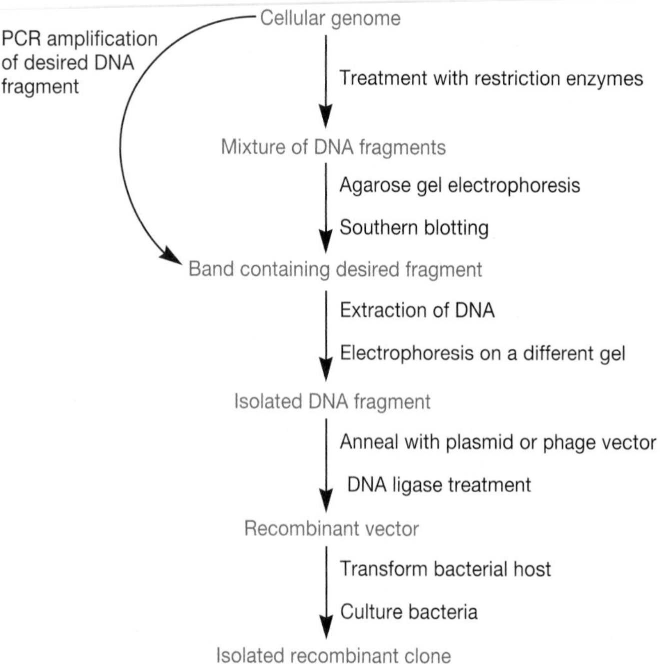

Figure 14.10 Cloning Cellular DNA Fragments. The preparation of a recombinant clone from isolated DNA fragments or DNA generated by PCR.

number of steps traditionally employed are no longer required. PCR is also used to generate DNA for nucleotide sequencing. Because PCR amplifies DNA even if it is present in very small, initial quantities, it is used in a number of diagnostic tests, including those for AIDS, Lyme disease, chlamydia, tuberculosis, hepatitis, the human papilloma virus, and other infectious agents and diseases. The tests are rapid, sensitive, and specific. PCR is particularly valuable in the detection of genetic diseases such as sickle cell anemia, phenylketonuria, and muscular dystrophy. The technique is also employed in forensic science where it is used in criminal cases as a part of DNA fingerprinting technology. It is possible to exclude or incriminate suspects using extremely small samples of biological material discovered at the crime scene.

14.4 Gel Electrophoresis

Agarose or polyacrylamide gels usually are used to separate DNA fragments electrophoretically. In **gel electrophoresis,** charged molecules are placed in an electrical field and allowed to migrate toward the positive and negative poles. The molecules separate because they move at different rates due to their differences in charge and size. Because DNA is negatively charged, it is loaded into wells at the negative pole of the gel and migrates toward the positive (**figure 14.11**). Each fragment's migration rate is inversely proportional to the log of its molecular weight. That is to say, the smaller a fragment is, the faster it moves through the gel. Migration rate is also a function of gel density. In practice, this means that higher concentrations of gel material (agarose or acrylamide) provide better resolution of small fragments and vice versa. DNA that has not been digested with restriction enzymes is usually supercoiled. For this and other reasons, DNA is usually cut with restriction endonucleases prior to electrophoresis. Small DNA molecules usually yield only a few bands because there are few restriction enzyme recognition sites. If the DNA fragment is large, or an entire chromosome is digested, many such sites are present and the DNA is cut in numerous places. When such DNA is electrophoresed, it produces a smear representing many thousands of DNA fragments of similar sizes that cannot be individually resolved. The region of the gel containing the desired DNA fragment must then be located using the Southern blot technique (figure 14.6). DNA from this region can be isolated from the gel material and electrophoresed again on a gel of another agarose or acrylamide concentration so that individual bands can be detected.

14.5 Cloning Vectors and Creating Recombinant DNA

Recombinant DNA technology depends on the propagation of many copies of the nucleotide sequence of choice. To accomplish this, genes or other genetic elements are inserted (i.e., cloned) into DNA vectors that replicate in a host organism. There are four major types of vectors: plasmids, bacteriophages and other viruses, cosmids, and artificial chromosomes (**table 14.3**). Each type has its own advantages, so the selection of the proper cloning vector is critical to the success of any cloning experiment. All engineered vectors share three features: an origin of replication; a region of DNA that bears unique restriction sites, called a multicloning site or polylinker; and a selectable marker. These elements are described in the discussion of plasmids, the most frequently used cloning vectors.

Plasmids

Plasmids make excellent cloning vectors because they replicate autonomously and are easy to purify. They can be introduced into microbes by conjugation and/or transformation. Many different plasmids are used in biotechnology, all derived from naturally occurring plasmids that have been genetically engineered (**figure 14.12**). Bacterial plasmids (section 3.6); Bacterial conjugation (section 13.7); Transformation (section 13.8)

Origin of Replication

The **origin of replication (*ori*)** allows the plasmid to replicate in the microbial host independently of the chromosome. pUC19, an *E. coli* plasmid, is said to have a high copy number because it replicates about 100 times in the course of one generation. That is to say, an *E. coli* cell with one chromosome can have as many as 100 copies of the plasmid. High copy number is often important because it facilitates plasmid purification and it can dramat-

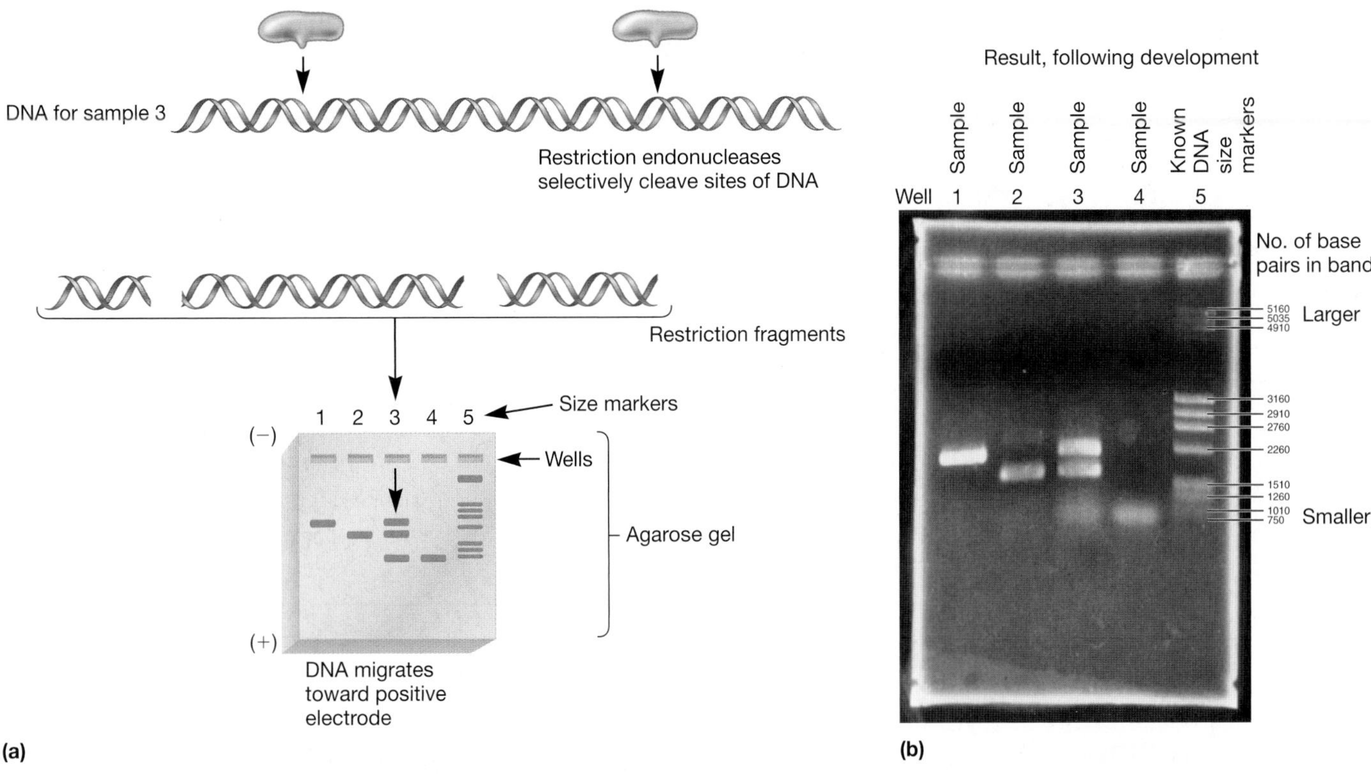

Figure 14.11 Gel Electrophoresis of DNA. **(a)** After cleavage into fragments, DNA is loaded into wells on one end of an agarose gel. When an electrical current is passed through the gel (from the negative pole to the positive pole), the DNA, being negatively charged, migrates toward the positive pole. The larger fragments, measured in numbers of base pairs, migrate more slowly and remain nearer the wells than the smaller (shorter) fragments. **(b)** An actual developed and stained gel reveals a separation pattern of the fragments of DNA. The size of a given DNA band can be determined by comparing it to a known set of molecular weight markers (lane 5) called a ladder.

ically increase the amount of cloned gene product produced by the cell. Some plasmids have two origins of replication, each recognized by different host organisms. These plasmids are called **shuttle vectors** because they can move or "shuttle" from one host to another. YEp24 is a shuttle vector that can replicate in yeast (*Saccharomyces cerevisiae*) and in *E. coli* because it has the 2μ circle yeast replication element and *E. coli* origin of replication, *ori* (figure 14.12).

Selectable Marker

Following the uptake of vector by host cells, one must be able to discriminate between those cells that successfully obtained vector from those that did not. Furthermore, one must be able to continue to select for the presence of plasmid, otherwise the host cell may stop replicating it. This is achieved by the presence of a gene that encodes a protein that is needed for the cell to survive under certain, selective conditions. Such a gene is called a **selectable marker.** In the case of pUC19, the selectable marker encodes the ampicillin resistance factor (*amp*^{R}, sometimes called *bla,* for *β-la*ctamase). The shuttle vector YEp24 bears both the *amp*^{R} gene for selection in *E. coli,* and *URA3,* which encodes a protein essential for uracil biosynthesis in yeast. Therefore when in *S. cerevisiae,* this plasmid must be maintained in uracil auxotrophs.

Multicloning Site (MCS) or Polylinker

A region of restriction enzyme cleavage sites found only once in the plasmid is essential for the insertion of foreign DNA. Cleavage at a unique restriction site generates a linear plasmid. Cleavage of the gene to be cloned with the same restriction enzyme results in compatible sticky ends, so that it may be inserted (ligated) into the **multicloning site (MCS).** Alternatively, two different, unique sites within the MCS may be cleaved and the DNA sequence between the two sites replaced with cloned DNA (**figure 14.13**). In either case, the plasmid and the DNA to be inserted are incubated in the presence of the enzyme **DNA ligase** so that when compatible sticky ends hydrogen bond, phosphodiester covalent bonds can be generated between the cloned DNA fragment and the vector. This requires the input of energy, thus ATP is added to this in vitro ligation reaction *(see figure 11.19).*

pUC19 has a number of unique restriction sites in its MCS (figure 14.12); this provides a number of cleavage options, making it easier to obtain the same, or compatible, sticky ends in both vector and the DNA to be inserted. In pUC19 the MCS is located within the 5′ end of the *lacZ* gene, which encodes β-galactosidase (β-Gal). This enzyme cuts the dissacharide lactose into galactose and glucose. When DNA has been cloned into the MCS, the *lacZ* gene is no longer intact, so a functional enzyme is not produced.

Table 14.3 Recombinant DNA Cloning Vectors

Vector	Insert Size (kb, 1 kb = 1,000 bp)	Example	Features
Plasmid	<20 kb	pBR322, pUC19	Replicates independently of microbial chromosome so many copies may be maintained in a single cell
Bacteriophage	9–25 kb	λ1059, λ gt11, M13mp18, EMBL3	Packaged into lambda phage particles; single-stranded DNA viruses like M13 have been modified (e.g., M13mp18) to generate either double- or single-stranded DNA in the host
Cosmids	30–47 kb	pJC720, pSupercos	Can be packaged into lambda phage particles for efficient introduction into bacteria, then replicates as a plasmid
PACs (P1 artificial chromosomes)	75–100 kb	pPAC	Based on the bacteriophage P1 packaging mechanism
BACs (bacterial artificial chromosomes)	75–300 kb	pBAC108L	Modified F plasmid that can carry large DNA inserts; very stable within the cell
YACs (yeast artificial chromosomes)	100–1,000 kb	pYAC	Can carry largest DNA inserts, replicates in *Saccharomyces cerevisiae*

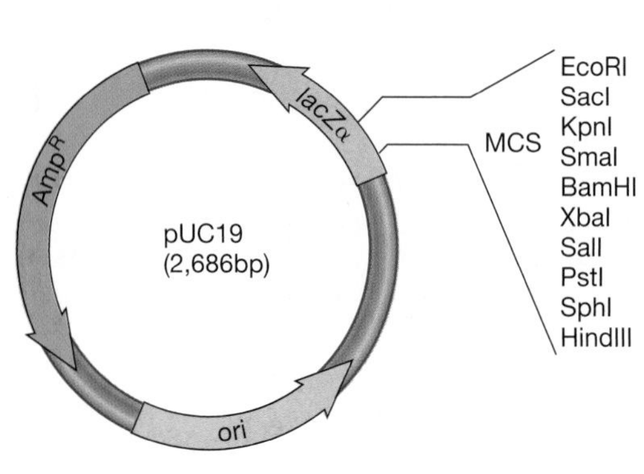

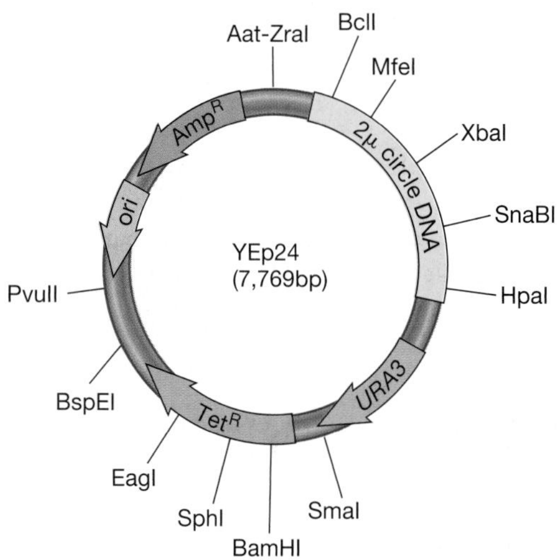

Figure 14.12 The Cloning Vectors pUC19 and YEp24. Restriction sites that are present only once in each vector are shown. pUC19 replicates only in *E. coli,* while YEp24 replicates in both *E. coli* and *S. cerevisiae.*

This can be detected by the color of colonies: cells turn blue when β-Gal splits the alternative substrate, X-Gal (5-bromo-4-chloro-3-indolyl-β-D-galactopyranoside), which is included in the medium (figure 14.13). This is important because the ligation of foreign DNA into a vector is never 100% efficient. Thus when the ligation mixture is introduced into host cells, one must be able to distinguish cells that carry just plasmid from those that carry plasmid into which DNA was successfully inserted. In the case of pUC19, all *E. coli* cells that take up plasmid (with or without insert) are selected by their resistance to ampicillin (Amp^R). Among these, colonies with plasmid lacking DNA insert will be blue (due to the presence of functional *lacZ* gene), while those that have pUC19 into which DNA was successfully cloned will be white. There are a number of other clever ways in which cells with vector versus those with vector with insert can be differentiated; the selection of blue versus white colonies is a common approach.

Phage Vectors

Phage vectors are engineered phage genomes that have been genetically modified to include useful restriction enzyme recognition sites for the insertion of foreign DNA. Once DNA has been inserted, the recombinant phage genome is packaged into viral capsids and used to infect host cells. The resulting phage lysate consists of thousands of phage particles that carry cloned DNA as well as the genes needed for host lysis. Two commonly used vectors are derived from the bacteriophages T7 and lambda (λ), both of which have double-stranded DNA genomes. Although these

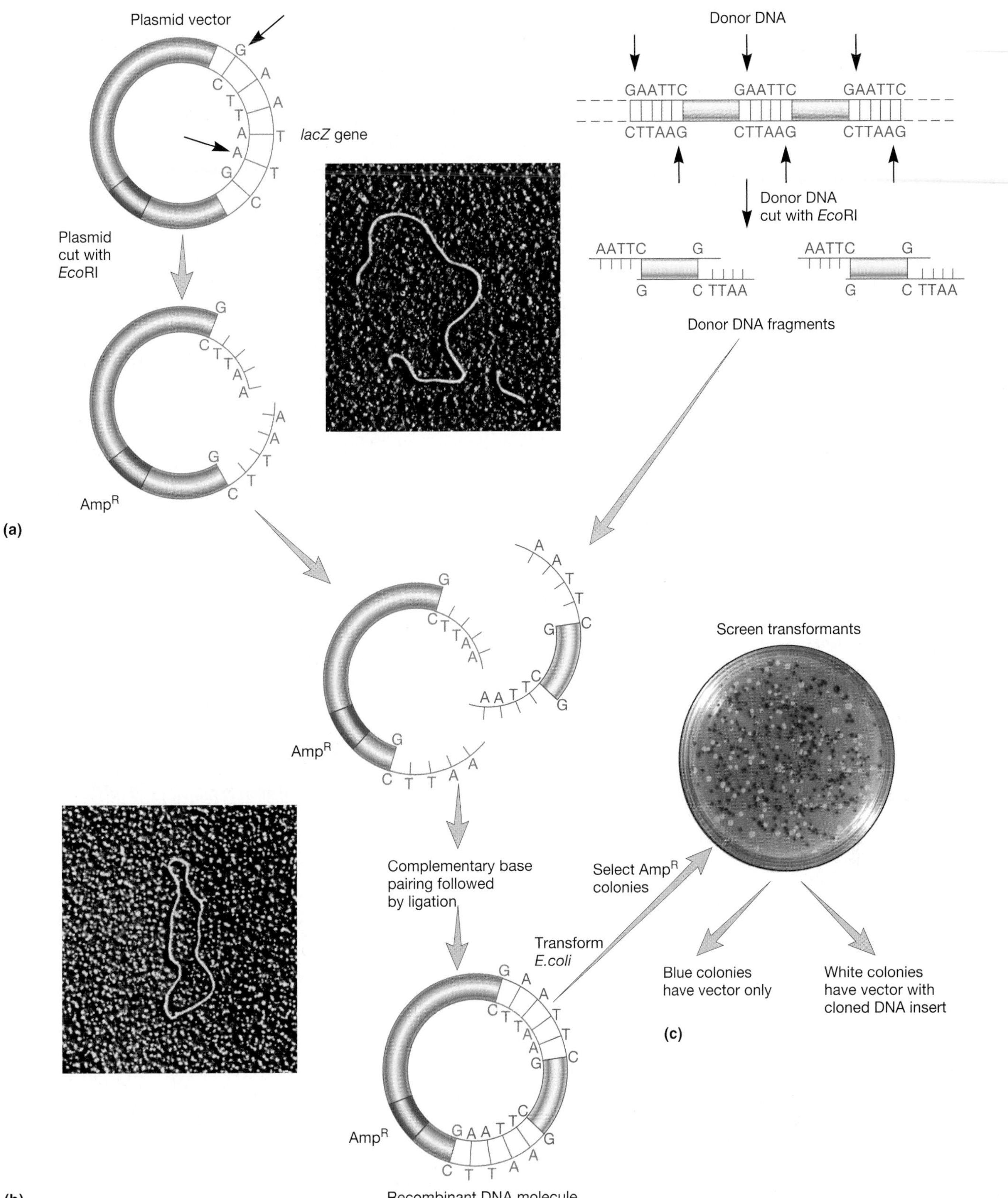

Figure 14.13 Recombinant Plasmid Construction and Cloning. The construction and cloning of a recombinant plasmid vector using an antibiotic resistance gene to select for the presence of the plasmid. The interruption of the *lacZ* gene by cloned DNA is used to select for vectors with insert. The scale of the sticky ends of the fragments and plasmid has been enlarged to illustrate complementary base pairing. **(a)** The electron micrograph shows a plasmid that has been cut by a restriction enzyme and a donor DNA fragment. **(b)** The micrograph shows a recombinant plasmid. **(c)** After transformation, *E. coli* cells are plated on medium containing ampicillin and X-Gal so that only ampicillin-resistant transformants grow; X-Gal enables the visualization of colonies that were transformed with recombinant vector (vector + insert, white colonies).

phages infect *E. coli,* phage vectors have been engineered for a number of different bacterial host species. Viruses of *Bacteria* and *Archaea* (chapter 17)

Cosmids

Cosmids were developed when it became clear that cloning vectors were needed that could tolerate larger fragments of cloned DNA (table 14.3). Unlike phages and plasmids, cosmids do not exist in nature. Instead, these engineered vectors have been constructed to contain features from both. Cosmids have a selectable marker and MCS from plasmids, and a *cos* site from λ phage. In λ phage, the *cos* site is where multiple copies of phage genome are linked prior to packaging. Cleavage at the *cos* sites yields single genomes that are the right size for packaging. Cosmids take advantage of the fact that the only requirement for λ phage heads to package DNA is two *cos* sites on a linear DNA molecule or a single *cos* site on a circular one. As long as the cosmid with its cloned DNA is the appropriate size (about 37 to 52 kb), it will be packaged. The phage is then used to introduce the recombinant DNA into *E. coli,* where it replicates as a plasmid. Temperate bacteriophages and lysogeny (section 17.5)

Artificial Chromosomes

Artificial chromosomes are special cloning vectors used when particularly large fragments of DNA must be cloned, as when constructing a genomic library or when sequencing an organism's entire genome. In fact, **bacterial artificial chromosomes (BACs)** were crucial to the timely completion of the human genome project. Like natural chromosomes, artificial chromosomes replicate only once per cell cycle. **Yeast artificial chromosomes (YACs)** were developed first and consist of a yeast telomere at each end (*TEL*), a centromere sequence (*CEN*), a yeast origin of replication (*ARS, a*utonomously, *r*eplicating *s*equence), a selectable marker such as URA3, and an MCS to facilitate the insertion of foreign DNA (**figure 14.14**). YACs are used when extraordinarily large DNA pieces (up to 1,000 kb; table 14.3) are to be cloned. BACs were developed, in part, because YACs tend to be unstable and may recombine with host chromosomes, thereby rearranging the cloned DNA. Although BACs accept smaller DNA inserts than do YACs (up to 300 kb), they are generally more stable. BACs are based on the F fertility factor of *E. coli.* The example shown in figure 14.14 is typical in that it includes genes that ensure a replication complex will be formed (*rep*E), as well as proper partitioning of one newly replicated BAC to each daughter cell (*sopA, sopB,* and *sopC*). It also includes features common to many plasmids such as an MCS within the *lac*Z gene for blue/white colony screening, and a selectable marker, in this case for resistance to the antibiotic chloramphenicol (Cm^R).

1. How are oligonucleotides synthesized? What is site-directed mutagenesis? How might site-directed mutagenesis be used to find the nucleotides that encode the active site of an enzyme?
2. Briefly describe the polymerase chain reaction. Explain the differences between reactions in which the products are collected after a defined number of cycles and real-time PCR. Suggest an application of each approach.
3. What is electrophoresis? How is it used in Southern blotting?
4. How are plasmids, cosmids, and artificial chromosomes different? What are some of the different purposes served by each?
5. Explain selection for antibiotic resistance followed by blue versus white screening of colonies containing recombinant plasmids. Why must both antibiotic selection and color screening be used? What would you conclude if, after transforming a ligation mixture into *E. coli*, only blue colonies were obtained?

14.6 Construction of Genomic Libraries

There are several ways in which the DNA to be cloned can be obtained. It can be synthesized by PCR or it can be located on the chromosome by Southern blotting. However, PCR amplification of a gene requires foreknowledge of its nucleotide sequence (or at least sequences flanking the gene), and a suitable probe must be obtained for Southern blotting. In both cases, once the DNA fragment is purified, it is cloned using a procedure like that described for recombinant plasmids and shown in figure 14.13. However, what if researchers wanted to clone a gene but they had no idea what its DNA sequence might be? A genomic library must then be constructed and screened.

The goal of **genomic library** construction is to have an organism's genome cloned as small fragments into separate vectors. Ideally the entire genome is represented; that is to say, the sum of the different fragments equals the whole genome. In this way specific groups of genes can be analyzed and isolated. The construction of a genomic library begins with cleaving the genome into small pieces by a restriction endonuclease (**figure 14.15**). These genomic

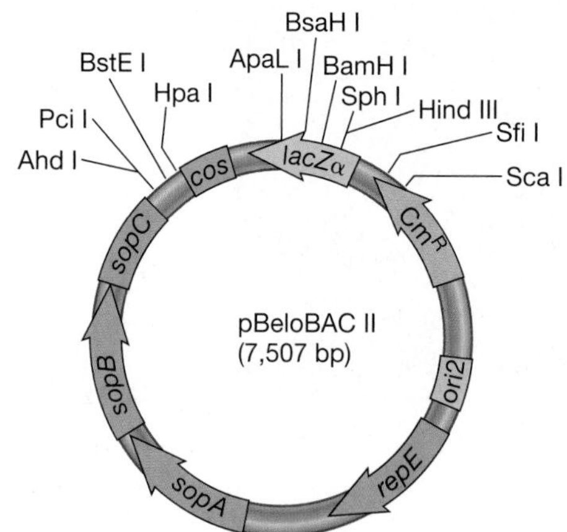

(a) Bacterial artificial chromosome (BAC)

TEL	TRP1	ARS	CEN	MCS	URA3	TEL

(b) Yeast artificial chromosome (YAC)

Figure 14.14 Artificial Chromosomes Can Be Used as Cloning Vectors. (a) The bacterial artificial chromosome pBelaBACIII and **(b)** a yeast artificial chromosome. See text for details.

fragments are then either cloned into vectors and introduced into a microbe or packaged into phage particles that are used to infect the host (**figure 14.16**). In either case, many thousands of different clones—each with a different genomic DNA insert—are created. To select the desired clone from the library, it is necessary to know something about the function of the target gene or genetic element. If the genomic library has been inserted into a microbe that expresses the foreign gene, it may be possible to assay each clone for a specific protein or phenotype. For example, if one is studying a newly isolated soil bacterium and wants to find the gene that encodes an enzyme needed for the biosynthesis of the amino acid alanine, the library could be expressed in an *E. coli* or *Bacillus subtilis* alanine auxotroph (figure 14.15). Recall that an alanine auxotroph requires the addition of this amino acid to the medium. Following introduction of the genomic library vector into host cells, those that now grow without alanine would be good candidates for the genomic library fragment that possesses the alanine biosynthetic gene or gene cluster. Success with this approach depends on the assumption that the function of the cloned gene product is similar in both organisms. If this is not the case, the host must be the same species from which the library was prepared. In this example, a soil bacterial mutant lacking the gene in question (e.g., an alanine auxotroph) is used as the genomic library host. The genetic complementation of a deficiency in the host cell is sometimes called **phenotypic rescue.** Detection and isolation of mutants (section 13.2; *see also figure 13.7*)

Alternatively, the library may be cloned into a phage vector (figure 14.16). The resulting plaques then contain phage particles whose genomes include the cloned DNA fragments. The plaques are screened by a technique based on hybridization of an oligonucleotide probe to the target DNA, much like in Southern blotting. However, in this case, DNA is transferred directly from the petri plate to the filter, which is then incubated with labeled probe. If the nucleic acid sequence of the DNA to be cloned is known, this approach can be used to screen transformed colonies as well.

If a genomic library is prepared from a eucaryote in an effort to isolate a structural gene, a cDNA library is usually constructed. In this way, introns are not present in the genomic library. Instead only the protein-coding regions of the genome are cloned. cDNA is prepared (figure 14.4) and cloned into a suitable vector. After the library is introduced into the host microbe, it may be screened by phenotypic rescue or by hybridization with an oligonucleotide as described earlier. In some cases, neither phenotypic rescue nor hybridization with a probe is possible. In such cases the researcher must develop a novel way that suits the particular set of circumstances to screen the genomic library.

14.7 Inserting Recombinant DNA into Host Cells

In cloning procedures, the selection of a host organism is as important as the choice of cloning vector. *E. coli* is the most frequent procaryotic host and *S. cerevisiae* is the most popular among eucaryotes. Host microbes that have been engineered to lack restriction enzymes and the recombination enzyme RecA make better hosts because it is less likely that the newly acquired DNA will be degraded and/or recombined with the host chromosome. There are several ways to introduce recombinant DNA into a host microbe. Transformation and electroporation are two commonly employed techniques. Often the host microbe does not have the capacity to be transformed naturally. This is the case with *E. coli* and most gram-negative bacteria as well as many gram-positive bacteria. In these cases, the host cells may be rendered competent by treatment with divalent cations and artificially transformed by heat shocking the cells. DNA transformation (section 13.8)

Electroporation is a technique that has gained popularity because of its simplicity and wide application to a number of host organisms, including plant and animal cells. In this procedure, cells are mixed with the recombinant DNA and exposed to a brief pulse of high-voltage electricity. The plasma membrane becomes temporarily permeable and DNA molecules are taken up by some of the cells. The cells are then grown on media that select for the presence of the cloning vector as previously described.

With the exception of yeast, inserting DNA into eucaryotic cells is often more difficult and less efficient. The most direct approach is **microinjection,** wherein genetic material is micropipetted into the host cell. The DNA is then taken up by the nucleus and stably incorporated into the host genome. When this is done in a fertilized mammalian egg, the egg is then transplanted into the uterus of the host animal where it will develop into a transgenic animal. Transgenic mice have become an essential tool for biomedical research.

One of the most effective techniques to insert DNA into eucaryotic cells is to shoot microprojectiles coated with DNA into plant and animal cells. The **gene gun,** first developed at Cornell University, operates somewhat like a shotgun. A blast of compressed gas shoots a spray of DNA-coated metallic microprojectiles into the cells. The device has been used to transform corn and produce fertile corn plants bearing foreign genes. Other guns use electrical discharges to propel the DNA-coated projectiles. These guns are sometimes called biolistic devices, a name derived from biological and ballistic. They have been used to transform microorganisms (yeast, the mold *Aspergillus,* and the protist *Chlamydomonas*), mammalian cells, and a variety of plant cells (corn, cotton, tobacco, onion, and poplar).

Plant cells can also be transformed with vectors derived from the bacterium *Agrobacterium* (p. 378). Viruses increasingly are used to insert desired genes into eucaryotic cells. For example, genes may be placed in a retrovirus, which then infects the target cell and integrates a DNA copy of its RNA genome into the host chromosome. Adenoviruses also can transfer genes to animal cells. Recombinant baculoviruses will infect insect cells and promote the production of many proteins. Eucaryotic viruses and other acellular infectious agents (chapter 18)

14.8 Expressing Foreign Genes in Host Cells

When a gene from one organism is cloned into another, it is said to be a **heterologous gene.** Heterologous genes are not always expressed in the host cell without further modification of the recombinant vector. To be transcribed, the recombinant gene must

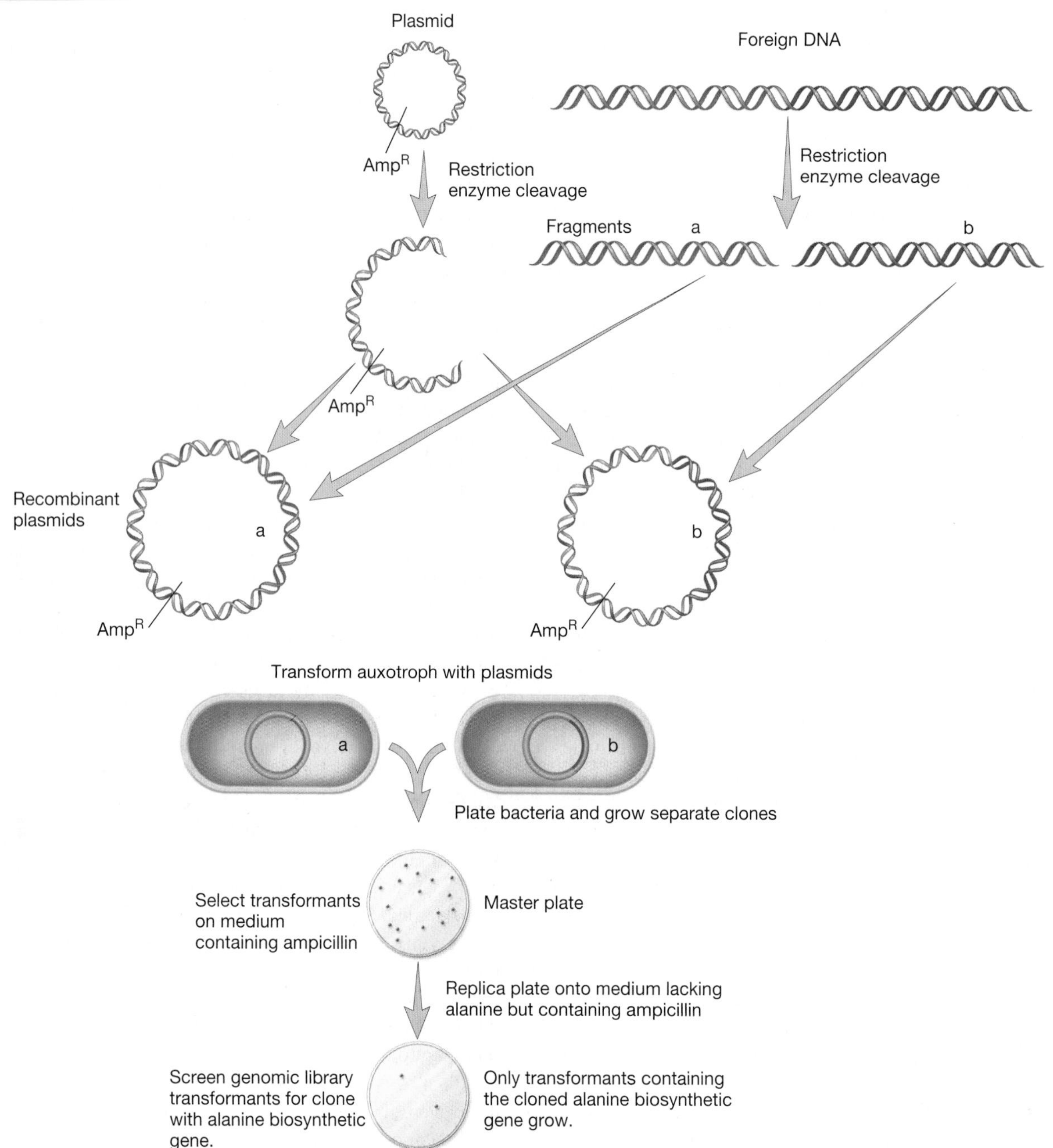
Plasmid
Foreign DNA
AmpR
Restriction enzyme cleavage
Restriction enzyme cleavage
Fragments a
b
AmpR
Recombinant plasmids
a
b
AmpR
AmpR
Transform auxotroph with plasmids
a
b
Plate bacteria and grow separate clones
Select transformants on medium containing ampicillin
Master plate
Replica plate onto medium lacking alanine but containing ampicillin
Screen genomic library transformants for clone with alanine biosynthetic gene.
Only transformants containing the cloned alanine biosynthetic gene grow.

Figure 14.15 Construction of a Genomic Library and Screening by Phenotypic Rescue. A genomic library is made by cloning fragments of an organism's entire genome into a vector. For simplicity, only two possible recombinant vectors are shown. In reality, a large mixture of clones is generated. This mixture of clones is then introduced into a suitable host. Phenotypic rescue is one way to screen the clones for the gene of interest. It involves using a host with a genetic defect that can be complemented or "rescued" by the expression of a specific gene that has been cloned.

have a promoter that is recognized by the host RNA polymerase. Translation of its mRNA depends on the presence of leader sequences and mRNA modifications that allow proper ribosome binding. These are quite different in eucaryotes and procaryotes. For instance, if the host is a procaryote and the gene has been cloned from a eucaryote, a procaryotic leader must be provided and introns removed.

The problems of expressing recombinant genes in host cells are largely overcome with the help of special cloning vectors called **expression vectors.** These vectors contain the necessary transcription and translation start signals in addition to convenient polylinker sites. Some expression vectors contain regulatory regions of the *lac* operon so that the expression of the cloned genes can be controlled in the same manner as the operon.

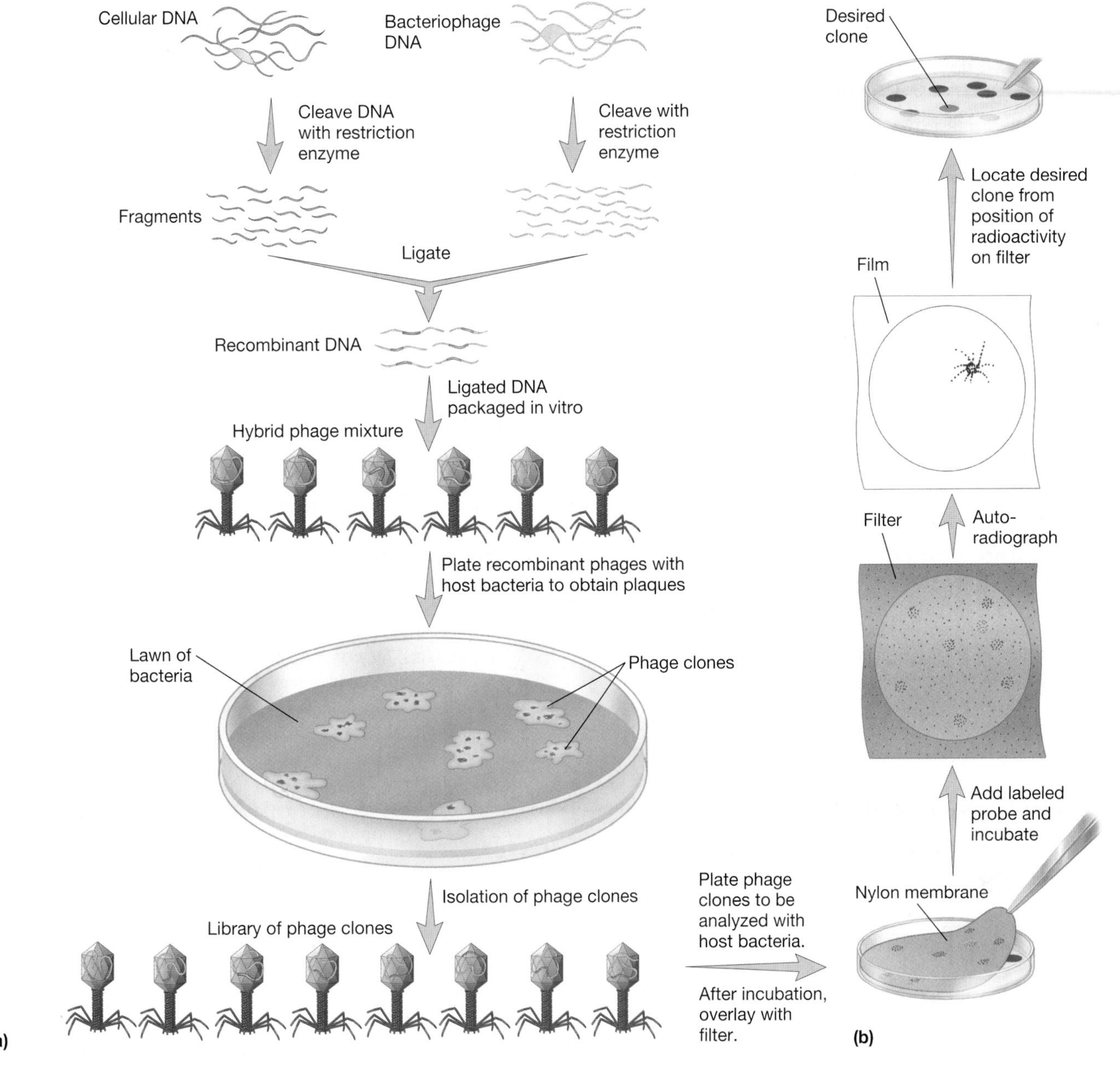

Figure 14.16 The Use of Lambda Phage as a Vector. **(a)** The preparation of a genomic library. Each plaque on the bacterial lawn contains a recombinant clone carrying a different DNA fragment. **(b)** Detection and cloning of the desired recombinant phage.

Somatostatin, the 14-residue hypothalamic peptide hormone that helps regulate human growth, provides an example of useful cloning and protein production. The gene for somatostatin was initially synthesized by chemical methods. Besides the 42 bases coding for somatostatin, the polynucleotide contained a codon for methionine at the 5′ end (which corresponds to the N-terminal end of the peptide) and two stop codons at the opposite end. To aid insertion into the plasmid vector, the 5′ ends of the synthetic gene were extended to form sticky ends complementary to those formed by the *Eco*RI and *Bam*HI restriction enzymes. A plasmid cloning vector was cut with both *Eco*RI and *Bam*HI to remove a part of the plasmid DNA. The synthetic gene was then inserted into the vector by taking advantage of its sticky ends (**figure 14.17**). Finally, a fragment containing the initial part of the *lac* operon (including the promoter, operator, ribosome binding site, and much of the β-galactosidase gene) was inserted upstream to, or at the 5′ end of, the somatostatin gene. The plasmid now contained the somatostatin gene fused in the proper orientation to the remaining portion of the β-galactosidase gene.

After introduction of this recombinant plasmid into *E. coli,* the somatostatin gene was transcribed with the β-galactosidase gene fragment to generate mRNA. Translation formed proteins consisting of the hormone peptide attached to the β-galactosidase fragment by a methionine residue. Cyanogen bromide cleaves

Techniques & Applications

14.1 Visualizing Proteins with Green Fluorescence

What does a jellyfish that lives in the cold waters of the northern Pacific have to do with biotechnology? It turns out, a lot. The jellyfish, *Aequorea victoria,* produces a protein called **green fluorescent protein (GFP)** that scientists have adopted to visualize gene expression and protein localization in living cells. GFP is encoded by a single gene that, when translated, undergoes self-catalyzed modification to generate a strong, green fluorescence. This means that it is easily cloned and expressed in any organism. And GFP isn't just green anymore; site-directed mutagenesis of the GFP gene has generated a variety of proteins that glow throughout the blue-green-yellow spectrum.

One widely used GFP application is to tag proteins so their cellular placement, or localization, can be seen with a light microscope. Formerly, electron microscopy (EM) was the only method by which the location of proteins could be visualized. Sample preparation for EM involves harsh treatment with solvents and cellular dehydration—procedures that can damage cell structures and lead to artifacts. In contrast, when a protein is tagged with GFP, the timing of protein production and its localization can be followed in living cells. To accomplish this, a structural gene is genetically fused to the GFP gene to create a chimeric protein—a protein that consists of two parts: the structural protein being studied and GFP. Of course, care must be taken to ensure that the protein fusion still functions like the original protein. One way to do this is to test for phenotypic rescue of a mutant lacking the structural gene of interest.

GFP protein fusion technology has been used to examine some of the most basic questions in biology. One example is cell division, or cytokinesis. Genetic evidence clearly shows that the tubulin-like protein FtsZ is key for new septum formation in most procaryotic cells. Similarly, the isolation and analysis of cell division mutants has identified other proteins, such as FtsA, MinC, MinD, and MinE, that are also needed for cytokinesis. Such studies have led to the development of models predicting the events required for cytokinesis. The procaryotic cell cycle (section 6.1)

A special type of cell division that occurs during the formation of an endospore has been intensively studied in *Bacillus subtilis*. In this case, septum formation does not generate two equally sized daughter cells. Instead, asymmetric division gives rise to two different progeny: the endospore and the mother cell. Asymmetric cell division means that rather than FtsZ assembling in the center of the cell, it is polymerized at the pole that gives rise to the endospore. Originally, it was hypothesized that the cell accomplishes this by preventing FtsZ ring assembly in the middle of the cell while simultaneously activating FtsZ polymerization sites near the poles (see **Box figure**).

The fusion of the *B. subtilis ftsZ* structural gene to the gene encoding GFP generated a surprising result: the deployment of FtsZ to the cell pole is a much more dynamic process. Time lapse photomicroscopy shows that when the cell switches from vegetative growth and binary fission to endospore formation, FtsZ forms a spiral-like filament that moves from the cell center to the poles. Careful analysis reveals that many cells appear to have FtsZ spirals that are more abundant in one half of the cell. This suggests that the accumulation of a critical level of FtsZ at one pole before another may determine endospore placement. Surely GFP fusions will help to resolve this and other questions regarding protein localization. Global regulatory systems: Sporulation in *Bacillus subtilis* (section 12.5)

Ben-Yehuda, S., and Losick, R. 2002. Asymmetic cell division in B. subtilis *involves a spiral-like intermediate of the cytokinetic protein FtsZ.* Cell. *109: 257–66.*

peptide bonds at methionine residues. Treatment of the fusion proteins with cyanogen bromide broke the peptide chains at the methionine and released the hormone (**figure 14.18**). Once free, the peptides were able to fold properly to become active. Because production of the fusion protein was under the control of the *lac* operon, it could be easily regulated. Many proteins have been produced since the synthesis of somatostatin. Examples include human growth hormone, interferons, and proteins used in vaccine production (**table 14.4**). In addition, the fusion of one protein to another has become a useful research tool (**Techniques & Applications 14.1**).

1. What is a genomic library? Describe two ways in which a genomic library might be screened for the clone of interest.
2. What is a transgenic animal? Describe how electroporation and gene guns are used to insert foreign genes into eucaryotic cells. What other approaches may be used? How are bacteria transformed?
3. How can one prevent recombinant DNA from undergoing recombination in a bacterial host cell?
4. List several reasons why a cloned gene might not be expressed in a host cell. What is an expression vector?

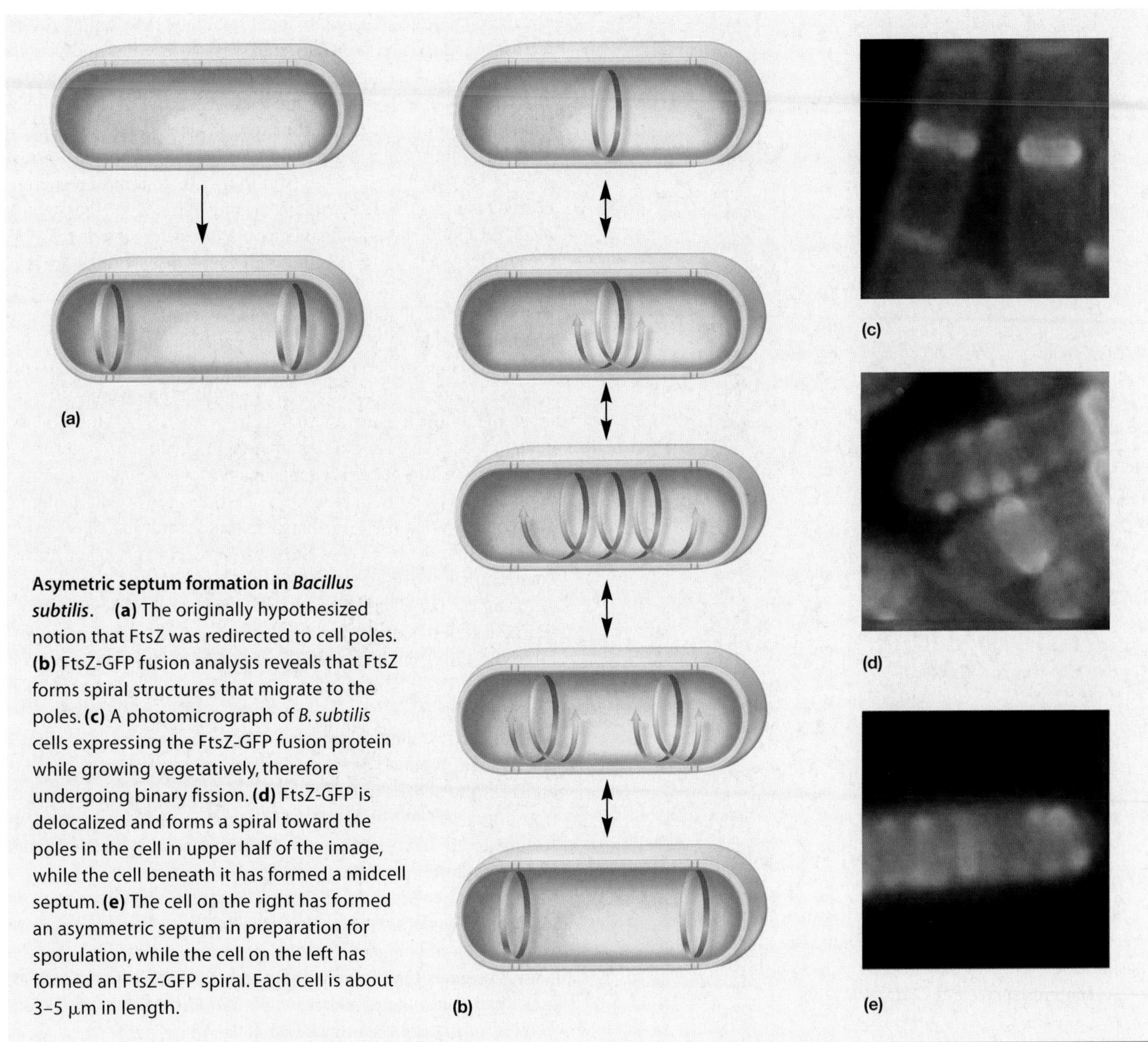

Asymetric septum formation in *Bacillus subtilis*. **(a)** The originally hypothesized notion that FtsZ was redirected to cell poles. **(b)** FtsZ-GFP fusion analysis reveals that FtsZ forms spiral structures that migrate to the poles. **(c)** A photomicrograph of *B. subtilis* cells expressing the FtsZ-GFP fusion protein while growing vegetatively, therefore undergoing binary fission. **(d)** FtsZ-GFP is delocalized and forms a spiral toward the poles in the cell in upper half of the image, while the cell beneath it has formed a midcell septum. **(e)** The cell on the right has formed an asymmetric septum in preparation for sporulation, while the cell on the left has formed an FtsZ-GFP spiral. Each cell is about 3–5 μm in length.

14.9 Applications of Genetic Engineering

Genetic engineering and biotechnology will continue to contribute in the future to medicine, industry, and agriculture, as well as to basic research. In this section some practical applications are briefly discussed.

Medical Applications

The production of medically useful proteins such as somatostatin, insulin, human growth hormone, and some interferons (signaling molecules of the immune system) is of great practical importance (table 14.4). This is particularly true of substances that previously only could be obtained from human tissues. For example, in the past, human growth hormone for treatment of pituitary dwarfism was extracted from pituitaries obtained during autopsies and was available only in limited amounts. Interleukin-2 (a protein that helps regulate the immune response) and blood-clotting factor VIII have been cloned, as well as a number of other important peptides and proteins. It also is possible to use genetically engineered plants to produce cloned gene products such as oral vaccines. Genetically engineered mice produce human monoclonal antibodies. Synthetic vaccines—for instance, vaccines for malaria and rabies—are also

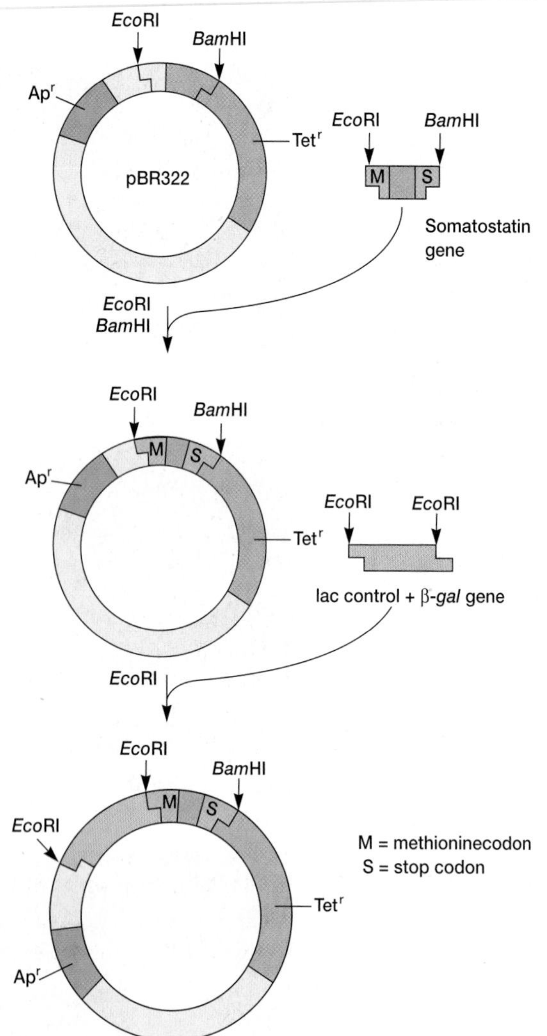

Figure 14.17 Cloning the Somatostatin Gene. An overview of the procedure used to synthesize a recombinant-plasmid containing the somatostatin gene. Ap^r, Tet^r, ampicillin- and tetracycline-resistance genes, respectively.

being developed with recombinant techniques. A recombinant hepatitis B vaccine is commercially available. Control of epidemics: Vaccines and immunization (section 36.8); Techniques & Applications 32.2: Monoclonal antibody technology

Livestock have become important in medical biotechnology through the use of an approach sometimes called molecular pharming. For instance, pig embryos injected with human hemoglobin genes develop into transgenic pigs that synthesize human hemoglobin. Current plans are to purify the hemoglobin and use it as a blood substitute. A pig could yield 20 units of blood substitute a year. Somewhat similar techniques have produced transgenic goats whose milk contains up to 3 grams of human tissue plasminogen activator (TPA) per liter. TPA dissolves blood clots and is used to treat cardiac patients.

However, genetic engineering has much more to offer medicine than the production of recombinant products. Probes are now being used to screen individuals and the blood supply for infectious agents such as AIDS and West Nile virus. Individuals can be screened for mutant genes; this is common in testing future parents as well as in prenatal testing. A growing area of medical intervention is **gene therapy,** which seeks to replace malfunctioning genes with wild-type genes. Here the normal gene is most commonly delivered in a virus that has been engineered to enter host cells without causing harm. Gene therapy has received a lot of attention in the popular press for several reasons. Gene therapy may cure debilitating genetic diseases that are currently not curable, offering hope to patients and their loved ones. But some clinical gene therapy trials have failed and a few have resulted in death or serious injury to patient volunteers. Also, the potential for gene therapy to be used in germline cells has generated much controversy. Germline gene therapy changes the genetic make-up of gametes or fertilized ova. Thus the genetic change would be passed on to subsequent generations—germline gene therapy introduces heritable changes. One can envision two applications of germline therapy: (1) to correct a genetic defect of an embryo that if left unchanged would result in a debilitating or lethal disease (e.g., Tay Sachs disease, a neurodegenerative condition with a 100% mortality rate by age 5); or (2) to produce a "designer baby" that bears some predetermined phenotypic trait. Fears of the latter have prompted many governments to prohibit germline gene therapy in humans. In contrast, somatic cell gene therapy is nonheritable, and treats only cells in the affected organ. For instance, gene therapy trials with cystic fibrosis (CF) patients have sought to deliver a functional gene to lung tissue. CF patients produce copious amounts of respiratory secretions and are plagued by recurrent *Pseudomonas aeruginosa* infections; the average lifespan of a CF patient is about 30 years. Unfortunately gene therapy has so far provided only localized and short-term relief for these patients.

The challenges to gene therapy are many; however, the potential benefits are so great that medical biotechnologists persevere. One of the biggest obstacles is the method of delivering the wild-type gene. One approach is to remove cells from the patient, genetically alter them, and grow them in vitro. Once a sufficient number of genetically engineered cells are available, these cells are returned to the patient. These cells can be either adult differentiated cells or stem cells. Following genetic manipulation, cells can be introduced directly into the diseased organ or infused into the patient's blood, from which they must migrate to the affected site. Currently, there are at least 15 clinical trials testing **adult stem cell therapy**—somatic gene therapy using adult stem cells. More controversial is the use of **embryonic stem cells.** These cells are collected from 5-day-old embryos obtained from fertility clinics. Unlike adult stem cells, embryonic stem cells are pluripotent—they can differentiate into any cell type (e.g., cardiac, muscle, blood). Scientists believe that these cells are key to understanding embryonic development and may perhaps yield important new therapies.

Finally, it is important to understand the difference between therapeutic cloning and reproductive cloning. **Therapeutic cloning** is what has been described—the use of genetic engineer-

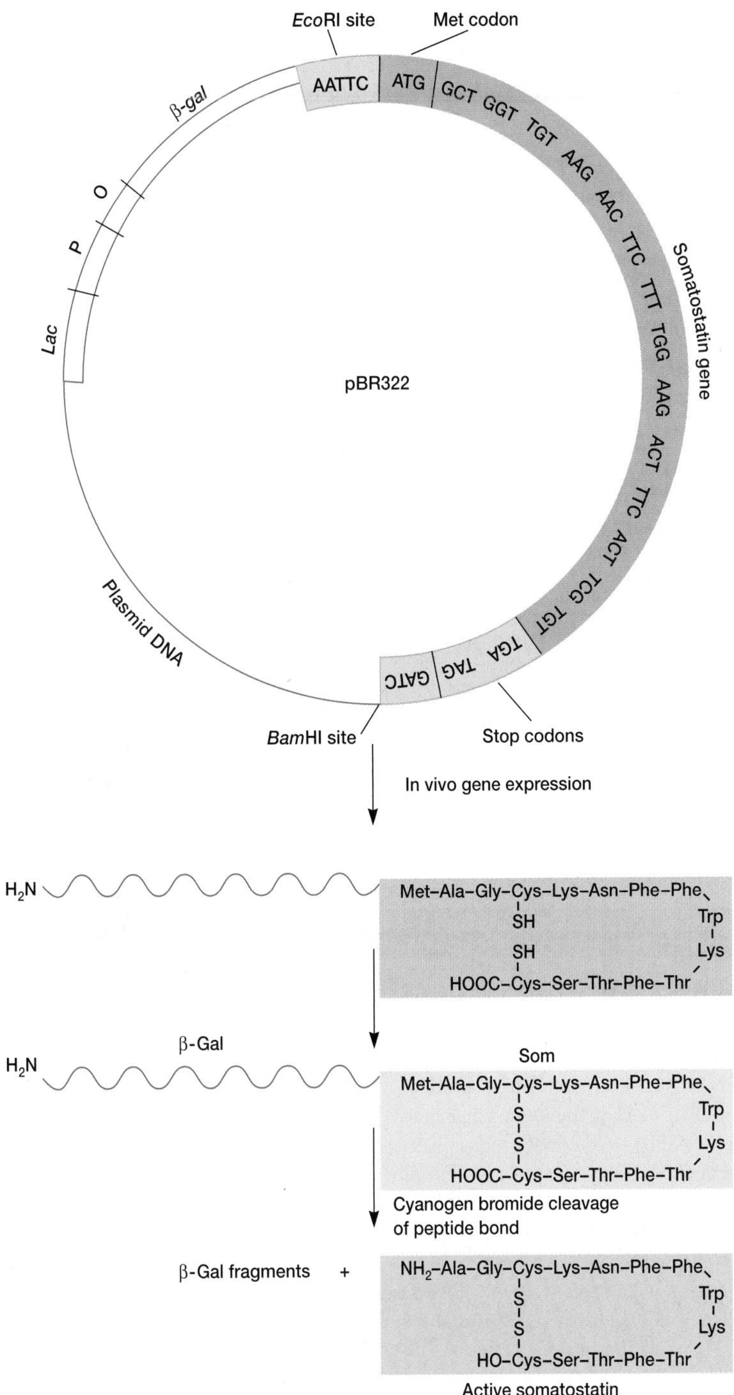

Figure 14.18 The Synthesis of Somatostatin by Recombinant *E. coli*. Cyanogen bromide cleavage at the methionine residue releases active hormone from the β-galactosidase fragment. The gene and associated sequences are shaded in color. Stop codons, the special methionine codon, and restriction enzyme sites are enclosed in boxes.

ing to alter a cell's genotype and return (or introduce) the modified cells to a patient. It is hoped that therapeutic cloning may one day provide treatments (if not cures) for conditions like Parkinson's disease, Alzheimer's disease, and spinal cord injuries. This differs from **reproductive cloning** in which an embryo is created by replacing the nucleus of an unfertilized egg with a nucleus from a somatic cell. This embryo is implanted into a female's uterus with the intention of creating a new life. Human reproductive cloning is legally forbidden throughout most of the world. The U.S. government does not fund therapeutic cloning that uses

Table 14.4 Some Human Peptides and Proteins Synthesized by Genetic Engineering

Peptide or Protein	Potential Use
α_1-antitrypsin	Treatment of emphysema
α-, β-, and γ-interferons	As antiviral, antitumor, and anti-inflammatory agents
Blood-clotting factor VIII	Treatment of hemophilia
Calcitonin	Treatment of osteomalacia
Epidermal growth factor	Treatment of wounds
Erythropoetin	Treatment of anemia
Growth hormone	Growth promotion
Insulin	Treatment of diabetes
Interleukins-1, 2, and 3	Treatment of immune disorders and tumors
Macrophage colony stimulating factor	Cancer treatment
Relaxin	Aid to childbirth
Serum albumin	Plasma supplement
Somatostatin	Treatment of acromegaly
Streptokinase	Anticoagulant
Tissue plasminogen activator	Anticoagulant
Tumor necrosis factor	Cancer treatment

embryonic stem cell technology, although some states do support such research.

Agricultural Applications

Cloned genes can be inserted into plant as well as animal cells. A popular way to insert genes into plants is with a recombinant **Ti plasmid** (tumor-inducing plasmid) obtained from the bacterium *Agrobacterium tumefaciens* (**Techniques & Applications 14.2**). It also is possible to donate genes by forming plant cell protoplasts, making them permeable to DNA, and then adding the desired recombinant DNA. The gene gun is also used in the production of transgenic plants. Microorganism associations with vascular plants: Agrobacterium (section 29.5)

Genetic engineering has made plants resistant to environmental stresses. For example, the genes for detoxification of glyphosate herbicides were isolated from *Salmonella,* cloned, and introduced into tobacco cells using the Ti plasmid. Plants regenerated from the recombinant cells were resistant to the herbicide. Herbicide-resistant varieties of cotton and fertile, transgenic corn also have been developed. This is of considerable importance because many crops suffer stress when treated with herbicides. Resistant plants are not stressed by the chemicals being used to control weeds.

U.S. farmers grow substantial amounts of genetically modified (GM) crops. About a third of the corn, half of the soybeans,

Techniques & Applications

14.2 Plant Tumors and Nature's Genetic Engineer

A plasmid from the plant pathogenic bacterium *Agrobacterium tumefaciens* is responsible for much success in the genetic engineering of plants. Infection of normal cells by the bacterium transforms them into tumor cells, and crown gall disease develops in dicotyledonous plants such as grapes and ornamental plants. Normally, the gall or tumor is located near the junction of the plant's root and stem. The tumor forms because of the insertion of genes into the plant cell genome, and only strains of *A. tumefaciens* possessing a large conjugative plasmid called the Ti plasmid are pathogenic (*see section 29.5* and *figure 29.19*). The Ti plasmid carries genes for virulence and the synthesis of substances involved in the regulation of plant growth. The genes that induce tumor formation reside between two 23 base pair direct-repeat sequences. This region is known as T-DNA and is very similar to a transposon. T-DNA contains genes for the synthesis of plant growth hormones (an auxin and a cytokinin) and an amino acid derivative called opine that serves as a nutrient source for the invading bacteria. In diseased plant cells, T-DNA is inserted into the chromosomes at various sites and is stably maintained in the cell nucleus.

When the molecular nature of crown gall disease was recognized, it became clear that the Ti plasmid and its T-DNA had great potential as a vector for the insertion of recombinant DNA into plant chromosomes. In one early experiment the yeast alcohol dehydrogenase gene was added to the T-DNA region of the Ti plasmid. Subsequent infection of cultured plant cells resulted in the transfer of the yeast gene. Since then, many modifications of the Ti plasmid have been made to improve its characteristics as a vector. Usually one or more antibiotic resistance genes are added, and the nonessential T-DNA, including the tumor inducing genes, is deleted. Those genes required for the actual infection of the plant cell by the plasmid are retained. T-DNA also has been inserted into *E. coli* plasmids to produce cloning vectors that can move between bacteria and plants (see **Box figure**). The gene or genes of interest are spliced into the T-DNA region between the direct repeats. Then the plasmid is returned to *A. tumefaciens,* plant culture cells are infected with the bacterium, and transformants are selected by screening for antibiotic resistance (or another trait encoded by T-DNA). Finally, whole plants are regenerated from the transformed cells. In this way several plants have been made herbicide resistant. Microbes as products: Biopesticides (section 41.8)

Unfortunately the *A. tumefaciens* Ti plasmid cannot be transferred naturally to monocotyledonous plants such as corn, wheat, and other grains. It has been used only to modify plants such as potato, tomato, celery, lettuce, and alfalfa. However, genetic modification of the Ti plasmid so that it can be productivity transferred to monocotyledonous plants is an active area of research. The creation of new procedures for inserting DNA into plant cells may well lead to the use of recombinant DNA techniques with many important crop plants.

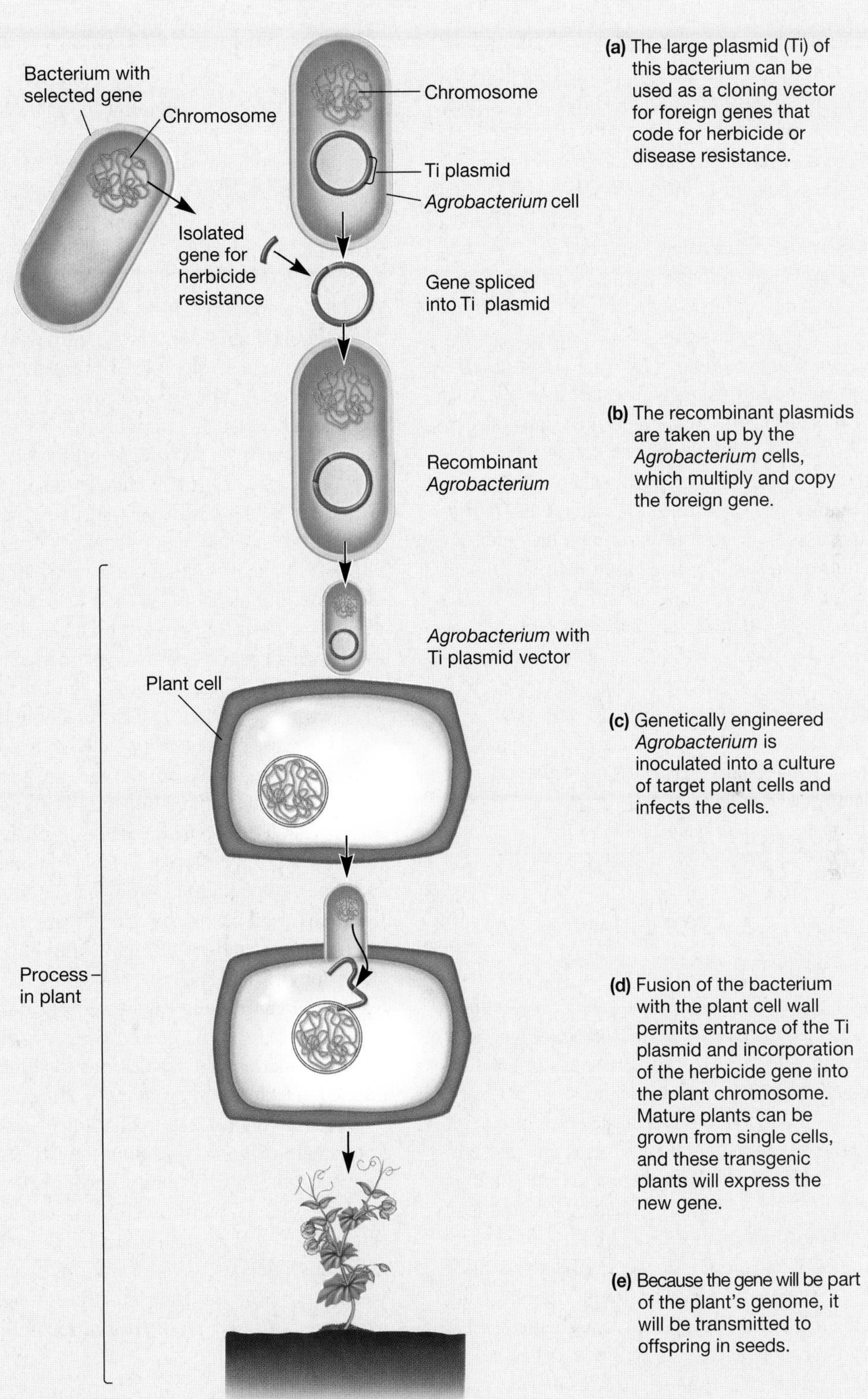
Bacterium with selected gene
Chromosome
Chromosome
Ti plasmid
Agrobacterium cell
(a) The large plasmid (Ti) of this bacterium can be used as a cloning vector for foreign genes that code for herbicide or disease resistance.
Isolated gene for herbicide resistance
Gene spliced into Ti plasmid
Recombinant Agrobacterium
(b) The recombinant plasmids are taken up by the Agrobacterium cells, which multiply and copy the foreign gene.
Agrobacterium with Ti plasmid vector
Plant cell
(c) Genetically engineered Agrobacterium is inoculated into a culture of target plant cells and infects the cells.
Process in plant
(d) Fusion of the bacterium with the plant cell wall permits entrance of the Ti plasmid and incorporation of the herbicide gene into the plant chromosome. Mature plants can be grown from single cells, and these transgenic plants will express the new gene.
(e) Because the gene will be part of the plant's genome, it will be transmitted to offspring in seeds.

Bioengineering of plants. Most techniques employ a genetically modified strain of a natural tumor-producing bacterium called *Agrobacterium tumefaciens*.

and a significant fraction of cotton crops are genetically modified. Cotton and corn are resistant to herbicides and insects. Soybeans have herbicide resistance and lowered saturated fat content. Other examples of genetically engineered commercial crops are canola, potato, squash, and tomato. Golden rice is a GM crop that may have an important impact in the developing world. Compared to unmodified rice, this strain stores about twice as much iron because it has been genetically modified to overexpress some of its own genes. In addition, it serves as a good source of beta-carotene (a vitamin A precursor) thanks to the introduction of genes from the petunia and bacteria.

Many new agricultural applications are being explored. Much effort is being devoted to defending plants against pests without the use of chemical pesticides. A strain of *Pseudomonas fluorescens* carrying the gene for the *Bacillus thuringiensis* toxin has been developed. This toxin destroys many insect pests such as the cabbage looper and the European corn borer. A variety of corn with the *B. thuringiensis* toxin gene has been created. Unfortunately, some insects seem able to develop resistance to the toxin. There is considerable interest in insect-killing viruses and particularly in the baculoviruses. A scorpion toxin gene has been inserted into the autographa californica multicapsid nuclear polyhedrosis virus (AcMNPV). The engineered AcMNPV kills cabbage looper more rapidly than the normal virus and reduces crop damage significantly. Finally, virus-resistant strains of soybeans, potatoes, squash, rice, and other plants are under development. Microbes as products: Biopesticides (section 41.8)

1. List several important present or future applications of genetic engineering in medicine, industry, and agriculture.
2. What is the Ti plasmid and why is it so important?

14.10 Social Impact of Recombinant DNA Technology

Despite the positive social impact of recombinant DNA technology, the potential to alter an organism genetically raises serious scientific and philosophical questions. These issues are the subjects of vigorous debate, as briefly reviewed here.

In contrast to the use of biotechnology in basic and applied science, the use of gene therapy in human beings raises pressing ethical and moral questions. These problems are not extreme as long as adult stem cells are used. However, as witnessed in American political discourse and legislation since 2001, the use of embryonic stem cells is problematic. Is it morally acceptable to sacrifice an embryo to obtain these cells? Proponents point out that adult stem cells are rare and are not truly pluripotent. They also note that these donated embryos are unwanted and will eventually be destroyed by the fertility clinics where they are frozen. Those opposed to embryonic stem cell research believe just as strongly that human life should not be destroyed, even at the very earliest stages. In the summer of 2001, President G. W. Bush banned the use of U.S. federal funds for the establishment of any additional embryonic stem cell cultures or lines. However, in 2004, California passed a ballot initiative releasing $3 billion for embryonic stem cell study in that state, further underscoring the controversial nature of this research. Other countries, most notably the United Kingdom, have actively pursued embryonic stem cell research.

Another area of considerable controversy involves the use of recombinant organisms in agriculture. Many ecologists are worried that the release of recombinant plants without careful prior risk assessment may severely disrupt the ecosystem. Another commonly cited concern is allergic reactions in humans consuming genetically modified (GM) foods, although no such responses have been reported. Others are anxious that the recombinant DNA from GM plants could be transferred to nearby wild plants. This could occur by the movement of pollen by wind and insects to neighboring fields. This would be problematic for a number of reasons including the possibility that GM crops bearing herbicide resistance genes may pollinate weeds, generating "super-weeds" that would require the application of more (rather than less) herbicide. In addition, farmers whose crops are certified organic fear genetic contamination; organic crops are tested for the presence of genetic modification, which if found excludes them from the organic market. Clearly, the rate of development of GM foods has outpaced the consideration of genetic pollution. In Europe, GM crops are not commonly produced by farmers or purchased by consumers. Some large U.S. food producers have responded to public concern and quit using GM crops.

As with any technology, the potential for abuse exists. A case in point is the use of genetic engineering in biological warfare and terrorism. Although international agreements limit research in this area to defense against other biological weapons, knowledge obtained in such research can easily be used in offensive biological warfare. Effective vaccines constructed using recombinant DNA technology can protect the attacker's troops and civilian population. It is relatively easy and inexpensive to prepare bacteria capable of producing massive quantities of toxins or to develop particularly virulent strains of viral and bacterial pathogens, so even small countries and terrorist organizations might acquire biological weapons. Since September 11, 2001 governments worldwide have established tighter regulations to control the availability of pathogens that might be used in a bioterrorism attack. Bioterrorism preparedness (section 36.9)

Recombinant DNA technology has greatly enhanced our knowledge of genes and how they function, and it promises to improve our lives in many ways. Yet, as this brief discussion shows, problems and concerns remain to be resolved. Past scientific advances have sometimes led to unanticipated and unfortunate consequences, such as environmental pollution and nuclear weapons. With prudence and forethought we may be able to avoid past mistakes in the use of recombinant DNA technology.

1. Describe four major areas of concern about the application of genetic engineering. In each case give both the arguments for and against the use of genetic engineering.

Summary

14.1 Historical Perspectives

a. Genetic engineering became possible after the discovery of restriction enzymes and reverse transcriptase, and the development of essential methods in nucleic acid chemistry such as the Southern blotting technique **(table 14.1)**.

14.2 Synthetic DNA

a. Oligonucleotides of any desired sequence can be synthesized by a DNA synthesizer machine. This made site-directed mutagenesis possible and is important for the synthesis of primers and DNA probes used in PCR and Southern blotting, respectively.

14.3 The Polymerase Chain Reaction

a. The polymerase chain reaction allows small amounts of specific DNA sequences to be increased in concentration thousands of times **(figure 14.9)**.

14.4 Gel Electrophoresis

a. Gel electrophoresis is used to separate molecules according to charge and size.

b. DNA fragments are separated on agarose and acrylamide gels. Because DNA is acidic it migrates from the negative to the positive end of a gel **(figure 14.11)**.

14.5 Cloning Vectors and Creating Recombinant DNA

a. There are four types of cloning vectors: plasmids, phages and viruses, cosmids, and artificial chromosomes. Cloning vectors generally have at least three components: an origin of replication, a selectable marker, and a multicloning site or polylinker **(table 14.3; figures 14.12** and **14.14)**.

b. The most common approach to cloning is to digest both vector and DNA to be inserted with the same restriction enzyme or enzymes so that compatible sticky ends are generated. The vector and DNA to be cloned are then incubated in vitro in the presence of DNA ligase, which catalyzes the covalent insertion of the DNA fragment into the vector **(figure 14.5)**.

c. Once the recombinant plasmid has been introduced into host cells, cells carrying vector must be selected. This is often accomplished by allowing the growth of only antibiotic-resistant cells because the vector bears the antibiotic-resistance gene. Cells that took up vector with inserted DNA can be distinguished from those that contain only vector in several ways. Often a blue versus white colony phenotype is used; this is based on the presence or absence of a functional *lacZ* gene, respectively **(figure 14.13)**.

14.6 Construction of Genomic Libraries

a. It is sometimes necessary to find a gene on a chromosome without the knowledge of the gene's DNA sequence. A genomic library is constructed by cleaving an organism's genome into many fragments, each of which is cloned into a vector to make a unique recombinant plasmid.

b. Genomic libraries can be screened for the gene of interest by either phenotypic rescue (genetic complementation) or DNA hybridization with an oligonucleotide probe. However, there are instances when a novel approach to screening the library for a specific gene must be devised **(figures 14.15** and **14.16)**.

14.7 Inserting Recombinant DNA into Host Cells

a. Bacteria and the yeast *S. cerevisiae* are the most common host species.

b. DNA can be introduced into microbes by transformation or electroporation.

c. A variety of techniques are used to introduce DNA into eucaryotic host cells including electroporation, microinjection, and the use of a gene gun. The Ti plasmid of the plant bacterial pathogen *Agrobacterium tumefaciens* has been engineered to transfer foreign DNA into plant genomes.

14.8 Expressing Foreign Genes in Host Cells

a. The recombinant vector often must be modified by the addition of promoters, leaders, and other elements. Eucaryotic gene introns also must be removed. An expression vector has the necessary features to express any recombinant gene it carries.

b. Many useful products, such as the hormone somatostatin, have been synthesized using recombinant DNA technology **(figures 14.17** and **14.18)**.

14.9 Applications of Genetic Engineering

a. Recombinant DNA technology will provide many benefits in medicine, industry, and agriculture.

14.10 Social Impact of Recombinant DNA Technology

a. Despite the great promise of genetic engineering, it also brings with it potential challenges in areas of safety, human experimentation, potential ecological disruption, and biological warfare.

Key Terms

adult stem cell therapy 376
autoradiography 359
bacterial artificial chromosome (BAC) 370
biotechnology 357
cloning 357
complementary DNA (cDNA) 358
cosmid 370
DNA ligase 367
electroporation 371
embryonic stem cells 376
expression vector 372
gel electrophoresis 366
gene gun 371
gene therapy 376
genetic engineering 357
genomic library 370
green fluorescent protein (GFP) 374
heterologous gene 371
microinjection 371
multicloning site (MCS) 367
oligonucleotides 361
origin of replication (*ori*) 366
phenotypic rescue 371
polymerase chain reaction (PCR) 362
primers 362
probe 358
real-time PCR (RT PCR) 363
recombinant DNA technology 357
reproductive cloning 377
restriction enzymes 357
reverse transcriptase (RT) 358
selectable marker 367
shuttle vector 367
site-directed mutagenesis 362
Southern blot procedure 358
sticky ends 358
therapeutic cloning 376
thermocycler 362
Ti plasmid 378
vectors 358
yeast artificial chromosome (YAC) 370

Critical Thinking Questions

1. Could the Southern blotting technique be applied to RNA? How might this be done?
2. Initial attempts to perform PCR were carried out using the DNA polymerase from *E. coli.* What was the major difficulty?
3. What advantage might there be in creating a genomic library first rather than directly isolating the desired DNA fragment?
4. You have cloned a structural gene required for riboflavin synthesis in *E. coli.* You find that an *E. coli* riboflavin auxotroph carrying the cloned gene on a vector makes less riboflavin than does the wild-type strain. Why might this be the case?
5. Suppose that you inserted a simple plasmid (one containing an antibiotic resistance gene and a separate restriction site) carrying a human interferon gene into *E. coli,* but none of the transformed bacteria produced interferon. Give as many plausible reasons as possible for this result.
6. What do you consider to be the greatest potential benefit of genetic engineering? Discuss possible ethical problems with this potential application?

Learn More

Ben-Ari, E. 2002. *Bacillus thuringiensis* as a paradigm for transgenic organisms. *ASM News* 68(12):597–602.

Cockerill, F. R., and Smith, T. F. 2002. Rapid-cycle real-time PCR: A revolution for clinical microbiology. *ASM News* 68(2):77–83.

Drlica, K. 2004. *Understanding DNA and gene cloning: A guide for the curious,* 4th ed. New York: John Wiley & Sons.

Falkow, S. 2001. I'll have the chopped liver please, or how I learned to love the clone. *ASM News* 67(11):555–59.

Glick, B. R., and Pasternak, J. J. 2003. *Molecular biotechnology: Principles and applications of recombination DNA,* 3d ed. Washington, D.C.: ASM Press.

Murray, N. E. 2000. DNA restriction and modification. In *Encyclopedia of microbiology,* 2d ed., vol. 2, J. Lederberg, editor-in-chief, 91–105. San Diego: Academic Press.

Rieger, M. A.; Lamond, M.; Preston, C.; Powles, S. B.; and Roush, R. T. 2002. Pollen-mediated movement of herbicide resistance between commercial canola fields. *Science* 296: 2386–88.

Wolfenbarger, L. L., and Phifer, P. R. 2000. The ecological risks and benefits of genetically engineered plants. *Science* 290:2088–93.

Please visit the Prescott website at www.mhhe.com/prescott7 for additional references.

15 Microbial Genomics

Each dot in the microarray pictured here consists of an oligonucleotide fragment of a single gene bound to a glass slide. Gene expression of two types of cells (for example, one wild-type and one mutant) can be compared by labeling the cDNA from each cell with either a red or green fluorescent label and allowing the cDNAs to bind to homologous sequences attached to the microarray. The color of each spot reveals the relative level of expression of each gene.

Preview

- Genomics is the study of the molecular organization of genomes, their information content, and the gene products they encode. It may be divided into a number of subdisciplines including structural genomics, functional genomics, and comparative genomics.
- Individual pieces of DNA can be sequenced using the Sanger method. The easiest way to analyze entire genomes is by whole-genome shotgun sequencing in which randomly produced fragments are sequenced individually and then aligned to give the complete genome.
- Newly sequenced genomes must be annotated to determine the location of genes and genetic elements such as promoters and ribosome-binding sites.
- Bioinformatics combines biology, mathematics, statistics, and computer science in the analysis of genomes and proteomes.
- Microarray technology enables the study of gene expression at the transcriptional level. In contrast, proteomics uses the entire collection of proteins produced at any given time by an organism to evaluate gene expression at the level of translation.
- Many microbial genomes have been sequenced and compared. The results tell us much about genome structure, microbial physiology, microbial phylogeny, and how pathogens cause disease. They will undoubtedly help in preparing new vaccines and drugs for the treatment of infectious disease.

Chapter 13 provides a brief introduction to microbial recombination, and chapter 14 describes the development of recombinant DNA technology. In this chapter, we carry these themes further with the discussion of the ongoing revolution in genomics. We begin with a general overview of the topic, followed by an introduction to DNA sequencing techniques. Next, the whole-genome shotgun sequencing method is briefly described. Genome function and the analysis of the transcripts and proteins produced by microbes are then explored. We focus on annotation, DNA microarrays, and the use of two-dimensional gel electrophoresis to study the cellular pool of proteins. The chapter concludes with a discussion of examples in which genomic analysis has extended our knowledge of microbial physiology, pathogenicity, evolution, and ecology.

15.1 Introduction

Genomics is the study of the molecular organization of genomes, their information content, and the gene products they encode. It is a broad discipline, which may be divided into at least three general areas. **Structural genomics** is the study of the physical nature of genomes. Its primary goal is to determine and analyze the DNA sequence of the genome. **Functional genomics** is concerned with the way in which the genome functions. That is, it examines the transcripts produced by the genome and the array of proteins they encode. The third area of study is **comparative genomics,** in which genomes from different organisms are compared to look for significant differences and similarities. This helps identify important, conserved regions of the genome in an effort to discern patterns in function and regulation. These data also provide much information about microbial evolution, particularly with respect to phenomena such as horizontal gene transfer.

Genomics is an exciting and growing field that has changed the ways in which key questions in microbial physiology, genetics, ecology, and evolution are pursued. Prior to the advent of genomics,

A prerequisite to understanding the complete biology of an organism is the determination of its entire genome sequence.

—*J. Craig Venter, et al.*

analysis of gene expression was limited to the identification of a small subset of transcripts and proteins. As we will see, genomic technologies enable scientists to study the cell in a holistic way by capturing a snapshot of changes in the entire pool of mRNA transcripts or proteins. The cell can thus be viewed as a network of interconnected circuits, not as a series of individual pathways. Further, genomics provides a window into entire microbial communities—microbial ecologists no longer need to confine their studies to the tiny fraction of microorganisms that have been cultivated. Finally, our understanding of the evolution of all organisms can be illuminated by the insights we gain in procaryotic evolution from genomic approaches. In these ways, and more, genomics has truly revolutionized biology.

15.2 Determining DNA Sequences

Techniques for sequencing DNA were developed in 1977 by two groups: Alan Maxam and Walter Gilbert, and Frederick Sanger. Sanger's method is most commonly used and is discussed here. This method involves the synthesis of a new strand of DNA using the DNA to be sequenced as a template. The reaction begins when single strands of template DNA are mixed with primer (a short piece of DNA complementary to the 5′ end of the region to be sequenced), DNA polymerase, the four deoxynucleoside triphosphates (dNTPs), and dideoxynucleoside triphosphates (ddNTPs). ddNTPs differ from dNTPs in that the 3′ carbon lacks a hydroxyl group (**figure 15.1**). In such a reaction mixture, DNA synthesis will proceed until a ddNTP, rather than a dNTP, is added to the growing chain. Without a 3′-OH group to attack the 5′-PO_4 of the next dNTP to be incorporated, synthesis stops (*see figure 11.14*). Indeed, Sanger's technique is frequently referred to as the **chain-termination DNA sequencing method.** In order to obtain sequence information, four separate synthesis reactions must be prepared, one for each ddNTP (**figure 15.2**). When each DNA synthesis reaction is stopped, a collection of DNA fragments of varying lengths has been generated. The reaction prepared with ddATP produces fragments ending with an A, those with ddTTP produce fragments with T termini, and so forth. If the DNA is to be manually sequenced, radioactive dNTPs are used and each reaction is electrophoresed in a separate lane on a polyacrylamide gel. The molecular weight of each fragment is determined by its

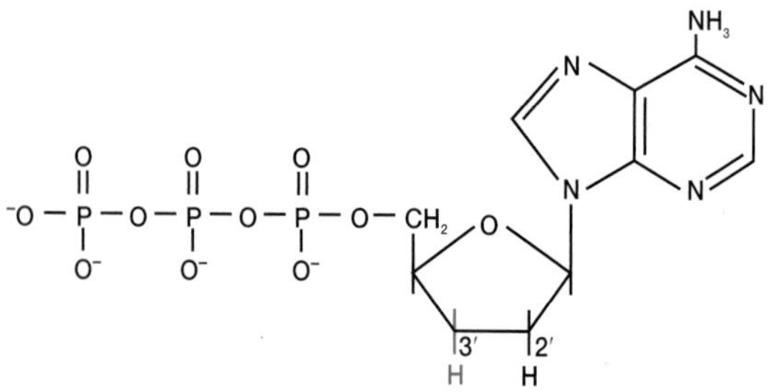

Figure 15.1 Dideoxyadenosine Triphosphate (ddATP). Note the lack of a hydroxyl group on the 3′ carbon, which prevents further chain elongation by DNA polymerase.

length, so shorter fragments migrate faster than larger fragments (*see figure 14.11*). Because synthesis proceeds with the addition of a nucleotide to the 3′-OH of the primer (i.e., in the 5′ to 3′ direction) the ddNTP at the end of the shortest fragment is assigned as the 5′ end of the DNA sequence, while the largest fragment is the 3′ end. In this way, the DNA sequence is read directly from the gel from the smallest to the largest fragment. DNA replication (section 11.4); Gel electrophoresis (section 14.4)

DNA is more often prepared for automated sequencing. Here the four reaction mixtures can be combined and loaded into a single lane of a gel because each ddNTP is labeled with a different colored fluorescent dye (figure 15.2*b*). These fragments are then electrophoresed on a polyacrylamide gel and a laser beam determines the order in which they exit the gel. A chromatogram is generated in which the amplitude of each spike represents the fluorescent intensity of each particular fragment (figure 15.2*c*). The corresponding DNA sequence is listed above the chromatogram.

Fully automated capillary electrophoresis DNA analyzers are required for large projects. These sequencing machines are very fast and can run for 24 hours without operator attention. As many as 96 samples can be sequenced simultaneously, making it possible to sequence as many as 1 million bases per day, per sequencer. This level of automation, involving many sequencers running at the same time, is needed for the completion of whole-genome sequences.

15.3 Whole-Genome Shotgun Sequencing

Although methods for sequencing relatively short regions of DNA have been in use for some time, efficient methods for sequencing whole genomes were not available until 1995, when J. Craig Venter, Hamilton Smith, and their collaborators developed **whole-genome shotgun sequencing** and the computer software needed to assemble sequence data into a complete genome. They used their new method to sequence the genomes of the bacteria *Haemophilus influenzae* and *Mycoplasma genitalium*. This was a significant accomplishment because prior to this only a few viral genomes had been fully sequenced. These are much smaller than the genome of *H. influenzae,* which contains about 1,743 genes and 1,830,137 base pairs, or about 1.8 Mb (*m*illion *b*ase pairs). Venter and Smith's contribution to biology has ushered in what has been called the genomic era. Within 10 years the number of complete genomes published grew from 2 to 249 with over 500 ongoing genome sequencing projects. **Table 15.1** lists just a few microbial genomes.

The process of whole-genome shotgun sequencing is fairly complex when considered in detail, and there are many procedures to ensure the accuracy of the results, but the following summary gives a general idea of the procedure. For simplicity, this approach may be broken into four stages: library construction, random sequencing, fragment alignment and gap closure, and editing.

1. *Library construction.* The DNA molecules are randomly broken into fairly small fragments using ultrasonic waves; the fragments are then purified (**figure 15.3**). These fragments are next attached to plasmid or cosmid vectors and plasmids or cosmids with a single insert are isolated. Special *Escherichia*

1 Isolated unknown DNA fragment

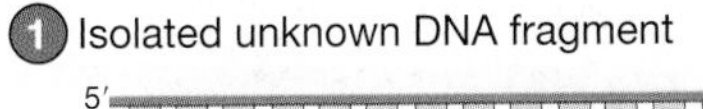

Original DNA to be sequenced

2 DNA is denatured to produce single template strand.

3′ G T G A A T C G G C T A G G 5′

3 Labeled specific primer molecule hybridizes to the DNA strand

Primer

C A C T T

3′ G T G A A T C G G C T A G G 5′

4 DNA polymerase and regular nucleotide mixture (dATP, dCTP, dGTP, and dTTP) are added; ddG, ddA, ddC, and ddT are placed in separate reaction tubes with the regular nucleotides. The dd nucleotides are labeled with some type of tracer, which allows them to be visualized.

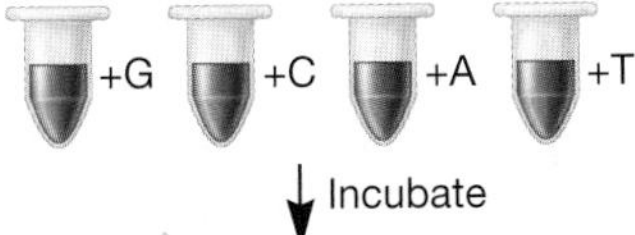

Incubate

5 Newly replicated strands are terminated at the point of addition of a dd nucleotide.

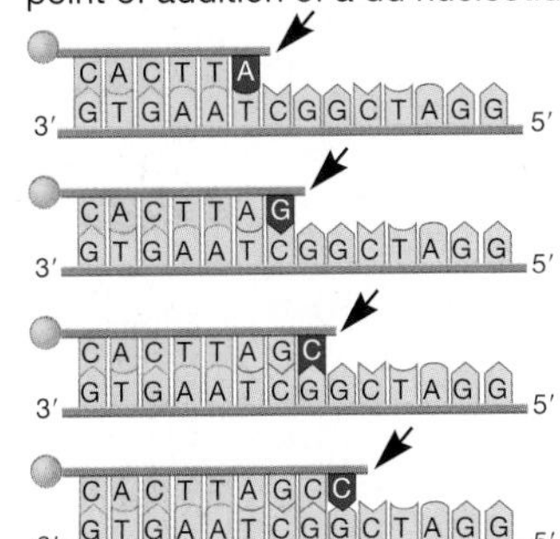

6 Schematic view of how all possible positions on the fragment are occupied by a labeled nucleotide

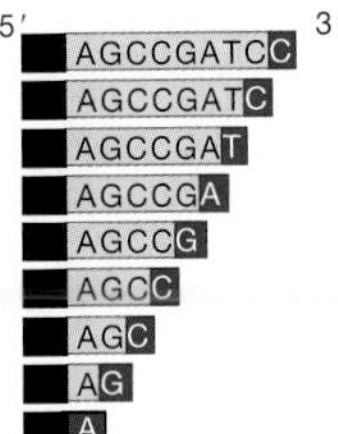

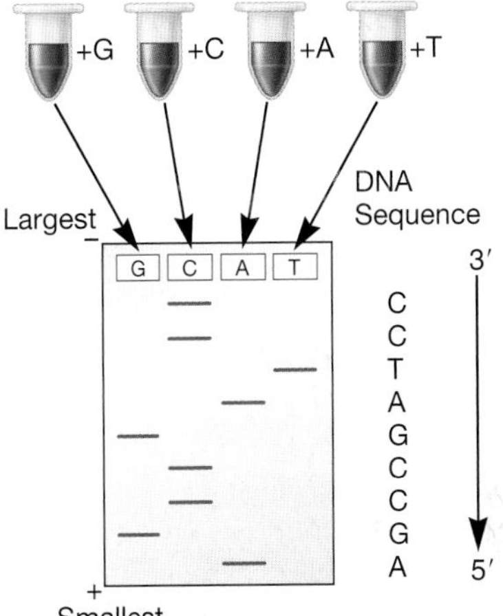

Smallest

7 Running the reaction tubes in four separate gel lanes separates them by size and nucleotide type. Reading from bottom to top, one base at a time, provides the correct DNA sequence.

(a)

—G C G A C A T—

+ddA +ddG +ddC +ddT

–G C G A C A –G C G –G C G A C –G C G A C A T

–G C G A –G –G C

Mix and electrophorese

T A C A G C G

(b)

Figure 15.2 The Sanger Method for DNA Sequencing. **(a)** Steps 1–6 are used for both manual and automated sequencing. Step (7) shows preparation of a gel for manual sequencing in which radiolabeled ddNTPs are used. **(b)** Part of an automated DNA sequencing run. Here the ddNTPs are labeled with fluorescent dyes. **(c)** Data generated during an automated DNA sequencing run. Bases 493 to 580 are shown.

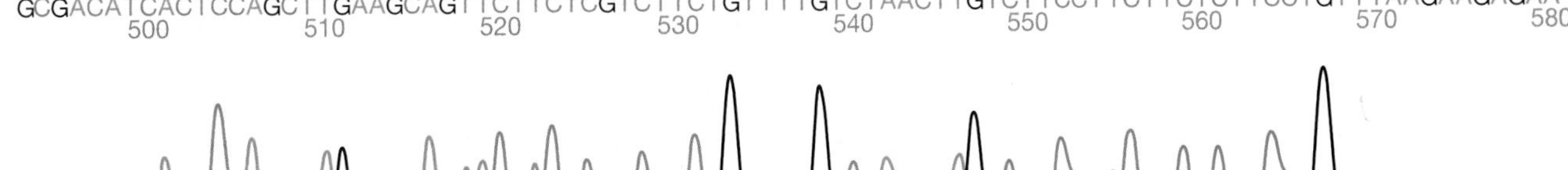

(c)

Table 15.1 **Examples of Complete Published Microbial Genomes**

Genome	Domain[a]	Number of Strains Sequenced	Size (Mb)	% G + C
Agrobacterium tumefaciens	B	2	4.92	60
Aquifex aeolicus	B	1	1.55	43
Archaeoglobus fulgidus	A	1	2.18	48
Bacillus anthracis	B	4	5.09–5.23	36
Bacillus subtilis	B	1	4.21	43
Borrelia burgdorferi	B	1	1.44	28
Campylobacter jejuni	B	1	1.64	31
Caulobacter crescentus	B	1	4.02	62–67
Chlamydia pneumoniae	B	2	1.23	40
Chlamydia trachomatis	B	2	1.05–1.07	41
Chlorobium tepidum	B	1	2.15	57
Clostridium perfringens	B	1	3.03	29
Corynebacterium glutamicum	B	1	3.31	55–58
Deinococcus radiodurans	B	1	3.06	67
Escherichia coli	B	6	4–5.45	50
Geobacter sulfurreducens PCA	B	1	3.81	61
Haemophilus influenzae Rd	B	1	1.83	39
Halobacterium sp. NRC-1	A	1	2.01	68
Helicobacter pylori	B	2	1.64–1.67	39
Listeria monocytogenes	B	2	2.9	37–39
Methanobacterium thermoautotrophicum	A	1	1.75	49
Methanocaldococcus jannaschii	A	1	1.66	31
Mycobacterium leprae	B	1	3.27	58
Mycobacterium tuberculosis	B	2	4.40	65
Mycoplasma genitalium	B	1	0.58	31
Mycoplasma pneumoniae	B	1	0.82	40
Nanobacterium equitans	A	1	0.49	32
Neisseria meningitidis	B	3	2.18–2.27	51
Prochlorococcus marinus	B	3	1.66–2.41	31–51
Pseudomonas aeruginosa	B	1	6.26	67
Pyrococcus abyssi	A	1	1.77	44
Pyrococcus horiksohii	A	1	1.74	42
Rhodopseudomonas palustris	B	1	5.46	65
Rickettsia prowazekii	B	1	1.11	29
Saccharomyces cerevisiae	E	1	12.14	38
Salmonella enterica serovar Typhimurium	B	1	4.86	50–53
Staphylococcus aureus	B	7	2.80–2.90	33
Streptococcus mutans	B	1	2.03	37
Streptococcus pneumoniae	B	2	2.16	40
Streptococcus pyogenes	B	6	1.84–1.90	39
Streptomyces coelicolor	B	1	8.67	72
Sulfolobus tokodaii	A	1	2.69	33
Synechocystis sp.	B	1	3.57	47
Thermoplasma acidophilum	A	1	1.56	46
Thermotoga maritima	B	1	1.86	46
Treponema pallidum	B	1	1.14	52
Vibrio cholerae	B	1	4.03	48
Yersinia pestis	B	3	4.60–4.65	48
Yersinia pseudotuberculosis	B	1	4.74	48

[a]The following abbreviations are used: A, *Archaea;* B, *Bacteria;* E, *Eucarya.*

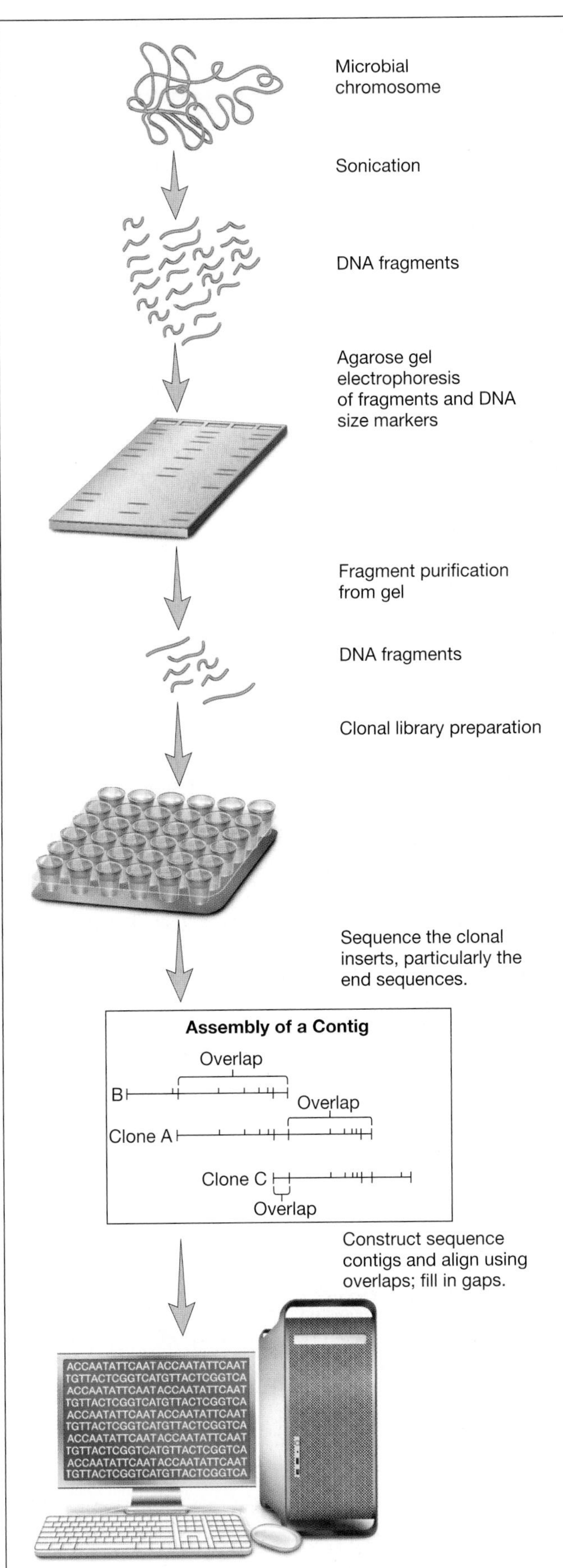

Figure 15.3 Whole-Genome Shotgun Sequencing.

coli strains lacking restriction enzymes are transformed with the plasmids to produce a library of the plasmid clones. Cloning vectors and creating recombinant DNA (section 14.5)

2. *Random sequencing.* After the clones are prepared and the DNA purified, thousands of DNA fragments are sequenced with automated sequencers, employing special dye-labeled primers. Thousands of templates are used, normally with primers that recognize the plasmid DNA sequences adjacent to the DNA insert. The nature of the process is such that almost all stretches of the genome are sequenced several times, and this increases the accuracy of the final results.
3. *Fragment alignment and gap closure.* Using special computer programs, the sequenced DNA fragment data are clustered and assembled into longer stretches of sequence by comparing nucleotide sequence overlaps between fragments. Two fragments are joined together to form a larger stretch of DNA if the sequences at their ends overlap and match. This overlap comparison process results in a set of larger, contiguous nucleotide sequences called contigs.

   Finally, the contigs are aligned in the proper order to form the complete genome sequence. If gaps exist between two contigs, sometimes fragments with ends in two adjacent contigs are available. These fragments are analyzed and the gaps filled in with their sequences. When this approach is not possible, a variety of other techniques are used to align contigs and fill in gaps. For example, λ phage libraries containing large DNA fragments can be constructed. The large fragments in these libraries overlap the previously sequenced contigs. These fragments are then combined with oligonucleotide probes that match the ends of the contigs to be aligned. If the probes bind to a λ library fragment, it can be used to prepare a stretch of DNA that represents the gap region. Overlaps in the sequence of this new fragment with two contigs allow them to be placed side-by-side and fill in the gap between them. Construction of genomic libraries (section 14.6)
4. *Editing.* The sequence is then carefully proofread in order to resolve any ambiguities in the sequence. Also the sequence must be checked for unwanted frameshift mutations and corrected if necessary.

Using this approach, it took less than 4 months to sequence the *M. genitalium* genome (about 500,000 base pairs in size). The shotgun technique also has been used successfully by Celera Genomics in the Human Genome Project and to sequence the *Drosophila* genome. Researchers at the Wellcome Trust Sanger Centre (UK) have also sequenced (and continue to sequence) the genomes of many important microbial pathogens, while the U.S. Department of Energy Joint Genome Institute supports the sequencing of many bacteria of environmental relevance. Once an organism's genome has been sequenced, the level of inquiry and the pace of research are greatly enhanced. The following sections describe only some of the ways a genome sequence can be used to learn more about an organism.

1. What is the goal of each of the three general areas of genomics?
2. Why is the Sanger technique of DNA sequencing also called the chain-termination method?
3. How would one recognize a gap in the genome sequence following whole-genome shotgun sequencing?

15.4 Bioinformatics

The analysis of entire genomes generates not only a tremendous amount of nucleotide sequence data but, as we shall see, a rapidly growing volume of information regarding genome content, structure, and arrangement, as well as data detailing protein structure and function. The only feasible way to organize and analyze these data is through the use of computers. This has led to the development of the field of **bioinformatics,** which combines biology, mathematics, computer science, and statistics. Determining the location and nature of genes or presumed genes on a newly sequenced genome is a complex process called **annotation.** Once genes have been identified, bioinformaticists can perform computer, or ***in silico,* analysis** to further examine the genome.

Genome Annotation

Obviously, obtaining genome sequence without any understanding of the location and nature of individual genes would be a pointless exercise. The process of genome annotation seeks to identify every potential (putative) protein-coding gene as well as each rRNA- and tRNA-coding gene. A protein-coding gene is usually first recognized as an **open reading frame (ORF);** to find all ORFs, both strands of DNA must be analyzed. A procaryotic ORF is generally defined as a sequence of at least 100 codons with the following three features: (1) it is not interrupted by a stop codon; (2) there is an apparent ribosomal binding site at the 5′ end; and (3) terminator sequences at the 3′ end. Only if these genetic elements are present is an ORF considered a putative gene (**figure 15.4**). ORFs that are presumed to encode proteins (as opposed to tRNA or rRNA) are called ***coding* sequences (CDS).** Bioinformaticists have developed algorithms to compare the sequence of predicted CDS with those in large databases containing nucleotide and amino acid sequences of known proteins. If an ORF matches one in the database, it is assumed to encode the same protein or type of protein. Although such comparisons are not without errors, they can provide tentative functional assignments for about 40 to 80% of the CDS in a given genome. The remaining fall into two classes: (1) genes that have matches in the database but no function has yet been assigned. These are said to encode conserved hypothetical proteins. (2) Genes whose translated products are unique to that organism. Such genes are said to encode proteins of unknown function. However, as more genomes are published, future comparisons may reveal a match in another organism. It is helpful to organize identified genes according to product function and/or location in the cell, such as ribosomal and transfer RNAs, lipid metabolism, energy metabolism, cell wall-associated, etc. (**table 15.2**). Clearly bioinformatics is a dynamic field and its continued development is crucial for further progress in structural, functional, and comparative genomics.

15.5 Functional Genomics

Functional genomics seeks to explain how genes and genomes operate. The base-by-base comparison of two or more gene sequences is called **alignment.** Alignment of genes on the same genome (i.e., from a single organism) may show that the nucleotide sequences are so alike that they most probably arose through gene duplication; such genes are called **paralogs.** Alignments of genes found in two or more different organisms may reveal that they are so strikingly similar that they are predicted to have the same function; these genes are called **orthologs.** Because the DNA code is degenerate, such alignments are generally performed after the gene's nucleotide sequence has been translated to amino acids.

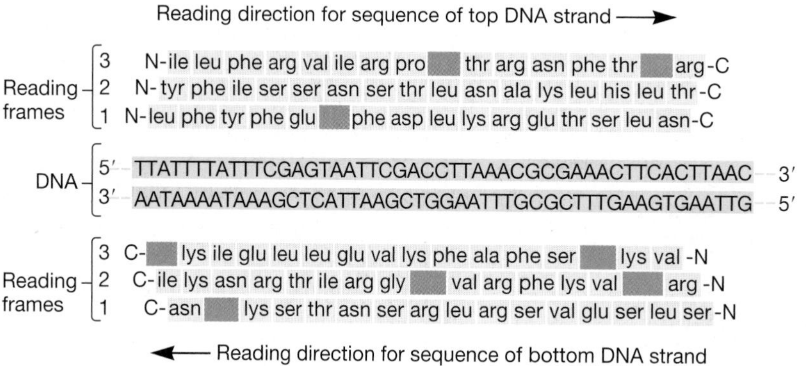

Figure 15.4 Finding Potential Protein Coding Genes. Annotation of genomic sequence requires that both strands of DNA be translated from the 5′ to 3′ direction in each of three possible reading frames. Stop codons are shown in green. See text for details.

Table 15.2 Estimated Number of Genes Involved in Various Cell Functions[a]

Gene Function	*Escherichia coli K12*	*Bacillus subtilis*	*Mycoplasma genitalium*	*Treponema pallidum*	*Rickettsia prowazekii*	*Chlamydia trachomatis*	*Mycobacterium tuberculosis*	*Methanocaldococcus jannaschii*	*Pyrococcus abyssi*
Approximate total number of genes[b]	4,289	4,100	484	1,040	834	894	4,425	1,728	1,765
Cellular processes[c]	190	374	6	77	40	46	132	26	64
Cell envelope components	172	185	29	53	74	45	152	25	106
Transport and binding proteins	315	400	33	59	49	58	168	56	140
DNA metabolism	102	122	30	51	63	48	68	53	63
Transcription	41	114	13	25	26	18	40	21	37
Protein synthesis	122	161	90	99	104	133	110	118	108
Regulatory functions	176	293	5	22	26	12	165	19	66
Energy metabolism[d]	368	439	33	54	89	56	234	158	180
Central intermediary metabolism[e]	73	96	7	6	19	12	293	19	79
Amino acid biosynthesis	114	143	0	7	13	18	91	64	76
Fatty acid and phospholipid metabolism	67	84	8	11	22	27	158	9	18
Purines, pyrimidines, nucleosides, and nucleotides	77	81	19	21	19	14	57	37	51
Biosynthesis of cofactors and prosthetic groups	100	113	5	15	24	27	109	50	53

[a]Data adapted from TIGR (The Institute for Genomic Research) databases.

[b]The number of genes with known or hypothetical functions.

[c]Genes involved in cell division, chemotaxis and motility, detoxification, transformation, toxin production and resistance, pathogenesis, adaptations to atypical conditions, etc.

[d]Genes involved in amino acid and sugar catabolism, polysaccharide degradation and biosynthesis, electron transport and oxidative phosphorylation, fermentation, glycolysis/gluconeogenesis, pentose phosphate pathway, Entner-Doudoroff, pyruvate dehydrogenase, TCA cycle, photosynthesis, chemoautotrophy, etc.

[e]Amino sugars, phosphorus compounds, polyamine biosynthesis, sulfur metabolism, nitrogen fixation, nitrogen metabolism, etc.

Bioinformaticists also analyze the translated amino acid sequence of presumed genes to gain an understanding of potential protein structure and function. Often a short pattern of amino acids, called a **motif,** will represent a functional unit within a protein, such as the active site of an enzyme. For instance, **figure 15.5** shows the C-terminal domain of the cell division protein MinD from a number of microbes. Because these amino acids are found in such a wide range of organisms, they are said to be phylogenetically well conserved. In this case, the conserved region is predicted to form a coil needed for proper localization of the protein to the membrane.

Comparing sequences with other microbes can also provide information about the physical structure of the genome, such as the presence of transposable elements, operons, and repeat elements. The fraction of the genome that has been acquired from another organism by horizontal or lateral gene transfer can be inferred by this type of comparative genomics. In addition, specific aspects of an organism's physiology can be deduced by the presence or absence of specific genes. Genomic analysis can also provide information about the phylogenetic relationships among microbes (**figure 15.6**). Examples of some of the insights derived from such analyses are discussed in section 15.6.

Evaluation of RNA-Level Gene Expression: Microarray Analysis

Once the identity and function of the genes that comprise a genome are established, the key question remains, "Which genes are expressed at any given time?" Prior to the genomic era, researchers could identify only a limited number of genes whose expression was altered under specific circumstances. However, the development of **DNA micorarrays (gene chips)** now allows scientists to look at the expression level of a vast collection of genes at once. DNA microarrays are solid supports, usually of glass or silicon, upon which DNA is attached in an organized grid fashion. Each spot of DNA, called a **probe,** represents a single gene or ORF. The location of each gene on the grid is carefully recorded so that when analyzed, the genetic identity of each spot is known.

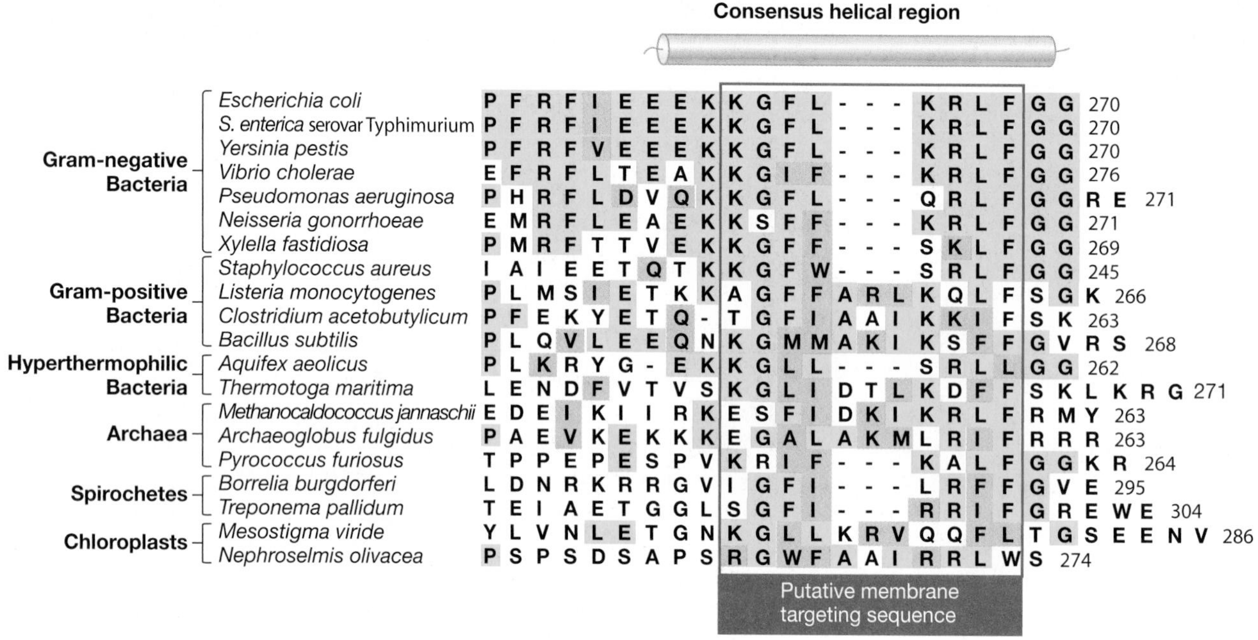

Figure 15.5 Analysis of Conserved Regions of Phylogenetically Well-Conserved Proteins. C-terminal amino acid residues of MinD from 20 organisms representing all three domains of life are aligned to show strong similarities. Amino acid residues identical to *E. coli* are boxed in yellow, and conservative substitutions (e.g., one hydrophobic residue for another) are boxed orange. The number of the last residue shown relative to the entire amino acid sequence is shown at the extreme right of each line.

Of the several ways in which microarrays can be constructed, two techniques are most commonly employed: **Spotted arrays** are prepared by the robotic application of probe to the chip. The probe may be a PCR product, cDNA, or a short DNA fragment within the gene or ORF, called an **oligonucleotide.** When eucaryotic genomes are to be analyzed, oligonucleotide probes are called ***e*xpressed *s*equence *t*ags (ESTs)** because each is derived from cDNA, which itself is the product of an expressed gene (*see figure 14.4*). The genes to be represented by DNA on spotted arrays are carefully selected, so this technology is most commonly used for custom-made microarrays.

Commercially prepared microarrays are usually prepared by a technique known as photolithography (**figure 15.7**). Here, a mask is laid over the chip, and light is used to control the synthesis of the oligonucleotide directly on the chip surface. Each hole in the mask will eventually hold many copies of a different oligonucleotide, and each mask can control the synthesis of several hundred thousand squares. Thus a single microarray can contain hundreds of thousands of different probes, and each probe is present in millions of copies. Commercial microarrays are available with probes for every expressed gene or ORF on the genomes of a number of microbes (**figure 15.8**). These include *E. coli* (about 4,200 ORFs), pathogens such as *Helicobacter pylori* (1,590 ORFs) and *Mycobacterium tuberculosis* (4,000 ORFs), as well as the yeast *Saccharomyces cerevisiae* (approximately 6,600 ORFs).

The analysis of gene expression using microarray technology, like many other molecular genetic techniques, is based on hybridization between the probe DNA and the nucleic acids to be analyzed, often called the targets, which may be either mRNA or single-stranded cDNA (**figure 15.9**). The target nucleotides are labeled with fluorescent dyes and incubated with the chip under conditions that ensure proper binding of target mRNA (or cDNA) to its complementary probe. Unbound target is washed off and the chip is then scanned with laser beams. Fluorescence at each spot or probe indicates that mRNA hybridized. Analysis of the color and intensity of each probe shows which genes were expressed.

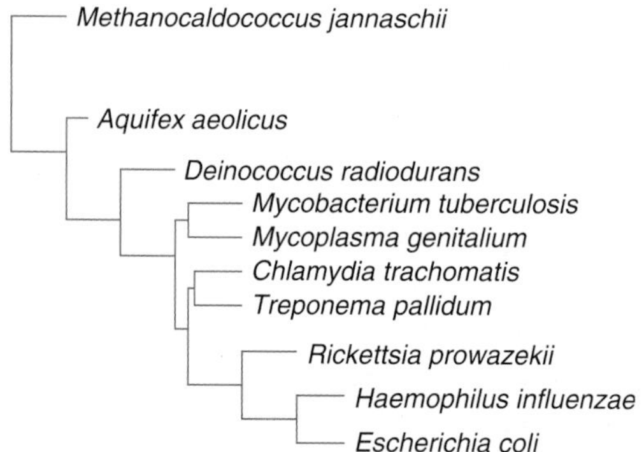

Figure 15.6 Phylogenetic Relationships of Some Procaryotes with Sequenced Genomes. These procaryotes are discussed in the text. *Methanocaldococcus jannaschii* is in the domain *Archaea*, the rest are members of the domain *Bacteria*. Genomes from a broad diversity of procaryotes have been sequenced and compared. *Source: The Ribosomal Database Project.*

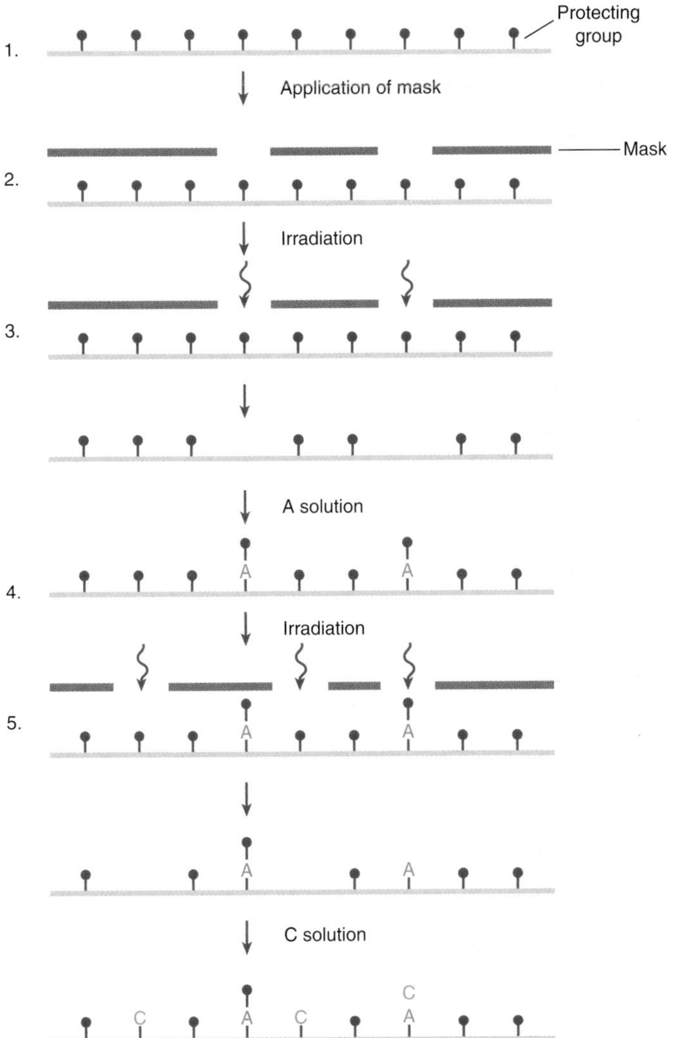

Figure 15.7 Construction of a DNA Chip with Attached Oligonucleotide Sequences. Only two cycles of synthesis are shown. The steps are as follows: (1) The glass support is coated with light-sensitive protecting groups that prevent random nucleoside attachment. (2) The surface is covered with a mask that has holes corresponding to the sites for attachment of the desired nucleosides. (3) Laser light passing through the mask holes removes the exposed protecting groups. (4) The chip is bathed in a solution containing the first nucleoside to be attached. The nucleoside will chemically couple to the light-activated sites. Each nucleoside has a light-removable protecting group to prevent addition of another nucleoside until the appropriate time. (5) Steps 2 through 4 are repeated using a new mask each time until all sequences on the chip have been completed.

DNA microarray analysis can be used to determine which genes are expressed during cellular differentiation or as the result of mutation. One common application of microarray technology in microbiology is to determine the genes whose expression is changed (either up or down regulated) in response to environmental changes (figure 15.9). In this sort of analysis, two different-colored fluorochromes are used. For instance, the pathogen *Helicobacter pylori* dwells in the stomach where it causes ulcers; its genome has been sequenced. One might want to know which *H. pylori* genes are repressed or induced upon exposure to acidic conditions. To determine this, total cellular mRNA from bacteria grown in neutral conditions is prepared and tagged with a green fluorochrome to serve as a control or reference, while mRNA from cells exposed to acidic pH is labeled red. The green (reference) and red (experimental) mRNA samples are then mixed and hybridized to the same microarray. After the unattached mRNA is washed off, the chip is scanned and the image is computer analyzed. A yellow spot or probe indicates that roughly equal numbers of green and red mRNA molecules (targets) were bound, so there was no change in the level of gene expression for that gene. If a target is red, more mRNA from bacteria grown under the experimental acidic conditions was present when the two mRNAs were mixed, thus this gene was induced. Conversely, a green target indicates that the gene was repressed upon exposure to acid stress. Careful image analysis is used to determine the relative intensity of each spot, so that the magnitude of induction or repression of each gene whose expression is altered can be approximated.

Such analysis leads to the detection of genes whose expression falls into specific patterns. For any given microbial genome, only about half to two-thirds of the ORFs are known genes, so such patterns may be used to help assign tentative functions to some unknown ORFs. This is because genes that are expressed under the same conditions may be co-regulated, suggesting that they share a common function or are involved in a common process. However, it is important to keep in mind that microarray results represent only mRNAs present at the time of preparation. Therefore, if a gene is transiently expressed, it may be missed. In addition, if a gene product is regulated posttranslationally, for instance by phosphorylation, its mRNA will be present but the protein may not be active.

15.6 Comparative Genomics

As we have seen, one approach to learning more about the genome of any given microbe is to compare it with that of others. This is the domain of comparative genomics. The publication of over 200 published microbial genomes has truly changed our understanding of microbial biology. One very striking insight is the fact that microbial genomes are not as static as once thought. In fact, microbial genomes are amazingly fluid with a substantial portion of the genome transferred *between* cells, not from parent to offspring. As discussed in chapter 13, this is called **lateral** or **horizontal gene transfer (HGT).** Broadly defined, HGT is the exchange of genetic material between organisms that need not be of similar evolutionary lineages. Genome analysis has revealed that HGT is frequently mediated by phages, and that lysogeny may be the rule, rather than the exception. In fact, some microbes carry multiple prophages. It has become clear that HGT is a major evolutionary force in short-term microbial evolution and long-term speciation. For instance, it is thought that *E. coli* may have acquired the lactose (*lac*) operon from another microbe and thus

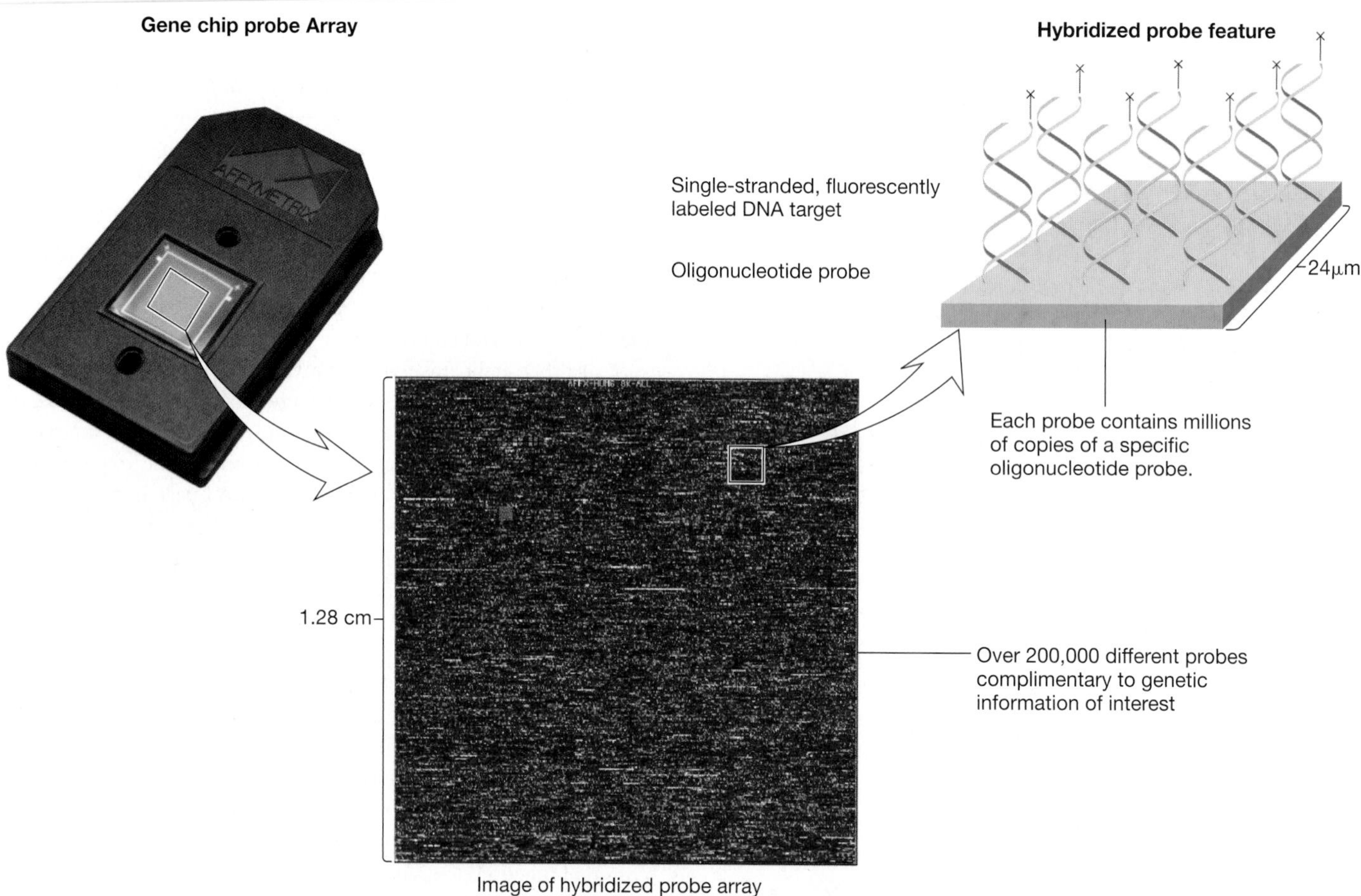

Figure 15.8 A DNA Microarray. The DNA chip manufactured by Affymetrix, Inc. contains probes designed to represent thousands or tens of thousands of genes.

became capable of colonizing the mammalian colon, where milk sugar is a common carbon source. On a larger scale, the methane-producing archaeon *Methanosarcina mazei* appears to have acquired about one-third of its genes from other procaryotes. It appears that microbes use HGT to quickly develop the capacity to colonize new ecological niches.

Comparative genomics can provide useful insights in trying to discern the origin and prevalence of particular phenotypic traits. An example is provided by the genome sequence of *Picrophilis torridus*. This archaeon grows optimally at a pH of 0.7 and a temperature of 65°C. It thrives in hot, acidic, sulfataric fields, a habitat it shares with other extremophilic archaea, like *Thermoplasma,* and the more distantly related *Sulfolobus* (**figure 15.10*a***). Understanding the mechanisms by which these microbes survive under these circumstances is not purely academic, as proteins engineered to withstand harsh treatment have industrial importance. In most cases, microbes living at low pH are able to maintain a neutral internal pH, but this is not the case for *P. torridus,* whose intracellular pH is around 4.6. This suggests it has evolved a unique strategy to prevent irreversible macromolecular damage. By comparing the genes and their products that are found only in these acidophiles and not in neutrophilic and alkalophilic microbes, researchers can focus on potential adaptive strategies (figure 15.10*b*).

Another area in which comparative genomics has recently been applied is vaccine development. The promise of new vaccines as a direct result of genome sequencing is covered extensively in the popular press. However, finding good targets for the development of new vaccines is complicated. The molecules (antigens) that form the basis of an effective vaccine must meet many requirements, including: (1) the antigen must be expressed by the pathogen during infection; (2) the antigen must be either secreted or found on the surface of the pathogen; (3) it must be found in all strains of the pathogen; (4) it must elicit a host immune response; and (5) the antigen must be essential for the survival of the pathogen, at least while it is in the host. Considering the thousands of ORFs that are revealed in each annotated genome, finding genes whose products meet all of these criteria might seem next to impossible. However, using comparative genomics, researchers can examine the genomes of multiple strains of a given pathogen to create a "short list" of candidates. This reduces the number of potential antigens to a manageable number. These antigens are then tested in a variety of assays to discover the best molecules for vaccine development. This

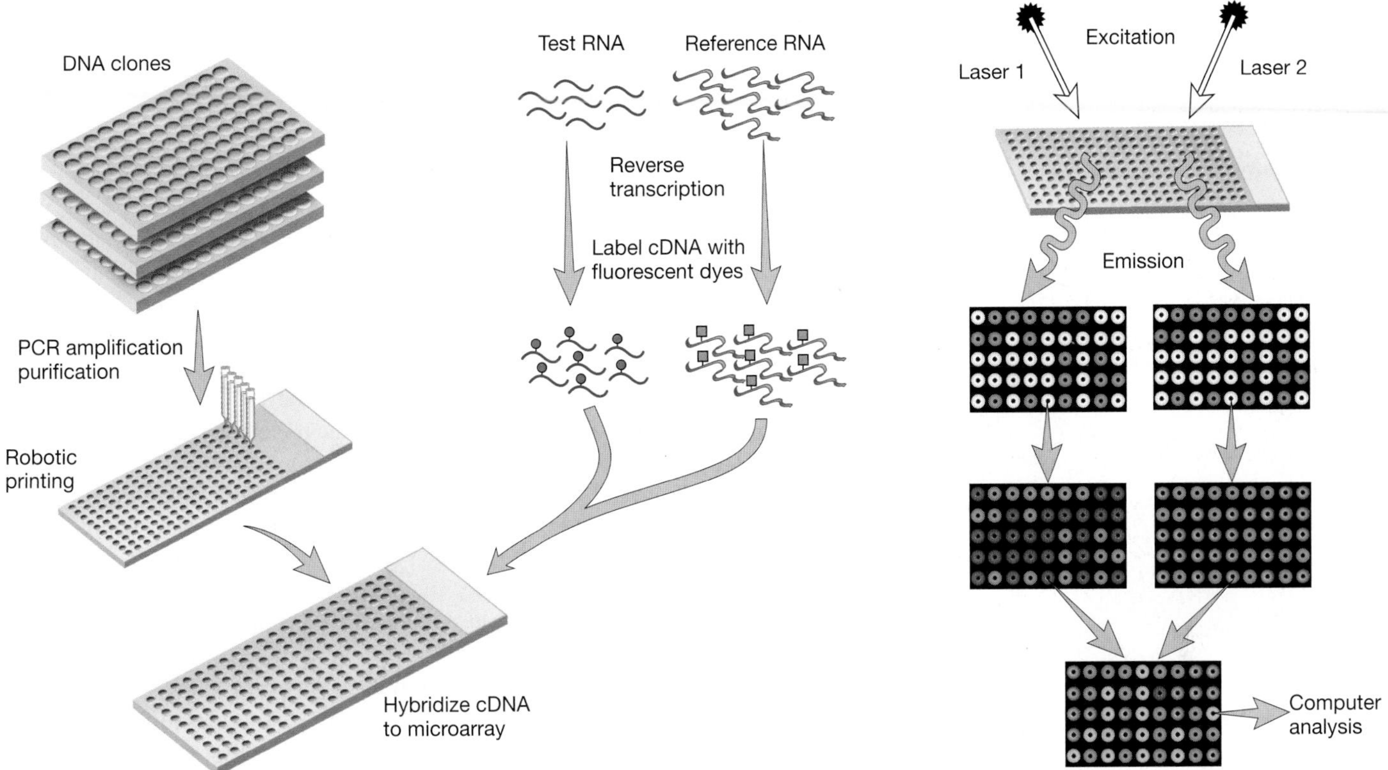

Figure 15.9 A Microarray System for Monitoring Gene Expression. Cloned genes from an organism are amplified by PCR, and after purification, samples are applied by a robotic printer to generate a spotted microarray. To monitor enzyme expression, mRNA from test and reference cultures are converted to cDNA by reverse transcriptase and labeled with two different fluorescent dyes. The labeled mixture is hybridized to the microarray and scanned using two lasers with different excitation wavelengths. The fluorescence responses are measured as normalized ratios that show whether the test gene response is higher or lower than that of the reference.

process was used for the development of an experimental Group B *Streptococcus* (GBS) vaccine. GBS causes serious infections in infants so a vaccine is highly desired. Eight genomes of GBS clinical isolates were compared and 312 surface antigens were tested for their ability to protect mice from infection. Four were found to be effective, and when combined into a single vaccine, they protected animals against infection by all known clinical strains. This demonstrates two important concepts: (1) genome analysis is a powerful mechanism for gaining information, but results must be experimentally verified; and (2) genomics will meet at least some of the expectations described by the popular press.

15.7 Proteomics

Genome function can be studied at the translation level as well as the transcription level. The entire collection of proteins that an organism produces is called its **proteome.** Thus **proteomics** is the study of the proteome or the array of proteins an organism can produce. It is an essential discipline because proteomics provides information about genome function that mRNA studies cannot. There is not always a direct correlation between mRNA and the pool of cellular proteins because of differing levels of mRNA and protein stability and posttranslational regulation. Measurement of mRNA levels can show the dynamics of gene expression and tell what might occur in the cell, whereas proteomics discovers what is actually happening. Much of the research in this area is referred to as **functional proteomics.** It is focused on determining the function of different cellular proteins, how they interact with one another, and the ways in which they are regulated.

Although new techniques in proteomics are currently being developed, we will focus briefly only on the most common approach, **two-dimensional gel electrophoresis.** In this procedure, a mixture of proteins is separated using two different electrophoretic procedures (dimensions). This permits the visualization of thousands of cellular proteins, which would not otherwise be separated in a single electrophoretic dimension. As shown in **figure 15.11,** the first dimension makes use of isoelectric focusing, in which proteins move electrophoretically through a pH gradient (e.g., pH 3 to 10). First the protein mixture is applied to an acrylamide gel in a tube with an immobilized pH gradient and electrophoresed. Each protein moves along the pH gradient until the protein's net charge is zero and the protein stops moving. The pH at this point is equal to the protein's isoelectric point. Thus the first dimension separates the proteins

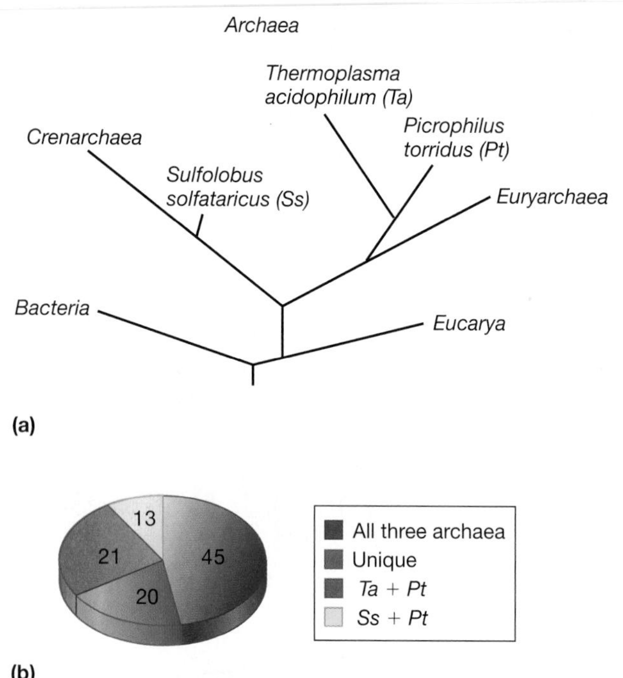

Figure 15.10 Comparative Genomics of *Picrophilus torridus*. **(a)** Simplified phylogenetic tree with the positions of the three acidiphilic, thermophilic archaea *P. torridus*, *T. acidophilum*, and *S. sulfataricus*. **(b)** ORFs found in *P. torridus*, *T. acidophilum*, and *S. sulfataricus*, ORFs shared between *P. torridus* and *T. acidophilum* (*Ta* + *Pt*), or *S. sulfataricus* (*Sa* + *Pt*), and those that are unique to *P. torridus* (Unique).

based on their content of ionizable amino acids. The second dimension is SDS *poly*acrylamide *g*el *e*lectrophoresis (SDS-PAGE). SDS (*s*odium *d*odecyl *s*ulfate) is an anionic detergent that denatures proteins and coats the polypeptides with a negative charge. After the isoelectric gel has been completed, it is soaked in SDS buffer and then placed at the edge of an SDS-PAGE gel. A voltage is then applied. Under these circumstances, polypeptides are separated according to their molecular weight—that is, the smallest polypeptide will travel fastest and farthest. Two-dimensional gel electrophesis can resolve thousands of proteins; each protein is visualized as a spot of varying intensity depending on its cellular abundance. Radiolabeled proteins are usually used, enabling greater sensitivity so that newly synthesized proteins can be distinguished and their rates of synthesis determined. Computer analysis is used to compare two-dimensional gels from microbes grown under different conditions or to compare wild-type and mutant strains. Websites have been developed for the deposition of such images, allowing researchers access to valuable and ever-growing databases. Gel electrophoresis (section 14.4)

Two-dimensional gel electrophoresis is even more powerful when coupled with mass spectrometry (MS). The unknown protein spot is cut from the gel and cleaved into fragments by treatment with proteolytic enzymes. Then the fragments are analyzed by a mass spectrometer and the mass of the fragments is plotted. This mass fingerprint can be used to estimate the probable amino acid composition of each fragment and tentatively identify the protein. Sometimes proteins or collections of fragments are run through two mass spectrometers in sequence, a process known as tandem MS (**figure 15.12**). The first spectrometer separates proteins and fragments, which are further fragmented. The second spectrometer then determines the amino acid sequence of each fragment produced in the first stage. The sequence of a whole protein often can be determined by analysis of such fragment sequence data. Alternatively, if the genome of the organism has been sequenced, only the sequence of the N-terminal amino acids need to be obtained. Computer analysis is then used to compare this partial amino acid sequence with the predicted translated sequences of all the annotated ORFs on the organism's genome. In this way, both the protein and the gene that encodes it can be identified. Further investigation of the protein may rely on one of the large databases of protein sequences that enable comparative analysis. Comparing amino acid sequences can provide information regarding the protein structure, function, and evolution.

Proteomics has been used to study the physiology of many microbes, including *E. coli*. Some areas of research have been the effect of phosphate limitation, proteome changes under anoxic conditions, heat-shock protein production, and the response to the toxicant 2,4-dinitrophenol. One particularly useful approach in studying genome function is to inactivate a specific gene and then look for changes in protein expression. Because changes in the whole proteome are followed, gene inactivation can tell much about gene function and the large-scale effects of gene activity. Gene-protein databases for a number of microbes have been established. These provide information about the conditions under which each protein is expressed and where it is located in the cell.

A second branch of proteomics is called **structural proteomics.** Here the focus is on determining the three-dimensional structures of many proteins and using these to predict the structures of other proteins and protein complexes. The assumption is that proteins fold into a limited number of shapes, and that proteins can be grouped into families of similar structures. When a number of protein structures are determined for a given family, the patterns of protein structure organization or protein-folding rules will be known. Then computational biologists (i.e., bioinformaticists) use this information and the amino acid sequence of a newly discovered protein to predict its final shape, a process known as **protein modeling.**

1. What is genome annotation? Why do you think it requires knowledge of mathematics and statistics as well as biology and computer science?
2. Describe how microarrays are constructed and used to analyze gene expression. How might the following scientists use DNA microarrays in their research? (a) An environmental microbiologist who is interested in how the soil microbe *Rhodopseudomonas palustris* degrades the toxic compound 3-chlorobenzene. (b) A medical microbiologist who wants to learn about how the pathogen *Salmonella* survives within a host cell.
3. Why does two-dimensional gel electrophoresis allow the visualization of many more cellular proteins than seen in electrophoresis in a single dimension?
4. What is the role of mass spectrometry in proteomics?
5. What is the difference between functional and structural proteomics? How do you think structural proteomics might be used in vaccine development?

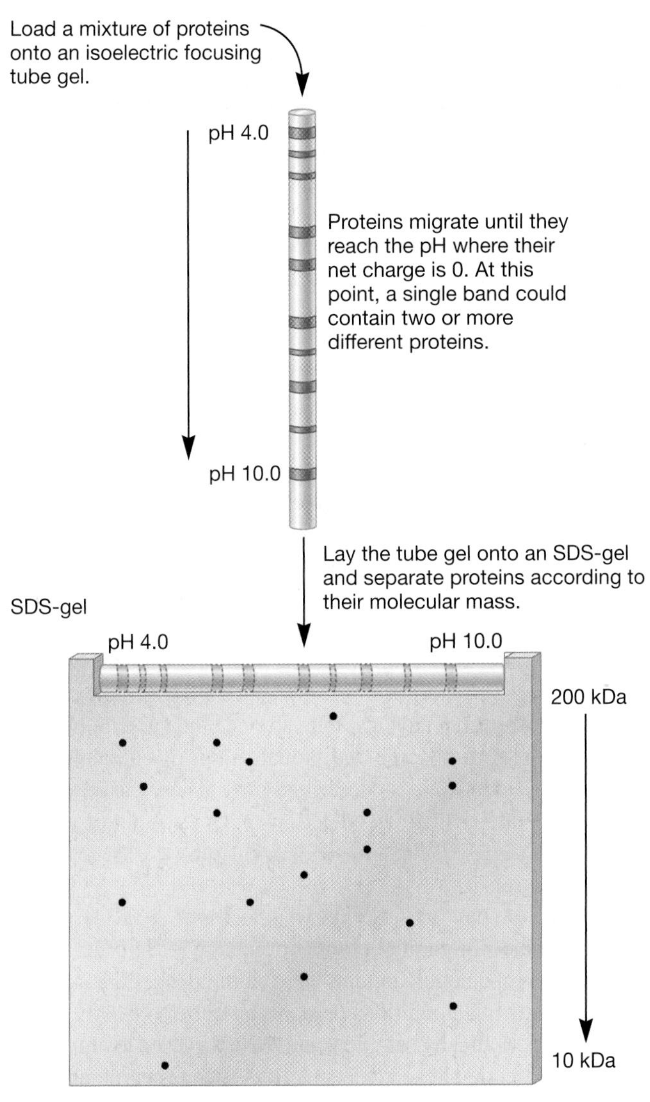

Load a mixture of proteins onto an isoelectric focusing tube gel.
pH 4.0
Proteins migrate until they reach the pH where their net charge is 0. At this point, a single band could contain two or more different proteins.
pH 10.0
Lay the tube gel onto an SDS-gel and separate proteins according to their molecular mass.
SDS-gel
pH 4.0
pH 10.0
200 kDa
10 kDa

(a) The technique of 2-dimensional gel electrophoresis

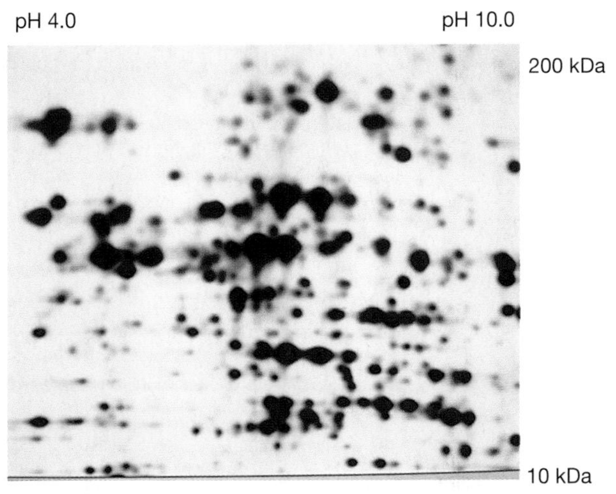

pH 4.0
pH 10.0
200 kDa
10 kDa

(b) An autoradiograph of a 2-dimensional gel. Each protein is a discrete spot.

Figure 15.11 Two-Dimensional Gel Electrophoresis. **(a)** The technique involves two electrophoresis steps. First, a mixture of proteins is separated on an isoelectric focusing gel that has the shape of a tube. Proteins migrate to the point where their net charge is zero. This tube gel is placed into a long well on top of an SDS-polyacrylamide gel. This second gel separates the proteins according to their mass. In this diagram, only a few spots are seen, but an actual experiment would involve a mixture of hundreds or thousands of different proteins. **(b)** An autoradiograph of a 2-D gel. Each spot represents a unique protein.

15.8 Insights from Microbial Genomes

The development of whole-genome shotgun sequencing and other genome sequencing techniques has led to the characterization of many microbial genomes in a relatively short time. These genomes represent great phylogenetic diversity and comparisons among them will contribute significantly to understanding evolutionary processes, deducing which genes are responsible for various cellular processes, dissecting the complexities of genetic regulation, and genome organization. In addition, genomics has become an important tool in developing new antimicrobial agents and may lead to new approaches to detoxify hazardous wastes and provide energy.

Examination of the genomes sequenced to date has provided valuable information, generated many questions, and stimulated new areas of research. In this section we learn how scientists are using experimental approaches to answer some unresolved questions that arise from genome analysis as well as how the genomic era has brought new insights into microbial physiology, ecology, and evolution.

Identification of Genes with Unknown Functions

A common result of genomic studies is the identification of genes for which no function can be assigned. For instance, genome analysis of the important pathogen *Neisseria meningitidis*, which is one causative agent of meningitis, is fairly typical. While just over half of the ORFs can be assigned biological roles based on similarity to proteins of known function, 16% of the ORFs match genes of unknown function in other organisms and about a quarter have no database matches at all. A number of approaches have been developed in an attempt to solve the identity of unknown genes. Perhaps the most comprehensive has been tested in Baker's yeast, otherwise known as *Saccharomyces cerevisiae*.

When the genome of *S. cerevisiae* was fully sequenced in 1996, only about one-third of the ORFs had a known function. At

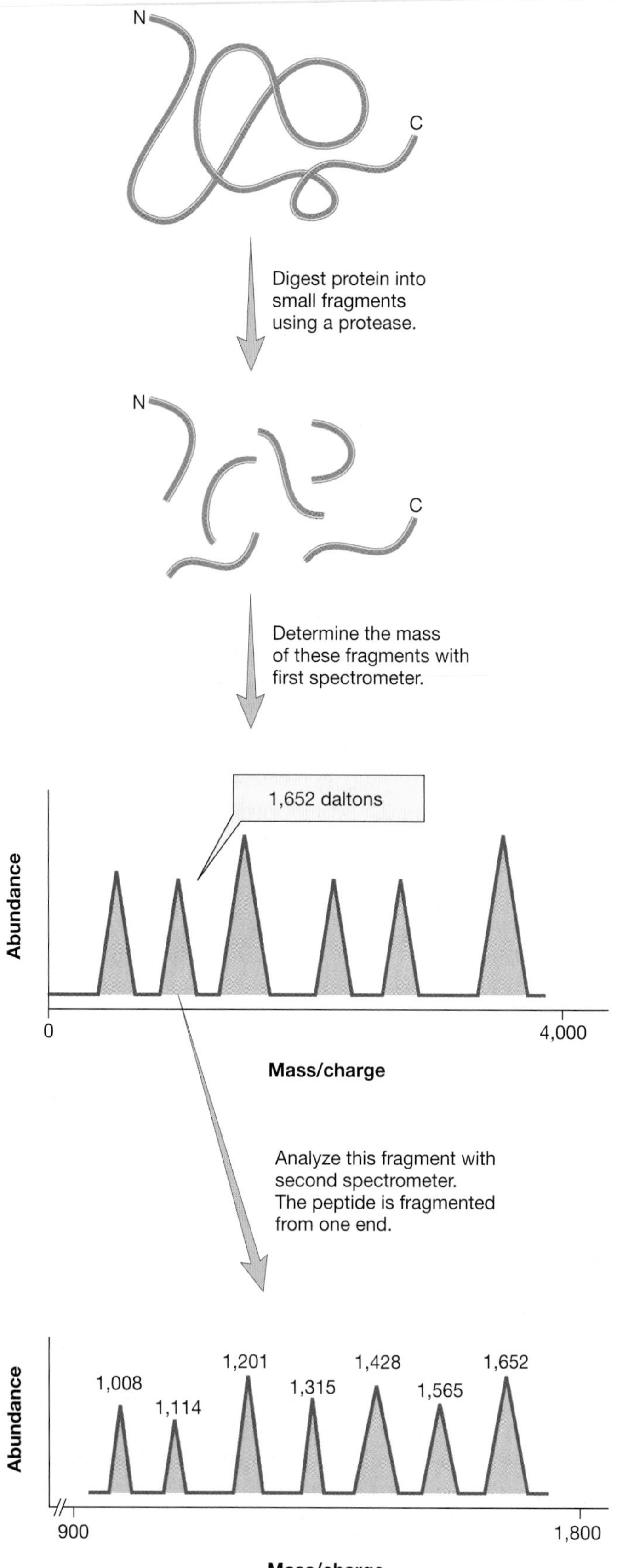

Figure 15.12 The Use of Tandem Mass Spectrometry to Determine the Amino Acid Sequence of a Peptide.

that time, an international consortium of scientists began a project to elucidate the role of ORFs to which no function could be assigned. They decided to create a collection of *S. cerevisiae* strains, each with a deletion in a specific ORF of unknown function. These mutants could then be used in studies designed to reveal their mutant phenotypes. The phenotype of each mutant would then be used to assign a tentative function for the gene.

The yeast deletions were created by a PCR-based gene strategy. In this approach, PCR was used to create DNA cassettes that could be introduced into yeast cells, and through recombination, replace a specific gene in the genome (**figure 15.13**). Each DNA cassette contained three major elements: (1) a gene encoding resistance to the antifungal agent geneticin (*KanMX*) that replaced the targeted ORF. Transformed cells were incubated in the presence of geneticin so only cells in which the targeted ORF had been replaced by the *KanMX* gene were able to grow. (2) Sequences flanking the start and stop codons of the target ORF were cloned on either side of the *KanMX* gene. Because these sequences matched those at the 5′ and 3′ ends of the target ORF, they were used to direct homologous recombination between the cassette and the chromosome, resulting in replacement of the ORF with the *KanMX* gene. (3) A unique DNA sequence called a "molecular bar code" was inserted in each cassette so once the cassette had integrated into the chromosome, individual strains in a mixed population could be identified, as described next. These cassettes were used to generate over 2,000 deletion mutant strains of *S. cerevisiae*. PCR (section 14.3); Creating genetic variability: Recombination at the molecular level (section 13.4)

The phenotypes of these "yeast knock-out" (YKO) mutants were determined in parallel competition assays. Here, large mixtures of different deletion mutants were cultured together under specific environmental conditions (e.g., high temperature or pH). The culture was periodically sampled, and PCR was used to amplify the molecular bar code of each deletant. Thus changes in mutant populations were followed over time. If the deleted gene was important for survival and growth under the test conditions, the deletant mutant missing that gene grew more slowly and the abundance of this mutant eventually declined. To assess the relative number of each mutant strain, a mixture of PCR-amplified molecular bar codes was hybridized to their complementary sequences on a microarray. The intensity of each spot on the microarray reflected the population size of each strain in the mixture. Mutations in genes whose products promote growth in the conditions being studied gave a less intense signal than did those whose products are not required.

Using this approach, many genes have been identified, including those that are involved in growth in rich and minimal media, at different pH values, and in the presence of certain drugs. Indeed, by the end of 2003, 80% of the annotated yeast genes had known functions (although some of these were revealed by analysis of newly identified orthologous genes) and all the genes on the genome are expected to be "known" sometime in 2007. Clearly, this approach to dissecting a genome provides important infor-

Figure 15.13 DNA Cassette Used in the Construction of Yeast-Deletion Mutants. The *KanMX* gene encodes resistance to the antifungal agent geneticin so yeast mutants can be selected and maintained. The blue sequences flanking the *KanMX* gene are the same as the sequences on either side of the ORF to be replaced. The molecular bar code is an additional genetic element inserted in each cassette so individual mutants can be identified by PCR amplification of genomic DNA.

mation, and similar projects are under way for *E. coli* and the endospore-forming bacterium *Bacillus subtilis*. Updates of these projects are posted on the Internet.

Genomic Analysis of Pathogenic Microbes

Analysis of the genomes of plant and animal pathogens provides considerable information about the evolution of virulence, host-pathogen interactions, potential treatment methods, and vaccine development. The genomes of numerous pathogens have been sequenced and compared to the genomes of phylogenetically related pathogens as well as close relatives that are not pathogenic.

The genome of *Mycoplasma genitalium* was one of the first to be sequenced. This microbe infects cells in the human genital and respiratory tracts. It has a genome of only 580 kilobases (0.58 Mb), one of the smallest genomes of any organism (**figure 15.14**). Thus the sequence data are of great interest because they help establish the minimal set of genes needed for free-living existence. There appear to be approximately 517 genes (480 protein-encoding genes and 37 genes for RNA species). About 90 proteins are involved in translation, and only around 29 proteins for DNA replication. Interestingly, 140 genes, or 29% of those in the genome, code for membrane proteins, and up to 4.5% of the genes seem to be involved in evasion of host immune responses. Only 5 genes have regulatory functions. Even in this small genome, about one-fifth of the genes do not match any known protein sequence. Comparison with the *M. pneumoniae* and *Ureaplasma urealyticum* genomes and studies of gene inactivation by transposon insertion suggest that about 108 to 121 *M. genitalium* genes may not be essential for survival. Thus for this microbe, the minimum gene set required for laboratory growth conditions seems to be approximately 300 genes; about 100 of these have unknown functions.

Haemophilus influenzae has a much larger genome, 1.8 megabases and 1,743 genes (**figure 15.15**). More than one-third of the genes have unknown functions. The bacterium lacks three Krebs cycle genes and thus a functional cycle. It devotes many genes (64 genes) to regulatory functions. *Haemophilus influenzae* is a species capable of natural transformation. The process must be very important to this bacterium because it contains 1,465 copies of the recognition sequence used in DNA uptake during transformation. DNA transformation (section 13.8)

One of the most alarming trends in the treatment of infectious disease is the rise of antibiotic-resistant bacteria. This is particularly true among the staphylococci. These gram-positive microbes cause an estimated 1 million serious infections each year. The two predominant opportunistic pathogens are *Staphylococcus epidermidis* and *S. aureus*. *S. epidermidis* is commonly found on the skin and has emerged as a serious pathogen only in recent years. It has been found to infect implanted medical devices like artificial heart valves. In contrast, *S. aureus* is more aggressive and can cause conditions that range from minor skin infections to life-threatening abscesses, heart infections, and toxic shock syndrome. During the 1960s, methicillin and other semi-synthetic penicillin antibiotics were frequently prescribed to treat staphylococcal infections, giving rise to the development of meticillin-resistant *S. aureus* (MRSA) and *S. epidermidis* (MRSE). By 2005, 60% of *S. aureus* clinical isolates were resistant to methicillin; some strains were resistant to up to 20 different antibiotics! The only drug effective against such multiply resistant strains was vancomycin, but *S. aureus* strains resistant to vancomycin have now been isolated. Antimicrobial chemotherapy (chapter 34)

Strains of *S. aureus* and *S. epidermidis* that were initially isolated between 1960 and 1998 vary in their resistance to a number of antibiotics and their virulence levels. The genomes of a number of such strains have been used in comparative genomic analysis to track the evolution of antibiotic resistance and virulence. The origin of about 1,700 genes that are shared by all strains of both species is difficult to ascertain. However, most genes that are strain-specific appear to have been introduced by prophages, transposons, insertion sequences, and plasmid-mediated gene transfer. For instance, this analysis reveals that an *Enterococcus faecalis* transposon introduced vancomycin resistance into *S. aureus*. Both *S. aureus* and *S. epidermidis* have acquired the genes that encode a capsule made of glutamate polymers from the gram-positive pathogen *Bacillus anthracis,* the causative agent of anthrax. In all three species, the polyglutamate capsule is a major virulence factor. Direct contact diseases: Staphylococcal diseases (section 38.3)

The genome of another pathogen, *Bacillus anthracis,* underwent intense scrutiny following a series of letter-based bioterrorism attacks in the fall of 2001. Although the genes that encode the components of the anthrax toxin are encoded on plasmids, the *B. anthracis* chromosome has a number of virulence-enhancing genes

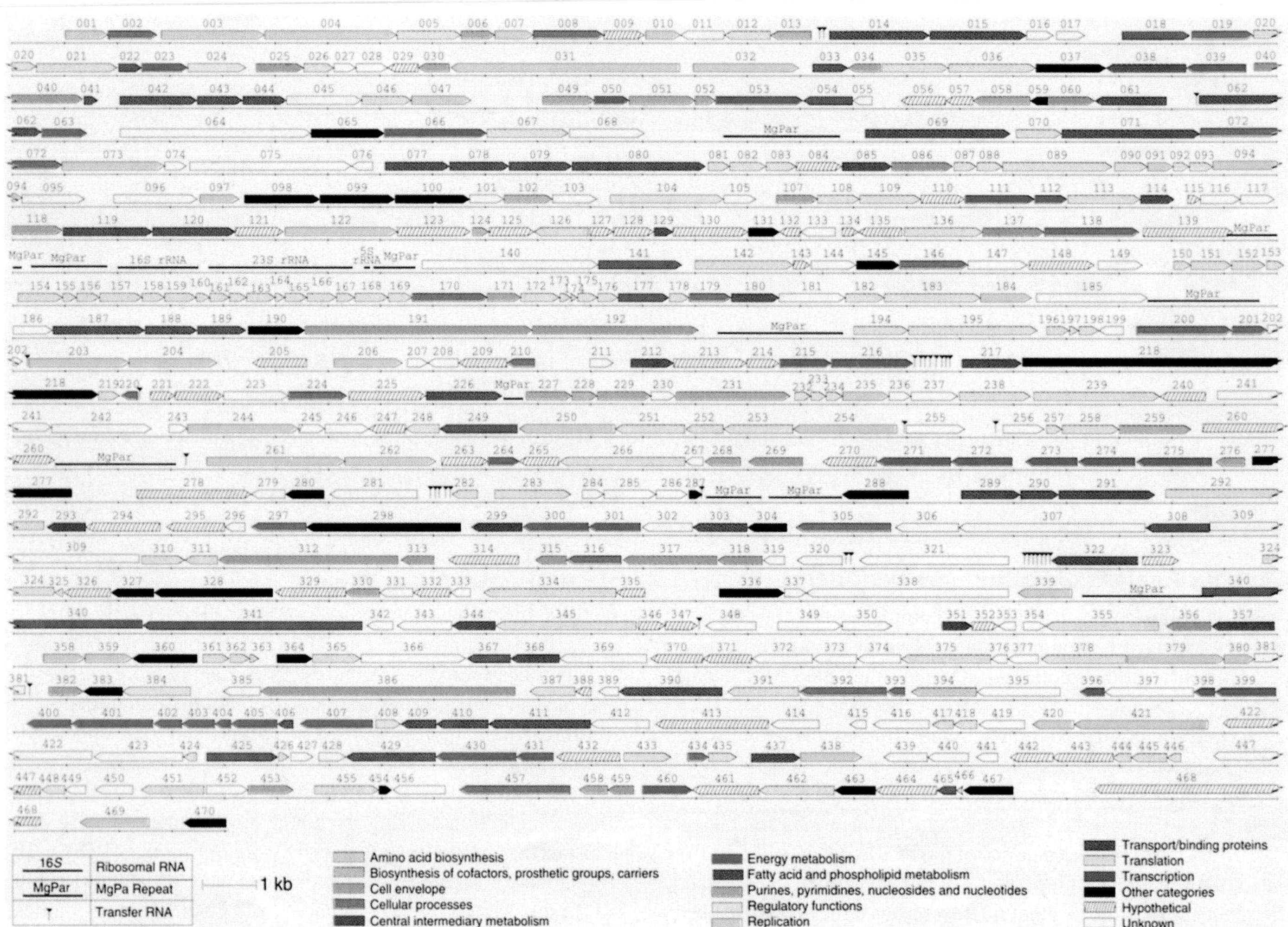

Figure 15.14 Map of the *Mycoplasma genitalium* Genome. The predicted coding regions are shown with the direction of transcription indicated by arrows. The genes are color coded by their functional role. The rRNA operon, tRNA genes, and adhesin protein operons (MgPa) are indicated. *Reprinted with permission from Fraser, C. M., et al. Copyright 1995. The minimal gene complement of* Mycoplasma genitalium. *Science 270:397–403. Figure 1, page 398 and The Institute for Genomic Research.*

with orthologs in the pathogens *B. cereus, B. thuringiensis,* and *Listeria monocytogenes*. Although the *B. anthracis* genome is most similar to that of *B. cereus* (a cause of food poisoning), the orthologs found on the *B. thuringiensis* genome are particularly interesting. *B. thuringiensis* is an insect pathogen and the source of Bt toxin, which is used as a commercial insecticide. Genomic evidence suggests that *B. anthracis* may be derived from an insect-infecting ancestor. *L. monocytogenes* is an intracellular pathogen and the genes found in both organisms may allow these pathogens to survive within host immune system cells called macrophages. Unlike the genomes of many other bacterial species, there is little variation in the nucleotide sequences among different strains of *B. anthracis*. However, careful comparative genome sequencing of several key strains provides clues regarding the origin of the *B. anthracis* strain used in the 2001 U.S. bioterrorism attacks (**figure 15.16**). Microbes as products: Biopesticides (section 41.8); Bioterrorism preparedness (section 36.9)

Genomics clearly helps us understand how antibiotic-resistant genes are shared and how new pathogens arise. It is also an important tool in the development of new antibiotics to combat the evergrowing threat of multiple drug resistant bacteria. Traditionally, new antimicrobial agents have been discovered by screening thousands of natural products for their capacity to inhibit bacterial growth. This is a labor-intensive process, and even when performed by robots, is not particularly efficient. However, genomics can be used to identify new structures or metabolites produced by specific pathogenic microbes. Scientists can then design new drugs based on these cellular components, which are called drug targets. This new approach of synthesizing drugs to interact with particular molecular targets is called **rational drug design** and as described previously for vaccine development, is greatly facilitated by the availability of annotated genomes.

Genomics has also promoted our understanding of pathogens that are difficult or impossible to culture and those with unusual life cycles. Chlamydiae are nonmotile, coccoid, gram-negative bacteria that reproduce only within cytoplasmic vesicles of eucaryotic cells by a unique life cycle. *Chlamydia trachomatis* infects humans and

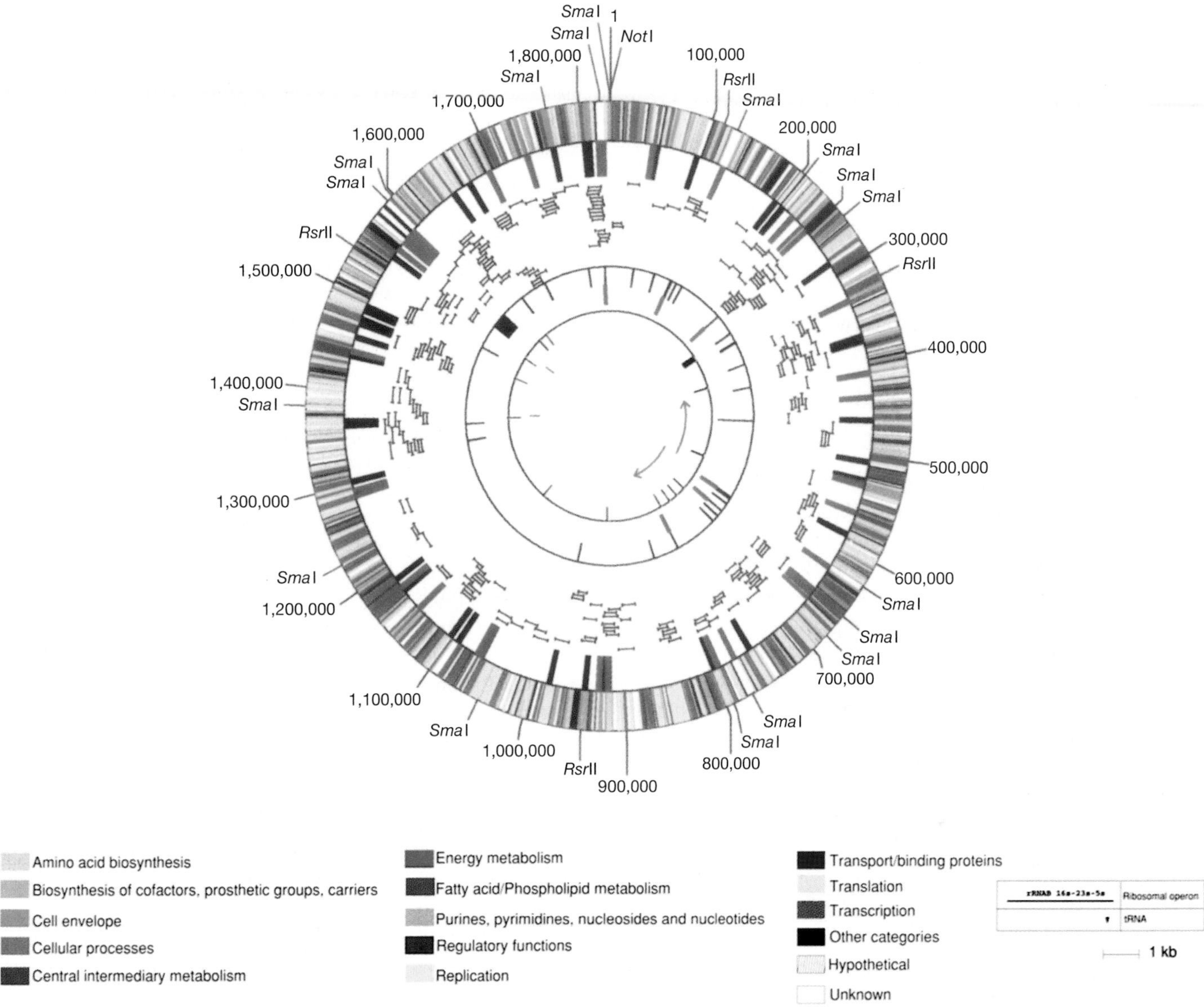

Figure 15.15 Map of the *Haemophilus influenzae* Genome. The predicted coding regions in the outer concentric circle are indicated with colors representing their functional roles. The outer perimeter shows the *NotI, RsrII,* and *SmaI* restriction sites. The inner concentric circle shows regions of high G + C content (red and blue) and high A + T content (black and green). The third circle shows the coverage by λ clones (blue). The fourth circle shows the locations of rRNA operons (green), tRNAs (black), and the mu-like prophage (blue). The fifth circle shows simple tandem repeats and the probable origin of replication (outward pointing green arrows). The red lines are potential termination sequences. *Reprinted with permission from Fleischman, R. D., et al. 1995. Whole-genome random sequencing and assembly of* Haemophilus influenzae *Rd.* Science *269:496–512. Figure 1, page 507 and The Institute of Genomic Research.*

causes the sexually transmitted disease nongonococcal urethritis, probably the most commonly transmitted sexual disease in the United States. It also is the leading cause of preventable blindness in the world. The bacterium's life cycle is so unusual that one would expect its genome to be somewhat atypical. Surprisingly, this is not the case. Microbiologists considered *Chlamydia* an "energy parasite" and believed that it obtained all its ATP from the host cell. The genome results show that *Chlamydia* has the genes to make at least some ATP on its own, although it also has genes for the transport of host ATP. The presence of enzymes for the synthesis of peptidoglycan was also unexpected because chlamydial cell walls lack peptidoglycan. Microbiologists had been unable to explain why the antibiotic penicillin, which disrupts peptidoglycan synthesis, is able to inhibit chlamydial growth. The presence of peptidoglycan biosynthetic enzymes helps account for the penicillin effect, but no one knows the purpose of peptidoglycan synthesis in this bacterium. Another major surprise is the absence of the *ftsZ* gene, which had been thought to be required by all *Bacteria* and *Archaea* for septum formation during cell division. The absence of this gene makes one wonder how *Chlamydia* divides. It may be that some of the genes

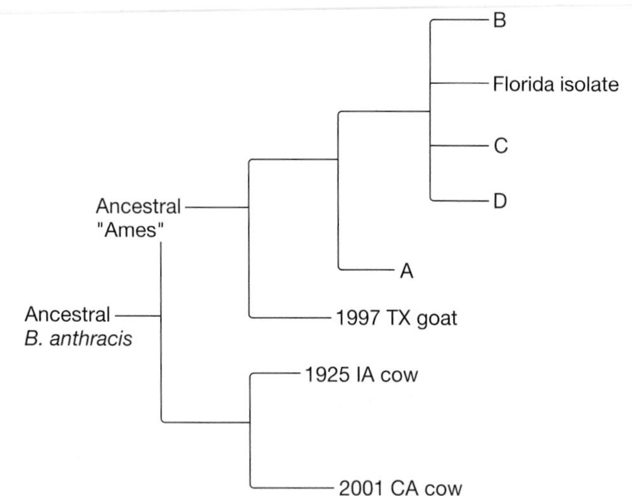

Figure 15.16 Proposed Phylogenetic Tree of the *B. anthracis* Ames Isolate Used in U.S. Bioterror Attacks in the Autumn of 2001. Isolates A–D are laboratory strains; other isolates are indicated by state and year of initial cultivation. Isolates B and C are identical to the isolate recovered from the first victim of the attacks (Florida isolate). Isolate D differs only by the insertion of a single adenine in a key region used for comparison. Laboratory strain A has two additional bases and has lost one of the toxin-encoding plasmids. This whole-genome analysis demonstrated the presence of four genetic elements on the genome that vary, despite previous analysis suggesting that these strains were identical or nearly identical. This approach is a powerful new technique for tracking infectious disease outbreaks.

with unknown functions play a major role in cell division. Perhaps *Chlamydia* employs a mechanism of cell division different from that of other procaryotes. Finally, the genome contains at least 20 genes that have been obtained from eucaryotic host cells (most bacteria have no more than 3 or 4 such genes). Some of these genes are plantlike; originally *Chlamydia* may have infected a plantlike host and then moved to animals. Phylum *Chlamydiae* (section 21.5); The procaryotic cell cycle: Cytokinesis (section 6.1)

One of the most difficult human pathogens to study is the causative agent of syphilis, *Treponema pallidum*. This is because it is not possible to grow *T. pallidum* outside the human body. We know little about its metabolism or the way it avoids host defenses, and no vaccine for syphilis has yet been developed. Naturally, the sequencing of the *T. pallidum* genome generated considerable excitement and hope. It turns out that *T. pallidum* is metabolically crippled. It can use carbohydrates as an energy source, but lacks the TCA cycle and oxidative phosphorylation enzymes (**figure 15.17**). *T. pallidum* also lacks many biosynthetic pathways (e.g., for enzyme cofactors, fatty acids, nucleotides, and some electron transport proteins) and must rely on molecules supplied by its host. In fact, about 5% of its genes code for transport proteins. Given the lack of several critical pathways, it is not surprising that the pathogen has not been cultured successfully. The genes for surface proteins are of particular interest. *T. pallidum* has a family of surface protein genes characterized by many repetitive sequences. Some have speculated that these genes might undergo recombination in order to generate new surface proteins, enabling the organism to avoid attack by the immune system. It may be possible to develop a vaccine for syphilis using some of the newly discovered surface proteins. We also may be able to identify strains of *T. pallidum* using these surface proteins, which would be of great importance in syphilis epidemiology. The genome results should ultimately help us understand how *T. pallidum* causes syphilis. About 40% of the genes have unknown functions. Possibly some of them are responsible for avoiding host defenses and for the production of toxins and other virulence factors. Direct contact diseases: Sexually transmitted diseases (section 38.3)

For centuries, tuberculosis has been one of the major scourges of humankind. About one-third of the human population is infected with the causative agent *Mycobacterium tuberculosis*. After establishing residence in immune system cells in the lung, it often remains in a dormant state until the host's immune system is compromised. The disease kills about 3 million people annually and is the direct cause of death for many AIDS patients. Predictably, *M. tuberculosis* is becoming ever more drug resistant. Genome studies could be of great importance in the fight to control the renewed spread of tuberculosis. The annotated *M. tuberculosis* genome was published in 1998; at that time it was predicted to have 3,974 genes. In 2002, its genome was reexamined and, based largely on inspection of small ORFs, an additional 82 genes were identified. The number of published genomes during the intervening four years is reflected by the change in the number of genes of unknown function: in 1998 there were 606 genes for which no function or ortholog could be assigned; by 2002, there were only 272 such genes. Significantly, the majority of these genes remain hypothetical because although orthologs were found, no functional assignment has yet been made. Undoubtedly, as more genomes are sequenced and proteomes examined, the number of genes of unknown function for all organisms whose genomes have been sequenced will decline.

What has the genome sequence of this ancient pathogen revealed? More than 250 genes are devoted to lipid metabolism (*E. coli* has only about 50 such genes), and *M. tuberculosis* may obtain much of its energy by degrading host lipids. There are a surprisingly large number of regulatory elements in the genome. This may mean that the infectious process is much more complex than previously thought. Two families of novel glycine-rich proteins with unknown functions are present and represent about 10% of the genome. They may be a source of antigenic variation involved in defense against the host immune system. One major medical problem has been the lack of a highly effective vaccine. A large number of proteins that are either secreted by the bacterium or on the bacterial surface have been identified from the genome sequence. It is hoped that some of these proteins can be used to develop better vaccines. This is particularly important in view of the spread of multiply drug resistant *M. tuberculosis*. *Mycobacterium tuberculosis* (sections 24.4 and 38.1)

The *M. tuberculosis* genome has been compared to the genomes of two relatives—*M. leprae,* which causes leprosy, and *M. bovis,* the causative agent of tuberculosis in a wide range of animals, including cows and humans. The *M. bovis* and *M. leprae* genomes differ from that of *M. tuberculosis* in some important ways. The genomes of *M. bovis* and *M. tuberculosis* are most similar—about 99.5% identical at the sequence level. However, the *M.*

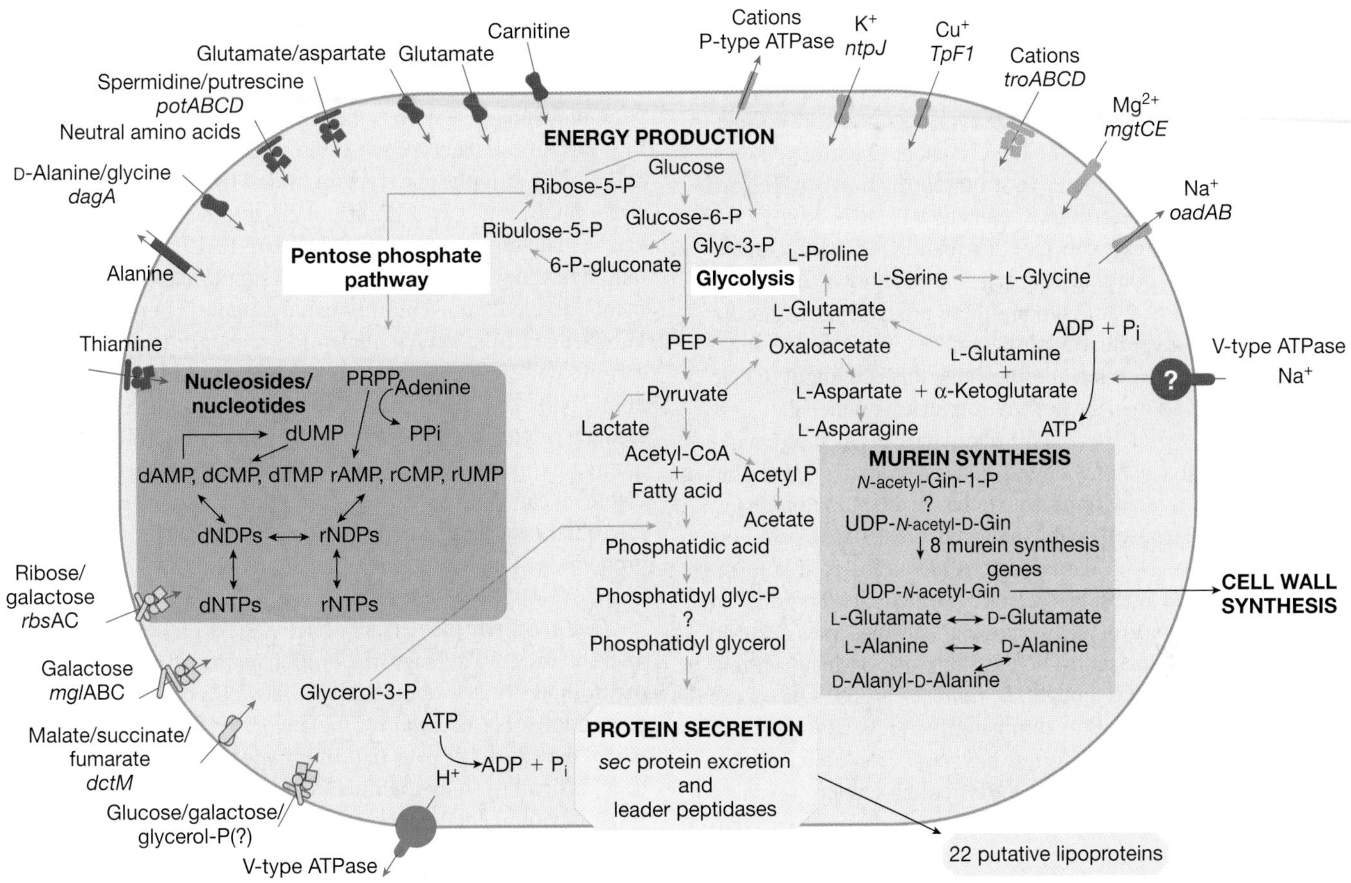

Figure 15.17 Metabolic Pathways and Transport Systems of *Treponema pallidum*. This depicts *T. pallidum* metabolism as deduced from genome annotation. Note the limited biosynthetic capabilities and extensive array of transporters. Although glycolysis is present, the TCA cycle and respiratory electron transport are lacking. Question marks indicate where uncertainties exist or expected activities have not been found.

bovis genome is missing 11 separate regions, making its genome slightly smaller (4.3 Mb vs. 4.4. Mb). The sequence dissimilarities involve the inactivation of some genes, leading to major differences in the way the two bacteria respond to environmental conditions. This may account for the host range differences between these two closely related pathogens. The divergence between *M. tuberculosis* and *M. leprae* is even more striking. The *M. leprae* genome is a third smaller than that of *M. tuberculosis.* About half the genome is devoid of functional genes. Instead, there are over 1,000 degraded, nonfunctional genes called pseudogenes. In total, *M. leprae* seems to have lost as many as 2,000 genes during its career as an intracellular parasite. It even lacks some of the enzymes required for energy production and DNA replication. This might explain why the bacterium has such a long doubling time, about two weeks in mice. One hope from genomics studies is that critical surface proteins can be discovered and used to develop a sensitive test for early detection of leprosy. This would allow immediate treatment of the disease before nerve damage occurs. Direct contact diseases: Leprosy (section 38.3)

Another microbe that has reduced the size of its genome is *Rickettsia prowazekii,* a member of the α-proteobacteria, and an obligate intracellular parasite of lice and humans. It is the causative agent of typhus, a disease that killed millions during the First and Second World Wars. It lacks genes for glycolysis and many genes for the biosynthesis of amino acids and nucleosides. Like *M. leprae,* its genome contains a number of pseudogenes. Gene inactivation and deletion have also been observed in plant pathogens. For instance, the plant pathogen *Phytoplasma asteris* is transmitted by insects and has an intracellular life cycle in both its insect vectors and plant hosts. It lacks many genes related to amino acid, nucleotide, and fatty acid biosynthesis. It also lacks many genes that function in energy metabolism. Presumably, microbes that obtain nutrients and ATP from their hosts lack the selective pressure needed to maintain the corresponding functional genes. Arthropod-borne diseases (section 38.2)

Genomic Analysis of Extremophiles

Microbes that live in harsh environments are called extremophiles. Genomic analysis of such microbes has been pursued with the goal of understanding how organisms can tolerate extreme environmental conditions. The genomes of over a dozen thermophiles and hyperthermophiles, including members of both the *Bacteria* and *Archaea,* have been analyzed. Originally, it was postulated that there would be a strong correlation between G+C content and optimal growth temperature because GC base pairs, which share three hydrogen bonds, are more stable at higher temperatures than AT base pairs, which share only two. However, no significant correlation has been discerned when the G+C content of entire genomes was compared. Curiously, there is a strong correlation between optimal growth temperature and G+C content of the tRNA and rRNA genes; however, the significance of this finding is not fully understood. It also seems odd that few genes are common to all thermophiles. One that is shared among these procaryotes is that for reverse gyrase. In hyperthermophilic archaea, this enzyme functions to relax the supercoiled chromosome; this is thought to help prevent its denaturation in the hot environments in which they live.

Deinococcus radiodurans is a bacterium that has the remarkable ability to survive not only desiccation and oxidizing agents, but γ-radiation at doses many times above that needed to kill humans. Ionizing radiation causes double-stranded breaks in DNA—the most lethal form of DNA damage. *D. radiodurans* is able to reassemble its genome after it has been fragmented into thousands of pieces. Its genome consists of two circular chromosomes (2.6 Mb and 0.4 Mb), a megaplasmid (177,466 bp), and a small plasmid (45,704 bp). Surely, it was thought, *D. radiodurans* must be superb in executing DNA repair. But, surprisingly, comparative genomic analysis shows that *D. radiodurans* has *fewer* DNA repair genes than does *E. coli.* To understand these findings, microbiologists used microarrays with about 94% of all *D. radiodurans* genes represented to examine the **transcriptome** (all the mRNA present) following radiation treatment. Such analysis generates information for each gene. This is organized by **hierarchical cluster analysis** wherein induced genes (red spots) are grouped separately from repressed genes (green spots); genes whose expression remains unaltered are shown in black (**figure 15.18**). These clusters are then scanned for genes that are known to have similar functions and then further grouped by relatedness (degree of relatedness is statistically quantified by a correlation coefficient, or "r value"). Such analysis has confirmed that the DNA repair gene *recA,* as well as genes involved in DNA replication and recombination, are dramatically upregulated following irradiation. In addition, genes whose products direct cell wall metabolism, cellular transport, and many genes whose protein products are unknown are also induced. These results make it clear that *D. radiodurans*' ability to survive such high levels of radiation is more complex than originally thought, requiring the coordination of a complex network of processes involving both DNA repair and metabolic activity. *Deinococcus-Thermus* (section 21.2)

15.9 Environmental Genomics

It is clear that the genomic era has ushered in new ways of answering questions. Perhaps nowhere is this more evident than in the growing field of **environmental genomics,** sometimes called **metagenomics.** While the dominant role of microorganisms in driving the nutrient cycles that support life on Earth has long been recognized, efforts to comprehensively understand microbial communities have been stymied by the fact that only about 1% of all procaryotes have been cultured in laboratory conditions. New genomic techniques offer cultivation-independent approaches to study microbial biodiversity. Environmental genomics can be used to take a census of microbial populations, as well as to discern the presence and abundance of certain classes of genes. That is to say, genomics can ask, "Who is there and what are they doing?" To do this, DNA fragments are extracted directly from the environment and cloned into plasmid vectors, much like genome library construction. In this way a library of environmental DNA fragments can be maintained and amplified (**figure 15.19**). Alternatively, certain genes may be obtained by PCR amplification of DNA fragments derived from environmental samples. For this approach, knowledge of the gene's nucleotide sequence is required; this is common when amplifying genes that encode 16S rRNA for taxonomic purposes. In either case, one produces a stable source of nucleotide sequences reflecting the diversity of microbes growing in nature, not just those that can be grown in the laboratory. The nucleotides can then be sequenced and analyzed, or expressed in a microbial host and screened for a specific function, such as the production of novel antimicrobial compounds. Techniques for determining microbial taxonomy and phylogeny (section 19.4)

One field which metagenomics has revolutionized is marine microbiology. An average of 1 million microbial cells can be found per milliliter of seawater. While it has long been recognized that marine microbes account for the majority of the oceans' biomass, where they perform over half of the global photosynthesis, it has been difficult to study their taxonomic and metabolic diversity. In 2000, environmental genomics led to the discovery of a new procaryotic gene that encodes a protein in the rhodopsin family. Rhodopsins convert light energy directly into an electrical gradient across a cell membrane. It had long been held that the *Archaea* were the only procaryotes to produce a protein in the rhodopsin family. When rhodopsin-like genes were amplified from a variety of procaryotic taxa, they were called proteorhodopsins because they were found in γ-proteobacteria (genes for proteorhodopsin have since been found in α-proteobacteria as well). The nature of microbial metabolic diversity in the sea is now being reconsidered as it is estimated that about 13% of marine bacteria may have the genes to encode rhodopsin-based, light-driven proton pumps. These pumps enable bacteria to produce a proton motive force that can fuel the production of ATP. Likewise, marine nitrogen budgets may need to be recalculated based on the discovery that the gene encoding nitrogenase, the enzyme that converts gaseous N_2 to ammonium, is present in far greater numbers among marine cyanobacteria than previously thought. Rhodopsin-based phototrophy (section 9.12); Phylum *Euryarchaeota:* The halobacteria (section 20.3); Biogeochemical cycling: The nitrogen cycle (section 27.2)

An ambitious metagenomics project was performed by J. Craig Venter, Hamilton Smith, and colleagues. They wanted to determine the procaryotic biodiversity of the Sargasso Sea, that portion of the Atlantic Ocean that surrounds Bermuda. They collected seawater and used filtration to exclude viruses ($<0.2\mu m$) and eucaryotes ($>0.8\mu m$). An environmental genomic library was prepared from DNA extracted from the remaining seawater. After sequencing over 1 *billion* base pairs, followed by manual and computer analysis to determine sequence relatedness, it was determined that at least 1,800 "genomic species," called **phylotypes,** were represented. Among these, about 145 phylotypes were previously unknown and

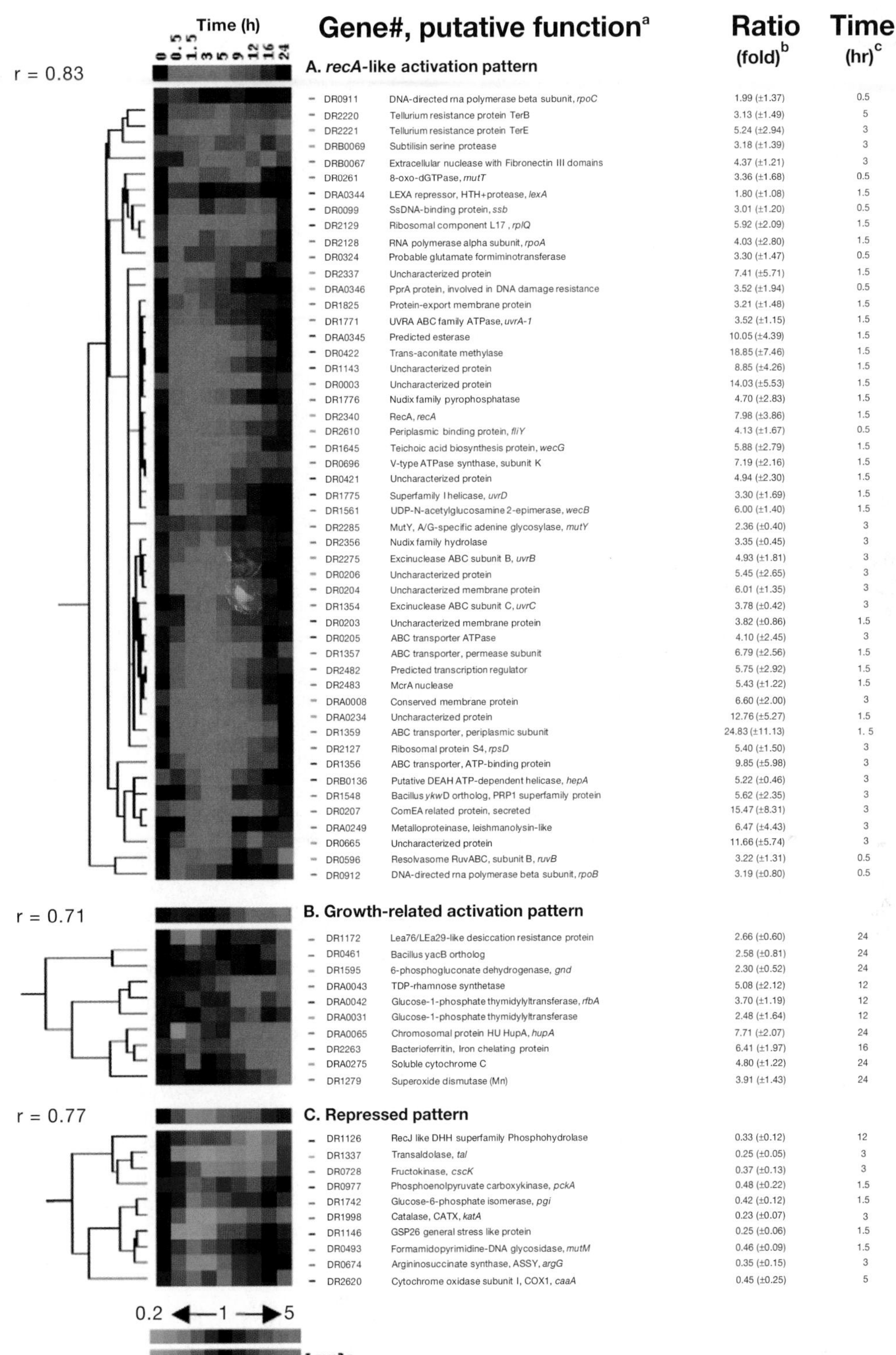

Time (h)
0 0.5 1.5 3 5 9 12 16 24
Gene#, putative function[a]
Ratio (fold)[b]
Time (hr)[c]
r = 0.83
A. recA-like activation pattern
DR0911 DNA-directed rna polymerase beta subunit, rpoC 1.99 (±1.37) 0.5
DR2220 Tellurium resistance protein TerB 3.13 (±1.49) 5
DR2221 Tellurium resistance protein TerE 5.24 (±2.94) 3
DRB0069 Subtilisin serine protease 3.18 (±1.39) 3
DRB0067 Extracellular nuclease with Fibronectin III domains 4.37 (±1.21) 3
DR0261 8-oxo-dGTPase, mutT 3.36 (±1.68) 0.5
DRA0344 LEXA repressor, HTH+protease, lexA 1.80 (±1.08) 1.5
DR0099 SsDNA-binding protein, ssb 3.01 (±1.20) 0.5
DR2129 Ribosomal component L17, rplQ 5.92 (±2.09) 1.5
DR2128 RNA polymerase alpha subunit, rpoA 4.03 (±2.80) 1.5
DR0324 Probable glutamate formiminotransferase 3.30 (±1.47) 0.5
DR2337 Uncharacterized protein 7.41 (±5.71) 1.5
DRA0346 PprA protein, involved in DNA damage resistance 3.52 (±1.94) 0.5
DR1825 Protein-export membrane protein 3.21 (±1.48) 1.5
DR1771 UVRA ABC family ATPase, uvrA-1 3.52 (±1.15) 1.5
DRA0345 Predicted esterase 10.05 (±4.39) 1.5
DR0422 Trans-aconitate methylase 18.85 (±7.46) 1.5
DR1143 Uncharacterized protein 8.85 (±4.26) 1.5
DR0003 Uncharacterized protein 14.03 (±5.53) 1.5
DR1776 Nudix family pyrophosphatase 4.70 (±2.83) 1.5
DR2340 RecA, recA 7.98 (±3.86) 1.5
DR2610 Periplasmic binding protein, fliY 4.13 (±1.67) 0.5
DR1645 Teichoic acid biosynthesis protein, wecG 5.88 (±2.79) 1.5
DR0696 V-type ATPase synthase, subunit K 7.19 (±2.16) 1.5
DR0421 Uncharacterized protein 4.94 (±2.30) 1.5
DR1775 Superfamily I helicase, uvrD 3.30 (±1.69) 1.5
DR1561 UDP-N-acetylglucosamine 2-epimerase, wecB 6.00 (±1.40) 1.5
DR2285 MutY, A/G-specific adenine glycosylase, mutY 2.36 (±0.40) 3
DR2356 Nudix family hydrolase 3.35 (±0.45) 3
DR2275 Excinuclease ABC subunit B, uvrB 4.93 (±1.81) 3
DR0206 Uncharacterized protein 5.45 (±2.65) 3
DR0204 Uncharacterized membrane protein 6.01 (±1.35) 3
DR1354 Excinuclease ABC subunit C, uvrC 3.78 (±0.42) 3
DR0203 Uncharacterized membrane protein 3.82 (±0.86) 1.5
DR0205 ABC transporter ATPase 4.10 (±2.45) 3
DR1357 ABC transporter, permease subunit 6.79 (±2.56) 1.5
DR2482 Predicted transcription regulator 5.75 (±2.92) 1.5
DR2483 McrA nuclease 5.43 (±1.22) 1.5
DRA0008 Conserved membrane protein 6.60 (±2.00) 3
DRA0234 Uncharacterized protein 12.76 (±5.27) 1.5
DR1359 ABC transporter, periplasmic subunit 24.83 (±11.13) 1. 5
DR2127 Ribosomal protein S4, rpsD 5.40 (±1.50) 3
DR1356 ABC transporter, ATP-binding protein 9.85 (±5.98) 3
DRB0136 Putative DEAH ATP-dependent helicase, hepA 5.22 (±0.46) 3
DR1548 Bacillus ykwD ortholog, PRP1 superfamily protein 5.62 (±2.35) 3
DR0207 ComEA related protein, secreted 15.47 (±8.31) 3
DRA0249 Metalloproteinase, leishmanolysin-like 6.47 (±4.43) 3
DR0665 Uncharacterized protein 11.66 (±5.74) 3
DR0596 Resolvasome RuvABC, subunit B, ruvB 3.22 (±1.31) 0.5
DR0912 DNA-directed rna polymerase beta subunit, rpoB 3.19 (±0.80) 0.5
r = 0.71
B. Growth-related activation pattern
DR1172 Lea76/LEa29-like desiccation resistance protein 2.66 (±0.60) 24
DR0461 Bacillus yacB ortholog 2.58 (±0.81) 24
DR1595 6-phosphogluconate dehydrogenase, gnd 2.30 (±0.52) 24
DRA0043 TDP-rhamnose synthetase 5.08 (±2.12) 12
DRA0042 Glucose-1-phosphate thymidylyltransferase, rfbA 3.70 (±1.19) 12
DRA0031 Glucose-1-phosphate thymidylyltransferase 2.48 (±1.64) 12
DRA0065 Chromosomal protein HU HupA, hupA 7.71 (±2.07) 24
DR2263 Bacterioferritin, Iron chelating protein 6.41 (±1.97) 16
DRA0275 Soluble cytochrome C 4.80 (±1.22) 24
DR1279 Superoxide dismutase (Mn) 3.91 (±1.43) 24
r = 0.77
C. Repressed pattern
DR1126 RecJ like DHH superfamily Phosphohydrolase 0.33 (±0.12) 12
DR1337 Transaldolase, tal 0.25 (±0.05) 3
DR0728 Fructokinase, cscK 0.37 (±0.13) 3
DR0977 Phosphoenolpyruvate carboxykinase, pckA 0.48 (±0.22) 1.5
DR1742 Glucose-6-phosphate isomerase, pgi 0.42 (±0.12) 1.5
DR1998 Catalase, CATX, katA 0.23 (±0.07) 3
DR1146 GSP26 general stress like protein 0.25 (±0.06) 1.5
DR0493 Formamidopyrimidine-DNA glycosidase, mutM 0.46 (±0.09) 1.5
DR0674 Argininosuccinate synthase, ASSY, argG 0.35 (±0.15) 3
DR2620 Cytochrome oxidase subunit I, COX1, caaA 0.45 (±0.25) 5
0.2 1 5
Levels

Figure 15.18 Hierarchical Cluster Analysis of Gene Expression of *D. radiodurans* Following Exposure to γ-Radiation. Each row of colored strips represents a single gene and the color indicates the level of expression over nine time intervals. The far-left column is the control and thus is black (at control levels of expression). The level of induction or repression relative to the control value is indicated as the Ratio (fold). The time indicates the number of hours after radiation exposure that the ratio was calculated. Each group of genes has been scored for relatedness and a "tree" has been generated on the far left of the clusters, with the indicated correlation coefficient (r value). A large number of genes encoding DNA repair, synthesis, and recombination proteins are induced upon radiation. These are grouped together. Fewer genes that encode proteins involved in metabolism are induced. Finally, genes involved in other aspects of metabolism are repressed.

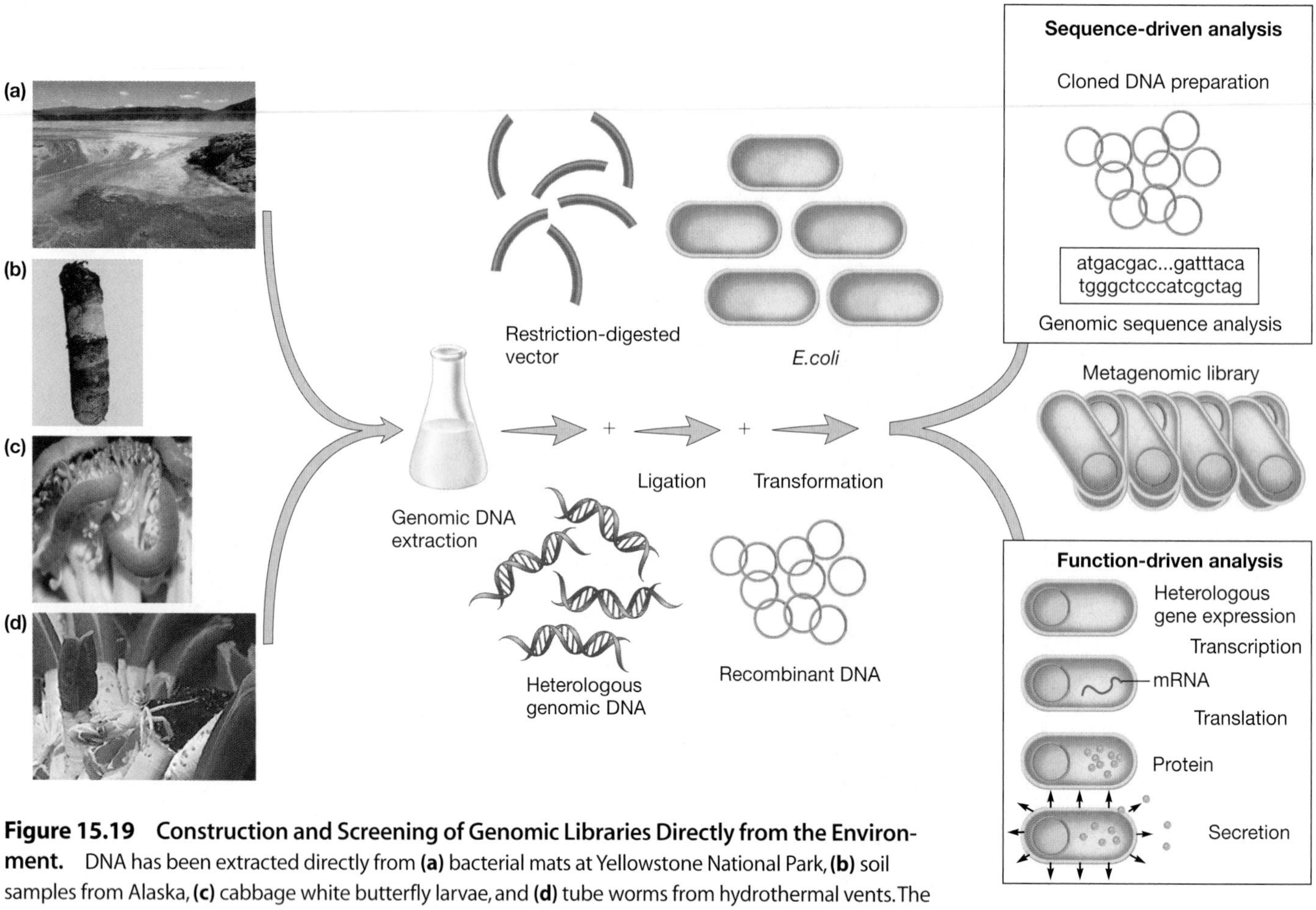

(a)

(b)

(c)

(d)

Restriction-digested vector

E.coli

Genomic DNA extraction

Ligation

Transformation

Heterologous genomic DNA

Recombinant DNA

Sequence-driven analysis

Cloned DNA preparation

atgacgac...gatttaca
tgggctcccatcgctag

Genomic sequence analysis

Metagenomic library

Function-driven analysis

Heterologous gene expression

Transcription

mRNA

Translation

Protein

Secretion

Figure 15.19 Construction and Screening of Genomic Libraries Directly from the Environment. DNA has been extracted directly from **(a)** bacterial mats at Yellowstone National Park, **(b)** soil samples from Alaska, **(c)** cabbage white butterfly larvae, and **(d)** tube worms from hydrothermal vents. The DNA is cloned into suitable vectors and transformed into a bacterial host. Sequences or gene products can then be analyzed.

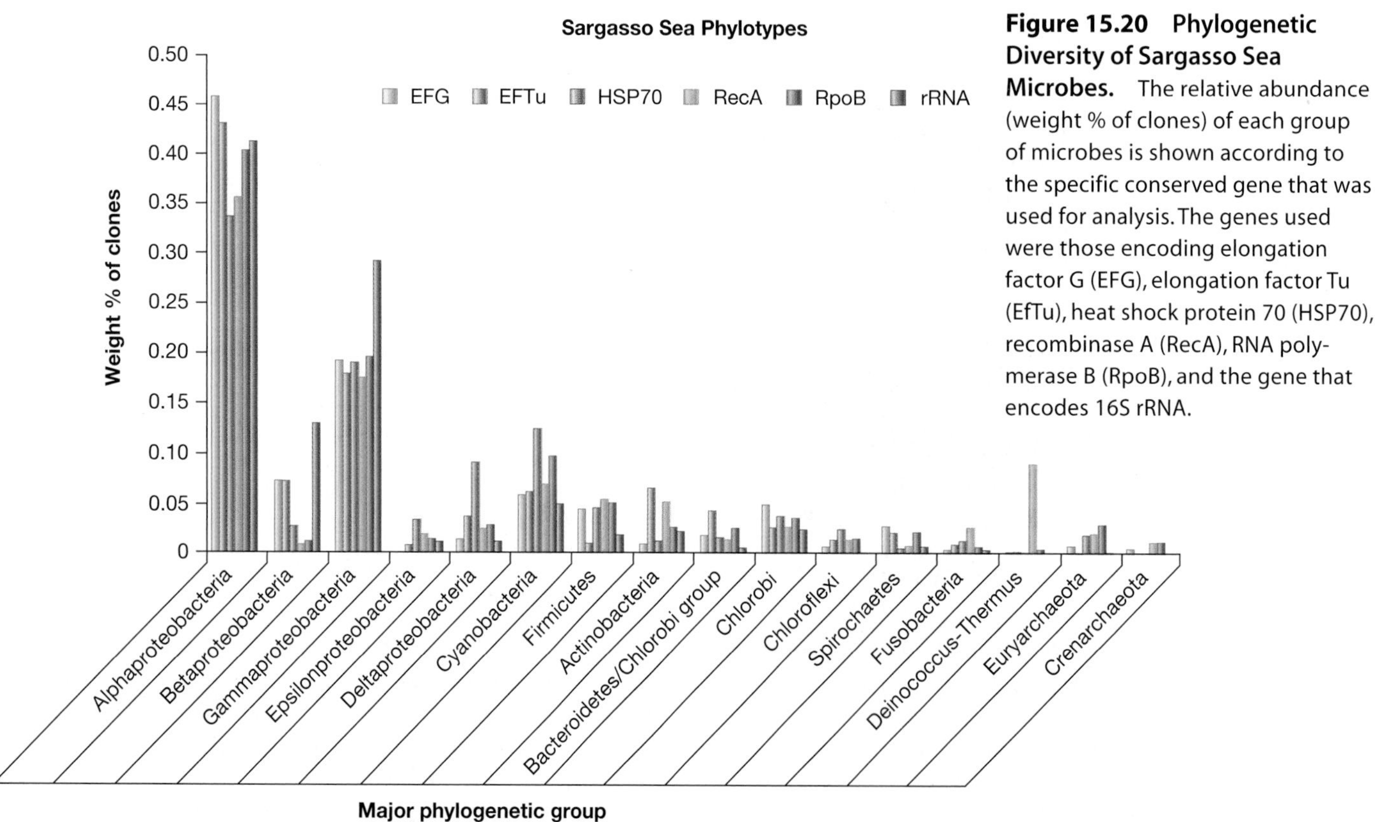

Figure 15.20 Phylogenetic Diversity of Sargasso Sea Microbes. The relative abundance (weight % of clones) of each group of microbes is shown according to the specific conserved gene that was used for analysis. The genes used were those encoding elongation factor G (EFG), elongation factor Tu (EfTu), heat shock protein 70 (HSP70), recombinase A (RecA), RNA polymerase B (RpoB), and the gene that encodes 16S rRNA.

most likely represent new species. Phylogenetic diversity was further evaluated by PCR amplification of genes, such as the recombinase gene *recA,* the 16S rRNA gene, and the gene that encodes RNA polymerase B (*rpoB*). The sequences for these genes are highly conserved, making them good candidates for assessing species diversity (**figure 15.20**). Amazingly, Venter, Smith, and their collaborators report the discovery of 1.2 million previously unknown genes (however, this number is controversial), including over 700 new proteorhodopsin-like photoreceptors from taxa not previously known to possess light-harvesting capabilities. Certainly, these results demonstrate the power of metagenomics and show that much more study is needed before we can fully appreciate the biological diversity in the world's oceans.

1. How can genomic sequencing be used to ask specific questions about the physiology of a given microbe? List three interesting or surprising results obtained from genomic analysis of pathogens.
2. Define lateral gene transfer. How might LGT be partly responsible for the rapid rise in antibiotic-resistant microbes?
3. For what types of microorganisms is extensive gene loss common? What is the most likely explanation for this phenomenon?
4. How might environmental genomics be used in expanding our knowledge of terrestrial microbial communities? How might environmental genomics be used by the biotechnology industry to develop new medically or industrially important natural products?

Summary

15.1 Introduction

a. Genomics is the study of the molecular organization of genomes, their information content, and the gene products they encode. It may be divided into three broad areas: structural genomics, functional genomics, and comparative genomics.

15.2 Determining DNA Sequences

a. DNA fragments are normally sequenced using dideoxynucleotides and the Sanger chain termination technique (**figure 15.2**).

15.3 Whole-Genome Shotgun Sequencing

a. Most often microbial genomes are sequenced using the whole-genome shotgun technique of Venter, Smith, and collaborators. Four stages are involved: library construction, sequencing of randomly produced fragments, fragment alignment and gap closure, and editing the final sequence (**figure 15.3**).

15.4 Bioinformatics

a. Analysis of vast amounts of genome data requires sophisticated computers and programs; these analytical procedures are a part of the discipline of bioinformatics.
b. Bioinformatic software enables the comparison of genes within genomes to identify paralogs, and genes between different organisms to identify orthologs.
c. Annotation of genomes can be used to identify many genes and their function, but the functional role of 35 to 50% of all ORFs on a given genome usually cannot be discerned (**figure 15.4**).

15.5 Functional Genomics

a. Functional genomics is used to reveal genome structure and function relationships (**figure 15.5**).
b. DNA microarrays (gene chips) can be used to assess gene expression as a measure of individual gene transcripts (mRNA). Gene expression can be determined for mutant versus wild-type strains or for a given organism under specific environmental conditions (**figure 15.9**).

15.6 Comparative Genomics

a. Comparing genome sequences reveals much information about genome structure and evolution, including the importance of lateral gene transfer.
b. Comparative genomics is an important tool in discerning how microbes have adapted to particular ecological niches and in the development of new therapeutic agents such as vaccines.

15.7 Proteomics

a. The entire collection of proteins that an organism can produce is its proteome, and its study is called proteomics.
b. The proteome is often analyzed by two-dimensional gel electrophoresis, in which the total cellular protein pool can be visualized. In many cases, the amino acid sequence of individual proteins is determined by mass spectrometry; if this is coupled to genomics, both a protein of interest and the gene that encodes it can be identified (**figures 15.11** and **15.12**).
c. Structural proteomics seeks to model the three-dimensional structure of proteins based on computer analysis of amino acid sequence data.

15.8 Insights from Microbial Genomes

a. One approach to identifying genes of unknown function is to construct a bank of mutants and study their competitive phenotypes. This has been done for the yeast *Saccharomyces cerevisiae,* and this approach is being pursued for procaryotes as well.
b. The complete genomes of many pathogens have been analyzed, providing information about virulence and evolution. In some cases, potential targets for new therapies and vaccines have been identified.
c. The genome of *Mycoplasma genitalium* is one of the smallest of any free-living organism. Analysis of this genome and others indicates that only about 265 to 350 genes are required for growth in the laboratory.
d. *Haemophilus influenzae* lacks a complete set of Krebs cycle genes and has 1,465 copies of the recognition sequence used in DNA uptake during transformation.
e. The genome of *Chlamydia trachomatis* has provided many surprises. For example, it appears able to make at least some ATP and peptidoglycan, despite the fact that it seems to obtain most ATP from the host and does not have a cell wall with peptidoglycan. The presence of plantlike genes indicates that it might have infected plantlike hosts before moving to animals.
f. *Treponema pallidum,* the causative agent of syphilis, has lost many of its metabolic genes, which may explain why it hasn't been cultivated outside a host.
g. *Mycobacterium tuberculosis* contains more than 250 genes for lipid metabolism and may obtain much of its energy from host lipids. Surface and secretory proteins have been identified and may help vaccine development.
h. The genomes of many extremophiles have been studied in an effort to gain a better understanding of the mechanisms by which these microbes survive in their harsh habitats. *Deinococcus radiodurans* survives high levels of gamma radiation, in part due to its ability to dramatically up-regulate DNA repair genes.

15.9 Environmental Genomics

a. Environmental genomics is a relatively new area of research that has been used to learn more about the biodiversity and metabolic potential of microbial communities (**figures 15.19** and **15.20**).

Key Terms

alignment 388
annotation 388
bioinformatics 388
chain-termination DNA sequencing method 384
coding sequence (CDS) 388
comparative genomics 383
DNA microarrays (gene chips) 389
environmental genomics 402
expressed sequence tag (EST) 390
functional genomics 383
functional proteomics 393
genomics 383
hierarchical cluster analysis 402
in silico analysis 388
lateral or horizontal gene transfer (LGT) 391
metagenomics 402
motif 389
oligonucleotide 390
open reading frame (ORF) 388
ortholog 388
paralog 388
phylotype 402
probe 389
protein modeling 394
proteome 393
proteomics 393
rational drug design 398
spotted arrays 390
structural genomics 383
structural proteomics 394
transcriptome 402
two-dimensional gel electrophoresis 393
whole-genome shotgun sequencing 384

Critical Thinking Questions

1. What impact might genome comparisons have on the current phylogenetic schemes for *Bacteria* and *Archaea* that are discussed in chapter 19?
2. Propose an experiment that can be done easily with a DNA microarray that would have required years before this new technology.
3. What are the pitfalls of searches for homologous genes and proteins?
4. You are developing a new vaccine for a pathogen. You want your vaccine to recognize specific cell-surface proteins. Explain how you will use genome analysis to identify potential protein targets. What functional genomics approaches will you use to determine which of these proteins is produced when the pathogen is in its host?

Learn More

Campbell, A. M., and Heyer, L. J. 2003. *Discovering genomics, proteomics, and bioinformatics*. San Francisco, CA: Benjamin Cunnings.

Camus, J. C.; Pryor, M. J., Médigue, C.; and Cole, S. T. 2002. Re-annotation of the genome sequence of *Mycobacterium tuberculosis* H37Rv. *Microbiol.* 148:2967–73.

Cole, S. T., et al., 1998. Deciphering the biology of *Mycobacterium tuberculosis* from the complete genome sequence. *Nature* 393:537–44.

de la Torre, J. R.; Christianson, L. M.; Béjà, O.; Suzuki, M. T.; Karl, D. M.; Heidelberg, J.; and DeLong, E. F. 2003. Proteorhodopsin genes are distributed among divergent marine bacterial taxa. *Proc. Natl. Acad. Sci.* 100:12830–35.

Fleischmann, R. D., et al. 1995. Whole-genome random sequencing and assembly of *Haemophilus influenzae* Rd. *Science* 269:496–512.

Gill, S. R., et al. 2005. Insights on evolution of virulence and resistance from the complete genome analysis of an early methicillin-resistant *Staphylococcus aureus* strain and a biofilm-producing methicillin-resistant *Staphylococcus epidermidis* strain. *J. Bacteriol.* 187:2426–38.

Graves, P. R., and Haystead, T. A. J. 2002. Molecular biologist's guide to proteomics. *Microbiol. Mol. Biol. Rev:* 66(1):39–63.

Handelsman, J. 2004. Metagenomics: Application of genomics to uncultured microorganisms. *Microbiol. Molec. Biol. Rev.* 68:669–85.

Hughes, T. R.; Robinson, M. D.; Mitsakakis, N.; and Johnston, M. 2004. The promise of functional genomics: Completing the encyclopedia of the cell. *Curr. Opin. Microbiol.* 7:546–54.

Jain, R.; Rivera, M.C.; and Lake, J. A. 1999. Horizontal gene transfer among genomes: The complexity hypothesis. *Proc. Natl. Acad. Sci.* 96:3801–6.

Knudson, S. 2004. *Guide to analysis of DNA microarray data,* 2nd ed. Hoboken, N. J.: John Wiley & Sons, Inc.

Krane, D. E., and Raymer, M. L. 2003. *Fundamental concepts of bioinformatics*. San Francisco, Calif.: Benjamin Cummings.

Lander, E. S., and Weinberg, R. A. 2000. Genomics: Journey to the center of biology. *Science* 287:1777–82.

Liu, Y. et al. 2003. Transcriptome dynamics of *Deinococcus radiodurans* recovering from ionizing radiation. *Proc. Natl. Acad. Sci.* 100:4191–96.

Meinke, A.; Henics, T.; and E. Nagy. 2004. Bacterial genomes pave the way to novel vaccines. *Curr. Opin. Microbiol.* 7:341–20.

Rhodius, V.; Van Dyk, T. K.; Gross, C.; and LaRossa, R. A. 2002. Impact of genomic technologies on studies of bacterial gene expression. *Annu. Rev. Microbiol.* 56:599–624.

Venter, J. C., et al. 2004. Environmental genome shotgun sequencing of the Sargasso Sea. *Science* 304:66–74.

Winzeler, E. A., et al. 1999. Functional characterization of the *S. cerevisiae* genome by gene deletion and parallel analysis. *Science* 285:901–6.

Please visit the Prescott website at www.mhhe.com/prescott7 for additional references.

16 The Viruses: Introduction and General Characteristics

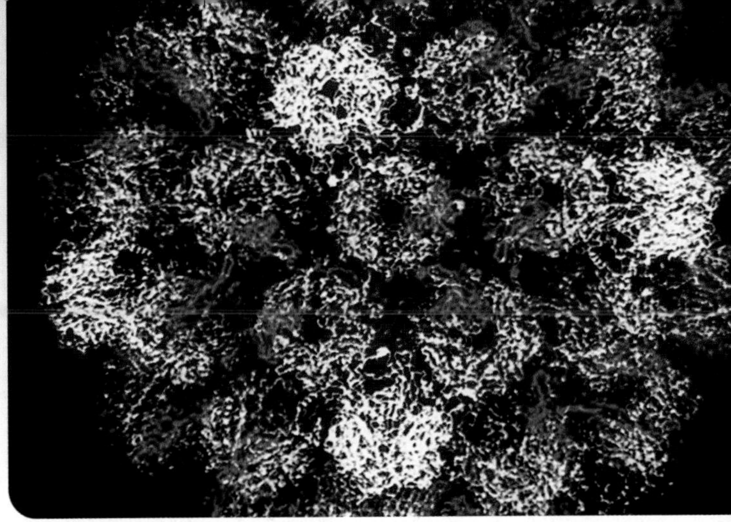

The simian virus 40 (SV-40) capsid shown here differs from most icosahedral capsids in containing only pentameric capsomers. SV-40 is a small double-stranded DNA polyomavirus with 72 capsomers. It may cause a central nervous system disease in rhesus monkeys and can produce tumors in hamsters. SV-40 was first discovered in cultures of monkey kidney cells during preparation of the poliovirus vaccine.

Preview

- Viruses are simple, acellular entities. They can reproduce only within living cells because they are obligate intracellular parasites.
- All viruses have a nucleocapsid composed of a nucleic acid genome surrounded by a protein capsid. Some viruses have a membranous envelope that lies outside the nucleocapsid. The nucleic acid of the virus can be RNA or DNA, single-stranded or double-stranded, linear or circular.
- Capsids may have helical, icosahedral, or complex symmetry. They are constructed of protomers that self-assemble through noncovalent bonds.
- Although each virus has unique aspects to its life cycle, a general pattern of replication is observable. The typical virus life cycle consists of five steps: attachment to the host cell, entry into the host cell, synthesis of viral nucleic acid and proteins within the host cell, self-assembly of virions within the host cell, and release of virions from the host cell.
- Viruses are cultured by inoculating living hosts or cell cultures with a virion preparation. Purification depends mainly on their large size relative to cell components, high protein content, and great stability. The virus concentration may be determined from the virion count or from the number of infectious units.
- Viruses are classified primarily on the basis of their nucleic acid's characteristics, reproductive strategy, capsid symmetry, and the presence or absence of an envelope.

In chapters 16, 17, and 18 we turn our attention to the viruses. These are infectious agents with fairly simple, acellular organization. Most possess only one type of nucleic acid, either DNA or RNA, and they only reproduce within living cells. Clearly viruses are quite different from procaryotic and eucaryotic microorganisms; they are studied by **virologists.**

Despite their simplicity, viruses are extremely important and deserve close attention. Many human viral diseases are known and more are discovered every year, as demonstrated by the appearance of SARS and avian influenza viruses. The study of viruses has contributed significantly to the discipline of molecular biology. In fact, the field of genetic engineering is based in large part upon discoveries in virology. Thus **virology** (the study of viruses) is a significant part of microbiology.

In this chapter we focus on the broader aspects of virology: its development as a scientific discipline, the general properties and structure of viruses, the ways in which viruses are cultured and studied, and viral taxonomy. In chapter 17 our concern is with viruses of the *Bacteria* and *Archaea,* and in chapter 18 we consider viruses of eucaryotes.

Viruses have had enormous impact on humans and other organisms, yet very little was known about their nature until fairly recently. A brief history of their discovery and recognition as uniquely different infectious agents can help clarify their nature.

16.1 Early Development of Virology

Although the ancients did not understand the nature of their illnesses, they were acquainted with diseases, such as rabies, that are now known to be viral in origin. In fact, there is some evidence that the great epidemics of A.D. 165 to 180 and A.D. 251 to 266, which severely weakened the Roman Empire and aided its decline, may have been caused by measles and smallpox viruses. Smallpox had an equally profound impact on the New World.

Great fleas have little fleas upon their backs to bite 'em
And little fleas have lesser fleas, and so on ad infinitum.

—*Augustus De Morgan*

Hernán Cortés's conquest of the Aztec Empire in Mexico was made possible by an epidemic that ravaged Mexico City. The virus was probably brought to Mexico in 1520 by the relief expedition sent to join Cortés. Before the smallpox epidemic subsided, it had killed the Aztec King Cuitlahuac (the nephew and son-in-law of the slain emperor, Montezuma II) and possibly 1/3 of the population. Since the Spaniards were not similarly afflicted, it appeared that God's wrath was reserved for Native Americans, and this disaster was viewed as divine support for the Spanish conquest (**Historical Highlights 16.1**).

Progress in preventing viral diseases began years before the discovery of viruses. Early in the eighteenth century, Lady Wortley Montagu, wife of the English ambassador to Turkey, observed that Turkish women inoculated their children against smallpox. The children came down with a mild case but subsequently were immune. Lady Montagu tried to educate the English public about the procedure but without great success. Later in the century an English country doctor, Edward Jenner, stimulated by a girl's claim that she could not catch smallpox because she had had cowpox, began inoculating humans with material from cowpox lesions. He published the results of 23 successful vaccinations in 1798. Although Jenner did not understand the nature of smallpox, he did manage to successfully protect his patients from the dreaded disease through exposure to the cowpox virus.

Until well into the nineteenth century, harmful agents were often grouped together and sometimes called viruses [Latin *virus,* poison or venom]. Even Louis Pasteur used the term virus for any living infectious disease agent. The development in 1884 of the porcelain bacterial filter by Charles Chamberland, one of Pasteur's collaborators and inventor of the autoclave, made possible the discovery of what are now called viruses. Tobacco mosaic disease was the first to be studied with Chamberland's filter. In 1892 Dimitri Ivanowski published studies showing that leaf extracts from infected plants would induce tobacco mosaic disease even after filtration removed all bacteria. However, he attributed this to the presence of a toxin. Martinus Beijerinck, working independently of Ivanowski, published the results of extensive studies on tobacco mosaic disease in 1898 and 1900. Because the filtered sap of diseased plants was still infectious, he proposed that the disease was caused by an entity different from bacteria, what he called a filterable virus. He observed that the virus would multiply only in living plant cells, but could survive for long periods in a dried state. At the same time Friedrich Loeffler and Paul Frosch in Germany found that the hoof-and-mouth disease of cattle was also caused by a virus rather than by a toxin. In 1900 Walter Reed began his study of the yellow fever disease whose incidence had been increasing in Cuba. Reed showed that this human disease was due to a virus that was transmitted by mosquitoes. Mosquito control soon reduced the severity of the yellow fever problem. Thus by the beginning of the 20th century, it had been established that viruses were different from bacteria and could cause diseases in plants, livestock, and humans.

Shortly after the turn of the century, Vilhelm Ellermann and Oluf Bang in Copenhagen reported that leukemia could be transmitted between chickens by cell-free filtrates and was probably caused by a virus. Three years later in 1911, Peyton Rous from the Rockefeller Institute in New York City reported that a virus,

Historical Highlights

16.1 Disease and the Early Colonization of America

There is considerable evidence that disease, and particularly smallpox, played a major role in reducing Indian resistance to the European colonization of North America. It has been estimated that Indian populations in Mexico declined about 90% within 100 years of initial contact with the Spanish. Smallpox and other diseases were a major factor in this decline, and there is no reason to suppose that North America was any different. As many as 10 to 12 million Indians may have lived north of the Rio Grande before contact with Europeans. In New England alone, there may have been over 72,000 in 1600; yet only around 8,600 remained in New England by 1674, and the decline continued in subsequent years.

Such an incredible catastrophe can be accounted for by consideration of the situation at the time of European contact with the Native Americans. The Europeans, having already suffered major epidemics in the preceding centuries, were relatively immune to the diseases they carried. On the other hand, the Native Americans had never been exposed to diseases like smallpox and were decimated by epidemics. In the sixteenth century, before any permanent English colonies had been established, many contacts were made by missionaries and explorers who undoubtedly brought disease with them and infected native populations. Indeed, the English noted at the end of the century that Indian populations had declined greatly but attributed it to armed conflict rather than to disease.

Establishment of colonies simply provided further opportunities for infection and outbreak of epidemics. For example, the Huron Indians decreased from a minimum of 32,000 people to 10,000 in 10 years. Between the time of initial English colonization and 1674, the Narraganset Indians declined from around 5,000 warriors to 1,000, and the Massachusetts Indians, from 3,000 to 300. Similar stories can be seen in other parts of the colonies. Some colonists interpreted these plagues as a sign of God's punishment of Indian resistance: the "Lord put an end to this quarrel by smiting them with smallpox. . . . Thus did the Lord allay their quarrelsome spirit and make room for the following part of his army."

It seems clear that epidemics of European diseases like smallpox decimated Native American populations and prepared the way for colonization of the North American continent. Many American cities—for example, Boston, Philadelphia, and Plymouth—grew upon sites of previous Indian villages.

now known as the Rous sarcoma virus, was responsible for a malignant muscle tumor in chickens. These studies established that at least some malignancies are caused by viruses. The Rous sarcoma virus is still extensively used in cancer research.

In 1915 Frederick Twort reported that bacteria also could be attacked by viruses. Twort isolated bacterial viruses that could attack and destroy micrococci and intestinal bacilli. Although he speculated that his preparations might contain viruses, Twort did not follow up on these observations. It remained for Felix d'Herelle to establish decisively the existence of bacterial viruses. d'Herelle isolated bacterial viruses from patients with dysentery, probably caused by *Shigella dysenteriae.* He noted that when a virus suspension was spread on a layer of bacteria growing on agar, clear circular areas containing viruses and lysed cells developed. A count of these clear zones allowed d'Herelle to estimate the number of viruses present. This procedure for enumerating viruses is now called a plaque assay; it is described in section 16.6. d'Herelle demonstrated that bacterial viruses could reproduce only in live bacteria; therefore he named them **bacteriophages** (or just **phages**) because they could eat holes in bacterial "lawns."

The chemical nature of viruses was established when Wendell Stanley announced in 1935 that he had crystallized the tobacco mosaic virus (TMV) and found it to be largely or completely protein. A short time later Frederick Bawden and Norman Pirie managed to separate the TMV virus particles into protein and nucleic acid. Thus by the late 1930s it was becoming clear that viruses are complexes of nucleic acids and proteins able to reproduce only in living cells.

16.2 General Properties of Viruses

Viruses are a unique group of infectious agents whose distinctiveness resides in their simple, acellular organization and pattern of reproduction. A complete virus particle or **virion** consists of one or more molecules of DNA or RNA enclosed in a coat of protein. Some viruses have additional layers that can be very complex and contain carbohydrates, lipids, and additional proteins (**figure 16.1**). Viruses can exist in two phases: extracellular and intracellular. Virions, the extracellular phase, possess few if any enzymes and cannot reproduce independent of living cells. In the intracellular phase, viruses exist primarily as replicating nucleic acids that induce host metabolism to synthesize virion components; eventually complete virus particles or virions are released.

In summary, viruses differ from living cells in at least three ways: (1) their simple, acellular organization; (2) the presence of either DNA or RNA, but not both, in almost all virions; and (3) their inability to reproduce independent of cells and carry out cell division as procaryotes and eucaryotes do.

1. Describe the major technical advances and important discoveries in the early development of virology. Why might virology have developed much more slowly without the use of Chamberland's filter?
2. Which scientists made important contributions to the development of virology? What were their contributions?
3. How are viruses similar to cellular organisms? How do they differ?

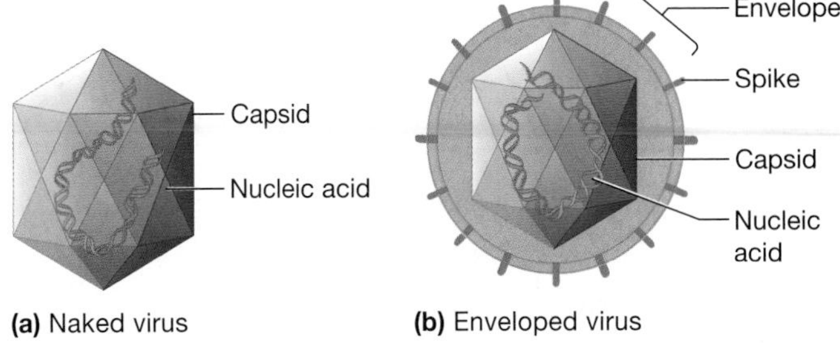

Figure 16.1 Generalized Structure of Viruses. **(a)** The simplest virus is a naked virus (nucleocapsid) consisting of a geometric capsid assembled around a nucleic acid strand. **(b)** An enveloped virus is composed of a nucleocapsid surrounded by a flexible membrane called an envelope. The envelope usually has viral proteins called spikes inserted into it.

16.3 The Structure of Viruses

Virus morphology has been intensely studied over the past decades because of the importance of viruses and the realization that virus structure was simple enough to be understood. Progress has come from the use of several different techniques: electron microscopy, X-ray diffraction, biochemical analysis, and immunology. Although our knowledge is incomplete due to the large number of different viruses, the general nature of virus structure is becoming clear.

Virion Size

Virions range in size from about 10 to 400 nm in diameter (**figure 16.2**). The smallest viruses are a little larger than ribosomes, whereas the poxviruses, which include vaccinia, are about the same size as the smallest bacteria and can be seen in the light microscope. Most viruses, however, are too small to be visible in the light microscope and must be viewed with scanning and transmission electron microscopes. Electron microscopy (section 2.4)

General Structural Properties

All virions, even if they possess other constituents, are constructed around a **nucleocapsid** core (indeed, some viruses consist only of a nucleocapsid). The nucleocapsid is composed of a nucleic acid, usually either DNA or RNA, held within a protein coat called the **capsid,** which protects viral genetic material and aids in its transfer between host cells.

Capsids are large macromolecular structures that self-assemble from many copies of one or a few types of proteins. The proteins used to build the capsid are called **protomers.** Probably the most important advantage of this design strategy is that the information stored in viral genetic material is used with maximum efficiency. For example, the tobacco mosaic virus (TMV) capsid is constructed using a single type of protomer that is 158 amino acids in length (**figure 16.3**). Therefore, of the 6,000 nucleotides in the TMV genome, only about 474 nucleotides are required to code

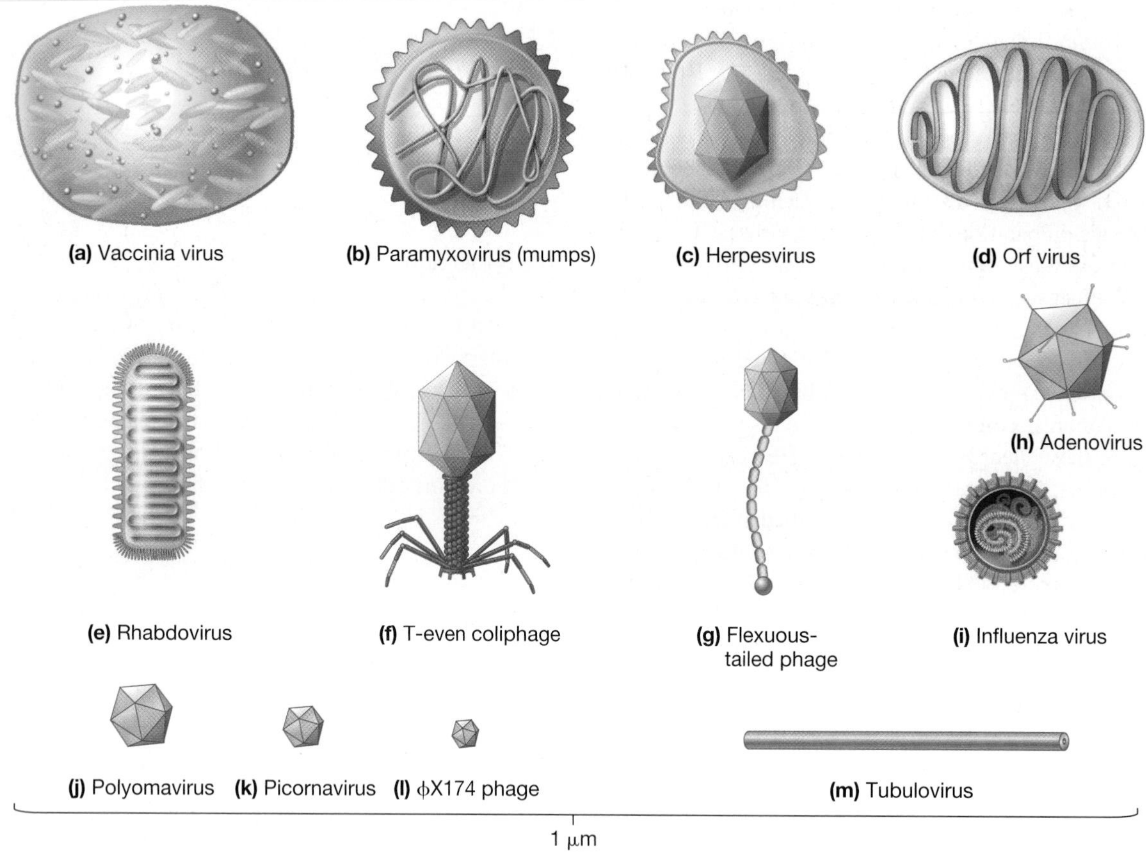

Figure 16.2 The Size and Morphology of Selected Viruses. The viruses are drawn to scale. A 1 μm line is provided at the bottom of the figure.

for the coat protein. Suppose, however, that the TMV capsid was composed of six different protomers all about 150 amino acids in length. If this were the case, about 2,900 of the 6,000 nucleotides in the TMV genome would be required just for capsid construction, and much less genetic material would be available for other purposes.

The various morphological types of viruses primarily result from the combination of a particular type of capsid symmetry with the presence or absence of an envelope, which is a lipid layer external to the nucleocapsid. There are three types of capsid symmetry: helical, icosahedral, and complex. Those virions having an envelope are called **enveloped viruses;** whereas those lacking an envelope are called **naked viruses** (figure 16.1).

Helical Capsids

Helical capsids are shaped like hollow tubes with protein walls. The tobacco mosaic virus provides a well-studied example of helical capsid structure (figure 16.3). In this virus, the self-assembly of protomers in a helical or spiral arrangement produces a long, rigid tube, 15 to 18 nm in diameter by 300 nm long. The capsid encloses an RNA genome, which is wound in a spiral and lies within a groove formed by the protein subunits. Not all helical capsids are as rigid as the TMV capsid. The influenza virus genome is enclosed in thin, flexible helical capsids that are folded within an envelope (**figure 16.4**).

The size of a helical capsid is influenced by both its protomers and the nucleic acid enclosed within the capsid. The diameter of the capsid is a function of the size, shape, and interactions of the protomers. The nucleic acid appears to determine helical capsid length because the capsid does not extend much beyond the end of the DNA or RNA.

Icosahedral Capsids

The icosahedron is a regular polyhedron with 20 equilateral triangular faces and 12 vertices (figure 16.2*h, j–l*). It is one of nature's favorite shapes. The **icosahedral capsid** is the most efficient way to enclose a space. A few genes, sometimes only one, can code for proteins that self-assemble to form the capsid. In this way a small number of genes can specify a large three-dimensional structure.

When icosahedral viruses are negatively stained and viewed in the transmission electron microscope, a complex structure is revealed (**figure 16.5**). The capsids are constructed from ring- or knob-shaped units called **capsomers,** each usually made of five

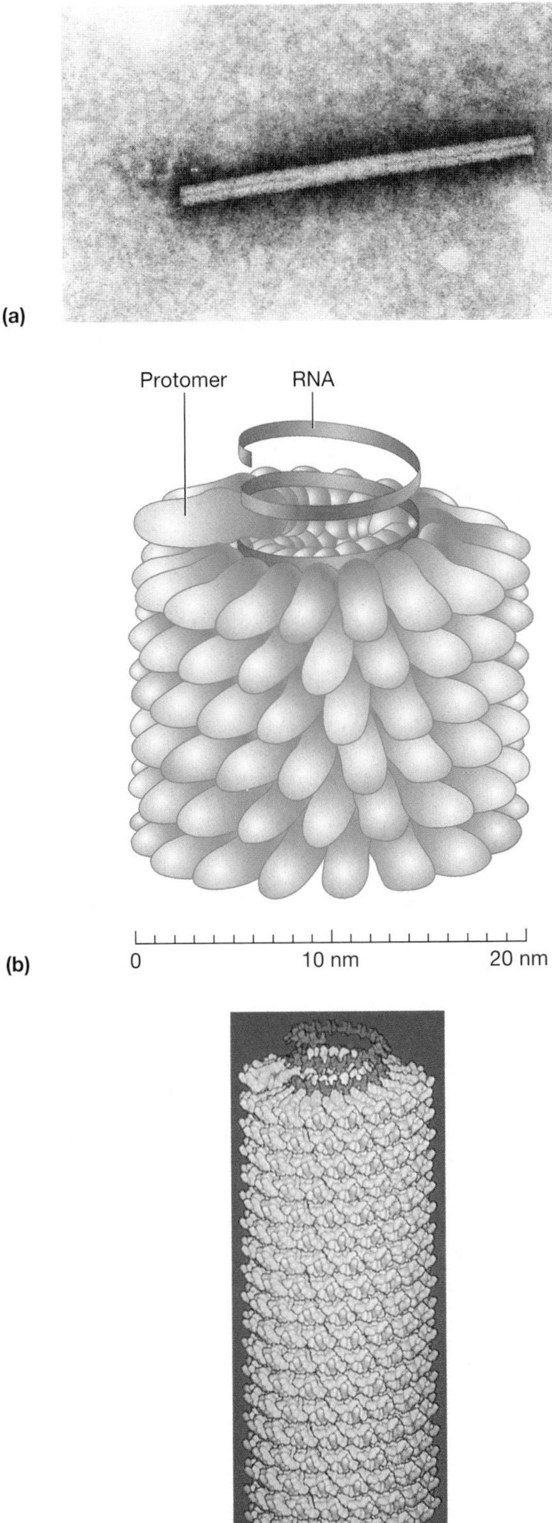

Figure 16.3 **Tobacco Mosaic Virus Structure.** **(a)** An electron micrograph of the negatively stained helical capsid (×400,000). **(b)** Illustration of TMV structure. Note that the nucleocapsid is composed of a helical array of protomers with the RNA spiraling on the inside. **(c)** A model of TMV.

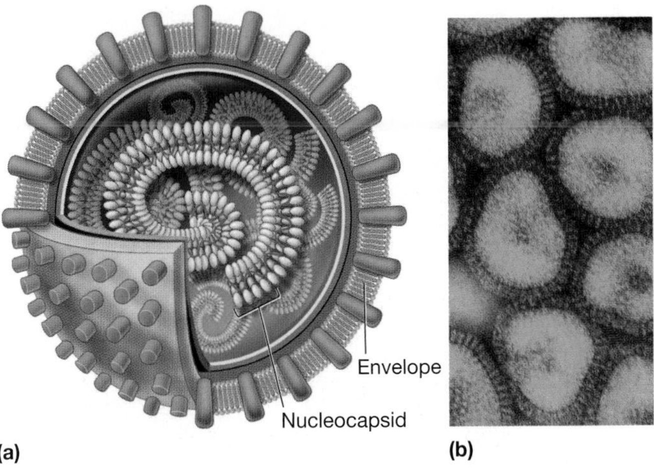

Figure 16.4 **Influenza Virus.** Influenza virus is an enveloped virus with a helical nucleocapsid. **(a)** Schematic view. Influenza viruses have segmented genomes consisting of 7 to 8 different RNA molecules. Each is coated by capsid proteins. **(b)** Because there are 7 to 8 flexible nucleocapsids enclosed by an envelope, the virions are pleomorphic. Electron micrograph (×350,000).

or six protomers. **Pentamers (pentons)** have five subunits; **hexamers (hexons)** possess six. Pentamers are usually at the vertices of the icosahedron, whereas hexamers generally form its edges and triangular faces (**figure 16.6**). The icosahedron in figure 16.6 is constructed of 42 capsomers; larger icosahedra are made if more hexamers are used to form the edges and faces (e.g., adenoviruses have a capsid with 252 capsomers as shown in figure 16.5*c,d*). In some RNA viruses, both the pentamers and hexamers of a capsid are constructed with only one type of subunit. In other viruses, pentamers are composed of different proteins than are the hexamers.

The self-assembly of capsids is a remarkable process that is not fully understood. Enzymatic activity is not required to link protomers together. However, noncapsid proteins may be involved. They usually provide a scaffolding upon which the protomers are assembled.

Although most icosahedral capsids appear to contain both pentamers and hexamers, simian virus 40 (SV-40), a small, double-stranded DNA virus, has only pentamers (**figure 16.7*a***). The virus is constructed of 72 cylindrical pentamers with hollow centers. Five flexible arms extend from the edge of each pentamer toward neighboring pentamers (figure 16.7*b,c*). The arms of adjacent pentamers twist around each other and act as ropes that tie the pentamers together.

Viruses with Capsids of Complex Symmetry

Although most viruses have either icosahedral or helical capsids, many viruses do not fit into either category. The poxviruses and large bacteriophages are two important examples.

The poxviruses are the largest of the animal viruses (about 400 × 240 × 200 nm in size) and can even be seen with a phase-contrast microscope or in stained preparations. They possess an

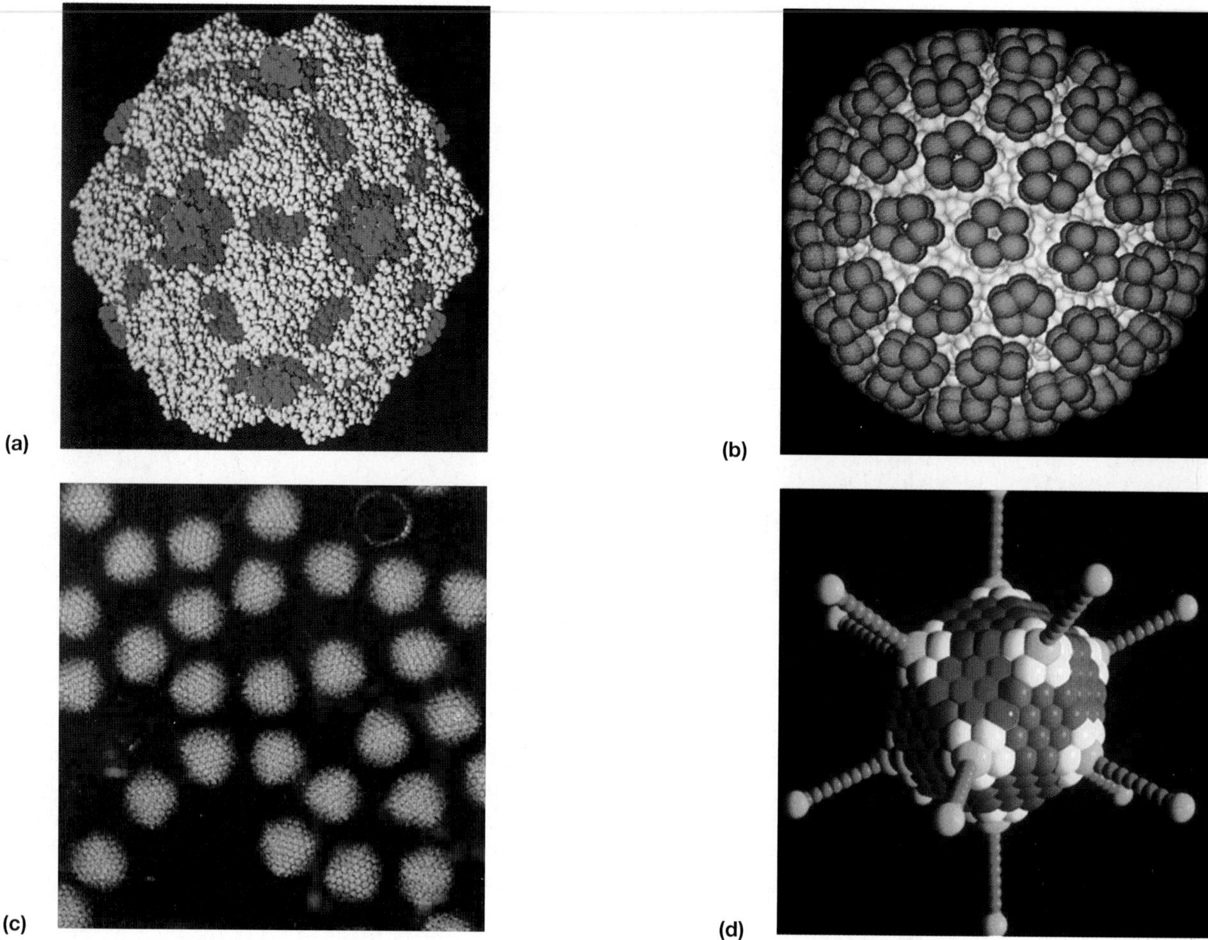

Figure 16.5 Examples of Icosahedral Capsids. **(a)** Canine parvovirus model, 12 capsomers. **(b)** Computer-simulated image of the polyomavirus (72 capsomers) that causes a rare demyelinating disease of the central nervous system. **(c)** Adenovirus, 252 capsomers (×171,000). **(d)** Computer-simulated model of adenovirus.

exceptionally complex internal structure with an ovoid- to brick-shaped exterior. **Figure 16.8** shows the morphology of vaccinia virus, a poxvirus. The double-stranded DNA is associated with proteins and contained in the nucleoid, a central structure shaped like a biconcave disk and surrounded by a membrane. Two elliptical or lateral bodies lie between the nucleoid and its outer envelope, a membrane and a thick layer covered by an array of tubules or fibers.

Some large bacteriophages are even more elaborate than the poxviruses. The T2, T4, and T6 phages (T-even phages) that infect *Escherichia coli* are said to have **binal symmetry** because they have a head that resembles an icosahedron and a tail that is helical. The icosahedral head is elongated by one or two rows of hexamers in the middle and contains the DNA genome (**figure 16.9**). The tail is composed of a collar joining it to the head, a central hollow tube, a sheath surrounding the tube, and a complex baseplate. The sheath is made of 144 copies of the gp18 protein arranged in 24 rings, each containing six copies. In T-even phages, the baseplate is hexagonal and has a pin and a jointed tail fiber at each corner.

There is considerable variation in structure among the large bacteriophages, even those infecting a single host. In contrast with the T-even phages, many other coliphages (phages that infect *E. coli*) have true icosahedral heads. T1, T5, and lambda phages have sheathless tails that lack a baseplate and terminate in rudimentary tail fibers. Coliphages T3 and T7 have short, noncontractile tails without tail fibers. Bacteriophages are discussed in more detail in chapter 17.

Viral Envelopes and Enzymes

Many animal viruses, some plant viruses, and at least one bacterial virus are bounded by an outer membranous layer called an **envelope** (**figure 16.10**). Animal virus envelopes usually arise from host cell nuclear or plasma membranes; their lipids and carbohydrates are normal host constituents. In contrast, envelope proteins are coded for by virus genes and may even project from the envelope surface as **spikes,** which are also called **peplomers.** In many cases, these spikes are involved in virus attachment to the host cell surface. Because they differ among viruses, they also

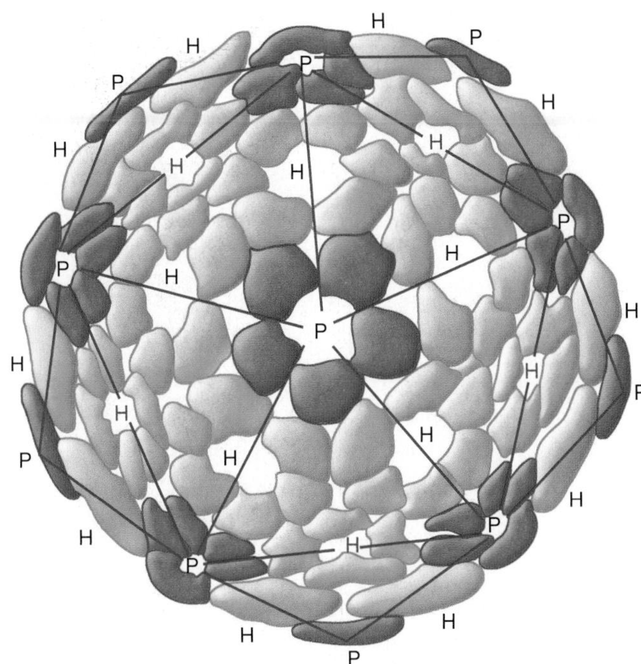

Figure 16.6 The Structure of an Icosahedral Capsid Formed from a Single Type of Protomer. The protomers associate to form either pentons (P), shown in red, or hexons (H), shown in gold. The blue lines define the triangular faces of the icosahedron. Notice that pentons are located at the vertices and that the hexons form the edges and faces of the icosahedron. This capsid contains 42 capsomers.

can be used to identify some viruses. The envelope is a flexible, membranous structure, so enveloped viruses frequently have a somewhat variable shape and are called pleomorphic. However, the envelopes of viruses like the bullet-shaped rabies virus are firmly attached to the underlying nucleocapsid and endow the virion with a constant, characteristic shape (figure 16.10*b*). In some viruses the envelope is disrupted by solvents like ether to such an extent that lipid-mediated activities are blocked or envelope proteins are denatured and rendered inactive. The virus is then said to be "ether sensitive."

Influenza virus (figure 16.10*a*) is a well-studied example of an enveloped virus. Spikes project about 10 nm from the surface at 7 to 8 nm intervals. Some spikes possess the enzyme neuraminidase, which functions in the release of mature virions from the host cell. Other spikes have hemagglutinin proteins, so named because they can bind the virions to red blood cell membranes and cause the red blood cells to clump together (agglutinate). This is called hemagglutination (*see figure 35.11*). Hemagglutinins participate in virion attachment to host cells. Proteins, like the spike proteins that are exposed on the outer envelope surface, are generally glycoproteins—that is, the proteins have carbohydrate attached to them. A nonglycosylated protein, the M or matrix protein, is found on the inner surface of the envelope and helps stabilize it.

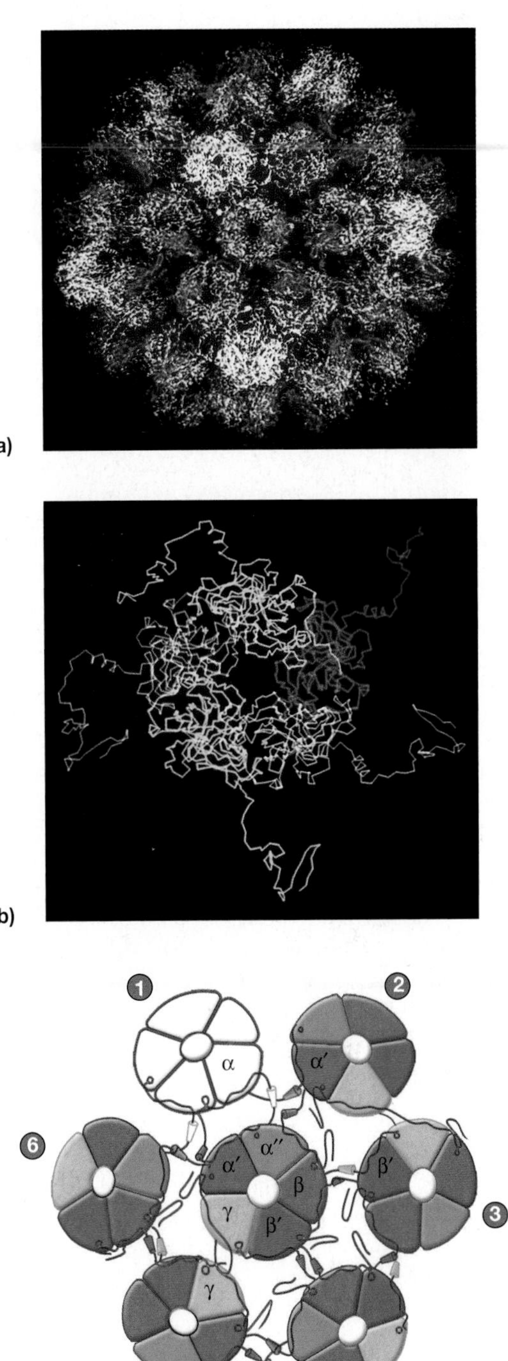

Figure 16.7 An Icosahedral Capsid Constructed of Pentamers. **(a)** The simian virus 40 capsid. The 12 pentamers at the icosahedron vertices are in white. The nonvertex pentamers are shown with each polypeptide chain in a different color. **(b)** A pentamer with extended arms. **(c)** A schematic diagram of the surface structure depicted in part a. The body of each pentamer is represented by a five-petaled flower design. Each arm is shown as a line or a line and cylinder (α-helix) with the same color as the rest of its protomer. The outer protomers are numbered clockwise beginning with the one at the vertex.

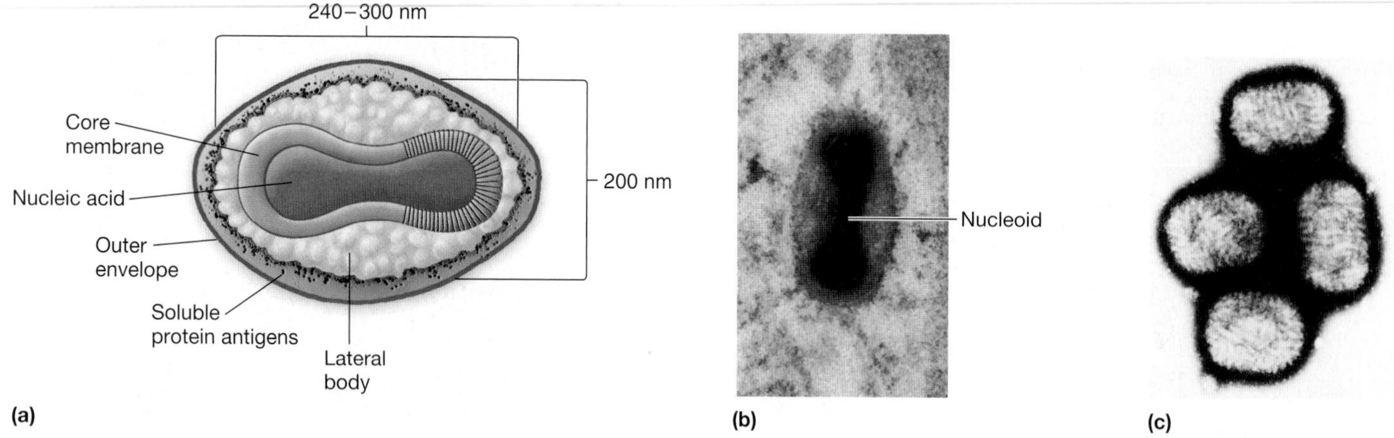

Figure 16.8 Vaccinia Virus Morphology. **(a)** Diagram of vaccinia structure. **(b)** Micrograph of the virion clearly showing the nucleoid (×200,000). **(c)** Vaccinia surface structure. An electron micrograph of four virions showing the thick array of surface fibers (×150,000).

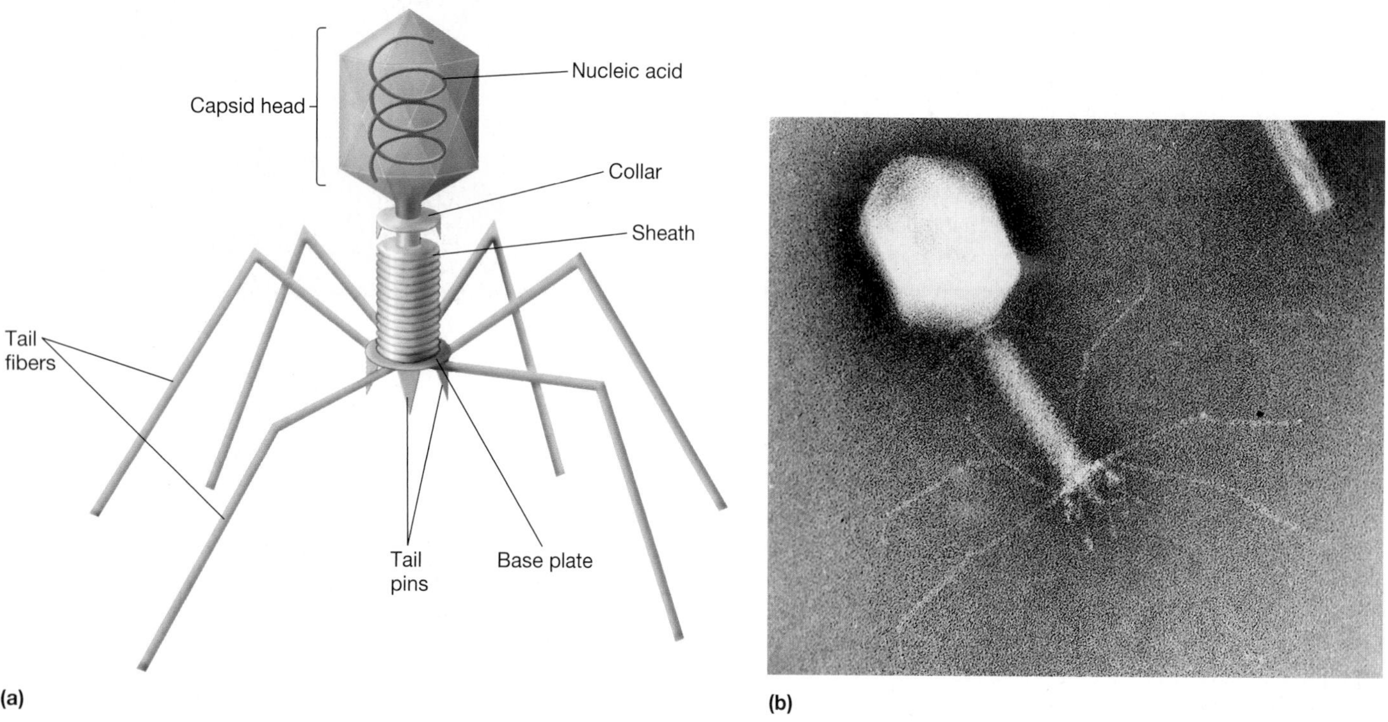

Figure 16.9 T-Even Coliphages. **(a)** The structure of the T4 bacteriophage. **(b)** The micrograph shows the phage before injection of its DNA.

It was originally thought that all virions lacked enzymes. However, as just illustrated in the discussion of influenza virus, this is not the case. In some instances, enzymes are associated with the envelope or capsid (e.g., influenza neuraminidase), but most viral enzymes are located within the capsid. Many of these are involved in nucleic acid replication. For example, the influenza virus uses RNA as its genetic material and carries an enzyme that synthesizes RNA using an RNA template. Such enzymes are called RNA-dependent RNA polymerases. Thus although viruses lack true metabolism and cannot reproduce independently of living cells, they may carry one or more enzymes essential to the completion of their life cycles.

Viral Genomes

Viruses are exceptionally flexible with respect to the nature of their genomes. They employ all four possible nucleic acid types: single-stranded DNA, double-stranded DNA, single-stranded

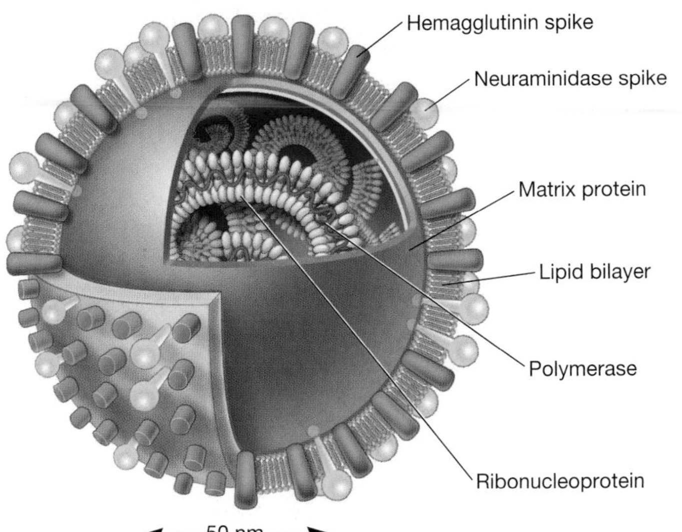

(a) Influenza virus

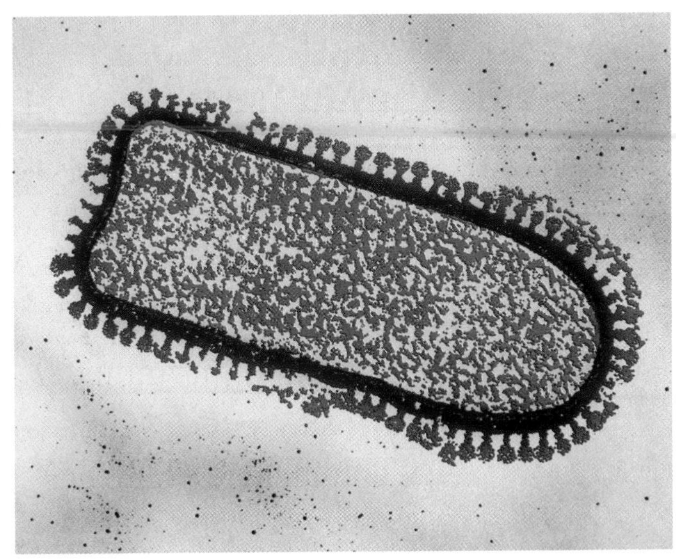

(b) Rabies virus

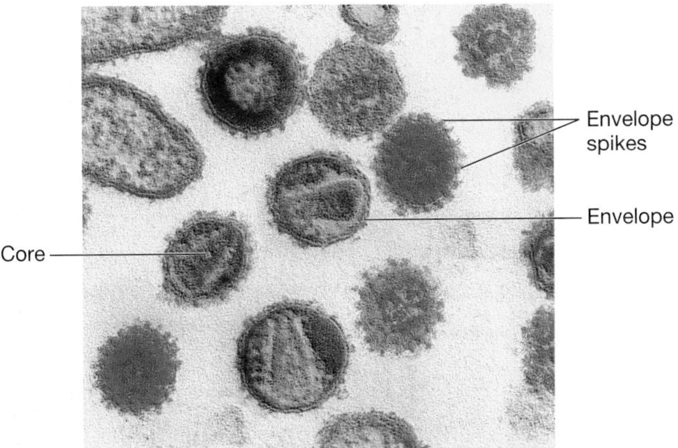

(c) HIV

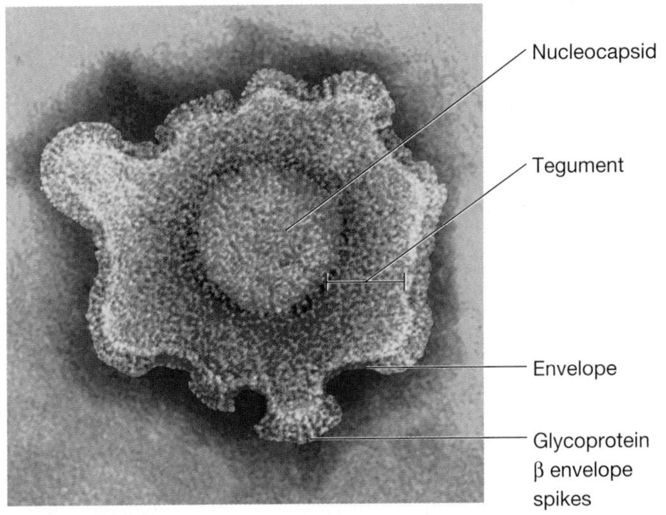

(d) Herpesvirus

(e) Semliki Forest virus

Figure 16.10 Examples of Enveloped Viruses. **(a)** Diagram of the influenza virion. **(b)** Negatively stained rabies virus. **(c)** Human immunodeficiency viruses. **(d)** Herpesviruses. **(e)** Computer image of the Semliki Forest virus, a virus that occasionally causes encephalitis in humans. Images (b), (c), and (d) are artificially colorized.

Table 16.1 Types of Viral Nucleic Acids

Nucleic Acid Type	Nucleic Acid Structure	Virus Examples
DNA		
Single Stranded	Linear, single-stranded DNA	Parvoviruses
	Circular, single-stranded DNA	ϕX174, M13, fd phages
Double Stranded	Linear, double-stranded DNA	Herpesviruses (herpes simplex viruses, cytomegalovirus, Epstein-Barr virus), adenoviruses, T coliphages, lambda phage, and other bacteriophages
	Linear, double-stranded DNA with single chain breaks	T5 coliphage
	Double-stranded DNA with cross-linked ends	Vaccinia, smallpox viruses
	Closed, circular, double-stranded DNA	Polyomaviruses (SV-40), papillomaviruses, PM2 phage, cauliflower mosaic virus
RNA		
Single-Stranded	Linear, single-stranded, positive-strand RNA	Picornaviruses (polio, rhinoviruses), togaviruses, RNA bacteriophages, TMV, and most plant viruses
	Linear, single-stranded, negative-strand RNA	Rhabdoviruses (rabies), paramyxoviruses (mumps, measles)
	Linear, single-stranded, segmented, positive-strand RNA	Brome mosaic virus (individual segments in separate virions)
	Linear, single-stranded, diploid (two identical single strands), positive-strand RNA	Retroviruses (Rous sarcoma virus, human immunodeficiency virus)
	Linear, single-stranded, segmented, negative-strand RNA	Paramyxoviruses, orthomyxoviruses (influenza)
Double-Stranded	Linear, double-stranded, segmented RNA	Reoviruses, wound-tumor virus of plants, cytoplasmic polyhedrosis virus of insects, phage ϕ6, many mycoviruses

Modified from S. E. Luria, et al., *General Virology*, 3d edition, 1983. John Wiley & Sons, Inc., New York, NY.

RNA, and double-stranded RNA. All four types are found in animal viruses. Most plant viruses have single-stranded RNA genomes, and most bacterial viruses contain double-stranded DNA. **Table 16.1** summarizes many variations seen in viral nucleic acids. The size of viral genetic material also varies greatly. The smallest genomes (those of the MS2 and Qβ viruses) are around 4,000 nucleotides, just large enough to code for three or four proteins. MS2, Qβ, and some other viruses even save space by using overlapping genes. At the other extreme, T-even bacteriophages, herpesvirus, and vaccinia virus have genomes of 1.0 to 2.0×10^5 nucleotides and may be able to direct the synthesis of over 100 proteins. In the following paragraphs the nature of each nucleic acid type is briefly summarized. Gene structure (section 11.5)

Most DNA viruses use double-stranded DNA (dsDNA) as their genetic material. However, some have single-stranded DNA (ssDNA) genomes. In both cases, the genomes can be either linear or circular (**figure 16.11**). Some DNA genomes can switch from one form to the other. For instance, the *E. coli* phage lambda has a genome that is linear in the capsid, but is converted into a circular form once the genome enters the host cell. Another important characteristic of DNA viruses is that their genomes often contain unusual nitrogenous bases. For example, the T-even phages of *E. coli* have 5-hydroxymethylcytosine (*see figure 17.9*) instead of cytosine, and the hydroxymethyl group is often modified by attachment of a glucose moiety.

RNA viruses also can be either double-stranded (dsRNA) or single-stranded (ssRNA). Although relatively few RNA viruses have dsRNA genomes, dsRNA viruses are known to infect animals, plants, fungi, and at least one bacterial species. More common are the viruses with ssRNA genomes. Some ssRNA genomes have a base sequence that is identical to that of viral mRNA, in which case the genomic RNA strand is called the **plus strand** or **positive strand.** In fact, plus strand RNAs can direct protein synthesis immediately after entering the cell. However, other viral RNA genomes are complementary rather than identical to viral mRNA, and are called **minus** or **negative strands.** Polio, tobacco mosaic, brome mosaic, and Rous sarcoma viruses are all positive strand RNA viruses; rabies, mumps, measles, and influenza

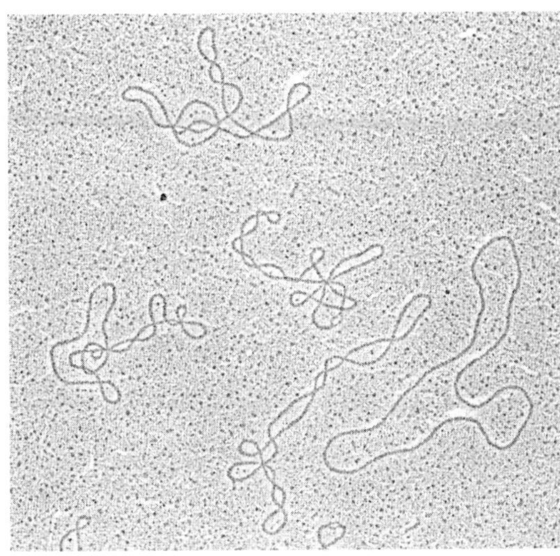

Figure 16.11 Circular Phage DNA. The closed circular DNA of the phage PM2 (×93,000). Note both the relaxed and highly twisted or supercoiled forms.

viruses are examples of negative strand RNA viruses. Many RNA viruses have **segmented genomes**—that is, the genome consists of more than one RNA strand or segment. In many cases, each segment codes for one protein. Usually all segments are enclosed in the same capsid even though some virus genomes may be composed of as many as 10 to 12 segments. However, it is not necessary that all segments be located in the same virion for successful reproduction. The genome of brome mosaic virus, a virus that infects certain grass species, is composed of four segments distributed among three different virus particles. All three of the largest segments are required for infectivity. Despite this complex and seemingly inefficient arrangement, the different brome mosaic virions manage to successfully infect the same host.

Plus strand viral RNA often resembles mRNA in more than the equivalence of its nucleotide sequence. Just as eucaryotic mRNA usually has a 5′ cap of 7-methylguanosine, many plant and animal viral RNA genomes are capped. In addition, most plus strand RNA animal viruses also have a poly-A sequence at the 3′ end of their genome, and thus closely resemble eucaryotic mRNA with respect to the structure of both ends. Strangely enough, a number of single-stranded plant viral RNAs have 3′ ends that resemble eucaryotic transfer RNA. Indeed, the genome of tobacco mosaic virus actually accepts amino acids. Transcription (section 11.6); Translation (section 11.8)

1. Define the following terms: nucleocapsid, capsid, icosahedral capsid, helical capsid, complex virus, binal symmetry, protomer, capsomer, pentamer or penton, and hexamer or hexon. How do pentamers and hexamers associate to form a complete icosahedron; what determines helical capsid length and diameter?
2. What is an envelope? What are spikes (peplomers)? Why are some enveloped viruses pleomorphic? Give two functions spikes might serve in the virus life cycle, and the proteins that the influenza virus uses in these processes.
3. All four nucleic acid forms can serve as virus genomes. Describe each, the types of virion possessing it, and any distinctive physical characteristics the nucleic acid can have. What are the following: plus strand, minus strand, and segmented genome?
4. What advantage would an RNA virus gain by having its genome resemble eucaryotic mRNA?

16.4 Virus Reproduction[1]

The differences in virus structure and viral genomes have important implications for the mechanism a virus uses to reproduce within its host cell. Indeed, even among viruses with similar structures and genomes, each can exhibit unique life cycles. However, despite these differences a general pattern of virus reproduction can be discerned. Because viruses need a host cell in which to reproduce, the first step in the life cycle of a virus is attachment to a host (**figure 16.12**). This is followed by entry of either the nucleocapsid or the viral nucleic acid into the host. If the nucleocapsid enters, uncoating of the genome usually occurs before further steps can occur. Once free in the cytoplasm, genes encoded by the viral genome are expressed. That is, the viral genes are transcribed and translated. This allows the virus to take control of the host cell's biosynthetic machinery so that new virions can be made. The viral genome is then replicated and viral proteins are synthesized. New virions are constructed by self-assembly of coat proteins with the nucleic acids, and finally the mature virions are released from the host. As discussed in chapters 17 and 18, the details of virus reproduction can vary dramatically. For instance, some viruses are released by lysing their hosts, whereas others bud from the host without lysis.

16.5 The Cultivation of Viruses

Because they are unable to reproduce independent of living cells, viruses cannot be cultured in the same way as procaryotic and eucaryotic microorganisms. For many years researchers have cultivated animal viruses by inoculating suitable host animals or embryonated eggs—fertilized chicken eggs incubated about 6 to 8 days after laying (**figure 16.13**). To prepare the egg for cultivation of viruses, the shell surface is first disinfected with iodine and penetrated with a small sterile drill. After inoculation, the drill hole is sealed with gelatin and the egg incubated. Some viruses reproduce only in certain parts of the embryo; consequently they must be injected into the proper region. For example, the myxoma virus grows well on the chorioallantoic membrane, whereas the mumps

[1]Virologists usually refer to the production of new virus particles within a host cell as virus replication. Indeed, many virologists state that viruses do not reproduce, they replicate. However, to avoid confusion about the meaning of the term replication, we will use the term reproduction when discussing the production of new virions, and use the term replication when discussing the synthesis of new copies of viral genomes.

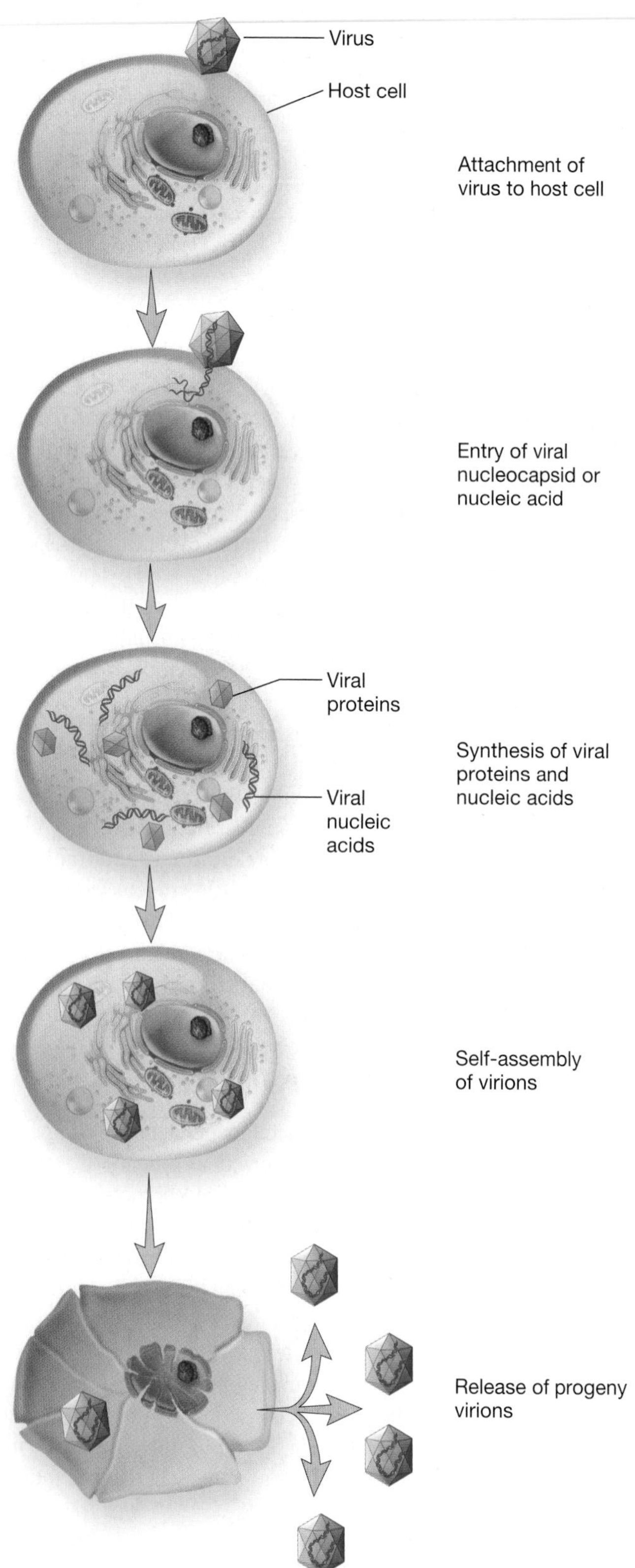

Figure 16.12 Generalized Illustration of Virus Reproduction. There is great variation in the details of virus reproduction for individual virus species.

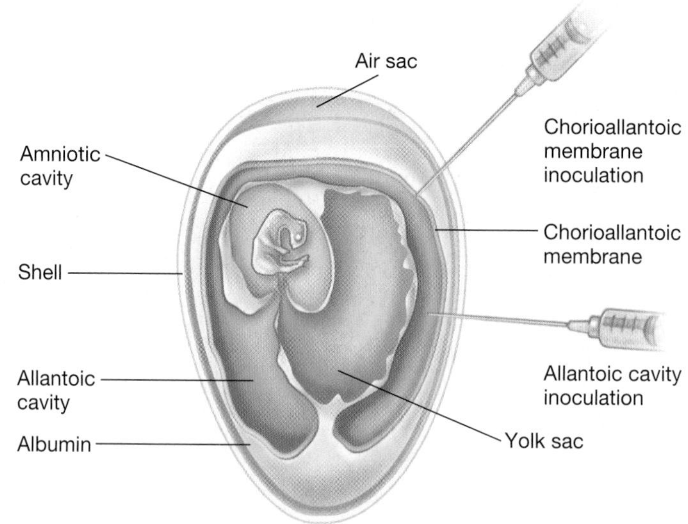

Figure 16.13 Cultivation of Viruses in an Embryonated Egg. Two sites that are often used to grow animal viruses are the chorioallantoic membrane and the allantoic cavity. The diagram shows a 9-day chicken embryo.

virus grows best in the allantoic cavity. The infection may produce a local tissue lesion known as a pock, whose appearance often is characteristic of the virus.

More recently animal viruses have been grown in tissue (cell) culture on monolayers of animal cells. This technique is made possible by the development of growth media for animal cells and by the use of antimicrobial agents that prevent bacterial and fungal contamination. Viruses are added to a layer of animal cells in a specially prepared petri dish and allowed time to attach to the cells. The cells are then covered with a thin layer of agar to limit virion spread so that only adjacent cells are infected by newly produced virions. As a result, localized areas of cellular destruction and lysis called **plaques** often are formed (**figure 16.14**) and may be detected if stained with dyes, such as neutral red or trypan blue, that can distinguish living from dead cells. Viral growth does not always result in the lysis of cells to form a plaque. Animal viruses, in particular, can cause microscopic or macroscopic degenerative changes or abnormalities in host cells and in tissues. These are called **cytopathic effects** (**figure 16.15**). Cytopathic effects may be lethal, but plaque formation from cell lysis does not always occur.

Bacterial and archaeal viruses are cultivated in either broth or agar cultures of young, actively growing cells. In some infected cultures, so many host cells are destroyed that turbid cultures clear rapidly because of cell lysis. Agar cultures are prepared by mixing viruses with cool, liquid agar and a suitable culture of host cells. The mixture is quickly poured into a petri dish containing a bottom layer of sterile agar. After hardening, cells in the layer of top agar grow and reproduce, forming a continuous, opaque layer or "lawn." Wherever a virion comes to rest in the top agar, the virus infects an adjacent cell and reproduces. Eventually, lysis of

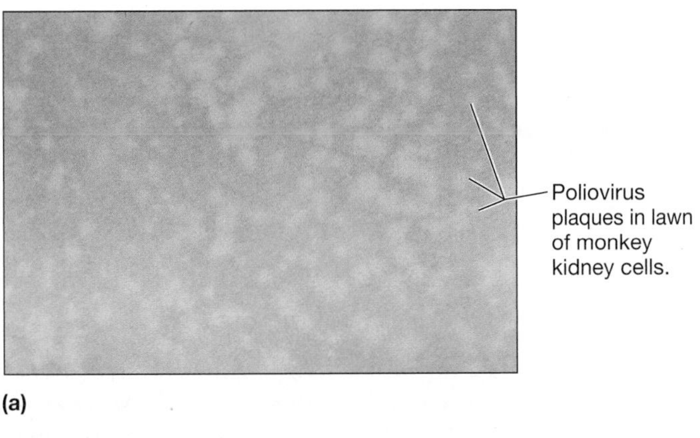

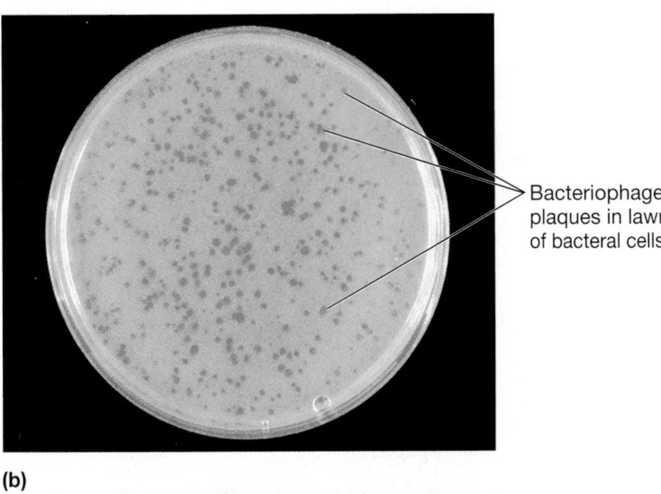

Figure 16.14 Virus Plaques. **(a)** Poliovirus plaques in a monkey kidney cell culture. **(b)** Plaques formed by bacteriophages growing on a lawn of bacterial cells.

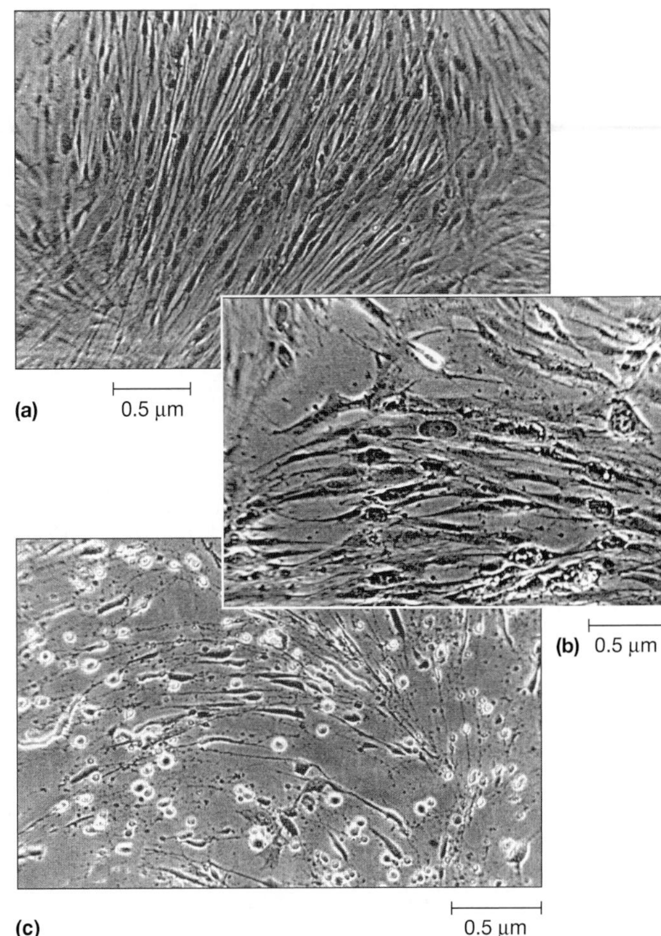

Figure 16.15 Cytopathic Effects of Viruses. **(a)** A monolayer of normal fibroblast cells from fetal tonsils. **(b)** Cytopathic effects caused by infection of fetal tonsil fibroblasts with adenovirus. **(c)** Cytopathic effects caused by infection of fetal tonsil fibroblasts with herpes simplex virus.

the cells generates a plaque or clearing in the lawn (figure 16.14*b* and **figure 16.16**). As shown in figure 16.16, plaque appearance often is characteristic of the virus being cultivated.

Plant viruses are cultivated in a variety of ways. Plant tissue cultures, cultures of separated cells, or cultures of protoplasts (cells lacking cell walls) may be used. Viruses also can be grown in whole plants. Leaves are mechanically inoculated when rubbed with a mixture of viruses and an abrasive. When the cell walls are broken by the abrasive, the viruses directly contact the plasma membrane and infect the exposed host cells. (In nature, the role of the abrasive is frequently filled by insects that suck or crush plant leaves and thus transmit viruses.) A localized **necrotic lesion** often develops due to the rapid death of cells in the infected area (**figure 16.17**). Even when lesions do not occur, the infected plant may show symptoms such as changes in pigmentation or leaf shape. Some plant viruses can be transmitted only if a diseased part is grafted onto a healthy plant.

16.6 Virus Purification and Assays

Virologists must be able to purify viruses and accurately determine their concentrations in order to study virus structure, reproduction, and other aspects of their biology. These methods are so important that the growth of virology as a modern discipline depended on their development.

Virus Purification

Purification makes use of several virus properties. Virions are very large relative to proteins, are often more stable than normal cell components, and have surface proteins. Because of these characteristics, many techniques useful for the isolation of proteins and organelles can be employed to isolate viruses. Four of the most widely used approaches are (1) differential and density gradient centrifugation, (2) precipitation of viruses,

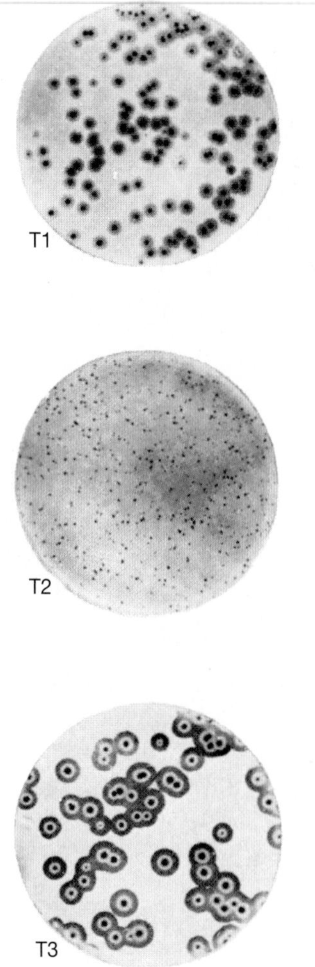

Figure 16.16 Phage Plaques. Plaques produced on a lawn of *E. coli* by some of the T coliphages (T1, T2, and T3 phages). Note the large differences in plaque appearance. The photographs are about 1/3 full size.

(a)

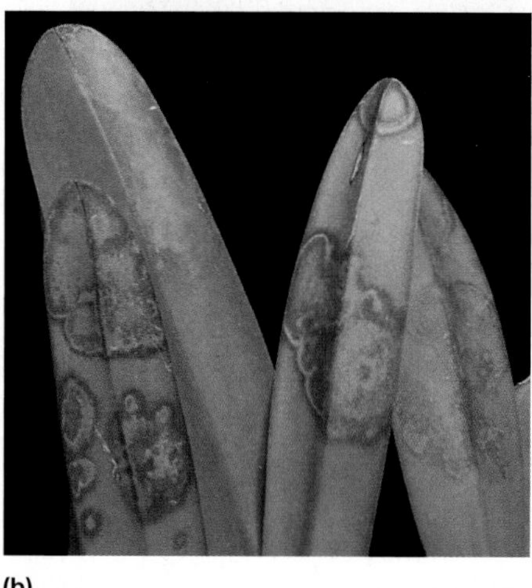

(b)

Figure 16.17 Necrotic Lesions on Plant Leaves. **(a)** Tobacco mosaic virus on *Nicotiana glutinosa.* **(b)** Tobacco mosaic virus infection of an orchid showing leaf color changes.

(3) denaturation of contaminants, and (4) enzymatic digestion of host cell constituents.

Differential and density gradient centrifugation often are used in the initial purification steps to separate virus particles from host cells. The process begins with host cells in later stages of infection because they contain mature virions. Infected cells are first disrupted in a buffer to produce an aqueous suspension or homogenate consisting of cell components and viruses. Viruses can then be isolated by **differential centrifugation,** the centrifugation of a suspension at various speeds to separate particles of different sizes (**figure 16.18**). Usually the homogenate is first centrifuged at high speed to sediment viruses and other large cellular particles. The supernatant, which contains the homogenate's soluble molecules, is discarded. The pellet is next resuspended and centrifuged at a low speed to remove substances heavier than viruses. Finally, higher speed centrifugation sediments the viruses. This process may be repeated to purify the virus particles further.

Additional purification of a virus preparation can be achieved by **gradient centrifugation** (**figure 16.19**). A sucrose solution is poured into a centrifuge tube so that its concentration smoothly and linearly increases from the top to the bottom of the tube. The virus preparation, often produced by differential centrifugation, is layered on top of the gradient and centrifuged. As shown in figure 16.19*a,* the particles settle under centrifugal force until they come to rest at the level where the virus and sucrose densities are equal (isopycnic gradient centrifugation). Viruses can be separated from other particles based on very small differences in density. Gradients also can separate viruses based on differences in their sedimentation rate (rate zonal gradient centrifugation). When this is done, particles are separated on the basis of both size and density; usually the largest virus will move most rapidly down the gradient. Figure 16.19*b* shows that viruses differ from one another and cell components with respect to either density (grams per milliliter) or sedimentation coefficient(s). Thus these

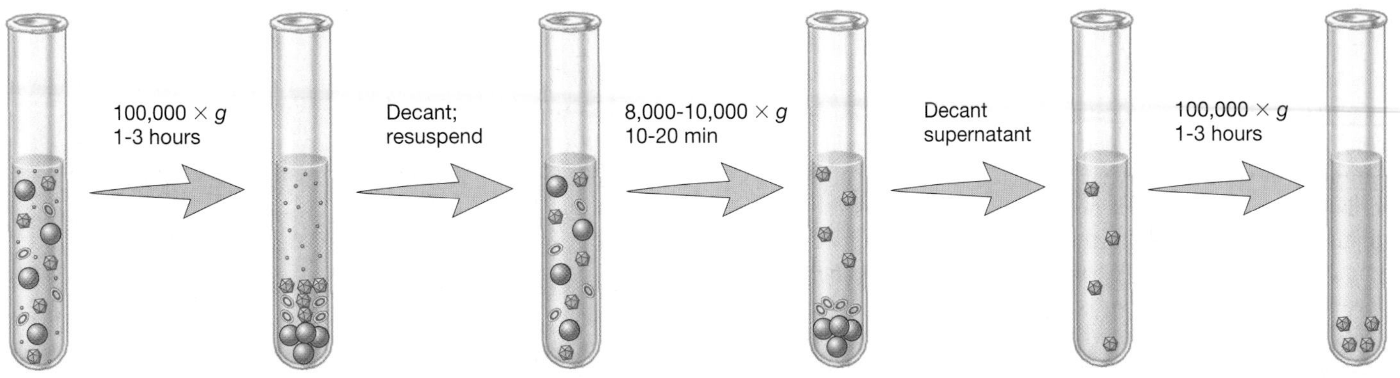

Figure 16.18 The Use of Differential Centrifugation to Purify a Virus. At the beginning the centrifuge tube contains homogenate and icosahedral viruses (in green). First, the viruses and heavier cell organelles are removed from smaller molecules. After resuspension, the mixture is centrifuged just fast enough to sediment cell organelles while leaving the smaller virus particles in suspension; the purified viruses are then collected. This process can be repeated several times to further purify the virions.

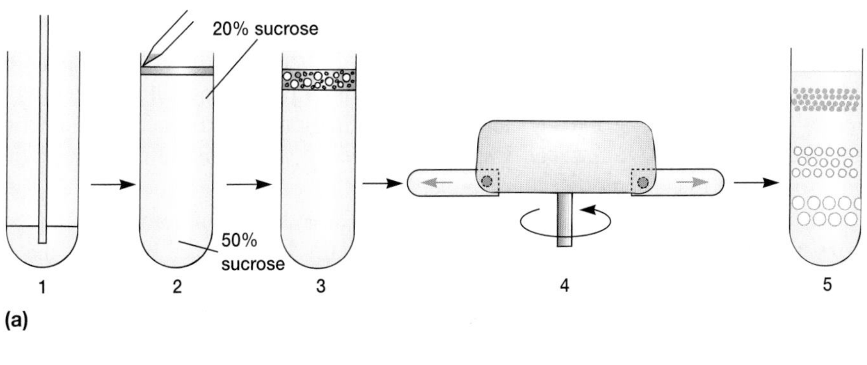

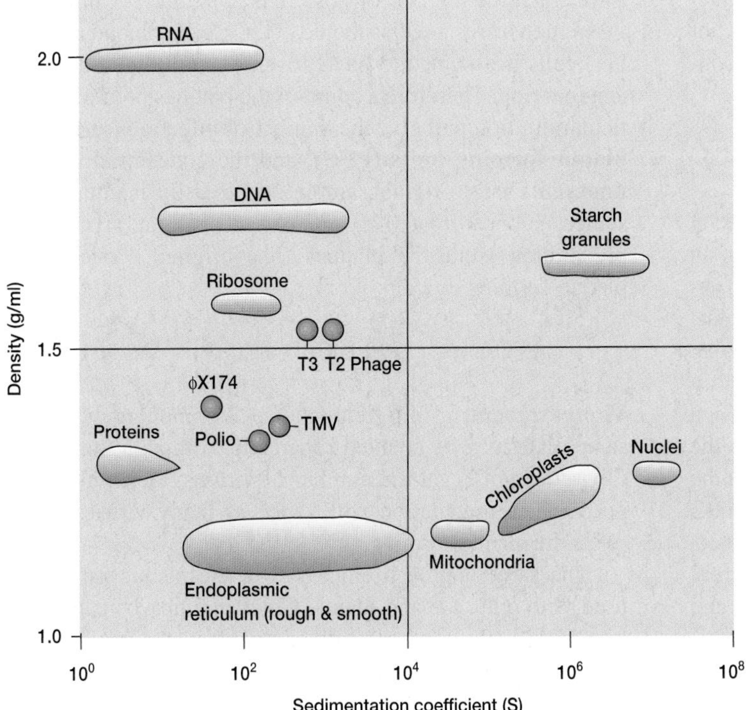

Figure 16.19 Gradient Centrifugation. **(a)** A linear sucrose gradient is prepared, *1*, and the particle mixture is layered on top, *2* and *3*. Centrifugation, *4*, separates the particles on the basis of their density and sedimentation coefficient, (the arrows in the centrifuge tubes indicate the direction of centrifugal force). *5*. In isopycnic gradient centrifugation, the bottom of the gradient is denser than any particle, and each particle comes to rest at a point in the gradient equal to its density. Rate zonal centrifugation separates particles based on their sedimentation coefficient, a function of both size and density, because the bottom of the gradient is less dense than the densest particles and centrifugation is carried out for a shorter time so that particles do not come to rest. The largest, most dense particles travel fastest. **(b)** The densities and sedimentation coefficients of representative viruses (shown in color) and other biological substances.

two types of gradient centrifugation are very effective in virus purification.

Although centrifugation procedures remove much cellular material, some cell components can remain in the virus preparation. Viruses can be separated from any remaining cellular contaminants by precipitation, denaturing, or enzymatic degradation of the contaminants. Many precipitation procedures use ammonium sulfate to precipitate the cellular contaminants. Initially, ammonium sulfate is added to a concentration just below that needed to precipitate the virus particles. Thus many cell components precipitate while the virus remains in solution. After any precipitated contaminants are removed, more ammonium sulfate is added and the precipitated viruses are collected by centrifugation. Viruses sensitive to ammonium sulfate often are purified by precipitation with polyethylene glycol.

Because viruses frequently are less sensitive to denaturing conditions than many cell components, exposure of a virus preparation to heat or a change in pH can be used in the final steps of virus purification. Furthermore, some viruses also tolerate treatment with organic solvents like butanol and chloroform. Thus solvent treatment can be used to both denature protein contaminants and extract any lipids in the preparation. The solvent is thoroughly mixed with the virus preparation, then allowed to stand and separate into organic and aqueous layers. The unaltered virus remains suspended in the aqueous phase while lipids dissolve in the organic phase. Substances denatured by organic solvents collect at the interface between the aqueous and organic phases.

One of the last steps used to purify viruses is enzymatic degradation of contaminants. This often is used to remove any remaining cellular proteins and nucleic acids in the virus preparation. Although viruses are composed of a protein coat surrounding a nucleic acid, they usually are more resistant to attack by nucleases and proteases than are free nucleic acids and proteins. For example, ribonuclease and trypsin often degrade cellular ribonucleic acids and proteins while leaving virions unaltered.

Virus Assays

The quantity of viruses in a sample can be determined either directly by counting particle numbers or indirectly by measurement of an observable effect of the virus. The values obtained by the two approaches often do not correlate closely; however, both are of value.

Virions can be counted directly with the electron microscope. In one procedure the virus-containing sample is mixed with a known concentration of small latex beads and sprayed on a coated specimen grid. The beads and virions are counted; the virus concentration is calculated from these counts and from the bead concentration (**figure 16.20**). This technique often works well with concentrated preparations of viruses of known morphology. Viruses can be concentrated by centrifugation before counting if the preparation is too dilute. However, if the beads and viruses are not evenly distributed (as sometimes happens), the final count will be inaccurate.

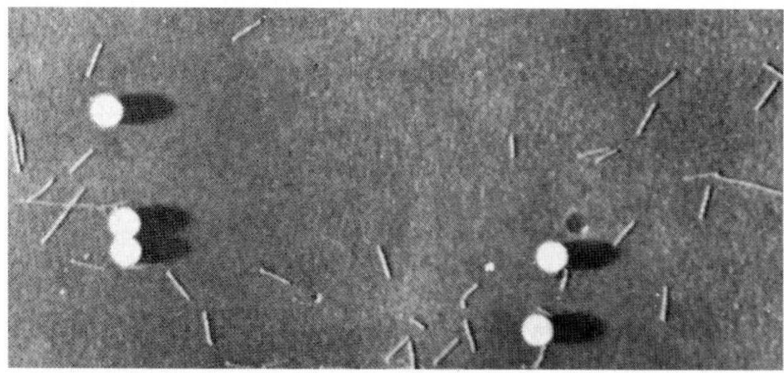

Figure 16.20 Tobacco Mosaic Virus. A tobacco mosaic virus preparation viewed in the transmission electron microscope. Latex beads 264 nm in diameter (white spheres) have been added.

An indirect method of counting virus particles is the **hemagglutination assay.** Many viruses can bind to the surface of red blood cells (*see figure 35.11*). If the ratio of viruses to cells is large enough, virus particles join the red blood cells together—that is, they agglutinate, forming a network that settles out of suspension. In practice, red blood cells are mixed with a series of virus dilutions and each mixture is examined. The hemagglutination titer is the highest dilution of virus (or the reciprocal of the dilution) that still causes hemagglutination. This assay is an accurate, rapid method for determining the relative quantity of viruses such as the influenza virus. If the actual number of viruses needed to cause hemagglutination is determined by another technique, the assay can be used to ascertain the number of virions present in a sample.

A variety of indirect assays determine virus numbers in terms of infectivity, and many of these are based on the same techniques used for virus cultivation. For example, in the **plaque assay** several dilutions of viruses are plated out with appropriate host cells. When the number of viruses plated are much lower than the number of host cells available for infection and when the viruses are distributed evenly, each plaque in a layer of host cells is assumed to have arisen from the reproduction of a single virion. Therefore a count of the plaques produced at a particular dilution will give the number of infectious virions, called **plaque-forming units (PFU),** and the concentration of infectious units in the original sample can be easily calculated. For instance, suppose that 0.10 ml of a 10^{-6} dilution of the virus preparation yields 75 plaques. The original concentration of plaque-forming units is

$$\text{PFU/ml} = (75\ \text{PFU}/0.10\ \text{ml})(10^6) = 7.5 \times 10^8.$$

Viruses producing different plaque morphology types on the same plate may be counted separately. Although the number of PFU does not equal the number of virions, their ratios are proportional: a preparation with twice as many viruses will have twice the plaque-forming units.

The same approach employed in the plaque assay may be used with embryos and plants. Chicken embryos can be inoculated with a diluted preparation or plant leaves rubbed with a mixture of diluted virus and abrasive. The number of pocks on embryonic membranes or necrotic lesions on leaves is multiplied by the dilution factor and divided by the inoculum volume to obtain the concentration of infectious units.

When biological effects are not readily quantified in these ways, the amount of virus required to cause disease or death can be determined by the endpoint method. Organisms or cell cultures are inoculated with serial dilutions of a virus suspension. The results are used to find the endpoint dilution at which 50% of the host cells or organisms are killed (**figure 16.21**). The **lethal dose (LD_{50})** is the dilution that contains a dose large enough to destroy 50% of the host cells or organisms. In a similar sense, the **infectious dose (ID_{50})** is the dose that, when given to a number of hosts, causes an infection of 50% of the hosts under the conditions employed.

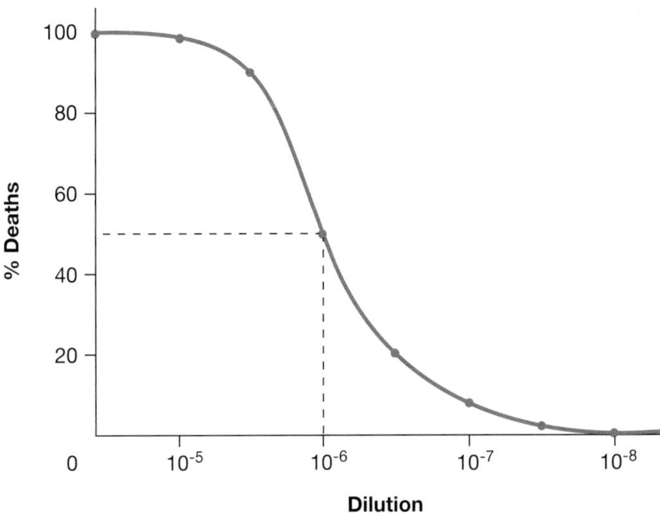

Figure 16.21 A Hypothetical Dose-Response Curve. The LD_{50} is indicated by the dashed line.

1. Discuss the ways that viruses can be cultivated. Define the terms pock, plaque, cytopathic effect, and necrotic lesion.
2. Give the four major approaches by which viruses may be purified, and describe how each works. Distinguish between differential and density gradient centrifugation in terms of how they are carried out.
3. How can one find the virus concentration, both directly and indirectly, by particle counts and measurement of infectious unit concentration? Define plaque-forming units, lethal dose, and infectious dose.

16.7 Principles of Virus Taxonomy

The classification of viruses is in a much less satisfactory state than that of cellular microorganisms. In part, this is due to a lack of knowledge of their origin and evolutionary history (**Microbial Tidbits 16.2**). In 1971, the International Committee for Taxonomy of Viruses (ICTV) developed a uniform classification system. Since then the number of viruses and taxonomic categories has continued to expand. In its eighth report, the ICTV described almost 2000 virus species and placed them in 3 orders, 73 families, 9 subfamilies, and 287 genera (**table 16.2**). The committee places greatest weight on specific properties to define families: nucleic acid type, nucleic acid strandedness, the sense (positive or negative) of ssRNA genomes, presence or absence of an envelope, symmetry of the capsid, and dimensions of the virion and capsid. Virus order

Microbial Tidbits

16.2 The Origin of Viruses

The origin and subsequent evolution of viruses are shrouded in mystery, in part because of the lack of a fossil record. However, recent advances in the understanding of virus structure and reproduction have made possible more informed speculation on virus origins. At present there are two major hypotheses entertained by virologists. It has been proposed that at least some of the more complex enveloped viruses, such as the poxviruses and herpesviruses, arose from small cells, probably procaryotic, that parasitized larger, more complex cells. These parasitic cells became ever simpler and more dependent on their hosts, much like multicellular parasites have done, in a process known as retrograde evolution. There are several problems with this hypothesis. Viruses are radically different from procaryotes, and it is difficult to envision the mechanisms by which such a transformation might have occurred or the selective pressures leading to it. In addition, one would expect to find some forms intermediate between procaryotes and at least the more complex enveloped viruses, but such forms have not been detected.

The second hypothesis is that viruses represent cellular nucleic acids that have become partially independent of the cell. Possibly a few mutations could convert nucleic acids, which are only synthesized at specific times, into infectious nucleic acids whose replication could not be controlled. This conjecture is supported by the observation that the nucleic acids of retroviruses (*see section 18.2*) and a number of other virions contain sequences quite similar to those of normal cells, plasmids, and transposons (*see chapter 13*). The small, infectious RNAs called viroids (*see section 18.9*) have base sequences complementary to transposons (*see section 13.5*), the regions around the boundary of mRNA introns, and portions of host DNA. This has led to speculation that they have arisen from introns or transposons. It has been proposed that cellular proteins spontaneously assembled into icosahedra around infectious nucleic acids to produce primitive virions.

It is possible that viruses have arisen by way of both mechanisms. Because viruses differ so greatly from one another, it seems likely that they have originated independently many times during the course of evolution. Many viruses have evolved from other viruses just as cellular organisms have arisen from specific predecessors. The question of virus origins is complex and quite speculative; future progress in understanding virus structure and reproduction may clarify this question.

names end in *virales;* virus family names in *viridae;* subfamily names, in *virinae;* and genus (and species) names, in *virus.* An example of this nomenclature scheme is shown in **figure 16.22.**

Although the ICTV committee reports are the official authority on viral taxonomy, many virologists find it useful to use an alternative classification scheme devised by Nobel laureate David Baltimore. The Baltimore system complements the ICTV system but focuses on the genome of the virus and the process used to synthesize viral mRNA. Baltimore's original system recognized six groups of viruses. Since then the system has been expanded to include seven groups. This was done in part by considering genome replication as well as mRNA synthesis in the classification scheme (**table 16.3**). As discussed in chapters 17 and 18, such a system helps virologists (and microbiology students) simplify the vast array of viral life cycles into a relatively small number of basic types.

Table 16.2 Some Common Virus Groups and Their Characteristics

ICTV Taxon (Baltimore System Group)[a]	Genome Size (kbp or kb)	Nucleic Acid	Strandedness	Capsid Symmetry[b]	Number of Capsomers	Presence of Envelope	Size of Capsid (nm)[c]	Host Range[d]
Picornaviridae (IV)	7–8	RNA	Single	I	32	–	22–30	A
Togaviridae (IV)	10–12		Single	I	32	+	40–70(e)	A
Retroviridae (VI)	7–12		Single	I?		+	100(e)	A
Orthomyxoviridae (V)	10–15		Single	H		+	9(h), 80–120(e)	A
Paramyxoviridae (V)	15		Single	H		+	18(h), 125–250(e)	A
Coronaviridae (IV)	27–31		Single	H		+	14–16(h), 80–160(e)	A
Rhabdoviridae (V)	11–15		Single	H		+	18(h), 70–80 × 130–240 (bullet shaped)	A
Bromoviridae (IV)	8–9		Single	I,B		–	26–35; 18–26 × 30–85	P
Tobamovirus (IV)	7		Single	H		–	18 × 300	P
Leviviridae [Qβ] *(IV)*	3–4		Single	I	32	–	26–27	B
Reoviridae (III)	19–32	RNA	Double	I	92	–	70–80	A,P
Cystoviridae (III)	13		Double	I		+	100(e)	B
Parvoviridae (II)	4–6	DNA	Single	I	12	–	20–25	A
Geminiviridae (II)	3–6		Single	I		–	18 × 30 (paired particles)	P
Microviridae (II)	4–6		Single	I		–	25–35	B
Inoviridae (II)	7–9		Single	H		–	6 × 900–1,900	B
Polyomaviridae (I)	5	DNA	Double	I	72	–	40	A
Papillomaviridae (I)	7–8		Double	I	72	–	55	A
Adenoviridae (I)	28–45		Double	I	252	–	60–90	A
Iridoviridae (I)	140–383		Double	I		–	130–180	A
Herpesviridae (I)	125–240		Double	I	162	+	100, 180–200(e)	A
Poxviridae (I)	130–375		Double	C		+	200–260 × 250–290(e)	A
Baculoviridae (I)	80–180		Double	H		+	40 × 300(e)	A
Hepadnaviridae (VII)	3		Double	C	42	+	28 (core), 42(e)	A
Caulimoviridae (I)	8		Double	I,B		–	50; 30 × 60–900	P
Corticoviridae (I)	9		Double	I		–	60	B
Myoviridae (I)	39–169		Double	Bi		–	80 × 110, 110[e]	B, Arch
Lipothrixviridae (I)	16		Double	H		+	38 × 410	Arch

[a]ICTV = International Committee on Virus Taxonomy. The ICTV and Baltimore Clarification Systems are discussed in section 16.7.

[b]Types of symmetry: I, icosahedral; H, helical; C, complex; Bi, binal; B, bacilliform.

[c]Diameter of helical capsid (h); diameter of enveloped virion (e).

[d]Host range: A, animal; P, plant; B, bacterium; Arch, archaeon.

[e]The first number is the head diameter; the second number, the tail length.

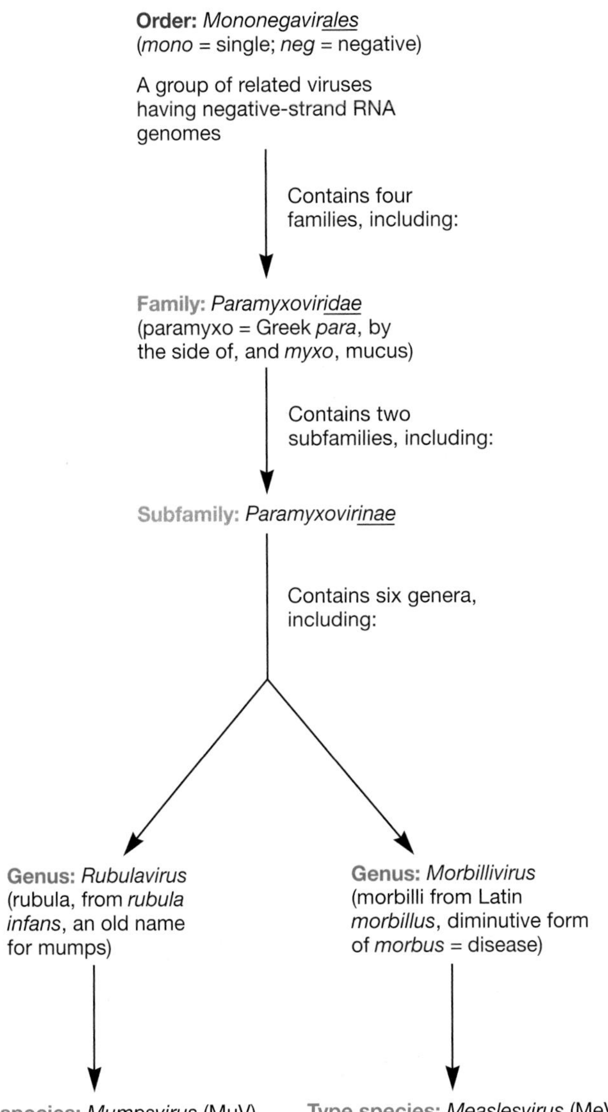
Order: Mononegavirales
(mono = single; neg = negative)
A group of related viruses having negative-strand RNA genomes
Contains four families, including:
Family: Paramyxoviridae
(paramyxo = Greek para, by the side of, and myxo, mucus)
Contains two subfamilies, including:
Subfamily: Paramyxovirinae
Contains six genera, including:
Genus: Rubulavirus
(rubula, from rubula infans, an old name for mumps)
Genus: Morbillivirus
(morbilli from Latin morbillus, diminutive form of morbus = disease)
Type species: Mumpsvirus (MuV)
Type species: Measlesvirus (MeV)

Figure 16.22 The Naming of Viruses. Because of the difficulty in establishing evolutionary relationships, most virus families have not been placed into an order. Virus names are derived from various aspects of their biology and history, including the features of their structure, diseases they cause, and locations where they were first identified or recognized.

Table 16.3 The Baltimore System of Virus Classification

Group	Description
I	Double-stranded DNA genome *genome replication: dsDNA → dsDNA* *mRNA synthesis: dsDNA → mRNA*
II	Single-stranded DNA genome *genome replication: ssDNA → dsDNA → ssDNA* *mRNA synthesis: ssDNA → dsDNA → mRNA*
III	Double-stranded RNA genome *replication: dsRNA → ssRNA → dsRNA* *mRNA synthesis: dsRNA → mRNA*
IV	Plus-strand RNA genome *replication: +RNA → −RNA → +RNA* *mRNA synthesis: +RNA = mRNA*
V	Negative-strand RNA genome *replication: −RNA → +RNA → −RNA* *mRNA synthesis: −RNA → mRNA*
VI	Single-stranded RNA genome *replication: ssRNA → dsDNA → ssRNA* *mRNA synthesis: ssRNA → dsDNA → mRNA*
VII	Double-stranded gapped DNA genome *replication: gapped dsDNA → dsDNA → +RNA → −DNA → gapped dsDNA* *mRNA synthesis: gapped dsDNA → dsDNA → mRNA*

As is the case with the taxonomy of cellular life forms, the taxonomy of viruses is rapidly changing as more and more viral genomes are sequenced. They have been useful in establishing evolutionary relationships among viruses, and have led to the creation of new virus families and genera.

1. List some characteristics used in classifying viruses. Which seem to be the most important?
2. What are the endings for the names of virus families, subfamilies, and genera or species?

Summary

16.1 Early Development of Virology

a. Europeans were first protected from a viral disease when Edward Jenner developed a smallpox vaccine in 1798.

b. Chamberland's invention of a porcelain filter that could remove bacteria from virus samples enabled microbiologists to show that viruses were different from bacteria.

c. In the late 1930s Stanley, Bawden, and Pirie crystallized the tobacco mosaic virus and demonstrated that it was composed only of protein and nucleic acid.

16.2 General Properties of Viruses

a. A virion is composed of either DNA or RNA enclosed in a coat of protein (and sometimes other substances as well). It cannot reproduce independently of living cells.

16.3 The Structure of Viruses

a. All virions have a nucleocapsid composed of a nucleic acid, usually either DNA or RNA, held within a protein capsid made of one or more types of protein subunits called protomers (**figure 16.1**).

b. There are four types of viral morphology: naked icosahedral, naked helical, enveloped icosahedral and helical, and complex.

c. Helical capsids resemble long hollow protein tubes and may be either rigid or quite flexible. The nucleic acid is coiled in a spiral on the inside of the cylinder (**figure 16.3*b***).

d. Icosahedral capsids are usually constructed from two types of capsomers: pentamers (pentons) at the vertices and hexamers (hexons) on the edges and faces of the icosahedron (**figure 16.6**).

e. Complex viruses (e.g., poxviruses and large phages) have complicated morphology not characterized by icosahedral and helical symmetry. Large phages often have binal symmetry: their heads are icosahedral and their tails, helical (**figure 16.9**).

f. Viruses can have a membranous envelope surrounding their nucleocapsid. The envelope lipids usually come from the host cell; in contrast, many envelope proteins are viral and may project from the envelope surface as spikes or peplomers.

g. Viral nucleic acids can be either single stranded or double stranded, DNA or RNA. Most DNA viruses have double-stranded DNA genomes that may be linear or closed circles (**table 16.1**).

h. RNA viruses usually have ssRNA that may be either plus (positive) or minus (negative) when compared with mRNA (positive). Many RNA genomes are segmented.

i. Although viruses lack true metabolism, some contain a few enzymes necessary for their reproduction.

16.4 Virus Reproduction

a. Virus reproduction can be divided into five steps: (1) attachment to host; (2) entry into host; (3) synthesis of viral nucleic acid and proteins; (4) self-assembly of virions and; (5) release from host (**figure 16.12**).

16.5 The Cultivation of Viruses

a. Viruses are cultivated using tissue cultures, embryonated eggs, bacterial cultures, and other living hosts (**figure 16.13**).

b. Sites of animal viral infection may be characterized by cytopathic effects such as pocks and plaques. Phages produce plaques in bacterial lawns. Plant viruses can cause localized necrotic lesions in plant tissues (**figures 16.14** to **16.17**).

16.6 Virus Purification and Assays

a. Viruses can be purified by techniques such as differential and gradient centrifugation, precipitation, and denaturation or digestion of contaminants (**figures 16.18** and **16.19**).

b. Virus particles can be counted directly with the transmission electron microscope or indirectly by the hemagglutination assay (**figure 16.20**).

c. Infectivity assays can be used to estimate virus numbers in terms of plaque-forming units, lethal dose (LD_{50}), or infectious dose (ID_{50}).

16.7 Principles of Virus Taxonomy

a. Currently viruses are classified with a taxonomic system placing primary emphasis on the type and strandedness of viral nucleic acids, and on the presence or absence of an envelope (**table 16.2**).

b. The Baltimore system of virus classification is used by many virologists to organize viruses based on their genome type and the mechanisms they use to synthesize mRNA and replicate their genomes (**table 16.3**).

Key Terms

bacteriophage 409
binal symmetry 412
capsid 409
capsomers 410
cytopathic effects 418
differential centrifugation 420
envelope 412
enveloped virus 410
gradient centrifugation 420
helical capsid 410
hemagglutination assay 422
hexamers (hexons) 411
icosahedral capsid 410
infectious dose (ID_{50}) 423
lethal dose (LD_{50}) 423
minus strand or negative strand 416
naked virus 410
necrotic lesion 419
nucleocapsid 409
pentamers (pentons) 411
peplomer or spike 412
phage 409
plaque 418
plaque assay 422
plaque-forming units (PFU) 422
plus strand or positive strand 416
protomers 409
segmented genome 417
virion 409
virologist 407
virology 407
virus 409

Critical Thinking Questions

1. Many classification schemes are used to identify bacteria. These start with Gram staining, progress to morphology/ arrangement characteristics, and include a battery of metabolic tests. Build an analogous scheme that could be used to identify viruses. You might start by considering the host, or you might start with viruses found in a particular environment, such as a marine filtrate.

2. Consider the different perspectives on the origin of viruses in Microbial Tidbits 16.2. Discuss whether you think viruses evolved before the first procaryote, or whether they have coevolved, and are perhaps still coevolving with their hosts.

Learn More

Diamond, J. 1999. *Guns, germs, and steel.* New York: W.W. Norton.

Flint, S. J.; Enquist, L. W.; Racaniello, V. R.; and Skalka, A. M. 2004. *Principles of virology,* 2d ed. Washington, D.C.: ASM Press.

Foster, K. R.; Jenkins, M. F.; and Toogood, A. C. 1998. The Philadelphia yellow fever epidemic of 1793. *Scientific American* (August):88–93.

Hull, R. 2002. *Matthews' plant virology,* 4th ed. San Diego: Academic Press.

Knipe, D. M., and Howley, P. M., editors-in-chief. 2001. *Fields virology,* 4th ed. New York: Lippincott Williams & Wilkins.

Mayo, M. A.; Maniloff, J., Desselberger, U.; Ball, L. A.; and Fauquet, C. M., editors. 2005. *Virus taxonomy: VIIIth report of the International Committee on Taxonomy of Viruses.* San Diego: Elservier Academic Press.

Nelson, D. 2004. Phage taxonomy: We agree to disagree. *J. Bacteriol.* 186(21):7029–31.

Oldstone, M. B. 1998. *Viruses, plagues, and history.* New York: Oxford University Press.

Zaitlin, M. 1999. Tobacco mosaic virus and its contributions to virology. *ASM News* 65(10):675–80.

Please visit the Prescott website at www.mhhe.com/prescott7 for additional references.

17 The Viruses: Viruses of *Bacteria* and *Archaea*

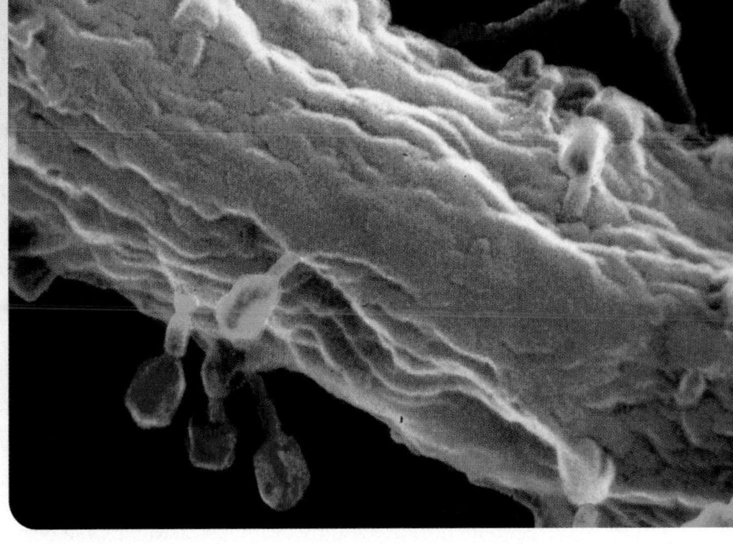

A scanning electron micrograph of T-even bacteriophages infecting *E. coli*. The phages are colored blue.

Preview

- Viruses that infect procaryotic cells, both bacterial and archaeal, have been identified. These viruses are diverse in their morphology and reproductive strategies.
- Some procaryotic viruses are virulent viruses. Shortly after infecting their host, they begin reproduction. At the completion of their life cycle, they lyse their host. This type of life cycle is called a lytic cycle.
- Some procaryotic viruses begin reproduction upon entering their host but do not kill the host by lysis. Some are extruded from the cell. Although many cellular processes are slowed, the cell remains viable and can release many virus progeny over time.
- Many procaryotic viruses are temperate viruses. The genomes of many temperate viruses can integrate into the host cell's chromosome, where they are replicated as the host chromosome is replicated. This relationship is called lysogeny and it can be maintained for long periods.
- Analysis of viral genomes is providing insight into the evolution of viruses and their hosts.

Chapter 16 introduces many of the facts and concepts underlying the field of virology, including information about the nature of viruses, their structure and taxonomy, and how they are cultivated and studied. Clearly the viruses are a complex, diverse, and fascinating group, the study of which has done much to advance disciplines such as genetics and molecular biology.

In this chapter we are concerned with viruses that infect procaryotic cells. Because the discovery of archaeal viruses is relatively recent and little is known about their biology, our focus will be on the **bacteriophages**—viruses that infect bacteria. Because of their crucial role in the history of genetics and molecular biology, it is tempting to think of bacteriophages as useful only for laboratory research in these disciplines. However, bacteriophages are significant members of terrestrial and aquatic ecosystems. Indeed, they may be the most abundant form of life on the planet and are major agents of microbial evolution. It has been estimated that numerous bacteriophage species infect each species of bacteria. *Escherichia coli,* for example, is subject to infection by more than 20 phage species. Bacteriophages are also important in industry and medicine. For example, many phages destroy the gram-positive lactic acid bacteria that are critical to the production of fermented milk products such as yogurt and cheese. Bacteriophages also can carry a variety of virulence factors that convert their bacterial hosts into potent pathogens. This is the case for major human pathogens such as *Streptococcus pyogenes, Staphylococcus aureus, Corynebacterium diphtheriae, Vibrio cholerae, E. coli* O157:H7, and *Salmonella enterica.* On the positive side, Russian physicians have used bacteriophages for years to treat bacterial diseases. Research indicates that phages may be effective in treating bacterial infections, including those caused by antibiotic-resistant bacteria.

You might wonder how such naive outsiders get to know about the existence of bacterial viruses. Quite by accident, I assure you. Let me illustrate by reference to an imaginary theoretical physicist, who knew little about biology in general, and nothing about bacterial viruses in particular. . . . Suppose now that our imaginary physicist, the student of Niels Bohr, is shown an experiment in which a virus particle enters a bacterial cell and 20 minutes later the bacterial cell is lysed and 100 virus particles are liberated. He will say: "How come, one particle has become 100 particles of the same kind in 20 minutes? That is very interesting. Let us find out how it happens!. . . Is this multiplying a trick of organic chemistry which the organic chemists have not yet discovered? Let us find out."

—*Max Delbrück*

17.1 Classification of Bacterial and Archaeal Viruses

Some of the families of bacterial and archaeal viruses are shown in **figure 17.1.** These families have been designated by the International Committee for the Taxonomy of Viruses (ICTV), the agency responsible for standardizing the classification of all viruses, including those that infect *Bacteria* and *Archaea*. Of the almost 2,000 virus species classified and catalogued by the ICTV, most are viruses of eucaryotes and bacteria. In fact, only about 40 archaeal viruses have been identified. Of those, only about eight have been assigned to virus taxa. Yet the discovery of archaeal viruses has had a significant impact on our understanding of viruses and on the ICTV classification scheme: archaeal viruses have led to the recognition of four new virus families, and at least three more families are awaiting ICTV approval **(Microbial Diversity and Ecology 17.1)**. This is due primarily to the unusual morphologies observed among the known archaeal viruses, including viruses that are spindle-shaped and droplet-shaped. Furthermore, many are enveloped viruses. Among the archaeal viruses, only *Methanobacterium* virus ψM1 and *Halobacterium* virus H are placed with bacterial viruses (families *Siphoviridae* and *Myoviridae*, respectively). This is because these two viruses have capsid and tail structures similar to the phages in these existing bacteriophage families. The remaining six assigned archaeal viruses define the new archaeal virus families *Fuselloviridae, Guttaviridae, Lipothrixviridae,* and *Rudiviridae*. All of the other families illustrated in figure 17.1 contain bacteriophages.

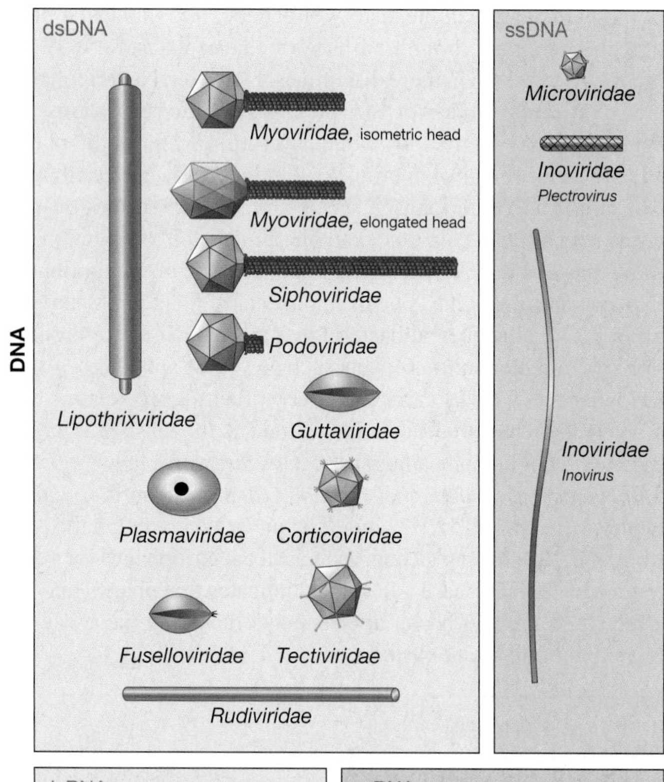

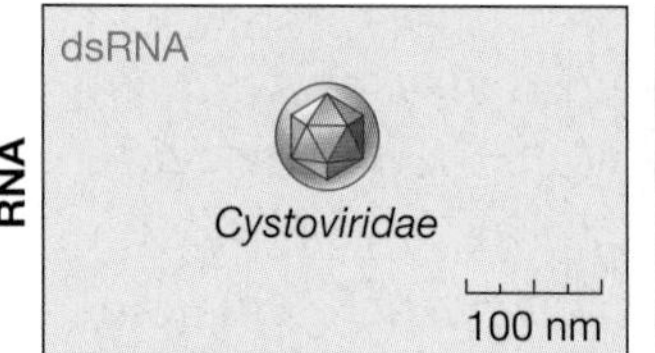

Figure 17.1 Families and Genera of Procaryotic Viruses. The *Myoviridae* are the only family with contractile tails. *Plasmaviridae* are pleomorphic. *Tectiviridae* have distinctive double capsids, whereas the *Corticoviridae* have complex capsids containing lipid.

Capsid structure is a critical phenotypic trait used to classify viruses of procaryotes. However, recent analyses of the genomes of bacteriophages have revealed problems with using phage morphology in taxonomy. This is especially true if one is trying to draw evolutionary relationships among the phages. As discussed in section 17.6, genetic modules appear to have been swapped across different species of viruses and even different families. This kind of lateral gene transfer makes the classification of bacteriophages difficult, and some virologists contend that current approaches to classifying bacteriophages need to be reconsidered and replaced by molecular approaches.

Bacteriophages exhibit a wide degree of diversity in terms of genome structure. Although most possess double-stranded DNA (dsDNA) genomes, phages with single-stranded DNA (ssDNA), single-stranded RNA (ssRNA), and double-stranded RNA (dsRNA) genomes have been identified and studied. Thus far, there is less diversity among the genomes of archaeal viruses. All are dsDNA viruses, either circular or linear. Because the nature of the genome correlates with the mechanisms used for synthesis of viral mRNA and with the replication of viral genomes, the discussion of bacteriophage reproduction that follows focuses primarily on these processes.

17.2 Virulent Double-Stranded DNA Phages

After DNA bacteriophages have reproduced within the host cell, many of them are released when the cell is destroyed by lysis. A phage life cycle that culminates with the host cell bursting and releasing virions is called a **lytic cycle,** and viruses that reproduce solely in this way are called **virulent viruses.** The events taking place during the lytic cycle are reviewed in this section, with the primary focus on the T-even phages of *E. coli,* which are some of the most complex viruses known *(see figure 16.9)*. T-even phages are double-stranded DNA bacteriophages with complex contractile tails. They are placed in the family *Myoviridae.*

The One-Step Growth Experiment

The development of the one-step growth experiment in 1939 by Max Delbrück and Emory Ellis marks the beginning of modern bacteriophage research. In a **one-step growth experiment,** the reproduction of a large phage population is synchronized so that the molecular events occurring during reproduction can be followed. A culture of susceptible bacteria such as *E. coli* is mixed with bacteriophage particles, and the phages are allowed a short interval to attach to their host cells. The culture is then greatly diluted so that any virions released upon host cell lysis do not immediately infect new

Microbial Diversity & Ecology

17.1 Host Independent Growth of an Archaeal Virus

The fact that viruses cannot replicate without first infecting a host cell has resulted in their classification as "acellular entities" or "forms"—they are not cells. So it was quite a surprise when an archaeal virus that develops long tails only when outside its host was discovered **(see Box figure)**. This archaeal virus was found in acidic hot springs (pH 1.5, 85–93°C) in Italy where it infects the hyperthermophilic archaeon *Acidianus convivator*. When the virus infects its host, lemon-shaped virions are assembled. Following host lysis, two "tails" begin to form at either end of the virion. These projections continue to assemble until they reach a length at least that of the viral capsid. Curiously, tails are only produced if virions are incubated at high temperature, leading to the hypothesis that they are part of a survival strategy. The virus is thus called ATV for *A*cidianus two-*t*ailed *v*irus.

To find out more about the structure of the tails, the ATV genome was sequenced. ATV is a double-stranded DNA virus that encodes only nine structural proteins. The tail protein is an 800 amino acid protein that bears homology to eucaryotic intermediate filament proteins. Both intermediate filaments and purified ATV tail protein assemble into filamentous structures without additional energy or cofactors. The cytoplasmic matrix, microfilaments, intermediate filaments, and microtubules (section 4.3)

It is suspected that the development of tails only at high temperatures may be a survival strategy for the virus when host cell density is low. So far, ATV is the only virus that infects procaryotes living in acidic hot springs and induces lysis rather than lysogeny. It would thus seem all other such viruses have evolved lysogeny as a means to survive these harsh conditions. Why this virus has evolved lysis and tail development is unknown, but it suggests that viruses may be more complicated than simple "entities."

Häring, M.; Vestergaard, G.; Rachel, R.; Chen, L.; Garret, R. A.; and Prangishvili, D. 2005. Independent virus development outside a host. Nature 436:1101–02.

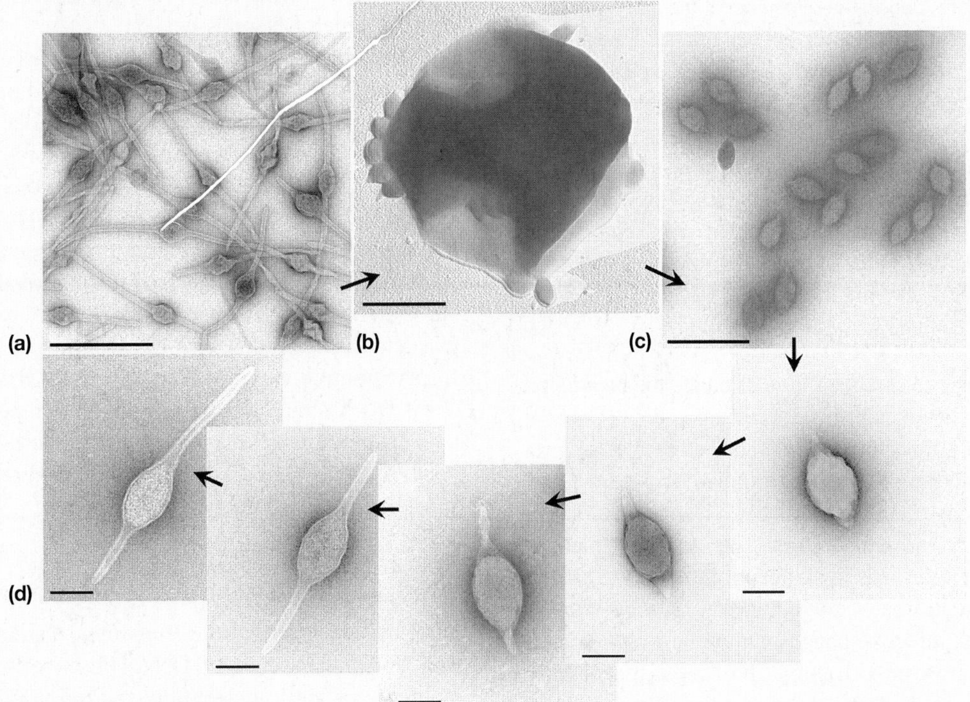

***Acidianus* Two-Tailed Virus, ATV.** **(a)** Virions collected from an acidic hot spring bear two long projections, or tails, at each end. **(b, c)** The hyperthermophilic archaeon *A. convivator* extrudes lemon-shaped virions. **(d)** These subsequently develop tail-like structures independent of the host and only when at high temperature. Scale bars: *a–c*, 0.5 μm; *d*, 0.1 μm.

cells. This strategy works because phages lack a means of seeking out host cells and must contact them during random movement through the solution. Thus phages are less likely to contact host cells in a dilute mixture. The number of infective phage particles released from bacteria is subsequently determined at various intervals by a plaque assay. Virus purification and assays (section 16.6)

A plot of the bacteriophages released from host cells versus time shows several distinct phases **(figure 17.2)**. During the **latent period,** which immediately follows phage addition, there is no release of virions. This is followed by the **rise period** (**burst**) when the host cells rapidly lyse and release infective phages. Finally, a plateau is reached and no more viruses are liberated. The total

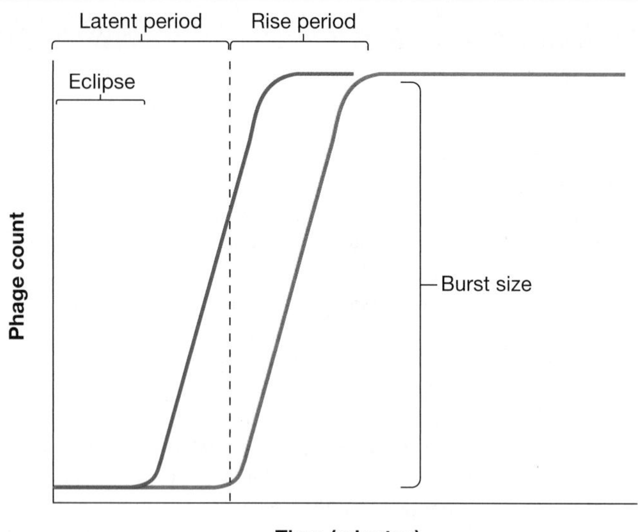

Figure 17.2 The One-Step Growth Curve. In the initial part of the latent period, the eclipse period, the host cells do not contain any complete, infective virions. During the remainder of the latent period, an increasing number of infective virions are present, but none are released. The latent period ends with host cell lysis and rapid release of virions during the rise period or burst. In this figure the blue line represents the total number of complete virions. The red line is the number of free viruses (the unadsorbed virions plus those released from host cells). When *E. coli* is infected with T2 phage at 37°C, the growth plateau is reached in about 30 minutes and the burst size is approximately 100 or more virions per cell. The eclipse period is 11–12 minutes, and the latent period is around 21–22 minutes.

number of phages released can be used to calculate the **burst size,** the number of viruses produced per infected cell.

The latent period is the shortest time required for virus reproduction and release. During the first part of this phase, host bacteria do not contain any complete, infective virions. This can be shown by lysing them with chloroform. This initial segment of the latent period is called the **eclipse period** because the virions detectable before infection are now concealed or eclipsed. The number of completed, infective phages within the host increases after the end of the eclipse period, and the host cell is prepared for lysis.

Between the beginning of a one-step growth experiment and the final burst, a carefully orchestrated series of events occurs **(figure 17.3)**. The major events—adsorption to the host cell, viral penetration of the host cell, synthesis of viral nucleic acids and proteins, assembly of phage particles and release of virions—are described next.

Adsorption and Penetration

Like all viruses, bacteriophages do not randomly attach to the surface of a host cell; rather, they fasten to specific surface structures called **receptors.** The nature of these receptors varies with the phage; cell wall lipopolysaccharides and proteins, teichoic acids, flagella, and pili can serve as receptors. The T-even phages of *E. coli* use cell wall lipopolysaccharides or proteins as receptors. Variation in receptor properties is at least partly responsible for phage host preferences.

T-even phage adsorption involves several tail structures. Phage attachment begins when a tail fiber contacts the appropriate receptor **(figure 17.4*a*)**. As more tail fibers make contact, the baseplate settles down on the surface (figure 17.4*b*). Binding is probably due to electrostatic interactions and is influenced by pH and the presence of ions such as Mg^{2+} and Ca^{2+}. After the baseplate is seated firmly on the cell surface, conformational changes occur in the baseplate and sheath, and the tail sheath reorganizes so that it shortens from a cylinder 24 rings long to one of 12 rings (figure 17.4*c, d*). That is, the sheath becomes shorter and wider, and the central tube or core is pushed through the bacterial wall. The baseplate contains the protein gp5, which has lysozyme activity. This aids in the penetration of the tube through the peptidoglycan layer. Finally, the linear DNA is extruded from the head, through the tail tube, and into the host cell (figure 17.4*e, f*). The tube may interact with the plasma membrane to form a pore through which DNA passes.

The penetration mechanisms of other bacteriophages can be different from that of the T-even phages, but most have not been studied in as much detail. An exception is the PRD1 phage of the family *Tectiviridae,* which has a membrane under its icosahedral capsid (figure 17.1). It infects pseudomonads and members of the *Enterobacteriaceae.* The PRD1 phage attaches to a surface receptor by a spike structure at one of its capsid vertices **(figure 17.5)**. This causes a conformational change in the capsid proteins. The attached spike-penton complex of the capsid dissociates from the virion, and the underlying membrane forms a tubular structure that penetrates the bacterial envelope. Virus DNA is then injected through the membrane tube into the bacterium. Penetration of the membrane tube is made possible partly by two enzymes, both of which break the same glycosidic bond that is attacked by lysozyme.

Synthesis of Phage Nucleic Acids and Proteins

With the exception of the *Hepadnaviridae* (*see tables 16.1 and 16.3*), all double-stranded DNA viruses follow a similar route for synthesis of viral nucleic acids and proteins **(figure 17.6)**. The DNA genome serves as the template for mRNA synthesis, and the mRNA molecules made are translated to yield viral proteins. Sometime after the onset of mRNA synthesis, DNA replication ensues and more viral genomes are made. The details of nucleic acid and protein synthesis vary from virus to virus, but all are designed to manipulate the host cell to the advantage of the virus. T4 bacteriophage will serve as our example.

Within 2 minutes after injection of T4 DNA into a host *E. coli* cell, the *E. coli* RNA polymerase starts synthesizing T4 mRNA (figure 17.3). This mRNA is called **early mRNA** because it is made before viral DNA is made. Early mRNA directs the synthesis of protein factors and enzymes required to take over the host

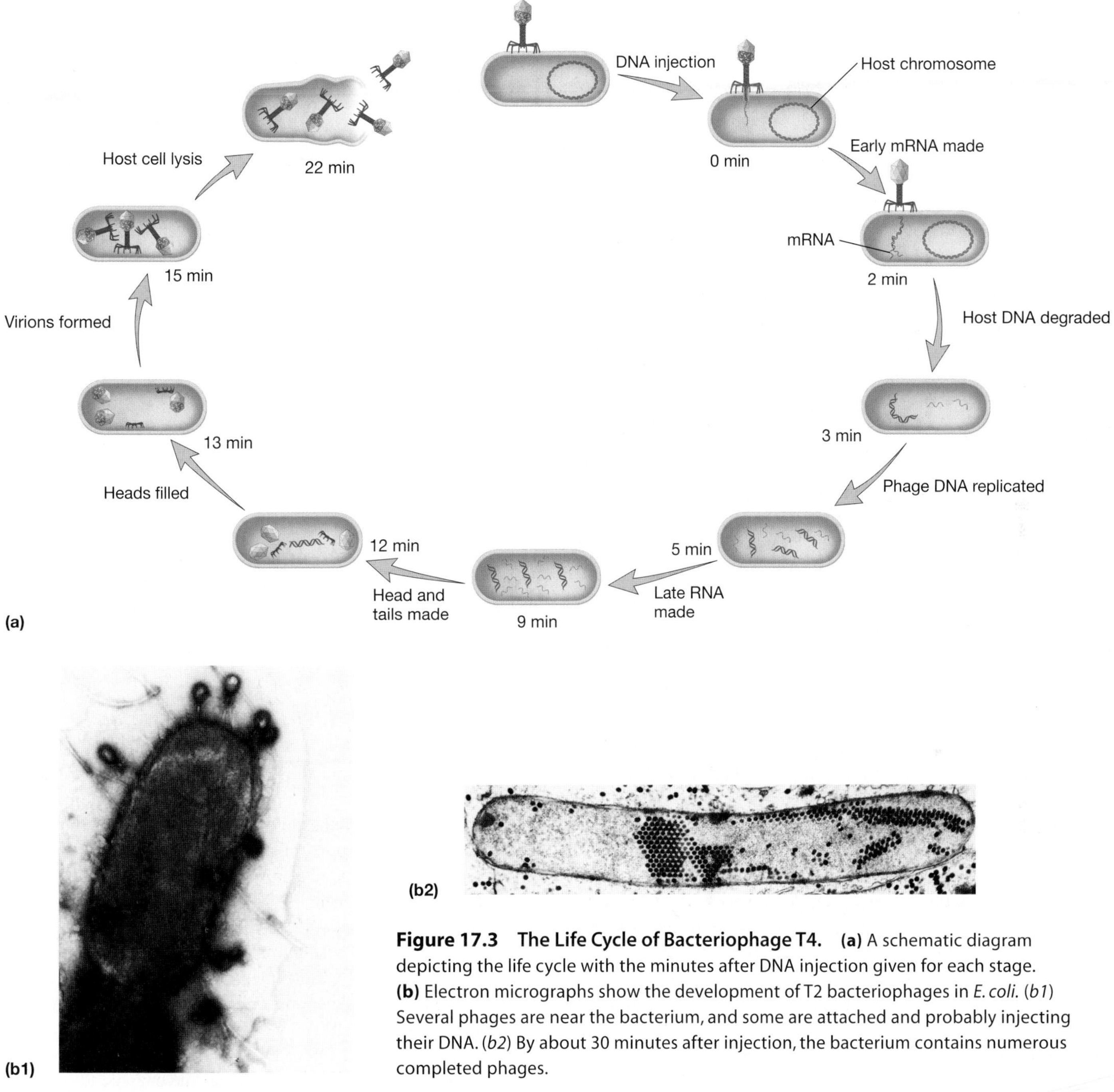

Figure 17.3 The Life Cycle of Bacteriophage T4. **(a)** A schematic diagram depicting the life cycle with the minutes after DNA injection given for each stage. **(b)** Electron micrographs show the development of T2 bacteriophages in *E. coli.* (*b1*) Several phages are near the bacterium, and some are attached and probably injecting their DNA. (*b2*) By about 30 minutes after injection, the bacterium contains numerous completed phages.

cell and force it to manufacture additional viral constituents. Some early virus-specific enzymes degrade host DNA to nucleotides, thereby simultaneously halting host gene expression and providing raw material for viral DNA synthesis. Within 5 minutes, viral DNA synthesis commences. DNA replication is initiated from several origins of replication and it proceeds bidirectionally from each. Viral DNA replication is followed by the synthesis of **late mRNAs,** which are important in later stages of the infection. Thus expression of viral genes is temporally ordered. How does T4 accomplish this?

T4 controls the expression of its genes by regulating the activity of the *E. coli* RNA polymerase. Initially T4 genes are transcribed by the regular host RNA polymerase and the sigma factor σ^{70} (*see table 12.3*). After a short interval, a virus enzyme catalyzes the transfer of the chemical group ADP-ribose from NAD to an α-subunit of RNA polymerase (*see figure 11.27*). This modification of the host enzyme helps inhibit the transcription of host genes and promotes virus gene expression. Later the second α-subunit receives an ADP-ribosyl group. This turns off some of the early T4 genes, but not before the product of one early gene,

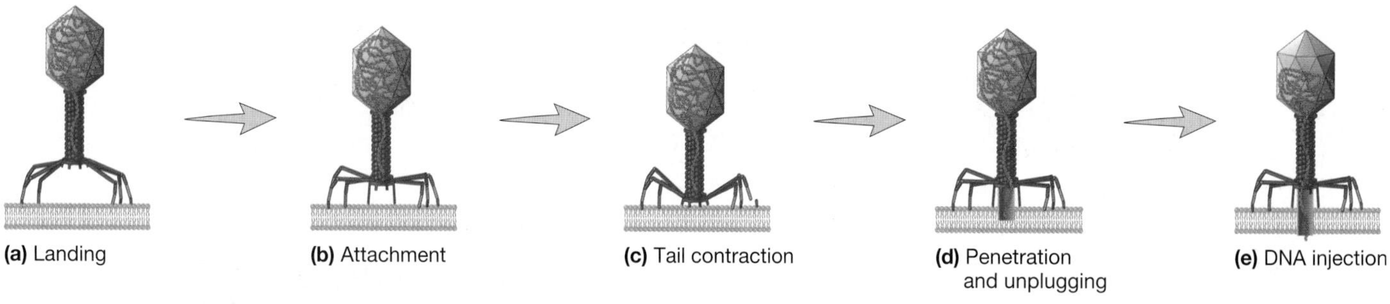

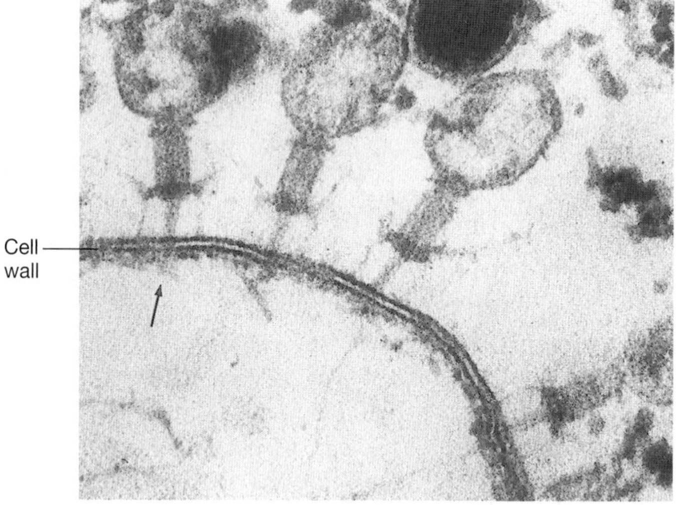

Figure 17.4 T4 Phage Adsorption and DNA Injection. **(a–e)** Adsorption and DNA injection is mediated by the phage's tail fibers and base plate as shown here. **(f)** An electron micrograph of an *E. coli* cell being infected by a T-even phage. These phages have injected their nucleic acid through the cell wall and now have empty capsids.

motA, stimulates transcription of somewhat later genes. One of these later genes encodes the sigma factor gp55. This sigma factor helps RNA polymerase bind to late promoters and transcribe late genes, which become active around 10 to 12 minutes after infection. Global regulatory systems (section 12.5)

The tight regulation of expression of T4 genes is aided by the organization of the T4 genome. As can be seen in **figure 17.7,** genes with related functions—such as the genes for phage head or tail fiber construction—are usually clustered together. Early and late genes also are clustered separately on the genome; they are even transcribed in different directions—early genes in the counterclockwise direction and late genes, clockwise.

Also apparent in figure 17.7 is that a considerable portion of the T4 genome encodes products needed for its replication, including all the protein subunits of its replisome and enzymes needed to prepare for synthesis of DNA **(figure 17.8)**. Some of these enzymes synthesize an important component of T4 DNA, **hydroxymethylcytosine (HMC) (figure 17.9)**. HMC is a modified nucleotide that replaces cytosine in T4 DNA. Once HMC is synthesized, replication ensues by a mechanism similar to that seen in bacteria. After T4 DNA has been synthesized, it is glucosylated by the addition of glucose to the HMC residues. Glucosylated HMC residues protect T4 DNA from attack by *E. coli* endonucleases called **restriction enzymes,** which would otherwise cleave the viral DNA at specific points and destroy it. This bacterial defense mechanism is called **restriction.** Other chemical groups also can be used to modify phage DNA and protect it against restriction enzymes. For example, methyl groups are added to the amino groups of adenine and cytosine in lambda phage DNA for the same reason.

The T4 genome is linear dsDNA and shows what is called terminal redundancy—that is, a base sequence is repeated at each end of the molecule. These two characteristics contribute to the formation of long DNA molecules called **concatemers,** which are composed of several genome units linked together in the same orientation **(figure 17.10)**. Why does this occur? As discussed in chapter 11, the ends of linear DNA molecules cannot be replicated without special machinery such as the enzyme telomerase observed in eucaryotes. T4 does not have telomerase activity. Therefore, each progeny DNA molecule has single-stranded 3′ ends. These ends participate in homologous recombination with double-stranded regions of other progeny DNA molecules, generating the concatemers. During assembly, concatemers are cleaved such that the genome packaged in the capsid is slightly longer than the T4 gene set. Thus each progeny virus has a genome unit that begins with a different gene. However, if each genome of the progeny viruses was circularized, the sequence of genes in each virion would be the same (figure 17.10). Therefore the T4 genome is said to be circularly permuted, and the genetic map of T4 is drawn as circular molecule.

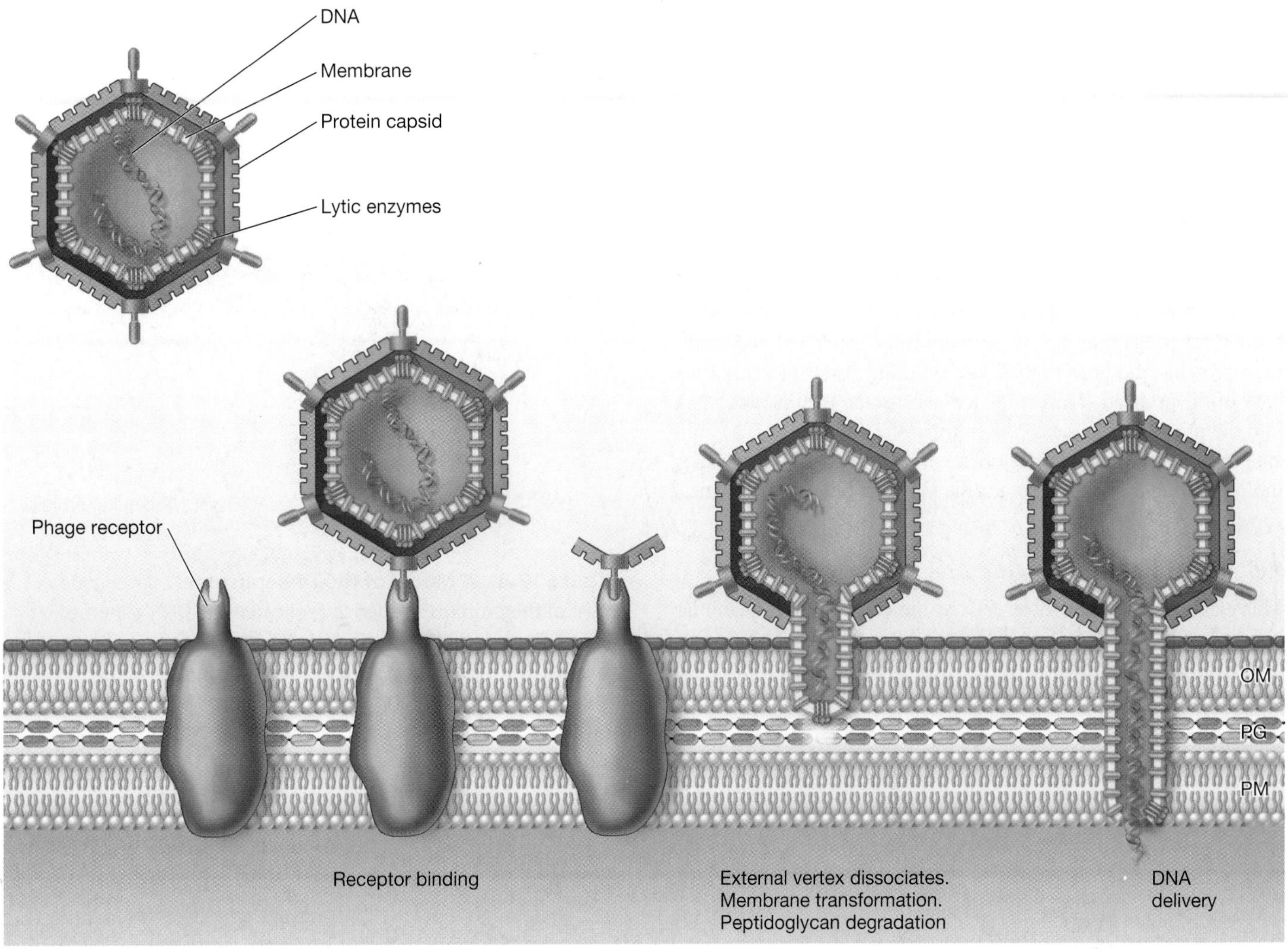

Figure 17.5 Attachment and Penetration of Host Cell by Phage PRD1. The gram-negative cell wall layers are indicated by OM (outer membrane), PG (peptidoglycan), and PM (plasma membrane). See text for details.

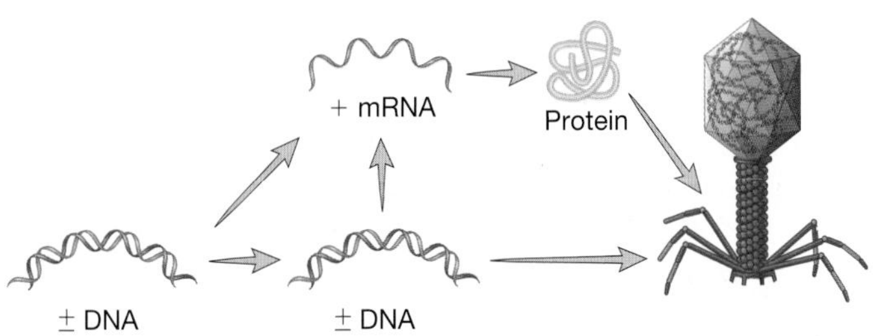

Figure 17.6 Replication Strategy Used by Double-Stranded DNA Viruses. Because the genome of double-stranded DNA viruses is similar to the host, the replication process closely resembles that of the host cell and can involve the use of host polymerases, viral polymerases, or both. The DNA serves as the template for DNA replication and mRNA synthesis. Translation of the mRNA by the host cell's translation machinery yields viral proteins, which are assembled with the viral DNA to make mature virions. These are eventually released from the host.

Assembly of Phage Particles

The assembly of T4 phage is an exceptionally complex self-assembly process that involves special virus proteins and some host cell factors. Late mRNA directs the synthesis of three kinds of proteins: (1) phage structural proteins, (2) proteins that help with phage assembly without becoming part of the virion structure, and (3) proteins involved in cell lysis and phage release. Late mRNA transcription begins about 9 minutes after T4 DNA injection into *E. coli.* All the phage proteins required for assembly are synthesized simultaneously and then used in four fairly independent subassembly lines **(figure 17.11)**. The baseplate is constructed of 16 gene products, which are assigned numbers rather than names **(figure 17.12)**. After the baseplate is finished, the tail tube is built on it and the sheath is assembled around the tube. The phage prohead (procapsid) is constructed of 10 proteins. The prohead is assembled with the aid of **scaffolding proteins** that are degraded or removed after construction is completed. A special portal protein is located at the base of the prohead where it connects to the tail. The portal protein participates in DNA packaging, yielding the mature head (figure 17.11). After completion of the head, it spontaneously combines with the tail assembly. Figure 17.12 shows the mature virion and the proteins from which it is constructed.

DNA packaging within the T4 prohead is accomplished by a complex of proteins sometimes called the "packasome." The packasome consists of the portal protein just mentioned. It also contains a set of proteins called the terminase complex, which generates double-stranded ends at the ends of the concatemers created when the viral genome was replicated. These double-stranded ends are needed for packaging the T4 genome. Once generated, the terminase proteins and phage DNA join with the portal protein at the base of the prohead. The completed packasome then moves DNA into the prohead, a process powered by ATP hydrolysis. The concatemer is cut when the phage head is filled with DNA—a DNA molecule roughly 3% longer than the length of one set of T4 genes.

Release of Phage Particles

Many phages lyse their host cells at the end of the intracellular phase. The lysis of *E. coli* by T4 takes places after about 150 virus particles have accumulated in the host cell **(figure 17.13)**. Two proteins are involved. One directs the synthesis of an enzyme that attacks peptidoglycan in the host's cell wall. It is sometimes called T4 lysozyme. Another T4 protein called holin creates holes in the *E. coli* plasma membrane, enabling T4 lysozyme to move from the cytoplasm to the peptidoglycan.

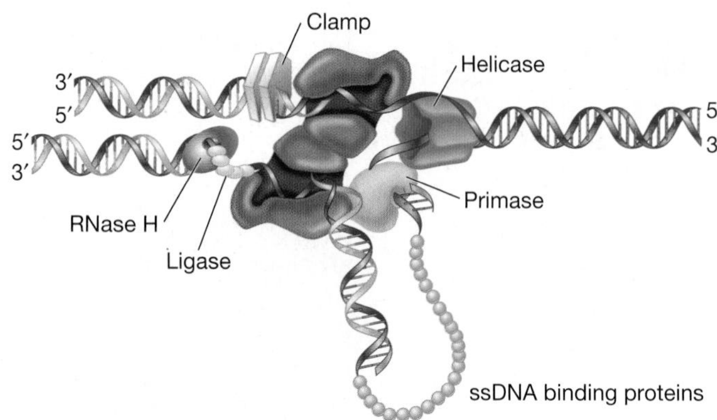

Figure 17.8 A Model of the T4 Replisome. T4 encodes most of the proteins needed to replicate its dsDNA genome, including components of the T4 replisome. This figure illustrates the viral replisome at the replication fork. Compare this model with the bacterial replisome shown in figure 11.16.

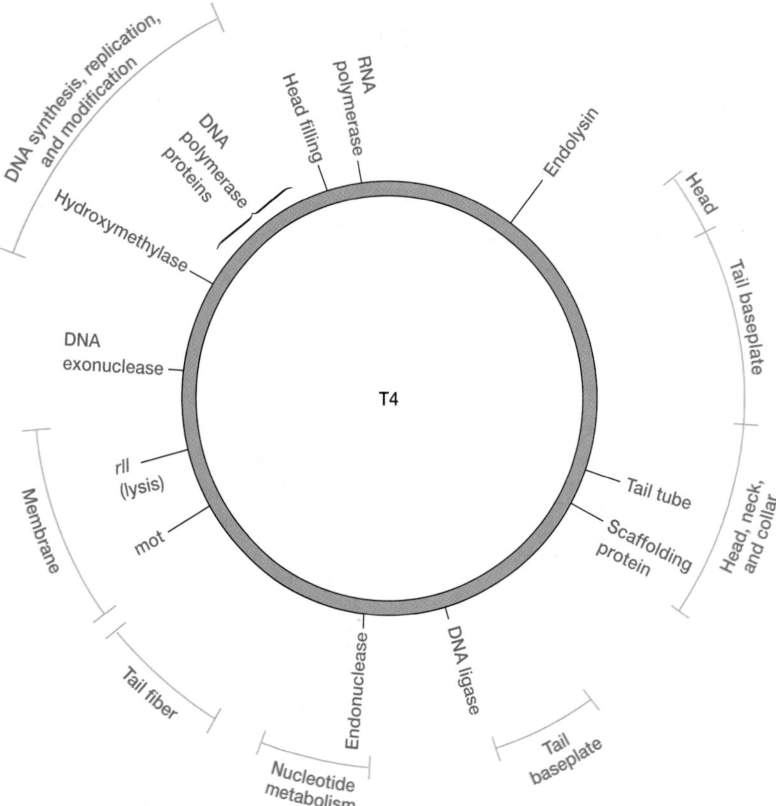

Figure 17.7 A Map of the T4 Genome. Some of its genes and their functions are shown. Genes with related functions tend to be clustered together.

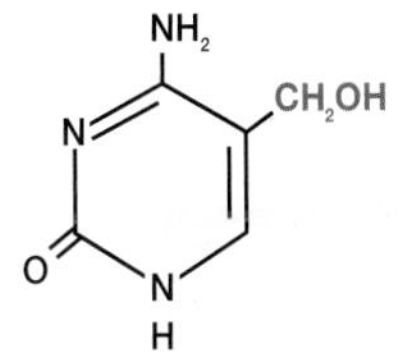

Figure 17.9 5-Hydroxymethylcytosine (HMC). In T4 DNA, the HMC often has glucose attached to its hydroxyl.

1. How is a one-step growth experiment carried out? Summarize what occurs in each phase. Define latent period, eclipse period, rise period (burst), and burst size.
2. Define the following terms: adsorption, penetration, phage assembly, phage release, lytic cycle, receptor site, early mRNA, late mRNA, hydroxymethylcytosine, restriction, restriction enzymes, concatemers, and scaffolding proteins. Use these terms to write a paragraph describing the life cycle of T4 bacteriophage.
3. Explain why the T4 phage genome is circularly permuted.

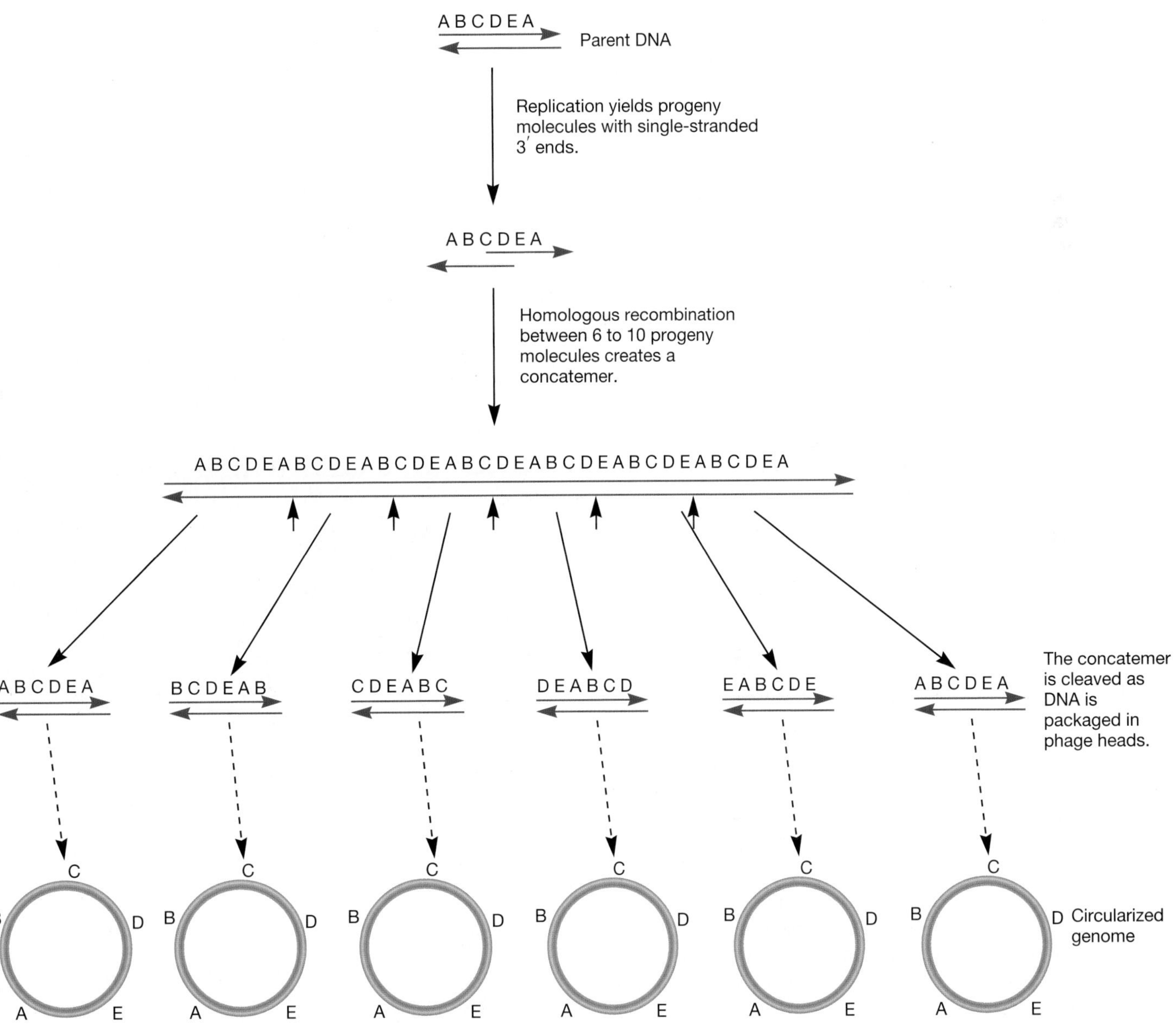

Figure 17.10 The Terminally Redundant, Circularly Permuted Genome of T4. The formation of concatemers during replication of the T4 genome is an important step in phage reproduction. During assembly of the virions, the phage head is filled with DNA cleaved from the concatemer. Because slightly more than one set of T4 genes is packaged in each head, each virion contains a different DNA fragment (note that the ends of the fragments are different). However, if each genome was circularized, the sequence of genes would be the same.

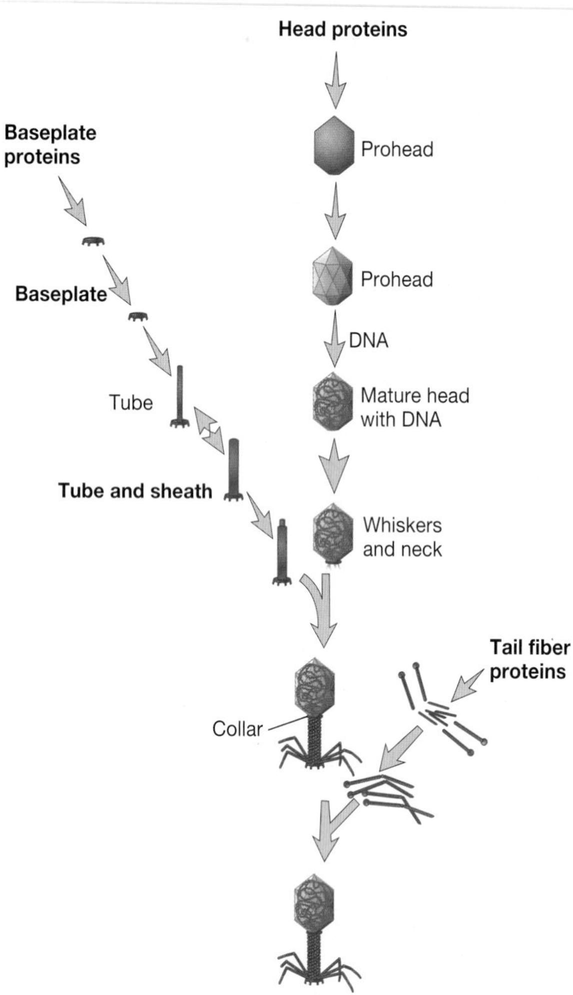

Figure 17.11 The Assembly of T4 Bacteriophage. Note the subassembly lines for the baseplate, tail tube and sheath, tail fibers, and head.

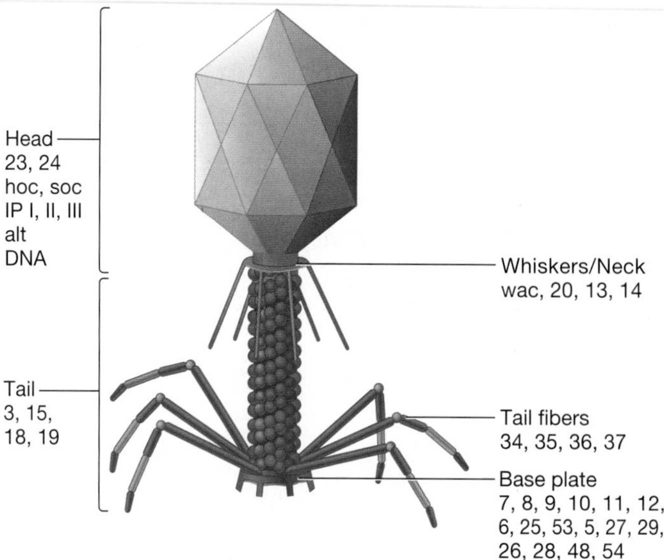

Figure 17.12 The Mature T4 Virion. T4 is composed of numerous proteins, most of which are designated with numbers rather than names. The proteins comprising each component of the phage are indicated. This image was reconstructed from electron micrographs. The phage is shown at 3 nm resolution.

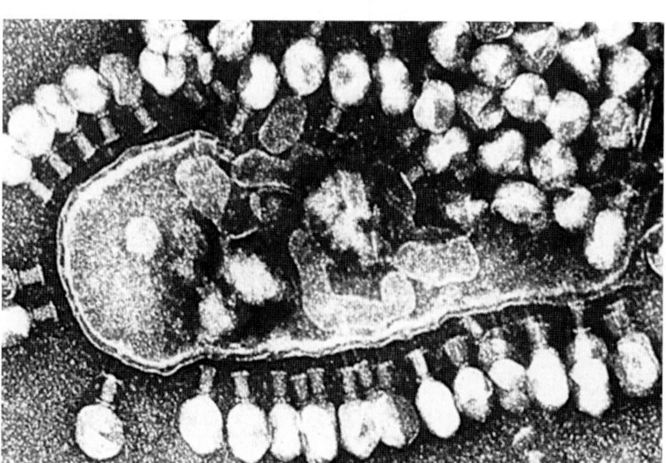

Figure 17.13 Release of T4 Bacteriophages by Lysis of the Host Cell. The host cell has been lysed (upper right portion of the cell) and virions have been released into the surroundings. Progeny virions also can be seen in the cytoplasm. In addition, empty capsids of the infecting phages coat the outside of the cell (X36,500).

17.3 Single-Stranded DNA Phages

Although many bacterial viruses are double-stranded DNA viruses, several single-stranded DNA (ssDNA) phages have been identified and their replication studied. Two ssDNA phages of *E. coli*—ϕX174, an icosahedral phage belonging to the *Microviridae,* and fd phage, a filamentous phage belonging to the *Inoviridae*—are the focus of this section.

The life cycle of ϕX174 begins with its attachment to the cell wall of its host. The circular ssDNA genome is injected into the cell, while the protein capsid remains outside the cell. The ϕX174 ssDNA genome has the same base sequence as viral mRNA and is therefore said to be plus-strand DNA. In order for either transcription or genome replication to occur, the phage DNA must be converted to a double-stranded form called the **replicative form (RF) (figure 17.14)**. This is accomplished by the bacterial DNA polymerase. The replicative form directs the synthesis of more RF copies and plus-strand DNA, both by rolling-circle replication. After assembly of virions, the phage is released by host lysis through a different method than that used by T4 phage. Rather than producing a lysozyme-like enzyme, as does T4, ϕX174 produces an enzyme (enzyme E) that blocks peptidoglycan synthesis. Enzyme E inhibits the activity of the bacterial protein MraY, which catalyzes the transfer of murein precursors to lipid carriers *(see figure 10.12)*. Blocking cell wall synthesis weakens the host cell wall, causing the cell to lyse and release the progeny virions.

DNA replication (section 11.4)

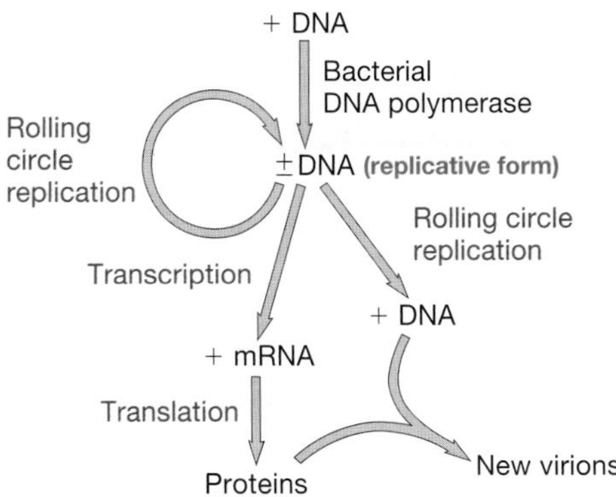

Figure 17.14 The Reproduction of ϕX174, a Plus-Strand DNA Phage. See text for details.

Although the fd phage also has a circular, positive-strand DNA genome, it behaves quite differently from ϕX174 in many respects. It is shaped like a long fiber about 6 nm in diameter by 900 to 1,900 nm in length (figure 17.1). Its ssDNA lies in the center of the filament and is surrounded by a tube made of a coat protein organized in a helical arrangement. The virus infects F^+, Hfr, and F′ *E. coli* cells by attaching to the tip of the pilus; the DNA enters the host along or possibly through the F factor-encoded sex pilus with the aid of a special adsorption protein. As with ϕX174, a replicative form is first synthesized and then transcribed. A phage-coded protein then aids in replication of the phage DNA by rolling-circle replication. Unlike either T4 or ϕX174, fd and other filamentous fd phages do not kill their host cell. Instead, they establish a relationship in which new virions are continually released by a secretory process. Filamentous phage coat proteins are first inserted into the membrane. The coat then assembles around the viral DNA as it is secreted through the host plasma membrane **(figure 17.15)**. Although the host cell is not lysed, it grows and divides at a slightly reduced rate.

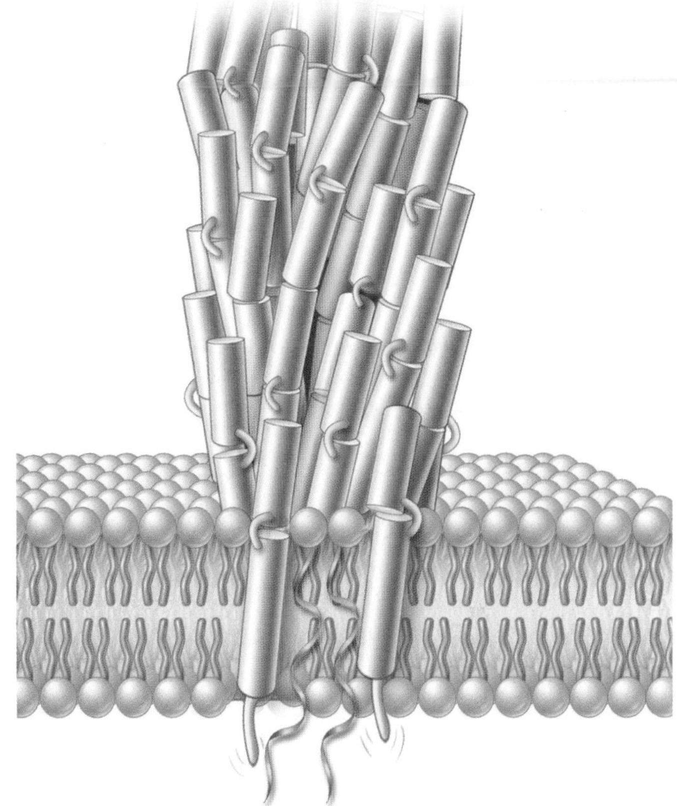

Figure 17.15 Release of Pf1 Phage. The Pf1 phage is a filamentous bacteriophage that is released from *Pseudomonas aeruginosa* without lysis. In this illustration the blue cylinders are hydrophobic α-helices that span the plasma membrane, and the red cylinders are amphipathic helices that lie on the membrane surface before virus assembly. In each protomer the two helices are connected by a short, flexible peptide loop (yellow). It is thought that the blue helix binds with circular, single-stranded viral DNA (green) as it is extruded through the membrane. The red helix simultaneously attaches to the growing virus coat that projects from the membrane surface. Eventually the blue helix leaves the membrane and also becomes part of the capsid.

17.4 RNA Phages

Although most bacteriophages are DNA viruses, numerous RNA phages have been identified and studied. Many of these are ssRNA viruses. Bacteriophages MS2 and Qβ, family *Leviviridae,* are small, tailless, icosahedral, plus-strand RNA viruses (figure 17.1). They attach to the side of the F pilus of their *E. coli* host. Retraction of the pilus brings the virions close to the outer membrane of the cell, from which they gain entry. Like many bacteriophages, the capsids of these viruses remain outside the cell and only the RNA genome enters. Because their genomes are plus stranded, the incoming RNA can act as messenger RNA and direct the synthesis of phage proteins. One of the first enzymes synthesized is a viral **RNA replicase,** an RNA-dependent RNA polymerase **(figure 17.16)**. The replicase then copies the plus strand to produce a double-stranded intermediate (±RNA), which is called the replicative form and is analogous to the ±DNA seen in the reproduction of ssDNA phages. The same replicase then uses this replicative form to synthesize thousands of copies of +RNA. Some of these plus strands are used to make more ±RNA in order to accelerate +RNA synthesis. Other +RNA strands act as mRNA and direct the synthesis of phage proteins. Finally, +RNA strands are incorporated into maturing virus particles. The mature virions are released by lysis. Release of virions has been studied for Qβ. It produces a protein that interferes with the bacterial protein MurA. MurA participates in the synthesis of UDP-NAM, a precursor for peptidoglycan synthesis. In the case of ϕX174, this interference weakens the cell wall and the host cell lyses. Synthesis of sugars and polysaccharides: Synthesis of peptidoglycan (section 10.4)

MS2 and Qβ have only three or four genes and are genetically the simplest phages known. In MS2, one protein is involved in phage adsorption to the host cell (and possibly also in virion construction or maturation). The other three genes code for a coat protein, RNA replicase, and a protein needed for cell lysis.

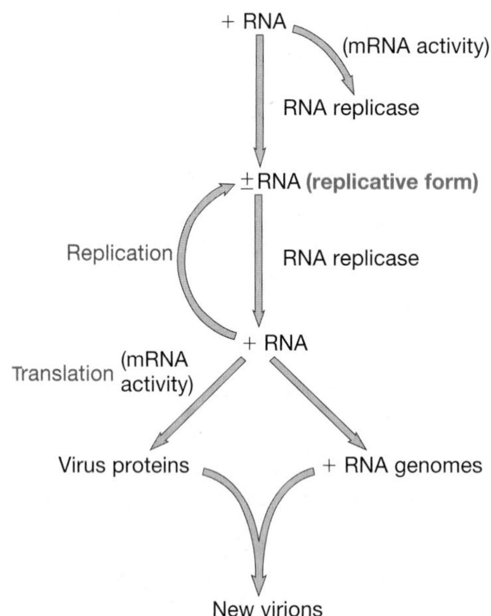

Figure 17.16 The Reproduction of Plus-Strand RNA Bacteriophages.

Several dsRNA phages have been discovered. The bacteriophage ϕ6 of *Pseudomonas syringae* pathovar phaseolicola (previously called *Pseudomonas phaseolicola*), a plant pathogen, is the best studied. It is an unusual bacteriophage for several reasons. One is that it is an enveloped virus. Within the envelope is a nucleocapsid containing a segmented genome consisting of three dsRNA segments and an RNA-dependent RNA polymerase. The life cycle of the virus also has unusual features. Like MS2 and Qβ, ϕ6 attaches to the side of a pilus. However, ϕ6 uses an envelope protein to facilitate adsorption. Retraction of the pilus brings the phage into contact with the outer membrane. The viral envelope then fuses with the cell's outer membrane, a process mediated by another envelope protein. Fusion of the two membranes delivers the nucleocapsid into the periplasmic space. Here, a protein associated with the nucleocapsid digests the peptidoglycan, allowing the nucleocapsid to cross that layer of the cell wall of its gram-negative host. Finally, the intact nucleocapsid enters the host cell by a process that resembles endocytosis. Because bacterial cells do not have the proteins and other factors needed for endocytosis, it is thought that viral proteins mediate this mechanism of entry. Once inside the host, the viral RNA polymerase acts as a **transcriptase,** catalyzing synthesis of viral mRNA from each dsRNA segment. The enzyme also acts as a replicase, synthesizing plus-strand RNA from each segment. These are enclosed within newly formed capsid proteins, where they serve as templates for the synthesis of the complementary negative strand, regenerating the dsRNA genome. Once the nucleocapsid is completed, a nonstructural viral protein called P12 functions in surrounding the nucleocapsid with a plasma membrane-derived envelope. Interestingly, the enveloped nucleocapsid is located within the cytoplasm. Finally, additional viral proteins are added to the envelope and the host cell is lysed, releasing the mature virions.

1. How does the reproduction of the ssDNA phages ϕX174 and fd differ from each other and from T4? How is their reproduction similar to the ssRNA phages? How does it differ?
2. What role does RNA replicase play in the reproduction of ssRNA phages? In dsRNA phages?
3. What is peculiar about the structure and life cycle of phage ϕ6?

17.5 Temperate Bacteriophages and Lysogeny

Thus far in our discussion of bacteriophages, we have focused primarily on virulent phages—those that only have one reproductive option: to begin replication immediately upon entering their host, followed by release from the host by lysis. However, many DNA phages are **temperate phages** that have two reproductive options: they can reproduce lytically as do the virulent phages or they can remain within the host cell without destroying it. Many temperate phages accomplish this by integrating their genome into the host cell's chromosome.

The relationship between a temperate phage and its host is called **lysogeny.** The form of the virus that remains within its host is called a **prophage,** and the infected bacteria are called **lysogens** or **lysogenic bacteria.** Lysogenic bacteria reproduce and in most other ways appear to be perfectly normal. However, they have two distinctive characteristics. The first is that they cannot be reinfected by the same virus—that is, they have immunity to superinfection. The second is that under appropriate conditions they lyse and release phage particles. This occurs when conditions within the cell cause the prophage to initiate synthesis of phage proteins and to assemble new virions, a process called **induction.** Induction leads to destruction of infected cells and release of new phages—that is, induction initiates the lytic cycle. Why and how these phenomena occur are discussed shortly.

Another important outcome of lysogeny is **lysogenic conversion.** This occurs when a temperate phage induces a change in the phenotype of its host. Lysogenic conversions often involve alterations in surface characteristics of the host. Many other lysogenic conversions give the host pathogenic properties. An example of the former is seen when *Salmonella* is infected by epsilon phage. The phage changes the activities of several enzymes involved in construction of the carbohydrate component of the bacterium's lipopolysaccharide, thereby altering the antigenic properties of the bacterium. Interestingly, these changes also eliminate the receptor for epsilon phage, and the bacterium becomes immune to infection by another epsilon phage. An example of altered pathogenic properties is observed when *Corynebacterium diphtheriae,* the cause of diphtheria, is infected with phage β. The phage β genome encodes diphtheria toxin, which is responsible for the symptoms of the disease. Thus only those strains of *C. diphtheriae* that are infected by the phage cause disease. Airborne diseases: Diphtheria (section 38.1)

Clearly, the infection of a bacterium by a temperate phage has significant impact on the host but why would viruses evolve this alternate life cycle? Two advantages of lysogeny have been recognized. The first is that lysogeny allows a virus to remain viable within a dormant host. Bacteria often become dormant due to nutrient deprivation and while in this state, they do not synthesize nucleic acids or proteins. In such situations, a prophage would survive but most virulent bacteriophages would not be replicated, as they require active cellular biosynthetic machinery. The second advantage arises when there are many more phages in an environment than there are host cells, a situation virologists refer to as a high multiplicity of infection (MOI). In these conditions, lysogeny allows for the survival of host cells so that the virus can continue to reproduce.

It should be apparent from this discussion that once a temperate phage infects its host, it must "decide" which reproductive cycle to follow. How does it make this choice? How temperate phages make this decision is best illustrated by bacteriophage lambda. Lambda is a double-stranded DNA phage that infects the K12 strain of *E. coli*. It has an icosahedral head 55 nm in diameter and a noncontractile tail with a thin tail fiber at its end **(figure 17.17)**. Its DNA genome is a linear molecule with cohesive ends—single-stranded stretches, 12 nucleotides long, that are complementary to each other and can base pair.

Like most bacteriophages, lambda attaches to its host and then injects its genome into the cytoplasm, leaving the capsid outside. Once inside the cell, the linear genome is circularized when the two cohesive ends base pair with each other; the breaks in the strands are sealed by the host cell's DNA ligase **(figure 17.18)**. The lambda genome has been carefully mapped, and over 40 genes have been located **(figure 17.19)**. Most genes are clustered according to their function, with separate groups involved in head synthesis, tail synthesis, lysogeny, DNA replication, and cell lysis. This organization is important because once the genome is circularized, a cascade of regulatory events occurs that determine if the phage pursues a lytic cycle or establishes lysogeny. Regulation of appropriate genes is facilitated by clustering and coordinated transcription from the same promoter. Below, we first provide an overview of lysogeny and the lytic cyle of lambda. This is followed by a more detailed examination of the decision-making process.

The cascade of events leading to either lysogeny or the lytic cycle involves a number of regulatory proteins that function as repressors or activators or both. Two regulatory proteins are of particular importance: the **lambda repressor** (product of the *cI* gene) and the **Cro protein** (product of the *cro* gene). The lambda repressor promotes lysogeny, and the Cro protein promotes the lytic cycle. In essence, the decision to pursue lysogeny or to pursue a lytic cycle is the result of a race between the production of these two proteins. If lambda repressor prevails, the production of Cro protein is inhibited and lysogeny occurs; if the Cro protein prevails, the production of lambda repressor is inhibited and the lytic cycle occurs. This is because the lambda repressor prevents transcription of viral genes, while Cro does just the opposite: it ensures viral gene expression.

The lambda repressor is 236 amino acids long and folds into a dumbbell shape with globular domains at each end **(figure 17.20)**. One domain binds DNA while the other binds another lambda repressor molecule to form a dimer. The dimer is the most active form of the repressor. Lambda repressor binds two operator sites, O_L and O_R, thereby blocking transcription of most viral genes **(figure 17.21** and **table 17.1)**. When bound at O_L, it represses transcription from the promoter P_L (promoter leftward). Likewise, when bound at O_R it represses transcription in the rightward direction from P_R (promoter rightward). However, it also activates transcription in the leftward direction from the *cI* promoter P_{RM} (RM stands for *r*epressor *m*aintenance). Recall that *cI* encodes the lambda repressor. Thus lambda repressor controls its own synthesis.

As noted earlier, if the lambda repressor wins the race with the Cro protein, lysogeny is established and the lambda genome is integrated into the host chromosome. Integration is catalyzed by the enzyme **integrase,** the product of the *int* gene, and takes place at a site in the host chromosome called the attachment site (*att*) (figure 17.19; *see also figure 13.40*). A homologous site is found on the phage genome, so the phage and bacterial *att* sites can base pair with each other. The bacterial site is located between the galactose (*gal*) and biotin (*bio*) operons, and as a result of integration, the circular lambda genome becomes a linear stretch of DNA located between these two host operons **(figure 17.22)**. The prophage can remain integrated indefinitely, being replicated as the bacterial genome is replicated.

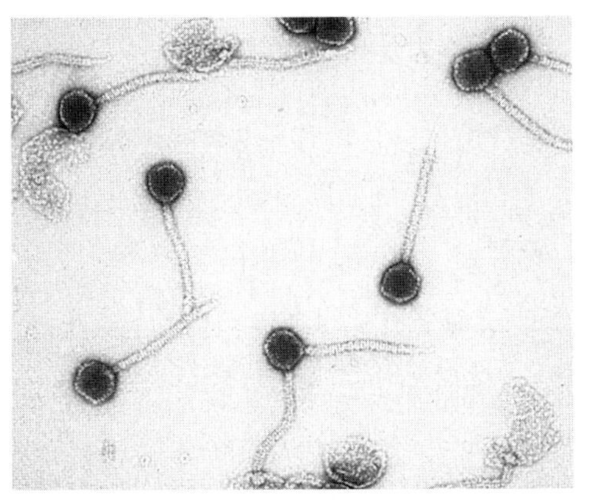

Figure 17.17 Bacteriophage Lambda.

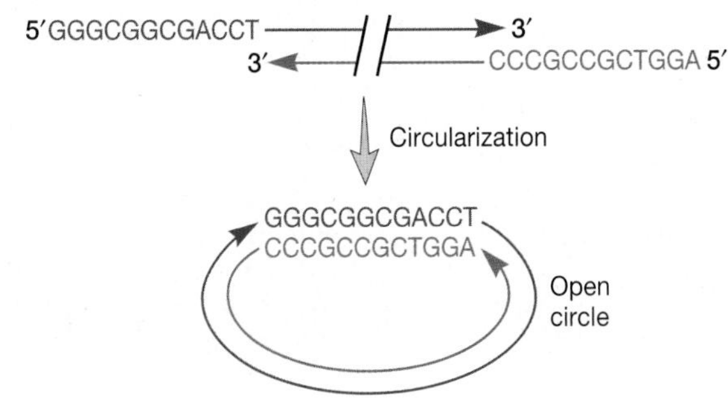

Figure 17.18 Lambda Phage DNA. A diagram of lambda phage DNA showing its 12 base, single-stranded cohesive ends (printed in purple) and the circularization their complementary base sequences make possible.

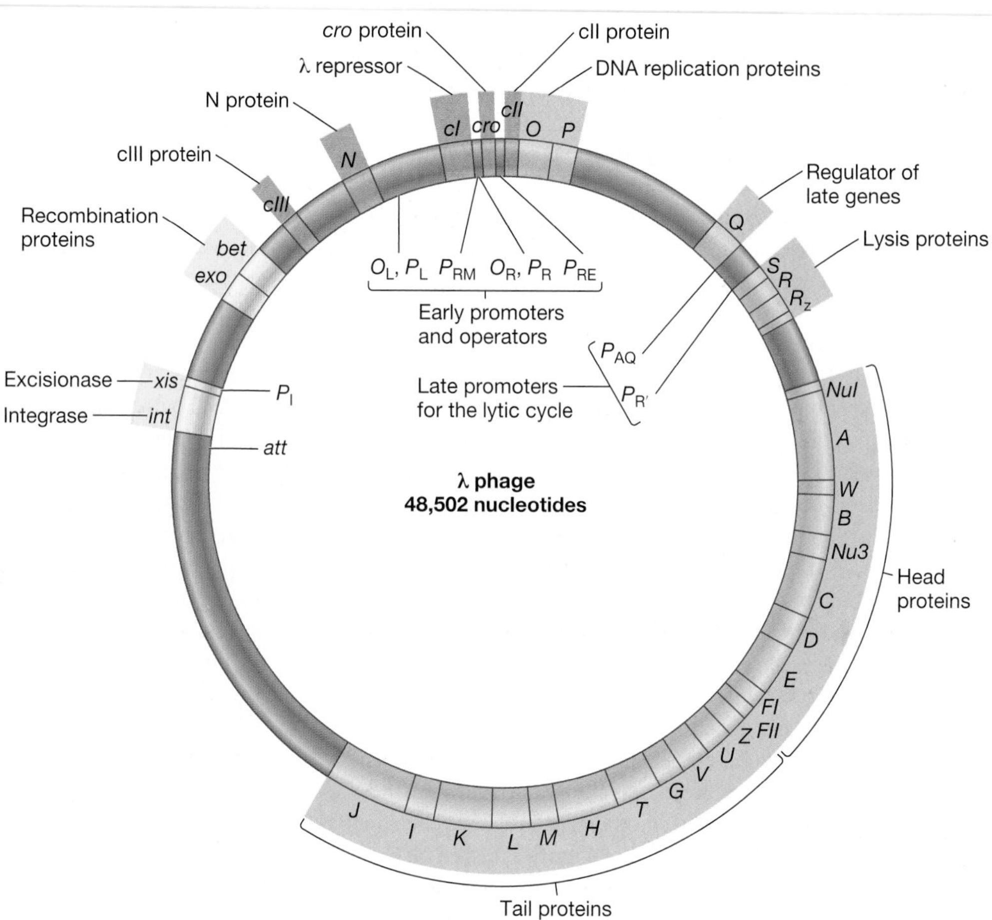

Figure 17.19 The Genome of Phage Lambda (λ). The genes are color-coded according to function. Those genes shaded in orange encode regulatory proteins that determine if the lytic or lysogenic cycle will be followed. Genes involved in establishing lysogeny are shaded in yellow. Those required for the lytic cycle are shaded in tan. Important promoters and operators are also noted.

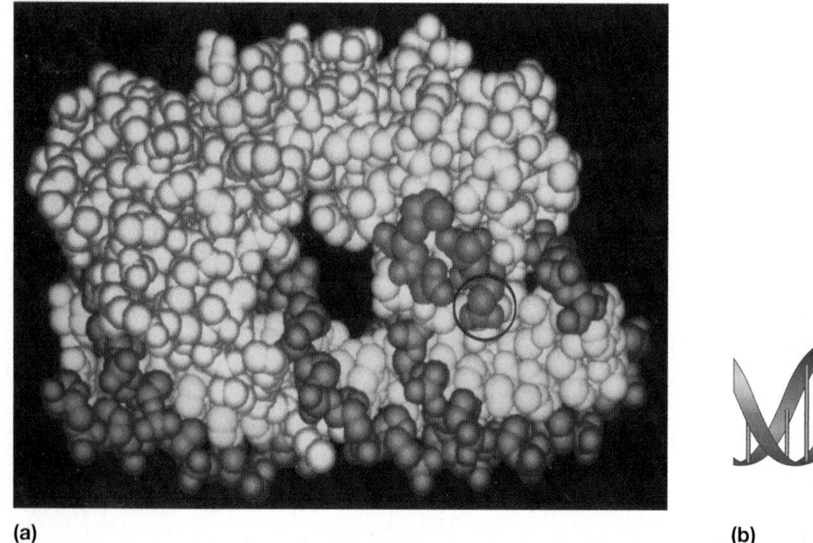
(a)

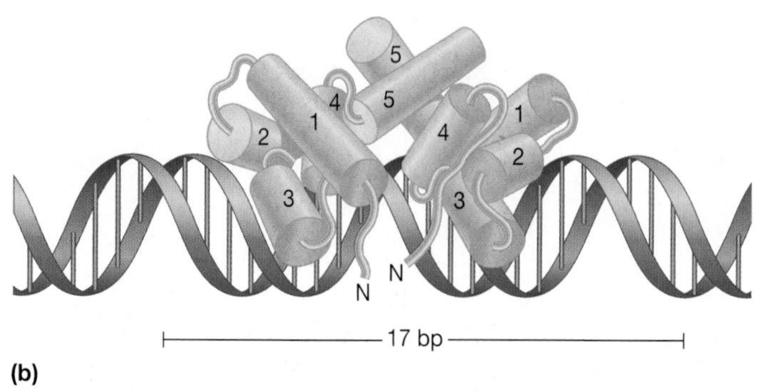

(b)

Figure 17.20 Lambda Repressor Binding. **(a)** A computer model of lambda repressor binding to the lambda operator. The lambda repressor dimer (brown and tan) is bound to DNA (blue and light blue). The arms of the dimer wrap around the major grooves of the double helix. **(b)** A diagram of the lambda repressor-DNA complex. The repressor binds to a 17 bp stretch of the operator. The α3-helices make closest contact with the major grooves of the operator (the helices are labeled in order, beginning at the N terminal of the chain).

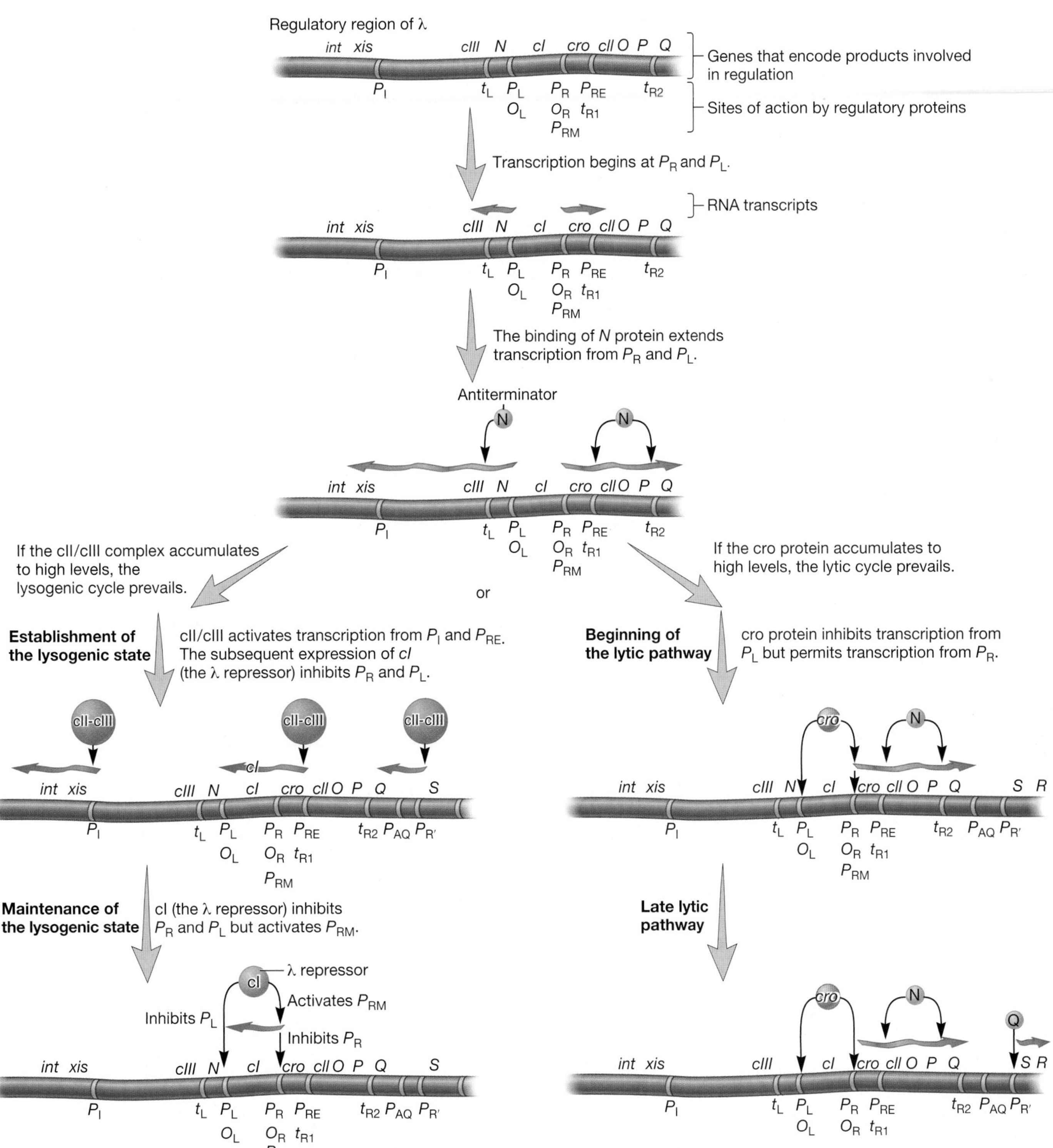

Figure 17.21 The Decision-Making Process for Establishing Lysogeny or the Lytic Pathway. The initial transcripts are synthesized by the host RNA polymerase. These encode the N protein and the Cro protein. The N protein is an antiterminator that allows transcription to proceed past the terminator sequences t_L, t_{R1}, and t_{R2}. This allows transcription of other regulatory genes as well as the *xis* and *int* genes. The latter genes encode the enzymes excisionase and integrase, respectively. The left side of the figure illustrates what occurs if lysogeny is established. The right side of the figure shows the lytic pathway.

Table 17.1 Functions of Lambda Promoters and Operators

Promoter or Operator	Name Derivation	Function
P_L	***P***romoter ***L***eftward	Promoter for transcription of *N, cIII, xis,* and *int* genes; important in establishing lysogeny
O_L	***O***perator ***L***eftward	Binding site for lambda repressor and Cro protein; binding by lambda repressor maintains lysogenic state; binding by Cro protein prevents establishment of lysogeny
P_R	***P***romoter ***R***ightward	Promoter for transcription of *cro, cII, O, P,* and *Q* genes; Cro, O, P, and Q proteins are needed for lytic cycle; CII protein helps establish lysogeny
O_R	***O***perator ***R***ightward	Binding site for lambda repressor and Cro protein; binding by lambda repressor maintains lysogenic state; binding by Cro allows transcription to occur
P_{RE}	***P***romoter for Lambda ***R***epressor ***E***stablishment	Promoter for *cI* gene (lambda repressor gene); recognized by CII protein, a transcriptional activator; important in establishing lysogeny
P_I	***P***romoter for ***I***ntegrase Gene	Transcription from P_I generates mRNA for integrase protein, but not excisionase; recognized by the transcriptional activator CII; important for establishing lysogeny
P_{AQ}	***P***romoter for ***A***nti-***Q*** mRNA	Transcription from P_{AQ} generates an antisense RNA that binds *Q* mRNA, preventing its translation; recognized by the transcriptional activator CII; important for establishing lysogeny
P_{RM}	***P***romoter for ***R***epressor ***M***aintenance	Promoter for transcription of lambda repressor gene (cI); activated by lambda repressor; important in maintaining lysogeny
$P_{R'}$	***P***romoter ***R***ightward′	Promoter for transcription of viral structural genes; activated by Q protein; important in lytic cycle

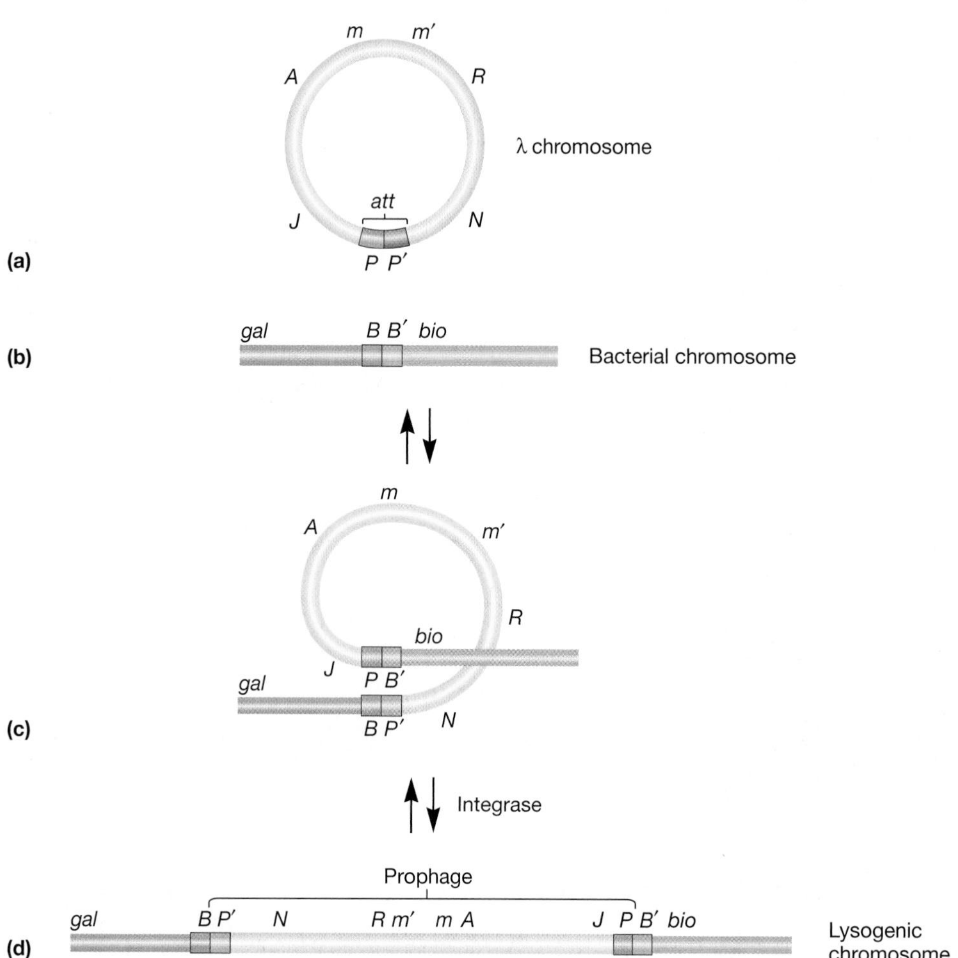

Figure 17.22 Reversible Insertion and Excision of Lambda Phage. After circularization, the *att* site P, P′ **(a)** lines up with a corresponding bacterial sequence B, B′ **(b)** and is integrated between the *gal* and *bio* operons to form the prophage, **(c)** and **(d).** If the process is reversed, the circular lambda chromosome will be restored and can then reproduce.

The Cro protein is composed of 66 amino acids and like lambda repressor, forms a dimer that binds the operator sites O_R and O_L, blocking transcription from the P_R and P_L promoters **(figure 17.23)**. If Cro protein wins the race with lambda repressor, it blocks synthesis of lambda repressor and prevents integration of the lambda genome into the host chromosome. By the time synthesis of lambda repressor is blocked, another regulatory protein called Q protein has accumulated. Q promotes transcription from a promoter called P_R', and in the presence of Q protein, the genes encoding viral structural proteins, as well other proteins needed for virus assembly and host lysis, are transcribed. Ultimately, the host is lysed and the new virions are released.

With this brief introduction and overview, we can now examine the decision-making process more closely. The sequence of events unfolds as follows. Once the lambda genome is circularized within the host cytoplasm, transcription is initiated by host RNA polymerase at promoters P_R and P_L (figure 17.21). Very early in the infection, only the *N* and *cro* genes are expressed, and transcription is terminated at the end of these two genes. However, once the N protein is synthesized, it functions as an antiterminator so that RNA polymerase continues transcription beyond the *N* and *cro* genes. Thus from P_R, the *cII* and DNA replication genes *O, P,* and *Q* are transcribed (the *Q* gene encodes the Q protein); from P_L, *cIII* and all the genes through the excision (*xis*) and integration (*int*) genes are transcribed. The synthesis of the CII and CIII proteins is critical to the next step in the infection—it is at this point where the choice to enter lysogeny or the lytic cycle is made.

CII is a transcriptional activator protein that recognizes the promoter P_{RE} (promoter for lambda *r*epressor *e*stablishment) and initiates transcription of the *cI* gene, which encodes the lambda repressor. It also recognizes the promoters P_I and P_{AQ}. Transcription leftward from the P_I promoter synthesizes mRNA encoding the enzyme integrase, which catalyzes the insertion of lambda DNA into the *E. coli* chromosome. Transcription leftward from P_{AQ} synthesizes an antisense RNA that is complementary to the mRNA encoding the Q protein. Recall that Q protein is needed for expression of viral structural and lysis genes. The activity of the CII protein is influenced by environmental factors. For instance, CII protein is particularly susceptible to proteases. When *E. coli* is in nutrient-rich conditions, many proteases are formed and CII is more likely to be degraded. The CIII protein's role is to protect CII from degradation. However, its protection is not complete and in certain conditions, CII is inactivated whether CIII is present or not.

At this point in the infection, several genes are being transcribed and their messages as well as the proteins they encode are accumulating within the host cell. These include Cro protein, lambda repressor, CII protein, CIII protein, integrase, N protein, and Q protein. It is now that the race between the lambda repressor and Cro protein begins in earnest. Recall that both Cro protein and lambda repressor bind the regulatory site O_R. If Cro binds O_R, it blocks synthesis of lambda repressor. If lambda repressor binds O_R, it promotes its own synthesis but blocks synthesis of the Cro protein. Because synthesis of Cro protein begins before synthesis of lambda repressor, initially the amount of Cro protein exceeds the amount of lambda repressor. However, Cro protein binds O_R less tightly than does lambda repressor. Thus it takes a higher concentration of Cro protein in the cell to bind O_R and block the binding of lambda repressor. If the CII protein is plentiful (i.e., it is not being degraded by host proteases), then the amount of lambda repressor in the cell will be sufficient to win the race for the O_R regulatory site. Thus integrase will catalyze integration of lambda genome into the host chromosome. Furthermore, antiQ RNA will be plentiful. Hybridization of this antisense molecule to the mRNA encoding Q protein leads to the destruction of Q protein message. Therefore, the action of CII, lambda repressor, and antiQ RNA represses the synthesis of all viral proteins except lambda repressor. If CII protein is not plentiful (i.e., the cell is in a nutrient-rich environment and many proteases are present within the cell, leading to degradation of CII), then Cro protein will accumulate to a sufficient level to outcompete lambda repressor for the O_L and O_R sites. Transcription of the lambda repressor genes and other genes that function to establish lysogeny will be blocked and the lytic cycle will proceed.

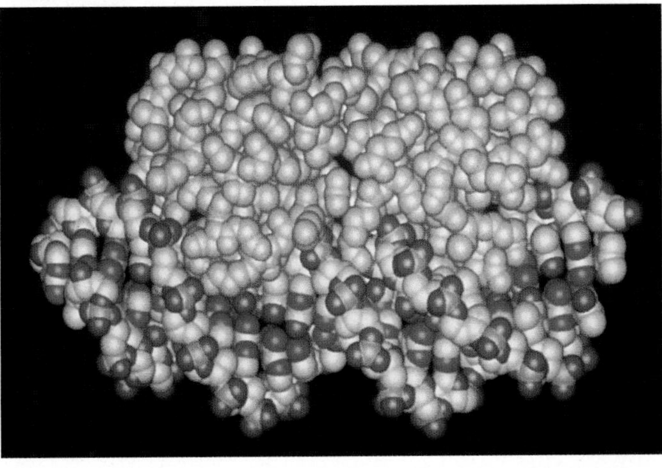

(a)

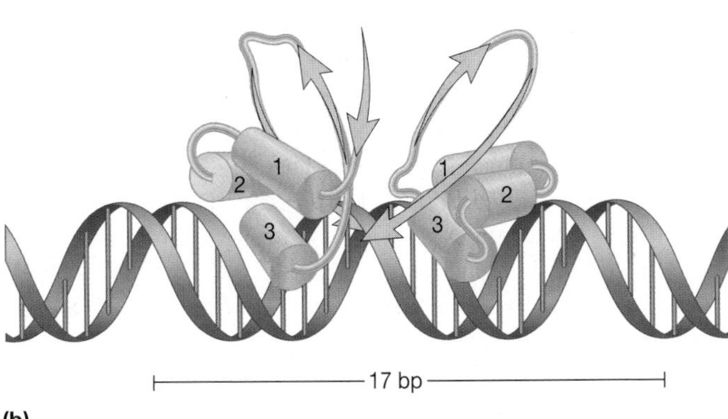

(b)

Figure 17.23 Cro Protein Binding. **(a)** A space-filling model of the Cro protein–DNA complex. The Cro protein is in yellow. **(b)** A diagram of the Cro protein dimer–DNA complex. Like the lambda repressor protein, the Cro protein functions as a dimer and binds to two adjacent DNA major grooves.

We have now considered the regulatory processes that dictate whether lysogeny is established or the lytic cycle is pursued. However, there are two additional phenomena related to the lysogenic state that we must also consider. The first is immunity to further infection. We have already seen how this is accomplished by the epsilon phage of *Salmonella*. Lambda phage uses a different mechanism. In a lambda lysogenic bacterium, the only viral protein synthesized is lambda repressor. Thus if a new lambda phage infects the cell, lambda repressor can bind the regulatory sites of the incoming viral genome immediately, and expression of all genes (and superinfection) is blocked. The second phenomenon is induction, which usually occurs in response to environmental factors such as UV light or chemical mutagens that damage DNA. This damage alters the activity of the RecA protein. As described in chapter 13, RecA plays important roles in recombination and DNA repair processes. When activated by DNA damage, RecA interacts with lambda repressor, causing the repressor to cleave itself. As more and more repressor proteins destroy themselves, transcription of the *cI* gene is decreased, further lowering the amount of lambda repressor in the cell. Eventually the level becomes so low that initiation of transcription of the *xis*, *int*, and *cro* genes occurs. The *xis* gene encodes the protein **excisionase.** It binds integrase, causing it to reverse the integration process, and the prophage is freed from the host chromosome. As lambda repressor levels decline, the Cro protein levels increase. Eventually, synthesis of lambda repressor is completely blocked and the lytic cycle proceeds to completion. Transduction: specialized transduction (section 13.9)

Our attention has been on lambda phage, but there are many other temperate phages. Most, like lambda, exist as integrated prophages in the lysogen. However, not all temperate phages integrate into the host chromosome at specific sites. Bacteriophage Mu uses a transposition mechanism to integrate randomly into the genome. It then expresses a repressor protein that inhibits lytic growth. Furthermore, integration is not an absolute requirement for lysogeny. The *E. coli* phage P1 is similar to lambda in that it circularizes after infection and begins to manufacture repressor. However, it remains as an independent circular DNA molecule in the lysogen and is replicated at the same time as the host chromosome. When *E. coli* divides, P1 DNA is apportioned between the daughter cells so that all lysogens contain one or two copies of the phage genome.

1. Define lysogeny, temperate phage, lysogen, prophage, immunity, and induction.
2. What advantages might a phage gain by being capable of lysogeny?
3. Describe lysogenic conversion and its significance.
4. Precisely how, in molecular terms, is a bacterial cell made lysogenic by a temperate phage like lambda?
5. How is a prophage induced to become active again?
6. Describe the roles of the lambda repressor, Cro protein, RecA protein, integrase, and excisionase in lysogeny and induction.
7. How do the temperate phages Mu and P1 differ from lambda phage?

17.6 Bacteriophage Genomes

Just as genomic analysis and bioinformatics are revolutionizing our understanding of cellular microbes, so too are they giving us a dramatic new picture of the biology and evolution of viruses. At present, the entire nucleotide sequences of the dsDNA genomes of over 150 tailed bacteriophages have been determined. This set of genomes includes those of many laboratory bacteriophage strains as well as new environmental isolates. Although this represents only a small fraction of the diversity of bacteriophages, comparison of available genomes has revealed surprising features and clear themes.

Perhaps the most significant discovery is that the genomes of bacteriophages are highly mosaic in character **(figure 17.24)**. Comparison of several tailed bacteriophages reveals blocks of related sequences shared among different pairs of genomes in a combinatorial fashion. For instance, a block of genes may be similar in phage A and phage B, but different from those in phage C. However, phage C may possess a second block of genes related to those in phage A, but not those in phage B. This mosaic nature is seen in a comparison of bacteriophages lambda, N15, and HK97. Lambda and N15 share similar genes for head and tail assembly but their lysogeny and lysis genes are unrelated. Lambda's lysogeny and lysis genes are more similar to those of HK97.

The mosaic nature has important implications for many aspects of phage biology. One is that it suggests bacteriophage genomes could not have diverged as a whole from a common ancestor; rather, each gene or block of genes appears to have a unique evolutionary history. Another implication is that genetic exchange to substitute whole blocks of genes may have occurred between nonhomologous sequences—sequences with little or no similarity. This mechanism of **nonhomologous recombination** is distinct from the types of recombination described in chapter 13. It is thought that nonhomologous recombination takes place relatively randomly throughout phage genomes, but genetic selection preserves only those recombinant genomes that produce viable bacteriophage progeny. The occurrence of nonhomologous recombination also allows bacteriophages to acquire genes from their bacterial hosts and to transfer genes into the host. Donation of genes by bacteriophages has clearly contributed to the evolution of pathogenic bacterial strains. The shiga-like toxins of enterohemorrhagic *E. coli* O157:H7 are encoded by a prophage absent from the genomes of *E. coli* strains that do not cause disease. Indeed, almost one-quarter of the *E. coli* 0157:H7 genome is not shared with the commensal strain *E. coli* K12, and fully half of this is conferred by prophages.

Our glimpse into bacteriophage genomics has provided new insights into the evolution of viruses and other life forms. It is expected that as more viral genomes are sequenced and analyzed, an even clearer understanding of their evolution and their contributions to the evolution of cellular organisms will follow.

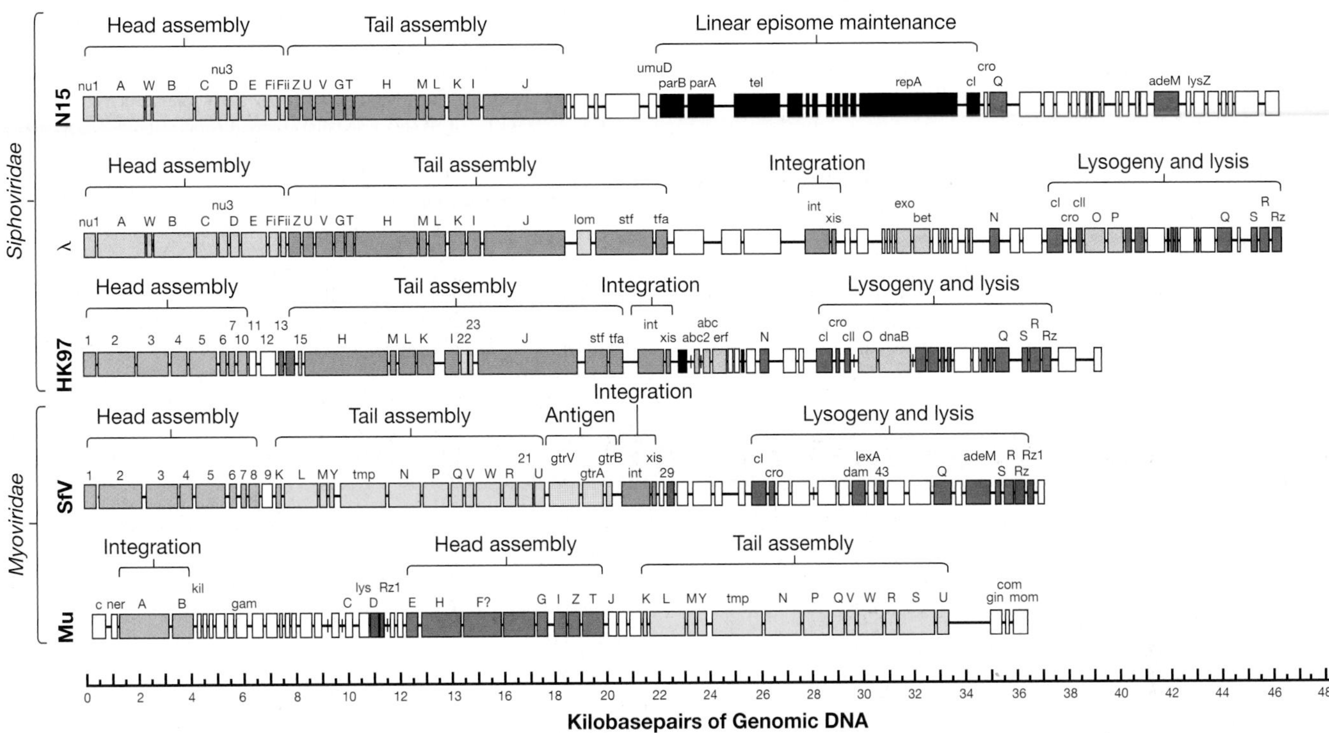

Figure 17.24 The Mosaic Nature of Bacteriophage Genomes. Homologous gene modules are similarly colored. Note that phage HK97 (family *Siphoviridae*) shares tail assembly genes with other members of its family (phages λ and N15), but shares head assembly genes with phage Sfv, a member of the *Myoviridae*.

Summary

17.1 Classification of Bacterial and Archaeal Viruses

a. The ICTV has recognized numerous virus families whose members infect procaryotic cells. Of these, four are new families that contain archaeal viruses having unusual capsid morphologies **(figure 17.1)**.

b. The classification of bacteriophages is made difficult by the discovery that considerable lateral gene transfer has occurred between viral species.

17.2 Virulent Double-Stranded DNA Phages

a. The lytic cycle of virulent bacteriophages is a life cycle that ends with host cell lysis and virion release.

b. The phage life cycle can be studied with a one-step growth experiment that is divided into an initial eclipse period within the latent period, and a rise period **(figure 17.2)**.

c. The life cycle of T4 phage, a dsDNA virus of *E. coli,* is composed of several phases. In the adsorption phase the phage attaches to a specific receptor site on the bacterial surface. This is followed by penetration of the cell wall and insertion of the viral nucleic acid into the cell **(figure 17.3)**.

d. Transcription of T4 DNA first produces early mRNA, which directs the synthesis of the protein factors and enzymes required to take control of the host and manufacture phage nucleic acids. Late mRNA is produced after DNA replication and directs the synthesis of capsid proteins, proteins involved in phage assembly, and those required for cell lysis and phage release.

e. T4 DNA contains hydroxymethylcytosine (HMC) in place of cytosine, and glucose is often added to the HMC to protect the phage DNA from attack by host restriction enzymes **(figure 17.9)**.

f. T4 DNA replication produces concatemers, long strands of several genome copies linked together **(figure 17.10)**.

g. Complete virions are assembled immediately after the separate components have been constructed. This is a self-assembly process, but requires participation of scaffolding proteins **(figure 17.11)**.

17.3 Single-Stranded DNA Phages

a. The replication of ssDNA phages proceeds through the formation of a double-stranded replicative form (RF) **(figure 17.14)**. The filamentous ssDNA phages are continually released without host cell lysis **(figure 17.15)**.

17.4 RNA Phages

a. When the RNA of plus-strand RNA phages enters a bacterial cell, it acts as a messenger and directs the synthesis of RNA replicase, which then produces double-stranded replicative forms and, subsequently, many +RNA copies **(figure 17.16)**.

b. The ϕ6 phage is a dsRNA phage. It is unusual in having a membranous envelope, an unusual endocytosis-like mechanism of entry into its host, and a segmented genome. The ϕ6 capsid contains an RNA-dependent RNA polymerase, which acts both as a transcriptase and as a replicase.

17.5 Temperate Bacteriophages and Lysogeny

a. Temperate phages, unlike virulent phages, often reproduce in synchrony with the host genome to yield a clone of virus-infected cells. This relationship is lysogeny, and the infected cell is called a lysogen. The latent form of the phage genome within the lysogen is the prophage.

b. Lysogeny is reversible, and the prophage can be induced to become active again and lyse its host.
c. A temperate phage may induce a change in the phenotype of its host cell that is not directly related to the completion of its life cycle. Such a change is called a conversion.
d. Two of the first proteins to appear after infection with lambda are the lambda repressor and the Cro protein. The lambda repressor blocks the transcription of the *cro* gene and other genes required for the lytic cycle, while the Cro protein inhibits transcription of the lambda repressor gene (**figure 17.21**).
e. There is a race between synthesis of lambda repressor and that of the Cro protein. If the Cro protein level rises high enough in time, lambda repressor synthesis is blocked and the lytic cycle initiated; otherwise, all genes other than the lambda repressor gene are repressed and the cell becomes a lysogen.
f. The final step in prophage formation is the insertion or integration of the lambda genome into the *E. coli* chromosome; this is catalyzed by the enzyme integrase (**figure 17.22**).
g. Several environmental factors can lower repressor levels and trigger induction. The prophage becomes active and makes an excisionase protein that causes the integrase to reverse integration, free the prophage, and initiate a lytic cycle.

17.6 Bacteriophage Genomes

a. The complete nucleotide sequences of over 150 tailed dsDNA bacteriophages have been determined.
b. Bacteriophage genomes appear mosaic in character, with short blocks of genes shared in different combinations. The mosaic nature of their genomes suggests that lateral gene transfer and nonhomologous recombination have contributed to the evolution of phages (**figure 17.24**).

Key Terms

bacteriophages 427
burst size 430
concatemer 432
Cro protein 439
early mRNA 430
eclipse period 430
excisionase 444
hydroxymethylcytosine (HMC) 432
induction 438
integrase 439
lambda repressor 439
late mRNA 431
latent period 429
lysogen 438
lysogenic bacteria 438
lysogenic conversion 438
lysogeny 438
lytic cycle 428
nonhomologous recombination 444
one-step growth experiment 428
prophage 438
receptor 430
replicative form (RF) 436
restriction 432
restriction enzyme 432
rise period or burst 429
RNA replicase 437
scaffolding proteins 433
temperate phage 438
transcriptase 438
virulent viruses 428

Critical Thinking Questions

1. Can you think of a way to simplify further the genomes of the ssRNA phages MS2 and Qβ? Would it be possible to eliminate one of their genes? If so, which one?
2. No temperate RNA phages have yet been discovered. How might this absence be explained?
3. How might a bacterial cell resist phage infections? Give those mechanisms mentioned in the chapter and speculate on other possible strategies.
4. The choice between lysogeny and lysis is influenced by many factors. How would external conditions such as starvation or crowding be "sensed" and communicated to the transcriptional machinery and influence this choice?
5. The most straightforward explanation as to why the endolysin of T4 is expressed so late in infection is that its promoter is recognized by the gp55 alternative sigma factor. Propose a different explanation.

Learn More

Calendar, R., editor. *The bacteriophages,* 2d ed. London: Oxford University Press.

Campbell, A. M. 2001. Bacteriophages. In *Fields Virology,* 4th ed., D. M. Knipe and P. M. Howley, editors-in-chief, 659–82. Philadelphia: Lippincott Williams & Wilkins.

Dyall-Smith, M.; Tang, S.-L.; and Bath, C. 2003. Haloarchaeal viruses: How diverse are they? *Res. Microbiol.* 154:309–13.

Fischetti, V. A. 2005. Bacteriophage lytic enzymes: Novel anti-infectives. *Trends Microbiol.* 13(10):491–96.

Flint, S. J.; Enquist, L. W.; Racaniello, V. R.; and Skalka, A. M. 2004. *Principles of virology,* 2d ed. Washington, D.C.: ASM Press.

Lawrence, J. G.; Hatfull, G. F.; and Hendrix, R. W. 2002. Imbroglios of viral taxonomy: Genetic exchange and failings of phenetic approaches. *J. Bact.* 184(17):4891–4905.

Mayo, M. A.; Maniloff, J.; Desselberger, U.; Ball, L. A.; and Fauquet, C. M., editors. 2005. *Virus taxonomy: VIIIth report of the International Committee on Taxonomy of Viruses*, San Diego: Elsevier Academic Press.

Miller, E. S.; Kutter, E.; Mosig, G.; Arisaka, F.; Kunisawa, T.; and Rüger, W. 2003. Bacteriophage T4 genome. *Microbiol. Mol. Biol. Rev.* 67(1):86–156.

Nechaev, S., and Severinov, K. 2003. Bacteriophage-induced modifications of host RNA polymerase. *Annu. Rev. Microbiol.* 57:301–22.

Poranen, M. M.; Daugelavičius, R.; and Bamford, D. H. 2002. Common principles in viral entry. *Annu. Rev. Microbiol.* 56:521–38.

Prangishvili, D., and Garrett, R. A. 2005. Viruses of hyperthermophilic crenarchaea. *Trends Microbiol.* 13(11):535–42.

Ptashne, M. 2004. *A genetic switch: Phage lambda revisited,* 3d ed. Cold Spring Harbor, New York: Cold Spring Harbor Laboratory Press.

Rydman, P. S., and Bamford, D. H. 2002. Phage enzymes digest peptidoglycan to deliver DNA. *ASM News* 68(7)330–5.

Snyder, J. C.; Stedman, K.; Rice, G.; Wiedenheft, B.; Spuhler, J.; and Young, M. J. 2003. Viruses of hyperthermophilic archaea. *Res. Microbiol.* 154:474–82.

Please visit the Prescott website at www.mhhe.com/prescott7 for additional references.

The Viruses:
Eucaryotic Viruses and Other Acellular Infectious Agents

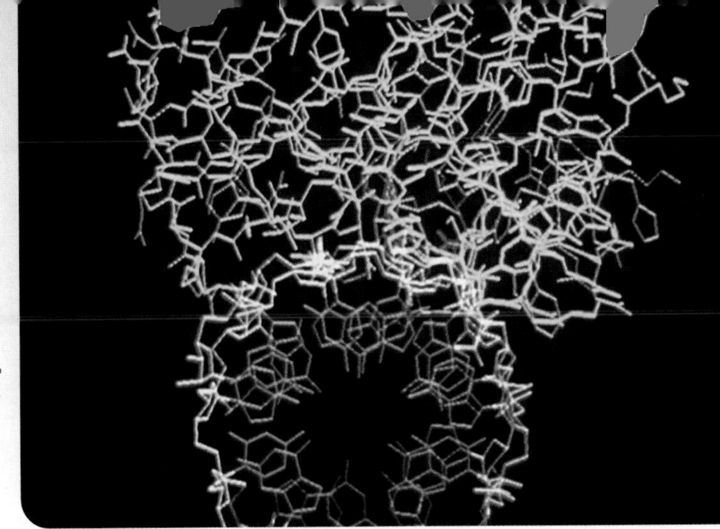

A model of the ribonuclease H component of reverse transcriptase that is complexed with an RNA-DNA hybrid (protein in yellow, DNA sugar-phosphate backbone in lavender, RNA backbone in pink, bases in blue).

Preview

- Eucaryotic viruses are extremely diverse, exhibiting an amazing variety of morphologies and life cycle strategies.
- Although the details differ, eucaryotic virus reproduction is similar to that of the bacteriophages in having the same series of phases: adsorption, penetration, replication of virus nucleic acids and synthesis of viral proteins, assembly of capsids, and virus release.
- Viruses may harm their host cells in a variety of ways, ranging from direct inhibition of DNA, RNA, and protein synthesis to the alteration of plasma membranes and formation of inclusion bodies.
- Some vertebrate virus infections have a rapid onset and relatively short duration. Others establish long-term chronic infections or are dormant for a while and then become active again. Slow virus infections may take years to develop.
- Cancer can be caused by a number of factors, including viruses. Viruses may bring oncogenes into a cell, carry transcription regulatory elements that stimulate a cellular proto-oncogene, or in other ways transform cells into tumor cells.
- Plant viruses are responsible for many important diseases but have not been as intensely studied due to technical challenges. Most are RNA viruses. Insects are the most important transmission agents, and some plant viruses multiply in insect tissues before being inoculated into another plant.
- Numerous viruses infect insects. Many of these infections are accompanied by the formation of characteristic inclusion bodies. A number of insect viruses show promise as biological control agents for insect pests.
- Infectious agents simpler than viruses also exist. Viroids are short strands of infectious RNA responsible for several plant diseases. Virusoids are infectious RNAs that require a helper virus to gain entry into a target cell.
- Prions are proteinaceous particles associated with certain degenerative neurological diseases in humans and livestock.

In chapter 17 we introduce bacterial and archaeal viruses in some detail because they are very important to the fields of molecular biology and genetics, as well as to virology. In this chapter we focus on viruses that use eucaryotic organisms as hosts. Although plant and insect viruses are discussed, we place particular emphasis on vertebrate viruses because they are very well studied and are the causative agents of so many important human diseases. The chapter closes with a brief summary of what is known about infectious agents that are even simpler in construction than viruses: the viroids, virusoids, and prions.

18.1 Taxonomy of Eucaryotic Viruses

Of the almost 6,000 viruses known, most infect eucaryotic organisms. They have been classified by the International Committee on the Taxonomy of Viruses (ICTV) into numerous families, based primarily on genome structure, replication strategy, morphology, and genetic relatedness. Some of the representative families of vertebrate viruses are shown in **figure 18.1.** Some representative genera and their descriptions are illustrated in **figures 18.2** and **18.3.** As illustrated in these figures, eucaryotic viruses can be naked (without an envelope) or enveloped. They also exhibit great diversity in their genomes, with all types observed. Their nucleic acid can be single stranded or double stranded, circular or linear. Some eucaryotic viruses have segmented genomes consisting of more than one distinct nucleic acid molecule.

The Virus. Observe this virus: think how small Its arsenal, and yet how loud its call; It took my cell, now takes your cell, And when it leaves will take our genes as well. Genes that are master keys to growth That turn it on, or turn it off, or both; Should it return to me or you It will own the skeleton keys to do A number on our tumblers; stage a coup.

—*Michael Newman*

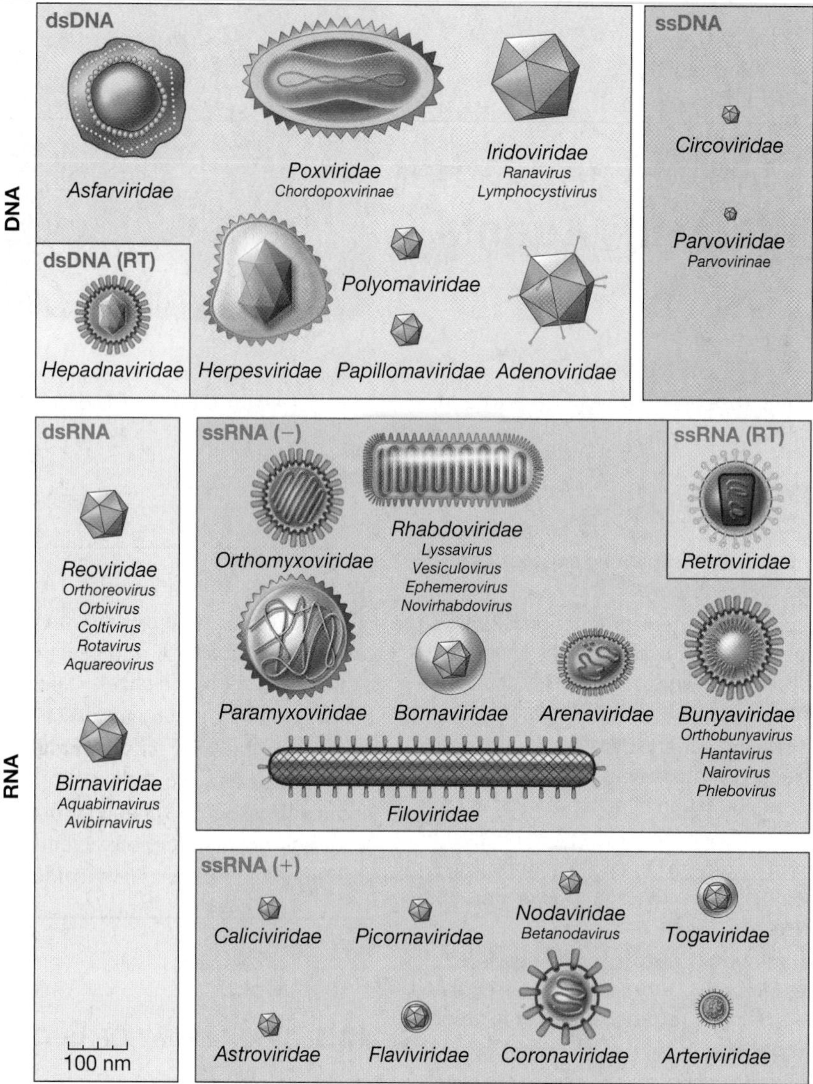

Figure 18.1 A Diagrammatic Description of the Families (-*idae*), subfamilies (-*inae*), and Genera (*virus*) of Viruses That Infect Vertebrates. RT stands for reverse transcriptase.

18.2 Reproduction of Vertebrate Viruses

The reproduction of vertebrate viruses is very similar in many ways to that of phages. Vertebrate virus reproduction may be divided into several stages: adsorption, penetration, replication of viral nucleic acids and synthesis of viral proteins, assembly of viral capsids, and release of mature viruses. Each of these stages is briefly described.

Adsorption of Virions

Encounters of the virus and the host cell surface are thought to occur through a random collision of the virion with a potential host. However, adsorption to the host is mediated by an interaction between receptors on the surface of the host cell and molecules on the surface of the virion. Because the virion attaches only to those host cells with the proper receptor, vertebrate viruses display **tropism.** That is, they only infect certain organisms, and in some cases only infect certain tissues within that host **(Microbial Diversity & Ecology 18.1)**. The receptors on the host cell have specific cellular functions. For instance, they may normally bind hormones or other molecules essential to the cell's function and role in the body **(table 18.1)**. Many host receptors are members of the immunoglobulin superfamily, a group of glycoproteins. Most members of the immunoglobulin superfamily are surface proteins that are involved in the immune response and cell-cell interactions. Examples are the human immunodeficiency virus (HIV) CD4 receptor, the poliovirus receptor, and the rhinovirus ICAM (*i*nter*c*ellular *a*dhesion *m*olecule) receptor. In some cases, it appears that two or more host cell receptors are involved in attachment. Herpes simplex virus interacts with a glycosaminoglycan and a member of the tumor necrosis factor/nerve growth factor receptor family. HIV uses

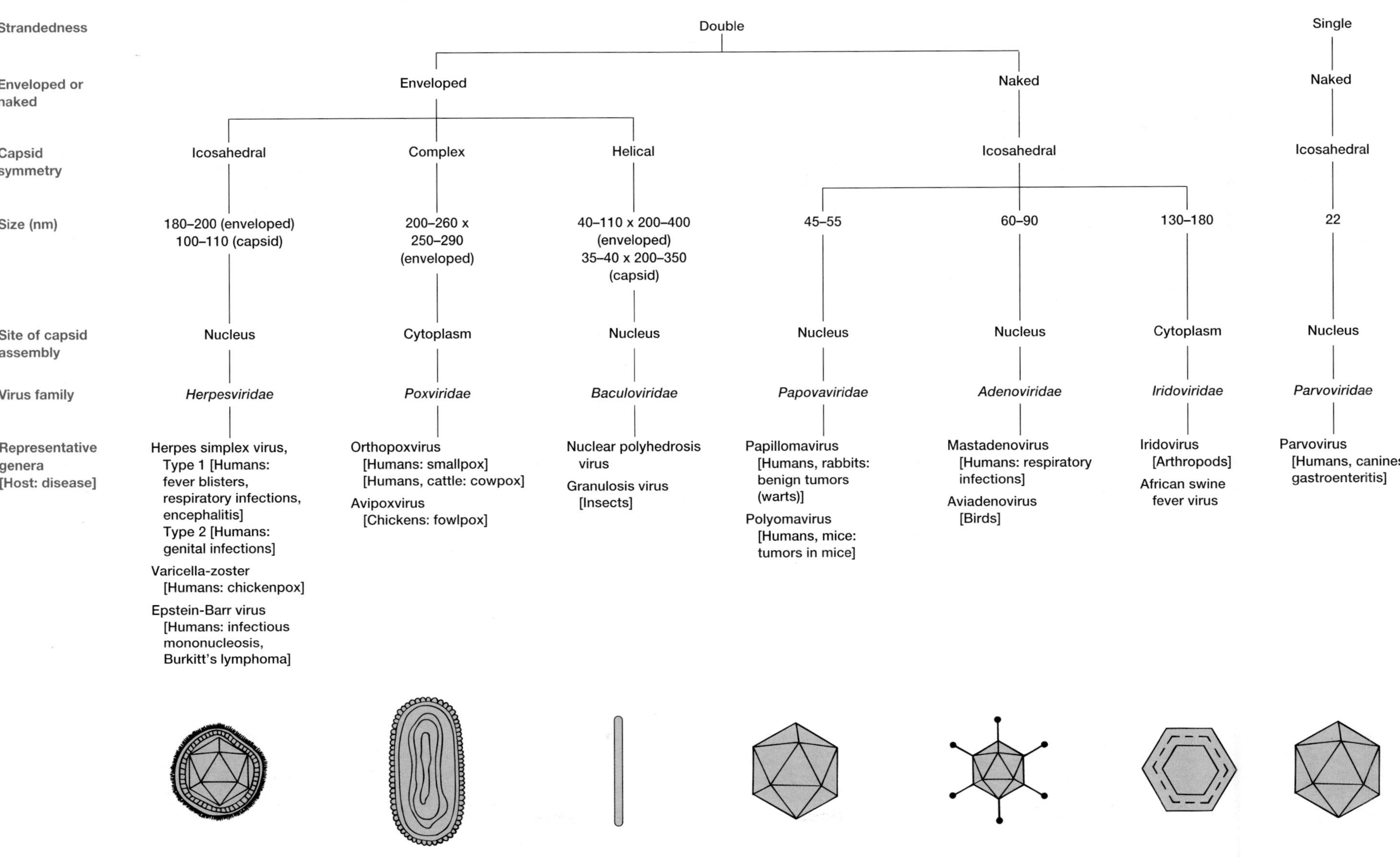
Strandedness
Double
Single
Enveloped or naked
Enveloped
Naked
Naked
Capsid symmetry
Icosahedral
Complex
Helical
Icosahedral
Icosahedral
Size (nm)
180–200 (enveloped) 100–110 (capsid)
200–260 x 250–290 (enveloped)
40–110 x 200–400 (enveloped) 35–40 x 200–350 (capsid)
45–55
60–90
130–180
22
Site of capsid assembly
Nucleus
Cytoplasm
Nucleus
Nucleus
Nucleus
Cytoplasm
Nucleus
Virus family
Herpesviridae
Poxviridae
Baculoviridae
Papovaviridae
Adenoviridae
Iridoviridae
Parvoviridae
Representative genera [Host: disease]
Herpes simplex virus, Type 1 [Humans: fever blisters, respiratory infections, encephalitis] Type 2 [Humans: genital infections]
Varicella-zoster [Humans: chickenpox]
Epstein-Barr virus [Humans: infectious mononucleosis, Burkitt's lymphoma]
Orthopoxvirus [Humans: smallpox] [Humans, cattle: cowpox]
Avipoxvirus [Chickens: fowlpox]
Nuclear polyhedrosis virus
Granulosis virus [Insects]
Papillomavirus [Humans, rabbits: benign tumors (warts)]
Polyomavirus [Humans, mice: tumors in mice]
Mastadenovirus [Humans: respiratory infections]
Aviadenovirus [Birds]
Iridovirus [Arthropods]
African swine fever virus
Parvovirus [Humans, canines: gastroenteritis]

Figure 18.2 The Taxonomy of DNA Animal Viruses.

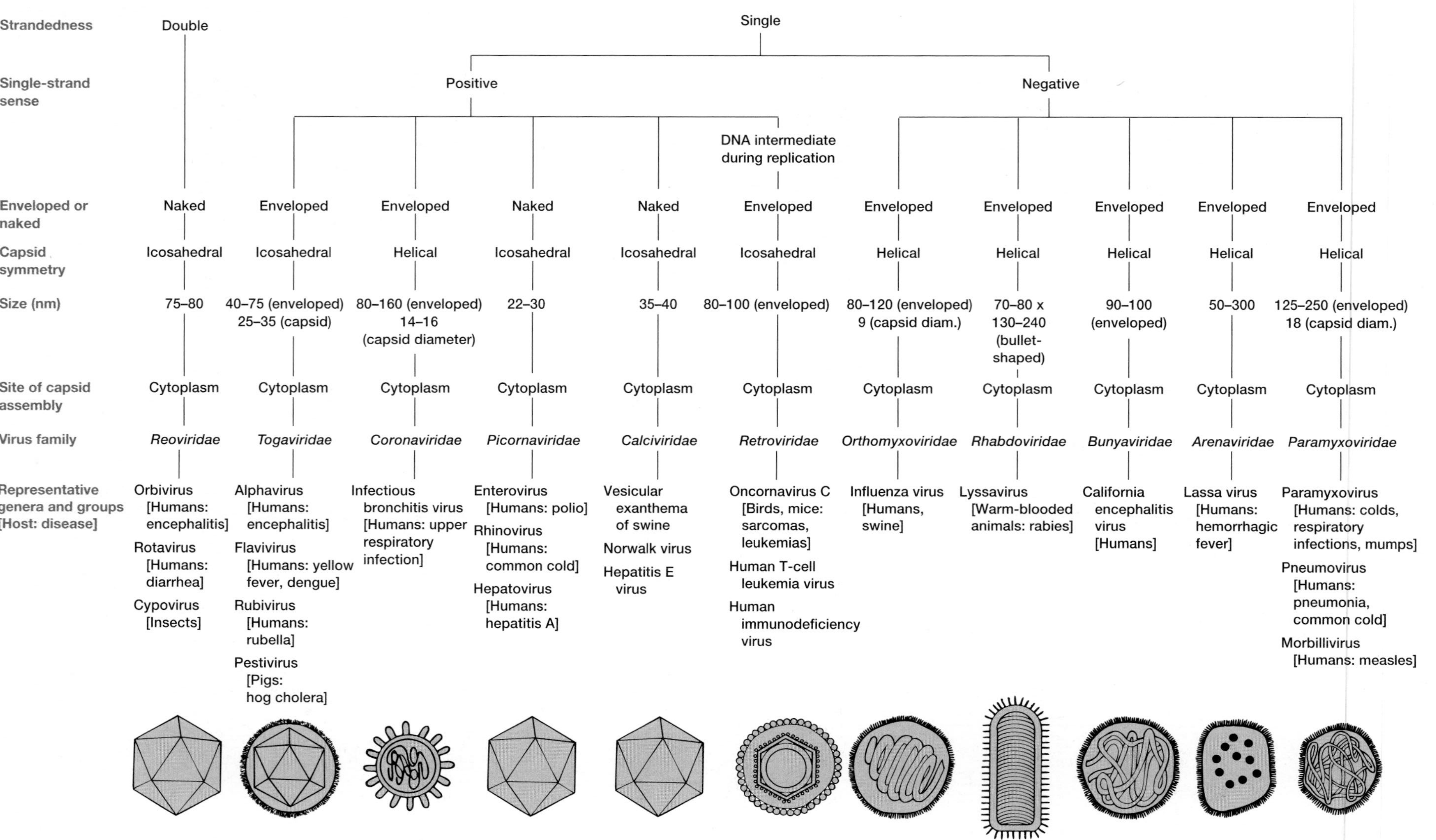

Strandedness
Single-strand sense
Enveloped or naked
Capsid symmetry
Size (nm)
Site of capsid assembly
Virus family
Representative genera and groups [Host: disease]
Double
Single
Positive
Negative
DNA intermediate during replication
Naked
Icosahedral
75–80
Cytoplasm
Reoviridae
Orbivirus [Humans: encephalitis]
Rotavirus [Humans: diarrhea]
Cypovirus [Insects]
Enveloped
Icosahedral
40–75 (enveloped) 25–35 (capsid)
Cytoplasm
Togaviridae
Alphavirus [Humans: encephalitis]
Flavivirus [Humans: yellow fever, dengue]
Rubivirus [Humans: rubella]
Pestivirus [Pigs: hog cholera]
Enveloped
Helical
80–160 (enveloped) 14–16 (capsid diameter)
Cytoplasm
Coronaviridae
Infectious bronchitis virus [Humans: upper respiratory infection]
Naked
Icosahedral
22–30
Cytoplasm
Picornaviridae
Enterovirus [Humans: polio]
Rhinovirus [Humans: common cold]
Hepatovirus [Humans: hepatitis A]
Naked
Icosahedral
35–40
Cytoplasm
Calciviridae
Vesicular exanthema of swine
Norwalk virus
Hepatitis E virus
Enveloped
Icosahedral
80–100 (enveloped)
Cytoplasm
Retroviridae
Oncornavirus C [Birds, mice: sarcomas, leukemias]
Human T-cell leukemia virus
Human immunodeficiency virus
Enveloped
Helical
80–120 (enveloped) 9 (capsid diam.)
Cytoplasm
Orthomyxoviridae
Influenza virus [Humans, swine]
Enveloped
Helical
70–80 x 130–240 (bullet-shaped)
Cytoplasm
Rhabdoviridae
Lyssavirus [Warm-blooded animals: rabies]
Enveloped
Helical
90–100 (enveloped)
Cytoplasm
Bunyaviridae
California encephalitis virus [Humans]
Enveloped
Helical
50–300
Cytoplasm
Arenaviridae
Lassa virus [Humans: hemorrhagic fever]
Enveloped
Helical
125–250 (enveloped) 18 (capsid diam.)
Cytoplasm
Paramyxoviridae
Paramyxovirus [Humans: colds, respiratory infections, mumps]
Pneumovirus [Humans: pneumonia, common cold]
Morbillivirus [Humans: measles]

Figure 18.3 The Taxonomy of RNA Animal Viruses.

Microbial Diversity & Ecology

18.1 SARS: Evolution of a Virus

In November 2002, a mysterious pneumonia was seen in the Guangdong Province of China, but the first case of this new type of pneumonia was not reported until February 2003. Thanks to the ease of global travel, it took only a couple of months for the pneumonia to spread to more than 25 countries in Asia, Europe, and North and South America. This newly emergent pneumonia was labeled Severe Acute Respiratory Syndrome (SARS) and its causative agent was identified as a previously unrecognized member of the coronavirus family, the SARS-CoV. Almost 10% of the roughly 8,000 people with SARS died. However, once the epidemic was contained, the virus appeared to "die out," and with the exception of a few mild, sporadic cases in 2004, no additional cases have been identified. From where does a newly emergent virus come? What does it mean when a virus "dies out"?

We can answer these questions thanks to the availability of the complete SARS-CoV genome sequence and the power of molecular modeling. Coronaviruses are large, enveloped viruses with positive-strand RNA genomes. They are known to infect a variety of mammals and birds. Researchers suspected that SARS-CoV had "jumped" from its animal host to humans, so samples of animals at open markets in Guangdong were taken for nucleotide sequencing. These studies revealed that cat-like animals called masked palm civits (*Paguma larvata*) harbored variants of the SARS-CoV. Although thousands of civits were then slaughtered, further studies failed to find widespread infection of domestic or wild civits. In addition, experimental infection of civits with human SARS-CoV strains made these animals ill, making the civit an unlikely candidate for the reservoir species. Such a species would be expected to harbor SARS-CoV without symptoms so that it could efficiently spread the virus.

Bats are reservoir hosts of several zoonotic viruses (viruses spread from animals to people) including the emerging Hendra and Nipah viruses that have been found in Australia and East Asia, respectively. Thus it was perhaps not too surprising when in 2005, two groups of international scientists independently demonstrated that Chinese horseshoe bats (genus *Rhinolophus*) are the natural reservoir of a SARS-like coronavirus. When the genomes of the human and bat SARS-CoV are aligned, 92% of the nucleotides are identical. More revealing is alignment of the translated amino acid sequences of the proteins encoded by each virus. The amino acid sequences are 96 to 100% identical for all proteins except the receptor-binding spike proteins, which are only 64% identical. The SARS-CoV spike protein mediates both host cell surface attachment and membrane fusion. Thus a mutation of the spike protein allowed the virus to "jump" from bat host cells to those of another species. It is not clear if the SARS-CoV was transmitted directly to humans (bats are eaten as a delicacy and bat feces are a traditional Asian cure for asthma) or if transmission to humans occurred through infected civits.

The relationship between the SARS-CoV found in civits and humans has also been studied in detail and offers insight into why there have been no additional cases of SARS since 2004 (at least as this book went to press). The region of the SARS-CoV spike protein that binds to the host receptor, angiotensin-converting enzyme-2 (ACE2), forms a shallow pocket into which ACE2 rests. The region of the spike protein that makes this pocket is called the receptor-binding domain (RBD). Of the approximately 220 amino acids within the RBD, only four differ between civit and human. Two of these amino acids appear to be critical. As shown in the **Box figure,** compared to the spike RBD in the SARS-CoV that caused the 2002–2003 epidemic, the civit spike has a serine (S) substituted for a threonine (T) at position 487 (T487S) and a lysine (K) at position 479 instead of asparagine (N), N479K. This causes a 1,000-fold decrease in the capacity of the virus to bind to human ACE2. Furthermore, the spike found in SARS-CoV isolated from patients in 2003 and 2004 also has a serine at position 487 as well as a proline (P) for leucine (L) substitution at position 472 (L472P). These amino acid substitutions could be responsible for the reduced virulence of the virus found in these more recent infections. In other words, these mutations could be the reason the SARS virus appears to have "died out."

Meanwhile a SARS vaccine based on the virulent 2002–2003 strain is being tested. This raises additional questions. Does the original virulent SARS-CoV strain still exist? Will the most recently identified SARS-CoV continue to evolve into less virulent forms? If not, will this vaccine be effective in preventing another highly infective SARS outbreak? Unfortunately, these questions cannot be easily answered.

Receptor Activity

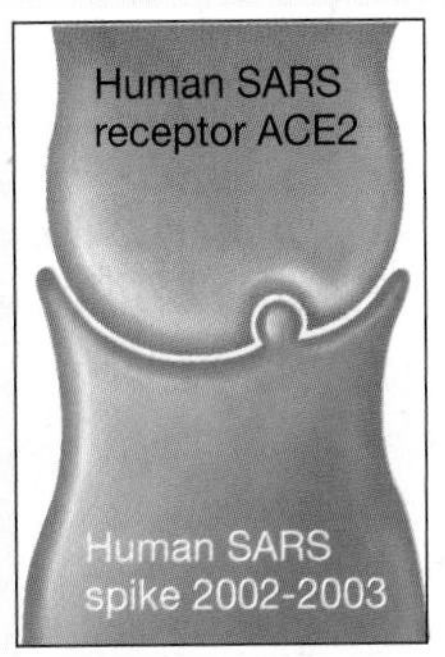

(a) Good

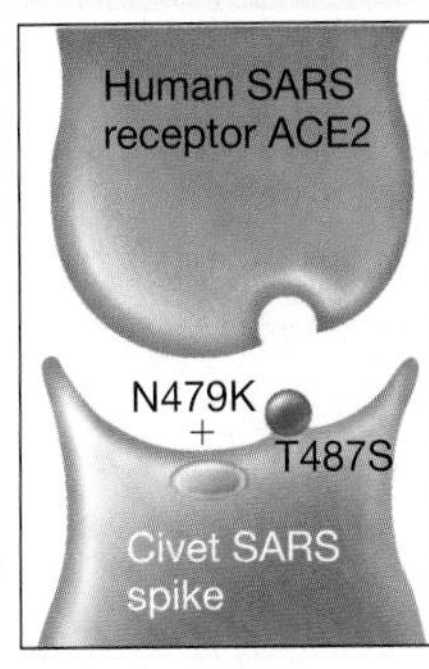

(b) Poor

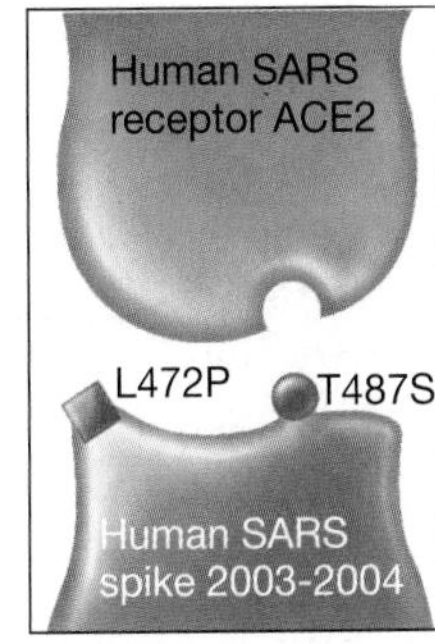

(c) Poor

Host Range of SARS-CoV Is Determined by Several Amino Acid Residues in the Spike Protein. **(a)** The spike protein of the SARS-CoV that caused the SARS epidemic in 2002–2003 fits tightly to the human host cell receptor ACE2. **(b)** The civit SARS-CoV has two different amino acids at positions 479 and 487. This spike protein binds very poorly to human ACE2, thus the receptor is only weakly activated. **(c)** The spike protein on the human SARS-CoV that was isolated from patients in 2003 and 2004 also differs from that seen in the epidemic-causing SARS-CoV by two amino acids. This SARS-CoV variant caused only mild, sporadic cases.

Li, F.; Li, W.; Farzan, M.; and Harrison, S. C. 2005. Structure of SARS coronavirus spike receptor-binding domain complexed with receptor. Science *309:1864–68.*

Li, W.; Shi, Z.; Yu, M.; Ren, W.; Smith, C.; Epstein, J H.; and Wang, H., et al. 2005. Bats are natural reservoirs of SARS-like coronavirus. Science *310:676–79.*

Table 18.1 Examples of Host Cell Surface Proteins That Serve as Virus Receptors

Virus	Cell Surface Receptor
Adenovirus	Coxsackie adenovirus receptor (CAR) protein
Epstein-Barr virus	Receptor for the C3d complement protein on human B lymphocytes
Hepatitis A virus	Alpha 2-macroglobulin
Herpes simplex virus, type 1	Heparan sulfate
Human immunodeficiency virus	CD4 protein on T-helper cells, macrophages, and monocytes; CXCR-4 or the CCR5 receptor
Influenza A virus	Sialic acid–containing glycoprotein
Measles virus	CD46 complement regulator protein
Poliovirus	Poliovirus receptor (PVR); Immunoglobulin-like molecule
Rabies virus	Acetylcholine receptor on neurons
Rhinovirus	Intercellular adhesion molecules (ICAMs) on the surface of respiratory epithelial cells
Rotavirus	$\alpha_2\beta_1$ and $\alpha_4\beta_1$ integrins
Vaccinia virus	Epidermal growth factor receptor

CD4 and CXCR-4 (fusin) or the CCR5 (CC-CKR-5) receptor. Both of these host molecules normally bind chemokines—signaling molecules produced by the immune system. Chemical mediators in nonspecific (innate) resistance: Cytokines (section 31.6)

The distribution of the host cell receptors to which viruses attach varies at both the cellular and tissue levels. Eucaryotic cell membranes have microdomains called lipid rafts that seem to be involved in both virion entrance and assembly. For example, the receptors for enveloped viruses such as HIV and Ebola are concentrated in lipid rafts. When the virus binds to these receptors, the host cell is tricked into endocytosing the virus. Distribution at the tissue level plays a crucial role in determining the tropism of the virus and the outcome of infection. For example, poliovirus receptors are found only in the human nasopharynx, gut, and anterior horn cells of the spinal cord. Therefore, it infects these tissues, causing gastrointestinal disease in its milder forms and paralytic disease in its more serious forms. In contrast, measles virus receptors are present in most tissues and disease is disseminated throughout the body, resulting in the widespread rash characteristic of measles. The plasma membrane and membrane structure (section 4.2)

The surface site on the virus that interacts with the host cell receptor can consist simply of a capsid structural protein or an array of such proteins. In some viruses—for example, the poliovirus and rhinoviruses—the binding site is at the bottom of a surface depression or valley. The site can bind to host cell surface projections but cannot be reached by host antibodies. Envelope glycoproteins also can be involved in the adsorption and penetration of enveloped viruses. For example, the herpes simplex virus has two envelope glycoproteins that are required for attachment, and at least four glycoproteins participate in penetration. In other cases the virus attaches to the host cell through special projections such as the fibers extending from the corners of adenovirus icosahedrons (*see figure 16.5d*) or the spikes of enveloped viruses. For instance, the influenza virus has hemagglutinin spikes that attach to host cells by interacting with sialic acid (*N*-acetylneuraminic acid) on the host cell surface.

Penetration and Uncoating

Viruses penetrate the plasma membrane and enter a host cell shortly after adsorption. During or soon after penetration, the viral nucleic acid is prepared for expression and replication. For some viruses, this involves shedding some or all capsid proteins, a process called uncoating, whereas other viruses remain encapsidated. Because penetration and uncoating are often coupled, we consider them together. The mechanisms of penetration and uncoating vary with the type of virus because viruses differ so greatly in structure and mode of reproduction. For example, enveloped viruses may enter cells in a different way than naked virions. Furthermore, some viruses inject only their nucleic acid, whereas others must ensure that a virus-associated RNA or DNA polymerase, or even an organized core, also enters the host cell. The entire process of adsorption and uncoating may take from minutes to several hours.

For many viruses, detailed mechanisms of penetration are unclear; it appears that one of two different modes of entry are employed by most viruses **(figure 18.4)**.

1. Fusion of the viral envelope with the host cell membrane—The envelopes of paramyxoviruses, the *Retroviridae,* and some other viruses fuse directly with the host cell plasma membrane (figure 18.4*a*). Fusion may involve envelope glycoproteins that bind to plasma membrane proteins. After attachment of paramyxoviruses, several things happen: membrane lipids rearrange, the adjacent halves of the contacting membranes merge, and a proteinaceous fusion pore forms. The nucleocapsid then enters the host cell cytoplasm, where a viral polymerase, associated with the nucleocapsid, begins transcribing the virus RNA while it is still within the capsid.
2. Entry by endocytosis—Nonenveloped viruses and some enveloped viruses enter cells by endocytosis. They may be engulfed by receptor-mediated endocytosis to form coated vesicles (figure 18.4*b*). The virions attach to clathrin-coated pits, and the pits then pinch off to form coated vesicles filled

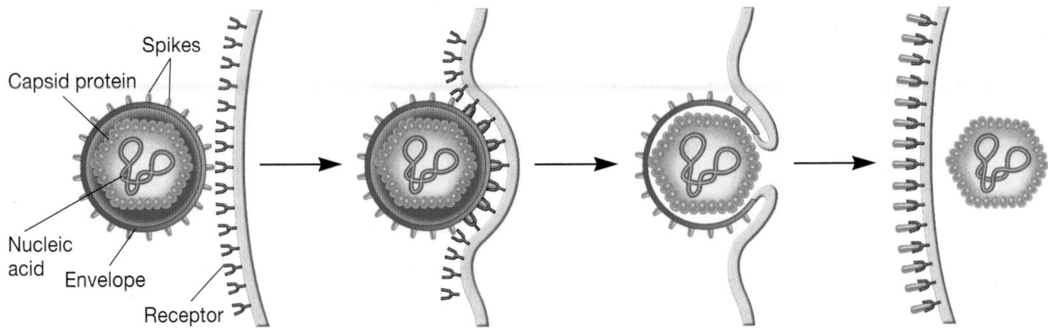

(a) Entry of enveloped virus by fusing with plasma membrane

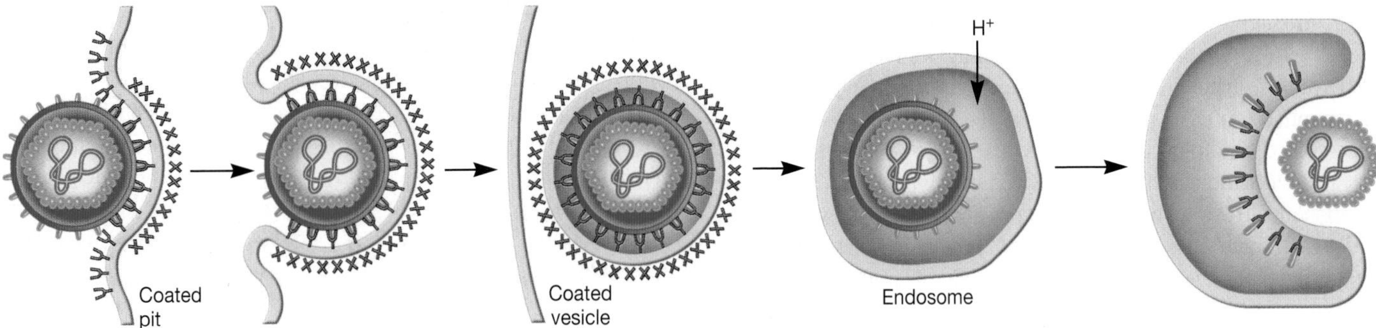

(b) Entry of enveloped virus by endocytosis

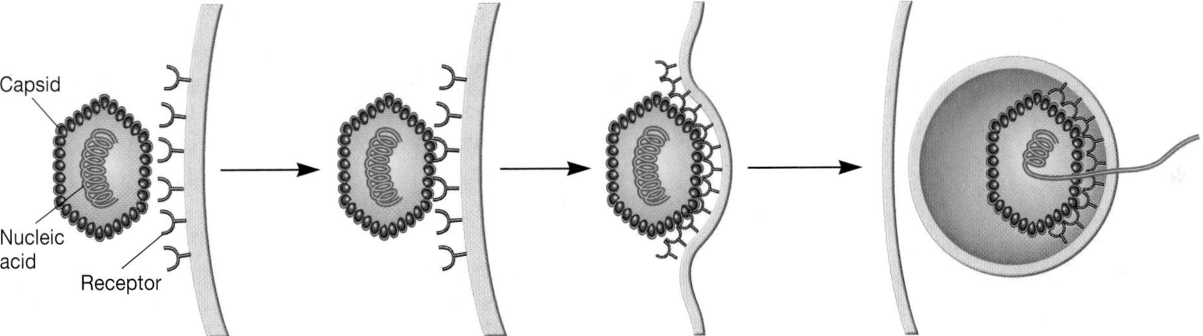

(c) Entry of naked virus by endocytosis

Figure 18.4 Animal Virus Entry. Examples of animal virus attachment and entry into host cells. Enveloped viruses can **(a)** enter after fusion of the envelope with the plasma membrane, or **(b)** escape from the vesicle after endocytosis. **(c)** Naked viruses such as poliovirus, a picornavirus, may be taken up by endocytosis and then insert their nucleic acid into the cytoplasm through the vesicle membrane. It also is possible that they insert the nucleic acid directly through the plasma membrane within a coated pit. See text for description of the entry modes.

with viruses. These vesicles fuse with endosomes after the clathrin has been removed; depending on the virus, escape of the nucleocapsid or its genome may occur either before or after vesicle fusion. Endosomal enzymes can aid in virus uncoating and low pHs often trigger the uncoating process. In at least some instances, the viral envelope fuses with the endosomal membrane, and the nucleocapsid is released into the cytoplasm (the capsid proteins may have been partially removed by endosomal enzymes). Once in the cytoplasm, viral nucleic acid may be released from the capsid upon completion of uncoating or may function while still attached to capsid components. Naked viruses lack an envelope and thus cannot employ the membrane fusion mechanism (figure 18.4*c*). In this case, it appears that vesicle acidification causes a capsid conformational change. The altered capsid contacts the vesicle membrane and either releases the viral nucleic acid into the cytoplasm through a membrane pore (picornaviruses) or ruptures the membrane to release the virion (adenovirus). Viruses may also enter the host cell by way of caveolae formation.

Organelles of the biosynthetic-secretory and endocytic pathways (section 4.4)

Genome Replication and Transcription in DNA Viruses

Although the details vary, genome replication and transcription in DNA viruses follow a similar course. The early part of the synthetic phase, governed by the **early genes,** is devoted to taking over the host cell and to the synthesis of viral DNA and RNA. Some animal viruses inhibit host cell DNA, RNA, and protein synthesis, though cellular DNA is not usually degraded. In contrast, other viruses may actually stimulate the synthesis of host macromolecules. DNA replication usually occurs in the host cell nucleus; poxviruses are exceptions since their genomes are replicated in the cytoplasm. Messenger RNA—at least early mRNA—is transcribed from DNA by host enzymes. Poxviruses are again the exception, as their early mRNA is synthesized by a viral polymerase. Some examples of DNA virus reproduction will help illustrate these generalizations.

Parvoviruses infect animals, including humans; one causes fifth's disease in children. They have a genome composed of one single-stranded (ss) DNA molecule of about 4,800 bases. Parvoviruses are among the simplest of the DNA viruses. The genome is so small that it directs the synthesis of only three polypeptides, all capsid components. Even so, the virus must resort to the use of overlapping genes to fit three genes into such a small molecule. That is, the base sequences that code for the three polypeptide chains overlap each other and are read using different reading frames. Since the genome does not code for any enzymes, the virus must use host cell enzymes for all biosynthetic processes. Thus viral DNA can only be replicated in the nucleus during the S phase of the cell cycle, when the host cell replicates its own DNA. Because the viral genome is single stranded, the host DNA polymerase must be tricked into copying it. By using a self-complementary sequence at the ends of the viral DNA, the parvovirus genome folds back on itself to form a primer for replication **(figure 18.5)**. This is recognized by the host DNA polymerase and DNA replication ensues. DNA replication (section 11.4)

Herpesviruses are a large group of icosahedral, enveloped, double-stranded (ds) DNA viruses responsible for many important human and animal diseases. The genome is a linear piece of DNA about 160,000 base pairs long and contains at least 50 to 100 genes. Immediately upon release of the viral DNA into the host nucleus, the DNA circularizes and is transcribed by host

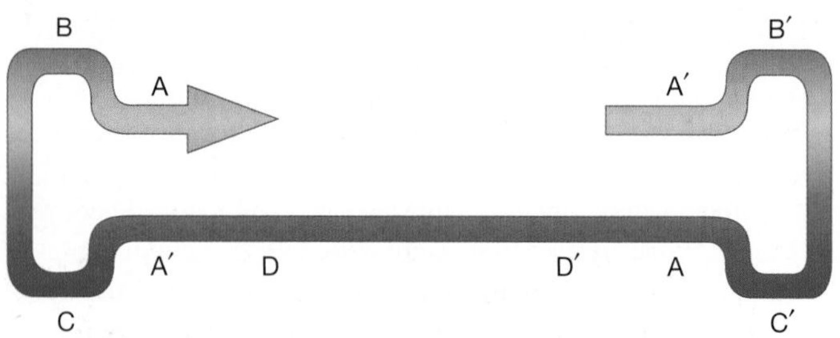

Figure 18.5 The Secondary Structure of the Parvovirus ssDNA Genome. The linear ssDNA genome of parvoviruses exhibits intrastrand base pairing that results in the formation of double-stranded regions at each end of the molecule. This provides a primer for DNA synthesis by the host DNA polymerase.

RNA polymerase to form mRNAs, which direct the synthesis of several early proteins. These are mostly regulatory proteins and the enzymes required for virus DNA replication **(figure 18.6,** steps 1 and 2). Replication of the gene with a virus-specific DNA polymerase begins in the cell nucleus within 4 hours after infection (step 3). Host DNA synthesis gradually slows during a lethal virus infection (not all herpes infections result in immediate cell death). Direct contact diseases: Genital herpes (section 37.3)

Poxviruses such as the vaccinia virus are among the largest viruses known and are morphologically complex. Their dsDNA possesses over 200 genes. These viruses enter through receptor-mediated endocytosis in coated vesicles; the central core escapes from the endosome and enters the cytoplasm. The core contains a DNA-dependent RNA polymerase that synthesizes early mRNAs, one of which directs the production of an enzyme that completes virus uncoating. DNA polymerase and other enzymes needed for DNA replication are also synthesized early in the reproductive cycle, and genome replication begins about 1.5 hours after infection. About the time DNA replication starts, transcription of late genes is initiated. Many late proteins are structural proteins used in capsid construction. The complete reproductive cycle in poxviruses takes about 24 hours. Airborne diseases: Smallpox (section 37.1)

The hepadnaviruses such as hepatitis B virus are quite different from other DNA viruses with respect to genome replication. They have circular dsDNA genomes that consist of one complete, but nicked, strand and a complementary strand that has a large gap—that is, it is incomplete **(figure 18.7)**. After infecting the cell, the virus's gapped DNA is released into the nucleus. There, host repair enzymes fill the gap and seal the nick, yielding a covalently closed, circular DNA. Transcription of viral genes occurs in the nucleus using host RNA polymerase and yields several mRNAs, including a large 3.4 kilobase RNA known as the pregenome (+RNAs). The RNAs move to the cytoplasm, and the mRNAs are translated to produce virus proteins including core proteins and a polymerase having three activities (DNA polymerase, reverse transcriptase, RNase H). Then the RNA pregenome associates with the polymerase and core protein to form an immature core particle. Reverse transcriptase subsequently reverse transcribes the RNA using a protein primer to form a minus-strand DNA from the pregenome +RNA. After almost all the pregenome RNA has been degraded by RNase H, the remaining RNA fragment serves as a primer for synthesis of the gapped dsDNA genome, using the minus-strand DNA as template. Finally, the nucleocapsid is completed and the progeny virions are released. Direct contact diseases: Viral hepatitides (section 37.3)

Genome Replication, Transcription, and Protein Synthesis in RNA Viruses

The RNA viruses are much more diverse in their reproductive strategies than are the DNA viruses. Most RNA viruses can be placed in one of four general groups based on their modes of genome replication and transcription, and their relationship to the host cell genome. **Figure 18.8** summarizes the reproductive cycles characteristic of these groups. Because of the unique features of retrovirus life cycles, we will consider them separately. For the re-

Figure 18.6 Generalized Life Cycle for Herpes Simplex Virus Type 1. Only one possible pathway for release of the virions is shown.

1. Circularization of genome and transcription of immediate-early genes
2. α-proteins, products of immediate-early genes, stimulate transcription of early genes.
3. β-proteins, products of early genes, function in DNA replication, yielding concatemeric DNA. Late genes are transcribed.
4. γ-proteins, products of late genes, participate in virion assembly.

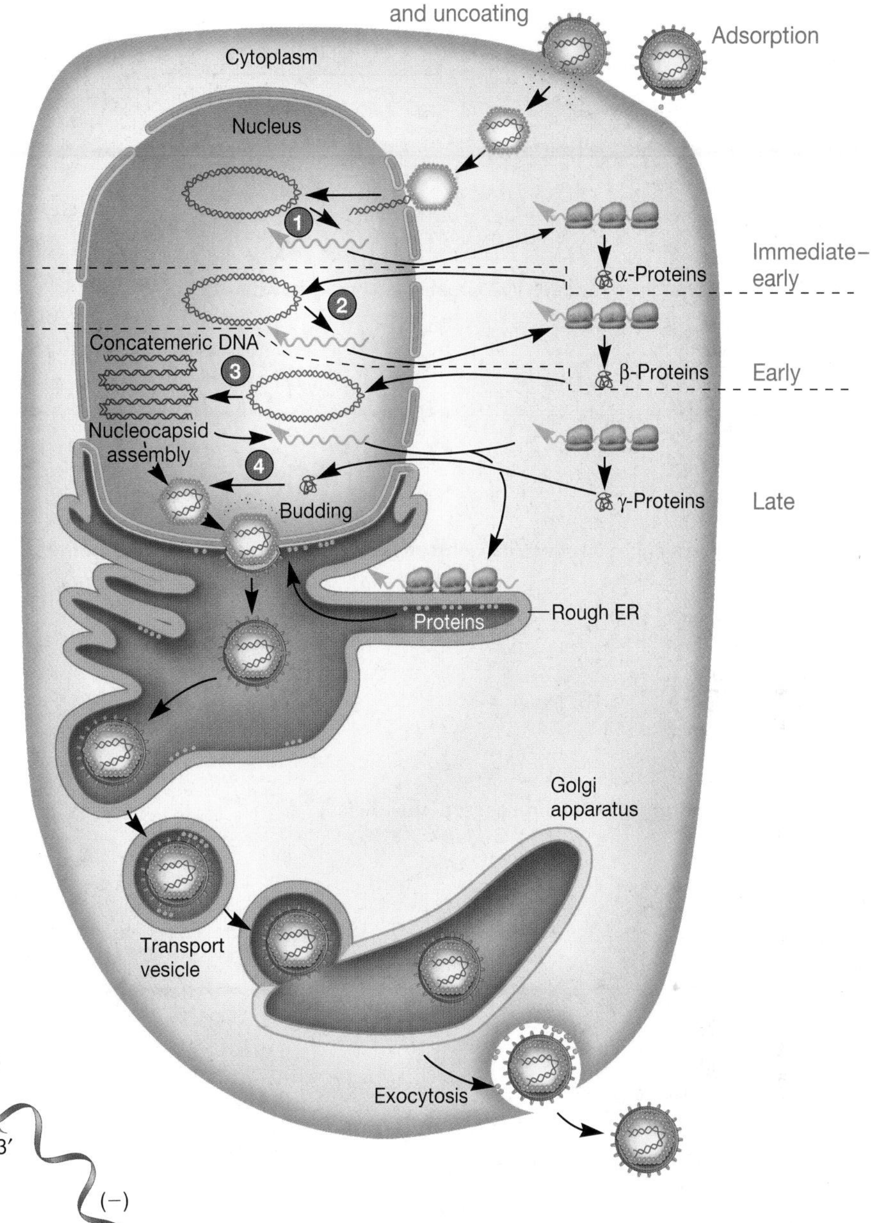

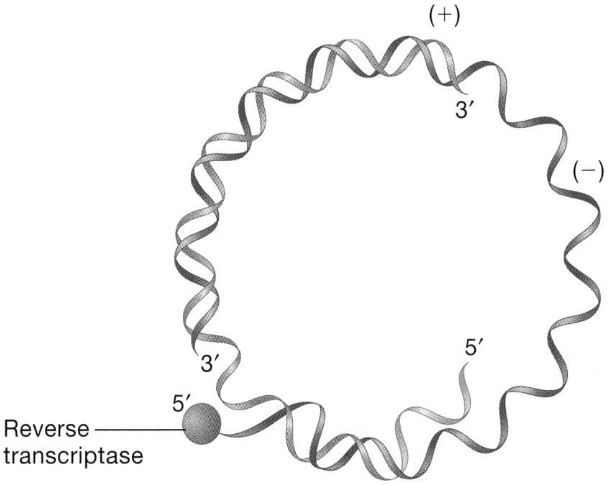

(+)
3′
(−)
3′
5′
5′
Reverse transcriptase

Figure 18.7 The Gapped Genome of Hepadnaviruses. The genomes of hepadnaviruses are unusual in several respects. The negative strand of the dsDNA molecule is complete, but nicked. The enzyme reverse transcriptase is attached to its 5′ end. The positive strand is gapped; that is, it is incomplete (shown in purple). A short stretch of RNA is attached to its 5′ end (shown in green). Upon entry into the host nucleus, the nick in the negative strand is sealed and the gap in the positive strand is filled, yielding a covalently closed, circular dsDNA molecule.

maining RNA viruses, one aspect is common to their life cycles: they must encode an RNA-dependent RNA polymerase, which is used to synthesize mRNA (**transcriptase** activity) or to replicate the RNA genome (**replicase** activity). Some ssRNA viruses use the identical enzyme to carry out both functions. However, in other viruses, one form of the polymerase synthesizes mRNA and another form replicates the genome. In all cases, mRNA is the initial product; later in the infection, a switch to genome replication occurs (figure 18.8*a–c*). We begin by examining plus-strand RNA viruses.

The picornaviruses such as poliovirus are the best studied positive-strand ssRNA viruses (**Techniques & Applications 18.2**). They use their positive-strand RNA genome as a giant mRNA, and host ribosomes synthesize an enormous peptide that is then cleaved (processed) by both host and virus-encoded enzymes to form the mature proteins (figure 18.8*a*). One of the proteins produced is an RNA-dependent RNA polymerase. It catalyzes the synthesis of a negative-strand RNA from the plus-strand genome,

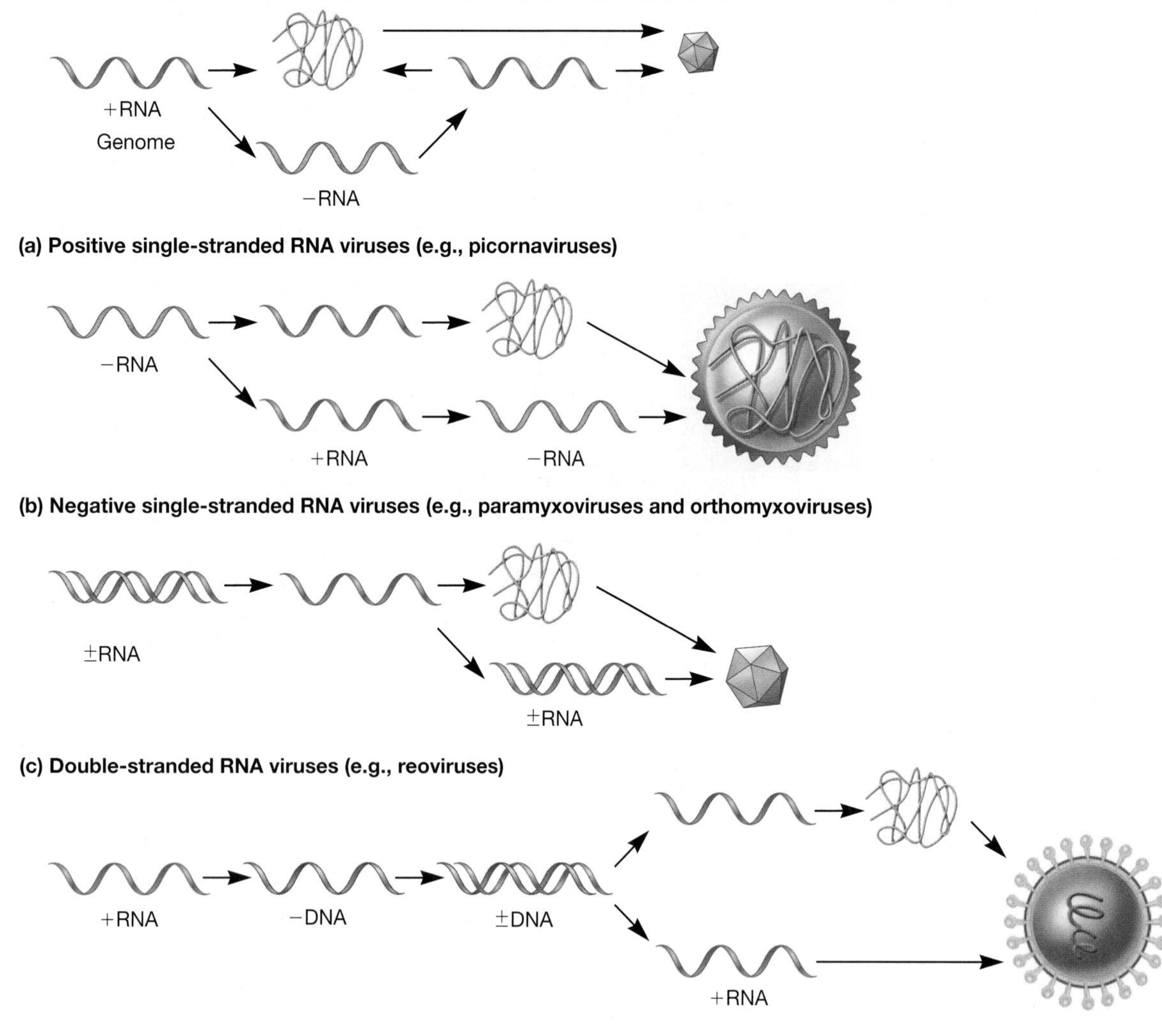

Figure 18.8 Reproductive Strategies of Vertebrate Viruses with RNA Genomes.

forming a dsRNA called the **replicative form (RF).** The replicative form is used as the template for synthesis of positive-strand RNA molecules, some of which serve as mRNA molecules. Other positive-strand RNA molecules are encapsulated and serve as progeny genomes. The formation of the replicative form creates a problem for the virus: dsRNA triggers certain host defenses, including interferon production, which inhibits viral replication. The virus avoids this problem by synthesizing only a small amount of negative-strand RNA molecules, thus limiting the number of double-stranded replicative forms present in the host cell.

Negative-strand RNA viruses must take a different approach to transcription and genome replication (figure 18.8*b*). Because their genome cannot function as an mRNA, these viruses must bring at least one RNA-dependent RNA polymerase into the host cell during entry. For instance, orthomyxoviruses such as the influenza A virus have segmented genomes (i.e., the genome consists of multiple pieces or segments), and the protein subunits that make up the polymerase, as well as other proteins, coat the RNA genome (**figure 18.9**). Thus the polymerase enters the host with the viral genome. The genome then serves as the template for mRNA synthesis (figure 18.9, step 1). Later, the virus switches from mRNA synthesis to genome replication. During this phase of the life cycle, the plus-strand RNA molecules synthesized from the minus-strand genome segments serve as templates for new negative-strand RNA genomes (figure 18.9, step 4).

The dsRNA viruses have additional challenges. The double-stranded nature of their genomes prevents translation of the plus strand within the RNA duplex. Furthermore, if the dsRNA genome is released into the host, it could trigger host defenses that might abort the infection process. Therefore these viruses not only bring into the cell an RNA-dependent RNA polymerase; they also keep their genomes enclosed in a subviral particle (figure 18.8*c*). It is within the subviral particle that synthesis of mRNA occurs. The mRNA molecules are released from the particle and are translated by the host into various viral proteins. Some of these are used in the assembly of additional particles containing positive-strand

1. The endonuclease activity of the PB1 protein cleaves the cap and about 10 nucleotides from the 5′ end of host mRNA (cap snatching). The fragment is used to prime viral mRNA synthesis by the RNA-dependent RNA polymerase activity of the PB1 protein.
2. Viral mRNA is translated. Early products include more NP and PB1 proteins.
3. RNA polymerase activity of the PB1 protein synthesizes +ssRNA from genomic −ssRNA molecules.
4. RNA polymerase activity of the PB1 protein synthesizes new copies of the genome using +ssRNA made in step 3 as templates. Some of these new genome segments serve as templates for the synthesis of more viral mRNA. Later in the infection, they will become progeny genomes.
5. Viral mRNA molecules transcribed from other genome segments encode structural proteins such as hemagglutinin (HA) and neuraminidase (NA). These messages are translated by ER-associated ribosomes and delivered to the cell membrane.
6. Viral genome segments are packaged as progeny virions bud from the host cell.

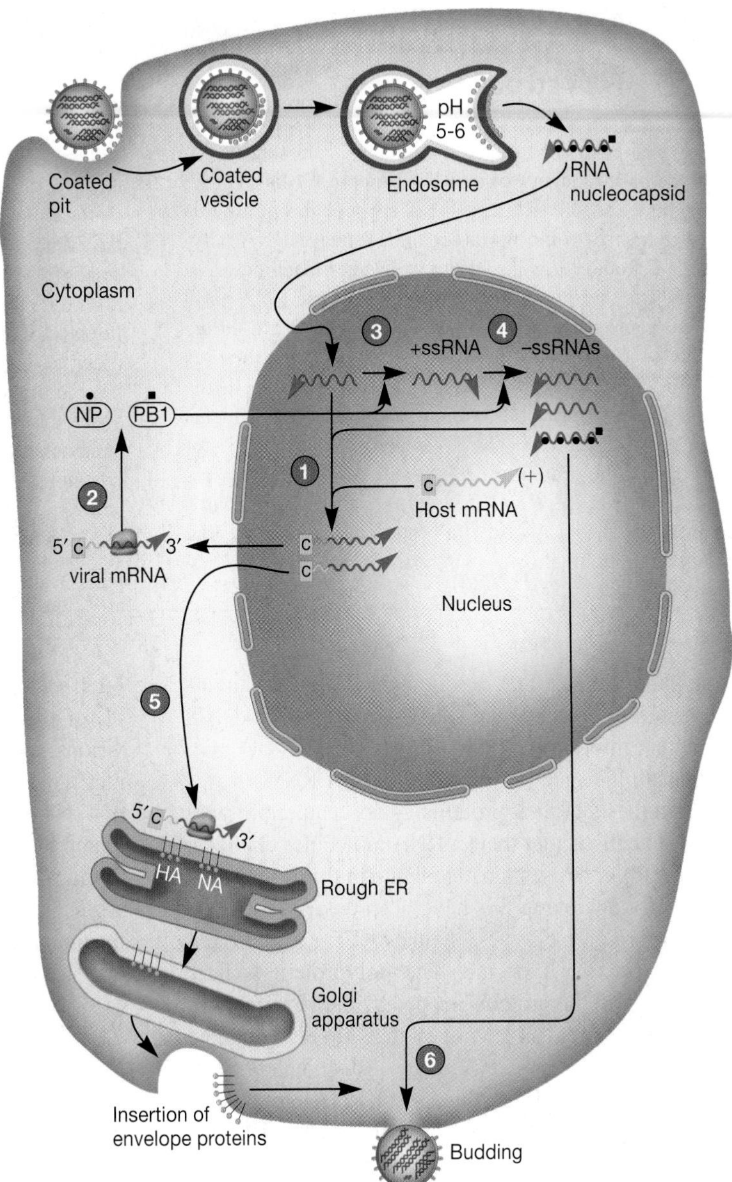

Figure 18.9 Simplified Life Cycle of Influenza Virus. Several steps have been eliminated for simplicity and clarity. After entry by receptor-mediated endocytosis, the virus envelope fuses with the endosome membrane, releasing the nucleocapsids into the cytoplasm (each genome segment is associated with nucleocapsid proteins [NP] and PB1 to form the nucleocapsid). The nucleocapsids enter the nucleus, where synthesis of viral mRNA and genomes occurs. Critical to the production of viral mRNA and genomes is the enzyme RNA-dependent RNA polymerase. This enzyme is one activity of the PB1 protein. The other is endonuclease activity, which is used to cleave the 5′ ends from host mRNA. Steps 1 through 6 illustrate the remaining steps of the virus's life cycle.

RNA. The positive-strand RNA is the template for the synthesis of negative-strand RNA, and the dsRNA genome is regenerated.

Retroviruses such as HIV possess ssRNA genomes but differ from other RNA viruses in that they must first synthesize DNA before transcribing mRNA and replicating their genome. The virus has an RNA-dependent DNA polymerase, commonly called **reverse transcriptase (RT),** that copies the +RNA genome to form a −DNA molecule (chapter opening figure and figure 18.8*d*). Interestingly, transfer RNA is carried by the virus and serves as the primer required for nucleic acid synthesis. The transformation of RNA into DNA takes place in two steps. First, reverse transcriptase uses the +RNA as a template to form a RNA-DNA hybrid. Then the **ribonuclease H (RNase H)** component of reverse transcriptase degrades the +RNA strand to leave −DNA. After synthesizing −DNA, the reverse transcriptase uses it as a template to produce a dsDNA called **proviral DNA.** The proviral DNA becomes integrated into the host cell genome. From there, host RNA polymerase can direct the synthesis of mRNA and

Techniques & Applications

18.2 Constructing a Virus

The poliovirus has been constructed completely from scratch beginning with the base sequence of its RNA genome. Because DNA is easier to synthesize than RNA, a DNA copy of the poliovirus RNA genome attached to the promoter for T7 phage RNA polymerase was first produced using DNA synthesizer machines. The synthetic DNA was then incubated with the T7 phage RNA polymerase and the appropriate nucleotides. The polymerase used the DNA template to form complete RNA copies of the poliovirus genome. The RNA genomes were then incubated with a cell-free cytoplasmic extract of HeLa cells, which supplied the constituents necessary for protein synthesis. The RNA directed the synthesis of poliovirus proteins; the proteins and RNA then spontaneously assembled into complete, infectious poliovirus virions. These particles could infect a special mouse strain and cause a disease that resembled human poliomyelitis.

Some scientists believe that it will be possible to make almost any virus once the genome sequence is known. This could be a particular threat if bioterrorists managed to create smallpox. Others believe that it will not be that easy to create a virulent pathogen like smallpox. The smallpox genome is much larger than that of poliovirus and would be harder to synthesize. Unlike the case with poliovirus, smallpox DNA is not infectious by itself and requires the presence of virion proteins such as polymerases. Thus one would have to find a way to supply these enzymes, perhaps by using helper viruses. A potentially greater threat is the synthesis of special hybrid strains that are very infectious and also quite lethal. On a more positive note, it may be possible to construct substantially weakened strains of viruses such as poliovirus for the production of more effective vaccines. Bioterrorism preparedness (section 36.9)

new +RNA genomes. Notice that during this process genetic information is transferred from RNA to DNA rather than from DNA to RNA as in cellular information flow. DNA replication (section 11.4)

Another interesting aspect of the life cycles of RNA viruses is related to the synthesis of their proteins. A key feature of translation in eucaryotic cells is that their mRNA molecules encode a single protein. Yet RNA viruses must generate multiple proteins. Several strategies for achieving this have been described. As already noted for the picornaviruses, synthesis of a polyprotein from a genomic-sized RNA occurs. The polyprotein is later cleaved to give rise to all the proteins needed by the virus to complete its life cycle. Those RNA viruses that have segmented genomes can use each segment to encode a different protein. Some RNA viruses have an RNA-dependent RNA polymerase that is able to initiate mRNA synthesis at internal sites along the template RNA. This generates **subgenomic mRNA,** which is mRNA that is smaller than the RNA genome. RNA splicing is used by some RNA viruses to generate different transcripts and therefore different proteins. Finally, translation itself may lead to synthesis of multiple proteins through **ribosomal frameshifting,** which causes translation of a different reading frame yielding a different protein with each frame read. Many RNA viruses use several of these strategies to produce a full array of viral proteins.

Assembly of Virus Capsids

Some **late genes** direct the synthesis of capsid proteins, and these spontaneously self-assemble to form the capsid just as in bacteriophage morphogenesis. Recently the self-assembly process has been dramatically demonstrated (Techniques & Applications 18.2). It appears that during icosahedral virus assembly empty **procapsids** are first formed; then the nucleic acid is inserted in some unknown way. The site of morphogenesis varies with the virus **(table 18.2)**. Large paracrystalline clusters of either complete virions or procapsids are often seen at the site of virus maturation **(figure 18.10)**. The assembly of enveloped virus capsids is generally similar to that of naked virions, except for poxviruses. These are assembled in the cytoplasm by a lengthy, complex process that begins with the enclosure of a portion of the cytoplasmic matrix through construction of a new membrane. Then newly synthesized DNA condenses, passes through the membrane, and moves to the center of the immature virus. Nucleoid and elliptical body construction takes place within the membrane.

Virion Release

Mechanisms of virion release differ between naked and enveloped viruses. Naked virions appear to be released most often by host cell lysis. In contrast, the formation of envelopes and the release of enveloped viruses are usually concurrent processes, and the host cell may continue virion release for some time. All viral envelopes are derived from host cell membranes by a multistep process. First, virus-encoded proteins are incorporated into the plasma membrane. Then the nucleocapsid is simultaneously released and the envelope formed by membrane budding **(figures 18.11** and **18.12)**. In several virus families, a matrix (M) protein attaches to the plasma membrane and aids in budding. Most envelopes arise from the plasma membrane. However, in herpesviruses, budding and envelope formation usually involve the nuclear membrane (table 18.2). The endoplasmic reticulum, Golgi apparatus, and other internal membranes also can be used to form envelopes.

Interestingly, it has been discovered that actin filaments can aid in virion release. Many viruses alter the actin microfilaments of the host cell cytoskeleton. For example, vaccinia virus appears to form long actin tails and use them to move intracellularly at up to 2.8 μm per minute. The actin filaments also propel vaccinia through the plasma membrane. In this way the virion escapes without destroying the host cell and infects adjacent cells. This is

Table 18.2 Intracellular Sites of Animal Virus Reproduction

Virus	Nucleic Acid Replication	Capsid Assembly	Membrane Used in Budding
DNA Viruses			
Adenoviruses	Nucleus	Nucleus	
Hepadnaviruses	Cytoplasm	Cytoplasm	Endoplasmic reticulum
Herpesviruses	Nucleus	At nuclear membrane	Nucleus
Papillomaviruses	Nucleus	Nucleus	
Parvoviruses	Nucleus	Nucleus	
Polyomaviruses	Nucleus	Nucleus	
Poxviruses	Cytoplasm	Cytoplasm	
RNA Viruses			
Coronaviruses	Cytoplasm	Cytoplasm	Golgi apparatus and endoplasmic reticulum
Orthomyxoviruses	Nucleus	Cytoplasm	Plasma membrane
Paramyxoviruses	Cytoplasm	Cytoplasm	Plasma membrane
Picornaviruses	Cytoplasm	Cytoplasm	
Reoviruses	Cytoplasm	Cytoplasm	
Retroviruses	Cytoplasm and nucleus	At plasma membrane	Plasma membrane
Rhabdoviruses	Cytoplasm	Cytoplasm	Plasma membrane, intracytoplasmic membranes
Togaviruses	Cytoplasm	Cytoplasm	Plasma membrane, intracytoplasmic membranes

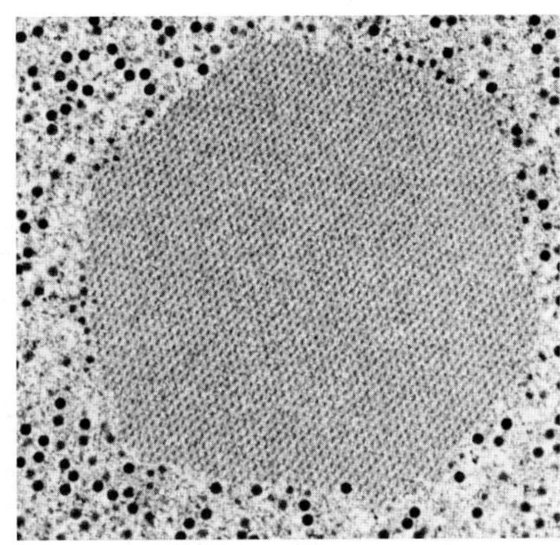

Figure 18.10 Paracrystalline Clusters. A crystalline array of adenoviruses within the cell nucleus (×35,000).

similar to the mechanism of pathogenesis used by the bacterium *Listeria monocytogenes.* The cytoplasmic matrix, microfilaments, intermediate filaments, and microtubules (section 4.3); Disease 4.1: Getting around

1. Compare and contrast each stage of vertebrate virus reproduction with those seen for bacteriophages.
2. What probably plays the most important role in determining the tissue and host specificity of vertebrate viruses? Give some specific examples.
3. In general, DNA viruses can be much more dependent on their host cells than can RNA viruses. Why is this so? What strategies are used by RNA viruses to complete their life cycles?
4. Outline the retrovirus life cycle. What is proviral DNA? How is it synthesized? How do you think the ability to form proviral DNA is related to the asymptomatic stage of HIV infection?
5. Compare the retrovirus life cycle with that of a lysogenic bacteriophage such as lambda. What advantage might a virus have in incorporating its genome into that of the host cell?

18.3 Cytocidal Infections and Cell Damage

An infection that results in cell death is a **cytocidal infection.** Vertebrate viruses can harm their host cells in many ways; often this leads to cell death. Microscopic or macroscopic degenerative changes or abnormalities in host cells and tissues are referred to as **cytopathic effects (CPEs).** Seven possible mechanisms of host cell damage are briefly described here. However, it should be emphasized that more than one of these mechanisms may be involved in a cytopathic effect.

1. Many viruses can inhibit host DNA, RNA, and protein synthesis. Cytocidal viruses (e.g., picornaviruses, herpesviruses, and adenoviruses) are particularly active in this regard. The mechanisms of inhibition are not yet clear.
2. Cell endosomes may be damaged, resulting in the release of hydrolytic enzymes and cell destruction.
3. Virus infection can drastically alter plasma membranes through the insertion of virus-specific proteins so that the infected cells are attacked by the immune system. When infected by viruses such as herpesviruses and measles virus, as many as 50 to 100 cells may fuse into one abnormal, giant,

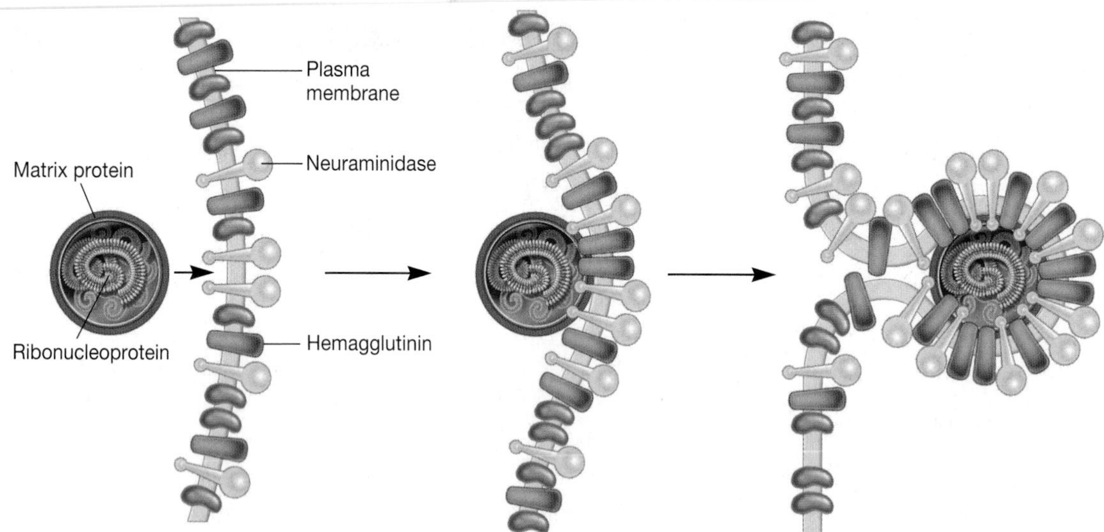

Figure 18.11 Release of Influenza Virus by Plasma Membrane Budding. First, viral envelope proteins (hemagglutinin and neuraminidase) are inserted into the host plasma membrane. Then the nucleocapsid approaches the inner surface of the membrane and binds to it. At the same time viral proteins collect at the site and host membrane proteins are excluded. Finally, the plasma membrane buds to simultaneously form the viral envelope and release the mature virion.

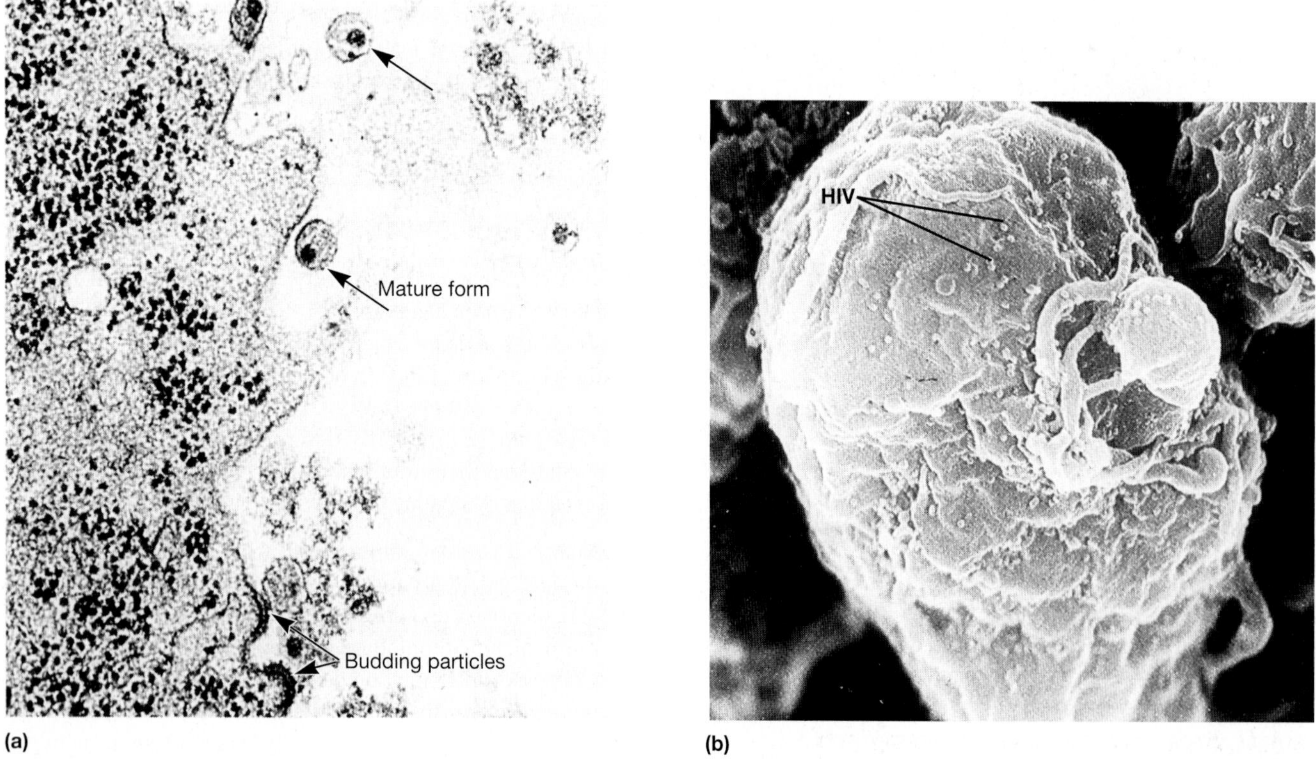

Figure 18.12 Human Immunodeficiency Virus (HIV) Release by Plasma Membrane Budding. **(a)** A transmission electron micrograph of HIV particles beginning to bud, as well as some mature particles. **(b)** A scanning electron micrograph view of HIV particles budding from a lymphocyte.

multinucleated cell called a **syncytium.** HIV appears to destroy $CD4^+$ T-helper cells at least partly through its effects on the plasma membrane.

4. High concentrations of proteins from several viruses (e.g., mumps virus and influenza virus) can have a direct toxic effect on cells and organisms.
5. Intracellular structures called **inclusion bodies** are formed during many virus infections. These may result from the clustering of subunits or virions within the nucleus or cytoplasm (e.g., the Negri bodies in rabies infections); they also may contain cell components such as ribosomes (arenavirus infections) or chromatin (herpesviruses). Regardless of their composition, these inclusion bodies can directly disrupt cell structure.
6. Chromosomal disruptions result from infections by herpesviruses and others.
7. Finally, the host cell may not be directly destroyed but transformed into a malignant cell. This is discussed in section 18.5.

18.4 Persistent, Latent, and Slow Virus Infections

Many virus infections (e.g., influenza) are **acute infections**—that is, they have a fairly rapid onset and last for a relatively short time (**figure 18.13**). However, some viruses can establish **persistent infections** lasting many years. There are several kinds of persistent infections. In **chronic virus infections,** the virus is almost always detectable and clinical symptoms may be either mild or absent for long periods. Examples are hepatitis B virus and HIV. In **latent virus infections** the virus stops reproducing and remains dormant for a period before becoming active again. During latency, no symptoms or viruses are detectable, although antibodies to the virus may be present at low levels. Examples are herpes simplex virus, varicella-zoster virus, cytomegalovirus, and Epstein-Barr virus, which causes mononucleosis. Herpes simplex type 1 virus often infects children and then becomes dormant within the nervous system ganglia; years later it can be activated to cause cold sores. The varicella-zoster virus causes chicken pox in children and then, after years of inactivity, may produce the skin disease shingles (initial adult infections result in chicken pox). These and other examples of persistent infections are discussed in more detail in chapter 37.

The causes of persistence and latency are probably multiple, although the precise mechanisms are still unclear. The virus genome may be integrated into the host genome thereby becoming a provirus. Viruses may become less antigenic and thus less susceptible to attack by the immune system. Often virus infections are latent in a site not subject to immune attack, such as the central nervous system. Viruses may also mutate to less virulent and slower reproducing forms. Sometimes a deletion mutation produces a ***d**efective **i**nterfering* **(DI) particle** that cannot reproduce but slows normal virus reproduction, thereby reducing host damage and establishing a chronic infection. Mutations and their chemical basis (section 13.1)

A small group of viruses causes extremely slowly developing infections, often called **slow virus diseases** or slow infections, in which symptoms may take years to emerge. Measles virus occasionally produces a slow infection. A child may have a normal case of measles, then 5 to 12 years later develop a degenerative brain disease called subacute sclerosing panencephalitis (SSPE). Lentiviruses such as HIV also cause slow diseases.

1. What is a cytocidal infection? What is a cytopathic effect? Outline the ways in which viruses can damage host cells during cytocidal infections.
2. Define the following: acute infection, persistent infection, chronic infection, latent virus infection, and slow virus disease.
3. Why might an infection be chronic or latent?

18.5 Viruses and Cancer

Cancer [Latin *cancer,* crab] is one of the most serious medical problems in developed nations. It is the focus of an immense amount of research. A **tumor** [Latin *tumere,* to swell] is a growth or lump of tissue resulting from **neoplasia,** abnormal new cell growth and reproduction due to loss of regulation. Tumor cells have aberrant shapes and altered plasma membranes that may contain distinctive tumor antigens. Their unregulated proliferation and loss of differentiation result in invasive growth that forms unorganized cell masses. This reversion to a more primitive or less differentiated state is called **anaplasia.**

There are two major types of tumors with respect to overall form or growth pattern. If the tumor cells remain in place to form a compact mass, the tumor is benign. In contrast, cells from malignant or cancerous tumors can actively spread throughout the body in a process known as **metastasis,** often by floating in the blood and establishing secondary tumors. Some cancers are not solid, but cell suspensions. For example, leukemias are composed of undifferentiated malignant white blood cells that circulate throughout the body. Indeed, dozens of kinds of cancers arise from a variety of cell types and afflict all kinds of organisms.

As one might expect from the wide diversity of cancers, there are many causes of cancer, only a few of which are directly related to viruses. Possibly as many as 30 to 60% of cancers may be related to diet and cigarette smoke. Many chemicals in our surroundings are carcinogenic and may cause cancer by inducing gene mutations or interfering with normal cell differentiation. However, it is important to note that many cancers are not linked to environmental risk factors.

Carcinogenesis is a complex, multistep process. It can be initiated by a chemical, usually a mutagen, but a cancer does not appear to develop until at least one more triggering event (possibly exposure to another chemical carcinogen or a virus) takes place. Cancer-causing genes, or **oncogenes,** are directly involved. Some oncogenes are contributed to a cell by viruses, as is discussed later. Others arise from genes within the cell called **proto-oncogenes.** Proto-oncogenes are cellular genes required for normal growth. If they are mutated or over-expressed, they may become oncogenes. That is, their products contribute to the malignant

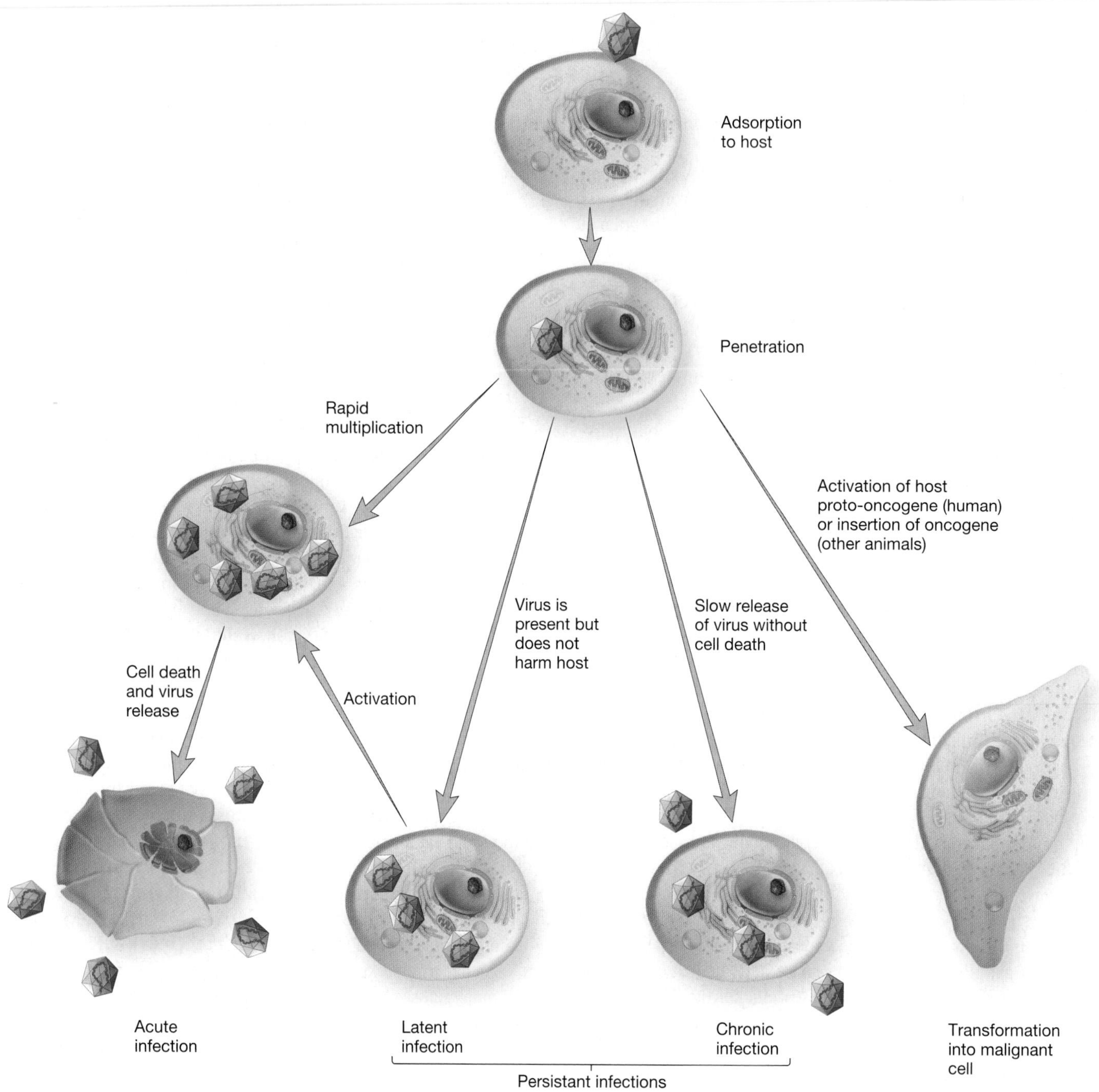

Figure 18.13 Types of Infections and Their Effects on Host Cells.

transformation of the cell. Many oncogenes are involved in the regulation of cell growth and signal transduction; for example, some code for growth factors that regulate cell reproduction. It may be that various cancers arise through different combinations of causes. Not surprisingly the chances of developing cancer rise with age because an older person will have had a longer time to accumulate the mutations needed for oncogenic transformation. Immune surveillance and destruction of cancer cells also may be less effective in older people.

Although viruses are known to cause many animal cancers, it is very difficult to prove that this is the case with human cancers because indirect methods of study must be used and Koch's postulates can't be applied completely. One tries to find virus particles and components within tumor cells, using techniques such as electron microscopy, immunologic tests, DNA-based assays, and enzyme assays. Attempts are also made to isolate suspected cancer viruses by cultivation in tissue culture or other animals. Sometimes a good correlation between the presence of a virus and cancer can be detected.

At present, viruses have been implicated in the genesis of at least eight human cancers. With the exception of a few retroviruses, these viruses have dsDNA genomes.

1. The Epstein-Barr virus (EBV) is one of the best-studied human cancer viruses. EBV is a herpesvirus and the cause of two cancers. Burkitt's lymphoma is a malignant tumor of the jaw and abdomen found in children of central and western Africa. EBV also causes nasopharyngeal carcinoma. Both EBV particles and genomes have been found within tumor cells; Burkitt's lymphoma patients also have high blood levels of antibodies to EBV. Interestingly there is some evidence that a person also must have had malaria to develop Burkitt's lymphoma. Environmental factors must play a role, because EBV does not cause much cancer in the United States despite its prevalence. This may be due to a low incidence of malaria in the United States.
2. Hepatitis B virus appears to be associated with one form of liver cancer (hepatocellular carcinoma) and can be integrated into the human genome.
3. Hepatitis C virus causes cirrhosis of the liver, which can lead to liver cancer.
4. Human herpesvirus 8 and HIV are associated with the development of Kaposi's sarcoma.
5. Some strains of human papillomaviruses have been linked to cervical cancer.
6. At least two retroviruses, the human T-cell lymphotropic virus I (HTLV-1) and HTLV-2, are associated with adult T-cell leukemia and hairy-cell leukemia, respectively. Other retrovirus-associated cancers may well be discovered in the future.

Viruses known to cause cancer are called **oncoviruses.** All known human dsDNA oncoviruses trigger cancerous transformation of cells by a similar mechanism. They encode proteins that bind to and thereby inactivate cellular proteins known as **tumor suppressors.** Tumor-suppressor proteins regulate cell cycling or monitor and/or repair DNA damage. Two tumor suppressors known to be targets of human oncovirus proteins are called Rb and p53. Rb has multiple functions in the nucleus, all of which are critical to normal cell cycling. When Rb molecules are rendered inactive by an oncoviral protein, cells undergo uncontrolled reproduction. Thus they are said to be hyperproliferative. The protein p53 is often referred to as "the guardian of the genome." This is because p53 normally initiates either cell cycle arrest or programmed cell death in response to DNA damage. However, when p53 is inactivated by an oncoviral protein, it cannot do so and genetic damage persists. From the point of view of the virus, hyperproliferation and the lack of programmed cell death are beneficial. However, for the cell, it can be catastrophic. Cells can rapidly accumulate the additional mutations needed for oncogenic transformation. The nucleus and cell division (section 4.8)

Retroviruses exert their oncogenic powers in a different manner. Some carry oncogenes captured from host cells many, many generations ago. Thus they transform the host cell by bringing the oncogene into the cell. For example, Rous sarcoma virus carries a mutated, oncogenic *src* gene that codes for an overactive tyrosine kinase. This enzyme is localized to the plasma membrane and phosphorylates the amino acid tyrosine in several cellular proteins. These Src-targeted proteins are essential in maintaining the cell's ability to respond to normal anti-growth signals from other cells and the extracellular matrix. When phosphorylated, they become active and override these signals, in effect signaling unregulated growth. The human retroviruses HTLV-1 and HTLV-2 transform a group of immune system cells called T cells by producing a regulatory protein that sometimes activates genes involved in cell division as well as stimulating virus reproduction. The second transformation mechanism used by retroviruses involves the integration of a viral genome into the host chromosome such that strong, viral regulatory elements are near a cellular proto-oncogene. This results in such a high level of expression of the cellular protein that the gene is now an oncogene. For example, some chicken retroviruses induce lymphomas when they are integrated next to the *c-myc* cellular proto-oncogene, which codes for a protein that is involved in the induction of either DNA or RNA synthesis.

1. What are the major characteristics of cancer?
2. How might viruses cause cancer? Are there other ways in which a malignancy might develop?
3. Define the following terms: tumor, neoplasia, anaplasia, metastasis, and oncogene.

18.6 Plant Viruses

Although it has long been recognized that viruses can infect plants and cause a variety of diseases, plant viruses generally have not been as well studied as bacteriophages and animal viruses. This is partly because they are often difficult to cultivate and purify. Some viruses, such as tobacco mosaic virus (TMV), can be grown in isolated protoplasts of plant cells just as phages and some animal viruses are cultivated in cell suspensions. However, many cannot grow in protoplast cultures and must be inoculated into whole plants or tissue preparations. This makes studying the life cycle difficult. Many plant viruses require insect vectors for transmission; some of these can be grown in monolayers of cell cultures derived from aphids, leafhoppers, or other insects.

The essentials of capsid morphology are outlined in chapter 16 and apply to plant viruses because they do not differ significantly in construction from their animal virus and phage relatives **(figure 18.14)**. Many have either rigid or flexible helical capsids

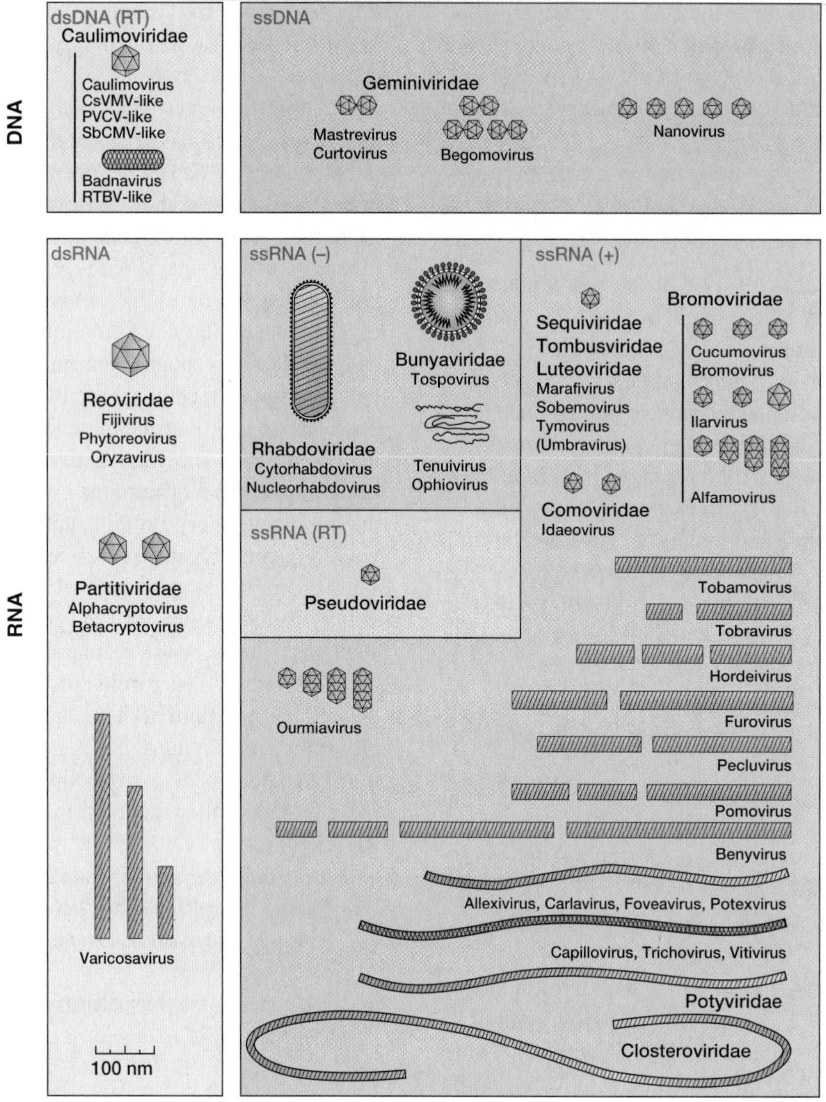

Figure 18.14 A Diagrammatic Description of Families and Genera of Viruses That Infect Plants. RT stands for reverse transcriptase.

(tobacco mosaic virus). Others are icosahedral or have modified the icosahedral pattern with the addition of extra capsomers (turnip yellow mosaic virus, **figure 18.15**). Most capsids are composed of one type of protein; no specialized attachment proteins have yet been detected. Almost all plant viruses are RNA viruses, either single stranded or double stranded. Caulimoviruses and geminiviruses with their DNA genomes are exceptions to this rule.

Like all viruses, plant viruses must penetrate a host cell before they can reproduce. However, penetration of plant cells is hampered by the fact that they are protected by complex outer layers, including cell walls. Entry of a plant virus into its host requires the presence of mechanical damage to the cell wall, and this is usually caused by insects or other animals that feed on plants. Particularly important are sucking insects such as aphids and leafhoppers. These insects not only create an entryway for the virus, they are often responsible for transmitting plant viruses from plant to plant. As the insect feeds on the plant, it can pick up a virus particle on its mouthparts. In some cases, the virus is stored in the insect's foregut as the insect moves to another plant. However, some plant viruses can reproduce within insect tissues—these viruses use both plants and insects as hosts. Whether passively transmitted or actively reproducing in the insect host, when the insect feeds on another plant, virus particles can be transferred to the new plant.

As noted earlier, most plant viruses are RNA viruses and, of these, plus-strand RNA viruses are most common. TMV is the best studied plus-strand RNA virus and is the focus of this discussion. Recall that the plus-strand genomes of RNA viruses can serve as mRNA. However, following entry into its host, the TMV RNA genome is not immediately translated. Rather the RNA is processed by a mechanism that is not understood. The resulting mRNAs encode several proteins, including the coat protein and an RNA-dependent RNA polymerase. Thus TMV can replicate its own genome. This is not the case with all positive-strand RNA plant viruses. Most plants contain an enzyme with RNA-dependent RNA polymerase activity. Thus it is possible that some plus-strand RNA plant viruses make use of the host enzyme.

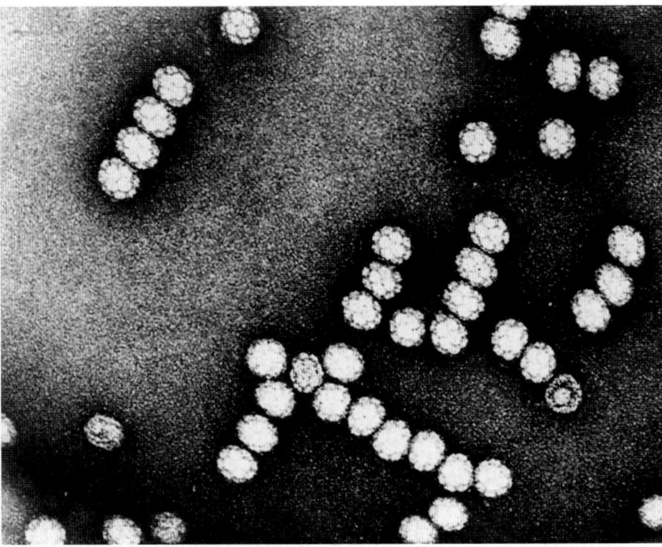

Figure 18.15 Turnip Yellow Mosaic Virus (TYMV). An RNA plant virus with icosahedral symmetry.

After the coat protein and RNA genome of TMV have been synthesized, they spontaneously assemble into complete TMV virions in a highly organized process **(figure 18.16)**. The protomers come together to form disks composed of two layers of protomers arranged in a helical spiral. Association of coat protein with TMV RNA begins at a specific assembly initiation site close to the 3′ end of the genome. The helical capsid grows by the addition of protomers, probably as disks, to the end of the rod. As the rod lengthens, the RNA passes through a channel in its center and forms a loop at the growing end. In this way the RNA can easily fit as a spiral into the interior of the helical capsid.

Reproduction within the host depends on the virus's ability to spread throughout the plant. Viruses can move long distances through the plant vasculature; usually they travel in the phloem. The spread of plant viruses in nonvascular tissue is hindered by the presence of tough cell walls. Nevertheless, a virus such as TMV spreads slowly, about 1 mm/day or less, moving from cell to cell through the plasmodesmata. These are slender cytoplasmic strands extending through holes in adjacent cell walls that join plant cells by narrow bridges. Viral "movement proteins" are required for transfer from cell to cell. The TMV movement protein accumulates in the plasmodesmata, but the way in which it promotes virus movement is not well understood.

Several cytological changes can take place in TMV-infected cells. Plant virus infections often produce microscopically visible intracellular inclusions, usually composed of virion aggregates.

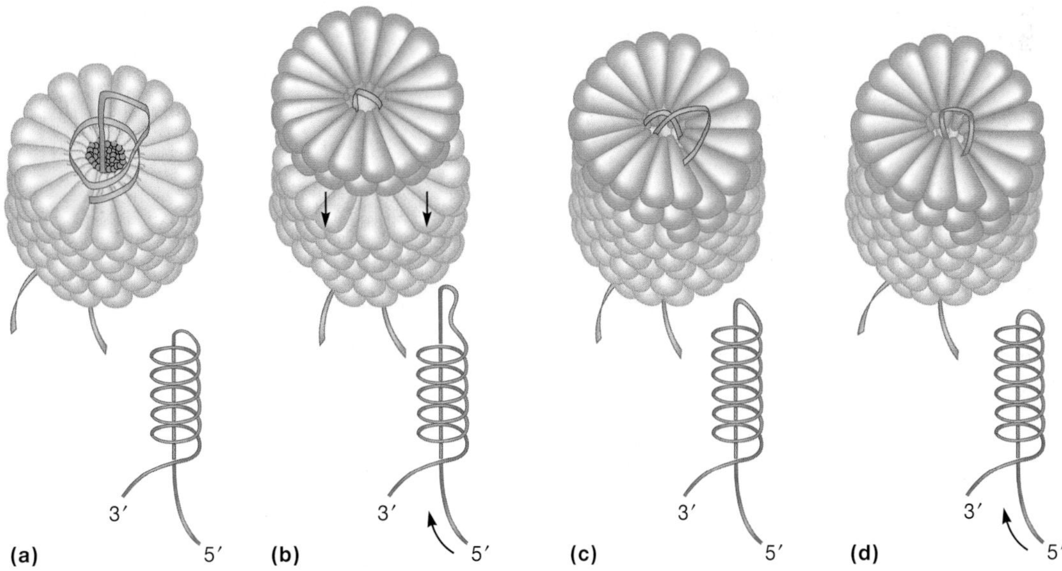

Figure 18.16 TMV Assembly. The elongation phase of tobacco mosaic virus nucleocapsid construction. The lengthening of the helical capsid through the addition of a protein disk to its end is shown in a sequence of four illustrations; line drawings depicting RNA behavior are included. The RNA genome inserts itself through the hole of an approaching disk and then binds to the groove in the disk as it locks into place at the end of the cylinder.

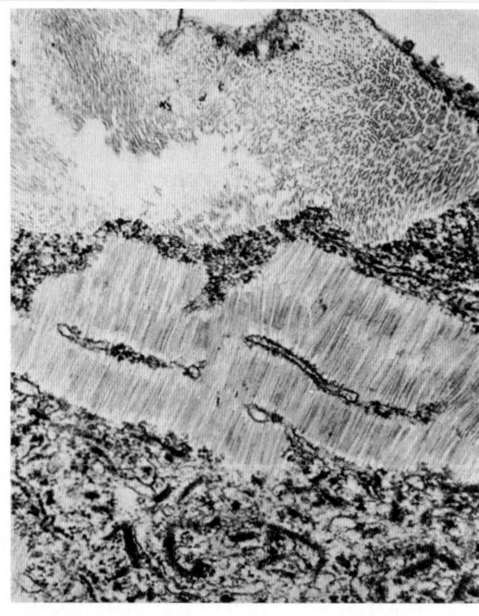

(a)

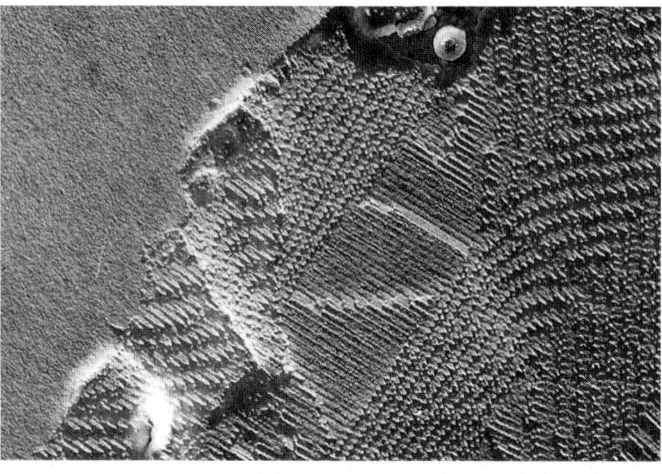

(b)

Figure 18.17 Intracellular TMV. **(a)** A crystalline mass of tobacco mosaic virions from a 10-day-old lesion in a *Chenopodium amaranticolor* leaf. **(b)** Freeze-fracture view of a crystalline mass of tobacco mosaic virions in an infected leaf cell. In both views the particles can be seen longitudinally and in cross section.

Hexagonal crystals of almost pure TMV virions sometimes develop in TMV-infected cells **(figure 18.17)**. The host cell chloroplasts become abnormal and often degenerate, while new chloroplast synthesis is inhibited.

1. Why have plant viruses not been as well studied as animal and bacterial viruses?
2. Describe in molecular terms the way in which TMV is reproduced.
3. How are plant viruses transmitted between hosts?
4. Compare plant and vertebrate viruses in terms of their entry into host cells. Why do they differ so greatly in this regard?

18.7 Viruses of Fungi and Protists

Most mycoviruses, viruses that infect fungi, have been isolated from higher fungi such as *Penicillium* and *Aspergillus*. They contain dsRNA and have isometric capsids (that is, their capsids are roughly spherical polyhedra), which are approximately 25 to 50 nm in diameter. Many appear to be latent viruses. Some mycoviruses induce disease symptoms in hosts such as the mushroom *Agaricus bisporus*, but cytopathic effects and toxic virus products have not yet been observed. The *Fungi* (chapter 26)

Much less is known about the viruses of lower fungi, and only a few have been examined in any detail. Both dsRNA and ssRNA genomes have been found; capsids usually are isometric or hexagonal and vary in size from 40 to over 200 nm. Unlike the viruses of higher fungi, virus reproduction in lower fungi is accompanied by host cell destruction and lysis.

Viruses that infect photosynthetic protists belong to the family *Phycodnaviridae*. Four genera are recognized by the ICTV. All have linear dsDNA genomes. Those algal viruses that have been studied have polyhedral capsids. One virus of *Uronema gigas* resembles many bacteriophages in having a tail. Another interesting group of viruses infects *Chlorella* strains that are endosymbionts of the ciliated protist *Paramecium busaria*. These viruses have very large genomes and encode proteins not usually encoded by viral genomes. They, along with the Mimivirus described below, have fostered considerable discussion about what features distinguish cellular organisms from acellular entities.

The viruses of only three genera of protozoa have been studied. It is known that *Giardia intestinalis* and *Leishmania* spp. are infected by members of the *Totiviridae*, a family of naked icosahedral viruses with dsRNA genomes. Finally, a giant dsDNA virus, named Mimivirus, has been discovered in the amoeba *Acanthamoeba polyphaga*. The virus is 400 nm in diameter and has a genome about 800 kilobase pairs in size; thus its genome is larger than that of some bacteria. Mimivirus is distantly related to the *Poxviridae* and *Phycodnaviridae*. The protists (chapter 25)

18.8 Insect Viruses

Members of many virus families are known to infect insects. Some use insects as agents for spreading to populations of susceptible animals and plants. These include *Flaviviridae* and *Togaviridae*, whose members cause yellow fever, West Nile disease, and several types of viral encephalitis. Other virus families use insects as primary hosts. Of these, probably the most important are the *Baculoviridae, Reoviridae, Iridoviridae,* and *Polydnaviridae*.

The *Iridoviridae* are icosahedral viruses with lipid in their capsids and a linear dsDNA genome. They are responsible for the iridescent virus diseases of the crane fly and some beetles. The group's name comes from the observation that larvae of infected insects can have an iridescent coloration due to the presence of crystallized virions in their fat bodies.

Many insect virus infections are accompanied by the formation of inclusion bodies within the infected cells. Granulosis viruses form granular protein inclusions, usually in the cytoplasm. Nuclear polyhedrosis and cytoplasmic polyhedrosis virus infections produce polyhedral inclusion bodies in the nucleus or the cytoplasm of affected cells. Although all three types of viruses generate inclusion bodies, they belong to two distinctly different families. The cytoplasmic polyhedrosis viruses are reoviruses; they are icosahedral with double shells and have dsRNA genomes. Nuclear polyhedrosis viruses and granulosis viruses are baculoviruses—rod-shaped, enveloped viruses of helical symmetry and with dsDNA.

The inclusion bodies, both polyhedral and granular, are protein in nature and enclose one or more virions **(figure 18.18)**. Insect larvae are infected when they feed on leaves contaminated with inclusion bodies. Polyhedral bodies protect the virions against heat, low pH, and many chemicals; the viruses can remain viable in the soil for years. However, when exposed to the alkaline contents of insect guts, the inclusion bodies dissolve to liberate the virions, which then infect midgut cells. Some viruses remain in the midgut while others spread throughout the insect. Just as with bacterial and vertebrate viruses, insect viruses can persist in a latent state within the host for generations while producing no disease symptoms. A reappearance of the disease may be induced by chemicals, thermal shock, or even a change in the insect's diet.

Much of the current interest in insect viruses arises from their promise as biological control agents for insect pests. Many people hope that some of these viruses may partially replace the use of toxic chemical pesticides. Baculoviruses have received the most attention for at least three reasons. First, they attack only invertebrates and have considerable host specificity; this means that they should be safe for nontarget organisms. Second, because they are encased in protective inclusion bodies, these viruses have a good shelf life and better viability when dispersed in the environment. Finally, they are well suited for commercial production because they often reach extremely high concentrations in larval tissue (as high as 10^{10} viruses per larva). The use of nuclear polyhedrosis viruses for the control of the cotton bollworm, Douglas fir tussock moth, gypsy moth, alfalfa looper, and European pine sawfly has either been approved by the U.S. Environmental Protection Agency or is being considered. The granulosis virus of the codling moth also is useful. Usually inclusion bodies are sprayed on foliage consumed by the target insects. More sensitive viruses are administered by releasing infected insects to spread the disease. As in the case with other pesticides, it is possible that resistance to these agents may develop in the future.

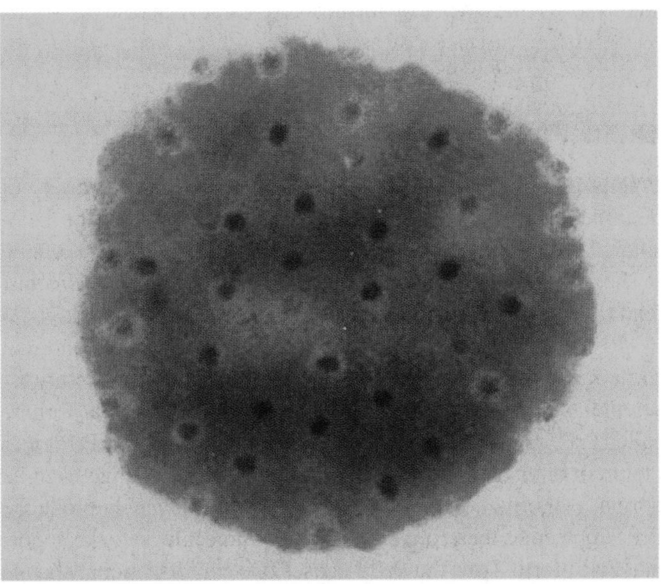

Figure 18.18 Inclusion Bodies. A section of a cytoplasmic polyhedron from a gypsy moth *(Lymantria dispar)*. The occluded virus particles with dense cores are clearly visible (×50,000).

1. Describe the major characteristics of the viruses that infect higher fungi, lower fungi, and protists. In what ways do they seem to differ from one another?
2. Summarize the nature of granulosis, nuclear polyhedrosis, and cytoplasmic polyhedrosis viruses and the way in which they are transmitted by inclusion bodies.
3. What are baculoviruses and why are they so promising as biological control agents for insect pests?

18.9 Viroids and Virusoids

Although some viruses are exceedingly small and simple, even simpler infectious agents exist. Viroids are infectious agents that consist only of RNA. Virusoids, formerly called satellite RNAs, are similar to viroids in that they also consist only of RNA. Finally, prions are infectious agents that consist only of protein. Prions are discussed in section 18.10.

Viroids cause over 20 different plant diseases, including potato spindle-tuber disease, exocortis disease of citrus trees, and chrysanthemum stunt disease. Viroids are covalently closed, circular, ssRNAs, about 250 to 370 nucleotides long **(figures 18.19** and **18.20)**. The circular RNA normally exists as a rodlike shape due to intrastrand base pairing, which forms double-stranded regions with single-stranded loops (figure 18.20). Some viroids are found in the nucleolus of infected host cells, where between 200 and 10,000 copies may be present. Others are located within chloroplasts. Interestingly, the RNA of viroids does not encode any gene products, so they cannot replicate themselves. Rather, it is thought that the viroid is replicated by one of the host cell's DNA-dependent RNA polymerases. The host polymerase evidently uses the viroid RNA as a template for RNA synthesis, rather than host DNA. The host polymerase synthesizes a complementary RNA molecule, a negative-strand RNA. This then serves as the template for the same host polymerase, and new viroid RNAs are synthesized. Both steps may occur by a rolling-circle-like mechanism.

A plant may be infected with a viroid without showing symptoms—that is, it may have a latent infection. However, the same viroid in another host species may cause severe disease.

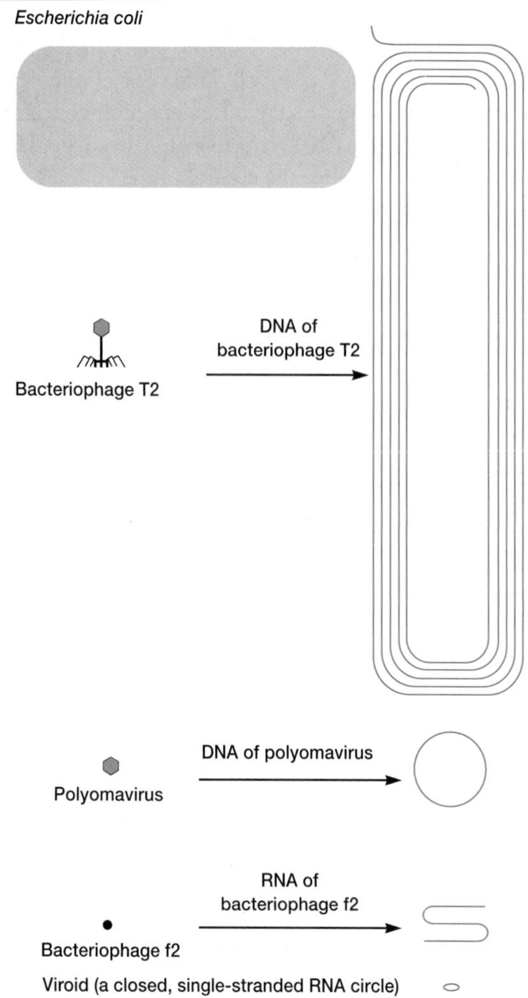

Figure 18.19 Viroids, Viruses, and Bacteria. A comparison of *Escherichia coli*, several viruses, and the potato spindle-tuber viroid with respect to size and the amount of nucleic acid possessed. (All dimensions are enlarged approximately ×40,000.)

The pathogenicity of viroids is not well understood, but it is known that particular regions of the RNA are required; studies have shown that removing these regions blocks the development of disease (figure 18.20). Some data suggest that viroids cause disease by triggering a eucaryotic response called **RNA silencing,** which normally functions to protect against infection by dsRNA viruses. During RNA silencing, the cell detects the presence of the dsRNA and selectively degrades it. Viroids may usurp this response by hybridizing to specific host mRNA molecules to which they have a complementary nucleotide sequence. Formation of the hybrid viroid:host mRNA double-stranded molecule is thought to elicit RNA silencing. This results in destruction of the host message and therefore silencing of the host gene. Failure to express a required host gene leads to disease in the host plant.

The potato spindle-tuber viroid (PSTV) is the most intensely studied viroid. Its RNA consists of about 359 nucleotides, much smaller than any virus genome. Several PSTV strains have been

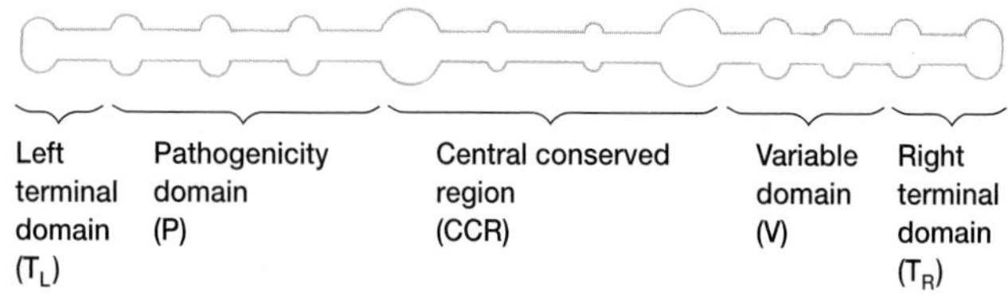

Figure 18.20 Viroid Structure. This schematic diagram shows the general organization of a viroid. The closed single-stranded RNA circle has extensive intrastrand base pairing and interspersed unpaired loops. Viroids have five domains. Most changes in viroid pathogenicity seem to arise from variations in the P and T_L domains.

isolated, ranging in virulence from those that cause only mild symptoms to lethal varieties. All variations in pathogenicity are due to a few nucleotide changes in two short regions on the viroid. It is believed that these sequence changes alter the shape of the rod and thus its ability to cause disease.

Virusoids are similar to viroids in that they are also covalently closed, circular, ssRNA molecules with regions capable of intrastrand base pairing. In contrast to viroids, they encode one or more gene products and they typically need a helper virus in order to infect host cells. The helper virus supplies gene products and other materials needed by the virusoid for completion of its replication cycle. The best-studied virusoid is the human hepatitis D virusoid, which is 1,700 nucleotides long. It uses the hepatitis B virus as its helper virus. If a host cell contains both the hepatitis B virus and the hepatitis D virusoid, the virusoid RNA and its gene product, called delta antigen, can be packaged within the envelope of the virus. These enveloped virusoids and delta antigens are capable of entering other host cells, where the virusoid RNA is transcribed by the host's RNA polymerase II. Direct contact diseases: Viral hepatitides (section 37.3)

18.10 Prions

Prions (for *pro*teinaceous *in*fectious particle) cause a variety of neurodegenerative diseases in humans and animals. The best-studied prion is the scrapie prion, which causes the disease scrapie in sheep. Afflicted animals lose coordination of their movements, tend to scrape or rub their skin, and eventually cannot walk.

Researchers have shown that scrapie is caused by an abnormal form of a cellular protein. The abnormal form is called PrP^{Sc} (for *sc*rapie-associated *pr*ion *p*rotein), and the normal cellular form is called PrP^{C}. Evidence supports a model in which entry of PrP^{Sc} into the brain of an animal causes the PrP^{C} protein to change from its normal conformation to the abnormal form. The newly produced PrP^{Sc} molecules then convert more PrP^{C} molecules into the abnormal PrP^{Sc} form. How the PrP^{Sc} causes this conformational change is unclear. However, the best-supported model is that the PrP^{Sc} directly interacts with PrP^{C}, causing the change. It is noteworthy that mice lacking the *PrP* gene cannot be infected with PrP^{Sc}. Although the evidence is strong that PrP^{Sc} causes PrP^{C} to fold abnormally, how this triggers neuron loss is poorly understood. Recent evidence suggests that the interaction of PrP^{Sc} with PrP^{C} serves to cross-link PrP^{C} molecules. The cross-linked PrP^{C} molecules trigger a series of events called apoptosis or programmed cell death. Thus the normal, but cross-linked, protein causes neuron loss, whereas the abnormal protein acts as the infectious agent.

In addition to scrapie, prions are responsible for bovine spongiform encephalopathy (BSE or "mad cow disease"), and the human diseases kuru, fatal familial insomnia, Creutzfeldt-Jakob disease (CJD), and Gerstmann-Strässler-Scheinker syndrome (GSS). All result in progressive degeneration of the brain and eventual death. At present, there is no effective treatment. Mad cow disease reached epidemic proportions in Great Britain in the 1990s and initially spread because cattle were fed meal made from all parts of cattle including brain tissue. It has now been shown that eating meat from cattle with BSE can cause a variant of Creutzfeldt-Jakob disease in humans (vCJD). More than 90 people have died in the United Kingdom and France from this source. Variant CJD differs from CJD in origin only: people acquire vCJD by eating contaminated meat, while CJD is an extremely rare condition caused by spontaneous mutation of the gene that encodes the prion protein. CJD and GSS are rare and cosmopolitan in distribution among middle-aged people, while kuru has been found only in the Fore, an eastern New Guinea tribe. This tribe had a custom of consuming dead kinsmen. Women and children were given the less desirable body parts to eat; this included the brain. Thus they and their children were infected. Cannibalism was stopped many years ago, and kuru has been eliminated.

1. What are viroids and why are they of great interest?
2. How does a viroid differ from a virus?
3. What is a prion? In what way does a prion appear to differ fundamentally from viruses and viroids?
4. Prions are difficult to detect in host tissues. Why do you think this is so? Why do you think we have not been able to develop effective treatments for these diseases?

Summary

18.1 Taxonomy of Eucaryotic Viruses

a. Eucaryotic viruses are classified according to many properties; the most important are their nucleic acids and replicative strategies (**figures 18.1–18.3**).

18.2 Reproduction of Vertebrate Viruses

a. The first step in the vertebrate virus reproductive cycle is adsorption of the virus to a target cell receptor site; often special capsid or envelope structures are involved in this process.

b. Entry of the virus into the host cell may be accompanied by capsid removal from the nucleic acid, a process called uncoating. Most often penetration occurs through either endocytotic engulfment to form coated vesicles or fusion of the envelope with the plasma membrane (**figure 18.4**).

c. In DNA viruses, early viral mRNA and proteins are involved in taking over the host cell and the synthesis of viral DNA and RNA. DNA replication often takes place in the host nucleus and mRNA is initially manufactured by host enzymes.

d. The parvoviruses are so small that they must conserve genome space by using overlapping genes and other similar mechanisms. Poxviruses differ from other DNA vertebrate viruses in that DNA replication takes place in host cytoplasm and they carry an RNA polymerase. Hepadnaviruses use reverse transcriptase to replicate their gapped dsDNA genome.

e. The genome of positive ssRNA viruses can act as an mRNA, whereas negative ssRNA virus genomes direct the synthesis of mRNA by a virus-associated transcriptase. Double-stranded RNA reoviruses use both virus-associated and newly synthesized transcriptases to make mRNA (**figure 18.8**).

f. RNA virus genomes are replicated in the host cell cytoplasm. Most ssRNA viruses use a viral replicase to synthesize a dsRNA replicative form that then directs the formation of new genomes.

g. Retroviruses use reverse transcriptase to synthesize a DNA copy of their RNA genome. After the double-stranded proviral DNA has been synthesized, it is integrated into the host genome and directs the formation of virus RNA and protein.

h. Late genes code for proteins needed in (1) capsid construction by a self-assembly process and (2) virus release.

i. Usually, naked virions are released upon cell lysis. In enveloped virus reproduction, virus release and envelope formation normally occur simultaneously after modification of the host plasma membrane, and the cell is not lysed (**figure 18.11**).

18.3 Cytocidal Infections and Cell Damage

a. Viruses can destroy host cells in many ways during cytocidal infections. These include such mechanisms as inhibition of host DNA, RNA, and protein synthesis; endosomal damage; alteration of host cell membranes; and the formation of inclusion bodies.

18.4 Persistent, Latent, and Slow Virus Infections

a. Although many virus infections are acute, having a rapid onset and short duration, some viruses can establish persistent infections lasting for years. Some infections are chronic. Viruses also can become dormant for a while and then resume activity in what is called a latent infection. Slow viruses may act so slowly that a disease develops over years (**figure 18.13**).

18.5 Viruses and Cancer

a. Cancer is characterized by the formation of a malignant tumor that metastasizes or invades other tissues and can spread through the body. Carcinogenesis is a complex, multistep process involving many factors.

b. Viruses cause cancer in several ways. For example, they may bring a cancer-causing gene, or oncogene, into a cell, or the virion may insert a transcription regulatory element next to a cellular proto-oncogene and stimulate the gene to greater activity. Alternatively, viral proteins may inactivate tumor-suppressor proteins, thereby promoting hyperproliferation and mutation.

18.6 Plant Viruses

a. Entry of plant viruses into their hosts is usually mediated by mechanical damage to the plant. This creates openings in the plant cell walls through which the virus can enter. Plant-feeding animals, especially insects, are often the cause of this damage.

b. Most plant viruses have an RNA genome and may be either helical or icosahedral. Depending on the virus the RNA genome may be replicated by either a host RNA-dependent RNA polymerase or a virus-specific RNA replicase.

c. The TMV nucleocapsid forms spontaneously by self-assembly when disks of coat protein protomers complex with the RNA.

18.7 Viruses of Fungi and Protists

a. Mycoviruses from higher fungi have isometric capsids and dsRNA, whereas the viruses of lower fungi may have either dsRNA or dsDNA genomes.

b. Only a few viruses of protists have been isolated and studied. Some are of special note because they have extremely large dsDNA genomes that include genes not usually found in viral genomes.

18.8 Insect Viruses

a. Members of several virus families infect insects, and many of these viruses produce inclusion bodies that aid in their transmission.

b. Baculoviruses and other viruses are finding uses as biological control agents for insect pests.

18.9 Viroids and Virusoids

a. Infectious agents simpler than viruses exist. For example, several plant diseases are caused by short strands of infectious RNA called viroids **(figure 18.20)**.

b. Virusoids are infectious RNAs that encode one or more gene products. They require a helper virus for replication.

18.10 Prions

a. Prions are small proteinaceous agents associated with at least six degenerative nervous system disorders: scrapie, bovine spongiform encephalopathy, kuru, fatal familial insomnia, the Gerstmann-Strässler-Scheinker syndrome, and Creutzfeldt-Jakob disease. The precise nature of prions is not yet clear.

b. Most evidence supports the hypothesis that prion proteins exist in two forms: the infectious, abnormally folded form and a normal cellular form. The interaction between the abnormal form and the cellular form converts the cellular form into the abnormal form.

Key Terms

acute infection 461
anaplasia 461
cancer 461
chronic virus infection 461
cytocidal infection 459
cytopathic effect (CPEs) 459
defective interfering (DI) particle 461
early genes 454
inclusion bodies 461
late genes 458
latent virus infection 461
metastasis 461
neoplasia 461
oncogene 461
oncovirus 463
persistent infection 461
prion 468
procapsids 458
proto-oncogene 461
proviral DNA 457
replicase 455
replicative form (RF) 455
retrovirus 457
reverse transcriptase (RT) 457
ribonuclease H 457
ribosomal frameshifting 458
RNA silencing 468
slow virus diseases 461
subgenomic mRNA 458
syncytium 461
transcriptase 455
tropism 448
tumor 461
tumor suppressor 463
viroid 467
virusoid 468

Critical Thinking Questions

1. Consider each of the following viral reproduction steps: adsorption, penetration, replication, and transcription. Suggest a strategy by which one could pharmacologically inhibit or discourage entry and propagation of viruses in animal cells. Can you explain host range using some of the same rationale?
2. Would it be advantageous for a virus to damage host cells? If it is not, why isn't damage to the host avoided? Is it possible that a virus might become less pathogenic when it has been associated with the host population for a longer time?
3. How does one prove that a virus is causing cancer? Try to think of approaches other than those discussed in the chapter. Give a major reason why it is so difficult to prove that a specific virus causes human cancer. Is it accurate to say that viruses cause cancer? Explain why or why not.
4. From what you know about cancer, is it likely that a single type of treatment can be used to cure it? What approaches might be effective in preventing cancer?
5. Propose some experiments that might be useful in determining what prions are and how they cause disease.

Learn More

Chien, P.; Weissman, J. S.; and DePace, A. H. 2004. Emerging principles of conformation-based prion inheritance. *Annu. Rev. Biochem.* 73:617–56.

Flint, S. J.; Enquist, L. W.; Racaniella, V. R.; and Skalka, A. M. 2004. *Principles of virology,* 2d ed. Washington, D.C.: ASM Press.

Gibbs, W. W. 2003. Untangling the roots of cancer. *Sci. Am.* 289(1):56–65.

Hull, R. 2002. *Matthew's Plant Virology,* 4th ed. San Diego: Academic Press.

Raoult, D.; Audic. S.; Robert, C; Abergel, C.; Renesto, P.; Ogata, H.; LaScola, B.; Suzan, M.; and Claverie, J.-M. 2004. The 1.2 megabase genome sequence of mimivirus. *Science* 306:1344–50.

Smith, A. E., and Helenius, A. 2004. How viruses enter animal cells. *Science* 304:237–41.

Wang, M. B.; Bian, X. Y.; Wu, L. M.; Liu, L. X.; Smith, N. A.; Isenegger, D.; Wu, R. M.; Masuta, C.; Vance, V. B.; Watson, J. M.; Rezaian, A.; Dennis, E. S.; and Waterhouse, P. M. 2004. On the role of RNA silencing in the pathogenicity and evolution of virioids and virus satellites. *Proc. Natl. Acad. Sci.* 101(9):3275–80.

Please visit the Prescott website at www.mhhe.com/prescott7 for additional references.

19 Microbial Evolution, Taxonomy, and Diversity

The stromatolites shown here are layered rocks formed by incorporation of minerals into microbial mats. Fossilized stromatolites indicate that microorganisms existed early in Earth's history.

Preview

- The evolutionary relationships among the three domains of life—*Bacteria, Archaea,* and *Eucarya*—are represented in the universal phylogenetic tree. The root, or origin, is placed early in the bacterial line of descent. This suggests that the *Archaea* and *Eucarya* share a common ancestry that is independent of the *Bacteria*. This long evolutionary history has generated a spectacular degree of microbial diversity.
- The RNA world hypothesis suggests that the first self-replicating entity was RNA and that this molecule formed the basis of the first primitive cell. Although it is unclear how the first eucaryotic nucleus arose, there is abundant evidence demonstrating that mitochondria and chloroplasts arose from endosymbiotic proteobacteria and cyanobacteria, respectively. Hydrogenosomes, found in some anaerobic protists, appear to share the same common ancestor as mitochondria.
- In order to make sense of the diversity of organisms, it is necessary to group similar organisms together and organize these groups in a nonoverlapping hierarchical arrangement. Taxonomy is the science of biological classification.
- A polyphasic approach is used to classify procaryotes. This combines information that is based on the analysis of microbial phenotypic, genotypic, and phylogenetic features. The results of these analyses are often summarized in treelike diagrams called dendrograms.
- Morphological, physiological, metabolic, ecological, genetic, and molecular characteristics are all useful in taxonomy because they reflect the organization and activity of the genome. Nucleic acid sequences are probably the best indicators of microbial phylogeny and relatedness because nucleic acids are either the genetic material itself or the products of gene transcription. Small subunit rRNAs and the genes that encode them display a number of features that make them useful in determining microbial phylogenies.
- Bacterial taxonomy is rapidly changing due to the acquisition of new data, particularly the use of molecular techniques such as the comparison of ribosomal RNA structure and chromosome sequences. This is leading to new phylogenetic classifications.
- The current edition of *Bergey's Manual of Systematic Bacteriology* is phylogenetically organized and distributes the *Bacteria* and *Archaea* among 25 phyla.

One of the most fascinating and attractive aspects of the microbial world is its extraordinary diversity. It seems that almost every possible shape, size, physiology, and lifestyle are represented. In this section of the text we focus on microbial diversity. Chapter 19 introduces the general principles of microbial evolution and taxonomy. This is followed by a five-chapter (20–24) survey of the most important procaryotic groups. Our survey of microbial diversity ends with an introduction to the major types of eucaryotic microorganisms: protists and fungi.

19.1 Microbial Evolution

Biological diversity is usually thought of in terms of plants and animals; yet, the assortment of microbial life forms is huge and largely uncharted. Consider the metabolic diversity of microorganisms—this alone suggests that the number of habitats occupied by microbes vastly exceeds that of all larger organisms. How has microbial life been able to radiate to such a bewildering level of diversity? To answer this question, one must consider microbial evolution. The field of microbial evolution, like any other scientific endeavor, is based on the formulation of hypotheses, the gathering of data, the analysis of the data, and the reformation of hypotheses based on newly acquired evidence. That is to say, the study of microbial evolution is based on the scientific method. To be sure, it is sometimes more difficult to amass evidence when considering events that occurred millions, and often billions, of

What's in a name? That which we call a rose by any other name would smell as sweet. . . .

—*W. Shakespeare*

years ago, but the advent of molecular biology has offered scientists a living record of life's ancient history. This chapter describes the outcome of this scientific research.

The Origin of Life

Dating meteorites through the use of radioisotopes places our planet at an estimated 4.5 to 4.6 billion years old. However, conditions on Earth for the first hundred million years or so were far too harsh to sustain any type of life. The first direct evidence of cellular life was discovered in 1977 in a geologic formation in South Africa known as the Swartkoppie chert, a granular type of silica. These microbial fossils as well as those from the Archaean Apex chert of Australia have been dated at about 3.5 billion years old (**figure 19.1**). Despite these findings, the microbial fossil record is understandably sparse. Thus to piece together the very early events that led to the origin of life, biologists must rely primarily on indirect evidence. Each piece of evidence must fit together like a jigsaw puzzle for a coherent picture to emerge.

The First Self-Replicating Entity: The RNA World

The origin of life rests on a single question: How did early cells arise? No one can say for certain; however, it seems likely that the first self-replicating entity was much simpler than even the most primitive modern, living cells. Before there was life, Earth was a cauldron of chemicals that reacted with one another, randomly "testing" the stability of the resulting molecules. This means that the first cells evolved when Earth was a very different place: hot and anoxic, with an atmosphere rich in gases like hydrogen, methane, carbon dioxide, nitrogen, and ammonia. To account for the evolution of life, one must consider the three essential cellular molecules: DNA, RNA, and proteins—one of these molecules presumably developed first and holds the key to understanding all that followed. Proteins are capable of performing cellular work but cannot replicate, while just the opposite is true of DNA. For life to evolve, a molecule was needed that could both replicate and perform cellular work. A possible solution to this problem was suggested in 1981 when Thomas Cech discovered self-splicing RNA in the eucaryotic microbe *Tetrahymena*. Three years later, Sidney Altman found that RNaseP in *Escherichia coli* is an RNA molecule that cleaves phosphodiester bonds. RNA molecules that possess catalytic activity are called **ribozymes** and to some, the ability of RNA to catalyze biochemical reactions suggests a precellular **RNA world,** a term coined by Walter Gilbert in 1986. This hypothesis suggests that the first self-replicating molecule was RNA, which is capable of storing, copying, and expressing genetic information, and possesses enzymatic activity as well. In this version of early life, various forms of molecules were assembled and destroyed over roughly half a billion years, until ultimately an entity something like modern RNA enclosed in a lipid vesicle was generated. Microbial Tidbits 11.2: Catalytic RNA (Ribozymes)

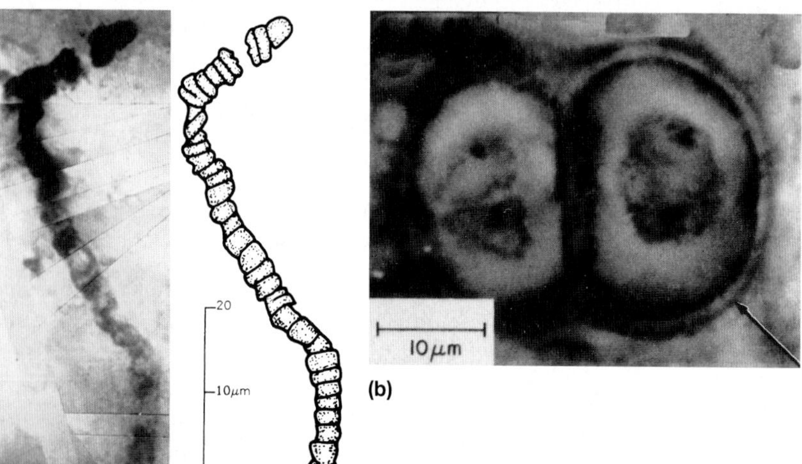

Figure 19.1 Fossilized Bacteria. Several microfossils resembling bacteria are shown, some with interpretive drawings. **(a)** Thin sections of Archean Apex chert from Western Australia; the fossilized remains of procaryotes are about 3.5 billion years old. **(b)** *Gloeodiniopsis,* about 1.5 billion years old, from carbonaceous chert in the Satka Formation of the southern Ural Mountains. The arrow points to the enclosing sheath. **(c)** *Palaeolyngbya,* about 950 million years old, from carbonaceous shale of the Lakhanda Formation of the Khabarovsk region in eastern Siberia.

Apart from its ability to replicate and perform enzymatic activities, the function of RNA suggests its ancient origin. Consider that much of the cellular pool of RNA in modern cells exists in the ribosome, a structure that consists largely of rRNA and it uses mRNA and tRNA to construct proteins. In fact, rRNA itself catalyzes peptide bond formation during protein synthesis. Thus RNA seems to be well poised for its importance in the development of proteins. Because RNA and DNA are structurally similar, RNA could have given rise to double-stranded DNA. It is posited that once DNA evolved it became the storage facility for cellular functions because it provided a more chemically stable structure. Two other pieces of evidence support the RNA world hypothesis: the fact that the energy currency of the cell, ATP, is a ribonucleotide, and the more recent discovery that RNA can regulate gene expression. So it would seem that proteins, genes, and cellular energy all can be traced back to RNA. Riboswitches (sections 12.3 and 12.4)

Others are skeptical of the RNA world hypothesis. They claim conditions on Earth 4 billion years ago would have prevented the stable formation of ribose, phosphate, purines, and pyrimidines—all needed to construct RNA. In fact, while purine bases have been generated abiotically in a heated mixture of hydrogen cyanide and ammonia, scientists have so far been unable to make pyrimidines in a similar fashion. Another problem with the RNA world hypothesis is the instability of RNA once it is assembled. In 1996, James Ferris and colleagues were able to overcome the problem of RNA degradation by adding the clay mineral montmorillonite to a solution of chemically charged nucleotides. They showed that the rate of RNA synthesis was faster than its degradation. Later they and others showed in similar experiments that amino acids in solution with the minerals hydroxyapatite and illite could also polymerize into polypeptides of about 50 amino acid residues. To some, these experiments provide experimental evidence for the biosynthesis of early organic polymers. However, one additional problem with the RNA world hypothesis concerns the ability of early RNA to self-replicate. Recall that in modern cells, RNA is synthesized by the enzyme RNA polymerase, a protein. Replication of early RNA without a protein was presumably accomplished by an ancient ribozyme. So far, no such ribozyme has been found, nor has it been generated experimentally. Thus although it is clear that microbial life ultimately emerged from a random mixture of chemicals, the actual mechanism by which the first cell-like entity arose is a controversy that may never be resolved. Transcription (section 11.6)

Early cellular life, although primitive compared to modern life, was still relatively complex. Cells had to derive energy from a harsh, anoxic environment. When scientists attempt to reconstruct the nature of very ancient life, they look to extant (living) microbes for clues. For instance, it is thought that the FeS-based metabolism seen in some hyperthermophilic archaea may be a remnant of the first form of chemiosmosis. Here it is suggested that the energy-yielding reaction $FeS + H_2S \rightarrow FeS_2 + H_2$ provided the reducing power (H_2) to produce a proton motive force. Photosynthesis also appears to have evolved early in Earth's history. There is fossil evidence to place the evolution of cyanobacteria and oxygenic photosynthesis at about 3 billion years ago. **Stromatolites** are layered rocks, often domed, that are formed by the incorporation of mineral sediments into microbial mats dominated by cyanobacteria (**figure 19.2**). Recent evidence has shown that some fossilized stromatolites formed in a similar fashion. The presence of oxygen was critical because it enabled the evolution of a wider variety of energy-capturing strategies, including aerobic respiration. Electron transport and oxidative phosphorylation (section 9.5); Anaerobic respiration (section 9.6)

Figure 19.2 Stromatolites. These are stromatolites at Shark Bay, Western Australia. Modern stromatolites are layered or stratified rocks formed by the incorporation of calcium sulfates, calcium carbonates, and other minerals into microbial mats. The mats are formed by cyanobacteria and other microorganisms.

Ironically, the study of the most ancient organisms is one of the youngest disciplines in the biological sciences. The ability to culture and examine microorganisms was developed only about 150 years ago. Almost immediately, early microbiologists attempted to classify microbes and organize them according to possible relationships to one another. Two important elements not understood until the late twentieth century made this especially difficult. First, only about 1% of all microbes have been cultured in the laboratory. Second, the most accurate assessment of evolutionary relationships between organisms is obtained by comparing nucleotide and amino acid sequences. Prior to the advent of sequence-based techniques, it was impossible to discern evolutionary relationships among microorganisms.

The Three Domains of Life

The most important sequence-based investigation was initiated by Carl Woese. In 1977, Woese and his collaborator George Fox used the nucleotide sequences of the **small subunit ribosomal RNAs (SSU rRNAs)** from a variety of organisms to determine that all living organisms belong to one of three domains: *Archaea, Bacteria,* and *Eucarya.* This initial observation has been further refined and substantiated by additional biochemical and genetic evidence. Recall from chapter 3 that most bacteria have cell wall peptidoglycan containing muramic acid and have membrane lipids with ester-linked, straight-chained fatty acids that resemble eucaryotic membrane lipids (**table 19.1**). The *Archaea* differ from the *Bacteria* in many respects and resemble the *Eucarya* in some ways. Although the *Archaea* are described more fully in chapter 20, it should be

Table 19.1 Comparison of *Bacteria, Archaea,* and *Eucarya*

Property	*Bacteria*	*Archaea*	*Eucarya*
Membrane-Enclosed Nucleus with Nucleolus	Absent	Absent	Present
Complex Internal Membranous Organelles	Absent	Absent	Present
Cell Wall	Almost always have peptidoglycan containing muramic acid	Variety of types, no muramic acid	No muramic acid
Membrane Lipid	Have ester-linked, straight-chained fatty acids	Have ether-linked, branched aliphatic chains	Have ester-linked, straight-chained fatty acids
Gas Vesicles	Present	Present	Absent
Transfer RNA	Thymine present in most tRNAs *N*-formylmethionine carried by initiator tRNA	No thymine in T or TψC arm of tRNA Methionine carried by initiator tRNA	Thymine present Methionine carried by initiator tRNA
Polycistronic mRNA	Present	Present	Absent
mRNA Introns	Absent	Absent	Present
mRNA Splicing, Capping, and Poly A Tailing	Absent	Absent	Present
Ribosomes			
Size	70S	70S	80S (cytoplasmic ribosomes)
Elongation factor 2 reaction with diphtheria toxin	Does not react	Reacts	Reacts
Sensitivity to chloramphenicol and kanamycin	Sensitive	Insensitive	Insensitive
Sensitivity to anisomycin	Insensitive	Sensitive	Sensitive
DNA-Dependent RNA Polymerase			
Number of enzymes	One	One	Three
Structure	Simple subunit pattern (6 subunits)	Complex subunit pattern similar to eucaryotic enzymes (8–12 subunits)	Complex subunit pattern (12–14 subunits)
Rifampicin sensitivity	Sensitive	Insensitive	Insensitive
Polymerase II Type Promoters	Absent	Present	Present
Metabolism			
Similar ATPase	No	Yes	Yes
Methanogenesis	Absent	Present	Absent
Nitrogen fixation	Present	Present	Absent
Chlorophyll-based photosynthesis	Present	Absent	Present[a]
Chemolithotrophy	Present	Present	Absent

[a]Present in chloroplasts (of bacterial origin).

noted that they differ from the *Bacteria* in lacking muramic acid in their cell walls and in possessing (1) membrane lipids with ether-linked branched aliphatic chains, (2) transfer RNAs without thymidine in the T or TψC arm, (3) distinctive RNA polymerase enzymes, and (4) ribosomes of different composition and shape. Thus although the *Archaea* and the *Bacteria* have a similar cell architecture, they vary considerably at the molecular level. Both groups differ from eucaryotes in their cell ultrastructure and many other properties. However, table 19.1 shows that both the *Bacteria* and the *Archaea* share some biochemical properties with eucaryotic cells. For example, *Bacteria* and eucaryotes have ester-linked membrane lipids; *Archaea* and eucaryotes are similar with respect to some components of the RNA and protein synthetic systems. The *Archaea* (chapter 20)

There are several views regarding the evolutionary history of microbes. We will first consider the **universal phylogenetic tree** as proposed by Norman Pace (**figure 19.3**). This analysis is based on SSU rRNA sequence analysis of organisms from all three domains of life. Importantly, a similar tree can be constructed from the nucleotide sequences of any gene whose product is involved in DNA replication, transcription, and translation, as long as that gene is found in all three domains. Although the details of phylogenetic tree construction and the use of SSU rRNAs to measure relatedness are discussed in more detail later (p. 489), the general concept is not difficult to understand. In this case, 16S and 18S rRNA sequences from a diverse collection of procaryotes and eucaryotes, respectively, are aligned from the 5' end to the 3' end and homologous residues are compared in a pair-wise fashion. Each nucleotide sequence difference is counted and serves to represent some evolutionary distance between the organisms. When data from a large number of organisms are compared, the evolutionary relatedness between organisms can be determined, but not the rate at which one organism diverged from another. Simply stated, rather than measuring time, the branches on such a tree measure the evolutionary distance between organisms. The concept is analogous to a map that accurately shows the distance between two cities but because of many factors, one cannot determine the time needed to travel that distance. Thus evolutionary distance is measured along each line: the longer the line, the more evolutionarily diverged are the two organisms (or types of organisms) at each end.

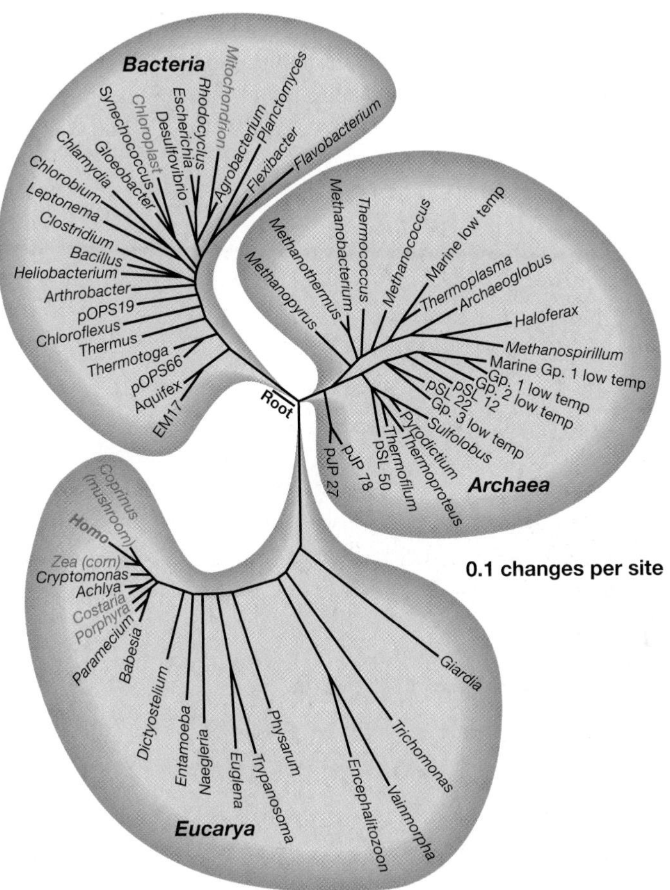

Figure 19.3 Universal Phylogenetic Tree. These evolutionary relationships are based on rRNA sequence comparisons. Length of branches indicates evolutionary relationships between organisms but not time. Microbes are printed in black.

What does the universal phylogenetic tree tell us about the origin of life? Close to the center is a line labeled "Root." This is where the data indicate the last common ancestor to all three domains should be placed (there are no branches here because there is no such extant organism). The root, or origin of modern life, is on the bacterial branch; it appears that the *Archaea* and the *Eucarya* evolved independently, separate from the *Bacteria*. Following the lines of descent away from the root, toward the *Archaea* and the *Eucarya,* it is evident that they shared common ancestry but diverged and became separate domains. The common evolution of these two forms of life is still evident in the manner in which the *Archaea* and the *Eucarya* process genetic information. For instance, the RNA polymerases of the *Eucarya* and the *Archaea* resemble each other, to the exclusion of the *Bacteria*. It follows that *Bacteria* use the sigma subunit of RNA polymerase to initiate gene transcription, while *Archaea* and eucaryotes use so-called TATA binding sites, as discussed in chapters 11 and 20. Thus the universal phylogenetic tree presents a picture whereby all life, regardless of eventual domain, arose from a single, common ancestor. One can envision the universal tree of life as a real tree that grows from a single seed.

To be sure, there are alternative hypotheses. One notion is that the *Bacteria* and *Eucarya* domains arose relatively independently, and the *Archaea* are a mosaic, having combined traits of the other two. This view is based largely on the observation that while the *Archaea* share large numbers of genes with *Eucarya,* they have an even larger number of genes in common with *Bacteria*. Another interpretation suggests that the first eucaryotes arose from two ancient procaryotic ancestors: an archaeon and a bacterium. This notion is further explored here.

As we have seen, genes from both the *Archaea* and the *Bacteria* can be found in eucaryotic chromosomes, but how did they get there? While the universal phylogenetic tree indicates that common genes reflect a single common ancestor, the **genome fusion hypothesis** attempts to explain the evolution of the nucleus. This hypothesis asserts that certain archaeal and bacterial genes were combined to form a single eucaryotic genome. It suggests that ancient archaeal cells were invaded by primitive gram-negative α-proteobacteria. The archaea are thought to have retained the bacteria because the latter performed some metabolic feat that conferred a survival advantage to their host. Eventually genes needed for independent living were lost from the bacterium while some essential genes were transferred to the host's proto-nucleus. The *Proteobacteria* (chapter 22)

The Endosymbiotic Origin of Mitochondria and Chloroplasts

In contrast to the unresolved origin of the nucleus, the **endosymbiotic hypothesis** is generally accepted as the origin of mitochondria and chloroplasts. That endosymbiosis was responsible for the development of these organelles (regardless of the exact mechanism) is supported by the fact that both organelles have bacterial-like ribosomes and most have a single, circular chromosome. Indeed, inspection of figure 19.3 shows that mitochondria and chloroplasts belong to the bacterial lineage. Important evidence for the origin of mitochondria comes from the genome sequence of the α-proteobacterium *Rickettsia prowazekii,* an obligate intracellular parasite. Its genome is more closely related to that of modern mitochondrial genomes than to any other bacterium. Mitochondria are believed to have descended from such an α-proteobacterium that became engulfed in a precursor cell and provided a function that was essential to the host cell. It may be that oxygen toxicity was eliminated because the intracellular bacterium used aerobic respiration to generate ATP. In return, the host provided nutrients and a safe place to live. These bacterial endosymbionts evolved to become mitochondria. This hypothesis also accounts for the evolution of chloroplasts from an endosymbiotic cyanobacterium. Presently the cyanobacterium *Prochloron* has become a favorite candidate as the extant relative of the endosymbiotic cyanobacterium that gave rise to green algae and plant chloroplasts. This microbe lives within marine invertebrates and is the only procaryote to have both chlorophyll *a* and *b,* but not phycobilins. This makes *Prochloron* most similar to chloroplasts. Photosynthetic bacteria: Phylum *Cyanobacteria* (section 21.3)

Recently, an additional endosymbiosis theory has been advanced. The **hydrogen hypothesis** asserts that the endosymbiont was an anaerobic α-proteobacterium that produced H_2 and CO_2 as end products of fermentation. In the absence of an external H_2 source, the host became dependent on the bacterium, which made ATP by substrate-level phosphorylation. Ultimately, the endosymbiont evolved into one of two organelles. If the endosymbiont developed the capacity to perform aerobic respiration, it evolved into a mitochondrion. However, in those cases where the endosymbiont did not acquire the ability to respire, it evolved into a **hydrogenosome**—an organelle found in some extant protists that produce ATP by fermentation (**figure 19.4**). While some be-

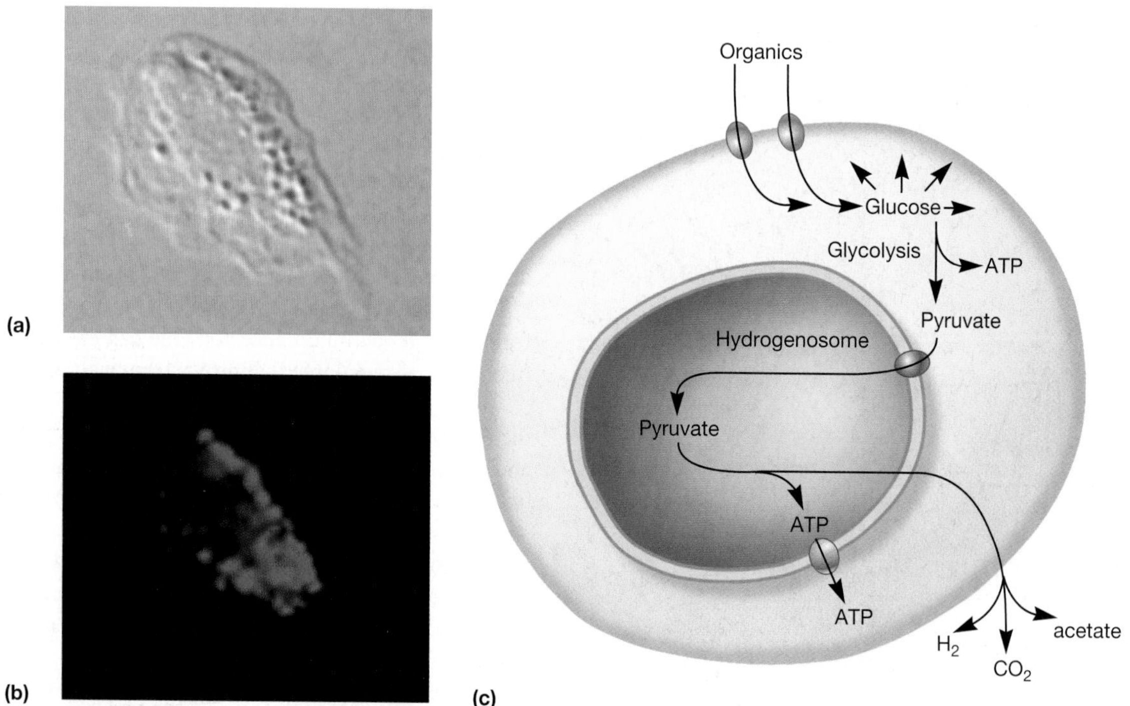

Figure 19.4 Hydrogenosomes of *Trichomonas vaginalis.* **(a)** Differential interference contrast image of the protist *Trichomonas vaginalis.* **(b)** Hydrogenosomes of the *T. vaginalis* cell shown in (a) labeled with red fluorescent antibody that recognizes the hydrogenosome-specific malic enzyme. **(c)** Schematic diagram of a hydrogenosome shows the function of substrate-level phosphorylation.

lieve that hydrogenosomes might be derived from mitochondria, three lines of evidence currently support the alternative idea that hydrogenosomes and mitochondria arose from the same ancestral organelle: (1) The heat-shock proteins of α-proteobacteria, mitochondria, and hydrogenosomes are closely related; (2) subunits of the mitochondrial enzyme NADH dehydrogenase are active in the hydrogenosomes of the protist *Trichomonas vaginalis;* and (3) the primitive genome found in the hydrogenosomes of the protist *Nyctotherus ovalis* encodes components of a mitochondrial electron transport chain. Taken together, these data suggest that mitochondria and hydrogenosomes are aerobic and anaerobic versions of the same ancestral organelle. The protists (chapter 25)

Finally, the endosymbiotic theory put forth by Lynn Margulis and her colleagues combines elements of the endosymbiotic origin of mitochondria with the genome fusion hypothesis. The **serial endosymbiotic theory (SET)** calls for the development of eucaryotes in a series of discrete endosymbiotic steps. This theory suggests that motility evolved first through endosymbiosis between anaerobic spirochetes and another anaerobe. Next, nuclei are thought to have formed by the development of internal membranes. These early nucleated forms would have been similar to modern protists with hydrogenosomes. The endosymbiotic events needed for the evolution of mitochondria are thought to have occurred later, giving rise to early fungi and animal cells, with subsequent endosymbiotic events leading to the development of chloroplasts and plants.

Evolutionary Processes

Clearly, figure 19.3 demonstrates the astounding level of microbial diversity reflecting hundreds of millions of years of evolution. The application of Darwin's theory of natural selection to microbial evolution requires special consideration. **Anagenesis,** also known as **microevolution,** refers to small, random genetic changes that occur over generations to slowly drive either speciation or extinction, both of which are forms of **macroevolution.** Neither microevolution nor macroevolution occur at a constant rate. Instead, the fossil record shows that the slow and steady pace of evolution is periodically interrupted by rapid bursts of speciation driven by abrupt changes in the environment (**figure 19.5**). This phenomenon is called **punctuated equilibria** and was introduced by Niles Eldredge and the late Steven Jay Gould. The theory of punctuated equilibria is one important reason why evolutionary distance, as measured by the similarity of genes in living organisms, provides little or no information regarding *when* evolutionary divergence occurred.

Procaryotic evolution results in the generation of microbial diversity upon which selective processes determine the development of new species. Recall that genetic diversity in the *Archaea* and *Bacteria* must occur asexually. Thus heritable genetic changes in these organisms are introduced principally by two mechanisms: mutation and lateral (horizontal) gene transfer (LGT). Genome sequencing has revealed that LGT, particularly in the form of transposition and phage-mediated gene transfer (transduction), appears to be more important than once thought. In addition, model studies designed to assess competition between microbial populations has led to some surprising observations. It had been thought that very small genetic differences between microbial populations of the same species were of little evolutionary significance. However, laboratory experiments demonstrate that when selection is applied, very small genetic differences can result in one population overtaking another. These recent analyses help illuminate the potential mechanisms by which the vast level of microbial diversity came about and will help guide future studies. Mechanisms of genetic diversity (chapter 13); Comparative genomics (section 15.6)

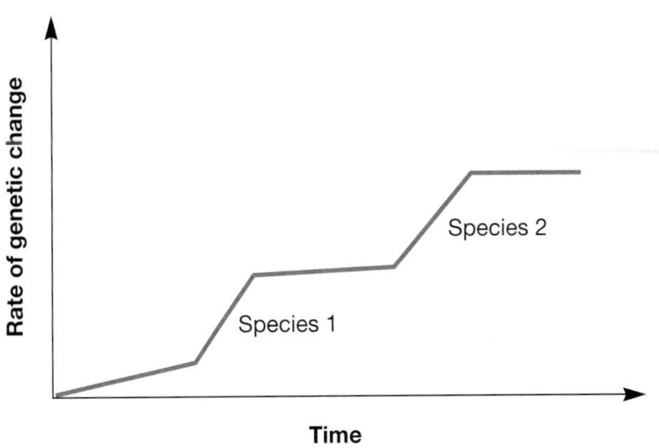

Figure 19.5 Punctuated Equilibria. This theory describes the rate of evolution as a result of periodic, abrupt changes in the environment. These changes interrupt the slow and steady pace of evolution, resulting in periods of relatively rapid speciation. In the diagram shown here, species 1 and species 2 arose following such dramatic environmental changes.

1. Why is RNA thought to be the first self-replicating biomolecule?
2. How is the evidence that the *Archaea* and the *Bacteria* share common genes interpreted in the universal tree of life theory? In the genome fusion hypothesis? Which theory do you favor? Why?
3. Explain the endosymbiotic hypothesis of the origin of mitochondria and chloroplasts. List two pieces of evidence that support this hypothesis.
4. What is a hydrogenosome? Why is it thought that mitochondria and hydrogenosomes arose from a single, common progenitor organelle? What is an alternative hypothesis?
5. What is the difference between macroevolution and microevolution? Define punctuated equilibria. Why does this theory help to explain why the universal phylogenetic tree cannot illustrate the rate at which the evolution of organisms occurred?

19.2 Introduction to Microbial Classification and Taxonomy

Microbiologists are faced with the daunting task of understanding the diversity of life forms that cannot be seen with the naked eye but can live seemingly anywhere on Earth. One of the first tools needed to survey this level of diversity is a reliable classification system. **Taxonomy** (Greek *taxis,* arrangement or order, and *nomos,* law, or *nemein,* to distribute or govern) is defined as the science of biological classification. In a broader sense it consists of three separate but interrelated parts: classification, nomenclature, and identification. Once a classification scheme is selected, it is used to arrange organisms into groups called **taxa** (s., **taxon**) based on mutual similarity. **Nomenclature** is the branch of taxonomy concerned with the assignment of names to taxonomic groups in agreement with published rules. **Identification** is the practical side of taxonomy, the process of determining if a particular isolate belongs to a recognized taxon. The term **systematics** is often used for taxonomy. However, many taxonomists define systematics in more general terms as "the scientific study of organisms with the ultimate object of characterizing and arranging them in an orderly manner." Any study of the nature of organisms, when the knowledge gained is used in taxonomy, is a part of systematics. Thus systematics encompasses disciplines such as morphology, ecology, epidemiology, biochemistry, molecular biology, and physiology.

One of the oldest classification systems, called a **natural classification,** arranges organisms into groups whose members share many characteristics and reflects as much as possible the biological nature of organisms. The Swedish botanist Carl von Linné, or Carolus Linnaeus as he often is called, developed the first natural classification, based largely on anatomical characteristics, in the middle of the eighteenth century. It was a great improvement over previously employed artificial systems because knowledge of an organism's position in the scheme provided information about many of its properties. For example, classification of humans as mammals denotes that they have hair, self-regulating body temperature, and milk-producing mammary glands in the female.

When natural classification is applied to higher organisms, evolutionary relationships become apparent simply because the morphology of a given structure (e.g., wings) in a variety of organisms (ducks, songbirds, hawks) suggests how that structure might have been modified to adapt to specific environments or behaviors. However, the taxonomic assignment of microbes is not necessarily rooted in evolutionary relatedness. For instance, bacterial pathogens and microbes of industrial importance were historically given names that described the diseases they cause or the processes they perform (i.e., *Vibrio cholerae, Clostridium tetani,* and *Lactococcus lactis*). Although these labels are of practical use, they do little to guide the taxonomist concerned with the vast majority of microbes that are neither pathogenic nor of industrial consequence. Our recent understanding of the evolutionary relationships among microbes now serves as the theoretical underpinning for taxonomic classification.

In practice, determination of the genus and species of a newly discovered procaryote is based on **polyphasic taxonomy.** This approach includes phenotypic, phylogenetic, and genotypic features. To understand how all of these data are incorporated into a coherent profile of taxonomic criteria, we must first consider the individual components and determine how they are assessed quantitatively through numerical taxonomy.

Phenetic Classification

For a very long time, microbial taxonomists relied exclusively on a **phenetic system,** which groups organisms together based on the mutual similarity of their phenotypic characteristics. This classification system succeeded in bringing order to biological diversity and clarified the function of morphological structures. For example, because motility and flagella are always associated in particular microorganisms, it is reasonable to suppose that flagella are involved in at least some types of motility. Although phenetic studies can reveal possible evolutionary relationships, they are not dependent on phylogenetic analysis. They compare many traits without assuming that any features are more phylogenetically important than others—that is, unweighted traits are employed in estimating general similarity. Obviously the best phenetic classification is one constructed by comparing as many attributes as possible. Organisms sharing many characteristics make up a single group or taxon.

Phylogenetic Classification

With the publication in 1859 of Darwin's *On the Origin of Species,* biologists began developing **phylogenetic** or **phyletic classification systems** that sought to compare organisms on the basis of evolutionary relationships. The term **phylogeny** (Greek *phylon,* tribe or race, and *genesis,* generation or origin) refers to the evolutionary development of a species. Scientists realized that when they observed differences and similarities between organisms as a result of evolutionary processes, they also gained insight into the history of life on Earth. However, for much of the twentieth century, microbiologists could not effectively employ phylogenetic classification systems, primarily because of the lack of a good fossil record. When Woese and Fox proposed using rRNA nucleotide sequences to assess evolutionary relationships among microorganisms, the door opened to the resolution of long-standing inquiries regarding the origin and evolution of the majority of life forms on Earth—the microbes. The validity of this approach is now widely accepted and there are currently over 200,000 different 16S and 18S rRNA sequences in the international databases GenBank and the Ribosomal Database Project (RDP-II). As discussed later (p. 485), the power of rRNA as a phylogenetic and taxonomic tool rests on the features of the rRNA molecule that make it a good indicator of evolutionary history and the ever-increasing size of the rRNA sequence database.

Genotypic Classification

There are currently many ways in which the genotype of a microbe can be evaluated in taxonomic terms. Some of these techniques are discussed later in this chapter (section 19.4). In general, **genotypic classification** seeks to compare the genetic similarity between organisms. Individual genes or whole genomes can be compared. Since the 1970s, it has been widely accepted that procaryotes whose genomes are at least 70% homologous belong to the same species. Unfortunately, this 70% threshold value was established to avoid disrupting existing species assignments; it is not based on theoretical considerations of species identity. Fortunately, the genetic data obtained using newer molecular approaches usually concur with these older assignments.

Numerical Taxonomy

The development of computers has made possible the quantitative approach known as **numerical taxonomy.** Peter H. A. Sneath and Robert Sokal have defined numerical taxonomy as "the grouping by numerical methods of taxonomic units into taxa on the basis of their character states." Information about the properties of organisms is converted into a form suitable for numerical analysis and then compared by means of a computer. The resulting classification is based on general similarity as judged by comparison of many characteristics, each given equal weight. This approach was not feasible before the advent of computers because of the large number of calculations involved.

The process begins with a determination of the presence or absence of selected characters in the group of organisms under study. A character usually is defined as an attribute about which a single statement can be made. Many characters, at least 50 and preferably several hundred, should be compared for an accurate and reliable classification. It is best to include many different kinds of data: morphological, biochemical, and physiological.

After character analysis, an association coefficient, a function that measures the agreement between characters possessed by two organisms, is calculated for each pair of organisms in the group. The **simple matching coefficient (S_{SM}),** the most commonly used coefficient in bacteriology, is the proportion of characters that match regardless of whether the attribute is present or absent (**table 19.2**). Sometimes the **Jaccard coefficient (S_J)** is calculated by ignoring any characters that both organisms lack (table 19.2). Both coefficients increase linearly in value from 0.0 (no matches) to 1.0 (100% matches).

The simple matching coefficients, or other association coefficients, are then arranged to form a **similarity matrix.** This is a matrix in which the rows and columns represent organisms, and each value is an association coefficient measuring the similarity of two different organisms so that each organism is compared to every other one in the table (**figure 19.6*a***). Organisms with great similarity are grouped together and separated from dissimilar organisms (figure 19.6*b*); such groups of organisms are called **phenons** (sometimes called phenoms).

The results of numerical taxonomic analysis are often summarized with a treelike diagram called a **dendrogram** (figure 19.6*c*). The diagram usually is placed on its side with the X-axis or abscissa graduated in units of similarity. Each branch point is at the similarity value relating the two branches. The organisms in the two branches share so many characteristics that the two groups are seen to be separate only after examination of association coefficients greater than the magnitude of the branch point value. Below the branch point value, the two groups appear to be one. The ordinate in such a dendrogram has no special significance, and the clusters may be arranged in any convenient order.

Table 19.2 The Calculation of Association Coefficients for Two Organisms

In this example, organisms A and B are compared in terms of the characters they do and do not share. The terms in the association coefficient equations are defined as follows:

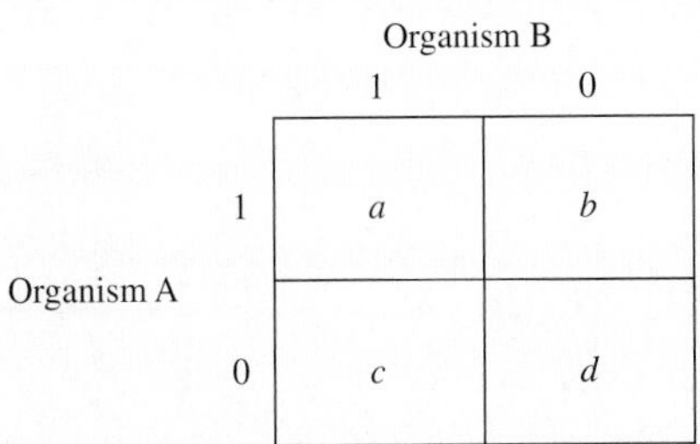

		Organism B 1	Organism B 0
Organism A	1	*a*	*b*
	0	*c*	*d*

a = number of characters coded as present (1) for both organisms

b and c = numbers of characters differing (1,0 or 0,1) between the two organisms

d = number of characters absent (0) in both organisms

Total number of characters compared = $a + b + c + d$

The simple matching coefficient (S_{SM}) = $\dfrac{a + d}{a + b + c + d}$

The Jaccard coefficient (S_J) = $\dfrac{a}{a + b + c}$

The significance of these clusters or phenons in traditional taxonomic terms is not always evident, and the similarity levels at which clusters are labeled species, genera, and so on, are a matter of judgment. Sometimes groups are simply called phenons and preceded by a number showing the similarity level above which they appear (e.g., a 70-phenon is a phenon with 70% or greater similarity among its constituents). Phenons formed at about 80% similarity often are equivalent to species.

Numerical taxonomy has proved to be a powerful tool in microbial taxonomy. Although it often has simply reconfirmed already existing classification schemes, sometimes accepted classifications are found wanting. Numerical taxonomic methods also can be used to compare sequences of macromolecules such as RNA and proteins.

1. What is a natural classification?
2. What is polyphasic taxonomy and what three types of data does it consider?
3. What is numerical taxonomy and why are computers so important to this approach?
4. Define the following terms: association coefficient, simple matching coefficient, Jaccard coefficient, similarity matrix, phenon, and dendrogram.
5. Which pair of species has more mutual similarity, a pair with an association coefficient of 0.9 or one with a coefficient of 0.6? Why?

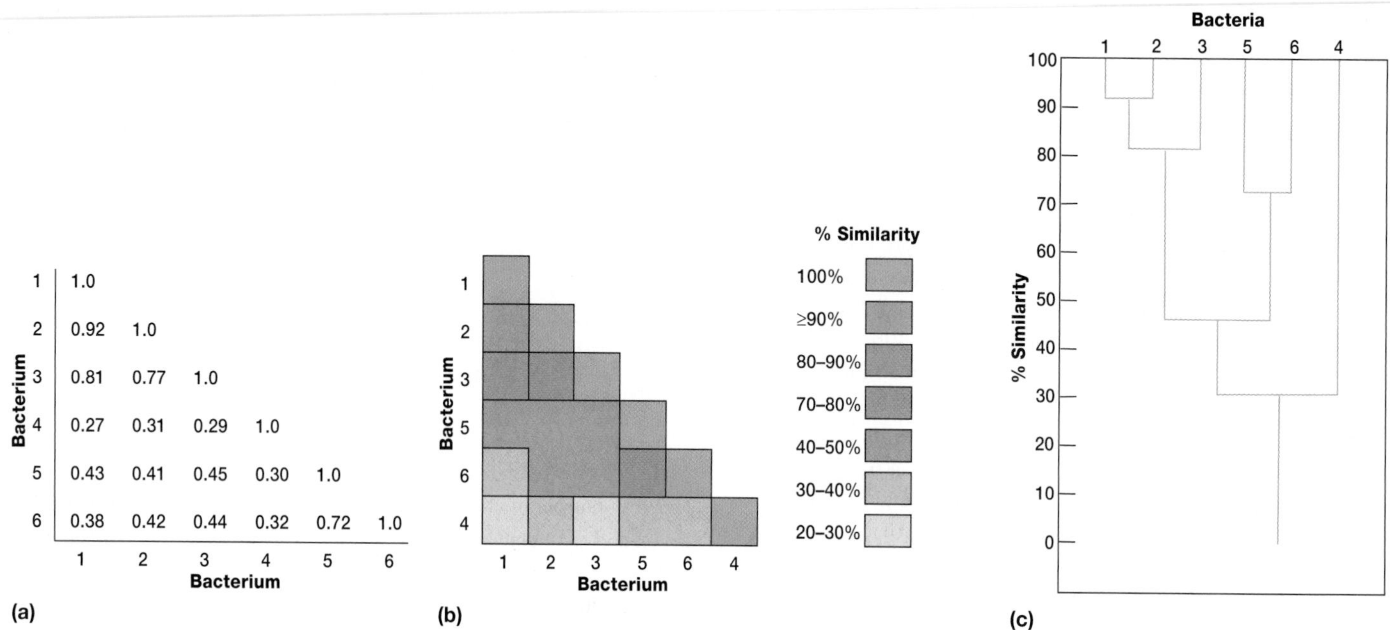

Figure 19.6 Clustering and Dendrograms in Numerical Taxonomy. **(a)** A small similarity matrix that compares six strains of bacteria. The degree of similarity ranges from none (0.0) to complete similarity (1.0). **(b)** The bacteria have been rearranged and joined to form clusters of similar strains. For example, strains 1 and 2 are the most similar. The cluster of 1 plus 2 is fairly similar to strain 3, but not at all to strain 4. **(c)** A dendrogram showing the results of the analysis in part (*b*). Strains 1 and 2 are members of a 90-phenon, and strains 1–3 form an 80-phenon. While strains 1–3 may be members of a single species, it is quite unlikely that strains 4–6 belong to the same species as 1–3.

19.3 Taxonomic Ranks

The classification of microbes involves placing them within hierarchical taxonomic levels. Microbes in each level or rank share a common set of specific features. The ranks are arranged in a nonoverlapping hierarchy so that each level includes not only the traits that define the rank above it, but a new set of more restrictive traits (**figure 19.7**). The highest rank is the domain, and all procaryotes belong to either the *Bacteria* or the *Archaea*. Within each domain, each microbe is assigned (in descending order) to a phylum, class, order, family, genus, and species (**table 19.3**). Some procaryotes are also given a subspecies designation. Microbial groups at each level have a specific suffix indicative of that rank or level. Microbiologists often use informal names in place of formal, hierarchical ones. Typical examples of such names are purple bacteria, spirochetes, methane-oxidizing bacteria, sulfate-reducing bacteria, and lactic acid bacteria. As we shall see, these informal names may not have taxonomic significance as they can include species from several phyla. A good example of this is the "sulfur bacteria."

The basic taxonomic group in microbial taxonomy is the **species.** Taxonomists working with higher organisms define the term species differently than do microbiologists. Species of higher organisms are groups of interbreeding or potentially interbreeding natural populations that are reproductively isolated from other groups. This is a satisfactory definition for organisms capable of sexual reproduction but fails with many microorganisms because they do not reproduce sexually. As we have discussed, procaryotic species are characterized by phenotypic, genotypic, and phylogenetic criteria. A **procaryotic species** is a collection of strains that share many stable properties and differ significantly from other groups of strains. A **strain** consists of the descendents of a single, pure microbial culture. The definition of a procaryotic species is subjective and can be interpreted in many ways. With an increasing amount of genome sequence data, some have argued that the definition of a procaryotic species needs further revision. Perhaps a species should be the collection of organisms that share the same sequences in their core housekeeping genes (genes coding for products that are required by all cells and which are usually continually expressed). It will take much more work to resolve this complex issue. Whatever the definition, ideally a species also should be phenotypically distinguishable from other similar species.

There are a number of different ways in which strains within a species may be described. **Biovars** are *var*iant strains characterized by biochemical or physiological differences, **morphovars** differ morphologically, and **serovars** have distinctive antigenic properties. For each species, one strain is designated as the **type strain.** It is usually one of the first strains studied and often is more fully characterized than other strains; however, it does not have to be the most representative member. The type strain for the species is called the type species and is the nomenclatural type or the holder of the species name. A nomenclatural type is a device to en-

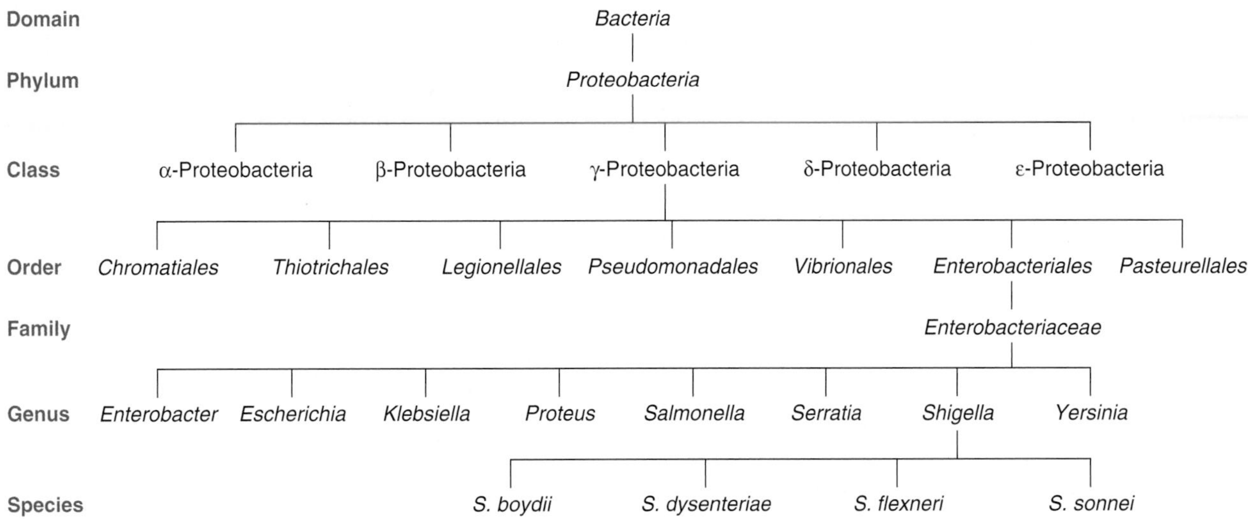

Figure 19.7 Hierarchical Arrangement in Taxonomy. In this example, members of the genus *Shigella* are placed within higher taxonomic ranks. Not all classification possibilities are given for each rank to simplify the diagram. Note that *-ales* denotes order and *-ceae* indicates family.

Table 19.3 An Example of Taxonomic Ranks and Names

Rank	Example
Domain	*Bacteria*
Phylum	*Proteobacteria*
Class	*Gammaproteobacteria*
Order	*Enterobacteriales*
Family	*Enterobacteriaceae*
Genus	*Shigella*
Species	*S. dysenteriae*

sure permanence of names when taxonomic rearrangements take place. When nomenclature revisions occur, the type species must remain within the genus of which it is the nomenclatural type. Only those strains very similar to the type strain or type species are included in a species. Each species is assigned to a genus, the next rank in the taxonomic hierarchy. A **genus** is a well-defined group of one or more species that is clearly separate from other genera. In practice there is considerable subjectivity in assigning species to a genus, and taxonomists may disagree about the composition of genera.

Microbiologists name microorganisms by using the **binomial system** of Linnaeus. The Latinized, italicized name consists of two parts. The first part, which is capitalized, is the generic name, and the second is the uncapitalized species name or specific epithet (e.g., *Escherichia coli*). The species name is stable; the oldest epithet for a particular organism takes precedence and must be used. In contrast, a generic name can change if the organism is assigned to another genus because of new information. For example, some members of the genus *Streptococcus* were placed into two new genera, *Enterococcus* and *Lactococcus,* based on rRNA analysis and other characteristics. Thus *Streptococcus faecalis* is now *Enterococcus faecalis.* Often the name will be shortened by abbreviating the genus name with a single capital letter, for example *E. coli.* A new procaryotic species cannot be recognized until it has been published in the *International Journal of Systematic and Evolutionary Microbiology;* until that time, the new species name will appear in quotation marks. *Bergey's Manual of Systematic Bacteriology* contains the currently accepted system of procaryotic taxonomy and is discussed later in section 19.8.

1. What is the difference between a procaryotic species and a strain?
2. Define morphovar, serovar, and type strain.
3. Which is the correct way to write this microbe's name: *bacillus subtilis, Bacillus subtilis, Bacillus Subtilis,* or Bacillus subtilis? Identify the genus name and the species name. Verify your answers by referring to the text.

19.4 Techniques for Determining Microbial Taxonomy and Phylogeny

Many different approaches are used in classifying and identifying microorganisms. For clarity, these have been divided into two groups: classical and molecular. Methods often employed in routine laboratory identification of bacteria are covered in the chapter on clinical microbiology (*see chapter 35*).

Classical Characteristics

Classical approaches to taxonomy make use of morphological, physiological, biochemical, ecological, and genetic characteristics. These characteristics have been employed in microbial taxonomy for many years. They are quite useful in routine identification and may provide phylogenetic information as well.

Morphological Characteristics

Morphological features are important in microbial taxonomy for many reasons. Morphology is easy to study and analyze, particularly in eucaryotic microorganisms and the more complex procaryotes. In addition, morphological comparisons are valuable because structural features depend on the expression of many genes, are usually genetically stable, and normally (at least in eucaryotes) do not vary greatly with environmental changes. Thus morphological similarity often is a good indication of phylogenetic relatedness.

Many different morphological features are employed in the classification and identification of microorganisms (**table 19.4**). Although the light microscope has always been a very important tool, its resolution limit of about 0.2 μm reduces its usefulness in viewing smaller microorganisms and structures. The transmission and scanning electron microscopes, with their greater resolution, have immensely aided the study of all microbial groups. Microscopy and specimen preparation (chapter 2)

Physiological and Metabolic Characteristics

Physiological and metabolic characteristics are very useful because they are directly related to the nature and activity of microbial enzymes and transport proteins. Since proteins are gene products, analysis of these characteristics provides an indirect comparison of microbial genomes. **Table 19.5** lists some of the most important of these properties.

Ecological Characteristics

The ability of a microorganism to colonize a specific environment is of taxonomic value. Some microbes may be very similar in many other respects but inhabit different ecological niches, suggesting they may not be as closely related as first suspected. Some examples of taxonomically important ecological properties are life cycle patterns; the nature of symbiotic relationships; the ability to cause disease in a particular host; and habitat preferences such as requirements for temperature, pH, oxygen, and osmotic concentration. Many growth requirements are considered physiological characteristics as well (table 19.5). Microbial interactions (section 30.1); The influence of environmental factors on growth (section 6.5)

Table 19.4 Some Morphological Features Used in Classification and Identification

Feature	Microbial Groups
Cell shape	All major groups[a]
Cell size	All major groups
Colonial morphology	All major groups
Ultrastructural characteristics	All major groups
Staining behavior	Bacteria, some fungi
Cilia and flagella	All major groups
Mechanism of motility	Gliding bacteria, spirochetes
Endospore shape and location	Endospore-forming bacteria
Spore morphology and location	Bacteria, protists, fungi
Cellular inclusions	All major groups
Color	All major groups

[a]Used in classifying and identifying at least some bacteria, fungi, and protists.

Table 19.5 Some Physiological and Metabolic Characteristics Used in Classification and Identification

Carbon and nitrogen sources
Cell wall constituents
Energy sources
Fermentation products
General nutritional type
Growth temperature optimum and range
Luminescence
Mechanisms of energy conversion
Motility
Osmotic tolerance
Oxygen relationships
pH optimum and growth range
Photosynthetic pigments
Salt requirements and tolerance
Secondary metabolites formed
Sensitivity to metabolic inhibitors and antibiotics
Storage inclusions

Genetic Analysis

Because most eucaryotes are able to reproduce sexually, genetic analysis has been quite useful in the classification of these organisms. As mentioned earlier, the species is defined in terms of sexual reproduction where possible. Although procaryotes do not reproduce sexually, the study of chromosomal gene exchange through transformation, conjugation, and transduction is sometimes useful in their classification.

Transformation can occur between different procaryotic species but only rarely between genera. The demonstration of transformation between two strains provides evidence of a close relationship since transformation cannot occur unless the genomes are fairly similar. Transformation studies have been carried out with several genera: *Bacillus, Micrococcus, Haemophilus, Rhizobium,* and others. Despite transformation's usefulness, its results are sometimes hard to interpret because an absence of transformation may result from factors other than major differences in DNA sequence. DNA transformation (section 13.8)

Conjugation studies also yield taxonomically useful data, particularly with the enteric bacteria. For example, *Escherichia* can undergo conjugation with the genera *Salmonella* and *Shigella* but not with *Proteus* and *Enterobacter.* These observations fit with other data showing that the first three of these genera are more closely related to one another than to *Proteus* and *Enterobacter.* Bacterial conjugation (section 13.7); Class *Gammaproteobacteria:* Order *Enterbacteriales* (section 22.3)

Plasmids are important taxonomically because they can confound the analysis of phenotypic traits. Most microbial genera carry plasmids and some plasmids are passed from one microbe to another with relative ease. When such plasmids encode a phenotypic trait (or traits) that is being used to develop a taxonomic scheme, the investigator may assume that the trait is encoded by chromosomal genes. Thus a microbe's phenetic characteristics are misunderstood and its relative degree of relatedness to another microbe may be overestimated. For example, hydrogen sulfide production and lactose fermentation are very important in the taxonomy of the enteric bacteria, yet genes for both traits can be borne on plasmids as well as bacterial chromosomes. One must take care to avoid errors as a result of plasmid-borne traits. Plasmids (section 3.5)

Molecular Characteristics

It is hard to overestimate how the study of the DNA, RNA, and proteins has advanced our understanding of microbial evolution and taxonomy. Evolutionary biologists studying plants and animals draw from a rich fossil record to assemble a history of morphological changes; in these cases, molecular approaches serve to supplement these data. In contrast, microorganisms have left almost no fossil record, so molecular analysis is the only feasible means of collecting a large and accurate data set from a number of microbes. When scientists are careful to make only valid comparisons, phylogenetic inferences based on molecular approaches provide the most robust analysis of microbial evolution.

Nucleic Acid Base Composition

Microbial genomes can be directly compared, and taxonomic similarity can be estimated in many ways. The first, and possibly the simplest, technique to be employed is the determination of DNA base composition. DNA contains four purine and pyrimidine bases: adenine (A), guanine (G), cytosine (C), and thymine (T). In double-stranded DNA, A pairs with T, and G pairs with C. Thus the (G + C)/(A + T) ratio or **G + C content,** the percent of G + C in DNA, reflects the base sequence and varies with sequence changes as follows:

$$\text{Mol\% G + C} = \frac{\text{G + C}}{\text{G + C + A + T}} \times 100$$

The base composition of DNA can be determined in several ways. Although the G + C content can be ascertained after hydrolysis of DNA and analysis of its bases with high-performance liquid chromatography (HPLC), physical methods are easier and more often used. The G + C content often is determined from the **melting temperature (T_m)** of DNA. In double-stranded DNA three hydrogen bonds join GC base pairs, and two bonds connect AT base pairs. As a result DNA with a greater G + C content have more hydrogen bonds, and its strands separate at higher temperatures—that is, it has a higher melting point. DNA melting can be easily followed spectrophotometrically because the absorbance of DNA at 260 nm (UV light) increases during strand separation. When a DNA sample is slowly heated, the absorbance

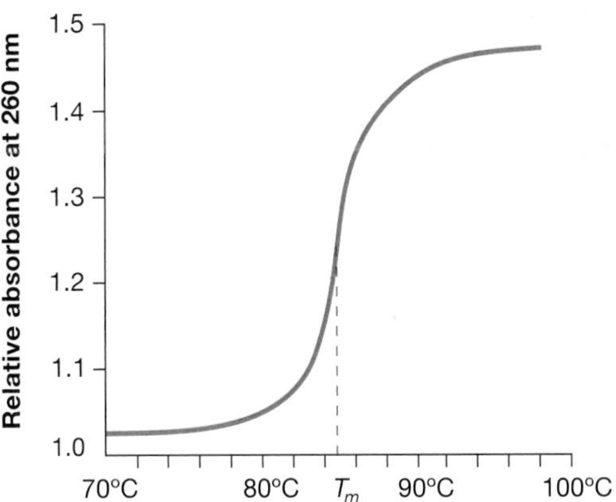

Figure 19.8 A DNA Melting Curve. The T_m is indicated.

increases as hydrogen bonds are broken and reaches a plateau when all the DNA has become single stranded (**figure 19.8**). The midpoint of the rising curve gives the melting temperature, a direct measure of the G + C content.

The G + C content of many microorganisms has been determined (**table 19.6**). The G + C content of DNA from animals and higher plants averages around 40% and ranges between 30 and 50%. In contrast, the DNA of both eucaryotic and procaryotic microorganisms varies greatly in G + C content; procaryotic G + C content is the most variable, ranging from around 25 to almost 80%. Despite such a wide range of variation, the G + C content of strains within a particular species is constant. If two organisms differ in their G + C content by more than about 10%, their genomes have quite different base sequences. On the other hand, it is not safe to assume that organisms with very similar G + C contents also have similar DNA base sequences because two very different base sequences can be constructed from the same proportions of AT and GC base pairs. Only if two microorganisms also are alike phenotypically does their similar G + C content suggest close relatedness.

G + C content data are valuable in at least two ways. First, they can confirm a taxonomic scheme developed using other data. If organisms in the same taxon are too dissimilar in G + C content, the taxon probably should be divided. Second, G + C content appears to be useful in characterizing procaryotic genera because the variation within a genus is usually less than 10% even though the content may vary greatly between genera. For example, *Staphylococcus* has a G + C content of 30 to 38%, whereas *Micrococcus* DNA has 64 to 75% G + C; yet these two genera of gram-positive cocci have many other features in common.

Nucleic Acid Hybridization

The similarity between genomes can be compared more directly by use of **nucleic acid hybridization** studies. If a mixture of single-stranded DNA (ssDNA) formed by heating double-stranded (ds) DNA is cooled and held at a temperature about 25°C below the T_m, strands with complementary base sequences will reassociate to form stable dsDNA, whereas noncomplementary strands will

Table 19.6 Representative G + C Content of Microorganisms

Organism	Percent G + C	Organism	Percent G + C	Organism	Percent G + C
Bacteria		*Rhodospirillum*	62–66	*Peridinium triquetrum*	53
Actinomyces	59–73	*Rickettsia*	29–33	*Physarum polycephalum*	38–42
Anabaena	39–44	*Salmonella*	50–53	*Plasmodium berghei*	41
Bacillus	32–62	*Spirillum*	38	*Scenedesmus*	52–64
Bacteroides	28–61	*Spirochaeta*	51–65	*Spirogyra*	39
Bdellovibrio	49.5–51	*Staphylococcus*	30–38	*Stentor polymorphus*	45
Caulobacter	62–65	*Streptococcus*	33–44	*Tetrahymena*	19–33
Chlamydia	41–44	*Streptomyces*	69–73	*Trichomonas*	29–34
Chlorobium	49–58	*Sulfolobus*	31–37	*Trypanosoma*	45–59
Chromatium	48–70	*Thermoplasma*	46	*Volvox carteri*	50
Clostridium	21–54	*Thiobacillus*	52–68		
Cytophaga	33–42	*Treponema*	25–53	**Fungi**	
Deinococcus	62–70			*Agaricus bisporus*	44
Escherichia	48–59	**Protists**		*Amanita muscaria*	57
Halobacterium	66–68	*Acanthamoeba castellanii*	56–58	*Aspergillus niger*	52
Hyphomicrobium	59–67	*Acetabularia mediterranea*	37–53	*Blastocladiella emersonii*	66
Methanobacterium	32–50	*Amoeba proteus*	66	*Candida albicans*	33–35
Micrococcus	64–75	*Chlamydomonas*	60–68	*Claviceps purpurea*	53
Mycobacterium	62–70	*Chlorella*	43–79	*Coprinus lagopus*	52–53
Mycoplasma	23–40	*Cyclotella cryptica*	41	*Fomes fraxineus*	56
Myxococcus	68–71	*Dictyostelium*	22–25	*Mucor rouxii*	38
Neisseria	48–56	*Euglena gracilis*	46–55	*Neurospora crassa*	52–54
Nitrobacter	59–62	*Lycogala*	42	*Penicillium notatum*	52
Oscillatoria	40–50	*Nitella*	49	*Polyporus palustris*	56
Prochloron	41	*Nitzschia angularis*	47	*Rhizopus nigricans*	47
Proteus	38–41	*Ochromonas danica*	48	*Saccharomyces cerevisiae*	36–42
Pseudomonas	58–69	*Paramecium* spp.	29–39	*Saprolegnia parasitica*	61

remain unpaired (**figure 19.9**). Because strands with similar, but not identical, sequences associate to form less temperature stable dsDNA hybrids, incubation of the mixture at 30 to 50°C below the T_m allows hybrids of more diverse ssDNAs to form. Incubation at 10 to 15°C below the T_m permits hybrid formation only with almost identical strands.

In one of the more widely used hybridization techniques, nylon filters with bound nonradioactive DNA strands are incubated at the appropriate temperature with ssDNA fragments made radioactive with ^{32}P, ^{3}H, or ^{14}C. After radioactive fragments are allowed to hybridize with the membrane-bound ssDNA, the membrane is washed to remove any nonhybridized ssDNA and its radioactivity is measured. The quantity of radioactivity bound to the filter reflects the amount of hybridization and thus the similarity of the DNA sequences. The degree of similarity or homology is expressed as the percent of experimental DNA radioactivity retained on the filter compared with the percent of homologous DNA radioactivity bound under the same conditions (**table 19.7** provides examples). Two strains whose DNAs show at least 70% relatedness under optimal hybridization conditions and less than a 5% difference in T_m often, but not always, are considered members of the same species.

If DNA molecules are very different in sequence, they will not form a stable, detectable hybrid. Therefore DNA-DNA hybridization is used to study only closely related microorganisms. More distantly related organisms can be compared by carrying out DNA-RNA hybridization experiments using radioactive ribosomal or transfer RNA. Distant relationships can be detected because rRNA and tRNA genes represent only a small portion of the total DNA genome and have not evolved as rapidly as most other microbial genes. The technique is similar to that employed for DNA-DNA hybridization: membrane-bound DNA is incubated with radioactive rRNA, washed, and counted. An even more accurate measurement of homology is obtained by finding the temperature required to dissociate and remove half the radioactive rRNA from the membrane; the higher this temperature, the stronger the rRNA-DNA complex and the more similar the sequences. Gene structure: Genes that code for tRNA and rRNA (section 11.5)

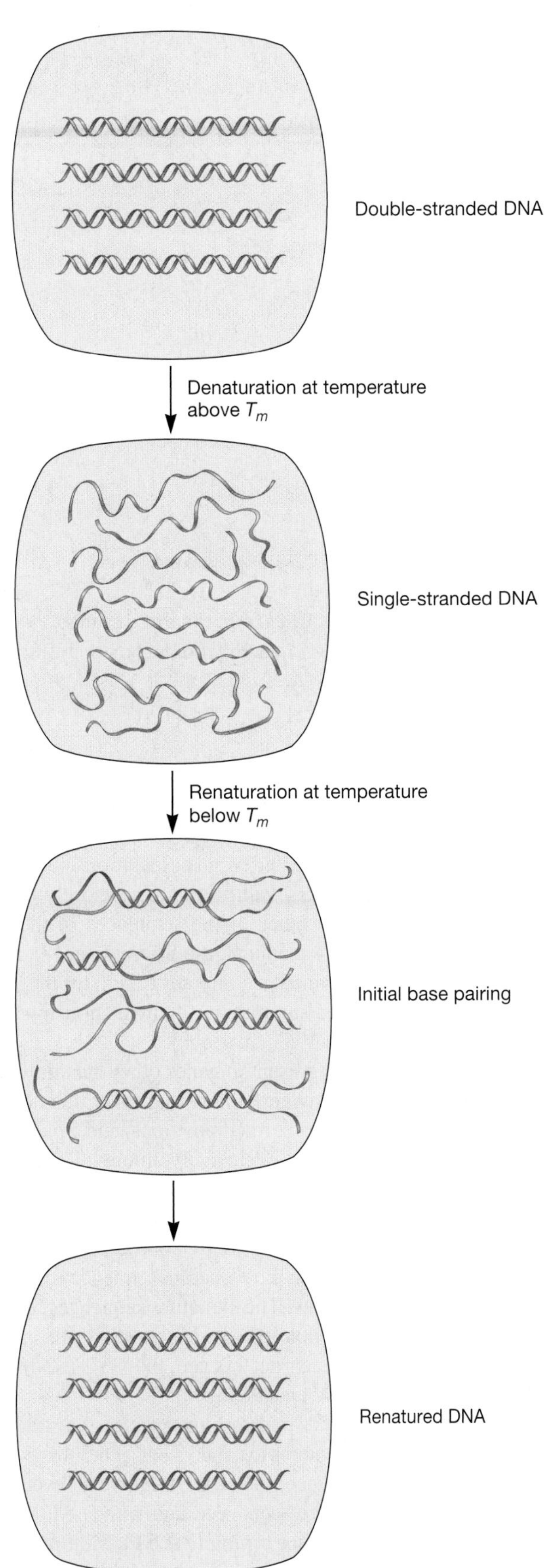

Figure 19.9 Nucleic Acid Melting and Hybridization. Complementary strands are shown in purple and blue.

Table 19.7 Comparison of *Neisseria* Species by DNA Hybridization Experiments

Membrane-Attached DNA[a]	Percent Homology[b]
Neisseria meningitidis	100
N. gonorrhoeae	78
N. sicca	45
N. flava	35

Source: Data from T. E. Staley and R. R. Colwell, "Applications of Molecular Genetics and Numerical Taxonomy to the Classification of Bacteria" in *Annual Review of Ecology and Systematics,* 8: 282, 1973.

[a]The experimental membrane-attached nonradioactive DNA from each species was incubated with radioactive *N. meningitidis* DNA, and the amount of radioactivity bound to the membrane was measured. The more radioactivity bound, the greater the homology between DNA sequences.

[b]*N.meningitidis* DNA bound to experimental DNA/Amount bound to membrane attached *N.meningitidis* DNA × 100

Nucleic Acid Sequencing

Despite the usefulness of G + C content determination and nucleic acid hybridization studies, rRNAs from small ribosomal subunits (16S and 18S rRNAs from procaryotes and eucaryotes, respectively) have become the molecules of choice for inferring microbial phylogenies and making taxonomic assignments at the genus level. The small subunit rRNAs (SSU rRNAs) are almost ideal for studies of microbial evolution and relatedness because they play the same role in all microorganisms. In addition, because the ribosome is absolutely necessary for survival and the SSU rRNAs are part of the complex ribosomal structure, the genes encoding SSU rRNAs cannot tolerate large mutations. Thus these genes change very slowly with time and do not appear to be subject to horizontal gene transfer, an important factor in comparing sequences from different phyla. The utility of SSU rRNAs is extended by the presence of certain sequences that are variable among organisms and other regions that are quite stable. The variable regions enable comparison between closely related microbes while the stable sequences allow the comparison of distantly related microorganisms.

The ability to amplify regions of rRNA genes (rDNA) by the polymerase chain reaction (PCR) and sequence the DNA using automated sequencing technology has greatly increased the efficiency by which SSU rRNA sequences can be obtained. PCR can be used to amplify rDNA from the genomes of organisms because conserved nucleotide sequences flank the regions of interest. In practice, this means that PCR primers are readily available or can be generated to amplify rDNA from both cultured and uncultured microbes. As noted, the Ribosome Database Project has sequences from over 200,000 microbes. PCR (section 14.3); Determining DNA sequences (section 15.2)

Comparative analysis of 16S rRNA sequences from thousands of organisms has demonstrated the presence of **oligonucleotide signature sequences** (**figure 19.10**). These are short, conserved nucleotide sequences that are specific for a phylogenetically defined group of organisms. Thus the signature sequences found in *Bacteria* are rarely or never found in *Archaea*

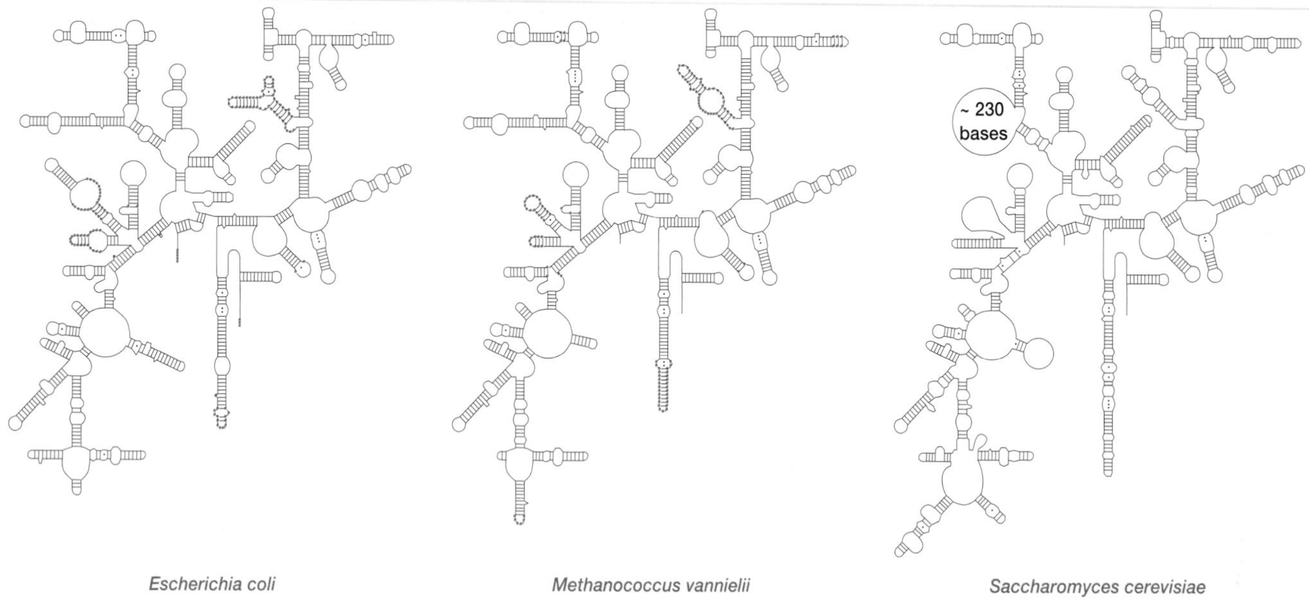

Figure 19.10 Small Ribosomal Subunit RNA. Representative examples of rRNA secondary structures from the three primary domains: *Bacteria* (*Escherichia coli*), *Archaea* (*Methanococcus vannielii*), and *Eucarya* (*Saccharomyces cerevisiae*). The red dots mark positions where *Bacteria* and *Archaea* normally differ. *Source: Data from C. P. Woese.* Microbiological Reviews, *51(2):221–227, 1987.*

Table 19.8 Selected 16S rRNA Signature Sequences for Some Bacterial Groups[a]

Position in rRNA	Consensus Composition	γ-Proteobacteria	Cyanobacteria	Spirochetes	*Bacteroides*	Green Sulfur	Green Nonsulfur	*Deinococcus*	Gram Positive (Low GC)	Gram Positive (High GC)	*Planctomyces*
47	C	+	+	U	+	+	+	+	+	+	G
53	A	+	+	G	+	+	G	+	+	+	G
570	G	+	+	+	U	+	+	+	+	+	U
812	G	c	+	+	+	+	+	C	+	+	+
906	G	Ag	+	+	+	+	A	+	+	A	+
955	U	+	+	+	+	+	+	+	+	A	C
1,207	G	+	C	+	+	+	+	+	C	C	+
1,234	C	+	+	a	U	A	+	+	+	+	+

[a]A plus sign in a column means that the group has the same base as the consensus sequence. If the letter is given in upper case, it is changed in more than 90% of the cases. A lowercase letter signifies a minor occurrence base (< 15% of the cases).

and vice versa (**table 19.8**). Likewise, the 18S rRNA of eucaryotes also bears signature sequences that are specific to the domain *Eucarya*. Either complete rRNAs or, more often, specific rRNA fragments can be compared. The proper alignment of SSU rRNA nucleotide sequences and the application of computer algorithms enable sequence comparison between any number of organisms. When comparing rRNA sequences between two microorganisms, their relatedness can be represented by an association coefficient, or S_{ab} value. The higher the S_{ab} values, the more closely the organisms are related to each other. If the sequences of the 16S rRNAs of two organisms are identical, the S_{ab} value is 1.0. After S_{ab} values have been determined, a computer calculates the relatedness of the organisms and summarizes their relationships in a tree or dendrogram (figures 19.3 and 19.6*c*).

Signature sequences are present in genes other than those encoding ribosomal RNA. Many genes have inserts or deletions of specific lengths and sequences at fixed positions, and a particular insert or deletion may be found exclusively among all members of one or more phyla. R. S. Gupta refers to these as conserved indels. These signature sequences are particularly useful in phylogenetic studies when they are flanked by conserved regions. In such cases, observed changes in the signature sequence cannot be due to sequence misalignments. The signature sequences located in some highly conserved housekeeping genes do not appear greatly affected by lateral gene transfers and, like SSU rRNA, can be employed in phylogenetic analysis.

The use of DNA sequences to determine species and strain (as opposed to genus) identity requires the analysis of genes that evolve more quickly than those that encode rRNA. Often five to seven conserved housekeeping genes are sequenced and compared, a technique called **multilocus sequence typing (MLST).** Multiple genes are usually examined to avoid misleading results that can arise through lateral gene transfer. MLST was originally developed to discriminate among pathogenic strains but has become more broadly applied to microbial taxonomy. Although MLST is helpful for differentiating isolates at the strain and species levels, the data become too difficult to interpret at higher taxonomic levels.

Genomic Fingerprinting

A group of techniques called **genomic fingerprinting** can also be used to classify microbes and help determine phylogenetic relationships. Unlike the molecular analyses so far discussed, genomic fingerprinting does not involve nucleotide sequencing. Instead, it employs the capacity of restriction endonucleases to recognize specific nucleotide sequences. Thus the pattern of DNA fragments generated by endonuclease cleavage (called restriction fragments) is a direct representation of nucleotide sequence (*see figure 14.3 and table 14.2*). The comparison of restriction fragments between species and strains is the basis of *r*estriction *f*ragment *l*ength *p*olymorphism (RFLP) analysis.

Another assay is based on highly conserved and repetitive DNA sequences present in many copies in the genomes of most gram-negative and some gram-positive bacteria. There are three families of repetitive sequences: the 154 bp BOX elements, the 124–127 bp *e*nterobacterial *r*epetitive *i*ntergenic *c*onsensus (ERIC) sequence, and 35–40 bp *r*epetitive *e*xtragenic *p*alindromic (REP) sequences. These sequences are generally found at distinct sites between genes—that is, they are intergenic. Because they are conserved among genera, oligonucleotide primers can be used to specifically amplify the repetitive sequences by PRC. Different primers are used for each type of repetitive element, and the results are classified as arising from BOX-PCR, ERIC-PCR, or REP-PCR (**figure 19.11**). In each case the amplified fragments from many microbial samples can be resolved and visualized on an agarose gel. Each lane of the gel corresponds to a single bacterial isolate, and the pattern created by many samples resembles a UPC bar code. The "bar code" is then computer analyzed using pattern recognition software as well as software that calculates phylogenetic relationships. Because DNA fingerprinting enables identification to the level of species, subspecies, and often strain, it is valuable not only in the study of microbial diversity, but in the identification of human, animal, and plant pathogens as well.

Amino Acid Sequencing

The amino acid sequences of proteins directly reflect mRNA sequences and therefore represent the genes coding for their synthesis. There are several ways to compare proteins. The most direct approach is to determine the amino acid sequence of proteins with the same function. The value of a given protein in taxonomic and phylogenetic studies varies. The sequences of proteins with dissimilar functions often change at different rates; some sequences change quite rapidly whereas others are very stable. Nevertheless, if the sequences of proteins with the same function are similar, the organisms possessing them may be

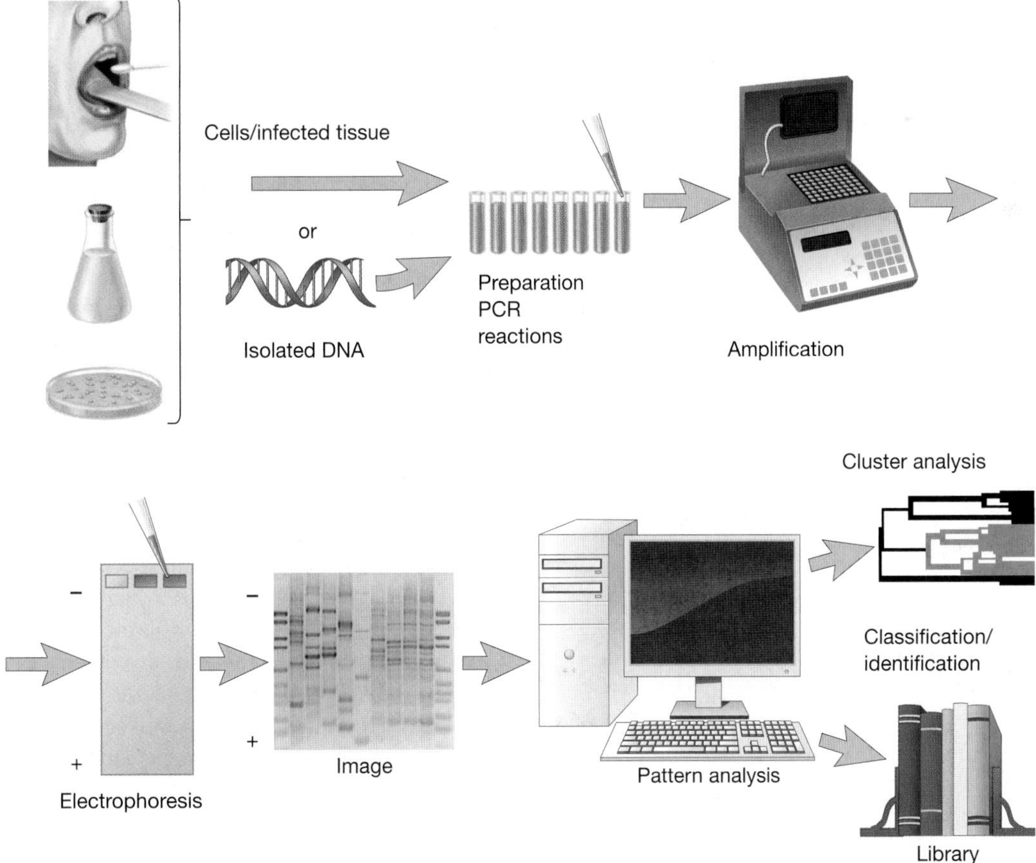

Figure 19.11 An Overview of the Genomic Fingerprinting Technique Based on Repetitive Nucleotide Sequences.

closely related. The sequences of cytochromes and other electron transport proteins, histones and heat-shock proteins, transcription and translation proteins, and a variety of metabolic enzymes have been used in taxonomic and phylogenetic studies. In contrast, rapidly evolving proteins, such as the outer surface proteins of the syphilis pathogen *Treponema pallidum,* are not appropriate for taxonomic or phylogenetic purposes. Thus not all proteins are suitable for studying large-scale changes that occur over long periods. However, suitable proteins may offer some advantages over rRNA comparisons. A sequence of 20 amino acids has more information per site than a sequence of four nucleotides. Protein sequences are less affected by organism-specific differences in G + C content than are DNA and RNA sequences. Finally, protein sequence alignment is easier because it is not dependent on secondary structure as is an rRNA sequence.

There are several ways to compare proteins. The most direct approach is to determine the amino acid sequence of proteins with the same function. Because protein sequencing is slow and expensive, more indirect methods of comparing proteins frequently have been employed. The electrophoretic mobility of proteins is useful in studying relationships at the species and subspecies levels. Antibodies can discriminate between very similar proteins, and immunologic techniques are used to compare proteins from different microorganisms. **Figure 19.12** shows the taxonomic utility of several kinds of molecular analyses including protein profiling; with the exception of genome sequencing, it is clear that a combination of approaches is best for identification at the species level or lower.

1. What are the advantages of using each major group of characteristics (morphological, physiological/metabolic, ecological, genetic, and molecular) in classification and identification? How is each group related to the nature and expression of the genome? Give examples of each type of characteristic.
2. What modes of genetic exchange in procaryotes have proved taxonomically useful? Why are plasmids of such importance in bacterial taxonomy?
3. What is the G + C content of DNA, and how can it be determined through melting temperature studies?
4. Why is it not safe to assume that two microorganisms with the same G + C content belong to the same species? In what two ways are G + C content data taxonomically valuable?
5. Describe how nucleic acid hybridization studies are carried out using membrane-bound DNA. Why might one wish to vary the incubation temperature during hybridization? What is the advantage of conducting DNA-RNA hybridization studies?
6. How are rRNA sequencing studies conducted, and why is rRNA so suitable for determining relatedness?
7. How is genomic fingerprinting similar to rRNA sequence analysis? How do the two techniques differ?
8. List some proteins used in phylogenetic and taxonomic studies. Why are they useful?

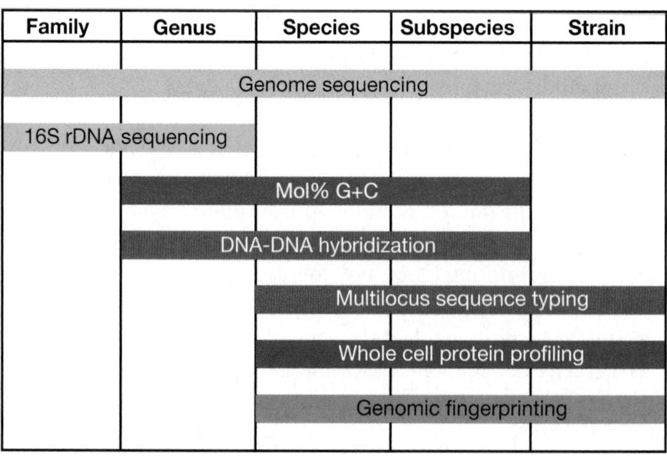

Figure 19.12 Relative Taxonomic Resolution of Various Molecular Techniques.

19.5 Assessing Microbial Phylogeny

Microbial taxonomy is changing rapidly. This is caused by ever-increasing knowledge of the biology of microorganisms, remarkable advances in computer technology, and the use of molecular characteristics to determine phylogenetic relationships between microorganisms. This section briefly describes some of the ways in which phylogenetic relationships are assessed.

Molecular Chronometers

The sequences of nucleic acids and proteins change with time and are considered to be **molecular chronometers.** This concept, first suggested by Zuckerkandl and Pauling (1965), is important in the use of molecular sequences in determining phylogenetic relationships and is based on the assumption that there is an evolutionary clock. According to this idea, the sequences of many rRNAs and proteins gradually change over time without destroying or severely altering their functions. One assumes that such changes are selectively neutral, occur fairly randomly, and increase linearly with time. When the sequences of similar molecules are quite different in two groups of organisms, the groups diverged from one another a long time ago. Phylogenetic analysis using molecular chronometers is somewhat complex and controversial because the rate of sequence change can vary. The swift and relatively infrequent speciation events described by punctuated equilibria (figure 19.5) call for periods characterized by especially rapid change. Furthermore, different molecules and various parts of the same molecule can change at different rates. Highly conserved molecules such as rRNAs are used to follow large-scale evolutionary changes, whereas rapidly changing molecules are employed to follow speciation. Squaring the notion of molecular chronometers with the fossil record that demonstrates punctuated equilibria is difficult. For this reason, some scientists prefer to speak of evolutionary relatedness rather than rates of evolution. Further studies will be required to establish the accuracy and usefulness of molecular chronometers.

Phylogenetic Trees

Phylogenetic relationships are illustrated in the form of branched diagrams or trees. A **phylogenetic tree** is a graph made of branches that connect nodes (**figure 19.13**). The nodes represent taxonomic units such as species or genes; the external nodes at the end of the branches represent living (extant) organisms. As in the universal phylogenetic tree (figure 19.3), the length of the branches represents the number of molecular changes that have taken place between the two nodes. Finally, a tree may be unrooted or rooted. An unrooted tree (figure 19.13*a*) simply represents phylogenetic relationships but does not provide an evolutionary path. Figure 19.13*a* shows that A is more closely related to C than it is to either B or D, but does not specify the common ancestor for the four species or the direction of change. In contrast, the rooted tree (figure 19.13*b*) gives a node that serves as the common ancestor and shows the development of the four species from this root. It is much more difficult to develop a rooted tree. For example, there are 15 possible rooted trees that connect four species, but only three possible unrooted trees.

Phylogenetic trees are developed by comparing nucleotide or amino acid sequences. To compare two molecules, their sequences must first be aligned so that similar parts match up. The object is to align and compare homologous sequences, ones that are similar because they had a common origin in the past. This is not an easy task, and computers and fairly complex mathematics must be employed to minimize the number of gaps and mismatches in the sequences being compared. Bioinformatics (section 15.4)

Once the molecules have been aligned, the number of positions that vary in the sequences are determined. These data are used to calculate a measure of the difference between the sequences. Often the difference is expressed as the **evolutionary distance.** This is simply a quantitative indication of the number of positions that differ between two aligned macromolecules. Statistical adjustments are made for back mutations and multiple substitutions that may have occurred. Organisms are then clustered together based on similarity in the sequences. The most similar organisms are clustered together, then compared with the remaining organisms to form a larger cluster associated together at a lower level of similarity or evolutionary distance. The process continues until all organisms are included in the tree.

Phylogenetic relationships also can be estimated by techniques such as **parsimony analysis.** In this approach, relationships are determined by estimating the minimum number of sequence changes required to give the final sequences being compared. It is presumed that evolutionary change occurs along the shortest pathway with the fewest changes or steps from an ancestor to the organism in question. The tree or pattern of relationships is favored that is simplest and requires the fewest assumptions.

1. What are molecular chronometers and upon what assumptions are they based? Compare these assumptions with the concept of punctuated equilibria.
2. Define phylogenetic tree and evolutionary distance. What is the difference between an unrooted and a rooted tree?

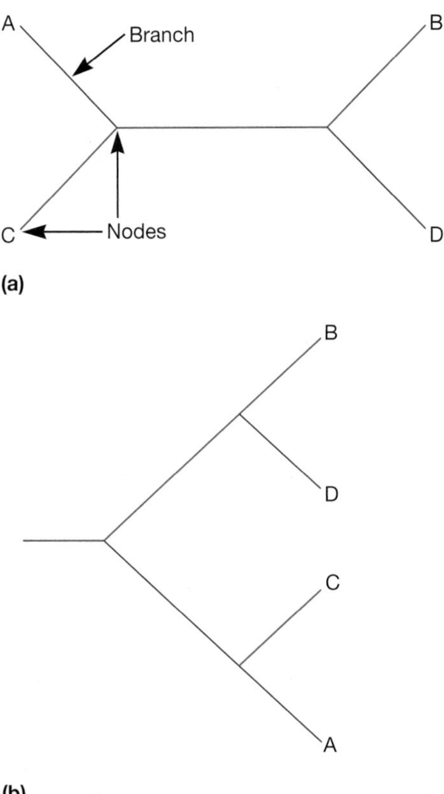

Figure 19.13 Examples of Phylogenetic Trees. **(a)** Unrooted tree joining four taxonomic units. **(b)** Rooted tree. See text for details.

19.6 The Major Divisions of Life

The division of all living organisms into three domains—*Archaea, Bacteria,* and *Eucarya*—has become widely accepted among microbiologists. Although in this text the universal phylogenetic tree as proposed by Norman Pace is emphasized (figure 19.3), some scientists contend that the *Bacteria* arose well before the *Archaea* and the *Eucarya,* as shown in **figure 19.14*a*.** Yet another interpretation is shown in the eocyte tree (figure 19.14*b*), which proposes that sulfur-dependent, extremely thermophilic procaryotes called eocytes [Greek *eo,* dawn and *cyta,* hollow vessel] form a separate group more closely related to eucaryotes than the *Archaea.* The existence of such organisms as a separate domain or group has been met with considerable skepticism. Finally, the idea that the *Eucarya* arose from a fusion of a bacterium and an archaeon is illustrated in figure 19.14*c*. There are many reasons why biologists are unable to agree on a single model. For instance, the selection of genes to be compared can have an impact on the resulting tree. When genes that encode proteins used for the storage and processing of genetic information are compared, trees that are consistent with those obtained by SSU rRNA analysis are generated (e.g., figure 19.3). On the other hand, when proteins involved in metabolism are compared, trees that place the *Bacteria* and the *Archaea* as closest relatives are generated. Yet the comparison of other "housekeeping" activities leads to trees that place

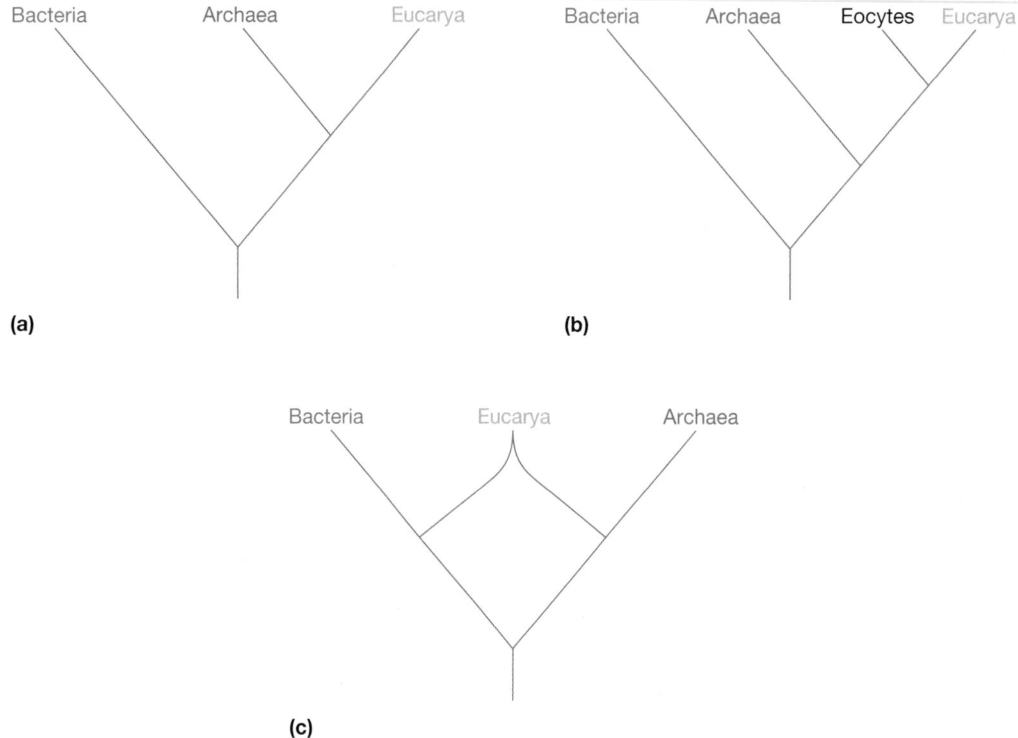

Figure 19.14 Variations in the Design of the "Tree of Life." These three alternative phylogenetic trees are discussed in the text.

the *Archaea* and the *Eucarya* most closely related (figure 19.14*a*). Other factors that can give rise to incongruent trees include unrecognized gene duplications that occurred before the domains formed, leading to confusing patterns. Unequal rates of evolution can distort the trees. Phylogenetically important information may have been lost in some molecular sequences. There may be significant sequence variation between the same molecules from different strains of the same species. Unless several strains are analyzed, false conclusions may be drawn. Thus inaccurate universal trees may result when only the sequences from a few molecules are employed.

One of the biggest challenges in constructing a satisfactory tree is widespread, frequent HGT. Genome sequence studies have shown that there is extensive HGT within and between domains. Eucaryotes possess genes from both bacteria and archaea, and there has been frequent gene swapping between the two procaryotic domains. It appears that at least some bacteria even have acquired eucaryotic genes. Although a variety of mechanisms may be responsible, it has been suggested that much of this gene movement occurs by way of virus-mediated transfer. Clearly the pattern of microbial evolution is not as linear and treelike as previously thought. This has prompted the development of trees that attempt to display HGT (**figure 19.15**). This tree resembles a web or network with many lateral branches linking various trunks, each branch representing the transfer of one or a few genes. Instead of having a single main trunk or common ancestor at its base, this tree has several trunks or groups of primitive cells that contribute to the original gene pool. Although there is extensive gene transfer between the *Archaea* and the *Bacteria* throughout their development, the *Eucarya* seldom participated in lateral gene transfer after the formation of fungi, plants, and animals. It is possible that eucaryotic cells originated in a complex process involving many gene transfers from both bacteria and archaea. This hypothesis is still consistent with the formation of mitochondria and chloroplasts by endosymbiosis with α-proteobacteria and cyanobacteria, respectively. Presumably the three domains remain separate because there are many more gene transfers within each domain than between them.

Molecular versus Organismal Trees

In this text, phylogenetic trees derived from 16S rRNA sequences are presented because these data are most extensive and are thought to be most accurate by the majority of microbiologists and evolutionary biologists. Phylogenetic trees based on the analysis of molecules like proteins or nucleic acids, are considered molecular phylogenetic trees. It should be remembered that these trees are based on individual genes, not whole organisms. Prior to the advent of molecular phylogeny, trees were constructed that classified eucaryotic organisms without significant consideration of the vast diversity of microbes or their evolutionary history. Such organismal trees generally reflect the organization requirements of zoologists and botanists at the expense of phylogenetics. A few are discussed here.

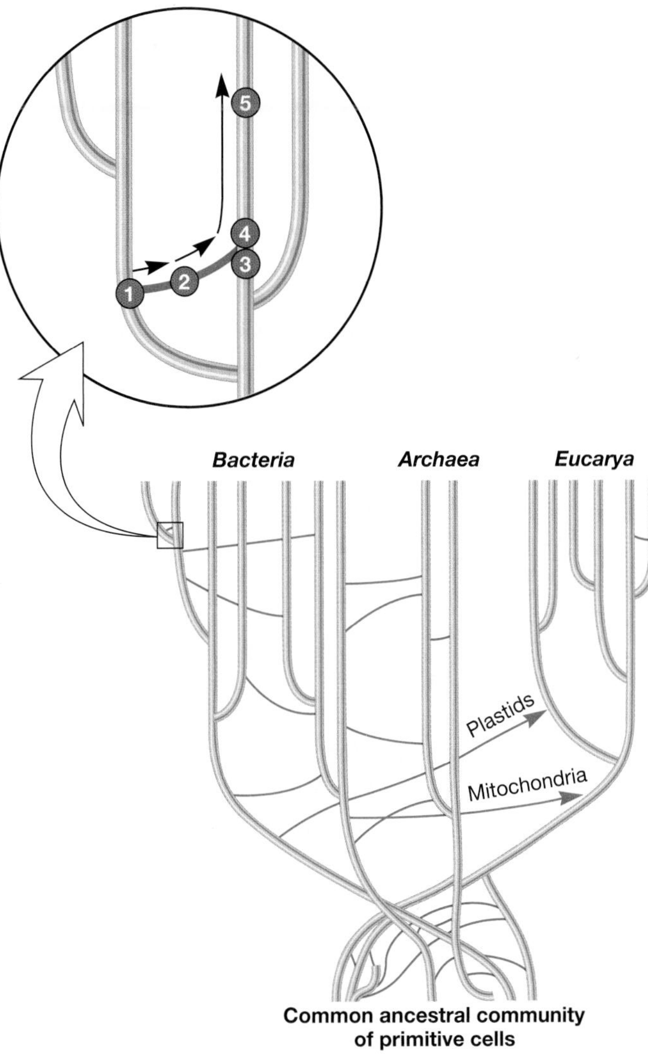

Figure 19.15 Universal Phylogenetic Tree with Lateral Gene Transfers. The effect of HGT on the evolution of life results in a tree with weblike interconnections that complicate the emergence of the three domains of life. The insert displays the series of events needed to give rise to the stable inheritance of a gene in a new organism.

Kingdoms

The first classification system to have gained popularity is the five-kingdom system first suggested by Robert Whittaker in 1969 (**figure 19.16*a***). Organisms are placed into five kingdoms based on at least three major criteria: (1) cell type—procaryotic or eucaryotic, (2) level of cellular organization—unicellular or multicellular, and (3) nutritional type. In this system the kingdom *Animalia* contains multicellular heterotrophs with wall-less eucaryotic cells and primarily ingestive nutrition, whereas the kingdom *Plantae* is composed of multicellular organisms with walled eucaryotic cells and primarily photoautotrophic nutrition. Microbiologists study members of the other three kingdoms. The kingdom *Monera* or *Procaryotae* contains all procaryotic organisms. The kingdom *Protista* is the least homogeneous and hardest to define. **Protists** are eucaryotes with unicellular organization, either in the form of solitary cells or colonies of cells lacking true tissues. They may have ingestive, absorptive, or photoautotrophic nutrition, and they include most of the microorganisms known as algae, protozoa, and many of the simpler fungi. The kingdom *Fungi* contains eucaryotic and predominately multinucleate organisms, with nuclei dispersed in a walled and often septate mycelium; their nutrition is absorptive. The taxonomy of the major protist and fungal phyla is discussed in more detail in chapters 25 and 26, respectively.

The five-kingdom system is no longer accepted by most biologists. A major problem is its lack of distinction between *Archaea* and *Bacteria.* Additionally, the kingdom *Protista* is too diverse to be taxonomically useful. Finally, the boundaries between the kingdoms *Protista, Plantae,* and *Fungi* are ill-defined. For example, the brown algae are probably not closely related to the plants even though the five-kingdom system places them in the *Plantae.*

It is thus not surprising that various alternatives have been suggested. The six-kingdom system is the simplest option; it divides the kingdom *Monera* or *Procaryotae* into two kingdoms, the *Eubacteria* and *Archaeobacteria* (figure 19.16*b*). Many attempts have been made to divide the protists into several better-defined kingdoms. The eight-kingdom system of Thomas Cavalier-Smith is a good example (figure 19.16*c*). Cavalier-Smith believes that differences in cellular structure and genetic organization are exceptionally important in determining phylogeny; thus he has used ultrastructural characteristics as well as rRNA sequences and other molecular data in developing his classification. He divides all organisms into two empires and eight kingdoms. The empire *Bacteria* contains two kingdoms, the *Eubacteria* and the *Archaeobacteria.* The second empire, the *Eucaryota,* contains six kingdoms of eucaryotic organisms. There are two new kingdoms of eucaryotes. The *Archezoa* are primitive eucaryotic unicellular organisms such as *Giardia* that have 70S ribosomes and lack Golgi apparatuses, mitochondria, chloroplasts, and peroxisomes. The kingdom *Chromista* contains mainly photosynthetic organisms that have their chloroplasts within the lumen of the rough endoplasmic reticulum rather than in the cytoplasmic matrix (as is the case in the kingdom *Plantae*). Diatoms, brown algae, cryptomonads, and oomycetes are all placed in the *Chromista.* The boundaries of the remaining four kingdoms—*Plantae, Fungi, Animalia,* and *Protozoa*—have been adjusted to better define each kingdom and distinguish it from the others.

Higher-Level Classification of the *Eucarya*

In 2005, the International Society of Protistologists, in collaboration with parasitologists, mycologists, and phycologists (scientists who study fungi and algae, respectively), proposed a classification scheme based on morphological, biochemical, and molecular phylogenetic analyses of eucaryotic microorganisms (procaryotes were not considered; however, the domains *Bacteria* and *Archaea* are recognized). Six phylogenetically coherent clusters were established but not placed in hierarchical order

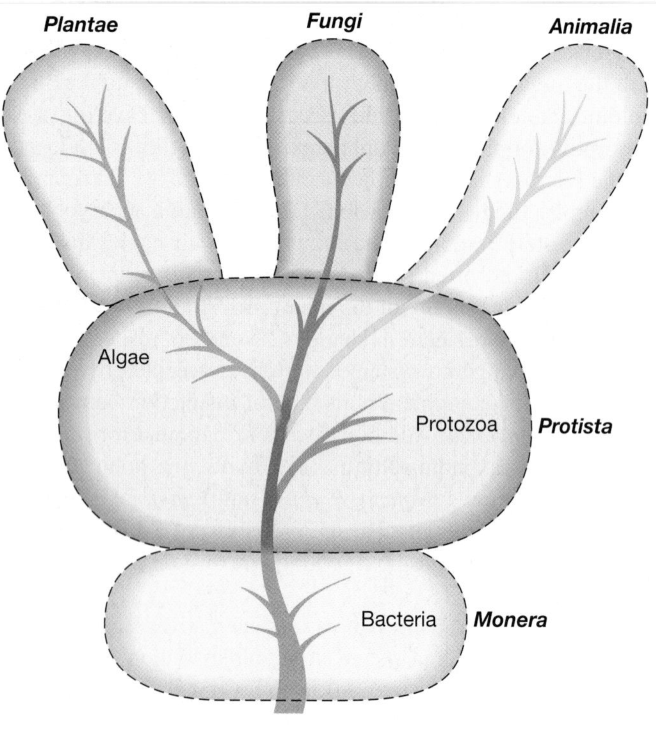

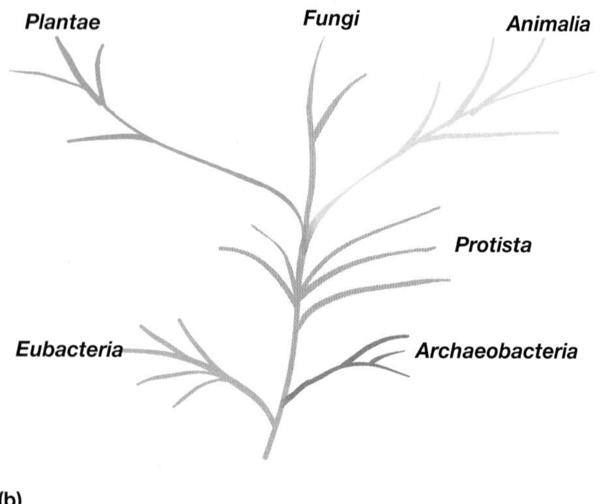

Figure 19.16 Systems of Eucaryotic and Procaryotic Phylogeny. Simplified schematic diagrams of the **(a)** five-kingdom system (Whittaker), **(b)** six-kingdom system, and **(c)** eight-kingdom system (Cavalier-Smith).

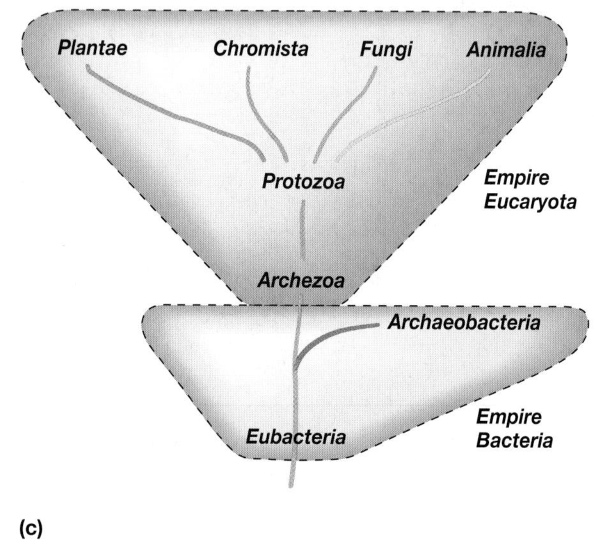

(**table 19.9**). Of specific interest, particularly when compared with the unrooted phylogenetic tree shown in figure 19.3, is the placement of animals and fungi in the same "super-group," the *Opisthokonta.* This implies that animals and fungi share a line of descent. Similarly, it is noted that higher plants (*Plantae*) arose from the green algae; both are placed in the super-group *Archaeplastida.* This classification scheme recognizes that kingdom-based classification schemes are not phylogenetically valid. For instance, to account for the molecular phylogenetic data, one would have to place kingdoms within kingdoms (e.g., the *Animalia* would have to be within the *Fungi* in Cavalier-Smith and Whittaker's schemes, and the *Plantae* would fall within the *Protista* according to Whittaker). This newest classification scheme is presented in further detail when protists and fungi are presented (chapters 25 and 26, respectively).

1. Describe the two major alternatives to Pace's universal phylogenetic tree (figure 19.3) that are depicted in figure 19.16*a–c*. Why have there been difficulties in developing an accurate tree? Discuss the effect of frequent horizontal gene transfer on phylogenetic trees.
2. With what three major criteria did Whittaker divide organisms into five kingdoms?

Table 19.9 Classification of the Eucarya as Proposed by the International Society of Protistologists

Super-Group	Unifying Features	First Rank	General Description
Opisthokonta		*Fungi* *Mesomycetozoa* *Metazoa* *Choanomonada*	Uni- or multicellular. During at least one stage of life cycle, cells have single posterior cilium without mastigonemes (hairlike projections on the flagella); possess kinetosomes (basal bodies) or centrioles; when unicellular, mitochondria have flat cristae. Includes yeast, fungi, and *Animalia*.
Archaeplastids		*Glaucophyta* *Rhodophyceae* *Chloroplastida*	Ancestral cyanobacterial endosymbiont has given rise to photosynthetic plastid with chlorophyll *a,* plastid later lost in some; uses starch as a storage product; usually with cell wall made of cellulose; mitochondria with flattened cristae. Includes the *Charophyta* (green algae) and higher plants.
Amoebozoa		*Tubulinea* *Flabellinea* *Stereomyxida* *Acanthamoebidae* *Entamoebida* *Mastigamoebidae* *Pelomyxa* *Eumycetozoa*	Amoeboid motility usually based on actino-myosin cytoskeleton with rounded pseudopodia (lobopodia); cells either naked or testate (having a shell); mitochondria usually have tubular cristae; usually uninucleate but sometimes multinucleate; cysts common; some lack mitochondria, peroxisomes, and hydrogenosomes (e.g., *Entamoeba*). Includes cellular and acellular slime molds.
Rhizaria		*Cercozoa* *Haplosporidia* *Foraminifera* *Gromia* *Radiolaria*	Possess thin pseudopodia (filopodia) that can be simple, branching, or supported by microtubules (axopodia); biciliate or amoeboid. Include plasmodial endoparasites of marine and freshwater animals.
Chromalveolata		*Cryptophyceae* *Haptophyta* *Stramenopiles* *Alveolata*	Auto-, mixo-, and heterotrophic forms; some bear ejectosomes (trichocysts—dartlike structures used for defense); plastid from secondary endosymbiosis with an ancestral archaeplastid; plastid then lost or reduced in some, or re-acquired in others. Includes diatoms, dinoflagellates, coccoliths, *Apixcomplexa* (e.g., *Plasmodium*), seaweeds/kelps, and *Ciliophora* (e.g., *Paramecium, Stentor*).
Excavata		*Fornicata* *Malawimonas* *Parabasalia* *Preaxostyla* *Jakobida* *Heterolobosea* *Euglenozoa*	Suspension-feeding groove (cytostome) present or thought to have been lost in some; feed by capturing particles from a flagellar-generated current. Includes *Giardia, Trichomonas,* and *Euglena*.

3. Briefly describe the six- and eight-kingdom systems. How do they differ from the five-kingdom system? What are the advantages and disadvantages of organismal trees?
4. How does the higher-level classification scheme differ from the kingdom-based approaches?

19.7 Bergey's Manual of Systematic Bacteriology

In 1923, David Bergey, professor of bacteriology at the University of Pennsylvania, and four colleagues published a classification of bacteria that could be used for identification of bacterial species, the *Bergey's Manual of Determinative Bacteriology.* This manual is now in its ninth edition. In 1984, the first edition of *Bergey's Manual of Systematic Bacteriology* was published. It contained descriptions of all procaryotic species then identified (**Microbial Diversity & Ecology 19.1**). The second edition will consist of five volumes; the first volume was published in 2001 and the second in 2005. Three additional volumes are due in 2007.

There has been enormous progress in procaryotic taxonomy since the first volume of *Bergey's Manual of Systematic Bacteriology* was published. In particular, the sequencing of rRNA, DNA, and proteins has made phylogenetic analysis of procaryotes feasible. Thus while microbial classification in the first edition was phenetic (based on phenotypic characterization), the second edition of *Bergey's Manual* is largely phylogenetic. Although gram-staining properties are generally considered phenetic characteristics, they also play a role in the phylogenetic

Microbial Diversity & Ecology

19.1 "Official" Nomenclature Lists—A Letter from Bergey's*

On a number of occasions lately, the impression has been given that the status of a bacterial taxon in *Bergey's Manual of Systematic Bacteriology* or *Bergey's Manual of Determinative Bacteriology* is in some sense official. Similar impressions are frequently given about the status of names in the *Approved List of Bacterial Names* and in the Validation Lists of newly proposed names that appear regularly in the *International Journal of Systematic Bacteriology*. It is therefore important to clarify these matters.

There is no such thing as an official classification. *Bergey's Manual* is not "official"—it is merely the best consensus at the time, and although great care has always been taken to obtain a sound and balanced view, there are also always regions in which data are lacking or confusing, resulting in differing opinions and taxonomic instability. When *Bergey's Manual* disavows that it is an official classification, many bacteriologists may feel that the solid earth is trembling. But many areas are in fact reasonably well established. Yet taxonomy is partly a matter of judgment and opinion, as is all science, and until new information is available, different bacteriologists may legitimately hold different views. They cannot be forced to agree to any "official classification." It must be remembered that, as yet, we know only a small percentage of the bacterial species in nature. Advances in technique also reveal new lights on bacterial relationships. Thus we must expect that existing boundaries of groups will have to be redrawn in the future, and it is expected that molecular biology, in particular, will imply a good deal of change over the next few decades.

The position with the Approved Lists and the Validation Lists is rather similar. When bacteriologists agreed to make a new start in bacteriological nomenclature, they were faced with tens of thousands of names in the literature of the past. The great majority were useless, because, except for about 2,500 names, it was impossible to tell exactly what bacteria they referred to. These 2,500 were therefore retained in the Approved Lists. The names are only approved in the sense that they were approved for retention in the new bacteriological nomenclature. The remainder lost standing in the nomenclature, which means they do not have to be considered when proposing new bacterial names (although names can be individually revived for good cause under special provisions).

The new International Code of Nomenclature of Bacteria requires all new names to be validly published to gain standing in the nomenclature, either by being published in papers in the *International Journal of Systematic Bacteriology* or, if published elsewhere, by being announced in the Validation Lists. The names in the Validation Lists are therefore valid only in the sense of being validly published (and therefore they must be taken account of in bacterial nomenclature). The names do not have to be adopted in all circumstances; if users believe the scientific case for the new taxa and validly published names is not strong enough, they need not adopt the names. For example, *Helicobacter pylori* was immediately accepted as a replacement for *Campylobacter pylori* by the scientific community, whereas *Tatlockia micdadei* had not generally been accepted as a replacement for *Legionella micdadei*. Taxonomy remains a matter of scientific judgment and general agreement.

From P. H. A. Sneath and D. J. Brenner, "Official Nomenclature Lists in ASM News, 58(4):175, 1992. Copyright © by the American Society for Microbiology. Reprinted by permission.

classification of microbes. Some of the major differences between gram-negative, gram-positive bacteria, and mycoplasmas (bacteria lacking a cell wall) are summarized in **table 19.10.**

In this section, the general features of the 2nd edition of *Bergey's Manual* are described. This edition has more ecological information about individual taxa. It does not group all the clinically important procaryotes together as the first edition did. Instead, pathogenic species are placed phylogenetically and thus scattered throughout the following five volumes.

- Volume 1—*The Archaea, and the Deeply Branching and Phototrophic Bacteria*
- Volume 2—*The Proteobacteria*
- Volume 3—*The Low G + C Gram-Positive Bacteria*
- Volume 4—*The High G + C Gram-Positive Bacteria*
- Volume 5—*The Planctomycetes, Spirochaetes, Fibrobacteres, Bacteroidetes, Fusobacteria, Chlamydiae, Acidobacteria, Verrumicrobia,* and *Dictyoglomus* (Volume 5 also will contain a section that updates descriptions and phylogenetic arrangements that have been revised since publication of volume 1.)

Table 19.11 summarizes the planned organization of the second edition and indicates where the discussion of a particular group may be found in this textbook.

19.8 A Survey of Procaryotic Phylogeny and Diversity

Before beginning a detailed introduction to procaryotic diversity, we will briefly survey the major groups as presented in the second edition of *Bergey's Manual*. This overview is meant only as a general survey of procaryotic diversity. The second edition places procaryotes into 25 phyla, only some of which will be mentioned here. Many of these groups are discussed in much more detail in chapters 20 through 24.

Volume 1 contains a wide diversity of procaryotes in the domains *Archaea* and *Bacteria*. At present the *Archaea* are divided into two phyla based on rRNA sequences (**figure 19.17**). The

Table 19.10 Some Characteristic Differences between Gram-Negative and Gram-Positive Bacteria

Property	Gram-Negative Bacteria	Gram-Positive Bacteria	Mycoplasmas
Cell wall	Gram-negative type wall with inner 2–7 nm peptidoglycan layer and outer membrane (7–8 nm thick) of lipid, protein, and lipopolysaccharide. (There may be a third outermost layer of protein.)	Gram-positive type wall with a homogeneous, thick cell wall (20–80 nm) composed mainly of peptidoglycan. Other polysaccharides and teichoic acids may be present.	Lack a cell wall and peptidoglycan precursors; enclosed by a plasma membrane
Cell shape	Spheres, ovals, straight or curved rods, helices or filaments; some have sheaths or capsules.	Spheres, rods, or filaments; may show true branching	Pleomorphic in shape; may be filamentous, can form branches
Reproduction	Binary fission, sometimes budding	Binary fission, filamentous forms grow by tip extension	Budding, fragmentation, and/or binary fission
Metabolism	Phototrophic, chemolithoautotrophic, or chemoorganoheterotrophic	Usually chemoorganoheterotrophic, a few phototrophic	Chemoorganoheterotrophic; most require cholesterol and long-chain fatty acids for growth.
Motility	Motile or nonmotile. Flagella placement can be varied—polar, lophotrichous, peritrichous. Motility may also result from the use of axial filaments (spirochetes) or gliding motility.	Most often nonmotile; have peritrichous flagella when motile	Usually nonmotile
Appendages	Can produce several types of appendages—pili and fimbriae, prosthecae, stalks	Usually lack appendages (may have spores on hyphae)	Lack appendages
Endospores	Cannot form endospores	Some groups	Cannot form endospores

Table 19.11 Organization of *Bergey's Manual of Systematic Bacteriology*

Taxonomic Rank	Representative Genera	Textbook Coverage
Volume 1. *The Archaea and the Deeply Branching and Phototrophic Bacteria*		
Domain *Archaea*		
Phylum *Crenarchaeota*		
Class I. *Thermoprotei*	*Thermoproteus, Pyrodictium, Sulfolobus*	pp. 507–8
Phylum *Euryarchaeota*		
Class I. *Methanobacteria*	*Methanobacterium*	pp. 508–13
Class II. *Methanococci*	*Methanococcus*	
Class III. *Methanomicrobia*	*Methanomicrobium*	
Class IV. *Halobacteria*	*Halobacterium, Halococcus*	pp. 514–16
Class V. *Thermoplasmata*	*Thermoplasma, Picrophilus, Ferroplasma*	pp. 516–17
Class VI. *Thermococci*	*Thermococcus, Pyrococcus*	p. 517
Class VII. *Archaeoglobi*	*Archaeoglobus*	p. 517
Class VIII. *Methanopyri*	*Methanopyrus*	pp. 510–12
Domain *Bacteria*		
Phylum *Aquificae*	*Aquifex, Hydrogenobacter*	p. 519
Phylum *Thermotogae*	*Thermotoga, Geotoga*	p. 520
Phylum *Thermodesulfobacteria*	*Thermodesulfobacterium*	
Phylum *Deinococcus-Thermus*	*Deinococcus, Thermus*	p. 520

(Continued)

Table 19.11 Organization of *Bergey's Manual of Systematic Bacteriology*, (Continued)

Domain *Bacteria, (Continued)*		
Phylum *Chrysiogenetes*	*Chrysogenes*	
Phylum *Chloroflexi*	*Chloroflexus, Herpetosiphon*	p. 523
Phylum *Thermomicrobia*	*Thermomicrobium*	
Phylum *Nitrospira*	*Nitrospira*	
Phylum *Deferribacteres*	*Geovibrio*	
Phylum *Cyanobacteria*	*Prochloron, Synechococcus, Pleurocapsa, Oscillatoria, Anabaena, Nostoc, Stigonema*	pp. 524–29
Phylum *Chlorobi*	*Chlorobium, Pelodictyon*	p. 523
Volume 2. *The Proteobacteria*		
Phylum *Proteobacteria*		
Class I. *Alphaproteobacteria*	*Rhodospirillum, Rickettsia, Caulobacter, Rhizobium, Brucella, Nitrobacter, Methylobacterium, Beijerinckia, Hyphomicrobium*	pp. 540–46
Class II. *Betaproteobacteria*	*Neisseria, Burkholderia, Alcaligenes, Comamonas, Nitrosomonas, Methylophilus, Thiobacillus*	pp. 546–51
Class III. *Gammaproteobacteria*	*Chromatium, Leucothrix, Legionella, Pseudomonas, Azotobacter, Vibrio, Escherichia, Klebsiella, Proteus, Salmonella, Shigella, Yersinia, Haemophilus*	pp. 551–61
Class IV. *Deltaproteobacteria*	*Desulfovibrio, Bdellovibrio, Myxococcus, Polyangium*	pp. 562–67
Class V. *Epsilonproteobacteria*	*Campylobacter, Helicobacter*	pp. 567–68
Volume 3. *The Low G + C Gram-Positive Bacteria*		
Phylum *Firmicutes*		
Class I. *Clostridia*	*Clostridium, Peptostreptococcus, Eubacterium, Desulfotomaculum, Heliobacterium, Veillonella*	pp. 576–78
Class II. *Mollicutes*	*Mycoplasma, Ureaplasma, Spiroplasma, Acholeplasma*	pp. 571–72
Class III. *Bacilli*	*Bacillus, Caryophanon, Paenibacillus, Thermoactinomyces, Lactobacillus, Streptococcus, Enterococcus, Listeria, Leuconostoc, Staphylococcus*	pp. 578–86
Volume 4. *The High G + C Gram-Positive Bacteria*		
Phylum *Actinobacteria*		
Class *Actinobacteria*	*Actinomyces, Micrococcus, Arthrobacter, Corynebacterium, Mycobacterium, Nocardia, Actinoplanes, Propionibacterium, Streptomyces, Thermomonospora, Frankia, Actinomadura, Bifidobacterium*	pp. 589–602
Volume 5. *The Planctomycetes, Spirochaetes, Fibrobacteres, Bacteriodetes, and Fusobacteria*		
Phylum *Planctomycetes*	*Planctomyces, Gemmata*	pp. 530–31
Phylum *Chlamydiae*	*Chlamydia*	pp. 531–32
Phylum *Spirochaetes*	*Spirochaeta, Borrelia, Treponema, Leptospira*	pp. 532–34
Phylum *Fibrobacteres*	*Fibrobacter*	
Phylum *Acidobacteria*	*Acidobacterium*	
Phylum *Bacteroidetes*	*Bacteroides, Porphyromonas, Prevotella, Flavobacterium, Sphingobacterium, Flexibacter, Cytophaga*	pp. 534–36
Phylum *Fusobacteria*	*Fusobacterium, Streptobacillus*	
Phylum *Verrucomicrobia*	*Verrucomicrobium*	
Phylum *Dictyoglomi*	*Dictyoglomus*	
Phylum *Gemmatimonadetes*	*Gemmatimonas*	

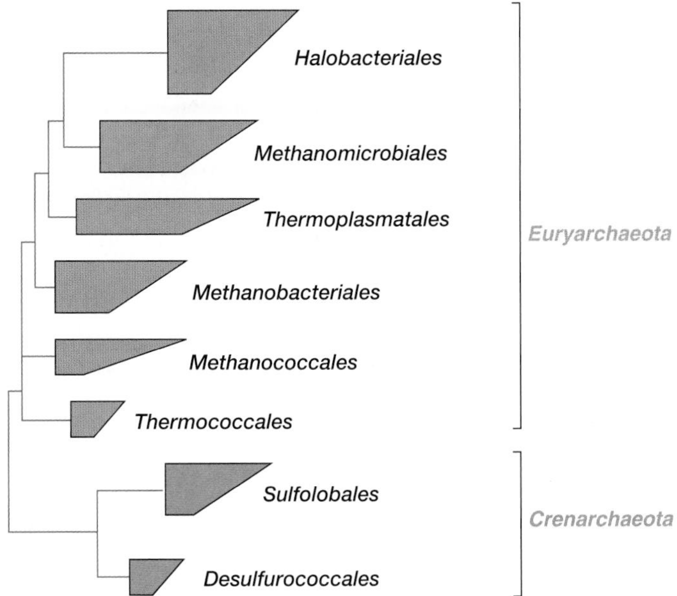

Figure 19.17 Phylogeny of the *Archaea*. The tree is based on 16S rRNA data and shows relationships between the better-studied orders. Each tetrahedron represents a group of related organisms; its horizontal edges indicate the shortest and longest branches in the group.

phylum *Crenarchaeota* contains thermophilic and hyperthermophilic sulfur-metabolizing organisms of the orders *Thermoproteales, Desulfurococcales,* and *Sulfolobales.* However, recently many other *Crenarchaeota* have been discovered. Some are inhibited by sulfur; others grow throughout the world's oceans as plankton. The phylum clearly is more diverse than first thought. The second phylum, the *Euryarchaeota,* contains primarily methanogenic procaryotes and halophilic procaryotes; thermophilic, sulfur-reducing organisms (the thermoplasmas and thermococci) also are in this phylum. The two phyla are divided into eight classes and 12 orders.

The *Bacteria* are an extraordinarily diverse assemblage of procaryotes that have been divided into 23 phyla (**figure 19.18**). Volume 1 covers deeply branching bacterial groups and all phototrophic bacteria except photosynthetic proteobacteria. The more important phyla are described in these sections.

1. Phylum *Aquificae.* The phylum *Aquificae* contains autotrophic bacteria such as *Aquifex* and *Hydrogenobacter* that can use hydrogen for energy production. *Aquifex* (meaning "water maker") actually produces water by using hydrogen to reduce oxygen. This group contains some of the most thermophilic bacteria known and is the deepest or earliest branch of the *Bacteria*.
2. Phylum *Thermotogae.* This phylum is composed of one class and six genera. *Thermotoga* and other members of the class *Thermotogae* are anaerobic, thermophilic, fermentative, gram-negative bacteria that have unusual fatty acids and resemble *Aquifex* with respect to their ether-linked lipids.
3. Phylum *Deinococcus-Thermus.* The order *Deinococcales* contains bacteria that are extraordinarily radiation resistant. The genus *Deinococcus* stains gram positive. It has unique lipids and a high concentration of carotenoid pigments, which may protect it from radiation.
4. Phylum *Chloroflexi.* The phylum *Chloroflexi* has one class and two orders. Many members of this gram-negative group are called green nonsulfur bacteria. *Chloroflexus* carries out anoxygenic photosynthesis and is a gliding bacterium; in contrast, *Herpetosiphon* is a nonphotosynthetic, respiratory gliding bacterium. Both genera have unusual peptidoglycans and lack lipopolysaccharides in their outer membranes.
5. Phylum *Cyanobacteria.* The oxygenic photosynthetic bacteria are placed in the phylum *Cyanobacteria,* which contains the class *Cyanobacteria* and five subsections based on morphology and life cycle. Cyanobacteria have chlorophyll *a* and almost all species possess phycobilins. These bacteria can be unicellular or filamentous, either branched or unbranched. Cyanobacteria incorporate CO_2 photosynthetically through use of the Calvin cycle just like plants and many purple photosynthetic bacteria.
6. Phylum *Chlorobi.* The phylum *Chlorobi* contains anoxygenic photosynthetic bacteria known as the green sulfur bacteria. They can incorporate CO_2 through the reductive tricarboxylic acid cycle rather than the Calvin cycle and oxidize sulfide to sulfur granules, which accumulate outside the cell.

Volume 2 is devoted completely to the gram-negative proteobacteria. The phylum *Proteobacteria* is a large and extremely complex group that currently contains over 2,000 species in 538 genera. Even though they are all related, the group is quite diverse in morphology, physiology, and life-style. All major nutritional types are represented: phototrophy, heterotrophy, and chemolithotrophy of several varieties. Many species are important in medicine, industry, and biological research. Prominent examples are the genera *Escherichia, Neisseria, Pseudomonas, Rhizobium, Rickettsia, Salmonella,* and *Vibrio.* The phylum is divided into five classes based on rRNA data. Because photosynthetic bacteria are found in the α, β, and γ classes of the proteobacteria, many believe that the whole phylum arose from a photosynthetic ancestor. Presumably many strains lost photosynthesis when adapting metabolically to new ecological niches.

1. Class I—*Alphaproteobacteria.* The α-proteobacteria include most of the oligotrophic forms (those capable of growing at low nutrient levels). *Rhodospirillum* and other purple nonsulfur bacteria are photosynthetic. Some genera have unusual metabolic modes: methylotrophy (e.g., *Methylobacterium*), chemolithotrophy (*Nitrobacter*), and nitrogen fixation (*Rhizobium*). *Rickettsia* and *Brucella* are important pathogens. About half of the microbes in this group have distinctive morphology such as prosthecae (*Caulobacter, Hyphomicrobium*).

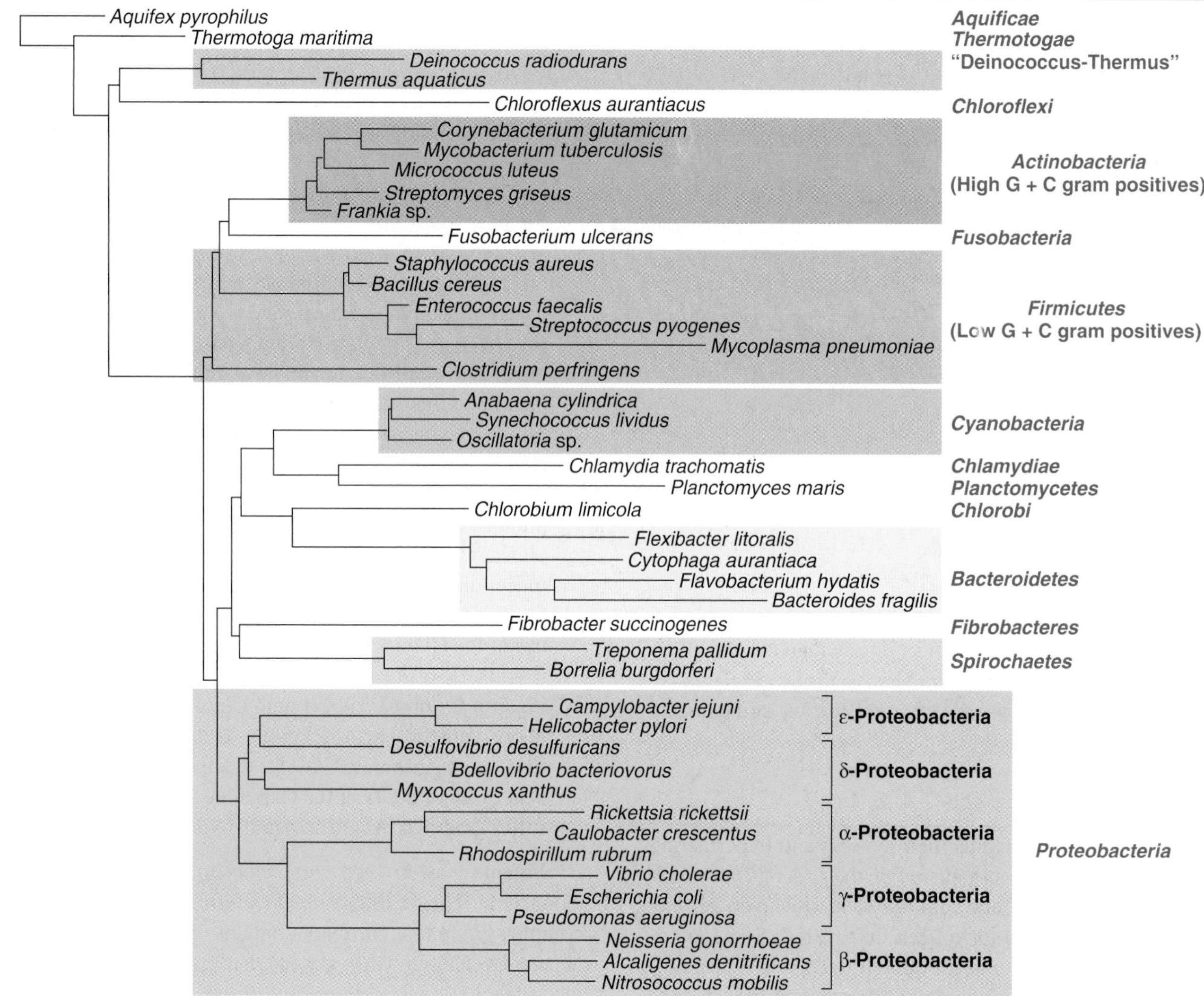

Figure 19.18 Phylogeny of the Bacteria. The tree is based on 16S rRNA comparisons. See text for discussion. *Source: The Ribosomal Database Project.*

2. Class II—*Betaproteobacteria.* The β-proteobacteria overlap the α subdivision metabolically. However, the β-proteobacteria tend to use substances that diffuse from organic decomposition in the anoxic zone of habitats. Some of these bacteria use such substances as hydrogen (*Alcaligenes*), ammonia (*Nitrosomonas*), methane (*Methylobacillus*), or volatile fatty acids (*Burkholderia*).
3. Class III—*Gammaproteobacteria.* The γ-proteobacteria compose a large and complex group of 14 orders and 28 families. Many are chemoorganotrophic, facultatively anaerobic, and fermentative. However, there is considerable diversity among the γ-proteobacteria with respect to energy metabolism. Some important families such as *Enterobacteriaceae, Vibrionaceae,* and *Pasteurellaceae* use the Embden-Meyerhof pathway and the pentose phosphate pathway. Others such as the *Pseudomonadaceae* and *Azotobacteriaceae* are aerobes and have the Entner-Doudoroff and pentose phosphate pathways. A few are photosynthetic (e.g., *Chromatium* and *Ectothiorhodospira*), methylotrophic (*Methylococcus*), or sulfur-oxidizing (*Beggiatoa*).
4. Class IV—*Deltaproteobacteria.* The δ-proteobacteria contain eight orders and 20 families. Many of these bacteria can be placed in one of three groups. Some are predators on other bacteria (e.g., *Bdellovibrio*). The order *Myxococcales* contains the fruiting myxobacteria such as *Myxococcus, Stigmatella,* and *Polyangium.* The myxobacteria often also prey on other bacteria. Finally, the class has a variety of anaerobes that generate sulfide from sulfate and sulfur while oxidizing organic nutrients (*Desulfovibrio*).
5. Class V—*Epsilonproteobacteria.* This class is composed of only one order, *Campylobacterales,* and three families. Despite its small size two important pathogenic genera are ϵ-proteobacteria: *Campylobacter* and *Helicobacter.*

Volume 3 of *Bergey's Manual* surveys the gram-positive bacteria with low G + C content in their DNA, which are members of the phylum *Firmicutes.* The dividing line is about 50% G + C; bacteria with a mol% lower than this value are in volume 3. Most of these bacteria stain gram positive and are heterotrophic. However, because of their close relationship to low G + C gram-positive bacteria, the mycoplasmas are placed here even though they lack cell walls and therefore stain gram negative. There is considerable variation in morphology: some are rods, others are cocci, and mycoplasmas are pleomorphic. Endospores may be present. The phylum contains three classes.

1. Class I—*Clostridia.* This class contains three orders and 11 families. Although they vary in morphology and size, the members tend to be anaerobic. Genera such as *Clostridium, Desulfotomaculum,* and *Sporohalobacter* form true bacterial endospores; many others do not. *Clostridium* is one of the largest bacterial genera.
2. Class II—*Mollicutes.* The class *Mollicutes* contains five orders and six families. Members of the class often are called mycoplasmas. These bacteria lack cell walls and cannot make peptidoglycan or its precursors. Because mycoplasmas are bounded only by the plasma membrane, they are pleomorphic and vary in shape from cocci to helical or branched filaments. They are normally nonmotile and stain gram negative because of the absence of a cell wall. In contrast with almost all other bacteria, most species require sterols for growth. The genera *Mycoplasma* and *Spiroplasma* contain several important animal and plant pathogens.
3. Class III—*Bacilli.* This large class comprises a wide variety of gram-positive, aerobic or facultatively anaerobic, rods and cocci. The class *Bacilli* has two orders, *Bacillales* and *Lactobacillales,* and 17 families. As with the members of the class *Clostridia,* some genera (e.g., *Bacillus, Sporosarcina, Paenibacillus,* and *Sporolactobacillus*) form true endospores. The class contains many medically and industrially important genera: *Bacillus, Lactobacillus, Streptococcus, Lactococcus, Enterococcus, Listeria,* and *Staphylococcus.*

Volume 4 is devoted to the high G + C gram positives, those bacteria with mol% values above 50 to 55%. All bacteria in this volume are placed in the phylum *Actinobacteria* and class *Actinobacteria.* There is enormous morphological variety among these procaryotes. Some are cocci, others are regular or irregular rods. High G + C gram positives called actinomycetes often form complex branching filaments called hyphae. Although none of these bacteria produce true endospores, many genera form asexual spores and some have complex life cycles. There is considerable variety in cell wall chemistry among the high G + C gram positives. For example, the composition of peptidoglycan varies greatly. Mycobacteria produce large mycolic acids that distinguish their cell walls from those of other bacteria.

The taxonomy of these bacteria is very complex. There are five subclasses, six orders, 14 suborders, and 44 families. Genera such as *Actinomyces, Arthrobacter, Corynebacterium, Micrococcus, Mycobacterium,* and *Propionibacterium* have recently been placed in the suborders *Actinomycineae, Micrococcineae, Corynebacterineae,* and *Propionibacterineae* because rRNA studies have shown them to be actinobacteria. The largest and most complex genus is *Streptomyces,* which contains about 150 species.

Volume 5 describes an assortment of ten phyla that are located here for convenience. The inclusion of these groups in volume 5 does not imply that they are directly related. Although they are all gram-negative bacteria, there is considerable variation in morphology, physiology, and life cycle pattern. Several genera are of considerable biological or medical importance. We briefly consider four of the 10 phyla.

1. Phylum *Planctomycetes.* The planctomycetes are related to the chlamydias according to their rRNA sequences. The phylum contains only one order, one family, and four genera. Planctomycetes are coccoid to ovoid or pear-shaped cells that lack peptidoglycan. Some have a membrane-enclosed nucleoid. Although they are normally unicellular, the genus *Isosphaera* will form chains. They divide by budding and may produce nonprosthecate appendages called stalks. Planctomycetes grow in aquatic habitats, and many move by flagella or gliding motility.
2. Phylum *Chlamydiae.* This small phylum contains one class, one order, and four families. The genus *Chlamydia* is by far the most important genus. *Chlamydia* are obligate intracellular parasites with a unique life cycle involving two distinctive stages: elementary bodies and reticulate bodies. These bacteria resemble planctomycetes in lacking peptidoglycan. They are small coccoid organisms with no appendages. Chlamydias are important pathogens and cause many human diseases.
3. Phylum *Spirochaetes.* This phylum contains helically shaped, motile, gram-negative bacteria characterized by a unique morphology and motility mechanism. The exterior boundary is a special outer membrane that surrounds the protoplasmic cylinder, which contains the cytoplasm and nucleoid. Periplasmic flagella lie between the protoplasmic cylinder and the outer membrane. The flagella rotate and move the cell even though they do not directly contact the external environment. These chemoheterotrophs can be free living, symbiotic, or parasitic. For example, the genera *Treponema* and *Borrelia* contain several important human pathogens. The phylum has one class, *Spirochaetes,* three families, and 13 genera.
4. Phylum *Bacteroidetes.* This phylum has three classes (*Bacteroides, Flavobacteria,* and *Sphingobacteria*), three orders, and 12 families. Some of the better-known genera are *Bacteroides, Flavobacterium, Flexibacter,* and *Cytophaga.* The gliding bacteria *Flexibacter* and *Cytophaga* are ecologically significant and are discussed later.

Because *Bergey's Manual* is the principal resource in procaryotic taxonomy used by microbiologists around the world, we follow *Bergey's Manual* in organizing the survey of procaryotic diversity, chapters 20 through 24. In so far as possible, the organization of the second edition of *Bergey's Manual* is employed. Chapter 20 is devoted to the *Archaea*. Chapter 21 covers the bacteria of volumes one and five except the *Archaea*. Chapter 22 is devoted to the proteobacteria. Chapters 23 and 24 deal with the low G + C and high G + C gram-positive bacteria, respectively. Chapter contents follow the overall phylogenetic scheme of *Bergey's Manual*. Phylogenetic and organizational details may well change somewhat before publication of each volume, but the general picture should adequately reflect the second edition.

Finally, it must be emphasized that procaryotic nomenclature is very much in flux. The names of families and genera are fairly well established and stable in the new system (at least in the absence of future discoveries); in fact, many family and genus names remain unchanged in the second edition of *Bergey's Manual*. In contrast, the names of orders and higher taxa are not always completely settled. Because the names of classes and orders are still changing, their use is kept to the minimum.

1. Briefly summarize the two phyla in the archaeal domain.
2. Give some ways in which the five classes of proteobacteria differ from each other.
3. In what phyla (and classes of *Proteobacteria* and *Firmicutes*) are the following placed: cyanobacteria, green nonsulfur bacteria, *Rickettsia*, the *Enterobacteriaceae*, *Campylobacter*, *Clostridium*, the mycoplasmas, *Bacillus*, *Streptomyces* and *Mycobacterium*, *Chlamydia*, *Treponema*, and *Cytophaga?*

Summary

19.1 Microbial Evolution

a. Precellular life may have been an "RNA world" because RNA has the capacity to both replicate and catalyze chemical reactions.

b. Living organisms can be divided into three domains: the *Eucarya*, the *Bacteria* and the *Archaea* (**table 19.1**).

c. The origin of eucaryotic cells is an unsettled question. The root of the universal phylogenetic tree suggests *Bacteria*, *Archaea*, and *Eucarya* have a single common ancestor but that the *Archaea* and *Eucarya* evolved independently of the *Bacteria* (**figure 19.3**).

d. The endosymbiotic theory asserts that mitochondria and chloroplasts evolved from an endosymbiotic α-proteobacterium and cyanobacterium, respectively. Hydrogenosomes and mitochondria are probably derived from a single, common ancestor (**figure 19.4**).

19.2 Introduction to Microbial Classification and Taxonomy

a. Taxonomy, the science of biological classification, is composed of three parts: classification, nomenclature, and identification.

b. A polyphasic approach is used to classify microbes. This incorporates information gleaned from genetic, phenotypic, and phylogenetic analysis.

c. Classifications may be constructed by means of numerical taxonomy, in which the general similarity of organisms is determined using computer software to calculate and analyze association coefficients (**figure 19.6**).

19.3 Taxonomic Ranks

a. Taxonomic ranks are arranged in a nonoverlapping hierarchy (**figure 19.7**).

b. The definition of species is different for sexually and asexually reproducing organisms. A procaryotic species is a collection of strains that have many stable properties in common and differ significantly from other groups of strains.

c. Microorganisms are named according to the binomial system.

19.4 Techniques for Determining Microbial Taxonomy and Phylogeny

a. The classical approach to determining microbial taxonomy and phylogeny includes the use of morphological, physiological, metabolic, ecological and genetic characteristics.

b. The study of transformation and conjugation in bacteria is sometimes taxonomically useful. Plasmid-borne traits can cause errors in bacterial taxonomy if care is not taken.

c. The G + C content of DNA is easily determined and taxonomically valuable because it is an indirect reflection of the base sequence (**table 19.6**).

d. Nucleic acid hybridization studies are used to compare DNA or RNA sequences and thus determine genetic relatedness (**figure 19.9**).

e. Nucleic acid sequencing is the most powerful and direct method for comparing genomes. The sequences of 16S and 18S rRNA are used most often in phylogenetic studies of procaryotic and eucaryotic microbes, respectively (**figure 19.10**). Complete microbial genomes are now being sequenced and compared.

f. Amino acid sequence of some proteins can be taxonomically and phylogenetically relevant, although the value of each protein must be assessed individually.

19.5 Assessing Microbial Phylogeny

a. Phylogenetic relationships often are shown in the form of branched diagrams called phylogenetic trees (**figure 19.13**). Trees may be either rooted or unrooted and are created in several different ways.

b. The sequences of rRNA, DNA, and proteins are used to produce phylogenetic trees. Often members of a group will have a unique characteristic rRNA sequence that distinguishes them from members of other taxonomic groups.

19.6 The Major Divisions of Life

a. Although most microbiologists favor the three-domain system, there are alternatives such as the five-, six-, and eight-kingdom systems (**figure 19.16**).

b. In 2005, the International Society of Protistologists proposed a higher-level classification scheme of the *Eucarya* that is phylogenetically based (**table 19.9**).

19.7 Bergey's Manual of Systematic Bacteriology

a. *Bergey's Manual of Systematic Bacteriology* gives the accepted system of procaryotic taxonomy.

b. The second edition of *Bergey's Manual* provides phylogenetic classifications. Procaryotes are divided between two domains and 25 phyla (**table 19.11,** and **figures 19.17** and **19.18**). Comparisons of nucleic acid sequences, particularly 16S rRNA sequences, are the foundation of this classification.

19.8 A Survey of Procaryotic Phylogeny and Diversity

a. The second edition of *Bergey's Manual* has five volumes. The general organization of the five volumes is summarized in table 19.10 and briefly outlined here.

(1) Volume 1: *The Archaea and the Deeply Branching and Phototrophic Bacteria.* This volume describes the *Archaea,* cyanobacteria, green sulfur and nonsulfur bacteria, deinococci, and other deeply branching groups.

(2) Volume 2: *The Proteobacteria.* All of the proteobacteria (purple bacteria) are placed in this volume and are divided into five major groups based on rRNA sequences and other characteristics: α-proteobacteria, β-proteobacteria, γ-proteobacteria, δ-proteobacteria, and ε-proteobacteria.

(3) Volume 3: *The Low G + C Gram-Positive Bacteria.* This volume contains gram-positive bacteria with G + C content below about 50%. Some of the major groups are the clostridia, bacilli, streptococci, and staphylococci. Mycoplasmas also are placed here.

(4) Volume 4: *The High G + C Gram-Positive Bacteria.* Gram-positive bacteria with G + C content above around 50 to 55% are in this volume. Such groups as *Corynebacterium, Mycobacterium, Nocardia,* and the actinomycetes are located here.

(5) Volume 5: *The Planctomycetes, Spirochaetes, Fibrobacteres, Bacteroidetes, and Fusobacteria.* Volume 5 has a variety of different gram-negative bacterial groups. The most practically important examples are the chlamydias and the spirochetes.

Key Terms

anagenesis 477
binomial system 480
biovar 480
dendrogram 479
endosymbiotic hypothesis 476
evolutionary distance 489
G + C content 483
genome fusion hypothesis 475
genomic fingerprinting 487
genotypic classification 478
genus 481
hydrogen hypothesis 476
hydrogenosome 476
identification 478
Jaccard coefficient (S_J) 479
macroevolution 477
melting temperature (T_m) 483
microevolution 477
molecular chronometers 488
morphovar 480
multilocus sequence typing (MLST) 486
natural classification 478
nomenclature 478
nucleic acid hybridization 483
numerical taxonomy 479
oligonucleotide signature sequences 485
parsimony analysis 489
phenetic system 478
phenons 479
phylogenetic or phyletic classification systems 478
phylogenetic tree 489
phylogeny 478
polyphasic taxonomy 478
procaryotic species 480
protists 491
punctuated equilibria 477
ribozymes 472
RNA world 472
serial endosymbiotic theory (SET) 477
serovar 480
similarity matrix 479
simple matching coefficient (S_{SM}) 479
small subunit ribosomal RNA (SSU rRNA) 474
species 480
strain 480
stromatolites 473
systematics 478
taxon 478
taxonomy 478
type strain 480
universal phylogenetic tree 475

Critical Thinking Questions

1. What experiments could be designed in a modern microbiology and/or chemistry lab to test the RNA world hypothesis?
2. Compare the findings of the universal phylogenetic tree and the genome fusion hypothesis. Debate the pros and cons of each.
3. Consider the fact that the use of 16S rRNA sequencing as a taxonomic and phylogenetic tool has resulted in tripling the number of procaryotic phyla. Why do you think the advent of this genetic technique has expanded the currently accepted number of microbial phyla?
4. Procaryotes were classified phenetically in the first edition of *Bergey's Manual of Systematic Bacteriology.* What do you think are the advantages and disadvantages of the phylogenetic classification used in the second edition?
5. Discuss the problems in developing an accurate phylogenetic tree. Is it possible to create a completely accurate universal phylogenetic tree?
6. Why is the current procaryotic classification system likely to change considerably? How would one select the best features to use in identification of unknown procaryotes and determination of relatedness?

Learn More

Adl, S. M.; Simpson, A. G. B.; Farmer, M. A.; Anderson, R. A.; Anderson, O. R.; Barta, J. R.; and Bowser, S. S., *et al.* 2005. The new higher level classification of Eukaryotes with emphasis on the taxonomy of protists. *J. Eukaryot. Microbiol.* 52:399–451.

Baquero, F. I.; Negri, M. C.; and Morosin, M. I. 1998. Selection of very small differences in bacterial evolution. *Internatl. Microbiol.* 1:295–300.

Boxma, B., et al. 2005. An anaerobic mitochondrion that produces hydrogen. *Nature* 434:74–79.

Ciccarelli, F. D.; Doerks, T.; von Mering, C.; Creevey, C. J.; Snel, B.; and Bork, P. 2006. Toward automatic reconstruction of a highly resolved tree of life. *Science.* 311:1283–1287.

Doolittle, W. F. 2000. Uprooting the tree of life. *Sci. Am.* 282(2):90–95.

Dyall, S. D.; Brown, M. T.; and Johnson, P. J. 2004. Ancient invasions: From endosymbionts to organelles. *Science* 304:253–57.

Garrity, G. M., editor-in-chief. 2001. *Bergey's manual of systematic bacteriology.* 2d ed., vol. 1, D. R. Boone and R. W. Castenholz, editors. New York: Springer-Verlag.

Garrity, G. M., editor-in-chief. 2005. *Bergey's manual of systematic bacteriology,* 2d ed., vol. 2, D. J. Brenner, N. R. Krieg, J. T. Staley, editors. New York: Springer-Verlag.

Gevers, D.; Cohan, F. M.; Lawrence, J. G.; Spratt, B. G.; Coenye, T.; Feil, E. J.; Stackenbrandt, E., et al. 2005. Re-evaluating prokaryotic species. *Nature Rev. Microbiol.* 3:733–39.

Hall, B. G. 2001. *Phylogenetic trees made easy: A how-to manual for molecular biologists.* Sunderland, Mass: Sinauer Associates.

Hrdy, I.; Hirt, R. P.; Dolezal, P.; Bardonová, L.; Foster, P. G.; Tachezy, J.; and Embley, T. M. 2004. *Trichomonas* hydrogenosomes contain the NADH dehydrogenase module of mitochondrial complex I. *Nature* 432:618–22.

Koch, A. L. 2003. Were Gram-positive rods the first bacteria? *Trends Microbial.* 11(4):166–70.

Martin, M., and Miklos, M. 1998. The hydrogen hypothesis for the first eukaryote. *Nature* 392:37–41.

Mayr, E. 1998. Two empires or three? *Proc. Natl. Acad. Sci.* 95:9720–23.

Pace, N. R. 1997. A molecular view of microbial diversity and the biosphere. *Science* 276:734–40.

Sapp, J. 2005. The prokaryote-eukaryote dichotomy: Meanings and mythology. *Microbiol. Mol. Bio. Rev.* 69:292–305.

Woese, C. R., and Fox, G. E. 1977. Phylogenetic structure of the prokaryotic domain: The primary kingdoms. *Proc. Natl. Acad. Sci. USA.* 74:5088–90.

Woese, C. R.; Kandler, O.; and Wheelis, M. L. 1990. Towards a natural system of organisms: Proposal for the domains *Archaea, Bacteria,* and *Eucarya. Proc. Natl. Acad. Sci.* 87:4576–79.

Please visit the Prescott website at www.mhhe.com/prescott7 for additional references.

20 The Archaea

Archaea are often found in extreme environments such as this geyser in Yellowstone National Park.

Preview

- The *Archaea* differ in many ways from both the *Bacteria* and the *Eucarya.* These include differences in cell wall structure and chemistry, membrane lipid structure, molecular biology, and metabolism.
- *Archaea* are best known for growing in a few restricted habitats (e.g., those that are hypersaline or high temperature). However, it is now evident that the *Archaea* are more widely distributed.
- The current edition of *Bergey's Manual* divides the *Archaea* into two phyla, the *Crenarchaeota* and *Euryarchaeota,* each with several orders.
- Many *Archaea* have special structural, chemical, and metabolic adaptations that enable them to grow in extreme environments.
- Methanogenic and sulfate-reducing archaea have unique cofactors that participate in methanogenesis.

In this chapter we begin with a general introduction to the *Archaea.* Then we briefly discuss the biology of each major archaeal group.

Chapters 20 through 24 survey the procaryotes described in *Bergey's Manual of Systematic Bacteriology.* Chapters 20 and 21 cover the material contained in volumes 1 and 5 of the second edition. Chapter 20 describes the *Archaea*; chapter 21 focuses on the bacterial groups in volumes 1 and 5. Chapter 22 covers the proteobacteria, which are located in volume 2. Volume 3 is devoted to the low G + C gram-positive bacteria, which we discuss in chapter 23. Finally, chapter 24 deals with the high G + C bacteria of volume 4.

Comparison of the sequences of rRNA from a great variety of organisms shows that organisms may be divided into three domains: *Bacteria, Archaea,* and *Eucarya (see figure 19.3).* Some of the most important features of these domains are summarized in table 19.1. Because the *Archaea* are different from both *Bacteria* and eucaryotes, their most distinctive properties are first described in some detail and compared with those of the latter two groups.

20.1 Introduction to the *Archaea*

The ***Archaea*** [Greek *archaios,* ancient] include microbes found in two phyla; the *Crenarchaeota,* and the *Euryarchaeota* (**figure 20.1**). Like the *Bacteria,* the *Archaea* are quite diverse, both in morphology and physiology. They can stain either gram positive or gram negative and may be spherical, rod-shaped, spiral, lobed, cuboidal, triangular, plate-shaped, irregularly shaped, or pleomorphic. Some are single cells, whereas others

As is often the case, epoch-making ideas carry with them implicit, unanalyzed assumptions that ultimately impede scientific progress until they are recognized for what they are. So it is with the prokaryote-eukaryote distinction. Our failure to understand its true nature set the stage for the sudden shattering of the concept when a "third form of life" was discovered in the late 1970s, a discovery that actually left many biologists incredulous. Archaebacteria, as this third form has come to be known, have revolutionized our notion of the prokaryote, have altered and refined the way in which we think about the relationship between prokaryotes and eukaryotes . . . and will influence strongly the view we develop of the ancestor that gave rise to all extant life.

—*C. R. Woese and R. S. Wolfe*

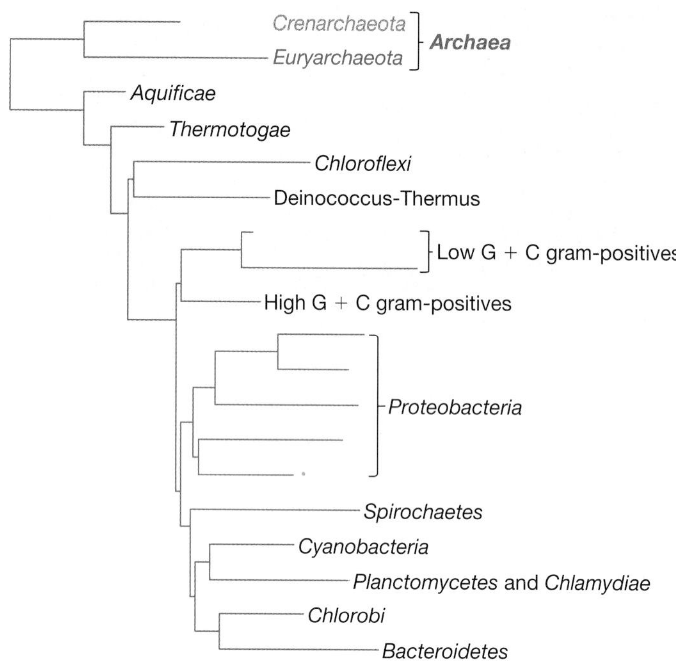

Figure 20.1 Phylogenetic Relationships Among Procaryotes. The *Archaea* are highlighted.

form filaments or aggregates. They range in diameter from 0.1 to over 15 μm, and some filaments can grow up to 200 μm in length. Multiplication may be by binary fission, budding, fragmentation, or other mechanisms. The *Archaea* are just as diverse physiologically. They can be aerobic, facultatively anaerobic, or strictly anaerobic. Nutritionally they range from chemolithoautotrophs to organotrophs. They include psychrophiles, mesophiles, and hyperthermophiles that can grow above 100°C.

Ecology

The types of environments where archaea have most often been found include areas with either very high or low temperatures or pH, concentrated salts, or completely anoxic. These are generally referred to as "extreme environments." However, terms such as extreme and hypersaline reflect a human perspective, meaning that they are situations where humans could not survive. On the contrary, most of the Earth (the oceans) is an "extreme environment" where it is very cold (about 4°C), dark, and under high pressure. Many archaea are well adapted to these environments, where they can grow to high numbers. For instance, archaea constitute at least 34% of the procaryotic biomass in at least some Antarctic coastal waters. In some hypersaline environments, their populations become so dense that the brine is red with archaeal pigments. Some archaea are symbionts in the digestive tracts of animals. Archaeal gene sequences have been found in soil and temperate and tropical ocean surface waters. Microorganisms in marine and freshwater environments (chapter 28)

Archaeal Cell Walls and Membranes

As discussed in chapter 3, archaea can stain either gram positive or gram negative, even though they lack the muramic acid and D-amino acids that make up peptidoglycan. Without the constraints of the conserved molecule peptidoglycan, archaeal cell walls can be quite diverse. For instance, some methanogenic archaea have **pseudomurein** (a peptidoglycan-like polymer that is cross-linked with L-amino acids), while others contain a complex polysaccharide similar to the chondroitin sulfate of animal connective tissue. Interestingly, some hyperthermophilic archaea and methanogens have protein walls. Archaeal cell walls (section 3.7)

One of the most distinctive archaeal features is their membrane lipids. As shown in table 19.1, the *Archaea* differ from both *Bacteria* and *Eucarya* in having branched chain hydrocarbons attached to glycerol by ether (rather than ester) linkages. Thermophilic archaea sometimes link two glycerol groups to form long tetraethers. Diether side chains are usually 20 carbons long, and tetraether chains contain 40 carbon atoms. However, cells can adjust chain lengths by cyclizing the chains to form pentacyclic rings. Such pentacyclic rings are used by thermophilic archaea to help maintain the delicate liquid crystalline balance of the membrane at high temperatures. Polar phospholipids, sulfolipids, and glycolipids are also found in archaeal membranes. Procaryotic cell membranes (section 3.2)

Genetics and Molecular Biology

Some features of archaeal genetics are similar to those in the *Bacteria,* while others more closely resemble the *Eucarya*. The genomes of some archaea are significantly smaller than those of many bacteria. For instance, while the genome of *Bacillus subtilis* is 4.20 million base pairs (Mb), the crenarchaeote *Pyrobaculum aerophilum* genome is 2.22 Mb and that of *Methanobacterium thermoautotrophicum,* a euryarchaeote, is 1.75 Mb. A sign of archaeal diversity is the variation in G + C content, from about 21% to 68%. To date, it appears that the *Archaea* have few plasmids.

Comparative genomics between the completely sequenced genomes of archaea, bacteria, and eucaryotes show several apparent trends. First, about 30% of all genes shared exclusively between archaea and eucaryotes encode proteins involved in transcription, translation, or DNA metabolism. In contrast, a large number of the genes shared only between *Bacteria* and *Archaea* are involved in metabolic pathways. In addition, there is evidence for horizontal gene transfer between these two domains, especially between thermophilic bacteria and archaea (*see figure 19.15*). The small number of genes found in all three domains does not seem to fit any specific pattern.

Archaeal DNA replication appears to be a complex mixture of eucaryotic and procaryotic features. Like *Bacteria,* most archaea have circular chromosomes with a single origin of replication, and replication appears to be bidirectional. However, in archaeal genomes that have been sequenced, the replication origin is flanked by genes encoding the eucaryotic-like initiation protein Cdc6/Orc1 and at least a few archaea have multiple origins. While it was originally thought that archaeal replication proteins were uniformly eucaryotic-like, further genome analysis reveals that some replication proteins are similar to those in *Bacteria,* while still others are uniquely archaeal. Some archaeal chromosomes differ from *Bacteria* in having eucaryotic-like histone proteins that bind DNA to form nucleosome-like structures.

Transcription in the *Archaea* likewise blends bacterial and eucaryotic features. Archaeal RNA polymerases consist of at least 10 subunits that are highly homologous to eucaryotic subunits. Also, like eucaryotic nuclear RNA polymerase, archaeal RNA polymerases do not efficiently recognize promoter regions without the aid of additional proteins. Instead, promoter recognition is dependent on at least two eucaryotic-like proteins: the *T*ATA-box-*b*inding *p*rotein (TBP) and *t*ranscription *f*actor *B* (TFB). It is therefore not surprising that many archaeal promoters are similar to certain eucaryotic promoters, possessing a TATA box (a 7-bp sequence found about 25 bp before the transcriptional start site) preceded by a purine-rich region called the *B r*esponsive *e*lement (BRE). In eucaryotes, the BRE is the site to which transcription factor IIB binds. It is thought that archaeal TFB and TBP bind the BRE region of DNA as a prerequisite for the assembly of RNA polymerase subunits prior to the initiation of transcription (**figure 20.2**). However, archaeal mRNA appears to be similar to bacterial mRNA in that it is polycistronic and there is no evidence for mRNA splicing. Transcription: Transcription in the *Archaea* (section 11.6)

Finally, the translational machinery in the *Archaea* is unique. Unlike both *Bacteria* and eucaryotes, the TψC arm of archaeal tRNA lacks thymine and contains pseudouridine or 1-methylpseudouridine. The archaeal initiator tRNA carries methionine as does the eucaryotic initiator tRNA. Although archaeal ribosomes are 70S, similar to bacterial ribosomes, electron microscopy studies show that their shape is quite variable and sometimes differs from that of both bacterial and eucaryotic ribosomes. They resemble eucaryotic ribosomes in their sensitivity to the antibiotic anisomycin and insensitivity to chloramphenicol and kanamycin. Furthermore, their elongation factor 2 reacts with diphtheria toxin like the eucaryotic EF-2 does. Translation (section 11.8)

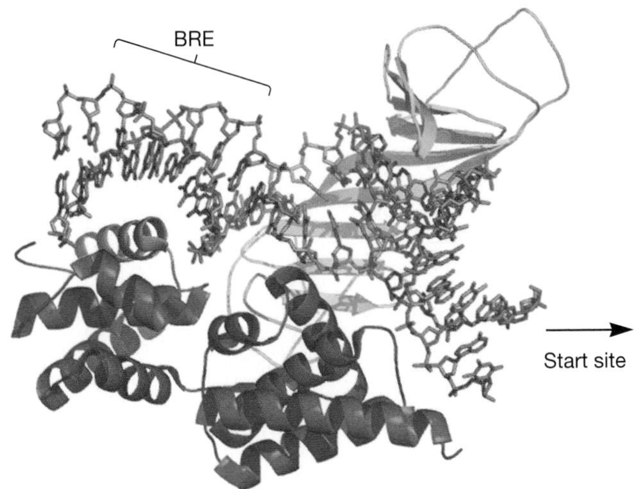

Figure 20.2 Archaeal Promoters Resemble Those of Eucaryotes. The crystal structure of the ternary complex between TBP, the carboxyl terminus of TFB, and a region of DNA containing a TATA box and BRE. DNA is shown in gray; TBP is the yellow ribbon structure; and TFB is magenta, with its recognition helix in turquoise.

Like archaeal protein synthesis, archaeal protein secretion has both bacterial and eucaryotic features. All three domains have signal recognition particles (SRPs) that target new proteins to translocation sites, but the archaeal SRP differs from those in the other two domains. As in the *Bacteria,* the archaeal SRP binds to the signal sequence of a preprotein and can direct it to the Sec-dependent protein secretion pathway for transport through the plasma membrane. The archaeal Sec-dependent pathway proteins, however, more closely resemble those of the eucaryotic pathway than the bacterial proteins. After the preprotein is moved across the membrane, its signal sequence is removed by a signal peptidase that resembles a subunit of the eucaryotic peptidase. Protein secretion in procaryotes (section 3.8)

Metabolism

Not surprisingly, in view of their variety of life-styles, archaeal metabolism varies greatly among the members of different groups. Some archaea are organotrophs; others are autotrophic. A few even carry out rhodopsin-based phototrophy.

Archaeal carbohydrate metabolism is best understood. The enzyme 6-phosphofructokinase has not been found in any archaea, and they do not appear to degrade glucose by way of the Embden-Meyerhof pathway. However, some hyperthermophiles appear to have a modified Embden-Meyerhof pathway that involves several novel enzymes, including an ADP-dependent phosphofructokinase. Extreme halophiles and thermophiles catabolize glucose using a modified form of the Entner-Doudoroff pathway in which the initial intermediates are not phosphorylated. All archaea that have been studied can oxidize pyruvate to acetyl-CoA. They lack the pyruvate dehydrogenase complex present in eucaryotes and respiratory bacteria and use the enzyme pyruvate oxidoreductase for this purpose. Halophiles and the extreme thermophile *Thermoplasma* seem to have a functional tricarboxylic acid cycle. Methanogens do not catabolyze glucose to any significant extent, and so it is not surprising that they lack a complete tricarboxylic acid cycle. Evidence for functional respiratory chains has been obtained in halophiles and thermophiles. The breakdown of glucose to pyruvate (section 9.3); The tricarboxylic acid cycle (section 9.4)

Very little is known in detail about biosynthetic pathways in the *Archaea.* Preliminary data suggest that the synthetic pathways for amino acids, purines, and pyrimidines are similar to those in other organisms. Some methanogens can fix atmospheric dinitrogen. Many archaea, including halophiles and methanogens, use a reversal of the Embden-Meyerhof pathway to synthesize glucose, and at least some methanogens and extreme thermophiles employ glycogen as their major reserve material. Synthesis of sugars and polysaccharides (section 10.4); Synthesis of amino acids (section 10.5)

Autotrophy is widespread among the methanogens and extreme thermophiles, and CO_2 fixation occurs in more than one way. *Thermoproteus* and possibly *Sulfolobus* incorporate CO_2 by the reductive tricarboxylic acid cycle (**figure 20.3*a***). This pathway is also present in the green sulfur bacteria. Methanogenic archaea and probably most extreme thermophiles incorporate CO_2 by the reductive acetyl-CoA pathway (figure 20.3*b*). A similar pathway also is present in acetogenic bacteria and autotrophic sulfate-reducing bacteria.

Archaeal Taxonomy

It should be clear by now that the *Archaea* are quite distinct from other living organisms. Within the domain, however, there is great diversity. As shown in **table 20.1,** the *Archaea* can be divided into five major groups based on physiological and morphological differences.

On the basis of phylogenetic evidence, *Bergey's Manual* divides the *Archaea* into the phyla *Euryarchaeota* [Greek *eurus,* wide, and Greek *archaios,* ancient or primitive] and *Crenarchaeota* [Greek *crene,* spring or fount, and *archaios*]. The euryarchaeotes are given this name because they occupy many different ecological niches and have a variety of metabolic patterns. The phylum *Euryarchaeota* is very diverse with eight classes (*Methanobacteria, Methanococci, Halobacteria, Thermoplasmata, Thermococci, Archaeglobi, Methanopyri,* and the recently added *Methanomicro-*

Figure 20.3 Mechanisms of Autotrophic CO_2 Fixation. **(a)** The reductive tricarboxylic acid cycle. The cycle is reversed with ATP and reducing equivalents [H] to form acetyl-CoA from CO_2. The acetyl-CoA may be carboxylated to yield pyruvate, which can then be converted to glucose and other compounds. This sequence appears to function in *Thermoproteus neutrophilus.* **(b)** The synthesis of acetyl-CoA and pyruvate from CO_2 in *Methanobacterium thermoautotrophicum.* One carbon comes from the reduction of CO_2 to a methyl group, and the second is produced by reducing CO_2 to carbon monoxide through the action of the enzyme CO dehydrogenase (E_1). The two carbons are then combined to form an acetyl group. Corrin-E_2 represents the cobamide-containing enzyme involved in methyl transfers. Special methanogen coenzymes and enzymes are described in figures 20.10 and 20.11.

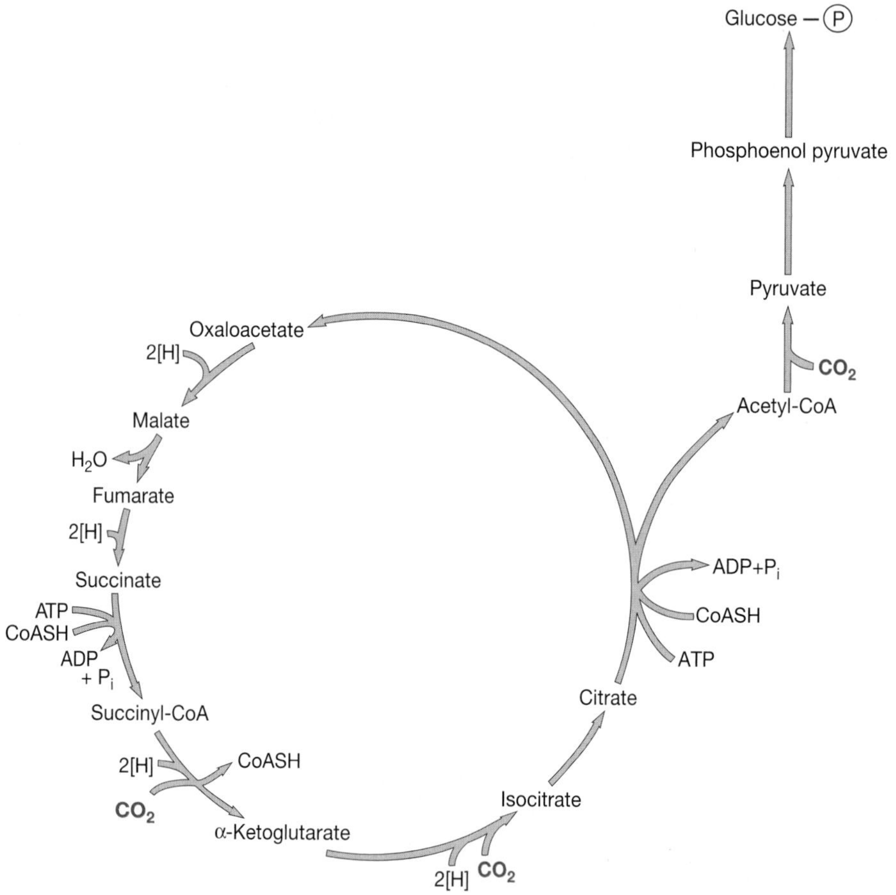

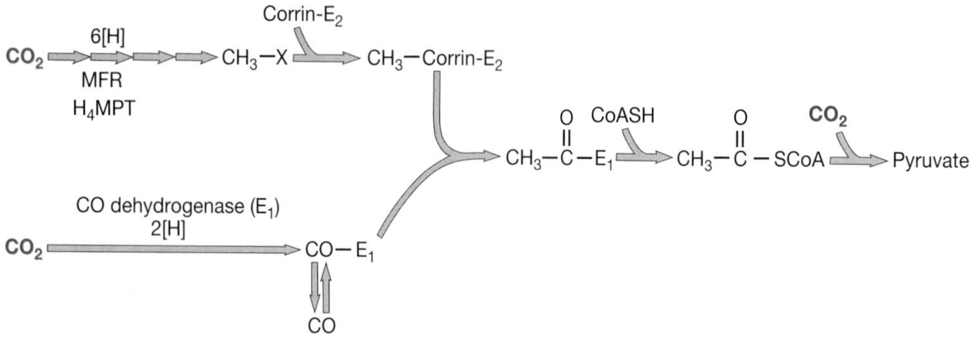

Table 20.1 Characteristics of the Major Archaeal Physiological Groups

Group	General Characteristics	Representative Genera
Methanogenic archaea	Strict anaerobes. Methane is the major metabolic end product. S^0 may be reduced to H_2S without yielding energy production. Cells possess coenzyme M, factors 420 and 430, and methanopterin.	*Methanobacterium* *Methanococcus* *Methanomicrobium* *Methanosarcina*
Archaeal sulfate reducers	Irregular gram-negative coccoid cells. H_2S formed from thiosulfate and sulfate. Autotrophic growth with thiosulfate and H_2. Can grow heterotrophically. Traces of methane also formed. Extremely thermophilic and strictly anaerobic. Possess factor 420 and methanopterin but not coenzyme M or factor 430.	*Archaeoglobus*
Extremely halophilic archaea	Rods, cocci, or irregular shaped cells, that may include pyramids or cubes. Stain gram negative or gram positive, but like all archaea lack peptidoglycan. Primarily chemoorganoheterotrophs. Most species require sodium chloride ≥1.5 M, but some survive in as little as 0.5 M. Most produce characteristic bright-red colonies; some are unpigmented. Neutrophilic to alkalophilic. Generally mesophilic; however, at least one species is known to grow at 55°C. Possess either bacteriorhodopsin or halorhodopsin and can use light energy to produce ATP.	*Halobacterium* *Halococcus* *Natronobacterium*
Cell wall-less archaea	Pleomorphic cells lacking a cell wall. Thermoacidophilic and chemoorganotrophic. Facultatively anaerobic. Plasma membrane contains a mannose-rich glycoprotein and a lipoglycan.	*Thermoplasma*
Extremely thermophilic S^0-metabolizers	Gram-negative rods, filaments, or cocci. Obligately thermophilic (optimum growth temperature between 70–110°C). Usually strict anaerobes but may be aerobic or facultative. Acidophilic or neutrophilic. Autotrophic or heterotrophic. Most are sulfur metabolizers. S^0 reduced to H_2S anaerobically; H_2S or S^0 oxidized to H_2SO_4 aerobically.	*Desulfurococcus* *Pyrodictium* *Pyrococcus* *Sulfolobus* *Thermococcus* *Thermoproteus*

bia), nine orders, and 16 families. The methanogens, extreme halophiles, sulfate reducers, and many extreme thermophiles with sulfur-dependent metabolism are located in the *Euryarchaeota*. Methanogens are the dominant physiological group.

The crenarchaeotes (**figure 20.4**) are thought to resemble the ancestor of the *Archaea*, and almost all the well-characterized species are thermophiles or hyperthermophiles. The phylum *Crenarchaeota* has only one class, *Thermoprotei*, which is divided into four orders and six families. The order *Thermoproteales* contains gram-negative-staining anaerobic to facultative, hyperthermophilic rods. They often grow chemolithoautotrophically by reducing sulfur to hydrogen sulfide. Members of the order *Sulfolobales* are coccus-shaped thermoacidophiles. The order *Desulfurococcales* contains gram-negative-staining coccoid or disk-shaped hyperthermophiles. They grow either chemolithotrophically by hydrogen oxidation or organotrophically by fermentation or respiration with sulfur as the electron acceptor. The order *Caldisphaerales* was recently added. It has only one genus, *Caldisphaera*, whose members are thermoacidophilic, aerobic, heterotrophic cocci. The taxonomy of both phyla will undoubtedly undergo further revisions as more organisms are discovered. This is particularly the case with the crenarchaeotes because of the discovery of mesophilic forms in the ocean; these crenarchaeotes may constitute a significant fraction of the oceanic picoplankton.

1. What are the *Archaea?* Briefly describe the major ways in which they differ from *Bacteria* and eucaryotes.
2. How do archaeal cell walls differ from those of the *Bacteria?* What is pseudomurein?
3. In what ways do archaeal membrane lipids differ from those of *Bacteria* and eucaryotes? How do these differences contribute to the survival of thermophilic and hyperthermophilic archaea?
4. List the differences between *Archaea* and other organisms with respect to DNA replication, transcription, and translation.
5. Briefly describe the way in which archaea degrade and synthesize glucose. In what two unusual ways do they incorporate CO_2?
6. How are the phyla *Euryarchaeota* and *Crenarchaeota* distinguished?

20.2 Phylum *Crenarchaeota*

As mentioned previously, most of the crenarchaeotes that have been cultured are extremely thermophilic, and many are acidophiles and sulfur dependent. The sulfur may be used either as an electron acceptor in anaerobic respiration or as an electron donor by lithotrophs. Many are strict anaerobes. They grow in geothermally heated water or soils that contain elemental sulfur. These environments are scattered all over the world. Examples are the

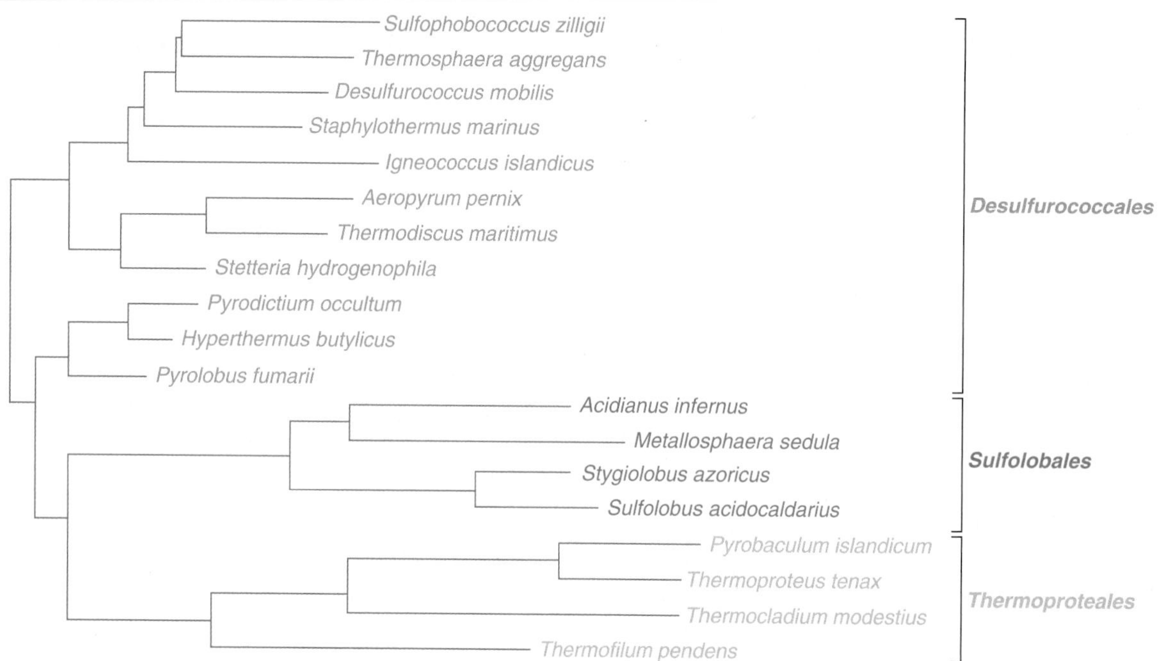

Figure 20.4 The Phylum *Crenarchaeota*. A phylogenetic tree developed with 16S rRNA data for crenarchaeotal-type species. Three orders are indicated.

sulfur-rich hot springs in Yellowstone National Park and the waters surrounding areas of submarine volcanic activity (**figure 20.5**). Such habitats are sometimes called solfatara. These archaea can be very thermophilic and often are classified as hyperthermophiles. The most extreme example was isolated from an active hydrothermal vent in the northeast Pacific Ocean. This is one of three novel isolates that constitute a new genus in the *Pyrodictiaceae* family. Its optimum growth rate is about 105°C, but even autoclaving this microbe at 121°C for one hour fails to kill it! It is strictly anaerobic, using Fe(III) as a terminal electron acceptor and H_2 or formate as electron donors (**figure 20.6**).

At present, the Crenarchaeota contains 25 genera; two of the better-studied genera are *Thermoproteus* and *Sulfolobus*. Members of the genus *Sulfolobus* stain gram negative, and are aerobic, irregularly lobed spherical archaea with a temperature optimum around 70 to 80°C and a pH optimum of 2 to 3 (**figure 20.7*a,b***). For this reason, they are **thermoacidophiles,** so called because they grow best at acid pH values and high temperatures. Their cell wall contains lipoprotein and carbohydrate. They grow lithotrophically on sulfur granules in hot acid springs and soils while oxidizing the sulfur to sulfuric acid (figures 20.5*b* and 20.7*b*). Oxygen is the normal electron acceptor, but ferric iron may be used. Sugars and amino acids such as glutamate also serve as carbon and energy sources.

Thermoproteus is a long, thin rod that can be bent or branched (figure 20.7*c*). Its cell wall is composed of glycoprotein. *Thermoproteus* is a strict anaerobe and grows at temperatures from 70 to 97°C and pH values between 2.5 and 6.5. It is found in hot springs and other hot aquatic habitats rich in sulfur. It can grow organotrophically and oxidize glucose, amino acids, alcohols, and organic acids with elemental sulfur as the electron acceptor. That is, *Thermoproteus* can carry out anaerobic respiration. It will also grow chemolithotrophically using H_2 and S^0. Carbon monoxide or CO_2 can serve as the sole carbon source.

Although the *Crenarchaeota* are notorious for their life at high temperatures and acidic pH, sequence analysis of DNA fragments derived directly from environmental samples reveals that this phylum is more widespread in nature. Recall that only a small fraction of microbes have been grown in culture, so the ability to analyze microbial communities using molecular techniques is an important way to truly understand microbial diversity. Such studies have revealed that the *Crenarchaeota* have significant populations in marine plankton from polar, temperate, and tropical waters. Crenarchaeotes also appear to inhabit rice paddies, soils, freshwater lake sediments, and at least two symbiotic species have been isolated, one from a cold water sea cucumber and another from a marine sponge. As more is learned about these microbes, our understanding of archaeal phylogeny will no doubt be enhanced and most likely modified (**Microbial Diversity & Ecology 20.1**). Techniques for determining microbial taxonomy and phylogeny: Molecular characteristics (section 19.4)

20.3 Phylum *Euryarchaeota*

The *Euryarchaeota* is a very diverse phylum with many genera (**figure 20.8**). Here, five major physiologic groups that comprise the euryarchaeotes are briefly discussed.

Figure 20.5 Habitats for Thermophilic Archaea. **(a)** The Pump Geyser in Yellowstone National Park. The orange color is due to the carotenoid pigments of thermophilic archaea. **(b)** The Sulfur Cauldron in Yellowstone National Park. The water is at its boiling point and very rich in sulfur. *Sulfolobus* grows well in such habitats.

(a)

(b)

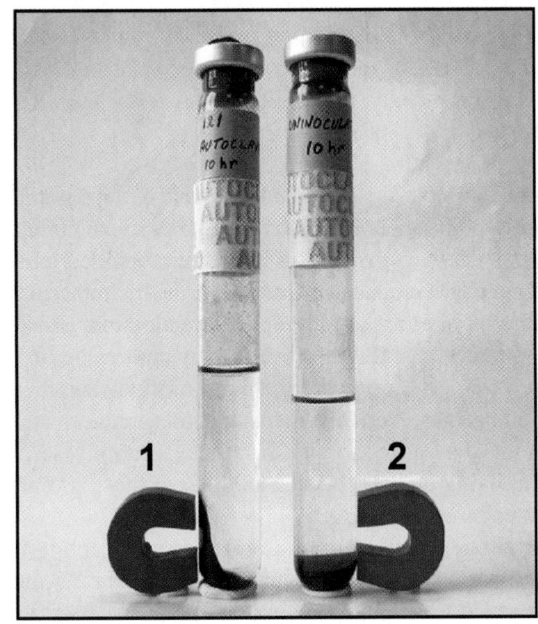

(a)

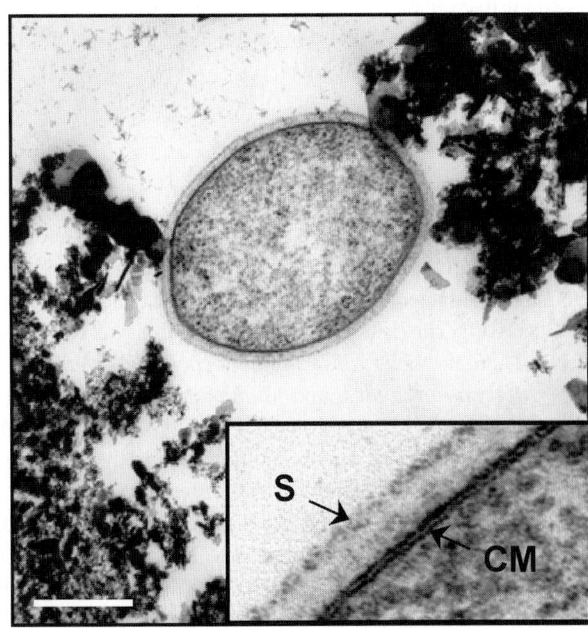

(b)

Figure 20.6 An Extremely Hyperthermophilic Crenarchaeote. **(a)** A member of the family *Pyrodictiaceae* grows following autoclaving at 121°C as shown by its ability to reduce Fe(III) to magnetite when incubated anaerobically. **(b)** A transmission electron micrograph shows the single layer cell envelope (S) and cytoplasmic membrane. Cell wall structure is one distinguishing feature of the *Archaea*. Scale bar = 1 μm.

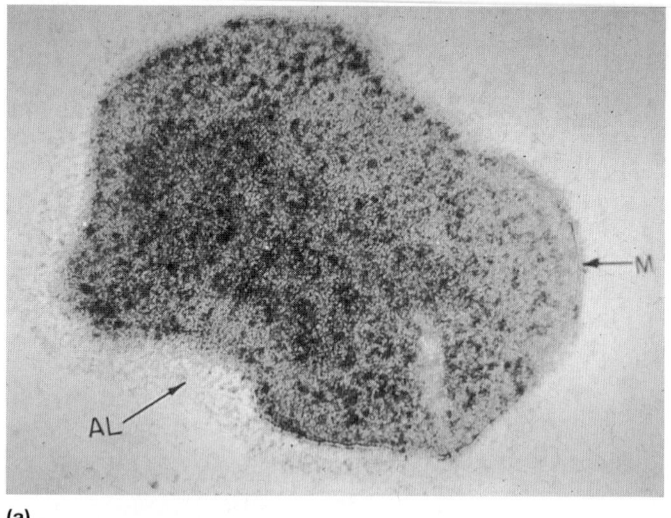

(a)

(b)

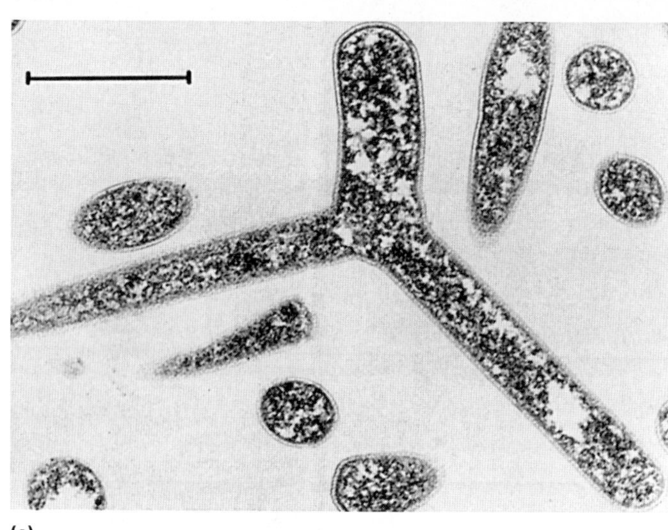

(c)

Figure 20.7 *Sulfolobus* and *Thermoproteus*. **(a)** A thin section of *Sulfolobus brierleyi*. The archaeon, about 1 μm in diameter, is surrounded by an amorphous layer (AL) instead of a well-defined cell wall; the plasma membrane (M) is also visible. **(b)** A scanning electron micrograph of a colony of *Sulfolobus* growing on the mineral molybdenite (MoS_2) at 60°C. At pH 1.5–3, the organism oxidizes the sulfide component of the mineral to sulfate and solubilizes molybdenum. **(c)** Electron micrograph of *Thermoproteus tenax*. Bar = 1 μm.

The Methanogens

Methanogens are strict anaerobes that obtain energy by converting CO_2, H_2, formate, methanol, acetate, and other compounds to either methane or methane and CO_2. They are autotrophic when growing on H_2 and CO_2. This is the largest group of archaea. There are five orders (*Methanobacteriales, Methanococcales, Methanomicrobiales, Methanosarcinales,* and *Methanopyrales*) and 26 genera, which differ greatly in overall shape, 16S rRNA sequence, cell wall chemistry and structure, membrane lipids, and other features. For example, methanogens construct three different types of cell walls. Several genera have walls with pseudomurein; other walls contain either proteins or heteropolysaccharides. The morphology of some methanogens is shown in **figure 20.9,** and selected properties of representative genera are presented in **table 20.2.** It should be noted that although almost all archaea in these orders are methanogens, methanotrophs (i.e., organisms that use methane as a carbon and energy source) have recently been discovered in the *Methanosarcinales* (**Microbial Diversity & Ecology 20.2**).

One of the most unusual methanogenic groups is the class *Methanopyri.* It has one order, *Methanopyrales,* one family and a single genus, *Methanopyrus.* This hyperthermophilic, rod-shaped methanogen has been isolated from a marine hydrothermal vent. *Methanopyrus kandleri* has a temperature minimum at 84°C and an optimum of 98°C; it will grow at temperatures up to 110°C. *Methanopyrus* occupies the deepest and most ancient branch of the euryarchaeotes. Perhaps methanogenic archaeal ancestors were among the earliest organisms. They certainly seem well adapted to living under conditions similar to those presumed to have existed on a young Earth.

As might be inferred from the methanogens' ability to produce methane anaerobically, their metabolism is unusual. These procaryotes contain several unique cofactors: tetrahydromethanopterin (H_4MPT), methanofuran (MFR), coenzyme M (2-mercaptoethanesulfonic acid), coenzyme F_{420}, and coenzyme F_{430} (**figure 20.10**). The first three cofactors bear the C_1 unit when CO_2 is reduced to CH_4. F_{420} carries electrons and protons, and F_{430}

Microbial Diversity & Ecology

20.1 Archaeal Phylogeny: More Than Just the *Crenarchaeota* and *Euryarchaeota?*

All known archaea currently in culture belong to either the *Crenarchaeota* or the *Euryarcheota*. However, an ever-growing collection of 16S rRNA nucleotide sequences cloned directly from the environment suggests that a third phylum of *Archaea* exists. Tentatively called the ***Korarchaeota*** (from the Greek word for "young man"), this phylum is gaining acceptance, although it is not clear if this designation will hold up to more complete analysis if and when any of these microbes are cultivated (**Box figure *a***).

Yet a fourth phylum was recently suggested by the discovery of the hyperthermophilic archaeon *Nanoarchaeum equitans* (**Box figure *b***). While investigating hyperthermophiles from submarine vents, scientists cultured a new hyperthermophilic crenarcheote belonging to the genus *Ingicoccus*. While examining this autotrophic, sulfur-reducing microbe, researchers noticed small cocci, about 0.4 μm in diameter, attached to its cell wall (**Box figure *c***). *N. equitans* has not been cultured axenically, so little is known about its physiology. However, because *N. equitans* and *Ignicoccus* vary in size, it is possible to collect isolated cells for genome analysis. The small size of the *N. equitans* genome (490,885 bp or 0.49 Mb) suggests that this microbe has lived in association with other organisms for a long time—long enough for it to have lost genes for lipid, nucleotide, amino acid, and enzyme cofactor biosynthesis. Not only has it eliminated genes, but its genome is very compact, with 95% of the DNA predicted to encode proteins or stable RNAs. The loss of essential biosynthetic genes and its limited catabolic capabilities indicates that *N. equitans* maintains a parasitic relationship with its host, making it the only known archaeal parasite.

So, how many archaeal phyla will eventually be accepted? No one knows for sure, but this state of flux in archaeal phylogeny demonstrates how dynamic and exciting taxonomy can be. The use of molecular probes to dissect microbial communities, combined with enhanced culture techniques, ensures that phylogenetic analysis will continue to evolve, just as the microbes do.

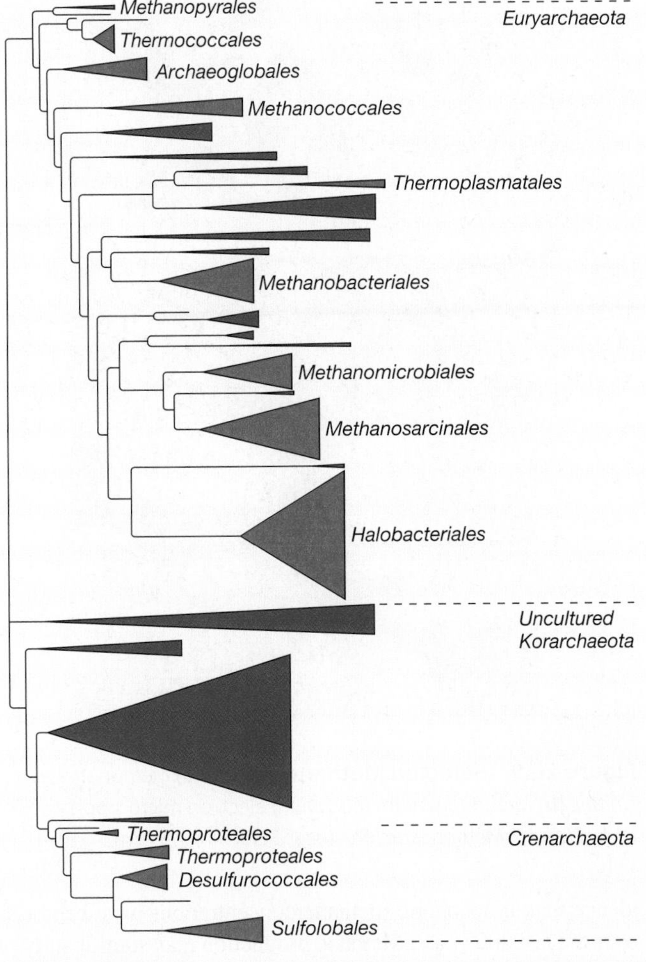

(a) Phylogeny of the Archaeal Domain. Based on 16S rRNA nucleotide sequence analysis of cultured species *(Crenarchaeota, Euryarchaeota)* and uncultured samples *(Korarchaeaota)*, the *Archaea* can be divided into at least three phyla.

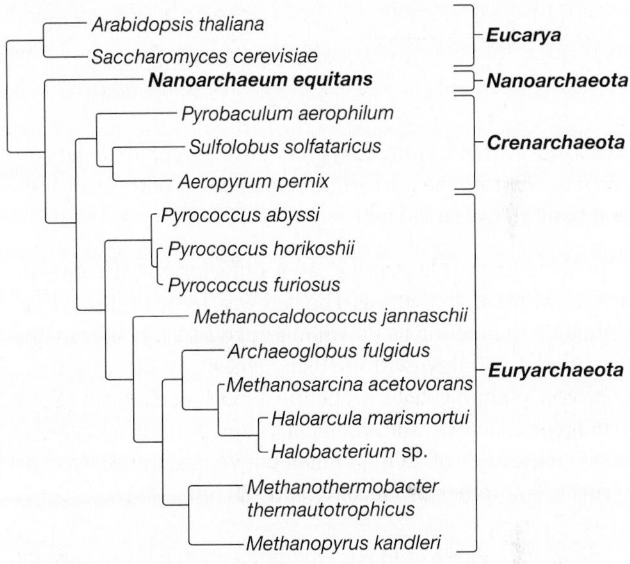

(b) Proposed phylogenetic position of *N. equitans* within the *Archaea*. Note that this tree omits the *Korarchaeota*.

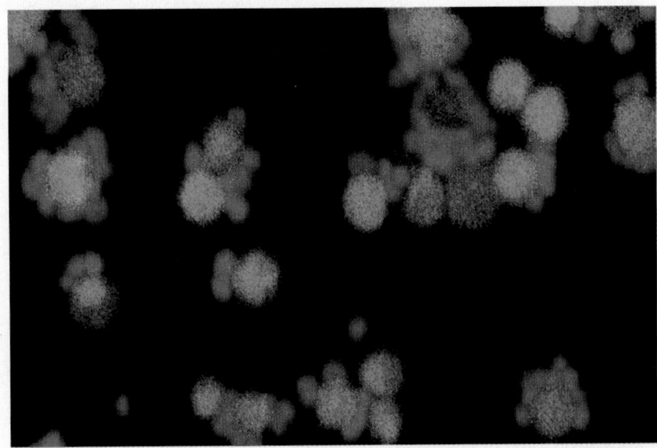

(c) Parasitic cells of *Nanoarchaeum equitans* attached to the surface of the crenarchaeote *Ingicoccus*. Confocal laser scanning micrograph in which *Ingicoccus* are stained green; *N. equitans* cells are red.

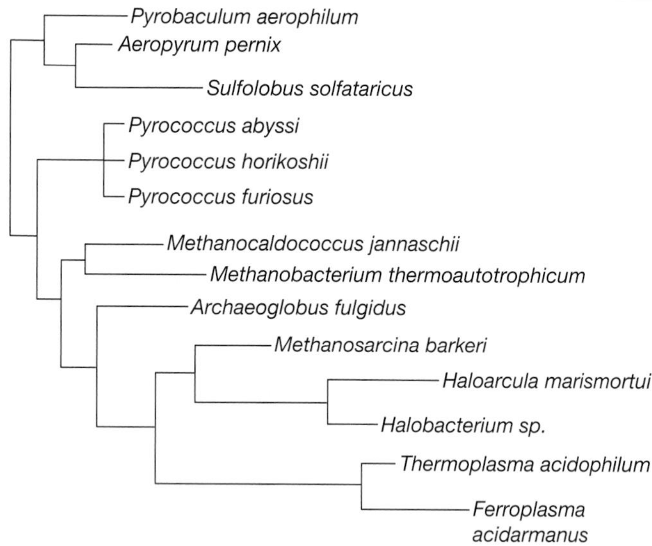

Figure 20.8 The Phylum *Euryarchaeota*. A phylogenetic tree developed from the nucleotide sequences of both small and large subunit rRNA sequences.

is a nickel tetrapyrrole serving as a cofactor for the enzyme methyl-CoM methylreductase. The pathway for methane synthesis is thought to function as shown in **figure 20.11.** It appears that ATP synthesis is linked with methanogenesis by electron transport, proton pumping, and a chemiosmotic mechanism. Some methanogens can live autotrophically by forming acetyl-CoA from two molecules of CO_2 and then converting the acetyl-CoA to pyruvate and other products (figure 20.3*b*). Electron transport and oxidative phosphorylation: The electron transport chain (section 9.5)

Methanogens thrive in anoxic environments rich in organic matter: the rumen and intestinal system of animals, freshwater and marine sediments, swamps and marshes, hot springs, anoxic sludge digesters, and even within anaerobic protozoa. Methanogens often are of ecological significance. The rate of methane production can be so great that bubbles of methane sometimes rise to the surface of a lake or pond. Rumen methanogens are so active that a cow can belch 200 to 400 liters of methane a day. Microbial interactions: The rumen ecosystem (section 30.1)

Methanogenic archaea are potentially of great practical importance since methane is a clean-burning fuel and an excellent energy source. For many years sewage treatment plants have been using the methane they produce as a source of energy for heat and electricity. Anaerobic digester microbes degrade particulate wastes such as sewage sludge to H_2, CO_2, and acetate. CO_2-reducing methanogens form CH_4 from CO_2 and H_2, while aceticlastic methanogens cleave acetate to CO_2 and CH_4 (about 2/3 of the methane produced by an anaerobic digester comes from acetate). A kilogram of organic matter can yield up to 600 liters of methane. It is quite likely that future research will greatly increase the efficiency of methane production and make methanogenesis an important source of pollution-free energy. Wastewater treatment (section 41.2)

Methanogenesis also can be an ecological problem. Methane absorbs infrared radiation and thus is a greenhouse gas. There is evidence that atmospheric methane concentrations have been rising over the last 200 years. Methane production may significantly promote future global warming. Recently it has been discovered that methanogens can oxidize Fe^0 and use it to produce methane and energy. This means that methanogens growing around buried or submerged iron pipes and other objects may contribute significantly to iron corrosion. Soil microorganisms and the atmosphere (section 29.6)

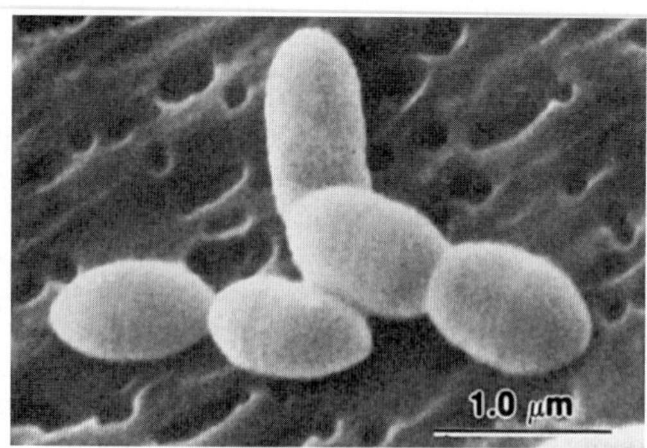

(a)

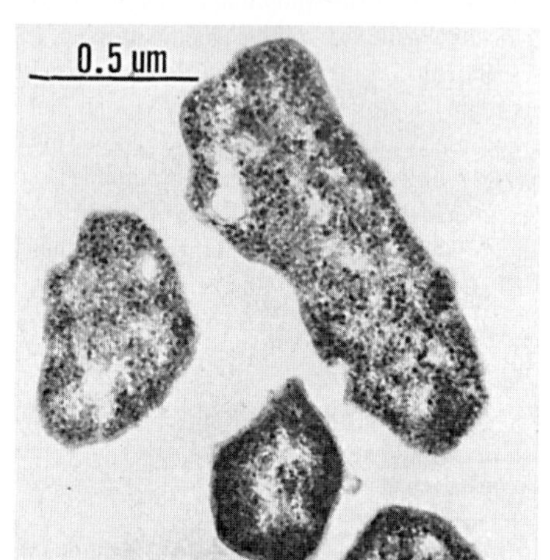

(b)

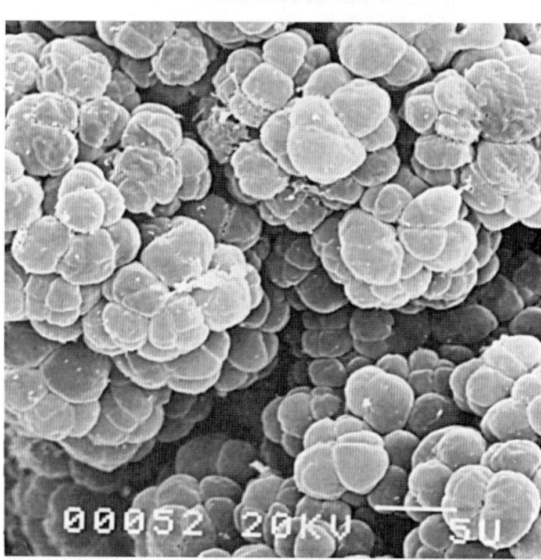

(c)

Figure 20.9 Selected Methanogens. **(a)** *Methanobrevibacter smithii.* **(b)** *Methanogenium marisnigri;* electron micrograph (×45,000). **(c)** *Methanosarcina mazei;* SEM. Bar = 5 μm.

Table 20.2 Selected Characteristics of Representative Genera of Methanogens

Genus	Morphology	% G + C	Wall Composition	Gram Reaction	Motility	Methanogenic Substrates Used
Order *Methanobacteriales*						
Methanobacterium	Long rods or filaments	32–61	Pseudomurein	+ to variable	–	$H_2 + CO_2$, formate
Methanothermus	Straight to slightly curved rods	33	Pseudomurein with an outer protein S-layer	+	+	$H_2 + CO_2$
Order *Methanococcales*						
Methanococcus	Irregular cocci	29–34	Protein	–	+	$H_2 + CO_2$, formate
Order *Methanomicrobiales*						
Methanomicrobium	Short curved rods	45–49	Protein	–	+	$H_2 + CO_2$, formate
Methanogenium	Irregular cocci	52–61	Protein or glycoprotein	–	–	$H_2 + CO_2$, formate
Methanospirillum	Curved rods or spirilla	47–52	Protein	–	+	$H_2 + CO_2$, formate
Order *Methanosarcinales*						
Methanosarcina	Irregular cocci, packets	36–43	Heteropolysaccharide or protein	+ to variable	–	$H_2 + CO_2$, methanol, methylamines, acetate

Microbial Diversity & Ecology

20.2 Methanotrophic Archaea

The marine environment may contain as much as 10,000 billion tons of methane hydrate buried in the ocean floor, around twice the amount of all known fossil fuel reserves. Although some methane rises toward the surface, it often is used before it escapes from the sediments in which it is buried. This is fortunate because methane is a much more powerful greenhouse gas than carbon dioxide. If the atmosphere were flooded with methane, the Earth could become too hot to support life as we know it. The reason for this disappearance of methane in sediments has been unclear until a recent discovery.

By using fluorescent probes for specific DNA sequences, an assemblage of archaea and bacteria has been discovered in anoxic, methane-rich sediments. These clusters of procaryotes contain a core of about 100 archaea from the order *Methanosarcinales* surrounded by a layer of sulfate-reducing bacteria related to the *Desulfosarcina* (see **Box figure**). These two groups appear to cooperate metabolically in such a way that methane is anaerobically oxidized and sulfate reduced; perhaps the bacteria use waste products of methane oxidation to derive energy from sulfate reduction. Isotope studies show that the archaea feed on methane and the bacteria get much of their carbon from the archaea. These methanotrophs may be crucial contributors to the Earth's carbon cycle because it is thought that they oxidize as much as 300 million tons of methane annually.

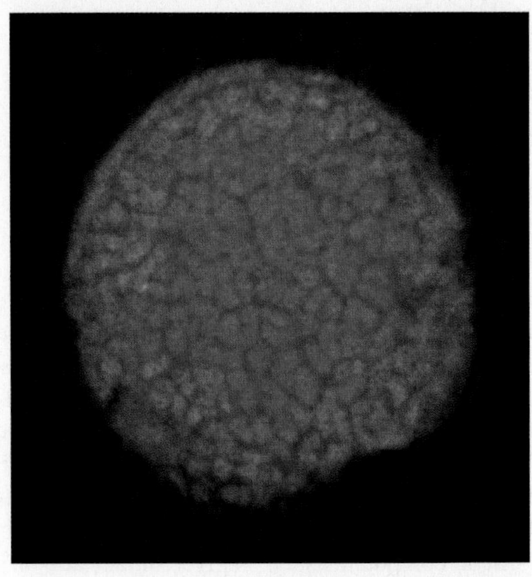

Methane-Consuming Archaea. A cluster of methanotrophic archaea, stained red by a specific fluorescent probe, surrounded by a layer of bacteria labeled by a green fluorescent probe.

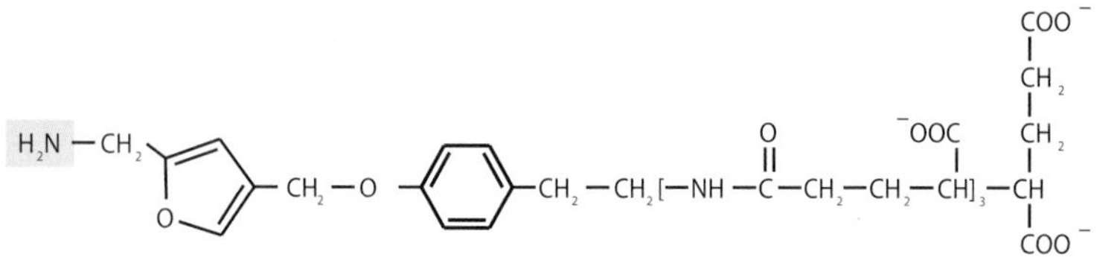

(a) Methanofuran (MFR)

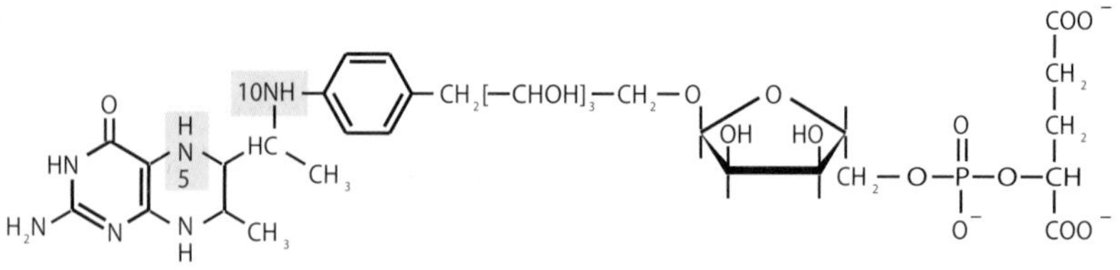

(b) Tetrahydromethanopterin (H_4MPT)

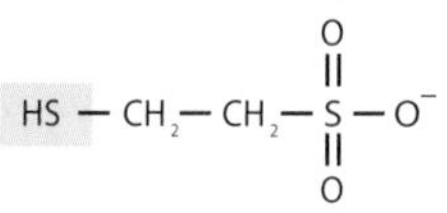

(c) Coenzyme M

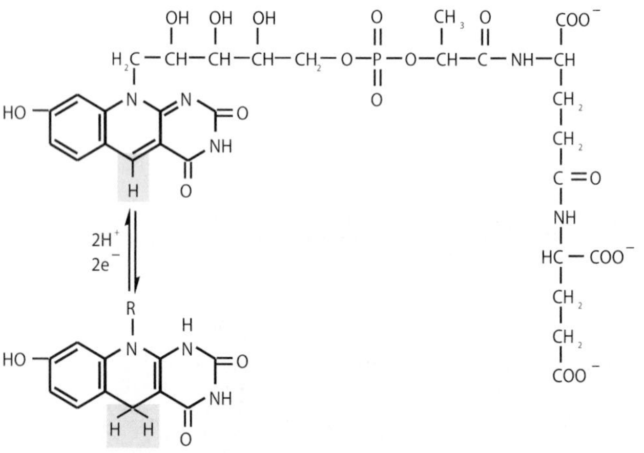

(d) Coenzyme F_{420}

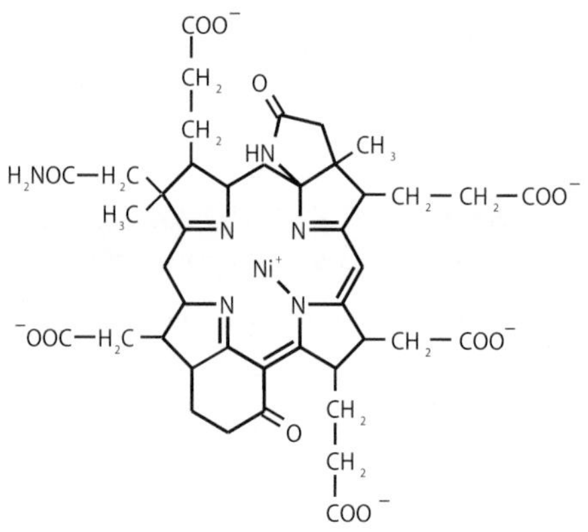

(e) Coenzyme F_{430}

Figure 20.10 Methanogen Coenzymes. **(a)** Coenzyme MFR, **(b)** H_4MPT, and **(c)** coenzyme M are used to carry one-carbon units during methanogenesis. MFR and a simpler form of H_4MPT called methantopterin (MPT; not shown) also participate in the synthesis of acetyl-CoA. The portions of the coenzymes that carry the one-carbon units are shown in blue. H_4MPT carries carbon units on nitrogens 5 and 10, like the more common enzyme tetrahydrofolate. **(d)** Coenzyme F_{420} participates in redox reactions. The part of the molecule that is reversibly oxidized and reduced is highlighted. **(e)** Coenzyme F_{430} participates in reactions catalyzed by the enzyme methyl-CoM methylreductase.

The Halobacteria

The **extreme halophiles** or **halobacteria,** order *Halobacteriales,* are another major archaeal group, currently with 17 genera in one family, the *Halobacteriaceae* (**figure 20.12**). Most are aerobic chemoorganotrophs with respiratory metabolism. Extreme halophiles demonstrate a wide variety of nutritional capabilities. The first halophiles were isolated from salted fish in the 1880s and required complex nutrients such as yeast extract. More recent isolates grow best in defined media, using carbohydrates or simple compounds such as glycerol, acetate, or pyruvate as their carbon source. Halophiles can be motile or nonmotile and are found in a variety of cell shapes. These include cubes and pyramids in addition to rods and cocci.

The most obvious distinguishing trait of this family is its absolute dependence on a high concentration of NaCl. These procaryotes require at least 1.5 M NaCl (about 8%, wt/vol), and usually have a growth optimum at about 3 to 4 M NaCl (17 to 23%). They will grow at salt concentrations approaching saturation (about 36%). The cell walls of most halobacteria are so dependent on the presence of NaCl that they disintegrate when the NaCl concentration drops below 1.5 M. Thus halobacteria only grow in high-salinity habitats such as marine salterns and salt lakes such as the Dead Sea between Israel and Jordan, and the Great Salt Lake in Utah. Halophiles are used in the production of many salted food products, including soy sauce. Halobacteria often have red-to-yellow pigmentation from carotenoids that are probably used as protection against strong sunlight. They can reach such high population levels that salt lakes, salterns, and salted fish actually turn red.

Probably the best-studied member of the family is *Halobacterium salinarium*. This archaeon is unusual because it produces a

Figure 20.11 Methane Synthesis. Pathway for CH_4 synthesis from CO_2 in *M. thermoautotrophicum.* Cofactor abbreviations: methanopterin (MPT), methanofuran (MFR), and 2-mercaptoethanesulfonic acid or coenzyme M (CoM). The nature of the carbon-containing intermediates leading from CO_2 to CH_4 are indicated in parentheses.

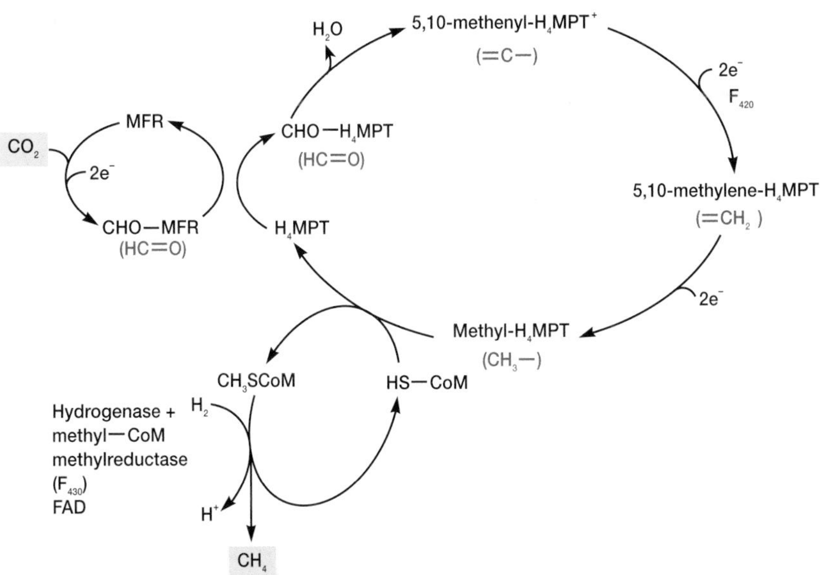

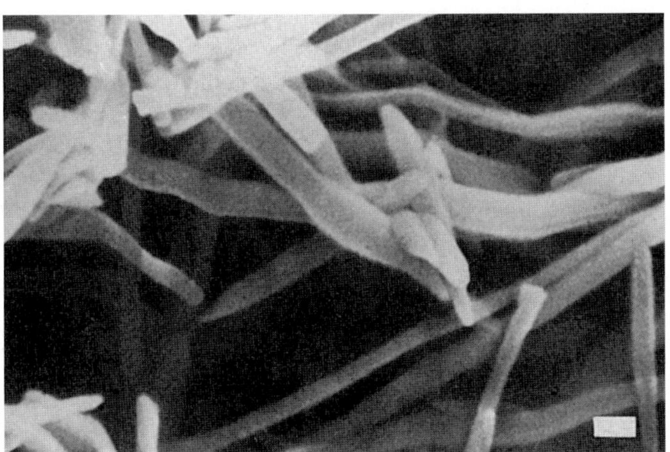

(a) *Halobacterium salinarium*

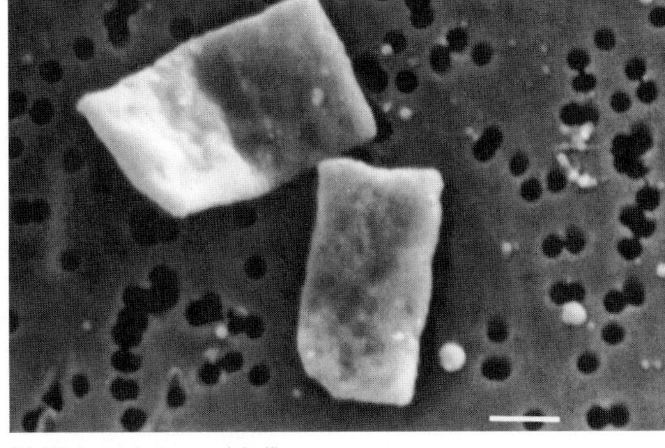

(b) *"Haloquadratum walsbyi"*

Figure 20.12 Examples of Halobacteria. **(a)** *Halobacterium.* A young culture that has formed long rods; SEM. Bar = 1 μm. **(b)** *"Haloquadratum walsbyi"*; SEM. Bar = 1 μm.

protein called **bacteriorhodopsin** that can trap light energy without the presence of chlorophyll. Structurally similar to the rhodopsin found in the mammalian eye, bacteriorhodopsin functions as a light-driven proton pump. Like all members of the rhodopsin family, bacteriorhodopsin has two distinct features: (1) a chromophore that is a derivative of retinal (an aldehyde of vitamin A), which is covalently attached to the protein by a Schiff base with the amino group of lysine (**figure 20.13**); and (2) seven membrane-spanning domains connected by loops on either side with the retinal resting within the membrane. Bacteriorhodpsin molecules form aggregates in a modified region of the cell membrane called **purple membrane.** When retinal absorbs light, the double bond between carbons 13 and 14 changes from a *trans* to a *cis* configuration and the Schiff base loses a proton. Protons move across the plasma membrane to the periplasmic space during these alterations, and the Schiff base changes are directly involved in this movement (figure 20.13). Bacteriorhodopsin undergoes several conformational changes during the photocycle. These conformational changes also are involved in proton transport. The light-driven proton pumping generates a pH gradient that can be used to power the synthesis of ATP by a chemiosmotic mechanism. Electron transport and oxidative phosphorylation (section 9.5); Phototrophy: Rhodopsin-based phototrophy (section 9.12)

Halobacterium has three additional rhodopsins, each with a different function. Halorhodopsin uses light energy to transport chloride ions into the cell and maintain a 4 to 5 M intracellular KCl concentration. Two more are called sensory rhodopsin I (SRI) and SRII. **Sensory rhodopsins** act as photoreceptors, in this case,

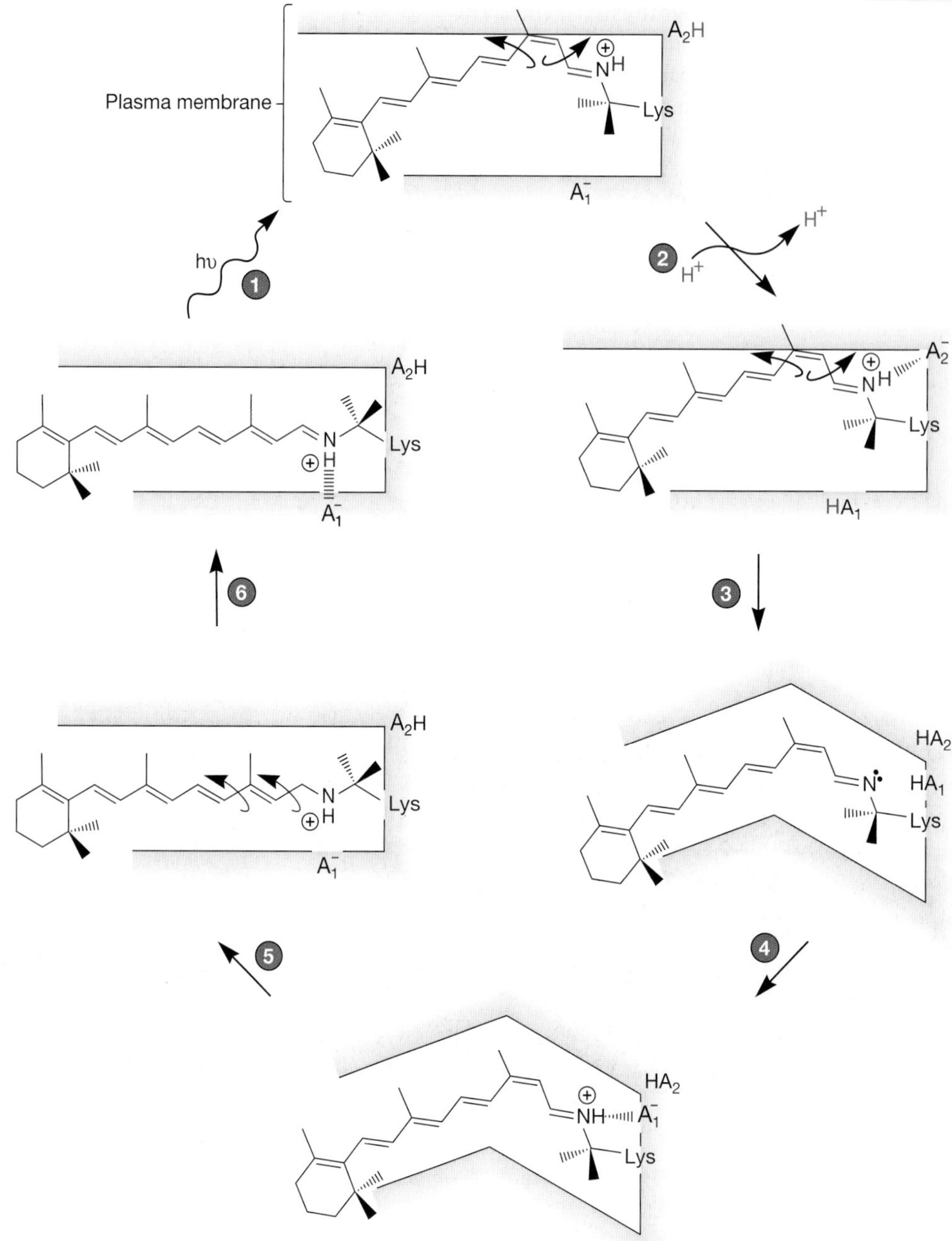

Figure 20.13 The Photocycle of Bacteriorhodopsin. In this hypothetical mechanism the retinal component of bacteriorhodopsin is buried in the membrane and retinal interacts with two amino acids, A_1 and A_2 (aspartates 96 and 85), that can reversibly accept and donate protons. A_2 is connected to the cell exterior, and A_1 is closer to the cell interior. Light absorption by retinal in **step 1** triggers an isomerization from 13-*trans*-retinal to 13-*cis*-retinal. The retinal then donates a proton to A_2 in **steps 2 and 3,** while A_1 is picking up another proton from the interior and A_2 is releasing a proton to the outside. In **steps 4 and 5,** retinal obtains a proton from A_1 and isomerizes back to the 13-*trans* form. The cycle is then ready to begin again after **step 6.**

one for red light and one for blue. They control flagellar activity to position the organism optimally in the water column. *Halobacterium* moves to a location of high light intensity, but one in which ultraviolet light is not sufficiently intense to be lethal.

Surprisingly, it now appears rhodopsin is widely distributed among procaryotes. DNA sequence analysis of uncultivated marine bacterioplankton reveals the presence of rhodopsin genes among both α- and γ-proteobacteria. This newly discovered rhodopsin is called proteorhodopsin. Cyanobacteria also have rhodopsin proteins, which like SRI and SRII of the halobacteria, sense the spectral quality of light. Thus the cyanobacterial molecules are also considered sensory rhodopsins. These procaryotic rhodopsins conserve the seven transmembrane helices through the cell membrane and the lysine residue that forms the Schiff base linkage with retinal. Microorganisms in marine environments: The photic zone of the open ocean (section 28.3)

The Thermoplasms

Procaryotes in the class *Thermoplasmata* are thermoacidophiles that lack cell walls. At present, three genera, *Thermoplasma, Picrophilus,* and *Ferroplasma* are known. They are sufficiently different from one another to be placed in separate families, *Thermoplasmataceae, Picrophilaceae,* and *Ferroplasmataceae.*

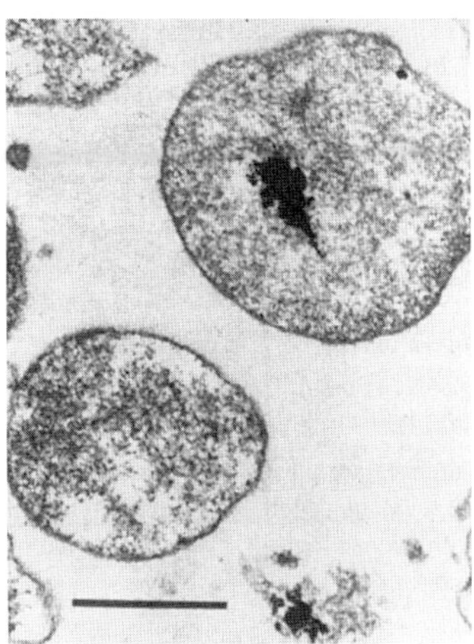

Figure 20.14 ***Thermoplasma.*** Transmission electron micrograph. Bar = 0.5 μm.

Thermoplasma grows in refuse piles of coal mines. These piles contain large amounts of iron pyrite (FeS), which is oxidized to sulfuric acid by chemolithotrophic bacteria. As a result the piles become very hot and acidic. This is an ideal habitat for *Thermoplasma* because it grows best at 55 to 59°C and pH 1 to 2. Although it lacks a cell wall, its plasma membrane is strengthened by large quantities of diglycerol tetraethers, lipid-containing polysaccharides, and glycoproteins. The organism's DNA is stabilized by association with archaeal histones that condense the DNA into structures resembling eucaryotic nucleosomes. At 59°C, *Thermoplasma* takes the form of an irregular filament, whereas at lower temperatures it is spherical (**figure 20.14**). The cells may be flagellated and motile.

Picrophilus is even more unusual than *Thermoplasma.* It originally was isolated from moderately hot solfataric fields in Japan. Although it lacks a regular cell wall, *Picrophilus* has an S-layer outside its plasma membrane. The cells grow as irregularly shaped cocci, around 1 to 1.5 μm in diameter, and have large cytoplasmic cavities that are not membrane bounded. *Picrophilus* is aerobic and grows between 47 and 65°C with an optimum of 60°C. It is most remarkable in its pH requirements: it grows only below pH 3.5 and has a growth optimum at pH 0.7. Growth even occurs at about pH 0!

Extremely Thermophilic S^0-Metabolizers

This physiological group contains the class *Thermococci,* with one order, *Thermococcales.* The *Thermococcales* are strictly anaerobic and can reduce sulfur to sulfide. They are motile by flagella and have optimum growth temperatures around 88 to 100°C. The order contains one family and three genera, *Thermococcus, Paleococcus,* and *Pyrococcus.*

Sulfate-Reducing *Euryarchaeota*

Euryarchaeal sulfate reducers are found in the class *Archaeoglobi* and the order *Archaeoglobales.* This order has only one family and three genera. *Archaeoglobus* contains gram-negative-staining, irregular coccoid cells with cell walls consisting of glycoprotein subunits. It can extract electrons from a variety of electron donors (e.g., H_2, lactate, glucose) and reduce sulfate, sulfite, or thiosulfate to sulfide. Elemental sulfur is not used as an acceptor. *Archaeoglobus* is extremely thermophilic (the optimum is about 83°C) and has been isolated from marine hydrothermal vents. The organism is not only unusual in being able to reduce sulfate, unlike other archaea, but it also possesses the methanogen coenzymes F_{420} and methanopterin.

1. What are thermoacidophiles and where do they grow? In what ways do they use sulfur in their metabolism? Briefly describe *Sulfolobus* and *Thermoproteus.*
2. Generally characterize methanogenic archaea and distinguish them from other groups.
3. Briefly describe how methanogens produce methane and the roles of their unique cofactors in this process.
4. Where does one find methanogens? Discuss their ecological and practical importance.
5. Where are the extreme halophiles found and what is unusual about their cell walls and growth requirements?
6. What is the purple membrane and what pigment does it contain?
7. How is *Thermoplasma* able to live in acidic, very hot coal refuse piles when it lacks a cell wall? How is its DNA stabilized? What is so remarkable about *Picrophilus?*
8. Characterize *Archaeoglobus.* In what way is it similar to the methanogens and how does it differ from other extreme thermophiles?

Summary

20.1 Introduction to the *Archaea*

a. The *Archaea* are highly diverse with respect to morphology, reproduction, physiology, and ecology. Although best known for their growth in anoxic, hypersaline, and high-temperature habitats they also inhabit marine arctic, temperate, and tropical waters.

b. Archaeal cell walls do not contain peptidoglycan and differ from bacterial walls in structure. They may be composed of pseudomurein, polysaccharides, or glycoproteins and other proteins.

c. Archaeal membrane lipids differ from those of other organisms in having branched chain hydrocarbons connected to glycerol by ether links. Bacterial and eucaryotic lipids have glycerol connected to fatty acids by ester bonds.

d. Their tRNA, ribosomes, elongation factors, RNA polymerases, and other components distinguish *Archaea* from *Bacteria* and eucaryotes.

e. Although much of archaeal metabolism appears similar to that of other organisms, the *Archaea* differ with respect to glucose catabolism, pathways for CO_2 fixation, and the ability of some to synthesize methane (**figure 20.3**).

f. *Archaea* may be divided into five groups: methanogenic archaea, sulfate reducers, extreme halophiles, cell wall-less archaea, and extremely thermophilic S^0-metabolizers (**table 20.1**).

g. The second edition of *Bergey's Manual* divides the *Archaea* into two phyla, the *Crenarchaeota* and *Euryarchaeota,* each with several orders (**figures 20.4** and **20.8**).

20.2 Phylum *Crenarchaeota*

a. The extremely thermophilic S^0-metabolizers in the phylum *Crenarchaeota* depend on sulfur for growth and are frequently acidophiles. The sulfur may be used as an electron acceptor in anaerobic respiration or as an electron donor by lithotrophs. They are almost always strict anaerobes and grow in geothermally heated soil and water that is rich in sulfur.

20.3 Phylum *Euryarchaeota*

a. The phylum *Euryarchaeota* contains five major groups: methanogens, halobacteria, the thermoplasms, extremely thermophilic S^0-metabolizers, and sulfate-reducing archaea.

b. Methanogenic archaea are strict anaerobes that can obtain energy through the synthesis of methane. They have several unusual cofactors that are involved in methanogenesis (**figures 20.10** and **20.11**).

c. Extreme halophiles or halobacteria are aerobic chemoheterotrophs that require at least 1.5 M NaCl for growth. They are found in habitats such as salterns, salt lakes, and salted fish.

d. *Halobacterium salinarum* can carry out phototrophy without chlorophyll or bacteriochlorophyll by using bacteriorhodopsin, which employs retinal to pump protons across the plasma membrane (**figure 20.13**).

e. The thermophilic archaeon *Thermoplasma* grows in hot, acidic coal refuse piles and survives despite the lack of a cell wall. Another thermoplasm, *Picrophilus,* can grow at pH 0.

f. The class *Thermococci* contains extremely thermophilic organisms that can reduce sulfur to sulfide.

g. Sulfate-reducing archaea are placed in the class *Archaeoglobi.* The extreme thermophile *Archaeoglobus* differs from other archaea in using a variety of electron donors to reduce sulfate. It also contains the methanogen cofactors F_{420} and methanopterin.

Key Terms

Archaea 503
bacteriorhodopsin 515
extreme halophiles 514
halobacteria 514
Korarchaeota 511
methanogens 510
pseudomurein 504
purple membrane 515
sensory rhodopsin 515
thermoacidophiles 508

Critical Thinking Questions

1. Do you think the *Archaea* should be separate from the *Bacteria* although both groups are procaryotic? Give your reasoning and evidence.
2. Explain why the fixation of CO_2 by *Thermoproteus* and possibly by *Sulfolobus* using a reductive reversal of the TCA cycle is *not* photosynthesis.
3. Often when the temperature increases, many procaryotes change their shapes from elongated rods into spheres. Suggest one reason for this change.
4. Why would ether linkages be more stable in membranes than ester lipids? How would the presence of tetraether linkages stabilize a thermophile's membrane?
5. Suppose you wished to isolate procaryotes from a hot spring in Yellowstone National Park. How would you go about it?

Learn More

Bell, S., and Jackson, S. P. 2001. Mechanism and regulation of transcription in archaea. *Curr. Opin. Microbiol.* 4:208–13.

Burggraf, S.; Huber, H.; and Stetter, K. O. 1997. Reclassification of the crenarchaeal orders and families in accordance with 16S rRNA sequence data. *Int. J. Syst. Bacteriol.* 47(3): 657–60.

Fuhrman, J. A., and Davis, A. A. 1997. Widespread archaea and novel bacteria from the deep sea as shown by 16S rRNA gene sequences. *Mar. Ecol. Prog. Ser.* 150: 275–85.

Gaasterland, T. 1999. Archaeal genomics. *Curr. Opin. Microbiol.* 2: 542–47.

Garrity, G. M., editor-in-chief. 2001. *Bergey's manual of systematic bacteriology,* 2d ed., vol. 1, D. R. Boone and R. W. Castenholz, editors. New York: Springer-Verlag.

Grabowski, B., and Kelman, Z. 2003. Archaeal DNA replication: Eukaryal proteins in a bacterial context. *Annu. Rev. Microbiol.* 57: 487–516.

Kashefi, K., and Lovley, D. 2004. Extending the upper temperature limit for life. *Science* 301: 934.

Kelman, L. M., and Kelman, A. 2003. *Archaea:* An archetype for replication initiation studies? *Mol. Microbiol.* 48: 605–15.

Oren, A. 1999. Bioenergetic aspects of halophilism. *Microbiol. Mol. Biol. Rev.* 63(2): 334–48.

Orphan, V. J.; House, C. H.; Hinrichs, K.-U.; McKeegan, K. D.; and DeLong, E. F. 2001. Methane-consuming archaea revealed by directly coupled isotopic and phylogenetic analysis. *Science* 293: 484–87.

Pereto, J.; Lopez-Garcia, P.; and Moreira, D. 2004. Ancestral lipid biosynthesis and early membrane evolution. *Trends Biochem. Sci.* 29: 469–77.

Sowers, K. R., and Schreier, H. J. 1999. Gene transfer systems for the *Archaea*. *Trends Microbiol.* 7(5): 212–19.

Walsby, A. E. 2005. *Archaea* with square cells. *Trends Microbiol.* 13: 193–95.

Waters, E., et al. 2003. The genome of *Nanobacterium equitans:* Insights into early *Archaeal* evolution and derived parasitism. *Proc. Natl. Acad. Sci. USA*. 100: 12984–88.

Please visit the Prescott website at www.mhhe.com/prescott7 for additional resources.

21 Bacteria: The Deinococci and Nonproteobacteria Gram Negatives

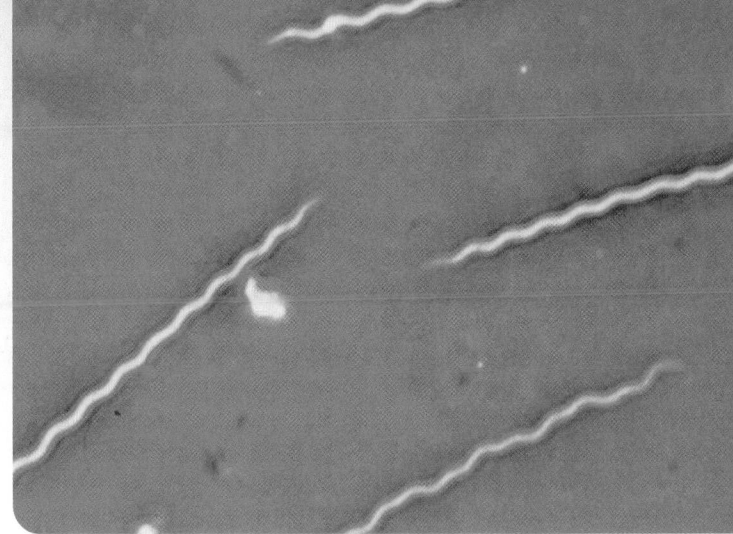

Spirochetes are distinguished by their structure and mechanism of motility. *Treponema pallidum* shown here causes syphilis.

Preview

- Some bacterial groups, such as those represented by the hyperthermophiles *Aquifex* and *Thermotoga,* are deeply branching and very old; other bacterial taxa have arisen more recently.
- All photosynthetic bacteria, including the cyanobacteria, were once considered a phenetically unified group. However, phylogenetic analysis reveals that the purple photosynthetic bacteria are proteobacteria, separating from them green sulfur and green nonsulfur bacteria. The cyanobacteria are separated from other photosynthetic bacteria because they resemble eucaryotic phototrophs in having both photosystems I and II and carrying out oxygenic photosynthesis. Their rRNA sequences also indicate that they are different from other photosynthetic bacteria.
- Bacteria, such as the chlamydiae, that are obligate intracellular parasites have relinquished some of their metabolic independence through loss of metabolic pathway genes. They use their host's energy supply and/or cell constituents.
- Gliding motility is widely distributed among bacteria and is very useful to organisms that digest insoluble nutrients or move over the surfaces of moist, solid substrata.

Chapter 20 surveys the *Archaea,* which are located in volume 1 of the second edition of *Bergey's Manual.* Volumes 1 and 5 of *Bergey's Manual* also describe a wide variety of other procaryotic groups that are members of the second domain: *Bacteria.* Chapter 21 is devoted to 10 of these bacterial phyla. Their phylogenetic locations are depicted in **figure 21.1.** We follow the general organization and perspective of the second edition of *Bergey's Manual* in most cases.

In describing each bacterial group, we include aspects such as distinguishing characteristics, morphology, reproduction, physiology, metabolism, and ecology. The taxonomy of each major group is summarized, and representative species are discussed. Students of microbiology should appreciate bacteria as living organisms rather than simply as agents of disease of little interest or importance in other contexts.

21.1 *Aquificae* and *Thermotogae*

Thermophilic microbes are found in both the bacterial and archaeal domains, but to date, all hyperthermophilic procaryotes (those with optimum growth temperatures above 85°C) belong to the *Archaea.* The phyla *Aquificiae* and *Thermotoga* are two examples of bacterial thermophiles.

The phylum *Aquificae* is thought to represent the deepest or oldest branch of *Bacteria* (*see figure 19.3*). It contains one class, one order, and eight genera. Two of the best-studied genera are *Aquifex* and *Hydrogenobacter. Aquifex pyrophilus* is a gram-negative, microaerophilic rod. It is thermophilic with a temperature optimum of 85°C and a maximum of 95°C. *Aquifex* is a chemolithoautotroph that captures energy by oxidizing hydrogen, thiosulfate, and sulfur with oxygen as the terminal electron acceptor. Because *Aquifex* and *Hydrogenobacter* are both thermophilic chemolithoautotrophs, it has been suggested that the original bacterial ancestor was probably thermophilic and chemolithoautotrophic. Chemolithotrophy (section 9.11)

There are wide areas of the bacteriological landscape in which we have so far detected only some of the highest peaks, while the rest of the beautiful mountain range is still hidden in the clouds and the morning fogs of ignorance. The goal is still lying on the ground, but we have to bend down to grasp it.

—*Preface to* The Prokaryotes

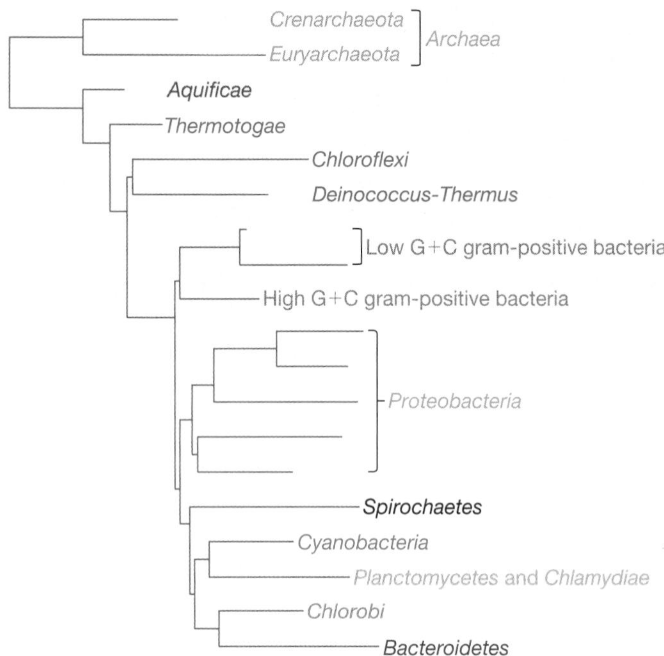

Figure 21.1 Phylogenetic Relationships Among Procaryotes. The *Deinococcus-Thermus* group and other nonproteobacterial gram-negatives are highlighted.

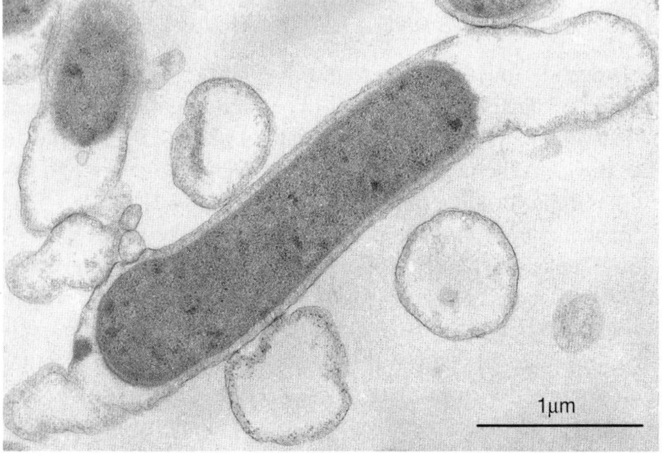

Figure 21.2 *Thermotoga maritima*. Note the loose sheath extending from each end of the cell.

The second oldest or deepest branch is the phylum *Thermotogae,* which has one class, one order, and six genera. The members of the genus *Thermotoga* [Greek *therme,* heat; Latin *toga,* outer garment], like *Aquifex,* are thermophiles with a growth optimum of 80°C and a maximum of 90°C. They are gram-negative rods with an outer sheathlike envelope (like a toga) that can extend or balloon out from the ends of the cell (**figure 21.2**). They grow in active geothermal areas found in marine hydrothermal systems and terrestrial solfataric springs. In contrast to *Aquifex, Thermotoga* is a chemoheterotroph with a functional glycolytic pathway that can grow anaerobically on carbohydrates and protein digests.

The genome of *Aquifex* is about a third the size of the *Escherichia coli* genome and, as expected, contains the genes required for chemolithoautotrophy. The *Thermotoga* genome is somewhat larger and has genes for sugar degradation. About 24% of its coding sequences are similar to archaeal genes; this proportion is greater than that of other bacteria, including *Aquifex* (16% similarity) and may be due to lateral (horizontal) gene transfer. Comparative genomics (section 15.6)

21.2 *Deinococcus-Thermus*

The phylum *Deinococcus-Thermus* contains the class *Deinococci* and the orders *Deinococcales* and *Thermales.* There are only three genera in the phylum; the genus *Deinococcus* is best studied. Deinococci are spherical or rod-shaped with distinctively different 16S rRNA. They often are associated in pairs or tetrads (**figure 21.3*a***) and are aerobic, mesophilic, and catalase positive; usually they can produce acid from only a few sugars. Although they stain gram positive, their cell wall is layered with an outer membrane like the gram-negative bacteria (figure 21.3*b*). They also differ from gram-positive cocci in having L-ornithine in their peptidoglycan, lacking teichoic acid, and having a plasma membrane with large amounts of palmitoleic acid rather than phosphatidylglycerol phospholipids. Almost all strains are extraordinarily resistant to both desiccation and radiation; they can survive as much as 3 to 5 million rad of radiation (an exposure of 100 rad can be lethal to humans).

Much remains to be discovered about the biology of these bacteria. Deinococci can be isolated from ground meat, feces, air, freshwater, and other sources, but their natural habitat is not yet known. Their great resistance to radiation results from their ability to repair a severely damaged genome, which consists of a circular chromosome, a megaplasmid, and a small plasmid. When exposed to high levels of γ radiation, the genome is broken into many fragments. Within 12 to 24 hours, the genome is pieced back together, ensuring viability. It is unclear how this is accomplished. Genome studies have shown that *D. radiodurans* has a very efficient DNA repair system; however, novel DNA repair genes have not yet been reported (*see figure 15.18*). It appears that the ability to accumulate high levels of Mn(II) may help protect the microbe from high levels of radiation-induced toxic oxygen species. Currently, the mechanism by which Mn(II) confers protection is unclear. Insights from microbial genomes: Genomic analysis of extremophiles (section 15.8)

21.3 Photosynthetic Bacteria

There are three groups of gram-negative photosynthetic bacteria: the purple bacteria, the green bacteria, and the cyanobacteria (**table 21.1**). The cyanobacteria differ most fundamentally from the green and purple photosynthetic bacteria in being able to carry out **oxygenic photosynthesis.** They have photosystems I

(a)

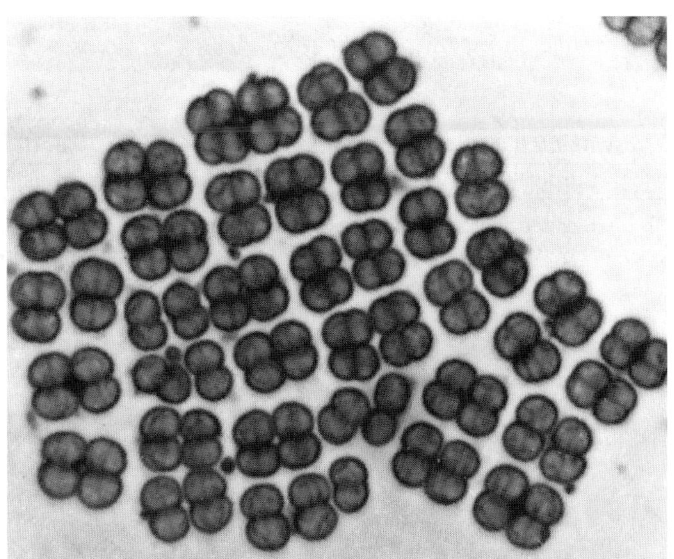

(b)

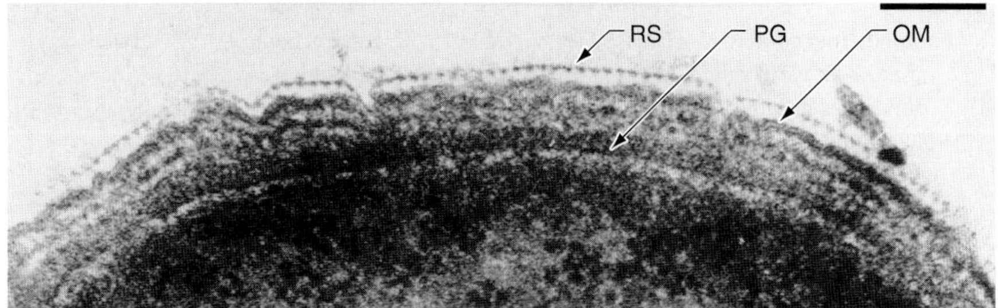

Figure 21.3 The Deinococci. **(a)** A *Deinococcus radiodurans* microcolony showing cocci arranged in tetrads (average cell diameter 2.5 μm). **(b)** The cell wall of *D. radiodurans* with a regular surface protein array (RS), a peptidoglycan layer (PG), and an outer membrane (OM). Bar = 100 nm.

and II, use water as an electron donor, and generate oxygen during photosynthesis. In contrast, purple and green bacteria have only one photosystem and use **anoxygenic photosynthesis.** Because they are unable to use water as an electron source, they employ reduced molecules such as hydrogen sulfide, sulfur, hydrogen, and organic matter as their electron source for the reduction of $NAD(P)^+$ to NAD(P)H. Consequently, many purple and green bacteria form sulfur granules. Purple sulfur bacteria accumulate granules within their cells, whereas green sulfur bacteria deposit the sulfur granules outside their cells. The purple nonsulfur bacteria normally use organic molecules as an electron source. There also are differences in photosynthetic pigments, the organization of photosynthetic membranes, nutritional requirements, and oxygen relationships. Phototrophy (section 9.12)

The differences in photosynthetic pigments and oxygen requirements among the photosynthetic bacteria have significant ecological consequences. As shown in **figure 21.4,** the chlorophylls, bacteriochlorophylls, and their associated accessory pigments have distinct absorption spectra. Oxygenic cyanobacteria and photosynthetic protists dominate the aerated upper layers of freshwater and marine microbial communities, where they absorb large amounts of red and blue light. Below these microbes, the anoxygenic purple and green photosynthetic bacteria inhabit the deeper anoxic zones that are rich in hydrogen sulfide and other reduced compounds that can be used as electron donors. Their bacteriochlorophyll and accessory pigments enable them to use light in the far-red spectrum that is not used by other photosynthetic organisms **(table 21.2)**. In addition, the bacteriochlorophyll absorption peaks at about 350 to 550 nm, enabling them to grow at greater depths because shorter wavelength light can penetrate water farther. As a result, when the water is sufficiently clear, a layer of green and purple bacteria develops in the anoxic, hydrogen sulfide-rich zone.

Bergey's Manual places photosynthetic bacteria into seven major groups distributed between five bacterial phyla. The phylum *Chloroflexi* contains the green nonsulfur bacteria, and the phylum *Chlorobi,* the green sulfur bacteria. The cyanobacteria are placed in their own phylum, *Cyanobacteria.* Purple bacteria are divided between three groups. Purple sulfur bacteria are placed in the γ-proteobacteria, families *Chromatiaceae* and *Ectothiorhodospiraceae.* The purple nonsulfur bacteria are distributed between the α-proteobacteria (five different families) and one family of the β-proteobacteria. Finally, the gram-positive heliobacteria in the phylum *Firmicutes* are also photosynthetic. There appears to have been considerable horizontal transfer of photosynthetic genes among the five phyla. At least 50 genes related to photosynthesis are common to all five. In this chapter, we describe the cyanobacteria and green bacteria; purple bacteria are discussed in chapter 22, while heliobacteria are featured in chapter 23. Class *Gammaproteobacteria:* The purple sulfur bacteria (section 22.3); Class *Clostridia* (section 23.4)

Table 21.1 Characteristics of the Major Groups of Gram-negative Photosynthetic Bacteria

Characteristic	Anoxygenic Photosynthetic Bacteria				Oxygenic Photosynthetic Bacteria
	Green Sulfur[a]	Green Nonsulfur[b]	Purple Sulfur	Purple Nonsulfur	Cyanobacteria
Major photosynthetic pigments	Bacteriochlorophylls *a* plus *c, d,* or *e* (the major pigment)	Bacteriochlorophylls *a* and *c*	Bacteriochlorophyll *a* or *b*	Bacteriochlorophyll *a* or *b*	Chlorophyll *a* plus phycobiliproteins *Prochlorococcus* has divinyl derivatives of chlorophyll *a* and *b*
Morphology of photosynthetic membranes	Photosynthetic system partly in chlorosomes that are independent of the plasma membrane	Chlorosomes present when grown anaerobically	Photosynthetic system contained in spherical or lamellar membrane complexes that are continuous with the plasma membrane	Photosynthetic system contained in spherical or lamellar membrane complexes that are continuous with the plasma membrane	Thylakoid membranes lined with phycobilisomes
Photosynthetic electron donors	H_2, H_2S, S	Photoheterotrophic donors—a variety of sugars, amino acids, and organic acids; photoautotrophic donors—H_2S, H_2	H_2, H_2S, S	Usually organic molecules: sometimes reduced sulfur compounds or H_2	H_2O
Sulfur deposition	Outside of the cell		Inside the cell[c]	Outside of the cell in a few cases	
Nature of photosynthesis	Anoxygenic	Anoxygenic	Anoxygenic	Anoxygenic	Oxygenic (sometimes facultatively anoxygenic)
General metabolic type	Obligately anaerobic photolithoautotrophs	Usually photoheterotrophic; sometimes photoautotrophic or chemoheterotrophic (when aerobic and in the dark)	Obligately anaerobic photolithoautotrophs	Usually anaerobic photoorgano-heterotrophs; some facultative photolithoautotrophs (in the dark, chemo-organoheterotrophs)	Aerobic photo-lithoautotrophs
Motility	Nonmotile; some have gas vesicles	Gliding	Motile with polar flagella; some are peritrichously flagellated	Motile with polar flagella or nonmotile; some have gas vesicles	Nonmotile, swimming motility without flagella or gliding motility; some have gas vesicles
Percent G + C	48–58	53–55	45–70	61–72	35–71
Phylum or class	*Chlorobi*	*Chloroflexi*	α-, β-, and γ-proteobacteria	α-proteobacteria β-proteobacteria (*Rhodocyclus*)	*Cyanobacteria*

[a]Characteristics of *Chlorobi*.
[b]Characteristics of *Chloroflexus*.
[c]With the exception of *Ectothiorhodospira*.

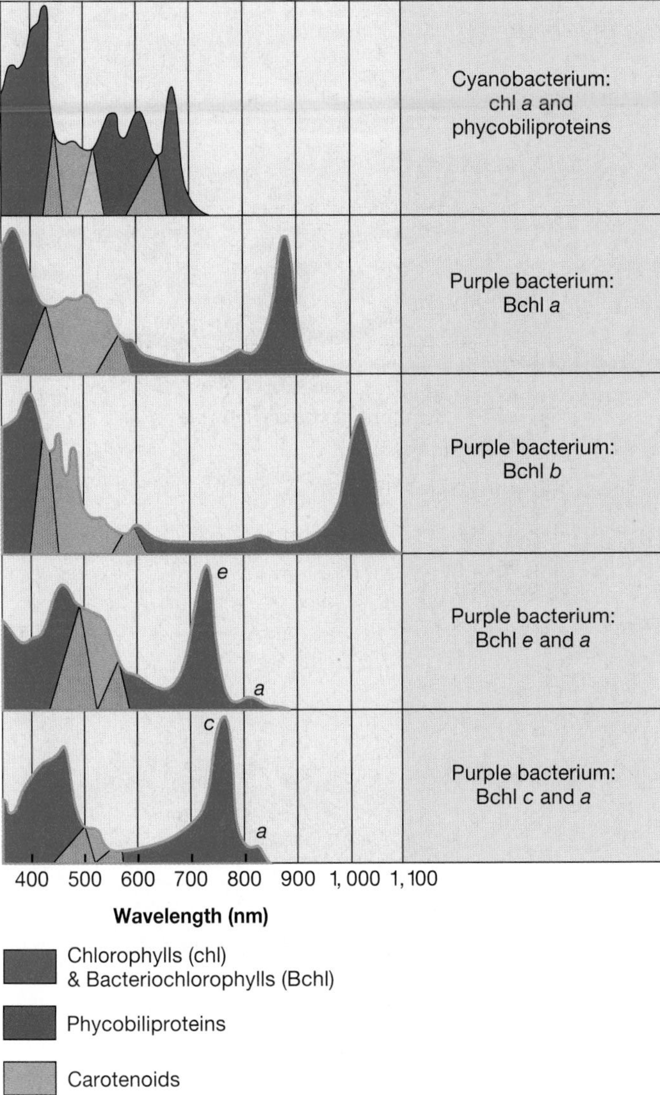

Figure 21.4 Photosynthetic Pigments. Absorption spectra of five photosynthetic bacteria showing the differences in absorption maxima and the contributions of various accessory pigments.

Phylum *Chlorobi*

The phylum *Chlorobi* has only one class (*Chlorobia*), order (*Chlorobiales*), and family (*Chlorobiaceae*). The **green sulfur bacteria** are a small group of obligately anaerobic photolithoautotrophs that use hydrogen sulfide, elemental sulfur, and hydrogen as electron sources. The elemental sulfur produced by sulfide oxidation is deposited outside the cell. Their photosynthetic pigments are located in ellipsoidal vesicles called **chlorosomes** or chlorobium vesicles, which are attached to the plasma membrane but are not continuous with it. Chlorosomes are the most efficient light-harvesting complexes found in nature. The chlorosome membrane is not a normal lipid bilayer. Instead bacteriochlorophyll molecules are grouped into lateral arrays held together by carotenoids and lipids. Chlorosomes contain accessory bacteriochlorophyll pigments, but the reaction center bacteriochlorophyll is located in the plasma membrane where it obtains energy from chlorosome pigments. These bacteria flourish in the anoxic, sulfide-rich zones of lakes. Although they lack flagella and are nonmotile, some species have gas vesicles **(figure 21.5*a*)** to adjust their depth for optimal light and hydrogen sulfide. Those forms without vesicles are found in sulfide-rich muds at the bottom of lakes and ponds.

The green sulfur bacteria are very diverse morphologically. They may be rods, cocci, or vibrios; some grow singly, and others form chains and clusters. They are either grass-green or chocolate-brown in color. Representative genera are *Chlorobium, Prosthecochloris,* and *Pelodictyon.*

Table 21.2 Procaryotic Bacteriochlorophyll and Chlorophyll Absorption Maxima

	Long Wavelength Maxima (nm)	
Pigment	In Ether or Acetone	Approximate Range of Values in Cells
Chlorophyll *a*	665	680–685
Bacteriochlorophyll *a*	775	850–910 (purple bacteria)[a]
Bacteriochlorophyll *b*	790	1,020–1,035
Bacteriochlorophyll *c*	660	745–760
Bacteriochlorophyll *d*	650	725–745
Bacteriochlorophyll *e*	647	715–725

[a]The spectrum of bacteriochlorophyll *a* in green bacteria has a different maximum, 805–810 nm.

Phylum *Chloroflexi*

The phylum *Chloroflexi* has both photosynthetic and nonphotosynthetic members. *Chloroflexus* is the major representative of the photosynthetic **green nonsulfur bacteria.** However, the term "green nonsulfur" is a misnomer because not all members of this group are green, and some use sulfur. *Chloroflexus* is a filamentous, gliding, thermophilic bacterium that often is isolated from neutral to alkaline hot springs where it grows in the form of orange-reddish mats, usually in association with cyanobacteria. It resembles the green sulfur bacteria with small chlorosomes and accessory bacteriochlorophyll *c*. However, like the purple bacteria, its light-harvesting complexes contain bacteriochlorophyll *a* and are in the plasma membrane. Finally, its metabolism is more similar to that of the purple nonsulfur bacteria. *Chloroflexus* can carry out anoxygenic photosynthesis with organic compounds as carbon sources or grow aerobically as a chemoheterotroph. It doesn't appear closely related to any other bacterial group based on 16S rRNA studies and is a deep and ancient branch of the bacterial tree (*see figure 19.3*). Genomic analysis of *Chloroflexus aurantiacus* should help elucidate the origin of photosynthesis.

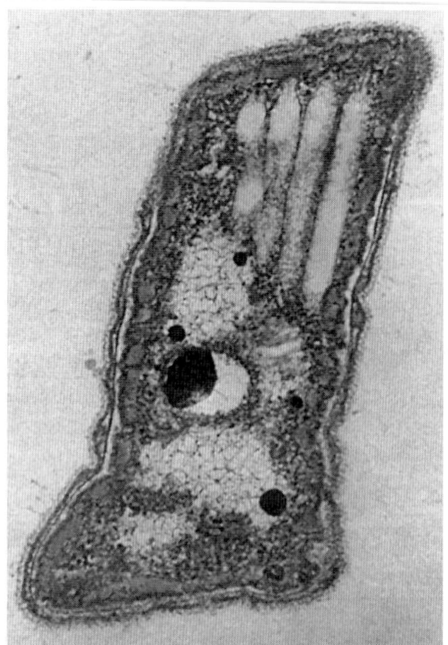
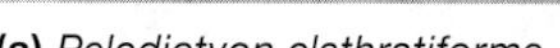

(a) *Pelodictyon clathratiforme*

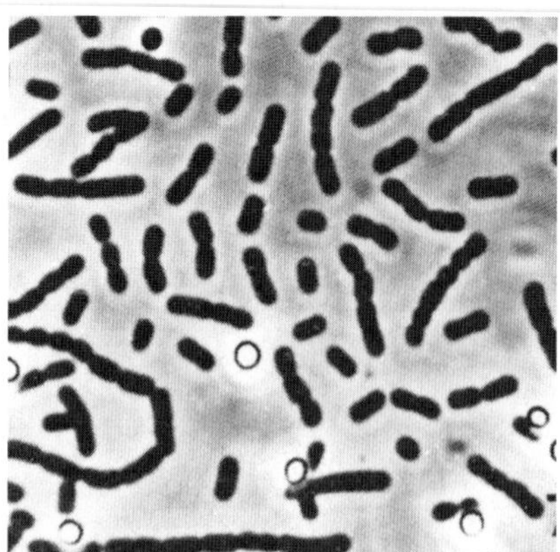

(b) *Chlorobium limicola*

Figure 21.5 Typical Green Sulfur Bacteria. **(a)** An electron micrograph of *Pelodictyon clathratiforme* (×105,000). Note the chlorosomes (dark gray areas) and gas vesicles (light gray areas with pointed ends). **(b)** *Chlorobium limicola* with extracellular sulfur granules.

The nonphotosynthetic, gliding, rod-shaped or filamentous bacterium *Herpetosiphon* also is included in this phylum. *Herpetosiphon* is an aerobic chemoorganotroph with respiratory metabolism that uses oxygen as the electron acceptor. It can be isolated from freshwater and soil habitats.

1. Give the major distinguishing characteristics of *Aquifex, Thermotoga,* and the deinococci. What is thought to contribute to the dessication and radiation resistance of the deinococci?
2. How do oxygenic and anoxygenic photosynthesis differ from each other and why? What is the ecological significance of these differences?
3. In general terms give the major characteristics of the following groups: purple sulfur bacteria, purple nonsulfur bacteria, and green sulfur bacteria. How do purple and green sulfur bacteria differ?
4. What are chlorosomes or chlorobium vesicles?
5. Compare the green nonsulfur (*Chloroflexi*) and green sulfur bacteria (*Chlorobi*).

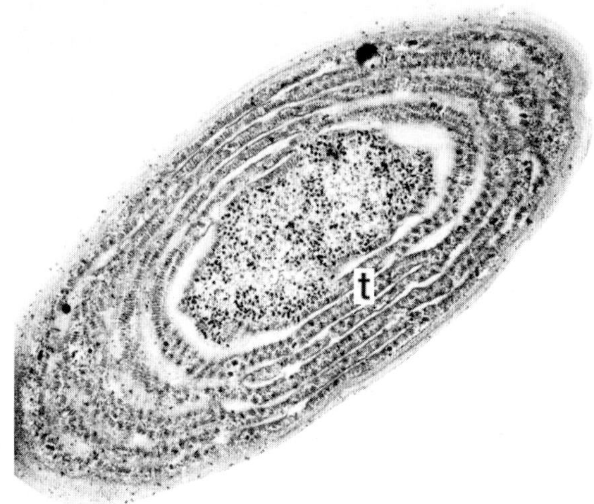

Figure 21.6 Cyanobacterial Thylakoids and Phycobilisomes. *Synechococcus lividus* with an extensive thylakoid system. The phycobilisomes lining these thylakoids are clearly visible as granules at location t (×85,000).

Phylum *Cyanobacteria*

The **cyanobacteria** are the largest and most diverse group of photosynthetic bacteria. There is little agreement about the number of cyanobacterial species. Older classifications had as many as 2,000 or more species. In one recent system this has been reduced to 62 species and 24 genera. *Bergey's Manual of Systematic Bacteriology* describes 56 genera. Cyanobacterial diversity is reflected in the G + C content of the group, which ranges from 35 to 71%. Although cyanobacteria are gram-negative bacteria, their photosynthetic system closely resembles that of the eucaryotes because they have chlorophyll *a* and photosystems I and II, thereby performing oxygenic photosynthesis. Indeed, the cyanobacteria were once known as "blue-green algae." Like the red algae, cyanobacteria use phycobiliproteins as accessory pigments. Photosynthetic pigments and electron transport chain components are located in thylakoid membranes lined with particles called **phycobilisomes (figure 21.6)**. These contain phycobilin pigments, particularly phycocyanin and phycoerythrin, that transfer energy to photosystem II. Carbon dioxide is assimilated through the Calvin cycle; the enzymes needed for this process are localized to internal structures called **carboxysomes.** The reserve carbohydrate is glycogen. Sometimes

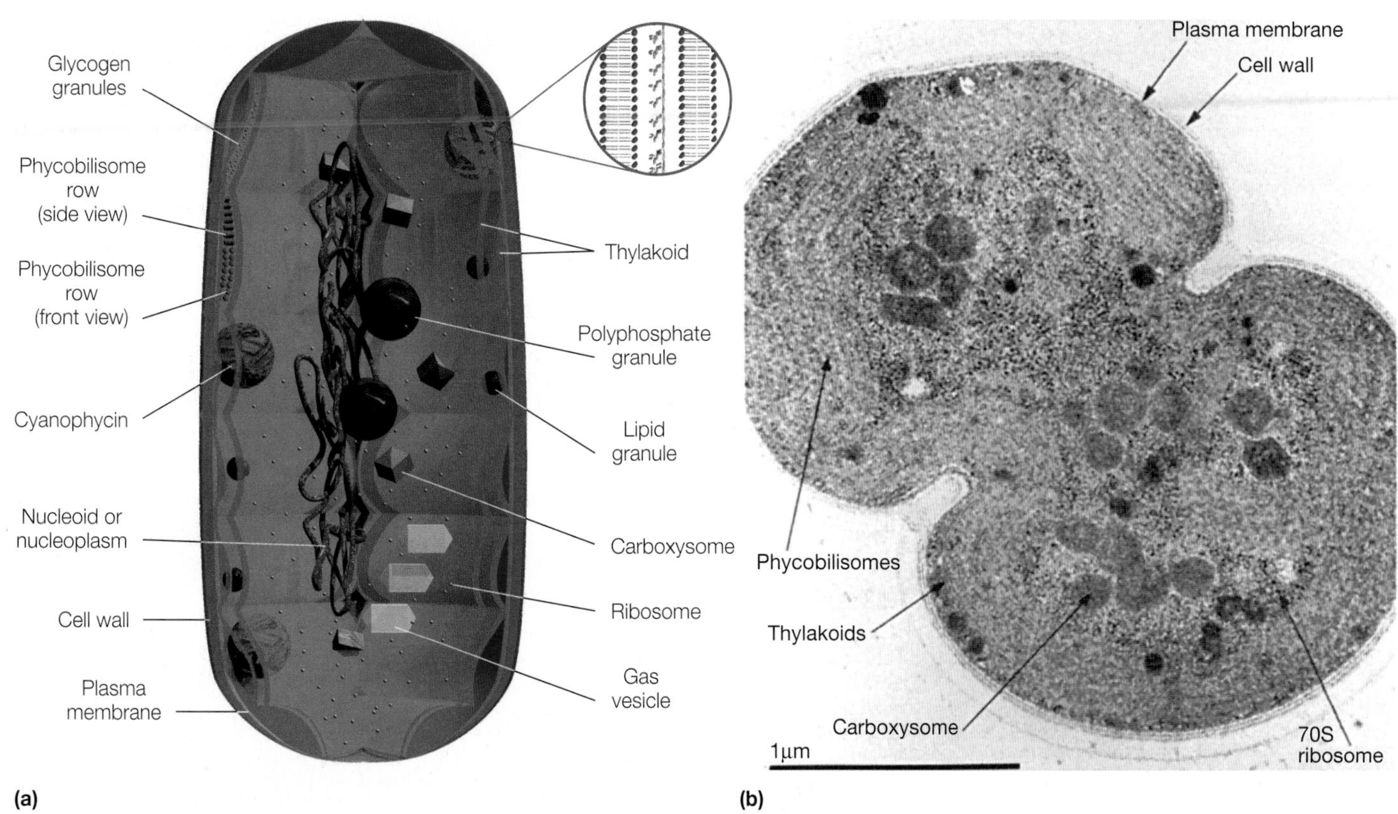

Figure 21.7 Cyanobacterial Cell Structure. **(a)** Schematic diagram of a vegetative cell. The insert shows an enlarged view of the envelope with its outer membrane and peptidoglycan. **(b)** Thin section of *Synechocystis* during division. Many structures are visible. (a) Illustration copyright © Hartwell T. Crim, 1998.

they store extra nitrogen as polymers of arginine or aspartic acid in **cyanophycin** granules. Phosphate is stored in polyphosphate granules (**figure 21.7**). Because cyanobacteria lack the enzyme α-ketoglutarate dehydrogenase, they do not have a fully functional citric acid cycle. The pentose phosphate pathway plays a central role in their carbohydrate metabolism. Although many cyanobacteria are obligate photolithoautotrophs, a few can grow slowly in the dark as chemoheterotrophs by oxidizing glucose and a few other sugars. Under anoxic conditions *Oscillatoria limnetica* oxidizes hydrogen sulfide instead of water and carries out anoxygenic photosynthesis much like the green photosynthetic bacteria. Obviously, cyanobacteria are capable of considerable metabolic flexibility.

Cyanobacteria also vary greatly in shape and appearance. They range in diameter from about 1 to 10 μm and may be unicellular, exist as colonies of many shapes, or form filaments called trichomes (**figure 21.8*c***). A **trichome** is a row of bacterial cells that are in close contact with one another over a large area. Although many appear blue-green because of phycocyanin, isolates from the open ocean are red or brown in color because of the pigment phycoerythrin. Cyanobacteria modulate the relative amounts of these pigments in a process known as **chromatic adaptation.** When orange light is perceived, phycocyanin production is stimulated, while blue and blue-green light promote the production of phycoerythrin. It is thought that sensory rhodopsins may play a role in signaling the spectral quality of light (*see p. 515*). Many cyanobacterial species use gas vacuoles to position themselves in optimum illumination in the water column—a form of **phototaxis. Gliding motility** is used by other cyanobacteria (**Microbial Diversity & Ecology 21.1**). Although cyanobacteria lack flagella, about one-third of the marine *Synechococcus* strains swim at rates up to 25 μm/sec by an unknown mechanism. Swimming motility is not used for phototaxis; instead, it appears to be used in chemotaxis toward simple nitrogenous compounds such as urea.

Cyanobacteria show great diversity with respect to reproduction and employ a variety of mechanisms: binary fission, budding, fragmentation, and multiple fission. In the last process a cell enlarges and then divides several times to produce many smaller progeny, which are released upon the rupture of the parental cell. Fragmentation of filamentous cyanobacteria can generate small, motile filaments called **hormogonia.** Some species develop **akinetes,** specialized, dormant, thick-walled resting cells that are resistant to desiccation (**figure 21.9*a***). Often these germinate to form new filaments.

Many filamentous cyanobacteria fix atmospheric nitrogen by means of special cells called **heterocysts** (figure 21.9). Around 5 to 10% of the cells develop into heterocysts when these cyanobacteria are deprived of both nitrate and ammonia, their preferred nitrogen sources. When individual cyanobacterial cells in

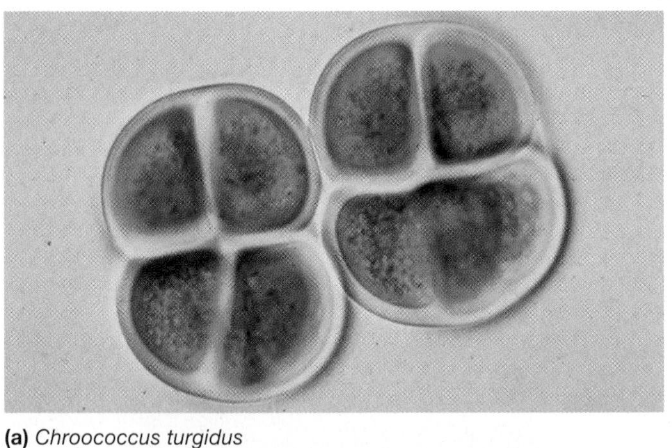

(a) *Chroococcus turgidus*

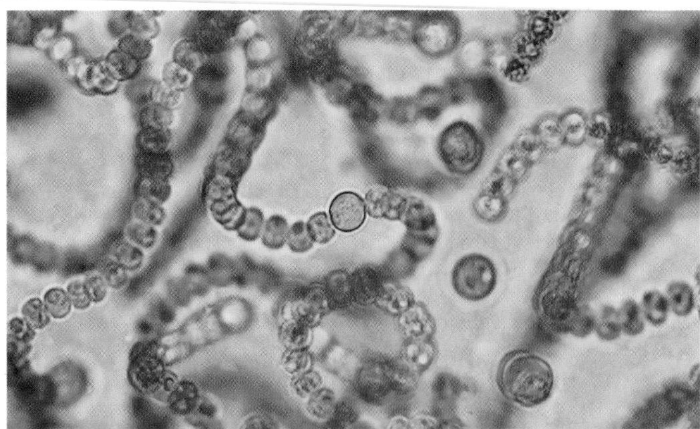

(b) *Nostoc*

(c) *Oscillatoria*

(d) *Anabaena spiroides* and *Microcystis aeruginosa*

Figure 21.8 Oxygenic Photosynthetic Bacteria. Representative cyanobacteria. **(a)** *Chroococcus turgidus,* two colonies of four cells each (×600). **(b)** *Nostoc* with heterocysts (×550). **(c)** *Oscillatoria* trichomes seen with Nomarski interference-contrast optics (×250). **(d)** The cyanobacteria *Anabaena spiroides* and *Microcystis aeruginosa.* The spiral *A. spiroides* is covered with a thick gelatinous sheath (×1,000).

a filamentous chain differentiate into heterocysts, the heterocysts develop a very thick cell wall. Within these specialized cells, photosynthetic membranes are reorganized and the proteins that make up photosystem II and phycobilisomes are degraded. Photosystem I remains functional to produce ATP, but no oxygen is generated. This inability to generate O_2 is critical because the enzyme nitrogenase is extremely oxygen sensitive. The thick heterocyst wall slows or prevents O_2 diffusion into the cell, and any O_2 present is consumed during respiration. Heterocyst structure and physiology ensure that it remains anaerobic; it is dedicated to nitrogen fixation and does not replicate. It obtains nutrients from adjacent vegetative cells and contributes fixed nitrogen in the form of amino acids. Nitrogen fixation also is carried out by some cyanobacteria that lack heterocysts. Some fix nitrogen under dark, anoxic conditions in microbial mats. Planktonic forms such as *Trichodesmium* fix nitrogen and contribute significantly to the marine nitrogen budget. Synthesis of amino acids: Nitrogen fixation (section 10.5); Biogeochemical cycling: Nitrogen cycle (section 27.2)

The classification of cyanobacteria is still unsettled. At present all taxonomic schemes must be considered tentative and while many genera have been assigned, species names have not yet been designated in most cases. *Bergey's Manual* divides the cyanobacteria into five subsections with 56 genera (**table 21.3**). These are distinguished using cell or filament morphology and reproductive patterns. Some other properties important in cyanobacterial characterization are ultrastructure, genetic characteristics, physiology and biochemistry, and habitat/ecology (preferred habitat and growth habit). Subsection I contains unicellular rods or cocci. Most are nonmotile and all reproduce by binary fission or budding. Organisms in subsection II are also unicellular, though several individual cells may be held together in an aggregate by an outer wall. Members of this group reproduce by multiple fission to form spherical, very small, reproductive cells called **baeocytes,** which escape when the outer wall ruptures. Some baeocytes disperse through gliding motility. The other three subsections contain filamentous cyanobacteria. Filaments are often surrounded by a sheath or slime layer. Cyanobacteria in subsection III form unbranched trichomes composed only of vegetative cells, whereas the other two subsections produce heterocysts in the absence of an adequate nitrogen source and also may

Microbial Diversity & Ecology

21.1 The Mechanism of Gliding Motility

Gliding motility always occurs on a solid surface, but it varies greatly in rate (from about 2 μm per minute to over 600 μm per minute) and in the nature of the motion. Although first observed over 100 years ago, the mechanism by which bacteria glide remains a mystery. Bacteria such as *Myxococcus* and *Flexibacter* glide along in a direction parallel to the longitudinal axis of their cells. Others (*Saprospira*) travel with a screwlike motion or even move in a direction perpendicular to the long axis of the cells in their trichome (*Simonsiella*). *Beggiatoa,* cyanobacteria, and some other bacteria rotate around their longitudinal axis while gliding, but this is not always seen. Many will flex or twitch as well as glide. Such diversity in gliding movement may indicate that more than one mechanism for motility exists. This conclusion is supported by the observation that some gliders (e.g., *Cytophaga, Flexibacter,* and *Flavobacterium*) move attached latex beads over their surface, whereas others such as *Myxococcus* either do not move beads or move them very slowly. (That is, not all gliding bacteria have rapidly moving cell-surface components.) Although slime is required for gliding, it does not appear to propel bacteria directly; rather, it probably attaches them to the substratum and lubricates the surface for more efficient movement.

A variety of mechanisms for gliding motility have been proposed. Cytoplasmic fibrils or filaments are associated with the envelope of many gliding bacteria. In *Oscillatoria* they seem to be contractile and may produce waves in the outer membrane, resulting in movement. Gliding motility may be best understood in the bacterium *Myxococcus xanthus.* This rod-shaped microbe glides in a pattern of reversals, so it has two motility "engines"—one at each pole. When gliding forward, type IV pili located at the front pole pull cells forward. This type of gliding motility is possible only when the cells are in a group and able to contact one another. Thus it is called social, or S, motility. The second motility engine consists of nozzlelike structures at the rear pole that eject slime to push the cells along the surface. Because slime production can occur in solitary cells, gliding driven by this mechanism is called adventurous, or A, motility. While the pili are always localized to the front of the cell, the nozzle structures are very dynamic and switch from one pole to another during gliding reversals. The presence of two different gliding mechanisms in a single bacterium suggests that further investigation may reveal additional diversity among gliding microorganisms. Class *Deltaproteobacteria:* Order *Myxococcales* (section 22.4)

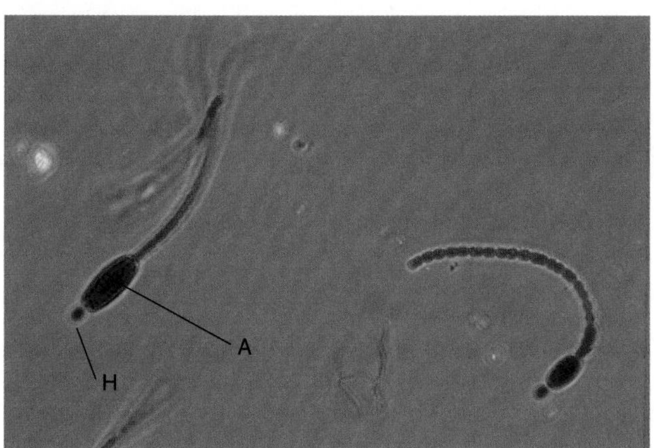

(a) *Cylindrospermum*

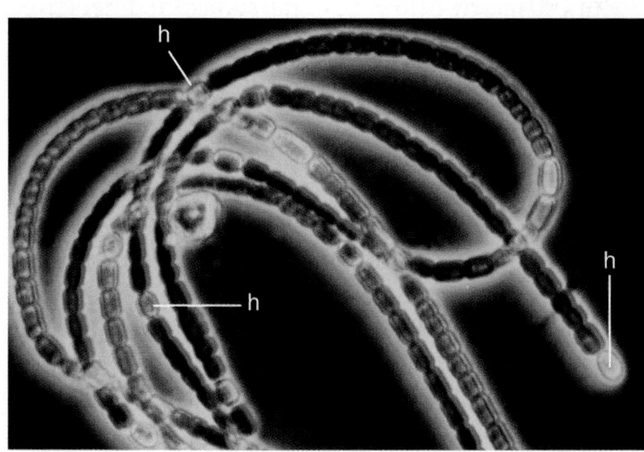

(b) *Anabaena*

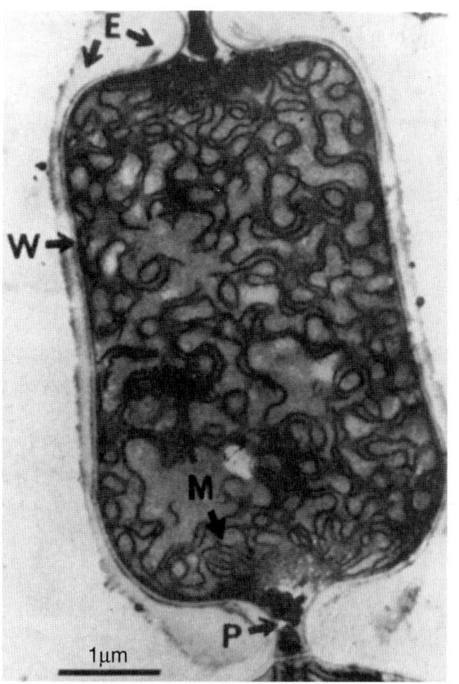

(c) *Anabaena* heterocyst

Figure 21.9 Examples of Heterocysts and Akinetes. **(a)** *Cylindrospermum* with terminal heterocysts (H) and subterminal akinetes (A) (×500). **(b)** *Anabaena,* with heterocysts. **(c)** An electron micrograph of an *Anabaena* heterocyst. Note the cell wall (W), additional outer walls (E), membrane system (M), and a pore channel to the adjacent cell (P).

Table 21.3 Characteristics of the Cyanobacterial Subsections

Subsection	General Shape	Reproduction and Growth	Heterocysts	% G + C	Other Properties	Representative Genera
I	Unicellular rods or cocci; nonfilamentous aggregates	Binary fission, budding	–	31–71	Nonmotile or swim without flagella	*Chroococcus* *Gloeothece* *Gleocapsa* *Prochlorococcus* *Prochloron* *Synechococcus*
II	Unicellular rods or cocci; may be held together in aggregates	Multiple fission to form baeocytes	–	40–46	Only some baeocytes are motile	*Pleurocapsa* *Dermocarpella* *Chroococcidiopsis*
III	Filamentous, unbranched trichome with only vegetative cells	Binary fission in a single plane, fragmentation	–	34–67	Usually motile	*Lyngbya* *Oscillatoria* *Prochlorothrix* *Spirulina* *Pseudanabaena*
IV	Filamentous, unbranched trichome may contain specialized cells	Binary fission in a single plane, fragmentation to form hormogonia	+	38–47	Often motile, may produce akinetes	*Anabaena* *Cylindrospermum* *Aphanizomenon* *Nostoc* *Scytonema* *Calothrix*
V	Filamentous trichomes either with branches or composed of more than one row of cells	Binary fission in more than one plane, hormogonia formed	+	42–44	May produce akinetes; greatest morphological complexity and differentiation in cyanobacteria	*Fischerella* *Stigonema* *Geitleria*

form akinetes. Heterocystous cyanobacteria are subdivided into those that form unbranched filaments (subsection IV) and cyanobacteria that divide in a second plane to produce branches or aggregates (subsection V).

The cyanobacteria that are collectively referred to as prochlorphytes (genera *Prochloron, Prochlorococcus,* and *Prochlorothrix*) are distinguished by the presence of both chlorophyll *a* and *b* and their lack of phycobilins. This pigment arrangement imparts a grass-green color to these microbes. As the only procaryotes to possess chlorophyll *b,* the ancestors of unicellular prochlorophytes are considered by some to be the best candidates as the endosymbionts that gave rise to chloroplasts. Considerable controversy has surrounded their phylogeny, but small subunit rRNA sequence analysis places them with the cyanobacteria. Microbial evolution: The endosymbiotic origin of mitochondria and chloroplasts (section 19.1)

The three recognized prochlorophyte genera are quite different from one another. *Prochloron* was first discovered as an extracellular symbiont growing either on the surface or within the cloacal cavity of marine colonial ascidian invertebrates (**figure 21.10**). These bacteria are single-celled, spherical, and from 8 to 30 μm in diameter. Their mol% of G + C is 31 to 41. *Prochlorothrix* is free living, has cylindrical cells that form filaments, and grows in freshwater. Its DNA has a higher G + C content (53 mol%).

Prochlorococcus marinus, which is less than 1 μm in diameter, flourishes about 100 meters below the ocean surface. It differs from other prochlorophytes in having divinyl chlorophyll *a* and *b* and α-carotene instead of chlorophyll *a* and β-carotene. During the summer, it reaches concentrations of 5×10^5 cells per milliliter. It is one of the most numerous of the marine plankton and may be the most abundant oxygenic photosynthetic organism on Earth.

Another indication of the vast diversity among the cyanobacteria is the wide range of habitats they occupy. Thermophilic species may grow at temperatures of up to 75°C in neutral to alkaline hot springs. Because these photoautotrophs are so hardy, they are primary colonizers of soils and surfaces that are devoid of plant growth. Some unicellular forms even grow in the fissures of desert rocks. In nutrient-rich warm ponds and lakes, surface cyanobacteria such as *Anacystis* and *Anabaena* can reproduce

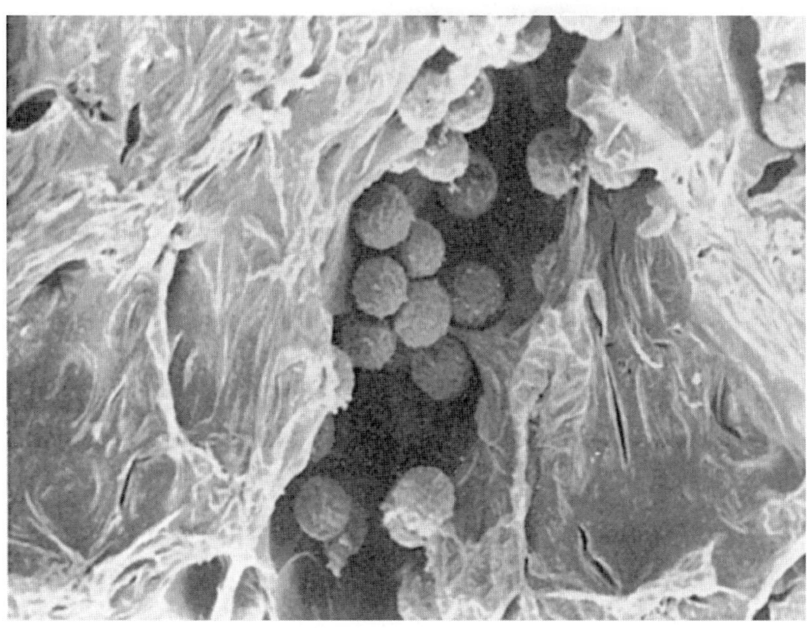
(a)

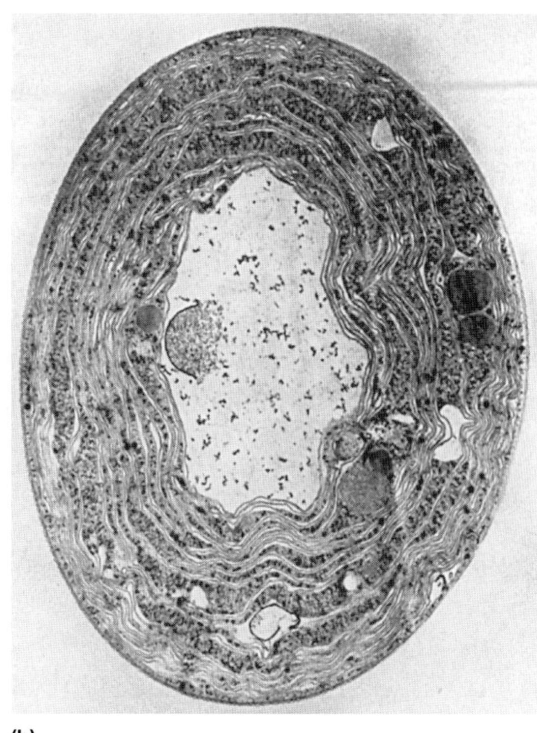
(b)

Figure 21.10 Prochloron. **(a)** A scanning electron micrograph of *Prochloron* cells on the surface of a *Didemnum candidum* colony. **(b)** *Prochloron didemni* thin section; transmission electron micrograph (×23,500).

Figure 21.11 Bloom of Cyanobacteria and Algae in a Eutrophic Pond.

rapidly to form blooms (**figure 21.11**). The release of large amounts of organic matter upon the death of the bloom microorganisms stimulates the growth of chemoheterotrophic bacteria. These microbes subsequently deplete available oxygen. This kills fish and other organisms. Some species can produce toxins that kill livestock and other animals that drink the water. Other cyanobacteria (e.g., *Oscillatoria*) are so pollution resistant and characteristic of freshwaters with high organic matter content that they are used as water pollution indicators. Disease 25.1: Harmful algal blooms; Microorganisms in marine and freshwater environments (chapter 28)

Cyanobacteria are particularly successful in establishing symbiotic relationships with other organisms. For example, they are the photosynthetic partner in most lichen associations. Cyanobacteria are symbionts with protozoa and fungi, and nitrogen-fixing species form associations with a variety of plants (liverworts, mosses, gymnosperms, and angiosperms). Microbial interactions (section 30.1)

1. Summarize the major characteristics of the cyanobacteria that distinguish them from other photosynthetic organisms.
2. Define or describe the following: phycobilisomes, hormogonia, akinetes, heterocysts, and baeocytes.
3. What is a trichome and how does it differ from a simple chain of cells?
4. Briefly discuss the ways in which cyanobacteria reproduce.
5. Describe how a vegetative cell, a heterocyst, and an akinete are different. How are heterocysts modified to carry out nitrogen fixation? Why do you think heterocysts are considered terminally differentiated cells?
6. Give the features of the five major cyanobacterial groups.
7. Compare the prochlorophytes with other cyanobacteria and chloroplasts. Why do you think the phylogeny of the prochlorophytes has been so difficult to establish?
8. List some important positive and negative impacts cyanobacteria have on humans and the environment.

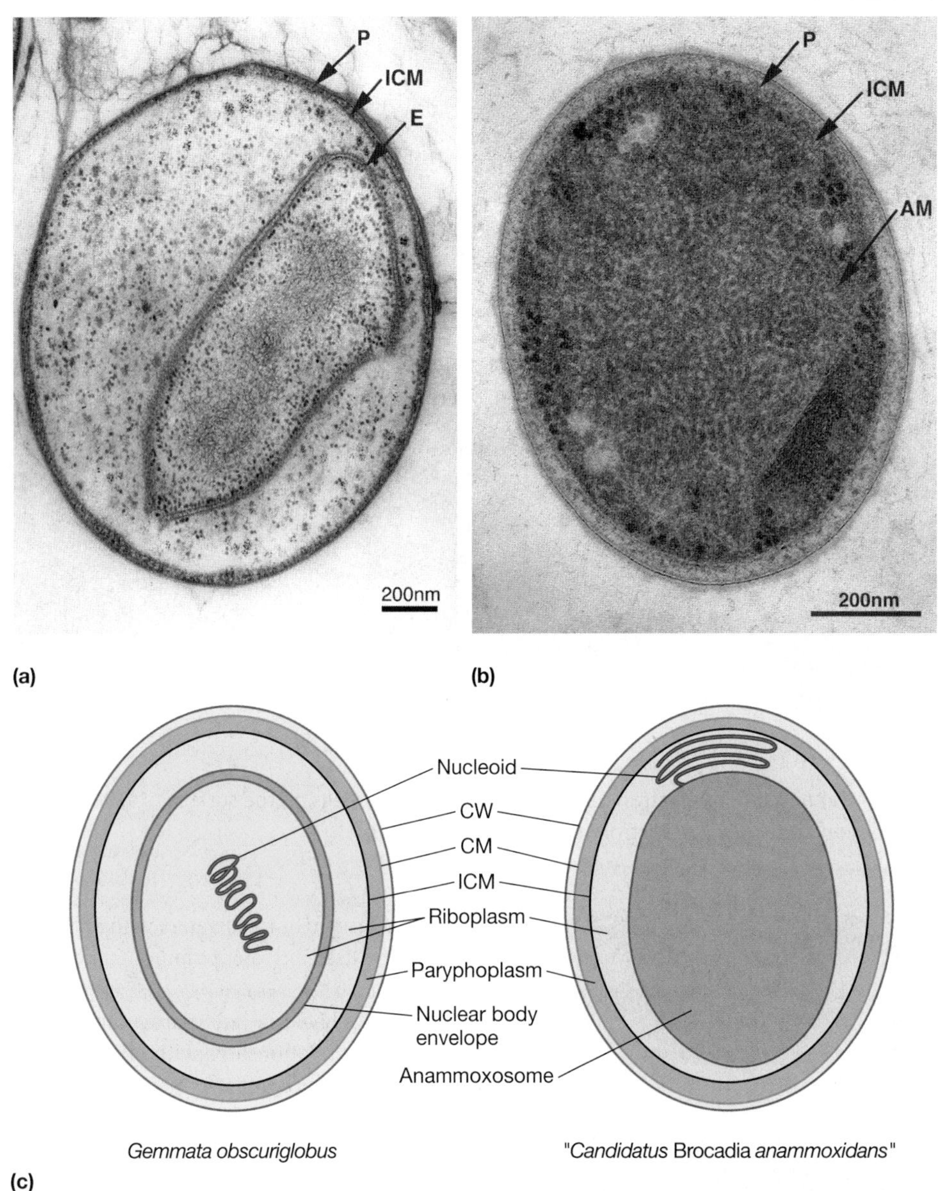

Figure 21.12 Planctomycete Cellular Compartmentalization. **(a)** An electron micrograph of *Gemmata obscuriglobus* showing the nuclear body envelope (E), the intracytoplasmic membrane (ICM), and the paryphoplasm (P). **(b)** An electron micrograph of the anaerobic ammonia-oxidizing planctomycete "*Candidatus Brocadia anammoxidans*" (quotation marks indicate unsettled nomenclature). The anamoxosome is labeled AM. **(c)** Schematic drawings corresponding to (a) and (b): cell wall (CW), cytoplasmic membrane (CM).

21.4 Phylum *Planctomycetes*

The phylum *Planctomycetes* contains one class, one order, and four genera. The planctomycetes are morphologically unique bacteria, having compartmentalized cells **(figure 21.12)**. Although each species is unique, all follow a basic cellular organization that includes a cytoplasmic membrane closely surrounded by the cell wall, which lacks peptidoglycan. The largest internal compartment, the intracytoplasmic membrane (ICM), is separated from the cytoplasmic membrane by a peripheral, ribosome-free region called the paryphoplasm. The nucleoid of the planctomycete *Gemmata obscuriglobus* is located in the nuclear body, which is enclosed in a double membrane. The species "*Candidatus Brocadia anammoxidans*" does not localize its DNA within a nuclear body; however, it has another compartment: the anammoxosome. This is the site of anaerobic ammonia oxidation, a unique and recently discovered form of chemolithotrophy in which ammonium ion (NH_4^+) serves as the electron donor and nitrite (NO_2^-) as the terminal electron acceptor; it is reduced to nitrogen gas (N_2). Note that the nomenclature of this microbe is still in flux so in some cases genus and species names are enclosed in quotation marks. Biogeochemical cycling: Nitrogen cycle (section 27.2)

The genus *Planctomyces* attaches to surfaces through a stalk and holdfast; the other genera in the order lack stalks. Most of these bacteria have life cycles in which sessile cells bud to produce motile swarmer cells. The swarmer cells are flagellated and swim for a while before settling down to attach and begin reproduction.

21.5 Phylum *Chlamydiae*

The gram-negative *Chlamydiae* are obligate intracellular parasites. That is, they must grow and reproduce within host cells. Although their ability to cause disease is widely recognized, many species grow within protists and animal cells without adverse affects. It is thought that these hosts represent a natural reservoir for the *Chlamydiae*.

The phylum *Chlamydiae* has one class, one order, four families, and only six genera. The genus *Chlamydia* is by far the most important and best studied; it will be the focus of our attention. **Chlamydiae** are nonmotile, coccoid bacteria, ranging in size from 0.2 to 1.5 μm. They reproduce only within cytoplasmic vesicles of host cells by a unique developmental cycle involving the formation of two cell types: elementary bodies and reticulate bodies. Although their envelope resembles that of other gram-negative bacteria, the cell wall differs in lacking muramic acid and a peptidoglycan layer. Elementary bodies achieve osmotic stability by cross-linking their outer membrane proteins, and possibly periplasmic proteins, with disulfide bonds. Chlamydiae are extremely limited metabolically, relying on their host cells for key metabolites. This is reflected in the size of their genome. It is relatively small at 1.0 to 1.3 Mb; the G + C content is 41 to 44%.

Chlamydial reproduction begins with the attachment of an **elementary body (EB)** to the cell surface (**figure 21.13**). Elementary bodies are 0.2 to 0.6 μm in diameter, contain electron-dense nuclear material and a rigid cell wall, and are infectious. The host cell phagocytoses the EB, which are held in inclusion bodies where the EB reorganizes itself to form a **reticulate body (RB)** or **initial body.** The RB is specialized for reproduction rather than infection. Reticulate bodies are 0.6 to 1.5 μm in diameter and have less dense nuclear material and more ribosomes than EBs; their walls are also more flexible. About 8 to 10 hours after infection, the reticulate body undergoes binary fission and RB reproduction continues until the host cell dies. A chlamydia-filled inclusion can become large enough to be seen

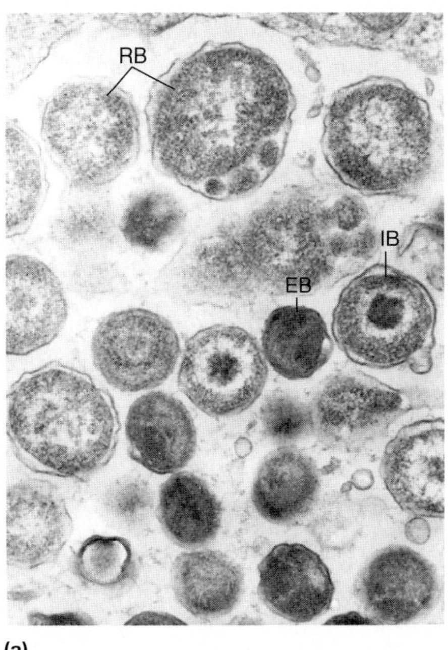

(a)

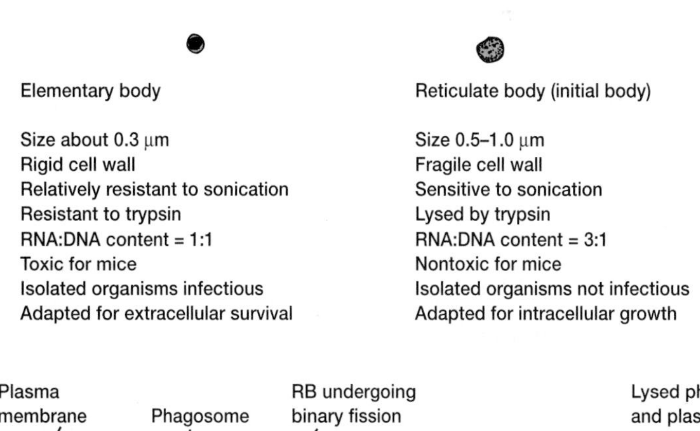

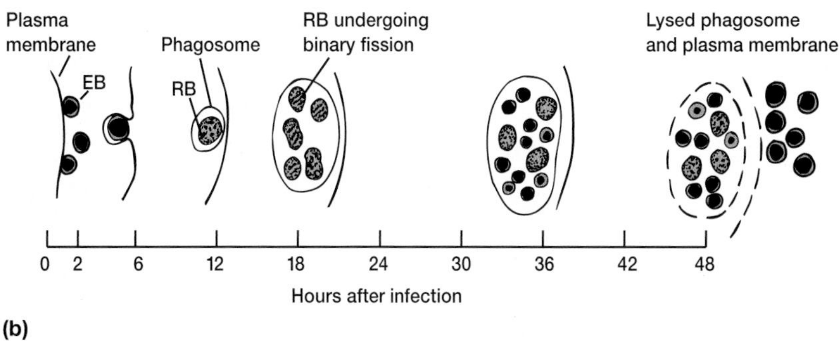

(b)

Figure 21.13 The Chlamydial Life Cycle. **(a)** An electron micrograph of an inclusion body containing a mixture of small, black elementary bodies (EB), larger reticulate bodies (RB), and an intermediate body (IB), a chlamydial cell intermediate in morphology between EB and RB. The RBs appear gray and granulated due to a high concentration of ribosomes. **(b)** A schematic representation of the infectious cycle of chlamydiae.

in a light microscope and even fill the host cytoplasm. After 20 to 25 hours, RBs begin to differentiate into infectious EBs and continue this process until the host cell lyses and releases the chlamydiae 48 to 72 hours after infection.

Chlamydial metabolism is very different from that of other gram-negative bacteria. It had been thought that chlamydiae cannot catabolize carbohydrates or other substances or synthesize ATP. *Chlamydia psittaci,* one of the best-studied species, lacks both flavoprotein and cytochrome electron transport chain carriers, but has a membrane translocase that acquires host ATP in exchange for ADP. Thus chlamydiae seem to be energy parasites that are completely dependent on their hosts for ATP. However, this might not be the complete story. The *C. trachomatis* genome sequence indicates that the bacterium may be able to synthesize at least some ATP. Although there are two genes for ATP/ADP translocases, there also are genes for substrate-level phosphorylation, electron transport, and oxidative phosphorylation. When supplied with precursors from the host, RBs can synthesize DNA, RNA, glycogen, lipids, and proteins. Presumably the RBs have porins and active membrane transport proteins, but little is known about these. They also can synthesize at least some amino acids and coenzymes. The EBs have minimal metabolic activity and cannot take in ATP or synthesize proteins. They are designed exclusively for transmission and infection. Insights from microbial genomes: Genomic analysis of pathogenic microbes (section 15.8)

Three chlamydial species are important pathogens of humans and other warm-blooded animals. *C. trachomatis* infects humans and mice. In humans it causes trachoma, nongonococcal urethritis, and other diseases. *C. psittaci* causes psittacosis in humans. However, unlike *C. trachomatis,* it also infects many other animals (e.g., parrots, turkeys, sheep, cattle, and cats) and invades the intestinal, respiratory, and genital tracts; the placenta and fetus; the eye; and the synovial fluid of joints. *Chlamydiophila pneumoniae* is a common cause of human pneumonia. Human diseases caused by bacteria (chapter 38)

21.6 Phylum *Spirochaetes*

The phylum *Spirochaetes* (Greek *spira,* a coil, and *chaete,* hair) contains gram-negative, chemoheterotrophic bacteria distinguished by their structure and mechanism of motility. They are slender, long bacteria (0.1 to 3.0 μm by 5 to 250 μm) with a flexible, helical shape (**figure 21.14**). Many species are so slim that they are only clearly visible in a light microscope by means of phase-contrast or dark-field optics. Spirochetes differ greatly from other bacteria with respect to motility and can move through very viscous solutions though they lack external rotating flagella. When in contact with a solid surface, they exhibit creeping or crawling movements. Their unique pattern of motility is due to an unusual morphological structure called the axial filament. The light microscope (section 2.2)

The distinctive features of spirochete morphology are evident in electron micrographs (**figure 21.15**). The central protoplasmic cylinder contains cytoplasm and the nucleoid, and is bounded by a plasma membrane and a gram-negative cell wall. Two to more than a hundred flagella, called **axial fibrils, periplasmic flagella,** or endoflagella, extend from both ends of the cylinder and often overlap one another in the center third of the cell (figure 21.16*c,d*). The whole complex of periplasmic flagella, the **axial filament,** lies inside a flexible outer sheath. The outer sheath contains lipid, protein, and carbohydrate and varies in structure between different genera. Its precise function is unknown, but the sheath is essential because spirochetes die if it is damaged or removed. The outer sheath of *Treponema pallidum* has few proteins exposed on

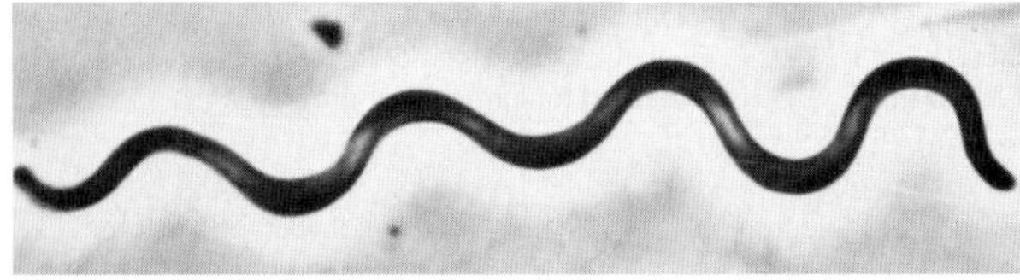

(a) *Cristispira*

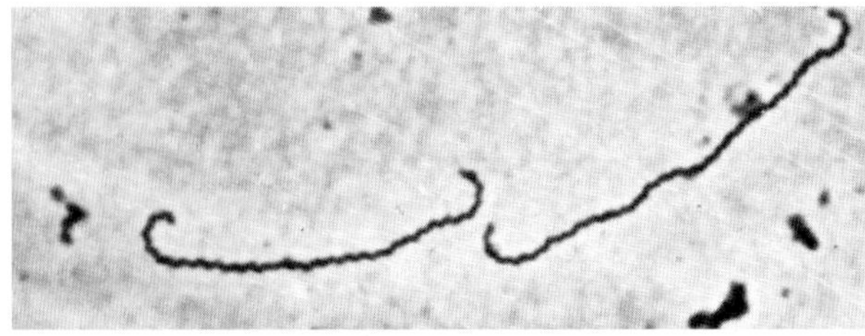

(c) *Leptospira interrogans*

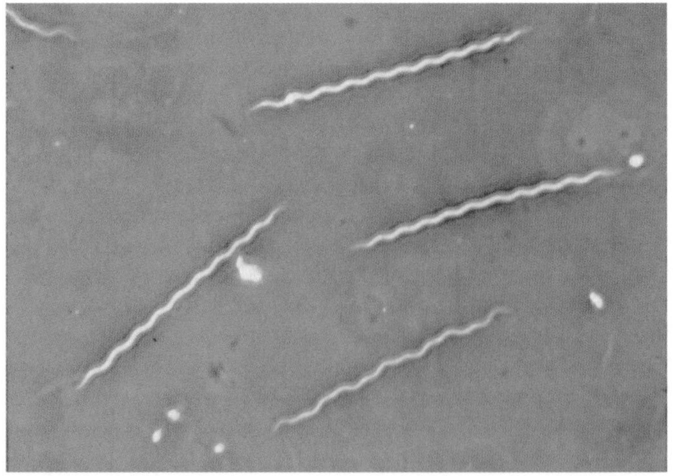

(b) *Treponema pallidum*

Figure 21.14 The Spirochetes. Representative examples. **(a)** *Cristispira* sp. from a clam; phase contrast (× 2,200). **(b)** *Treponema pallidum* (×1,000). **(c)** *Leptospira interrogans* (×2,200).

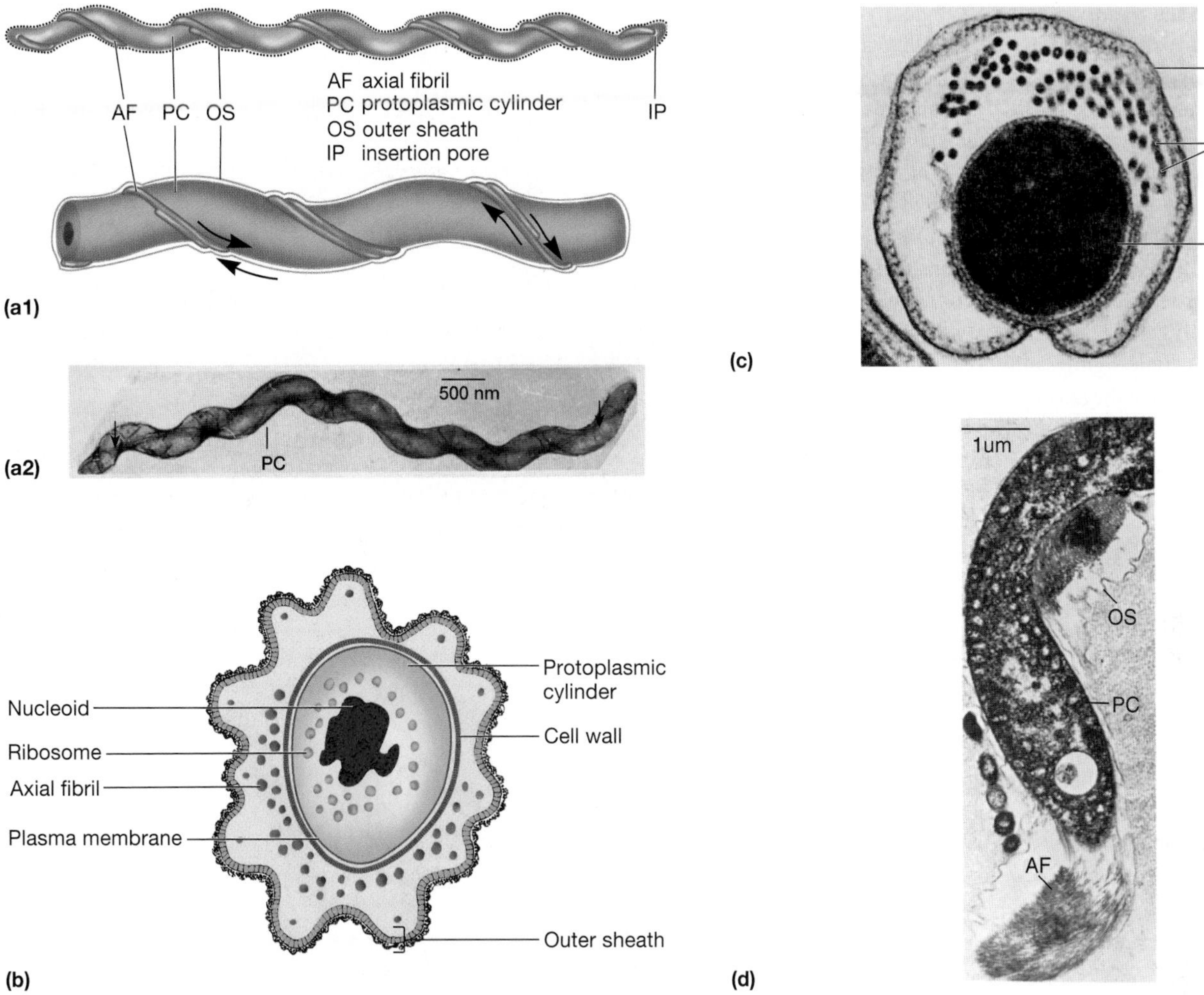

Figure 21.15 Spirochete Morphology. **(a1)** A surface view of spirochete structure as interpreted from electron micrographs. **(a2)** A longitudinal view of *Treponema zuelzerae* with axial fibrils extending most of the cell length. **(b)** A cross section of a typical spirochete showing morphological details. **(c)** Electron micrograph of a cross section of *Clevelandina* from the termite *Reticulitermes flavipes* showing the outer sheath, protoplasmic cylinder, and axial fibrils (×70,000). **(d)** Longitudinal section of *Cristispira* showing the outer sheath (OS), the protoplasmic cylinder (PC), and the axial fibrils (AF).

its surface. This allows the syphilis spirochete to avoid attack by host antibodies.

The way in which periplasmic flagella propel the cell has not been fully established. Mutants with straight rather than curved flagella are nonmotile. The periplasmic flagella rotate like the external flagella of other bacteria. This causes the corkscrew-shaped outer sheath to rotate and move the cell through the surrounding liquid (**figure 21.16**). Flagellar rotation may also flex or bend the cell and account for the crawling movement seen on solid surfaces.

Spirochetes can be anaerobic, facultatively anaerobic, or aerobic. Carbohydrates, amino acids, long-chain fatty acids, and long-chain fatty alcohols may serve as carbon and energy sources.

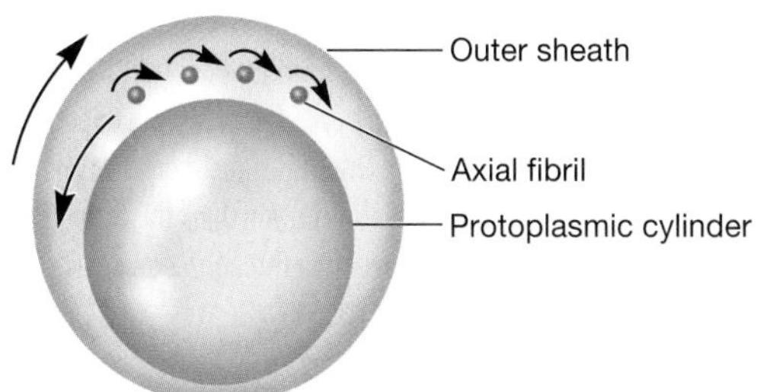

Figure 21.16 Spirochete Motility. A hypothetical mechanism for spirochete motility. See text for details.

The group is exceptionally diverse ecologically and grows in habitats ranging from mud to the human mouth. Members of the genus *Spirochaeta* are free-living and often grow in anoxic and sulfide-rich freshwater and marine environments. Some species of the genus *Leptospira* grow in oxic water and moist soil. Some spirochetes form symbiotic associations with other organisms and are found in a variety of locations: the hindguts of termites and wood-eating roaches, the digestive tracts of mollusks (*Cristispira*) and mammals, and the oral cavities of animals (*Treponema denticola, T. oralis*). Spirochetes from termite hindguts and freshwater sediments have nitrogenase and can fix nitrogen. There is evidence that they contribute significantly to the nitrogen nutrition of termites. Spirochetes coat the surfaces of many protozoa from termite and wood-eating roach hindguts (**figure 21.17**). For example, the flagellate *Myxotricha paradoxa* is covered with slender spirochetes (0.15 by 10 μm in length) that are firmly attached and help move the protozoan. Some members of the genera *Treponema, Borrelia,* and *Leptospira* are important pathogens; for example, *Treponema pallidum* causes syphilis, and *Borrelia burgdorferi* is responsible for Lyme disease. The study of *T. pallidum* and its role in syphilis has been hindered by the inability to culture the spirochete outside its human host. The *T. pallidum* genome sequence shows that this spirochete is metabolically crippled and quite dependent on its host. The *B. burgdorferi* genome consists of a linear chromosome of 910,725 base pairs and at least 17 linear and circular plasmids, which constitute another 533,000 base pairs. The plasmids have some genes that are normally found on chromosomes, and plasmid proteins seem to be involved in bacterial virulence. Insights from microbial genomes: Genomic analysis of pathogenic microbes (section 15.8); Arthropod-borne diseases: Lyme disease (section 38.2); Direct contact diseases: Sexually transmitted diseases (section 38.3)

Bergey's Manual divides the phylum *Spirochaetes* into one class, one order (*Spirochaetales*), and three families (*Spirochaetaceae, Serpulinaceae,* and *Leptospiraceae*). At present, there are 13 genera in the phylum. **Table 21.4** summarizes some of the more distinctive properties of selected genera.

1. Describe the *Planctomycetes* and their more distinctive properties.
2. Give the major characteristics of the phylum *Chlamydiae.* What are elementary and reticulate bodies? Briefly describe the steps in the chlamydial life cycle. Why do you think *Chlamydiae* differentiate into specialized cell types for infection and reproduction?
3. How does chlamydial metabolism differ from that of other bacteria?
4. Define the following terms: protoplasmic cylinder, axial fibrils or periplasmic flagella, axial filament, and outer sheath. Draw and label a diagram of spirochete morphology, locating these structures. Why do you think this form of motility might be especially well suited for movement through viscous fluids?

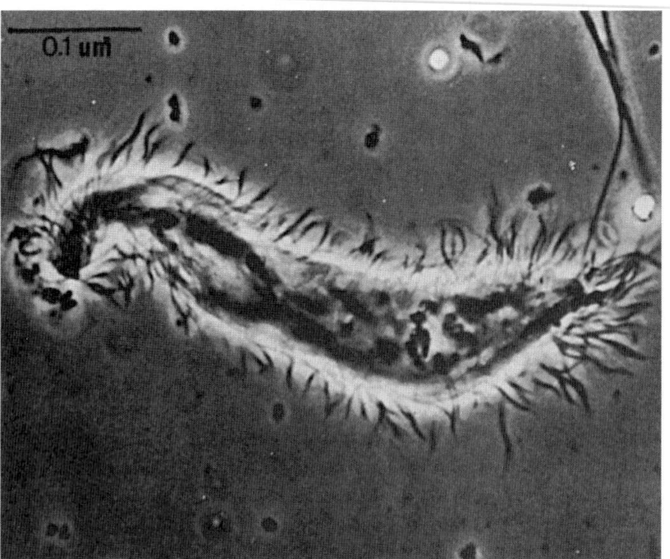

(a)

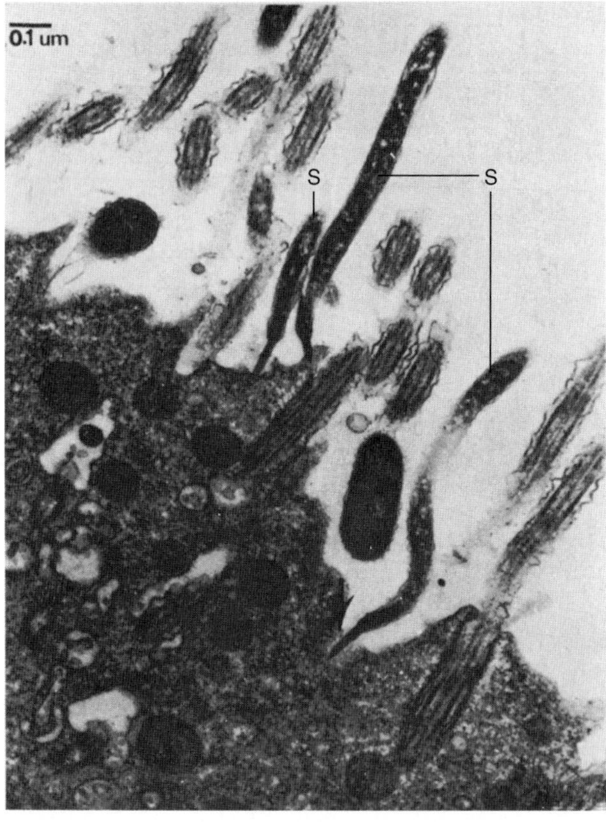

(b)

Figure 21.17 Spirochete-Protozoan Associations. The surface spirochetes serve as organs of motility for protozoa. **(a)** The spirochete-*Personympha* association with the spirochetes projecting from the protist's surface. **(b)** Electron micrograph of small spirochetes (S) attached to the membrane of the flagellate protozoan *Barbulanympha.*

21.7 Phylum *Bacteroidetes*

The phylum *Bacteroidetes* is very diverse and seems most closely related to the phylum *Chlorobi.* The phylum has three classes (*Bacteroides, Flavobacteria,* and *Sphingobacteria*), 12 families, and 63 genera.

Table 21.4 Characteristics of Spirochete Genera

Genus	Dimensions (μm) and Flagella	G + C Content (mol%)	Oxygen Relationship	Carbon + Energy Source	Habitats
Spirochaeta	0.2–0.75 × 5–250; 2–40 periplasmic flagella (almost always 2)	51–65	Facultatively anaerobic or anaerobic	Carbohydrates	Aquatic and free-living
Cristispira	0.5–3.0 × 30–180; ≥ 100 periplasmic flagella	N.A.[a]	Facultatively anaerobic?	N.A.	Mollusk digestive tract
Treponema	0.1–0.4 × 5–20: 2–16 periplasmic flagella	25–53	Anaerobic or microaerophilic	Carbohydrates or amino acids	Mouth, intestinal tract, and genital areas of animals; some are pathogenic (syphilis, yaws)
Borrelia	0.2–0.5 × 3–20; 14–60 periplasmic flagella	27–32	Anaerobic or microaerophilic	Carbohydrates	Mammals and arthropods; pathogens (relapsing fever, Lyme disease)
Leptospira	0.1 × 6–24; 2 periplasmic flagella	35–53	Aerobic	Fatty acids and alcohols	Free-living or pathogens of mammals, usually located in the kidney (leptospirosis)
Leptonema	0.1 × 6–20; 2 periplasmic flagella	51–53	Aerobic	Fatty acids	Mammals
Brachyspira	0.2 × 1.7–6.0; 8 periplasmic flagella	25–27	Anaerobic	N.A.	Mammalian intestinal tract
Serpulina	0.3–0.4 × 7–9; 16–18 periplasmic flagella	25–26	Anaerobic	Carbohydrates and amino acids	Mammalian intestinal tract

[a]N.A., information not available

The class *Bacteroides* contains anaerobic, gram-negative, nonsporing, motile or nonmotile rods of various shapes. These bacteria are chemoheterotrophic and usually produce a mixture of organic acids as fermentation end products. They do not reduce sulfate or other sulfur compounds. The genera are identified using properties such as general shape, motility and flagellation pattern, and fermentation end products. These bacteria grow in habitats such as the oral cavity and intestinal tract of vertebrates and the rumen of ruminants. Microbial interactions: The rumen ecosystem (section 30.1); Normal microbiota of the human body: Large intestine (section 30.3)

Although difficulty culturing these anaerobes has hindered our understanding of them; they are clearly widespread and important. Often they benefit their host. *Bacteroides ruminicola* is a major component of the rumen flora; it ferments starch, pectin, and other carbohydrates. About 30% of the bacteria isolated from human feces are members of the genus *Bacteroides,* and these organisms may provide extra nutrition by degrading cellulose, pectins, and other complex carbohydrates (*see figure 30.18*). The family also is involved in human disease. Members of the genus *Bacteroides* are associated with diseases of major organ systems, ranging from the central nervous system to the skeletal system. *B. fragilis* is a particularly common anaerobic pathogen found in abdominal, pelvic, pulmonary, and blood infections.

Another important group in the *Bacteroidetes* is the class *Sphingobacteria.* Besides the similarity in their 16S rRNA sequences, sphingobacteria often have sphingolipids in their cell walls. Some genera in this class are *Sphingobacterium, Saprospira, Flexibacter, Cytophaga, Sporocytophaga,* and *Crenothrix.*

The genera *Cytophaga, Sporocytophaga,* and *Flexibacter* differ from each other in morphology, life cycle, and physiology. Bacteria of the genus *Cytophaga* are slender rods, often with pointed ends (**figure 21.18*a***). *Sporocytophaga* is similar to *Cytophaga* but forms spherical resting cells called microcysts (figure 21.18*b,c*). *Flexibacter* produces long, flexible threadlike cells when young (figure 21.18*d*) and is unable to use complex polysaccharides. Often colonies of these bacteria are yellow to orange because of carotenoid or flexirubin pigments. Some of the flexirubins are chlorinated, which is unusual for biological molecules.

Members of the genera *Cytophaga* and *Sporocytophaga* are aerobes that actively degrade complex polysaccharides. Soil cytophagas digest cellulose; both soil and marine forms attack

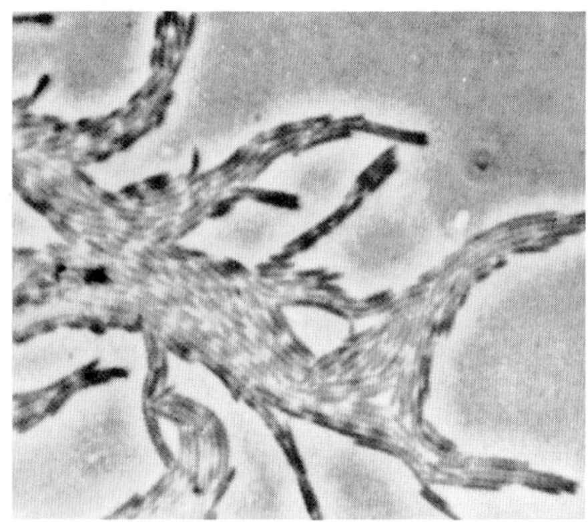

(a) *Cytophaga*

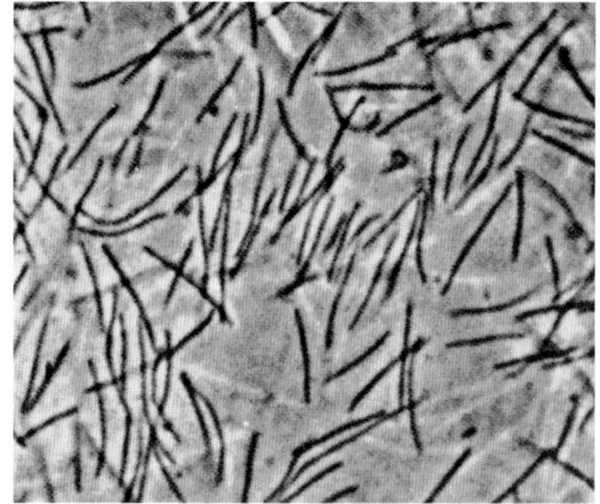

(b) *Sporocytophaga myxococcoides*

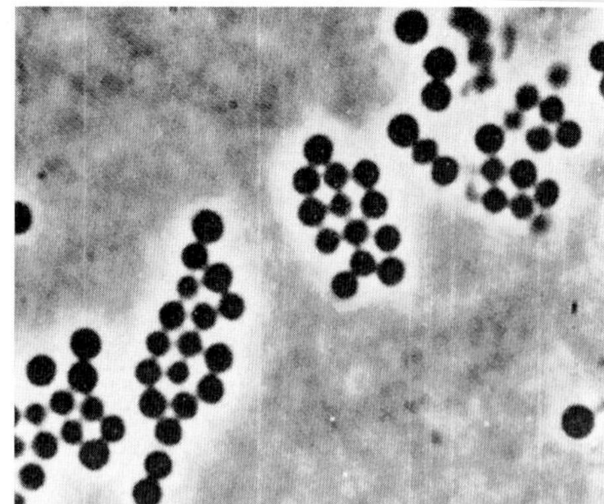

(c) *S. myxococcoides* microcysts

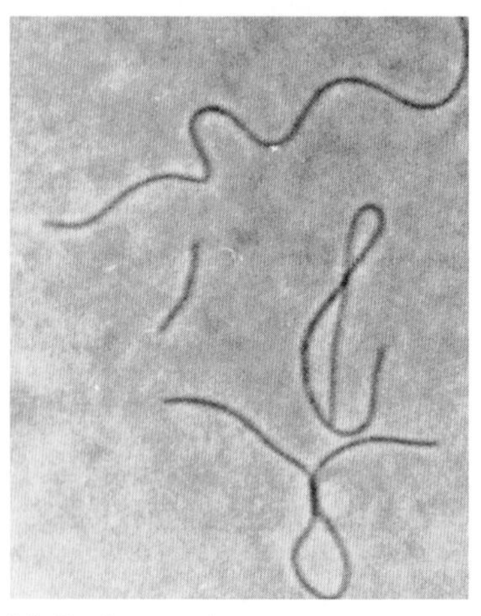

(d) *Flexibacter elegans*

Figure 21.18 Nonphotosynthetic, Nonfruiting, Gliding Bacteria. Representative members of the order *Cytophagales.* **(a)** *Cytophaga* sp. (×1,150). **(b)** *Sporocytophaga myxococcoides,* vegetative cells on agar (×1,170). **(c)** *Sporocytophaga myxococcoides,* mature microcysts (×1,750). **(d)** Long thread cells of *Flexibacter elegans* (×1,100).

chitin, pectin, and keratin. Some marine species even degrade agar, a component of seaweed. Cytophagas play a major role in the mineralization of organic matter and can cause great damage to exposed fishing gear and wooden structures. They also are a major component of the bacterial population in sewage treatment plants and presumably contribute significantly to the waste treatment process. Wastewater treatment (section 41.2)

Although most cytophagas are free-living, some can be isolated from vertebrate hosts and are pathogenic. *Cytophaga columnaris* and others cause diseases such as columnaris disease, cold water disease, and fin rot in freshwater and marine fish.

The gliding motility so characteristic of these organisms is quite different from flagellar motility. Gliding motility is present in a wide diversity of taxa: fruiting and nonfruiting aerobic chemoheterotrophs, cyanobacteria, green nonsulfur bacteria, and at least two gram-positive genera (*Heliobacterium* and *Desulfonema*). Gliding bacteria lack flagella and are stationary while suspended in liquid medium. When in contact with a surface, they glide along, leaving a slime trail. Movement can be very rapid; some cytophagas travel 150 μm in a minute, whereas filamentous gliding bacteria may reach speeds of more than 600 μm/minute. Young organisms are the most motile, and motility often is lost with age. Low nutrient levels usually stimulate gliding. The gliding mechanism is not well understood (Microbial Diversity & Ecology 21.1).

Gliding motility gives a bacterium many advantages. Many aerobic chemoheterotrophic gliding bacteria actively digest insoluble macromolecular substrates such as cellulose and chitin, and gliding motility is ideal for searching these out. Because many of the digestive enzymes are cell wall associated, the bacteria must be in contact with insoluble nutrient sources; gliding motility makes this possible. Gliding movement is well adapted to drier habitats and to movement within solid masses such as soil, sediments, and rotting wood that are permeated by small channels. Finally, gliding bacteria, like flagellated bacteria, can position themselves at optimal conditions of light intensity, oxygen, hydrogen sulfide, temperature, and other factors that influence growth and survival.

1. Give the major properties of the class *Bacteroides*.
2. How do these bacteria benefit and harm their hosts?
3. List three advantages of gliding motility.
4. Briefly describe the following genera: *Cytophaga, Sporocytophaga,* and *Flexibacter.*
5. Why are the cytophagas ecologically important?

Summary

21.1 *Aquificae* and *Thermotogae*

a. *Aquifex* and *Thermotoga* are hyperthermophilic gram-negative rods that represent the two deepest or oldest phylogenetic branches of the *Bacteria*.

21.2 *Deinococcus-Thermus*

a. Members of the order *Deinococcales* are aerobic, gram-positive cocci and rods that are distinctive in their unusually great resistance to desiccation and radiation.

21.3 Photosynthetic Bacteria

a. Cyanobacteria carry out oxygenic photosynthesis; purple and green bacteria use anoxygenic photosynthesis.
b. The four most important groups of purple and green photosynthetic bacteria are the purple sulfur bacteria, the purple nonsulfur bacteria, the green sulfur bacteria, and the green nonsulfur bacteria (**table 21.1**).
c. The bacteriochlorophyll pigments of purple and green bacteria enable them to live in deeper, anoxic zones of aquatic habitats.
d. The phylum *Chlorobi* includes the green sulfur bacteria—obligately anaerobic photolithoautotrophs that use hydrogen sulfide, elemental sulfur, and hydrogen as electron sources.
e. Green nonsulfur bacteria such as *Chloroflexus* are placed in the phylum *Chloroflexi. Chloroflexus* is a filamentous, gliding thermophilic bacterium that is metabolically similar to the purple nonsulfur bacteria.
f. Cyanobacteria carry out oxygenic photosynthesis by means of a photosynthetic apparatus similar to that of the eucaryotes. Phycobilisomes contain the light-harvesting pigments phycocyanin and phycoerythrin (**figure 21.7**).
g. Cyanobacteria reproduce by binary fission, budding, multiple fission, and fragmentation by filaments to form hormogonia. Some produce a dormant akinete.
h. Many nitrogen-fixing cyanobacteria form heterocysts, specialized cells in which nitrogen fixation occurs (**figure 21.9*b,c***).
i. *Bergey's Manual* divides the cyanobacteria into five subsections and includes the prochlorophytes in the phylum *Cyanobacteria* (**table 21.3**).

21.4 Phylum *Planctomycetes*

a. The phyla *Planctomycetes* and *Chlamydiae* lack peptidoglycan in their walls. The *Planctomycetes* have unusual cellular compartmentalization (**figure 21.12**).

21.5 Phylum *Chlamydiae*

a. Chlamydiae are nonmotile, coccoid, gram-negative bacteria that reproduce within the cytoplasmic vacuoles of host cells by a life cycle involving elementary bodies (EBs) and reticulate bodies (RBs) (**figure 21.13**). They are energy parasites.

21.6 Phylum *Spirochaetes*

a. The spirochetes are slender, long, helical, gram-negative bacteria that are motile because of the axial filament underlying an outer sheath or outer membrane (**figure 21.15**).

21.7 Phylum *Bacteroidetes*

a. Members of the class *Bacteroides* are obligately anaerobic, chemoheterotrophic, nonsporing, motile or nonmotile rods of various shapes. Some are important rumen and intestinal symbionts, others can cause disease.
b. Gliding motility is present in a diversity of bacteria, including the sphingobacteria.
c. Cytophagas degrade proteins and complex polysaccharides and are active in the mineralization of organic matter.

Key Terms

akinetes 525
anoxygenic photosynthesis 521
axial fibrils 532
axial filament 532
baeocytes 526
carboxysome 524
chlamydiae 531
chlorosomes 523
chromatic adaptation 525
cyanobacteria 524
cyanophycin 525
elementary body (EB) 531
gliding motility 525
green nonsulfur bacteria 523
green sulfur bacteria 523
heterocysts 525
hormogonia 525
initial body 531
oxygenic photosynthesis 520
periplasmic flagella 532
phototaxis 525
phycobilisomes 524
reticulate body (RB) 531
trichome 525

Critical Thinking Questions

1. The cyanobacterium *Anabaena* grows well in liquid medium that contains nitrate as the sole nitrogen source. Suppose you transfer some of these filaments to the same medium except it lacks nitrate and other nitrogen sources. Describe the morphological and physiologial changes you observe.
2. Many types of movement are employed by bacteria in these phyla. Review them and propose mechanisms by which energy (ATP or proton gradients) might drive the locomotion.
3. Propose two experimental approaches you might use to examine the mechanism by which *Cytophaga* glides.

Learn More

Daly, M. J.; Gaidamakova, E. K.; Matrosova, V. Y.; Valienko, A.; Zhai, M.; Venkateswaran, A.; Hess, M.; Omelchenko, M. V.; Kostandarithes, H. M.; Makarova, K. S.; Wackett, L. P.; Fredrickson, J. K.; and Ghosal, D. 2004. Accumulation of Mn(II) in *Deinococcus radiodurans* facilitates gamma-radiation resistance. *Science* 306:1025–28.

Everett, K. D. 2000. *Chlamydia* and *Chlamydiales:* More than meets the eye. *Vet. Microbiol.* 75:109–26.

Golden, J. W., and Yoon, H-S. 2003. Heterocyst development in *Anabaena. Curr. Opin. Microbiol.* 6:557–63.

Honda, D.; Yokota, A.; and Sugiyama, J. 1999. Detection of seven major evolutionary lineages in cyanobacteria based on the 16S rRNA gene sequence analysis with new sequences of five marine *Synechococcus* strains. *Mol. Evol.* 48: 723–39.

Lilburn, T. G.; Kim, K. S.; Ostrom, N. E.; Byzek, K. R.; Leadbetter, J. R.; and Breznak, J. A. 2001. Nitrogen fixation by symbiotic and free-living spirochetes. *Science* 292:2495–98.

Lindsay, M. R.; Webb, R. I.; Strous, M.; Jetten, M. S.; Butler, M. K.; Forde, R. J.; and Fuerst, J. A. 2001. Cell compartmentalization in planctomycetes: Novel types of structural organization for the bacterial cell. *Arch. Microbiol.* 175:413–29.

Litvaitis, M. K. 2002. A molecular test of cyanobacterial phylogeny: Inferences from constraint analysis. *Hydrobiologia* 468:135–45.

McBride, M. J. 2001. Bacterial gliding motility: Multiple mechanisms for cell movement over surfaces. *Annu. Rev. Microbiol.* 55:49–75.

Olson, I.; Paster, B. J.; and Dewhirst, F. E. 2000. Taxonomy of spirochetes. *Anaerobe* 6:39–57.

Partensky, F.; Hess, W. R.; and Vaulot, D. 1999. *Prochlorococcus,* a marine photosynthetic prokaryote of global significance. *Microbiol. Mol. Biol. Rev.* 63(1): 106–27.

Raymond, J.; Zhaxybayeva, O.; Gogarten, J. P.; Gerdes, S. Y.; and Blankenship, R. E. 2002. Whole-genome analysis of photosynthetic prokaryotes. *Science* 298:1616–20.

Strous, M.; Fuerst, J. A.; Kramer, E. H. M.; Logemann, S.; Muyzer, G.; van de Pas-Schoonen, K.; Webb, R.; Kuenen, J. G.; and Jetten, M. S. M. 1999. Missing lithotroph identified as new planctomycete. *Nature* 400:446–49.

Ting, C. S.; Rocap, G.; King, J.; and Chisholm, S. W. 2002. Cyanobacterial photosynthesis in the oceans: The origins and significance of divergent light-harvesting strategies. *Trends Microbiol.* 10(3):134–42.

Please visit the Prescott website at www.mhhe.com/prescott7 for additional references.

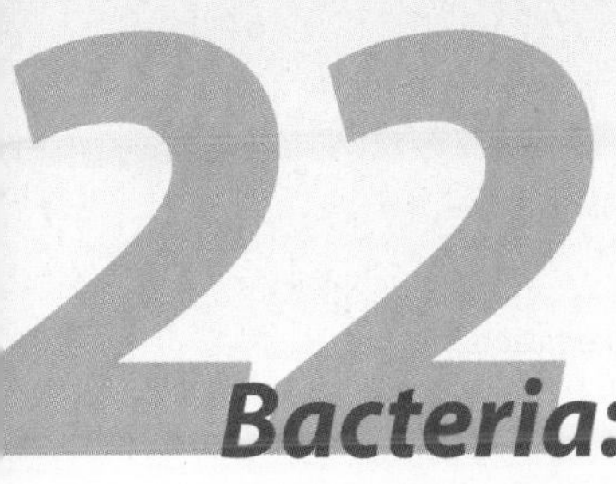

Bacteria: The Proteobacteria

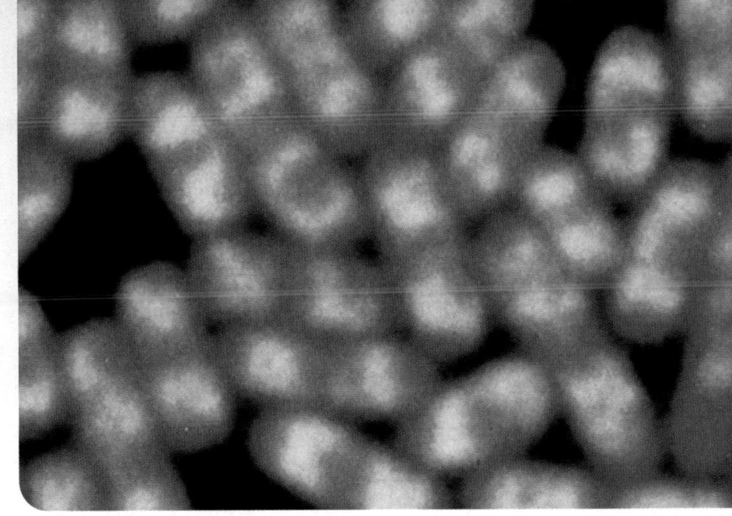

Salmonella enterica serovar Typhi, stained here with the fluorescent dye acridine orange, is a significant human pathogen. It causes typhoid fever.

Preview

- The phylum *Proteobacteria* is the largest, phylogenetically coherent bacterial group with over 2,000 species assigned to more than 500 genera.
- Many of these gram-negative bacteria are of considerable importance, either as disease agents or because of their contributions to ecosystems. In addition, bacteria such as *Escherichia coli,* are major experimental organisms studied in many laboratories.
- These bacteria are very diverse in their metabolism and life-styles, which range from obligate intracellular parasitism to a free-living existence in soil and aquatic habitats.
- Chemolithotrophic bacteria obtain energy and electrons by oxidizing inorganic compounds rather than the organic nutrients employed by many bacteria. They often have substantial ecological impact because of their ability to oxidize many forms of inorganic nitrogen and sulfur.
- Some *Proteobacteria* produce specialized structures such as prosthecae, stalks, buds, sheaths, or complex fruiting bodies.
- Many bacteria that specialize in predatory or parasitic modes of existence, such as *Bdellovibrio* and the rickettsias, have relinquished some of their metabolic independence through the loss of metabolic pathways. They depend on the prey's or host's energy supply and/or cell constituents.

In chapters 20 and 21 we describe many of the groups found in volumes 1 and 5 of the second edition of *Bergey's Manual.* We now introduce the bacteria that are covered in volume 2 of the second edition of *Bergey's Manual.* We provide an overview of the major biological features of each group and a few selected representative bacteria of particular interest.

Volume 2 of the second edition of *Bergey's Manual* is devoted entirely to the **proteobacteria.** This is the largest and most diverse group of bacteria; currently there are over 500 genera. Although 16S rRNA studies show that they are phylogenetically related, proteobacteria vary markedly in many respects. The morphology of these gram-negative bacteria ranges from simple rods and cocci to genera with prosthecae, buds, and even fruiting bodies. Physiologically they are just as diverse. Photoautotrophs, chemolithotrophs, and chemoheterotrophs are all well represented. There is no obvious overall pattern in metabolism, morphology, or reproductive strategy that characterizes proteobacteria.

Comparison of 16S rRNA sequences has revealed five lineages of descent within the phylum *Proteobacteria*: *Alphaproteobacteria, Betaproteobacteria, Gammaproteobacteria, Deltaproteobacteria,* and *Epsilonproteobacteria* (**figure 22.1**). However, new sequence data suggest that the separation between the *Betaproteobacteria* and *Gammaproteobacteria* is less distinct than once thought. If further analysis supports this, the *Betaproteobacteria* may be considered a subgroup of the *Gammaproteobacteria.* Members of the purple photosynthetic bacteria are found among the α-, β-, and γ-proteobacteria. This has led to the proposal that the proteobacteria arose from a single photosynthetic ancestor, presumably similar to the purple bacteria. That is to say, the phylum is considered monophyletic, although this has recently been questioned. Subsequently photosynthesis would have been lost by various lines, and new metabolic capacities were acquired as these bacteria adapted to different ecological niches.

Microbes is a vigitable, an'ivry man is like a conservatory full iv millyons iv these potted plants.

—Finley Peter Dunne

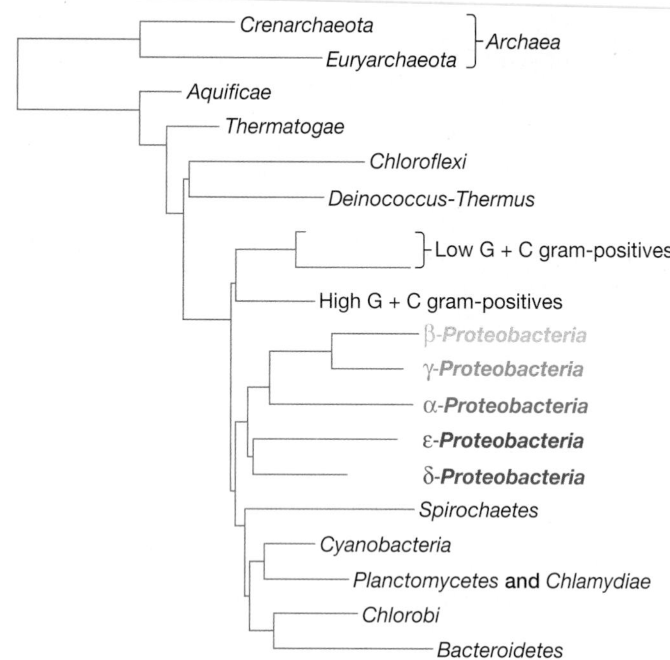

Figure 22.1 Phylogenetic Relationships Among the Procaryotes. The *Proteobacteria* are highlighted.

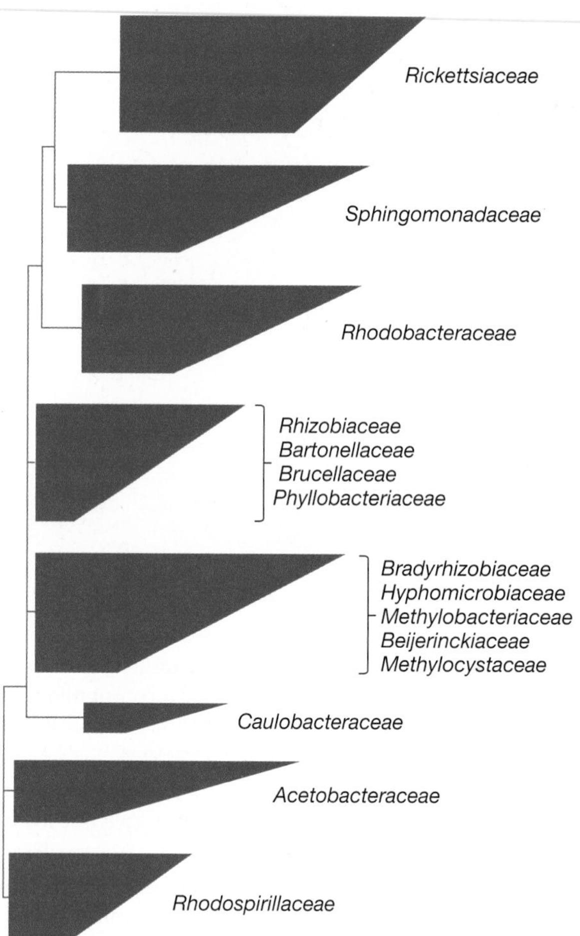

Figure 22.2 Phylogenetic Relationships Among Major Families Within the α-Proteobacteria. The relationships are based on 16S rRNA sequence data.

22.1 Class *Alphaproteobacteria*

The **α-proteobacteria** include most of the oligotrophic proteobacteria (those capable of growing at low nutrient levels). Some have unusual metabolic modes such as methylotrophy (*Methylobacterium*), chemolithotrophy (*Nitrobacter*), and the ability to fix nitrogen (*Rhizobium*). Members of genera such as *Rickettsia* and *Brucella* are important pathogens; in fact, *Rickettsia* is an obligate intracellular parasite. Many genera are characterized by distinctive morphology such as prosthecae.

The class *Alphaproteobacteria* has seven orders and 20 families. **Figure 22.2** illustrates the phylogenetic relationships among major groups within the α-proteobacteria, and **table 22.1** summarizes the general characteristics of many of the bacteria discussed in the following sections.

The Purple Nonsulfur Bacteria

All the purple bacteria use anoxygenic photosynthesis, possess bacteriochlorophylls *a* or *b,* and have their photosynthetic apparatus in membrane systems that are continuous with the plasma membrane. Most are motile by polar flagella. All purple nonsulfur bacteria are α-proteobacteria, with the exception of *Rhodocyclus* (β-proteobacteria). Photosynthetic bacteria (section 21.3)

The **purple nonsulfur bacteria** are exceptionally flexible in their choice of an energy source. Normally they grow anaerobically as photoorganoheterotrophs; they trap light energy and employ organic molecules as both electron and carbon sources (*see table 21.1*). Although they are called nonsulfur bacteria, some species can oxidize very low, nontoxic levels of sulfide to sulfate, but they do not oxidize elemental sulfur to sulfate. In the absence of light, most purple nonsulfur bacteria can grow aerobically as chemoorganoheterotrophs, but some species carry out fermentations anaerobically. Oxygen inhibits bacteriochlorophyll and carotenoid synthesis so that cultures growing aerobically in the dark are colorless.

Purple nonsulfur bacteria vary considerably in morphology (**figure 22.3**). They may be spirals (*Rhodospirillum*), rods (*Rhodopseudomonas*), half circles or circles (*Rhodocyclus*), or they may even form prosthecae and buds (*Rhodomicrobium*). Because of their metabolism, they are most prevalent in the mud and water of lakes and ponds with abundant organic matter and low sulfide levels. There also are marine species.

Rhodospirillum and *Azospirillum* (both in the family *Rhodospirillaceae*) are among several bacterial genera capable of forming cysts. These resting cells differ from the well-characterized endospores made by gram-positive bacteria such as *Bacillus* and *Clostridium.* Like spores, cysts are very resistant to desicca-

Table 22.1 Characteristics of Selected α-Proteobacteria

Genus	Dimensions (μm) and Morphology	G + C Content (mol%)	Genome size (Mb)	Oxygen Requirement	Other Distinctive Characteristics
Agrobacterium	0.6–1.0 × 1.5–3.0; motile, nonsporing rods with peritrichous flagella	57–63	2.5	Aerobic	Chemoorganotroph that can invade plants and cause tumors
Caulobacter	0.4–0.6 × 1–2; rod- or vibrioid-shaped with a flagellum and prostheca and holdfast	62–65	4.0	Aerobic	Heterotrophic and oligotrophic; asymmetric cell division
Hyphomicrobium	0.3–1.2 × 1–3; rod-shaped or oval with polar prosthecae	59–65	Nd*	Aerobic	Reproduces by budding; methylotrophic
Nitrobacter	0.5–0.9 × 1.0–2.0; rod- or pear-shaped, sometimes motile by flagella	59–62	3.4	Aerobic	Chemolithotroph, oxidizes nitrite to nitrate
Rhizobium	0.5–1.0 × 1.2–3.0; motile rods with flagella	57–66	5.1	Aerobic	Invades leguminous plants to produce nitrogen-fixing root nodules
Rhodospirillum	0.7–1.5 wide; spiral cells with polar flagella	62–64	4.4	Anaerobic, microaerobic, aerobic	Photoheterotroph under anoxic conditions
Rickettsia	0.3–0.5 × 0.8–2.0; short nonmotile rods	29–33	1.1–1.3	Aerobic	Obligate intracellular parasite

*Nd: Not determined; genome not yet sequenced

tion but are less tolerant of other environmental stresses like heat and UV light. Cysts are made in response to nutrient limitation. They have a thick outer coat and store an abundance of poly-β-hydroxybutyrate (PHB). Cyst-forming bacteria are not limited to α-proteobacteria; for instance *Azotobacter*, a γ-proteobacterium, also forms cysts. The cytoplasmic matrix: Inclusion bodies (section 3.3); The bacterial endospore (section 3.11)

Rickettsia and *Coxiella*

The genus *Rickettsia* is placed in the order *Rickettsiales* and family *Rickettsiaceae* of the α-proteobacteria, whereas *Coxiella* is in the order *Legionellales* and family *Coxiellaceae* of the γ-proteobacteria. However, here we discuss *Rickettsia* and *Coxiella* together because of their similarity in life-style, despite their apparent phylogenetic distance.

These bacteria are rod-shaped, coccoid, or pleomorphic with typical gram-negative walls and no flagella. Although their size varies, they tend to be very small. For example, *Rickettsia* is 0.3 to 0.5 μm in diameter and 0.8 to 2.0 μm long; *Coxiella* is 0.2 to 0.4 μm by 0.4 to 1.0 μm. All species are parasitic or mutualistic. The parasitic forms grow in vertebrate erythrocytes, macrophages, and vascular endothelial cells. Often they also live in blood-sucking arthropods such as fleas, ticks, mites, or lice, which serve as vectors or as primary hosts. Microbial interactions (section 30.1)

Because these genera include important human pathogens, their reproduction and metabolism have been intensively studied. Rickettsias enter the host cell by inducing phagocytosis. Members of the genus *Rickettsia* immediately escape the phagosome and reproduce by binary fission in the cytoplasm (**figure 22.4**). In contrast, *Coxiella* remains within the phagosome after it has fused with a lysosome and actually reproduces within the phagolysosome. Eventually the host cell bursts, releasing new organisms. Besides incurring damage from cell lysis, the host is harmed by the toxic effects of rickettsial cell walls (wall toxicity appears related to the mechanism of penetration into host cells). Phagocytosis (section 31.3)

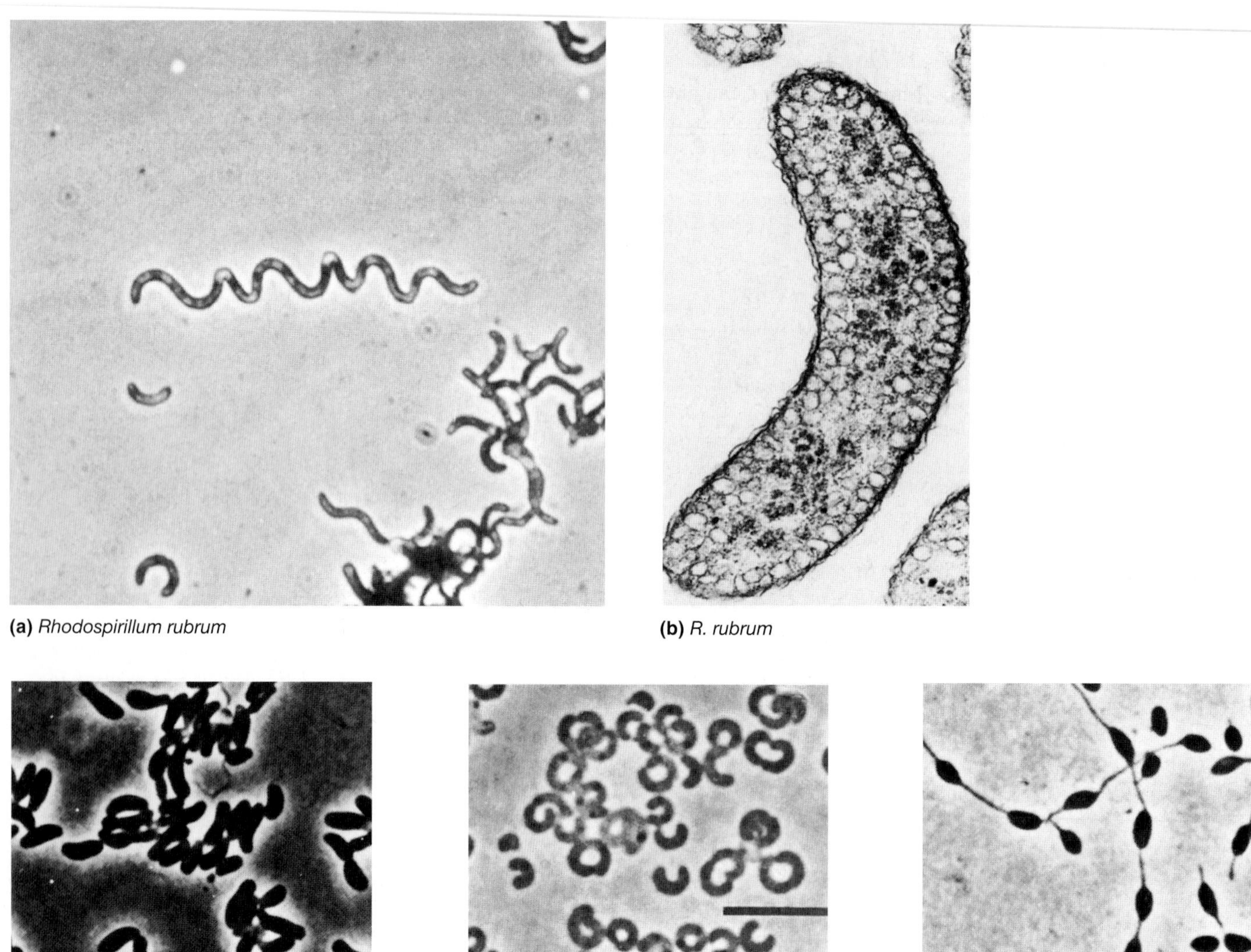

(a) *Rhodospirillum rubrum* (b) *R. rubrum*

(c) *Rhodopseudomonas acidophila* (d) *Rhodocyclus purpureus* (e) *Rhodomicrobium vannielii*

Figure 22.3 Typical Purple Nonsulfur Bacteria. **(a)** *Rhodospirillum rubrum*; phase contrast (×410). **(b)** *R. rubrum* grown anaerobically in the light. Note vesicular invaginations of the cytoplasmic membranes; transmission electron micrograph (×51,000). **(c)** *Rhodopseudomonas acidophila*; phase contrast. **(d)** *Rhodocyclus purpureus*; phase contrast. Bar = 10 μm. **(e)** *Rhodomicrobium vannielii* with vegetative cells and buds; phase contrast.

Rickettsias are very different from most other bacteria in physiology and metabolism. They lack glycolytic pathways and do not use glucose as an energy source, but rather oxidize glutamate and tricarboxylic acid cycle intermediates such as succinate. The rickettsial plasma membrane has carrier-mediated transport systems, and host cell nutrients and coenzymes are absorbed and directly used. For example, rickettsias take up both NAD^+ and uridine diphosphate glucose. Their membrane also has an adenylate exchange carrier that exchanges ADP for external ATP. Thus host ATP may provide much of the energy needed for growth. This metabolic dependence explains why many of these organisms must be cultivated in the yolk sacs of chick embryos or in tissue culture cells. Genome sequencing shows that *R. prowazekii* is similar in many ways to mitochondria. Possibly mitochondria arose from an endosymbiotic association with an ancestor of *Rickettsia*. Microbial evolution: Endosymbiotic origin of mitochondria and chloroplasts (section 19.1)

These orders contain many important pathogens. *Rickettsia prowazekii* and *R. typhi* are associated with typhus fever, and *R. rickettsii*, with Rocky Mountain spotted fever. *Coxiella burnetii* causes Q fever in humans. These diseases are discussed in some detail in chapter 38. Rickettsias are also important pathogens of domestic animals such as dogs, horses, sheep, and cattle.

The *Caulobacteraceae* and *Hyphomicrobiaceae*

A number of the proteobacteria are not simple rods or cocci but have some sort of appendage. These bacteria have interesting life

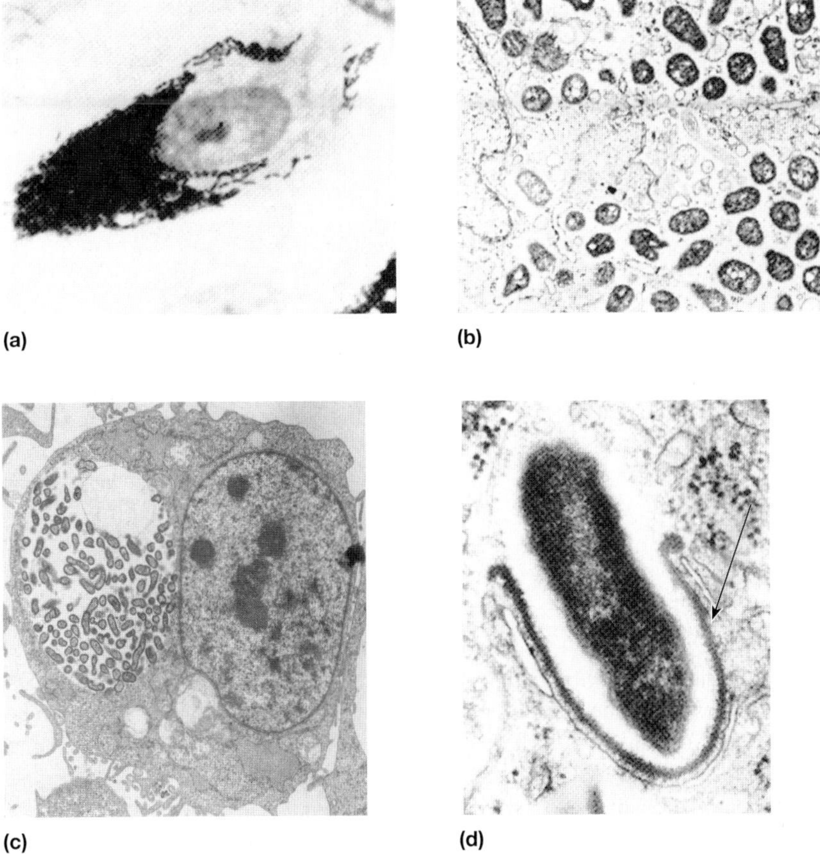

Figure 22.4 ***Rickettsia*** **and** ***Coxiella.*** Rickettsial morphology and reproduction. **(a)** A human fibroblast filled with *Rickettsia prowazekii* (×1,200). **(b)** A chicken embryo fibroblast late in infection with free cytoplasmic *R. prowazekii* (×13,600). **(c)** *Coxiella burnetti* growing within fibroblast vacuoles (×9,000). **(d)** *R. prowazekii* leaving a disrupted phagosome (arrow) and entering the cytoplasmic matrix (× 46,000).

cycles that feature a prostheca or reproduction by budding. A **prostheca** (pl., prosthecae), also called a **stalk,** is an extension of the cell, including the plasma membrane and cell wall, that is narrower than the mature cell. **Budding** is distinctly different from the **binary fission** normally used by bacteria. The bud first appears as a small protrusion at a single point and enlarges to form a mature cell. Most or all of the bud's cell envelope is newly synthesized. In contrast, portions of the parental cell envelope are shared with the progeny cells during binary fission. Finally, the parental cell retains its identity during budding, and the new cell is often smaller than its parent. In binary fission the parental cell disappears as it forms progeny of equal size. The families *Caulobacteraceae* and *Hyphomicrobiaceae* of the α-proteobacteria contain two of the best studied prosthecate genera: *Caulobacter* and *Hyphomicrobium.* The procaryotic cell cycle (section 6.1)

The genus *Hyphomicrobium* contains chemoheterotrophic, aerobic, budding bacteria that frequently attach to solid objects in freshwater, marine, and terrestrial environments. (They even grow in laboratory water baths.) The vegetative cell measures about 0.5 to 1.0 by 1 to 3 μm (**figure 22.5**). At the beginning of the reproductive cycle, the mature cell produces a prostheca (also called a hypha), 0.2 to 0.3 μm in diameter, that grows to several μm in length (**figure 22.6**). The nucleoid divides, and a copy moves into the hypha while a bud forms at its end. As the bud matures, it produces one to three flagella, and a septum divides the bud from the hypha. The bud is finally released as an oval- to pear-shaped swarmer cell, which swims off, then settles down and begins budding. The mother cell may bud several times at the tip of its hypha.

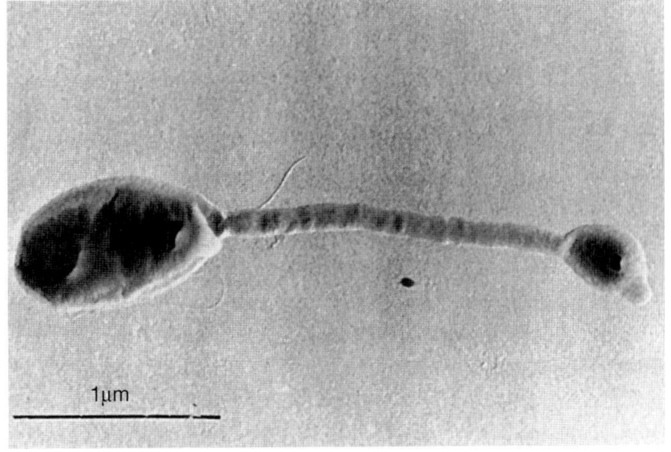

Figure 22.5 Prosthecate, Budding Bacteria. *Hyphomicrobium facilis* with hypha and young bud.

Hyphomicrobium also has distinctive physiology and nutrition. Sugars and most amino acids do not support abundant

Figure 22.6 The Life Cycle of *Hyphomicrobium*.

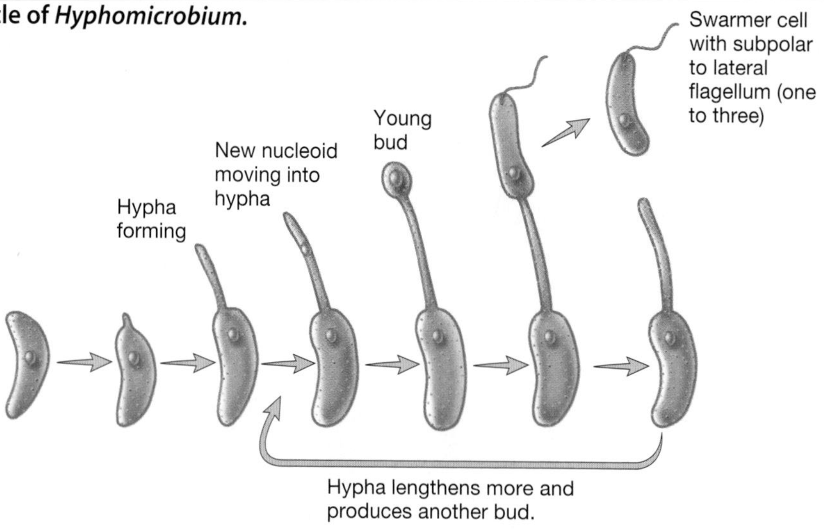

growth; instead *Hyphomicrobium* grows on ethanol and acetate and flourishes with one-carbon compounds such as methanol, formate, and formaldehyde. That is, it is a facultative **methylotroph** and can derive both energy and carbon from reduced one-carbon compounds. It is so efficient at acquiring one-carbon molecules that it can grow in a medium without an added carbon source (presumably the medium absorbs sufficient atmospheric carbon compounds). *Hyphomicrobium* may comprise up to 25% of the total bacterial population in oligotrophic or nutrient-poor freshwater habitats.

Bacteria in the genus *Caulobacter* alternate between polarly flagellated rods and cells that possess a prostheca and **holdfast,** by which they attach to solid substrata (**figure 22.7**). Incredibly, the material secreted at the end of the *Caulobacter cresentus* holdfast is the strongest biological adhesion molecule known—sort of a bacterial superglue. Caulobacters are usually isolated from freshwater and marine habitats with low nutrient levels, but they also are present in the soil. They often adhere to bacteria, photosynthetic protists, and other microorganisms and may absorb nutrients released by their hosts. The prostheca differs from that of *Hyphomicrobium* in that it lacks cytoplasmic components and is composed almost totally of the plasma membrane and cell wall. It grows longer in nutrient-poor media and can reach more than 10 times the length of the cell body. The prostheca may improve the efficiency of nutrient uptake from dilute habitats by increasing surface area; it also gives the cell extra buoyancy.

The life cycle of *Caulobacter* is unusual (**figure 22.8**). When ready to reproduce, the cell elongates and a single polar flagellum forms at the end opposite the prostheca. The cell then undergoes asymmetric transverse binary fission to produce a flagellated swarmer cell that swims away. The swarmer, which cannot reproduce, comes to rest, ejects its flagellum, and forms a new prostheca on the formerly flagellated end. The new stalked cell then starts the cycle anew. This process takes about two hours to complete. The species *C. cresentus* has become an important model organism in the study of microbial development and the bacterial cell cycle.

Family *Rhizobiaceae*

The order *Rhizobiales* of the α-proteobacteria contains 11 families with a great variety of phenotypes. This includes the family *Hyphomicrobiaceae,* which has already been discussed. An important family in this order is *Rhizobiaceae,* which includes the aerobic genera *Rhizobium* and *Agrobacterium.*

Members of the genus *Rhizobium* are 0.5 to 0.9 by 1.2 to 3.0 μm motile rods that become pleomorphic under adverse conditions (**figure 22.9**). Cells often contain poly-β-hydroxybutyrate inclusions. They grow symbiotically within root nodule cells of legumes as nitrogen-fixing bacteroids (figure 22.9*b, also see figure 29.13*). In fact, the *Leguminosae* is the most successful plant family on Earth, with over 18,000 species. Their proliferation reflects their capacity to establish symbiotic relationships with bacteria that form nodules on their roots. Within the nodules the microbes reduce or fix atmospheric nitrogen into ammonium, making it directly available to the plant host. The process by which bacteria perform this fascinating and important symbiosis is discussed in chapter 29. Synthesis of amino acids: Nitrogen fixation (section 10.5); Microorganism associations with vascular plants: The Rhizobia (section 29.5)

The genus *Agrobacterium* is placed in the family *Rhizobiaceae* but differs from *Rhizobium* in not stimulating root nodule formation or fixing nitrogen. Instead agrobacteria invade the crown, roots, and stems of many plants and transform plant cells into autonomously proliferating tumor cells. Most of the genes that encode distinguishing characteristics are carried on plas-

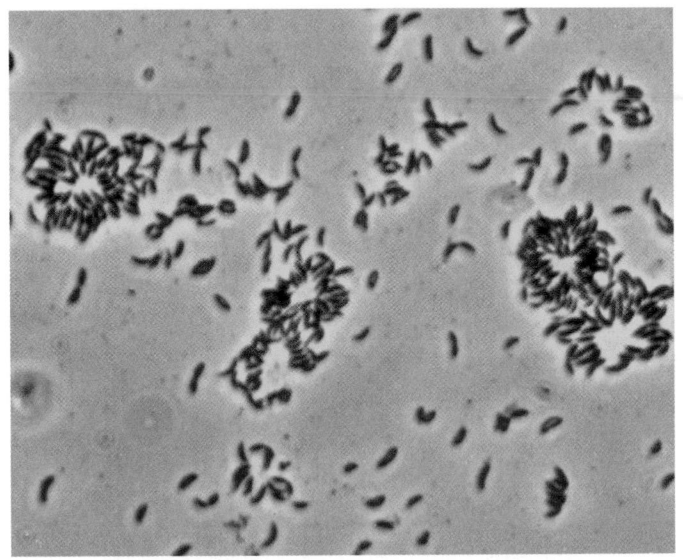

(a)

(d)

(b)

(c)

Figure 22.7 ***Caulobacter*** **Morphology and Reproduction.** **(a)** Rosettes of cells adhering to each other by their prosthecae; phase contrast (×600). **(b)** A cell dividing to produce a swarmer (×6,030). Note prostheca and flagellum. **(c)** A stalked cell and a flagellated swarmer cell (×6,030). **(d)** A rosette of cells as seen in the electron microscope.

mids (*see figure 29.19*). The best-studied species is *A. tumefaciens,* which enters many broad-leaved plants through wounds and causes crown gall disease (**figure 22.10**). The ability to produce tumors depends on the presence of a large Ti (for *t*umor-*i*nducing) plasmid. Tumor production by *Agrobacterium* is discussed in greater detail in Techniques & Applications 14.2 and section 29.5. Plasmids (section 3.5)

Nitrifying Bacteria

The taxonomy of the aerobic chemolithotrophic bacteria, those bacteria that derive energy and electrons from reduced inorganic compounds, is quite complex. Normally these bacteria employ CO_2 as their carbon source and thus are chemolithoautotrophs, but some can function as chemolithoheterotrophs and use reduced organic carbon sources. In *Bergey's Manual,* the chemolithotrophic bacteria are distributed between the α-, β-, and γ-proteobacteria. The nitrifying bacteria are found in all three classes. Chemolithotrophy (section 9.11)

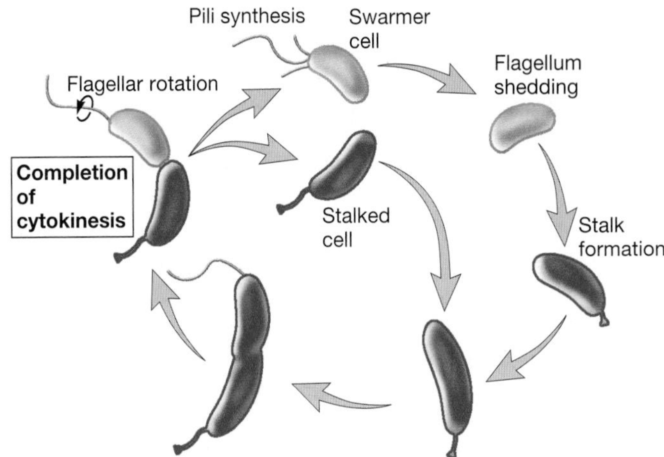

Figure 22.8 Caulobacter Life Cycle. Stalked cells attached to a substrate undergo asymmetric binary fission producing a stalked and a flagellated cell, called a swarmer cell. The swarmer cell swims freely and makes pili until it settles, ejects its flagella, and forms a stalk. Only stalked cells can divide.

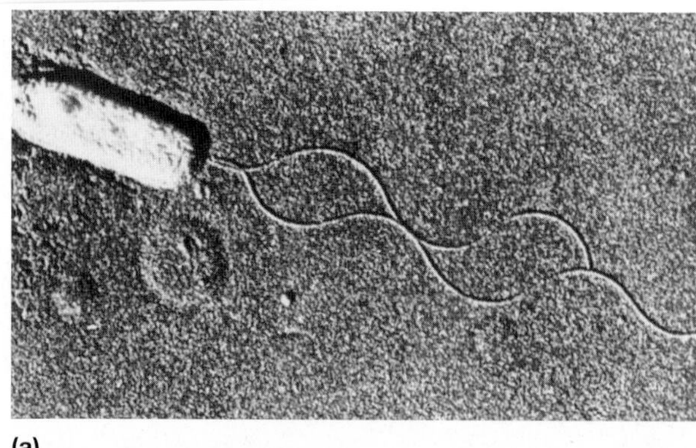

(a)

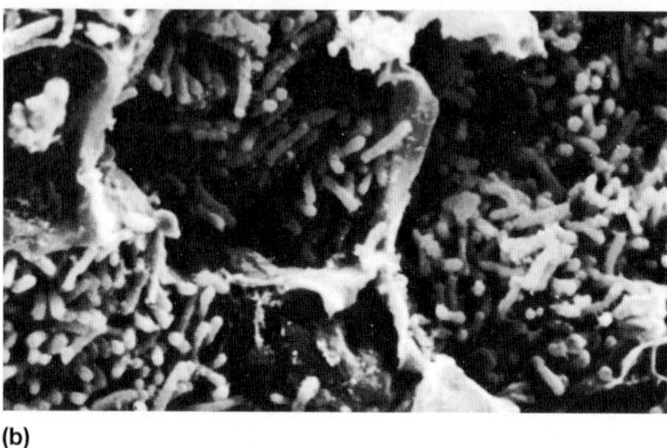

(b)

Figure 22.9 Rhizobium. **(a)** *Rhizobium leguminosarum* with two polar flagella (×14,000). **(b)** Scanning electron micrograph of bacteroids in alfalfa root (×640).

Figure 22.10 Agrobacterium. Crown gall tumor of a tomato plant caused by *Agrobacterium tumefaciens.*

The **nitrifying bacteria** are a very diverse collection of bacteria. *Bergey's Manual* places nitrifying genera in three classes and several families: *Nitrobacter* in the *Bradyrhizobiaceae,* α-proteobacteria; *Nitrosomonas* and *Nitrosospira* in the *Nitrosomonadaceae,* γ-proteobacteria; *Nitrococcus* in the *Ectothiorhodospiraceae,* γ-proteobacteria; and *Nitrosococcus* in the *Chromatiaceae,* γ-proteobacteria. All are aerobic, gram-negative organisms with the ability to capture energy from the oxidation of either ammonia or nitrite. However, they differ considerably in other properties (**table 22.2**). Nitrifiers may be rod-shaped, ellipsoidal, spherical, spirillar or lobate, and they may possess either polar or peritrichous flagella (**figure 22.11**). Often they have extensive membrane complexes in their cytoplasm. Identification is based on properties such as their preference for nitrite or ammonia, their general shape, and the nature of any cytomembranes present.

Nitrifying bacteria make important contributions to the nitrogen cycle. In soil, sewage disposal systems, and freshwater and marine habitats, the β-proteobacteria *Nitrosomonas* and *Nitrospira* and the γ-proteobacterium *Nitrosococcus* oxidize ammonia to nitrite. In the same niches, members of the γ-proteobacterial genus *Nitrococcus* then oxidize nitrite to nitrate. The whole process of converting ammonia to nitrite to nitrate is called **nitrification** and it occurs rapidly in oxic soil treated with fertilizers containing ammonium salts. Nitrate is readily used by plants, but it is also rapidly lost through leaching of water-soluble nitrates and by denitrification to nitrogen gas, so the benefits gained from nitrification can be fleeting. Biogeochemical cycling: Nitrogen cycle (section 27.2)

1. Describe the general properties of the α-proteobacteria.
2. Discuss the characteristics and physiology of the purple nonsulfur bacteria. Where would one expect to find them growing?
3. Briefly describe the characteristics and life cycle of the genus *Rickettsia.*
4. In what way does the physiology and metabolism of the rickettsias differ from that of other bacteria?
5. Name some important rickettsial diseases.
6. Define the following terms: prostheca, stalk, budding, swarmer cell, methylotroph, and holdfast.
7. Briefly describe the morphology and life cycles of *Hyphomicrobium* and *Caulobacter.*
8. What is unusual about the physiology of *Hyphomicrobium?* How does this influence its ecological distribution?
9. How do *Agrobacterium* and *Rhizobium* differ in life-style? What effect does *Agrobacterium* have on plant hosts?
10. What are chemolithotrophic bacteria?
11. Give the major characteristics of the nitrifying bacteria and discuss their ecological importance. How does the metabolism of *Nitrobacter* differ from that of *Nitrosomonas?*

22.2 Class *Betaproteobacteria*

The **β-proteobacteria** overlap the α-proteobacteria metabolically but tend to use substances that diffuse from organic decomposition in the anoxic zone of habitats. Some of these bacteria use hydrogen,

Table 22.2 Selected Characteristics of Representative Nitrifying Bacteria

Species	Cell Morphology and Size (μm)	Reproduction	Motility	Cytomembranes	G + C Content (mol%)	Habitat
Ammonia-Oxidizing Bacteria						
Nitrosomonas europaea (β-proteobacteria)	Rod; 0.8–1.1 × 1.0–1.7	Binary fission	–	Peripheral, lamellar	50.6–51.4	Soil, sewage, freshwater, marine
Nitrosococcus oceani (γ-proteobacteria)	Coccoid; 1.8–2.2 in diameter	Binary fission	+; 1 or more subpolar flagella	Centrally located parallel bundle, lamellar	50.5	Obligately marine
Nitrosospira briensis (β-proteobacteria)	Spiral; 0.3–0.4 in diameter	Binary fission	+ or –; 1 to 6 peritrichous flagella	Rare	53.8–54.1	Soil
Nitrite-Oxidizing Bacteria						
Nitrobacter winogradskyi (α-proteobacteria)	Rod, often pear-shaped; 0.5–0.9 × 1.0–2.0	Budding	+ or –; 1 polar flagellum	Polar cap of flattened vesicles in peripheral region of the cell	61.7	Soil, freshwater, marine
Nitrococcus mobilis (γ-proteobacteria)	Coccoid; 1.5–1.8 in diameter	Binary fission	+; 1 or 2 subpolar flagella	Tubular cytomembranes randomly arranged in cytoplasm	61.3 (1 strain)	Marine

From Brenner, D. J.; Krieg, N. R.; and Staley, J. T. Eds. 2005. Bergey's Manual to Systemic Bacteriology 2nd ed. Vol. 2: *The Proteobacteria.* Garrity, G. M. Ed-in-Chief. New York: Springer.

ammonia, methane, volatile fatty acids, and similar substances. As with the α-proteobacteria, there is considerable metabolic diversity; the β-proteobacteria may be chemoheterotrophs, photolithotrophs, methylotrophs, and chemolithotrophs.

The class *Betaproteobacteria* has seven orders and 12 families. **Figure 22.12** shows the phylogenetic relationships among major groups within the β-proteobacteria, and **table 22.3** summarizes the general characteristics of many of the bacteria discussed in this section. Here we discuss two genera with important human pathogens: *Neisseria* and *Bordetella*.

Order *Neisseriales*

The order *Neisseriales* has one family, *Neisseriaceae,* with 15 genera. The best-known and most intensely studied genus is *Neisseria*. Members of this genus are nonmotile, aerobic, gram-negative cocci that most often occur in pairs with adjacent sides flattened. They may have capsules and fimbriae. The genus is chemoorganotrophic, and produces the enzymes oxidase and catalase (thus they are said to be oxidase positive and catalase positive). They are inhabitants of the mucous membranes of mammals, and some are human pathogens. *Neisseria gonorrhoeae* is the causative agent of gonorrhea; *Neisseria meningitidis* is responsible for some cases of bacterial meningitis. Direct contact diseases: Sexually transmitted diseases (section 38.3)

Order *Burkholderiales*

The order contains four families, three of them with well-known genera. The genus *Burkholderia* is placed in the family *Burkholderiaceae.* This genus was established when *Pseudomonas* was divided into at least seven genera based on rRNA data: *Acidovorax, Aminobacter, Burkholderia, Comamonas, Deleya, Hydrogenophaga,* and *Methylobacterium.* Members of the genus *Burkholderia* are gram-negative, aerobic, nonfermentative, nonsporing, mesophilic straight rods. With the exception of one species, all are motile with a single polar flagellum or a tuft of polar flagella. Catalase is produced and they often are oxidase positive. Most species use PHB as their carbon reserve. One of the most important species is *B. cepacia,* which can degrade over 100 different organic molecules and is very active in recycling organic materials in nature. Originally described as the plant pathogen that causes onion rot, it

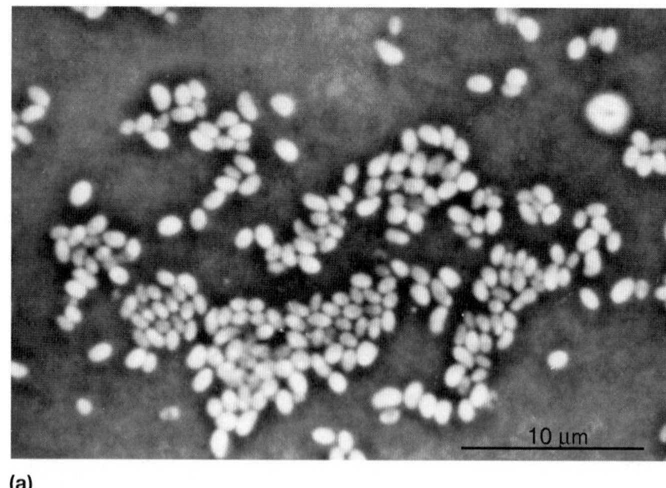

(a)

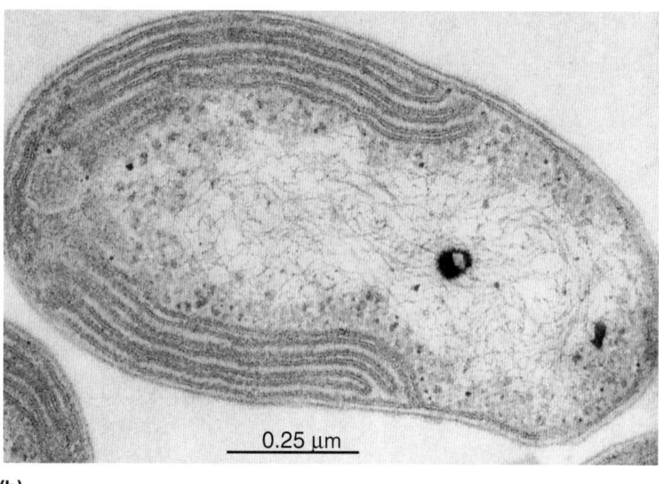

(b)

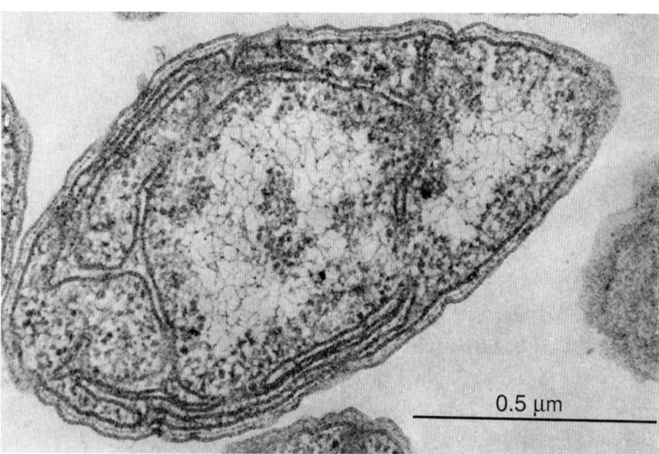

(c)

Figure 22.11 Representative Nitrifying Bacteria. **(a)** *Nitrobacter winogradskyi*; phase contrast (×2,500). **(b)** *N. winogradskyi*. Note the polar cap of cytomembranes (×213,000). **(c)** *N. europaea* with extensive cytoplasmic membranes (×81,700).

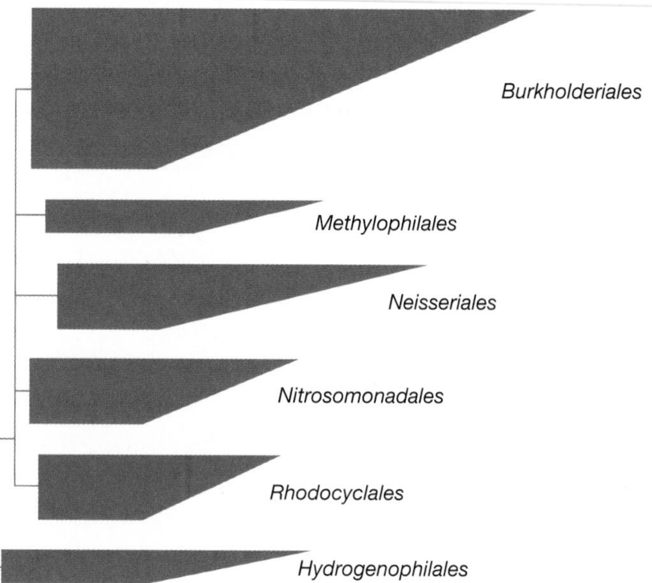

Figure 22.12 Phylogenetic Relationships Among Major Groups Within the β-Proteobacteria. The relationships are based on 16S rRNA sequence data.

has emerged in the last 20 years as a major nosocomial pathogen. It is a particular problem for cystic fibrosis patients. Two other species, *B. mallai* and *B. pseudomallai* are human pathogens that could be misused as bioterrorism agents. Bioterrorism preparedness (section 36.9)

Surprisingly, two genera within the *Burkholderiaceae* family are capable of forming nitrogen-fixing symbioses with legumes much like the rhizobia that belong to the α-proteobacteria. Genome analysis of the nitrogen-fixing β-proteobacteria *Burkholderia* and *Ralstonia* reveals the presence of nodulation (*nod*) genes that are very similar to those of the rhizobia. This suggests a common genetic origin. It is thought the β-proteobacteria gained the capacity to form symbiotic, nitrogen-fixing nodules with legumes through lateral gene transfer.

The family *Alcaligenaceae* contains the genus *Bordetella*. This genus is composed of gram-negative, aerobic coccobacilli, about 0.2 to 0.5 by 0.5 to 2.0 μm in size. *Bordetella* is a chemoorganotroph with respiratory metabolism that requires organic sulfur and nitrogen (amino acids) for growth. It is a mammalian parasite that multiplies in respiratory epithelial cells. *Bordetella pertussis* is a nonmotile, encapsulated species that causes whooping cough. Airborne diseases: Diphtheria (section 38.1)

Some genera in the order have a **sheath**—a hollow, tubelike structure surrounding a chain of cells. Sheaths often are close fitting, but they are never in intimate contact with the cells they enclose and may contain ferric or manganic oxides. They have at least two functions. Sheaths help bacteria attach to solid surfaces and acquire nutrients from slowly running water as it flows past, even if it is nutrient-poor. Sheaths also protect against predators such as protozoa and the δ-proteobacterium *Bdellovibrio* (p. 563).

Table 22.3 Characteristics of Selected β-Proteobacteria

Genus	Dimensions (μm) and Morphology	G + C Content (mol%)	Genome Size (Mb)	Oxygen Requirement	Other Distinctive Characteristics
Bordetella	0.2–0.5 × 0.5–2.0; nonmotile coccobacillus	66–70	3.7–5.3	Aerobic	Requires organic sulfur and nitrogen; mammalian parasite
Burkholderia	0.5–1.0 × 1.5–4; straight rods with single flagella or a tuft at the pole	59–69.5	4.1–7.2	Aerobic, some capable of anaerobic respiration with NO_3^-	Poly-β-hydroxybutyrate as reserve; can be pathogenic
Leptothrix	0.6–1.5 × 2.5–15; straight rods in chains with sheath, free cells flagellated	68–71	Nd*	Aerobic	Sheaths encrusted with iron and manganese oxides
Neisseria	0.6–1.9; cocci in pairs with flattened adjacent sides	48–56	2.2–2.3	Aerobic	Inhabitant of mucous membranes of mammals
Nitrosomonas	Size varies with strain; spherical to ellipsoidal cells with intracytoplasmic membranes	45–54	2.8	Aerobic	Chemolithotroph that oxidizes ammonia to nitrite
Sphaerotilus	1.2–2.5 × 2–10; single chains of cells with sheaths, may have holdfasts	70	Nd	Aerobic	Sheaths not encrusted with iron and manganese oxides
Thiobacillus	0.3–0.5 × 0.9–4; rods, often with polar flagella	52–68	Nd	Aerobic	All chemolithotrophic, oxidizes reduced sulfur compounds to sulfate, some also chemoorganotrophic

*Nd: Not determined; genome not yet sequenced

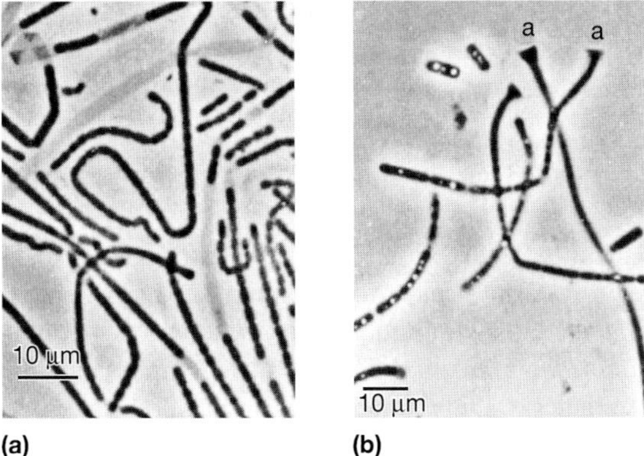

Figure 22.13 Sheathed Bacteria, *Sphaerotilus natans*. **(a)** Sheathed chains of cells and empty sheaths. **(b)** Chains with holdfasts (indicated by the letter a) and individual cells containing poly-β-hydroxybutyrate granules.

Two well-studied sheathed genera are *Sphaerotilus* and *Leptothrix*. *Sphaerotilus* forms long sheathed chains of rods, 0.7 to 2.4 by 3 to 10 μm, attached to submerged plants, rocks, and other solid objects, often by a holdfast (**figure 22.13**). Single swarmer cells with a bundle of subpolar flagella escape the filament and form a new chain after attaching to a solid object at another site. *Sphaerotilus* prefers slowly running freshwater polluted with sewage or industrial waste. It grows so well in activated sewage sludge that it sometimes forms tangled masses of filaments and interferes with the proper settling of sludge. *Leptothrix* characteristically deposits large amounts of iron and manganese oxides in its sheath (**figure 22.14**). This seems to protect it and allow *Leptothrix* to grow in the presence of high concentrations of soluble iron compounds.

Order *Nitrosomonadales*

A number of chemolithotrophs are found in the order *Nitrosomonadales.* Two genera of nitrifying bacteria (*Nitrosomonas* and *Nitrosospira*) are members of the family *Nitrosomonadaceae*

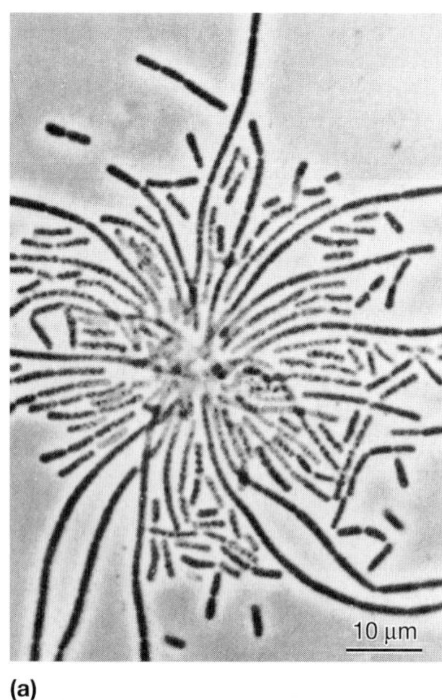

(a)

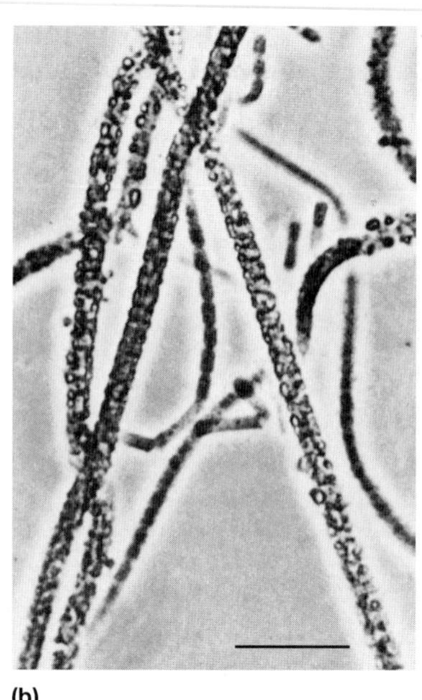

(b)

Figure 22.14 Sheathed Bacteria, *Leptothrix* Morphology. **(a)** *L. lopholea* trichomes radiating from a collection of holdfasts. **(b)** *L. cholodnii* sheaths encrusted with MnO_2.

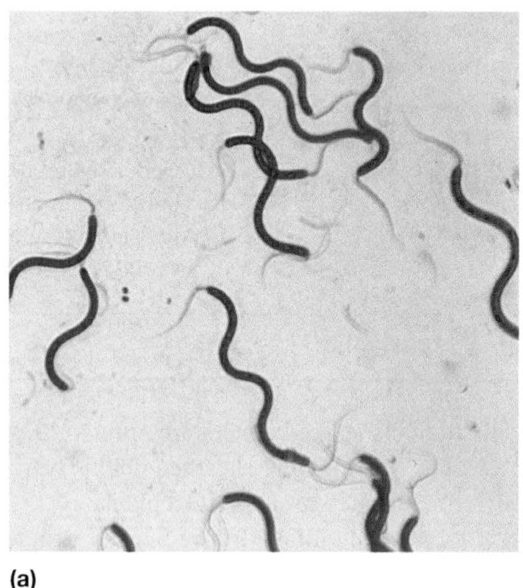

(a)

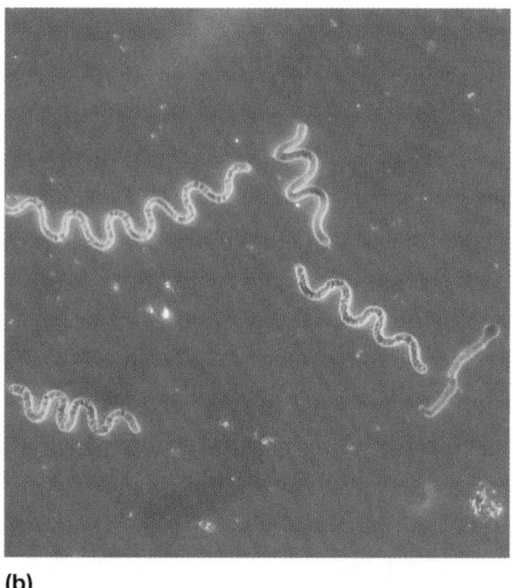

(b)

Figure 22.15 The Genus *Spirillum*. **(a)** *Spirillum volutans* with bipolar flagella visible (×450). **(b)** *Spirillum volutans*; phase contrast (×550).

but were discussed earlier with other genera of nitrifying bacteria (pp. 545–546). The stalked chemolithotroph *Gallionella* is in this order. The family *Spirillaceae* has one genus, *Spirillum* (**figure 22.15**).

Order *Hydrogenophilales*

This small order contains *Thiobacillus*, one of the best-studied chemolithotrophs and most prominent of the colorless sulfur bacteria. Like the nitrifying bacteria, **colorless sulfur bacteria** are a highly diverse group. Many are unicellular rod-shaped or spiral sulfur-oxidizing bacteria that are nonmotile or motile by flagella (**table 22.4**). *Bergey's Manual* disperses these bacteria between two classes; for example, *Thiobacillus* and *Macromonas* are β-proteobacteria, whereas *Thiomicrospira, Thiobacterium, Thiospira, Thiothrix, Beggiatoa,* and others are γ-proteobacteria. Only some of these bacteria have been isolated and studied in pure culture. Most is known about the genera *Thiobacillus* and *Thiomicrospira. Thiobacillus* is a gram-negative rod, and *Thiomicrospira* is a long spiral cell (**figure 22.16**); both have polar flagella. They differ from many of the nitrifying bacteria in that they lack extensive internal membrane systems.

Table 22.4 Colorless Sulfur-Oxidizing Genera

Genus	Cell Shape	Motility; Flagella	G + C Content (mol%)	Location of Sulfur Deposit[a]	Nutritional Type
Thiobacillus	Rods	+; polar	62–67	Extracellular	Obligate or facultative chemolithotroph
Thiomicrospira	Spirals, comma, or rod shaped	− or +; polar	39.6–49.9	Extracellular	Obligate chemolithotroph
Thiobacterium	Rods embedded in gelatinous masses	−	N.A.[b]	Intracellular	Probably chemoorganoheterotroph
Thiospira	Spiral rods, usually with pointed ends	+; polar (single or in tufts)	N.A.	Intracellular	Unknown
Macromonas	Rods, cylindrical or bean shaped	+; polar tuft	67	Intracellular	Probably chemoorganoheterotroph

[a]When hydrogen sulfide is oxidized to elemental sulfur.

[b]N.A., data not available.

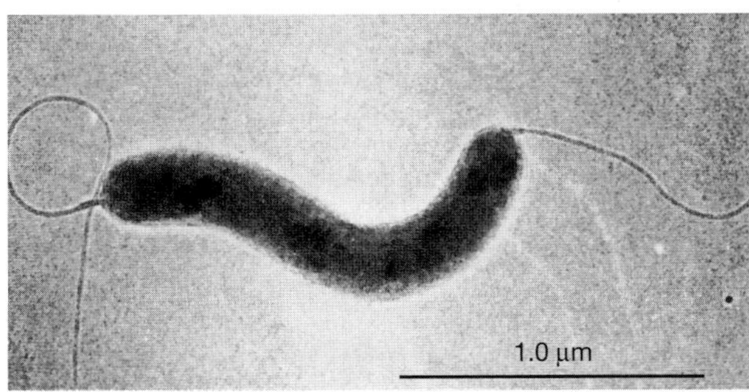

Figure 22.16 Colorless Sulfur Bacteria. *Thiomicrospira pelophila,* a γ-proteobacterium, with polar flagella.

The metabolism of *Thiobacillus* has been intensely studied. It grows aerobically by oxidizing a variety of inorganic sulfur compounds (elemental sulfur, hydrogen sulfide, thiosulfate) to sulfate. ATP is produced by a combination of oxidative phosphorylation and substrate-level phosphorylation by means of adenosine 5′-phosphosulfate. Although *Thiobacillus* normally uses CO_2 as its major carbon source, *T. novellus* and a few other strains can grow heterotrophically. Some species are very flexible metabolically. For example, *Thiobacillus ferrooxidans* also uses ferrous iron as an electron donor and produces ferric iron as well as sulfuric acid. *T. denitrificans* even grows anaerobically by reducing nitrate to nitrogen gas. Interestingly, some other sulfur-oxidizing bacteria such as *Thiobacterium* and *Macromonas* probably do not derive energy from sulfur oxidation. They may use the process to detoxify metabolically produced hydrogen peroxide.

Sulfur-oxidizing bacteria have a wide distribution and great practical importance. *Thiobacillus* grows in soil and aquatic habitats, both freshwater and marine. In marine habitats *Thiomicrospira* is more important than *Thiobacillus.* Because of their great acid tolerance (*T. thiooxidans* grows at pH 0.5 and cannot grow above pH 6), these bacteria prosper in habitats they have acidified by sulfuric acid production, even though most other organisms cannot. The production of large quantities of sulfuric acid and ferric iron by *T. ferrooxidans* corrodes concrete and pipe structures. Thiobacilli often cause extensive acid and metal pollution when they release metals from mine wastes. However, sulfur-oxidizing bacteria also are beneficial. They may increase soil fertility when they release elemental sulfur by oxidizing it to sulfate. Thiobacilli are used in processing low-grade metal ores because of their ability to leach metals from ore. Biogeochemical cycling: Sulfur cycle (section 27.2)

1. Describe the general properties of the β-proteobacteria.
2. Briefly describe the following genera and their practical importance: *Neisseria, Burkholderia,* and *Bordetella.*
3. What is a sheath and of what advantage is it?
4. How does *Sphaerotilus* maintain its position in running water? How does it reproduce and disperse its progeny?
5. Characterize the colorless sulfur bacteria and discuss their placement in the second edition of *Bergey's Manual.*
6. How do colorless sulfur bacteria obtain energy by oxidizing sulfur compounds? What is adenosine 5′-phosphosulfate?
7. List several positive and negative impacts sulfur-oxidizing bacteria have on the environment and human activities.

22.3 Class *Gammaproteobacteria*

The **γ-proteobacteria** constitute the largest subgroup of proteobacteria with an extraordinary variety of physiological types. Many important genera are chemoorganotrophic and facultatively anaerobic. Other genera contain aerobic chemoorganotrophs, photolithotrophs, chemolithotrophs, or methylotrophs. According to some DNA-rRNA hybridization studies, the γ-proteobacteria are composed of several deeply branching groups. One consists of the purple sulfur bacteria; a second includes the intracellular parasites *Legionella* and *Coxiella.* The two largest groups contain a wide variety of nonphotosynthetic genera. Ribosomal RNA superfamily I is represented by the families *Vibrionaceae, Enterobacteriaceae,* and *Pasteurellaceae.* These bacteria use the Embden-Meyerhof and pentose phosphate pathways to catabolize carbohydrates. Most are facultative anaerobes. Ribosomal RNA superfamily II contains mostly aerobes that often use the Entner-Doudoroff and pentose phosphate pathways to catabolize many different kinds of organic molecules. The genera *Pseudomonas, Azotobacter, Moraxella, Xanthomonas,* and *Acinetobacter* belong to this superfamily.

The exceptional diversity of these bacteria is evident from the fact that *Bergey's Manual* divides the class *Gammaproteobacteria* into 14 orders and 28 families. **Figure 22.17** illustrates the phylogenetic relationships among major groups and selected γ-proteobacteria, and **table 22.5** outlines the general characteristics of some of the bacteria discussed in this section.

The Purple Sulfur Bacteria

As mentioned previously, the purple photosynthetic bacteria are distributed between three subgroups of the proteobacteria. Despite the diversity of these organisms, they share some general characteristics, which are summarized in table 21.1 (*see p. 522*). Most of the purple nonsulfur bacteria are α-proteobacteria and were discussed earlier in this chapter (pp. 540–541). Because the purple sulfur bacteria are γ-proteobacteria, they are described here.

Bergey's Manual divides the purple sulfur bacteria into two families: the *Chromatiaceae* and *Ectothiorhodospiraceae* in the

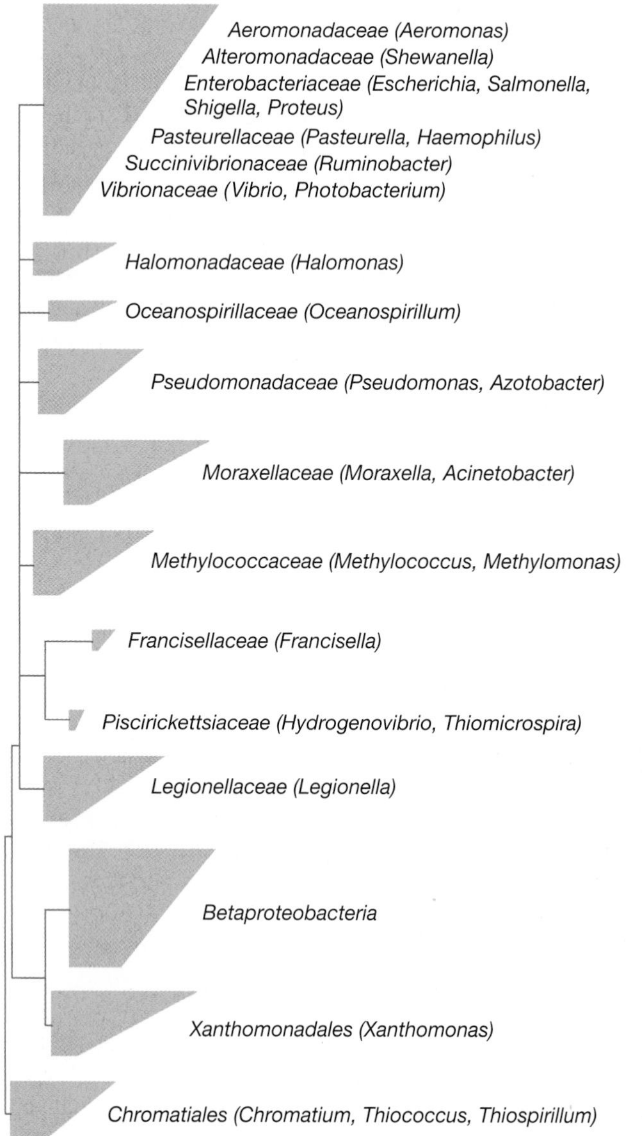

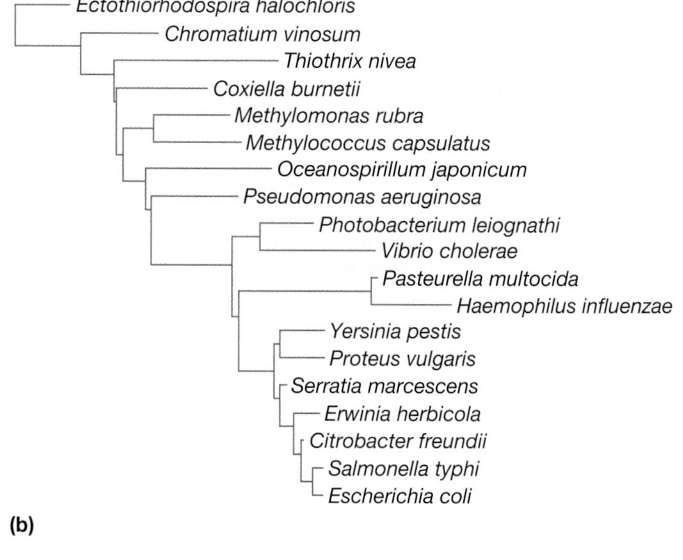

Figure 22.17 Phylogenetic Relationships Among γ-Proteobacteria. **(a)** The major phylogenetic groups based on 16S rRNA sequence comparisons. Representative genera are given in parentheses. Each tetrahedron in the tree represents a group of related organisms; its horizontal edges show the shortest and longest branches in the group. Multiple branching at the same level indicates that the relative branching order of the groups cannot be determined from the data. **(b)** The relationships of a few species based on 16S rRNA sequence data. *Source: The Ribosomal Database Project.*

Table 22.5 Characteristics of Selected γ-Proteobacteria

Genus	Dimensions (μm) and Morphology	G + C Content (mol %)	Oxygen Requirement	Other Distinctive Characteristics
Azotobacter	1.5–2.0; ovoid cells, pleomorphic, peritrichous flagella or nonmotile	63.2–67.5	Aerobic	Can form cysts; fix nitrogen nonsymbiotically
Beggiatoa	1–200 × 2–10; colorless cells form filaments, either single or in colonies	35–39	Aerobic or microaerophilic	Gliding motility; can form sulfur inclusions with hydrogen sulfide present
Chromatium	1–6 × 1.5–16; rod-shaped or ovoid, straight or slightly curved, polar flagella	48–50	Anaerobic	Photolithoautotroph that can use sulfide; sulfur stored within the cell
Ectothiorhodospira	0.7–1.5 in diameter; vibrioid- or rod-shaped, polar flagella	61.4–68.4	Anaerobic, some aerobic or microaerophilic	Internal lamellar stacks of membranes; deposits sulfur granules outside cells
Escherichia	1.1–1.5 × 2–6; straight rods, peritrichous flagella or nonmotile	48–59	Facultatively anaerobic	Mixed acid fermenter; formic acid converted to H_2 and CO_2, lactose fermented, citrate not used
Haemophilus	<1.0 in width, variable lengths; coccobacilli or rods, nonmotile	37–44	Facultative or aerobic	Fermentative; requires growth factors present in blood; parasites on mucous membranes
Leucothrix	Long filaments of short cylindrical cells, usually holdfast is present	46–51	Aerobic	Dispersal by gonidia, filaments don't glide; rosettes formed; heterotrophic
Methylococcus	0.8–1.5 × 1.0–1.5; cocci with capsules, nonmotile	59–65	Aerobic	Can form a cyst; uses methane, methanol, and formaldehyde as sole carbon and energy sources
Photobacterium	0.8–1.3 × 1.8–2.4; straight, plump rods with polar flagella	39–44	Facultatively anaerobic	Two species can emit blue-green light; Na^+ needed for growth
Pseudomonas	0.5–1.0 × 1.5–5.0; straight or slightly curved rods, polar flagella	58–69	Aerobic or facultatively anaerobic	Respiratory metabolism with oxygen or nitrate as acceptor; some use H_2 or CO as energy source
Vibrio	0.5–0.8 × 1.4–2.6; straight or curved rods with sheathed polar flagella	38–51	Facultatively anaerobic	Fermentative or respiratory metabolism; sodium ions stimulate or are needed for growth; oxidase positive

order *Chromatiales.* The family *Ectothiorhodospiraceae* contains eight genera. *Ectothiorhodospira* has red, spiral-shaped, polarly flagellated cells that deposit sulfur globules externally (**figure 22.18**). Internal photosynthetic membranes are organized as lamellar stacks. The typical purple sulfur bacteria are located in the family *Chromatiaceae,* which is much larger and contains 26 genera.

The **purple sulfur bacteria** are strict anaerobes and usually photolithoautotrophs. They oxidize hydrogen sulfide to sulfur and deposit it internally as sulfur granules (usually within invaginated pockets of the plasma membrane); often they eventually oxidize the sulfur to sulfate. Hydrogen also may serve as an electron donor. *Thiospirillum, Thiocapsa,* and *Chromatium* are typical purple sulfur bacteria (**figure 22.19**). They are found in anoxic, sulfide-rich zones of lakes, bogs, and lagoons where large blooms can occur under certain conditions (**figure 22.20**).

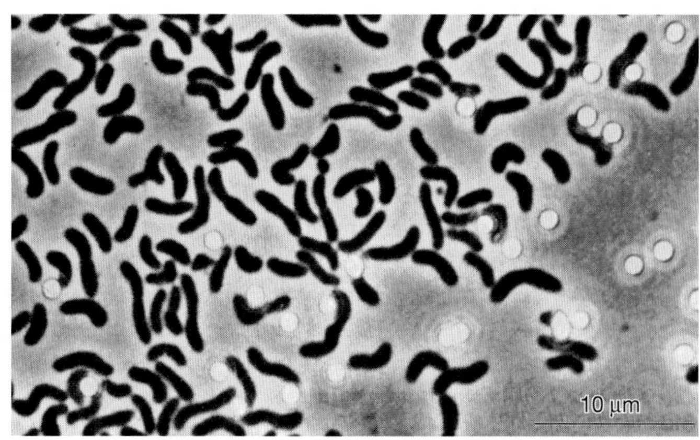

Figure 22.18 Purple Bacteria. *Ectothiorhodospira mobilis;* light micrograph.

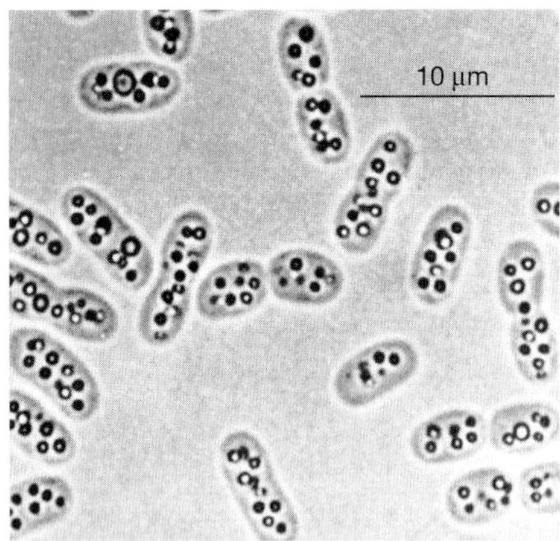

(a) *Chromatium vinosum*

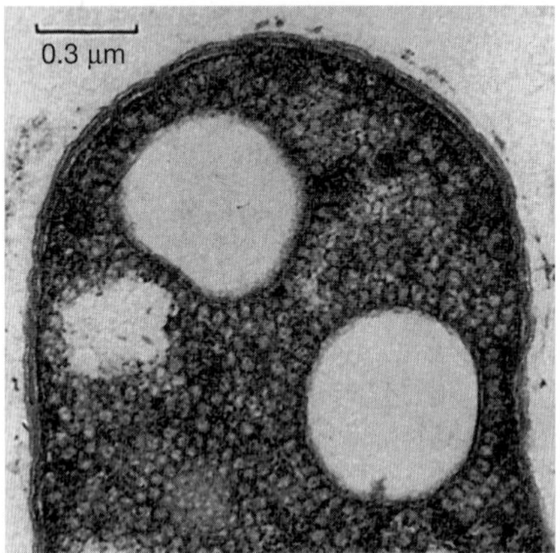

(b) *C. vinosum*

Figure 22.19 Typical Purple Sulfur Bacteria. **(a)** *Chromatium vinosum* with intracellular sulfur granules. **(b)** Electron micrograph of *C. vinosum*. Note the intracytoplasmic vesicular membrane system. The large white areas are the former sites of sulfur globules.

(a)

(b)

Figure 22.20 Purple Photosynthetic Sulfur Bacteria. **(a)** Purple photosynthetic sulfur bacteria growing in a bog. **(b)** A sewage lagoon with a bloom of purple photosynthetic bacteria.

Order *Thiotrichales*

The order *Thiotrichales* contains three families, the largest of which is the family *Thiotrichaceae*. This family has several genera that oxidize sulfur compounds (see the colorless sulfur bacteria [p. 550] and chapter 9 for sulfur oxidation and chemolithotrophy). Morphologically both rods and filamentous forms are present.

Two of the best-studied gliding genera in this family are *Beggiatoa* and *Leucothrix* (**figures 22.21** and **22.22**). *Beggiatoa* is microaerophilic and grows in sulfide-rich habitats such as sulfur springs, freshwater with decaying plant material, rice paddies, salt marshes, and marine sediments. Its filaments contain short, disklike cells and lack a sheath. *Beggiatoa* is very versatile metabolically. It oxidizes hydrogen sulfide to form large sulfur grains located in pockets formed by invaginations of the plasma membrane. *Beggiatoa* can subsequently oxidize the sulfur to sulfate. The electrons are used by the electron transport chain in energy production. Many strains also can grow heterotrophically with acetate as a carbon source, and some incorporate CO_2 autotrophically.

Leucothrix (figure 22.22) is an aerobic chemoorganotroph that forms filaments or trichomes up to 400 μm long. It is usually marine and is attached to solid substrates by a holdfast. *Leucothrix* has a complex life cycle in which it is dispersed by the formation of gonidia. Rosette formation often is seen in culture (figure 22.22*d*). *Thiothrix* is a related genus that forms sheathed filaments and releases gonidia from the open end of the sheath (**figure 22.23**). In contrast with *Leucothrix, Thiothrix* is a chemolithotroph that oxidizes hydrogen sulfide and deposits sulfur granules internally. It also requires an organic compound for growth (i.e., it is a mixotroph). *Thiothrix* grows in sulfide-rich flowing water and activated sludge sewage systems.

Figure 22.21 ***Beggiatoa alba.*** A light micrograph showing part of a colony (×400). Note the dark sulfur granules within many of the filaments.

Order *Methylococcales*

The single family in ths order is *Methylococcaceae.* It contains rods, vibrios, and cocci that use methane, methanol, and other reduced one-carbon compounds as their sole carbon and energy sources under aerobic or microaerobic (low oxygen) conditions. That is, they are methylotrophs distinguishing them from bacteria that use methane exclusively as their carbon and energy source, which are called methanotrophs. The family contains seven genera, two of which are *Methylococcus* (spherical, nonmotile cells) and *Methylomonas* (straight, curved, or branched rods with a single, polar flagellum). When oxidizing methane, the bacteria contain complex arrays of intracellular membranes. Almost all are capable of forming cysts. Methanogenesis from substrates such as H_2 and CO_2 is widespread in anoxic soil and water, and methylotrophic bacteria grow above anoxic habitats all over the world.

Methane-oxidizing bacteria use methane as a source of both energy and carbon. Methane is first oxidized to methanol by the enzyme methane monooxygenase. The methanol is then oxidized to formaldehyde by methanol dehydrogenase, and the electrons

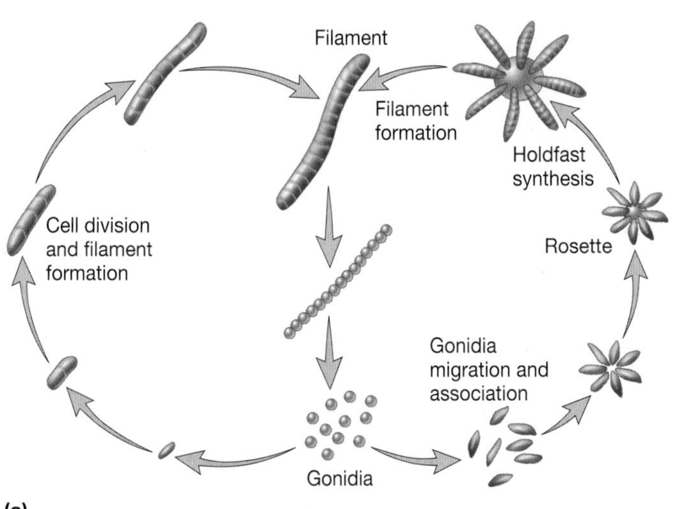

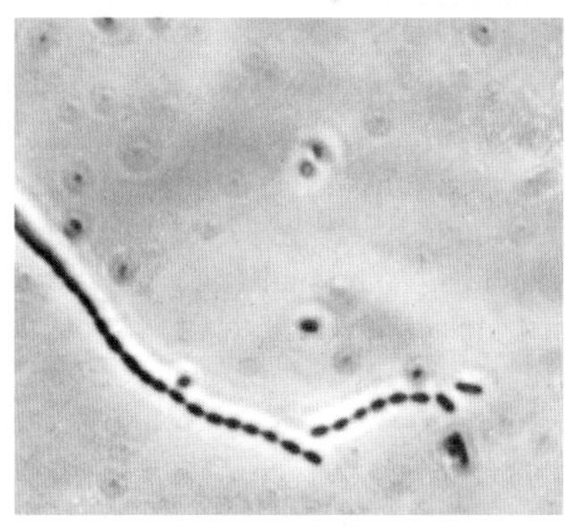

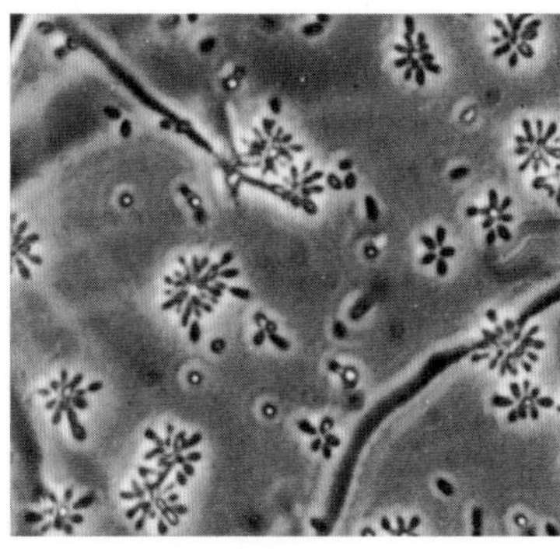

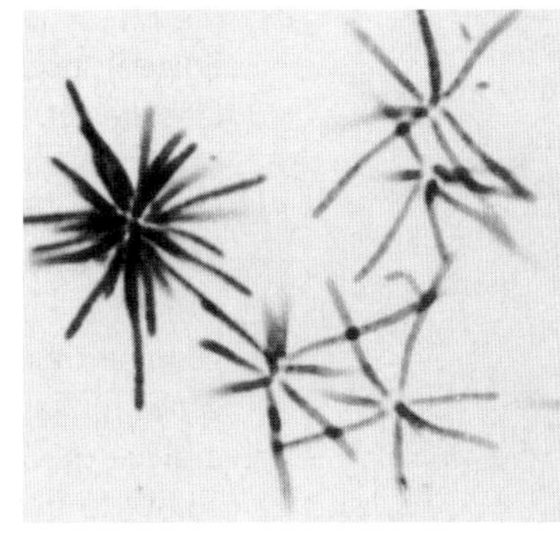

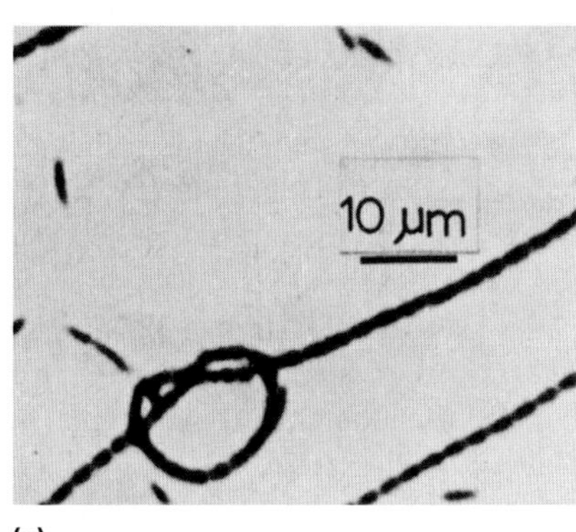

Figure 22.22 Morphology and Reproduction of *Leucothrix mucor.* **(a)** Life cycle of *L. mucor.* **(b)** Separation of gonidia from the tip of mature filament; phase contrast (×1,400). **(c)** Gonidia aggregating to form rosettes; phase contrast (×950). **(d)** Young developing rosettes (×1,500). **(e)** A knot formed by a *Leucothrix* filament.

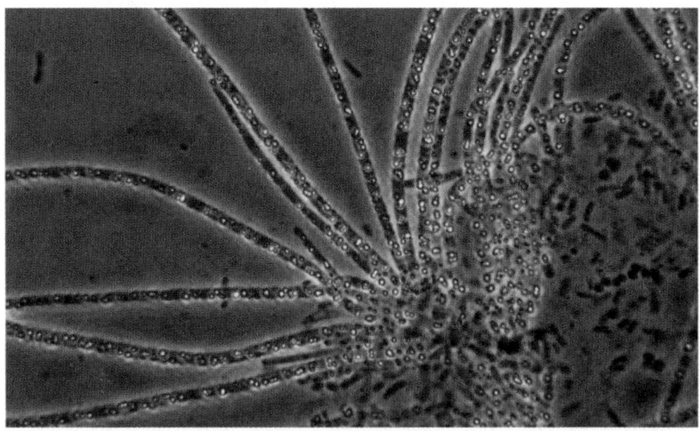

Figure 22.23 ***Thiothrix.*** A *Thiothrix* colony viewed with phase-contrast microscopy (×1,000).

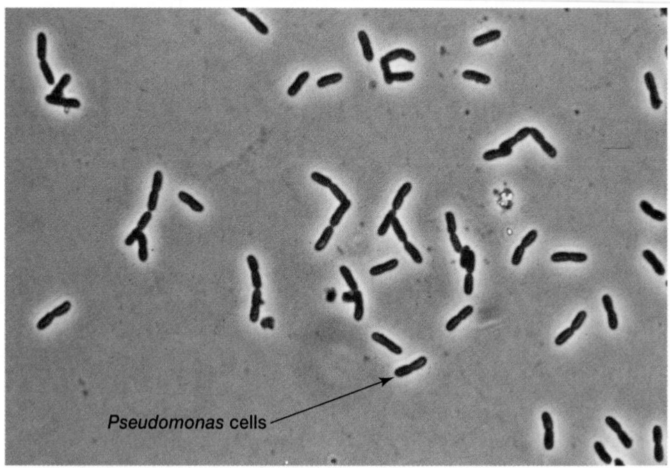

(a)

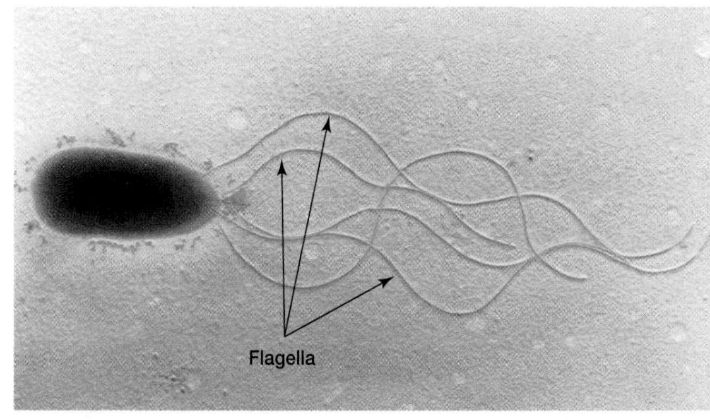

(b)

Figure 22.24 The Genus *Pseudomonas.* **(a)** A phase-contrast micrograph of *Pseudomonas* cells containing PHB (poly-β-hydroxybutyrate) granules. **(b)** A transmission electron micrograph of *Pseudomonas putida* with five polar flagella, each flagellum about 5–7 μm in length.

from this oxidation are donated to an electron transport chain for ATP synthesis. Formaldehyde can be assimilated into cell material by the activity of either of two pathways, one involving the formation of the amino acid serine and the other proceeding through the synthesis of sugars such as fructose 6-phosphate and ribulose 5-phosphate.

Order *Pseudomonadales*

Pseudomonas is the most important genus in the order *Pseudomonadales,* the family *Pseudomonaceae*. These bacteria are straight or slightly curved rods, 0.5 to 1.0 μm by 1.5 to 5.0 μm in length and are motile by one or several polar flagella (**figure 22.24;** *see figure 3.32a*). These chemoheterotrophs usually carry out aerobic respiration. Sometimes nitrate is used as the terminal electron acceptor in anaerobic respiration. All pseudomonads have a functional tricarboxylic acid cycle and can oxidize substrates completely to CO_2. Most hexoses are degraded by the Entner-Doudoroff pathway rather than the Embden-Meyerhof pathway. The breakdown of glucose to pyruvate (section 9.3); The tricarboxylic acid cycle (section 9.4); also see appendix II.

The genus *Pseudomonas* is an exceptionally heterogeneous taxon currently composed of about 60 species. Many can be placed in one of seven rRNA homology groups. The three best characterized groups are subdivided according to properties such as the presence of poly-β-hydroxybutyrate (PHB), the production of a fluorescent pigment, pathogenicity, the presence of arginine dihydrolase, and glucose utilization. For example, the fluorescent subgroup does not accumulate PHB and produces a diffusible, water-soluble, yellow-green pigment that fluoresces under UV radiation (**figure 22.25**). *Pseudomonas aeruginosa, P. fluorescens, P. putida,* and *P. syringae* are members of this group.

The pseudomonads have a great practical impact in several ways, including these:

1. Many can degrade an exceptionally wide variety of organic molecules. Thus they are very important in the **mineralization** process (the microbial breakdown of organic materials to inorganic substances) in nature and in sewage treatment. The fluorescent pseudomonads can use approximately 80 different substances as their carbon and energy sources. Microorganisms in the soil environment (section 29.3)
2. Several species (e.g., *P. aeruginosa*) are important experimental subjects. Many advances in microbial physiology and biochemistry have come from their study. For example, the study of *P. aeruginosa* has significantly advanced our understanding of how bacteria form biofilms and the role of extracellular signaling in bacterial communities and pathogenesis. The genome of *P. aeruginosa* has an unusually large number of genes for catabolism, nutrient transport, the efflux of organic molecules, and metabolic regulation. This may explain its ability to grow in many environments and resist antibiotics. Microbial growth in natural environments: Biofilms (section 6.6); Global regulatory systems: Quorum sensing (section 12.5)

3. Some pseudomonads are major animal and plant pathogens. *P. aeruginosa* infects people with low resistance such as cystic fibrosis patients. It also invades burns, and causes urinary tract infections. *P. syringae* is an important plant pathogen.
4. Pseudomonads such as *P. fluorescens* are involved in the spoilage of refrigerated milk, meat, eggs, and seafood because they grow at 4°C and degrade lipids and proteins.

The genus *Azotobacter* also is in the family *Pseudomonadaceae.* The genus contains large, ovoid bacteria, 1.5 to 2.0 μm in diameter, that may be motile by peritrichous flagella. The cells are often pleomorphic, ranging from rods to coccoid shapes, and form cysts as the culture ages (**figure 22.26**). The genus is aerobic, catalase positive, and fixes nitrogen nonsymbiotically. *Azotobacter* is widespread in soil and water.

Order *Vibrionales*

Three closely related orders of the γ-proteobacteria contain a number of important bacterial genera. Each order has only one family of facultatively anaerobic gram-negative rods. **Table 22.6** summarizes the distinguishing properties of the families *Enterobacteriaceae, Vibrionaceae,* and *Pasteurellaceae* from the orders *Enterbacteriales, Vibrionales,* and *Pasteurellales,* respectively.

The order *Vibrionales* contains only one family, the *Vibrionaceae.* Members of the family *Vibrionaceae* are gram-negative, straight or curved rods with polar flagella (**figure 22.27**). Most are oxidase positive, and all use D-glucose as their sole or primary carbon and energy source (table 22.6). The majority are aquatic microorganisms, widespread in freshwater and the sea. There are eight genera in the family: *Vibrio, Photobacterium, Salinivibrio, Listonella, Allomonas, Enterovibrio, Catencoccus,* and *Grimontia.*

Several vibrios are important pathogens. *Vibrio cholerae* causes cholera, and *V. parahaemolyticus* can cause gastroenteritis in humans following consumption of contaminated seafood. *V. anguillarum* and others are responsible for fish diseases. Food-borne and waterborne diseases: Cholera (section 38.4)

The *Vibrio cholerae* genome contains about 3,800 open reading frames distributed between two circular chromosomes, chromosome 1 (2.96 million base pairs) and chromosome 2 (1.07 million bp). The larger chromosome primarily has genes for essential cell functions such as DNA replication, transcription, and protein synthesis. It also has most of the virulence genes. For example, the cholera toxin gene is located in an integrated CTX phage on chromosome 1. Chromosome 2 also has essential genes such as transport genes and ribosomal protein genes. Copies of some genes are present on both chromosomes. Perhaps *V. cholerae* achieves faster genome duplication and cell division by distributing its genes between two chromosomes.

Several members of the family are unusual in being bioluminescent. *Vibrio fischeri*, *V. harveyi*, and at least two species of *Photobacterium* are among the few marine bacteria capable of **bioluminescence** and emit a blue-green light because of the activity of the enzyme luciferase (**Microbial Diversity & Ecology 22.1**). The peak emission of light is usually between 472 and 505 nm, but one strain of *V. fischeri* emits yellow light with a major peak at 545 nm. Although many of these bacteria are free-living, *V. fischeri, V. harveyi, P. phosphoreum,* and *P. leiognathi* live symbiotically in the luminous organs of fish (**figure 22.28**) and squid (*see figure 6.30*).

Order *Enterobacteriales*

The family *Enterobacteriaceae* is the largest of the families listed in table 22.6. It contains gram-negative, peritrichously flagellated or nonmotile, facultatively anaerobic, straight rods with simple

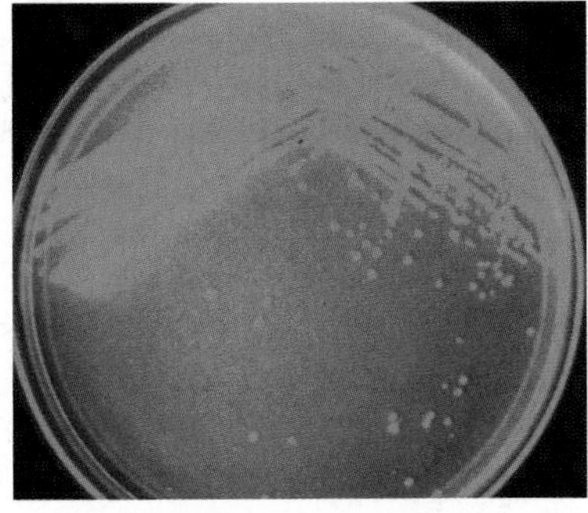

Figure 22.25 ***Pseudomonas* Fluorescence.** *Pseudomonas aeruginosa* colonies fluorescing under ultraviolet light.

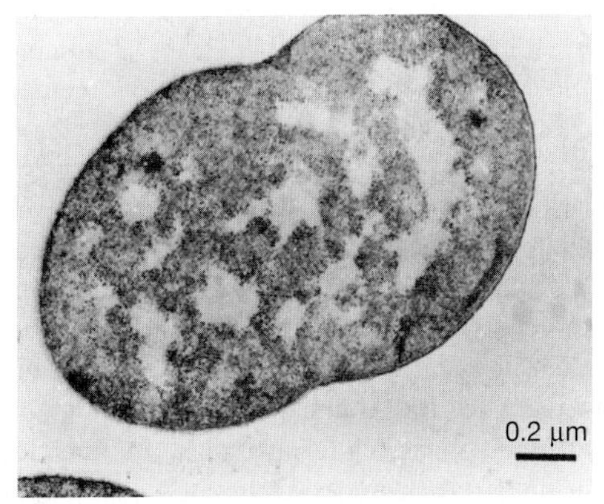

(a)

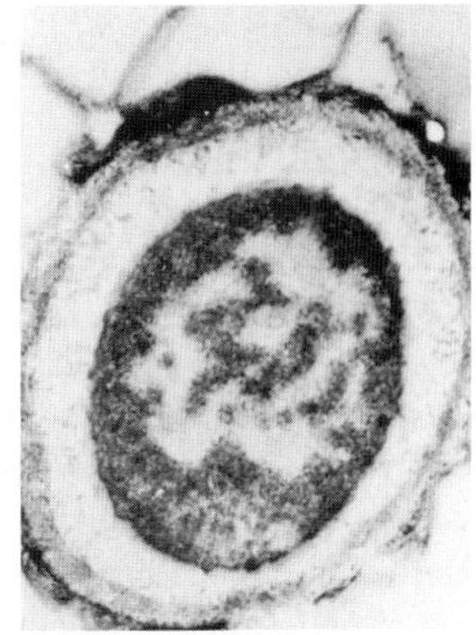

(b)

Figure 22.26 ***Azotobacter.*** **(a)** Electron micrograph of vegetative *A. chroococcum.* **(b)** *Azotobacter* cyst.

Table 22.6 Characteristics of Families of Facultatively Anaerobic Gram-Negative Rods

Characteristics	*Enterobacteriaceae*	*Vibrionaceae*	*Pasteurellaceae*
Cell dimensions	0.3–1.0 × 1.0–6.0 μm	0.3–1.3 × 1.0–3.5 μm	0.2–0.4 × 0.4–2.0 μm
Morphology	Straight rods; peritrichous flagella or nonmotile	Straight or curved rods; polar flagella; lateral flagella may be produced on solid media	Coccoid to rod-shaped cells, sometimes pleomorphic; nonmotile
Physiology	Oxidase negative	Oxidase positive; all can use D-glucose as sole or principal carbon source	Oxidase positive; heme and/or NAD often required for growth; organic nitrogen source required
G + C content	38–60%	38–51%	38–47%
Symbiotic relationships	Some parasitic on mammals and birds; some species plant pathogens	Most not pathogens; several inhabit light organs of marine organisms	Parasites of mammals and birds
Representative genera	*Escherichia, Shigella, Salmonella, Citrobacter, Klebsiella, Enterobacter, Erwinia, Serratia, Proteus, Yersinia*	*Vibrio, Photobacterium*	*Pasteurella, Haemophilus*

From G. M. Garrity editor-in-chief. *Bergey's Manual of Systematic Bacteriology,* vol. 2. Copyright © 2005 New York: Springer. Reprinted by permission.

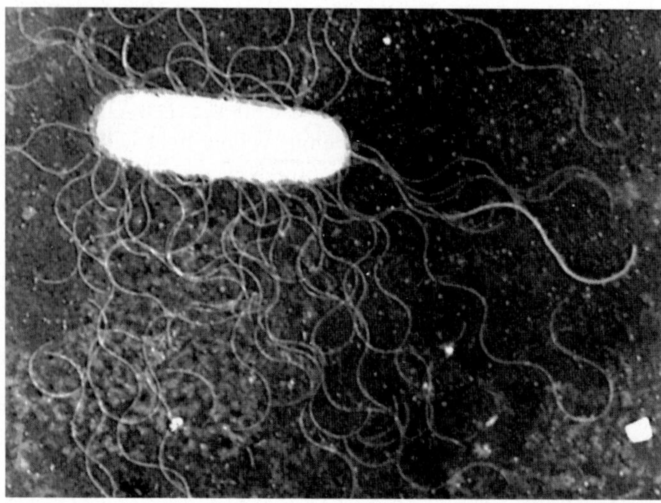

Figure 22.27 The *Vibrionaceae.* Electron micrograph of *Vibrio alginolyticus* grown on agar, showing a sheathed polar flagellum and unsheathed lateral flagella (×18,000).

nutritional requirements. The order *Enterobacteriales* has only one family, *Enterobacteriaceae,* with 44 genera. The relationship between *Enterobacteriales* and the orders *Vibrionales* and *Pasteurellales* can be seen by inspecting figure 22.17.

The metabolic properties of the *Enterobacteriaceae* are very useful in characterizing its constituent genera. Members of the family, often called **enterobacteria** or **enteric bacteria** [Greek *enterikos,* pertaining to the intestine], all degrade sugars by means of the Embden-Meyerhof pathway and cleave pyruvic acid to yield formic acid in formic acid fermentations. Those enteric bacteria that produce large amounts of gas during sugar fermentation, such as *Escherichia* spp., have the formic hydrogenlyase complex that degrades formic acid to H_2 and CO_2. The family can be divided into two groups based on their fermentation products. The majority (e.g., the genera *Escherichia, Proteus, Salmonella,* and *Shigella*) carry out mixed acid fermentation and produce mainly lactate, acetate, succinate, formate (or H_2 and CO_2), and ethanol. In contrast, *Enterobacter, Serratia, Erwinia,* and *Klebsiella* are butanediol fermenters. The major products of butanediol fermentation are butanediol, ethanol, and carbon dioxide. The two types of formic acid fermentation are distinguished by the methyl red and Voges-Proskauer tests. Fermentations (section 9.7)

Because the enteric bacteria are so similar in morphology, biochemical tests are normally used to identify them after a preliminary examination of their morphology, motility, and growth responses (**figure 22.29** provides a simple example). Some more commonly used tests are those for the type of formic acid fermentation, lactose and citrate utilization, indole production from tryptophan, urea hydrolysis, and hydrogen sulfide production. For example, lactose fermentation occurs in *Escherichia* and *Enterobacter* but not in *Shigella, Salmonella,* or *Proteus.* **Table 22.7** summarizes a few of the biochemical properties useful in distinguishing between genera of enteric bacteria. The mixed acid fermenters are located on the left in this table and the butanediol fermenters on the right. The usefulness of biochemical tests in identifying enteric bacteria is shown by the popularity of commercial identification systems, such as the Enterotube and API 20-E systems, that are based on these tests. Identification of microorganisms from specimens (section 35.2)

Microbial Diversity & Ecology

22.1 Bacterial Bioluminescence

Several species in the genera *Vibrio* and *Photobacterium* can emit light of a blue-green color. The enzyme luciferase catalyzes the reaction and uses reduced flavin mononucleotide, molecular oxygen, and a long-chain aldehyde as substrates.

$$FMNH_2 + O_2 + RCHO \xrightarrow{\text{luciferase}} FMH + H_2O + RCOOH + \text{light}$$

The evidence suggests that an enzyme-bound, excited flavin intermediate is the direct source of luminescence. Because the electrons used in light generation are probably diverted from the electron transport chain and ATP synthesis, the bacteria expend considerable energy on luminescence. Luminescence activity is regulated and can be turned off or on under the proper conditions.

There is much speculation about the role of bacterial luminescence and its value to bacteria, particularly because it is such an energetically expensive process. Luminescent bacteria occupying the luminous organs of fish do not emit light when they grow as free-living organisms in the seawater. Free-living luminescent bacteria can reproduce and infect young fish. Once settled in a fish's luminous organ, the quorum-sensing molecule autoinducer produced by the bacteria stimulates the emission of light. Other luminescent bacteria growing on potential food items such as small crustacea may use light to attract fish to the food source. After ingestion, they could establish a symbiotic relationship in the host's gut.

The mechanism by which autoinducer regulates light production in these marine bacteria is an important model for understanding quorum sensing in many gram-negative bacteria, including a number of pathogens. Global regulatory systems: Quorum sensing (section 12.5)

Figure 22.28 Bioluminescence. **(a)** A photograph of the Atlantic flashlight fish *Kryptophanaron alfredi.* The light area under the eye is the fish's luminous organ, which can be covered by a lid of tissue. **(b)** The masses of photobacteria in the SEM view are separated by thin epithelial cells. **(c)** Ultrathin section of the luminous organ of a fish, *Equulities novaehollandiae,* with the bioluminescent bacterium *Photobacterium leiognathi,* PL.

(a)

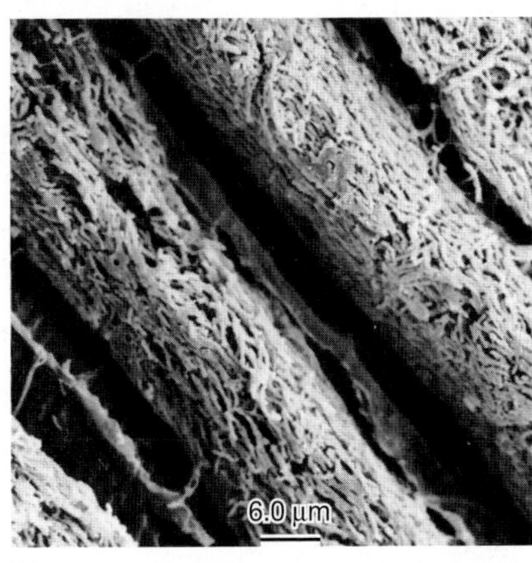

(b)

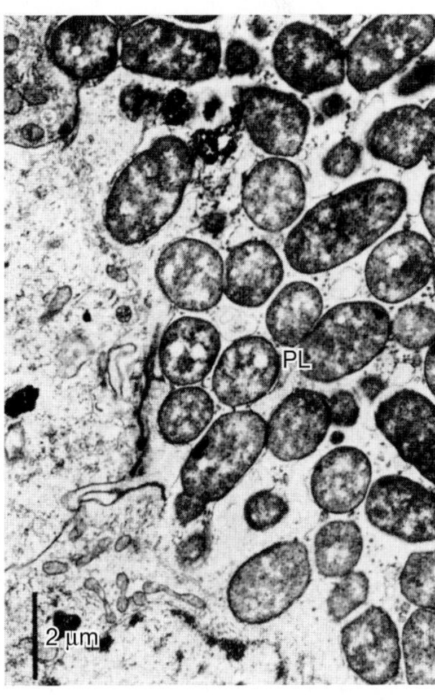

(c)

Members of the *Enterobacteriaceae* are so common, widespread, and important that they are probably more often seen in most laboratories than any other bacteria. *Escherichia coli* is undoubtedly the best-studied bacterium and the experimental organism of choice for many microbiologists. It is an inhabitant of the colon of humans and other warm-blooded animals, and it is quite useful in the analysis of water for fecal contamination. Some strains cause gastroenteritis or urinary tract infections. Several genera contain very important human pathogens responsible for a variety of diseases: *Salmonella* (**figure 22.30**), typhoid fever and gastroenteritis; *Shigella,* bacillary dysentery; *Klebsiella,* pneumonia; *Yersinia,* plague. Members of the genus *Erwinia* are major pathogens of crop plants and cause blights, wilts, and several other plant diseases. These and other members of the family are discussed in more detail in chapter 38. Water purification and sanitary analysis (section 41.1)

Order *Pasteurellales*

The family *Pasteurellaceae* in the order *Pasteurellales* differs from the *Vibrionales* and the *Enterobacteriales* in several ways (table 22.6). Most notably, they are small (0.2 to 0.4 µm in diameter) and nonmotile, normally oxidase positive, have complex nutritional requirements of various kinds, and are parasitic in

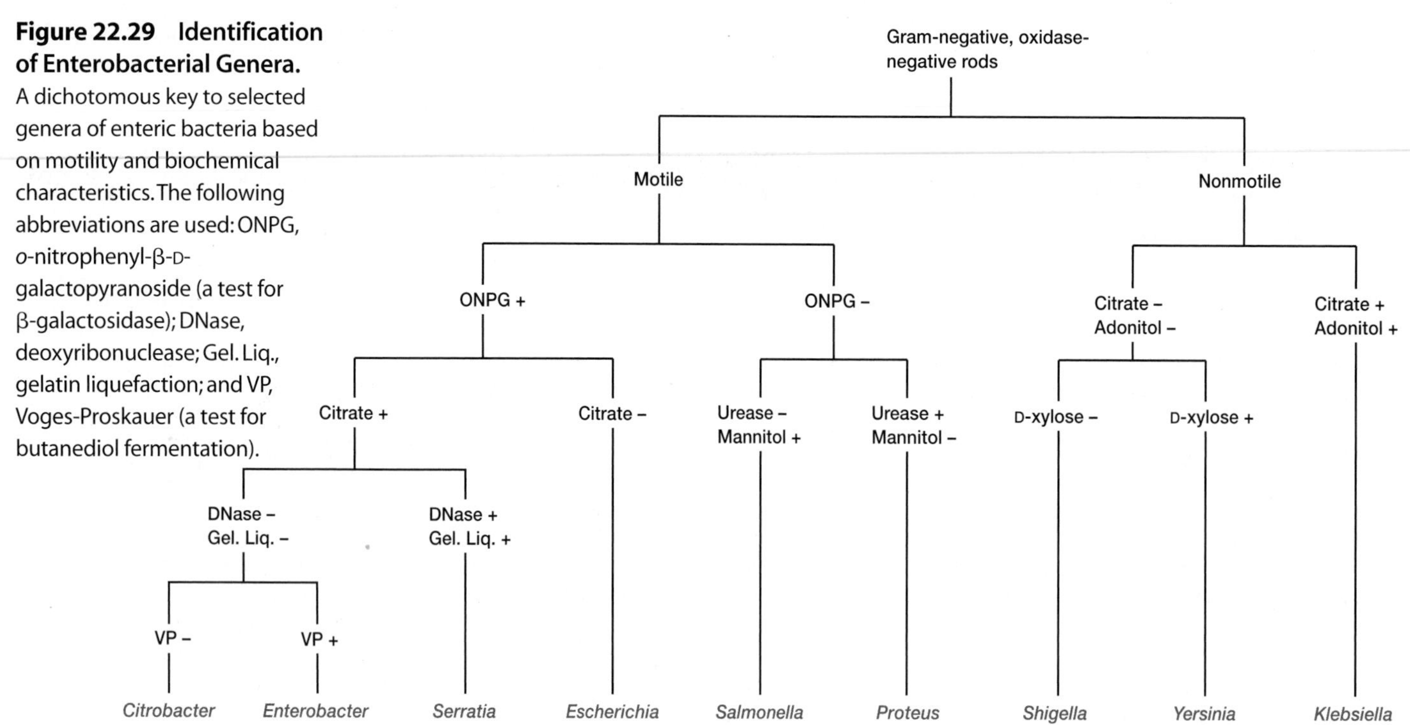

Figure 22.29 Identification of Enterobacterial Genera. A dichotomous key to selected genera of enteric bacteria based on motility and biochemical characteristics. The following abbreviations are used: ONPG, *o*-nitrophenyl-β-D-galactopyranoside (a test for β-galactosidase); DNase, deoxyribonuclease; Gel. Liq., gelatin liquefaction; and VP, Voges-Proskauer (a test for butanediol fermentation).

Table 22.7 Some Characteristics of Selected Genera in the *Enterobacteriaceae*

Characteristics	*Escherichia*	*Shigella*	*Salmonella*	*Citrobacter*	*Proteus*
Methyl red	+	+	+	+	+
Voges-Proskauer	–	–	–	–	d
Indole production	(+)	d	–	d	d
Citrate use	–	–	(+)	+	d
H_2S production	–	–	(+)	d	(+)
Urease	–	–	–	(+)	+
β-galactosidase	(+)	d	d	+	–
Gas from glucose	+	–	(+)	+	+
Acid from lactose	+	–	(–)	d	–
Phenylalanine deaminase	–	–	–	–	+
Lysine decarboxylase	(+)	–	(+)	–	–
Ornithine decarboxylase	(+)	d	(+)	(+)	d
Motility	d	–	(+)	+	+
Gelatin liquifaction (22°C)	–	–	–	–	+
% G + C	48–59	49–53	50–53	50–52	38–41
Genome size (Mb)	4.6–5.5	4.6	4.5–4.9	Nd[d]	Nd
Other characteristics	1.1–1.5 × 2.0–6.0 μm; peritrichous when motile	No gas from sugars	0.7–1.5 × 2–5 μm; peritrichous flagella	1.0 × 2.0–6.0 μm; peritrichous	0.4–0.8 × 1.0–3.0 μm; peritrichous

[a](+) usually present
[b](–) usually absent
[c]d, strains or species vary in possession of characteristic
[d]Nd: Not determined; genome not yet sequenced

vertebrates. The family contains seven genera: *Pasteurella, Haemophilus, Actinobacillus, Lonepinella, Mannheimia, Phocoenobacter,* and *Gallibacterium.*

As might be expected, members of this family are best known for the diseases they cause in humans and many animals. *Pasteurella multocida* and *P. haemolytica* are important animal pathogens. *P. multocida* is responsible for fowl cholera, which kills many chickens, turkeys, ducks, and geese each year. *P. haemolytica* is at least partly responsible for pneumonia in cattle, sheep, and goats (e.g., "shipping fever" in cattle). *H. influenzae* type b is a major human pathogen that causes a variety of diseases, including meningitis in children. Airborne diseases: Meningitis (section 38.1)

1. Describe the general properties of the γ-proteobacteria.
2. What are the major characteristics of the purple sulfur bacteria? Contrast the families *Chromatiaceae* and *Ectothiorhodospiraceae.*
3. Describe the genera *Beggiatoa, Leucothrix,* and *Thiothrix.*
4. In what habitats would one expect to see the *Methylococcaceae* growing and why?
5. What is a methylotroph? How do methane-oxidizing bacteria use methane as both an energy source and a carbon source?
6. Give the major distinctive properties of the genera *Pseudomonas* and *Azotobacter.* Briefly discuss the taxonomic changes that have occurred in the genus *Pseudomonas.*
7. Why are the pseudomonads such important bacteria? What is mineralization?
8. List the major distinguishing traits of the families *Vibrionaceae, Enterobacteriaceae,* and *Pasteurellaceae.*
9. What major human disease is associated with the *Vibrionaceae,* and what species causes it?
10. Briefly describe bioluminescence and the way it is produced.
11. Into what two groups can the enteric bacteria be placed based on their fermentation patterns?
12. Give two reasons why the enterobacteria are so important.

Yersinia	***Klebsiella***	***Enterobacter***	***Erwinia***	***Serratia***
+	(+)[a]	(−)[b]	+	d[c]
− (37°C)	(+)	+	(+)	+
d	d	−	(−)	(−)
(−)	(+)	+	(+)	+
−	−	−	(+)	−
d	(+)	(−)	−	−
+	(+)	+	+	+
(−)	(+)	(+)	(−)	d
(−)	(+)	(+)	d	d
−	−	(−)	(−)	−
(−)	(+)	d	−	d
d	−	(+)	−	d
− (37°C)	−	+	+	+
(−)	−	d	d	(+)
46–50	53–58	52–60	50–54	52–60
4.6	Nd	Nd	5.1	5.1
0.5–0.8 × 1.0–3.0 μm; peritrichous when motile	0.3–1.0 × 0.6–6.0 μm; capsulated	0.6–1.0 × 1.2–3.0 μm; peritrichous	0.5–1.0 × 1.0–3.0μm; peritrichous; plant pathogens and saprophytes	0.5–0.8 × 0.9–2.0 μm; peritrichous; colonies often pigmented

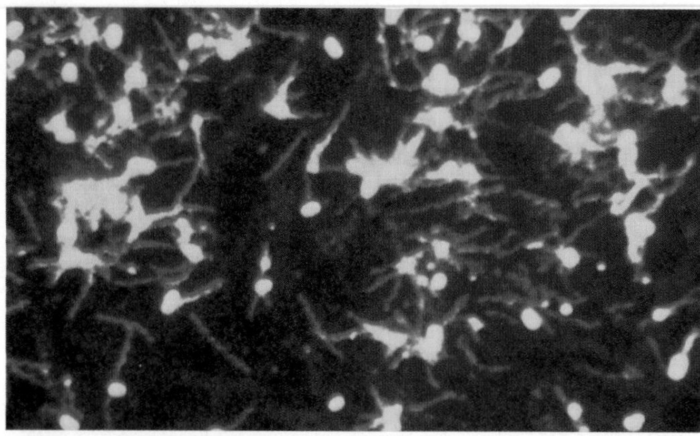

(a) *Salmonella enteritidis*

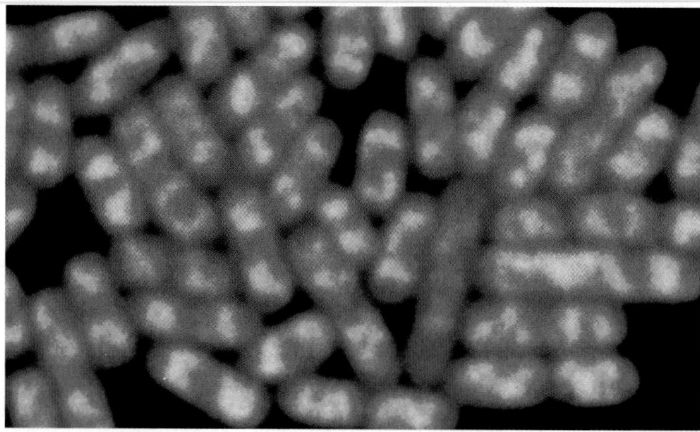

(b) *S. typhi*

Figure 22.30 The *Enterobacteriaceae*. *Salmonella* treated with fluorescent stains. **(a)** *Salmonella enterica* serovar Enteritidis with peritrichous flagella (×500). *S. enteritidis* is associated with gastroenteritis. **(b)** *S. enterica* serovar Typhi with acridine orange stain (×2,000). *S. typhi* causes typhoid fever.

22.4 Class *Deltaproteobacteria*

Although the **δ-proteobacteria** are not a large assemblage of genera, they show considerable morphological and physiological diversity. These bacteria can be divided into two general groups, all of them chemoorganotrophs. Some genera are predators such as the bdellovibrios and myxobacteria. Others are anaerobes that generate sulfide from sulfate and sulfur while oxidizing organic nutrients. The class has eight orders and 20 families. **Figure 22.31** illustrates the phylogenetic relationships among major groups within the δ-proteobacteria, and **table 22.8** summarizes the general properties of some representative genera.

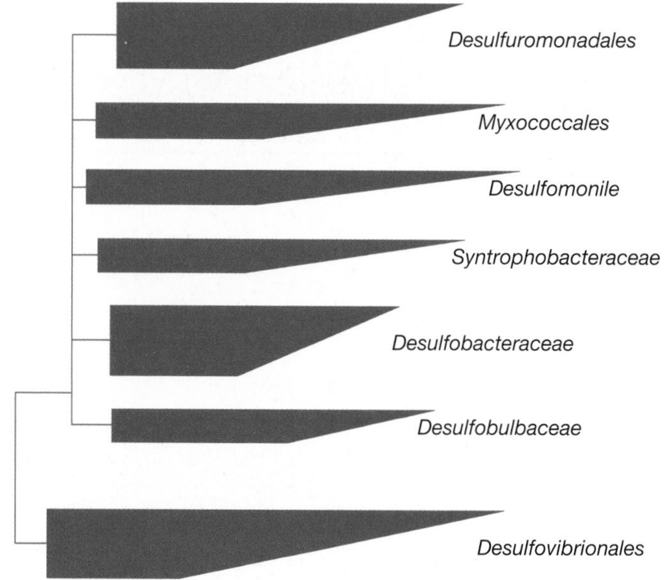

Figure 22.31 Phylogenetic Relationships Among Major Groups Within the δ-Proteobacteria. The relationships are based on 16S rRNA sequence.

Orders *Desulfovibrionales, Desulfobacterales,* and *Desulfuromonadales*

These sulfate- or sulfur-reducing bacteria are a diverse group united by their anaerobic nature and the ability to use elemental sulfur or sulfate and other oxidized sulfur compounds as electron acceptors during anaerobic respiration (**figure 22.32**). An electron transport chain reduces sulfur and sulfate to hydrogen sulfide and generates the proton motive force that drives the synthesis of ATP. The best-studied sulfate-reducing genus is *Desulfovibrio; Desulfuromonas* uses only elemental sulfur as an acceptor. Anaerobic respiration (section 9.6)

These bacteria are very important in the cycling of sulfur within the ecosystem. Because significant amounts of sulfate are present in almost all aquatic and terrestrial habitats, sulfate-reducing bacteria are widespread and active in locations made anoxic by microbial digestion of organic materials. *Desulfovibrio* and other sulfate-reducing bacteria thrive in habitats such as muds and sediments of polluted lakes and streams, sewage lagoons and digesters, and waterlogged soils. *Desulfuromonas* is most prevalent in anoxic marine and estuarine sediments. It also can be isolated from methane digesters and anoxic hydrogen sulfide-rich muds of freshwater habitats. It uses elemental sulfur, but not sulfate, as its electron acceptor. Often sulfate and sulfur reduction are apparent from the smell of hydrogen sulfide and the blackening of water and sediment by iron sulfide. Hydrogen sulfide production in waterlogged soils can kill animals, plants, and microorganisms. Sulfate-reducing bacteria negatively impact industry because of their primary role in the anaerobic corrosion of

Table 22.8 Characteristics of Selected δ- and ϵ-Proteobacteria

Class Genus	Dimensions (μm) and Morphology	G + C Content (mol%)	Oxygen Requirement	Other Distinctive Characteristics
δ-Proteobacteria				
Bdellovibrio	0.2–0.5 × 0.5–1.4; comma-shaped rods with a sheathed polar flagellum	49.5–51	Aerobic	Preys on other gram-negative bacteria where it grows in the periplasm, alternates between predatory and intracellular reproductive phases
Desulfovibrio	0.5–1.5 × 2.5–10; curved or sometimes straight rods, motile by polar flagella	46.1–61.2	Anaerobic	Oxidizes organic compounds to acetate and reduces sulfate or sulfur to H_2S
Desulfuromonas	0.4–0.9 × 1.0–4.0; straight or slightly curved or ovoid rods, lateral or subpolar flagella	54–62	Anaerobic	Reduces sulfur to H_2S, oxidizes acetate to CO_2, forms pink or peach-colored colonies
Myxococcus	0.4–0.7 × 2–8; slender rods with tapering ends, gliding motility	68–71	Aerobic	Forms fruiting bodies with microcysts not enclosed in a sporangium
Stigmatella	0.7–0.8 × 4–8; straight rods with tapered ends, gliding motility	67–68	Aerobic	Stalked fruiting bodies with sporangioles containing myxospores (0.9–1.2 × 2–4 μm)
ϵ-Proteobacteria				
Campylobacter	0.2–0.8 × 0.5–5; spirally curved cells with a single polar flagellum at one or both ends	29–47	Microaerophilic	Carbohydrates not fermented or oxidized; oxidase positive and urease negative; found in intestinal tract, reproductive organs, and oral cavity of animals
Helicobacter	0.2–1.2 × 1.5–10; helical, curved, or straight cells with rounded ends; multiple, sheathed flagella	24–48	Microaerophilic	Catalase and oxidase positive; urea rapidly hydrolyzed; found in the gastric mucosa of humans and other animals

iron in pipelines, heating systems, and other structures. Biogeochemical cycling: Sulfur cycle (section 27.2)

Order *Bdellovibrionales*

The order has only the family *Bdellovibrionaceae* and four genera. The genus *Bdellovibrio* [Greek *bdella,* leech] contains aerobic gram-negative, curved rods with polar flagella (**figure 22.33**). The flagellum is unusually thick due to the presence of a sheath that is continuous with the cell wall. *Bdellovibrio* has a distinctive life-style: it preys on other gram-negative bacteria and alternates between a nongrowing predatory phase and an intracellular reproductive phase.

The life cycle of *Bdellovibrio* is complex although it requires only 1 to 3 hours for completion (**figure 22.34**). The free bacterium swims along very rapidly (about 100 cell lengths per second) until it collides violently with its prey. It attaches to the bacterial surface, begins to rotate as fast as 100 revolutions per second, and bores a hole through the host cell wall in 5 to 20 minutes with the aid of several hydrolytic enzymes that it releases. Its flagellum is lost during penetration of the cell.

After entry, *Bdellovibrio* takes control of the host cell and grows in the periplasmic space (between the cell wall and plasma membrane) while the host cell loses its shape and rounds up. The predator inhibits host DNA, RNA, and protein synthesis within minutes and disrupts the host's plasma membrane so that cytoplasmic constituents leak out of the cell. The growing bacterium uses host amino acids as its carbon, nitrogen, and energy source. It employs fatty acids and nucleotides directly in biosynthesis, thus saving carbon and energy. The bacterium rapidly grows into a long filament under the cell wall and then divides into many smaller, flagellated progeny, which escape upon host cell lysis. Such multiple fission is rare in procaryotes.

The *Bdellovibrio* life cycle resembles that of bacteriophages in many ways. Not surprisingly, if a *Bdellovibrio* culture is plated out on agar with host bacteria, plaques will form in the bacterial lawn. This technique is used to isolate pure strains and count the number of viable organisms just as with phages.

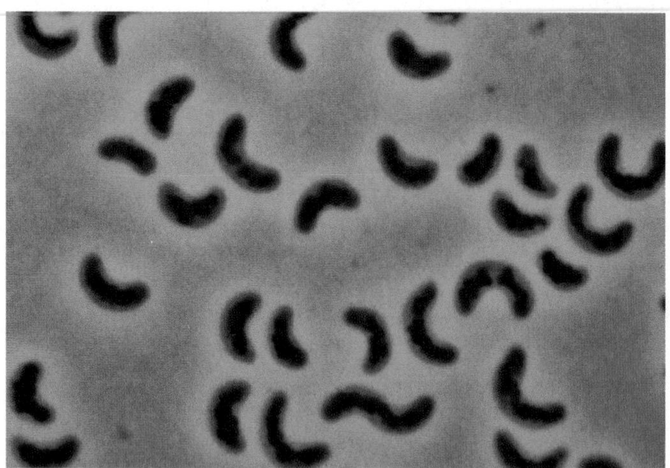

(a) *Desulfovibrio saprovorans*

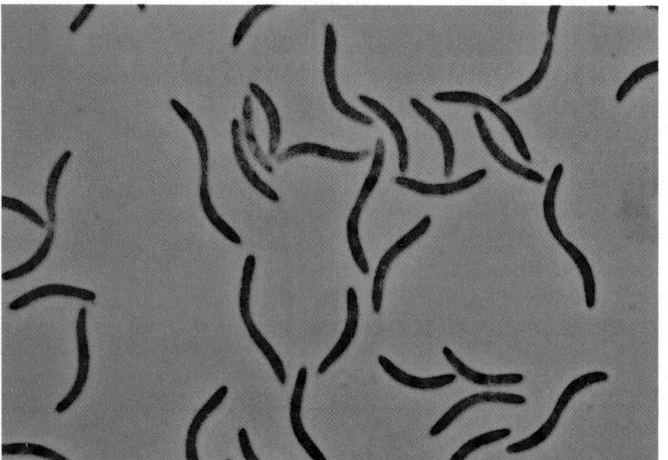

(b) *Desulfovibrio gigas*

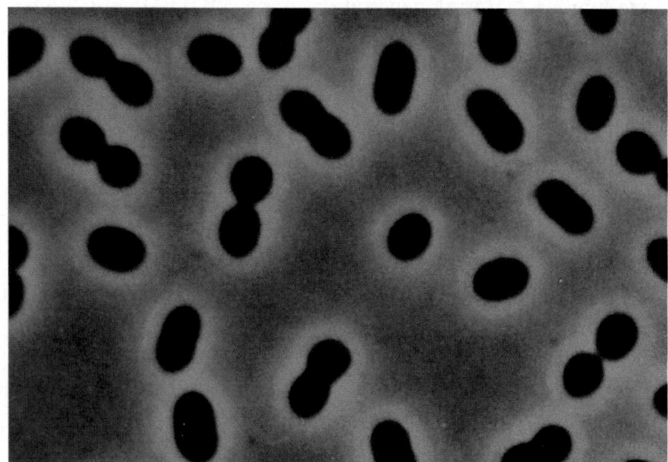

(c) *Desulfobacter postgatei*

Figure 22.32 The Dissimilatory Sulfate- or Sulfur-Reducing Bacteria. Representative examples. **(a)** Phase-contrast micrograph of *Desulfovibrio saprovorans* with PHB inclusions (×2,000). **(b)** *Desulfovibrio gigas;* phase contrast (×2,000). **(c)** *Desulfobacter postgatei;* phase contrast (×2,000).

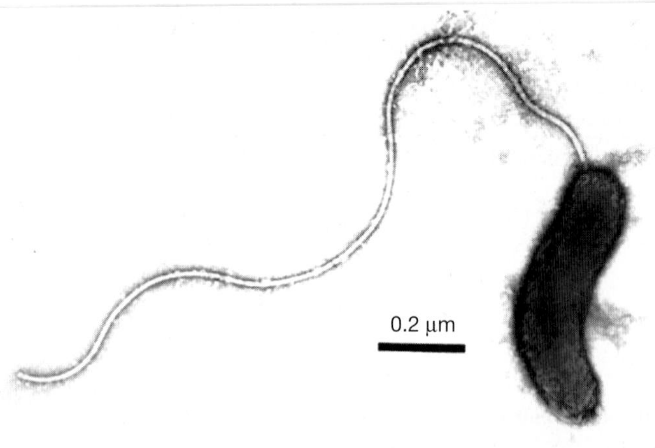

Figure 22.33 *Bdellovibrio* Morphology. Negatively stained *Bdellovibrio bacteriovorus* with its sheathed polar flagellum.

Order *Myxococcales*

The **myxobacteria** are gram-negative, aerobic soil bacteria characterized by gliding motility, a complex life cycle with the production of fruiting bodies, and the formation of dormant myxospores. In addition, their G + C content is around 67 to 71%, significantly higher than that of most gliding bacteria. Myxobacterial cells are rods, about 0.4 to 0.7 by 2 to 8 μm long, and may be either slender with tapered ends or stout with rounded, blunt ends (**figure 22.35**). The order *Myxococcales* is divided into six families based on the shape of vegetative cells, myxospores, and sporangia. Microbial diversity & ecology 21.1: The mechanism of gliding motility

Most myxobacteria are micropredators or scavengers. They secrete an array of digestive enzymes that lyse bacteria and yeasts. Many myxobacteria also secrete antibiotics, which may kill their prey. The digestion products, primarily small peptides, are absorbed. Most myxobacteria use amino acids as their major source of carbon, nitrogen, and energy. All are chemoheterotrophs with respiratory metabolism.

The myxobacterial life cycle is quite distinctive and in many ways resembles that of the cellular slime molds (**figure 22.36**). In the presence of a food supply, myxobacteria migrate along a solid surface, feeding and leaving slime trails. During this stage the cells often form a swarm and move in a coordinated fashion. Some species congregate to produce a sheet of cells that moves rhythmically to generate waves or ripples. When their nutrient supply is exhausted, the myxobacteria aggregate and differentiate into a **fruiting body.** Protist classification: *Eumycetozoa* (section 25.6)

The life cycle of the species *Myxococcus xanthus* has been well studied. Development in this microbe is induced by nutrient limitation and involves the exchange of at least five different extracellular signaling molecules that allow the cells to communicate with one another. Two of these signals have been characterized. Both the A factor, a mixture of peptides and amino acids, and the protein C factor are released and help trigger the process. Fruiting body development also requires gliding motil-

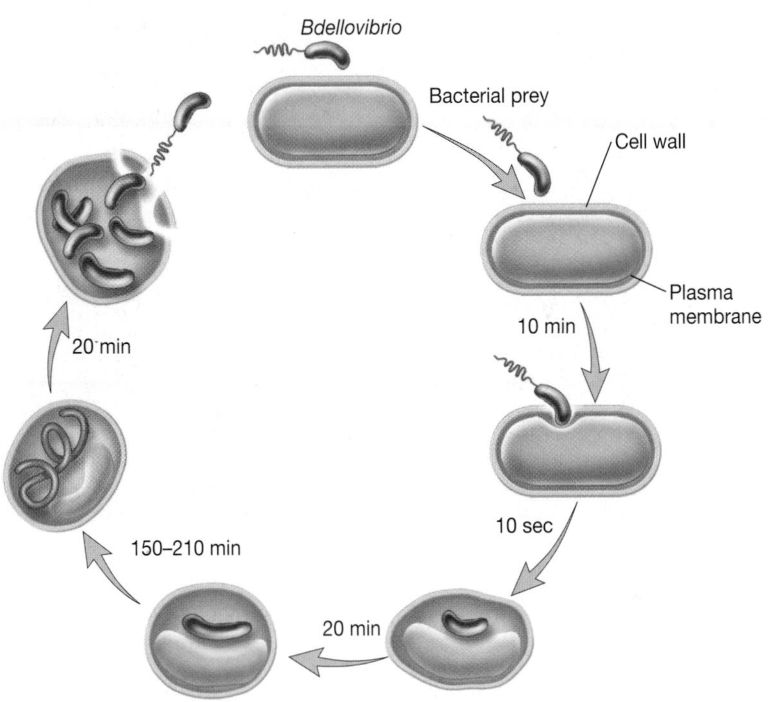

(a)

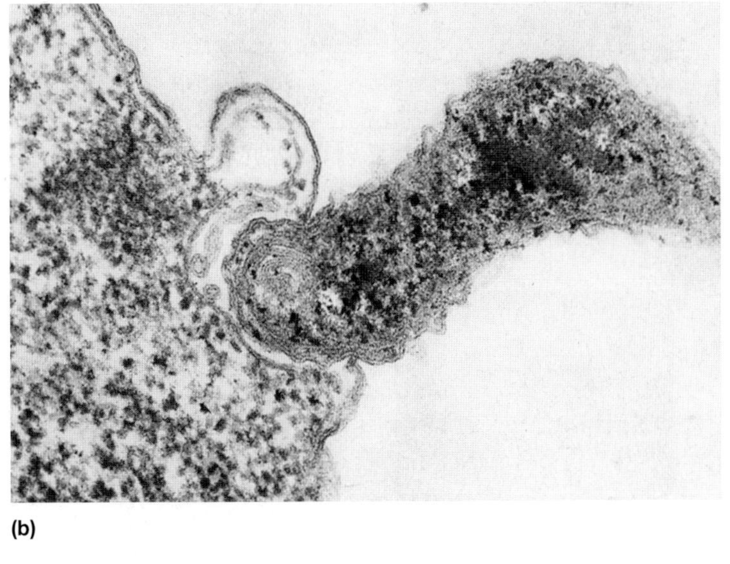

(b)

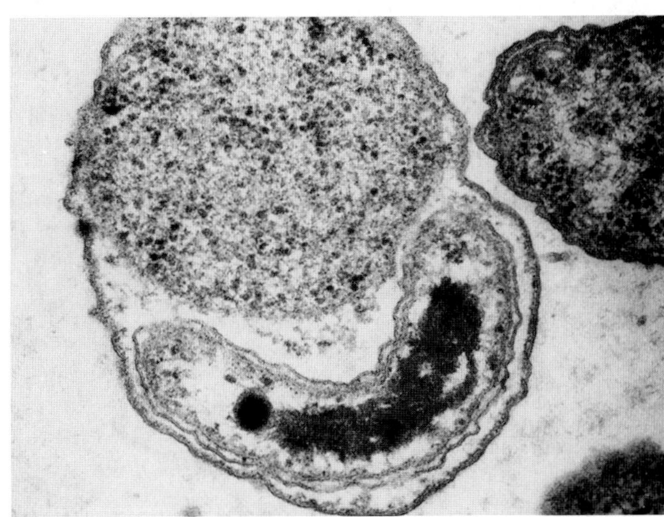

(c)

Figure 22.34 The Life Cycle of *Bdellovibrio*. **(a)** A general diagram showing the complete life cycle. **(b)** *Bdellovibrio bacteriovorus* penetrating the cell wall of *E. coli* (×55,000). **(c)** A *Bdellovibrio* encapsulated between the cell wall and plasma membrane of *E. coli* (×60,800).

ity. Two types of gliding have been characterized in *M. xanthus:* adventurous (A) motility is propelled by the extrusion of a gel-like material from the rear pole; social (S) motility is governed by the production of retractable pili from the front end of the cell. When the pili retract, the cell creeps forward. This type of motility was originally called social motility because it is only observed in cells that are close together. It is now known that cell-to-cell contact is required for S motility because cells share outer membrane lipoproteins involved in pili secretion.

A variety of new proteins are synthesized during fruiting body formation. Fruiting bodies range in height from 50 to 500 μm and often are colored red, yellow, or brown by carotenoid pigments. They vary in complexity from simple globular objects made of about 100,000 cells (*Myxococcus*) to the elaborate, branching, treelike structures formed by *Stigmatella* and *Chondromyces* (**figure 22.37**). Some cells develop into dormant **myxospores** that frequently are enclosed in walled structures called sporangioles or sporangia. Each species forms a characteristic fruiting body.

Myxospores are not only dormant but desiccation-resistant, and they may survive up to 10 years under adverse conditions. They enable myxobacteria to survive long periods of dryness and nutrient deprivation. The use of fruiting bodies provides further protection for the myxospores and assists in their dispersal. (The myxospores often are suspended above the soil surface.) Because myxospores are kept together within the fruiting body, a colony

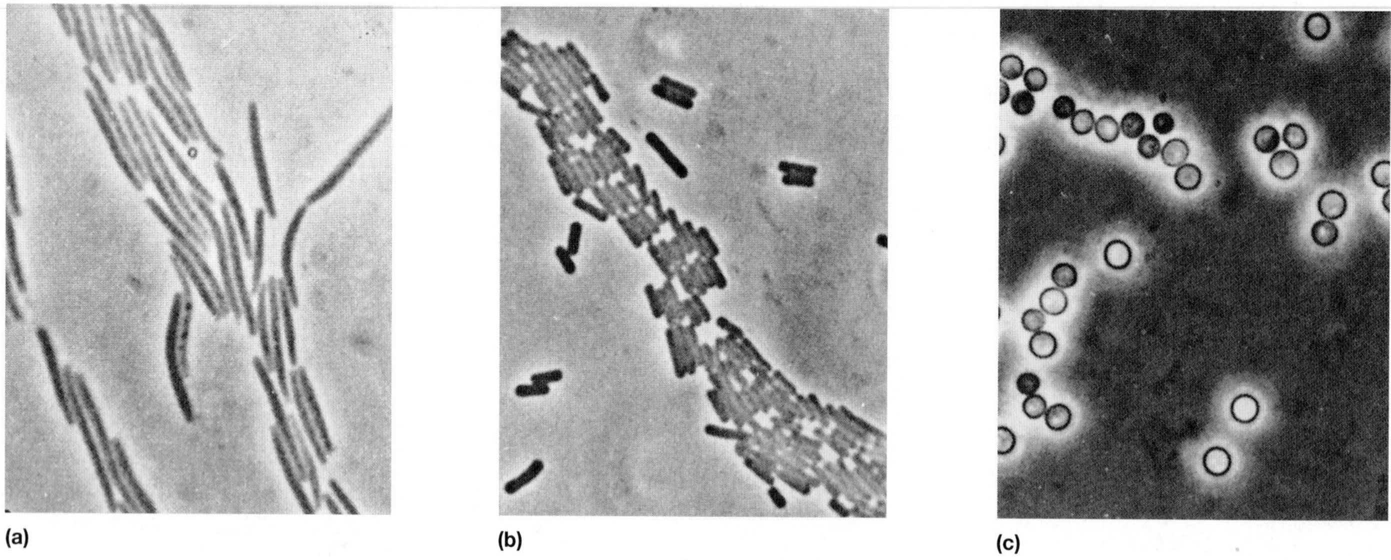

Figure 22.35 Gliding, Fruiting Bacteria (Myxobacteria). Myxobacterial cells and myxospores. **(a)** *Stigmatella aurantiaca* (×1,200). **(b)** *Chondromyces crocatus* (×950). **(c)** Myxospores of *Myxococcus xanthus* (×1,100). All photographs taken with a phase-contrast microscope.

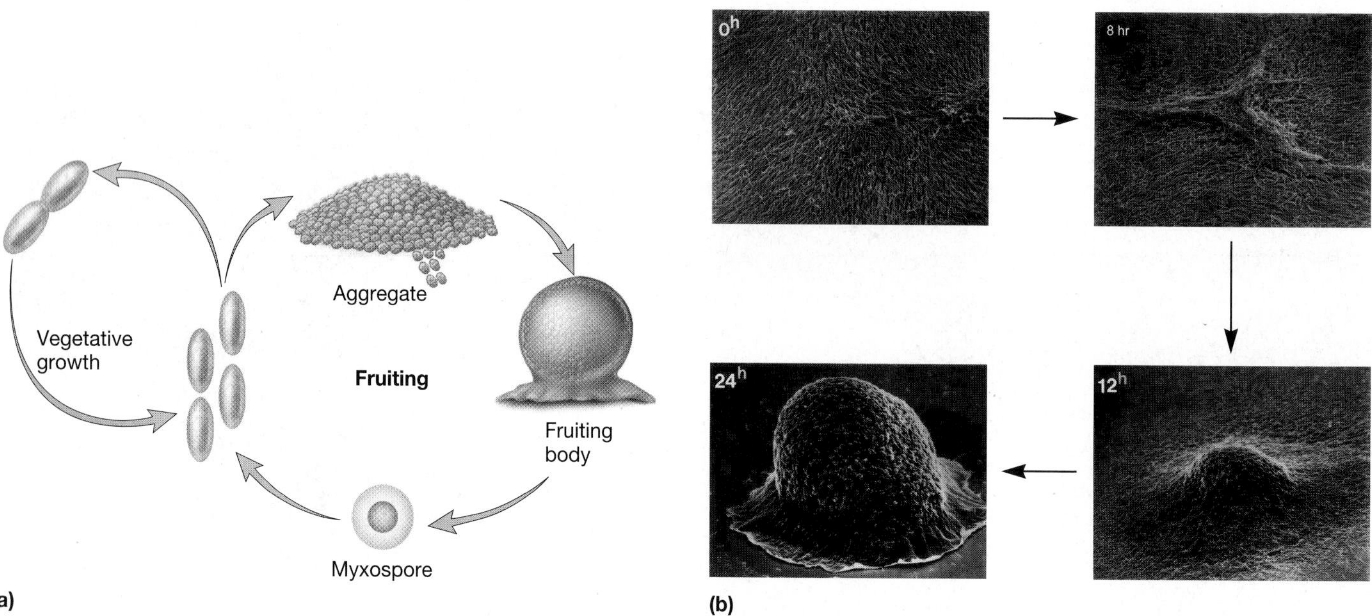

Figure 22.36 Life cycle of *Myxococcus xanthus*. **(a)** When nutrients are plentiful, *M. xanthus* grows vegetatively. However, when nutrients are depleted, a complex exchange of extracellular signaling molecules triggers the cells to aggregate and form fruiting bodies. Most of the cells within a fruiting body will become resting myxospores that will not germinate until nutrients are available. **(b)** Scanning electron micrographs taken during aggregate (0–12 hours) and fruiting body (24 hours) formation.

of myxobacteria automatically develops when the myxospores are released and germinate. This communal organization may be advantageous because myxobacteria obtain nutrients by secreting hydrolytic enzymes and absorbing soluble digestive products. A mass of myxobacteria can produce enzyme concentrations sufficient to digest their prey more easily than can an individual cell. Extracellular enzymes diffuse away from their source, and an individual cell will have more difficulty overcoming diffusional losses than a swarm of cells.

Myxobacteria are found in soils worldwide. They are most commonly isolated from neutral soils or decaying plant material such as leaves and tree bark, and from animal dung. Although they grow in habitats as diverse as tropical rain forests and the arctic tundra, they are most abundant in warm areas.

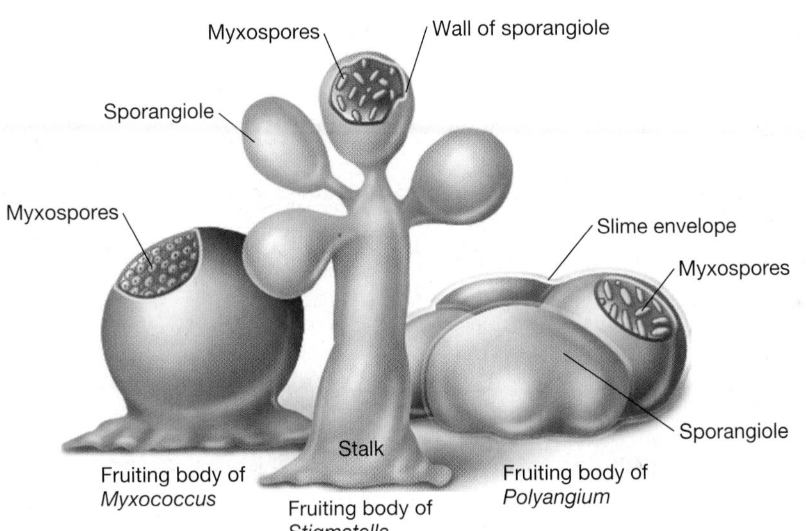

(b) *Myxococcus fulvus*

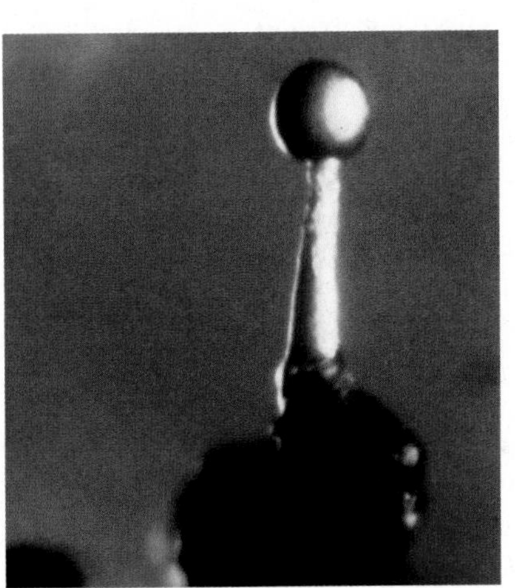
(c) *Myxococcus stipitatus*

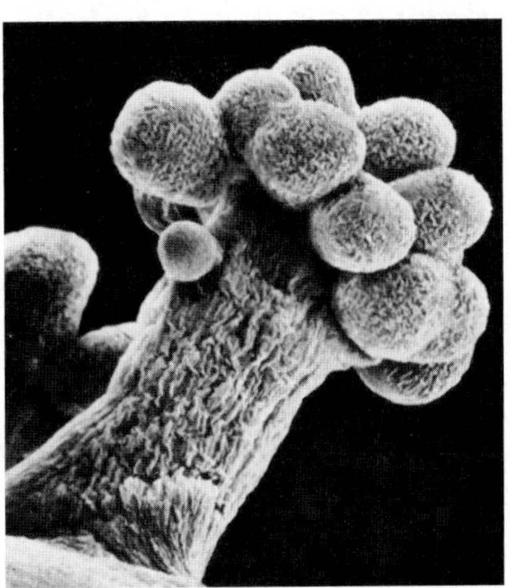
(d) *Chondromyces crocatus*

Figure 22.37 Myxobacterial Fruiting Bodies. **(a)** An illustration of typical fruiting body structure. **(b)** *Myxococcus fulvus*. Fruiting bodies are about 150–400 μm high. **(c)** *Myxococcus stipitatus*. The stalk is as tall as 200 μm. **(d)** *Chondromyces crocatus* viewed with the SEM. The stalk may reach 700 μm or more in height.

1. Briefly characterize the δ-proteobacteria.
2. Describe the metabolic specialization of the dissimilatory sulfate- or sulfur-reducing bacteria. Why are they important?
3. Characterize the genus *Bdellovibrio* and outline its life cycle in detail.
4. Give the major distinguishing characteristics of the myxobacteria. How do they obtain most of their nutrients?
5. Briefly describe the myxobacterial life cycle. What are fruiting bodies and myxospores?

22.5 Class *Epsilonproteobacteria*

The **ε-proteobacteria** are the smallest of the five proteobacterial classes. They all are slender gram-negative rods, which can be straight, curved, or helical. The ε-proteobacteria have one order, *Campylobacterales,* and three families: *Campylobacteraceae, Helicobacteraceae,* and the recently added *Nautiliaceae*. Two pathogenic genera, *Campylobacter* and *Helicobacter,* are microaerophilic, motile, helical or vibrioid, gram-negative rods. Table 22.8 summarizes some of the characteristics of these two genera.

The genus *Campylobacter* contains both nonpathogens and species pathogenic for humans and other animals. *C. fetus* causes reproductive disease and abortions in cattle and sheep. It is associated with a variety of conditions in humans ranging from **septicemia** (pathogens or their toxins in the blood) to **enteritis** (inflammation of the intestinal tract). *C. jejuni* causes abortion in sheep and enteritis diarrhea in humans. Food-borne and waterborne diseases: *Campylobacter jejuni* gastroenteritis (section 38.4)

There are at least 23 species of *Helicobacter,* all isolated from the stomachs and upper intestines of humans, dogs, cats, and other mammals. In developing countries 70 to 90% of the population is infected; developed countries range from 25 to 50%. Most infections are probably acquired during childhood, but the precise mode of transmission is unclear. The major human pathogen is *Helicobacter pylori,* which causes gastritis

(a)

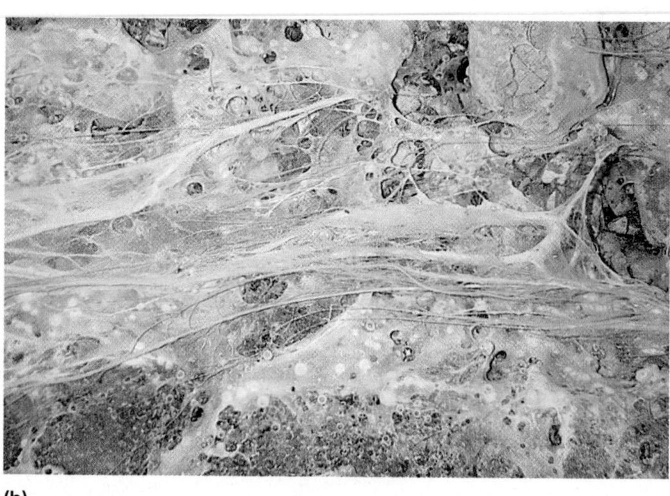
(b)

Figure 22.38 ε-Proteobacteria Dominate Filamentous Microbial Mats in a Wyoming Sulfidic Cave Spring. **(a)** A channel formed by the spring appears white due to the high density of filamentous ε-proteobacteria. The stake in the center of photo (arrow) is about 25 cm high. **(b)** Filamentous ε-proteobacteria within the springs.

and peptic ulcer disease. *H. pylori* produces large quantities of urease, and urea hydrolysis appears to be associated with its virulence. Direct-contact diseases: Peptic ulcer disease and gastritis (section 38.3)

The genomes of *C. jejuni* and *H. pylori* (both about 1.6 million base pairs in size) have been sequenced. They are now being studied and compared in order to understand the life styles and pathogenicity of these bacteria.

The ε-proteobacteria are now recognized to be more metabolically and ecologically diverse than previously thought. For instance, filamentous microbial mats in anoxic, sulfide-rich cave springs are dominated by members of the ε-proteobacteria (**figure 22.38**). The inclusion of a new family, the *Nautiliaceae,* in the second edition of *Bergey's Manual* is a result of the recent isolation of two genera of moderately thermophilic (optimum growth temperature about 55°C) chemolithoautotrophs from deep-sea hydrothermal vents. Members of the genera *Nautilia* and *Caminibacter* are strict anaerobes that oxidize H_2 and use sulfur as an electron acceptor. Species are found as either freely living or as symbionts of vent macrofauna. Microbial interactions: Sulfide-based mutualisms (section 30.1)

1. Briefly describe the properties of the ε-proteobacteria.
2. Give the general characteristics of *Campylobacter* and *Helicobacter.* What is their public health significance?

Summary

The proteobacteria are the largest and most diverse group of bacteria. On the basis of rRNA sequence data, they are divided into five classes: the α-, β-, γ-, δ-, and ε-proteobacteria.

22.1 Class *Alphaproteobacteria*

a. The purple nonsulfur bacteria can grow anaerobically as photoorganoheterotrophs and often aerobically as chemoorganoheterotrophs (**figure 22.3**). They are found in aquatic habitats with abundant organic matter and low sulfide levels.

b. Rickettsias are obligately intracellular parasites responsible for many diseases (**figure 22.4**). They have numerous transport proteins in their plasma membranes and make extensive use of host cell nutrients, coenzymes, and ATP.

c. Many proteobacteria have prosthecae, stalks, or reproduction by budding. Most of these bacteria are placed among the α-proteobacteria.

d. Two examples of budding and/or appendaged bacteria are *Hyphomicrobium* (budding bacteria that produce swarmer cells) and *Caulobacter* (bacteria with prosthecae and holdfasts) (**figures 22.5–22.8**).

e. *Rhizobium* carries out nitrogen fixation, whereas *Agrobacterium* causes the development of plant tumors. Both are in the family *Rhizobiaceae* of the α-proteobacteria (**figures 22.9** and **22.10**).

f. Chemolithotrophic bacteria derive energy and electrons from reduced inorganic compounds. Nitrifying bacteria are aerobes that oxidize either ammonia or nitrite to nitrate and are responsible for nitrification (**table 22.2**).

22.2 Class *Betaproteobacteria*

a. The genus *Neisseria* contains nonmotile, aerobic, gram-negative cocci that usually occur in pairs. They colonize mucous membranes and cause several human diseases.

b. *Sphaerotilus, Leptothrix,* and several other genera have sheaths, hollow tube-like structures that surround chains of cells without being in intimate contact with the cells (**figure 22.14**).

c. The colorless sulfur bacteria such as *Thiobacillus* oxidize elemental sulfur, hydrogen sulfide, and thiosulfate to sulfate while generating energy chemolithotrophically.

22.3 Class *Gammaproteobacteria*

a. The γ-proteobacteria are the largest subgroup of proteobacteria with great variety in physiological types (**table 22.5** and **figure 22.17**).

b. The purple sulfur bacteria are anaerobes and usually photolithoautotrophs. They oxidize hydrogen sulfide to sulfur and deposit the granules internally (**figure 22.19**).

c. Bacteria like *Beggiatoa* and *Leucothrix* grow in long filaments or trichomes (**figures 22.21** and **22.22**). Both genera have gliding motility. *Beggiatoa* is primarily a chemolithotroph and *Leucothrix,* a chemoorganotroph.

d. The *Methylococcaceae* are methylotrophs; they use methane, methanol, and other reduced one-carbon compounds as their sole carbon and energy sources.

e. The genus *Pseudomonas* contains straight or slightly curved, gram-negative, aerobic rods that are motile by one or several polar flagella and do not have prosthecae or sheaths (**figure 22.24**).

f. The pseudomonads participate in natural mineralization processes, are major experimental subjects, cause many diseases, and often spoil refrigerated food.

g. The most important facultatively anaerobic, gram-negative rods are found in three families: *Vibrionaceae, Enterobacteriaceae,* and *Pasteurellaceae* (**table 22.6**).

h. The *Enterobacteriaceae,* often called enterobacteria or enteric bacteria, are gram-negative, peritrichously flagellated or nonmotile, facultatively anaerobic, straight rods with simple nutritional requirements.

i. The enteric bacteria are usually identified by a variety of physiological tests and are very important experimental organisms and pathogens of plants and animals (**table 22.7** and **figure 22.29**).

22.4 Class *Deltaproteobacteria*

a. The δ-proteobacteria contain chemorganotrophic gram-negative bacteria that are anaerobic and can use elemental sulfur and oxidized sulfur compounds as electron acceptors in anaerobic respiration (**table 22.8**). They are very important in sulfur cycling in the ecosystem. Other δ-proteobacteria are predatory aerobes.

b. *Bdellovibrio* is an aerobic curved rod with sheathed polar flagellum that preys on other gram-negative bacteria and grows within their periplasmic space (**figures 22.33** and **22.34**).

c. Myxobacteria are gram-negative, aerobic soil bacteria with gliding motility and a complex life cycle that leads to the production of dormant myxospores held within fruiting bodies (**figures 22.35–22.37**).

22.5 Class *Epsilonproteobacteria*

a. The ϵ-proteobacteria are the smallest of the proteobacterial classes and contain two important pathogenic genera: *Campylobacter* and *Helicobacter.* These are microaerophilic, motile, helical or vibrioid, gram-negative rods.

b. Recently a new family, the *Nautiliaceae,* has been added. Many of these bacteria are chemolithoautotrophs from deep-sea hydrothermal vents ecosystems.

Key Terms

α-proteobacteria 540
β-proteobacteria 546
binary fission 543
bioluminescence 557
budding 543
colorless sulfur bacteria 550
δ-proteobacteria 562
enteric bacteria (enterobacteria) 558
enteritis 567
ϵ-proteobacteria 567
fruiting body 564
γ-proteobacteria 551
holdfast 544
methylotroph 544
mineralization 556
myxobacteria 564
myxospores 565
nitrification 546
nitrifying bacteria 546
prostheca 543
proteobacteria 539
purple nonsulfur bacteria 540
purple sulfur bacteria 553
septicemia 567
sheath 548
stalk 543

Critical Thinking Questions

1. *Helicobacter pylori* produces large quantities of urease. Urease catalyzes the reaction:

$$H_2N{-}\overset{\overset{\displaystyle O}{\|}}{C}{-}NH_2 \rightarrow CO_2 + 2NH_3$$

Suggest why this allows *H. pylori* to inhabit the acidic habitat of the gastric mucosa.

2. Methylotrophs oxidize methane to methanol, then to formaldehyde, and finally into acetate. Suggest mechanisms by which the bacterium protects itself from the toxic effects of the intermediates, methanol and formaldehyde.

3. *Bdellovibrio* is an intracellular predator. Once it invades the periplasm, it manages to inhibit many aspects of host metabolism. Suggest a mechanism by which this inhibition could occur so rapidly.

4. Why might the ability to form dormant cysts be of great advantage to *Agrobacterium* but not as much to *Rhizobium?*

5. Why are gliding and budding and/or appendaged bacteria distributed among so many different sections in *Bergey's Manual?*

Learn More

Balows, A.; Trüper, H. G.; Dworkin, M.; Harder, W.; and Schleifer, K.-H. 1992. *The prokaryotes,* 2d ed. New York: Springer-Verlag.

Engel, A., S.; Lee, N.; Porter, M. L.; Stern, L. A.; Bennett, P. C.; and Wagner, M. 2003. Filamentous "*Epsilonproteobacteria*" dominate microbial mats from sulfidic cave springs. *Appl. Environ. Microbiol.* 69:5503–11.

Eremeeva, M. E., and Dasch, G.A. 2000. Rickettsiae. In *Encyclopedia of microbiology,* 2d ed., vol. 4, J. Lederberg, editor-in-chief, 140–80. San Diego: Academic Press.

Garrity, G. M., editor-in-chief. 2005. *Bergey's Manual of Systematic Bacteriology,* 2d ed., vol. 2, D. J. Brenner, N. R. Krieg, and J. T. Staley, editors. New York: Springer-Verlag.

Kaplan, H. B. 2003. Multicellular development and gliding motility in *Myxococcus xanthus. Curr. Opin. Microbiol.* 6:572–77.

McGrath, P. T.; Vollier, P.; and McAdams, H. H. 2004. Setting the pace: Mechanisms tying *Caulobacter* cell-cycle progression to macroscopic cellular events. *Curr. Opin. Microbiol.* 7:192–97.

Miroshnichenko, M. L.; Haridon, S. L.; Schumann, P.; Spring, S.; Bonch-Osmolovskaya, E. A.; Jeanthon, C.; and Stackenbrandt, E. 2004. *Caminibacter profundus* sp. nov., a novel thermophile of *Nautiliales* ord. nov. within the class "*Epsilonproteobacteria,*" isolated from a deep-sea hydrothermal vent. *Int. J. Sys. Evol. Microbiol.* 54:41–5.

Miroshnichenko, M. L.; Kostrikina, N. A.; Haridon, S. L.; Jeathon, C.; Hippe, H.; Stackenbrandt, E.; and Bonch-Osmolovskaya, E. A. 2002. *Nautilia lithotrophica* gen. nov., sp. nov. a thermophilic sulfur-reducing ε-proteobacterium isolated from a deep-sea hydrothermal vent. *Int. J. Sys. Evol. Microbiol.* 52: 1299–1304.

Moulin, L. O.; Muinve, A.; Dreyfus, B.; and Boivin-Masson, C. 2001. Nodulation of legumes by members of the β-subclass of Proteobacteria. *Nature* 411: 948–50.

Parkhill, J., et al. 2000. The genome sequence of the food-borne pathogen *Campylobacter jejuni* reveals hypervariable sequences. *Nature* 403:655–68.

Shapiro, L.; McAdams, H. H.; and Losick, R. 2002. Generating and exploiting polarity in bacteria. *Science* 298:1942–46.

Zhu, J.; Oger, P. M.; Schrammeijer, B.; Hooykaas, P. J.; Farrand, S. K.; and Winans, S. C. 2000. The basis of crown gall tumorigenesis. *J. Bacteriol.* 182: 3885–95.

Please visit the Prescott website at www.mhhe.com/prescott7 for additional references.

Bacteria: The Low G + C Gram Positives

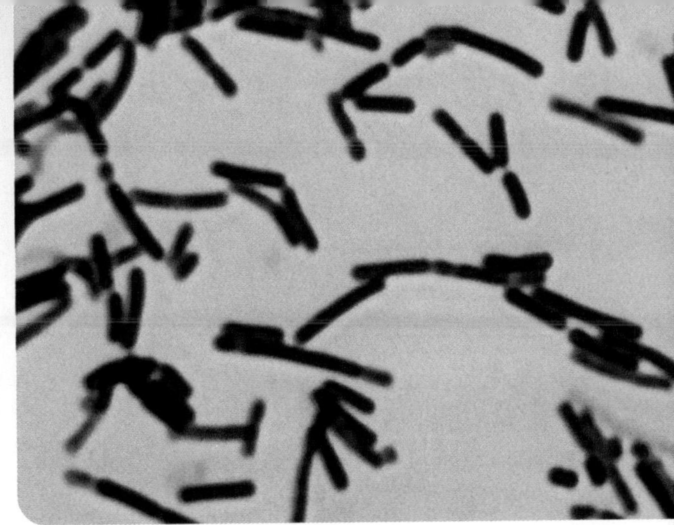

Lactobacilli are indispensable to the food and dairy industry. They are not considered pathogens.

Preview

- *Bergey's Manual* groups the gram-positive bacteria phylogenetically into two major groups: the low G + C gram-positive bacteria and the high G + C gram-positive bacteria. This classification is based primarily on nucleic acid sequences rather than phenotypic similarity.
- The low G + C gram positives contain (1) clostridia and relatives, (2) the mycoplasmas, and (3) the bacilli and lactobacilli. Endospore formers, cocci, and rods are found among the clostridia and bacilli groups. Thus common possession of a complex structure such as an endospore does not necessarily indicate close relatedness between the genera.
- Peptidoglycan structure varies among different groups in ways that are often useful in their identification.
- The majority of gram-positive bacteria are harmless, free-living saprophytes, but most major groups include pathogens of humans, other animals, and plants. Some gram-positive bacteria are very important in the food and dairy industries.

This chapter surveys many of the bacteria found in volume 3 of the second edition of *Bergey's Manual of Systematic Bacteriology.* This edition of *Bergey's Manual* divides the gram-positive bacteria phylogenetically between volumes 3 and 4. Here we focus on the mycoplasmas, *Clostridium* and its relatives, and the bacilli and lactobacilli.

23.1 General Introduction

Gram-positive bacteria were historically grouped on the basis of their general shape (e.g., rods, cocci, or irregular) and their ability to form endospores. However, analysis of the phylogenetic relationships within the gram-positive bacteria by comparison of 16S rRNA sequences shows that they are divided into a low G + C group and high G + C, or actinobacterial, group (**figure 23.1**). The most recent edition of *Bergey's Manual of Systematic Bacteriology* places the low G + C gram-positive bacteria in volume 3. This volume describes over 1,300 species placed in 255 genera. The bacterial cell wall: Gram-positive cell walls (section 3.6)

The low G + C gram-positive bacteria are placed in the phylum *Firmicutes* and divided into three classes: *Clostridia, Mollicutes,* and *Bacilli.* The phylum *Firmicutes* is large and complex; it has 10 orders and 34 families. The mycoplasmas, class *Mollicutes,* are also considered low G + C gram positives despite their lack of a cell wall. Ribosomal RNA data indicate that the mycoplasmas are closely related to the lactobacilli. **Figure 23.2** shows the phylogenetic relationships among some of the bacteria in this chapter.

23.2 Class *Mollicutes* (The Mycoplasmas)

The class *Mollicutes* has five orders and six families. The best-studied genera are found in the orders *Mycoplasmatales (Mycoplasma, Ureaplasma), Entomoplasmatales (Entomoplasma,*

We noted, after having grown the bacterium through a series of such cultures, each fresh culture being inoculated with a droplet from the previous culture, that the last culture of the series was able to multiply and act in the body of animals in such a way that the animals developed anthrax with all the symptoms typical of this affection.

Such is the proof, which we consider flawless, that anthrax is caused by this bacterium.

—Louis Pasteur

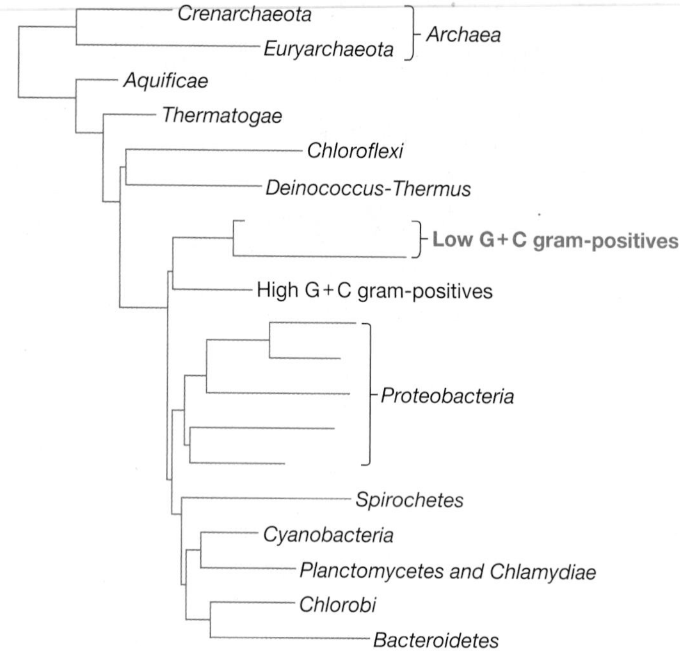

Figure 23.1 Phylogenetic Relationships Among the Procaryotes. The low G + C gram-positive bacteria are highlighted.

Mesoplasma, Spiroplasma), *Acholeplasmatales* (*Acholeplasma*), and *Anaeroplasmatales* (*Anaeroplasma, Asteroleplasma*). **Table 23.1** summarizes some of the major characteristics of these genera.

Members of the class *Mollicutes* are commonly called **mycoplasmas.** Although they evolved from ancestors with gram-positive cell walls, they now lack cell walls and cannot synthesize peptidoglycan precursors. Thus they are penicillin resistant but susceptible to lysis by osmotic shock and detergent treatment. Because they are bounded only by a plasma membrane, these procaryotes are pleomorphic and vary in shape from spherical or pear-shaped organisms, about 0.3 to 0.8 μm in diameter, to branched or helical filaments (**figure 23.3**). Some mycoplasmas (e.g., *M. genitalium*) have a specialized terminal structure that projects from the cell and gives them a flask or pear shape. This structure aids in attachment to eucaryotic cells. They are among the smallest bacteria capable of self-reproduction. Although most are nonmotile, some can glide along liquid-covered surfaces. Most species differ from the vast majority of bacteria in requiring sterols for growth, which are incorporated into the plasma membrane. Here sterols may facilitate osmotic stability. Most are facultative anaerobes, but a few are obligate anaerobes. When growing on agar, most species form colonies with a "fried-egg" appearance because they grow into the agar surface at the center while spreading outward on the surface at the colony edges (**figure 23.4**). Their genomes are among the smallest found in procaryotes, ranging from 0.7 to 1.7 Mb (table 23.1); the G + C content ranges from 23 to 41%. The complete genomes of the human pathogens *Mycoplasma genitalium, M. pneumoniae,* and *Ureaplasma urealyticum* have been sequenced. These genomes are characteristically small with less than 1,000 genes; it seems that not many genes are required to sustain a free-living existence. Mycoplasmas can be saprophytes, commensals, or parasites, and many are pathogens of plants, animals, or insects. Insights from microbial genomes: Genomic analysis of pathogenic microbes (section 15.8)

Metabolically, the mycoplasmas are incapable of synthesizing a number of macromolecules. In addition to requiring sterols, they also need fatty acids, vitamins, amino acids, purines, and pyrimidines. Some produce ATP by the Embden-Meyerhof pathway and lactic acid fermentation. Others catabolize arginine or urea to generate ATP. The pentose phosphate pathway seems to be functional in at least some mycoplasmas; none appear to have the complete tricarboxylic acid cycle.

Mycoplasmas are remarkably widespread and can be isolated from animals, plants, the soil, and even compost piles. Although their complex growth requirements can make their growth in pure (axenic) cultures difficult, about 10% of the mammalian cell cultures in use are probably contaminated with mycoplasmas. This seriously interferes with tissue culture experiments. In animals, mycoplasmas colonize mucous membranes and joints and often are associated with diseases of the respiratory and urogenital tracts. Mycoplasmas cause several major diseases in livestock, for example, contagious bovine pleuropneumonia in cattle (*M. mycoides*), chronic respiratory disease in chickens (*M. gallisepticum*), and pneumonia in swine (*M. hyopneumoniae*). *M. pneumoniae* causes primary atypical pneumonia in humans. *Ureaplasma urealyticum* is commonly found in the human urogenital tract. It is now known to be associated with premature delivery of newborns, as well as neonatal meningitis and pneumonia. Spiroplasmas have been isolated from insects, ticks, and a variety of plants. They cause disease in citrus plants, cabbage, broccoli, corn, honey bees, and other hosts. Arthropods may often act as vectors and carry the spiroplasmas between plants. It is likely that more pathogenic mollicutes will be discovered as techniques for their detection, isolation, and study improve.

1. What morphological feature distinguishes the mycoplasmas? In what class are they found? Why have they been placed with the low G + C gram-positive bacteria?
2. What might mycoplasmas use sterols for?
3. What do you think the relationship between *Mycoplasma* genome size and growth requirements might be?
4. Where are mycoplasmas found in animals? List several animal and human diseases caused by them. What kinds of organisms do spiroplasmas usually infect?

23.3 Peptidoglycan and Endospore Structure

The gram-positive bacteria have traditionally been classified largely on the basis of observable characteristics such as cell shape, the clustering and arrangement of cells, the presence or absence of endospores, oxygen relationships, fermentation patterns, and peptidoglycan chemistry. Because of the importance of pep-

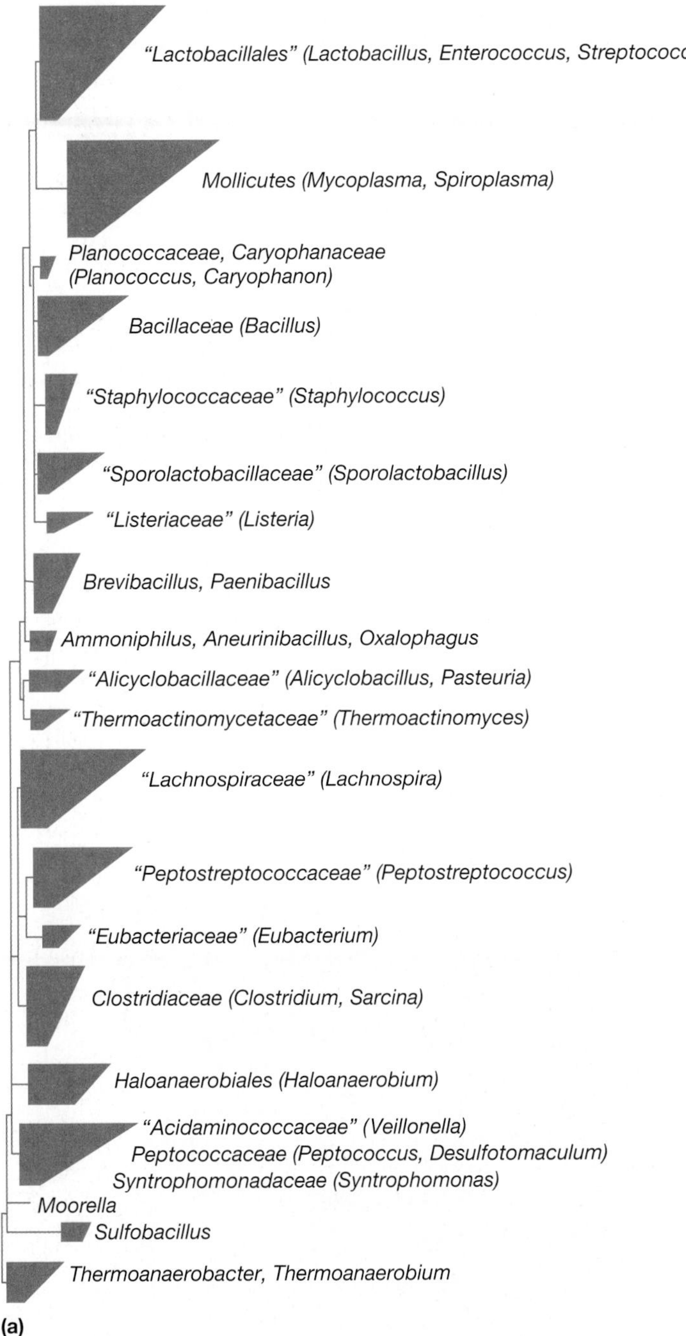

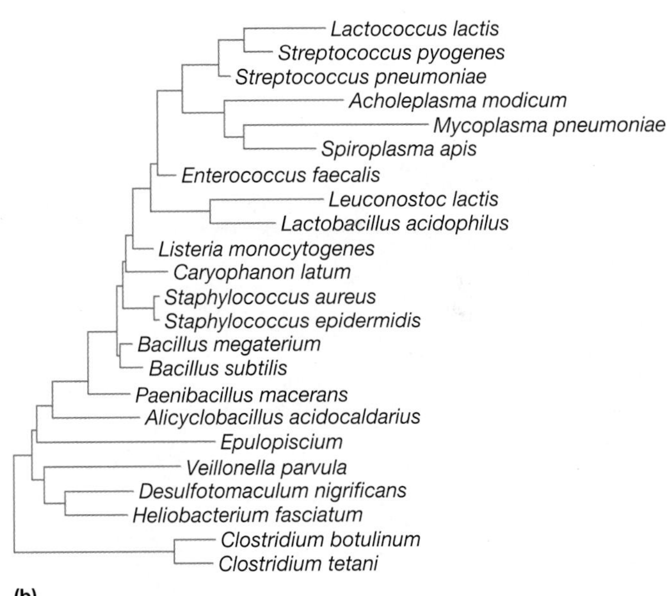

Figure 23.2 Phylogenetic Relationships in the Phylum *Firmicutes* (Low G + C Gram Positives). **(a)** The major phylogenetic groups with representative genera in parentheses. Each tetrahedron in the tree represents a group of related organisms; its horizontal edges show the shortest and longest branches in the group. Multiple branching at the same level indicates that the relative branching order of the groups cannot be determined from the data. The quotation marks around some names indicate that they are not formally approved taxonomic names. **(b)** The relationships of a few species based on 16S rRNA sequence date. *Source: The Ribosomal Database Project.*

tidoglycan and endospores in these bacteria, we briefly discuss these two important components.

Peptidoglycan structure varies considerably among different gram-positive groups. Many gram-positive bacteria (and most gram-negatives) have a peptidoglycan structure in which *meso*-diaminopimelic acid in position 3 is directly linked through its free amino group with the free carboxyl of the terminal D-alanine of an adjacent peptide chain (**figure 23.5*a***). These include *Bacillus, Clostridium, Corynebacterium, Mycobacterium,* and *Nocardia.* In other gram-positive bacteria, lysine is substituted for diaminopimelic acid in position 3, and the peptide subunits of the glycan chains are cross-linked by interpeptide bridges containing monocarboxylic L-amino acids or glycine, or both (figure 23.5*b*). Many genera including *Staphylococcus, Streptococcus, Micrococcus, Lactobacillus,* and *Leuconostoc* have this type of peptidoglycan. The high G + C gram-positive genus *Streptomyces* and several other actinobacterial genera have replaced *meso*-diaminopimelic acid with L,L-diaminopimelic acid in position 3 and have one glycine residue as the interpeptide bridge. The plant pathogenic corynebacteria (also a high G + C gram-positive) provide another example of peptidoglycan variation. In some of these bacteria, the interpeptide bridge connects positions 2 and 4 of the peptide

Table 23.1 Properties of Some Members of the Class *Mollicutes*

Genus	No. of Recognized Species	G + C Content (mol%)	Genome Size (Mb)	Sterol Requirement	Habitat	Other Distinctive Features
Acholeplasma	13	26–36	1.50–1.65	No	Vertebrates, some plants and insects	Optimum growth at 30–37°C
Anaeroplasma	4	29–34	1.50–1.60	Yes	Bovine or ovine rumen	Oxygen-sensitive anaerobes
Asteroleplasma	1	40	1.50	No	Bovine or ovine rumen	Oxygen-sensitive anaerobes
Entomoplasma	5	27–29	0.79–1.14	Yes	Insects, plants	Optimum growth, 30°C
Mesoplasma	12	27–30	0.87–1.10	No	Insects, plants	Optimum growth, 30°C; sustained growth in serum-free medium only with 0.04% Tween 80
Mycoplasma	104	23–40	0.60–1.35	Yes	Humans, animals	Optimum growth usually at 37°C
Spiroplasma	22	25–30	0.94–2.20	Yes	Insects, plants	Helical filaments; optimum growth at 30–37°C
Ureaplasma	6	27–30	0.75–1.20	Yes	Humans, animals	Urea hydrolysis

Adapted from J. G. Tully, et al., "Revised Taxonomy of the Class *Mollicutes*" in *International Journal of Systematic Bacteriology,* 43(2):378–85. Copyright © 1993 American Society for Microbiology, Washington, D.C. Reprinted by permission.

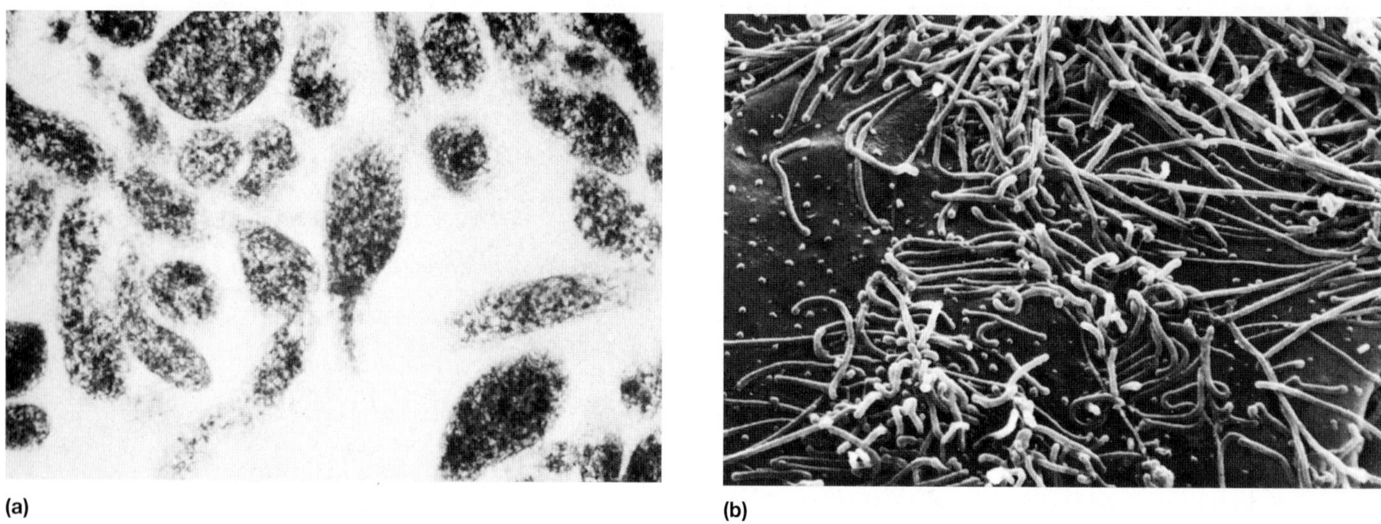

Figure 23.3 The *Mycoplasmas*. Electron micrographs of *Mycoplasma pneumoniae* showing its pleomorphic nature. **(a)** A transmission electron micrograph of several cells (× 47,880). The central cell appears flask or pear-shaped because of its terminal structure. **(b)** A scanning electron micrograph (× 26,000).

subunits rather than 3 and 4 (figure 23.5*c*). Because the interpeptide bridge connects the carboxyl groups of glutamic acid and alanine, a diamino acid such as ornithine is used in the bridge. Many other variations in peptidoglycan structure are found, including other interbridge structures and large differences in the frequency of cross-linking between glycan chains. Bacilli and most gram-negative bacteria have fewer cross-links between chains than do gram-positive bacteria such as *Staphylococcus aureus* in which almost every muramic acid is cross-linked to another. These structural variants are often characteristic of particular groups and are therefore taxonomically useful. Synthesis of sugars and polysaccharides: Synthesis of peptidoglycan (section 10.4)

Bacterial endospores are assembled within the differentiating mother cell, which lyses to release the free spore. Mature spores

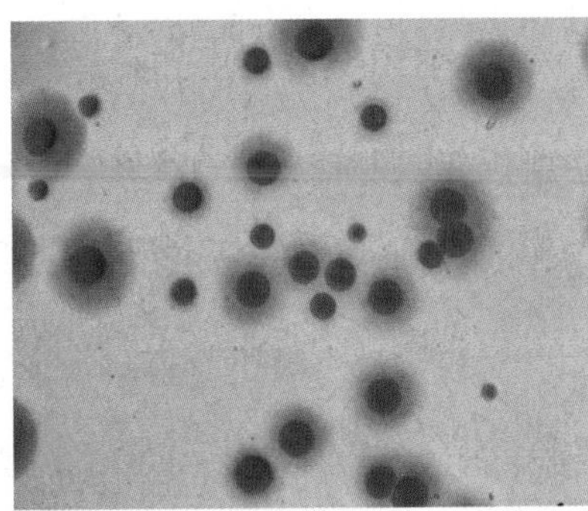

Figure 23.4 ***Mycoplasma* Colonies.** Note the "fried-egg" appearance, colonies stained before photographing (× 100).

have a complex structure with an outer coat, cortex, and inner spore membrane surrounding the protoplast (**figure 23.6**). They contain dipicolinic acid, are very heat resistant, and can remain dormant and viable for very long periods (**Microbial Tidbits 23.1**). Although endospore-forming bacteria are distributed widely, they are primarily soil inhabitants. Soil conditions are often extremely variable, and endospores are an obvious advantage in surviving periods of dryness or nutrient deprivation. In one well-documented experiment, spores remained viable for about 70 years. It has also been reported that viable spores have been recovered from Dominican bees that were encased in 25- to 40-million year-old amber. If this result is confirmed, spores from an ancestor of *Bacillus sphaericus* have survived for more than 25 million years! Similar

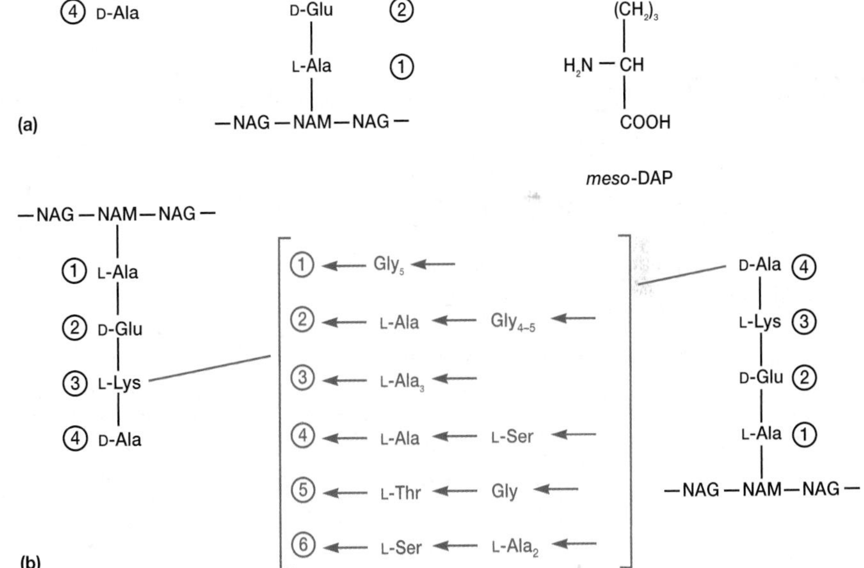

Figure 23.5 Representative Examples of Peptidoglycan Structure. **(a)** The peptidoglycan with a direct cross-linkage between positions 3 and 4 of the peptide subunits, which is present in most gram-negative and many gram-positive bacteria. **(b)** Peptidoglycan with lysine in position 3 and an interpeptide bridge. The bracket contains six typical bridges: (1) *Staphylococcus aureus,* (2) *S. epidermidis,* (3) *Micrococcus roseus* and *Streptococcus thermophilus,* (4) *Lactobacillus viridescens,* (5) *Streptococcus salvarius,* and (6) *Leuconostoc cremoris.* The arrows indicate the polarity of peptide bonds running in the C to N direction. **(c)** An example of the cross-bridge extending between positions 2 and 4 from *Corynebacterium poinsettiae.* The interbridge contains a D-diamino acid like ornithine, and L-homoserine (L-Hsr) is in position 3. The abbreviations and structures of amino acids in the figure are found in appendix I.

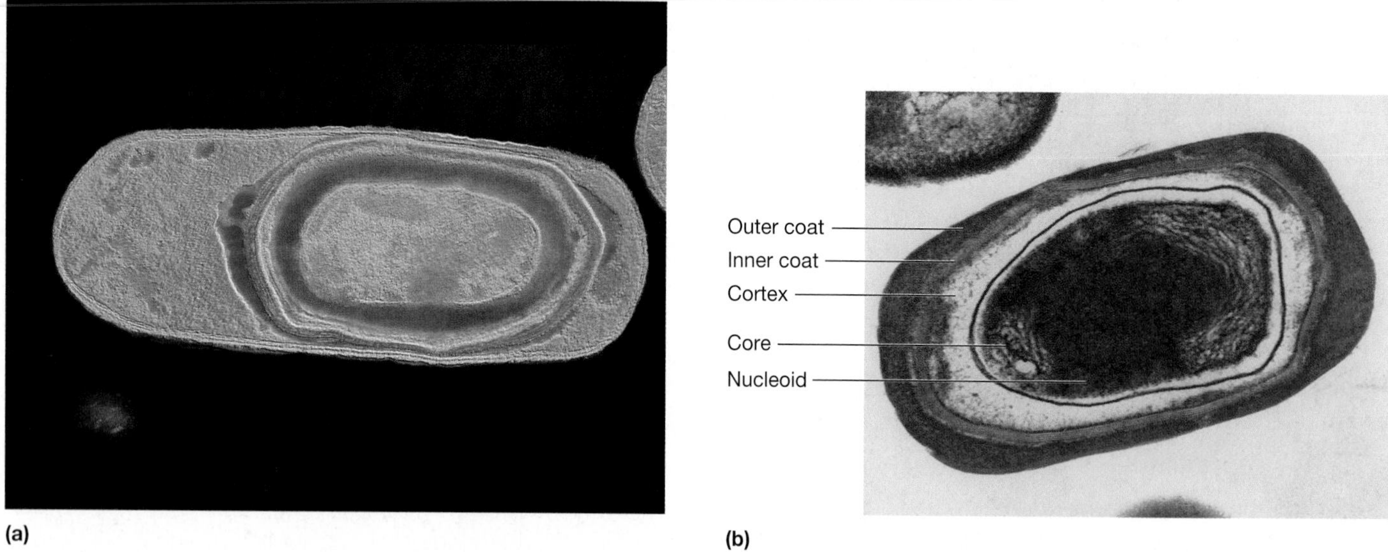

Figure 23.6 Bacterial Endospores. **(a)** A colorized cross section of a *Bacillus subtilis* cell undergoing sporulation. The oval in the center is an endospore that is almost mature; when it reaches maturity, the mother cell will lyse to release it. **(b)** A cross section of a mature *B. subtilis* spore showing the cortex and spore coat layers that surround the core. The endospore in (a) is 1.3 μm; the spore in (b) is 1.2 μm.

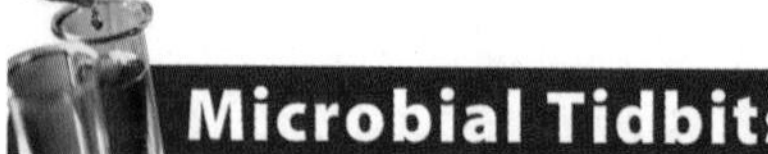

Microbial Tidbits

23.1 Spores in Space

During the nineteenth-century argument over the question of the evolution of life, the panspermia hypothesis became popular. According to this hypothesis, life did not evolve from inorganic matter on Earth but arrived as viable bacterial spores that had escaped from another planet. More recently, the British astronomer Fred Hoyle has revived the hypothesis based on his study of the absorption of radiation by interstellar dust. Hoyle maintains that dust grains were initially viable bacterial cells that have been degraded, and that the beginning of life on Earth was due to the arrival of bacterial spores that had survived their trip through space.

Even more recently Peter Weber and J. Mayo Greenberg from the University of Leiden in the Netherlands have studied the effect of very high vacuum, low temperature, and UV radiation on the survival of *Bacillus subtilis* spores. Their data suggest that spores within an interstellar molecular cloud might be able to survive between 4.5 to 45 million years. Molecular clouds move through space at speeds sufficient to transport spores between solar systems in this length of time. Although these results do not prove the panspermia hypothesis, they are consistent with the possibility that bacteria might be able to travel between planets capable of supporting life.

reports have been made subsequently; all these studies will have to be reconfirmed. Usually endospores are observed either in the light microscope after spore staining or by phase-contrast microscopy of unstained cells. They also can be detected by heating a culture at 70 to 80°C for 10 minutes followed by incubation in the proper growth medium. Because only endospores and some thermophiles would survive such heating, bacterial growth tentatively confirms their presence. The bacterial endospore (section 3.11); Preparation and staining of specimens (section 2.3)

There are two classes of low G + C endospore-forming bacteria: *Clostridia* (the clostridia and relatives), and *Bacilli* (the bacilli and lactobacilli) (figure 23.2). Each class includes both rods and cocci and are discussed here.

23.4 Class *Clostridia*

The class *Clostridia* has a very wide variety of gram-positive bacteria distributed into three orders and 11 families. The characteristics of some of the more important genera are summarized in **table 23.2.** Phylogenetic relationships are shown in figure 23.2.

By far the largest genus is *Clostridium.* It includes obligately anaerobic, fermentative, gram-positive bacteria that form endospores. The genus contains well over 100 species in several distinct phylogenetic clusters. The genus *Clostridium* may be subdivided into several genera in the future.

Members of the genus *Clostridium* have great practical impact. Because they are anaerobic and form heat-resistant en-

Table 23.2 Characteristics of Clostridia and Relatives

Genus	Dimensions (μm) and Morphology	G + C Content (mol%)	Oxygen Relationship	Other Distinctive Characteristics
Clostridium	0.3–2.0 × 1.5–20; rod-shaped, often pleomorphic, nonmotile or peritrichous	22–55	Anaerobic	Does not carry out dissimilatory sulfate reduction; usually chemoorganotrophic, fermentative, and catalase negative; forms oval or spherical endospores
Desulfotomaculum	0.3–1.5 × 3–9; straight or curved rods, peritrichous or polar flagella	37–50	Anaerobic	Reduces sulfate to H_2S, forms subterminal to terminal endospores; stains gram negative but has gram-positive wall, catalase negative
Heliobacterium	1.0 × 4–10; rods that are frequently bent, gliding motility	52–55	Anaerobic	Photoheterotrophic with bacteriochlorophyll *g;* stains gram negative but has gram-positive wall, some form endospores
Veillonella	0.3–0.5; cocci in pairs, short chains, and masses; nonmotile	36–43	Anaerobic	Stains gram negative; pyruvate and lactate fermented, but not carbohydrates; acetate, propionate, CO_2, and H_2 produced from lactate; parasitic in mouths, intestines, and respiratory tracts of animals

dospores, they are responsible for many cases of food spoilage, even in canned foods. *C. botulinum* is the causative agent of botulism. Clostridia often can ferment amino acids to produce ATP by oxidizing one amino acid and using another as an electron acceptor in a process called the Stickland reaction (*see figure 9.19*). This reaction generates ammonia, hydrogen sulfide, fatty acids, and amines during the anaerobic decomposition of proteins. These products are responsible for many unpleasant odors arising during putrefaction. Food-borne and waterborne diseases: Botulism (section 38.4)

Several clostridia produce toxins and are major disease agents. *C. tetani* (**figure 23.7**) is the causative agent of tetanus, and *C. perfringens,* of gas gangrene and food poisoning. *C. perfringens* genome sequence analysis reveals that the microbe possesses the genes for fermentation with gas production but lacks genes encoding enzymes for the TCA cycle or a respiratory chain. Nonetheless, *C. perfringens* has an extraordinary doubling time of only 8 to 10 minutes when in the human host. Clostridia also are industrially valuable; for example, *C. acetobutylicum* is used to manufacture butanol in some countries.

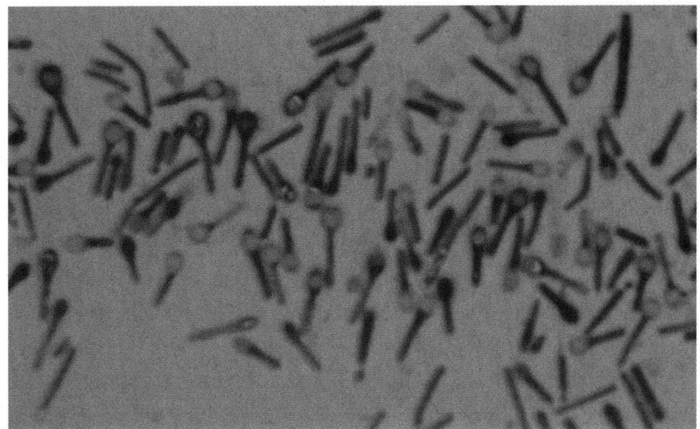

Figure 23.7 ***Clostridium tetani*** **with Spores That Are Round and Terminal.**

Desulfotomaculum is another anaerobic, endospore-forming genus. Unlike *Clostridium,* it reduces sulfate and sulfite to hydrogen sulfide during anaerobic respiration (**figure 23.8**). Although it stains gram negative, electron microscopic studies have shown that *Desulfotomaculum* has a gram-positive type cell wall. This concurs with phylogenetic studies that place it with the low G + C gram positives.

The heliobacteria are an excellent example of the diversity in this class. The genera *Heliobacterium* and *Heliophilum* are a group of unusual anaerobic, photosynthetic bacteria characterized by the presence of bacteriochlorophyll *g*. They have a photosystem I type reaction center like the green sulfur bacteria, but have no intracytoplasmic photosynthetic membranes; pigments are contained in the plasma membrane. Like *Desulfotomaculum,* they have a gram-positive type cell wall with lower than normal peptidoglycan content, and they stain gram negative. Some heliobacteria form endospores. Phototrophy (section 9.12)

The phylogenetic placement of the genus *Veillonella* bears mentioning. Although these bacteria stain gram-negative, *Bergey's Manual* places them in the family *Acidominococcaceae,* in the order *Clostridiales.* Members of the genus *Veillonella* are anaerobic,

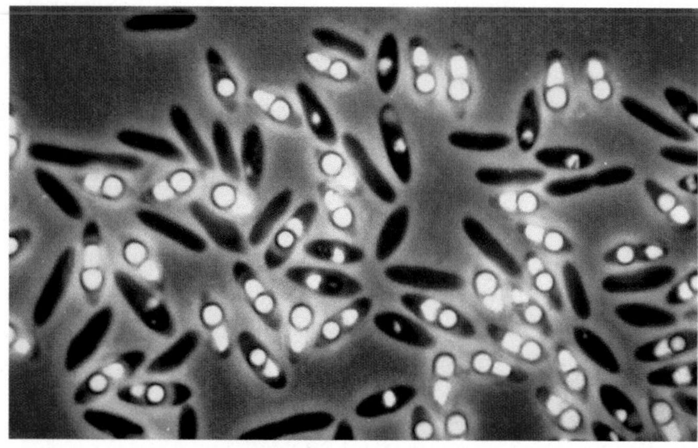

Figure 23.8 ***Desulfotomaculum.*** *Desulfotomaculum acetoxidans* with spores; phase contrast (×2,000).

chemoheterotrophic cocci ranging in diameter from about 0.3 to 2.5 μm. Usually they are diplococci (often with their adjacent sides flattened), but they may exist as single cells, clusters, or chains. All have complex nutritional requirements and ferment substances such as carbohydrates, lactate and other organic acids, and amino acids to produce gas (CO_2 and often H_2) plus a mixture of volatile fatty acids. They are parasites of homeothermic (warm-blooded) animals.

Like many groups of anaerobic bacteria, members of this genus have not been thoroughly studied. Some species are part of the normal biota of the mouth, the gastrointestinal tract, and the urogenital tract of humans and other animals. For example, *Veillonella* is plentiful on the tongue surface and dental plaque of humans and it can be isolated from the vagina. *Veillonella* is unusual in growing well on organic acids such as lactate, pyruvate, and malate while being unable to ferment glucose and other carbohydrates. It is well adapted to the oral environment because it can use the lactic acid produced from carbohydrates by the streptococci and other oral bacteria. Veillonella are found in infections of the head, lungs, and the female genital tract, but their precise role in such infections is unclear.

1. Describe, in a diagram, the chemical composition and structure of the peptidoglycan found in gram-negative bacteria and many gram-positive genera.
2. How do the cell walls of bacilli and most gram-negative bacteria differ from those of gram-positive bacteria such as *S. aureus* with respect to cross-linking frequency?
3. What is a bacterial endospore? Give its most important properties and two ways to demonstrate its presence. Why do you think reports of viable spores from ancient source have been met with skepticism?
4. Give the general characteristics of *Clostridium, Desulfotomaculum*, the heliobacteria, and *Veillonella*. Briefly discuss why each is interesting or of practical importance.
5. What do you think is the evolutionary significance of the discovery of a photosynthetic genus within the low G + C gram-positive bacteria?

23.5 Class *Bacilli*

The second edition of *Bergey's Manual* gathers a large variety of gram-positive bacteria into one class, *Bacilli,* and two orders, *Bacillales* and *Lactobacillales.* These orders contain 17 families and over 70 gram-positive genera representing cocci, endospore-forming rods and cocci, and nonsporing rods. The biology of some members of the order *Bacillales* will be described first; then important representatives of the order *Lactobacillales* will be considered. The phylogenetic relationships between some of these organisms are pictured in figure 23.1, and the characteristics of selected genera are summarized in **table 23.3.**

Order Bacillales

The genus *Bacillus,* family *Bacillaceae,* is the largest in the order (**figure 23.9**). The genus contains gram-positive, endospore-forming, chemoheterotrophic rods that are usually motile with peritrichous flagella. It is aerobic, or sometimes facultative, and catalase positive. Many species once included in this genus have been placed in other families and genera based on rRNA sequence data. For example, the genus *Alicyclobacillus* contains acidophilic, sporing, gram-positive or gram-variable rods that have ω-alicyclic fatty acids with 6- or 7-carbon rings in their membranes. Members are aerobic or facultative and have a G + C content of about 51 to 60%. Another genus, *Paenibacillus* [Latin *paene,* almost, and bacillus], contains gram-positive rods that are facultative, motile by peritrichous flagella, have ellipsoidal endospores and swollen sporangia, produce acid and sometimes gas from glucose and various sugars, and have a G + C content of 40 to 54%. Some examples of organisms that were formerly in the genus *Bacillus* are *Paenibacillus alvei, P. macerans,* and *P. polymyxa.*

Bacillus subtilis, the type species for the genus, is the most well-studied gram-positive bacterium. It is a useful model organism for the study of gene regulation, cell division, quorum sensing, and cellular differentiation. Its 4.2-Mb genome was one of the first genomes to be completely sequenced. Genome sequencing reveals a number of interesting elements. For instance, several families of genes have been expanded by gene duplication; the largest such family encodes ABC transporters, which are the most frequent type of protein in *B. subtilis.* There are 18 genes that encode sigma factors. Recall that the use of alternative sigma subunits of RNA polymerase is one way in which bacteria regulate gene expression. In this case, many of the sigma factors govern sporulation and other responses to stressful conditions. The genome contains genes for the catabolism of many diverse carbon sources and antibiotic synthesis. There are at least 10 integrated prophages or remnants of prophages. Protein secretion in procaryotes (section 3.8); Global regulations systems: Sporulation in *Bacillus subtilis* (section 12.5); Comparative genomics (section 15.6)

Many species of *Bacillus* are of considerable importance. Some produce the antibiotics bacitracin, gramicidin, and polymyxin. *B. cereus* (figure 23.9*b*) causes some forms of food

Table 23.3 Characteristics of Members of the Class *Bacilli*

Genus	Dimensions (μm) and Morphology	G + C Content (mol%)	Genome Size (Mb)	Oxygen Relationship	Other Distinctive Characteristics
Bacillus	0.5–2.5 × 1.2–10; straight rods, peritrichous	32–69	4.2–5.4	Aerobic or facultative	Forms endospores; catalase positive; chemoorganotrophic
Caryophanon	1.5–3.0 × 10–20; multicellular rods with rounded ends, peritrichous	41–46	Nd*	Aerobic	Acetate only major carbon source; catalase positive; trichome cells have greater width than length, trichomes can be in short chains
Enterococcus	0.6–2.0 × 0.6–2.5; spherical or ovoid cells in pairs or short chains, nonsporing, sometimes motile	34–42	3.2	Facultative	Ferments carbohydrates to lactate with no gas; complex nutritional requirements; catalase negative; occurs widely, particularly in fecal material
Lactobacillus	0.5–1.2 × 1.0–10; usually long, regular rods, nonsporing, rarely motile	32–53	1.9–3.3	Facultative or microaerophilic	Fermentative, at least half the end-product is lactate; requires rich, complex media; catalase and cytochrome negative
Lactococcus	0.5–1.2 × 0.5–1.5; spherical or ovoid cells in pairs or short chains, nonsporing, nonmotile	38–40	2.4	Facultative	Chemoorganotrophic with fermentative metabolism; lactate without gas produced; catalase negative; complex nutritional requirements; in dairy and plant products
Leuconostoc	0.5–0.7 × 0.7–1.2; cells spherical or ovoid, in pairs or chains; nonmotile and nonsporing	38–44	Nd	Facultative	Requires fermentable carbohydrate and nutritionally rich medium for growth; fermentation produces lactate, ethanol, and gas; catalase and cytochrome negative
Staphylococcus	0.9–1.3; spherical cells occurring singly and in irregular clusters, nonmotile and nonsporing	30–39	2.5–2.8	Facultative	Chemoorganotrophic with both respiratory and fermentative metabolism, usually catalase positive, associated with skin and mucous membranes of vertebrates
Streptococcus	0.5–2.0; spherical or ovoid cells in pairs or chains, nonmotile and nonsporing	34–46	1.8–2.2	Facultative	Fermentative, producing mainly lactate and no gas; catalase negative; commonly attack red blood cells (α- or β-hemolysis); complex nutritional requirements; commensals or parasites on animals
Thermoactinomyces	0.4–1.0 in diameter; branched, septate mycelium resembles those of actinomycetes	52.0–54.8	Nd	Aerobic	Usually thermophilic; true endospores form singly on hyphae; numerous in decaying hay, vegetable matter, and compost

*Nd: Not determined; genome not yet sequenced.

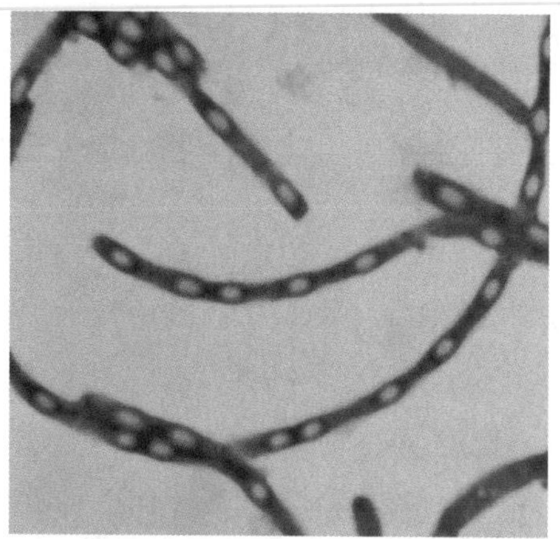

(a) *Bacillus anthracis*

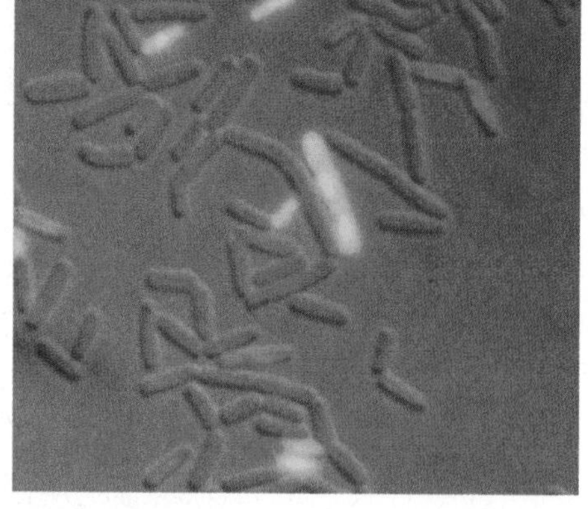

(b) *B. cereus*

Figure 23.9 Bacillus. **(a)** *B. anthracis,* spores elliptical and central (×1,600). **(b)** *B. cereus* stained with SYTOX Green nucleic acid stain and viewed by epifluorescence and differential interference contrast microscopy. The cells that glow green are dead.

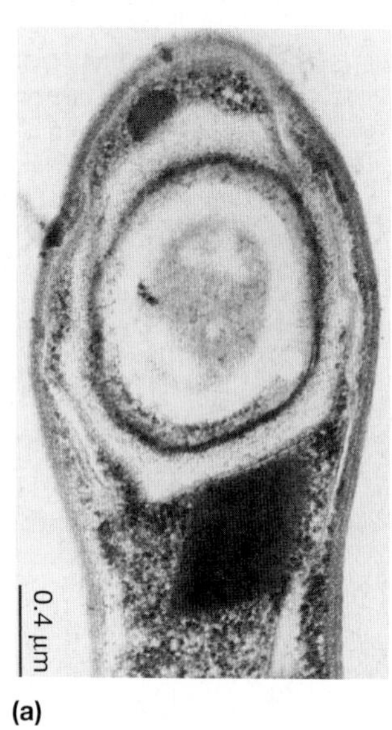

(a)

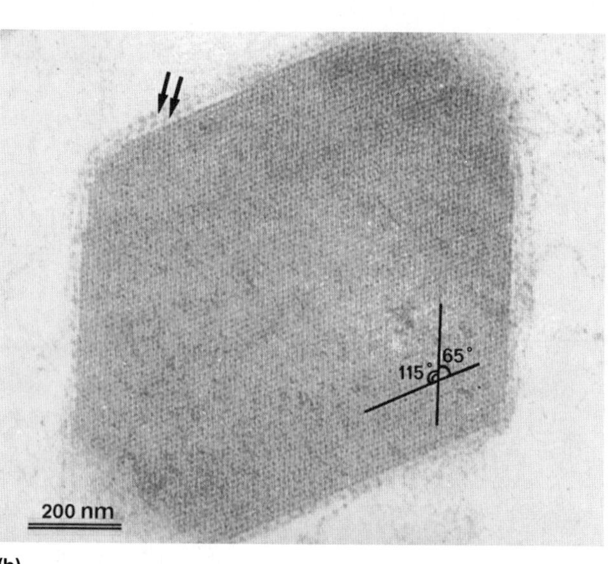

(b)

Figure 23.10 The Parasporal Body. **(a)** An electron micrograph of a *B. sphaericus* sporulating cell containing a parasporal body just beneath the endospore. **(b)** The crystalline parasporal body at a higher magnification. The crystal is surrounded by a two-layered envelope (arrows).

poisoning and can infect humans. *B. anthracis* is the causative agent of the disease anthrax, which can affect both farm animals and humans. Several species are used as insecticides. For example, *B. thuringiensis* and *B. sphaericus* form a solid protein crystal, the **parasporal body,** next to their endospores during spore formation (**figure 23.10**). The *B. thuringiensis* parasporal body contains protein toxins that kill over 100 species of moths by dissolving in the alkaline gut of caterpillars and destroying the epithelium. The solubilized toxin proteins are cleaved by midgut proteases to smaller toxic polypeptides that attack the gut epithelial cells. The alkaline gut contents escape into the blood, causing paralysis and death. One of these toxins has been shown to form pores in the plasma membrane of the target insect's cells. These channels allow monovalent cations such as potassium to pass through. *B. thuringiensis* toxin genes have been engineered to make a variety of pest-resistant, genetically modified plants. The *B. sphaericus* parasporal body contains proteins toxic for mosquito larvae and may be useful in controlling the mosquitos that carry the malaria parasite *Plasmodium.* Zoonotic diseases: Anthrax (section 38.6); Microbes as products: Biopesticides (section 41.8)

The genus *Thermoactinomyces* has historically been classified as an actinomycete because it forms filaments that differentiate from soil-associated substrate hyphae into upwardly growing, aerial hyphae. However, phylogenetic analysis places it with the low G + C microbes in the order *Bacillales,* family *Thermoactinomycetaceae.* Its G + C content (52–55 mol%) is considerably lower than that of the *Actinobacteria.* This genus is thermophilic and grows between 45 and 60°C; it forms single spores on both its aerial and substrate mycelia (**figure 23.11**) and its 16S rRNA sequence suggests a relationship with the genus *Bacillus. Thermoactinomyces* is commonly found in damp haystacks, compost piles, and other high-temperature habitats. Unlike actinomycete exospores, *Thermoactinomyces* (figure 23.11*b*) forms true endospores that are

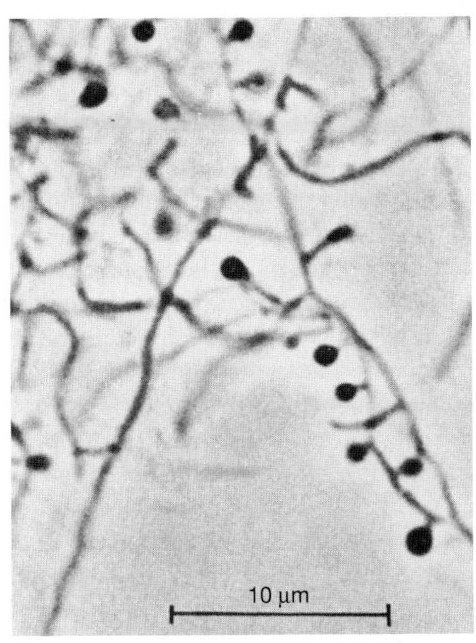

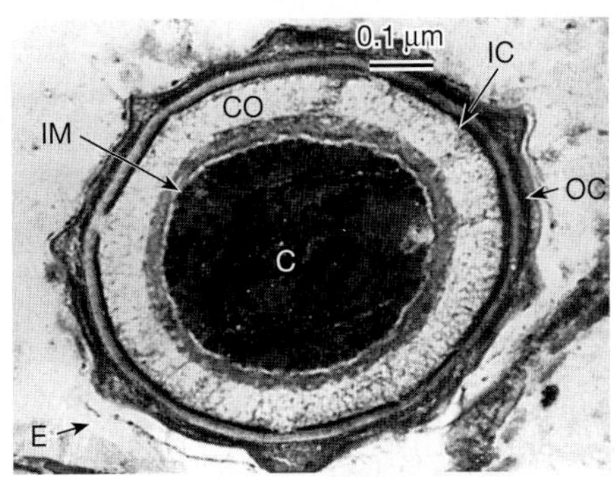

Figure 23.11 ***Thermoactinomyces.*** **(a)** *Thermoactinomyces vulgaris* aerial mycelium with developing endospores at tips of hyphae. **(b)** Thin section of a *T. sacchari* endospore. E, exosporium; OC, outer spore coat; IC, inner spore coat; CO, cortex; IM, inner forespore membrane; C, core.

heat resistant; they can survive at 90°C for 30 minutes. They are formed within hyphae and appear to have typical endospore structure, including the presence of calcium and dipicolinic acid. *Thermoactinomyces vulgaris* (figure 23.11*a*) from haystacks, grain storage silos, and compost piles is a causative agent of farmer's lung, an allergic disease of the respiratory system in agricultural workers. Recently spores from *Thermoactinomyces vulgaris* were recovered from the mud of a Minnesota lake and found to be viable after about 7,500 years of dormancy.

One of the more unusual genera in this order is *Caryophanon.* This gram-positive bacterium is strictly aerobic, catalase positive, and motile by peritrichous flagella. Its normal habitat is cow dung. *Caryophanon* morphology is distinctive. Individual cells are disk-shaped (1.5 to 2.0 μm wide by 0.5 to 1.0 μm long) and joined to form filaments called trichomes that are about 10 to 20 μm long (**figure 23.12**).

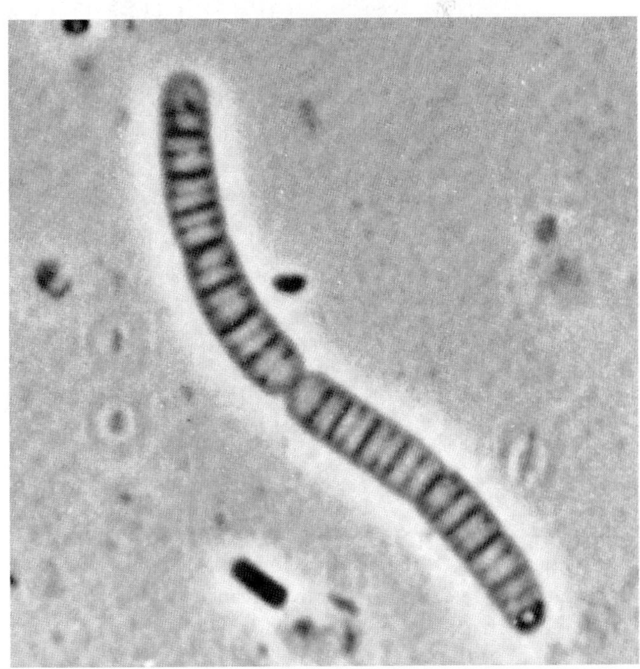

Figure 23.12 ***Caryophanon*** **Morphology.** *Caryophanon latum* in a trichome chain. Note the disk-shaped cells stacked side by side; phase contrast (×3,450).

The family *Staphylococcaceae* contains four genera, the most important of which is the genus *Staphylococcus*. Members of this genus are facultatively anaerobic, nonmotile, gram-positive cocci that usually form irregular clusters (**figure 23.13**). They are catalase positive, oxidase negative, ferment glucose, and have teichoic acid in their cell walls. Staphylococci are normally associated with the skin, skin glands, and mucous membranes of warm-blooded animals.

Staphylococci are responsible for many human diseases. *S. epidermidis* is a common skin resident that is sometimes responsible for endocarditis and infections of patients with lowered resistance (e.g., wound infections, surgical infections, urinary tract infections, body piercing). *S. aureus* is the most important human staphylococcal pathogen and causes boils, abscesses, wound infections, pneumonia, toxic shock syndrome, and other diseases. Strains of **meticillin-resistant *Staphylococcus aureus* (MRSA;** formerly methicillin) and vancomycin-resistant *S. aureus* are among the most threatening antibiotic-resistant pathogens known. Vancomycin is considered the "drug of last resort" and infections caused by vancomycin-resistant *S. aureus* generally cannot be treated by antibiotic therapy. How did *S. aureus* so quickly evolve the capacity to defeat a diverse array of antibiotics? Comparative genome analysis of MRSA strains with antibiotic-sensitive *S. aureus* strains shows that this microbe is very adept at

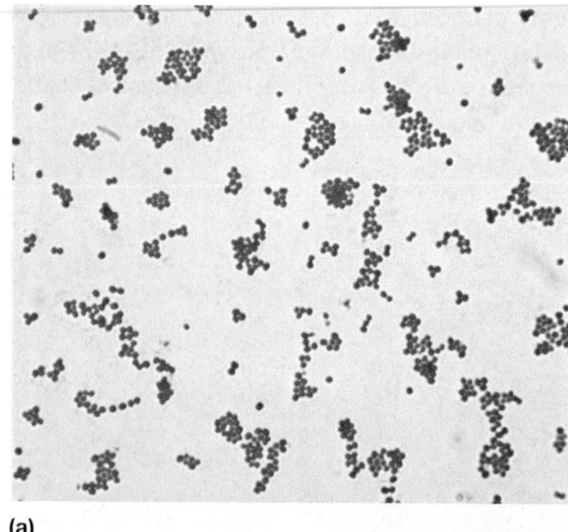

(a)

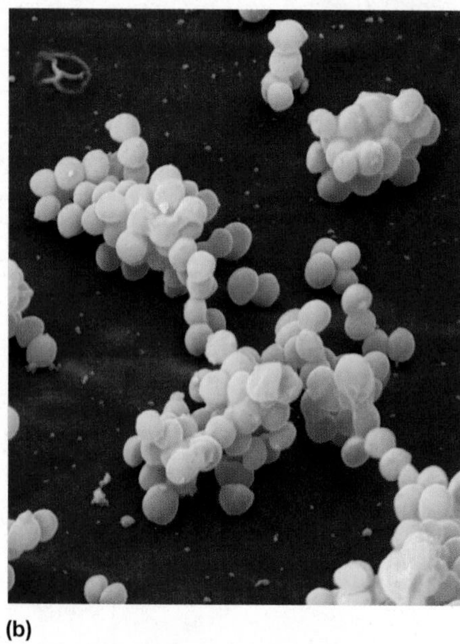

(b)

Figure 23.13 ***Staphylococcus.*** **(a)** *Staphylococcus aureus,* Gram-stained smear (X1,500). **(b)** *Staphylococcus aureus* cocci arranged like clusters of grapes; color-enhanced scanning electron micrograph. Each cell is about 1 μm in diameter.

acquiring genetic elements from other bacteria. In fact, large, mobile genetic elements appear to encode both antibiotic-resistance factors and proteins that increase virulence. Insights from microbial genomes: Genomic analysis of pathogenic microbes (section 15.8)

One of the virulence factors produced by *S. aureus* is the enzyme **coagulase,** which causes blood plasma to clot. Growth and hemolysis patterns on blood agar are also useful in identifying these staphylococci (figure 23.18). The structure of staphylococcal α-hemolysin has been determined. The toxin lyses a cell by forming solvent-filled channels in its plasma membrane. Water-soluble toxin monomers bind to the cell surface and associate with each other to form pores. The hydrophilic channels then allow free passage of water, ions, and small molecules. *S. aureus* usually grows on the nasal membranes and skin; it also is found in the gastrointestinal and urinary tracts of warm-blooded animals. Direct contact diseases: Staphylococcal diseases (section 38.3)

S. aureus is a major cause of food poisoning. For instance several years ago in Texas, 1,364 elementary school children were sickened by tainted chicken. A food service worker responsible for deboning the chicken was the most likely source of contamination. After the chicken was deboned, it was cooled to room temperature before further processing and refrigeration. This provided sufficient time for the bacterium to grow and produce toxins. This case is not uncommon. Poultry accounts for about one-quarter of all bacterial food poisoning cases in which the source of poisoning is known, and it must be cooked and handled carefully.

Listeria, family *Listeriaceae,* is another medically important genus in this order. The genus contains short rods that are aerobic or facultative, catalase positive, and motile by peritrichous flagella. It is widely distributed in nature, particularly in decaying matter. *Listeria monocytogenes* is a pathogen of humans and other animals and causes listeriosis, an important food infection. Food-borne and waterborne diseases: Listeriosis (section 38.4)

Order *Lactobacillales*

Many members of the order *Lactobacillales* produce lactic acid as their major or sole fermentation product and are sometimes collectively called **lactic acid bacteria (LAB)**. *Streptococcus, Enterococcus, Lactococcus, Lactobacillus,* and *Leuconostoc* are all members of this group. Lactic acid bacteria are nonsporing and usually nonmotile. They normally depend on sugar fermentation for energy. They lack cytochromes and obtain energy by substrate-level phosphorylation rather than by electron transport and oxidative phosphorylation. Nutritionally, they are fastidious and many vitamins, amino acids, purines, and pyrimidines must be supplied because of their limited biosynthetic capabilities. Lactic acid bacteria usually are categorized as facultative anaerobes, but some classify them as aerotolerant anaerobes. The influence of environmental factors on growth: Oxygen concentration (section 6.5); Fermentations (section 9.7)

The largest genus in this order is *Lactobacillus* with around 100 species. *Lactobacillus* contains nonsporing rods and sometimes coccobacilli that lack catalase and cytochromes, are usually facultative anaerobic or microaerophilic, produce lactic acid as their main or sole fermentation product, and have complex nutritional requirements (**figure 23.14**). Lactobacilli carry out either a homolactic fermentation using the Embden-Meyerhof pathway or a heterolactic fermentation with the pentose phosphate pathway. They grow optimally under slightly acidic conditions, when the pH is between 4.5 to 6.4. The genus is found on plant surfaces and in dairy products, meat, water, sewage, beer, fruits, and many other materials. Lactobacilli also are part of the normal flora of the human body in the mouth, intestinal tract, and vagina. They usually are not pathogenic.

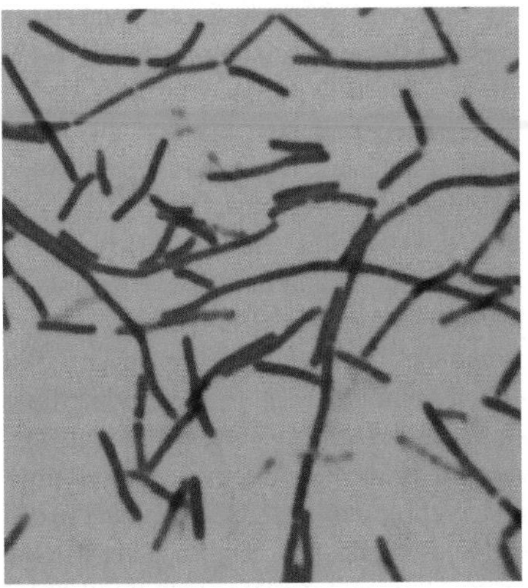
(a) *Lactobacillus acidophilus*

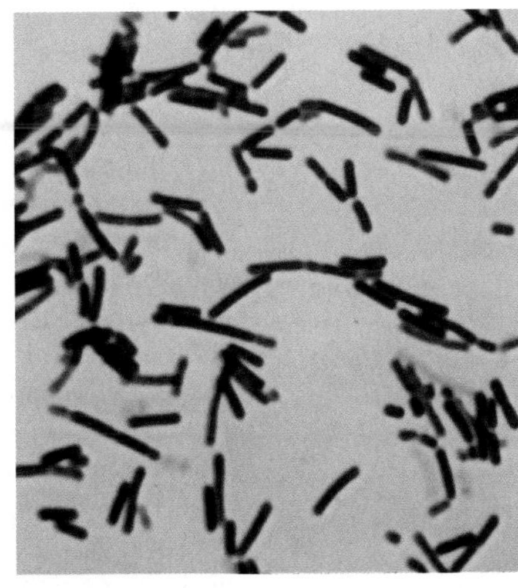
(b) *L. lactis*

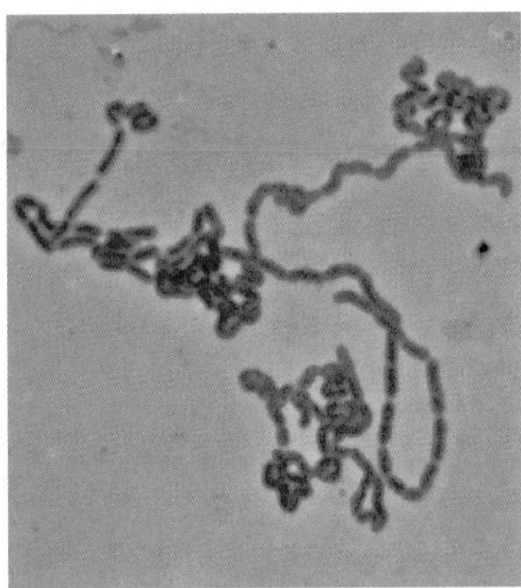
(c) *L. bulgaricus*

Figure 23.14 ***Lactobacillus.*** **(a)** *L. acidophilus* (×1,000). **(b)** *L. lactis.* Gram stain (×1,000). **(c)** *L. bulgaricus;* phase contrast (×600).

Lactobacillus is indispensable to the food and dairy industry. Lactobacilli are used in the production of fermented vegetable foods (sauerkraut, pickles, silage), beverages (beer, wine, juices), sour dough bread, Swiss and other hard cheeses, yogurt, and sausage. Yogurt is probably the most popular fermented milk product in the United States. In commercial production, nonfat or low-fat milk is pasteurized, cooled to 43°C or lower, inoculated with *Streptococcus thermophilus* and *Lactobacillus bulgaricus. S. thermophilus* grows more rapidly at first and renders the milk anoxic and weakly acidic. *L. bulgaricus* then acidifies the milk even more. Acting together, the two species ferment almost all the lactose to lactic acid and flavor the yogurt with diacetyl (*S. thermophilus*) and acetaldehyde (*L. bulgaricus*). Fruits or fruit flavors are pasteurized separately and then combined with the yogurt. Microbiology of fermented foods: Fermented milks (section 40.6)

At least one species, *L. plantarum,* is sold commercially as a probiotic agent that may provide some health benefits for the consumer. On the other hand some lactobacilli also create problems. They sometimes are responsible for spoilage of beer, milk, and meat because their metabolic end products contribute undesirable flavors and odors.

Leuconostoc, family *Leuconostocaceae,* contains facultative gram-positive cocci, which may be elongated or elliptical and arranged in pairs or chains (**figure 23.15**). Leuconostocs lack catalase and cytochromes and carry out **heterolactic fermentation** by converting glucose to D-lactate and ethanol or acetic acid by means of the phosphoketolase pathway (**figure 23.16**). They can be isolated from plants, silage, and milk. The genus is used in wine production, in the fermentation of vegetables such as cabbage (sauerkraut; *see figure 40.19*) and cucumbers (pickles), and in the manufacture of buttermilk, butter, and cheese. *L. mesenteroides* synthesizes dextrans from sucrose and is important in industrial dextran production. *Leuconostoc* species are involved in food spoilage and tolerate high sugar concentrations so well that they grow in syrup and are a major problem in sugar refineries.

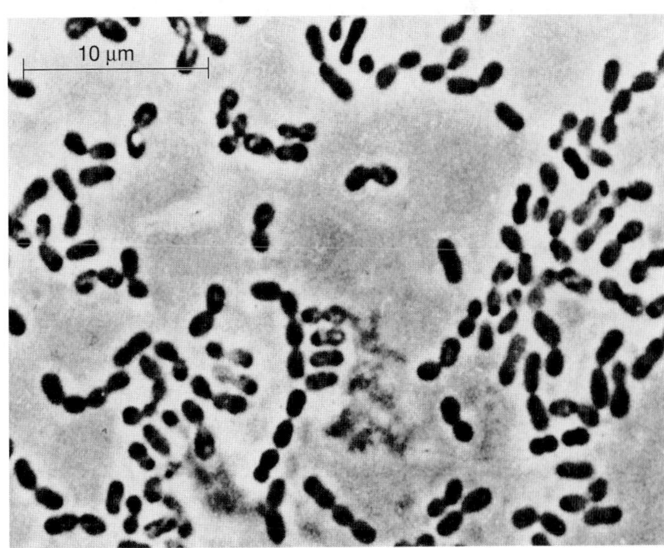

Figure 23.15 ***Leuconostoc.*** *Leuconostoc mesenteroides;* phase-contrast micrograph.

Enterococcaceae (Enterococcus) and *Streptococcaceae (Streptococcus, Lactococcus)* are important families of chemoheterotrophic, mesophilic, nonsporing, gram-positive cocci. In practice, they are often distinguished primarily based on phenotypic properties such as oxygen relationships, cell arrangement, the presence of catalase and cytochromes, and peptidoglycan structure. The most important of these genera is *Streptococcus,* which is facultatively anaerobic and catalase negative. The

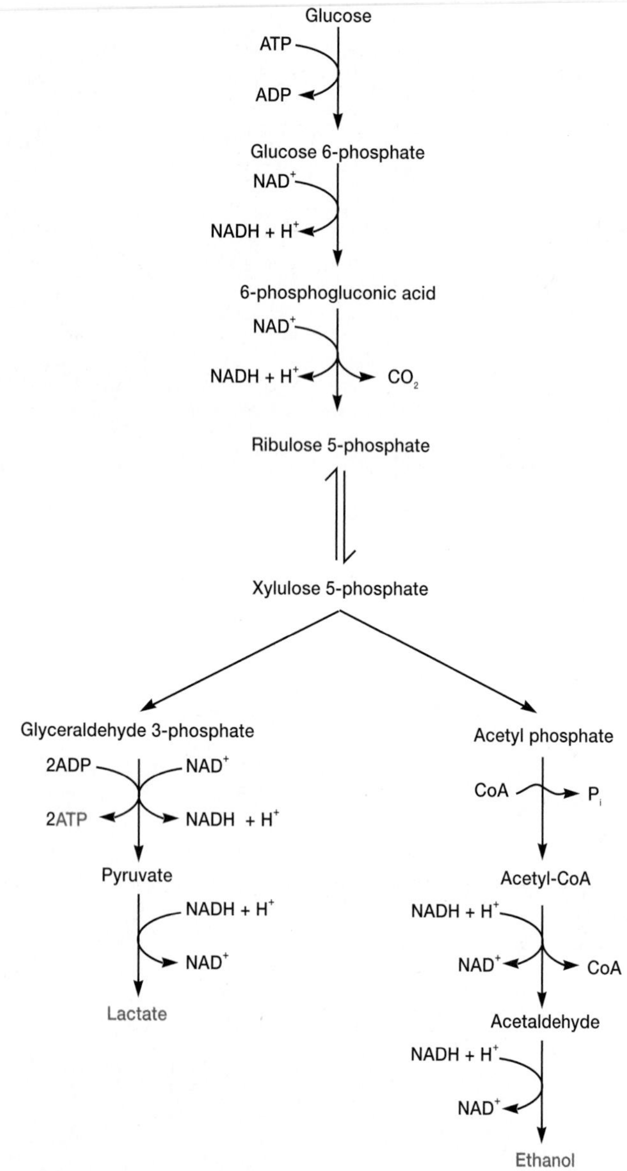

Figure 23.16 Heterolactic Fermentation and the Phosphoketolase Pathway. The phosphoketolase pathway converts glucose to lactate, ethanol, and CO_2.

streptococci and their close relatives, the enterococci and lactococci, occur in pairs or chains when grown in liquid media (**figure 23.17**), do not form endospores, and usually are nonmotile. They all ferment sugars with lactic acid, but no gas, as the major product—that is, they carry out homolactic fermentation. A few species are anaerobic rather than facultative.

The genus *Streptococcus* is large and complex. These bacteria have been clustered into three groups: pyogenic streptococci, oral streptococci, and other streptococci. Many bacteria originally placed within the genus have been moved to two other genera, *Enterococcus* and *Lactococcus*. Some major characteristics of these three closely related genera are summarized in **table 23.4. Table 23.5** lists a few properties of selected genera.

Many characteristics are used to identify these cocci. One of their most important taxonomic characteristics is the ability to lyse erythrocytes when growing on blood agar, an agar medium containing 5% sheep or horse blood (**figure 23.18**). In **α-hemolysis,** a 1 to 3 mm greenish zone of incomplete hemolysis forms around the colony; **β-hemolysis** is characterized by a zone of clearing or complete lysis without a marked color change. In addition, other hemolytic patterns are sometimes seen. Serological studies are also very important in identification because these genera often have distinctive cell wall antigens. Polysaccharide and teichoic acid antigens found in the cell wall or between the wall and the plasma membrane are used to identify these cocci, particularly pathogenic β-hemolytic streptococci, by the **Lancefield grouping system.** Biochemical and physiological tests are essential in identification (e.g., growth temperature preferences, carbohydrate fermentation patterns, acetoin production, reduction of litmus milk, sodium chloride and bile salt tolerance, and the ability to hydrolyze arginine, esculin, hippurate, and starch). Sensitivity to bacitracin, sulfa drugs, and optochin (ethylhydrocuprein) also are used to identify particular species. Some of these techniques are being replaced by molecular genetic approaches such as multilocus sequence typing (MSLT). Clinical microbiology and immunology (chapter 35); Techniques for determining microbial taxonomy and phylogeny: Molecular characteristics (section 19.4)

Members of the three genera have considerable practical importance. Pyogenic streptococci usually are pathogens and associated with pus formation (pyogenic means pus producing). Most species produce β-hemolysis on blood agar and form chains of cells. The major human pathogen in this group is *S. pyogenes,* which causes streptococcal sore throat, acute glomerulonephritis, and rheumatic fever. The normal habitat of oral streptococci is the oral cavity and upper respiratory tract of humans and other animals. In other respects oral streptococci are not necessarily similar. *S. pneumoniae* is α-hemolytic and grows as pairs of cocci (figures 23.17*c* and 23.18*b*). It is associated with lobar pneumonia and otitis media (inflammation of the middle ear). *S. mutans* is associated with the formation of dental caries. The enterococci such as *E. faecalis* are normal residents of the intestinal tracts of humans and most other animals. *E. faecalis* is an opportunistic pathogen that can cause urinary tract infections and endocarditis. Unlike the streptococci the enterococci will grow in 6.5% sodium chloride. They are major agents in the horizontal transfer of antibiotic-resistance genes. The lactococci ferment sugars to lactic acid and can grow at 10°C but not at 45°C. *L. lactis* is widely used in the production of buttermilk and cheese because it can curdle milk and add flavor through the synthesis of diacetyl and other products. Airborne diseases: Streptococcal diseases (section 38.1); Dental infections (section 38.7); Microbiology of fermented foods (section 40.6)

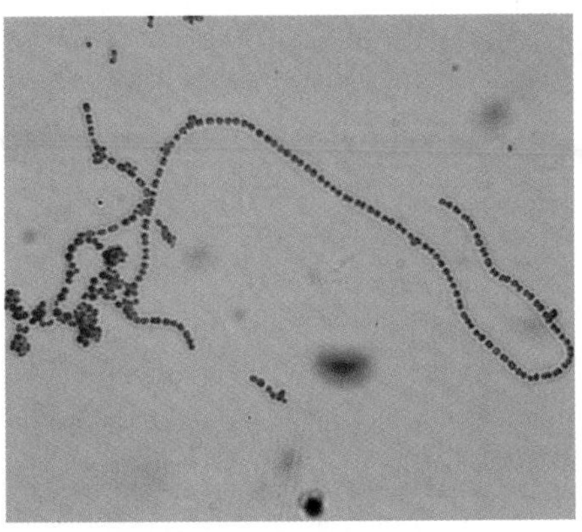

(a) *Streptococcus pyogenes*

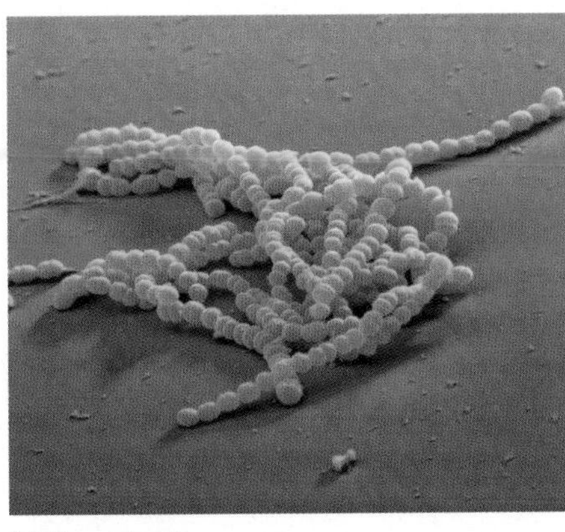

(b) *S. agalactiae*

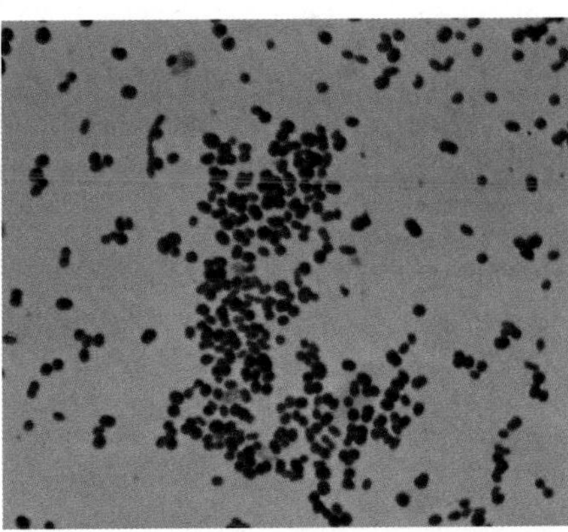

(c) *S. pneumoniae*

Figure 23.17 ***Streptococcus.*** **(a)** *Streptococcus pyogenes* (×900). **(b)** *Streptococcus agalactiae,* the cause of Group B streptococcal infections. Note the long chains of cells; color-enhanced scanning electron micrograph (×4,800). **(c)** *Streptococcus pneumoniae* (×900).

Table 23.4 Classification of the Streptococci, Entercocci, and Lactococci

Characteristics	*Streptococcus*	*Enterococcus*	*Lactococcus*
Predominant arrangement (most common first)	Chains, pairs	Pairs, chains	Pairs, short chains
Capsule/slime layer	+	−	−
Habitat	Mouth, respiratory tract	Gastrointestinal tract	Dairy products
Growth at 45°C	Variable	+	−
Growth at 10°C	Variable	Usually +	+
Growth at 6.5% NaCl broth	Variable	+	−
Growth at pH 9.6	Variable	+	−
Hemolysis	Usually β (pyogenic) or α (oral)	α, β, −	Usually −
Serological group (Lancefield)	Variable (A–O)	Usually D	Usually N
Mol% G + C	34–46	34–42	38–40
Representative species	Pyogenic streptococci	*E. faecalis*	*L. lactis*
	S. agalactiae	*E. faecium*	*L. raffinolactis*
	S. pyogenes	*E. avium*	*L. plantarum*
	S. equi	*E. durans*	
	S. dysgalactiae	*E. gallinarum*	
	Oral streptococci		
	S. gordonii		
	S. salvarius		
	S. sanguis		
	S. oralis		
	S. pneumoniae		
	S. mitis		
	S. mutans		
	Other streptococci		
	S. bovis		
	S. thermophilus		

Table 23.5 Properties of Selected Streptococci and Relatives

	Pyogenic Streptococci	Oral Streptococci			Enterococci	Lactococci
Characteristics	***S. pyogenes***	***S. pneumoniae***	***S. sanguis***	***S. mutans***	***E. faecalis***	***L. lactis***
Growth at 10°C	–[a]	–	–	–	+	+
Growth at 45°C	–	–	d	d	+	–
Growth at 6.5% NaCl	–	–	–	–	+	–
Growth at pH 9.6	–	–	–	–	+	–
Growth with 40% bile	–	–	d	d	+	+
α-hemolysis	–	+	+	–	–	d
β-hemolysis	+	–	–	–	+	–
Arginine hydrolysis	+	+	+	–	+	d
Hippurate hydrolysis	–	–	–	–	+	d
Mol% G + C of DNA	35–39	30–39	40–46	36–38	34–38	39
Genome size (Mb)	1.8	2.2	Nd*	2.0	3.2	2.3

Modified from *Bergey's Manual of Systematic Bacteriology,* Vol. 2, edited by P. H. A. Sneath, et al. Copyright © 1986 Williams and Wilkins, Baltimore, MD. Reprinted by permission.

[a]Symbols: +, 90% or more of strains positive; –, 10% or less of strains positive; d, 11–89% of strains are positive.

*Nd: Not determined; genome not yet sequenced

(a)

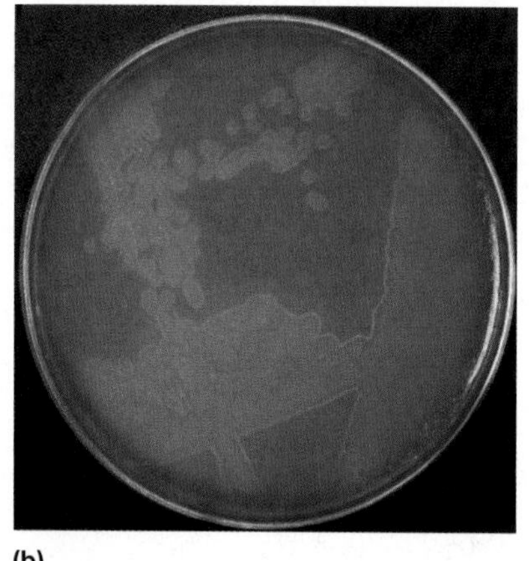
(b)

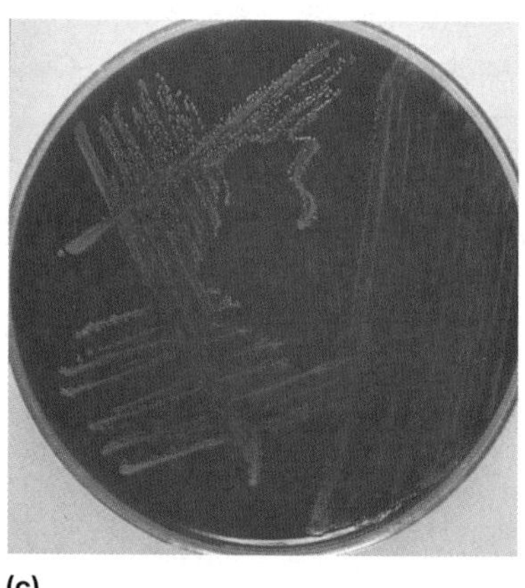
(c)

Figure 23.18 Streptococcal and Staphylococcal Hemolytic Patterns. **(a)** *Streptococcus pyogenes* on blood agar, illustrating β-hemolysis. **(b)** *Streptococcus pneumoniae* on blood agar, illustrating α-hemolysis. **(c)** *Staphylococcus epidermidis* on blood agar with no hemolysis.

1. List the major properties of the genus *Bacillus.* What practical impacts does it have on society? Define parasporal body.
2. Briefly describe the genus *Thermoactinomyces,* with particular emphasis on its unique features. What disease does it cause?
3. What is distinctive about the morphology of *Caryophanon?* Can you think of a selective advantage that may be conferred by the morphology of *Thermoactinomyces* or *Caryophanon?*
4. Describe the genus *Staphylococcus.* How does the pathogen *S. aureus* differ from the common skin resident *S. epidermidis,* and where is it normally found?
5. List the major properties of the genus *Lactobacillus.* Why is it important in the food and dairy industries?
6. Describe the major distinguishing characteristics of the following taxa: *Streptococcus, Enterococcus, Lactococcus,* and *Leuconostoc.*
7. Of what practical importance is *Leuconostoc?* What are lactic acid bacteria?
8. What are α-hemolysis, β-hemolysis, and the Lancefield grouping system?
9. What is the difference between pyogenic and oral streptococci?

Summary

23.1 General Introduction

a. Traditionally, gram-positive bacteria were classified according to characteristics such as general shape, peptidoglycan structure, the possession of endospores, and their response to acid-fast staining.

b. Based on 16S rRNA analysis, gram-positive bacteria are now divided into low G + C and high G + C groups; this system is used in the current edition of *Bergey's Manual.*

c. The second edition of *Bergey's Manual* places the low G + C gram positives in the phylum *Firmicutes,* which contains three classes: *Clostridia, Mollicutes,* and *Bacilli* (**tables 23.1–23.3** and **figure 23.2**).

23.2 Class *Mollicutes* (The Mycoplasmas)

a. Mycoplasmas stain gram-negative because they lack cell walls and cannot synthesize peptidoglycan precursors. Many species require sterols for growth. They are one of the smallest bacteria capable of self-reproduction and usually grow on agar to give colonies a "fried-egg" appearance (**figure 23.4**).

23.3 Peptidoglycan and Endospore Structure

a. Peptidoglycan structure often differs between groups in taxonomically useful ways. Most variations are in amino acid 3 of the peptide subunit or in the interpeptide bridge (**figure 23.5**).

b. Endospores resist desiccation and heat; they are used by bacteria to survive harsh conditions, especially in the soil (**figure 23.6**).

23.4 Class *Clostridia*

a. Members of the genus *Clostridium* are anaerobic gram-positive rods that form endospores and don't carry out dissimilatory sulfate reduction (**figure 23.7**). They are responsible for botulism, tetanus, food spoilage, and putrefaction.

b. *Desulfotomaculum* is an anaerobic, endospore-forming genus that reduces sulfate to sulfide during anaerobic respiration (**figure 23.8**).

c. The heliobacteria are anaerobic, photosynthetic bacteria with bacteriochlorophyll *g*. Some form endospores.

d. The family *Veillonellaceae* contains anaerobic cocci that stain gram-negative. Some are parasites of vertebrates.

23.5 Class *Bacilli*

a. The class *Bacilli* is divided into two orders: *Bacillales* and *Lactobacillales.*

b. The genus *Bacillus* contains aerobic and facultative, catalase-positive, endospore-forming, chemoheterotrophic, gram-positive rods that are usually motile and have peritrichous flagella (**figure 23.9**). Species of *Bacillus* synthesize antibiotics and insecticides, and cause food poisoning and anthrax.

c. *Thermoactinomyces* is a gram-positive thermophile that forms a mycelium and true endospores (**figure 23.11**). It causes allergic reactions and leads to farmer's lung.

d. Members of the genus *Staphylococcus* are facultatively anaerobic, nonmotile, gram-positive cocci that form irregular clusters (**figure 23.13**). They grow on the skin and mucous membranes of warm-blood animals, and some are important human pathogens.

e. Several important genera such as *Lactobacillus, Listeria,* and *Caryophanon* contain regular, nonsporing, gram-positive rods. *Lactobacillus* carries out lactic acid fermentation and is extensively used in the food and dairy industries.

f. *Leuconostoc* carries out heterolactic fermentation using the phosphoketolase pathway (**figure 23.16**) and is involved in the production of fermented vegetable products, buttermilk, butter, and cheese.

g. The genera *Streptococcus, Enterococcus,* and *Lactococcus* contain gram-positive cocci arranged in pairs and chains that are usually facultative and carry out homolactic fermentation (**tables 23.4** and **23.5**). Some important species are the pyogenic coccus *S. pyogenes,* the oral streptococci *S. pneumoniae* and *S. mutans,* the enterococcus *E. faecalis* and lactococcus *L. lactis* (**figure 23.17**).

Key Terms

α-hemolysis 584
β-hemolysis 584
coagulase 582
heterolactic fermentation 583
lactic acid bacteria (LAB) 582
Lancefield grouping system 584
meticillin-resistant *Staphylococcus aureus* (MRSA) 581
mycoplasmas 572
parasporal body 580

Critical Thinking Questions

1. Many low G + C bacteria are parasitic. The dependence on a host might be a consequence of the low G + C content. Elaborate on this concept.
2. How might one go about determining whether the genome of *M. genitalium* is the smallest one compatible with a free-living existence?
3. Account for the ease with which anaerobic clostridia can be isolated from soil and other generally aerobic niches.

Learn More

Amesz, J. 1995. The heliobacteria, a new group of photosynthetic bacteria. *J. Photochem. Photobiol. B* 30:89–96.

Cunningham, M. W. 2000. Pathogenesis of group A streptococcal infections: *Clin. Microbiol. Rev.* 13(3):470–511.

Glass, J. I.; Lefkowitz, E. J.; Glass, J. S.; Heiner, C. R.; Chen, E. Y.; and Cassell, G. H. 2000. The complete sequence of the mucosal pathogen *Ureaplasma urealyticum. Nature* 470: 757–62.

Himmelreich, R.; Hilbert, H.; Plagens, H.; Pirkl, E.; Li, B.-C.; and Hermann, R. 1996. Complete sequence analysis of the genome of the bacterium *Mycoplasma pneumoniae. Nucleic Acids Res.* 24(22):4420–49.

Holden, M. T.; Feil, E. J.; et al., 2004. Complete genomes of two clinical *Staphylococcus aureus* strains: Evidence for the rapid evolution of virulence and drug resistance. *Proc. Natl. Acad. Sci. USA.* 101: 9786–91.

Holt, J. G., editor-in-chief. 1994. *Bergey's Manual of Determinative Bacteriology.* 9th ed. Baltimore, Md: Williams & Wilkins.

Johnson, E. A. 2000. Clostridia. In *Encyclopedia of microbiology,* 2d ed., vol. 1, J. Lederberg, editor-in-chief, 834–39. San Diego: Academic Press.

Karlin, S.; Theriot, J.; and Mrazek, J. 2004. Comparative analysis of gene expression among low G + C positive genomes. *Proc. Natl. Acad. Sci. USA.* 101: 6182–87.

Kleerevezem, M., et al. 2003. Complete genome sequence of *Lactobacillus plantarum* WCFS1. *Proc. Natl. Acad. Sci. USA.* 100: 1990–95.

Kunst, F., et al. 1997. The complete genome sequence of the gram-positive bacterium *Bacillus subtilis. Nature* 390:249–56.

Lambert, B., and Peferoen, M. 1992. Insecticidal promise of *Bacillus thuringiensis:* Facts and mysteries about a successful biopesticide. *BioScience* 42(2):112–22.

Nicholson, W. L.; Munakata, N.; Horneck, G.; Melosh, H. J.; and Setlow, P. 2000. Resistance of *Bacillus* endospores to extreme terrestrial and extraterrestrial environments. *Micro. Mol. Biol. Rev.* 64(3):548–72.

Vreeland, R. H.; Rosenweig, W. D.; and Powers, D. W. 2000. Isolation of a 250 million-year-old halotolerant bacterium from a primary salt crystal. *Nature* 407: 897–900.

Please visit the Prescott website at www.mhhe.com/prescott7 for additional references.

---

24 Bacteria: The High G + C Gram Positives

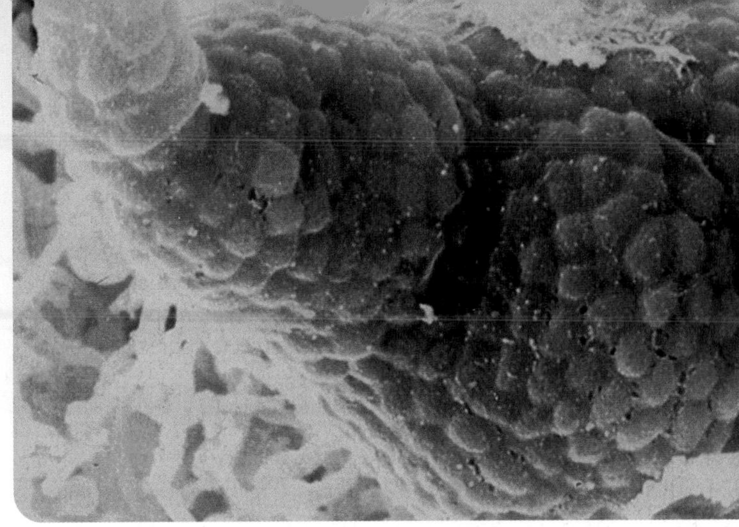

Frankia forms nonmotile spores and is symbiotic with a number of higher plants such as alder trees.

Preview

- *Bergey's Manual* classifies the actinomycetes and other high G + C gram-positive bacteria using 16S rRNA data. They are placed in the phylum *Actinobacteria,* which is a large, complex grouping with one class and six orders.
- The morphology and arrangement of spores, cell wall chemistry, and the types of sugars present in cell extracts are particularly important in actinomycete taxonomy and are used to divide these bacteria into different groups.
- Actinomycetes have considerable practical impact because they play a major role in the mineralization of organic matter in the soil and are the primary source of most naturally synthesized antibiotics. The genera *Corynebacterium, Mycobacterium,* and *Nocardia* include important human pathogens.

Chapter 24, the last of the survey chapters on bacteria, describes the high G + C gram-positive bacteria (**figure 24.1**). They are found in volume 4 of the second edition of *Bergey's Manual*. Many of these bacteria are called actinomycetes. Actinomycetes are gram-positive, aerobic bacteria, but are distinctive because they have filamentous hyphae that differentiate to produce asexual spores. Many closely resemble fungi in overall morphology. Presumably this resemblance results partly from adaptation to the same habitats. In this chapter, we first summarize the general characteristics of the actinomycetes. Representatives are described next, with emphasis on morphology, taxonomy, reproduction, and general importance. The **actinomycetes** [s., actinomycete] are a diverse group, but they share many properties.

24.1 General Properties of the Actinomycetes

The actinomycetes are a fascinating group of microorganisms. They are the source of most of the antibiotics used in medicine today. They also produce metabolites that are used as anticancer drugs, antihelminthics (for instance ivermectin, which is given to dogs to prevent heart worm), and drugs that suppress the immune system in patients who have received organ transplants. This practical aspect of the actinomycetes is linked very closely to their mode of growth. Like the myxobacteria, the prosthecate bacteria, and several other microbes described in previous chapters, the actinomycetes undergo a complex life cycle. The life cycle of many actinomycetes includes the development of filamentous cells, called **hyphae,** and spores. When growing on a solid substratum such as soil or agar, the actinomycetes develop a branching network of hyphae. The hyphae grow both on the surface of the substratum and into it to form a dense mat of hyphae termed a **substrate mycelium.** Septae usually divide the hyphae into long cells (20 μm and longer) containing several nucleoids. In many actinomycetes, substrate hyphae differentiate into upwardly growing hyphae to form an **aerial mycelium** that extends above the substratum. It is at this time that medically useful compounds are formed. Because the physiology of the actinomycete has switched from actively growing vegetative cells into this special cell type, these compounds are often called **secondary metabolites.**

The aerial hyphae form thin-walled spores upon septation (**figure 24.2**). These spores are considered exospores because they do not develop within a mother cell like the endospores of

Actinomycetes are very important from a medical point of view. . . . They may be a nuisance, as when they decompose rubber products, grow in aviation fuel, produce odorous substances that pollute water supplies, or grow in sewage-treatment plants where they form thick clogging foams. . . . In contrast, actinomycetes are the producers of most of the antibiotics.

—*H. A. Lechevalier and M. P. Lechevalier*

Bacillus and *Clostridium.* If the spores are located in a sporangium, they may be called **sporangiospores.** The spores can vary greatly in shape (**figure 24.3**). Like spore formation in other bacteria, actinomycete sporulation is usually in response to nutrient deprivation. However, most actinomycete spores are not particularly heat resistant but withstand desiccation well, so they have considerable adaptive value. Most actinomycetes are not motile, and spores are dispersed by wind or adhering to animals; in this way they may find a new habitat that will provide needed nutrients. In the few motile genera, motility is confined to flagellated spores.

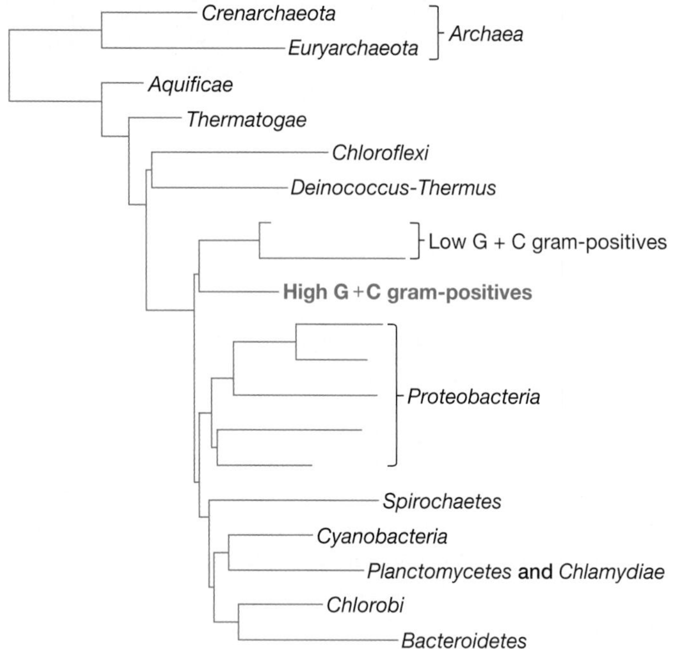

Figure 24.1 Phylogenetic Relationships Among Procaryotes. The high G + C gram-positive bacteria are highlighted.

Actinomycete cell wall composition varies greatly among different groups and is of considerable taxonomic importance. Four major cell wall types can be distinguished according to three features of peptidoglycan composition and structure: the amino acid in tetrapeptide side chain position 3, the presence of glycine in interpeptide bridges, and peptidoglycan sugar content (**table 24.1;** *see also figure 23.5*). Whole cell extracts of actinomycetes with wall types II, III, and IV also contain characteristic sugars that are useful in identification (**table 24.2**). Some other taxonomically valuable properties are the morphology and color of the mycelium and sporangia, the surface features and arrangement of spores, the percent G + C in DNA, the phospholipid composition of cell membranes, and spore heat resistance. Of course 16S rRNA sequences has proven valuable. Several actinomycete genomes have been sequenced including *Mycobacterium tuberculosis, M. leprae, Streptomyces coelicolor,* and *S. avermitilis.* The bacterial cell wall (section 3.6); Peptidoglycan and endospore structure (section 23.3)

Actinomycetes also have great ecological significance. They are primarily soil inhabitants and are very widely distributed. They can degrade an enormous number and variety of organic compounds and are extremely important in the mineralization of organic matter. Although most actinomycetes are free-living microorganisms, a few are pathogens of humans, other animals, and some plants.

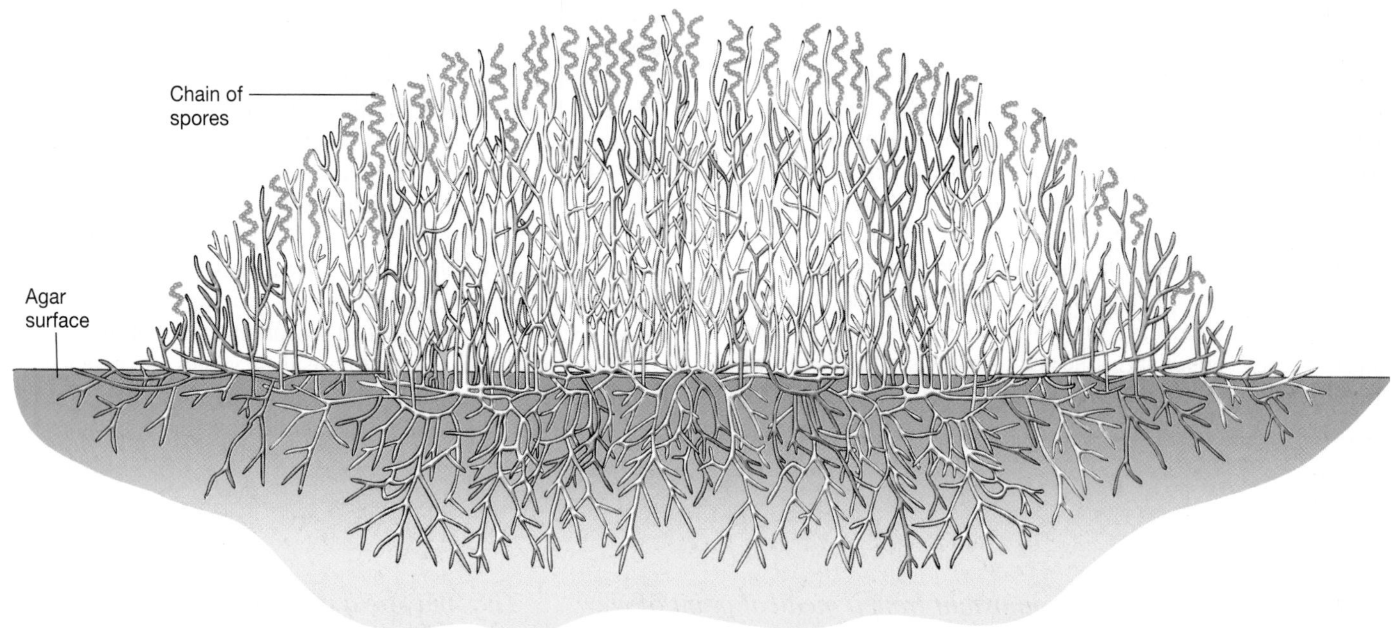

Figure 24.2 An Actinomycete Colony. The cross section of an actinomycete colony with living (blue-green) and dead (white) hyphae. The substrate mycelium and aerial mycelium with chains of spores are shown.

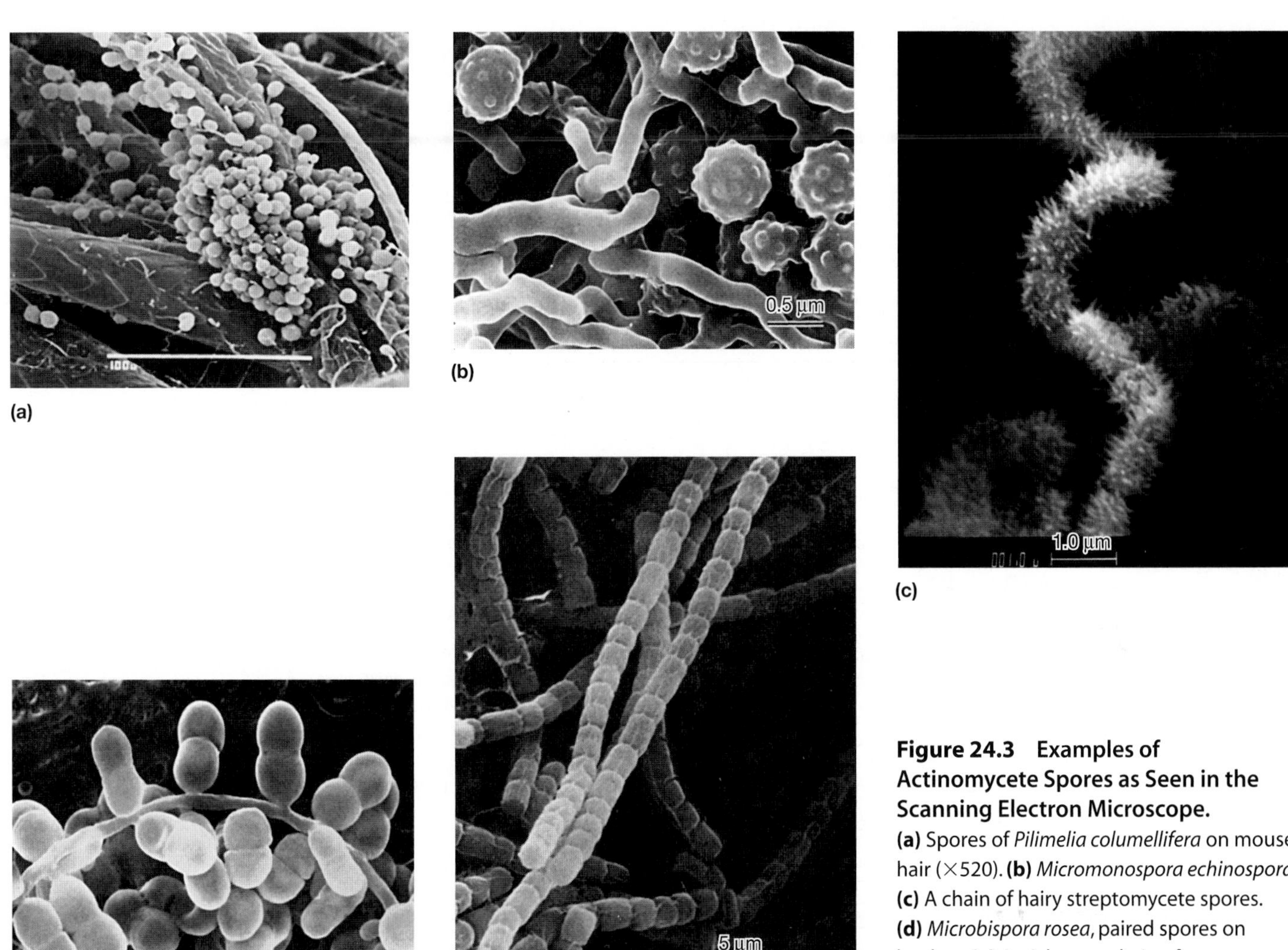

(a) (b) (c) (d) (e)

Figure 24.3 Examples of Actinomycete Spores as Seen in the Scanning Electron Microscope. **(a)** Spores of *Pilimelia columellifera* on mouse hair (×520). **(b)** *Micromonospora echinospora.* **(c)** A chain of hairy streptomycete spores. **(d)** *Microbispora rosea,* paired spores on hyphae. **(e)** Aerial spore chain of *Kitasatosporia setae.*

Table 24.1 Actinomycete Cell Wall Types

Cell Wall Type	Diaminopimelic Acid Isomer	Glycine in Interpeptide Bridge	Characteristic Sugars	Representative Genera
I	L, L	+	NA[a]	*Nocardioides, Streptomyces*
II	*meso*	+	NA	*Micromonospora, Pilimelia, Actinoplanes*
III	*meso*	−	NA	*Actinomadura, Frankia*
IV	*meso*	−	Arabinose, galactose	*Saccharomonospora, Nocardia*

[a]NA, either not applicable or no diagnostic sugars.

It has been clear for some time that some of the phenotypic traits traditionally used to determine actinomycete taxonomy do not always fit with 16S rRNA sequence data. Phylogenetic analyses based on 16S rRNA sequences are used to classify the high G + C gram-positive bacteria (i.e., those gram-positive bacteria with a DNA base composition above approximately 50 mol% G + C). All are placed in the phylum *Actinobacteria* and classified as shown in **figure 24.4.** The phylum is large and very complex; it contains one class (*Actinobacteria*), five subclasses, six orders, 14 suborders, and 44 families. In this system

Table 24.2 Actinomycete Whole Cell Sugar Patterns

Sugar Pattern Types[a]	Characteristic Sugars	Representative Genera
A	Arabinose, galactose	*Nocardia, Rhodococcus, Saccharomonospora*
B	Madurose[b]	*Actinomadura, Streptosporangium, Dermatophilus*
C	None	*Thermomonospora, Actinosynnema, Geodermatophilus*
D	Arabinose, xylose	*Micromonospora, Actinoplanes*

[a]Characteristic sugar patterns are present only in wall types II-IV, those actinomycetes with *meso*-diaminopimelic acid.

[b] Madurose is 3-*O*-methyl-D-galactose.

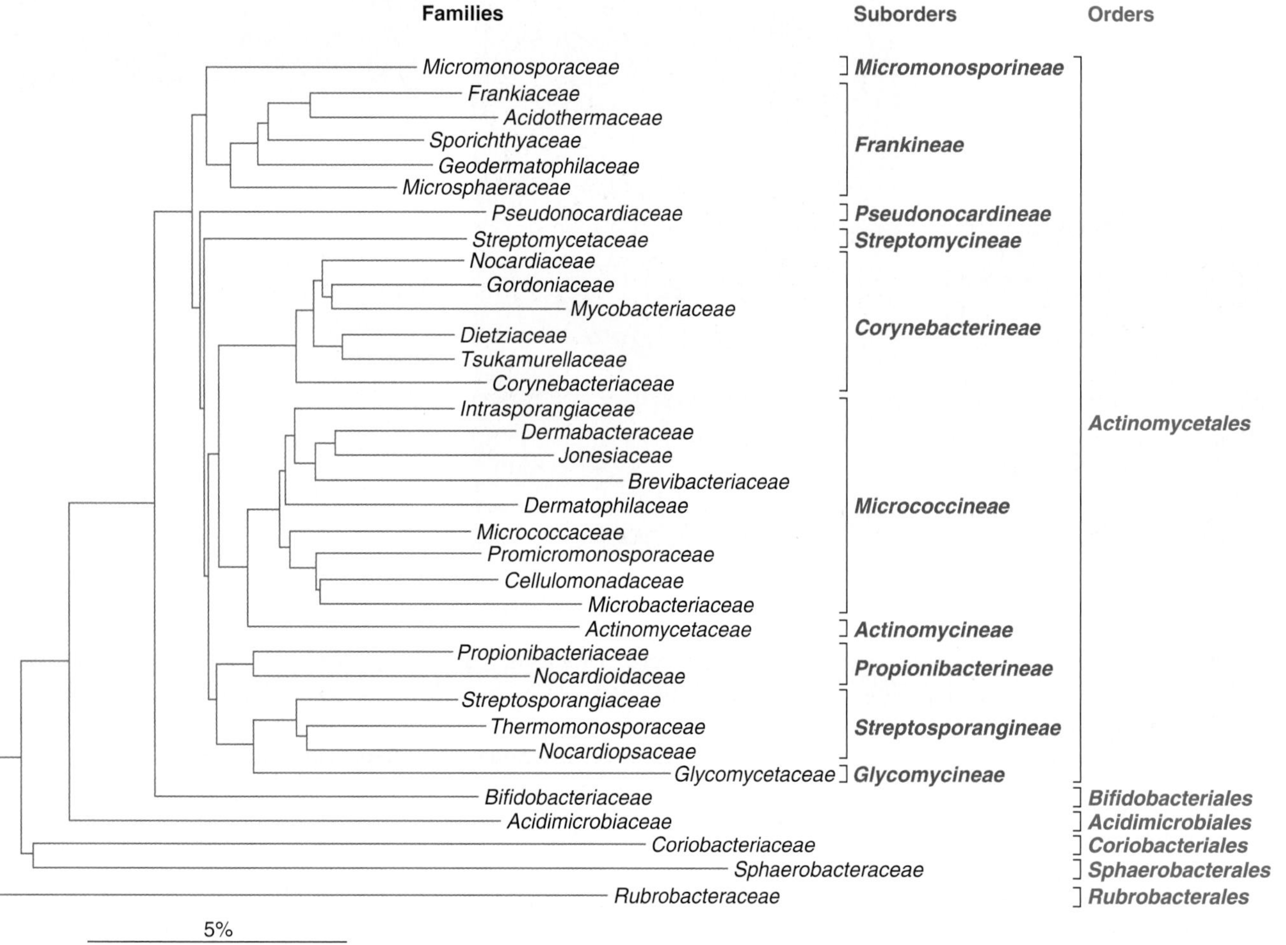

(a)

Figure 24.4 Classification of the Phylum *Actinobacteria*. **(a)** The phylogenetic relationships between orders, suborders, and families based on 16S rRNA data are shown. The bar represents 5 nucleotide substitutions per 100 nucleotides. *Source: (a) E. Stackebrandt, F. A. Rainey, and N. L. Ward-Rainey. Proposal for a new hierarchic classification system,* Actinobacteria, *classis nov.* Int. J. Syst. Bacteriol. *47(2):479–491, 1997. Figure 3, p. 482.*

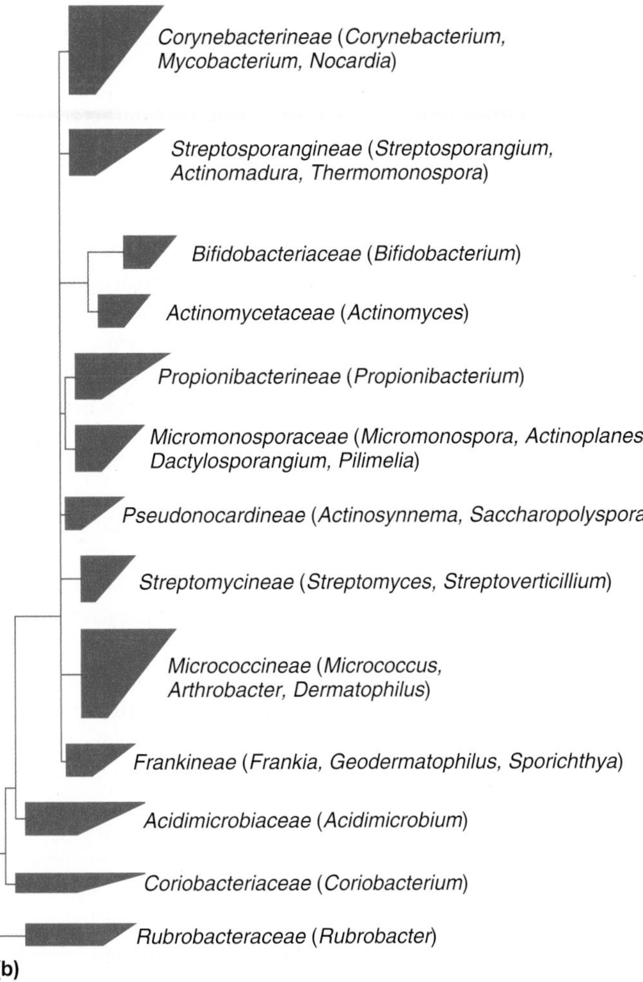

Figure 24.4 ***(continued).*** **(b)** Phylogenetic relationships among major actinobacterial groups based on 16S rRNA data. Representative genera are given in parentheses. Each tetrahedron in the tree represents a group of related organisms; its horizontal edges show the shortest and longest branches in the group. Multiple branching at the same level indicates that the relative branching order of the groups cannot be determined from the data.

the **actinobacteria** are composed of the actinomycetes and their high G + C relatives. **Figure 24.5** shows the phylogenetic relationships among selected representatives of the high G + C gram positives, and **table 24.3** summarizes the characteristics of some of the genera discussed in this chapter. Techniques for determining microbial taxonomy and phylogeny: Molecular characteristics (section 19.4)

Most of the genera discussed in the following survey are in the subclass *Actinobacteridae* and order *Actinomycetales* that is divided into 10 suborders. The survey focuses on several of these suborders. The order *Bifidobacteriales* also is briefly described.

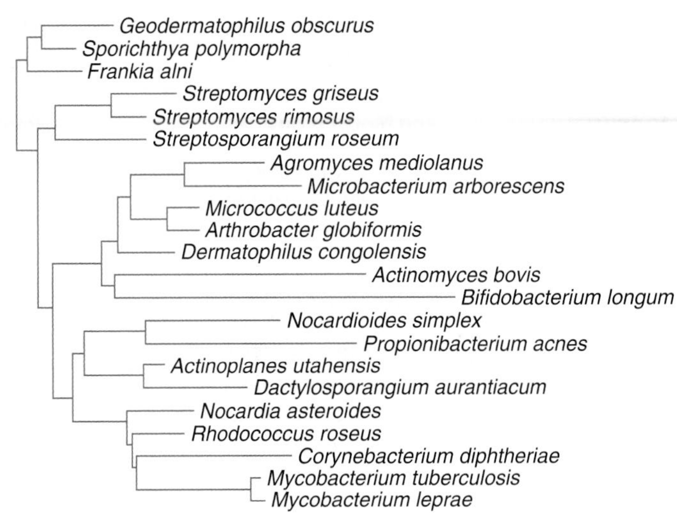

Figure 24.5 Phylogenetic Relationships Among Selected High G + C Gram-Positive Bacteria. The relationships among a few species based on 16S rRNA sequence data are shown. *Source: The Ribosomal Database Project.*

24.2 Suborder *Actinomycineae*

There is one family with five genera in the suborder *Actinomycineae*. These include *Actinomyces, Actinobaculum, Arcanobacterium, Mobiluncus* and *Varibaculum*. Most are irregularly shaped, nonsporing, gram-positive rods with aerobic or faculative metabolism. The rods may be straight or slightly curved and usually have swellings, club shapes, or other deviations from normal rod-shape morphology.

Members of the genus *Actinomyces* are either straight or slightly curved rods that vary considerably in shape or slender filaments with true branching (**figure 24.6**). The rods and filaments may have swollen or clubbed ends. They are either facultative or strict anaerobes that require CO_2 for optimal growth. The cell walls contain lysine but not diaminopimelic acid or glycine. *Actinomyces* species are normal inhabitants of mucosal surfaces of humans and other warm-blooded animals; the oral cavity is their preferred habitat. *A. bovis* causes lumpy jaw in cattle. *Actinomyces* is responsible for actinomycoses, ocular infections, and periodontal disease in humans. The most important human pathogen is *A. israelii.*

24.3 Suborder *Micrococcineae*

The suborder *Micrococcineae* has 14 families and a wide variety of genera. Two of the best-known genera are *Micrococcus* and *Arthrobacter*.

The genus *Micrococcus* contains aerobic, catalase-positive cocci that occur mainly in pairs, tetrads, or irregular clusters and are usually nonmotile (**figure 24.7**). Unlike many other actinomycetes,

Table 24.3 Characteristics of Actinobacteria

Genus	Dimensions (μm) and Morphology	G + C Content (mol%)	Oxygen Relationship	Other Distinctive Characteristics
Actinoplanes	Nonfragmenting, branching mycelium with little aerial growth; sporangia formed; motile spores with polar flagella	72–73	Aerobic	Hyphae often in palisade arrangement; highly colored; type II cell walls; found in soil and decaying plant material
Arthrobacter	0.8–1.2 × 1.0–8.0; young cells are irregular rods, older cells are small cocci; usually nonmotile	59–70	Aerobic	Have rod-coccus growth cycle; metabolism respiratory; catalase positive; mainly in soil
Bifidobacterium	0.5–1.3 × 1.5–8; rods of varied shape, usually curved; nonmotile	55–67	Anaerobic	Cells can be clubbed or branched, pairs often in V arrangement; ferment carbohydrates to acetate and lactate, but no CO_2; catalase negative
Corynebacterium	0.3–0.8 × 1.5–8.0; straight or slightly curved rods with tapered or clubbed ends; nonmotile	51–63	Facultatively anaerobic	Cells often arranged in a V formation or in palisades of parallel cells; catalase positive and fermentative; metachromatic granules
Frankia	0.5–2.0 in diameter; vegetative hyphae with limited to extensive branching and no aerial mycelium; multilocular sporangia formed	66–71	Aerobic to microaerophilic	Sporangiospores nonmotile; usually fixes nitrogen; type III cell walls; most strains are symbiotic with angiosperm plants and induce nodules
Micrococcus	0.5–2.0 diameter; cocci in pairs, tetrads, or irregular clusters; usually nonmotile	64–75	Aerobic	Colonies usually yellow or red; catalase positive with respiratory metabolism; primarily on mammalian skin and in soil
Mycobacterium	0.2–0.6 × 1.0–10; straight or slightly curved rods, sometimes branched; acid-fast; nonmotile and nonsporing	62–70	Aerobic	Catalase positive; can form filaments that are readily fragmented; walls have high lipid content; in soil and water; some parasitic
Nocardia	0.5–1.2 in diameter; rudimentary to extensive vegetative hyphae that can fragment into rod-shaped and coccoid forms	64–72	Aerobic	Aerial hyphae formed; catalase positive; type IV cell wall; widely distributed in soil
Propionibacterium	0.5–0.8 × 1–5; pleomorphic nonmotile rods, may be forked or branched; nonsporing	53–67	Anaerobic to aerotolerant	Fermentation produces propionate and acetate, and often gas; catalase positive
Streptomyces	0.5–2.0 in diameter; vegetative mycelium extensively branched; aerial mycelium forms chains of three to many spores	69–78	Aerobic	Form discrete lichenoid or leathery colonies that often are pigmented; use many different organic compounds as nutrients; soil organisms

Micrococcus does not undergo morphological differentiation. Micrococci colonies often are yellow, orange, or red. They are widespread in soil, water, and on mammalian skin, which may be their normal habitat.

The genus *Arthrobacter* contains aerobic, catalase-positive rods with respiratory metabolism and lysine in its peptidoglycan. Its most distinctive feature is a rod-coccus growth cycle (**figure 24.8**). When *Arthrobacter* grows in exponential phase, the bacteria are irregular, branched rods that may reproduce by a process called **snapping division.** As they enter stationary phase, the cells change to a coccoid form. Upon transfer to fresh medium, the coccoid cells differentiate to form actively growing rods. Although arthrobacters often are isolated from fish, sewage, and plant surfaces, their most important habitat is the soil, where they constitute a significant

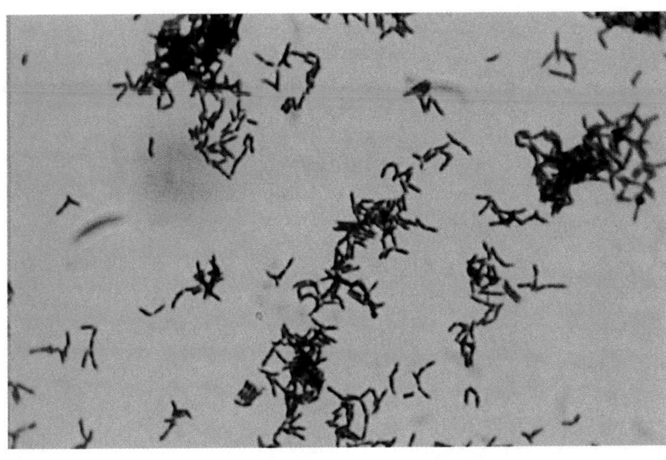

(a)

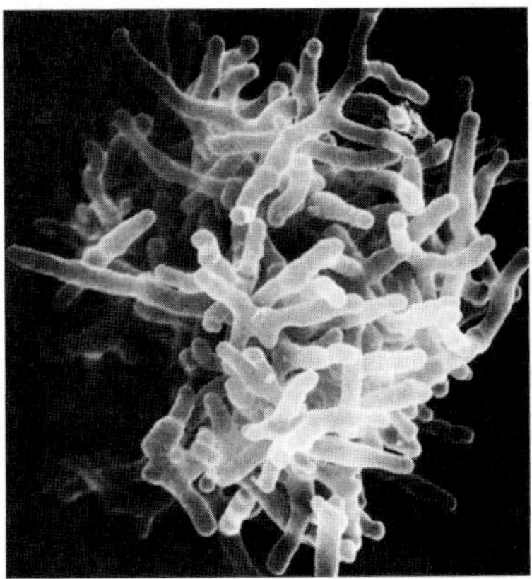

(b)

Figure 24.6 Representatives of the Genus *Actinomyces*. **(a)** *A. naeslundii*; Gram stain (×1,000). **(b)** *Actinomyces*; scanning electron micrograph (×18,000). Note filamentous nature of the colony.

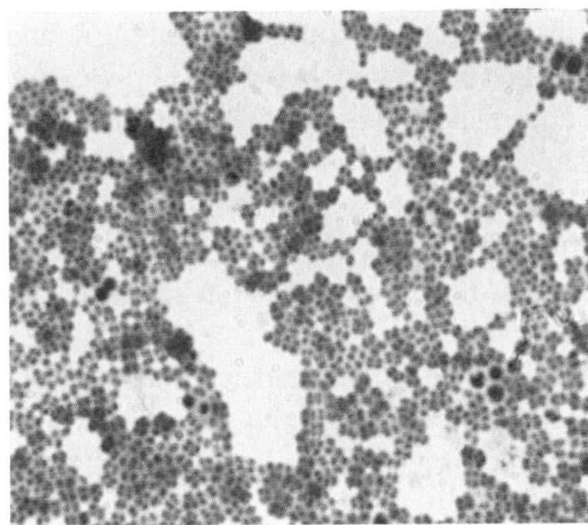

Figure 24.7 Micrococcus. *Micrococcus luteus,* methylene blue stain (×1,000).

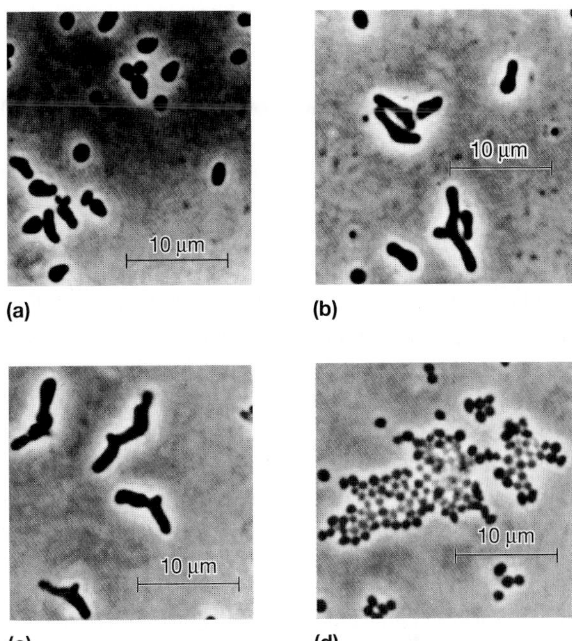

Figure 24.8 The Rod-Coccus Growth Cycle. The rod-coccus cycle of *Arthrobacter globiformis* when grown at 25°C. **(a)** Rods are outgrowing from cocci 6 hours after inoculation. **(b)** Rods after 12 hours of incubation. **(c)** Bacteria after 24 hours. **(d)** Cells after reaching stationary phase (3 day incubation). The cells used for inoculation resembled these stationary-phase cocci.

component of the microbial flora. They are well adapted to this niche because they are very resistant to desiccation and nutrient deprivation. This genus is unusually flexible nutritionally and can even degrade some herbicides and pesticides.

The mechanism of snapping division has been studied in *Arthrobacter.* These bacteria have a two-layered cell wall, and only the inner layer grows inward to generate a transverse wall dividing the new cells. The completed transverse wall or septum next thickens and puts tension on the outer wall layer, which still holds the two cells together. Eventually, increasing tension ruptures the outer layer at its weakest point, and a snapping movement tears the outer layer apart around most of its circumference. The new cells now rest at an angle to each other and are held together by the remaining portion of the outer layer that acts as a hinge.

A third genus in this suborder is *Dermatophilus. Dermatophilus* (type IIIB) also forms packets of motile spores with tufts of flagella, but it is a facultative anaerobe and a parasite of mammals responsible for the skin infection streptothricosis.

24.4 Suborder *Corynebacterineae*

This suborder contains seven families with several well-known genera. Three of the most important genera are *Corynebacterium, Mycobacterium,* and *Nocardia.*

The family *Corynebacteriaceae* has one genus, *Corynebacterium,* which contains aerobic and facultative, catalase-positive, straight to slightly curved rods, often with tapered ends. Club-shaped forms are also seen. The bacteria often remain partially attached after snapping division, resulting in angular arrangements of the cells, or a palisade arrangement in which rows of cells are lined up side by side (**figure 24.9**). Corynebacteria form metachromatic granules, and their walls have *meso*-diaminopimelic acid. Although some species are harmless saprophytes, many corynebacteria are plant or animal pathogens. For example, *C. diphtheriae* is the causative agent of diphtheria in humans. Airborne diseases: Diphtheria (section 38.1)

The family *Mycobacteriaceae* contains the genus *Mycobacterium,* which is composed of slightly curved or straight rods that sometimes branch or form filaments (**figure 24.10**). Mycobacterial filaments differ from those of actinomycetes in readily fragmenting into rods and coccoid bodies. They are aerobic and catalase positive. Mycobacteria grow very slowly and must be incubated for 2 to 40 days after inoculation of a solidified complex medium to form a visible colony. Their cell walls have a very high lipid content and contain waxes with 60 to 90 carbon **mycolic acids.** These are complex fatty acids with a hydroxyl group on the β-carbon and an aliphatic chain attached to the α-carbon (**figure 24.11**). The presence of mycolic acids and other lipids outside the peptidoglycan layer makes mycobacteria **acid-fast** (basic fuchsin dye cannot be removed from the cell by acid alcohol treatment). Extraction of wall lipid with alkaline ethanol destroys acid-fastness. Preparation and staining of specimens: Differential staining (section 2.3)

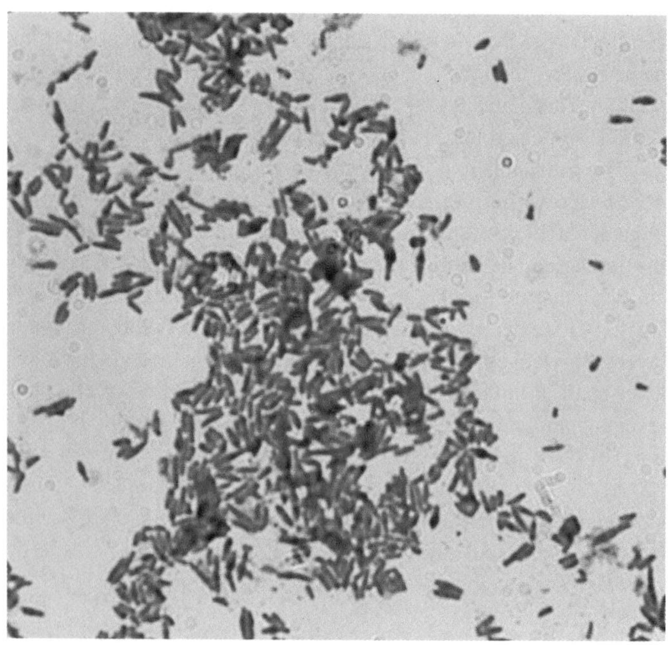

Figure 24.9 ***Corynebacterium diphtheriae.*** Note the irregular shapes of individual cells, the angular associations of pairs of cells, and palisade arrangements (×1,000). These gram-positive rods do not form endospores.

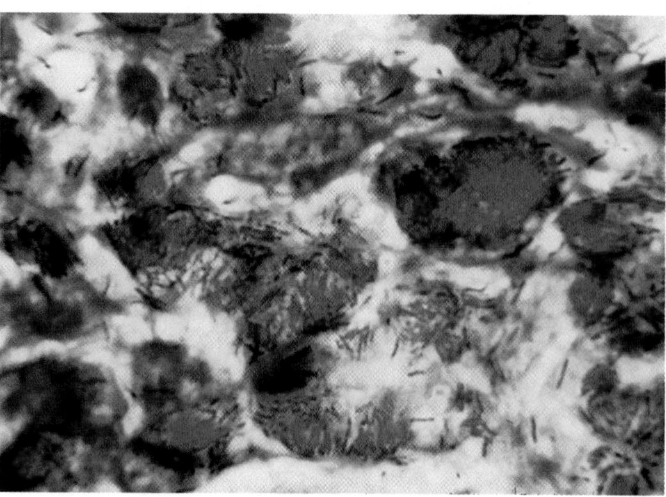

Figure 24.10 The Mycobacteria. *Mycobacterium leprae.* Acid-fast stain (×400). Note the masses of red mycobacteria within blue-green host cells.

Although some mycobacteria are free-living saprophytes, they are best known as animal pathogens. *M. bovis* causes tuberculosis in cattle, other ruminants, and primates. Because this bacterium can produce tuberculosis in humans, dairy cattle are tested for the disease yearly; milk pasteurization kills the pathogen. Prior to widespread milk pasteurization, contaminated milk was a problematic source of transmission. Currently, *M. tuberculosis* is the chief source of tuberculosis in humans. The other major mycobacterial human disease is leprosy, caused by *M. leprae.* Airborne diseases: *Mycobacterium avium-M. intercellulare* and *M. tuberculosis* pulmonary infections (section 38.1); Insights from microbial genomes: Genomic analysis of pathogenic microbes (section 15.8)

The family *Nocardiaceae* is composed of two genera, *Nocardia* and *Rhodococcus.* These bacteria develop a substrate mycelium that readily breaks into rods and coccoid elements (**figure 24.12**). They also form an aerial mycelium that rises above the substratum and may produce conidia. Almost all are strict aerobes. Most species have peptidoglycan with *meso*-diaminopimelic acid and no peptide interbridge. The wall usually contains a carbohydrate composed of arabinose and galactose; mycolic acids are present in *Nocardia* and *Rhodococcus.* Because these and related genera resemble members of the genus *Nocardia* (named after Edmond Nocard [1850–1903], French bacteriologist and veterinary pathologist), they are collectively called **nocardioforms.**

Nocardia is distributed worldwide in soil and aquatic habitats. Nocardiae are involved in the degradation of hydrocarbons and waxes and can contribute to the biodeterioration of rubber joints in water and sewage pipes. Although most are free-living saprophytes, some species, particularly *N. asteroides,* are opportunistic pathogens that cause nocardiosis in humans and other animals. People with low resistance due to other health problems, such as individuals with HIV-AIDS, are most at risk. The lungs are most often infected, but the central nervous system and other organs may be invaded.

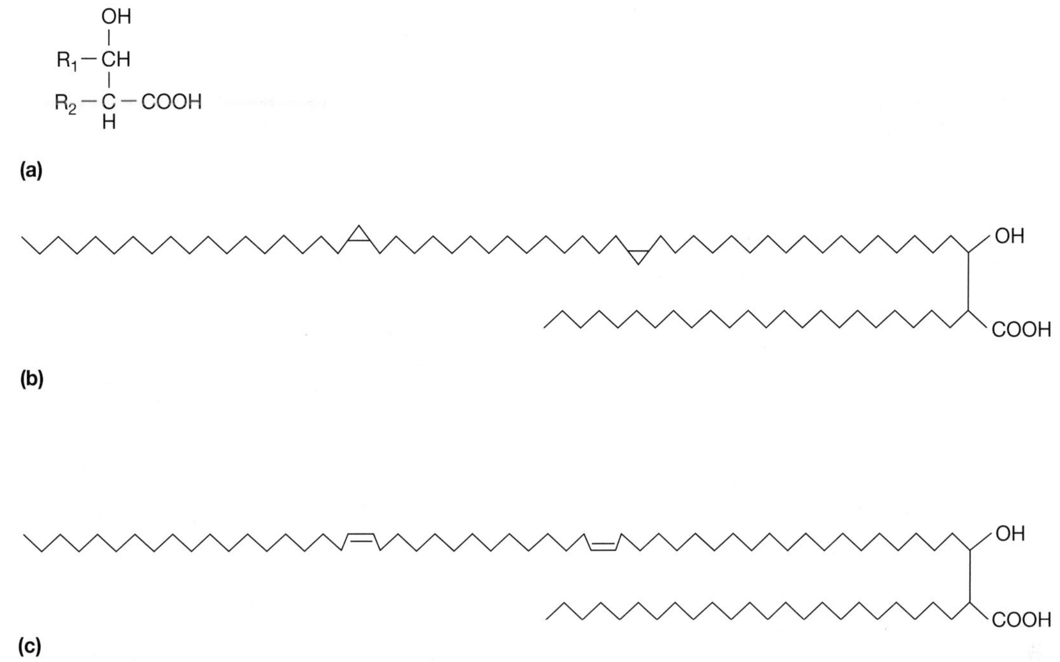

OH
|
R_1 — CH
|
R_2 — C — COOH
H

(a)

OH
COOH

(b)

OH
COOH

(c)

Figure 24.11 Mycolic Acid Structure. **(a)** The generic structure of mycolic acids, a family that includes over 500 different types. **(b)** A mycolic acid with two cyclopropane rings and **(c)** an unsaturated mycolic acid.

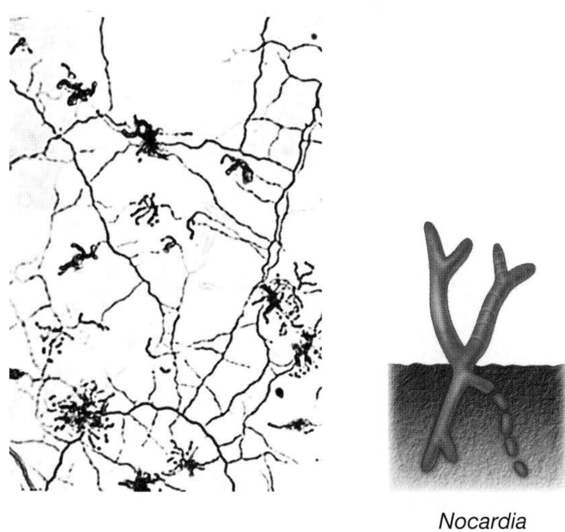

Figure 24.12 *Nocardia.* *Nocardia asteroides,* substrate mycelium and aerial mycelia with conidia illustration and light micrograph (×1,250).

Rhodococcus is widely distributed in soils and aquatic habitats. It is of considerable interest because members of the genus can degrade an enormous variety of molecules such as petroleum hydrocarbons, detergents, benzene, polychlorinated biphenyls (PCBs), and various pesticides. It may be possible to use rhodococci to remove sulfur from fuels, thus reducing air pollution by sulfur oxide emissions.

1. Define actinomycete, substrate mycelium, aerial mycelium, and exospore. Explain how these structures confer a survival advantage.
2. Describe how cell wall structure and sugar content are used to classify the actinomycetes. Include a brief description of the four major wall types.
3. Why are the actinomycetes of such practical interest?
4. Describe the phylum *Actinobacteria* and its relationship to the actinomycetes.
5. Describe the major characteristics of the following genera: *Actinomyces, Micrococcus, Arthrobacter,* and *Corynebacterium.* Include comments on their normal habitat and importance.
6. What is snapping division? the rod-coccus growth cycle?
7. Give the distinctive properties of the genus *Mycobacterium.*
8. List two human mycobacterial diseases and their causative agents. Which pathogen causes tuberculosis in cattle?
9. What is a nocardioform, and how can the group be distinguished from other actinomycetes?
10. Where is *Nocardia* found, and what problems may it cause? Consider both environmental and public health concerns.

24.5 Suborder *Micromonosporineae*

The suborder *Micromonosporineae* contains only one family, *Micromonosporaceae.* Genera include *Micromonospora, Dactylosporangium, Pilimelia,* and *Actinoplanes.* Often the family is collectively referred to as the actinoplanetes [Greek *actinos,* a ray or beam, and *planes,* a wanderer]. They have an extensive substrate mycelium and are type IID cells. Often the hyphae are highly colored and diffusible pigments may be produced. Normally an aerial mycelium is absent or rudimentary. Spores are

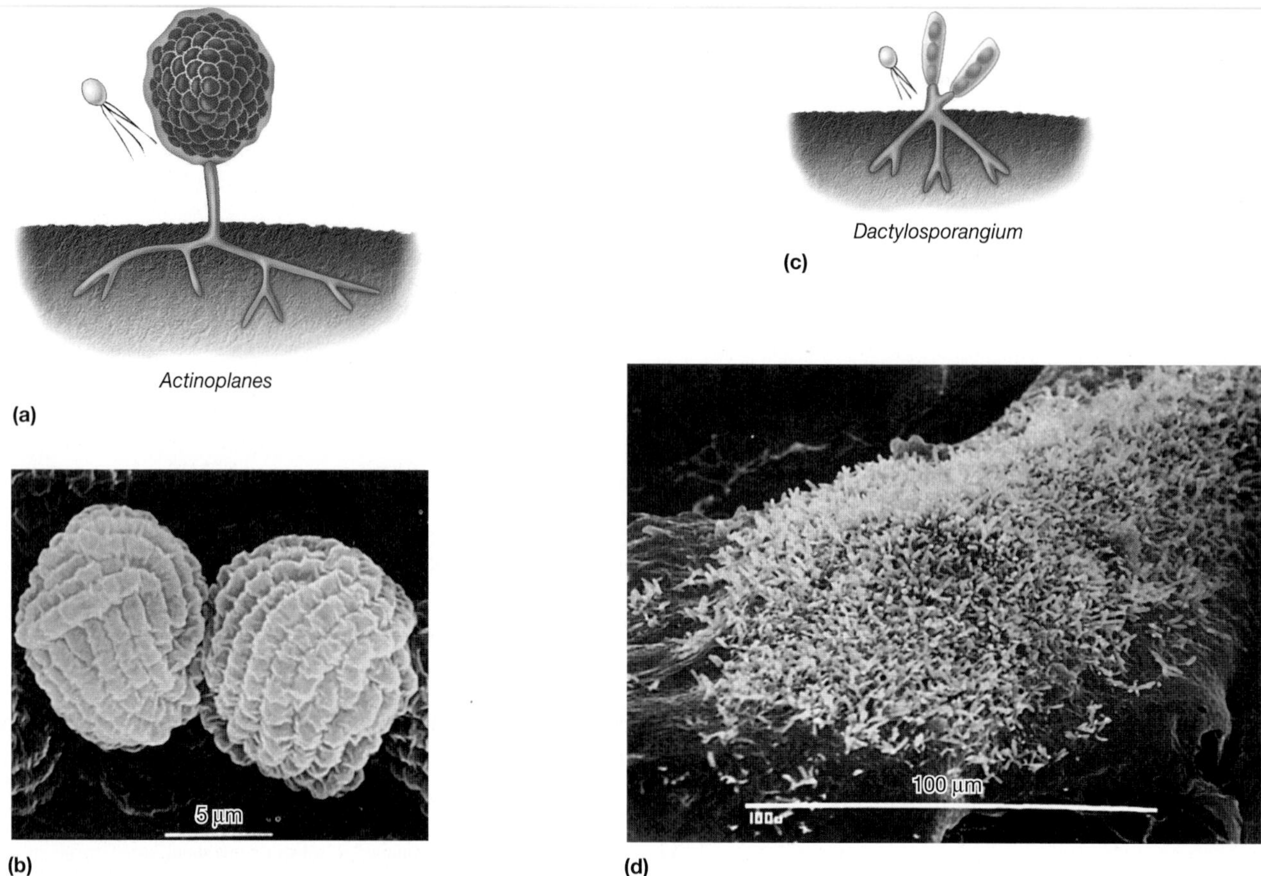

Figure 24.13 Family *Micromonosporaceae*. Actinoplanete morphology. **(a)** *Actinoplanes* structure. **(b)** A scanning electron micrograph of mature *Actinoplanes* sporangia. **(c)** *Dactylosporangium* structure. **(d)** A *Dactylosporangium* colony covered with sporangia.

usually formed within a sporangium raised above the surface of the substratum at the end of a special hypha called a sporangiophore. The spores can be either motile or nonmotile. These bacteria vary in the arrangement and development of their spores. Some genera (*Actinoplanes, Pilimelia*) have spherical, cylindrical, or irregular sporangia with a few to several thousand spores per sporangium (figure 24.3*a* and **figure 24.13**). The sporangium develops above the substratum at the tip of a sporangiophore; the spores are arranged in coiled or parallel chains (**figure 24.14**). *Dactylosporangium* forms club-shaped, fingerlike, or pyriform sporangia with one to six spores (figure 24.13*c,d*). *Micromonospora* bears single spores, which often occur in branched clusters of sporophores (figure 24.3*b*).

Actinoplanetes grow in almost all soil habitats, ranging from forest litter to beach sand. They also flourish in freshwater, particularly in streams and rivers (probably because of abundant oxygen and plant debris). Some have been isolated from the ocean. The soil-dwelling species may have an important role in the decomposition of plant and animal material. *Pilimelia* grows in association with keratin. *Micromonospora* actively degrades chitin and cellulose, and produces antibiotics such as gentamicin.

24.6 Suborder *Propionibacterineae*

This suborder contains two families and 14 genera. The genus *Propionibacterium* contains pleomorphic, nonmotile, nonsporing rods that are often club-shaped with one end tapered and the other end rounded. Cells also may be coccoid or branched. They can be single, in short chains, or in clumps. The genus is facultatively anaerobic or aerotolerant; lactate and sugars are fermented to produce large quantities of propionic and acetic acids, and often carbon dioxide. *Propionibacterium* is usually catalase positive. The genus is found growing on the skin and in the digestive tract of animals, and in dairy products such as cheese. *Propionibacterium* contributes substantially to the production of Swiss cheese. *P. acnes* is involved with the development of body odor and acne vulgaris. Microbiology of feremented foods (section 40.6)

24.7 Suborder *Streptomycineae*

The suborder *Streptomycineae* has only one family, *Streptomycetaceae,* and three genera, the most important of which is *Streptomyces.* These bacteria have aerial hyphae that divide in a single

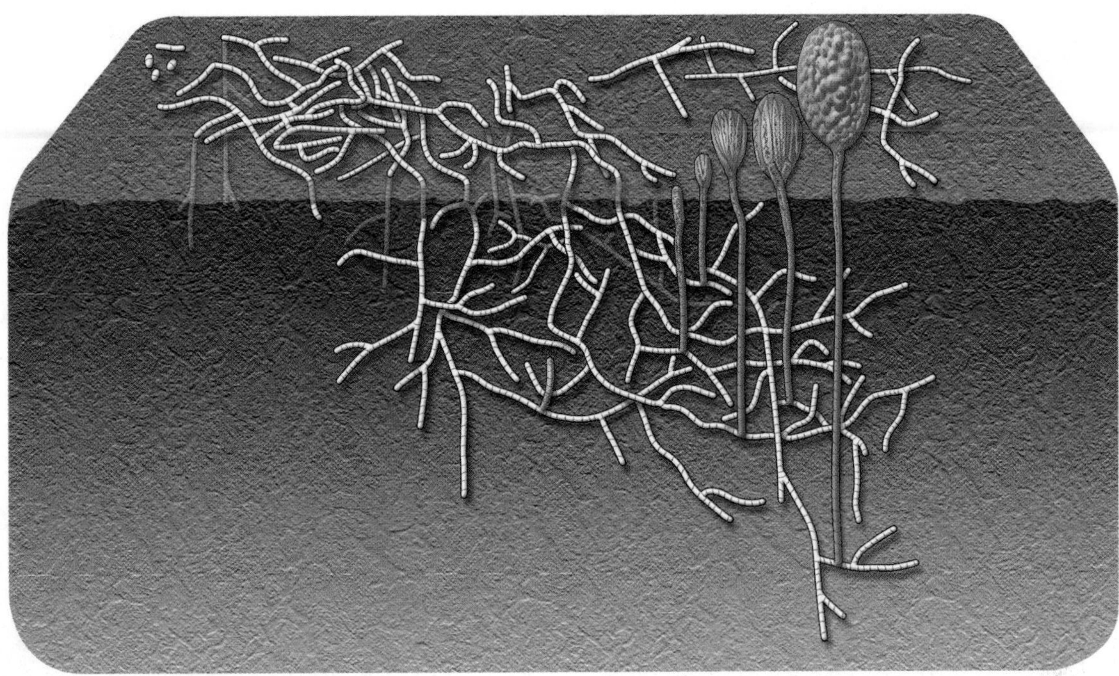

Figure 24.14 Sporangium Development in an Actinoplanete. The developing sporangium is shown in purple with more mature stages to the right.

plane to form chains of 3 to 50 or more nonmotile spores with surface texture ranging from smooth to spiny and warty (figure 24.3*e;* **figures 24.15** and **24.16**). All have a type I cell wall and a G + C content of 69 to 78%. Filaments grow by tip extension rather than by fragmentation. Members of this family and similar bacteria are often called **streptomycetes** [Greek *streptos,* bent or twisted, and *myces,* fungus].

Streptomyces is a large genus; there are around 150 species. Members of the genus are strict aerobes, have cell wall type I, and form chains of nonmotile spores (figure 24.15*c* and 24.16). The three to many spores in each chain are often pigmented and can be smooth, hairy, or spiny in texture. *Streptomyces* species are determined by means of a mixture of morphological and physiological characteristics, including the following: the color of the aerial and substrate mycelia, spore arrangement, surface features of individual spores, carbohydrate use, antibiotic production, melanin synthesis, nitrate reduction, and the hydrolysis of urea and hippuric acid.

Streptomycetes are very important, both ecologically and medically. The natural habitat of most streptomycetes is the soil, where they may constitute from 1 to 20% of the culturable population. In fact, the odor of moist earth is largely the result of streptomycete production of volatile substances such as **geosmin.** Streptomycetes play a major role in mineralization. They are flexible nutritionally and can aerobically degrade resistant substances such as pectin, lignin, chitin, keratin, latex, agar, and aromatic compounds. Streptomycetes are best known for their synthesis of a vast array of antibiotics.

Stanley Waksman's discovery that *S. griseus* (**figure 24.17*a***) produces streptomycin was an enormously important contribution to science and public health. Streptomycin was the first drug to effectively combat tuberculosis, and in 1952 Waksman earned the Nobel Prize. In addition, this discovery set off a massive search resulting in the isolation of new *Streptomyces* species that produce other compounds of medicinal importance. In fact, since that time, the streptomycetes have been found to produce over 10,000 bioactive compounds. Hundreds of these natural products are now used in medicine and industry; about two-thirds of the antimicrobial agents used in human and veterinary medicine are derived from the streptomycetes. Examples include amphotericin B, chloramphenicol, erythromycin, neomycin, nystatin, and tetracycline. Some *Streptomyces* species produce more than one antibiotic. Antibiotic-producing bacteria have genes that encode proteins that make them resistant to such compounds. Antimicrobial chemotherapy (chapter 34)

The genome of *Streptomyces coelicolor,* which produces four antibiotics and serves as a model species for research, has been sequenced. At 8.67 Mbp, it is one of the largest procaryotic genomes. Its large number of genes (7,825) no doubt reflects the number of proteins required to undergo a complex life cycle. Many genes are devoted to regulation, with an astonishing 65 predicted RNA polymerase sigma subunits and over 50 two-component regulatory systems. The ability to exploit a variety of soil nutrients is also demonstrated by the presence of a large number of ABC transporters, the Sec protein translocation system, and secreted degradative enzymes. Finally, genes were discovered that are thought to encode an additional 18 secondary metabolites. Protein secretion in procaryotes (section 3.8); Regulation of transcription initiation: Two-component regulatory systems (section 12.2); Global regulatory systems (section 12.5)

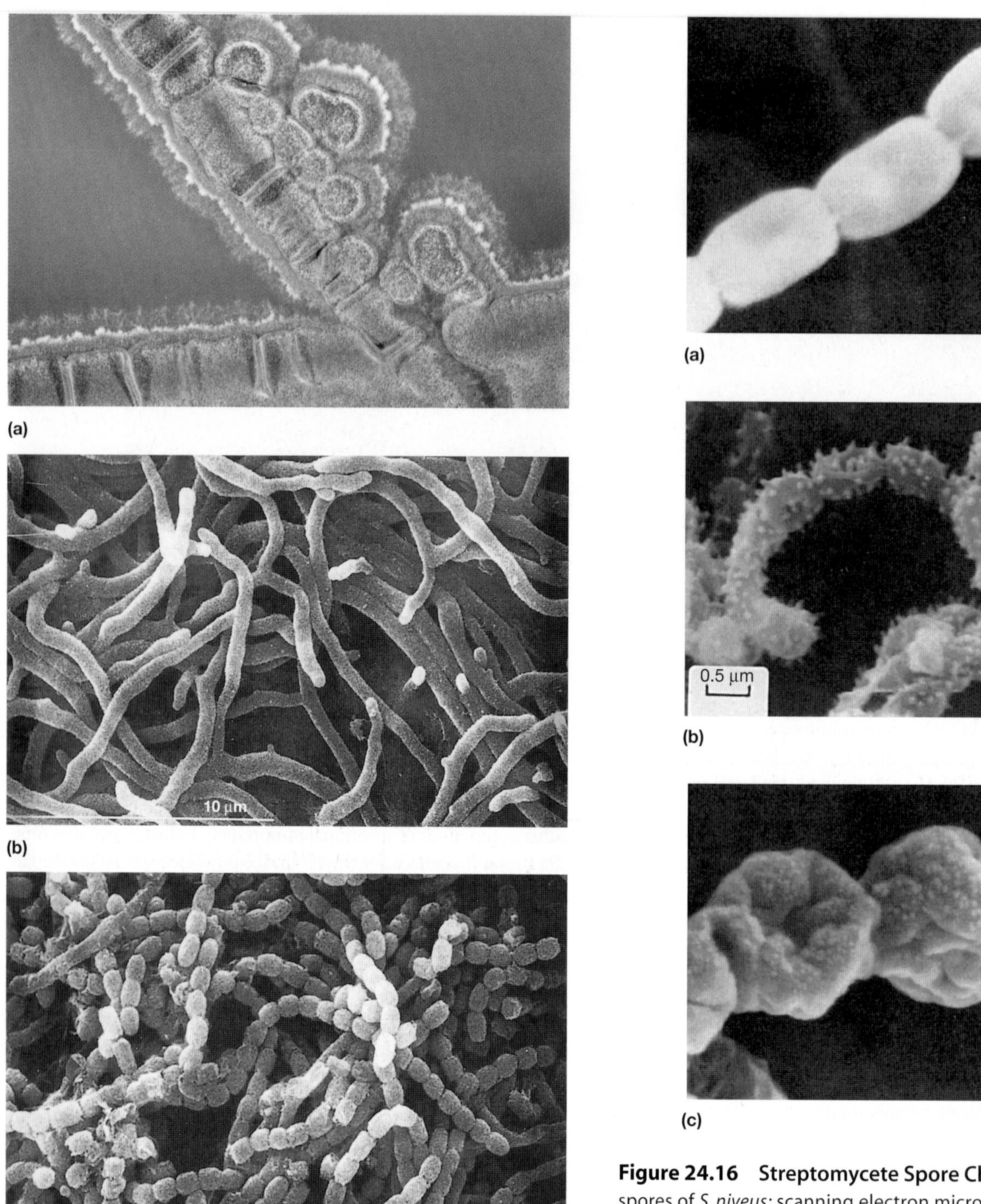

Figure 24.15 Streptomyces Development. **(a)** Growth of *Streptomyces coelicolor* on a solid substrate results in the formation of vegetative hyphae that differentiate into aerial hyphae to form an aerial mycelium, which imparts a white, fuzzy appearance to the colony surface. The blue background is caused by the diffusion of the pigmented polyketide antibiotic actinorhodin into the agar. **(b)** *S. coelicolor* vegetative hyphae are straight with branches. **(c)** Chains of *S. coelicolor* spores that will eventually pinch off and be released into the environment.

Figure 24.16 Streptomycete Spore Chains. **(a)** Smooth spores of *S. niveus;* scanning electron micrograph. **(b)** Spiney spores of *S. viridochromogenes.* **(c)** Warty spores of *S. pulcher.*

Although most streptomycetes are nonpathogenic saprophytes, a few are associated with plant and animal diseases. *Streptomyces scabies* causes scab disease in potatoes and beets (figure 24.17*b*). *S. somaliensis* is the only streptomycete known to be pathogenic to humans. It is associated with **actinomycetoma,** an infection of subcutaneous tissues that produces lesions and leads to swelling, abscesses, and even bone destruction if untreated. *S. albus* and other species have been isolated from patients with various ailments and may be pathogenic.

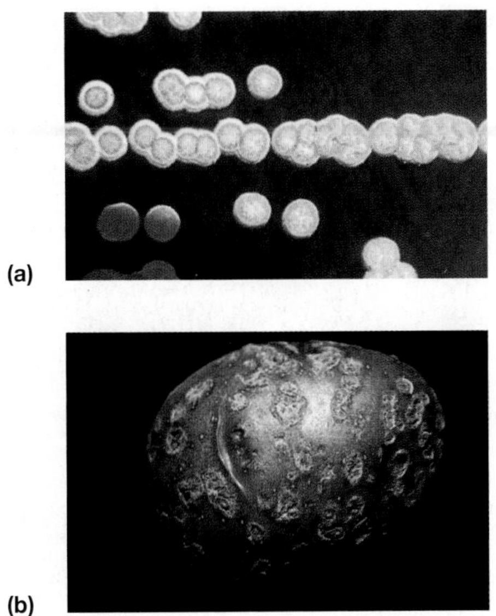

Figure 24.17 Streptomycetes of Practical Importance. **(a)** *Streptomyces griseus.* Colonies of the actinomycete that produces streptomycin. **(b)** *S. scabies* growing on a potato.

24.8 Suborder *Streptosporangineae*

The suborder *Streptosporangineae* contains three families and 16 genera. The family includes the maduromycetes, which have type III cell walls and the sugar derivative **madurose** (3-*O*-methyl-D-galactose) in whole cell homogenates. Their G + C content is 64 to 74 mol%. Aerial hyphae bear pairs or short chains of spores, and the substrate hyphae are branched (**figure 24.18**). Some genera form sporangia; spores are not heat resistant. Like *S. somaliensis, Actinomadura* is another actinomycete associated with the disease actinomycetoma. *Thermomonospora* produces single spores on the aerial mycelium or on both the aerial and substrate mycelia. It has been isolated from moderately high-temperature habitats such as compost piles and hay; it can grow at 40 to 48°C.

24.9 Suborder *Frankineae*

The suborder *Frankineae* includes the genera *Frankia* and *Geodermatophilus*. Both form multilocular sporangia, characterized by clusters of spores when a hypha divides both transversely and longitudinally. (Multilocular means having many cells or compartments.) They have type III cell walls (table 24.1), although the cell extract sugar patterns differ. The G + C content varies from 57 to 75 mol%. *Geodermatophilus* (type IIIC) has motile spores and is an aerobic soil organism. *Frankia* (type IIID) forms nonmotile spores in a sporogenous body (**figure 24.19**). It grows in symbiotic association with the roots of at least eight families of higher nonleguminous plants (e.g., alder trees) and is a microaerophile that can fix atmospheric nitrogen.

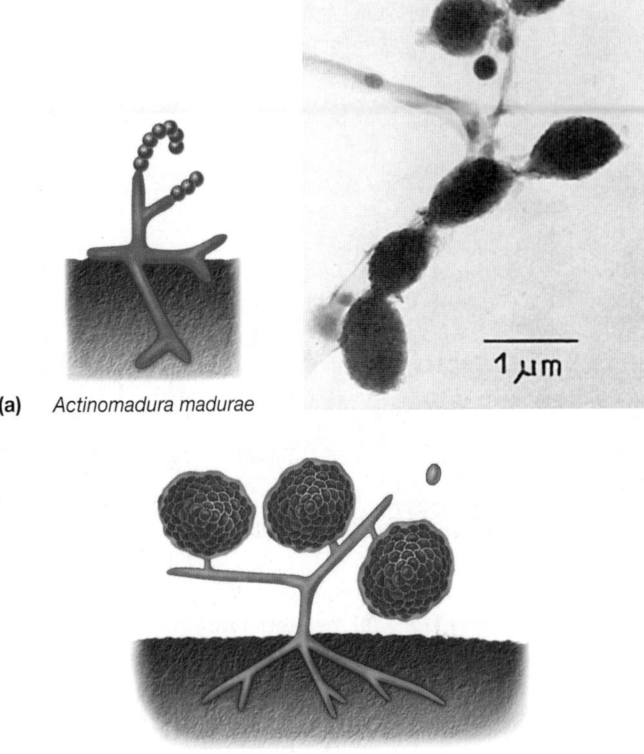

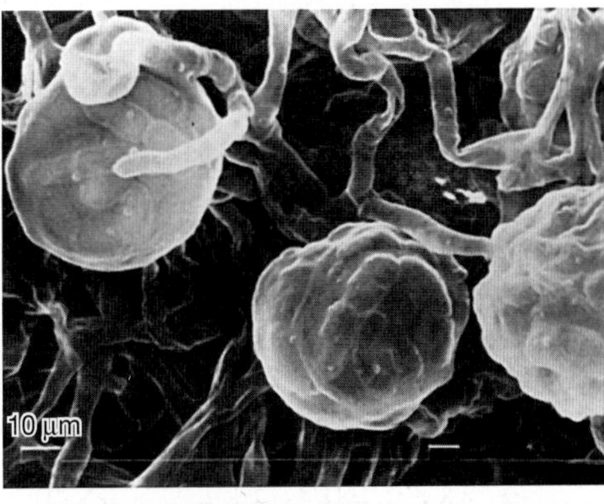

Figure 24.18 Maduromycetes. **(a)** *Actinomadura madurae* morphology; illustration and electron micrograph of a spore chain (×16,500). **(b)** *Streptosporangium* morphology; illustration and micrograph of *S. album* on oatmeal agar with sporangia and hyphae; SEM.

The roots of infected plants develop nodules that fix nitrogen so efficiently that a plant such as an alder can grow in the absence of combined nitrogen (e.g., NO_3^-) when nodulated. Within the nodule cells, *Frankia* forms branching hyphae with globular vesicles at their ends (figure 24.19*c*). These vesicles may be the sites of nitrogen fixation. The nitrogen-fixation process resembles that of *Rhizobium* in that it is oxygen sensitive and requires molybdenum and cobalt. Microorganisms associated with vascular plants: Nitrogen fixation (section 29.5)

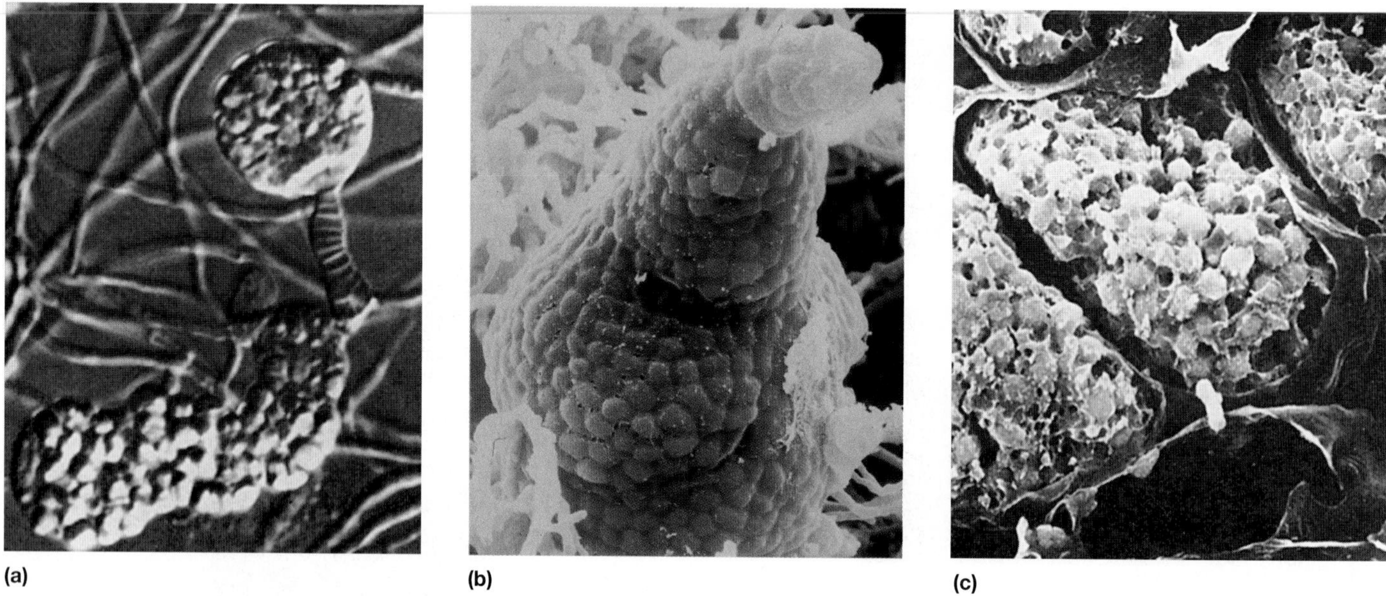

Figure 24.19 ***Frankia.*** **(a)** An interference contrast micrograph showing hyphae, multilocular sporangia, and spores. **(b)** A colorized scanning electron micrograph of a sporangium surrounded by hyphae. **(c)** A nodule of the alder *Alnus rubra* showing cells filled with vesicles of *Frankia.* Scanning electron microscopy.

Another genus in this suborder, *Sporichthya,* is one of the strangest of the actinomycetes. It lacks a substrate mycelium. The hyphae remain attached to the substratum by holdfasts and grow upward to form aerial mycelia that release motile, flagellate spores in the presence of water.

24.10 Order *Bifidobacteriales*

The order *Bifidobacteriales* has one family, *Bifidobacteriaceae,* and 10 genera (five of which have unknown affiliation). *Falcivibrio* and *Gardnerella* are found in the human genital/urinary tract; *Gardnerella* is thought to be a major cause of bacterial vaginitis. *Bifidobacterium* probably is the best-studied genus. Bifidobacteria are nonmotile, nonsporing, gram-positive rods of varied shapes that are slightly curved and clubbed; often they are branched (**figure 24.20**). The rods can be single or in clusters and V-shaped pairs. *Bifidobacterium* is anaerobic and actively ferments carbohydrates to produce acetic and lactic acids, but no carbon dioxide. It is found in the mouth and intestinal tract of warm-blooded vertebrates, in sewage, and in insects. *B. bifidum* is a pioneer colonizer of the human intestinal tract, particularly when babies are breast fed. A few *Bifidobacterium* infections have been reported in humans, but the genus does not appear to be a major cause of disease. Direct contact diseases: Sexually transmitted diseases (section 38.3)

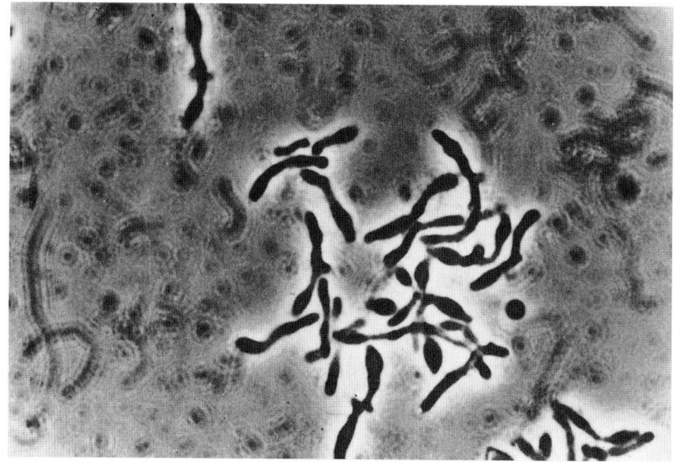

Figure 24.20 ***Bifidobacterium.*** *Bifidobacterium bifidum;* phase-contrast photomicrograph (×1,500).

1. Give the distinguishing properties of the actinoplanetes.
2. Describe the genus *Propionibacterium* and comment on its practical importance.
3. Describe the major properties of the genus *Streptomyces.*
4. Describe three ways in which *Streptomyces* is of ecological importance. Why do you think *Streptomyces* spp. produce antibiotics?
5. Briefly describe the genera of the suborder *Streptosporangineae.* What is madurose? Why is *Actinomadura* important?
6. Describe *Frankia* and discuss its importance.
7. Characterize the genus *Bifidobacterium.* Where is it found and why is it significant?

Summary

24.1 General Properties of the Actinomycetes

a. Actinomycetes are aerobic, gram-positive bacteria that form branching, usually nonfragmenting, hyphae and asexual spores (**figure 24.2**).

b. The asexual spores borne on aerial mycelia are called spores if they are at the tip of hyphae or sporangiospores if they are within sporangia.

c. Actinomycetes have several distinctively different types of cell walls and often also vary in terms of the sugars present in cell extracts. Properties such as color and morphology are also taxonomically useful (**tables 24.1, 24.2,** and **24.3**).

d. The second edition of *Bergey's Manual* classifies the high G + C bacteria phylogenetically using 16S rRNA data. The phylum *Actinobacteria* contains the actinomycetes and their high G + C relatives (**figure 24.4**).

24.2 Suborder *Actinomycineae*

a. The suborder *Actinomycineae* contains the genus *Actinomyces,* members of which are irregularly shaped, nonsporing rods that can cause disease in cattle and humans (**figure 24.6**).

24.3 Suborder *Micrococcineae*

a. The suborder *Micrococcineae* includes the genera *Micrococcus, Arthrobacter,* and *Dermatophilus. Arthrobacter* has an unusual rod-coccus growth cycle and carries out snapping division (**figures 24.7** and **24.8**).

24.4 Suborder *Corynebacterineae*

a. The genera *Corynebacterium, Mycobacterium,* and *Nocardia* are placed in the suborder *Corynebacterineae.* Mycobacteria form either rods or filaments that readily fragment. Their cell walls have a high lipid content and mycolic acids; the presence of these lipids makes them acid-fast (**figures 24.9–24.11**). The genera *Corynebacterium* and *Mycobacterium* contain several very important human pathogens.

b. Nocardioform actinomycetes have hyphae that readily fragment into rods and coccoid elements, and often form aerial mycelia with spores (**figure 24.12**).

24.5 Suborder *Micromonosporineae*

a. The suborder *Micromonosporineae* has genera that include *Micromonospora* and *Actinoplanes.* These actinomycetes have an extensive substrate mycelium and form special aerial sporangia (**figure 24.14**). They are present in soil, freshwater, and the ocean. The soil forms are probably important in decomposition.

24.6 Suborder *Propionibacterineae*

a. The genus *Propionibacterium* is in the suborder *Propionibacterineae*. Members of this genus are common skin and intestinal inhabitants and are important in cheese manufacture and the development of acne vulgaris.

24.7 Suborder *Streptomycineae*

a. The suborder *Streptomycineae* includes the genus *Streptomyces.* Members of this genus have type I cell walls and aerial hyphae bearing chains of 3 to 50 or more nonmotile spores (**figures 24.15** and **24.16**).

b. Streptomycetes are important in the degradation of more resistant organic material in the soil and produce many useful antibiotics. A few cause diseases in plants and animals.

24.8 Suborder *Streptosporangineae*

a. Many genera in suborder *Streptosporangineae* have the sugar derivative madurose and type III cell walls.

24.9 Suborder *Frankineae*

a. The genera *Frankia* and *Geodermatophilus* are placed in the suborder *Frankineae.* They produce clusters of spores at hyphal tips and have type III cell walls. *Frankia* grows in symbiotic association with nonleguminous plants and fixes nitrogen (**figure 24.19**).

24.10 Order *Bifidobacteriales*

a. The genus *Bifidobacterium* is placed in the order *Bifidobacteriales*. This irregular, anaerobic rod is one of the first colonizers of the intestinal tract in nursing babies.

Key Terms

acid-fast 596
actinobacteria 593
actinomycete 589
actinomycetoma 600
aerial mycelium 589
geosmin 599
hypha(e) 589
madurose 601
mycolic acids 596
nocardioforms 596
secondary metabolite 589
snapping division 594
sporangiospores 590
streptomycetes 599
substrate mycelium 589

Critical Thinking Questions

1. Even though these are "high G + C" organisms, there are regions of the genome that must be more AT-rich. Suggest a few such regions and explain why they must be more AT-rich.

2. Choose two different species in the phylum *Actinobacteria* and investigate their physiology and ecology. Compare and contrast these two organisms. Can you determine why having a high G + C genomic content might confer an evolutionary advantage?

3. *Streptomyces coelicolor* is studied as a model system for cellular differentiation. Some of the genes involved in sporulation contain a rare codon not used in vegetative genes. Suggest how *Streptomyces* might use the rare codon to regulate sporulation.

4. Suppose that you discovered a nodulated plant that could fix atmospheric nitrogen. How might you show that a bacterial symbiont was involved and that *Frankia* rather than *Rhizobium* was responsible?

Learn More

Anderson, A. S., and Wellington, E. M. H. 2001. The taxonomy of *Streptomyces* and related genera. *Int. J. Syst. Evol. Microbiol.* 51: 797–814.

Beaman, B. L.; Saubolle, M. A.; and Wallace R. J. 1995. *Nocardia, Rhodococcus, Streptomyces, Oerskovia,* and other aerobic actinomycetes of medical importance. In *Manual of Clinical Microbiology,* 6th ed., P. R. Murray, editor-in-chief, 379–99. Washington, D.C.: American Society for Microbiology.

Benson, D. R., and Silvester, W. B. 1993. Biology of *Frankia* strains, actinomycete symbionts of actinorhizal plants. *Microbiol. Rev.* 57(2): 293–319.

Bentley, S. D.; Chater, K. F.; Cerdeño-Tárraga, A.-M.; Challis, G. L.; Tomson, N. R.; James, K. D.; Harris, D. E.; Quail, M. A.; Kieser, H.; Harper, D.; Bateman, A.; Brown, S.; Chandra, G.; Chen, C. W.; Collins, M.; Cronin, A.; Fraser, A.; Goble, A.; Hidalgo, J.; Hornsby, T.; Howarth, S.; Huang, C.-H.; Kieser, T.; Larke, L.; Murphy, L.; Oliver, K.; O'Neil, S.; Rabbinowitsch, E.; Rajandream, M.-A.; Rutherford, K.; Rutter, S.; Segger, K.; Saunders, D.; Sharp, S.; Squares, R.; Squares, S.; Taylor, K.; Warren, T.; Wietzorrek, A.; Woodward, J.; Barrell, B. G.; Parkhill, J.; and Hopwood, D. A. 2002. Complete genome sequence of the model actinomycete *Streptomyces coelicolor* A3(2). *Nature* 417: 141–47.

Clawson, M. L.; Bourret, A.; and Benson, D. R. 2004. Assessing the phylogeny of *Frankia*-actinorhizal plant nitrogen-fixing root nodule symbioses with *Frankia* 16S rRNA and glutamine synthetase gene sequences. *Mol. Phylogen. Evol.* 31: 131–38.

Dyson, P. 2000. *Streptomyces* genetics. In *Encyclopedia of microbiology,* 2d ed., vol. 4, J. Lederberg, editor-in-chief, 451–66. San Diego: Academic Press.

Parenti, F., and Coronelli, C. 1979. Members of the genus *Actinoplanes* and their antibiotics. *Annu. Rev. Microbiol.* 33:389–411.

Stackebrandt, E.; Rainey, F. A.; and Ward-Rainey, N. L. 1997. Proposal for a new hierarchic classification system. *Actinobacteria* classis nov. *Int. J. Syst. Bacteriol.* 47(2):479–91.

Please visit the Prescott website at www.mhhe.com/prescott7 for additional references.

---

25 The Protists

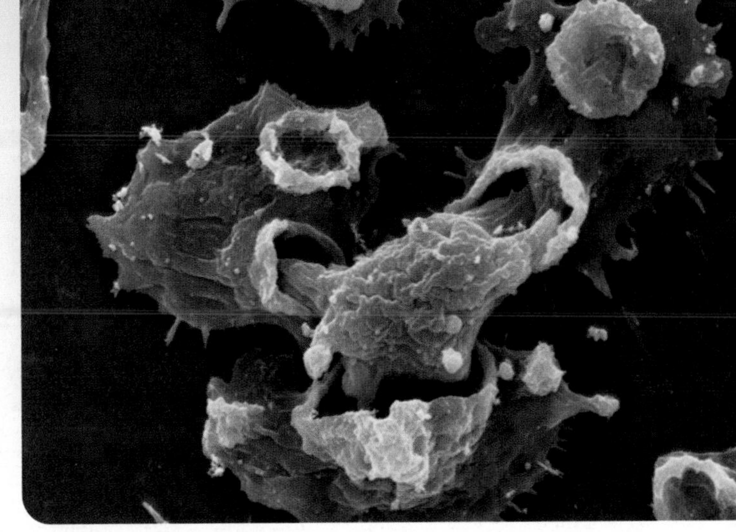

This is a scanning electron micrograph (×2,160) of the protozan *Naegleria fowleri*. Three *N. fowleri* from an axenic culture, attacking and beginning to devour or engulf a fourth, presumably dead amoeba, with their amoebastomes (suckerlike structures that function in phagocytosis). This amoeba is the major cause of the disease in humans called primary amebic meningoencephalitis.

Preview

- Protists are a polyphyletic collection of organisms. Most are unicellular and lack the level of tissue organization seen in higher eucaryotes.
- Protists are ubiquitous—they are found wherever there is adequate moisture. They are prominent members of marine planktonic communities where they contribute to both the amount of fixed CO_2 as well as the recycling of nutrients. Aquatic and terrestrial protists are also important in nutrient regeneration.
- Protists can be photoautotrophic, chemoorganotrophic, or mixotrophic. They can assimilate organic nutrients by saprotrophy or holotrophy, feeding on bacteria and other particulate forms of carbon. Many protists live in association with other organisms as mutualists, commensals, or parasites.
- Protists usually reproduce asexually by binary fission; some undergo multiple fission or budding. Many also have sexual cycles that involve meiosis and the fusion of gametes or gametic nuclei, resulting in a diploid zygote. The zygote is often a thick-walled, resistant, resting cell called a cyst.
- All protists have one or more nuclei; some have a macronucleus and a micronucleus. Energy metabolism occurs in mitochondria, hydrogenosomes, and chloroplasts, although some protists lack these organelles.
- The systematics and taxonomic classification of the protists are areas of active research and debate.

Chapter 25 presents the major biological features of the protists. We introduce protists of medical and ecological significance and follow the higher-level classification scheme of eucaryotes as proposed by the International Society of Protistologists. The kingdom *Protista,* as defined by Whittaker's five-kingdom scheme, is an artificial grouping of over 64,000 different single-celled life forms that lack common evolutionary heritage; that is to say, they are polyphyletic (**figure 25.1**). In fact, the protists are unified only by what they lack: absent is the level of tissue organization found in fungi, plants, and animals. The term **protozoa** [s., protozoan; Greek *protos,* first, and *zoon,* animal] has traditionally referred to chemoorganotrophic protists, and **protozoology** generally refers to the study of protozoa. The term **algae** can be used to describe photosynthetic protists. "Algae" was originally used to refer to all "simple aquatic plants," but this term has no phylogenetic utility. The study of photosynthetic protists (algae) is often referred to as **phycology** and is the realm of both botanists and protistologists. The study of all protists, regardless of their metabolic type, is called **protistology.**

For many years the protozoa were classified into four major groups based on their means of locomotion: flagellates (*Mastigophora*), ciliates (*Infusoria* or *Ciliophora*), amoebae (*Sarcodina*), and stationary forms (*Sporozoa*). Although nonprotistologists still use these terms, these divisions have no bearing on evolutionary relationships and should be avoided. It is now agreed that the old classification system is best abandoned, but for many years there was little agreement on what should take its place. Recent morphological, biochemical, and phylogenetic analyses have resulted in the development of a higher-level classification system for the eucaryotes. This scheme, as proposed by the International Society of Protistologists, is introduced in chapter 19 (*see table 19.9*) and is followed here. However, first we introduce some major ecological, morphological, and physiological elements that describe most protists.

And a pleasant sight they are indeed. Their shapes range from teardrops to bells, barrels, cups, cornucopias, stars, snowflakes, and radiating suns, to the common amoebas, which have no real shape at all. Some live in baskets that look as if they were fashioned of exquisitely carved ivory filigree. Others use colored bits of silica to make themselves bright mosaic domes. Some even form graceful transparent containers shaped like vases or wine glasses of fine crystal in which they make their homes.

—Helena Curtis

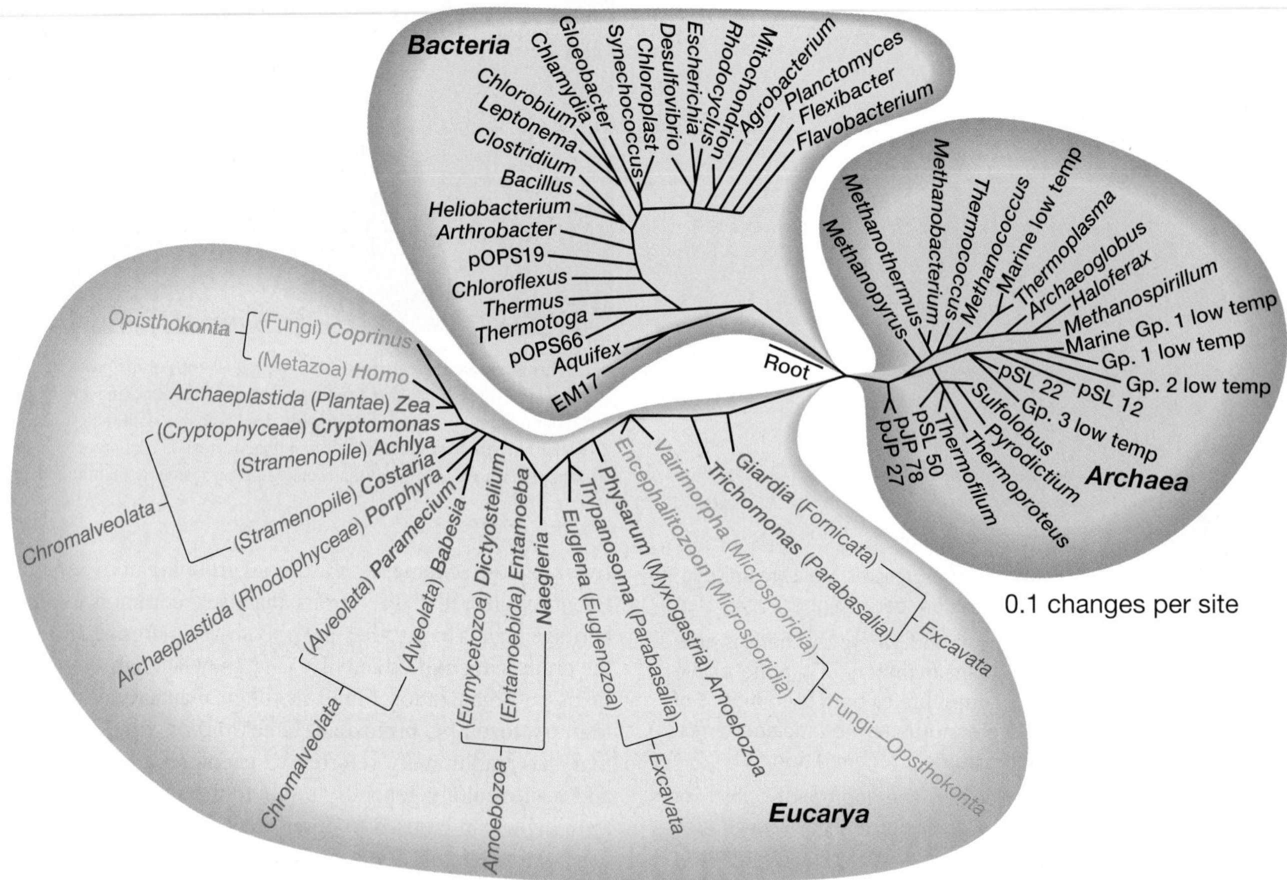

Figure 25.1 Universal Phylogenetic Tree, Highlighting Protists. Recent molecular phylogenetic evidence in combination with morphological and biochemical data has resulted in the establishment of super groups among the *Eucarya* (each super group is highlighted with a different color). This new, higher-order classification scheme is not entirely congruent with the placement of the protists on the Universal Tree of Life; the latter is based entirely on analysis of SSU rRNA. Nonetheless, it demonstrates that the protists are a highly polyphyletic group. Because the protists are not a monophyletic group, the term "protist" cannot be used to represent evolutionary histories. Instead the term "protist" as used in this chapter denotes a group of eucaryotic organisms that share some morphological, reproductive, ecological, and biochemical characteristics.

25.1 Distribution

Protists grow in a wide variety of moist habitats. Moisture is absolutely necessary for their existence because they are susceptible to desiccation. Most protists are free living and inhabit freshwater or marine environments. Many terrestrial chemoorganotrophic forms can be found in decaying organic matter and in soil, where they are important in recycling the essential elements nitrogen and phosphorus. Others are **planktonic**—floating free in lakes and oceans. Planktonic microbes (both procaryotic and eucaryotic) are responsible for a majority of the nutrient cycling that occurs in these ecosystems.

Every major group of protists includes species that live in association with other organisms. For instance, some photosynthetic protists associate with fungi to form lichens while others live with corals where they provide fixed carbon to the coral animal. Thousands of others are parasites and cause important diseases in humans and domesticated animals (**table 25.1**). Finally, protists are useful in biochemical and molecular biological research. Not only do they display an amazing array of unique adaptations, but many biochemical pathways used by protists are found in other eucaryotes, making them useful model organisms.

25.2 Nutrition

Photosynthetic protists are exclusively aerobic. Like cyanobacteria and plants, they possess both photosystems I and II and perform oxygenic photosynthesis. Most photosynthetic forms are photoautotrophic, obtaining energy from light and fixing CO_2 to meet their carbon requirements. However, some are photoheterotrophic, using organic carbon rather than CO_2. Chemoheterotrophic protists can be holozoic or saprozoic. In **holozoic nutrition,** solid nutrients such as bacteria are acquired by phago-

Table 25.1 Pathogenic Protists That Cause Major Diseases of Domestic Animals

Super Group (subrank)	Genus	Host	Preferred Site of Infection	Disease
Amoebozoa	*Entamoeba*	Mammals	Intestine	Amebiasis
	Iodamoeba	Swine	Intestine	Enteritis
Chromalveolata	*Babesia*	Cattle	Blood cells	Babesiosis
(Apicomplexa)	*Therileria*	Cattle, sheep, goats	Blood cells	Therileriasis
	Sarcocystis	Mammals, birds	Muscles	Sarcosporidiosis
	Taxoplasma	Cats	Intestine	Toxoplasmosis
	Isospora	Dogs	Intestine	Coccidiosis
	Eimeria	Cattle, cats, chickens, swine	Intestine	Coccidiosis
	Plasmodium	Many animals	Red blood cells, liver	Malaria
	Leucocytozoon	Birds	Spleen, lungs, blood	Leucocytozoonosis
	Cryptosporidium	Mammals	Intestine	Cryptosporidiosis
(Ciliophora)	*Balantidium*	Swine	Large intestine	Balantidiasis
Excavata (Kinetoplasta)	*Leishmania*	Dogs, cats, horses, sheep, cattle	Spleen, bone marrow, mucous membranes	Leishmaniasis
	Trypanosoma	Most animals	Blood	Chagas disease Sleeping sickness
(Parabasalia)	*Trichomonas*	Horses, cattle	Genital tract	Trichmoniasis (abortion)
	Histomonas	Birds	Intestine	Blackhead disease
(Fornicata)	*Giardia*	Mammals	Intestine	Giardiasis

cytosis and the subsequent formation of phagocytic vacuoles (**figure 25.2**). In **saprozoic nutrition,** soluble nutrients such as amino acids and sugars cross the plasma membrane by endocytosis, diffusion, or carrier-mediated transport (facilitated diffusion or active transport). The mechanisms by which soluble nutrients are assimilated are sometimes collectively called **osmotrophy.**

It is difficult to classify the nutritional strategies of some protists. Certain protists can derive energy through the oxidation of inorganic substrates but assimilate organic carbon compounds. Some simultaneously use both organic and inorganic forms of carbon; this is sometimes referred to as **mixotrophy.** This metabolic flexibility is clearly evident in those forms like the dinoflagellates that can perform photosynthesis and holozoic feeding concurrently.

25.3 Morphology

Before we describe specific morphological features, it is helpful to define the terms amoeba and ciliate. Older, now discredited classification schemes placed all amoeboid and ciliated protists in the "*Sarcodina*" and "*Infusoria,*" respectively. While these terms may remain useful in describing protists that demonstrate amoeboid and ciliated motility, they do not describe monophyletic groups. Thus in this chapter, "amoebae" and "ciliates" are not taxonomic terms—rather, they are adjectives, much like we might describe flagellated archaea or vibroid bacteria.

Because protists are eucaryotic cells, in many respects their morphology and physiology are the same as the cells of multicellular plants and animals. However, because all of life's various functions must be performed within a single cell, complexity arises at the level of specialized organelles rather than at the tissue level. Protists must remain relatively small because without multicellularity to facilitate the exchange of nutrients and metabolites, they need a high ratio of cell surface to intracellular volume. This is because the distance from the cell membrane to the center of the cell cannot exceed the distance a molecule can diffuse. Even the largest algae (the seaweeds) are long, thin, and often flattened—these are shapes that maximize the surface-to-volume ratio.

The protist cell membrane, called the **plasmalemma,** is identical to that of multicellular organisms. In some protists the cytoplasm immediately under the plasmalemma is divided into an outer gelatinous region called the **ectoplasm,** and an inner fluid region, the **endoplasm.** The ectoplasm imparts rigidity to the cell body. The plasmalemma and structures immediately beneath it are called the **pellicle.** One or more vacuoles are usually present in the cytoplasm of protozoa. These are differentiated into contractile, secretory, and food vacuoles. **Contractile vacuoles** function as osmoregulatory organelles in those protists that live in hypotonic environments, such as freshwater lakes. Osmotic balance is maintained by continuous water expulsion. Most marine and parasitic species are isotonic to their environment and lack

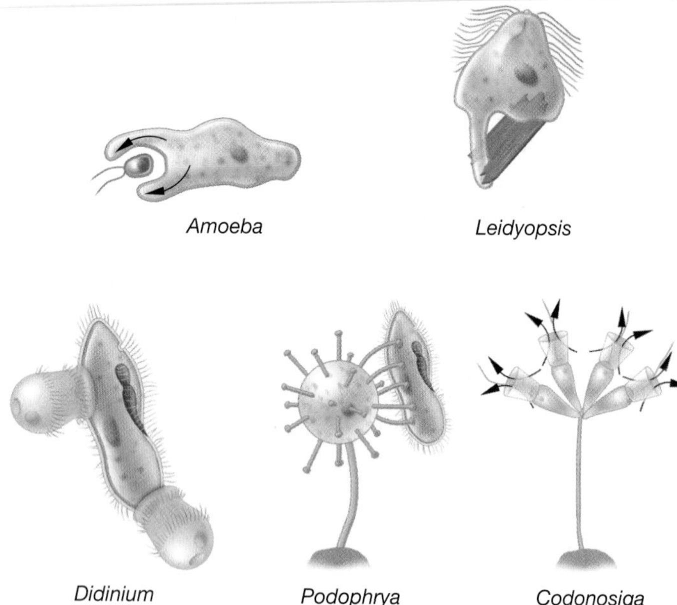

Figure 25.2 Holozoic Feeding Methods Among the Protists. *Amoeba* uses pseudopodia to surround bacteria. *Leidyopsis* lives in the guts of termites, where it uses pseudopodia to entrap wood particles. The ciliate *Didinium* eats only *Paramecium* (a larger ciliate) through a temporary cytostome. *Podophyra* uses tentacles to first attach to its prey and then siphon prey cytoplasm into its body where it is digested in food vacuoles. The sessile protist *Codonosiga* has a collar of microvilli; flagella help sweep food particles suspended in water into the collar. Despite this variety, all of these feeding methods are considered forms of pseudopodia.

such vacuoles. **Phagocytic vacuoles** are conspicuous in holozoic and parasitic species and are the sites of food digestion. In some organisms, they may occur anywhere on the cell surface, while others have a specialized structure for phagocytosis called the **cytostome** (cell mouth). When digestion commences, the phagocytic vacuole is acidic, but as it proceeds, the vacuolar pH increases and the membrane forms small blebs. These pinch off and carry nutrients throughout the cytoplasm. The undigested contents of the original phagocytic vacuole are expelled from the cell either at a random site on the cell membrane (as is the case with amoebae) or at a designated position called the **cytoproct.** Organelles of the biosynthetic-secretory and endocytic pathways (section 4.4)

Several energy-conserving organelles are observed in protists. Most aerobic chemoorganotrophic forms have mitochondria with cristae that are characterized by their physical appearance: they can be discoid, tubular, or lamellar; cristae morphology is of taxonomic value. The majority of anaerobic protists (such as *Trichonympha,* which lives in the gut of termites) lack mitochondria and cytochromes, and have an incomplete tricarboxylic acid cycle. However, some have small, membrane-bound organelles termed **hydrogenosomes.** Within these organelles, pyruvate derived from glycolysis is oxidized and decarboxylated to form CO_2, H_2, and the high-energy molecule acetyl-CoA. The conversion of each acetyl-CoA to acetate yields an ATP (*see figure 19.4*). Because hydrogenosomes lack a complete tricarboxylic acid cycle, acetate must be excreted. Fermentation end products may be consumed by symbiotic procaryotes living within the protist. In some cases, methanogenic archaea consume the CO_2 and H_2 and generate methane. Recent evidence suggests that hydrogenosomes and mitochondria evolved from the same endosymbiotically derived organelle. Photosynthetic protists have chloroplasts featuring thylakoid membranes. A dense proteinaceous area, the **pyrenoid,** which is associated with the synthesis and storage of starch, may be present in the chloroplasts. Microbial evolution (section 19.1)

Many protists feature cilia or flagella at some point in their life cycle. Their formation is associated with a basal bodylike organelle called the **kinetosome,** which is similar in structure to the centriole. In addition to aiding in motility, these organelles may be used to generate water currents for feeding and respiration. Because there is no precise morphological difference between flagella and cilia, some scientists prefer to call them both undulipodia (meaning "wave foot"). Cilia and flagella (section 4.10)

1. How do the terms protozoa and algae differ from the term protist?
2. In what habitats can protists be found?
3. What roles do protists play in the trophic structure of their communities and in the organisms with which they associate?
4. What is a hydrogenosome? How does it differ from a mitochondrion?
5. Trace the path of a food item from the phagocytic vacuole to the cytoproct.

25.4 Encystment and Excystment

Many protists are capable of **encystment.** During encystment, the organism de-differentiates (becomes simpler in morphology) and develops into a resting stage called a **cyst.** The cyst is a dormant form marked by the presence of a cell wall and by very low metabolic activity. Cyst formation is particularly common among aquatic, free-living protists and parasitic forms. Cysts serve three major functions: (1) they protect against adverse changes in the environment, such as nutrient deficiency, desiccation, adverse pH, and low partial pressure of O_2; (2) they are sites for nuclear reorganization and cell division (reproductive cysts); and (3) they serve as a means of transfer between hosts in parasitic species. Although the exact stimulus for **excystment** (escape from the cysts) is unknown, it is generally triggered by a return to favorable environmental conditions. For example, cysts of parasitic species excyst after ingestion by the host and form the vegetative form called the **trophozoite.**

25.5 Reproduction

Most protists reproduce asexually and some also carry out sexual reproduction. The most common method of asexual reproduction is **binary fission.** During this process the nucleus first undergoes mitosis, then the cytoplasm divides by cytokinesis to form two identical individuals (**figure 25.3**). Multiple fission is also com-

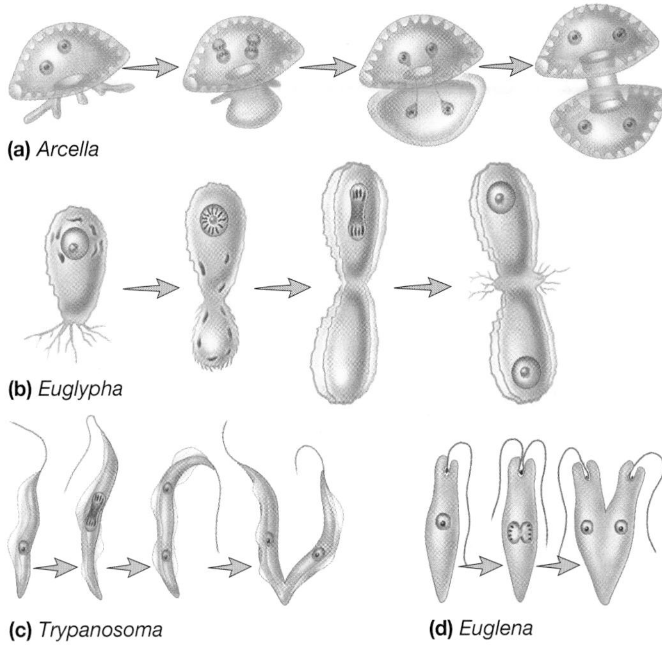

Figure 25.3 Binary Fission in Protists. (a) The two nuclei of the testate (shelled) amoeba *Arcella* divide as some of its cytoplasm is extruded and a new test for the daughter cell is secreted. **(b)** In another testate amoeba, *Euglypha*, secretion of new platelets is begun before cytoplasm begins to move out of the aperture. The nucleus divides while the platelets are used to construct the test for the daughter cell. For many protists, all organelles must be replicated before the cell divides. This is the case with **(c)** *Trypanosoma* and **(d)** *Euglena.*

mon, as is budding. Some filamentous, photosynthetic protists undergo fragmentation so that each piece of the broken filament grows independently.

Most protists also undergo sexual reproduction at some point in their life cycle. Protist cells that produce gametes are termed **gamonts.** The fusion of haploid gametes is called **syngamy.** Among protists, syngamy can involve the fusion of two morphologically similar gametes (**isogamy**) or the fusion of morphologically different types (**anisogamy**). Meiosis may occur either before the formation and union of gametes, as in most animals, or just after fertilization, as is also the case with lower plants. Furthermore, the exchange of nuclear material may occur in the familiar fashion—between two different individuals (**conjugation**)—or by the development of a genetically distinct nucleus within a single individual (**autogamy**).

With this level of reproductive complexity, perhaps it is not surprising that the nuclei among protists show considerable diversity. Most commonly, a **vesicular nucleus** is present. This is characterized by a nucleus that is 1 to 10 μm in diameter, spherical, with a distinct nucleolus and uncondensed chromosomes. **Ovular nuclei** are up to 10 times this size and possess many peripheral nucleoli. Still others have **chromosomal nuclei,** in which the chromosomes remain condensed throughout interphase with a single nucleolus associated with one chromosome. Finally, the *Ciliophora* have two types of nuclei: a large **macronucleus** with distinct nucleoli and condensed chromatin and a smaller, diploid **micronucleus** with dispersed chromatin but lacking nucleoli. Macronuclei are engaged in trophic activities and regeneration processes while micronuclei are involved only in genetic recombination during sexual reproduction and the regeneration of the macronucleus.

1. What functions do cysts serve for a typical protist? What causes excystment to occur?
2. What is a gamont? What is the difference between anisogamy and syngamy?
3. How does conjugation differ from autogamy?
4. Describe vesicular, ovular, and chromosomal nuclei. Which is most like the nuclei found in higher eucaryotes?

25.6 Protist Classification

Ever since Antony van Leeuwenhoek described the first protozoan "animalcule" in 1674, the taxonomic classification of these microbes has remained in flux. During the 20th century, classification schemes based on functional morphology rather than evolutionary relationships were used. The application of molecular techniques is providing new insights into protist systematics and, in most cases, modern morphological and biochemical analyses are in agreement with molecular phylogenetic data. The 2005 International Society of Protistologists' publication of a higher-level classification scheme of all eucaryotes with an emphasis on the protists provides a long-needed consensus (**table 25.2**). Note that this scheme does not use formal hierarchical rank designations such as class and order, reflecting the fact that protist taxonomy remains an area of active research.

Super Group *Excavata*

The *Excavata* includes some of the most primitive, or deeply branching, eucaryotes (figure 25.1). Most possess a cytostome characterized by a suspension-feeding groove. This apparatus features a posteriorly directed flagellum to generate a current that enables the capture of small particles from a feeding current. Those that lack this morphological feature are presumed to have had it at one time during their evolution—that is to say, it is thought to have been secondarily lost.

Fornicata

Ever curious, Anton van Leeuwenhoek described *Giardia intestinalis* (**figure 25.4**) from his own diarrheic feces. Over 300 years later, this species continues to be a public health concern. Today it most often infects campers and other individuals who unwittingly consume contaminated water. Members of the *Fornicata* bear flagella and lack mitochondria. Unlike *Giardia,* most are harmless symbionts. The few free-living forms are most often found in waters that are heavily polluted with organic nutrients (a

Table 25.2 Classification of the Protists as Proposed by the International Society of Protistologists[a]

Super Group	Unifying Features	First Rank	General Description
Opisthokonta	During at least one stage of life cycle, cells have single posterior cilium without mastigonemes (hair-like projections on the flagella); possess kinetosomes or centrioles; when unicellular, mitochondria have flat cristae	*Mesomycetozoa*	Most have mitochondria with flat cristae; one or more life cycle stages feature spherical cells; one posterior flagellum or amoeboid; some have parasitic, nonflagellated stages and endospores; some have trophic stages that feature a cell wall. Includes *Aphelidium, Dermocystidium, Ichthyophonus,* and *Nuclearia.*
		Choanomonada	Radially symmetric; phagotrophic with collar of microvilli encircling a single flagellum; mitochondria have flat cristae; solitary or colonial. Includes *Monosiga, Salpingoeca,* and *Stenphanoeca.*
Archaeplastids	Photosynthetic plastid with chlorophyll *a* from ancestral cyanobacterial endosymbiont, plastid later lost in some; uses starch as a storage product; usually have cell wall made of cellulose	*Glaucophyta*	Plastid is a cyanelle, which unlike chloroplasts has peptidoglycan between the two membranes; stacked thylakoids with chlorophyll *a* and phycobiliproteins and other pigments; includes flagellated and nonflagellated species or life cycle stages. Includes *Cyanophora, Glaucocystis,* and *Gloeochaeta.*
		Rhodophycae	Lack flagellated stages, kinetosome, and centrioles, unstacked thylakoids; chloroplast without external endoplasmic reticulum; also called red algae, although traditional subgroups are no longer considered valid. Includes *Ceramium, Porphyra,* and *Sphaerococcus.*
		Chloroplastida	Pyrenoid often within plastid, which features chlorophylls *a* and *b;* cellulose-containing cell wall typical. Includes the *Charophyta* (green algae), *Chara, Nitella,* and *Volvox.*
Amoebozoa	Amoeboid motility with lobopodia; naked or testate; mitochondria with tubular cristae; usually uninucleate, sometimes multinucleate; cysts common	*Tubulinea*	Naked or testate with tubular pseudopodia; lacks centrosomes; motility based on actinomyosin cytoskeleton; cytoplasmic microtubules rare; lacks flagellated stages. Includes *Amoeba, Hydramoeba, Flabellula,* and *Rhizamoeba.*
		Flabellinea	Flattened amoebae lacking tubular pseudopodia; motility based on actinomyosin cytoskeleton; lacks centrosome and flagellated stages. Includes *Dermamoeba, Podostoma, Sappinia,* and *Thecamoeba.*
		Stereomyxida	Branched or reticulate plasmodial organisms; trilaminate centrosome. Includes *Corallomyxe* and *Stereomyxa.*
		Acanthamoebidae	Thin glycocalyx; prominent subpseudopodia that narrow to a blunt or fine tip (acanthopodia); single nucleus; cysts usually have a double wall; motility based on actinomyosin cytoskeleton; centriole-like body observed. Includes *Acanthamoeba, Balamuthia,* and *Protacanthamoeba.*
		Entamoebida	No flagella or centrioles; lacks mitochondria, hydrogenosomes, and peroxisomes. Includes *Entamoeba.*
		Mastigamoebidae	Amoeboid with several pseudopodia; single flagellum projecting forward; single kinetosome and nucleus, although some multinucleate; lacks mitochondria; inhabits low-oxygen to anoxic, nutrient-rich environments. Includes *Mastigella* and *Mastigamoeba.*
		Pelomyxa	Multiple cilia; anaerobic; lack mitochondria, hydrogenosomes, and peroxisomes; polymorphic life cycle with multinucleate stages; some are symbionts. Includes *Pelomyxa paulstris.*
		Eumycetozoa	Amoeboid organisms that produce fruiting bodylike structures; both cellular and acellular slime molds. Includes *Dictyostelium, Physarum, Hemitichia,* and *Stemonitis.*
Rhizaria	Possess thin pseudopodia (filopodia)	*Cercozoa*	Biciliated and/or amoeboid; most have mitochondria with tubular cristae; many encyst; kinetosomes connected to nucleus with cytoskeleton. Includes *Cercomonas, Katabia, Medusetta,* and *Sagosphaera.*

		Haplosporidia	Plasmodial endoparasites of marine and freshwater animals; distinct mechanism of spore formation; mitochondria with tubular cristae. Includes *Haplosporidium, Minchinia,* and *Urosporidum.*
		Foraminifera	Filopodia with granular cytoplasm that forms a complex network of reticulopodia. Simplest forms are open tubes or hollow spheres; in others, shells, called tests, are divided into chambers added during growth. Tests made of organic compounds or inorganic particles cemented together, or crystalline calcite. Includes *Allogromia, Carpenteria, Globigerinella, Lana,* and *Textularia.*
		Gromia	Organic tests; branched filipodia; nongranular cytoplasm; flagellated dispersal cells or gametes. Includes *Gromia.*
		Radiolaria	Many species exhibit radial symmetry, from which the name is derived. All have a porous capsular cell wall through which axopodia project. Skeletons, when present, made of amorphous silica (opal) or strontium sulfate; morphology can be simple to ornate; mitochondria with tubular cristae; axopodia supported by internal microtubules. Includes *Acanthometra, Lophospyris,* and *Staurocon.*
Chromalveolata	Plastid from secondary endosymbiosis with an ancestral archaeplastid; plastid then lost in some; reacquired in others	*Cryptophyceae*	Auto-, mixo-, and heterotrophic forms with ejectosomes or trichocysts (dartlike structures used for defense); mitochondria with flat cristae; biflagellated; tubular channels and/or longitudinal groove lined with ejectosomes; chlorophyll *a* and c_2 if chloroplasts present. Includes *Campylomonas, Cryptomonas, Goniomonas,* and *Rhodomonas.*
		Haptophyta	Scales cover cell; solitary or colonial; motile cells biflagellate and usually have a haptonema (thin appendage between flagella used for prey capture and attachment to substrates); outer nuclear membrane continuous with chloroplast membrane; auto-, mixo-, and heterotrophic forms. Includes the coccoliths and *Diacronema, Isochrysis,* and *Phaeocystis.*
		Stramenopiles	Motile cells usually with two flagella, heterokont flagellation typical. Includes diatoms and labyrinthulids.
		Alveolata	Mixo- or heterotrophic. Includes dinoflagellates, *Apixcomplexa* (e.g., *Plasmodium*), and *Ciliphora* (e.g., *Paramecium* and *Stentor*).
Excavata	Suspension feeding groove (cytostome) present or presumed to have been lost; feed by a flagella-generated current	*Fornicata*	Lack typical mitochondria; uninucleate; usually have a feeding groove. Includes *Giardia.*
		Malawimonas	Has mitochondria, two kinetosomes, and a single ventral flagellar vane. Includes *Malawinas.*
		Parabasalia	Have parabasal structure; striated parabasal fibers connect Golgi to flagellar apparatus; up to thousands of flagella; hydrogenosomes present. Includes *Calonympha, Holomastigotes, Spirotrichosoma,* and *Trichomonas.*
		Preaxostyla	Unicellular; flagellated; no mitochondria; heterotrophic. Includes *Dinenympha, Polymastix,* and *Streblomastix.*
		Jakobida	Two flagella placed at top of wide ventral feeding groove. Includes *Jakoba* and *Histiona.*
		Heterolobosea	Heterotrophic amoebae with eruptive pseudopodia; if flagellated, has 2 or 4 parallel flagella. Includes *Acrasis, Gruberella,* and *Rosculus.*
	Euglenozoa	One or two flagella inserted into apical or subapical pocket; usually with tubular feeding apparatus; two kinetosomes; mitochondria have discoid cristae. Includes *Dinema, Euglena, Leishmania, Trypanoplasma* and *Trypanosoma.*	

[a]Adapted from: Adl, S. M.; Simpson, A. G. B.; Farmer, M. A.; Anderson, R. A.; Anderson, O. R.; Barta, J. R.; Browser, S. S.; *et al.* 2005. The new higher level classification of Eukaryotes wtih emphasis on the taxonomy of protists. *J. Eukaryot. Microbiol.* 52:399–451.

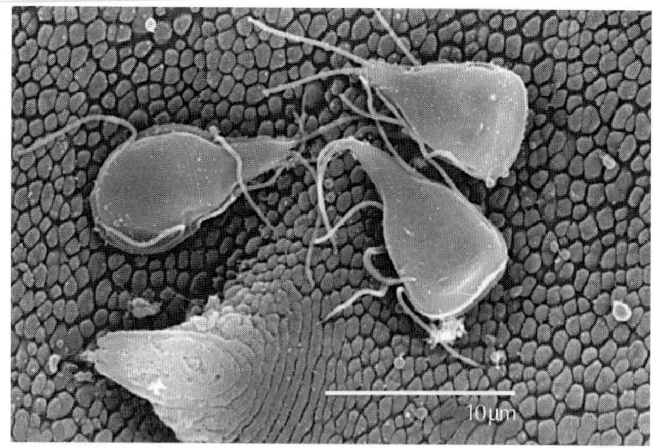

Figure 25.4 Scanning Electron Micrograph of *Giardia* sp. This pathogenic protist is shown attached to the parasitic flatworm *Echinostoma caproni.*

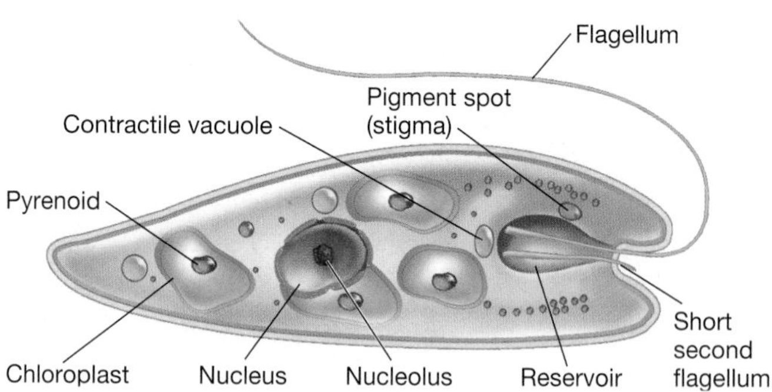

Figure 25.5 *Euglena.* A Diagram Illustrating the Principle Structures Found in this Euglenoid. Notice that a short second flagellum does not emerge from the anterior invagination. In some euglenoids both flagella are emergent.

condition known as eutrophication). Only asexual reproduction by binary fission has been observed. Other pathogenic species include *Hexamida salmonis,* a troublesome fish parasite found in hatcheries and fish farms, and *H. meleagridis,* a turkey pathogen that is responsible for the annual loss of millions of dollars in poultry revenue. Food and waterborne diseases: *Giardiasis* (section 39.5)

Parabasalia

Members of the *Parabasalia* are flagellated; most are endosymbionts of animals. Without a distinct cytostome, they use phagocytosis to engulf food items. Here we consider two subgroups: the *Trichonymphida* and the *Trichomonadida. Trichonymphida* are obligate mutualists in the digestive tracts of wood-eating insects such as termites and wood roaches, where they secrete the enzyme cellulase needed for the digestion of wood. In fact, these protists use pseudopodia to entrap wood particles (figure 25.2). One species, *Trichonympha campanula,* can account for up to one-third of the biomass of an individual termite. This species is particularly large for a protist (several hundred micrometers) and can bear several thousand flagella. Although asexual reproduction is the norm, a hormone called ecdysone produced by the host when molting triggers sexual reproduction.

Trichomonadida, or simply the trichomonads, are symbionts of the digestive, reproductive, and respiratory tracts of many vertebrates, including humans. Four species infect humans: *Dientamoeba fragilis, Pentatrichomonas hominis, Trichomonas tenax,* and *T. vaginalis. D. fragilis* has recently been recognized as a cause of diarrhea, while *T. vaginalis* has long been known to be pathogenic. Found in the genitourinary tract of both men and women, most *T. vaginalis* strains are either not pathogenic or only mildly so. However, the sexual transmission of pathogenic strains results in painful inflammation associated with a whitish-green discharge. Microscopic examination of this discharge reveals very high populations of the protist. *Tritrichomonas foetus* is a cattle parasite and an important cause of spontaneous abortion in these animals. Trichomonads do not require oxygen and possess hydrogenosomes rather than mitochondria. They undergo asexual reproduction only. Direct contact diseases: Trichomoniasis (section 39.4)

Euglenozoa

These protists are commonly found in freshwater, although a few species are marine. About one-third of euglenids are photoautotrophic; the remaining are free-living chemoorganotrophs (principally saprotrophic) although a few parasitic species have been described. The representative genus is the photoautotroph *Euglena.* A typical *Euglena* cell (**figure 25.5**) is elongated and bounded by a plasmalemma. The pellicle consists of proteinaceous strips and microtubules; it is elastic enough to enable turning and flexing of the cell, yet rigid enough to prevent excessive alterations in shape. Photosynthetic protists can be characterized by the pigments and carbohydrates they possess. *Euglena* contains chlorophylls *a* and *b* together with carotenoids. The large nucleus contains a prominent nucleolus. The primary storage product is paramylon (a polysaccharide composed of β-1,3 linked glucose molecules), which is unique to euglenoids. A red eye spot or **stigma** helps the organism orient to light and is located near an anterior reservoir. A large contractile vacuole near the reservoir continuously collects water from the cell and empties it into the reservoir, thus regulating the osmotic pressure within the organism. Two flagella arise from the base of the reservoir, although only one emerges from the canal and actively beats to move the cell. Reproduction in euglenoids is by longitudinal mitotic cell division (figure 25.3*d*).

Several protists of medical relevance belong to the *Euglenozoa.* These include the trypanosomes, which exist only as parasites of plants and animals. Trypanosome diseases have global significance. Members of the genus *Leishmania* cause a group of conditions, collectively termed **leishmaniasis,** that include systemic and skin/mucous membrane afflictions affecting some 12 million people. Chagas' disease is caused by *Trypanosoma cruzi,* which is transmitted by "kissing bugs" (*Triatominae*), so called because they bite the face of sleeping victims (*see figure 39.10*). Two to three million citizens of South and Central America show the central and peripheral nervous system dysfunction that is characteristic of this disease; of these, about 45,000 die of the disease each year. *Try-*

panosoma gambiense and *T. rhodesiense* (often considered a subspecies of *T. brucei*) cause African sleeping sickness (**figure 25.6**). Ingestion of these parasites by the blood-sucking tsetse fly triggers a complex cycle of development and reproduction, first in the fly's gut and then in its salivary glands. From there it is easily transferred to a vertebrate host, where it often causes a fatal infection. It is estimated that about 65,000 people die annually of sleeping sickness. The presence of this dangerous parasite prevents the use of about 11 million square kilometers of African grazing land. Arthropod-borne diseases: Leishmaniasis and trypanosomiasis (section 39.3)

African trypanosomes have a thick glycoprotein layer coating the cell wall surface. The chemical composition of the glycoprotein layer is switched cyclically, expressing only one of 1,000 to 2,000 variable antigens at any given time. This process, known as antigenic variation, enables the parasite's escape from host immune surveillance. It is therefore not surprising that there are no vaccines for either Chagas' disease or African sleeping sickness and the few drugs available for treatment are not particularly effective. However, the annotated genome sequences of *T. cruzi* and *T. brucei* were reported in 2005. The release of these genomes allows scientists to compare the parasitic features and the mechanisms by which these protists so successfully evade the host's immune system. This may help identify new drug and vaccine targets. Comparative genomics (section 15.6)

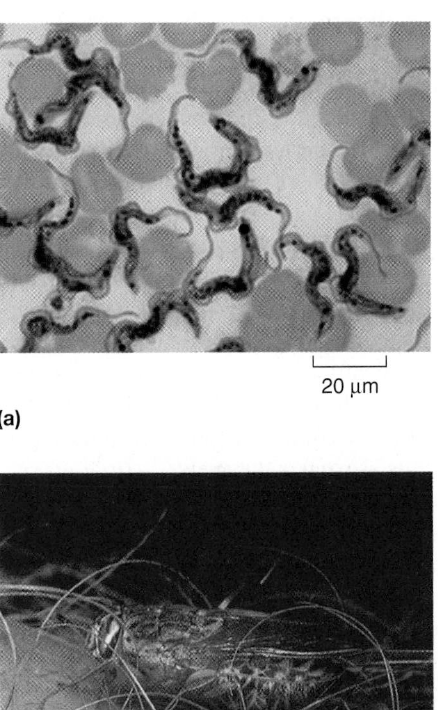

Figure 25.6 The Euglenozoan *Trypanosoma* and Its Insect Host. **(a)** *Trypanosoma* among red blood cells. Note the dark-staining nuclei, anterior flagella, and undulating changeable shape (×500). **(b)** The tsetse fly, shown here sucking blood from a human arm, is an important vector of the *Trypanosoma* species that causes African sleeping sickness.

1. Why is the classification of chemoorganotrophic protists according to their mode of locomotion no longer considered valid?
2. What are some features that distinguish the *Parabasalia* from the *Euglenozoa*?
3. What is the function of the stigma in *Euglena*? How does this protist maintain osmotic balance?
4. What *Euglenozoa* genera cause disease? What adaptations make these protists successful pathogens?

Super Group *Amoebozoa*

It is clear that the amoeboid form arose independently numerous times from various flagellated ancestors. Thus we see that some ameboid forms are placed in the super group *Amoebozoa* while others are placed in the *Rhizaria*. One of the morphological hallmarks of amoeboid motility is the use of **pseudopodia** (meaning "false feet") for both locomotion and feeding (**figure 25.7**). Pseudopodia can be rounded (**lobopodia**), long and narrow (**filopodia**), or form a netlike mesh (**reticulopodia**). Amoebae that lack a cell wall or other supporting structures and are surrounded

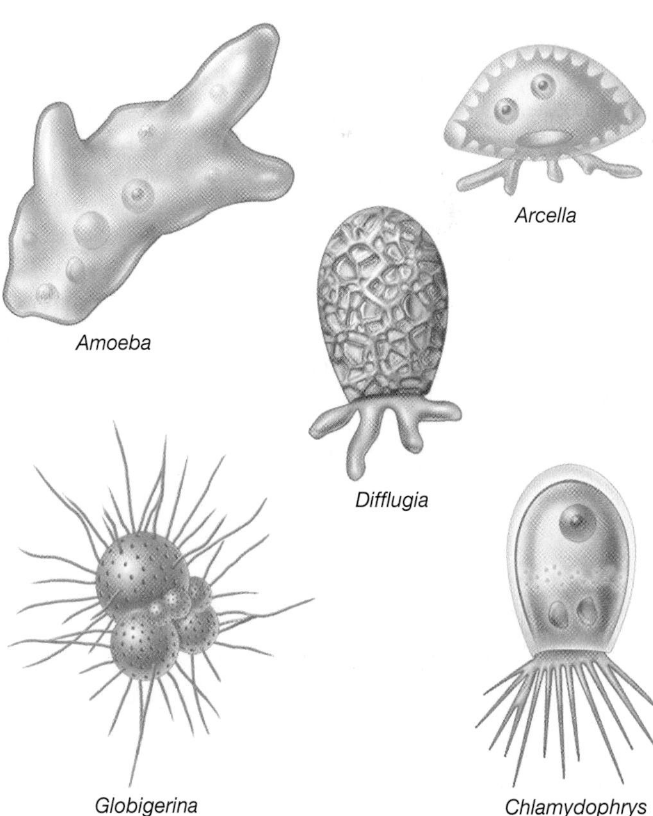

Figure 25.7 Pseudopodia. *Amoeba, Difflugia,* and *Arcella* have lobopodia; *Chlamydophrys* bears filopodia; the foraminiferan *Globigerina* has reticulopodia.

only by a plasma membrane are called **naked amoebae.** In contrast, the plasma membrane of the **testate amoebae** is covered by material that is either made by the protist itself or collected by the organism from the environment. Binary fission is the usual means of asexual division, although some amoebae form cysts that undergo multiple fission (figure 25.3*a,b*).

Tubulinea

These protists inhabit almost any environment where they will remain moist; this includes glacial meltwater, marine plankton, tidepools, lakes, and streams. Free-living forms are known to dwell in ventilation ducts and cooling towers where they feed on microbial biofilms. Others are endosymbionts, commensals, or parasites of invertebrates, fishes, and mammals. Some harbor intracellular symbionts, including algae, bacteria, and viruses, but the nature of these relationships is not well understood. *Amoeba proteus* (**figure 25.8**), a favorite among introductory biology laboratory instructors, is included in this group.

Entamoebida

Amoebic dysentery is the third leading cause of parasitic death worldwide and is caused by *Entamoeba histolytica.* Individuals acquire this pathogen by eating feces-contaminated food or by drinking water contaminated by *E. histolytica* cysts. These pass unharmed through the stomach and undergo multiple fission when introduced to the alkaline conditions in the intestines. There they not only graze on bacteria, but produce a suite of digestive enzymes that degrade gut epithelial cells. *E. histolytica* can penetrate into the bloodstream and migrate to the liver, lungs, and/or skin. Cysts in feces remain viable for weeks, but are killed by heat greater than 40°C. Food and waterborne diseases: Amebiasis (section 39.5)

Eumycetozoa

Since their first description in the 1880s, the *Eumycetozoa* or "slime molds" have been classified as plants, animals, and fungi. As we examine their morphology and behavior, the source of this confusion should become apparent. Analysis of certain proteins (for example, elongation factor EF-1α, β-tubulin, and actin) as well as physiological, behavioral, biochemical, and developmental data point to a monophyletic group (figure 25.1). The *Eumycetozoa* includes the *Myxogastria* and *Dictyostelia.* The **acellular slime mold,** or ***Myxogastria,*** life cycle includes a distinctive stage when the organisms exist as streaming masses of colorful protoplasm that creep along in amoeboid fashion over moist, rotting logs, leaves, and other organic matter, which they degrade (**figure 25.9**). Their name derives from the lack of individual cell membranes from which a large, multinucleate mass called a **plasmodium** is formed; there can be as many as 10,000 synchronously dividing nuclei within a single plasmodium (figure 25.9*b*). Feeding is by endocytosis. When starved or dried, the plasmodium develops ornate fruiting bodies. As these mature, they form stalks with cellulose walls that are resistant to environmental stressors (figure 25.9*c,d,e*). When conditions improve, spores germinate and release haploid amoeboflagellates. These fuse and as the resulting zygotes feed, nuclear division and synchronous mitotic divisions give rise to the multinucleate plasmodium.

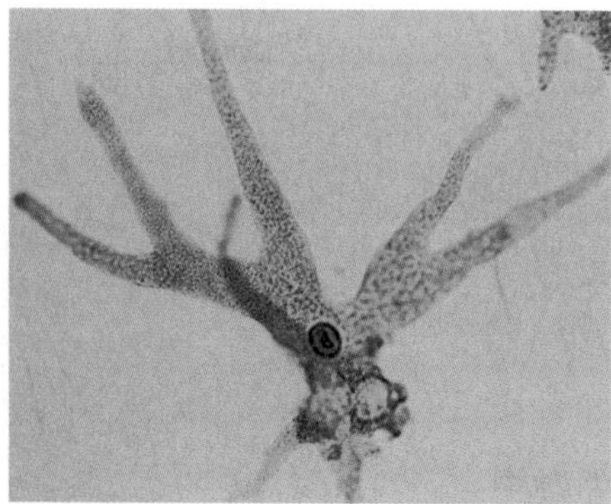

Figure 25.8 ***Amoeba proteus.*** The lobopodia are seen as long projections; the dark-staining nucleus is in the central region of the cell.

The **cellular slime molds (*Dictyostelia*)** are strictly amoeboid and feed endocytically on bacteria and yeasts. Their complex life cycle involves true multicellularity, despite their primitive evolutionary status (**figure 25.10*a***). The species *Dictyostelium discoideum* is an attractive model organism. The vegetative cells move as a mass, sometimes called a pseudoplasmodium, because individual cells retain their cell membranes. When starved, cells release cyclic AMP and a specific glycoprotein, which serve as molecular signals. Other cells sense these compounds and respond by forming an aggregate around the signal-producing cells (figure 25.10*b*). In this way large, motile, multicellular slugs develop and serve as precursors to fruiting body formation (figure 25.10*c*). Fruiting body morphogenesis commences when the slug stops and cells pile on top of each other. Cells at the bottom of this vertically oriented structure form a stalk by secreting cellulose, while cells at the tip differentiate into spores (figure 25.10*d,e*). Germinated spores become vegetative amoebae to start this asexual cycle anew.

Sexual reproduction in *D. discoideum* involves the formation of special spores call macrocysts. These arise by a form of conjugation that has some unusual features. First, a group of amoebae become enclosed within a wall of cellulose. Following conjugation a single, large amoeba forms and cannibalizes the remaining amoebae. The now giant amoeba matures into a macrocyst. Macrocysts can remain dormant within their cellulose walls for extended periods of time. Vegetative growth resumes after the diploid nucleus undergoes meiosis to generate haploid amoebae.

The 33.8-Mb genome of *D. discoideum* has been sequenced and annotated. Analysis of a number of proteins supports the notion that these soil-dwelling microbes are more primitive than the fungi. For instance, *D. discoideum* has 14 different histidine kinase receptor proteins; these proteins are generally thought to be distinctly procaryotic. Also of note is the presence of 40 genes that appear to be involved in cellulose biosynthesis or degradation. These genes could be involved in producing the cellulose the microbe needs during morphological differentiation and/or for degrading cellulose-containing microbes.

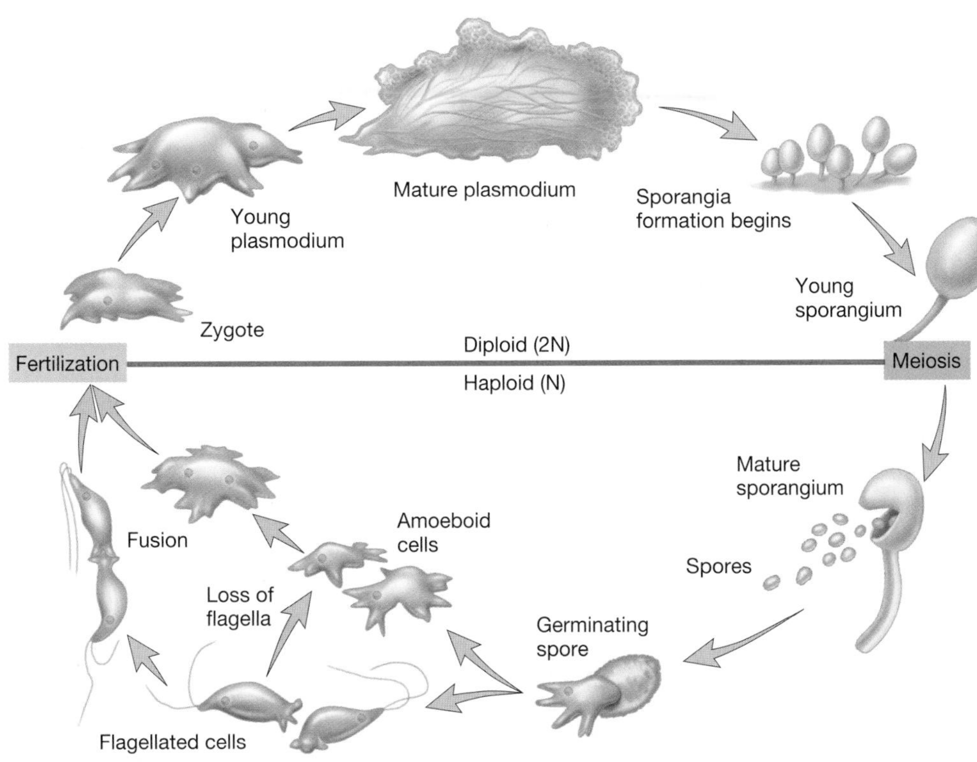

(a)

(b) *Physarum* sp.

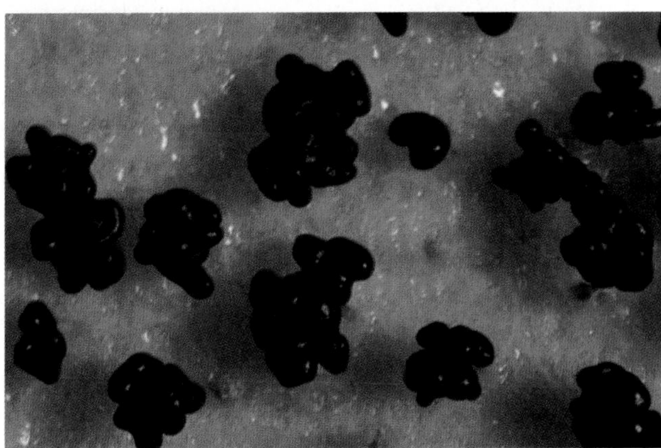
(c) *Physarum polycephalum*

(d) *Hemitrichia*

(e) *Stemonitis*

Figure 25.9 Acellular Slime Molds. **(a)** The life cycle of a plasmodial slime mold includes sexual reproduction; when conditions are favorable for growth the adult diploid forms sporangia. Following meiosis, the haploid spores germinate, releasing haploid amoeboid or flagellated cells that fuse. **(b)** Plasmodium of the slime mold *Physarum* sp. (×175). Sporangia of **(c)** *Physarum polycephalum*, **(d)** *Hemitrichia*, and **(e)** *Stemonitis*.

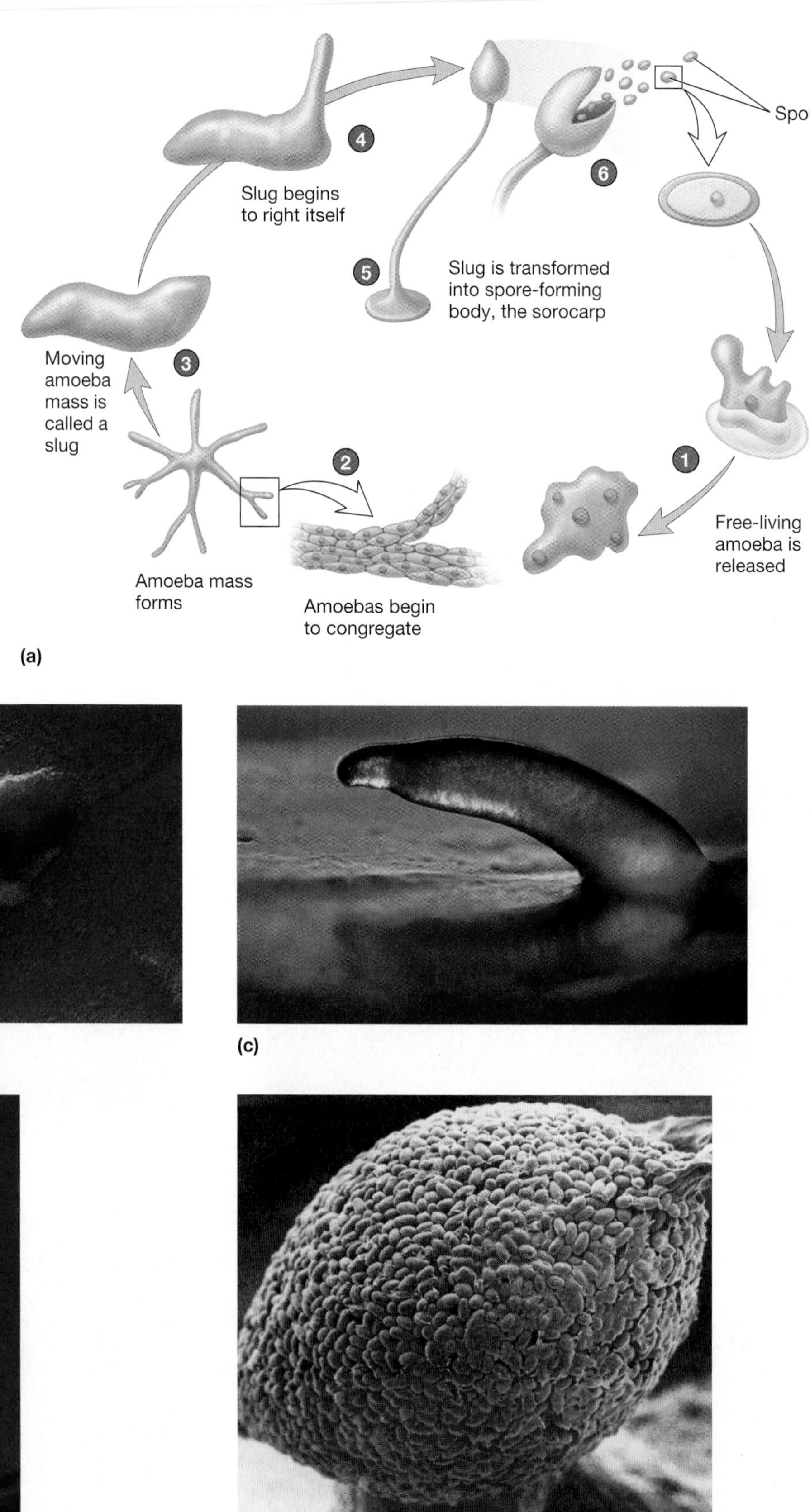

Figure 25.10 Development of *Dictyostelium discoideum*, a Cellular Slime Mold. **(a)** Life Cycle. **(b)** Aggregating *D. discoideum* become polar and begin to move in an oriented direction in response to the molecular signal cAMP. **(c)** Slug begins to right itself and **(d)** forms a spore-forming body called a sorocarp. **(e)** Electron micrograph of a sorocarp showing individual spores (×1,800).

Super Group *Rhizaria*

These protists are amoeboid in morphology and thus were historically grouped with the members of what we now call the *Amoebozoa*. However, molecular phylogenetic analysis makes it clear that the *Amoebozoa* and *Rhizaria* are not monophyletic. Morphololgically, the *Rhizaria* can be distinguished by their fine pseudopodia (filopodia), which can be simple, branched, or connected. Filopodia supported by microtubules are known as an **axopodia.** Axopodia protrude from a central region of the cell called the axoplast and are primarily used in feeding (**figure 25.11**).

Radiolaria

Most *Radiolaria* have an internal skeleton made of siliceous material; however, members of the subgroup *Acantharia* have endoskeletons consisting of strontium sulfate. A few genera have an exoskeleton of siliceous spines or scales, while some lack a skeleton completely. Skeletal morphology is highly variable and often includes radiating spines that help the organisms float, as does the storage of oils and other low-density fluids (**figure 25.12**). The skeletal material of ancient radiolarians that settled on the ocean floor millions of years ago remains preserved and is useful to scientists. The strontium sulfate skeletons of acantharians are used to measure the relative amounts of natural versus anthropogenic radioactivity in marine sediments. Siliceous skeletons of radiolaria and diatoms (see *Stramenopiles*, p. 621) form deposits called siliceous ooze, which serve as ancient or paleoenvironmental indicators.

The *Radiolaria* feed by endocytosis using mucus-coated filopodia to entrap prey including bacteria, other protists, and even small invertebrates. Large prey items are partially digested extracellularly before becoming encased in a food vacuole. Many surface-dwelling radiolarians have algal symbionts thought to enhance their net carbon assimilation. Asexual reproduction is found in some forms; the acantharians reproduce only sexually by

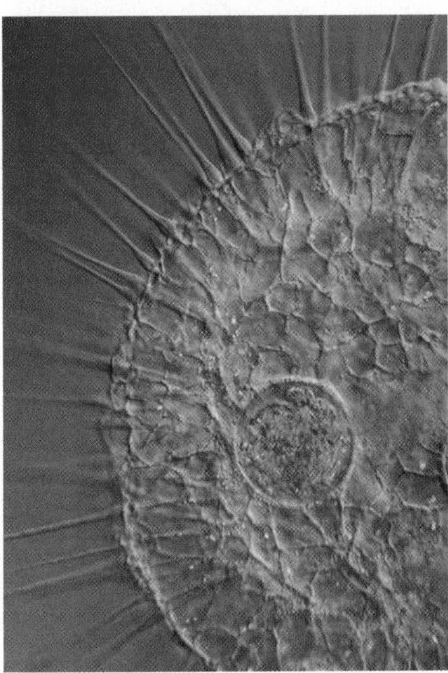

Figure 25.11 Axopodia. The rhizarian *Actinosphaerium* has needlelike axopodia.

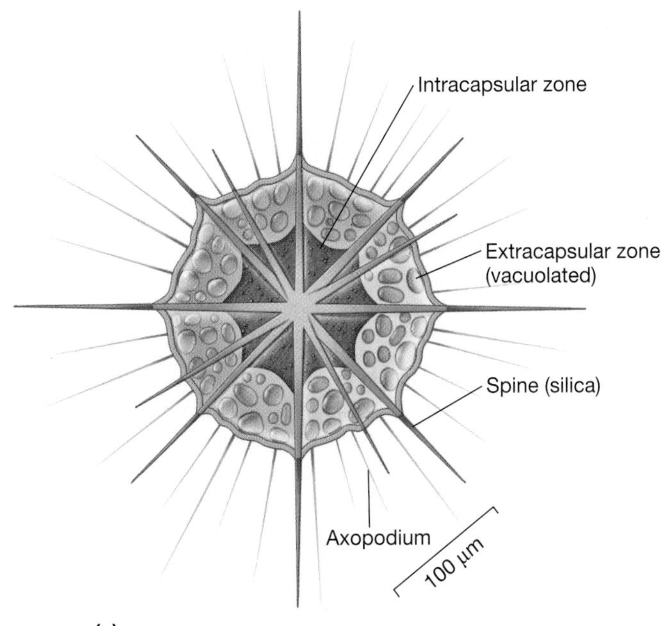

Figure 25.12 Radiolaria. **(a)** The radiolarian *Acanthometra elasticum* demonstrates the internal skeleton. **(b)** Radiolarian shells made of silica.

consecutive mitotic and meiotic divisions that release hundreds of biciliated isogametic cells. In others, asexual reproduction (binary or multiple fission, or budding) is most common, but sexual reproduction can be triggered by either nutrient limitation or a heavy feeding. In this case, two haploid nuclei fuse to form a diploid zygote encased in a cyst from which it is released when survival conditions improve.

Foraminifera

The *Foraminifera* (also simply called forams) range in size from roughly 20 μm to several centimeters. Their filopodia are arranged in a branching network called reticulopodia. They have characteristic tests arranged in multiple chambers that are sequentially added as the protist grows (**figure 25.13**).

Reticulopodia bear vesicles at their tips that secrete a sticky substance used to trap prey. Many species harbor endosymbiotic algae that can migrate out of reticulopodia (without being eaten) to expose themselves to more sunlight. It has been shown experimentally that individual forams with algal symbionts grow to larger sizes than those of the same species without symbionts, lending credibility to the notion that these algae contribute significantly to foram nutrition.

Foraminiferan life cycles can be complex. While some smaller species reproduce only asexually by budding and/or multiple fission, larger forms frequently alternate between sexual and asexual phases. During the sexual phase, flagellated gametes pair, fuse, and generate asexual individuals (agamonts). Meiotic division of the agamonts gives rise to haploid gamonts. There are several mechanisms by which gamonts return to the diploid condition. For instance, a variety of forams release flagellated gametes that become fertilized in the open water. In others, two or more gamonts attach to one another, enabling gametes to fuse within the chambers of the paired tests. When the shells separate, newly formed agamonts are released. True autogamy has been observed in at least one genus (*Rotaliella*): each gamont produces gametes that pair and fuse within a single test and the zygote is then released as an agamont.

Foraminifera are found in marine and estuarine habitats. Some forms are planktonic, but most are benthic. Foraminiferin tests accumulate on the sea floor where they constitute a fossil record dating back to the Early Cambrian (543 million years ago), which is helpful in oil exploration. Their remains, or ooze, can be up to hundreds of meters deep in some tropical regions. In Indonesia, the ancient tests are collected and used as paving material. Foram tests make up most of modern-day chalk, limestone, and marble, and are familiar to most as the White Cliffs of Dover in England (**figure 25.14**). They also formed the stones used to build the great pyramids.

1. Describe filopodia, lobopodia, and reticulopodia form and function.
2. List at least one mechanism used by *Entamoeba histolytica* to successfully infect a human host.
3. Why do you think the slime molds have been so hard to classify?
4. What is a plasmodium? How does it differ between the acellular and cellular slime molds?

Figure 25.13 A Foraminiferan. The reticulopodia are seen projecting through pores in the calcareous test, or shell, of this protist.

Figure 25.14 White Cliffs of Dover. The limestone that forms these cliffs is composed almost entirely of fossil shells of protists, including foraminifera.

5. Describe the life cycle of *Dictyostelium discoideum*. Why is this organism a good model for the study of cellular differentiation, coordinated cell movement, and chemotaxis?
6. What adaptations do the planktonic radiolaria have to help them float?
7. Compare the means by which radiolaria use axopodia with the way foraminifera use reticulopodia.
8. Describe the forms of sexual reproduction in the *Foraminifera*.

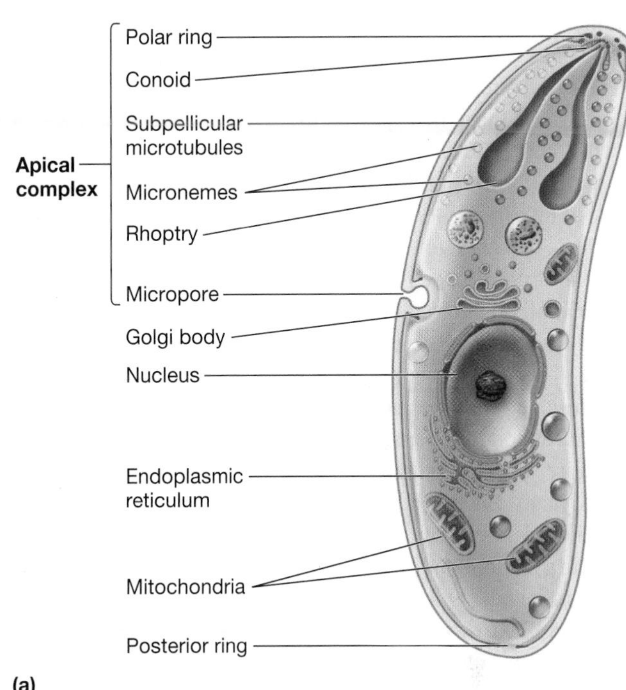

Figure 25.15 The Apicomplexan Cell. **(a)** The vegetative cell, or merozoite, illustrating the apical complex, which consists of the polar ring, conoid, rhoptries, subpellicular microtubules, and micropore. **(b)** The infective oocyte of *Eimeria*. The oocyst is the resistant stage and has undergone multiple fission after zygote formation (sporogony).

Super Group *Chromalveolata*

The *Chromalveolata* are diverse and include autotrophic, mixotrophic, and heterotrophic protists. They are united in plastid origin, which appears to have been acquired by endosymbiosis with an ancestral archaeplastid (which itself acquired plastids from endosymbiotic cyanobacteria). Here we introduce three subgroups: the *Alveolata, Stramenopiles,* and *Haptophyta,* some members of which have been previously considered to be orders or super groups.

Alveolata

The *Alveolata* is a large group that includes the *Apicomplexa, Dinoflagellata* (dinoflagellates), and the *Ciliophora*. We begin our discussion with the *Apicompexa*. All **apicomplexans** are either intra- or intercellular parasites of animals and are distinguished by a unique arrangement of fibrils, microtubules, vacuoles, and other organelles, collectively called the apical complex, which is located at one end of the cell (**figure 25.15*a***). This unique combination of organelles is designed to penetrate host cells. Motility (flagellated or amoeboid) is confined to the gametes and zygotes of a few species.

Apicomplexans have complex life cycles in which certain stages sometimes occur in one host and other stages occur in a different host. The life cycle has both asexual (clonal) and sexual phases and is characterized by an alternation of haploid and diploid generations. The clonal and sexual stages are haploid, except for the zygote. The motile, infective stage is called the **sporozoite.** When this haploid form infects a host, it differentiates into a gamont; male and female gamonts pair and undergo multiple fission, which produces many gametes. Released gametes pair, fuse, and form zygotes. Each zygote secretes a protective covering and is then considered a spore. Within the spore, the nucleus undergoes meiosis (restoring the haploid condition) followed by mitosis to generate eight sporozoites ready to infect a new host (figure 25.15*b*).

A number of apicomplexans are important infectious agents. The most significant is *Plasmodium,* which causes malaria in some 500 million people annually, with a yearly death toll of 1 to 3 million (*see figure 39.5*). *Eimera* is the causative agent of cecal coccidiosis in chickens, a condition that costs hundreds of millions of dollars in lost animals each year in the United States. Toxoplasmosis, caused by members of the genus *Toxoplasma,* is transmitted either by consumption of undercooked meat or by fecal contamination from a cat's litterbox. Cryptosporidia are responsible for cryptosporidiosis, an infection that begins in the intestines but can disseminate to other parts of the body. Cryptosporidiosis has become problematic for AIDS patients and other immunocompromised individuals. Arthropod borne diseases: Malaria (section 39.3); Direct contact diseases: Toxoplasmosis (section 39.4); Food and waterborne diseases: Cryptosporidiosis (section 39.5)

Recently the genomes of the apicomplexans *Theilaria parva* and *Theilaria annulata* were sequenced and annotated. These tick-borne parasites cause diseases marked by rapid proliferation of white blood cells (lymphoproliferation). *T. parva* infects cattle and causes a fatal disease called East Coast Fever. This disease kills over a million cattle each year in sub-Saharan Africa, costing over $200 million and targeting farmers who can least afford such economic losses. The nuclear genomes of these two parasites are similar in size (about 8.3 Mb); both have four chromosomes that range in size from 1.9 to 2.6 Mb. Among the roughly 4,000 identified genes are those that may be involved in altering mitosis in host cells, thereby resulting in hyperproliferation. In addition, putative secreted polypeptides are present. These may enhance the protists' ability to evade host immune responses. It is hoped that experiments designed to test these hypotheses may help in the development of more effective treatment and a vaccine.

The **dinoflagellates (*Dinoflagellata*)** are a large group most commonly found in marine plankton, where some species are responsible for the phosphorescence sometimes seen in seawater. Their nutrition is complex; about half of all dinoflagellates are photosynthetic, although photoautotrophy is rare. Most are saprotrophic (either entirely, or as facultative chemoorganotrophs), but some also use endocytosis. Each cell bears two, distinctively placed flagella: one is wrapped around a transverse groove (the girdle) and the other is draped in a longitudinal groove (the sulcus; **figure 25.16**). The orientation and beating patterns of these flagella cause the cell to spin as it is propelled forward; the name dinoflagellate is derived from the Greek *dinein,* "to whirl." Many dinoflagellates are covered with cellulose plates that are secreted within alveolar sacs that lie just under the plasma membrane. These forms are said to be thecate or armored; those with empty alveoli are called athecate or naked and include the luminescent genus *Noctiluca*. Most dinoflagellates are free living, although some form important associations with other organisms. Endosymbiotic dinoflagellates that live as undifferentiated cells occasionally send out motile cells called **zooxanthellae.** The most well-known zooxanthella belongs to the genus *Symbiodinium*. These are photosynthetic endosymbionts of reef-building coral. They provide fixed carbon to the coral animal and help maintain the internal chemical environment needed for the coral to secrete its calcium carbonate exoskeleton. Dinoflagellates are also responsible for toxic "red tides" that harm other organisms, including humans (**Disease 25.1**).

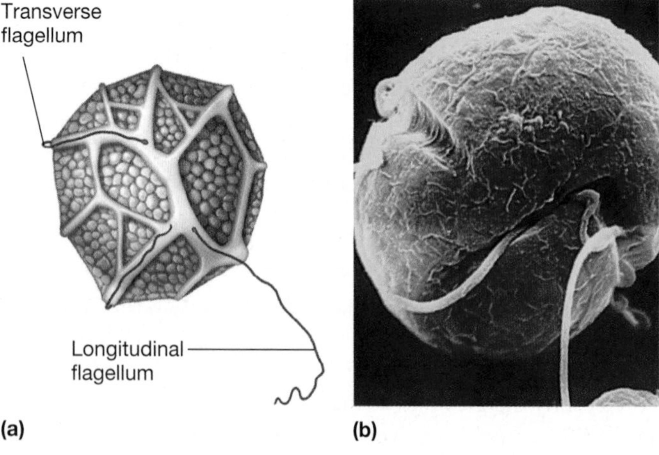

Figure 25.16 Dinoflagellates. **(a)** *Ceratium.* **(b)** Scanning electron micrograph of *Gymnodinium* (×4,000). Notice the plates of cellulose and two flagella: one in the transverse groove and the other projecting outward.

The **ciliates (*Ciliophora*)** include about 12,000 species. All are chemoorganotrophic and range from about 10 μm to 4.5 mm long. They inhabit both benthic and planktonic communities in marine and freshwater systems, as well as moist soils. As their name implies, *Ciliophora* employ many cilia as locomotory and feeding organelles. The cilia are generally arranged either in longitudinal rows (**figure 25.17**) or in spirals around the body of the organism. They beat with an oblique stroke; therefore, the protist revolves as it swims. They coordinate ciliary beating so precisely that they can go both forward and backward. There is great variation in shape, and most ciliates do not look like the slipper-shaped *Paramecium*. Some species including *Vorticella,* attach to substrates by a long stalk. *Stentor* attaches to substrates and stretches out in a trumpet shape to feed (figure 25.17*a*). A few species have tentacles for the capture of prey. Some can discharge toxic, threadlike darts called toxicysts, which are used in capturing prey. A striking feature of the *Ciliophora* is their ability to capture many particles in a short time by the action of the cilia around the buccal cavity. Food first enters the cytostome and passes into phagocytic vacuoles that fuse with lysosomes after detachment from the cytostome. The ciliate digests the vacuole's contents when the vacuole is acidified and lysosomes release digestive enzymes into it. After the digested material has been absorbed into the cytoplasm, the vacuole fuses with the cytoproct and waste material is expelled.

Most ciliates have two types of nuclei: a large macronucleus and a smaller micronucleus. The micronucleus is diploid and contains the normal somatic chromosomes. It divides by mitosis and transmits genetic information through meiosis and sexual reproduction. Macronuclei are derived from micronuclei by a complex series of steps. Within the macronucleus are many chromatin bodies, each containing many copies of only one or two genes. Macronuclei are thus polyploid and divide by elongating and then constricting. They produce mRNA to direct protein synthesis, maintain routine cellular functions, and control normal cell metabolism.

Some ciliates reproduce asexually by transverse binary fission, forming two equal daughter cells. The most common means of sexual reproduction among ciliates is conjugation. In this process there is an exchange of gametes between paired cells of complementary mating types (**conjugants**). A well-studied example is *Paramecium caudatum* (**figure 25.18**). At the beginning of conjugation, two ciliates unite, fusing their pellicles at the contact point. The macronucleus in each is degraded. The individual micronuclei undergo meiosis to form four haploid pronuclei, three of which disintegrate. The remaining pronucleus divides again mitotically to form two gametic nuclei—a stationary one and a migratory one. The migratory nuclei pass into the respective conjugants. Then the ciliates separate, the gametic nuclei fuse, and the resulting diploid zygote nucleus undergoes three rounds of mitosis. The eight resulting nuclei have different fates: one nucleus is retained as a micronucleus; three others are destroyed; and the four remaining nuclei develop into macronuclei. Each separated conjugant now undergoes cell division. Eventually progeny with one macronucleus and one micronucleus are formed.

Disease

25.1 Harmful Algal Blooms (HABs)

The Bible reports that the first plague Moses visited on the Egyptians was a blood-red tide that killed fish and fouled water. The Red Sea probably is named after these toxic algal blooms. Thousands of years later we still have problems with this plague.

The poisonous and destructive red tides that occur frequently in coastal areas often are associated with population explosions, or "blooms," of dinoflagellates. *Gymnodinium* and *Gonyaulax* species are the dinoflagellates most often involved. The pigments in the dinoflagellate cells are responsible for the red color of the water. Under these bloom conditions, the dinoflagellates produce a powerful neurotoxin called saxitoxin. The toxin paralyzes the striated respiratory muscles in many vertebrates by inhibiting sodium transport, which is essential to the function of their nerve cells. The toxin does not harm the shellfish that feed on the dinoflagellates. However, the shellfish accumulate the toxin and are themselves highly poisonous to organisms, such as humans, who consume the shellfish, resulting in a condition known as paralytic shellfish poisoning or neurotoxic shellfish poisoning. Paralytic shellfish poisoning is characterized by numbness of the mouth, lips, face, and extremities. Duration of the illness ranges from a few hours to a few days and usually is not fatal.

Another type of poisoning in humans is called ciguatera. It results from eating marine fishes (e.g., grouper, snapper) that have consumed the dinoflagellate *Gambierdiscus toxicus*. The protist's toxin, called ciguatoxin, accumulates in the flesh of fish. This is one of the most powerful toxins known and remains in the flesh even after it has been cooked. Unfortunately it cannot be detected in fish and they are not visibly affected. In humans the toxin may cause gastrointestinal disturbances, profuse diarrhea, central nervous system involvement, and respiratory failure.

In 1988 a red tide that has long plagued the Gulf Coast of Florida spread northward to North Carolina. The dinoflagellates released a neurotoxin called brevetoxin, and this prompted state health authorities to shut down all shellfishing for three months. In 1987, in Prince Edward Island, Canada, several people died and hundreds became sick from eating mussels contaminated with domoic acid. The domoic acid was traced to a bloom of diatoms, *Stramenopiles* once thought to be innocent of all toxicity. The resulting disease, called amnesic shellfish poisoning, produces short-term memory loss in its victims. In 1998 over 400 sea lions eating anchovies off the California coast died from domoic acid poisoning. In 1993 saxitoxin was found for the first time in crabs from Alaska. Unfortunately there are no treatments for these types of poisonings. Supportive measures are the only therapy.

Overall, toxic algal blooms are on the rise. For example, in 1997 *Pfiesteria piscicida* (Latin for "fish killer") and other *Pfiesteria*-like dinoflagellates caused large fish kills along the coast of Maryland and Virginia. Similar fish kills have been occurring along the Atlantic coast at least since the 1980s. The flagellated form of the ambush-predator dinoflagellate swims toward the fish and attacks the prey, which it can then feed on. No one is certain why these toxic blooms are becoming more frequent, but most believe that the blooms are caused by the continuous pumping of nutrients such as nitrogen and phosphorus into coastal waters. Sewage and agricultural runoff are probably the major sources. Another possibility is world trade: oceangoing ships are unintentionally trafficking in harmful algae, giving the protist a free ride to foreign ports and new habitats in which they can flourish. With people eating more seafood, this toxic menace in the world's oceans will become increasingly more common in the future. Microogranisms in the marine environment: Coastal marine systems (section 28.3)

Although most ciliates are free living, symbiotic forms do exist. Some live as harmless commensals—for example, *Entodinium* is found in the rumen of cattle and *Nyctotherus* occurs in the colon of frogs. Other ciliates are strict parasites—for example, *Balantidium coli* lives in the intestine of mammals, including humans, where it can produce dysentery. *Ichthyophthirius* lives in freshwater where it can attack many species of fish, producing a disease known as "ick."

1. What is the apical complex seen in apicomplexans?
2. Why is the life cycle of apicomplexans so difficult to study? What do you think the implications of the complicated *Plasmodium* life cycle are for the development of a malaria cure or vaccine?
3. What are zooxanthellae? Describe the relationship between *Symbiodinium* and its coral host.
4. What is the morphology of typical *Ciliophora?* Why do you think these are the fastest-moving protists?
5. Describe conjugation as it occurs in the *Ciliophora*. What is the fate of the micronucleus and the macronucleus during this process?

Stramenopiles

This is a large and diverse group that includes photosynthetic protists such as the diatoms, brown and golden algae (the *Chrysophyceae*), and chemoorganotrophic (saprophytic) genera such as the öomycetes (*Peronosporomycetes*), labyrinthulids (slime nets), and the *Hyphochytriales*. The *Stramenopiles* also include brown seaweeds and kelp that form large, rigid structures and macroscopic forms that were once considered fungi and plants. One unifying feature of this very diverse taxon is the possession of **heterokont flagella** at some point in the life cycle. This is characterized by two flagella—one extending anteriorly and the other posteriorly. These flagella bear small hairs with a unique, three-part morphology; the name stramenopila means "straw hair."

The **diatoms (*Bacillariophyta*)** possess chlorophylls *a* and c_1/c_2, and the carotenoid fucoxanthin. When fucoxanthin is the dominant pigment, the cells have a golden-brown color. Their major carbohydrate reserve is chrysolaminarin (a polysaccharide storage product composed principally of β (1→3) linked glucose residues). Diatoms have a distinctive, two-piece cell wall of silica

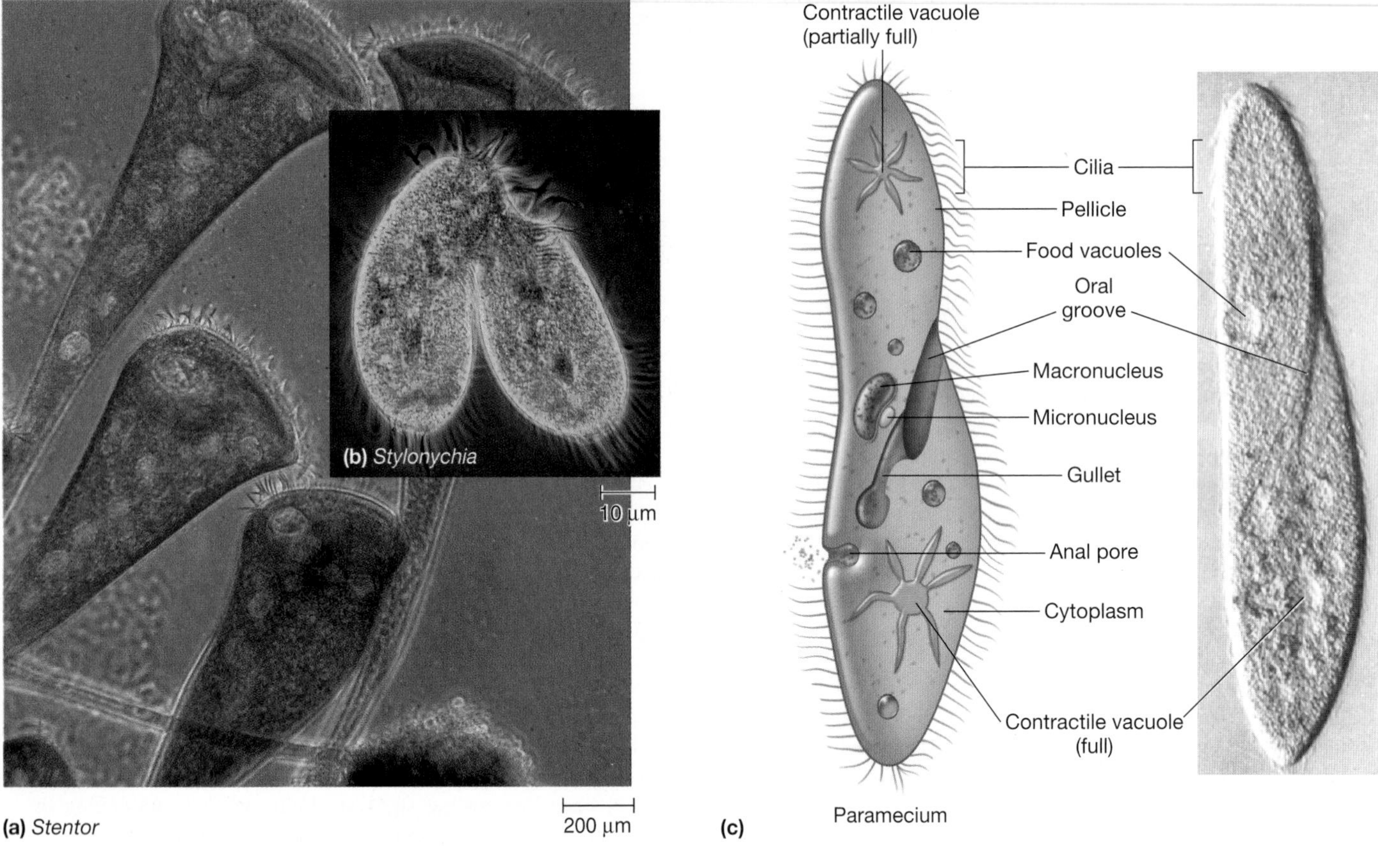

Figure 25.17 The *Ciliophora*. **(a)** *Stentor,* a large, vase-shaped, freshwater protozoan. **(b)** Two *Stylonychia* conjugating. **(c)** Structure of *Paramecium*, adjacent to an electron micrograph.

called a **frustule.** Diatom frustules are composed of two halves or thecae that overlap like a petri dish (**figure 25.19*a***). The larger half is the **epitheca,** and the smaller half is the **hypotheca.** Diatom frustules are composed of crystallized silica [$Si(OH)_4$] with very fine markings (figure 25.19*b*). They have distinctive, and often exceptionally beautiful, patterns that are unique for each species. Frustule morphology is very useful in diatom identification, and diatom frustules have a number of practical applications (**Techniques & Applications 25.2**). The fine detail and precise morphology of these frustules has made them attractive for nanotechnology. Microbes as products: Nanotechnology (section 41.8)

Although the majority of diatoms are strictly photoautotrophic, some are facultative chemoorganotrophs, absorbing carbon-containing molecules through the holes in their walls. The vegetative cells of diatoms are diploid and can be unicellular, colonial, or filamentous. They lack flagella and have a single, large nucleus and smaller plastids. Reproduction consists of the organism dividing asexually, with each half then constructing a new theca within the old one. Because the epitheca and hypotheca are of different sizes, each time the hypotheca is used as a template to construct a new theca, the diatom gets smaller. However, when a cell has diminished to about 30% of its original size, sexual reproduction is usually triggered. The diploid vegetative cells undergo meiosis to form gametes, which then fuse to produce a zygote. The zygote develops into an auxospore, which increases in size again and forms a new wall. The mature auxospore eventually divides mitotically to produce vegetative cells with frustules of the original size.

Diatoms are found in freshwater lakes, ponds, streams, and throughout the world's oceans. Marine planktonic diatoms produce 40 to 50% of the organic carbon in the ocean; they are therefore very important in global carbon cycling. In fact, marine diatoms are thought to contribute as much fixed carbon as all rain forests combined. Biogeochemical cycling: Carbon cycle (section 27.2)

The complete genome of the marine planktonic diatom *Thalassiosira pseudonana* was recently sequenced. Its genome consists of 24 chromosomes (34 Mb), a plastid genome (0.139 Mb), and a mitochondrial genome (0.044 Mb). It contains novel genes for silicic acid transport and silica-based cell wall synthesis. Also discovered were genes for scavenging iron, multiple nitrate and ammonium transporters, and the enzymes for urea metabolism. The annotation of this genome will help scientists discover how this microbe has adapted so successfully to its nutrient-limited environment.

A group of protists once considered true fungi and traditionally called **öomycetes,** meaning "egg fungi," were recently assigned the name *Peronosporomycetes*. They differ from true fungi in a num-

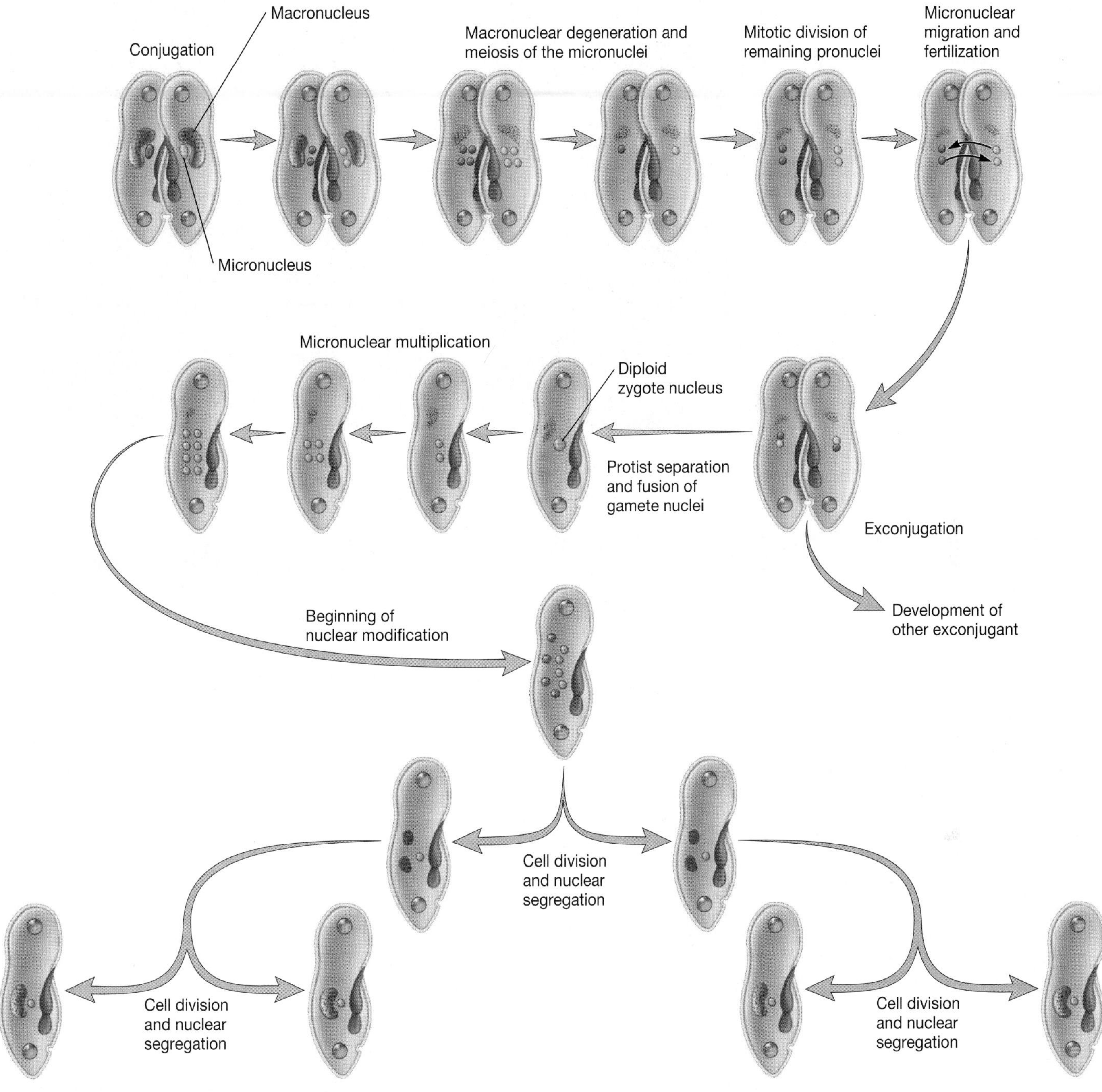

Figure 25.18 Conjugation in *Paramecium caudatum*. After the conjugants separate, only one of the exconjugants is shown; however, a total of eight new protists result from each conjugation.

ber of features including their cell wall composition (cellulose and β-glucan instead of chitin) and the fact that they are diploid throughout their life cycle. When undergoing sexual reproduction, they form a relatively large egg cell (öogonium) that is fertilized by either a sperm cell or a smaller gametic cell (called an antheridium) to produce a zygote. When the zygote germinates, the asexual zoospores display heterokont flagellation. The *Fungi:* Reproduction (section 26.5)

Peronosporomycetes such as *Saprolegnia* and *Achlya* are saprophytes that grow as cottony masses on dead algae and animals, mainly in freshwater environments. Some öomycetes are parasitic on the gills of fish. *Peronospora hyoscyami* is responsible for "blue mold" on tobacco plants and grape downy mildew is caused by *Plasmopara viticola*. Certainly the most famous öomycete is *Phytophthora infestans,* which attacked the European potato crop in the mid-1840s, spawning the Irish famine.

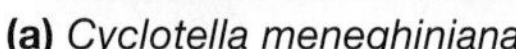

Figure 25.19 Diatoms. **(a)** The silaceous epitheca and hypotheca of the diatom *Cyclotella meneghiniana* fit together like a petri dish. **(b)** A variety of diatoms show the intricate structure of the silica cell wall.

(a) *Cyclotella meneghiniana* (b)

Techniques & Applications

25.2 Practical Importance of Diatoms

Diatoms have both direct and indirect economic significance for humans. Because diatoms make up most of the phytoplankton of the cooler parts of the ocean, they are the most important ultimate source of food for fish and other marine animals in these regions. It is not unusual for 1 liter of seawater to contain almost a million diatoms. They are also extremely important for the biochemical cycling of silica and as contributors to global fixed carbon.

When diatoms die, their frustules sink to the bottom. Because the siliceous part of the frustule is not affected by the death of the cell, diatom frustules tend to accumulate at the bottom of aquatic environments. These form deposits of material called diatomaceous earth. This material is used as an active ingredient in many commercial preparations, including detergents, fine abrasive polishes, paint removers, decolorizing and deodorizing oils, and fertilizers. Diatomaceous earth also is used extensively as a filtering agent, as a component in insulating (firebrick) and soundproofing products, and as an additive to paint to increase the night visibility of signs and license plates.

The use of diatoms as indicators of water quality and of pollution tolerance is becoming increasingly important. Specific tolerances for given species to various environmental parameters (concentrations of salts, pH, nutrients, nitrogen, temperature) have been compiled.

Diatomaceous earth can also be used to control insects. Insects have their soft body parts exposed but covered by a waxy film to prevent dehydration. When they contact the diatoms in diatomaceous earth the silica frustules break the waxy film on the insects, causing them to dehydrate and die. Insects cannot build up resistance to diatomaceous earth, and it can be fed to poultry, livestock, and pets with no ill effects.

The original classification of *P. infestans* as a fungus was misleading and for decades farmers attempted to control its growth with fungicide, to which it is (of course) resistant. This protist continues to take its toll; potato blight costs some $5 billion annually worldwide.

Labyrinthulids also have a complex taxonomic history: like the *Peronosporomycetes* they were formerly considered fungi. However, molecular phylogenetic evidence combined with the observation that they form heterokont flagellated zoospores places them among the *Stramenopiles*. The more familiar, nonflagellated stage of the life cycle features spindle-shaped cells that form complex colonies that glide rapidly along an ectoplasmic net made by the organism. This net is actually an external network of calcium-dependent contractile fibers made up of actinlike proteins that facilitate the movement of cells. Their feeding mechanism is like that of fungi: osmotrophy aided by the production of extracellular degradative enzymes. In marine habitats, the genus *Labyrinthula* grows on plants and algae and is thought to play a role in the "wasting disease" of eelgrass, an important intertidal plant. The *Fungi:* Nutrition and metabolism (section 26.4)

Haptophyta

One interesting subgroup of the haptophyta is the *Coccolithales*. These photosynthetic protists bear ornate calcite scales called coccoliths (**figure 25.20**). Together with the *Foraminifera,* the **coccolithophores** precipitate calcium carbonate ($CaCO_3$) in the open ocean, thereby influencing Earth's carbon budget. Cells are usually biflagellated and possess a unique organelle called a haptonema, which is somewhat similar to a flagellum but differs in microtubule arrangement. One species, *Emiliania huxleyi,* has been studied extensively. Like all coccolithophores, it is planktonic. High concentrations or blooms of *E. huxleyi* can significantly alter nutrient flux by emitting sulfur (as dimethyl sulfide) to the atmosphere and sequestering calcium carbonate in the sediments. Other coccolithophore species are known to cause toxic blooms. Marine and freshwater environments: Nutrient cycling (section 28.1)

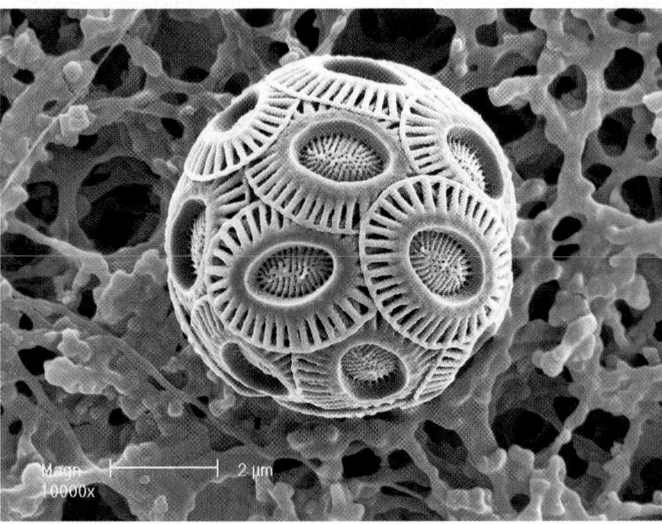

Figure 25.20 The Haptophyte *Emiliana huxleyi.* Note the ornamental scales made of calcite.

Super Group *Archaeplastida*

The *Archaeplastida* includes all organisms with a photosynthetic plastid that arose through an ancient endosymbiosis with a cyanobacterium. It thus includes all higher plants as well as many protist species. Microbial evolution: Endosymbiotic origin of mitochondria and chloroplasts (section 19.1)

Chloroplastida

The *Chloroplastida* are often referred to as green algae [Greek *chloros,* green]. These phototrophs grow in fresh and salt water, in soil, on other organisms, and within other organisms. They have chlorophylls *a* and *b* along with specific carotenoids, and they store carbohydrates such as starch. Many have cell walls made of cellulose. They exhibit a wide diversity of body forms, ranging from unicellular to colonial, filamentous, membranous or sheetlike, and tubular types (**figure 25.21**). Some species have a holdfast structure that anchors them to the substratum. Both asexual and sexual reproduction are observed.

Chlamydomonas is a member of the subgroup *Chlorophyta* (**figure 25.22**). Individuals have two flagella of equal length at the anterior end by which they move rapidly in water. Each cell has a single haploid nucleus, a large chloroplast, a conspicuous pyrenoid, and a stigma (eyespot) that aids the cell in phototactic responses. Two small contractile vacuoles at the base of the flagella function as osmoregulatory organelles. *Chlamydomonas* reproduces asexually by producing zoospores through cell division. Sexual reproduction occurs when some products of cell division act as gametes and fuse to form a four-flagellated, diploid zygote that ultimately loses its flagella and enters a resting phase. Meiosis occurs at the end of this resting phase and produces four haploid cells that give rise to adults.

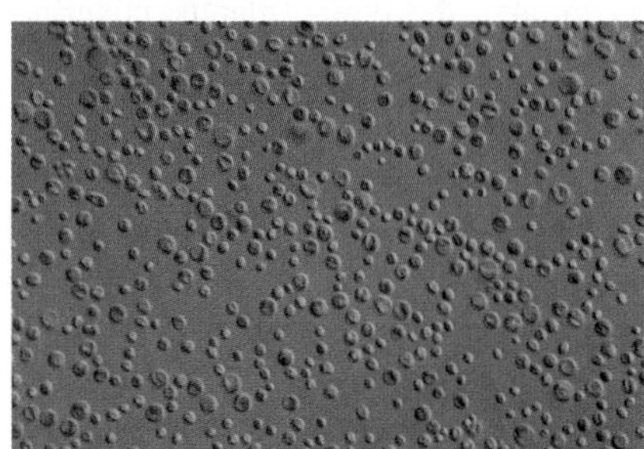

(a) *Chlorella*

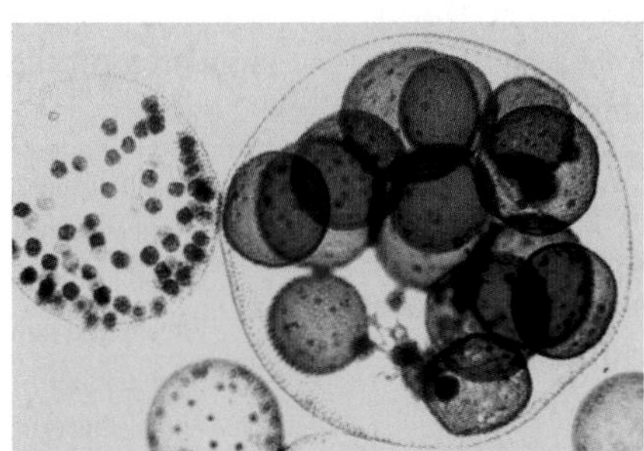

(b) *Volvox*

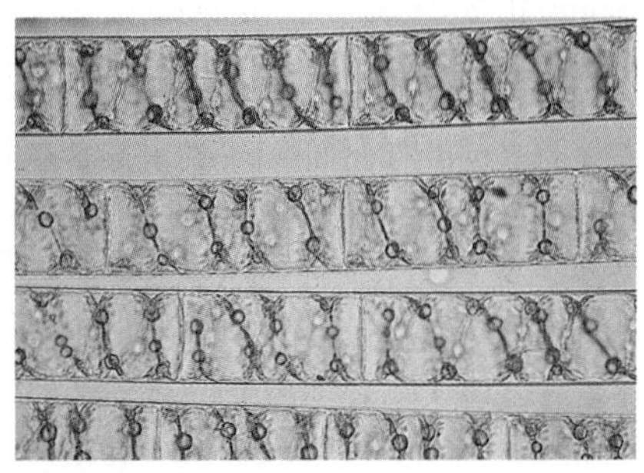

(c) *Spirogyra*

(d) *Acetabularia*

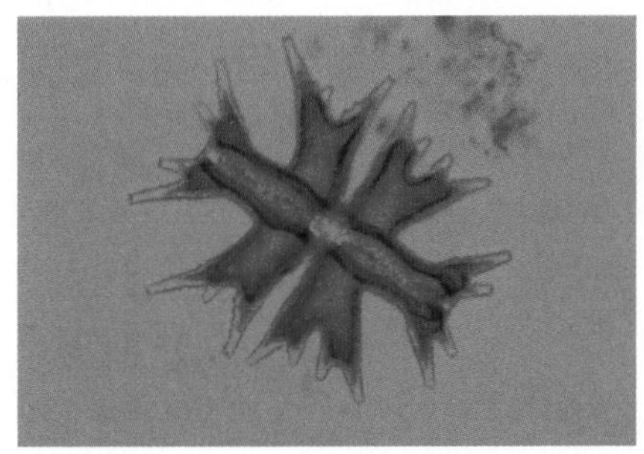

(e) *Micrasterias*

Figure 25.21 *Chlorophyta* (Green Algae); Light Micrographs. **(a)** *Chlorella,* a unicellular nonmotile Chlorophyte (×160). **(b)** *Volvox,* which demonstrates colonial growth (×450). **(c)** *Spirogyra* (×100). Four filaments are shown. Note the ribbonlike, spiral chloroplasts within each filament. **(d)** *Acetabularia,* the mermaid's wine goblet. **(e)** *Micrasterias* (×150).

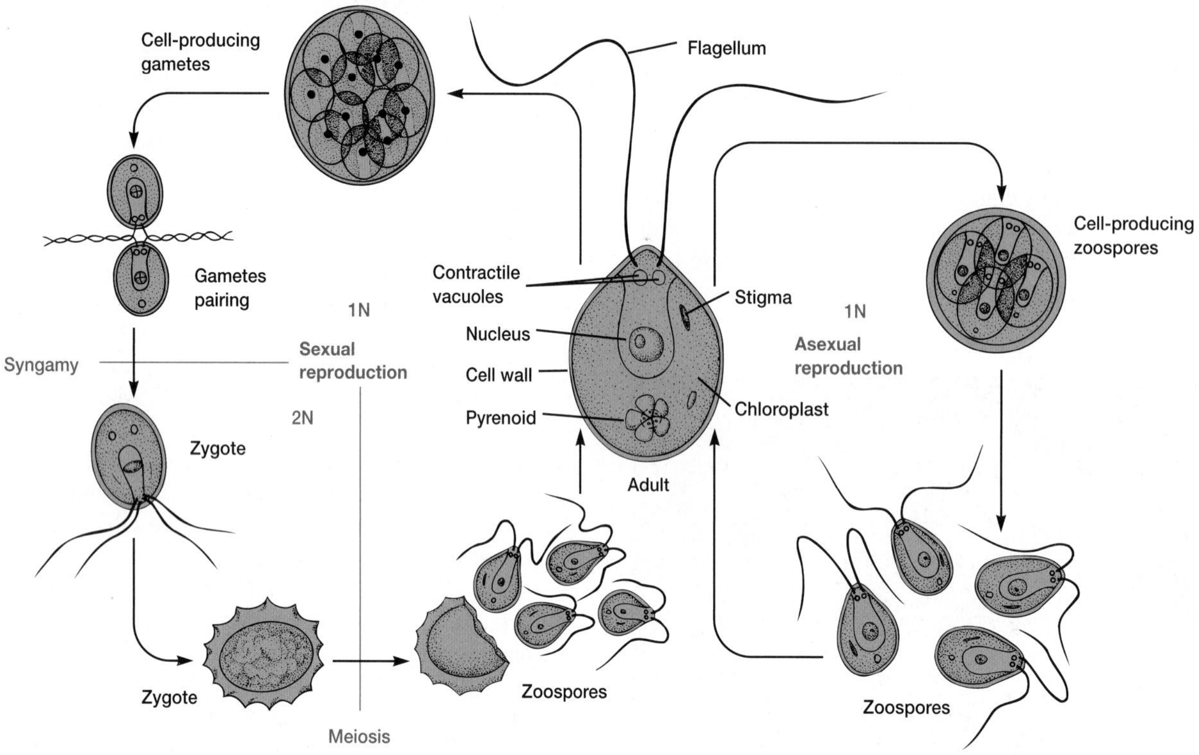

Figure 25.22 ***Chlamydomonas:*** **The Structure and Life Cycle of this Motile Green Alga.** During asexual reproduction, all structures are haploid; during reproduction, only the zygote is diploid.

Another *Chlorophyta, Chlorella* (figure 25.21*a*), is widespread both in fresh- and saltwater and also in soil. It only reproduces asexually and lacks flagella, eyespots, and contractile vacuoles; the nucleus is very small. Motile, colonial organisms such as *Volvox* represent another line of evolutionary specialization. A *Volvox* colony (figure 25.21*b*) is a hollow sphere made up of a single layer of 500 to 60,000 individual cells, each resembling a *Chlamydomonas* cell. The flagella of all the cells beat in a coordinated way to rotate the colony in a clockwise direction as it moves through the water. Only a few cells are reproductive and these are located at the posterior end of the colony. Some divide asexually and produce new colonies; others produce gametes. After fertilization, the zygote divides to form a daughter colony. In both cases the daughter colonies stay within the parental colony until it ruptures.

The chlorophyte *Prototheca moriformis* causes the disease protothecosis in humans and animals. *Prototheca* cells are fairly common in the soil and it is from this site that most infections occur. Severe systemic infections, such as massive invasion of the bloodstream, have been reported in animals. The subcutaneous type of infection is more common in humans. It starts as a small lesion and spreads slowly through the lymph glands, covering large areas of the body.

1. Explain the unique structural features of the diatoms. How does their morphology play a role in the alternation between asexual and sexual reproduction?
2. Why do you think the öomycetes and the *Labyrinthulids* were formerly considered fungi?
3. What is the ecological importance of the coccolithophores?
4. Compare the morphology and swimming behavior of *Chlamydomonas* and *Volvox*.

Summary

25.1 Distribution

a. Protists are found wherever other eucaryotic organisms exist.
b. They are important components of many terrestrial, aquatic, and marine ecosystems where they contribute to nutrient cycling. Many are parasitic in humans and animals (**table 25.1**) and some have become very useful in the study of molecular biology.

25.2 Nutrition

a. Photosynthetic protists are aerobic and perform oxygenic photosynthesis. They can be either photoautotrophic or photoheterotrophic.
b. Chemoorganotrophic protists (protozoa) can be either holozoic or saprozoic. Holozoic forms use phagocytosis to entrap and consume food particles (**figure 25.2**).

25.3 Morphology

a. Because protists are eucaryotic cells, in many respects their morphology and physiology resemble those of multicellular plants and animals. However, because all of their functions must be performed within the individual protist, many morphological and physiological features are unique.

b. The protistan cell membrane is called the plasmalemma and the cytoplasm can be divided into the ectoplasm and endoplasm. The cytoplasm contains contractile, secretory, and food (phagocytic) vacuoles (**figures 25.5** and **25.15**).

c. Energy metabolism occurs within mitochondria, hydrogenosomes, or chloroplasts. Some primitive chemoorganotrophic protists are fermentative, lacking mitochondria and hydrogenosomes.

25.4 Encystment and Excystment

a. Some protists can secrete a resistant covering and go into a resting stage (encystment) called a cyst. Cysts protect the organism against adverse environments, function as a site for nuclear reorganization, and serve as a means of transmission in parasitic species.

25.5 Reproduction

a. Most protists reproduce asexually by binary or multiple fission or budding (**figure 25.3**).

b. Some also use sexual reproduction via a variety of sexual reproductive strategies including conjugation, syngamy, and autogamy (**figure 25.18**).

25.6 Protist Classification

a. Protist phylogeny is the subject of active research and debate. The classification scheme presented here is that of the International Society of Protistologists (**table 25.2**).

b. The super group *Excavata* includes the *Fornicata,* the *Parabasalia,* and the *Euglenozoa.* Most have a cytoproct and use a flagellum for suspension feeding.

c. The human pathogen *Giardia* is a member of the *Fornicata.* It is considered one of the most primitive eucaryotes (**figure 25.4**).

d. Most *Parabasalia* are flagellated endosymbionts of animals. They include the obligate mutualists of wood-eating insects such as *Trichonympha* and the human pathogen *Trichomonas.*

e. Many of the members of the *Euglenozoa* are photoautotrophic (**figure 25.5**). The remainder are chemoorganotrophs, of which most are saprotrophic. Important human pathogens include members of the genus *Trypanosoma.* Trypanosomes cause a number of important human diseases including leishmaniasis, Chagas' disease, and African sleeping sickness (**figure 25.6**).

f. Amoeboid forms use pseudopodia, which can be lobopodia, filopodia, or reticulopodia (**figure 25.7**). Amoebae that bear external plates are called testate; those without plates are called naked or atestate amoebae.

g. Members of the *Amoebozoa* subclasses *Tubulinea* and *Entamoebida* include a number of endosymbiotic and parasitic protists, including *Entamoeba histolytica*, a major cause of parasitic death worldwide.

h. The *Amoebozoa* subclass *Eumycetozoa* includes the acellular and cellular slime molds, which were previously classified as plants, animals, or fungi. The acellular slime molds form a large mass of protoplasm, called a plasmodium, in which individual cells lack a cell membrane (**figure 25.9**). The cellular slime molds produce a pseudoplasmodium and each cell within has a cell wall. *Dictyostelium discoideum* is a cellular slime mold that is used as a model organism in the study of chemotaxis, cellular development, and behavior (**figure 25.10**).

i. The super group *Rhizaria* are amoeboid forms that include the *Radiolaria,* which have filopodia, and the *Foraminifera,* which bear netlike reticulopodia and tests that can be ornate. Most foraminifera are benthic and their tests accumulate on the ocean floor where they are useful in oil exploration (**figures 25.11** and **25.13**).

j. The super group *Chromalveolata* is diverse. It includes the *Alveolata,* which consists of the apicomplexans, dinoflagellates, the stramenopiles and the ciliophora.

k. Apicomplexans are parasitic with complex life cycles. The motile, infective stage is called the sporozoite (**figure 25.15**). The most important apicomplexan is *Plasmodium,* which causes malaria.

l. The dinoflagellates are a large group of nutritionally complex protists. Most are marine and planktonic. They are known for their phosphorescence and for causing toxic blooms. Symbiotic forms live in association with reef-building corals (**figure 25.16**).

m. The *Ciliophora* are chemorganotrophic protists that use cilia for locomotion and feeding (**figure 25.17**). In addition to asexual reproduction, conjugation is used in sexual reproduction (**figure 25.18**).

n. The *Stramenopila,* is extremely diverse and includes diatoms, golden and brown algae, the öomycetes, and labyrinthulids. Diatoms are found in fresh- and saltwater and are important components of marine plankton (**figure 25.19**). The öomycetes and labyrinthulids were once thought to be fungi.

o. The haptophytes include the coccolithophores, planktonic photosynthetic protists that contribute to the global carbon budget by precipitating calcium carbonate for their ornate scales (**figure 25.20**).

p. The *Archaeplastida* include the *Chlorophyta,* also known as green algae. All are photosynthetic with chlorophylls *a* and *b* along with specific carotenoids. They exhibit a wide range of morphologies (**figure 25.21**).

Key terms

acellular slime mold (Myxogastria) 614
algae 605
anisogamy 609
apicomplexan 619
autogamy 609
axopodia 617
binary fission 608
cellular slime mold (*Dictyostelia*) 614
chromosomal nuclei 609
ciliates (*Ciliophora*) 620
coccolithophore 624
conjugant 620
conjugation 609
contractile vacuole 607
cyst 608
cytoproct 608
cytostome 608
diatoms (*Bacillariophyta*) 621
dinoflagellates (*Dinoflagellata*) 620
ectoplasm 607
encystment 608
endoplasm 607
epitheca 622
excystment 608
filopodia 613
frustule 622
gamonts 609
heterokont flagella 621
holozoic nutrition 606
hydrogenosome 608
hypotheca 622
isogamy 609
kinetosome 608
labyrinthulid 624
leishmaniasis 612
lobopodia 613
macronucleus 609
micronucleus 609
mixotrophy 607
naked amoebae 614
öomycete 622
osmotrophy 607
ovular nuclei 609
pellicle 607
phagocytic vacuole 608
phycology 605
planktonic 606
plasmalemma 607
plasmodium 614
protistology 605
protozoa 605
protozoology 605
pseudopodia 613
pyrenoid 608
reticulopodia 613
saprozoic nutrition 607
sporozoite 619
stigma 612
syngamy 609
testate amoebae 614
trophozoite 608
vesicular nucleus 609
zooxanthellae 620

Critical Thinking Questions

1. Why do you think our knowledge of the biology of protists has lagged so far behind that of procaryotes, fungi, and higher eucaryotes?
2. Encystment is usually triggered by changes in the environment. How do you think protists perceive these changes? How might this be similar or different from endospore formation in gram-positive bacteria?
3. Suggest why, in some protists, the cytoplasmic material (ectoplasm) just under the plasma membrane is so rigid?
4. Which of the protists discussed in this chapter do you think are the most evolutionarily advanced or derived? Explain your reasoning.
5. Vaccine development for diseases caused by protists (e.g., malaria, Chagas disease) has been much less successful than that for bacterial diseases. Discuss one biological reason and one geopolitical reason for this fact.

Learn more

Adl, S. M.; Simpson, A. G. B.; Farmer, M. A.; Anderson, R. A.; Anderson, O. R.; Barta, J. R.; Bowser, S. S.; *et al.* 2005. The new higher level classification of Eukaryotes with emphasis on the taxonomy of protists. *J. Eukaryot. Microbiol.* 52: 399–451.

Armbrust, E. V.; Berges, J. A.; Bowler, C.; *et al.* 2004. The genome of the diatom *Thalassiosira pseudonana:* Ecology, evolution, and metabolism. *Science* 306: 79–86.

Baldauf, S. L.; Roger, A. J.; Senk-Siefert, I.; and Doolittle, W. F. 2000. A kingdom-level phylogeny of eukaryotes based on combined protein data. *Science* 290: 972–76.

Biron, D.; Libros, P.; Sagi, D.; Mirelman, D.; and Moses, E. 2001. "Midwives" assist dividing amoebae. *Nature* 410: 973–77.

Falkowski, P. G.; Katz, M. E.; Knoll, A. H.; Quigg, A.; Raven, J. A.; Schofield, O.; and Taylor, F. J. R. 2004. The evolution of modern eukaryotic phytoplankton. *Science* 305: 354–60.

Gardner, M. J.; Bishop, R.; Shah, T.; *et al.* 2005. Genome sequence of *Theileria parva,* a bovine pathogen that transforms lymphocytes. *Science* 309: 134–37.

Grell, K. B. 1973. *Protozoology.* New York: Springer-Verlag.

Lee, J. J.; Leedale, G. F.; and Bradbury, P., editors. 2000. *An illustrated guide to the protozoa,* 2d ed. Lawrence, Kansas: Society of Protozoologists.

Lipscomb, D. L.; Farris, J. S.; Kallersjo, M.; and Tehler, A. 1998. Support, ribosomal sequences and the phylogeny of the eukaryotes. *Cladistics* 14: 303–38.

Roberts, L. S., and Javovy, J. J. 2005. *Foundations in parasitology,* 7th ed. Dubuque, Iowa. McGraw-Hill Higher Education.

Sogin, M. L., and Silberman, J. D. 1998. Evolution of the protists and protistan parasites from the perspective of molecular systematics. *Int. J. Parasitol.* 28: 11–20.

Please visit the Prescott website at www.mhhe.com/prescott7 for additional references.

26 The *Fungi* (*Eumycota*)

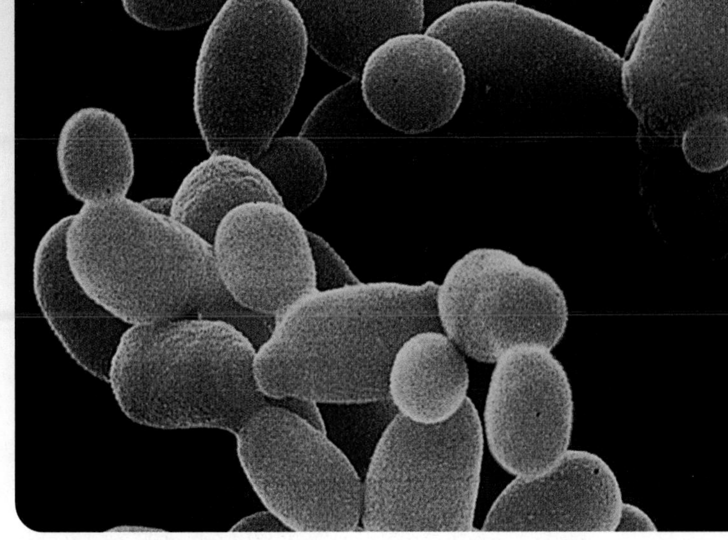

This is a scanning electron micrograph of the microscopic, unicellular yeast, *Saccharomyces cerevisiae* (×21,000). *S. cerevisiae* is the most thoroughly investigated eucaryotic microorganism. This has led to a better understanding of the biology of the eucaryotic cell. Today it serves as a widely used biotechnological production organism as well as a eucaryotic model system.

Preview

- Fungi are widely distributed and are found wherever moisture is present. They are of great importance to humans in both beneficial and harmful ways.
- Fungi exist primarily as filamentous hyphae. A mass of hyphae is called a mycelium.
- Like some bacteria and protists, fungi digest insoluble organic matter by secreting exoenzymes, then absorbing the solubilized nutrients.
- Two reproductive structures occur in fungi: (1) sporangia form asexual spores, and (2) gametangia form sexual gametes.
- Like the study of protists, fungal systematics is an area of active research. Eight fungal subdivisions are presented here, including the *Chytridiomycetes, Zygomycota, Ascomycota, Basidiomycota, Urediniomycetes, Ustilaginomycetes, Glomeromycota*, and *Microsporidia*.
- The *Chytridiomycetes* are a group of terrestrial and aquatic fungi that reproduce by motile zoospores with single, posterior, whiplash flagella.
- The *Zygomycota* are characterized by resting structures called zygospores—cells in which zygotes are formed.
- The *Ascomycota* form zygotes within a characteristic saclike structure, the ascus. The ascus contains two or more ascospores.
- Yeasts are unicellular fungi—most are ascomycetes.
- Basidiomycetes possess dikaryotic hyphae, one of each mating type. The hyphae divide uniquely, forming basidiocarps within which club-shaped basidia can be found. The basidia bear two or more basidiospores.
- The *Ustilaginomycetes* and *Urediniomycetes* include important plant pathogens, whereas the *Glomeromycota* form important associations with vascular plants and enhance plant nutrient uptake. Some members of the *Microsporidia* are considered emerging pathogens of humans.

In this chapter we introduce the *Fungi*. Like protists, fungi have a long and confused taxonomic history. Their relatively simple morphology, wide diversity, and lack of a fossil record limit the value of traditional taxonomic approaches. More recently, the application of molecular techniques including sequence comparisons of small subunit rRNA and conserved proteins has offered new insights into fungal evolution. For example, the division Deuteromycetes (also known as *fungi imperfecti*) is no longer recognized. Here we present eight fungal groups: the *Chytridiomycetes, Zygomycota, Ascomycota, Basidiomycota, Urediniomycetes, Ustilaginomycetes, Glomeromycota,* and *Microsporidia* (**figure 26.1**). The *Urediniomycetes* and the *Ustilaginomycetes* are commonly considered *Basidiomycota*. However, recent evidence suggests that they may be taxonomically distinct.

Microbiologists use the term **fungus** [pl., fungi; Latin *fungus,* mushroom] to describe eucaryotic organisms that are spore-bearing, have absorptive nutrition, lack chlorophyll, and reproduce sexually and asexually. Scientists who study fungi are **mycologists** [Greek *mykes,* mushroom, and *logos,* science], and the scientific discipline devoted to fungi is called **mycology.** The study of fungal toxins and their effects is called **mycotoxicology,** and the diseases caused by fungi in animals are known as **mycoses** (s., mycosis). According to the universal phylogenetic tree, fungi are members of the domain *Eucarya* (figure 26.1*a*). Morphological, biochemical and molecular phylogenetic analyses demonstrate

Yeasts, molds, mushrooms, mildews, and the other fungi pervade our world. They work great good and terrible evil. Upon them, indeed, hangs the balance of life; for without their presence in the cycle of decay and regeneration, neither man nor any other living thing could survive.

—*Lucy Kavaler*

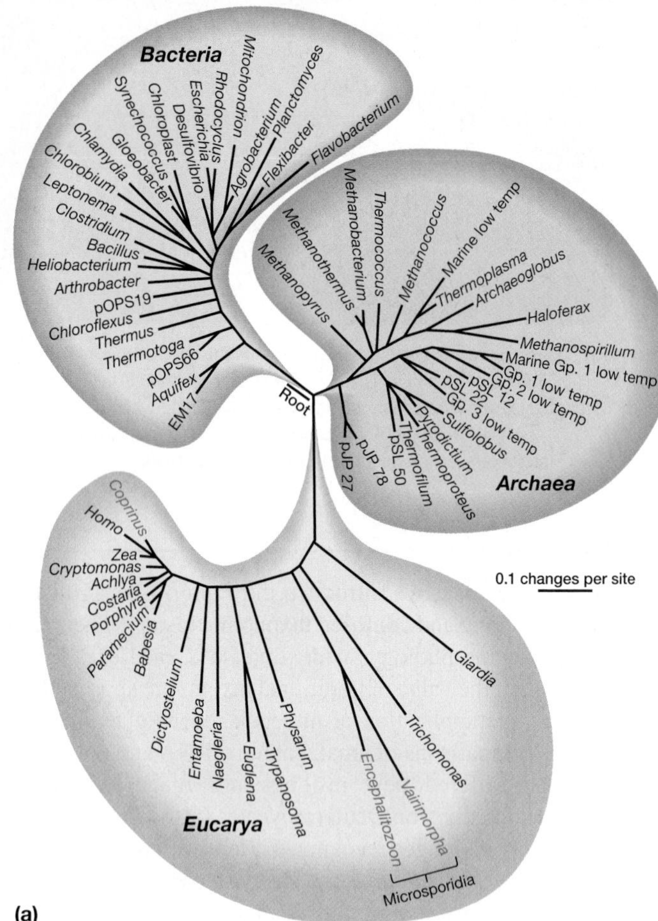

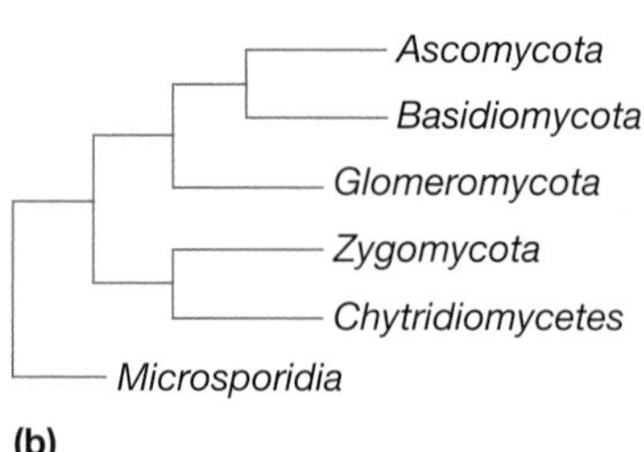

Figure 26.1 The *Fungi*. **(a)** The Universal Tree of Life, an unrooted tree, divides fungi so that with the exception of the *Microsporidia,* all are closely related to the *Metazoa* (*Homo*). *Coprinus* is shown as the representative fungal group. The *Microsporidia* have a confused taxonomic history but are now considered by most to be fungi. **(b)** A rooted tree showing most of the fungal groups discussed in this chapter. Unlike the unrooted three (a), here the *Microsporidia* are shown to be closely related to other fungal groups. Many consider *Urediniomycetes* and the *Ustilaginomycetes* to be *Basidiomycota*.

that the *Fungi* constitute a monophyletic group. They are sometimes referred to as the true fungi or ***Eumycota*** [Greek *eu,* true, and *mykes,* fungus].

26.1 Distribution

Fungi are primarily terrestrial organisms, although a few are freshwater or marine. They have a global distribution from polar to tropical regions. Many are pathogenic and infect plants and animals. Fungi also form beneficial relationships with other organisms. For example, the vast majority of vascular plant roots form associations (called mycorrhizae) with fungi. Fungi also are found in the upper portions of many plants. These endophytic fungi affect plant reproduction and palatability to herbivores. Lichens are associations of fungi and photosynthetic protists or cyanobacteria. Microorganism associations with vascular plants: Mycorrhizae (section 29.5); Microbial interactions (section 30.1)

26.2 Importance

About 90,000 fungal species have been described; however, some estimates suggest that 1.5 million species may exist. Fungi are important to humans in both beneficial and harmful ways. With bacteria and a few other groups of chemoorganotrophic organisms, fungi act as decomposers, a role of enormous significance. They degrade complex organic materials in the environment to simple organic compounds and inorganic molecules. In this way carbon, nitrogen, phosphorus, and other critical constituents of dead organisms are released and made available for living organisms. Microorganisms in the soil environment (section 29.3)

On the other hand, fungi are a major cause of disease. Plants are particularly vulnerable to fungal diseases because fungi can invade leaves through their stomates (**figure 26.2**). Over 5,000 species attack economically valuable crops, garden plants, and many wild plants. Fungi also cause many diseases of animals (**table 26.1**) and humans. In fact, about 20 new human fungal pathogens are documented each year.

Fungi, especially the yeasts, are essential to many industrial processes involving fermentation. Examples include the making of bread, wine, and beer. Fungi also play a major role in the preparation of some cheeses, soy sauce, and sufu; in the commercial production of many organic acids (citric, gallic) and certain drugs (ergometrine, cortisone); and in the manufacture of many antibiotics (penicillin, griseofulvin) and the immunosuppressive drug cyclosporin. These topics are discussed more fully in chapters 34 and 40.

Finally, fungi are important research tools in the study of fundamental biological processes. Cytologists, geneticists, biochemists, biophysicists, and microbiologists regularly use fungi in their research. The yeast *Saccharomyces cerevisiae* is the best understood eucaryotic cell. It has been a valuable model organism in the study of cell biology, genetics, and cancer.

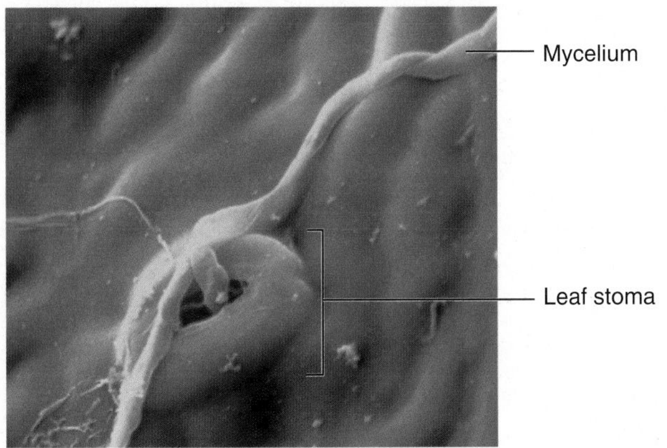

Figure 26.2 Fungal Pathogens of Plants. Scanning electron micrograph of a young mycelial aggregate forming over a leaf stoma (×1,000).

26.3 Structure

The body or vegetative structure of a fungus is called a **thallus** [pl., thalli]. It varies in complexity and size, ranging from the single-cell microscopic yeasts to multicellular molds, macroscopic puffballs, and mushrooms (**figure 26.3**). The fungal cell usually is encased in a cell wall of **chitin.** Chitin is a strong but flexible nitrogen-containing polysaccharide consisting of *N*-acetylglucosamine residues.

A **yeast** is a unicellular fungus that has a single nucleus and reproduces either asexually by budding and transverse division or sexually through spore formation. Each bud that separates can grow into a new yeast, and some group together to form colonies. Generally yeast cells are larger than bacteria, vary considerably in size, and are commonly spherical to egg shaped. They lack flagella but possess most of the other eucaryotic organelles (**figure 26.4**).

The thallus of a **mold** consists of long, branched, threadlike filaments of cells called **hyphae** [s., hypha; Greek *hyphe,* web] that form a **mycelium** (pl., mycelia), a tangled mass or tissuelike aggregation of hyphae (**figure 26.5**). In some fungi, protoplasm streams through hyphae, uninterrupted by cross walls. These hyphae are

Table 26.1 Some Mycotoxicoses[a] Produced by Fungal Mycotoxins in Domestic Animals

Disease	Fungus	Mycotoxin	Contaminated Foodstuff	Animals Affected
Aflatoxicosis	*Aspergillus flavus*	Aflatoxins	Rice, corn, sorghum, cereals, peanuts, soybeans	Poultry, swine, cattle, sheep, dogs
Ergotism	*Claviceps purpurea*	Ergot alkaloids	Seedheads of many grasses, grains	Cattle, horses, swine, poultry
Mushroom poisoning	*Amanita verna*	Amanitins	Eaten from pastures	Cattle
Poultry hemorrhagic syndrome	*Aspergillus flavus* and others	Aflatoxins	Toxic grain and meal	Chickens
Slobbers	*Rhizoctonia*	Alkaloid slaframine	Red clover	Sheep, cattle
Tall fescue toxicosis	*Acremonium coenophialum* (an endophytic fungus)	Ergot alkaloids	Endophyte-infected tall fescue plants	Cattle, horses

[a]A mycotoxicosis [pl., mycotoxicoses] is a poisoning caused by a fungal toxin.

(a) *Penicillium*

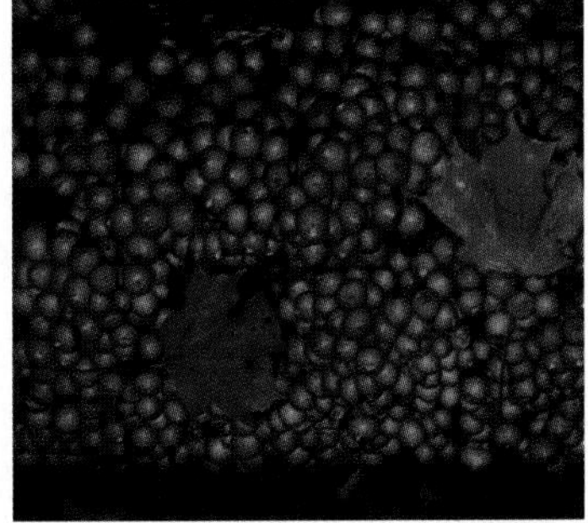
(b) *Lycoperdon*

(c) A mushroom

Figure 26.3 Fungal Thalli. (a) The multicellular common mold, *Penicillium,* growing on an apple. **(b)** A large group of puffballs, *Lycoperdon,* growing on a log. **(c)** A mushroom is made of densely packed hyphae that form the mycelium or visible structure (thallus).

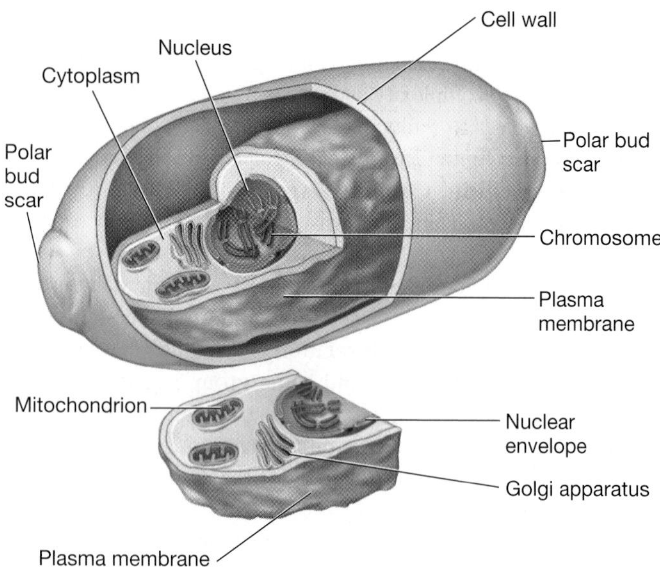

Figure 26.4 A Yeast. Diagrammatic drawing of a yeast cell showing typical morphology. For clarity, the plasma membrane has been drawn separated from the cell wall. In a living cell the plasma membrane adheres tightly to the cell wall.

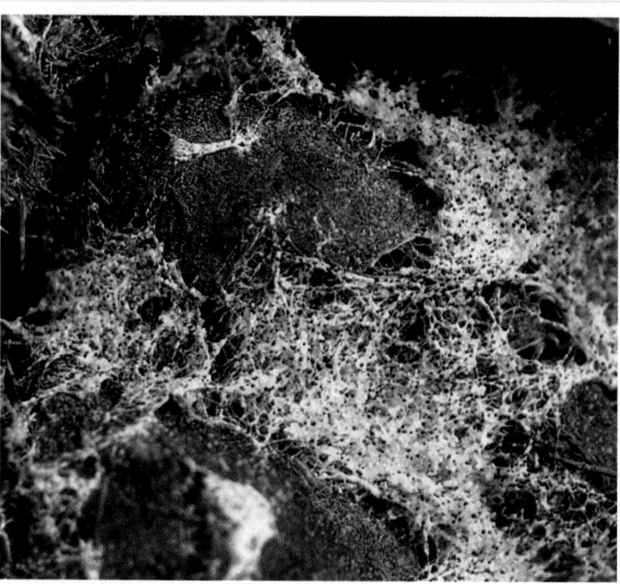

Figure 26.5 Mold Mycelia. The hyphae that compose the fungal mycelium can form a macroscopic mass, as shown by this basidiomycete growing in and on soil.

called **coenocytic** or aseptate (**figure 26.6*a***). The hyphae of other fungi (figure 26.6*b*) have cross walls called **septa** (s., septum) with either a single pore (figure 26.6*c*) or multiple pores (figure 26.6*d*) that enable cytoplasmic streaming. These hyphae are termed **septate.**

Hyphae are composed of an outer cell wall and an inner lumen, which contains the cytosol and organelles (**figure 26.7**). A plasma membrane surrounds the cytoplasm and lies next to the cell wall. The filamentous nature of hyphae results in a large surface area relative to the volume of cytoplasm. This makes adequate nutrient absorption possible.

Many fungi, especially those that cause diseases in humans and animals, are dimorphic (**table 26.2**)—that is, they have two forms. Dimorphic fungi can change from the yeast (Y) form in the animal to the mold or mycelial form (M) in the external environment in response to changes in various environmental factors (nutrients, CO_2 tension, oxidation-reduction potentials, temperature). This shift is called the **YM shift.** In plant-associated fungi the opposite type of dimorphism exists: the mycelial form occurs in the plant and the yeast form in the external environment.

1. How can a fungus be defined?
2. With what organisms do fungi associate? What does the global distribution of fungi imply about the diversity of this group? Explain your answer.
3. What does the term coenocytic mean? Consider the fact that the genomes of coenocytic fungi are not separated by septa. How might this affect clonal growth?
4. What are some forms represented by different fungal thalli?
5. What organelles would you expect to find in the cytoplasm of a typical fungus?
6. What are the major differences between a yeast and a mold?

26.4 Nutrition and Metabolism

Fungi grow best in dark, moist habitats where there is little danger of desiccation, but they are found wherever organic material is available. Most fungi are **saprophytes,** securing their nutrients from dead organic material. Like many bacteria and protists, fungi release hydrolytic exoenzymes that digest external substrates. They then absorb the soluble products—a process sometimes called **osmotrophy.** They are chemoorganoheterotrophs and use organic compounds as a source of carbon, electrons, and energy. Glycogen is the primary storage polysaccharide in fungi. Most fungi use carbohydrates (preferably glucose or maltose) and nitrogenous compounds to synthesize their own amino acids and proteins.

Fungi usually are aerobic. Some yeasts, however, are facultatively anaerobic and can obtain energy by fermentation. Many fungal fermentations are of industrial importance, such as the production of ethyl alcohol in the manufacture of beer and wine. Obligately anaerobic fungi are found in the rumen of cattle.

26.5 Reproduction

Reproduction in fungi can be either asexual or sexual. Asexual reproduction is accomplished in several ways:

1. A parent cell can undergo mitosis and divide into two daughter cells by a central constriction and formation of a new cell wall (**figure 26.8*a***).
2. Mitosis in vegetative cells may be concurrent with budding to produce a daughter cell. This is very common in the yeasts.
3. The most common method of asexual reproduction is spore production. Asexual spore formation occurs in an individual

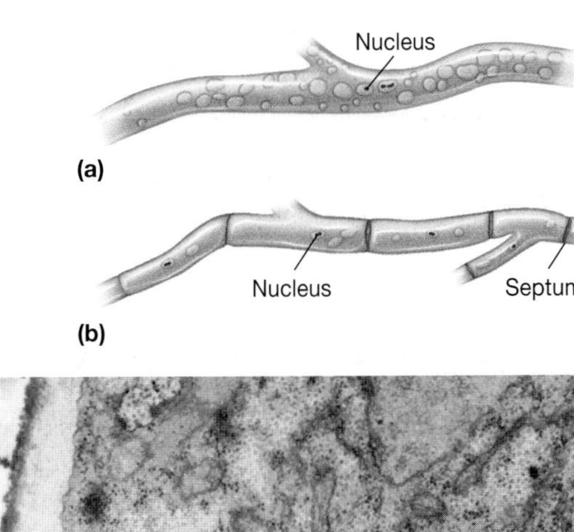

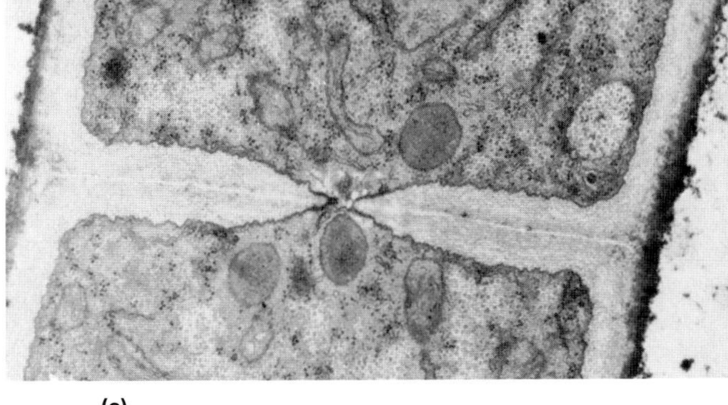
(c)

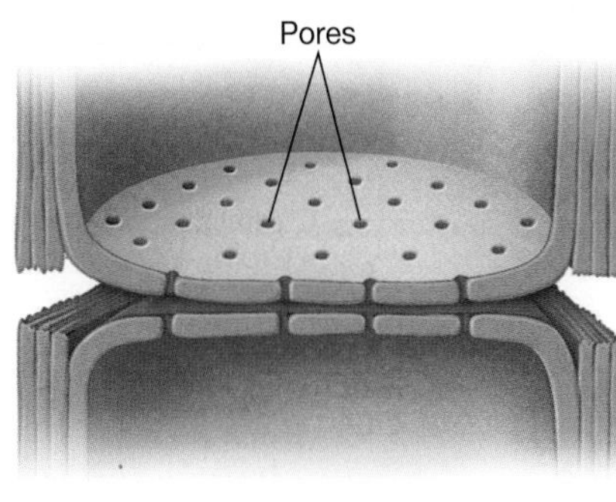

(d)

Figure 26.6 Hyphae. Drawings of **(a)** coenocytic hyphae (aseptate) and **(b)** hyphae divided into cells by septa. **(c)** Electron micrograph (×40,000) of a section of *Drechslera sorokiniana* showing wall differentiation and a single pore. **(d)** Drawing of a multiperforate septal structure.

Table 26.2 Some Medically Important Dimorphic Fungi

Fungus	Disease[a]
Blastomyces dermatitidis	Blastomycosis
Candida albicans	Candidiasis (Thrush)
Coccidioides immitis	Coccidioidomycosis
Histoplasma capsulatum	Histoplasmosis
Sporothrix schenckii	Sporotrichosis
Paracoccidioides brasiliensis	Paracoccidioidomycosis

[a]See chapter 40 for a discussion of each of these diseases.

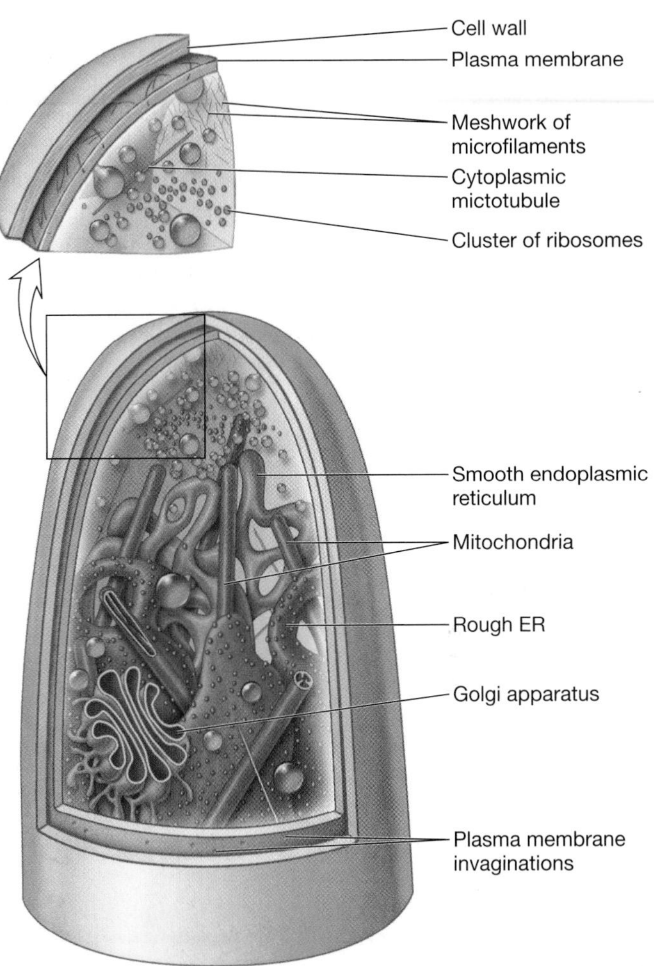

Figure 26.7 Hyphal Morphology. Diagrammatic representation of a hyphal tip showing typical organelles and other structures.

fungus through mitosis and subsequent cell division. There are several types of asexual spores, each with its own name:

a. A hypha can fragment (by the separation of hyphae through splitting of the cell wall or septum) to form cells that behave as spores. These cells are called **arthroconidia** or **arthrospores** (figure 26.8*b*).
b. If the cells are surrounded by a thick wall before separation, they are called **chlamydospores** (figure 26.8*c*).
c. If the spores develop within a sac (**sporangium;** pl., sporangia) at a hyphal tip, they are called **sporangiospores** (figure 26.8*d*).
d. If the spores are not enclosed in a sac but produced at the tips or sides of the hypha, they are termed **conidiospores** (figures 26.8*e* and 26.13).
e. Spores produced from a vegetative mother cell by budding (figure 26.8*f*) are called **blastospores.**

Sexual reproduction in fungi involves the fusion of compatible nuclei. Homothallic fungal species are self-fertilizing and produce sexually compatible gametes on the same mycelium.

Heterothallic species require outcrossing between different but sexually compatible mycelia. It has long been held that sexual reproduction must occur between mycelia of opposite **mating types (MAT).** However, one instance of same-sex mating was discovered following an outbreak of the pathogenic yeast *Crytococcus gatti* in Canada. Depending on the species, sexual fusion may occur between haploid gametes, gamete-producing bodies called **gametangia,** or hyphae. Sometimes both the cytoplasm and haploid nuclei fuse immediately to produce the diploid zygote. Usually, however, there is a delay between cytoplasmic and nuclear fusion. This produces a **dikaryotic stage** in which cells contain two separate haploid nuclei (N + N), one from each parent (**figure 26.9**). After a period of dikaryotic existence, the two nuclei fuse and undergo meiosis to yield spores. For example, in the zygomycetes the zygote develops into a **zygospore** (**figure 26.10**); in the ascomycetes, an **ascospore** (figure 26.14); and in the basidomycetes; a **basidiospore** (figure 26.15).

Fungal spores are important for several reasons. The spores enable fungi to survive environmental stresses such as desiccation, nutrient limitation, and extreme temperatures, although they are not as stress resistant as bacterial endospores. Because they are often small and light, spores can remain suspended in air for long periods. Thus they frequently aid in fungal dissemination, a significant factor that helps explain the wide distribution of many fungi. Fungal spores often spread by adhering to the bodies of insects and other animals. The bright colors and fluffy textures of many molds often are due to their aerial hyphae and spores. The

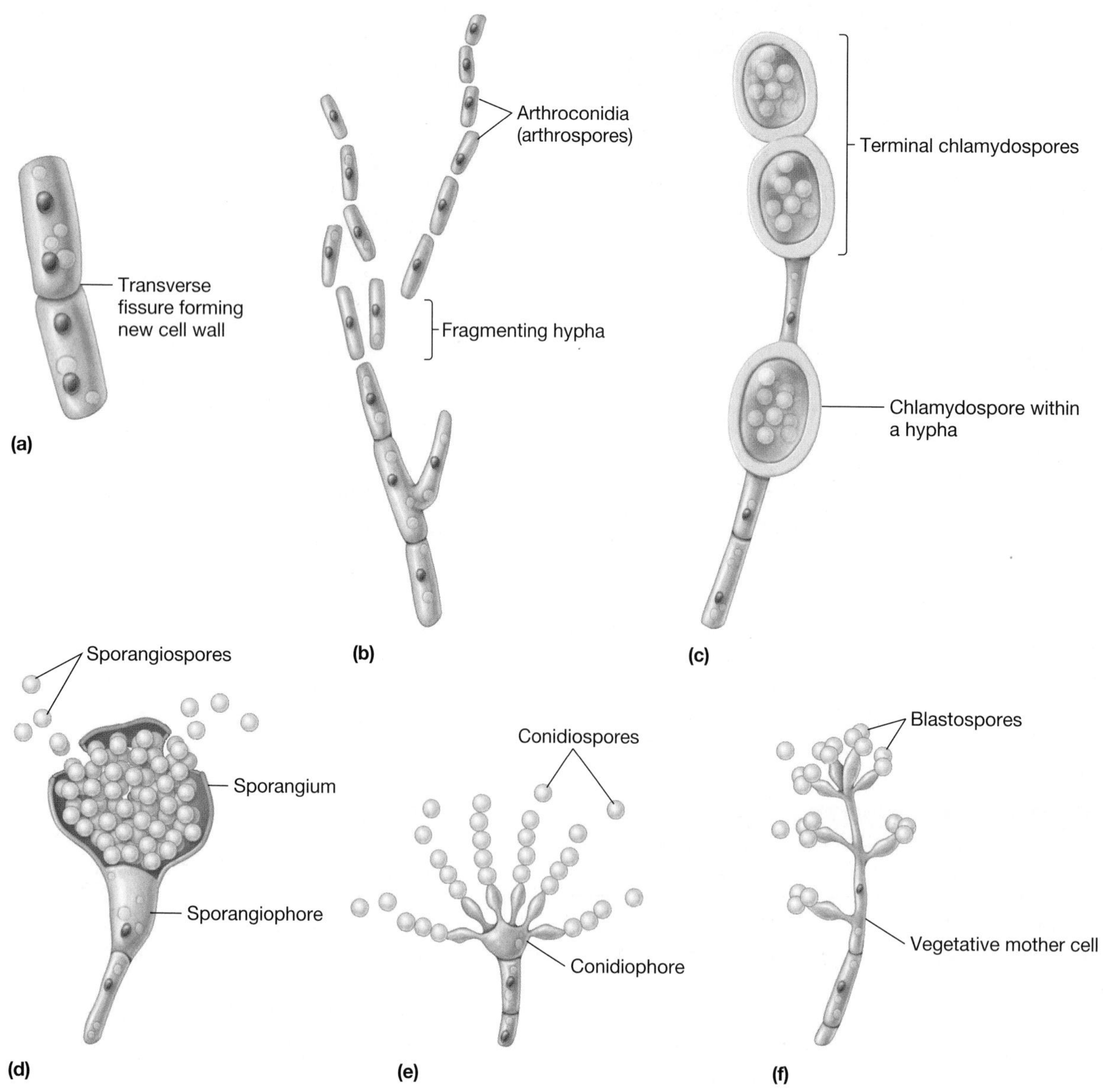

Figure 26.8 Diagrammatic Representation of Asexual Reproduction in the Fungi and Some Representative Spores. **(a)** Transverse fission. **(b)** Hyphal fragmentation resulting in arthroconidia (arthrospores) and **(c)** chlamydospores. **(d)** Sporangiospores in a sporangium. **(e)** Conidiospores arranged in chains at the end of a conidiophore. **(f)** Blastospores are formed from buds off of the parent cell.

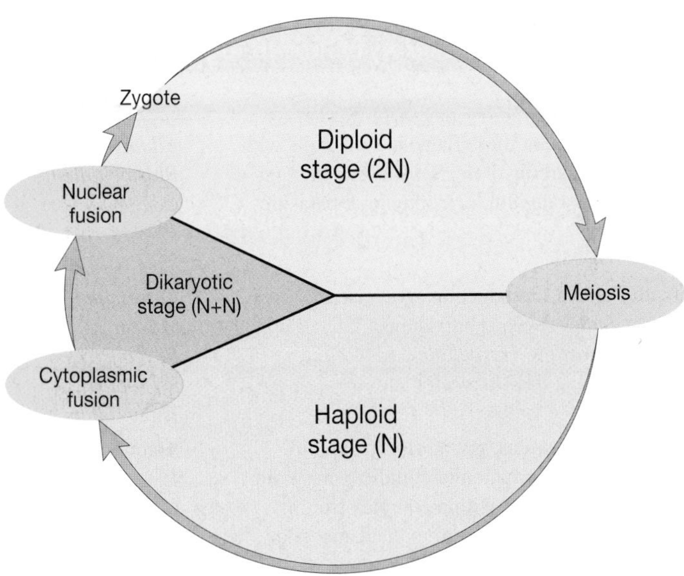

Figure 26.9 Reproduction in Fungi. A drawing of the generalized life cycle for fungi showing the alternation of haploid and diploid stages. Some fungal species do not pass through the dikaryotic stage indicated in this drawing. The asexual (haploid) stage is used to produce spores that aid in the dissemination of the species. The sexual (diploid) stage involves the formation of spores that survive adverse environmental conditions (e.g., cold, dryness, heat).

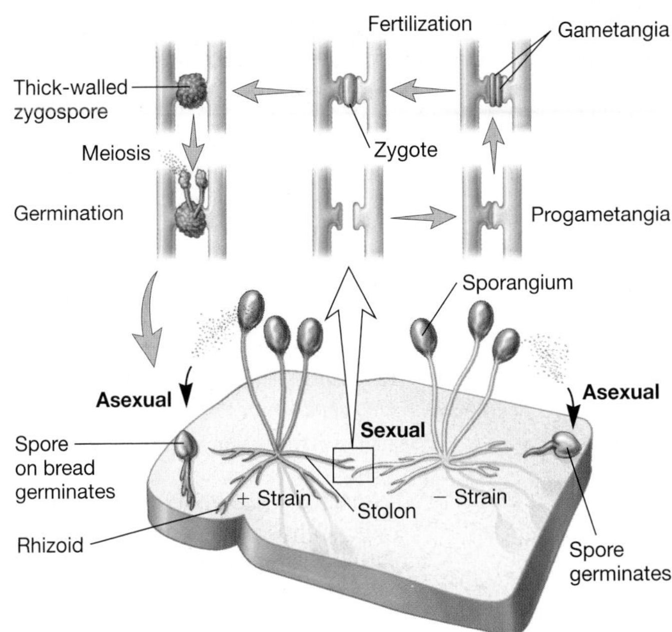

Figure 26.10 The *Zygomycota*. Diagrammatic representation of the life cycle of *Rhizopus stolonifer.* Both the sexual and asexual phases are illustrated.

size, shape, color, and number of spores are useful in the identification of fungal species.

1. What are saprophytes? How do fungi usually obtain their nutrients?
2. Fungi were originally classified as plants. Why do you think early 20th-century biologists made this mistake?
3. How is asexual reproduction accomplished in the fungi? sexual reproduction?
4. Describe each of the following types of asexual fungal spores: sporangiospore, conidiospore, and blastospore.
5. How are fungi dispersed in the environment?

26.6 Characteristics of the Fungal Divisions

The long evolutionary history of fungi is rich in examples of **convergent** and **divergent evolution.** That is to say, many structurally and functionally similar structures evolved independently (converged) while other structures became dissimilar (diverged) over time. Without an extensive fossil record, the use of morphology in phylogenetic analysis is limited. However, sequence analysis of 18S rRNA and certain protein-coding genes has shown that the *Fungi* comprise a monophyletic group with eight subdivisions. Four of these—the *Chytridiomycetes, Zygomycota, Ascomycota,* and *Basidiomytota*—have been recognized as separate groups for some time. The other four—the *Urediniomycetes, Ustilaginomycetes, Glomeromycota,* and *Microsporidia*—have only recently been proposed as separate groups. **Table 26.3** surveys all eight groups; we focus most of our discussion on the better-known groups.

Chytridiomycota

The simplest of the fungi belong to the *Chytridiomycota,* or **chytrids.** They are unique among fungi in the production of a zoospore with a single, posterior, whiplash flagellum. This is considered a primitive feature that was lost in more evolved fungi. Free-living members of this taxon are saprotrophic—living on plant or animal matter in freshwater, mud, or soil. Parasitic forms infect aquatic plants and animals including insects. A few are found in the anoxic rumen of herbivores. Based on zoospore morphology, the subdivisions within the chytridiomycetes include the *Blastocladiales, Monoblepharidales, Neocallimastigaceae, Spizellomycetales,* and the *Chytridiales.*

Chytridiomycota display a variety of life cycles involving both asexual and sexual reproduction. Members of this group are microscopic in size and may consist of a single cell, a small multinucleate mass, or a true mycelium with hyphae capable of penetrating porous substrates. Sexual reproduction results in the release of sporangiospores from sporangia (figure 26.8*d*) at the surface. Many are capable of degrading cellulose and even keratin, which enables the degradation of crustacean exoskeletons. The genus *Allomyces* is used to study morphogenesis.

Zygomycota

The *Zygomycota* contains fungi called **zygomycetes.** Most live on decaying plant and animal matter in the soil; a few are parasites of

Table 26.3 Abbreviated Classification of the Fungi as proposed by International Society of Protistologists[a]

Subclass	Characteristics	Examples
Chytridiomycetes	Flagellated cells in at least one stage of life cycle; may have one or more flagella. Cell walls with chitin and β-1,3-1,6-glucan; glycogen is used as a storage carbohydrate. Sexual reproduction often results in a zygote that becomes a resting spore or sporangium; saprophytic or parasitic. Currently six major subdivisions.	*Allomyces* *Blastocladiella* *Coelomomyces* *Physoderma* *Synchytrium*
Zygomycota	Thalli usually filamentous and nonseptate, without cilia; sexual reproduction gives rise to thick-walled zygospores that are often ornamented. Includes seven subdivisions: *Basidiobolus, Dimargaritales, Endogonales, Entomophthorales, Harpellales, Kickxellales, Mucorales,* and *Zoopagales.* Human pathogens found among the *Mucorales* and *Entomophthorales.*	*Amoebophilus* *Mucor* *Phycomyces* *Rhizopus* *Thamnidium*
Ascomycota	Sexual reproduction involves meiosis of a diploid nucleus in an ascus, giving rise to haploid ascospores; most also undergo asexual reproduction with the formation of conidiospores with specialized aerial hyphae called conidiophores. Many produce asci within complex fruiting bodies called ascocarps. Includes saprophytic, parasitic forms; many form mutualisms with phototrophic microbes to form lichens. Four monophyletic subdivisions including: *Saccharomycetes, Pezizomycotina, Taphrinomycotina,* and *Neolecta.*	*Ascobolus* *Aspergillis* *Candida* *Crinula* *Neurospora* *Penicillium* *Pneumocystis* *Saccharomyces*
Basidiomycota	Includes many common mushrooms and shelf fungi. Sexual reproduction involves formation of a basidium (small, club-shaped structure that typically forms spores at the ends of tiny projections) within which haploid basidiospores are formed. Usually 4 spores per basidium, but can range from 1 to 8. Sexual reproduction involves fusion with opposite mating type resulting in a dikaryotic mycelium with parental nuclei paired but not initially fused. No subdivisions recognized.	*Agaricus* *Boletes* *Dacrymyces* *Lycoperdon* *Polyporus* *Russula* *Tremella*
Urediniomycetes	Mycelial or yeast forms. Sexual reproduction involves fusion of parental nuclei in probasidium followed by meiosis in a separate compartment. Many are plant pathogens called rusts, animal pathogens, nonpathogenic endophytes, and rhizosphere species. No subdivisions recognized.	*Caeoma* *Melampsora* *Uromyces*
Ustilaginomycetes	Plant parasites that cause rusts and smuts. Mycelial in parasitic phase; meiospores formed on septate or aseptate basidia; cell wall principally composed of glucose. No subdivisions recognized.	*Malassezia* *Tilletia* *Ustilago*
Glomeromycota	Filamentous, most are endomycorrhizal, arbuscular; lack cilium; form asexual spores outside of host plant; lack centrioles, conidia, and aerial spores. No subdivisions recognized.	*Acaulospora* *Entrophospora* *Glomus*
Microsporidia	Obligate intracellular parasites usually of animals. Lack mitochondria, peroxisomes, kinetosomes, cilia, and centrioles; spores have an inner chitin wall and outer wall of protein; produce a tube for host penetration. Subdivisions currently uncertain.	*Amblyospora* *Encephalitozoon* *Enterocytozoon* *Nosema*

[a]Adapted from: Adl, S. M.; Simpson, A. G. B.; Farmer, M. A.; Anderson, R. A.; Anderson, O. R.; Barta, J. R.; Bowser, S. S.; et al. 2005. The new higher level classification of Eukaryotes with emphasis on the taxonomy of protists. *J. Eukaryot. Microbiol.* 52:399–451.

plants, insects, other animals, and humans. The hyphae of zygomycetes are coenocytic, with many haploid nuclei. Asexual spores, usually wind dispersed, develop in sporangia at the tips of aerial hyphae. Sexual reproduction produces tough, thick-walled zygotes called zygospores that can remain dormant when the environment is too harsh for growth of the fungus.

The bread mold, *Rhizopus stolonifer,* is a very common member of this division. This fungus grows on the surface of moist, carbohydrate-rich foods, such as breads, fruits, and vegetables. On breads, for example, *Rhizopus*'s hyphae rapidly cover the surface. Hyphae called rhizoids extend into the bread, and absorb nutrients (figure 26.10). Other hyphae (stolons) become erect, then arch back into the substratum forming new rhizoids. Still others remain erect and produce at their tips asexual sporangia filled with the black spores, giving the mold its characteristic color. Each spore, when liberated, can germinate to start a new mycelium.

Rhizopus usually reproduces asexually, but if food becomes scarce or environmental conditions unfavorable, it begins sexual reproduction. Sexual reproduction requires compatible strains of opposite mating types (figure 26.10). These have traditionally been labeled + and − strains because they are not morphologically distinguishable as male and female. When the two mating strains are close, hormones are produced that cause their hyphae to form projections called **progametangia** [Greek *pro,* before], and then mature gametangia. After fusion of the gametangia, the nuclei of the two gametes fuse, forming a zygote. The zygote develops a thick, rough, black coat and becomes a dormant zygospore. Meiosis often occurs at the time of germination; the zygospore then splits open and produces a hypha that bears an asexual sporangium and the cycle begins anew.

The genus *Rhizopus* is also important because it is involved in the rice disease known as seedling blight. If one considers that rice feeds more people on Earth than any other crop, the implications of this disease are obvious. It was thought that *Rhizopus* secreted a toxin that kills rice seedlings, so scientists set about isolating the toxin and the genes that produce it. Much to everyone's surprise, a β-proteobacterium, *Burkholderia,* found growing within the fungus produces the toxin. Although the evolutionary history of the *Rhizopus-Burkholderia* association remains to be clarified, there is at least one interesting twist to this story: the same toxin has been shown to stop cell division in some human cancer cells and is now being investigated as an antitumor agent.

The zygomycetes also contribute to human welfare. For example, one species of *Rhizopus* is used in Indonesia to produce a food called tempeh from boiled, skinless soybeans. Another zygomycete (*Mucor* spp.) is used with soybeans in Asia to make a curd called sufu. Others are employed in the commercial preparation of some anesthetics, birth control agents, industrial alcohols, meat tenderizers, and the yellow coloring used in margarine and butter substitutes. Microbiology of food (chapter 40)

Ascomycota

The *Ascomycota* contain fungi called **ascomycetes,** commonly known as sac fungi. Ascomycetes are ecologically important in freshwater, marine, and terrestrial habitats because they degrade many chemically stable organic compounds including lignin, cellulose, and collagen. Many species are quite familiar and economically important (**figure 26.11**). For example, most of the red, brown, and blue-green molds that cause food spoilage are ascomycetes. The powdery mildews that attack plant leaves and the fungi that cause chestnut blight and Dutch elm disease are ascomycetes. Many yeasts as well as edible morels and truffles are ascomycetes. The pink bread mold *Neurospora crassa,* also an ascomycete, is an extremely important research tool in genetics and biochemistry.

Many ascomycetes are parasites on higher plants. *Claviceps purpurea* parasitizes rye and other grasses, causing the plant disease **ergot. Ergotism,** the toxic condition in humans and animals who eat grain infected with the fungus, is often accompanied by gangrene, psychotic delusions, nervous spasms, abortion, and convulsions. During the Middle Ages ergotism, then known as St. Anthony's fire, killed thousands of people. For example, over 40,000 deaths from ergot poisoning were recorded in France in the year 943. It has been suggested that the widespread accusations of witchcraft in Salem Village and other New England communities in the late 1690s may have resulted from outbreaks of ergotism. The pharmacological activities are due to an active ingredient, lysergic acid diethylamide (LSD). In controlled dosages other active compounds can be used to induce labor, lower blood pressure, and ease migraine headaches.

The ascomycetes are named for their characteristic reproductive structure, the saclike **ascus** [pl., asci; Greek *askos,* sac]. Many ascomycetes are yeast. The term yeast is used to refer to unicellular fungi that reproduce asexually by either budding or binary fission (**figure 26.12*a***); the life cycle of the yeast *Saccharomyces cerevisiae* is well understood. This ascomycete alternates between haploid and diploid states (figure 26.12*b*). As long as nutrients remain plentiful, haploid and diploid cells undergo mitosis to produce haploid and diploid daughter cells, respectively. Each daughter cell leaves a scar on the mother cell as it separates, and daughter cells bud only from unscarred regions of the cell wall. When a mother cell has no more unscarred cell wall remaining, it can no longer reproduce and will senesce (die). When nutrients are limited, diploid *S. cerevisiae* cells undergo meiosis to produce four haploid cells that remain bound within a common cell wall, the acsus. Upon the addition of nutrients, if two haploid cells of opposite mating types (*a* and α) come into contact, they will fuse to create a diploid. Typically only cells of opposite mating types can fuse; this process is tightly regulated by the action of pheromones.

Filamentous ascomycetes form septate hyphae. Asexual reproduction is common in these ascomycetes and takes place by means of conidiospores (**figure 26.13**). Sexual reproduction also

(a) *Morchella esculenta*

(b) *Sarcoscypha coccinea*

(c) *Tuber brumale*

Figure 26.11 The *Ascomycota*. **(a)** The common morel, *Morchella esculenta,* is one of the choicest edible fungi. It fruits in the spring. **(b)** Scarlet cups, *Sarcoscypha coccinea,* with open ascocarps (apothecia). **(c)** The black truffle, *Tuber brumale,* is highly prized for its flavor by gourmet cooks. Truffles are mycorrhizal associations on oak trees.

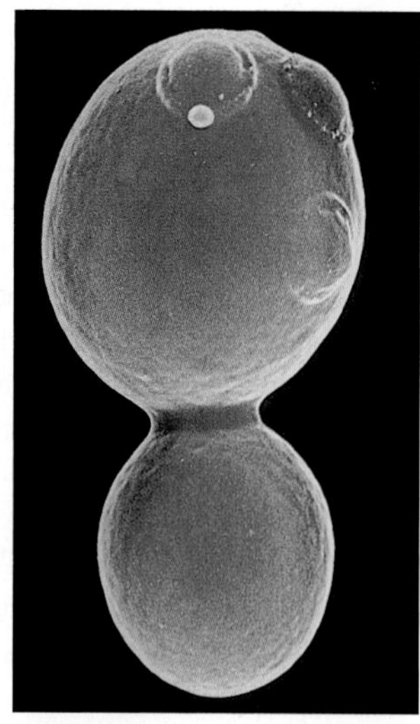

(a) *Saccharomyces cerevisiae:* budding division

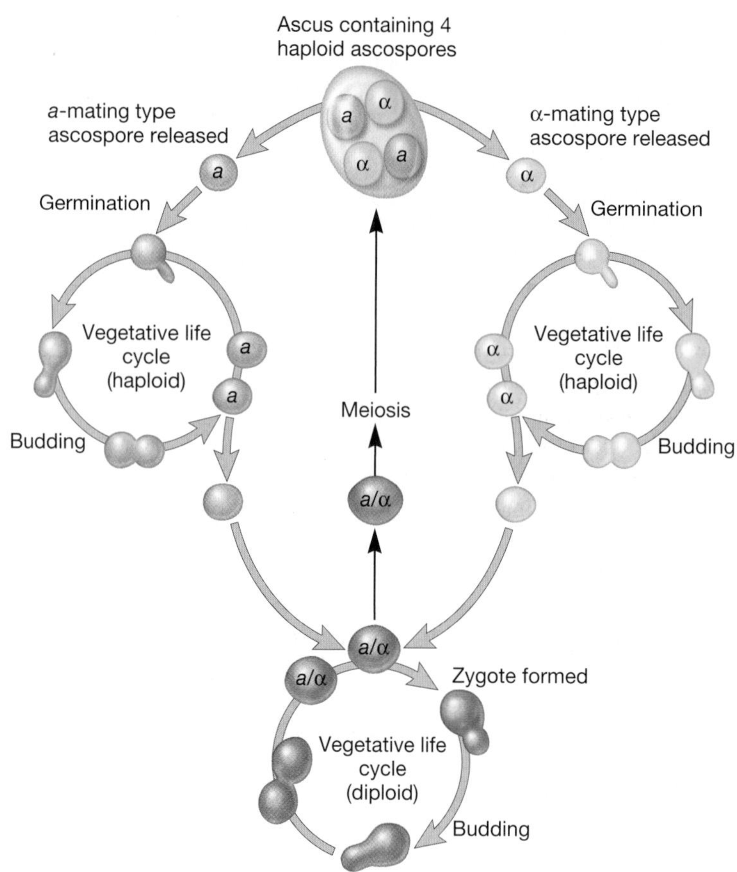

(b) *S.cereviseae* life cycle

Figure 26.12 The Life Cycle of the Yeast *Saccharomyces cerevisiae.* **(a)** Budding division results in asymmetric septation and the formation of a smaller daughter cell. **(b)** When nutrients are abundant, haploid and diploid cells undergo mitosis and grow vegetatively. When nutrients are limited, diploid *S. cerevisiae* cells undergo meiosis to produce four haploid cells that remain bound within a common cell wall, the ascus. Upon the addition of nutrients, two haploid cells of opposite mating types (*a* and α) fuse to create a diploid cell.

Figure 26.13 Asexual Reproduction in *Ascomyota.* Characteristic conidiospores of *Aspergillus* as viewed with the electron microscope (×1,200).

involves ascus formation, with each ascus usually bearing eight haploid ascospores, although some species can produce over 1,000 (**figure 26.14*a***). In the more complex ascomycetes, ascus formation is preceded by the development of special **ascogenous hyphae** into which pairs of nuclei migrate (figure 26.14*b*). One nucleus of each pair originates from a "male" mycelium (**antheridium**) or cell and the other from a "female" organ or cell (**ascogonium**) that has fused with it. As the ascogenous hyphae grow, the paired nuclei divide so that there is one pair of nuclei in each cell. After the ascogenous hyphae have matured, nuclear fusion occurs at the hyphal tips in the ascus mother cells. The diploid zygote nucleus then undergoes meiosis, and the resulting four haploid nuclei divide mitotically again to produce a row of eight nuclei in each developing ascus. These nuclei are walled off from one another. Thousands of asci may be packed together in a cup- or flask-shaped fruiting body called an **ascocarp** (figure 26.14*b*). When the ascospores mature, they often are released from the asci with great force. If the mature ascocarp is jarred, it may appear to belch puffs of "smoke" consisting of thousands of ascospores. Upon reaching a suitable environment, the ascospores germinate and start the cycle anew.

The genomes of three *Aspergillus* species have been sequenced, annotated, and compared. This ascomycete genus is important for a number of reasons. *A. fumigatus* is ubiquitous, commonly found in homes and in the environment. It is known to trigger allergic responses and is implicated in the increased incidence in severe asthma and sinusitis. It is also pathogenic, infecting immunocompromised individuals with a mortality rate of nearly 50%. Its 29.4-Mb genome consists of eight chromosomes and about 10,000 genes. *A. nidulans* is a model organism that is

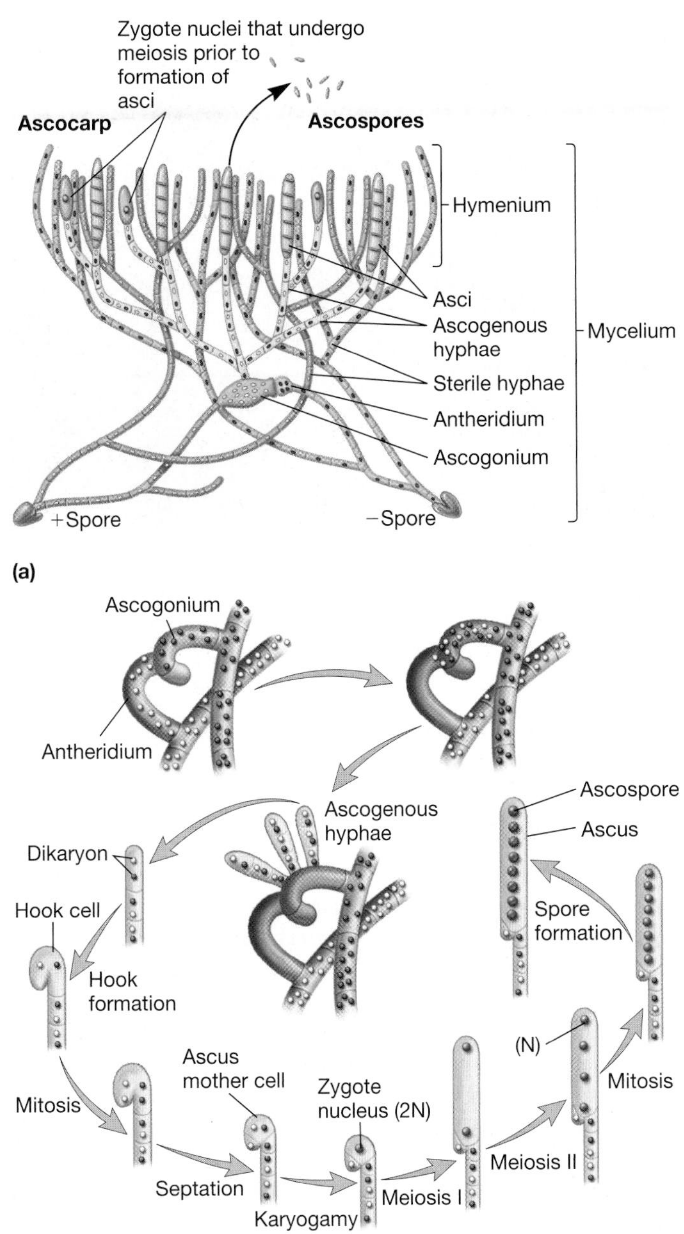

Figure 26.14 The Typical Life Cycle of a Filamentous Ascomycete. Sexual reproduction involves the formation of asci and ascospores. Within the ascus, karyogamy is followed by meiosis to produce the ascospores. **(a)** Sexual reproduction and ascocarp morphology of a cup fungus. **(b)** The details of sexual reproduction in ascogenous hyphae. The nuclei of the two mating types are represented by unfilled and filled circles.

used to study questions of eucaryotic cell and developmental biology. *A. oryzae* is used in the production of traditional fermented foods and beverages in Japan including saki and soy sauce. Because it secretes many industrially useful proteins and can be genetically manipulated, it has become an important organism in biotechnology. Its metabolic and genetic versatility is reflected in the size of its genome—at 37 Mb, it is considerably larger than either *A. fumigatus* or *A. nidulans*. Much of this additional sequence is involved in the production of secreted hydrolytic enzymes, nutrient transport systems, and secondary metabolites.

Comparative analysis of the three genomes reveals over 5,000 nonprotein-coding regions that are conserved in all three species. A variety of regulatory elements are present, including a riboswitch and other forms of translational control. These genome sequences will be useful to scientists seeking to understand the interaction between *Aspergillus* and the immune system, its role in food and industrial microbiology, and eucaryotic evolution. Regulation at the level of translation: Riboswitches (section 12.4); Comparative genomics (section 15.6); Microbiology of fermented foods (section 40.6)

Basidiomycota

The *Basidiomycota* includes the **basidiomycetes,** commonly known as club fungi. Examples include jelly fungi, rusts, shelf fungi, stinkhorns, puffballs, toadstools, mushrooms, and bird's nest fungi. Basidiomycetes are named for their characteristic structure or cell, the **basidium,** which is involved in sexual reproduction (**figure 26.15**). A basidium [Greek *basidion,* small base] is produced at the tip of hyphae and normally is club shaped. Two or more basidiospores are produced by the basidium, and basidia may be held within fruiting bodies called **basidiocarps.**

The basidiomycetes affect humans in many ways. Most are saprophytes that decompose plant debris, especially cellulose and lignin. For example, the common fungus *Polyporus squamosus* forms large, shelflike structures that project from the lower portion of dead trees, which they help decompose. The fruiting body can reach 2 feet in diameter and has many pores (hence the name *Polyporus*), each lined with basidia that produce basidiospores. Thus a single fruiting body can produce millions of spores. Many mushrooms are used as food throughout the world. The cultivation of the mushroom *Agaricus campestris* is a multimillion-dollar business (*see figure 40.20*). Of course not all mushrooms are edible; as suggested by its name, ingestion of *Russula emetica* induces vomiting.

Many mushrooms produce specific alkaloids that act as either poisons or hallucinogens. One such example is the "death angel" mushroom, *Amanita phalloides.* Two toxins isolated from this species are phalloidin and α-amanitin. Phalloidin primarily attacks liver cells where it binds to plasma membranes, causing them to rupture and leak their contents. Alpha-amanitin attacks the cells lining the stomach and small intestine and is responsible for the severe gastrointestinal symptoms associated with mushroom poisoning.

The basidiomycete *Cryptococcus neoformans* is an important human and animal pathogen. It produces the disease called cryptococcosis, a systemic infection primarily involving the lungs and central nervous system. The production of an elaborate capsule is an important virulence factor for the microbe. Analysis of the *C. neoformans* 19-Mb genome has uncovered over 30 new genes involved in capsule biosynthesis. This may help researchers develop new antifungal agents. Airborne diseases (section 39.2)

Urediniomycetes and *Ustilaginomycetes*

Often considered *Basidiomycota,* both the *Urediniomycetes* and the *Ustilaginomycetes* include important plant pathogens causing

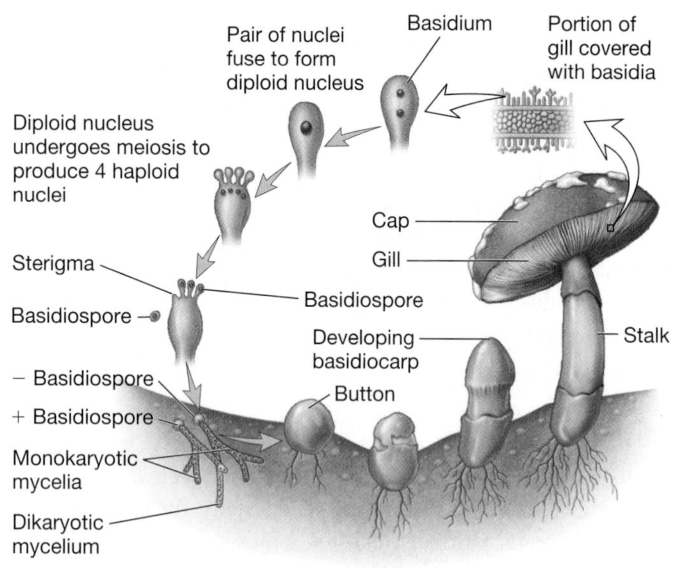

Figure 26.15 The *Basidiomycota*. The life cycle of a typical soil basidiomycete starts with a basidiospore germinating to produce a monokaryotic mycelium (one with a single nucleus in each septate cell). The mycelium quickly grows and spreads throughout the soil. When this primary mycelium meets another monokaryotic mycelium of a different mating type, the two fuse to initiate a new dikaryotic secondary mycelium. The secondary mycelium is divided by septa into cells, each of which contains two nuclei, one of each mating type. This dikaryotic mycelium is eventually stimulated to produce basidiocarps. A solid mass of hyphae forms a button that pushes through the soil, elongates, and develops a cap. The cap contains many platelike gills, each of which is coated with basidia. The two nuclei in the tip of each basidium fuse to form a diploid zygote nucleus, which immediately undergoes meiosis to form four haploid nuclei. These nuclei push their way into the developing basidiospores, which are then released at maturity.

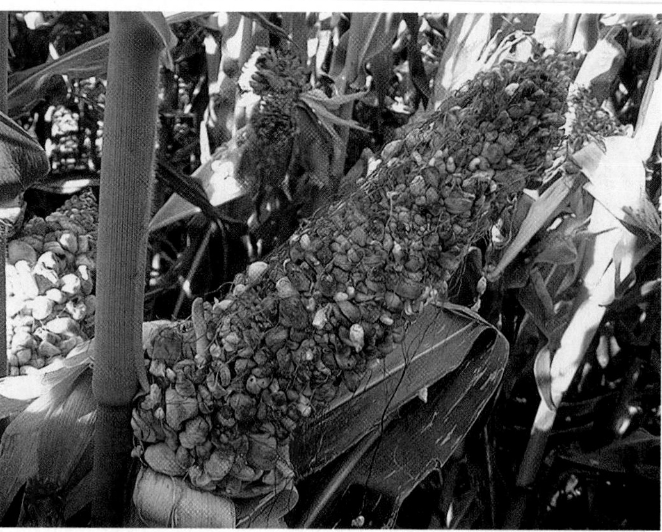

Figure 26.16 *Ustilago maydis*. This pathogen causes tumor formation in corn. Note the enlarged tumors releasing dark teliospores in place of normal corn kernals.

"rusts" and "smuts." In addition, some *Urediniomycetes* include human pathogens. These fungi are virulent plant pathogens that cause extensive damage to cereal crops; millions of dollars worth of crops are destroyed annually. In contrast to the basidiomycetes, the *Urediniomycetes* and *Ustilaginomycetes* do not form large basidiocarps. Instead small basidia arise from hyphae at the surface of the host plant. The hyphae grow either intra- or extracellularly in plant tissue.

The ustilaginomycete *Ustilago maydis* is a common corn pathogen that has become a model organism for plant smuts (**figure 26.16**). It is dimorphic and the yeastlike saprophytic form can be easily grown in the laboratory. In nature, the yeast form must mate to produce infectious, filamentous dikaryons that depend on the host plant for continued development. Once a plant is infected, *U. maydis* forms specialized flat hyphae called **appressoria** (s., appressorium) that enable penetration and subsequent reproduction within the host. This triggers the plant to form tumors, in which the fungus proliferates and eventually produces diploid spores called teliospores. Upon germination, cells undergo meiosis and haploid sporidia are released, causing the infection to spread from plant to plant.

Glomeromycota

Considered zygomycetes by some, the *Glomeromycota* are of critical ecological importance because most are endomycorrhizal symbionts of vascular plants. **Mycorrhizal fungi** form important associations with the roots of almost all herbaceous plants and tropical trees. As described in section 30.1, this is considered a mutualistic relationship because both the host plant and the fungus benefit: the fungus helps protect its host from stress and delivers soil nutrients to the plant, which in turn provides carbohydrate to the fungus. Microorganism associations with vascular plants: Mycorrhizae (section 29.5)

Only asexual reproduction is known to occur in the *Glomeromycota*. Spores are produced and germinate when in contact with the roots of a suitable host plant. An appressoria is formed and the outgrowth of hyphae forms a new mycelial symbiosis. Propagation can also occur by fragmentation and colonization of hyphae from the soil or a nearby plant.

Microsporidia

Of all the fungi, perhaps the *Microsporidia* have had the most confused taxonomic history. They have been considered protists and are sometimes still cited as such. However, molecular analysis of ribosomal RNA and specific proteins such as α- and β-tubulin shows that they are most closely related to fungi; they are perhaps "curious fungi." Unlike other fungi, they lack mitochonidria, peroxisomes, and centrioles. Importantly, they are obligate intracellular parasites that infect insects, fish, and humans. In particular, they infect immunocompromised individuals, especially those with HIV/AIDS. Common human pathogens include *Enterocystozoon bieneusi,* which causes diarrhea and pneumonia (depending on whether it was acquired through ingestion or inhalation, respectively) and *Encephaolitozoon cuniculi,* which causes encephalitis and nephritis (kidney disease).

Microsporidia morphology is also unique among eucaryotes. Small spores of 1 to 40 μm are viable outside the host. Depending on the species, spores may be spherical, rod, egg- or crescent-shaped. Spore germination is triggered by a signal from the host cells and results in the expulsion of a tightly packed organelle called the polar tube or filament (**figure 26.17**). The polar tube is ejected with enough force to pierce the host cell membrane, which permits the parasite's entry. Once inside the host cell, the microsporidian undergoes a developmental cycle that differs among the various microsporidian species. However, in all cases more spores are produced and eventually take over the host cell. Although this can have catastrophic consequences for humans, the use of microsporidia that infect destructive and pathogen-bearing insects is under development as a biocontrol measure.

1. What are the *Chytridiomycetes?* How do they differ from other fungi?
2. Describe how a typical zygomycete reproduces. Give some beneficial uses for zygomycetes.
3. Describe the ascomycete life cycle. How are the ascomycetes important to humans?
4. How do yeasts reproduce sexually? asexually? Why do you think *Saccharomyces cerevisiae* has become such an important model organism?
5. Describe the life cycle of a typical basidiomycete. Discuss their importance.
6. How does *Ustilago maydis* trigger tumor formation? How is tumor formation advantageous to the fungi?
7. What are mycorrhizae and why are they important?
8. Why do you think *Microsporidia* are still sometimes considered protists? Describe the germination of a microsporidian spore.

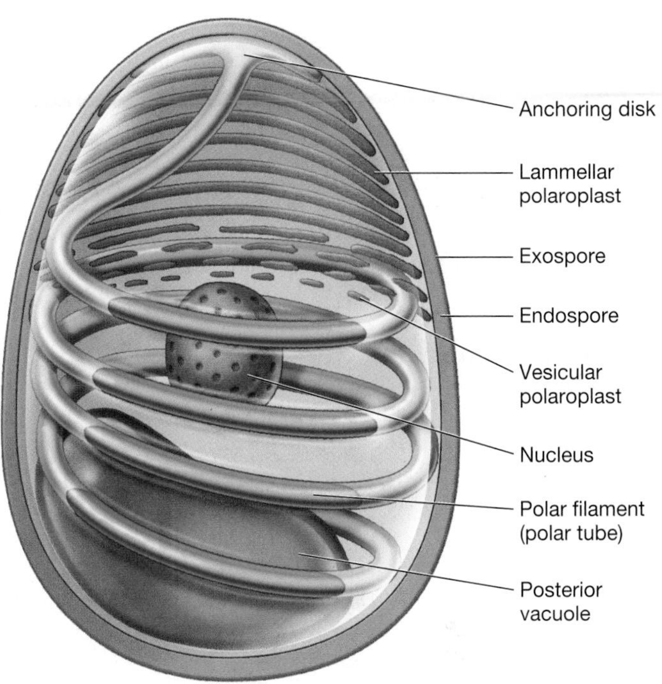

Figure 26.17 The Unique Structure of a Microsporidian Spore. Upon germination in a host cell, the polar tube is ejected with enough force to pierce the host cell membrane, allowing the fungus to gain entry.

Summary

26.1 Distribution

a. Fungi are widespread in the environment, found wherever water, suitable organic nutrients, and an appropriate temperature occur. They secrete enzymes outside their body structure and absorb the digested food.

26.2 Importance

a. Fungi are important decomposers that break down organic matter; live as parasites on animals, humans, and plants; play a role in many industrial processes; and are used as research tools in the study of fundamental biological processes.

26.3 Structure

a. The body or vegetative structure of a fungus is called a thallus. Fungi may be grouped into molds or yeasts based on the development of the thallus (**figure 26.3**).

b. A fungus is a eucaryotic, spore-bearing organism that has absorptive nutrition and lacks chlorophyll; that reproduces asexually, sexually, or by both methods; and that normally has a cell wall containing chitin.

c. Yeasts are unicellular fungi that have a single nucleus and reproduce either asexually by budding and transverse division or sexually through spore formation (**figure 26.4**).

d. A mold consists of long, branched, threadlike filaments of cells, the hyphae, that form a tangled mass called a mycelium. Hyphae may be either septate or coenocytic (aseptate). The mycelium can produce reproductive structures (**figures 26.5** and **26.6*a***).

e. Some fungi are dimorphic—they alternate between a yeast and a mold form (**table 26.2**).

26.4 Nutrition and Metabolism

a. Most fungi are saprophytes and grow best in moist, dark habitats. These chemoorganoheterotrophs are usually aerobic; some yeasts are fermentative.

26.5 Reproduction

a. Asexual reproduction often occurs in fungi by the production of specific types of spores that are easily dispersed (**figure 26.8**).

b. Sexual reproduction is initiated in fungi by the fusion of hyphae or cells of different mating types. In some fungi the parental nuclei immediately combine to form a zygote. In others the two genetically distinct nuclei remain separate, forming pairs that divide synchronously. Eventually some nuclei fuse (**figure 26.9**).

26.6 Characteristics of the Fungal Divisions

a. *Chytridiomycota* produce motile spores. Most are saprophytic; some reside in the rumen of herbivores.

b. Zygomycetes are coenocytic. Most are saprophytic. One example is the common bread mold, *Rhizopus stolonifer.* Sexual reproduction occurs through a form of conjugation involving + and − strains (**figure 26.10**).

c. *Ascomycota* are known as the sac fungi because they form a sac-shaped reproductive structure called an ascus (**figure 26.11**). In asexual reproduction they produce characteristic conidia (**figure 26.12**). Sexual reproduction involves strains of different mating types (**figure 26.13**).

d. *Basidiomycota* are the club fungi. They are named after their basidium that produces basidiospores (**figure 26.15**).

e. Urediniomycetes and Ustilaginomycetes are often considered *Basidiomycota.* Genera from both groups include important plant pathogens (**figure 26.16**).

f. *Glomeromycota* include the mycorrhizal fungi that grow in association with plant roots. They serve to increase plant nutrient uptake.

g. *Microsporidia* have a unique morphology and are still sometimes considered protists. They include virulent human pathogens; some infect other vertebrates and insects (**figure 26.17**).

Key Terms

antheridium 638
appressorium 640
arthroconidia 633
arthrospore 633
ascocarp 638
ascogenous hypha 638
ascogonium 638
ascomycetes 637
ascospore 634
ascus 637
basidiocarp 639
basidiomycetes 639
basidiospore 634
basidium 639
blastospore 633
chitin 631
chlamydospore 633
chytrids 635
coenocytic 631
conidiospore 633
convergent evolution 635
dikaryotic stage 634
divergent evolution 635
ergot 637
ergotism 637
Eumycota 630
fungus 629
gametangium 634
hypha 631
mating type (MAT) 634
mold 631
mycelium 631
mycologist 629
mycology 629
mycosis 629
mycorrhizal fungi 640
mycotoxicology 629
osmotrophy 632
progametangium 636
saprophyte 632
septa 632
septate 632
sporangiospore 633
sporangium 633
thallus 631
yeast 631
YM shift 632
zygomycetes 635
zygospore 634

Critical Thinking Questions

1. What are some logical targets to exploit in treating animals or plants suffering from fungal infections? Are they different from the targets you would use when treating infections caused by bacteria? By viruses? Explain.
2. Fungi tend to reproduce asexually when nutrients are plentiful and conditions are favorable for growth, but reproduce sexually when environmental or nutrient conditions are not favorable. Why is this an evolutionarily important and successful strategy?
3. Because asexual spores are such a rapid way of reproducing for some fungi, what adaptive "use" is there for an additional sexual phase?
4. Some fungi can be viewed as coenocytic organisms that exhibit little differentiation. When differentiation does occur, such as in the formation of reproductive structures, it is preceded by septum formation. Why does this occur?
5. Both bacteria and fungi are major environmental decomposers. Obviously competition exists in a given environment, but fungi usually have an advantage. What characteristics specific to the fungi provide this advantage?

Learn More

Adl, S. M.; Simpson, A. G. B.; Farmer, M. A.; Anderson, R. A.; Anderson, O. R.; Barta, J. R.; Bowser, S. S.; *et al.* 2005. The new higher level classification of Eukaryotes with emphasis on the taxonomy of protists. *J. Eukaryot. Microbiol.* 52: 399–451.

Alexopoulos, C. J.; Mims, C. W.; Blackwell, M. 1996. *Introductory mycology,* 4th ed. New York: John Wiley and Sons.

Barr, D. J. S. 1992. Evolution and kingdoms of organisms from the perspective of a mycologist *Mycologia* 84:1–11.

Carlile, M. J., and Gooday, G. W. 2001. The *Fungi,* 2d ed. New York: Academic Press.

Feofilova, E. P. 2001. The Kingdom Fungi: Heterogeneity of physiological and biochemical properties and relationships with plants, animals, and prokaryotes (Review). *Appl. Biochem. Microbiol.* 37:124–37.

Griffin, D. H. 1996. *Fungal physiology,* 2d ed. New York: Wiley-Liss.

Loftus, B. J.; Fung, E.; Roncaglia, P.; *et al.* 2005. The genome of the basidiomycetous yeast and human pathogen *Cryptococcus neoformans. Science* 307: 1321–24.

Mueller, G. M.; Bills, G. F.; and Foster, M. S., editors. 2004. *Biodiversity of Fungi: Inventory and monitoring methods.* New York: Elsevier Academic Press.

Ostergaard, S.; Olsson, L.; and Nielson, J. 2000. Metabolic engineering of *Saccharomyces cerevisiae. Microbiol. Mol. Biol. Rev.* 64(1):34–50.

Partida-Martinez, L. P., and Herweck, C. 2005. Pathogenic fungus harbours endosymbiotic bacteria for toxin production. *Nature* 437: 884–88.

Scholte, E.-J.; Ng'habi, K.; Kihonda, J.; Takken, W.; Paaijmans, K.; Abdulla, S.; Killeen, G. F.; and Knols, B. G. J. 2005. An entomopathogenic fungus for control of adult African malaria mosquitoes. *Science* 308: 1641–42.

Please visit the Prescott website at www.mhhe.com/prescott7 for additional references.

27 Biogeochemical Cycling and Introductory Microbial Ecology

Microorganisms living in environments where most known organisms cannot survive are important for understanding microbial diversity. These strands of iron-oxidizing *Ferroplasma* were discovered growing at pH 0 in an abandoned mine near Redding, California. This hardy microorganism has only a plasma membrane to protect itself from the rigors of this harsh environment.

Preview

- Life on Earth would not be possible without microbes. Despite their importance, less than 1% of all microbial species have been cultured, identified, and studied.
- Microbial ecology is the study of community dynamics and the interaction of microbes (both procaryotic and eucaryotic) with each other and with plants, animals, and the environment in which they live.
- Energy, electrons, and nutrients must be available in a suitable physical environment for microorganisms to function. Microbes interact with their environment to obtain energy (from light or chemical sources), electrons, and nutrients, which leads to a process called biogeochemical cycling. Microorganisms change the physical state and mobility of many nutrients.
- Biogeochemical cycling refers to the biological and chemical processes that elements such as carbon, nitrogen, sulfur, iron, and magnesium undergo during microbial metabolism. Life on Earth would not be possible without these microbial activities.
- Microorganisms are an important part of ecosystems—self-regulating biological communities and their physical environments. The microbial loop describes the many functions microbes play in the cycling of nutrients.
- Environments that seem extreme from a human point of view support a wide diversity of bacterial and archaeal species that offer insights into the adaptive capabilities of cells and the dynamics of community structure.
- Microbial ecologists employ a variety of diverse analytical techniques to understand the critical role of microbes in specific ecosystems and in maintaining life on Earth.

In previous chapters microorganisms are usually considered as isolated entities. However, microorganisms exist in communities. Recent estimates suggest that most microbial communities have between 10^{10} to 10^{17} individuals representing at least 10^7 different taxa. How can such huge populations exist and, moreover, survive together in a productive fashion? To answer this question, one must know which microbes are present and how they interact—that is, one must study **microbial ecology.** In this chapter we begin our consideration of this multidisciplinary field by discussing the process of elemental cycling—the microbe-mediated exchange of nutrients that makes life on Earth possible. This is sometimes called **environmental microbiology** (**Microbial Diversity & Ecology 27.1**). We then present an overview of some of the physical and biological features that define microbial habitats. Finally, an overview of some of the more important tools and techniques used to study microbial ecology is presented. This chapter thus provides the foundation for a more detailed review of microbial communities in marine and freshwater environments (chapter 28) and terrestrial ecosystems (chapter 29).

27.1 Foundations in Microbial Diversity and Ecology

Microorganisms function as **populations** or assemblages of similar organisms, and as **communities,** or mixtures of different microbial populations. These microorganisms have evolved while interacting with each other, with higher organisms, and with the inorganic world. They largely play beneficial and vital roles; disease-causing organisms are only a minor component of the microbial world. Microorganisms, as they interact with other organisms and their environment, also contribute to the functioning of **ecosystems**—self-regulating biological communities and their physical environment. Knowledge of these interactions is

Everything is everywhere, the environment selects.

—*M. W. Beijerinck*

Microbial Diversity & Ecology

27.1 Microbial Ecology Versus Environmental Microbiology

The term microbial ecology is now used in a general way to describe the presence and contributions of microorganisms, through their activities, to the places where they are found. Students of microbiology should be aware that much of the information on microbial presence and contributions to soils, waters, and associations with plants, now described by this term, would have been considered "environmental microbiology" in the past. Thomas Brock, the discoverer of *Thermus aquaticus,* which is known the world over as the source of *Taq* polymerase for the polymerase chain reaction (PCR), has given a definition of microbial ecology that may be useful: "Microbial ecology is the study of the behavior and activities of microorganisms in their natural environments." The important operator in this sentence is *their* environment instead of *the* environment. To emphasize this point, Brock has noted that "microbes are small; their environments also are small." In these small environments or "microenvironments," other kinds of microorganisms (and macroorganisms) often also are present, a critical point that was emphasized by Sergei Winogradsky in 1947.

Environmental microbiology, in comparison, relates primarily to all-over microbial processes that occur in soil, water, or food, as examples. It is not concerned with the particular "microenvironment" where the microorganisms actually are functioning, but with the broader-scale effects of microbial presence and activities. One can study these microbially mediated processes and their possible global impacts at the scale of "environmental microbiology" without knowing about the specific microenvironment (and the organisms functioning there) where these processes actually take place. However, it is critical to be aware that microbes function in their localized environments and affect ecosystems at greater scales, including causing global-level effects. In the last decades the term microbial ecology largely has lost its original meaning, and recently the statement has been made that microbial ecology has become a "catch-all" term. As you read various textbooks and scientific papers, possible differences between microbial ecology and environmental microbiology should be kept in mind.

important in understanding both microbial contributions to the natural world and microbial roles in disease processes.

A major problem in understanding microbial interactions is that only about 1% of microorganisms have been grown in the laboratory. The differences between observable and culturable microorganisms have been noted for at least 70 years; however, molecular analyses has demonstrated the magnitude of the problem. This problem was perhaps first discussed in a 1932 textbook on soil microbiology written by Selman Waksman, who discovered streptomycin. Molecular techniques and sequence data provide valuable information on these uncultured microorganisms, and attempting to grow them remains a central challenge in microbial ecology.

27.2 Biogeochemical Cycling

Microorganisms, in the course of their growth and metabolism, interact with each other in the cycling of nutrients, including carbon, nitrogen, phosphorus, sulfur, iron, and manganese. This nutrient cycling, called **biogeochemical cycling** when applied to the environment, involves both biological and chemical processes and is of global importance. Nutrients are transformed and cycled, often by oxidation-reduction reactions that can change the chemical and physical characteristics of the nutrients. All of the biogeochemical cycles are linked (**figure 27.1**), and the metabolism-related transformations of these nutrients make life on Earth possible. Oxidation-reduction reactions, electron carriers, and electron transport systems (section 8.6)

The major reduced and oxidized forms of the most important elements are noted in **table 27.1,** together with their valence states. Significant gaseous components occur in the carbon and nitrogen cycles and, to a lesser extent, in the sulfur cycles. Soil, aquatic, and marine microorganisms often can fix gaseous forms of carbon and nitrogen compounds. In the "sedimentary" cycles, such as that for phosphorus and iron, there is no gaseous component.

Carbon Cycle

Carbon is present in reduced forms, such as methane (CH_4) and organic matter, and in more oxidized forms, such as carbon monoxide (CO) and carbon dioxide (CO_2). The major pools in an integrated, simplified carbon cycle are shown in **figure 27.2.** Electron donors (e.g., hydrogen, which is a strong reductant) and electron acceptors (e.g., O_2) influence the course of biological and chemical reactions involving carbon. Hydrogen can be produced when organic matter is degraded, especially under anoxic conditions when fermentation occurs. Although carbon cycles continuously from one form to another, for the sake of clarity, we shall say that the cycle "begins" with carbon fixation—the conversion of CO_2 into organic matter. Plants like trees and crops are often thought of as the principal CO_2-fixing organisms, but at least half the carbon on Earth is fixed by microbes, particularly marine photosynthetic procaryotes and protists (e.g., the cyanobacteria *Prochlorococcus* and *Synechococcus,* and diatoms, respectively). Carbon is also fixed by chemolithoautotrophic microbes. All fixed carbon enters a common pool of organic matter that can then be oxidized back to CO_2 through aerobic or anaerobic respiration and fermentation. Microorganisms in marine environments: The photic zone of the open ocean (section 28.3)

Alternatively, inorganic (CO_2) and organic carbon can be reduced anaerobically to methane (CH_4). Methane is produced by

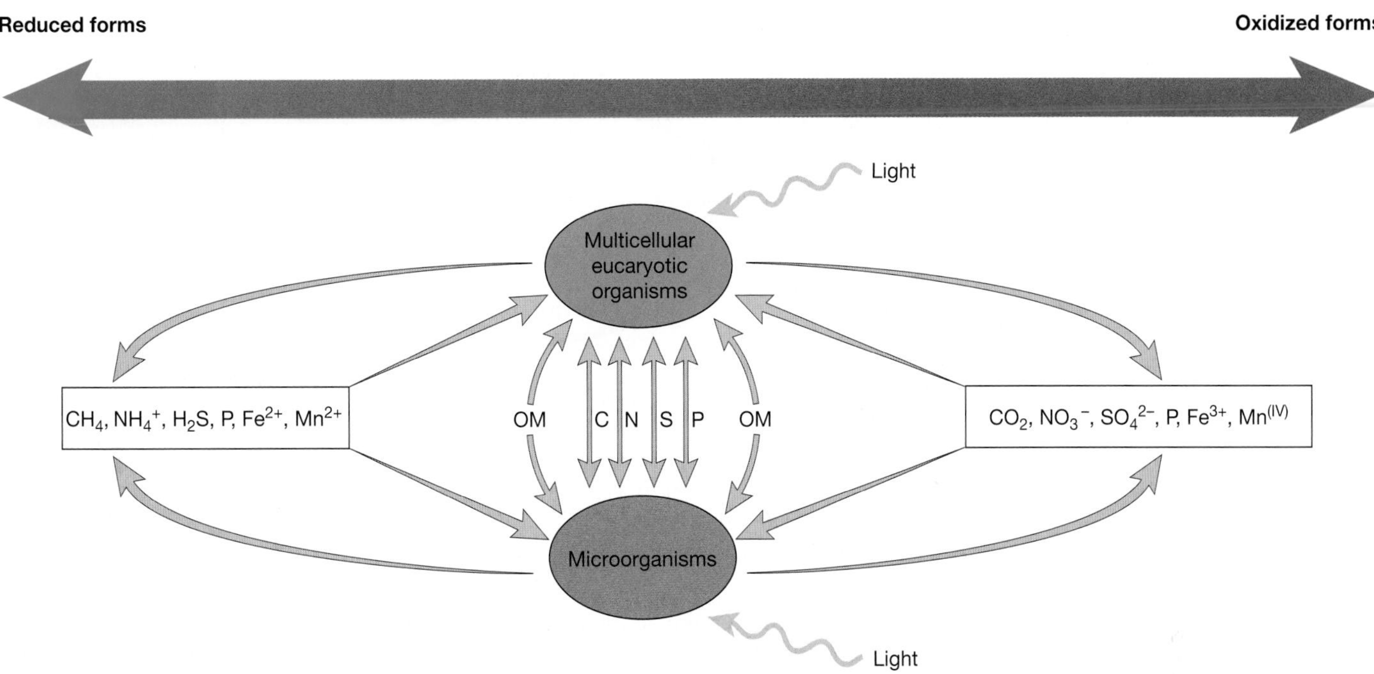

Figure 27.1 Macrobiogeochemistry: A Cosmic View of Mineral Cycling by Microorganisms, Higher Organisms, and the Abiotic Chemical World. All biogeochemical cycles are linked, with energy obtained from light and pairs of reduced and oxidized compounds. Only major flows are shown. The forms that move between the microorganisms and multicellular organisms can vary. The biotic components include both living forms and those that have died/senesced and are being processed. Flows from lithogenic sources are important for phosphorus cycling. Organic matter (OM).

Table 27.1 The Major Forms of Carbon, Nitrogen, Sulfur, and Iron Important in Biogeochemical Cycling

		Major Forms and Valences				
Cycle	**Significant Gaseous Component Present?**	**Reduced Forms**	**Intermediate Oxidation State Forms**			**Oxidized Forms**
C	Yes	Methane: CH_4 (−4)	Carbon monoxide CO (+2)			CO_2 (+4)
N	Yes	Ammonium: NH_4^+, organic N (−3)	Nitrogen gas: N_2 (0)	Nitrous oxide N_2O (+1)	Nitrite: NO_2^- (+3)	Nitrate: NO_3^- (+5)
S	Yes	Hydrogen sulfide: H_2S, SH groups in organic matter (−2)	Elemental sulfur: S^0 (0)	Thiosulfate: $S_2O_3^{2-}$ (+2)	Sulfite: SO_3^{2-} (+4)	Sulfate: SO_4^{2-} (+6)
Fe	No	Ferrous iron: Fe^{2+}(+2)				Ferric Iron: Fe^{3+}(+3)

Note: The carbon, nitrogen, and sulfur cycles have significant gaseous components, and these are described as gaseous nutrient cycles. The iron cycle does not have a gaseous component, and this is described as a sedimentary nutrient cycle. Major reduced, intermediate oxidation state, and oxidized forms are noted, together with valences.

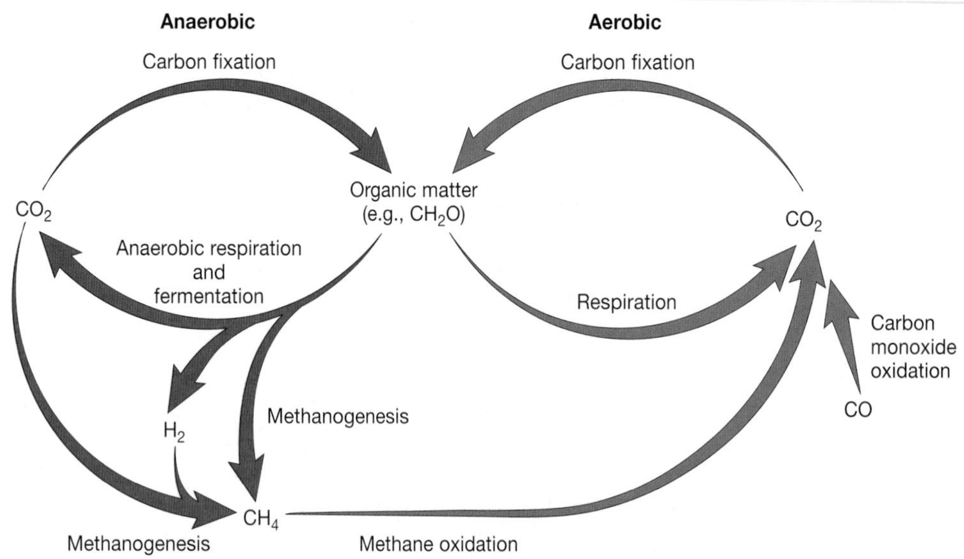

Figure 27.2 The Basic Carbon Cycle in the Environment. Carbon fixation can occur through the activities of photoautotrophic and chemoautotrophic microorganisms. Methane can be produced from inorganic substrates ($CO_2 + H_2$) or from organic matter. Carbon monoxide (CO)—produced by sources such as automobiles and industry—is returned to the carbon cycle by CO-oxidizing bacteria. Aerobic processes are noted with blue arrows, and anaerobic processes are shown with red arrows.

archaea in anoxic habitats. In a water column, the anoxic zone where methane is produced is often below an oxic zone. Therefore, as methane moves up the water column, it is oxidized before reaching the atmosphere. However, in many situations, such as in rice paddies without an overlying oxic water zone, the methane is released directly to the atmosphere, thus contributing to global atmospheric methane increases. Rice paddies, ruminants, coal mines, sewage treatment plants, landfills, and marshes are important sources of methane. Archaea such as *Methanobrevibacter* in the guts of termites also contribute to methane production. Phylum *Euryarchaeota:* Methanogens (section 20.3)

In the carbon cycle depicted in figure 27.2, no distinction is made between the different types of organic matter formed and degraded. This is a marked oversimplification because organic matter varies widely in physical characteristics and in the biochemistry of its synthesis and degradation. Organic matter varies in terms of elemental composition, structure of basic repeating units, linkages between repeating units, and physical and chemical characteristics. Its degradation is influenced by a series of factors. These include (1) nutrient availability; (2) abiotic conditions (e.g., pH, oxidation-reduction potential, O_2, osmotic conditions), and (3) the microbial community present.

The major complex organic substrates used by microorganisms are summarized in **table 27.2.** Because microorganisms require each macronutrient, if an environment is enriched in one nutrient but relatively deficient in another, the nutrients may not be completely recycled into living biomass. For instance, chitin, protein, and nucleic acids contain nitrogen in large amounts. If these substrates are used for growth, the excess nitrogen and other minerals that are not used in the formation of new microbial biomass are released to the environment in the process of **mineralization.** This is the process by which organic matter is decomposed to release simpler, inorganic compounds (e.g., CO_2, NH_4^+, CH_4, H_2).

The other complex substrates listed in table 27.2 contain only carbon, hydrogen, and oxygen. If microorganisms are to grow by using these substrates, they must acquire the remaining nutrients they need for biomass synthesis from the environment. This is often very difficult, as the concentration of nitrogen and phosphorus may be very low. The inability to assimilate sufficient levels of a macronutrient will then limit the growth of a given population. For instance, in open-ocean microbial communities, growth of many autotrophic microbes is often nitrogen limited. In other words, if higher concentrations of usable nitrogen (NO_3, NH_4^+) were available, the rate of growth of individual microbes would increase, as would their overall population size. Those nutrients that are converted into biomass become temporarily "tied up" from nutrient cycling; this is sometimes called nutrient **immobilization.** Microorganisms in marine environments: The photic zone of the open ocean (section 28.3)

Most carbon substrates can be degraded easily with or without oxygen present, but this is not always the case. The exceptions are hydrocarbons and lignin. Hydrocarbons are unique in that microbial degradation most often involves the initial addition of molecular O_2. However, anaerobic degradation of hydrocarbons with sulfate or nitrate as electron acceptors can occur. With sulfate present, bacteria of the genus *Desulfovibrio* are active. Anaerobic degradation of hydrocarbons proceeds more slowly

Table 27.2 Complex Organic substrate Characteristics That Influence Decomposition and Degradability

Substrate	Basic Subunit	Linkages (if Critical)	Elements Present in Large Quantity					Degradation	
			C	H	O	N	P	With O_2	Without O_2
Starch	Glucose	$\alpha(1\rightarrow4)$ $\alpha(1\rightarrow6)$	+	+	+	−	−	+	+
Cellulose	Glucose	$\beta(1\rightarrow4)$	+	+	+	−	−	+	+
Hemicellulose	C6 and C5 monosaccharides	$\beta(1\rightarrow4)$, $\beta(1\rightarrow3)$, $\beta(1\rightarrow6)$	+	+	+	−	−	+	+
Lignin	Phenylpropene	C−C, C−O bonds	+	+	+	−	−	+	−
Chitin	*N*-acetylglucosamine	$\beta(1\rightarrow4)$	+	+	+	+	−	+	+
Protein	Amino acids	Peptide bonds	+	+	+	+	−	+	+
Hydrocarbon	Aliphatic, cyclic, aromatic		+	+	−	−	−	+	+/−
Lipids	Glycerol, fatty acids; some contain phosphate and nitrogen	Esters, ethers	+	+	+	+	+	+	+
Microbial biomass		Varied	+	+	+	+	+	+	+
Nucleic acids	Purine and pyrimidine bases, sugars, phosphate	Phosphodiester and *N*-glycosidic bonds	+	+	+	+	+	+	+

and only in microbial communities that have been exposed to these compounds for extended periods. Biodegradation and bioremediation by natural communities (section 41.6)

Lignin, an important structural component in mature plant materials, is a complex amorphous polymer linked by carbon-carbon and carbon-ether bonds. It makes up approximately 1/3 of the weight of wood. Lignin is degraded by filamentous fungi, a process that occurs under oxic conditions. Lignin's diminished biodegradability under anoxic conditions results in accumulation of lignified materials, including the formation of peat bogs and muck soils. The absence of lignin degradation under anoxic conditions also is important in construction. Large masonry structures often are built on swampy sites by driving in wood pilings below the water table and placing the building footings on the pilings. As long as the foundations remain water-saturated and anoxic, the structure is stable. If the water table drops, however, the pilings will begin to rot and the structure will be threatened. Similarly, the cleanup of harbors can lead to decomposition of costly docks built with wooden pilings due to increased degradation of wood by aerobic filamentous fungi. Soils, plants, and nutrients (section 29.2)

Oxygen availability also affects the final products that accumulate when organic substrates have been processed and mineralized by microorganisms. Under oxic conditions, oxidized products such as nitrate, sulfate, and carbon dioxide will result from microbial degradation of complex organic matter (**figure 27.3**). In comparison, under anoxic conditions reduced end products tend to accumulate, including ammonium ion, sulfide, and methane.

The carbon cycle has come under intense scrutiny in the last decade or so. This is because CO_2 levels in the atmosphere have

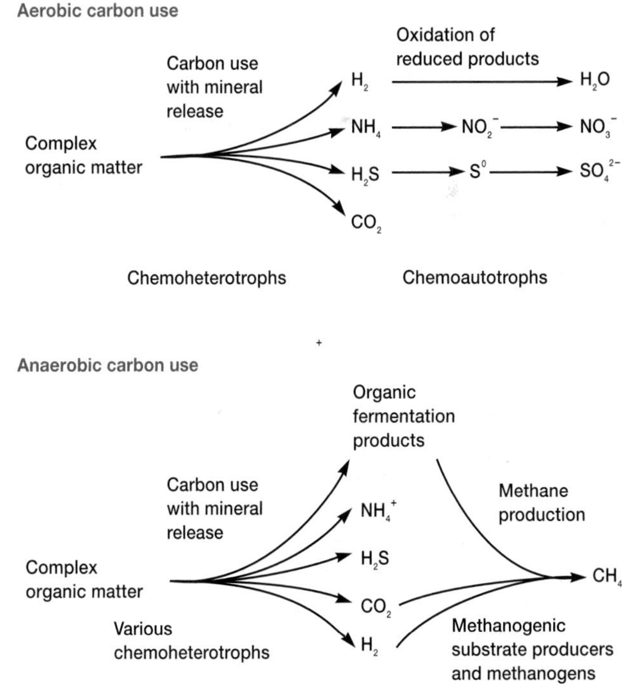

Figure 27.3 The Influence of Oxygen on Organic Matter Decomposition. Microorganisms form different products when breaking down complex organic matter aerobically than they do under anoxic conditions. Under oxic conditions oxidized products accumulate, while reduced products accumulate anaerobically. These reactions also illustrate that the waste products of one group of microorganisms may be used by a second type of microorganism.

risen from their preindustrial concentration of about 280 μmol per mol to 376 μmol per mol in 2003. This represents an increase of about one-third, and CO_2 levels continue to rise. Like CO_2, methane is also a **greenhouse gas** and its atmospheric concentration is likewise increasing about 1% per year, from 0.7 to 1.7 ppm (volume) since the early 1700s. These changes are clearly the result of the combustion of fossil fuels and altered land use. The term greenhouse gas describes the ability of these gasses to trap heat within Earth's atmosphere, leading to a documented increase in the planet's mean temperature. Indeed, over the past 100 years, Earth's average temperature has increased by 0.6°C and continues to rise at a rapid rate. As discussed more fully in chapters 28 and 29, the increase in CO_2 levels would be even more dramatic if it were not for the removal of large quantities of CO_2 from the atmosphere by carbon fixation in both marine and terrestrial ecosystems.

1. What is biogeochemical cycling?
2. Which organic polymers discussed in this section do and do not contain nitrogen?
3. Define mineralization and immobilization and give examples.
4. What is unique about lignin and its degradation?
5. What C, N, and S forms will accumulate after anaerobic degradation of organic matter?

Nitrogen Cycle

Like the carbon cycle, cycling of nitrogenous materials makes life on Earth possible. A simplified nitrogen cycle is presented in **figure 27.4.** Again, we begin our discussion of this cycle with the fixation of the inorganic element (N_2) to its organic form (NH_4^+, amino acids). **Nitrogen fixation** is a uniquely procaryotic process; apart from a limited amount of nitrogen fixation that occurs during lightning strikes, all organic nitrogen is of procaryotic origin. Nitrogen fixation can be carried out under oxic and anoxic conditions. Microbes such as *Azotobacter* and the cyanobacterium *Trichodesmium* fix nitrogen aerobically, while free-living anaerobes such as members of the genus *Clostridium* fix nitrogen anaerobically. Perhaps the best-studied nitrogen-fixing microbes are the bacterial symbionts of leguminous plants, including *Rhizobium,* its α-proteobacterial relatives, and some recently discovered β-proteobacteria (e.g., *Burkholderia* and *Ralstonia* spp.). However, other bacterial symbionts fix nitrogen. For instance, the actinomycete *Frankia* fixes nitrogen while colonizing many types of woody shrubs, and the heterocystous cyanobacterium *Anabaenea* fixes nitrogen when in association with the water fern *Azolla.* Synthesis of amino acids: Nitrogen fixation (section 10.5); Photosynthetic bacteria: Phylum *Cyanobacteria* (section 21.3); Microorganisms in association with vascular plants: Nitrogen fixation (section 29.5)

The product of N_2 fixation is ammonia (NH_3); it is immediately incorporated into organic matter as an amine. The addition of eight electrons per N atom requires a great deal of energy and reducing power. The nitrogenase enzyme is thus very sensitive to O_2 and must be protected from oxidizing conditions. Aerobic and microaerophilic nitrogen-fixing bacteria employ a number of strategies to protect their nitrogenase enzymes. For example, heterocystous cyanobacteria physically separate nitrogen fixation from oxygenic photosynthesis by confining the process to special cells called heterocysts, while other cyanobacteria fix nitrogen only at night when photosynthesis is impossible. These strategies are discussed in more detail in sections 21.3 and 29.5.

Ammonia made by N_2 fixation is immediately incorporated into organic matter as amines. These amine N-atoms are eventually introduced into proteins, nucleic acids, and other biomolecules. The N cycle continues with the degradation of these molecules into ammonium (NH_4^+) within mixed assemblages of microbes. One important fate of this ammonium is its conversion to nitrate (NO_3^-), a process called **nitrification.** This is a two-step process whereby ammonium ion (NH_4^+) is first oxidized to nitrite (NO_2^-), which is then oxidized to nitrate.

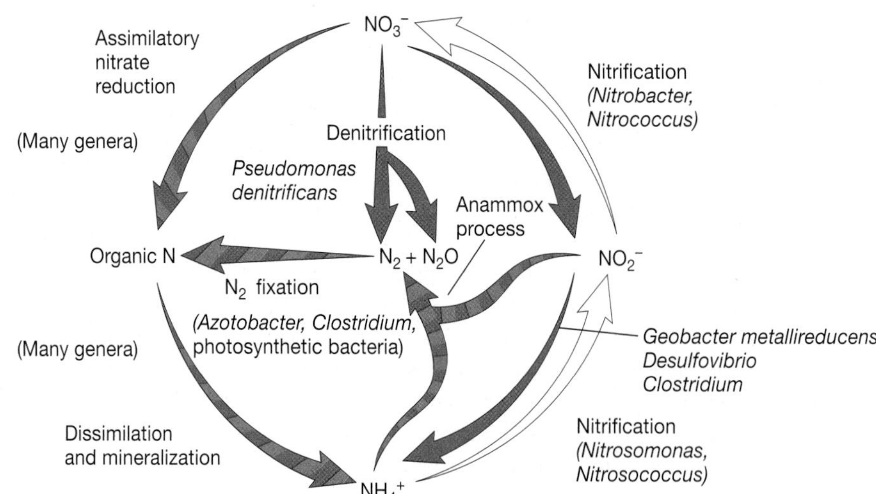

Figure 27.4 A Simplified Nitrogen Cycle. Flows that occur predominantly under oxic conditions are noted with open arrows. Anaerobic processes are noted with solid bold arrows. Processes occurring under both oxic and anoxic conditions are marked with cross-barred arrows. The anammox reaction of NO_2^- and NH_4^+ to yield N_2 is shown. Important genera contributing to the nitrogen cycle are given as examples.

Bacteria of the genera *Nitrosomonas* and *Nitrosococcus,* for example, play important roles in the first step, and *Nitrobacter* and related chemolithoautotrophic bacteria carry out the second step. In addition, *Nitrosomonas eutropha* has been found to oxidize ammonium ion anaerobically to nitrite and nitric oxide (NO) using nitrogen dioxide (NO_2) as an acceptor in a denitrification-related reaction.

The production of nitrate is important because it can be reduced and incorporated into organic nitrogen; this process is known as assimilatory nitrate reduction. The use of nitrate as a source of organic nitrogen is an example of **assimilatory reduction.** Because assimilatory reduction of nitrate to ammonium is energetically expensive, nitrate sometimes accumulates as a transient intermediate. Alternatively, for some microbes nitrate serves as a terminal electron acceptor during anaerobic respiration; this is a form of **dissimilatory reduction.** In this case, nitrate is removed from the ecosystem and returned to the atmosphere as dinitrogen gas (N_2) through a series of reactions that are collectively known as **denitrification.** This dissimilatory process, in which nitrate is used as an electron acceptor in anaerobic respiration, usually involves heterotrophs such as *Pseudomonas denitrificans.* The major products of denitrification include nitrogen gas (N_2) and nitrous oxide (N_2O), although nitrite (NO_2^-) also can accumulate. Nitrite is of environmental concern because it can contribute to the formation of carcinogenic nitrosamines. Finally, nitrate can be transformed to ammonia in dissimilatory reduction by a variety of bacteria, including *Geobacter metallireducens, Desulfovibrio* spp., and *Clostridium* spp. Anaerobic respiration (section 9.6); Synthesis of amino acids: Nitrogen assimilation (section 10.5)

A recently identified form of nitrogen conversion is called the **anammox reaction** (*an*oxic *amm*onium *ox*idation). In this anaerobic reaction, chemolithotrophs use ammonium ion (NH_4^+) as the electron donor and nitrite (NO_2^-) as the terminal electron acceptor; it is reduced to nitrogen gas (N_2). In effect, the anammox reaction is a shortcut to N_2, proceeding directly from ammonium and nitrite without having to cycle first through nitrate (figure 27.4). Although this reaction was known to be energetically possible, microbes capable of performing the anammox reaction were only recently documented. The discovery that marine bacteria perform the anammox reaction in the anoxic waters just below oxygenated regions in the open ocean solved a long-standing mystery. For many years microbiologists wondered where the "missing" NH_4^+ could be—mass calculations did not agree with experimentally derived nitrogen measurements. The discovery that planctomycete bacteria oxidize measurable amounts of NH_4^+ to N_2, thereby removing it from the marine ecosystem, has necessitated a reevaluation of nitrogen cycling in the open ocean. Phylum *Planctomycetes* (section 21.4)

1. Under what circumstances does nitrogen fixation occur? Describe some microbes that are capbable of nitrogen fixation. How does the process of nitrogen fixation make their life-styles unique?
2. Describe the two-step process that makes up nitrification. Why do you think nitrification requires two different types of bacteria?
3. What is the difference between assimilatory nitrate reduction and denitrification? Which reaction is performed by most microbes and which is a more specialized metabolic capability?
4. Describe the anammox reaction. Why do you think it was so difficult for microbiologists to discover the microbes that perform this reaction?

Phosphorus Cycle

Unlike the carbon and nitrogen cycles, the phosphorus cycle has no gaseous component (**figure 27.5**). Biogeochemical cycling of phosphorus is important for a number of reasons. All living cells require phosphorus for nucleic acids, lipids, and some polysaccharides. However, most environmental phosphorus is present in low concentrations, locked within the Earth's crust. Thus it is frequently the nutrient that limits growth.

Unlike carbon and nitrogen, which can be obtained from the atmosphere, phosphorus is derived solely from the weathering of phosphate-containing rocks. Therefore in soil, phosphorus exists in both inorganic and organic forms. Organic phosphorus includes not only that found in biomass, but in materials like humus and other organic compounds. The phosphorus in these organic materials is recycled by microbial activity. Inorganic phosphorus is negatively charged, so it complexes readily with cations in the environment, such as iron, aluminum, and calcium. These compounds are relatively insoluble, and their dissolution is pH dependent such that it is most available to plants and microbes between pH 6 and 7. Under such conditions, these organisms rapidly convert phosphate to its organic form so that it becomes available to animals. The microbial transformation of phosphorus features the transformation of simple orthophosphate (PO_4^-), which bears phosphorus in the +5 valence state, to more complex forms. These include the polyphosphates seen in metachromatic granules as well as more familiar macromolecules. Note that the phosphorus in all of these organic forms remains in the +5 valence state. Synthesis of purines, pyrimadines, and nucleotides: Phosphorus assimilation (section 10.6)

Sulfur Cycle

Microorganisms contribute greatly to the sulfur cycle; a simplified version is shown in **figure 27.6.** Recall that sulfide can serve as an electron source for both photosynthetic microorganisms and chemolithoautotrophs such as *Thiobacillus;* it is converted to elemental sulfur and sulfate. When sulfate diffuses into reduced habitats, it provides an opportunity for different groups of microorganisms to carry out **sulfate reduction.** For example, when a usable organic electron donor is present, *Desulfovibrio* uses sulfate as its terminal electron acceptor during anaerobic respiration. Dissimilatory sulfate reduction (i.e., the use of sulfate as an external electron acceptor) results in sulfide accumulation in the environment. In comparison, the reduction of

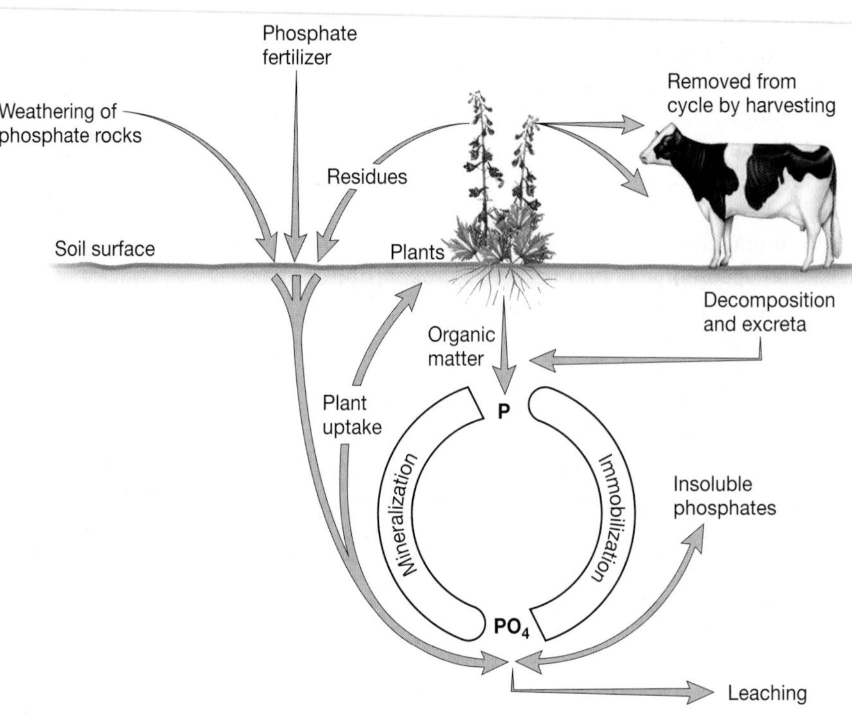

Figure 27.5 A Simplified Phosphorus Cycle. Phosphorus enters soil and water through the weathering of rocks, phosphate fertilizer, and surface residue of plant degradation. Plants and microbes rapidly convert inorganic phosphorus to its organic form, causing immobilization. However, much of the soil phosphorus can leach great distances or complex with cations to form relatively insoluble compounds.

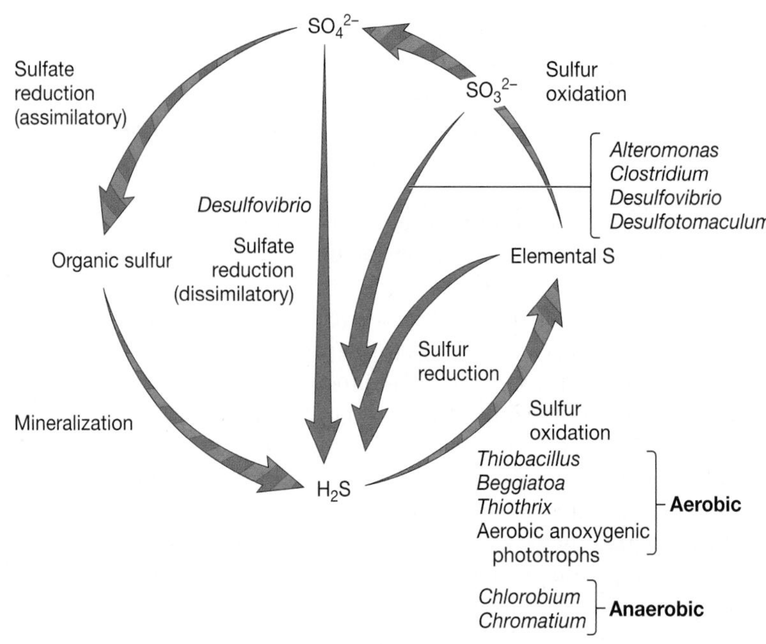

Figure 27.6 A Simplified Sulfur Cycle. Photosynthetic and chemosynthetic microorganisms contribute to the environmental sulfur cycle. Sulfate and sulfite reductions carried out by *Desulfovibrio* and related microorganisms, noted with purple arrows, are dissimilatory processes. Sulfate reduction also can occur in assimilatory reactions, resulting in organic sulfur forms. Elemental sulfur reduction to sulfide is carried out by *Desulfuromonas*, thermophilic archaea, or cyanobacteria in hypersaline sediments. Sulfur oxidation can be carried out by a wide range of aerobic chemotrophs and by aerobic and anaerobic phototrophs.

sulfate for use in amino acid and protein biosynthesis is described as assimilatory sulfate reduction. Other microorganisms have been found to carry out dissimilatory elemental sulfur (S°) reduction. These include *Desulfuromonas,* thermophilic archaea, and also cyanobacteria in hypersaline sediments. Sulfite (SO_3^{2-}) is another critical intermediate that can be reduced to sulfide by a wide variety of microorganisms, including *Alteromonas* and *Clostridium,* as well as *Desulfovibrio* and *Desulfotomaculum. Desulfovibrio* is usually considered an obligate anaerobe. Research, however, has shown that this interesting organism also respires using oxygen under microaerobic conditions (dissolved oxygen level of 0.04%).

In addition to the very important photolithotrophic sulfur oxidizers such as *Chromatium* and *Chlorobium,* which function under strict anoxic conditions in deep water columns, a large and varied group of bacteria carry out **aerobic anoxygenic photosynthesis.** These phototrophs use bacteriochlorophyll *a* and carotenoid pigments and are found in marine and freshwater environments; they are often components of microbial mat communities. Important genera include *Erythromonas, Roseococcus, Porphyrobacter,* and *Roseobacter.*

Minor compounds in the sulfur cycle play major roles in biology. An excellent example is dimethylsulfoniopropionate (DMSP), which is used by bacterioplankton (floating bacteria) as a sulfur source for protein synthesis, and which is transformed to dimethylsulfide (DMS), a volatile sulfur form that can affect atmospheric processes.

When pH and oxidation-reduction conditions are favorable, several key transformations in the sulfur cycle also occur as the result of chemical reactions in the absence of microorganisms. An important example of such an abiotic process is the oxidation of sulfide to elemental sulfur. This takes place rapidly at a neutral pH, with a half-life of approximately 10 minutes for sulfide at room temperature.

Iron Cycle

The iron cycle principally features the interchange of ferrous iron (Fe^{2+}) to ferric iron (Fe^{3+}) (**figure 27.7**). Iron oxidation can be carried out at neutral pH by *Gallionella* spp. and in acidic conditions by *Thiobacillus ferrooxidans* and the thermphile *Sulfolobus*. Much of the earlier literature suggested that additional genera could oxidize iron, including *Sphaerotilus* and *Leptothrix*. These two genera are still termed "iron bacteria" by many nonmicrobiologists. Confusion about the role of these genera resulted from the occurrence of the chemical oxidation of ferrous ion to ferric ion (forming insoluble iron precipitates) at neutral pH values, where these microorganisms grow on organic substrates. These microorganisms are now classified as chemoorganotrophs.

Some microbes oxidize Fe^{2+} using nitrate as an electron acceptor. These include the interesting microorganism *Dechlorosoma suillum,* a mixotroph that can also use chlorate and perchlorate as electron acceptors. Because perchlorate is a major component of explosives and rocket propellents, it is a frequent contaminant at retired munitions facilities and military bases. Thus *D. suillum* may be used in the bioremediation (biological clean-up) of such sites. This process also occurs in aquatic sediments with depressed levels of oxygen and may be another route by which large zones of oxidized iron have accumulated in environments with lower oxygen levels. Banded iron formation that occurred when atmospheric oxygen levels were beginning to increase at the end of the Precambrian era may be evidence of increased iron bacteria activity. Biodegradation and bioremediation by natural communities (section 41.6); Microbial evolution (section 19.1)

Iron reduction occurs under anoxic conditions resulting in the accumulation of ferrous ion. Although many microorganisms can reduce small amounts of iron during their metabolism, most iron reduction is carried out by specialized iron-respiring microorganisms such as *Geobacter metallireducens, Geobacter sulfurreducens, Ferribacterium limneticum,* and *Shewanella putrefaciens,* which can obtain energy for growth on organic matter using ferric iron as the election acceptor. Anaerobic respiration (section 9.6)

In addition to these relatively simple reductions to ferrous ion, some magnetotactic bacteria such as *Aquaspirillum magnetotacticum* transform extracellular iron to the mixed valence iron oxide mineral magnetite (Fe_3O_4) and construct intracellular magnetic compasses. Furthermore, dissimilatory iron-reducing bacteria accumulate magnetite as an extracellular product. The cytoplasmic matrix: Inclusion bodies (section 3.3)

Magnetite has been detected in sediments, where it is present in particles similar to those found in bacteria, indicating a long-term contribution of bacteria to iron cycling processes. Genes for magnetite synthesis have been cloned into other organisms, creating

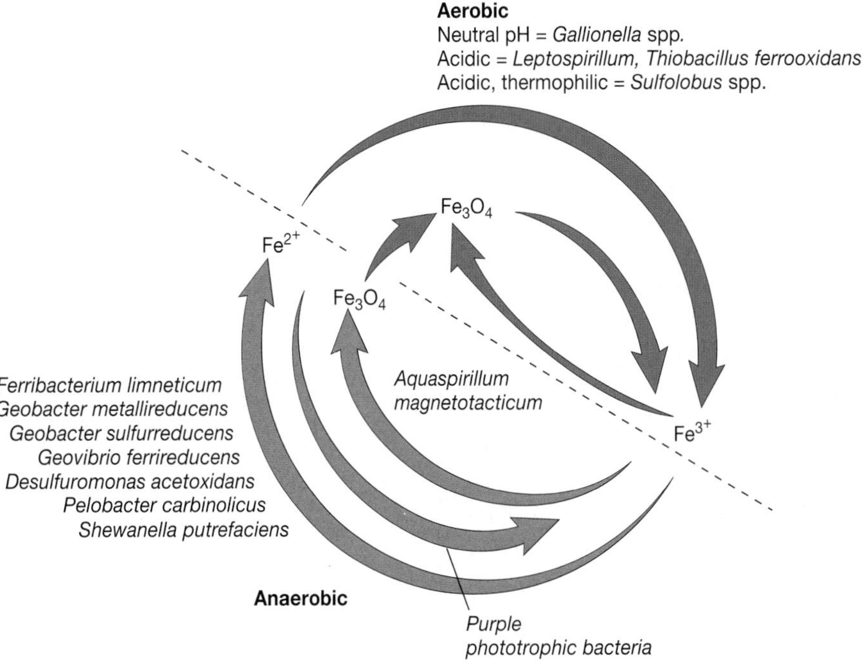

Figure 27.7 The Iron Cycle. A simplified iron cycle with examples of microorganisms contributing to these oxidation and reduction processes. In addition to ferrous ion (Fe^{2+}) oxidation and ferric ion (Fe^{3+}) reduction, magnetite (Fe_3O_4), a mixed valence iron compound formed by magnetotactic bacteria, is important in the iron cycle. Different microbial groups carry out the oxidation of ferrous ion depending on environmental conditions.

new magnetically sensitive microorganisms. Magnetotactic bacteria may be described as **magneto-aerotactic bacteria** because they are thought to use magnetic fields to migrate to the position in a bog or swamp where the oxygen level is best suited for their functioning. Microorganisms have also been described that use ferrous ion as an electron donor in anoxygenic photosynthesis. Thus with production of ferric ion in lighted anoxic zones by iron-oxidizing bacteria, the stage is set for subsequent chemotrophic-based iron reduction, such as that done by *Geobacter* and *Shewanella,* creating a strictly anoxic oxidation/reduction cycle for iron.

Manganese Cycle

The importance of microorganisms in manganese cycling is becoming much better appreciated. The manganese cycle involves the transformation of manganous ion (Mn^{2+}) to MnO_2 (equivalent to manganic ion [Mn^{4+}]), which occurs in hydrothermal vents and bogs (**figure 27.8**). *Leptothrix, Arthrobacter,* and *Pedomicrobium,* are important in Mn^{2+} oxidation. *Shewanella, Geobacter,* and other chemoorganotrophs can carry out the complementary manganese reduction process.

1. Trace the fate of a single phosphorus atom from a rock to a stream and back to the earth.
2. Describe the difference between assimilatory and dissimilatory sulfate reduction.
3. Why do you think some bacteria can reduce $Fe_3^{\ +}$ only in acidic conditions?
4. What are some important microbial genera that contribute to manganese cycling?

Microorganisms and Metal Toxicity

In addition to metals such as iron and manganese, which are largely nontoxic to microorganisms and animals, there are a series of metals that have varied toxic effects on microorganisms and homeothermic animals. Microorganisms play important roles in modifying the toxicity of these metals (**table 27.3**).

The metals can be considered in broad categories. The noble metals have distinct effects on microorganisms, including growth inhibition. The second group includes metals or metalloids that microorganisms can methylate to form more mobile products called organometals. Organometals contain carbon-metal bonds. Some organometals can cross the blood-brain barrier and affect the central nervous system and organ function of vertebrates.

The mercury cycle is of particular interest and illustrates many characteristics of those metals that can be methylated. Mercury compounds were widely used in industrial processes over the centuries. Indeed Lewis Carroll alluded to this problem when he wrote of the Mad Hatter in *Alice in Wonderland.* At that time mercury was used in the shaping of felt hats. Microorganisms methylated some of the mercury, thus rendering it more toxic to the hatmakers.

A devastating situation developed in southwestern Japan when large-scale mercury poisoning occurred in the Minamata Bay region because of industrial mercury released into the marine environment. Inorganic mercury that accumulated in bottom muds of the bay was methylated by anaerobic bacteria of the genus *Desulfovibrio* (**figure 27.9**). Such methylated mercury forms are volatile and lipid soluble, and the mercury concentrations increased in the food chain by the process of biomagnification (the progressive accumulation of refractile compounds by successive trophic levels). The mercury was ultimately ingested by the human population, the "top consumers," through their primary food source—fish—leading to severe neurological disorders.

The third group of metals occurs in ionic forms directly toxic to microorganisms and more complex organisms. In higher organisms, plasma proteins react with the ionic forms of

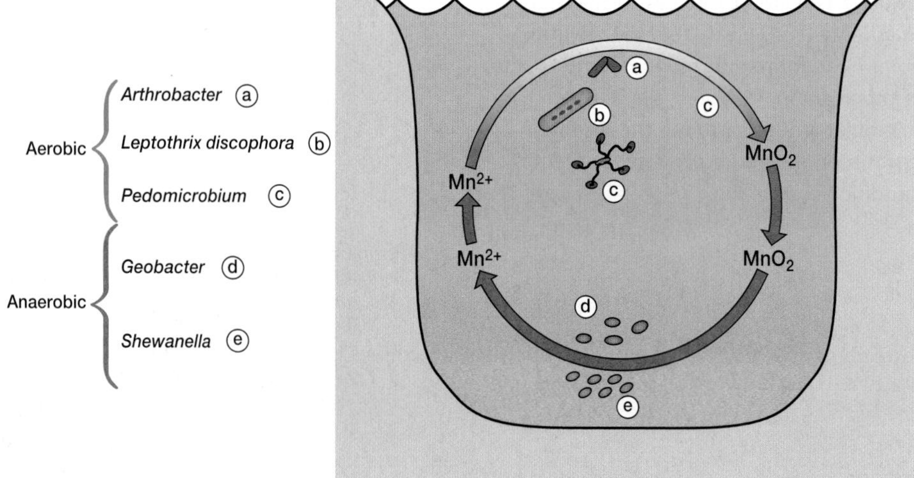

Figure 27.8 The Manganese Cycle, Illustrated in a Stratified Lake. Microorganisms make many important contributions to the manganese cycle. After diffusing from anoxic (pink) to oxic (blue) zones, manganous ion (Mn^{2+}) is oxidized chemically and by many morphologically distinct microorganisms in the oxic water column to manganic oxide—$MnO_2^{(IV)}$, valence equivalent to 4+. When the $MnO_2^{(IV)}$ diffuses into the anoxic zone, bacteria such as *Geobacter* and *Shewanella* carry out the complementary reduction process. Similar processes occur across oxic/anoxic transitions in soils, muds, and other environments.

Table 27.3 Examples of Microorganism-Metal Interactions and Relations to Effects on Microorganisms and Homeothermic Animals

Metal Group	Metal		Interactions and Transformations: Microorganisms	Interactions and Transformations: Homeothermic Animals
Noble metals	Ag Au Pt	Silver Gold Platinum	Microorganisms can reduce ionic forms to the elemental state. Low levels of ionized metals released to the environment have antimicrobial activity.	Many of these metals can be reduced to elemental forms and do not tend to cross the blood-brain barrier. Silver reduction can lead to inert deposits in the skin.
Metals that form stable carbon metal bonds	As Hg Se	Arsenic Mercury Selenium	Microorganisms can transform inorganic and organic forms to methylated forms, some of which tend to bioaccumulate in higher trophic levels.	Methylated forms of some metals can cross the blood-brain barrier, resulting in neurological effects or death.
Other metals	Cu Zn Co	Copper Zinc Cobalt	In the ionized form, at higher concentrations, these metals can directly inhibit microorganisms. They are often required at lower concentrations as trace elements.	At higher levels, clearance from higher organisms occurs by reaction with plasma proteins and other mechanisms. Many of these metals serve as trace elements at lower concentrations.

these metals and aid in their excretion, unless excessive long-term contact and ingestion occur. Relatively high doses of these metals are required to cause lethal effects. At lower concentrations many of these metals serve as required trace elements. The use of chemical agents in control: Heavy metals (section 7.5)

1. What are examples of the three groups of metals in terms of their toxicity to microorganisms and homeothermic animals?
2. How can microbial activity render some metals more or less toxic to warm-blooded animals?
3. Why do metals such as mercury have such major effects on higher organisms?

27.3 The Physical Environment

Microorganisms, as they interact with each other and with other organisms in biogeochemical cycling, also are influenced by their immediate physical environment, whether this is soil, water, the deep marine environment, or a plant or animal host. It is important to consider the specific environments where microorganisms interact with each other, other organisms, and the physical environment.

The Microenvironment and Niche

The specific physical location of a microorganism is its **microenvironment.** In this physical microenvironment, the flux of required electron donors and acceptors, and nutrients to the actual location of the microorganism can be limited. At the same time, waste products may not be able to diffuse away from the microorganism at rates sufficient to avoid growth inhibition by high waste product concentrations. These fluxes and gradients create a unique **niche,** which includes the microorganism, its physical habitat, the time of resource use, and the resources available for microbial growth and function (**figure 27.10**).

This physically structured environment also can limit the predatory activities of protozoa. If the microenvironment has pores with diameters of 3 to 6 μm, bacteria in the pores will be protected from predation, while allowing diffusion of nutrients and waste products. If the pores are larger, perhaps greater than 6 μm in diameter, protozoa may be able to feed on the bacteria. It is important to emphasize that microorganisms can create their own microenvironments and niches. For example, microorganisms in the interior of a colony have markedly different microenvironments and niches than those of the same microbial populations located on the surface or edge of the colony. Microorganisms also can associate with clays and form inert microhabitats called "clay hutches" for protection. Microbial growth in soils is discussed more fully in chapter 29.

Biofilms and Microbial Mats

One way microorganisms create their own microenvironments and niches is by forming **biofilms.** These are organized microbial systems consisting of layers of microbial cells associated with surfaces (**figure 27.11*a***). Biofilms are an important factor in almost all areas of microbiology. Microbial growth in natural environments: Biofilms (section 6.6); Global regulatory systems: Quorum sensing (section 12.5)

Simple biofilms develop when microorganisms attach and form a monolayer of cells. Depending on the particular microbial growth environment (e.g., light, nutrients present, and diffusion rates), biofilms can become more complex with layers of organisms of different types (figure 27.11*b*). A typical example would involve photosynthetic organisms on the surface, facultative chemoorganotrophs in the middle, and possibly sulfate-reducing microorganisms on the bottom.

Figure 27.9 The Mercury Cycle. Interactions between the atmosphere, oxic water, and anoxic sediment are critical. Microorganisms in anoxic sediments, primarily *Desulfovibrio*, can transform mercury to methylated forms that can be transported to water and the atmosphere. These methylated forms also undergo biomagnification. The production of volatile elemental mercury (Hg^0) releases this metal to waters and the atmosphere. Sulfide, if present in the anoxic sediment, can react with ionic mercury to produce less soluble HgS.

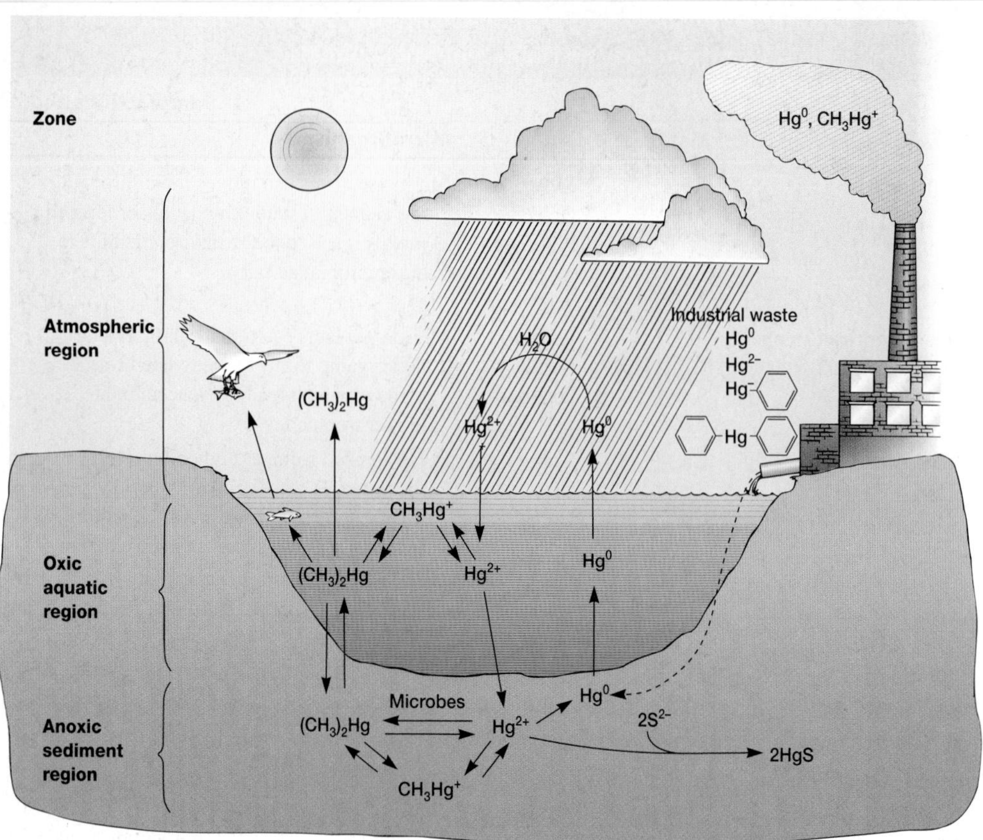

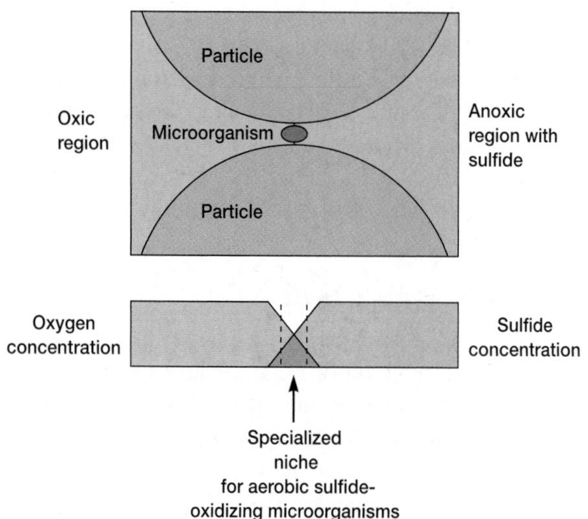

Figure 27.10 The Creation of a Niche from a Microenvironment. As shown in this illustration, two nearby particles create a physical microenvironment for possible use by microorganisms. Chemical gradients, as with oxygen from the oxic region, and sulfide from the anoxic region, create a unique niche. This niche is the physical environment and the resources available for use by specialized aerobic sulfide-oxidizing bacteria.

More complex biofilms can develop to form a four-dimensional structure (X, Y, Z, and time) with cell aggregates, interstitial pores, and conduit channels (figure 27.11*c*). This developmental process involves the growth of attached microorganisms, resulting in accumulation of additional cells on the surface, together with the continuous trapping and immobilization of free-floating microorganisms that move over the expanding biofilm. This structure allows nutrients to reach the biomass, and the channels are shaped by protozoa that graze on bacteria.

These more complex biofilms, in which microorganisms create unique environments, can be observed by the use of confocal scanning laser microscopy (CSLM) as discussed in chapter 2 (*see figure 2.25*). The diversity of nonliving and living surfaces that can be exploited by biofilm-forming microorganisms include surfaces in catheters and dialysis units, which have intimate contact with human body fluids. Control of such microorganisms and their establishment in these sensitive medical devices is an important part of modern hospital care.

Biofilms also can protect pathogens from disinfectants, create a focus for later occurrence of disease, or release microorganisms and microbial products that may affect the immune system of a susceptible host. Biofilms are critical in ocular diseases because *Chlamydia, Staphylococcus,* and other pathogens survive in ocular devices such as contact lenses and in cleaning solutions.

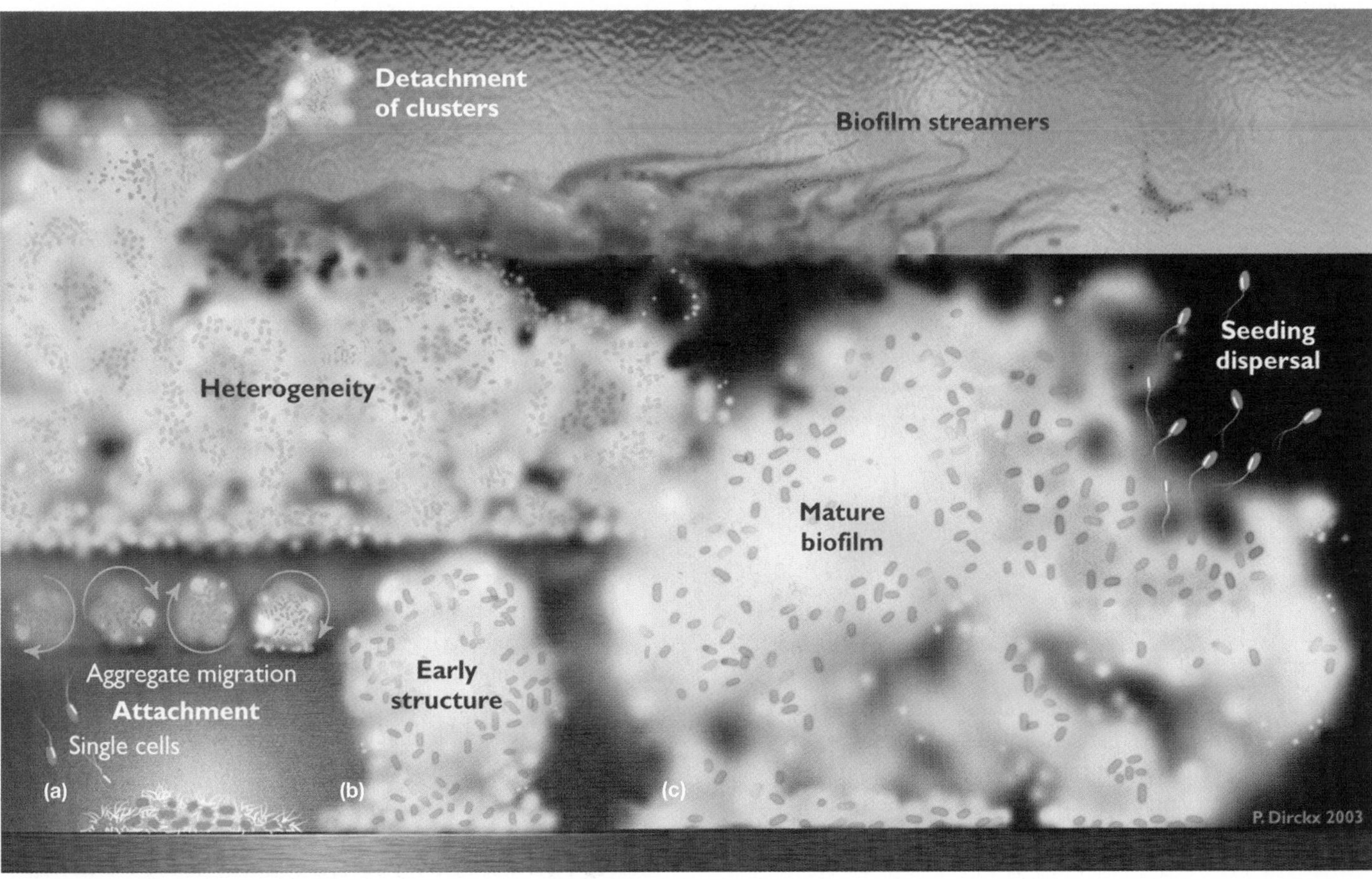

Figure 27.11 The Growth of Biofilms. Biofilms, or microbial growths on surfaces such as in freshwater and marine environments, can develop and become extremely complex, depending on the energy sources that are available. **(a)** Initial colonization by a single type of bacterium. **(b)** Development of a more complex biofilm with layered microorganisms of different types. **(c)** A mature biofilm with cell aggregates, interstitial pores, and conduits.

Depending on environmental conditions, biofilms can become so large that they are visible and have macroscopic dimensions. Bands of microorganisms of different colors are evident as shown in **figure 27.12.** These thick biofilms, called **microbial mats,** are found in many freshwater and marine environments. These mats are complex layered microbial communities that can form at the surface of rocks or sediments in hypersaline and freshwater lakes, lagoons, hot springs, and beach areas. They consist of filamentous microbes, including cyanobacteria. A major characteristic of mats is the extreme gradients that are present. Visible light only penetrates approximately 1 mm into these communities, and below this photosynthetic zone, anoxic conditions occur and sulfate-reducing bacteria play a major role. The sulfide that these organisms produce diffuses to the anoxic lighted region, allowing sulfur-dependent photosynthetic microorganisms to grow. Some believe that microbial mats could have allowed the formation of terrestrial ecosystems prior to the development of vascular plants, and fossil microbial mats, called stromatolites, have been dated at over 3.5 billion years old *(see figure 19.2).* Molecular techniques and stable isotope measurements (see section 27.4) are being used to better understand these unique microbial communities.

1. What are the similarities and differences between a microenvironment and a niche?
2. Why might pores in soils, waters, and animals be important for survival of bacteria if protozoa are present?
3. Why might conditions vary for a bacterium on the edge of a colony in comparison with the center of the colony?
4. What are biofilms? What types of surfaces on living organisms can provide a site for biofilm formation?
5. Why are biofilms important in human health?
6. What are microbial mats, and where are they found?

Microorganisms and Ecosystems

Microorganisms, as they interact with each other and other organisms, and influence nutrient cycling in their specific microenvironments and niches, also contribute to the functioning of ecosystems.

Figure 27.12 Microbial Mats. Microorganisms, through their metabolic activities, can create environmental gradients resulting in layered ecosystems. A vertical section of a hot spring (55°C) microbial mat, showing the various layers of microorganisms.

Ecosystems have been defined as "communities of organisms and their physical and chemical environments that function as self-regulating units." These self-regulating biological units respond to environmental changes by modifying their structure and function.

Microorganisms in ecosystems can have two complementary roles: (1) the synthesis of new organic matter from CO_2 and other inorganic compounds during **primary production** and (2) decomposition of accumulated organic matter. The general relationship between **primary producers** that synthesize organic matter and chemoorganotrophic **decomposers** was once thought to be quite simple. It was held that different organisms performed these nonoverlapping processes. We now understand that microbial communities are more complicated. Because aquatic systems are easier to investigate than terrestrial microbial communities, much of our current understanding of the relative contributions of microbes to ecosystem function comes from studies of these environments. As shown in **figure 27.13*a*,** the traditional food chain is only part of the picture. Larger plants and animals contribute to a common pool of **dissolved organic matter (DOM)** that is consumed by a variety of procaryotic and eucaryotic microbes. These microbes then return some of this DOM to larger animals in the form of particulate material—that is to say, protists eat bacteria and primary consumers eat protists (figure 27.13*b,c*). In addition, the metabolism and death of these microbes recycles some organic matter back to the general pool of DOM. This complex web of interactions is called the **microbial loop.** In terrestrial systems, the roles of microbes appear to be similar, with the exception of primary production, which is performed chiefly by vascular plants instead of microbes. The microbial loop is discussed in further detail in chapter 28 (*see sections 28.1 and 28.3; figure 28.15*).

Microorganisms thus carry out many important functions as they interact in ecosystems, including:

1. Contributing to the formation of organic matter through photosynthetic and chemosynthetic processes.
2. Decomposing organic matter, often with the release of inorganic compounds (e.g., CO_2, NH_4^+, CH_4, H_2) in mineralization processes.
3. Serving as a nutrient-rich food source for other chemoheterotrophic microorganisms, including protozoa and animals.
4. Modifying substrates and nutrients used in symbiotic growth processes and interactions, thereby contributing to biogeochemical cycling.
5. Changing the amounts of materials in soluble and gaseous forms. This occurs either directly by metabolic processes or indirectly by modifying the environment (e.g., altering the pH).
6. Producing inhibitory compounds that decrease microbial activity or limit the survival and functioning of plants and animals.
7. Contributing to the functioning of plants and animals through positive and negative symbiotic interactions, as discussed in chapter 30.

Microorganism Movement between Ecosystems

Microorganisms are moved constantly between ecosystems. This often happens naturally in many ways: (1) soil is transported around the Earth by windstorms and falls on land areas and waters far from its origin; (2) rivers transport eroded materials, sewage plant effluents, and urban wastes to the ocean; and (3) insects and animals release urine, feces, and other wastes to environments as they migrate around the Earth. When plants and animals die after moving to a new environment, they decompose and their specially adapted and coevolved microorganisms (and their nucleic acids) are released. The fecal-oral route of disease transmission, often involving foods and waters, and the acquisition of diseases in hospitals (nosocomial infections) are important examples of pathogen movement between ecosystems. Each time a person coughs or sneezes, microorganisms also are being transported to new ecosystems.

Humans also both deliberately and unintentionally move microorganisms between different ecosystems. This occurs when microbes are added to environments to speed up microbially mediated degradation processes or when a plant-associated inoculum, such as *Rhizobium*, is added to soil to increase the formation of nitrogen-fixing nodules on legumes. One of the most important accidental modes of microbial movement is the use of modern transport vehicles such as automobiles, trains, ships, and airplanes. These often rapidly move microorganisms long distances. Biodegradation and bioremediation in natural communi-

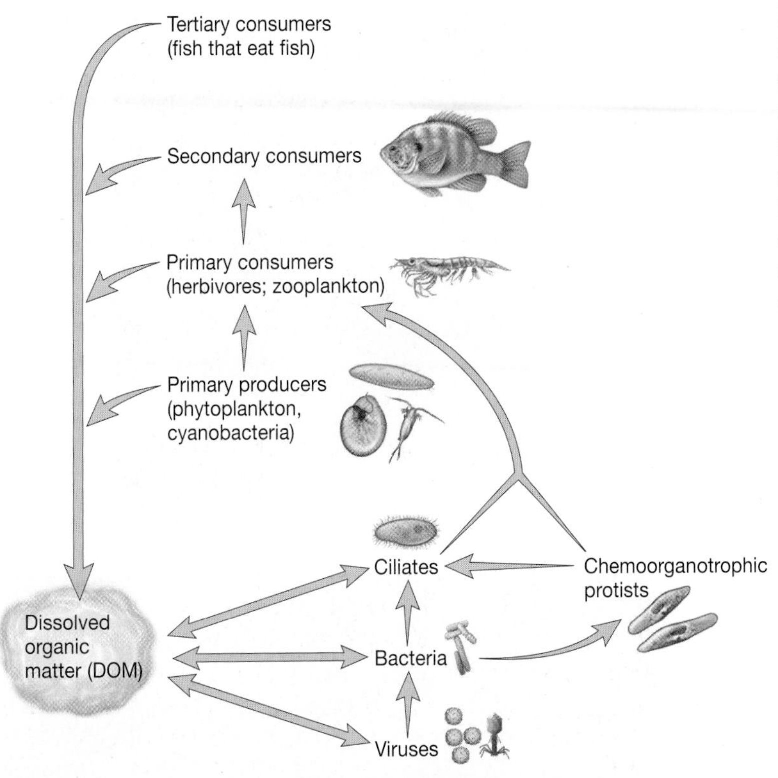

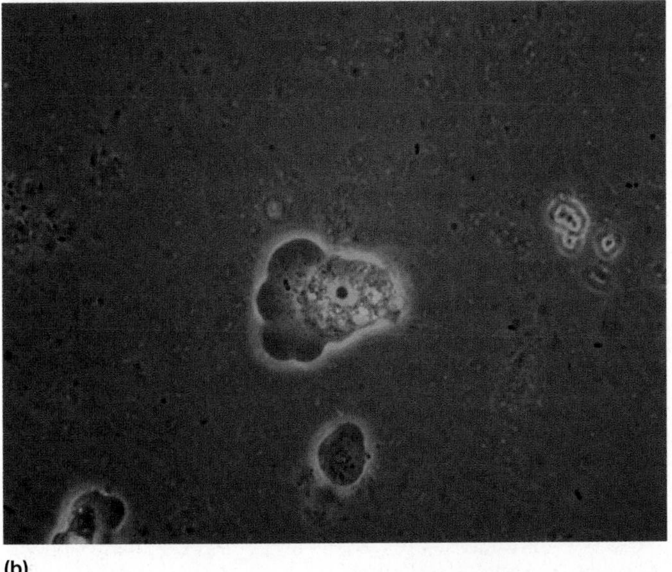
(b)

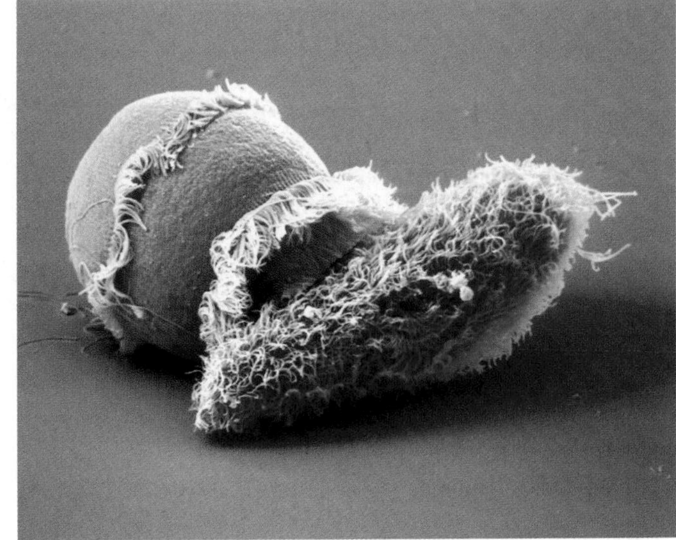
(c)

Figure 27.13 The Microbial Loop. **(a)** Microorganisms play vital roles in ecosystems as primary producers, decomposers, and primary consumers. The microbial loop describes the exchange of dissolved organic matter (DOM) between organisms so that it is recycled many times and not immediately lost to the system. **(b)** Protists consume bacteria; in this case a naked amoeba is consuming the cyanobacterium *Synechococcus* which fluoresce red. **(c)** Protists consume protists; here the ciliate *Didinium* (gold) is preying upon another ciliate, *Paramecium*.

ties (section 41.6); Microorganism associations with vascular plants: The rhizobia (section 29.5)

The fate of microorganisms placed in environments where they normally do not live, or of microorganisms returned to their original environments, is of theoretical and practical importance. Pathogens that are normally associated with an animal host largely have lost their ability to compete effectively for nutrients with microorganisms indigenous to other environments. Upon moving to a new environment, the population of viable and culturable pathogens gradually decreases. However, more sensitive viability assessment procedures, particularly molecular techniques, indicate that **vibiable but nonculturable**

(VBNC) microorganisms may play critical roles in disease occurrence (p. 660). The growth curve: Senescence and death (section 6.2)

1. Define the following terms: ecosystem, primary production, decomposer.
2. List important functions of higher consumers in natural environments.
3. What are the important functions of microorganisms in ecosystems?
4. How can microorganisms move between different ecosystems?

Extreme Environments

Microorganisms are found in a wide range of environments that differ in pH, temperature, atmospheric pressure, salinity, water availability, and ionizing radiation. In some environments, these conditions can be at either end of a continuum (e.g., very alkaline or acidic; extremely hot or cold). Such environments are called **extreme environments** (**figure 27.14**), and some of the characteristics of such ecosystems are summarized in **table 27.4.** The microorganisms that survive in such environments are described as **extremophiles.** Although extreme environments are usually considered to have decreased microbial diversity, as judged by the microorganisms that can be cultured, with the increased use of molecular detection techniques, it appears that many contain a surprising diversity of microorganisms. Work to establish relationships between the microorganisms that can be observed and detected by molecular techniques and culturable microorganisms is ongoing. The influence of environmental factors on growth (section 6.5)

Many microbial genera have specific requirements for survival in extreme environments. For example, a high sodium ion concentration is required to maintain membrane integrity in many halophilic procaryotes, including members of the genus *Halobacterium.* Halobacteria require a sodium ion concentration of at least 1.5 M, and about 3 to 4 M for optimum growth. Phylum *Euryarchaeota:* The halobacteria (section 20.3)

Some bacteria found in deep-sea environments provide another example of extremophiles. These bacteria can be described as **baro-** or **piezotolerant bacteria** (growth from approximately 1 to 500 atm), **moderately barophilic bacteria** (growth optimum 5,000 meters, and still able to grow at 1 atm), and **extreme barophilic bacteria,** which require approximately 400 atm or higher for growth. Microorganisms in marine environments: Benthic marine environments (section 28.3)

Intriguing changes in basic physiological processes occur in microorganisms functioning under extreme acidic or alkaline conditions. These acidophilic and alkalophilic microorganisms have markedly different problems in maintaining a more neutral internal pH and chemiosmotic processes. Obligately acidophilic microorganisms can grow at a pH of 3.0 or lower, and major pH differences can exist between the interior and exterior of the cell. These acidophiles include members of the genera *Thiobacillus, Sulfolobus,* and *Thermoplasma.* The higher relative internal pH is maintained by a net outward translocation of protons. This may occur as the result of unique membrane lipids, hydrogen ion re-

(a)

(b)

(c)

Figure 27.14 Microorganisms Growing in Extreme Environments. Many microorganisms are especially suited to survive in extreme environments. **(a)** Salterns turned red by halophilic algae and halobacteria. **(b)** A hot spring colored green and blue by cyanobacterial growth. **(c)** A source of acid drainage from a mine into a stream. The soil and water have turned red due to the presence of precipitated iron oxides caused by the activity of bacteria such as *Thiobacillus.*

Table 27.4 Characteristics of Extreme Environments in Which Microorganisms Grow

Stress	Environmental Conditions	Microorganisms Observed
High temperature	121°C	*Geogemma barossii*
	110–113°C, deep marine trenches	*Pyrolobus fumarii* *Methanopyrus kandleri* *Pyrodictium abyssi*
	67–102°C, marine basins	*Pyrococcus abyssi*
	85°C, hot springs	*Thermus* *Sulfolobus*
	75°C, sulfur hot springs	*Thermothrix thiopara*
Low temperature	−12°C, antarctic ice	*Psychromonas ingrahamii*
Osmotic stress	13–15% NaCl	*Chlamydomonas*
	25% NaCl	*Halobacterium* *Halococcus*
pH	pH 10.0 or above	*Bacillus*
	pH 3.0 or lower	*Saccharomyces* *Thiobacillus*
	pH 0.5	*Picrophilus oshimae*
	pH 0.0	*Ferroplasma acidarmanus*
Low water availability	a_w = 0.6–0.65	*Torulopsis* *Candida*
Temperature and low pH	85°C, pH 1.0	*Cyanidium* *Sulfolobus acidocaldarum*
Pressure	500–1,035 atm	*Colwellia hadaliensis*
Radiation	1.5 million rads	*Deinococcus radiodurans*

moval during reduction of oxygen to water, or the pH-dependent characteristics of membrane-bound enzymes. An archaeal iron-oxidizing acidophile, *Ferroplasma acidarmanus,* capable of growth at pH 0, has been isolated from a sulfide ore body in California. This unique procaryote, capable of massive surface growth in flowing waters in the subsurface, possesses a single peripheral cytoplasmic membrane and no cell wall (**figure 27.15**).

The extreme alkalophilic microorganisms grow at pH values of 10.0 and higher and must maintain a net inward flux of protons. These obligate alkalophiles cannot grow below a pH of 8.5 and are often members of the genus *Bacillus; Micrococcus* and *Exiguobacterium* representatives have also been reported. Some cyanobacteria also have similar characteristics. Increased internal proton concentrations may be maintained by means of coordinated hydrogen and sodium ion fluxes.

Figure 27.15 Massive Growth of the Extreme Acidophile *Ferroplasma* in a California Mine. Slime streamers of *Ferroplasma acidarmanus,* an archaeon, which have developed within pyritic sediments at and near pH 0. This unique procaryote has a plasma membrane and no cell wall.

Observations of microbial growth at 121°C in thermal vent areas, or of **hyperthermophiles** (**Techniques & Applications 27.2**), indicate that this area will continue to be a fertile field for investigation. For some successful microorganisms, an extreme environment may not be "extreme" but required and even, perhaps, ideal. Phylum *Euryarchaeota* (section 20.3)

1. What factors can create extreme environments?
2. Define extremophile and discuss an example of adaptation to extremes alkalinity or acidity.
3. Why are molecular techniques changing our view of extreme environments?
4. What is unique about *Ferroplasma acidarmanus?*

27.4 Microbial Ecology and Its Methods: An Overview

Microbial ecologists study natural microbial communities that may exist in soils, waters, or in association with other organisms, including humans. Regardless of the habitat they study, these scientists seek to answer several fundamental questions. First they are interested in the microbial population: How many microbes are present? What genera and species are represented in the ecosystem? Once these questions have been addressed, community dynamics can be explored: How do these microorganisms interact with one another, with higher eucaryotes, and with the abiotic features found in the environment? The multidisciplinary nature of microbial ecology has generated an enormous assortment of methods including microscopic, cultural, physical,

Techniques & Applications

27.2 Themophilic Microorganisms and Modern Biotechnology

There is great interest in the characteristics of procaryotes isolated from the outflow mixing regions above deep hydrothermal vents that release water at 250 to 350°C. This is because these procaryotes can grow at temperatures as high as 121°C. The problems in growing these microorganisms, often archaea, are formidable. For example, to grow some of them, it is necessary to use special culturing chambers and other specialized equipment to maintain water in the liquid state at these high temperatures.

Such microorganisms, termed hyperthermophiles, with optimum growth temperatures of 85°C or above, confront unique challenges in nutrient acquisition, metabolism, nucleic acid replication, and growth. Many of these are anaerobes that depend on elemental sulfur as an electron acceptor and reduce it to sulfide. Enzyme stability is critical. Some DNA polymerases are inherently stable at 140°C, whereas many other enzymes are stabilized in vivo with unique thermoprotectants. When these enzymes are separated from their protectant, they lose their unique thermostability.

These enzymes may have important applications in methane production, metal leaching and recovery, and for use in immobilized enzyme systems. In addition, the possibility of selective stereochemical modification of compounds normally not in solution at lower temperatures may provide new routes for directed chemical syntheses. This is an exciting and expanding area of the modern biological sciences to which microbiologists can make significant contributions.

chemical, and particularly molecular techniques. Some of these techniques are now discussed.

Examination of Microbial Populations

It is now well understood that the vast majority of microbes have not been grown under laboratory conditions and the isolation and growth of microorganisms in pure or axenic culture remains of fundamental importance. A variety of standardized growth media is used in viable count procedures, which are based on colony formation and enumeration. These methods are inherently biased, as most microbes are unable to grow under a particular set of conditions. If it is necessary to isolate specific groups of microbes, or to attempt to search for organisms with new capabilities, enrichment techniques are used. These techniques are based on expansion of the microenvironment to allow massive growth of an organism formerly restricted to a small ecological niche. This approach still is valuable in studies of microbial ecology and plays a central role in finding new and undescribed microbes. It also can be used in most probable number approaches to estimate populations of specific physiological groups in an environment. Measurement of microbial growth (section 6.3)

With enrichment techniques, however, microorganisms must be able to grow under the test conditions that are used. Too often microbes can be observed in environments, but enrichment approaches and other cultural techniques do not work. There are two alternative explanations: the observed organisms truly are nonviable, or the right conditions for their growth in the lab haven't been created. This has led to the description of such potentially viable microbes as being "nonculturable." A microbe is deemed viable but nonculturable (VBNC) if it shows "signs of life" (e.g., motility or the presence of dividing cells) when directly observed or is known to grow in a natural environment (**figure 27.16**). For example, some *Vibrio* spp. may not grow in the laboratory but will grow when they infect a susceptible host. Thus it may be difficult or impossible to use culture techniques to monitor a pathogen's survival in waters and foods such as shellfish. This has spurred the development of kits based on molecular approaches for food safety analysis. The growth curve (section 6.2)

Another critical problem in growing and characterizing microorganisms, particularly protists and cyanobacteria, is that many of these microorganisms exist in microbial assemblages. Often cultures of these types of microbes are not axenic; they can have surface-associated commensal partners, and phagotrophs such as protozoa can "trap" other organisms. When such microorganisms are finally studied as axenic cultures, their morphological and physiological characteristics may change due to the lack of growth factors and vitamins formerly provided by the commensal organisms.

The importance of pure cultures, however, cannot be underestimated. Significant advances in developing new growth media and conditions continue to be made. For instance, it is now possible to grow the root-associated, nitrogen-fixing actinomycete *Frankia,* although this required about 70 years of effort. Another area where ingenuity and creativity have contributed to science is the establishment of pure cultures of barophiles that are also thermophilic.

It is now well understood that even with the use of special media and incubation conditions, only about 1% of microbial species present in any given community have been coaxed into growing in the laboratory. Today bulk nucleic acids are routinely recovered by direct extraction techniques. However, it has been found that the DNA obtained can vary depending on the method employed. This makes it difficult to state with certainty that the microbial community has specific characteristics based on the use of a single DNA extraction procedure. A second problem, especially with muds and soils, is that one has no knowledge of the source of the bulk-extracted DNA. The DNA recovered by this approach may not even be derived from living organisms. Third, the DNA may have been recovered from nonfunctional propagules such as

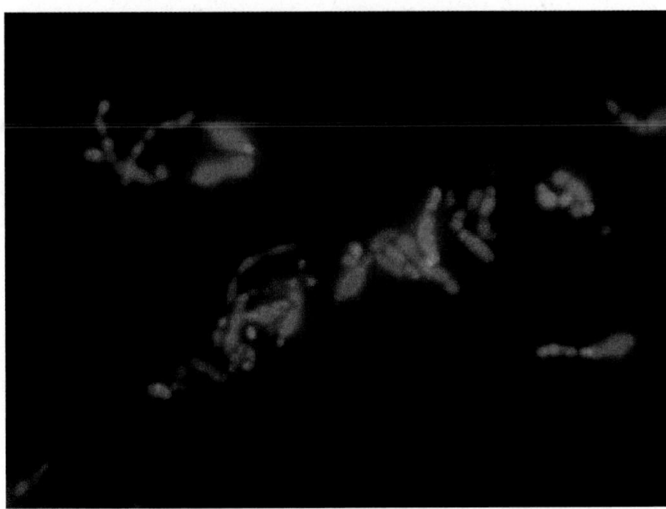

Figure 27.16 Assessment of Microbial Viability by Use of Direct Staining. By using differential staining methods, it is possible to estimate the portion of cells in a given population that are viable. In the LIVE/DEAD *Bac*Light Bacterial Viability procedure, two stains are used: a membrane-permeable green fluorescent, nucleic acid stain, and propidium iodide (red) that penetrates only cells with damaged membranes. A *Bifidobacterium* culture is shown here. Living cells stain green, while dead and dying cells stain red.

fungal or bacterial spores, or other resting structures. Thus the genera identified by cloned DNA libraries may have minimal relevance to the microbial community actually functioning in the particular environment. Environmental genomics (section 15.9)

It is important to note that all methods have inherent limitations; a critical challenge is to recognize these limitations and to understand what information a particular method will (and will not) provide. Generally, it is best to use more than one method to obtain complementary information on different aspects of a microbial community. Also, some methods are better suited to the study of one type of environment than another. For example, because waters have fewer interfering inert particles than muds or soils, it is easier to view microbes or to extract cellular constituents for use in molecular studies. Ideally, one should use approaches suitable for the study of all microbial components of a habitat (viruses, bacteria, archaea, and eucaryotes). Otherwise, one may miss critical relationships that arise because of interactions between different groups.

With these caveats in mind, microbial ecologists regularly use **small subunit (SSU) ribosomal RNA** analysis (16S for procaryotes, 18S for eucaryotes) to determine the identity of microbes that populate a community. SSU rRNA can be amplified by the polymerase chain reaction (PCR) directly from samples of soil, water, or other natural material (for instance sputum or blood in a clinical setting). The use of specific primers that target either archaeal, bacterial, or eucaryotic SSU rRNA genes enables researchers to use PCR to obtain a sufficient number of nucleic acid fragments for DNA sequencing. However, recall that if one is using a specific primer to amplify SSU rRNA from a population of genomes, one will generate a population of PCR products, most of which have a very similar molecular weight (and thus appear as a single band on an agarose gel). Techniques for determining microbial taxonomy and phylogeny: Nucleic acid sequencing (section 19.4); PCR (section 14.3)

How can one separate such a collection of DNA fragments for further analysis? The answer lies in the fact that although all the fragments are about the same size, they differ in nucleotide sequence. Commonly, **denaturing gradient gel electrophoresis (DGGE)** is used (**figure 27.17**). Recall from chapter 19 that the temperature at which double-stranded DNA is denatured varies with the G + C content; that is, it varies with DNA sequence. DGGE is based on the fact that DNA of different nucleotide sequences will denature at varying rates, although a gradient of chemicals that denature the DNA (usually urea and formamide), rather than temperature, is used. In this technique, a mixture of DNA fragments is placed in a single well in a gradient gel. As electrophoresis proceeds, fragments will migrate until they become denatured. What appeared to be a single DNA fragment (band) on nongradient agarose gel will resolve into separate fragments (multiple bands) by DGGE. These individual fragments can be cut from the gel, purified, and cloned for DNA sequencing. Gel electrophoresis (section 14.4); Determining DNA sequences (section 15.2)

Once nucleotide sequences are obtained, they can be compared with sequences from the SSU rRNA genes isolated from other microbes using several different databases. In this way, microbial ecologists can get a reasonable idea of the identity of the microbes that occupy a specific niche. Because whole organisms are not isolated and studied, it is said that specific **phylotypes,** or unique SSU rRNA genes have been identified. Sometimes other genes besides SSU rRNA analysis are used. For example, a structural gene that confers a specific metabolic capability might be used as a means of determining not only the number of phylotypes but community function. Regardless of the gene that is chosen for study, it is important to emphasize that this approach is only quantitative if real-time PCR is performed (*see section 14.3*). Thus endpoint PCR of SSU rRNA can answer the question "who is there?" but only real-time PCR can address the question "how many are there?" Environmental genomics (section 15.9)

The importance of microbial communities has led to vigorous debate regarding the magnitude of procaryotic diversity in recent years. SSU rRNA provides some insight regarding species diversity, but it is generally agreed that this technique is not well suited to determine the true number of species present in a given ecosystem. Nonetheless, it is clear that a culture-independent technique is the only valid strategy. One approach is to "count" the number of genomes. This is accomplished through **DNA reassociation.** As explained in chapter 19, DNA can be rendered single-stranded by heating, and it will spontaneously reanneal (become double-stranded) when allowed to cool. The rate at which DNA reanneals is dependent on its size: the larger the fragment (or chromosome), the longer it takes. Determining procaryotic diversity based on rates of DNA reassociation rests on the notion that DNA extracted

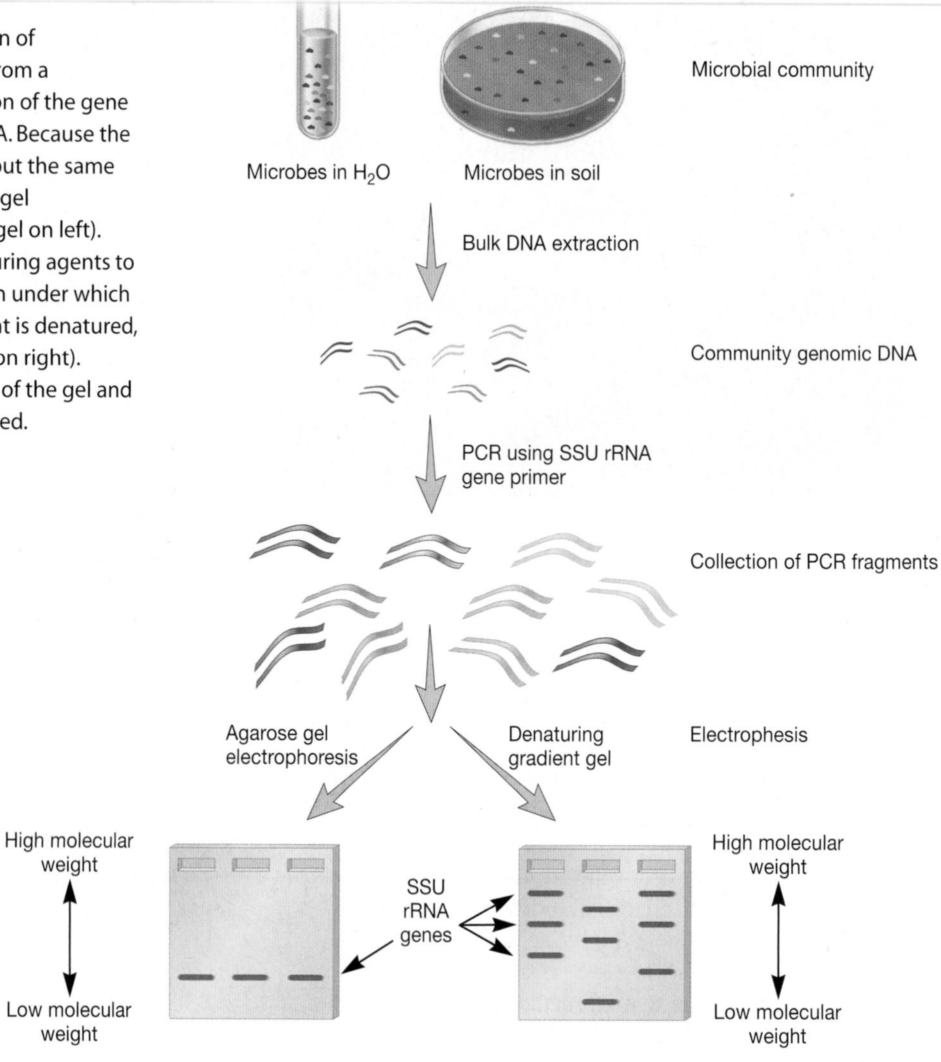

Figure 27.17 Denaturing Gradient Gel Electrophoresis (DGGE). The identification of phylotypes starts with the extraction of DNA from a microbial community and the PCR amplification of the gene of choice, usually that which encodes SSU rRNA. Because the majority of amplified DNA fragments have about the same molecular weight, when visualized by agarose gel electrophoresis they appear as a single band (gel on left). However, DGGE uses a gradient of DNA denaturing agents to separate the fragments based on the condition under which they become single-stranded. When a fragment is denatured, it stops migrating through the gel matrix (gel on right). Individual DNA fragments can then be cut out of the gel and cloned and the nucleotide sequence determined.

from the environment can be viewed as a giant genome—the length of time it takes for it to reanneal can be divided by the length of time it takes for the average procaryotic genome to reanneal. This then gives researchers a general idea of how many genomes are present. The use of this technique has revealed that on the whole, microbial communities in soil are more diverse than those in aquatic and marine ecosystems, and that unspoiled environments have more microbial species than those that are polluted. Most recently, the application of mathematical models to analyze new and previously reported data indicates that the extent of microbial diversity may be even greater than previously imagined, with over a million different species in a single pristine soil community.

Examination of Microbial Community Structure

The most direct way to assess microbial community structure is to observe complex microbial communities in nature. This can be carried out in situ using immersed slides or electron microscope grids placed in a location of interest, which are then recovered later for observation. Samples taken from an environment also are examined in the laboratory using classical cellular stains, fluorescent stains, or fluorescent molecular probes.

By combining direct observational and molecular techniques, microorganisms and their physical relationships can be studied, as shown for the *Nanoarchaeum equitans*–*Ignicoccus* coculture (**figure 27.18**). Using specific molecular probes, a unique archaeon, 400 nm in diameter, was found growing on a larger archaeon, *Ignicoccus,* in a special relationship. *Nanoarchaeum* has one of the smallest archaeal genomes found to date: only 0.5 megabases. Thus by using direct observation and molecular probes it was possible to document the life-style of this unusual archaeal symbiont, which appears to depend on its larger host for survival.

Microbial communities also can be described in terms of their structure and the nutrients contained in the community. Aquatic microorganisms can be recovered directly using filtration; the volume, dry weight, or chemical content of the microorganisms can then be measured. If it is not possible to directly measure the cell density,

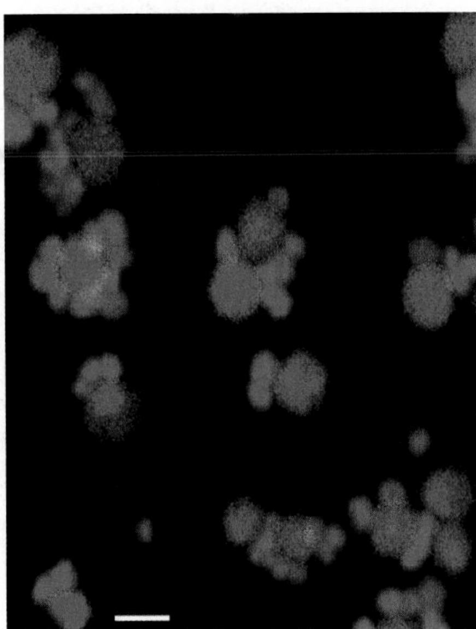

Figure 27.18 The Use of Differential Molecular Probes to Study Microbial Community Characteristics. In this study, specially designed molecular probes were used to study the physical relationships between two archaea: a larger host, a member of the genus *Ignicoccus* (*green*), and a nanosized (~400 nm) hyperthermophilic symbiont named *Nanoarchaeum equitans* (*red*). Bar = 1 μm.

carbon, nitrogen, phosphorus, or an organic constituent of the cells (such as lipids or ergosterol) can be determined. This will give a single-value estimate of the microbial community. Such chemical measurements may be used either directly or expressed as microbial biomass. Inferences can then be made based on mass balance calculations.

Single-value estimates of microbial constituents or biomass, although valuable, provide no information about the physical structure of microbes or of the microbial community. For example, in nature most filamentous fungi consist primarily of empty hyphae. A small amount of cytoplasm moves within the tubular network as the organism penetrates and exploits new substrates. Such microbes, without distinct edges and boundaries, do not have predictable volume-biomass ratios, and are described as indeterminate or **nondiscrete microorganisms.** With such microorganisms, a single-value biomass measurement is of limited value, as the organisms only can be described by direct observation of their physical structure.

The advent of environmental genomics, also called **metagenomics,** has opened vast, new possibilities in the analysis of microbial community dynamics. As discussed previously and in chapter 15, microbial populations can be viewed as a community of genomes. The exchange of genes within these communities is more widespread that once thought. This implies that members of any given microbial community co-evolve with one another and any census offers just a snapshot of its structure at that moment in time. Metagenomics starts with pooled DNA or RNA from a given microbial community and uses either shotgun sequencing of randomly cloned fragments or targeted sequencing of specific genomic regions that have been amplified by the PCR (figure 27.17). The goal is to define the function of the gene pool under a variety of conditions. Ideally, samples are taken at different times and under different circumstances, enabling some understanding of expression patterns, which provides more insight into community function. As one might imagine, there are a number of technical and computing challenges that must be overcome to produce and interpret such a vast dataset so that it can be considered valid. However, it is predicted that neither the molecular techniques nor computing power will limit the growth of this powerful new field. Rather it will be the number of biologists trained in the necessary fields of microbiology, ecology, mathematics, and bioinformatics that will slow the progress of metagenomics. Whole-genome shotgun sequencing (section 15.3); Bioinformatics (section 15.4); Environmental genomics (section 15.9)

Microbial Activity and Turnover

Measurement of microbial activity seeks to ask not simply "who is there?" but, "what are they doing?" Such measurements can be made over various time intervals, ranging from the essentially instantaneous responses of samples containing active microbes to long-term geological process-related measurements. A few examples of activity measurements are described here.

Specific processes, such as nitrification, denitrification, and sulfate reduction are studied by the use of direct chemical measurements. Microarrays can be used to measure gene expression and activity in complex microbial communities. Again, the results depend on the source and quality of the nucleic acids that are recovered from a particular microbial community and its physical environment. Similarly reverse transcriptase, real-time PCR can be used to determine what genes are expressed by community members. Stable isotope measurements can indicate whether carbon, nitrogen, or other elements have been processed by organisms. Microbes generally prefer the lighter of two stable isotopes, such as ^{12}C over ^{13}C. When a microbe uses carbon dioxide or an organic substrate, cells and their metabolic products often have lower concentrations of the heavy isotope than does the original substrate. The application of stable isotope analysis is discussed in more detail in chapter 29. Functional genomics: Evaluation of RNA-level gene expression: Microarray analysis (section 15.5); PCR (section 14.3)

Microbial growth rates in complex systems also can be measured directly. Colonization of surfaces can be observed using microscope slides or other materials. Changes in microbial numbers are followed over time, and the frequency of dividing cells (FDC) is also used to estimate production. This approach is especially valuable in studies of aquatic microorganisms. Finally, the incorporation of radiolabelled components such as thymidine (a DNA constituent) into microbial biomass provides information about growth rates and microbial turnover.

Recovery or Addition of Individual Microbes

The direct observation of microorganisms in their environments is central to the methodology used in any study of microbial ecology. In recent years, valuable new experimental approaches have been developed to recover and study individual microorganisms from an environment.

Such single-cell isolations can be carried out using **optical tweezers** (a laser beam used to drag a microbe away from its neighbors) and by micromanipulation. With a **micromanipulator,** a desired cell or cellular organelle is drawn up into a micropipette after direct observation. Once the microbe is isolated, PCR amplification of the DNA from the individual cell or cell organelle provides sequence data for use in phylogenetic analysis. For example, it has been possible to establish the phylogenetic relationship of a mycoplasma recovered from the flagellate *Koruga bonita* by micromanipulation (**figure 27.19**).

Consideration of microbial ecology on the scale of the individual cell has led to important ecological insights. It is now evident that there is surprising heterogeneity in what have been assumed to be homogenous microbial populations. Cells of a genetically uniform population do not have similar phenotypic attributes, the phenomenon of **phenotypic** or **population heterogeneity.** This is important in understanding responses of microorganisms in complex environments, and particularly in disease processes.

Microbial ecologists also use "reporter"microbes to characterize the physical microenvironment on the scale of an individual bacterium (around 1 to 3 μm). This is done by constructing cells with **reporter genes,** often based on green fluorescent protein (GFP) that change their fluorescence in response to environmental and physiological alterations. Such "reporter" microbes are used to measure oxygen availability, UV radiation dose, pollutant or toxic chemical effects, and stress. For example, when microbes that contain a moisture stress reporter gene have less available water, there is an increase in GFP-based fluorescence.
Techniques & Applications 14.1: Visualizing proteins with green fluorescence

In summary, the direct observation of microorganisms in their natural environments, combined with carefully selected classical culture, chemical, and molecular techniques, is leading to new views of how microorganisms interact with each other, with other organisms such as plants and animals, and with their abiotic environment. Important new advances continue to make microbial ecology one of the most exciting areas of modern science.

1. Why are "classic" microscopic and physical methods still used for the study of microorganisms when molecular techniques are available?
2. Describe the techniques that can be used to assess the identity of microbes versus their function in a given microbial community.
3. What time scales can be used when studying the activity of microorganisms?
4. What important advances have been made in microbial ecology, based on the recovery of individual microbes from complex environmental samples, or by addition of microbes that contain reporter genes?

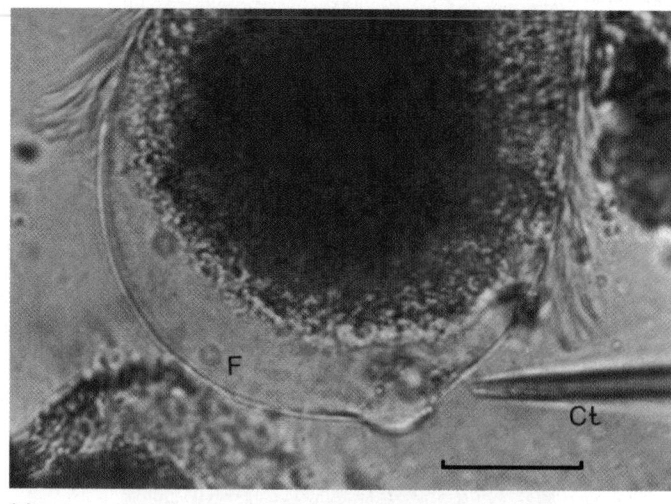

(a)

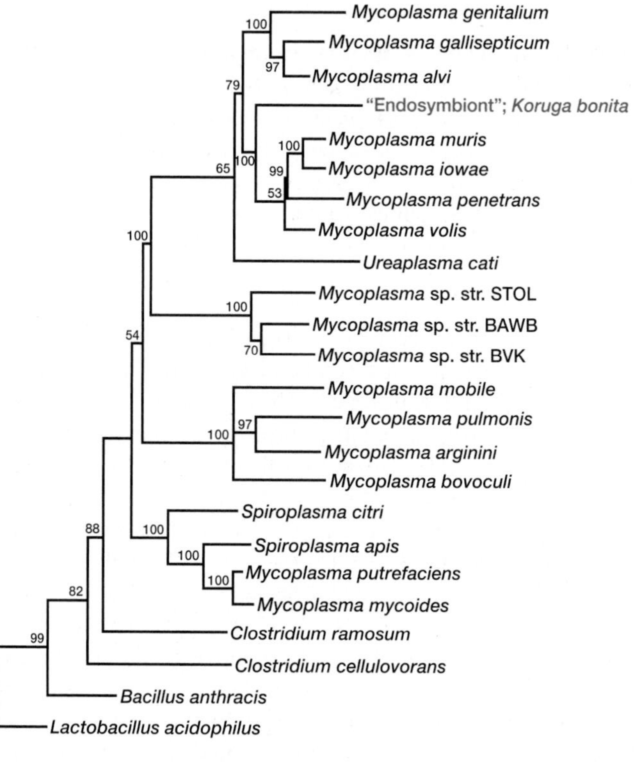

(b)

Figure 27.19 Combining Micromanipulation for Isolation of Single Cells or Organelles with the Polymerase Chain Reaction (PCR). **(a)** Recovery of an endosymbiotic mycoplasma from single cell of the flagellate *Koruga bonita* by micromanipulation (bar = 10 μm) and **(b)** phylogenetic analysis of the recovered mycoplasma following PCR amplification and sequencing of the PCR products, with the bar indicating 10% estimated sequence divergence. *Lactobacillus acidophilus* is the outgroup reference. This approach makes it possible to link a specific microorganism or organelle, isolated from a natural environment, to its molecular sequence and phylogenetic information. Flagellate (F), capillary tube (Ct).

Summary

27.1 Foundations in Microbial Diversity and Ecology

a. Only about 1% of the microorganisms that can be observed in complex natural assemblages under a microscope have been grown at the present time. Molecular techniques are making it possible to obtain information on uncultured microorganisms.

b. Microbial ecology is the study of microorganisms' interactions with their living and nonliving environments.

27.2 Biogeochemical Cycling

a. Microorganisms—functioning with plants, animals, and the environment—play important roles in nutrient cycling, which is also termed biogeochemical cycling. Assimilatory processes involve incorporation of nutrients into the organism's biomass during metabolism; dissimilatory processes, in comparison, involve the release of nutrients to the environment after metabolism.

b. Biogeochemical cycling involves oxidation and reduction processes, and changes in the concentrations of gaseous cycle components, such as carbon, nitrogen, phosphorus, and sulfur can result from microbial activity (**figures 27.2, 27.4–27.9**).

c. Major organic compounds used by microorganisms differ in structure, linkage, elemental composition, and susceptibility to degradation under oxic and anoxic conditions. Lignin is degraded only under oxic conditions, a fact that has important implications in terms of carbon retention in the biosphere.

d. In terms of effects on humans, metals can be considered in three broad groups: (1) the noble metals, which have antimicrobial properties but which do not have negative effects on humans; (2) metals such as mercury and lead, from which toxic organometallic compounds can be formed; and (3) certain other metals, which are antimicrobial in ionic form, such as copper and zinc. The second of these groups is of particular concern (**table 27.3**).

27.3 The Physical Environment

a. A microorganism functions in a physical location that can be described as its microenvironment. The resources available in a microenvironment and their time of use by a microorganism describe the niche. Pores are important microenvironments that can protect bacteria from predation.

b. Biofilms, or organized layers of microorganisms, are widespread and are formed on a wide variety of living and nonliving surfaces. These are important in disease occurrence and the survival of pathogens. Biofilms can develop to form complex layered communities called microbial mats (**figure 27.11**).

c. Microorganisms serve as primary producers that accumulate organic matter. Energy sources include hydrogen, sulfide, and methane. In addition, many chemoheterotrophs decompose the organic matter that primary producers accumulate and carry out mineralization, the release of inorganic nutrients from organic matter. The multiple and overlapping roles played by microorganisms in nutrient cycling is called the microbial loop (**figure 27.13**).

d. Decreased species diversity usually occurs in extreme environments, and many microorganisms that can function in such habitats, called extremophiles, have specialized growth requirements. For them, extreme environmental conditions can be required.

27.4 Microbial Ecology and Its Methods: An Overview

a. Many approaches can be used to study microorganisms in the environment. These include analyses of nutrient cycling, biomass, population size and activity, and community structure.

b. Methods presently being used make it possible to study presence, types, and activities of microorganisms in their natural environments (including soils, waters, plants, and animals). Although the vast majority of microorganisms that can be observed cannot yet be grown in the laboratory, molecular techniques make it possible to obtain information about these noncultured microorganisms.

c. The construction of DNA libraries from microbial communities from which SSU rRNA genes or other genes of interest can be amplified by PCR and sequenced has revealed that microbial populations and communities are more diverse and complex than traditionally thought (**figure 27.17**).

d. Optical tweezers and micromanipulators can be used to recover individual cells or cell organelles from complex microbial communities. This makes it possible to obtain genomic and phylogenetic information from specific individual microbial cells for use in studies of microbial ecology (**figure 27.19**).

Key Terms

aerobic anoxygenic photosynthesis 650
anammox reaction 649
assimilatory reduction 649
barotolerant or piezotolerant bacteria 658
biofilm 653
biogeochemical cycling 644
community 643
decomposer 656
denaturing gradient gel electrophoresis (DGGE) 661
denitrification 649
dissimilatory reduction 649
dissolved organic matter (DOM) 656
DNA reassociation 661
ecosystem 643
environmental microbiology 643
extreme barophilic bacteria 658
extreme environment 658
extremophile 658
greenhouse gas 648
hyperthermophile 659
immobilization 646
magneto-aerotactic bacteria 652
metagenomics 663
microbial ecology 643
microbial loop 656
microbial mat 655
microenvironment 653
micromanipulator 664
mineralization 646
moderately barophilic bacteria 658
niche 653
nitrification 648
nitrogen fixation 648
nondiscrete microorganism 663
optical tweezers 664
phenotypic or population heterogeneity 664
phylotype 661
population 643
primary producer 656
primary production 656
reporter genes 664
small subunit (SSU) ribosomal rRNA 661
sulfate reduction 649
viable but nonculturable (VBNC) microorganisms 657

Critical Thinking Questions

1. Compare and contrast diversity among microorganisms with diversity among macroorganisms.
2. Describe a naturally occurring niche on this planet that you believe is inhospitable to microbial life. Explain, in light of what is known about extremophiles, why you believe this environment will not support microbial life.
3. How might you show that a microorganism found in a particular extreme environment is actually growing there?
4. How might you attempt to grow a microorganism in the laboratory to increase its chances of being a strong competitor when placed back in a natural habitat?
5. Considering the possibility of microorganisms functioning at temperatures approaching 120°C, what do you think the limiting factor for microbial growth at higher temperatures will be and why?
6. Considering the intensive searches for unique microorganisms that have been carried out all over the world, where can we look for new microbes?

Learn More

Acinas, S. G.; Klepac-Ceraj, V.; Hunt, D. E.; Pharino, C.; Ceraj, I.; Distel, D. L.; and Polz, M. F. 2004. Fine-scale phylogenetic architecture of a complex bacterial community. *Nature* 430:551–54.

Colwell, R. R., and Grimes, D. J. 2000. *Nonculturable microorganisms in the environment*. Washington, D.C.: ASM Press.

Curtis, T. P., and Sloan, W. T. 2005. Exploring microbial diversity—A vast below. *Science* 309:1331–33.

Curtis, T. P., and Sloan, W. T. 2004. Prokaryotic diversity and its limits: Microbial community structure in nature and its implications for microbial ecology. *Curr. Opin. Microbiol.* 7:221–26.

DeLong, E. F. 2002. Microbial populations genomics and ecology. *Curr. Opin. Microbiol.* 5:520–24.

Fenchel, T.; King, G. M.; and Blackburn, T. H. 1998. *Bacterial biogeochemistry: The ecophysiology of mineral cycling*, 2d ed. New York: Academic Press.

Forney, L. J.; Zhou, X.; and Brown, C. J. 2004. Molecular microbial ecology: Land of the one-eyed king. *Curr. Opin. Microbiol.* 7:210–20.

Overbeck, J., and Chróst, R. J., editors. 1999. *Aquatic microbial ecology, biochemical and molecular approaches.* New York: Springer-Verlag.

Radajewski, S.; Ineson, P.; Parekh, N. R.; and Murrell, J. C. 2000. Stable-isotope probing as a tool in microbial ecology. *Nature* 403:646–49.

Rappae, M. S., and Giovannoni, S. J. 2003. The uncultured microbial majority. *Annu. Rev. Microbiol.* 57:369–94.

Stevenson, B. S.; Eichorst, S. A.; Wertz, J. T.; Schmidt, T. M., and Breznak, J. A. 2004. New strategies for cultivation and detection of previously uncultured microorganisms. *Appl. Environ. Microbiol.* 70:4748–55.

Tyson, G. W.; Chapman, J.; Hugenholtz, P.; Allen, E. E.; Ram, R. J.; Richardson, P. M.; Solovyev, V. V.; Rubin, E. M.; Rokhsar, D. S.; and Banfield, J. F. 2004. Community structure and metabolism through reconstruction of microbial genomes from the environment. *Nature* 428:37–43.

Zengler, K.; Toledo, G.; Rappé, M.; Elkins, J.; Mathur, E. J.; Short, J. M.; and Keller, M. 2002. Cultivating the uncultured. *Proc. Natl. Acad. Sci.* 99(24):15681–86.

Please visit the Prescott website at www.mhhe.com/prescott7 for additional references.

28 Microorganisms in Marine and Freshwater Environments

New procaryotes are discovered at locations where reduced and oxidized nutrients mix. This giant bacterium, *Thiomargarita namibiensis,* about 100 to 300 μm in diameter, accumulates sulfur from sediments in its refractive sulfur granules and nitrate from overlying waters to support its growth. *Thiomargarita,* which resembles a string of pearls, is found off the coast of Namibia in West Africa.

Preview

- Water bodies all over the planet support large and diverse microbial populations. In addition, microbial communities in marine and freshwater sediments make a significant contribution to the Earth's total biomass.
- The microbial communities found in freshwater and marine ecosystems are greatly influenced by complex interactions between dissolved gases and nutrient flux. Gas solubility, especially that of oxygen, has a profound impact on microbial activities. The marine environment is well buffered by the carbonate equilibrium system.
- The microbial loop has been best characterized in marine microbial ecosystems. Microbes play diverse roles in nutrient cycling, and the microbial community recycles and retains most nutrients.
- Marine microbial environments include nearshore systems such as estuaries and salt marshes, the open ocean, and benthic communities deep within the sediments.
- Autotrophic microbes in the open ocean are responsible for about one-half of the primary production on Earth. Benthic microbes have been under-explored but appear to represent a significant percentage of global microbial biomass.
- The ability of microbes to cycle vast amounts of nutrients has implications for the Earth's carbon budget, and thus is of concern as atmospheric CO_2 levels continue to increase as global warming persists.
- Freshwater microbial systems can be found in diverse habitats including glaciers, streams and rivers, and lakes. Each ecosystem presents unique physical and biological challenges to microbes.

All microbes are aquatic—even those that live on land. Procaryotes, protists, and most fungi require at least a thin film of liquid for replication. In this chapter, we turn our attention to those microbes that inhabit marine and freshwater ecosystems. The oceans cover over two-thirds of the planet and contain all but 3% of its water. Freshwater environments are the source of our drinking water and are thus required for terrestrial life. Microbiologists studying these systems are examining some of the most important and life-sustaining microbial communities on Earth. We begin with a general description of some of the physical factors that microbial populations encounter in marine and aquatic environments and some of the strategies they have evolved to meet these challenges. We then consider marine environments, including coastal and open-ocean ecosystems. Several freshwater systems are then discussed.

28.1 Marine and Freshwater Environments

Marine and freshwater environments have varied surface areas and volumes. They are found in locations as diverse as the human body; beverages; and the usual places one would expect—rivers, lakes, and the oceans. They also occur in water-saturated zones in materials we usually describe as soils. These environments can range from alkaline to extremely acidic, with temperatures from −5 to −15°C at the lower range, to at least 121°C in geothermal areas. Some of the most intriguing microbes have come from the study of high-temperature environments, including the now-classic studies of Thomas Brock and his coworkers at Yellowstone National Park. Their work led to the discovery of *Thermus aquaticus,* the source of the temperature-stable DNA polymerase, which makes PCR possible.

Water is a very good servant, but it is a cruel master.

—*John Bullein*

In addition to temperature, the penetration of sunlight and the mixing of nutrients, O_2, and waste products that occur in freshwater and marine environments are dominant factors controlling the microbial community. For example, in deep lakes or oceans, organic matter from the surface can sink to great depths, creating nutrient-rich zones where decomposition takes place. Gases and soluble wastes produced by microorganisms in these deep zones can move into overlying waters and stimulate the activity of other microbial groups.

Microbiologists who study marine and freshwater microbes and their habitats seek to understand the enormous diversity of microbes that contribute to these communities and their interactions with the natural environments they inhabit. It is an exciting time for aquatic microbiology—the development of molecular biology, novel culturing techniques, remote sensing, and deep-sea exploration has propelled this discipline to a new age of discovery. Recent reports have revealed a level of microbial diversity not previously imagined as well as the importance of microbes in maintaining a balanced ecosystem. We now realize the role microbes play in addressing problems such as global warming, disease, and pollution (**Disease 28.1**). In addition, these new technologies have advanced our understanding of the food webs that govern the world's fisheries. Thus the microbiology of lakes, streams, and oceans is of enormous interest and importance.

Water as a Microbial Habitat

The nature of water as a microbial habitat depends on a number of physical factors such as temperature, pH, and light penetration. One of the most important of these is dissolved oxygen content. The flux rate of oxygen through water is about 1/10,000 times less than its rate through air. However, in some marine habitats the limits to oxygen diffusion can be offset by the increased solubility of oxygen at colder temperatures and increasing atmospheric pressures. Thus for the very deep ocean, the oxygen concentration actually increases with depth, even though the air/water interface is literally miles away (**figure 28.1**). On the other hand, tropical lakes and summertime-temperate lakes may become oxygen limited only meters below the surface. In this case, aerobic microbes consume the surface-associated oxygen faster than it can be replenished. This frequently leads to the formation of **hypoxic** or **anoxic** zones in these aquatic environments. This enables specialized anaerobic microbes, both chemotrophic and phototrophic, to grow in the lower regions of lakes where light can penetrate.

The second major gas in water, CO_2, plays many important roles in chemical and biological processes. The pH of distilled water, which is not buffered, is determined by the dissolved CO_2 in equilibrium with the air and is approximately 5.0 to 5.5. The pH of freshwater systems such as lakes and streams, which are usually only weakly buffered, is therefore controlled by the nature of terrestrial input (for instance, minerals that may be either acidic or alkaline) and the rate at which CO_2 is removed by photosynthesis. When autotrophic organisms such as diatoms use CO_2, the pH of the water is often increased.

In contrast, seawater is strongly buffered by the balance of CO_2, bicarbonate (HCO_3^-), and carbonate (CO_3^{2-}). Atmospheric CO_2 enters the oceans where it is either converted to organic carbon by photosynthesis or it reacts with seawater to form carbonic acid (H_2CO_3), which quickly dissociates to form bicarbonate and carbonate (**figure 28.2**):

$$CO_2 + H_2O \rightleftharpoons H_2CO_3 \rightleftharpoons H^+ + HCO_3^- \rightleftharpoons 2H^+ + CO_3^{2-}$$

The oceans are effectively buffered between pH 7.6 and 8.2 by this **carbonate equilibrium system.** Much like the buffer one might use in a chemistry experiment, the pH of seawater is determined by the relative concentrations of the weak acids bicarbonate and carbonate. The equilibrium of these reactions has taken on new importance. Some oceanographers predict that the pH of the ocean will drop by 0.35 units by 2100 unless effective means of limiting greenhouse gas emissions (in particular CO_2) are implemented. Recall that the pH scale is logarithmic; it is unclear what the implications of this change in carbonate equilibrium will mean for the marine environment, and indeed all of life on Earth.

Disease

28.1 New Agents in Medicine—The Sea as the New Frontier

Most currently available antibiotics have been derived from soil microorganisms, primarily from the actinomycetes, but also from nonfilamentous gram-positive bacteria and fungi. Hundreds of these natural products are in use as antibiotics, antitumor agents, and agrochemicals.

In recent years, with the need for additional compounds for use in medicine, marine microorganisms are receiving increased attention. Some of the newer chemicals that have been discovered are microalgal metabolites. There also is an interest in the culture of symbiotic marine microorganisms, including *Prochloron,* which are associated with macroscopic hosts. A variety of interesting compounds of unknown origin have been discovered. Many are assumed to be of microbial origin, but more work will be needed to establish this. Many biologists feel that marine microorganisms may provide unique bioactive compounds, including marine toxins, which do not occur in terrestrial microorganisms. There is a worldwide effort to better characterize the marine microbial community and to harness these often poorly studied microorganisms for modern medicine.

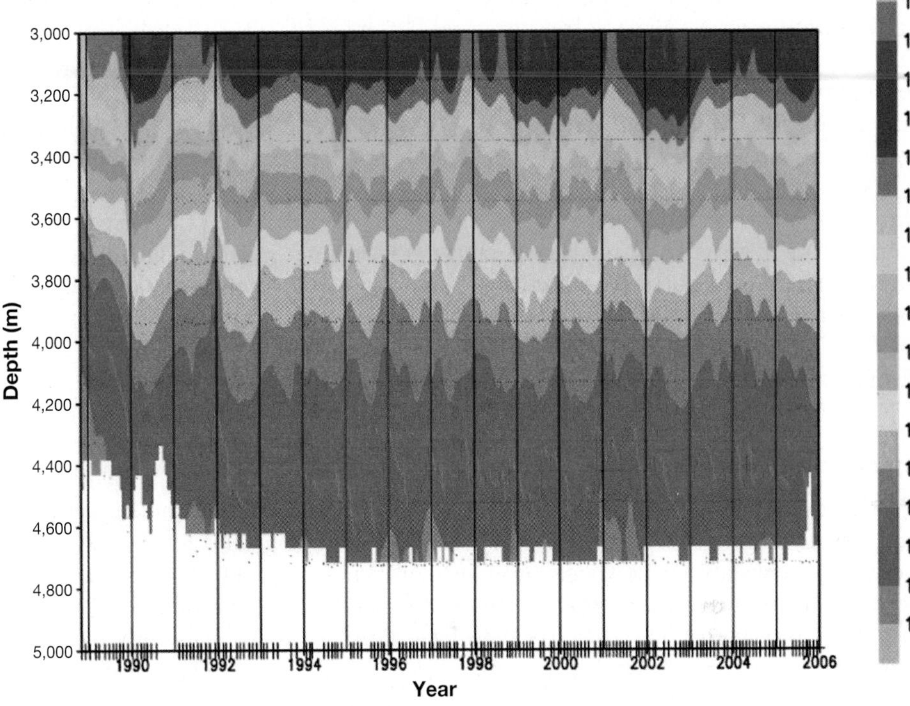

Figure 28.1 Oxygen vs. Depth in the Deep Ocean. Dissolved oxygen measured in water samples between 3,000 and 5,000 meters at Station Aloha, in the Hawaii Ocean Time series (HOT) program. Note that oxygen concentration increases with depth due to increased oxygen solubility in cold waters and at high pressure.

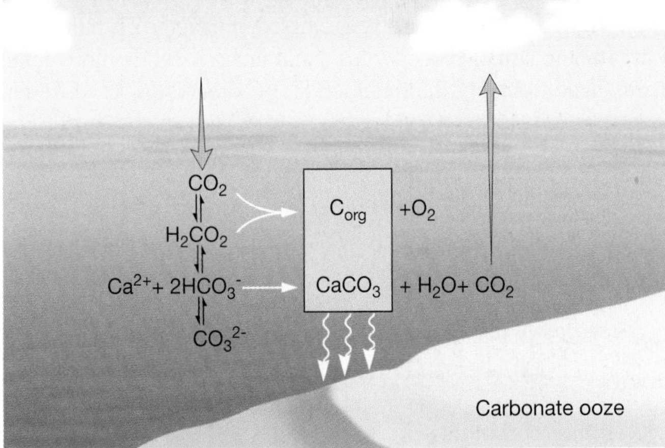

Figure 28.2 The Carbonate Equilibrium System. Atmospheric CO_2 enters seawater and is converted to organic carbon (C_{org}) or is converted to carbonic acid (H_2CO_3) that rapidly dissociates into the weak acids bicarbonate (HCO_3^-) and carbonate (CO_3^{2-}). Calcium carbonate ($CaCO_3$), a solid, precipitates to the seafloor where it helps form a carbonate ooze. This system keeps seawater buffered at about pH 8.0.

Other gases also are important in aquatic environments. These include nitrogen gas, used as a nitrogen source by nitrogen fixers; hydrogen, which is both a waste product and a vital substrate; and methane (CH_4). These gases vary in their water solubility; methane is the least soluble of the three. Under certain conditions, methane can be an ideal microbial waste product: once it is produced under anoxic conditions, it leaves the microorganism's environment by diffusing up in the water column where it can be oxidized by methanotrophs or released to the atmosphere. This eliminates the problem of toxic waste accumulation that occurs with many microbial metabolic products, such as organic acids and ammonium ion.

Light is also critical for the health of marine and freshwater ecosystems. Like all life on Earth, microbes in these environments depend on **primary producers**—autotrophic organisms—to provide organic carbon. In streams, lakes, and coastal systems, much of the carbon is fixed by macroscopic algae and plants, and organic carbon enters these systems in terrestrial runoff. The situation is very different in the open ocean where all organic carbon is the product of microbial autotrophy. In fact, about one-half of all the organic carbon on Earth is the result of this microbial (eucaryotic and procaryotic) carbon fixation. Water from the surface to the depth to which light penetrates with sufficient intensity to support these important autotrophs is called the **photic zone.** We see a marked difference in the depth of the photic zone when we compare nearshore waters with the open ocean. In lakes and estuaries where the water is turbid, the photic zone may be only a meter or two in depth. This is in sharp contrast to nutrient-depleted areas such as the open ocean and many tropical areas where the water seems "crystal clear." In these regions the photic zone ranges from 150 to 200 meters deep.

Solar radiation warms the water and this can lead to thermal stratification. Warmer water is less dense than cool water, so as the sun warms the surface in tropical and temperate waters, a **thermocline** develops. A thermocline can be thought of as a mass of warmer water "floating" on top of cooler water. These two water masses remain separate until there is either a substantial mixing event, such as a severe storm, or in temperate climates, the onset of autumn. As the weather cools, the upper layer of warm water becomes cooled and the two water masses mix. This is often associated with a pulse of nutrients from the lower, darker waters to the surface. This can trigger a sudden and rapid increase in the population of some microbes and a bloom may develop. This is discussed more fully in section 28.3.

1. What factors influence oxygen solubility? How is this important in considering marine and aquatic environments?
2. Describe the buffering system that regulates the pH of seawater. What might be the implications of this stable buffering on microbial evolution?
3. What is the photic zone? How and why does it differ in lakes and coastal ecosystems versus the open ocean?
4. What is a thermocline?

Nutrient Cycling in Marine and Freshwater Environments

There are obviously many differences between nearshore and open-ocean environments. From a microbial point of view, lakes, estuaries, and other coastal regions have relatively high rates of primary production. They therefore must have a higher influx and turnover (or re-use) of essential nutrients. In these regions, nitrogen and phosphorus are most essential in terms of limiting growth. Nearby agricultural and urban activity frequently generates runoff that provides substantial nutrient inputs to these environments. In contrast, nutrient levels are very low in the open ocean, which is unaffected by rivers, streams, and terrestrial runoff. Here nitrogen, phosphorus, iron, and even silica, which diatoms need to construct their frustules, can be limiting. Protist classification: *Stramenopiles* (section 25.6)

The major source of organic matter in illuminated surface waters is photosynthetic activity, primarily from **phytoplankton** [Greek *phyto,* plant and *planktos,* wandering], autotrophic organisms that float in the photic zone. Common planktonic cyanobacterial genera are *Prochlorococcus* and *Synechococcus,* which can reach densities of 10^4 to 10^5 cells per milliliter at the ocean surface. These **picoplankton** (planktonic microbes between 0.2 and 2.0 μm in size) can represent 20 to 80% of the total phytoplankton biomass. Eucaryotic autotrophs, especially diatoms, also contribute a significant fraction of fixed carbon to these ecosystems.

As they grow and fix carbon dioxide to form organic matter, phytoplankton acquire needed nitrogen and phosphorus from the surrounding water. In marine systems, the nutrient composition of the water affects the final carbon-nitrogen-phosphorus (C:N:P) ratio of the phytoplankton, which is termed the **Redfield ratio,** named for the oceanographer Alfred Redfield. A commonly used value for this ratio is 106 parts C, 16 parts N, and 1 part P. This ratio is important for following nutrient dynamics, especially mineralization and immobilization processes, and for studying the sensitivity of oceanic photosynthesis to atmospheric additions of CO_2 nitrogen, sulfur, and iron. Microbial ecology and its methods: Microbial activity and turnover (section 27.4)

In addition to their role as primary producers, microbes play an essential role in cycling other nutrients as well. The **microbial loop** (**figure 28.3**) was briefly discussed in chapter 27, but it is so important to aquatic ecosystems that it is discussed in more detail here. Traditionally, the interaction of organisms at different

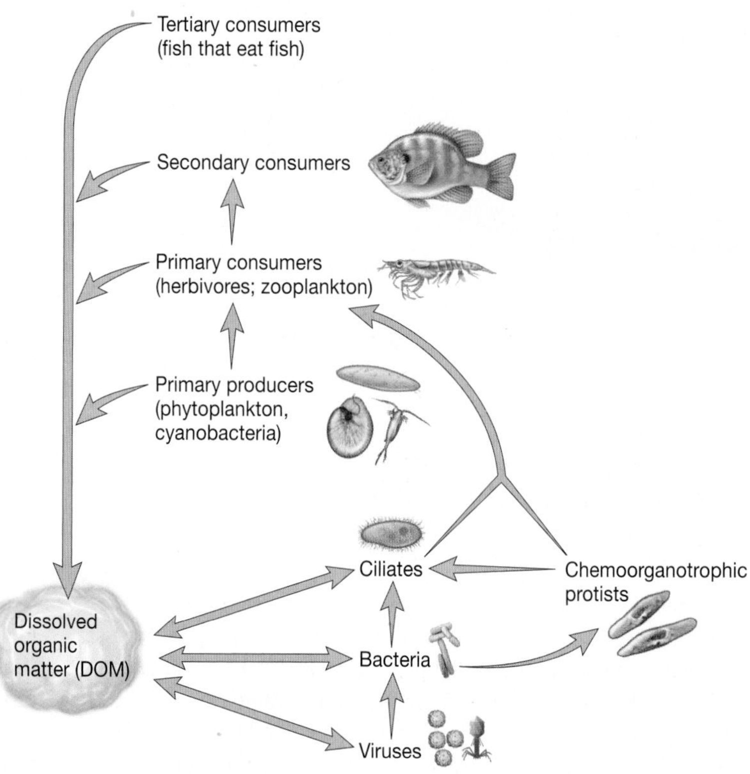

Figure 28.3 The Microbial Loop. Microorganisms play vital roles in ecosystems as primary producers, decomposers, and primary consumers. All organisms contribute to a common pool of dissolved organic matter (DOM) that is consumed by microbes. Viruses contribute DOM by lysing their hosts, and procaryotes are consumed by protists, which also consume other protists. These microbes are then consumed by herbivores that often select their food items by size, thereby ingesting both heterotrophic and autotrophic microbes. Thus nutrient cycling is a complex system driven in large part by microbes.

trophic levels has been depicted as a food chain in which primary producers are most numerous. They must provide all the organic carbon consumed by herbivores. Herbivores are then consumed by carnivores, which may occupy several trophic levels. Such diagrams generally show microorganisms functioning strictly as decomposers, mineralizing most of the waste products produced in the ecosystem. However, this simplified version of trophic interactions does not adequately describe the important and varied roles of microbes. Consider that microbial ecologists estimate there are at least 6×10^{30} microbial cells residing on the planet at any given time. This unseen biomass far exceeds that of all other organismal groups combined. If microbes functioned only to degrade and mineralize organic material to their inorganic forms, these essential elements would be in danger of being irreversibly removed from the ecosystem. Instead, microbes interact with several trophic levels, serving to recycle nutrients many times within the community before any given element is either mineralized or sinks to the sediments below.

The microbial loop describes the many roles that microbes serve. For example, the metabolic flexibility of eucaryotic and procaryotic microbes allows them to consume **dissolved organic matter (DOM)** that larger organisms cannot degrade. Sources of DOM include the liquid waste of zooplankton and material that leaks from the phytoplankton, sometimes called **photosynthate.** Viruses are also a source of DOM. Marine viruses can be present at concentrations up to 10^8 per milliliter; the lysis of their host cells contributes significantly to the return of nutrients back into the microbial loop (see page 679). Protists, including flagellates and ciliates, consume these smaller microbes, which can be thought of as **particulate organic matter (POM).** Because these organisms are then consumed by zooplankton, both DOM and POM are recycled back into the food web for use at a number of trophic levels.

1. Describe the differences in nutrient input in coastal ecosystems as compared to open ocean.
2. What is picoplankton and why is it important?
3. How does the microbial loop differ from a food chain?

28.2 Microbial Adaptations to Marine and Freshwater Environments

Water provides an environment in which a wide variety of microorganisms survive and function. Microbial diversity depends on available nutrients, their varied concentrations (ranging from extremely low to very high levels), the transitions from oxic to anoxic zones, and the mixing of electron donors and acceptors in this dynamic environment. In addition, the penetration of light into many anoxic zones creates environments for certain types of photosynthetic microorganisms. Here we discuss adaptations of specific microbes to some particular aquatic environments.

One adaptation that has taken many marine microbiologists by surprise is just how small most oceanic microbes are. In fact, they are so small it was not until the development of very fine filtration systems (less than 0.2 μm) and the application of direct counting methods such as epifluorescence microscopy that the abundance of **ultramicrobacteria** was discovered. How is small size an adaptation? Recall that microbial cells must assimilate all nutrients across their plasma membranes. Cells with a large surface area relative to their total intracellular volume are able to maximize nutrient uptake, and can therefore grow more quickly than their larger neighbors. Thus the majority of microbes growing in nutrient-limited, or **oligotrophic,** open oceans are between 0.3 μm and 0.6 μm. The question as to whether small size is a response to oligotrophy or an adaptation has been difficult to answer because most microbes have not be cultivated. However, the fact that some cultured ultramicrobacteria do not become larger when nutrients are added suggests that, at least in these cases, the microbes have evolved to maximize their surface area to volume ratio to oligotrophic conditions.

At the other extreme is an unusual marine microbe found off the coast of Namibia in western Africa. *Thiomargarita namibiensis,* which means the "sulfur pearl of Namibia," is considered to be the world's largest bacterium (**figure 28.4**). Individual cells are

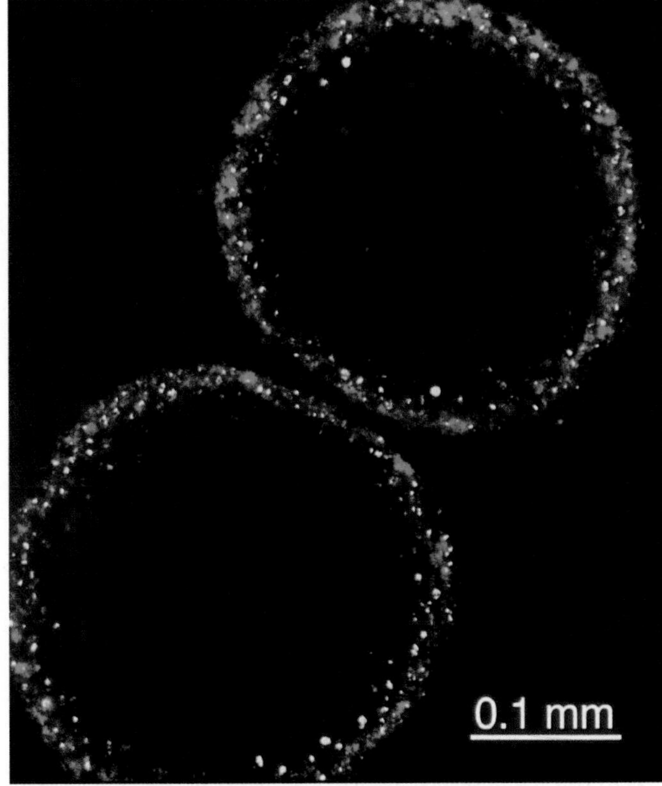

Figure 28.4 ***Thiomargarita namibiensis,*** **the World's Largest Known Bacterium.** This procaryote, usually 100 to 300 μm in diameter as shown here, occasionally reaches a size of 750 μm (larger than a period on this page), 100 times the size of a common bacterium. This unique bacterium uses sulfide from bottom sediments as an energy source and nitrate, which is found in the overlying waters, as an electron acceptor.

(a)

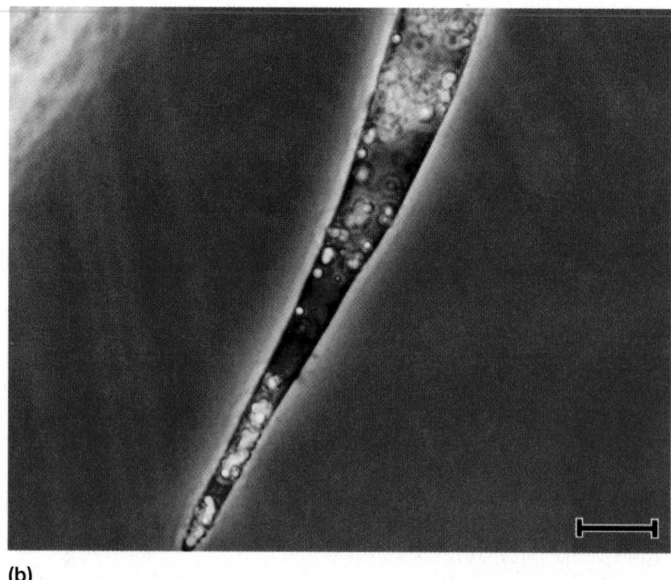

(b)

Figure 28.5 ***Thioploca*** **the "Spaghetti Bacterium."** *Thioploca* ("sulfur braid") is an unusual microorganism that links separated resources of sulfide from the mud and nitrate from the overlying water. **(a)** Bundles that join the oxic surface and the lower anoxic mud. **(b)** An individual *Thioploca,* showing the elemental sulfur globules and tapered ends. Bar = 40 μm.

usually 100 to 300 μm in diameter (750 μm cells occasionally occur with a biovolume of 200,000,000 μm^3). Sulfide and nitrate are used as the electron donor and acceptor, respectively. In this case nitrate, from the overlying seawater, penetrates the anoxic sulfide-containing muds only during storms. When this short-term mixing occurs, *Thiomargarita* takes up and stores the nitrate in a huge internal vacuole, which may occupy 98% of the organism's volume. The vacuolar nitrate can approach a concentration of 800 mM. The elemental sulfur granules appear near the cell edge in a thin layer of cytoplasm. Between storms, the organism lives using the stored nitrate as an electron acceptor. These unique bacteria are important in sulfur and nitrogen cycling in these environments. Microbial Diversity & Ecology 3.1: Monstrous microbes

A critical adaptation of microorganisms in aquatic systems is the ability to link and use resources that are in separate locations, or that are available at the same location only for short intervals such as during storms. One of the most interesting bacteria linking widely separated resources is *Thioploca* spp., which lives in bundles surrounded by a common sheath (**figure 28.5**). These microbes are found in upwelling areas along the coast of Chile, where oxygen-poor but nitrate-rich waters are in contact with sulfide-rich bottom muds (much like the environment off the coast of western Africa). The individual cells are 15 to 40 μm in diameter and many centimeters long, making them, like *T. namibiensis,* one of the largest bacteria known. They form filamentous sheathed structures, and the individual cells can glide 5 to 15 cm deep into the sulfide-rich sediments. These unique microorganisms are found in expanses off the coast of Chile.

In addition to living a planktonic existence, many microorganisms take advantage of surfaces. These include sessile microorganisms of the genera *Sphaerotilus* and *Leucothrix* and the prosthecate and budding bacteria of the genera *Caulobacter* and *Hyphomicrobium.* There are also a wide range of aerobic gliding bacteria such as the genera *Flexithrix* and *Flexibacter,* which move over surfaces where organic matter is adsorbed. These organisms are characterized by their exploitation of surfaces and nutrient gradients. They are obligate aerobes, although sometimes they can carry out denitrification, as occurs in the genus *Hyphomicrobium.* In addition, bacteria may be primary colonizers of submerged surfaces, allowing subsequent development of a complex biofilm. Biofilms (sections 6.6 and 27.3); Class *Alphaproteobacteria:* The *Caulobacter* and *Hyphomicrobium* (section 22.1); Class *Gammaproteobacteria:* Order *Thiotrichales* (section 22.3); Microbial Diversity & Ecology 21.1: The mechanism of gliding motility

Microscopic fungi, which usually are thought to be terrestrial organisms living in soils and on fruits and other foods, also grow in freshwater and marine environments. Zoosporic organisms adapted to life in the water include the chytrids, which have motile asexual reproductive spores with a single whiplash flagellum. Chytrids are important because of their role in decomposing dead organic matter. In addition, many chytrids attack algae (**figure 28.6**). Characteristics of fungal divisions: *Chytridiomycota* (sections 26.6)

Another important group includes filamentous fungi that can sporulate under water. These hyphomycetes include the **Ingoldian fungi,** named after C. T. Ingold. In 1942 Ingold discovered fungi that produce unique tetraradiate forms (**figure 28.7**). The ecology

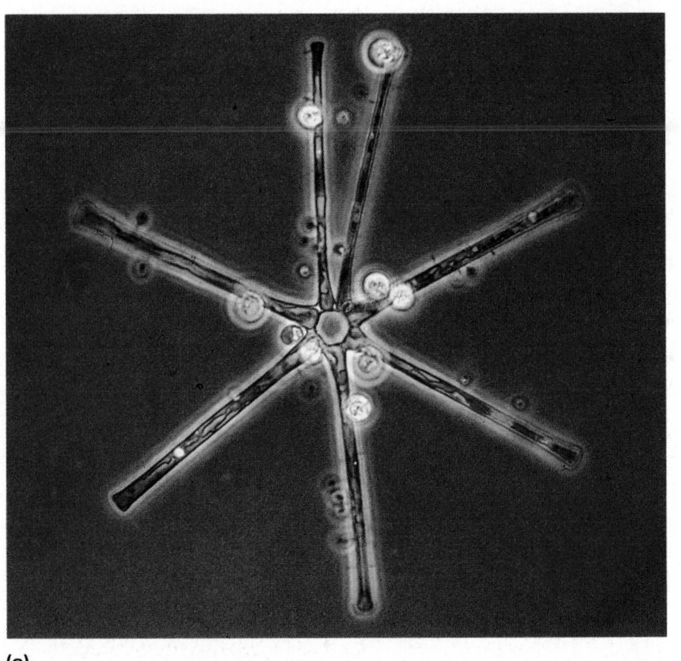

(a)

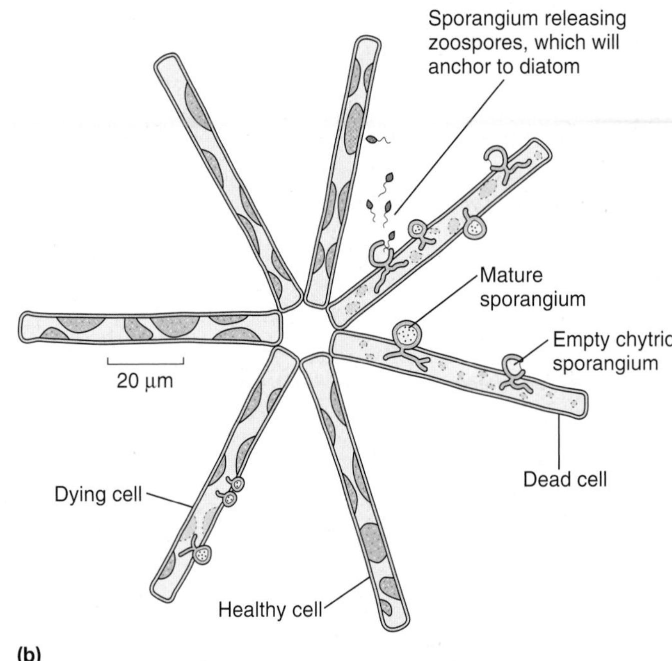

(b)

Figure 28.6 Chytrids and Aquatic Environments. Chytrids play important roles in aquatic environments. **(a)** The infection of the diatom by the chytrid *Rhizophydium* is shown in the photograph, and **(b)** in the illustration showing details of the parasitic process.

of these aquatic fungi is very interesting. The tetraradiate conidium forms on a vegetative mycelium, which grows inside decomposing leaves. When the vegetative hyphae differentiate into an aerial mycelium, conidia are released into the water. Released conidia are then transported and often are present in surface foam. When they contact leaves, the conidia attach and establish new centers of growth. These uniquely adapted fungi contribute significantly to the processing of organic matter, especially leaves. Often aquatic insects will feed only on leaves that contain fungi.

1. What are ultramicrobacteria? What is the evidence that they have adapted genetically to oligotrophic environments as opposed to simply responding to nutrient limitation?
2. Compare the environments in which *Thiomargarita namibiensis* and *Thioploca* spp. occur. How are these two microbes similar?
3. Describe the life cycle of Ingoldian fungi.

28.3 Microorganisms in Marine Environments

As terrestrial organisms, we must remind ourselves that 97% of the Earth's water is in marine environments. Although much of this is in the deep sea, from a microbiological perspective, the surface waters have been most intensely studied. This is where the photosynthesis that drives all the marine ecosystems occurs. Only recently have scientists had the capacity to probe the deep-sea sediments and the subsurface (the benthos), and investigations of this kind are revealing a number of surprises. We begin our discussion of marine ecosystems with estuaries, and then discuss the microbial communities that inhabit the open ocean and finally the dark, cold, high-pressure benthos.

Coastal Marine Systems: Estuaries and Salt Marshes

An estuary is a semi-enclosed coastal region where a river meets the sea. Estuaries are defined by tidal mixing between freshwater and salt water. They feature a characteristic salinity profile called a **salt wedge** (**figure 28.8**). Salt wedges are formed because saltwater is denser than freshwater, so seawater sinks below overlying freshwater. As the contribution from the incoming river increases and that of the ocean decreases, the relative amount of seawater declines with the estuary's increased distance from the sea. The distance the salt wedge intrudes up the estuary is not static. Most estuaries undergo large-scale tidal flushing; this forces organisms to adapt to changing salt concentrations on a daily basis. Microbes that live under such conditions combat the resulting osmotic stress by adjusting their intracellular osmolarity to limit the difference with that of the surrounding water. Most protists and fungi produce osmotically active carbohydrates for this purpose, whereas procaryotic microbes regulate internal concentrations of potassium or special amino acids such as ecoine and betaine. Thus most microbes that inhabit estuaries are **halotolerant,** which is distinct from halophilic. Halotolerant microbes can withstand significant changes in salinity; halophilic microorganisms have an absolute requirement for high salt concentrations.

Figure 28.7 Ingoldian Fungi. These aquatic fungi are capable of sporulation under water. They play important roles in the processing of complex organic matter, such as leaves, which fall into streams and lakes. These microbes include types with tetraradiate conidia. **(a)** Fungal hyphae grow inside the decomposing leaf and give rise to tetraradiate conidia. These aerial structures project from the leaf surface into the water column. **(b)** The new tetraradiate conidium then can be released and attach to a leaf surface, repeating the process and accelerating leaf decomposition and the release of nutrients.

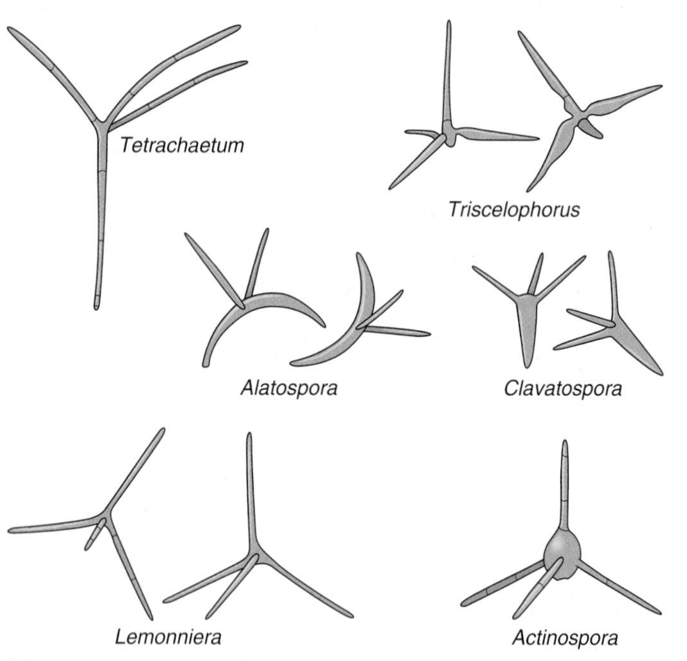

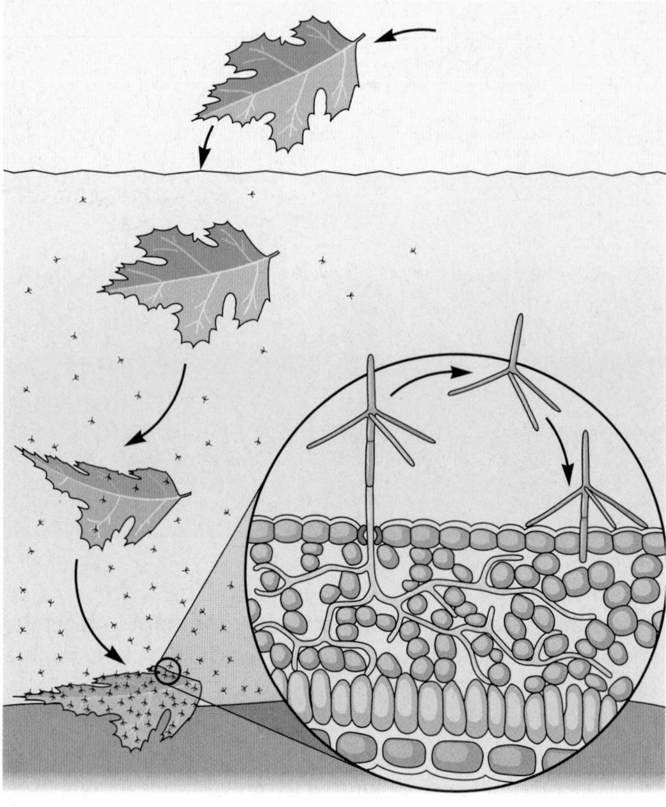

Figure 28.8 A Salt Wedge. An estuary contains both freshwater and saltwater. Because seawater is denser than freshwater, the water masses do not mix. Rather, the seawater remains below the freshwater with the relative amount of seawater decreasing in the upper reaches of the estuary.

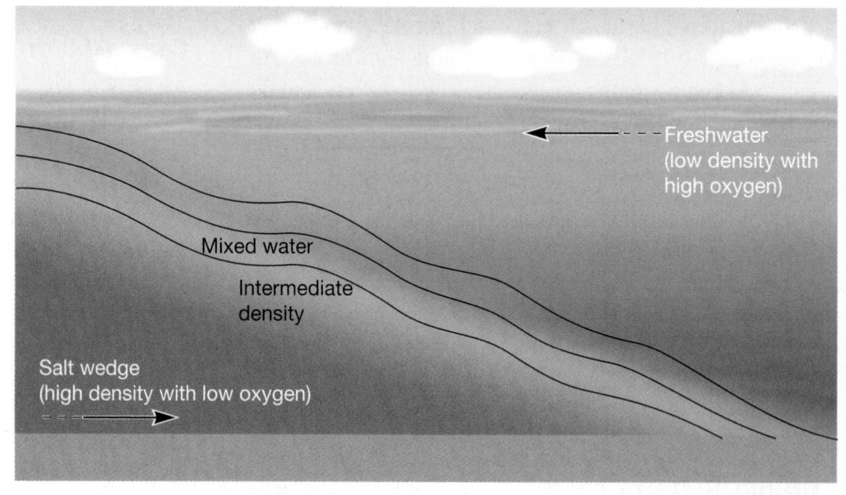

Estuaries are unique in many respects. Their calm, nutrient-rich waters serve as nurseries for juvenile forms of many commercially important fish and invertebrates. However, despite their importance to the commercial fishing industry, estuaries are among the most polluted marine environments. They are the receptacles of waste that is dumped in rivers and pollutants that have been discharged during industrial processes. Recall that from the 17th century through most of the 20th century, industries dumped their wastes without fear of punitive consequences. The cleanup of rivers and estuaries contaminated with industrial wastes such as polychlorinated biphenyls (PCBs) continues to this day. Often industrial pollution includes organic materials that can be used as nutrients. This further magnifies the problem because chemoorganotrophic microbes consume available oxygen, forming anoxic dead zones. Such anoxic regions, which are devoid of almost all macroscopic life, are now present in the Gulf of Mexico, Chesapeake Bay, and fjords entering the North Sea.

Organic pollution can also create the opposite problem: too much growth. However, in this case, a single microbial species, either algal or cyanobacterial, grows at the expense of all other organisms in the community. This phenomenon, called a bloom, often results from the introduction of nutrients combined with mixing sediments. If the microbes produce a toxic product or are in themselves toxic to other organisms such as shellfish or fish, the term **harmful algal bloom (HAB)** is used. Some HABs are called red tides because the microbial density is so great that the water becomes red or pink (the color of the algae). The number of HABs has dramatically increased in the last decade or so. HABs are sometimes responsible for killing large numbers of fish or marine mammals. For instance, off the coast of California, an HAB was responsible for sudden, large-scale deaths among sea lions. In this case, the bloom species was a diatom of the genus *Pseudonitzschia*. Anchovies consumed the diatoms and the potent neurotoxin, domoic acid, accumulated in the fish. The mammals were poisoned after they ingested large quantities of the fish, an important component of the sea lion diet. Disease 25.1: Harmful algal blooms (HABs)

HABs are often caused by dinoflagellates. Some bloom-causing dinoflagellates produce a potent neurotoxin, called a **brevetoxin.** Dinoflagellates of the genus *Alexandrium* produce a brevetoxin responsible for most of the *p*aralytic *s*hellfish *p*oisoning (PSP), which affects humans as well as other animals in coastal, temperate North America. The brevetoxin produced by the dinoflagellate *Karinia brevis* killed a number of endangered manatees and bottlenose dolphins in 2002 in a bloom in Florida. Another dinoflagellate, *Pfiesteria piscicida,* has become a problem in the Chesapeake Bay and regions south. This protist produces lethal lesions in fish (**figure 28.9**) and has had a devastating effect on the local fishery industry. Exposure to this microbe also causes neurological damage to humans, including short-term memory loss. Our understanding of this microbe has been limited by a number of factors, including a debate about its life cycle and problems in isolating its toxin. Protist classification: *Alveolata* and *Stramenopiles* (section 25.6)

Figure 28.9 ***Pfiesteria* Lesions.** Lesions on menhaden resulting from parasitism by the dinoflagellate *Pfiesteria piscicida*.

Salt marshes generally differ from estuaries in that they lack freshwater input from a single source. Smaller streams enter salt marshes, which are flatter and have a wider expanse of sediment and plant life exposed at low tide. For this reason, salt marshes are sometimes called salt meadows. The microbial communities within salt marsh sediments are very dynamic. These ecosystems can be modeled in **Winogradsky columns** (**figure 28.10**), named after the pioneering microbial ecologist Sergei Winogradsky (1856–1953). A Winogradsky column is easily constructed using a glass cylinder into which either marine or freshwater sediment is placed and then overlaid with saltwater or freshwater, respectively. Winogradsky columns prepared with saltwater and salt marsh sediments contain a considerable amount of sulfur because sulfate is found in seawater. Anaerobic microbes use the sulfate as a terminal electron acceptor and produce hydrogen sulfide. When the top of the column is sealed, much of the cylinder will eventually become anoxic. The addition of shredded newspaper introduces a source of cellulose that is degraded to fermentation products by the genus *Clostridium*. With these fermentation products available as electron donors and using sulfate as an acceptor, *Desulfovibrio* produces hydrogen sulfide. The hydrogen sulfide diffuses upward toward the oxygenated zone, creating a stable hydrogen sulfide gradient. In this gradient the photoautotrophs *Chlorobium* and *Chromatium* develop as visible olive green and purple zones. These microorganisms use hydrogen sulfide as an electron source, and CO_2, from sodium carbonate, as a carbon source. Above this region the purple nonsulfur bacteria of the genera *Rhodospirillum* and *Rhodopseudomonas* can grow. These photoorganotrophs use organic matter as an electron donor under anoxic conditions and function in a zone where the sulfide level is lower. Both O_2 and hydrogen sulfide may be present higher in the column, allowing specially adapted microorganisms to function. These include the chemolithotrophs *Beggiatoa* and *Thiothrix,* which use reduced sulfur compounds as electron donors and O_2 as an acceptor. In the upper portion of the column, diatoms and cyanobacteria may be visible. Phototrophy (section 9.12); Photosynthetic bacteria (section 21.3); Class *Alphaproteobacteria:* Purple nonsulfur bacteria (section 22.1); Class *Gammaproteobacteria* (section 22.3); Class *Clostridia* (section 23.4)

1. Define salt wedge and explain its influence on estuarine microbial communities.
2. Consider the fact that during droughts, rivers that normally flow into an estuary have a significantly diminished flow. Examine figure 28.8 and explain how this could result in salt intrusion into a public water supply. What do you think the consequences would be on estuarine microbial communities?
3. Name two groups of protists known to cause HABs in marine ecosystems.
4. Describe the ecosystem that develops within a Winogradsky column. How is this similar to the microbial community you would expect to find in a salt marsh? How might it be different?

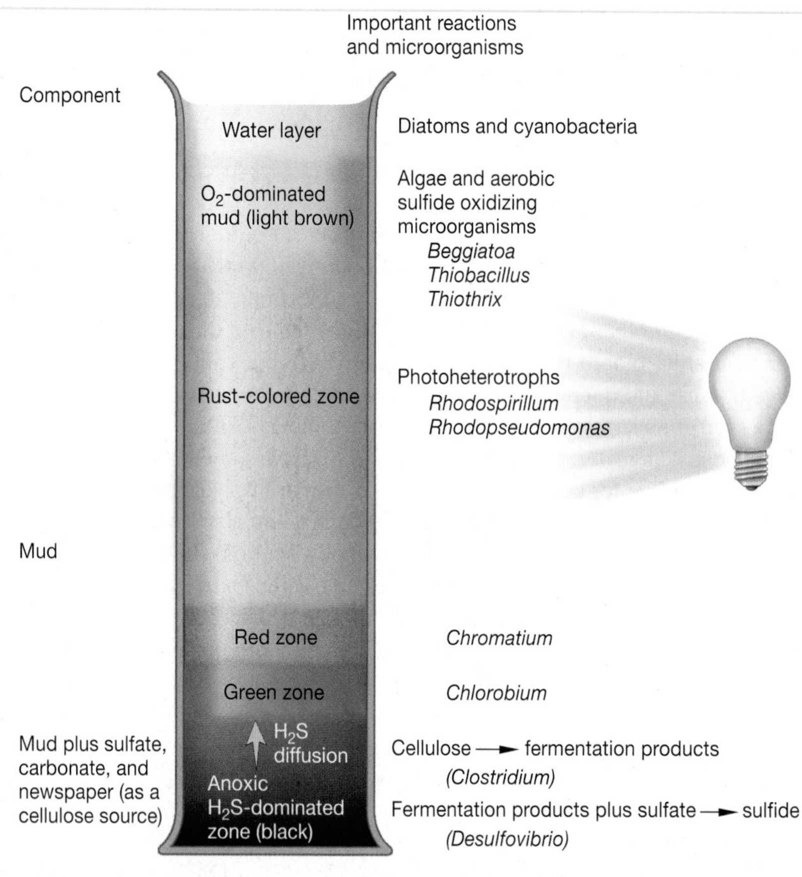

Figure 28.10 The Winogradsky Column. A microcosm in which microorganisms and nutrients interact over a vertical gradient. Fermentation products and sulfide migrate up from the reduced lower zone, and oxygen penetrates from the surface. This creates conditions similar to those in a lake or salt marsh with nutrient-rich sediments. Light is provided to simulate the penetration of sunlight into the anoxic lower region, which allows photosynthetic microorganisms to develop.

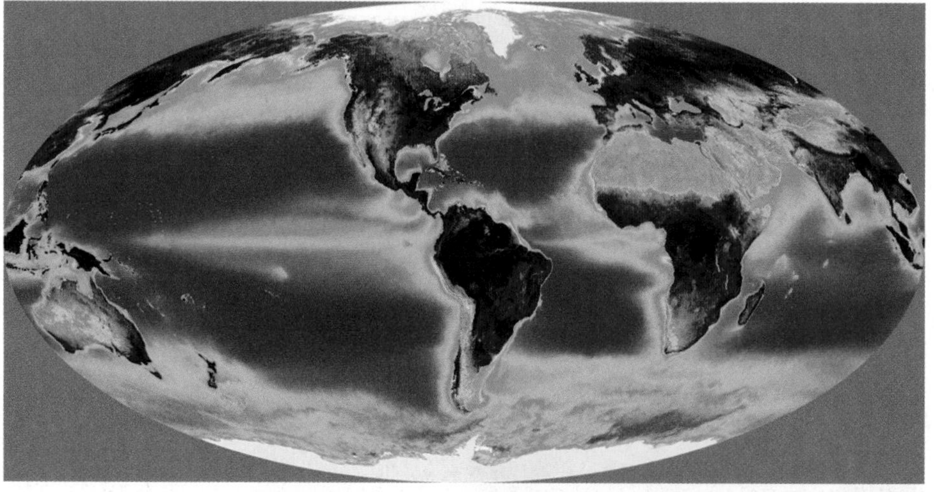

Figure 28.11 Mean Annual Surface Chlorophyll Levels. Global chlorophyll as measured by ocean satellites. The dark-blue regions correspond to subtropical gyres (massive spirals of water) where nutrient levels are especially low; note scale below image. Image courtesy of NASA-Goddard Space Flight Center and the Orbital Sciences Corporation.

The Photic Zone of the Open Ocean

Sometimes called "the invisible rain forest," the upper 200 to 300 meters of the open ocean is home to a diverse collection of photosynthetic microbes. Open ocean regions are also called **pelagic.** The use of satellite imagery to measure chlorophyll (**figure 28.11**) shows that although chlorophyll levels are lower in the sea than on land, the sheer volume of the oceans accounts for the fact that about half the world's photosynthesis is performed by marine microbes. The open ocean is an oligotrophic environment—that is to say, nutrient levels are very low. Recall that the influx of new nutrients is limited, so primary productivity (and thus the whole ecosystem) depends on rapid recycling of nutrients. Unlike terrestrial ecosys-

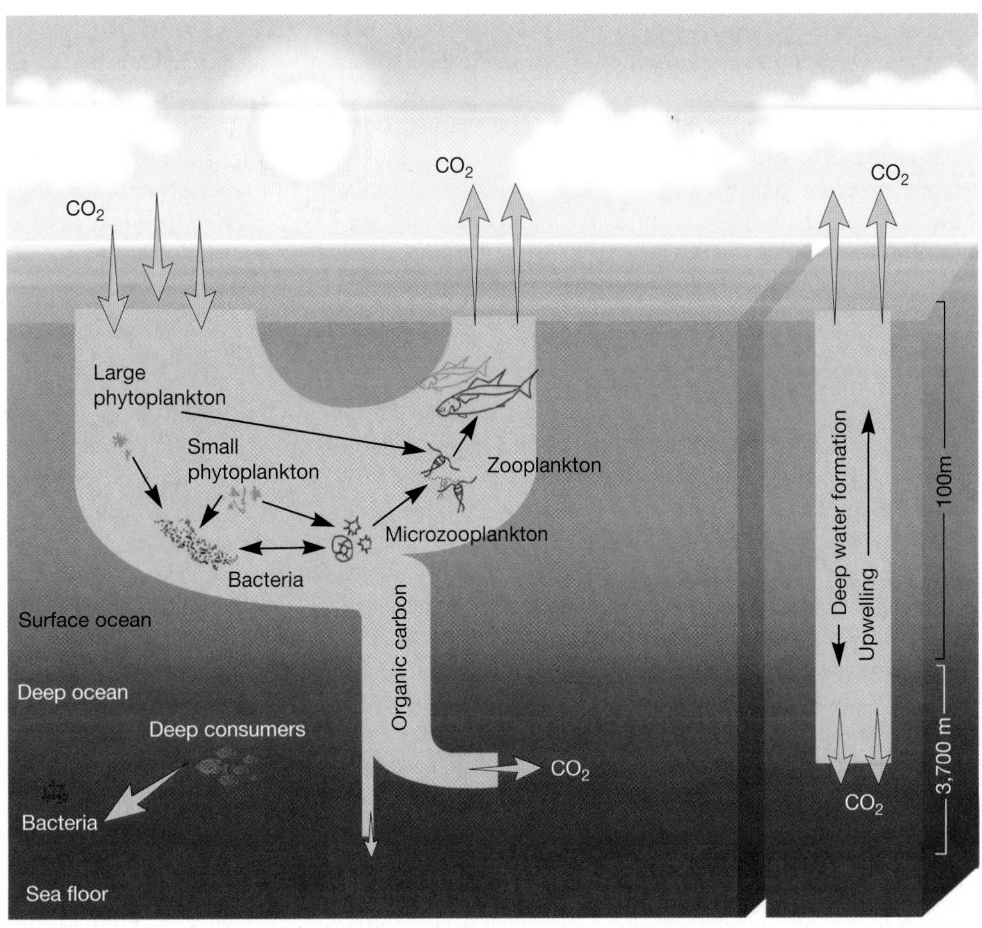

Figure 28.12 The Biological Carbon Pump. The vast majority of the carbon fixed by microbial autotrophs in the open ocean remains in the photic zone. However, a small fraction is "pumped" to the seafloor, which in turn returns much of it to the surface in regions where deep and surface waters mix (upwelling regions). The flux of CO_2 in and out of the world's oceans is in equilibrium, but this equilibrium may be upset by the increase in atmospheric CO_2, the hallmark of global warming. Note depth scale on right.

tems, at least 90% of the nutrients required by the microbial community are recycled within the microbial loop (figure 28.3).

Global warming has focused intense scrutiny on determining how much CO_2 phytoplankton can "draw down" out of the atmosphere and sequester in the benthos. As shown in **figure 28.12,** there is a constant exchange of CO_2 at the ocean surface, as well as limited export of carbon to the seafloor. Organic matter escapes the photic zone and falls through the depths as **marine snow** (**figure 28.13**). This material gets its name from its appearance, sometimes seen in video images of mid- and deep-ocean water, where drifting flocculate particles look much like snow. Marine snow consists primarily of fecal pellets, diatom frustules, and other materials that are not easily degraded. During its fall to the bottom, marine snow is colonized by a community of microbes so that mineralization continues. By the time it reaches the seafloor, less than 1% of photosynthetically derived materials remains unaltered. Once there, only a tiny fraction will be buried in the sediments; most will return to the surface in upwelling regions (figure 28.12). An urgent question is whether or not the oceans can draw down more CO_2 as atmospheric concentrations continue to increase. One approach to reversing (or at least slowing) global warming is to fertilize specific regions of the oceans. These regions, known as high-nutrient, low-chlorophyll (HNLC)

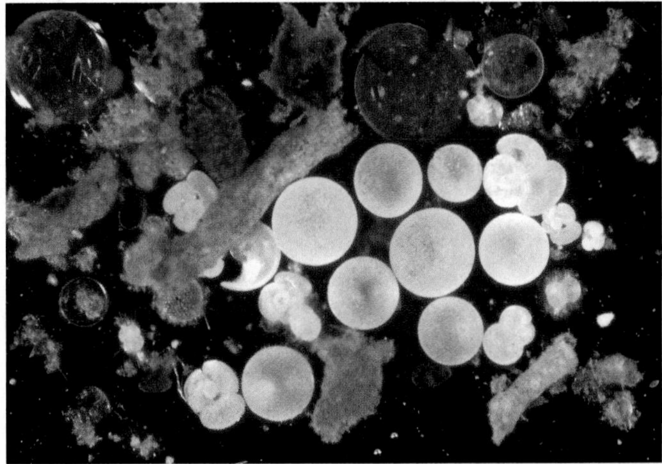

Figure 28.13 Marine Snow. The export of organic matter out of the photic zone to deeper water occurs through the sinking of marine snow. Shown here is material collected in a sediment trap at 5,367 meters on the Sohm Abyssial Plain in the Sargasso Sea. It includes cylindrical fecal pellets, planktonic tests (round white objects), transparent snail-like pteropod shells, radiolarians, and diatoms.

areas, are limited by iron. A number of experiments have been performed wherein large transects of the southern Pacific were fertilized with iron to trigger diatom blooms. As might be expected, the notion of fertilizing vast regions of the ocean to rid the atmosphere of CO_2 is controversial for a number of reasons. First, it is not apparent that this will be effective—while all studies show a temporary increase in primary production, only some studies show a measurable loss of carbon from the photic zone; others report no increase in CO_2 draw-down. Second, many scientists are skeptical about the long-term effects of altering such a large and important ecosystem.

The cycling of other elements, in addition to carbon, occurs within the photic zone. In general, the open ocean is limited by nitrogen, not iron. Two recent discoveries have led marine microbiologists and oceanographers to reexamine the traditionally accepted nitrogen cycle. First, the filamentous cyanobacterium *Trichodesmium* is fixing far more N_2 than previously thought. Second, unicellular cyanobacteria collected at sea are also fixing N_2. These microbes were known to fix nitrogen under laboratory conditions but researchers assumed they did not do so in the open ocean. Thus there appears to be at least two recently identified sources of "new" organic nitrogen. The other recent discovery that has revolutionized our understanding of nitrogen cycling is the presence of bacteria that perform the anammox reaction below the photic zone where oxygen concentrations reach a minimum. In nitrogen-limited areas, it is generally understood that there is a net loss of ammonium, nitrate, and nitrite. A large fraction of this loss has been attributed to denitrification (the anaerobic reduction of nitrate to N_2) occurring at the oxygen-minimum zones. It now appears that consortia of bacteria capable of the anammox reaction—the anaerobic oxidation of NH_4^+ to N_2—are responsible for much of the loss of nitrogen that could otherwise support life. This consortia includes members of the interesting phylum *Planctomycetes*. Once again, the urgency to understand the global oceanic nitrogen budget reflects concern over increasing levels of atmospheric CO_2, global warming, and the capacity of the world's oceans to remove more CO_2 while maintaining overall ecosystem equilibrium. Photosynthetic bacteria: Phylum *Cyanobacteria* (section 21.3); Phylum *Planctomycetes* (section 21.4); Biogeochemical cycling: Nitrogen cycle (section 27.2)

The pelagic cyanobacterium *Trichodesmium* has long been of interest not only because it can fix nitrogen, but because it forms extensive blooms in the open ocean. These blooms, which look similar to floating straw, can cover up to 300,000 square kilometers. Considering the fact that nutrient levels, especially nitrogen and phosphorus, are extremely low in the open ocean, it has been clear for a long time that the cyanobacterium must have evolved clever mechanisms to survive. As a nitrogen-fixing cyanobacterium, neither carbon nor nitrogen would be limiting. It appears that *Trichodesmium* must be able to outcompete other phytoplankton for phosphorus. Recent experiments that combine genomics with real-time PCR show how *Trichodesmium* has mastered phosphorus limitation. Annotation of the *Trichodesmium* genome reveals genes predicted to encode proteins associated with high-affinity transport and assimiliation of phosphonate, compounds with a C—P bond. Phosphonate is an important part of DOM, but most photosynthetic planktonic microbes are unable to use it. Researchers at the Woods Hole Oceanographic Institution in Massachusetts used real-time PCR to show that the genes thought to be involved in phosphonate assimilation (in cultured and Sargasso Sea field populations) were expressed only under phosphate-limiting conditions. Thus *Trichodesmium* may be so successful in the oligotrophic oceans because it uses a source of phosphorus not available to most other phytoplankton. Biogeochemical cycling: Phosphorus cycle (section 27.2)

It is clear that when researchers consider microbial activities in the open ocean, they are assessing global processes. So perhaps it is not surprising that the most abundant group of monophyletic organisms on Earth is marine. Members of the α-proteobacterial clade called **SAR11** have been detected by rRNA gene cloning from almost all open-ocean samples taken worldwide. In addition, it has been found at depths of 3,000 meters as well as in coastal waters and even in some freshwater lakes. Using a technique called ***f*luorescence *i*n *s*itu *h*ybridization (FISH),** in which specific DNA fragments are labeled with fluorescent dye, SAR11 has been found to constitute 25 to 50% of the total procaryotic community in the surface waters in both nearshore and open-ocean samples. Indeed SAR11 bacteria are estimated to constitute about 25% of all microbial life on the planet.

But what exactly are SAR11 microbes and what are they doing? The answer to that question took about a decade to emerge because SAR11 (named after the *Sar*gasso Sea, where it was first detected) could not be coaxed into growing in the laboratory. In 2002 Stephen Giovannoni, the scientist who first discovered SAR11, and his graduate students isolated several members of the SAR11 clade in pure culture, and named the microbe *Pelagibacter ubique*. It grows only in seawater cultures with very low nutrients, conditions much like the oligotrophic ocean. *P. ubique* is vibrioid and only 0.4 to 0.9 μm in length, making it one of the smallest known, free-living microbes. Unlike most microbes in culture, which grow to densities exceeding 10^8 cells per milliliter, SAR11 isolates stop replicating after reaching about 10^6 cells per milliliter. Curiously, this is the density at which they are generally found in nature, suggesting that natural factors in seawater somehow control population growth. By measuring how much and how fast SAR11 microbial isolates assimilate radiolabeled amino acids, glucose, and complex biomolecules, scientists have begun to get a picture of their role in the microbial loop. These microbes contribute as much as 50% to the bacterial biomass production and DOM flux in some marine environments, and it appears that they may selectively degrade the kinds of complex biomolecules that comprise marine snow. Microbial ecology and its methods: An overview (section 27.4)

The genome of *P. ubique* was recently sequenced and annotated. At 1.31 Mb, it is the smallest genome of any independently replicating cell sequenced to date. Unlike the small genomes of parasitic bacteria that have lost genes for energy capture and other essentials that can be obtained from the host, SAR11 accomplishes this feat by eliminating "genomic waste." There are no pseudogenes, phage genes, or recent gene duplications. In addition, the number of nucleotides between coding regions is also very limited.

Nonetheless, *P. ubique* has the genes necessary for the Entner-Douderoff pathway, the TCA cycle, and a complete electron transport chain. It has adapted to life in the oligotrophic ocean by encoding a number of high-affinity nutrient transport systems and maintaining its small size, thereby optimizing its surface area to volume ratio. In addition, it uses both respiration and a proteorhodopsin proton pump to capture energy. Insights from microbial genomes (section 15.8); Phylum *Euryarchaeota:* The *Halobacteria* (section 20.3)

The genome sequence of *Silicobacter pomeroyi,* another member of the bacterioplankton community, demonstrates that we have much to learn about the strategies bacteria use to cope in the oligotrophic ocean. Like *P. ubique, S. pomeroyi* is an α-proteobacterium but is part of the marine *Roseobacter* clade, which constitutes 10 to 20% of coastal and oceanic bacterioplankton. Originally thought to be a chemoorganotroph, the microbe depends on chemolithoheterotrophy, supplementing its heterotrophic existence with the use of inorganic compounds such as sulfide and carbon monoxide (CO) as sources of energy and electrons. Sulfide is found on marine snow and the *S. pomeroyi* genome encodes two putative quorum-sensing systems that would be useful to the microbe when at high cell densities, as might occur on heavily colonized pieces of marine snow. Two operons encode CO dehydrogenases; these enzymes catalyze the oxidation of CO to CO_2. CO is abundant in seawater due to photooxidation of DOM. However, because *S. pomeroyi* lacks the genes for carbon fixation, CO does not appear to be a carbon source. Rather CO oxidation is used to donate electrons for energy production. Once again, this has implications for the global carbon budget as *Silicobacter*-like consumption of CO may remove as much as 10 to 60 *trillion* grams of C annually, thereby buffering the partial pressure of CO_2 in the ocean.

One of the most interesting discoveries in recent years is the widespread presence of large numbers of marine archaea, once thought to inhabit only extreme environments. In fact, the *Archaea* are ubiquitous in the marine environment—they have been found in polar and tropical regions, and in estuarine, planktonic, and deep-sea communities. Archaea are also found in freshwater environments. 16S rRNA analysis and direct visualization of archaea using epifluorescence has revealed that at least 20% of oceanic picoplankton are *Crenarchaeota*. Many of these populations are closely related, suggesting that a shared collection of adaptations has been fine-tuned to meet the needs of the different niches within specific picoplankton communities. The distribution of archaea relative to that of bacteria differs widely with the particular ecosystem. **Figure 28.14** shows typical results for the open ocean. Bacteria are most numerous in the upper 150 to 200 meters (i.e., the photic zone), but archaea increase in relative abundance with depth until they approximate, or sometimes even exceed, bacteria.

As mentioned in our discussion of the microbial loop, viruses are important members of marine and freshwater microbial communities. In fact, **virioplankton** are the most numerous members of marine ecosystems. It was not until the 1990s that the abundance of marine viruses was recognized; their study has become one of the most active and exciting areas of marine microbiology.

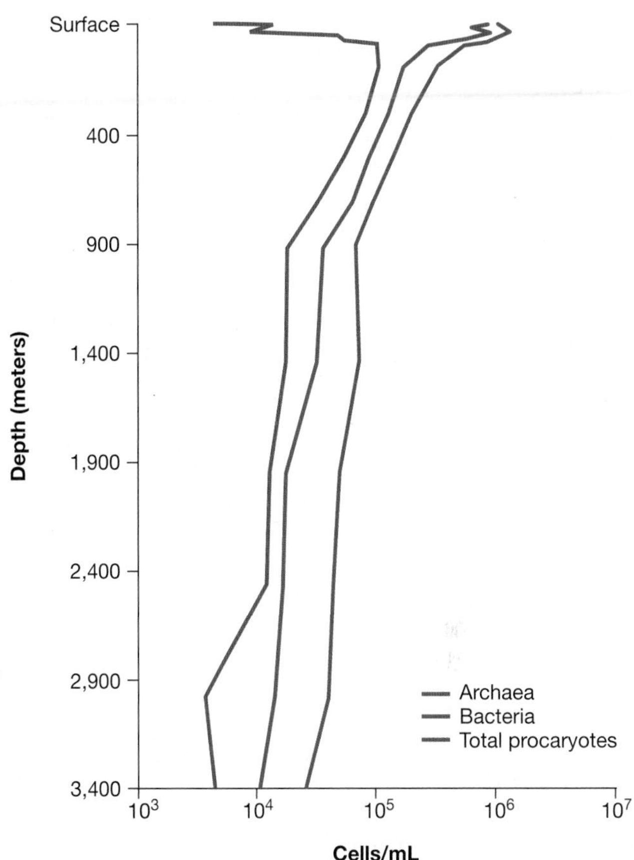

Figure 28.14 Archaea Are Plentiful in Ocean Depths. The distribution of archaea and bacteria, together with an estimate of total procaryotes, over a depth of 3,400 meters is shown at a Pacific Ocean location. These results indicate that archaea make up a significant part of the observable picoplankton below the surface zone.

Quantifying viruses is tricky: the traditional method of plaque formation requires knowledge of virus and host, as well as the ability to grow the host in the laboratory. Recall that only about 1% of all microbes have been cultured; this often prevents the measurement of virus diversity by examining actual virus infection. Instead, virus particles may be visualized directly. This requires the concentration of many tens of liters of water for direct examination by transmission electron microscopy. Because this does not prove that any given virus can actually infect a host cell, viruses enumerated in this way are called ***virus*l*ike particles* (VLPs).** Using this approach, the average VLP density in seawater is between 10^6 to 10^7 per milliliter (although it may be closer to 10^8 per milliliter); their numbers decline to roughly 10^6 below about 250 meters. Marine viruses are so abundant that virus particles are now recognized as the most abundant life forms on Earth. As might be expected by their numbers, these viruses are very diverse, including single- and double-stranded RNA and DNA viruses that infect archaea, bacteria, and protists. Measurement of viral lysis of bacteria and archaea in the field is difficult

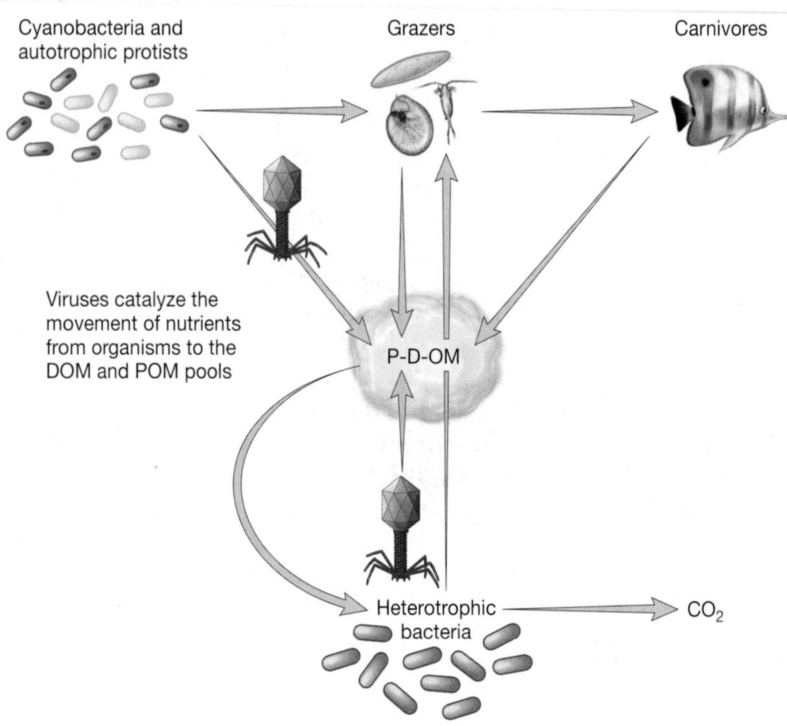

Figure 28.15 The Role of Viruses in the Microbial Loop. Viral lysis of autotrophic and heterotrophic microbes accelerates the rate at which these microbes are converted to particulate and dissolved organic matter (P-D-OM). This is thought to increase net community respiration and decrease the efficiency of nutrient transfer to higher trophic levels.

and yields variable estimates, but appears to account for between 10 and 50% of the total mortality of the procaryotic members of the marine microbial community. Computer modeling and model experiments indicate that viruses contribute to nutrient cycling by accelerating the rate at which their microbial hosts are converted to POM and DOM, thereby "feeding" other microorganisms without first making them available for protists and other bacteriovores. This "short-circuits" the microbial loop (**figure 28.15**).

Genome sequencing of DNA cloned directly from marine ecosystems suggests that phages are important vectors for horizontal gene transfer. In fact, it has been calculated that in the oceans, phage-mediated gene transfer occurs at an astounding rate of 20 billion times per second. Recently, the importance of phage-mediated lateral gene transfer was demonstrated when it was discovered that cyanophages thought to infect the cyanobacteria *Synechococcus* and *Prochlorococcus* carry the structural gene for a photosynthetic reaction center protein. These phages are presumably shuttling this essential gene between the two genera of cyanobacteria, and may thereby play a critical role in the evolution of these important primary producers. Transmission electron microscopy (section 2.4); Viruses of *Bacteria* and *Archaea* (chapter 17); Transduction (section 13.9); Environmental genomics (section 15.9); Microbial ecology and its methods: An overview (section 27.4)

Benthic Marine Environments

The majority of the Earth's crust is under the sea, which means that of all the world's microbial ecosystems, we know the least about the largest. However, the combination of deep-ocean drilling projects and the exploration of geologically active sites, such as submarine volcanoes and hydrothermal vents, has revealed that the study of ocean sediments, or **benthos,** can be rewarding and surprising. Marine sediments range from the very shallow to the deepest trenches, from dimly illuminated to completely dark, and from the newest sediment on Earth to material that is millions of years old. The temperature of such sediments depends on the proximity of geologically active areas. The exciting discovery of hydrothermal vent communities with large and diverse invertebrates, some of which depend on endosymbiotic chemolithotrophic bacteria, has been intensely investigated since their discovery in the late 1970s. These microbes are discussed in chapter 30. Because the vast majority of Earth's crust lies at great depth far from geothermally active regions, most benthic marine microbes live under high pressure, without light, and at temperatures between 1°C to 4°C.

Deep-ocean sediments were once thought to be devoid of all life and were therefore not considered worth the considerable effort it takes to study them. In fact, it took an international consortium of scientists to organize the Ocean Drilling Project in 1985, which has now been expanded to the Integrated Ocean Drilling Program through 2013. Researchers aboard the research vessel *JOIDES Resolution* drill cores of sediments from water depths up to 8,200 meters (at its deepest, the ocean is about 11,000 m). Microbiologists now know that far from being sterile, it appears the total subsurface (intraterrestrial) microbial biomass equals that of all terrestrial and marine plants. This is possible because benthic marine microbes inhabit not just the surface of the sea floor, but within sediments to a depth of least 0.6 km. To survive at these depths, microorganisms must be able to tolerate atmospheric pressures up to 1,100 atm (pressure increases about 1 atm/10 meters depth). Such microbes are said to be **barophilic** (Greek, *baro,* weight, and *philein,* to love). Some are obligate barophiles and must be cultured in special hyperbaric incubation chambers. In fact, scientists have yet to find the subterranean depth limit of microbial growth. However, it appears that it will not be governed by pressure; rather, it will be determined by temperature. It seems unlikely that the maximum temperature at which life can be sustained has been identified.

One outcome of deep-ocean sediment exploration has been the discovery of methane hydrates. These pools of trapped methane are produced by methanogenic archaea that convert acetate to methane. This accumulates in lattice-like cages of crystalline water 500 meters or more below the sediment surface in many regions of the world's oceans. The formation of methane hydrates requires both cold temperatures and high pressure. This discovery is very significant because there may be up to 10^{13} metric tons of methane hydrate worldwide—80,000 times the world's current known natural gas reserve.

It is also exciting that recent deep-sea sediment drilling has turned our understanding of bacterial energetics literally upside down. As discussed in chapter 9, it is generally understood that anaerobic respiration occurs such that there is preferential use of available terminal electron acceptors. That which yields the most energy (ΔG) from the oxidation of NADH or an inorganic reduced compound (e.g., H_2, H_2S) will be used before those electron acceptors that produce a smaller ΔG (*see table 8.1 and figure 8.8*). Thus following oxygen depletion, nitrate will be reduced; then manganese, iron, sulfate, and finally carbon dioxide. When sediment cores up to 420 meters deep were collected off the coast of Peru, this predictable profile of electron acceptors and their microbial-derived reduced products were observed within the upper strata of the sediments (**figure 28.16**). However, when researchers measured these signature chemical compounds at great depth, the profile was upside down. This suggests the presence of unknown sources of these electron acceptors at subsurface depths of more than 420 meters. In addition, contrary to our long-held notion of thermodynamic limits, methane formation (methanogenesis) and iron and manganese reduction seem to be co-occurring in sediments with high sulfate concentrations. Although the identity of the microbes that make up this community awaits further study, it is clear that with densities of 10^8 cells per gram of sediment at the seafloor surface and 10^4 cells per gram in deep subsurface sediments, these communities are important. Free energy and reactions (section 8.4); Oxidation-reduction reactions, electron carriers, and electron transport systems (section 8.6); Anaerobic respiration (section 9.6)

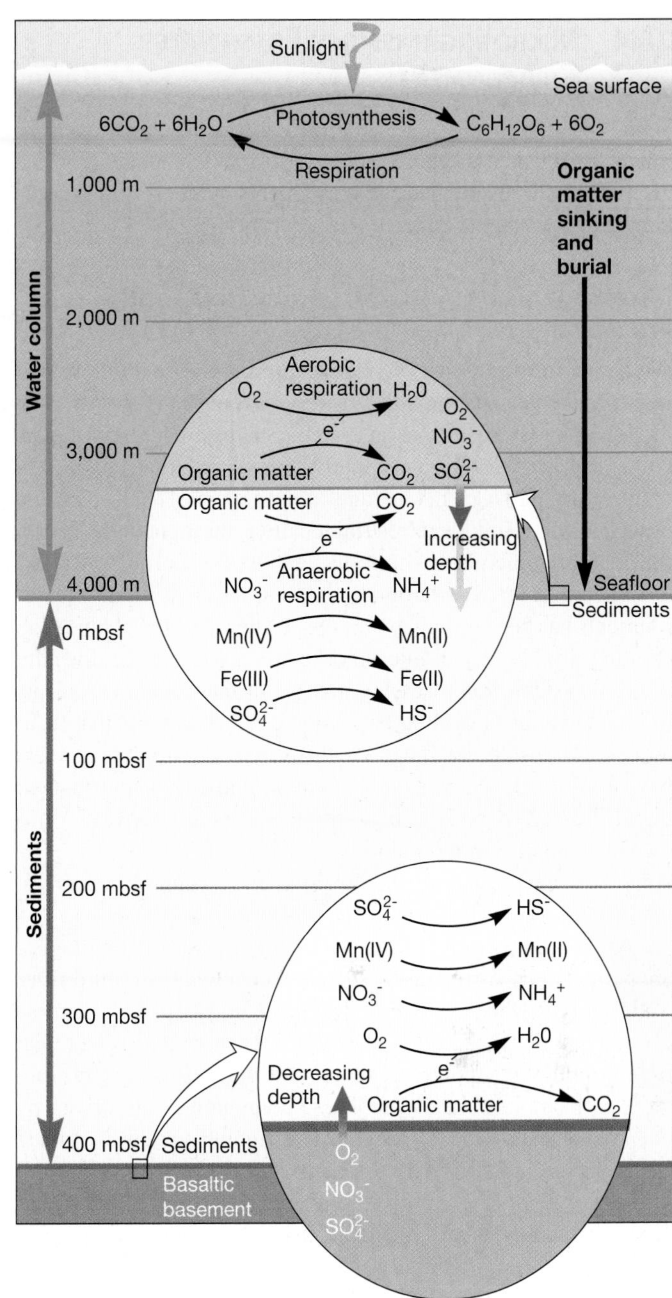

Figure 28.16 Microbial Activity in Deep-Ocean Sediments. At the surface of the seafloor, reduction of oxidized substrates that serve as electron acceptors in anaerobic respiration follows a predictable stratification based on thermodynamic considerations. This sequence is just the opposite in very deep subsurface sediments, suggesting a source of electron acceptors from deep within Earth's crust. Meters below seafloor, mbsf.

1. What is marine snow? Why is it important in CO_2 draw-down?
2. Name two sources of organic nitrogen in the open ocean that have only recently been recognized.
3. Draw a diagram of the anammox reaction and explain its importance to the global carbon and nitrogen budgets.
4. Why do you think that despite its great abundance, SAR11 was not discovered until the late 20th century?
5. Why do you think marine viruses are difficult to study?
6. Describe the role of marine viruses in the microbial loop.
7. What are methane hydrates?
8. Explain what is meant by "upside-down microbial energetics" as described for deep subsurface ocean sediments.

28.4 Microorganisms in Freshwater Environments

While the vast majority of water on Earth is in marine environments, freshwater is crucial to our terrestrial existence. Here we discuss freshwater (aquatic) environments and describe them as habitats for a diverse collection of microbes.

Microorganisms in Glaciers and Permanently Frozen Lakes

We begin our discussion of freshwater microbes with those that reside in ice that has remained frozen for thousands of years. Although this may seem like an extreme environment, it is important to note that a majority of the Earth's surface never exceeds a temperature of 5°C. This includes polar regions, the deep ocean, as well as high-altitude terrestrial locations throughout the world. Surprisingly, microbes within glaciers are not dormant. Rather, evidence that active microbial communities exist in these environments has been growing over the last decade. In fact, this is an exciting time in glacial microbiology; determining the diversity in these systems and assessing the role of these microorganisms in biogeochemical cycling can involve novel and creative techniques. The results may be of great consequence because glaciers have traditionally been regarded as areas that do not contribute to the global carbon budget. In addition, since the discovery of ice on Mars and on Jupiter's moon Europa, astrobiologists have become very interested in ice-dwelling psychrophilic microbes. The influence of environmental factors on growth: Temperature (section 6.5)

Among the frozen landscapes of interest to microbiologists are permanently frozen lakes, such as Antarctica's McMurdo Dry Valley Lakes where the ice is 3 to 6 meters deep. Life in these ecosystems depends on the photosynthetic activity of microbial psychrophiles. In contrast, lakes that lie below glaciers are blocked from solar radiation. These communities are driven by chemosynthesis. One of the most intriguing and well-studied Antarctic habitats is Lake Vostok, one of 68 lakes located 3 to 4 kilometers below the East Antarctic Ice Sheet (**figure 28.17**). Geothermal heating, pressure, and the insulation of the overlying ice keep these lakes in a liquid state. It is thought that Lake Vostok was formed approximately 420,000 years ago and that its water is about a million years old. This stable environment supports a number of microbes including gram-negative proteobacteria and gram-positive actinomycetes. The overlying ice also harbors similar microbes, although at lower population densities. To find out if these microbial populations are active, radiolabeled substrates, including ^{14}C-acetate and ^{14}C-glucose, were added to samples incubated at an Antarctic laboratory. Indeed, the respiration of these compounds, measured as $^{14}C\text{-}CO_2$, demonstrates that these communities are dynamic and of great interest.

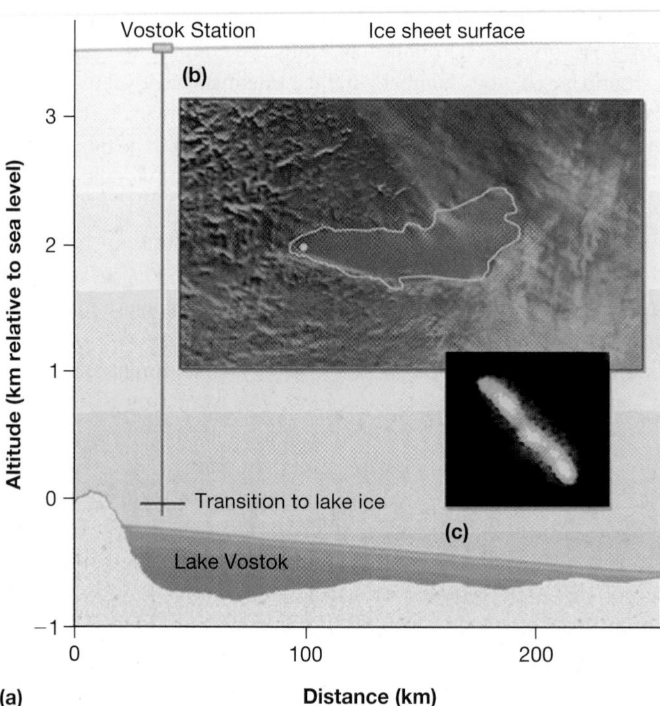

Figure 28.17 The East Antarctic Ice Sheet and Lake Vostok. **(a)** Lake Vostok lies beneath several kilometers of the ice sheet. **(b)** The deep drilling station from which ice and lake water samples were obtained is indicated by the yellow dot. The yellow outline identifies a smooth plateau of snow and ice that floats on top of Lake Vostok. The surrounding rough ice is formed when the ice sheet moves over bedrock rather than liquid water. **(c)** Microbes have been found in both the overlying ice and the lake water, as seen in this epifluorescent image.

Microorganisms in Streams and Rivers

As freshwater glaciers melt, their waters enter streams and rivers. This marks a departure from an environment that is stable on a geologic time scale to one that is extremely changeable. The continuous flow of water in streams and all but the largest rivers prevents the development of significant planktonic communities. Instead, most of the microbial biomass is attached to surfaces. Depending on the size of the stream or river, the source of nutrients may vary. The source may be in-stream, called **autochthonous** production based on photosynthetic microorganisms (**figure 28.18*a***). Nutrients also may come from outside the stream, including runoff sediment from riparian areas (the edge of a river), or leaves and other organic matter falling directly into the water (figure 28.18*b*). Such nutrients are called **allochthonous.** Chemoorganotrophic microorganisms metabolize the available organic material, recycling nutrients within the ecosystem. Autotrophic microorganisms grow using the minerals released from the organic matter. This leads to the production of O_2 during the daylight hours; respiration occurs at night farther down the river, resulting in diurnal oxygen shifts. Eventually the O_2 level approaches saturation, completing a self-purification process. Thus when the amount of organic matter added to streams and rivers does not exceed the system's oxidative capacity, productive and aesthetically pleasing streams and rivers are maintained.

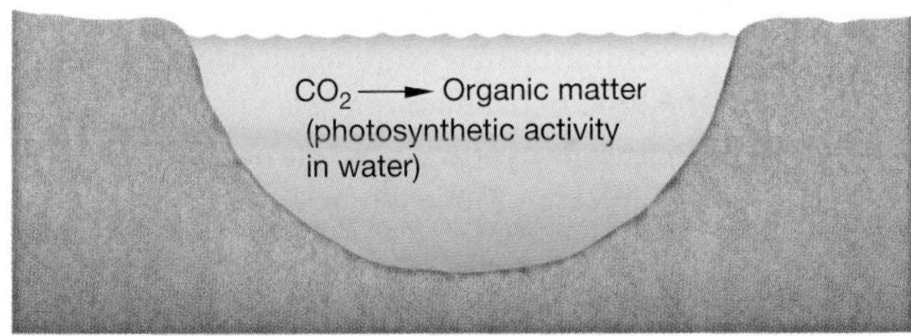

(a)

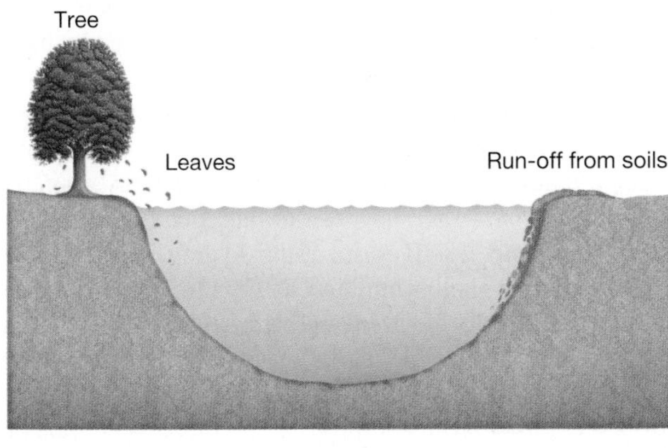

(b)

Figure 28.18 Organic Matter Sources for Lakes and Rivers. Organic matter used by microorganisms in lakes and rivers can be synthesized in the water (autochthonous), or can be added to the water from outside sources (allochthonous). **(a)** Autochthonous sources of organic matter, primarily photosynthesis, and **(b)** allochthonous sources of organic matter.

The capacity of streams and rivers to process added organic matter is limited. If too much organic matter is added, the system is said to be **eutrophic** and the rate of respiration exceeds that of photosynthesis. Thus the water may become anoxic. This is especially the case with urban and agricultural areas located adjacent to streams and rivers. The release of inadequately treated municipal wastes and other materials from a specific location along a river or stream represents **point source pollution.** Such point source additions of organic matter can produce distinct and predictable changes in the microbial community and available oxygen, creating an oxygen sag curve (**figure 28.19**). Runoff from agriculturally active fields and feedlots is an example of **nonpoint source pollution.** This can cause disequilibrium in the microbial community leading to algal or cyanobacterial blooms.

Along with the stresses of added nutrients, removal of silica from rivers by the construction of dams and trapping of sediments causes major ecological disturbances. For example, construction of the dam at the "iron gates" on the Danube (600 miles above the Black Sea) has led to a decrease in silica to 1/60th of the previous concentration. This decreased silica availability inhibits the growth of diatoms because the ratio of silica to nitrate has been altered (silica is required for diatom frustule formation). With this shift in resources, Black Sea diatoms are not able to grow and immobilize nutrients. The result has been a 600-fold increase in nitrate levels and a massive development of toxic algae. This example shows that the delicate balance of river ecosystems can be altered in unexpected ways by dams changing microbiological processes.

Although there are over 36,000 dams worldwide, many nations have recognized the devastating consequences of unrestricted or improper dam construction and efforts are under way to remove

Figure 28.19 The Dissolved Oxygen Sag Curve. Microorganisms and their activities can create gradients over distance and time when nutrients are added to rivers. An excellent example is the dissolved oxygen sag curve, caused when organic wastes are added to a clean river system. During the later stages of self-purification, the phototrophic community will again become dominant, resulting in diurnal changes in river oxygen levels.

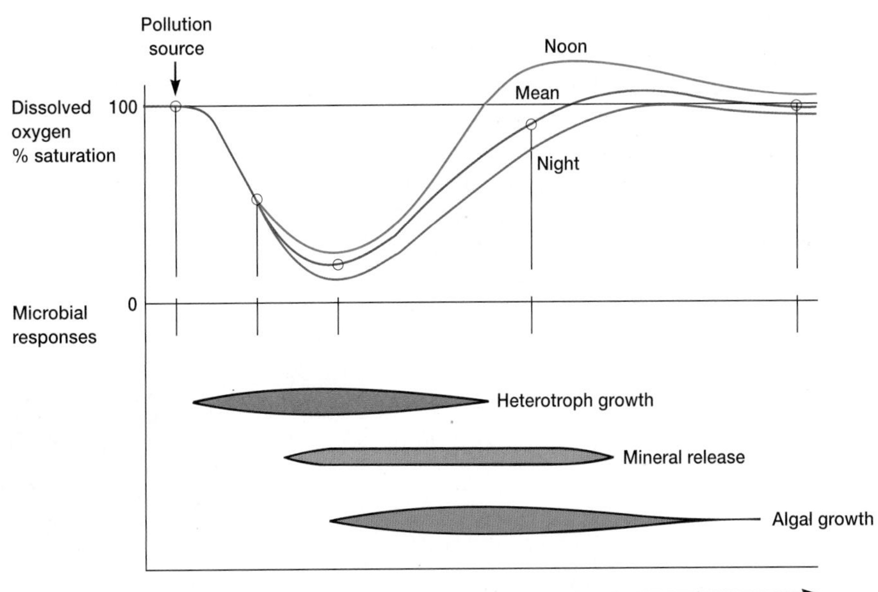

some of the worst dams. These structures must be breached to restore normal water flow and enable important commercial fish renewed access to spawning grounds. Clearly this will be an area requiring microbiological expertise for many years to come.

1. Describe the Lake Vostok ecosystem. Why is this a chemosynthesis based, rather than a photosynthetically based, microbial community?
2. What is an oxygen sag curve? What changes in a river cause these effects?
3. What are point and nonpoint souce pollution? Can you think of examples in your community?
4. Why might dams influence microorganisms and microbial processes in rivers?

Microorganisms in Lakes

Lakes offer a completely different set of physical and biological features for microbial communities, although not all lake environments are the same. Lakes vary in nutrient status. Some are oligotrophic (**figure 28.20*a***), others are eutrophic (figure 28.20*b*). Nutrient-poor lakes remain oxic throughout the year, and seasonal temperature shifts usually do not result in distinct oxygen stratification. In contrast, eutrophic lakes usually have bottom sediments that are rich in organic matter. In thermally stratified lakes the **epilimnion** (warmer, upper layer) is oxic, while the **hypolimnion** (colder, bottom layer) often is anoxic (particularly if the lake is nutrient-rich). The epilimnion and hypolimnion are separated by a thermocline.

Because there is little mixing between the epilimnion and the hypolimnion, the bottom waters may become deprived of oxygen. This is a permanent situation in tropical eutrophic lakes and occurs in the summer in temperate eutrophic lakes. If nutrient levels are high, this bottom, or benthic, zone becomes dominated by anaerobic microbial activity. In very warm eutrophic lakes, anaerobic microbes release gases such as H_2S into the water. In addition, human activities (e.g., septic and agricultural runoff) may add high levels of nitrogen and phosphorus to the lake waters. This can result in a bloom of algae, plants, and/or bacteria in the epilimnion. During autumn cooling, temperate lakes lose their thermocline because surface waters increase in density and storms mix the two layers. This sometimes happens within a 24-hour period; if bottom water has become filled with anaerobic byproducts, the sudden upwelling causes fish kills.

In oligotrophic lakes, phosphorus is often the limiting nutrient. If added to oligotrophic freshwater, cyanobacteria capable of nitrogen fixation may bloom. Several genera, notably *Anabaena, Nostoc,* and *Cylindrospermum,* can fix nitrogen under oxic conditions. The genus *Oscillatoria,* using hydrogen sulfide as an electron donor for photosynthesis, can fix nitrogen under anoxic conditions. If both nitrogen and phosphorus are present, cyanobacteria compete with algae. Cyanobacteria function more efficiently in alkaline waters (8.5 to 9.5) and higher temperatures (30 to 35°C). Photosynthetic protists, in comparison, generally grow best at neutral pH and have lower optimum temperatures. By using CO_2 at rapid rates, cyanobacteria also increase the pH, making the environment less suitable for protists. Photosynthetic bacteria: Phylum *Cyanobacteria* (section 21.3)

Cyanobacteria have additional competitive advantages. Many produce hydroxamates, which bind iron, making this important

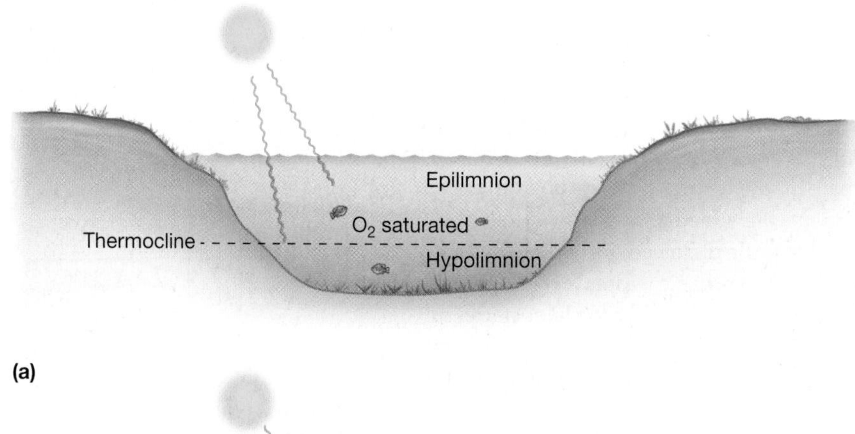

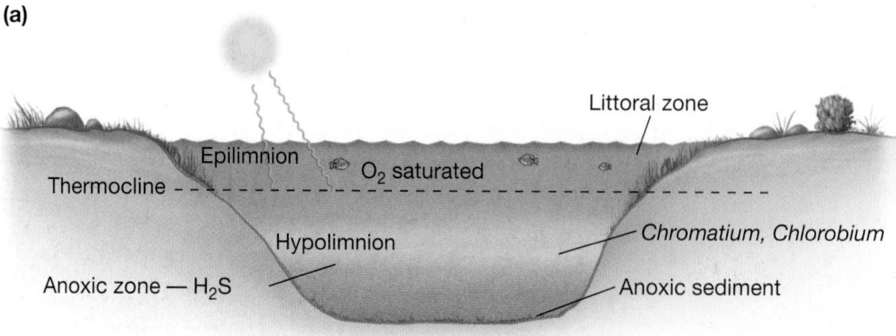

Figure 28.20 Oligotrophic and Eutrophic Lakes. Lakes can have different levels of nutrients, ranging from low nutrient to extremely high nutrient systems. The comparison of **(a)** an oligotrophic (nutrient-poor) lake, which is oxygen saturated and has a low microbial population, with **(b)** a eutrophic (nutrient-rich) lake. The eutrophic lake has a bottom sediment layer and can have an anoxic hypolimnion. As microbial biomass increases with nutrient levels, light penetration is diminished. Thus the bottom of eutrophic lakes may be become dark, anoxic, and even poisonous from H_2S production.

trace nutrient less available for protists. Some cyanobacteria also resist predation because they produce toxins. In addition, some synthesize odor-producing compounds that affect the quality of drinking water. However, both cyanobacteria and protists can contribute to massive blooms in strongly eutrophied lakes. Lake management can improve the situation by removing or sealing bottom sediments or adding coagulating agents to speed sedimentation.

1. What terms can be used to describe the different parts of a lake?
2. What are some important effects of eutrophication on lakes?
3. Why are cyanobacteria so important in waters that have been polluted by phosphorus additions?

Summary

28.1 Marine and Freshwater Environments

a. Most water on the Earth is marine (97%). The majority of this is cold (2 to 5°C) and at high pressure. Fresh water is a minor but important part of Earth's biosphere.

b. Oxygen solubility and diffusion rates in surface waters are limited; waters are low oxygen diffusion rate environments, in comparison with soils. Carbon dioxide, nitrogen, hydrogen, and methane are also important gases for microbial activity in waters.

c. The carbonate equilibrium system keeps the oceans buffered at pH 7.6 to 8.3 (**figure 28.2**).

d. The penetration of light into the surface water determines the depth of the photic zone. Warming the surface waters can lead to the development of a thermocline.

e. The nutrient composition of the ocean influences the C:N:P ratio of the phytoplankton, which is called the Redfield ratio. This ratio is important for predicting nutrient cycling in oceans. Atmospheric additions of minerals, including iron and nitrogen, affect this ratio and global-level oceanic processes.

f. The microbial loop describes the transfer of nutrients between trophic levels while taking into account the multiple contributions of microbes to recycling nutrients. Nutrients are recycled so efficiently, the majority remain in the photic zone (**figure 28.3**).

28.2 Microbial Adaptations to Marine and Freshwater Environments

a. The marine microbial community is dominated, in terms of numbers and biomass, by ultramicrobacteria.

b. Many unusual microbial groups are found in waters, especially when oxidants and reductants can be linked. These include *Thioploca* and *Thiomargarita,* both of which are found in coastal areas where nutrient mixing occurs. *Thiomargarita* is the world's largest known bacterium (**figure 28.4**).

c. Aquatic fungi are important members of the aquatic microbial community. These include the chytrids, with a motile zoosporic stage, and the Ingoldian fungi, which often have tetraradiate structures. Both of these are uniquely adapted to an aquatic existence, and the chytrids may contribute to disease in amphibians (**figures 28.6** and **28.7**).

28.3 Microorganisms in Marine Environments

a. Tidal mixing in estuaries, as characterized by a salt wedge, is osmotically stressful to microbes in this habitat. Thus they have evolved mechanisms to cope with rapid changes in salinity (**figure 28.8**).

b. Coastal regions like estuaries and salt marshes can be the sites of harmful algal blooms such as those caused by the diatom *Pseudonitzchia* and the dinoflagellates *Alexanderium* and *Pfiesteria* (**figure 28.9**).

c. Autotrophic microbes in the photic zone within the open ocean account for about one-half of all the carbon fixation on Earth.

d. The carbon and nitrogen budgets of the open-ocean photic zone are intensely studied because of their implications for controlling global warming (**figure 28.12**).

e. Two members of the α-proteobacteria—SAR11 and *Silicobacter pomeroyi*—demonstrate unique adaptations to life in the oligotrophic open ocean. SAR11 is the most numerous organism on Earth.

f. Archaea are important components of the microbial community. Viruses are present at high concentrations in many waters, and occur at 10-fold higher levels than the bacteria. In marine systems they may play a major role in controlling cyanobacterial development and nutrient turnover (**figures 28.14** and **28.15**).

g. Sediments deep beneath the ocean's surface are home to about one-half of the world's procaryotic biomass.

h. Methane hydrates, the result of psychrophilic archaeal methanogenesis under extreme atmospheric pressure, may contain more natural gas than is currently found in known reserves.

i. Deep sediment drilling reveals that microbes within this habitat may employ unique energetic strategies (**figure 28.16**).

28.4 Microorganisms in Freshwater Environments

a. Glaciers and permanently frozen lakes are sites of active microbial communities. The East Antarctic Ice Sheet and Lake Vostok, which lies beneath it, are productive study sites (**figure 28.17**).

b. Nutrient sources for streams and rivers may be autochthonous or allochthonous. Often allochthonous inputs include urban, industrial, and agricultural runoff (**figure 28.18**).

c. Lakes can be oligotrophic or eutrophic. Eutrophication can cause increased growth of chemoorganotrophic microbes and the system may become anoxic (**figure 28.20**).

Key Terms

allochthonous 682
anoxic 668
autochthonous 682
barophilic 681
benthos 680
brevetoxin 675
carbonate equilibrium system 668
dissolved organic matter (DOM) 671
epilimnion 684
eutrophic 683
*f*luorescence *in s*itu *h*ybridization (FISH) 678
halotolerant 673
harmful algal bloom (HAB) 675
hypolimnion 684
hypoxic 668

Ingoldian fungi 672
marine snow 677
microbial loop 670
nonpoint source pollution 683
oligotrophic 671
particulate organic matter (POM) 671
pelagic 676
photic zone 669
photosynthate 671
phytoplankton 670
picoplankton 670
point source pollution 683
primary producers 669
Redfield ratio 670
salt wedge 673
SAR11 678
thermocline 669
ultramicrobacteria 671
virioplankton 679
viruslike particles (VLPs) 679
Winogradsky column 675

Critical Thinking Questions

1. In what habitats might you find microbes like *Thioploca* and *Thiomargarita?* What culture conditions would you use to purify such microbes from their natural habitat?
2. How might it be possible to cleanse an aging eutrophic lake? Consider chemical, biological, and physical approaches as you formulate your plan.
3. *Clostridium botulinum,* the causative agent of botulism, sometimes causes fish kills in lakes where people swim. Currently there are few, if any, monitoring procedures for this potential source of disease transmission to humans. Do you think a monitoring program is needed and, if so, how would you implement such a program?
4. Do you think fertilization of the ocean with iron to increase CO_2 drawdown is a good approach to controlling global warming? Why or why not? Why do you think similar experiments yield different results?

Learn More

Arrigo, K. R. 2005. Marine micro-organisms and global nutrient cycles. *Nature* 437:349–55.

Capone, D. G. 2001. Marine nitrogen fixation: What's all the fuss? *Curr. Opin. Microbiol.* 4:341–48.

DeLong, E. F., and Karl, D. M. 2005. Genomic perspectives in microbial oceanography. *Nature* 437:336–42.

Dyhrman, S. T.; Chapell, P. D.; Haley, S. T.; Moffett, J. W.; Orchard, E. D.; Waterbury, J. B.; and Webb, E. A. 2006. Phosphonate utilization by the globally important marine diazotrophic *Trichodesmium. Nature* 439:68–71.

Edwards, K. J.; Bach, W.; and McCollom, T. M. 2005. Geomicrobiology in oceanography: Microbe-mineral interactions at and below the seafloor. *Trends Microbiol.* 13:449–56.

Giovannoni, S. J., and Stingl, U. 2005. Molecular diversity and ecology of microbial plankton. *Nature* 437:343–48.

Karl, D. M.; Bird, D. F.; Björkman, K.; Houlihan, T.; Shackelford, R.; and Tupas, L. 1999. Microorganisms in the accreted ice of Lake Vostok, Antarctica. *Science* 286:2144–47.

Kuypers, M. M.; Lavik, G.; Woebken, D.; Schmid, M.; Fuchs, B. M.; Amann, R.; Jørgensen, B. B.; and Jetten, M. S. M. 2005. Massive nitrogen loss from the Benguela upwelling system through anaerobic ammonium oxidation. *Proc. Natl. Acad. Sci. USA* 102:6478–83.

Moran, M. A.; Buchan, A.; González, J. M.; Heidelberg, J. F.; Whitman, W. B.; Kiene, R. P., et al. 2004. Genome sequence of *Silicobacter pomeryi* reveals adaptations to the marine environment. *Nature* 432:910–13.

Rappé, M. S.; Connon, S. A.; Vergin, K. L.; and Giovinonni, S. J. 2002. Cultivation of the ubiquitous SAR11 marine bacterioplankton clade. *Nature* 418: 630–33.

Schippers, A.; Neritin, L. N.; Kallmeyer, J.; Ferdelman, T. G.; Cragg, B. A.; Parkes, R. J., and Jørgensen, B. B. 2005. Prokaryotic cells of the deep sub-seafloor biosphere identified as living bacteria. *Nature* 433:861–64.

Smatacek, V., and Nicol, S. 2005. Polar ocean ecosystems in a changing world. *Nature* 437:362–68.

Suttle, C. A. 2005. Viruses in the sea. *Nature* 437:356–61.

Please visit the Prescott website at www.mhhe.com/prescott7 for additional references.

29 Microorganisms in Terrestrial Environments

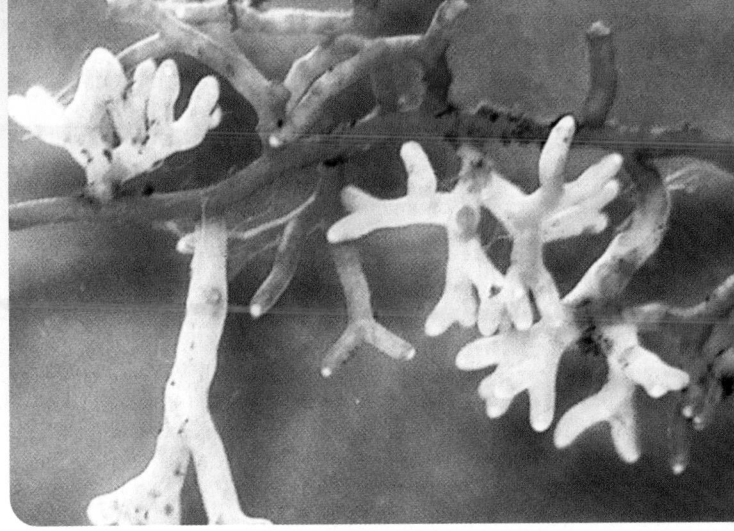

Terrestrial plants and filamentous fungi have developed long-term relationships that benefit both partners. The tips of pine tree roots are usually surrounded by dense fungal sheaths that are part of a hyphal network which extends out into the soil. The plant supplies organic matter to maintain the fungus, and the fungus, in turn, provides the plant with nutrients and water.

Preview

- Soil is a complex environment offering a variety of microhabitats. This is one reason why microbial diversity in soils is much greater than that found in aquatic environments.
- In terrestrial ecosystems, primary production is performed by plants, but the nutrient recycling that occurs though a microbial loop is also essential. Each climate and soil type has a community of microbes specifically adapted to that particular microhabitat.
- Many microbes inhabit the pores between soil particles; others live in association with plants. The plant root surface (rhizoplane) and the region close to plant roots (the rhizosphere) are important sites for microbial growth.
- Mycorrhizal fungi associate with most plants. In this relationship, the fungi provide their plant partner with essential nutrients like nitrogen and phosphorus, while the plant supplies organic carbon to the fungi.
- Rhizobia include α- and β-proteobacteria that form nodules within the roots of leguminous plants, where they fix nitrogen. This process has been best studied in the genus *Rhizobium* and its relatives; it involves a complex plant-microbe communication system and the differentiation of the bacterium into a form that can fix nitrogen. A variety of other bacteria, including the actinomycete *Frankia,* also fix nitrogen while interacting with plants.
- The plant pathogen *Agrobacterium* also relies on an intercellular communication system with host plants, in which it causes tumors called galls. These arise following the insertion of a fragment of bacterial DNA, called T DNA, into the plant cell's chromosome.
- Subsurface microbiology is a relatively new and exciting field that explores vast microbial communities living deep beneath the topsoil. Recent studies show that the biomass within this microbial world equals at least one-third of that living above ground.

Chapter 28 introduces aquatic and marine environments; we now turn our attention to land. The microbiology of soil is important for a variety of reasons. These include the contribution terrestrial microbes make to global biogeochemical cycles and the essential role soil microbes play in agriculture and in maintaining environmental quality. These are just some of the reasons that the microbial ecology of soil is a dynamic and growing field. In the past, culture-based investigations limited scientists' understanding to an estimated 1% of the soil microbes in any given community. As we shall see, the ability to study these complex communities without relying on the direct isolation and growth of individual species has had a profound impact on the appreciation of the soil as a complex and vital environment.

We begin by describing the soil habitat and how microbes contribute to the development of soils. This is followed by a discussion of specific soil microbial communities and the interaction of microbes with vascular plants. We pay particular attention to two very important relationships: that between fungi and plants and between nitrogen-fixing bacteria and leguminous plants. Finally, the new and exciting field of deep subsurface microbiology is introduced.

29.1 Soils as an Environment for Microorganisms

A soil scientist would describe soil as weathered rock combined with organic matter and nutrients. An agronomist would point out that soil supports plant life. However, a microbial ecologist knows

They [the leaves] that waved so loftily, how contentedly, they return to dust again and are laid low, resigned to lie and decay at the foot of the tree and afford nourishment to new generations of their kind, as well as to flutter on high!

—Henry D. Thoreau

that the formation of organic matter and the growth of plants depend on the microbial community within the soil. Historically, the complexity of soil as a habitat has been a challenge to understanding soil microbial ecology. Soil is very dynamic and is formed in a wide variety of environments. These environments range from Arctic tundra regions, where approximately 11% of the Earth's soil carbon pool is stored, to Antarctic dry valleys, where there are no vascular plants. In addition, deeper subsurface zones, where plant roots and their products cannot penetrate, also have surprisingly large microbial communities. Microbial activities in these environments can lead to the formation of minerals such as dolomite; microbial activity also occurs in deep continental oil reservoirs, in stones, and even in rocky outcrops. These microbes are dependent on energy sources from photosynthetic protists and nutrients in rainfall and dust.

Most soils are dominated by inorganic geological materials, which are modified by the biotic community, including microorganisms and plants, to form soils. The spaces between soil particles

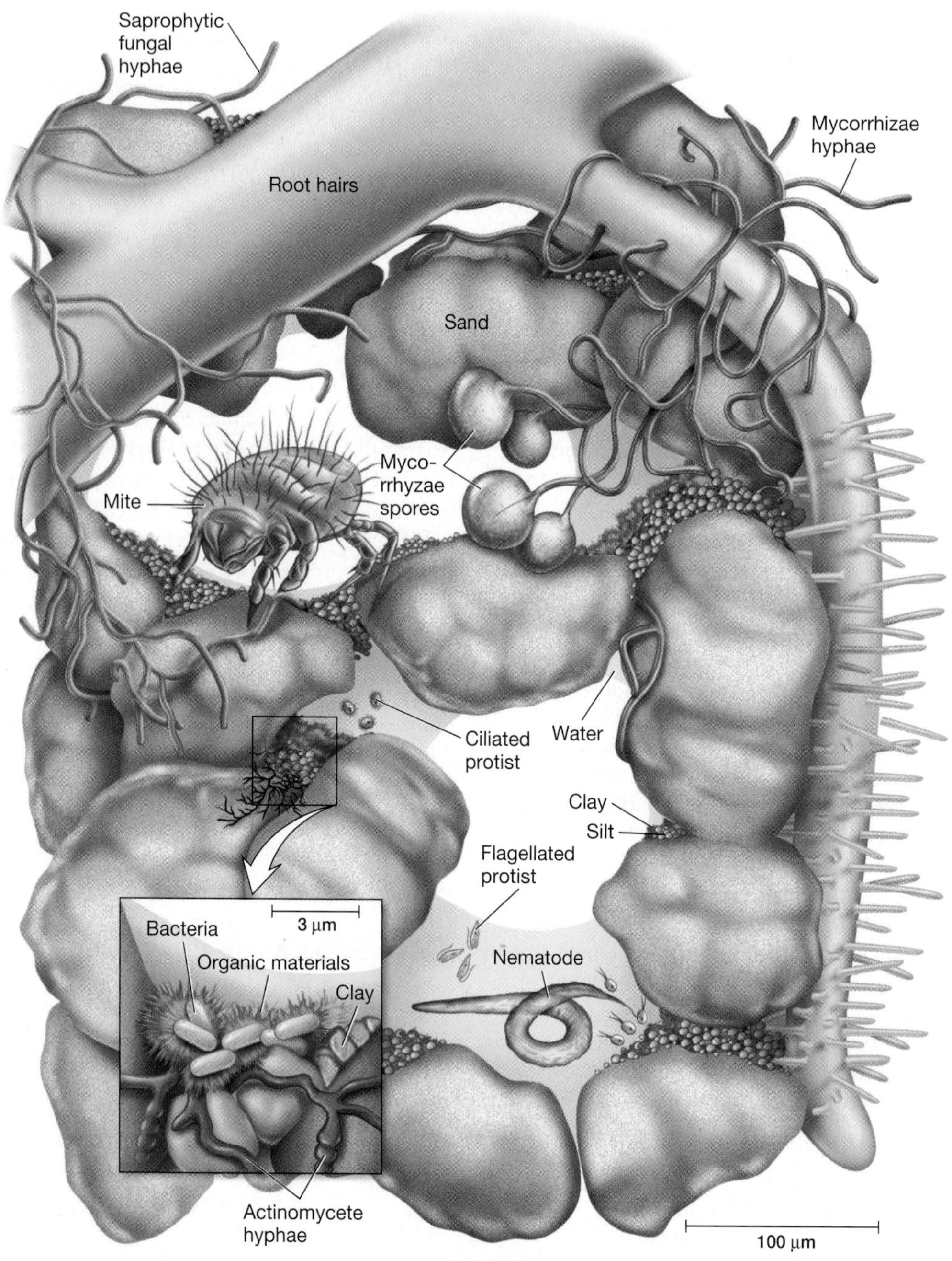

Figure 29.1 The Soil Habitat. A typical soil habitat contains a mixture of clay, silt, and sand along with soil organic matter. Roots, animals (e.g., nematodes and mites), as well as chemoorganotrophic bacteria consume oxygen, which is rapidly replaced by diffusion within the soil pores where the microbes live. Note that two types of fungi are present: mycorrhizal fungi, which derive their organic carbon from their symbiotic partners—plant roots; and saprophytic fungi, which contribute to the degradation of organic material.

are critical for the movement of water and gases (**figure 29.1**). Total pore space, and thus gas diffusion, is determined by the texture of the soil. For instance, sandy soils have larger pore spaces than do clay soils, so sandy soils tend to drain quickly. Pores are also critical because they provide the optimum environment for microbial growth. Here the microbes are within thin water films on the particle surfaces where oxygen is present at high levels and can be easily replenished by diffusion. Oxygen diffusion through air in the soil occurs about 10,000 times faster than it does through water (**figure 29.2**). The oxygen concentrations and flux rates in pores and channels is high, whereas within water-filled zones the oxygen flux rate is much lower. As an example, particles as small as about 2.0 mm can be oxic on the outside and anoxic on the inside.

Depending on the physical characteristics of the soil, rainfall or irrigation may rapidly change a soil from being well aerated to an environment with isolated pockets of water, which are "mini-aquatic" habitats. If this process of flooding continues, a water-logged soil can be created that is more like a lake sediment. If oxygen consumption exceeds that of oxygen diffusion, water-logged soils can become anoxic.

Shifts in water content and gas fluxes also affect the concentrations of CO_2, CO, and other gases present in the soil atmosphere, as noted in **table 29.1.** These changes are accentuated in the smaller pores where many bacteria are found. The roots of plants growing in aerated soils also consume oxygen and release CO_2, influencing the concentrations of these gases in the root environment.

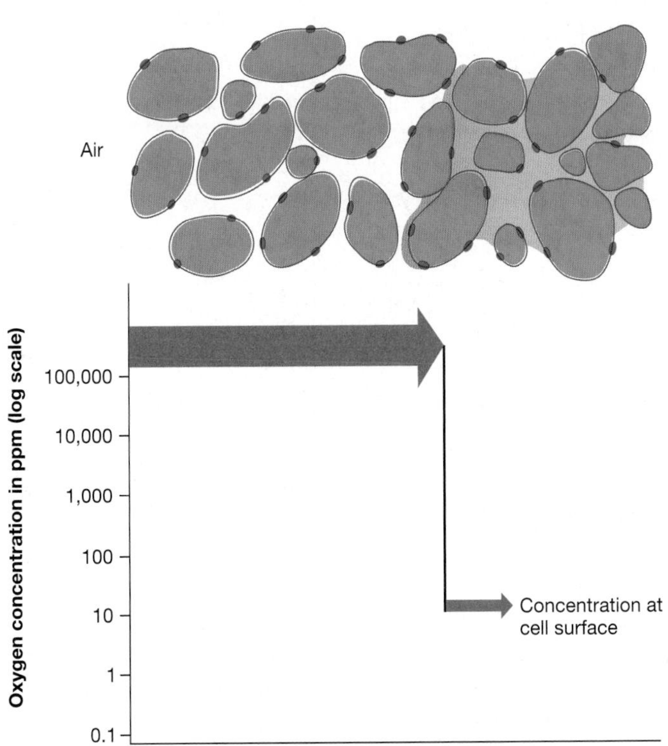

Figure 29.2 Oxygen Concentrations and Fluxes in a Soil. Microorganisms in thin water films on the surface of soil particles have ample access to oxygen. In comparison, microbes in isolated water volumes have limited oxygen fluxes, creating miniaquatic environments.

Table 29.1 Concentrations of Oxygen and Carbon Dioxide in the Atmosphere of a Tropical Soil under Wet and Dry Conditions

	Oxygen Content (%)		Carbon Dioxide Content (%)	
Soil Depth (cm)	**Wet**	**Dry**	**Wet**	**Dry**
10	13.7	20.7	6.5	0.5
25	12.7	19.8	8.5	1.2
45	12.2	18.8	9.7	2.1
90	7.6	17.3	10.0	3.7
120	7.8	16.4	9.6	5.1

From E. W. Russell, *Soil Conditions and Plant Growth,* 10th edition. Copyright © 1973 Longman Group Limited, Essex, United Kingdom. Reprinted by permission.

Note: Normal air contains approximately 21% oxygen and 0.035% carbon dioxide.

1. What is the importance of soil pores?
2. Contrast differences in oxygen flux rates and concentrations in a soil with those in a miniaquatic environment.
3. How do the concentrations of oxygen and carbon dioxide differ between the atmosphere and the soil interior?

29.2 Soils, Plants, and Nutrients

Soils can be divided into two general categories: A **mineral soil** contains less than 20% organic carbon whereas an **organic soil** possesses at least this amount. By this definition, the vast majority of Earth's soils are mineral. The importance of organic matter within soils cannot be underestimated. **Soil organic matter (SOM)** helps to retain nutrients, maintain soil structure, and hold water for plant use. SOM is subject to gains and losses, depending on changes in environmental conditions and agricultural management practices. Plowing and other disturbances expose SOM to more oxygen, leading to extensive microbiological degradation of organic matter. Irrigation causes periodic wetting and drying, which can also lead to increased degradation of SOM, especially at higher temperatures.

Microbial degradation of plant material results in the evolution of CO_2 and the incorporation of the plant carbon into additional microbial biomass. However, a small fraction of the decomposed plant material remains in the soil as SOM. When considering this material, it is convenient to divide the SOM into humic and nonhumic fractions (**table 29.2**). **Nonhumic SOM** has not undergone significant biochemical degradation. It can represent up to about 20% of the soil organic matter. **Humic SOM,** or **humus,** is dark brown to black. It results when the products of microbial metabolism have undergone chemical transformation within the soil. Although there is no precise

Table 29.2 Fractions of Soil Organic Matter

SOM Fraction	Definition	Physical Appearance
Humic substances	High-molecular-weight organic material produced by secondary synthesis reactions	Dark brown to black
Nonhumic substances	Unaltered remains of plants, animals, and microbes, from which macromolecules have not yet been extracted	Light brown
Humic acid	Organic matter extracted from soils by various reagents (often dilute alkali treatment) that is then precipitated by acidification	Dark brown to black
Fulvic acid	Soluble organic matter that remains after humic acid extraction	Yellow
Humin	SOM that cannot be extracted from soil with dilute alkali	

chemical composition of humus, it can be described as a complex blend of phenolic compounds, polysaccharides, and proteins. The recalcitrant nature of this material to degradation is evident by ^{14}C dating: the average age of most SOM ranges from 150 to 1,500 years.

The degradation of plant material and the development of SOM can be thought of as a three-step process. First, easily degraded compounds such as soluble carbohydrates and proteins are broken down. About half the carbon is respired as CO_2 and the remainder is rapidly incorporated into new biomass. During the second stage, complex carbohydrates, such as the plant structural polysaccharide **cellulose,** are degraded. Fungi and members of the bacterial genera *Streptomyces, Pseudomonas,* and *Bacillus* produce extracellular cellulase enzymes that break down cellulose into two to three glucose units called cellobiose and cellotriose, respectively. These smaller compounds are readily degraded and assimilated as glucose monomers. Finally, very resistant material, in particular lignin, is attacked. **Lignin** is an important structural component of woody plants. While its exact structure differs among species, the common building block is the phenylpropene unit. This consists of a hydroxylated six-carbon aromatic benzene ring and a three-carbon linear side chain (**figure 29.3**). A single lignin molecule can consist of up to 600 cross-linked phenylpropene units. It is therefore not surprising that lignin degradation is much slower than that of cellulose. Basidiomycete fungi and actinomycetes (e.g., *Streptomyces* spp.) are capable of extracellular lignin degradation. These microbes produce extracellular phenoloxidase enzymes needed for aerobic lignin degradation. Lignin decomposition is also limited by the physical nature of the material. For example, healthy woody plants are saturated with sap, which limits oxygen diffusion. In addition, high ethylene and CO_2 levels and the presence of phenolic and terpenoid compounds retard the growth of lignin-degrading fungi. It follows that no more than 10% of the carbon found in lignin is recycled into new microbial biomass. Lignin can be degraded anaerobically but this process is very slow, so lignin tends to accumulate in wet, poorly oxygenated soils, like peat bogs.

SOM represents only a small part of the total soil volume, but exerts a disproportionate influence on the biological, chemical, and physical dynamics of soil. Therefore, steps are frequently taken to maximize the SOM content of soil. The practice of reduced tillage has been employed to retain the amount of crop residue that contributes to SOM formation and water retention in the soil, which in turn helps prevent erosion.

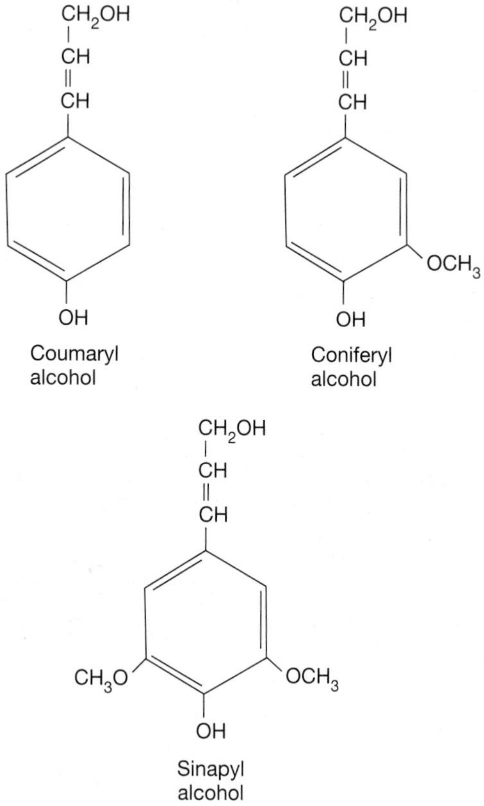

Figure 29.3 Example Phenylpropene Units. These molecules are polymerized to form lignin.

Nitrogen is another important element of the soil ecosystem. Many soils are nitrogen limited; this is why each year tons of nitrogen fertilizer are added to agricultural soils. Nitrogen in soil is often considered in relation to the soil carbon content as the organic **carbon to nitrogen ratio (C/N ratio).** A C/N ratio of 20 or more (i.e., much more carbon than nitrogen) results in loss of soluble nitrogen from the system. Because the addition of nitrogen under such conditions does not stimulate growth, it is said that the soil has reached its **nitrogen saturation point.** Conversely, ratios below 20 enable microbes to convert ammonium and nitrate to biomass (e.g., proteins and nucleic acids).

Throughout the world, soils are increasingly being impacted by mineral nitrogen releases resulting from human activities. This nitrogen has two major sources: (1) agricultural fertilizers containing chemically synthesized nitrogen, and (2) fossil fuel combustion. Fossil fuel-based releases occur especially in eastern North America, Europe, eastern China, and Japan. The major types of nitrogen fertilizers used in agriculture are liquid ammonia and ammonium nitrate. Ammonium ion usually is added because it will be attracted to the negatively charged clays in a soil and be retained on the clay surfaces until used as a nutrient by the plants. However, the nitrifier populations in a soil can oxidize the ammonium ion to nitrite and nitrate, and these anions can be leached from the plant environment and enter surface waters and groundwaters. Biogeochemical cycling: Nitrogen cycle (section 27.2)

An inevitable consequence of the application of nitrogen fertilizers has been higher nitrate levels in waters, which can contribute to infant respiratory problems, and possibly to the production of nitrites and the formation of nitrosamine carcinogens. In addition, plants grown in high nitrate-concentration soils may accumulate nitrate to a level that is harmful for animals. Cereal grains, many weeds, and grass hay contain high nitrate levels when grown in such soils. Nitrogen fertilizers also can affect microbial community structure and function. The use of nitrogen fertilizers has led to decreases in filamentous fungal development in a wide variety of soils. The loss of fungi affects the soil structure because fungi are important conduits of nitrogen and phosphorus to most plants. With a weakened and decreased fungal community, many plants become more susceptible to stresses such as drought and toxic metals.

Phosphorus in fertilizers also is critical. The binding of this anionic fertilizer component to soils is dependent on the cation exchange capacity (CEC) and soil pH. As the soil phosphorus sorption capacity is reached, the excess, together with phosphorus that moves to lakes, streams, and estuaries with soil erosion can stimulate the growth of freshwater organisms, particularly cyanobacteria, in the process of eutrophication. The cyanobacteria are then able to fix more nitrogen, because in most freshwater systems, phosphorus is the critical limiting element. As noted in **Microbial Tidbits 29.1**, excessive fertilizer can have unexpected global consequences. Microorganisms in marine environment: Coastal marine systems (section 28.3)

1. List three reasons why soil organic matter (SOM) is important.
2. What is the difference between humic and nonhumic SOM? Which is more abundant in most soils? Why?
3. Describe the three phases of plant degradation and SOM formation.
4. What are possible effects of nitrogen-containing fertilizers on microbial communities?
5. Most nitrogen fertilizer is added as ammonium ion. Why is this preferred over nitrate?
6. Why is nitrate of concern when it reaches rivers, lakes, and groundwaters?
7. Why might enrichment of fresh waters with phosphorus be even more critical than nitrogen enrichment?

Microbial Tidbits

29.1 An Unintended Global-Scale Nitrogen Experiment

Technological advances can have many unexpected consequences. An excellent example is the discovery of the Haber-Bosch Process at the beginning of the twentieth century that made possible the synthesis of ammonium nitrogen from inert dinitrogen gas. This led to the low-cost availability of mineral nitrogen for crop fertilization, which until then was largely dependent on animal-derived nitrogen sources (such as guano, primarily from Chile, and animal manure) and crop rotation schemes, often involving planting nitrogen-fixing legumes. With the availability of ample ammonium ion as fertilizer from the Haber-Bosch Process, there was no further need to continue such "inefficient" agricultural systems; a green cash crop could be generated each year without the tiresome use of manure and crop rotations. This, however, had unexpected consequences; without the addition of manures, organic matter was lost from soils in these more intensive cropping systems, and there were changes in plant and microbial communities, as well as increased water pollution with nitrates.

Nitrogen production by the Haber-Bosch Process, together with burning fossil fuels, adds mineral nitrogen to the atmosphere with other unexpected consequences. Studies of Northern Hemisphere forests indicate that nitrogen is released from soils and drainage waters primarily as nitrate. It has been assumed that this reflected global-level natural processes as the phenomenon was so widespread. However, studies of forests in South America, which are far from sources of mineral nitrogen, suggest a very different view. Steven Perakis and Lars Hedin of Cornell University have shown that remote South American forests release only small amounts of nitrate, and instead that most nitrogen is released in organic forms. What appears to have happened? By having large-scale releases of mineral nitrogen in the Northern Hemisphere for almost a century, human activities have dramatically changed the nitrogen cycle of soils and waters in most areas of the populated world. It appears that the entire Earth has been turned into a giant nitrogen-amendment experiment. Agricultural and industrial pollution has strongly impacted the Northern Hemisphere, and is now affecting the Southern Hemisphere as well. Most researchers and policymakers have not been aware of this global-level nitrogen impact. The problem is that we cannot return to the beginning and modify such a global-scale experiment.

29.3 Microorganisms in the Soil Environment

If we look at a soil in greater detail (figure 29.1), we find that bacteria, archaea fungi, and protists use different functional strategies to take advantage of this complex physical matrix. Most soil procaryotes are located on the surfaces of soil particles and require water and nutrients that must be located in their immediate vicinity. Procaryotes are found most frequently on surfaces within smaller soil pores (2 to 6 μm in diameter). Here they are probably less liable to be eaten by protozoa, unlike those located on the exposed outer surface of a sand grain or organic matter particle.

Terrestrial filamentous fungi, in comparison, bridge open areas between soil particles or aggregates, and are exposed to high levels of oxygen. These fungi will tend to darken and form oxygen-impermeable structures called sclerotia and hyphal cords. This is particularly important for basidiomycetes, which form such structures as an oxygen-sealing mechanism. Within these structures, the filamentous fungi move nutrients and water over great distances, including across air spaces, a unique part of their functional strategy. These oxidatively polymerized, oxygen-impermeable hyphal boundaries do not usually occur in fungi growing in aquatic environments. Characteristics of the fungal divisions: *Basidiomycota* (section 26.6)

The microbial populations in soils can be very high. In a surface soil the bacterial population can approach 10^9 to 10^{10} cells per gram dry weight of soil as measured microscopically. Fungi can be present at up to several hundred meters of hyphae per gram of soil. We tend to think of soil fungi as small structures like the mushrooms sprouting from our lawns. However, the vast majority of fungal biomass is below ground. For instance, an individual clone of the fungus *Armillaria bulbosa,* which lives associated with tree roots in hardwood forests, was discovered that covers about 30 acres in the Upper Peninsula of Michigan. It is estimated to weigh a minimum of 100 tons (an adult blue whale weighs about 150 tons) and be at least 1,500 years old. Thus some fungal mycelia are among the largest and most ancient living organisms on Earth.

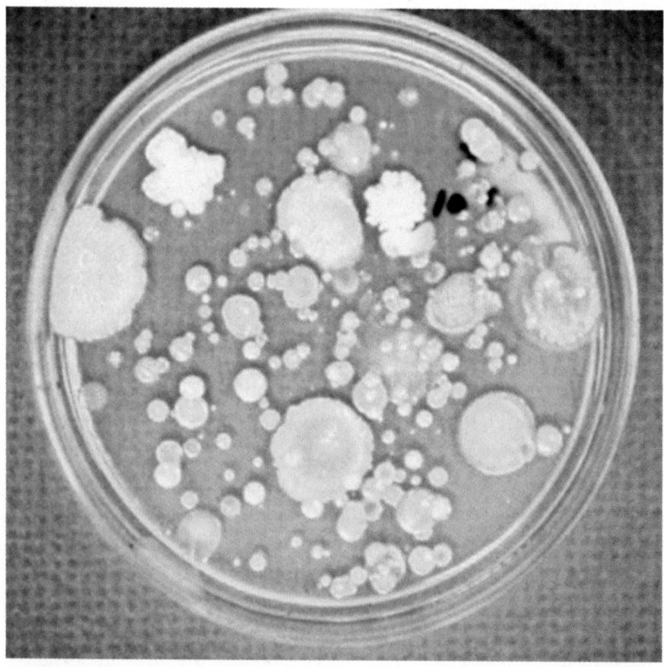

Figure 29.4 Cultured Soil Microbes. Although only about 1% of microorganisms have been cultured, those that grow in the laboratory show morphological diversity.

Soil microbial communities appear to be more diverse than those found in most freshwater and marine environments. Recall that less than 1% of all soil microbes have been cultured (**figure 29.4**), so molecular techniques have been key to gaining an understanding of these complex ecosystems. Small subunit rRNA analysis and the re-association rate of denatured (single-stranded) DNA extracted directly from environmental samples have been used to measure community diversity. DNA reassociation measures the rate at which denatured DNA returns to the double-stranded state; this depends on the number of homologous chromosomes in the sample. Procaryotic community genome size is then calculated based on the assumption that the average procaryotic genome is about the same size as that of *E. coli* (4.1 Mb). Community genome size thus reflects the genomic diversity within the sample. Such analyses show that diversity is highest in pristine organic soils, with as many as 11,000 different genomes per cubic cm. The lowest diversity is found in extreme environments such as hypersaline ecosystems (**table 29.3**). Microbial ecology and its methods: Examination of microbial community structure (section 27.4)

Molecular techniques have been key to identifying microbes not previously thought to inhabit soils. The *Crenarchaeota* have been discovered in soil and ocean sediments by extracting microbial DNA and amplifying it with the polymerase chain reaction (*see figure 27.17*). Examination of soils from different areas of the world continues to yield surprises. Microorganisms are present and prolific in subsurface environments, including oil reservoirs. Hyperthermophilic archaea have been found in such harsh sub-

Table 29.3 Procaryotic Diversity Determined by Direct Cell Counts Using Fluorescence Microscopy (Abundance) and the Re-association Rate of DNA Isolated from Each Community Listed (Genome Equivalents)

DNA Source	Abundance (cells/cm^3)	Genome Equivalents[1]
Forest soil	4.8×10^9	6,000
Pasture soil	1.8×10^{10}	3,500–8,800
Arable soil	2.1×10^{10}	140–350
Marine fish farm	7.7×10^9	50
Hypersaline pond (22% salinity)	6.0×10^9	7

From V. Torsvik, L. Ovraes, and T. F. Thingstad. 2002. *Science* 296:1064–66.

[1] Genome equivalents are based on the assumption that the average procaryotic genome is about the size of the *E. coli* genome (4.1 Mb).

surface environments and are probably indigenous to these poorly understood regions of our world. *Crenarchaeota* (section 20.2)

The coryneforms, the nocardioforms, and the true filamentous bacteria (the streptomycetes) (**table 29.4**) are an important part of the soil microbial community. These gram-positive bacteria play a major role in the degradation of hydrocarbons, older plant materials, and soil humus. In addition, some members of these groups actively degrade pesticides. The filamentous actinomycetes, primarily of the genus *Streptomyces,* produce an odor-causing compound called **geosmin,** which gives soils their characteristic earthy odor. Polyprosthecate bacteria such as the genera *Verrucomicrobium, Pedomicrobium,* and *Prosthecobacter* are present in soils at high levels. With their small size, and the difficulties involved in culturing them, they have been largely overlooked in assessments of soil microbial diversity. High G + C gram-positive bacteria (chapter 24)

As discussed in section 29.6, the microbial community in soil makes important contributions to the carbon, nitrogen, sulfur, iron, and manganese biogeochemical cycles. Because the soil is primarily an oxidized environment, the inorganic forms of these elements tend to be in the oxidized state. If there are localized water-saturated, lower oxygen-flux environments, the biogeochemical cycles shift toward reduced species. Biogeochemical cycling (section 27.2)

Soil insects and other animals such as nematodes and earthworms also contribute to organic matter transformations in soils. These organisms carry out decomposition, often leading to the release of minerals, and physically "reducing" the size of organic particles such as plant litter. This increases the surface area and makes organic materials more available for use by bacteria and fungi. Earthworms also mix substrates with their internal gut microflora and enzymes; this contributes substantially to decomposition and has major effects on soil structure and the soil microbial community.

Nutrients are regenerated in soils through a microbial loop that differs from that which operates in the photic zone of the open ocean. A major distinction is that plants rather than microbes account for most primary production in nearly all terrestrial systems. But much like the microbial loop in marine waters, microbes in soils rapidly recycle the organic material derived from plants and animals, including the many nematodes and insects. In turn, microbes themselves are preyed upon by soil protists, whose numbers can reach 100,000 per gram of soil. This makes microbial organic matter available to other trophic levels. Another difference between marine and terrestrial microbial loops reflects the physical and biological properties of soil. Degradative enzymes released by plants, insects, and other animals do not rapidly diffuse away; instead, they represent a significant contribution to the biological activity in soil ecosystems. In fact, these free enzymes contribute to many hydrolytic degradation reactions, such as proteolysis; catalase and peroxidase activities also have been detected.

Table 29.4 Easily Cultured Gram-Positive Irregular Branching and Filamentous Bacteria Common in Soils

Bacterial Group	Representative Genera	Comments and Characteristics
Coryneforms	*Arthrobacter*	Rod-coccus cycle
	Cellulomonas	Important in degradation of cellulose
	Corynebacterium	Club-shaped cells
Mycobacteria	*Mycobacterium*	Acid-fast
Nocardioforms	*Nocardia*	Rudimentary branching
Streptomycetes	*Streptomyces*	Aerobic filamentous bacteria
Bacilli	*Thermoactinomyces*	Higher temperature growth

1. What are the differences in preferred soil habitats between bacteria and filamentous fungi?
2. What types of archaea have been detected in soils?
3. How can earthworms, nematodes, and insects influence microbial communities?
4. How does the microbial loop function in soils?

29.4 Microorganisms and the Formation of Different Soils

Soils are formed when geologic materials are exposed. This may occur after a dramatic event such as a volcanic eruption or from a simple disturbance. Soil formation is the result of the combined action of weathering and colonization of geologic material by microbes. The microbial community that initially colonizes a newly disturbed environment will bring changes to the local environment by degrading and recycling organic material. For example, if extreme soil erosion has removed topsoil, phosphorus may be present, but nitrogen and carbon must be imported by physical or biological processes. Under these circumstances, autotrophic, N_2-fixing cyanobacteria are active in pioneer-stage nutrient accumulation. These primary microflora will eventually be replaced by a new community of microorganisms that further alter the ecosystem. This sets in motion successive waves of microbial communities until a community that is sufficiently diverse and physiologically well-suited is established and a stable climax ecosystem is developed. The major types of plant-soil systems are shown in **figure 29.5** and are discussed here.

Tropical and Temperate Region Soils

In warm, moist tropical soils, organic matter is decomposed very quickly and the mobile inorganic nutrients can be leached out of the surface soil environment, causing a rapid loss of fertility. To limit nutrient loss, many tropical plant root systems penetrate the rapidly decomposing litter layer. As soon as organic material and minerals are released during decomposition, the roots take them

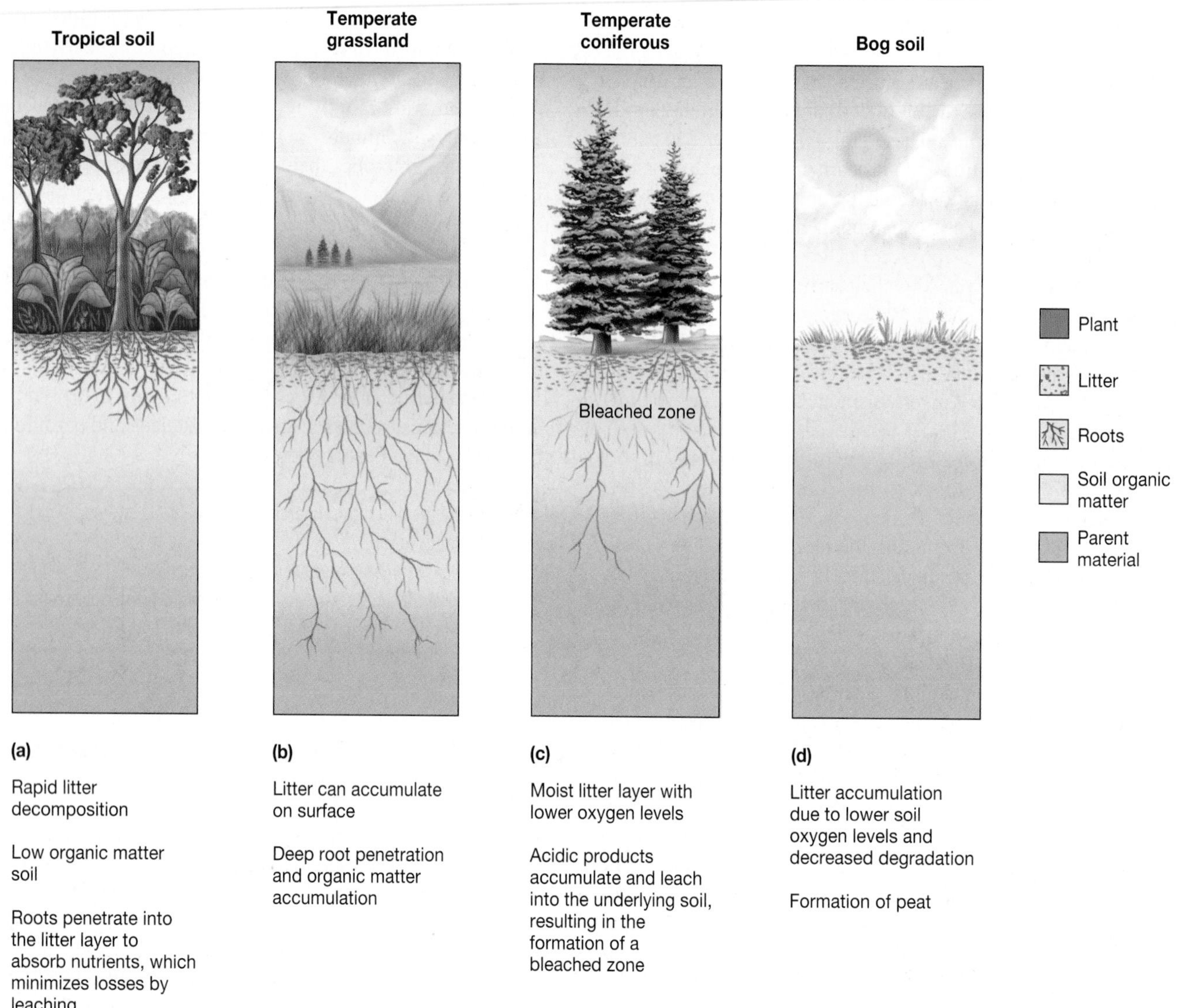
Tropical soil
Temperate grassland
Temperate coniferous
Bog soil
Bleached zone
Plant
Litter
Roots
Soil organic matter
Parent material
(a)
Rapid litter decomposition
Low organic matter soil
Roots penetrate into the litter layer to absorb nutrients, which minimizes losses by leaching
(b)
Litter can accumulate on surface
Deep root penetration and organic matter accumulation
(c)
Moist litter layer with lower oxygen levels
Acidic products accumulate and leach into the underlying soil, resulting in the formation of a bleached zone
(d)
Litter accumulation due to lower soil oxygen levels and decreased degradation
Formation of peat

Figure 29.5 Examples of Tropical and Temperate Region Plant-Soil Biomes. Climate, parent material, plants, topography, and microorganisms interact over time to form different plant-soil systems. In these figures the characteristics of **(a)** tropical, **(b)** temperate grassland, **(c)** temperate coniferous forest, and **(d)** bog soils are illustrated.

up to avoid losses in leaching. Thus it is possible to recycle nutrients before they are lost due to water movement through the soil (figure 29.5*a*). With deforestation, there is no leaf litter so nutrients are not recycled, leading to their loss from the soil and decreased soil fertility.

Tropical plant-soil communities are often used in **slash-and-burn agriculture.** The vegetation on a site is chopped down and burned to release the trapped nutrients. For a few years, until the minerals are washed from the soils, crops can be grown. When the minerals are lost from these low organic matter soils, the farmer must move to a new area and start over by again cutting and burning the native plant community. This cycle of slash-and-burn agriculture is stable if there is sufficient time for the plant community to regenerate before it is again cut and burned. If the cycle is too short, rapid and almost irreversible degradation of the soil occurs.

In many temperate region soils, in contrast, the decomposition rates are less than that of primary production, leading to litter accumulation. Deep root penetration in temperate grasslands results in the formation of fertile soils, which provide a valuable resource for the growth of crops in intensive agriculture (figure 29.5*b*).

The soils in many cooler coniferous forest environments suffer from an excessive accumulation of organic matter as plant litter (figure 29.5*c*). In winter, when moisture is available, the soils are cool, and this limits decomposition. In summer, when the soils are warm, water is not as available for decomposition. Organic acids are produced in the cool, moist litter layer, and they leach into the underlying soil. These acids solubilize soil components such as aluminum and iron, and a bleached zone may form. Litter continues to accumulate, and fire becomes the major means by which nutrient cycling is maintained. Controlled burns have become part of the environmental management in this type of plant-soil system.

Bog soils provide a unique set of conditions for microbial communities (figure 29.5*d*). In these soils, decomposition is slowed by the waterlogged, predominantly anoxic conditions, which lead to peat accumulation. When such areas are drained, they become more oxic and SOM is degraded, resulting in soil subsidence. Under oxic conditions the lignin-cellulose complexes of the accumulated organic matter are more susceptible to decomposition by filamentous fungi.

Cold Moist Soils

Soils in cold environments, whether in Arctic, Antarctic, or alpine regions, are of extreme interest because of their wide distribution and impacts on global-level processes. The colder mean soil temperatures at these sites decrease the rates of both decomposition and plant growth. In these cases SOM accumulates, and plant growth can become limited due to the immobilization of nutrients. Often, below the plant growth zone, these soils are permanently frozen. These permafrost soils hold about 11% of the Earth's soil carbon and 95% of its organically bound nutrients. These soils are very sensitive to physical disturbance and pollution, and the widespread exploration of such areas for oil and minerals can have long-term effects on their structure and function.

In water-saturated bog areas, oxygen limitation means that bacteria are more important than fungi in decomposition processes, and there is decreased degradation of lignified materials. As in other soils, the nutrient cycling processes of nitrification, denitrification, nitrogen fixation, and methane synthesis and utilization, although occurring at slower rates, can have major impacts on global gaseous cycles.

Desert Soils

Soils of hot and cold arid and semiarid deserts are dependent on periodic and infrequent rainfall. When it rains, water can puddle in low areas and be retained on the soil surface by microbial communities called **desert crusts.** These consist of cyanobacteria and associated microbes, including *Anabaena, Microcoleus, Nostoc,* and *Scytonema*. The depth of the photosynthetic layer is perhaps 1 mm, and the cyanobacterial filaments and slime link the sand particles (**figure 29.6**), which change the surface soil albedo (the amount of sunlight reflected), water infiltration rate, and susceptibility to erosion. These crusts are quite fragile, and vehicle damage can be evident for decades. After a rainfall, nitrogen fixation begins within approximately 25 to 30 hours, and when the rain evaporates or drains, the crust dries up and nitrogen is released for use by other microorganisms and the plant community. Photosynthetic bacteria: *Phylum Cyanobacteria* (section 21.3)

Figure 29.6 A Desert Crust as Observed with the Scanning Electron Microscope. The crust has been disturbed to show the extracellular sheaths and filaments of the cyanobacterium *Microcoleus vaginatus*. The sand grains are linked by these filamentous growths, creating a unique ecological structure.

Geologically Heated Hyperthermal Soils

Geologically heated soils are found in such areas as Iceland, the Kamchatka peninsula in eastern Russia, Yellowstone National Park, and at many mining waste sites. These soils are populated by bacteria and archaea, many of which are chemolithoautotrophs. A wide variety of chemoorganotrophic genera also are found in these environments; these include the aerobes *Thermomicrobium, Thermoleophilum,* and also the anaerobes *Thermosipho* and *Thermotoga*. An important microorganism found in heated mining wastes is *Thermoplasma*. Such geothermal soils have been of great interest as a source of new microbes to use in biotechnology, and the search for new, unique microorganisms in these areas is intensifying all over the world. Phylum *Euryarchaeota: Thermoplasms* (section 20.3)

1. Characterize each major soil type discussed in this section in terms of the balance between primary production and organic matter decomposition.
2. What is slash-and-burn agriculture? Describe the roles of microorganisms in this process.
3. What is unique about bogs in terms of organic matter degradation?
4. Describe desert crusts. What types of microorganisms function in these unique environments?
5. What unique microbial genera are found in geothermally heated soils?

29.5 Microorganism Associations with Vascular Plants

The vast majority of soil microbes are heterotrophic, so it should come as no surprise that many have evolved close relationships with plants, the major source of terrestrial primary production. Many microbe-plant interactions do no harm to the plant, whereas the microbe gains some advantage. Such relationships, in which one partner benefits but the other is neither hurt nor helped, is called commensalism. Many other important interactions are beneficial to both the microorganism and the plant (i.e., are mutualistic). Finally, other microbes are plant pathogens and parasitize their plant hosts. In all cases, the microbe and the plant have established the capacity to communicate with each other. The microbe detects and responds to plant-produced chemical signaling molecules. This generally triggers the release of microbial compounds that are in turn recognized by the plant, thereby beginning a two-way "conversation" that employs a molecular lexicon. Indeed, once a microbe-plant relationship is initiated, microbes and plants continue to monitor the physiology of their partner and adjust their own activities accordingly. The nature of the signaling molecules and the mechanisms by which both plants and microbes respond has become an exciting multidisciplinary focus of soil microbiology research, as it encompasses ecology, molecular biology, genetics, and biochemistry. Microbial interactions (section 30.1)

Microbe-plant interactions can be broadly divided into two classes: microbes that live on the surface of plants are called **epiphytes;** those that colonize internal plant tissues are called **endophytes.** Further, we can consider those microbes that live in the above-ground, or aerial, surfaces of plants separately from those that inhabit below-ground plant tissues. We begin our discussion of microbe-plant interactions by first introducing the microbial communities associated with aerial regions of plants. We then turn our attention to two important microbe-root symbioses—the mycorrhizal fungi and the nitrogen-fixing rhizobia. Finally, we consider several microbial plant pathogens.

Phyllosphere Microorganisms

The environment of the aerial portion of a plant, called the **phyllosphere,** was once thought to be too hostile to support a stable microbial community. Leaves and stems undergo frequent and rapid changes in humidity, UV exposure, and temperature. This in turn results in fluctuations in the leaching of organic material (primarily simple sugars) that could support a microbial population. It is now known that the phyllosphere is home to a diverse assortment of microbes including bacteria, filamentous fungi, yeasts, and photosynthetic and heterotrophic protists. Numerically, it appears that the γ-proteobacteria *Pseudomonas syringae* and *Erwinia,* and *Pantoea* spp. are most important. Another abundant bacterial genus, *Sphingomonas,* produces pigments that function like suncreen so it can survive the high levels of UV irradiation occurring on these plant surfaces. This bacterium, also common in soils and waters, can occur at 10^8 cells per gram of plant tissue. *Sphingomonas* often represents a majority of the culturable species.

Rhizosphere and Rhizoplane Microorganisms

Plant roots receive between 30 to 60% of the net photosynthesized carbon. Of this, an estimated 40 to 90% enters the soil as a wide variety of materials including alcohols, ethylene, sugars, amino and organic acids, vitamins, nucleotides, polysaccharides, and enzymes as shown in **table 29.5.** These materials create a unique environment for soil microorganisms called the **rhizosphere.** The plant root surface, termed the **rhizoplane,** also provides a unique environment for microorganisms, as these gaseous, soluble, and particulate materials move from the plant to the soil. Rhizosphere and rhizoplane microorganisms increase their numbers when these newly available substrates become available; their composition and function also change. In addition, rhizosphere and rhizoplane microorganisms serve as labile sources of nutrients for other organisms, creating a soil microbial loop and thereby playing critical roles in organic matter synthesis and degradation.

A wide range of microbes in the rhizosphere can promote plant growth, orchestrated by their ability to communicate with plants using complex chemical signals. Some of these chemical signal compounds include auxins, gibberellins, glycolipids, and cytokinins, and are beginning to be fully appreciated in terms of their biotechnological potential. Plant growth-promoting rhizobacteria include the genera *Pseudomonas* and *Achromobacter.* These can be added to the plant, even in the seed stage, if the bacteria have the required surface attachment proteins. The genes that control the expression of these attachment proteins are of great interest to agricultural biotechnologists.

A critical process that occurs on the surface of the plant, and particularly in the root zone, is **associative nitrogen fixation,** in which nitrogen-fixing microorganisms are on the surface of the plant root, the rhizoplane (**figure 29.7**), as well as in the rhizosphere. This process is carried out by representatives of the genera *Azotobacter, Azospirillum,* and *Acetobacter.* These bacteria contribute to nitrogen accumulation by tropical grasses. Evidence suggests that their major contribution may not be nitrogen fixation but the production of growth-promoting hormones that increase root hair development,

Table 29.5 Compounds Excreted by Microorganism-Free Wheat Roots

Volatile Compounds	Low-Molecular-Weight Compounds	High-Molecular-Weight Compounds
CO_2	Sugars	Polysaccharides
Ethanol	Amino acids	Enzymes
Isobutanol	Vitamins	
Isoamyl alcohol	Organic acids	
Acetoin	Nucleotides	
Isob6utyric acid		
Ethylene		

From J. W. Woldendorp, "The Rhizosphere as Part of the Plant-Soil System" in *Structure and Functioning of Plant Populations* (Amsterdam, Holland: Proceedings, Royal Dutch Academy of Sciences, Natural Sciences Section: 2d Series, 1978) 70:243.

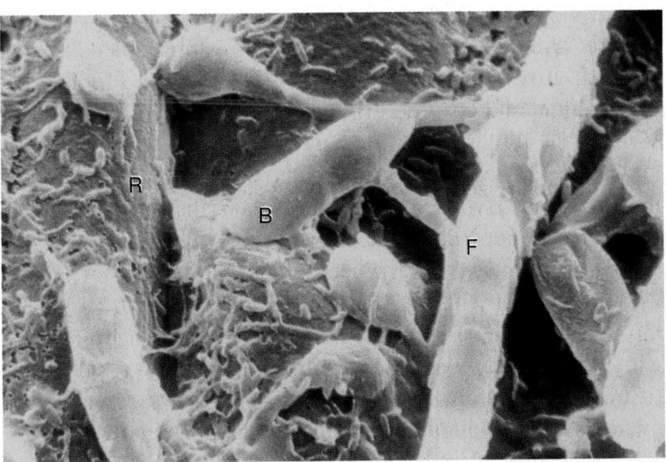

Figure 29.7 Root Surface Microorganisms. Plant roots release nutrients that allow intensive development of bacteria and fungi on and near the plant root surface, the rhizoplane. A scanning electron micrograph shows bacteria and fungi growing on a root surface. R = root surface; B = bacterium; F = fungal hypha.

thereby enhancing plant nutrient uptake. This is an area of research that is particularly important in tropical agricultural areas.

Recently methanogenic archaea have been identified in the rhizosphere of rice. Rice paddies are semi-submerged; the limited flux of O_2 into the soil creates an anoxic zone just below the surface. This, combined with the vast amount of land devoted to rice cultivation, makes rice fields a major source of the greenhouse gas methane (CH_4).

A technique called **stable isotope probing** has been employed to explore methanogenesis in rice fields. Stable isotopes are those forms of elements that differ in atomic weight because they bear different numbers of neutrons but are not radioactive. For example, ^{12}C and ^{13}C both occur in nature, but ^{12}C is found in great abundance whereas ^{13}C is rare. Organisms discriminate between stable isotopes and for any given element, the lighter (in this case ^{12}C) is preferentially incorporated into biomass. To study the rice paddy soil ecosystem, researchers built microcosms to simulate the natural rice paddies and introduced $^{13}CO_2$. Gaseous $^{13}CH_4$ was collected and RNA was isolated from the soil. ^{12}C-containing RNA was separated from the ^{13}C-RNA on the basis of differing buoyant densities. The ^{13}C-containing RNA, which could only be synthesized by those bacteria that assimilated the $^{13}CO_2$, was then used to identify the microbes.

Recall that methanogenic microbes are archaea; here they were identified as members of "Rice Cluster-I" (RC-I). These methanogens rely on other microbes to ferment photosynthetically derived compounds excreted from plant roots. The H_2 produced by these fermenters is then used by the RC-I archaea to reduce CO_2 to CH_4. The fact that RC-I archaea have not yet been grown in culture demonstrates the value and importance of culture-independent approaches to understanding microbial community ecology and physiology. Phylum *Euryarchaeota:* Methanogens (section 20.3); Biogeochemical cycling: Carbon cycle (section 27.2)

1. Define the following terms: rhizosphere, rhizoplane, and associative nitrogen fixation.
2. What unique stresses face a microorganism on a leaf but not in the soil?
3. What is the importance of plant growth-promoting bacteria?
4. What important genera are involved in associative nitrogen fixation?
5. Describe stable isotope probing. Explain why this technique can simultaneously reveal microbial activity and identity without the need to culture individual microbes.

Mycorrhizae

Mycorrhizae (derived from the Greek "fungus root") are mutualistic relationships that develop between most plants and a limited number of fungal species. Both partners in mutualistic relationships are dependent on the activities of the other and as such have coevolved (**Microbial Diversity & Ecology 29.2**). In this case, fungi colonize the roots of about 80% of all higher plants as well as ferns and mosses. Unlike most fungi, mycorrhizal fungi are not saprophytic—that is, they do not obtain organic carbon from the degradation of organic material. Instead, they use photosynthetically derived carbohydrate provided by their host. In return, they provide a number of services for their plant hosts, including enhanced nutrient uptake. The importance of mycorrhizae cannot be

Microbial Diversity & Ecology

29.2 Mycorrhizae and the Evolution of Vascular Plants

Fossil evidence shows that endomycorrhizal symbioses were as frequent in vascular plants during the Devonian period, about 400 million years ago, as they are today. As a result some botanists have suggested that the evolution of this type of association may have been a critical step in allowing colonization of land by plants. During this period soils were poorly developed, and as a result mycorrhizal fungi were probably significant in aiding the uptake of phosphorus and other nutrients. Even now, those plants that start to colonize extremely nutrient-poor soils survive much better if they have endomycorrhizae. Thus it may have been a symbiotic association of plants and fungi that initially colonized the land and led to our modern vascular plants.

underestimated; as described by the plant pathologist Stephen Wilhelm, "in agricultural field conditions, plants do not, strictly speaking have roots, they have mycorrhizae." Microbial interactions: Mutualism (section 30.1)

Mycorrhizae can be broadly classified as **endomycorrhizae**—those with fungi that enter the root cells, or as **ectomycorrhizae**—those that remain extracellular, forming a sheath of interconnecting filaments (hyphae) around the roots. Although all six types of mycorrhizae are detailed in **table 29.6,** we confine most of our discussion to the two most important types: ectomycorrhizae and the endomycorrhizae called arbuscular mycorrhizae. Characteristics of fungal divisions (section 26.6)

The ectomycorrhizae (ECM) are formed by both ascomycete and basidiomycete fungi. The latter are best known for their fruiting bodies, which include toadstools and puffballs. ECM colonize almost all trees in cooler climates. Their importance arises from their ability to transfer essential nutrients, especially phosphorus and nitrogen, to the root. The development of an ECM starts with the growth of a fungal mycelium around the root. As the mycelium thickens, it forms a sheath or mantle so that the entire root may be covered by the fungal mycelium (**figure 29.8*d***). Most ECM produce signaling molecules that limit the growth of root hairs, thus ECM-colonized roots often appear blunt and covered in fungi (**figure 29.9**). From the root surface, the fungi extend hyphae into the soil; these filaments may aggregate to form **rhizomorphs,** which are often visible to the naked eye. Hyphae on the inner side of the sheath penetrate between (but not within) the cortical root cells, forming a characteristic meshwork of hyphae called the **Hartig net** (figure 29.8*d*). Soil nutrients taken up by rhizomorphs must first pass through the hyphal sheath and then into the Hartig net filaments, which form numerous contacts with root cells. This results in efficient two-way transfer of soil nutrients to the plant and carbohydrates to the fungus. This relationship has evolved to the point that some plants synthesize sugars such as mannitol and trehelose that cannot be used by the plants and can only be assimilated by their fungal symbionts.

Arbuscular mycorrhizae (AM) are the most common type of mycorrhizae. They can be found in association with many tropical plants and, importantly, with most crop plants. AM fungi be-

Table 29.6 Mycorrhizal Associations

Mycorrhizal Classification	Fungi Involved	Plants Colonized	Fungal Structural Features	Fungal Function
Ectomycorrhizae	Basidiomycetes including those with large fruiting bodies (e.g., toadstools) Some ascomycetes	~90% of trees and woody plants in temperate regions; fungal/plant colonization is often species specific	Hartig net Mantle or sheath Rhizomorphs Root hair development is usually limited	Nutrient (N and P) uptake and transfer
Arbuscular	Glomeromycetes, in particular six genera of the order *Glomales*	Wild and crop plants, tropical trees; fungal/plant colonization is not highly specific	Arbuscules: hyphae-filled invaginations of cortical root cell	Nutrient (N and P) uptake and transfer Facilitate soil aggregation Promote seed production Reduce pest and nematode infection Increase drought and disease resistance
Ericaceous	Ascomycetes Basidiomycetes	Low evergreen shrubs, heathers	Some intracellular, some extracellular	Mineralization of organic matter
Orchidaceous	Basidiomycetes	Orchids	Hyphal coils called pelotons within host tissue	Some orchids are non-photosynthetic and others produce chlorophyll when mature; these organisms are almost completely dependent on mycorrhizae for organic carbon and nutrients
Ectendo-mycorrhizae	Ascomycetes	Conifers	Hartig net with some intracellular hyphae	Nutrient uptake and mineralization of organic matter
Monotropoid mycorrhizae	Ascomycetes Basidiomycetes	Flowering plants that lack chlorophyll (*Monotropaceae;* e.g., Indian pipe)	Hartig net one cell deep in the root cortex	Nutrient uptake and transfer

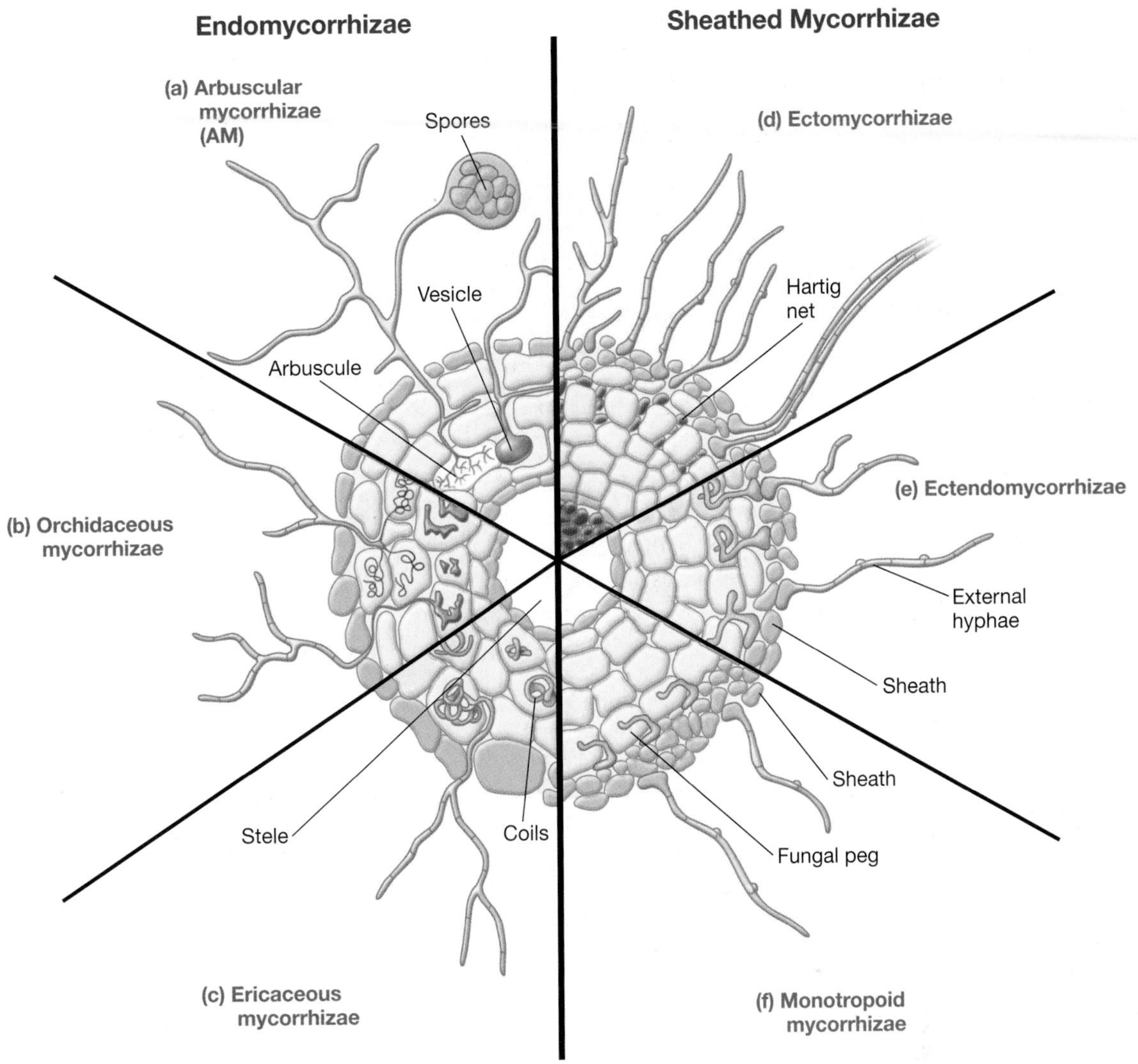

Endomycorrhizae
Sheathed Mycorrhizae
(a) Arbuscular mycorrhizae (AM)
Spores
Vesicle
Arbuscule
(b) Orchidaceous mycorrhizae
Stele
Coils
(c) Ericaceous mycorrhizae
(d) Ectomycorrhizae
Hartig net
(e) Ectendomycorrhizae
External hyphae
Sheath
Sheath
Fungal peg
(f) Monotropoid mycorrhizae

Figure 29.8 Mycorrhizae. Fungi can establish mutually beneficial relationships with plant roots, called mycorrhizae. Root cross sections illustrate different mycorrhizal relationships.

Figure 29.9 Ectomycorrhizae as Found on Roots of a Pine Tree. Typical irregular branching of the white, smooth mycorrhizae is evident.

long to the division *Glomeromycota;* however, they have not yet been grown in pure culture without their plant hosts. These microbes enter root cells between the plant cell wall and invaginations in the plasma membrane (figure 29.8*a*). So, although AM are endomycorrhizae, they do not breach the root cell membrane. Instead, treelike hyphal networks called **arbuscules** develop within the folds of the plasma membrane (**figure 29.10**). Individual arbuscules are transient; they last at the most two weeks. AM can be vigorous colonizers: a 5-cm segment of root can support the growth of as many as eight species and hyphae from a single germinated spore can simultaneously colonize multiple roots from unrelated plant species.

AM are believed to provide a number of services to their plant hosts including protection from disease, drought, nematodes, and other pests. Their capacity to transfer phosphorus to roots has been well documented and recently the nature of their transfer of nitrogen has been explored in detail. Stable isotope experiments

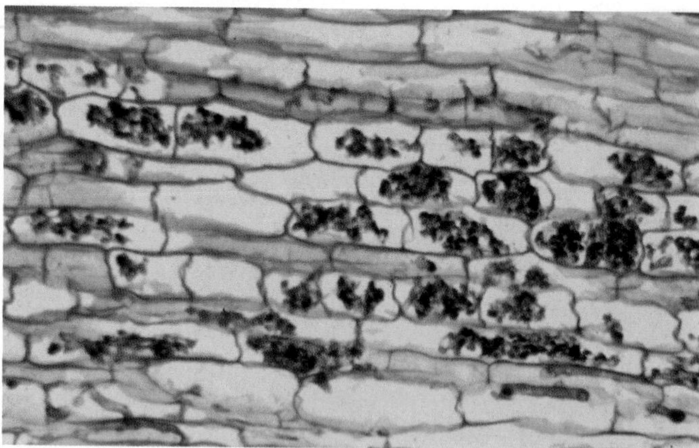

Figure 29.10 Endomycorrhizae. Endomycorrhizae, or arbuscular mycorrhizae, form characteristic structures within roots. These can be observed with a microscope after the roots are stained. The arbuscules of *Gigaspora margarita* can be seen inside the root cortex cells of cotton.

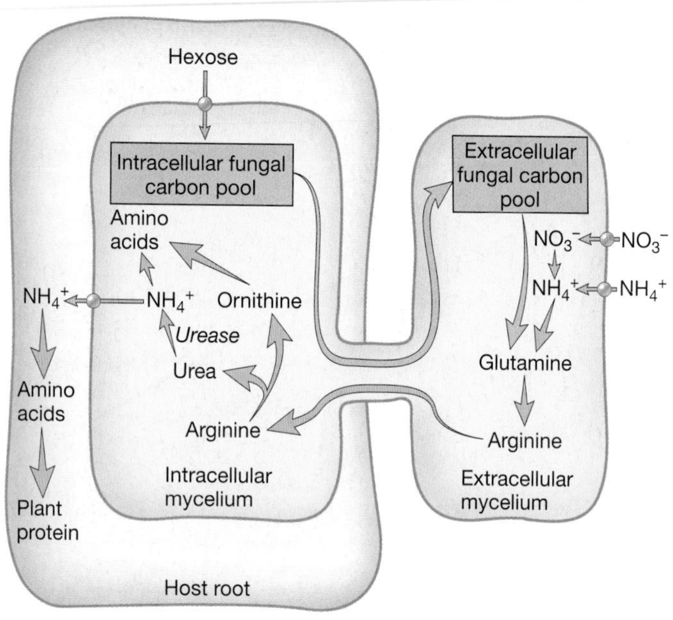

Figure 29.11 Nitrogen Exchange between Arbuscular Mycorrhizal Fungi and Host Plant. Nitrate and ammonium are taken up by the fungal mycelium that is outside the host plant cell (extracellular mycelium) and converted to arginine. This amino acid is transferred to the mycelium within the host plant cell and broken down so that only the ammonium enters the plant.

were performed in which AM-colonized plants (and the appropriate negative controls lacking mycorrhizae) were treated with $^{15}NO_3^-$ and $^{15}NH_4^+$ (**figure 29.11**). These forms of nitrogen are incorporated into fungal tissue through the glutamine synthetase-glutamate synthase pathway (GS-GOGAT). Thus ^{15}N-containing glutamine (which is converted to arginine) is recovered. However, prior to transferring the nitrogen to host cells, intracellular fungal hyphae degrade the amino acids, transferring only the $^{15}NH_4^+$. Thus while the fungus provides its host with much needed ammonium, it retains the carbon skeleton it needs for nitrogen uptake. Synthesis of amino acids: Nitrogen assimilation (section 10.5)

In addition to these two most abundant mycorrhizal types, the orchidaceous mycorrhizae are of particular interest (figure 29.8*b*). Orchids are unusual plants in that many never produce chlorophyll, while others only do so after they have matured past the seedling stage. Therefore all orchids have an absolute dependence on their endomycorrhizal partners for at least part of their lives. Indeed, orchid seeds will not germinate unless first colonized by a basidiomycete orchid mycorrhiza. Because the host orchid cannot produce photosynthetically derived fixed carbon (or produces very little), unlike AM and ECM fungi, orchid mycorrhizal fungi are saprophytic. They must degrade organic matter to obtain carbon, which the orchids then also consume. In this case, the orchid functions as a parasite.

Depending on the environment of the plant, mycorrhizae can increase a plant's competitiveness. In wet environments they increase the availability of nutrients, especially phosphorus. In arid environments, where nutrients do not limit plant functioning to the same degree, the mycorrhizae aid in water uptake, allowing increased transpiration rates in comparison with nonmycorrhizal plants. These benefits have distinct energy costs for the plant in the form of photosynthate required to support the plant's "mycorrhizal habit." Based on the ubiquity of mycorrhizae, most plants are apparently willing to trade photosynthate—produced with the increased water acquisition—for water.

Bacteria are also associated with the mycorrhizal fungi. As the external hyphal network radiates out into the soil, a **mycorrhizosphere** is formed due to the flow of carbon from the plant into the mycorrhizal hyphal network and then into the surrounding soil. In addition, "mycorrhization helper bacteria" can play a role in the development of mycorrhizal relationships with ectomycorrhizal fungi. Bacterial symbionts also are found in the cytoplasm of AM fungi, as shown in **figure 29.12.** Such bacteria-like organisms (BLOs) appear to be related to *Burkholderia cepacia*. It has been suggested that these "trapped" bacteria contribute to the nitrogen metabolism of the plant-fungal complex by assisting with the synthesis of essential amino acids.

1. Describe the two-way relationship between mycorrhizal fungi and the plant host.
2. List three major differences between arbuscular mycorrhizae and ectomycorrhizae.
3. What is the function of the rhizomorph and Hartig's net?
4. Describe the uptake and transfer of ammonium by arbuscular mycorrhizae to the plant host. Why do you think only the ammonium is transferred?
5. What makes orchid mycorrhizae different from other mycorrhizal associations?
6. Propose two potential functions for mycorrhization helper bacteria.

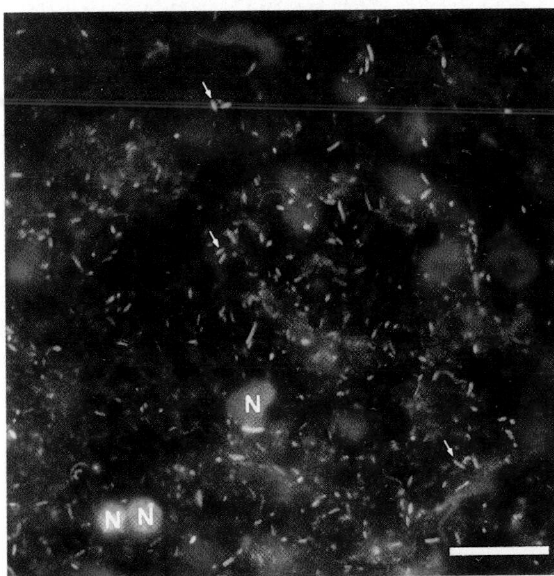

Figure 29.12 Mycorrhization Helper Bacteria. Stained bacterial endosymbionts in unfixed spores of the AM fungus *Gigaspora margarita*. Living bacteria fluoresce bright yellow-green; lipids and fungal nuclei (N) appear as diffuse masses. Bar = 7 μm.

Nitrogen fixation

The enzymatic conversion of gaseous nitrogen (N_2) to ammonia (NH_3) often occurs as part of a symbiotic relationship between bacteria and plants. These symbioses produce more than 100 million metric tons of fixed nitrogen annually and are a vital part of the global nitrogen cycle. In addition, symbiotic nitrogen fixation accounts for more than half of the nitrogen used in agriculture. The provision of fixed nitrogen enables the growth of host plants in soils that would otherwise be nitrogen limiting, and simultaneously reduces loss of nitrogen by denitrification and leaching. Indeed, with over 18,000 species, legumes are the most successful plants on Earth, probably due to their symbiotic relationships with nitrogen-fixing bacteria. For these reasons nitrogen fixation, particularly that by members of the genus *Rhizobium* and related α-proteobacteria in association with their leguminous host plants have been the subject of intense investigation.

The Rhizobia

Several microbial genera are able to form nitrogen-fixing nodules with legumes. These include the α-proteobacteria *Allorhizobium, Azorhizobium, Bradyrhizobium, Mesorhizobium, Sinorhizobium,* and *Rhizobium*. Collectively these genera are often called the **rhizobia.** Recently the phylogenetic diversity of the rhizobia has been extended by the discovery that the β-proteobacteria *Burkholderia caribensis* and *Ralstonia taiwanensis* also form nitrogen-fixing nodules on legumes. Here, the general process of nodulation is presented followed by molecular details that have been revealed largely though studies of the genus *Rhizobium*. Class *Alphaproteobacteria* (section 22.1); Class *Betaproteobacteria* (section 22.2)

Rhizobia live freely in the soil, but when they approach the plant root, they are assumed to be an alien invader. The plant responds with an **oxidative burst,** producing a mixture that can contain superoxide radicals, hydrogen peroxide, and N_2O. This redox-based oxidative burst, involving glutathione and homoglutathione, is critical for determining the fate of the infection process and influences whether further steps in the early infection process will occur. The rhizobia, if they are to be effective colonizers, must use antioxidant defenses to survive and continue the infection process. Only rhizobia and related genera with sufficient antioxidant abilities are able to proceed to the next steps in the infection process.

The plant roots also release **flavonoid inducer molecules** that stimulate rhizobial colonization of the root surfaces (**figure 29.13*a***). In response to this molecular message, rhizobia produce their own signaling compounds called **Nod factors.** The precise structure of individual Nod factors depends on the bacterial species, but all consist of four to five units of β-1,4 linked N-acetyl-D-glucosamine bearing an acyl chain at the nonreducing terminal residue, and a sulfate attached to the reducing end (figure 29.13*c*). Upon receipt of the Nod factor message, gene expression in the outer (epidermal) cells of the roots is altered so that the root hairs become deformed. In some cases the root hairs will curl to resemble a shepherd's crook, entrapping bacteria (figure 29.13*c,d*). In these regions, the plant cell wall is locally modified, the plant plasma membrane invaginates, and new plant material is laid down. These modifications lead to the development of a bacteria-filled, tubelike structure called the **infection thread** (29.13*e,f*). The infection thread grows toward the base of the root hair cell to a region called the primordium. Division of these root cells ultimately gives rise to the nodule. When bacteria are released from the infection thread into the primodium, they remain surrounded by a plant cell membrane called the peribacteroid membrane (figure 29.13*h*). It is here that each bacterial cell differentiates into the nitrogen-fixing form called a **bacteroid.** Bacteroids are terminally differentiated—they can neither divide nor revert back to the nondifferentiated state. Further growth and differentiation leads to the development of a structure called a **symbiosome.** Recall that the nitrogenase enzyme is very sensitive to oxygen. To help protect the nitrogenase, a protein called **leghemoglobin,** which binds to oxygen and helps maintain microaerobic conditions within the mature nodule, is produced (figure 29.13*i, j*). This protein is similar in structure to myo- and hemoglobins found in animals; however, it has a higher affinity for oxygen. Interestingly, the protein moiety is encoded by plant genes whereas the heme group is the product of bacterial genes. Synthesis of amino acids: Nitrogen fixation (section 10.5)

The symbiosomes within mature **root nodules** are the site of nitrogen fixation. Within these nodules, the differentiated bacteriods reduce atmospheric N_2 to ammonium, and in return they receive carbon and energy in the form of dicarboxylic acids from their host legume. It had long been thought the principal form of nitrogen that was transferred to the host plant was ammonium, but more recent evidence shows that a more complex cycling of amino acids occurs. Apparently, the plant provides amino acids to the bacteroids so that

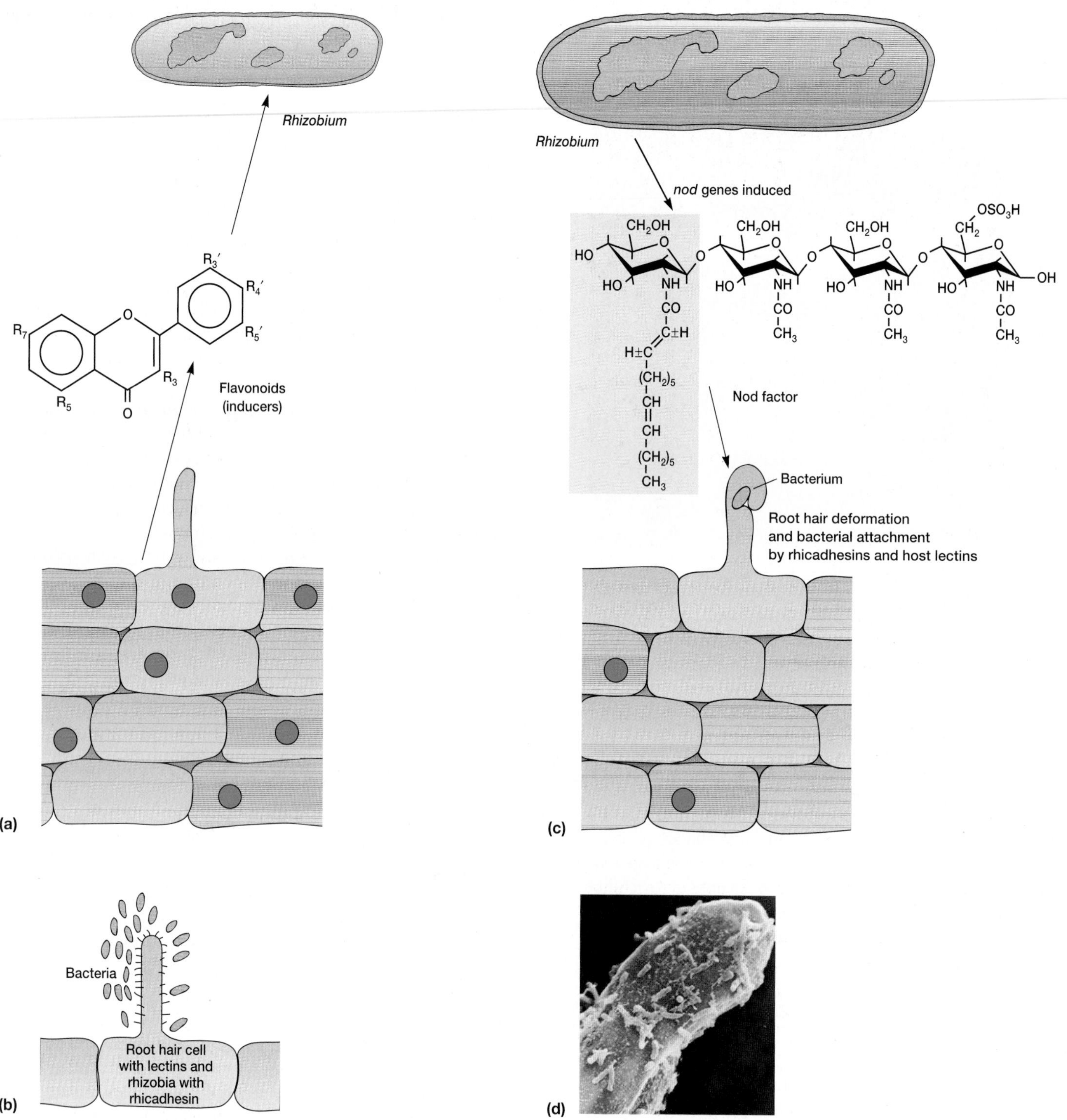

Figure 29.13 Root Nodule Formation by *Rhizobium*. Root nodule formation on legumes by *Rhizobium* is a complex process that produces the nitrogen-fixing symbiosis. **(a)** The plant root releases flavonoids that stimulate the production of various Nod metabolites by *Rhizobium*. There are many different Nod factors that control infection specificity. **(b)** Attachment of *Rhizobium* to root hairs involves specific bacterial proteins called rhicadhesins and host plant lectins that affect the pattern of attachment and *nod* gene expression. **(c)** Structure of a typical Nod factor that promotes root hair curling and plant cortical cell division. The bioactive portion (nonreducing *N*-fatty acyl glucosamine) is highlighted. These Nod factors enter root hairs and migrate to their nuclei. **(d)** A plant root hair covered with *Rhizobium* and undergoing curling. **(e)** Initiation of bacterial penetration into the root hair cell and infection thread growth coordinated by the plant nucleus "N." **(f)** A branched infection thread shown in an electron micrograph. **(g)** Cell-to-cell spread of *Rhizobium* through transcellular infection threads followed by release of rhizobia and infection of host cells. **(h)** Formation of bacteroids surrounded by plant-derived peribacteroid membranes and differentiation of bacteroids into nitrogen-fixing symbiosomes. The bacteria change morphologically and enlarge around 7 to 10 times in volume. The symbiosome contains the nitrogen-fixing bacteroid, a peribacteroid space, and the peribacteroid membrane. **(i)** Light micrograph of two nodules that develop by cell division (×5). This section is oriented to show the nodules in longitudinal axis and the root in cross section. **(j)** *Sinorhizobium meliloti* nitrogen-fixing nodules on roots of white sweet clover *(Melilotus alba)*.

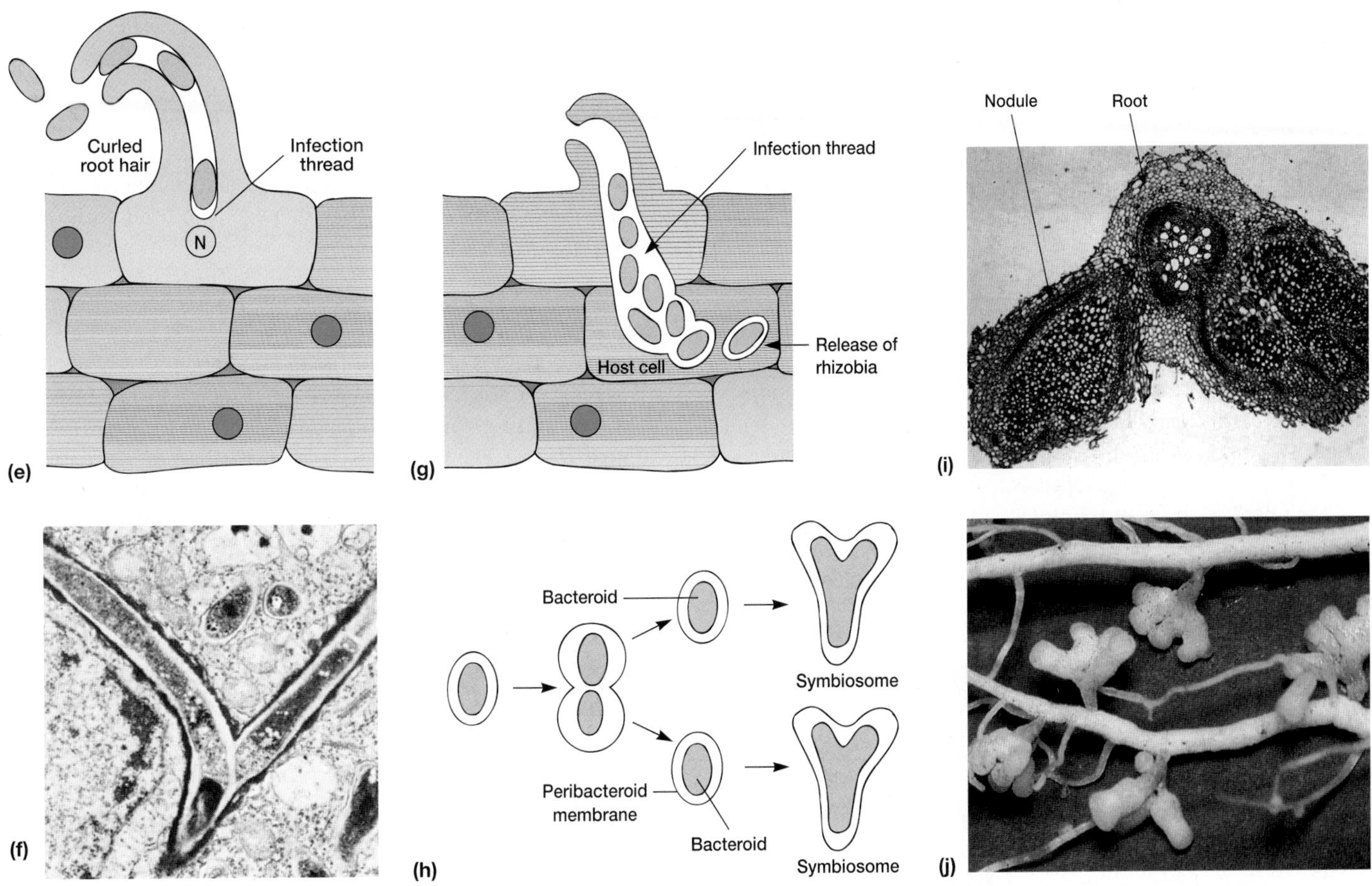

Figure 29.13 *(Continued).*

they do not need to assimilate ammonium. In return, the bacteroids shuttle amino acids (which bear the newly fixed nitrogen) back to the plant. This creates an interdependent relationship, providing selective pressure for the evolution of mutualism.

The genes essential for nodulation (*nod, nol,* and *noe* genes) and nitrogen fixation (*nif* and *fix* genes) show homology among the α- and β-proteobacterial rhizobia. However, the arrangement of these genes on the bacterial genome is not conserved. In some cases, these genes are clustered; in others they are not. For instance, in *Rhizobium meliloti,* which forms nodules on alfalfa, clusters of symbiosis genes are encoded on huge megaplasmids that are over a million base pairs long. In *Bradyrhizobium japonicum* (which forms a symbiosis with soybean plants), the *nod* genes are located on the chromosome. Evidence for transfer of a symbiosis gene island has been demonstrated for *Mesorhizobium loti,* which infects the model legume *Lotus japonicus* (bird's-foot trefoil). This genomic diversity has led to taxonomic confusion in the reclassification of the rhizobia that were previously identified only by phenotypic features.

Interestingly, plants use the same initial response to the establishment of productive nitrogen-fixing symbionts as they use in establishing arbuscular mycorrhizae associations. For example, the plant genetic locus called DMI (*d*oesn't *m*ake *i*nfections—recall that genes are frequently named for their mutant phenotype) is induced upon the initial colonization of rhizobia or arbuscular mycorrhizal fungi. Thus some plant mutants unable to form nodules are also unable to interact with arbuscular mycorrhizae.

The molecular mechanisms by which both the legume host and the rhizobial symbionts establish productive nitrogen-fixing bacteriods within nodules continues to be an intense area of research. A major goal of biotechnology is to introduce nitrogen-fixation genes into plants that do not normally form such associations. It has been possible to produce modified lateral roots on nonlegumes such as rice, wheat, and oilseed rape; the roots are invaded by nitrogen-fixing bacteria. It appears that infection begins with bacterial attachment to the root tips. Although these modified root structures have not yet been found to fix useful amounts of nitrogen, they do enhance rice production and intense work is expected to continue in this area.

1. List several bacteria that are considered rhizobia.
2. Describe the communication system between a rhizobia bacterium and its legume host.
3. What is the function of the infection thread? Why do you think it is important that the bacteria do not enter plant cell cytoplasm until they reach the primordium?

4. What does the term "terminally differentiated" mean? Can you think of other cells that are also terminally differentiated?
5. How is leghemoglobin made and what is its function?

Actinorhizae

Another example of symbiotic nitrogen fixation occurs between the actinomycete *Frankia* and eight nonleguminous host plant families (**table 29.7**). These bacterial associations with plant roots are called **actinorhizae** or actinorhizal relationships (**figure 29.14**). *Frankia* fixes nitrogen and is important, particularly in trees and shrubs. As examples, these associations occur in areas where Douglas fir forests have been clear-cut, and in bog and heath environments where bayberries and alders are dominant. The nodules of some plants (*Alnus, Ceanothus*) are as large as baseballs. The nodules of *Casuarina* (Australian pine) approach soccer ball size.

Members of the genus *Frankia* are slow-growing and were impossible to culture apart from the plant until 1978. Since then, this actinomycete has been grown on specialized media supplemented with metabolic intermediates such as pyruvate. Major advances in understanding the physiology, genetics, and molecular biology of these microorganisms are now taking place.

As in all plant-microbe associations, the actinorhizal relationship costs the plant energy. However, the plant benefits and is better able to compete in nature. This association provides a unique opportunity for microbial management to improve plant growth processes.

Stem-Nodulating Rhizobia

Other associations of nitrogen-fixing microorganisms with plants also occur. A particularly interesting association is caused by **stem-nodulating rhizobia,** found primarily in tropical legumes (**figure 29.15**). These nodules form at the base of adventitious roots branching out of the stem just above the soil surface and, because they contain oxygen-producing photosynthetic tissues, they have unique mechanisms to protect the oxygen-sensitive nitrogen fixation enzymes. One microorganism that forms such

Table 29.7 Nonleguminous Nodule-Bearing Plants with *Frankia* Symbioses[a]

Family	Genus	*Frankia* Isolated?	Isolated Strains Infective?
Casuarinaceae	*Allocasuarina*	+	+
	Casuarina	+	+
	Ceuthostoma	–	–
	Gymnostoma	+	+
Coriariaceae	*Coriaria*	+	+
Datiscaceae	*Datisca*	+	–
Betulaceae	*Alnus*	+	+
Myricaceae	*Comptonia*	+	+
	Myrica	+	+
Elaeagnaceae	*Elaeagnus*	+	+
	Hippophae	+	+
	Shepherdia	+	+
Rhamnaceae	*Adolphia*	–	–
	Ceanothus	+	–
	Colletia	+	+
	Discaria	+	+
	Kentrothamnus	–	–
	Retanilla	+	+
	Talguenea	+	+
	Trevoa	+	+
Rosaceae	*Cercocarpus*	+	–
	Chaemabatia	–	–
	Cowania	+	–
	Dryas	–	–
	Purshia	+	–

Source: Data from Dr. D. Baker, MDS Panlabs and Dr. J. Dawson, University of Illinois. Personal communications.

[a]*Frankia* isolation from nodules and ability of these isolated strains to initiate nodulation are also noted.

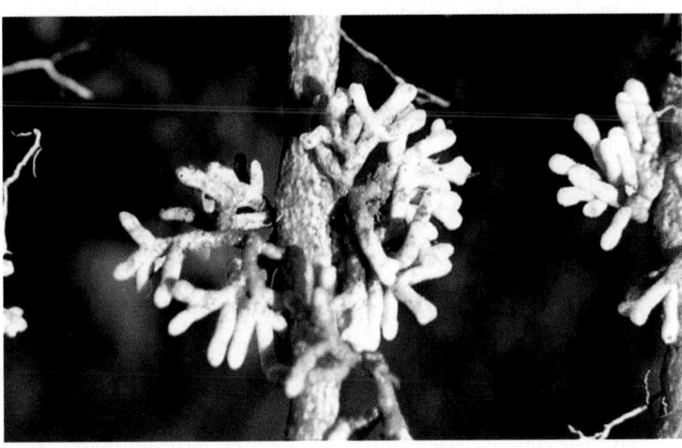

Figure 29.14 Actinorhizae. *Frankia*-induced actinorhizal nodules in *Ceanothus* (Buckbrush).

Figure 29.15 Stem-Nodulating Rhizobia. Nitrogen-fixing microorganisms also can form nodules on stems of some tropical legumes. Nodules formed on the stem of a tropical legume by a stem-nodulating *Rhizobium*.

root and stem nodules is *Azorhizobium caulinodans,* which forms nodules on the tropical legume *Sesbania rostrata*. It has been shown that some of these stem-nodulating rhizobia are photosynthetic. Thus they can obtain their energy not only from the plant's organic compounds, but also from the light.

Fungal and Bacterial Endophytes

Specialized fungi and bacteria can live within some plants as endophytes and may be beneficial. Specialized clavicipitaceous fungi form systemic fungal infections in which the endophyte grows between the plant's cortex cells (**figure 29.16**). Plants infected with these endophytes may be less susceptible to attack by various chewing insects due to the production of alkaloids, a form of "chemical defense." *Rhizobium leguminosarum* bv. *trifolii,* which forms a nitrogen-fixing symbiosis with clover, can

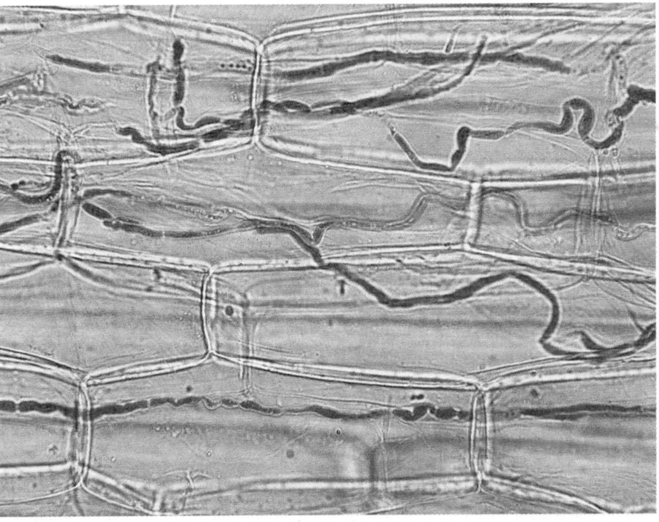

Figure 29.16 Fungal Endophytes. Fungi can invade the upper parts of some plants. A fungal endophyte growing inside the leaf sheath of a grass, tall fescue, is shown.

Figure 29.17 Parasitic Castration of Plants by Endophytic Fungi. Stroma of the fungus *Atkinsonella hypoxylon* infecting *Danthonia compressa* and causing abortion of the terminal spikelets.

also form a natural endophytic association with rice roots. This interaction, first observed in the Nile Delta, is supported by the rotation of clover with the rice. The association promotes rice root and shoot growth, resulting in increased rice grain yield at maturity. The rice/*Rhizobium*/clover association provides approximately one-quarter of the nitrogen needed by the rice crop.

Not all endophytic relationships are mutualistic; some are parasitic. Parasitic fungal endophytes can actually reduce the genetic variability of the plant by sterilizing their host (**figure 29.17**). This "parasitic castration of plants" by systemic fungi, which promotes increased fungal spread in a less variable plant community, is suggested to be of major importance in the co-evolution of plants and fungi.

Endophytic bacteria have been discovered in sugar cane, cotton, pears, and potatoes. Some are plant pathogens that can survive for extended periods in a quiescent state. The majority have no known positive or deleterious effect on plant growth or development. The use of these bacteria as microbial delivery systems in agriculture is a current topic in agricultural biotechnology. It also has been possible to establish *Azorhizobium,* a root and stem nodulating bacterium of *Sesbana rostrata,* in the lateral roots of wheat plants, leading to a possible increased plant dry weight and nitrogen content.

1. What is the major contribution of *Frankia* to plant functioning? Which types of plants are infected?
2. What are stem-nodulating rhizobia?
3. Describe some possible effects of endophytic fungi on plants.

Agrobacterium

Clearly, some microbe-plant interactions are beneficial for both partners. However, others involve microbial pathogens that harm or even kill their host. *Agrobacterium tumefaciens* is an α-proteobacterium that has been studied intensely for several decades. Initially, research focused on how this microbe causes **crown gall disease,** which results in the formation of tumorlike growths in a wide variety of plants (**figure 29.18**). Within the last 15 years or so, however, *A. tumefaciens* has become one of biotechnology's most important tools. The molecular genetics by which this pathogen infects its host is the basis for plant genetic engineering. Techniques & Applications 14.2: Plant tumors and nature's genetic engineer; Class *Alphaproteobacteria* (section 22.1)

The genes for plant infection and virulence are encoded on an *A. tumefaciens* plasmid, called the **Ti (*t*umor-*i*nducing) plasmid.** These genes include 21 *vir* genes (*vir* stands for *vir*ulence), which are found in six separate operons. Two of these genes, *virD1* and *virD2,* encode proteins that cleave a separate region of the Ti plasmid, called the **T DNA.** After excision, this T DNA fragment is integrated into the host plant's genome. Once incorporated into a plant cell's genome, T DNA directs the overproduction of phytohormones that cause unregulated growth and reproduction of plant cells, thereby generating a tumor or gall in the plant.

Figure 29.18 ***Agrobacterium.*** *Agrobacterium*-caused tumor on a *Kalanchoe* sp. plant.

The *vir* genes are not expressed when *A. tumefaciens* is living saprotrophically in the soil. Instead, they are induced by the presence of plant phenolics and monosaccharides present in an acidic (pH 5.2–5.7) and cool (below 30°C) environment (**figure 29.19*a***). The microbe usually infects its host through a wound. Upon reception of the plant signal, a two-component signal transduction system is activated: VirA is a sensor kinase that, in the presence of a phenolic signal, phosphorylates the response regulator VirG. Activated VirG then induces transcription of the other *vir* genes. This enables the bacterial cell to become adequately positioned relative to the plant cell, at which point the *virB* operon expresses the apparatus that will transfer the T DNA. This transfer is similar to bacterial conjugation and involves a type IV secretion system. After VirD1 and VirD2 excise the T DNA from the Ti plasmid, the T DNA, with the VirD2 protein attached to the 5′ end, is delivered to the plant cell cytoplasm. The protein VirE2 is also transferred and, together with VirD2, the T DNA is shepherded to the plant cell nucleus where it is integrated into the host's genome. Here the T DNA has two specific functions. First, it directs the host cell to overproduce phytohormones that cause tumor formation. Second, it stimulates the plant to produce special amino acid and sugar derivatives called opines (figure 29.19*b*). Opines are not metabolized by the plant but *A. tumefaciens* is attracted to opines; chemotaxis of bacteria from the surrounding soil population will further advance the infection because the bacterium can use opines as sources of carbon, energy, nitrogen, and phosphorus. Regulation of transcription initiation: Two-component signal transduction systems (section 12.2), Bacterial conjugation (section 13.7)

Other Plant Pathogens

In addition to *Agrobacterium,* a variety of other bacteria cause an array of spots, blights, wilts, rots, cankers, and galls, as shown in **table 29.8.** The soft rots caused by the enterobacteria *Erwinia chrysanthemi* and *E. carotovora* have significant economic impact. These bacteria digest plant tissue by producing extracellular enzymes that degrade pectin, cellulose, and proteins. These exoenzymes are secreted by type II and type I secretion systems; mutants lacking these secretion systems are no longer pathogenic. Similarly, proteobacteria belonging to the genera *Ralstonia, Pseudomonas, Pantoea,* and *Xanthomonas* rely on type III secretion systems to deliver virulence proteins. Although these microbes cause a diverse collection of plant diseases, all colonize the spaces between plant cells to kill their hosts. Another group of important plant pathogens includes the wall-less phytoplasms that infect vegetable and fruit crops such as sweet potatoes, corn, and citrus. Protein secretion in procaryotes (section 3.8)

As discussed in chapters 25 and 26, respectively, protists and fungi can be devastating plant pathogens. Examples are the fungus *Puccinia graminis,* which causes wheat rust, and the öomycete

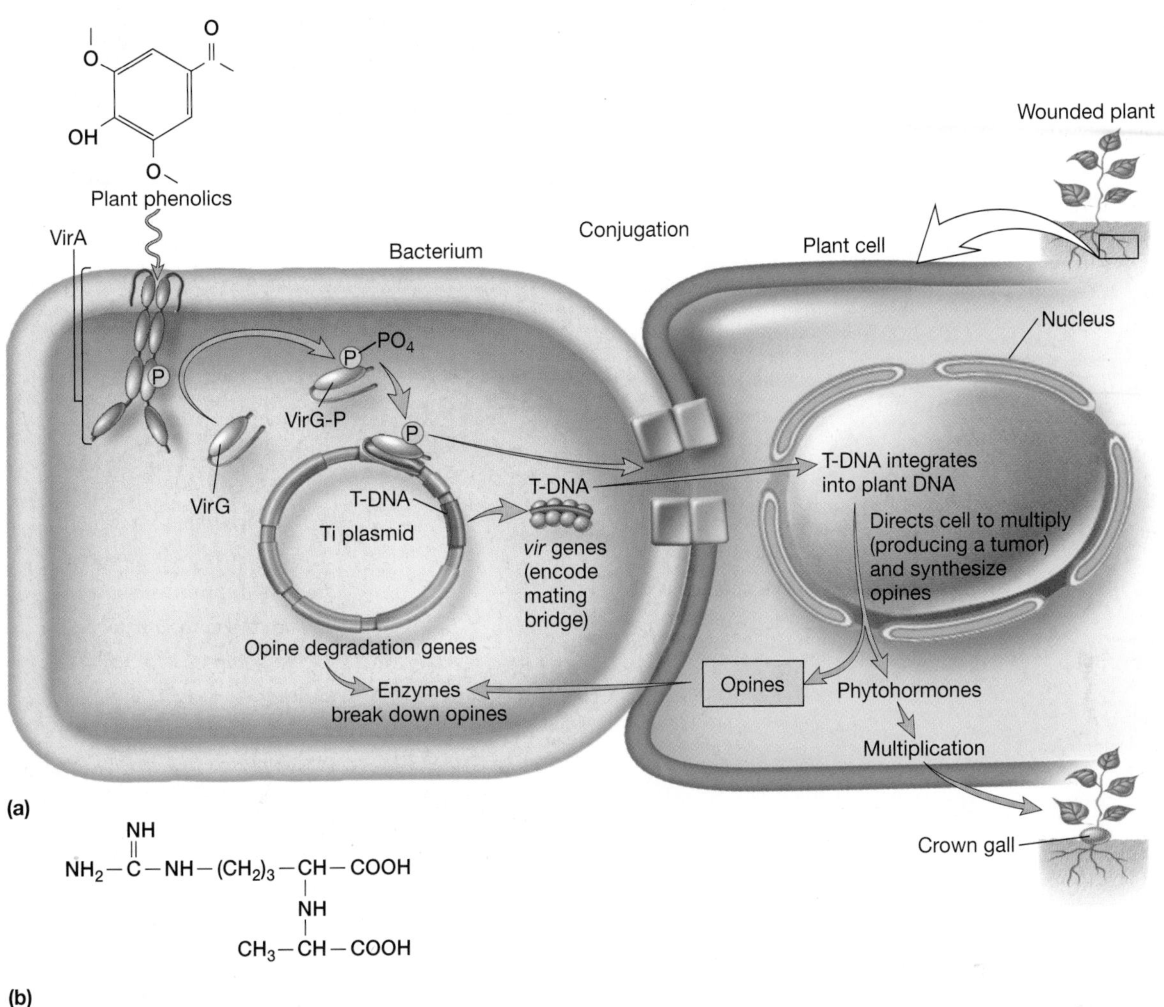

Figure 29.19 Functions of Genes Carried on the *Agrobacterium* Ti Plasmid. **(a)** Genes carried on the Ti plasmid of *Agrobacterium* control tumor formation by a two-component regulatory system that stimulates formation of the mating bridge and excision of the T-DNA. The T-DNA is moved by transfer genes, which lead to integration of the T-DNA into the plant nucleus. T-DNA encodes plant hormones that cause the plant cells to divide, producing the tumor. The tumor cells produce opines (shown in **b**) that can serve as a carbon source for the infecting *Agrobacterium*. Ultimately a crown gall is formed on the stem of the wounded plant above the soil surface.

Phytophthora infestans, which was responsible for the Irish potato famine.

A wide range of viruses and virusoids infect plants, as described in sections 18.6 and 18.9 respectively, and are of worldwide importance in terms of plant disease and economic losses. These include tobacco mosaic virus (TMV), the first virus to be characterized. A virus of particular interest in terms of plant-pathogen interactions is one member of the hypoviruses that infects the fungus *Cryphonectria parasitica,* the cause of chestnut blight. Based on pioneering studies carried out in Italy and France, workers in Connecticut and West Virginia noted that if they infected the fungus with the hypovirus, the rate and occurrence of blight was decreased. They are hoping to treat trees with the less lethal virus strains and eventually transform the indigenous lethal strains of *Cryphonectria* into more benign fungi.

Tripartite and Tetrapartite Associations

An additional set of interactions occurs when the same plant develops relationships with two or three different types of microorganisms. These more complex interactions are important to a variety of plant types in both temperate and tropical agricultural systems. First described in 1896, these symbiotic associations involve the interaction of the plant-associated microorganisms with each other and the host plant. Several **tripartite associations** are known to occur: the plant plus (1) endomycorrhizae plus rhizobia, including *Rhizobium* and *Bradyrhizobium;* (2) endomycorrhizae and actinorhizae; and (3) ectomycorrhizae and actinorhizae. Nodulated and mycorrhizal plants are better suited for coping with nutrient-deficient environments. **Tetrapartite associations** also occur. These consist of endomycorrhizae, ectomycorrhizae, *Frankia,* and the host plant. These complex associations, in spite

Table 29.8 Major Plant Diseases Caused by Bacteria

Symptoms	Examples	Pathogen
Spots and blights	Wildfire (tobacco)	*Pseudomonas syringae* pv.[a] *tabaci*
	Haloblight (bean)	*P. syringae* pv. *phaseolica*
	Citrus blast	*P. syringae* pv. *syringae*
	Leaf spot (bean)	*P. syringae* pv. *syringae*
	Blight (rice)	*Xanthomonas campestris* pv. *oryzae*
	Blight (cereals)	*X. campestris* pv. *translucens*
	Spot (tomato, pepper)	*X. campestris* pv. *vesicatoria*
	Ring rot (potato)	*Clavibacter michiganensis* pv. *sepedonicum*
Vascular wilts	Wilt (tomato)	*C. michiganensis* pv. *michiganensis*
	Stewart's wilt (corn)	*Erwina stewartii*
	Fire blight (apples)	*E. amylovora*
	Moko disease (banana)	*P. solanacearum*
Soft rots	Black rot (crucifers)	*X. campestris* pv. *campestris*
	Soft rots (numerous)	*E. carotovora* pv. *carotovora*
	Black leg (potato)	*E. carotovora* pv. *atroseptica*
	Pink eye (potato)	*P. marginalis*
	Sour skin (onion)	*P. cepacia*
Canker	Canker (stone fruit)	*P. syringae* pv. *syringae*
	Canker (citrus)	*X. campestris* pv. *citri*
Galls	Crown galls (numerous)	*Agrobacterium tumefaciens*
	Hairy root	*A. rhizogenes*
	Olive knot	*P. syringae* pv. *savastonoi*

Source: From J. W. Lengler, G. Drews, H. G. Schlegel. 1999. *Biology of the prokaryotes*, Blackwell Science, Malden, Mass., table 34.4.

[a]pv., pathover, a variety of microorganisms with phytopathogenic properties.

of their additional energy costs, provide important benefits for the plant.

1. Discuss the nature and importance of the Ti plasmid.
2. What functions do the members of the two-component system play in infection of a plant by *Agrobacterium?* What are the roles of phenolics and opines in this infection process?
3. What kinds of exoenzymes are produced by some plant pathogens.
4. How are plant pathologists attempting to control chestnut blight?
5. What are tripartite and tetrapartite associations?

29.6 Soil Microorganisms and the Atmosphere

Soil microorganisms, like marine microbes, can have major effects on global fluxes of a variety of gases. These gases can be considered as those that are "relatively stable" and those that are "reactive gases." Relatively stable gases that are influenced by microbial activities include carbon dioxide, nitrous oxide, nitric oxide, and methane. Microorganisms also contribute to the flow of reactive gases such as ammonia, hydrogen sulfide, and dimethylsulfide. These reactive gases tend to be produced in more waterlogged environments. Biogeochemical cycling (section 27.2)

Atmospheric gases such as carbon dioxide, nitrous oxide, nitric oxide, and methane are greenhouse gases**.** The production and consumption of greenhouse gases can be influenced by a range of human activities. These include plant fertilization and automobile use, conversion of soils to agricultural use, and landfills.

Just as marine primary production helps prevent an even more rapid increase in atmospheric CO_2, terrestrial forests are a tremendous CO_2 "sink." During the 1990s, terrestial ecosystems sequestered about three gigatons of carbon per year (a gigaton is one billion metric tons—that's a lot of carbon). There has been much speculation regarding the rate of plant growth in response to increasing atmospheric CO_2. Some believe that plants will assimilate the extra CO_2, resulting in an accelerated average rate of growth. This optimistic view holds that increases in CO_2 will be buffered by forest ecosystems (and apparently ignores the current rate of deforestation). An international team of scientists recently addressed this question by growing a variety of trees for several years under higher concentrations of CO_2. Although they found that indeed there was plant growth stimulation, it was coupled to an increase in respiration by soil microbes. Recall that as respiration rates increase, so does the volume of CO_2 released. If these findings can be extrapolated to a world where CO_2 levels continue to increase, forest ecosystems may sequester less carbon than predicted.

Methane is a greenhouse gas of increasing concern that can be derived from a variety of sources. These include ruminants, rice paddies, and landfills. Landfills, especially, can release methane to the atmosphere over longer terms (decades, centuries). Another interesting source of atmospheric methane is microorganisms that inhabit the gut of wood-eating termites. As noted in **Microbial Diversity & Ecology 29.3,** with increased deforestation and the accumulation of plant residues, populations of termites (and termite gut-inhabiting methanogenic microorganisms) are increasing.

Methane levels are influenced by microorganisms and their functioning in the environment. Well-drained, oxic soils are capable of methane oxidation by **methanotrophs** (aerobic bacteria that oxidize methane), whereas in water-saturated sites of soils and in bogs and wetlands, methane may be produced faster than it can be used by methanotrophs. Based on analyses of gas bubbles in glacier ice cores, the levels of methane in the atmosphere remained essentially constant until about 400 years ago. Since then the methane level has increased 2.5-fold to the present level of 1.7 parts per million (ppm) by volume. Considering these trends, there is a worldwide interest in understanding the factors that control methane synthesis and use by microorganisms.

The processes of methane synthesis and use occur on a variety of scales in upland soils and in wetlands, as shown in **figure 29.20.** In nominally oxic upland soils, there may be anoxic hot spots (water-saturated local areas) where methane is produced. If these areas are surrounded by oxic soils, methanotrophs degrade most of the methane before it can be released to the atmosphere. In more waterlogged areas such as lowland soils, methanogenesis can proceed and there is less opportunity for methanotrophs to function. In spite of this, most of the methane will be degraded. Aquatic plants transport oxygen to their rhizospheres and thus facilitate methane oxidation in these localized oxic hot spots. Methane oxidation also is sensitive to nitrogen-containing fertilizers and increases in atmospheric CO_2 levels. The balance between methane synthesis and degradation is very important. It is estimated that soils of all types provide 60% of the total atmospheric methane; wet areas, water-saturated hot spots, and rice paddies are particularly significant contributors.

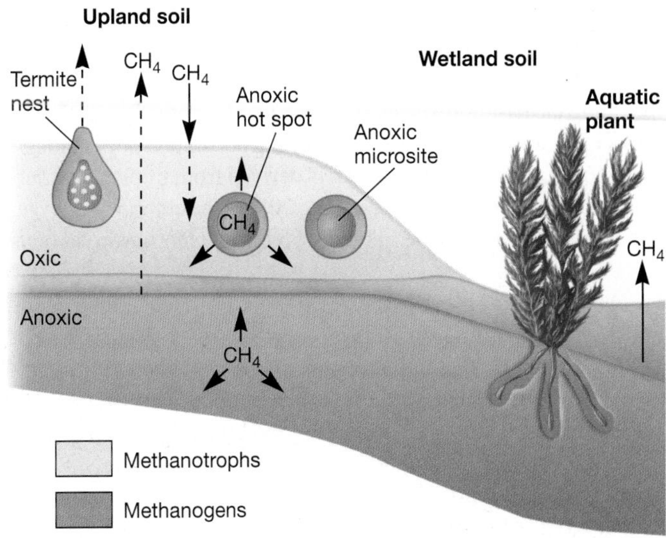

Figure 29.20 Methane Production and Use in Soils. Methane production and degradation can occur in closely located oxic and anoxic zones. In upland soils, methane synthesis can take place in localized anoxic hot spots and termite nests. Much of this methane can be degraded in the surrounding oxic soils. In wetland areas, methane production dominates in the waterlogged areas, and methane has a greater chance of being released to the atmosphere. If aquatic plants are present and translocate oxygen into anoxic zones, localized oxic hot spots in the rhizosphere for methane oxidation are created.

Microbial Diversity & Ecology

29.3 Soils, Termites, Intestinal Microbes, and Atmospheric Methane

Termites are important components of tropical ecosystems, where their use of cellulose plant materials allows rapid—sometimes too rapid—recycling of plant materials. Termites harbor significant populations of archaea that use products of cellulose digestion, including CO_2 and hydrogen, to produce methane.

Termites occur on 2/3 of the Earth's land surface, and, based on laboratory studies, 0.77% of the carbon ingested by termites can be released as methane. In tropical wet savannas and cultivated areas, termite populations are increasing rapidly. This increase is being accelerated by the destruction of tropical forests, which results in the accumulation of dead plant materials on the soil surface. This provides an ideal environment for the growth of termites. Termites are estimated to be contributing annually at least 1.5×10^{14} grams of methane, together with hydrogen and CO_2, to the atmosphere. This is believed to be contributing to measurable increases in the atmospheric methane level. Thus unseen termites and their associated gut microorganisms may be affecting global warming.

Greenhouse gases are also produced by fungi in the process of woody plant decomposition. Large amounts of chloromethane (CH_3Cl), an important greenhouse gas, are produced by many fungi, including the basidiomycetes *Phellinus* and *Inonotus*. The global input of CH_3Cl to the atmosphere from plant decomposition is thought to be 160,000 tons, 75% of which is derived from tropical and subtropical soils. It is estimated that 15 to 20% of the chlorine-catalyzed ozone destruction is due to naturally produced chlorohydrocarbons. Characteristics of fungal divisions: *Basidiomycota* (section 26.6)

As noted in **table 29.9,** terrestrial environments, the oceans, and biomass burning are all important sources of atmospheric chloromethane. Large amounts also have been detected in some basidiomycetes. These include concentrations of 74 to 2,400 mg/kg in the basidia of some agarics and bracket fungi. It is of interest that the maximum permissible chlorophenol levels in soils are only 1 to 10 mg/kg.

Addition of nitrogen-containing fertilizers also affects atmospheric gas exchange processes in a soil. Nitrogen additions stimulate the production of the nitrification intermediates NO and N_2O, which are critical greenhouse gases. NO also appears to be required for *Nitrosomonas eutropha* to carry out nitrification. The oxidation of NH_4^+ involves the formation of hydroxylamine (NH_2OH) and NO as intermediates. NO reacts with oxygen to give NO_2, which then can repeat the process in this reaction:

$$NH_4^+ + NO_2 \rightarrow NH_2OH + NO$$
$$2NO + O_2 \rightarrow NO_2 \text{ (nitrogen dioxide)}$$

In this sequence molecular oxygen does not react with NH_4^+ but with NO. If NO is absent the reaction will not proceed.

Soil microorganisms also influence the atmosphere by degrading airborne pollutants such as methane, hydrogen, CO, benzene, trichloroethylene (TCE), and formaldehyde. They can substantially improve the air in closed buildings (**Techniques & Applications 29.4**). Although soil microorganisms cannot completely eliminate air-borne pollutants, they can decrease these to equilibrium levels of approximately 1 to 2 ppm.

Microbes also respond to greenhouse gases. In fact, scientists have been questioning how global warming may change patterns of infectious disease outbreaks in humans and other animals. Recently, a significant clue was provided by ecologists studying the extinction of 67% of the 110 species of harlequin frogs (*Atelopus*) native to tropical America, which has occurred over the last 20 years (**figure 29.21**). They have strong evidence to support the hypothesis that these frogs have succumbed to a pathogenic chytrid fungus (*Batrachochytrium dendrobatidis*) whose range has expanded in response to warmer temperatures. This is not the first account of a pathogen responding to climate change. Pine blister rust, caused by the fungus *Cronartium ribicola,* is on the rise in mountainous regions of North America because its vector, the mountain pine beetle (*Dendroctonus poderosae*), is now able to complete its life cycle in one, rather than two, years because of warmer temperatures. Epidemiologists are currently monitoring the geography of infectious disease outbreaks in an effort to model potential patterns on a warmer planet. The *Fungi* (chapter 26)

Cyanide is another chemical of widespread concern produced by fungi, especially by basidiomycetes and ascomycetes. This cyanide may be derived from the S-methyl group of L- and D-methionine, or its production may be linked to methyl benzoate synthesis. The best studied cyanide-producing fungi are *Marasmius oreades* (the cause of fairy ring disease) and the snow mold basid-

Table 29.9 Global Emissions of Chloromethane

Source	Inputs to Atmosphere (10^5 tons/year)[a]
Natural Sources	
Terrestrial processes	0–20
Ocean fluxes	3–20
Biomass burning	4–14
Anthropogenic Sources	0–3

Source: From R. Watling and D. B. Harper. 1998. *Mycol. Res.* 102(7):769–87.

[a]Estimated atmospheric inputs from natural and anthropogenic sources.

Techniques & Applications

29.4 Keeping Inside Air Fresh with Soil Microorganisms

A major problem in the development of more energy-efficient homes and office buildings is the potential effect of such closed environments on human health. With many people spending much of their lives in such enclosed environments, the "sick building syndrome" is an increasing concern. While saving energy, these "sick buildings" have higher levels of many volatile compounds, including benzene, trichloroethylene (TCE), formaldehyde, phenolics, and solvents. These are released from rugs, furniture, plastic flooring, paints, and office machines such as photocopiers and printers. An important but still largely unappreciated means of improving the air in such "sick buildings" is through plants and their associated soil microorganisms. Plants not only produce oxygen, but the soil microorganisms degrade many airborne pollutants. It is recommended that one plant be used per 100 square feet of living area. As noted by B. C. Wolverton, "The ultimate solution to the indoor air pollution problem must involve plants, the plant soils and their associated microorganisms." Soil microorganisms, especially in association with plants, can help keep air in closed environments fresher and more healthful (plants are also nice to look at).

Figure 29.21 Chytrid Fungi Appear to Be Responsible for the Extinction of Many Species of Harlequin Frogs. This species of Panamanian golden frog can still be seen but many tree frog species have been eliminated as result of fungal infection. The fungus *(Batrachochytrium dendrobatidis)* has been able to expand its range in response to warmer temperatures.

iomycetes. Some bacteria also produce cyanide. Cyanide synthesis involves the oxidative decarboxylation of glycine, which is stimulated by methionine or other methyl-group donors in this reaction:

$$NH_2CH_2COOH \rightarrow HCN + CO_2 + 4[H]$$

Cyanide can inhibit respiration. It also can serve as a carbon and nitrogen source for microorganisms, including cyanogenic fungi such as *Marasmius* and *Pholiota,* and some actinomycetes. This illustrates the adaptability of microorganisms in the use of a nominally toxic metabolic product.

1. List some greenhouse gases. Discuss their origins.
2. Discuss the possible role of forests in the control of CO_2.
3. What microbial processes occur in soils to both produce and degrade methane?
4. Describe factors that might lead to the formation of localized hot spots for the production and consumption of greenhouse gases.
5. Describe the role of fungi in cycling chloromethanes and cyanide.

29.7 The Subsurface Biosphere

For many years it was thought that life could exist only in the thin veneer on Earth's surface and that any microbes recovered from sediments hundreds of meters deep were contaminants obtained during sampling. This view was drastically altered in the 1980s when the U.S. Department of Energy started looking for novel ways to clean up toxic waste. The agency began funding studies that applied modern technologies to sample the deep **subsurface biosphere.** Subsequent reports of microbes at great depth gained credibility and international teams of geologists and microbiologists have since recovered microbes from thousands of meters below Earth's surface. The application of culture-independent approaches to quantify the numbers and diversity of microbes has revealed that not only are subsurface microbes present, but that they constitute about *one-third* of Earth's living biomass. This realization has made deep subsurface microbiology an exciting and dynamic field.

Microbial processes take place in different subsurface regions, including (1) the shallow subsurface where water flowing from the surface moves below the plant root zone; (2) subsurface regions where organic matter, originating from the Earth's surface in times past, has been transformed by chemical and biological processes to yield coal (from land plants), kerogens (from marine and freshwater microorganisms), and oil and gas; and (3) zones where methane is being synthesized as a result of microbial activity.

In the shallow subsurface, surface waters often move through **aquifers,** porous geological structures below the plant root zone. In a pristine system with an oxic surface zone, the electron acceptors used in catabolism are distributed from the most oxidized and energetically favorable (oxygen) near the surface to the least favorable (in which CO_2 is used in methanogenesis) in lower zones (**figure 29.22**).

In subsurface regions where organic matter from the Earth's surface has been buried and processed by thermal and possibly biological processes, kerogen and coals break down to yield gas and oil (**figure 29.23**). After their generation, these mobile products, predominantly hydrocarbons, move upward into the more porous geological structures where microorganisms can be active. Chemical signature molecules from plant and microbial biomass are present in these petroleum hydrocarbons.

Below these zones lie vast regions where methane is present in geological structures (**figure 29.24**); this methane is continuously being released to the overlying strata. Based on studies of stable carbon isotopes, in the "biogenic" zone, methane has less

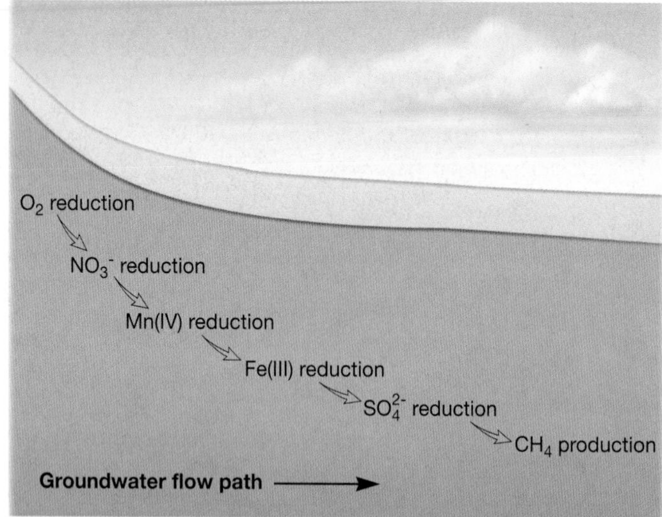

Figure 20.22 The Shallow Subsurface Biosphere. The shallow subsurface, as in a stable sediment, showing the distribution with depth of electron acceptors that can occur in an oxic pristine aquifer. In oxic sediments, the electron acceptors will be distributed with the most energetically favorable (oxygen) near the surface and the least energetically favorable at the lower zones of the geological structure. *Source: Lovley, D. K., 1991. Dissimilatory Fe (III) and Mn (IV) reduction. Microbiol. Rev. 55:259–87.*

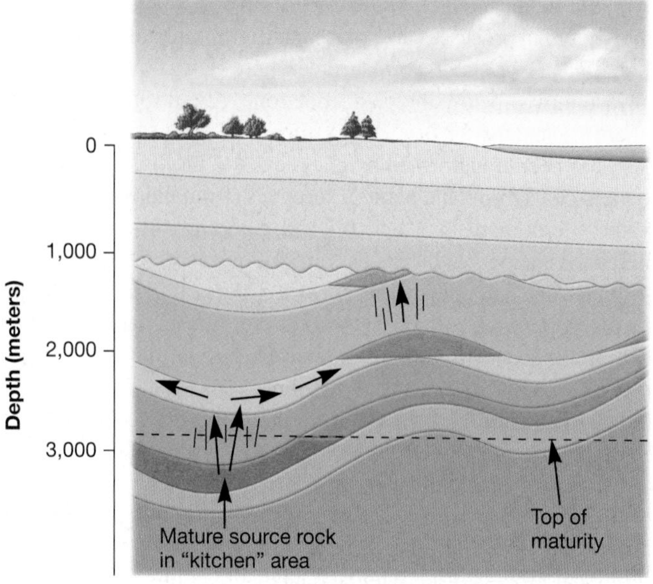

Figure 20.23 The Oil and Gas Region of the Subsurface. Organic matter, originating from the Earth's surface, is transformed to oil, gas, and coal by chemical, thermal, and biological processes. Above the higher-temperature "kitchen" area, where chemical changes occur, microorganisms can contribute to the processing of these organic materials. Hydrocarbons migrate through porous strata and fractures, finally accumulating in porous, overlying geological structures. Lines indicate fractures, and arrows indicate hydrocarbon flows.

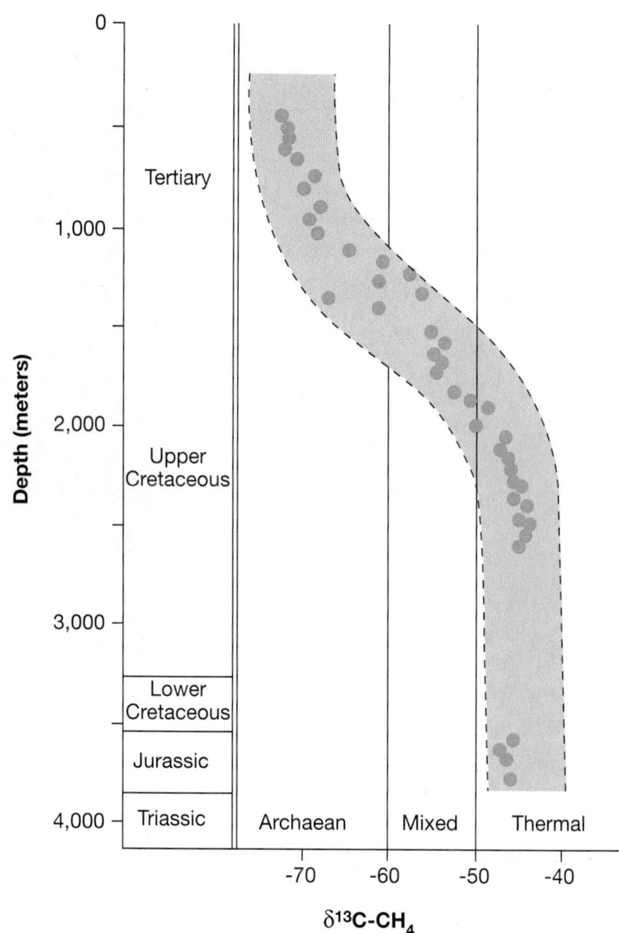

Figure 29.24 Methane Synthesis by Archaea in the Subsurface. Stable isotope techniques have shown that microbially mediated production of methane occurs in the subsurface. The decrease in the occurrence of the ^{13}C isotope of carbon indicates that the methane was produced by microorganisms down to a depth of 1,500 meters under the floor of the North Sea. Below a depth of 2,000 meters, the methane does not have a lower frequency of the ^{13}C isotope of carbon, indicating that it was formed by abiotic processes. The δ ^{13}C value gives an indication of the relative proportion of ^{13}C to ^{12}C in the sample. The more negative scale values signify a decreased presence of the heavier isotope in the upper zone at this location.

^{13}C isotope, indicating that it was derived from microorganisms using H_2 as an energy source and CO_2 as a carbon source and electron acceptor. Recall that microbes tend to use the lighter of two isotopes—in this case ^{12}C over the ^{13}C isotope-containing CO_2. In the underlying hotter abiogenic zone, methane is not depleted of the heavier carbon isotope, indicating that this is of chemical and thermal origin.

What is the origin of hydrogen for deep subsurface methanogenesis? Suggested sources are geological and may include (1) the reaction of water with basaltic rocks, (2) possible radioactive interactions, and (3) so-called magma degassing processes that occur along geological faults. The actual origin of this almost limitless supply of hydrogen, upon which deep surface methanogenesis depends, has not been determined with certainty.

Microorganisms also appear to be growing in intermediate depth oil-bearing structures. Recent studies indicate that active procaryotic assemblages are present in high-temperature (60 to 90°C) oil reservoirs, including such genera as *Thermotoga, Thermoanaerobacter,* and *Thermococcus*. The archaeal genera are dominated by methanogens. Thus microbial activities may be occurring above or in the "deep hot biosphere," a term suggested by Thomas Gold to describe this poorly understood region.

The discovery of deep subsurface microbes has a variety of implications. Evidence suggests that these microbes have been trapped in this environment for at least 80 million years, perhaps as long as 160 million years. They have evolved to exist in a stable environment that is anoxic and without sunlight; some chemolithoautotrophs survive only on H_2, CO_2, and water. The presence of primary production below ground on this planet suggests that it might also be possible in the subsurface sediments of Mars. Exploring these possibilities, along with demonstrating just how metabolically active Earth's deep subsurface microbes are, remains the next challenge.

1. What types of microbial activities have been observed in the deep subsurface?
2. What happens in terms of microbiological processes when organic matter leaches from the surface into the subsurface?
3. Why are stable isotope analyses so important in studies of microbe-geological interactions?
4. What microbial genera have been observed in oil field materials?

29.8 Soil Microorganisms and Human Health

Humans are in constant contact with soils. This occurs directly when children or adults play or work in the "dirt," or even when raw leafy and root vegetables are eaten. In most cases the contact with soil is harmless. However, soils contain a wide variety of pathogenic organisms. What is needed is an entry point and favorable conditions within or on the human body. A wide variety of anaerobes, including *Clostridium,* are present in soils. Unless there is a deep puncture wound that provides the anoxic environment required for their growth, anaerobes are of little concern. However, puncture wounds that occur in warfare and accidents can lead to gangrene. Soils contain other pathogens. Organisms such as the protist *Acanthamoeba,* which can be inhaled from dust, may cause primary amebic meningoencephalitis. When soils are used for surface disposal of human wastes without sewage treatment, the transmission of a wide variety of pathogens, including protozoa such as *Acanthamoeba* and *Cyclospora,* can occur. Wastewater treatment (section 41.2)

Soil and soil-related microorganisms also are of concern when they grow in buildings (**figure 29.25**). This increasingly common problem, often linked to the flooding of houses located in low-lying districts or to moisture accumulation in sink and bathroom areas has led to major health problems. This is particularly severe when water penetrates into house walls and insulation materials. The problem reached a critical point in the aftermath of Hurricane Katrina. However, throughout the US, it has been estimated that as many as 50% of homes have mold problems, a major source of chronic sinus infections. These molds also have been related to increases in asthma rates. The major responsible fungi are *Stachybotrys chartarum, Eurotium herbariorum,* and *Aspergillus versicolor*. Fungal growth results in a black slime; when this fungal growth dries, a dry dusty layer remains and the spores can be dispersed into the air. These spores are particularly dangerous for infants, whose lungs are less developed. *Stachybotrys* infection can result in pulmonary hemosiderosis, which causes bleeding of the lungs and sometimes death. Rapid drying of water-damaged buildings is required to control this problem. The *Fungi* (chapter 26)

Wallboards can contain baseline bioburdens of fungi, which need only high humidity to trigger additional growth. In addition, *Mycobacterium komossense* and gram-negative endotoxin-producing bacteria have been isolated. There are limited

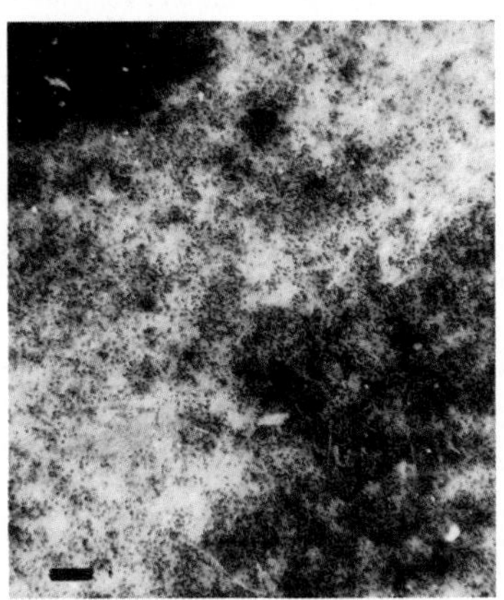

(a)

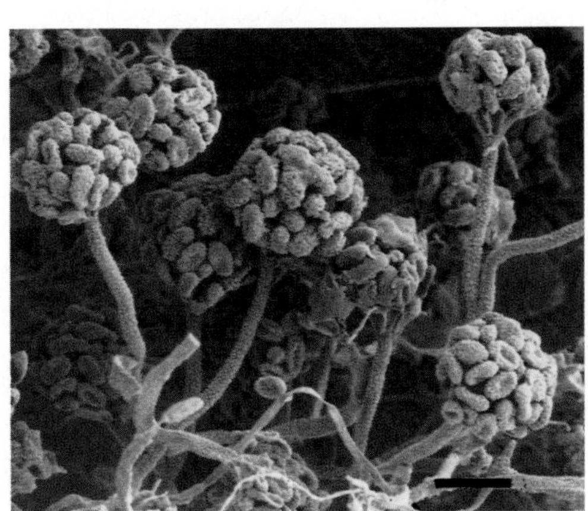

(b)

Figure 29.25 Fungal Growth in a Building. Fungal growth on sheet rock removed from a water-damaged building. **(a)** Stereo microscopic view showing black discolorations. Bar = 500 μm. **(b)** Scanning electron micrograph of dense mycelia and conidiophores characteristic of *Stachybotrys*. Bar = 10 μm.

alternatives for controlling and removing such dangerous pathogens; the most important are the removal/disinfection of moldy materials and keeping a house dry.

1. Surface soil spreading of untreated human wastes is carried out in many parts of the world. What are some of the possible effects of this practice?
2. How can the growth of fungi affect human health inside of a home?
3. What important fungal genera are involved in mold problems in homes?

Summary

29.1 Soils as an Environment for Microorganisms

a. Terrestrial environments are dominated by the solid phase, consisting of organic and inorganic components.

b. In an ideal soil, microorganisms function in thin water films that have close contact with air. Miniaquatic environments can form within soils (**figure 29.1**).

29.2 Soils, Plants, and Nutrients

a. Soil organic matter (SOM) helps retain nutrients and water and maintain soil structure. It can be divided into humic and nonhumic material (**table 29.2**).

b. Microbial degradation of SOM occurs in three phases, starting with the degradation and consumption of soluble compounds, followed by the extracellular attack of more resistant material such a cellulose, and finally the slow degradation of structurally complex molecules such as lignin.

29.3 Microorganisms in the Soil Environment

a. Most microorganisms in these environments are associated with surfaces, and these surfaces influence microbial use of nutrients and interactions with plants and other living organisms (**figure 30.2**).

b. Bacteria and fungi in soils have different functional strategies. Fungi tend to develop on the surfaces of aggregates, whereas microcolonies of bacteria are commonly associated with smaller pores.

c. Insects, earthworms, and other soil animals are also important parts of the soil. These decomposer-reducers interact with the microorganisms to influence nutrient cycling and other processes.

29.4 Microorganisms and the Formation of Different Soils

a. Soils form under many conditions. In all cases organic matter accumulation occurs through the direct activities of primary producers or by the import of preformed organic materials. Soils can be formed in regions such as the antarctic where there are no vascular plants (**figure 29.5**).

29.5 Microorganism Associations with Vascular Plants

a. Plants develop associations with many types of microorganisms. These include important associations involving mycorrhizae, rhizobia, and actinorhizae.

b. Mycorrhizal relationships (plant-fungal associations) are varied and complex. Six basic types can be observed including endomycorrhizal and sheathed/ectomycorrhizal types. Specialized monotropoid fungi make it possible for achlorophyllous plants to survive using carbon fixed by green plants. The hyphal network of the mycobiont can lead to the formation of a mycorrhizosphere (**figure 29.8; table 29.6**).

c. The mycorrhizal relationship often is established with the assistance of mycorrhization helper bacteria. In addition, bacteria may occur inside of the mycorrhizal fungus. These bacteria apparently contribute to the nitrogen cycling of the plant-fungus complex.

d. The *Rhizobium*-legume symbiosis is one of the best-studied examples of plant-microorganism interactions. This interaction is mediated by complex chemicals that serve as communication signals (**figure 29.13**).

e. The actinomycete *Frankia* forms nitrogen-fixing symbioses with some trees and shrubs.

f. *Agrobacterium* establishes a complex communication system with its plant host into which it transfers a fragment of DNA. Genes on this DNA encode proteins that result in the formation of plant tumors or galls (**figures 29.18** and **29.19**).

g. More complex microbe-plant interactions include tripartite and tetrapartite associations which often involve the plant, mycorrhizal fungi and bacteria.

29.6 Soil Microorganisms and the Atmosphere

a. Microorganisms can play major roles in the dynamics of greenhouse gases such as carbon dioxide, nitrous oxide, nitric oxide, and methane. Microorganisms can contribute to both the production and consumption of these gases (**figure 29.20**).

b. Fungi, especially, can produce chemicals that are normally considered as anthropogenic pollutants. These include chloromethane and cyanide.

29.7 The Subsurface Biosphere

a. The subsurface includes at least three zones: the shallow subsurface; the zone where gas, oil, and coal have accumulated; and the deep subsurface, where methane synthesis occurs (**figures 29.22, 29.23** and **29.24**).

29.8 Soil Microorganisms and Human Health

a. Microorganisms, particularly the fungi, can develop in moist areas in houses and cause major health problems for humans, including asthma. An important fungus involved in these problems is *Stachybotrys* (**figure 29.25**).

Key Terms

actinorhizae 704
aquifer 711
arbuscular mycorrhizae (AM) 698
arbuscule 699
associative nitrogen fixation 696
bacteroid 701
carbon to nitrogen (C/N) ratio 690
cellulose 690
crown gall disease 706
desert crust 695
ectomycorrhiza 698
endomycorrhiza 698
endophyte 696
epiphyte 696
flavonoid inducer molecules 701
geosmin 693
Hartig net 698
humic SOM (humus) 689
infection thread 701
leghemoglobin 701
lignin 690
methanotroph 709
mineral soil 689
mycorrhizosphere 700
nitrogen saturation point 690
Nod factors 701
nonhumic SOM 689
organic soil 689
oxidative burst 701
phyllosphere 696
rhizobia 701
rhizomorph 698
rhizoplane 696
rhizosphere 696
root nodule 701
slash-and-burn agriculture 694
soil organic matter (SOM) 689
stable isotope probing 697
stem-nodulating rhizobia 704
subsurface biosphere 711
symbiosome 701
T DNA 706
tetrapartite association 707
Ti (tumor-inducing) plasmid 706
tripartite association 707

Critical Thinking Questions

1. Tropical soils throughout the world are under intense pressure in terms of agricultural development. What land use and microbial approaches might be employed to better maintain this valuable resource?
2. Why might vascular plants have developed relationships with so many types of microorganisms? What do these molecular-level interactions, which show so many similarities when microbe-plant and microbe-human interactions are considered, suggest concerning possible common evolutionary relationships?
3. How might you maintain organisms from the deep hot subsurface under their in situ conditions? Compare this problem with that of working with microorganisms from deep marine environments.
4. Soil bacteria such as *Streptomyces* produce the bulk of known antibiotics. Look up the competitors for *Streptomyces,* the types of antibiotics these bacteria produce, and how the compounds are effective against competitors (what are the physiological targets?). Would you expect aquatic/marine bacteria to be major producers of antibiotics? Why or why not?

Learn More

Allen, M. F. 2000. Mycorrhizae. In *Encyclopedia of microbiology,* 2d ed., vol. 3, J. Lederberg, editor-in-chief, 328–36. San Diego: Academic Press.

Bala, H. P.; Prithiviraj, B.; Jha, A. K.; Ausubel, F.; and Vivanco, J. M. 2005. Mediation of pathogen resistance by exudation of antimicrobials from roots. *Nature* 434:217–21.

Bencic, A., and Winans, S. 2005. Detection and response to signals involved in host-microbe interactions by plant-associated bacteria. *Microbiol. Molec. Biol. Rev.* 69:155–94.

Bidartondo, M. I.; Redecker, D.; Hijri, I.; Wiemken, A.; Bruns, T. D.; Dominguez, L.; Sersic, D. J.; Leake, J. R.; and Read, D. J. 2002. Epiparasitic plants specialized on arbuscular mycorrhizal fungi. *Nature* 419:389–92.

Chapelle, F. H.; O'Neill, K.; Bradley, P. M.; Methé, B. A.; Clufo, S. A.; Knobel, L. L.; and Lovely, D. R. 2002. A hydrogen-based subsurface microbial community dominated by methanogens. *Nature* 415:312–14.

Dulla, G.; Marco, M.; Quinones, B.; and Lindow, S. 2005. A closer look at *Pseudomonas syringae* as a leaf colonist. *ASM News* 71:469–75.

Gao, R., and Lynn, D. G. 2005. Environmental pH sensing: Resolving the VirA/VirG two-component system inputs for *Agrobacterium* pathogenesis. *J. Bacteriol.* 187:2182–89.

Geurts, R., and Bisseling, T. 2002. *Rhizobium* Nod factor perception and signaling. *Plant Cell* 14:S239–49.

Govindarajulu, M.; Pfeffer, P. E.; Jin, H.; Abubaker, J.; Douds, D. D.; Allen, J. W.; Bücking, H.; Lammers, P. J.; and Shacher-Hill, Y. 2005. Nitrogen transfer in the arbuscular mycorrhizal symbiosis. *Nature* 435:819–23.

Heath, J.; Ayers, E.; Possell, M.; Bardgett, R. D.; Black, H. I. J.; Grant, H.; Ineson, P.; and Kerstiens, G. 2005. Rising atmospheric CO_2 reduces sequestration of root-derived soil carbon. *Science* 309:1711–13.

Klein, D. A. 2000. The rhizosphere. In *Encyclopedia of microbiology,* 2d ed., vol. 4, J. Lederberg, editor-in-chief, 117–26. San Diego: Academic Press.

Klironomos, J. N. 2002. Feedback with soil biota contributes to plant rarity and invasiveness in communities. *Nature* 417:67–70.

Kuhn, D. M., and Ghannoum, M. A. 2003. Indoor mold, toxigenic fungi, and *Stachybotrys chartarum:* Infectious disease perspective. *Clin. Microbial. Rev.* 16(1):144–72.

Lu, Y., and Conrad, R. 2005. In situ stable isotope probing of methanogenic *Archaea* in the rice rhizosphere. *Science* 309: 1088–90.

Pounds, J. A.; Bustamante, M. R.; Coloma, L. A.; Consuegra, J. A.; Fogden, M. P. L.; Foster, P. N.; La Marca, E., et al. 2006. Widespread amphibian extinctions from epidemic disease driven by global warming. *Nature* 439:161–67.

Torsvik, V.; Øvreås, L.; and Thingstad, T. F. 2002. Prokarytotic diversity—Magnitude, dynamics, and controlling factors. *Science* 296:1064–66.

Vandenkoornhuyse, P.; Baldauf, S. L.; Leyval, C.; Straczek, J.; and Young, J. P. W. 2002. Extensive fungal diversity in plant roots. *Science* 295:2051–61.

Please visit the Prescott website at www.mhhe.com/prescott7 for additional references.

30 Microbial Interactions

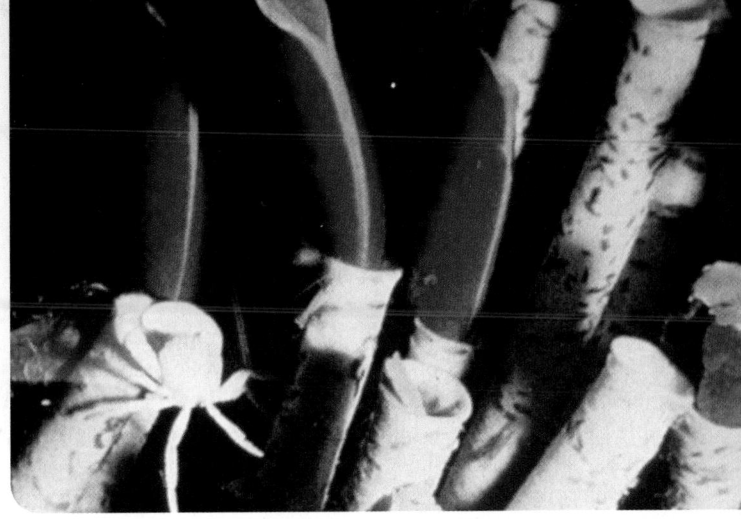

A "garden" of tube worms (*Riftia pachyptila*) at the Galápagos Rift hydrothermal vent site (depth 2,550 m). Each worm grows to more than a meter in length thanks to endosymbiotic chemolithoautotrophic bacteria, which provide carbohydrate to these gutless worms. The bacteria use the Calvin cycle to fix CO_2 and H_2S as an electron donor.

Preview

- The term symbiosis, or "together-life," can be used to describe many of the interactions between microorganisms, and also microbial interactions with higher organisms, including plants and animals. These interactions may be positive or negative.
- Symbiotic interactions include mutualism, cooperation, commensalism, parasitism, predation, amensalism, and competition. These interactions are important in natural processes and in the occurrence of disease. The interactions can vary depending on the environment and changes in the interacting organisms.
- Microorganisms, as they interact, can form complex physical assemblages that include biofilms. These form on living and inert surfaces and have major impacts on microbial survival and the occurrence of disease.
- Microorganisms also interact by the use of chemical signal molecules, which allow the microbial population to respond to increased population density. Such responses include quorum sensing, which controls a wide variety of microbial activities.
- Gnotobiotic refers to a microbiologically monitored environment or animal in which the identities of all microorganisms present are known or to an environment or animal that is germfree.
- Most microorganisms associated with the human body are bacteria; they normally colonize specific sites. There are both positive and negative aspects of these normal microbial associations. Sometimes they compete with pathogens, other times they are capable of producing opportunistic infections.

Our discussion of microbial ecology has so far considered microbial communities in complex ecosystems. However, the ecology of microorganisms also involves the physiology and behavior of microbes as they interact with one another and with higher organisms. In this chapter, we begin by defining types of microbial interactions and present a number of illustrative examples. We conclude this chapter with a consideration of perhaps the most intimate, yet still relatively unexplored, microbial habitat: the human body.

Our discussion of interactions between microbes and between microbes and eucaryotes focuses on the nature of symbioses. Although **symbiosis** is often used in a nonscientific sense to mean a mutually beneficial relationship, here we use the term in its original broadest sense, as an association of two or more different species of organisms, as suggested by H. A. deBary in 1879.

30.1 Microbial Interactions

Microorganisms can associate physically with other organisms in a variety of ways. One organism can be located on the surface of another, as an **ectosymbiont.** In this case, the ectosymbiont usually is a smaller organism located on the surface of a larger organism. In contrast, one organism can be located within another organism as an **endosymbiont.** While the simplest microbial interactions involve two members, a symbiont and its host, a number of interesting organisms host more than one symbiont. The term **consortium** can be used to describe this physical relationship. For example, *Thiothrix* species, a sulfur-using bacterium, is attached to the surface of a mayfly larva and itself

. . . every organic being is related, in the most essential yet often hidden manner, to that of all other organic beings, with which it comes into competition for food or residence, or from which it has to escape, or on which it preys.

Charles Darwin

contains a parasitic bacterium. Fungi associated with plant roots (mycorrhizal fungi) often contain endosymbiotic bacteria, as well as having bacteria living on their surfaces. Microorganism associations with vascular plants: Mycorrhizae (section 29.5)

These physical associations can be intermittent and cyclic or permanent. Examples of intermittent and cyclic associations of microorganisms with plants and marine animals are shown in **table 30.1.** Important human diseases, including listeriosis, malaria, leptospirosis, legionellosis, and vaginosis also involve such intermittent and cyclic symbioses. These diseases are discussed in chapters 38 and 39. Interesting permanent relationships also occur between bacteria and animals, as shown in **table 30.2.** Hosts include squid, leeches, aphids, nematodes, and mollusks. In each of these cases, an important characteristic of the host animal is conferred by the permanent bacterial symbiont.

Although it is possible to observe microorganisms in these varied physical associations with other organisms, the fact that there is some type of physical contact provides no information on the nature of the interactions that might be occurring. These interactions include mutualism, cooperation, commensalism, predation, parasitism, amensalism, and competition (**figure 30.1**). These interactions are now discussed.

Table 30.1 Intermittent and Cyclical Symbioses of Microorganisms with Plants and Marine Animals

Symbiosis	Host	Cyclical Symbiont
Plant-bacterial	*Gunnera* (tropical angiosperm)	*Nostoc* (cyanobacterium)
	Azolla (rice paddy fern)	*Anabaena* (cyanobacterium)
	Phaseolus (bean)	*Rhizobium* (N_2 fixer)
	Ardisia (angiosperm)	*Protobacterium*
Marine animals	Coral coelenterates	*Symbiodinium* (dinoflagellate)
	Luminous fish	*Vibrio, Photobacterium*

Adapted from L. Margulis and M. J. Chapman, 1998. Endosymbioses: Cyclical and permanent in evolution. *Trends in Microbiology* 6(9):342–46, tables 1, 2, and 3.

1. Define the term symbiosis.
2. List several important diseases that involve cyclic and intermittent symbioses. Compare these with permanent relationships.

Mutualism

Mutualism [Latin *mutuus,* borrowed or reciprocal] defines the relationship in which some reciprocal benefit accrues to both partners. This is an obligatory relationship in which the **mutualist** and the host are dependent on each other. When separated, in many cases, the individual organisms will not survive. Several examples of mutualism are presented next.

Microorganism-Insect Mutualisms

Mutualistic associations are common in the insects. This is related to the foods used by insects, which often include plant sap or animal fluids lacking in essential vitamins and amino acids. The required vitamins and amino acids are provided by bacterial symbionts in exchange for a secure habitat and ample nutrients (**Microbial Diversity & Ecology 30.1).** The aphid is an excellent example of this mutualistic relationship. This insect harbors the γ-proteobacterium *Buchnera aphidicola* in its cytoplasm, and a mature insect contains literally millions of these bacteria in its body. The *Buchnera* provides its host with 10 essential amino acids, and if the insect is treated with antibiotics, it dies. Likewise, *B. aphidicola* is an obligate mutualistic symbiont. The inability of either partner to grow without the other indicates that the two organisms demonstrate coevolution, or have evolved together. It is estimated that the *B. aphidicola*-aphid endosymbiosis was established about 150 million years ago. The genomes of two different *B. aphidicola* strains have been sequenced and annotated to reveal extreme genomic stability. These strains diverged 50 to 70 million years ago, and since that time there have

Table 30.2 Examples of Permanent Bacterial-Animal Symbioses and the Characteristics Contributed by the Bacterium to the Symbiosis

Animal Host	Symbiont	Symbiont Contribution
Sepiolid squid (*Euprymna scolopes*)	Luminous bacterium (*Vibrio fischeri*)	Luminescence
Medicinal leech (*Hirudo medicinalis*)	Enteric bacterium (*Aeromonas veronii*)	Blood digestion
Aphid (*Schizaphis graminum*)	Bacterium (*Buchnera aphidicola*)	Amino acid synthesis
Nematode worm (*Heterorhabditis* spp.)	Luminous bacterium (*Photorhabdus luminescens*)	Predation and antibiotic synthesis
Shipworm mollusk (*Lyrodus pedicellatus*)	Gill cell bacterium	Cellulose digestion and nitrogen fixation

Source: From E. G. Ruby, 1999. Ecology of a benign "infection": Colonization of the squid luminous organ by *Vibrio fischeri.* In *Microbial ecology and infectious disease,* E. Rosenberg, editor, American Society for Microbiology, Washington, D.C., 217–31, table 1.

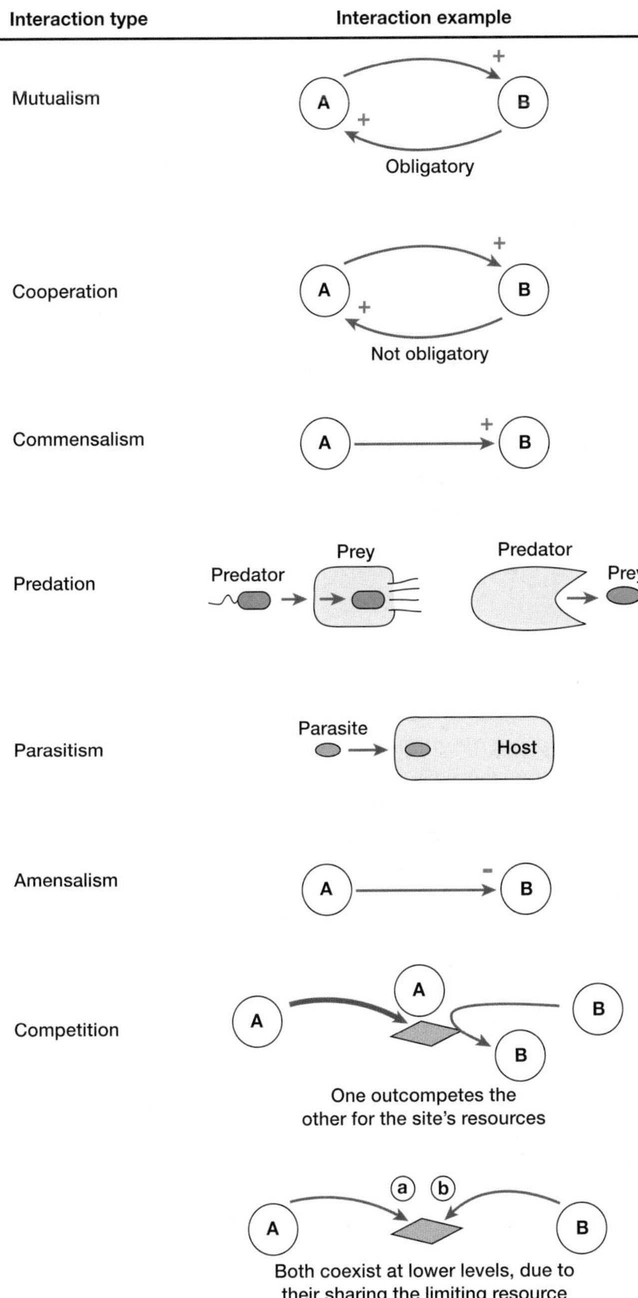

Figure 30.1 Microbial Interactions. Basic characteristics of symbiotic interactions that can occur between different organisms.

been no gene duplications, translocations, inversions, or genes acquired by horizontal transfer. The genomes are small, only 0.64 Mb each with 93% of their genes common to both strains. Furthermore, only two genes have no orthologs in their close relative *E. coli*. This tremendous degree of stability implies that although the initial acquisition of the endosymbiont by ancestral aphids enabled their use of an otherwise deficient food source (sap), the bacteria have not continued to expand the ecological niche of their insect host through the acquisition of new traits that might be advantageous. Bioinformatics: Genome annotation (section 15.4); Microbial evolution (section 19.1)

The protozoan-termite relationship is another classic example of mutualism in which the flagellated protozoa live in the gut of termites and wood roaches (**figure 30.2*a***). These flagellates exist on a diet of carbohydrates, acquired as cellulose ingested by their host (figure 28.2*b*). The protozoa engulf wood particles, digest the cellulose, and metabolize it to acetate and other products. Termites oxidize the acetate released by their flagellates. Because the host is almost always incapable of synthesizing cellulases (enzymes that catalyse the hydrolysis of cellulose), it is dependent on the mutualistic protozoa for its existence. This mutualistic relationship can be readily tested in the laboratory if wood roaches are placed in a bell jar containing wood chips and a high concentration of O_2. Because O_2 is toxic to the flagellates, they die. The wood roaches are unaffected by the high O_2 concentration and continue to ingest wood, but they soon die of starvation due to a lack of cellulases.

Zooxanthellae

Many marine invertebrates (sponges, jellyfish, sea anemones, corals, ciliates) harbor endosymbiotic dinoflagellates called **zooxanthellae** within their tissue (**figure 30.3*a***). Because the degree of host dependency on the mutualistic protist is somewhat variable, only one well-known example is presented. Protist classification: *Alveolata* (section 25.6)

The hermatypic (reef-building) corals (figure 30.3*b*) satisfy most of their energy requirements using their zooxanthellae, which are found at densities between 5×10^5 and 5×10^6 cells per square centimeter of coral animal. In exchange for up to 95% of their photosynthate (fixed carbon), zooxanthellae receive nitrogenous compounds, phosphates, CO_2, and protection from UV light from their hosts. This efficient form of nutrient cycling and tight coupling of trophic levels accounts for the stunning success of reef-building corals in developing vibrant ecosystems. However, during the past several decades, the number of coral bleaching events has increased dramatically. **Coral bleaching** is defined as a loss of either the photosynthetic pigments from the corals or expulsion of the zooxanthallae. It has been determined that damage to photosystem II of the zooxanthellae generates reactive oxygen species (ROS); it is these ROS that appear to be the direct cause of damage (recall that photosystem II uses water as the electron source resulting in the evolution of oxygen). Coral bleaching appears to be caused by a variety of stressors, but it has been experimentally determined, as well as observed in field sites, that temperature increases as small as 2°C above the average summer maxima can cause coral bleaching. Sadly, evidence suggests that many corals will be unable to evolve quickly enough to keep pace with the observed and predicted increases in ocean temperatures if global warming continues unchecked. Phototrophy: Light reaction in oxygenic photosynthesis (section 9.12)

Sulfide-Based Mutualisms

Tube worm-bacterial relationships exist several thousand meters below the surface of the ocean, where the Earth's crustal plates are spreading apart (**figure 30.4**). Vent fluids are anoxic, contain

Microbial Diversity & Ecology

30.1 *Wolbachia pipientis:* The World's Most Infectious Microbe?

Most people have never heard of the bacterium *Wolbachia pipientis,* but this rickettsia infects more organisms than does any other microbe. It is known to infect a variety of crustaceans, spiders, mites, millipedes, and parasitic worms and may infect more than a million insect species worldwide. *Wolbachia* inhabits the cytoplasm of these animals where it apparently does no harm. To what does *Wolbachia* owe its extraordinary success? Quite simply, this endosymbiont is a master at manipulating its hosts' reproductive biology. In some cases it can even change the sex of the infected organism.

Wolbachia is transferred from one generation of host to the next through the eggs of infected female hosts. So in order to survive, this microbe must ensure the fertilization and viability of infected eggs while decreasing the likelihood that uninfected eggs survive. In the 1970s, scientists noticed that if a male wasp infected with *Wolbachia* mated with an uninfected female, few if any of these uninfected offspring survived. However, if infected females of the same wasp species mated with either infected or uninfected males, all of the eggs were viable—and infected with *Wolbachia.* Although scientists had no clear understanding of the cellular or molecular mechanisms involved, they suggested that "cytoplasmic incompatibility" might be responsible. They proposed that the cytoplasm of infected sperm was toxic to uninfected eggs, but that eggs carrying *Wolbachia* produced an antidote to the hypothetical poison.

It wasn't until the 1990s that the mechanism of cytoplasmic incompatibility became clear. Researchers at the University of California, Santa Cruz used an assortment of dyes so they could visualize fertilization of eggs of the wasp *Nasonia vitripennis.* They discovered cytoplasmic incompatibility involved timing, not toxins. Specifically, when infected sperm fertilize uninfected eggs, the sperm chromosomes try to align with those of the egg while the egg's chromosomes are still confined to the pronucleus. These eggs ultimately divide as if never fertilized and develop into males. However, chromosomes behave normally when a male and infected female mate. This yields a normal sex distribution and all progeny are infected with *Wolbachia.*

Another important breakthrough in the 1990s was an understanding of how prevalent *Wolbachia* infection is in the insect world. John Werren, a microbiologist at the University of Rochester in New York, and his colleagues used the polymerase chain reaction (PCR) to survey insects from Panama, England, and the United States, and found that roughly 20% of all species they sampled harbored *Wolbachia.* Other researchers found infection rates to be as high as 75% in smaller geographic locations. These high rates of global infection result because once a few individuals harbor *Wolbachia,* it spreads quickly through a population. In studying these newly identified infected species, scientists discovered that in addition to cytoplasmic incompatibility, the endosymbiont has developed other means to ensure its endurance (**Box figure a**). In some insect species, *Wolbachia* simply kills all the male offspring and induces parthenogenesis of infected females—that is, the mothers simply clone themselves. This limits genetic diversity, but allows 100% transmission of *Wolbachia* to the next generation. In other species, the microbe allows the birth of males but then modifies their hormones so that the males become feminized and produce eggs.

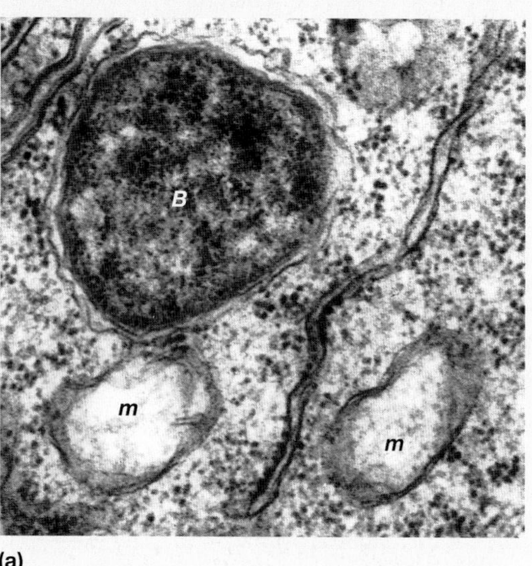

(a)

(a) *Wolbachia pipientis* Within the Egg Cytoplasm of the Ant *Gnamptogenys menadensis*. In this insect, *Wolbachia* is maternally transmitted, so the bacterium has evolved mechanisms to manipulate the sex distribution of the offspring so that the host produces mostly females. The *Wolbachia* cell is indicated by (B) and host mitochondria are labeled m.

Wolbachia's ability to manipulate the reproduction of its hosts has brought it to the attention of evolutionary biologists. These scientists are interested in the bacterium's potential role in speciation. It was noticed that two North American wasp species (*N. giraulti* and *N. longicornis*), each carrying a different strain of *Wolbachia,* appeared to be morphologically, behaviorally, and most importantly genetically similar. Predictably, when the two species mate with each other, there are no viable offspring. But much to the surprise of the scientific community, when wasps are treated with an antibiotic to cure them of their *Wolbachia* infections, the two wasp species mate and produce viable, fertile offspring. Could it be that while there are no genetic barriers to reproduction, the presence of two different *Wolbachia* endosymbionts forms a reproductive wall that could ultimately drive the evolution of new species?

Finally, *Wolbachia* may be driving more than speciation. They are required for the embryogenesis of filarial nemotodes, which cause diseases like elephantitis and river blindness. *Onchocerca volvulus* is the filarial nematode that causes river blindness. It is

(continued)

transmitted by blackflies in Africa, Latin America, and Yemen with a worldwide incidence of about 18 million people. When a fly bites a person, the nematode establishes its home in small nodules beneath the skin, where it can survive for as long as 14 years. During that time, it releases millions of larvae, many of which migrate to the eye. Eventually the host mounts an inflammatory response that results in progressive vision loss (**Box figure b**). It is now recognized that this inflammatory response is principally directed at the *Wolbachia* infecting the nematodes, not the worms themselves. German researchers discovered that in patients treated with a single course of antibiotic to kill the endosymbiont, nematode reproduction stopped. While inflammatory damage cannot be reversed, disease progression is halted. This may prove a more effective and cost-efficient means of treatment than the current anti-parasitic method that must be repeated every six months.

Our understanding of the distribution of *Wolbachia* and the mechanistically clever ways it has evolved to assure its survival will continue to grow. The microbe may ultimately become a valuable tool in investigating the complexities of speciation as well as the key to curing devastating diseases that strike those in developing countries.

Source: A. St. Andre, N. M. Blackwell, L. R. Hall, A. Hoerauf, N, W. Brattig, L. Volkman, M. J. Taylor, L. Ford, A. G. Hise, J. H. Lass, E. Diaconu, and E. Pearlman. 2002. The role of endosymbiotic Wolbachia *bacteria in the pathogenesis of river blindness.* Science *295: 1892–95.*

C. Zimmer. 2001. Wolbachia: *A tale of sex and survival.* Science *292:1093–95.*

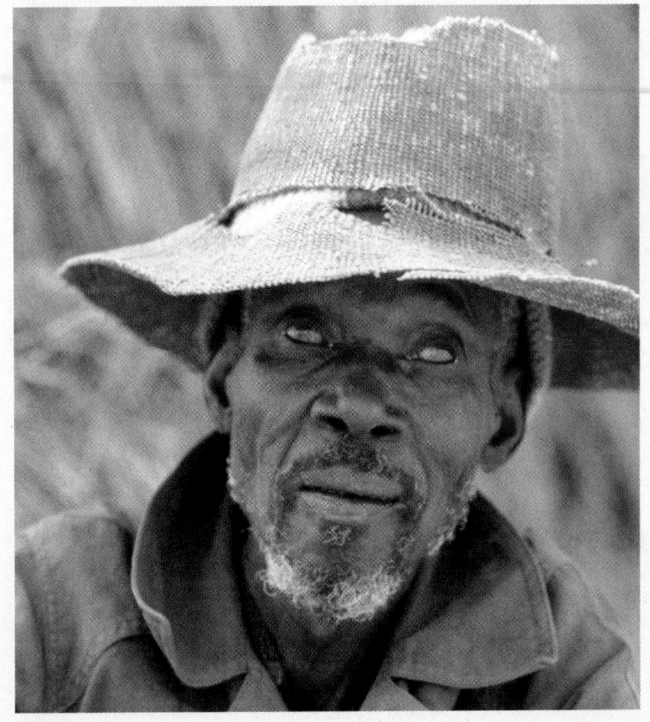

(b)

(b) River blindness. This is the second-leading cause of blindness worldwide. Evidence suggests that it is not the nematode but its endosymbiont, *Wolbachia pipientis*, that causes the severe inflammatory response that leaves many, like the man shown here, blind.

(a)

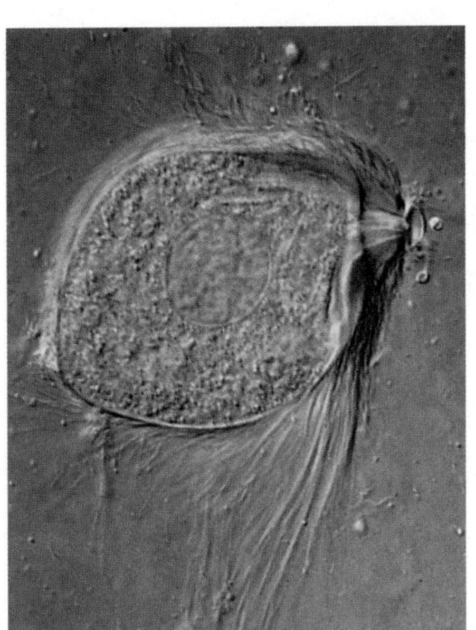

(b)

Figure 30.2 Mutualism. Light micrographs of **(a)** a worker termite of the genus *Reticulitermes* eating wood (×10), and **(b)** *Trichonympha*, a multiflagellated protozoan from the termite's gut (×135). Notice the many flagella that occur over most of its length. The ability of *Trichonympha* to break down cellulose allows termites to use wood as a food source.

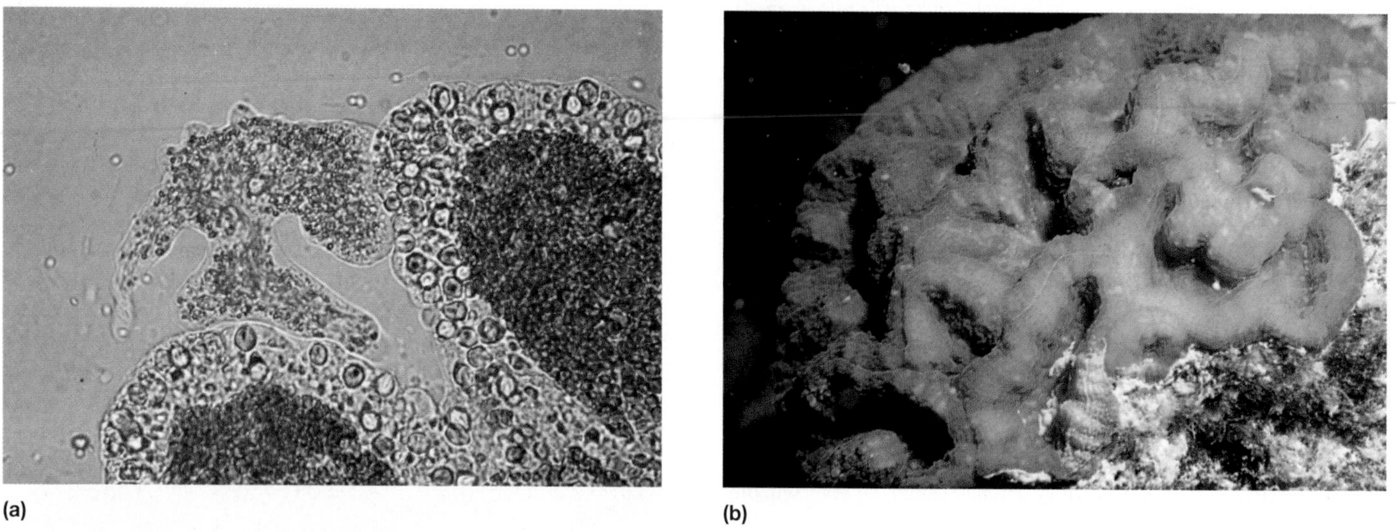

(a) (b)

Figure 30.3 Zooxanthellae. **(a)** Zooxanthellae (green) within the tip of a hydra tentacle (×150). **(b)** The green color of this rose coral (*Manilina*) is due to the abundant zooxanthellae within its tissues.

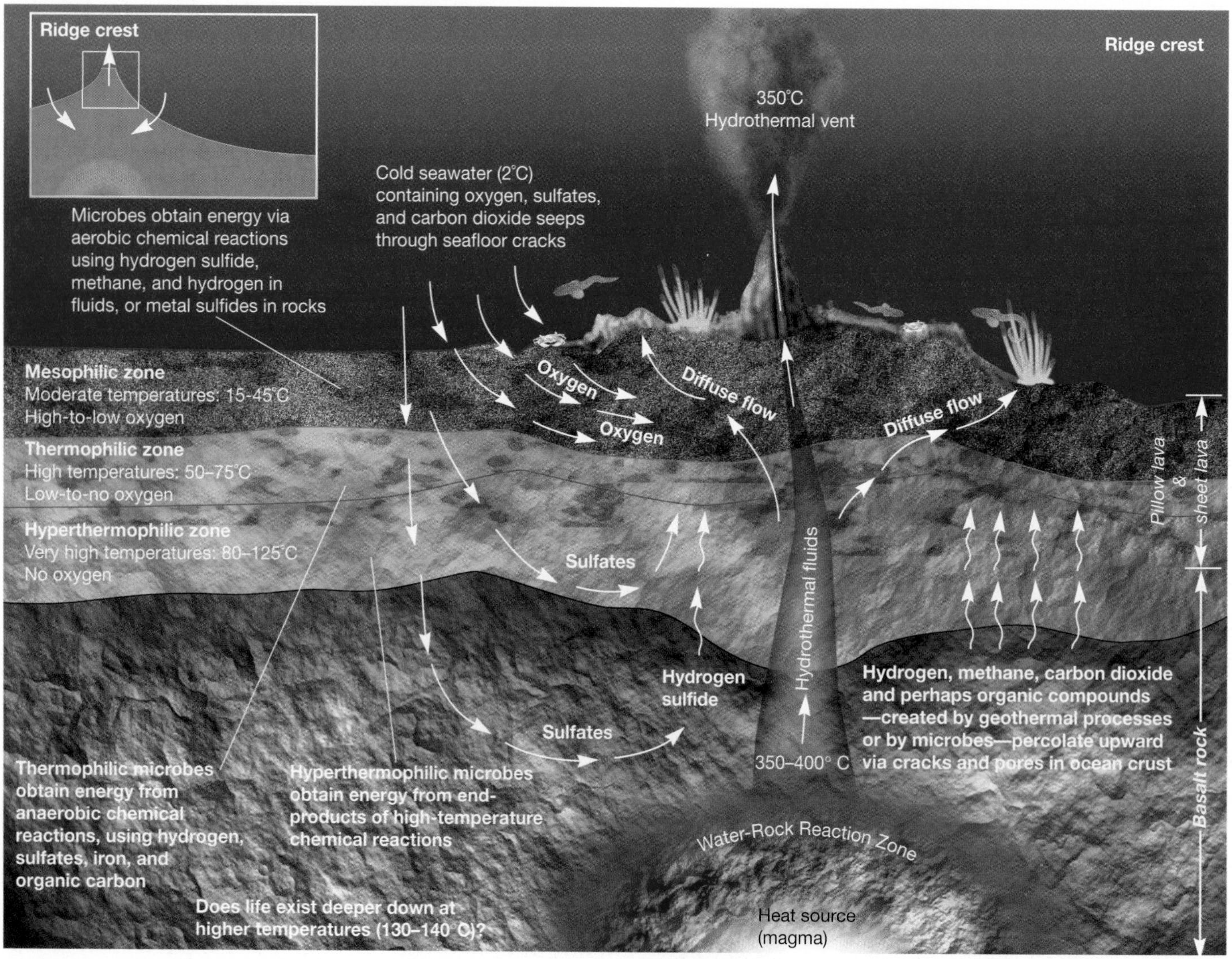

Ridge crest

Ridge crest

350°C
Hydrothermal vent

Microbes obtain energy via aerobic chemical reactions using hydrogen sulfide, methane, and hydrogen in fluids, or metal sulfides in rocks

Cold seawater (2°C) containing oxygen, sulfates, and carbon dioxide seeps through seafloor cracks

Mesophilic zone
Moderate temperatures: 15–45°C
High-to-low oxygen

Thermophilic zone
High temperatures: 50–75°C
Low-to-no oxygen

Hyperthermophilic zone
Very high temperatures: 80–125°C
No oxygen

Oxygen

Oxygen

Diffuse flow

Diffuse flow

Pillow lava & sheet lava

Sulfates

Hydrothermal fluids

Hydrogen sulfide

Hydrogen, methane, carbon dioxide and perhaps organic compounds —created by geothermal processes or by microbes—percolate upward via cracks and pores in ocean crust

Sulfates

350–400° C

Basalt rock

Thermophilic microbes obtain energy from anaerobic chemical reactions, using hydrogen, sulfates, iron, and organic carbon

Hyperthermophilic microbes obtain energy from end-products of high-temperature chemical reactions

Water-Rock Reaction Zone

Does life exist deeper down at higher temperatures (130–140 C)?

Heat source (magma)

Figure 30.4 Hydrothermal Vents and Related Geological Activity. The chemical reactions between seawater and rocks that occur over a range of temperatures on the seafloor supplies the carbon and energy that support a diverse collection of microbial communities in specific niches within the vent system.

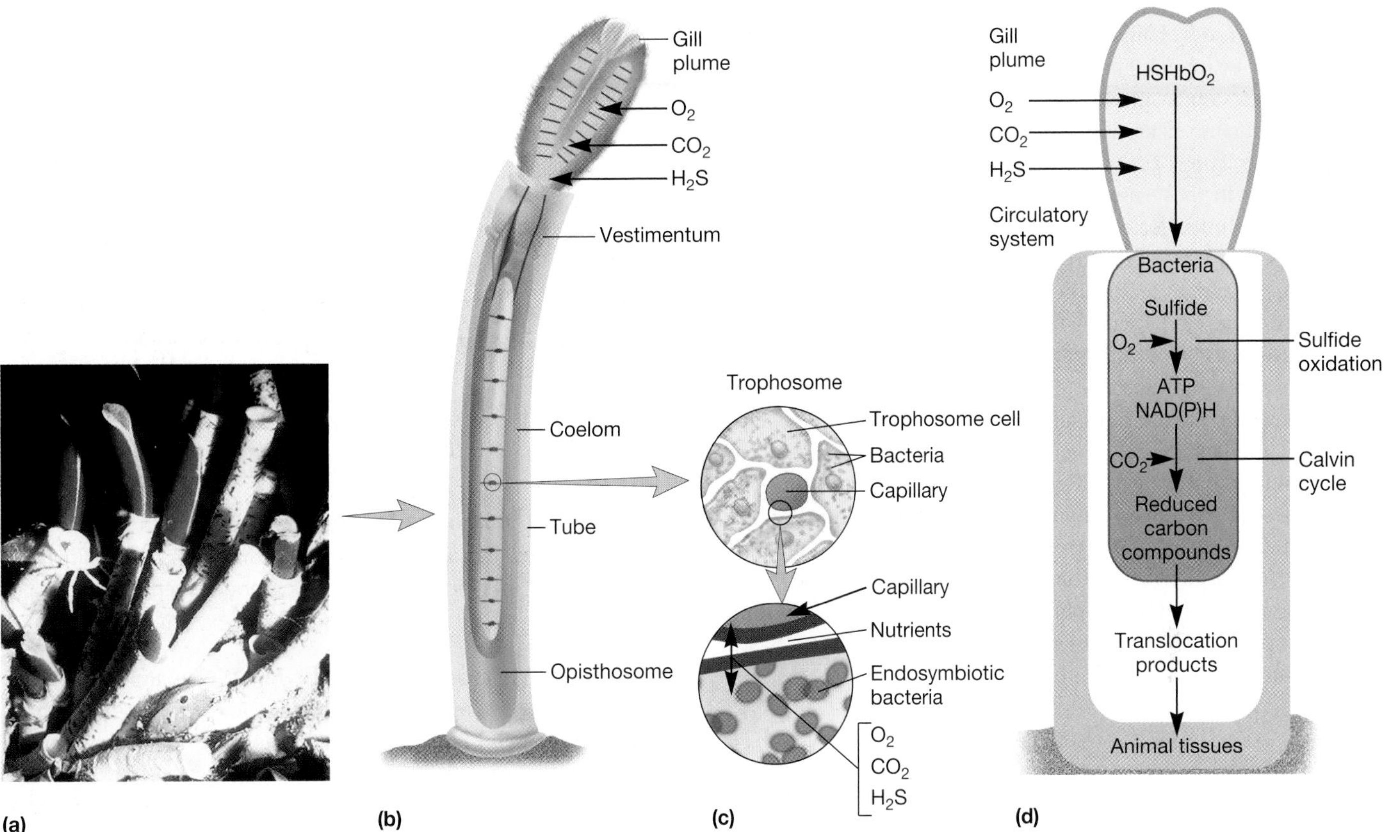

Figure 30.5 The Tube Worm–Bacterial Relationship. **(a)** A community of tube worms (*Riftia pachyptila*) at the Galápagos Rift hydrothermal vent site (depth 2,550 m). Each worm is more than a meter in length and has a 20 cm gill plume. **(b, c)** Schematic illustration of the anatomical and physiological organization of the tube worm. The animal is anchored inside its protective tube by the vestimentum. At its anterior end is a respiratory gill plume. Inside the trunk of the worm is a trophosome consisting primarily of endosymbiotic bacteria, associated cells, and blood vessels. At the posterior end of the animal is the opisthosome, which anchors the worm in its tube. **(d)** Oxygen, carbon dioxide, and hydrogen sulfide are absorbed through the gill plume and transported to the blood cells of the trophosome. Hydrogen sulfide is bound to the worm's hemoglobin ($HSHbO_2$) and carried to the endosymbiont bacteria. The bacteria oxidize the hydrogen sulfide and use some of the released energy to fix CO_2 in the Calvin cycle. Some fraction of the reduced carbon compounds synthesized by the endosymbiont is translocated to the animal's tissues.

high concentrations of hydrogen sulfide, and can reach a temperature of 350°C. However, because of increased atmospheric pressure, the water does not boil. The seawater surrounding these vents has sulfide concentrations around 250 μM and temperatures 10 to 20°C above the ambient seawater temperature of about 2°C.

Giant (>1 m in length), red, gutless tube worms (*Riftia* spp.; **figure 30.5*a***) near these hydrothermal vents provide an example of a unique form of mutualism and animal nutrition in which chemolithotrophic bacterial endosymbionts are maintained within specialized cells of the tube worm host (figure 30.5 *b,c,d*). The *Riftia* tube worms live at the interface between the hot, anoxic fluids of the vents and the cold, oxygen-containing seawater. Here, reduced sulfides from the vents react rapidly and spontaneously with oxygen in the seawater. In order to provide both reduced sulfur and oxygen to their bacterial endosymbionts, *Riftia's* blood contains a unique kind of hemoglobin, which accounts for the bright-red plume extending out of their tubes. Hydrogen sulfide (H_2S) and O_2 are removed from the seawater by the worm's hemoglobin, and delivered to a special organ called the trophosome. The trophosome is packed with chemolithotrophic bacterial endosymbionts that fix CO_2 using the Calvin cycle (*see figure 10.4*) with electrons provided by H_2S. The CO_2 is carried to the endosymbionts in three ways: (1) freely in the bloodstream, (2) bound to hemoglobin, and (3) as organic acids such as malate and succinate. When these acids are decarboxylated, they release CO_2. This process is similar to carbon fixation by plants and cyanobacteria, but it occurs in the deepest, darkest reaches of the ocean. This mutualism is enormously successful: not only do the worms grow to an astounding size in densely packed communities, but the bacteria (which have not yet been cultured in the laboratory) reach densities of up to 10^{11} cells per gram of worm tissue.

Methane-Based Mutualisms

Other unique food chains involve methane-fixing microorganisms. By converting methane to carbohydrate, these bacteria perform the first step in providing organic matter for consumers. Methanotrophs are bacteria capable of using methane as a sole carbon source. They occur as intracellular symbionts of methane-vent mussels. In these mussels the thick, fleshy gills are filled with bacteria. In the Barbados Trench, methanotrophic carnivorous sponges have been discovered in a mud volcano at a depth of 4,943 m. Abundant methanotrophic symbionts were confirmed by the presence of enzymes related to methane oxidation in sponge tissues. These sponges are not satisfied with just bacteria; they also trap swimming prey to give variety to their diet.

Methanotrophic microorganisms are important in ecosystems outside of methane vents. For example, a methanotrophic endosymbiont was recently discovered to reduce the flux of methane from peat bogs; wetlands are the largest natural source of this greenhouse gas. *Sphagnum* moss, the principal plant in peat bogs (and a favorite among florists), can grow when submerged in water. Methanotrophic α-proteobacteria living within the outer cortex cells of *Sphagnum* stems oxidize methane as it diffuses through the water column:

$$CH_4 + 2O_2 \rightarrow CO_2 + H_2O$$

The resulting CO_2 is then readily fixed by the plant, which uses the Calvin cycle:

$$2CO_2 + 2H_2O \rightarrow 2CH_2O + 2O_2$$

This enables extremely efficient carbon recycling within this ecosystem:

$$CH_4 + 2CO_2 \rightarrow 2CH_2O$$

The Rumen Ecosystem

Ruminants are the most successful and diverse group of mammals on Earth today. Examples include cattle, deer, elk, bison, water buffalo, camels, sheep, goats, giraffes, and caribou. These animals spend vast amounts of time chewing their cud—a small ball of partially digested grasses that the animal has consumed but not yet completely digested. It is thought that the ruminants evolved an "eat now, digest later" strategy because their grazing can often be interrupted by predator attacks.

These herbivorous animals have stomachs that are divided into four chambers (**figure 30.6**). The upper part of the ruminant stomach is expanded to form a large pouch called the **rumen** and a smaller, honeycomb-like region, the reticulum. The lower portion is divided into an antechamber, the omasum, followed by the "true" stomach, the abomasum. The rumen is a highly muscular, anaerobic fermentation chamber where huge amounts of grasses eaten by the animal are digested by a diverse microbial community that includes bacteria, archaea, fungi, and protists. This microbial community is large—about 10^{12} organisms per milliliter of digestive fluid. When the animal eats plant material, it is mixed with saliva and swallowed without chewing to enter the rumen. Here the material is churned and thoroughly mixed. Eventually, microbial attack and mixing coats the grass with microbes, reducing it to a pulpy, partially digested, mass. At this point the mass moves into the reticulum where it is regurgitated as cud, chewed, and re-swallowed by the animal. As this process proceeds, the grass becomes progressively more liquefied and flows out of the rumen into the omasum and then the abomasum. Here the nutrient-enriched grass material meets the animal's digestive enzymes and soluble organic and fatty acids are absorbed into the animal's bloodstream.

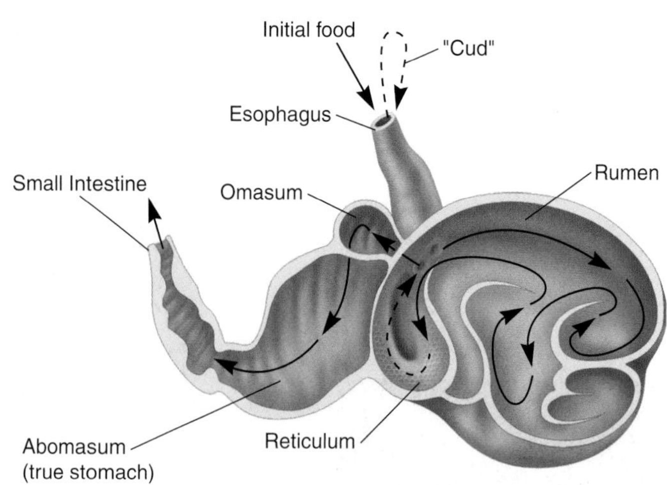

Figure 30.6 Ruminant Stomach. The stomach compartments of a cow. The microorganisms are active mainly in the rumen. Arrows indicate direction of food movement.

The microbial community in the rumen is extremely dynamic. The rumen is slightly warmer than the rest of the animal and with a redox potential of about -30 mV, all resident microorganisms must carry out anaerobic metabolism. Not only does the animal have a mutualistic relationship with the microbial community, but within the microbial community there are very specific interactions. One population of bacteria produce extracellular cellulases that cleave the $\beta(1 \rightarrow 4)$ linkages between the successive D-glucose molecules that form plant cellulose. The D-glucose is then fermented to organic acids such as acetate, butyrate, and propionate. These organic acids, as well as fatty acids are the true energy source for the animal. In some ruminants, the processing of organic matter stops at this stage (**Microbial Diversity & Ecology 30.2**). In others, such as cows, acetate, CO_2, and H_2 are used by methanogenic archaea to generate methane (CH_4), a greenhouse gas. In fact, a single cow can produce as much as 200 to 400 liters of CH_4 per day. The animal releases this CH_4 by a process called eructation (Latin *eructare,* to belch). Although the methanogens consume acetate that could be used by their animal hosts, they provide most of the vitamins needed by the ruminant. In fact, rumen microbes are so effective in fortifying the grass consumed by the animal, unlike humans, most ruminants have no required dietary amino acids. Fermentations (section 9.7)

Microbial Diversity & Ecology

30.2 Coevolution of Animals and Their Gut Microbial Communities

Organisms with digestive tracts had to make an interesting evolutionary choice: will the microbial community produce methane or not? The use of plant materials as a major food source does not always lead to methane production in the digestive tract. For example, kangaroos do not produce methane, whereas sheep and cattle do. The kangaroo has a distinct advantage in terms of nutrition. When complex plant materials are degraded only to organic acids such as acetate, the animal's digestive system can directly absorb the acids. In sheep and cattle, the microbial community is more complex and converts acetate-level substrates to methane and carbon dioxide, leading to nutrient loss from the original plant material. This loss can be substantial, with 10 to 15% of the organic matter in the feed lost to the atmosphere as methane.

An examination of over 250 reptiles, birds, and mammals showed that their maintenance of methanogenic microorganisms and methane production is under phylogenetic and not dietary control (see **Box figure**). Although low levels of methanogens can be detected in vertebrates that do not produce much of this important greenhouse gas, the lack of methane production seems to result from absence of methanogen receptor sites in the digestive system. As shown in this figure, the ability to maintain methanogens often has been lost. A similar situation occurs with arthropods: methane is produced by only a few organisms, including tropical millipedes, cockroaches, termites, and scarab beetles.

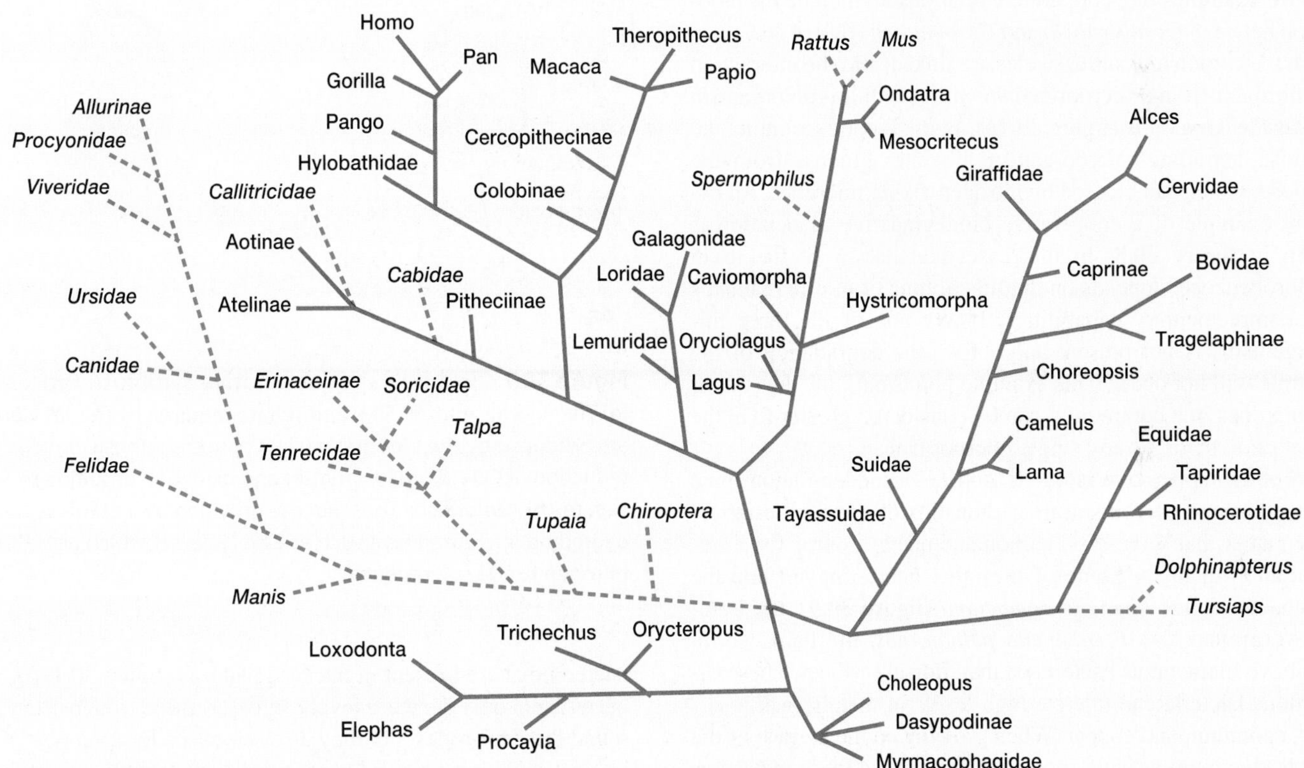

Coevolution of animals and their gut microbial communities: the methane choice. Methane-producing vertebrates are noted with solid red lines and Roman letters; nonmethane producers are noted with blue dotted lines and italics.

1. What is the critical characteristic of a mutualistic relationship?
2. How might one test to see if an insect-microbe relationship is mutualistic?
3. What is the role of *Riftia* hemoglobin to the success of the tube worm-endosymbiont mutualistic relationship?
4. How is the *Riftia* endosymbiont similar to cyanobacteria? How is it different?
5. Describe how the Sphagnum–methanotroph mutualism results in efficient carbon cycling.
6. What structural features of the rumen make it suitable for an herbivorous diet? Why do ruminants chew their cud?
7. Why is it important that the rumen is a reducing environment?

Cooperation

Cooperation and commensalism are two positive but not obligatory types of symbioses found widely in the microbial world. These involve syntrophic relationships. **Syntrophism** [Greek *syn,* together, and *trophe,* nourishment] is an association in which the growth of one organism either depends on or is improved by growth factors, nutrients, or substrates provided by another organism growing nearby. Sometimes both organisms benefit.

Cooperation benefits both organisms (figure 30.1). A cooperative relationship is not obligatory and, for most microbial ecologists, this nonobligatory aspect differentiates cooperation from mutualism. Unfortunately, it is often difficult to distinguish obligatory from nonobligatory because that which is obligatory in one habitat may not be in another (e.g., the laboratory). Nonetheless, the most useful distinction between cooperation and mutualism is the observation that cooperating organisms can be separated from one another and remain viable, although they may not function as well.

Two examples of a cooperative relationship include the association between *Desulfovibrio* and *Chromatium* (**figure 30.7*a***), in which the carbon and sulfur cycles are linked, and the interaction of a nitrogen-fixing microorganism with a cellulolytic organism such as *Cellulomonas* (figure 30.7*b*). In the second example, the cellulose-degrading microorganism liberates glucose from the cellulose, which can be used by nitrogen-fixing microbes. An excellent example of a cooperative biodegradative association is shown in **figure 30.8.** In this case degradation of the toxin 3-chlorobenzoate depends on the functioning of microorganisms with complementary capabilities. If any one of the three microorganisms is not present and active, the degradation of the substrate will not occur. This example points out how the sum of the microbes in a community can be considered greater than the contribution made by any single microorganism.

In other cooperative relations, sulfide-dependent autotrophic filamentous microorganisms fix carbon dioxide and synthesize organic matter that serves as a carbon and energy source for a heterotrophic organism. Some of the most interesting include the polychaete worms *Alvinella pompejana* (**figure 30.9**), the Pompeii worm, and also *Paralvinella palmiformis,* the Palm worm. Both have filamentous bacteria on their dorsal surfaces. These filamentous bacteria can tolerate high levels of metals such as arsenic, cadmium, and copper. When growing on the surface of the animal, they may provide protection from these toxic metals, as well as thermal protection; in addition, they appear to be used as a food source. A deep-sea crustacean has been discovered that uses sulfur-oxidizing autotrophic bacteria as its food source. This shrimp, *Rimicaris exoculata* (**figure 30.10**) has filamentous sulfur-oxidizing bacteria growing on its surface (figure 30.10*b*). When these are dislodged the shrimp ingests them. This nominally "blind" shrimp use a reflective organ to respond to the glow emitted by geothermally active black smoker chimneys. The organ is sensitive to a light wavelength that is not detectable by humans.

Another interesting example of bacterial epigrowth is shown by nematodes, including *Eubostrichus parasitiferus,* that live at the interface between oxic and anoxic sulfide-containing marine sediments (**figure 30.11*a,b***). These animals are covered by sulfide-oxidizing bacteria that are present in intricate patterns (figure 30.11*b*). The bacteria not only decrease levels of toxic sulfide, which often surround the nematodes, but they also serve as a food supply.

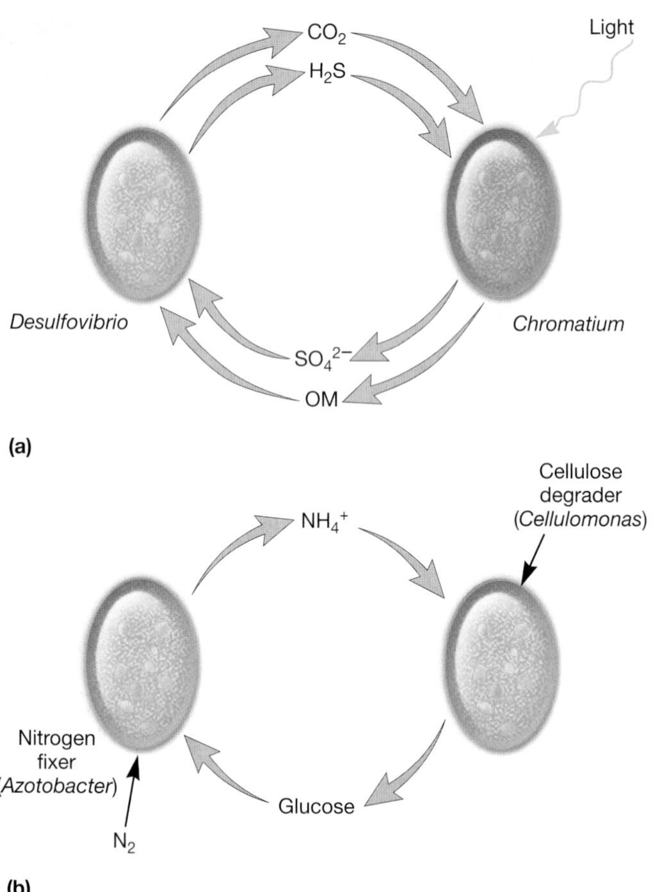

Figure 30.7 Examples of Cooperative Symbiotic Processes. **(a)** The organic matter (OM) and sulfate required by *Desulfovibrio* are produced by the *Chromatium* in its photosynthesis-driven reduction of CO_2 to organic matter and oxidation of sulfide to sulfate. **(b)** *Azotobacter* uses glucose provided by a cellulose-degrading microorganism such as *Cellulomonas,* which uses the nitrogen fixed by *Azotobacter.*

In 1990, hydrothermal vents were discovered in a freshwater environment at the bottom of Lake Baikal, the oldest (25 million years old) and deepest lake in the world. This lake is located in the far east of Russia (**figure 30.12*a,b***) and has the largest volume of any freshwater lake (not the largest area—which is Lake Superior). Microbial mats featuring long, white strands are in the center of the vent field where the highest temperatures are found. At the edge of the vent field, where the water temperature is lower, the microbial mats end, and sponges, gastropods, and other organisms, which use the sulfur-oxidizing bacteria as a food source, are present (figure 30.12*b*). Similar although less developed areas have been found in Yellowstone Lake, Wyoming.

A form of cooperation also occurs when a population of similar microorganisms monitors its own density—the process of

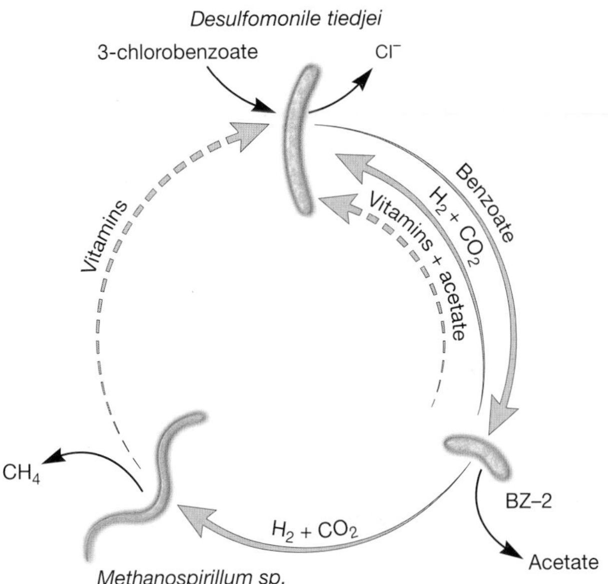

Figure 30.8 Associations in a Defined Three-Membered Cooperative and Commensalistic Community That Can Degrade 3-Chlorobenzoate. If any member is missing, degradation will not take place. The solid arrows demonstrate nutrient flows, and the dashed lines represent hypothesized flows.

Figure 30.9 A Marine Worm-Bacterial Cooperative Relationship. *Alvinella pompejana*, a 10 cm long worm, forms a cooperative relationship with bacteria that grow as long threads on the worm's surface. The bacteria and *Alvinella* are found near the black smoker-heated water fonts.

quorum sensing, which was discussed in sections 6.6 and 12.5. The microorganisms produce specific autoinducer compounds, and as the population increases and the concentration of these compounds reaches critical levels, specific genes are expressed.

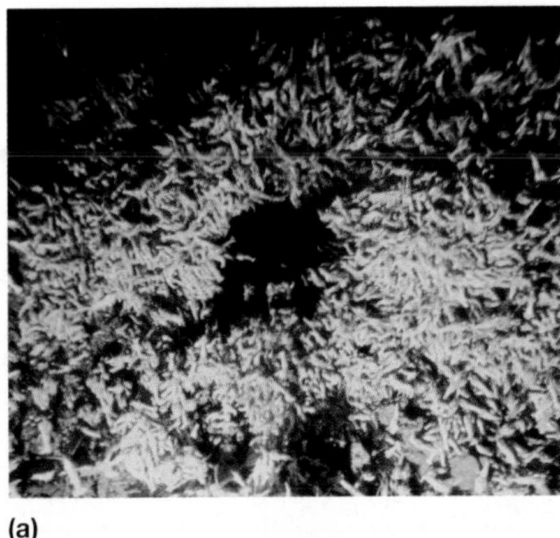

(a)

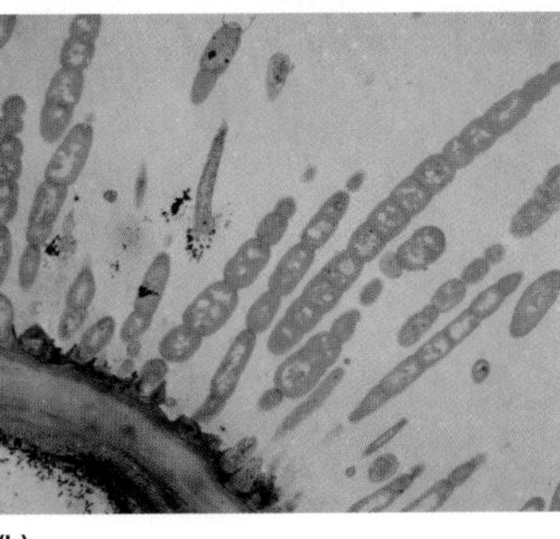

(b)

Figure 30.10 A Marine Crustacean-Bacterial Cooperative Relationship. **(a)** A picture of the marine shrimp *Rimicaris exoculata* clustered around a hydrothermal vent area, showing the massive development of these crustaceans in the area where chemolithotrophic bacteria grow using sulfide as an electron and energy source. The bacteria, which grow on the vent openings and also on the surface of the crustaceans, fix carbon, and serve as the nutrient for the shrimp. **(b)** An electron micrograph of a thin section across the leg of the marine crustacean *Rimicaris exoculata*, showing the chemolithotrophic bacteria that cover the surface of the shrimp. The filamentous nature of these bacteria, upon which this commensalistic relationship is based, is evident in this thin section.

These responses are important for microorganisms that form associations with each other, plants, and animals. Intercellular communication is critical for the establishment of biofilms and for colonization of hosts by pathogens.

1. How does cooperation differ from mutualism? What might be some of the evolutionary implications of both types of symbioses?
2. What is syntrophism? Is physical contact required for this relationship?
3. Why are *Alvinella, Rimicaris,* and *Eubostrichus* good examples of cooperative microorganism-animal interactions?
4. Where is Lake Baikal located, and why is it unique in terms of its microbial communities?

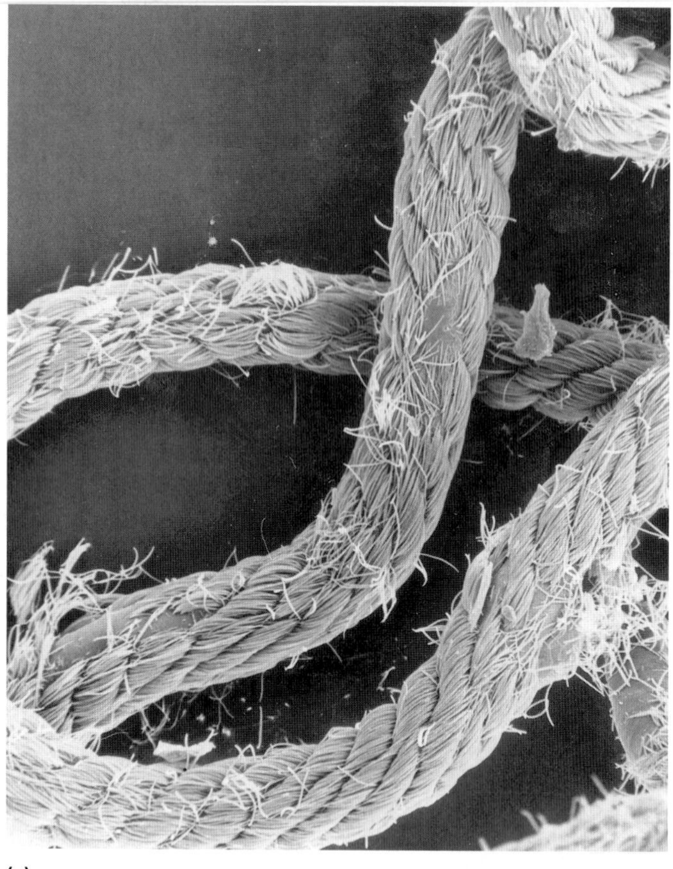
(a)

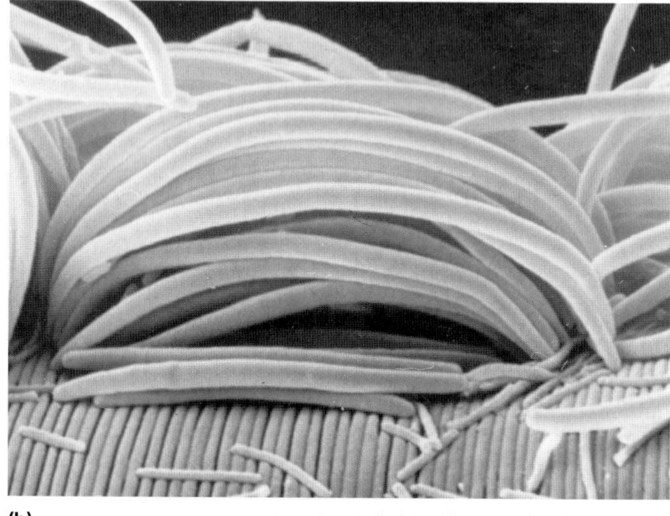
(b)

Figure 30.11 A Marine Nematode-Bacterial Cooperative Relationship. Marine free-living nematodes, which grow at the oxidized-reduced interface where sulfide and oxygen are present, are covered by sulfide-oxidizing bacteria. The bacteria protect the nematode by decreasing sulfide concentrations near the worm, and the worm uses the bacteria as a food source. **(a)** The marine nematode *Eubostrichus parasitiferus* with bacteria arranged in a characteristic helix pattern. **(b)** The chemolithotrophic bacteria attached to the cuticle of the marine nematode *Eubostrichus parasitiferus*. Cells are fixed to the nematode surface at both ends.

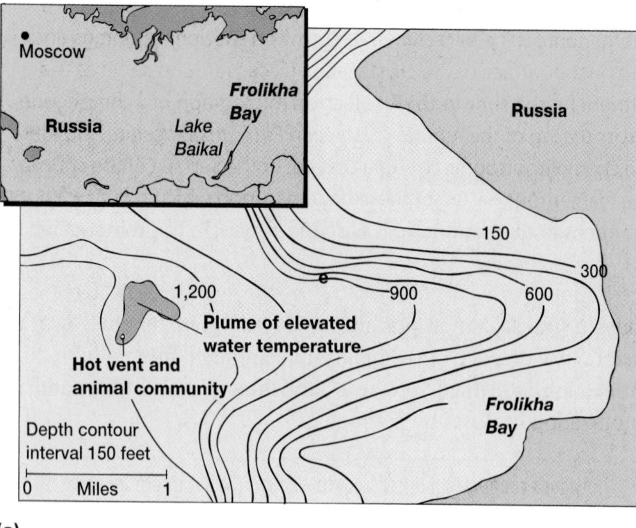

(a)

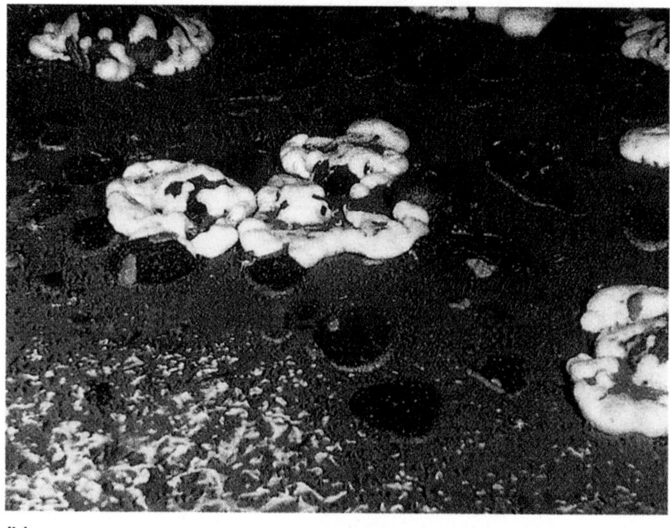
(b)

Figure 30.12 Hydrothermal Vent Ecosystems in Freshwater Environments. Lake Baikal has been found to have low temperature hydrothermal vents. **(a)** Location of Lake Baikal, site of the hydrothermal vent field. **(b)** Bacterial filaments and sponges at the edge of the vent field. *(a) Source: Data from the National Geographic Society.*

Commensalism

Commensalism [Latin *com,* together, and *mensa,* table] is a relationship in which one symbiont, the **commensal,** benefits while the other (sometimes called the host) is neither harmed nor helped, as shown in figure 30.1. This is a unidirectional process. Often both the host and the commensal "eat at the same table." The spatial proximity of the two partners permits the commensal to feed on substances captured or ingested by the host, and the commensal often obtains shelter by living either on or in the host. The commensal is not directly dependent on the host metabolically and causes it no particular harm. When the commensal is separated from its host experimentally, it can survive without the addition of factors of host origin.

Commensalistic relationships between microorganisms include situations in which the waste product of one microorganism is a substrate for another species. One good example is nitrification, the oxidation of ammonium ion to nitrite by microorganisms such as *Nitrosomonas,* and the subsequent oxidation of the nitrite to nitrate by *Nitrobacter* and similar bacteria. *Nitrobacter* benefits from its association with *Nitrosomonas* because it uses nitrite to obtain energy for growth. A second example of this type of relationship is found in anoxic methanogenic ecosystems such as sludge digesters, anoxic freshwater aquatic sediments, and flooded soils. In these environments, fatty acids can be degraded to produce H_2 and methane by the interaction of two different bacterial groups. Methane production by methanogens depends on **interspecies hydrogen transfer.** A fermentative bacterium generates hydrogen gas, and the methanogen uses it quickly as a substrate for methane gas production.

Various fermentative bacteria produce low molecular weight fatty acids that can be degraded by anaerobic bacteria such as *Syntrophobacter* to produce H_2 as follows:

$$\text{Propionic acid} \rightarrow \text{acetate} + CO_2 + H_2$$

Syntrophobacter uses protons ($H^+ + H^+ \rightarrow H_2$) as terminal electron acceptors in ATP synthesis. The bacterium gains sufficient energy for growth only when the H_2 it generates is consumed. The products H_2 and CO_2 are used by methanogenic archaea such as *Methanospirillum* as follows:

$$4H_2 + CO_2 \rightarrow CH_4 + 2H_2O$$

By synthesizing methane, *Methanospirillum* maintains a low H_2 concentration in the immediate environment of both microbes. Continuous removal of H_2 promotes further fatty acid fermentation and H_2 production. Because increased H_2 production and consumption stimulate the growth rates of *Syntrophobacter* and *Methanospirillum,* both participants in the relationship benefit.

Commensalistic associations also occur when one microbial group modifies the environment to make it more suited for another organism. For example, common, nonpathogenic strains of *Escherichia coli* live in the human colon, but also grow quite well outside the host, and thus are typical commensals. When oxygen is used up by facultatively anaerobic *E. coli,* obligate anaerobes such as *Bacteroides* are able to grow in the colon. The anaerobes benefit from their association with the host and *E. coli,* but *E. coli* derives no obvious benefit from the anaerobes. In this case the commensal *E. coli* contributes to the welfare of other symbionts. Commensalism can involve other environmental modifications. The synthesis of acidic waste products during fermentation stimulate the proliferation of more acid-tolerant microorganisms, which are only a minor part of the microbial community at neutral pH. A good example is the succession of microorganisms during milk spoilage. Biofilm formation provides another example. The colonization of a newly exposed surface by one type of microorganism (an initial colonizer) makes it possible for other microorganisms to attach to the microbially modified surface. Microbial growth in natural environments: Biofilms (section 6.6); The physical environment: Biofilms and microbial mats (section 27.3)

Commensalism also is important in the colonization of the human body and the surfaces of other animals and plants. The microorganisms associated with an animal's skin and body orifices can use volatile, soluble, and particulate organic compounds from the host as nutrients. Under most conditions these microbes do not cause harm, other than possibly contributing to body odor. Sometimes when the host organism is stressed or the skin is punctured, these normally commensal microorganisms may become pathogenic by entering a different environment. These interactions are discussed in more detail in section 30.2.

1. How does commensalism differ from cooperation?
2. Why is nitrification a good example of a commensalistic process?
3. What is interspecies hydrogen transfer, and why can this be beneficial to both producers and consumers of hydrogen?
4. Why are commensalistic microorganisms important for humans? Where are they found in relation to the human body?

Predation

As is the case with larger organisms, **predation** among microbes involves a predator species that attacks and usually kills its prey. Over the last several decades, microbiologists have discovered a number of fascinating bacteria that survive by their ability to prey upon other microbes. Several of the best examples are *Bdellovibrio, Vampirococcus,* and *Daptobacter* (**figure 30.13**).

Bdellovibrio is an active hunter that is vigorously motile, swimming about looking for susceptible gram-negative bacterial prey. Upon sensing such a cell, *Bdellovibrio* swims faster until it collides with the prey cell. It then bores a hole through the outer membrane of its prey and enters the periplasmic space. As it grows, it forms a long filament that eventually septates to produce progeny bacteria. Lysis of the prey cell releases new *Bdellovibrio* cells. *Bdellovibrio* will not attack mammalian cells, and gram-negative prey bacteria have never been observed to acquire resistance to *Bdellovibrio* attack. This has raised interest in the use of *Bdellovibrio* as a "probiotic" to treat infected wounds. Although this is has not yet been tried, one can imagine

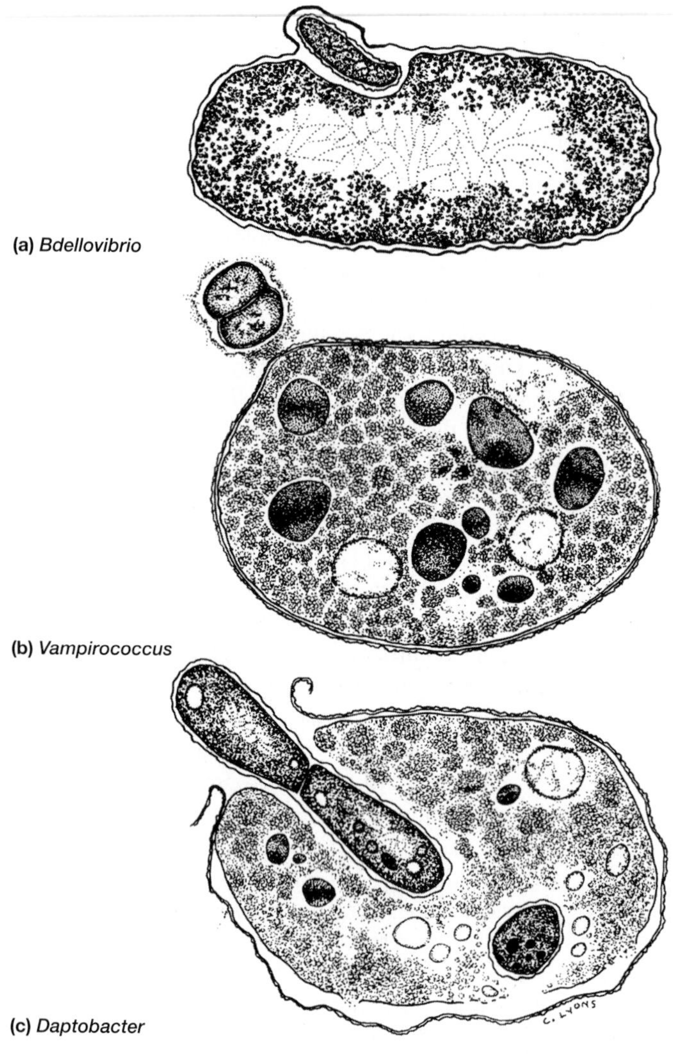

Figure 30.13 Examples of Predatory Bacteria Found in Nature. **(a)** *Bdellovibrio*, a periplasmic predator that penetrates the cell wall and grows outside the plasma membrane, **(b)** *Vampirococcus* with its unique epibiotic mode of attacking a prey bacterium, and **(c)** *Daptobacter* showing its cytoplasmic location as it attacks a susceptible bacterium.

that with the rise in antibiotic-resistant pathogens, such forms of treatments may become viable alternatives. Class *Deltaproteobacteria* (section 22.4); Microbiology of fermented foods: Probiotics (section 40.6)

Although *Vampirococcus* and *Daptobacter* also kill their prey, they gain entry in a less-dramatic fashion. *Vampirococcus* attaches itself as an epibiont to the outer membrane of its prey. It then secretes degradative enzymes that result in the release of the prey's cytoplasmic contents. In contrast, *Daptobacter* penetrates the prey cell and consumes the cytoplasmic contents directly.

A surprising finding is that predation has many beneficial effects, especially when one considers interactive populations of predators and prey, as summarized in **table 30.3.** Simple ingestion and assimilation of a prey bacterium can lead to increased rates of nutrient cycling, critical for the functioning of the microbial loop *(see figure 27.13)*. In this process, organic matter produced through photosynthetic and chemotrophic activity is mineralized before it reaches higher consumers, allowing the minerals to be made available to the primary producers. Ingestion and short-term retention of bacteria also are critical for ciliate functioning in the rumen, where methanogenic bacteria contribute to the health of the ciliates by decreasing toxic hydrogen levels by using H_2 to produce methane, which then is passed from the rumen. The physical environmnet: Microorganisms and ecosystems (section 27.3)

Predation also can provide a protective, high-nutrient environment for particular prey. Ciliates ingest the gram-positive bacterium *Legionella* and protect this important pathogen from chlorine, which often is used in an attempt to control *Legionella* in cooling towers and air-conditioning units. The ciliate serves as a reservoir host. *Legionella pneumophila* also has been found to have a greater potential to invade macrophages and epithelial cells after predation, indicating that ingestion not only provides protection but also may enhance pathogenicity. A similar phenomenon of survival in protozoa has been observed for *Mycobacterium avium,* a pathogen of worldwide concern. These protective aspects of predation have major implications for survival and control of disease-causing microorganisms in the biofilms present in water supplies and air-conditioning systems. Airborne diseases (section 38.1)

Fungi often show interesting predatory skills. Some fungi can trap protozoa by the use of sticky hyphae or knobs, sticky networks of hyphae, or constricting or nonconstricting rings. A classic example is *Arthrobotrys,* which traps nematodes by use of constricting rings. After the nematode is trapped, hyphae grow into the immobilized prey and the cytoplasm is used as a nutrient. Other fungi have conidia that, after ingestion by an unsuspecting predator, grow and attack the susceptible host from inside the intestinal tract. In this situation the fungus penetrates the host cells in a complex interactive process.

Clearly predation in the microbial world is not straightforward. It often has a fatal and final outcome for an individual prey organism but it can have a wide range of beneficial effects on prey populations. Predation is clearly critical in the functioning of natural environments.

Parasitism

Parasitism is one of the most complex microbial interactions; the line between parasitism and predation is difficult to define (figure 30.1). This is a relationship between two organisms in which one benefits from the other, and the host is usually harmed. This can involve nutrient acquisition and/or physical maintenance in or on the host. In parasitism there is always some co-existence between host and parasite. Successful parasites have evolved to co-exist in equilibrium with their hosts. This is because a host that dies immediately after parasite invasion may prevent the microbe from reproducing to sufficient numbers to ensure colonization of a new host. But what happens if the host-parasite equilibrium is upset? If the balance favors the host (perhaps by a strong immune defense or

Table 30.3 The Many Faces of Predation

Predation Result	Example
Digestion	The microbial loop. Soluble organic matter from primary producers is normally used by bacteria, which become a particulate food source for higher consumers. Flagellates and ciliates prey on these bacteria and digest them, making the nutrients they contain available again in mineral form for use in primary production. In this way a large portion of the carbon fixed by the photosynthetic microbes is mineralized and recycled and does not reach the higher trophic levels of the ecosystem (*see figure 27.13*).
	Predation also can reduce the density-dependent stress factors in prey populations, allowing more rapid growth and turnover of the prey than would occur if the predator were not active.
Retention	Bacteria retained within the predator serve a useful purpose, as in the transformation of toxic hydrogen produced by ciliates in the rumen to methane. Also, trapping of chloroplasts (kleptochloroplasty) by protozoa provides the predator with photosynthate.
Protection and increased fitness	The intracellular survival of *Legionella* ingested by ciliates protects it from stresses such as heating and chlorination. Ingestion also results in increased pathogenicity when the prey is again released to the external environment, and this may be required for infection of humans. The predator serves as a reservoir host.
	Nanoplankton may be ingested by zooplankton and grow in the zooplankton digestive system. They are then released to the environment in a more fit state. Dissemination to new locations also occurs.

antimicrobial therapy), the parasite loses its habitat and may be unable to survive. On the other hand, if the equilibrium is shifted to favor the parasite, the host becomes ill, and depending on the specific host-parasite relationship, may die. One good example is the disease typhus. This disease is caused by the rickettsia *Rickettsia typhi,* which is harbored in fleas that live on rats. It is transmitted to humans who are bitten by such fleas, so in order to contract typhus, one must be in close proximity to rats. Humans often live in association with rats, and in such communities there is always a small number of people with typhus—that is to say, typhus is endemic. However, during times of war or when people are forced to become refugees, lack of sanitation and overcrowding result in an increased number of rat-human interactions. Typhus can then reach epidemic proportions. During the Crimean War (1853–1856), about 213,000 men were killed or wounded in combat while over 850,000 were sickened or killed by typhus.

On the other hand, a controlled parasite-host relationship can be maintained for long periods of time. For example, **lichens (figure 30.14)** are the association between specific ascomycetes (a fungus) and certain genera of either green algae or cyanobacteria. In a lichen, the fungal partner is termed the **mycobiont** and the algal or cyanobacterial partner, the **phycobiont.** In the past the lichen symbiosis was considered to be a mutualistic interaction. It recently has been found that a lichen forms only when the two potential partners are nutritionally deprived. In nutrient-limited environments, the relationship between the fungus and its photosynthetic partner has coevolved to the point where lichen morphology and metabolic relationships are extremely stable. In fact, lichens are assigned generic and species names. The characteristic morphology of a given lichen is a property of the association and is not exhibited by either symbiont individually.

Characteristics of fungal divisions: *Ascomycota* (section 26.6); Photosynthetic bacteria: Phylum *Cyanobacteria* (section 21.3)

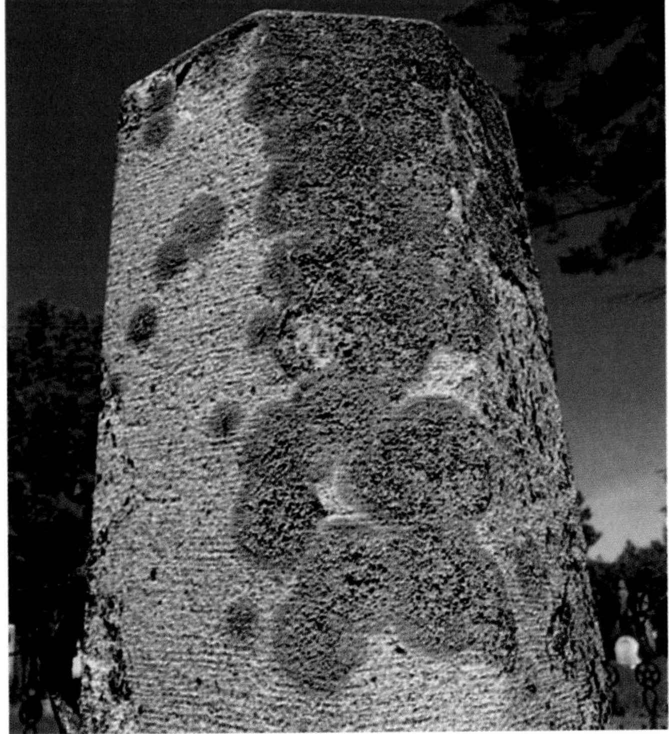

Figure 30.14 Lichens. Crustose (encrusting) lichens growing on a granite post.

Because the phycobiont is a photoautotroph—dependent only on light, carbon dioxide, and mineral nutrients—the fungus can get its organic carbon directly from the alga or cyanobacterium. The fungus often obtains nutrients from its partner by projections of fungal hyphae called haustoria, which penetrate the phycobiont cell wall. It also uses the O_2 produced during phycobiont photophosphorylation in carrying out respiration. In turn the fungus protects the phycobiont from high light intensities, provides water and minerals to it, and creates a firm substratum within which the phycobiont can grow protected from environmental stress. The invasive nature of the fungal partner is why lichens are considered parasitic relationships.

An important aspect of many symbiotic relationships, including parasitism, is that over time, the symbiont, once it has established a relationship with the host, will tend to discard excess, unused genomic information, a process called **genomic reduction.** This is clearly the case with the aphid endosymbiont *Buchnera aphidicola* (p. 718) and it has also occurred with the parasite *Mycobacterium leprae,* and with the microsporidium *Encephalitozoon cuniculi.* The latter organism, which parasitizes a wide range of animals, including humans, now can only survive inside the host cell. Insights from microbial genomes (section 15.8)

1. Define predation and parasitism. How are these similar and different?
2. How can a predator confer positive benefits on its prey? Think of the responses of individual organisms versus populations as you consider this question.
3. What are examples of parasites that are important in microbiology?
4. What is a lichen? Discuss the benefits the phycobiont and mycobiont provide each other.

Amensalism

Amensalism (from the Latin for *not* at the same table) describes the adverse effect that one organism has on another organism (figure 30.1). This is a unidirectional process based on the release of a specific compound by one organism which has a negative effect on another organism. A classic example of amensalism is the production of antibiotics that can inhibit or kill a susceptible microorganism (**figure 30.15*a***). Community complexity is demonstrated by the capacity of attine ants (ants belonging to a New World tribe) to take advantage of an amensalistic relationship between an actinomycete and the parasitic fungi *Escovopsis.* This amensalistic relationship enables the ant to maintain a mutualism with another fungal species, *Leucocoprini.* Amazingly, these ants cultivate a garden of *Leucocoprini* for their own nourishment (figure 30.15b). To prevent the parasitic fungus *Escovopsis* from decimating their fungal garden, the ants also promote the growth of an actinomycete of the genus *Pseudonocardia,* which produces an antimicrobial compound that inhibits the growth of *Escovopsis.* This unique amensalistic process appears to have evolved 50 to 65 million years ago in South America. Thus this relationship has been subject to millions of years of coevolution, such that particular groups of ants cultivate specific strains of fungi that are then subject to different groups of *Escovopsis* parasites. In addition, the ants have developed intricate crypts within their exoskeletons for the growth of the antibiotic-producing *Pseudonocardia.* As shown in figure 30.15*c,* these crypts have been modified throughout the ants' evolutionary history. The most primitive "paleo-attine" ants carry the bacterium on their forelegs, "lower" and "higher" attines have evolved special plates on their ventral surfaces, while the entire surface of the most recent attines, leaf-cutter ants of the genus *Acromyrmex,* are covered with the bacterium. Related ants that do not cultivate fungal gardens (e.g., *Atta* sp.) do not host *Pseudonocardia.* This unique multipartner relationship has enabled scientists to explore the behavioral, physiological, and structural aspects of the organisms involved.

Other important amensalistic relationships involve microbial production of specific organic compounds that disrupt cell wall or plasma membrane integrity of target microorganisms. These include the bacteriocins. The bacteriocin nisin has been used as an additive for controlling the growth of undesired pathogens in dairy products for over 40 years. Antibacterial peptides also can be released by the host in the intestine and other sites. These molecules, called cecropins in insects and defensins in mammals, are effector molecules that play significant roles in innate immunity. In vertebrates these molecules are released by phagocytes and intestinal cells, and have powerful antimicrobial activity. Human sweat is also antimicrobial. Sweat glands produce an antimicrobial peptide called dermicidin. The skin also produces similar compounds including an antimicrobial peptide called cathelicidin. Finally, metabolic products, such as organic acids formed in fermentation, can produce amensalistic effects. These compounds inhibit growth by changing the environmental pH, for example, during natural milk spoilage. Chemical mediators in nonspecific (innate) resistance: Antimicrobial peptides (section 31.6)

Competition

Competition arises when different organisms within a population or community try to acquire the same resource, whether this is a physical location or a particular limiting nutrient (figure 30.1). If one of the two competing organisms can dominate the environment, whether by occupying the physical habitat or by consuming a limiting nutrient, it will outgrow the other organism. This phenomenon was studied by E. F. Gause, who in 1934 described it as the **competitive exclusion principle.** He found that if two competing ciliates overlapped too much in terms of their resource use, one of the two protozoan populations was excluded. In chemostats, competition for a limiting nutrient may occur among microorganisms with transport systems of differing affinity. This can lead to the exclusion of the slower-growing population under a particular set of conditions. If the dilution rate is changed, the previously slower-growing population may become dominant. Often two microbial populations that appear to be similar nevertheless coexist. In this case, they share the limiting resource (space, a limiting nutrient) and coexist while surviving at lower population levels. The continuous culture of microorganisms: The chemostat (section 6.4)

1. What is the origin of the term amensalism?
2. The production of antimicrobial agents and their effects on target species has been called a "microbial arms race." Explain what you think this phrase means and its implications for the attine ant system.
3. What are bacteriocins?
4. What is the competitive exclusion principle?

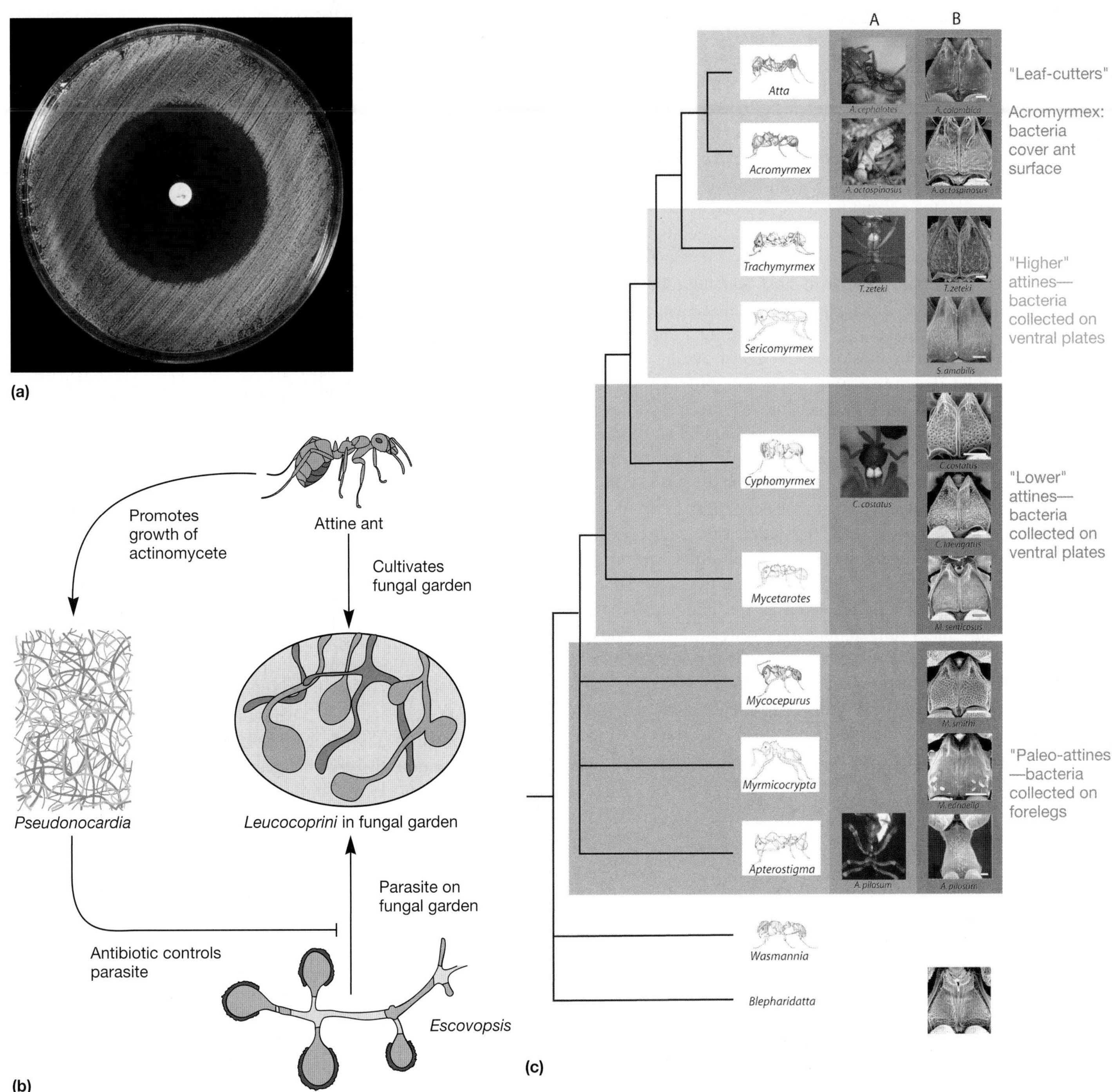

Figure 30.15 Amensalism: An Adverse Microbe-Microbe Interaction. **(a)** Antibiotic production and inhibition of growth of a susceptible bacterium on an agar medium. **(b)** A schematic diagram describing the use of antibiotic-producing streptomycetes by ants to control fungal parasites in their fungal garden. **(c)** Coevolution of attine ants and the antibiotic-producing *Pseudonocardia* has resulted in specialized localization of the bacterium on the ant. This rooted tree illustrates the phylogeny of fungus-growing ants; column A shows the placement of the bacteria on the ants' body, column B presents scanning electron micrographs of the areas colonized by the microbe.

30.2 Human-Microbe Interactions

As we have seen, many microorganisms live much of their lives in a special ecological relationship: an important part of their environment is a member of another species. Here we discuss microorganisms normally associated with the human body, the **normal microbial flora** or **microbiota.** If we consider the human body as a diverse environment in and on which specific niches are formed, the normal flora may be discussed as the microbial ecology of a human. The application of ecological principles can assist in our understanding of host-microbe relationships. Interactions between host and microbe are dynamic, permitting niche fulfillment that maximizes benefit to the microbe and, in some cases, the host. Tolerating a normal flora likewise suggests that the host derives benefit. Acquisition of a normal microbial flora represents a selective process, where a niche may be defined by cellular receptors, surface properties, or secreted products. It should be noted that microbial niche variations are also related to age, gender, diet, nutrition, and developmental stage of the host. In general, the adult human microbial flora is relatively constant and can thus be mapped.

The survival of a host, such as a human, depends upon an elaborate network of defenses that keeps harmful microorganisms and other foreign material from entering the body. Should they gain access, additional host defenses are summoned to prevent them from establishing another type of relationship, one of parasitism or **pathogenicity.** Pathogenicity is the ability to produce pathologic changes or disease. A **pathogen** [Greek *patho,* and *gennan,* to produce] is any disease-producing microorganism. Here we introduce the normal human microbiota, which function not as pathogens but as symbionts that are part of the host's first line of defense against harmful infectious agents.

Gnotobiotic Animals

To determine the role of the normal microorganisms associated with a host and evaluate the consequences of colonization, it is possible to deliver an animal by cesarean section and raise that animal in the absence of microorganisms—that is, germfree. These microorganism-free animals provide suitable experimental models for investigating the interactions of animals and their microorganisms. Comparing animals possessing normal microorganisms (conventional animals) with germfree animals permits the elucidation of many complex relationships between microorganisms, hosts, and specific environments. Germfree experiments also extend and challenge the microbiologist's "pure culture concept" to in vivo research.

The term **gnotobiotic** [Greek *gnotos,* known, and *biota,* the flora and fauna of a region] has been defined in two ways. Some think of a gnotobiotic environment or animal as one in which all the microbiota are known; they distinguish it from one that is truly germfree. We shall use the term in a more inclusive sense. Gnotobiotic refers to a microbiologically monitored environment or animal that is germfree (**axenic** [*a,* neg, and Greek *Xenos,* a stranger]) or in which the identities of all microbiota are known.

Development of a lifelong symbiotic relationship with microbes begins during birth. The infant's exposure to the vaginal mucosa, skin, hair, food, and other nonsterile objects quickly results in the acquisition of a predominantly commensal normal flora. The human fetus in utero (as is the case in most mammals) is usually free from microorganisms. As an infant begins to acquire a normal microbiota, the microbial population stabilizes during the first week or two of life. Colonization of the newborn varies with respect to its environment. The newborn likely acquires external flora from those who provide its care. Likewise, internal flora are acquired through its diet. Bifidobacteria represent more than 90% of the total intestinal bacteria in breast-fed infants, with *Enterobacteriaceae* and enterococci in smaller proportions. This suggests that human milk may act as a selective medium for nonpathogenic bacteria, as bottle-fed babies appear to have a much smaller proportion of intestinal bifidobacteria. Switching to cow's milk or solid food (mostly polysaccharide) appears to result in the loss of bifidobacteria predominance, as *Enterobacteriaceae,* enterococci, bacteroides, lactobaccili, and clostridia increase in number. Bacterial chemotaxis and trophism may explain the high frequency of bacterial-tissue associations. Additionally, the host may partly direct microbe-tissue associations as seen in the selective destruction of gram-positive bacteria by the antimicrobial peptide angiogenin-4 secreted by special intestinal cells called **Paneth cells.**

Louis Pasteur first suggested that animals could not live in the absence of microorganisms. Attempts between 1899 and 1908 to grow germfree chickens had limited success because the birds died within a month. Thus it was believed that intestinal bacteria were essential for the adequate nutrition and health of the chickens. It was not until 1912 that germfree chickens were shown to be as healthy as normal birds when they were fed an adequate diet. Since then, gnotobiotic animals and systems have become commonplace in research laboratories (**figure 30.16**).

What have we learned from germfree animals? Germfree animals are usually more susceptible to pathogens. With the normal commensal microbiota absent, foreign and pathogenic microorganisms establish themselves very easily. The number of microorganisms necessary to infect a germfree animal and produce a diseased state is much smaller. Conversely, germfree animals are almost completely resistant to the intestinal protozoan *Entamoeba histolytica* that causes amebic dysentery. This resistance results from the absence of the bacteria that *E. histolytica* uses as a food source. Germfree animals also do not show any dental caries or plaque formation. However, if they are inoculated with cariogenic (caries or cavity-causing) streptococci of the *Streptococcus mutans–Streptococcus gordonii* group and fed a high-sucrose diet, they develop caries. Protist classification: Super group *Amoebozoa* (section 25.6); Food- and water-borne diseases (sections 37.4, 38.4, and 39.5)

1. Define gnotobiotic.
2. Compare a germfree mouse to a normal one with regard to overall susceptibility to pathogens. What benefits does an animal gain from its microbiota?

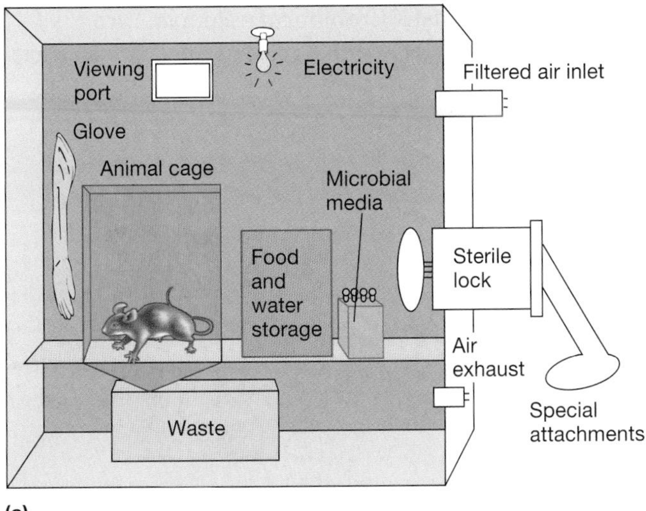

(a)

(b)

Figure 30.16 Raising Gnotobiotic Animals. **(a)** Schematic of a gnotobiotic isolator. The microbiological culture media monitor the sterile environment. If growth occurs on any of the cultures, gnotobiotic conditions do not exist. **(b)** Gnotobiotic isolators for rearing colonies of small mammals.

30.3 Normal Microbiota of the Human Body

In a healthy human the internal tissues (e.g., brain, blood, cerebrospinal fluid, muscles) are normally free of microorganisms. Conversely, the surface tissues (e.g., skin and mucous membranes) are constantly in contact with environmental microorganisms and become readily colonized by various microbial species. The mixture of microorganisms regularly found at any anatomical site is referred to as the normal microbiota, the indigenous microbial population, the microflora, or the normal flora. For consistency, the term normal microbiota is used in this chapter. An overview of the microbiota native to different regions of the body is presented next (**figure 30.17**). Because bacteria make up most of the normal microbiota, they are emphasized over the fungi (mainly yeasts) and protists.

There are many reasons to acquire knowledge of the normal human microbiota. Three specific examples include:

1. An understanding of the different microorganisms at particular locations provides greater insight into the possible infections that might result from injury to these body sites.
2. A knowledge of the normal microbiota helps the physician-investigator understand the causes and consequences of colonization and growth by microorganisms normally absent at a specific body site.
3. An increased awareness of the role that these normal microbiota play in stimulating the host immune response can be gained. This awareness is important because the immune system provides protection against potential pathogens.

As noted previously, three of the most important types of symbiotic relationships are commensalism, mutualism, and parasitism. Within each category the association may be either ectosymbiotic or endosymbiotic. In the following subsections, examples are presented of both ecto- and endosymbiotic relationships. Both commensalistic and mutualistic relationships are also considered. Parasitism and pathogenicity are presented in chapter 33.

Skin

The adult human is covered with approximately 2 square meters of skin. It has been estimated that this surface area supports about 10^{12} bacteria. Recall that commensalism is a symbiotic relationship in which one species benefits and the other is unharmed. Commensal microorganisms living on or in the skin can be either resident (normal) or transient microbiota. Resident organisms normally grow on or in the skin. Their presence becomes fixed in well-defined distribution patterns. Those that are temporarily present are transient microorganisms. Transients usually do not become firmly entrenched and are unable to multiply.

It should be emphasized that the skin is a mechanically strong barrier to microbial invasion. Few microorganisms can penetrate the skin because its outer layer consists of thick, closely packed cells called keratinocytes. In addition to direct resistance to penetration, continuous shedding of the outer epithelial cells removes many of those microorganisms adhering to the skin surface.

The anatomy and physiology of the skin vary from one part of the body to another, and the normal resident microbiota reflect these variations. The skin surface or epidermis is not a favorable environment for microbial colonization. In addition to a slightly acidic pH, a high concentration of sodium chloride, and a lack of moisture in many areas, certain inhibitory substances (bactericidal and/or bacteriostatic) on the skin help control microbial colonization. For example, the sweat glands release lysozyme (muramidase), an enzyme that lyses *Staphylococcus epidermidis* and other gram-positive bacteria by hydrolyzing the $\beta(1 \rightarrow 4)$ glycosidic bond connecting *N*-acetylmuramic acid and *N*-acetylglucosamine

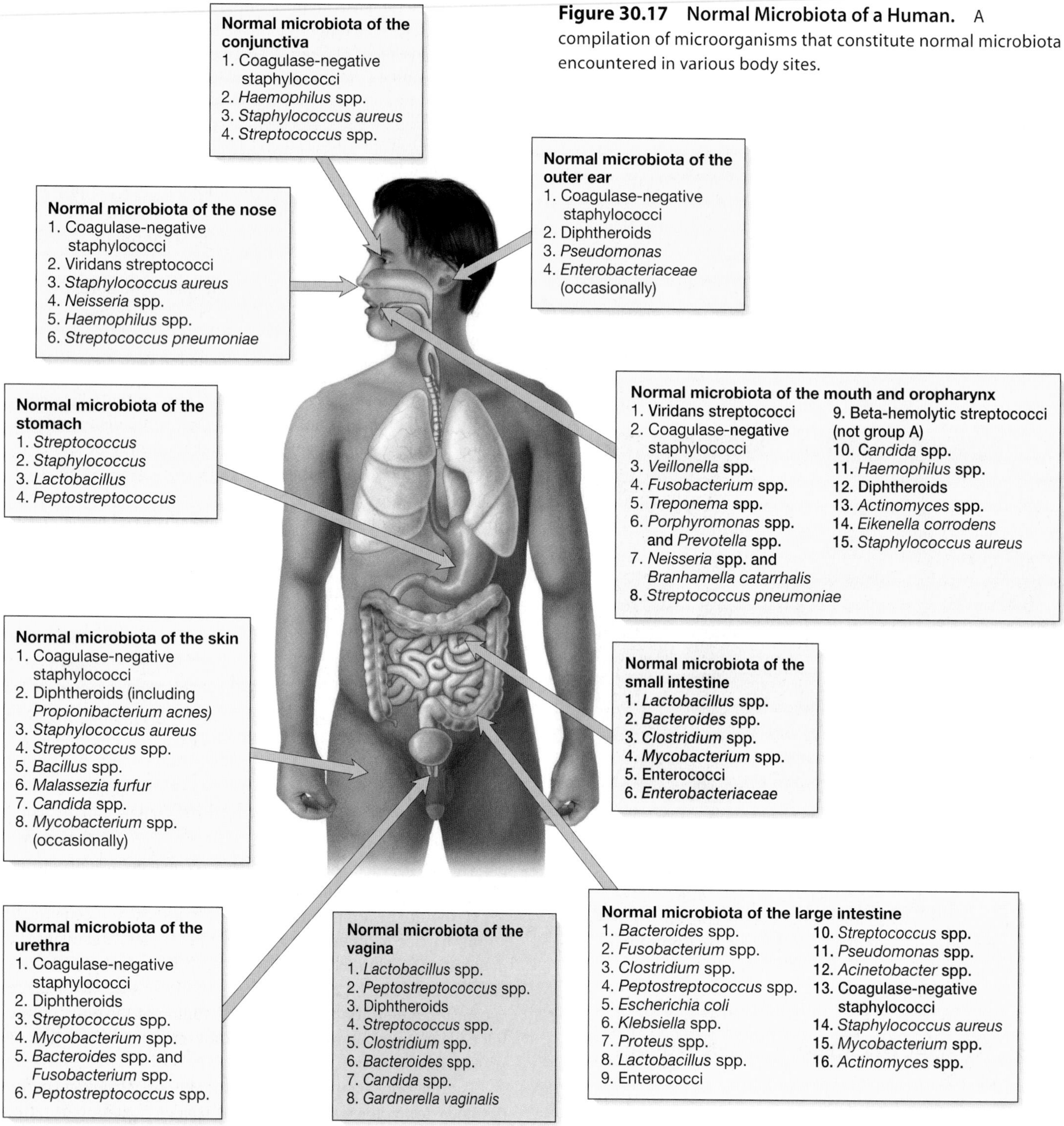

Figure 30.17 Normal Microbiota of a Human. A compilation of microorganisms that constitute normal microbiota encountered in various body sites.

in the bacterial cell wall peptidoglycan (*see figure 31.17*). Sweat glands also produce antimicrobial peptides called **cathelicidins** (Latin *catharticus,* to purge, and *cida,* to kill) that help protect against infectious agents by forming pores in bacterial plasma membranes. The bacterial cell wall (section, 3.6); Chemical mediators in nonspecific (innate) resistance (section 31.6)

The oil glands secrete complex lipids that may be partially degraded by the enzymes from certain gram-positive bacteria (e.g., *Propionibacterium acnes*). These bacteria can change the secreted lipids to unsaturated fatty acids such as oleic acid that have strong antimicrobial activity against gram-negative bacteria and some fungi. Some of these fatty acids are volatile and may be associated with a strong odor. Therefore many deodorants contain antibacterial substances that act selectively against gram-positive bacteria to reduce the production of volatile unsaturated fatty acids and body odor.

Most skin bacteria are found on superficial cells, colonizing dead cells, or closely associated with the oil and sweat glands. Secretions from these glands provide the water, amino acids, urea, electrolytes, and specific fatty acids that serve as nutrients primarily for *S. epidermidis* and aerobic corynebacteria. Gram-negative bacteria generally are found in the moister regions. The yeasts *Pityrosporum ovale* and *P. orbiculare* normally occur on the scalp.

The most prevalent bacterium in the oil glands is the gram-positive, anaerobic, lipophilic rod *Propionibacterium acnes*. This bacterium usually is harmless; however, it is associated with the skin disease acne vulgaris. Acne commonly occurs during adolescence when the endocrine system is very active. Hormonal activity stimulates an overproduction of **sebum,** a fluid secreted by the oil glands. A large volume of sebum accumulates within the glands and provides an ideal microenvironment for *P. acnes*. In some individuals this accumulation triggers an inflammatory response that causes redness and swelling of the gland's duct and produces a **comedo** [pl., comedones], a plug of sebum and keratin in the duct. Inflammatory lesions (papules, pustules, nodules) commonly called "blackheads" or "pimples" can result when pores or ducts clog with cebum or bacteria. *P. acnes* produces lipases that hydrolyse the sebum triglycerides into free fatty acids. Free fatty acids are especially irritating because they can enter the dermis and promote inflammation. Because *P. acnes* is extremely sensitive to tetracycline, this antibiotic may aid acne sufferers. Retin A and accutane, synthetic forms of vitamin A, are also used.

1. What are three reasons why knowledge of the normal human microbiota is important?
2. Why is the skin not usually a favorable microenvironment for colonization by bacteria?
3. How do microorganisms contribute to body odor?
4. What physiological role does *Propionibacterium acnes* play in the establishment of acne vulgaris?

Nose and Nasopharynx

The normal microbiota of the nose is found just inside the nostrils. *Staphylococcus aureus* and *S. epidermidis* are the predominant bacteria present and are found in approximately the same numbers as on the skin of the face.

The nasopharynx, that part of the pharynx lying above the level of the soft palate, may contain small numbers of potentially pathogenic bacteria such as *Streptococcus pneumoniae, Neisseria meningitidis,* and *Haemophilus influenzae*. Diphtheroids, a large group of nonpathogenic gram-positive bacteria that resemble *Corynebacterium,* are commonly found in both the nose and nasopharynx.

Oropharynx

The oropharynx is that division of the pharynx lying between the soft palate and the upper edge of the epiglottis. The most important bacteria found in the oropharynx are the various alpha-hemolytic streptococci (*S. oralis, S. milleri, S. gordonii, S. salivarius*); large numbers of diphtheroids; *Branhamella catarrhalis;* and small gram-negative cocci related to *N. meningitidis*. The palatine and pharyngeal tonsils harbor a similar microbiota, except within the tonsillar crypts, where there is an increase in *Micrococcus* and the anaerobes *Porphyromonas, Prevotella,* and *Fusobacterium*.

Respiratory Tract

The upper and lower respiratory tracts (trachea, bronchi, bronchioles, alveoli) do not have a normal microbiota. This is because microorganisms are removed in at least three ways. First, a continuous stream of mucus is generated by the goblet cells. This entraps microorganisms, and the ciliated epithelial cells continually move the entrapped microorganisms out of the respiratory tract. Second, alveolar macrophages phagocytize and destroy microorganisms. Finally, a bactericidal effect is exerted by the enzyme lysozyme, which is present in the nasal mucus.

Eye

At birth and throughout human life, a small number of bacterial commensals are found on the conjunctiva of the eye. The predominant bacterium is *S. epidermidis* followed by *S. aureus, Haemophilus* spp., and *S. pneumoniae*.

External Ear

The normal microbiota of the external ear resemble those of the skin, with coagulase-negative staphylococci and *Corynebacterium* predominating. Mycological studies show the following fungi to be normal microbiota: *Aspergillus, Alternaria, Penicillium, Candida,* and *Saccharomyces*.

Mouth

The normal microbiota of the mouth or oral cavity contains organisms that resist mechanical removal by adhering to surfaces like the gums and teeth. Those that cannot attach are removed by the mechanical flushing of the oral cavity contents to the stomach where they are destroyed by hydrochloric acid. The continuous desquamation (shedding) of epithelial cells also removes microorganisms. Those microorganisms able to colonize the mouth find a very comfortable environment due to the availability of water and nutrients, the suitability of pH and temperature, and the presence of many other growth factors. Biofilms (sections 6.6 and 27.3)

The oral cavity is colonized by microorganisms from the surrounding environment within hours after a human is born. Initially the microbiota consists mostly of the genera *Streptococcus, Neisseria, Actinomyces, Veillonella,* and *Lactobacillus*. Some yeasts also are present. Most microorganisms that invade the oral cavity initially are aerobes and obligate anaerobes. When the first teeth erupt, anaerobes (*Porphyromonas, Prevotella,* and *Fusobacterium*) become dominant due to the anoxic nature of the space between the teeth and gums. As the teeth grow, *Streptococcus parasanguis* and *S. mutans* attach to their enamel surfaces; *S. salivarius* attaches to the buccal and gingival epithelial surfaces and colonizes the saliva. These streptococci produce a glycocalyx and various other adherence factors that enable them to attach to oral surfaces. The presence of these bacteria contributes to the eventual formation of dental plaque, caries, gingivitis, and periodontal disease. Dental infections (section 38.7)

Stomach

As noted earlier, many microorganisms are washed from the mouth into the stomach. Owing to the very acidic pH values (2 to 3) of the gastric contents, most microorganisms are killed. As a result the stomach usually contains less than 10 viable bacteria per milliliter of gastric fluid. These are mainly *Streptococcus, Staphylococcus, Lactobacillus, Peptostreptococcus,* and yeasts such as *Candida* spp. Microorganisms may survive if they pass rapidly through the stomach or if the organisms ingested with food are particularly resistant to gastric pH (e.g., mycobacteria).

Small Intestine

The small intestine is divided into three anatomical areas: the duodenum, jejunum, and ileum. The duodenum (the first 25 cm of the small intestine) contains few microorganisms because of the combined influence of the stomach's acidic juices and the inhibitory action of bile and pancreatic secretions that are added here. Of the bacteria present, gram-positive cocci and rods comprise most of the microbiota. *Enterococcus faecalis, lactobacilli,* diphtheroids, and the yeast *Candida albicans* are occasionally found in the jejunum. In the distal portion of the small intestine (ileum), the microbiota begin to take on the characteristics of the colon microbiota. It is within the ileum that the pH becomes more alkaline. As a result anaerobic gram-negative bacteria and members of the family *Enterobacteriaceae* become established.

Large Intestine (Colon)

The large intestine or colon has the largest microbial community in the body. Microscopic counts of feces approach 10^{12} organisms per gram wet weight. Over 400 different species have been isolated from human feces. The microbiota consist primarily of anaerobic, gram-negative bacteria and gram-positive, spore-forming, and nonsporing rods. Not only are the vast majority of microorganisms anaerobic, but many different species are present in large numbers. Several studies have shown that the ratio of anaerobic to facultative anaerobic bacteria is approximately 300 to 1. Besides the many bacteria in the large intestine, the yeast *Candida albicans* and certain protozoa may occur as harmless commensals. *Trichomonas hominis, Entamoeba hartmanni, Endolimax nana,* and *Iodamoeba butschlii* are common inhabitants.

The importance of the microbes living within the human colon, which can be likened to an anaerobic bioreactor, has prompted a number of investigations using culture-independent molecular approaches. Recent 16S rRNA analysis of microbes shed in feces, as well as microbes collected from gut epithelium, reveals that the majority of procaryotes are currently uncultivated. However, one bacterium, *Bacteroides thetaiontaomicron,* has been the focus of recent interest. This microbe is well suited for survival in the gut, where it is able to degrade complex dietary polysaccharides. Genome analysis reveals that *B. thetaiontaomicron* has a large collection of genes that encode proteins needed for the acquisition and metabolism of carbohydrates. It resides in a specific microenvironment: rather than adhering to the intestinal epithelium, it produces substrate-specific binding proteins that allow it to colonize exfoliated host cells, food particles, and even sloughed mucus (**figure 30.18**). It is thought that such attachment helps retain the microbes in the gut and, once bound, the induced expression of extracellular hydrolases enables efficient digestion. Of course, the diversity and density of microbes within the colon suggests that such "nutrient rafts" are colonized

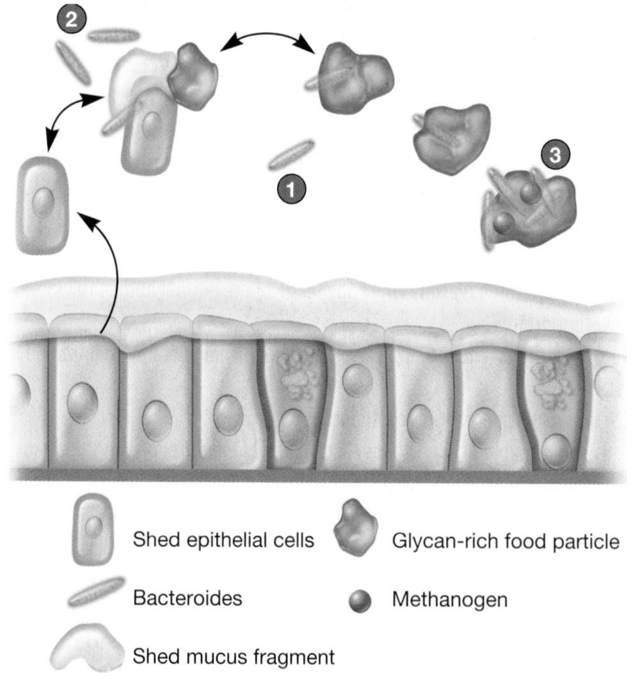

Figure 30.18 ***Bacteriodes thetaiontaomicron* as a Model for Colon Microbial Physiology and Community Dynamics.** (1) *B. thetaiontaomicron* is rapidly eliminated if it remains planktonic; however, (2) it efficiently adheres to substrates within the lumen of the gut rather than the gut itself (3) where, together with other microbes including methanogenic archaea, it degrades complex carbohydrates.

by a community of bacteria. For example, methanogenic bacteria are thought to remove the products of fermentation by converting H_2 and CO_2 to methane, just as they do in the rumen microbial community.

Various physiological processes move the microbiota through the colon so an adult eliminates about 3×10^{13} microorganisms daily. These processes include peristalsis and desquamation of the surface epithelial cells to which microorganisms are attached, and continuous flow of mucus that carries adhering microorganisms with it. To maintain homeostasis of the microbiota, the body must continually replace lost microorganisms. The bacterial population in the human colon usually doubles once or twice a day. Under normal conditions the resident microbial community is self-regulating. Competition and mutualism between different microorganisms and between the microorganisms and their host serve to maintain a status quo. However, if the intestinal environment is disturbed, the normal microbiota may change greatly. Disruptive factors include stress, altitude changes, starvation, parasitic organisms, diarrhea, and use of antibiotics or probiotics (**Techniques & Applications 30.3**). Finally, it should be emphasized that the actual proportions of the individual bacterial populations within the indigenous microbiota depend largely on a person's diet.

Genitourinary Tract

The upper genitourinary tract (kidneys, ureters, and urinary bladder) is usually free of microorganisms. In both the male and female, a few bacteria (*S. epidermidis, E. faecalis,* and *Corynebacterium* spp.) usually are present in the distal portion of the urethra.

In contrast, the adult female genital tract, because of its large surface area and mucous secretions, has a complex microbiota that constantly changes with the female's menstrual cycle. The major microorganisms are the acid-tolerant lactobacilli, primarily *Lactobacillus acidophilus,* often called Döderlein's bacillus. They ferment the glycogen produced by the vaginal epithelium, forming lactic acid. As a result the pH of the vagina and cervix is maintained between 4.4 and 4.6, inhibiting other microorganisms.

1. What are the most common microorganisms found in the nose? The oropharynx? The nasopharynx? The tonsillar crypts? The lower respiratory tract? The mouth? The eye? The external ear? The stomach? The small intestine? The colon? The genitourinary tract?
2. Why is the colon considered a large fermentation vessel?
3. What physiological processes move the microbiota through the gastrointestinal tract?
4. Describe the microbiota of the upper and lower female genitourinary tract.

Techniques & Applications

30.3 Probiotics for Humans and Animals

The large intestine of humans and animals contains a very complex and balanced microbiota. These microorganisms normally prevent infection and have a positive effect on nutrition. Any abrupt change in diet, stress, or antibiotic therapy can upset this microbial balance, making the host susceptible to disease and decreasing the efficiency of food use. **Probiotics** [Greek *pro,* for and *bios,* life], the oral administration of either living microorganisms or substances to promote health and growth, has the potential to reestablish the natural balance and return the host to normal health and nutrition.

Probiotic microorganisms are host specific; thus a strain selected as a probiotic in one animal may not be suitable in another species. Furthermore, microorganisms selected for probiotic use should exhibit these characteristics:

1. Adhere to the intestinal mucosa of the host
2. Be easily cultured
3. Be nontoxic and nonpathogenic to the host
4. Exert a beneficial effect on the host
5. Produce useful enzymes or physiological end products that the host can use
6. Remain viable for a long time
7. Withstand HCl in the host's stomach and bile salts in the small intestine

There are several possible explanations of how probiotic microorganisms displace pathogens and enhance the development and stability of the microbial balance in the large intestine. These include:

1. Competition with pathogens for nutrients and adhesion sites
2. Inactivation of pathogenic bacterial toxins or metabolites
3. Production of substances that inhibit pathogen growth
4. Stimulation of nonspecific immunity

A wide variety of probiotic preparations have been patented for cattle, goats, horses, pigs, poultry, sheep, domestic animals, and most recently for humans. Most of these preparations contain lactobacilli and/or streptococci; a few contain bifidobacteria.

Evidence has accumulated that certain probiotic microorganisms also offer considerable health benefits for humans. Potential benefits include:

1. Anticarcinogenic activity
2. Control of intestinal pathogens
3. Improvement of lactose use in individuals who have lactose intolerance
4. Reduction in the serum cholesterol concentration

Although the use of probiotics has only recently begun, a better understanding of the normal microbiota in the large intestine of both animals and humans will be forthcoming as more microbiologists investigate probiotic activity. Microbiology of fermented foods: Probiotics (section 40.6)

The Relationship between Normal Microbiota and the Host

The interaction between a host and a microorganism is a dynamic process in which each partner acts to maximize its survival. In some instances, after a microorganism enters or contacts a host, a positive mutually beneficial relationship occurs that becomes integral to the health of the host. These microorganisms become the normal microbiota. In other instances, the microorganism produces or induces deleterious effects on the host; the end result may be disease or even death of the host. Pathogenecity of microorganisms (chapter 33)

Our environment is teeming with microorganisms and we come in contact with many of them every day. Some of these microorganisms are pathogenic—that is, they cause disease. Yet these pathogens are at times prevented from causing disease by competition provided by the normal microbiota. In general, the normal microbiota use space, resources, and nutrients needed by pathogens. In addition, they may produce chemicals that repel invading pathogens. These normal microbiota prevent colonization by pathogens and possible disease through "bacterial interference." For instance, the lactobacilli in the female genital tract maintain a low pH and inhibit colonization by pathogenic bacteria and yeast, and the corynebacteria on the skin produce fatty acids that inhibit colonization by pathogenic bacteria. This is an excellent example of amensalism.

Products made by colonic bacteria (such as vitamins B and K) also benefit the host. Interestingly, studies using germfree animals suggest a strong correlation between the establishment of a stable microbial flora and the induction of immune compentency. For example, the introduction of normal fecal flora to germfree rodents stimulates the production and secretion of angiogenin-4, an antimicrobial peptide of intestinal Paneth cells. Furthermore, the reconstitution of germfree rodents with flora from conventionally raised siblings causes the abnormal gut-associated lymphoid (GALT) tissue and intestinal lamina propria to resemble that of the conventional animals (i.e., their lymphoid tissues and immunity are normalized). Even cell wall fragments from gram-positive bacteria can induce these changes. This normalization also includes an increase in the local lymphocyte populations and increased mucosal antibody production. Cells, tissues, and organs of the innate immune system (section 31.2)

Although normal microbiota offer some protection from invading pathogens, they may themselves become pathogenic and produce disease under certain circumstances, and then are termed **opportunistic microorganisms** or **pathogens.** These opportunistic microorganisms are adapted to the noninvasive mode of life defined by the limitations of the environment in which they are living. If removed from these environmental restrictions and introduced into the bloodstream or tissues, disease can result. For example, streptococci of the viridans group are the most common resident bacteria of the mouth and oropharynx. If they are introduced into the bloodstream in large numbers (e.g., following tooth extraction or a tonsillectomy), they may settle on deformed or prosthetic heart valves and cause endocarditis.

Opportunistic microorganisms often cause disease in compromised hosts. A **compromised host** is seriously debilitated and has a lowered resistance to infection. There are many causes of this condition including malnutrition, alcoholism, cancer, diabetes, leukemia, another infectious disease, trauma from surgery or an injury, an altered normal microbiota from the prolonged use of antibiotics, and immunosuppression by various factors (e.g., drugs, viruses [HIV], hormones, and genetic deficiencies). For example, *Bacteroides* species are one of the most common residents in the large intestine (figure 30.18) and are quite harmless in that location. If introduced into the peritoneal cavity or into the pelvic tissues as a result of trauma, they cause suppuration (the formation of pus) and bacteremia (the presence of bacteria in the blood). Many other examples of opportunistic infections will be presented in chapters 37, 38, and 39. The important point here is that the normal microbiota are harmless and are often beneficial in their normal location in the host and in the absence of coincident abnormalities. However, they can produce disease if introduced into foreign locations or compromised hosts.

1. Give two examples of the normal microbiota benefiting a host.
2. Provide two examples of how the normal host microbiota prevent the establishment of a pathogen.
3. How would you define an opportunistic microorganism or pathogen? A compromised host?

Summary

30.1 Microbial Interactions

a. Symbiotic interactions include mutualism (mutually beneficial and obligatory), cooperation (mutually beneficial, not obligatory), and commensalism (product of one organism can be used beneficially by another organism). Predation involves one organism (the predator) ingesting/killing a larger or smaller prey, parasitism (a longer-term internal maintenance of another organism or acellular infectious agent), and amensalism (a microbial product can inhibit another organism). Competition involves organisms competing for space or a limiting nutrient. This can lead to dominance of one organism, or coexistence of both at lower populations (**figure 30.1**).

b. A consortium is a physical association of organisms that have a mutually beneficial relationship based on positive interactions.

c. Mutual advantage is central to many organism-organism interactions. These interactions can be based on material transfers related to energetics, cell-to-cell communication, or physical protection. With several important mutualistic interactions, chemolithotrophic microorganisms play a critical role in making organic matter available for use by an associated organism (e.g., endosymbionts in *Riftia*).

d. The rumen is an excellent example of a mutualistic interaction between an animal and a complex microbial community. In this microbial community, complex plant materials are broken down to simple organic compounds that can be absorbed by the ruminant, as well as forming waste gases such as methane that are released to the environment (**figure 30.6**).

e. Syntrophism simply means growth together. It does not require physical contact and involves a mutually positive transfer of materials, such as interspecies hydrogen transfer.

f. Cooperative interactions are beneficial for both organisms but are not obligatory. Important examples are marine animals, including *Alvinella, Rimicarus,* and *Eubostrichus,* that involve interactions with hydrogen sulfide-oxidizing chemotrophs (**figure 30.7**).

g. Predation and parasitism are closely related. Predation has many beneficial effects on populations of predators and prey. These include the microbial loop (returning minerals immobilized in organic matter to mineral forms for reuse by chemotrophic and photosynthetic primary producers), protection of prey from heat and damaging chemicals, and possibly aiding pathogenicity, as with *Legionella* (**figure 30.13**).

30.2 Human-Microbe Interactions

a. Animals and environments that are germfree or have one or more known microorganisms are termed gnotobiotic. Gnotobiotic animals and techniques provide good experimental systems with which to investigate the interactions of animals and specific species or microorganisms (**figure 30.16**).

b. Various microbes have adapted to specific niches found on the human host. These niches are uniquely able to support microbe growth by maintaining a relatively constant environment.

30.3 Normal Microbiota of the Human Body

a. Commensal microorganisms living on or in the skin can be characterized as either transients or residents (**figure 30.17**).

b. The normal microbiota of the oral cavity is composed of those microorganisms able to resist mechanical removal.

c. The stomach contains very few microorganisms due to its acidic pH.

d. The distal portion of the small intestine and the entire large intestine have the largest microbial community in the body. Over 400 species have been identified, the vast majority of them anaerobic.

e. The upper genitourinary tract is usually free of microorganisms. In contrast, the adult female genital tract has a complex microbiota.

f. In some instances, after a microorganism contacts or enters a host, a positive mutually beneficial relationship occurs and becomes integral to the health of the host. In other instances, the microorganism may produce disease or even death of the host.

g. Many of the normal host microbiota compete with pathogenic microorganisms.

h. An opportunistic microorganism is generally harmless in its normal environment but may become pathogenic when moved to a different body location or in a compromised host.

Key Terms

amensalism 732
axenic 734
cathelicidins 736
comedo 737
commensal 729
commensalism 729
competition 732
competitive exclusion principle 732
compromised host 740
consortium 717
cooperation 726
coral bleaching 719
ectosymbiont 717
endosymbiont 717
genomic reduction 732
gnotobiotic 734
interspecies hydrogen transfer 729
lichen 731
mutualism 718
mutualist 718
mycobiont 731
normal microbial flora or microbiota 734
opportunistic microorganism or pathogen 740
Paneth cell 734
parasitism 730
pathogen 734
pathogenicity 734
phycobiont 731
predation 729
probiotics 739
rumen 724
sebum 737
symbiosis 717
syntrophism 726
zooxanthellae 719

Critical Thinking Questions

1. Describe an experimental approach to determine if a plant-associated microbe is a commensal or a mutualist.
2. Some patients who take antibiotics for acne develop yeast infections. Explain.
3. How does knowing the anatomical location of commensal flora help clinicians diagnose infection?
4. Compare and contrast the microbial communities that reside in a ruminant with those in a human gut.

Learn More

Bäckhed, F.; Ley, R. E.; Sonnenburg, J. L.; Peterson, D. A.; and Gordon, J. I. 2005. Host-bacterial mutualisms in the human intestine. *Science* 307: 1915–20.

Currie, C. R.; Poulsen, M.; Mendenhall, J.; Boomsma, J. J.; and Billen, J. 2005. Coevolved crypts and exocrine glands support mutualist bacteria in fungus-growing ants. *Science* 311:81–83.

Ekburg, P. B.; Bik, E. M.; Bernstein, C. N.; Purdom, E.; Dethlefsen, L.; Sargent, M.; Gill, S. R.; Nelson, K. E.; and Relman, D. A. 2005. Diversity of the human intestinal microbial flora. Science 308:1635–38.

Hentschel, U., and Steinert, M. 2001. Symbiosis and pathogenesis: Common themes, different outcomes. *Trends Microbiol.* 9(12):585.

Potera, C. 2002. Antimicrobial peptides from skin effective in killing pathogens. *ASM News* 68:108–109.

Raghoebarsing, A. A.; Smolders, A. J. P.; Schmid, M. C.; Rijpstra, W. I. C.; Wolters-Arts, M.; Derksen, J.; Jetten, M. S. M.; Schouten, S.; Damste, J. S. S.; Lamers, L. P. M.; Roelofs, J. G. M.; Op den Camp, H. J. M.; and Strous, M. 2005. Methylotrophic symbionts provide carbon for photosynthesis in peat bogs. *Nature* 436:1153–56.

Smith, D. J.; Suggett, D. J.; and Baker, N. R. 2005. Is photoinhibition of zooxanthellae photosynthesis the primary cause of thermal bleaching in corals? *Global Change Biol.* 11:1–11.

Stewart, F. J.; Newton, I. L. G.; and Cavanaugh, C. M. 2005. Chemosynthetic endosymbioses: Adaptations to oxic-anoxic interfaces. *Trends Microbiol.* 13: 439–48.

Thomas, I.; Klasson, L.; Canbäck, B.; Näslund, A. K.; Eridsson, A. S.; Wenegreen, J. J.; Sandström, J. P.; Moran, N. A.; and Andersson, S. G. E. 2002. Fifty million years of genomic stasis in endosymbiotic bacteria. *Science* 296:2376–79.

Velicer, G. J. 2003. Social strife in the microbial world. *Trends Microbiol.* 11(7):330–37.

Xi, Z.; Khoo, C. C. H.; and Dobson, S. L. 2005. *Wolbachia* establishment and invasion in an *Aedes aegypti* laboratory population. *Science* 310:326–28.

Xu, J.; Bjursell, M. K.; Himrod, J.; Deng, S.; Carmichael, L. K.; Chiang, H. C.; Hooper, L. V.; and Gordon, J. I. 2003. A genomic view of the human—*Bacteroides thetaiotaomicron* symbiosis. *Science* 299:2074–76.

Please visit the Prescott website at www.mhhe.com/prescott7 for additional references.

31 Nonspecific (Innate) Host Resistance

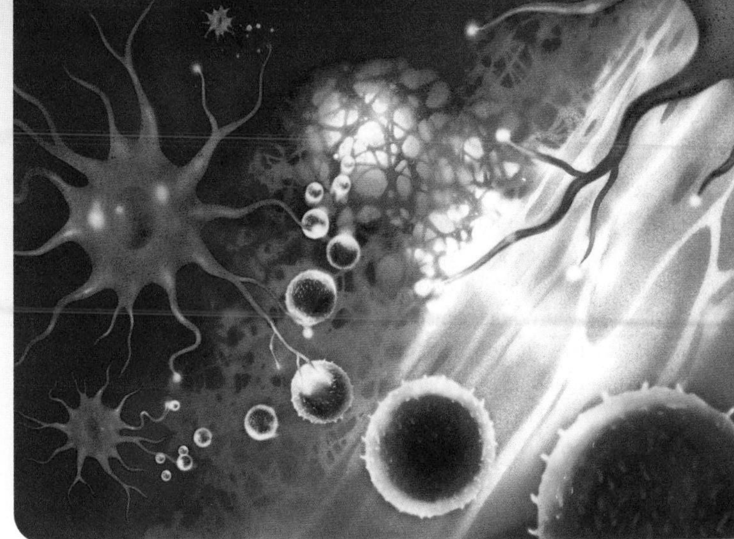

The immune system is a myriad of cells, tissues, and soluble factors that cooperate to defend against foreign invaders.

Preview

- The host's ability to resist infection depends on a constant defense against microbial invasion. Resistance arises from both nonspecific (innate) and specific (adaptive) body defense mechanisms.
- Nonspecific host defenses are those innate mechanisms that are constitutively expressed within a host. Examples include specialized cells, tissues, organs, cellular processes, physical barriers, and chemical mediators.
- Physical and mechanical barriers include components of the skin, mucous membranes, the respiratory system, gastrointestinal tract, genitourinary tract, and the eye. All are formidable impediments to microbial invasion. Other nonspecific defenses include chemical mediators such as cationic peptides and acute-phase proteins.
- Inflammation, the complement pathways, phagocytosis, various cytokines, fever, and natural killer cells are other examples of nonspecific defenses that help protect the host against microorganisms and cancer.

The integrity of any eucaryotic organism depends not only on the proper expression of its genes but also on its freedom from invading microorganisms. Commensal relationships aside, when microorganisms inhabit a multicellular host, there is competition for resources at the cellular level. In previous chapters we explore the various types of symbiotic relationships that two organisms may have. In future chapters we examine the diseases that result when bacteria, viruses, fungi, or protists access cells and tissues within human hosts and successfully compete for nutrients and/or produce factors that inhibit or kill host cells. First however, we explore the mechanisms by which humans (and other mammalian hosts) defend themselves against such microbial invasion.

31.1 Overview of Host Resistance

To establish an infection, an invading microorganism must first overcome many surface barriers, such as skin, degradative enzymes, and mucus, that have either direct antimicrobial activity or inhibit attachment of the microorganism to the host. Because neither the surface of the skin nor the mucus-lined body cavities are ideal environments for the vast majority of microorganisms, most **pathogens** must breach these barriers to reach underlying tissues. Any microorganism that penetrates these barriers encounters two levels of resistance: other nonspecific resistance mechanisms and the specific immune response.

Vertebrates (including humans) are continuously exposed to microorganisms and their metabolic products that can cause disease. Fortunately these animals are equipped with an immune system that protects them against adverse consequences of this exposure. The **immune system** is composed of widely distributed cells, tissues, and organs that recognize foreign substances, including microorganisms. Together they act to neutralize or destroy them.

Immunity

The term immunity [Latin *immunis,* free of burden] refers to the general ability of a host to resist a particular infection or disease. **Immunology** is the science that is concerned with immune responses to foreign challenge and how these responses are used to resist infection. It includes the distinction between "self" and "nonself" and all the biological, chemical, and physical aspects of the immune response.

There are two fundamentally different types of immune responses to an invading microorganism and/or foreign material. The **nonspecific immune response** is also known as **nonspecific resistance** and **innate** or **natural immunity;** it offers resistance to any

Half the secret of resistance is cleanliness, the other half is dirtiness.

—*Anonymous*

microorganism or foreign material encountered by the vertebrate host. It includes general mechanisms inherited as part of the innate structure and function of each animal (such as skin, mucus, and constituitively produced antimicrobial mediators like lysozyme), and acts as a first line of defense. The nonspecific immune response defends against foreign particles equally and lacks immunological memory—that is, nonspecific responses occur to the same extent each time a microorganism or foreign body is encountered.

In contrast, the **specific immune responses,** also known as **acquired, adaptive,** or **specific immunity,** resist a particular foreign agent. Moreover, the effectiveness of specific immune responses increases on repeated exposure to foreign agents such as viruses, bacteria, or toxins; that is to say, specific responses have "memory." Substances that are recognized as foreign and provoke immune responses are called **antigens** (contraction of the words *anti*body and *gen*erators). The antigens cause specific cells to replicate and manufacture a variety of proteins that function to protect the host. One such cell, the B cell, produces and secretes glycoproteins called antibodies. **Antibodies** bind to specific antigens and inactivate them or contribute to their elimination. Other immune cells become activated to destroy cells harboring intracellular pathogens. The nonspecific and specific responses usually work together to eliminate pathogenic microorganisms and other foreign agents (**figure 31.1**).

The distinction between the innate and adaptive systems is, to a degree, artificial. Although innate systems predominate immediately upon initial exposure to foreign substances, multiple bridges occur between innate and adaptive immune system components (figure 31.1). Importantly, there are a variety of cells that function in both innate and adaptive immunity. These cells are known as the **white blood cells,** or leukocytes (as they were first identified as part of a white "buffy coat" between red blood cells and the plasma of centrifuged blood). Blood cell development occurs in the bone marrow of mammals during the process of **hematopoesis.** White blood cells (WBCs) are divided between the innate and adaptive immune systems according to specific cellular behaviors. Cells that typically mature prior to leaving the bone marrow and provide the same physiological response regardless of antigen are assigned to the innate immune system (e.g., macrophages and dendritic cells). Cells that are not completely functional after leaving the bone marrow but differentiate in response to various types of antigens are assigned to the adaptive immune system (e.g., B and T cells). The WBCs form the basis for immune responses to invading microbes and foreign substances. Many of these cells reside in specific tissues and/or organs. Some tissues and organs provide supportive functions in nurturing the cells so that they can mature and/or respond correctly to antigens.

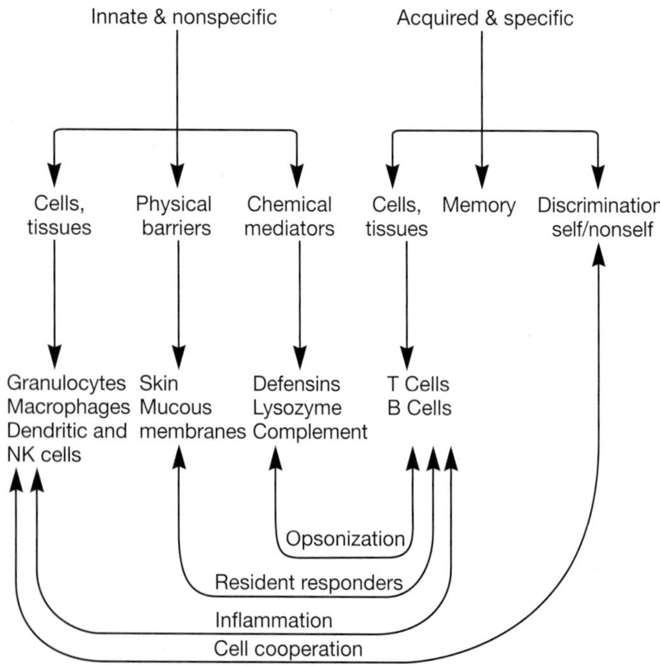

Figure 31.1 Some of the Major Components that Make Up the Mammalian Immune System. Double-headed arrows indicate potential bridging events that unite innate and acquired forms of immunity.

The bridges that interconnect the innate and specific immune responses are numerous and present a difficulty for any author trying to describe immunity. We begin our discussion with an introduction to the various cells, tissues, and organs of the innate defenses. In this discussion, reference will be made to processes and chemical mediators that are described in detail either later in this chapter or in chapter 32 (specific defenses). For your convenience, cross-references to the detailed discussion are provided. Thus the discussion of innate immunity proceeds from the general to the ever-more specific and detailed.

1. Define each of the following terms: immune system, immunity, immunology, antigen, antibody.
2. Compare and contrast the specific and nonspecific immune responses.
3. Describe how the activity of white blood cells is often used to assign their role in immunity.

31.2 Cells, Tissues, and Organs of the Immune System

As indicated, the immune system is an organization of molecules, cells, tissues, and organs, each with a specialized role in defending against viruses, microorganisms, cancer cells, and nonself proteins (e.g., organ transplants). Immune system cells and tissue are now considered.

Cells of the Immune System

The cells responsible for both nonspecific and specific immunity are the **leukocytes** [Greek *leukos,* white, and *kytos,* cell], or white blood cells. All of the leukocytes originate from pluripotent stem cells in the fetal liver and in the bone marrow of the animal host (**figure 31.2**). Pluripotency indicates that these stem cells have

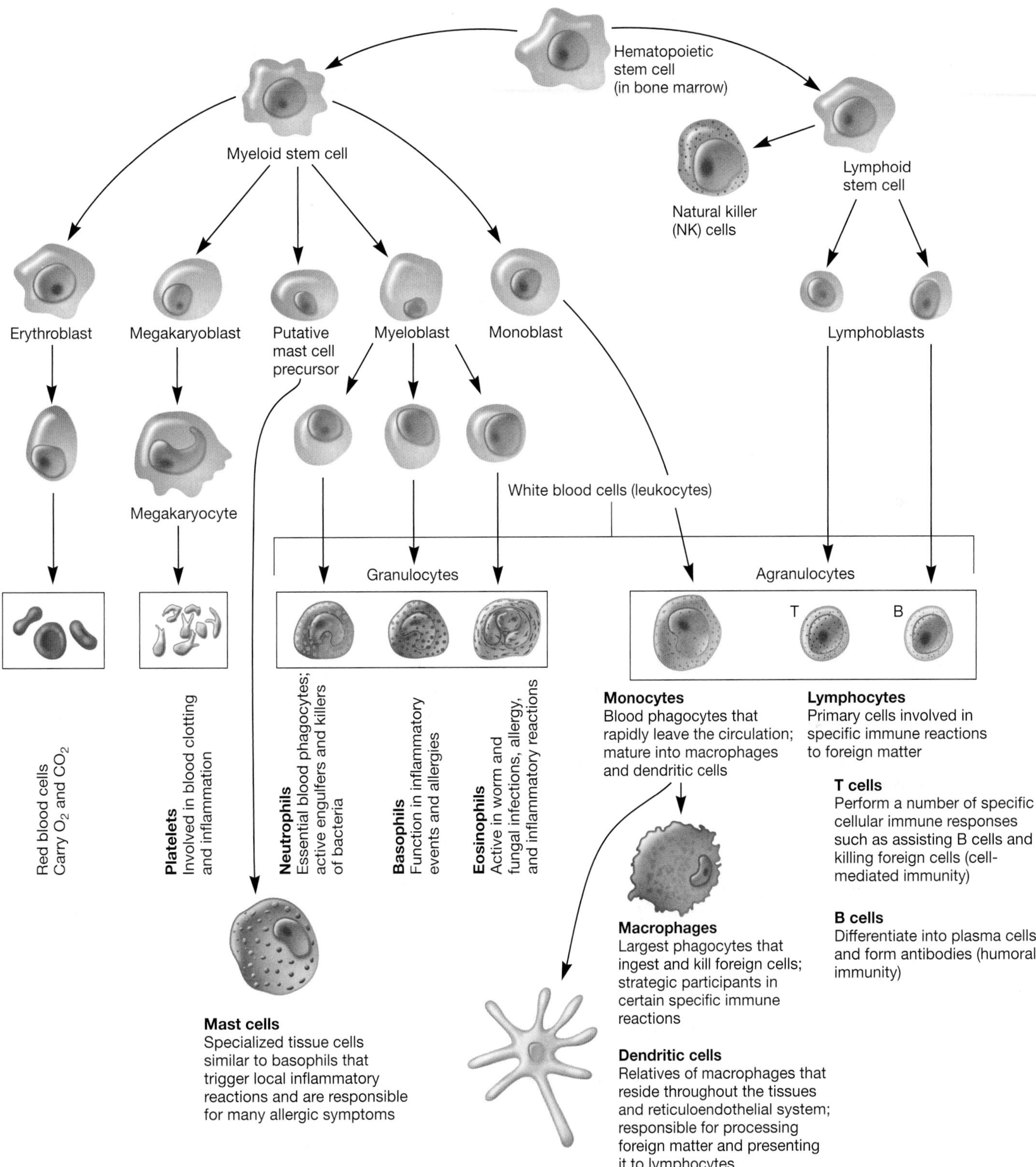

Figure 31.2 The Different Types of Human Blood Cells. Pluripotent stem cells in the bone marrow divide to form two blood cell lineages: (1) the lymphoid stem cell gives rise to B cells that become antibody-secreting plasma cells, T cells that become activated T cells, and natural killer cells. (2) The common myeloid progenitor cell gives rise to the granulocytes (neutrophils, eosinophils, basophils), monocytes that give rise to macrophages and dendritic cells, an unknown precursor that gives rise to mast cells, megakaryocytes that produce platelets, and the erythroblast that produces erythrocytes (red blood cells).

Table 31.1 Normal Adult Blood Count

Cell Type	Cells/mm^3	Percent WBC
Red blood cells	5,000,000	
Platelets	250,000	
White blood cells	7,200	100
Neutrophils	4,320	60
Lymphocytes	2,160	30
Monocytes	430	6
Eosinophils	215	3
Basophils	70	1

not yet committed to differentiating into one specific cell type. When they migrate to other body sites, some differentiate into hematopoietic stem cells that are destined to become leukocytes. When stimulated to undergo further development, some leukocytes become residents within tissues, where they respond to local trauma. These cells may sound the alarm that signals invasion by foreign organisms. Other leukocytes circulate in body fluids and are recruited to the sites of infection after the alarm has been raised. The average adult has approximately 7,400 leukocytes per mm^3 of blood (**table 31.1**). This average value shifts substantially during infectious and allergic responses of the host. Thus the complete blood count (CBC) and the phenotype differential (DIFF), which determines the relative number of each type of blood cell, are used clinically to assist in the diagnosis of disease and infection. In defending the host against pathogenic microorganisms, leukocytes cooperate with each other first to recognize the pathogen as an invader and then to destroy it. These different leukocytes are now briefly examined.

Monocytes and Macrophages

Monocytes and macrophages are highly phagocytic and make up the **monocyte-macrophage system** (**figure 31.3**). Although the specifics of phagocytosis will be discussed shortly, recall that it involves the engulfment of large particles and microorganisms that are then enclosed in a phagocytic vacuole or phagosome. **Monocytes** [Greek *monos,* single, and *cyte,* cell] are mononuclear leukocytes with an ovoid- or kidney-shaped nucleus and granules in the cytoplasm that stain gray-blue with basic dyes (figure 31.2). They are produced in the bone marrow and enter the blood, circulate for about eight hours, enlarge, migrate to the tissues, and mature into macrophages or dendritic cells.

Because **macrophages** [Greek *macros,* large, and *phagein,* to eat] are derived from monocytes, they are also classified as mononuclear phagocytic leukocytes. However, they are larger than monocytes, contain more organelles that are critical for phagocytosis, and have a plasma membrane covered with microvilli (**figure 31.4**). Macrophages have surface molecules that function as receptors to nonspecifically recognize common components of pathogens. These receptors include mannose and fucose receptors and a special class of molecules called toll-like receptors (p. 753), which bind lipopolysaccharide (LPS), peptidoglycan, fungal cell wall components called zymosan, viral nucleic acids, and foreign DNA. In addition, macrophages possess specialized scavenger receptors including a surface protein called CD14, which also binds LPS. Macrophages also have receptors for antibodies and serum glycoproteins known as complement (p. 763). Both antibody and complement proteins can coat microorganisms or foreign material and enhance their phagocytosis. This enhancement is termed opsonization and is discussed in further detail in section 31.6. Macrophages spread throughout the animal body and take up residence in specific tissues where they are given special names (figure 31.3). Since macrophages are highly phagocytic, their function in nonspecific resistance is discussed in more detail in the context of phagocytosis. Antigens: Cluster of differentiation molecules (section 32.2); Antibodies (section 32.7)

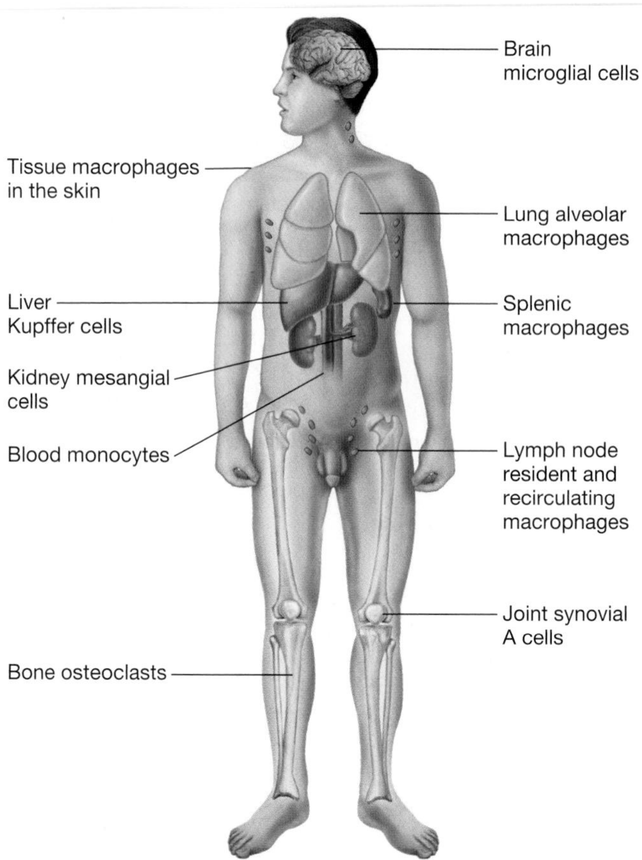

Figure 31.3 The Monocyte-Macrophage System. This system consists of tissue (such as found within the liver, spleen, and lymph nodes) containing "fixed" or immobile phagocytes that have specific names depending on their location.

Granulocytes

Granulocytes have irregular-shaped nuclei with two to five lobes, and as such, are often called **polymorphonuclear leukocytes.** Their cytoplasmic matrix has granules that contain reactive substances that kill microorganisms and enhance inflammation (figure 31.2). Three types of granulocytes exist: basophils, eosinophils, and neutrophils. Because of the irregular-shaped nuclei, neutrophils are also called **polymorphonuclear neutrophils** or **PMNs.**

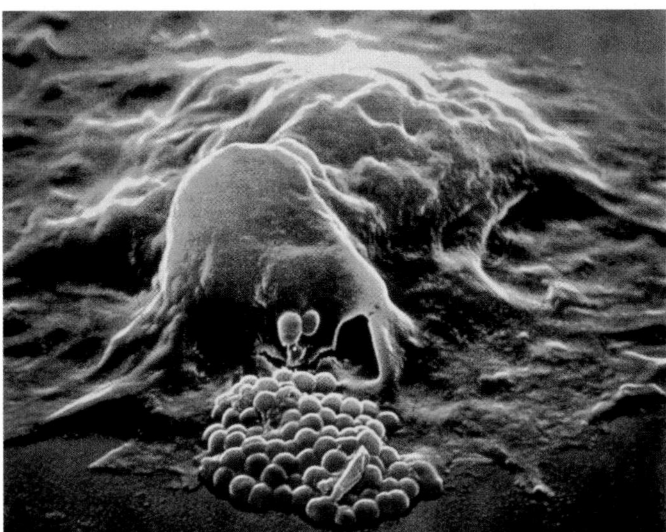

Figure 31.4 Phagocytosis by a Macrophage. One type of nonspecific host resistance involves white blood cells called macrophages and the process of phagocytosis. This scanning electron micrograph (×3,000) shows a macrophage devouring a colony of bacteria. Phagocytosis is one of many nonspecific defenses humans and other animals use to combat microbial pathogens.

Basophils [Greek *basis,* base, and *philein,* to love] have an irregular-shaped nucleus with two lobes, and granules that stain bluish-black with basic dyes (figure 31.2). Basophils are nonphagocytic cells that function by releasing specific compounds from their cytoplasmic granules in response to certain types of stimulation. These molecules include histamine, prostaglandins, serotonin, and leukotrienes. Because these physiological mediators influence the tone and diameter of blood vessels, they are termed vasoactive mediators. Basophils (and mast cells) possess high-affinity receptors for one type of antibody, known as immunoglobulin E (IgE), that is associated with allergic responses. When these cells become coated with IgE antibodies, binding of antigen to the IgE can trigger the secretion of vasoactive mediators. As discussed in chapter 32, these inflammatory mediators play a major role in certain allergic responses such as eczema, hay fever, and asthma. Antibodies (section 32.7); Immune disorders: Hypersensitivities (section 32.11)

Eosinophils [Greek *eos,* dawn, and *philien*] have a two-lobed nucleus connected by a slender thread of chromatin, and granules that stain red with acidic dyes (figure 31.2). Unlike basophils, eosinophils migrate from the bloodstream into tissue spaces, especially mucous membranes. Their role is important in the defense against protozoan and helminth parasites, mainly by releasing cationic peptides (p. 762) and reactive oxygen intermediates (p.755) into the extracellular fluid. These molecules damage the parasite plasma membrane, destroying it. Eosinophil numbers are often increased during allergic reactions, especially type 1 hypersensitivities. Eosinophils also play a role in allergic reactions, as they have granules containing histaminase and aryl sulphatase, down-regulators of the inflammatory mediators histamine and leukotrienes, respectively.

Neutrophils [Latin *neuter,* neither, and *philien*] stain readily at a neutral pH, have a nucleus with three to five lobes connected by slender threads of chromatin, and contain inconspicuous organelles known as the primary and secondary granules. Lytic enzymes and bactericidal substances are contained within larger primary and smaller secondary granules. Primary granules contain peroxidase, lysozyme, defensins, and various hydrolytic enzymes, whereas secondary granules have collagenase, lactoferrin, cathelicidins, and lysozyme. Both of these granules help accomplish intracellular digestion of foreign material after it is phagocytosed. Neutrophils also use oxygen-dependent and oxygen-independent pathways that generate additional antimicrobial substances to kill ingested microorganisms. Like macrophages, neutrophils have receptors for antibodies and complement proteins and are highly phagocytic. However, unlike macrophages, neutrophils do not reside in healthy tissue but circulate in blood so as to rapidly migrate to the site of tissue damage and infection, where they become the principal phagocytic and microbicidal cells. Neutrophils and their antimicrobial compounds are described in more detail in the contexts of the inflammatory response (section 31.4) and phagocytosis (section 31.3).

1. Describe the structure and function of each of the following blood cells: monocyte, macrophage, basophil, eosinophil, and neutrophil.
2. What is the significance of the respective blood cell percentages in blood?

Mast Cells

Mast cells are bone marrow-derived cells that differentiate in the blood and connective tissue. Although they contain granules with histamine and other pharmacologically active substances similar to those in basophils, they arise from a different cellular lineage (figure 31.2). Mast cells, along with basophils, play an important role in the development of allergies and hypersensitivities.

Dendritic Cells

Dendritic cells constitute only 0.2% of white blood cells in the blood but play an important role in nonspecific resistance (**figure 31.5**). They are present in the skin and mucous membranes of the nose, lungs, and intestines where they readily contact invading pathogens, phagocytose and process antigens, and display foreign antigens on their surface. This is a process known as **antigen presentation,** which is discussed in more detail in section 32.2.

Dendritic cells recognize specific *p*athogen-*a*ssociated *m*olecular *p*atterns (PAMPs) on microorganisms (p. 753). These molecular patterns enable dendritic cells to distinguish between potentially harmful microbes and other host molecules. After the pathogen is recognized, the dendritic cell's *p*attern *r*ecognition *r*eceptors (PRRs) bind the pathogen and phagocytose it. The dendritic cells then migrate to lymphoid tissues where, as activated cells, they present antigen to T cells. Antigen presentation triggers the activation of T cells, which are critical for the initiation and regulation of an effective specific immune response. Thus not only do dendritic cells destroy invading pathogens as part of the innate response, but they also help trigger specific immune responses. T cell biology (section 32.5)

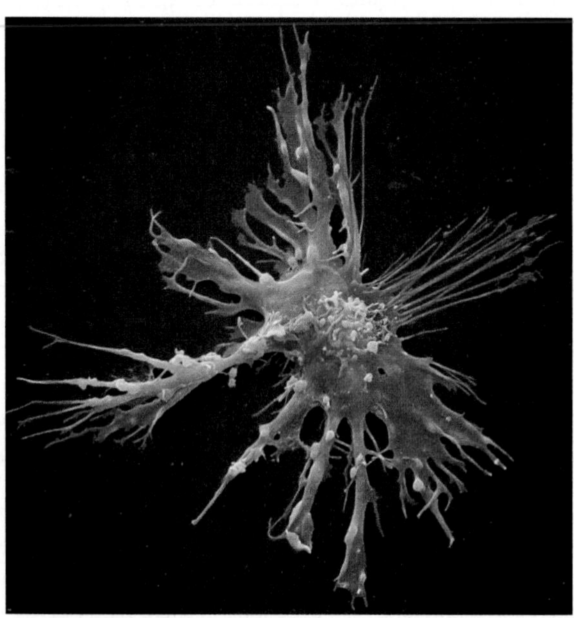

Figure 31.5 The Dendritic Cell. The dendritic cell was named for its cellular extensions, which resemble the dendrites of nerve cells. Dendritic cells reside in most tissue sites where they survey their local environments for pathogens and altered host cells.

Lymphocytes

Lymphocytes [Latin *lympha,* water, and *cyte,* cell] are the major cells of the specific immune system. Lymphocytes can be divided into three populations: T cells, B cells, and null cells (which include special cells called natural killer, or NK, cells). Lymphocytes leave the bone marrow in a kind of cellular stasis. In other words, they are not actively replicating like other somatic cells. B and T lymphocytes differentiate from their respective lymphoid stem cell precursor cells and are then blocked from exiting the G_0 phase of the cell cycle. This is important because it ensures that their gene products are made only when they are needed. In general, lymphocytes require a specific antigen to bind to a surface receptor (the B-cell receptor on B cells or the T-cell receptor on T cells) so that cellular activation can occur. Activation stimulates the cell to enter mitosis. Once activated, the lymphocytes continue to replicate as they circulate throughout the host, leaving several clones of activated lymphocytes to populate various lymphoid tissues. In addition to activated cells, some of the replicated lymphocytes are inhibited from further replication, waiting to be activated by the same antigen sometime later in the life of the host. These cells are called **memory cells.** The nucleus and cell division: Mitosis and meiosis (section 4.8)

After **B lymphocytes** or **B cells** reach maturity within the bone marrow, they circulate in the blood and disperse into various lymphoid organs, where they become activated. The activated B cell becomes more ovoid. Its nuclear chromatin condenses and numerous folds of endoplasmic reticulum become more visible. A mature, activated B cell is called a **plasma cell.** The phenotypic changes reflect functional changes occurring within the plasma cell as it begins to secrete large quantities of glycoproteins called antibodies. Some of these antibodies can directly neutralize toxins and viruses, and are important in stimulating an efficient phagocytic response (**figure 31.6**). Actions of antibodies (section 32.8)

Lymphocytes destined to become **T lymphocytes** or **T cells** leave the bone marrow to mature in the thymus gland; they can remain in the thymus, circulate in the blood, or reside in lymphoid organs such as the lymph nodes and spleen, like B cells do. Also like B cells, T cells require a specific antigen to bind to their receptor to signal the continuation of replication. Unlike B cells, however, T cells do not secrete antibodies. Activated T cells produce and secrete proteins called **cytokines** (figure 31.6 and p. 766). Cytokines can have various effects on other cells including other T cells, B cells, granulocytes, and other somatic cells. In some cases the cytokines stimulate cells to mature and differentiate, produce new effector products, and even cause some cells to die. Because B and T cells must be activated by specific antigens, they are included in the adaptive or specific immune system. B cells and T cells are discussed further in chapter 32.

Natural killer (NK) cells are a small population of large, nonphagocytic granular lymphocytes that play an important role in innate immunity (figure 31.2). The major NK cell function is to destroy malignant cells and cells infected with microorganisms. They recognize their targets in one of two ways. They can bind to antibodies that coat infected or malignant cells; thus the antibody bridges the two cell types. This process is called **antibody-dependent cell-mediated cytotoxicity (ADCC) (figure 31.7)** and can result in the death of the target cell. The second way that NK cells recognize infected cells and cancer cells relies on the presence of specialized proteins on the surface of all nucleated host cells, known as the class I *m*ajor *h*isto*c*ompatibility (MHC) antigen. If a host cell loses this MHC protein, as when some viruses or cancers overtake the cell, the NK cell kills it by releasing pore-forming proteins and cytotoxic enzymes called granzymes. Together the pore-forming proteins and the granzymes cause the target cell to lyse (**figure 31.8**). Recognition of foreignness (section 32.4)

1. How might mast cells, lymphocytes, and dendritic cells work together in innate immunity?
2. Discuss the role of NK cells in protecting the host.
3. What is the purpose of antibody-dependent cell-mediated cytotoxicity?

Organs and Tissues of the Immune System

Based on function, the organs and tissues of the immune system can be divided into primary or secondary lymphoid organs and tissues (**figure 31.9**). The primary organs and tissues are where immature lymphocytes mature and differentiate into antigen-sensitive B and T cells. The thymus is the primary lymphoid organ for T cells, and the bone marrow is the primary lymphoid tissue for B cells. The secondary organs and tissues serve as areas where lymphocytes may encounter and bind antigen, whereupon they proliferate and differentiate into fully active, antigen-specific effector

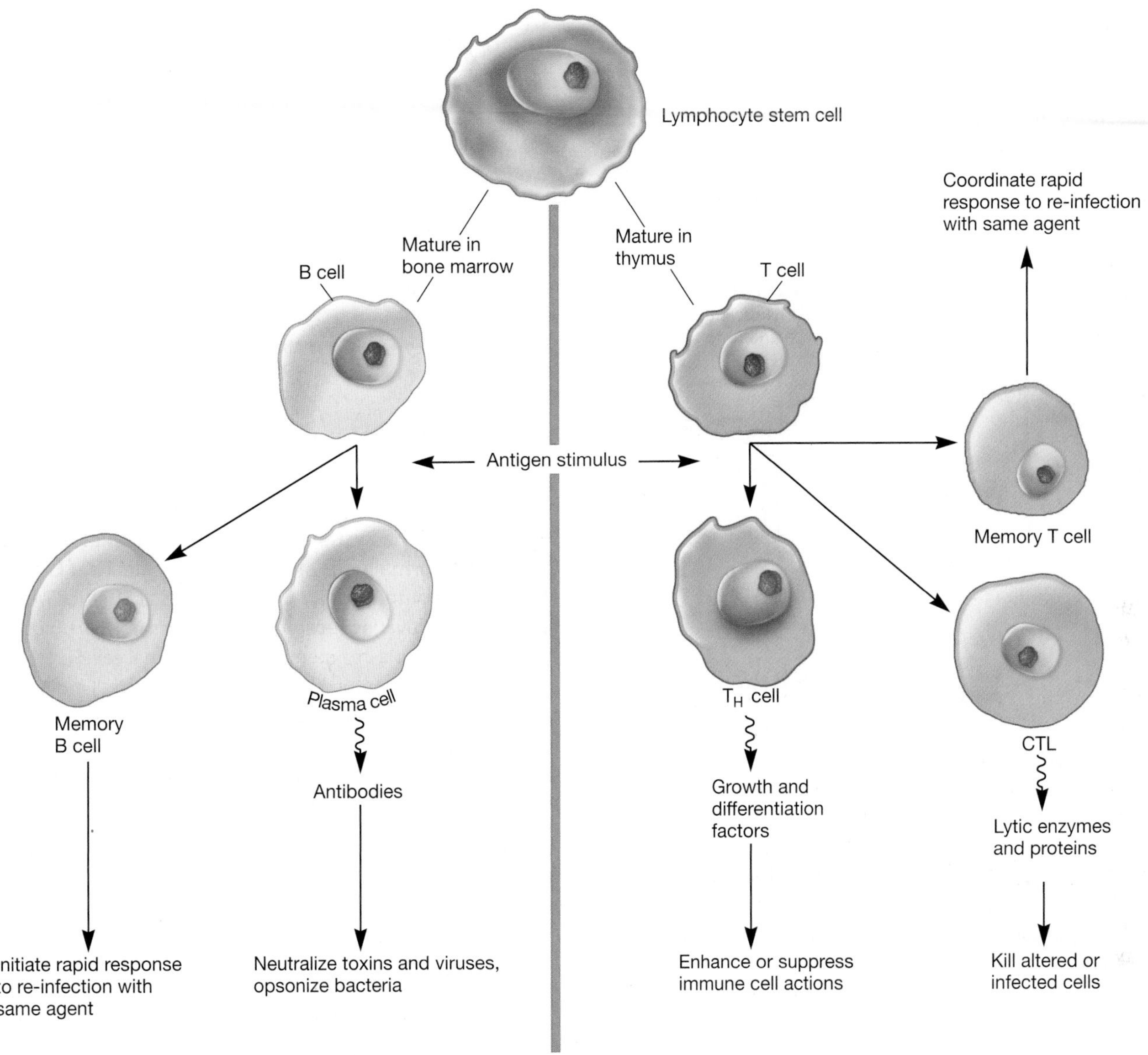

Figure 31.6 The Development and Function of B and T Lymphocytes. B cells and T cells arise from the same cell lineage but diverge into two different functional types. Immature B cells and T cells are indistinguishable by histological staining. However, they express different proteins on their surfaces that can be detected by immunohistochemistry. Additionally, the final secreted products of mature B and T cells can be used to identify the cell type.

cells. The spleen is a secondary lymphoid organ and the lymph nodes and mucus-associated tissues (GALT, gut-associated lymphoid tissue and SALT, skin-associated lymphoid tissues) are the secondary lymphoid tissues. The thymus, bone marrow, lymph nodes, and spleen are now discussed in more detail. GALT and SALT are described in section 31.5 as part of the host's physical and mechanical barriers.

Primary Lymphoid Organs and Tissues

Immature undifferentiated lymphocytes are generated in the bone marrow. They mature and become committed to a specific antigen within the primary lymphoid organ/tissues. The two most important primary sites in mammals are the thymus and bone marrow.

The **thymus** is a highly organized lymphoid organ located above the heart. Precursor cells from the bone marrow migrate into the outer cortex of the thymus where they proliferate. As they mature and acquire T-cell surface markers, approximately 90% die. This is due to a selection process in which T cells that recognize host (self) antigens are destroyed. The remaining 10% move into the medulla of the thymus, become mature T cells, and subsequently enter the bloodstream (figure 31.9*a*).

In mammals, the bone marrow (figure 31.9*b*) is the site of B-cell maturation. Like thymic selection during T-cell maturation,

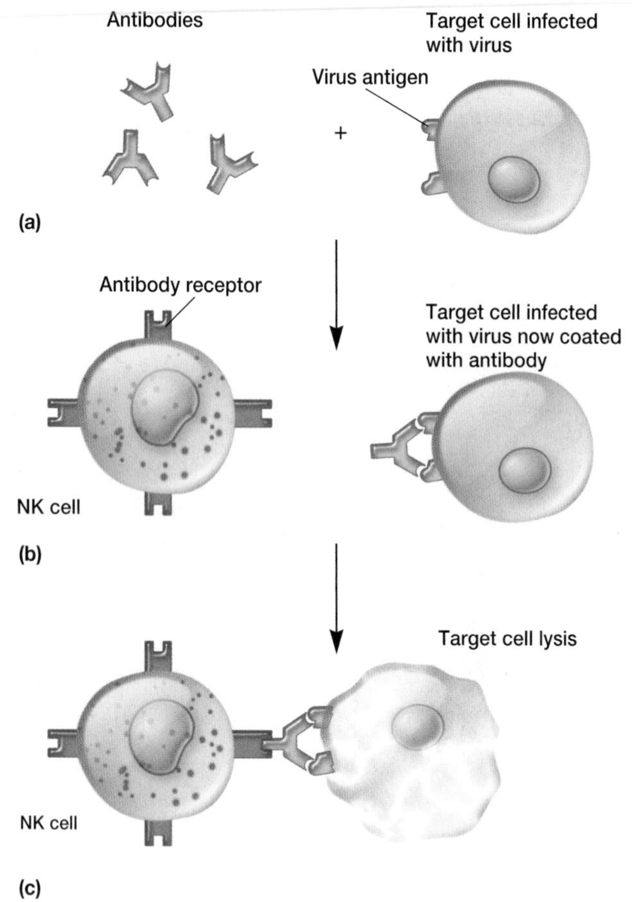

Figure 31.7 Antibody-Dependent Cell-Mediated Cytotoxicity. **(a)** In this mechanism, IgG antibodies bind to a target cell infected with a virus. **(b)** NK cells have specific antibody receptors on their surface. **(c)** When the NK cells encounter virus, infected cells coated with antibody, they kill the target cell.

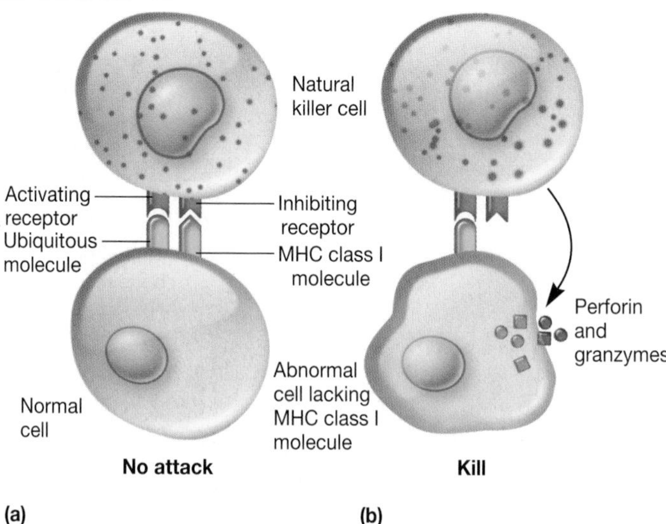

Figure 31.8 The System Used by Natural Killer Cells to Recognize Normal Cells and Abnormal Cells That Lack the Major Histocompatability Complex Class I Surface Molecule. **(a)** The killer-activating receptor recognizes a normal ubiquitous molecule on the plasma membrane of a normal cell. Since the killer-inhibitory receptor recognizes the MHC class I molecule, there is no attack. **(b)** In the absence of the inhibitory signal, the receptor issues an order to the NK cell to attack and kill the abnormal cell. The cytotoxic granules of the NK cell contain perforin and granzymes. With no inhibitory signal, the granules release their contents, killing the abnormal cell.

a selection process within the bone marrow eliminates B cells bearing self-reactive antigen receptors. In birds, undifferentiated lymphocytes move from the bone marrow to the **bursa of Fabricius** where B cells mature; this is where B cells were first identified and how they came to be known as "B" (for bursa) cells.

Secondary Lymphoid Organs and Tissues

The spleen is the most highly organized secondary lymphoid organ. The **spleen** is a large organ located in the abdominal cavity (figure 31.9). It specializes in filtering the blood and trapping blood-borne microorganisms and antigens. Once trapped by splenic macrophages or dendritic cells, the pathogen is phagocytosed, killed, and digested. The resulting peptide (protein fragment consisting of less than about 50 amino acid residues) antigens are delivered to the macrophage or dendritic cell surface where they are presented to B and T cells. This is the most common means by which lymphocytes become activated to carry out their immune functions.

Lymph nodes lie at the junctions of lymphatic vessels where they filter out harmful microorganisms and antigens from the lymph; pathogens and antigens are trapped by phagocytic macrophages and dendritic cells (figure 31.9*c*). They then phagocytose the foreign material and present antigen to lymphocytes. It is within the lymph nodes that B cells differentiate into memory cells and antibody-secreting plasma cells. This involves specialized T cells, called T-helper cells, which are also found here. Dendritic cells and macrophages (antigen-presenting cells) present antigens to the T-helper cells, which subsequently secrete cytokines that promote B-cell immune responses. B cell biology (section 32.6)

Lymphoid tissues are found throughout the body and act as regional centers of antigen sampling and processing (figure 31.9). Lymphoid tissues are found as highly organized or loosely associated cellular complexes. Some lymphoid cells are closely associated with specific tissues such as skin (*s*kin-*a*ssociated *l*ymphoid *t*issue, or SALT) and mucous membranes (*m*ucus-*a*ssociated *l*ymphoid *t*issue, or MALT). SALT and MALT are good examples of highly organized lymphoid tissues, typically seen histologically as macrophages surrounded by specific areas of B and T lymphocytes, and sometimes dendritic cells. Loosely associated lymphoid tissue is best represented by the *b*ronchial-

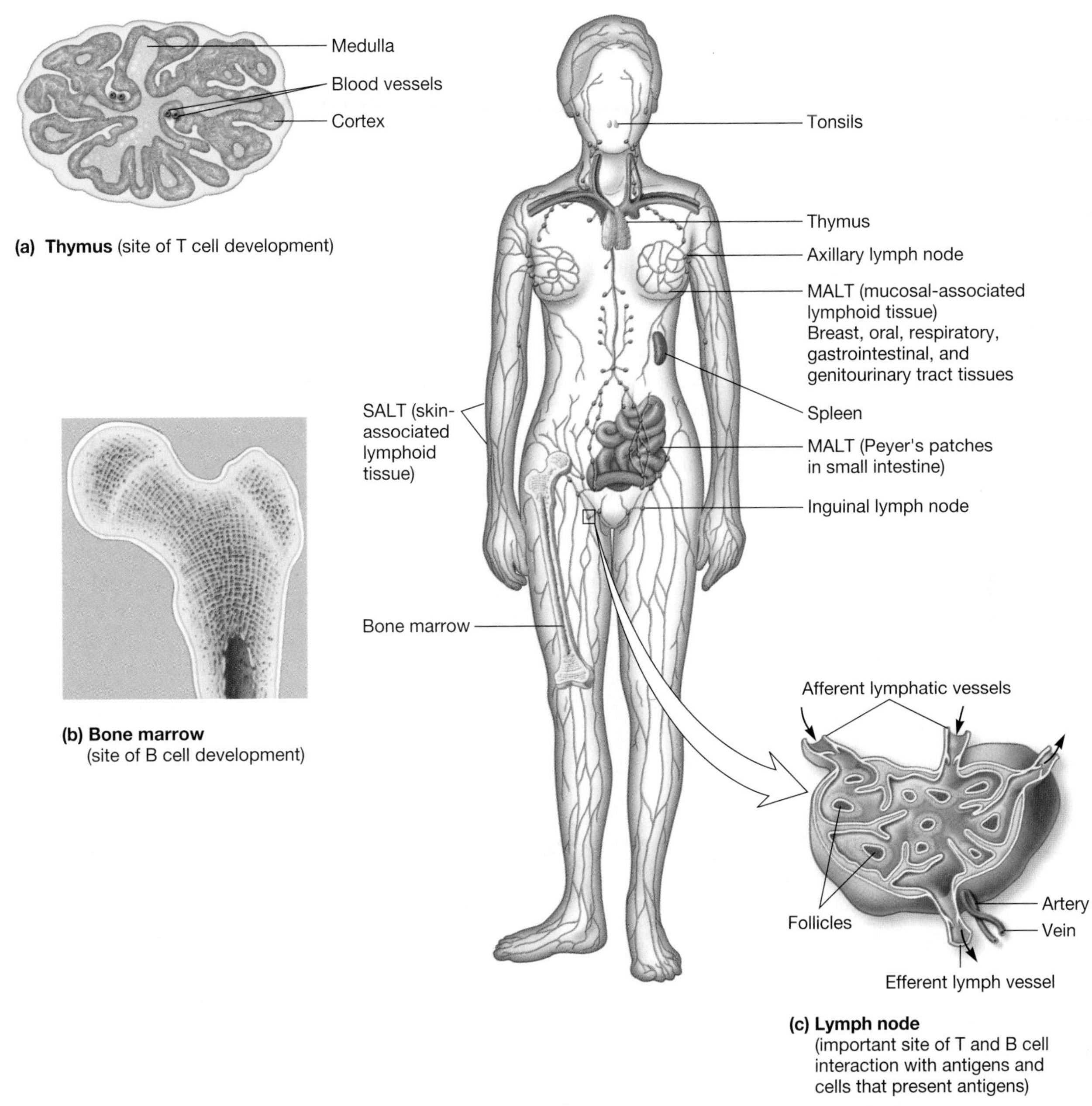

Figure 31.9 Anatomy of the Lymphoid System. Lymph is distributed through a system of lymphatic vessels, passing through many lymph nodes and lymphoid tissues. For example, **(a)** the thymus is involved in T-cell development and atrophies with age; **(b)** the bone marrow is the site of B-cell development; and **(c)** lymph enters a lymph node through the afferent lymph vessels, percolates through and around the follicles in the node, and leaves through the efferent lymphatic vessels. The lymphoid follicles are the site of cellular interactions and extensive immunologic activity.

*a*ssociated *l*ymphoid *t*issue, or BALT, characterized by the lack of cellular partitioning. The primary role of these lymphoid tissues is to efficiently organize leukocytes to increase interaction between the innate and the acquired arms of the immune response. In other words, the lymphoid tissues serve as the interface between the innate and acquired immunity of a host. Thus as we shall see, a microbe attempting to invade a potential host is greeted by nonspecific, physical, chemical, and granulocyte barriers that are designed to kill the invader, digest the carcass into small antigens, and assist the lymphocytes in formulating long-term protection against the next invasion. We now examine the phagocytic processes in more detail and then consider how the host integrates many of the innate immune activities into a substantial barrier to microbial invasion, known as the inflammatory response.

1. Briefly describe each of the primary lymphoid organs and tissues.
2. What is the function of the spleen? A lymph node? The thymus?
3. Injury to the spleen can lead to its removal. What impact would this have on host defenses?

31.3 Phagocytosis

During their lifetimes, humans and other vertebrates encounter many microbial species, but only a few of these species can grow and cause serious disease in otherwise healthy hosts. Phagocytic cells (monocytes, tissue macrophages, dendritic cells, and neutrophils) are an important early defense against invading microorganisms. These phagocytic cells recognize, ingest, and kill many extracellular microbial species by the process called **phagocytosis** [Greek *phagein,* to eat, *cyte,* cell, and *osis,* a process]. The concept of phagocytosis is briefly introduced in section 4.4 within the discussion of the endocytic pathway for obtaining nutrients. Phagocytosis is now considered in more detail in the context of nonspecific host resistance (**figure 31.10**).

Phagocytic cells use two basic molecular mechanisms for the recognition of microorganisms: (1) opsonin-independent (nonopsonic) recognition and (2) opsonin-dependent (opsonic) recognition. The phagocytic process can be greatly enhanced by opsonization. We discuss nonopsonic recognition here as it augments our study of phagocytosis. We reserve our discussion of opsonic recognition for section 31.6.

Pathogen Recognition

The opsonin-independent mechanism is a receptor-based system wherein components common to many different pathogens are recognized to activate phagocytes (figure 31.10*a*). Phagocytic cells recognize pathogens by several means but appear to exploit a common signaling system to respond (**table 31.2**). One recognition mode, termed lectin phagocytosis, is based on the binding of a mi-

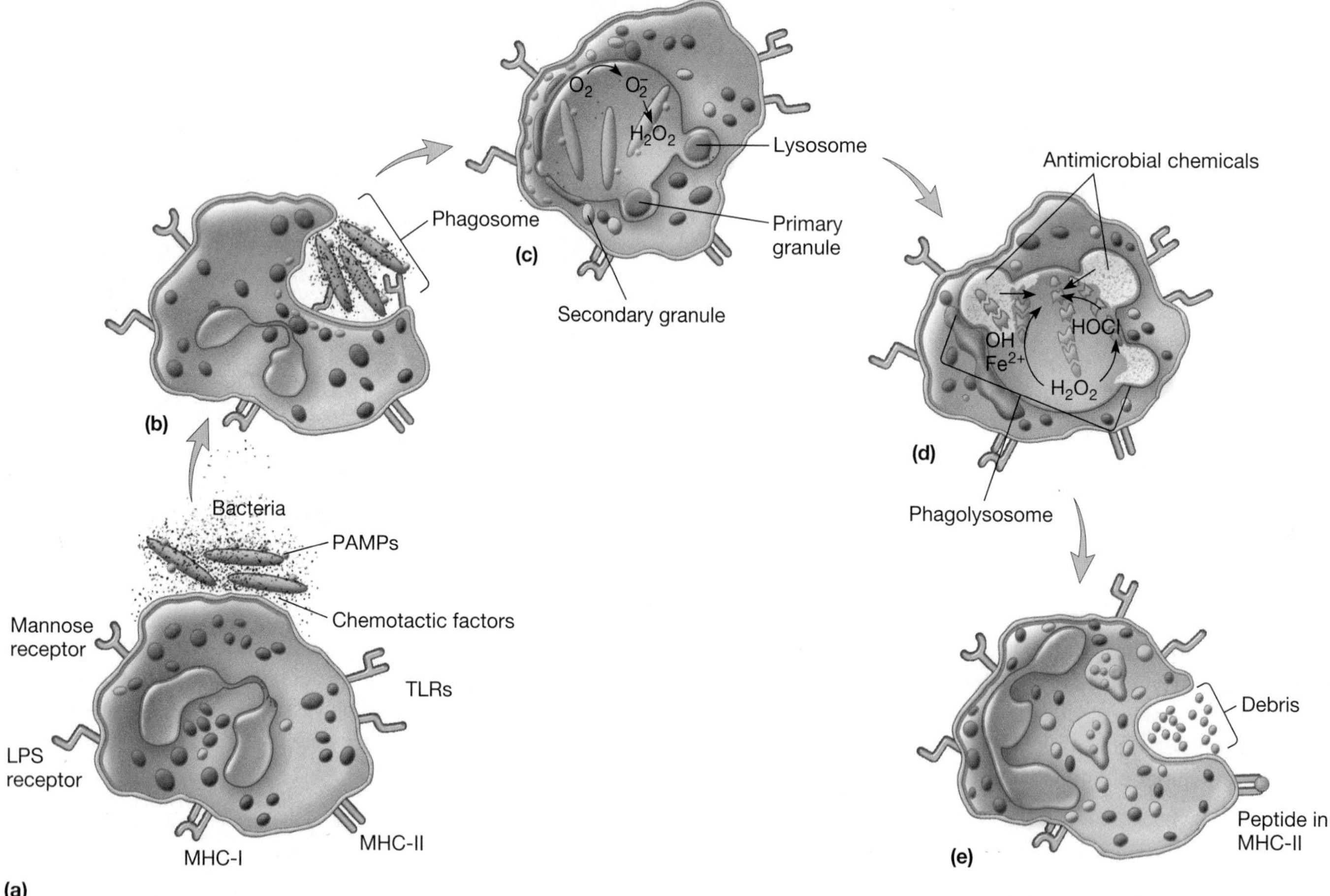

Figure 31.10 Phagocytosis. **(a)** Drawing shows receptors on a phagocytic cell, such as a macrophage, and the corresponding PAMPs participating in phagocytosis. The schematic depicts the process of phagocytosis showing **(b)** ingestion, **(c)** participation of primary and secondary granules, and O_2-dependent killing events, **(d)** intracellular digestion, and **(e)** exocytosis. LPS receptor: lipopolysaccharide receptor; TLRs: toll-like receptors; MHCI: class I major histocompatibility protein; MHCII: class II major histocompatibility protein; PAMPs: pathogen-associated molecular patterns.

Table 31.2 Nonopsonic Modes of Recognition and Signaling by Phagocytes

Type of Interaction	Bacterial Ligand (and Example)	Phagocytic Receptor (and Example)
Lectin-carbohydrate	Lectin (type I fimbriae)	Glycoprotein (integrins)
	Polysaccharide (capsule)	Lectin (Man/GlcNAc receptors)
Protein-protein	Arginine-glycine-aspartic acid (RGD)-containing proteins (filamentous hemagglutinin)	RGD receptor (integrins)
Hydrophobic protein	Glycolipid (lipoteichoic acid)	Lipid receptors (integrins)
Signaling	Bacterial lipopeptides	TLR1[1]/TLR2
	G^+ lipoteichoic acid and zymosan	TLR2/TLR6
	Peptidoglycan	TLR2
	Double-stranded viral RNA	TLR3
	LPS, Heat-shock proteins	TLR4/TLR4
	Flagellin	TLR5
	U-rich single-stranded (ss) viral RNA	TLR7
	ss viral RNA	TLR8
	Unmethylated CpG of bacterial & viral DNA	TLR9

[1]TLR: toll-like receptor

crobial lectin [latin *legere,* to select or choose], a protein that specifically binds or cross-links carbohydrates to a carbohydrate moiety of a cell receptor (**figure 31.11**). A second mode results from protein-protein interactions between the peptide sequence arginine-glycine-aspartic acid (RGD) on the cell surface of microorganisms and RGD receptors found on all phagocytes (figure 31.11). Third, hydrophobic interactions between bacteria and phagocytic cells also promote phagocytosis. A particular microbial species can express multiple binding sites, each recognized by a distinct receptor present on phagocytic cells (figure 31.11).

A fourth type of interaction also involves the recognition of microbial antigen and plays a crucial role in nonspecific host resistance. This recognition strategy is based on the detection of conserved molecular structures that occur in patterns and are the essential products of normal microbial physiology. These invariant structures are called **pathogen-*a*ssociated *m*olecular *p*atterns (PAMPs).** PAMPs are unique to microorganisms, invariant among microorganisms of a given class, and not produced by the host. The most well-known examples of PAMPs are the lipopolysaccharide (LPS) of gram-negative bacteria and the peptidoglycan of gram-positive bacteria. These and other PAMPs are recognized by receptors on phagocytic cells called **pattern recognition receptors (PRRs).** Because PAMPs are produced only by microorganisms, they are perceived by the phagocytic cells of the innate immune system as molecular signatures of infection. This is one way the innate immune system distinguishes self from microbial nonself.

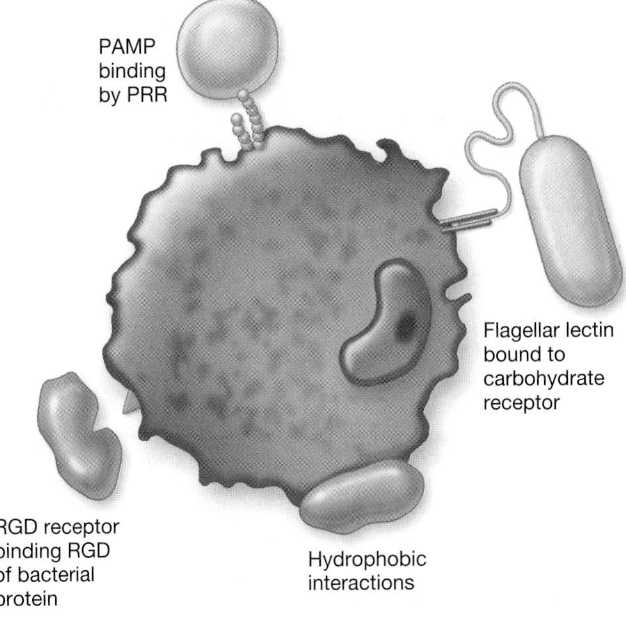

Figure 31.11 Some of the Possible Mechanisms by which a Macrophage can Recognize and Capture Microbes. See text for details.

Toll-like Receptors

Several structurally and functionally distinct classes of PRRs evolved in phagocytic cells to recognize PAMPs and to induce various host defensive pathways. For example, secreted PRRs bind to microbial cells and mark them for destruction by either the complement system or phagocytosis. Another class of PRRs function exclusively as signaling receptors. These receptors are known as **toll-like receptors (TLRs)** (**figure 31.12**). TLRs recognize and bind unique PAMPs of different classes of pathogens (viruses, bacteria, or fungi) and subsequently communicate that binding to the

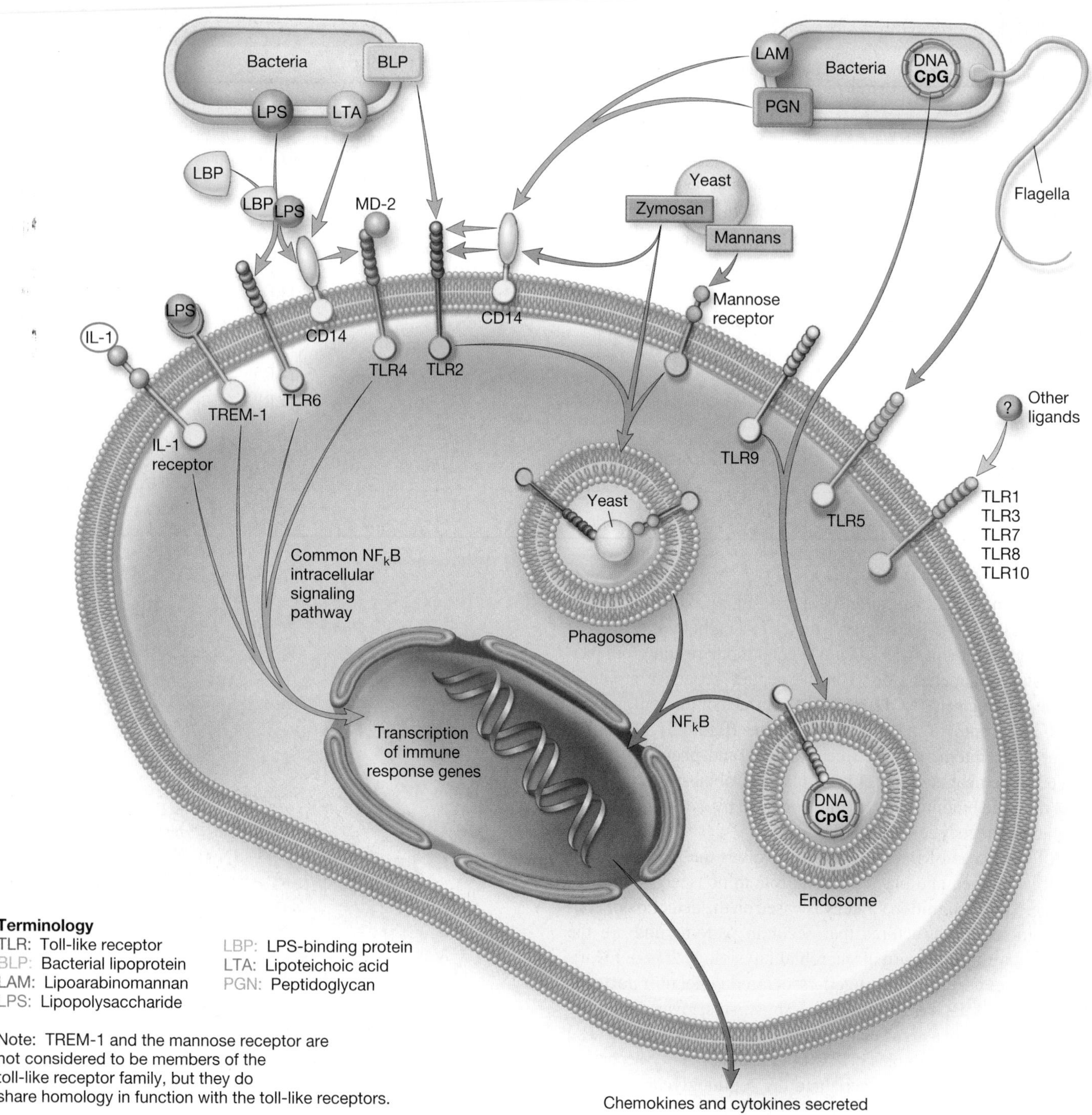

Figure 31.12 Recognition of Pathogen-Associated Molecular Patterns (PAMPs) by Toll-like Receptors (TLRs). PAMP binding of TLR results in a signaling process that upregulates gene expression. A common NF$_k$B signal transduction pathway is used.

host cell nucleus to initiate appropriate gene expression and host response. There are at least 10 distinct proteins in this family of mammalian receptors. For example, TLR-4 signals the presence of bacterial lipopolysaccharide (LPS) and heat-shock proteins (table 31.2). TLR-9 signals the dinucleotide CpG motif present on DNA released by dying bacteria. TLR-2 signals the presence of bacterial lipoproteins and peptidoglycans. Binding of TLRs triggers an evolutionarily ancient signaling pathway that activates transcription factor NF$_k$B by degrading its inhibitor, I$_k$B. This induces expression of a variety of genes, including genes for cytokines, chemokines, and costimulatory molecules that play essential roles in calling forth and directing the adaptive immune response later in an infection. Thus binding of specific microbial components to phagocyte receptors is an important first step in phagocytosis. Once bound, the microbe and/or its components can be internalized as part of a **phagosome** that is then united with a lysosome to facilitate microbial killing and digestion.

Intracellular Digestion

Once ingested by phagocytosis, microorganisms in membrane-enclosed vesicles are delivered to a lysosome by fusion of the phagocytic vesicle, called a phagosome, with the lysosome membrane, forming a new vacuole called a **phagolysosome** (figure 31.10 *c,d*). This is when the killing begins because lysosomes deliver a variety of hydrolases such as lysozyme, phospholipase A_2, ribonuclease, deoxyribonuclease, and proteases. The activity of these degradative enzymes is enhanced by the acidic vacuolar pH. Collectively, these enzymes participate in the destruction of the entrapped microorganisms. In addition to these oxygen-independent lysosomal hydrolases, macrophage and neutrophil lysosomes contain oxygen-dependent enzymes that produce toxic **reactive oxygen intermediates (ROIs)** such as the superoxide radical ($O_2^{-}\cdot$), hydrogen peroxide (H_2O_2), singlet oxygen ($^1O_2$), and hydroxyl radical (OH^-). The NADPH required for this process is supplied by a large increase in pentose phosphate pathway activity (*see figure 9.6*). Neutrophils also contain the heme-protein myeloperoxidase, which catalyzes the production of hypochlorous acid. Some reactions that form ROIs are shown in **table 31.3.** These reactions result from the **respiratory burst** that accompanies the increased oxygen consumption and ATP generation needed for phagocytosis. These reactions occur within the lysosome as soon as the phagosome is formed; lysosome fusion is not necessary for the respiratory burst. ROIs are effective in killing invading microorganisms. The influence of environmental factors on growth: Oxygen concentration (section 6.5)

Macrophages, neutrophils, and mast cells have also been shown to form **reactive nitrogen intermediates (RNIs).** These molecules include nitric oxide (NO) and its oxidized forms, nitrite (NO_2^-) and nitrate (NO_3^-). The RNIs are very potent cytotoxic agents and may be either released from cells or generated within cell vacuoles. Nitric oxide is probably the most effective RNI. Macrophages produce it from the amino acid arginine. Nitric oxide can block cellular respiration by complexing with the iron in electron transport proteins. Macrophages use RNIs in the destruction of a variety of infectious agents including the herpes simplex virus, the protozoa *Toxoplasma gondii* and *Leishmania major,* the opportunistic fungus *Cryptococcus neoformans,* and the metazoan pathogen *Schistosoma mansoni.* RNIs are also used to kill tumor cells.

Neutrophil granules contain a variety of other microbicidal substances such as several cationic peptides, the bactericidal permeability-increasing protein (BPI), and the family of broad-spectrum antimicrobial peptides including defensins. There are four human defensins produced by neutrophils called (HNPs): HNP-1, 2, 3, and 4. These defensins are synthesized by myeloid (mononuclear granulocyte) precursor cells during their sojourn in the bone marrow, and are then stored in the cytoplasmic granules of mature neutrophils. This compartmentalization strategically locates defensins (and other antimicrobial products) for extracellular secretion or delivery to phagocytic vacuoles. Susceptible microbial targets include a variety of gram-positive and gram-negative bacteria, yeasts and molds, and some viruses. Defensins act against bacteria and fungi by permeabilizing cell membranes. They form voltage-dependent membrane channels that allow ionic efflux. Antiviral activity involves direct neutralization of enveloped viruses, so they can no longer bind host cell receptors; nonenveloped viruses are not affected by defensins. Defensins are described in more detail in section 31.6.

Table 31.3 Formation of Reactive Oxygen Intermediates

Oxygen Intermediate	Reaction
Superoxide ($O_2^{-}\cdot$)	$NADPH + 2O_2 \xrightarrow{\text{NADPH oxidase}} 2O_2^{-}\cdot + H^+ + NADP^+$
Hydrogen peroxide (H_2O_2)	$O_2^{-}\cdot + 2H^+ \xrightarrow{\text{Superoxide dismutase}} H_2O_2 + O_2$
Hypochlorous acid (HOCl)	$H_2O_2 + Cl^- \xrightarrow{\text{Myeloperoxidase}} HOCl + OH^+$
Singlet oxygen ($^1O_2$)	$ClO^- + H_2O_2 \xrightarrow{\text{Peroxidase}} {}^1O_2 + Cl^- + H_2O$
Hydroxyl radical (OH^-)	$O_2^{-}\cdot + H_2O_2 \xrightarrow{\text{Peroxidase}} OH^- + OH^- + O_2$

Exocytosis

Once the microbial invaders have been killed and digested into small antigenic fragments, the phagocyte may do one of two things. Neutrophils tend to expel the microbial fragments by the process of exocytosis. This is essentially a reverse of the phagocytic process whereby the phagolysosome unites with the cell membrane resulting in the extracellular release of the microbial fragments. Other phagocytic cells, such as macrophages and dendritic cells, continue to process the microbial fragments by passing them from the phagolysosome to the endoplasmic reticulum. Here the peptide components of the fragments are united with glycoproteins destined for the cell membrane. The glycoproteins bind the peptides within their extracellular domain so that they are presented outward from the cell once the glycoprotein is secured in the cell membrane. This so-called antigen presentation is critical because it is the event that permits wandering lymphocytes to evaluate killed microbes (as antigens) and be activated. Thus antigen presentation links a nonspecific immune response to a specific immune response. Recognition of foreignness (section 32.4)

1. What is the role of opsonin-independent phagocytosis?
2. Once a phagolysosome forms, how is the entrapped microorganism destroyed?
3. What is the purpose of the respiratory burst that occurs within macrophages and other phagocytic cells? Describe the nature and function of reactive oxygen and nitrogen intermediates.

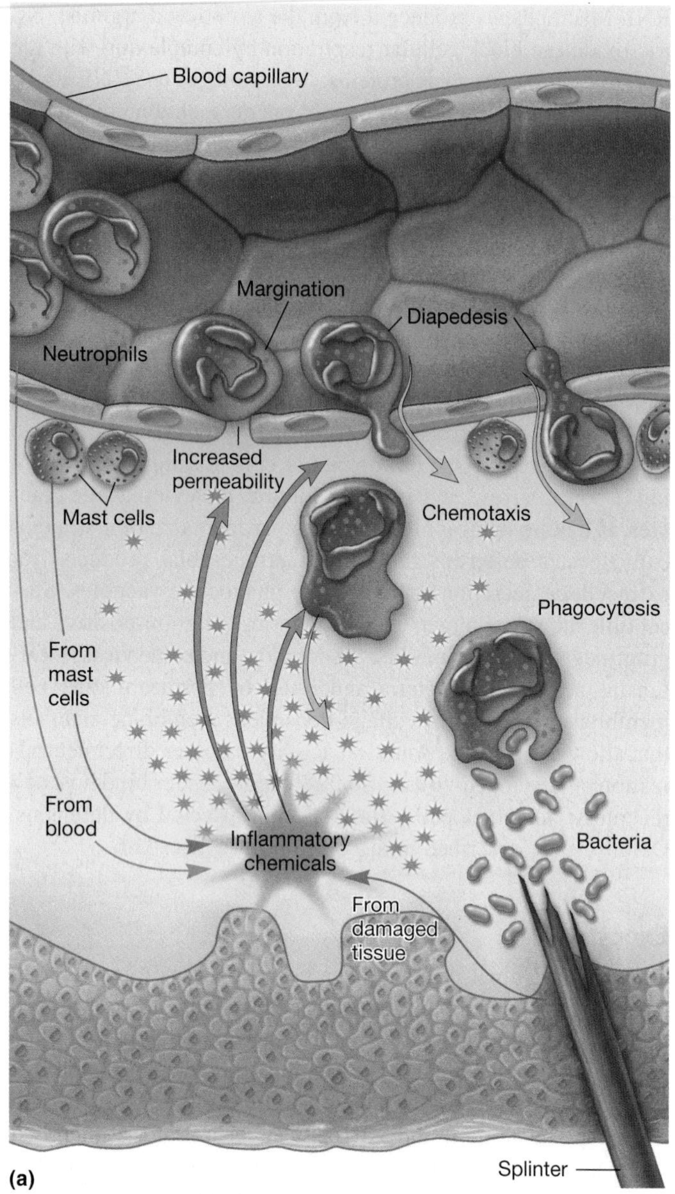

31.4 Inflammation

So far we have discussed the cellular, tissue, and organ components of innate immunity, along with an explanation of how pathogens are recognized by the innate immune system, destroyed, and their presence is communicated to the acquired immune system. But how do the innate immune cells perceive an impending invasion by pathogens so as to be recruited for host defense? One part of the answer lies in the process known as inflammation. **Inflammation** [Latin, *inflammatio,* to set on fire] is an important nonspecific defense reaction to tissue injury, such as that caused by a pathogen or wound. Acute inflammation is the immediate response of the body to injury or cell death. The gross features were described over 2,000 years ago and are still known as the cardinal signs of inflammation. These signs include redness (*rubor*), warmth (*calor*), pain (*dolor*), swelling (*tumor*), and altered function (*functio laesa*).

The acute inflammatory response begins when injured tissue cells release chemical signals (chemokines) that activate the inner lining (endothelium) of nearby capillaries (**figure 31.13**). Within the capillaries, **selectins** (a family of cell adhesion molecules) are displayed on the activated endothelial cells. These adhesion molecules attract and attach wandering neutrophils to the endothelial cells. This slows the neutrophils and causes them to roll along the endothelium where they encounter the inflammatory chemicals that act as activating signals (figure 31.13*b*). These signals activate **integrins** (adhesion receptors) on the neutrophils. The neutrophil integrins then attach tightly to the selectins. This causes the neutrophils to stick to the endothelium and stop rolling (margination). The neutrophils now undergo dramatic shape changes, squeeze through the endothelial wall (diapedesis) into the interstitial tissue fluid, migrate to the site of injury (extravasion), and attack the pathogen or other cause of the tissue damage. Neutrophils and other leukocytes are attracted to the infection site by chemotactic factors, which are also called chemotaxins. They include substances released by bacteria, endothelial cells, mast cells, and tissue breakdown products. Depending on the severity and nature of tissue damage, other types

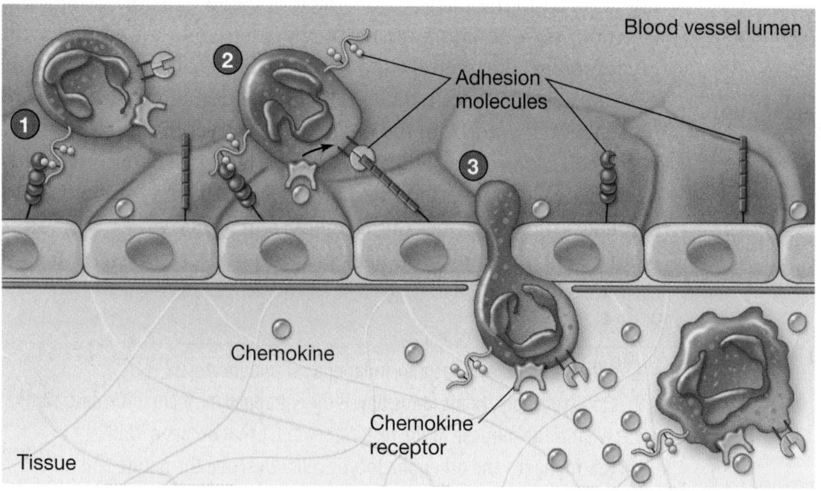

Figure 31.13 Physiological Events of the Acute Inflammatory Response. **(a)** At the site of injury (splinter), chemical messengers are released from the damaged tissue, mast cells, and the blood plasma. These inflammatory chemicals stimulate neutrophil migration, diapedesis, chemotaxis, and phagocytosis. **(b)** Neutrophil integrins interact with endothelial selectins (1) to facilitate margination (2) and diapedesis (3).

of leukocytes (e.g., lymphocytes, monocytes, and macrophages) may follow the neutrophils.

The release of inflammatory mediators from injured tissue cells sets into motion a cascade of events that result in the development of the signs of inflammation. The mediators increase the acidity in the surrounding extracellular fluid, which activates the extracellular enzyme **kallikrein (figure 31.14)**. Kallikrein cleavage releases the peptide bradykinin from its long precursor chain. Bradykinin then binds to receptors on the capillary wall, opening the junctions between cells and allowing fluid and infection-fighting leukocytes to leave the capillary and enter the infected tissue. Simultaneously, bradykinin binds to mast cells in the connective tissue associated with most small blood vessels. This activates the mast cells by causing an influx of calcium ions, which leads to degranulation and release of preformed mediators such as histamine. If nerves in the infected area are damaged, they release substance P, which also binds to mast cells, boosting preformed-mediator release. Histamine in turn makes the intercellular junctions in the capillary wall wider so that more fluid, leukocytes, kallikrein, and bradykinin move out, causing swelling or edema. Bradykinin then binds to nearby capillary cells and stimulates the production of prostaglandins (PGE_2 and $PGF_{2\alpha}$) to promote tissue swelling in the infected area. Prostaglandins also bind to free nerve endings, making them fire and start a pain impulse.

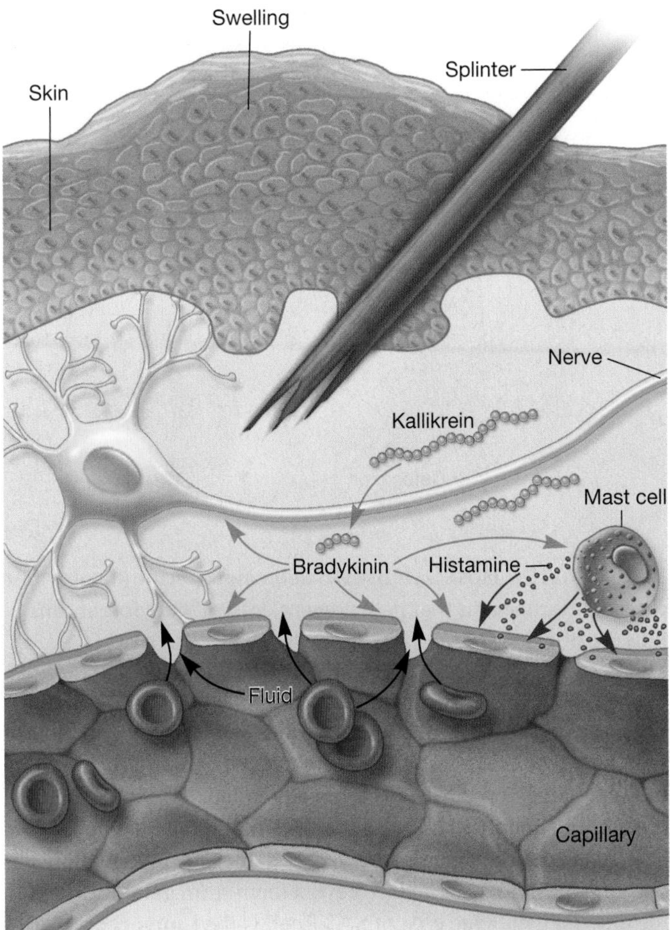

Figure 31.14 Tissue Injury Results in the Recruitment of Kallikrein, From Which Bradykinin Is Released. Bradykinin acts on endothelial and nerve cells resulting in edema and pain, respectively. It also stimulates mast cells to release histamine. Histamine also acts on endothelial cells, further increasing fluid leakage into injured tissue sites.

Activated mast cells also release a small molecule called arachidonic acid, the product of a reaction catalyzed by phospholipase A_2. Arachidonic acid is metabolized by the mast cell to form potent mediators including prostaglandins E_2 and $F_{2\alpha}$, thromboxane A_2, slow-reacting substance (SRS), and leukotrienes (LTC_4 and LTD_4). All of these mediators play specific roles in the inflammatory response. During acute inflammation, the offending pathogen is neutralized and eliminated by a series of important events:

1. The increase in blood flow and capillary dilation bring into the area more antimicrobial factors and leukocytes that destroy the pathogen. Dead host cells also release antimicrobial factors.
2. Blood leakage into tissue spaces increases the temperature and further stimulates the inflammatory response and may inhibit microbial growth.
3. A fibrin clot often forms and may limit the spread of the invaders so that they remain localized.
4. Phagocytes collect in the inflamed area and phagocytose the pathogen. In addition, chemicals stimulate the bone marrow to release neutrophils and increase the rate of granulocyte production.

Chronic Inflammation

In contrast to acute inflammation, which is a rapid and transient process, chronic inflammation is a slow process characterized by the formation of new connective tissue, and it usually causes permanent tissue damage. Regardless of the cause, chronic inflammation lasts two weeks or longer. Chronic inflammation can occur as a distinct process without much acute inflammation. The persistence of bacteria by a variety of mechanisms can stimulate chronic inflammation. For example, mycobacteria, which include species that cause tuberculosis and leprosy, have cell walls with a very high lipid and wax content, making them relatively resistant to phagocytosis and intracellular killing. These bacteria and a number of other pathogens can survive within the macrophage, as do some protozoan pathogens such as *Leishmania*. In addition, some bacteria produce toxins that stimulate tissue-damaging reactions even after bacterial death.

Suborder *Corynebacterinaea* (section 24.4); Protist classification: *Excavata* (section 25.6)

Chronic inflammation is characterized by a dense infiltration of lymphocytes and macrophages. If the macrophages are unable to protect the host from tissue damage, the body attempts to wall off and isolate the site by forming a **granuloma** [Latin, *granulum,* a small particle; Greek, *oma,* to form]. Granulomas are formed when neutrophils and macrophages are unable to destroy the microorganism during inflammation. Infections caused by some bacteria (listeriosis, brucellosis), fungi (histoplasmosis, coccidioidomycosis), helminth parasites (schistosomiasis), protozoa (leishmaniasis), and large antibody-antigen complexes (rheumatoid arthritis) result in granuloma formation and chronic inflammation. These infectious diseases are discussed in chapters 38 and 39.

1. What major events occur during an inflammatory reaction, and how do they contribute to pathogen destruction?
2. How does chronic inflammation differ from acute inflammation?

31.5 Physical Barriers in Nonspecific (Innate) Resistance

With few exceptions, a potential microbial pathogen invading a human host immediately confronts a vast array of nonspecific (innate) defense mechanisms (**figure 31.15**). Although the effectiveness of some individual mechanisms is not great, collectively their defense is formidable. Many direct factors (nutrition, physiology, fever, age, genetics) and equally as many indirect factors (personal hygiene, socioeconomic status, living conditions) influence all host-microbe relationships. At times they favor the establishment of the microorganism within the host; at other times they provide some measure of defense to the host. For example, when the host is either very young or very old, susceptibility to infection increases. Babies are at a particular risk after their maternal immunity has waned and before their own immune systems have matured. Very old persons experience a decline in the immune system itself and in the homeostatic functioning of many organs, which reduce host defenses. In addition to these direct and indirect factors, a vertebrate host has some specific physical and mechanical barriers.

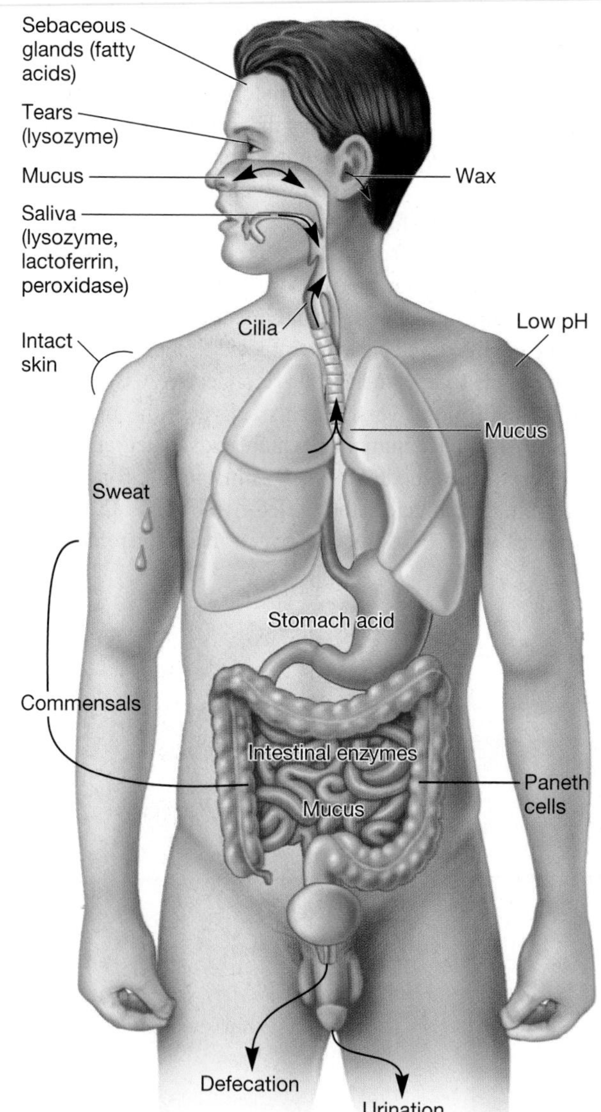

Figure 31.15 Host Defenses. Some nonspecific (innate) host defense mechanisms that help prevent entry of microorganisms into the host's tissues.

Physical and Mechanical Barriers

Physical and mechanical barriers, along with the host's secretions (flushing mechanisms), are the first line of defense against microorganisms. Protection of the most important body surfaces by these mechanisms is discussed next.

Skin

The intact skin contributes greatly to nonspecific host resistance. It forms a very effective mechanical barrier to microbial invasion. Its outer layer consists of thick, closely packed cells called keratinocytes, which produce keratins. Keratins are scleroproteins (i.e., insoluble proteins) that make up the main components of hair, nails, and the outer skin cells. These outer skin cells shed continuously, removing any grime or microorganisms that manage to adhere to their surface. The skin is slightly acidic (around pH 5 to 6) due to skin oil, secretions from sweat glands, and organic acids produced by commensal staphylococci. It also contains a high concentration of sodium chloride and is subject to periodic drying.

Despite the skin's defenses, at times some pathogenic microorganisms gain access to the tissue under the skin surface. Here they encounter a specialized set of cells called the ***s*kin-*a*ssociated *l*ymphoid *t*issue (SALT)** (**figure 31.16**). The major function of SALT is to confine microbial invaders to the area immediately underlying the epidermis and to prevent them from gaining access to the bloodstream. One type of SALT cell is the **Langerhans cell,** a specialized myeloid cell that can phagocytose antigens. Once the Langerhans cell has internalized the antigen, it migrates from the epidermis to nearby lymph nodes where it differentiates into a mature dendritic cell. Recall that dendritic cells can present antigens and activate nearby lymphocytes to induce the acquired immune system. This dendritic cell-lymphocyte interaction illustrates another bridge between the innate and acquired immune systems.

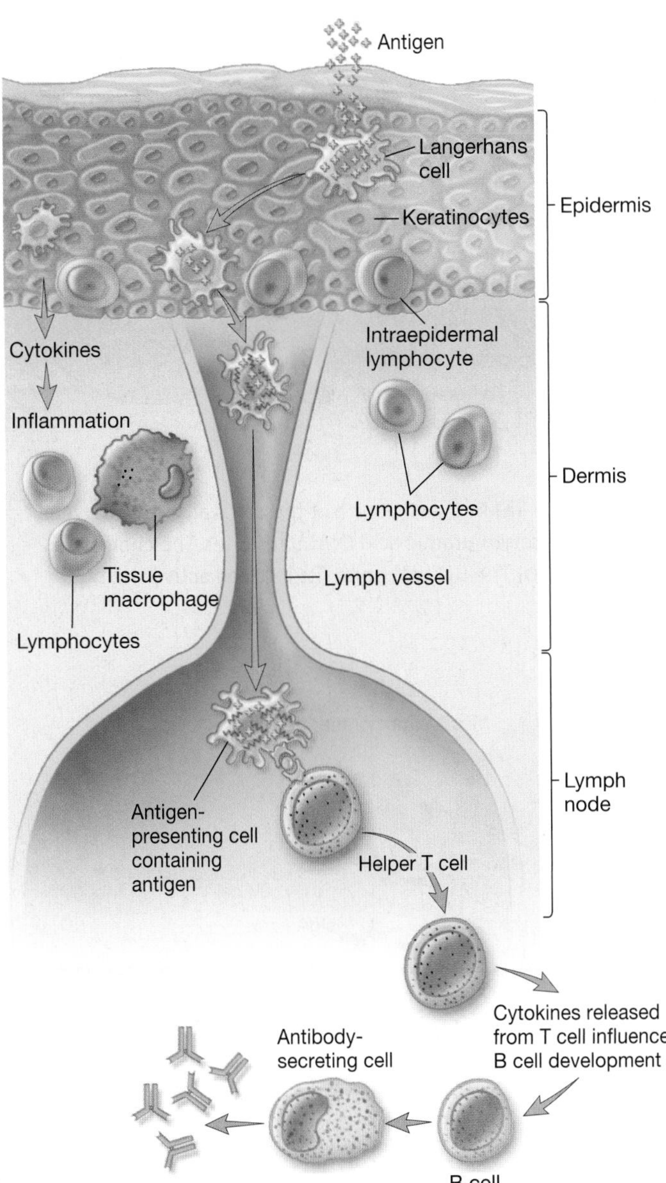

Figure 31.16 Skin-Associated Lymphoid Tissue (SALT). Keratinocytes make up 90% of the epidermis. They are capable of secreting cytokines that cause an inflammatory response to invading pathogens. Langerhans cells internalize antigen and move to a lymph node where they differentiate into dendritic cells that present antigen to helper T cells. The intraepidermal lymphocytes may function as T cells that can activate B cells to induce an antibody response.

The epidermis also contains another type of SALT cell called the **intraepidermal lymphocyte** (figure 31.16). These cells are strategically located in the skin so that they can intercept any antigens that breach the first line of defense. Most of these specialized SALT cells are T cells. Unlike other T cells, they have limited receptor diversity and have likely evolved to recognize common skin pathogen patterns. A large number of tissue macrophages (figure 31.3) are also located in the dermal layer of the skin and phagocytose most microorganisms they encounter (figure 31.4).

Mucous Membranes

The mucous membranes of the eye (conjunctiva) and the respiratory, digestive, and urogenital systems withstand microbial invasion because the intact stratified squamous epithelium and mucous secretions form a protective covering that resists penetration and traps many microorganisms. This mechanism contributes to nonspecific immunity. Furthermore, many mucosal surfaces are bathed in specific antimicrobial secretions. For example, cervical mucus, prostatic fluid, and tears are toxic to many bacteria. One antibacterial substance in these secretions is **lysozyme** (muramidase), an enzyme that lyses bacteria by hydrolyzing the $\beta(1 \rightarrow 4)$ bond connecting *N*-acetylmuramic acid and *N*-acetylglucosamine of the bacterial cell wall peptidoglycan—especially in gram-positive bacteria (**figure 31.17**). These mucous secretions also contain specific immune proteins that help prevent the attachment of microorganisms. They also contain significant amounts of the iron-binding protein, lactoferrin. **Lactoferrin** is released by activated macrophages and polymorphonuclear leukocytes (PMNs). It sequesters iron from the plasma, reducing the amount of iron available to invading microbial pathogens and limiting their ability to multiply. Finally, mucous membranes produce lactoperoxidase, an enzyme that catalyzes the production of superoxide radicals, a reactive oxygen intermediate that is toxic to many microorganisms (table 31.3). The bacterial cell wall: Peptidoglycan structure (section 3.6)

Like the skin, mucous membranes also have a specialized immune barrier called ***m*ucus-*a*ssociated *l*ymphoid *t*issue (MALT).** There are several types of MALT. The system most studied is the **gut-*a*ssociated *l*ymphoid *t*issue (GALT).** GALT includes the tonsils, adenoids, diffuse lymphoid areas along the gut, and specialized regions in the intestine called Peyer's patches. Less well-organized MALT also occurs in the respiratory system and is called ***b*ronchial *a*ssociated *l*ymphoid *t*issue (BALT);** the diffuse MALT in the urogenital system does not have a specific name. MALT can operate by two basic mechanisms. First, when an antigen arrives at the mucosal surface, it contacts a type of cell called the **M cell** (**figure 31.18*a***). The M cell does not have the brush border or microvilli found on adjacent columnar epithelial cells. Instead it has a large pocket containing B cells, T cells, and macrophages. When an antigen contacts the M cell, it is endocytosed and released into the pocket. Macrophages engulf the antigen or pathogen and try to destroy it. An M cell also can endocytose an antigen and transport it to a cluster of cells called an organized lymphoid follicle (figure 31.18*b*). The B cells within this follicle recognize the antigen and mature into antibody-producing plasma cells. The plasma cells leave the follicle and secrete a class of mucous membrane-associated antibody called secretory (s) IgA. sIgA is then transported into the lumen of the gut where it interacts with the antigen that caused its production. Similar to the SALT, GALT intra- and inter-epithelial lymphocytes are strategically distributed so the likelihood of antigen detection is increased should the intestinal membrane be breached. Antibodies (section 32.7)

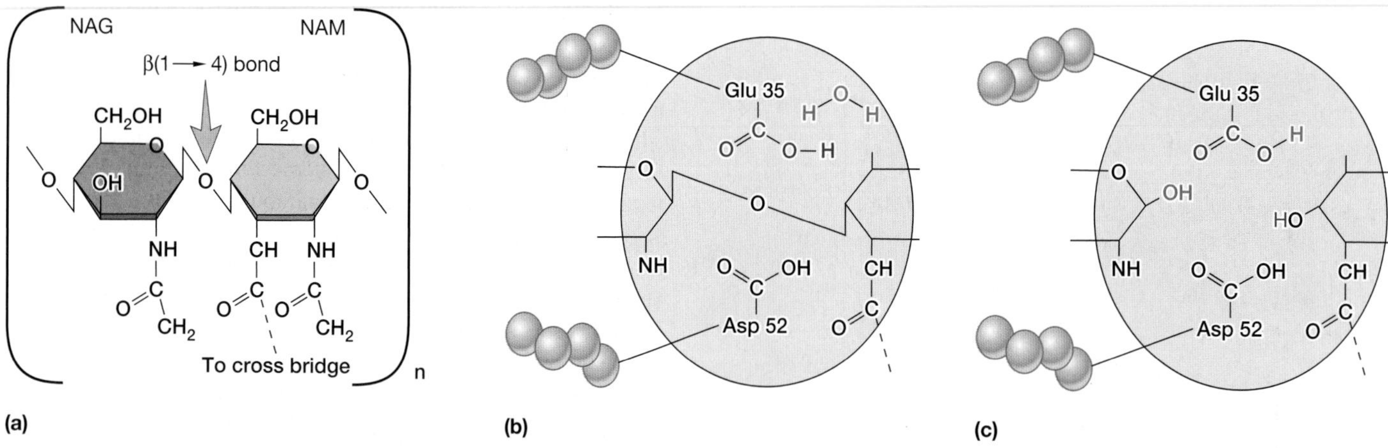

Figure 31.17 Action of Lysozyme on the Cell Wall of Gram-Positive Bacteria. **(a)** In the structure of the cell wall peptidoglycan backbone, the β(1→4) bonds connect alternating *N*-acetylglucosamine (NAG) and *N*-acetylmuramic acid (NAM) residues. The chains are linked through cross-bridges. Lysozyme splits the molecule as indicated by the arrow. **(b)** The β(1→4) bond fits into the active site of lysozyme (shaded area) facilitating **(c)** bond hydrolysis.

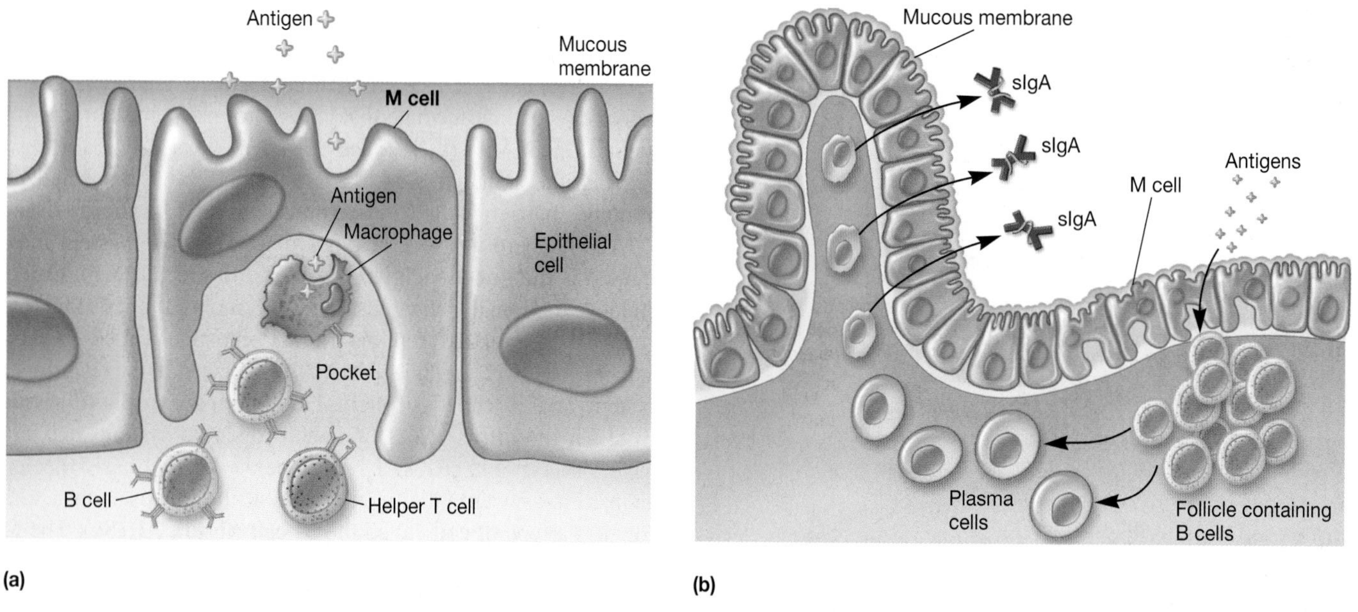

Figure 31.18 Function of M Cells in Mucosal-Associated Immunity. **(a)** Structure of an M cell located between two epithelial cells in a mucous membrane. The M cell endocytoses the pathogen and releases it into the pocket containing helper T cells, B cells, and macrophages. It is within the pocket that the pathogen often is destroyed. **(b)** The antigen is transported by the M cell to the organized lymphoid follicle containing B cells. The activated B cells mature into plasma cells, which produce secretory IgA and release it into the lumen where it reacts with the antigen that caused its production.

1. Why is the skin such a good first line of defense against pathogenic microorganisms?
2. How do intact mucous membranes resist microbial invasion of the host?
3. Describe SALT function in the immune response.
4. How do M cells function in MALT?

Respiratory System

The mammalian respiratory system has formidable defense mechanisms. The average person inhales at least eight microorganisms a minute, or 10,000 each day. Once inhaled, a microorganism must first survive and penetrate the air-filtration system of the upper and lower respiratory tracts. Because the airflow in these tracts is very turbulent, microorganisms are deposited on the moist, sticky mucosal surfaces. Microbes larger than 10 μm usually are trapped by hairs and cilia lining the nasal cavity. The cilia in the nasal cavity beat toward the pharynx, so that mucus with its trapped microorganisms is moved toward the mouth and expelled (**figure 31.19**). Humidification of the air within the nasal cavity causes many hygroscopic (attracting moisture from the air) microorganisms to swell, and this aids phagocytosis. Microbes smaller than 10 μm often pass through the nasal cavity and are trapped by the **mucociliary blanket** that coats the mucosal surfaces of lower portions of the respiratory system. The trapped microbes are transported by ciliary action (**mucociliary escalator**) that moves them away from the lungs. Coughing and sneezing reflexes clear the respiratory system of microorganisms by expelling air forcefully from the lungs through the mouth and nose, respectively. Salivation also washes microorganisms from the mouth and nasopharyngeal areas into the stomach. Microorganisms that succeed in reaching the alveoli of the lungs encounter a population of fixed phagocytic cells called **alveolar macrophages** (figure 31.3). These cells can ingest and kill most bacteria by phagocytosis.

Gastrointestinal Tract

Most microorganisms that reach the stomach are killed by gastric juice (a mixture of hydrochloric acid, proteolytic enzymes, and mucus). The very acidic gastric juice (pH 2 to 3) is sufficient to destroy most organisms and their toxins, although exceptions exist (protozoan cysts, *Helicobacter pylori, Clostridium* and *Staphylococcus* toxins). However, organisms embedded in food particles are protected from gastric juice and reach the small intestine. Once in the small intestine, microorganisms often are damaged by various pancreatic enzymes, bile, enzymes in intestinal secretions, and the GALT system. **Peristalsis** [Greek *peri,* around, and *stalsis,* contraction] and the normal loss of columnar epithelial cells act in concert to purge intestinal microorganisms. In addition, the normal microbiota of the large intestine (*see figure 30.17*) is extremely important in preventing the establishment of pathogenic organisms. For example, many normal commensals in the intestinal tract produce metabolic products, such as fatty acids, that prevent unwanted microorganisms from becoming established. Other normal microbiota outcompete potential pathogens for attachment sites and nutrients. The mucous membranes of the intestinal tract contain cells called **Paneth cells.** These cells produce lysozyme (figure 31.17) and a set of peptides called **cryptins.** Cryptins are toxic for some bacteria, although their mode of action is not known.

Genitourinary Tract

Under normal circumstances the kidneys, ureters, and urinary bladder of mammals are sterile. Urine within the urinary bladder also is sterile. However, in both the male and female, a few

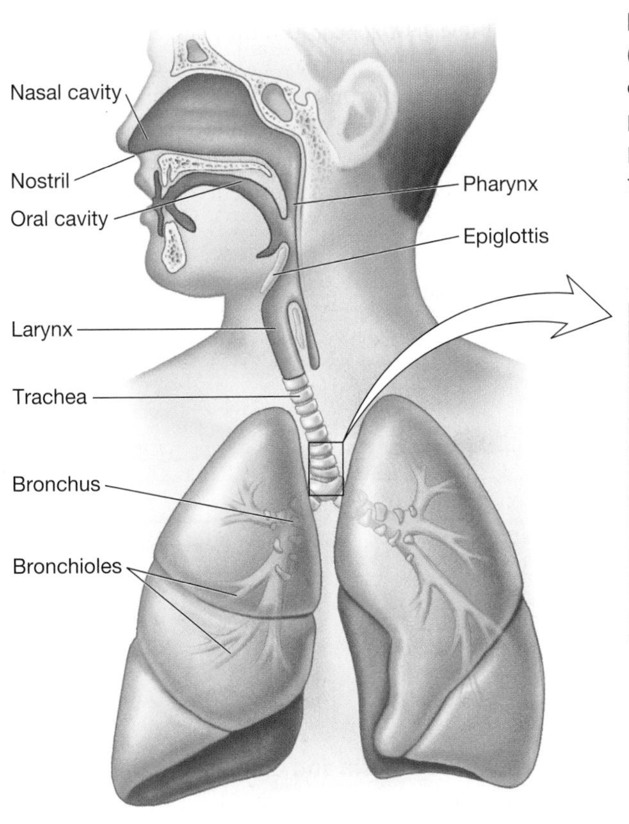

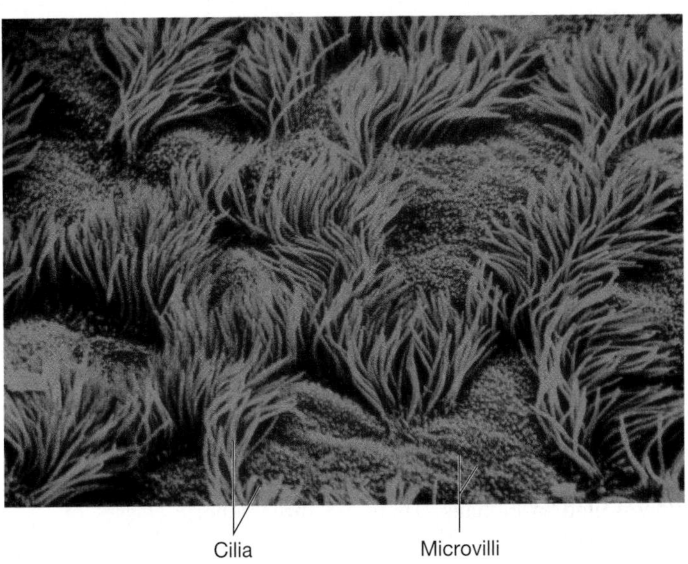

Figure 31.19 The Bronchial-Associated Lymphoid Tissue (BALT). **(a)** The respiratory tract is lined with a mucous membrane of ciliated epithelial cells. The circuitous passages of the nasal cavity prevent large particles from entering deeper into the respiratory tract. Mucus traps particles along the tract and the **(b)** cilia (×5,000) sweep them upward toward the throat to be expectorated.

bacteria are usually present in the distal portion of the urethra (*see figure 30.17*). The factors responsible for this sterility are complex. In addition to removing microbes by flushing action, urine kills some bacteria due to its low pH and the presence of urea and other metabolic end products (uric acid, hippuric acid, indican, fatty acids, mucin, enzymes). The kidney medulla is so hypertonic that few organisms can survive there. In males, the anatomical length of the urethra (20 cm) provides a distance barrier that excludes microorganisms from the urinary bladder. Conversely, the short urethra (5 cm) in females is more readily traversed by microorganisms; this explains why urinary tract infections are 14 times more common in females than in males.

The vagina has another unique defense. Under the influence of estrogens, the vaginal epithelium produces increased amounts of glycogen that acid-tolerant *Lactobacillus acidophilus* bacteria degrade to form lactic acid. Normal vaginal secretions contain up to 10^8 of these bacilli per ml. Thus an acidic environment (pH 3 to 5) unfavorable to most organisms is established. Cervical mucus also has some antibacterial activity.

The Eye

The conjunctiva is a specialized, mucus-secreting epithelial membrane that lines the interior surface of each eyelid and the exposed surface of the eyeball. It is kept moist by the continuous flushing action of tears (lacrimal fluid) from the lacrimal glands. Tears contain large amounts of lysozyme, lactoferrin, and sIgA and thus provide chemical as well as physical protection.

1. Describe the different antimicrobial defense mechanisms that operate within the respiratory system of mammals.
2. What factors operate within the gastrointestinal system that help prevent the establishment of pathogenic microorganisms?
3. Except for the anterior portion of the urethra, why is the genitourinary tract a sterile environment?

31.6 Chemical Mediators in Nonspecific (Innate) Resistance

Mammalian hosts have a chemical arsenal with which to combat the continuous onslaught of microorganisms. Some of these chemicals (gastric juices, salivary glycoproteins, lysozyme, oleic acid on the skin, urea) have already been discussed with respect to the specific body site(s) they protect. In addition, blood, lymph, and other body fluids contain a potpourri of defensive chemicals such as defensins and other polypeptides.

Antimicrobial Peptides

Cationic Peptides

Antimicrobial cationic peptides appear to be highly conserved through evolution (**figure 31.20**). We will only discuss those peptides found in humans. There are three generic classes of cationic peptides whose biological activity is related to their ability to damage bacterial plasma membranes. This is accomplished by electrostatic interactions with membranes—the formation of ionic pores and/or transient gaps thereby altering membrane permeability.

The first group of cationic peptides includes those that are linear, alpha-helical peptides that lack cysteine amino acid residues. An important example is **cathelicidin,** a peptide that arises from a precursor protein having a C-terminus bearing the mature peptide of some 12 to 80 amino acids. Cathelicidins are produced by a variety of cells (e.g., neutrophils, respiratory epithelial cells, and alveolar macrophages) and there is substantial heterogeneity between cathelicidins made by various cells. Protein structure (appendix I)

A second group, the **defensins,** is composed of peptides that are open-ended, rich in arginine and cysteine, and disulfide linked. The group is composed of various structural motifs with an approximate average molecular weight of 4,000 Daltons. In mammals, defensins have anti-parallel beta sheet structures with beta hairpin loops containing cationic amino acids. Two types of defensins have been reported in humans—alpha and beta. Alpha defensins tend to be peptides of 29 to 35 amino acid residues while beta defensins are usually 36 to 42 amino acids in length and are found in the primary granules of neutrophils, intestinal Paneth cells, and in intestinal and respiratory epithelial cells.

A third group contains larger peptides that are enriched for specific amino acids and exhibit regular structural repeats. **Histatin,** one such peptide isolated from human saliva, has antifungal activity. Histatin is a 24 to 38 amino acid peptide, heavily enriched with histidine, that does not appear to form ionic channels, but rather translocates to the fungal cytoplasm where it targets mitochondria.

Other natural antimicrobial products include fragments from (1) histone proteins, (2) lactoferrin, and (3) chemokines. A number of antibacterial peptides are produced by bacteria as well. The most notable of these are the bacteriocins.

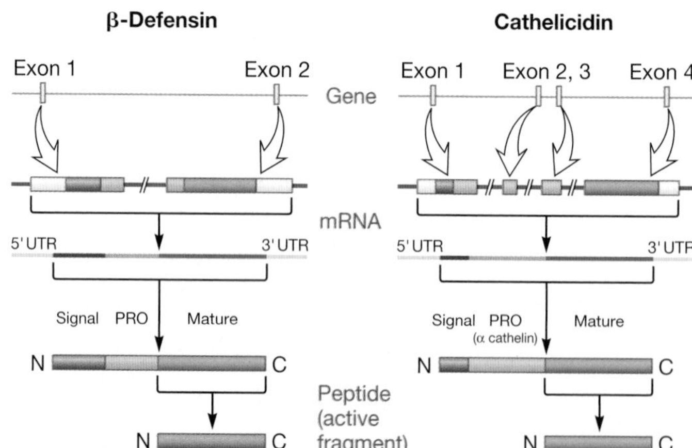

Figure 31.20 β-Defensin and Cathelicidin DNA, Messenger RNA and Peptides. Note that both exhibit biological activity only when smaller peptide fragments are cleaved from the native peptide.

Bacteriocins

As noted previously, the first line of defense against microorganisms is the host's anatomical barrier, consisting of the skin and mucous membranes. These surfaces are colonized by normal microbiota, which by themselves provide a biological barrier against uncontrolled proliferation of foreign microorganisms. Many of the bacteria that are part of the normal microflora synthesize and release toxic proteins (e.g., colicin, staphylococcin) called **bacteriocins** that are lethal to other strains of the same species as well as other bacterial species. Bacteriocin peptides range from 900 to 5,800 Daltons and can be cationic, neutral, or anionic. Bacteriocins may give their producers, which are naturally immune to antibacterial products they make, an adaptive advantage against other bacteria. Ironically, they sometimes increase bacterial virulence by damaging host cells such as mononuclear phagocytes. Bacteriocins are produced by gram-negative and gram-positive bacteria. For example, *E. coli* synthesizes bacteriocins called **colicins,** which are encoded by genes on several different plasmids (ColB, ColE1, ColE2, ColI, and ColV). Some colicins bind to specific receptors on the cell envelope of sensitive target bacteria and cause cell lysis, attack specific intracellular sites such as ribosomes, or disrupt energy production. Other examples include the lantibiotics produced by the genera *Streptococcus, Bacillus, Lactococcus* and *Staphylococcus*. It is now widely recognized that these antimicrobial peptides act as defensive effector molecules protecting the bacterial flora and its human host. Normal microbiota of the human body (section 30.3)

1. How do cationic peptides function against gram-positive bacteria?
2. How do bacteriocins function?

Complement

Complement was discovered many years ago as a heat-labile component of human blood plasma that augments phagocytosis. This activity was said to "complement" the antibacterial activity of antibody; hence, the name complement. It is now known that the **complement system** is composed of over 30 serum proteins that have a complex (and somewhat confusing) nomeclature. This system has three major physiological activities: (1) defending against bacterial infections by facilitating and enhancing phagocytosis (through opsonization, chemotaxis, activation of leukocytes, and lysis of bacterial cell walls); (2) bridging innate and adaptive immunity (augmentation of antibody responses, enhancement of immunologic memory); and (3) disposing of wastes (immune complexes, the products of inflammatory injury, clearance of dead host cells).

To achieve these activities, we need to revisit the idea of opsonization, first discussed in section 31.3. **Opsonization** [Greek *opson,* to prepare victims for] is a process in which microorganisms or other particles are coated by serum components (antibodies, mannose-binding proteins, and/or the complement glycoprotein C3b) thereby preparing them for recognition and ingestion by phagocytic cells. Molecules that function in this capacity are collectively known as opsonins. In the opsonin-dependent recognition mechanism, the host serum components function as a bridge between the microorganism and the phagocyte. They act by binding to the surface of the microorganism at one end and to specific receptors on the phagocyte surface at the other (**figure 31.21**). Some of the complement proteins are opsonins in that they bind to microbial cells, coating them for recognition by phagocytes (figure 31.21*b*). Additionally, other complement proteins are strong chemotactic signals that recruit phagocytes to the site of their activation. Still other complement proteins puncture cell membranes to cause lysis. Interestingly, one of the several triggers that can activate the complement process is the recognition of specific antibody on a target cell. All together, the complement activities unite the nonspecific and specific arms of the immune system to assist in the killing and removal of invading pathogens.

Complement proteins are produced in an inactive form; they become active following enzymatic cleavage. There are three pathways of complement activation: the alternative, lectin, and classical pathways (**figure 31.22**). Although they employ similar mechanisms, specific proteins are unique to the first part of each pathway (**table 31.4**). Each complement pathway is activated in a cascade fashion: the activation of one component results in the activation of the next. Thus complement proteins are poised for

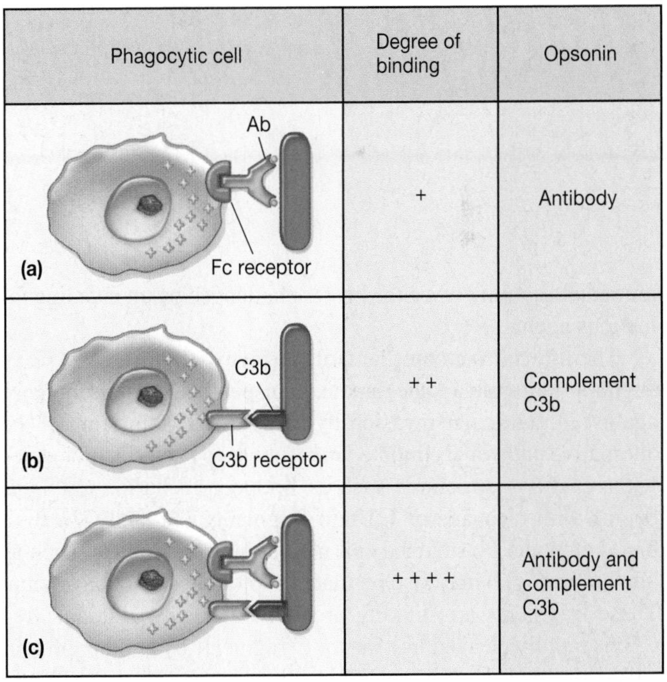

Phagocytic cell	Degree of binding	Opsonin
(a) Ab, Fc receptor	+	Antibody
(b) C3b, C3b receptor	+ +	Complement C3b
(c)	+ + + +	Antibody and complement C3b

Figure 31.21 Opsonization. **(a)** The intrinsic ability of a phagocyte to bind to a microorganism is enhanced if the microorganism elicits the formation of antibodies (Ab) that act as a bridge to attach the microorganism to the Fc receptor on the phagocytic cell. **(b)** If the microorganism has activated complement (C3b), the degree of binding is further enhanced by the C3b receptor. **(c)** If both antibody and C3b opsonize, binding is greatly enhanced.

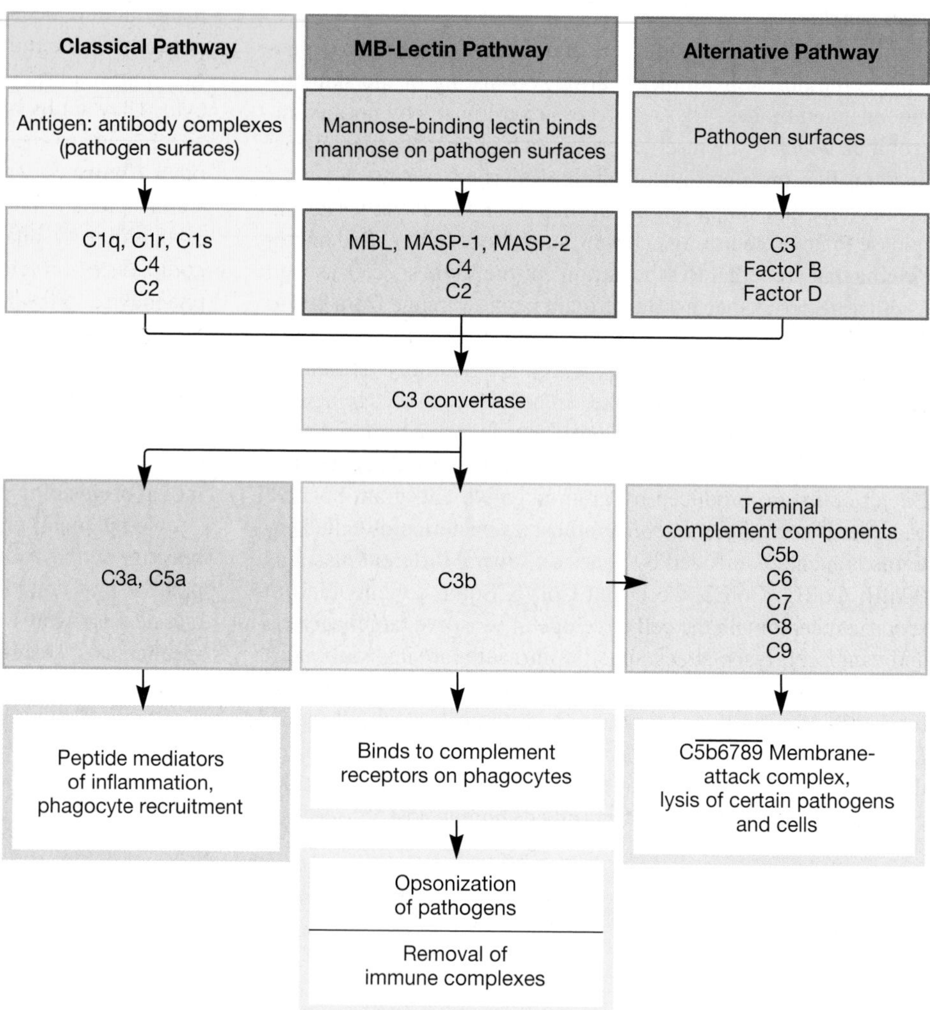

Figure 31.22 The Main Components and Actions of Complement. Complement activation involves a series of enzymatic reactions that culminate in the formation of C3 convertase, which cleaves complement component C3 into C3b and C3a. The production of the C3 convertase is where the three pathways converge. C3a is a peptide mediator of local inflammation. C3b binds covalently to the bacterial cell membrane and opsonizes the bacteria, enabling phagocytes to internalize them. C5a and C5b are generated by the cleavage of C5 by a C5 convertase. C5a is also a powerful peptide mediator of inflammation. C5b promotes the terminal components of complement to assemble into a membrane-attack complex.

immediate activity when the host is challenged by an invading infectious agent.

The **alternative complement pathway** (figure 31.22) plays an important role in the innate, nonspecific immune defense against intravascular invasion by bacteria and some fungi. The alternative pathway is initiated in response to bacterial molecules with repetitive structures such as lipopolysaccharide (LPS). It begins with cleavage of C3 into fragments C3a and C3b by a blood enzyme. Plasma and cell membrane components (such as human Factor H) may also regulate C3 proteolysis or conversion. These fragments are initially produced at a slow rate and free C3b is rapidly cleaved into inactive fragments by another protein called Factor I. However, C3b becomes stable when it binds to the LPS of gram-negative bacterial cell walls, or to aggregates of antibodies in the classes IgA or IgE. A protein in blood termed Factor B adsorbs to bound C3b and is cleaved into two fragments by Factor $\overline{\text{D}}$ (the bar indicates an activated enzyme complex), leading to the formation of active enzyme $\overline{\text{C3bBb}}$. This complex is called the C3 convertase of the alternative pathway because it cleaves more C3 to C3a and C3b thereby increasing the rate at which C3 is converted. $\overline{\text{C3bBb}}$ is further stabilized by a second blood protein, properdin, which allows another addition of C3b forming C5 convertase ($\overline{\text{C3bBb3b}}$). This convertase then cleaves C5 to C5a and C5b. The two proteins C6 and C7 rapidly bind to C5b, forming a $\overline{\text{C5b67}}$ complex that possesses an unstable membrane-binding site; once bound to a membrane, this complex is stable. C8 and C9 then bind, forming the **membrane attack complex** ($\overline{\text{C5b6789}}$), which creates a pore in the plasma membrane of the target cell (**figure 31.23**). If the cell is eucaryotic, Na^+ and H_2O enter through the pore and the cell lyses. Lysozyme can pass through pores in the outer membrane of gram-negative cell walls and digest the peptidoglycan cell wall, thus weakening the wall and aiding lysis. In contrast, gram-positive bacteria resist the cytolytic action of the membrane attack complex because they lack an exposed outer membrane and have a thick peptidoglycan protecting the plasma membrane. Unfortunately, host cell membranes are also susceptible to attack by complement proteins and bystander lysis is a potential consequence of complement activation.

The generation of complement fragments C3a and C5a leads to several important inflammatory effects. For example, binding of C3a and C5a to their cellular receptors induces some cells to

Table 31.4 Some Important Proteins of the Complement Cascade

Protein	Fragment	Function
Recognition Unit		
C1	q	Binds to the Fc portion of antigen-antibody complexes
	r	Activates C1s
	s	Cleaves C4 and C2 due to its enzymatic activity
Activation Unit		
C2		Causes viral neutralization
C3	a	Anaphylatoxin, immunoregulatory
	b	Key component of the alternative pathway and major opsonin in serum
	e	Induces leukocytosis
C4	a	Anaphylatoxin
	b	Causes viral neutralization; opsonin
Membrane Attack Unit		
C5	a	Anaphylatoxin; principal chemotactic factor in serum; induces neutrophil attachment to blood vessel walls
	b	Initiates membrane attack
C6, C7, C8, C9		Participate with C5b in formation of the membrane attack complex that lyses targeted cells
Alternative Pathway		
Factor B		Causes macrophage spreading on surfaces; precursor of C3 convertase
Factor $\overline{D}$		Cleaves Factor B to form active C3b$\overline{Bb}$ in alternative pathway
Properdin		Stabilizes alternative pathway C3 convertase
Regulatory Proteins		
Factor H		Promotes C3b breakdown and regulates alternative pathway
Factor I		Degrades C3b and regulates alternative pathway
C4b binding protein		Inhibits assembly and accelerates decay of C$\overline{4bC2a}$
C1 INH complex		Binds to and dissociates C1r and C1s from C1
S protein		Binds fluid-phase $\overline{C5b67}$; prevents membrane attachment

release other biological mediators. These mediators amplify the inflammatory signals of C3a and C5a by dilating vessels, increasing permeability, stimulating nerves, and recruiting phagocytic cells. C5a induces a directed, chemotactic migration of neutrophils to the site of complement activation. Macrophages in the area can synthesize even more complement components to interact with the bacteria. All of these defensive events promote the ingestion and ultimate destruction of the bacteria by neutrophils and macrophages.

The **lectin complement pathway** (also called the mannan-binding lectin pathway) also begins with the activation of C3 convertase. However, in this case a lectin, a special protein that binds to specific carbohydrates, initiates the proteolytic cascade. When macrophages ingest viruses, bacteria, or other foreign material, they release chemicals that stimulate liver cells to secrete acute phase proteins such as *m*annose-*b*inding *p*rotein (MBP). Because mannose, in certain three-dimensional configurations, is a major component of bacterial cell walls and of some virus envelopes and antigen-antibody complexes, MBP binds to these components. MBP enhances phagocytosis and is therefore an opsonin. When MBP is bound to the *M*BP-*a*ssociated *s*erine esterase (MASP), it activates the same C3 convertase found in the alternative complement pathway. Thus the lectin pathway activates the same complement cascade that the classical and alternative pathways do. However, it uses a mechanism that is independent of antibody-antigen interactions (the classical pathway), and it does not require interaction of complement with pathogen surfaces (alternative pathway).

This overview of the alternative and lectin complement pathways (figure 31.22) provides a basis for consideration of the function of complement as an integrated system during an animal's defensive effort. Bacteria arriving at a local tissue site will interact with components of the alternative pathway, resulting in the generation of biologically active fragments, opsonization of the

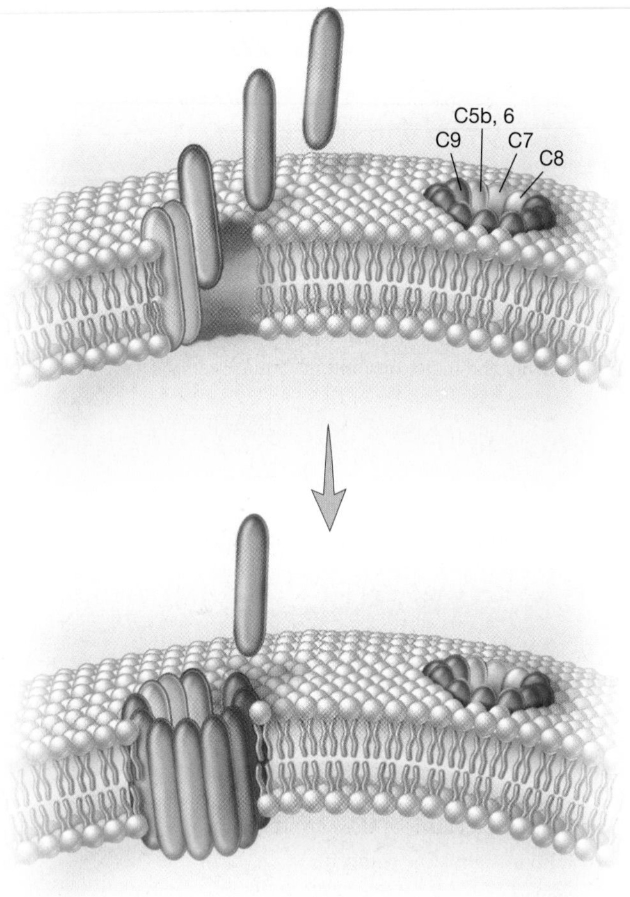

(a)

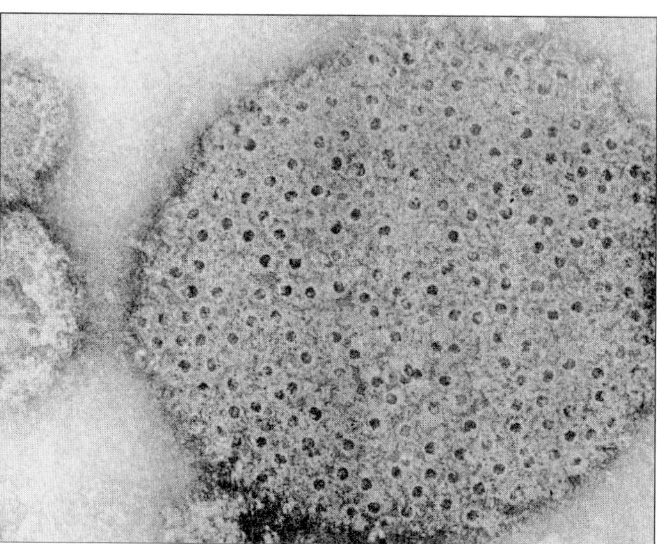

(b)

Figure 31.23 The Membrane Attack Complex. The membrane attack complex (MAC) is a tubular structure that forms a transmembrane pore in the target cell's plasma membrane. (**a**) This representation shows the subunit architecture of the membrane attack complex. The transmembrane channel is formed by a $\overline{\text{C5b678}}$ complex and 10 to 16 polymerized molecules of C9. (**b**) MAC pores appear as craters or donuts by electron microscopy.

bacteria, and initiation of the lytic sequence. If the bacteria persist or if they invade the animal a second time, antibody responses also will activate the classical complement pathway.

Activation of the **classical complement pathway** can occur in response to some microbial products (lipid A, staphylococcal protein A, etc.). However, it is usually initiated by the interaction of antibodies with an antigen (figure 31.22). Antibody secretion is part of an acquired immune response (and is discussed in chapter 32). It is important to emphasize that antibodies are glycoproteins that bind to specific antigens. This binding triggers the C1 complement component, composed of three proteins (q, r, and s), to attach to the antibody through its C1q subcomponent. In the presence of calcium ions, a trimolecular complex (C1qrs • antigen • antibody) with esterase activity is rapidly formed. The activated C1s subcomponent attacks and cleaves its natural substrates in serum (C2 and C4). This leads to binding of a portion of each molecule (C2a and C4b) to the antigen-antibody-complement complex with the release of C4a and C2b fragments. With the binding of C2a to C4b, an enzyme with trypsinlike proteolytic activity is generated. The natural substrate for this enzyme is C3; thus C2a4b is a C3 convertase. Just as we saw with the lectin and alternative pathways, the C3 convertase cleaves C3 into a bound subcomponent C3b and a C3a soluble component. This sets in motion the activation of the complement cascade, which leads to the formation of the membrane attack complex, opsonins, and the release of mediators that influence inflammation. Thus the three complement pathways have three different initiating processes. However, their common outcomes of opsonization, stimulation of inflammatory mediators, and lysis of membrane-bound microorganisms achieve the goal of innate host defense against foreign invaders.

1. What effect does the formation of the membrane attack complex have on eucaryotic cells? procaryotic cells?
2. How is the alternative pathway activated? the lectin pathway?
3. What role do complement fragments C3a and C5a play in an animal's defense against gram-negative bacteria?
4. How is the classical complement pathway activated?

Cytokines

Defense against viruses, microorganisms and their products, parasites, and cancer cells is mediated by both nonspecific immunity and specific immunity. Cytokines are required for immunoregulation of both of these immune responses. The term **cytokine** [Greek *cyto,* cell, and *kinesis,* movement] is a generic term for any soluble protein or glycoprotein released by one cell population that acts as an intercellular (between cells) mediator or signaling molecule. When released from mononuclear phagocytes, these proteins are called **monokines;** when released from T lymphocytes they are called **lymphokines;** when produced by a leukocyte and the action is on another leukocyte, they are **interleukins;** and if their effect is to stimulate the growth and differentiation of immature leukocytes in the bone marrow, they are called **colony-stimulating factors (CSFs).** Cytokines have been grouped into the following categories or families: chemokines, hematopoietins, interleukins, and members of the **tumor necrosis factor (TNF)** family. Some examples of these cytokine families are listed in **table 31.5.** Cytokines can affect the same cell responsible for their production (an autocrine function) or nearby cells (a paracrine function), or they can be distributed by the circulatory system to distant target cells (an endocrine function). Their production is induced by nonspecific stimuli such as a viral, bacterial, or parasitic infection; cancer; inflammation; or the interaction between a T cell and antigen. Some cytokines also can induce the production of other cytokines.

Interest in the biological actions of cytokines has grown enormously over the past three decades. This is due in part to their incredible potency and range of effects on eucaryotic cells, and to the fact that cytokines are involved in all aspects of disease. Cytokines produce biological actions only when they bind to specific, high-affinity receptors on the surface of target cells. An extracellular molecule that binds a specific receptor is called a ligand. The affinity of cytokine receptors for their cytokine ligands is very high, and consequently cytokines are effective at very low concentrations.

Most cells have hundreds to a few thousand cytokine receptors, but a maximal cellular response results even when only a small number of these are occupied by a cytokine. This binding activates specific intracellular signaling pathways that switch on genes encoding proteins essential to the appropriate target cell functions. For example, cytokine binding may result in the target cell's production of other cytokines, cell-to-cell adhesion receptors, proteases, lipid-synthesizing enzymes, and nitric oxide synthase (the production of nitric oxide has potent antimicrobial activity). In addition, cytokines can activate cell proliferation and/or cell differentiation (**figure 31.24**). They also can inhibit cell division and cause apoptosis (programmed cell death). Chemokines, one type of

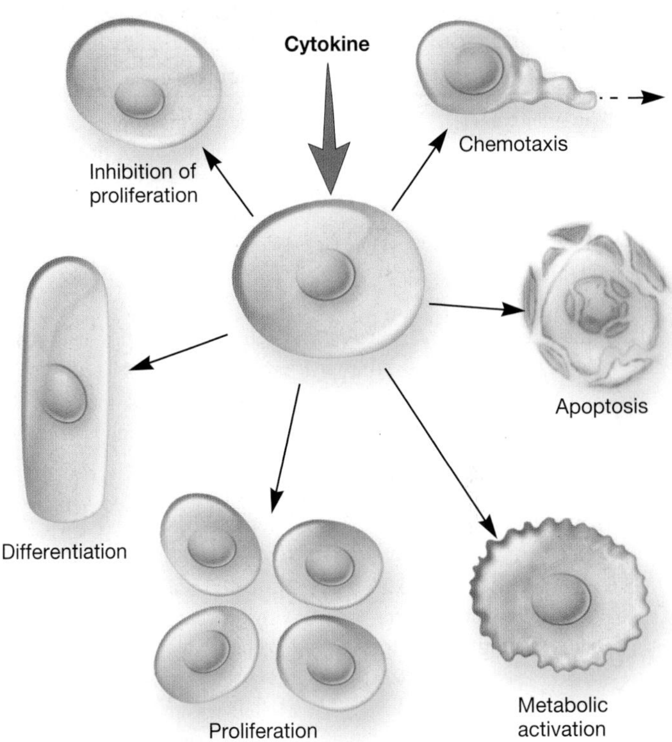

Figure 31.24 Range of Biological Actions That Cytokines Have on Eucaryotic Cells. Chemokines are one family of cytokines that induce leukocyte chemotaxis and migration. Other cytokines activate cell metabolism and synthesis. This can lead to the synthesis of a wide range of proteins including cyclooxygenase II, proteolytic enzymes, NO synthase, and various adhesion receptors. In addition, other cytokines can cause proliferation, inhibition of cell proliferation, or apoptosis.

Table 31.5 The Four Cytokine Families

Family	Examples	Functions
Chemokines	IL-8, RANTES[a], MIP (macrophage inflammatory protein)	Cytokines that are chemotactic and chemokinetic for leukocytes. They stimulate cell migration and attract phagocytic cells and lymphocytes. Chemokines play a central role in the inflammatory response.
Hematopoietins	Epo (erythropoietin), various colony-stimulating factors	Cytokines that stimulate and regulate the growth and differentiation processes involved in blood cell formation (hematopoiesis).
Interleukins	IL-1 to IL-18	Cytokines produced by lymphocytes and monocytes that regulate the growth and differentiation of other cells, primarily lymphocytes and hematopoietic stem cells. They often also have other biological effects.
Tumor necrosis factor (TNF) family	TNF-α, TNF-β, Fas ligand	Cytokines that are cytotoxic for tumor cells and have many other effects such as promoting inflammation, fever, and shock; some can induce apoptosis.

[a]RANTES: *R*egulated on *a*ctivation, *n*ormal *T* *e*xpressed and *s*ecreted; also called CCL5; member of the IL-8 cytokine superfamily.

cytokine, stimulate chemotaxis and chemokinesis (i.e., they direct cell movement) and thus play an important role in the acute inflammatory response (figure 31.13). Some examples of important cytokines and their functions are given in **table 31.6.**

Interferons (IFNs) are a group of related low molecular weight, regulatory cytokines produced by certain eucaryotic cells in response to a viral infection. Besides defending against viruses, they also help regulate the immune response. Interferons usually

Table 31.6 Some of the Cytokines That Mediate Immune Responses

Cytokine	Cell Source	Functions
IL-1 (interleukin-1)	Monocytes/macrophages, endothelial cells, fibroblasts, neuronal cells, glial cells, keratinocytes, epithelial cells	Produces a wide variety of effects on the differentiation and function of cells involved in inflammatory and immune effector responses; also affects central nervous and endocrine systems; it is an endogenous pyrogen
IL-2 (interleukin-2, T-cell growth factor)	T cells (T_H1)	Stimulates T-cell proliferation and differentiation; enhances cytolytic activity of NK cells; promotes proliferation and immunoglobin secretion of activated B cells
IL-3 (interleukin-3)	T cells, keratinocytes, neuronal cells, mast cells	Stimulates the production and differentiation of macrophages, neutrophils, eosinophils, basophils, and mast cells
IL-4 (interleukin-4, B-cell growth factor-1 [BCGF-1], B-cell stimulatory factor-1 [BCSF-1])	T cells (T_H2), macrophages, mast cells, basophils, B cells	Induces the differentiation of naive $CD4^+$ T cells into T-helper cells; induces the proliferation and differentiation of B cells; exhibits diverse effects on T cells, monocytes, granulocytes, fibroblasts, and endothelial cells
IL-5 (interleukin-5)	T cells (T_H2)	Growth and activation of B cells and eosinophils; chemotactic for eosinophils
IL-6 (interleukin-6, cytotoxic T-cell differentiation factor, B-cell differentiation factor)	T_H2 cells, monocytes/macrophages, fibroblasts, hepatocytes, endothelial cells, neuronal cells	Activates hematopoietic cells; induces growth of T cells, B cells, hepatocytes, keratinocytes, and nerve cells; stimulates the production of acute-phase proteins
IL-8 (interleukin-8)	Monocytes, endothelial cells, fibroblasts, alveolar epithelium, T cells, keratinocytes, neutrophils, hepatocytes	Chemoattractant for PMNs and T cells; causes PMN degranulation and expression of receptors; inhibits adhesion of PMNs to cytokine-activated endothelium; promotes migration of PMNs through nonactivated endothelium
IL-10 (interleukin-10)	T cells (T_H2), B cells, macrophages, keratinocytes	Reduces the production of IFN-γ, IL-1, TNF-α, and IL-6 by macrophages; in combination with IL-3 and IL-4, causes mast cell growth; in combination with IL-2, causes growth of cytotoxic T cells and differentiation of $CD8^+$ cells
IFNs α/β (interferons α/β)	T cells, B cells, monocytes/macrophages, fibroblasts	Antiviral activity, antiproliferative; stimulates macrophage activity; increases MHC class I protein expression on cells; regulates the development of the specific immune response
IFN-γ (interferon-γ)	T cells (T_H1, CTLs), NK cells	Activation of T cells, macrophages, neutrophils, and NK cells; antiviral and antiproliferative activities; increases class I and II MHC molecule expression on various cells
TNF-α (tumor necrosis factor-α [cachectin])	T cells, macrophages and NK cells	A wide variety of effects due to its ability to mediate expression of genes for growth factors and cytokines, transcription factors, receptors, inflammatory mediators, and acute-phase proteins; plays a role in host resistance to infection by serving as an immunostimulant and mediator of the inflammatory response; cytotoxic for tumor cells
TNF-β (tumor necrosis factor-β [lymphotoxin])	T cells, B cells	Same as TNF-α
G-CSF (granulocyte colony-stimulating factor)	T cells, macrophages, neutrophils	Enhances the differentiation and activation of neutrophils
M-CSF (macrophage colony-stimulating factor)	T cells, neutrophils, macrophages, fibroblasts, endothelial cells	Stimulates various functions of monocytes and macrophages, promotes the growth and development of macrophage colonies from undifferentiated precursors

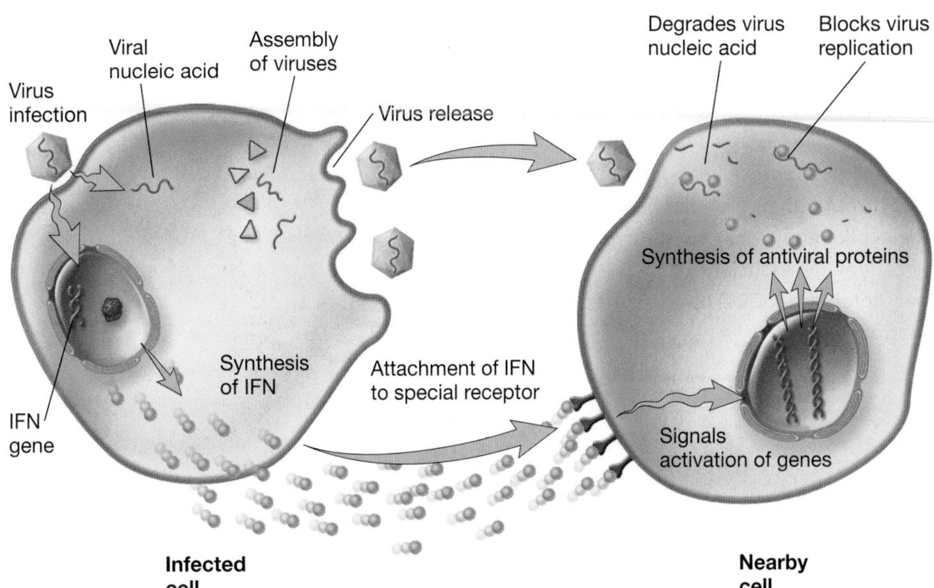

Figure 31.25 The Antiviral Action of Interferon. Interferon (IFN) synthesis and release is often induced by a virus infection or double-stranded RNA (dsRNA). Interferon binds to a ganglioside receptor on the plasma membrane of a second cell and triggers the production of enzymes that render the cell resistant to virus infection. The two most important such enzymes are oligo(A) synthetase and a special protein kinase. When an interferon-stimulated cell is infected, viral protein synthesis is inhibited by an active endoribonuclease that degrades viral RNA. An active protein kinase phosphorylates and inactivates the initiation factor eIF-2 required for viral protein synthesis.

are species specific but virus nonspecific. Several classes of interferons are recognized: IFN-γ is a family of 20 different molecules that can be synthesized by virus-infected leukocytes, antigen-stimulated T cells, and natural killer cells (table 31.5). IFN-α/β is derived from virus-infected fibroblasts. Although interferons do not prevent virus entry into host cells, they prevent viral replication and assembly thereby preventing further amplification of the viral infection (**figure 31.25**).

Another group of noteworthy cytokines are **endogenous pyrogens,** which elicit fever in the host. From a physiological point of view, fever results from disturbances in hypothalamic thermoregulatory activity, leading to an increase of the thermal "set point." In adult humans **fever** is defined as an oral temperature above 98.6°F (37°C) or a rectal temperature above 99.5°F (37.5°C). The most common cause of a fever is a viral or bacterial infection (or bacterial toxins). Examples of these pyrogens include interleukin-1 (IL-1), IL-6, and tissue necrosis factor (TNF); all are produced by host macrophages in response to pathogenic microorganisms. After their release, these pyrogens circulate to the hypothalamus and induce neurons to secrete prostaglandins. Prostaglandins reset the hypothalamic thermostat at a higher temperature, and temperature-regulating reflex mechanisms then act to bring the core body temperature up to this new setting. IL-1 also causes proliferation, maturation, and activation of T and B cells, which in turn augment the immune response of the host to the pathogen.

The fever induced by a microorganism augments the host's defenses by three complementary pathways: (1) it stimulates leukocytes so that they can destroy the microorganism; (2) it enhances the specific activity of the immune system; and (3) it enhances microbiostasis (growth inhibition) by decreasing available iron to the microorganism. Evidence suggests that some hosts are able to redistribute the iron during a fever in an attempt to withhold it from the microorganism (**hypoferremia**). Conversely, the virulence of many microorganisms is enhanced with increased iron availability (**hyperferremia**). For example, gonococci, the causative agent of gonorrhea, spread most often during menstruation, a time in which there is an increased concentration of free iron available to these bacteria.

Acute-Phase Proteins

Macrophages release cytokines (IL-1, IL-6, IL-8, TNF-α, etc.) upon activation by bacteria, which stimulate the liver to rapidly produce acute-phase proteins. These include C-reactive protein (CRP), mannose-binding lectin (MBL), and surfactant proteins A (SP-A) and D (SP-D), all of which can bind bacterial surfaces and act as opsonins. CRP can interact with C1q to activate the classical complement pathway. MBL activates the alternative complement pathway. SP-A, SP-D, and C1q are "collectins"—proteins composed of a "*col*lagen"-like motif connected by α-helices to globular "*lectin*" binding sites (**figure 31.26**). Thus these proteins (along with others) police host tissues by binding to and assisting in the removal of bacteria.

1. Describe the role of cytokines and interferons in innate immunity.
2. How do interferons render cells resistant to viruses?
3. How might acute phase reactants assist in pathogen removal?
4. How can a fever be beneficial to a host?
5. What is the role of collectins in innate immunity?

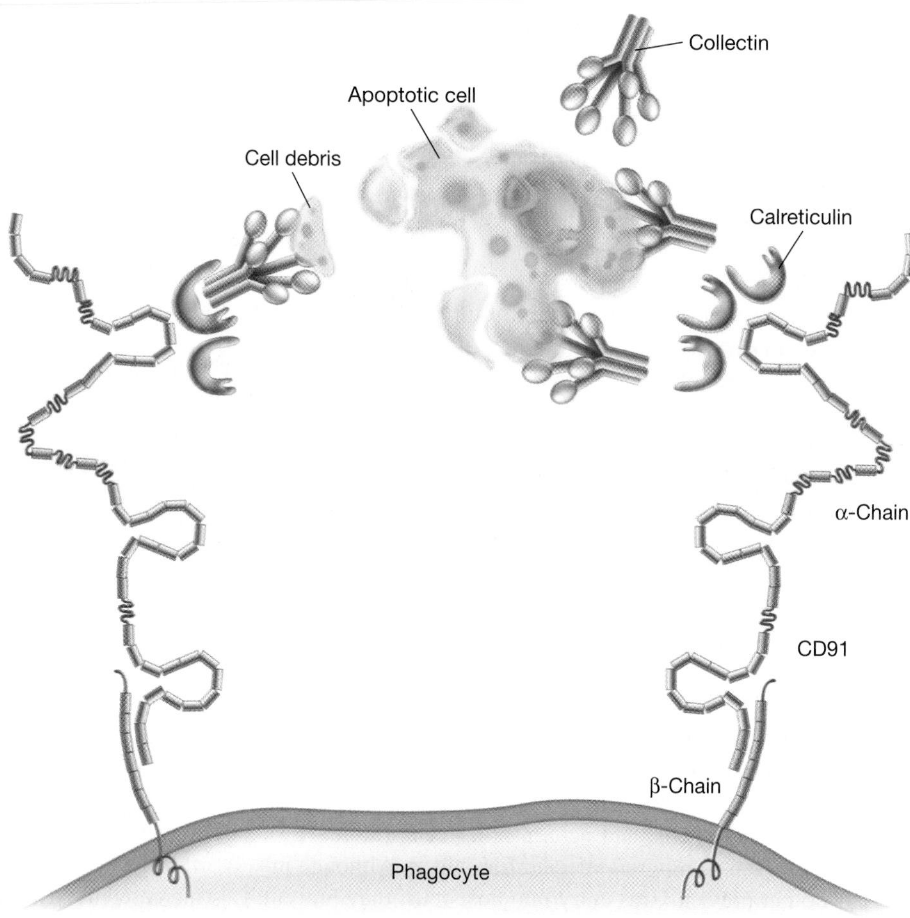

Figure 31.26 Collectins are Molecular Scavengers. This schematic depicts the binding of cellular debris and apoptotic cells by collectins. Collectins (also known as defense collagens) are a family of similar proteins that bind cellular debris and dying cells through their globular head groups. Their collagenous tails are then recognized by calreticulin associated with α-2 macroglobulin (CD91) on the surface of phagocytes.

Summary

31.1 Overview of Host Resistance

a. There are two fundamentally different types of immune responses to invading microorganisms and foreign material. The nonspecific (innate) response offers resistance to any microorganism or foreign material. It includes general mechanisms that are a part of the animal's innate structure and function. The nonspecific system has no immunological memory—that is, nonspecific responses occur to the same extent each time. In contrast, the specific (adaptive) response resists a particular foreign agent; moreover, specific immune responses improve on repeated exposure to the agent (**figure 31.1**).

31.2 Cells, Tissues, and Organs of the Immune System

a. The cells responsible for both nonspecific and specific immunity are the white blood cells called leukocytes (**figure 31.2**). Examples include monocytes and macrophages, dendritic cells, granulocytes, and mast cells (**figures 31.3** and **31.5**).

b. Immature undifferentiated lymphocytes generated in the bone marrow mature and become committed to a particular antigenic specificity within the primary lymphoid organs and tissues. In mammals, T cells mature in the thymus and B cells in the bone marrow. The thymus is the primary lymphoid organ; the bone marrow is the primary lymphoid tissue (**figure 31.6**).

c. Natural killer cells are a small population of large, nonphagocytic lymphocytes that destroy cancer cells and cells infected with microorganisms (**figures 31.7** and **31.8**).

d. The secondary lymphoid organs and tissues serve as areas where lymphocytes may encounter and bind antigens, then they proliferate and differentiate into fully mature, antigen specific effector cells. The spleen is a secondary lymphoid organ and the lymph nodes and mucosal-associated tissues (GALT and SALT) are the secondary lymphoid tissue (**figure 31.9**).

31.3 Phagocytosis

a. Phagocytosis involves the recognition, ingestion, and destruction of pathogens by lysosomal enzymes, superoxide radicals, hydrogen peroxide, defensins, RNIs, and metallic ions. Phagocytic cells use two basic mechanisms for the recognition of microorganisms: opsonin-dependent and opsonin-independent (**figure 31.10**).

31.4 Inflammation

a. Inflammation is one of the host's nonspecific defense mechanisms to tissue injury that may be caused by a pathogen. Inflammation can either be acute or chronic (**figures 31.13** and **31.14**).

31.5 Physical Barriers in Nonspecific (Innate) Resistance

a. Many direct factors (age, nutrition) or general barriers contribute in some degree to all host-microbe relationships. At times they favor the establishment of the microorganism; at other times they provide some measure of general defense to the host.

b. Physical and mechanical barriers along with host secretions are the host's first line of defense against pathogens. Examples include the skin and mucous membranes; the epithelia of the respiratory, gastrointestinal, and genitourinary systems (**figures 31.15, 31.16, 31.18,** and **31.19**).

31.6 Chemical Mediators in Nonspecific (Innate) Resistance

a. Mammalian hosts have specific chemical barriers that help combat the continuous onslaught of pathogens. Examples include general chemicals, cationic peptides, bacteriocins, cytokines, interferons, pyrogens, acute-phase proteins, and complement.

b. The complement system is composed of a large number of serum proteins that play a major role in the animal's defensive immune response. There are three pathways of complement activation: the classical, alternative, and lectin pathways (**figure 31.22** and **table 31.4**).

c. Cytokines are required for immunoregulation of both the nonspecific and specific immune responses. Cytokines have a broad range of actions on eucaryotic cells (**figure 31.24** and **tables 31.5** and **31.6**).

d. Interferons are a group of cytokines that respond in a defensive way to viral infections, double-stranded RNA, endotoxins, antigenic stimuli, mitogenic agents, and many pathogens capable of intracellular growth (**figure 31.25**).

e. Fever induced by a microorganism augments the host's defenses in three ways: it stimulates leukocytes so they can destroy the invading microorganism; it enhances microbiostasis by decreasing iron available to the microorganism; and it enhances the specific activity of the immune system.

Key Terms

alternative complement pathway 764
alveolar macrophage 761
antibody 744
antibody-dependent cell-mediated cytotoxicity (ADCC) 748
antigen 744
antigen presentation 747
bacteriocin 763
basophil 747
B cell 748
B lymphocyte 748
bronchial-associated lymphoid tissue (BALT) 759
bursa of Fabricius 750
cathelicidin 762
classical complement pathway 766
colicin 763
colony-stimulating factor (CSF) 767
complement system 763
cryptin 761
cytokine 748, 766
defensin 762
dendritic cell 747
endogenous pyrogen 769
eosinophil 747
fever 769
granulocyte 746
granuloma 757
gut-associated lymphoid tissue (GALT) 759
hematopoesis 744
histatin 762
hyperferremia 769
hypoferremia 769
immune system 743
immunology 743
inflammation 756
innate or natural immunity 743
integrin 756
interferon (IFN) 768
interleukin 767
intraepidermal lymphocyte 759
kallikrein 757
lactoferrin 759
Langerhans cell 758
lectin complement pathway 765
leukocyte 744
lymph node 750
lymphocyte 748
lymphokine 767
lysozyme 759
M cell 759
macrophage 746
mast cell 747
membrane attack complex 764
memory cells 748
monocyte 746
monocyte-macrophage system 746
monokine 767
mucociliary blanket 761
mucociliary escalator 761
mucus-associated lymphoid tissue (MALT) 759
natural killer (NK) cell 748
neutrophil 747
nonspecific immune response 743
nonspecific resistance 743
opsonization 763
Paneth cell 761
pathogen 743
pathogen-associated molecular pattern (PAMP) 753
pattern recognition receptor (PRR) 753
peristalsis 761
phagocytosis 752
phagolysosome 755
phagosome 755
plasma cell 748
polymorphonuclear leukocyte 746
polymorphonuclear neutrophil (PMN) 746
reactive nitrogen intermediate (RNI) 755
reactive oxygen intermediate (ROI) 755
respiratory burst 755
selectin 756
skin-associated lymphoid tissue (SALT) 758
specific immune response (acquired, adaptive, or specific immunity) 744
spleen 750
T cell 748
T lymphocyte 748
thymus 749
toll-like receptor (TLR) 753
tumor necrosis factor (TNF) 767
white blood cell 744

Critical Thinking Questions

1. Some pathogens invade cells, others invade tissue spaces. Explain how the nonspecific immune response differs for both types of pathogens.
2. How might the various antimicrobial chemical factors be developed into new methods to control infectious disease?
3. How might a scientist use selective gene "knock-outs" to test the role of the toll-like receptor proteins?

Learn More

Bals, R., 2000. Epithelial antimicrobial peptides in host defense against infection. *Resp. Res.* 1:141–50.

Banchereau, J.; and Steinman, R. 1998. Dendritic cells and the control of immunity. *Nature* 392:245–52.

Beutler, B. 2004. Inferences, questions, and possibilities in toll-like receptor signaling. *Nature* 430:257–63.

Biragyn, A.; Ruffini, P. A.; Leifer, C. A.; et al. 2002. Toll-like receptor 4-dependent activation of dendritic cells by beta-defensin 2. *Science* 298:1025–29.

Degli-Esposti, M.; and Smyth, M. 2005. Close encounters of a different kind: Dendritic cells and NK cells take center stage. *Nature Rev. Immunol.* 5:112–24.

Djaldetti, M.; Salman, H.; Bergman, M.; Djaldetti, R.; and Bessler, H. 2002. Phagocytosis—the mighty weapon. *Microsc. Res. Tech.* 57:421–31.

Goldberg, A. L. 2000. Probing the proteosome pathway. *Nature Biotechnol.* 18:494–96.

Iwasaki, A., and Medzhitov, R. 2004. Toll-like receptor control of the adaptive immune response. *Nature Immunol.* 5:987–95.

Marshall, S. H., and Arenas, G. 2003. Antimicrobial peptides: A natural alternative to chemical antibiotics and a potential for applied biotechnology. *Electronic J. Biotechnol.* 6: 271–84.

Medzhitov, R.; Preston-Hurlburt, P.; and Janeway, C. A. 1997. A human analogue of the *Drosophila* Toll protein signals activation of adaptive immunity. *Nature* 388: 394–97.

Roy, C., and Sansonetti, P. 2004. Host-microbe interactions: Bacteria, host-pathogen interactions: Interpreting the dialog. *Curr. Opin. Microbiol.* 7:1–3.

Please visit the Prescott website at www.mhhe.com/prescott7 for additional references.

32 Specific (Adaptive) Immunity

Nude (athymic) mice have a genetic defect (*nu* mutation) that affects thymus gland development. Thus T cells do not form. They, however, do have a B-cell component. This unique deficiency provides animals in which to study B/T-cell dichotomy and environmental influences on the maturation and differentiation of T cells, as well as many different immune disorders.

Preview

- The major function of the specific immune response in vertebrates is to provide protection (immunity) against harmful microorganisms, toxins, and abnormal (cancer) cells through the recognition of foreign (nonself) antigens.
- Antigens are substances, such as proteins, nucleoproteins, polysaccharides, and some glycolipids, to which lymphocytes respond.
- Specific (adaptive) immunity has two branches: humoral and cell-mediated. Cell-mediated immunity involves specialized white blood cells called T cells that act against microbe-infected cells and foreign tissues. They also regulate the activation and proliferation of other immune system cells such as macrophages, B cells, and other T cells. Humoral immunity, or antibody-mediated immunity, involves the production of glycoprotein antibodies by plasma cells derived from B cells.
- In order to function, B and T cells must be exposed to a specific antigen and other required signals. This results in their proliferation and differentiation in a process known as clonal selection.
- Activated T cells produce cytokines that assist in the development and function of other immunocytes.
- Activated B cells produce antibody, also called immunoglobulin (Ig). Two distinguishing characteristics of Ig are their diversity and specificity. There are five human Ig classes based on physicochemical and biological properties: IgG, IgM, IgD, IgA, and IgE.
- The binding of antigen to antibody initiates the participation of other elements that determine the fate of the antigen. For example, the classical complement pathway can be activated, leading to cell lysis or phagocytosis. Other defensive mechanisms include toxin neutralization, adherance inhibition, and opsonization.
- Immune disorders range from mild conditions like hay fever to life-threatening diseases. They can be categorized as hypersensitivities, autoimmune diseases, transplantation (tissue) rejection, and immunodeficiencies.

The immune system is responsible for the ability of a host to resist foreign invaders. It includes an array of cells and molecules with specialized roles in defending the host against the continuous onslaught of microbes, toxins, and cancer cells. Chapter 31 discusses nonspecific host resistance and the innate mechanisms by which the host protects itself from invading microorganisms. Recall that the innate resistance system responds to a foreign substance in the same manner and to the same magnitude each time, and that its activation can assist in the formation of specific immune responses.

In chapter 32 we continue our discussion of the immune response by describing the system of specific (adaptive) responses used to protect the host. Specific immunity is acquired and requires sufficient time to fully develop. However, upon subsequent exposure to the same substance, activation of a specific immune response is significantly faster and stronger than that of the initial response. Thus the cooperation between the host's innate resistance mechanisms and its specific killing responses prevents pathogen invasion and maintains host integrity through the combination of immediate, nonspecific resistance and delayed, specific responses.

After a brief overview of how the specific immune responses work, we launch into chapter 32 by first defining the nature of molecules (antigens) that elicit specific immune reactions, including the methods by which specific immunity can be induced, and the role recognition of "self" plays in a host's detection of invaders. We then elaborate on the structure and function of T and B cells, along with their effector products, that we began in chapter 31. Finally, we end the chapter with a brief overview of why the host shouldn't attack itself (tolerance) and the disorders that occur when the host does.

The remarkable capacity of the immune system to respond to many thousands of different substances with exquisite specificity saves us all from certain death by infection

—*Martin C. Raff*

32.1 Overview of Specific (Adaptive) Immunity

The specific (adaptive) immune system of vertebrates has three major functions: (1) to recognize anything that is foreign to the body ("nonself"); (2) to respond to this foreign material; and (3) to remember the foreign invader. The recognition response is highly specific. The immune system is able to distinguish one pathogen from another, to identify cancer cells, and to discriminate the body's own "self" proteins and cells as different from "nonself" proteins, cells, tissues, and organs. After recognition of an invader has occurred, the specific immune system responds by amplifying and activating specific lymphocytes to attack it. This is called an **effector response.** A successful effector response either eliminates the foreign material or renders it harmless to the host, thus preventing disease. If the same invader is encountered at a later time, the immune system is prepared to mount a more intense and rapid memory or **anamnestic response** that eliminates the invader once again and protects the host from disease. Four characteristics distinguish specific (adaptive) immunity from nonspecific (innate) resistance:

1. Discrimination between self and nonself. The (adaptive) specific immune system almost always responds selectively to nonself and produces specific responses against the stimulus.
2. Diversity. The system is able to generate an enormous diversity of molecules such as antibodies that recognize trillions of different foreign substances.
3. Specificity. Immunity is also selective in that it can be directed against one particular pathogen or foreign substance (among trillions); the immunity to this one pathogen or substance usually does not confer immunity to others.
4. Memory. When re-exposed to the same pathogen or substance, the body reacts so quickly that there is usually no noticeable pathogenesis. By contrast, the reaction time for inflammation and other nonspecific (innate) defenses is just as long for a later exposure to a given antigen as it was for the initial one.

The recognition of foreign substances by a mammalian host occurs because host cells express a unique protein on their surface, marking them as residents of that host, or as "self." Thus the introduction of materials lacking that unique self-marker results in their destruction by the host. There are, of course, exceptions to this process. For example, some materials are highly conserved throughout evolution—that is, their three-dimensional structures are identical, or very similar, regardless of the host that produces them, and as such, they are not distinguished as foreign.

Two branches or arms of specific (adaptive) immunity are recognized (**figure 32.1**): humoral (antibody-mediated) immunity and cellular (cell-mediated) immunity. **Humoral (antibody-mediated) immunity,** named for the fluids or "humors" of the body, is based on the action of soluble glycoproteins called antibodies that occur in body fluids and on the plasma membranes of B lymphocytes. Circulating antibodies bind to microorganisms, toxins, and extracellular viruses, neutralizing them or "tagging or marking" them for destruction by mechanisms as described in section 31.2. **Cellular (cell-mediated) immunity** is based on the action of specific kinds of T lymphocytes that directly attack cells infected with viruses or parasites, transplanted cells or organs, and cancer cells. T cells can lyse these cells or release chemicals (cytokines) that enhance specific immunity and nonspecific (innate) defenses such as phagocytosis and inflammation. Because the activity of the acquired immune response is so potent, it is imperative that T and B cells consistently discriminate between self and nonself with great accuracy. How they accomplish this is discussed next.

32.2 Antigens

The immune system distinguishes between "self" and "nonself" through an elaborate recognition process. During their development, B and T cells that would recognize components of their host (self-determinants) are induced to undergo apoptosis (programmed cell death). This is critical for the removal of effector lymphocytes that could react against their host. This ensures that lymphocytes can produce specific immunologic reactions only against foreign materials and organisms, leading to their removal. Self and nonself substances that elicit an immune response and react with the products of that response are often called **antigens.** Antigens include molecules such as proteins, nucleoproteins, polysaccharides, and some glycolipids. While the term "immunogen" (*immun*ity *gen*erator) is a more precise descriptor for a substance that elicits a specific immune response, "antigen" is used more frequently. Most antigens are large, complex molecules with a molecular weight generally greater than 10,000 Daltons (Da). The ability of a molecule to function as an antigen depends on its size, structural complexity, chemical nature, and degree of foreignness to the host. Cells, tissues, and organs of the immune system: Lymphocytes (section 31.2)

Each antigen can have several **antigenic determinant sites,** or **epitopes** (**figure 32.2**). Epitopes are the regions or sites in the antigen that bind to a specific antibody or T-cell receptor through an antigen-binding site. Antibodies are formed most readily in response to epitopes that project from the foreign molecule or to terminal residues of a specific polymer chain. Chemically, epitopes include sugars, organic acids and bases, amino acid side chains, hydrocarbons, and aromatic groups. The number of epitopes on the surface of an antigen is its **valence.** The valence determines the number of antibody molecules that can combine with the antigen at one time. If one determinant site is present, the antigen is monovalent. Most antigens, however, have more than one copy of the same epitope and are termed multivalent. Multivalent antigens generally elicit a stronger immune response than do monovalent antigens. As we will see, each antibody molecule has at least two antigen-binding sites so multivalent antigens can be "cross-linked" by antibodies, a phenomenon that can result in precipitation or agglutination of antigen. **Antibody affinity** relates to the strength with which an antibody binds to its antigen at a given antigen-binding site. Affinity tends to

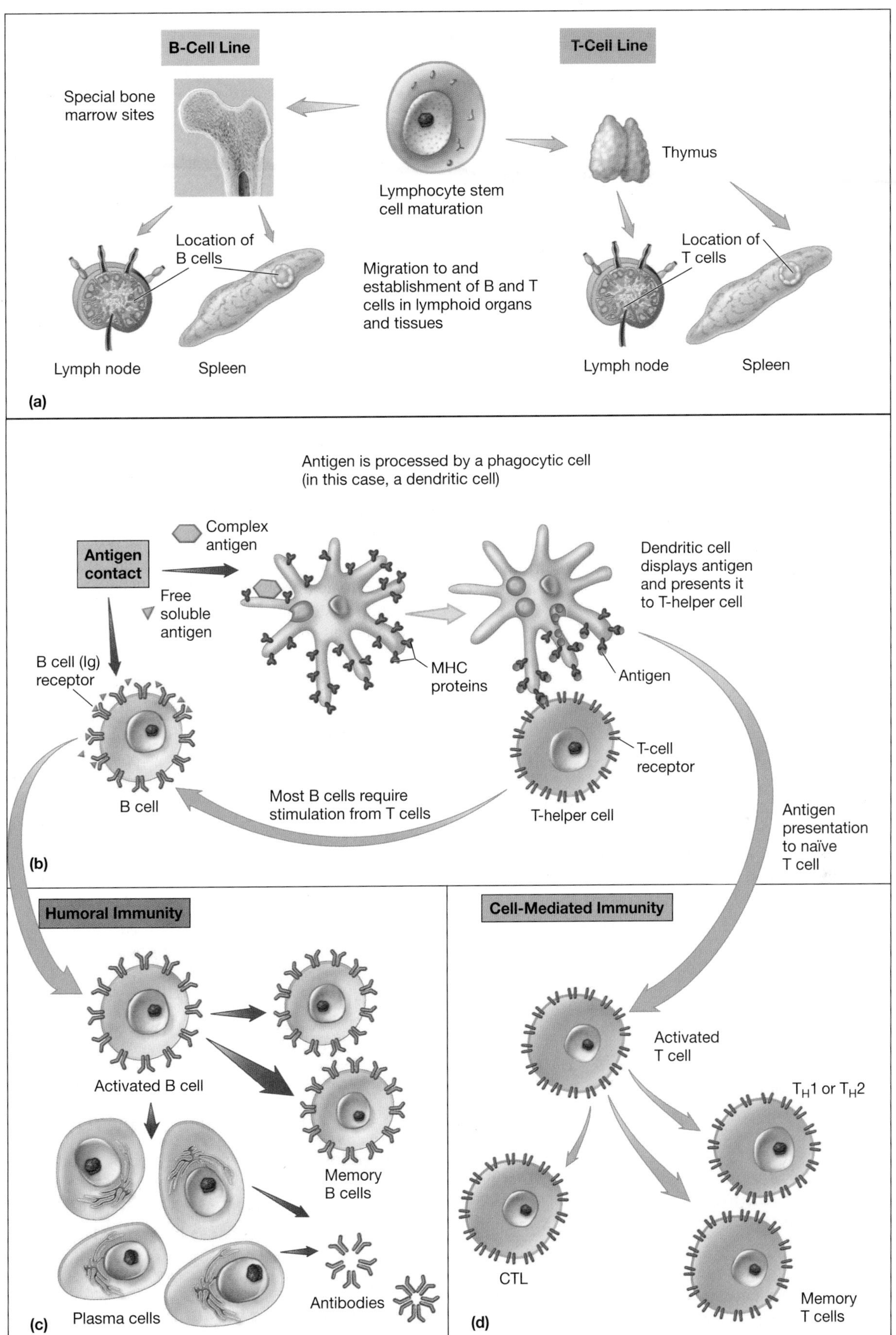

Figure 32.1 Acquired Immune System Development. **(a)** Lymphocyte stem cells develop into B- and T-cell precursors that migrate to the bone marrow or thymus, respectively. Mature B and T cells seed secondary lymphoid tissues. **(b)** Lymphocyte receptor binding of antigen activates B and T cells to become effector cells. **(c)** B lymphocytes develop into memory cells and antibody-secreting plasma cells. **(d)** T cells develop into memory cells, helper T cells, and cytotoxic T cells.

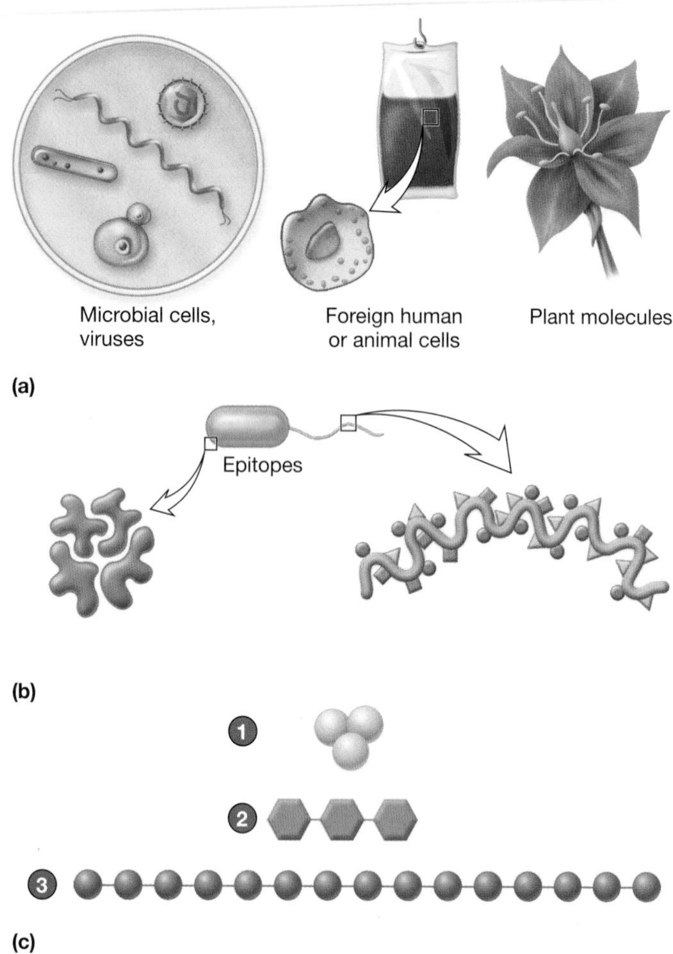

Figure 32.2 Antigen Characteristics Are Numerous and Diverse. **(a)** Whole cells and viruses make good immunogens. **(b)** Complex molecules with several epitopes make good immunogens. **(c)** Poor immunogens include small molecules not attached to a carrier molecule (*1*), simple molecules (*2*), and large but repetitive molecules (*3*).

increase during the course of an immune response and is discussed in section 32.7. The **avidity** of an antibody relates to its overall ability to bind antigen at all antigen-binding sites.

Haptens

Many small organic molecules are not antigenic by themselves but become antigenic if they bond to a larger carrier molecule such as a protein. These small antigens are called **haptens** [Latin *haptein,* to grasp]. When lymphocytes are stimulated by the combined hapten-carrier molecule, they can react to either the hapten or the larger carrier molecule. This occurs because the hapten functions as one epitope of the carrier. When the carrier is processed (section 31.2) and presented to T cells, responses to both the hapten and the carrier protein can be elicited. As a result, both hapten-specific and carrier-specific antibodies can be made. One example of a hapten is penicillin. By itself penicillin is a small molecule and is not antigenic. However, when it is combined with certain serum proteins of sensitive individuals, the resulting molecule becomes immunogenic, activates lymphocytes, and initiates a severe and sometimes fatal allergic immune reaction. In these instances the hapten is acting as an antigenic determinant on the carrier molecule.

Cluster of Differentiation Molecules (CDs)

Recall that antigens are molecules that elicit an immune response. It is important to recognize that the molecules of a given host can elicit an immune reaction when exposed to lymphocytes obtained from a different host; they are indeed, antigens. This is the basis for identifying and studying cell receptors. One practical example of this is seen in the identification of cell surface proteins that have specific roles in intercellular communication. For example, lymphocytes and other cells of the immune system bear particular membrane proteins called **cluster of *d*ifferentiation (CDs) molecules or antigens.** CDs are cell surface proteins and many are receptors. CDs can be measured in situ and from peripheral blood, biopsy samples, or other body fluids. They often are used in a classification system to differentiate between leukocyte subpopulations. To date, over 300 CDs have been characterized. **Table 32.1** summarizes some of the functions of several CDs.

CDs have both biological and diagnostic significance. The presence of various CDs on the cell's surface can be used to determine the cell's identity. For example, it has been established that the CD4 molecule is a cell surface receptor for human immunodeficiency virus (HIV; the virus that causes AIDS), and CD34 is the cell surface indicator of stem cells. As we will see, using the CD antigen system to name cells is more efficient than describing all of a cell's functions. We also use this approach in naming specific cell types as we discuss their relative functions in immunity.

1. Distinguish between "self" and "nonself" substances.
2. Define and give several examples of an antigen. What is an antigenic determinant site or epitope?
3. Define hapten and CD antigens.
4. Give some examples of the biological significance of cluster of differentiation molecules (CDs).

32.3 Types of Specific (Adaptive) Immunity

Acquired immunity refers to the type of specific (adaptive) immunity a host develops after exposure to foreign substances, or after transfer of antibodies or lymphocytes from an immune donor. Acquired immunity can be obtained actively or passively by natural or artificial means (**figure 32.3**).

Naturally Acquired Immunity

Naturally acquired active immunity occurs when an individual's immune system contacts a foreign stimulus (antigen) such as a pathogen that causes an infection. The immune system responds by producing antibodies and activated lymphocytes that inactivate or

Table 32.1 Functions of Some Cluster of Differentiation (CD) Molecules

Molecule	Function
CD1 a,b,c	MHC class I-like receptor used for lipid antigen presentation
CD3 δ,ϵ,γ	T-cell antigen receptor
CD4	MHC class II co-receptor on T cells, monocytes, and macrophages; HIV-1 and HIV-2 (gp120) receptor
CD8	MHC class I co-receptor on cytotoxic T cells
CD11 a,b,c,d	α-subunits of integrin found on various myeloid and lymphoid cells; used for binding to cell adhesion molecules
CD19	B-cell antigen co-receptor
CD25	IL-2 α chain on activated T cells, B cells, and monocytes
CD34	Stem cell protein that binds to sialic acid residues
CD45	Tyrosine phosphatase common to all hematopoeitic cells
CD56	NK cell and neural cell adhesion molecule
CD97 α/β	Igα/Igβ receptor on B cells
CD106	Endothelial cell vascular cell adhesion molecule-1
CD209	Dendritic cell-specific c-type lectin

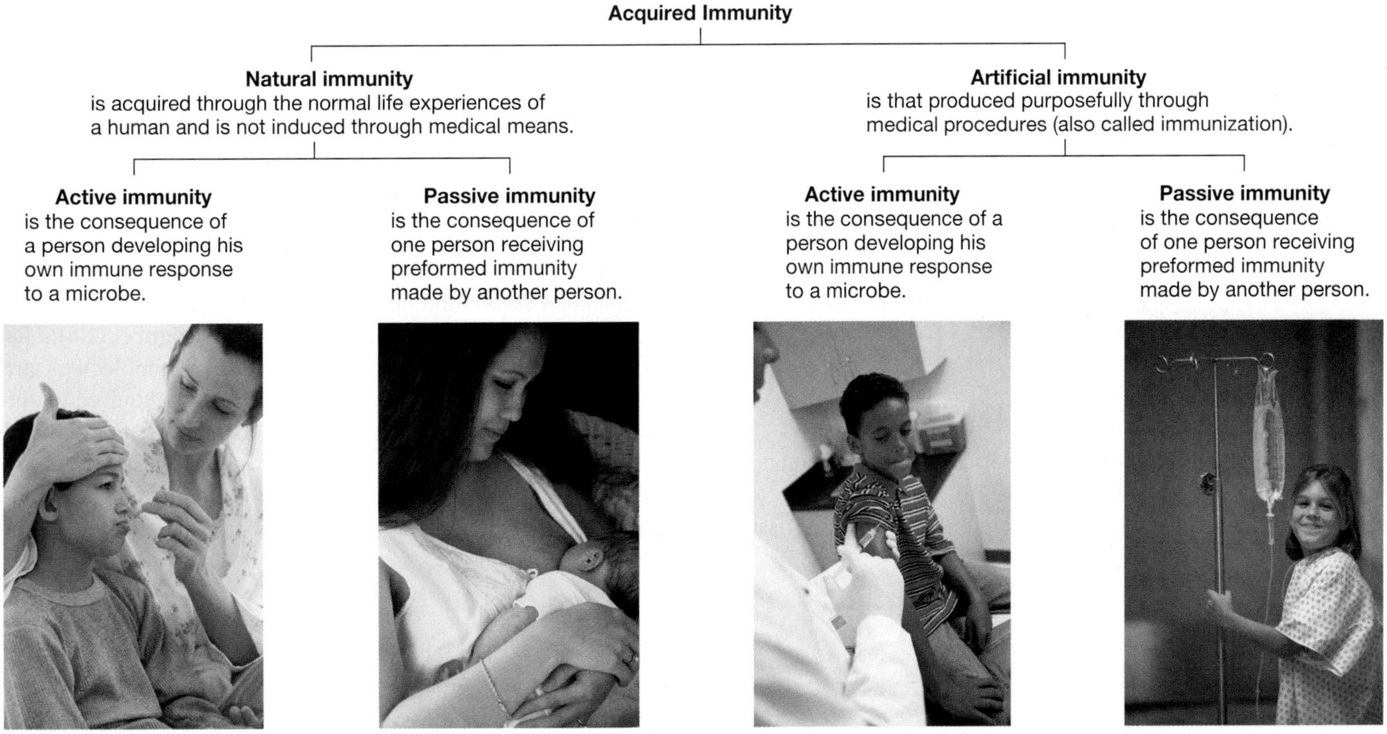

Acquired Immunity

Natural immunity is acquired through the normal life experiences of a human and is not induced through medical means.

Active immunity is the consequence of a person developing his own immune response to a microbe.

Passive immunity is the consequence of one person receiving preformed immunity made by another person.

Artificial immunity is that produced purposefully through medical procedures (also called immunization).

Active immunity is the consequence of a person developing his own immune response to a microbe.

Passive immunity is the consequence of one person receiving preformed immunity made by another person.

Figure 32.3 Immunity Can Be Acquired by Various Means. Naturally acquired immunity can be either active or passive. Artificially acquired immunity can also be either active or passive.

destroy the pathogen. The immunity produced can be either lifelong, as with measles or chickenpox, or last for only a few years, as with influenza. **Naturally acquired passive immunity** involves the transfer of antibodies from one host to another. For example, some of a pregnant woman's antibodies pass across the placenta to her fetus. If the female is immune to diseases such as polio or diphtheria, this placental transfer also gives her fetus and newborn temporary immunity to these diseases. Certain other antibodies can pass from the female to her offspring in the first secretions (called colostrum) from the mammary glands. These maternal antibodies are essential for providing immunity to the newborn for the first few weeks or months of life, until its own immune system matures. Protection of the newborn by antibodies from colostrum is especially important in certain animal species (such as cattle and horses), which have less antibody transfer across the placenta than do primates. Naturally acquired passive immunity generally lasts only a short time (weeks to months, at most).

Artificially Acquired Immunity

Artificially acquired active immunity results when an animal is intentionally exposed to a foreign material and induced to form antibodies and activated lymphocytes. This foreign material is called a vaccine and the procedure is vaccination (immunization). A vaccine may consist of a preparation of killed microorganisms; living, weakened (attenuated) microorganisms; genetically engineered organisms; or inactivated bacterial toxins (toxoids) that are administered to induce immunity artificially. Vaccines and immunizations are discussed in detail in section 36.8.

Artificially acquired passive immunity results when antibodies or lymphocytes that have been produced outside the host are introduced into a host. Although this type of immunity is immediate, it is short lived, lasting only a few weeks to a few months. An example would be botulinum antitoxin produced in a horse and given to a human suffering from botulism food poisoning, or a bone marrow transplant given to a patient with genetic immunodeficiency.

1. What are the three related activities mediated by the specific (adaptive) immune systems?
2. What distinguishes specific immunity from nonspecific (innate) resistance?
3. What are the two arms of specific (adaptive) immunity?
4. How does naturally acquired immunity occur? Contrast active and passive immunity.

32.4 Recognition of Foreignness

Distinguishing between self and nonself is essential in maintaining host integrity. This distinction is highly specific and selective so as to eliminate invading pathogens but not destroy host tissue. An important extension of this exquisite survival mechanism is seen in the modern use of tissue transplantation, where organs, tissues, and cells from unrelated persons are carefully matched to the self-recognition markers on the recipient's cells when used in disease treatment. While entire textbooks are devoted to this subject, we only discuss the aspects of foreignness recognition that assist us in understanding why and how lymphocytes respond. Without this ability to recognize foreign materials, lymphocytes have no reason to differentiate into effector cells.

Recall that each cell of a particular host needs to be identified as a member of that host so it can be distinguished from foreign invaders. To accomplish this, each cell must express gene products that mark it as a resident of that host. Furthermore, it is not enough to simply identify resident (self) cells—in addition, effective cooperation between cells must occur so that efficient information sharing and selective effector activities occur. Such a system has evolved in mammals and is encoded in the major histocompatibility gene complex.

The **major histocompatibility complex (MHC)** is a collection of genes on chromosome 6 in humans and chromosome 17 in mice. This term is derived from the Greek word for tissue [*histo*] and the ability to get along [compatibility]. The MHC is called the **human leukocyte antigen (HLA)** complex in humans and the H-2 complex in mice. Almost all human cells contain HLA molecules on their plasma membranes. HLA molecules can be divided into three classes: class I molecules are found on all types of nucleated body cells; class II molecules appear only on cells that can process nonself materials and present antigens to other cells (i.e., macrophages, dendritic cells, and B cells); and class III molecules include various secreted proteins that have immune functions (**figure 32.4**). Unlike class I and II MHC molecules, class III molecules are mostly secreted products whose presence is not required to discriminate between self and nonself. Furthermore, the class III MHC molecules are not membrane proteins, are not related to class I or II molecules, and have no role in antigen presentation. Thus they are not discussed further.

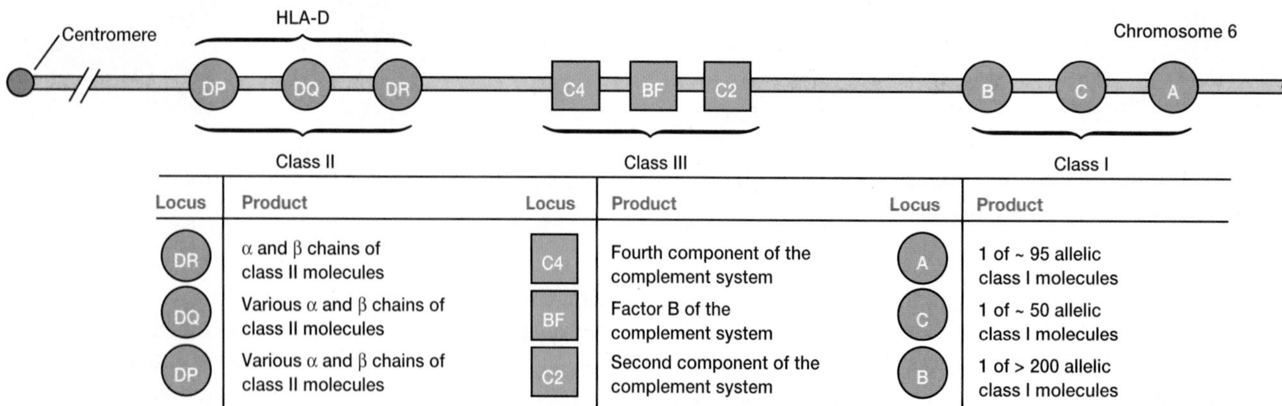

Figure 32.4 Major Histocompatibility Complex. The MHC region of human chromosome 6 and the gene products associated with each locus.

Each individual has two sets of MHC genes—one from each parent, and both are expressed (i.e., they are codominant). Thus a person expresses many different HLA products (figure 32.4). The HLA proteins differ among individuals; the closer two people are related, the more similar are their HLA molecules. In addition, many forms of each HLA gene exist. This is because multiple alleles of each gene have arisen by high gene mutation rates, gene recombination, and other mechanisms (i.e., each gene locus is polymorphic). The differences in the HLA products expressed by individuals appear to account for some of the variation in infectious disease susceptibility. Class I molecules comprise HLA types A, B, and C and serve to identify almost all cells of the body as "self." They also stimulate antibody production when introduced from one host into another host with different class I molecules. This is the basis for MHC typing when a patient is being prepared for an organ transplant (**Techniques & Applications 32.1**).

Class II HLA molecules are produced only by certain white blood cells, such as activated macrophages, dendritic cells, mature B cells, some T cells, and certain cells of other tissues. Class II molecules are required for T-cell communication with macrophages, dendritic cells, and B cells. As discussed later in section 32.7, part of the T-cell receptor must recognize a peptide within a class II molecule on the antigen-presenting cell before the T cell can secrete cytokines necessary for the immune response.

Class I MHC molecules consist of a complex of two protein chains, one with a mass of 45,000 Daltons (Da), known as the alpha chain, and the other with a mass of 12,000 Da (β2-microglobulin) (**figure 32.5*a***). The alpha chain can be divided into three functional domains, designated α_1, α_2, and α_3. The α_3 domain is attached to the plasma membrane by a short amino acid sequence that extends into the cell interior, while the rest of the protein protrudes to the outside. The β_2-microglobulin (β_2m) protein and α_3 segment of the alpha chain are noncovalently associated with one another and are close to the plasma membrane. The α_1 and α_2 domains lie to the outside and form the antigen-binding pocket. Only non-nucleated cells (red blood cells) lack class I MHC molecules.

Class II MHC molecules are also transmembrane proteins consisting of α and β chains of mass 34,000 Da and 28,000 Da, respectively (figure 32.5*b*). Both chains combine to form a three-dimensional protein pocket, the antigen-binding pocket, into

Techniques & Applications

32.1 Donor Selection for Tissue or Organ Transplants

The successful transplantation of tissues began in 1908 when Alexis Carrel reported survival of nine cats with functional kidneys transplanted among them. Since then a number of different tissues and organs have been successfully transplanted between humans. Tissues that are not rejected and do not induce an immunological response when transplanted are said to be "histocompatible." Thus when the genes encoding the cellular proteins responsible for recognition of self and nonself were identified, they were labeled the major histocompatibility complex. Over the years many of the major histocompatibility complex (MHC) molecules have been identified and diagnostic antibodies made against them.

Successful tissue transplantation requires that the ABO blood group and the MHC molecules of the donor and the recipient be as closely matched as possible. Other than identical twins, no two persons have the same MHC (called human leukocyte antigens [HLAs] in humans) proteins on their cells. Thus it is necessary to identify these antigens so that a donated tissue or organ is less likely to be recognized as foreign by the recipient's immune system.

The likelihood of tissue or organ transplant acceptance can be increased by using persons who have a great degree of genetic similarity. The greater the similarity, the more likely the persons will have similar MHC complexes. The reason for this is that 77 histocompatibility alleles are determined by just four human histocompatibility genes—A, B, C, and D (which includes DQ, DR, and DP). These genes are transmitted from parents to offspring according to Mendelian genetics. For example, two siblings (brother and sister) may have approximately a 25% chance of being B identical, a 50% chance of sharing one B allele, and only a 25% chance of having completely different B alleles. Therefore, when tissue or organ transplants are performed, a deliberate attempt is made to use tissues from siblings or other genetically related, histocompatible people.

Family members are the logical first choice in the search for close HLA matches; the HLA genes are usually inherited as a complete set from parents. After matching the blood types of donor and recipient, identification of HLA proteins is performed. Potential donors are first screened using a modification of the complement fixation assay. In this assay donor lymphocytes are incubated with recipient serum (containing antibodies) in the presence of complement proteins. If the donor lymphocytes are killed (lysed by complement), the recipient serum contains antibodies against donor HLA proteins on the lymphocytes. This rules out that potential donor. If the donor lymphocytes survive exposure to the recipient serum and complement, HLA typing ensues. Clinical immunology: Complement fixation (section 35.3)

HLA typing is typically accomplished using the microcytotoxicity test. In this test, potential donor and recipient leukocytes are placed separately into wells of a microtiter plate. Antibodies specific for various class I and class II MHC (HLA-A, B, C, or D) proteins are added to different cell-filled wells, along with complement. Cytotoxicity (cell damage and death) occurs if cells express the HLA protein recognized by the antibody. These dead cells are readily measured using a specific dye. Once enough HLA molecules are identified, the donor and recipient HLA profiles are compared to determine compatibility. Additional testing can determine the degree of antibody recognition as a potential measure of rejection severity.

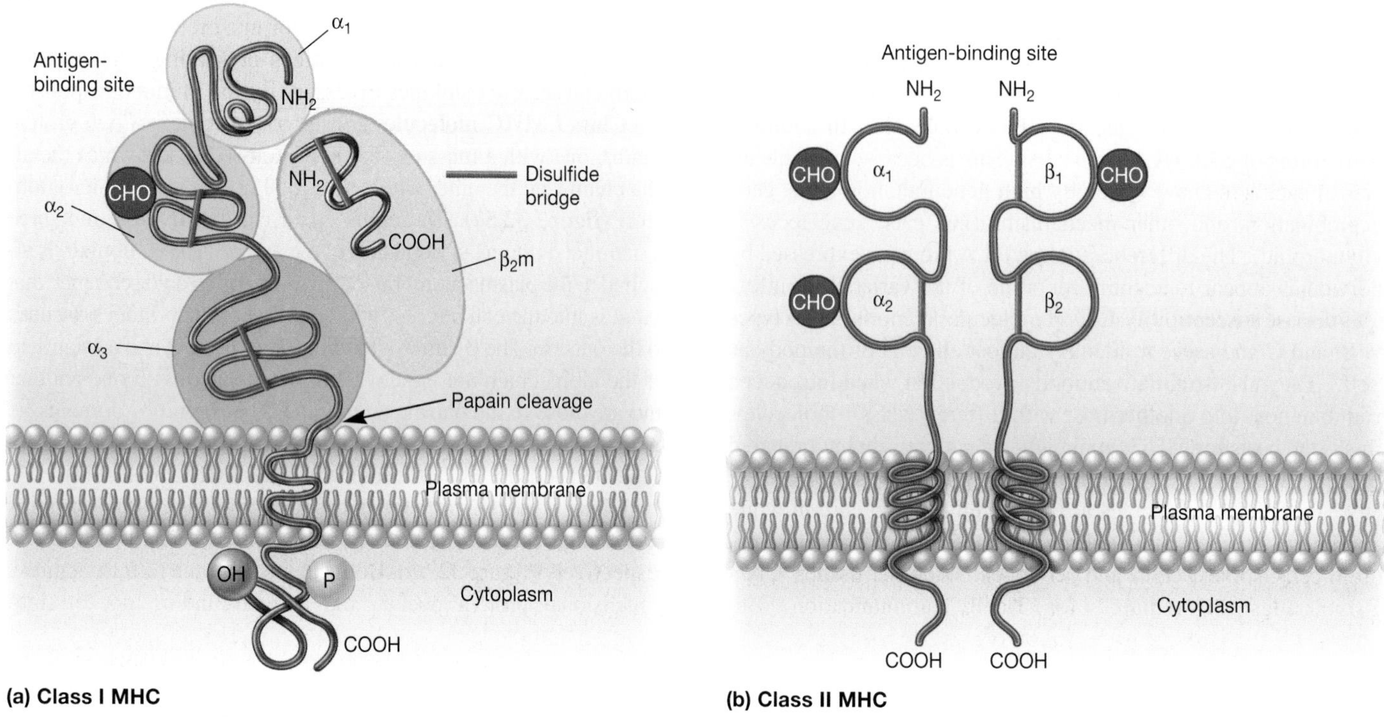

(c) (d)

Figure 32.5 The Membrane-Bound Class I and Class II Major Histocompatibility Complex Molecules. **(a)** The class I molecule is a heterodimer composed of the alpha protein, which is divided into three domains: α_1, α_2, and α_3, and the protein β_2 microglobulin. **(b)** The class II molecule is a heterodimer composed of two distinct proteins called alpha and beta. Each is divided into two domains, α_1, α_2, and β_1, β_2, respectively. **(c)** This space-filling model of a class I MHC protein illustrates that it holds shorter peptide antigens (blue) than does a class II MHC. **(d)** The difference is because the peptide binding site of the class I molecule is closed off, whereas the binding site of the class II molecules is open on both ends.

which a nonself peptide fragment can be captured for presentation to other cells of the immune system (immunocytes). Although MHC class I and class II molecules are structurally distinct, both fold into similar shapes. Each MHC molecule has a deep groove into which a short peptide derived from a foreign substance can bind (figure 32.5*c,d*). Because this peptide is not part of the MHC molecule, it can vary from one MHC molecule to the next. Foreign peptides (antigen fragments) in the MHC groove must be present to activate T cells, which in turn activate other immunocytes.

By binding and presenting foreign peptides, class I and class II molecules inform the immune system of the presence of nonself. These peptides arise in different places within cells as the result of a process known as **antigen processing.** Class I molecules bind to peptides that originate in the cytoplasm. Foreign peptides within the cytoplasm of mammalian cells come from replicating viruses or other intracellular pathogens, or are the result of cancerous transformation. These intracellular antigenic proteins are digested by a cytoplasmic structure called the proteasome (*see figure 4.9*), as part of the natural process by which a cell continually renews its protein contents. The resulting short peptide fragments are pumped by specific transport proteins from the cytoplasm into the endoplasmic reticulum. Within the endoplasmic reticulum the class I MHC alpha chain is synthesized and associates with β_2-microglobulin. This dimer appears to bind antigen as soon as the foreign peptide enters the endoplasmic reticulum. The class I MHC molecule and antigenic peptide are then carried to, and anchored in, the plasma membrane. In this way, the host cell presents the antigen to a subset of T cells called $CD8^+$, or cytotoxic, T lymphocytes. $CD8^+$ T cells bear a receptor that is specific for class I MHC molecules that present antigen; as will be discussed in section 32.5, these T cells bind and ultimately kill infected host cells.

Class II MHC molecules bind to fragments that initially come from antigens outside the cell. This pathway functions with bacteria, viruses, and toxins that have been taken up by endocytosis. An **antigen-presenting cell (APC),** such as a macrophage, dendritic cell, or B cell, takes in the antigen or pathogen by receptor-mediated endocytosis or phagocytosis, and produces antigen fragments by digestion in the phagolysosome. Fragments then combine with preformed class II MHC molecules and are delivered to the cell surface. It is here that the peptide is recognized by $CD4^+$ T-helper cells. Unlike $CD8^+$ T cells, $CD4^+$ T cells do not directly kill target cells. Instead they respond in two distinct ways. One is to proliferate, thereby increasing the number of $CD4^+$ cells that can react to the antigen. Some of these cells will become memory T cells, which can respond to subsequent exposures to the same antigen. The second response is to secrete cytokines (e.g., interleukin-2) that either directly inhibit the pathogen that produced the antigen or recruit and stimulate other cells to join in the immune response. Phagocytosis (section 31.3); Chemical mediators in nonspecific resistance: Cytokines (section 31.6)

32.5 T Cell Biology

In order for acquired immunity to develop, T cells and B cells must be activated. T cells are major players in the cell-mediated immune response (figure 32.1), and also have a major role in B-cell activation. They are immunologically specific, can carry a vast repertoire of immunologic memory, and can function in a variety of regulatory and effector ways. Because of their paramount importance, we discuss them first.

T-Cell Receptors

T cells have specific **T-cell receptors (TCRs)** for antigens on their plasma membrane surface. The receptor site is composed of two parts: an alpha polypeptide chain and a beta polypeptide chain (**figure 32.6*a***). Each chain is stabilized by disulfide bonds. The receptor is anchored in the plasma membrane and parts of the α and β chains extend into the cytoplasm. The recognition sites of the T-cell receptor extend out from the membrane and have a terminal variable section complementary to antigen fragments. T-cells respond to antigen fragments presented in the MHC molecules.

Types of T Cells

T cells originate from stem cells in the bone marrow, but T cell precursors migrate to the thymus for further differentiation. This includes destruction of T cells that recognize self antigens (so-called self-reactive T cells). Like all lymphocytes, T cells that have survived the process of development are called mature cells, but they are also considered to be "naïve" cells because they have not yet been activated by a specific MHC-antigen peptide combination. This activation of T cells involves specific molecular signaling events inside the cell, which is discussed in more detail shortly. Once activation occurs, T cells proliferate to form memory cells, as well as activated or "effector" cells, which carry out specific functions to protect the host against the invading antigen. The two major types of T cells, the T-helper (T_H) cells and the cytotoxic T lymphocytes (CTLs) (figure 32.3), are discussed first, then some of the details of T-cell activation are examined (**table 32.2**).

T-Helper Cells

T-helper (T_H) cells, also known as **$CD4^+$ T cells,** are activated by antigen presented by class II MHC molecules on APCs. They can be further subdivided into T_H0 cells, T_H1 cells, and T_H2 cells.

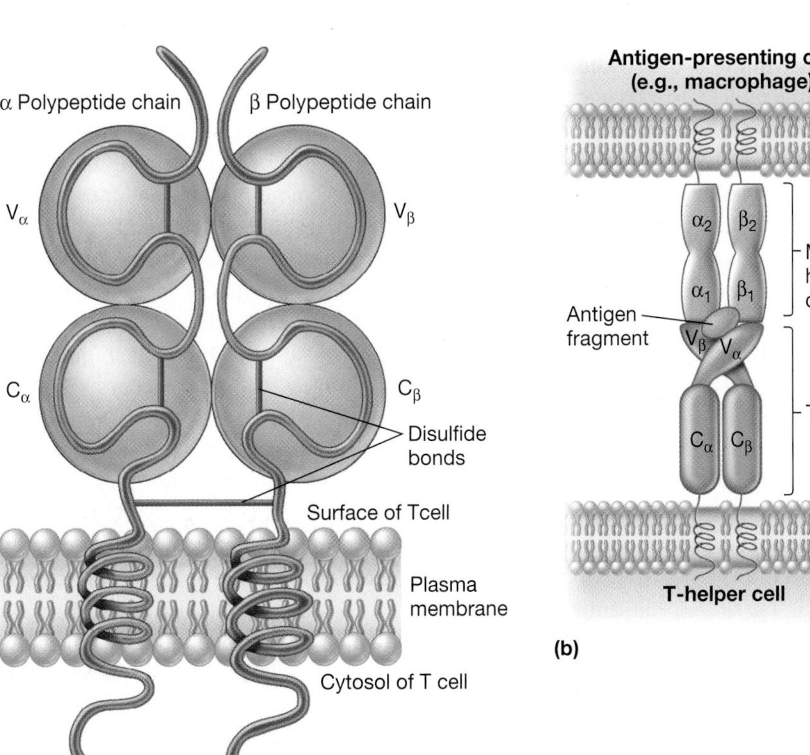

Figure 32.6 The Role of the T-cell Receptor Protein in T-helper Cell Activation. (a) A schematic illustration of the proposed overall structure of the antigen receptor site on a T-cell plasma membrane. **(b)** An antigen-presenting cell begins the activation process by displaying an antigen fragment (e.g., peptide) on its surface as part of a complex with the histocompatibility molecules. A T-helper cell is activated after the variable region of its receptor (designated V_α and V_β) reacts with the antigen fragment in a class II MHC molecule on the presenting cell surface.

Table 32.2 Some Functional Examples of T Cell Subpopulations

Type	Function
$CD4^+$ T-helper (T_H0) cell	Precursor cell activated by specific antigen presented on class II MHC to differentiate into T_H1 or T_H2 cells
T_H1 cell	Produces IL-2, IFN-γ and TNF-α, which activate macrophages and CTLs to promote cellular immune responses
T_H2 cell	Produces IL-4, IL-5, IL-6, IL-10, and IL-13 to promote B cell maturation and humoral immune responses
$CD8^+$ T cell	Precursor cell activated by specific antigen presented on MHC-I to differentiate into cytotoxic lymphocyte
Cytotoxic T lymphocyte (CTL), also called cytotoxic T cell	Kills cells expressing foreign specific antigen on class I MHC by perforin and granzyme release or induction of apoptosis by CD95L (Fas ligand)

T_H0 cells are simply undifferentiated precursors of T_H1 cells and T_H2 cells, while T_H1 and T_H2 cells are distinguished by the different types of cytokines they produce (**figure 32.7**). Activated **T_H1 cells** promote cytotoxic T lymphocyte (CTL) activity, activate macrophages, and mediate inflammation by producing interleukin (IL)-2, interferon (IFN)-γ, and tumor necrosis factor (TNF)-α. These cytokines are also responsible for delayed-type (type IV) hypersensitivity reactions, in which host cells and tissues are damaged nonspecifically by activated T cells (pp. 807–809). **T_H2 cells** tend to stimulate antibody responses in general, and defend against helminth parasites by producing cytokines such as IL-4, IL-5, IL-6, IL-10, and IL-13. An overabundance of T_H2 type responses may also be involved in promoting allergic reactions. Allergic and hypersensitivity reactions are discussed in section 32.11. Chemical mediators in nonspecific (innate) resistance: Cytokines (section 31.6)

Cytotoxic T Lymphocytes

Cytotoxic T lymphocytes (CTLs) are **$CD8^+$ T cells** that function to destroy host cells that have been infected by an intracellular pathogen, such as a virus. Activation of CTLs can be thought of as a two step process. First, naïve $CD8^+$ cells must interact with an APC (i.e., a macrophage or dendritic cell) that has processed the antigen and presents it to the immature CTL on its class I MHC molecule (**figure 32.8**). Note that unlike $CD4^+$ cells, $CD8^+$ cells interact with APCs through their class I MHCs. This is important because $CD8^+$ cells then mature into CTLs that can respond to the same antigen as presented in the class I MHC of any host cells that have been infected by the intracellular pathogen (recall that such cells do not possess class II MHCs). All host cells that present the same antigen thus become target cells. Once activated, these CTLs kill target cells in at least two ways: the perforin pathway and the CD95 pathway.

In the **perforin pathway,** binding of the CTL to the target cell triggers movement of cytoplasmic granules toward the part of the plasma membrane that is in contact with the target cell. These CTL granules fuse with the plasma membrane, releasing molecules called perforin and granzymes into the intercellular space. Perforin, which has considerable homology to the C9 component of complement that forms the membrane attack complex, polymerizes in the target cell's membrane to form pores. These allow the granzymes to enter the target cell, where they induce programmed cell death (apoptosis). Chemical mediators of innate resistance: Complement (section 31.6)

In the **CD95** or **Fas-FasL pathway,** the activated CTL increases expression of a protein called Fas ligand (FasL; also known as CD95L) on its surface. FasL can interact with the transmembrane Fas protein receptor found on the target cell surface. This induces the target cell to undergo apoptosis. By inducing target cell apoptosis rather than cell lysis, both the perforin and the CD95 pathways stimulate membrane changes that are thought to trigger phagocytosis and slow destruction of the apoptotic cell by macrophages. Thus any infectious agent (e.g., a virus) that caused the cell to be initially attacked by the CTL is also destroyed. If the target cell were simply to be lysed, any infectious agent it harbored could potentially be released unharmed and infect surrounding cells.

T-Cell Activation

In order to respond to a foreign substance, lymphocytes must be activated by the antigen to which they ultimately respond. Lymphocyte activation is a complex process that is still not completely understood. However, in general, the binding of an antigen within the lymphocyte receptor initiates a signaling cascade involving other membrane-bound proteins and intracellular messengers. Lymphocyte proliferation, differentiation, and expression of specific cytokine genes, however, occurs only when a second signal is communicated along with the antigen. A general discussion of this process in T cells now follows.

All naïve T cells, whether $CD4^+$ or $CD8^+$ cells, require two signals to be activated by antigen. Signal 1 occurs when an antigen fragment, presented in a MHC molecule of an antigen-presenting cell (APC), fills the appropriate T-cell receptor. Signal 1 differs for T_H and CTL cells. In the case of T_H cells, an exogenous antigen is taken into an APC by phagocytosis or endocytosis, processed, and presented to the T_H cell by class II MHC molecules. CD4 co-receptors on the T_H cell interact with the antigen-bound MHC molecule to assist with signal 1 recognition. For CTLs to be activated, an endogenous (cytoplasmic) antigen is processed by the APC proteasome and presented on class I MHC molecules. In this case, the CD8 co-receptor on the CTL interacts with the antigen-bound MHC molecule on the APC, assisting with signal 1 recognition. Organelles of the biosynthetic-secretory and endocytic pathways (section 4.4)

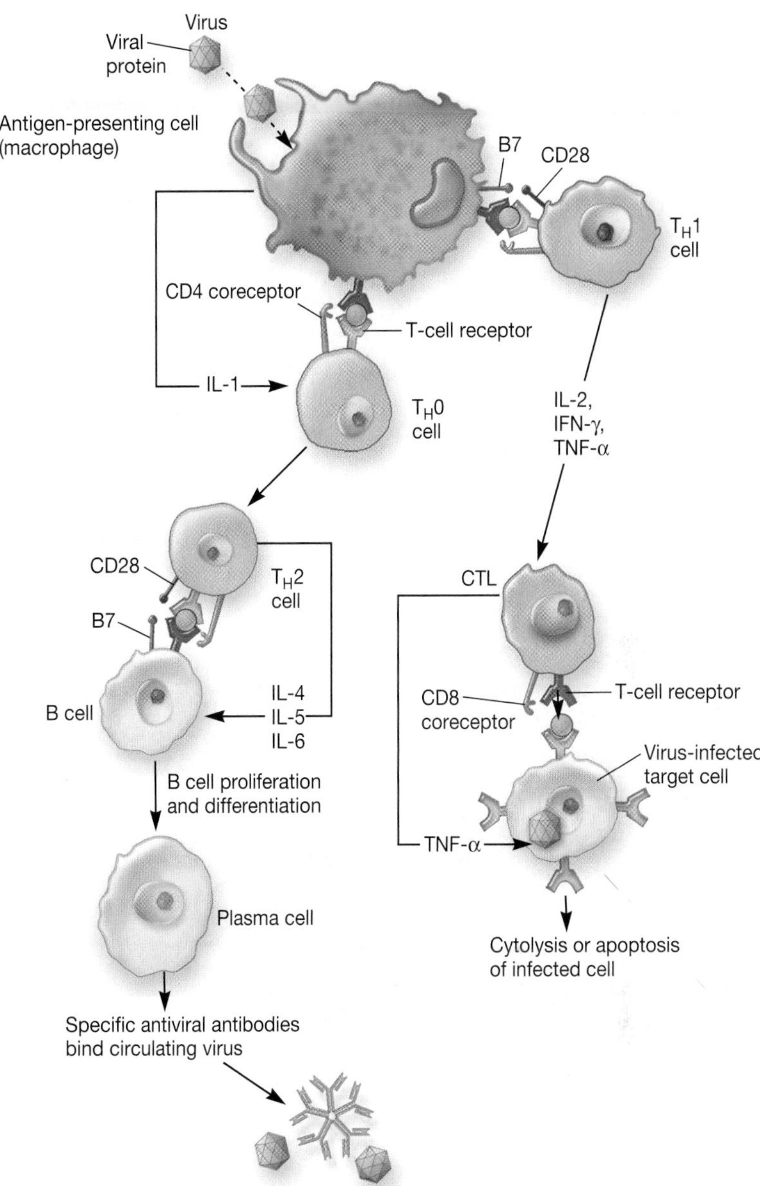

Figure 32.7 T-Cell Responses. A virus is phagocytosed by a macrophage and a small antigen fragment (peptide) presented to naïve T_H cells in association with class II MHC molecules. Once activated (activation signals 1 and 2), the T_H0 cell may differentiate into a T_H2 cell that secretes the cytokines IL-4, IL-5, and IL-6, followed by IL-10, and IL-13 (not shown) which causes B cell proliferation and subsequent secretion of specific antiviral antibodies. Alternatively they differentiate into T_H1 cells that secrete IL-2, IFN-γ, and TNF-α. IL-2 regulates the proliferation of cytotoxic T cells (CTLs). Once a CTL proliferates and differentiates into an activated effector cell, it attacks and causes lysis or programmed cell death (apoptosis) of a virus-infected cell by either the perforin or Fas pathways, respectively.

In addition to signal 1, both naïve cell types require a second, co-stimulatory signal (signal 2) to become activated. Without signal 2, a T cell receiving signal 1 only will often become **anergic,** or unresponsive to that antigen. There may be more than one factor that contributes to signal 2, but the most important seems to be the B7 (CD80) protein on the surface of an APC. One type of APC that is particularly good at stimulating naïve T cells is the dendritic cell. This type of phagocytic cell expresses high levels of B7 constitutively (at all times). Thus the combination of signals 1 and 2 provided by a mature dendritic cell presenting the antigen fragment stimulates molecular events inside the T cell, which causes it to proliferate and differentiate.

In T_H cells, signal 1 stimulates a signal transduction pathway that results in the production of key cytokines. First, signal 1 activates a tyrosine kinase located in the cytoplasm (**figure 32.9**). Tyrosine kinases add phosphate groups to the amino acid tyrosine in proteins. In this case, phosphorylation of the enzyme phospholipase $C_\gamma1$ stimulates it to cleave the molecule phosphatidylinositol bisphosphate located in the T-helper cell plasma membrane. Two cleavage products are formed and each contributes to a separate pathway within the cell. One of the cleavage products, diacylglycerol, activates protein kinase C (PKC). PKC moves into the nucleus where it catalyzes the formation of a protein complex called AP-1 from separate components. The other cleavage product, inositol triphosphate, causes a calcium channel to open; calcium ions then rush into the cytoplasm, leading to further enzymatic activity and the activation of calmodulin, calcineurin, and nuclear factor of activated T_H1 cells (NF-AT). NF-AT then migrates into the nucleoplasm where it binds to the newly formed AP-1, forming a NF-AT/AP-1 complex. This functions as a transcription factor, causing interleukin-2 mRNA to be expressed. Interleukin-2 (IL-2) mRNA moves out of the nucleus to the ribosomes where the IL-2

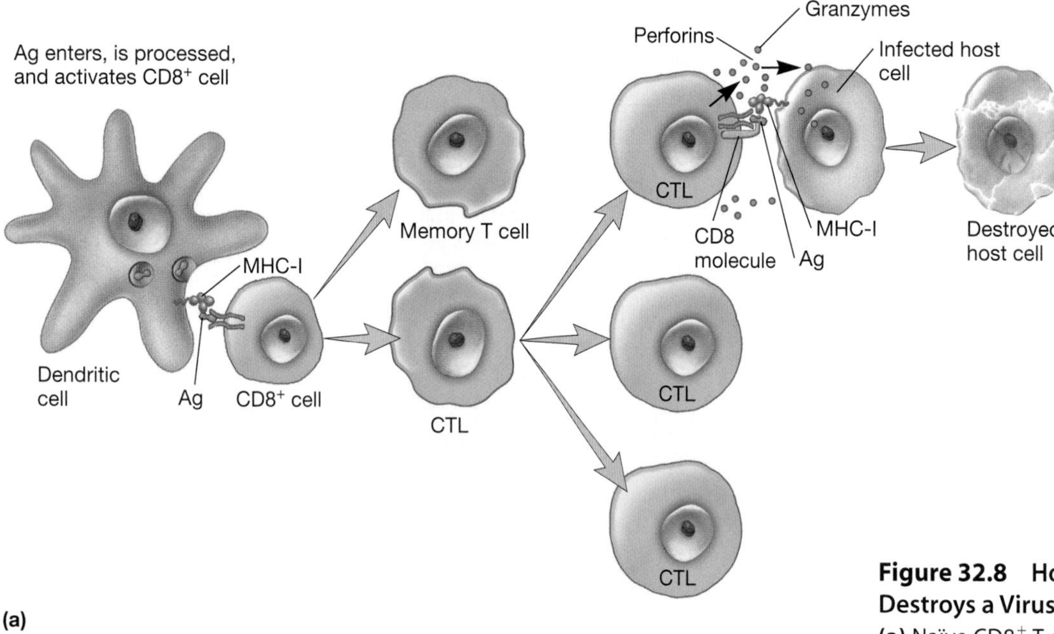

(a)

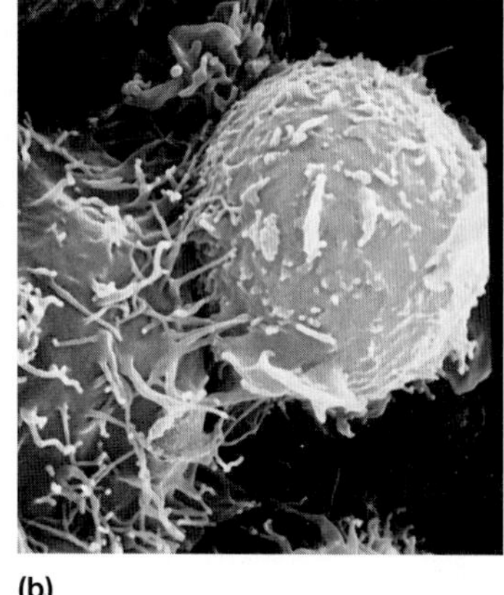

(b)

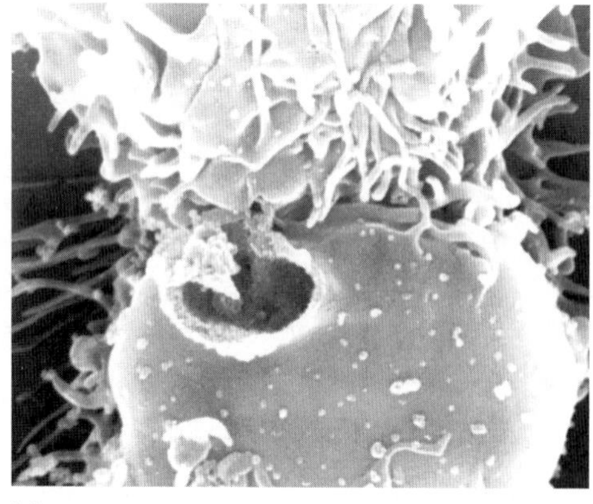

(c)

Figure 32.8 How an Effector Cytotoxic T Cell Destroys a Virus-Infected Target Cell. **(a)** Naïve $CD8^+$ T cells are activated when they are exposed to antigen within a class I MHC molecule on an antigen-presenting cell. Antigen activation leads to development of effector CTL and memory cells. Effector CTLs and their memory cells subsequently react with antigen expressed in class I MHC molecules of any host cell to destroy it. T cell cytotoxicity often involves the perforin pathway and leads to apoptosis and cytolysis. **(b)** A cytotoxic T cell (left) contacts a target cell (right) (×5,700). **(c)** The T cell secretes perforin that forms pores in the target cell's plasma membrane. These pores allow the contents of the target cell to leak out and granzymes to enter and induce apoptosis. Ag, antigen.

protein is produced. IL-2 is a T-cell growth factor. Its production stimulates T cell differentiation in an autocrine-like manner. Thus signal 1 triggers the production of this important cytokine that is required for further T cell development.

Signal 2, mediated by the CD28 receptor and B7 molecule, activates a different tyrosine kinase, causing the formation of the transcription factor CD28RC (figure 32.9). It stabilizes the mRNA transcribed from the IL-2 gene, thereby further increasing production of IL-2. T_H1 cells that are activated by these two signals secrete large amounts of IL-2, which can also activate nearby cytotoxic T cells in a paracrine-like fashion (figure 32.7).

$CD8^+$ cell activation occurs in a manner similar to that of T_H cells. Recall that signal 1 is a specific antigenic fragment; it is presented in the class I MHC receptor of an APC to the T-cell receptor, in association with the CD8 co-receptor. Although $CD8^+$ cells detect antigen that is presented in the class I MHC, the B7 on the APC also costimulates the $CD8^+$ cell, just as it does for $CD4^+$ cells. Once signals 1 and 2 trigger $CD8^+$ T cells to differentiate into CTLs, they can respond to antigen presented by the class I MHC of other target cells (typically virus infected or cancerous cells) without the stringent B7 co-stimulation required for their intial activation. The binding of CD28 to B7 is also signal 2 for the CTL. The B7 of a mature dendritic cell is sufficient to convey signal 2 to the $CD8^+$ cell. However, other APCs may not produce sufficient B7 to communicate signal 2. In these cases, activated T_H cells are required to stimulate the APC to make another co-stimulatory signal (a different signal 2). For example, APCs can also be stimulated by T_H cells to produce 4-IBBL (CD137 ligand). 4-IBBL binds to the 4-IBB receptor of signal 1-stimulated $CD8^+$ cell to complete their activation. The activated $CD8^+$ cell then synthesizes and secretes IL-2 to drive its own proliferation and differentiation. Overall, once $CD8^+$ cells have been activated by two signals, they differentiate into CTLs, which can rapidly respond to host cells infected with an intracellular pathogen (i.e., a target cell) by simply recognizing foreign antigen fragments within the target

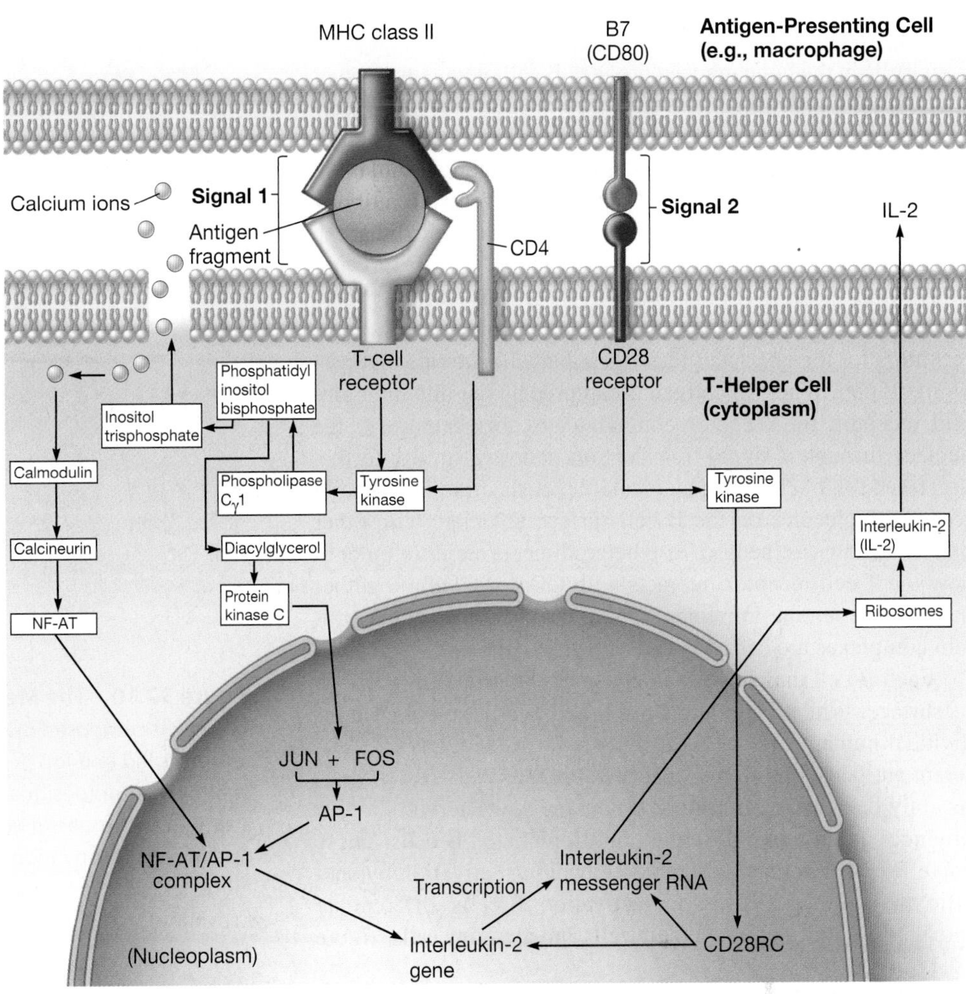

Figure 32.9 Two Signals (Costimulation) Are Essential for T-helper Cell Activation. The first signal is the presentation of the antigen fragment by a macrophage or other antigen-presenting cell along with the MHC class II molecule to the T-helper cell receptor and CD4 protein. The second signal occurs when the macrophage presents the B7 (CD80) protein to the T-helper cell with its CD28 protein receptor. Both signals send information into the cytoplasm of the T-helper cell. The first signal causes interleukin-2 mRNA to be produced. The second signal boosts the production of interleukin-2 to effective concentrations. It now is known that the gene for interleukin-2 is under very tight regulation. It cannot be transcribed unless the NF-AT/AP-1 complex, and other transcription factors (e.g., CD28RC) are present. All these factors must be produced anew or activated when the T-helper cell is stimulated through its antigen-specific receptor.

cell's class I MHC protein and docking on the MHC-antigen complex.

Superantigens

Several bacterial and viral proteins can provoke a drastic and harmful response when they are exposed to T cells. These proteins are known as **superantigens** because they activate a large number of cells as if they have been exposed to a specific antigen; they proliferate and produce cytokines. However, superantigens do this by "tricking" T cells into activation when no specific antigen has triggered them. The mechanism by which superantigens do this is by bridging class II MHC molecules on APCs to T-cell receptors (TCRs) in the absence of a specific antigen in the MHC binding site. This interaction allows many different T cells with different antigen specificities to become activated. The consequence of this nonspecific activation is the release of massive quantities of cytokines from $CD4^+$ T cells, leading to organ failure and suppression of specific immune responses. Thus numerous T cells (nearly 30%) can be activated by superantigen to over-produce cytokines such as TNF-α and interleukins 1 and 6, resulting in endothelial damage, circulatory shock, and multiorgan failure. Superantigens can be considered virulence factors whose effects contribute to microbial pathogenicity. Examples of superantigens include the staphylococcal enterotoxins (which can cause food poisoning), the toxin that causes toxic shock syndrome, mouse tumor virus superantigen, and perhaps proteins from Epstein-Barr and rabies viruses. Because of these activities, staphylococcal enterotoxin B has been added to the Select Agent List of the U.S. government as a potential agent of terrorism. The devastating effects superantigens have on the host serve to emphasize the importance of tightly regulating a normal immune response.

1. What is the function of an antigen-presenting cell? What is a T-cell receptor and how is it involved in T-cell activation?
2. What are MHCs and HLAs? Describe the roles of the three MHC classes.
3. Describe antigen processing. How does this process differ for endogenous and exogenous antigens?
4. Briefly describe the cytotoxic T cell, its general role, how it is activated, and the two ways in which it destroys target cells.
5. Outline the functions of a T-helper cell. How do T_H1 and T_H2 cells differ in function? Briefly describe how T_H cells are activated by co-stimulation versus superantigen.

32.6 B Cell Biology

Stem cells in the bone marrow produce B cell precursors (figure 32.1). Like T cells, B cells must be activated by a specific antigen to continue mitosis, replicate, and differentiate into antibody-secreting plasma cells. Prior to antigen stimulation, B cells express genes that code for a special form of antibody that is attached to their cell membrane and oriented so that the part of the antibody that binds to antigen is facing outward, away from the cell (**figure 32.10**). These cell-surface, transmembrane antibodies (also known as immunoglobulins) act as receptors for the one specific antigen that will activate that particular B cell. When an antigen is captured by the immunoglobulin receptor, the receptor communicates this capture to the nucleus through a signal transduction pathway similar to that described for T cells. On a molecular level, the immunoglobulin receptor molecules on the B cell surface associate with other proteins known as the Ig-α/Ig-β heterodimer proteins (similar to how the T-cell receptor interacts with CD4 or CD8). Together the transmembrane immunoglobulin and the heterodimer protein complexes are called **B-cell receptors (BCRs).**

Each B cell may have as many as 50,000 of these BCRs on its surface. While the total mature B-cell population of each individual human carries BCRs specific for as many as 10^{13} different antigens, each individual B cell possesses BCRs specific for only one particular epitope on an antigen. Therefore a host produces at least 10^{13} different, undifferentiated B cells. These naïve B cells circulate in the blood awaiting activation by specific antigenic epitopes. Upon activation, B cells differentiate into antibody-producing **plasma cells** and **memory cells** (figure 32.1).

So far we have introduced the mechanism by which B cells are activated and that once activated they secrete antibody. However, it is important to note that B cells also internalize the antigen-bound receptor to share its three-dimensional configuration with other cells—that is, BCRs that have captured their antigenic epitope are able to trigger endocytosis of that antigen leading to antigen processing inside the B cell. As is the case with macrophages and dendritic cells, a small antigen fragment is then presented on the surface of the B cell in association with class II MHC molecules. Thus B cells have two immunological roles: (1) they proliferate and differentiate into memory cells and plasma cells, which respond to antigens by making antibodies, and at the same time (2) they can act as antigen-presenting cells.

B Cell Activation

In general, an activated B cell still requires growth and differentiation factors supplied by other cells. Recall that T-helper cells produce cytokines, some of which act on B cells to assist in their growth and differentiation. Although activation of the B cell is typically antigen-specific, it can additionally be T-cell dependent or T-cell independent. This distinction reflects additional supportive activity provided by T-helper cells beyond the release of initial cytokine growth factors.

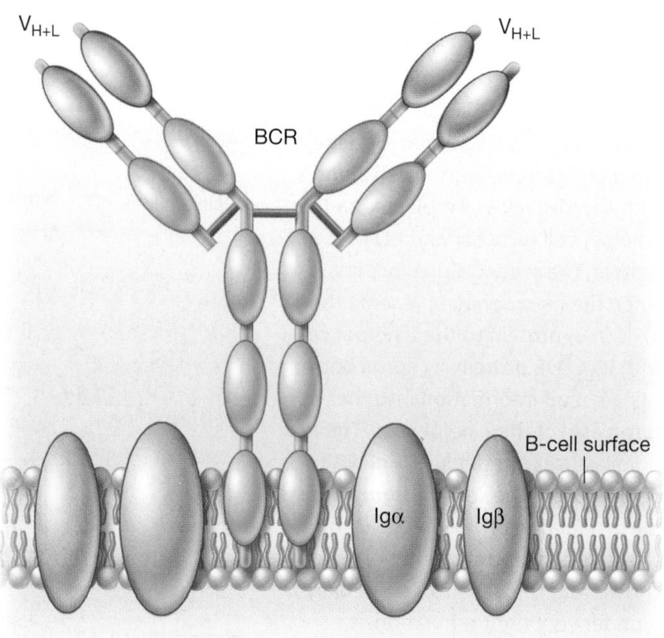

Figure 32.10 The Membrane-Bound B-Cell Receptor (BCR). The BCR is composed of monomeric IgM antibody and the co-receptors Igα and Igβ. A B cell is activated after the variable region of one receptor (designated V_{H+L}) binds an antigen fragment attached to the class II MHC molecule of an activated T_H2 cell (T-dependent activation), or after the variable regions of (two or more) receptors are bridged by an antigen (T-independent activation).

T-Dependent Antigen Triggering

As noted previously, most antigens have more than one type of antigenic determinant site (epitope) on each molecule (figure 32.2). In **T-dependent antigen triggering,** B cells that are specific for a given epitope on the antigen (e.g., epitope X) cannot develop into plasma cells that secrete antibody (anti-X) without the collaboration of T-helper cells. In other words, binding of epitope X to the B cell may be necessary, but it is not usually sufficient for B-cell activation. Antigens that elicit a response with the aid of T-helper cells are called **T-dependent antigens.** Examples include bacteria, foreign red blood cells, certain proteins, and hapten-carrier combinations.

The basic mechanism for T-dependent antigen triggering of a B cell is illustrated in **figure 32.11** and involves three cells: (1) a macrophage or other APC to process and present the antigen; (2) a T-helper cell able to recognize the antigen and respond to it; and (3) a B cell specific for the antigen. When all of these cells and the antigen are present, the following sequence takes place: (1) The APC (e.g., macrophage) presents part of the antigen in its class II MHC to the T-helper cell (signal #1 to the T cell). (2) Co-stimulation is provided by the B7-CD28 interaction (signal #2) between the APC and the T cell, resulting in cytokine production. (3) These T-helper cells then directly associate with B cells that display the same antigen-MHC complex that was presented on the APC. This promotes the secretion of additional cy-

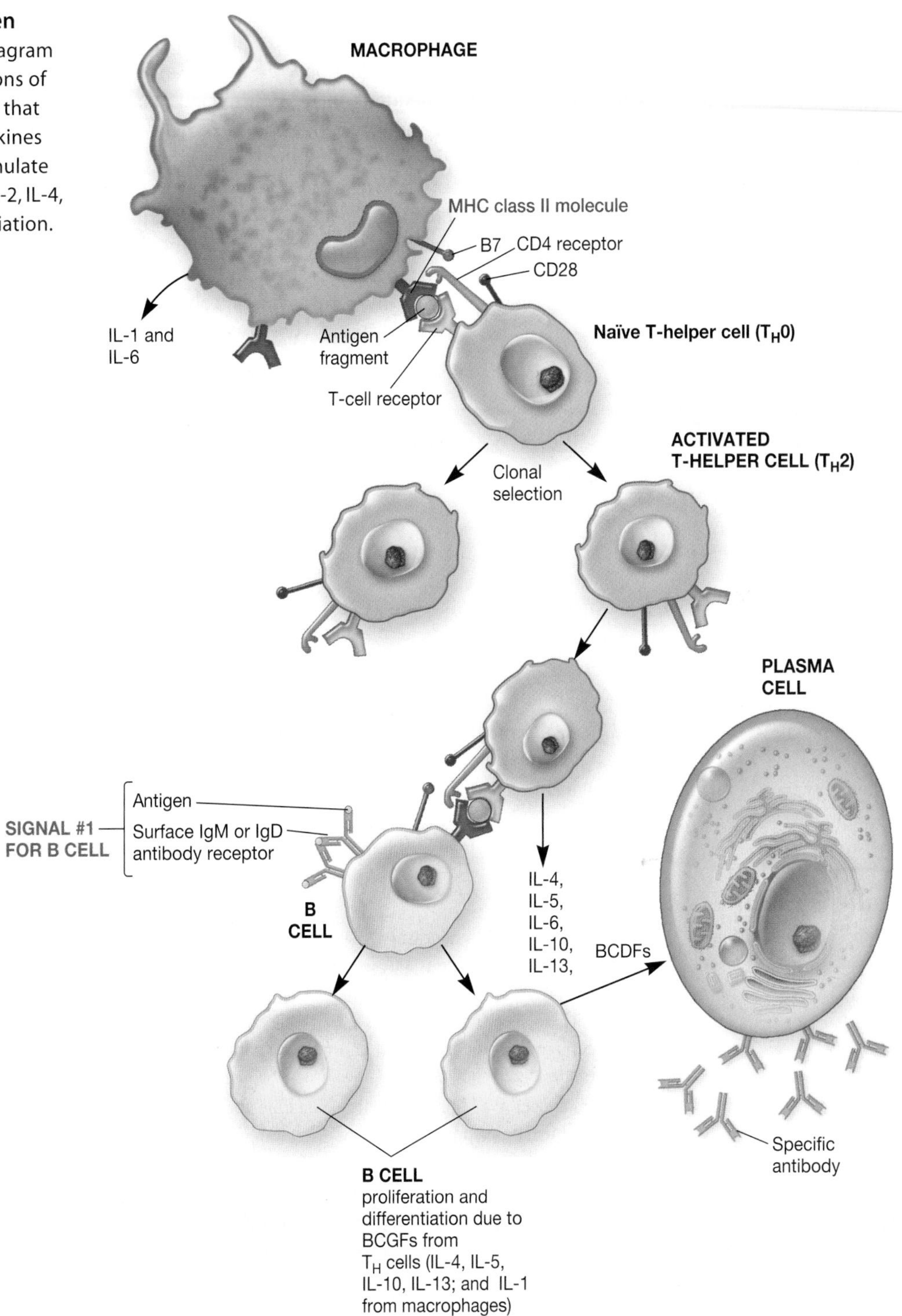

Figure 32.11 T-Dependent Antigen Triggering of a B Cell. Schematic diagram of the events occurring in the interactions of macrophages, T-helper cells, and B cells that produce humoral immunity. Many cytokines (e.g., IL-1, IL-4, IL-5, IL-6, IL-10, IL-13) stimulate B-cell proliferation. Cytokines such as IL-2, IL-4, IL-6, and IL-13 stimulate B-cell differentiation.

tokines by the T cell. (4) This interaction causes the B cells to proliferate and differentiate into plasma cells, which start producing antibodies.

However, B cell activation requires more than just T-helper cell activity. The B cell also requires the antigen, recognized through its BCR (signal #1 for the B cell), to help trigger B cell proliferation and differentiation into a plasma cell. Thus B cells, like T cells, require two signals: antigen-BCR interaction (signal 1) and T cell cytokines (signal 2). This is a very effective process: one plasma cell can synthesize more than 10 million antibody molecules per hour! It should be noted that once the antigen binds to surface BCRs, B cells become more effective antigen presenters than macrophages especially at low antigen concentrations. Here B cells bind antigen, take it up by receptor-mediated endocytosis, and present it to T-helper cells to activate them. In this situation B cells and T-helper cells activate each other. We will see later that ultimately, B cells use T-dependent antigen triggering to alter their antibody production, a process called class switching (p. 795).

T-Independent Antigen Triggering

Not all antibody responses require direct T cell help. There are a few specific antigens that trigger B cells into antibody production without T cell cooperation. These are called **T-independent antigens** and their stimulation of B cells is known as **T-independent antigen triggering.** Examples include bacterial lipopolysaccharides, certain tumor-promoting agents, antibodies against other antibodies, and antibodies to certain B cell differentiation antigens. The T-independent antigens tend to be polymeric—that is, they are composed of repeating polysaccharide or protein subunits. The resulting antibody generally has a relatively low affinity for antigen.

The mechanism for activation by T-independent antigens probably depends on their polymeric structure. Large molecules present a large array of identical epitopes to a B cell specific for that determinant. The repeating epitopes cross-link membrane-bound BCRs such that cell activation occurs and antibody is secreted. Because there is no T cell help, the B cell cannot alter its antibody production, and no memory B cells are formed. Thus T-independent B cell activation is less effective than T-dependent B cell activation: the antibodies produced have a low affinity for antigen and no immunologic memory is formed. **Tables 32.3** and **32.4** summarize and compare many of the important properties of lymphocytes that we have discussed.

Table 32.3 Comparison of Lymphocytes Involved in the Immune Response

Property	T Cells	B Cells
Origin	Bone marrow	Bone marrow
Maturation and expression of antigen receptors	Thymus	Bone marrow; bursa of Fabricius in birds
Differentiation	Lymphoid tissue	Lymphoid tissue
Mobility	Great	Very little (some stages circulate)
Complement receptors	Absent	Present
Surface immunoglobulins	Absent	Present
Proliferation	Upon antigenic stimulation, proliferate and differentiate into effector and memory cells	Upon antigenic stimulation, proliferate and differentiate into plasma and memory cells
Immunity type	Cell mediated and humoral (B cell activation by T_H cells)	Humoral
Distribution	High in blood, lymph, and lymphoid tissue	High in spleen, lymph nodes, bone marrow, and other lymphoid tissue; low in blood
Secretory product	Cytokines	Antibodies
Subsets and functions	T-helper (T_H) cell: necessary for B-cell activation by T-dependent antigens and T-effector cells. There are three types of T-helper cells: T_H1, T_H2, and T_H0.	Plasma cell: a cell arising from a B cell that manufactures specific antibodies
	Cytotoxic T cell: differentiates into a CTL that lyses cells recognized as nonself and virus or parasite-infected cells	Memory Cell: a long-lived cell responsible for the anamnestic response
	Memory cells: a long-lived cell responsible for the anamnestic response	

Table 32.4 Antigen Recognition by T and B Cells

Characteristic	T Cells	B Cells
Binds soluble antigen	No	Yes
Biochemistry of the antigens	Mostly proteins, but some glycolipids presented on MHC molecules	Proteins, glycolipids, polysaccharides
Antigen recognition	Antigens processed internally and presented as linear peptides bound to MHC molecules	Accessible areas of protein structure containing sequential amino acids and nonsequential amino acids
	Involve two cells: T-cell receptor, with processed antigen in MHC molecule on APC	Immunoglobulin receptor binds antigen in native conformation
		Involves two partners: antigen and membrane immunoglobulin

1. What are B cell receptors? How are they involved in B cell activation?
2. Briefly compare and contrast B cells and T cells with respect to their formation, structure, and roles in the immune response.
3. How does antigen-antibody binding occur? What is the basis for antibody specificity?
4. How does T-independent antigen triggering of B cells differ from T-dependent triggering?

32.7 Antibodies

An **antibody** or **immunoglobulin (Ig)** is a glycoprotein that is made by activated B cells called plasma cells. Certain antibodies serve as the antigen receptor (BCR) on the B cell surface. After antigen capture by the membrane-bound immunoglobulin and B cell activation, B cells differentiate into plasma cells, which secrete a soluble form of antibody that circulates through the bloodstream to recognize and bind the antigen that induced its synthesis. Antibodies are present in the blood serum, tissue fluids, and mucosal surfaces of vertebrate animals. There are five classes of human antibodies. **Table 32.5** summarizes some of the more important physiochemical properties of the human immunoglobulin classes. The classes differ from each other in molecular size, structure, charge, amino acid composition, and carbohydrate content. Before we examine each of the immunoglobulin types, a thorough evaluation of immunoglobulin structure will help us appreciate the differences among immunoglobulin classes.

Table 32.5 Physicochemical Properties of Human Immunoglobulin Classes

	Immunoglobulin Classes				
Property	**IgG**[a]	**IgM**	**IgA**[b]	**IgD**	**IgE**
Heavy chain	γ_1	μ	α_1	δ	ε
Mean serum concentration (mg/ml)	9	1.5	3.0	0.03	0.00005
Percent of total serum antibody	80–85	5–10	5–15	<1	0.002–0.05
Valency	2	5(10)	2(4)	2	2
Mass of heavy chain (kDa)	51	65	56	70	72
Mass of entire molecule (kDa)	146	970	160[c]	184	188
Placental transfer	+	–	–	–	–
Half-life in serum (days)[d]	23	5	6	3	2
Complement activation					
Classical pathway	++	+++	–	–	–
Alternative pathway	–	–	+	–	–
Induces mast cell degranulation	–	–	–	–	+
Major characteristics	Most abundant Ig in body fluids; neutralizes toxins; opsonizes bacteria; activates complement; transplacental antibody	First to appear after antigen stimulation; very effective agglutinator; expressed as membrane-bound antibody on B cells	Secretory antibody; protects external surfaces	Present on B-cell surface; B-cell recognition of antigen	Anaphylactic-mediating antibody; resistance to helminths
% carbohydrate	3	7–10	7	12	11

[a] Properties of IgG subclass 1.

[b] Properties of IgA subclass 1.

[c] sIgA = 360 – 400 kDa

[d] Time required for half of the antibodies to disappear.

Immunoglobulin Structure

Recall that antigen is captured by its respective antibody through the antigen-binding or more appropriately, antigen-combining, site. Each B cell has numerous surface-bound (transmembrane) antibodies, which have two combining sites—that is, they are bivalent. Some bivalent antibody molecules can combine to form multimeric antibodies that have up to 10 combining sites. All immunoglobulin molecules have a basic structure composed of four polypeptide chains: two identical heavy and two identical light chains connected to each other by disulfide bonds (**figure 32.12**). Each light chain polypeptide usually consists of about 220 amino acids and has a mass of approximately 25,000 Da. Each heavy chain consists of about 440 amino acids and has a mass of about 50,000 to 70,000 Da. The heavy chains are structurally distinct for each immunoglobulin class or subclass. Both light (L) and heavy (H) chains contain two different regions. **Constant (C) regions** (**C_L** and **C_H**) have amino acid sequences that do not vary significantly between antibodies of the same class. The **variable (V) regions (V_L and V_H)** from different antibodies have different amino acid sequences. It is the variable regions (V_L and V_H) that, when folded together, form the antigen-binding sites.

The four chains are arranged in the form of a flexible "Y" with a hinge region. This hinge allows the antibody molecule to be more flexible, adjusting to the different spatial arrangements of epitopes or antigenic determinants of antigens. The stalk of the Y is termed the **crystallizable fragment (Fc)** and binds to a cell by interacting with the cell surface Fc receptor. The top of the Y consists of two **antigen-binding fragments (Fab)** that bind with compatible epitopes. The Fc fragments are composed only of constant regions, whereas the Fab fragments have both constant and variable regions. Both the heavy and light chains contain several homologous units of about 100 to 110 amino acids. Within each unit, called a **domain,** disulfide bonds form a loop of approximately 60 amino

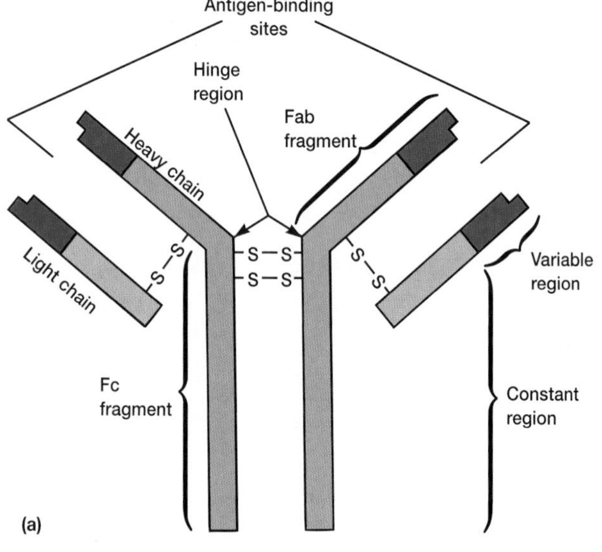

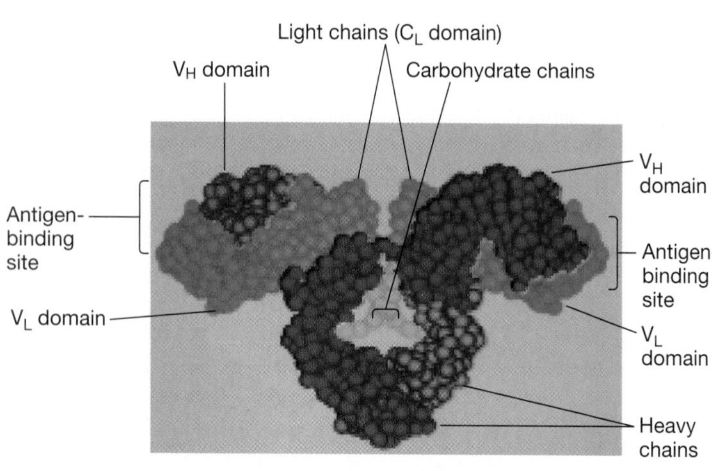

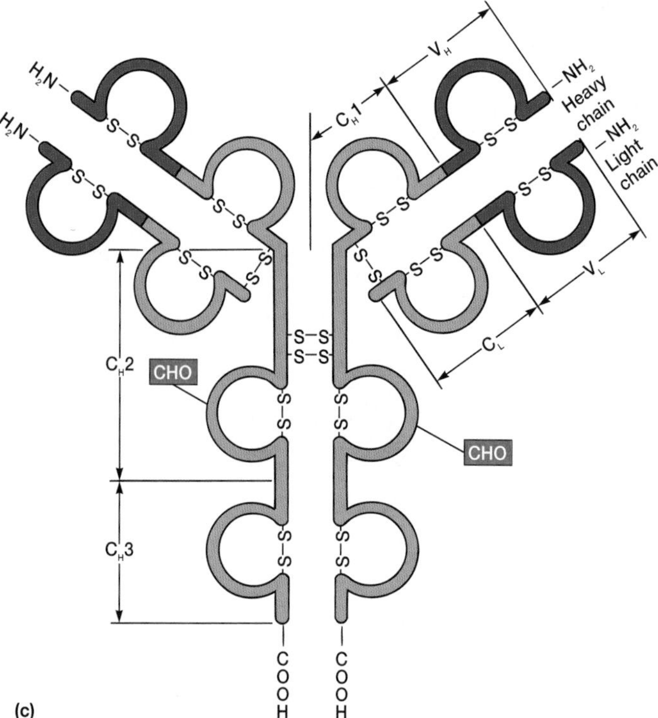

Figure 32.12 Immunoglobulin (Antibody) Structure. **(a)** An immunoglobulin molecule. The molecule consists of two identical light chains and two identical heavy chains held together by disulfide bonds. **(b)** A computer-generated model of antibody structure showing the arrangement of the four polypeptide chains. **(c)** Within the immunoglobulin unit structure, intrachain disulfide bonds create loops that form domains. All light chains contain a single variable domain (V_L) and a single constant domain (C_L). Heavy chains contain a variable domain (V_H) and either three or four constant domains (C_H1, C_H2, C_H3, and C_H4). The variable regions (V_H, V_L), when folded together in three-dimensions, form the antigen-binding sites.

acids (figure 32.12*c*). Interchain disulfide bonds also link heavy and light chains together.

The light chain may be either of two distinct forms called kappa (κ) and lambda (λ). These can be distinguished by the amino acid sequence of the constant (C) portion of the chain (**figure 32.13*a***). In humans, the constant regions of all κ chains are identical. However, there are four similar λ protein sequences possible, reflecting four, slightly different λ subtypes. Regardless of immunoglobulin class, each antibody molecule produced by a sole B cell will contain either κ or λ light chains, but never both. Within the light chain variable (V) domain are hypervariable regions, or **complementarity-determining regions (CDRs),** that differ in amino acid sequence more frequently than the rest of the variable domain. These figure prominently in determining antigen specificity.

The amino-terminal domain of the heavy chain has a pattern of variability similar to that of the variable (V) region of the kappa chain (V_κ) and the variable region of the lambda chain (V_λ) domains, and is termed the V_H domain. The other domains of the heavy chains are termed constant (C) domains (figure 32.13*b*). The constant domains of the heavy chain form the constant (C_H) region. The amino acid sequence of this region determines the classes of heavy chains. In humans there are five classes of heavy chains designated by lowercase Greek letters: gamma (γ), alpha (α), mu (μ), delta (δ), and epsilon (ε), and sometimes written as G, A, M, D or E. The properties of these heavy chains determine, respectively, the five immunoglobulin (Ig) classes—IgG, IgA, IgM, IgD, and IgE (table 32.5). Each immunoglobulin class differs in its general properties, half-life, distribution in the body, and interaction with other components of the host's defensive systems. There are variants of immunoglobulins that can be classified as: (1) **Isotypes** are the variations in the heavy chain constant regions associated with the different classes that are normally present in all individuals (**figure 32.14*a***). Therefore there are five isotypes corresponding to the five antibody classes. (2) **Allotypes** are the genetically controlled, allelic forms of immunoglobulin molecules (figure 32.14*b*) that are not present in all individuals. They arise by genetic recombination (pp. 796–798). (3) **Idiotypes** are individual, specific immunoglobulin molecules that differ in the hypervariable region of the Fab portion due to mutations that occur during B cell development (figure 32.14*c*). These variations of immunoglobulin structure reflect the diversity of antibodies generated by the immune response.

1. What is the variable region of an antibody? The hypervariable or complementarity-determining region? The constant region?
2. What is the function of the Fc region of an antibody? The Fab region?
3. Name the two types of antibody light chains.
4. What determines the class of heavy chain of an antibody?
5. Name the five immunoglobulin classes.
6. Distinguish among isotype, allotype, and idiotype.

Immunoglobulin Function

Each end of the immunoglobulin molecule has a unique role. The Fab region is concerned with binding to antigen, whereas the Fc region mediates binding to Fc receptors found on various cells of the immune system, or the first component of the classical complement system. The binding of an antibody to an antigen usually

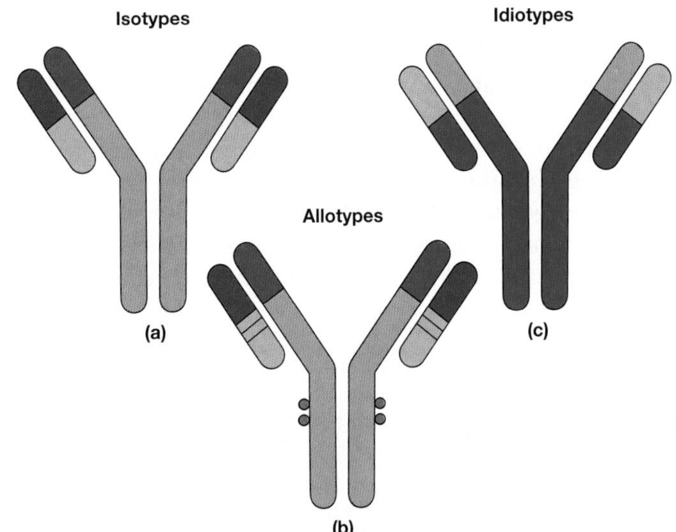

Figure 32.14 Variants of Immunoglobulins. **(a)** Isotypes represent variants present in serum of a normal individual. **(b)** Allotypes represent alternative forms coded for by different alleles and so are not present in all individuals of a population. **(c)** Idiotypes are individually specific to each immunoglobulin molecule. Carbohydrate side chains are in red.

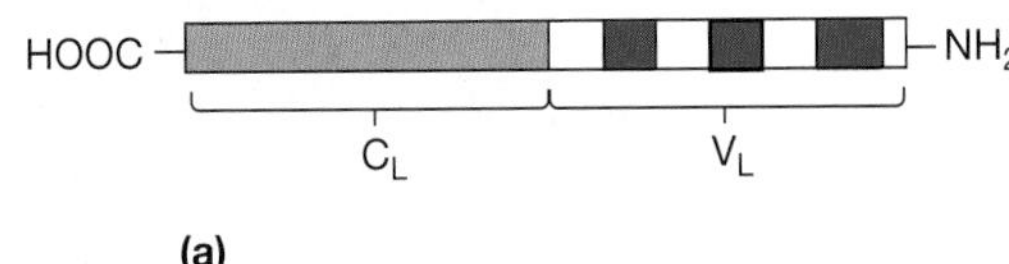

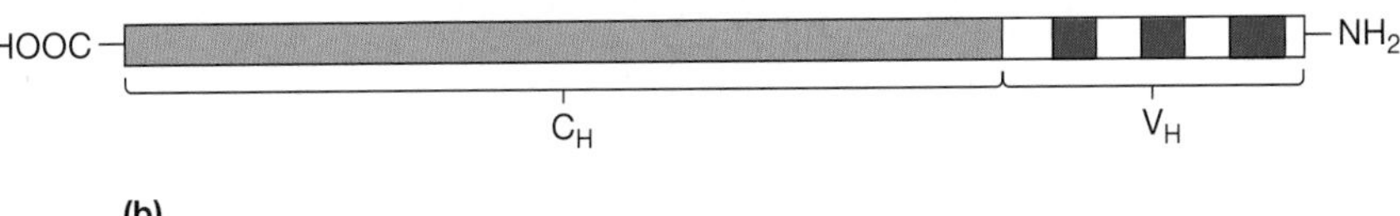

Figure 32.13 Constant and Variable Domains. Location of constant (C) and variable (V) domains within **(a)** light chains and **(b)** heavy chains. The dark blue bands represent hypervariable regions or complementarity-determining regions within the variable domains.

does not cause destruction of the antigen or of the microorganism, cell, or agent to which it is attached. Rather the antibody serves to mark and identify the nonself agent as a target for immunological attack and to activate nonspecific immune responses that can destroy the target.

An antigen binds to an antibody at the antigen-binding site within the Fab region of the antibody. More specifically, a pocket is formed by the folding of the V_H and V_L regions (figure 32.12). At this site, specific amino acids contact the antigen's epitope or haptenic groups and form multiple noncovalent bonds between the antigen and amino acids of the binding site (**figure 32.15**). Because binding is due to weak, noncovalent bonds such as hydrogen bonds and electrostatic attractions, the antigen's shape must exactly match that of the antigen-binding site. If the shape of the epitope and binding site are not truly complementary, the antibody will not effectively bind the antigen. Although a lock-and-key mechanism normally may operate, in at least one case, the antigen-binding site does change shape when it complexes with the antigen (an induced-fit mechanism). Regardless of the precise mechanism, antibody specificity results from the nature of antibody-antigen binding.

Phagocytes have Fc receptors for immunoglobulin on their surface, so bacteria that are covered with antibodies are better targets for phagocytosis by neutrophils and macrophages. This is termed **opsonization.** Other cells may kill antibody-coated cells through a process called antibody-dependent cell-mediated cytotoxicity. Immune destruction also is promoted by antibody-induced activation of the classical complement system. Chemical mediators of innate immunity: Complement (section 31.6)

Immunoglobulin Classes

Immunoglobulin γ, or **IgG,** is the major immunoglobulin in human serum, accounting for 80% of the immunoglobulin pool (**figure 32.16*a***). IgG is present in blood plasma and tissue fluids. The IgG class acts against bacteria and viruses by opsonizing the invaders and neutralizing toxins. It is also one of the two immunoglobulin classes that activate complement by the classical pathway. IgG is the only immunoglobulin molecule able to cross the placenta and provide natural immunity in utero and to the neonate at birth.

There are four human IgG subclasses (IgG1, IgG2, IgG3, and IgG4) that vary chemically in their heavy chain composition and the number and arrangement of interchain disulfide bonds (figure 32.16*b*). About 65% of the total serum IgG is IgG1, and 23% is IgG2. Differences in biological function have been noted in these subclasses. For example, IgG2 antibodies are opsonic and develop in response to toxins. IgG1 and IgG3, upon recognition of their specific antigens, bind to Fc receptors expressed on neutrophils and macrophages. This increases phagocytosis by these cells. The IgG4 antibodies function as skin-sensitizing immunoglobulins.

Immunoglobulin μ, or **IgM,** accounts for about 10% of the immunoglobulin pool. It is usually a polymer of five monomeric units (pentamer), each composed of two heavy chains and two light chains (**figure 32.17**). The monomers are arranged in a pinwheel array with the Fc ends in the center, held together by disulfide bonds and a special **J** (joining) **chain.** IgM is the first immunoglobulin made during B-cell maturation and individual IgM monomers are expressed on B cells, serving as the antibody

Fc
Hinge
Elbow
Fab
2-fold axis
Canyon

(a)

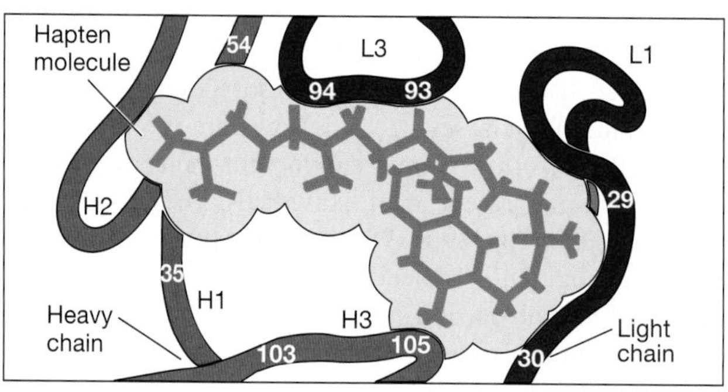

(b)

Figure 32.15 Antigen-Antibody Binding. (a) An example of antigen-antibody binding is represented in this model of the monoclonal antibody mAb17-IA bound to the surface of the human rhinovirus. The heavy chains are red, light chains are blue, and the antigen capsid protein is yellow. The RNA interior of the virus would be toward the bottom of the diagram. The antibody is bound bivalently across icosahedral twofold axes of the virus. **(b)** Based on X-ray crystallography, the hapten molecule nestles in a pocket formed by the antibody combining site. In the illustration the hapten makes contact with only 10 to 12 amino acids in the hypervariable regions of the light and heavy chains. The numbers represent contact amino acids.

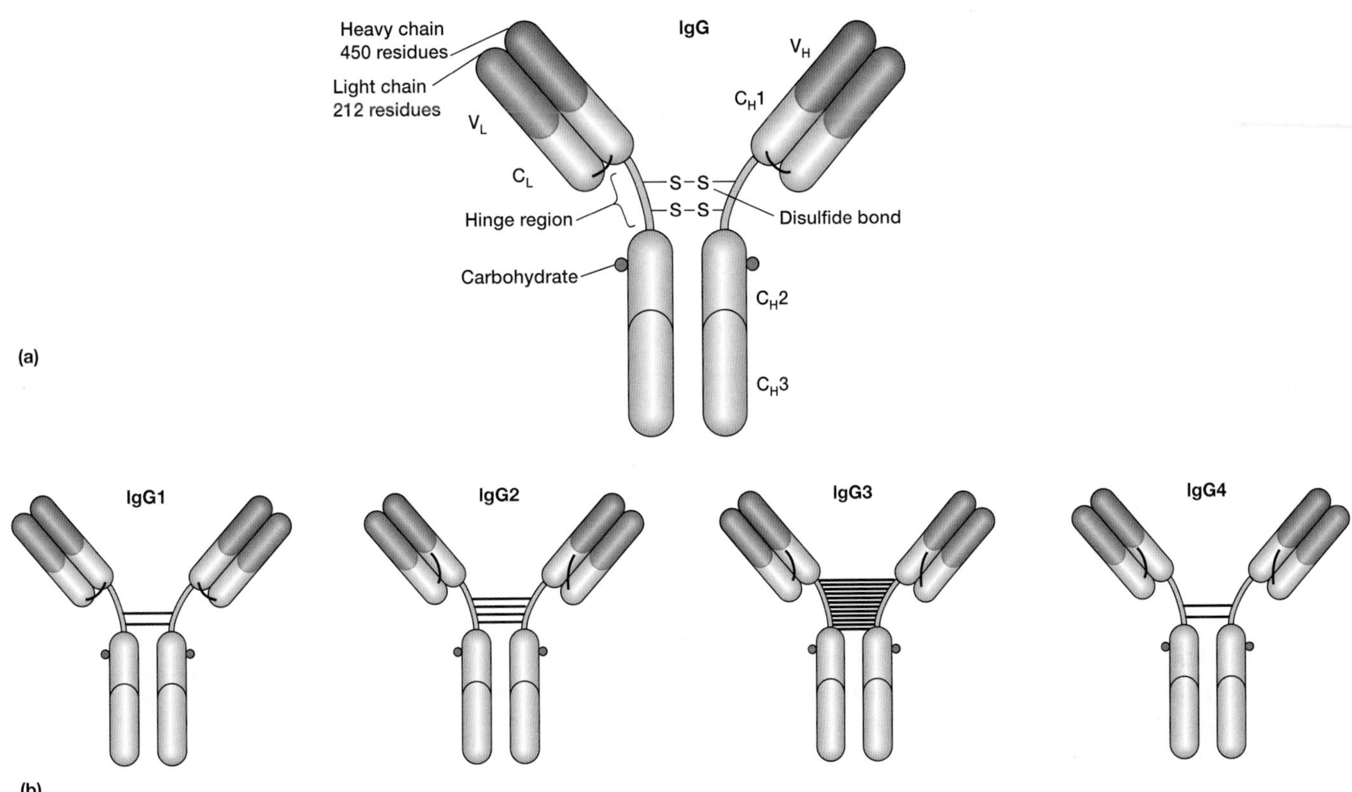

Figure 32.16 Immunoglobulin G. **(a)** The basic structure of human IgG. **(b)** The structure of the four human IgG subclasses. Note the arrangement and numbers of disulfide bonds (shown as thin black lines). Carbohydrate side chains are shown in red.

component of the BCR. Pentameric IgM is secreted into serum during a primary antibody response (as is discussed shortly). IgM tends to remain in the bloodstream where it agglutinates (or clumps) bacteria, activates complement by the classical pathway, and enhances the ingestion of pathogens by phagocytic cells.

Although most IgM appears to be pentameric, around 5% or less of human serum IgM exists in a hexameric form. This molecule contains six monomeric units but seems to lack a J chain. Hexameric IgM activates complement up to 20-fold more effectively than does the pentameric form. It has been suggested that bacterial cell wall antigens such as gram-negative lipopolysaccharides may directly stimulate B cells to form hexameric IgM without a J chain. If this is the case, the immunoglobulins formed during primary immune responses are less homogeneous than previously thought.

Immunoglobulin α, or **IgA,** accounts for about 15% of the immunoglobulin pool. Some IgA is present in the serum as a monomer. However, IgA is most abundant in mucous secretions where it is a dimer held together by a J chain (**figure 32.18**). IgA has special features that are associated with secretory mucosal surfaces. IgA, when transported from the mucus-associated lymphoid tissue to mucosal surfaces, acquires a protein called the secretory component. **Secretory IgA (sIgA),** as the modified molecule is called, is the primary immunoglobulin of mucus-

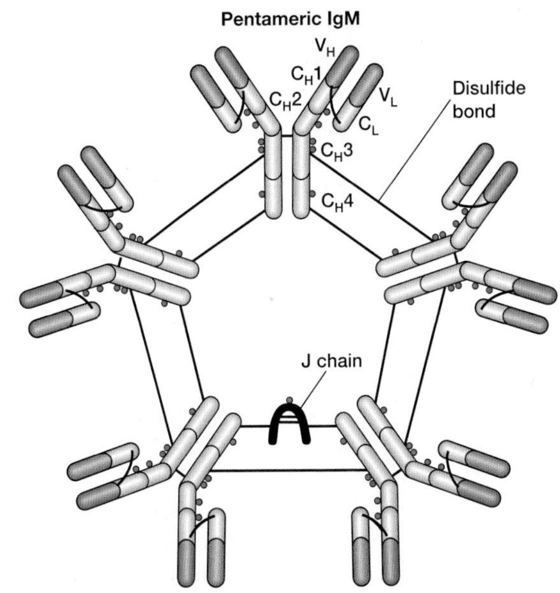

Figure 32.17 Immunoglobulin M. The pentameric structure of human IgM. The disulfide bonds linking peptide chains are shown in black; carbohydrate side chains are in red. Note that 10 antigen-binding sites are present.

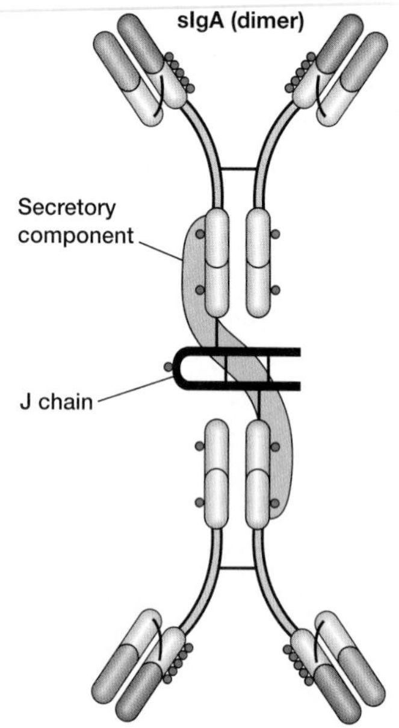

Figure 32.18 Immunoglobulin A. The dimeric structure of human secretory IgA. Notice the secretory component (tan) wound around the IgA dimer and attached to the constant domain of each IgA monomer. Carbohydrate side chains are shown in red.

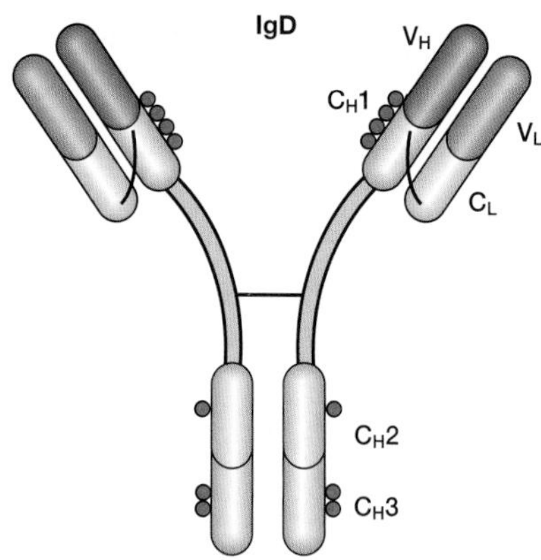

Figure 32.19 Immunoglobulin D. The structure of human IgD. The disulfide bonds linking protein chains are shown in black; carbohydrate side chains are in red.

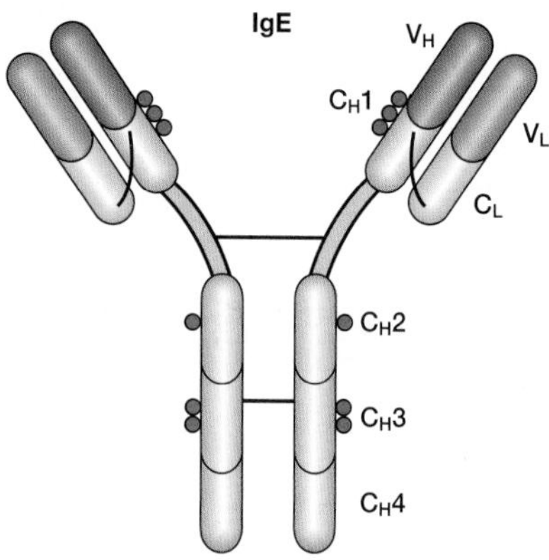

Figure 32.20 Immunoglobulin E. The structure of human IgE.

associated lymphoid tissue. Secretory IgA is also found in saliva, tears, and breast milk. In these fluids and related body areas, sIgA plays a major role in protecting surface tissues against infectious microorganisms by the formation of an immune barrier. For example, in breast milk sIgA helps protect nursing newborns. In the intestine, sIgA attaches to viruses, bacteria, and protozoan parasites such as *Entamoeba histolytica.* This prevents pathogen adherence to mucosal surfaces and invasion of host tissues, a phenomenon known as immune exclusion. In addition, sIgA binds to antigens within the mucosal layer of the small intestine; subsequently the antigen-sIgA complexes are excreted through the adjacent epithelium into the gut lumen. This rids the body of locally formed immune complexes and decreases their access to the circulatory system. Secretory IgA also plays a role in the alternative complement pathway. Chemical mediators of innate immunity: Complement (section 31.6)

Immunoglobulin δ, or **IgD,** is an immunoglobulin found in trace amounts in blood serum. It has a monomeric structure (**figure 32.19**) similar to that of IgG. IgD antibodies do not fix complement and cannot cross the placenta, but they are abundant in combination with IgM on the surface of B cells and thus are part of the B cell receptor complex. Therefore their function is to signal the B cell to start antibody production upon antigen binding.

Immunoglobulin ε, or **IgE** (**figure 32.20**), makes up only a small percent of the total immunoglobulin pool. The classic skin-sensitizing and anaphylactic antibodies belong to this class. The Fc portion of IgE can bind to special Fc_{ε} receptors on mast cells, eosinophils, and basophils. When two IgE molecules on the surface of these cells are cross-linked by binding to the same antigen, the cells degranulate. This degranulation releases histamine and other pharmacological mediators of inflammation. It also stimulates eosinophilia (an excessive number of eosinophils in the blood) and gut hypermotility (increased rate of movement of the intestinal contents), which aid in the elimination of helminthic parasites. Thus although IgE is present in small amounts, this class of antibodies has potent biological capabilities, as is discussed in section 32.11.

1. Explain the different functions of antibody when it is bound to B cells versus when it is soluble in serum.
2. Describe the major functions of each immunoglobulin class.
3. Why is the structure of IgG considered the model for all five immunoglobulin classes?
4. Which immunoglobulin can cross the placenta?
5. Which immunoglobulin is most prevalent in the immunoglobulin pool? The least prevalent?

Antibody Kinetics

The synthesis and secretion of antibody can also be evaluated with respect to time. Monomeric IgM serves as the B-cell receptor for antigen, and pentameric IgM is secreted after B-cell activation. Furthermore, under the influence of T-helper cells (responding to other stimuli), the IgM-secreting plasma cells may stop producing and secreting IgM in favor of another antibody class (IgG, IgA, IgE, for example). This is known as **class switching.** These events take time to unfold.

The Primary Antibody Response

When an individual is exposed to an antigen (for example, an infection or vaccine), there is an initial lag phase, or latent period, of several days to weeks before an antibody response is mounted. During this latent period no antibody can be detected in the blood (**figure 32.21**). Once B cells have differentiated into plasma cells, antibody is secreted and can be detected. This explains why antibody-based HIV tests, for example, are not accurate until weeks after exposure. The **antibody titer,** which is a measurement of serum antibody concentration (the reciprocal of the highest dilution of an antiserum that gives a positive reaction in the test being used), then rises logarithmically to a plateau during the second, or log, phase. In the plateau phase the antibody titer stabilizes. This is followed by a decline phase, during which antibodies are naturally metabolized or bound to the antigen and cleared from the circulation. During the primary antibody response, IgM appears first, then IgG, or another antibody class. The affinity of the antibodies for the antigen's determinants is low to moderate during this primary antibody response.

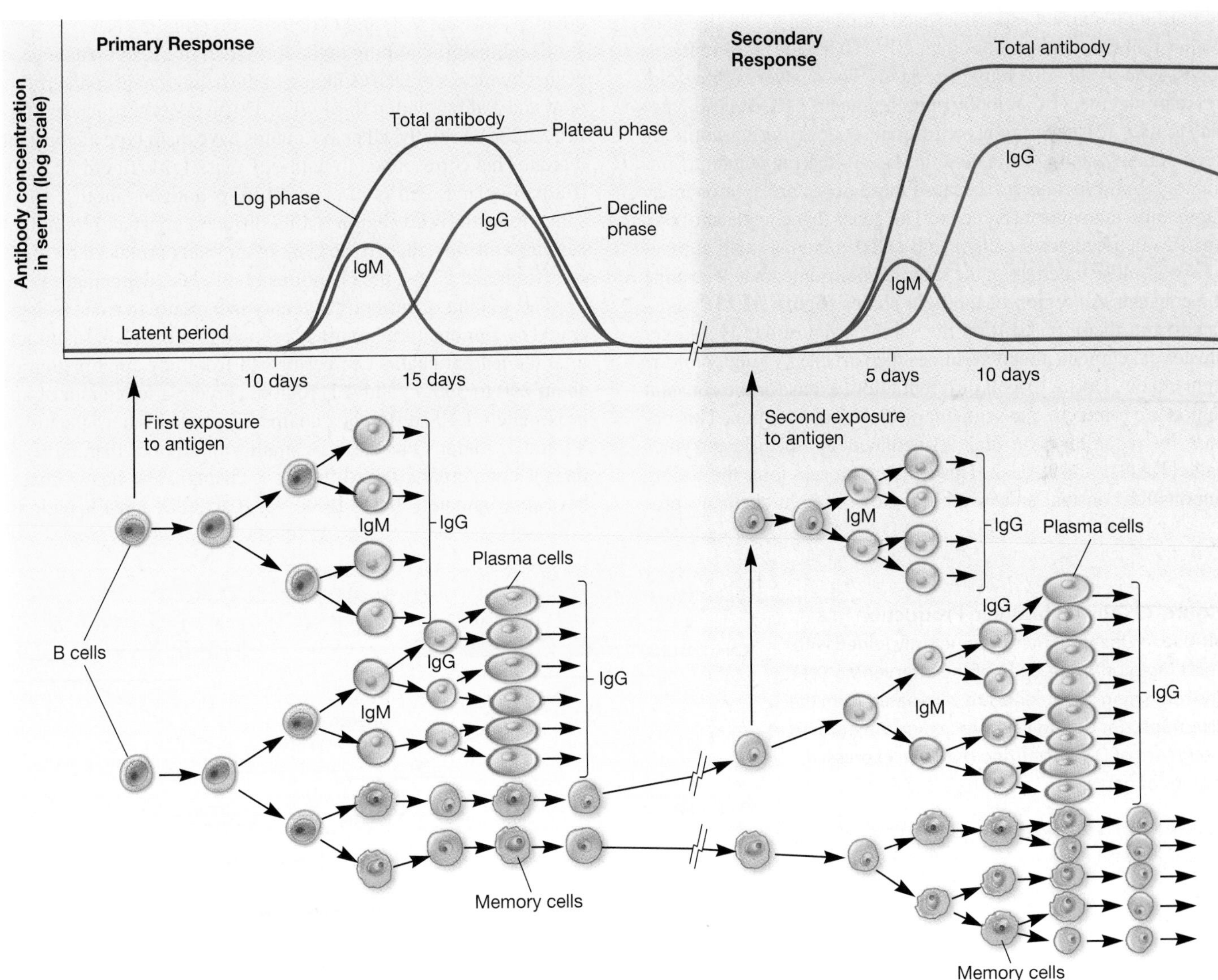

Figure 32.21 Antibody Production and Kinetics. The four phases of a primary antibody response correlate to the clonal expansion of the activated B cell, differentiation into plasma cells, and secretion of the antibody protein. The secondary response is much more rapid and total antibody production is nearly 1,000 times greater than that of the primary response.

The Secondary Antibody Response

The primary antibody response primes the immune system so that it possesses specific immunological memory through its clones of memory B cells. Upon secondary antigen challenge, as occurs when an individual is re-exposed to a pathogen or receives a vaccine booster, B cells mount a heightened secondary, or anamnestic [Greek *anamnesis,* remembrance], response to the same antigen (figure 32.21). Compared to the primary antibody response, the secondary antibody response has a shorter lag phase and a more rapid log phase, persists for a longer plateau period, attains a higher IgG titer, and produces antibodies with a higher affinity for the antigen.

Diversity of Antibodies

One unique property of antibodies is their remarkable diversity. According to current estimates, each human can synthesize antibodies that can bind to more than 10^{13} (10 trillion) different epitopes. How is this diversity generated? The answer is threefold: (1) rearrangement of antibody gene segments, called combinatorial joining (2) generation of different codons during antibody gene splicing, and (3) somatic mutations. Rearrangement of immunoglobulin loci occurs because these genes are split or interrupted into many gene segments. The genes that encode antibody proteins in precursor B cells (Pro B cells) contain a small number of exons, close together on the same chromosome, that determine the constant (C) region of the light chains (**figure 32.22**). Separated from them, but still on the same chromosome, is a larger cluster of segments that determines the variable (V) region of the light chains. During B-cell differentiation, exons for the constant region are joined to one segment of the variable region. This occurs by recombination and is mediated by specific enzymes called RAG-1 and RAG-2. This splicing process joins the coding regions for constant and variable regions of light chains to produce a complete light chain of an antibody. A similar splicing produces a complete heavy-chain antibody gene (figure 32.22).

Because the light-chain genes actually consist of three parts, and the heavy-chain genes consist of four, the formation of a finished antibody molecule is slightly more complicated than previously outlined. The germ line DNA for the light-chain gene contains multiple coding sequences called V and J (joining) regions. During the development of a B cell in the bone marrow, the RAG enzymes join one V gene segment with one J segment. This DNA joining process is termed combinatorial joining because it can create many combinations of the V and J regions. In addition, an enzyme called *t*erminal *d*eoxynucleotidyl *t*ransferase (tdt) inserts nucleotides at the V-J junction, creating additional diversity. When the light-chain gene is transcribed, transcription continues through the DNA region that encodes the constant portion of the gene. RNA splicing subsequently joins the V, J, and C regions, creating mRNA.

Combinatorial joining in the formation of a heavy-chain gene occurs by means of DNA splicing of the heavy-chain counterparts of V and J along with a third set of D (diversity) sequences (**figure 32.23*a***). Initially, all heavy chains have the μ type of constant region. This corresponds to antibody class IgM (figure 32.23*b*). If a particular B cell is re-exposed to its antigen, another DNA splice joins the VDJ region with a different constant region that can subsequently change the class of antibody produced by the B cell (figure 32.22*c*)—the phenomenon of class switching.

The amount of antibody diversity in a mouse that can be generated by combinatorial joining is shown in **table 32.6.** In this animal the κ light chains can be formed from any combination of about 250 to 350 V_κ and 4 J_κ regions, giving a maximum of approximately 1,400 different κ chains. The λ chains have their own V_λ and J_λ regions but they are smaller in number than those of their κ counterparts (six different λ chains). The heavy chains have approximately 250 to 1,000 V_H, 10 to 30 D, and 4 J_H regions,

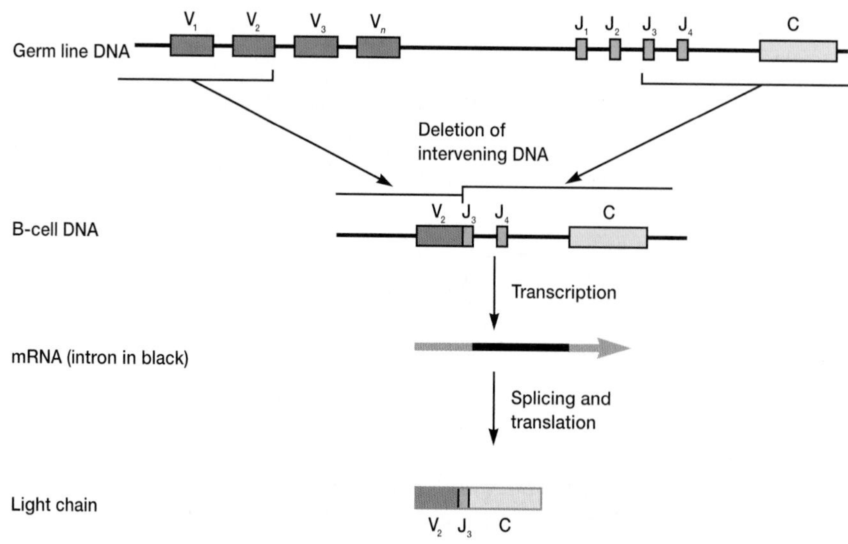

Figure 32.22 Light Chain Production in a Mouse. One V segment is randomly joined with one J-C region by deletion of the intervening DNA. The remaining J segments are eliminated from the RNA transcript during RNA processing. An intron is a segment of DNA occurring between expressed regions of genes.

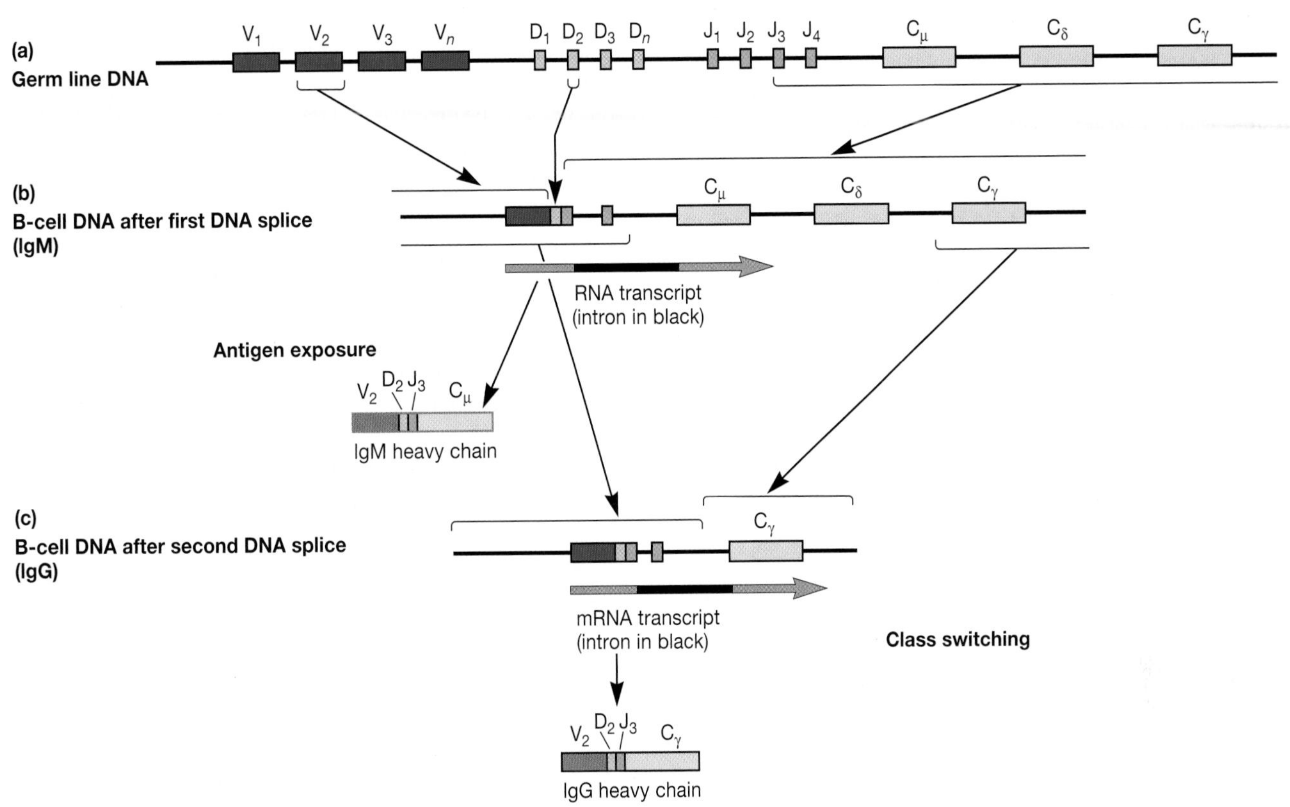

Figure 32.23 The Formation of a Gene for the Heavy Chain of an Antibody Molecule. See text for further details.

Table 32.6 Number of Antibodies Possible through the Combinatorial Joining of Mouse Germ Line Genes[a]

λ light chains	V regions = 2
	J regions = 3
	Combinations = 2 × 3 = 6
κ light chains	V_κ regions = 250−350
	J_κ regions = 4
	Combinations = 250 × 4 = 1,000
	= 350 × 4 = 1,400
Heavy chains	V_H = 250−1,000
	D = 10–30
	J_H = 4
	Combinations = 250 × 10 × 4 = 10,000
	= 1,000 × 30 × 4 = 120,000
Diversity of antibodies	κ-containing: 1,000 × 10,000 = 10^7
	1,400 × 120,000 = 2×10^8
	λ-containing: 6 × 10,000 = 6×10^4
	6 × 120,000 = 7×10^5

[a]Approximate values.

giving a maximum of 120,000 different combinations. Because any light chain can combine with any heavy chain, there will be a maximum of 2×10^8 possible κ chain antibody types. However, this value is actually an underestimate because antibody diversity is further augmented by two processes:

1. Splice-site variability: The junction for either VJ or VDJ splicing in combinatorial joining can occur between different nucleotides and thus generate different codons in the spliced gene. In addition, the activation of tdt can greatly increase the variability of the nucleotide sequence at the VJ or VDJ junctions during the splicing process. For example, one VJ splicing event can join the V sequence CCTCCC with the J sequence TGGTGG in two ways: CCTCCC + TGGTGG = CCGTGG, which codes for proline and tryptophan. Alternatively, the V_J splicing event can give rise to the sequence CCTCGG, which codes for proline and arginine. Thus the same V_J joining could produce polypeptides differing in a single amino acid.
2. Somatic mutation of V regions: The V regions of germ-line DNA are susceptible to a high rate of somatic mutation during B-cell development in response to an antigen challenge. These mutations allow B-cell clones to produce antibodies with somewhat different polypeptide sequences.

1. How many chromosomes encode for antibody production in humans?
2. What is the name of each part of the gene that encodes for the different regions of antibody chains?
3. Describe what is meant by combinatorial joining of V, D, and J gene segments.
4. In addition to combinatorial joining, what other two processes play a role in antibody diversity?

Clonal Selection

As noted previously, combinatorial joinings, somatic mutations, and variations in the splicing process generate the great variety of antibodies produced by mature B cells. From a large, diverse B-cell pool, specific cells are stimulated by antigens to reproduce and form B-cell clones containing the same genetic information. This is known as **clonal selection;** this explains immunological specificity and memory.

The clonal selection theory has four components or tenets. The first tenet is that there exists a pool of lymphocytes that can bind to a tremendous range of antigenic epitopes (**figure 32.24**). The process of how antibody diversity is generated for B cells is well understood. Because some of the B cells generated by this process will produce antibodies that can react with self-epitopes, the second tenet of the theory is that these self-reactive cells are eliminated at an early stage of development. Indeed, this has been shown to be true for developing B cells (in the bone marrow) and self-reacting T cells (in the thymus). The third tenet is that once a lymphocyte has been released into the body and is exposed to its specific antigen, it proliferates to form a **clone** (a population of identical cells derived from a single parent cell). Note that this clone has been "selected" by exposure to specific antigen, hence the name of the theory. The final tenet states that all clonal cells react with the same antigenic epitope that stimulated its formation. However, the cells may differentiate to have somewhat different functions. Figure 32.24 shows this process for a B cell, which, after proliferating in response to antigen exposure, forms two different cell populations: antibody-producing plasma cells and memory cells.

Plasma cells are literally protein factories that produce about 2,000 antibodies per second in their brief five- to seven-day life span. Memory B cells can initiate the antibody-mediated immune

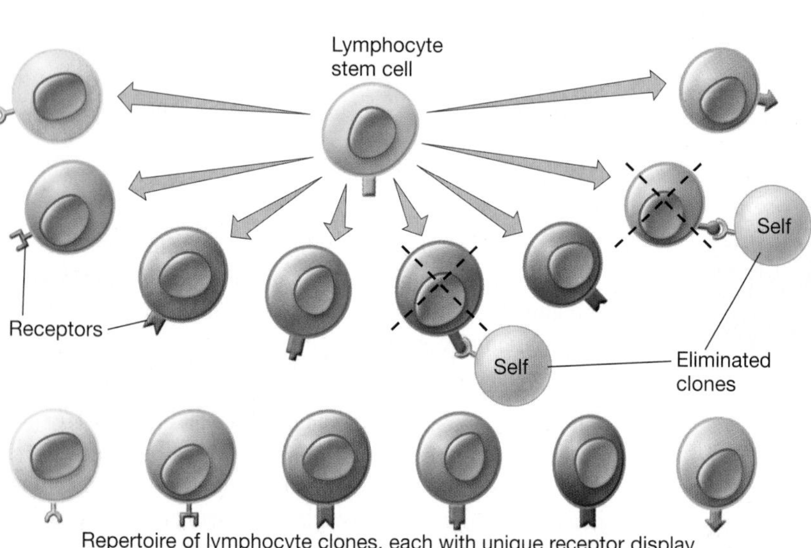

(a) Antigen-Independent Period

1 During development of early lymphocytes from stem cells, a given stem cell undergoes rapid cell division to form numerous progeny.

During this period of cell differentiation, random rearrangements of the genes that code for cell surface protein receptors occur. The result is a large array of genetically distinct cells, called clones, each clone bearing a different receptor that is specific to react with only a single type of foreign molecule or antigen.

2 At the same time, any lymphocyte clones that develop a specificity for self molecules and could be harmful are eliminated from the pool of diversity. This is called immune tolerance.

3 The specificity for a single antigen molecule is programmed into the lymphocyte and is set for the life of a given clone. The end result is an enormous pool of mature but naïve lymphocytes that are ready to further differentiate under the influence of certain organs and immune stimuli.

(b) Antigen-Dependent Period

4 Lymphocytes come to populate the lymphatic organs, where they will finally encounter antigens. These antigens will become the stimulus for the lymphocytes' final activation and immune function. Entry of a specific antigen selects only the lymphocyte clone or clones that carry matching surface receptors. This will trigger an immune response, which varies according to the type of lymphocyte involved.

Figure 32.24 Lymphocyte Clonal Expansion. **(a)** Cell populations expand and are restricted based on their MHC expression. **(b)** They are further expanded when activated by a specific antigen.

response upon detecting the particular antigen specific for their B cell receptors (BCRs). These memory cells circulate more actively from blood to lymph and live much longer (years or even decades) than do plasma cells. Memory cells are responsible for the immune system's rapid secondary antibody response (figure 32.21) to the same antigen. Finally, memory B cells and plasma cells are usually not produced unless the B cell has interacted with, and received cytokine signals from, activated T-helper cells (figure 32.11). In addition to providing a theoretical basis for understanding how the adaptive immune system can amplify its responses to specific antigens, clonal selection is now widely accepted as the explanation for the differences between primary and secondary antibody responses. It has also led to the development of hybridoma (**monoclonal antibody [mAb]**) technology (**Techniques & Applications 32.2**).

32.8 Action of Antibodies

The antigen-antibody interaction is a bimolecular association that exhibits exquisite specificity. The in vivo interactions that occur in vertebrate animals are absolutely essential in protecting the animal against the continuous onslaught of viruses, other microorganisms and their products, and cancer cells. This occurs partly because the antibody coats the invading foreign material, marking it for enhanced recognition by other components of the innate and adaptive immune systems. The mechanism by which antibodies achieve this are now discussed.

Neutralization

Some bacteria produce extracellular toxins that contribute to their pathogenic effects. Immunity to such a disease (e.g., diphtheria or anthrax) depends on the production of specific antibodies that inactivate the toxins produced by the bacteria. This process is called **toxin neutralization** (**figure 32.25**). Once neutralized, the toxin-antibody complex is either unable to attach to receptor sites on host target cells and is unable to enter the cell, or it is ingested by macrophages. For example, diphtheria toxin, a heterodimer, inhibits protein synthesis after binding to the cell surface by its B subunit and subsequent passage of its active A subunit into the cytoplasm of the target cell. The antibody blocks the toxic effect by inhibiting the entry of the A subunit or the binding of the B fragment. An antibody capable of neutralizing a toxin or antiserum containing neutralizing antibody against a toxin is called **antitoxin.**

Toxigenicity: AB toxins (section 33.4)

IgG, IgM, and IgA antibodies can bind to some viruses during their extracellular phase and inactivate them. This antibody-mediated viral inactivation is called **viral neutralization.** Fixation of the classical pathway complement component C3b to a virus aids the neutralization process. Viral neutralization prevents a viral infection due to the inability of the virus to bind to and enter its target cell.

The capacity of bacteria to colonize the mucosal surfaces of mammalian hosts is dependent in part on their ability to adhere to mucosal epithelial cells. Secretory IgA (sIgA) antibodies inhibit certain bacterial adherence-promoting factors. Thus sIgA can protect the host against infection by some pathogenic bacteria, and perhaps by other microorganisms on mucosal surfaces by neutralizing their adherence to host cells.

Immune reactions against protozoan and helminthic parasites are only partially understood. Parasites that have a tissue-invasive phase in their life cycle often are associated with both eosinophilia and elevated IgE levels. Evidence suggests that, in the presence of elevated IgE, eosinophils can bind to the parasites and discharge their granules. Degranulation releases lytic and inflammatory mediators that neutralize and even kill parasites.

Opsonization

Phagocytes have an intrinsic ability to bind directly to microorganisms by nonspecific cell surface receptors, engulf the microorganisms, form phagosomes, and digest the microorganisms. This phagocytic process can be greatly enhanced by opsonization. As noted in section 31.6, opsonization is the process by which microorganisms or other foreign particles are coated with antibody and/or complement, and thus prepared for "recognition" and ingestion by phagocytic cells. Opsonizing antibodies, especially IgG1 and IgG3, bind to Fc receptors on the surface of macrophages and neutrophils. This binding provides the phagocyte with a method for the specific capture of antigens. In other words, the antibody forms a bridge between the phagocyte and the antigen thereby increasing the likelihood of its phagocytosis (*see figure 31.21*).

Immune Complex Formation

Because antibodies have at least two antigen-binding sites and most antigens have at least two antigenic determinants, cross-linking can occur, producing large aggregates termed **immune complexes** (figure 32.25). If the antigens are soluble molecules and the complex becomes large enough to settle out of solution, a **precipitation** [Latin *praecipitare,* to cast down] or **precipitin reaction** occurs, and is caused by a **precipitin** antibody. When the immune complex involves the cross-linking of cells or particles, an **agglutination reaction** occurs and the responsible antibody is an **agglutinin.** These immune complexes are more rapidly phagocytosed in vivo than are free antigens. The extent of immune complex formation, whether within an animal or in vitro, depends on the relative concentrations of the precipitin antibody and antigen. If there is a large excess of antibody, separate antibody molecules usually bind to each antigenic determinant and a less insoluble network or lattice forms.

When antigen is present in excess, two separate antigen molecules tend to bind to each antibody and network development or cross-linking is inhibited. The ratio of antibody and antigen is said to be in the equivalence zone when their concentration is optimal for the formation of a large network of interconnected antibody and antigen molecules. All antibody and antigen molecules precipitate or agglutinate as an insoluble complex. Precipitin reactions can occur in both solutions and agar gel media. In either case, antibody-antigen equivalence is required for optimal results.

Techniques & Applications

32.2 Monoclonal Antibody Technology

The value of antibodies as tools for locating or identifying antigens is well established. For many years, antiserum extracted from human or animal blood was the main source of antibodies for tests and therapy, but most antiserum has a basic problem. It contains polyclonal antibodies, meaning it is a mixture of different antibodies because it reflects dozens of immune reactions from a wide variety of B-cell clones. This characteristic is to be expected, because several immune reactions may be occurring simultaneously, and even a single species of microbe can stimulate several different types of antibodies. Certain applications in immunology require a pure preparation of monoclonal antibodies (mAbs) that originate from a single clone and have a single specificity for antigen.

The technology for producing monoclonal antibodies is possible by hybridizing cancer cells and activated B cells in vitro. This technique began with the discovery that tumors isolated from multiple myelomas in mice consist of identical plasma cells. These monoclonal plasma cells secrete a strikingly pure form of antibody with a single specificity and continue to divide indefinitely. Immunologists recognized the potential in these plasma cells and devised a hybridoma approach to creating mAb. The basic idea behind this approach is to hybridize or fuse a myeloma cell with a normal plasma cell from a mouse spleen to create an immortal cell that secretes a supply of functional antibodies with a single specificity.

The introduction of this technology has the potential for numerous biomedical applications. Monoclonal antibodies have provided immunologists with excellent standardized tools for studying the immune system and for expanding disease diagnosis and treatment. Most of the successful applications thus far use mAbs in in vitro diagnostic testing and research. Although injecting monoclonal antibodies to treat human disease is an exciting prospect, this therapy has been stymied because most mAbs are of mouse origin, and many humans will develop hypersensitivity to them. However, using genetic engineering, human antibody constant regions are cloned to mouse antibody-binding regions to create a hybrid antibody that is highly specific but less likely to cause hypersensitivity reactions.

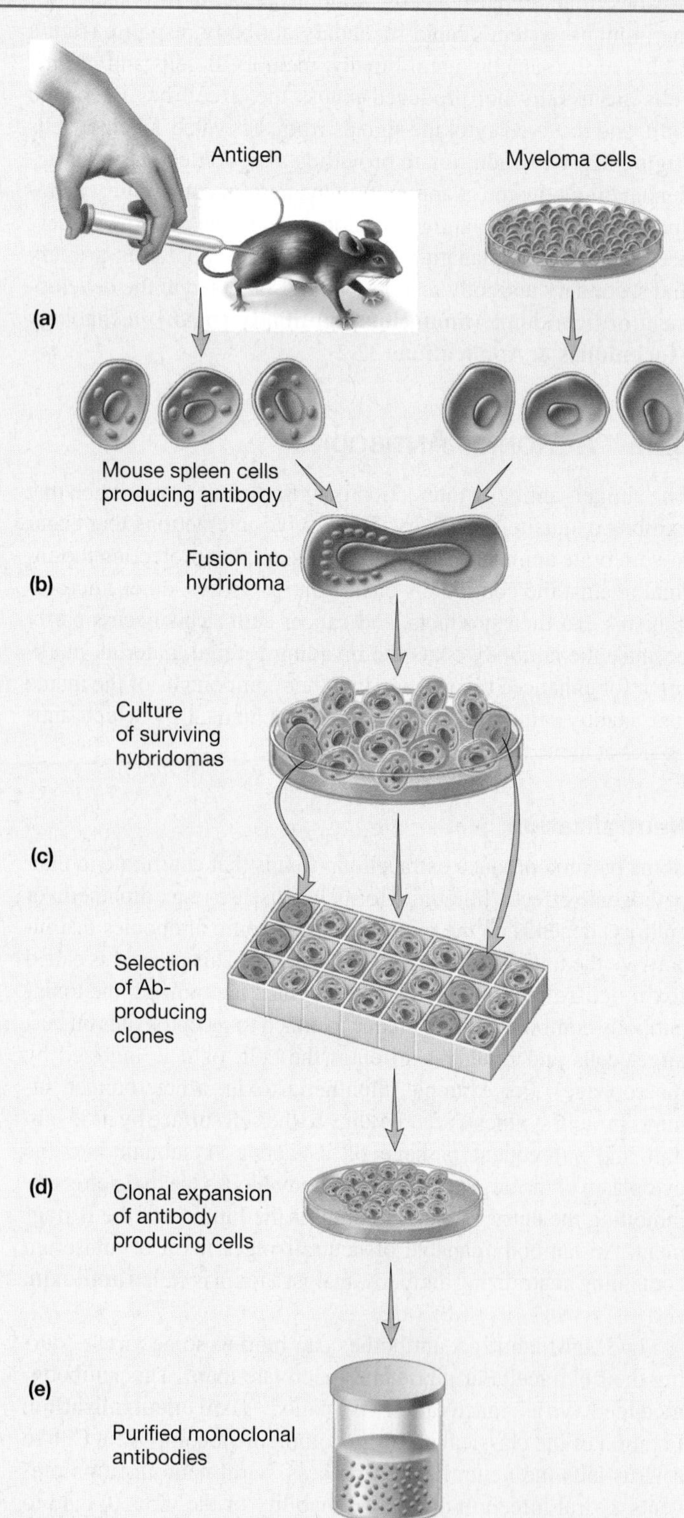

Monoclonal Antibody Formation. **(a)** A mouse is inoculated with an antigen having the desired specificity, and activated cells are isolated from its spleen. A special strain of mouse provides the myeloma cells. **(b)** The two cell populations are mixed with polyethylene glycol, which causes some cells in the mixture to fuse and form hybridomas. **(c)** Surviving cells are cultured and separated into individual wells. **(d)** Tests are performed on each hybridoma to determine specificity of the antibody (Ab) it secretes. **(e)** A hybridoma with the desired specificity is grown in tissue culture; antibody is then isolated and purified.

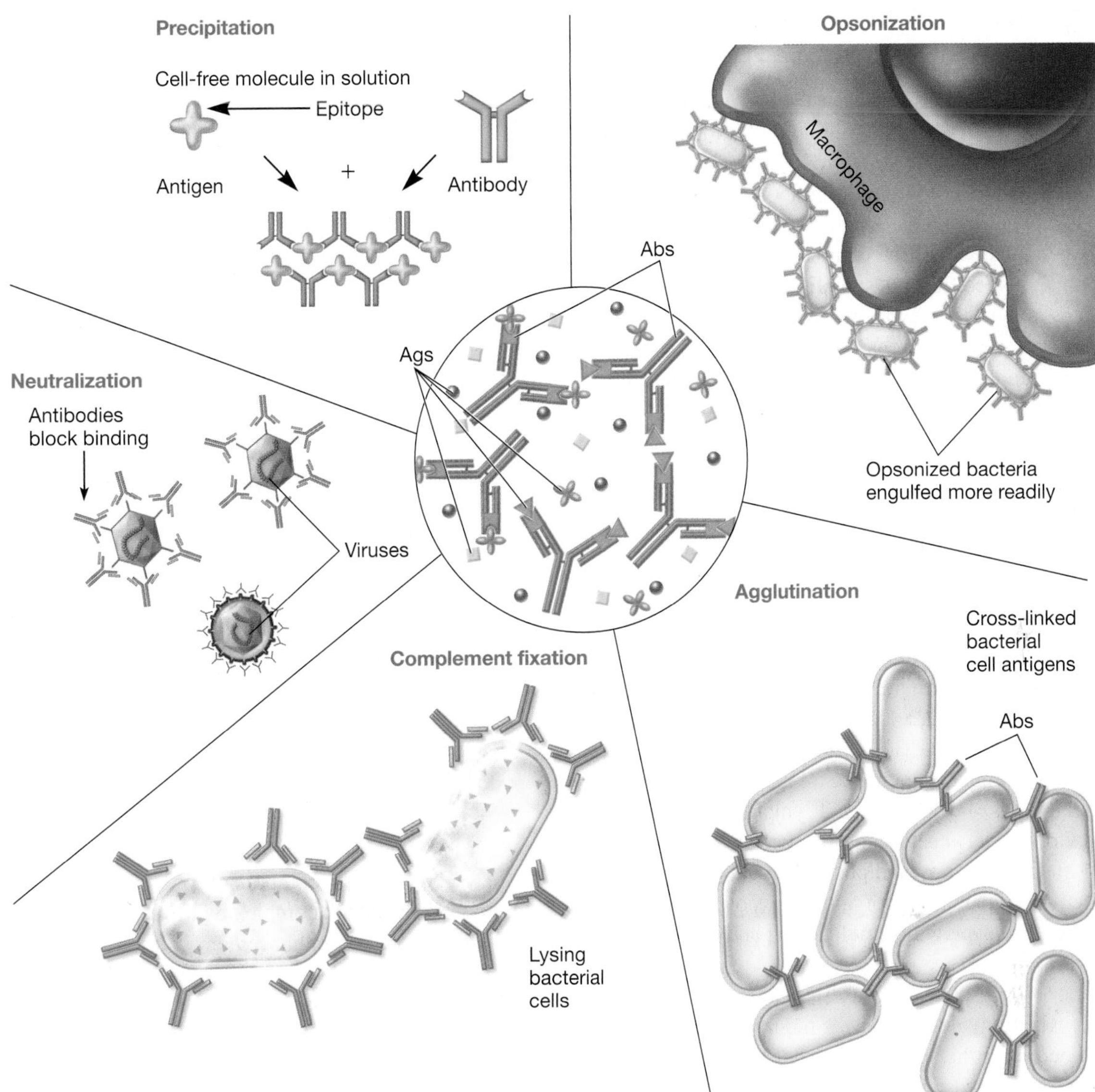

Figure 32.25 Consequences of Antigen-Antibody Binding. Immune complexes can form when soluble antigens (Ag) bind soluble antibody (Ab), resulting in precipitation. Opsonization occurs when antibody binds to antigens on larger molecules or cells to be recognized by phagocytic cells. Agglutination results when insoluble antigens (like viral or bacterial cells) are cross-linked by antibody. The classical complement cascade can be activated by immune complexes. Neutralization results when antibody binds to antigens, preventing the antigen from binding to host cells.

Outside the animal body (in vitro), this same specificity has led to the development of a variety of immunological assays that can detect the presence of either antibody or antigen. These assays are important in the diagnosis of diseases; in the identification of specific viruses, bacteria, and parasites; in monitoring the level of the humoral response and immunologic problems; and in identifying molecules of medical and biological interest. Immunological assays differ in their speed and sensitivity; some are qualitative whereas others are quantitative. Clinical immunology: Immunoprecipitation (section 35.3)

1. How does toxin neutralization occur? Viral neutralization?
2. How does adherence inhibition occur?
3. Describe an immune complex. What two types are formed?

32.9 Summary: The Role of Antibodies and Lymphocytes in Immune Defense

Chapter 31 presents nonspecific (innate) host immunity and this chapter describes specific (adaptive) immunity. Although both the humoral and cellular arms of the specific immune response have been considered separately, it is important to understand that the host response to any particular pathogen may involve complex interactions between the host and the pathogen, as well as the components of both nonspecific and specific immunity. The next section summarizes the defense mechanisms vertebrate hosts use against viral and bacterial pathogens. At times, however, these defenses are not enough to protect the host because pathogens have evolved mechanisms to circumvent many of the host's defenses. This is the subject of chapter 33.

Immunity to Viral Infections

Resistance to viral infections involves humoral immunity, interferon sensitization of host cells, and cell-mediated immunity.

1. Interferons are important in resistance in the early stage of viral infection, as in the case of colds and influenza. Interferon-stimulated cells shut down viral protein synthesis and destroy viral mRNA (*see figure 31.25*). Some interferons also stimulate the activity of T cells (figure 32.7) and natural killer cells, thus accelerating the immune response to a viral infection. Natural killer cells (NK cells) are non-B, non-T lymphocytes capable of destroying virus-infected cells and cancer cells. They are components of innate immunity and possess no antigen receptors. They are active without any prior antigen exposure. Interferon and antibodies, however, will stimulate their activity. Cells, tissues, and organs of the immune system (section 31.2)
2. Cell-mediated immunity to viruses is a major resistance mechanism when enveloped viruses modify host cell membranes and bud from the surface (e.g., herpesvirus, poxvirus, influenza, mumps, measles, rabies, and rubella viruses). Activated lymphocytes can recognize and destroy virus-infected cells by detecting changes in surface molecules. Cytotoxic T lymphocytes (CTLs) destroy virus-infected cells by inducing apoptosis of the infected cell through the release of the CD95L peptide (FasL), and the production of granzymes and perforin, which form channels through the plasma membrane of infected cells, resulting in cytolysis (figure 32.8). The class I MHC proteins are involved in T-cell recognition of infected cells (figure 32.7). Cells displaying both viral antigens and the proper class I MHC will be destroyed. CTLs are also involved in the destruction of cancer cells, a process known as **immune surveillance.**
3. At some point in the infection, viruses are released and can be detected by macrophages and other immune system cells. These extracellular viruses can trigger humoral responses. Binding of antibodies to virus particles has two important outcomes. Antibody binding can neutralize viruses, thereby interfering with their adsorption and entrance into host cells (figure 32.5). This limits spread of the infection. Antibodies also act as opsonins (*see figure 31.21*) and enhance phagocytosis of the viruses. Phagocytosis (section 31.3); Chemical mediators in nonspecific (innate) resistance: Complement (section 31.6)

Immunity to Bacterial Infections

If a bacterium successfully breaches the physical barriers (skin and mucous membranes) that serve as the host's first line of defense, then other innate defenses, as well as specific immune responses are elicited. Inflammation, complement activation, and humoral immunity are more important than cell-mediated immunity for those bacteria that remain outside host cells. However, for intracellular bacterial pathogens, cell-mediated responses are also important.

1. The inflammatory response helps destroy bacterial pathogens. In addition, it recruits macrophages to the site of bacterial invasion. These APCs not only ingest the bacteria, but also signal its presence to T-helper lymphocytes.
2. T_H2 cells are formed. They help activate B cells, triggering humoral responses. When antibodies bind the bacteria, several outcomes are possible: (a) opsonization, (b) agglutination, (c) neutralization, and (d) complement activation. IgG is an opsonin that aids in the phagocytgosis of bacteria by macrophages and granulocytes. IgM and IgG agglutinate bacterial pathogens, thus limiting their spread and enhancing the efficiency of phagocytosis (figure 32.25). Some antibodies act as antitoxins and neutralize bacterial exotoxins—toxins that are secreted by bacteria. Activation of complement by the classical pathway can result in opsonization of the bacteria by the C3b and C4b components of the system, formation of the membrane attack complex by the C5b-9 components (*see figure 31.25*), and enhancement of the inflammatory response by C3a, C5a, and $\overline{C5b67}$. These complement proteins attract neutrophils and macrophages to the site of the infection.
3. Cell-mediated responses by activated macrophages and T cells (figure 32.7) are also important, particularly in resisting intracellular bacterial pathogens. Activated T cells secrete several cytokines that have a variety of effects. Among these (a) interferon-gamma (IFN-γ) is a major factor that stimulates macrophages to become "angry" and more effectively phagocytose and destroy pathogens; (b) the macrophage chemotactic factor and migration inhibition factor attract more macrophages and keep them in the area of infection after arrival; and (c) interleukin-2 (IL-2) stimulates the proliferation of activated T cells to increase the population of cells involved in the cell-mediated immune response. It also increases the effectiveness of cytotoxic T cells and NK cells by promoting the synthesis of other cytokines by T cells.

32.10 Acquired Immune Tolerance

Acquired immune tolerance is the body's ability to produce T cells and antibodies against nonself antigens such as microbial antigens, while "tolerating" (not responding to) self-antigens. Some of this tolerance arises early in embryonic life when immunologic competence is being established. Three general tolerance mechanisms have been proposed: (1) negative selection by clonal deletion, (2) the induction of anergy, and (3) inhibition of the immune response by T cells with suppressor/regulatory function.

Negative selection is one mechanism that produces immunologic tolerance. Negative selection by clonal deletion removes from the immune system lymphocytes that recognize any self antigens that are present. These cells are eliminated by apoptosis, or programmed cell death. T-cell tolerance induced in the thymus and B-cell tolerance in the bone marrow is called **central tolerance.** However, another mechanism is needed to prevent immune reactions against self-antigens, termed autoimmunity, because many antigens are tissue-specific and are not present in the thymus or bone marrow.

Mechanisms occurring elsewhere in the body are collectively referred to as **peripheral tolerance.** These supplement central tolerance. Peripheral tolerance is thought to be based largely on incomplete activation signals given to the lymphocyte when it encounters self-antigens in the periphery of the body. This mechanism leads to a state of unresponsiveness called **anergy** (immunologists describe an inactive lymphocyte as "anergic" from the Greek *an,* negative, and *ergon,* work), which is associated with impaired intracellular signaling and apoptosis. Many autoreactive B cells undergo clonal deletion or become anergic as they mature in the bone marrow. Negative selection occurs in the bone marrow if the B cells encounter large amounts of self-antigen, either in the soluble phase or as cell membrane constituents. The deletion of self-reactive B cells also takes place in secondary lymphoid tissue such as the spleen and lymph nodes. Since B cells recognize native antigen, there is no need for the participation of MHC molecules in these processes. For those self-antigens present at relatively low concentrations, immunologic tolerance is often maintained only within the T-cell population. This is sufficient to sustain tolerance because it denies the help essential for antibody production by self-reactive B cells. T cells with suppressor activity have been defined as cells that can specifically inhibit responses of other T cells in an antigen-specific manner, although their existence has not been conclusively proven.

1. Describe the three ways acquired immune tolerance develops in the vertebrate host.
2. How would you define anergy?

32.11 Immune Disorders

Like any system in a vertebrate animal, disorders (malfunctions) also occur in the immune system. Immune disorders can be categorized as hypersensitivities, autoimmune diseases, transplantation (tissue) rejection, and immunodeficiencies. Each of these immune disorders is now discussed.

Hypersensitivities

Hypersensitivity is an exaggerated immune response that results in tissue damage and is manifested in the individual on a second or subsequent contact with an antigen. Hypersensitivity reactions can be classified as either immediate or delayed. Obviously immediate reactions appear faster than delayed ones, but the main difference between them is the nature of the immune response to the antigen. Realizing this fact in 1963, Peter Gell and Robert Coombs developed a classification system for reactions responsible for hypersensitivities. Their system correlates clinical symptoms with information about immunologic events that occur during hypersensitivity reactions. The Gell-Coombs classification system divides hypersensitivity into four types: I, II, III, and IV.

Type I Hypersensitivity

An **allergy** [Greek *allos,* other and *ergon,* work] is one kind of **type I hypersensitivity** reaction. Allergic reactions occur when an individual who has produced IgE antibody in response to the initial exposure to an antigen (**allergen**) subsequently encounters the same allergen. Upon initial exposure to a soluble allergen, B cells are stimulated to differentiate into plasma cells and produce specific IgE antibodies with the help of T_H cells (**figure 32.26**). This IgE is sometimes called a **reagin,** and the individual has a hereditary predisposition for its production. Once synthesized, IgE binds to the Fc receptors of mast cells (basophils and eosinophils can also be bound) and sensitizes these cells, making the individual sensitized to the allergen. When a subsequent exposure to the allergen occurs, the allergen attaches to the surface-bound IgE on the sensitized mast cells, causing mast cell degranulation.

Degranulation releases physiological mediators such as histamine, leukotrienes, heparin, prostaglandins, PAF (platelet-activating factor), ECF-A (eosinophil chemotactic factor of anaphylaxis), and proteolytic enzymes. These mediators trigger smooth muscle contractions, vasodilation, increased vascular permeability, and mucus secretion (figure 32.26). The inclusive term for these responses is **anaphylaxis** [Greek *ana,* up, back again, and *phylaxis,* protection]. Anaphylaxis can be divided into systemic and localized reactions.

Systemic anaphylaxis is a generalized response that occurs when an individual sensitized to an allergen receives a subsequent exposure to it. The reaction is immediate due to a sudden burst in the release of mast cell mediators. Usually there is respiratory impairment caused by smooth muscle constriction in the bronchioles. The arterioles dilate, which greatly reduces arterial blood pressure and increases capillary permeability with rapid loss of fluid into the tissue spaces. These physiological changes can be rapid and severe enough to be fatal within a few minutes from reduced venous return, asphyxiation, reduced blood pressure, and circulatory shock. Common examples of allergens that can produce systemic anaphylaxis include drugs (penicillin), passively administered antisera, peanuts, and insect venom from the stings or bites of wasps, hornets, or bees.

Localized anaphylaxis is called an **atopic** ("out of place") **reaction.** The symptoms that develop depend primarily on the route by which the allergen enters the body. **Hay fever** (allergic rhinitis) is a good example of atopy involving the upper respiratory tract. Initial exposure involves airborne allergens—such as plant pollen, fungal spores, animal hair and dander, and house dust mites—that sensitize mast cells located within the mucous membranes of the respiratory tract. Re-exposure to the allergen causes

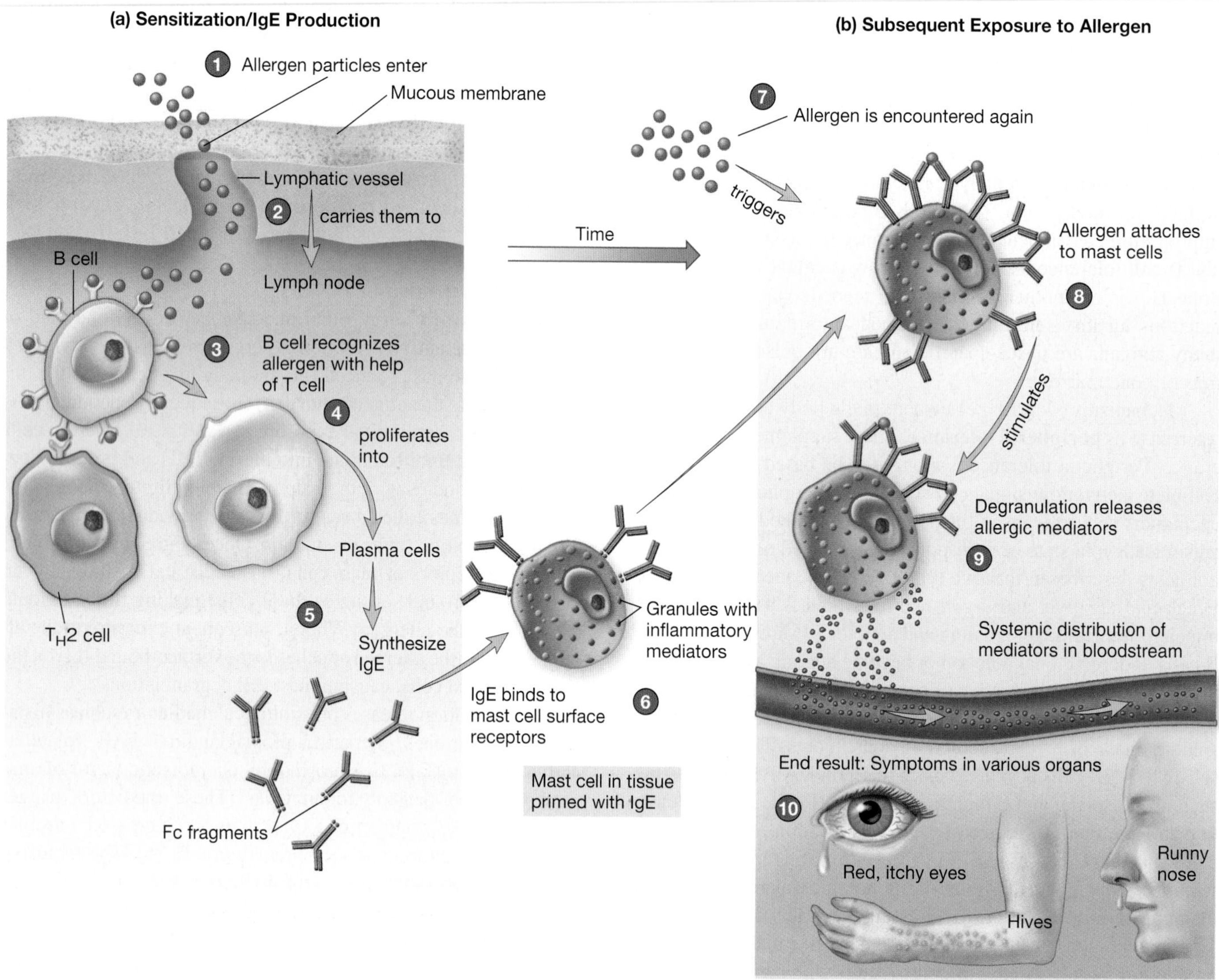

Figure 32.26 Type I Hypersensitivity (Allergic Response). **(a)** The initial contact (sensitization) of lymphocytes by small-protein allergens at mucous membranes results in T_H2 cell-assisted antibody class switching; plasma cells secrete IgE antibody. The IgE binds to its receptor on tissue mast cells (1–6). **(b)** Subsequent exposure to the same allergens results in their capture by the cell-bound IgE (7), triggering mast cell degranulation (8, 9). Characteristic signs and symptoms (hives, swelling, itching, etc.) of allergy ensue (10).

a localized anaphylactic response: itchy and tearing eyes, congested nasal passages, coughing, and sneezing. Antihistamine drugs are used to help alleviate these symptoms.

Bronchial asthma is an example of atopy involving the lower respiratory tract. Common allergens can be the same as for hay fever. In bronchial asthma, however, the air sacs (alveoli) become over-distended and fill with fluid and mucus; the smooth muscle contracts and narrows the walls of the bronchi. Bronchial constriction may produce a wheezing or whistling sound during exhalation. Symptomatic relief is obtained from bronchodilators, which help to relax the bronchial muscles, and from liquefacients and expectorants, which dissolve and expel mucus plugs that accumulate, respectively.

Allergens that enter the body through the digestive system may cause food allergies. **Hives** (eruptions of the skin) are a good diagnostic sign of a true food allergy. Once established, type I food allergies are usually permanent but can be partially controlled with antihistamines or by avoidance of the allergen. Skin testing can be used to identify the antigen responsible for allergies. These tests involve inoculating small amounts of suspect allergen(s) into the skin. Sensitivity to the antigen is shown by a rapid inflammatory reaction characterized by redness, swelling, and itching at the site of inoculation (**figure 32.27**). The affected area in which the allergen-mast cell reaction takes place is called a wheal and flare reaction site. Once the responsible allergen has been identified, the individual should avoid contact with it. If this is not possible, **desen-**

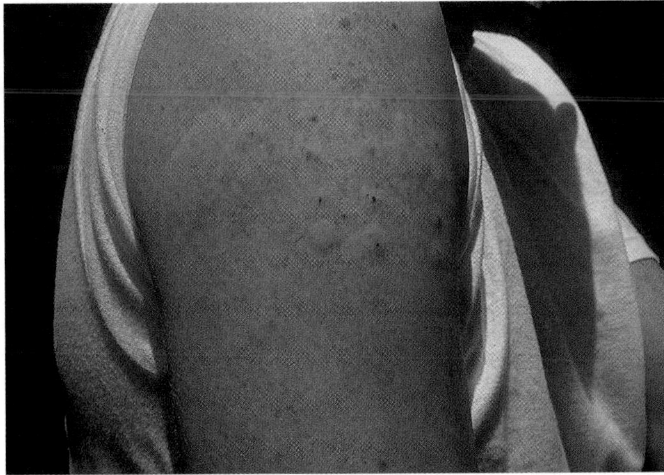

(a)

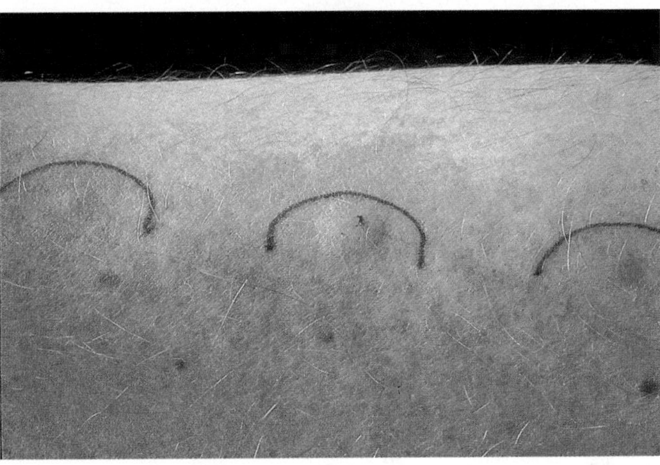

(b)

Figure 32.27 In Vivo Skin Testing. **(a)** Skin prick tests with grass pollen in a person with summer hay fever. Notice the various reactions with increasing dosages (from top to bottom). **(b)** Skin patch test. The surface of the skin (left) is abraded and the suspect allergic extract placed on the skin. After 48 hours (center) it is eczematous and positive for the suspect antigen.

sitization is warranted. This procedure consists of a series of allergen doses injected beneath the skin to stimulate the production of IgG antibodies rather than IgE antibodies. The circulating IgG antibodies can then act to intercept and neutralize allergens before they have time to react with mast cell-bound IgE. Desensitizations are about 65 to 75% effective in individuals whose allergies are caused by inhaled allergens.

Type II Hypersensitivity

Type II hypersensitivity is generally called a cytolytic or cytotoxic reaction because it results in the destruction of host cells, either by lysis or toxic mediators. In type II hypersensitivity, IgG or IgM antibodies are inappropriately directed against cell surface or tissue-associated antigens. They usually stimulate the classical complement pathway and a variety of effector cells (**figure 32.28**). The antibodies interact with complement (Clq) and the effector cells through their Fc regions. The damage mechanisms are a reflection of the normal physiological processes involved in interaction of the immune system with pathogens. Classical examples of type II hypersensitivity reactions are the response exhibited by a person who receives a transfusion with blood from a donor with a different blood group and erythroblastosis fetalis.

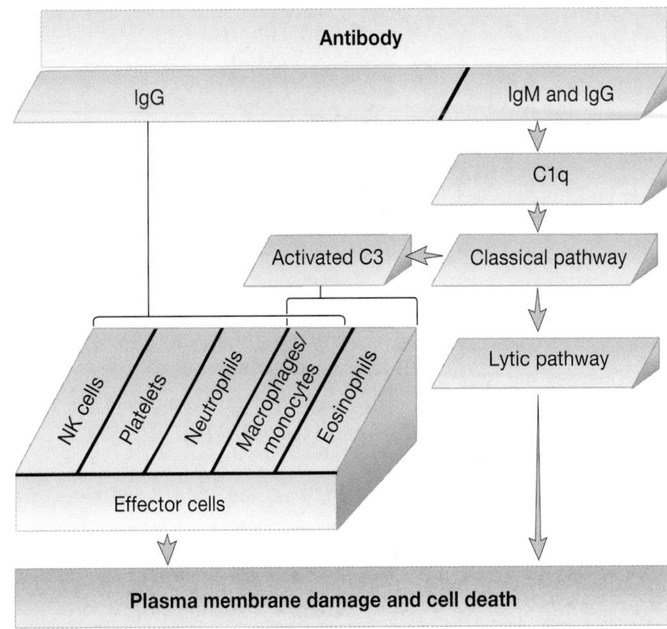

Figure 32.28 Type II Hypersensitivity. The action of antibody occurs through effector cells or the membrane attack complex, which damages target cell plasma membranes, causing cell destruction.

Blood transfusion was often fatal prior to the discovery of distinct blood types by Karl Landsteiner in 1904. Landsteiner's observation that sera from one person could agglutinate the blood cells of another person led to his identification of four distinct types of human blood. The red blood cell types were subsequently determined to result from cell surface glycoproteins, now called the ABO blood groups (**figure 32.29*a***). The four types are genetically inherited as two (out of three alternative) alleles, so called A, B, or O alleles respectively encoding the A- or B-type glycoprotein, or no glycoprotein at all. Thus homozygous expression of A or B alleles results in type A or type B blood, respectively. Heterozygous expression of the co-dominant A and B alleles results in type AB blood. Heterozygous expression of the dominant A or B alleles with the O allele results in type A or type B blood, respectively. Homozygous expression of the O allele results in type O blood. AB glycoproteins are self-antigens. Thus AB reactive lymphocytes of the developing host are destroyed during the negative selection process. However, the lymphocytes specific for AB glycoproteins not expressed by the host remain to be activated upon exposure to those specific antigens (which are ubiquitously distributed throughout nature). Consequently, type A

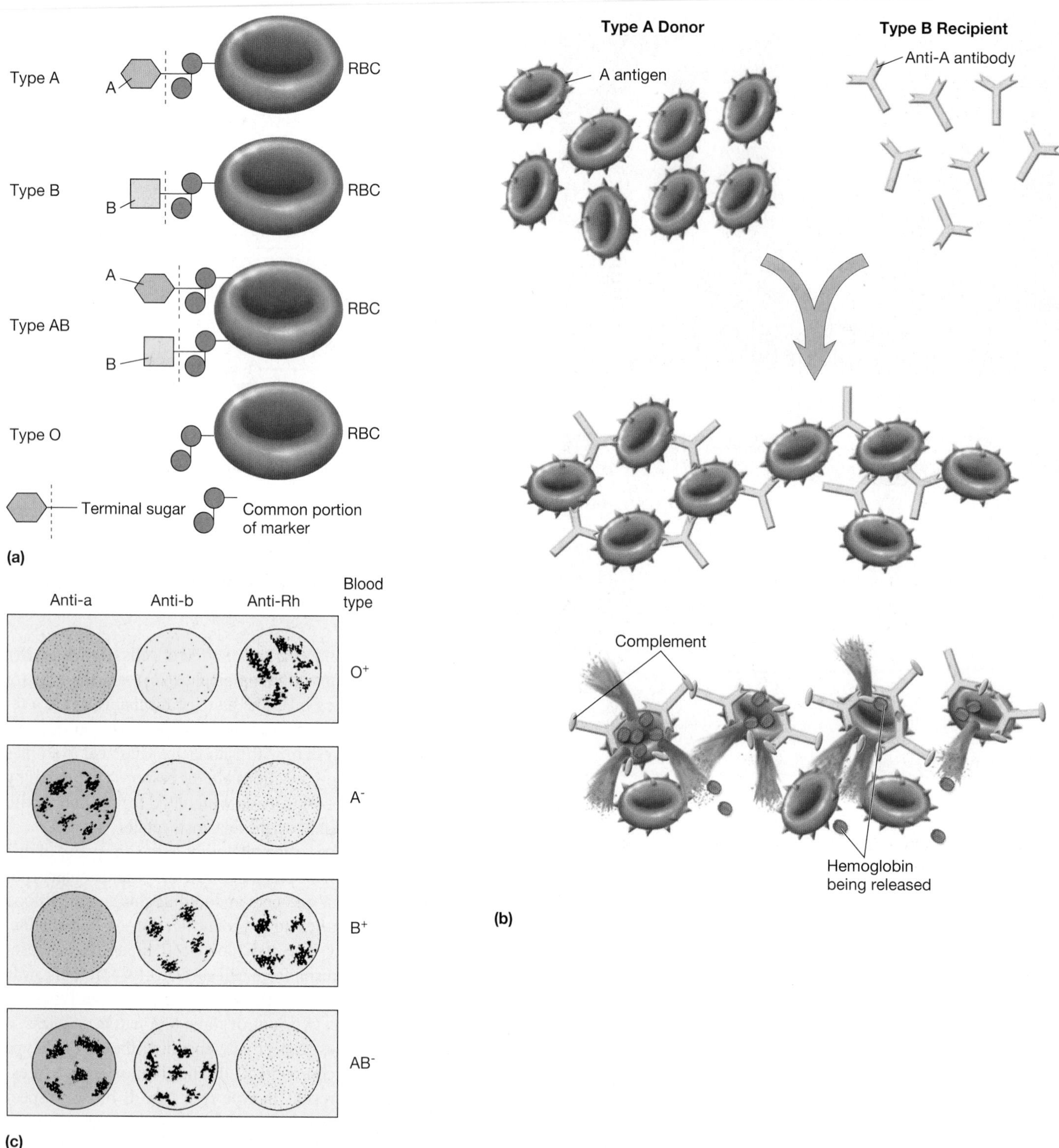

Figure 32.29 Immunohematology is the Study of Immune Reactions Associated With Blood. **(a)** Red blood cells (RBCs) can have genetically inherited carbohydrate antigens (two possible sugar residues) on their surface. The presence of the antigen(s) determines the blood type. Some individuals may have one, both, or no antigen, resulting in the A or B, AB, or O blood types, respectively. **(b)** A host does not make antibodies to its own blood antigen(s); antibodies to the blood antigens not found in a host are made. Exposure of blood to antibody specific for its carbohydrate type results in RBC agglutination. RBC lysis can then occur if complement is activated by the antibody-agglutinated cells. **(c)** RBC agglutination with specific antibody is the basis for blood typing. Another molecule, the Rhesus (Rh) factor, is another major RBC antigen that is typed to determine blood compatibility.

hosts produce anti-B antibodies, type B hosts produce anti-A antibodies, and type O hosts produce both anti-A and anti-B antibodies. Type O individuals are considered "universal donors" because their red blood cells lack A and B surface antigens. Conversely, type AB hosts produce neither anti-A nor anti-B antibodies so such individuals are called "universal recipients." The type II hypersensitivity reaction seen in blood transfusion occurs as complement is activated by cross-linking antibodies (figure 32.29*b*).

Blood typing can be accomplished by a slightly more sophisticated method of Landsteiner's process whereby blood from one host is mixed with antibodies specific for type A or type B blood. Agglutination of red cells by antibody (figure 32.29*c*) is used as a diagnostic tool to determine blood type. Another red blood cell antigen often reported with the ABO type was discovered during experiments with *Rh*esus monkeys. The so-called Rh factor (or D antigen) is determined by the expression of two alleles, one dominant (coding for the factor) and one recessive (not coding for the factor). Thus homozygous or heterozygous expression of the Rh allele confers the antigen (indicated as Rh^+). Expression of two recessive alleles results in the designation of Rh^- (no Rh factor). Incompatibility between Rh^- mothers and their Rh^+ fetus can result in maternal anti-Rh antibodies destroying fetal blood cells (**figure 32.30**). This type II hypersensitivity is called erythroblastosis fetalis. Control of this potentially fatal hemolytic disease of the newborn can be mitigated if the mother is passively immunized with anti-Rh factor antibodies, or RhoGam.

Type III Hypersensitivity

Type III hypersensitivity involves the formation of immune complexes (**figure 32.31**). Normally these complexes are phagocytosed effectively by the fixed monocytes and macrophages of the monocyte-macrophage system. In the presence of excess amounts of some soluble antigens, the antigen-antibody complexes may not be efficiently removed. Their accumulation can lead to a hypersensitivity reaction from complement that triggers a variety of inflammatory processes. The antibodies of type III reactions are primarily IgG. The inflammation caused by immune complexes and cells responding to such inflammation can result in significant damage, especially of blood vessels (vasculitis), kidney glomerular basement membranes (glomerulonephritis), joints (arthritis), and skin (systemic lupus erythematosus).

Type IV Hypersensitivity

Type IV hypersensitivity involves delayed, cell-mediated immune reactions. A major factor in the type IV reaction is the time required for T cells to migrate to and accumulate near the antigens. Both T_H and CTL cells can elicit type IV hypersensitivity reactions depending on the pathway in which the antigen is processed and presented. These events usually take a day or more to plateau.

In general, type IV reactions occur when antigens, especially those binding to serum proteins or tissue cells, are processed and presented to T cells. If the antigen is phagocytosed, it will be presented to T_H cells by the class II MHC molecules on the APC.

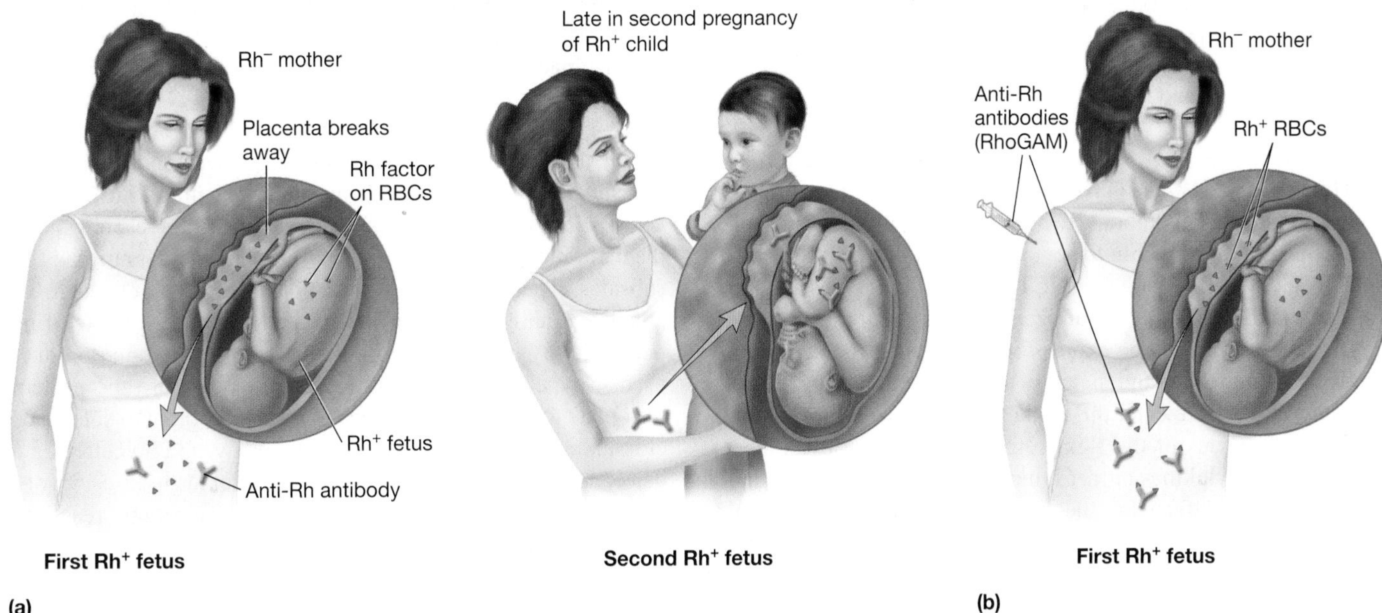

Figure 32.30 Rh Factor Incompatibility Can Result in RBC Lysis. **(a)** A naturally occurring blood cell incompatibility results when a Rh^+ fetus develops within a Rh^- mother. Initial sensitization of the maternal immune system occurs when fetal blood passes the placental barrier. In most cases, the fetus develops normally. However, a subsequent pregnancy with a Rh^+ fetus results in a severe, fetal hemolysis. **(b)** Anti-Rh antibody (RhoGAM) can be administered to Rh^- mothers during pregnancy to help bind, inactivate, and remove any Rh factor that may be transferred from the fetus. In some cases, RhoGAM is administered before sensitization occurs.

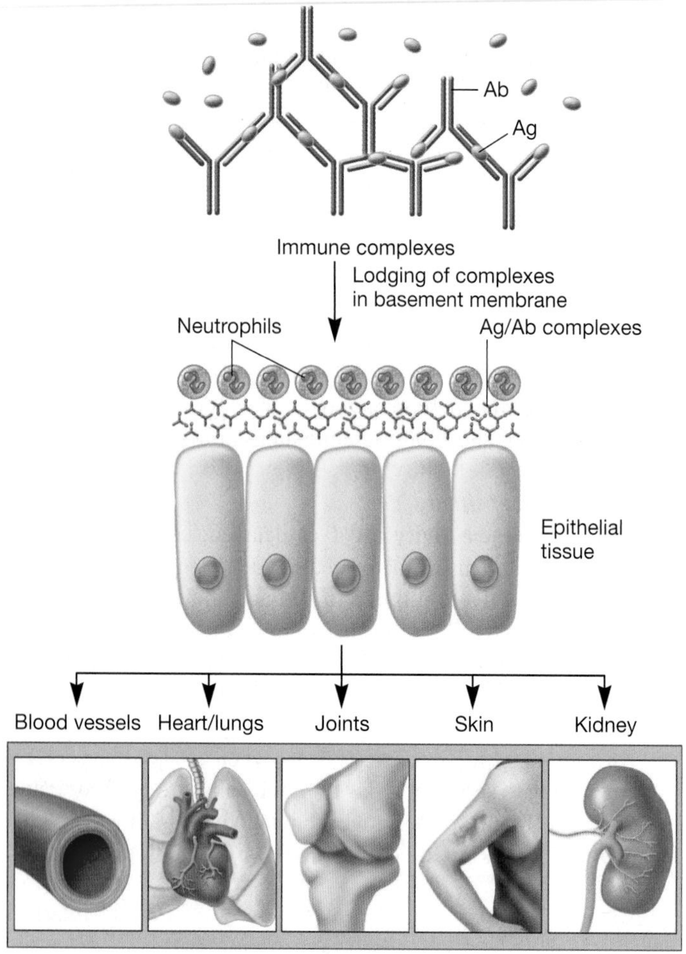

Major organs that can be targets of immune complex deposition

Steps:

1. Antibody combines with excess soluble antigen, forming large quantities of Ab/Ag complexes.
2. Circulating immune complexes become lodged in the basement membrane of epithelia in sites such as kidney, lungs, joints, skin.
3. Fragments of complement cause release of histamine and other mediator substances.
4. Neutrophils migrate to the site of immune complex deposition and release enzymes that cause severe damage to the tissues and organs involved.

Figure 32.31 Type III Hypersensitivity. Circulating immune complexes may lodge at various tissue sites where they activate complement and subsequently cause tissue cell lysis. Complement activation also results in granulocyte recruitment and release of their mediators, causing further injury to the tissue. Increased vascular permeability, resulting from granulocyte mediators, allows immune complexes to be deposited deeper into tissue sites where platelet recruitment leads to microthrombi (blood clots), which impair blood flow, resulting in additional tissue damage.

This activates the T_H1 cell, causing it to proliferate and secrete cytokines like IFN-γ and TNF-α. If the antigen is lipid soluble, it can cross the cell membrane to be processed within the cytosol. Antigens processed within the cytosol will be presented to CTL cells by class I MHC molecules. CTLs secrete cytokines and kill the cell that is presenting the antigen. Regardless of the cytokine source, they stimulate the expression of adhesion molecules on the local endothelium and increase vascular permeability allowing fluid and cells to enter the tissue space. Cytokines also stimulate keratinocytes of the skin to release their own cytokines. Together, the cytokines attract lymphocytes, macrophages, and basophils to the affected tissue, exacerbating the inflammation. Extensive tissue damage may result. Examples of type IV hypersensitivities include tuberculin hypersensitivity (the **TB skin test; figure 32.32**), allergic contact dermatitis, some autoimmune diseases, transplantation rejection, and killing of cancer cells.

In tuberculin hypersensitivity, a partially purified protein called tuberculin, which is obtained from the bacterium that causes tuberculosis, is injected into the skin of the forearm (figure 32.32). The response in a tuberculin-positive individual begins in about 8 hours, and a reddened area surrounding the injection site becomes indurated (firm and hard) within 12 to 24 hours. The T_H1 cells that migrate to the injection site are responsible for the induration. The reaction reaches its peak in 48 hours and then subsides. The size of the induration is directly related to the amount of antigen that was introduced and to the degree of hypersensitivity of the tested individual. Other microbial products used in type IV skin testing to detect disease are the proteins histoplasmin to detect histoplasmosis, coccidioidin to detect coccidioidomycosis, lepromin to detect leprosy, and brucellergen to detect brucellosis. Several important chronic diseases involve cell and tissue destruction by type IV hypersensitivity reactions. These diseases are caused by viruses, mycobacteria, protozoa, and fungi that produce chronic infections in which the macrophages and T cells are continually stimulated. Examples are leprosy, tuberculosis, leishmaniasis, candidiasis, and herpes simplex lesions. These infectious diseases are discussed in chapters 37–39.

Allergic contact dermatitis is a type IV reaction caused by haptens that combine with proteins in the skin to form the allergen

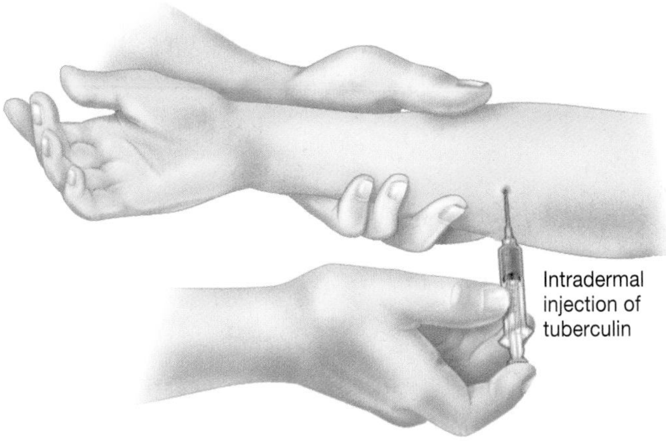

(a)

(b)

Figure 32.32 Type IV (or Delayed-type) Hypersensitivity. The mechanism of type IV hypersensitivity is illustrated by the tuberculin skin test, used to determine exposure to *M. tuberculosis*. Injection of the tuberculin antigens into the skin of individuals previously sensitized by *M. tuberculosis* results in the localized recruitment of macrophages and T_H1 cells, over 12 to 48 hours. The T_H1 cells are activated by the antigens presented by the class II MHC molecules on the macrophages. T_H1 cells then secrete inflammatory cytokines, which increase vascular permeability and recruit other immune cells, resulting in a visible swelling at the injection site.

that elicits the immune response (**figure 32.33**). The haptens are the antigenic determinants, and the skin proteins are the carrier molecules for the haptens. Examples of these haptens include cosmetics, plant materials (catechol molecules from poison ivy and poison oak; **figure 32.34**), topical chemotherapeutic agents, metals, and jewelry (especially jewelry containing nickel).

1. Discuss the mechanism of type I hypersensitivity reactions and how these can lead to systemic and localized anaphylaxis.
2. What causes a wheal and flare reaction site?
3. Why are type II hypersensitivity reactions called cytolytic or cytotoxic?
4. What characterizes a type III hypersensitivity reaction? Give an example.
5. Characterize a type IV hypersensitivity reaction.

Autoimmune Diseases

As discussed earlier, the body is normally able to distinguish its own self-antigens from foreign nonself antigens and does not mount an immunologic attack against itself. This phenomenon is called immune tolerance. At times the body loses tolerance and mounts an abnormal immune attack, either with antibodies or T cells, against a person's own self antigens.

It is important to distinguish between autoimmunity and autoimmune disease. Autoimmunity often is benign, whereas autoimmune disease often is fatal. **Autoimmunity** is characterized only by the presence of serum antibodies that react with self-antigens. These antibodies are called autoantibodies. The formation of autoantibodies is a normal consequence of aging; is readily inducible by infectious agents, organisms, or drugs; and is potentially reversible (it disappears when the offending "agent" is removed or eradicated). **Autoimmune disease** results from the activation of self-reactive T and B cells that, following stimulation by genetic or environmental triggers, cause actual tissue damage (**table 32.7**). Examples include rheumatoid arthritis and Type I diabetes mellitus. Four factors influence the development of autoimmune disease. Two major factors are genetic and viral. The third factor is endocrine—the effect of hormones. The fourth factor is psycho-neuro-immunological —the influence of stress and neurochemicals on the immune response. Overall, all four of these factors can affect gene expression, which directly or indirectly interferes with important immunoregulatory actions. Although their causal mechanism is not well known, these diseases may involve viral or bacterial infections. Some investigators believe that the release of abnormally large quantities of antigens may occur when the infectious agent causes tissue damage. The same agents also may cause body proteins to change into forms that stimulate antibody production or T-cell activation. Simultaneously, the activity of T cells with suppressor activity, which normally limits this type of reaction, seems to be repressed. Many autoimmune diseases have a genetic component. For example, there is a well-documented association between an individual's susceptibility to Graves' disease (which causes hyperthyroidism) and the neuro-degenerative disease multiple sclerosis and specific determinants on the major histocompatibility complex.

Transplantation (Tissue) Rejection

Tissue transplant rejection is the third area (after hypersensitivity and autoimmunity) in which the immune system can act detrimentally. Transplants between genetically different individuals

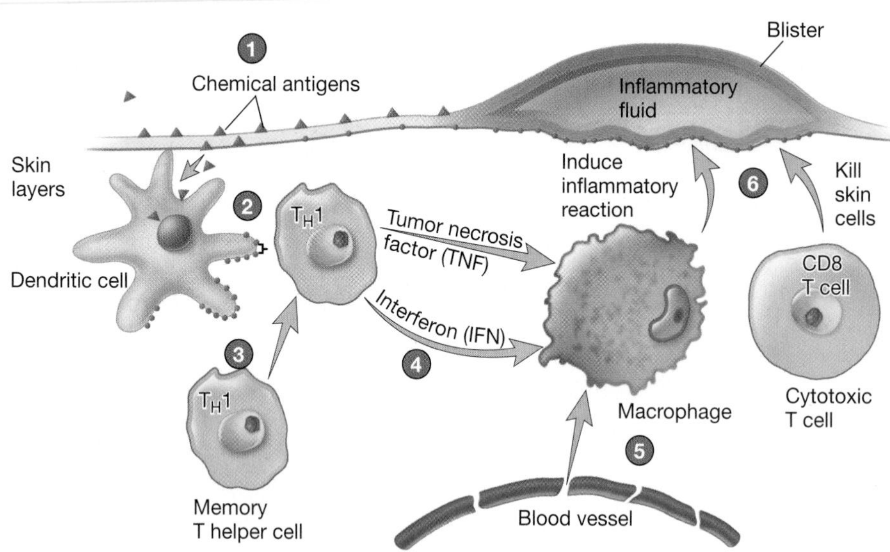

1 Lipid-soluble catechols are absorbed by the skin.

2 Dendritic cells close to the epithelium pick up the allergen, process it, and display it on MHC receptors.

3 Previously sensitized T_H1 ($CD4^+$) cells recognize the presented allergen.

4 Sensitized T_H1 cells are activated and secrete cytokines (IFN, TNF).

5 These cytokines attract macrophages and cytotoxic T cells to the site.

6 Macrophages release mediators that stimulate a strong, local inflammatory reaction. Cytotoxic T cells directly kill cells and damage the skin. Fluid-filled blisters result.

Figure 32.33 Contact Dermatitis. In contact dermatitis to poison ivy, a person initially becomes exposed to the allergen, predominantly 3-n-pentadecyl-catechol found in the resinous sap material uroshiol, which is produced by the leaves, fruit, stems, and bark of the poison ivy plant. The catechol molecules, acting as haptens, combine with high molecular weight skin proteins. After 7 to 10 days sensitized T cells are produced and give rise to memory T cells. Upon second contact, the catechols bind to the same skin proteins, and the memory T cells become activated in only 1 to 2 days, leading to an inflammatory reaction (contact dermatitis).

within a species are termed **allografts** [Greek *allos,* other]. Some transplanted tissues do not stimulate an immune response. For example, a transplanted cornea is rarely rejected because lymphocytes do not circulate into the anterior chamber of the eye. This site is considered an immunologically privileged site. Another example of a privileged tissue is the heart valve, which in fact, can be transplanted from a pig to a human without stimulating an immune response. Such a graft between different species is termed a **xenograft** [Greek *xenos,* strayed].

Transplanting tissue that is not immunologically privileged generates the possibility that the recipient's cells will recognize the donor's tissues as foreign. This triggers the recipient's immune mechanisms, which may destroy the donor tissue. Such a response is called a tissue rejection reaction. A tissue rejection reaction can occur by two different mechanisms. First, foreign class II MHC molecules on transplanted tissue, or the "graft," are recognized by host T-helper cells, which aid cytotoxic T cells in graft destruction (**figure 32.35**). Cytotoxic T cells then recognize the graft through the foreign class I MHC molecules. This response is much like the activation of CTLs by virally infected host cells. A second mechanism involves the T-helper cells reacting to the graft (transplanted tissue) and releasing cytokines. The cytokines stimulate macrophages to enter, accumulate within the graft, and destroy it. The MHC molecules play a dominant role in tissue rejection reactions because of their unique association with the recognition system of T cells. Unlike antibodies, T cells cannot recognize or react directly with non-MHC molecules (viruses, allergens). They recognize these molecules only in association with, or complexed to, an MHC molecule.

Because class I MHC molecules are present on every nucleated cell in the body they are important targets of the rejection reaction. The greater the antigenic difference between class I molecules of the recipient and donor tissues, the more rapid and severe the rejection reaction is likely to be. However, the reaction can sometimes be minimized if recipient and donor tissues are matched as closely as possible. Most recipients are not 100% matched to their donors, so immunosuppressing drugs are used to prevent host-mediated rejection of the graft.

Organ transplant recipients also can develop **graft-versus-host disease.** This occurs when the transplanted tissue contains immunocompetent cells that recognize host antigens and attack the host. The immunosuppressed recipient cannot control the response of the grafted tissue. Graft-versus-host disease is a common problem in allogenic bone marrow transplants. The transplanted bone marrow contains many mature, post-thymic T cells. These cells recognize the host MHC antigens and attack the immunosuppressed recipient's normal tissue cells. Currently one way to prevent graft-versus-host disease is to deplete the bone marrow of mature T cells by using immunosuppressive techniques. Examples include drugs that attack T cells (azathioprine, methotrexate, and cyclophosphamide), immunosuppressive drugs (cyclosporin, tacrolimus, and rapamycin), anti-inflammatory drugs (corticosteroids), irradiation of the lymphoid tissue, and antibodies directed against T-cell antigens.

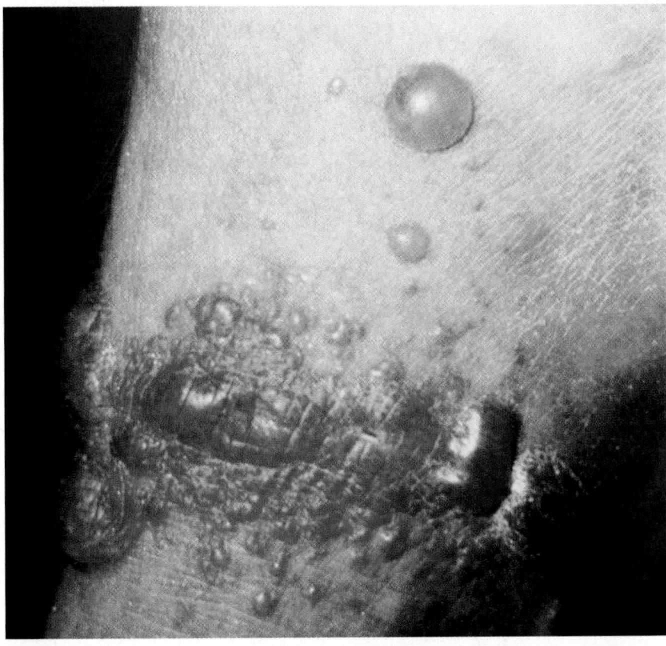

Figure 32.34 Contact Dermatitis from Poison Oak. The various stages of dermatitis (blister, scales and thickened skin patches) caused by skin contact with poison oak result from $CD8^+$ T cells, responding to plant antigens, processed and presented by class I MHC molecules. The dermatitis ensues from the inflammatory reaction resulting from CTL cytokines and cellular damage.

Immunodeficiencies

Defects in one or more components of the immune system can result in its failing to recognize and respond properly to antigens. Such **immunodeficiencies** can make a person more prone to infection than those people capable of a complete and active immune response. Despite the increase in knowledge of functional derangements and cellular abnormalities in the various immunodeficiency disorders, the fundamental biological errors responsible for them remain largely unknown. To date, most genetic errors associated with these immunodeficiencies are located on the X chromosome and produce primary or congenital immunodeficiencies (**table 32.8**). Other immunodeficiencies can be acquired because of infections by immunosuppressive microorganisms, such as HIV. Direct contact diseases: Acquired immune deficiency syndrome (AIDS) (section 37.3)

1. What is an autoimmune disease and how might it develop?
2. What is an immunologically privileged site and how is it related to transplantation success?
3. How does a tissue rejection reaction occur?
4. Describe an immunodeficiency. How might immunodeficiencies arise?

Table 32.7 Some Autoimmune Diseases in Humans

Disease	Autoantigen	Pathophysiology
Acute rheumatic fever	Streptococcal cell wall antigens; antibodies cross-react with cardiomyocytes	Arthritis, scarring of heart valves, myocarditis
Autoimmune hemolytic anemia	Rh blood group, I antigen	Red blood cells are destroyed by complement and phagocytosis, anemia
Autoimmune thrombocytopenia purpura	Platelet integrin	Perfuse bleeding
Goodpasture's syndrome	Basement membrane collagen	Glomerulonephritis, pulmonary hemorrhage
Graves' disease	Thyroid-stimulating hormone receptor	Hyperthyroidism
Multiple sclerosis	Myelin basic protein	Demyelination of axons
Myasthenia gravis	Acetycholine receptor	Progressive muscular weakness
Pemphigus vulgaris	Cadherin in epidermis	Skin blisters
Rheumatoid arthritis	Unknown synovial joint antigen	Joint inflammation and destruction
Systemic lupus erythematosus	DNA, histones, ribosomes	Arthritis, glomerulonephritis, vasculitis, rash
Type 1 diabetes mellitus	Pancreatic beta cell antigen	Beta cell destruction

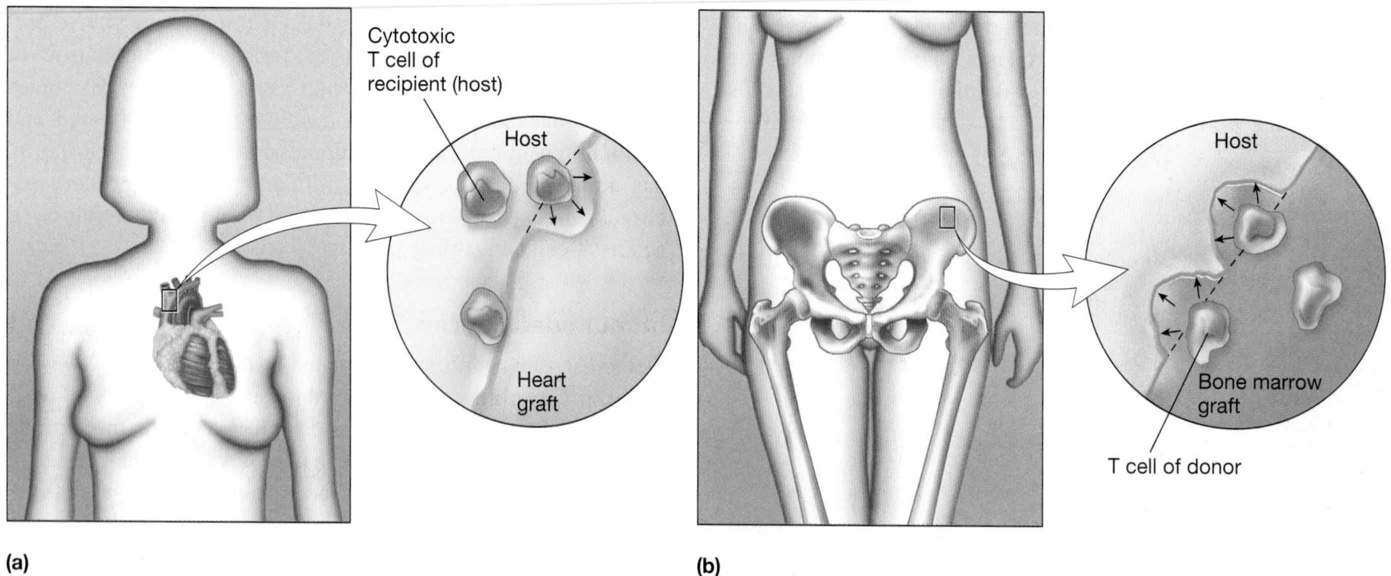

Figure 32.35 Potential Transplantation Reactions. **(a)** Donated tissues, not from an identical twin, contain cellular MHC proteins that are recognized as foreign by the recipient host (host-versus-graft disease). The tissue is then attacked by host CTLs, resulting in its damage and rejection. **(b)** Donated tissue may also contain immune cells that react against host antigens. Recognition of a foreign host by donor CTLs results in graft-versus-host disease.

Table 32.8 Some Congenital Immune Deficiencies in Humans

Condition	Symptoms	Cause
Chronic granulomatous disease	Defective monocytes and neutrophils leading to recurrent bacterial and fungal infections	Failure to produce reactive oxygen intermediates due to defective NADPH oxidase
X-linked agammaglobulinemia	Plasma cell or B-cell deficiency and inability to produce adequate specific antibodies	Defective B-cell differentiation due to loss of tyrosine kinase
DiGeorge syndrome	T-cell deficiency and very poor cell-mediated immunity	Lack of thymus or a poorly developed thymus
Severe combined immunodeficiency disease (SCID)	Both antibody production and cell-mediated immunity impaired due to a great reduction of B- and T-cell levels	Various mechanisms (e.g., defective B- and T-cell maturation because of X-linked gene mutation; absence of adenosine deaminase in lymphocytes)

Summary

32.1 Overview of Specific (Adaptive) Immunity

a. The specific (adaptive) immune response system consists of lymphocytes that can recognize foreign molecules (antigens) and respond to them. Two branches or arms of immunity are recognized: humoral (antibody-mediated) immunity and cellular (cell-mediated) immunity (**figure 32.1**).

b. Acquired immunity refers to the type of specific (adaptive) immunity that a host develops after exposure to a suitable antigen. It can be obtained actively or passively by natural or artificial means.

32.2 Antigens

a. An antigen is a substance that stimulates an immune response and reacts with the products of that response. Each antigen can have several antigenic determinant sites or epitopes that stimulate production of and combine with specific antibodies (**figure 32.2**).

b. Haptens are small organic molecules that are not antigenic by themselves but can become antigenic if bound to a larger carrier molecule.

32.3 Types of Specific (Adaptive) Immunity

a. Immunity can be acquired by natural means—actively through infection or passively through receipt of preformed antibodies, as through colostrum (**figure 32.3**).

b. Immunity can be acquired by artificial means—actively through immunization or passively through receipt of preformed antibodies, as with anti-sera.

32.4 Recognition of Foreignness

a. MHC molecules are cell surface proteins coded by a group of genes termed the major histocompatibility complex. Class I MHC proteins are found on all nucleated cells of mammals. Class II MHC proteins are only expressed on cells that can phagocytose foreign materials and organisms. The human MHC gene products are called the human leukocyte antigens (HLA) (**figures 32.4** and **32.5**).

b. Class I MHC proteins collect foreign peptides processed by the proteasome and present them to cytotoxic T cells.

c. Class II MHC proteins collect foreign peptides processed by the phagosome and present them to helper T cells.

32.5 T Cell Biology

a. T cells are pivotal elements of the immune response. T cells have antigen-specific receptor proteins (**figure 32.6**).

b. Antigen-presenting cells—most of which are macrophages, dendritic cells, and B cells—take in foreign antigens or pathogens, process them, and present antigenic fragments complexed with MHC molecules to T-helper cells (**figure 32.7**).

c. Cytotoxic T lymphocytes recognize target cells such as virus-infected cells that have foreign antigens and class I MHC molecules on their surface. The CTLs then attack and destroy the target cells using the CD95 pathway and/or the perforin pathway (**figure 32.8**).

d. T cells control the development of other cells, including effector B and T cells. T-helper cells ($CD4^+$) regulate cell behavior and cytotoxic T cells ($CD8^+$) also regulate cell behavior, but in addition, they can kill altered host cells directly. There are three subsets of T-helper cells: T_H1, T_H2, and T_H0. T_H1 cells produce various cytokines and are involved in cellular immunity (**figure 32.7**). The T_H2 cells also produce various cytokines but are involved in humoral immunity (**figure 32.11**). T_H0 cells are simply undifferentiated precursors of T_H1 and T_H2 cells.

32.6 B Cell Biology

a. B cells defend against antigens by differentiating into plasma cells that secrete antibodies into the blood and lymph, providing humoral or antibody-mediated immunity.

b. B cells can be stimulated to divide and/or differentiate to secrete antibody when triggered by the appropriate signals.

c. B cells have receptor immunoglobulins on their plasma membrane surface that are specific for given antigenic determinants. Contact with the antigenic determinant is required for the B cell to divide and differentiate into plasma cells and memory cells (**figures 32.10** and **32.11**).

32.7 Antibodies

a. Antibodies (immunoglobulins) are a group of glycoproteins present in the blood, tissue fluids, and mucous membranes of vertebrates. All immunoglobulins have a basic structure composed of four polypeptide chains (two light and two heavy) connected to each other by disulfide bonds (**figure 32.12**). In humans, five immunoglobulin classes exist: IgG, IgA, IgM, IgD, and IgE (**figures 32.16–32.20** and **table 32.5**).

b. The primary antibody response in a host occurs following initial exposure to the antigen. This response has lag, log, plateau, and decline phases. Upon secondary antigen challenge, the B cells mount a heightened and accelerated anamnestic response (**figure 32.21**).

c. Antibody diversity results from the rearrangement and splicing of the individual gene segments on the antibody-coding chromosomes, somatic mutations, the generation of different codons during splicing, and the independent assortment of light- and heavy-chain genes (**figures 32.22** and **32.23**).

d. Immunologic specificity and memory is partly explained by the clonal selection theory (**figure 32.24**).

e. Hybridomas result from the fusion of lymphocytes with myeloma cells. These cells produce a single monoclonal antibody. Monoclonal antibodies have many uses (**Techniques & Applications 32.2**).

32.8 Action of Antibodies

a. Various types of antigen-antibody reactions occur in vertebrates and initiate the participation of other body processes that determine the ultimate fate of the antigen. For example, the complement system can be activated, leading to cell lysis, phagocytosis, chemotaxis, or stimulation of the inflammatory response. Other defensive antigen-antibody interactions include toxin neutralization, viral neutralization, adherence inhibition, opsonization, and immune complex formation (**figure 32.25**).

32.9 Summary: The Role of Antibodies and Lymphocytes in Immune Defense

a. Although both the humoral and cellular arms of the specific (adaptive) immune response have been considered separately, it is important to understand that the response of a vertebrate host to any particular pathogen may involve a complex set of responses. Both humoral and cellular immune responses can join nonspecific (innate) defenses to ensure a maximal survival advantage against viral and bacterial pathogens.

32.10 Acquired Immune Tolerance

a. Acquired immune tolerance is the ability of a host to react against nonself antigens while tolerating self-antigens. It can be induced in several ways.

32.11 Immune Disorders

a. When the immune response occurs in an exaggerated form and results in tissue damage to the individual, the term hypersensitivity is applied. There are four types of hypersensitivity reactions, designated as types I through IV (**figures 33.26–33.34**).

b. Autoimmune diseases result when self-reactive T and B cells attack the body and cause tissue damage. A variety of factors can influence the development of autoimmune disease.

c. The immune system can act detrimentally and reject tissue transplants. There are different types of transplants. Xenografts involve transplants of privileged tissue between different species, and allografts are transplants between genetically different individuals of the same species.

d. Immunodeficiency diseases are a diverse group of conditions in which an individual's susceptibility to various infections is increased; several severe diseases can arise because of one or more defects in the specific (adaptive) or nonspecific (innate) immune response.

Key Terms

acquired immune tolerance 802
acquired immunity 776
agglutination reaction 799
agglutinin 799
allergen 803
allergic contact dermatitis 808
allergy 803
allografts 810
allotypes 791
anamnestic response 774
anaphylaxis 803
anergic 783

Critical Thinking Questions

1. Why do you think antibodies are proteins rather than polysaccharides or lipids? List all properties of proteins that make them suitable molecules from which to make antibodies.
2. How did the clonal selection theory inspire the development of monoclonal antibody techniques?
3. What is the difference in the kinetics of antibody formation in response to a first and second exposure to the same antigen?
4. Why do MHC, TCR, and BCR molecules require accessory proteins or co-receptors for a signal to be sent within the cell?
5. Why do you think two signals are required for B- and T-cell activation, but only one signal is required for activation of an APC?
6. Most immunizations require multiple exposures to the vaccine (i.e., boosters). Why is this the case?

Learn More

Ansel, K. M.; Harris, R. B. S.; and Cyster, J. G. 2002. CXCL13 is required for B1 cell homing, natural antibody production and body cavity immunity. *Immunity* 16:67–76.

Dempsey, P.; Allicson, M.; Akkaraju, S.; Goodnow C.; and Fearon, D. 1996. C3d of complement as a molecular adjuvant: Bridging innate and acquired immunity. *Science* 271:348–50.

Fruend, J., and McDermott, K. 1942. Sensitization to horse serum by means of adjuvants. *Proc. Soc. Exp. Biol. Med.* 49:548–53.

Horton, R.; Wilming, L.; and Rand, V. *et al.* 2004. Gene map of the extended human MHC. *Nature Rev. Genetics* 5:889–99.

Janeway, C. 2004. *Immunobiology*, 6th ed. New York: Garland Science.

Kondilis, H. D.; Kor, B.; Steckman, B.; and Kargel, M. 2005. Regulation of T-cell receptor β allelic exclusion at a level beyond accessibility. *Nature Immunol.* 8:189–97.

Linton, P., and Dorskind, K. 2004. Age-related changes in lymphocyte development and function. *Nature Immunol.* 5:133–39.

Liu, C. *et al.* 2005. The role of CCL2 in recruitment of T-precursors to fetal thymi. *Blood* 105: 31–39.

Modlin, R., and Sieling, P. 2005. Now presenting: γδ T cells. *Science* 309:252–53.

Sallusto, F.; Cella, M.; Danieli C.; and Lanzavecchia, A. 1995. Dendritic cells use macropinocytosis and the mannose receptor to concentrate macromolecules in the major histocompatibility complex class II compartment: down regulation by cytokines and bacterial products. *J. Exp. Med.* 182:389–400.

Silva-Santos, B.; Pennington, D.; and Heyday, A. 2005. Lymphotoxin-mediated regulation of γδ cell differentiation by αβ T cell progenitors. *Science* 307: 925–28.

Please visit the Prescott website at www.mhhe.com/prescott7 for additional references.

33 Pathogenicity of Microorganisms

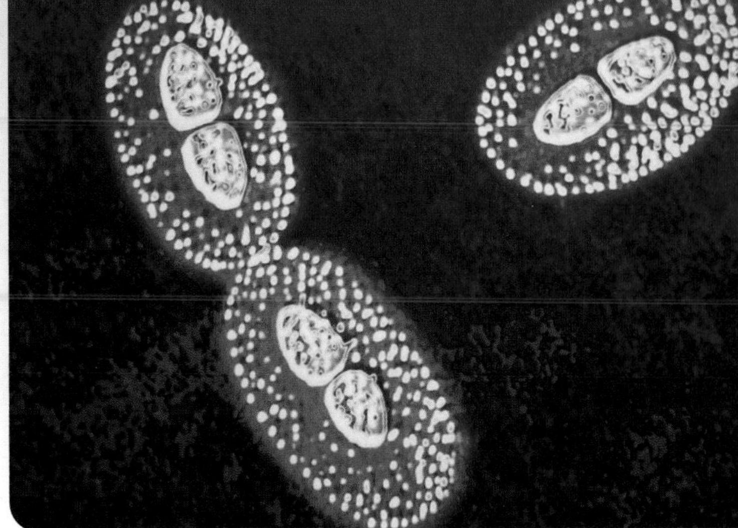

Three *Streptococcus pneumoniae,* each surrounded by a slippery mucoid capsule (shown as a layer of white spheres around the diplococcus bacteria). The polysaccharide capsule is vital to the pathogenicity of this bacterium because it prevents phagocytic cells from accomplishing phagocytosis.

Preview

- If a microorganism (symbiont) either harms or lives at the expense of another organism, it is called a parasitic organism and the relationship is termed parasitism. In this relationship the infected organism is referred to as the host.
- Those organisms capable of causing disease are called pathogens. Disease is any change in the host from a healthy to an unhealthy, abnormal state in which part or all of the host's body is not capable of carrying on its normal functions.
- The steps for the infectious process involving viral diseases include the following: entry into a potential host, attachment to a susceptible cell, penetration of viral nucleic acid, replication of virus particles within the host cell, and ultimate release from the host cell. Newly replicated virus particles are available to infect other susceptible cells. Viral infection can result in cellular injury, stimulation of immune responses, or evasion of the virus from immune detection resulting in chronic infection.
- The steps of the infectious process involving bacterial diseases usually include the following: the bacterium is transmitted to a suitable host, attaches to and/or colonizes the host, grows and multiplies within or on the host, and interferes with or impairs the normal physiological activities of the host.
- During coevolution with human hosts, pathogenic bacteria have evolved complex signal transduction pathways to regulate the genes necessary for virulence.
- The genes that encode virulence factors are often located on large segments of DNA within the bacterial genome, called pathogenicity islands, that carry genes responsible for virulence.
- Two distinct categories of disease can be recognized based on the role bacteria play in the disease causing process: infections (invasion and growth) and intoxications.
- Toxins produced by pathogenic bacteria are either exotoxins or endotoxins.
- Viruses and bacteria are continuously evolving and producing unique mechanisms that enable them to escape the host's arsenal of defenses.

Chapter 30 introduces the concept of symbiosis and deals with several of its subordinate categories, including commensalism and mutualism. In this chapter the process of parasitism is presented along with one of its possible consequences—pathogenicity. The parasitic way of life is so successful, that it has evolved independently in nearly all groups of organisms. In recent years concerted efforts to understand organisms and their relationships with their hosts have developed within the disciplines of virology, bacteriology, mycology, parasitology (protozoology and helminthology), entomology, and zoology. This chapter examines the parasitic way of life in terms of health and disease and emphasizes viral and bacterial disease mechanisms. We conclude the chapter with some viral and bacterial mechanisms used to evade host defenses.

33.1 Host-Parasite Relationships

Relationships between two organisms can be very complex. A larger organism that supports the survival and growth of a smaller organism is called the host. The interaction of the two is symbiotic. **Symbiosis** refers to the "living together" of organisms and includes

Pathogenicity is not the rule. Indeed, it occurs so infrequently and involves such a relatively small number of species, considering the huge population of bacteria on earth, that it has a freakish aspect. Disease usually results from inconclusive negotiations for symbiosis, an overstepping of the line by one side or the other, a biological misinterpretation of borders.

—Lewis Thomas

commensalism, mutualism, and parasitism (*see figure 30.1*). A commensalistic relationship is demonstrated by the microflora of the cecum of mammals. The mammal provides food and shelter for the microflora, while the microorganisms enzymatically break down complex nutrients to be utilized by the mammal. In addition to relationships between two living organisms, many microorganisms are saprophytic. These organisms obtain nutrients from dead or decaying organic matter. Although some saprophytes are capable of causing disease, most are not parasites; rather they can be thought of as scavengers. Technically, **parasites** are those organisms that live on or within a host organism and are metabolically dependent on the host. Unfortunately, the term parasite has other meanings. It is often used to mean a protozoan or helminth organism living within a host. However, any organism that causes disease is a parasite. Microbiologists can also define infectious disease by the **host-parasite relationship.** A small number of microorganisms can exist as either saprophytes or parasites. Furthermore, commensals, like those associated with the gut, can become parasites when they are present in a location within the host other than the site they normally colonize. These organisms are often referred to as opportunists.

Several types of parasitism are recognized. If an organism lives on the surface of its host, it is an **ectoparasite;** if it lives internally, it is an **endoparasite.** Some parasites, especially those with complex life cycles, inhabit multiple hosts. The host on or in which the parasitic organism either attains sexual maturity or reproduces is the **final host.** A host that serves as a temporary but essential environment for some stages of development is an **intermediate host.** In contrast, a **transfer host** is not necessary for the completion of the organism's life cycle but is used as a vehicle for reaching a final host. A host infected with a parasitic organism that also can infect humans is called a **reservoir host.**

The host-parasite relationship is complex and dynamic. When a parasite is growing and multiplying within or on a host, the host is said to have an **infection.** The nature of an infection can vary widely with respect to severity, location, and number of organisms involved (**table 33.1**). An infection may or may not result in overt disease. An **infectious disease** is any change from a state of health in which part or all of the host body is not capable of carrying on its normal functions due to the presence of a parasite or its products. Any organism or agent that produces such a disease is also known as a **pathogen** [Greek *patho,* disease, and *gennan,* to produce]. Its ability to cause disease is called **pathogenicity.** A **primary (frank) pathogen** is any organism that causes disease in a healthy host by direct interaction. Conversely, an **opportunistic pathogen** refers to an organism that is part of the host's normal microbiota, but is able to cause disease when the host is immunocompromised or when it has gained access to other tissue sites.

At times an infectious organism can enter a latent state in which there is no shedding of the organism (that is, the organism is not infectious at that time) and no symptoms present within the host. This latency can be either intermittent or quiescent. Intermittent latency is exemplified by the herpesvirus that causes cold sores (fever blisters). After an initial infection, the symptoms subside. However, the virus remains in nerve tissue and can be cyclically activated weeks or years later by factors such as stress or sunlight. In a quiescent latency the organism persists but remains inactive for long periods of time, usually for years. For example, the varicella-zoster virus causes chickenpox in children and remains after the disease has subsided. In adulthood, under certain conditions, the same virus may erupt into a disease called shingles. Direct contact diseases: Cold sores (section 37.3); Airborne diseases: Chickenpox and shingles (section 37.1)

The outcome of most host-parasite relationships is dependent on three main factors: (1) the number of microorganisms infecting the host, (2) the degree of pathogenicity (or virulence) of the organism, and (3) the host's defenses or degree of resistance (**figure 33.1**). Usually the greater the number of organisms within a given host, the greater the likelihood of disease. However, a few organisms can cause disease if they are extremely virulent or if the host's resistance is low. Such infections can be a serious problem among hospitalized patients with very low resistance.

The term **virulence** [Latin *virulentia,* from *virus,* poison] refers to the degree or intensity of pathogenicity. As mentioned previously, pathogenicity is a general term that refers to an organism's potential to cause disease. Various physical and chemical characteristics (such as structures that facilitate attachment and molecules that bypass host defenses) contribute to pathogenicity, and thus virulence. Individual characteristics that confer virulence are called **virulence factors** (e.g., capsules, pili, toxins). Virulence is determined by three characteristics of the pathogen: invasiveness, infectivity, and pathogenic potential. **Invasiveness** is the ability of the organism to spread to adjacent or other tissues. **Infectivity** is the ability of the organism to establish a focal point of infection. **Pathogenic potential** refers to the degree that the pathogen causes damage. A major aspect of pathogenic potential is toxigenicity. **Toxigenicity** is the pathogen's ability to produce toxins, chemical substances that damage the host and produce disease. Virulence is often meas-

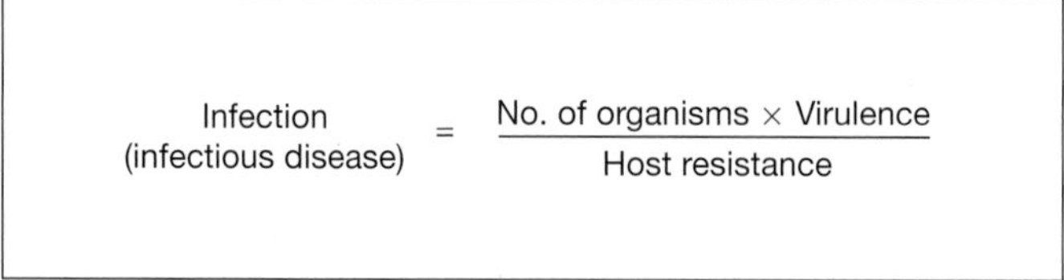

Figure 33.1 Mathematical Expression of Infection. As a mathematical expression, infection or infectious disease can be evaluated by determining the relative contributions of the number of organisms, their virulence, and the host resistance. Organism number reflects the infectious dose and the rate at which the organism can reproduce. Virulence reflects the total number of virulence factors encoded by the genome and expressed in the host. Host resistance is a function of immune status (immunizations, nutrition, previous exposure, etc.) or the effects of chemotherapeutic intervention.

Table 33.1 Various Types of Infections Associated with Parasitic Organisms

Type	Definition
Abscess	A localized infection with a collection of pus surrounded by an inflamed area
Acute	Short but severe
Bacteremia	Presence of viable bacteria in the blood
Chronic	Persisting over a long time
Covert	Subclinical, with no symptoms
Cross	Transmitted between hosts infected with different organisms
Focal	Existing in circumscribed areas
Fulminating	Infectious agent multiplying with great intensity
Iatrogenic	Caused as a result of health care
Latent	Persisting in tissues for long periods, during most of which there are no symptoms
Localized	Restricted to a limited region or to one or more anatomical areas
Nosocomial	Developed during a stay at a hospital or other clinical care facility
Opportunistic	Resulting from endogenous microbiota, especially when host resistance is very low
Overt	Symptomatic
Phytogenic	Caused by plant pathogens
Polymicrobial	More than one organism present simultaneously
Primary	First infection that often allows other organisms to invade host at that site
Pyogenic	Resulting in pus formation
Secondary	Caused by an organism following an initial or primary infection
Sepsis	(1) The condition resulting from the presence of bacteria or their toxins in blood or tissues; the presence of pathogens or their toxins in the blood or other tissues (2) Systemic response to infection; this systemic response is manifested by two or more of the following conditions as a result of infection: temperature, >38 or <36°C; heart rate, >90 beats per min; respiratory rate, >20 breaths per min, or pCO_2, <32 mm Hg; leukocyte count, >12,000 cells per ml^3, or >10% immature (band) forms
Septicemia	Blood poisoning associated with persistence of pathogenic organisms or their toxins in the blood
Septic shock	Sepsis with hypotension despite adequate fluid resuscitation, along with the presence of perfusion abnormalities that may include, but are not limited to, lactic acidosis, oliguria, or an acute alteration in mental status
Severe sepsis	Sepsis associated with organ dysfunction, hypoperfusion, or hypotension; hypoperfusion and perfusion abnormalities may include, but are not limited to, lactic acidosis, oliguria, or an acute alteration in mental status
Sporadic	Occurring only occasionally
Subclinical	No detectable symptoms or manifestations occurring (covert)
Systemic	Spread throughout the body
Toxemia	Condition arising from toxins in the blood
Zoonotic	Caused by a parasitic organism that is normally found in animals other than humans

ured experimentally by determining the **lethal dose 50 (LD_{50})** or the **infectious dose 50 (ID_{50})**. These values refer to the dose or number of pathogens that either kill or infect, respectively, 50% of an experimental group of hosts within a specified period (**figure 33.2**).

It should be noted that disease can result from causes other than toxin production. Sometimes a host triggers exaggerated immunological responses (**immunopathology**) upon a second exposure or chronic exposure to a microbial antigen. These hypersensitivity reactions damage the host even though the pathogen doesn't produce a toxin. Tuberculosis is a good example of the involvement of hypersensitivity reactions in disease. Some diseases also might be due to autoimmune responses. For instance, a viral or bacterial pathogen may stimulate the immune system to attack host tissues because it carries antigens that resembled those of the host, a phenomenon known as molecular mimicry. Streptococcal infections may cause rheumatic fever in this way. Immune disorders: Hypersensitivities (section 32.11)

1. Define parasitic organism, parasitism, infection, infectious disease, pathogenicity, virulence, invasiveness, infectivity, pathogenic potential, and toxigenicity.
2. What factors determine the outcome of most host-parasite relationships?

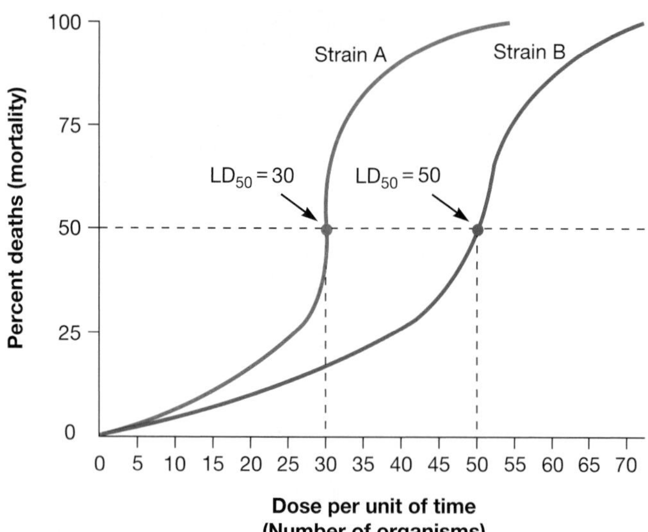

Figure 33.2 Determination of the LD_{50} of a Pathogenic Microorganism. Various doses of a specific pathogen are introduced into experimental host animals. Deaths are recorded and a graph constructed. In this example, the graph represents the susceptibility of host animals to two different strains of a pathogen—strain A and strain B. For strain A the LD_{50} is 30, and for strain B it is 50. Hence strain A is more virulent than strain B.

33.2 Pathogenesis of Viral Diseases

The fundamental process of viral infection is the expression of the viral replicative cycle (*see section 18.2*) in a host cell. The steps for the infectious process involving viruses usually include the following:

1. Maintain a reservoir
2. Enter a host
3. Contact and enter susceptible cells
4. Replicate within the cells
5. Release from host cells (immediate or delayed)
6. Spread to adjacent cells
7. Virus-host interactions engender host immune response
8. Be either cleared from the body of the host, establish a persistent infection, or kill the host
9. Be shed back into the environment

The determinants of pathogenicity are now discussed in more detail. Reproduction of verebrate viruses (section 18.2)

Maintaining a Reservoir

Like other infectious agents, viruses must reside somewhere before they are transmitted to a specific host or tissue site. Because most viruses are limited to the type of host that they can infect (animal viruses infect animals, plant viruses infect plants, and so on), they must gain access to a susceptible host so that they can replicate. Thus the most common **reservoirs** of human viruses are humans and other animals. Some viruses are acquired early in the life of a host, only to cause disease at some later time. More often, however, viruses are transmitted from reservoir (a human host) to host (another human), to cause noticable infection in a relatively short time frame. Because viruses require viable host cells in which to replicate, the source and/or reservoir may harbor large numbers of viral particles that can infect equally large numbers of new hosts upon their release. However, some viruses may not be able to leave their reservoirs or may leave at a very slow rate. Because the source and/or reservoir of a pathogen are part of the infectious disease cycle, this aspect of pathogenicity is discussed in detail in chapter 36, which covers the epidemiology of infectious diseases. Microbial Diversity & Ecology 18.1: SARS: Evolution of a virus

Contact, Entry, and Primary Replication

The first step in the infectious process is the attachment and entrance of the virus into a susceptible host and the host's cells. Entrance may be accomplished through one of the body surfaces (skin, respiratory system, gastrointestinal system, urogenital system, or the conjunctiva of the eye). Other viruses enter the host by sexual contact, needle sticks, blood transfusions and organ transplants, or by insect **vectors** (organisms that transmit the pathogen from one host to another).

Regardless of the method of entry into the host organism, viral infection begins when the viral particle penetrates a host cell to gain access to the cell's replicative machinery. This process is called adsorption, or the attachment to the cell surface. Recall that adsorption occurs because viruses produce specific protein ligands that bind to host cell receptors embedded within their plasma membranes. Host specificity for the virus is a function of viral gene expression; the virus must express the ligand so as to dock with a specific host cell. Protein ligands are usually positioned on the virus to maximize contact with the cell. Enveloped viruses use spikes—viral proteins that protrude from their membrane. Naked viruses have their ligands as part of their capsid proteins.

Each viral ligand only binds to a complementary receptor on the host cell surface. Binding of a virus to its receptor typically results in penetration of the cell or the delivery of virus nucleic acid to the cytoplasm of the cell. In the case of human viruses, nucleic acid enters the host cell by (1) direct entry of just the nucleic acid, as with poliovirus; (2) endocytosis and the release of nucleic acid from the capsid (uncoating), as with the poxviruses; or (3) fusion of the viral envelope with the cell membrane and subsequent uncoating, as with influenza virus (*see figure 18.4*).

Some viruses replicate at the site of entry, cause disease at the same site (e.g., respiratory and gastrointestinal infections), and do not spread throughout the body. Others spread to sites distant from the point of entry and replicate at these sites. For example, the poliovirus enters through the gastrointestinal tract but produces disease in the central nervous system. Food-borne and waterborne diseases: Poliomyelitis (section 37.4)

Release from Host Cells

The details by which various viruses exit their host cells are described in chapter 18. Briefly, there are two distinct release mechanisms that viruses use. The first mechanism is very dramatic and results in relatively large numbers of virions leaving the host cell at the same time and host cell death. This mechanism is called host cell lysis. Replication of viral particles increases within the host cell until the cell membrane can no longer contain all that is within its boundaries. The cell simply expands beyond a size that the cell membrane can maintain its integrity—it lyses. The virions are then free to infect other susceptible cells.

The second general release mechanism is called budding or "blebbing." Here a newly formed nucleocapsid pushes against the host cell membrane until the membrane evaginates and pinches off behind the virus. The released virus is coated with host cell membrane, now called the viral envelope. Release of viral particles by budding is a slower process than lysis; exiting viral particles take relatively small amounts of host cell membrane. The host cell can replenish its membrane permitting continued virus release over the life of the infected cell (*see figure 18.11*).

Viral Spread and Cell Tropism

Mechanisms of viral spread vary, but the most common routes are the bloodstream and lymphatic system. The presence of viruses in the blood is called **viremia.** In some instances, spread is by way of nerves (e.g., rabies, herpes simplex, and varicella-zoster viruses).

Viruses exhibit cell, tissue, and organ specificities. These specificities are called **tropisms** (Greek *trope,* turning). A tropism by a specific virus usually reflects the presence of specific cell surface receptors on the eucaryotic host cell for that virus (*see figure 37.14*).

Virus-Host Interactions

The interaction between a virus and its host cell can result in a variety of effects. Viruses can be either cytopathic or noncytopathic. **Cytopathic viruses** are those that ultimately kill the host cell; the result is often local necrosis. Alternatively, cytopathic viruses can trigger **apoptosis,** or programmed cell death, which culminates in the death of the host cell, often before viral replication can occur (*see figure 16.15*). Although both involve death of host cells, necrosis and apoptosis are very different phenomena. Apoptosis is a normal process in multicellular organisms. It is used during development to remove cells or tissues that are longer needed. It involves nuclear degeneration, the partial digestion of many cell proteins by proteolytic enzymes called capsases, and the formation of apoptotic bodies (membrane-enclosed cell components), which are subsequently phagocytosed by macrophages. Unlike necrosis, the apoptotic cell does not lyse and release its contents. Rather, apoptosis is a controlled dismantling of the cell that results in cell death. In the case of viral infection, apoptosis also prevents virus replication and spread. Some viruses stimulate apoptosis but use special viral proteins to prolong the process long enough to complete viral replication. This ensures that the virus can continue to infect new host cells.

Noncytopathic viruses do not immediately produce cell death and result in latent or persistent infections. As a result, noncytopathic viruses can be subdivided into productive and nonproductive. Noncytopathic viruses that produce persistent infection with the release of only a few new viral particles at a time are said to be productive. Noncytopathic viruses that do not actively make virus at detectable levels for a period of time (latent infection) are considered nonproductive. However, these viruses may be triggered to a reactivated (productive) state by environmental stressors or other factors. Persistent, latent, and slow virus infections (section 18.4)

As anyone who has ever had the flu or a cold knows, clinical illness may be a result of virus-host cell interactions. Some tissues, such as intestinal epithelium, can quickly regenerate when damaged by viruses. Thus they are easily repaired following cellular damage. In contrast, tissues of the nervous system are limited in their ability to regenerate and are thus difficult to repair following damage by viruses. Moreover, infection of cells with some viruses can result in the integration of viral DNA. In a few cases, this can cause them to transform into cancerous cells. This is the result of viral DNA interference with host DNA growth cycle regulation. Viruses and cancer (section 18.5)

Host Immune Response

Both humoral and cellular components of the immune response are involved in the control of viral infections and are discussed in detail in chapters 31 and 32 and summerized in section 32.9.

Recovery from Infection

The host will either succumb to or recover from a viral infection. Recovery processes involve nonspecific defense mechanisms and specific humoral and cellular immunity. The relative importance of each of these factors varies with the virus and the disease, and is covered in chapter 37.

Virus Shedding

The last step in the infectious process is shedding of the virus back into the environment. This is necessary to maintain a source of viruses in a population of hosts. Shedding often occurs from the same body surface used for entry. During this period, an infected host is infectious (contagious) and can spread the virus. In some viral infections, such as a rabies infection, the infected human is the final host because virus shedding does not occur.

1. For a virus to cause disease, certain steps are usually accomplished. Briefly describe each of these steps.
2. If you were to design an antiviral drug, which step or steps in the viral life cycle would you target? Explain your answer.
3. What are the four most common patterns of viral infections? Describe apoptosis and its role in viral infections.

33.3 Overview of Bacterial Pathogenesis

The steps for infections by pathogenic bacteria usually include the following:

1. Maintain a reservoir. A reservoir is a place to live and multiply before and after causing an infection.
2. Initial transport to and entry into the host.
3. Adhere to, colonize, and/or invade host cells or tissues.
4. Evade host defense mechanisms.
5. Multiply (grow) or complete its life cycle on or in the host or the host's cells.
6. Damage the host.
7. Leave the host and return to the reservoir or enter a new host.

The first five factors influence the degree of infectivity and invasiveness. Toxigenicity plays a major role in the sixth. These determinants are now discussed in more detail.

Maintaining a Reservoir of the Bacterial Pathogen

All bacterial pathogens must have at least one reservoir. The most common reservoirs for human pathogens are other humans, animals, and the environment. Since the source and/or reservoir of the pathogen is part of the infectious disease cycle, this aspect of pathogenicity is discussed in detail in chapter 36, which covers the epidemiology of infectious diseases. The infectious disease cycle (section 36.5)

Transport of the Bacterial Pathogen to the Host

An essential feature in the development of an infectious disease is the initial transport of the bacterial pathogen to the host. The most obvious means is direct contact—from host to host (coughing, sneezing, body contact). Bacteria also are transmitted indirectly in a variety of ways. Infected hosts shed bacteria into their surroundings. Once in the environment bacteria can be deposited on various surfaces, from which they can be either resuspended into the air or indirectly transmitted to a host. Soil, water, and food are indirect vehicles that harbor and transmit bacteria to hosts. Arthropod vectors and **fomites** (inanimate objects that harbor and transmit pathogens) also are involved in the spread of many bacteria.

Attachment and Colonization by the Bacterial Pathogen

After being transmitted to an appropriate host, the bacterial pathogen must be able to adhere to and colonize host cells or tissues. In this context **colonization** means the establishment of a site of microbial reproduction on or within a host. It does not necessarily result in tissue invasion or damage. Colonization depends on the ability of the bacteria to survive in the new (host) environment and to compete successfully with the host's normal microbiota for essential nutrients. Specialized structures that allow bacteria to compete for surface attachment sites also are necessary for colonization.

Bacterial pathogens adhere with a high degree of specificity to particular tissues. Adherence structures such as pili and fimbriae (**table 33.2**), and specialized adhesion molecules on the bacterium's cell surface that bind to complementary receptor sites on the host cell surface (**figure 33.3**), facilitate bacterial attachment to host cells. They are one type of virulence factor. Recall that virulence factors are bacterial products or structural components (e.g., capsules and adhesins) that contribute to virulence or pathogenicity. Components external to the cell wall (section 3.9)

Invasion of Host Tissues

Entry into host cells and tissues is a specialized strategy used by many bacterial pathogens for survival and multiplication. Pathogens often actively penetrate the host's mucous membranes and epithelium after attachment to the epithelial surface. This may be accomplished through production of lytic substances that alter the host tissue by (1) attacking the extracellular matrix and basement membranes of integuments and intestinal linings, (2) degrading carbohydrate-protein complexes between cells or on the cell surface (the glycocalyx), or (3) disrupting the cell surface.

At times a bacterial pathogen can penetrate the epithelial surface by passive mechanisms not related to the pathogen itself. Examples include (1) small breaks, lesions, or ulcers in a mucous membrane that permit initial entry; (2) wounds, abrasions, or burns on the skin's surface; (3) arthropod vectors that create small wounds while feeding; (4) tissue damage caused by other organisms; (e.g., a dog bite) and (5) existing eucaryotic internalization pathways (e.g., endocytosis and phagocytosis. Phagocytosis (section 31.3)

Table 33.2 Bacterial Adherence Factors That Play a Role in Infectious Diseases

Adherence Factor	Description
Fimbriae	Filamentous structures that help attach bacteria to other bacteria or to solid surfaces
Glycocalyx or capsule	A layer of exopolysaccharide fibers with a distinct outer margin that surrounds many cells; it inhibits phagocytosis and aids in adherence; when the layer is well organized and not easily washed off it is called a capsule
Pili	Filamentous structures that bind procaryotes together for the transfer of genetic material
S layer	The outermost regularly structured layer of cell envelopes of some bacteria that may promote adherence to surfaces
Slime layer	A bacterial film that is less compact than a capsule and is removed easily
Teichoic and lipoteichoic acids	Cell wall components in gram-positive bacteria that aid in adhesion

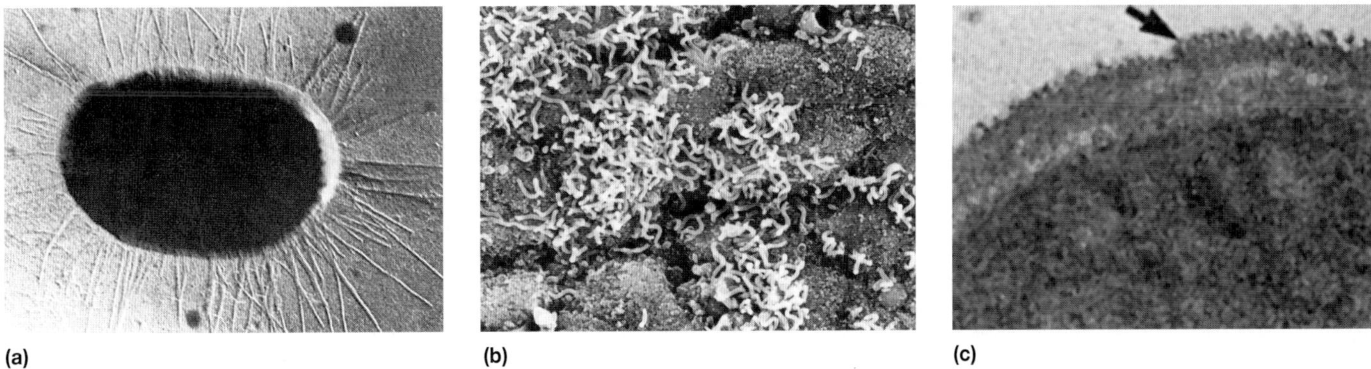

Figure 33.3 Microbial Adherence. **(a)** Transmission electron micrograph of fimbriated *Escherichia coli* (×16,625). **(b)** Scanning electron micrograph of epithelial cells with adhering vibrios (×1,200). **(c)** *Candida albicans* fimbriae (arrow) are used to attach the fungus to vaginal epithelial cells.

Once under the mucous membrane, the bacterial pathogen may penetrate to deeper tissues and continue disseminating throughout the body of the host. One way the pathogen accomplishes this is by producing specific structures and/or enzymes that promote spreading (**table 33.3**). These products represent other types of virulence factors. Bacteria may also enter the small terminal lymphatic capillaries that surround epithelial cells. These capillaries merge into large lymphatic vessels that eventually drain into the circulatory system. Once the circulatory system is reached, the bacteria have access to all organs and systems of the host.

Bacterial invasiveness varies greatly among pathogens. For example, *Clostridium tetani* (cause of tetanus) produces a variety of virulence factors but is considered noninvasive. *Bacillus anthracis* (cause of anthrax) and *Yersinia pestis* (cause of plague) also produce substantial virulence factors and are highly invasive. Members of the genus *Streptococcus* span the spectrum of virulence factors and invasiveness.

Growth and Multiplication of the Bacterial Pathogen

For a bacterial pathogen to be successful in growth and reproduction (colonization), it must find an appropriate environment (e.g., nutrients, pH, temperature, redox potential) within the host. Those areas of the host's body that provide the most favorable conditions will harbor the pathogen and allow it to grow and multiply to produce an infection. Some bacteria can actively grow and multiply in the blood plasma. The presence of viable bacteria in the bloodstream is called **bacteremia.** The presence of bacteria or their toxins in the blood often is termed **septicemia** [Greek *septikos*, produced by putrefaction, and *haima*, blood].

Some bacteria are able to grow and multiply within various cells of a host. Organisms with this ability to live intracellularly are subdivided into two groups. Facultative intracellular pathogens are those organisms that can reside within the cells of the host or in the environment. An example of a facultative intracellular pathogen is *Brucella abortus*, which is capable of growth and replication within macrophages, neutrophils, and trophoblast cells. However, facultative intracellular pathogens can also be grown in pure culture without host cell support. In contrast, obligate intracellular pathogens are incapable of growth and multiplication outside a host cell. Examples of obligate intracellular pathogens include viruses and the rickettsia. These microbes cannot be grown in the laboratory outside of their host cells.

Leaving the Host

The last determinant of a successful bacterial pathogen is its ability to leave the host and enter either a new host or a reservoir. Unless a successful escape occurs, the disease cycle will be interrupted and the microorganism will not be perpetuated. Most bacteria employ passive escape mechanisms. Passive escape occurs when a pathogen or its progeny leave the host in feces, urine, droplets, saliva, or desquamated cells.

Regulation of Bacterial Virulence Factor Expression

As noted in many chapters, some pathogenic bacteria have adapted to both the free-living state and to an environment within a human host. In the adaptive process, these pathogens have evolved complex signal transduction pathways to regulate the genes necessary for virulence. A virulence factor may be present simply because the bacterium has been infected by a phage—that is, the genes for virulence factors reside on a lysogenic phage genome (prophage). Often environmental factors control the expression of the virulence genes. Common signals include temperature, osmolality, available iron, pH, specific ions, and other nutrient factors. Several examples are now presented.

The gene for diphtheria toxin from *Corynebacterium diphtheriae* (the pathogen that causes diphtheria) is carried on the temperate bacteriophage β, and its expression is regulated by iron. The toxin is produced only by strains lysogenized by the phage. Expression of the virulence genes of *Bordetella pertussis* (the pathogen that causes whooping cough) is enhanced when the bacteria grow at body temperature (37°C) and suppressed when grown at a lower temperature. Finally, the virulence factors of *Vibrio cholerae* (the pathogen that causes cholera) are carried on a temperate phage and regulated at various levels by many environmental factors. Expression of the cholera toxin is higher at pH 6 than at pH 8 and higher at 30 than at 37°C.

Table 33.3 Microbial Products (Virulence Factors) Involved in Bacterial Pathogen Dissemination Throughout a Mammalian Host

Product	Organism Involved	Mechanism of Action
Coagulase	*Staphylococcus aureus*	Coagulates (clots) the fibrinogen in plasma. The clot protects the pathogen from phagocytosis and isolates it from other host defenses.
Collagenase	*Clostridium* spp.	Breaks down collagen that forms the framework of connective tissues; allows the pathogen to spread.
Deoxyribonuclease (along with calcium and magnesium)	Group A streptococci, staphylococci, *Clostridium perfringens*	Lowers viscosity of exudates, giving the pathogen more mobility.
Elastase and alkaline protease	*Pseudomonas aeruginosa*	Cleaves laminin associated with basement membranes.
Hemolysins	Staphylococci, streptococci, *Escherichia coli, Clostridium perfringens*	Lyse erythrocytes; make iron available for microbial growth.
Hyaluronidase	Groups A, B, C, and G streptococci, staphylococci, clostridia	Hydrolyzes hyaluronic acid, a constituent of the extracellular matrix that cements cells together and renders the intercellular spaces amenable to passage by the pathogen.
Hydrogen peroxide (H_2O_2) and ammonia (NH_3)	*Mycoplasma* spp., *Ureaplasma* spp.	Are produced as metabolic wastes. These are toxic and damage epithelia in respiratory and urogenital systems.
Immunoglobulin A protease	*Streptococcus pneumoniae*	Cleaves immunoglobulin A into Fab and Fc fragments.
Lecithinase or phospholipase	*Clostridium* spp.	Destroys the lecithin (phosphatidylcholine) component of plasma membranes, allowing pathogen to spread.
Leukocidins	Staphylococci, pneumococci, streptococci	Pore-forming exotoxins that kill leukocytes; cause degranulation of lysosomes within leukocytes, which decreases host resistance.
Porins	*Salmonella enterica* serovar Typhimurium	Inhibit leukocyte phagocytosis by activating the adenylate cyclase system.
Protein A Protein G	*Staphylococcus aureus* *Streptococcus pyogenes*	Located on cell wall. Immunoglobulin G (IgG) binds to protein A by its Fc end, thereby preventing complement from interacting with bound IgG.
Pyrogenic exotoxin B (cysteine protease)	Group A streptococci, (*Streptococcus pyogenes*)	Degrades proteins.
Streptokinase (fibrinolysin, staphylokinase)	Group A, C, and G streptococci, staphylococci	A protein that binds to plasminogen and activates the production of plasmin, thus digesting fibrin clots; this allows the pathogen to move from the clotted area.

Pathogenicity Islands

The genes that encode major virulence factors in many bacteria (e.g., *Yersinia* spp., *Pseudomonas aeruginosa, Shigella flexneri, Salmonella,* enteropathogenic *E. coli*) are found on large segments of DNA, called **pathogenicity islands,** which carry genes responsible for virulence. Pathogenicity islands have been acquired during evolution by horizontal gene transfer. A pathogen may have more than one pathogenicity island. They have several common sequence characteristics. The 3′ and 5′ ends of the islands contain insertion-like elements, suggesting their promiscuity as mobile genetic elements. The G + C nucleotide content of pathogenicity islands differs significantly from the G + C content of the remaining bacterial genome. The pathogenicity island DNA also exhibits several open reading frames, suggesting other putative genes. Interestingly, pathogenicity islands are typically associated with genes that encode tRNA. An excellent example of virulence genes carried in a pathogenicity island are those involved in protein secretion. So far, five pathways of protein secretion (types I to V) have been described in gram-negative bacteria. A set of approximately 25 genes encodes a pathogenicity mechanism termed the **type III secretion system (TTSS)** that enables gram-negative bacteria to secrete and inject virulence proteins into the cytoplasm of eucaryotic host cells. Protein secretion in procaryotes (section 3.8)

Many gram-negative bacteria that live in close relationships with host organisms are able to modulate host activities by secreting proteins directly into the interior of the host cell using the TTSS. Perhaps the best studied TTSS is that of *Yersinia pestis* and *Y. enterocolitica,* which cause bubonic plague and gastroenteritis, respectively. Both bacteria use the same plasmid-encoded TTSS consisting of the Yop (*Yersinia* *o*uter *p*rotein) secretion (Ysc) injectisome and secreted Yop products (**figure 33.4*a***). The TTSS injectisome is composed of a basal body and a needle. The

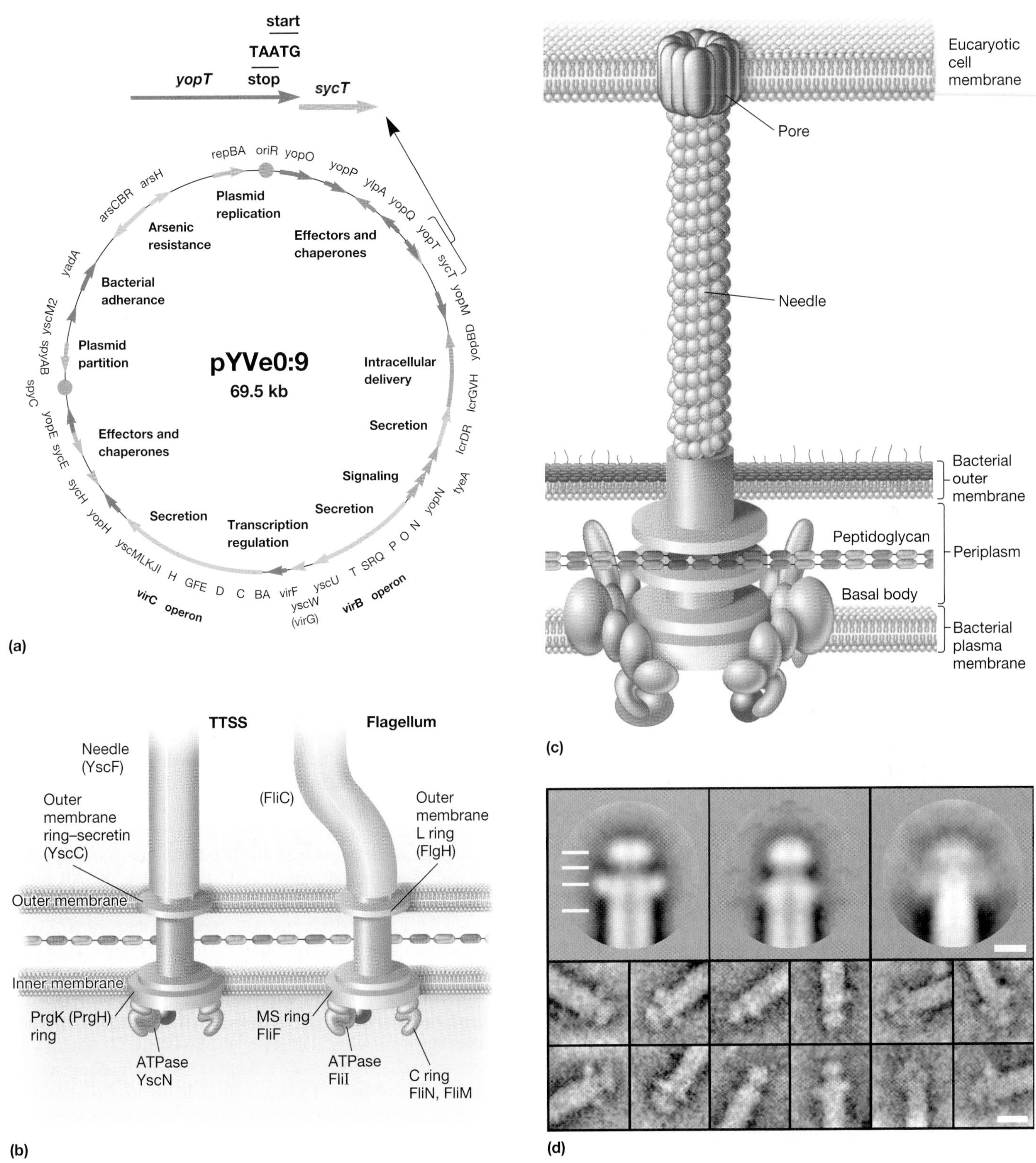

Figure 33.4 Type III Secretion System. **(a)** The type III secretion system (TTSS) and other virulence genes of *Yersinia* are encoded on the pYV plasmid. The TTSS genes encoding the *Yersinia* outer proteins (Yop) are homologous to many of the genes encoding flagellar proteins. **(b)** Both the TTSS injectisome and the flagella are anchored in the plasma membrane by similar basal body structures. **(c)** X-ray fiber diffraction resolves the injectisome as a helical structure. **(d)** Scanning tunneling electron microscopy reveals the injectisome tip, indicating how it may lock into the translocator pore on the target cell.

basal body is made from a number of proteins that are homologous (having similar amino acid sequences) to proteins that make up the basal body of bacterial flagella. This suggests that the injectisome is held in the bacterial envelope, similar to the ring system that holds a flagellum (figure 33.4*b*). The injectisome employs an ATPase that "energizes" the transport of other TTSS proteins, called "effectors," through "translocator pores" (formed by YopB and YopD proteins), into the host cell. Another protein, LcrV (also known as V antigen), is required for the correct assembly of the translocator pores. X-ray fiber diffraction of TTSS of *Shigella* demonstrates that the needle component of the TTSS has a helical arrangement (figure 33.4*c*). Scanning tunneling electron microscopy of the injectisome needle reveals a characteristic tip (head, neck, and base) through which a central channel is seen (figure 34.4*d*). LcrV localizes at the tip of the needle, which may explain its critical role in facilitating transport of TTSS-mediated proteins and their role in *Yersinia* virulence.

Unlike other bacterial secretory systems, the type III system is triggered specifically by contact with host cells, which helps avoid inappropriate activation of host defenses. Secretion of these virulence proteins into a host cell allows the pathogen to subvert the host cell's normal signal transduction system. Redirection of cellular signal transduction can disarm host immune responses or reorganize the cytoskeleton, thus establishing subcellular niches for bacterial colonization and facilitating "stealth and interdiction" of host defense communication lines.

Pathogenicity islands generally increase microbial virulence and are absent in nonpathogenic members of the same genus or species. One specific example is found in *E. coli.* The enteropathogenic *E. coli* strains possess large DNA fragments, 35 to 170 kilobases in size, that contain several virulence genes absent from commensal *E. coli* strains. Some of these genes code for proteins that alter actin microfilaments within a host intestinal cell. As a consequence, the host cell surface bulges and develops into a cuplike pedestal to which the bacterium tightly binds.

1. What seven steps are involved in the infection process and pathogenesis of bacterial diseases?
2. What are some ways in which bacterial pathogens are transmitted to their hosts? Define vector and fomite.
3. Describe several specific adhesins by which bacterial pathogens attach to host cells.
4. Once under the mucus and epithelial surfaces, what are some mechanisms that bacterial pathogens possess to promote their dissemination throughout the body of a host?
5. What are virulence factors? Pathogenicity islands?

33.4 Toxigenicity

Two distinct categories of disease can be recognized based on the role of the bacteria in the disease-causing process: infections and intoxications. An infectious disease results partly from the pathogen's growth and reproduction (or invasiveness) that often produce tissue alterations.

Intoxications are diseases that result from a specific toxin (e.g., botulinum toxin) produced by bacteria. Some toxins are only produced during host infection. Toxins can even induce disease in the absence of the organism that produced them. A **toxin** [Latin *toxicum,* poison] is a substance, such as a metabolic product of the organism, that alters the normal metabolism of host cells with deleterious effects on the host. The term **toxemia** refers to the condition caused by toxins that have entered the blood of the host. Some toxins are so potent that even if the bacteria that produced them are eliminated (for instance, by antibiotic chemotherapy), the disease conditions persist. Toxins produced by bacteria can be divided into two main categories: exotoxins and endotoxins. The primary characteristics of the two groups are compared in **table 33.4.**

Exotoxins

Exotoxins are soluble, heat-labile, proteins (a few are enzymes) that usually are released into the surroundings as the bacterial pathogen grows. In general, exotoxins are produced by gram-positive bacteria, although some gram-negative bacteria also make exotoxins. Often exotoxins may travel from the site of infection to other body tissues or target cells in which they exert their effects.

Exotoxins are usually synthesized by specific bacteria that often have plasmids or prophages bearing the toxin genes. They are associated with specific diseases and often are named for the disease they produce (e.g., the diphtheria toxin). Exotoxins are among the most lethal substances known; they are toxic in microgram-per-kilogram concentrations (e.g., botulinum toxin), but are typically heat-labile (inactivated at 60 to 80°C). Exotoxins are proteins that exert their biological activity by specific mechanisms. As proteins, the toxins are highly immunogenic and can stimulate the production of neutralizing antibodies called **antitoxins.** The toxin proteins can also be inactivated by formaldehyde, iodine, and other chemicals to form immunogenic **toxoids** (tetanus toxoid, for example). In fact, the tetanus vaccine is a solution of tetanus toxoid.

Exotoxins can be grouped into four types based on their structure and physiological activities. (1) One type is the AB toxin, which gets its name from the fact that the portion of the toxin (B) that binds to a host cell receptor is separate from the portion (A) that has the enzyme activity that causes the toxicity (**figure 33.5*a***). (2) A second type, which also may be an AB toxin, consists of those toxins that affect a specific host site (nervous tissue [neurotoxins], the intestines [enterotoxins], general tissues [cytotoxins]) by acting extracellularly or intracellularly on the host cells. (3) A third type does not have separable A and B portions and acts by disorganizing host cell membranes. Examples include the leukocidins, hemolysins, and phospholipases. (4) A fourth type is the superantigen that acts by stimulating T cells directly to release cytokines. Examples of these types are now discussed. The general properties of some AB exotoxins are presented in **table 33.5.**

AB Toxins

AB toxins are composed of an enzymatic subunit (A) that is responsible for the toxic effect once inside the host cell and a binding subunit (B) (figure 33.5). Isolated A subunits are enzymatically active but lack binding and cell entry capability, whereas isolated

Table 33.4 Characteristics of Exotoxins and Endotoxins

Characteristic	Exotoxins	Endotoxins
Chemical composition	Protein, often with two components (A and B)	Lipopolysaccharide complex on outer membrane; lipid A portion is toxic
Disease examples	Botulism, diphtheria, tetanus	Gram-negative infections, meningococcemia
Effect on host	Highly variable between different toxins	Similar for all endotoxins
Fever	Usually do not produce fever	Produce fever by induction of interleukin-1 and TNF
Genetics	Frequently carried by extrachromosomal genes such as plasmids	Synthesized directly by chromosomal genes
Heat stability	Most are heat sensitive and inactivated at 60-80°C	Heat stable to 250°C
Immune response	Antitoxins provide host immunity; highly antigenic	Weakly immunogenic; immunogenicity associated with polysaccharide
Location	Usually excreted outside the living cell	Part of outer membrane of gram-negative bacteria
Production	Produced by both gram-positive and gram-negative bacteria	Found only in gram-negative bacteria; Released on bacterial death and some liberated during growth
Toxicity	Highly toxic and fatal in nanogram quantities	Less potent and less specific than exotoxin; causes septic shock
Toxoid production	Converted to antigenic, nontoxic toxoids; toxoids are used to immunize (e.g., tetanus toxoid)	Toxoids cannot be made

B subunits bind to target cells but are nontoxic and biologically inactive. The B subunit interacts with specific receptors on the target cell or tissue such as the gangliosides GM1 for cholera toxin, GT1 and/or GD1 for tetanus toxin, and SV2 for botulinum toxin. The B subunit therefore determines what cell type the toxin will affect.

Several mechanisms for the entry of A subunits or fragments into target cells have been proposed. In one mechanism the B subunit inserts into the plasma membrane and creates a pore through which the A subunit enters (figure 33.5*a*). In another mechanism entry is by receptor-mediated endocytosis (figure 33.5*b*).

The mechanism of action of an AB toxin can be quite complex, as shown by the example of diphtheria toxin (figure 33.5*b*). The diphtheria toxin is a protein of about 62,000 Daltons. It binds to cell surface receptors by the B subunit and is taken into the cell through the formation of a clathrin-coated vesicle. The toxin then enters the vesicle membrane and the two subunits are separated; the A subunit escapes into the cytosol. The A subunit is an enzyme that catalyzes the addition of an ADP-ribose group to the eucaryotic elongation factor EF2 that aids in translocation during protein synthesis. The substrate for this reaction is the coenzyme NAD^+.

$$NAD^+ + EF2 \rightarrow ADP\text{-ribosyl-}EF2 + \text{nicotinamide}$$

The modified EF2 protein cannot participate in the elongation cycle of protein synthesis, and the cell dies because it can no longer synthesize proteins. ADP-ribosylation is a common mechanism for the A subunit of a number of toxins; however, the specific host molecule to which the ADP-ribose group is attached differs. AB exotoxins vary widely in their relative contribution to the disease process with which they are associated.

A variation of this AB toxin is the cytolethal distending toxin (CDT) produced by *Campylobacter* spp. Discovered in 1987, CDT is a tripartite holotoxin complex encoded by three tandem genes; *cdtA, cdtB, and cdtC.* CDT binding and internalization appear to be encoded by the *cdtA* and *cdtC* genes, while the active component of the holotoxin is encoded within the *cdtB* gene. The predicted amino acid sequence of CdtB is homologous to proteins having deoxyribonuclease (DNase) I activity. However, the mechanism of CDT in disease is unclear. *C. jejuni* has all three genes. In culture with epithelial cells, the CDT of *C. jejuni* induces a progressive epithelial cell distension resulting from an irreversible blockage of the cell cycle at the G2/M phase. This leads to oversized cells without cell division (distension) and cell death.

Specific Host Site Exotoxins

The second type of exotoxin is categorized on the basis of the site affected: **neurotoxins** (nerve tissue), **enterotoxins** (intestinal mucosa), and **cytotoxins** (general tissues). Some of the bacterial pathogens that produce these exotoxins are presented in table 33.5: neurotoxins (botulinum toxin and tetanus toxin), enterotoxins (cholera toxin, *E. coli* heat labile toxins), and cytotoxins (diphtheria toxin, Shiga toxin). Note that many AB toxins are also host site specific, thus these categories are not mutually exclusive.

Neurotoxins usually are ingested as preformed toxins that affect the nervous system and indirectly cause enteric (pertaining to the small intestine) symptoms. Examples include staphylococcal enterotoxin B, *Bacillus cereus* emetic toxin [Greek *emetos,* vomiting], and botulinum toxin.

True enterotoxins [Greek *enter,* intestine] have a direct effect on the intestinal mucosa and elicit profuse fluid secretion (diarrhea). The classic enterotoxin, cholera toxin (choleragen), has been studied extensively. It is an AB toxin. The B subunit is made of five parts arranged as a donut-shaped ring. The B subunit ring anchors itself to the epithelial cell's plasma membrane and then inserts the smaller A subunit into the cell. The A subunit ADP-ribosylates and thereby activates tissue adenylate cyclase to increase intestinal cyclic AMP (cAMP) concentrations. High concentrations of cAMP provoke the movement of massive quantities of water and electrolytes across the intestinal cells into

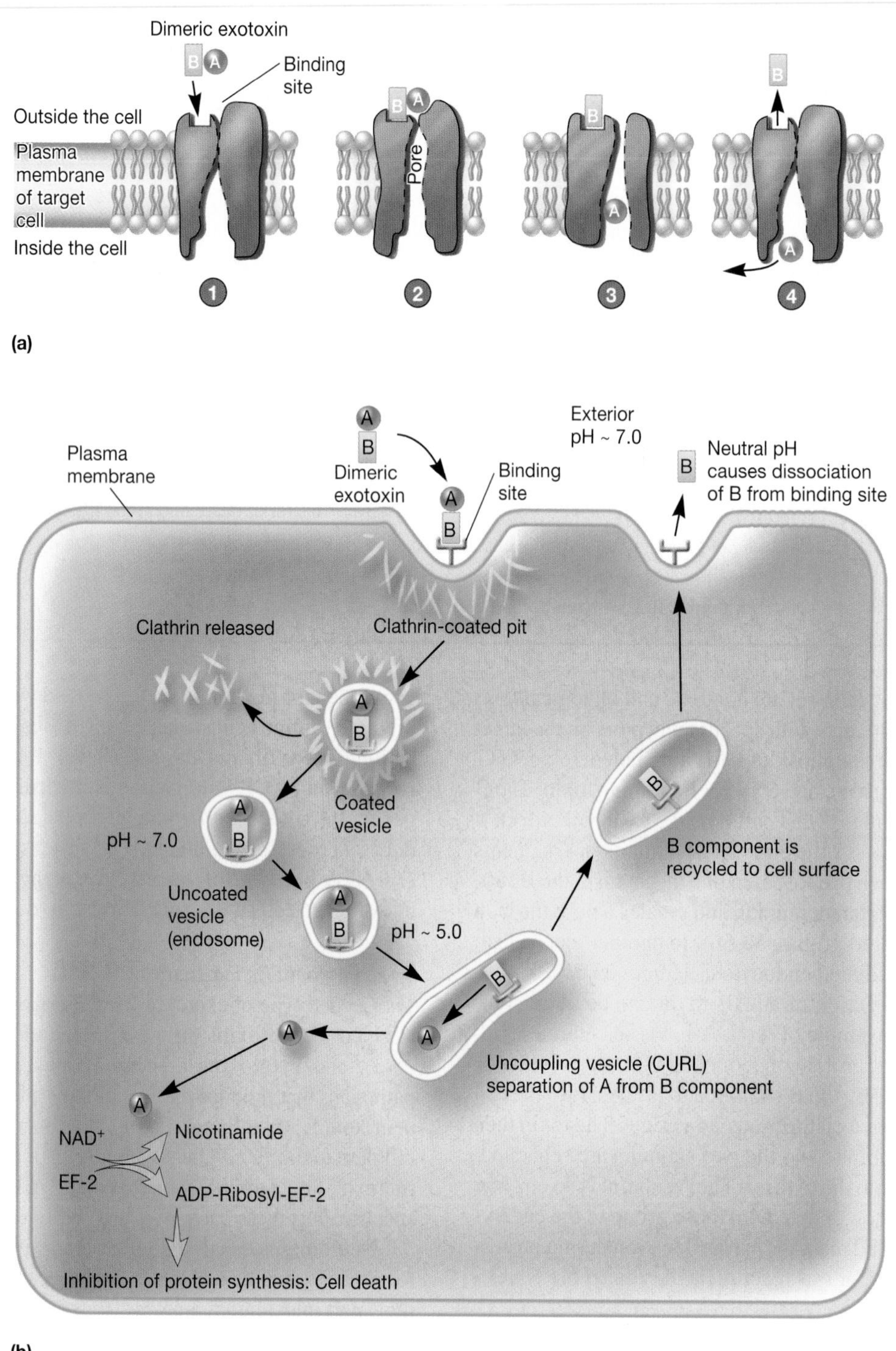

Figure 33.5 Two AB Exotoxin Transport Mechanisms **(a)** Subunit B of the dimeric exotoxin (AB) binds to a specific membrane receptor of a target cell [1]. A conformational change [2] generates a pore [3] through which the A subunit crosses the membrane and enters the cytosol, followed by re-creation [4] of the binding site. **(b)** Receptor-mediated endocytosis of the diphtheria toxin involves the dimeric exotoxin binding to a receptor-ligand complex that is internalized in a clathrin-coated pit that pinches off to become a coated vesicle. The clathrin coat depolymerizes resulting in an uncoated endosome vesicle. The pH in the endosome decreases due to the H^+-ATPase activity. The low pH causes A and B components to separate. An endosome in which this separation occurs is sometimes called a CURL (*c*ompartment of *u*ncoupling of *r*eceptor and *l*igand). The B subunit is then recycled to the cell surface. The A subunit moves through the cytosol, catalyzes the ADP-ribosylation of EF-2 (elongation factor 2) and inhibits protein synthesis, leading to cell death.

Table 33.5	Properties of Some AB Model Bacterial Exotoxins					
Toxin	**Organism**	**Gene Location**	**Subunit Structure**	**Target Cell Receptor**	**Enzymatic Activity**	**Biologic Effects**
Anthrax toxins	*Bacillus anthracis*	Plasmid	Three separate proteins (EF, LF, PA)[a]	Capillary morphogenesis protein 2 (CMP-2) and tumor endothelium marker 8 (TEM8)	EF is a calmodulin-dependent adenylate cyclase; LF is a zinc-dependent protease that cleaves a host signal transduction molecule (MAPKK)	EF + PA: increase in target cell cAMP level, localized edema; LF + PA: altered cell signaling; death of target cells
Bordetella adenylate cyclase toxin	*Bordetella* spp.	Chromosomal	A-B[b]	CR3 intergrin (CD11–CD18)	Calmodulin-activated adenylate cyclase	Increase in target cell cAMP level; decrease ATP production; modified cell function or cell death
Botulinum toxin	*Clostridium botulinum*	Phage	A-B[c]	Synaptic vesicle 2 (SV2)	Zinc-dependent endoprotease cleavage of presynaptic protein (SNARE)	Decrease in peripheral, presynaptic acetylcholine release; flaccid paralysis
Cholera toxin	*Vibro cholera*	Phage	A-5B[d]	Ganglioside (GM_1)	ADP ribosylation of adenylate cyclase regulatory protein, G_S	Activation of adenylate cyclase, increase in cAMP level; secretory diarrhea
Diphtheria toxin	*Corynebacterium diphtheriae*	Phage	A-B[e]	Heparin-binding, EGF-like growth factor precursor	ADP ribosylation of elongation factor 2	Inhibition of protein synthesis; cell death
Heat-labile enterotoxins[f]	*E. coli*	Plasmid	—— Similar or Identical to Cholera Toxin ——			
Pertussis toxin	*Bordetella pertussis*	Chromosomal	A-5B[g]	Asparagine-linked oligosaccharide and lactosylceramide sequences	ADP ribosylation of signal-transducing G proteins	Block of signal transduction mediated by target G proteins
Pseudomonas exotoxin A	*P. aeruginosa*	Chromosomal	A-B	α_2-Macroglobulin/LDL receptor	— Similar or Identical to Diphtheria Toxin —	
Shiga toxin	*Shigella dysenteriae*	Chromosomal	A-5B[h]	Globotriaosylceramide (Gb_3)	RNA *N*-glycosidase	Inhibition of protein synthesis, cell death
Shiga-like toxin 1	*Shigella* spp., *E. coli*	Phage	—— Similar or Identical to Shiga Toxin ——			
Tetanus toxin	*C. tetani*	Plasmid	A-B[c]	Ganglioside (GT_1 and/or GD_{1b})	Zinc-dependent endopeptidase cleavage of synaptobrevin	Decrease in neurotransmitter release from inhibitory neurons; spastic paralysis

[a]The binding component (known as protective antigen [PA]) catalyzes/facilitates the entry of either edema factor (EF) or lethal factor (LF).

[b]Apparently synthesized as a single polypeptide with binding and catalytic (adenylate cyclase) domains.

[c]Holotoxin is apparently synthesized as a single polypeptide and cleaved proteolytically as diphtheria toxin; subunits are referred to as L: light chain, A equivalent; H: heavy chain, B equivalent.

[d]The A subunit is proteolytically cleaved into A1 and A2, with A1 possessing the ADP-ribosyl transferase activity; the binding component is made up of five identical B units.

[e]Holotoxin is synthesized as a single polypeptide and cleaved proteolytically into A and B components held together by disulfide bonds.

[f]The heat-labile enterotoxins of *E. coli* are now recognized to be a family of related molecules with identical mechanisms of action.

[g]The binding portion is made up of two dissimilar heterodimers labeled S2-S3 and S2-S4 that are held together by a bridging peptide, SS.

[h]Subunit composition and structure similar to cholera toxin.

the lumen of the gut. To maintain osmotic homeostasis, the cell then relases this water; this results in severe dirarrhea (cholera victims can lose 20% of their water per day). The genes for this enterotoxigenicity are encoded on a filamentous phage within *Vibrio cholerae.* Food-borne and waterborne diseases: Cholera (section 38.4)

Cytotoxins have a specific toxic action upon cells/tissues of special organs and are named according to the type of cell/tissue or organ for which they are specific. Examples include nephrotoxin (kidney), hepatotoxin (liver), and cardiotoxin (heart).

Membrane-Disrupting Exotoxins

The third type of exotoxin lyses host cells by disrupting the integrity of the plasma membrane. There are two subtypes of **membrane-disrupting exotoxins.** The first, is a protein that binds to the cholesterol portion of the host cell plasma membrane, inserts itself into the membrane, and forms a channel (pore) (**figure 33.6*a***). This causes the cytoplasmic contents to leak out. Also, because the osmolality of the cytoplasm is higher than the extracellular fluid, this causes a sudden influx of water into the cell, causing it to swell and rupture. Two specific examples of this type of membrane-disrupting exotoxin are now presented.

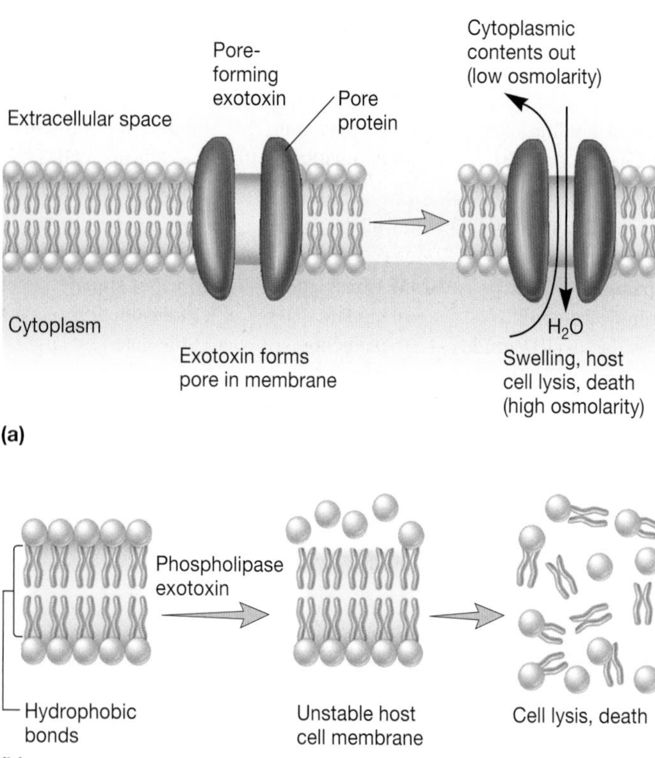

Figure 33.6 Two Subtypes of Membrane-Disrupting Exotoxins. **(a)** A channel-forming (pore-forming) type of exotoxin inserts itself into the normal host cell membrane and makes an open channel (pore). Formation of multiple pores causes cytoplasmic contents to leave the cell and water to move in, leading to cellular lysis and death of the host cell. **(b)** A phospholipid-hydrolyzing phospholipase exotoxin destroys membrane integrity. The exotoxin removes the charged polar head groups from the phospholipid part of the host cell membrane. This destabilizes the membrane and causes the host cell to lyse.

Some pathogens produce membrane-disrupting toxins that kill phagocytic leukocytes. These are termed **leukocidins** [*leuko*cyte and Latin *caedere,* to kill]. Most leukocidins are produced by pneumococci, streptococci, and staphylococci. Since the pore-forming exotoxin produced by these bacteria destroys leukocytes, this in turn decreases host resistance. Other toxins, called **hemolysins** [*haima,* blood, and Greek *lysis,* dissolution], also can be secreted by pathogenic bacteria. Many hemolysins probably form pores in the plasma membrane of erythrocytes through which hemoglobin and/or ions are released (the erythrocytes lyse or, more specifically, hemolyze). **Streptolysin-O (SLO)** is a hemolysin, produced by *Streptococcus pyogenes,* that is inactivated by O_2 (hence the "O" in its name). SLO causes beta hemolysis of erythrocytes on agar plates incubated anaerobically. A complete zone of clearing around the bacterial colony growing on blood agar is called **beta hemolysis,** and a partial clearing of the blood (leaving a greenish halo of hemoglobin) is called **alpha hemolysis. Streptolysin-S (SLS)** is also produced by *S. pyogenes* but is insoluble and bound to the bacterial cell. It is O_2 stable (hence the "S" in its name) and causes beta hemolysis on aerobically incubated blood-agar plates. In addition to hemolysins, SLO and SLS are also leukocidins and kill leukocytes. It should also be noted that hemolysins attack the plasma membranes of many cells, not just erythrocytes and leukocytes.

The second subtype of membrane-disrupting toxins are the **phospholipase** enzymes. Phospholipases remove the charged head group (figure 33.6*b*) from the lipid portion of the phospholipids in the host-cell plasma membrane. This destabilizes the membrane so that the cell lyses and dies. One example of the pathogenesis caused by phospholipases is observed in the disease gas gangrene. In this disease, the *Clostridium perfringens* α-toxin almost completely destroys the local population of white blood cells (that are drawn in by inflammation to fight the infection) through phospholipase activity.

Superantigens

As discussed in chapter 32, superantigens are bacterial and viral proteins that can provoke as many as 30% of a person's T cells to release massive concentrations of cytokines. The best-studied superantigen is also a staphylococcal enterotoxin. Staphylococcal entertoxin B (SEB) exhibits biological activity as a superantigen at nanogram concentrations and is therefore classified as a select agent; it has the potential to be misused as a bioterror agent. SEB exerts its superantigen activity by bridging the unfilled class II MHC molecules of antigen-presenting cells to T-cell receptors. Because no processed antigen is involved, many T cells are activated at once. This activation of T cells results in normal cytokine release; however, the sum total of the combined cytokines overwhelms cells and tissues. Cytokines stimulate endothelial damage, circulatory shock, and multiorgan failure. T-cell biology: Superantigens (section 32.5)

Roles of Exotoxins in Disease

Humans are exposed to bacterial exotoxins in three main ways: (1) ingestion of preformed exotoxin, (2) colonization of a mucosal surface followed by exotoxin production, and (3) colonization of a wound or abscess followed by local exotoxin production. Each of these is now briefly discussed.

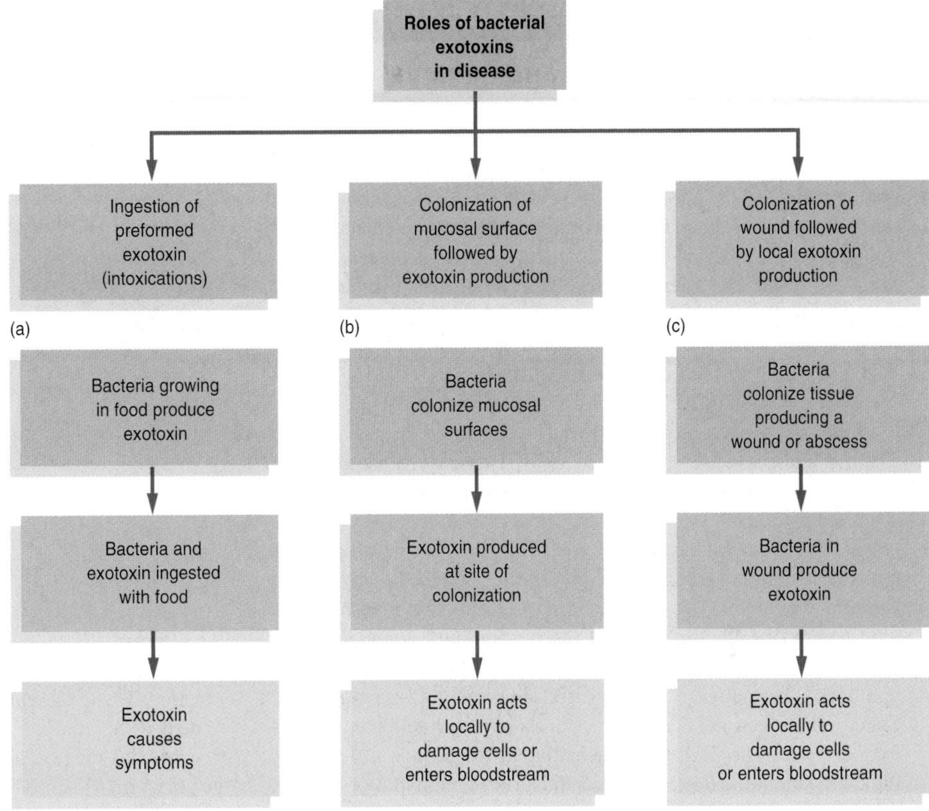

Figure 33.7 Roles of Exotoxins in Disease Pathogenesis. Three ways **(a, b, c)** in which bacterial exotoxins can contribute to the progression of disease in a human.

In the first example (**figure 33.7*a***), the exotoxin is produced by bacteria growing in food. When food is consumed, the preformed exotoxin is also consumed. The classical example is staphylococcal food poisoning caused solely by the ingestion of preformed enterotoxin. Since the bacteria (*Staphylococcus aureus*) cannot colonize the gut, they pass through the body without producing any more exotoxin; thus, this type of bacterial disease is self-limiting.

In the second example (figure 33.7*b*), bacteria colonize a mucosal surface but do not invade underlying tissue or enter the bloodstream. The toxin either causes disease locally or enters the bloodstream and is distributed systemically where it can cause disease at distant sites. The classical example here is the disease cholera caused by *Vibrio cholerae*. Once the bacteria enter the body, they adhere to the intestinal mucosa. They are not invasive but secrete the cholera toxin. As a result, cholera toxin stimulates hypersecretion of water and chloride ions and the patient loses massive quantities of water through the gastrointestinal tract.

The third example of exotoxins in disease pathogenesis occurs when bacteria grow in a wound or abscess (figure 33.7*c*). The exotoxin causes local tissue damage or kills phagocytes that enter the infected area. A disease of this type is gas gangrene in which the exotoxin (α-toxin) of *Clostridium perfringens* lyses red blood cells, induces edema, and causes tissue destruction in the wound.

1. What is the difference between an infectious disease and an intoxication? Define toxemia.
2. Describe some general characteristics of exotoxins.
3. How do exotoxins get into host cells?
4. Describe the biological effects of several bacterial exotoxins.
5. Discuss the mechanisms by which exotoxins can damage cells.
6. What are the four types of exotoxins?
7. What is the mode of action of a leukocidin? Of a hemolysin?
8. Name two specific hemolysins.
9. What are the three main roles exotoxins have in human disease pathogenesis?

Endotoxins

Gram-negative bacteria have lipopolysaccharide (LPS) in the outer membrane of their cell wall that, under certain circumstances, is toxic to specific hosts. This LPS is called an **endotoxin** because it is bound to the bacterium and is released when the microorganism lyses (**Techniques & Applications 33.1**). Some is also released during bacterial multiplication. The toxic component of the LPS is the lipid portion, called lipid A. Lipid A is not a single macromolecular structure but appears to be a complex array of lipid residues. The lipid A component exhibits all the properties associated with endotoxicity and gram-negative bacteremia. The bacterial cell wall: Gram-negative cell walls (section 3.6)

Besides the preceding characteristics, bacterial endotoxins are

1. Heat stable
2. Toxic (nanogram amounts)
3. Weakly immunogenic
4. Generally similar, despite source
5. Usually capable of producing general systematic effects: fever (are pyrogenic), shock, blood coagulation, weakness, diarrhea, inflammation, intestinal hemorrhage, and fibrinolysis (enzymatic breakdown of fibrin, the major protein component of blood clots)

Techniques & Applications

33.1 Detection and Removal of Endotoxins

Bacterial endotoxins plagued the pharmaceutical industry and medical device producers for years. For example, administration of drugs contaminated with endotoxins resulted in complications—even death—to patients. In addition, endotoxins can be problematic for individuals and firms working with cell cultures and genetic engineering. The result has been the development of sensitive tests and methods to identify and remove these endotoxins. The procedures must be very sensitive to trace amounts of endotoxins. Most firms have set a limit of 0.25 **endotoxin units (E.U.),** 0.025 ng/ml, or less as a release standard for their drugs, media, or products.

One of the most accurate tests for endotoxins is the in vitro *Limulus* amoebocyte lysate (LAL) assay. The assay is based on the observation that when an endotoxin contacts the clot protein from circulating amoebocytes of the horseshoe crab (*Limulus*), a gel-clot forms. The assay kits available today contain calcium, proclotting enzyme, and procoagulogen. The proclotting enzyme is activated by bacterial endotoxin (lipopolysaccharide) and calcium to form active clotting enzyme (see **Box figure**). Active clotting enzyme then catalyzes the cleavage of procoagulogen into polypeptide subunits (coagulogen). The subunits join by disulfide bonds to form a gel-clot. Spectrophotometry is then used to measure the protein precipitated by the lysate. The LAL test is sensitive at the nanogram level but must be standardized against Food and Drug Administration Bureau of Biologics endotoxin reference standards. Results are reported in endotoxin units per milliliter and reference made to the particular reference standards used.

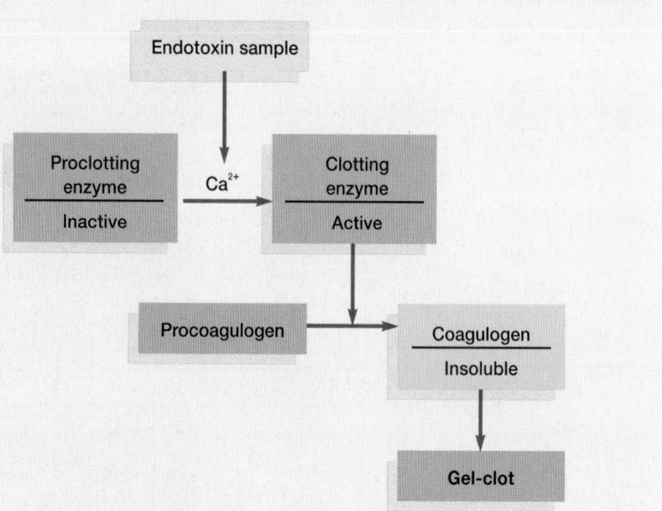

Removal of endotoxins presents more of a problem than their detection. Those present on glassware or medical devices can be inactivated if the equipment is heated at 250°C for 30 minutes. Soluble endotoxins range in size from 20 kDa to large aggregates with diameters up to 0.1 μm. Thus they cannot be removed by conventional filtration systems. Manufacturers have developed special filtration systems and filtration cartridges that retain these endotoxins and help alleviate contamination problems.

The characteristics of endotoxins and exotoxins are contrasted in table 33.4. The main biological effect of lipid A is an indirect one, being mediated by host molecules and systems rather than by lipid A directly. For example, endotoxins can initially activate Hageman Factor (blood clotting factor XII), which in turn activates up to four humoral systems: coagulation, complement, fibrinolytic, and kininogen systems (*see figure 38.26*).

Gram-negative endotoxins also indirectly induce a fever in the host by causing macrophages to release **endogenous pyrogens** that reset the hypothalamic thermostat. One important endogenous pyrogen is the cytokine interleukin-1 (IL-1). Other cytokines released by macrophages, such as the tumor necrosis factor, also produce fever.

Evidence indicates that LPS affects macrophages, monocytes, and neutrophils by binding to the soluble pattern-recognition receptor (formerly LPS-binding protein) for transfer to the membrane-bound CD14 on these cells. LPS-bound CD14 then complexes with toll-like receptor (TLR) 4 to initiate a signaling process that upregulates the phagocyte response to LPS. Part of this response is the synthesis and release of cytokines IL-1, IL-6, IL-8, tumor necrosis factor α, and platelet-activating factor. These and other proinflammatory mediators signal target cells resulting in fever, complement activation, prostaglandin synthesis, and activation of the coagulation cascade. Phagocytosis: Toll-like receptors (section 31.3); Chemical mediators in nonspecific resistance: Cytokines (section 31.6)

1. Describe the chemical structure of the LPS endotoxin.
2. List some general characteristics of endotoxins.
3. How do gram-negative endotoxins induce fever in a mammalian host?

33.5 Host Defense Against Microbial Invasion

In host-pathogen relationships, the balance between host integrity and resource utilization results in the co-evolution of survival strategies (**figure 33.8**). Competition, harsh environments, and cellular biowarfare have demanded unique solutions from host and pathogen alike. The complex, albeit "sneaky," methods employed by microbes to gain access to host resources have invariably been met with equally complex countermeasures. The host-pathogen relationship is indeed defined by ecological principles. The host provides a myriad of niches for microbes that have adapted to various temperature optima, nutrient content, oxygen concentration, and tolerance for host, as well as other microorganisms (antagonism).

Chapters 31 and 32 detail the innate and adaptive responses, respectively, available to the human host in preventing or limiting infection. A variety of physical, chemical, and biological barriers establish a formidable defense against microbial invasion.

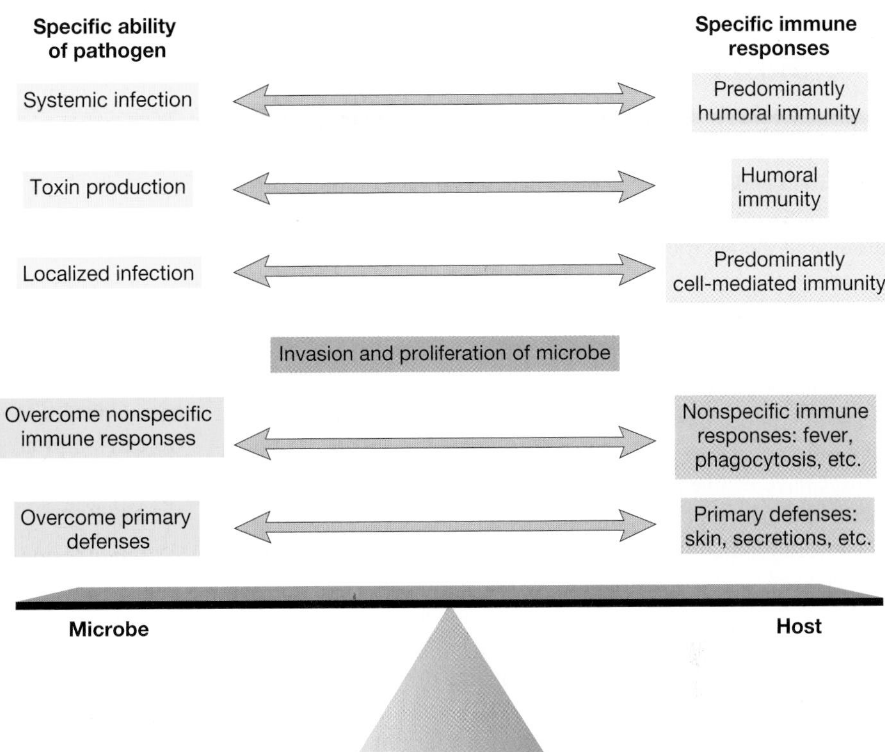

Figure 33.8 The Balance Between Microorganism Activity and Host Immunity. The interaction between the two determines the final host-parasite balance.

Nonetheless, microorganisms often breech these sophisticated barriers, occasionally prevailing. However, more recent evidence suggests that host cells are equipped with sensitive inter- and intracellular surveillance systems that recognize unique pathogen-associated patterns. As in medical diagnostics, early immune detection usually leads to early clearance of disease. Coupled with the ability to produce highly specific receptors and antibodies, the host prevents many more microbial invasions than those that are noticed.

Primary Defenses

Recall that the host has evolved several strategies to prevent pathogen entry. This is intuitively the most logical mechanism to develop. In other words, there would be no need for secondary defenses if the primary systems were 100% effective. An impenetrable suit of armor might restrict microbial entry, but limits other essential functions. Thus a multilayered skin, speckled with glands producing antimicrobial substances and an army of formidable microbial allies, is a sound compromise. Mucous membranes with cilia, pH regulation, flushing mechanisms, and additional antimicrobial products are efficient transition sites where the host tissues interface with their environment.

Secondary Defenses

Because the defenses of the skin and mucous membranes can be overcome, a secondary system has evolved. Composed of soluble antimicrobial products and cells capable of sensing and responding to invading microbes, the secondary system is quite effective in controlling infection. The host blood and interstitial fluids contain evolutionarily conserved, antimicrobial proteins that are very efficacious. These proteins include a variety of low-molecular-weight, "pore-forming" peptides and the ubiquitous lysozyme. Additionally, host cells have evolved strategies to exploit pathogen sensitivities to toxic oxygen radicals. Finally, sophisticated, soluble receptors police the host in search of microbial ligands. Binding of such ligands to their receptors initiates processes designed to amplify microbial detection and destruction. Examples of these processes include complement activation, inflammation, fever, and phagocytosis.

Factors Influencing Host Defenses

There are a variety of factors that influence the primary and secondary immune responses of the host. Age, stress, nutritional deficiencies, and genetic background all play substantial roles in host defense. The very young and the very old tend to be more susceptible to diseases. Some individuals are inherently (genetically) more resistant or more susceptible to particular diseases. Nutrition also influences these factors. In fact, historians have discovered a link between times of famine and times of disease. For example, nutritional deficiencies can decrease epithelial integrity, weaken antibody responses, and facilitate changes in the normal flora. Stress too can inhibit the immune response by stimulating the production of corticosteroids, which depress immune responses. Lastly, long-term exposure to environmental pollutants, drug abuse, or certain prescribed medicines may also inhibit normal host defenses against infection.

33.6 Microbial Mechanisms for Escaping Host Defenses

So far, we have discussed some of the ways viral and bacterial pathogens cause disease in a host. During the course of microbe and human evolution, these same pathogens have evolved ways for evading host defenses. Many of these mechanisms are found throughout the microbial world and several are now discussed.

Evasion of Host Defenses by Viruses

As noted earlier in this chapter, the pathology arising from a viral infection is due to either (1) the host's immune response, which attacks virus-infected cells or produces hypersensitivity reactions, or (2) the direct consequence of viral multiplication within host cells. Viruses have evolved a variety of ways to suppress or evade the host's immune response. These mechanisms are now becoming recognized through genomics and the functional analysis of specific gene products. Immune disorders: Hypersensitivities (section 32.11)

Some viruses may mutate and change antigenic sites (antigenic drift) on the virion proteins (e.g., the influenza virus) or may down-regulate the level of expression of viral cell surface proteins (e.g., the herpesvirus). Other viruses (HIV) may infect cells (T cells) of the immune system and diminish their function. HIV as well as the measles virus and cytomegalovirus cause the fusion of host cells. This allows these viruses to move from an infected cell to an uninfected cell without exposure to the antibody-containing fluids of the host. The herpesvirus may infect neurons that express little or no major histocompatibility complex molecules. The adenovirus produces proteins that inhibit major histocompatibility complex function. Finally, hepatitis B virus-infected cells produce large amounts of antigens not associated with the complete virus. These antigens bind the available neutralizing antibody so that there is insufficient free antibody to bind with the complete virion. Airborne diseases: Influenza (section 37.1); Recognition of foreignness (section 32.4)

Evasion of Host Defenses by Bacteria

Bacteria also have evolved many mechanisms to evade host defenses. Because bacteria would not be well served either by the death of their host or their own death, their survival strategy is protection against host defenses rather than host destruction.

Evading the Complement System

To evade the activity of complement, some bacteria have capsules (see chapter opening figure) that prevent complement activation. Some gram-negative bacteria can lengthen the O chains in their lipopolysaccharide to prevent complement activation. Others such as *Neisseria gonorrhoeae* generate **serum resistance.** These bacteria have modified lipooligosaccharides on their surface that interfere with proper formation of the membrane attack complex (*see figure 31.23*) during the complement cascade. The virulent forms of *N. gonorrhoeae* that possess serum resistance are able to spread throughout the body of the host and cause systemic disease, whereas those *N. gonorrhoeae* that lack serum resistance remain localized in the genital tract. Chemical mediators in nonspecific resistance: Complement (section 31.6)

Resisting Phagocytosis

As noted previously, before a phagocytic cell can engulf a bacterium, it must first directly contact the bacterium's surface. Some bacteria such as *Streptococcus pneumoniae, Neisseria meningitidis,* and *Haemophilus influenzae* can produce a slippery mucoid capsule that prevents the phagocyte from effectively contacting the bacterium. Other bacteria evade phagocytosis by producing specialized surface proteins such as the M protein on *S. pyogenes.* Like capsules, these proteins interfere with adherence between a phagocytic cell and the bacterium.

Bacterial pathogens use other mechanisms to resist phagocytosis. For example, *Staphylococcus* produces leukocidins that destroy phagocytes before phagocytosis can occur. *S. pyogenes* releases a protease that cleaves the C5a complement factor and thus inhibits complement's ability to attract phagocytes to the infected area.

Survival Inside Phagocytic Cells

Some bacteria have evolved the ability to survive inside neutrophils, monocytes, and macrophages. They are very pathogenic because they are impervious to a most important host protective mechanism. One method of evasion is to escape from the phagosome before it merges with the lysosome, as seen with *Listeria monocytogenes, Shigella,* and *Rickettsia.* These bacteria use actin-based motility to move within mammalian host cells and spread between them. Upon lysing the phagosome, they gain access to the cytoplasm. Each bacterium then recruits to its surface host cell actin and other cytoskeletal proteins and activates the assembly of an actin tail (**figure 33.9*a***). The actin tails propel the bacteria through the cytoplasm of the infected cell to its surface where they push out against the plasma membrane and form protrusions (figure 33.9*b*). The protrusions are engulfed by adjacent cells, and the bacteria once again enter phagosomes and escape into the cytoplasm. In this way the infection spreads to adjacent cells. The lysosomes never have a chance to merge with the phagosomes. Another approach is to resist the toxic products released into the phagolysosome after fusion occurs. A good example of a bacterium that is resistant to the lysosomal enzymes is *Mycobacterium tuberculosis,* probably at least partly because of its waxy external layer. Still other bacteria prevent fusion of phagosomes with lysosomes (e.g., *Chlamydia*). Phagocytosis (section 31.3)

Evading the Specific Immune Response

To evade the specific immune response, some bacteria (e.g., *S. pyogenes*) produce capsules that are not antigenic because they resemble host tissue components. *N. gonorrhoeae* can evade the specific immune response by two mechanisms: (1) it makes genetic variations in its pili (phase variation) so that specific antibodies are useless against the new pili and adherence to host tissue occurs, and (2) it produces IgA proteases that destroy secretory IgA and allow adherence. Finally, some bacteria produce proteins (such as staphylococcal protein A and protein G of *S. pyogenes*) that interfere with antibody-mediated opsonization by binding to the Fc portion of immunoglobulins.

1. What are some mechanisms viruses use to evade host defenses?
2. How do bacteria evade each of the following host defenses: the complement system, phagocytosis, and the specific immune response?

Figure 33.9 Formation of Actin Tails by Intracellular Bacterial Pathogens. **(a)** Transmission electron micrograph of *Lysteria monocytogenes* in a host macrophage. The bacterium has polymerized host actin into a long tail that it uses for intracellular propulsion and to move from one host cell to another. **(b)** *Burkholderia pseudomallei* (stained red) also forms actin tails (stained dark green) as shown in this confocal micrograph. Note that the actin tails enable the bacterial cells to be propelled out of the host cell.

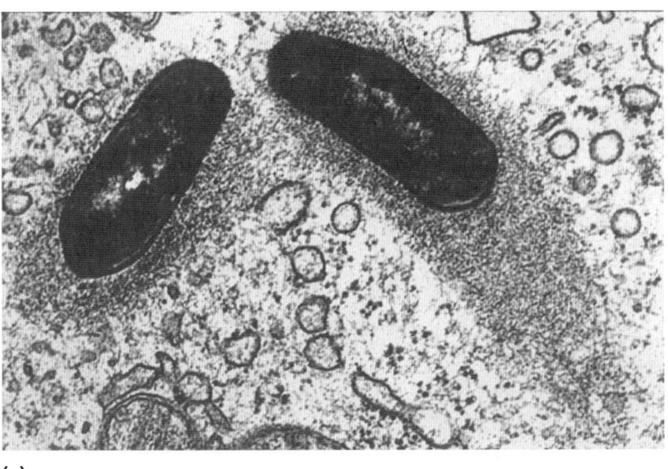

(a)

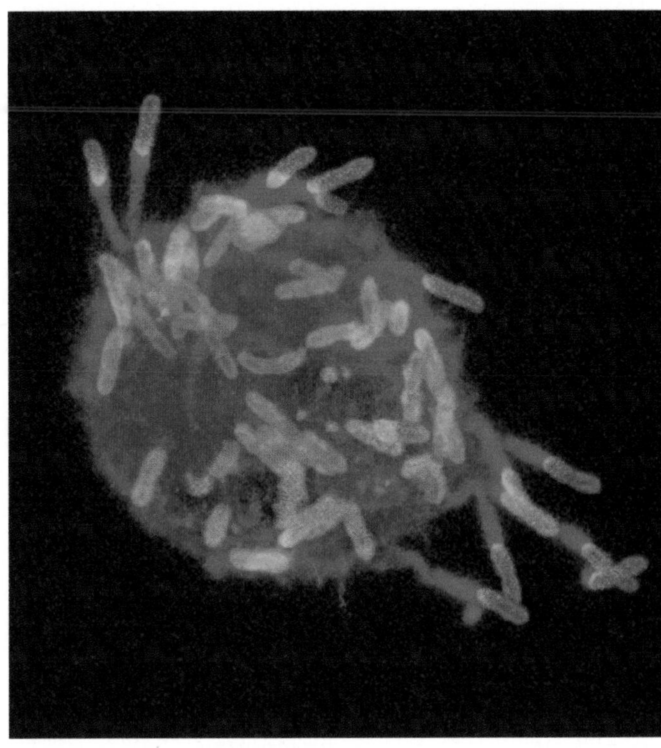

(b)

Summary

33.1 Host-Parasite Relationships

a. Parasitism is a type of symbiosis between two species in which the smaller organism is physiologically dependent on the larger one, termed the host. The parasitic organism usually harms its host in some way.

b. An infection is the colonization of the host by a parasitic organism. An infectious disease is the result of the interaction between the parasitic organism and its host, causing the host to change from a state of health to one of a diseased state. Any organism that produces such a disease is a pathogen (**figure 33.1**).

c. Pathogenicity refers to the quality or ability of an organism to produce pathological changes or disease. Virulence refers to the degree or intensity of pathogenicity of an organism and is measured experimentally by the LD_{50} or ID_{50} (**figure 33.2**).

33.2 Pathogenesis of Viral Diseases

a. The fundamental process of viral infection is the expression of the viral replicative cycle in a host cell. To produce disease a virus enters a host, comes into contact with susceptible cells, and reproduces.

b. Viruses spread to adjacent cells when they are released by either host cell lysis or budding.

c. Host cell damage caused by viruses stimulates a host immune response involving neutralizing antibodies for free virions and activated killer cells for intracellular viruses.

d. Recovery from infection results when the virus has either been cleared from the body of the host, establishes a persistent infection, or kills the host.

e. The viral infection cycle is complete when the virus is shed back into the environment, to be acquired by another host.

33.3 Overiew of Bacterial Pathogenesis

a. Pathogens or their products can be transmitted to a host by either direct or indirect means. Transmissibility is the initial requisite in the establishment of an infectious disease.

b. Special adherence factors allow pathogens to bind to specific receptor sites on host cells and colonize the host (**table 33.2** and **figure 33.3**).

c. Pathogens can enter host cells by both active and passive mechanisms. Once inside, they can produce specific products and/or enzymes that promote dissemination throughout the body of the host. These are termed virulence factors (**table 33.3**).

d. The pathogen generally is found in the area of the host's body that provides the most favorable conditions for its growth and multiplication.

e. During coevolution with human hosts, some pathogenic bacteria have evolved complex signal transduction pathways to regulate the genes necessary for virulence.

f. Many bacteria are pathogenic because they have large segments of DNA called pathogenicity islands that carry genes responsible for virulence.

33.4 Toxigenicity

a. Intoxications are diseases that result from the entrance of a specific toxin into a host. The toxin can induce the disease in the absence of the toxin-producing organism. Toxins produced by pathogens can be divided into two main categories: exotoxins and endotoxins (**table 33.4**).

b. Exotoxins are soluble, heat-labile, potent, toxic proteins produced by the pathogen. They have very specific effects and can be categorized as neurotoxins, cytotoxins, or enterotoxins. Most exotoxins conform to the AB model in which the A subunit or fragment is enzymatic and the B subunit or fragment, the binding portion (**table 33.5**). Several mechanisms exist by which the A component enters target cells (**figure 33.5**).

c. Exotoxins can be divided into four types: (1) the AB toxins, (2) specific host site toxins (neurotoxins, enterotoxins, cytotoxins), (3) toxins that disrupt plasma membranes of host cells (leukocidins, hemolysins, and phospholipases), and (4) superantigens.

d. Bacterial exotoxins cause disease in a human host in three main ways: (1) ingestion of preformed exotoxin, (2) colonization of a mucosal surface followed by exotoxin production, and (3) colonization of a wound followed by local exotoxin production.

e. Endotoxins are heat-stable, toxic substances that are part of the cell wall lipopolysaccharide of some gram-negative bacteria. Most endotoxins function by initially activating Hageman Factor, which in turn activates one to four humoral systems.

33.5 Host Defense Against Microbial Invasion

a. Hosts have primary and secondary defenses against microbial invasion. Primary defenses include the skin, mucous membranes, and antimicrobial chemicals. Secondary defenses include specialized cells that are antigen specific and have memory, lysozyme, and soluble receptors.

b. Factors influencing host defenses against microbial invasion include age, nutrition, genetics, and stress.

33.6 Microbial Mechanisms for Escaping Host Defenses

a. During the course of microbe and human evolution, some pathogens have evolved ways for escaping host defenses. Viruses have mechanisms that either suppress or evade the host's immune response. Bacteria have evolved mechanisms to evade the complement system, phagocytosis, and the specific immune response.

Key Terms

AB toxins 824
alpha hemolysis 828
antitoxin 824
apoptosis 819
bacteremia 821
beta hemolysis 828
colonization 820
cytopathic viruses 819
cytotoxin 825
ectoparasite 816
endogenous pyrogen 830
endoparasite 816
endotoxin 829
endotoxin unit (E.U.) 830
enterotoxin 825
exotoxin 824
final host 816
fomite 820
hemolysin 828
host-parasite relationship 816
immunopathology 817
infection 816
infectious disease 816
infectious dose 50 (ID_{50}) 817
infectivity 816
intermediate host 816
intoxication 824
invasiveness 816
lethal dose 50 (LD_{50}) 817
leukocidin 828
membrane-disrupting exotoxin 828
neurotoxin 825
noncytopathic viruses 819
opportunistic pathogen 816
parasite 816
pathogen 816
pathogenicity 816
pathogenicity island 822
pathogenic potential 816
phospholipase 828
primary (frank) pathogen 816
reservoir 818
reservoir host 816
septicemia 821
serum resistance 832
streptolysin-O (SLO) 828
streptolysin-S (SLS) 828
symbiosis 815
toxemia 824
toxigenicity 816
toxin 824
toxoid 824
transfer host 816
tropism 819
type III secretion system (TTSS) 822
vector 818
viremia 819
virulence 816
virulence factor 816

Critical Thinking Questions

1. Why does a parasitic organism not have to be a parasite?
2. In general, infectious diseases that are commonly fatal are newly evolved relationships between the parasitic organism and the host. Why is this so?
3. Explain the observation that different pathogens infect different parts of the host.
4. Intracellular bacterial infections present a particular difficulty for the host. Why is it harder to defend against these infections than against viral infections and extracellular bacterial infections?

Learn More

Abrami, L.; Reig, N.; and van der Goot, F. G. 2005. Anthrax toxin: The long and winding road that leads to the kill. *Trends Microbiol.* 13:72–78.

Aktories, K., and Barbieri, J. T. 2005. Bacterial cytotoxins: Targeting eukaryotic switches. *Nature Rev. Microbiol.* 3:397–410.

Blanke, S. L. 2006. Portals and pathways: Principles of bacterial toxin entry into host cells. *Microbe.* 1:26–410.

Casadevall, A., and Pirofski, L. A. 2003. The damage-response framework of microbial pathogenesis. *Nature Rev. Microbiol.* 1:17–24.

Day, T. 2002. The evolution of virulence in vector-borne and directly transmitted parasites. *Theor. Popul. Biol.* 62:199–213.

Ewald, P. W. 2004. Evolution of virulence. *Infect. Dis. Clin. North Am.* 18:1–15.

Foster, T. J. 2005. Immune evasion by staphylococci. *Nature Rev. Microbiol.* 3: 948–58.

Gandon, S. 2004. Evolution of multihost parasites. *Int. J. Org. Evol.* 58:455–69.

Ghosh, P. 2004. Process of protein transport by the type III secretion system. *Microbiol. Mol. Biol. Rev.* 68:771–95.

Lara-Tejero, M., and Galan, J. E. 2002. Cytolethal distending toxin: Limited damage as a strategy to modulate cellular functions. *Trends Microbiol.* 10:147–52.

Stevens, J. M.; Galyov, E. E.; and Stevens, M. P. 2006. Actin-dependent movement of bacterial pathogens. *Nature Rev. Microbiol.* 4:91–101.

Waldvogel, F. A. 2004. Infectious diseases in the 21st century: Old challenges and new opportunities. *Int. J. Infect. Dis.* 8:5–12.

Please visit the Prescott website at www.mhhe.com/prescott7 for additional references.

34 Antimicrobial Chemotherapy

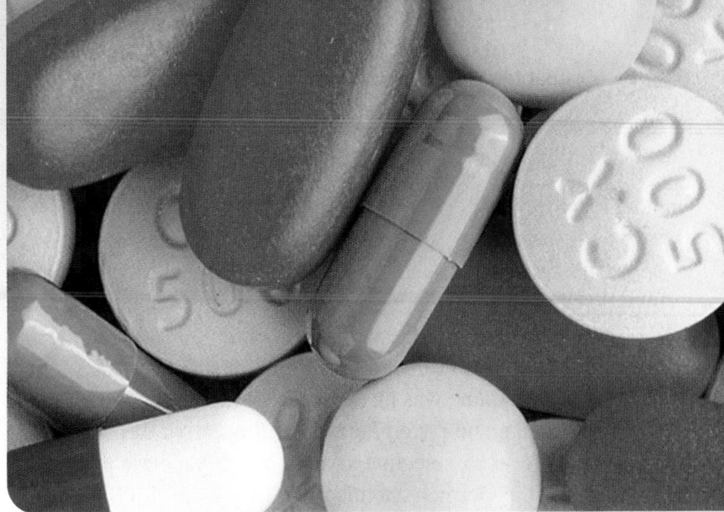

Many antimicrobial medications are available to combat infections. Nonetheless, they fall into a limited number of classes based on their modes of action.

Preview

- Many infectious diseases are treated with chemotherapeutic agents, such as antibiotics, that inhibit or kill the pathogen while harming the host as little as possible.
- Ideally, antimicrobial agents disrupt microbial processes or structures that differ from those of the host. They may damage pathogens by hampering cell wall synthesis, inhibiting microbial protein and nucleic acid synthesis, disrupting microbial membrane structure and function, or blocking metabolic pathways through inhibition of key enzymes.
- The effectiveness of chemotherapeutic agents depends on many factors: the route of administration and location of the infection, the presence of interfering substances, the concentration of the drug in the body, the nature of the pathogen, the presence of drug allergies, and the resistance of microorganisms to the drug.
- The increasing number and variety of drug-resistant pathogens is a serious public health problem.
- Although antibacterial chemotherapy is more advanced, drugs for the treatment of fungal, protozoan, and viral infections are also becoming increasingly available.

The control of microorganisms is critical for the prevention and treatment of disease. Chapter 7 is concerned principally with the chemical and physical agents used to treat inanimate objects in order to destroy microorganisms or inhibit their growth. Microorganisms also grow on and within other organisms, and microbial colonization can lead to disease, disability, and death. Thus the control or destruction of microorganisms residing within the bodies of humans and other animals is of great importance.

When disinfecting or sterilizing an inanimate object, one naturally must use procedures that do not damage the object itself. The same is true for the treatment of living hosts. The most successful drugs interfere with vital processes that differ between the pathogen and host, thereby seriously damaging the target microorganism while harming its host as little as possible. This chapter introduces the principles of antimicrobial chemotherapy and briefly reviews the characteristics of selected antibacterial, antifungal, and antiprotozoan antiviral drugs.

Modern medicine is dependent on **chemotherapeutic agents,** chemical agents that are used to treat disease. Ideally, chemotherapeutic agents used to treat infectious disease destroy pathogenic microorganisms or inhibit their growth at concentrations low enough to avoid undesirable damage to the host. Most of these agents are **antibiotics** [Greek *anti,* against, and *bios,* life], microbial products or their derivatives that can kill susceptible microorganisms or inhibit their growth. Drugs such as the sulfonamides are sometimes called antibiotics although they are synthetic chemotherapeutic agents, not microbially synthesized.

34.1 The Development of Chemotherapy

The modern era of chemotherapy began with the work of the German physician Paul Ehrlich (1854–1915). Ehrlich was fascinated with dyes that specifically bind to and stain microbial cells. He reasoned that one of the dyes could be a chemical that would selectively destroy pathogens without harming human cells—a

It was the knowledge of the great abundance and wide distribution of actinomycetes, which dated back nearly three decades, and the recognition of the marked activity of this group of organisms against other organisms that led me in 1939 to undertake a systematic study of their ability to produce antibiotics.

—*Selman A. Waksman*

"magic bullet." By 1904 Ehrlich found that the dye trypan red was active against the trypanosome that causes African sleeping sickness (*see figure 25.6*) and could be used therapeutically. Subsequently Ehrlich and a young Japanese scientist named Sahachiro Hata tested a variety of arsenicals on syphilis-infected rabbits and found that arsphenamine was active against the syphilis spirochete. Arsphenamine was made available in 1910 under the trade name Salvarsan, and paved the way to the testing of hundreds of compounds for their selective toxicity and therapeutic potential.

In 1927, the German chemical industry giant, I. G. Farbenindustrie, began a long-term search for chemotherapeutic agents under the direction of Gerhard Domagk. Domagk had screened a vast number of chemicals for other "magic bullets" and discovered that Prontosil Red, a new dye for staining leather, protected mice completely against pathogenic streptococci and staphylococci without apparent toxicity. Jacques and Therese Trefouel later showed that the body metabolized the dye to sulfanilamide. Domagk received the 1939 Nobel Prize in Physiology or Medicine for his discovery of sulfonamides, or sulfa drugs.

In the 1920s, Alexander Fleming, a Scottish physician, found that human tears contained a naturally occurring antibacterial substance that he termed "lysozyme." This substance unfortunately had little therapeutic value because it could not be isolated in large quantities and was not effective against many microorganisms. However, it prepared Fleming for the discovery of penicillin, the first true antibiotic to be used therapeutically.

Penicillin was actually discovered in 1896 by a 21-year-old French medical student named Ernest Duchesne. His work was forgotten until Fleming's accidental rediscovery of the antibiotic in September 1928. After returning from a weekend vacation, Fleming noticed that a petri plate of *Staphylococcus* also had a mold growing on it and, like the lysozyme he had discovered years before, there were no *Staphylococcus* colonies surrounding it (**figure 34.1**). Although the precise events are still unclear, it has been suggested that a *Penicillium notatum* spore had made its way onto the petri dish before it had been inoculated with the staphylococci. The mold apparently grew before the bacteria and produced penicillin. The bacteria nearest the fungus were lysed. Fleming correctly deduced that the mold contaminant produced a diffusible substance, which he called penicillin. In subsequent studies he showed that this substance could diffuse through agar so that even small amounts of it extracted from broth cultures could kill several pathogenic bacteria, including *S. aureus*. Unfortunately, Fleming could not demonstrate that penicillin remained active in vivo long enough to destroy pathogens and thus dropped the research.

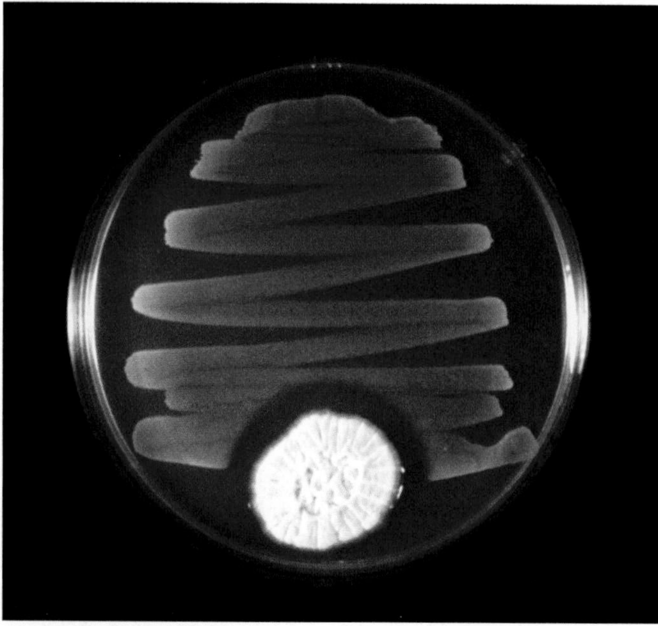

Figure 34.1 Bacteriocidal Action of Penicillin. The *Penicillium* mold colony secretes penicillin that kills *Staphylococcus aureus* that was streaked nearby.

In 1939 Howard Florey, a professor of pathology at Oxford University, was in the midst of testing the bactericidal activity of many substances, including lysozyme and the sulfonamides. After reading Fleming's paper on penicillin, one of Florey's coworkers, Ernst Chain, obtained the *Penicillium* culture from Fleming and set about culturing it and purifying penicillin. Florey and Chain were greatly aided in this by the biochemist Norman Heatley. Heatley devised the original assay, culture, and purification techniques needed to produce crude penicillin for further experimentation. When purified penicillin was injected into mice infected with streptococci or staphylococci, practically all the mice survived. Florey and Chain's success was reported in 1940, and subsequent human trials were equally successful. Fleming, Florey, and Chain received the Nobel Prize in 1945 for the discovery and production of penicillin.

The discovery of penicillin stimulated the search for other antibiotics. Selman Waksman announced in 1944 that he and his associates had found a new antibiotic, streptomycin, produced by the actinomycete *Streptomyces griseus*. This discovery arose from the careful screening of about 10,000 strains of soil bacteria and fungi. The importance of streptomycin cannot be understated, as it was the first drug that could successfully treat tuberculosis. Waksman received the Nobel Prize in 1952, and his success led to a worldwide search for other antibiotic-producing soil microorganisms. Microorganisms producing chloramphenicol, neomycin, terramycin, and tetracycline were isolated by 1953.

The discovery of chemotherapeutic agents and the development of newer, more powerful drugs has transformed modern medicine and greatly alleviated human suffering. Furthermore, antibiotics have proven exceptionally useful in microbiological research (**Techniques & Applications 34.1**).

1. What are chemotherapeutic agents? Antibiotics?
2. What contributions to chemotherapy were made by Ehrlich, Domagk, Fleming, Florey and Chain, and Waksman?

Techniques & Applications

34.1 The Use of Antibiotics in Microbiological Research

Although the use of antibiotics in the treatment of disease is emphasized in this chapter, it should be noted that antibiotics are extremely important research tools. For example, they aid the cultivation of viruses by preventing bacterial contamination. When eggs are inoculated with a virus sample, antibiotics often are included in the inoculum to maintain sterility. Usually a mixture of antibiotics (e.g., penicillin, amphotericin, and streptomycin) also is added to tissue cultures used for virus cultivation and other purposes.

Researchers often use antibiotics as instruments to dissect metabolic processes by inhibiting or blocking specific steps and observing the consequences. Although selective toxicity is critical when antibiotics are employed therapeutically, specific toxicity is more important in this context: the antibiotic must act by a specific and precisely understood mechanism. A clinically useful antimicrobial agent such as ampicillin sometimes may be employed in research, but often an agent with specific toxicity and excellent research potential is too toxic for therapeutic use. The actinomycins, discovered in 1940 by Selman Waksman, are a case in point. They are so toxic to higher organisms that it was suggested they be used as rat poison. Today actinomycin D is a standard research tool specifically used to block RNA synthesis. Other examples of antibiotics useful in research, with the process inhibited, are the following: chloramphenicol (bacterial protein synthesis), cycloserine (peptidoglycan synthesis), nalidixic acid and novobiocin (bacterial DNA synthesis), rifampin (bacterial RNA synthesis), cycloheximide (eucaryotic protein synthesis), daunomycin (fungal RNA synthesis), mitomycin C (eucaryotic DNA synthesis), polyoxin D (fungal cell wall chitin synthesis), and cerulenin (fatty acid synthesis).

In practice, the antibiotic is administered and changes in cell function are monitored. If one desired to study the dependence of bacterial flagella synthesis on RNA transcription, the flagella could be removed by high-speed mixing in a blender, followed by actinomycin D addition to the incubation mixture. The bacterial culture would then be observed for flagella regeneration in the absence of RNA synthesis. The results of such experiments must be interpreted with caution. Flagella synthesis may have been blocked because actinomycin D inhibited some other process, thus affecting flagella regeneration indirectly rather than simply inhibiting transcription of a gene required for flagella synthesis. Furthermore, not all microorganisms respond in the same way to a particular drug.

34.2 General Characteristics of Antimicrobial Drugs

As Ehrlich so clearly saw, to be successful a chemotherapeutic agent must have **selective toxicity:** it must kill or inhibit the microbial pathogen while damaging the host as little as possible. The degree of selective toxicity may be expressed in terms of (1) the therapeutic dose, the drug level required for clinical treatment of a particular infection, and (2) the toxic dose, the drug level at which the agent becomes too toxic for the host. The **therapeutic index** is the ratio of the toxic dose to the therapeutic dose. The larger the therapeutic index, the better the chemotherapeutic agent (all other things being equal).

A drug that disrupts a microbial function not found in eucaryotic animal cells often has a greater selective toxicity and a higher therapeutic index. For example, penicillin inhibits bacterial cell wall peptidoglycan synthesis but has little effect on host cells because they lack cell walls; therefore penicillin's therapeutic index is high. A drug may have a low therapeutic index because it inhibits the same process in host cells or damages the host in other ways. The undesirable effects on the host, or side effects, are of many kinds and may involve almost any organ system (**table 34.1**). Because side effects can be severe, chemotherapeutic agents should be administered with great care.

Some bacteria and fungi are able to naturally produce many of the commonly employed antibiotics (**table 34.2**). In contrast, several important chemotherapeutic agents, such as sulfonamides, trimethoprim, chloramephenicol, ciprofloxacin, isoniazid, and dapsone, are synthetic—that is, manufactured by chemical procedures independent of microbial activity (table 34.1). An increasing number of antibiotics are semisynthetic—they are natural antibiotics that have been structurally modified by the addition of chemical groups to make them less susceptible to inactivation by pathogens (e.g., ampicillin, carbenicillin, and methicillin). In addition, many semisynthetic drugs have a broader spectrum of antibiotic activity than does their parent molecule. This is particularly true of the semisynthetic penicillins (e.g., ampicillin, amoxycillin) versus the naturally produced penicillin G and penicillin V. It is likely that the manufacture of newer semisynthetic antimicrobials will increase in the coming years as the rise in microbes resistant to existing antibiotics continues to grow and newer drugs must be introduced.

Drugs vary considerably in their range of effectiveness. Many are **narrow-spectrum drugs**—that is, they are effective only against a limited variety of pathogens (table 34.1). Others are **broad-spectrum drugs** that attack many different kinds of pathogens. Drugs may also be classified based on the general microbial group they act against: antibacterial, antifungal, antiprotozoan, and antiviral. Some agents can be used against more than one group; for example, sulfonamides are active against bacteria and some protozoa. Chemotherapeutic agents, like disinfectants, can be either **cidal** or **static.** Static agents reversibly inhibit growth; if the agent is removed, the microorganisms will recover and grow again. The pattern of microbial death (section 7.2)

Although a cidal agent kills the target pathogen, its activity is concentration dependent and the agent may be only static at low

Table 34.1 Properties of Some Common Antibacterial Drugs

Antibiotic Group	Primary Effect	Mechanism of Action	Members	Spectrum	Common Side Effects
Cell Wall Synthesis Inhibition					
Penicillins	Cidal	Inhibit transpeptidation enzymes involved in cross-linking the polysaccharide chains of the bacterial cell wall peptidoglycan. Activate cell wall lytic enzymes.	Penicillin G, penicillin V, methicillin Ampicillin, carbenicillin	Narrow (gram-positive) Broad (gram-positive, some gram-negative)	Allergic responses (diarrhea, anemia, hives, nausea, renal toxicity)
Cephalosporins	Cidal	Same as above	Cephalothin, cefoxitin, cefaperazone, ceftriaxone	Broad (gram-positive, some gram-negative)	Allergic responses, thrombophlebitis, renal injury
Vancomycin	Cidal	Prevents transpeptidation of peptidoglycan subunits by binding to D-Ala-D-Ala amino acids at the end of peptide cross-bridges. Thus it has a different binding site than that of the penicillins.	Vancomycin	Narrow (gram-positive)	Ototoxic (tinnitus and deafness), nephrotoxic, allergic reactions
Protein Synthesis Inhibition					
Aminoglycosides	Cidal	Bind to small ribosomal subunit (30S) and interfere with protein synthesis by directly inhibiting synthesis and causing misreading of mRNA	Neomycin, kanamycin, gentamicin Streptomycin	Broad (gram-negative, mycobacteria) Narrow (aerobic gram-negative)	Deafness, renal damage, loss of balance, nausea, allergic responses Same as above
Tetracyclines	Static	Same as above	Oxytetracycline, chlortetracycline	Broad (gram-positive and -negative, rickettsia and chlamydia)	Gastrointestinal upset, teeth discoloration, renal, hepatic injury
Macrolides	Static	Bind to 23S rRNA of large ribosomal subunit (50S) to inhibit peptide chain elongation during protein synthesis	Erythromycin, clindamycin	Broad (aerobic and anaerobic gram-positive, some gram-negative)	Gastrointestinal upset, hepatic injury, anemia, allergic responses
Chloramphenicol	Static	Same as above	Chloramphenicol	Broad (gram-positive and -negative, rickettsia and chlamydia)	Depressed bone marrow function, allergic reactions
Nucleic Acid Synthesis Inhibition					
Quinolones and Fluoroquinolones	Cidal	Inhibit DNA gyrase and topoisomerase IV, thereby blocking DNA replication and transcription	Norfloxacin, ciprofloxacin, Levofloxacin	Narrow (gram-negatives better than gram-positives) Broad spectrum	Tendonitis, headache, lightheadedness, convulsions, allergic reactions

Rifampin	Cidal	Inhibits bacterial DNA-dependent RNA polymerase	R-Cin, rifacilin, rifamycin, rimactane, rimpin, siticox	*Mycobacterium* infections and some gram-negative such as *Neisseria meningitidis* and *Haemophilus influenzae* b	Nausea, vomiting, diarrhea, fatigue, anemia, drowsiness, headache, mouth ulceration, liver damage
Cell Membrane Disruption					
Polymyxin B	Cidal	Binds to plasma membrane and disrupts its structure and permeability properties	Polymyxin B, polymyxin topical ointment	Narrow—gram-negatives only	Can cause severe kidney damage, drowsiness, dizziness
Antimetabolites					
Sulfonamides	Static	Inhibits folic acid synthesis by competing with ρ-aminobenzoic acid (PABA)	Silver sulfadiazine, sodium sulfacetamide, sulfamethoxazole, sulfanilamide, sulfasalazine, sulfisoxazole	Broad spectrum	Nausea, vomiting, and diarrhea; hypersensitivity reactions such as rashes, photosensitivity
Trimethoprim	Static	Blocks folic acid synthesis by inhibiting the enzyme tetrahydrofolate reductase	Trimethoprim (in combination with a sulfamethoxazole [1:5])	Broad spectrum	Same as sulfonamides, but less frequent
Dapsone	Static	Thought to interfere with folic acid synthesis	Dapsone	Narrow—mycobacterial infections, principally leprosy	Back, leg, or stomach pains; discolored fingernails, lips, or skin; breathing difficulties fever, loss of appetite, skin rash, fatigue
Isoniazid	Cidal if bacteria are actively growing, static if bacteria are dormant	Exact mechanism is unclear, but it is thought to inhibit lipid synthesis (especially mycolic acid); putative enoyl-reductase inhabitor	Isoniazid	Narrow—mycobacterial infections, principally tuberculosis	Nausea, vomiting, liver damage, seizures, "pins and needles" in extremities (peripheral neuropathy)

Table 34.2 Microbial Sources of Some Antibiotics

Microorganism	Antibiotic
Bacteria	
Streptomyces spp.	Amphotericin B
	Chloramphenicol (also synthetic)
	Kanamycin
	Neomycin
	Nystatin
	Rifampin
	Streptomycin
	Tetracyclines
	Vancomycin
Micromonospora spp.	Gentamicin
Bacillus spp.	Bacitracin
	Polymyxins
Fungi	
Penicillium spp.	Griseofulvin
	Penicillin
Cephalosporium spp.	Cephalosporins

levels. The effect of an agent also varies with the target species: an agent may be cidal for one species and static for another. Because static agents do not directly destroy the pathogen, elimination of the infection depends on the host's own resistance mechanisms. A static agent may not be effective if the host's resistance is too low.

Some idea of the effectiveness of a chemotherapeutic agent against a pathogen can be obtained from the **minimal inhibitory concentration (MIC).** The MIC is the lowest concentration of a drug that prevents growth of a particular pathogen. On the other hand, the **minimal lethal concentration (MLC)** is the lowest drug concentration that kills the pathogen. A cidal drug generally kills pathogens at levels only two to four times the MIC, whereas a static agent kills at much higher concentrations (if at all).

1. Define the following terms: selective toxicity, therapeutic index, side effect, narrow-spectrum drug, broad-spectrum drug, synthetic and semisynthetic antibiotics, cidal and static agents, minimal inhibitory concentration (MIC), and minimal lethal concentration (MLC).
2. Why is it necessary to make synthetic and semisynthetic antibiotics?
3. Use the MIC and MLC concepts to distinguish between cidal and static agents.

34.3 Determining the Level of Antimicrobial Activity

Determination of antimicrobial effectiveness against specific pathogens is essential to proper therapy. Testing can show which agents are most effective against a pathogen and give an estimate of the proper therapeutic dose.

Dilution Susceptibility Tests

Dilution susceptibility tests can be used to determine MIC and MLC values. Antibiotic dilution tests can be done in both agar and broth. In the broth dilution test, a series of broth tubes (usually Mueller-Hinton broth) containing antibiotic concentrations in the range of 0.1 to 128 μg/ml (2-fold dilutions) is prepared and inoculated with a standard density of the test organism. The lowest concentration of the antibiotic resulting in no growth after 16 to 20 hours of incubation is the MIC. The MLC can be ascertained if the tubes showing no growth are subcultured into fresh medium lacking antibiotic. The lowest antibiotic concentration from which the microorganisms do not grow when transferred to fresh medium is the MLC. The agar dilution test is very similar to the broth dilution test. Plates containing Mueller-Hinton agar and various amounts of antibiotic are inoculated and examined for growth. Several automated systems for susceptibility testing and MIC determination with broth or agar cultures have been developed.

Disk Diffusion Tests

If a rapidly growing aerobic or facultative pathogen like *Staphylococcus* or *Pseudomonas* is being tested, a disk diffusion technique may be used to save time and media. The principle behind the assay technique is fairly simple. When an antibiotic-impregnated disk is placed on agar previously inoculated with the test bacterium, the antibiotic diffuses radially outward through the agar, producing an antibiotic concentration gradient. The antibiotic is present at high concentrations near the disk and affects even minimally susceptible microorganisms (resistant organisms will grow up to the disk). As the distance from the disk increases, the antibiotic concentration decreases and only more susceptible pathogens are harmed. A clear zone or ring is present around an antibiotic disk after incubation if the agent inhibits bacterial growth. The wider the zone surrounding a disk, the more susceptible the pathogen is. Zone width also is a function of the antibiotic's initial concentration, its solubility, and its diffusion rate through agar. Thus zone width cannot be used to compare directly the effectiveness of two different antibiotics.

Currently the disk diffusion test most often used is the **Kirby-Bauer method,** which was developed in the early 1960s at the University of Washington Medical School by William Kirby, A.W. Bauer, and their colleagues. An inoculating loop or needle is touched to four or five isolated colonies of the pathogen growing on agar and then used to inoculate a tube of culture broth. The culture is incubated for a few hours at 35°C until it becomes slightly turbid and is diluted to match a turbidity standard. A sterile cotton swab is dipped into the standardized bacterial test suspension and used to evenly inoculate the entire surface of a Mueller-Hinton agar plate. After the agar surface has dried for about 5 minutes, the appropriate antibiotic test disks are placed on it, either with sterilized forceps or with a multiple applicator device (**figure 34.2**). The plate is immediately placed in a 35°C incubator. After 16 to 18 hours of incubation, the diameters of the zones of inhibition are measured to the nearest mm.

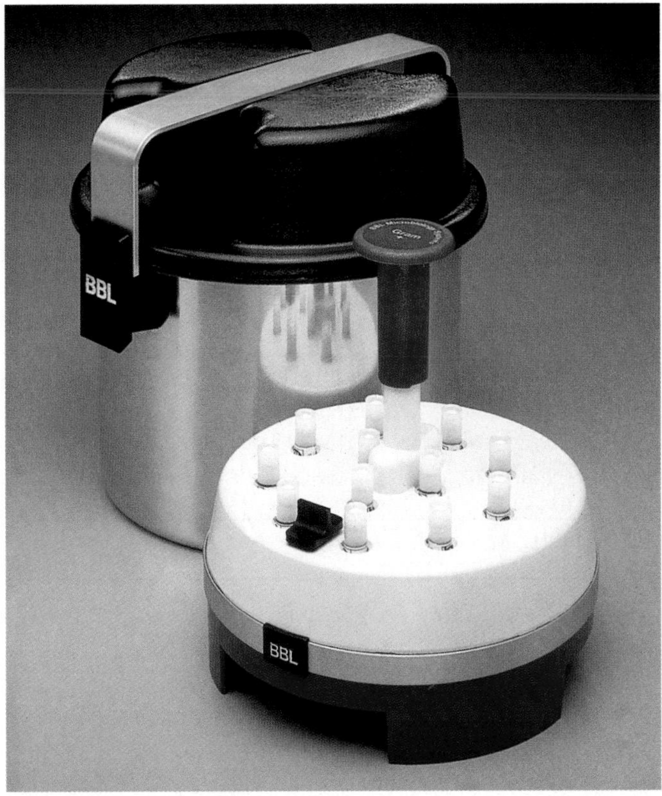

(a)

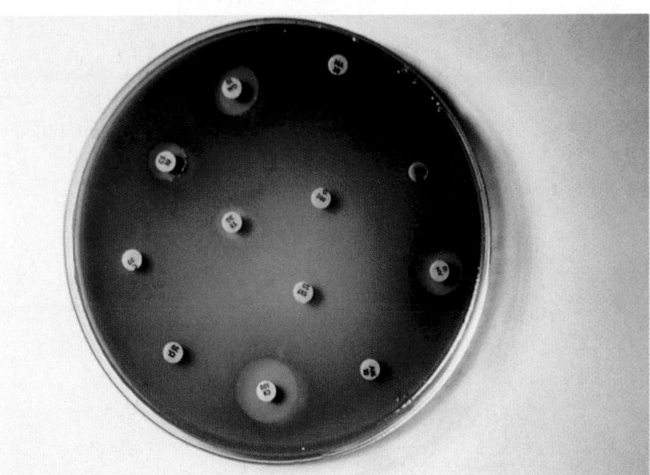

(b)

Figure 34.2 The Kirby-Bauer Method. **(a)** A multiple antibiotic disk dispenser and **(b)** disk diffusion test results.

Kirby-Bauer test results are interpreted using a table that relates zone diameter to the degree of microbial resistance (**table 34.3**). The values in table 34.3 were derived by finding the MIC values and zone diameters for many different microbial strains. A plot of MIC (on a logarithmic scale) versus zone inhibition diameter (arithmetic scale) is prepared for each antibiotic (**figure 34.3**). These plots are then used to find the zone diameters corresponding to the drug concentrations actually reached in the body. If the zone diameter for the lowest level reached in the body is smaller than that seen with the test pathogen, the pathogen should have an MIC value low enough to be destroyed by the drug. A pathogen with too high a MIC value (too small a zone diameter) is resistant to the agent at normal body concentrations.

The Etest

The Etest from AB Biodisk may be used in sensitivity testing under some conditions. It is particularly convenient for use with anaerobic pathogens. A petri dish of the proper agar is streaked in three different directions with the test organism and special plastic Etest® strips are placed on the surface so that they extend out radially from the center (**figure 34.4**). Each strip contains a gradient of an antibiotic and is labeled with a scale of minimal inhibitory concentration values. The lowest concentration in the strip lies at the center of the plate. After 24 to 48 hours of incubation, an elliptical zone of inhibition appears. As shown in the figure, MICs are determined from the point of intersection between the inhibition zone and the strip's scale of MIC values.

Measurement of Drug Concentrations in the Blood

A drug must reach a concentration at the site of infection above the pathogen's MIC to be effective. In cases of severe, life-threatening disease, it often is necessary to monitor the concentration of drugs in the blood and other body fluids. This may be achieved by microbiological, chemical, immunologic, enzymatic, or chromatographic assays. Extra care is needed to also evaluate antibiotic binding to serum proteins and are thus unavailable for measurement by common antibiotic assays.

1. How can dilution susceptibility tests and disk diffusion tests be used to determine microbial drug sensitivity?
2. Briefly describe the Kirby-Bauer test and its purpose.
3. How is the Etest carried out?

34.4 Antibacterial Drugs

Since Fleming's discovery of penicillin, natural antibiotics (table 34.2) have been found that can damage pathogens in several ways. A few antibacterial drugs are described here and summarized in table 34.1, with emphasis on their mechanisms of action.

Inhibitors of Cell Wall Synthesis

The most selective antibiotics are those that interfere with bacterial cell wall synthesis. Drugs like penicillins, cephalosporins, vancomycin, and bacitracin have a high therapeutic index because they target structures not found in eukaryotic cells. The bacterial cell wall (section 3.6)

Penicillins

Most **penicillins** (e.g., penicillin G or benzylpenicillin) are derivatives of 6-aminopenicillanic acid and differ from one another with

Table 34.3 Inhibition Zone Diameter of Selected Chemotherapeutic Drugs

Chemotherapeutic Drug	Disk Content	Zone Diameter (Nearest mm) Resistant	Intermediate	Susceptible
Carbenicillin (with *Proteus* spp. and *E. coli*)	100 μg	≤17	18–22	≥23
Carbenicillin (with *Pseudomonas aeruginosa*)	100 μg	≤13	14–16	≥17
Ceftriaxone	30 μg	≤13	14–20	≥21
Chloramphenicol	30 μg	≤12	13–17	≥18
Erythromycin	15 μg	≤13	14–17	≥18
Penicillin G (with staphylococci)	10 U[a]	≤20	21–28	≥29
Penicillin G (with other microorganisms)	10 U	≤11	12–21	≥22
Streptomycin	10 μg	≤11	12–14	≥15
Sulfonamides	250 or 300 μg	≤12	13–16	≥17
Tetracycline	30 μg	≤14	15–18	≥19

[a]One milligram of penicillin G sodium = 1,600 units (U).

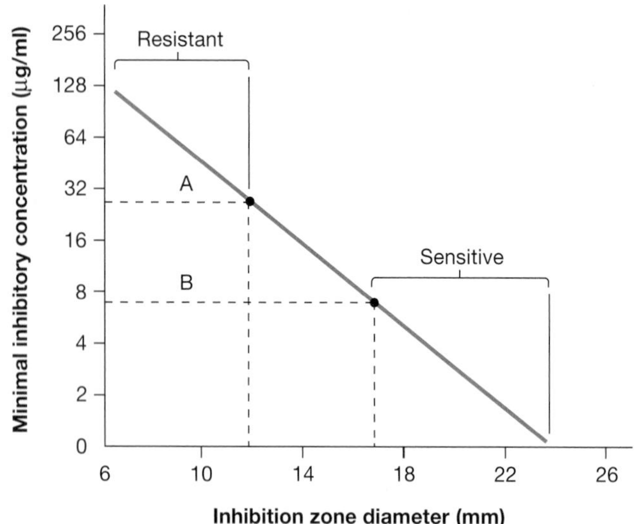

Figure 34.3 Interpretation of Kirby-Bauer Test Results. The relationship between the minimal inhibitory concentrations of a hypothetical drug and the size of the zone around a disk in which microbial growth is inhibited. As the sensitivity of microorganisms to the drug increases, the MIC value decreases and the inhibition zone grows larger. Suppose that this drug varies from 7–28 μg/ml in the body during treatment. Dashed line A shows that any pathogen with a zone of inhibition less than 12 mm in diameter will have an MIC value greater than 28 μg/ml and will be resistant to drug treatment. A pathogen with a zone diameter greater than 17 mm will have an MIC less than 7 μg/ml and will be sensitive to the drug (see line B). Zone diameters between 12 and 17 mm indicate intermediate sensitivity and usually signify resistance.

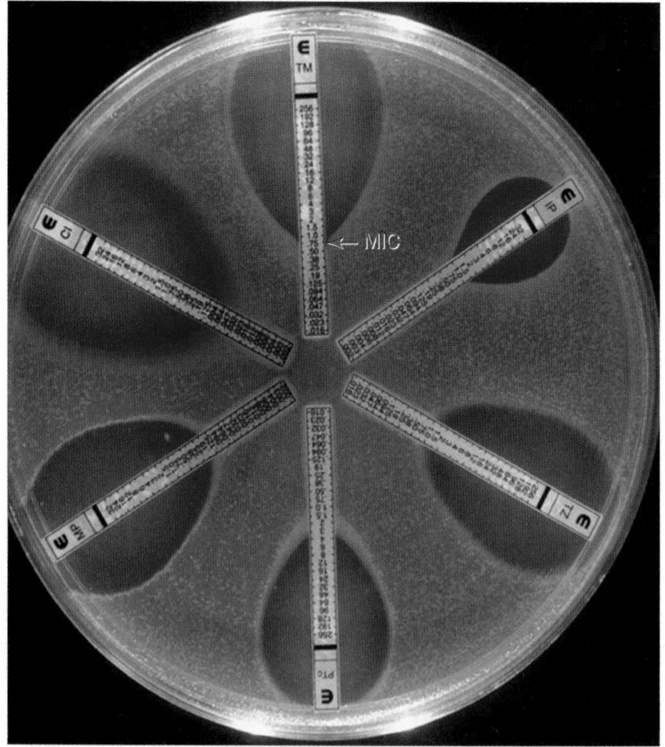

Figure 34.4 The Etest®. An example of a bacterial culture plate with Etest® strips arranged radially on it. The strips are arranged so that the lowest antibiotic concentration in each is at the center. The MIC concentration is read from the scale at the point it intersects the zone of inhibition as shown by the arrow in this example. Etest® is a registered trademark of AB BIODISK and patented in all major markets.

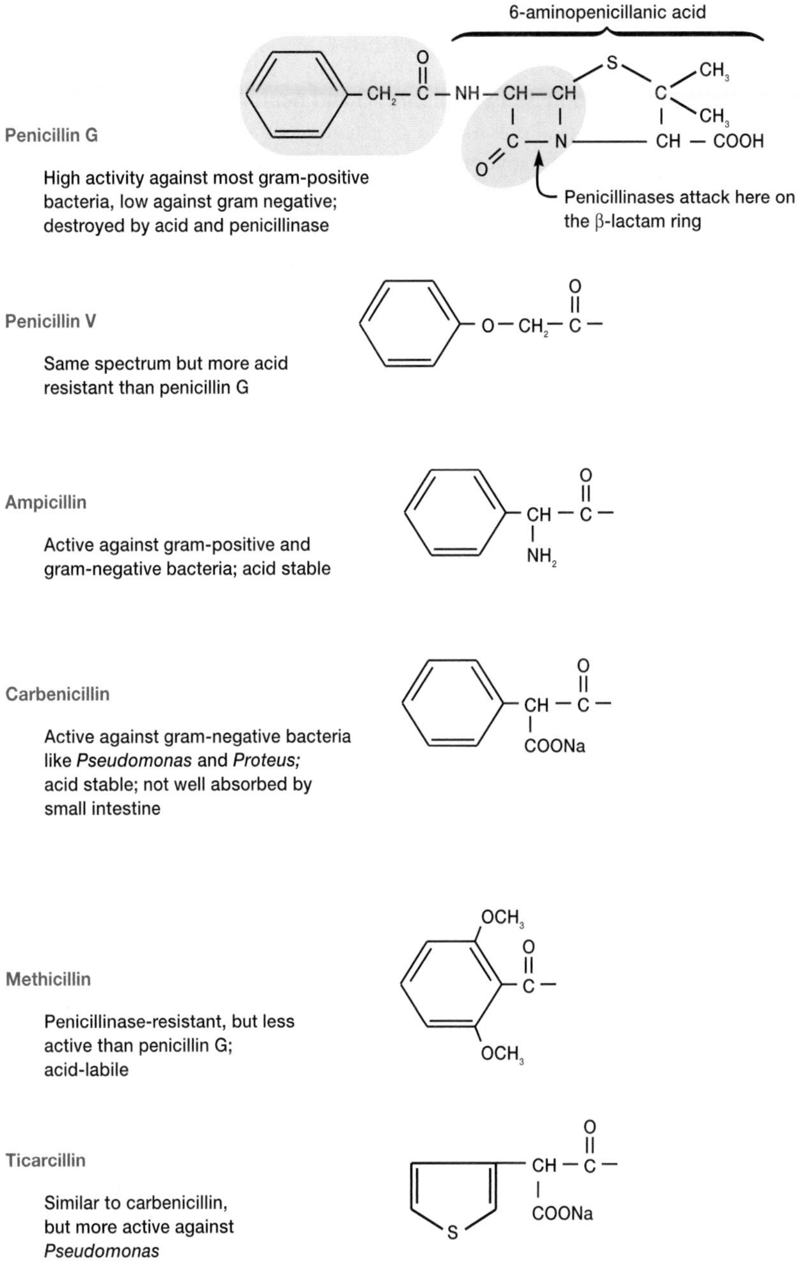

Figure 34.5 Penicillins. The structures and characteristics of representative penicillins. All are derivatives of 6-aminopenicillanic acid; in each case the shaded portion of penicillin G is replaced by the side chain indicated. The β-lactam ring is also shaded (blue), and an arrow points to the bond that is hydrolyzed by penicillinase.

6-aminopenicillanic acid

Penicillins attack here on the β-lactam ring

Penicillin G

High activity against most gram-positive bacteria, low against gram negative; destroyed by acid and penicillinase

Penicillin V

Same spectrum but more acid resistant than penicillin G

Ampicillin

Active against gram-positive and gram-negative bacteria; acid stable

Carbenicillin

Active against gram-negative bacteria like *Pseudomonas* and *Proteus;* acid stable; not well absorbed by small intestine

Methicillin

Penicillinase-resistant, but less active than penicillin G; acid-labile

Ticarcillin

Similar to carbenicillin, but more active against *Pseudomonas*

respect to the side chain attached to its amino group (**figure 34.5**). The most crucial feature of the molecule is the **β-lactam ring,** which is essential for bioactivity. Many penicillin-resistant bacteria produce **penicillinase** (also called **β-lactamase**), an enzyme that inactivates the antibiotic by hydrolyzing a bond in the β-lactam ring.

Although the complete mechanism of action of penicillins is still not completely known, their structures resemble the terminal D-alanyl-D-alanine found on the peptide side chain of the peptidoglycan subunit. It has been proposed that this structural similarity blocks the enzyme catalyzing the transpeptidation reaction that forms the peptidoglycan cross-links (*see figure 10.12*). Thus formation of a complete cell wall is blocked, leading to osmotic lysis. This mechanism is consistent with the observation that penicillins act only on growing bacteria that are synthesizing new peptidoglycan.

Evidence has indicated that the mechanism of penicillin action is even more complex than previously imagined. It has been discovered that penicillins bind to several periplasmic proteins (penicillin-binding proteins, or PBPs) and may also destroy bacteria by activating their own autolytic enzymes. However, there is also some evidence that penicillin kills bacteria even in the absence of autolysins or murein hydrolases. Lysis could occur after bacterial viability has already been lost. Penicillin may stimulate special proteins called bacterial holins to form holes or lesions in the plasma membrane, leading directly to membrane leakage and death. Murein hydrolases also could move through the holes, disrupt the peptidoglycan, and lyse the cell.

Penicillins differ from each other in several ways. The two naturally occurring penicillins, penicillin G and penicillin V, are narrow-spectrum drugs. Penicillin G is effective against gonococci, meningococci, and several gram-positive pathogens such as streptococci and staphylococci. However, it must be administered by injection (parenterally) because it is destroyed by stomach acid. Penicillin V (figure 34.5) is similar to penicillin G in spectrum of activity, but can be given orally because it is more resistant to acid. The semisynthetic penicillins, on the other hand, have a broader spectrum of activity. Ampicillin can be administered orally and is effective against gram-negative bacteria such as *Haemophilus, Salmonella,* and *Shigella.* Carbenicillin and ticarcillin are potent against *Pseudomonas* and *Proteus.*

An increasing number of bacteria have become resistant to natural penicillins and many of the semisynthetic analogs. Physicians frequently employ specific semisynthetic penicillins that are not destroyed by β-lactamases to combat antibiotic-resistant pathogens. These include methicillin (figure 34.5), nafcillin, and oxacillin. However, this practice has been confounded by the emergence of meticillin-resistant bacteria.

Although penicillins are the least toxic of the antibiotics, about 1 to 5% of the adults in the United States are allergic to them. Occasionally, a person will die of a violent allergic response; therefore, patients should be questioned about penicillin allergies before treatment is begun. Immune disorders: Hypersensitivities (section 32.11)

Cephalosporins

Cephalosporins are a family of antibiotics originally isolated in 1948 from the fungus *Cephalosporium*. They contain a **β-lactam** structure that is very similar to that of the penicillins (**figure 34.6**). As might be expected from their structural similarities to penicillins, cephalosporins also inhibit the transpeptidation reaction during peptidoglycan synthesis. They are broad-spectrum drugs frequently given to patients with penicillin allergies (although about 10% of patients allergic to penicillin are also allergic to cephalosporins).

Many cephalosporins are in use. Cephalosporins are broadly categorized into four generations (groups of drugs that are sequentially developed) based on their spectrum of activity. First-generation cephalosporins are more effective against gram-positive pathogens than gram-negatives. Second-generation drugs, developed after the first generation, have improved effects on gram-

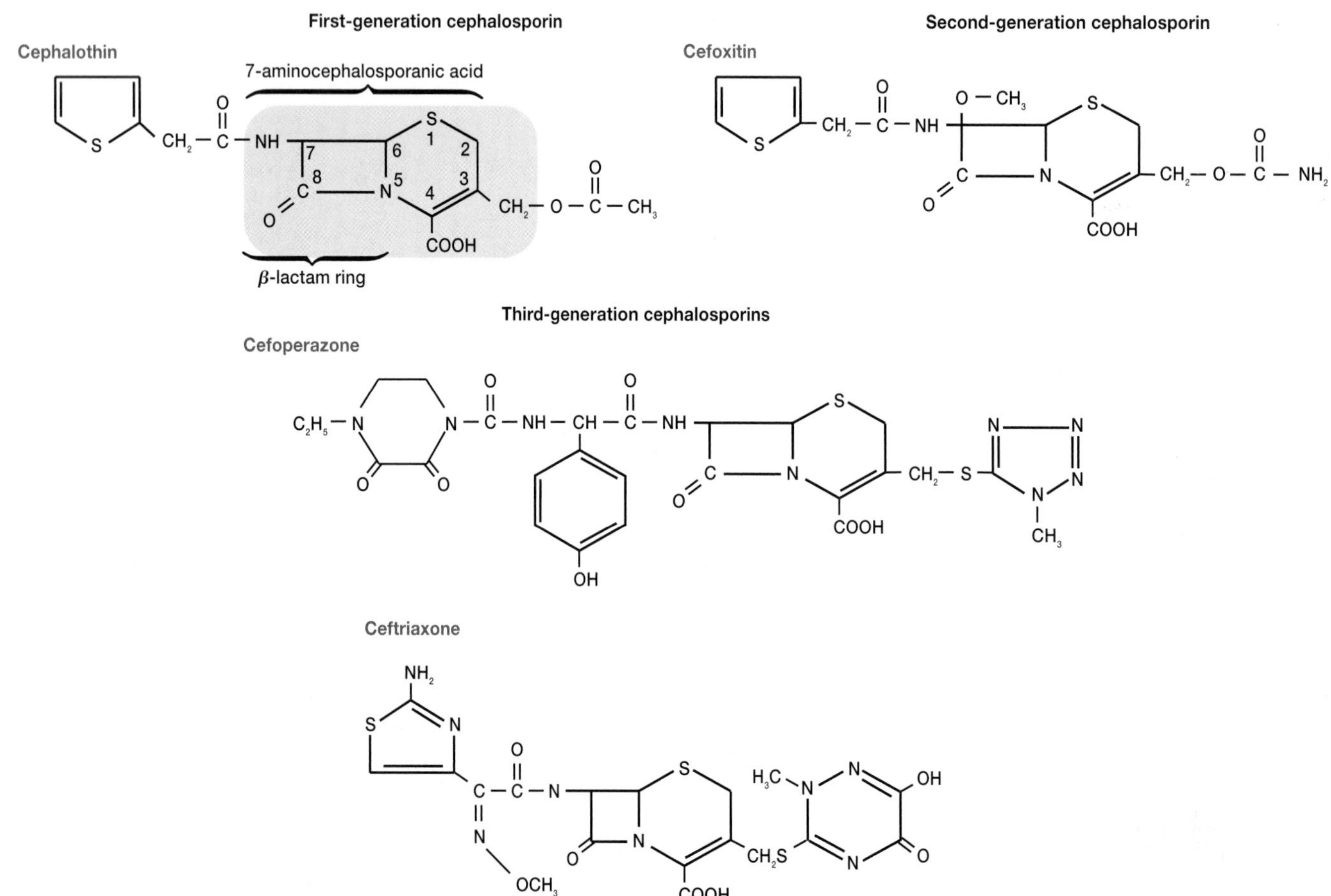

Figure 34.6 Cephalosporin Antibiotics. These drugs are derivatives of 7-aminocephalosporanic acid and contain a β-lactam ring.

negative bacteria with some anaerobe coverage. Third-generation drugs are particularly effective against gram-negative pathogens, and some reach the central nervous system. This is of particular note because many antimicrobial agents do not cross the blood-brain barrier. Finally, fourth-generation cephalosporins are broad spectrum with excellent gram-positive and gram-negative coverage and, like their third-generation predecessors, inhibit the growth of the difficult opportunistic pathogen *Pseudomonas aeruginosa*.

Vancomycin and Teicoplanin

Vancomycin is a glycopeptide antibiotic produced by *Streptomyces oreintalis*. It is a cup-shaped molecule composed of a peptide linked to a disaccharide. Vancomycin's peptide portion blocks the transpeptidation reaction by binding specifically to the D-alanine-D-alanine terminal sequence on the pentapeptide portion of peptidoglycan. The antibiotic is bactericidal for *Staphylococcus* and some members of the genera *Clostridium, Bacillus, Streptococcus,* and *Enterococcus*. It is given both orally and intravenously and has been particularly important in the treatment of antibiotic-resistant staphylococcal and enterococcal infections. However, vancomycin-resistant strains of *Enterococcus* have become widespread and cases of resistant *Staphylococcus aureus* have appeared. This poses a serious public health threat—vancomycin has been considered the "drug of last resort" in cases of antibiotic-resistant *S. aureus*. Clearly newer drugs must be developed.

Teicoplanin is a glycopeptide antibiotic from the actinomycete *Actinoplanes teichomyceticus* that is similar in structure and mechanism of action to vancomycin, but has fewer side effects. It is active against staphylococci, enterococci, streptococci, clostridia, *Listeria,* and many other gram-positive pathogens.

Protein Synthesis Inhibitors

Many antibiotics inhibit protein synthesis by binding with the procaryotic ribosome. Because these drugs discriminate between procaryotic and eucaryotic ribosomes, their therapeutic index is fairly high, but not as high as that of cell wall inhibitors. Some drugs bind to the 30S (small) ribosomal subunit, while others attach to the 50S (large) subunit. Several different steps in protein synthesis can be affected: aminoacyl-tRNA binding, peptide bond formation, mRNA reading, and translocation. Translation (section 11.8)

Aminoglycosides

Although there is considerable variation in structure among several important **aminoglycoside antibiotics,** all contain a cyclohexane ring and amino sugars (**figure 34.7**). **Streptomycin,** kanamycin, neomycin, and tobramycin are synthesized by different species of the genus *Streptomyces,* whereas gentamicin comes from another actinomycete, *Micromonospora purpurea*. Aminoglycosides bind to the 30S (small) ribosomal subunit and interfere with protein synthesis by directly inhibiting the synthesis process and also by causing misreading of the mRNA.

These antibiotics are bactericidal and tend to be most effective against gram-negative pathogens. Streptomycin's usefulness has decreased greatly due to widespread drug resistance, but it is still effective in treating tuberculosis and plague. Gentamicin is used to treat *Proteus, Escherichia, Klebsiella,* and *Seratia* infections. Aminoglycosides can be quite toxic, however, and can cause deafness, renal damage, loss of balance, nausea, and allergic responses.

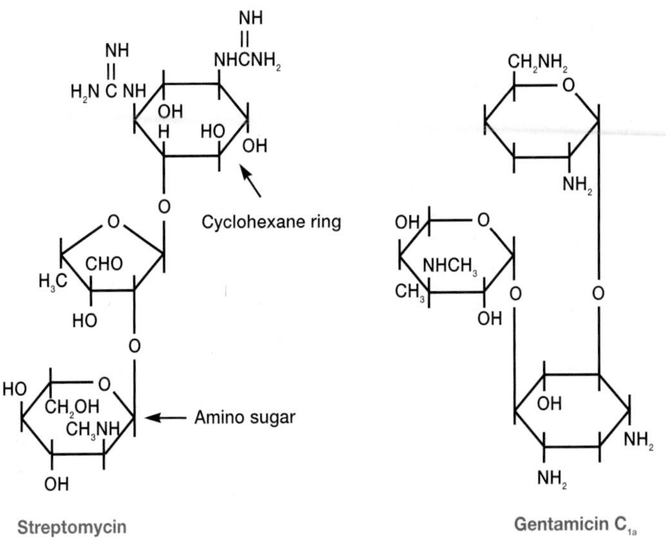

Streptomycin — Cyclohexane ring — Amino sugar — Gentamicin C_{1a}

Figure 34.7 Representative Aminoglycoside Antibiotics.

Tetracycline (chlortetracycline, doxycycline)

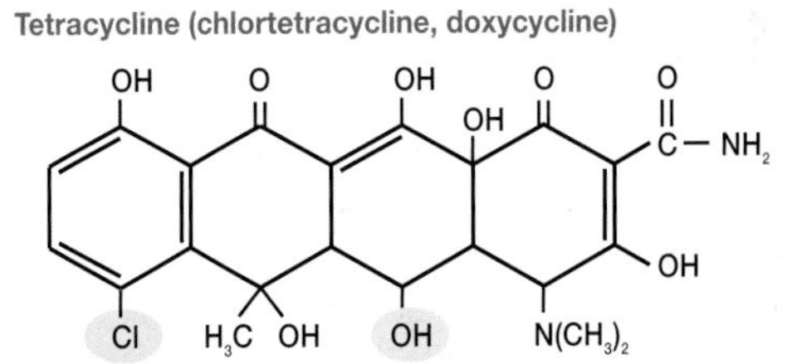

Figure 34.8 Tetracyclines. Three members of the tetracycline family. Tetracycline lacks both of the groups that are shaded. Chlortetracycline (aureomycin) differs from tetracycline in having a chlorine atom (blue): doxycycline consists of tetracycline with an extra hydroxyl (light blue).

Tetracyclines

The **tetracyclines** are a family of antibiotics with a common four-ring structure to which a variety of side chains are attached (**figure 34.8**). Oxytetracycline and chlortetracycline are produced naturally by *Streptomyces* species while others are semisynthetic drugs. These antibiotics are similar to the aminoglycosides and combine with the 30S (small) subunit of the ribosome. This inhibits the binding of aminoacyl-tRNA molecules to the A site of the ribosome. Because their action is only bacteriostatic, the effectiveness of treatment depends on active host resistance to the pathogen.

Tetracyclines are broad-spectrum antibiotics that are active against gram-negative, as well as gram-positive, bacteria, rickettsias, chlamydiae, and mycoplasmas. High doses may result in nausea, diarrhea, yellowing of teeth in children, and damage to the liver and kidneys. Although their use has declined in recent years, they are still sometimes used to treat acne.

Macrolides

The **macrolide antibiotics** contain 12- to 22-carbon lactone rings linked to one or more sugars (**figure 34.9**). **Erythromycin** is usually bacteriostatic and binds to the 23S rRNA of the 50S (large) ribosomal subunit to inhibit peptide chain elongation during protein synthesis. Erythromycin is a relatively broad-spectrum antibiotic effective against gram-positive bacteria, mycoplasmas, and a few gram-negative bacteria. It is used with patients who are allergic to penicillins and in the treatment of whooping cough, diphtheria, diarrhea caused by *Campylobacter,* and pneumonia from *Legionella* or *Mycoplasma* infections. Clindamycin is effective against a variety of bacteria including staphylococci, and anaerobes such as *Bacteroides*. Azithromycin, which has surpassed erythromycin in use, is particularly effective against *Chlamydia trachomatis*.

Chloramphenicol

Chloramphenicol (figure 34.10) was first produced from cultures of *Streptomyces venezuelae* but it is now synthesized chemically. Like erythromycin, this antibiotic binds to 23S rRNA on the 50S ribosomal subunit to inhibit the peptidyl transferase reaction. It has a very broad spectrum of activity but, unfortunately, is quite toxic. One may see allergic responses or neurotoxic reactions. The most common side effect is depression of bone marrow function, leading to aplastic anemia and a decreased number of white blood cells. Consequently, this antibiotic is used only in life-threatening situations when no other drug is adequate.

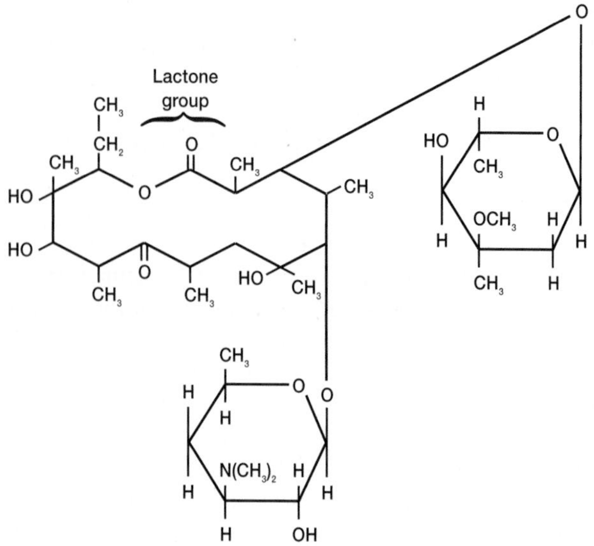

Figure 34.9 Erythromycin, a Macrolide Antibiotic. The 14-member lactone ring is connected to two sugars.

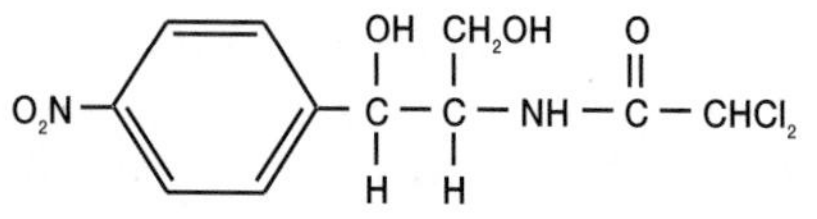

Figure 34.10 Chloramphenicol.

Metabolic Antagonists

Several valuable drugs act as **antimetabolites**—they antagonize, or block, the functioning of metabolic pathways by competitively inhibiting the use of metabolites by key enzymes. These drugs can act as structural analogs, molecules that are structurally similar to naturally occurring metabolic intermediates. These analogs compete with intermediates in metabolic processes because of their similarity, but are just different enough so that they prevent normal cellular metabolism. As such they are bacteriostatic but broad spectrum.

Sulfonamides or Sulfa Drugs

The first antimetabolites to be used successfully as chemotherapeutic agents were the sulfonamides, discovered by G. Domagk. **Sulfonamides,** or sulfa drugs, are structurally related to sulfanilamide, an analog of *p*-aminobenzoic acid, or PABA (**figures 34.11** and **34.12**). PABA is used in the synthesis of the cofactor folic acid (folate). When sulfanilamide or another sulfonamide enters a bacterial cell, it competes with PABA for the active site of an enzyme involved in folic acid synthesis, causing a decline in folate concentration. This decline is detrimental to the bacterium because folic acid is a precursor of purines and pyrimidines, the bases used in the construction of DNA, RNA, and other important cell constituents. The resulting inhibition of purine and pyrimidine synthesis leads to cessation of protein synthesis and DNA replication, thus the pathogen dies. Sulfonamides are selectively toxic for many pathogens because these bacteria manufacture their own fo-

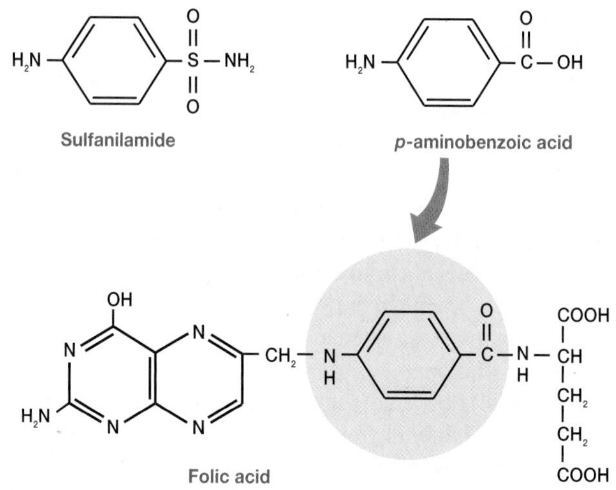

Figure 34.11 Sulfanilamide. Sulfanilamide and its relationship to the structure of folic acid.

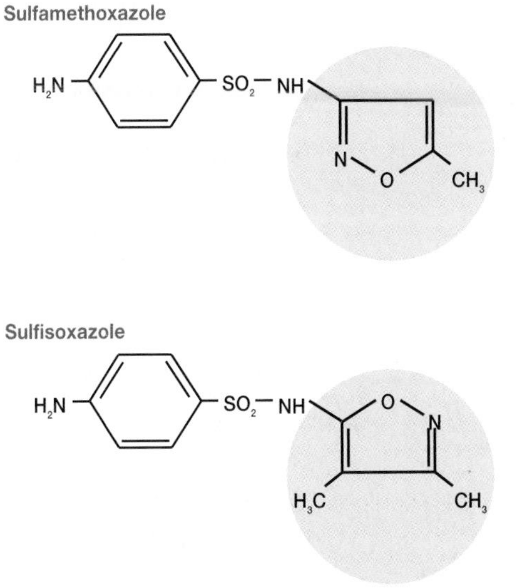

Sulfamethoxazole

Sulfisoxazole

Figure 34.12 Two Sulfonamide Drugs. The blue shaded areas are side chains substituted for a hydrogen in sulfanilamide (figure 35.4).

late and cannot effectively take up this cofactor, whereas humans do not synthesize folate (we must obtain it in our diet). Sulfonamides thus have a high therapeutic index.

The increasing resistance of many bacteria to sulfa drugs limits their effectiveness. Furthermore, as many as 5% of the patients receiving sulfa drugs experience adverse side effects, chiefly allergic responses such as fever, hives, and rashes.

Trimethoprim

Trimethoprim is a synthetic antibiotic that also interferes with the production of folic acid. It does so by binding to dihydrofolate reductase (DHFR), the enzyme responsible for converting dihydrofolic acid to tetrahydrofolic acid, competing against the dihydrofolic acid substrate (**figure 34.13**). It is a broad-spectrum antibiotic often used to treat respiratory and middle ear infections, urinary tract infections, and traveler's diarrhea. It can be combined with sulfa drugs to increase efficacy of treatment by blocking two key steps in the folic acid pathway (**figure 34.14**). The inhibition of two successive steps in a single biochemical pathway means that less of each drug is needed in combination than when used alone. This is termed a synergistic drug interaction.

The most common side effects associated with trimethoprim are abdominal pain, abnormal taste, diarrhea, loss of appetite, nausea, swelling of the tongue, and vomiting. Taking trimethoprim with food may reduce some of these side effects. Some patients are allergic to trimethoprim, exhibiting rash and itching. Some patients develop photosensitivity reactions (i.e., rashes due to sun exposure).

Nucleic Acid Synthesis Inhibition

The antibacterial drugs that inhibit nucleic acid synthesis function by inhibiting DNA polymerase and DNA helicase or RNA polymerase, thus blocking processes of replication or transcription, respectively. These drugs are not as selectively toxic as other antibiotics because procaryotes and eucaryotes do not differ greatly with respect to nucleic acid synthesis.

(a) Dihydrofolic acid (DFA)

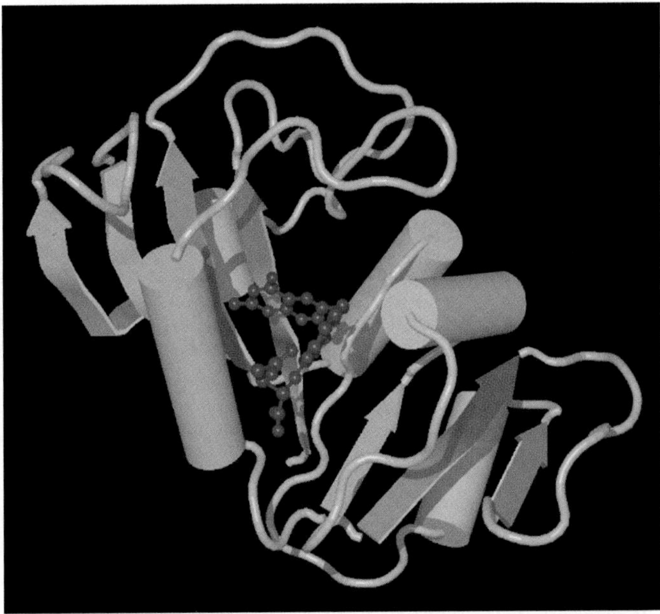

(b) Dihydrofolate reductase

(c) Trimethoprim

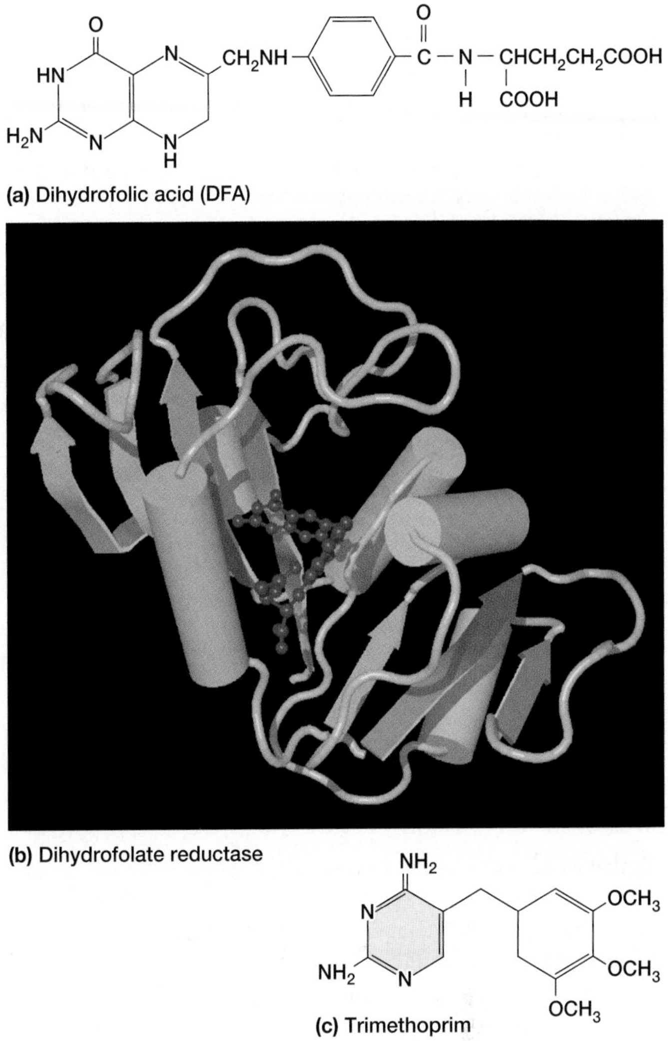

Figure 34.13 Competitive Inhibition of Dihydrofolate Reductase (DHFR) by Trimethoprim. **(a)** Dihydrofolic acid (DFA) is the natural substrate for the DHFR enzyme of the folic acid pathway. **(b)** DHFR structure and its interaction with DFA (red). Note the chemical structure and how it fits into the active site of the enzyme. **(c)** Trimethoprim mimics the structural orientation of the DFA and thus competes for the active site of the enzyme. The consequence of this is delayed or absent folic acid synthesis because the DFA cannot be converted to tetrahydrofolic acid when trimethoprim occupies the DHFR active site.

Quinolones

The **quinolones** are synthetic drugs that contain the 4-quinolone ring. The quinolones are important antimicrobial agents that inhibit nucleic acid synthesis. They are increasingly used to treat a wide variety of infections. The first quinolone, nalidixic acid

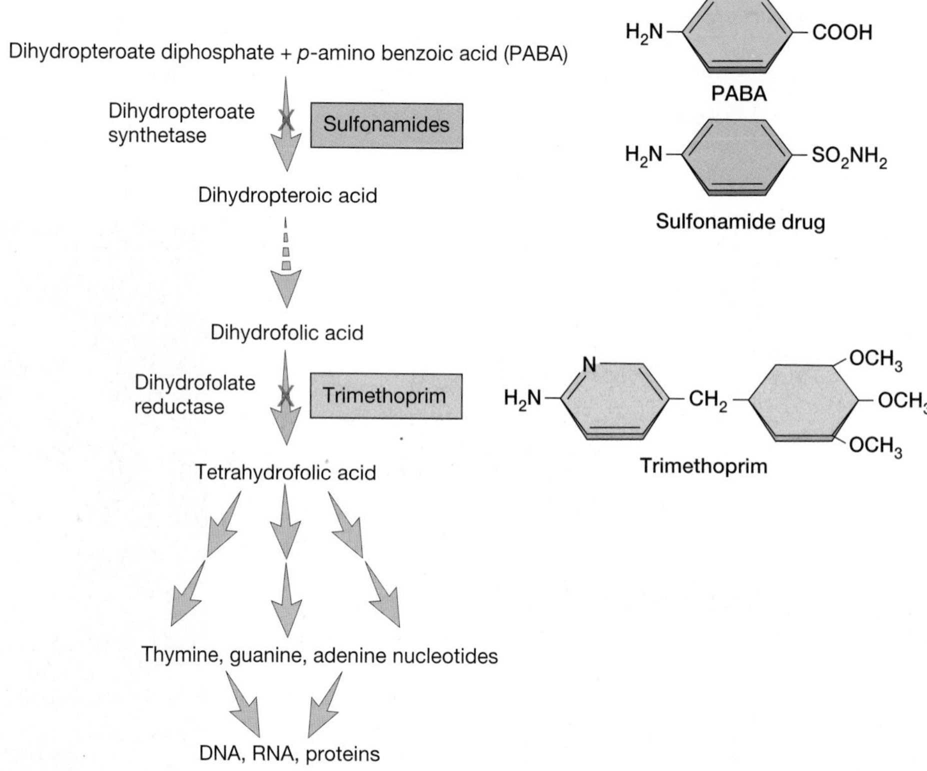

Figure 34.14 Synergistic Drug Interaction Between the Sulfonamides and Trimethoprim. Two successive steps in the biochemical pathway for folic acid synthesis are blocked by these drugs. Thus the efficacy of the drug combination is greater than that of either drug used alone.

(**figure 34.15**), was synthesized in 1962. Since that time, generations of fluoroquinolones have been produced. Three of these—ciprofloxacin, norfloxacin, and ofloxacin—are currently used in the United States, and more fluoroquinolones are being synthesized and tested. Ciprofloxacin (Cipro) gained notoriety during the 2001 bioterror attacks in the United States as one treatment for anthrax. Bioterrorism preparedness (section 36.9)

Quinolones act by inhibiting the bacterial DNA gyrase and topoisomerase II. DNA gyrase introduces negative twist in DNA and helps separate its strands (**figure 34.16**). Inhibition of DNA gyrase disrupts DNA replication and repair, bacterial chromosome separation during division, and other cell processes involving DNA. Fluoroquinolones also inhibit topoisomerase II, another enzyme that untangles DNA during replication. It is not surprising that quinolones are bactericidal. DNA replication (section 11.4)

The quinolones are broad-spectrum antibiotics. They are highly effective against enteric bacteria such as *E. coli* and *Klebsiella pneumoniae*. They can be used with *Haemophilus, Neisseria, P. aeruginosa,* and other gram-negative pathogens. The quinolones also are active against gram-positive bacteria such as *S. aureus, Streptococcus pyogenes,* and *Mycobacterium tuberculosis*. Currently, they are used in treating urinary tract infections, sexually transmitted diseases caused by *Neisseria* and *Chlamydia,* gastrointestinal infections, respiratory infections, skin infections, and osteomyelitis (bone infection). Quinolones are effective when administered orally but can sometimes cause diverse side effects, particularly gastrointestinal upset.

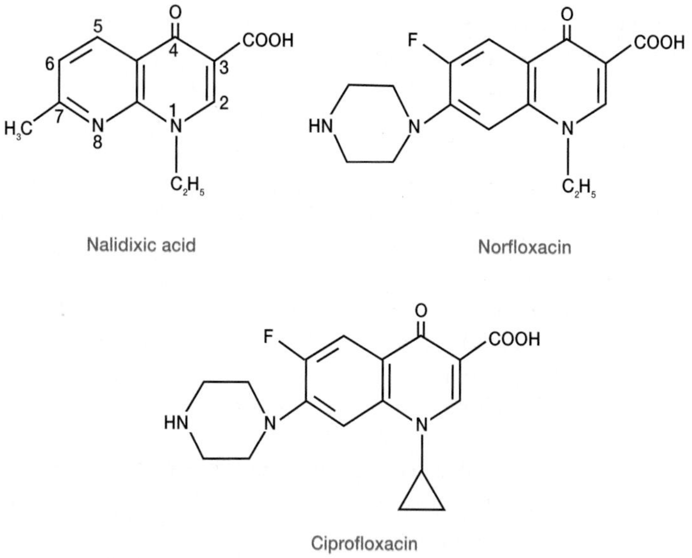

Nalidixic acid

Norfloxacin

Ciprofloxacin

Figure 34.15 Quinolone Antimicrobial Agents. Ciprofloxacin and norfloxacin are newer generation fluoroquinolones. The 4-quinolone ring in nalidixic acid has been numbered.

1. Explain five ways in which chemotherapeutic agents kill or damage bacterial pathogens.
2. Why do penicillins and cephalosporins have a higher therapeutic index than most other antibiotics?

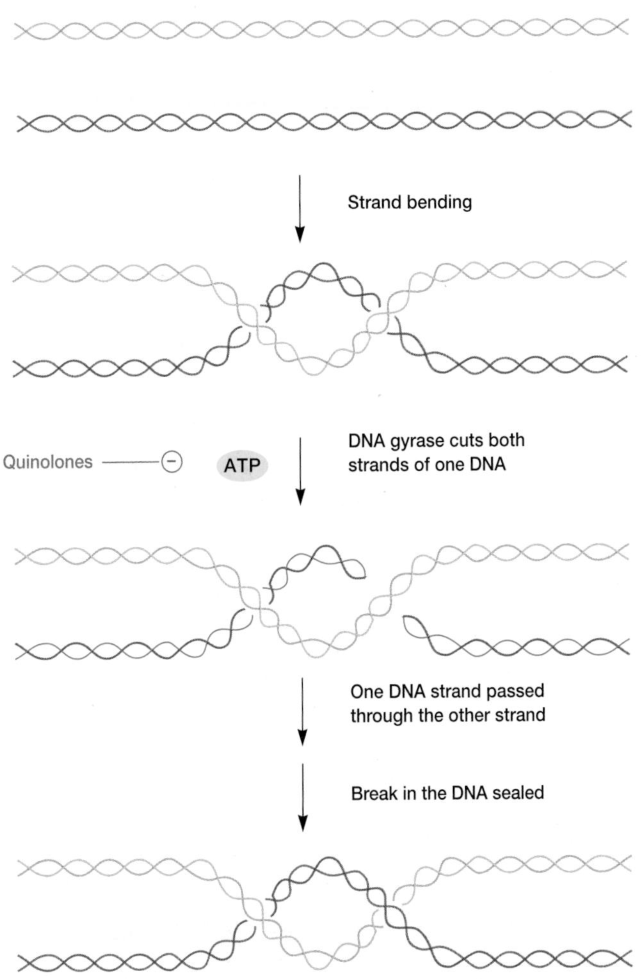

Figure 34.16 DNA Gyrase Action and Quinolone Inhibition.

3. Would there be any advantage to administering a bacteriostatic agent along with penicillins? Any disadvantage?
4. What are antimetabolites?
5. Why are some antibiotics toxic?

34.5 Factors Influencing Antimicrobial Drug Effectiveness

It is crucial to recognize that drug therapy is not a simple matter. Drugs may be administered in several different ways, and they do not always spread rapidly throughout the body or immediately kill all invading pathogens. A complex array of factors influence the effectiveness of drugs.

First, the drug must actually be able to reach the site of infection. Understanding the factors that control drug activity, stability, and metabolism in vivo are essential in drug formulation. For example, the mode of administration plays an important role. A drug such as penicillin G is not suitable for oral administration because it is relatively unstable in stomach acid. Some antibiotics—for example, gentamicin and other aminoglycosides—are not well absorbed from the intestinal tract and must be injected intramuscularly or given intravenously. Other antibiotics (neomycin, bacitracin) are so toxic that they can only be applied topically to skin lesions. Non-oral routes of administration often are called **parenteral routes.** Even when an agent is administered properly, it may be excluded from the site of infection. For example, blood clots or necrotic tissue can protect bacteria from a drug, either because body fluids containing the agent may not easily reach the pathogens or because the agent is absorbed by materials surrounding it.

Second, the pathogen must be susceptible to the drug. Bacteria in biofilms or abscesses may be replicating very slowly and are therefore resistant to chemotherapy, because many agents affect pathogens only if they are actively growing and dividing. A pathogen, even though growing, may simply not be susceptible to a particular agent. For example, penicillins and cephalosporins, which inhibit cell wall synthesis (table 34.1), do not harm mycoplasmas, which lack cell walls. To control resistance, drug cocktails can be used to treat some infections. A notable example of this is the use of clavulonic acid (to inactivate penicillinase) combined with ampicillin (Augmentin) to treat penicillin-resistant bacteria.

Third, the chemotherapeutic agent must exceed the pathogen's MIC value if it is going to be effective. The concentration reached will depend on the amount of drug administered, the route of administration and speed of uptake, and the rate at which the drug is cleared or eliminated from the body. It makes sense that a drug will remain at high concentrations longer if it is absorbed over an extended period and excreted slowly.

Finally, chemotherapy has been rendered less effective and much more complex by the spread of drug-resistance genes.

1. Briefly discuss the factors that influence the effectiveness of antimicrobial drugs.
2. What is parenteral administration of a drug?

34.6 Drug Resistance

The spread of drug-resistant pathogens is one of the most serious threats to public health in the 21st century (**Disease 34.2**). This section describes the ways in which bacteria acquire drug resistance and how resistance spreads within a bacterial population.

Mechanisms of Drug Resistance

The long-awaited "superbug" arrived in the summer of 2002. *S. aureus,* a common but sometimes deadly bacterium, had acquired a new antibiotic-resistance gene. The new strain was isolated from foot ulcers on a diabetic patient in Detroit, Michigan. Meticillin-resistant (formerly methicillin-resistant) *S. aureus* (MRSA) had been well known as the bane of hospitals. This newer strain had developed resistance to vancomycin, one of the few antibiotics that was still able to control *S. aureus*. This new vancomycin-resistant

34.2 Antibiotic Misuse and Drug Resistance

The sale of antimicrobial drugs is big business. In the United States millions of pounds of antibiotics valued at billions of dollars are produced annually. As much as 70% of these antibiotics are added to livestock feed.

Because of the massive quantities of antibiotics being prepared and used, an increasing number of diseases are resisting treatment due to the spread of drug resistance. A good example is *Neisseria gonorrhoeae,* the causative agent of gonorrhea. Gonorrhea was first treated successfully with sulfonamides in 1936, but by 1942 most strains were resistant and physicians turned to penicillin. Within 16 years a penicillin-resistant strain emerged in Asia. A penicillinase-producing gonococcus reached the United States in 1976 and is still spreading in this country. Thus penicillin is no longer used to treat gonorrhea.

In late 1968 an epidemic of dysentery caused by *Shigella* broke out in Guatemala and affected at least 112,000 persons; 12,500 deaths resulted. The strains responsible for this devastation carried an R plasmid conferring resistance to chloramphenicol, tetracycline, streptomycin, and sulfonamide. In 1972 a typhoid epidemic swept through Mexico producing 100,000 infections and 14,000 deaths. It was due to a *Salmonella* strain with the same multiple-drug-resistance pattern seen in the previous *Shigella* outbreak.

Haemophilus influenzae type b is responsible for many cases of childhood pneumonia and middle ear infections, as well as respiratory infections and meningitis. It is now becoming increasingly resistant to tetracyclines, ampicillin, and chloramphenicol. Similarly, the worldwide rate of penicillin-nonsusceptible (i.e., resistant) *Streptococcus pneumoniae* (PNSP) continues to increase. There is a direct correlation between the daily use of antibiotics (expressed as defined daily dose [DDD] per day) and the percent of PNSP isolates cultured (**Box figure**). This dramatic correlation is alarming. More alarming is the continued indiscriminant use of antibiotics in light of these data.

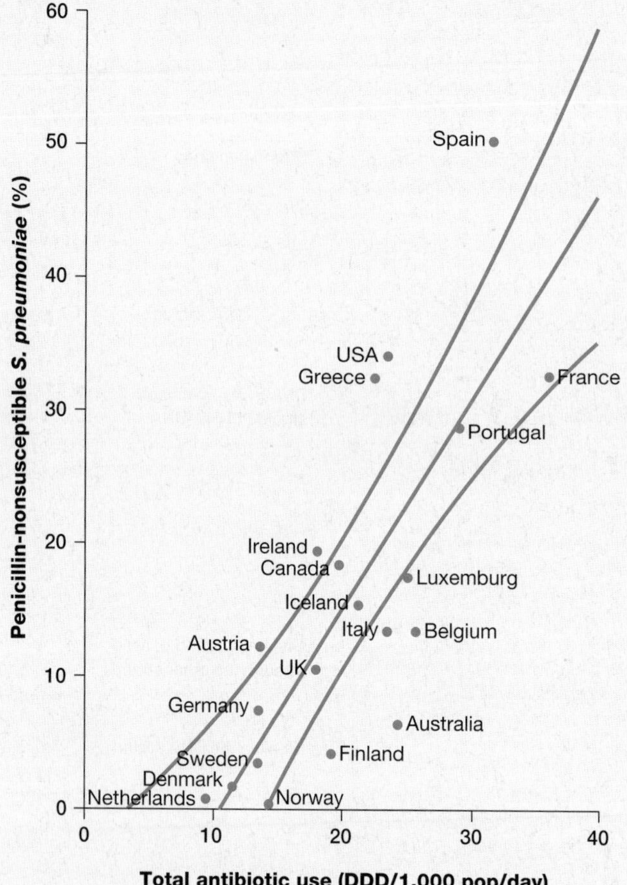

In 1946 almost all strains of *Staphylococcus* were penicillin sensitive. Today most hospital strains are resistant to penicillin G, and some are now also resistant to methicillin and/or gentamicin and only can be treated with vancomycin. Some strains of *Enterococcus* have become resistant to most antibiotics, including vancomycin. Recently a few cases of vancomycin-resistant *S. aureus* have been reported in the United States and Japan. At present these strains are only intermediately resistant to vancomycin. If full vancomycin resistance spreads, *S. aureus* may become untreatable.

It is clear from these and other examples (e.g., multiresistant *Mycobacterium tuberculosis*) that drug resistance is an extremely serious public health problem. Much of the difficulty arises from drug misuse. Drugs frequently have been overused in the past. It has been estimated that over 50% of the antibiotic prescriptions in hospitals are given without clear evidence of infection or adequate medical indication. Many physicians have administered antibacterial drugs to patients with colds, influenza, viral pneumonia, and other viral diseases. A recent study showed that over 50% of the patients diagnosed with colds and upper respiratory infections and 66% of those with chest colds (bronchitis) are given antibiotics, even though over 90% of these cases are caused by viruses. Frequently antibiotics are prescribed without culturing and identifying the pathogen or without determining bacterial sensitivity to the drug. Toxic, broad-spectrum antibiotics are sometimes given in place of narrow-spectrum drugs as a substitute for culture and sensitivity testing, with the consequent risk of dangerous side effects, opportunistic infections, and the selection of drug-resistant mutants. The situation is made worse by patients not completing their course of medication. When antibiotic treatment is ended too early, drug-resistant mutants may survive. Drugs are available without prescription to the public in many countries; people may practice self-administration of antibiotics and further increase the prevalence of drug-resistant strains.

The use of antibiotics in animal feeds is undoubtedly another contributing factor to increasing drug resistance. The addition of low levels of antibiotics to livestock feeds raises the efficiency and rate of weight gain in cattle, pigs, and chickens (partially because of infection control in overcrowded animal populations). However, this also increases the number of drug-resistant bacteria in animal intestinal tracts. There is evidence for the spread of bacteria such as *Salmonella* from animals to human populations. In 1983, 18 people in four midwestern states were infected with a multiple-drug-resistant strain of *Salmonella newport.* Eleven were hospitalized for salmonellosis and one died. All 18 patients had recently been infected by eating hamburger from beef cattle fed subtherapeutic doses of chlortetracycline for growth promotion. Resistance to some antibiotics has been traced to the use of specific farmyard antibiotics. Avoparcin resembles vancomycin in structure, and virginiamycin resembles Synercid. There is good circumstantial evidence that extensive use of these two antibiotics in animal feed has led to an increase in vancomycin and Synercid resistance among enterococci. The use of the quinolone antibiotic enrofloxacin in swine herds appears to have promoted ciprofloxacin resistance in pathogenic strains of *Salmonella.* In 2005, the use of fluoroquinolones in U.S. poultry farming was banned in recognition of this public health threat.

The spread of antibiotic resistance can be due to quite subtle factors. For example, products such as soap and deodorants often now contain triclosan and other germicides. There is increasing evidence that the widespread use of triclosan actually favors an increase in antibiotic resistance (*see section 7.5*).

S. aureus (VRSA) strain also resisted most other antibiotics including ciprofloxacin, methicillin, and penicillin. Isolated from the same patient was another dread of hospitals—vancomycin-resistant enterococci (VRE). Genetic analyses revealed that the patient's own vancomycin-sensitive *S. aureus* had acquired the vancomycin-resistance gene, *vanA*, from VRE through conjugation. So was born a new threat to the health of the human race. Bacterial conjugation (section 13.7); Bacterial plasmids (sections 3.5 and 13.6)

Bacteria often become resistant in several different ways (**figure 34.17**). Unfortunately, a particular type of resistance mechanism is not confined to a single class of drugs (**figure 34.18**). Two bacteria may use different resistance mechanisms to withstand the same chemotherapeutic agent. Furthermore, resistant mutants arise spontaneously and are then selected for in the presence of the drug.

Pathogens often become resistant simply by preventing entrance of the drug. Many gram-negative bacteria are unaffected by penicillin G because it cannot penetrate the envelope's outer membrane. Genetic mutations that lead to changes in penicillin binding proteins also render a cell resistant. A decrease in permeability can lead to sulfonamide resistance. Mycobacteria resist many drugs because of the high content of mycolic acids (*see figure 24.11*) in a complex lipid layer outside their peptidoglycan. This layer is impermeable to most water soluble drugs. Suborder *Corynebacterineae:* Genus *Mycobacterium* (section 24.4)

A second resistance strategy is to pump the drug out of the cell after it has entered. Some pathogens have plasma membrane translocases, often called efflux pumps, that expel drugs. Because they are relatively nonspecific and can pump many different drugs, these transport proteins often are called multidrug-resistance pumps. Many are drug/proton antiporters—that is, protons enter the cell as the drug leaves. Such systems are present in *E. coli, P. aeruginosa,* and *S. aureus* to name a few.

Many bacterial pathogens resist attack by inactivating drugs through chemical modification. The best-known example is the hydrolysis of the β-lactam ring of penicillins by the enzyme penicillinase. Drugs also are inactivated by the addition of chemical groups. For example, chloramphenicol contains two hydroxyl groups (figure 34.10) that can be acetylated in a reaction catalyzed by the enzyme chloramphenicol acyltransferase with acetyl CoA as the donor. Aminoglycosides (figure 34.7) can be modified and inactivated in several ways. Acetyltransferases catalyze the acetylation of amino groups. Some aminoglycoside-modifying enzymes catalyze the addition to hydroxyl groups of either phosphates (phosphotransferases) or adenyl groups (adenyltransferases).

Because each chemotherapeutic agent acts on a specific target, resistance arises when the target enzyme or cellular structure is modified so that it is no longer susceptible to the drug.

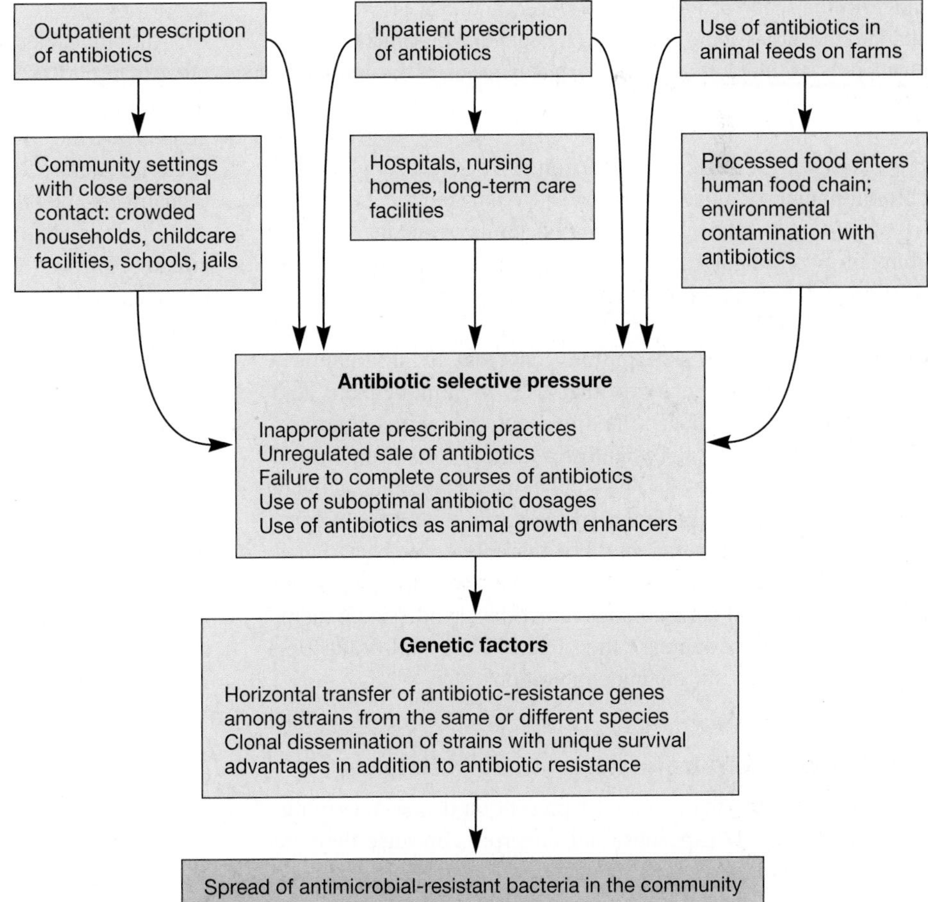

Figure 34.17 Antibiotic Resistance Has Many Sources. Incomplete and indiscriminant use of antibiotics in people and animals leads to increased selective pressure on bacteria. Bacteria capable of resisting antibiotics survive and spread these traits by horizontal gene transfer.

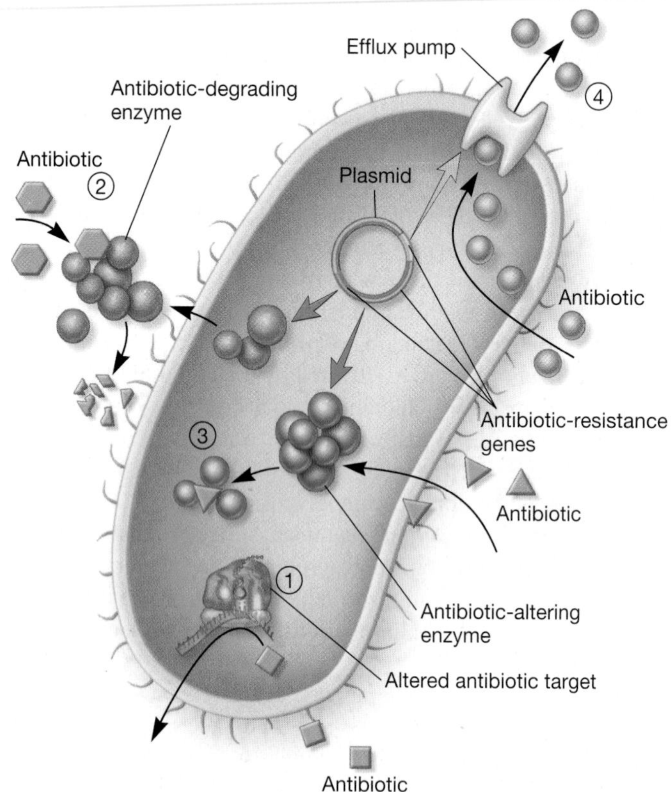

Figure 34.18 Antibiotic Resistance Mechanisms. Bacteria can resist the action of antibiotics by (1) preventing access to (or altering) the target of the antibiotic, (2) degrading the antibiotic, (3) altering the antibiotic, and/or (4) rapid extrusion of the antibiotic.

For example, the affinity of ribosomes for erythromycin and chloramphenicol can be decreased by a change in the 23S rRNA to which they bind. Enterococci become resistant to vancomycin by changing the terminal D-alanine-D-alanine in their peptidoglycan to a D-alanine-D-lactate. This drastically reduces antibiotic binding. Antimetabolite action may be resisted through alteration of susceptible enzymes. In sulfonamide-resistant bacteria the enzyme that uses *p*-aminobenzoic acid during folic acid synthesis (the dihydropteroic acid synthetase; figure 34.14) often has a much lower affinity for sulfonamides.

Finally, resistant bacteria may either use an alternate pathway to bypass the sequence inhibited by the agent or increase the production of the target metabolite. For example, some bacteria are resistant to sulfonamides simply because they use preformed folic acid from their surroundings rather than synthesize it themselves. Other strains increase their rate of folic acid production and thus counteract sulfonamide inhibition.

The Origin and Transmission of Drug Resistance

Genes for drug resistance may be present on bacterial chromosomes, plasmids, transposons, and integrons. Because they are often found on mobile genetic elements, they can freely exchange between bacteria. Spontaneous mutations in the bacterial chromosome, although they do not occur very often, can make bacteria drug resistant. Usually such mutations result in a change in the drug target; therefore the antibiotic cannot bind and inhibit growth (e.g., the protein target to which streptomycin binds on bacterial ribosomes). Many mutants are probably destroyed by natural host resistance mechanisms. However, when a patient is being treated extensively with antibiotics, some resistant mutants may survive and flourish because of their competitive advantage over nonresistant strains.

Frequently a bacterial pathogen is drug resistant because it has a plasmid bearing one or more resistance genes; such plasmids are called **R plasmids** (*r*esistance plasmids). Plasmid resistance genes often code for enzymes that destroy or modify drugs; for example, the hydrolysis of penicillin or the acetylation of chloramphenicol and aminoglycoside drugs. Plasmid-associated genes have been implicated in resistance to the aminoglycosides, choramphenicol, penicillins and cephalosporins, erythromycin, tetracyclines, sulfonamides, and others. Once a bacterial cell possesses an R plasmid, the plasmid (or its genes) may be transferred to other cells quite rapidly through normal gene exchange processes such as conjugation, transduction, and transformation (**figure 34.19**). Because a single plasmid may carry genes for resistance to several drugs, a pathogen population can become resistant to several antibiotics simultaneously, even though the infected patient is being treated with only one drug. Bacterial conjugation (section 13.7); Transduction (section 13.9); DNA transformation (section 13.8); Bacterial plasmids (sections 3.5 and 13.6)

Antibiotic resistance genes can be located on genetic elements other than plasmids. Many composite transposons contain genes for antibiotic resistance, and some bear more than one resistance gene. They are found in both gram-negative and gram-positive bacteria. Some examples and their resistance markers are Tn*5* (kanamycin, bleomycin, streptomycin), Tn*9* (chloramphenicol), Tn*10* (tetracycline), Tn*21* (streptomycin, spectinomycin, sulfonamide), Tn*551* (erythromycin), and Tn*4001* (gentamicin, tobramycin, kanamycin). Resistance genes on composite transposons can move rapidly between plasmids and through a bacterial population. Often several resistance genes are carried together as gene cassettes in association with a genetic element known as an integron. An **integron** is composed of an integrase gene and sequences for site-specific recombination. Thus integrons can capture genes and gene cassettes. **Gene cassettes** are genetic elements that may exist as circular nonreplicating DNA when moving from one site to another, but which normally are a linear part of a transposon, plasmid, or bacterial chromosome. Cassettes usually carry one or two genes and a recombination site. Several cassettes can be integrated sequentially in an integron. Thus integrons also are important in spreading resistance genes. Finally, conjugative transposons, like composite transposons, can carry resistance genes. Because they are capable of moving between bacteria by conjugation, they are also effective in spreading resistance. Transposable elements (section 13.5)

Extensive drug treatment favors the development and spread of antibiotic-resistant strains because the antibiotic destroys

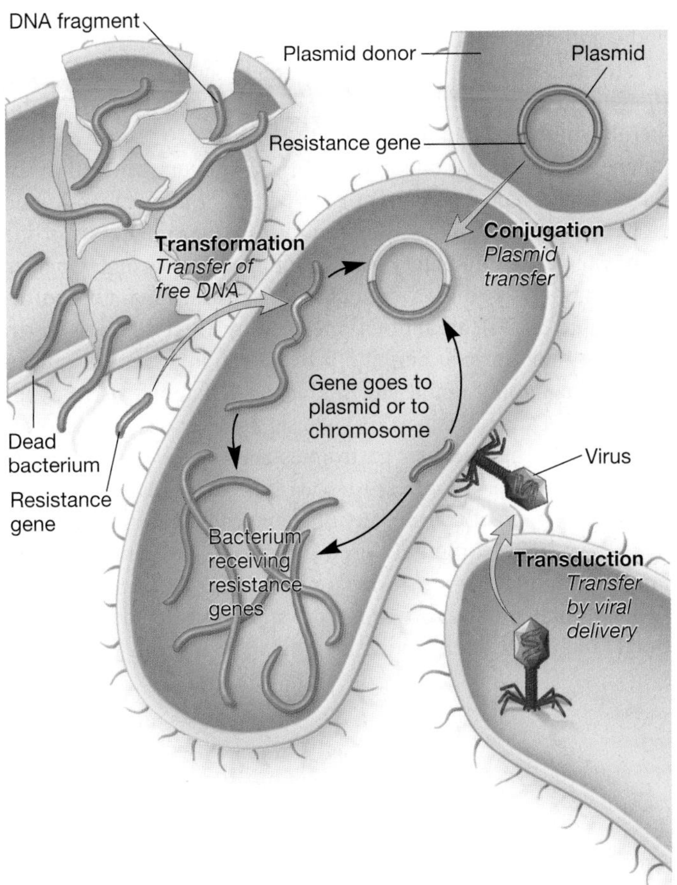

Figure 34.19 Horizontal Gene Exchange. Bacteria exchange genetic information, like antibiotic resistance genes, through conjugation, transformation, and/or transduction.

susceptible bacteria that would usually compete with drug-resistant strains. The result may be the emergence of drug-resistant pathogens. Several strategies can be employed to discourage the emergence of drug resistance. The drug can be given in a high enough concentration to destroy susceptible bacteria and most spontaneous mutants that might arise during treatment. Sometimes two or even three different drugs can be administered simultaneously with the hope that each drug will prevent the emergence of resistance to the other. This approach is used in treating tuberculosis and malaria. Finally, chemotherapeutic drugs, particularly broad-spectrum drugs, should be used only when definitely necessary. If possible, the pathogen should be identified, drug sensitivity tests run, and the proper narrow-spectrum drug employed.

Despite efforts to control the emergence and spread of drug resistance, the situation continues to worsen. Of course, antibiotics should be used in ways that reduce the development of resistance. Another approach is to search for new antibiotics that microorganisms have never encountered. Pharmaceutical companies collect and analyze samples from around the world in a search for completely new antimicrobial agents. Structure-based or rational drug design is a third option. If the three-dimensional structure of a susceptible target molecule such as an enzyme essential to microbial function is known, computer programs can be used to design drugs that precisely fit the target molecule. These drugs might be able to bind to the target and disrupt its function sufficiently to destroy the pathogen. Pharmaceutical companies are using this approach to attempt to develop drugs for the treatment of AIDS, cancer, septicemia caused by lipopolysaccharide (LPS), and the common cold. At least one company is developing "enhancers." These are cationic peptides that disrupt bacterial membranes by displacing their magnesium ions. Antibiotics then penetrate and rapidly exert their effects. Other pharmaceutical companies are developing efflux-pump inhibitors to administer with antibiotics and prevent their expulsion by the resistant pathogen.

There has been some progress in developing new antibiotics that are effective against drug-resistant pathogens. Two new drugs are fairly effective against vancomycin-resistant enterococci. Synercid is a mixture of the streptogramin antibiotics quinupristin and dalfopristin that inhibits protein synthesis. A second drug, linezolid (Zyvox), is the first drug in a new family of antibiotics, the oxazolidinones. It inhibits protein synthesis and is active against both vancomycin-resistant enterococci and meticillin-resistant *Staphylococcus aureus*.

Information that is coming from the sequencing and analysis of pathogen genomes is also useful in identifying new targets for antimicrobial drugs. For example, genomics studies are providing data for research on inhibitors of both aminoacyl-tRNA synthetases and the enzyme that removes the formyl group from the N-terminal methionine during bacterial protein synthesis. Bacteria must synthesize the fatty acids they require for growth rather than acquiring the acids from their environment. The drug susceptibility of enzymes in the fatty acid synthesis system is being analyzed by screening pathogens for potential targets. Genomics (chapter 15)

A most interesting response to the current crisis is the renewed interest in an idea first proposed early in the twentieth century by Felix d'Herelle, one of the discoverers of bacterial viruses or bacteriophages. d'Herelle proposed that bacteriophages could be used to treat bacterial diseases. Although most microbiologists did not pursue his proposal actively due to technical difficulties and the advent of antibiotics, Russian scientists developed the medical use of bacteriophages. Currently Russian physicians use bacteriophages to treat many bacterial infections. Bandages are saturated with phage solutions, phage mixtures are administered orally, and phage preparations are given intravenously to treat *Staphylococcus* infections. Three American companies are actively conducting research on phage therapy and preparing to carry out clinical trials. Viruses of *Bacteria* and *Archaea* (chapter 17)

1. Briefly describe the five major ways in which bacteria become resistant to drugs and give an example of each.
2. Define plasmid, R plasmid, integron, and gene cassette. How are R plasmids involved in the spread of drug resistance?
3. List several ways in which the development of antibiotic-resistant pathogens can be slowed or prevented.

34.7 Antifungal Drugs

Treatment of fungal infections generally has been less successful than that of bacterial infections largely because eucaryotic fungal cells are much more similar to human cells than are bacteria. Many drugs that inhibit or kill fungi are therefore quite toxic for humans. In addition, most fungi have a detoxification system that modifies many antifungal agents. As a result the added antibiotics are fungistatic only as long as repeated application maintains high levels of unmodified antibiotic. Despite their relatively low therapeutic index, a few drugs are useful in treating many major fungal diseases. Effective antifungal agents frequently either extract membrane sterols or prevent their synthesis. Similarly, because animal cells do not have cell walls, the enzyme chitin synthase is the target for fungal-active antibiotics such as polyoxin D and nikkomycin.

Fungal infections are often subdivided into infections of superficial tissues or superficial mycoses and systemic mycoses. Treatment for these two types of disease is very different. Several drugs are used to treat superficial mycoses. Three drugs containing imidazole—miconazole, ketoconazole (**figure 34.20**), and clotrimazole—are broad-spectrum agents available as creams and solutions for the treatment of dermatophyte infections such as athlete's foot, and oral and vaginal candidiasis. They are thought to disrupt fungal membrane permeability and inhibit sterol synthesis. Tolnaftate is used topically for the treatment of cutaneous infections, but is not as effective against infections of the skin and hair. **Nystatin** (figure 34.20), a polyene antibiotic from *Streptomyces,* is used to control *Candida* infections of the skin, vagina, or alimentary tract. It binds to sterols and damages the membrane, leading to fungal membrane leakage. **Griseofulvin** (figure 34.20), an antibiotic formed by *Penicillium,* is given orally to treat chronic dermatophyte infections. It is thought to disrupt the mitotic spindle and inhibit cell division; it also may inhibit protein and nucleic acid synthesis. Side effects of griseofulvin include headaches, gastrointestinal upset, and allergic reactions. Human diseases caused by fungi and protists (chapter 39)

Systemic infections are very difficult to control and can be fatal. Three drugs commonly used against systemic mycoses are **amphotericin B,** 5-flucytosine, and fluconazole (figure 34.20). Amphotericin B from *Streptomyces* spp. binds to the sterols in fungal membranes, disrupting membrane permeability and caus-

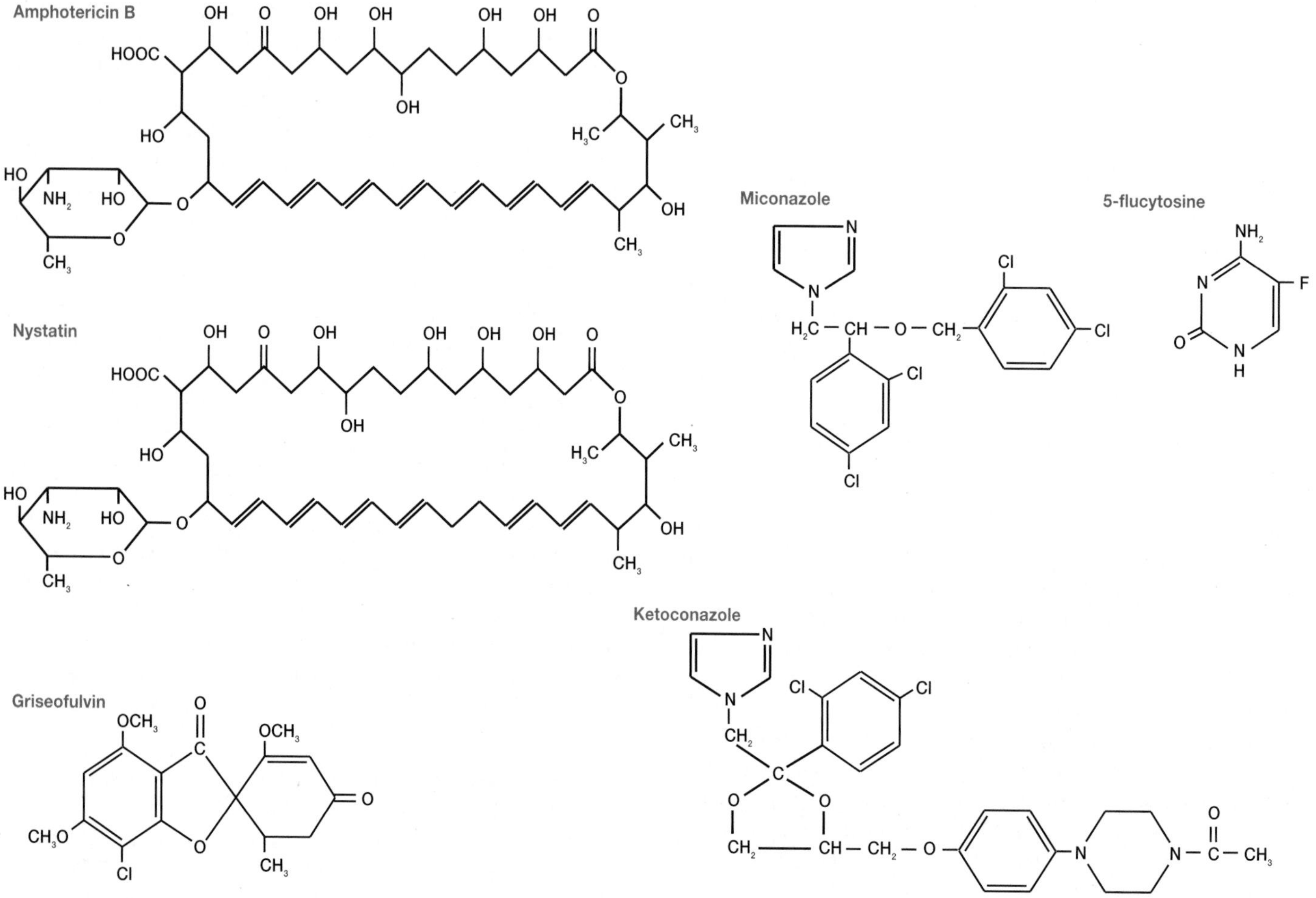

Figure 34.20 Antifungal Drugs. Six commonly used drugs are shown here.

ing leakage of cell constituents. It is quite toxic and used only for serious, life-threatening infections. The synthetic oral antimycotic agent 5-flucytosine (5-fluorocytosine) is effective against most systemic fungi, although drug resistance often develops rapidly. The drug is converted to 5-fluorouracil by the fungi, incorporated into RNA in place of uracil, and disrupts RNA function. Its side effects include skin rashes, diarrhea, nausea, aplastic anemia, and liver damage. Fluconazole is used in the treatment of candidiasis, cryptococcal meningitis, and coccidioidal meningitis. Because adverse effects to fluconazole are relatively uncommon, it is used prophylactically to prevent life-threatening fungal infections in AIDS patients and other individuals who are severely immunosuppressed.

1. Summarize the mechanism of action and the therapeutic use of the following antifungal drugs: miconazole, nystatin, griseofulvin, amphotericin B, and 5-flucytosine.

34.8 Antiviral Drugs

For many years the possibility of treating viral infections with drugs appeared remote because viruses enter host cells and make use of host cell enzymes and constituents. A drug that would block virus reproduction also was thought to be toxic for the host. Inhibitors of virus-specific enzymes and life cycle processes have now been discovered, and several drugs are used therapeutically. Some important examples are shown in **figure 34.21.** Reproduction of vertebrate viruses (section 18.2)

Most antiviral drugs disrupt either critical stages in the virus life cycle or the synthesis of virus-specific nucleic acids. **Amantadine** and rimantadine can be used to prevent influenza A infections. When given early in the infection (in the first 48 hours), they reduce the incidence of influenza by 50 to 70% in an exposed population. Amantadine blocks the penetration and uncoating of influenza virus particles. **Adenine arabinoside** or **vidarabine**

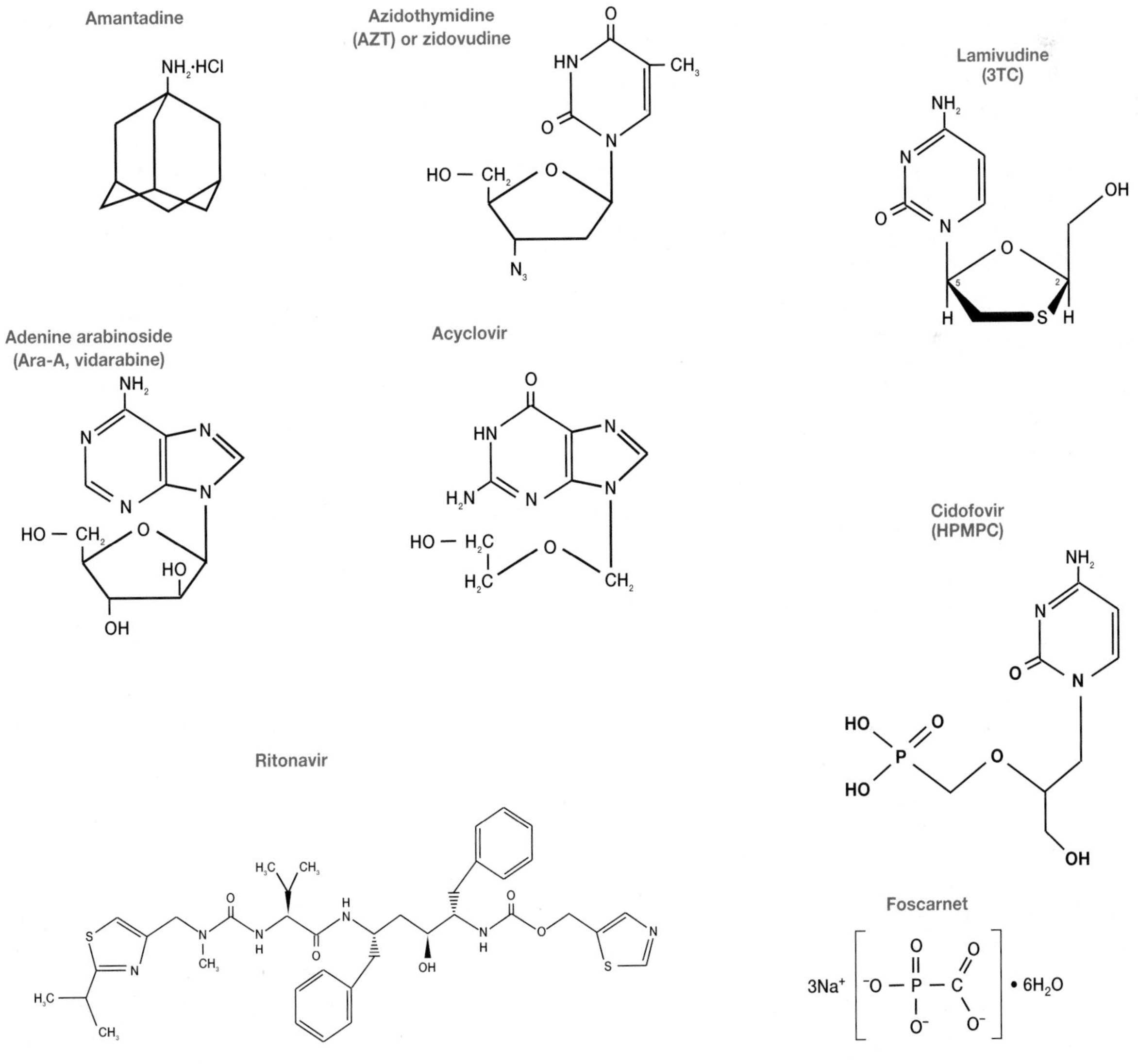

Amantadine

Azidothymidine (AZT) or zidovudine

Lamivudine (3TC)

Adenine arabinoside (Ara-A, vidarabine)

Acyclovir

Cidofovir (HPMPC)

Ritonavir

Foscarnet

Figure 34.21 Representative Antiviral Drugs.

disrupts the activity of DNA polymerase and several other enzymes involved in DNA and RNA synthesis and function. It is given intravenously or applied as an ointment to treat herpes infections. A third drug, **acyclovir,** is also used in the treatment of herpes infections. Upon phosphorylation, acyclovir resembles deoxy-GTP and inhibits the virus DNA polymerase. Unfortunately acyclovir-resistant strains of herpes are already developing. Effective acyclovir derivatives and relatives are now available. Valacyclovir is an orally administered prodrug form of acyclovir. Prodrugs are inactive until metabolized. Ganciclovir, penciclovir, and its oral form famciclovir are effective in treatment of herpesviruses. Another kind of drug, foscarnet, inhibits the virus DNA polymerase in a different way. Foscarnet is an organic analog of pyrophosphate (figure 34.21) that binds to the polymerase active site and blocks the cleavage of pyrophosphate from nucleoside triphosphate substrates. It is used in treating herpes and cytomegalovirus infections. Airborne diseases: Influenza (section 37.1)

Several broad-spectrum anti-DNA virus drugs have been developed. A good example is the drug HPMPC or cidofovir (figure 34.21). It is effective against papovaviruses, adenoviruses, herpesviruses, iridoviruses, and poxviruses. The drug acts on the viral DNA polymerase as a competitive inhibitor and alternative substrate of dCTP. It has been used primarily against cytomegalovirus but also against herpes simplex and human papillomavirus infections.

Research on anti-HIV drugs has been particularly active. Many of the first drugs to be developed were reverse transcriptase inhibitors such as **azidothymidine (AZT)** or **zidovudine,** lamivudine (3TC), didanosine (ddI), zalcitabine (ddC), and stavudine (d4T) (figure 34.21). These interfere with reverse transcriptase activity and therefore block HIV reproduction. More recently **HIV protease inhibitors** have also been developed. Three of the most used are saquinvir, indinavir, and ritonavir (figure 34.21). Protease inhibitors are effective because HIV, like many viruses, translates multiple proteins as a single polypeptide. This polypeptide must then be cleaved into individual proteins required for virus replication. Protease inhibitors mimic the peptide bond that is normally attacked by the protease. The most successful treatment regimen involves a cocktail of agents given at high dosages to prevent the development of drug resistance. For example, the combination of AZT, 3TC, and ritonavir is very effective in reducing HIV plasma concentrations almost to zero. However, the treatment does not eliminate latent proviral HIV DNA that still resides in memory T cells, and possibly elsewhere. Reproduction of vertebrate viruses: Genome replication, transcription, and protein synthesis in RNA viruses (section 18.2); Direct contact diseases: Acquired immune deficiency syndrome (AIDS) (section 37.3)

Probably the most publicized antiviral agent has been Tamiflu (generically, oseltamivir phosphate). Tamiflu is a neuraminidase inhibitor that has received much attention in light of 21st-century predictions of an influenza pandemic, including avian influenza ("bird flu"). While Tamiflu is not a cure for neurominidase-expressing viruses, two clinical trials showed that patients who took Tamiflu were relieved of flu symptoms 1.3 days faster than patients who did not take Tamiflu. Prophylactic use has resulted in viral resistance to Tamiflu. Tamiflu is not a substitute for yearly flu vaccination and frequent hand-washing.

34.9 Antiprotozoan Drugs

The mechanism of drug action for most antiprotozoan drugs is not completely elucidated. Drugs such as chloroquine, atovaquone, mefloquine, iodoquinol, metronidazole, nitazoxanide, and pentamidine, for example, have potent antiprotozoan action but a clear mechanism of action for each class of protozoa is unknown. It may be that each drug has more than one activity and that the relative role of each mechanism to the overall antiprotozoan activity may be different for the various species of protozoa. However, most antiprotozoan drugs appear to act on protozoan nucleic acid or some metabolic event.

Chloroquine is used to treat malaria. Several mechanisms of action have been reported. It can raise the internal pH, clump the plasmodial pigment, and intercalate into plasmodial DNA. Chloroquine also inhibits heme polymerase, an enzyme that converts toxic heme into nontoxic hemazoin. Inhibition of this enzyme leads to a buildup of toxic heme. Mefloquine is also used to treat malaria and has been found to swell the *Plasmodium falciparum* food vacuoles, where it may act by forming toxic complexes that damage membranes and other plasmodial components.

Metronidazole is used to treat *Entamoeba* infections. Anaerobic organisms readily reduce it to the active metabolite within the cytoplasm. Aerobic organisms appear to reduce it using ferrodoxin (a protein of the electron transport system). Reduced metronidazole interacts with DNA altering its helical structure and causing DNA fragmentation; it prevents normal nucleic acid synthesis, resulting in cell death.

A number of antibiotics that inhibit bacterial protein synthesis are also used to treat protozoan infection. These include the aminoglycosides clindamycin, and paromomycin. Aminoglycosides can be considered polycationic molecules that have a high affinity for nucleic acids. Specifically, aminoglycosides possess high affinities for RNAs. Different aminoglycoside antibiotics bind to different sites on RNAs. RNA binding interferes with the normal expression and function of the RNA, resulting in cell death.

Interference of eucaryotic electron transport is one common activity of some antiprotozoan drugs. Atovaquone is used to treat *Pneumocystis jiroveci* (formerly called *P. carinii*) and *Toxoplasma gondii*. It is an analog of ubiquinone, an integral component of the eucaryotic electron transport system. As an analog of ubiquinone, atovaquone can act as a competitive inhibitor and thus suppress electron transport. The ultimate metabolic effects of electron transport blockade include inhibited or delayed synthesis of nucleic acids and ATP. Another drug that interferes with electron transport is nitazoxanide, which used to treat cryptosporidiosis. It appears to exert its effect through interference with the pyruvate:ferredoxin oxidoreductase. It has also been reported to form toxic free radicals once the nitro group is reduced intracellularly.

Pentamidine is used to treat *Pneumocystis* infection. Some reports indicate that it interferes with protozoan metabolism, although the drug only moderately inhibits glucose metabolism, protein synthesis, RNA synthesis, and intracellular amino acid transport in vitro. Pyrimethamine, used to treat *Toxoplasma* infection, and dapsone, used for *Pneumocystis* infection, appear to act in the same way as trimethoprim—interfering with folic acid synthesis by inhibition of dihydrofolate reductase.

As with other antimicrobial therapies, traditional drug development starts by identifying a unique target to which a drug can bind and thus prevent some vital function. A second consideration is often related to drug spectrum—how many different species have that target so that the proposed drug can be used broadly as a chemotherapeutic agent. This is also true for use of agents needed to remove protozoan parasites from their hosts. However, because protozoa are eucaryotes, the potential for drug action on host cells and tissues is greater than it is when targeting procaryotes. Most of the drugs used to treat protozoan infection have significant side effects; nonetheless, the side effects are usually acceptable when weighed against the parasitic alternative.

1. Why do you think drugs that inhibit bacterial protein synthesis are also effective against some protists?
2. Why do you think malaria, like tuberculosis, is now treated with several drugs simultaneously?
3. What special considerations must be taken into account when treating infections caused by protozoan parasites?

Summary

Chemotherapeutic agents are compounds that destroy pathogenic microorganisms or inhibit their growth and are used in the treatment of disease. Most are antibiotics: microbial products or their derivatives that can kill susceptible microorganisms or inhibit their growth.

34.1 The Development of Chemotherapy

a. The modern era of chemotherapy began with Paul Ehrlich's work on drugs against African sleeping sickness and syphilis. Other early pioneers were Gerhard Domagk, Alexander Fleming, Howard Florey, Ernst Chain, Norman Heatley, and Selman Waksman.

34.2 General Characteristics of Antimicrobial Drugs

a. An effective chemotherapeutic agent must have selective toxicity. A drug with great selective toxicity has a high therapeutic index and usually disrupts a structure or process unique to the pathogen. It has fewer side effects.

b. Antibiotics can be classified in terms of the range of target microorganisms (narrow spectrum versus broad spectrum); their source (natural, semisynthetic, or synthetic); and their general effect (static versus cidal) (**table 34.1**).

34.3 Determining the Level of Antimicrobial Activity

a. Antibiotic effectiveness can be estimated through the determination of the minimal inhibitory concentration and the minimal lethal concentration with dilution susceptibility tests. Tests like the Kirby-Bauer test (a disk diffusion test) and the Etest are often used to estimate a pathogen's susceptibility to drugs quickly (**figures 34.2, 34.3,** and **34.4**).

34.4 Antibacterial Drugs

a. Members of the penicillin family contain a β-lactam ring and disrupt bacterial cell wall synthesis, resulting in cell lysis (**figure 34.5**). Some, like penicillin G, are usually administered by injection and are most effective against gram-positive bacteria. Others can be given orally (penicillin V), are broad spectrum (ampicillin, carbenicillin), or are penicillinase resistant (methicillin).

b. Cephalosporins are similar to penicillins, and are given to patients with penicillin allergies (**figure 34.6**).

c. Vancomycin is a glycopeptide antibiotic that inhibits the transpeptidation reaction during peptidoglycan synthesis. It is used against drug-resistant staphylococci, enterococci, and clostridia.

d. Aminoglycoside antibiotics like streptomycin and gentamicin bind to the small ribosomal subunit, inhibit protein synthesis, and are bactericidal (**figure 34.7**).

e. Tetracyclines are broad-spectrum antibiotics having a four-ring nucleus with attached groups (**figure 34.8**). They bind to the small ribosomal subunit and inhibit protein synthesis.

f. Erythromycin is a bacteriostatic macrolide antibiotic that binds to the large ribosomal subunit and inhibits protein synthesis (**figure 34.9**).

g. Chloramphenicol is a broad-spectrum, bacteriostatic antibiotic that inhibits protein synthesis (**figure 34.10**). It is quite toxic and used only for very serious infections.

h. Sulfonamides or sulfa drugs resemble *p*-aminobenzoic acid and competitively inhibit folic acid synthesis (**figure 34.12**).

i. Trimethoprim is a synthetic antibiotic that inhibits the dihydrofolate reductase, which is required by organisms in the manufacture of folic acid (**figure 34.13**).

j. Quinolones are a family of bactericidal synthetic drugs that inhibit DNA gyrase and thus inhibit such processes as DNA replication (**figure 34.15**).

34.5 Factors Influencing Antimicrobial Drug Effectiveness

a. A variety of factors can greatly influence the effectiveness of antimicrobial drugs during actual use.

34.6 Drug Resistance

a. Bacteria can become resistant to a drug by excluding it from the cell, pumping the drug out of the cell, enzymatically altering it, modifying the target enzyme or organelle to make it less drug sensitive, as examples. The genes for drug resistance may be found on the bacterial chromosome, a plasmid called an R plasmid, or other genetic elements such as transposons (**figures 34.18** and **34.19**).

b. Chemotherapeutic agent misuse fosters the increase and spread of drug resistance, and may lead to superinfections.

34.7 Antifungal Drugs

a. Because fungi are more similar to human cells than bacteria, antifungal drugs generally have lower therapeutic indexes than antibacterial agents and produce more side effects.

b. Superficial mycoses can be treated with miconazole, ketoconazole, clotrimazole, tolnaftate, nystatin, and griseofulvin (**figure 34.20**). Amphotericin B, 5-flucytosine, and fluconazole are used for systemic mycoses.

34.8 Antiviral Drugs

a. Antiviral drugs interfere with critical stages in the virus life cycle (amantadine, rimantadine, and ritonavir) or inhibit the synthesis of virus-specific nucleic acids (zidovudine, adenine arabinoside, acyclovir) (**figure 34.21**).

34.9 Antiprotozoan Drugs

a. The mechanisms by which most drugs used to treat protozoan infection are unknown.

b. Antiprotozoan drugs interfere with critical steps in nucleic acid synthesis, protein synthesis, and electron transport of folic acid synthesis.

Key Terms

acyclovir 856
adenine arabinoside or vidarabine 855
amantadine 855
aminoglycoside antibiotic 845
amphotericin B 854
antibiotic 835
antimetabolites 846
azidothymidine (AZT) 856
β-lactam 844
β-lactam ring 843
β-lactamase 843
broad-spectrum drugs 837
cephalosporin 844
chemotherapeutic agent 835
chloramphenicol 846
cidal 837
dilution susceptibility tests 840
erythromycin 846
gene cassette 852
griseofulvin 854
HIV protease inhibitors 856
integron 852
Kirby-Bauer method 840
macrolide antibiotic 846
minimal inhibitory concentration (MIC) 840
minimal lethal concentration (MLC) 840
narrow-spectrum drugs 837
nystatin 854
parenteral route 849
penicillinase 843
penicillins 841
quinolones 847
R plasmids 852
selective toxicity 837
static 837
streptomycin 845
sulfonamide 846
tetracycline 845
therapeutic index 837
trimethoprim 847
vancomycin 845
zidovudine 856

Critical Thinking Questions

1. What advantage might soil bacteria and fungi gain from the synthesis of antibiotics?
2. Why might it be desirable to prepare a variety of semisynthetic antibiotics?
3. Why is it so difficult to find or synthesize effective antiviral drugs?
4. How might the use of antibiotics as growth promoters in livestock contribute to antibiotic resistance among human pathogens?
5. A recent study found that 480 *Streptomyces* strains freshly isolated from the soil are resistant to at least six different antibiotics. In fact, some isolates are resistant to 20 different antibiotic drugs. Why do you think these bacteria, which are neither pathogenic nor exposed to human use of antibiotics, are resistant to so many drugs? What might be the implications for human bacterial pathogens?
6. You are a pediatrician treating a child with an upper respiratory infection that is clearly caused by a virus. The child's mother insists that you prescribe antibiotics—she's not leaving without them! How do you convince the child's mother that antibiotics will do more harm than good?

Learn More

DeClercq, E. 2005. Antivirals and antiviral strategies. *Nature Rev. Microbiol.* 2:704–20.

D'Costa, V. M.; McGrann, K. M.; Hughes, D. W.; and Wright, G. D. 2006. Sampling the antibiotic resistome. *Science* 311:374–77.

Fischetti, V.A. 2005. Bacteriophage lytic enzymes: Novel anti-infectives. *Trends Microbiol.* 13:491–96.

Furuya, E. Y., and Lowy, F. D. 2006. Antimicrobial-resistant bacteria in the community setting. *Nature Rev. Microbiol.* 4:36–45.

Harbarth, S., and Samore, M. H. 2005. Antimicrobial resistance determinants and future control. *Emerg. Infect. Dis.* 11:794–801.

Klugman, K. P., and Lonks, J. R. 2005. Hidden epidemic of macrolide-resistant pneumococci. *Emerg. Infect. Dis.* 11:802–7.

Payne, D., and Tomasz, A. 2004. The challenge of antibiotic-resistant bacterial pathogens: The medical need, the market and prospects for new antimicrobial agents. *Curr. Opin. Microbiol.* 7:435–38.

Schmid, M. 2005. Seeing is believing: The impact of structural genomics on antimicrobial drug discovery. *Nature Rev. Microbiol.* 2:739–46.

Walsh, C. 2003. Where will the new antibiotics come from? *Nature Rev. Microbiol.* 1:65–79.

Walsh, F. M., and Amyes, S. G. B. 2004. Microbiology and drug resistance mechanisms of fully resistant pathogens. *Curr. Opin. Microbiol.* 7:439–44.

White, D. G.; Zhao, S.; Singh, R.; and McDermott, P. F. 2004. Antimicrobial resistance among gram-negative foodborne bacterial pathogens associated with foods of animal origin. *Foodborne Pathol. Dis.* 1:137–52.

Please visit the Prescott website at www.mhhe.com/prescott7 for additional references.

35 Clinical Microbiology and Immunology

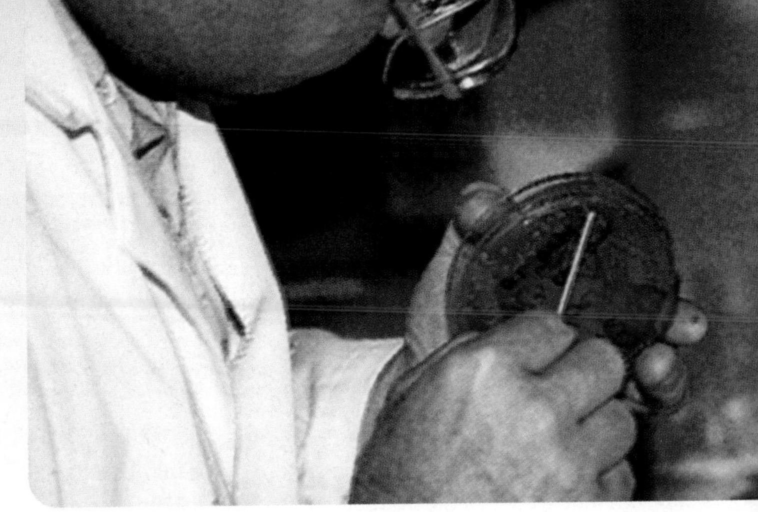

The major objective of the clinical microbiologist is to isolate and identify pathogens from clinical specimens rapidly. In this illustration, a clinical microbiologist is picking up suspect bacterial colonies for biochemical, immunologic, or molecular testing.

Preview

- Clinical microbiologists perform many services related to the identification and control of pathogens and the detection of immune dysfunction.
- Success in clinical microbiology depends on (1) using the proper aseptic technique; (2) correctly obtaining the clinical specimen from the infected patient by swabs, needle aspiration, intubation, or catheters; (3) correctly handling the specimen; and (4) quickly transporting the specimen to the laboratory.
- Once the clinical specimen reaches the laboratory, it is cultured to identify the infecting pathogens. Identification techniques include microscopy; growth on enrichment, selective, differential, or characteristic media; specific biochemical tests; rapid test methods; immunologic techniques; bacteriophage typing; and molecular methods such as nucleic acid-based hybridization techniques, gas-liquid chromatography, and genomic fingerprinting.
- After the microorganism has been isolated, cultured, and/or identified, samples are used in susceptibility tests to find which method of control will be most effective.
- Clinical specimens can also be tested for the presence and/or concentration of either antigen or antibody. These immunological tests use principles of antigen-antibody binding. Various methods of reporting the binding events are used including colorimetric, enzyme-substrate reactions, radionucleotide detection, and precipitation in agar.

Pathogens, particularly bacteria and yeasts, coexist with harmless microorganisms on or in the host. These pathogens must be properly identified as the actual cause of infectious diseases. This is the purpose of clinical microbiology and immunology. The clinical microbiologist identifies agents and organisms based on morphological, biochemical, immunologic, and molecular procedures. Time is a significant factor in the identification process, especially in life-threatening situations. Advances in technology for rapid identification have greatly aided the clinical microbiologist. Molecular methods allow identification of microorganisms based on highly specific genomic and biochemical properties. Once isolated and identified, the microorganism can then be subjected to antimicrobial sensitivity tests. Even in the absence of a culture, immunologic tests can detect pathogens by measuring antigens or antibodies in the specimen. In the final analysis the patient's well-being and health can benefit significantly from information provided by the clinical laboratory—the subject of this chapter.

35.1 Specimens

Infection is the invasion and multiplication in body tissues by bacteria, fungi, viruses, protozoa or helminths that often results in localized cellular injury due to competition for nutrients, toxin production, and/or intracellular replication. The major goal of the **clinical microbiologist** is to isolate and identify pathogenic microorganisms from clinical specimens rapidly. The purpose of the clinical laboratory is to provide the physician with information concerning the presence or absence of microorganisms that may be involved in the infectious disease process (**figure 35.1**). These individuals and facilities also determine the susceptibility of microorganisms to antimicrobial agents. Clinical microbiology makes use of information obtained from research on such diverse topics as microbial biochemistry and physiology, immunology, molecular biology, genomics, and the host-parasite relationships

The specimen is the beginning. All diagnostic information from the laboratory depends upon the knowledge by which specimens are chosen and the care with which they are collected and transported.

—Cynthia A. Needham

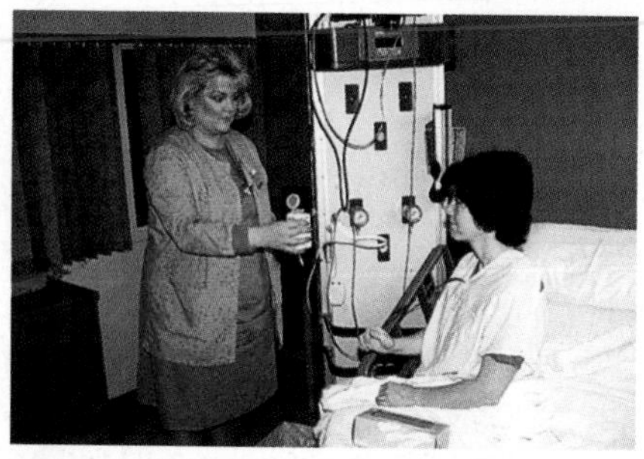

(a) The identification of the microorganism begins at the patient's beside. The nurse is giving instructions to the patient on how to obtain a sputum specimen.

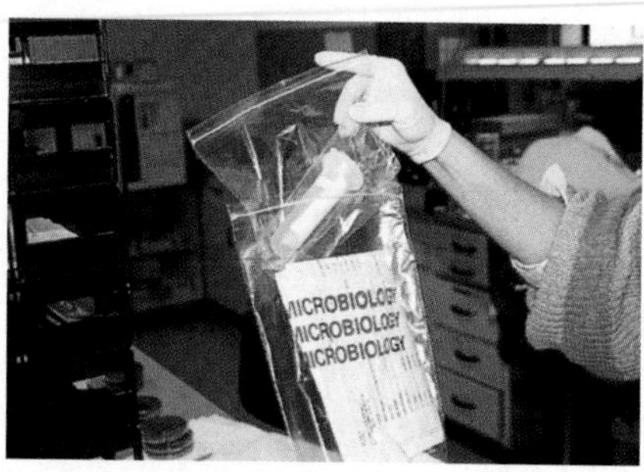

(b) The specimen is sent to the laboratory to be processed. Notice that the specimen and worksheet are in different bags.

(c) Specimens such as sputum are plated on various types of media under a laminar airflow hood. This is to prevent specimen aerosols from coming in contact with the microbiologist.

(d) Sputum and other specimens are usually Gram stained to determine whether or not bacteria are present and to obtain preliminary results on the nature of any bacteria found.

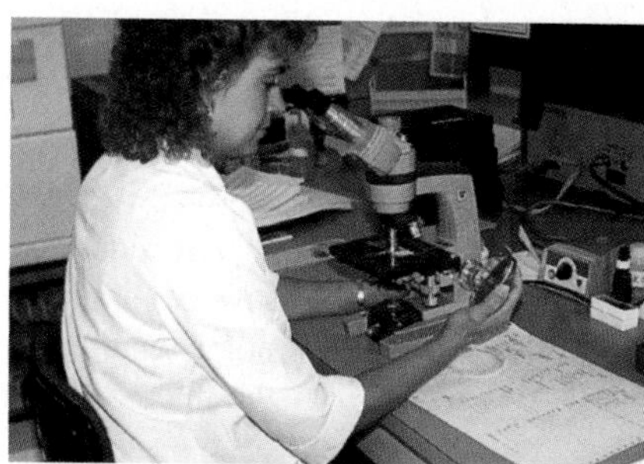

(e) After incubation, the plates are examined for significant isolates. The Gram stain may be repeated for correlation.

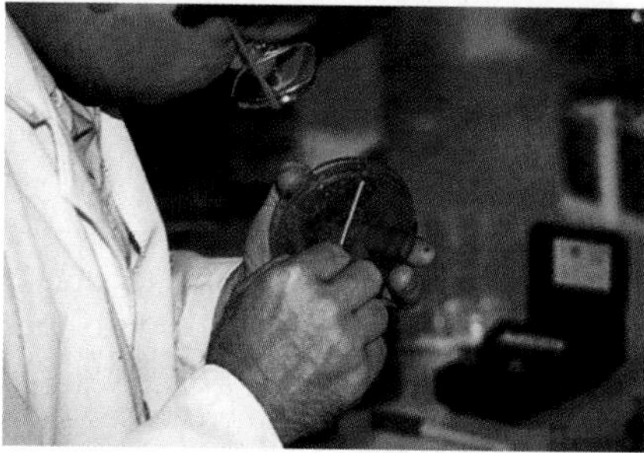

(f) Suspect colonies are picked for biochemical, immunologic, or molecular testing.

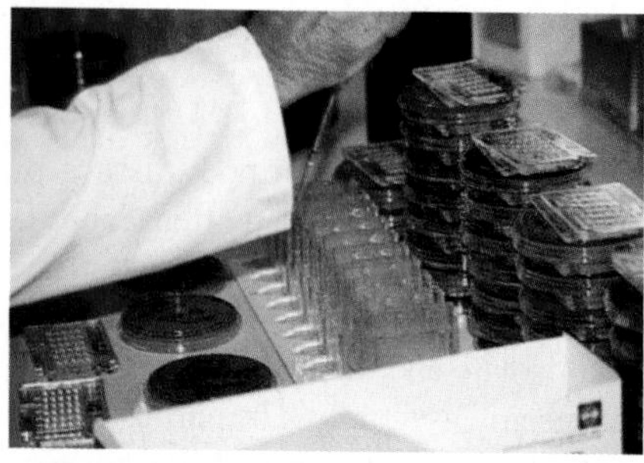

(g) Colonies are prepared for identification by rapid test systems.

(h) In a short time, sometimes 4 hours, computer-generated information is obtained that consist of biochemical identification and antibiotic susceptibility results.

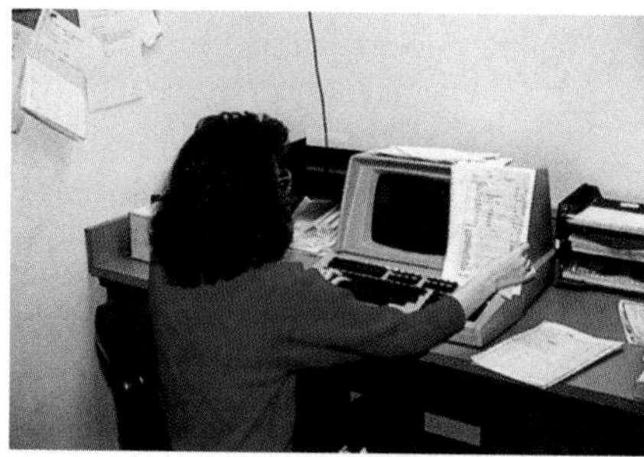

(I) All information about the specimen is now entered into a computer and the data are transmitted directly to the hospital ward.

Figure 35.1 Isolation and Identification of Microorganisms in a Clinical Laboratory.

involved in the infectious disease process. Importantly, tests developed to exploit the antigen-antibody binding capabilities, the focus of clinical immunology, can often detect microorganisms in specimens by identifying microbial antigens and/or quantifying the type and amount of responding antibody.

In clinical microbiology a clinical specimen represents a portion or quantity of human material that is tested, examined, or studied to determine the presence or absence of particular microorganisms. Safety for the patients, hospital, and laboratory staff is of utmost importance. The guidelines presented in **Tech-**

Techniques & Applications

35.1 Standard Microbiological Practices

The identification of potentially fatal, blood-borne infectious agents (HIV, hepatitis B virus, and others) spurred the codification of standard microbiological practices to limit exposure to such agents. These standard microbiological practices are *minimum* guidelines that should be supplemented with other precautions based on the potential exposure risks and biosafety level regulations for the lab. Briefly:

1. Eating, drinking, manipulation of contact lenses, and the use of cosmetics, gum, and tobacco products are strictly prohibited in the lab.
2. Hair longer than shoulder length should be tied back. Hands should be kept away from face at all times. Items (e.g., pencils) should not be placed in the mouth while in the lab. Protective clothing (lab coat, smock, etc.) is recommended while in the lab. Exposed wounds should be covered and protected.
3. Lab personnel should know how to use the emergency eyewash and/or shower stations.
4. Work space should be disinfected at the beginning and completion of lab time. Hands should be washed thoroughly after any exposure and before leaving the lab.
5. Precautions should be taken to prevent injuries caused by sharp objects (needles, scalpels, etc.). Sharp instruments should be discarded for disposal in specially marked containers.

Recommended guidelines for additional precautions should reflect the laboratory's biosafety level (BSL). The following table defines the BSL for the four categories of biological agents and suggested practices.

BSL	Agents	Practices
1	Not known to consistently cause disease in healthy adults (e.g., *Lactobacillus casei, Vibrio fischeri*)	Standard Microbiological Practices
2	Associated with human disease, potential hazard if percutaneous injury, ingestion, mucous membrane exposure occurs (e.g., *Salmonella typhi, E. coli* O157:H7, *Staphylococcus aureus*)	BSL-1 practice plus: • Limited access • Biohazard warning signs • "Sharps" precautions • Biosafety manual defining any needed waste decontamination or medical surveillance policies
3	Indigenous or exotic agents with potential for aerosol transmission; disease may have serious or lethal consequences (e.g., *Coxiella burnetti, Yersinia pestis,* Herpes viruses)	BSL-2 practice plus: • Controlled access • Decontamination of all waste • Decontamination of lab clothing before laundering • Baseline serum values determined in workers using BSL-3 agents
4	Dangerous/exotic agents that pose high risk of life-threatening disease, aerosol-transmitted lab infections; or related agents with unknown risk of transmission (e.g., *Variola major* (smallpox virus), *Ebola* virus, hemorrhagic fever viruses)	BSL-3 practices plus: • Clothing change before entering • Shower on exit • All material decontaminated on exit from facility

niques & Applications 35.1 were established by the Centers for Disease Control and Prevention (CDC) to address areas of personal protection and specimen handling. Other important concerns regarding specimens need emphasis:

1. The specimen selected should adequately represent the diseased area and also may include additional sites (e.g., urine and blood specimens) in order to isolate and identify potential agents of the particular disease process.
2. A quantity of specimen adequate to allow a variety of diagnostic testing should be obtained.
3. Attention must be given to specimen collection in order to avoid contamination from the many varieties of microorganisms indigenous to the skin and mucous membranes (*see figure 30.17*).
4. The specimen should be collected in appropriate containers and forwarded promptly to the clinical laboratory.
5. If possible, the specimen should be obtained before antimicrobial agents have been administered to the patient.

Collection

Overall, the results obtained in the clinical laboratory are only as good as the quality of the specimen collected for analysis. Specimens may be collected by several methods using aseptic technique. In this case, aseptic technique refers to specific procedures used to prevent unwanted microorganisms from contaminating the clinical specimen. Each method is designed to ensure that only the proper material is sent to the clinical laboratory.

The most common method used to collect specimens from the anterior nares or throat is the sterile **swab.** A sterile swab is a rayon-, calcium alginate, or dacron-tipped polystyrene applicator. Manufacturers of swabs have their own unique container design and instructions for proper use. For example, many commercially manufactured swabs contain a transport medium designed to preserve any microorganisms and to prevent multiplication of rapidly growing members of the population (**figure 35.2*a***). However, with the exception of the nares or throat, the use of swabs for the collection of specimens is of little value and should be discouraged for two major reasons: swabs are associated with a greater risk of contamination with surface and subsurface microorganisms, and they have a limited volume capacity (<0.1 ml).

Needle aspiration is used to collect specimens aseptically (e.g., anaerobic bacteria) from cerebrospinal fluid, pus, and blood. Stringent antiseptic techniques are used to avoid skin contamination. To prevent blood from clotting and entrapping microorganisms, various anticoagulants (e.g., heparin, sodium citrate) are included within the specimen bottle or tube (figure 35.2*b*).

Intubation [Latin *in,* into, and *tuba,* tube] is the insertion of a tube into a body canal or hollow organ. For example, intubation can be used to collect specimens from the stomach. In this procedure a long sterile tube is attached to a syringe, and the tube is either swallowed by the patient or passed through a nostril (figure 35.2*c*) into the patient's stomach. Specimens are then withdrawn periodically into the sterile syringe. The most common intubation tube is the Levin tube.

A **catheter** is a tubular instrument used for withdrawing or introducing fluids from or into a body cavity. For example, urine specimens may be collected with catheters to detect urinary tract infections caused by bacteria and from newborns and neonates who cannot give a voluntary urinary specimen. Three types are commonly used for urine. The hard catheter is used when the urethra is very narrow or has strictures. The French catheter is a soft tube used to obtain a single specimen sample. If multiple samples are required over a prolonged period, a Foley catheter is used (figure 35.2*d*).

The most common method used for the collection of urine is the clean-catch method. After the patient has cleansed the urethral meatus (opening), a small container is used to collect the urine. The optimal time to use the clean-catch method is early morning because the urine contains more microorganisms as a result of being in the bladder overnight. In the clean-catch midstream method, the first urine voided is not collected because it becomes contaminated with those transient microorganisms normally found in the lower portion of the urethra. Only the midstream portion is collected since it most likely will contain those microorganisms found in the urinary bladder. If warranted for some patients, needle aspirations also are done directly into the urinary bladder.

Sputum is the most common specimen collected in suspected cases of lower respiratory tract infections. **Sputum** is the mucous secretion expectorated from the lungs, bronchi, and trachea through the mouth, in contrast to saliva, which is the secretion of the salivary glands that contains oral microflora. Sputum is collected in specially designed sputum cups (figure 35.2*e*).

Handling

Immediately after collection the specimen must be properly labeled and handled. The person collecting the specimen is responsible for ensuring that the name, hospital, registration number, location in the hospital, diagnosis, current antimicrobial therapy, name of attending physician, admission date, and type of specimen are correctly and legibly written or imprinted on the culture request form. This information must correspond to that written or imprinted on a label affixed to the specimen container. The type or source of the sample and the choice of tests to be performed also must be specified on the request form.

Transport

Speed in transporting the specimen to the clinical laboratory after it has been obtained from the patient is of prime importance. Some laboratories refuse to accept specimens if they have been in transit too long.

Microbiological specimens may be transported to the laboratory by various means (figure 35.1*b*). For example, certain specimens should be transported in a medium that preserves the microorganisms and helps maintain the ratio of one organism to another. This is especially important for specimens in which normal microorganisms may be mixed with microorganisms foreign to the body location.

Under some circumstances, transport media may require supplementation to support microbial survival or to inhibit normal flora microbes that may be in the specimen. For example, the use of 50,000 U of penicillin, 10 mg of streptomycin, or 0.2 mg of chloramphenicol can be added per mL of specimen to ensure recovery of fungi. Alternatively, polyvinyl alcohol-based preservatives can be used for fixation of ova and parasites in clinical specimens. Importantly, the Select Agents legislation (Federal Register 12/20/02 and 4/18/05) governs policy for the possession, use, and transport (outside of the clinical collection point) of potential biothreat agents. Thus cultivation, storage, and transport of clinical and/or environmental samples known to contain select agents (microbes of potential bioterrorism threat) are now regulated by the Select Agents Program. In most cases, specific packaging and approvals for transport (including postal) are required for specimens containing these organisms.

Special treatment is required for specimens when the microorganism is thought to be anaerobic. The material is aspirated with a needle and syringe. Most of the time it is practical to remove the needle, cap the syringe with its original seal, and bring the specimen directly to the clinical laboratory. Transport of these specimens should take no more than 10 minutes; otherwise, the specimen must be injected immediately into an anaerobic transport vial (**figure 35.3**). Vials should contain a transport medium with an indicator, such as resazurin, to show that the interior of the vial is anoxic at the time the specimen is introduced. Swabs for anaerobic culture usually are less satisfactory than aspirates or tissues, even if they are transported in an anaerobic vial, because of greater risk of contamination and poorer recovery of anaerobes.

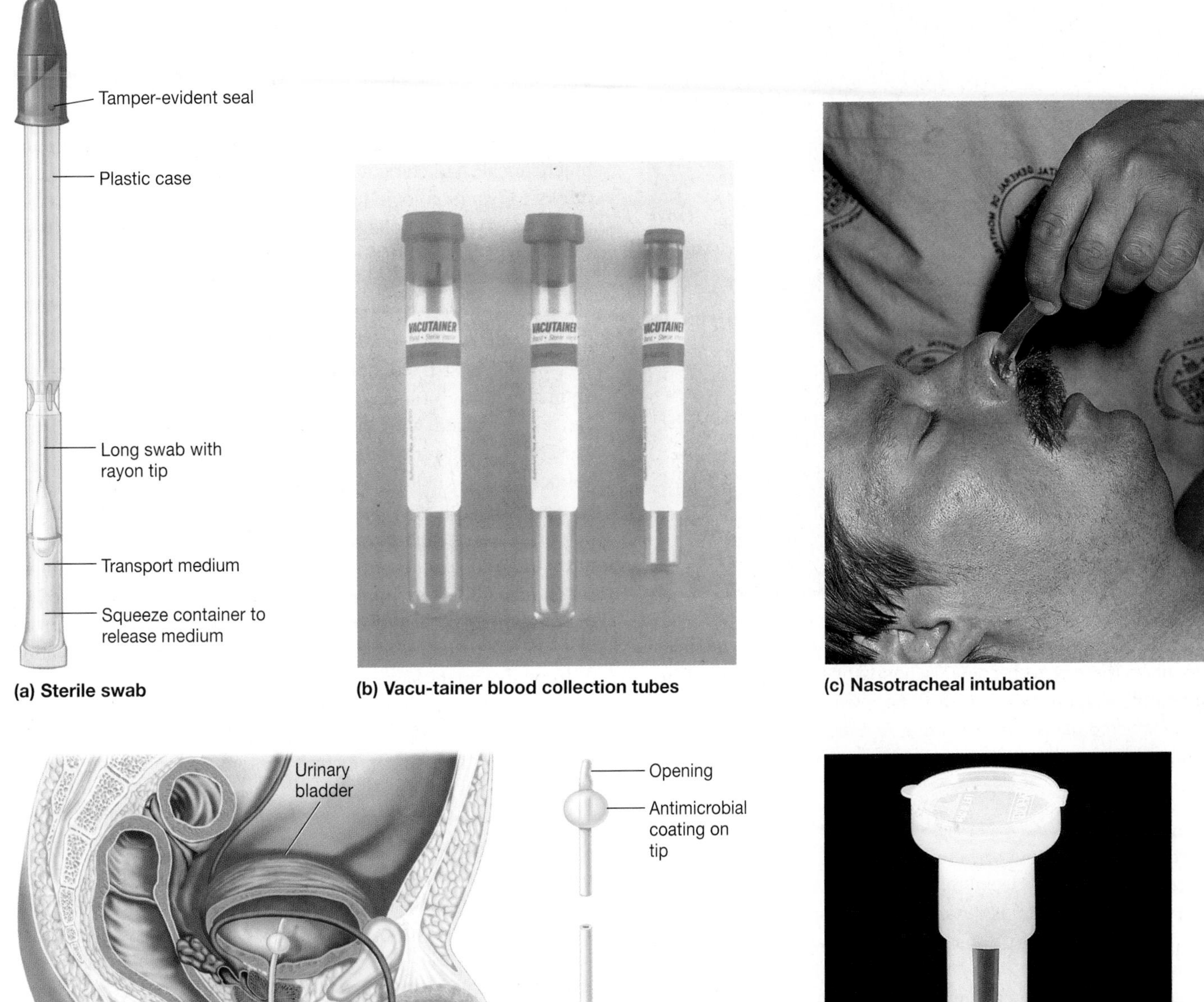

Figure 35.2 Collection of Clinical Specimens. **(a)** A drawing of a sterile swab containing a specific transport medium. **(b)** Sterile Vacutainer tubes for the collection of blood. **(c)** Nasotracheal intubation. **(d)** A drawing of a Foley catheter. Notice that three separate lumens are incorporated within the round shaft of the catheter for drainage of urine, inflation, and introducing irrigating solutions into the urinary bladder. After the Foley catheter has been introduced into the urinary bladder, the tip is inflated to prevent it from being expelled. **(e)** This specially designed sputum cup allows the patient to expectorate a clinical specimen directly into the cup. In the laboratory, the cup can be opened from the bottom to reduce the chance of contamination from extraneous pathogens.

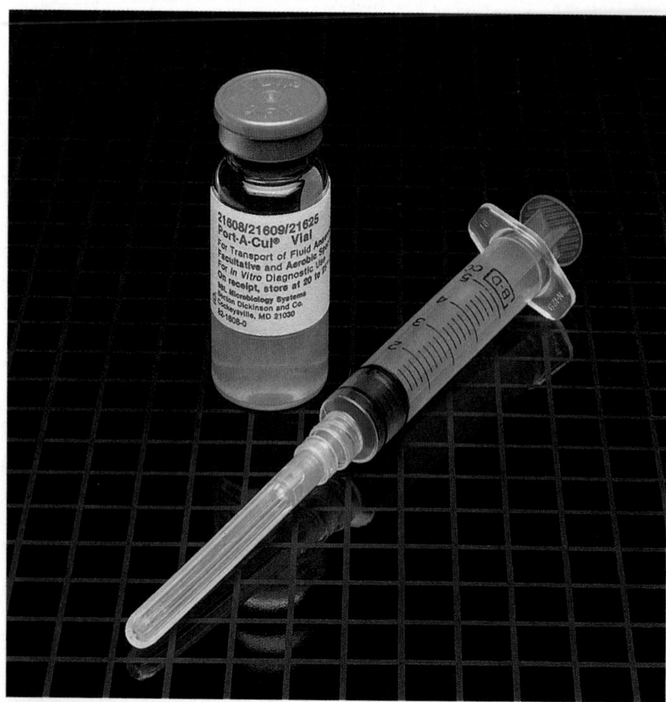

Figure 35.3 Some Anaerobic Transport Systems. A vial and syringe. These systems contain a nonnutritive transport medium that retards diffusion of oxygen after specimen addition and helps maintain microorganism viability up to 72 hours. A built-in color indicator is clear and turns lavender in the presence of oxygen.

Many clinical laboratories insist that stool specimens for culture be transported in special buffered preservatives. Preparation of these transport media is described in various manuals.

Transport of urine specimens to the clinical laboratory must be done as soon as possible. No more than 1 hour should elapse between the time the specimen is obtained and the time it is examined. If this time schedule cannot be followed, the urine sample must be refrigerated immediately.

Cerebrospinal fluid (CSF) from patients suspected of having meningitis should be examined immediately by personnel in the clinical microbiology laboratory. CSF is obtained by lumbar puncture under conditions of strict asepsis, and the sample is transported to the laboratory within 15 minutes. Specimens for the isolation of viruses are iced before transport, and can be kept at 4°C for up to 72 hours; if the sample will be stored longer than 72 hours, it should be frozen at −72°C.

1. What is the function of the clinical microbiologist? the clinical microbiology laboratory?
2. What general guidelines should be followed in collecting and handling clinical specimens?
3. Define the following terms: specimen, swab, catheter, and sputum.
4. What are some transport problems associated with stool specimens? anaerobic cultures? urine specimens?

35.2 Identification of Microorganisms from Specimens

The clinical microbiology laboratory can provide preliminary or definitive identification of microorganisms based on (1) microscopic examination of specimens, (2) study of the growth and biochemical characteristics of isolated microorganisms (pure cultures), (3) immunologic tests that detect antibodies or microbial antigens, (4) bacteriophage typing (restricted to research settings and the CDC), and (5) molecular methods.

Microscopy

Wet-mount, heat-fixed, or chemically fixed specimens can be examined with an ordinary bright-field microscope. Examination can be enhanced with either phase-contrast or dark-field microscopy. The latter is the procedure of choice for the detection of spirochetes in skin lesions associated with early syphilis or Lyme disease. The fluorescence microscope can be used to identify certain acid-fast microorganisms (*Mycobacterium tuberculosis*) after they are stained with fluorochromes such as auramine-rhodamine. Some morphological and genetic features used in classification and identification of microorganisms are presented in section 19.4 and in table 19.4. Direct microscopic examination of most specimens suspected of containing fungi can be made as well. Identification of hyphae in clinical specimens is a presumptive positive result for fungal infection. Definitive identification of most fungi is based on the morphology of reproductive structures (spores). The light microscope (section 2.2)

Many stains that can be used to examine specimens for specific microorganisms have been described. Two of the more widely used bacterial stains are the Gram stain and the acid-fast stain. Because these stains are based on the chemical composition of cell walls, they are not useful in identifying bacteria without cell walls (e.g., mycoplasmas). Lactophenol aniline (cotton) blue is typically used to stain fungi from cultures. Fungal infections (i.e., mold and yeast infections) often are diagnosed by direct microscopic examination of specimens using fluorescence. For example, the identification of molds often can be made if a portion of the specimen is mixed with a drop of 10% Calcofluor White stain on a glass slide. Concentrated wet mounts of blood, stool, or urine specimens can be examined microscopically for the presence of eggs, cysts, larvae, or vegetative cells of parasites. D'Antoni's iodine (1%) is often used to stain internal structures of parasites. Blood smears for apicomplexan (malaria) and flagellate (trypanosome) parasites are stained with Giemsa. Refer to standard references, such as the *Manual of Clinical Microbiology* published by the American Society for Microbiology, for details about other reagents and staining procedures.

The field of clinical microbiology changed dramatically in the 1980s when immunologists created hybrid cells (hybridomas) that secrete antibodies and live a very long time (*see Techniques & Applications 32.2*). Recall that each **hybridoma** cell and its progeny normally produce a **monoclonal antibody (mAb)** of a single specificity. Thus antibodies recognizing a single epitope are produced and used for diagnostics. One such method, known as immunofluorescence or immunohistochemistry, results from the chemical attachment of fluorescent molecules to mAbs; the mAb binds to a single epitope and the fluorescent molecule "reports" that binding. The technique is used to "stain" microorganisms, or clinical specimens thought to contain microorganisms. In the clinical microbiology laboratory, fluorescently labeled mAbs to viral or bacterial antigens have replaced polyclonal antisera for use in culture confirmation when accurate, rapid identification is required. With the use of sensitive techniques such as fluorescence microscopy, it is possible to perform antibody-based microbial identifications with improved accuracy, speed, and fewer organisms. Antibodies (section 32.7), Action of antibodies: Immune complex formation (section 32.8)

Immunofluorescence is a process in which fluorochromes (fluorescent dyes) are exposed to UV, violet, or blue light to make them fluoresce or emit visible light. Dyes such as rhodamine B or fluorescein isothiocyanate (FITC) can be coupled to antibody molecules without changing the antibody's capacity to bind to a specific antigen. Fluorochrome dyes also can be attached to antigens. There are two main kinds of fluorescent antibody assays: direct and indirect.

Direct immunofluorescence involves fixing the specimen (cell or microorganism) containing the antigen of interest onto a slide (**figure 35.4*a***). Fluorescein-labeled antibodies are then added to the slide and incubated. The slide is washed to remove any unbound antibody and examined with the fluorescence microscope (*see figure 2.12*) for a yellow-green fluorescence. The pattern of fluorescence reveals the antigen's location. Direct immunofluorescence is used

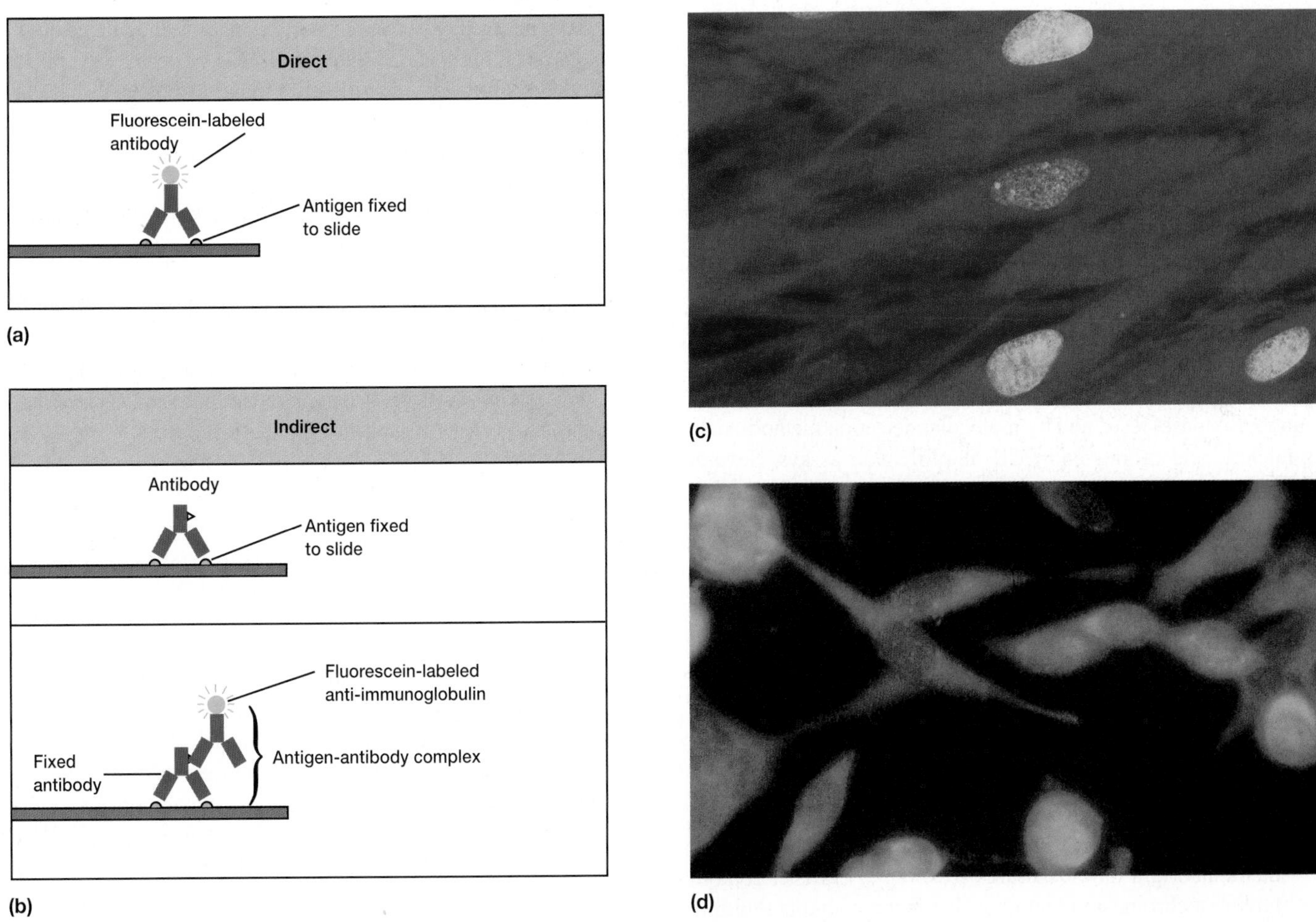

Figure 35.4 Direct and Indirect Immunofluorescence. **(a)** In the direct fluorescent-antibody (DFA) technique, the specimen containing antigen is fixed to a slide. Fluorescently labeled antibodies that recognize the antigen are then added, and the specimen is examined with a fluorescence microscope for yellow-green fluorescence. **(b)** The indirect fluorescent-antibody technique (IFA) detects antigen on a slide as it reacts with an antibody directed against it. The antigen-antibody complex is located with a fluorescent antibody that recognizes antibodies. **(c)** Three infected nuclei in a cytomegalovirus (CMV) positive tissue culture. **(d)** Several infected cells in a herpes simplex virus positive tissue culture.

to identify antigens such as those found on the surface of group A streptococci and to diagnose enteropathogenic *Escherichia coli, Neisseria meningitidis, Salmonella* spp., *Shigella sonnei, Listeria monocytogenes, Haemophilus influenzae* type b, and the rabies virus. The light microscope: The fluorescence microscope (section 2.2)

Indirect immunofluorescence (figure 35.4*b*) is used to detect the presence of antibodies in serum following an individual's exposure to microorganisms. In this technique a known antigen is fixed onto a slide. The test antiserum is then added, and if the specific antibody is present, it reacts with antigen to form a complex. When fluorescein-labeled antibodies are added, they react with the fixed antibody. After incubation and washing, the slide is examined with the fluorescence microscope. The occurrence of fluorescence shows that antibody specific to the test antigen is present in the serum. Indirect immunofluorescence is used to identify the presence of *Treponema pallidum* antibodies in the diagnosis of syphilis (treponemal antibody absorption, FTA-ABS), as well as antibodies produced in response to other microorganisms.

Growth and Biochemical Characteristics

Typically microorganisms have been identified by their particular growth patterns and biochemical characteristics. These characteristics vary depending on whether the clinical microbiologist is dealing with viruses, fungi (yeasts, molds), parasites (protozoa, helminths), common gram-positive or gram-negative bacteria, rickettsias, chlamydiae, or mycoplasmas.

Viruses

Viruses are identified by isolation in conventional cell (tissue) culture, by immunodiagnosis (fluorescent antibody, enzyme immunoassay, radioimmunoassay, latex agglutination, and immunoperoxidase) tests, and by molecular detection methods such as nucleic acid probes and PCR amplification assays. Several types of systems are available for virus cultivation: cell cultures, embryonated hen's eggs, and experimental animals; these are discussed further in section 35.3.

Cell cultures are divided into three general classes:

1. Primary cultures consist of cells derived directly from tissues such as monkey kidney and mink lung cells that have undergone one or two passages (subcultures) since harvesting.
2. Semicontinuous cell cultures or low-passage cell lines are obtained from subcultures of a primary culture and usually consist of diploid fibroblasts that undergo a finite number of divisions.
3. Continuous or immortalized cell cultures, such as HEp-2 cells, are derived from transformed cells that are generally epithelial in origin. These cultures grow rapidly, are heteroploid (having a chromosome number that is not a simple multiple of the haploid number), and can be subcultured indefinitely.

Each type of cell culture favors the growth of a different array of viruses, just as bacterial culture media have differing selective and restrictive properties for growth of bacteria. Viral replication in cell cultures is detected in two ways: (1) by observing the presence or absence of cytopathic effects (CPEs), and (2) by hemadsorption. A **cytopathic effect** is an observable morphological change that occurs in cells because of viral replication. Examples include ballooning, binding together, clustering, or even death of the culture cells (*see figure 16.15*). During the incubation period of a cell culture, red blood cells can be added. Several viruses alter the plasma membrane of infected culture cells so that red blood cells adhere firmly to them. This phenomenon is called **hemadsorption.** Virus reproduction (section 16.4)

Embryonated hen's eggs can be used for virus isolation. There are three main routes of egg inoculation for virus isolation as different viruses grow best on different cell types. (1) the allantoic cavity, (2) the amniotic cavity, and (3) the chorioallantoic membrane (*see figure 16.13*). Egg tissues are inoculated with clinical specimens to determine the presence of virus; virus is revealed by the development of pocks on the chorioallantoic membrane, by the development of hemagglutinins in the allantoic and amniotic fluid, and by death of the embryo.

Embryonated chicken eggs and laboratory animals, especially suckling mice, may be used for virus isolation. Inoculated animals are observed for specific signs of disease or death. Several serological tests for viral identification make use of mAb-based immunofluorescence. These tests (figure 35.4*c,d*) detect viruses such as the cytomegalovirus and herpes simplex virus in tissue cultures.

---

1. Name two specimens for which microscopy would be used in the initial diagnosis of an infectious disease.
2. Name three general classes of cell cultures.
3. Explain two ways by which the presence of viral replication is detected in cell culture.
4. What are the three main routes of egg inoculation for virus isolation?
5. What are the advantages of using monoclonal antibody (mAb) immunofluorescence in the identification of viruses?

Fungi

Fungal cultures remain the standard for the recovery of fungi from patient specimens; however, the time needed to culture fungi varies anywhere from a few days to several weeks depending on the organism. For this reason, fungal cultures demonstrating no growth should be maintained for a minimum of 30 days before they are discarded as a negative result. Cultures should be evaluated for rate and appearance of growth on at least one selective and one nonselective agar medium, with careful examination of colonial morphology, color, and dimorphism. Fungal serology (e.g., complement fixation and immunodiffusion) is designed to detect serum antibody but is limited to a few fungi (e.g., *Blastomyces dermatitidis, Coccidioides immitis, Histoplasma capsulatum*). The cryptococcal latex antigen test is routinely used for the direct detection of *Cryptococcus neoformans* in serum and cerebrospinal fluid. In the clinical laboratory, nonautomated and automated methods for rapid identification (minutes to hours) are used to detect most yeasts. Any biochemical methods used to detect fungi should always be accompanied by morphological studies examining for pseudohyphae, yeast cell structure, chlamydospores, and so on.

Parasites

Culture of parasites from clinical specimens is not routine. Identification and characterization of ova, trophozoites, and cysts in the specimen result in the definitive diagnosis of parasitic infection. This is typically accomplished by direct microscopic evaluation of the clinical specimen. Typical histological staining of blood, negative staining of other body fluids, and immunofluorescence staining are routinely used in the identification of parasites. Some serological tests also are available.

Bacteria

Isolation and growth of bacteria are required before many diagnostic tests can be used to confirm the identification of the pathogen. The presence of bacterial growth usually can be recognized by the development of colonies on solid media or turbidity in liquid media. The time for visible growth to occur is an important variable in the clinical laboratory. For example, most pathogenic bacteria require only a few hours to produce visible growth, whereas it may take weeks for colonies of mycobacteria or mycoplasmas to become evident. The clinical microbiologist as well as the clinician should be aware of reasonable reporting times for various cultures.

The initial identity of a bacterial organism may be suggested by (1) the source of the culture specimen; (2) its microscopic appearance and gram reaction; (3) its pattern of growth on selective, differential, or metabolism-determining media (**table 35.1**); and (4) its hemolytic, metabolic, and fermentative properties on the various media (table 35.1; *see also table 19.5*). After the microscopic and growth characteristics of a pure culture of bacteria are examined, specific biochemical tests can be performed. Some of the most common biochemical tests used to identify bacterial isolates are listed in **table 35.2.** Classic dichotomous keys are coupled with the biochemical tests for the identification of bacteria from specimens. Generally, fewer than 20 tests are required to identify clinical bacterial isolates to the species level (**figure 35.5**). Microbial nutrition (chapter 5)

Rickettsias

Although rickettsias, chlamydiae, and mycoplasmas are bacteria, they differ from other bacterial pathogens in a variety of ways. Therefore the identification of these three groups is discussed separately. Rickettsias can be diagnosed by immunoassays or by isolation of the microorganism. Because isolation is both hazardous to the clinical microbiologist and expensive, immunological methods are preferred. Isolation of rickettsias and diagnosis of rickettsial diseases are generally confined to reference and specialized research laboratories.

Chlamydiae

Chlamydiae can be demonstrated in tissues and cell scrapings with Giemsa staining, which detects the characteristic intracellular inclusion bodies (*see figure 21.14*). Immunofluorescent staining of tissues and cells with monoclonal antibody reagents is a more sensitive and specific means of diagnosis. The most sensitive methods for demonstrating chlamydiae in clinical specimens involve nucleic acid sequencing and PCR-based methods. Techniques for determining microbial taxonomy and phylogeny: Molecular characteristics (section 19.4)

Mycoplasmas

The most routinely used techniques for identification of the mycoplasmas are immunological (hemagglutinin), complement-fixing antigen-antibody reactions using the patient's sera, and PCR (depending on the lab). These microorganisms are slow growing; therefore positive results from isolation procedures are rarely available before 30 days—a long delay with an approach that offers little advantage over standard techniques. DNA probes are also used for the detection of *Mycoplasma pneumoniae* in clinical specimens.

---

1. How can fungi and parasites be detected in a clinical specimen? Rickettsias? Chlamydiae? Mycoplasmas?
2. Why must the clinical microbiologist know what are reasonable reporting times for various microbial specimens?
3. How can a clinical microbiologist determine the initial identity of a bacterium?
4. Describe a dichotomous key that could be used to identify a bacterium.

Rapid Methods of Identification

Clinical microbiology has benefited greatly from technological advances in equipment, computer software and data bases, molecular biology, and immunochemistry (**Microbial Tidbits 35.2**). With respect to the detection of microorganisms in specimens, there has been a shift from the multistep methods previously discussed to unitary procedures and systems that incorporate standardization, speed, reproducibility, miniaturization, mechanization, and automation. These rapid identification methods can be divided into three categories: (1) manual biochemical "Kit" systems, (2) mechanized/automated systems, and (3) immunologic systems.

One example of a "kit approach" biochemical system for the identification of members of the family *Enterobacteriaceae* and other gram-negative bacteria is the API 20E system. It consists of a plastic strip with 20 microtubes containing dehydrated biochemical substrates that can detect certain biochemical characteristics (**figure 35.6**). The biochemical substrates in the 20 microtubes are inoculated with a pure culture of bacteria evenly suspended in sterile physiological saline. After 5 hours or overnight incubation, the 20 test results are converted to a seven- or nine-digit profile (**figure 35.7**). This profile number can be used with a computer or a book called the *API Profile Index* to identify the bacterium.

Clinical laboratory scientists (medical technologists) are the trained and certified workforce that is the front line in laboratory-based disease detection. They staff the sentinel laboratories that receive patient specimens. The production of faster and more specific detection technologies has allowed them to rapidly and

Table 35.1 Isolation of Pure Bacterial Cultures from Specimens

Selective Media

A selective medium is prepared by the addition of specific substances to a culture medium that will permit growth of one group of bacteria while inhibiting growth of some other groups. These are examples:

Salmonella-Shigella agar (SS) is used to isolate *Salmonella* and *Shigella* species. Its bile salt mixture inhibits many groups of coliforms. Both *Salmonella* and *Shigella* species produce colorless colonies because they are unable to ferment lactose. Lactose-fermenting bacteria will produce pink colonies.

Mannitol salt agar (MS) is used for the isolation of staphylococci. The selectivity is obtained by the high (7.5%) salt concentration that inhibits growth of many groups of bacteria. The mannitol in this medium helps in differentiating the pathogenic from the nonpathogenic staphylococci, as the former ferment mannitol to form acid while the latter do not. Thus this medium is also differential.

Bismuth sulfite agar (BS) is used for the isolation of *Salmonella enterica* serovar Typhi, especially from stool and food specimens. *S. enterica* serovar Typhi reduces the sulfite to sulfide, resulting in black colonies with a metallic sheen.

The addition of blood, serum, or extracts to tryptic soy agar or broth will support the growth of many fastidious bacteria. These media are used primarily to isolate bacteria from cerebrospinal fluid, pleural fluid, sputum, and wound abscesses.

Differential Media

The incorporation of certain chemicals into a medium may result in diagnostically useful growth or visible change in the medium after incubation. These are examples:

Eosin methylene blue agar (EMB) differentiates between lactose fermenters and nonlactose fermenters. EMB contains lactose, salts, and two dyes—eosin and methylene blue. *E. coli,* which is a lactose fermenter, will produce a dark colony or one that has a metallic sheen. *S. enterica* serovar Typhi, a nonlactose fermenter, will appear colorless.

MacConkey agar is used for the selection and recovery of *Enterobacteriaceae* and related gram-negative rods. The bile salts and crystal violet in this medium inhibit the growth of gram-positive bacteria and some fastidious gram-negative bacteria. Because lactose is the sole carbohydrate, lactose-fermenting bacteria produce colonies that are various shades of red, whereas nonlactose fermenters produce colorless colonies.

Hektoen enteric agar is used to increase the yield of *Salmonella* and *Shigella* species relative to other microbiota. The high bile salt concentration inhibits the growth of gram-positive bacteria and retards the growth of many coliform strains.

Blood agar: addition of citrated blood to tryptic soy agar makes possible variable hemolysis, which permits differentiation of some species of bacteria. Three hemolytic patterns can be observed on blood agar.

1. α-hemolysis—greenish to brownish halo around the colony (e.g., *Streptococcus gordonii, Streptococcus pneumoniae*).
2. β-hemolysis—complete lysis of blood cells resulting in a clearing effect around growth of the colony (e.g., *Staphylococcus aureus* and *Streptococcus pyogenes*).
3. Nonhemolytic—no change in medium (e.g., *Staphylococcus epidermidis* and *Staphylococcus saprophyticus*).

Media to Determine Biochemical Reactions

Some media are used to test bacteria for particular metabolic activities, products, or requirements. These are examples:

Urea broth is used to detect the enzyme urease. Some enteric bacteria are able to break down urea, using urease, into ammonia and CO_2.

Triple sugar iron (TSI) agar contains lactose, sucrose, and glucose plus ferrous ammonium sulfate and sodium thiosulfate. TSI is used for the identification of enteric organisms based on their ability to attack glucose, lactose, or sucrose and to liberate sulfides from ammonium sulfate or sodium thiosulfate.

Citrate agar contains sodium citrate, which serves as the sole source of carbon, and ammonium phosphate, the sole source of nitrogen. Citrate agar is used to differentiate enteric bacteria on the basis of citrate utilization.

Lysine iron agar (LIA) is used to differentiate bacteria that can either deaminate or decarboxylate the amino acid lysine. LIA contains lysine, which permits enzyme detection, and ferric ammonium citrate for the detection of H_2S production.

Sulfide, indole, motility (SIM) medium is used for three different tests. One can observe the production of sulfides, formation of indole (a metabolic product from tryptophan utilization), and motility. This medium is generally used for the differentiation of enteric organisms.

accurately identify agents of disease. However, the bioterror incidents of 2001 spawned a renewed demand for "better, faster, and smarter" microbial detection and identification technologies. While nucleic acid-based detection systems, like PCR, have garnered much attention as the basis of newer detection systems, antibody-based identification technologies are still considered more flexible and easier to modify. Traditional antibody-based detection technologies are being linked to sophisticated reporting systems that provide "med techs" with an ever-increasing cadre of cutting-edge technology. Examples of more recent microbial identification technologies include biosensors based on: (1) microfluidic antigen sensors, (2) real time (20-minute) PCR, (3) highly

Table 35.2 Some Common Biochemical Tests Used by Clinical Microbiologists in the Diagnosis of Bacteria from a Patient's Specimen

Biochemical Test	Description	Laboratory Application
Carbohydrate fermentation	Acid and/or gas are produced during fermentative growth with sugars or sugar alcohols.	Fermentation of specific sugars used to differentiate enteric bacteria as well as other genera or species.
Casein hydrolysis	Detects the presence of caseinase, an enzyme able to hydrolyze milk protein casein. Bacteria that use casein appear as colonies surrounded by a clear zone.	Used to cultivate and differentiate aerobic actinomycetes based on casein utilization. For example, *Streptomyces* uses casein and *Nocardia* does not.
Catalase	Detects the presence of catalase, which converts hydrogen peroxide to water and O_2.	Used to differentiate *Streptococcus* (−) from *Staphylococcus* (+) and *Bacillus* (+) from *Clostridium* (−).
Citrate utilization	When citrate is used as the sole carbon source, this results in alkalinization of the medium.	Used in the identification of enteric bacteria. *Klebsiella* (+), *Enterobacter* (+), *Salmonella* (often +); *Escherichia* (−), *Edwardsiella* (−).
Coagulase	Detects the presence of coagulase. Coagulase causes plasma to clot.	This is an important test to differentiate *Staphylococcus aureus* (+) from *S. epidermidis* (−).
Decarboxylases (arginine, lysine, ornithine)	The decarboxylation of amino acids releases CO_2 and amine.	Used in the identification of enteric bacteria.
Esculin hydrolysis	Tests for the cleavage of a glycoside.	Used in the differentiation of *Staphylococcus aureus, Streptococcus mitis,* and others (−) from *S. bovis, S. mutans,* and enterococci (+).
β-galactosidase (ONPG) test	Demonstrates the presence of an enzyme that cleaves lactose to glucose and galactose.	Used to separate enterics (*Citrobacter* +, *Salmonella* −) and to identify pseudomonads.
Gelatin liquefaction	Detects whether or not a bacterium can produce proteases that hydrolyze gelatin and liquify solid gelatin medium.	Used in the identification of *Clostridium, Serratia, Pseudomonas,* and *Flavobacterium.*
Hydrogen sulfide (H_2S)	Detects the formation of hydrogen sulfide from the amino acid cysteine due to cysteine desulfurase.	Important in the identification of *Edwardsiella, Proteus,* and *Salmonella.*
IMViC (indole; methyl red; Voges-Proskauer; citrate)	The indole test detects the production of indole from the amino acid tryptophan. Methyl red is a pH indicator to determine whether the bacterium carries out mixed acid fermentation. VP (Voges-Proskauer) detects the production of acetoin. The citrate test determines whether or not the bacterium can use sodium citrate as a sole source of carbon.	Used to separate *Escherichia* (MR+, VP−, indole+) from *Enterobacter* (MR−, VP+, indole−) and *Klebsiella pneumoniae* (MR−, VP+, indole−); also used to characterize members of the genus *Bacillus.*
Lipid hydrolysis	Detects the presence of lipase, which breaks down lipids into simple fatty acids and glycerol.	Used in the separation of clostridia.
Nitrate reduction	Detects whether a bacterium can use nitrate as an electron acceptor.	Used in the identification of enteric bacteria which are usually +.
Oxidase	Detects the presence of cytochrome *c* oxidase that is able to reduce O_2 and artificial electron acceptors.	Important in distinguishing *Neisseria* and *Moraxella* spp. (+) from *Acinetobacter* (−), and enterics (all −) from pseudomonads (+).
Phenylalanine deaminase	Deamination of phenylalanine produces phenylpyruvic acid, which can be detected colorimetrically.	Used in the characterization of the genera *Proteus* and *Providencia.*
Starch hydrolysis	Detects the presence of the enzyme amylase, which hydrolyzes starch.	Used to identify typical starch hydrolyzers such as *Bacillus* spp.
Urease	Detects the enzyme that splits urea to NH_3 and CO_2.	Used to distinguish *Proteus, Providencia rettgeri,* and *Klebsiella pneumoniae* (+) from *Salmonella, Shigella* and *Escherichia* (−).

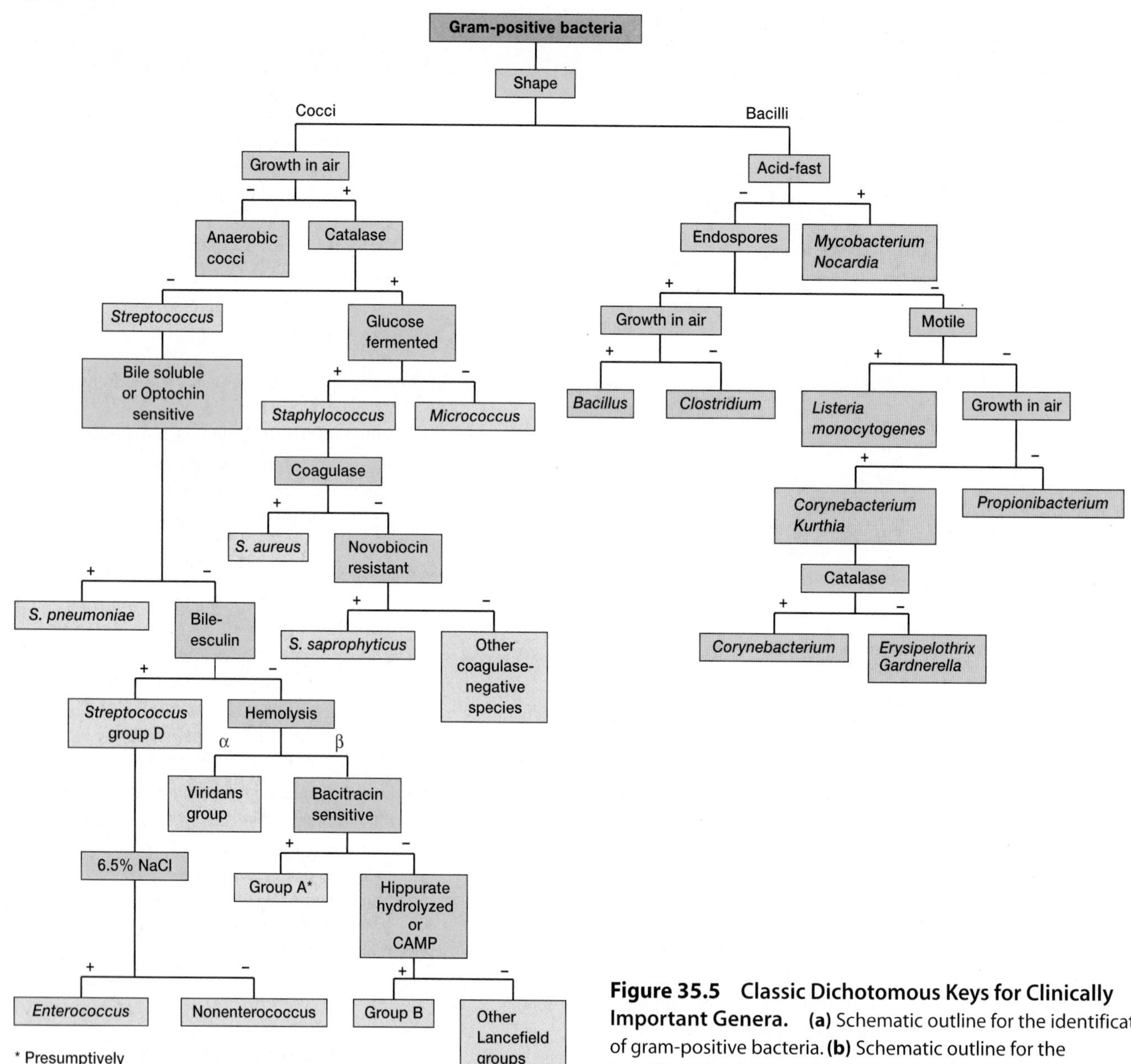

Figure 35.5 Classic Dichotomous Keys for Clinically Important Genera. **(a)** Schematic outline for the identification of gram-positive bacteria. **(b)** Schematic outline for the identification of gram-negative bacteria.

sensitive spectroscopy systems, and (4) liquid crystal amplification of microbial immune complexes. Some of these technologies are being used as part of military sentinel detection programs; others are awaiting approval by various licensing agencies before deployment into clinical laboratories. Additional technologies are expected as the demand for immediate, highly sensitive microbial detection increases globally. Thus the rapidly growing discipline of immunology has greatly aided the clinical microbiologist. Numerous technologies now exist that exploit the specificity and sensitivity of monoclonal antibodies to detect and identify microorganisms. These are briefly discussed here and expanded upon in section 35.3.

Recall that hybridomas have the antibody-producing capacity of their original plasma cells and are long-lived. The monoclonal antibodies (mAbs) they produce have many applications. For example, they are routinely used in the typing of tissue; in the identification and epidemiological study of infectious microorganisms, tumors, and other surface antigens; in the classification of leukemias; in the identification of functional populations of different types of T cells; and in the identification and mapping of antigenic determinants (epitopes) on proteins. Importantly, mAbs can be conjugated with molecules that provide colorimetric, fluorometric, or enzymatic activity to report the binding of the mAb to specific microbial antigens. Numerous microbial detection kits are available to screen clinical specimens for the presence of specific microorganisms (**table 35.3**). mAbs have been produced against a wide variety of bacteria, viruses, fungi, and protozoans and have been made as cross-species or cross-genus reactivity to be used as an adjunct

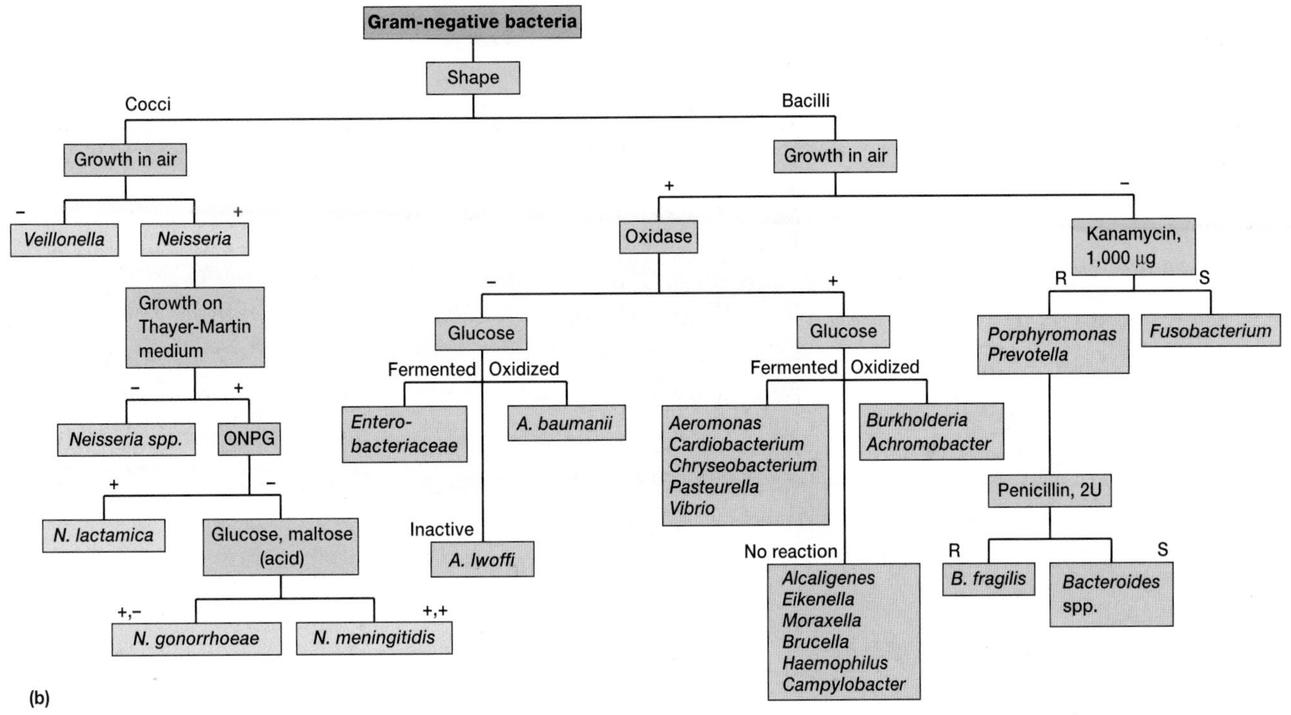

Figure 35.5 *continued*

Microbial Tidbits

35.2 Biosensors: The Future Is Now

The 120-plus-year-old pathogen detection systems based on culture and biochemical phenotyping are being challenged. Fueled by the release of anthrax spores in the U.S. postal system, government agencies have been calling for newer technologies for the near-immediate detection and identification of microbes. In the past, detection technologies have traded speed for cost and complexity. The agar plate technique, refined by Robert Koch and his contemporaries in the 1880s, is a trusted and highly efficient method for the isolation of bacteria into pure cultures. Subsequent phenotyping biochemical methods, often using differential media in a manner similar to that used in the isolation step, then identifies common bacterial pathogens. Unfortunately, reliable results from this process often take several days. More rapid versions of the phenotyping systems can be very efficient, yet still require pure culture inoculations. The rapid immunological tests offer faster detection responses but may sacrifice sensitivity. Even DNA sequence comparisons, which are extremely accurate, may require significant time for DNA amplification and significant cost for reagents and sensitive readers. As usual, necessity has begat invention.

The more recent microbial detection systems, many of which are still untested in the clinical arena, sound like science fiction gizmos, yet promise a new age for near-immediate detection and identification of pathogens. These technologies are collectively referred to as "biosensors," and if the biosensor is integrated with a computer microchip for information management, it is then called a "biochip." Biosensors should ideally be capable of highly specific recognition so as to discriminate between nearest relatives, and "communicate" detection through some type of transducing system. Biosensors that detect specific DNA sequences, expressed proteins, and metabolic products have been developed that use optical (mostly fluorescence), electrochemical, or even mass displacement, to report detection. The high degree of recognition required to reduce false-positive results has demanded the uniquely specific, receptor-like capture that is associated with nucleic acid hybridization and antibody binding. Several microbial biosensors employ single-stranded DNA or RNA sequences, or antibody, for the detection component. The transducing or sensing component of biosensors may be markedly different, however. For example, microcantilever systems detect the increased mass of the receptor-bound ligand; the surface acoustic wave device detects change in specific gravity; the bulk quartz resonator monitors fluid density and viscosity; the quartz crystal microbalance measures frequency change in proportion to the mass of material deposited on the crystal; the micromirror sensor uses an optical fiber waveguide that changes reflectivity; and the liquid crystal-based system reports the reorientation of polarized light. Thus the specific capture of a ligand is reflected in the net change measured by each system and results in a signal that announces the initial capture event. Microchip control of the primary and subsequent secondary signals has resulted in automation of the detection process. The reliable detection of pathogens in complex specimens will be the real test as each of these technologies continues to compete for a place in the clinical laboratory.

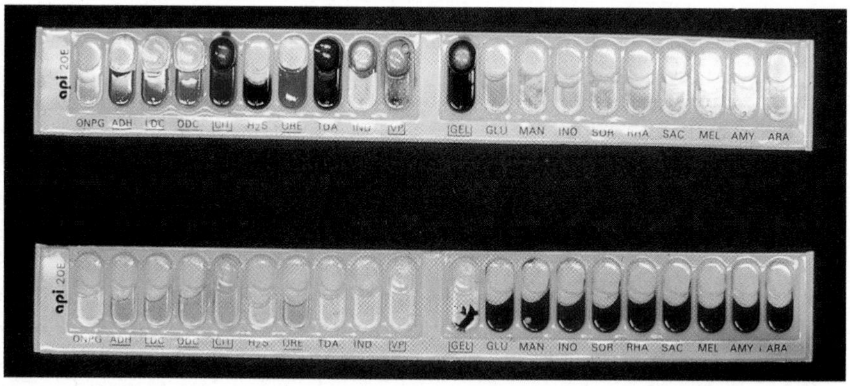

Figure 35.6 A "Kit Approach" to Bacterial Identification. The API 20E manual biochemical system for microbial identification. **(a)** Positive and **(b)** negative results.

(a)

(b)

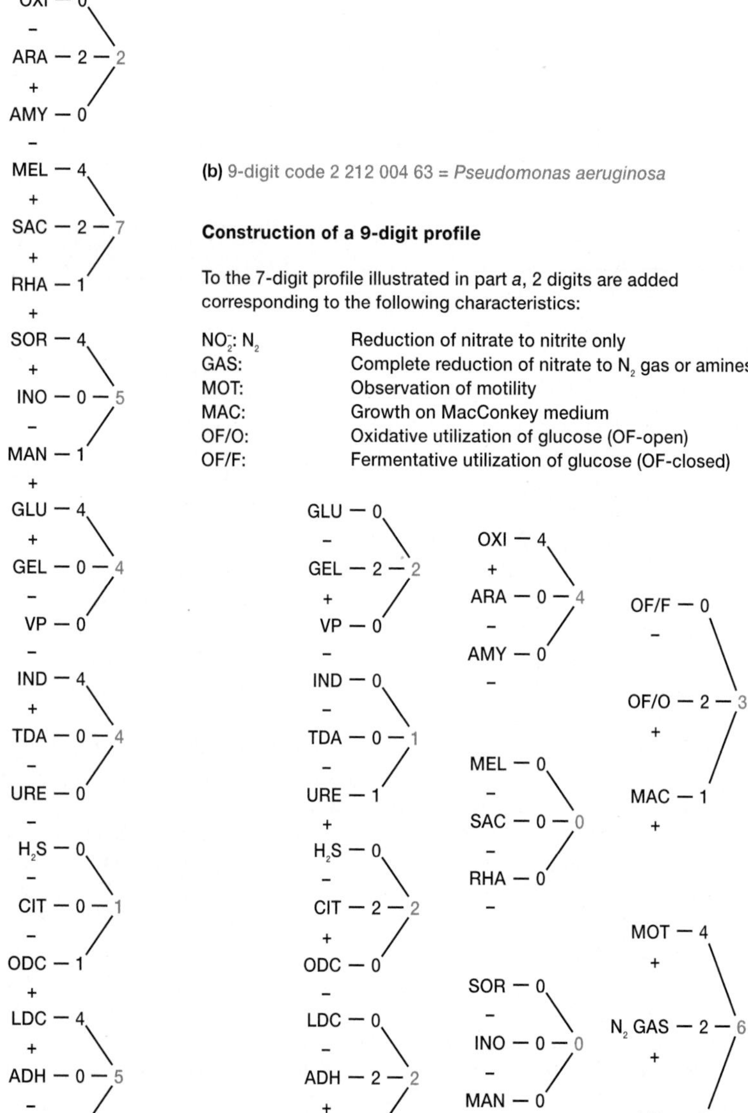

Figure 35.7 The API 20E Profile Number. The conversion of API 20E test results to the codes used in identification of unknown bacteria. The test results read top to bottom (and right to left in part *b*) correspond to the 7- and 9-digit codes when read in the right-to-left order. The tests required for obtaining a 7-digit code take an 18–24 hour incubation and will identify most members of the *Enterobacteriaceae.* The longer procedure that yields a 9-digit code is required to identify many gram-negative nonfermenting bacteria. The following tests are common to both procedures: ONPG (β-galactosidase); ADH (arginine dihydrolase); LDC (lysine decarboxylase); ODC (ornithine decarboxylase); CIT (citrate utilization); H_2S (hydrogen sulfide production); URE (urease); TDA (tryptophane deaminase); IND (indole production); VP(Voges-Proskauer test for acetoin); GEL (gelatin liquefaction); the fermentation of glucose (GLU), mannitol (MAN), inositol (INO), sorbitol (SOR), rhamnose (RHA), sucrose (SAC), melibiose (MEL), amygdalin (AMY), and arabinose (ARA); and OXI (oxidase test).

Table 35.3 Some Common Rapid Immunologic Test Kits for the Detection of Bacteria and Viruses in Clinical Specimens

Bactigen (Wampole Laboratories, Cranburg, N.J.)

The Bactigen kit is used for the detection of *Streptococcus pneumoniae, Haemophilus influenzae* type b, and *Neisseria meningitidis* groups A, B, C, and Y from cerebrospinal fluid, serum, and urine.

Culturette Group A Strep ID Kit (Marion Scientific, Kansas City, Mo.)

The Culturette kit is used for the detection of group A streptococci from throat swabs.

Directigen (Hynson, Wescott, and Dunning, Baltimore, Md.)

The Directigen Meningitis Test kit is used to detect *H. influenzae* type b, *S. pneumoniae,* and *N. meningitidis* groups A and C.

The Directigen Group A Strep Test kit is used for the direct detection of group A streptococci from throat swabs.

Gono Gen (Micro-Media Systems, San Jose, Calif.)

The Gono Gen kit detects *Neisseria gonorrhoeae.*

Ora Quick (OraAure Technologies, Bethleham, Pa.) Detects HIV in saliva in 10 minutes.

QuickVue *H. pylori* Test (Quidel, San Diego, Calif.)

A 7-minute test for detection of IgG antibodies against *Helicobacter pylori* in human serum or plasma.

Staphaurex (Wellcome Diagnostics, Research Triangle Park, N.C.)

Staphaurex screens and confirms *Staphylococcus aureus* in 30 seconds.

Directigen RSV (Becton Dickinson Microbiology Systems, Cockeysville, Md.)

By using a nasopharyngeal swab, the respiratory syncytial virus can be detected in 15 minutes.

SureCell Herpes (HSV) Test (Kodak, Rochester, N.Y.)

Detects the herpes (HSV) 1 and 2 viruses in minutes.

SUDS HIV-1 Test (Murex Corporation, Norcross, Ga.)

Detects antibodies to HIV-1 antigens in about 10 minutes.

method in the taxonomic identification of microorganisms. Those monoclonal antibodies that define species-specific antigens are extremely valuable in diagnostic reagents. Monoclonal antibodies that exhibit more restrictive specificity can be used to identify strains of biotypes within a species and in epidemiological studies involving the matching of microbial strains. Coupling sensitive visualization technologies such as fluorescence or scanning tunneling microscopy to mAb detection systems makes it possible to perform microbial identifications with improved accuracy, speed, and fewer organisms. Techniques & Applications 32.1: Donor selection for tissue or organ transplants

1. Describe in general how biochemical tests are used in the API 20E system to identify bacteria.
2. Why might cultures for some microorganisms be unavailable?

Bacteriophage Typing

Bacteriophages are viruses that attack members of a particular bacterial species, or strains within a species. **Bacteriophage (phage) typing** is based on the specificity of phage surface receptors for cell surface receptors. Only those bacteriophages that can attach to these surface receptors can infect bacteria and cause lysis. On a petri dish culture, lytic bacteriophages cause plaques on lawns of sensitive bacteria. These plaques represent infection by the virus (*see figure 16.14*). Viruses of *Bacteria* and *Archaea* (chapter 17)

In bacteriophage typing the clinical microbiologist inoculates the bacterium to be tested onto a petri plate. The plate is heavily and uniformly inoculated so that the bacteria will grow to form a solid sheet or lawn of cells. The plate is then marked off into squares (15 to 20 mm per side), and each square is inoculated with a drop of suspension from the different phages available for typing. After the plate is incubated for 24 hours, it is observed for plaques. The phage type is reported as a specific genus and species followed by the types that can infect the bacterium. For example, the series 10/16/24 indicates that this bacterium is sensitive to phages 10, 16, and 24, and belongs to a collection of strains, called a **phagovar,** that have this particular phage sensitivity. Bacteriophage typing remains a tool of the research and reference laboratory.

Molecular Methods and Analysis of Metabolic Products

The application of molecular technology enables the analysis of molecular characteristics of microorganisms in the clinical laboratory. Some of the most accurate approaches to microbial identification are through the analysis of proteins and nucleic acids. Examples include comparison of proteins; physical, kinetic, and regulatory properties of microbial enzymes; nucleic acid–base composition; nucleic acid hybridization; and nucleic acid sequencing (*see figure 19.12*). Three other molecular methods being widely used are nucleic acid probes, gas-liquid chromatography, and DNA fingerprinting. Techniques for determining microbial taxonomy and phylogeny (section 19.4)

Nucleic Acid–Based Detection Methods

Nucleic acid–based diagnostic methods for the detection and identification of microorganisms have become routine in many clinical microbiology laboratories. For example, DNA hybridization technology can identify a microorganism by probing its genetic composition. The use of cloned DNA as a probe is based upon the capacity of single-stranded DNA to bind (hybridize) with a complementary nucleic acid sequence present in test specimens to form a double-stranded DNA hybrid (**figure 35.8**). Thus DNA derived from one microorganism (the probe) is used to search for others containing the same sequence. Hybridization reactions may be applied to purified DNA preparations, to bacterial colonies, or to clinical specimens such as tissue, serum, sputum, and pus. DNA probes have been developed that bind to complementary strands of ribosomal RNA. These DNA:rRNA hybrids are more sensitive than conventional DNA probes, give

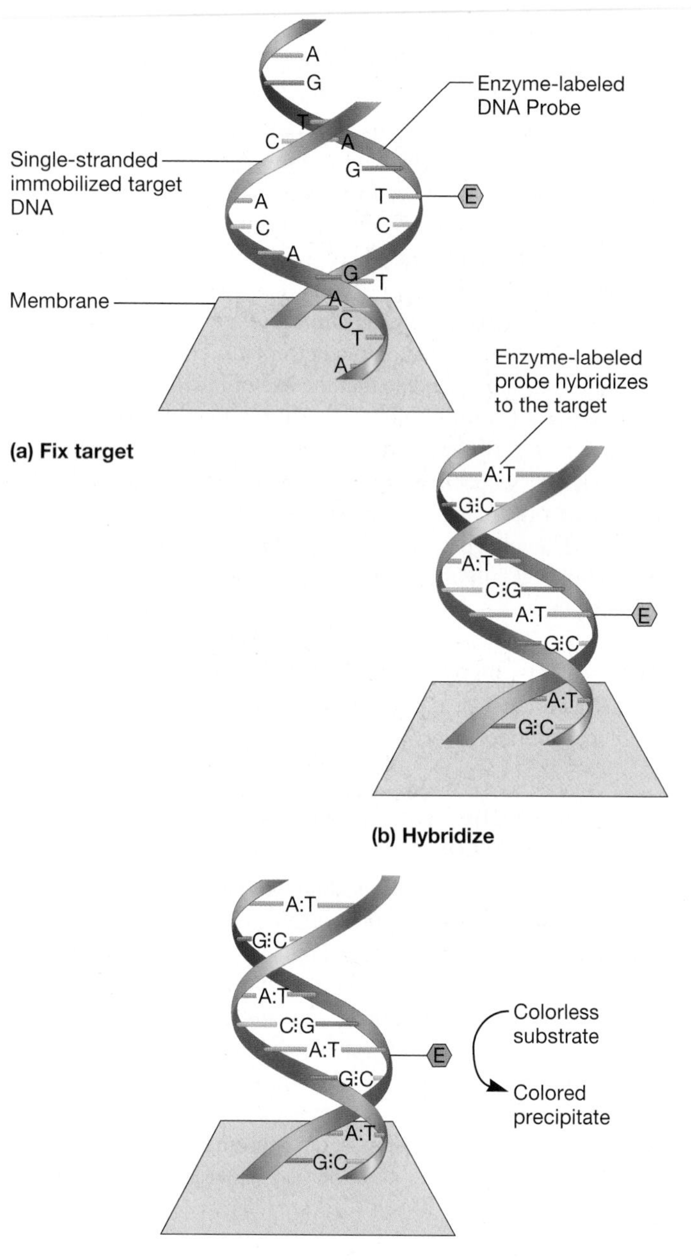

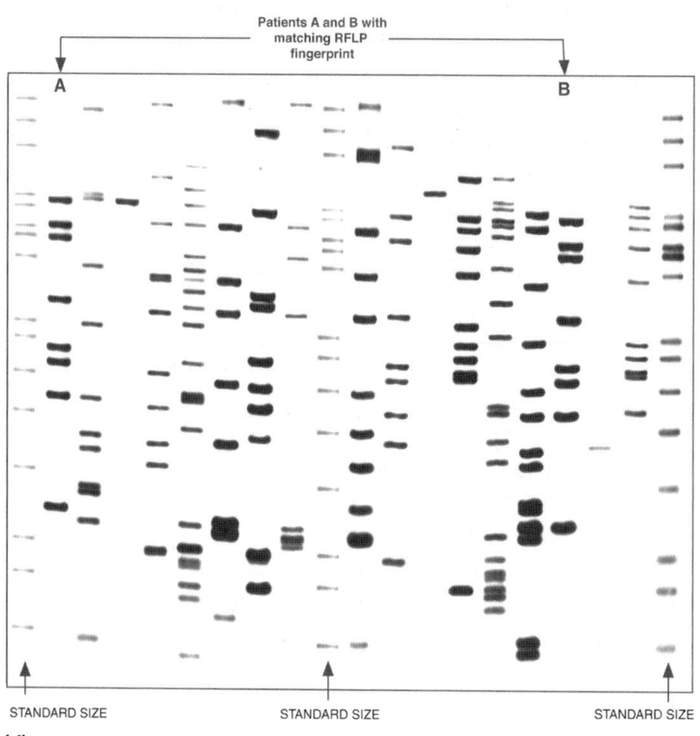

Figure 35.8 Basic Steps in a DNA Probe Hybridization Assay. **(a)** Single-stranded target nucleic acid is bound to a membrane. A DNA probe with attached enzyme (E) also is employed. **(b)** The probe is added to the membrane. If the probe hybridizes to the target DNA, a double-stranded DNA hybrid is formed. **(c)** A colorless substrate is added. The enzyme attached to the probe converts the substrate to a colored precipitate. This detection system is semiquantitative, in that color intensity is proportional to the quantity of hybridized target nucleic acid present. **(d)** The pattern shows results for various strains isolated from 17 patients. Bands were developed by DNA hybridization using probes specific to genes from several *M. tuberculosis* strains. Lanes 1, 10, and 20 provide size markers for reference. Patients A and B are infected with the same common strain of the pathogen.

results in 2 hours or less, and require the presence of fewer microorganisms. DNA probe sensitivity can be increased by over one million-fold if the target DNA is first amplified using PCR. DNA:rRNA probes are available or are currently being developed for most clinically important microorganisms. The polymerase chain reaction (section 14.3)

The nucleotide sequence of small subunit ribosomal RNA (rRNA) can be used to identify bacterial genera (*see figure 19.10*). Usually, the rRNA encoding gene or gene fragment is amplified by PCR. After nucleotide sequencing, the rRNA gene is compared with those in the international database maintained by the National Center for Biotechnology (NCBI). This method of bacterial identification, called **ribotyping,** is based on the high level of 16s rRNA conservation among bacteria. Another approach, genomic fingerprinting, is also used in identifying pathogens. This does not involve nucleotide sequencing; rather, it compares the similarity of specific DNA fragments generated by restriction endonuclease digestion. BOX-, ERIC-, and REP PCR are also described in section 19.4 (*see figure 19.11*).

Gas-Liquid Chromatography

During chromatography a chemical mixture carried by a liquid or gas is separated into its individual components due to processes such as adsorption, ion-exchange, and partitioning between different solvent phases. In gas-liquid chromatography (GLC), specific microbial metabolites, cellular fatty acids, and products from the pyrolysis (a chemical change caused by heat) of whole bacterial cells are analyzed and identified. These compounds are easily removed from growth media by extraction with an organic solvent such as ether. The ether extract is then injected into the GLC system. Both volatile and nonvolatile acids can be identified. Based on the pattern of fatty acid production, common bacteria isolated from clinical specimens can be identified.

The reliability, precision, and accuracy of GLC have been improved significantly with continued advances in instrumentation; the introduction of instruments for high-performance liquid chromatography; and the use of mass spectrometry, nuclear magnetic resonance spectroscopy, and associated analytical techniques for the identification of components separated by the chromatographic process. These combined techniques can be used to discover specific chemical markers of various infectious disease agents by direct analysis of body fluids.

Plasmid Fingerprinting

As presented in section 3.5, a plasmid is an autonomously replicating extrachromosomal molecule of DNA. **Plasmid fingerprinting** identifies microbial isolates of the same or similar strains; related strains often contain the same number of plasmids with the same molecular weights and similar phenotypes. In contrast, microbial isolates that are phenotypically distinct have different plasmid fingerprints. Plasmid fingerprinting of many *E. coli, Salmonella, Campylobacter,* and *Pseudomonas* strains and species has demonstrated that this method often is more accurate than other phenotyping methods such as biotyping, antibiotic resistance patterns, phage typing, and serotyping.

The technique of plasmid fingerprinting involves five steps:

1. The bacterial strains are grown in broth or on agar plates.
2. The cells are harvested and lysed with a detergent.
3. The plasmid DNA is separated from the chromosomal DNA and then cut with specific restriction endonucleases.
4. The plasmid DNA is applied to agarose gels and electrophoretically separated.
5. The gel is stained with ethidium bromide, which binds to DNA, causing it to fluoresce under UV light. The plasmid DNA bands are then located.

Because the migration rate of plasmid DNA in agarose is inversely proportional to the molecular weight, plasmids of a different size appear as distinct bands in the stained gel. The molecular weight of each plasmid species can then be determined from a plot of the distance that each species has migrated versus the log of the molecular weights of plasmid markers of known size that have been electrophoresed simultaneously in the same gel (**figure 35.9**).

1. What is the basis for bacteriophage typing?
2. How can nucleic acid–based detection methods be used by the clinical microbiologist? Gas-liquid chromatography?
3. How can a suspect bacterium be plasmid fingerprinted?

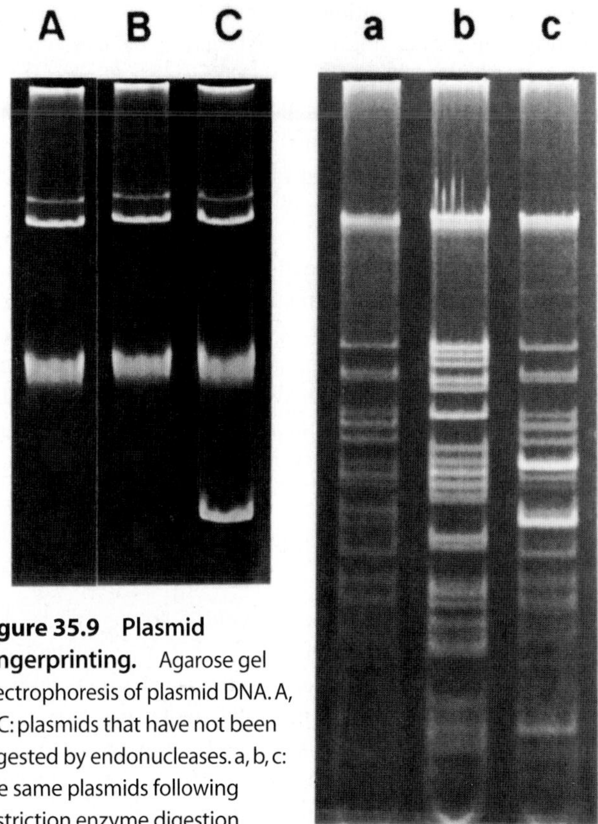

Figure 35.9 Plasmid Fingerprinting. Agarose gel electrophoresis of plasmid DNA. A, B, C: plasmids that have not been digested by endonucleases. a, b, c: the same plasmids following restriction enzyme digestion.

35.3 Clinical Immunology

The culturing of certain viruses, bacteria, fungi, and protozoa from clinical specimens may not be possible because the methodology remains undeveloped (e.g., *Treponema pallidum;* hepatitis A, B, C; and Epstein-Barr virus), is unsafe (rickettsias and HIV), or is impractical for all but a few microbiology laboratories (e.g., mycobacteria, strict anaerobes, *Borrelia*). Cultures also may be negative because of prior antimicrobial therapy. (This is why it is so important to obtain a reliable sample prior to starting antimicrobial chemotherapy.) Under these circumstances, detection of antibodies or antigens may be quite valuable diagnostically.

Immunologic systems for the detection and identification of pathogens from clinical specimens are easy to use, give relatively rapid reaction endpoints, and are sensitive and specific (they give a low percentage of false positives and negatives). Some of the more popular immunologic rapid test kits for viruses and bacteria are presented in table 35.3.

Due to dramatic advances in clinical immunology, there has been a marked increase in the number, sensitivity, and specificity of serological tests. This increase reflects a better understanding of (1) immune cell surface antigens (CD antigens), (2) lymphocyte biology, (3) the production of monoclonal antibodies, and (4) the development of sensitive antibody-binding reporter systems. Furthermore, each individual's immunologic response to a microorganism is quite variable. As a result the interpretation of immunologic tests is sometimes difficult. For example, a single, elevated IgM titer does not distinguish between active and past infections. Rather an elevated IgG titer typically indicates an active infection, especially when subsidence of symptoms correlates with a four-fold (or greater) decrease in antibody titer. Furthermore, a lack of a measurable antibody titer may reflect an organism's lack of immunogenicity or an insufficient time for an antibody response to develop following the onset of the infectious disease. Some patients are also immunosuppressed due to other disease processes and/or treatment procedures (e.g., cancer and AIDS patients) and therefore do not respond. For these reasons, test selection and timing of specimen collection are essential to the proper interpretation of immunologic tests. In this section some of the more common antibody-based techniques that are employed in the diagnosis of microbial and immunological diseases are discussed.

Serotyping

Serum is the liquid portion of blood (devoid of clotting factors) that contains many different components, especially the immunoglobulins or antibodies. **Serotyping** refers to the use of serum (antibodies) to specifically detect and identify other molecules. Serotyping can be used to identify specific white blood cells or the proteins on cell surfaces (*see Techniques & Applications 32.1*). Serotyping can also be used to differentiate strains (serovars or serotypes) of microorganisms that differ in the antigenic composition of a structure or product (**Techniques & Applications 35.3**). The serological identification of a pathogenic strain has diagnostic value. Often the symptoms of infections depend on the nature of the cell products released by the pathogen. Therefore it is sometimes possible to identify a pathogen serologically by testing for cell wall antigens. For example, there are 90 different strains of *Streptococcus pneumoniae,* each unique in the nature of its capsular material. These differences can be detected by antibody-induced capsular swelling (termed the **Quellung reaction**) when the appropriate antiserum for a specific capsular type is used (**figure 35.10**).

Agglutination

As noted in figure 32.25, when an immune complex is formed by cross-linking cells or particles with specific antibodies, it is called an agglutination reaction. Agglutination reactions usually form visible aggregates or clumps (**agglutinates**) that can be seen with the unaided eye. Direct agglutination reactions are very useful in the diagnosis of certain diseases. For example, the **Widal test** is a reaction involving the agglutination of typhoid bacilli when they are mixed with serum containing typhoid antibodies from an individual who has typhoid fever. Action of antibodies: Immune complex formation (section 32.8)

Techniques have also been developed that employ microscopic synthetic latex spheres coated with antigens. These coated microspheres are extremely useful in diagnostic agglutination reactions. For example, the modern pregnancy test detects the elevated level of human chorionic gonadotropin (hCG) hormone that occurs in a woman's urine and blood early in pregnancy. Latex agglutination tests are also used to detect antibodies that develop during certain mycotic, helminthic, and bacterial infections, and in drug testing.

Hemagglutination usually results from antibodies cross-linking red blood cells through attachment to surface antigens and is routinely used in blood typing. In addition, some viruses can accomplish **viral hemagglutination.** For example, if a person has a certain viral disease, such as measles, antibodies will be present in the serum to react with the measles viruses and neutralize them. Normally, hemagglutination occurs when measles viruses and red blood cells are mixed. However, red blood cells may be mixed first with a person's serum followed by the addition of virions. If no hemagglutination occurs, the serum antibodies have neutralized the measles viruses. This is considered a positive test result for the presence of virus-specific antibodies (**figure 35.11**).

Techniques & Applications

35.3 History and Importance of Serotyping

In the early 1930s Rebecca Lancefield (1895–1981) recognized the importance of serological tests. She developed a classification system for the streptococci based on the antigenic nature of cell wall carbohydrates. Her system is now known as the **Lancefield system** in which each different serotype is a Lancefield group and identified by a letter (A through T). This scheme is based on specific antibody agglutination reactions with cell wall carbohydrate antigens (C polysaccharides) extracted from the streptococci. Lancefield also showed that further subdividing of the group A streptococci into specific serological types was possible, based on the presence of type-specific M (protein) antigens.

Escherichia coli, Salmonella, and other bacteria are routinely serotyped with specific antigen-antibody reactions involving flagella (H) antigens, capsular (K) antigens, and lipopolysaccharide (O) antigens. Among *E. coli* there are over 167 different O antigens.

The current value of serotyping may be seen in the fact that *E. coli* O55, O111, and O127 serotypes are the ones most frequently associated with infantile diarrhea and *E. coli* O157:H7 is largely to blame for many life-threatening *E. coli* outbreaks. Thus the serotype of *E. coli* from stool samples is of diagnostic value and aids in identifying the source of the infection.

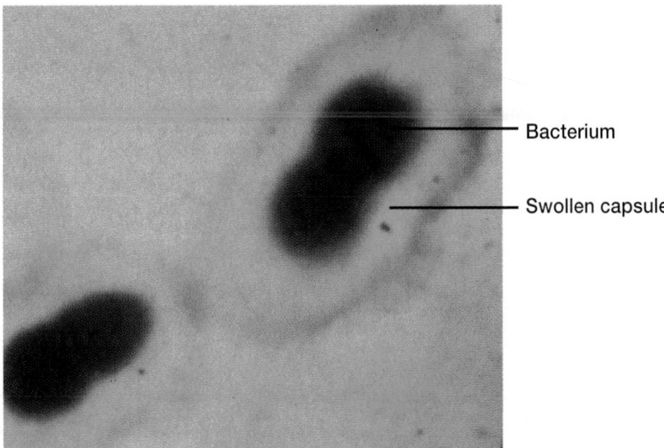

Figure 35.10 Serotyping. *Streptococcus pneumoniae* has reacted with a specific pneumococcal antiserum leading to capsular swelling (the Quelling reaction). The capsules seen around the bacteria indicated potential virulence.

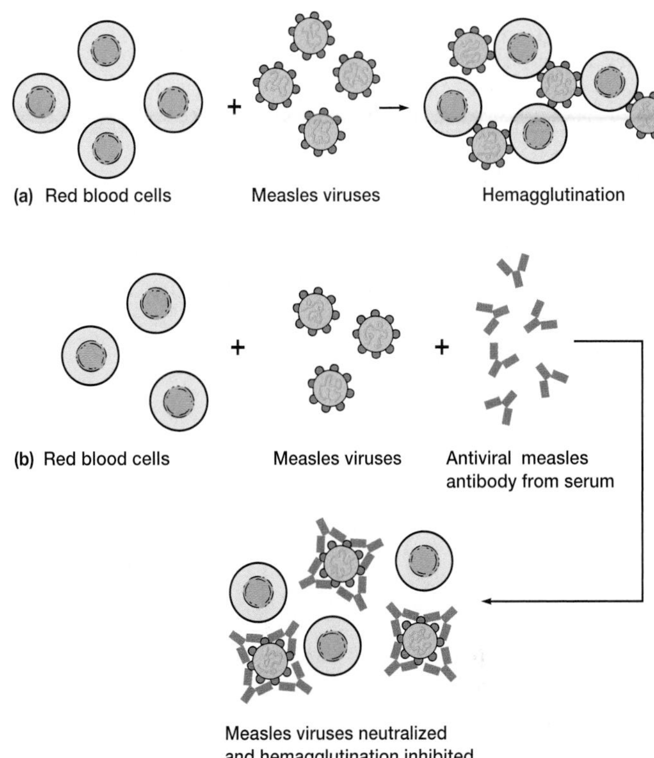

Figure 35.11 Viral Hemagglutination. **(a)** Certain viruses can bind to red blood cells causing hemagglutination. **(b)** If serum containing specific antibodies to the virus is mixed with the red blood cells, the antibodies will neutralize the virus and inhibit hemagglutination (a positive test).

This hemagglutination inhibition test is widely used to diagnose influenza, measles, mumps, mononucleosis, and other viral infections.

Agglutination tests are also used to measure antibody titer. In the tube or well agglutination test, a specific amount of antigen is added to a series of tubes or shallow wells in a microtiter plate (**figure 35.12**). Serial dilutions of serum (1/20, 1/40, 1/80, 1/160, etc.) containing the antibody are then added to each tube or well. The greatest dilution of serum showing an agglutination reaction is determined, and the reciprocal of this dilution is the serum antibody titer.

1. What is serology?
2. When would you use the Widal test?
3. Why does hemagglutination occur and how can it be used in the clinical laboratory?

Complement Fixation

When complement binds to an antigen-antibody complex, it becomes "fixed" and "used up." Complement fixation tests are very sensitive and can be used to detect extremely small amounts of an antibody for a suspect microorganism in an individual's serum. A known antigen is mixed with test serum lacking complement (**figure 35.13*a***). When immune complexes have had time to form, complement is added (figure 35.13*b*) to the mixture. If immune complexes are present, they will fix and consume complement. Afterward, sensitized indicator cells, usually sheep red blood cells previously coated with complement-fixing antibodies, are added to the mixture. Lysis of the indicator cells (figure 35.13*c*) results if immune complexes do not form in part *a* of the test because the antibodies are not present in the test serum. In the absence of antibodies, complement remains and lyses the indicator cells. On the other hand, if specific antibodies are present in the test serum and complement is consumed by the immune complexes, insufficient amounts of complement will be available to lyse the indicator cells. Absence of lysis shows that specific antibodies are present in the test serum. Complement fixation was once used in the diagnosis of syphilis (the Wassermann test) and is still used as a rapid, inexpensive screening method in the diagnosis of certain viral, fungal, rickettsial, chlamydial, and protozoan diseases.

Enzyme-Linked Immunosorbent Assay

The ***e**nzyme-**l**inked **i**mmuno**s**orbent **a**ssay* **(ELISA)** has become one of the most widely used serological tests for antibody or antigen detection. This test involves the linking of various "label" enzymes to either antigens or antibodies. Two basic methods are used: the direct (also called the double antibody sandwich assay) and the indirect immunosorbent assay.

The double antibody sandwich assay is used for the detection of antigens (**figure 35.14*a***). In this assay, specific antibody is placed in wells of a microtiter plate (or it may be attached to a membrane). The antibody is absorbed onto the walls, coating the plate. A test antigen (in serum, urine, etc.) is then added to each well. If the antigen reacts with the antibody, the antigen is retained

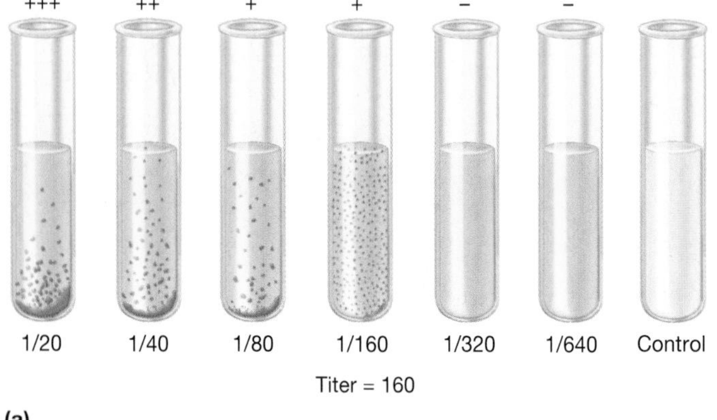

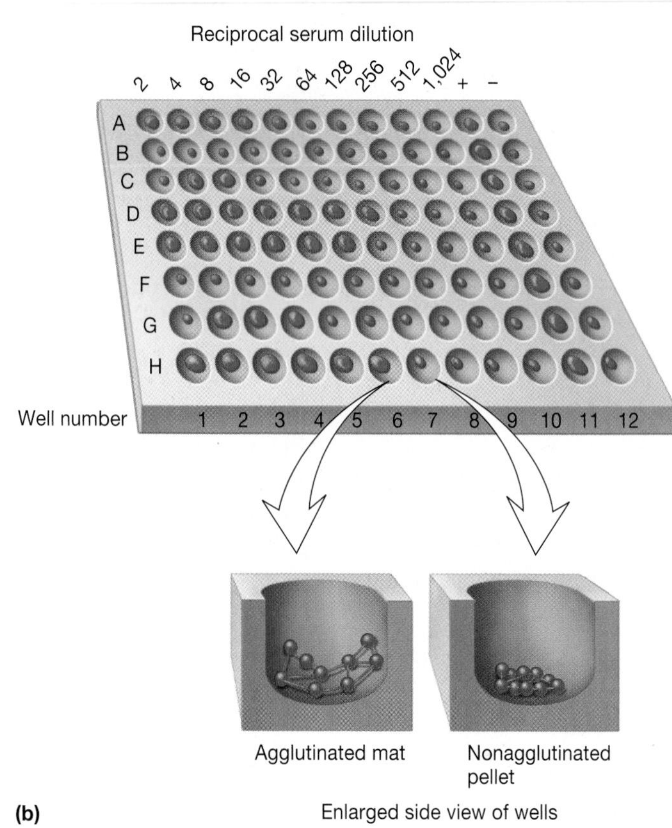

Figure 35.12 Agglutination Tests. **(a)** Tube agglutination test for determining antibody titer. The titer in this example is 160 because there is no agglutination in the next tube in the dilution series (1/320). The blue in the dilution tubes indicates the presence of the patient's serum. **(b)** A microtiter plate illustrating hemagglutination. The antibody is placed in the wells (1–10). Positive controls (row 11) and negative controls (row 12) are included. Red blood cells are added to each well. If sufficient antibody is present to agglutinate the cells, they sink as a mat to the bottom of the well. If insufficient antibody is present, they form a pellet at the bottom. Can you read the different titers in rows A–H?

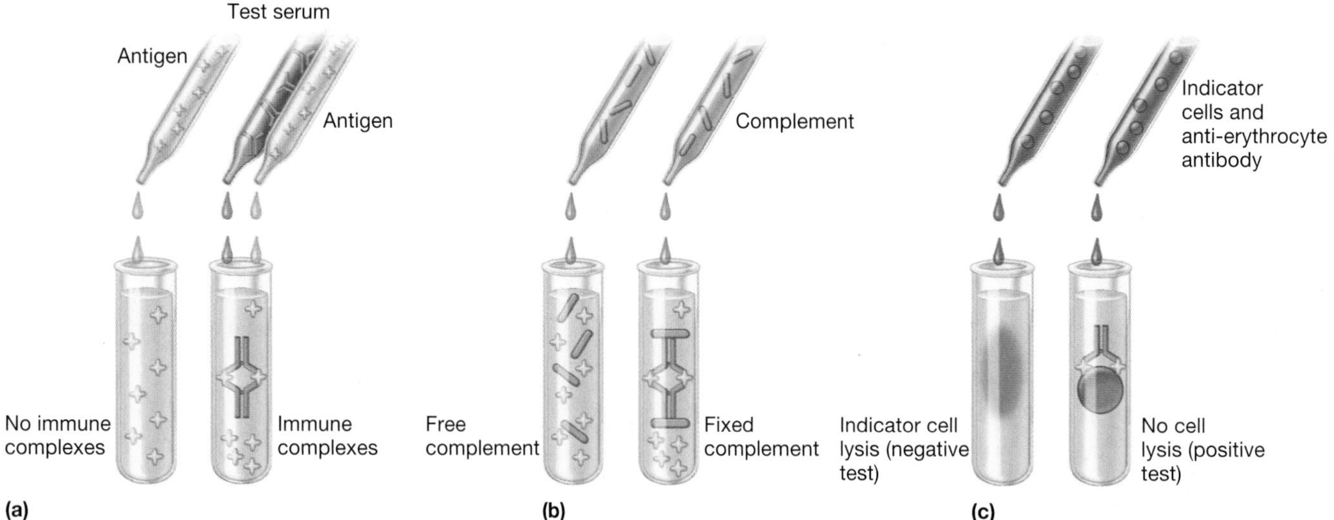

Figure 35.13 Complement Fixation. **(a)** Test serum is added to one test tube. A fixed amount of antigen is then added to both tubes. If antibody is present in the test serum, immune complexes form. **(b)** When complement is added, if complexes are present, they fix complement and consume it. **(c)** Indicator cells and a small amount of anti-erythrocyte antibody are added to the two tubes. If there is complement present, the indicator cells will lyse (a negative test): if the complement is consumed, no lysis will occur (a positive test).

when the well is washed to remove unbound antigen. A commercially prepared antibody-enzyme conjugate specific for the antigen is then added to each well. The final complex formed is an outer antibody-enzyme, middle antigen, and inner antibody—that is, it is a layered (Ab-Ag-Ab) sandwich. A substrate that the enzyme will convert to a colored product is then added, and any resulting product is quantitatively measured by optical density scanning of the plate. If the antigen has reacted with the absorbed

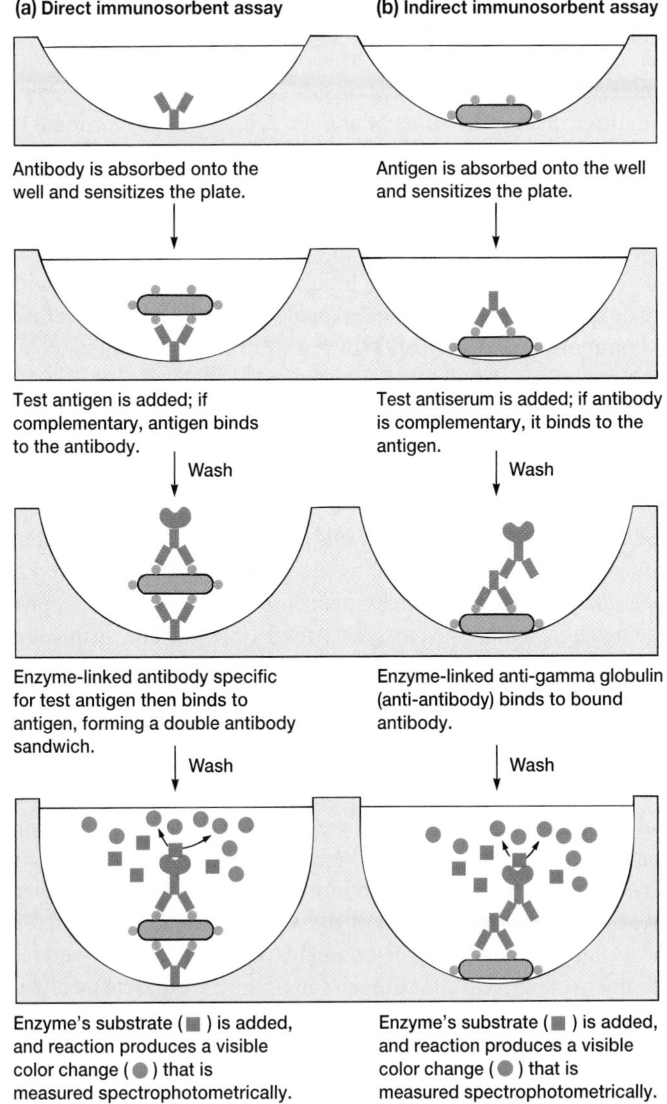

(a) Direct immunosorbent assay
(b) Indirect immunosorbent assay
Antibody is absorbed onto the well and sensitizes the plate.
Antigen is absorbed onto the well and sensitizes the plate.
Test antigen is added; if complementary, antigen binds to the antibody.
Test antiserum is added; if antibody is complementary, it binds to the antigen.
Wash
Wash
Enzyme-linked antibody specific for test antigen then binds to antigen, forming a double antibody sandwich.
Enzyme-linked anti-gamma globulin (anti-antibody) binds to bound antibody.
Wash
Wash
Enzyme's substrate ( ) is added, and reaction produces a visible color change ( ) that is measured spectrophotometrically.
Enzyme's substrate ( ) is added, and reaction produces a visible color change ( ) that is measured spectrophotometrically.

Figure 35.14 The ELISA or EIA Test. **(a)** The direct or double antibody sandwich method for the detection of antigens. **(b)** The indirect immunosorbent assay for detecting antibodies. See text for details.

antibodies in the first step, the ELISA test is positive (i.e., it is colored). If the antigen is not recognized by the absorbed antibody, the ELISA test is negative because the unattached antigen has been washed away, and no antibody-enzyme is bound (it is colorless). This assay is currently being used for the detection of *Helicobacter pylori* infections, brucellosis, salmonellosis, and cholera. Many other antigens also can be detected by the sandwich method. For example, there are ELISA kits on the market that can test for many different food allergens.

The indirect immunosorbent assay detects antibodies rather than antigens. In this assay, antigen in appropriate buffer is incubated in the wells of a microtiter plate (figure 35.14*b*) and is absorbed onto the walls of the wells. Free antigen is washed away. Test antiserum is added, and if specific antibody is present, it binds to the antigen. Unbound antibody is washed away. Alternatively the test sample can be incubated with a suspension of latex beads that have the desired antigen attached to their surface. After allowing time for antibody-antigen complex formation, the beads are trapped on a filter and unbound antibody is washed away. An anti-antibody that has been covalently coupled to an enzyme, such as horseradish peroxides, is added next. The antibody-enzyme complex (the conjugate) binds to the test antibody, and after unbound conjugate is washed away, the attached ligand is visualized by the addition of a chromogen. A **chromogen** is a colorless substrate acted on by the enzyme portion of the ligand to produce a colored product. The amount of test antibody is quantitated in the same way as an antigen is in the double antibody sandwich method. The indirect immunosorbent assay currently is being used to test for antibodies to human immunodeficiency virus (HIV) and rubella virus (German measles), and to detect certain drugs in serum. For example, antigen-coated latex beads are used in the SUDS HIV-1 test to detect HIV serum antibodies in about 10 minutes.

Immunoblotting (Western Blot)

Another immunologic technique used in the clinical microbiology laboratory is **immunoblotting.** Immunoblotting involves polyacrylamide gel electrophoresis of a protein specimen followed by transfer of the separated proteins to nitrocellulose sheets. Protein bands are then visualized by treating the nitrocellulose sheets with solutions of enzyme-tagged antibodies. This procedure demonstrates the presence of common and specific proteins among different strains of microorganisms (**figure 35.15**). Immunoblotting also can be used to show strain-specific immune responses to microorganisms, to serve as an important diagnostic indicator of a recent infection with a particular strain of microorganism, and to allow for prognostic implications with severe infectious diseases.

Immunoprecipitation

The **immunoprecipitation** technique detects soluble antigens that react with antibodies called precipitins. The precipitin reaction occurs when bivalent or multivalent antibodies and antigens are mixed in the proper proportions. The antibodies link the antigen to form a large antibody-antigen network or lattice that settles out of solution when it becomes sufficiently large (**figure 35.16*a***). Immunoprecipitation reactions occur only at the equivalence zone when there is an optimal ratio of antigen to antibody so that an insoluble lattice forms. If the precipitin reaction takes place in a test tube (figure 35.16*b*), a precipitation ring forms in the area in which the optimal ratio or equivalence zone develops. Action of antibodies: Immune complex formation (section 32.8)

Immunodiffusion

Immunodiffusion refers to a precipitation reaction that occurs between an antibody and antigen in an agar gel medium. Two techniques are routinely used: single radial immunodiffusion and double diffusion in agar.

The **single *r*adial *i*mmuno*d*iffusion (RID) assay** or Mancini technique quantitates antigens. Monospecific antibody is added to agar, then the mixture is poured onto slides and allowed to set. Wells are cut in the agar and known amounts of standard antigen added. The unknown test antigen is added to a separate well (**figure 35.17*a***). The slide is left for 24 hours or until equilibrium has been reached, during which time the antigen diffuses out of the wells to form insoluble complexes with anitbodies. The size of the resulting precipitation ring surrounding various dilutions of antigen selected is proportional to the amount of antigen in the well (the wider the ring, the greater the antigen concentration). This is because the antigen's concentration drops as it diffuses farther out into the agar. The antigen forms a precipitin ring in the agar when its level has decreased sufficiently to reach equivalence and combine with the antibody to produce a large, insoluble network. This method is commonly used to quantitate serum immunoglobulins, complement proteins, and other substances.

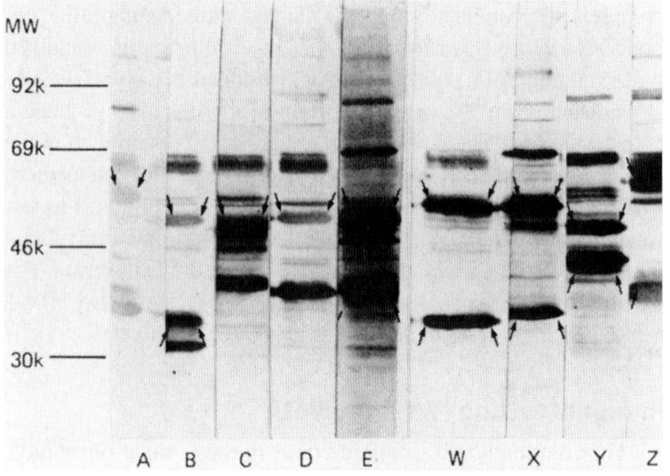

Figure 35.15 Immunoblotting (Western Blot). Immunoblot of the standard strains of *Clostridium difficile*. Arrows indicate strain-specific bands (specific molecular weights) of the various proteins in different lanes (A–E). The molecular weight of the protein is indicated on the left.

The **double diffusion agar assay (Öuchterlony technique)** is based on the principle that diffusion of both antibody and antigen (hence, double diffusion) through agar can form stable and easily observable immune complexes. Test solutions of antigen and antibody are added to the separate wells punched in agar. The solutions diffuse outward, and when antigen and the appropriate antibody meet, they combine and precipitate at the equivalence zone, producing an indicator line (or lines) (figure 35.17*b*). The visible line of precipitation permits a comparison of antigens for identity (same antigenic determinants), partial identity (cross-reactivity), or non-identity against a given selected antibody. For example, if a V-shaped line of precipitation forms, this demonstrates that the antibodies bind to the same antigenic determinants in each antigen sample and are identical. If one well is filled with a different antigen that shares some but not all determinants with the first antigen, a Y-shaped line of precipitation forms, demonstrating partial identity. In this reaction the stem of the Y, called a spur, is formed if those antigen or antigenic determinants absent in the first well but present in the second one (antigen a in figure 35.17*b*) react with the diffusing antibodies. If two completely unrelated antigens are

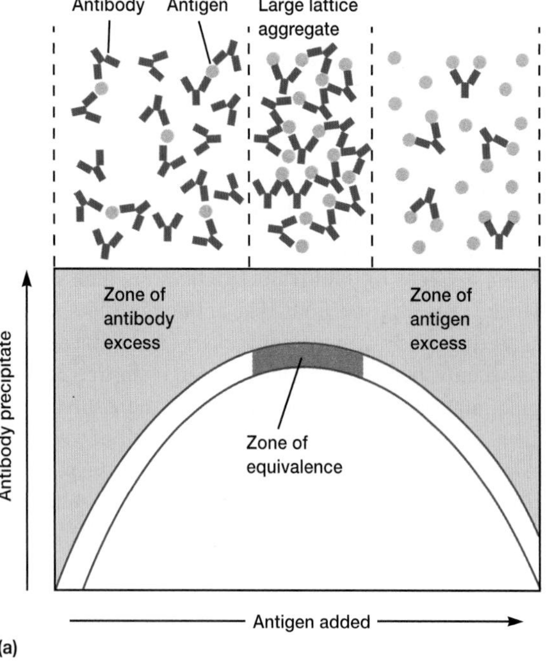

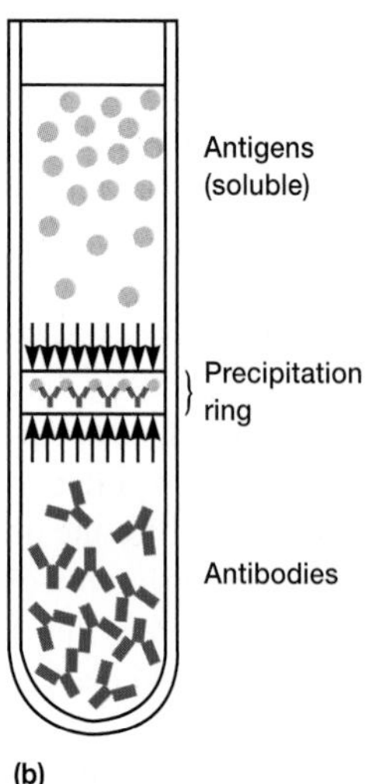

Figure 35.16 Immunoprecipitation. **(a)** Graph showing that a precipitation curve is based on the ratio of antigen to antibody. The zone of equivalence represents the optimal ratio for precipitation. **(b)** A precipitation ring test. Antibodies and antigens diffuse toward each other in a test tube. A precipitation ring is formed at the zone of equivalence.

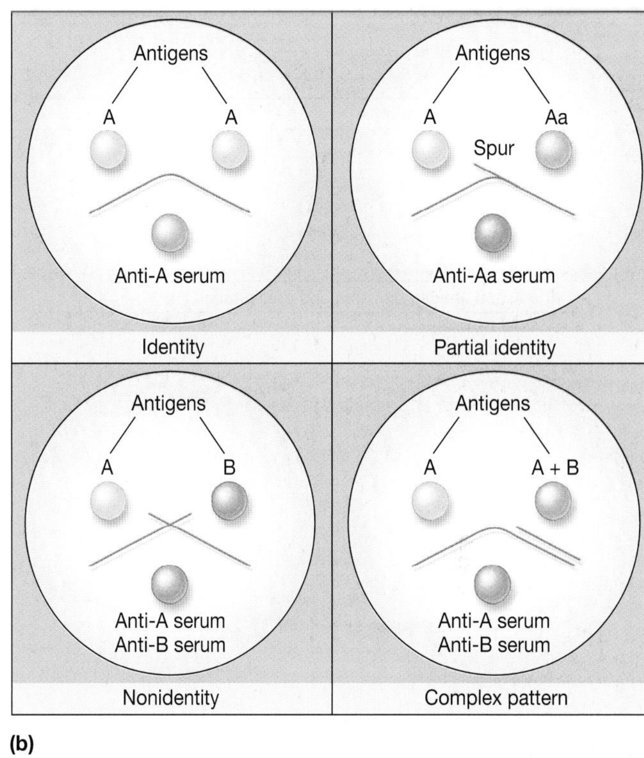

Figure 35.17 Immunodiffusion. **(a)** Single radial immunodiffusion assay. Three standard solutions of different antigen concentrations (Ag1, Ag2, Ag3) and an unknown (AgX) are placed on agar. After equilibration the ring diameters are measured. Usually the square of the diameter of the standard rings is plotted on the x-axis and the antigen concentration on the y-axis. From this standard curve, the concentration of an unknown can be determined. **(b)** Double diffusion agar assay showing characteristics of identity (top left), reaction of nonidentity (bottom left), partial identity (top right), and a complex pattern (bottom right).

added to the wells, either a single straight line of precipitation forms between the two wells, or two separate lines of precipitation form, creating an X-shaped pattern, a reaction of nonidentity.

Immunoelectrophoresis

Some antigen mixtures are too complex to be resolved by simple diffusion and precipitation. Greater resolution is obtained by the technique of classical **immunoelectrophoresis** in which antigens are first separated based on their electrical charge, then visualized by the precipitation reaction. In this procedure antigens are separated by electrophoresis in an agar gel. Positively charged proteins move to the negative electrode, and negatively charged proteins move to the positive electrode (**figure 35.18*a***). A trough is then cut next to the wells (figure 35.18*b*) and filled with antibody. The plate is incubated, the antibodies and antigens will diffuse and form precipitation bands or arcs (figure 35.18*c*) that can be better visualized by staining (figure 35.18*d*). This assay is used to separate the major blood proteins in serum for certain diagnostic tests. Gel electrophoresis (section 14.4)

1. What does a negative complement fixation test show? a positive test?
2. What are the two types of ELISA methods and how do they work? What is a chromogen?
3. Specifically, when do immunoprecipitation reactions occur?
4. Name two types of immunodiffusion tests and describe how they operate.
5. Describe the classical immunoelectrophoresis technique.

Flow Cytometry

Flow cytometry allows single- or multiple-microorganism detection in clinical samples in an easy, reliable, fast way. In flow cytometry microorganisms are identified on the basis of their unique cytometric parameters or by means of certain dyes called fluorochromes that can be used either independently or bound to specific antibodies or oligonucleotides. The flow cytometer forces a suspension of cells through a laser beam and measures the light they scatter or the florescence the cells emit as they pass through the beam. For example, cells can be tagged with a fluorescent antibody directed against a specific surface antigen. As

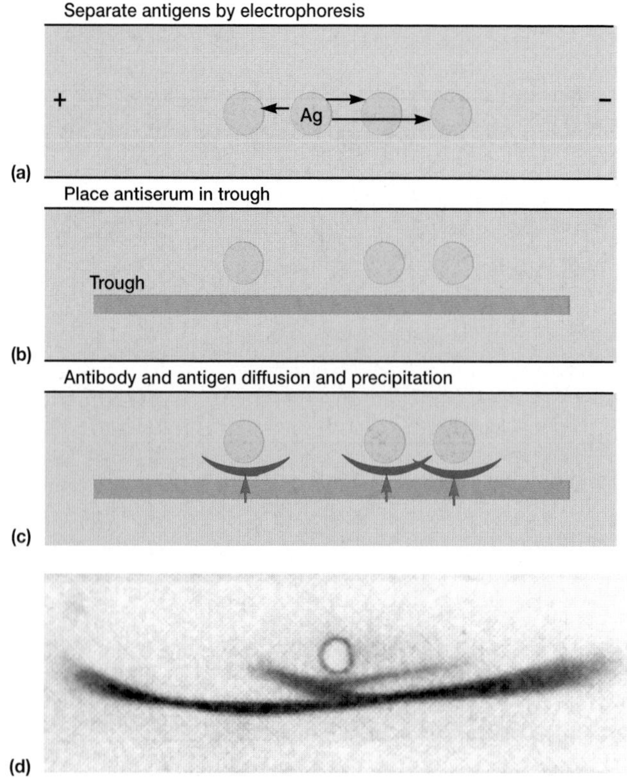

Figure 35.18 Classical Immunoelectrophoresis. **(a)** Antigens are separated in an agar gel by an electrical charge. **(b)** Antibody (antiserum) is then placed in a trough cut parallel to the direction of the antigen migration. **(c)** The antigens and antibodies diffuse through the agar and form precipitin arcs. **(d)** After staining, better visualization is possible.

the stream of cells flows past the laser beam, each fluorescent cell can be detected, counted, and even separated from the other cells in the suspension. The cytometer also can measure a cell's shape, size, and content of DNA and RNA. This technique has enabled the development of quantitative methods to assess antimicrobial susceptibility and drug cytotoxicity in a rapid, accurate, and highly reproducible way. The most outstanding contribution of this technique is the ability to detect the presence of heterogenous microbial populations with different responses to antimicrobial treatments.

Radioimmunoassay

The ***radioimmunoassay* (RIA)** technique has become an extremely important tool in biomedical research and clinical practice (e.g., in cardiology, blood banking, diagnosis of allergies, and endocrinology). Indeed, Rosalyn Yalow won the 1977 Nobel Prize in Physiology or Medicine for its development. RIA uses a purified antigen that is radioisotope-labeled and competes for antibody with unlabeled standard antigen or test antigen in experimental samples. The radioactivity associated with the antibody is then detected by means of radioisotope analyzers and autoradiography (photographic emulsions that show areas of radioactivity). If there is much antigen in an experimental sample, it will compete with the radioisotope-labeled antigen for antigen-binding sites on the antibody, and little radioactivity will be bound. A large amount of bound radioactivity indicates that there is little antigen present in the experimental sample.

1. Explain how flow cytometry is both qualitative and quantitative (i.e., it can determine both the identity and the number of a specific cell type).
2. Describe the RIA technique.

35.4 Susceptibility Testing

Many clinical microbiologists believe that determining the susceptibility of a microorganism to specific antibiotics is one of the most important tests performed in the clinical microbiology laboratory. Results can show the antibiotics to which a microorganism is most susceptible and the proper therapeutic dose needed to treat the infectious disease (*see figures 34.2b and 34.4*). Dilution susceptibility tests, disk diffusion tests (Kirby-Bauer method), the Etest, and drug concentration measurements in the blood are discussed in detail in section 34.3.

35.5 Computers in Clinical Microbiology

In the United States and other developed nations, computer systems in the clinical microbiology laboratory have replaced the handwritten mode of information acquisition and transmission. Computers improve the efficiency of the laboratory operation and increase the speed and clarity with which results can be reported. From a work-flow standpoint, the major functions involving the computer are test ordering, result entry, analysis of results, and report preparation.

Besides reporting laboratory tests, computers manage specimen logs, reports of overdue tests, quality control statistics, antimicrobial susceptibility probabilities, hospital epidemiological data, and many other items. The computer can be interfaced with various automated instruments for rapid and accurate calculation and transfer of clinical data. Direct entry into electronic notebooks is replacing written laboratory records. The personal digital assistant (PDA) has found a home in patient management, reporting histories, making diagnoses, and ordering prescriptions. The more stringent requirement for security and the increased applications will require the electronic laboratory notebook to be larger and more versatile than current PDAs, however.

1. Why is susceptibility testing so important in clinical microbiology?
2. What are some different ways in which computers can be used in the clinical microbiology laboratory? What are their major functions from the standpoint of work flow?

Summary

35.1 Specimens

a. The major focus of the clinical microbiologist is to isolate and identify microorganisms from clinical specimens accurately and rapidly. A clinical specimen represents a portion or quantity of biological material that is tested, examined, or studied to determine the presence or absence of specific microorganisms (**figure 35.1**).

b. Specimens may be collected by various methods that include swabs, needle aspiration, intubation, catheters, and clean-catch techniques. Each method is designed to ensure that only the proper material will be sent to the clinical laboratory (**figures 35.2** and **35.3**).

c. Immediately after collection the specimen must be properly handled and labeled. Speed in transporting the specimen to the clinical laboratory after it has been collected is of prime importance.

35.2 Identification of Microorganisms from Specimens

a. The clinical microbiology laboratory can provide preliminary or definitive identification of microorganisms based on (1) microscopic examination of specimens; (2) growth and biochemical characteristics of microorganisms isolated from cultures; and (3) immunologic techniques that detect antibodies or microbial antigens.

b. Viruses are identified by isolation in living cells or immunologic tests. Several types of living cells are available: cell culture, embryonated hen's eggs, and experimental animals. Rickettsial disease can be diagnosed immunologically or by isolation of the organism. Chlamydiae can be demonstrated in tissue and cell scrapings with Giemsa stain, which detects the characteristic intracellular inclusion bodies. The most routinely used techniques for identification of the mycoplasmas are immunologic. Identification of fungi often can be made if a portion of the specimen is mixed with a drop of 10% Calcofluor White stain. Wet mounts of stool specimens or urine can be examined microscopically for the presence of parasites.

c. Immunofluorescence is a process in which fluorochromes are irradiated with UV, violet, or blue light to make them fluoresce. These dyes can be coupled to an antibody. There are two main kinds of fluorescent antibody assays: direct and indirect (**figure 35.4**).

d. The initial identity of a bacterial organism may be suggested by (1) the source of the culture specimen; (2) its microscopic appearance; (3) its pattern of growth on selective, differential, enrichment, or characteristic media; and (4) its hemolytic, metabolic, and fermentative properties.

e. Rapid methods for microbial identification can be divided into three categories: (1) manual biochemical systems (**figure 35.6**), (2) mechanized/automated systems, and (3) immunologic systems.

f. Bacteriophage typing for bacterial identification is based on the fact that phage surface receptors bind to specific cell surface receptors. On a petri plate culture, bacteriophages cause plaques on lawns of bacteria with the proper receptors.

g. Various molecular methods and analyses of metabolic products also can be used to identify microorganisms. Examples include nucleic acid-based detection, gas-liquid chromatography, and plasmid fingerprinting.

35.3 Clinical Immunology

a. Serotyping refers to serological procedures used to differentiate strains (serovars or serotypes) of microorganisms that have differences in the antigenic composition of a structure or product (**figure 35.10**).

b. Agglutination reactions in vitro usually form aggregates or clumps (agglutinates) that are visible with the naked eye. Tests have been developed, such as the Widal test, latex microsphere agglutination reaction, hemagglutination, and viral hemagglutination, to detect antigen as well as to determine antibody titer (**figures 35.11** and **35.12**).

c. The complement fixation test can be used to detect a specific antibody for a suspect microorganism in an individual's serum (**figure 35.13**).

d. The enzyme-linked immunosorbent assay (ELISA) involves linking various enzymes to either antigens or antibodies. Two basic methods are involved: the double antibody sandwich method and the indirect immunosorbent assay (**figure 35.14**). The first method detects antigens and the latter, antibodies.

e. Immunoblotting involves polyacrylamide gel electrophoresis of a protein specimen followed by transfer of the separated proteins to nitrocellulose sheets and identification of specific bands by labeled antibodies (**figure 35.15**).

f. Immunoprecipitation reactions occur only when there is an optimal ratio of antigen and antibody to produce a lattice at the zone of equivalence, which is evidenced by a visible precipitate (**figure 35.16**).

g. Immunodiffusion refers to a precipitation reaction that occurs between antibody and antigen in an agar gel medium. Two techniques are routinely used: double diffusion in agar and single radial diffusion (**figure 35.17**).

h. In classical immunoelectrophoresis antigens are separated based on their electrical charge, then visualized by precipitation and staining (**figure 35.18**).

i. Flow cytometry and fluorescence allow single- or multiple-microorganism detection based on their cytometric parameters or by means of certain dyes called fluorochromes.

35.4 Susceptibility Testing

a. After the microorganism has been isolated, cultured, and/or identified, samples are used in susceptibility tests to find which method of control will be most effective. The results are provided to the physician as quickly as possible.

35.5 Computers in Clinical Microbiology

a. Computer systems in clinical microbiology are designed to replace handwritten information exchange and to speed data evaluation and report preparation.

Key Terms

agglutinates 876
bacteriophage (phage) typing 873
catheter 862
chromogen 879
clinical microbiologist 859
cytopathic effect 866
double diffusion agar assay (Öuchterlony technique) 880
enzyme-linked immunosorbent assay (ELISA) 877
flow cytometry 881
hemadsorption 866
hybridoma 864
immunoblotting 879
immunodiffusion 879
immunoelectrophoresis 881
immunofluorescence 865
immunoprecipitation 879
intubation 862
Lancefield system 876
monoclonal antibody (mAb) 864
needle aspiration 862
phagovar 873
plasmid fingerprinting 875
Quellung reaction 876
radioimmunoassay (RIA) 882
ribotyping 874
serotyping 876
single radial immunodiffusion (RID) assay 880
sputum 862
swab 862
viral hemagglutination 876
Widal test 876

Critical Thinking Questions

1. As more new ways of identifying the characteristics of microorganisms emerge, the number of distinguishable microbial strains also seems to increase. Why do you think this is the case?
2. Why are miniaturized identification systems used in clinical microbiology? Describe one such system and its advantage over classic dichotomous keys.
3. It has been speculated that as good as nucleic acid tests are in identifying viruses, antibody-based identification tests will be just as specific but easier and faster to develop and get to the marketplace. Do you agree or disagree? Defend your answer.
4. ELISA tests usually use a primary and secondary antibody. Why? What are the necessary controls one would need to perform to ensure that the antibody specificities are valid (that is, no false-positive or false-negative reactions)?

Learn More

de Las Rivas B.; Marcobal, A.; and Munoz, R. 2006. Development of a multilocus sequence typing method for analysis of *Lactobacillus plantarum* strains. *Microbiol.* 152:85–93.

Heikens, E.; Fleer, A.; Paauw, A.; Florijn, A.; and Fluit, A. C. 2005. Comparison of genotypic and phenotypic methods for species-level identification of clinical isolates of coagulase-negative staphylococci. *J. Clin. Microbiol.* 43:2286–90.

Jardi, R.; Rodriguez, F.; Buti, M.; Costa, X.; Cotrina, M.; Valdes, A.; Galimany, R.; Esteban, R.; and Guardia, J. 2001. Quantitative detection of hepatitis B virus DNA in serum by a new rapid real-time fluorescence PCR assay. *J. Virol. Hepat.* 8:465–71.

Tumpey, T. M.; Basler, C. F.; Aguilar, P. V.; Zeng, H.; Solórzano, A.; Swayne, D. E.; Cox, N. J.; Katz, J. M.; Taubenberger, J. K.; Palese, P.; and García-Sastre, A. 2005. Characterization of the reconstructed 1918 Spanish influenza pandemic virus. *Science* 310:77–80.

Please visit the Prescott website at www.mhhe.com/prescott7 for additional references.

36 The Epidemiology of Infectious Disease

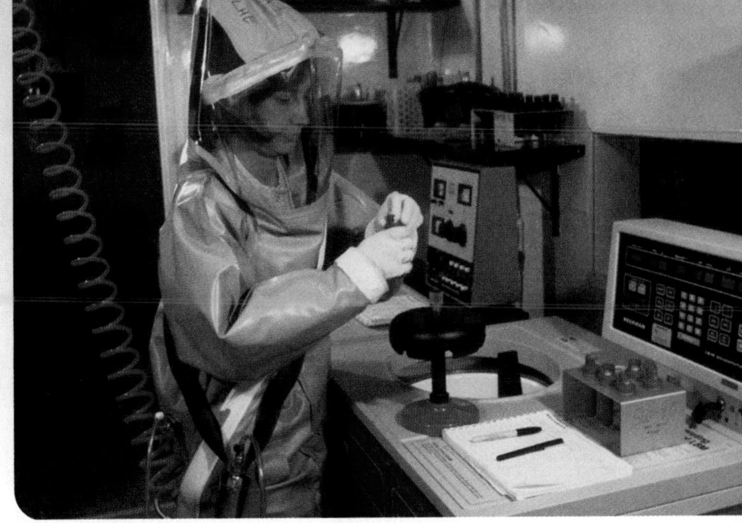

This laboratory worker at the Centers for Disease Control and Prevention (CDC) is in the highest level of isolation (Level 4) to avoid contact with microorganisms and to prevent their escape into the environment.

Preview

- The science of epidemiology deals with the occurrence and distribution of disease within a given population. Infectious disease epidemiology is concerned with organisms or agents responsible for the spread of infectious diseases in human and other animal populations.
- Statistics, an important working tool in this discipline, are used to determine morbidity and mortality rates.
- To trace the origin and manner of spread of an infectious disease outbreak, it is necessary to learn what pathogen is responsible.
- Epidemiologists identify the pathogen and investigate five links in the infectious disease cycle: (1) characteristics of the pathogen, (2) source and/or reservoir of the pathogen, (3) mode of transmission, (4) susceptibility of the host, and (5) exit mechanisms.
- Emerging and reemerging diseases and pathogens are a major global concern, as is the threat of bioterrorism.
- Bioterrorism will likely appear as a sudden increase in infectious disease—an epidemic. Newer laws now regulate use and transport of "select" agents that could be misused as agents of biocrimes or bioterrorism.
- Global travel requires global health considerations.
- The control of nosocomial (hospital acquired) infections has received increasing attention in recent years because of the number of individuals involved, increasing costs, and the length of hospital stays.

In this chapter we describe the epidemiological parameters that are studied in the infectious disease cycle. The practical goal of epidemiology is to establish effective recognition, control, prevention, and eradication measures within a given population. Because emerging and reemerging diseases and pathogens, as well as the threat of bioterrorism, are worldwide concerns, these topics are covered here. Global travel requires global health considerations that are continually monitored by epidemiologists. Nosocomial (hospital) acquired infections have increased in recent years, and a brief synopsis of their epidemiology also is presented.

The science of epidemiology originated and evolved in response to the great epidemic diseases such as cholera, typhoid fever, smallpox, and yellow fever (**Historical Highlights 36.1**). More recent epidemics of ebola, HIV/AIDS, cryptosporidiosis, enteropathogenic *Escherichia coli,* SARS, and avian influenza have underscored the importance of epidemiology in preventing global catastrophies caused by infectious diseases. Today its scope encompasses all diseases: infectious diseases, genetic abnormalities, metabolic dysfunction, malnutrition, neoplasms, psychiatric disorders, and aging. This chapter emphasizes only infectious disease epidemiology.

By definition, **epidemiology** [Greek *epi,* upon, and *demos,* people or population, and *logy,* study] is the science that evaluates the occurrence, determinants, distribution, and control of health and disease in a defined human population (**figure 36.1**). **Health** is the condition in which the organism (and all of its parts) performs its vital functions normally or properly. It is a state of physical and mental well-being and not merely the absence of disease. A **disease** [French *des,* from, and *aise,* ease] is an impairment of the normal state of an organism or any of its components that hinders the performance of vital functions. It is a response to environmental factors (e.g., malnutrition, industrial hazards, climate),

Epidemics of infectious disease are often compared with forest fires. Once fire has spread through an area, it does not return until new trees have grown up. Epidemics in humans develop when a large population of susceptible individuals is present. If most individuals are immune, then an epidemic will not occur.

—*Andrew Cliff and Peter Haggett*

Historical Highlights

36.1 John Snow—The First Epidemiologist

Much of what we know today about the epidemiology of cholera is based on the classic studies conducted by the British physician John Snow between 1849 and 1854. During this period a series of cholera outbreaks occurred in London, England, and Snow set out to find the source of the disease. Some years earlier when he was still a medical apprentice, Snow had been sent to help during an outbreak of cholera among coal miners. His observations convinced him that the disease was usually spread by unwashed hands and shared food, not by "bad" air or casual direct contact.

Thus when the outbreak of 1849 occurred, Snow believed that cholera was spread among the poor in the same way as among the coal miners. He suspected that water, and not unwashed hands and shared food, was the source of the cholera infection among the wealthier residents. Snow examined official death records and discovered that most of the victims in the Broad Street area had lived close to the Broad Street pump or had been in the habit of drinking from it. He concluded that cholera was spread by drinking water from the Broad Street pump, which was contaminated with raw sewage containing the disease agent. When the pump handle was removed, the number of cholera cases dropped dramatically.

In 1854 another cholera outbreak struck London. Part of the city's water supply came from two different suppliers: the Southwark and Vauxhall Company and the Lambeth Company. Snow interviewed cholera patients and found that most of them purchased their drinking water from the Southwark and Vauxhall Company. He also discovered that this company obtained its water from the Thames River below locations where Londoners had discharged their sewage. In contrast, the Lambeth Company took its water from the Thames before the river reached the city. The death rate from cholera was over eightfold lower in households supplied with Lambeth Company water. Water contaminated by sewage was transmitting the disease. Finally, Snow concluded that the cause of the disease must be able to multiply in water. Thus he nearly recognized that cholera was caused by a microorganism, though Robert Koch didn't discover the causative bacterium (*Vibrio cholerae*) until 1883.

To commemorate these achievements, the John Snow Pub now stands at the site of the old Broad Street pump. Those who complete the Epidemiologic Intelligence Program at the Centers for Disease Control and Prevention receive an emblem bearing a replica of a barrel of Whatney's Ale—the brew dispensed at the John Snow Pub.

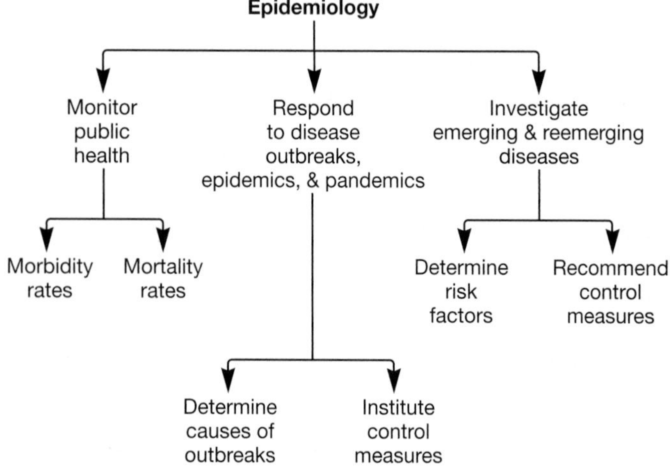

Figure 36.1 Epidemiology. Epidemiology is a multifacited science that investigates diseases to discover their origin, evaluates diseases to assess their risk, and controls diseases to prevent future outbreaks.

specific infective agents (e.g., viruses, bacteria, fungi, protozoa, helminths), inherent defects of the body (e.g., various genetic or immunologic anomalies), or combinations of these.

Any individual who practices epidemiology is an **epidemiologist.** Epidemiologists are, in effect, disease detectives. Their major concerns are the discovery of the factors essential to disease occurrence and the development of methods for disease prevention. In the United States, the Centers for Disease Control and Prevention (CDC, headquartered in Atlanta, GA) serves as the national agency for developing and applying disease prevention and control, environmental health, and health promotion and education activities. Its worldwide counterpart is the World Health Organization (WHO) located in Geneva, Switzerland.

36.1 Epidemiological Terminology

When a disease occurs occasionally, and at irregular intervals in a human population, it is a **sporadic disease** (e.g., bacterial meningitis). When it maintains a steady, low-level frequency at a moderately regular interval, it is an **endemic** [Greek *endemos,* dwelling in the same people] **disease** (e.g., the common cold). **Hyperendemic diseases** gradually increase in occurrence frequency beyond the endemic level but not to the epidemic level (e.g., the common cold during winter months). An **outbreak** is the sudden, unexpected occurrence of a disease, usually focally or in a limited segment of a population (e.g., Legionnaires' disease). An **epidemic** [Greek *epidemios,* upon the people], on the other hand, is an outbreak affecting many people at once (i.e., there is a sudden increase in the occurrence of a disease above the expected level) (**figure 36.2**). Influenza is an example of a disease that may occur suddenly and unexpectedly in a family and often achieves epidemic status in a community. The first case in an epi-

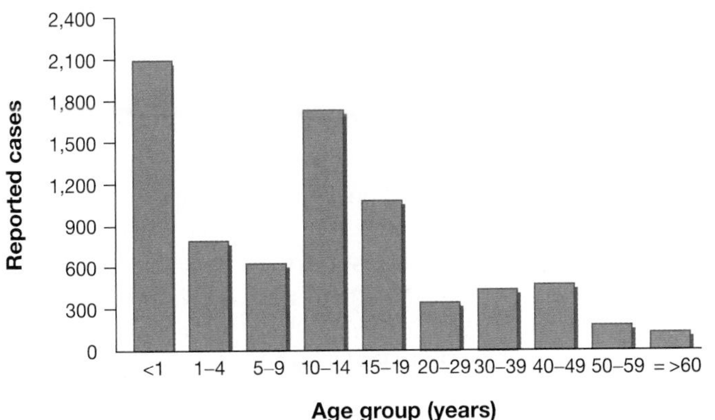

(a) Pertussis—Reported cases by age group, United States, 2000

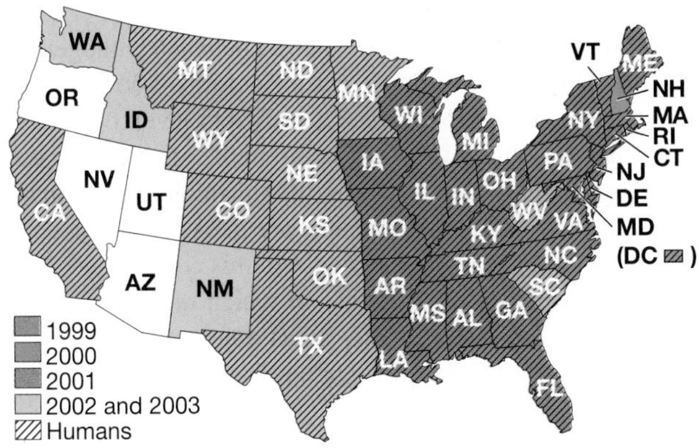

(b) West Nile virus cases were first reported in 1999 among a variety of animals and humans. Colors track its rapid progress across the United States in just 4 years.

Figure 36.2 Graphic Representation of Epidemiological Data. The Centers for Disease Control and Prevention collects and evaluates a number of parameters related to human health and disease. The data are then presented in various formats including **(a)** age and **(b)** geographic region.

demic is called the **index case.** Finally, a **pandemic** [Greek *pan,* all] is an increase in disease occurrence within a large population over a very wide region (usually the world). Usually, pandemic diseases spread among continents. The global H1N1 influenza outbreak of 1918 is a good example.

36.2 Measuring Frequency: The Epidemiologist's Tools

In order to determine if an outbreak, epidemic, or pandemic is occurring, epidemiologists must measure disease frequency at single time points and over time. The epidemiologist then uses statistics to analyze the data and determine risk factors and other factors associated with disease. **Statistics** is the branch of mathematics dealing with the collection, organization, and interpretation of numerical data. As a science particularly concerned with rates and the comparison of rates, epidemiology was the first medical field in which statistical methods were extensively used.

Measures of frequency usually are expressed as fractions. The numerator is the number of individuals experiencing the event—infection or other problem—and the denominator is the number of individuals in whom the event could have occurred, that is, the population at risk. The fraction is a proportion or ratio but is commonly called a rate because a time period is always specified. (A rate also can be expressed as a percentage.) In population statistics, rates usually are stated per 1,000 individuals, although other powers of 10 may be used for particular diseases (e.g., per 100 for very common diseases and per 10,000 or 100,000 for uncommon diseases).

A **morbidity rate** measures the number of individuals that become ill due to a specific disease within a susceptible population during a specific time interval. It is an incidence rate and reflects the number of new cases in a period. The rate is commonly determined when the number of new cases of illness in the general population is known from clinical reports. It is calculated as follows:

$$\text{Morbidity rate} = \frac{\text{number of new cases of a disease during a specified period}}{\text{number of individuals in the population}}$$

For example, if in one month there were 700 new cases of influenza per 100,000 individuals, then the morbidity rate would be expressed as 700 per 100,000 or 0.7%.

The **prevalence rate** refers to the total number of individuals infected in a population at any one time no matter when the disease began. The prevalence rate depends on both the incidence rate and the duration of the illness.

The **mortality rate** is the relationship of the number of deaths from a given disease to the total number of cases of the disease. The mortality rate is a simple statement of the proportion of all deaths that are assigned to a single cause. It is calculated as follows:

$$\text{Mortality rate} = \frac{\text{number of deaths due to a given disease}}{\text{size of the total population with the same disease}}$$

For example, if there were 15,000 deaths due to AIDS in a year, and the total number of people infected was 30,000, the mortality rate would be 15,000 per 30,000 or 1 per 2 or 50%.

The determination of morbidity, prevalence, and mortality rates aids public health personnel in directing health-care efforts to control the spread of infectious diseases. For example, a sudden increase in the morbidity rate of a particular disease may indicate a need for the implementation of preventive measures designed to reduce mortality.

1. What is epidemiology?
2. How would you define a disease? health?
3. What terms are used to describe the occurrence of a disease in a human population? in an animal population?
4. Define morbidity rate, prevalence rate, and mortality rate.

36.3 Recognition of an Infectious Disease in a Population

An infectious disease is a disease resulting from an infection by microbial agents such as viruses, bacteria, fungi, protozoa, and helminths. A **communicable disease** is an infectious disease that can be transmitted from person to person (not all infectious diseases are communicable; for example, rabies is an infectious disease acquired only through the bite of a rabid animal). The manifestations of an infectious or communicable disease can range from mild to severe to deadly depending on the agent and host. An epidemiologist studying an infectious disease is concerned with the causative agent, the source and/or reservoir of the disease agent (section 36.6), how it was transmitted, what host and environmental factors could have aided development of the disease within a defined population, and how best to control or eliminate the disease. These factors describe the natural history or cycle of an infectious disease.

Epidemiologists recognize an infectious disease in a population by using various surveillance methods. Surveillance is a dynamic activity that includes gathering information on the development and occurrence of a disease, collating and analyzing the data, summarizing the findings, and using the information to select control methods. Some combination of these surveillance methods is used most often for the:

1. Generation of morbidity data from case reports
2. Collection of mortality data from death certificates
3. Investigation of actual cases
4. Collection of data from reported epidemics
5. Field investigation of epidemics
6. Review of laboratory results: surveys of a population for antibodies against the agent and specific microbial serotypes, skin tests, cultures, stool analyses, etc.
7. Population surveys using valid statistical sampling to determine who has the disease
8. Use of animal and vector disease data
9. Collection of information on the use of specific biologics—antibiotics, antitoxins, vaccines, and other prophylactic measures
10. Use of demographic data on population characteristics such as human movements during a specific time of the year
11. Use of remote sensing and geographic information systems

As noted, surveillance may not always require the direct examination of cases. However, to accurately interpret surveillance data and study the course of a disease in individuals, epidemiologists and other medical professionals must be aware of the pattern of infectious diseases. Often infectious diseases have characteristic signs and symptoms. **Signs** are objective changes in the body, such as a fever or rash, that can be directly observed. **Symptoms** are subjective changes, such as pain and loss of appetite, that are personally experienced by the patient. The term symptom is often used in a broader scope to include the clinical signs. A **disease syndrome** is a set of signs and symptoms that are characteristic of the disease. Frequently additional laboratory tests are required for an accurate diagnosis because symptoms and readily observable signs are not sufficient for diagnosis.

The course of an infectious disease usually has a characteristic pattern and can be divided into several phases (**figure 36.3**). Knowledge of the pattern is essential in accurately diagnosing the disease.

1. The **incubation period** is the period between pathogen entry and the expression of signs and symptoms. The pathogen is spreading but has not reached a sufficient level to cause clinical manifestations. This period's length varies with disease.
2. The **prodromal stage** is the period in which there is an onset of signs and symptoms, but they are not yet specific enough to make a diagnosis. The patient often is contagious.
3. The illness period is the phase in which the disease is most severe and has characteristic signs and symptoms. The immune response has been triggered; B and T cells are becoming active.
4. In the period of decline, the signs and symptoms begin to disappear. The recovery stage often is referred to as **convalescence.**

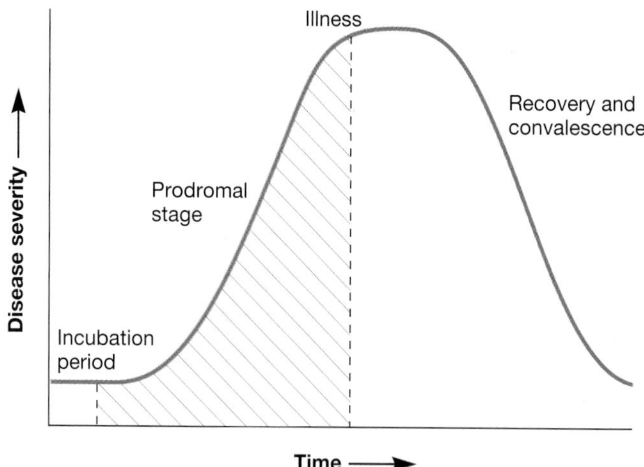

Figure 36.3 The Course of an Infectious Disease. Most infectious diseases occur in four stages. The duration of each stage is a characteristic feature of each disease. This is one of the major reasons the Centers for Disease Control and Prevention study the course of infectious diseases. The shaded area represents when a disease is typically communicable.

Remote Sensing and Geographic Information Systems: Charting Infectious Diseases

Remote sensing and geographic information systems are map-based tools that can be used to study the distribution, dynamics, and environmental correlates of microbial diseases. **Remote sensing (RS)** is the gathering of digital images of the Earth's surface from satellites and transforming the data into maps. A **geographic information system (GIS)** is a data management system that organizes and displays digital map data from RS and facilitates the analysis of relationships between mapped features. Statistical relationships often exist between mapped features and diseases in natural host or human populations. Examples include the location of the habitats of the malaria parasite and mosquito vectors in Mexico and Asia, Rift Valley fever in Kenya, Lyme disease in the United States, and African trypanosomiasis and schistosomiasis in both humans and livestock in the southeastern United States. RS and GIS may also permit the assessment of human risk from pathogens such as Sin Nombre virus (the virus that causes hantavirus pulmonary syndrome in North America). RS and GIS are most useful if disease dynamics and distributions are clearly related to mapped environmental variables. For example, if a microbial disease is associated with certain vegetation types or physical characteristics (e.g., elevation, precipitation), RS and GIS can identify regions where risk is relatively high.

Correlation with a Single Causative Agent

After an infectious disease has been recognized in a population, epidemiologists correlate the disease outbreak with a specific organism—its exact cause must be discovered (**Historical Highlights 36.2**). At this point the clinical or diagnostic microbiology laboratory enters the investigation. Its purpose is to isolate and identify the organism responsible for the disease.

36.4 Recognition of an Epidemic

As previously noted, an infectious disease epidemic is usually a short-term increase in the occurrence of the disease in a particular population. Two major types of epidemic are recognized: common source and propagated.

A **common-source epidemic** is characterized as having reached a peak level within a short period of time (1 to 2 weeks) followed by a moderately rapid decline in the number of infected patients (**figure 36.4*a***). This type of epidemic usually results from a single common contaminated source such as food (food poisoning) or water (Legionnaires' disease).

A **propagated epidemic** is characterized by a relatively slow and prolonged rise and then a gradual decline in the number of individuals infected (figure 36.4*b*). This type of epidemic usually results from the introduction of a single infected individual into a susceptible population. The initial infection is then propagated to others in a gradual fashion until many individuals within the population are infected. An example is the increase in strep throat cases that coincides with new populations of sensitive children who arrive in classrooms. Only one infected child is necessary to initiate the epidemic.

To understand how epidemics are propagated, consider **figure 36.5.** At time 0, all individuals in this population are susceptible to a hypothetical pathogen. The introduction of an infected individual initiates the epidemic outbreak (lower curve), which spreads and reaches a peak (day 15). As individuals recover from the disease, they become immune and no longer transmit the pathogen (upper curve). The number of susceptible individuals therefore decreases. The decline in the number of susceptibles to the threshold density (the minimum number of individuals necessary to continue propagating the disease) coincides with the peak of the epidemic wave, and the incidence of new cases declines because the pathogen cannot propagate itself.

Historical Highlights

36.2 "Typhoid Mary"

In the early 1900s there were thousands of typhoid fever cases, and many died of the disease. Most of these cases arose when people drank water contaminated with sewage or ate food handled by or prepared by individuals who were shedding the typhoid fever bacterium (*Salmonella enterica* serovar Typhi). The most famous carrier of the typhoid bacterium was Mary Mallon.

Between 1896 and 1906 Mary Mallon worked as a cook in seven homes in New York City. Twenty-eight cases of typhoid fever occurred in these homes while she worked in them. As a result the New York City Health Department had Mary arrested and admitted to an isolation hospital on North Brother Island in New York's East River. Examination of Mary's stools showed that she was shedding large numbers of typhoid bacteria though she exhibited no external symptoms of the disease. An article published in 1908 in the *Journal of the American Medical Association* referred to her as "Typhoid Mary," an epithet by which she is still known today. After being released when she pledged not to cook for others or serve food to them, Mary changed her name and began to work as a cook again. For five years she managed to avoid capture while continuing to spread typhoid fever. Eventually the authorities tracked her down. She was held in custody for 23 years until she died in 1938. As a lifetime carrier, Mary Mallon was positively linked with 10 outbreaks of typhoid fever, 53 cases, and 3 deaths.

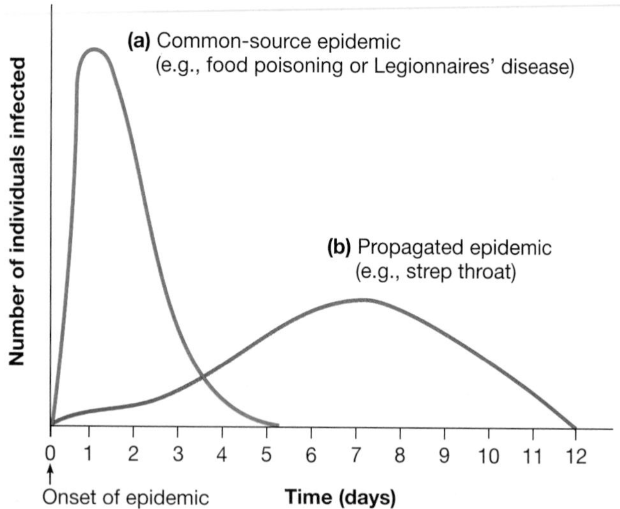

Figure 36.4 Epidemic Curves. **(a)** In a common-source epidemic, there is a rapid increase up to a peak in the number of individuals infected and then a rapid but more gradual decline. Cases usually are reported for a period that equals approximately one incubation period of the disease. **(b)** In a propagated epidemic the curve has a gradual rise and then a gradual decline. Cases usually are reported over a time interval equivalent to several incubation periods of the disease.

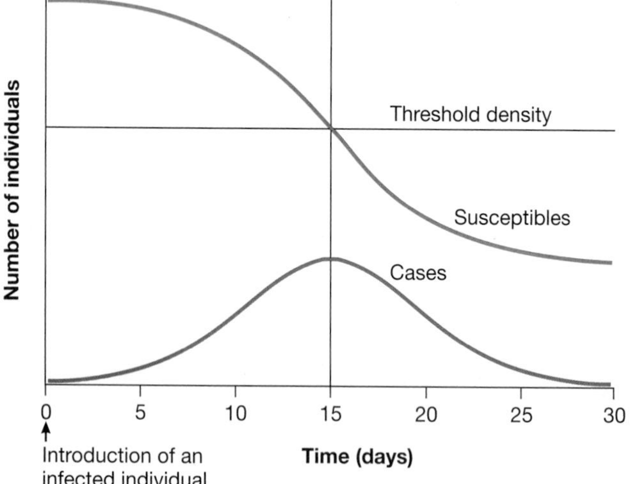

Figure 36.5 The Spread of an Imaginary Propagated Epidemic. The lower curve represents the number of cases and the upper curve the number of susceptible individuals. Notice the coincidence of the peak of the epidemic wave with the threshold density of susceptible people.

Herd immunity is the resistance of a population to infection and pathogen spread because of the immunity of a large percentage of the population. The larger the proportion of those immune, the smaller the probability of effective contact between infective and susceptible individuals—that is, many contacts will be immune, and thus the population will exhibit a group resistance. A susceptible member of such an immune population enjoys an immunity that is not of his or her own making (not self-made) but instead arises because of membership in the group.

At times public health officials immunize large portions of the susceptible population in an attempt to maintain a high level of herd immunity. Any increase in the number of susceptible individuals may result in an endemic disease becoming epidemic. The proportion of immune to susceptible individuals must be constantly monitored because new susceptible individuals continually enter a population through migration and birth. In addition, pathogens can change through processes such as antigenic shift (see next paragraph) whereby immune individuals become susceptible again.

Pathogens cause endemic diseases because infected humans continually transfer them to others (e.g., sexually transmitted diseases) or because they continually reenter the human population from animal reservoirs (e.g., rabies). Other pathogens continue to evolve and may produce epidemics (e.g., AIDS, influenza virus [A strain], and *Legionella* bacteria). One way in which a pathogen changes is by **antigenic shift,** a major genetically determined change in the antigenic character of a pathogen (**figure 36.6**). An antigenic shift can be so extensive that the pathogen is no longer recognized by the host's immune system. For example, influenza viruses frequently change by recombination from one antigenic type to another. Antigenic shift also occurs through the hybridization of different influenza virus serovars; two serovars of a virus intermingle to form a new antigenic type. Hybridization may occur between an animal strain and a human strain of the virus. Even though resistance in the human population becomes so high that the virus can no longer spread (herd immunity), it can be transmitted to animals, where the hybridization takes place. Smaller antigenic changes also can take place by mutations in pathogen strains that help the pathogen avoid host immune responses. These smaller changes are called **antigenic drift.**

Whenever antigenic shift or drift occurs, the population of susceptible individuals increases because the immune system has not been exposed to the new mutant strain. If the percentage of susceptible people is above the threshold density (figure 36.5), the level of protection provided by herd immunity will decrease and the morbidity rate will increase. For example, the morbidity rates of influenza among school children may reach epidemic levels if the number of susceptible people rises above 30% for the whole population. As a result the goal of public health agencies is to make sure that at least 70% of the population is immunized against these diseases to provide the herd immunity necessary for protection of those who are not immunized.

1. How can epidemiologists recognize an infectious disease in a population? Define sign, symptom, and disease syndrome. What are the four phases seen during the course of an infection?
2. How can remote sensing and geographic information systems chart infectious diseases?
3. Differentiate between common-source and propagated epidemics.
4. Explain herd immunity. How does this protect the community?
5. What is the significance of antigenic shift and drift in epidemiology?

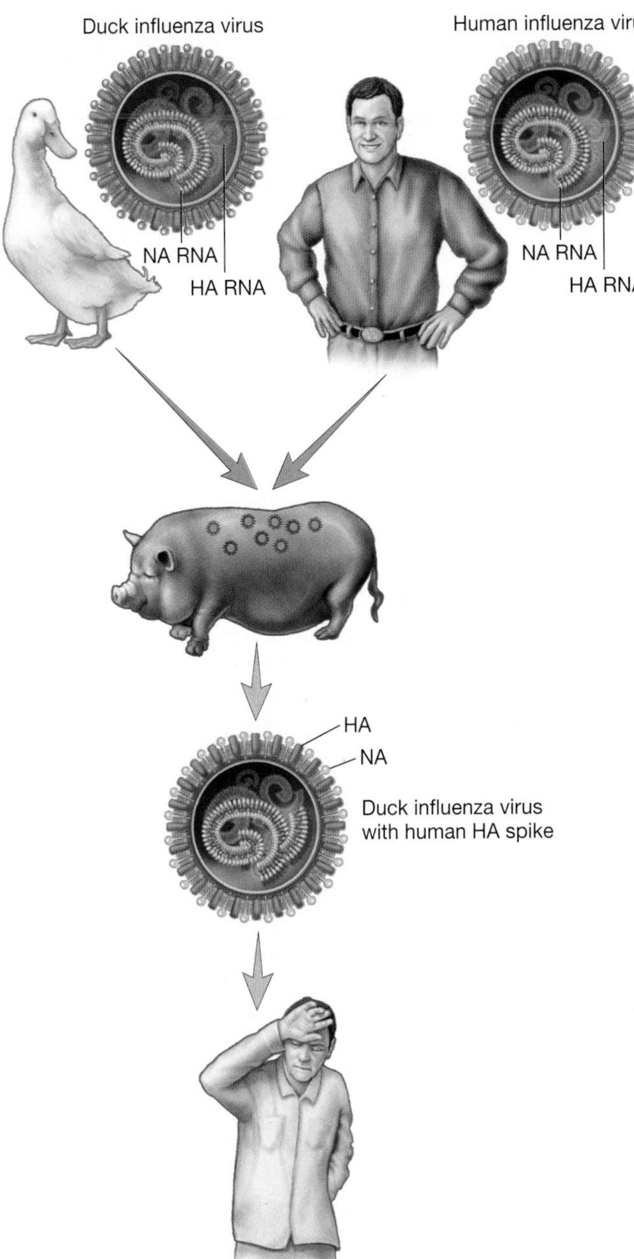

Figure 36.6 Antigenic Shift in Influenza Viruses. Close habitation of humans, pigs, and ducks permits the multiple infection of the pigs with both human and duck influenza viruses. Exchange of genetic information then results in a new strain of influenza that is new to humans; humans have no immunity to the new strain.

36.5 The Infectious Disease Cycle: Story of a Disease

To continue to exist, a pathogen must reproduce and be disseminated among its hosts. Thus an important aspect of infectious disease epidemiology is a consideration of how reproduction and dissemination occur. The **infectious disease cycle** or **chain of infection** represents these events in the form of an intriguing epidemiological mystery story (**figure 36.7**).

What Pathogen Caused the Disease?

The first link in the infectious disease cycle is the pathogen. After an infectious disease has been recognized in a population, epidemiologists must correlate the disease outbreak with a specific pathogen. The disease's exact cause must be discovered. This is where Koch's postulates, and modifications of them, are used to determine the etiology or cause of an infectious disease. At this point the clinical or diagnostic microbiology laboratory enters the investigation. Its purpose is to isolate and identify the pathogen that caused the disease and to determine the pathogen's susceptibility to antimicrobial agents or methods that may assist in its eradication. The golden age of microbiology: Koch's postulates (section 1.4); Clinical microbiology and immunology (chapter 35)

Many pathogens can cause infectious diseases in humans and are discussed in detail in chapters 37 to 39. Often these pathogens are transmissible from one individual to another resulting in a communicable disease. Pathogens have the potential to produce disease (pathogenicity); this potential is a function of such factors as the number of pathogens, their virulence, and the nature and magnitude of host defenses.

What Was the Source and/or Reservoir of the Pathogen?

The source and/or reservoir of a pathogen is the second link in the infectious disease cycle. Identifying the source and/or reservoir is an important aspect of epidemiology. If the source or reservoir of the infection can be eliminated or controlled, the infectious disease cycle itself will be interrupted and transmission of the pathogen will be prevented (Historical Highlights 36.1 and 36.2).

A **source** is the location from which the pathogen is immediately transmitted to the host, either directly through the environment or indirectly through an intermediate agent. The source can be either animate (e.g., humans or animals) or inanimate (e.g., water, soil, or food). The **period of infectivity** is the time during which the source is infectious or is disseminating the pathogen.

The **reservoir** is the site or natural environmental location in which the pathogen normally resides. It is also the site from which a source acquires the pathogen and/or where direct infection of the host can occur. Thus a reservoir sometimes functions as a source. Reservoirs also can be animate or inanimate.

Much of the time, human hosts are the most important animate sources of the pathogen and are called carriers. A **carrier** is an infected individual who is a potential source of infection for others. Carriers play an important role in the epidemiology of disease. Four types of carriers are recognized:

1. An **active carrier** is an individual who has an overt clinical case of the disease.
2. A **convalescent carrier** is an individual who has recovered from the infectious disease but continues to harbor large numbers of the pathogen.

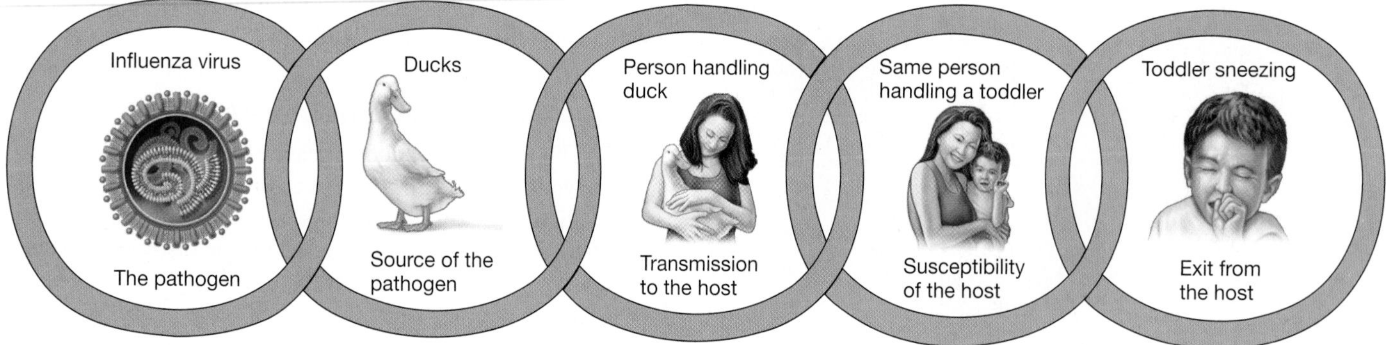

Figure 36.7 Chain of Infectious Disease.

3. A **healthy carrier** is an individual who harbors the pathogen but is not ill.
4. An **incubatory carrier** is an individual who is incubating the pathogen in large numbers but is not yet ill.

Convalescent, healthy, and incubatory carriers may harbor the pathogen for only a brief period (hours, days, or weeks) and then are called **casual, acute,** or **transient carriers.** If they harbor the pathogen for long periods (months, years, or life), they are called **chronic carriers.**

Animal diseases that can be transmitted to humans are termed **zoonoses** (Greek *zoon,* animal, and *nosos,* disease); thus animals also can serve as reservoirs. Humans contract the pathogen by several mechanisms: coming into direct contact with diseased animal flesh (e.g., tularemia); drinking contaminated cow's milk (e.g., tuberculosis and brucellosis); inhaling dust particles contaminated by animal excreta or products (e.g., Q fever, hantavirus pulmonary infection, anthrax); or eating insufficiently cooked infected flesh (e.g., anthrax, trichinosis). In addition, being bitten by arthropod **vectors** (organisms that spread disease from one host to another) such as mosquitoes, ticks, fleas, mites, or biting flies (e.g., equine encephalomyelitis, malaria, Lyme disease, Rocky Mountain spotted fever, plague, scrub typhus, and tularemia); or being bitten by a diseased animal (e.g., rabies) can lead to infection.

Table 36.1 lists some common zoonoses found in the Western Hemisphere. This table is noninclusive in scope; it merely abbreviates the enormous spectrum of zoonotic diseases that are relevant to human epidemiology. Domestic animals are the most common source of zoonoses because they live in greater proximity to humans than do wild animals. Diseases of wild animals that are transmitted to humans tend to occur sporadically because close contact is infrequent. Other major reservoirs of pathogens are water, soil, and food. These reservoirs are discussed in detail in chapters 40 and 41.

How Was the Pathogen Transmitted?

To maintain an active infectious disease in a human population, the pathogen must be transmitted from one host or source to another. Transmission is the third link in the infectious disease cycle and occurs by four main routes: airborne, contact, vehicle, and vector-borne (**figure 36.8**).

Airborne Transmission

Because air is not a suitable medium for the growth of pathogens, any pathogen that is airborne must have originated from a source such as humans, other animals, plants, soil, food, or water. In **airborne transmission** the pathogen is truly suspended in the air and travels over a meter or more from the source to the host. The pathogen can be contained within droplet nuclei or dust. **Droplet nuclei** can be small particles, 1 to 4 μm in diameter, that result from the evaporation of larger particles (10 μm or more in diameter) called droplets. Droplet nuclei can remain airborne for hours or days and travel long distances. Chicken pox and measles are examples of droplet-spread diseases.

When animals or humans are the source of the airborne pathogen, it usually is propelled from the respiratory tract into the air by an individual's coughing, sneezing, or vocalization. For example, enormous numbers of moisture droplets are aerosolized during a typical sneeze (**figure 36.9**). Each droplet is about 10 μm in diameter and initially moves about 100 m/second or more than 200 mi/hour!

Dust also is an important route of airborne transmission. At times a pathogen adheres to dust particles and contributes to the number of airborne pathogens when the dust is resuspended by some disturbance. A pathogen that can survive for relatively long periods in or on dust creates an epidemiological problem, particularly in hospitals, where dust can be the source of hospital acquired infections. **Table 36.2** summarizes some human airborne pathogens and the diseases they cause.

Contact Transmission

Contact transmission implies the coming together or touching of the source or reservoir of the pathogen and the host (**Historical Highlights 36.3**). Contact can be direct or indirect. Direct contact implies an actual physical interaction with the infectious source (figure 36.8). This route is frequently called person-to-person contact. Person-to-person transmission occurs primarily by touching, kissing, or sexual contact (sexually transmitted diseases); by contact with oral secretions or body lesions (e.g., herpes and boils);

Table 36.1 Infectious Organisms in Nonhuman Reservoirs That May Be Transmitted to Humans

Disease	Etiologic Agent	Usual or Suspected Nonhuman Host	Usual Method of Human Infection
Anthrax	*Bacillus anthracis*	Cattle, horses, sheep, swine, goats, dogs, cats, wild animals, birds	Inhalation or ingestion of spores; direct contact
Babesiosis	*Babesia bovis, B. divergens, B. microti, B. equi*	*Ixodes* ticks of various species	Bite of infected tick
Brucellosis (undulant fever)	*Brucella melitensis, B. abortus, B. suis*	Cattle, goats, swine, sheep, horses, mules, dogs, cats, fowl, deer, rabbits	Milk; direct or indirect contact
Campylobacteriosis	*Campylobacter fetus, C. jejuni*	Cattle, sheep, poultry, swine, pets	Contaminated water and food
Cat-scratch disease	*Bartonella henselae*	Cats, dogs	Cat or dog scratch
Colorado tick fever	*Coltivirus*	Squirrels, chipmunks, mice, deer	Tick bite
Cowpox	Cowpox virus	Cattle, horses	Skin abrasions
Cryptosporidiosis	*Cryptosporidium* spp.	Farm animals, pets	Contaminated water
Encephalitis (California)	Arbovirus	Rats, squirrels, horses, deer, hares, cows	Mosquito
Encephalitis (St. Louis)	Arbovirus	Birds	Mosquito
Encephalomyelitis (Eastern equine)	Arbovirus	Birds, ducks, fowl, horses	Mosquito
Encephalomyelitis (Venezuelan equine)	Arbovirus	Rodents, horses	Mosquito
Encephalomyelitis (Western equine)	Arbovirus	Birds, snakes, squirrels, horses	Mosquito
Giardiasis	*Giardia intestinalis*	Rodents, deer, cattle, dogs, cats	Contaminated water
Glanders	*Burkholderia mallei*	Horses	Skin contact; inhalation
Hantavirus pulmonary syndrome	Pulmonary syndrome hantavirus	Deer mice	Contact with the saliva, urine, or feces of deer mice; aerosolized viruses
Herpes B viral encephalitis	*Herpesvirus simiae*	Monkeys	Monkey bite; contact with material from monkeys
Influenza	Influenza virus	Water fowl, pigs	Direct contact or inhalation
Leptospirosis	*Leptospira interrogans*	Dogs, rodents, wild animals	Direct contact with urine, infected tissue, and contaminated water
Listeriosis	*Listeria monocytogenes*	Sheep, cattle, goats, guinea pigs, chickens, horses, rodents, birds, crustaceans	Food-borne
Lyme disease	*Borrelia burgdorferi*	Ticks (*Ixodes scapularis* or related ticks)	Bite of infected tick
Lymphocytic choriomeningitis	Arenavirus	Mice, rats, dogs, monkeys, guinea pigs	Inhalation of contaminated dust; ingestion of contaminated food
Mediterranean fever (boutonneuse fever, African tick typhus)	*Rickettsia conorii*	Dogs	Tick bite
Melioidosis	*Burkholderia pseudomallei*	Rats, mice, rabbits, dogs, cats	Arthropod vectors, water, food
Orf (contagious ecthyma)	*Parapoxvirus*	Sheep, goats	Through skin abrasions
Pasteurellosis	*Pasteurella multocida*	Fowl, cattle, sheep, swine, goats, mice, rats, rabbits	Animal bite

(continued)

Table 36.1 Infectious Organisms in Nonhuman Reservoirs That May Be Transmitted to Humans, *(Continued)*

Plague (bubonic)	*Yersinia pestis*	Domestic rats, many wild rodents	Flea bite
Psittacosis	*Chlamydia psittaci*	Birds	Direct contact, respiratory aerosols
Q fever	*Coxiella burnetii*	Cattle, sheep, goats	Inhalation of contaminated soil and dust
Rabies	Rabies virus	Dogs, bats, opposums, skunks, raccoons, foxes, cats, cattle	Bite of rabid animal
Rat bite fever	*Spirillum minus*	Rats, mice, cats	Rat bite
	Streptobacillus moniliformis	Rats, mice, squirrels, weasels, turkeys, contaminated food	Rat bite
Relapsing fever (borreliosis)	*Borrelia* spp.	Rodents, porcupines, opposums, armadillos, ticks, lice	Tick or louse bite
Rickettsialpox	*Rickettsia akari*	Mice	Mite bite
Rocky Mountain spotted fever	*Rickettsia rickettsii*	Rabbits, squirrels, rats, mice, groundhogs	Tick bite
Salmonellosis	*Salmonella* spp. (except *S. typhosa*)	Fowl, swine, sheep, cattle, horses, dogs, cats, rodents, reptiles, birds, turtles	Direct contact; food
SARS	SARS coronavirus	Bats, civits	Contact with infected animal or person
Scrub typhus	*Rickettsia tsutsugamushi*	Wild rodents, rats	Mite bite
Tuberculosis	*Mycobacterium bovis, M. tuberculosis*	Cattle, horses, cats, dogs	Milk; direct contact
Tularemia	*Francisella tularensis*	Wild rabbits, most other wild and domestic animals	Direct contact with infected carcass, usually rabbit; tick bite, biting flies
Typhus fever (endemic)	*Rickettsia mooseri*	Rats	Flea bite
Vesicular stomatitis	Vesicular stomatitis virus	Cattle, swine, horses	Direct contact
Weil's disease (leptospirosis)	*Leptospira interrogans*	Rats, mice, skunks, opposums, wildcats, foxes, raccoons, shrews, bandicoots, dogs, cattle, swine	Through skin, drinking water, eating food
Yellow fever (jungle)	Yellow fever virus	Monkeys, marmosets, lemurs, mosquitoes	Mosquito

by nursing mothers (e.g., staphylococcal infections); and through the placenta (e.g., AIDS, syphilis). Some infectious pathogens also can be transmitted by direct contact with animals or animal products (e.g., *Salmonella* and *Campylobacter*).

Indirect contact refers to the transmission of the pathogen from the source to the host through an intermediary—most often an inanimate object. The intermediary is usually contaminated by an animate source. Common examples of intermediary inanimate objects include thermometers, eating utensils, drinking cups, stethoscopes, and neckties. *Pseudomonas* bacteria are easily transmitted by this route. This mode of transmission is often also considered a form of vehicle transmission (see next section).

In droplet spread the pathogen is carried on particles smaller than 5 μm. The route is through the air but only for a very short distance—usually less than a meter. As a result droplet transmission of a pathogen depends on the proximity of the source and the host. Contact with oral secretions may also result when droplet nuclei contaminate body surfaces that touch mucous membranes (e.g., respiratory secretions on hands that contact eyes).

Vehicle Transmission

Inanimate materials or objects involved in pathogen transmission are called **vehicles.** In **common vehicle transmission** a single inanimate vehicle or source serves to spread the pathogen to multiple hosts but does not support its reproduction. Examples include surgical instruments, bedding, and eating utensils. In epidemiology these common vehicles are called **fomites** [s., fomes or fomite]. A single source containing pathogens (e.g., blood, drugs, IV fluids) can contaminate a common vehicle that causes multiple infections. Food and water are important common vehicles for many human diseases.

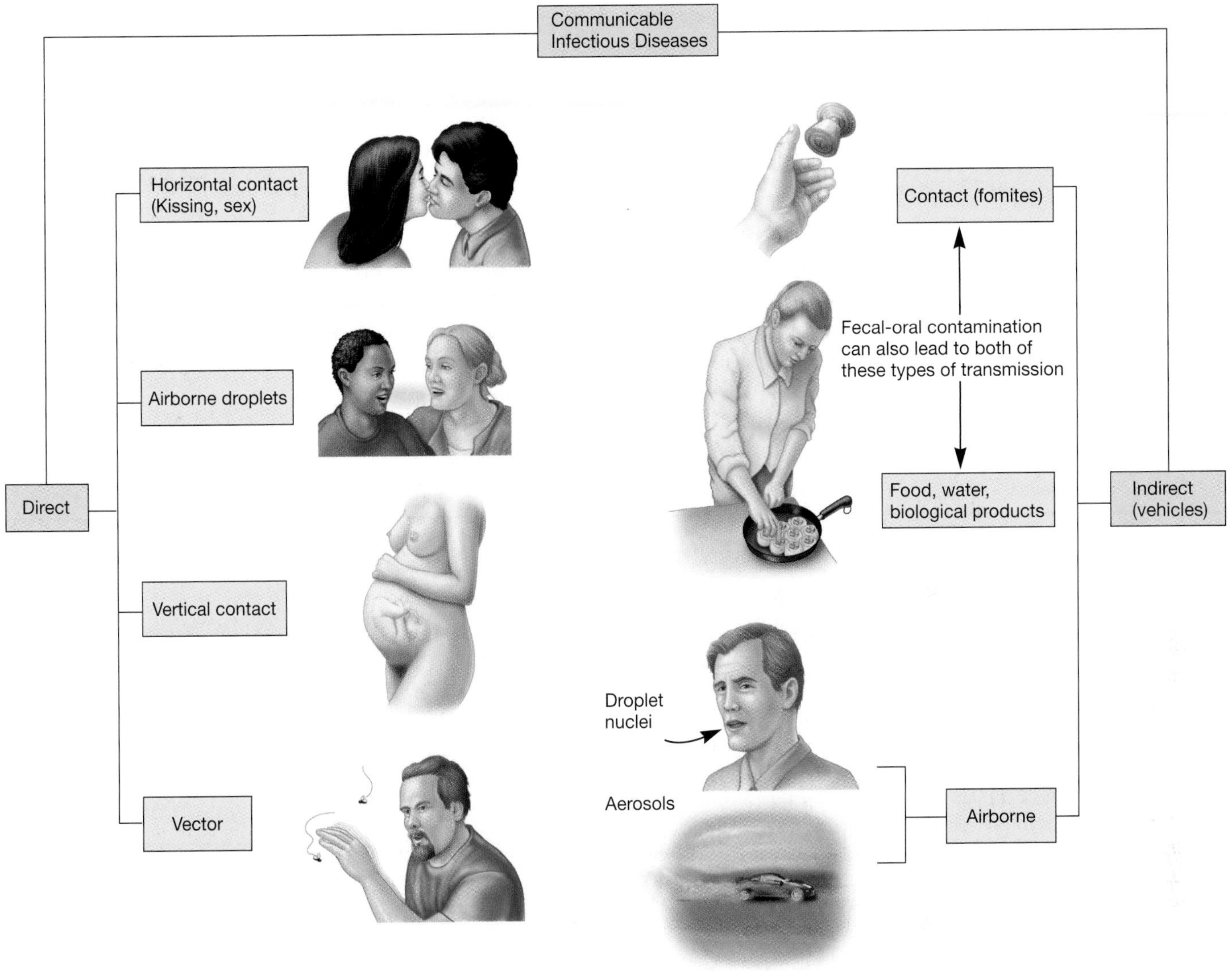

Communicable Infectious Diseases

Direct

Horizontal contact (Kissing, sex)

Airborne droplets

Vertical contact

Vector

Indirect (vehicles)

Contact (fomites)

Fecal-oral contamination can also lead to both of these types of transmission

Food, water, biological products

Droplet nuclei

Aerosols

Airborne

Figure 36.8 Transmission of Communicable Disease. Infectious diseases can be transmitted from person to person by various direct and indirect methods.

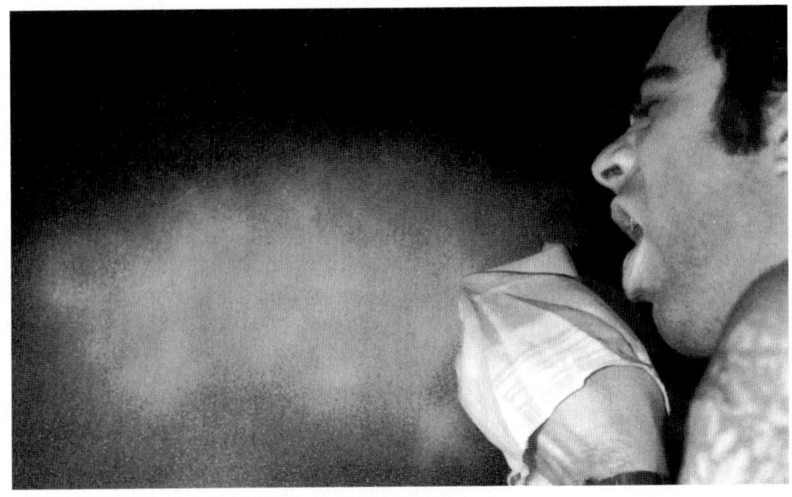

Figure 36.9 A Sneeze. High-speed photograph of an aerosol generated by an unstifled sneeze. The particles seen are comprised of saliva and mucus laden with microorganisms. These airborne particles may be infectious when inhaled by a susceptible host. Even a surgical mask will not prevent the spread of all particles.

Table 36.2 Some Airborne Pathogens and the Diseases They Cause in Humans

Microorganisms	Disease	Microorganism	Disease
Viruses		**Bacteria**	
Varicella	Chickenpox	*Actinomyces* spp.	Lung infections
Influenza	Flu (Influenza)	*Bordetella pertussis*	Whooping cough
Rubeola	Measles	*Chlamydia psittaci*	Psittacosis
Rubella	German measles	*Corynebacterium diphtheriae*	Diphtheria
Mumps	Mumps	*Mycoplasma pneumoniae*	Pneumonia
Polio	Poliomyelitis	*Mycobacterium tuberculosis*	Tuberculosis
Acute respiratory viruses	Viral pneumonia	*Neisseria meningitidis*	Meningitis
Pulmonary syndrome hantavirus	Hantavirus pulmonary syndrome	*Streptococcus* spp.	Pneumonia, sore throat
Variola	Smallpox	**Fungi**	
		Blastomyces spp.	Lung infections
		Coccidioides spp.	Coccidioidomycosis
		Histoplasma capsulatum	Histoplasmosis

Historical Highlights

36.3 The First Indications of Person-to-Person Spread of an Infectious Disease

In 1773 Charles White, an English surgeon and obstetrician, published his "Treatise on the Management of Pregnant and Lying-In Women." In it, he appealed for surgical cleanliness to combat childbed or puerperal fever. (Puerperal fever is an acute febrile condition that can follow childbirth and is caused by streptococcal infection of the uterus and/or adjacent regions.) In 1795 Alexander Gordon, a Scottish obstetrician, published his "Treatise on the Epidemic Puerperal Fever of Aberdeen," which demonstrated for the first time the contagiousness of the disease. In 1843 Oliver Wendell Holmes, a noted physician and anatomist in the United States, published a paper entitled "On the Contagiousness of Puerperal Fever" and also appealed for surgical cleanliness to combat this disease.

However, the first person to realize that a pathogen could be transmitted from one person to another was the Hungarian physician Ignaz Phillip Semmelweis. Between 1847 and 1849 Semmelweis observed that women who had their babies at the hospital with the help of medical students and physicians were four times as likely to contract puerperal fever as those who gave birth with the help of midwives. He concluded that the physicians and students were infecting women with material remaining on their hands after autopsies and other activities. Semmelweis thus began washing his hands with a calcium chloride solution before examining patients or delivering babies. This simple procedure led to a dramatic decrease in the number of cases of puerperal fever and saved the lives of many women. As a result Semmelweis is credited with being the pioneer of antisepsis in obstetrics. Unfortunately, in his own time, most of the medical establishment refused to acknowledge his contribution and adopt his procedures. After years of rejection Semmelweis had a nervous breakdown in 1865. He died a short time later of a wound infection. It is very probable that it was a streptococcal infection, arising from the same pathogen he had struggled against his whole professional life.

Vector-Borne Transmission

As noted earlier, living transmitters of a pathogen are called vectors. Most vectors are arthropods (e.g., insects, ticks, mites, fleas) or vertebrates (e.g., dogs, cats, skunks, bats). **Vector-borne transmission** can be either external or internal. In external (mechanical) transmission the pathogen is carried on the body surface of a vector. Carriage is passive, with no growth of the pathogen during transmission. An example would be flies carrying *Shigella* organisms on their feet from a fecal source to a plate of food that a person is eating.

In internal transmission the pathogen is carried within the vector. Here it can go into either a harborage or biologic transmission phase. In **harborage transmission** the pathogen does not undergo morphological or physiological changes within the vector. An example would be the transmission of *Yersinia pestis* (the etiologic agent of plague) by the rat flea from rat to human. **Biologic transmission** implies that the pathogen does go through a morphological or physiological change within the vector. An example would be the developmental sequence of the malarial parasite inside its mosquito vector. Arthropod-borne diseases: Malaria (section 39.3)

Why Was the Host Susceptible to the Pathogen?

The fourth link in the infectious disease cycle is the host. The susceptibility of the host to a pathogen depends on both the pathogenicity of the organism and the nonspecific and specific defense mechanisms of the host. These susceptibility factors are the basis for chapters 31 and 32. In addition to host defense mechanisms, issues of nutrition, genetic predisposition, and stress are also germaine when considering host susceptibility to infection.

How Did the Pathogen Leave the Host?

The fifth and last link in the infectious disease cycle is release or exit of the pathogen from the host. From the point of view of the pathogen, successful escape from the host is just as important as its initial entry. Unless a successful escape occurs, the disease cycle will be interrupted and the pathogenic species will not be perpetuated. Escape can be active or passive, although often a combination of the two occurs. Active escape takes place when a pathogen actively moves to a portal of exit and leaves the host. Examples include the many parasitic helminths that migrate through the body of their host, eventually reaching the surface and exiting. Passive escape occurs when a pathogen or its progeny leaves the host in feces, urine, droplets, saliva, or desquamated cells. Microorganisms usually employ passive escape mechanisms.

1. What are some epidemiologically important characteristics of a pathogen? What is a communicable disease?
2. Define source, reservoir, period of infectivity, and carrier.
3. What types of infectious disease carriers does epidemiology recognize?
4. Describe the four main types of infectious disease transmission and give examples of each.
5. Define the terms droplet nuclei, vehicle, fomite, and vector.

36.6 Virulence and the Mode of Transmission

There is evidence that a pathogen's virulence may be strongly influenced by its mode of transmission and ability to live outside its host. When the pathogen uses a mode of transmission such as direct contact, it cannot afford to make the host so ill that it will not be transmitted effectively. This is the case with the common cold, which is caused by rhinoviruses and several other respiratory viruses. If the virus reproduced too rapidly and damaged its host extensively, the person would be bedridden and not contact others. The efficiency of transmission would drop because rhinoviruses shed from the cold sufferer could not contact new hosts and would be inactivated by exposure. Cold sufferers must be able to move about and directly contact others. Thus virulence is low and people are not incapacitated by the common cold.

On the other hand, if a pathogen uses a mode of transmission not dependent on host health and mobility, then the person's health will not be a critical matter. The pathogen might be quite successful—that is, transmitted to many new hosts even though it kills its host relatively quickly. Host death will mean the end of any resident pathogens, but the species as a whole can spread and flourish as long as the increased transmission rate outbalances the loss due to host death. This situation may arise in several ways.

When a pathogen is transmitted by a vector, it will benefit by extensive reproduction and spread within the host. If pathogen levels are very high in the host, a vector such as a biting insect has a better chance of picking up the pathogen and transferring it to a new host. Indeed, pathogens transmitted by biting arthropods often are very virulent (e.g., malaria, typhus, sleeping sickness). It is important that such pathogens do not harm their vectors, and the vector generally remains healthy, at least long enough for pathogen transmission.

Virulence also is often directly correlated with a pathogen's ability to survive in the external environment. If a pathogen cannot survive well outside its host and does not use a vector, it depends on host survival and will tend to be less virulent. When a pathogen can survive for long periods outside its host, it can afford to leave the host and simply wait for a new one to come along. This seems to promote increased virulence. Host health is not critical, but extensive multiplication within the host will increase the efficiency of transmission. Good examples are tuberculosis and diphtheria. *Mycobacterium tuberculosis* and *Corynebacterium diphtheriae* survive for a long time, at least weeks to months, outside human hosts.

Human cultural patterns and behavior almost certainly also affect pathogen virulence. Waterborne pathogens such as *Vibrio cholerae* (which causes cholera) are transmitted through drinking water systems. They can be virulent because immobile hosts still shed pathogens, which frequently reach the water. This is why the establishment of an uncontaminated drinking water supply is critical in limiting a cholera outbreak, particularly in refugee camps. The same appears to be true of *Shigella* and dysentery. Often one of the best ways to reduce virulence may be to reduce the frequency of transmission.

36.7 Emerging and Reemerging Infectious Diseases and Pathogens

Only a few decades ago, a grateful public trusted that science had triumphed over infectious diseases by building a fortress of health protection. Antibiotics, vaccines, and aggressive public health campaigns had yielded a string of victories over old enemies like whooping cough, pneumonia, polio, and smallpox. In developed countries, people were lulled into believing that microbial threats were a thing of the past. Trends in the number of deaths caused by infectious diseases in the United States from 1900 through 1982 supported this conclusion (**figure 36.10**). However, this downward trend ended in 1982 and the death rate has since risen. The world has seen the global spread of AIDS, the resurgence of tuberculosis, and the appearance of new enemies like hantavirus pulmonary syndrome, hepatitis C and E, Ebola virus, Lyme disease, cryptosporidiosis, and *E. coli* O157:H7. In addition, during this same time period:

- A "bird flu" virus that had never before attacked humans began to kill people in southeast Asia.
- A new variant of a fatal prion disease of the brain, Creutzfeldt-Jakob disease was identified in the United

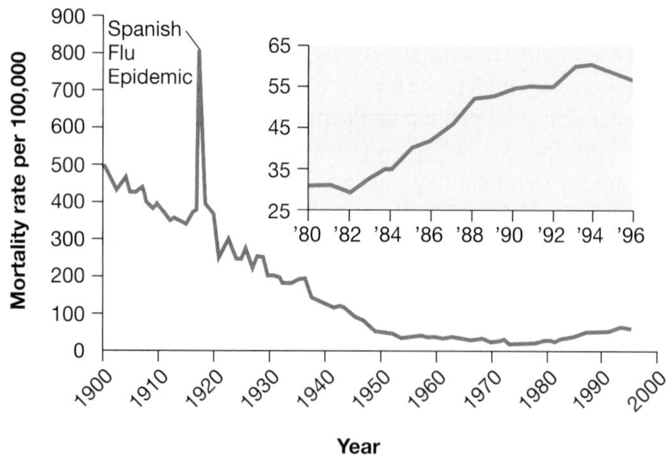

Figure 36.10 Infectious Disease Mortality in the United States Decreased Greatly during Most of the Twentieth Century. The insert is an enlargement of the right-hand portion of the graph and shows that the death rate from infectious diseases increased between 1980 and 1994.

Kingdom, transmitted by beef from animals with "mad cow disease."

- *Staphylococcus* bacteria with resistance to methicillin and vancomycin, long the antibiotics of first choice and last resort, respectively, were seen for the first time.
- Several major multistate foodborne outbreaks occurred in the United States, including those caused by parasites on raspberries, viruses on strawberries, and bacteria in produce, ground beef, cold cuts, and breakfast cereal.
- A new strain of tuberculosis that is resistant to many drugs, and occurs most often in people infected with HIV, arose in the city of New York and other large cities.

By the 1990s, the idea that infectious diseases no longer posed a serious threat to human health was obsolete. In the 21st century, it is clear that globally, humans will continually be faced with both new infectious diseases and the reemergence of older diseases once thought to be conquered (e.g., tuberculosis, dengue hemorrhagic fever, yellow fever). William McNeill, in *Plagues and Peoples* (1976), addresses this problem as follows: "Ingenuity, knowledge, and organization alter but cannot cancel humanity's vulnerability to invasion by parasitic forms of life. Infectious disease which antedated the emergence of humankind will last as long as humanity itself, and will surely remain, as it has been hitherto, one of the fundamental parameters and determinants of human history."

The Centers for Disease Control and Prevention has defined these diseases as "new, reemerging, or drug-resistant infections whose incidence in humans has increased within the past three decades or whose incidence threatens to increase in the near future." Some of the most recent examples of these diseases are shown in **figure 36.11**.The increased importance of emerging and reemerging infectious diseases has stimulated the establishment of a field called **systematic epidemiology,** which focuses on the ecological and social factors that influence the development of these diseases.

After a century marked by dramatic advances in medical research and drug discovery, technology development, and sanitation, why are viruses, bacteria, fungi, and parasites posing such a problem and challenge? Many factors characteristic of the modern world undoubtedly favor the development and spread of these microorganisms and their diseases. Examples include:

1. Unprecedented worldwide population growth, population shifts (demographics), and urbanization
2. Increased international travel
3. Increased worldwide transport (commerce), migration, and relocation of animals and food products
4. Changes in food processing, handling, and agricultural practices
5. Changes in human behavior, technology, and industry
6. Human encroachment on wilderness habitats that are reservoirs for insects and animals that harbor infectious agents
7. Microbial evolution (e.g., selective pressure) and the development of resistance to antibiotics and other antimicrobial drugs (e.g., penicillin-resistant *Streptococcus pneumoniae,* meticillin-resistant *Staphylococcus aureus,* and vancomycin-resistant enterococci)
8. Changes in ecology and climate
9. Modern medicine (e.g., immunosuppression)
10. Inadequacy of public infrastructure and vaccination programs
11. Social unrest and civil wars
12. Bioterrorism (section 36.9)
13. Virulence-enhancing mechanisms of pathogens (e.g., the mobile genetic elements—bacteriophages, plasmids, transposons)

As population density increases in cities, the dynamics of microbial exposure and evolution increase in humans themselves. Urbanization often crowds humans and increases exposure to microorganisms. Crowding leads to unsanitary conditions and hinders the effective implementation of adequate medical care, enabling more widespread transmission and propagation of pathogens. In modern societies, crowded workplaces, community-living settings, day-care centers, large hospitals, and public transportation all facilitate microbial transmission. Furthermore, land development and the exploration and destruction of natural habitats have increased the likelihood of human exposure to new pathogens and may put selective pressures on pathogens to adapt to new hosts and changing environments. The introduction of pathogens to a new environment or host can alter transmission and exposure patterns, leading to sudden proliferation of disease. For example, the spread of Lyme disease in New England probably was due partly to ecological disruption that eliminated predators of deer. An increase in deer and the deer tick populations provided a favorable situation for pathogen spread to humans. Whenever there is alteration of the

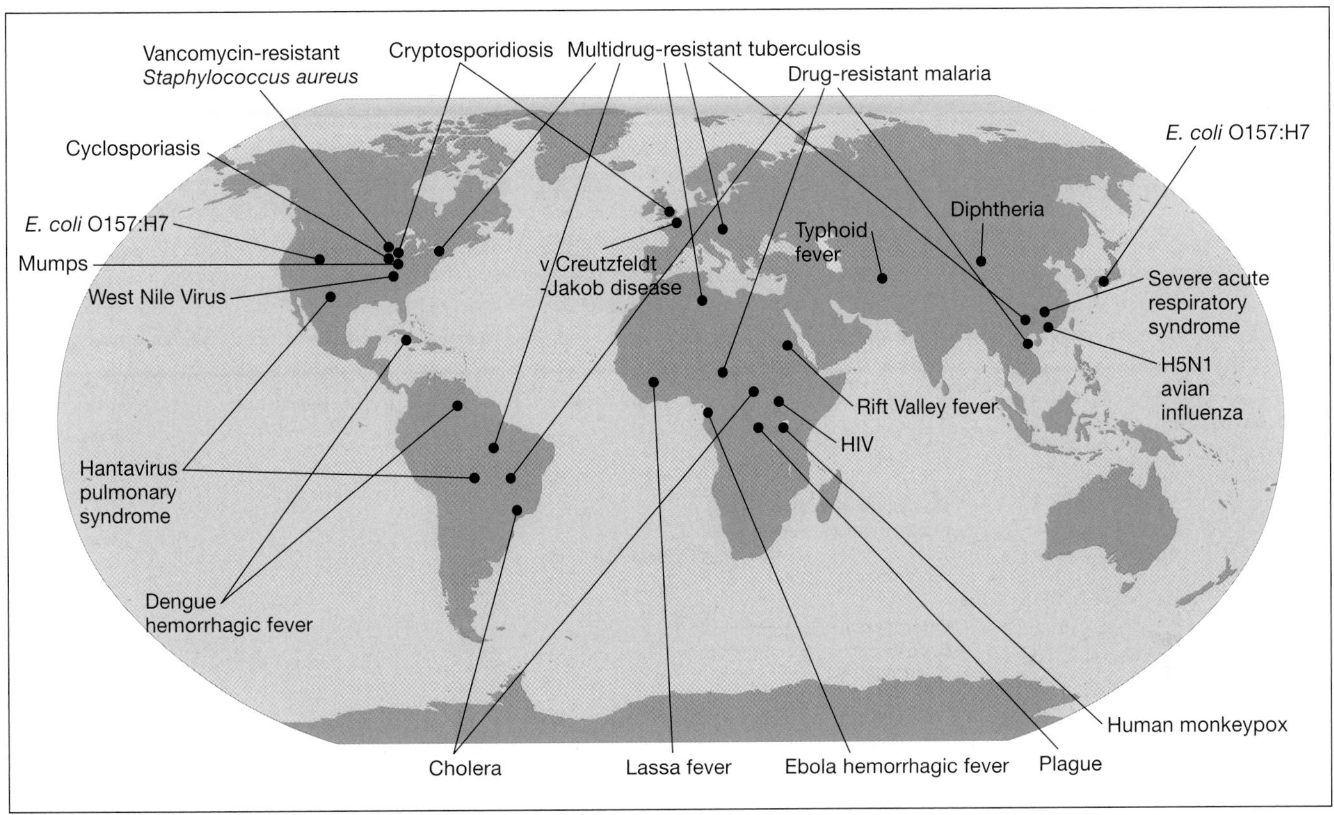

Figure 36.11 Some Examples of Emerging and Reemerging Infectious Diseases. Although diseases such as HIV are indicated in only one or two significant locations, they are very widespread and a threat in many regions.

environment and new environments are created, this may not only confer a survival advantage, but may also increase a pathogen's virulence and alter its drug susceptibility profile. When there are changes in climate or ecology, it should not be surprising to find changes in both beneficial and detrimental microorganisms. Global warming also affects microorganism selection and survival. Finally, mass migrations of refugees, workers, and displaced persons have led to a steady growth of urban centers at the expense of rural areas.

Microbiologists are all too familiar with the development of resistance to antibiotics used in human medicine. The distribution of nosocomial pathogens (**figure 36.12**) has changed throughout the antibiotic era. Hospital acquired infections were dominated early on by staphylococci, which initially responded to penicillin. During subsequent years, the emergence of meticillin-resistant *Staphylococcus aureus* (MRSA) increased dramatically; similar patterns are emerging for penicillin-resistant *Streptococcus pneumoniae.* Vancomycin-resistant *S. aureus* infections are now a major nosocomial threat. Glycopeptide-resistant *Enterococcus faecium* was first reported in the late 1980s; vancomycin-resistant enterococci (VRE) are now common in U.S. hospitals. Newly recognized nosocomial, gram-positive species include *Corynebacterium jeikeium* and *Rhodococcus equi.* The incidences of infections by the gram-negative pathogens *Pseudomonas aeruginosa* and *Acinetobacter* ssp. and the recently renamed gram-negative bacterial pathogens *Burkholderia cepacia* and *Stenotrophomonas maltophilia* have increased. With respect to the extended-spectrum β-lactam-resistant gram-negative bacilli, bacteria such as *Klebsiella pneumoniae, E. coli,* other *Klebsiella* spp., *Proteus* spp., *Morganella* spp., *Citrobacter* spp., *Salmonella* spp., and *Serratia marcescens* are resistant to penicillins; first-generation cephalosporins; and some third-generation cephalosporins such as cefotaxime (Claforan), ceftriaxone (Rocephin), ceftazidime, and aztreonam (Azactam).

Without a doubt, the key factors responsible for the rise in drug-resistant pathogens have been the excessive or inappropriate use of antimicrobial therapy and the sometimes indiscriminant use of broad-spectrum antibiotics. Although the CDC acknowledges that it is too late to solve the resistance problem simply by using antimicrobial agents more prudently, it is no less true that the problem of drug resistance will continue to worsen if this does not occur. Also needed (especially in underdeveloped countries) is a renewed emphasis on alternate prevention and control strategies that prevailed in the years before antimicrobial chemotherapy. These include improved sanitation and hygiene, isolation of infected persons, antisepsis, and vaccination. Drug resistance (section 34.6)

In this new millennium, the speed and volume of international travel are major factors contributing to the global emergence of

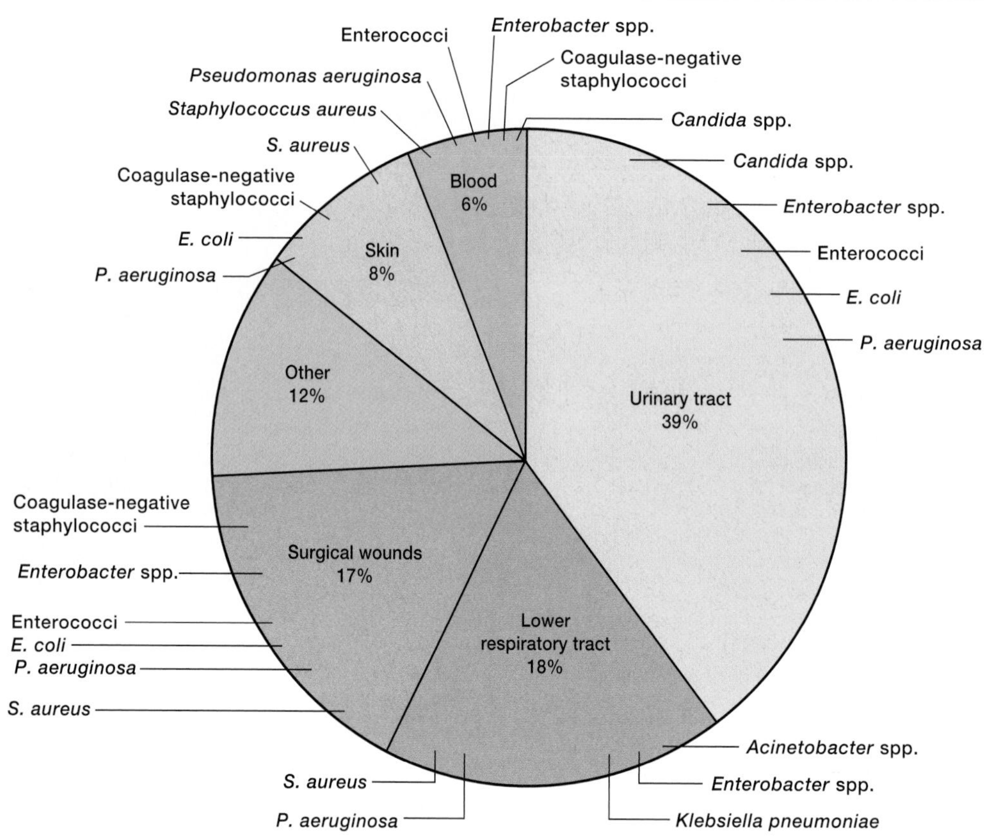

Figure 36.12 Nosocomial Infections. Relative frequency by body site. These data are from the National Nosocomial Infections Surveillance, which is conducted by the Centers for Disease Control and Prevention (CDC).

infectious diseases. The spread of a new disease often used to be limited by the travel time needed to reach a new host population. If the travel time was sufficiently long, as when a ship crossed the ocean, the infected travelers would either recover or die before reaching a new population. Because travel by air has obliterated time between exposure and disease outbreak, a traveler can spread virtually any human disease in a matter of hours. Vehicles of human transport, such as aircraft and ships, also transport the infectious agents and their vectors.

It is probably best to view emerging and reemerging pathogens and their diseases as the outcome of many different factors. Because the world is now so interconnected, we cannot isolate ourselves from other countries and continents. Changes in the disease status of one part of the world may well affect the health of the remainder. As Nobel laureate Joshua Lederberg has so eloquently stated, "The microbe that felled one child in a distant continent yesterday can reach your child today and seed a global pandemic tomorrow."

1. Describe how virulence and the mode of transmission may be related. What might cause the development of new human diseases?
2. How would you define emerging or reemerging infectious diseases?
3. What are some of the factors responsible for the emergence or reemergence of pathogens?
4. What are some key factors responsible for the rise in drug-resistant bacteria?

36.8 Control of Epidemics

The development of an infectious disease is a complex process involving many factors, as is the design of specific epidemiological control measures. Epidemiologists must consider available resources and time constraints, adverse effects of potential control measures, and human activities that might influence the spread of the infection. Many times control activities reflect compromises among alternatives. To proceed intelligently, one must identify components of the infectious disease cycle that are primarily responsible for a particular epidemic. Control measures should be directed toward that part of the cycle that is most susceptible to control—the weakest link in the chain (figure 36.7).

There are three types of control measures. The first type is directed toward reducing or eliminating the source or reservoir of infection:

1. Quarantine and isolation of carriers
2. Destruction of an animal reservoir of infection
3. Treatment of water sewage to reduce contamination *(see chapter 41)*
4. Therapy that reduces or eliminates infectivity of the individual

The second type of control measure is designed to break the connection between the source of the infection and susceptible individuals. Examples include general sanitation measures:

1. Chlorination of water supplies
2. Pasteurization of milk and other beverages
3. Supervision and inspection of food and food handlers
4. Destruction of vectors

The third type of control measure reduces the number of susceptible individuals and raises the general level of herd immunity by immunization. Examples include:

1. Passive immunization to give a temporary immunity following exposure to a pathogen or when a disease threatens to take an epidemic form
2. Active immunization to protect the individual from the pathogen and the host population from the epidemic

Vaccines and Immunization

A **vaccine** [Latin *vacca,* cow] is a preparation of one or more microbial antigens used to induce protective immunity. It may consist of killed microorganisms, living, weakened microorganisms (**attenuated vaccine**); inactivated bacterial toxins (**toxoids**); purified cellular subunits; recombinant vectors (e.g., modified polio vaccine); or DNA. **Immunization** is the result achieved by the successful delivery of vaccines; it stimulates immunity. Vaccination attempts to induce antibodies and activated T cells to protect a host from future infection. Many epidemics have been stayed by mass prophylactic immunization. Vaccines have eradicated smallpox, pushed polio to the brink of extinction, and spared countless individuals from hepatitis A and B, influenza, measles, rotavirus disease, tetanus, typhus, and other dangerous diseases. **Vaccinomics,** the application of genomics and bioinformatics to vaccine development, is bringing a fresh approach to the Herculean problem of making vaccines against various microorganisms and parasites.

To promote a more efficient immune response, antigens in vaccines can be mixed with an **adjuvant** [Latin *adjuvans,* aiding], which enhances the rate and degree of immunization. Adjuvants can be any nontoxic material that prolongs antigen interaction with immune cells, assists in the antigen-presenting cell (APC) processing of antigens, or otherwise nonspecifically stimulates the immune response to the antigen. Several types of adjuvants can be used. Oil in water emulsions (Freund's incomplete adjuvant), aluminum hydroxide salts (alum), beeswax, and various combinations of bacteria (live or killed) are used in vaccine adjuvants. In most cases, the adjuvant materials trap the antigen, thereby promoting a sustained release as APCs digest and degrade the preparation. In other cases, the adjuvant activates APCs so that antigen recognition, processing, and presentation are more efficient.

The modern era of vaccines and immunization began in 1798 with Edward Jenner's use of cowpox as a vaccine against smallpox (**Historical Highlights 36.4**) and in 1881 with Louis Pasteur's anthrax vaccine. Vaccines for other diseases did not emerge until later in the 19th century, when largely through a process of trial and error, methods for inactivating and attenuating microorganisms were improved and vaccines were produced. Vaccines were eventually developed against most of the epidemic diseases that plagued western Europe and North America (diphtheria, measles, mumps, pertussis, German measles and polio, for example). Indeed, toward the end of the of the 20th century it seemed that the combination of vaccines and antibiotics would temper the problem of microbial infections. Such optimism was cut short by the emergence of new and previously unrecognized pathogens, and antibiotic resistance among existing pathogens. Nevertheless, vaccination is still one of the most cost-effective weapons for prevention of microbial disease.

Vaccination of most children should begin at about age 2 months (**table 36.3**). Before that age, they are protected by passive natural immunity from maternal antibodies. Further vaccination of teens and adults depends on their relative risk for exposure to infectious disease. Individuals living in close quarters (for instance, college students in residence halls, military personnel), the elderly, and individuals with reduced immunity (e.g., those with chronic and metabolic diseases), should receive vaccines for influenza, meningitis, and pneumonia. International travelers may be immunized against cholera, hepatitis A, plague, polio, typhoid, typhus, and yellow fever, depending on the country visited. Veterinarians, forest rangers, and others whose jobs involve contact with animals may be vaccinated against rabies, plague, and anthrax. Health-care workers are typically immunized against hepatitis B virus. The role of immunization as a protective therapy cannot be overstated; immunizations save lives.

Whole-Cell Vaccines

Many of the current vaccines used for humans that are effective against viral and bacterial diseases consist of whole microorganisms that are either inactivated (killed) or attenuated (live but avirulent) (**table 36.4**). These are termed **whole-cell vaccines.** The major characteristics of these vaccines are compared in **table 36.5.** Inactivated vaccines are effective, but they are less immunogenic so often require several boosters and normally do not adequately stimulate cell-mediated immunity or secretory IgA production. In contrast, attenuated vaccines usually are given in a single dose and stimulate both humoral and cell-mediated immunity.

Even though whole-cell vaccines are considered the "gold standard" of existing vaccines, they can be problematic in their own way. For example, whole-organism vaccines fail to shield against some diseases. Attenuated vaccines that work can also cause full-blown illness in an individual whose immune system is compromised (e.g., AIDS patients, cancer patients undergoing chemotherapy, the elderly). These same individuals may also contract the disease from healthy people who have been vaccinated recently. Moreover, attenuated viruses can at times mutate in ways that restore virulence, as has happened in some monkeys given an attenuated simian form of the AIDS virus.

Acellular or Subunit Vaccines

A few of the common risks associated with whole-cell vaccines can be avoided by using only specific, purified macromolecules

Historical Highlights

36.4 The First Immunizations

Since the time of the ancient Greeks, it has been recognized that people who have recovered from plague, smallpox, yellow fever, and various other infectious diseases rarely contract the diseases again. The first scientific attempts at artificial immunizations were made in the late eighteenth century by Edward Jenner (1749–1823), who was a country doctor from Berkley, Gloucestershire, England. Jenner investigated the basis for the widespread belief of the English peasants that anyone who had vaccinia (cowpox) never contracted smallpox. Smallpox was often fatal—10 to 40% of the victims died—and those who recovered had disfiguring pockmarks. Yet most English milkmaids, who were readily infected with cowpox, had clear skin because cowpox was a relatively mild infection that left no scars.

It was on May 14, 1796, that Jenner extracted the contents of a pustule from the arm of a cowpox-infected milkmaid, Sarah Nelmes, and injected it into the arm of eight-year-old James Phipps. As Jenner expected, immunization with the cowpox virus caused only mild symptoms in the boy. When he subsequently inoculated the boy with smallpox virus, the boy showed no symptoms of the disease.

Jenner then inoculated large numbers of his patients with cowpox pus, as did other physicians in England and on the European continent (see **Box figure**). By 1800 the practice known as variolation had begun in America, and by 1805 Napoleon Bonaparte had ordered all French soldiers to be vaccinated.

Further work on immunization was carried out by Louis Pasteur (1822–1895). Pasteur discovered that if cultures of chicken cholera bacteria were allowed to age for two or three months the bacteria produced only a mild attack of cholera when inoculated into chickens. Somehow the old cultures had become less pathogenic (attenuated) for the chickens. He then found that fresh cultures of the bacteria failed to produce cholera in chickens that had been previously inoculated with old, attenuated cultures. To honor Jenner's work with cowpox, Pasteur gave the name vaccine to any preparation of a weakened pathogen that was used (as was Jenner's "vaccine virus") to immunize against infectious disease.

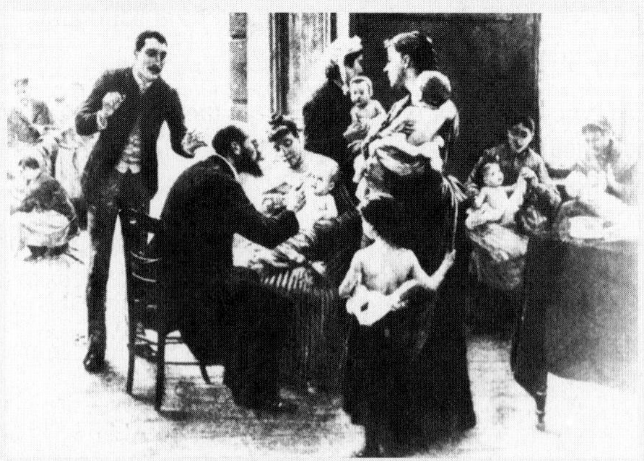

Nineteenth-Century Physicians Performing Vaccinations on Children

Table 36.3 Recommended Childhood and Adolescent Immunization Schedule, United States, 2006

Vaccine ▼ / Age ▶	Birth	1 Month	2 Months	4 Months	6 Months	12 Months	15 Months	18 Months	24 Months	4–6 Years	11–12 Years	13–14 Years	15 Years	16–18 Years
Hepatitis B	HepB	HepB		HepB		HepB					HepB Series			
Diptheria, Tetanus, Pertussis			DTaP	DTaP	DTaP		DTaP			DTaP	Tdap		Tdap	
Haemophilus influenzae type b			Hib	Hib	Hib	Hib								
Inactivated Poliovirus			IPV	IPV		IPV				IPV				
Measles, Mumps, Rubella						MMR				MMR		MMR		
Varicella							Varicella				Varicella			
Meningococcal							Vaccines within broken line are for selected populations		MPSV4		MCV4		MCV4 MPSV4	
Pneumococcal			PCV	PCV	PCV	PCV			PCV			PPV		
Influenza							Influenza (yearly)				Influenza (yearly)			
Hepatitis A										HepA Series				

This schedule indicates the recommended ages for routine administration of currently licensed childhood vaccines, as of December 1, 2005, for children through age 18 years. Any dose not administered at the recommended age should be administered at any subsequent visit when indicated and feasible. ▬ Indicates age groups that warrant special effort to administer those vaccines not previously administered. Additional vaccines may be licensed and recommended during the year. Licensed combination vaccines may be used whenever any components of the combination are indicated and other components of the vaccine are not contraindicated and if approved by the Food and Drug Administration for that dose of the series. Providers should consult the respective Advisory Council on Immunization Practices (ACIP) statement for detailed recommendations. (www.cdc.gov/nip/acip). Clinically significant adverse events that follow immunization should be reported to the Vaccine Adverse Event Reporting System (VAERS).

▬ Range of recommended ages ▬ Catch-up immunization ▬ 11–12 year old assessment

Table 36.4 Examples of Vaccines to Prevent Viral and Bacterial Diseases in Humans

Disease	Vaccine	Booster	Recommendation
Viral Diseases			
Brain infection	Inactivated Japanese encephalitis virus	None	Residents of and travelers to areas of endemic disease
Chickenpox	Attenuated Oka strain (Varivax)	None	Children 12–18 months: older children who have not had chickenpox
Hepatitis A	Inactivated virus (Havrix)	6–12 months	International travelers
Hepatitis B	HB viral antigen (Engerix-B, Recombivax HB)	None	High-risk medical personnel: children, birth to 18 months and 11–12 years of age
Influenza A	Inactivated virus or live attenuated	Yearly	All persons
Measles, Mumps, Rubella	Attenuated viruses (combination MMR vaccine)	None	Children 15–19 months old
Poliomyelitis	Attenuated (oral poliomyelitis vaccine, OPV) or inactivated vaccine	Adults as needed	Children 2–3 years old
Rabies	Inactivated virus	None	For individual in contact with wildlife, animal control personnel, veterinarians
Respiratory disease	Live attenuated adenovirus	None	Military personnel
Smallpox	Live attenuated vaccinia virus	None	Laboratory, health-care, and military personnel
Yellow fever	Attenuated virus	10 years	Military personnel and individuals traveling to endemic areas
Bacterial Diseases			
Anthrax	Extracellular components of unencapsulated *B. anthracis*	None	Agricultural workers, veterinary, and military personnel
Cholera	Fraction of *Vibrio cholerae*	6 months	Individuals in endemic areas, travelers
Diphtheria, Pertussis, Tetanus	Diphtheria toxoid, killed *Bordetella pertussis,* tetanus toxoid (DPT vaccine) or with acellular *pertussis* (DtaP); or tetanus toxoid, reduced diphtheria toxoid, and acellular pertussis vaccine, adsorbed (Tdap)	10 years	Children from 2–3 months old to 12 years, and adults; children 10–18 years, at least 5 years after DPT series, should receive Tdap
Haemophilus influenzae type b	Polysaccharide-protein conjugate (HbCV) or bacterial polysaccharide (HbPV)	None	Children under 5 years of age
Meningococcal infections	Bacterial polysaccharides of serotypes A/C/Y/W-135	None	Military; high-risk individuals; college students living in dorma-tories; elderly in nursing homes
Plague	Fraction of *Yersinia pestis*	Yearly	Individuals in contact with rodents in endemic areas
Pneumococcal pneumonia	Purified *S. pneumoniae* polysaccharide of 23 pneumococcal types	None	Adults over 50 with chronic disease
Q fever	Inactivated *Coxiella burnetii*	None	Workers in slaughter houses and meat-processing plants
Tuberculosis	Attenuated *Mycobacterium bovis* (BCG vaccine)	3–4 years	Individuals exposed to TB for prolonged periods of time; not licensed for use in the U.S.
Typhoid fever	Ty21a (live attenuated, polysaccharide)	None	Resident of and travelers to areas of endemic disease
Typhus fever	Killed *Rickettsia prowazekii*	Yearly	Scientists and medical personnel in areas where typhus is endemic

Table 36.5 A Comparison of Inactivated (Killed) and Attenuated (Live) Vaccines

Major Characteristic	Inactivated Vaccine	Attenuated Vaccine
Booster shots	Multiple boosters required	Only a single booster typically required
Production	Virulent microorganism inactivated by chemicals or irradiation	Virulent microorganism grown under adverse conditions or passed through different hosts until avirulent
Reversion tendency	None	May revert to a virulent form
Stability	Very stable, even where refrigeration is unavailable	Less stable
Type of immunity induced	Humoral	Humoral and cell-mediated

Source: Adapted from Goldsby, T. J. Kindt, and B. A. Osborne, *Kuby Immunology*. 2003, New York: W. H. Freeman

Table 36.6 Purified Acellular or Subunit Vaccines Currently Available for Human Use

Type of Subunit (Disease or Microorganism)	Form of Vaccine
Capsular polysaccharide	
Haemophilus influenzae type b	Polysaccharide-protein conjugate (HbCV) or bacterial polysaccharide (HbPV)
Neisseria meningitidis	Polysaccharides of serotypes A/C/Y/W-135
Streptococcus pneumoniae	23 distinct capsular polysaccharides
Surface antigen	
Hepatitis B	Recombinant surface antigen (HbsAg)
Toxoids	
Diphtheria	Inactivated exotoxin
Tetanus	Inactivated exotoxin

derived from pathogenic microorganisms. There are three general forms of **subunit vaccines:** (1) capsular polysaccharides, (2) recombinant surface antigens, and (3) inactivated exotoxins called toxoids. The purified microbial subunits or their secreted products can be prepared as nontoxic antigens to be used in the formulation of vaccines (**table 36.6**).

Recombinant-Vector and DNA Vaccines

Genes isolated from a pathogen that encode major antigens can be inserted into nonvirulent viruses or bacteria. Such recombinant microorganisms serve as vectors, replicating within the host and expressing the gene product of the pathogen-encoded antigenic proteins. The antigens elicit humoral immunity (i.e., antibody production) when they escape from the vector, and they also elicit cellular immunity when they are broken down and properly displayed on the cell surface (just as occurs when host cells harbor an active pathogen). Several microorganisms, such as adenovirus and attenuated *Salmonella,* have been used in the production of these **recombinant-vector vaccines.**

On the other hand, **DNA vaccines** introduce DNA directly into the host cell (often via an air pressure or gene gun). When injected into muscle cells, the DNA is taken into the nucleus and the pathogen's DNA fragment is expressed, generating foreign proteins to which the host immune system responds. DNA vaccines are very stable; refrigeration is often unnecessary. At present, there are human trials underway with several different DNA vaccines against malaria, AIDS, influenza, hepatitis B, and herpesvirus. Vaccines against a number of cancers (such as lymphomas, prostate, colon) are also being tested.

The Role of the Public Health System: Epidemiological Guardian

The control of an infectious disease relies heavily on a well-defined network of clinical microbiologists, nurses, physicians, epidemiologists, and infection control personnel who supply epidemiological information to a network of local, state, national, and international organizations. These individuals and organizations comprise the public health system. For example, each state has a public health laboratory that is involved in infection surveillance and control. The communicable disease section of a state laboratory includes specialized laboratory services for the examination of specimens or cultures submitted by physicians, the local health department, hospital laboratories, sanitarians, epidemiologists, and others. These groups share their findings with other health agencies in the state, the Centers for Disease Control and Prevention, and the World Health Organization.

36.9 Bioterrorism Preparedness

Bioterrorism (Greek *bios,* life, and terrorism, the systematic use of terror to demoralize, intimidate, and subjugate) is defined as "the intentional or threatened use of viruses, bacteria, fungi, or toxins from living organisms to produce death or disease in humans, animals, and plants." The use of biological agents to effect personal or political outcome is not new (**Historical Highlights 36.5**), and the modern use of biological agents is a reality. The most notable intentional uses of biological agents for criminal or terror intent are (1) the use of *Salmonella enterica* serovar Typhimurium in 10 restaurant salad bars (Rajneeshee religious cult in The Dalles, OR, 1984); (2) the intentional release of *Shigella dysentariae* in a hospital laboratory break room (Texas, 1996); and (3) the use of weaponized *Bacillus anthracis* spores delivered through the U.S. postal system (perpetrator[s] still unknown, seven eastern U.S. states, 2001). The *Salmonella*-contaminated salads resulted in 751 documented cases and 45 hospitalizations due to salmonellosis. The *Shigella* release resulted in eight confirmed cases and four hospitalizations for shigellosis. The *Bacillus* spores infected 22 people (11 cases of inhalation anthrax and 11 cases of cutaneous anthrax) and were the cause of five deaths. The list of biological agents that could pose the greatest public health risk in the event of a bioterrorist attack is relatively short and includes viruses, bacteria, parasites, and toxins (**table 36.7**). Although short, the list includes agents that, if acquired and properly disseminated, could become a difficult public health challenge in terms of limiting the numbers of casualties and controlling panic. The agents are catagorized (A, B, or C) based on (1) ease of dissemination, (2) communicability, and (3) morbidity and mortality.

Biological agents are likely to be chosen as a means of localized attack (**biocrime**) or mass casualty (bioterrorism) for several reasons. They are mostly invisible, odorless, tasteless, and difficult to detect. Use of biological agents for terrorism also means that perpetrators may escape undetected as it may take hours to days before signs and symptoms of their use become evident. Additionally, the general public is not likely to be protected immunologically against agents that are thought to be used in bioterrorism. Furthermore, the use of biological agents in terrorism results in fear, panic and chaos.

There are several key indicators of a bioterrorism event. These include sudden increased numbers of sick individuals, especially with unusual (nonendemic) diseases (for that place and/or time of year). Also, sudden increased numbers of zoonoses, diseased animals, or vehicle-borne illnesses may indicate bioterrorism. Among weapons of mass destruction, biological weapons can be more destructive than chemical weapons, including nerve gas. In certain circumstances, biological weapons can be as devastating as a nuclear explosion—a few kilograms of anthrax could kill as many people as a Hiroshima-size nuclear bomb.

In 1998, the U.S. government launched the first national effort to create a biological weapons defense. The initiatives included

Historical Highlights

36.5 1346—The First Recorded Biological Warfare Attack

The Black Death *(see pp. 962–963),* which swept through Europe, Asia, and North Africa in the mid-fourteenth century, was probably the greatest public health disaster in recorded history. Europe, for example, lost an estimated quarter to third of its population. This is not only of great historical interest but also relevant to current efforts to evaluate the threat of military or terrorist use of biological weapons.

Some believe that evidence for the origin of the Black Death in Europe is found in a memoir by the Genoese Gabriele de' Mussi. According to this fourteenth-century memoir, the Black Death reached Europe from the Crimea (a region of the Ukraine) in 1346 as a result of a biological warfare attack. The Mongol army hurled plague-infected cadavers into the besieged Crimean city of Caffa (now Feodosija, Ukraine), thereby transmitting the disease to the inhabitants; fleeing survivors then spread the plague from Caffa to the Mediterranean Basin. Such transmission was especially likely at Caffa where cadavers would have been badly mangled by being hurled, and the defenders probably often had cut or abraded hands from coping with the bombardment. Because many cadavers were involved, the opportunity for disease transmission was greatly increased. Disposal of victims' bodies in a major disease outbreak is always a problem, and the Mongol army used their hurling machines as a solution to limited mortuary facilities. It is possible that thousands of cadavers were disposed of this way; de' Mussi's description of "mountains of dead" might have been quite literally true. Indeed, Caffa could be the site of the most spectacular incident of biological warfare ever, with the Black Death as its disastrous consequence. It is a powerful reminder of the horrific consequences that can result when disease is successfully used as a weapon.

Table 36.7 Pathogens and Toxins Defined by the CDC as Select Agents

Category	Definition	Disease (Agent)
A	Easily disseminated or transmitted from person to person; high mortality rates; potential for major public health impact; cause public panic and social disruption; require special action for public health preparedness	**Anthrax** (*Bacillus anthracis*) **Botulism** (*Clostridium botulinum* toxin) **Plague** (*Yersinia pestis*) **Smallpox** (*Variola major*) **Tularemia** (*Francisella tularensis*) **Viral hemorrhagic fever** (filoviruses and arenaviruses)
B	Moderately easy to disseminate, moderate morbidity and mortality rates; require specific enhancements of CDC's diagnostic capacity and enhanced disease surveillance	**Brucellosis** (*Brucella* species) **Glanders** (*Burkholderia mallei*) **Melioidosis** (*Burkholderia pseudomallei*) **Psittacosis** (*Chlamydia psittaci*) **Q fever** (*Coxiella burnetii*) **Typhus fever** (*Rickettsia prowazekii*) **Viral encephalitis** (alphaviruses) **Toxemia** Ricin from castor beans Staphylococcal enterotoxin B Epsilon toxin *Clostridium perfringens* **Other** Water safety threats (e.g., *Vibrio cholerae, Cryptosporidium parvum*) Food safety threats (e.g., *Salmonella* spp., *E. coli* O157:H7, *Shigella* spp.)
C	Emerging pathogens that could be engineered for mass dissemination; potential for high morbidity and mortality rates; major health impact potential	Nipah virus Hantaviruses Tickborne hemorrhagic fever viruses Tickborne encephalitis viruses Yellow fever virus Multidrug-resistant *Mycobacterium tuberculosis*

(1) the first-ever procurement of specialized vaccines and medicines for a national civilian protection stockpile; (2) invigoration of research and development in the science of biodefense; (3) investment of more time and money in genome sequencing, new vaccine research, and new therapeutic research; (4) development of improved detection and diagnostic systems; and (5) preparation of clinical microbiologists and the clinical microbiology laboratory as members of the "first responder" team to respond in a timely manner to acts of bioterrorism. In 2002, the U.S. government passed the Public Health Security and Bioterrorism Preparedness and Response Act, which identified "select" agents whose use is now tightly regulated. A final rule implementing the provisions of the act that govern the possession, use, and transport of biological agents that are considered likely to be used for biocrimes or bioterrorism was issued in 2005.

In 2003, the U.S. government established the Department of Homeland Security to coordinate the defense of the United States against terrorist attacks. As one of many duties, the Secretary of Homeland Security is responsible for developing and maintaining a National Incident Management System (NIMS) to monitor large-scale hazards. Bioterrorism and other public health incidents are managed within this system. The Department of Health and Human Services has the initial responsibility for the national public health and will deploy assets as needed within the areas of its statutory responsibility (e.g., the Public Health Service Act and the Federal Food Drug and Cosmetic Act) while keeping the Secretary of Homeland Security apprised during an incident and the nature of the response. The Secretary of Health and Human Services directs the CDC to effect the necessary integration of public health activities.

The events of September and October 2001 in the United States have changed the world. Global efforts to prevent terrorism, especially using biological agents, are evolving from cautious planning to proactive preparedness. In the United States, the CDC has partnered with academic institutions across the country to educate, train, and drill public health employees, traditional first responders, and numerous environmental and health-care providers. Centers for Public Health Preparedness were established to bolster the overall response capability to bioterrorism. Another CDC-managed program that began in 1999, the Laboratory Response Network (LRN), serves to ensure an effective laboratory response to bioterrorism by helping to improve the nation's public health laboratory infrastructure, through its partnership with the FBI and the Association of Public Health Laboratories (APHL). The LRN maintains an integrated network that links state and local public health, federal, military, and international laboratories so that a rapid and coordinated response to bioterrorism or other public health emergencies (including veterinary, agriculture, military, and water- and food- related) can occur.

In the absence of overt terrorist threats and without the ability to rapidly detect bioterrorism agents, it is likely that an act of bioterrorism will be defined by sudden spikes in illnesses reported to the public health system. Thus important guidelines have been prepared for all sentinel (local hospital, contract, clinic, etc.) laboratories to assist in the management of clinical specimens containing select agents. These guidelines were developed by the American Society for Microbiology in coordination with the CDC and the APHL to offer standardized protocols to assist microbiologists in ruling out critical agents so specimens containing select agents can then be referred to the public health reference laboratories (of the LRN) for final confirmation. A summary of the rule-out tests for six bacterial agents is presented in **table 36.8.** The disease and microbiology associated with specific select agents are discussed in chapters 37, 38, and 39.

1. In what three general ways can epidemics be controlled? Give one or two specific examples of each type of control measure.
2. Name some of the microorganisms that can be used to commit biocrimes. From this list, which pose the greatest risk for causing large numbers of casualties?
3. Why are biological weapons more destructive than chemical weapons?
4. What is the Public Health Security and Bioterrorism Preparedness and Response Act designed to do?

36.10 Global Travel and Health Considerations

From a global health perspective, developed countries such as Australia, the European countries, Israel, New Zealand, and the United States have highly effective public health systems. About 25% of the over 6 billion people on our planet Earth live in these countries. As a result, of the approximately 12 million deaths in these countries per year, only about 500,000 (about 4%) are due to infectious diseases. Less developed areas such as Africa, Central and South America, India, the parts of eastern Europe, and Asia have less developed public health systems and represent 75% of the human population. It is in these regions that infectious diseases are the major cause of death; for example, of approximately 38.5 million deaths per year, about 18 million (about 47%) are attributed to infectious microbial diseases. Despite the efforts of many governmental and nongovernmental agencies, it will take many years before all people have access to clean water, decent sanitation, and a basic health care infrastructure.

The high incidence of infectious diseases in less developed countries must be of great concern for people traveling to these destinations. Each year 1 billion passengers travel by air, and over

Table 36.8 Criteria for Presumptive Identification of Six Bacterial Select Agents

Pathogen	Gram Morphology	Colonial Morphology	Biochemical Results
Bacillus anthracis	Gram positive endospore-forming rod	Grey, flat, "Medusa head" irregularity, nonhemolytic on sheep's blood agar	Catalase positive, oxidase positive, urea negative, Voges-Proskauer (VP) positive, phenylalanine (Phe) deaminase negative, NO_3^- to NO_2^- positive
Brucella suis	Gram negative rod (tiny)	Nonpigmented, convex-raised, pin-point after 48 hr, nonhemolytic	Catalase positive, oxidase variable, urea positive, VP negative
Burkholderia mallei	Gram negative straight or slightly curved cocobacilli, bundles	Grey, smooth, translucent after 48 hr, nonhemolytic	Catalase positive, oxidase variable, indole negative, Arginine dihydrolase positive, NO_3^- to NO_2^- positive
Clostridium botulinum	Gram positive endospore-forming rod	Creamy, irregular, rough, broad, nonhemolytic	Catalase negative, urea negative, gelatinase positive, indole negative, VP negative, Phe deaminase negative, NO_3^- to NO_2^- negative
Francisella tularensis	Gram negative rod (tiny)	Grey-white, shiny, convex, pin-point after 72 hr, nonhemolytic	Catalase positive (weak), oxidase negative, β-lactamase positive, urea negative
Yersinia pestis	Gram negative rod (bipolar staining)	Grey-white, "fried-egg" irregularity, nonhemolytic, grows faster and larger at 28°C	Catalase positive, oxidase negative, urea negative, indole negative, VP negative, Phe deaminase negative

50 million people from developed countries visit less developed countries. Furthermore, the time required to circumnavigate the globe has decreased from 365 days to fewer than 2 days.

Several kinds of precautions can be taken by individuals to prevent travel-related infectious diseases. Examples include:

1. Wash hands with soap and water frequently, especially before each meal.
2. Get or update vaccinations appropriate for specific destinations. Check the CDC travel advisory web site (www.cdc.gov/travel) for precautions regarding specific locations.
3. Avoid uncooked food, nonbottled water and beverages, and unpasteurized dairy products. Use bottled water for drinking, making ice cubes, and brushing teeth.
4. Use barrier protection if engaging in sexual activity.
5. Minimize skin exposure and use repellents to prevent arthropod-borne illnesses (e.g., malaria, dengue, yellow fever, Japanese encephalitis).
6. Avoid skin-perforating procedures (e.g., acupuncture, body piercing, tattooing, venipuncture, sharing of razors).
7. Do not pet or feed animals, especially dogs and monkeys.
8. Avoid swimming or wading in nonchlorinated fresh water.

Vaccinations are one of the most important strategies of prophylaxis in travel medicine. A medical consultation before travel is an excellent opportunity to update routine immunizations. Selection of immunizations should be based on requirements and risk of infection at the travel destination. According to International Health Regulations, many countries require proof of yellow fever vaccination on the International Certificate of Vaccination. Additionally, a few countries still require proof of vaccination against cholera, diphtheria, and meningococcal disease. The basic CDC immunization recommendations for those traveling abroad are presented in **table 36.9.** In the not-too-distant future, it is hoped that travelers will be offered a variety of oral vaccines against the microorganisms causing dengue fever and travelers' diarrhea (e.g., enterotoxigenic *E. coli, Campylobacter, Shigella*); vaccines against malaria and AIDS, the infections causing most deaths in travelers, are much farther in the future.

In addition, a traveler should:

1. Read carefully CDC information about the destination and follow recommendations
2. Begin the vaccination process early
3. Find a travel clinic for information and specialized immunizations
4. Plan ahead when traveling with children or if there are special needs
5. Learn about safe food and water (contaminated food and water are the major sources of stomach or intestinal illness while traveling), protection against insects, and other precautions

Table 36.9 Vaccine Recommendations for Travelers*

Category	Vaccine
Routine recommended vaccination**	Diphtheria/Tetanus/Pertussis (DPT)#†
	Hepatitis B (HBV)#
	Haemophilus influenzae type b (Hib)#
	Influenza
	Measles (MMR)#
	Poliomyelitis (IPV)#
	Varicella#
Selective vaccination based on exposure risk	Cholera
	Hepatitis A (HAV)
	Japanese Encephalitis
	Meningococcal (polysaccharide)
	Pneumococcal (polysaccharide)
	Rabies
	Tick-borne Encephalitis
	Typhoid Fever
	Yellow Fever
Mandatory vaccination for entry§	Meningococcal (polysaccharide)
	Yellow Fever

* Based on the CDC and WHO 2005 recommendations
** For travelers aged 2 years or older
# Normally administered during childhood; should be updated if indicated for travel
† Adult booster of Tetanus/diphtheria (Td) vaccine should be every 10 years
§ Meningococcal vaccination to enter Saudi Arabia; Yellow Fever vaccination to enter various countries in the endemic zone of South America and Africa

1. Give some examples where population movements affect microbial disease transmission.
2. In addition to vaccinations, what are some additional precautions global travelers should take?

36.11 Nosocomial Infections

Nosocomial infections [Greek *nosos,* disease, and *komeion,* to take care of] result from pathogens that develop within a hospital or other type of clinical care facility and are acquired by patients while they are in the facility (figure 36.12). Besides harming patients, nosocomial infections can affect nurses, physicians, aides, visitors, salespeople, delivery personnel, custodians, and anyone who has contact with the hospital. Most nosocomial infections become clinically apparent while patients are still hospitalized; however, disease onset can occur after patients have been discharged. Infections that are incubating when patients are admitted to a hospital are not nosocomial; they are community acquired. However, because such infections can serve as a ready source or reservoir of pathogens for other patients or personnel, they are also considered in the total epidemiology of nosocomial infections.

The CDC estimates that about 10% of all hospital patients acquire some type of nosocomial infection. Because approximately 40 million people are admitted to hospitals annually, about 2 to 4 million people may develop an infection they did not have upon entering the hospital. Thus nosocomial infections represent a significant proportion of all infectious diseases acquired by humans.

Nosocomial diseases are usually caused by bacteria, most of which are noninvasive and part of the normal microbiota; viruses, protozoa, and fungi are rarely involved. Figure 36.12 summarizes the most common types of nosocomial infections and the most common nosocomial pathogens.

Source

The nosocomial pathogens that cause diseases come from either endogenous or exogenous sources. Endogenous sources are the patient's own microbiota; exogenous sources are microbiota other than the patient's. Endogenous pathogens are either brought into the hospital by the patient or are acquired when the patient becomes colonized after admission. In either case the pathogen colonizing the patient may subsequently cause a nosocomial disease (e.g., when the pathogen is transported to another part of the body or when the host's resistance drops). If it cannot be determined that the specific pathogen responsible for a nosocomial disease is exogenous or endogenous, then the term autogenous is used. An **autogenous infection** is one that is caused by an agent derived from the microbiota of the patient, despite whether it became part of the patient's microbiota following his or her admission to the hospital.

There are many potential exogenous sources in a hospital. Animate sources are the hospital staff, other patients, and visitors. Some examples of inanimate exogenous sources are food, computer keyboards, urinary catheters, intravenous and respiratory therapy equipment, and water systems (e.g., softeners, dialysis units, and hydrotherapy equipment).

Control, Prevention, and Surveillance

In the United States nosocomial infections prolong hospital stays by 4 to 13 days, result in over 4.5 billion dollars a year in direct hospital charges, and lead to over 20,000 direct and 60,000 indirect deaths annually. The enormity of this problem has led most hospitals to allocate substantial resources to the development of methods and programs for the surveillance, prevention, and control of nosocomial infections.

All personnel involved in the care of patients should be familiar with basic infection control measures such as isolation policies of the hospital; aseptic techniques; proper handling of equipment, supplies, food, and excreta; and surgical wound care and dressings. To adequately protect their patients, hospital personnel must practice proper aseptic technique and handwashing procedures, and must wear gloves when contacting mucous membranes and secretions. Patients should be monitored with respect to the frequency, distribution, symptomatology, and other characteristics common to nosocomial infections. A dynamic control and surveillance program can be invaluable in preventing many nosocomial infections, patient discomfort, extended stays, and further expense.

The Hospital Epidemiologist

Because of nosocomial infections, all hospitals desiring accreditation by the Joint Commission on Accreditation of Healthcare Organizations (JCAHO) must have a designated individual directly responsible for developing and implementing policies governing control of infections and communicable diseases. This individual is often a registered nurse known as a hospital epidemiologist, nurse epidemiologist, infection control nurse, infection control practitioner, or a clinical microbiologist/ technologist. In larger hospitals, a physician is the hospital epidemiologist and should be trained in infectious diseases. He or she oversees a staff that includes nurse epidemiologists, quality assurance specialists, fellows in infectious disease/hospital epidemiology, and epidemiology technicians. The hospital epidemiologist must meet with an infection control committee composed of various professionals who have expertise in the different aspects of infection control and hospital operation. The infection control committee periodically evaluates laboratory reports, patients' charts, and surveys done by the hospital epidemiologist to determine whether there has been any increase in the frequency of particular infectious diseases or potential pathogens.

Overall, the services provided by the hospital epidemiologist should include the following:

1. Research in infection control
2. Evaluation of disinfectants, rapid test systems, and other products
3. Efforts to encourage appropriate legislation related to infection control, particularly at the state level
4. Efforts to contain hospital operating costs, especially those related to fixed expenses such as the DRGs (diagnosis-related groups)
5. Surveillance and comparison of endemic and epidemic infection frequencies
6. Direct participation in a variety of hospital activities relating to infection control and maintenance of employee health

7. Education of hospital personnel in communicable disease control and disinfection and sterilization procedures
8. Establishment and maintenance of a system for identifying, reporting, investigating, and controlling infections and communicable diseases of patients and hospital personnel
9. Maintenance of a log of incidents related to infections and communicable diseases
10. Monitoring trends in the antimicrobial drug resistance of infectious agents

Computer software packages are available to aid the infection control practitioner. Such packages generate standard reports, cause-and-effect tabulations, and graphics for the daily epidemiological monitoring that must be done.

1. Describe a nosocomial infection.
2. What two general sources are responsible for nosocomial infections? Give some specific examples of each general source.
3. Why are nosocomial infections important?
4. What does a hospital epidemiologist do to control nosocomial infections?

Summary

36.1 Epidemiological Terminology

a. Epidemiology is the science that evaluates the determinants, occurrence, distribution, and control of health and disease in a defined population.

b. Specific epidemiological terminology is used to communicate disease incidence in a given population. Frequently used terms include sporadic disease, endemic disease, hyperendemic disease, epidemic, index case, outbreak, and pandemic.

36.2 Measuring Frequency: The Epidemiologist's Tools

a. Statistics is an important tool used in the study of modern epidemiology.

b. Epidemiological data can be obtained from such factors as morbidity, prevalence, and mortality rates.

36.3 Recognition of an Infectious Disease in a Population

a. An infectious disease is caused by microbial agents such as viruses, bacteria, fungi, protozoa, and helminths. A communicable disease can be transmitted from person to person.

b. The manifestations of an infectious disease can range from mild to severe to deadly, depending on the agent and host.

c. Surveillance is necessary for recognizing a specific infectious disease within a given population. This consists of gathering data on the occurrence of the disease, collating and analyzing the data, summarizing the findings, and applying the information to control measures.

d. It is important to recognize the signs and symptoms that compose a disease syndrome. This includes the characteristic course of the disease, such aspects as the incubation period and prodromal stage (**figure 36.3**).

e. Remote sensing and geographic information systems can be used to gather epidemiological data on the environment.

36.4 Recognition of an Epidemic

a. A common-source epidemic is characterized by a sharp rise to a peak and then a rapid, but not as pronounced, decline in the number of individuals infected (**figure 36.4**). A propagated epidemic is characterized by a relatively slow and prolonged rise and then a gradual decline in the number of individuals infected (**figure 36.5**).

b. Herd immunity is the resistance of a population to infection and pathogen spread because of the immunity of a large percentage of the individuals within the population.

36.5 The Infectious Disease Cycle: Story of a Disease

a. The infectious disease cycle or chain involves the characteristics of the pathogen, the source and/or reservoir of the pathogen, the transmission of the pathogen, the susceptibility of the host, the exit mechanism of the pathogen from the body of the host, and its spread to a new reservoir or host (**figure 36.7**).

b. There are four major modes of transmission: airborne, contact, vehicle, and vector-borne (**figure 36.8**).

36.6 Virulence and the Mode of Transmission

a. The degree of virulence may be influenced by the pathogen's preferred mode of transmission. New human diseases may arise and spread because of ecosystem disruption, rapid transportation, human behavior, and other factors.

36.7 Emerging and Reemerging Infectious Diseases and Pathogens

a. It is now clear that globally, humans will continually be faced with both new infectious diseases and the reemergence of older diseases once thought to be conquered.

b. CDC has defined these diseases as "new, reemerging, or drug-resistant infections whose incidence in humans has increased within the past two decades or whose incidence threatens to increase in the near future" (**figure 36.11**).

c. Many factors characteristic of the modern world undoubtedly favor the development and spread of these microorganisms and their diseases.

36.8 Control of Epidemics

a. The public health system consists of individuals and organizations that function in the control of infectious diseases and epidemics.

b. Vaccination is one of the most cost-effective weapons for microbial disease prevention, and vaccines constitute one of the greatest achievements of modern medicine.

c. Artificially acquired immunity to pathogens can be accomplished by either active or passive immunization.

d. Many of the current vaccines in use for humans (**table 36.4**) consists of whole organisms that are either inactivated (killed) or attenuated (live but avirulent).

e. Some of the risks associated with whole-cell vaccines can be avoided by using only specific purified macromolecules derived from pathogenic microorganisms. Currently, there are three general forms of subunit or acellular vaccines: capsular polysaccharides, recombinant surface antigens, and inactivated exotoxins (toxoids) (**table 36.6**).

f. A number of microorganisms have been used for recombinant-vector vaccines. The attenuated microorganism serves as a vector, replicating within the host, and expressing the gene product of the pathogen-encoded antigenic proteins. The proteins can elicit humoral immunity when the proteins escape from the cells and cellular immunity when they are broken down and properly displayed on the cell surface.

g. Other genetic vaccines are termed DNA vaccines. DNA vaccines elicit protective immunity against a pathogen by activating both branches of the immune system: humoral and cellular.

h. Epidemiological control measures can be directed toward reducing or eliminating infection sources, breaking the connection between sources and susceptible individuals, or isolating the susceptible individuals and raising the general level of herd immunity by immunization.

36.9 Bioterrorism Preparedness

a. Today bioterrorism is a reality. Terrorist incidents and hoaxes involving toxic or infectious agents have been on the rise.

b. Among weapons of mass destruction, biological weapons are more destructive than chemical weapons. The list of biological agents that could pose the greatest public health risk in the event of a bioterrorist attack is short and includes viruses, bacteria, parasites, and toxins (**table 36.7**).

36.10 Global Travel and Health Considerations

a. Certain precautions and health considerations should be taken into consideration when traveling globally.

b. Vaccinations are one of the most important strategies of prophylaxis in travel medicine (**table 36.9**).

36.11 Nosocomial Infections

a. Nosocomial infections are infections acquired during hospitalization and are produced by a pathogen acquired during a patient's stay. These infections come from either endogenous or exogenous sources (**figure 36.12**).

b. Hospitals must designate an individual to be responsible for identifying and controlling nosocomial infections. This person is known as a hospital epidemiologist, nurse epidemiologist, infection control nurse, or infection control practitioner.

Key Terms

active carrier 891
acute carrier 892
adjuvant 901
airborne transmission 892
antigenic drift 890
antigenic shift 890
attenuated vaccine 901
autogenous infection 909
biocrime 905
biologic transmission 896
bioterrorism 905
carrier 891
casual carrier 892
chronic carrier 892
common-source epidemic 889
common vehicle transmission 894
communicable disease 888
contact transmission 892
convalescence 888
convalescent carrier 891
disease 885
disease syndrome 888
DNA vaccine 904
droplet nuclei 892
endemic disease 886
epidemic 886
epidemiologist 886
epidemiology 885
fomite 894
geographic information system (GIS) 889
harborage transmission 896
health 885
healthy carrier 892
herd immunity 890
hyperendemic disease 886
immunization 901
incubation period 888
incubatory carrier 892
index case 887
infectious disease cycle (chain of infection) 891
morbidity rate 887
mortality rate 887
nosocomial infection 908
outbreak 886
pandemic 887
period of infectivity 891
prevalence rate 887
prodromal stage 888
propagated epidemic 889
recombinant-vector vaccine 904
remote sensing (RS) 889
reservoir 891
signs 888
source 891
sporadic disease 886
statistics 887
subunit vaccine 904
symptoms 888
systematic epidemiology 898
toxoid 901
transient carrier 892
vaccine 901
vaccinomics 901
vector 892
vector-borne transmission 896
vehicle 894
whole-cell vaccines 901
zoonoses 892

Critical Thinking Questions

1. Why is international cooperation a necessity in the field of epidemiology? What specific problem can you envision if there were no such organizations?
2. What common sources of infectious disease are found in your community? How can the etiologic agents of such infectious diseases spread from their source or reservoir to members of your community?
3. How could you prove that an epidemic of a given infectious disease was occurring?
4. How can changes in herd immunity contribute to an outbreak of a disease on an island?
5. College dormitories are notorious for outbreaks of flu and other infectious diseases. This is particularly prevalent during final exam weeks. Using your knowledge of the immune response and epidemiology, suggest practices that could be adopted to minimize the risks at such a critical time of the term.
6. Why does an inactivated vaccine induce only a humoral response, whereas an attenuated vaccine induces both humoral and cell-mediated responses?
7. Why is a DNA vaccine delivered intramuscularly and not by intravenous or oral routes?

Learn More

Buckland, B. C. 2005. The process development challenge for a new vaccine. *Nature Medicine* 11:S16–S19.

Curtis, R. 2002. Bacterial infectious disease control by vaccine development. *J. Clin. Invest.* 110:1061–66.

Desrosiers, R. 2004. Prospects for an AIDS vaccine. *Nature Medicine* 10:221–23.

Diamond, J. M. 1997. *Guns, germs, and steel.* New York: W.W. Norton & Company.

Friedman, D. S.; Heisey-Grove, D.; Argyros, F.; Berl, E.; Nsubuga, J.; Stiles, T.; Fontana, J.; Board, R. S.; Monroe, S.; McGrath, M. E.; Sutherby, H.; Dicker, R. C.; DeMaria, A.; and Matyas, B. T. 2005. An outbreak of norovirus gastroenteritis associated with wedding cakes. *Epidemiology and Infection* 133: 1057–63.

Soares, C. 2005. Cooping up avian flu: Buying time to arm for a pandemic is possible—maybe. *Sci. Am.* April 25, 2005.

Wayt Gibbs, W., and Soares, C. 2005. Preparing for a pandemic. *Sci. Am.* Nov 24, 2005.

Please visit the Prescott website at www.mhhe.com/prescott7 for additional references.

---

37 Human Diseases Caused by Viruses and Prions

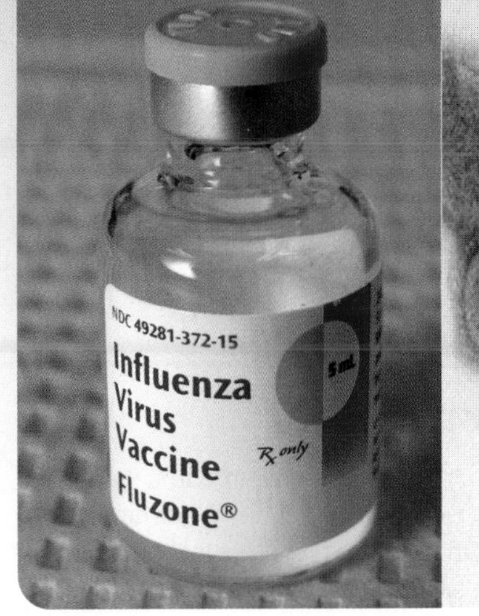

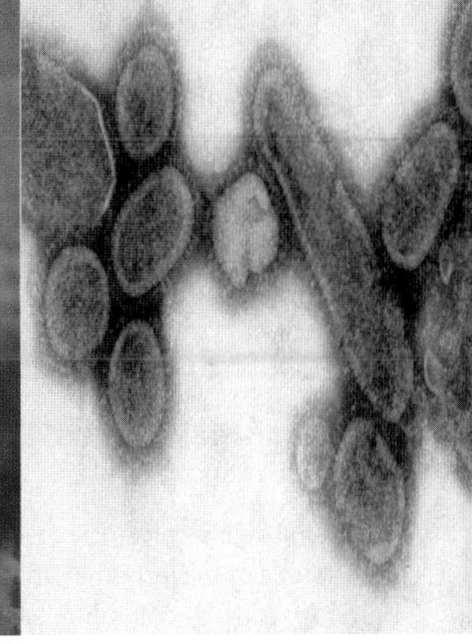

Numerous global epidemics are attributed to influenza viruses. The rapid mutation of their surface antigens results in sudden antigenic shifts that lead to widespread human infections. More frequent mutations leading to antigenic drifting can be anticipated such that yearly immunization with influenza vaccine may prevent infection. The H5N1 strain may be the agent of the next influenza pandemic.

Preview

- Hundreds of viruses are known to cause illness in humans. There are currently very few treatment options to control viruses; thus viruses are the source of many deadly human diseases.
- Some viruses can be transmitted through the air and directly or indirectly involve the respiratory system. Most of these viruses are highly communicable and cause diseases such as chickenpox, influenza, measles, mumps, respiratory syndromes and viral pneumonia, rubella, and hantavirus pulmonary syndrome.
- The arthropod-borne diseases are transmitted by arthropod vectors from human to human or animal to human. Examples include the various encephalitides, Colorado tick fever, West Nile fever, and historically important yellow fever.
- Some viruses are so sensitive to environmental influences that they are unable to survive for significant periods outside their hosts. These viruses are transmitted from host to host by direct contact and cause diseases such as HIV-AIDS, cold sores, the common cold, cytomegalovirus inclusion disease, genital herpes, human herpesvirus 6 infections, human parvovirus B19 infections, certain leukemias, infectious mononucleosis, human papillomavirus, viral hepatitides and warts.
- Viruses can be transmitted by food and wate. They usually either grow in or pass through the intestinal system and are acquired through fecal-oral transmission. Examples of such diseases include viral gastroenteritis, hepatitis A and E, and poliomyelitis.
- Many highly contagious and often fatal viral infections are initially contracted from animals; they are then spread from human to human. Some of the agents responsible for these zoonotic diseases, such as Ebola hemorrhagic and Lassa fevers, are potential agents of bioterrorism.
- The transmissible spongiform encephalopathies are caused by prions that remain clinically silent during a prolonged period of months or years, after which progressive disease becomes apparent, usually ending months later in profound disability or death. Examples include new variant Creutzfeldt-Jakob disease, kuru, and Gerstmann-Straussler-Scheinker disease.

Chapters 16, 17, and 18 review the general biology of viruses and introduce basic virology. In chapter 37 we continue this coverage by discussing some of the most important viruses that are pathogenic to humans. The viral diseases are grouped according to their mode of acquisition and transmission; viral diseases that occur in the United States are emphasized. Diseases caused by viruses that are listed as Select Agents are identified within the chapter by two asterisks (**).

More than 400 different viruses can infect humans. Human diseases caused by viruses are particularly interesting, considering the small amount of genetic information introduced into a host cell. This apparent simplicity belies the severe pathological features and clinical consequences that result from many viral diseases. With few exceptions, only prophylactic or supportive treatment is available. Collectively these diseases are some of the most common and yet most puzzling of all infectious diseases. The resulting frustration is compounded when year after year familiar diseases

Only once in human history have we witnessed the total eradication of a dreaded disease, and that was smallpox more than two decades ago. Now humanity stands on the brink of a second: the global eradication of polio.

—From UNICEF's Polio Website

of unknown etiology or new diseases become linked to virus infections (**table 37.1**).

37.1 Airborne Diseases

Because air does not support virus growth, any virus that is airborne must have originated from a living source. When humans are the source of the airborne virus, it usually is propelled from the respiratory tract by coughing, sneezing, or vocalizing.

Chickenpox (Varicella) and Shingles (Herpes Zoster)

Chickenpox (varicella) is a highly contagious skin disease primarily of children 2 to 7 years of age. Humans are the reservoir and the source for this virus, which is acquired by droplet inhalation into the respiratory system. The virus is highly infectious with secondary infection rates in susceptible household contacts of 65% to 86%. In the pre-vaccine era, about 4 million cases of chickenpox occurred annually in the United States, resulting in approximately 11,000 hospitalizations and 100 deaths.

The causative agent is the enveloped, DNA varicella-zoster virus (VZV), a member of the family *Herpesviridae.* The virus produces at least six glycoproteins that play a role in viral attachment to specific receptors on respiratory epithelial cells. Their recognition by the human immune system results in humoral and cellular immunity. This virus has been shown to inhibit the expression of MHC molecules by infected cells; however, this inhibition only temporarily interferes with immune recognition of the virus, perhaps as a way of increasing its transmission. Following an incubation period of 10 to 23 days, small vesicles erupt on the face or upper trunk, fill with pus, rupture, and become covered by scabs (**figure 37.1**). Healing of the vesicles occurs in about 10 days. During this time intense itching often occurs.

Laboratory confirmation of varicella virus is by detection of varicella-zoster immunoglobulin M (IgM) antibody; detection of VZV, demonstration of VZV antigen by direct fluorescent antibody and by polymerase chain reaction in clinical specimens or a significant rise in serum IgG antibody level to VZV. However, laboratory testing for VZV is not normally required, as the diagnosis of chickenpox is typically made by clinical assessment. Laboratory confirmation is recommended, though, to confirm the diagnosis of severe or unusual cases of chickenpox. As the incidence of chickenpox continues to decline due to vaccination, fewer cases are seen clinically, resulting in the likelihood of misdiagnosis. Furthermore, in persons who have previously received varicella vaccination, the disease is usually mild or atypical, and can pose particular challenges for clinical diagnosis. Therefore, laboratory confirmation of varicella cases is becoming more important. Chickenpox can be prevented or the infection shortened with an attenuated varicella vaccine (Varivax; *see table 36.4*) or the drug acyclovir (Zovirax or Valtrex). It should be noted that Valtrex (valacyclovir) is an orally administered prodrug of Zovirax or acyclovir (*see figure 34.21*). Valtrex is the valyl ester of acyclovir and is rapidly hydrolyzed to acyclovir in the body.

Table 37.1 Some Examples of Human Viral Diseases Recognized Since 1967

Year	Virus	Disease
1967	Marburg virus	Hemorrhagic fever
1973	Rotavirus	Major cause of infantile diarrhea worldwide
1974	Parvovirus B19	Aplastic crisis in chronic hemolytic anemia
1977	Ebola virus	Ebola hemorrhagic fever
1977	Hantavirus	Hemorrhagic fever with renal syndrome
1980	Human T-cell lymphotrophic virus 1 (HTLV-1)	Adult T-cell leukemia
1982	Human T-cell lymphotrophic virus 2 (HTLV-2)	Hairy-cell leukemia
1983	Human immunodeficiency virus (HIV)	Acquired immunodeficiency syndrome (AIDS)
1988	Human herpesvirus 6 (HHV-6)	Sixth disease (roseola subitum); may be associated with multiple sclerosis
1988	Hepatitis E	Enterically transmitted non-A, non-B hepatitis
1989	Hepatitis C	Parenterally transmitted non-A, non-B liver infection
1991	Guanarito virus	Venezuelan hemorrhagic fever
1992	Lymphocytic choriomeningitis virus	Central nervous system infection often leading to meningitis, encephalomyelitis, or other diseases
1993	Sin Nombre virus	Hantavirus respiratory syndrome
1994	Sabia virus	Brazilian hemorrhagic fever
1994	Ross River virus	Ross River viral disease (Australia)
1994	Human herpesvirus 8 (HHV-8)	Associated with Kaposi's sarcoma in AIDS patients
1996	O'nyoung-nyong virus	Epidemic O'nyong fever
1997	Deer tick virus	Enzootic tick-borne encephalitis
1997	Avian flu (H5N1) virus	Avian influenza illness
1997	Transfusion-transmitted virus (TTV)	Hepatitis
1999	Australian bat lyssavirus (ABL)	ABL infection
2000	Hepatitis G	Chronic liver inflammation
2002	Metapneumovirus	Respiratory tract infections
2003	SARS-Coronavirus	Severe acute respiratory syndrome (SARS)

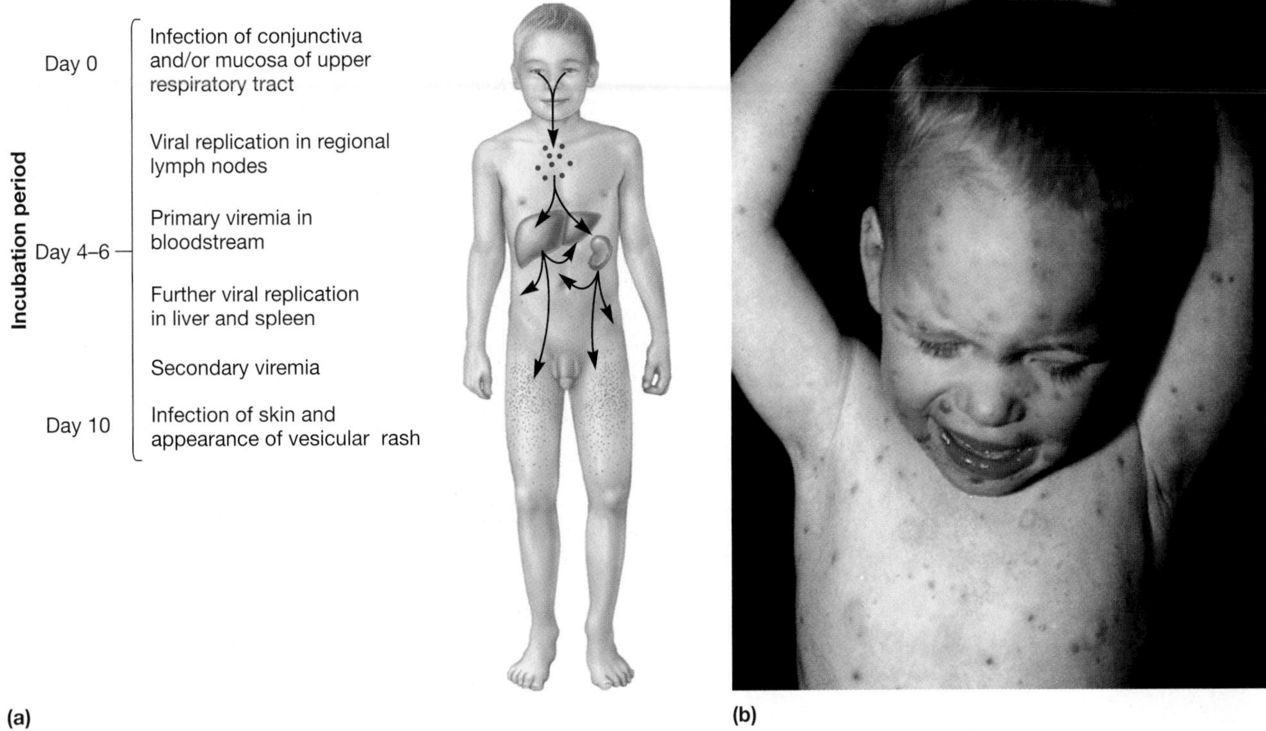

Figure 37.1 Chickenpox (Varicella). **(a)** Course of infection. **(b)** Typical vesicular skin rash. This rash occurs all over the body, but is heaviest on the trunk and diminishes in intensity toward the periphery.

Individuals who recover from chickenpox are subsequently immune to this disease; however, they are not free of the virus, as viral DNA resides in a dormant (latent) state within the nuclei of cranial nerves and sensory neurons in the dorsal root ganglia. This viral DNA is maintained in infected cells but virions cannot be detected (**figure 37.2*a***). When the infected person becomes immunocompromised by such factors as age, neoplastic diseases, organ transplants, AIDS, or psychological or physiological stress, the viruses may become activated (figure 37.2*b*). They migrate down sensory nerves, initiate viral replication, and produce painful vesicles because of sensory nerve damage (figure 37.2*c*). This syndrome is called **postherpetic neuralgia.** To manage the intense pain, corticosteroids or the drug gabapentin (Neurontin) can be prescribed. The reactivated form of chickenpox is called **shingles (herpes zoster).** Most cases occur in people over 50 years of age. Except for the pain of postherpetic neuralgia, shingles does not require specific therapy; however, in immunocompromised individuals, acyclovir, valacyclovir, vidarabine (Vira-A), or famciclovir (Famvir) are recommended. More than 500,000 cases of herpes zoster occur annually in the United States.

Influenza (Flu)

Influenza [Italian, to be influenced by the stars—*un influenza di freddo*], or the **flu,** is a respiratory system disease caused by negative-strand RNA viruses that belong to the family *Orthomyxoviridae*. There are four groups: influenza A, influenza B, influenza C, and Thogoto viruses. They contain 7 to 8 segments of linear RNA, with a genome length between 12,000 to 15,000 nucleotides. The enveloped virion can be spherical (50–120 nm in diameter) or filamentous (200–300 nm long, 20 nm in diameter) (*see figure 16.4*). Influenza A infections are responsible for the majority of clinical influenza cases, with influenza B accounting for approximately 3% of flu in the United States. Influenza A infections usually peak in the winter and involve 10% or more of the population, with rates as high as 50 to 75% in school-age children. Influenza A viruses are widely distributed, infect a variety of mammal and bird hosts, and are further classified into subtypes (or strains) based on their membrane surface glycoproteins, hemagglutinin (HA), and neuraminidase (NA). HA and NA function in viral attachment and virulence. There are 16 HA and 9 NA antigenic forms known; they can recombine to produce various HA/NA subtypes of influenza. All subtype combinations infect birds. Influenza A viruses having H1, H2, and H3 HA antigens, along with N1 and N2 NA antigens, are predominant in nature, infecting humans since the early 1900s. H1N1 viruses appeared in 1918 and were replaced in 1957 by H2N2 subtypes as the predominant subtype. The H2N2 viruses were subsequently replaced by H3N2 as the principle subtypes in 1968. The H1N1 subtype reappeared in 1977 and co-circulates today with H2N1, H3N2, H5N2, H7N2, H7N3, H7N7, H9N2, H10N7, and H5N1 viruses.

The H5N1 subtype (also know as bird flu) appears to be the most likely candidate to initiate another influenza pandemic (global epidemic). The H5N1 subtype was responsible for six

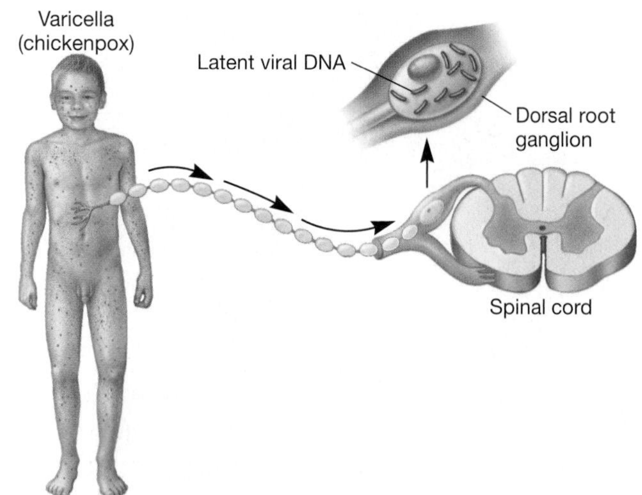

(a) Primary infection—Chickenpox

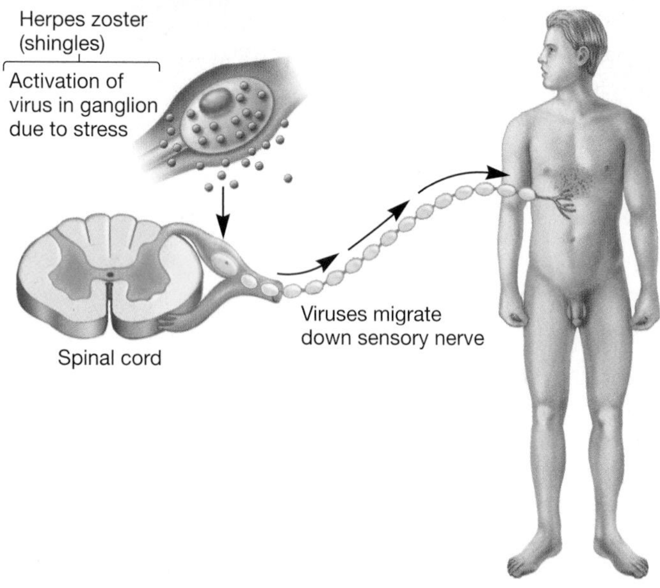

(b) Recurrence—Shingles

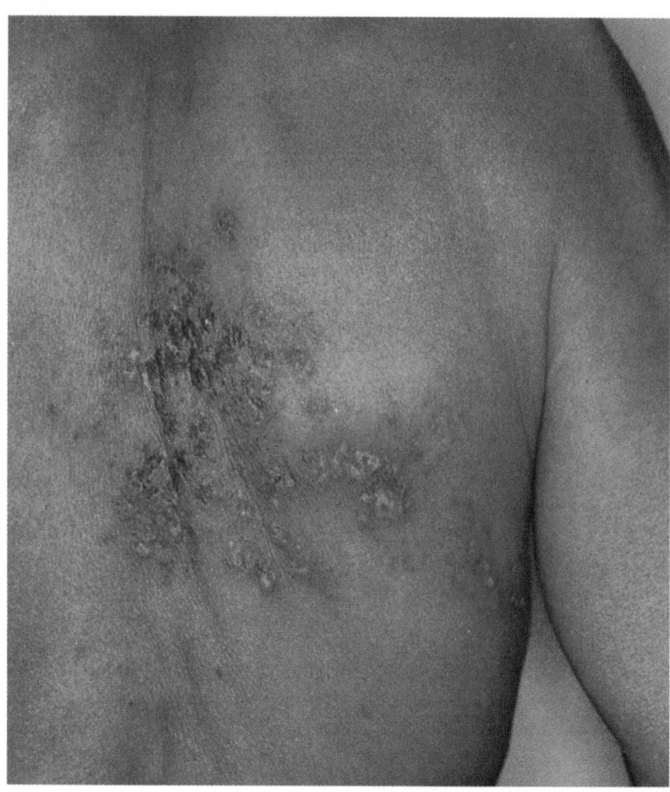

(c)

Figure 37.2 Pathogenesis of the Varicella-Zoster Virus. **(a)** After an initial infection with varicella (chickenpox), the viruses migrate up sensory peripheral nerves to their dorsal root ganglia, producing a latent infection. **(b)** When a person becomes immunocompromised or is under psychological or physiological stress, the viruses may be activated. **(c)** They migrate down sensory nerve axons, initiate viral replication, and produce painful vesicles. Since these vesicles usually appear around the trunk of the body, the name *zoster* (Greek for girdle) was used.

deaths in 1997, although it had been sporadically detected prior to that. Throughout 2003 to 2006, however, H5N1 was responsible for a substantial number of bird infections, resulting in the culling of millions of birds. The birds were destroyed to help prevent transmission of the virus to susceptible humans. Yet, by September 2006, H5N1 was responsible for 241 confirmed cases of influenza and 141 deaths in six countries.

One of the most important features of the influenza viruses is the frequency with which changes in antigenicity occur. If the variation is small, it is called **antigenic drift** (*see figure 36.6*). Antigenic drift results from the accumulation of mutations of HA and NA in a single strain of flu virus within a geographic region. This usually occurs every 2 to 3 years, causing local increases in the number of flu infections. **Antigenic shift** is a large antigenic change resulting from the reassortment of genomes when two different strains of flu viruses (from both animals and humans) infect the same host cell and are incorporated into a single new capsid (*see figure 36.6*). Because there is a greater change with antigenic shift than antigenic drift, antigenic shifts can yield major epidemics and pandemics. Antigenic variation occurs almost yearly with influenza A virus, less frequently with the B virus, and has not been demonstrated in the C virus. Recognition of an epidemic (section 36.4)

The standard nomenclature system for influenza virus subtypes includes the following information: group (A, B, or C), host of origin, geographic location, strain number, and year of original isolation. Antigenic descriptions of the HA and NA are given in parentheses for type A. The host of origin is not indicated for human isolates—for example, A/Hong Kong/03/68 (H3N2). However, the host origin is given for others—for example, A/swine/Iowa/15/30 (H1N1).

Animal reservoirs are critical to the epidemiology of human influenza. For example, rural China is one region of the world where chickens, pigs, and humans live in close, crowded conditions. Influenza is widespread in chickens; although chickens can't usually transmit the virus to humans, they can transfer it to pigs. Pigs can transfer it to humans, and humans back to pigs. Recombination between human and avian strains thus occurs in pigs, leading to major antigenic shifts. This explains why influenza continues to be a major epidemic disease and frequently produces worldwide pandemics. Hippocrates described influenza in 412 B.C. The first well-documented global epidemic of influenza-like disease occurred in 1580. Since then, 31 possible influenza pandemics have been documented, with four occurring in the twentieth century. The worst pandemic on record occurred in 1918 and killed between 20 million and 50 million people. This disaster, traced to the Spanish influenza virus (*see figure 36.10*), was followed by pandemics of Asian flu (1957), Hong Kong flu (1968), and Russian flu (1977). (The names reflect popular impressions of where the episodes began, although all are now thought to have originated in China.)

The 2003–2004 U.S. influenza season began earlier than most and was moderately severe. Influenza A (H1N1, H1N2, and H3N2) and influenza B viruses co-circulated, with the predominant strain being influenza A (H3N2). In 2005, influenza activity in the United States peaked early in February and then declined, surprising many who predicted a severe flu season. However, widespread outbreaks of avian influenza A (H5N1) continued from late 2003 and were reported in Southeast Asia in 2004 through 2006, predominantly among poultry. In a number of Asian countries, though, these outbreaks were associated with severe human illnesses and deaths. Because of potentially severe consequences of pandemic H5N1 infection of humans, an international effort coordinated by the World Health Organization monitors virus surveillance, epidemiology, and control efforts. The National Institutes of Health have contracted for the manufacture of a vaccine to H5N1. It began clinical trials in 2005.

The virus (*see figure 18.9*) is acquired by inhalation or ingestion of virus-contaminated respiratory secretions. During an incubation period of 1 to 2 days, the virus adheres to the epithelium of the respiratory system (the neuraminidase present in envelope spikes may hydrolyze the mucus that covers the epithelium). The virus attaches to the epithelial cell by its hemagglutinin spike protein, causing part of the cell's plasma membrane to bulge inward, seal off, and form a vesicle (receptor-mediated endocytosis). This encloses the virus in an endosome (*see figure 18.11*). The hemagglutinin molecule in the virus envelope undergoes a dramatic conformational change when the endosomal pH decreases.The hydrophobic ends of the hemagglutinin spring outward and extend toward the endosomal membrane. After they contact the membrane, fusion occurs and the RNA nucleocapsid is released into the cytoplasmic matrix.

Influenza is characterized by chills, fever (usually > 102°F, 39°C), headache, malaise, cough, sore throat, and general muscular aches and pains. These symptoms arise from the death of respiratory epithelial cells, probably due to attacks by activated T cells. These symptoms are more debilitating than are symptoms of the common cold. Recovery usually occurs in 3 to 7 days, during which coldlike symptoms appear as the fever subsides. Influenza alone usually is not fatal. However, death may result from pneumonia caused by secondary bacterial invaders such as *Staphylococcus aureus, Streptococcus pneumoniae,* and *Haemophilus influenzae.* A commercially available identification technique is Directigen FLU-A (an enzyme immunoassay [EIA] rapid test). This test can detect influenza A virus in clinical specimens in less than 15 minutes.

As with many other viral diseases, only the symptoms of influenza usually are treated. Amantadine (Symmetrel) (*see figure 34.21*), rimantadine (Flumadine), zanamivir (Relenza), and oseltamivir (Tamiflu) have been shown to reduce the duration and symptoms of type A influenza if administered during the first two days of illness. Unfortunately, 91% of the virus samples (representing the predominant influenza strain) tested by the CDC in December 2005 were resistant to rimantidine and amantidine, compared to 11% in the previous year. Amantadine and rimantadine are chemically related, antiviral drugs known as adamantanes. These usually have activity against influenza A viruses but not influenza B viruses. Amantadine and rimantadine are thought to interfere with influenza A virus M2 protein, a membrane ion channel protein. They also inhibit virus uncoating, which inhibits virus replication, resulting in decreased viral shedding. Zanamivir and oseltamivir are chemically related antiviral drugs known as neuraminidase inhibitors that have activity against both influenza A and B viruses. Neuraminidase inhibitors attack the virus directly by plugging the catalytic site of the enzyme neuraminidase. With the enzyme inactivated, viral particles can't travel from cell to cell. Importantly, aspirin (salicylic acid) should be avoided in children younger than 14 years to reduce the risk of Reye's syndrome (**Disease 37.1**).

The mainstay for prevention of influenza since the late 1940s has been inactivated virus vaccines, (*see table 36.4*), especially for the chronically ill, individuals over age 65, residents of nursing homes, and health-care workers in close contact with people at risk. Clinical disease in these patients is most likely to be severe. Because of influenza's high genetic variability, efforts are made each year to incorporate new virus subtypes into the vaccine. Even when no new subtypes are identified in a given year, annual immunization is still recommended because immunity using the inactivated virus vaccine typically lasts only 1 to 2 years.

Antiviral drugs (section 34.8)

Measles (Rubeola)

Measles [**rubeola:** Latin *rubeus,* red] is a highly contagious skin disease that is endemic throughout most of the world. In March 2000, a group of experts convened by the CDC concluded that fortunately, measles is no longer endemic in the United States. It seems that all cases of measles (less than 200 per year since 1998) in the United States were imported from other countries, usually Europe and Asia. We discuss measles because it remains of global importance. It is a negative-strand, enveloped RNA virus, in the genus *Morbillivirus* and the family *Paramyxoviridae.* The measles virus is monotypic, but small variations at the epitope level have been described.

Disease

37.1 Reye's and Guillain-Barré Syndromes

An occasional complication of influenza, chickenpox, and a few other viral diseases in children under 14 years of age is **Reye's syndrome.** After the initial infection has disappeared, the child suddenly begins to vomit persistently and experiences convulsions followed by delirium and a coma. Pathologically, the brain swells with injury to the neuronal mitochondria, fatty infiltration into the liver occurs, blood ammonia is elevated, and both serum glutamic oxaloacetic transaminase (SGOT) and serum glutamic pyruvic transaminase (SGPT) are elevated in the blood. Diagnosis is made by the measurement of the levels of these enzymes and ammonia.

The relationship between the initial viral infection and the brain and liver damage is unknown. Treatment is nonspecific and directed toward reducing intracranial pressure and correcting metabolic and electrolyte abnormalities. Some children who recover have residual neurological deficits—impaired mental capacity, seizures, and hemiplegia (paralysis on one side of the body). Mortality ranges from 10 to 40%. It is suspected that the use of aspirin or salicylate-containing products to lower the initial viral fever increases a child's chances of acquiring Reye's syndrome.

Another condition that involves the central nervous system and is associated with influenza infections is **Guillain-Barré syndrome** (sometimes called **French polio**). In this disorder the individual suffers a delayed reaction (usually within 8 weeks) either to the actual virus infection or to vaccines against influenza. The virus or viral antigen in the vaccine damages the Schwann cells that myelinate the peripheral nerves and thus causes demyelination. As a result a prominent feature of this syndrome is a symmetric weakness of the extremities and sensory loss. Fortunately recovery usually is complete because the remaining undamaged Schwann cells eventually proliferate and wrap around the demyelinated nerves.

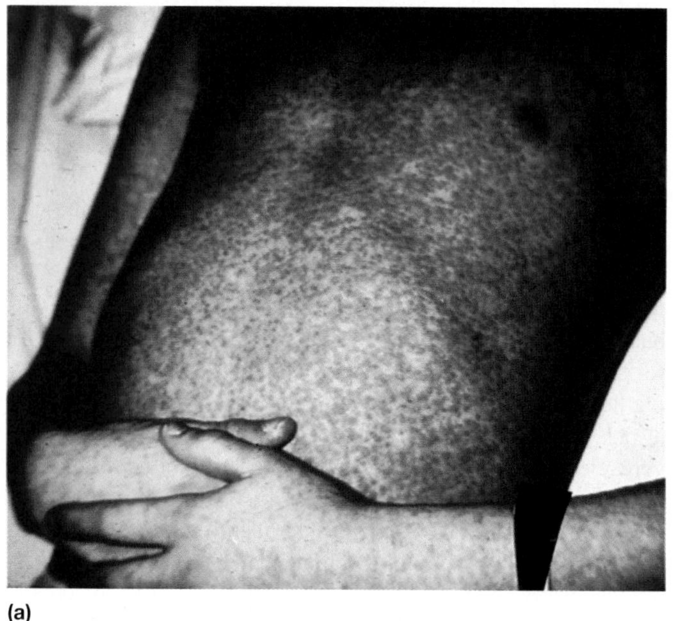

(a)

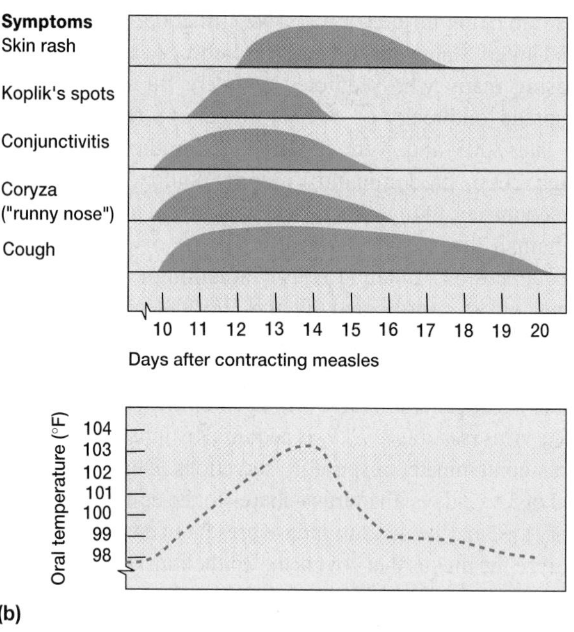

(b)

Figure 37.3 Measles (Rubeola). **(a)** The rash of small, raised spots is typical of measles. The rash usually begins on the face and moves downward to the trunk. **(b)** Signs and symptoms of a measles infection.

The variations are based on genetic variability in the virus genes. Such variations, however, have no effect on protective function since a measles infection still provides a lifelong immunity against reinfection. The virus enters the body through the respiratory tract or the conjunctiva of the eyes. The receptor for the measles virus is the complement regulator CD46, also known as membrane cofactor protein.

The incubation period is usually 10 to 21 days, and the first symptoms begin about the tenth day with a nasal discharge, cough, fever, headache, and conjunctivitis. Within 3 to 5 days skin eruptions occur as faintly pink maculopapular lesions that are at first discrete, but gradually become confluent (**figure 37.3**). The rash normally lasts about 5 to 10 days. Lesions of the oral cavity include the diagnostically useful bright-red **Koplik's spots** with a bluish-white speck in the center of each. Koplik's spots represent a viral exanthem (a skin eruption) occurring in the form of macules or papules as a result of the viral infection. Very infrequently a progressive degeneration of the central nervous system called **subacute sclerosing panencephalitis** occurs. No specific treatment is available for measles. The use of attenuated measles vaccine (Attenuvax) or in combination (MMR vaccine; *m*easles, *m*umps, *r*ubella) is recommended for all children (*see table 36.4*). Since public health immunization programs began in 1963, there has been near eradication of measles in the United States. In less well-developed countries, however, the morbidity and mortality in young children from measles infection remain high. It has been estimated that measles infects 50 million people and kills about 4 million a year worldwide. Serious outbreaks of measles are still reported in North America and Europe, especially among college students.

Mumps

Mumps is an acute, generalized disease that occurs primarily in school-age children. The number of mumps cases in the United States has decreased 99% since the widespread use of the MMR vaccine, with fewer than 300 cases reported annually. Occasional outbreaks occur in nonimmunized populations. The mumps virus is a member of the genus *Rubulavirus* in the family *Paramyxoviridae.* This virus is a pleomorphic, enveloped virus that contains a helical nucleocapsid containing negative-strand RNA. The virus is transmitted in saliva and respiratory droplets. The portal of entry is the respiratory tract. Mumps is about as contagious as influenza and rubella, but less so than measles or chickenpox. The virus replicates in the nasopharynx and lymph nodes of an infected person. Viral transmission is airborne or through direct contact with contaminated droplets or saliva. The most prominent manifestations of mumps are swelling and tenderness of the salivary (parotid) glands 16 to 18 days after infection of the host by the virus (**figure 37.4**). The swelling usually lasts for 1 to 2 weeks and is accompanied by a low-grade fever. Severe complications of mumps are rare, however, meningitis and inflammation of the epididymis and testes (**orchitis**) can be important complications associated with this disease—especially in the postpubescent male. Therapy of mumps is limited to symptomatic and supportive measures. A live, attenuated mumps virus vaccine is available. It usually is given as part of the trivalent MMR vaccine (*see table 36.4*).

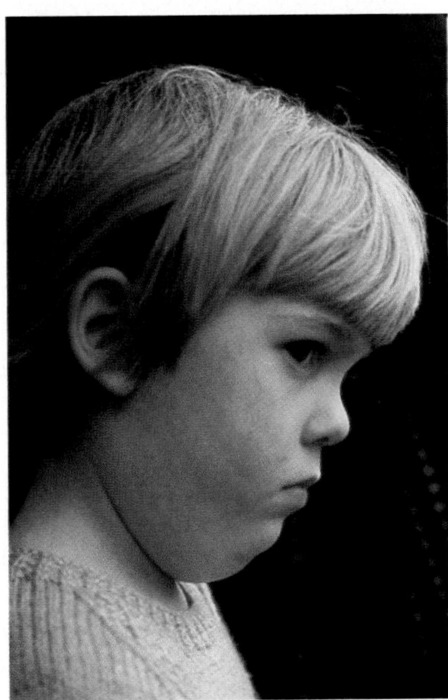

Figure 37.4 Mumps. Child with diffuse swelling of the salivary (parotid) glands due to the mumps virus.

Respiratory Syndromes and Viral Pneumonia

Acute viral infections of the respiratory system are among the most common causes of human disease. The infectious agents are called the acute respiratory viruses and collectively produce a variety of clinical manifestations, including rhinitis (inflammation of the mucous membranes of the nose), tonsillitis, laryngitis, and bronchitis. The adenoviruses, coxsackievirus A, coxsackievirus B, echovirus, influenza viruses, parainfluenza viruses, poliovirus, respiratory syncytial virus, and reovirus are thought to be responsible. It should be emphasized that for most of these viruses there is a lack of specific correlation between the agent and the clinical manifestation—hence the term syndrome. Immunity is not complete, and reinfection is common. The best treatment is rest. Taxonomy of eucaryotic viruses (section 18.1)

In those cases of pneumonia for which no cause can be identified, viral pneumonia may be assumed if mycoplasmal pneumonia has been ruled out. The clinical picture is nonspecific. Symptoms may be mild, or there may be severe illness and death.

Respiratory syncytial virus (RSV) often is described as the most dangerous cause of lower respiratory infections in young children. In the United States over 90,000 infants are hospitalized each year and at least 4,000 die. RSV is a member of the negative-strand RNA virus family *Paramyxoviridae.* The virion is variable in shape and size (average diameter of between 120 and 300 nm). It is enveloped with two virally specific glycoproteins as part of the structure. One of the two glycoproteins, G, is responsible for the binding of the virus to the host cell; the other, the fusion protein or F, permits fusion of the viral envelope with the host cell plasma membrane, leading to entry of the virus. The F protein also induces the fusion of the plasma membranes of infected cells. RSV thus gets its name from the resulting formation of a syncytium or multinucleated mass of fused cells. The multinucleated syncytia are responsible for inflammation, alveolar thickening, and the filling of alveolar spaces with fluid. The source of the RSV is hand contact and respiratory secretions of humans. The virus is unstable in the environment (surviving only a few hours on environmental surfaces), and is readily inactivated with soap and water and disinfectants.

Clinical manifestations consist of an acute onset of fever, cough, rhinitis, and nasal congestion. In infants and young children, this often progresses to severe bronchitis and viral pneumonia. Diagnosis is by either Directigen RSV or Test-Pack RSV rapid test kits. The virus is found worldwide and causes seasonal (November to March) outbreaks lasting several months. Treatment is with inhaled ribavirin (Virazole). A series of antibody (RSV-immune globulin) injections has been shown to reduce the severity of this disease in infants by 75%. Prevention and control consists of isolation for RSV-infected individuals, use of nursing barrier protection, and strict attention to good handwashing practices.

Rubella (German Measles)

Rubella [Latin *rubellus,* reddish] was first described in Germany in the 1800s and was subsequently called **German measles.** It is a moderately contagious disease that occurs primarily in children 5 to 9 years of age. It is caused by the rubella virus, an enveloped, single, positive-stranded RNA virus that is a member of the family *Togaviridae.* Rubella is worldwide in distribution and occurs more frequently during the winter and spring months. This virus is spread in droplets that are shed from the respiratory secretions of infected individuals. Once the virus is inside the body, the incubation period ranges from 12 to 23 days. A rash of small red spots (**figure 37.5**), usually lasting no more than 3 days, and a light fever are the normal symptoms. (This is why rubella is sometimes referred to as the "three-day measles.") The rash appears as immunity develops and the virus disappears from the blood, suggesting that the rash is immunologically mediated and not caused by the virus infecting skin cells.

Rubella can be a disastrous disease in the first trimester of pregnancy (**congenital rubella syndrome**) and can lead to fetal death, premature delivery, or a wide array of congenital defects that affect the heart, eyes, and ears. However, because rubella is usually such a mild infection, no treatment is indicated. All children and women of childbearing age who have not been previously exposed to rubella should be vaccinated. The live attenuated rubella vaccine (part of MMR, *see table 36.4*) is recommended. Because routine vaccination began in the United States in 1969, fewer than 1,000 cases of rubella and 10 cases of congenital rubella currently occur annually.

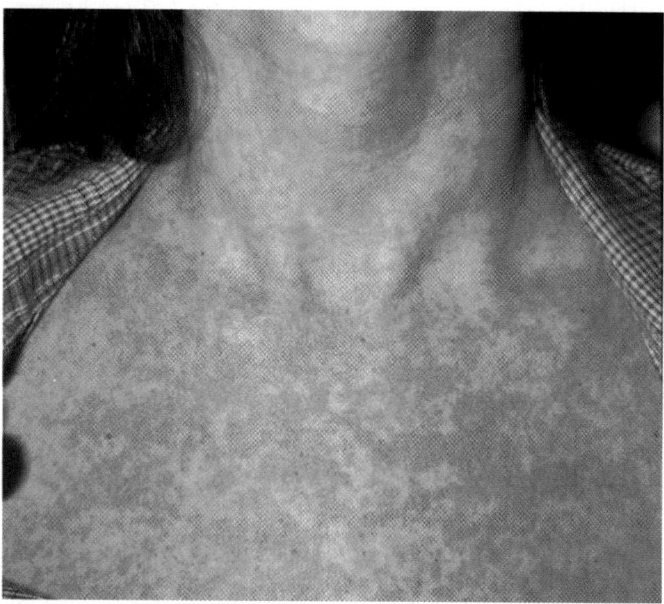

Figure 37.5 German Measles (Rubella). This disease is characterized by a rash of red spots. Notice that the spots are not raised above the surrounding skin as in measles (rubeola; see figure 37.3).

Severe Acute Respiratory Syndrome (SARS)

Severe acute respiratory syndrome (SARS) is a highly contagious viral disease caused by a novel coronavirus, known as the SARS-associate coronavirus (SARS-CoV). Coronaviruses are positive-strand RNA viruses. Coronaviruses are relatively large (120–150 nm), composed of RNA within a helical nucleocapsid, surrounded by a membranous envelope. Large peplomers (spikes) protrude from the envelope to aid in attachment and entry into host cells. The protruding peplomers extend from the oval-to-spherical virion to give the illusion of a halo, or corona, around the virus. The virus causes a febrile (>100.4°F or >38°C) lower respiratory track illness. Sudden, severe illness in otherwise healthy individuals is a hallmark of the disease. Other symptoms may include headache; mild, flu-like discomfort; and body aches. SARS patients may develop a dry cough after a few days, and most will develop pneumonia. About 10 to 20% of patients have diarrhea. If not detected early, this disease can be fatal even with supportive care. SARS is transmitted by close contact with respiratory secretions (droplet spread).

The initial outbreak of SARS appears to have originated in China in late 2002 and it spread rapidly to at least 29 other countries by summer 2003. The outbreak resulted in 8,098 persons with possible SARS, including 744 deaths being reported by the World Health Organization. There were 373 possible SARS cases in the United States; however, SARS-CoV identification has been confirmed in only 8 of them. Seven of the eight cases were likely due to exposure during international travel and the eighth case was probably due to exposure to one of the other seven. The 2003 SARS epidemic demonstrated to the world the ease with which a viral infectious agent can spread. Rapid detection and prevention measures were pursued during and after the outbreak. Diligent screening for signs of fever or respiratory disease at airports and the initiation of SARS-CoV vaccine trials are two examples of protective measures. No specific treatment is currently approved. Microbial diversity & ecology 18.1: SARS: Evolution of a virus

**Smallpox (Variola)

Smallpox (variola) is a highly contagious illness of humans caused by the orthopoxvirus, variola major. Characteristic symptoms of infection include acute onset of fever ≥101°F (38.3°C) followed by a rash that features firm, deep-seated vesicles or pustules in the same stage of development without other apparent cause. The variola virus belongs to the family *Poxviridae,* which includes vaccinia (cowpox and also the smallpox vaccine virus), monkeypox virus, and molluscum contagiosum virus. Humans are the only natural hosts of variola. The virion is large, brick-shaped, and contains a dumbell-shaped core. Interestingly, the size of the smallpox virus (300 by 250 to 200 nm) is slightly larger than that of some of the smallest bacteria—for example, *Chlamydia*. The genome inside the core consists of a single, linear molecule of double-stranded DNA and replicates in the host cell's cytoplasm. Smallpox was once one of the most prevalent of all diseases. It was a universally dreaded scourge for more than 3 millennia, with case fatality rates of 20 to 50%. First subjected to some control by variolation in tenth-century India and China, it was gradually suppressed in the industrialized world after Edward Jenner's 1796 landmark discovery that infection with the harmless cowpox (vaccinia) virus renders humans immune to the smallpox virus. Historical highlights 36.4: The first immunizations

Since the advent of immunization with the vaccinia virus, and because of concerted efforts by the World Health Organization, smallpox has been eradicated throughout the world—the greatest public health achievement ever. (The last case from a natural infection occurred in Somalia in 1977.) Eradication was possible because smallpox has obvious clinical features, virtually no asymptomatic carriers, only human hosts as reservoirs, and a short period of infectivity (3 to 4 weeks). The disease was successfully eliminated by a global immunization effort to prevent the spread of smallpox until no new cases developed.

There are two clinical forms of smallpox. Disease caused by Variola major is more severe and the most common form of smallpox, with a more extensive rash and higher fever. Historically, this form of smallpox had an overall fatality rate of about 33%; with significant morbidity in those who did not die. Variola minor is a less common form of smallpox, with much less severe disease and death rates of 1% or less. Variola is generally transmitted by direct and fairly prolonged face-to-face contact. It also can be spread through direct contact with infected bodily fluids or contaminated objects such as bedding or clothing. Smallpox has been reported to spread through the air in enclosed settings such as buildings, buses, and trains. Smallpox is not known to be transmitted by insects or animals.

The virus enters the respiratory tract, seeding the mucous membranes and passing rapidly into regional lymph nodes. The average incubation period is 12 to 14 days but can range from 7 to 17 days. During this time, the virus multiplies in the monocyte-macrophage system (*see figure 31.3*), but the host is not contagious. Another brief period of viremia precedes the prodromal phase (*see figure 36.3*). The prodromal phase lasts for 2 to 4 days and is characterized by malaise, severe head and body aches, occasional vomiting, and fever (over 40°C), all beginning abruptly. During the prodromal phase, the mucous membranes in the mouth and pharynx become infected, resulting is a rash of small, red spots. These spots develop into open sores that spread large amounts of the virus into the mouth and throat. At this time, the person is highly contagious. The virus then invades the capillary epithelium of the skin, leading to the development of the following sequence of lesions in/on the skin: eruptions, papules, vesicles, pustules, crusts, and desquamation (**figure 37.6**). The rash appears on the face, spreads to the arms and legs, and then the hands and feet. Usually the rash spreads over the body within 24 hours. By the third day of the rash, it forms raised bumps. By the fourth day, the bumps fill with a thick, opaque fluid and often have a depression in the center that looks like a bellybutton. (This is a major distinguishing characteristic of smallpox, as compared to other diseases that exhibit rashes.) The fever usually declines as the rash appears but rises and is sustained from the time the vesicles form until they crust over. Oropharyngeal and skin lesions contain abundant virus particles, particularly early in the disease process. Death from smallpox is due to toxemia associated with immune-mediated blood clots and elevated blood pressure.

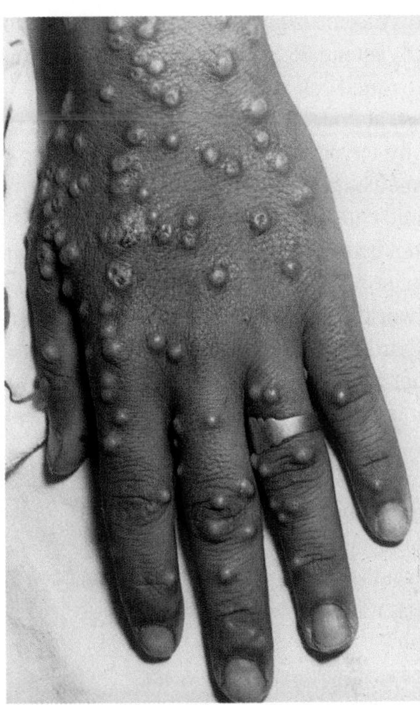

Figure 37.6 Smallpox. Back of hand showing single crop of smallpox vesicles.

Protection from smallpox is through vaccination (once referred to as variolation). The smallpox vaccine is a live virus immunization using the related vaccina virus. The vaccine is given using a bifurcated (two-pronged) needle that is dipped into the vaccine. The bifurcated needle retains a droplet of the vaccine so that pricking the skin allows vaccinia entry into the skin. This immunization practice causes a sore spot and one or two droplets of blood to form. If the vaccination is successful, a red and itchy bump develops at the vaccine site that becomes a large blister, filled with pus. The blister will dry and form a scab that falls off, leaving a small scar. People vaccinated for the first time have a stronger reaction than those who are revaccinated. Routine immunization for smallpox was discontinued in the United States once global eradication was confirmed. Today, smallpox vaccination is controversial in light of its unknown efficacy in bioterrorism prevention and potential side-effects. There is no FDA-approved treatment for smallpox, although several antiviral agents have been suggested as adjunct therapies.

A suspect case of smallpox should be managed in a negative-pressure room, if possible, and the patient should be vaccinated, particularly if the disease is in the early stage. Unlike most vaccines, when smallpox vaccine is given very early in the incubation period, it can markedly attenuate or even prevent clinical manifestations. Strict respiratory and contact isolation is important.

There is great concern that the smallpox virus could be used as a weapon by terrorists. An accidental or deliberate release of smallpox virus would be catastrophic in an unimmunized population and could cause a major pandemic. Because smallpox vaccination has not been performed routinely since about 1972, there is now a large population of susceptible persons. Currently, less than half the world's population has been exposed to either smallpox (variola virus) or to the vaccine. Thus if an outbreak occurred, prompt recognition and institution of control measures would be paramount. Historical Highlights 16.1: Disease and the early colonization of America

1. Why are chickenpox and shingles discussed together? What is their relationship?
2. Briefly describe the course of an influenza infection and how the virus causes the symptoms associated with the flu. Why has it been difficult to develop a single flu vaccine?
3. What are some common symptoms of measles?
4. What are Koplik's spots?
5. What is one side effect that mumps can cause in a young postpubescent male?
6. Describe some clinical manifestations caused by the acute respiratory viruses.
7. Is viral pneumonia a specific disease? Explain.
8. When is a German measles infection most dangerous and why?

37.2 Arthropod-Borne Diseases

The *ar*thropod-*bo*rne viruses (arboviruses) are transmitted by bloodsucking arthropods from one vertebrate host to another. They multiply in the tissues of the arthropod without producing disease, and the vector acquires a lifelong infection. Approximately 150 of the recognized arboviruses cause illness in humans. Diseases produced by the arboviruses can be divided into three clinical syndromes: (1) fevers of an undifferentiated type with or without a rash; (2) encephalitis (inflammation of the brain), often with a high fatality rate; and (3) hemorrhagic fevers, also frequently severe and fatal (**Disease 37.2**). **Table 37.2** summarizes the six major human arbovirus diseases that occur in the United States. For all of these diseases, immunity is believed to be permanent after a single infection. No vaccines are available for the human arthropod-borne diseases listed in table 37.2, although supportive treatment is beneficial.

**Equine Encephalitis

Equine encephalitis is caused by viruses in the genus *Alphavirus*, family *Togaviridae*. They are positive-strand, enveloped RNA viruses. In humans, the disease can present as a spectrum from fever and headache to (aseptic) meningitis and encephalitis. The disease can progress to include seizures, paralysis, coma, and death. The virus is transmitted to humans by *Aedes* and *Culex* spp. mosquitoes. Various geographic descriptors are used to define the disease caused by genetically distinct strains of these arboviruses: Eastern equine encephalitis (EEE) occurs along the eastern Atlantic coast from Canada to South America; Western equine encephalitis (WEE) occurs from Canada to South America along the western coast; and Venezuelan equine encephalitis occurs in central and southern parts of the United States into South America. Between 1964 and 2000, 182 cases of EEE and 640 cases of WEE were reported to the CDC. Reservoir hosts are important in the replication, maintenance, and dissemination of these arboviruses. Treatment consists of the supportive care of symptoms. The equine hosts generally show little or no disease after infection. Currently no vaccine is available to prevent disease. Preventative measures rely on common mosquito precautions.

**Tick-Borne Encephalitis

Tick-borne encephalitis (TBE) is a viral infection of the central nervous system caused by the TBE virus (TBEV) transmitted by ticks. TBEV is a positive-strand RNA virus in the genus *Flavivirus*, family *Flaviviridae*. Human infections are transmitted through bites from infected *Ixodes ricinus* ticks. Humans can also acquire the infection by consuming unpasteurized dairy products from infected cows, goats, or sheep. The disease most often manifests as meningitis, encephalitis, or meningoencephalitis. Long-lasting or permanent neuropsychiatric sequelae are observed in 10 to 20% of TBEV-infected patients. TBE is an important infectious disease in Europe and Asia. These are regions where the ixodid tick reservoir is found. The annual number of cases varies from year to year, with several thousand reported annually and many more unreported.

The incubation period for TBE is usually between 7 to 14 days. Shorter incubation times have been reported after milk-borne exposure. Following an initial asymptomatic phase, fever develops and lasts 2 to 4 days, corresponding to viremia. Other symptoms include malaise, anorexia, muscle aches, headache, nausea, and/or vomiting. A second phase of the disease occurs in 20 to 30% of patients after about 8 days of remission and involves central nervous system symptoms of meningitis (e.g., fever, headache, and a stiff neck) or encephalitis (e.g., drowsiness, confusion, sensory disturbances, and/or motor abnormalities such as paralysis) or meningoencephalitis. TBE is more severe in adults than in children. The virus can be isolated from the blood during the first phase of the disease. Specific diagnosis usually depends on detection of specific IgM in either blood or cerebral spinal fluid, usually appearing later, during the second phase of the disease. Prevention of TBE is achieved through use of vaccines (available in Europe and in Canada). U.S. data do not support routine immunization except for those at high risk for infection. Prevention involves common tick precautions.

**Rift Valley Fever

Rift Valley fever (RVF) is an acute, febrile disease caused by a negative-strand RNA virus in the genus *Phlebovirus*, family *Bunyaviridae*. RVF was first reported in the early 1900s as a disease

Disease

37.2 Viral Hemorrhagic Fevers—A Microbial History Lesson

Scientists know of several viruses lurking in the tropics that—with a little help from nature—could wreak far more loss of life than will likely result from the AIDS pandemic. Collectively these viruses produce what is known as the **hemorrhagic fevers.** The viruses are passed among wild vertebrates, which serve as reservoir hosts. Arthropods transmit the viruses among vertebrates, and humans are infected when they invade the environment of the natural host. These diseases, distributed throughout the world, are known by over 70 names, usually denoting the geographic area where they were first described.

Viral hemorrhagic fevers can be fatal. Patients suffer headache, muscle pain, flushing of the skin, massive hemorrhaging either locally or throughout the body, circulatory shock, and death.

The first documented cases of hemorrhagic fever occurred in the late 1960s when dozens of scientists in West Germany fell seriously ill, and several died. Victims suffered from a breakdown of liver function and a bizarre combination of bleeding and blood clots. The World Health Organization traced the outbreak to a batch of fresh monkey cells the scientists had used to grow polioviruses. The cells from the imported Ugandan monkeys were infected with the lethal tropical Marburg virus (see **Box figure**) and the scientists suffered from Marburg viral hemorrhagic fever.

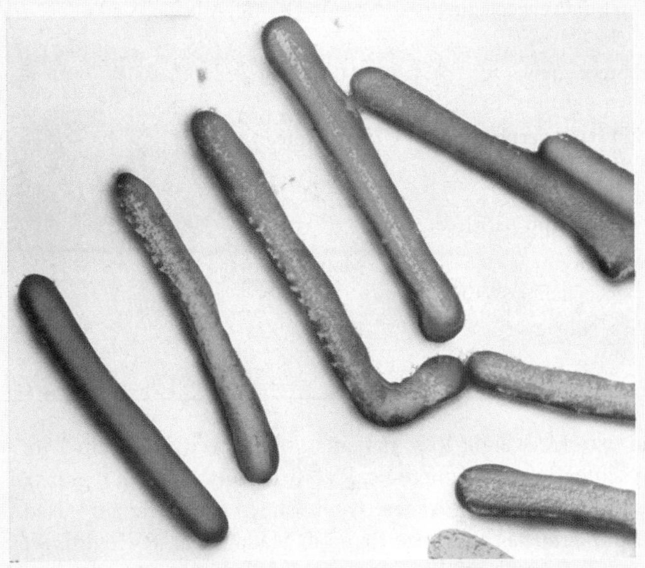

Sinister Foe. The deadly Marburg virus was first isolated in 1967 at the Institute for Hygiene and Microbiology in Marburg, Germany.

In 1977 the *Plebovirus* causing Rift Valley Fever in sheep and cattle moved from these animals into the South African population. The virus, which causes severe weakness, incapacitating headaches, damage to the retina, and hemorrhaging, then made its way to Egypt, where millions of humans became infected and thousands died.

Among the most frightening hemorrhagic outbreak was that of the Ebola virus hemorrhagic fever in Zaire and Sudan in 1976. This disease infected more than 1,000 people and left over 500 dead. It became concentrated in hospitals, where it killed many of the Belgian physicians and nurses treating infected patients. Since that time there have been numerous additional outbreaks in which hundreds of patients and health care workers have died. All of these outbreaks have so far occurred in Africa.

In the United States in 1989, epidemiologists provided new evidence that rats infected with a potentially deadly hemorrhagic virus are prevalent in Baltimore slums. The virus appears to be taking a previously unrecognized toll on the urban poor by causing Korean hemorrhagic fever.

In the summer of 1993, reports appeared in the news media about a mysterious illness that had caused over 30 deaths among the Navajo Nation in the Four-Corners area of the southwestern United States. The CDC finally determined the causative agent to be a hantavirus, a negative-strand RNA virus that is a member of the family *Bunyaviridae.* Hantaviruses are endemic in rodents, such as deer mice, in many areas of the world. Deer mice shed the virus in their saliva, feces, and urine. Humans contract the disease when they inhale aerosolized particles containing the excreted virus. Throughout Asia and central Europe, hantaviruses cause hemorrhagic fever with renal syndrome in humans. But the type of virus found in the Southwest had not been previously recognized, and no hantavirus anywhere in the world had been associated with the clinical syndrome initially seen among the Navajo; namely, the hantavirus pulmonary syndrome in which the virus destroys the lungs. In 1993 the CDC named this virus **pulmonary syndrome hantavirus** (sometimes called the Sin Nombre or no-name virus) and isolated cases have since been reported from almost every state. Prevention involves wearing gloves when handling mice and spraying the feces and urine of all mice with a disinfectant.

Although to date, these epidemics have not become global, they provide a humbling vision of humankind's vulnerability. History shows that the life-threatening viral hemorrhagic outbreaks often arise when humans move into unexplored terrain or when living conditions deteriorate in ways that generate new viral hosts. In each case medical and scientific resources have been reactive, not proactive.

of livestock in Kenya. The disease is transmitted by mosquitoes and infects large numbers of livestock and domestic animals; that is, it is an **epizoonotic** disease. Data also suggest that the blood of infected animals and other biting insects can transmit the virus.

Infection with RVF virus typically presents with no symptoms or as a mild, febrile illness associated with abnormal liver function. However, in some patients the illness can progress to a hemorrhagic fever, encephalitis, or disease of the eye. Symptoms usually include fever, back pain, dizziness, generalized weakness,

Table 37.2 Summary of the Six Major Human Arbovirus Diseases That Occur in the United States

Disease	Distribution	Vectors	Natural Host	Mortality Rate
California encephalitis (La Crosse)	North Central, Atlantic, South	Mosquitoes (*Aedes* spp.)	Birds	Fatalities rare
Eastern equine encephalitis (EEE)	Atlantic, Southern Coast	Mosquitoes (*Aedes* spp.)	Birds	50–70%
St. Louis encephalitis (SLE)	Widespread	Mosquitoes (*Culex* spp.)	Birds	10–30%
Venezuelan equine encephalitis (VEE)	Southern United States	Mosquitoes (*Aedes* spp. and *Culex* spp.)	Rodents, horses	20–30% (children) <10% (adults)
Western equine encephalitis (WEE)	Mountains of the West	Mosquitoes (*Culex* spp.)	Birds	3–7%
West Nile fever	Widespread	Mosquitoes (various spp.)	Birds	50–70% (elderly) <14% (all others)

and extreme weight loss. Patients usually recover within 2 to 7 days after onset of symptoms. RVF is usually found in regions of eastern and southern Africa (where sheep and cattle are raised), but the virus has also been found in Madagascar. RVF virus was found in Egypt in 1977 and caused a large outbreak among animals and people. The first human epidemic of RVF in West Africa was reported in 1987 and was linked to flooding that forced close interactions between humans and animals. In September 2000, a RVF outbreak was reported in Saudi Arabia and then in Yemen. These cases represent the first Rift Valley fever cases identified outside Africa. There is no established course of treatment for patients infected with RVF virus. However, studies in monkeys and other animals have shown promise for ribavirin, an antiviral drug, for future use in humans. Additional studies suggest that interferon, immune modulators, and convalescent-phase plasma may also help in the treatment of patients with RVF. Preventative measures rely on common mosquito precautions.

West Nile Fever (Encephalitis)

West Nile fever (encephalitis) is caused by a positive-strand RNA flavivirus that occurs primarily in the Middle East, Africa, and Southwest Asia. The disease was first discovered in 1937 in the West Nile district of Uganda. In 1999, the virus appeared unexpectedly in the United States (New York), causing seven deaths among 62 confirmed human encephalitis cases and extensive mortality in a variety of domestic and exotic birds. It probably crossed the Atlantic in an infected bird, mosquito, or human traveler.

By 2003, 46 U.S. states reported West Nile virus (WNV) infections in over 9,800 people, resulting in 264 deaths. By the start of 2006, every state in the continental United States reported West Nile virus in either animals or humans. WNV is transmitted predominately to humans by *Culex* spp. mosquitoes that feed on infected birds (crows and sparrows). Mosquitoes harbor the greatest concentration of virus in the early fall; there is a peak of disease in late August to early September. The risk of disease then decreases as the mosquitoes die when the weather becomes colder. Although many people are bitten by WNV-infected mosquitoes, most do not know they have been exposed. Most infected individuals remain asymptomatic or exhibit only mild, flu-like symptoms. Data from the outbreak in Queens, New York suggests that 2.6% of the population was infected, 20% of the infected people developed mild illness, and only 0.7% of the infected people developed meningitis or encephalitis.

Human-to-human transmission has been reported through blood and organ donation. However, the risk of acquiring WNV infection from donated blood or organs has greatly diminished since the introduction of a PCR-based detection assay in 2003. There are no data to suggest that WNV transmission to humans occurs from handling infected birds (live or dead), but barrier protection is suggested in handling potentially infected animals. The virus can be recovered from *Culex* mosquitoes, birds, and blood taken in the acute stage of a human infection. Diagnosis is by a rise in neutralizing antibody in a patient's serum. Only one antigenic type exists and immunity is presumed permanent. An ELISA test for IgM anti-WNV antibody is the FDA-approved diagnostic test. There is no treatment other than hospitalization and intravenous fluids. Clinical immunology (section 35.3)

There is no human vaccine to prevent WNV infection as of yet. Mosquito abatement and the use of repellents such as DEET appear to be the only control measures.

**Yellow Fever

Yellow fever is less lethal than other viral diseases caused by a flavivirus, and no longer occurs in the U.S. It is featured here because it remains a public health problem in Africa and South America, causing over 200,000 infections and 30,000 deaths each year. In addition it is a potential bioweapon. In the early years of the United States when trading with the West Indies was vital (1600–1800), yellow fever was dreaded because of its sudden appearance, and debilitating symptoms. Arriving at port cities, yellow fever epidemics would often ravage communities without respecting the previously observed disease barriers of affluence or social status. The epidemic of 1798, for example, claimed more than 5,000 victims between Boston and Philadelphia. Yellow fever was first identified by Benjamin Rush, who catalogued its signs and symptoms in attempt to discover its cause and cure. Yellow fever holds the distinction as the first human disease found to be caused by a virus (Walter Reed discovered this in 1901). It also provided the first confirmation (by Carlos Juan Finley) that an insect could transmit a virus.

The disease received its first name, yellow jack, because jaundice is a prominent sign in severe cases. The jaundice is due to the deposition of bile pigments in the skin and mucous membranes because of liver damage. The disease is spread through a population in two epidemiological patterns. In the urban cycle, human-to-human transmission is by *Aedes aegypti* mosquitoes. In the sylvatic cycle the mosquitoes transmit the virus between monkeys and from monkeys to humans (sylvatic means in the woods or affecting wild animals).

Once inside a person the virus spreads to local lymph nodes and multiplies; from this site it moves to the liver, spleen, kidneys, and heart, where it can persist for days. Yellow fever has an abrupt onset after an incubation period of 3 to 6 days, and usually includes fever, prostration, headache, sensitivity to light, low-back pain, extremity pain, epigastric pain, anorexia, and vomiting. The illness can progress to liver and renal failure, and hemorrhagic symptoms and signs caused by thrombocytopenia (low platelet count) and abnormal clotting and coagulation can occur. The fatality rate of severe yellow fever is approximately 20%.

Diagnosis of yellow fever is made by culture of virus from blood or tissue specimens or by identification of viral antigen or nucleic acid in tissues using *i*mmuno*h*isto*c*hemistry (IHC), ELISA antigen capture, or PCR. There is no specific treatment for yellow fever. An active immunity to yellow fever results from initial infection or from vaccines containing the attenuated yellow fever 17D strain or the Dakar strain virus. Prevention and control of this disease involves vaccination (*see table 36.4*) and control of the insect vector.

37.3 Direct Contact Diseases

Acquired Immune Deficiency Syndrome (AIDS)

It is now recognized that **AIDS (acquired immune deficiency syndrome)** was the great pandemic of the second half of the twentieth century. First described in 1981, AIDS is the result of an infection by the **human immunodeficiency virus (HIV),** a positive-strand, enveloped RNA virus within the family *Retroviridae.* Although the disease has been studied intensively for the past two and a half decades, its origin only now seems clear. Molecular epidemiology data indicate that HIV-1 arose from the SIVcpz retrovirus harbored by the chimpanzee, *Pan troglodytes troglodytes* (Ptt). The deduced evolutionary sequence suggests that SIVcpz ancestors recombined when they crossed between several species of nonhuman primates, then into chimpanzees, and finally into humans. Stable viral infections in humans occurred on at least three occasions with several groups of mutated SIVcpz viruses adapting to humans residing within chimp habitats. Full length genome analyses indicate that HIV-1 groups M, N, and O are most similiar to SIVcpz viruses of Ptt, evolving from separate SIVcpz lineages. However, only the SIVcpz strain now referred to as the group M HIV-1 gave rise to the virus causing the global AIDS pandemic. Notably, group M HIV-1 has diverged into several subtypes (clades) indicated as A-K. Subtype B was the first to appear in the United States and it remains the predominant type (>80%) through the Americas.

Epidemiologically, AIDS occurs worldwide although the HIV-2 strain predominates in West Africa (**figure 37.7**). HIV is acquired and may be passed from one person to another when infected blood, semen, or vaginal secretions come in contact with an uninfected person's broken skin or mucous membranes. In the developing world, AIDS affects men and women alike, with many women getting AIDS from their husbands who have multiple sex partners. In the United States, the groups most at risk for acquiring AIDS are (in descending order) men who have sex with other men; intravenous drug users; heterosexuals who have sex with drug users and prostitutes; children born of infected mothers, as well as their breast-fed infants; transfusion patients; and transplant recipients. Transmission of HIV in the latter risk groups is exceedingly rare due to extensive testing of blood products before use. The mortality rate from AIDS is almost 100% if it is not treated. The use of combined antiviral medications has significantly reduced the morbidity and mortality of AIDS in developed nations.

In the United States, AIDS is caused primarily by HIV-1 (some cases result from an HIV-2 infection). This virus is closely related to HTLV-1, the cause of adult T-cell leukemia, and HTLV-2, which has been isolated from individuals with hairy-cell leukemia. HIV-1 has a cylindrical core inside its capsid (**figure 37.8*a***). The

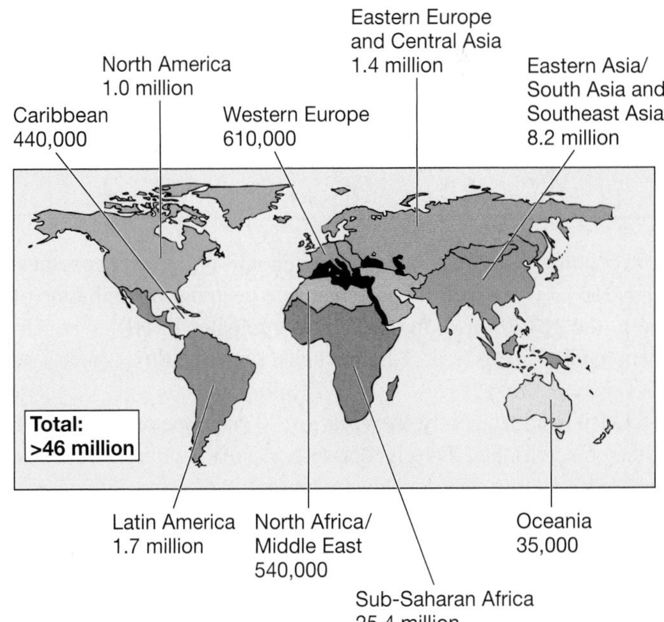

Figure 37.7 Distribution of HIV/AIDS in Adults by Continent or Region. The figure shows data from a 2003 United Nations report. According to recent estimates by the UN, the number of HIV/AIDS cases may be over 46 million. *Source of data: UNAIDS.*

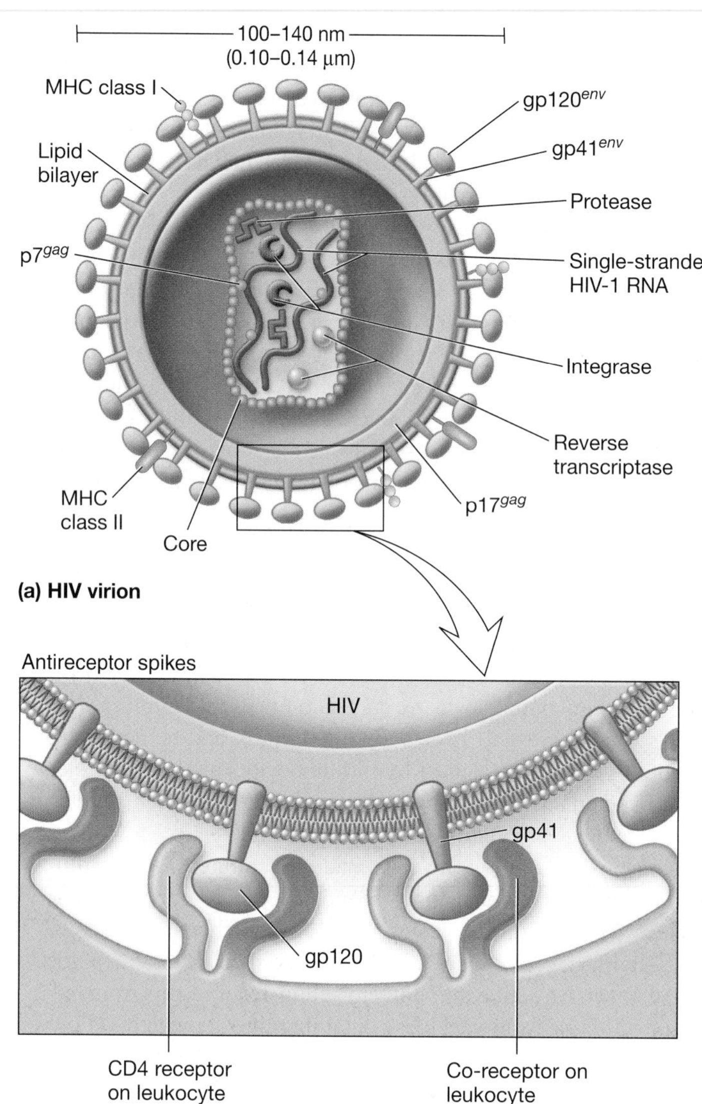

Figure 37.8 Schematic Diagram of the HIV-1 Virion. **(a)** The HIV-1 virion is an enveloped structure containing 72 external spikes. These spikes are formed by the two major viral-envelope proteins, gp120 and gp41. (gp stands for glycoprotein—proteins linked to sugars—and the number refers to the mass of the protein, in thousands of daltons.) The HIV-1 lipid bilayer is also studded with various host proteins, including class I and class II major histocompatibility complex molecules, acquired during virion budding. The cone-shaped core of HIV-1 contains four nucleocapsid proteins (p24, p9, p7) each of which is proteolytically cleaved from a 53 kDa *gag* precursor by the HIV-1 protease. The phosphorylated p24 polypeptide forms the chief component of the inner shell of the nucleocapsid, whereas the p17 protein is associated with the inner surface of the lipid bilayer and stabilizes the exterior and interior components of the virion. The p7 protein binds directly to the genomic RNA through a zinc finger structural motif and together with p9 forms the nucleoid core. The retroviral core contains two copies of the single stranded HIV-1 genomic RNA that is associated with the various preformed viral enzymes, including the reverse transcriptase, integrase, ribonuclease, and protease. **(b)** The snug attachment of HIV glycoprotein molecules (gp41 and gp120) to their specific receptors on a human cell membrane. These receptors are CD4 and a co-receptor called CXCR-5 (fusin) that permit docking with the host cell and fusion with the cell membrane.

core contains two copies of its RNA genome and several enzymes. Thus far 10 virus-specific proteins have been discovered. One of them, the gp120 envelope protein, participates in HIV-1 attachment to $CD4^+$ cells (e.g., T-helper cells; figure 37.8*b*). Viruses and cancer (section 18.5)

Once inside the body, the virus gp120 envelope protein (figure 37.8*b*) binds to the CD4 glycoprotein plasma membrane receptor on $CD4^+$ T cells, macrophages, dendritic cells, and monocytes. Dendritic cells are present throughout the body's mucosal surfaces and bear the CD4 protein. Thus it is possible that these are the first cells infected by HIV in sexual transmission. The virus requires a coreceptor in addition to the CD4 receptor. Macrophage-tropic strains, which seem to predominate early in the disease and infect both macrophages and T cells, require the CCR5 (CC-CKR-5) chemokine receptor protein as well as CD4. A second chemokine coreceptor, called CXCR-4 or fusin, is used by T-cell-tropic strains that are active at later stages of infection. These strains induce the formation of syncytia. Individuals with two defective copies of the *CCR5* gene do not seem to develop AIDS; apparently the virus cannot infect their T cells. People with one good copy of the *CCR5* gene do get AIDS but survive several years longer than those with no mutation. Reproduction of vertebrate viruses (section 18.2)

Entry into the host cell begins when the envelope fuses with the plasma membrane, and the virus releases its core containing two RNA strands into the cytoplasm (**figure 37.9*a***). Inside the infected cell, the core protein remains associated with the RNA as it is copied into a single strand of DNA by the RNA-dependent DNA polymerase activity of the reverse transcriptase enzyme. The RNA is next degraded by another reverse transcriptase component, ribonuclease H, and the DNA strand is duplicated to form a double-stranded DNA copy of the original RNA genome. A complex of the double-stranded DNA (the provirus) and the integrase enzyme moves into the nucleus. Then the proviral DNA is integrated into the cell's DNA through a complex sequence of reactions catalyzed

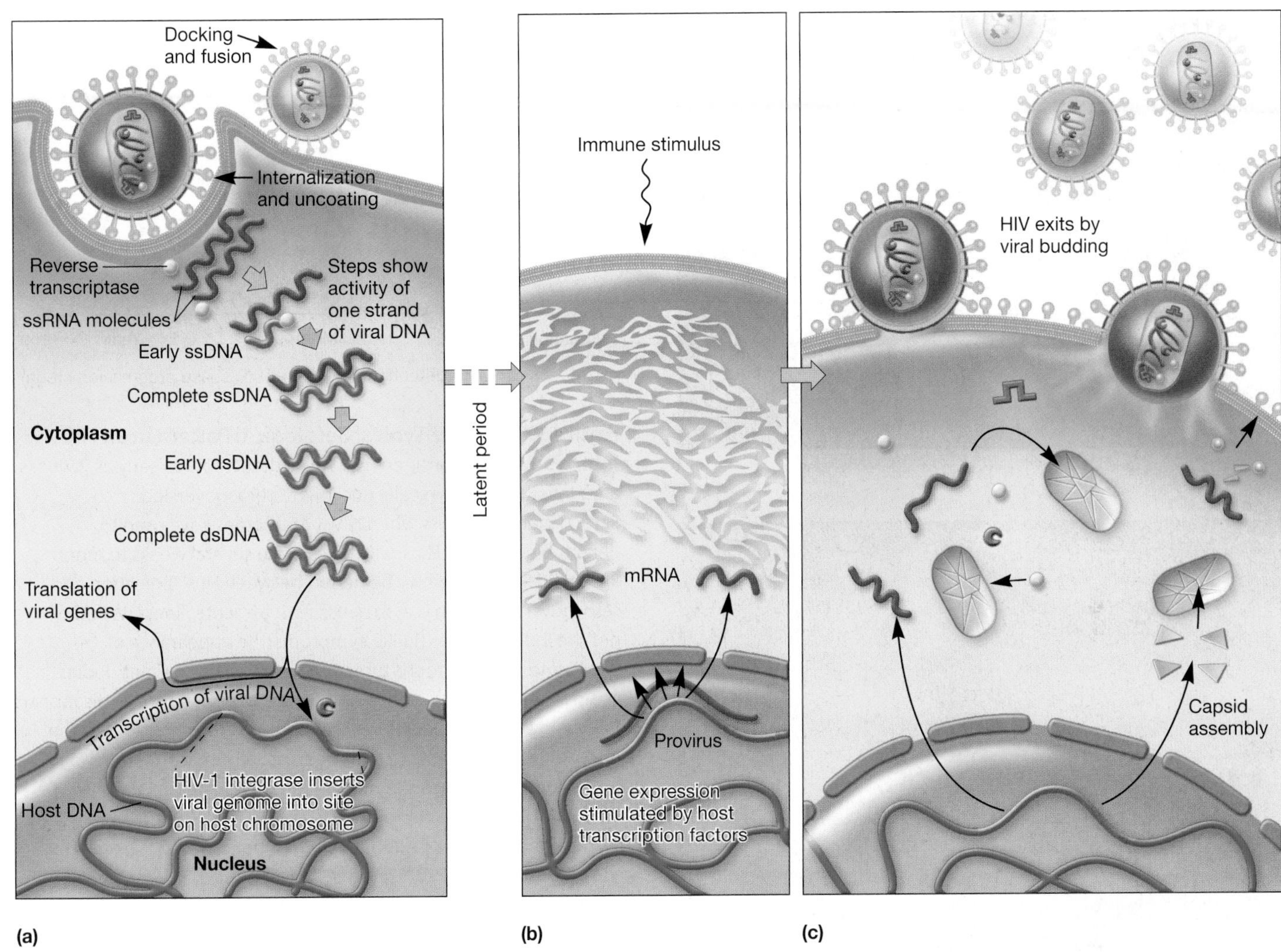

Figure 37.9 HIV Life Cycle. **(a)** The virus is adsorbed onto the $CD4^+$ host cell and endocytosed. The twin RNAs are uncoated and the reverse transcriptase catalyzes the synthesis of a single complementary strand of DNA. This DNA serves as a template for synthesis of double-stranded DNA. The dsDNA can be inserted into the host chromosome as a provirus (latency). **(b)** The provirus genes are transcribed. **(c)** Viral mRNA is translated into virus components (capsid, reverse transcriptase, spike proteins), and the virus is assembled. Mature viral particles bud from the host cell, taking host membrane as their envelope.

by the integrase. The integrated provirus can remain latent, giving no clinical sign of its presence. Alternatively the provirus can force the cell to synthesize viral mRNA (figure 37.9*b*). Some of the RNA is translated to produce viral proteins by the cell's ribosomes. Some of the proteins have been shown to affect host cell function; for example, HIV NEF decreases MHC class I expression and may prevent apoptosis at some infection stages. Viral proteins and the complete HIV-1 RNA genome are then assembled into new virions that bud from the infected host cell (figure 37.9*c*). Eventually the host cell lyses. Recognition of foreignness: Major histocompatibility complex (section 32.4)

Once a person becomes infected with HIV, the course of disease may vary greatly. Some rapid progressors may develop clinical AIDS and die within 2 to 3 years. A small percentage of long-term nonprogressors remain relatively healthy for at least 10 years after infection. For the majority of HIV-infected individuals, HIV infection progresses to AIDS in 8 to 10 years. The CDC has developed a classification system for the stages of HIV-related conditions: acute, asymptomatic, chronic symptomatic, and AIDS.

The acute infection stage occurs 2 to 8 weeks after HIV infection. About 70% of individuals in this stage experience a brief illness referred to as acute retroviral syndrome, with symptoms that may include fever, malaise, headache, macular rash, weight loss, lymph node enlargement (lymphadenopathy), and oral candidiasis (**figure 37.10*a***). During this stage the virus multiplies rapidly and disseminates to lymphoid tissues throughout the body, until an acquired immune response (antibodies and cytotoxic T cells; **figure 37.11**) can be generated to bring virus replication under control. During the acute infection stage, levels of HIV may reach 10^5 to 10^6 copies of viral RNA per mL of plasma. It is believed

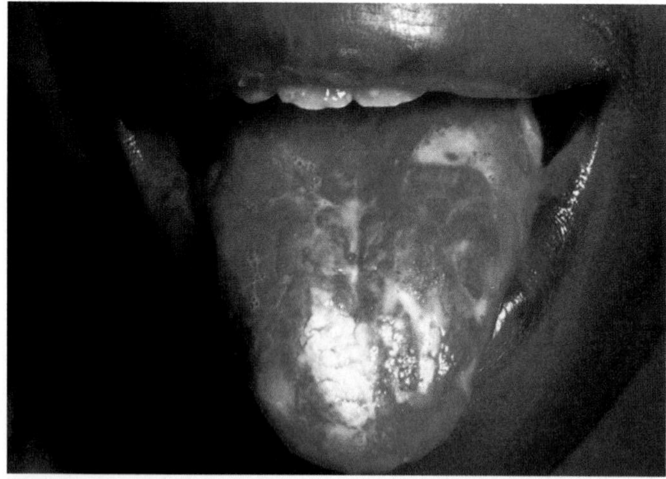

(a)

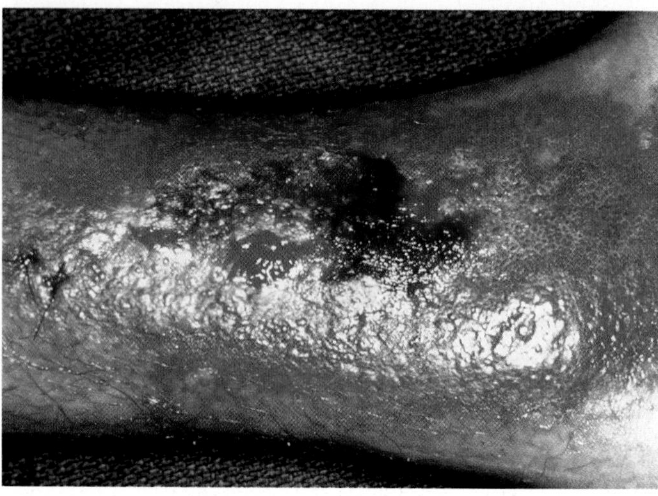

(b)

Figure 37.10 Some Diseases Associated with AIDS. **(a)** Candidiasis of the oral cavity and tongue (thrush) caused by *Candida albicans.* **(b)** Kaposi's sarcoma on the arm of an AIDS patient. The flat purple tumors can occur in almost any tissue and are frequently multiple.

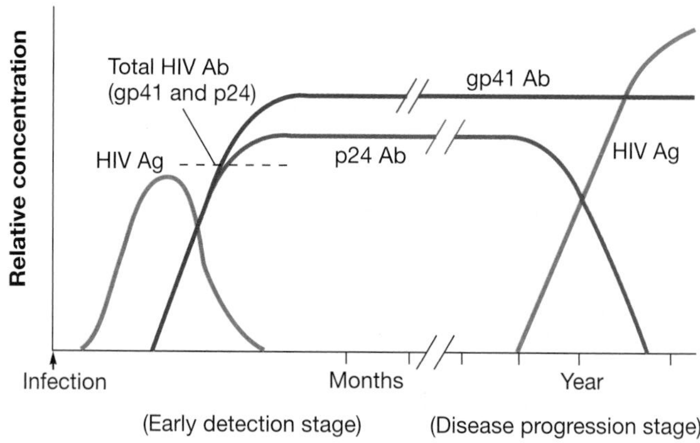

Figure 37.11 The Typical Serological Pattern in an HIV-1 Infection. HIV-1 antigen (HIV Ag) is detectable as early as 2 weeks after infection and typically declines at seroconversion. Seroconversion occurs when HIV-1 antibodies have risen to detectable levels. This usually takes place several weeks to months after the HIV-1 infection. The period between HIV-1 infection and seroconversion often is associated with an acute illness. Whether or not the individual has flulike symptoms, the appearance of circulating HIV-1 antigens typically occurs before IgG antibodies against gp41 and p24 develop. HIV-1 antigen then usually disappears following seroconversion but reappears in the latter stages of the disease. The reappearance of antigen usually indicates impending clinical deterioration. An asymptomatic HIV-1 antigen-positive individual is six times more likely to develop AIDS within 3 years than a similar individual who is HIV-1 antigen negative. Thus testing for the presence of the HIV-1 antigen assists clinicians in monitoring the progression of the disease.

that the extent to which the immune response is able to control this initial burst of virus replication may determine the amount of time required for progression to the next clinical stages.

The asymptomatic stage of HIV infection may last from 6 months to 10 years or more in some individuals. During this stage the levels of detectable HIV in the blood decrease (figure 37.11), but the virus continues to replicate, particularly in lymphoid tissues. Even before any changes in $CD4^+$ T cells can be detected, the virus may affect certain immune functions, such as memory cell responses to common antigens like tetanus toxoid or *Candida albicans*.

During the chronic symptomatic stage (formerly called AIDS-related complex or ARC), which can last for months to years, virus replication continues and the number of $CD4^+$ T cells in the blood begins to significantly decrease. Because these T-helper cells are critically important in the generation of acquired immunity, individuals at this stage develop a variety of symptoms including fever, weight loss, malaise, fatigue, anorexia, abdominal pain, diarrhea, headaches, and lymphadenopathy (enlarged lymph nodes). Paradoxically, some patients develop increased serum antibody production during this stage, perhaps as a result of generalized immune dysfunction. These antibodies, however, do little to protect the host from infection. As $CD4^+$ T cell numbers continue to decline, some patients develop opportunistic infections such as oral candidiasis (figure 37.10*a*) or Kaposi's sarcoma (figure 37.10*b*).

There is experimental evidence to support several potential mechanisms of $CD4^+$ T cell depletion by HIV, although the most important one remains enigmatic. The mechanisms include: (1) direct cytopathic effects of HIV on T cells, (2) formation of syncytia, (3) immune-mediated destruction of HIV-infected cells, and (4) effects of viral products (such as gp120) on uninfected cells. The cytopathic effect may be due to the disruption of plasma membrane integrity and function by excessive budding of virus. Insertion of HIV proviral DNA into the host cell's DNA can disrupt cell function, destroying the host T cells. Expression of

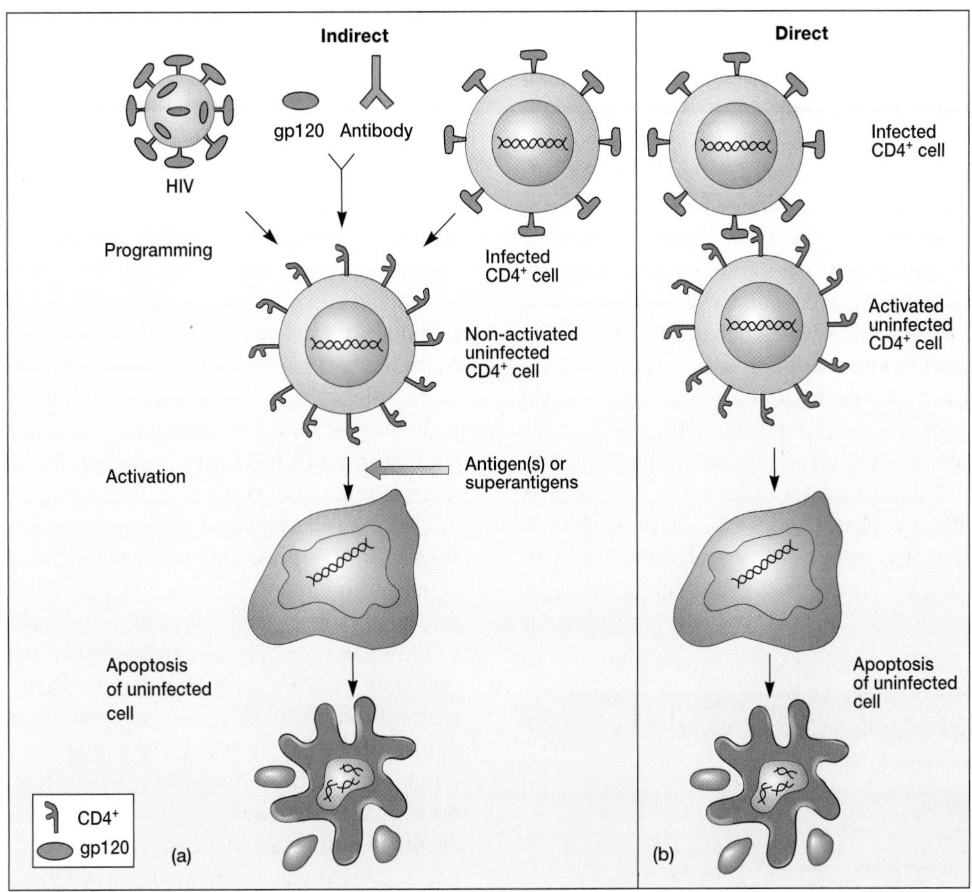

Figure 37.12 Apoptosis and AIDS. Apoptosis is a homeostatic physiological suicide mechanism in which cell death occurs naturally during normal tissue turnover. Cells undergoing apoptosis display profound structural changes such as a decrease in cell volume, blebbing of the plasma membrane, and nuclear fragmentation. The nuclear DNA is cleaved into short lengths. The dying cell sheds small membrane-bound apoptotic bodies, which are phagocytosed and digested. **(a)** There may be several ways in which an HIV infection can indirectly trigger apoptosis. In all cases an initial event would program or prime the target cell so that apoptosis would be triggered by the binding of antigens to the cell's T-cell receptors. Possibly the external gp120 envelope glycoprotein of the HIV virion binds to the CD4 protein on lymphocytes and programs the lymphocyte. A combination of free gp120 and antibodies to gp120 also could stimulate programmed cell death. First, the gp120 would bind to CD4 receptors. Then antibodies would attach to the gp120 and cause clustering of the receptors, thus priming the uninfected $CD4^+$ cell. It also is possible that binding of the infected cell's surface gp120 proteins to the CD4 receptors on an uninfected cell will program the uninfected cell for apoptosis in response to antigens. **(b)** Apoptosis may be directly triggered in an uninfected cell. The gp120 envelope proteins on the surface of an infected cell may combine with the CD4 proteins of an uninfected cell and directly stimulate programmed cell death without activation by antigens.

gp120 on virally infected host cells may interact with T-cell CD4 receptors, causing them to fuse and form multinucleated syncytia that eventually die. Moreover, free gp120 proteins released from infected cells may bind to CD4 on uninfected cells and induce those cells to undergo apoptosis (programmed cell death). Other potential methods used by HIV to trigger apoptosis are illustrated in **figure 37.12.** Finally, multiple components of the immune system (cytotoxic or $CD8^+$ T cells, NK cells, complement, and antibody-dependent cellular cytotoxicity) may contribute to the continuing destruction of virus-infected $CD4^+$ T cells, resulting in acquired immune deficiency. Other factors besides $CD4^+$ cell destruction may also contribute to AIDS pathogenesis. For example, HIV may also inhibit or destroy dendritic (antigen-presenting) cells. HIV mutates exceptionally rapidly and thus could evade and eventually overwhelm the host immune system. HIV may disrupt the balance between the various types of T cells also altering immune system integrity. Viral replication eventually outpaces the host's attempts to control it, resulting in clinical AIDS.

Many considerations may be incorporated into a diagnosis of AIDS, the fourth stage of HIV infection. At this point the host immune system is no longer able to defend against the virus, as it has largely depleted the potential effector cells. In 1993, the CDC revised the definition of AIDS to include all HIV-infected individuals who have fewer than 200 $CD4^+$ T cells per microliter of blood (or a $CD4^+$ T cell percentage of total lymphocytes of less than 14). The reason for this definition is that the development of particular opportunisitic infections is related to the $CD4^+$ T cell concentration. Healthy persons have about 1,000 $CD4^+$ T cells per microliter of blood. This number declines by an average of 40 to 80 cells per microliter per year in HIV-infected individuals. The first opportunistic infections and disease processes typically occur once the $CD4^+$ T cell count declines to 200 to 400 per microliter, despite antimicrobial therapies (**table 37.3**). Examples of these infections and diseases include *Pneumocystis* pneumonia, *Mycobacterium avium-intracellulare* pneumonia, toxoplasmosis, herpes zoster infection, chronic diarrhea caused by *Cyclospora,* cryptococcal meningitis, and *Histoplasma capsulatum* infection.

In addition to its devastating effects on the immune system, HIV infection can also lead to disease of the central nervous system because virus-infected macrophages can cross the blood-brain barrier. The classical symptoms of central nervous system disease in AIDS patients are headaches, fevers, subtle cognitive changes, abnormal reflexes, and ataxia (irregularity of muscular action). Dementia and severe sensory and motor changes characterize more advanced stages of the disease. Autoimmune neuropathies, cerebrovascular disease, and brain tumors also are common. Histological changes include inflammation of neurons, nodule formation, and demyelination. Evidence indicates that these neurological changes are correlated with higher levels of HIV-1 antigen (figure 37.11) and/or the HIV-1 genome in central nervous system cells. In AIDS dementia, macrophages and glial cells (supporting cells of the nervous system) are primarily infected and bud new viruses. However, it is unlikely that direct infection of these cells by HIV-1 is responsible for the symptoms. More likely the symptoms arise through either the secretion of viral proteins or viral induction of cytokines that bind to glial cells and neurons. HIV-1 induction of interleukin-1 and tumor necrosis factor-α (TNF-α) may stimulate further viral reproduction and the induction of other cytokines (e.g., interleukin-6 (IL-6), granulocyte-macrophage colony-stimulating factor [GMCSF]). IL-1 and TNF-α in combination with IL-6 and GMCSF could account for the many clinical and histopathological findings in the central nervous system of AIDS individuals.

Another potential complication of HIV infection is cancer. Individuals infected with HIV-1 have an increased risk of three types of tumors: (1) Kaposi's sarcoma (figure 37.10*b*), (2) carcinomas of the mouth and rectum, and (3) B-cell lymphomas or lymphoproliferative disorders. It seems likely that the depression of the initial immune response enables secondary tumor-causing agents to initiate the cancers.

In 1994 it was discovered that Kaposi's sarcoma–associated herpesvirus (KSHV, also known as human herpesvirus 8, or HHV-8) is a virus that is consistently present in Kaposi's sarcoma and in primary effusion (body cavity–based) lymphomas. These cancers occur most frequently in AIDS patients. KSHV is a gamma herpesvirus with homology to herpesvirus saimiri and Epstein-Barr virus, both of which can transform lymphocytes. It is now known that this virus is spread in saliva through kissing.

In contrast, a harmless virus (GB virus C) discovered in 1995 and carried by tens of millions of people worldwide appears to prolong the lives of those who are infected with HIV. This virus decreases mortality, slows damage to the immune system, and boosts the effects of AIDS drugs. How the GB virus C performs its beneficial effects is not known but two possibilities exist. The virus could directly suppress HIV replication or it could stimulate the immune system. There is some evidence that it attaches to the same CD4 T-cell receptors used by HIV. This would block virion entrance.

The laboratory diagnosis of HIV infection can be by viral isolation and culture or by using assays for viral reverse transcriptase activity or viral antigens (figure 37.11). However, diagnosis is most commonly accomplished through the detection of specific anti-HIV antibodies in the blood. For routine screening purposes, an ELISA is commonly used because it is sensitive and relatively inexpensive. However, false-positive results can occur with this method, requiring positive samples to be retested using a more specific Western Blot technique. The most sensitive HIV assay employs the polymerase chain reaction. PCR can be used to am-

Table 37.3 Disease Processes Associated with AIDS
Candidiasis of bronchi, trachea, or lungs
Candidiasis, esophageal
Cervical cancer, invasive
Coccidioidomycosis, disseminated or extrapulmonary
Cryptosporidiosis, chronic intestinal (>1 month's duration)
Cyclospora, diarrheal disease
Cytomegalovirus disease (other than liver, spleen, or lymph nodes)
Cytomegalovirus retinitis (with loss of vision)
Encephalopathy, HIV-related
Herpes simplex: chronic ulcer(s) (>1 month's duration); or bronchitis, pneumonitis, or esophagitis
Histoplasmosis, disseminated or extrapulmonary
Isosporiasis, chronic intestinal (>1 month's duration)
Kaposi's sarcoma
Lymphoma, Burkitt's (or equivalent term)
Lymphoma, immunoblastic (or equivalent term)
Lymphoma, primary, of brain
Mycobacterium avium complex or *M. kansasii*
Mycobacterium tuberculosis, any site
Mycobacterium, other species or unidentified species
Pneumocystis pneumonia
Pneumonia, recurrent
Progressive multifocal leukoencephalopathy
Salmonella septicemia, recurrent
Toxoplasmosis of brain
Wasting syndrome due to AIDS

Source: Data from *MMWR* 41 (No. RR17). 1993 Revised Classification System for HIV Infection and Expanded Surveillance Case Definition for AIDS Among Adolescents and Adults.

plify and detect tiny amounts of viral RNA and cDNA in virions or infected host cells. Quantitative PCR assays will provide an estimate of a patient's viral load. This is particularly significant because the level of virions in the blood, as well as the concentration of $CD4^+$ cells, is very predictive of the clinical course of the infection. The probable time to development of AIDS can be estimated from the patient's blood virion level and $CD4^+$ cell count.

Polymerase chain reaction (section 14.3); Clinical immunology (section 35.3)

At present there is no cure for AIDS. Primary treatment is directed at reducing the viral load and disease symptoms, and at treating opportunistic infections and malignancies. The antiviral drugs currently approved for use in HIV disease are of four types. (1) *nu*cleoside *r*everse *t*ranscriptase *i*nhibitors (NRTIs) are analogues that inhibit the enzyme reverse transcriptase as it synthesizes viral DNA. Examples include zidovudine (AZT or Retrovir), didanosine (Videx), zalcitabine (ddC or HIVID), tenofovir (Viread), emtricitabine (Emtriva or Coviracil), stavudine (Zerit), and lamivudine (Epivir or 3TC). (2) The *non*nucleoside *r*everse *t*ranscriptase *i*nhibitors (NNRTIs) include delavirdine (Rescriptor), efavirenz (Sustiva), and nevirapine (Viramune). (3) The *p*rotease *i*nhibitors (PIs) work by blocking the activity of the HIV protease and thus interfere with virion assembly. Examples include indinavir (Crixivan), ritonavir (Norvir), nelfinavir (Viracept), lopinavir (Kaletra), Fosamprenavir (Lexiva), atazanavir (Reyataz), and saquinavir (Invirase). (4) The fusion inhibitors (FIs) are a newer category of drugs that prevent HIV entry into cells. This category is represented by enfuvirtide (Fuzeon). The most successful treatment approach in combating HIV/AIDS is to use drug combinations. An effective combination is a cocktail of various NRTIs, NNRTIs, PIs, and the FI. Such drug combination use is referred to as HAART (*h*ighly *a*ctive *a*nti-*r*etroviral *t*herapy). However, the specific combination of drugs is a function of a number of factors including time from exposure, symptoms, viral load, pregnancy, and many others. In many patients the virus disappears from the patient's blood with proper treatment and drug-resistant strains do not seem to arise. HIV can remain dormant in memory T cells, survive drug cocktails, and reactivate. Thus patients are not completely cured with drug treatment. It should be noted that side effects can be very severe, and treatment is prohibitively expensive for those without medical insurance.

The development of a vaccine has been a long-sought research goal. Such a vaccine would ideally (1) stimulate the production of neutralizing antibodies, which can bind to the virus envelope and prevent it from entering host cells; and (2) promote the formation of cytotoxic T cells (CTLs), which can destroy cells infected with virus. Among the many problems encountered in developing an HIV vaccine is the fact that the envelope proteins of the virus continually change their antigenic properties.

Many HIV researchers continue to take great interest in HIV-infected persons who are long-term nonprogressors. These individuals maintain $CD4^+$ T cell counts of at least 600 per microliter of blood, have less than 5,000 copies of HIV RNA per milliliter of blood, and have remained this way for more than 10 years after documented infection (even in the absence of antiviral agents). At least three explanations of this phenomenon have been proposed: long-term nonprogressors (1) may react with a more effective immune response (CTL and neutralizing antibody) to relatively conserved proteins; (2) may have been initially infected with an attenuated strain; or (3) may have predisposing genetic differences that prevent or inhibit infection, as in the example of the CCR5 mutation already discussed.

Prevention and control of AIDS is achieved primarily through education. Understanding risk factors and practicing strategies to reduce risk are essential in the fight against AIDS. Barrier protection from blood and body fluids greatly limits risk of HIV infection. Education to prevent the sharing of intravenous needles and syringes is also very important. Additionally, prevention includes the continued screening of blood and blood products.

Cold Sores

Cold sores or **fever blisters (herpes labialis)** are caused by the herpes simplex virus type 1 (HSV-1). Like all herpesviruses, it is a double-stranded DNA virus with an enveloped, icosahedral capsid. The term herpes is derived from the Greek word meaning "to creep," and clinical descriptions of herpes labialis (lips) date back to the time of Hippocrates (circa 400 B.C.). Transmission is through direct contact of epithelial tissue surfaces with the virus (*see figure 18.6*). A blister(s) develops at the inoculation site because of host- and viral-mediated tissue destruction (**figure 37.13**). Most blisters involve the epidermis and surface mucous membranes of the lips, mouth, and gums (**gingivostomatitis**). The blisters generally heal within a week. However, after a primary infection, the virus travels to the trigeminal nerve ganglion, where it remains in a latent state for the lifetime of the infected person. Stressful stimuli such as excessive sunlight, fever, trauma, chilling, emotional stress, and hormonal changes can reactivate the virus. Once reactivated, the virus moves from the trigeminal ganglion down a peripheral nerve to the border of the lip or other parts of the face to produce another fever blister. Primary and recurring infections also may occur in the eyes, causing

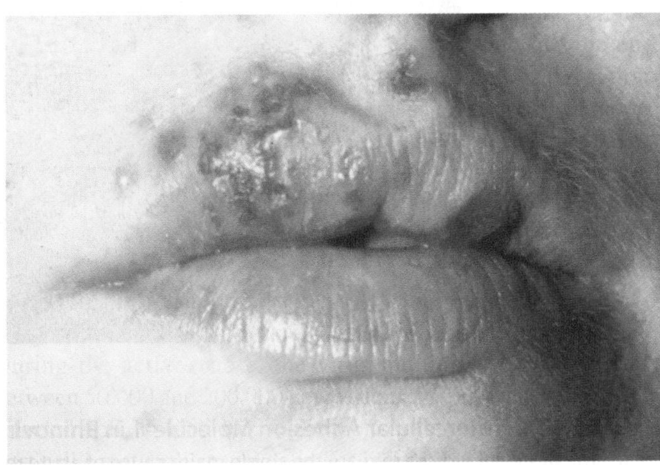

Figure 37.13 Cold Sores. Herpes simplex fever blisters on the lip, caused by herpes simplex type 1 virus.

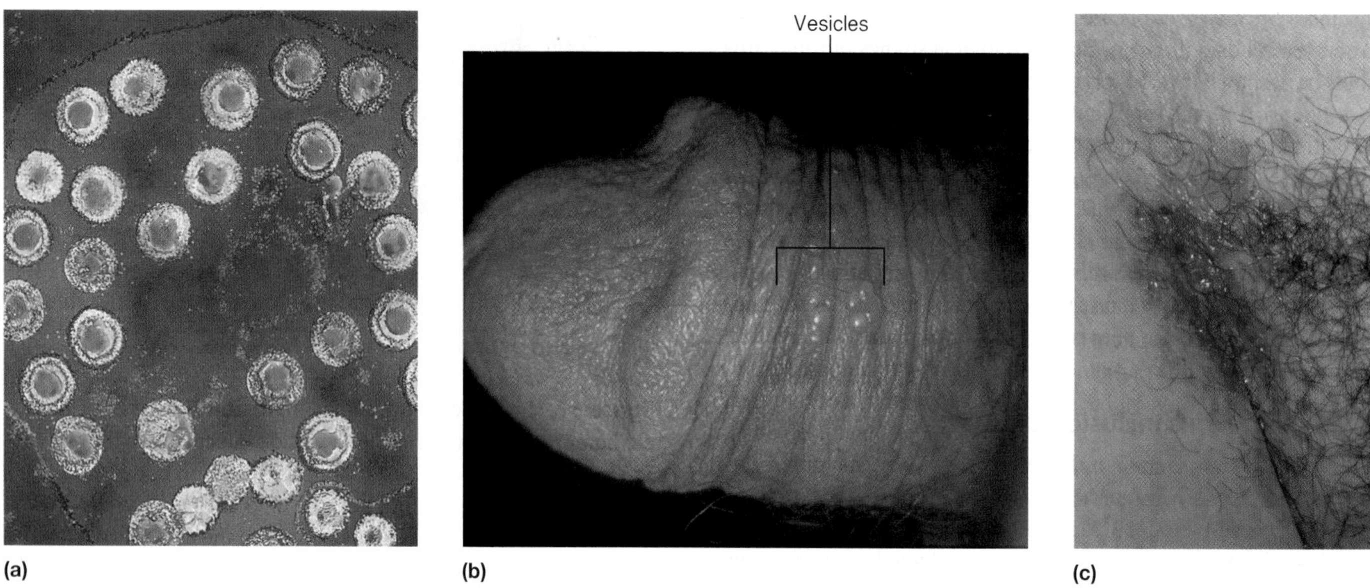

Figure 37.16 Genital Herpes. **(a)** Herpes simplex virus type 2 (yellow and green) inside an infected cell (*see also figure 16.10d*). **(b)** Herpes vesicles on the penis. **(c)** Herpes vesicles and blisters on the vaginal labia. The vesicles contain fluid that is infectious.

inflammatory response; they contain fluid and infectious viruses. A fever, headache, muscle aches and pains, a burning sensation, and genital soreness are frequently present during the active phase. Although blisters generally heal spontaneously in a few weeks, the viruses retreat to nerve cells in the sacral plexus of the spinal cord, where they remain in a latent form. During the latent phase, the viral genome becomes incorporated into the chromosome of the host cell. During this phase, the host cell does not die. Because viral genes are not expressed, the infected person is symptom-free. However, periodically the viruses multiply and migrate down nerve fibers to the skin or mucous membranes that the nerve supplies, where they produce new blisters. Activation may be due to sunlight, sexual activity, illness accompanied by fever, hormones, or stress. It should be noted that both primary infection and reactivation can occur without any symptoms and apparently healthy people can transmit HSV-2 to their sexual partners or their newborns.

Besides being transmitted by sexual contact, herpes can be spread to an infant during vaginal delivery, leading to **congenital (neonatal) herpes.** Congenital herpes is one of the most life-threatening of all infections in newborns, affecting approximately 1,500 to 2,200 babies per year in the United States. It can result in neurological involvement as well as blindness. As a result any pregnant female who has had genital herpes should have a caesarean section instead of delivering vaginally. For unknown reasons, the HSV-2 virus is also associated with a higher-than-normal rate of cervical cancer and miscarriages. Diagnosis of HSV-2 infection is by ELISA screening of blood or serum, direct fluorescent antibody testing of tissue, and/or by PCR.

Although there is no cure for genital herpes, oral use of the antiviral drugs acyclovir (Zovirax or Valtrex) and famciclovir (Famvir) has proven to be effective in ameliorating the recurring blister outbreaks. Topical acyclovir is also effective in reducing virus shedding, the time until the crusting of blisters occurs, and new lesion formation. Idoxuridine and trifluridine are used to treat herpes infections of the eye.

In the United States the incidence of genital herpes has increased so much during the past several decades that it is now a very common sexually transmitted disease. It is estimated that over 25 million Americans (20% of adults) are infected with the herpes simplex virus type 2.

Human Herpesvirus 6 Infection

Human herpesvirus 6 (HHV-6) is the etiologic agent of **exanthem subitum** [Greek *exanthema,* rash] in infants. HHV-6 is a unique member of the family *Herpesviridae* that is distinct serologically and genetically from the other herpesviruses. The virus envelope encloses an icosahedral capsid and a core containing double-stranded DNA. The disease caused by HHV-6 was originally termed **roseola infantum** and then given the ordinal designation **sixth disease** to differentiate it from other exanthems and roseolas. Exanthem subitum is a short-lived disease characterized by a high fever of 3 to 4 days' duration, after which the temperature suddenly drops to normal and a macular rash appears on the trunk and then spreads to other areas of the body. HHV-6 infects over 95% of the United States infant population, and most children are seropositive for HHV-6 by 3 years of age. $CD4^+$ T cells are the main site of viral replication, whereas monocytes are in an infected, latent state. The tropism of HHV-6 appears to be wide, including $CD8^+$ T cells, natural killer cells, and probably epithelial cells. In adults HHV-6 is commonly found in peripheral-blood mononuclear cells and saliva, suggesting that the infection is lifelong. Since the salivary glands are the major site of latent infection, transmission is probably by way of saliva.

HHV-6 also produces latent and chronic infections and is occasionally reactivated in immunocompromised hosts leading to pneumonitis. Furthermore, HHV-6 has been implicated in several other diseases (lymphadenitis, multiple sclerosis, and infectious mononucleosis-like syndrome or chronic fatigue syndrome) in immunocompetent adults. Diagnosis is by immunofluorescence or enzyme immunoassay, or PCR. To date, there is no antiviral therapy or prevention.

Human Parvovirus B19 Infection

Since its discovery in 1974, **human parvovirus B19** (family *Parvoviridae,* genus *Parvovirus*) has emerged as a significant human pathogen. B19 virions are uniform, icosahedral, naked particles approximately 23 nm in diameter. The genome of B19 is a single-stranded (ss) DNA. There is a spectrum of disease caused by parvovirus B19 infection, ranging from mild symptoms (fever, headache, chills, malaise) in normal persons, **erythema infectiosum** in children (**fifth disease**), and a joint disease syndrome in adults. More serious diseases include aplastic crisis in persons with sickle cell disease and autoimmune hemolytic anemia, and pure red cell aplasia due to persistent B19 virus infection in immunocompromised individuals. The B19 parvovirus can also infect the fetus, resulting in anemia, fetal hydrops (the accumulation of fluid in tissues), and spontaneous abortion. This has prompted a grassroots awareness of B19 complications among school teachers who may be pregnant. It is assumed that the natural mode of infection is by the respiratory route. The average incubation period is 4 to 14 days. Approximately 20% of infected individuals are asymptomatic and a smaller percentage of infected individuals have symptoms for up to three weeks. Infection typically results in a life-long immunity to B19. A variety of techniques are available for the detection of the B19 virus. Antiviral antibodies appear to represent the principal means of defense against B19 parvovirus infection and disease. The treatment of individuals suffering from acute and persistent B19 infections with commercial immunoglobins containing anti-B19 and human monoclonal antibodies to B19 is an effective therapy. As with other diseases spread by contact with respiratory secretions, frequent handwashing is the best prevention of the disease.

Leukemia (Virus-induced)

Certain **leukemias** in humans are caused by two retroviruses: human T-cell lymphotropic virus I (HTLV-I) and HTLV-II. HTLV-I and HTLV-II are members of the family *Retroviridae.* They have a nuclear core containing two positive-strand RNA genomes. Once a cell is infected, the RNA genome is converted by reverse transcriptase to DNA and integrates into the host's chromosome. The viruses are transmitted among drug addicts sharing needles, by sexual contact, across the placenta, from the mother's milk, or by mosquitoes. Viruses and cancer (section 18.5)

HTLV-1 causes **adult T-cell leukemia.** Once within the body the HTLV-I virus enters white blood cells and integrates into the cellular genome, where it activates growth-promoting genes. For example, in infected cells, a viral protein called TAX increases expression of the gene that encodes interleukin-2 (IL-2), an important T-cell growth factor. The transformed cell proliferates extensively, and death generally results from the explosive proliferation of the leukemia cells or from opportunistic infections. To date, no effective treatment exists.

In 1982 the second human retrovirus (HTLV-II) was shown to be the agent responsible for hairy-cell leukemia. This virus shares the same disease-causing mechanism as HTLV-I. Hairy-cell leukemia gets its name from the many membrane-derived protrusions that give white blood cells the appearance of being "hairy" (**figure 37.17**). This leukemia is a chronic, progressive lymphoproliferative disease. The malignancy is believed to originate in a stage of B-cell development. The bone marrow, spleen, and liver become infiltrated with malignant cells. This lowers the person's immunity. The primary cause of mortality is bacterial and other opportunistic infections. IFN-α n3 (Alferon N) has shown some promise for treatment in certain cases.

Mononucleosis (Infectious)

The Epstein-Barr virus (EBV) is a member of the family *Herpesviridae.* EBV exhibits the characteristic herpes morphology—all herpesviruses consist of an icosahedral capsid (approximately 125 nm in diameter) surrounded by a membrane envelope. The capsid contains the viral double-stranded (ds) DNA. Its ds DNA exists as a linear form in the mature viron and a circular episomal form in latently infected cells. EBV is the etiologic agent of **infectious mononucleosis (mono)**, a disease whose symptoms closely resemble those of cytomegalovirus-induced mononucleosis.

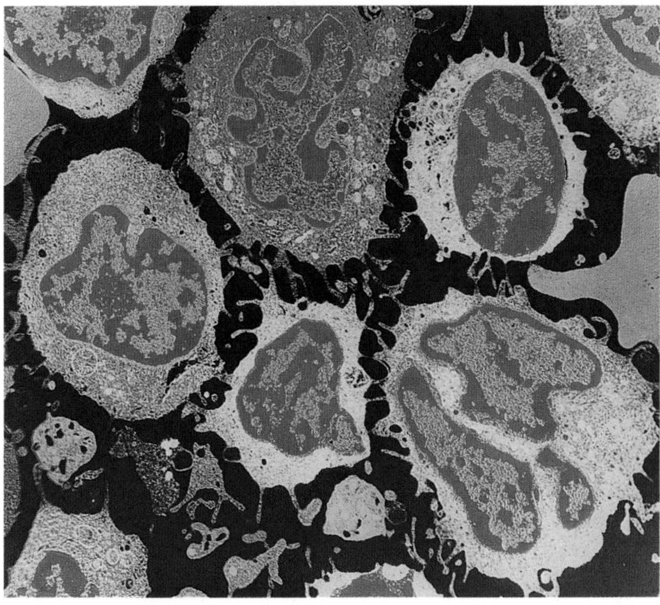

Figure 37.17 Hairy-Cell Leukemia. False-color transmission electron micrograph (×3,100) of abnormal B lymphocytes. Notice that the lymphocytes are covered with characteristic hairlike membrane-derived protrusions.

Because the Epstein-Barr virus occurs in oropharyngeal secretions, it can be spread by mouth-to-mouth contact (hence the terminology infectious and kissing disease) or shared drinking bottles and glasses. A person gets infected when the virus from someone else's saliva makes its way into epithelial cells lining the throat. After a brief bout of replication in the epithelial cells, the new viruses are shed and infect memory B cells. Infected B cells rapidly proliferate and take on an atypical appearance (Downey cells) that is useful in diagnosis (**figure 37.18**). The disease is manifested by enlargement of the lymph nodes and spleen, sore throat, headache, nausea, general weakness and tiredness, and a mild fever that usually peaks in the early evening. The disease lasts for 1 to 6 weeks and is self-limited. Like other herpesviruses, EBV becomes latent in its host.

Treatment of mononucleosis is largely supportive and includes plenty of rest. Diagnosis of mononucleosis is usually confirmed by demonstration of an increase in circulating mononuclear cells, along with a serological test for nonspecific (heterophile) antibodies, specific viral antibodies, or identification of viral nucleic acid. Several rapid tests are on the market.

The peak incidence of mononucleosis occurs in people 15 to 25 years of age. Collegiate populations, particularly those in the upper-socioeconomic class, have a high incidence of the disease. About 50% of college students have no immunity, and approximately 15% of these can be expected to contract mononucleosis. People in lower-socioeconomic classes tend to acquire immunity to the disease because of early childhood infection. The Epstein-Barr virus may well be the most common virus in humans as it infects 80 to 90% of all adults worldwide.

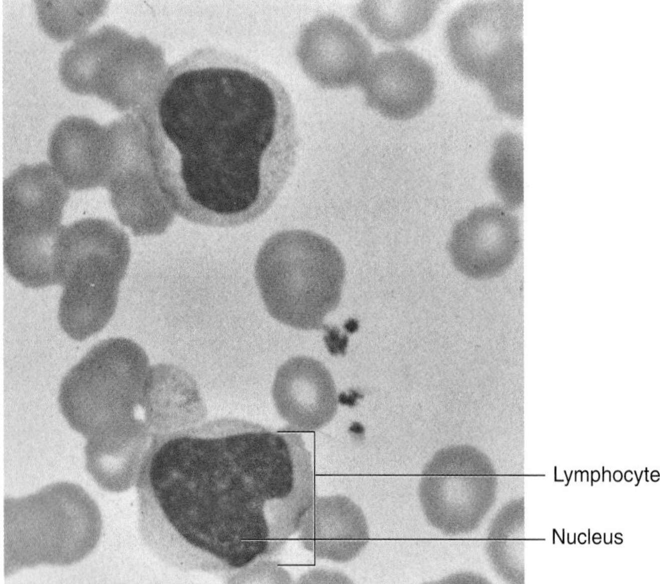

Figure 37.18 Evidence of Epstein-Barr Infection in the Blood Smear of a Patient with Infectious Mononucleosis. Note the abnormally large lymphocytes containing indented nuclei with light discolorations.

EBV infections are associated with chronic fatigue syndrome and the cancers Burkitt's lymphoma in tropical Africa and nasopharyngeal carcinoma in Southeast Asia, East and North Africa, and in Inuit populations.

Viral Hepatitides

Inflammation of the liver is called **hepatitis** [pl., hepatitides] [Greek *hepaticus,* liver]. Currently 11 viruses are recognized as causing hepatitis. Two are herpesviruses (cytomegalovirus [CMV] and Epstein-Barr virus [EBV]) and 9 are hepatotropic viruses that specifically target liver hepatocytes.

EBV and CMV cause mild, self-resolving forms of hepatitis with no permanent hepatic damage. Both viruses cause the typical infectious mononucleosis syndrome of fatigue, nausea, and malaise.

Of the nine human hepatotropic viruses, only five are well characterized; hepatitis G (**table 37.4**) and TTV (transfusion-transmitted virus) are more recently discovered viruses. Hepatitis A (sometimes called infectious hepatitis) and hepatitis E are transmitted by fecal-oral contamination and discussed in the section on food- and water-borne diseases (section 37.4). The other major types include hepatitis B (sometimes called serum hepatitis), hepatitis C (formerly non-A, non-B hepatitis), and hepatitis D (a virusoid formerly called delta hepatitis).

Hepatitis B (serum hepatitis) is caused by the hepatitis B virus (HBV), an enveloped, double-stranded circular DNA virus of complex structure. HBV is classified as an *Orthohepadnavirus* within the family *Hepadnaviridae.* Serum from individuals infected with hepatitis B contains three distinct antigenic particles: a spherical 22 nm particle, a 42 nm spherical particle (containing DNA and DNA polymerase) called the **Dane particle,** and tubular or filamentous particles that vary in length (**figure 37.19**). The viral genome is 3.2 kb in length, consisting of four partially overlapping, open-reading frames that encode viral proteins. Viral replication takes place predominantly in hepatocytes. The infecting virus encases its double-shelled Dane particles within membrane envelopes coated with hepatitis B surface antigen (HBsAg). The inner nucleocapsid core antigen (HBcAg) encloses a single molecule of double-stranded HBV DNA and an active DNA polymerase. HBsAg in body fluids is (1) an indicator of hepatitis B infection, (2) used in the large-scale screening of blood for the hepatitis B virus, and (3) the basis for the first vaccine for human use developed by recombinant DNA technology. Diagnosis of HBV is made by detection of HBsAg in unimmunized individuals or HBcAg antibody, or detection of HBV nucleic acid by PCR.

The hepatitis B virus is normally transmitted through blood or other body fluids (saliva, sweat, semen, breast milk, urine, feces) and body-fluid-contaminated equipment (including shared intravenous needles). The virus can also pass through the placenta to the fetus of an infected mother. The number of new HBV cases in the United States declined by 60% between 1985 and 1995, and was only 78,000 in 2001 (down from 260,000 in the 1980s). It is estimated, however, that there are currently 1.25 million chroni-

Table 37.4 Characteristics of Hepatitides Caused by Hepatotropic Viruses[a]

Disease (Virus)	Genome	Classification	Transmission	Outcome	Prevention
Hepatitis A (hepatitis A)	RNA	*Picornaviridae, Hepatovirus*	Fecal-oral	Subclinical, acute infection	Killed HAV (Havrix vaccine)
Hepatitis B (hepatitis B)	DNA	*Hepadnaviridae, Orthohepadnavirus*	Blood, needles, body secretions, placenta, sexually	Subclinical, acute chronic infection; cirrhosis; primary hepatocarcinoma	Recombinant HBV vaccines
Hepatitis C (hepatitis C)	RNA	*Flaviviridae, Pestivirus, or Flavivirus* (?)	Blood, sexually	Subclinical, acute chronic infection; primary hepatocarcinoma	Routine screening of blood
Hepatitis D (hepatitis D)	RNA	Virusoid	Blood, sexually	Superinfection or coinfection with HBV	HBV vaccine
Hepatitis E (hepatitis E)	RNA	*Caliciviridae* (?)	Fecal-oral	Subclinical, acute infection (but high mortality in pregnant women)	Improve sanitary conditions
Hepatitis G	RNA	*Flaviviridae*	Sexually, parenterally	Chronic liver inflammation	HBV vaccine

[a] Hepatitis TTV has been discovered but not well characterized. Thus it is not included in this table.

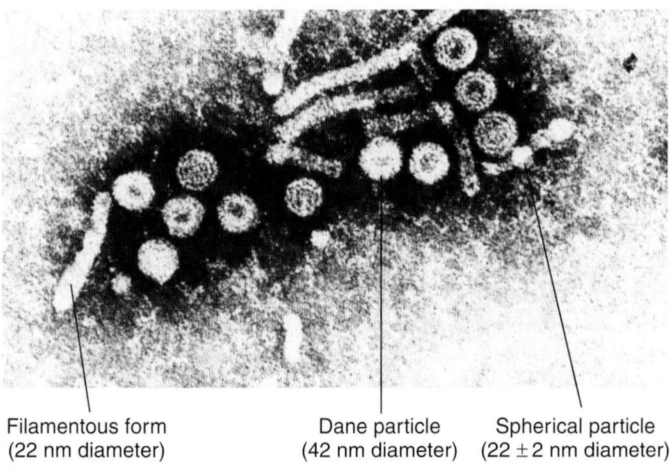

Figure 37.19 Hepatitis B Virus in Serum. Electron micrograph (×210,000) showing the three distinct types of hepatitis B antigenic particles. The spherical particles and filamentous forms are small spheres or long filaments without an internal structure, and only two of the three characteristic viral envelope proteins appear on their surface. Dane particles are the complete, infectious virion.

cally infected Americans. In the United States, about 5,000 persons die yearly from hepatitis-related cirrhosis and about 1,000 die from HBV-related liver cancer. (HBV is second only to tobacco as a known cause of human cancer.) Worldwide, HBV infects over 200 million people.

The clinical signs of hepatitis B vary widely. Most cases are asymptomatic. However, sometimes fever, loss of appetite, abdominal discomfort, nausea, fatigue, and other symptoms gradually appear following an incubation period of 1 to 3 months. The virus infects liver hepatic cells and causes liver tissue degeneration and the release of liver-associated enzymes (transaminases) into the bloodstream. This is followed by jaundice, the accumulation of bilirubin (a breakdown product of hemoglobin) in the skin and other tissues with a resulting yellow appearance. Chronic hepatitis B infection also causes the development of primary liver cancer, known as hepatocellular carcinoma.

General measures for prevention and control involve (1) excluding contact with HBV-infected blood and secretions, and minimizing accidental needle-sticks; (2) passive prophylaxis with intramuscular injection of hepatitis B immune globulin within 7 days of exposure; and (3) active prophylaxis with recombinant vaccines: Energix-B, Recombivax HB, Pediatrix, and Twinrix. These vaccines are widely used and are recommended for routine prevention of HBV in infants to 18-year-olds, and risk groups of all ages (for example, household contacts of HBV carriers, healthcare and public safety professionals, men who have sex with other men, international travelers, hemodialysis patients). Recommended treatments for HBV include Adefovir dipivoxil, alpha-interferon, and lamivudine.

Hepatitis C is caused by the enveloped **hepatitis C** virus (HCV), which has an 80 nm diameter, a lipid coat, contains a single strand of linear RNA. The hepatitis C virus is a member of the family *Flaviviridae.* HCV is classified into multiple genotypes.

This virus is transmitted by contact with virus-contaminated blood, by the fecal-oral route, by in utero transmission from mother to fetus, sexually, or through organ transplantation. Diagnosis is made by enzyme-linked immunosorbent assay (ELISA), which detects serum antibody to a recombinant antigen of HCV, and nucleic acid detection by PCR. HCV is found worldwide. Prior to routine screening, HCV accounted for more than 90% of hepatitis cases developed after a blood transfusion. Worldwide, hepatitis C has reached epidemic proportions, with more than 1 million new cases reported annually. In the United States, nearly 4 million persons are infected and 25,000 new cases occur annually. Currently, HCV is responsible for about 8,000 deaths annually in the United States. Furthermore, HCV is the leading reason for liver transplantation in the United States. Treatment is with Ribovirin and pegylated (coupled to polyethylene glycol) recombinant interferon-alpha (Intron A, Roferon-A). This combination therapy can rid the virus in 50% of those infected with genotype 1 and in 80% of those infected with genotype 2 or 3.

In 1977 a cytopathic hepatitis agent termed the **Delta agent** was discovered. Later it was called the hepatitis D virus (HDV) and the disease **hepatitis D** was designated. HDV is a unique agent in that it is dependent on the hepatitis B virus to provide the envelope protein (HBsAg) for its RNA genome. Thus HDV only replicates in liver cells co-infected with HBV. Both must be actively replicating. Furthermore, the RNA of the HDV is smaller than the RNA of the smallest picornaviruses and its circular conformation differs from the linear structure typical of animal RNA viruses. Thus its similarity to the plant virusoids has led some to also call this agent a virusoid. HDV is spread only to persons who are already infected with HBV (superinfection) or to individuals who get HBV and the virusoid at once (coinfection). The primary laboratory tools for the diagnosis of an HDV infection are serological tests for anti-delta antibodies. Treatment of patients with chronic HDV remains difficult. Some positive results can be obtained with alpha interferon treatment for 3 months to 1 year. Liver transplantation is the only alternative to chemotherapy. Worldwide, there are approximately 300 million HBV carriers, and available data indicate that no fewer than 5% of these are infected with HDV. Thus because of the propensity of HDV to cause fulminant as well as chronic liver disease, continued incursion of HDV into areas of the world where persistent hepatitis B infection is endemic has serious implications. Prevention and control involves the widespread use of the hepatitis B vaccine.

Two other forms of hepatitis have been identified: **hepatitis F** (causing fulminant, posttransfusion hepatitis) and a syncytial giant-cell hepatitis (hepatitis G) with viruslike particles resembling the measles virus. **Hepatitis G** (HGV) is a member of the *Flaviviridae* family. It has been cloned and is widely distributed in humans. HGV can be transmitted through needles or sexually. The significance of HGV infections in liver disease is not yet clear. However, infection causes chronic liver inflammation with its associated sequelae. Further virologic, epidemiological, and molecular efforts to characterize these new agents and their diseases are currently being undertaken.

Warts

Warts, or verrucae [Latin *verruca,* wart], are horny projections on the skin caused by the human papillomaviruses. The papillomaviruses are placed in the family *Papillomaviridae* (formerly they were in the *Papovaviridae*). These viruses have naked icosahedral capsids with a double-stranded, supercoiled, circular DNA genome. At least eight distinct genotypes produce benign epithelial tumors that vary in respect to their location, clinical appearance, and histopathologic features. Warts occur principally in children and young adults and are limited to the skin and mucous membranes. The viruses are spread between people by direct contact; autoinoculation occurs through scratching. Four major kinds of warts are **plantar warts, verrucae vulgaris, flat** or **plane warts,** and **anogenital condylomata (venereal warts)** (**figure 37.20**). Treatment includes physical destruction of the wart(s) by electrosurgery, cryosurgery with liquid nitrogen or solid CO_2, laser fulguration (drying), direct application of the drug podophyllum to the wart(s), or injection of IFN-α (Intron A, Alferon N). Anogenital condylomata (venereal warts) are sexually transmitted and caused by types 6, 11, and 42 human papillomavirus (HPV). Once the virus enters the body, the incubation period is 1 to 6 months. The warts (figure 37.20*d*) are soft, pink, cauliflowerlike growths that occur on the external genitalia, in the vagina, on the cervix, or in the rectum. They often are multiple and vary in size. In addition to being a common sexually transmitted disease, genital infection with HPV is of considerable importance because specific types of genital HPV play a major role in the pathogenesis of epithelial cancers of the male and female genital tracts. Over the last decade, many studies have convincingly demonstrated that specific types of HPV are the causal agents of at least 90% of cervical cancers. The most common types conferring a high risk for cervical cancer include HPV types 16, 18, 31, 33, 35, 45, 51, 52, and 56. There is also a possible link between papillomaviruses and nonmelanoma squamous and basal cell cancers. Thirty percent of humans with a rare syndrome of persistent warts (not common warts, but a particular type of warty growth) eventually develop skin cancer, and HPV viral DNA is found in the malignant cells. The epidemiology, molecular biology, and role of HPV in the development of such cancers is an area of active research. Recently a vaccine has been licensed for use in the U.S. that is highly effective in preventing cervical cancer caused by HPVs 16 and 18. These two viruses cause an estimated 80% of all such cancers.

1. Describe the AIDS virus and how it cripples the immune system. How is the virus transmitted? What types of pathological changes can result?
2. Why do people periodically get cold sores? Describe the causative agent.
3. Why do people get the common cold so frequently? How are cold viruses spread?
4. Give two major ways in which herpes simplex virus type 2 is spread. Why do herpes infections become active periodically?
5. What two types of leukemias are caused by viruses?
6. Describe the causative agent and some symptoms of mononucleosis and exanthem subitum.

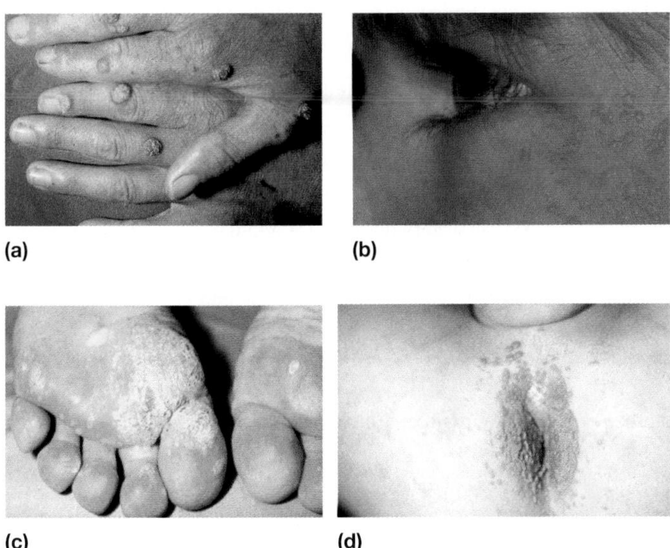

Figure 37.20 Warts. **(a)** Common warts on fingers. **(b)** Flat warts on the face. **(c)** Plantar warts on the feet. **(d)** Perianal condyloma acuminata.

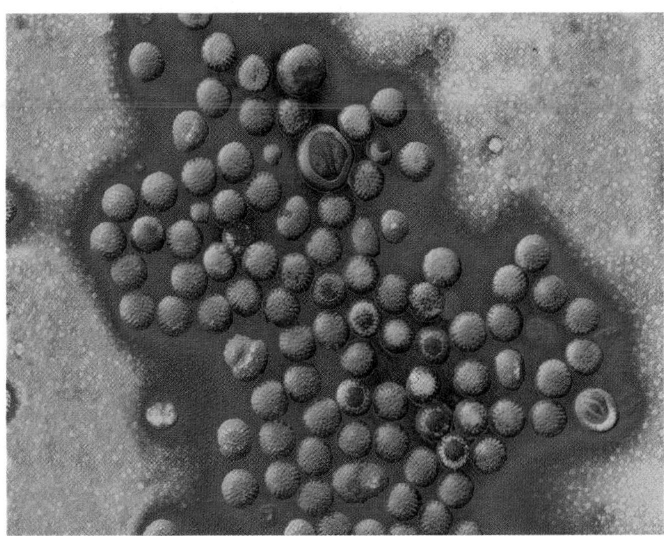

Figure 37.21 Viral Gastroenteritis. Electron micrograph of rotaviruses (reoviruses) in a human gastroenteritis stool filtrate (×90,000). Note the spokelike appearance of the icosahedral capsids that surround double-stranded RNA within each virion.

7. What are the different causative viruses of hepatitis and how do they differ from one another? How can one avoid hepatitis? Do you know anyone who is a good candidate for infection with these viruses?
8. What kind of viruses cause the formation of warts? Describe the formation of venereal warts.

37.4 Food-Borne and Waterborne Diseases

Food and water have been recognized as potential carriers (vehicles) of disease since the beginning of recorded history. Collectively more infectious diseases occur by these two routes than any other. A few of the many human viral diseases that are food- and waterborne are now discussed. Food-borne diseases (section 40.4); Water purification and sanitary analysis (section 41.1)

Gastroenteritis (Viral)

Acute viral gastroenteritis (inflammation of the stomach or intestines) is caused by six major categories of viruses: rotaviruses (**figure 37.21**), adenoviruses, caliciviruses, astroviruses, Norwalk virus, and a group of Noroviruses (previously known as Norwalk-like viruses). The medical importance of these viruses is summarized in **table 37.5.**

The viruses responsible for gastroenteritis are transmitted by the fecal-oral route. Infection with rotaviruses and astroviruses is most common during the cooler months, while infection with adenovirus occurs year-round. Bacteria-caused diarrheal diseases usually occur in the warmer months of the year. The average incubation period for most of these diseases is 1 to 2 days. The clinical manifestations typically range from asymptomatic to a relatively mild diarrhea with headache and fever; to a severe, watery, nonbloody diarrhea with abdominal cramps. Fatal dehydration is most common in young children. Vomiting is almost always present, especially in children. Viral gastroenteritis is usually self-limited. Treatment is designed to provide relief through the use of oral fluid replacement with isotonic liquids, analgesics, and antiperistaltic agents. Symptoms usually last for 1 to 5 days and recovery often results in protective immunity to subsequent infection.

Diarrheal diseases are the leading cause of childhood deaths (5 to 10 million deaths per year) in developing countries where malnutrition is common. Current estimates are that viral gastroenteritis produces 30 to 40% of the cases of infectious diarrhea in the United States, far outnumbering documented cases of bacterial and protozoan diarrhea (the cause of approximately 40% of presumed cases of diarrhea remains unknown). In the United States, rotavirus is responsible for the hospitalization of approximately 55,000 children each year and, worldwide, the death of over 600,000 children annually. Norovirus, on the other hand, is estimated to cause 23 million cases of acute gastroenteritis representing at least 50% of all food-borne outbreaks of gastroenteritis. Viral gastroenteritis is seen most frequently in infants 1 to 11 months of age, where the virus attacks the epithelial cells of the upper intestinal villi, causing malabsorption, impairment of sodium transport, and diarrhea.

Hepatitis A

Hepatitis A (infectious hepatitis) usually is transmitted by fecal-oral contamination of food, drink, or shellfish that live in contaminated water and contain the virus in their digestive system. The

Table 37.5 Medically Important Gastroenteritis Viruses

Virus	Epidemiological Characteristics	Clinical Characteristics
Rotaviruses		
Group A	Endemic diarrhea in infants worldwide	Dehydrating diarrhea for 5–7 days; fever, abdominal cramps, nausea, and vomiting common
Group B	Large outbreaks in adults and children in China	Severe watery diarrhea for 3–5 days
Group C	Sporadic cases in children in Japan	Similar to group A
Norwalk virus and Norovirus	Epidemics of vomiting and diarrhea in older children and adults; occurs in families, communities, and nursing homes; often associated with shellfish, other food, or water and infected food handlers	Acute vomiting, fever, myalgia, and headache lasting 1–2 days, diarrhea
Caliciviruses other than the Norwalk group	Pediatric diarrhea; associated with shellfish and other foods in adults	Rotavirus-like illness in children; Norwalk-like in adults
Astroviruses	Pediatric diarrhea; reported in nursing homes	Watery diarrhea for 1–3 days

disease is caused by the hepatitis A virus (HAV) of the genus *Hepatovirus* in the family *Picornaviridae.* The hepatitis A virus is an icosahedral, linear, positive-strand RNA virus that lacks an envelope. Once in the digestive system, the viruses multiply within the intestinal epithelium. Usually only mild intestinal symptoms result. Occasionally viremia (the presence of viruses in the blood) occurs and the viruses may spread to the liver. The viruses reproduce in the liver, enter the bile, and are released into the small intestine. This explains why feces are so infectious. Symptoms last from 2 to 20 days and include anorexia, general malaise, nausea, diarrhea, fever, and chills. If the liver becomes infected, jaundice ensues. Laboratory diagnosis is by detection of anti-hepatitis A antibody. During epidemic years, about 30,000 cases were reported annually in the United States. The number of new cases has been dramatically reduced since the introduction of the hepatitis A vaccine in the 1990s. Fortunately the mortality rate is low (less than 1%), and infections in children are usually asymptomatic. Most cases resolve in 4 to 6 weeks and yield a strong immunity. Approximately 40 to 80% of the United States population have serum antibodies though few have been aware of the disease. Control of infection is by simple hygienic measures, the sanitary disposal of excreta, and the killed HAV vaccine (Havrix). This vaccine is recommended for travelers (*see table 36.9*) going to regions with high evidence rates of hepatitis A.

Hepatitis E

Hepatitis E is implicated in many epidemics in certain developing countries in Asia, Africa, and Central and South America. It is uncommon in the United States but is occasionally imported by infected travelers. The monopartite, positive-strand, RNA viral genome (7,900 nucleotides) is linear. The virion is spherical, nonenveloped, and 32 to 34 nm in diameter. Based on biologic and physicochemical properties, HEV has been provisionally classified in the *Caliciviridae* family; however, the organization of the HEV genome is substantially different from that of other caliciviruses and, therefore, HEV may eventually be classified in a separate family.

Infection usually is associated with feces-contaminated drinking water. Presumably HEV enters the blood from the gastrointestinal tract, replicates in the liver, is released from hepatocytes into the bile, and is subsequently excreted in the feces. Like hepatitis A, an HEV infection usually runs a benign course and is self-limiting. The incubation period varies from 15 to 60 days, with an average of 40 days. The disease is most often recorded in patients that are 15 to 40 years of age. Children are typically asymptomatic or present mild signs and symptoms, similar to those of other types of viral hepatitis, including abdominal pain, anorexia, dark urine, fever, hepatomegaly, jaundice, malaise, nausea, and vomiting. Case fatality rates are low (1 to 3%) except for pregnant women (15 to 25%), who may die from fulminant hepatic failure. Diagnosis of HEV is by ELISA (IgM or IgG to recombinant HEV) or reverse transcriptase PCR. There are no specific measures for preventing HEV infections other than those aimed at improving the level of health and sanitation in affected areas.

Poliomyelitis

Poliomyelitis [Greek *polios,* gray, and *myelos,* marrow or spinal cord], **polio,** or **infantile paralysis** is caused by the poliovirus, a member of the family *Picornaviridae* (**Historical Highlights 37.3**). The poliovirus is a naked, positive-strand RNA virus with three different serotypes—P1, P2, and P3. The virus is very stable, especially at acidic pH, and can remain infectious for relatively long periods in food and water—its main routes of transmission. The average incubation period is 6 to 20 days. Once ingested, the virus multiplies in the mucosa of the throat and/or small intestine. From these sites the virus invades the tonsils and lymph nodes of the neck and terminal portion of the small intestine. Generally, there are either no symptoms or a brief illness characterized by fever, headache, sore throat, vomiting, and loss of appetite. The virus sometimes enters the bloodstream and causes a viremia. In most cases (more than 99%), the viremia is transient and clinical disease does not result. In

Historical Highlights

37.3 A Brief History of Polio

Like many other infectious diseases, polio is probably of ancient origin. Various Egyptian hieroglyphics dated approximately 2000 B.C. depict individuals with wasting, withered legs and arms (see **Box figure**). In 1840 the German orthopedist Jacob von Heine described the clinical features of poliomyelitis and identified the spinal cord as the problem area. Little further progress was made until 1890, when Oskar Medin, a Swedish pediatrician, portrayed the natural history of the disease as epidemic in form. He also recognized that a systemic phase, characterized by minor symptoms and fever, occurred early and was complicated by paralysis only occasionally. Major progress occurred in 1908, when Karl Landsteiner and William Popper successfully transmitted the disease to monkeys. In the 1930s much public interest in polio occurred because of the polio experienced by Franklin D. Roosevelt. This led to the founding of the March of Dimes campaign in 1938; the sole purpose of the March of Dimes was to collect money for research on polio. In 1949 John Enders, Thomas Weller, and Frederick Robbins discovered that the poliovirus could be propagated in vitro in cultures of human embryonic tissues of nonneural origin. This was the keystone that later led to the development of vaccines.

In 1952 David Bodian recognized that there were three distinct serotypes of the poliovirus. Jonas Salk successfully immunized humans with formalin-inactivated poliovirus in 1952, and this vaccine (IPV) was licensed in 1955. The live attenuated poliovirus vaccine (oral polio vaccine, OPV) developed by Albert Sabin and others had been employed in Europe since 1960 and was licensed for U.S. use in 1962. Both the Salk and Sabin vaccines led to a dramatic decline of paralytic poliomyelitis in most developed countries and, as such, have been rightfully hailed as two of the great accomplishments of medical science.

Ancient Egyptian with Polio. Note the withered leg.

the minority of cases (less than 1%), the viremia persists and the virus enters the central nervous system and causes paralytic polio. The virus has a high affinity for anterior horn motor nerve cells of the spinal cord. Once inside these cells, it multiplies and destroys the cells; this results in motor and muscle paralysis. Since the licensing of the formalin-inactivated Salk vaccine (1955) and the attenuated virus Sabin vaccine (1962), the incidence of polio has decreased markedly. No endogenous reservoir of polioviruses exists in the United States. An on-going global effort to eliminate polio has been very successful. However, sporadic cases were reported in 2005, mostly in areas where religious views and misinformation diminish vaccination efforts. Nonetheless, it is likely that polio will be the next human disease to be completely eradicated.

1. What two virus groups are associated with acute viral gastroenteritis? How do they cause the disease's symptoms?
2. Describe some symptoms of hepatitis A.
3. Why was hepatitis A called infectious hepatitis?
4. At what specific sites within the body can the poliomyelitis virus multiply? What is the usual outcome of an infection?

37.5 Zoonotic Diseases

The diseases discussed here are caused by viruses that are normally zoonotic (animal-borne). The RNA virus families *Arenaviridae, Bunyaviridae, Flaviviridae, Filoviridae,* and *Picornoviridae* represent notable examples of human viral infections found in animal reservoirs before transmission to and between humans. Some of these viruses are exotic and rare; others are being irradicated by public health efforts. Some of the virus types are found in relatively small geographic areas; others are distributed across continents. Several of these viruses cause diseases with substantial morbidity and mortality. It is for these reasons that many are placed on the Select Agents list as potential bioweapons, as indicated by double astericks.

**Ebola and Marburg Hemorrhagic Fevers

Viral hemorrhagic fever (VHF) is the term used to describe a severe, multisystem syndrome caused by several distinct viruses. Characteristically, the overall host vascular system is damaged, resulting in vascular leakage (hemorrhage) and dysfunction (coagulopathy). **Ebola hemorrhagic fever** is caused by the Ebola virus, first recognized near the Ebola River in the Democratic Republic of the Congo in Africa. The virus is a member of a family of negative-strand RNA viruses called the *Filoviridae*. Four Ebola subtypes are known. Ebola-Zaire, Ebola-Sudan, and Ebola-Ivory Coast cause disease in humans. The fourth, Ebola-Reston, appears to only cause disease in nonhuman primates, and unlike the others, is spread by aerosol transmission as well as by contact with body fluids. Infection with Ebola is severe and approximately 80% fatal. The incubation period for the hemorrhagic fever caused by Ebola ranges from 2 to 21 days and is characterized by abrupt fever, headache, joint and muscle aches, sore throat, and weakness, followed by diarrhea, vomiting, and stomach pain. Signs of infection include rash, red eyes, bleeding, and hiccups, while symptoms alert of internal hemorrhage. The reservoir of Ebola virus appears to be at least three species of fruit bats that are native only to Africa. Contact with an infected animal (bats, apes, or other primates) and subsequent transmission to other humans likely initiates an outbreak. Transmission can be from direct contact with the blood and/or secretions from an infected person or clinical samples. Exposure can also occur through contact with bodies of Ebola victims. There is no standard treatment for Ebola infection. Patients receive supportive therapy. This consists of balancing patients' fluids and electrolytes, maintaining their oxygen status and blood pressure, and treating them for any complicating infections. Experimental vaccines are currently being evaluated and show promise in nonhuman primate models.

Marburg hemorrhagic fever is caused by a genetically unique RNA virus in the *Filoviridae* family. Marburg fever is a rare, severe type of hemorrhagic fever that affects both humans and nonhuman primates. Recognition of this virus led to the creation of the *Filoviridae* family. Marburg virus was first recognized in 1967, when outbreaks of hemorrhagic fever occurred simultaneously in laboratories in Marburg and Frankfurt, Germany and in Belgrade, Yugoslavia (now Serbia). A total of 32 people became ill; they included laboratory workers as well as several medical personnel and family members who had cared for them. The first people infected had been exposed to African green monkeys or their tissues. In Marburg, the monkeys had been imported for research and to prepare polio vaccine. Marburg virus is indigenous to Africa, but its specific origin is yet unknown. A definitive animal host is also unknown. The average incubation period for Marburg hemorrhagic fever is 5 to 10 days. The disease symptoms are abrupt, marked by fever, chills, headache, and myalgia. A maculopapular rash, most prominent on the chest, back, and stomach, typically appears around the fifth day after the onset. Nausea, vomiting, chest pain, sore throat, abdominal pain, and diarrhea may also occur in infected patients. Symptoms become increasingly severe and may include jaundice, delirium, liver failure, pancreatitis, severe weight loss, shock, and multi-organ dysfunction. A specific treatment for this disease is unknown. However, supportive hospital therapy should be utilized. This includes balancing the patient's fluids and electrolytes, maintaining oxygen status and blood pressure, replacing lost blood and clotting factors, and treating for other complicating infections (*see Disease 37.2*).

**Hantavirus Pulmonary Syndrome

Hantavirus pulmonary syndrome (HPS) is a deadly disease caused by a negative-strand RNA virus of the *Bunyaviridae*. HPS is typically transmitted to humans by inhalation of viral particles shed in urine, feces, or saliva of infected rodents. HPS was first recognized in 1993 and has since been identified throughout the United States. Although rare, HPS is potentially deadly. Rodent control in and around the home remains the primary strategy for preventing hantavirus infection. HPS in the United States is not transmitted from person to person, nor is it known to be transmitted by rodents purchased from pet stores. Hantaviruses have lipid envelopes that are susceptible to most disinfectants. The length of time hantaviruses can remain infectious in the environment is variable and depends on environmental conditions.

Temperature, humidity, exposure to sunlight, and even the rodent's diet (affecting the chemistry of rodent urine) strongly influence viral survival. Viability of dried virus has been reported at room temperature for 2 to 3 days. Hantaviruses are shed in body fluids but do not appear to cause disease in their reservoir hosts. Data indicate that viral transfer then may occur through biting as field studies suggest that viral transmission in rodent populations occurs horizontally and more frequently between males. A specific treatment for HPS is unknown. Supportive therapy is used to treat symptoms, including balancing the patient's fluids and electrolytes, maintaining oxygen status and blood pressure, replacing lost blood and clotting factors, and treating for other complicating infections.

**Lassa Fever

Lassa fever is an acute illness caused by a negative-stand RNA virus in the family *Arenaviridae*. The illness was discovered in 1969 in Nigeria, West Africa. Lassa fever can be mild and has no observable symptoms in about 80% of people infected. The remaining 20% have severe multisystem disease. Case-fatality rate can reach 50% during epidemics; thus Lassa fever can be a significant cause of morbidity and mortality. Lassa virus appears to be harbored by Old World rats and mice (family *Muridae,* subfamily *Murinae*). These rodents become chronically infected with arenaviruses, yet the viruses do not appear to cause disease in their hosts. The viruses are shed by infected rodents in urine or feces. Lassa virus can be transmitted from person to person; airborne and contact transmission have been reported. No vaccine is yet available to prevent Lassa fever. However, ribavirin has been approved for use as a preventative therapy.

Lymphocytic Choriomeningitis

Lymphocytic choriomeningitis (LCM) is another rodent-borne viral infection caused by the lymphocytic choriomeningitis virus (LCMV), a negative-strand RNA virus. LCMV is a member of the family *Arenaviridae* and often presents as aseptic meningitis, encephalitis, or meningoencephalitis. Asymptomatic infection or mild febrile illnesses are also common clinical manifestations of LCMV. Infection in utero may result in spontaneous abortion, congenital hydrocephalus and chorioretinitis, and mental retardation. LCMV is known to be transmissible through organ transplantation. Recently, the death of three organ transplant patients prompted an investigation by the Rhode Island Department of Health, the Massachusetts Department of Public Health, the CDC, the New England Organ Bank, and the transplant centers involved. The three deceased patients and one other living patient received organs from a common donor. The CDC confirmed that all four patients were infected with LCMV. The donor's blood and tissue were found to be negative for LCMV upon testing. However, it is speculated that the donor acquired the LCMV from a pet hamster bought three weeks prior to his death. LCMV is associated with Old World rats and mice, which are now found worldwide. Human infection with arenaviruses occurs upon contact with the excretions, or materials contaminated with the excretions, of an infected rodent. Infection can also occur by aerosol transmission upon inhalation of virus-contaminated rodent urine or saliva.

**Nipah Virus

Nipah virus is a member of the family *Paramyxoviridae,* negative-strand RNA viruses similar to the measles and mumps viruses. Infection with Nipah virus initiates 3 to 14 days of fever and headache, followed by drowsiness and mental confusion that can lead to coma within 24 to 48 hours. About 40% of the patients with serious neurological disease during an outbreak in 1998–1999, died from the illness. Some patients present with history of a respiratory illness during the early part of their infections. Nipah virus was initially isolated in 1999 during an outbreak of encephalitis and respiratory illness among adult men (associated with infected pigs) in Malaysia and Singapore. The natural reservoir for Nipah virus is unknown; however, data suggest that bats of the genus *Pteropus* can harbor Nipah virus.

Rabies

Rabies [Latin *rabere,* rage or madness] is caused by a number of different strains of highly neurotropic viruses. Most belong to a single serotype in the genus *Lyssavirus* [Greek *lyssa,* rage or rabies], family *Rhabdoviridae.* The bullet-shaped virion contains a negative-strand RNA genome (**figure 37.22*a***). Rabies has been the object of human fascination, torment, and fear since the disease was first recognized. Prior to Pasteur's development of an antirabies vaccine, few words were more terrifying than the cry of "mad dog!" Improvements in prevention during the past 50 years have led to almost complete elimination of indigenously acquired rabies in the United States where rabies is primarily a disease of feral animals. Most wild animals can become infected with rabies, but susceptibility varies according to species. Foxes, coyotes, and wolves are the most susceptible; intermediate are skunks, raccoons, insectivorous bats, and bobcats; while opossums are quite resistant (figure 37.22*b*). Worldwide, almost all cases of human rabies are attributed to dog bites. In developing countries where canine rabies is still endemic, rabies accounts for up to 40,000 deaths per year. Occasionally, other domestic animals are responsible for

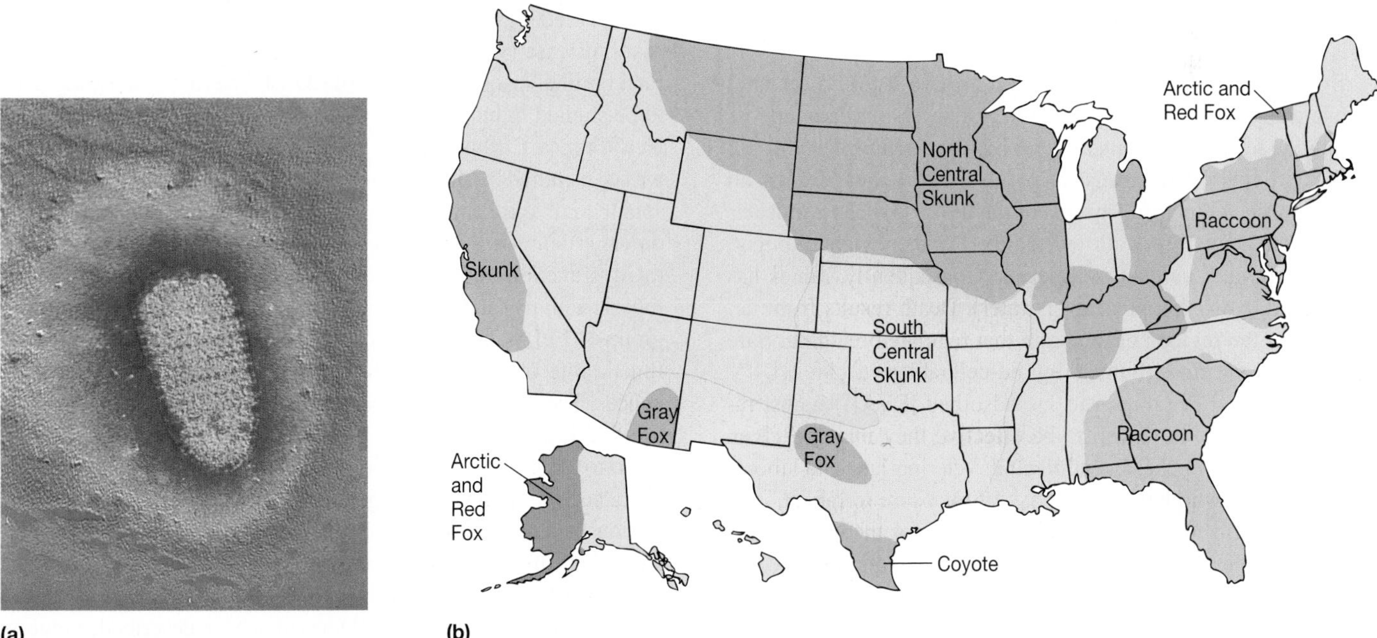

Figure 37.22 Rabies. **(a)** Electron micrograph of the rabies virus (yellow) (×36,700). Note the bullet shape. The external surface of the virus contains spikelike glycoprotein projections that bind specifically to cellular receptors. **(b)** In the United States rabies is found in terrestrial animals in 10 distinct geographic areas. In each area a particular species is the reservoir and one of five antigenic variants of the virus predominates as illustrated by the five different colors. Although not shown, another eight viral variants are found in insectivorous bats and cause sporadic cases of rabies in terrestrial animals throughout the country. Absence of a strain does not imply absence of rabies.

transmission of rabies to humans. It should be noted, however, that not all rabid animals exhibit signs of agitation and aggression (known as furious rabies). In fact, paralysis (dumb rabies) is the more common sign exhibited by rabid animals.

The virus multiplies in the salivary glands of an infected host. It is transmitted to humans or other animals by the bite of an infected animal whose saliva contains the virus; by aerosols of the virus that can be spread in caves where bats dwell; or by contamination of scratches, abrasions, open wounds, and mucous membranes with saliva from an infected animal. After inoculation, a region of the virions' glycoprotein envelope spike attaches to the plasma membrane of nearby skeletal muscle cells, the virus enters the cells, and multiplication of the virus occurs. When the concentration of the virus in the muscle is sufficient, the virus enters the nervous system through unmyelinated sensory and motor terminals; the binding site is the nicotinic acetylcholine receptor.

The virus spreads by retrograde axonal flow at 8 to 20 mm per day until it reaches the spinal cord, when the first specific symptoms of the disease—pain or paresthesia at the wound site—may occur. A rapidly progressive encephalitis develops as the virus quickly disseminates through the central nervous system. The virus then spreads throughout the body along the peripheral nerves, including those in the salivary glands, where it is shed in the saliva.

Within brain neurons the virus produces characteristic **Negri bodies,** masses of viruses or unassembled viral subunits that are visible in the light microscope. In the past, diagnosis of rabies consisted solely of examining nervous tissue for the presence of these bodies. Today diagnosis is based on *d*irect immuno*f*luorescent *a*ntibody (DFA) of brain tissue, virus isolation, detection of Negri bodies, and a rapid rabies enzyme-mediated immunodiagnosis test.

Symptoms of rabies usually begin 2 to 16 weeks after viral exposure and include anxiety, irritability, depression, fatigue, loss of appetite, fever, and a sensitivity to light and sound. The disease quickly progresses to a stage of paralysis. In about 50% of all cases, intense and painful spasms of the throat and chest muscles occur when the victim swallows liquids. The mere sight, thought, or smell of water can set off spasms. Consequently, rabies has been called hydrophobia (fear of water). Death results from destruction of the regions of the brain that regulate breathing. Safe and effective vaccines (human diploid-cell rabies vaccine HDCV [Imovax Rabies] or rabies vaccine adsorbed [RVA]) against rabies are available; however, to be effective they must be given soon after the person has been infected. Veterinarians and laboratory personnel, who have a high risk of exposure to rabies, usually are immunized every 2 years and tested for the presence of suitable antibody titer. About 30,000 people annually receive this treatment. In the United States fewer than 10 cases of rabies occur yearly in humans, although about 8,000 cases of animal rabies are reported each year from various sources (figure 37.22*b*). Prevention and control involves pre-exposure vaccination of dogs and cats, postexposure vaccination of humans, and pre-exposure vaccination of humans at special risk (persons spending a month or more in countries where rabies is common in dogs).

Some states and countries (Hawaii and Great Britain, for example) retain their rabies-free status by imposing quarantine periods on any entering dog or cat. If an asymptomatic, unvaccinated dog or cat bites a human, the animal is typically confined and observed by a veterinarian for at least 10 days. If the animal shows no signs of rabies in that time, it is determined to be uninfected. Animals demonstrating signs of rabies are killed and brain tissue submitted for rabies testing. Postexposure prophylaxis—rabies immune globulin for passive immunity and rabies vaccine for active immunity—is initiated to exploit the relatively long incubation period of the virus. This is usually recommended for anyone bitten by one of the common reservoir species (raccoons, skunks, foxes, and bats), unless it is proven that the animal was uninfected. Once symptoms of rabies develop in a human, death usually occurs.

1. Why are Ebola and Marburg hemorrhagic fever diseases so deadly?
2. What precautions can be taken to prevent Hantavirus and Lassa virus transmission to humans?
3. How does the rabies virus cause death in humans?

37.6 Prion Diseases

Prion diseases, also called **transmissible spongiform encephalopathies (TSEs)**, are fatal neurodegenerative disorders that have attracted enormous attention not only for their unique biological features but also for their impact on public health. Prions (protein infectious particles) are thought to consist of abnormally folded proteins (PrP^{SC}), which can induce normal forms of the protein (PrP^{C}) to fold abnormally. This group of diseases includes kuru, Creutzfeldt-Jakob disease (CJD), new variant Creutzfeldt-Jakob disease (vCJD), Gerstmann-Sträussler disease (GSD), and fatal familial insomnia (FFI; **table 37.6**). The first of these diseases to be studied in humans was kuru, discovered in the Fore tribe of New Guinea. Carlton Gadjusek and others showed that the disease was transmitted by ritual cannibalism (especially where brains and spinal cords were eaten). The primary symptom of the human disorders is dementia, usually accompanied by manifestations of motor dysfunction such as cerebral ataxia (inability to coordinate muscle activity) and myoclonus (shocklike contractions of muscle groups). FFI is also characterized by dysautonomia (abnormal functioning of the autonomic nervous system) and sleep disturbances. These symptoms appear insidiously in middle to late adult life and last from months (CJD, FFI, and kuru) to years (GSD) prior to death. Neuropathologically, these disorders produce a characteristic spongiform degeneration of the brain, as well as deposition of amyloid plaques. Prion diseases thus share important clinical, neuropathological, and cell biological features with another, more common cerebral amyloidosis, Alzheimer's disease. A familial (inherited) form of CJD has also been described, suggesting that certain genetic mutations cause the PrP^{C} protein to more easily assume the PrP^{SC} conformation. Prions (section 18.10)

New variant CJD is transmitted from cattle that have bovine spongiform encephalopathy (BSE, or mad cow disease) as described in section 40.4. There have been two confirmed cases of

Table 37.6 Prion Diseases of Humans

Disease	Incubation Period	Nature of Disease
Creutzfeldt-Jakob disease (CJD) (sporadic, iatrogenic, familial, new-variant)	Months to years	Spongiform encephalopathy (degenerative changes in the central nervous system)
Kuru	Months to years	Spongiform encephalopathy
Gerstmann-Sträussler-Scheinker disease (GSD)	Months to years	Genetic neurodegenerative disease
Fatal familial insomnia (FFI)	Months to years	Genetic neurodegenerative disease with progressive, untreatable insomnia

BSE in the United States (Washington state, 2004 and Texas, 2005) compared to approximately 40,000 BSE cases in the United Kingdom. Cattle experimentally infected by the oral route have tested positive for the BSE agent in the brain, spinal cord, retina, dorsal root ganglia, distal ileum, and bone marrow, suggesting that (1) the BSE agent survives passage along the GI tract, (2) the BSE agent is neurotropic, and (3) these tissues represent a source of infectious material that may be transmitted to humans and other animals. In fact, much evidence has accumulated to suggest that human vCJD can be acquired by individuals who eat meat products (especially if they contain brain and spinal cord) prepared from infected cattle. Data from the U.K. report 107 confirmed and 42 probable deaths attributed to vCJD; compared to 260 total worldwide deaths attributed to vCJD (as of September 2006). Estimates of the final total vCJD cases (extrapolated from analysis of positive results from a tonsil and appendix tissue bank) expected in the United Kingdom by 2080 range from a few hundred to 140,000. Of additional concern is the report of four vCJD cases associated with blood transfusion in the United Kingdom. Iatrogenic CJD is induced by a physician or surgeon, a medical treatment, or diagnostic procedures. It has been transmitted by prion-contaminated human growth hormone, corneal grafts, and grafts of dura mater (tissue surrounding the brain). Donor screening and more thorough testing of grafts have decreased the frequency of prion transmission.

1. How are prions different from viruses; how are they similar?
2. In what way are spongiform encephalopathies commonly acquired?

Summary

37.1 Airborne Diseases

a. More than 400 different viruses can infect humans. These viruses can be grouped and discussed according to their mode of transmission and acquisition.

b. Most airborne viral diseases involve either directly or indirectly the respiratory system. Examples include chickenpox (varicella, **figure 37.1**), shingles (herpes zoster, **figure 37.2**), rubella (German measles, **figure 37.5**), influenza (flu), measles (rubeola, **figure 37.3**), mumps (**figure 37.4**), the acute respiratory viruses such as the respiratory syncytial virus, the eradicated smallpox (variola, **figure 37.6**), and viral pneumonia.

37.2 Arthropod-Borne Diseases

a. The arthropod-borne viral diseases are transmitted by arthropod vectors from human to human or animal to human (**table 37.2**). Examples include Rift Valley fever; St. Louis encephalitis; eastern, western, and Venezuelan equine encephalitis; West Nile fever; and yellow fever. All these diseases are characterized by fever, headache, nausea, vomiting, and characteristic encephalitis.

37.3 Direct Contact Diseases

a. Person-to-person contact is another way of acquiring or transmitting a viral disease. Examples of such diseases include AIDS (**figures 37.7–37.12**), cold sores (**figure 37.13**), the common cold, cytomegalovirus inclusion disease, genital herpes (**figure 37.16*b,c***), human herpesvirus 6 infections, human parvovirus B19 infection, certain leukemias, infectious mononucleosis, and hepatitis (**table 37.4**): hepatitis B (serum hepatitis); hepatitis C; hepatitis D (delta hepatitis); hepatitis F; and hepatitis G.

37.4 Food-Borne and Waterborne Diseases

a. The viruses that are transmitted in food and water usually grow in the intestinal system and leave the body in the feces (**table 37.5**). Acquisition is generally by the oral route. Examples of diseases caused by these viruses include acute viral gastroenteritis (rotavirus and others), infectious hepatitis A, hepatitis E, and poliomyelitis.

37.5 Zoonotic Diseases

a. Diseases transmitted from animals are zoonotic. Several animal viruses can cause disease in humans. Examples of viral zoonoses include Ebola and Marburg fevers, hantavirus pulmonary syndrome, Lassa fever, and rabies.

37.6 Prion Diseases

a. A prion disease is a pathological process caused by a transmissible agent (a prion) that remains clinically silent for a prolonged period, after which the clinical disease becomes apparent. Examples include (**table 37.6**) Creutzfeldt-Jakob disease, kuru, Gerstmann-Sträussler-Scheinker disease, and fatal familial insomnia. These diseases are chronic infections of the central nervous system that result in progressive degenerative changes and eventual death.

Key Terms

acute viral gastroenteritis 939
adult T-cell leukemia 935
AIDS (acquired immune deficiency syndrome) 925
anogenital condylomata (venereal warts) 938
antigenic drift 916
antigenic shift 916
chickenpox (varicella) 914
cold sore 931
common cold 932
congenital (neonatal) herpes 934
congenital rubella syndrome 920
cytomegalovirus inclusion disease 933
Dane particle 936
Delta agent 938
Ebola hemorrhagic fever 942
epizoonotic 923
equine encephalitis 922
erythema infectiosum 935
exanthem subitum 934
fever blister 931
fifth disease 935
flat or plane warts 938
genital herpes 933
gingivostomatitis 931
Guillain-Barré syndrome (French polio) 918
hantavirus pulmonary syndrome (HPS) 942
hemorrhagic fevers 923
hepatitis 936
hepatitis A 939
hepatitis B 936
hepatitis C 937
hepatitis D 938
hepatitis E 940
hepatitis F 938
hepatitis G 938
herpes labialis 931
herpetic keratitis 932
human herpesvirus 6 934
human immunodeficiency virus (HIV) 925
human parvovirus B19 935
infantile paralysis 940
infectious mononucleosis (mono) 935
influenza or flu 915
intranuclear inclusion body 933
Koplik's spots 918
Lassa fever 942
leukemia 935
Lymphocytic choriomeningitis (LCM) 942
Marburg viral hemorrhagic fever 942
measles (rubeola) 917
mumps 919
Negri bodies 944
Nipah virus 943
orchitis 919
plantar warts 938
polio 940
poliomyelitis 940
postherpetic neuralgia 915
pulmonary syndrome hantavirus 923
rabies 943
respiratory syncytial virus (RSV) 919
Reye's syndrome 918
Rift Valley fever (RVF) 922
roseola infantum 934
rubella (German measles) 920
severe acute respiratory syndrome (SARS) 920
shingles (herpes zoster) 915
sixth disease 934
smallpox (variola) 920
subacute sclerosing panencephalitis 918
tick-borne encephalitis (TBE) 922
transmissible spongiform encephalopathies (TSE) 944
verrucae vulgaris 938
viral hemorrhagic fever (VHF) 941
wart 938
West Nile fever (encephalitis) 924
yellow fever 924

Critical Thinking Questions

1. Explain why antibiotics are not effective against viral infections. Advise a person about what can be done to relieve symptoms of a viral infection and recover most quickly. Address your advice to (a) someone who has had only a basic course in high school biology, and (b) a third-grade student.
2. Several characteristics of AIDS render it particularly difficult to detect, prevent, and treat effectively. Discuss two of them. Contrast the disease with polio and smallpox.
3. From an epidemiological perspective, why are most arthropod-borne viral diseases hard to control?
4. In terms of molecular genetics, why is the common cold such a prevalent viral infection in humans?
5. Will it be possible to eradicate many viral diseases in the same way as smallpox? Why or why not?

Learn More

Jahrling, P. B.; Fritz, E. A.; and Hensley, L. E. 2005. Countermeasures to the bioterrorist threat of smallpox. *Curr. Mol. Med.* 5:817–26.

Kerr, J. 2005. Pathogenesis of parvovirus B19 infection: Host gene variability, and possible means and effects of virus persistence. *J. Vet. Med. B. Infect. Dis. Vet. Public Health* 52:335–9.

Preston, R. 1989. *The hot zone: A terrifying true story.* New York: Anchor Books.

Tarkowski, T. A.; Koumans, H. E.; Sawyer, M.; Pierce, A.; Black, C. M.; Papp, J. R.; Markowitz, L.; and Unger, E. R. 2004. Epidemiology of human papillomavirus infection and abnormal cytologic test results in an urban adolescent population. *J. Infect. Dis.* 189:46–50.

Taubenberger, J. K., and Morens, D. M. 2006. 1918 influenza: The mother of all pandemics. *Emerging Infect. Dis.* 12:15–22.

Webster, R. G.; Peiris, M.; Chen, H.; and Guan, Y. 2006. H5N1 outbreaks and enzootic influenza. *Emerging Infect. Dis.* 12:3–8.

Wright, L. 2003. To vanquish a virus. *Sci. Amer.* July 21.

Please visit the Prescott website at www.mhhe.com/prescott7 for additional references.

38 Human Diseases Caused by Bacteria

The toll of tetanus. The bacterial genus *Clostridium* contains many pathogenic species, including the species responsible for tetanus (*C. tetani*). Sir Charles Bell's portrait (c. 1821) of a soldier wounded in the Peninsular War in Spain shows the suffering from generalized tetanus.

Preview

- Most of the airborne diseases caused by bacteria involve the respiratory system. Examples include diphtheria, Legionnaires' disease and Pontiac fever, *Mycobacterium avium–M. intracellulare* and *M. tuberculosis* infections, pertussis, streptococcal diseases, and mycoplasmal pneumonia. Other airborne bacteria can cause skin diseases, including cellulitis and erysipelas, or systemic diseases such as meningitis, glomerulonephritis, and rheumatic fever.
- Although arthropod-borne bacterial diseases are generally rare, they are of interest either historically (plague) or because they have interesting clinical presentation (Lyme disease). Most of the rickettsial diseases are arthropod-borne. The rickettsias found in the United States can be divided into the typhus group (epidemic typhus caused by *R. prowazekii* and murine typhus caused by *R. typhi*) and the spotted fever group (Rocky Mountain spotted fever caused by *R. rickettsii* and ehrlichiosis caused by *Ehrlichia chaffeensis*). Q fever (caused by *Coxiella burnetti*) is an exception. It is not a rickettsia; it forms endospore-like structures and does not have to use an insect vector.
- Most of the direct contact bacterial diseases involve the skin, mucous membranes, or underlying tissues. Examples include bacterial vaginosis, chancroid, gas gangrene, leprosy, peptic ulcer disease and gastritis, staphylococcal diseases, and syphilis. Others can become disseminated throughout specific regions of the body—for example, gonorrhea, staphylococcal diseases, syphilis, and tetanus. Two chlamydial species cause direct contact disease: *Chlamydophila (Chlamydia) trachomatis* causes inclusion conjunctivitis, lymphogranuloma venereum, nongonococcal urethritis, and chlamydial pneumonia. At least three species of mycoplasmas are human pathogens: *Mycoplasma hominis* and *Ureaplasma urealyticum* cause genitourinary tract disease.
- Humans contract the food-borne and waterborne bacterial diseases when they ingest contaminated food or water. These diseases are essentially of two types: infections and intoxications. An infection occurs when a pathogen enters the gastrointestinal tract and multiplies. Examples include *Campylobacter* gastroenteritis, salmonellosis, listerosis, shigellosis, *Escherichia coli* infections, and typhoid fever. An intoxication occurs because of the ingestion of a toxin. Examples include botulism, cholera, and staphylococcal food poisoning.
- Some microbial diseases and their effects cannot be related to a specific mode of transmission. Two important examples are sepsis and septic shock. Gram-positive bacteria, fungi, and endotoxin-containing gram-negative bacteria can initiate the pathogenic cascade of sepsis leading to septic shock.
- Many bacterial diseases can be acquired directly from animals. These zoonotic diseases include anthrax, brucellosis, psitticosis, and tularemia.
- Several bacterial odonto-pathogens are responsible for the most common bacterial diseases in humans—tooth decay and periodontal disease. Both are the result of biofilm formation and the production of lactic and acetic acids by the odonto-pathogens.

In this chapter we continue our discussion of infectious disease by turning our attention to bacterial pathogens. These include bacteria that cause localized and systemic infec-

Soldiers have rarely won wars. They more often mop up after the barrage of epidemics. And typhus, with its brothers and sisters—plague, cholera, typhoid, dysentery—has decided more campaigns than Caesar, Hannibal, Napoleon, and all the . . . generals of history. The epidemics get the blame for the defeat, the generals the credit for victory. It ought to be the other way around. . . .

—Hans Zinsser

tions. The microorganisms involved in dental infections are also described. Diseases caused by bacteria that are now listed as Select Agents (potential bioterror agents) are identified within the chapter by two asterisks (**).

Of all the known bacterial species, only a few are pathogenic to humans. Some human diseases have been only recently recognized **(table 38.1)**; others have been known since antiquity. In the following sections the more important disease-causing bacteria are discussed according to their mode of acquisition/transmission.

Table 38.1 Some Examples of Human Bacterial Diseases Recognized Since 1977

Year	Bacterium	Disease
1977	*Legionella pneumophila*	Legionnaires' disease
1977	*Campylobacter jejuni*	Enteric disease (gastroenteritis)
1981	*Staphylococcus aureus*	Toxic shock syndrome
1982	*Escherichia coli* O157:H7	Hemorrhagic colitis; hemolytic uremic syndrome (HUS)
1982	*Borrelia burgdorferi*	Lyme disease
1982	*Helicobacter pylori*	Peptic ulcer disease
1984	Meticillin-resistant *Staphylococcus aureus*	Epidemic nosocomial infections
1986	*Ehrlichia chaffeensis*	Human ehrlichiosis
1988	*Chlamydia pneumoniae*	Atherosclerosis
1988	*Salmonella enteritidis* F14	Egg-borne salmonellosis
1989	*Enterococcus faecium;* vancomycin-resistant enterococci	Colitis and enteritis
1990	*Streptococcus pyogenes*	"Flesh-eating"and streptococcal toxic shock
1992	*Vibrio cholerae* O139	New strain associated with epidemic cholera in Asia
1992	*Bartonella henselae*	Cat-scratch disease; bacillary angiomatosis
1994	*Ehrlichia* spp.	Human granulocytic ehrlichiosis
1995	*Neisseria meningitidis*	Meningococcal supraglottitis
1996	Vancomycin-resistant *S. aureus*	Nosocomial infections
1997	*Kingella kingae*	Pediatric infections
2000	*Tropheryma whipplei*	Whipple's disease

38.1 Airborne Diseases

Most airborne diseases caused by bacteria involve the respiratory system. Other airborne bacteria can cause skin diseases. Some of the better known of these diseases are now discussed.

Chlamydial Pneumonia

Chlamydial pneumonia is caused by *Chlamydophila (Chlamydia) pneumoniae.* Clinically, infections are generally mild; pharyngitis, bronchitis, and sinusitis commonly accompany some lower respiratory tract involvement. Symptoms include fever, a productive cough (respiratory secretion brought up by coughing), sore throat, hoarseness, and pain on swallowing. Infections with *C. pneumoniae* are common but sporadic; about 50% of adults have antibody to the chlamydiae. Evidence suggests that *C. pneumoniae* is primarily a human pathogen directly transmitted from human to human by droplet (respiratory) secretions. Diagnosis of chlamydial pneumonia is based on symptoms and a microimmunofluorescence test. Tetracycline and erythromycin are routinely used for treatment. In seroepidemiological studies, *C. pneumoniae* infections have been linked with coronary artery disease as well as vascular disease at other sites. Following a demonstration of *C. pneumoniae*-like particles in atherloscerotic plaque tissue by electron microscopy, *C. pneumoniae* genes and antigens have been detected in artery plaque. Rarely, however, has the microorganism been recovered in cultures of atheromatous tissue (i.e., artery plaque). As a result of these findings, the possible etiologic role of *C. pneumoniae* in coronary artery disease and systemic atherosclerosis is under intense scrutiny. Phylum *Chlamydiae* (section 21.5)

Diphtheria

Diphtheria [Greek *diphthera,* membrane, and *ia,* condition] is an acute, contagious disease caused by the gram-positive bacterium *Corynebacterium diphtheriae* (*see figure 24.9*). *C. diphtheriae* is well-adapted to airborne transmission by way of nasopharyngeal secretions because it is very resistant to drying. Diphtheria mainly affects unvaccinated, poor people living in crowded conditions. Once within the respiratory system, bacteria that carry the prophage β containing the *tox* gene produce diphtheria toxin; tox^+ phage infection of *C. diphtheriae* is required for toxin production. This toxin is an exotoxin that causes an inflammatory response and the formation of a grayish pseudomembrane on the pharynx and respiratory mucosa **(figure 38.1)**. The pseudomembrane consists of dead host cells and cells of *C. diphtheriae.* Diphtheria toxin is absorbed into the circulatory system and distributed throughout the body, where it may cause destruction of cardiac, kidney, and nervous tissues by inhibiting protein synthesis. The toxin is composed of two polypeptide subunits: A and B. The A subunit consists of the catalytic domain; the B subunit is composed of the receptor and transmembrane domains (*see figure 33.5*). The receptor domain binds to the heparin-binding epidermal growth factor receptor on the surface of various eucaryotic cells. Once bound, the toxin enters the cytoplasm by

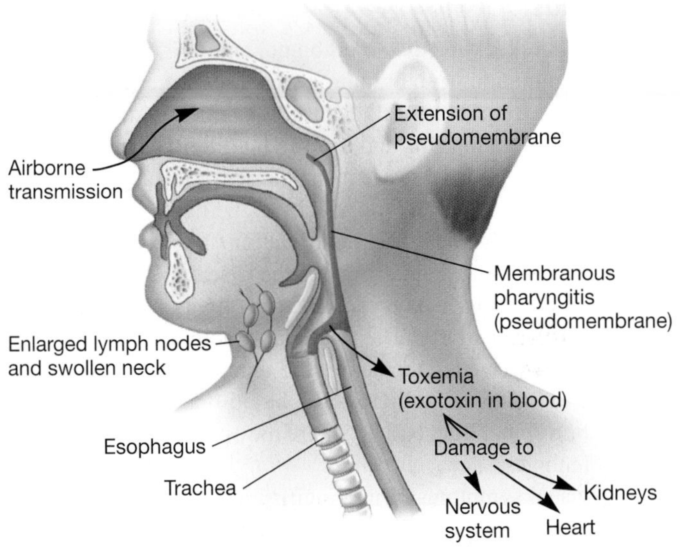

(a)

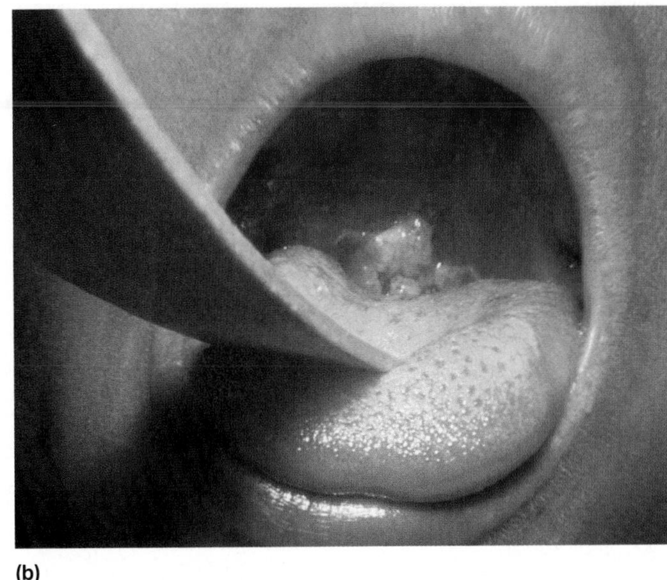

(b)

Figure 38.1 Diphtheria Pathogenesis. **(a)** Diphtheria is a well-known, exotoxin-mediated infectious disease caused by *Corynebacterium diphtheriae*. The disease is an acute, contagious, febrile illness characterized by local oropharyngeal inflammation and pseudomembrane formation. If the exotoxin gets into the blood, it is disseminated and can damage the peripheral nerves, heart, and kidneys. **(b)** The clinical appearance includes gross inflammation of the pharynx and tonsils marked by grayish patches (a pseudomembrane) and swelling of the entire area.

endocytosis. The transmembrane domain of the toxin embeds itself into the target cell membrane causing the catalytic domain to be cleaved and translocated into the cytoplasm. The cleaved catalytic domain becomes an active enzyme, catalyzing the attachment of ADP-ribose (from NAD^+) to elongation factor-2 (EF-2). A single enzyme (i.e., catalytic domain) can exhaust the entire supply of cellular EF-2 within hours, resulting in protein synthesis inhibition and cell death. Suborder *Corynebacterineae* (section 24.4); Toxigenicity: AB toxins (section 33.4)

Typical symptoms of diphtheria include a thick mucopurulent (containing both mucus and pus) nasal discharge, pharyngitis, fever, cough, paralysis, and death. (*C. diphtheriae* can also infect the skin, usually at a wound or skin lesion, causing a slow-healing ulceration termed **cutaneous diphtheria.**) Diagnosis is made by observation of the pseudomembrane in the throat and by bacterial culture. Diphtheria antitoxin is given to neutralize any unabsorbed exotoxin in the patient's tissues; penicillin and erythromycin are used to treat the infection. Prevention is by active immunization with **DPT** (*d*iphtheria-*p*ertussis-*t*etnus) **vaccine;** and then boosted with DTap (*d*iphtheria *t*oxoid, *t*etanus toxoid, *a*cellular *B. pertussis* vaccine); or Tdap (*t*etanus toxoid, reduced *d*iphtheria toxoid, *a*cellular *p*ertussis vaccine, adsorbed), approved in 2005 (*see table 36.4*). Most cases involve people over 30 years of age who have a weakened immunity to the diphtheria toxin and live in tropical areas. Since 1980, fewer than six diphtheria cases have been reported annually in the United States, and most occur in nonimmunized individuals. Control of epidemics: Vaccines and immunization (section 36.8)

Legionnaire's Disease and Pontiac Fever

In 1976 the term **Legionnaires' disease,** or **legionellosis,** was coined to describe an outbreak of pneumonia that occurred at the Pennsylvania State American Legion Convention in Philadelphia. The bacterium responsible for the outbreak was *Legionella pneumophila,* a nutritionally fastidious, aerobic, gram-negative rod (**figure 38.2**). It is now known that this bacterium is part of the natural microbial community of soil and freshwater ecosystems, and it has been found in large numbers in air-conditioning systems and shower stalls. Class *Gammaproteobacteria* (section 22.3)

An increasing body of evidence suggests that environmental protozoa are the most important factor for the survival and growth of *Legionella* in nature. A variety of free-living amoebae and ciliated protozoa that contain *Legionella* spp. have been isolated from water sites suspected as sources of *Legionella* infections. *Legionella* spp. multiply intracellularly within the amoebae, just as they do within human monocytes and macrophages. This might explain why there is no human-to-human spread of legionellosis.

Infection with *L. pneumophila* and other *Legionella* spp. results from the airborne spread of bacteria from an environmental reservoir to the human respiratory system. Males over 50 years of age most commonly contract the disease, especially if their immune system is compromised by heavy smoking, alcoholism, or chronic illness. The bacteria reside within the phagosomes of alveolar macrophages, where they multiply and produce localized tissue destruction through export of a cytotoxic exoprotease. Symptoms start 2 to 10 days after exposure and include a high

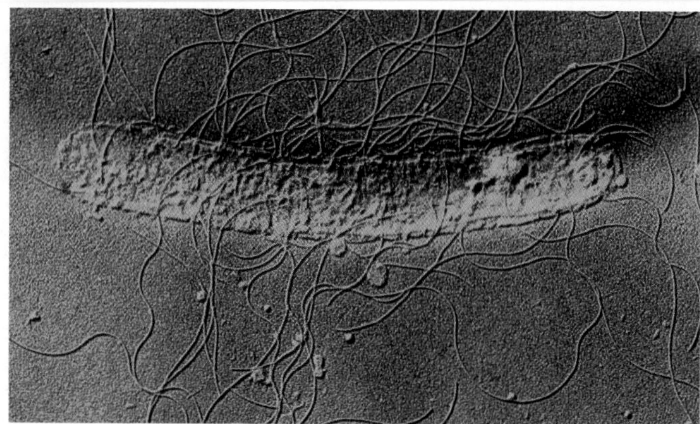

Figure 38.2 Legionnaires' Disease. *Legionella pneumophila*, the causative agent of Legionnaires' disease, with many lateral flagella; SEM (× 10,000).

fever, nonproductive cough (respiratory secretions are not brought up during coughing), headache, neurological manifestations, and severe bronchopneumonia. Diagnosis depends on isolation of the bacterium, documentation of a rise in antibody titer over time, or the presence of *Legionella* antigens in the urine as detected by a rapid test kit. Treatment begins with supportive measures and the administration of erythromycin or rifampin. Death occurs in 10 to 15% of cases.

Prevention of Legionnaires' disease depends on the identification and elimination of the environmental source of *L. pneumophila*. Chlorination, the heating of water, and the cleaning of water-containing devices can help control the multiplication and spread of *Legionella*. These control measures are effective because the pathogen does not appear to be spread from person to person. Since the initial outbreak of this disease in 1976, many outbreaks during summer months have been recognized in all parts of the United States. About 1,000 to 1,600 cases are diagnosed each year, and about 18,000 or more additional mild or subclinical cases are thought to occur. It is estimated that 23% of all nosocomial pneumonias are due to *L. pneumophila*, especially among immunocompromised patients.

L. pneumophila also causes a milder illness called **Pontiac fever.** This disease, which resembles an allergic disease more than an infection, is characterized by an abrupt onset of fever, headache, dizziness, and muscle pains. It is indistinguishable clinically from the various respiratory syndromes caused by viruses. Pneumonia does not occur. The disease resolves spontaneously within 2 to 5 days. No deaths from Pontiac fever have been reported.

Pontiac fever was first described from an outbreak in a county health department in Pontiac, Michigan. Ninety-five percent of the employees became ill and eventually showed elevated serum titers against *L. pneumophila*. These bacteria were later isolated from the lungs of guinea pigs exposed to the air of the building. The likely source was water from a defective air conditioner.

Meningitis

Meningitis [Greek *meninx*, membrane, and *–itis*, inflammation] is an inflammation of the brain or spinal cord meninges (membranes). Based on the specific cause, it can be divided into **bacterial (septic) meningitis** and **aseptic meningitis syndrome.** As shown in **table 38.2,** there are many causes of the aseptic meningitis syndrome, only some of which can be treated with antimicrobial agents. Thus accurate identification of the causative agent is essential for proper treatment of the disease. The immediate sources of the bacteria responsible for meningitis are respiratory secretions from carriers. The bacteria initially colonize the nasopharynx, after which they cross the mucosal barrier. They can enter the bloodstream and cross the blood-brain barrier to enter the cerebral spinal fluid (CSF), where they produce inflammation of the meninges.

The usual symptoms of meningitis include an initial respiratory illness or sore throat interrupted by one of the meningeal syndromes: vomiting, headache, lethargy, confusion, and stiffness in the neck and back. Bacterial meningitis can be diagnosed by a Gram stain and culture or rapid tests of the bacteria from CSF. Once bacterial meningitis is suspected, specific antibiotics (penicillin, chloramphenicol, cefotaxime, ceftriazone, ofloxacin) are administered immediately. In fact, antibiotics are often administered prophylactically to patient contacts.

Bacterial meningitis can be caused by various gram-positive and gram-negative bacteria. However, three organisms tend to be

Table 38.2 Causative Agents of Meningitis by Diagnostic Category

Type of Meningitis	Causative Agent
Bacterial (Septic) Meningitis	
	Streptococcus pneumoniae
	Neisseria meningitidis
	Haemophilus influenzae type b
	Group B streptococci
	Listeria monocytogenes
	Mycobacterium tuberculosis
	Nocardia asteroides
	Staphylococcus aureus
	Staphylococcus epidermidis
Aseptic Meningitis Syndrome	
Agents Requiring Antimicrobials	Fungi
	Amoebae
	Treponema pallidum
	Mycoplasmas
	Leptospires
Agents Requiring Other Treatments	Viruses
	Cancers
	Parasitic cysts
	Chemicals

associated with meningitis more frequently than others: *Streptococcus pneumoniae, Neisseria meningitidis,* and *Haemophilus influenzae* (serotype b). *S. pneumoniae* is discussed with other streptococcal diseases later in this chapter. *N. meningiditis,* often referred to as the meningococcus, is a normal inhabitant of the human nasopharynx (5 to 15% of humans carry the nonpathogenic serotypes). Most disease-causing *N. meningitidis* strains belong to serotypes A, B, C, Y and W-135. In general, serotype A strains are the cause of epidemic disease in developing countries, while serotype C and W-135 strains are responsible for meningitis outbreaks in the United States. Infection results from airborne transmission of the bacteria, typically through close contact with a primary carrier. The disease process is initiated by pili-mediated colonization of the nasopharynx by pathogenic bacteria. The bacteria cross the nasopharyngeal epithelium (typically through endocytosis) and invade the bloodstream (meningococcemia) where they proliferate. Symptoms caused by *N. meningitidis* are variable depending on the degree of bacterial dissemination. Infection of the CSF leads to meningitis; untreated meningitis is fatal.

Control of *N. meningitidis* infection is with vaccination and antibiotics. There are currently two vaccines available: the meningococcal polysaccharide (MPSV4) and the meningococcal conjugate vaccine (MCV4). Both vaccines are effective against serotypes A, C, Y and W-135. Vaccination is recommended for all college students living in residence halls. MCV4 is recommended for preteen children, teens, and adults less than 55 years of age. MPSV4 should be used for children 2 to 10 years of age and adults over 55 who are at risk.

Another agent of meningitis is *H. influenzae,* a small, gram-negative bacterium. Transmission is by inhalation of droplet nuclei shed by infectious individuals or carriers. *H. influenzae* serotype b can infect mucous membranes, resulting in sinusitis, pneumonia, and bronchitis. It can disseminate to the bloodstream and cause a bacteremia. *H. influenzae* serotype b can cross into the CSF, resulting in inflammation of the meninges (meningitis). *H. influenzae* disease (including pneumonia and meningitis) is primarily observed in children less than 5 years of age. It is estimated that *H. influenzae* serotype b causes at least 3 million cases of serious disease, and several hundreds of thousands of deaths each year.

A sharp reduction in the incidence of *H. influenzae* serotype b infections began in the mid-1980s due to administration of the *H. influenzae* type b conjugate vaccine, rifampin prophylaxis of disease contacts, and the availability of more efficacious therapeutic agents. From 1987 through 1999, the incidence of invasive infection among U.S. children under 5 years of age declined by 95%. Three to six percent of all *H. influenzae* infections are fatal. Furthermore, up to 20% of surviving patients have permanent hearing loss or other long-term sequelae. Currently, all children should be vaccinated with the *H. influenzae* type b conjugate vaccine at the age of 2 months.

Complicating diagnostic practices is the fact that a person may have meningitis symptoms but show no microbial agent in gram-stained specimens, and have negative cultures. In such a case the diagnosis often is called aseptic meningitis syndrome. Aseptic meningitis is typically more difficult to treat.

Mycobacterium avium-M. intracellulare and *M. tuberculosis* Pulmonary Diseases

M. avium-M. intracellulare Infections

An extremely large group of mycobacteria are normal inhabitants of soil, water, and house dust. Two of these are noteworthy pathogens in the United States—the two, *Mycobacterium avium* and *Mycobacterium intracellulare,* are so closely related that they are referred to as the *M. avium* complex (MAC). Globally, *M. tuberculosis* has remained more prevalent in developing countries, where as MAC has become the most common cause of mycobacterial infections in the United States. Suborder *Corynebacterineae* (section 24.4)

These mycobacteria are found worldwide and infect a variety of insects, birds, and animals. Both the respiratory and the gastrointestinal tracts have been proposed as entry portals for the *M. avium* complex; however, person-to-person transmission is not very efficient. The gastrointestinal tract is thought to be the most common site of colonization and dissemination in AIDS patients. MAC causes a pulmonary infection in humans similar to that caused by *M. tuberculosis*. Pulmonary MAC is more common in non-AIDS patients, particularly in elderly persons with preexisting pulmonary disease.

Shortly after the recognition of AIDS and its associated opportunistic infections, it became apparent that one of the more common AIDS-related infections was caused by MAC. In the United States, disseminated infection with MAC occurs in 15 to 40% of AIDS patients with $CD4^+$ cell counts of less than 100 per cubic millimeter. Disseminated infection with MAC produces disabling symptoms, including fever, malaise, weight loss, night sweats, and diarrhea. Carefully controlled epidemiological studies have shown that MAC shortens survival by 5 to 7 months among persons with AIDS. With more effective antiviral therapy for AIDS and with prolonged survival, the number of cases of disseminated MAC is likely to increase substantially, and its contribution to AIDS mortality will increase. Direct contact diseases: AIDS (section 37.3)

MAC can be isolated from sputum, blood, and aspirates of bone marrow. Acid-fast stains are of value in making a diagnosis. The most sensitive method for detection is the commercially available lysis-centrifugation blood culture system (Wampole Laboratories). Although no drugs are currently approved by the FDA for the therapy of MAC, every regimen should contain either azithromycin or clarithromycin and ethambutol as a second drug. One or more of the following can be added: closfazimine, rifabutin, rifampin, ciprofloxacin, and amikacin.

Mycobacterium tuberculosis Infections

Over a century ago Robert Koch identified *Mycobacterium tuberculosis* as the causative agent of **tuberculosis (TB)**. At the time, TB was rampant, causing one-seventh of all deaths in

Europe and one-third of deaths among productive young adults. Today TB remains a global health problem of enormous dimension. It is estimated that one-third of the world's human population is infected, with 9 million new cases and 2 million deaths per year (**figure 38.3*a***).

In the United States, this disease occurs most commonly among the homeless, elderly, and malnourished, or among alcoholic males, minorities, immigrants, prison populations, and Native Americans. Between 1999 and 2005, the incidence of tuberculosis in the United States steadily declined to about 14,000

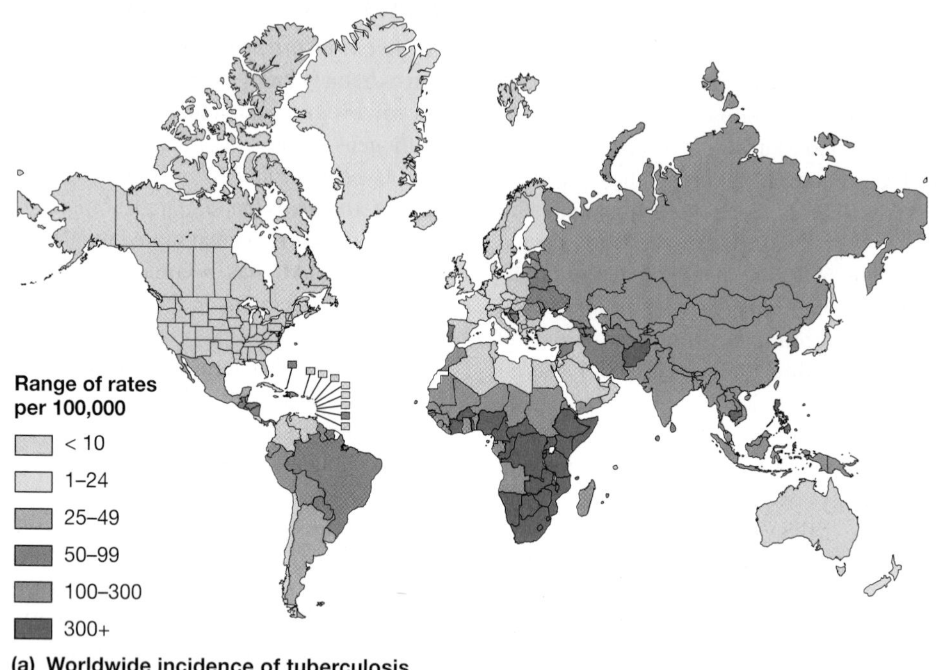

(a) Worldwide incidence of tuberculosis

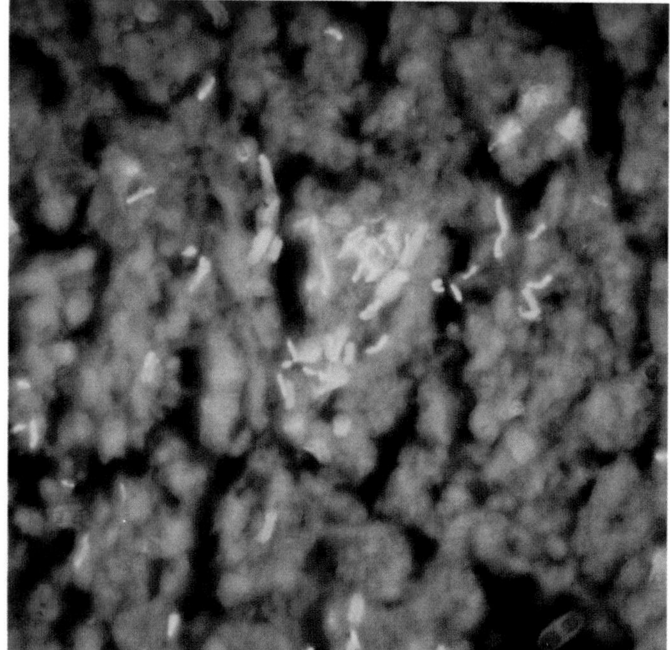

(b) *M. tuberculosis* in sputum

Multinucleate giant cell
Epithelioid cells
Caseous necrosis (tubercle bacilli at center)
Granuloma (fibroblast) cells

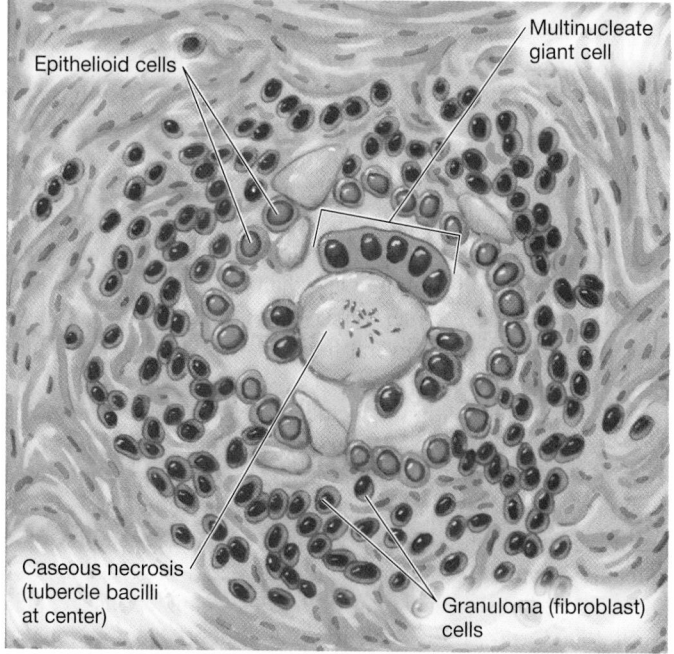

(c) A tubercle

Figure 38.3 Tuberculosis. **(a)** Tuberculosis is a significant global disease. **(b)** *Mycobacteria* are recovered in the sputum of tuberculosis patients and can be identified using a fluorescent acid-fast stain. **(c)** In the lungs, tuberculosis is identified by the tubercle, a massive granuloma of white blood cells, bacteria, fibroblasts, and epithelioid cells. The center of the tubercle contains caseous (cheesy) pus and bacteria. **(d)** The natural history of mycobacterial infection leading to tuberculosis demonstrates its public health threat.

cases; about 1,000 deaths are reported each year. During 2005, a total of 14,093 confirmed TB cases were reported in the United States, representing a 3.8% decline in the rate from 2004. Slightly more than half (53.7%) of these U.S. cases were in foreign-born persons. Most cases in the United States are acquired from other humans through droplet nuclei and the respiratory route (figure 38.3*d*). It appears that about one-fourth to one-third of active TB cases in the United States may be due to recent transmission. The majority of active cases result from the reactivation of old, dormant infections.

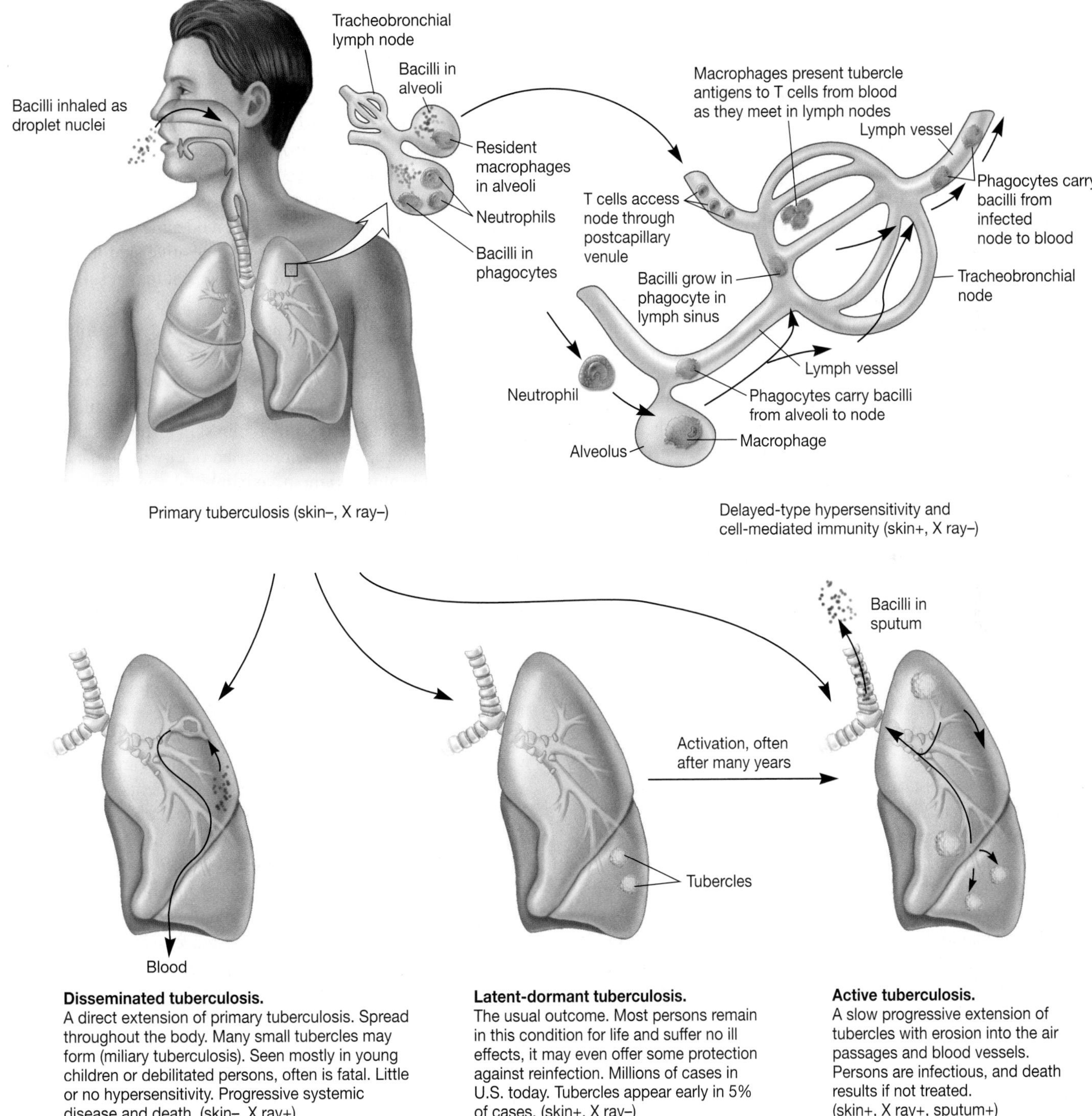

(d) ***Mycobacterium tuberculosis*** **infection and its possible outcomes.**

Figure 38.3 *continued.*

Worldwide, TB is caused by *M. bovis* and *M. africanum,* in addition to *M. tuberculosis*. This is likely due to closer interactions between people and livestock, another host for the organisms. Transmission to humans from susceptible animal species and their products (e.g., milk) is also possible. With the advent of the AIDS epidemic, there has been a steady yearly increase in the number of global TB cases. Available statistics indicate that a close association exists between AIDS and TB. Therefore, further spread of HIV infection among populations with a high prevalence of TB infection is resulting in dramatic increases in TB. However, the *Mycobacterium avium* complex has become the most common mycobacterial disease in U.S. AIDS patients.

The bacteria are phagocytosed by macrophages in the lungs, where they survive the normal antimicrobial processes (figure 38.3*b*). In fact, macrophages that have phagocytosed mycobacteria often die attempting to destroy them. Other immune effector cells are recruited to the site of infection by cytokines released from the responding macrophages. Together, and in response to several mycobacterial products, a hypersensitivity response results in the formation of small, hard nodules called **tubercles** composed of bacteria, macrophages, T cells, and various human proteins (figure 38.3*c*). Tubercles are characteristic of tuberculosis and give the disease its name. The disease process usually stops at this stage but the bacteria often remain alive within macrophage phagosomes. However, in some cases, the disease may become active, even after many years of latency. The incubation period is about 4 to 12 weeks, and the disease develops slowly. The symptoms of tuberculosis are fever, fatigue, and weight loss. A cough, which is characteristic of pulmonary involvement, may result in expectoration of bloody sputum.

M. tuberculosis (Mtb) does not produce classic virulence factors such as toxins, capsules, and fimbriae. Instead, Mtb has some rather unique products and properties that contribute to its virulence. The cell envelope of Mtb differs substantially from that of gram-positive and gram-negative bacteria in that it contains several unique lipids and glycolipids. These include mycolic acids, lipoarabinomannan, trehalose dimycolate, and phthiocerol dimycocerosate (*see figure 24.11*). These materials are directly toxic to eucaryotic cells and create a hydrophobic barrier around the bacterium that facilitates impermeability and resistance to antimicrobial agents, resistance to killing by acidic and alkaline compounds, resistance to osmotic lysis, and resistance to lysozyme. Cell wall glycolipids also associate with mannose giving Mtb control over entry into macrophages, exploiting the macrophage mannose receptors. Once inside, Mtb can inhibit phagosome-lysosome fusion by altering the phagosome membrane. Resistance to oxidative killing, inhibition of phagosome-lysosome fusion, and inhibition of diffusion of lysosomal enzymes are just some of the mechanisms that help explain the survival of *M. tuberulosis* inside macrophages.

In time, the tubercle may change to a cheeselike consistency and is then called a **caseous lesion** (figure 38.3*b*). If such lesions calcify, they are called **Ghon complexes,** which show up prominently in a chest X-ray. (Often the primary lesion is called the Ghon's tubercle or Ghon's focus.) Sometimes the tubercle lesions liquefy and form air-filled **tuberculous cavities.** From these cavities the bacteria can spread to new foci of infection throughout the body. This spreading is often called **miliary tuberculosis** due to the many tubercles the size of millet seeds that are formed in the infected tissue. It also may be called **reactivation tuberculosis** because the bacteria have been reactivated in the initial site of infection.

Persons infected with *M. tuberculosis* develop cell-mediated immunity because the bacteria are phagocytosed by macrophages (i.e., it is an intracellular pathogen). This immunity involves sensitized T cells and is the basis for the tuberculin skin test. In this test a purified protein derivative (PPD) of *M. tuberculosis* is injected intracutaneously (the Mantoux test). If the person has had tuberculosis, or was exposed to *M. tuberculosis,* sensitized T cells react with these proteins and a delayed hypersensitivity reaction occurs within 48 hours (*see figure 32.32*). This positive skin reaction appears as an induration (hardening) and reddening of the area around the injection site. In a young person, a positive skin test could indicate active tuberculosis. In older persons, it may result from previous disease, vaccination, or a false-positive test. In both cases, X-rays and bacterial isolation are completed to confirm the diagnosis. Immune disorders: Type IV hypersensitivy (section 32.11)

Laboratory diagnosis of tuberculosis is by visualization of the acid-fast bacterium, chest X-ray, commercially available DNA probes, and the Mantoux or tuberculin skin test. Both chemotherapy and chemoprophylaxis are carried out by administering isoniazid (INH), plus rifampin, ethambutol, and pyrazinamide. These drugs are administered simultaneously for 6 to 9 months as a way of decreasing the possibility that the bacterium develops drug resistance.

Multidrug-resistant strains of tuberculosis (MDR-TB) have developed and are spreading. A multidrug-resistant strain is defined as *M. tuberculosis* that is resistant to isoniazid and rifampin, with or without resistance to other drugs. Between 2003 and 2004 in the United States, MDR-TB increased 13.3%, the largest yearly increase in over a decade. These MDR-TB cases represent 1.2% of cases for which drug-susceptibility test results were reported. Inadequate therapy is the most common means by which resistant bacteria are acquired, and patients who have previously undergone therapy are presumed to harbor MDR-TB until proven otherwise. MDR-TB can be fatal.

MDR-TB arises because tubercle bacilli have spontaneous, predictable rates of chromosomally born mutations that confer resistance to drugs. These mutations are unlinked; hence, resistance to one drug is not associated with resistance to an unrelated drug. The emergence of drug resistance represents the survival of random preexisting mutations, not a change caused by exposure to the drug—that the mutations are not linked is the cardinal principle underlying TB chemotherapy. For example, mutations that cause resistance to isoniazid or rifampin occur roughly in 1 in 10^8 replications of *M. tuberculosis*. The likelihood of spontaneous mutations causing resistance to both isoniazid and rifampin is the product of these probabilities, or 1 in 10^{16}. However, these biological mechanisms of resistance break down when chemotherapy is inadequate. In the circumstances of monotherapy, erratic drug ingestion, omission of one or more drugs, suboptimal dosage, poor drug absorption, or an insufficient number of active drugs in a regimen, a susceptible strain of *M. tuberculosis* may become resistant to multiple drugs within a matter of months.

Prevention and control of tuberculosis requires rapid, specific therapy to interrupt infectious spread. Retreatment of patients who have multidrug-resistant tuberculosis should be carried out in programs with comprehensive microbiological, pharmacokinetic, psychosocial, and nutritional support systems. In many countries (not the United States), individuals, especially infants and children, are vaccinated with **bacille Calmette-Guerin (BCG)** vaccine to prevent complications such as meningitis. The BCG vaccine confers a positive PPD skin test result but appears to protect only about half of those inoculated. Tuberculosis rates also can be lowered by better public health measures and social conditions—for example, a reduction in homelessness and drug abuse.

1. How do humans contract chlamydial pneumoniae?
2. What causes the typical symptoms of diphtheria? How are individuals protected against this disease?
3. What is the environmental source of the bacterium that causes Legionnaires' disease? Pontiac fever?
4. What are the three major types of meningitis? Why is it so important to determine which type a person has?
5. How is tuberculosis diagnosed? Describe the various types of lesions and how they are formed. How do multidrug-resistant strains of tuberculosis develop?

Pertussis

Pertussis [Latin *per,* intensive, and *tussis,* cough], sometimes called "whooping cough," is caused by the gram-negative bacterium *Bordetella pertussis*. (*B. parapertussis* is a closely related species that causes a milder form of the disease.) Pertussis bacteria colonize the respiratory epithelium to produce a disease (whooping cough) characterized by fever, malaise, uncontrollable cough, and cyanosis (bluish skin color resulting from inadequate tissue oxygenation). The disease gets its name from the characteristically prolonged and paroxysmal coughing that ends in an inspiratory gasp, or whoop. Pertussis is a highly contagious, vaccine-preventable disease that primarily affects children. It has been estimated that over 95% of the world's population has experienced either mild or severe symptoms of the disease. Around 300,000 die from the disease each year. However, there are less than around 7,000 cases and less than 10 deaths annually in the United States. Class *Betaproteobacteria:* Order *Burkholdariales* (section 22.2)

Transmission occurs by inhalation of the bacterium in droplets released from an infectious person. The incubation period is 7 to 14 days. Once inside the upper respiratory tract, the bacteria colonize the cilia of the mammalian respiratory epithelium through fimbrial-like structures, called filamentous hemagglutinins, that bind to phagocyte complement receptors. Additionally, some of the components of one of the *B. pertussis* toxins (S2 and S3 subunits of the PTx toxin) assist in adherence to cilia by bridging the bacterial and host cells; S2 binds to the cilial glycolipid lactosylceramide, and S3 binds to phagocyte glycoproteins. Thus attachment is an important virulence factor in the initiation of the disease.

B. pertussis produces several toxins. The most important is the pertussis toxin (PTx). PTx is a two-component, AB exotoxin (*see figure 33.5*). The A subunit (S1) is an ADP ribosyl transferase, similar to the diphtheria toxin. The B subunit is composed of five polypeptides (S2–S5, there are two S4 subunits) that bind to specific carbohydrates on cell surfaces. The B subunit binds to host cells, transporting the A subunit to the cell membrane where it is inserted and released into the cytoplasm. As an enzyme, the A subunit transfers the ADP ribosyl moiety of NAD^+ to a membrane-bound, regulatory G protein, G_i. G_i normally inhibits eucaryotic adenylate cyclase, which catalyzes the conversion of ATP to cyclic AMP (cAMP). Thus the net effect of PTx on a cell is an increase in intracellular levels of cAMP. *B. pertussis* also produces an extra-cytoplasmic invasive adenylate cyclase, tracheal cytotoxin, and dermonecrotic toxin, which destroy epithelial tissue. Working together, the tracheal cytotoxin and pertussis toxin also provoke the secretory cells in the respiratory tract to produce nitric oxide, which kills nearby ciliated cells, inhibiting removal of bacteria and mucus. The secretion of a thick mucus also impedes ciliary action.

Pertussis is divided into three stages: (1) the catarrhal stage, so named because of the mucous membrane inflammation, which is insidious and resembles the common cold; (2) the paroxysmal stage, which is characterized by prolonged coughing sieges. During this stage the infected person tries to cough up the mucous secretions by making 5 to 15 rapidly consecutive coughs followed by the characteristic whoop—a hurried, deep inspiration. The catarrhal and paroxysmal stages last about 6 weeks. (3) The convalescent stage, when final recovery may take several months.

Laboratory diagnosis of pertussis is by culture of the bacterium, fluorescent antibody staining of smears from nasopharyngeal swabs, other antibody-based detection tests and PCR. The development of a strong, lasting immunity takes place after an initial infection. Treatment is with erythromycin, tetracycline, or chloramphenicol. Treatment ameliorates clinical illness when begun during the catarrhal phase and may also reduce the severity of the disease when begun within 2 weeks of the onset of the paroxysmal cough. Prevention is with the DPT vaccine in children when they are 2 to 3 months old; and with the tetanus toxoid, reduced diphtheria toxoid and acellular pertussis (Tdap) vaccine (approved in 2005; replaces the older tetanus and reduced diphtheria toxoid [Td] booster shots) for older children and adults (*see table 36.3*).

Mycoplasmal Pneumonia

Typical pneumonia has a bacterial origin (most frequently *Streptococcus pneumoniae*) with fairly consistent signs and symptoms. If the symptoms of pneumonia are different from what is typically observed, the disease is often called **atypical pneumonia.** One cause of atypical pneumonia is *Mycoplasma pneumoniae*, a mycoplasma with worldwide distribution. Spread involves close contact and airborne droplets. The disease is fairly common and mild in infants and small children; serious disease is seen principally in older children and young adults. Class *Mollicutes* (section 23.2)

M. pneumoniae usually infects the upper respiratory tract and subsequently moves to the lower respiratory tract, where it attaches to respiratory mucosal cells. It then produces peroxide, which may be a toxic factor, but the exact mechanism of pathogenesis is unknown. A change in mucosal cell nucleic acid synthesis has been observed. The manifestations of this disease vary in severity from asymptomatic to a serious pneumonia. The latter is accompanied by death of the surface mucosal cells, lung infiltration, and congestion. Initial symptoms include headache, weakness, a low-grade fever, and a predominant, characteristic cough. The disease and its symptoms usually persist for weeks. The mortality rate is less than 1%.

Several rapid tests using latex-bead agglutination for *M. pneumoniae* antibodies are available for diagnosis of mycoplasmal pneumonia. When isolated from respiratory secretions, some mycoplasmas form distinct colonies with a "fried-egg" appearance on agar (*see figure 23.4*). During the acute stage of the disease, diagnosis must be made by clinical observations. Tetracyclines or erythromycin are effective in treatment. There are no preventive measures.

Streptococcal Diseases

Streptococci, commonly called strep, are a heterogeneous group of gram-positive bacteria. In this group, *Streptococcus pyogenes* (group A β-hemolytic streptococci) is one of the most important bacterial pathogens. The different serotypes of **group A streptococci (GAS)** produce (1) extracellular enzymes that break down host molecules; (2) streptokinases, enzymes that activate a host-blood factor that dissolves blood clots; (3) the cytolysins streptolysin O and streptolysin S, which kill host leukocytes; and (4) capsules and M protein, which help to retard phagocytosis **(figure 38.4)**. M protein, a filamentous protein anchored in the streptococcal cell membrane, facilitates attachment to host cells and prevents opsonization by complement protein C3b. It is the major virulence factor of GAS. M protein types 1, 3, 12, and 28 are commonly found in patients with streptococcal toxic shock and multi-organ failure. Chemical mediators in nonspecific (innate) resistance: Complement (section 31.6)

S. pyogenes is widely distributed among humans; some become asymptomatic carriers. Individuals with acute infections may spread the pathogen, and transmission can occur through respiratory droplets, as direct or indirect contact. When highly virulent strains appear in schools, they can cause acute outbreaks of sore throats. Due to the cumulative buildup of antibodies to many different *S. pyogenes* serotypes over the years, outbreaks among adults are less frequent. Class *Bacilli:* Order *Lactobacillales* (section 23.5)

Diagnosis of a streptococcal infection is based on both clinical and laboratory findings. Several rapid tests are available. Treatment is with penicillin or macrolide antibiotics. Vaccines are not available for streptococcal diseases other than streptococcal pneumonia because of the large number of serotypes. The best control measure is prevention of transmission. Individuals with a known infection should be isolated and treated. Personnel working with infected patients should follow standard aseptic procedures. In the following sections some of the more important human streptococcal diseases are discussed.

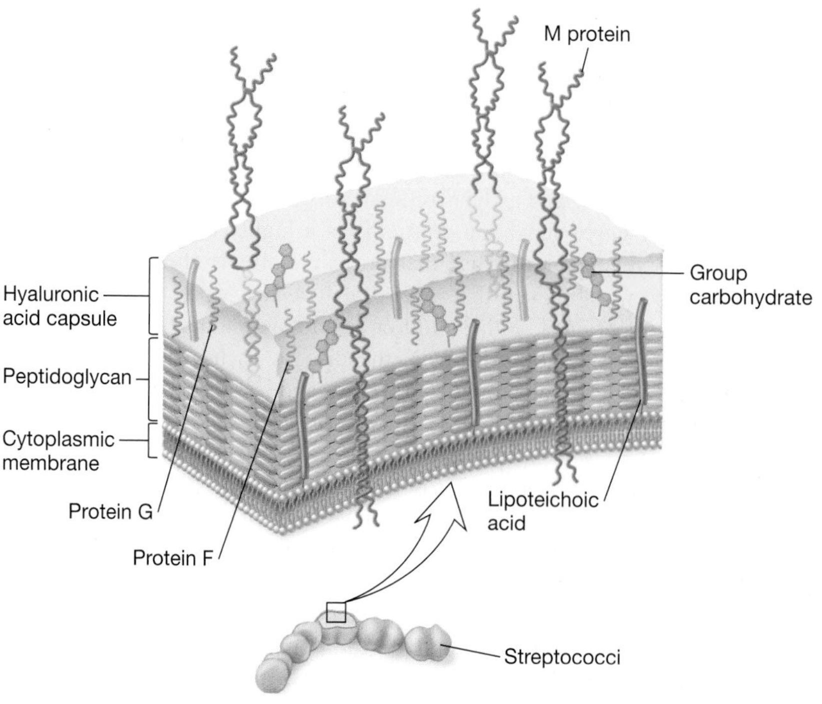

Figure 38.4 Streptococcal Cell Envelope. The M protein is a major virulence factor for streptococci. It facilitates bacterial attachment to host cells and has antiphagocytic activity. Protein G prevents attack by antibodies because it binds to the Fc portion, preventing the antigen-binding site from bacterial capture. Protein F is also an epithelial cell attachment factor.

Cellulitis, Impetigo, and Erysipelas

Cellulitis is a diffuse, spreading infection of subcutaneous skin tissue. The resulting inflammation is characterized by a defined area of redness (erythema) and the accumulation of fluid (edema). A number of different bacteria can cause cellulitis.

The most frequently diagnosed skin infection caused by *S. pyogenes* is **impetigo** (impetigo also can be caused by *Staphylococcus aureus*). Impetigo is a superficial cutaneous infection, most commonly seen in children, usually located on the face, and characterized by crusty lesions and vesicles surrounded by a red border. Impetigo is most common in late summer and early fall. The drug of choice for impetigo is penicillin; erythromycin is prescribed for those individuals who are allergic to penicillin.

Erysipelas [Greek *erythros,* red, and *pella,* skin] is an acute infection and inflammation of the dermal layer of the skin. It occurs primarily in infants and people over 30 years of age with a history of streptococcal sore throat. The skin often develops painful reddish patches that enlarge and thicken with a sharply defined edge **(figure 38.5)**. Recovery usually takes a week or longer if no treatment is given. The drugs of choice for the treatment of erysipelas are erythromycin and penicillin. Erysipelas may recur periodically at the same body site for years.

Invasive Streptococcus A Infections

In the 19th century, invasive *Streptococcus pyogenes* infections were a major cause of morbidity and mortality. However, during the 20th century the incidence of severe group A streptococcal infections declined, especially with the arrival of antibiotic therapy. In the mid-1980s there was a worldwide increase in group A streptococcal sepsis; clusters of rheumatic fever were reported from locations within the United States, and a streptococcal toxic shocklike syndrome emerged. In fact, a virulent "strep A" infection killed *Sesame Street* muppeteer Jim Henson in 1990, and in 1994 the bacterium made headlines with articles on "the flesh-eating invasive disease."

The development of invasive strep A disease appears to depend on the presence of specific virulent strains (M-1 and M-3 serotypes, for example) and predisposing host factors (surgical or nonsurgical wounds, diabetes, and other underlying medical problems). A life-threatening infection begins when invasive strep A strains penetrate a mucous membrane or take up residence in a wound such as a bruise. This infection can quickly lead either to **necrotizing fasciitis** [Greek *nekrosis,* deadness, Latin *fascis,* band or bandage, and *itis,* inflammation], which destroys the sheath covering skeletal muscles, or to **myositis** [Greek *myos,* muscle, and *itis*], the inflammation and destruction of skeletal muscle and fat tissue **(figure 38.6)**. Because necrotizing fasciitis and myositis arise and spread so quickly, they have been colloquially called "galloping gangrene."

Rapid treatment is necessary to reduce the risk of death, and penicillin G remains the treatment of choice. In addition, surgical removal of dead and dying tissue usually is needed in more advanced cases of necrotizing fasciitis. It is estimated that approximately 10,000 cases of invasive strep A infections occur annually in the United States, and between 5 and 10% of them are associated with necrotizing conditions.

One reason why invasive strep A strains are so deadly is that about 85% of them carry the genes for the production of *s*trepto*c*occal *p*yrogenic *e*xotoxins A and B (Spe exotoxins). Exotoxin A acts as a superantigen, a nonspecific T-cell activator. This superantigen quickly stimulates T cells to begin producing abnormally large quantities of cytokines. These cytokines damage the endothelial cells that line blood vessels, causing fluid loss and rapid tissue death from a lack of oxygen. Another pathogenic mechanism involves the secretion of exotoxin B, a cysteine protease (a

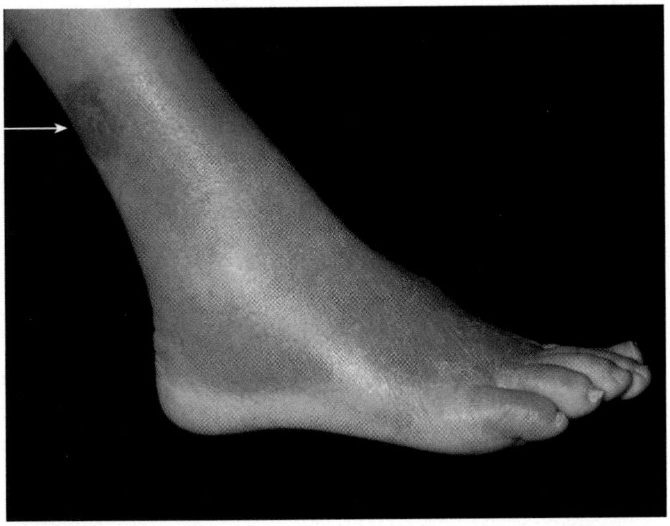

Figure 38.5 Erysipelas. Notice the bright, raised, rubbery, lesion at the site of initial entry (white arrow) and the spread of the inflammation to the foot. The reddening is caused by toxins produced by the streptococci as they invade new tissue.

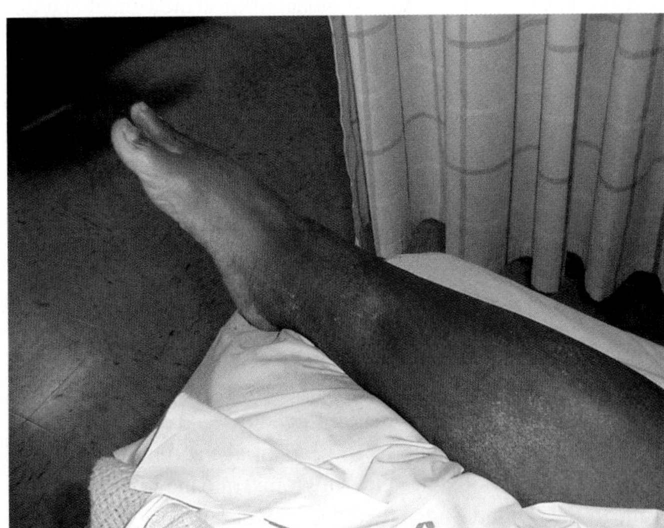

Figure 38.6 Necrotizing Fasciitis. Rapidly advancing streptococcal disease can lead to large, necrotic sites, sometimes with blisters that rupture and expose the dying tissue. This is often called flesh-eating disease or necrotizing fasciitis.

proteolytic enzyme that has a cysteine residue in the active site). This protease rapidly destroys tissue by breaking down proteins. T-cell biology: Superantigens (section 32.5)

Since 1986 it has been recognized that invasive strep A infections can also trigger a ***t*oxic *s*hock*l*ike *s*yndrome (TSLS)**, characterized by a precipitous drop in blood pressure, failure of multiple organs, and a very high fever. TSLS is caused by an invasive strep A that produces one or more of the streptococcal pyrogenic exotoxins. TSLS has a mortality rate of over 30%.

Because group A streptococci are less contagious than cold or flu viruses, infected individuals do not pose a major threat to people around them. The best preventive measures are simple ones such as covering food, washing hands, and cleansing and medicating wounds.

Poststreptococcal Diseases

The poststreptococcal diseases are glomerulonephritis and rheumatic fever. They occur 1 to 4 weeks after an acute streptococcal infection (hence the prefix *post*). These two nonsupporative (nonpus-producing) diseases are the most serious problems associated with streptococcal infections in the United States.

Glomerulonephritis, also called **Bright's disease,** is an inflammatory disease of the renal glomeruli—membranous structures within the kidney where blood is filtered. Damage probably results from the deposition of antigen-antibody complexes, possibly involving the streptococcal M protein, in the glomeruli. Thus the disease arises from a type III hypersensitivity reaction in the kidney. The complexes cause destruction of the glomerular membrane, allowing proteins and blood to leak into the urine. Clinically the affected person exhibits edema, fever, hypertension, and hematuria (blood in the urine). The disease occurs primarily among schoolage children. Diagnosis is based on the clinical history, physical findings, and confirmatory evidence of prior streptococcal infection. The incidence of glomerulonephritis in the United States is less than 0.5% of streptococcal infections. Penicillin G or erythromycin can be given for any residual streptococci. However, there is no specific therapy once kidney damage has occurred. About 80 to 90% of all cases undergo slow, spontaneous healing of the damaged glomeruli, whereas the others develop a chronic form of the disease. The latter may require a kidney transplant or lifelong renal dialysis. Immune disorders: Type III hypersensitivity (section 32.11)

Rheumatic fever is an autoimmune disease characterized by inflammatory lesions involving the heart valves, joints, subcutaneous tissues, and central nervous system. It usually results from a prior streptococcal pharyngitis. The exact mechanism of rheumatic fever development remains unknown. However, it has been associated with specific M strains. The disease occurs most frequently among children 6 to 15 years of age and manifests itself through a variety of signs and symptoms, making diagnosis difficult. In the United States rheumatic fever has become very rare (less than 0.05% of streptococcal infections). It occurs 100 times more frequently in tropical countries. Therapy is directed at decreasing inflammation and fever, and controlling cardiac failure. Salicylates and corticosteroids are the mainstays of treatment. Although rheumatic fever is rare, it is still the most common cause of permanent heart valve damage in children.

Streptococcal Pharyngitis

Streptococcal pharyngitis is one of the most common bacterial infections of humans and is commonly called strep throat. The β-hemolytic, group A streptococci are spread by droplets of saliva or nasal secretions. The incubation period in humans is 2 to 4 days. The incidence of sore throat is greater during the winter and spring months.

The action of the strep bacteria in the throat (**pharyngitis**) or on the tonsils (**tonsillitis**) stimulates an inflammatory response and the lysis of leukocytes and erythrocytes. An inflammatory exudate consisting of cells and fluid is released from the blood vessels and deposited in the surrounding tissue, although only about 50% of patients with strep pharyngitis present with an exudate. This is accompanied by a general feeling of discomfort or malaise, fever (usually above 101°F), and headache. Prominent physical manifestations include redness, edema, and lymph node enlargement in the throat. Signs and symptoms alone are not diagnostic because viral infections have a similar presentation. Several common rapid test kits are available for diagnosing strep throat. In the absence of complications, the disease can be self-limiting and may disappear within a week. However, treatment with penicillin G benzathine (or erythromycin for penicillin-allergic people) can shorten the infection and clinical syndromes, and is especially important in children for the prevention of complications such as rheumatic fever and glomerulonephritis. Infections in older children and adults tend to be milder and less frequent due in part to the immunity they have developed against the many serotypes encountered in early childhood. Prevention and control measures include proper disposal or cleansing of objects (e.g., facial tissue, handkerchiefs) contaminated by discharges from the infected individual.

Streptococcal Pneumonia

Streptococcal pneumonia is now considered an **opportunistic infection**—that is, it is contracted from one's own normal microbiota. It is caused by the gram-positive *Streptococcus pneumoniae,* normally found in the upper respiratory tract **(figure 38.7*a*)**. However, disease usually occurs only in those individuals with predisposing factors such as viral infections of the respiratory tract, physical injury to the tract, alcoholism, or diabetes. About 60 to 80% of all respiratory diseases known as pneumonia are caused by *S. pneumoniae*. An estimated 150,000 to 300,000 people in the United States contract this form of pneumonia annually, and between 13,000 to 66,000 deaths result.

The primary virulence factor of *S. pneumoniae* is its capsular polysaccharide, which is composed of hyaluronic acid (*see figure 35.10*). The production of large amounts of hyaluronic capsular polysaccharide plays an important role in protecting the organism from ingestion and killing by phagocytes. The pathogenesis is due to the rapid multiplication of the bacteria in alveolar spaces (figure 38.7*b*). The bacteria also produce the toxin pneumolysin, which destroys host cells. The alveoli fill with blood cells and fluid and become inflamed. The sputum is often rust-colored because of blood coughed up from the lungs. The onset of clinical symptoms is usually abrupt, with chills; hard, labored breathing; and chest pain. Diagnosis is by chest X ray, Gram stain, culture, and tests for

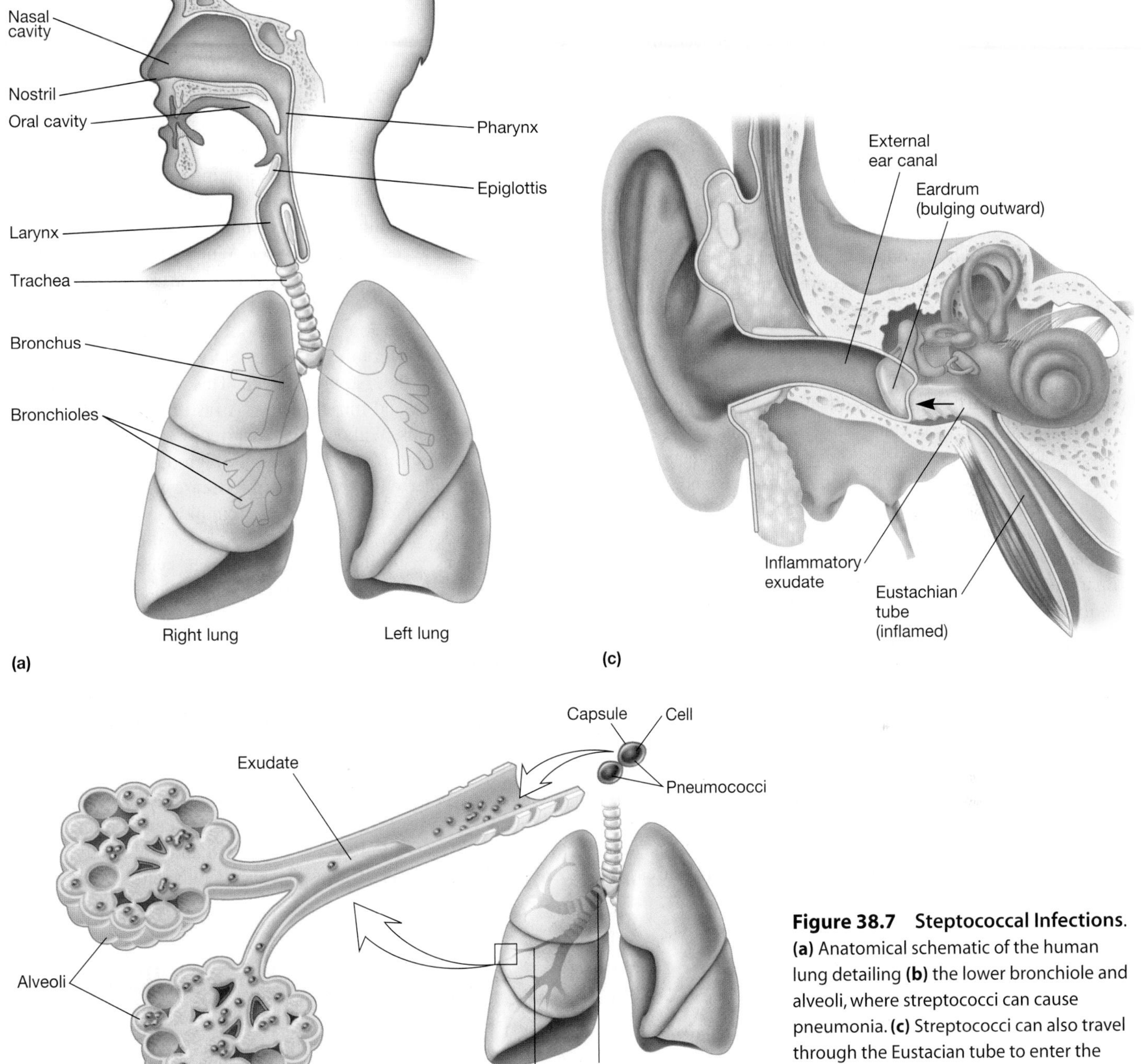

Figure 38.7 Steptococcal Infections. **(a)** Anatomical schematic of the human lung detailing **(b)** the lower bronchiole and alveoli, where streptococci can cause pneumonia. **(c)** Streptococci can also travel through the Eustacian tube to enter the middle ear, leading to infection (otitis media).

metabolic products. Cefotaxime, ofloxacin, and ceftriaxone have contributed to a greatly reduced mortality rate. For individuals who are sensitive to penicillin, erythromycin or tetracycline can be used.

S. pneumoniae is also associated with sinusitis, conjunctivitis, and otitis media (figure 38.7*c*). It is an important cause of bacteremia (blood infections) and meningitis. Penicillin- and tetracycline-resistant strains of *S. pneumoniae* are now in the United States. Pneumococcal vaccines (Pneumovax 23, Pnu-Imune 23) are available for people who are at greater risk for exposure (e.g., college students and people in chronic-care facilities). The Pneumovax vaccines (pooled collections of 23 different *S. pneumoniae* capsular polysaccharides) are effective because they generate antibodies to the capsule. When these antibodies are deposited on the surface of the capsule, they become opsonic and enhance phagocytosis (*see figure 31.21*). In 2000, a pediatric vaccine containing seven different capsular serotypes was approved for use in the United States. Preventive and control measures include immunization and adequate treatment of infected persons.

1. Name the three stages of pertussis.
2. Describe the pneumonia caused by *M. pneumoniae*.
3. Name the most important human diseases caused by *Streptococcus pyogenes* and *S. pneumoniae*. How do they differ from one another?

38.2 Arthropod-Borne Diseases

Although arthropod-borne bacterial diseases are generally rare, they are of interest either historically (plague, typhus), or because they have been newly introduced into humans (ehrlichiosis, Q fever, Lyme disease). In the next sections, only diseases that occur in the United States are discussed.

Ehrlichiosis

The first case of **ehrlichiosis** was diagnosed in the United States in 1986. The disease was caused by a relatively new bacterial species (table 38.1)—*Ehrlichia chaffeensis*. Members of the genus *Ehrlichia* are related to the genus *Rickettsia* and placed in the order *Rickettsiales* of the α-proteobacteria. More than 200 cases of ehrlichiosis are reported in the United States annually. *E. chaffeensis* is transmitted from dogs and white-tailed deer, the primary reservoirs, to humans, primarily by the Lone Star tick (*Amblyomma americanum*). Once inside the human body, *E. chaffeensis* infects circulating monocytes, causing a nonspecific febrile illness (*h*uman *m*onocytic *e*hrlichiosis, HME) that resembles Rocky Mountain spotted fever. Diagnosis is by serological tests and DNA probes. Doxycycline is the drug of choice.

A new form of ehrlichiosis was discovered in 1994. *H*uman *g*ranulocytic *e*hrlichiosis (HGE) is transmitted by deer ticks (*Ixodes scapularis*) and possibly dog ticks (*Dermacentor variabilis*), and has been found in 30 states, particularly in the southeastern, northern, and central United States. The causative agents are *Ehrlichia* species different from *E. chaffeensis*. The disease is characterized by the rapid onset of fever, chills, headaches, and muscle aches. Treatment is also with doxycycline.

**Epidemic (Louse-Borne) Typhus

Epidemic (louse-borne) typhus is caused by *Rickettsia prowazekii*, which is transmitted from person to person by the body louse (**Historical Highlights 38.1**). In the United States, a reservoir of *R. prowazekii* also exists in the southern flying squirrel. When a louse feeds on an infected rickettsemic person, the rickettsias infect the insect's gut and multiply, and large numbers of organisms appear in the insect's feces in about a week. When a louse takes a blood meal, it defecates. The irritation causes the affected individual to scratch the site and contaminate the bite wound with rickettsias. The rickettsias then spread by way of the bloodstream and infect the endothelial cells of the blood vessels, causing a **vasculitis** (inflammation of the blood vessels). This produces an abrupt headache, fever, and muscle aches. A rash begins on the upper trunk and spreads. Without treatment, recovery can take about 2 weeks, though mortality rates are very high (around 50%), especially in the elderly. Recovery from the disease gives a vigorous immunity and also protects the person from murine typhus (see next section). Class *Alphaproteobacteria: Rickettsia* and *Coxiella* (section 22.1)

Diagnosis is by the characteristic rash, symptoms, and immunofluorescence testing. Chloramphenicol and tetracycline are effective against typhus. Control of the human body louse (*Pediculus humanus corporis*) and the conditions that foster its proliferation are mainstays in the prevention of epidemic typhus, although a typhus vaccine is available for high-risk individuals. The importance of louse control and good public hygiene is shown by the prevalence of typhus epidemics during times of war and famine when there is crowding and little attention to the maintenance of proper sanitation. (For example, around 30 million cases of typhus fever and 3 million deaths occurred in the Soviet Union and Eastern Europe between 1918 and 1922. The bacteriologist Hans Zinsser believes that Napoleon's retreat from Russia

Historical Highlights

38.1 The Hazards of Microbiological Research

The investigation of human pathogens often is a very dangerous matter, and several microbiologists have been killed by the microorganisms they were studying. The study of typhus fever provides a classic example. In 1906 Howard T. Ricketts (1871–1910), an associate professor of pathology at the University of Chicago, became interested in Rocky Mountain spotted fever, a disease that had decimated the Nez Percé and Flathead Indians of Montana. By infecting guinea pigs, he established that a small bacterium was the disease agent and was transmitted by ticks. In late 1909 Ricketts traveled to Mexico to study Mexican typhus.

He discovered that a microorganism similar to the Rocky Mountain spotted fever bacterium could cause the disease in monkeys and be transmitted by lice. Despite his careful technique, he was bitten while transferring lice in his laboratory and died of typhus fever on May 3, 1910. The causative agent of typhus fever was fully described in 1916 by the Brazilian scientist H. da Roche-Lima and named *Rickettsia prowazekii* in honor of Ricketts and Stanislaus von Prowazek, a Czechoslovakian microbiologist who died in 1915 while studying typhus.

Today modern equipment to control microorganisms, such as laminar airflow hoods, have greatly reduced the risks of research on microbial pathogens.

in 1812 may have been partially provoked by typhus and dysentery epidemics that ravaged the French army.) Fewer than 25 cases of epidemic typhus are reported in the United States each year.

Endemic (Murine) Typhus

The etiologic agent of **endemic (murine) typhus** is *Rickettsia typhi*. It occurs in isolated areas around the world, including southeastern and Gulf Coast states, especially Texas. The disease occurs sporadically in individuals who come into contact with rats and their fleas (*Xenopsylla cheopi*). The disease is nonfatal in the rat and is transmitted from rat to rat by fleas. When an infected flea takes a human blood meal, it defecates. Its feces are heavily laden with rickettsias, which infect humans by contaminating the bite wound.

The clinical manifestations of murine [Latin *mus, muris,* mouse or rat] typhus are similar to those of epidemic typhus except that they are milder in degree and the mortality rate is much lower—less than 5%. Diagnosis and treatment are also the same. Rat control and avoidance of rats are preventive measures for the disease. Few cases of endemic typhus are reported in the United States each year, resulting in its removal from the reportable disease list.

Lyme Disease

Lyme disease (LD, Lyme borreliosis) was first observed and described in 1975 among people of Old Lyme, Connecticut. It has become the most common tick-borne zoonosis in the United States, with about 17,000 cases reported annually. The disease is also present in Europe and Asia.

The Lyme spirochetes responsible for this disease comprise at least three species, currently designated *Borrelia burgdorferi* (**figure 38.8*a***), *B. garinii,* and *B. afzelii*. Deer and field mice are the natural hosts. In the northeastern United States, *B. burgdorferi* is transmitted to humans by the bite of the infected deer tick (*Ixodes scapularis;* figure 38.8*b*). On the Pacific Coast, especially in California, the reservoir is a dusky-footed woodrat, and the tick, *I. pacificus*. Phylum *Spirochaetes* (section 21.6)

Clinically, Lyme disease is a complex illness with three major stages. The initial, localized stage occurs a week to 10 days after an infectious tick bite. The illness often begins with an expanding, ring-shaped, skin lesion with a red outer border and partial central clearing called erythema migrans (figure 38.8*c*). This often is accompanied by flulike symptoms (malaise and fatigue, headache, fever, and chills). Often the tick bite is unnoticed, and the skin lesion may be missed due to skin coloration or its obscure location, such as on the scalp. Thus treatment, which is usually effective at this stage, may not be given because the illness is assumed to be "just the flu."

The second, disseminated stage may appear weeks or months after the initial infection. It consists of several symptoms such as neurological abnormalities, heart inflammation, and bouts of arthritis (usually in the major joints such as the elbows or knees). Current research indicates that Lyme arthritis might be an autoimmune response to major histocompatibility (MHC) molecules on cells in the synovium (joint) that are similar to the bacterial antigens. Inflammation that produces organ damage is initiated and possibly perpetuated by the immune response to one or more spirochetal proteins. Finally, like the progression of syphilis—another disease caused by a spirochete—the late stage may appear years later. Infected individuals may develop neuron demyelination with symptoms resembling Alzheimer's disease and multiple sclerosis. Behavioral changes can also occur. Recognition of foreignness: Major histocompatability complex (section 32.4)

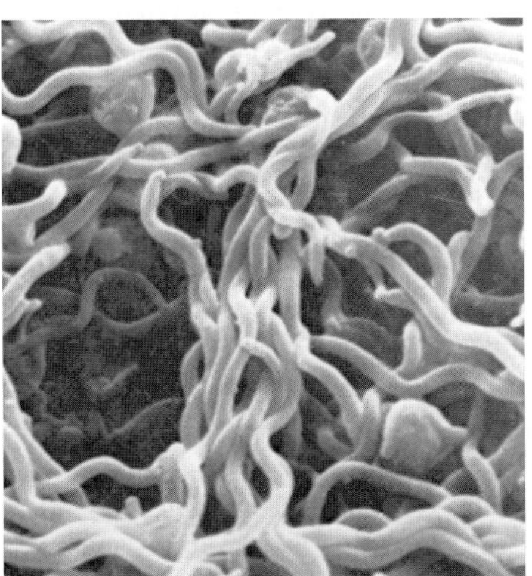
(a)

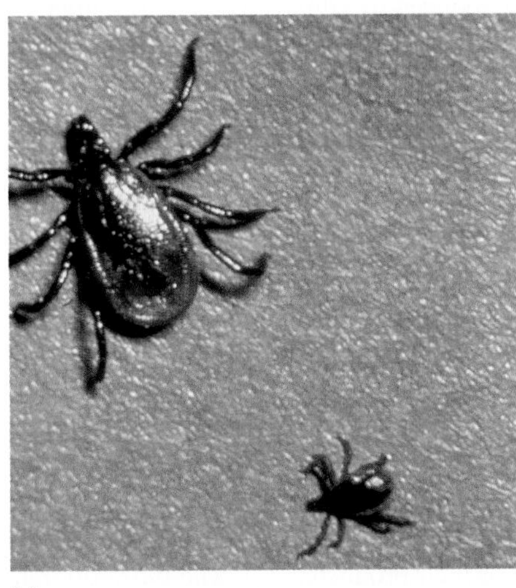
(b)

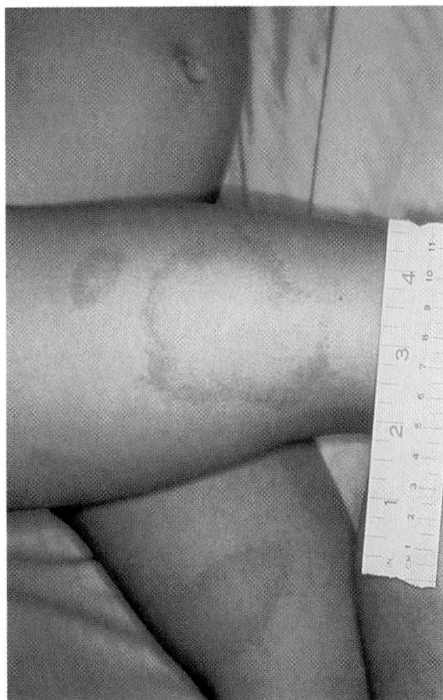
(c)

Figure 38.8 Lyme Disease. **(a)** One etiological agent is the spirochete *Borrelia burgdorferi;* SEM. **(b)** The vector in the northeastern U.S. is the tick *Ixodes scapularis*. An unengorged adult (bottom) is about the size of pinhead, and an engorged adult (top) can reach the size of a jelly bean. **(c)** The typical rash (erythema migrans) showing concentric rings around the initial site of the tick bite.

Table 38.3 Lyme Disease Prevention Strategies

Persons who are active in an area where Lyme disease or other tick-borne zoonoses occur should keep in mind the following points.

1. It takes a minimum of 24 hours of attachment and feeding for bacterial transmission to occur; thus prompt removal of attached ticks will greatly reduce the risk of infection. To remove an embedded tick, use tweezers to grasp the tick as close as possible to the skin and then pull with slow, steady pressure in a direction perpendicular to the skin.
2. Because each deer tick life cycle stage usually occurs at a certain time, there are periods when an individual should be most aware of the risk of infection. The most dangerous times are May through July, when the majority of nymphal deer ticks are present and the risk of transmission is greatest.
3. In the woods, wear light-colored pants and good shoes. Tuck the cuffs of your pants into long socks to deny ticks easy entry under your clothes. After coming out of the woods, check all clothes for ticks.
4. Repellents containing high concentrations of DEET (diethyltoluamide) or permanone are available over the counter and are very noxious to ticks. Premethrin kills ticks on contact but is approved only for use on clothing.
5. Immediately after being in a high-risk area, examine your body for ticks, bite marks, swellings, and redness. Taking a shower and using lots of soap aids in this examination. Areas such as the scalp, armpits, navel, and groin are difficult to examine effectively but are preferred sites for tick attachment. Pay special attention to these parts of the body.

Laboratory diagnosis of LD is based on (1) serological testing (Lyme ELISA or Western Blot) for IgM or IgG antibodies to the pathogen, (2) detection of *Borrelia* DNA in patient specimens (especially synovial fluid) after amplification by PCR, and (3) recovery of the spirochete from patient specimens, although cultures are laborious with modest success. Treatment with amoxicillin or tetracycline early in the illness results in prompt recovery and prevents arthritis and other complications. If nervous system involvement is suspected, ceftriaxone is used because it can cross the blood-brain barrier. PCR (section 14.3)

Prevention and control of LD involves environmental modification (clearing and burning tick habitat) and the application of acaricidal compounds (agents that destroy mites and ticks). An individual's risk of acquiring LD may be greatly reduced by education and personal protection **(table 38.3)**.

**Plague

In the southwestern part of the United States, **plague** [Latin *plaga*, pest] occurs primarily in wild ground squirrels, chipmunks, mice, and prairie dogs. However, massive human epidemics occurred in Europe during the Middle Ages, where the disease was known as the Black Death due to black-colored, subcutaneous hemorrhages. Infections now occur in humans only sporadically or in limited outbreaks. In the United States, 10 to 20 cases are reported annually; the mortality rate is about 14%.

The disease is caused by the gram-negative bacterium *Yersinia pestis*. It is transmitted from rodent to human by the bite of an infected flea, direct contact with infected animals or their products, or inhalation of contaminated airborne droplets **(figure 38.9*a*)**. Initially spread by contact with flea-infested animals, *Y. pestis* can spread among people by airborne transmission. Once in the human body, the bacteria multiply in the blood and lymph. An important factor in the virulence of *Y. pestis* is its ability to survive and proliferate inside phagocytic cells rather than being killed by them. One of the ways this is accomplished is by its modification of host cell behavior. Like other gram-negative extracellular pathogens, *Y. pestis* uses type III secretion system to deliver effector proteins. *Y. pestis* secretes plasmid-encoded *y*ersinal *o*uter membrane *p*roteins (YOPS) into phagocytic cells to counteract natural defense mechanisms and help the bacteria multiply and disseminate in the host (*see figure 33.4*). Class *Gammaproteobacteria:* Order *Enterobacteriales* (section 22.3); Overview of bacterial pathogenesis: Pathogenicity islands (section 33.3)

Symptoms—besides the subcutaneous hemorrhages—include fever, chills, headache, extreme exhaustion, and the appearance of enlarged lymph nodes called **buboes** (hence another old name, **bubonic plague**) (figure 38.9*c*). In 50 to 70% of the untreated cases, death follows in 3 to 5 days from toxic conditions caused by the large number of bacilli in the blood.

Laboratory diagnosis of plague is made in reference labs where direct microscopic examination, culture of the bacterium, and serological tests are used. Because the plague bacillus is listed as a Select Agent, the sentinel laboratory identification of *Y. pestis* is restricted to these few tests. Identity confirmation of *Y. pestis* by PCR, advanced serological studies, and other approved practices is restricted to national reference laboratories. Treatment is with streptomycin, chloramphenicol, or tetracycline, and recovery from the disease gives a good immunity.

Pneumonic plague arises (1) from primary exposure to infectious respiratory droplets from a person or animal with respiratory plague, or (2) secondary to hematogenous spread in a patient with bubonic or septicemic plague. Pneumonic plague can also arise from accidental inhalation of *Y. pestis* in the laboratory. The mortality rate for this kind of plague is almost 100% if it is not recognized within 12 to 24 hours. Obviously great care must be taken to prevent the spread of airborne infections to personnel who care for pneumonic plague patients.

Prevention and control involve flea and rodent control, isolation of human patients, prophylaxis or abortive therapy of exposed persons, and vaccination (USP Plague vaccine) of persons at high risk.

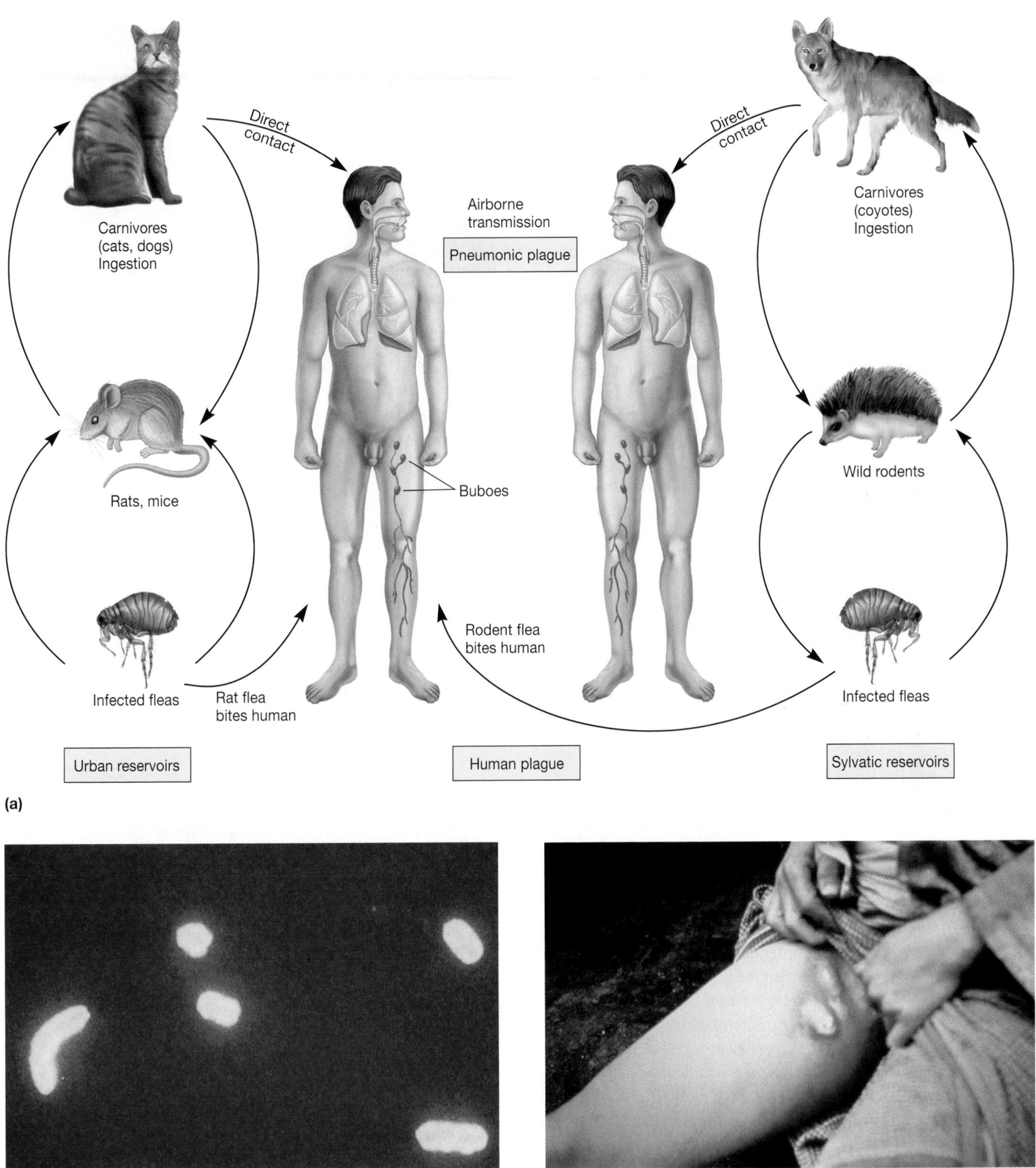

Figure 38.9 Plague. **(a)** Plague is spread to humans through (1) the urban cycle and rat fleas. (2) the sylvatic cycle centers on wild rodents and their fleas, or (3) by airborne transmission from an infected person leading to pneumonic plague. Dogs, cats, and coyotes can also acquire the bacterium by ingestion of infected animals. **(b)** *Yersinia pestis* can be stained with fluorescent antibodies for identification. **(c)** Enlarged lymph nodes called buboes are characteristic of *Yersinia* infection.

**Q Fever

Q fever (Q for query because the cause of the fever was not known for some time) is an acute zoonotic disease caused by the γ-proteobacterium *Coxiella burnetii,* an intracellular, gram-negative bacterium. *C. burnetii* can survive outside host cells by forming a resistant, endospore-like body. It does not need an arthropod vector for transmission. This bacterium infects both wild animals and livestock. Cattle, sheep, and goats are the primary reservoirs. In animals, ticks (many species) transmit *C. burnetii,* whereas human transmission is primarily by inhalation of dust contaminated with bacteria from dried animal feces or urine, or consumption of unpasteurized milk. The disease can occur in epidemic form among slaughterhouse workers and sporadically among farmers and veterinarians. Each year, fewer than 30 cases of Q fever are reported in the United States. Class *Alphaproteobacteria: Rickettsia* and *Coxiella* (section 22.1)

In humans, Q fever is an acute illness characterized by the sudden onset of severe headache, malaise, confusion, sore throat, chills, sweats, nausea, chest pain, myalgia (muscle pain), and high fever. It is rarely fatal, but endocarditis—inflammation of the heart muscle—occurs in about 10% of the cases. Five to ten years may elapse between the initial infection and the appearance of the endocarditis. During this interval the bacteria reside in the liver and often cause hepatitis.

Diagnosis is most commonly made in national reference labs using an indirect immunofluorescence assay or PCR. Treatment is with doxycycline or quinolone antibiotics. Prevention and control consists of vaccinating researchers and others at high occupational risk and in areas of endemic Q fever; cow and sheep milk should be pasteurized before consumption.

Rocky Mountain Spotted Fever

Rocky Mountain spotted fever is caused by *Rickettsia rickettsii.* Although this disease was originally detected in the Rocky Mountain area, most cases now occur east of the Mississippi River. The disease is transmitted by ticks and usually occurs in people who are or have been in tick-infested areas. There are two principal vectors: *Dermacentor andersoni,* the wood tick, is distributed in the Rocky Mountain states and is active during the spring and early summer. *D. variabilis,* the dog tick, has assumed greater importance and is almost exclusively confined to the eastern half of the United States. Unlike other rickettsias, *R. rickettsii* can pass from generation to generation of ticks through their eggs in a process known as **transovarian passage.** No humans or mammals are needed as reservoirs for the continued propagation of this rickettsia in the environment.

When humans contact infected ticks, the rickettsias are either deposited on the skin (if the tick defecates after feeding) and then subsequently rubbed or scratched into the skin, or the rickettsias are deposited into the skin as the tick feeds. Once inside the skin, the rickettsias enter the endothelial cells of small blood vessels, where they multiply and produce a characteristic vasculitis.

The disease is characterized by the sudden onset of a headache, high fever, chills, and a skin rash **(figure 38.10)** that initially appears on the ankles and wrists and then spreads to the trunk of the body. If the disease is not treated, the rickettsias can destroy the blood vessels in the heart, lungs, or kidneys and cause death. Usually, however, severe pathological changes are avoided by antibiotic therapy (chloramphenicol, chlortetracycline), the development of immune resistance, and supportive therapy. Diagnosis is made through observation of symptoms and signs such as the characteristic rash and by serological tests. The best means of prevention remains the avoidance of tick-infested habitats and animals. There are roughly 400 to 800 reported cases of Rocky Mountain spotted fever annually in the United States.

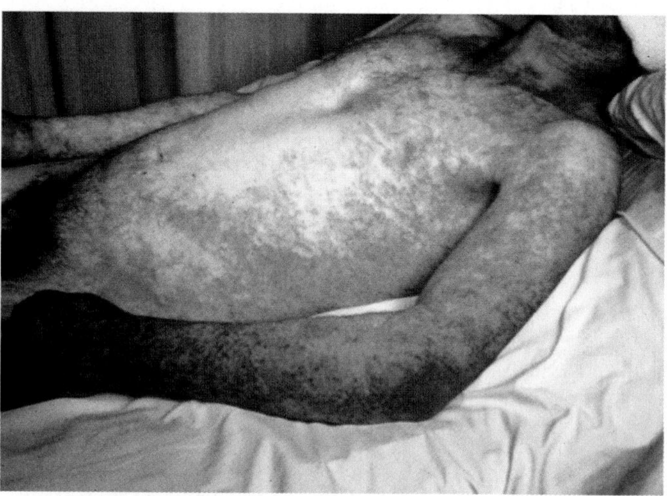

Figure 38.10 Rocky Mountain Spotted Fever. Typical rash occurring on the arms and chest consists of generally distributed, sharply defined macules.

1. How is epidemic typhus spread? Ehrlichiosis? Murine typhus? What are their symptoms?
2. Why are two antibiotics used to treat most rickettsial infections?
3. What is the causative agent of Lyme disease and how is it transmitted to humans? How does the illness begin? Describe the three stages of Lyme disease.
4. Why is plague sometimes called the Black Death? How is it transmitted? Distinguish between bubonic and pneumonic plague.
5. Describe the symptoms of Rocky Mountain spotted fever.
6. How does transovarian passage occur?

38.3 Direct Contact Diseases

Most of the direct contact bacterial diseases involve the skin or underlying tissues. Others can become disseminated through specific regions of the body. Some of the better-known of these diseases are now discussed.

Gas Gangrene or Clostridial Myonecrosis

Clostridium perfringens, C. novyi, and *C. septicum* are gram-positive, endospore-forming rods termed the histotoxic clostridia. They can produce a necrotizing infection of skeletal muscle called

gas gangrene [Greek *gangraina,* an eating sore] or **clostridial myonecrosis** [*myo,* muscles, and *necrosis,* death]; however, *C. perfringens* is the most common cause. Class *Clostridia* (section 23.4)

Histotoxic clostridia occur in the soil worldwide and also are part of the normal endogenous microflora of the human large intestine. Contamination of injured tissue with spores from soil containing histotoxic clostridia or bowel flora is the usual means of transmission. Infections are commonly associated with wounds resulting from abortions, automobile accidents, military combat, or frostbite. If the spores germinate in anoxic tissue, the bacteria grow and secrete α-toxin, which breaks down muscle tissue. Growth often results in the accumulation of gas (mainly hydrogen as a result of carbohydrate fermentation), and of the toxic breakdown products of skeletal muscle tissue **(figure 38.11)**. Toxigenicity: Exotoxins (section 33.4)

Clinical manifestations include severe pain, edema, drainage, and muscle necrosis. The pathology arises from progressive skeletal muscle necrosis due to the effects of α-toxin (a lecithinase). Lecithinase disrupts cell membranes leading to cell lysis. Other enzymes produced by the bacteria degrade collagen and tissue, facilitating spread of the disease.

Gas gangrene is a medical emergency. Laboratory diagnosis is through recovery of the appropriate species of clostridia accompanied by the characteristic disease symptoms. Treatment is extensive surgical debridement (removal of all dead tissue), the administration of antitoxin, and antimicrobial therapy with penicillin and tetracycline. Hyperbaric oxygen therapy (the use of high concentrations of oxygen at elevated pressures) also is considered effective. The oxygen saturates the infected tissue and thereby prevents the growth of the obligately anaerobic clostridia. Prevention and control include debridement of contaminated traumatic wounds plus antimicrobial prophylaxis and prompt treatment of all wound infections. Amputation of limbs often is necessary to prevent further spread of the disease.

Group B Streptococcal Disease

Streptococcus agalactiae, or **Group B streptococcus (GBS)**, is a gram-positive bacterium that causes illness in newborn babies, pregnant women, the elderly, and adults compromised by other severe illness. GBS is the most common cause of sepsis and meningitis in newborns, a frequent cause of newborn pneumonia, and a common cause of life-threatening, neonatal infections. GBS in neonates is more common than rubella, congenital syphilis, and respiratory syncytial virus, for example. Nearly 75% of infected newborns present symptoms in the first week of life; premature babies are more susceptible. Most GBS cases are apparent within hours after birth. This is called "early-onset disease." GBS can also develop in infants who are one week to several months old ("late-onset disease"), however this is very rare. Meningitis is more common with late-onset disease. Interestingly, about half of the infants with late-onset GBS disease are associated with a mother who is a GBS carrier; the source of infection for the other newborns with late-onset GBS disease is unknown. Infant mortality due to GBS disease is about 5%. Babies that survive GBS disease, particularly those with meningitis, may have permanent disabilities, such as hearing or vision loss or developmental disabilities.

GBS is transmitted directly from person to person. Many people are asymptomatic GBS carriers—they are colonized by GBS but do not become ill from it. Adults can carry GBS in the bowel, vagina, bladder, or throat. Twenty to 25% of pregnant women carry GBS in the rectum or vagina. Thus a fetus may be exposed to GBS before or during birth if the mother is a carrier. GBS carriers appear to be so temporarily—that is, they do not harbor the bacteria for life.

GBS disease is diagnosed when the gram-positive, beta-hemolytic, streptococcal bacterium is grown from cultures of otherwise sterile body fluids, such as blood or spinal fluid. Confirmation of GBS is by latex agglutination immunoassay. GBS cultures may take 24 to 72 hours for strong growth to appear. GBS infections in both newborns and adults are usually treated with penicillin or ampicillin, unless resistance or hypersensitivity is indicated.

GBS carriage can be detected during pregnancy by the presence of the bacterium in culture specimens of both the vagina and the rectum. An FDA-approved DNA test for GBS is also available if rapid diagnosis of disease is necessary or culture results

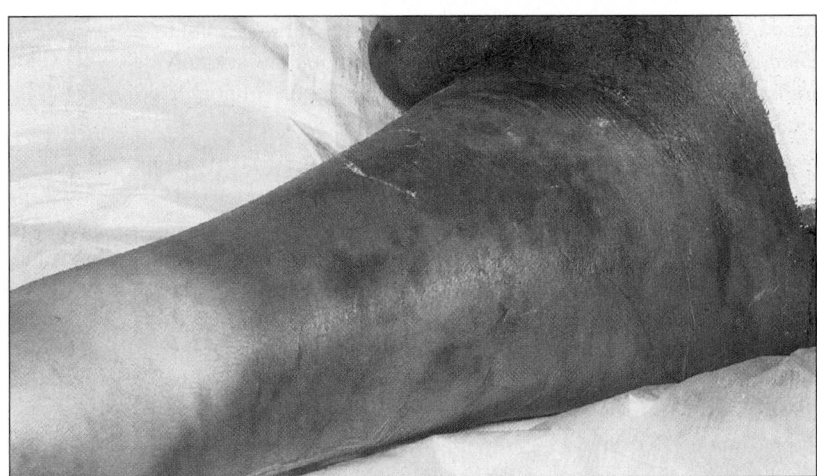

Figure 38.11 Gas Gangrene (Clostridial Myonecrosis). Necrosis of muscle and other tissues results from the numerous toxins produced by *Clostridium perfringens.*

are equivocal. The CDC recommends that pregnant women have vaginal and rectal specimens cultured for GBS in late pregnancy (35 to 37 weeks' gestation) to accurately predict whether GBS colonization needs to be treated prior to delivery. Antibiotic chemotherapy is mostly effective. There is no preventative vaccine.

A positive culture result means that the mother carries GBS—not that she or her baby will definitely become ill. Women who carry GBS are not given oral antibiotics before labor because antibiotic treatment at this time does not prevent GBS disease in newborns. An exception to this is when GBS is identified in urine during pregnancy. GBS in the urine is treated at the time it is diagnosed—it indicates an active infection in the mother. Carriage of GBS, in either the vagina or rectum, becomes important at the time of labor and delivery, or any time the placental membrane ruptures. At this time, antibiotics are effective in preventing the spread of GBS from mother to baby. GBS carriers at highest risk are those with any of the following conditions: fever during labor, rupture of membranes 18 hours or more before delivery, labor or membrane rupture before 37 weeks.

Inclusion Conjunctivitis

Inclusion conjunctivitis is an acute, infectious disease caused by *Chlamydia trachomatis* serotypes D–K, and it occurs throughout the world. It is characterized by a copious mucous discharge from the eye, an inflamed and swollen conjunctiva, and the presence of large inclusion bodies in the host cell cytoplasm. In inclusion conjunctivitis of the newborn, the chlamydiae are acquired during passage through an infected birth canal. The disease appears 7 to 12 days after birth. Erythromycin eyedrops given prophylactically to prevent ophthalmia neonatorum also prevents chlamydial inclusion conjunctivitis. If the chlamydiae colonize an infant's nasopharynx and tracheobronchial tree, pneumonia may result. Adult inclusion conjunctivitis is acquired by contact with infective genital tract discharges. Phylum *Chlamydiae* (section 21.5)

Without treatment, recovery usually occurs spontaneously over several weeks or months. Therapy involves treatment with tetracycline, erythromycin, or a sulfonamide. The specific diagnosis of *C. trachomatis* can be made by direct immunofluorescence, Giemsa stain, nucleic acid probes, and culture. Genital chlamydial infections and inclusion conjunctivitis are sexually transmitted diseases (pp. 975–76). Prevention depends upon diagnosis and treatment of all infected individuals.

Leprosy

Leprosy [Greek *lepros,* scaly, scabby, rough], or **Hansen's disease,** is a severely disfiguring skin disease caused by *Mycobacterium leprae* (*see figure 24.10*). The only reservoirs of proven significance are humans. The disease most often occurs in tropical countries, where there are more than 11 million cases. An estimated 4,000 cases exist in the United States, with approximately 100 new cases reported annually. Transmission of leprosy is most likely to occur when individuals are exposed for prolonged periods to infected individuals who shed large numbers of *M. leprae.* Nasal secretions probably are the infectious material for family contacts. Suborder *Corynebacterineae* (section 24.4)

The incubation period is about 3 to 5 years but may be much longer, and the disease progresses slowly. The bacterium invades peripheral nerve and skin cells and becomes an obligately intracellular parasite. It is most frequently found in the Schwann cells that surround peripheral nerve axons and in mononuclear phagocytes. The earliest symptom of leprosy is usually a slightly pigmented skin eruption several centimeters in diameter. Approximately 75% of all individuals with this early, solitary lesion heal spontaneously because of the cell-mediated immune response to *M. leprae.* However, in some individuals this immune response may be so weak that one of two distinct forms of the disease occurs: tuberculoid or lepromatous leprosy **(figure 38.12)**.

Tuberculoid (neural) leprosy is a mild, nonprogressive form of leprosy associated with a delayed-type hypersensitivity reaction to antigens on the surface of *M. leprae.* It is characterized by damaged nerves and regions of the skin that have lost sensation and are surrounded by a border of nodules **(figure 38.13)**. Afflicted individuals who do not develop hypersensitivity have a relentlessly progressive form of the disease, called **lepromatous (progressive) leprosy,** in which large numbers of *M. leprae* develop in the skin cells. The bacteria kill skin tissue, leading to a progressive loss of facial features, fingers, toes, and other structures. Moreover, disfiguring nodules form all over the body. Nerves are also infected, but usually are less damaged than occurs in tuberculoid leprosy. Immune disorders: Type IV hypersensitivity (section 32.11)

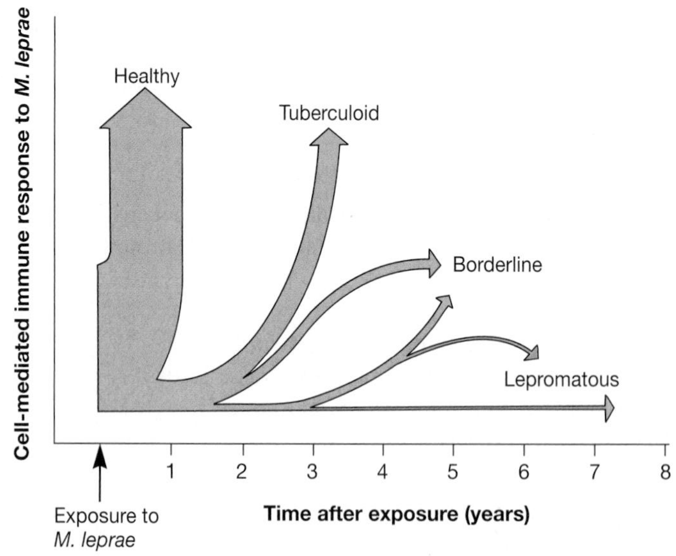

Figure 38.12 Development of Leprosy. A schematic representation of a hypothesis explaining the relationship between development of subclinical infection and various types of leprosy to the time of onset of cell-mediated immune response after initial exposure. The thickness of the lines indicates the proportion of individuals from the exposed population that is likely to fall into each category.

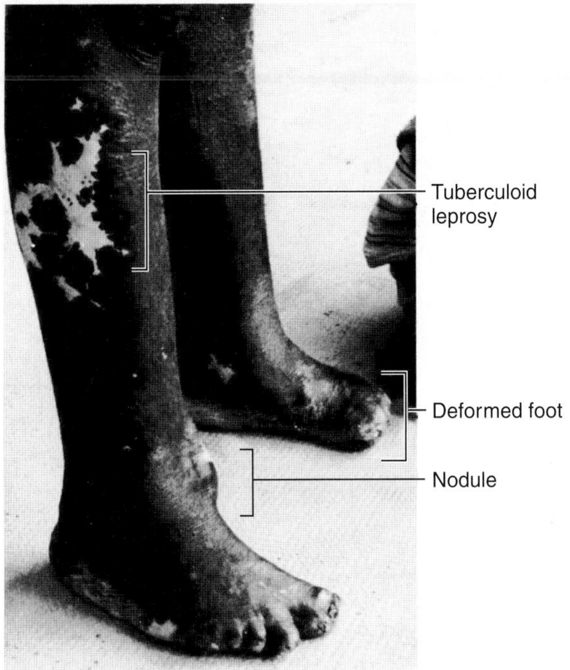

Figure 38.13 Leprosy. In tuberculoid leprosy the skin within the nodule is completely without sensation. Note the disfiguring nodule on the ankle. The deformed foot is associated with lepromatous leprosy.

Because the leprosy bacillus cannot be cultured in vitro, laboratory diagnosis is supported by the demonstration of the bacterium in biopsy specimens using direct fluorescent antibody staining. Serodiagnostic methods, such as the fluorescent leprosy antibody absorption test, DNA amplification, and ELISA have been developed.

Treatment is long-term with the sulfone drug, diacetyl dapsone (which acts by inhibiting the synthesis of dihydrofolic acid through competition with para-amino-benzoic acid; *see figure 34.11*), and rifampin, with or without clofazimine. Alternative drugs are ethionamide or protionamide. Use of *Mycobacterium* vaccine in conjunction with the drugs shortens the duration of drug therapy and speeds recovery from the disease.

There is good evidence that the nine-banded armadillo is an animal reservoir for the leprosy bacillus in the United States but plays no role in transmission of leprosy to humans. Identification and treatment of patients with leprosy is the key to control. Children of presumably contagious parents should be given chemoprophylactic drugs until treatment of their parents has made them noninfectious.

1. How can humans acquire gas gangrene? Group B strep? Leprosy? Describe the major symptoms of each.
2. How does an infant acquire inclusion conjunctivitis?
3. Define the following terms: tuberculoid and lepromatous leprosy.

Peptic Ulcer Disease and Gastritis

A gram-negative, microaerophilic, spiral bacillus found in gastric biopsy specimens from patients with histologic **gastritis** [Greek *gaster,* stomach, and *itis,* inflammation] was successfully cultured in Perth, Australia, in 1982 and named *Campylobacter pylori.* In 1993 its name was changed to *Helicobacter pylori*. Barry Marshall and Robin Warren discovered that this bacterium is responsible for most cases of chronic gastritis not associated with another known primary cause (e.g., autoimmune gastritis or eosinophilic gastritis), and it is the leading factor in the pathogenesis of **peptic ulcer disease**. In addition, there are strong, positive correlations between gastric cancer rates and *H. pylori* infection rates in certain populations, and it has been classified as a Class I carcinogen by the World Health Organization. Marshall and Warren were awarded the Nobel Prize in Physiology or Medicine in 2005 for their work. Class *Epsilonproteobacteria* (section 22.5)

H. pylori colonizes only gastric mucus-secreting cells, beneath the gastric mucous layers, and surface fimbriae are believed to be one of the adhesins associated with this process (**figure 38.14**). *H. pylori* binds to Lewis B antigens (which are part of the blood group antigens that determine blood group O) and to the monosaccharide sialic acid, also found in the glycoproteins on the surface of gastric epithelial cells. The bacterium moves into the mucous layer to attach to mucus-secreting cells. Movement into the mucous layer may be aided by the fact that *H. pylori* is a strong producer of urease. Urease activity is thought to create a localized alkaline environment when hydrolysis of urea produces ammonia. The increased pH may protect the bacterium

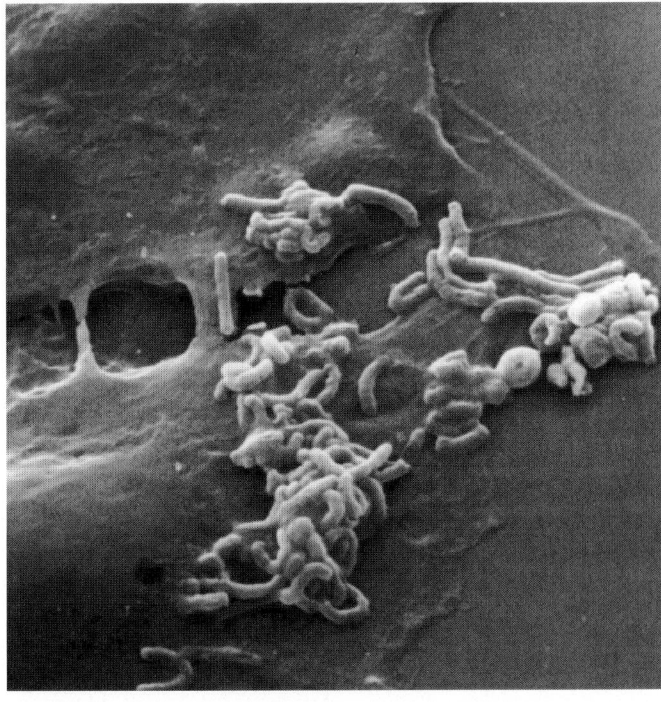

Figure 38.14 Peptic Ulcer Disease. Scanning electron micrograph (×3,441) of *Helicobacter pylori* adhering to gastric cells.

from gastric acid until it is able to grow under the layer of mucus in the stomach. The potential virulence factors responsible for epithelial cell damage and inflammation probably include proteases, phospholipases, cytokines, and cytotoxins.

Approximately 50% of the world's population is estimated to be infected with *H. pylori. H. pylori* is most likely transmitted from person to person (usually acquired in childhood), although infection from a common exogenous source cannot be completely ruled out, and some think that it is spread by food or water. Support for the person-to-person transmission comes from evidence of clustering within families and from reports of higher than expected prevalences in residents of custodial institutions and nursing homes.

Laboratory identification of *H. pylori* is by culture of gastric biopsy specimens, examination of stained biopsies for the presence of bacteria, detection of serum IgG (Pyloriset EIA-G, Malakit *Helicobacter pylori*), the urea breath test, urinary excretion of [^{15}N] ammonia, stool antigen assays, or detection of urease activity in the biopsies. The goal of *H. pylori* treatment is the complete elimination of the organism. Treatment is two-pronged: use of drugs to decrease stomach acid and antibiotics to kill the bacteria. Three regimes are commonly used: bismuth subsalicylate (Pepto-Bismol) combined with metronidazole and either tetracycline or amoxicillin; clarithromycin (Biaxin), ranitidine, and bismuth citrate; or clarithromycin, amoxicillin, and lansoprazole (Prevacid).

Staphylococcal Diseases

The genus *Staphylococcus* consists of gram-positive cocci, 0.5 to 1.5 μm in diameter, occurring singly, in pairs, and in tetrads, and characteristically dividing in more than one plane to form irregular clusters (*see figure 23.13*). The cell wall contains peptidoglycan and teichoic acid. Staphylococci are facultative anaerobes and usually catalase positive. Class *Bacilli:* Order *Bacillales* (section 23.5)

Staphylococci are among the most important bacteria that cause disease in humans. They are normal inhabitants of the upper respiratory tract, skin, intestine, and vagina. Staphylococci, with pneumococci (*S. pneumoniae*) and streptococci, are members of a group of invasive gram-positive bacteria known as the pyogenic (or pus-producing) cocci. These bacteria cause various suppurative, or pus-forming diseases in humans (e.g., boils, carbuncles, folliculitis, impetigo contagiosa, scalded-skin syndrome).

Staphylococci can be divided into pathogenic and relatively nonpathogenic strains based on the synthesis of the enzyme coagulase. The coagulase-positive species *S. aureus* is the most important human pathogen in this genus. Coagulase-negative staphylococci (CoNS) such as *S. epidermidis* do not produce coagulase, are nonpigmented, and are generally less invasive. However, they have increasingly been associated (as opportunistic pathogens) with serious nosocomial infections. Insights from microbial genomes: Genomic analysis of pathogenic microbes (section 15.8)

Staphylococci are further classified into *s*lime *p*roducers (SP) and *n*on-*s*lime *p*roducers (NSP). The ability to produce slime has been proposed as a marker for pathogenic strains of staphylococci **(figure 38.15a)**. **Slime** is a viscous, extracellular glycoconjugate that allows these bacteria to adhere to smooth surfaces such as prosthetic medical devices and catheters. Scanning electron microscopy has clearly demonstrated that biofilms (figure 38.15*b*) consisting of staphylococci encased in a slimy matrix are formed in association with biomaterial-related infections **(Disease 38.2)**. Slime also appears to inhibit neutrophil chemotaxis, phagocytosis, and the antimicrobial agents vancomycin and teicoplanin.

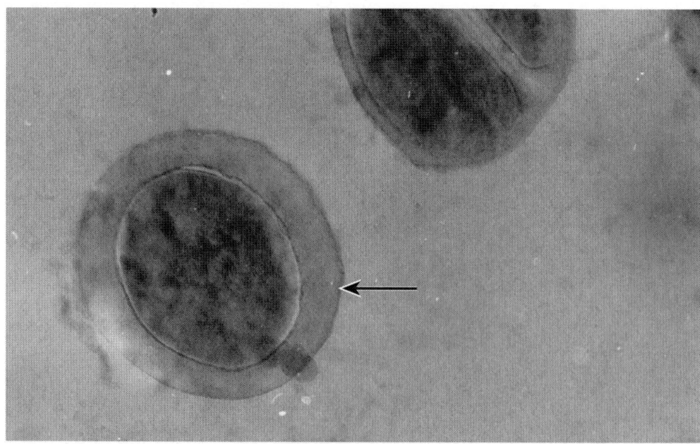

(a)

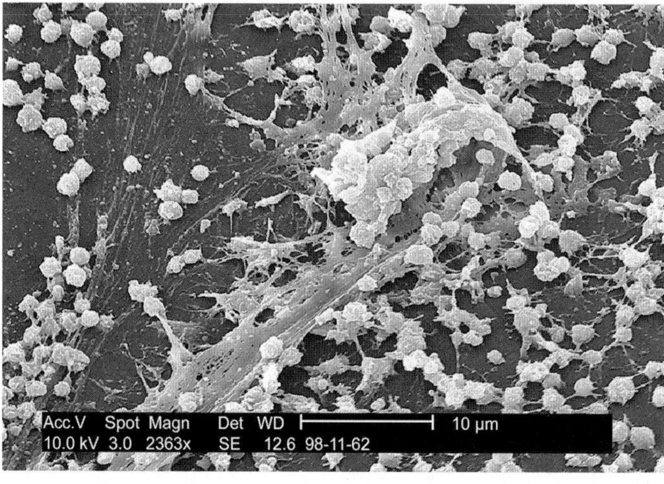

(b)

Figure 38.15 Slime and Biofilms. *S. aureus* and certain coagulase-negative staphylococci produce a viscous extracellular glyco conjugate called slime. **(a)** Cells of *S. aureus*, one of which produces a slime layer (arrowhead; transmission electron microscopy, ×10,000). **(b)** A biofilm on the inner surface of an intravenous catheter. Extracellular polymeric substances, mostly polysaccharides, surround and encase the staphylococci (scanning electron micrograph, ×2,363).

Staphylococci, harbored by either an asymptomatic carrier or a person with the disease (i.e., an active carrier), can be spread by the hands, expelled from the respiratory tract, or transported in or on animate and inanimate objects. Staphylococci can produce disease in almost every organ and tissue of the body **(figure 38.16)**. For the most part, however, staphyloccal disease occurs in people whose defensive mechanisms have been compromised, such as those in hospitals.

Disease

38.2 Biofilms

Biofilms consist of microorganisms immobilized at a substratum surface and typically embedded in an organic polymer matrix of microbial origin. They develop on virtually all surfaces immersed in natural aqueous environments, including both biological (aquatic plants and animals) and abiological (concrete, metal, plastics, stones). Biofilms form particularly rapidly in flowing aqueous systems where a regular nutrient supply is provided to the microorganisms (*see figure 27.11*). Extensive microbial growth, accompanied by excretion of copious amounts of extracellular organic polymers, leads to the formation of visible slimy layers (biofilms) on solid surfaces. Microbial growth in natural environments: Biofilms (section 6.6)

Most of the human gastrointestinal tract is colonized by specific microbial groups (the normal indigenous microbiota) that give rise to natural biofilms. At times, these natural biofilms provide protection for pathogenic species, allowing them to colonize the host. Normal microbiota of the human body: Large intestine (section 30.3)

Insertion of a prosthetic device into the human body often leads to the formation of biofilms on the surface of the device (*see figure 6.27*). The microorganisms primarily involved are *Staphylococcus epidermidis,* other coagulase-negative staphylococci, and gram-negative bacteria. These normal skin inhabitants possess the ability to tenaciously adhere to the surfaces of inanimate prosthetic devices. Within the biofilms they are protected from the body's normal defense mechanisms and also from antibiotics; thus the biofilm also provides a source of infection for other parts of the body as bacteria detach during biofilm sloughing.

Some examples of biofilms of medical importance include:

1. The deaths following massive infections of patients receiving Jarvik 7 artificial hearts
2. Cystic fibrosis patients harboring great numbers of *Pseudomonas aeruginosa* that produce large amounts of alginate polymers, which inhibit the diffusion of antibiotics
3. Teeth, where biofilm forms plaque that leads to tooth decay (figure 38.28)
4. Contact lenses, where bacteria may produce severe eye irritation, inflammation, and infection
5. Air-conditioning and other water retention systems where potentially pathogenic bacteria, such as *Legionella* species, may be protected by biofilms from the effects of chlorination.

Staphylococci produce disease through their ability to multiply and spread widely in tissues and through their production of many virulence factors **(table 38.4)**. Some of these factors are exotoxins, and others are enzymes thought to be involved in staphylococcal invasiveness. Many toxin genes are carried on plasmids; in some cases genes responsible for pathogenicity reside on both a plasmid and the host chromosome. The pathogenic capacity of a particular *S. aureus* strain is due to the combined effect of extracellular factors and toxins, together with the invasive properties of the strain. At one end of the disease spectrum is staphylococcal food poisoning, caused solely by the ingestion of preformed enterotoxin (table 38.4). At the other end of the spectrum are staphylococcal bacteremia and disseminated abscesses in most organs of the body.

The classic example of a staphylococcal lesion is the localized abscess **(figure 38.17*a,b*)**. When *S. aureus* becomes established in a hair follicle, tissue necrosis results. Coagulase is produced and forms a fibrin wall around the lesion that limits the spread. Within the center of the lesion, liquefaction of necrotic tissue occurs, and the abscess spreads in the direction of least resistance. The abscess may be either a furuncle (boil) (figure 38.17*c*) or a carbuncle (figure 38.17*d*). The central necrotic tissue drains, and healing eventually occurs. However, the bacteria may spread from any focus by the lymphatics and bloodstream to other parts of the body.

Newborn infants and children can develop a superficial skin infection characterized by the presence of encrusted pustules (figure 38.17*e*). This disease, called impetigo contagiosa, is caused by *S. aureus* or group A streptococci. It is contagious and can spread rapidly through a nursery or school. It usually occurs in areas where sanitation and personal hygiene are poor.

***T*oxic *s*hock *s*yndrome (TSS)** is a staphylococcal disease with potentially serious consequences. Most cases of this syndrome have occurred in females who used superabsorbent tampons during menstruation. However, the toxin associated with this syndrome is also produced in men and in nonmenstruating women by *S. aureus* present at sites other than the genital area (e.g., in surgical wound infections). Toxic shock syndrome is characterized by low blood pressure, fever, diarrhea, an extensive skin rash, and shedding of the skin. TSS results from the massive overproduction of cytokines by T cells induced by the TSST-1 protein (or to staphylococcal enterotoxins B and C1). TSST-1 binds both class II MHC receptors and T-cell receptors, stimulating T-cell responses in the absence of specific antigen. TSST-1 and other proteins having this property are called **superantigens.** Superantigens activate 5 to 30% of the total T-cell population, whereas specific antigens activate only 0.01 to 0.1% of the T-cell population. The net effect of cytokine overproduction is circulatory collapse leading to shock and multiorgan failure. Tumor necrosis factor α (TNF-α) and interleukins (IL) 1 and 6 are strongly associated with superantigen-induced shock. Mortality rates for TSS are 30 to 70% and morbidity due to surgical debridement and amputation is very high. Approximately 150 cases of toxic shock syndrome are reported annually in the United States. For these reasons, staphylococcal and streptococcal superantigens are categorized as Select Agents; their production and use are restricted, as they may be used as bioterror agents. T-cell biology: Superantigens (section 32.5)

***S*taphylococcal *s*calded *s*kin *s*yndrome (SSSS)** is a third example of a common staphylococcal disease (figure 38.17*f*). SSSS

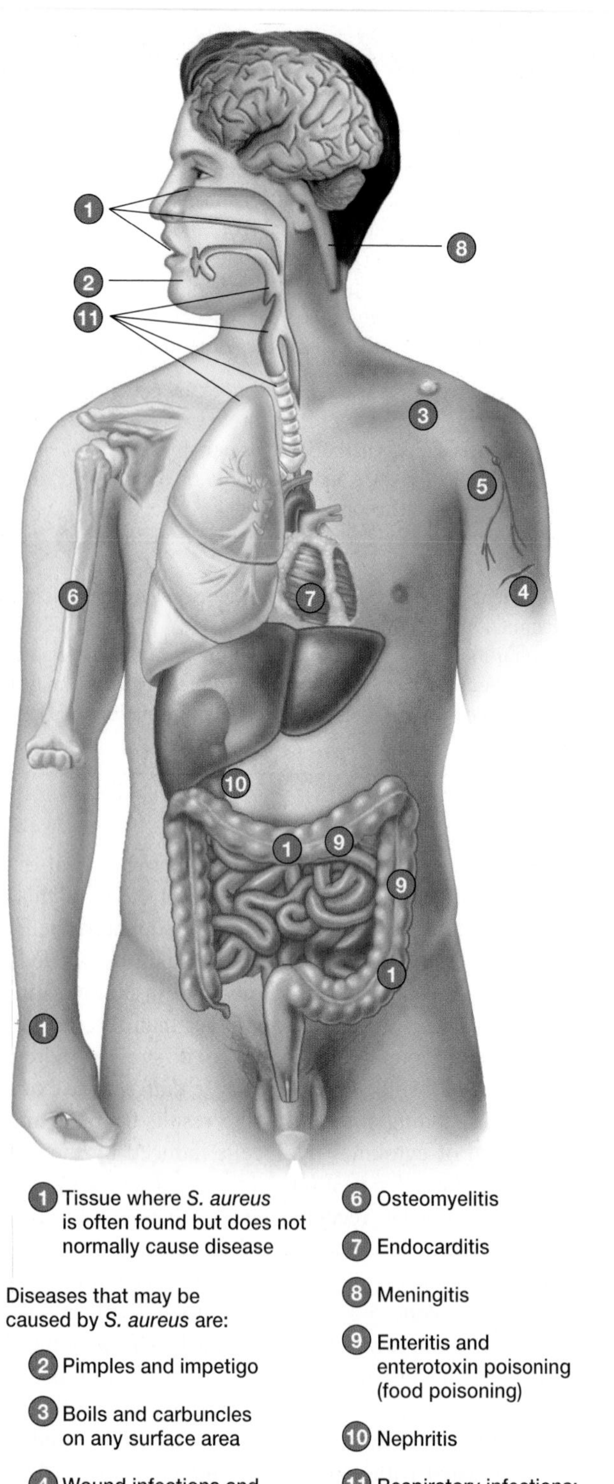

Figure 38.16 Staphylococcal Diseases. The anatomical sites of the major staphylococcal infections of humans are indicated by the corresponding numbers.

is caused by strains of *S. aureus* that produce the **exfoliative toxin,** or **exfoliatin.** This protein is usually plasmid encoded, although in some strains the toxin gene is on the bacterial chromosome. In this disease the epidermis peels off to reveal a red area underneath—thus the name of the disease. SSSS is seen most commonly in infants and children, and neonatal nurseries occasionally suffer large outbreaks of the disease.

The definitive diagnosis of staphylococcal disease can be made only by isolation and identification of the staphylococcus involved. This requires culture, catalase, coagulase, and other biochemical tests. Commercial rapid test kits also are available. There is no specific prevention for staphylococcal disease. The mainstay of treatment is the administration of specific antibiotics: penicillin, cloxacillin, methicillin, vancomycin, oxacillin, cefotaxime, ceftriaxone, a cephalosporin, or rifampin and others. Because of the prevalence of drug-resistant strains (e.g., meticillin-resistant staph), all staphylococcal isolates should be tested for antimicrobial susceptibility **(Disease 38.3)**. Cleanliness, good hygiene, and aseptic management of lesions are the best means of control. Drug resistance (section 34.6)

Sexually Transmitted Diseases

Sexually transmitted diseases (STDs) represent a worldwide public health problem. The various viruses that cause STDs are presented in chapter 37, the responsible bacteria in this chapter, and the yeasts and protozoa in chapter 39. **Table 38.5** lists the various microorganisms that can be sexually transmitted and the diseases they cause.

The spread of most sexually transmitted diseases is currently out of control. The World Health Organization estimates that 300 million new cases of sexually transmitted diseases occur annually, with the predominant number of infections in 15 to 30-year-old individuals. In the United States, sexually transmitted diseases remain a major public health challenge. The 2004 STD Surveillance Report published by the CDC indicated that while substantial progress has been made in preventing, diagnosing, and treating certain STDs, 19 million new infections occur each year, nearly half of them among people aged 15 to 24. Sexually transmitted diseases also exact a tremendous economic toll—with direct medical costs estimated at $13 billion annually in the United States alone. Many cases of STDs go undiagnosed; others, like human papillomavirus and genital herpes, are not reported at all. Some STDs can also lead to infertility or cancer.

STDs were formerly called venereal diseases (from Venus, the Roman goddess of love), and may sometimes be referred to as sexually transmitted infections (STIs). They occur most frequently in the most sexually active age group—15 to 30 years of age—but anyone who has sexual contact with an infected individual is at increased risk. In general, the more sexual partners a person has, the more likely the person will acquire an STD **(Disease 38.4)**.

As noted in previous chapters, some of the microorganisms that cause STDs can also be transmitted by nonsexual means. Examples include transmission by contaminated hypodermic nee-

Table 38.4 Various Enzymes and Toxins Produced by Staphylococci

Product	Physiological Action
β-lactamase	Breaks down penicillin
Catalase	Converts hydrogen peroxide into water and oxygen and reduces killing by phagocytosis
Coagulase	Reacts with prothrombin to form a complex that can cleave fibrinogen and cause the formation of a fibrin clot; fibrin may also be deposited on the surface of staphylococci, which may protect them from destruction by phagocytic cells; coagulase production is synonymous with invasive pathogenic potential
DNase	Destroys DNA
Enterotoxins	Are divided into heat-stable toxins of six known types (A, B, C1, C2, D, E); responsible for the gastrointestinal upset typical of food poisoning
Exfoliative toxins A and B (superantigens)	Causes loss of the surface layers of the skin in scalded-skin syndrome
Hemolysins	Alpha hemolysin destroys erythrocytes and causes skin destruction Beta hemolysin destroys erythrocytes and sphingomyelin around nerves
Hyaluronidase	Also known as spreading factor; breaks down hyaluronic acid located between cells, allowing for penetration and spread of bacteria
Panton-Valentine leukocidin	Inhibits phagocytosis by granulocytes and can destroy these cells by forming pores in their phagosomal membranes
Lipases	Break down lipids
Nuclease	Breaks down nucleic acids
Protein A	Is antiphagocytic by competing with neutrophils for the Fc portion of specific opsonins
Proteases	Break down proteins
Toxic shock syndrome toxin-1 (a superantigen)	Is associated with the fever, shock, and multisystem involvement of toxic shock syndrome

dles and syringes shared among intravenous drug users, and transmission from infected mothers to their infants. Some STDs can be cured quite easily, but others, especially those caused by viruses, are presently difficult or impossible to cure. Because treatments are often inadequate, prevention is essential. Preventive measures are based mainly on better education of the total population and when possible, control of the sources of infection and treatment of infected individuals with chemotherapeutic agents.

1. Describe how *H. pylori* can survive the acidic conditions of the stomach.
2. Compare and contrast the diseases caused by the TSST and SSSS proteins of *S. aureus*.
3. Describe several diseases caused by the staphylococci.
4. Name four ways in which a person may contract an STD.

Bacterial Vaginosis

Bacterial vaginosis is considered a sexually transmitted disease (table 38.5). It has a polymicrobial etiology that includes *Gardnerella vaginalis* (a gram-positive to gram-variable, pleomorphic, nonmotile rod), *Mobiluncus* spp., and various anaerobic bacteria. The finding that these microorganisms inhabit the vagina and rectum of 20 to 40% of healthy women indicates a potential source of autoinfection in addition to sexual transmission. Although it is a mild disease, it is a risk factor for obstetric infections, various adverse outcomes of pregnancy, and pelvic inflammatory disease. Vaginosis is characterized by a copious, frothy, fishy-smelling discharge with varying degrees of pain or itching. Diagnosis is based on this fishy odor and the microscopic observation of clue cells in the discharge. **Clue cells** are sloughed-off vaginal epithelial cells covered with bacteria, mostly *G. vaginalis.* Treatment for bacterial vaginosis is with metronidazole (Flagyl, MetroGel-Vaginal), a drug that kills anaerobic streptococci and the *Mobiluncus* spp. that appear to be needed for the continuation of the disease.

Chancroid

Chancroid [French *chancre,* a destructive sore, and Greek *eidos,* to form], also known as **genital ulcer disease,** is a sexually transmitted disease caused by the pleomorphic gram-negative bacillus *Haemophilus ducreyi* (table 38.5). The bacterium enters the skin through a break in the epithelium. After an incubation period of 4 to 7 days, a papular lesion develops within the epithelium, causing swelling and white blood cell infiltration. Within several days a pustule forms and ruptures, producing a painful, circumscribed ulcer with a ragged edge; hence the term genital ulcer disease. Most of the ulcers in males are on the penis and, in females, at the

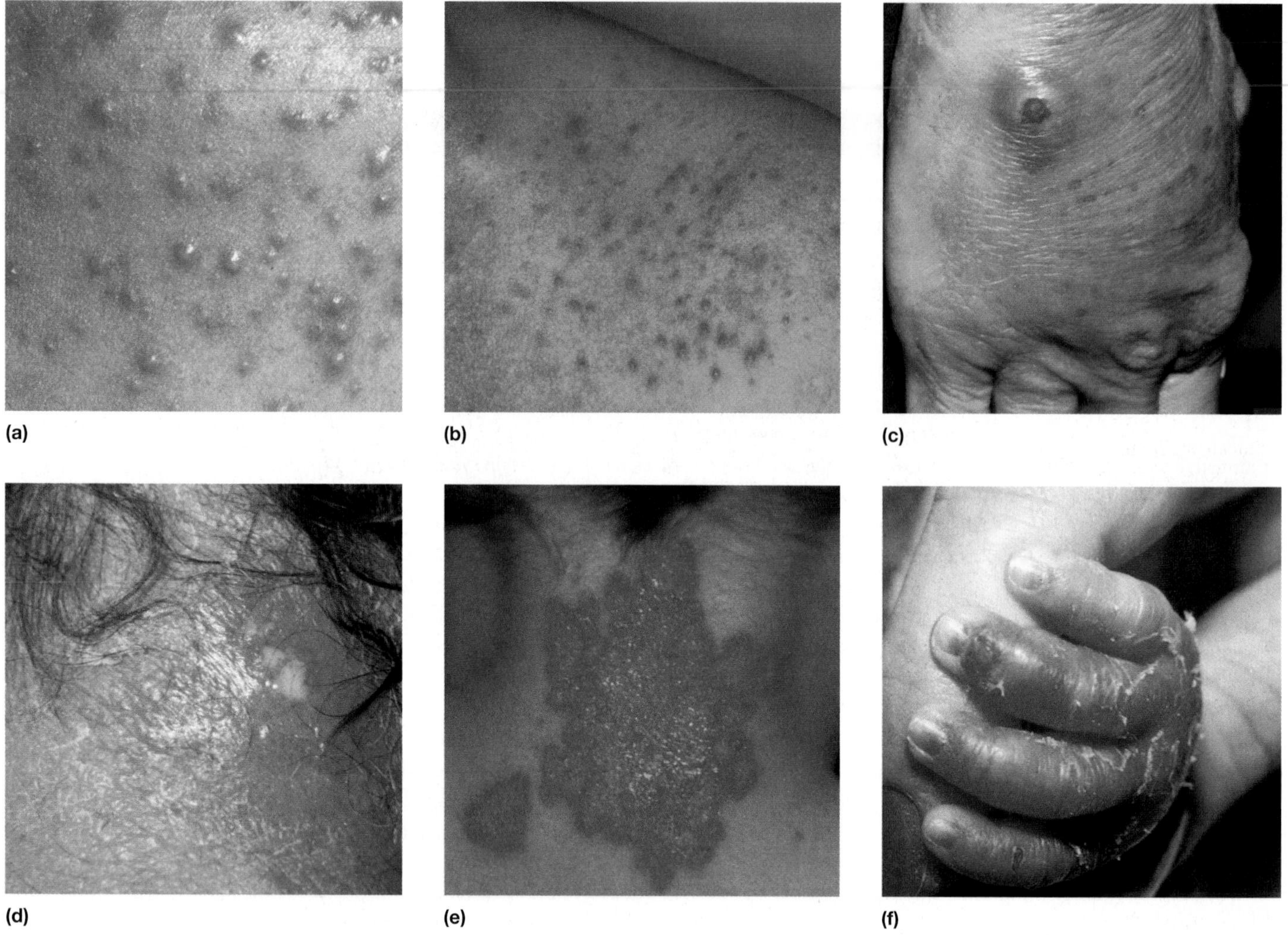

Figure 38.17 Staphylococcal Skin Infections. **(a)** Superficial folliculitis in which raised, domed pustules form around hair follicles. **(b)** In deep folliculitis the microorganism invades the deep portion of the follicle and dermis. **(c)** A furuncle arises when a large abscess forms around a hair follicle. **(d)** A carbuncle consists of a multilocular abscess around several hair follicles. **(e)** Impetigo on the neck of 2-year-old male. **(f)** Scalded skin syndrome in a 1-week-old premature male infant. Reddened areas of skin peel off, leaving "scalded"-looking moist areas.

Disease

38.3 Antibiotic-Resistant Staphylococci

During the late 1950s and early 1960s, *Staphylococcus aureus* caused considerable morbidity and mortality as a nosocomial, or hospital-acquired, pathogen. Penicillinase-resistant, semisynthetic penicillins have been successful antimicrobial agents in the treatment of staphylococcal infections. Unfortunately *m*eticillin-*r*esistant *S. aureus* (MRSA) strains have emerged as a major nosocomial problem. One way in which staphylococci become resistant is through acquisition of a chromosomal gene (*mecA*) that encodes an alternate target protein which is not inactivated by methicillin and its relatives, all belonging to the metacillin family of β-lactam drugs. The majority of the strains are also resistant to several of the most commonly used antimicrobial agents, including macrolides, aminoglycosides, and other beta-lactam antibiotics, including the latest generation of cephalosporins. Serious infections by meticillin-resistant strains have been most often successfully treated with an older, potentially toxic antibiotic, vancomycin. However, strains of *Enterococcus* and *Staphylococcus* recently have become resistant to vancomycin.

Meticillin-resistant *S. epidermidis* strains also have emerged as a nosocomial problem, especially in individuals with prosthetic heart valves or in people who have undergone other forms of cardiac surgery. Resistance to methicillin also may extend to the cephalosporin antibiotics. Difficulties in performing in vitro tests that adequately recognize cephalosporin resistance of these strains continue to exist. Most serious infections due to meticillin-resistant *S. epidermidis* have been successfully treated with combination therapy, including vancomycin plus rifampin or an aminoglycoside.

Table 38.5 Summary of the Major Sexually Transmitted Diseases (STDs)

Microorganism	Disease	Comments	Treatment
Viruses			
Human immunodeficiency virus (HIV)	Acquired immune deficiency syndrome (AIDS)	Pandemic in many parts of the world	A cocktail of reverse transcriptase and protease inhibitors
Herpes simplex virus (HSV-2)	Genital herpes	Painful blisters; enters latent stage, with reactivation due to stress; also oral, pharyngeal, and rectal herpes; no cure; very prevalent in the U.S.	Acyclovir and similar drugs alleviate the symptoms
Human papillomavirus (HPV) various serotypes	Condyloma acuminata (genital warts)	Predisposes to cervical cancer; no cure; very common in the U.S. HPV vaccine recently developed	Removal by various mechanical and chemical means; interferon injection
Hepatitis B virus (HBV)	Hepatitis B (serum hepatitis)	Transmitted in body fluids; cirrhosis, primary hepatocarcinoma	No treatment; recombinant HBV vaccine for prevention
Cytomegalovirus (CMV)	Congenital cytomegalic inclusion disease	Avoid sexual contact with an infected person	Ganciclovir and cidofovir for high-risk patients
Molluscum contagiosum	Genital molluscum contagiosum	Localized wartlike skin lesions	None
Bacteria			
Calymmatobacterium granulomatis	Granuloma inguinale (donovanosis)	Rare in the U.S.; draining ulcers can persist for years	Tetracycline, erythromycin, newer quinolones
Chlamydia trachomatis	Nongonococcal urethritis (NGU); cervicitis, pelvic inflammatory disease (PID), lymphogranuloma venereum	Serovars D-K cause most of the STDs in the U.S.; lymphogranuloma venereum rare in the U.S.	Tetracyclines, erythromycin, doxycycline, ceftriaxone
Gardnerella vaginalis	Bacterial vaginosis	Clue cells present	Metronidazole
Haemophilus ducreyi	Chancroid ("soft chancre")	Open sores on the genitals can lead to scarring without treatment; on the rise in the U.S.	Erythromycin or ceftriaxone
Helicobacter cinaedi, H. fennelliae	Diarrhea and rectal inflammation in homosexual men	Common in immunocompromised individuals	Metronidazole, macrolides
Mycoplasma genitalium	Implicated in some cases of NGU	Only recently described as an STD	Tetracyclines or erythromycin
Mycoplasma hominis	Implicated in some cases of PID	Widespread, often asymptomatic but can cause PID in women	Tetracyclines or erythromycin
Neisseria gonorrhoeae	Gonorrhea, PID	Most commonly reported STD in the U.S.; usually symptomatic in men and asymptomatic in women; antibiotic-resistant strains	Third-generation cephalosporins and/or quinolones
Treponema pallidum subsp. *pallidum*	Syphilis, congenital syphilis	Manifests many clinical syndromes	Benzathine penicillin G
Ureaplasma urealyticum	Urethritis	Widespread, often asymptomatic but can cause PID in women and NGU in men; premature birth	Tetracyclines or erythromycin
Yeasts			
Candida albicans	Candidiasis (moniliasis)	Produces a thick white vaginal discharge and severe itching	Nystatin, terconazole
Protozoa			
Trichomonas vaginalis	Trichomoniasis	Produces a frothy vaginal discharge; very common in the U.S.	Oral metronidazole

Disease

38.4 A Brief History of Syphilis

Syphilis was first recognized in Europe near the end of the fifteenth century. During this time the disease reached epidemic proportions in the Mediterranean areas. According to one hypothesis, syphilis is of New World origin and Christopher Columbus (1451–1506) and his crew acquired it in the West Indies and introduced it into Spain after returning from their historic voyage. Another hypothesis is that syphilis had been endemic for centuries in Africa and may have been transported to Europe at the same time that vast migrations of the civilian population were occurring (1500). Others believe that the Vikings, who reached the New World well before Columbus, were the original carriers.

Syphilis was initially called the Italian disease, the French disease, and the great pox as distinguished from smallpox. In 1530 the Italian physician and poet Girolamo Fracastoro wrote *Syphilis sive Morbus Gallicus* (Syphilis or the French Disease). In this poem a Spanish shepherd named Syphilis is punished for being disrespectful to the gods by being cursed with the disease. Several years later Fracastoro published a series of papers in which he described the possible mode of transmission of the "seeds"of syphilis through sexual contact.

Its venereal transmission was not definitely shown until the eighteenth century. The term venereal is derived from the name Venus, the Roman goddess of love. Recognition of the different stages of syphilis was demonstrated in 1838 by Philippe Ricord, who reported his observations on more than 2,500 human inoculations. In 1905 Fritz Schaudinn and Erich Hoffmann discovered the causative bacterium, and in 1906 August von Wassermann introduced the diagnostic test that bears his name. In 1909 Paul Ehrlich introduced an arsenic derivative, arsphenamine or salvarsan, as therapy. During this period, an anonymous limerick aptly described the course of this disease:

There was a young man from Black Bay
Who thought syphilis just went away
He believed that a chancre
Was only a canker
That healed in a week and a day.

But now he has "acne vulgaris"—
(Or whatever they call it in Paris);
On his skin it has spread
From his feet to his head,
And his friends want to know where his hair is.

There's more to his terrible plight:
His pupils won't close in the light
His heart is cavorting,
His wife is aborting,
And he squints through his gun-barrel sight.

Arthralgia cuts into his slumber;
His aorta is in need of a plumber;
But now he has tabes,
And saber-shinned babies,
While of gummas he has quite a number.

He's been treated in every known way,
But his spirochetes grow day by day;
He's developed paresis,
Has long talks with Jesus,
And thinks he's the Queen of the May.

entrance of the vagina. The disease is frequently accompanied by very swollen lymph nodes in the groin. Genital ulcer disease occurs commonly in the tropics; however, in the past decade there have been major outbreaks in the United States. Worldwide, genital ulcer disease is an important cofactor in the transmission of the AIDS virus. Diagnosis is by isolating *H. ducreyi* from the ulcers; treatment is with azithromycin or ceftriaxone. No vaccine is available. Control is the same as for other STDs: avoid contact with infected tissues by the use of barrier protection or abstinence.

Genitourinary Mycoplasmal Diseases

The mycoplasmas *Ureaplasma urealyticum* and *Mycoplasma hominis* are common parasitic microorganisms of the genital tract and their transmission is related to sexual activity (table 38.5). Both mycoplasmas can opportunistically cause inflammation of the reproductive organs of males and females. Because mycoplasmas are not usually cultured by clinical microbiologists, management and treatment of these infections depend on recognition of clinical syndromes and provision for adequate therapy. Tetracyclines are active against most strains; resistant organisms can be treated with erythromycin. Class *Mollicutes* (section 23.2)

Gonorrhea

Gonorrhea [Greek *gono,* seed, and *rhein,* to flow] is an acute, infectious, sexually transmitted disease of the mucous membranes of the genitourinary tract, eye, rectum, and throat (table 38.5). It is caused by the gram-negative, oxidase-positive, diplococcus, *Neisseria gonorrhoeae*. These bacteria are also referred to as **gonococci** [pl. of gonococcus; Greek *gono,* seed, and *coccus,* berry] and have a worldwide distribution. Over 500,000 cases are reported annually in the United States; the actual incidence is significantly higher. Class *Betaproteobacteria:* Order *Neisseriales* (section 22.2)

Once inside the body the gonococci attach to the microvilli of mucosal cells by means of pili and protein II, which function as adhesins. This attachment prevents the bacteria from being washed away by normal cervical and vaginal discharges or by the flow of urine. They are then phagocytosed by the mucosal cells and may even be transported through the cells to the intercellular spaces and subepithelial tissue. Phagocytes, such as neutrophils, also may contain gonococci inside vesicles **(figure 38.18)**. Because the gonococci are intracellular at this time, the host's defenses have little effect on the bacteria. Following penetration of the bacteria, the host tissue responds locally by the infiltration of mast cells, more

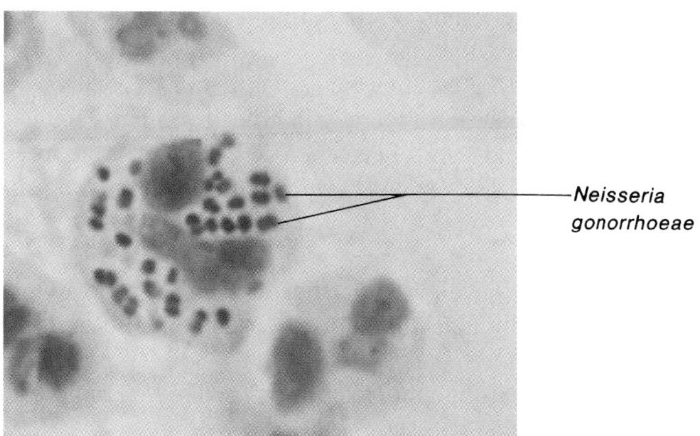

Figure 38.18 Gonorrhea. Gram stain of male urethral exudate showing *Neisseria gonorrhoeae* (diplococci) inside a PMN; light micrograph (×500). Although the presence of gram-negative diplococci in exudates is a probable indication of gonorrhea, the bacterium should be isolated and identified.

PMNs, and anitbody-secreting plasma cells. These cells are later replaced by fibrous tissue that may lead to urethral closing, or stricture, in males. Cells, tissues, and organs of the immune system (section 31.2); Phagocytosis (section 31.3)

In males the incubation period is 2 to 8 days. The onset consists of a urethral discharge of yellow, creamy pus, and frequent, painful urination that is accompanied by a burning sensation. In females, the cervix is the principal site infected. The disease is more insidious in females and few individuals are aware of any symptoms. However, some symptoms may begin 7 to 21 days after infection. These are generally mild; some vaginal discharge may occur. The gonococci also can infect the Fallopian tubes and surrounding tissues, leading to ***p*elvic *i*nflammatory *d*isease (PID)**. This occurs in 10 to 20% of infected females. Gonococcal PID is a major cause of sterility and ectopic pregnancies because of scar formation in the Fallopian tubes. Gonococci disseminate most often during menstruation, a time in which there is an increased concentration of free iron available to the bacteria. In both genders, disseminated gonococcal infection with bacteremia may occur. This can lead to involvement of the joints (gonorrheal arthritis), heart (gonorrheal endocarditis), or pharynx (gonorrheal pharyngitis). Gonorrheal eye infections can occur in newborns as they pass through an infected birth canal. The resulting disease is called **ophthalmia neonatorum,** or **conjunctivitis of the newborn.** This was once a leading cause of blindness in many parts of the world. To prevent this disease, tetracycline, erythromycin, povidone-iodine, or silver nitrate in dilute solution is placed in the eyes of newborns. This type of treatment is required by law in the United States and many other nations.

Laboratory diagnosis of gonorrhea relies on the successful growth of *N. gonorrhoeae* in culture to determine oxidase reaction, Gram stain reaction, and colony and cell morphology. The performance of confirmation tests also is necessary. Because the gonococci are very sensitive to adverse environmental conditions and survive poorly outside the body, specimens should be plated directly; when this is not possible, special transport media are necessary. A DNA probe (Gen-Probe Pace) for *N. gonorrhoeae* has been developed and is used to supplement other diagnostic techniques.

The Centers for Disease Control and Prevention recommends as treatment five single doses of cefixime, cefriaxone, ciprofloxacin, ofloxacin, and levofloxacin to eradicate the infection. Penicillin-resistant strains of gonococci occur worldwide. Most of these strains carry a plasmid that directs the formation of penicillinase, a β-lactamase enzyme that inactivates penicillin G and ampicillin. Since 1980, strains of *N. gonorrhoeae* with chromosomally mediated penicillin resistance have developed. Instead of producing a penicillinase, these strains have altered penicillin-binding proteins. Since 1986, tetracycline-resistant *N. gonorrhoeae* also have developed. Therefore, neither penicillins nor tetracyclines are recommended for treating gonococcal infections. Recently, the CDC has recommended discontinuation of quinolone use to treat gonorrhea in men who have sex with other men, as quinolone-resistant strains are increasing in this group. Instead, they recommend ceftriaxone or spectinomycin as an alternative. Drug resistance (section 34.6)

The most effective method for control of this sexually transmitted disease is public education, diagnosing and treating the asymptomatic patient, barrier protection, and treating infected individuals quickly to prevent further spread of the disease. More than 60% of all cases occur in the 15- to 24-year-old age group. Repeated gonococcal infections are common. Protective immunity to reinfection does not arise because of antigenic variation in which a single strain changes its pilin gene by a recombinational event and alters the expression of the various protein II genes by slipped strand mispairing. This can thus be viewed as a programmed evasion technique employed by the bacterium rather than a mere reflection of strain variation.

Lymphogranuloma Venereum

Lymphogranuloma venereum (LGV) is a sexually transmitted disease (table 38.5) caused by *Chlamydia trachomatis* serotypes L1–L3. It has a worldwide distribution but is more common in tropical climates. LGV proceeds through three phases. (1) In the primary phase a small ulcer appears several days to several weeks after a person is exposed to the chlamydiae. The ulcer may appear on the penis in males or on the labia or vagina in females. The ulcer heals quickly and leaves no scar. (2) The secondary phase begins 2 to 6 weeks after exposure, when the chlamydiae infect lymphoid cells, causing the regional lymph nodes to become enlarged and tender; such nodes are called buboes **(figure 38.19)**. Systemic symptoms such as fever, chills, and anorexia are common. (3) If the disease is not treated, a late phase ensues. This results from fibrotic changes and abnormal lymphatic drainage that produces fistulas (abnormal passages leading from an abscess or a hollow organ to the body surface or from one hollow organ to another), and urethral or rectal strictures (a decrease in size). An untreatable fluid accumulation in the penis, scrotum, or vaginal area may result. Phylum *Chlamydiae* (section 21.5)

The disease is detected by staining infected cells with iodine to observe inclusions (chlamydia-filled vacuoles), culture of the

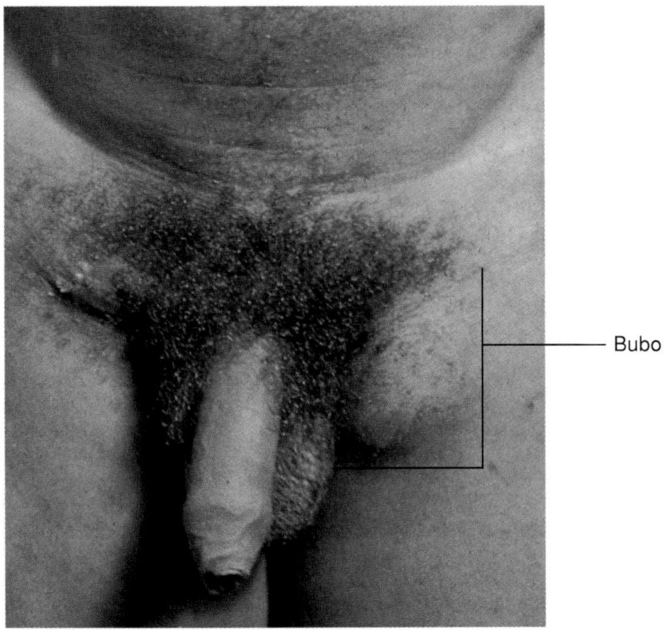

Figure 38.19 Lymphogranuloma Venereum. The bubo in the left inguinal area is draining.

chlamydiae from a bubo, nucleic acid probes, or by the detection of a high antibody titer to *C. trachomatis*. Treatment in the early phases consists of aspiration of the buboes and administration of the drugs azithromycin, ceftriaxone, erythromycin, or ciprofloxacin. The late phase may require surgery. The methods used for the control of LGV are the same as for other sexually transmitted diseases: abstinence, barrier protection, and early diagnosis and treatment of infected individuals. About 100 cases of LGV occur annually in the United States.

Nongonococcal Urethritis

Nongonococcal urethritis (NGU) is any inflammation of the urethra not due to the bacterium *Neisseria gonorrhoeae*. This condition is caused both by nonmicrobial factors such as catheters and drugs and by infectious microorganisms. The most important causative agents are *C. trachomatis, Ureaplasma urealyticum, Mycoplasma hominis, Trichomonas vaginalis, Candida albicans,* and herpes simplex viruses. Most infections are acquired sexually (table 38.5), and of these, approximately 50% are *Chlamydia* infections. NGU is endemic throughout the world, with an estimated 10 million Americans infected.

Symptoms of NGU vary widely. Males may have few or no manifestations of disease; however, complications can exist. These include a urethral discharge, itching, and inflammation of the male reproductive structures. Females may be asymptomatic or have a severe infection leading to PID, which often leads to sterility. *Chlamydia* may account for as many as 200,000 to 400,000 cases of PID annually in the United States. In the pregnant female, a chlamydial infection is especially serious because it is directly related to miscarriage, stillbirth, inclusion conjunctivitis, and infant pneumonia.

Diagnosis of NGU requires the demonstration of a leukocyte exudate and exclusion of urethral gonorrhea by Gram stain and culture. Several rapid tests for detecting *Chlamydia* in urine specimens are also available. Treatment varies with the causative agent.

Syphilis

Venereal syphilis [Greek *syn,* together, and *philein,* to love] is a contagious, sexually transmitted disease (table 38.5) caused by the spirochete *Treponema pallidum* subsp. *pallidum* (*T. pallidum; see figure 21.14b*). **Congenital syphilis** is the disease acquired in utero from the mother. Phylum *Spirochaetes* (section 21.6)

T. pallidum enters the body through mucous membranes or minor breaks or abrasions of the skin. It migrates to the regional lymph nodes and rapidly spreads throughout the body. The disease is not highly contagious, and there is only about a 1 in 10 chance of acquiring it from a single exposure to an infected sex partner.

Three recognizable stages of syphilis occur in untreated adults. In the primary stage, after an incubation period of about 10 days to 3 weeks or more, the initial symptom is a small, painless, reddened ulcer, or **chancre** [French *canker,* a destructive sore] with a hard ridge that appears at the infection site and contains spirochetes **(figure 38.20*a*)**. Contact with the chancre during sexual contact may result in disease transmission. In about one-third of the cases, the disease does not progress further and the chancre disappears. Serological tests are positive in about 80% of the individuals during this stage **(figure 38.21)**. The spirochetes typically enter the bloodstream and are distributed throughout the body.

Within 2 to 10 weeks after the primary lesion appears, the disease may enter the secondary stage, which is characterized by a highly variable skin rash (figure 38.20*b*). By this time 100% of the individuals are serologically positive. Other symptoms during this stage include the loss of hair patches, malaise, and fever. Both the chancre and the rash lesions are infectious.

After several weeks the disease becomes latent. During the latent period the disease is not normally infectious, except for possible transmission from mother to fetus (congenital syphilis). After many years a tertiary stage develops in about 40% of untreated individuals with secondary syphilis. During this stage degenerative lesions called **gummas** (figure 38.20*c*) form in the skin, bone, and nervous system as the result of hypersensitivity reactions. This stage also is characterized by a great reduction in the number of spirochetes in the body. Involvement of the central nervous system may result in tissue loss that can lead to cognitive deficits, blindness, a "shuffle"walk (tabes), or insanity. Many of these symptoms have been associated with such well-known people as Al Capone, Francisco Goya, Henry VIII, Adolf Hitler, Scott Joplin, Friedrich Nietzsche, Franz Schubert, Oscar Wilde, and Kaiser Wilhelm (Disease 38.4). Immune disorders: Hypersensitivity (section 32.11)

Diagnosis of syphilis is through a clinical history, physical examination, and dark-field and immunofluorescence examination of lesion fluids (except oral lesions) for typical motile spirochetes. Because humans respond to *T. pallidum* with the formation of anti-

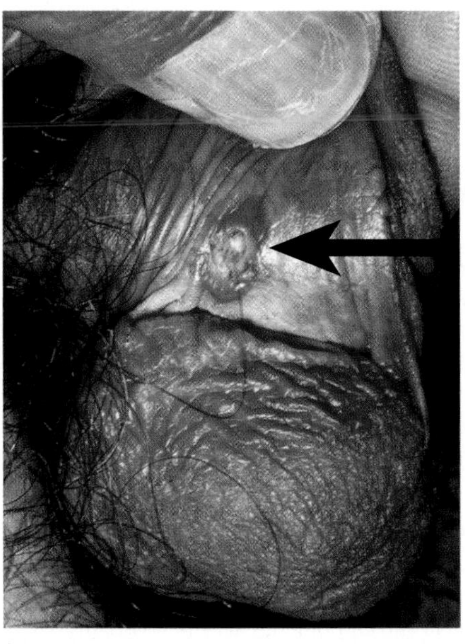
(a)

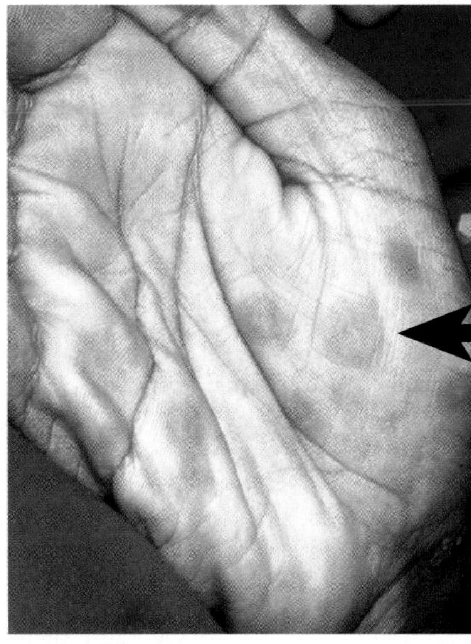
(b)

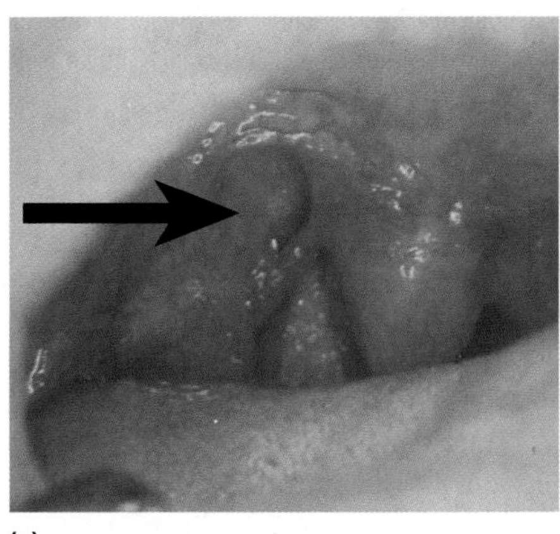
(c)

Figure 38.20 Syphilis. **(a)** Primary syphilitic chancre of the penis. **(b)** Palmar lesions of secondary syphilis. **(c)** Ruptured gumma and ulcer of upper hard palate of the mouth.

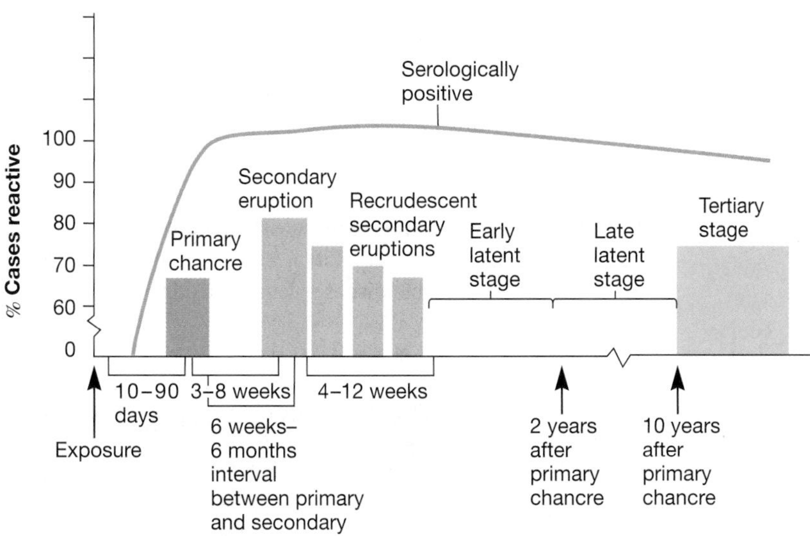

Figure 38.21 The Course of Untreated Syphilis.

treponemal antibody and a complement-fixing reagin, serological tests are very informative. Examples include tests for nontreponemal antigens (VDRL, *V*enereal *D*isease *R*esearch *L*aboratories test; RPR, *R*apid *P*lasma *R*eagin test) and treponemal antibodies (FTA-ABS, *f*luorescent *t*reponemal *a*ntibody-*abs*orption test; TP-PA, *T. pallidum* particle *a*gglutination). Action of antibodies: Immune complex formation (section 32.8)

Treatment in the early stages of the disease is easily accomplished with long-acting benzathine penicillin G or aqueous procaine penicillin. Later stages of syphilis are more difficult to treat with drugs and require much larger doses over a longer period. For example, in neurosyphilis cases, treponemes occasionally survive such drug treatment. Immunity to syphilis is not complete, and subsequent infections can occur.

Prevention and control of syphilis depends on (1) public education (2) prompt and adequate treatment of all new cases, (3) follow-up on sources of infection and contact so they can be treated, and (4) prophylaxis (barrier protection) to prevent exposure. At present, the incidence of syphilis, as well as other sexually transmitted diseases, is rising in most parts of the world. An estimated 12 million new cases occur each year. In the United States, less than 10,000 cases of primary and secondary syphilis in the civilian population and about 500 cases of congenital syphilis are reported annually. The highest incidence is among those 20 to 39 years of age.

1. Why is bacterial vaginosis considered a sexually transmitted disease?
2. What is ophthalmia neonatorum and how is it transmitted?
3. Describe the lesions of chancroid and syphilis. How are they the same? Different?
4. Besides the bacterial cause, what distinguishes gonorrhea from NGU?
5. Describe the symptoms of LGV in both the male and female host.
6. Describe the progression of syphilis if treatment is not obtained.

Tetanus

Tetanus [Greek *tetanos,* to stretch] is caused by *Clostridium tetani,* an anaerobic, gram-positive, endospore-forming rod. The spores of *C. tetani* are commonly found in hospital environments, in soil and dust, and in the feces of many farm animals and humans. Class *Clostridia* (section 23.4)

Transmission to humans is associated with skin wounds. Any break in the skin can allow *C. tetani* spores to enter. If the oxygen tension is low enough, the spores germinate and release the neurotoxin tetanospasmin. **Tetanospasmin** is an endopeptidase that selectively cleaves the synaptic vesicle membrane protein synaptobrevin. This prevents exocytosis (release from the cell) and release of inhibitory neurotransmitters (gamma-aminobutyric acid and glycine) at synapses within the spinal cord motor nerves. The result is uncontrolled stimulation of skeletal muscles (spastic paralysis). A second toxin, **tetanolysin,** is a hemolysin that aids in tissue destruction.

Early in the course of the disease, tetanospasmin causes tension or cramping and twisting in skeletal muscles surrounding the wound and tightness of the jaw muscles. With more advanced disease, there is trismus ("lockjaw"), an inability to open the mouth because of the spasm of the masseter muscles. Facial muscles may go into spasms, producing the characteristic expression known as risus sardonicus. Spasms or contractions of the trunk and extremity muscles may be so severe that there is boardlike rigidity, painful tonic convulsions, and opisthotonos (backward bowing of the back so that the heels and back approach each other [see chapter opening figure]). Death usually results from spasms of the diaphragm and intercostal respiratory muscles.

Testing for tetanus is suggested whenever an individual has a history of wound infection and muscle stiffness. Prevention of tetanus involves the use of the tetanus toxoid. The toxoid, which incorporates an adjuvant (aluminum salts) to increase its immunizing potency, is given routinely with diphtheria toxoid and pertussis vaccine. An initial dose is normally administered a few months after birth, a second dose 4 to 6 months later, and finally a reinforcing dose 6 to 12 months after the second injection. Another booster is given between the ages of 4 to 6 years (*see table 36.3*). For many years, booster doses of tetanus toxoid were administered every 3 to 5 years. However, that practice has been discontinued since it has been shown that a single booster dose can provide protection for 10 to 20 years. Serious hypersensitivity reactions have occurred when too many doses of toxoid were administered over a period of years. Booster doses today are generally given only when an individual has sustained a wound infection and only if it has been 10 or more years since the previous dose.

Control measures for tetanus are not possible because of the wide dissemination of the bacterium in the soil and the long survival of its spores. The case fatality rate in generalized tetanus ranges from 30 to 90% because tetanus treatment is not very effective. Therefore prevention is all important and depends on (1) active immunization with toxoid, (2) proper care of wounds contaminated with soil, (3) prophylactic use of antitoxin, and (4) administration of penicillin. Around 50 cases of tetanus are reported annually in the United States—the majority of which are in intravenous drug users.

Trachoma

Trachoma [Greek *trachoma,* roughness] is a contagious disease caused by *Chlamydia trachomatis* serotypes A–C. It is one of the oldest known infectious diseases of humans and is the greatest single cause of blindness throughout the world. Probably over 500 million people are infected and 20 million blinded each year by this chlamydia. In endemic areas, most children are chronically infected within a few years of birth. Active disease in adults over age 20 is three times as frequent in females as in males because of mother-child contact. Although trachoma is uncommon in the United States, except among Native Americans in the Southwest, it is widespread in Asia, Africa, and South America.

Trachoma is transmitted by contact with inanimate objects such as soap and towels, by hand-to-hand contact that carries *C. trachomatis* from an infected eye to an uninfected eye, or by flies. The disease begins abruptly with an inflamed conjunctiva. This leads to an inflammatory cell exudate and necrotic eyelash follicles (**figure 38.22**). The disease usually heals spontaneously.

Figure 38.22 Trachoma. An active infection showing marked follicular hypertrophy of both eyelids. The inflammatory nodules cover the thickened conjunctive of the eye.

However, with reinfection, secondary infections, vascularization of the cornea, or **pannus** formation, scarring of the conjunctiva can occur. If scar tissue accumulates over the cornea, blindness results.

Diagnosis and treatment of trachoma are the same as for inclusion conjunctivitis (previously discussed). However, prevention and control of trachoma depends more on health education and personal hygiene—such as access to clean water for washing—than on treatment.

1. Explain why tetanus is potentially life-threatening. How is the disease tetanus acquired? What are its symptoms and how do they arise?
2. How does *C. trachomatis*, serotypes A–C, cause trachoma? Describe how it is transmitted, and the way in which blindness may result. What happens if it is left untreated?

38.4 Food-Borne and Waterborne Diseases

Many microorganisms that contaminate food and water can cause acute gastroenteritis—inflammation of the stomach and intestinal lining. When food is the source of the pathogen, the condition is often called **food poisoning.** Gastroenteritis can arise in two ways. The microorganisms may actually produce a **food-borne infection.** That is, they may first colonize the gastrointestinal tract and grow within it, then either invade host tissues or secrete exotoxins. Alternatively, the pathogen may secrete an exotoxin that contaminates the food and is then ingested by the host. This is sometimes referred to as a **food intoxication** because the toxin is ingested and the presence of living microorganisms is not required. Because these toxins disrupt the functioning of the intestinal mucosa, they are called **enterotoxins.** Common symptoms of enterotoxin poisoning are nausea, vomiting, and diarrhea.

Worldwide, diarrheal diseases are second only to respiratory diseases as a cause of adult death; they are the leading cause of childhood death, and in some parts of the world they are responsible for more years of potential life lost than all other causes combined. For example, each year around 5 million children (more than 13,600 a day) die from diarrheal diseases in Asia, Africa, and South America. In the United States estimates exceed 10,000 deaths per year from diarrhea, and an average of 500 childhood deaths are reported.

This section describes several of the more common bacteria associated with gastrointestinal infections, food intoxications, and waterborne diseases. **Table 38.6** summarizes many of the bacterial pathogens responsible for food poisoning and **table 38.7** lists many important water-based bacterial pathogens. The protozoa responsible for food- and waterborne diseases are covered in chapter 39.
Controlling food spoilage (section 40.3); Food-borne diseases (section 40.4)

**Botulism

Food-borne **botulism** [Latin *botulus*, sausage] is a form of food poisoning caused by an exotoxin produced by *Clostridium botulinum*, an obligately anaerobic, endospore-forming, gram-positive rod found in soil and aquatic sediments. The most common source of infection is home-canned food that has not been heated sufficiently to kill contaminating *C. botulinum* spores. The spores then germinate, and a toxin is produced during vegetative growth. If the food is later eaten without adequate cooking, the active toxin results in disease.
Class *Clostridia* (section 23.4); The bacterial endospore (section 3.11)

The botulinum toxin is a neurotoxin that binds to the synapses of motor neurons **(figure 38.23)**. It selectively cleaves the synaptic vesicle membrane protein synaptobrevin, thus preventing exocytosis and release of the neurotransmitter acetylcholine. As a consequence, muscles do not contract in response to motor neuron activity, and flaccid paralysis results **(Techniques & Applications 38.5)**. Symptoms of botulism occur within 12 to 72 hours of toxin ingestion and include blurred vision, difficulty in swallowing and speaking, muscle weakness, nausea, and vomiting. Without adequate treatment, one-third of the patients may die of either respiratory or cardiac failure within a few days.

Laboratory diagnosis is restricted to Laboratory Response Network facilities and is by demonstration of the toxin in the patient's serum, stools, or vomitus. In addition, recovery of *C. botulinum* in stool cultures is diagnostic. Treatment relies on supportive care and polyvalent antitoxin. Fewer than 100 cases of botulism occur in the United States annually.

Infant botulism is the most common form of botulism in the United States and is confined to infants under a year of age. Approximately 100 cases are reported each year. It appears that ingested spores, which may be naturally present in honey or house dust, germinate in the infant's intestine. *C. botulinum* then multiplies and produces the toxin. The infant becomes constipated, listless, generally weak, and eats poorly. Death may result from respiratory failure.

Prevention and control of botulism involves (1) strict adherence to safe food-processing practices by the food industry, (2) educating the public on safe home-preserving (canning) methods for foods, and (3) not feeding honey to infants younger than 1 year of age.

Campylobacter jejuni Gastroenteritis

Campylobacter jejuni is a slender, gram-negative, motile, curved rod found in the intestinal tract of animals. *Campylobacter* infections cause more diarrhea in the United States than *Salmonella* and *Shigella* combined. Studies with chickens, turkeys, and cattle have shown that as much as 50 to 100% of a flock or herd of these birds or animals excrete *C. jejuni.* These bacteria also can be isolated in high numbers from surface waters. They are transmitted to humans by contaminated food and water, contact with infected animals, or anal-oral sexual activity. *C. jejuni* causes an estimated 2 million cases of *Campylobacter* **gastroenteritis**—inflammation of the intestine—or **campylobacteriosis** and subsequent diarrhea in the United States each year.
Class *Epsilonproteobacteria* (section 22.5)

The incubation period is 2 to 10 days. *C. jejuni* invades the epithelium of the small intestine, causing inflammation, and also secretes an exotoxin that is antigenically similar to the cholera toxin.

Table 38.6 Bacteria That Cause Acute Bacterial Diarrhea and Food Poisoning

Organism	Incubation Period (Hours)	Vomiting	Diarrhea	Fever	Epidemiology
Staphylococcus aureus	1–8 (rarely, up to 18)	+++	+	−	Staphylococci grow in meats, dairy and bakery products and produce enterotoxins.
Bacillus cereus	2–16	+++	++	−	Reheated fried rice causes vomiting or diarrhea.
Clostridium perfringens	8–16	±	+++	−	Clostridia grow in rewarmed meat dishes.
Clostridium botulinum	18–24	±	Rare	−	Clostridia grow in anoxic foods and produce toxin.
Escherichia coli (enterohemorrhagic)	3–5 days	±	++	±	Generally associated with ingestion of undercooked ground beef, and unpasteurized fruit juices and cider.
Escherichia coli (enterotoxigenic strain)	24–72	±	++	−	Organisms grow in gut and are a major cause of traveler's diarrhea.
Vibrio parahaemolyticus	6–96	+	++	±	Organisms grow in seafood and in gut and produce toxin, or invade.
Vibrio cholerae	24–72	+	+++	−	Organisms grow in gut and produce toxin.
Shigella spp.(mild cases)	24–72	±	++	+	Organisms grow in superficial gut epithelium. *S. dysenteriae* produces toxin.
Salmonella spp. (gastroenteritis)	8–48	±	++	+	Organisms grow in gut.
Salmonella enterica serovar Typhi (typhoid fever)	10–14 days	±	±	++	Bacteria invade the gut epithelium and reach the lymph nodes, liver, spleen, and gallbladder.
Clostridium difficile	Days to weeks after antibiotic therapy	−	+++	+	Antibiotic-associated colitis
Campylobacter jejuni	2–10 days	−	+++	++	Infection by oral route from foods, pets. Organism grows in small intestine.
Yersinia enterocolitica	4–7days	±	++	+	Fecal-oral transmission, food-borne, animals infected

Pathogenesis	Clinical Features
Enterotoxins act on gut receptors that transmit impulses to medullary centers; may also act as superantigens.	Abrupt onset, intense vomiting for up to 24 hours, recovery in 24–48 hours. Occurs in persons eating the same food. No treatment usually necessary except to restore fluids and electrolytes. With incubation period of 2–8 hours, mainly vomiting.
Enterotoxins formed in food or in gut from growth of *B. cereus*.	With incubation period of 8–16 hours, mainly diarrhea.
Enterotoxins produced during sporulation in gut, causes hypersecretion.	Abrupt onset of profuse diarrhea; vomiting occasionally. Recovery usual without treatment in 1–4 days. Many clostridia in cultures of food and feces of patients.
Toxin absorbed from gut and blocks acetylcholine release at neuromuscular junction.	Diplopia, dysphagia, dysphonia, difficulty breathing. Treatment requires clearing the airway, ventilation, and intravenous polyvalent antitoxin. Exotoxin present in food and serum. Mortality rate high.
Toxins cause epithelial necrosis in colon; mild to severe complications.	Symptoms vary from mild to severe bloody diarrhea. The toxin can be absorbed, becoming systemic and producing hemolytic uremic syndrome, most frequently in children.
Heat-labile (LT) and heat-stable (ST) enterotoxins cause hypersecretion in small intestine.	Usually abrupt onset of diarrhea; vomiting rare. A serious infection in newborns. In adults, "traveler's diarrhea" is usually self-limited in 1–3 days.
Toxin causes hypersecretion; vibrios invade epithelium; stools may be bloody.	Abrupt onset of diarrhea in groups consuming the same food, especially crabs and other seafood. Recovery is usually complete in 1–3 days. Food and stool cultures are positive.
Toxin causes hypersecretion in small intestine. Infective dose $>10^5$ vibrios.	Abrupt onset of liquid diarrhea in endemic area. Needs prompt replacement of fluids and electrolytes IV or orally. Tetracyclines shorten excretion of vibrios. Stool cultures positive.
Organisms invade epithelial cells; blood, mucus, and neutrophils in stools. Infective dose $<10^3$ organisms.	Abrupt onset of diarrhea, often with blood and pus in stools, cramps, tenesmus, and lethargy. Stool cultures are positive. Trimethoprim sulfamethoxazole, ampicillin, or chloramphenicol given in severe cases. Do not give opiates. Often mild and self-limited. Restore fluids.
Superficial infection of gut, little invasion. Infective dose $>10^5$ organisms.	Gradual or abrupt onset of diarrhea and low-grade fever. Nausea, headache, and muscle aches common. Administer no antimicrobials unless systemic dissemination is suspected. Stool cultures are positive. Prolonged carriage is frequent.
Symptoms probably due to endotoxins and tissue inflammation; infective dose $\geq 10^7$ organisms.	Initially fever, headache, malaise, anorexia, and muscle pains. Fever may reach 40°C by the end of the first week of illness and lasts for 2 or more weeks. Diarrhea often occurs, and abdominal pain, cough, and sore throat may be prominent. Antibiotic therapy shortens duration of the illness.
Toxins causes epithelial necrosis in colon; psuedomembranous colitis.	Especially after abdominal surgery, abrupt bloody diarrhea and fever. Toxins in stool. Oral vancomycin useful in therapy.
Invasion of mucous membrane; toxin production uncertain	Fever, diarrhea; PMNs and fresh blood in stool, especially in children. Usually self-limited. Special media needed for culture at 43°C. Erythromycin given in severe cases with invasion. Usual recovery in 5–8 days.
Gastroenteritis or mesenteric adenitis; occasional bacteremia; toxin produced occasionally.	Severe abdominal pain, diarrhea, fever; PMNs and blood in stool; polyarthritis, erythema nodosum, especially in children. Gentamicin used in severe cases. Keep stool specimen at 4°C before culture.

Table 38.7 Water-Borne Bacterial Pathogens

Organism	Reservoir	Comments
Aeromonas hydrophila	Free-living	Sometimes associated with gastroenteritis, cellulitis, and other diseases
Campylobacter	Bird and animal reservoirs	Major cause of diarrhea; common in processed poultry; a microaerophile
Helicobacter pylori	Free-living	Can cause type B gastritis, peptic ulcers, gastric adenocarcinomas
Legionella pneumophila	Free-living and associated with protozoa	Found in cooling towers, evaporators, condensers, showers, and other water sources
Leptospira	Infected animals	Hemorrhagic effects, jaundice
Mycobacterium	Infected animals and free-living	Complex recovery procedure required
Pseudomonas aeruginosa	Free-living	Swimmer's ear and related infections
Salmonella enteriditis	Animal intestinal tracts	Common in many waters
Vibrio cholerae	Free-living	Found in many waters including estuaries
Vibrio parahaemolyticus	Free-living in coastal waters	Causes diarrhea in shellfish consumers
Yersinia enterocolitica	Frequent in animals and in the environment	Waterborne gastroenteritis

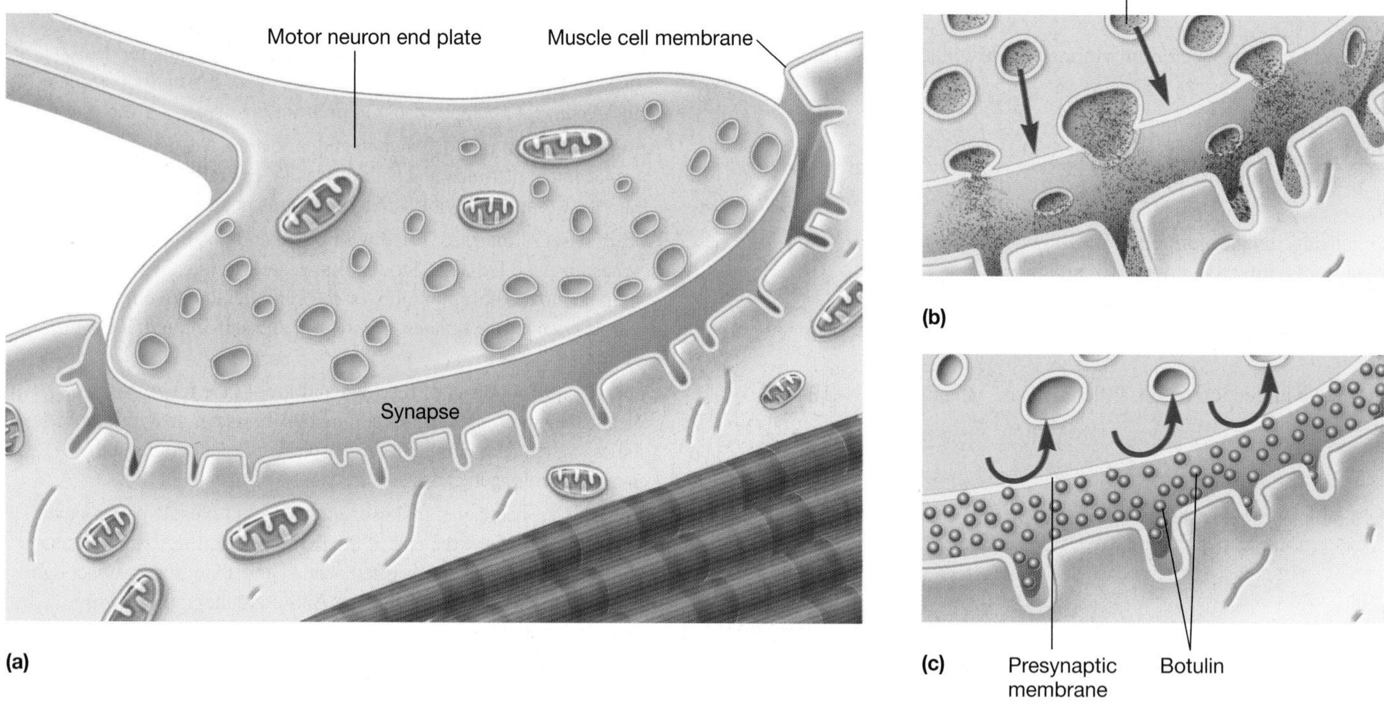

Figure 38.23 The Physiological Effects of Botulism Toxin. **(a)** The relationship between the motor neuron and the muscle at the neuromuscular junction. **(b)** In the normal state, acetylcholine released at the synapse crosses to the muscle and creates an impulse that stimulates muscle contraction. **(c)** In botulism, the toxin enters the motor end plate and attaches to the presynaptic membrane, where it blocks release of the chemical. This prevents impulse transmission, and keeps the muscle from contracting.

Symptoms include diarrhea, high fever, severe inflammation of the intestine along with ulceration, and bloody stools. *C. jejuni* infection has also been linked to Guillain-Barre syndrome, a disorder in which the body's immune system attacks peripheral nerves, resulting in life-threatening paralysis. Toxigenicity: Exotoxins (section 33.4)

Laboratory diagnosis is by culture in an atmosphere with reduced O_2 and added CO_2. The disease is usually self-limited, and treatment is supportive; fluids, electrolyte replacement, and erythromycin may be used in severe cases. Recovery usually takes from 5 to 8 days. Prevention and control involve good per-

Techniques & Applications

38.5 Clostridial Toxins as Therapeutic Agents—Benefits of Nature's Most Toxic Proteins

Some toxins are currently being used for the treatment of human disease. Specifically, botulinum toxin, the most poisonous biological substance known, is being used for the treatment of specific neuromuscular disorders characterized by involuntary muscle contractions. Since approval of type-A botulinum toxin (Botox) by the FDA in 1989 for three disorders (strabismus [crossing of the eyes], blepharospasm [spasmotic contractions of the eye muscles], and hemifacial spasm [contractions of one side of the face]), the number of neuromuscular problems being treated has increased to include other tremors, migraine and tension headaches, and other maladies. In 2000, dermatologists and plastic surgeons began using Botox to eradicate wrinkles caused by repeated muscle contractions as we laugh, smile, or frown. The remarkable therapeutic utility of botulinum toxin lies in its ability to specifically and potently inhibit involuntary muscle activity for an extended duration. Overall, the clostridia (currently one of the largest and most diverse genera of bacteria containing about 130 species) produce more protein toxins than any other known bacterial genus and are a rich reservoir of toxins for research and medicinal uses. For example, research is underway to use clostridial toxins or toxin domains for drug delivery, prevention of food poisoning, and the treatment of cancer and other diseases. The remarkable success of botulinum toxin as a therapeutic agent has thus created a new field of investigation in microbiology.

sonal hygiene and food-handling precautions, including pasteurization of milk and thorough cooking of poultry.

Cholera

Throughout recorded history, **cholera** [Greek *chole,* bile] has caused seven pandemics in various areas of the world, especially in Asia, the Middle East, and Africa. The disease has been rare in the United States since the 1800s, but an endemic focus is believed to exist on the Gulf Coast of Louisiana and Texas.

Cholera is caused by the comma-shaped, gram-negative *Vibrio cholerae* bacterium of the family *Vibrionaceae* **(figure 38.24)**. *V. cholerae* is actively motile by way of its single, polar flagellum. Although there are many serogroups, only O1 and O139 have exhibited the ability to cause epidemics. *V. cholerae* O1 is divided into two serotypes, Inaba and Ogawa, and two biotypes, classic and El Tor. Class *Gammaproteobacteria:* Order *Vibrionales* (section 22.3)

Individuals acquire cholera by ingesting food or water contaminated by fecal material from patients or carriers. Shellfish are natural reservoirs. In 1961 the El Tor biotype emerged as an important cause of cholera pandemics, and in 1992 the newly identified strain *V. cholerae* O139 emerged in Asia. This novel toxigenic strain does not agglutinate with O1 antiserum but possesses epidemic and pandemic potential. In Calcutta, India, serogroup O139 of *V. cholerae* has displaced El Tor *V. cholerae* serogroup O1, an event that has never before happened in the recorded history of cholera.

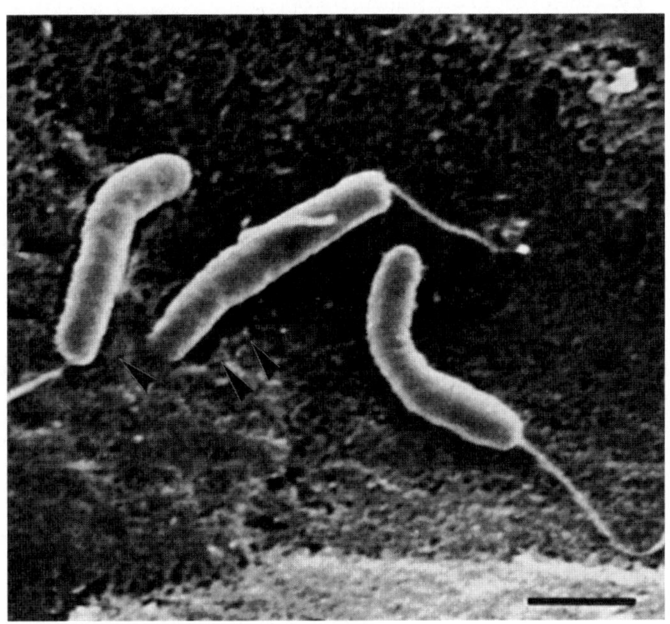

Figure 38.24 Cholera. *Vibrio cholerae* adhering to intestinal epithelium; scanning electron micrograph (×12,000). Notice that the bacteria is slightly curved with a single polar flagellum.

Once the bacteria enter the body, the incubation period is 12 to 72 hours. The bacteria adhere to the intestinal mucosa of the small intestine, where they are not invasive but secrete **choleragen,** a cholera toxin. Choleragen is an AB toxin composed of two functional subunits—an enzymatic A subunit (the toxic component) and an intestinal receptor-binding B subunit (*see figure 33.5*). The A subunit enters the intestinal epithelial cells and activates the enzyme adenylate cyclase by the addition of an ADP-ribosyl group in a way similar to that employed by diphtheria toxin. As a result, choleragen stimulates hypersecretion of water and chloride ions while inhibiting absorption of sodium ions. The patient loses massive quantities of fluid and electrolytes, causing abdominal muscle cramps, vomiting, fever, and watery diarrhea. The voided fluid is often referred to as "rice-water-stool" because of the flecks of mucus floating in it. The diarrhea can be so profuse that a person can lose 10 to 15 liters of fluid during the infection. Death may result from the elevated concentrations of blood proteins, caused by reduced fluid levels, which leads to circulatory shock and collapse. The cholera toxin gene is carried by the CTX filamentous bacteriophage. The phage binds to the pilus used to colonize the host's gut, enters the bacterium, and incorporates its genes into the bacterial chromosome.

Evidence indicates that passage through the human host enhances infectivity, although the exact mechanism is unclear. Before *V. cholerae* exits the body in watery stools, some unknown aspect of the intestinal environment stimulates the activity of certain bacterial genes. These genes, in turn, seem to prepare the bacteria for ever more effective colonization of their next victims, possibly fueling epidemics. *V. cholerae* can also be free-living in warm, alkaline, and saline environments.

Laboratory diagnosis is by culture of the bacterium from feces and subsequent identification by agglutination reactions with specific antisera. Treatment is by oral rehydration therapy with NaCl plus glucose to stimulate water uptake by the intestine; the antibiotics of choice are tetracycline, trimethoprim-sulfamethoxazole, or ciprofloxacin. The most reliable control methods are based on proper sanitation, especially of water supplies. The mortality rate without treatment is often over 50%; with treatment and supportive care, it is less than 1%. Fewer than 20 cases of cholera are reported each year in the United States.

Listeriosis

Listeria monocytogenes is a gram-positive rod that can be isolated from soil, vegetation, and many animal reservoirs. Human disease due to *L. monocytogenes* generally occurs in pregnancy or in people who are immunosuppressed due to illness or medication. Recent evidence suggests that a substantial number of cases of human **listeriosis** are attributable to the food-borne transmission of *L. monocytogenes. Listeria* outbreaks have been traced to sources such as contaminated milk, soft cheeses, vegetables, and meat. Unlike many of the food-borne pathogens, which cause primarily gastrointestinal illness, *L. monocytogenes* causes invasive syndromes such as meningitis, sepsis, and stillbirth. Class *Bacilli:* Order *Bacillales* (section 23.5)

L. monocytogenes is an intracellular pathogen, a characteristic consistent with its predilection for causing illness in persons with deficient cell-mediated immunity. This bacterium can be found as part of the normal gastrointestinal microbiota in healthy individuals. In immunosuppressed individuals, invasion, intracellular multiplication, and cell-to-cell spread of the bacterium appears to be mediated through proteins such as internalin, the hemolysin listeriolysin O, and phospholipase C. *Listeria* also uses host cell actin filaments to move within and between cells (*see figure 33.9*). The increased risk of infection in pregnant women may be due to both systemic and local immunological changes associated with pregnancy. For example, local immunosuppression at the maternal-fetal interface of the placenta may facilitate intrauterine infection following transient maternal bacteremia.

Diagnosis of listeriosis is by culture of the bacterium. Treatment is intravenous administration of either ampicillin or penicillin. Because *L. monocytogenes* is frequently isolated from food, the USDA (U.S. Department of Agriculture) and manufacturers are pursuing measures to reduce the contamination of food products by this bacterium. Controlling food spoilage (section 40.3)

Salmonellosis

Salmonellosis (*Salmonella* gastroenteritis) is caused by over 2,000 *Salmonella* serovars (*sero*logical *var*iations, or strains). Based on DNA homology studies, all known *Salmonella* are thought to belong to a single species, *S. enterica,* although the taxonomy of this bacterium remains controversial. The most frequently isolated serovars from humans are Typhimurium and Enteritidis. (Serovar names are not italicized, and the first letter is capitalized.) The salmonellae are gram-negative, motile, nonspore-forming rods. Class *Gammaproteobacteria:* Order *Enterobacteriales* (section 22.3)

The initial source of the bacterium is the intestinal tracts of birds and other animals. Humans acquire the bacteria from contaminated foods such as beef products, poultry, eggs, egg products, or water. Around 45,000 cases a year are reported in the United States, but there actually may be as many as 2 to 3 million cases annually.

Once the bacteria are in the body, the incubation time is only about 8 to 48 hours. The disease results from a true food-borne infection because the bacteria multiply and invade the intestinal mucosa, where they produce an enterotoxin and a cytotoxin that destroy the epithelial cells. Abdominal pain, cramps, diarrhea, nausea, vomiting, and fever are the most prominent symptoms, which usually persist for 2 to 5 days but can last for several weeks. During the acute phase of the disease, as many as 1 billion *Salmonella* can be found per gram of feces. Most adult patients recover, but the loss of fluids can cause problems for children and elderly people.

Laboratory diagnosis is by isolation of the bacterium from food or from patients' stools. Treatment is with fluid and electrolyte replacement. Prevention depends on good food-processing practices, proper refrigeration, and adequate cooking.

Typhoid Fever

Typhoid [Greek *typhodes,* smoke] **fever** is caused by *Salmonella enterica* serovar Typhi and is acquired by ingestion of food or water contaminated by feces of infected humans or person-to-person contact. In earlier centuries the disease occurred in great epidemics. A milder form of the disease, paratyphoid fever, is caused by serovars Paratyphi A, B, and C of *Salmonella enterica* subspecies *enterica.* Class *Gammaproteobacteria:* Order *Enterbacteriales* (section 22.3)

In the small intestine, the incubation period is about 10 to 14 days. The bacteria colonize the small intestine, penetrate the epithelium, and spread to the lymphoid tissue, blood, liver, and gallbladder. Symptoms include fever, headache, abdominal pain, anorexia, and malaise, which last several weeks. Bacteria then reinfect the gastrointestinal tract, producing abdominal pain and diarrhea. After approximately 3 months, most individuals stop shedding bacteria in their feces. However, a few individuals continue to shed *S.* Typhi for extended periods but show no symptoms. In these carriers, the bacteria continue to grow in the gallbladder and reach the intestine through the bile duct. Historical Highlights 36.2: "Typhoid Mary"

Laboratory diagnosis of typhoid fever is by demonstration of typhoid bacilli in the blood, urine, or stools and serology (the Widal test). Treatment with ceftriaxone or ciprofloxacin has reduced the mortality rate to less than 1%. Recovery from typhoid confers a permanent immunity. Purification of drinking water, prevention of food handling by carriers, and complete isolation of patients are the most successful prophylactic measures. There is a vaccine for high-risk individuals. About 300 to 400 cases of typhoid fever occur annually in the United States.

Shigellosis

Shigellosis, or bacillary dysentery, is a diarrheal illness resulting from an acute inflammatory reaction of the intestinal tract caused by the four species of the genus *Shigella* (gram-negative, nonmotile, nonspore-forming, facultative rods). About 20,000 to 25,000 cases a year are reported in the United States, and around 600,000 deaths a year worldwide are due to bacillary dysentery. Class *Gammaproteobacteria:* Order *Enterobacteriales* (section 22.3)

Shigella is restricted to human hosts. *S. sonnei* is the usual pathogen in the United States and Britain, but *S. flexneri* is also fairly common. The organism is transmitted by the fecal-oral route—primarily by food, fingers, feces, and flies (the four "F's")—and is most prevalent among children, especially 1- to 4-year-olds. The infectious dose is only around 10 to 100 bacteria. In the United States, shigellosis is a particular problem in daycare centers and custodial institutions where there is crowding.

The shigellae are facultatively anaerobic, intracellular parasites that multiply within the villus cells of the colon epithelium. The bacteria induce the Peyer's patch cells to phagocytose them. After being ingested, the bacteria then disrupt the phagosome membrane and are released into the cytoplasm where they reproduce. They then invade adjacent mucosal cells. *Shigella* initiate an inflammatory reaction in the mucosa. Both endotoxins and exotoxins may participate in disease progression, but the bacteria do not usually spread beyond the colonic epithelium. The watery stools often contain blood, mucus, and pus. In severe cases the colon can become ulcerated. Virulent *Shigella* produce a heat-labile AB exotoxin (Sxt) known as the shiga-toxin (formerly verotoxin). The complete toxin molecule is composed of one A protein surrounded by five B proteins. The B proteins attach to host vascular cells, stimulating the internalization of the whole toxin. The A subunit protein is subsequently released from the B protein units and binds to host ribosomes, inhibiting protein synthesis. A specific target of the B protein seems to be the glomerular endothelium; toxin action on these cells leads to kidney failure. Phagocytosis (section 31.3); Toxigenicity (section 33.4)

Pathogenic shigellae also use a type III secretion system to deliver virulence factors to target epithelial cells as well. The *Shigella* type III secretion machinery is responsible for delivering to host cells the specific protein components required for its invasion. Recall that the type III bacterial secretion system is specialized for the direct export of virulence factors into target host cells. It is comprised of 20 to 30 different proteins. Some of the proteins form a needlelike structure connected to a basal body. The needle penetrates the target cell membrane and delivers other proteins through it *(see figure 33.4b–d)*. The virulence factors typically subvert normal host cell functions so as to benefit the invading bacterium, such as the ability to acquire iron, adhere to host cells, or invade them. Specifically, *Shigella* uses Spa proteins for structural components like the needle; IpaB protein appears to be used to invade the host—it is homologous with many pore-forming toxins. Protein secretion in procaryotes (section 3.8)

The incubation period usually ranges from 1 to 3 days and the organisms are shed over a period of 1 to 2 weeks. Identification of isolates is based on biochemical characteristics and serology. The disease normally is self-limiting in adults and lasts an average of 4 to 7 days; in infants and young children it may be fatal. Usually fluid and electrolyte replacement are sufficient, and antibiotics may not be required in mild cases although they can shorten the duration of symptoms and transmission to family members. Sometimes, particularly in malnourished infants and children, neurological complications and kidney failure result. When necessary, treatment is with trimethoprim-sulfamethoxazole *(see figure 34.14)* or fluoroquinolones. Antibiotic-resistant strains are becoming a problem. Prevention is a matter of good personal hygiene and the maintenance of a clean water supply.

Staphylococcal Food Poisoning

Staphylococcal food poisoning is the major type of food intoxication in the United States. It is caused by ingestion of improperly stored or cooked food (particularly foods such as ham, processed meats, chicken salad, pastries, ice cream, and hollandaise sauce) in which *Staphylococcus aureus* has grown. Class *Bacilli:* Order *Bacillales* (section 23.5)

S. aureus (a gram-positive coccus) is very resistant to heat, drying, and radiation; it is found in the nasal passages and on the skin of humans and other mammals worldwide. From these sources it can readily enter food. If the bacteria are allowed to incubate in certain foods, they produce heat-stable enterotoxins that render the food dangerous even though it appears normal. Once the bacteria have produced the toxin, the food can be extensively and properly cooked, killing the bacteria without destroying the toxin. Intoxication can therefore result from food that has been thoroughly cooked. Thirteen different enterotoxins have been identified; enterotoxins A, B, C1, C2, D, and E are the most common. (Recall that enterotoxins A and B are superantigens.) These toxins appear to act as neurotoxins that stimulate vomiting through the vagus nerve. Toxigenicity (section 33.4)

Typical symptoms include severe abdominal pain, cramps, diarrhea, vomiting, and nausea. The onset of symptoms is rapid (usually 1 to 8 hours) and of short duration (usually less than 24 hours). The mortality rate of staphylococcal food poisoning is negligible among healthy individuals. Diagnosis is based on symptoms or laboratory identification of the bacteria from foods. Enterotoxins may be detected in foods by animal toxicity tests or antibody-based methods. Treatment is with fluid and electrolyte replacement. Prevention and control involve avoidance of food contamination, and control of personnel responsible for food preparation and distribution.

Traveler's Diarrhea and *Escherichia coli* Infections

Millions of people travel yearly from country to country. Unfortunately, a large percentage of these travelers acquire a rapidly acting, dehydrating condition called **traveler's diarrhea.** This diarrhea results from an encounter with certain viruses, bacteria, or protozoa usually absent from the traveler's normal environment. One of the major causative agents is *E. coli.* This bacterium circulates in the resident population, typically without causing symptoms due to the immunity afforded by previous exposure. Because many of these bacteria are needed to initiate infection, contaminated food and water are the major means by which the bacteria are spread. This is the basis for the popular warnings to international travelers: "Don't drink the local water" and "Boil it, peel it, cook it, or forget it."

Although the vast majority of *E. coli* strains are nonpathogenic members of the normal intestinal flora, some strains may cause diarrheal disease by several mechanisms. Six categories or strains of diarrheagenic *E. coli* are now recognized **(figure 38.25)**: enterotoxigenic *E. coli* (ETEC), enteroinvasive *E. coli* (EIEC), enterohemorrhagic *E. coli* (EHEC), enteropathogenic *E. coli* (EPEC), enteroaggregative *E. coli* (EAggEC), and diffusely adhering *E. coli* (DAEC). Class *Gammaproteobacteria*: Order *Enterbacteriales* (section 22.3)

The **enterotoxigenic *E. coli* (ETEC)** strains produce one or both of two distinct enterotoxins, which are responsible for the diarrhea and distinguished by their heat stability: heat-stable enterotoxin (ST) and heat-labile enterotoxin (LT) (figure 38.25*a*). The genes for ST and LT production and for colonization factors are usually plasmid-borne and acquired by horizontal gene transfer. ST binds to a glycoprotein receptor that is coupled to guanylate cyclase on the surface of intestinal epithelial cells. Activation of guanylate cyclase stimulates the production of cyclic guanosine monophosphate (cGMP), which leads to the secretion of electrolytes and water into the lumen of the small intestine, manifested as the watery diarrhea characteristic of an ETEC infection. LT binds to specific gangliosides on the epithelial cells and activates membrane-bound adenylate cyclase, which leads to increased production of cyclic adenosine monophosphate (cAMP) through the same mechanism employed by cholera toxin. Again, the result is hypersecretion of electrolytes and water into the intestinal lumen.

The **enteroinvasive *E. coli* (EIEC)** strains cause diarrhea by penetrating and multiplying within the intestinal epithelial cells (figure 38.25*b*). The ability to invade epithelial cells is associated with the presence of a large plasmid; EIEC may also produce a cytotoxin and an enterotoxin.

The **enteropathogenic *E. coli* (EPEC)** strains attach to the brush border of intestinal epithelial cells and cause a specific type of cell damage called effacing lesions (figure 38.25*c*). **Effacing lesions** or attaching-effacing (AE) lesions represent destruction of brush border microvilli adjacent to adhering bacteria. This cell destruction leads to the subsequent diarrhea. As a result of this pathology, the term AE *E. coli* is used to describe true EPEC strains. It is now known that AE *E. coli* is an

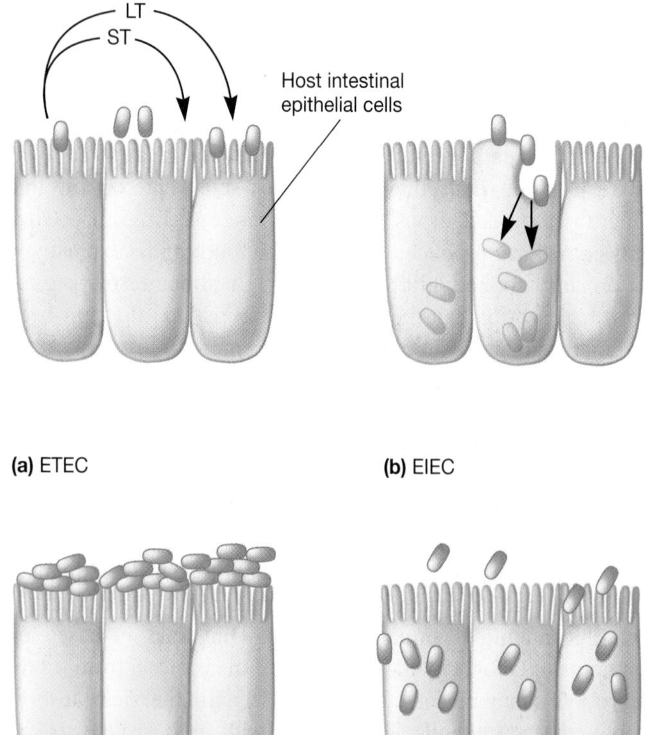

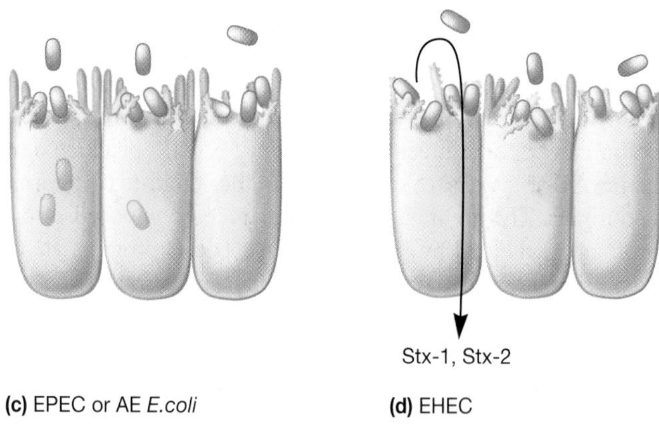

Figure 38.25 Six Classes of Diarrheagenic *E. coli.* Each class of diarrhea-causing *E. coli* can be classified by the nature of its interaction with host intenstinal epithelial cells.

important cause of diarrhea in children residing in developing countries.

The **enterohemorrhagic *E. coli* (EHEC)** strains carry the bacteriophage-encoded genetic determinants for shiga-like toxin (Stx-1 and Stx-2 proteins; figure 38.25*d*). EHEC also produce AE lesions causing hemorrhagic colitis with severe abdominal pain and cramps followed by bloody diarrhea. Stx-1 and Stx-2 (previously verotoxins 1 and 2) have also been implicated in the extra-intestinal disease, **hemolytic uremic syndrome,** a severe hemolytic anemia that leads to kidney failure. It is believed these toxins kill vascular endothelial cells. A major form of EHEC is *E. coli* O157:H7, which has caused many outbreaks of hemorrhagic colitis in the United States since it was first recognized in 1982. Currently there are an estimated 73,000 *E. coli* O157:H7 cases in the United States each year, resulting in 60 deaths. Other serotypes of *E. coli* can cause similar disease, but they do not typically carry the genes for shiga-like toxin. However, most laboratories do not test for non-O157 strains, so the actual incidence of EHEC is under-reported.

The **enteroaggregative *E. coli* (EAggEC)** strains adhere to epithelial cells in localized regions, forming clumps of bacteria with a "stacked brick"appearance (figure 38.25*e*). Conventional extracellular toxins have not been detected in EAggEC, but unique lesions are seen in epithelial cells, suggesting the involvement of toxins.

The **diffusely adhering *E. coli* (DAEC)** strains adhere over the entire surface of epithelial cells and usually cause disease in immunologically naive or malnourished children (figure 38.25*f*). It has been suggested that DAEC may have an as-yet undefined virulence factor.

Diagnosis of traveler's diarrhea caused by *E. coli* is based on past travel history and symptoms. Laboratory diagnosis is by isolation of the specific type of *E. coli* from feces and identification using DNA probes, the determination of virulence factors, and the polymerase chain reaction. Treatment is with fluid and electrolytes plus doxycycline and trimethroprim-sulfamethoxazole. Recovery can be without complications except in EHEC damage to kidneys. Prevention and control involve avoiding contaminated food and water.

1. Define food intoxication, food poisoning, and food-borne infection. What is an enterotoxin?
2. How does one acquire botulism? Describe how botulinum toxin causes flaccid paralysis.
3. Why is cholera the most severe form of gastroenteritis?
4. What is a common source of *Listeria* infections? How is the intracellular growth of *Listeria* related to the symptoms it produces and the observation that immunocompromised individuals are most at risk?
5. What is the usual source of the bacterium responsible for salmonellosis? Shigellosis? Where and how does *Shigella* infect people?
6. Describe a typhoid carrier. How does one become a carrier?
7. Describe the most common type of food intoxication in the United States and explain how it arises.
8. What are some specific causes of traveler's diarrhea? Briefly describe the six major types of pathogenic *E. coli.*

38.5 Sepsis and Septic Shock

Some microbial diseases and their effects cannot be categorized under a specific mode of transmission. Two important examples are sepsis and septic shock. Septic shock is the most common cause of death in intensive care units and the thirteenth most common cause of death in the United States. Unfortunately, the incidence of these two disorders continues to rise: 400,000 cases of sepsis and 200,000 episodes of septic shock are estimated to occur annually in the United States, resulting in more than 100,000 deaths.

Sepsis has been redefined by physicians as the systemic response to a microbial infection. This response is manifested by two or more of the following conditions: temperature above 38°C or below 36°C; heart rate above 90 beats per minute; respiratory rate above 20 breaths per minute or a pCO_2 below 32 mmHg; leukocyte count above 12,000 cells per ml^3 or below 4,000 cells per ml^3. **Septic shock** is sepsis associated with severe hypotension (low blood pressure) despite adequate fluid replacement. Gram-positive bacteria, fungi, and endotoxin-containing gram-negative bacteria can initiate the pathogenic cascade of sepsis leading to septic shock. Gram-negative sepsis is most commonly caused by *E. coli*, *Klebsiella* spp., *Enterobacter* spp., or *Pseudomonas aeruginosa*. Endotoxin, or more specifically, the lipid A moiety of lipopolysaccharide (LPS), an integral component of the outer membrane of gram-negative bacteria, has been implicated as a primary initiator of the pathogenesis of gram-negative septic shock. The bacterial cell wall: Gram-negative cell walls (section 3.6)

The pathogenesis of sepsis and septic shock begins with the proliferation of the microorganism at the infection site **(figure 38.26)**. The microorganism may invade the bloodstream directly or may proliferate locally and release various products into the bloodstream. These products include both structural components of the microorganisms (endotoxin, teichoic acid antigen) and exotoxins synthesized by the microorganism. All of these products can stimulate the release of the endogenous mediators of sepsis from endothelial cells, plasma cells (monocytes, macrophages, neutrophils), and plasma cell precursors.

The endogenous mediators have profound physiological effects on the heart, vasculature, and other body organs. Because there is no drug therapy (despite vigorous research efforts), the consequences of septic shock are either recovery or death. Death usually ensues if one or more organ systems fail completely.

38.6 Zoonotic Diseases

Diseases transmitted from animals to humans are called zoonotic diseases. A number of important human pathogens begin as normal flora or parasites of animals and can often adapt to cause disease in humans. Here we highlight a few of the more notable diseases and the agents that cause them.

**Anthrax

Anthrax (Greek *anthrax,* coal) is a highly infectious animal disease that can be transmitted to humans by direct contact with infected animals (cattle, goats, sheep) or their products, especially

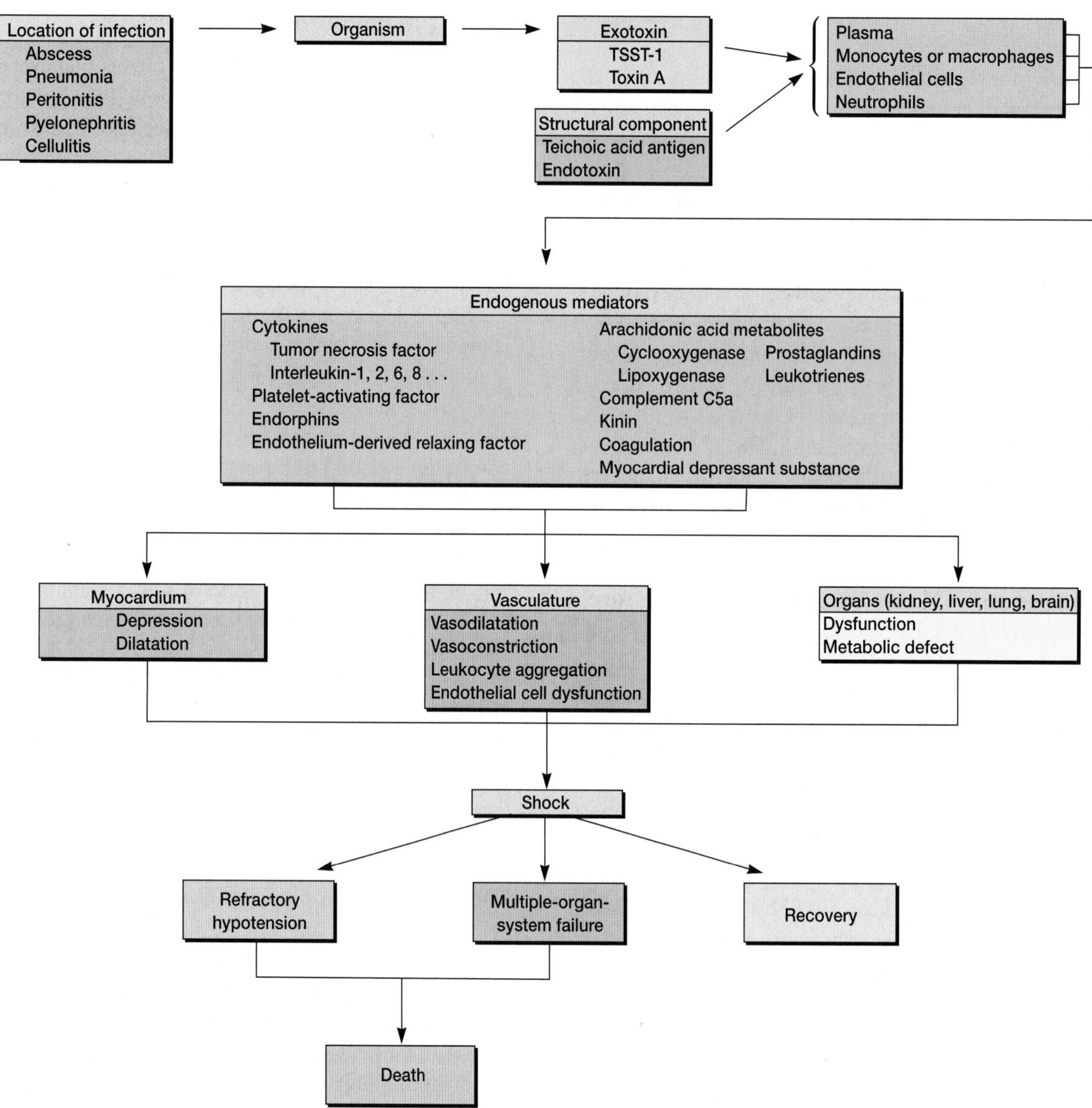

Figure 38.26 The Septic Shock Cascade. Microbial exotoxins (toxic shock syndrome toxin-I [TSST-1], *Pseudomonas aeruginosa* toxin A [toxin A]) and structural components of the microorganism (teichoic acid antigen, endotoxin) trigger the biochemical events that lead to such serious complications as shock, adult respiratory distress syndrome, and disseminated intravascular coagulation.

hides. The causative bacterium is the relatively large, gram-positive, aerobic, endospore-forming *Bacillus anthracis,* which has a nearly worldwide distribution. Its spores can remain viable in soil and animal products for decades (*see figure 3.47*). Although *B. anthracis* is one of the most molecularly monomorphic (of one shape) bacteria, it is now possible to separate all known strains into five categories (providing some clues to their geographic sites of origin) based on the number of tandem repeats in various genes (*see figure 15.16*). Techniques for determining microbial taxonomy and phylogeny: Molecular characteristics (section 19.4); Class *Bacilli:* Order *Bacillales* (section 23.5)

Human infection is usually through a cut or abrasion of the skin, resulting in **cutaneous anthrax;** however, inhaling spores may result in **pulmonary anthrax,** also known as woolsorter's disease. If spores reach the gastrointestinal tract, **gastrointestinal anthrax** may result. *B. anthracis* bacteremia can develop from any form of anthrax. The principal virulence factors of *B. anthracis* are encoded on two plasmids—one involved in the synthesis of a polyglutamyl capsule that inhibits phagocytosis and the other bearing the genes for the synthesis of its exotoxins (a complex exotoxin system composed of three proteins: protective antigen [PA], edema factor [EF], and lethal factor [LF]).

For a successful infection, *B. anthracis* must evade the host's innate immune system by killing macrophages. Macrophages have many anthrax toxin receptors (capillary morphogenesis protein-2) on their plasma membranes to which the PA portion of the exotoxin

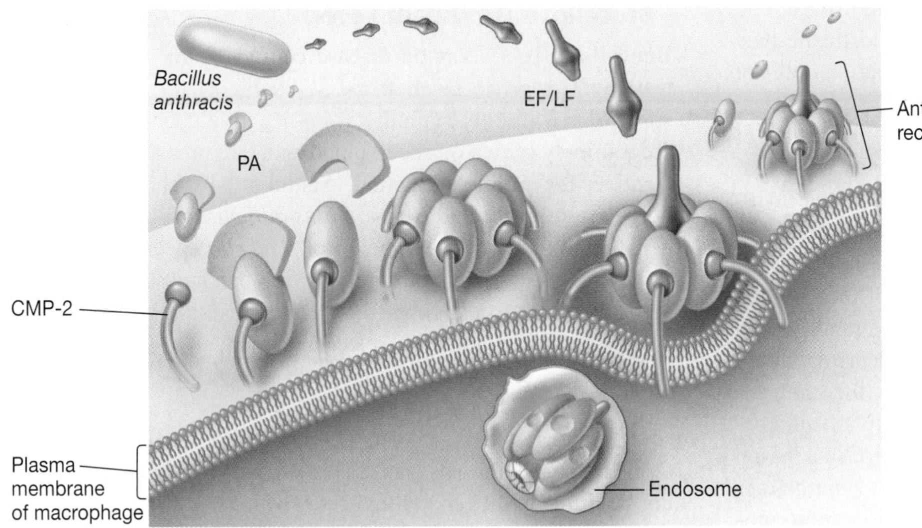

(a)

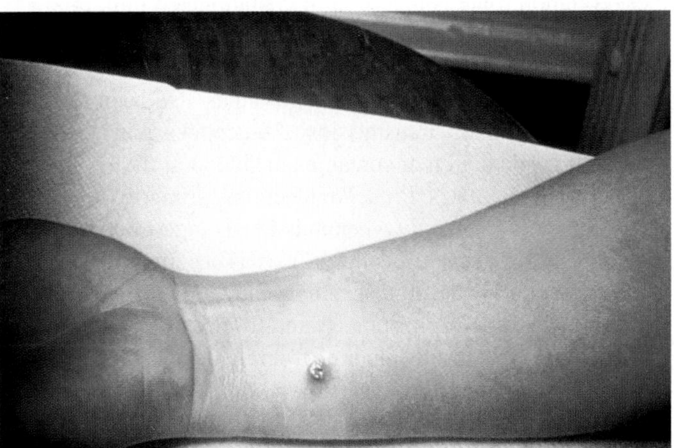

(b)

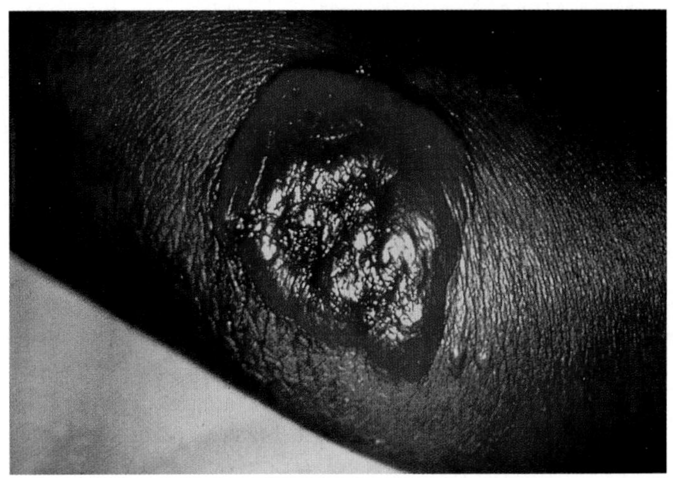

(c)

Figure 38.27 Anthrax.
(a) A protein called protective antigen (PA) delivers two other proteins, edema factor (EF) and lethal factor (LF), to the capillary morphogenesis protein-2 (CMP-2) receptor on the cell membrane of a target macrophage where PA, EF, and LF are transported to an endosome. PA then delivers EF and LF from the endosome into the cytoplasm of the macrophage where they exert their toxic effects. **(b)** A cutaneous anthrax papule will ulcerate and necrose into **(c)** an eschar.

system attaches. Attachment continues until seven PA-ATR complexes gather in a doughnut-shaped ring **(figure 38.27*a*)**. The ring acts like a syringe, boring through the plasma membrane of the macrophage. The ring then binds EF and LF, after which the entire complex is engulfed by the macrophages' plasma membrane and shuttled to an endosome inside the cell (*see figure 4.10*). Once there, the PA molecules form a special pore that pierces the endosome's membrane and lets EF and LF out into the cytoplasm. EF has adenylate cyclase activity, similar to diphtheria and pertussis toxins; increasing intracellular cAMP. Toxin activity results in fluid release, or the formation of edema. Additionally, LF interferes with a transcription factor, nuclear factor κB (NFκB), which regulates numerous cytokine and other immunity genes, promoting macrophage survival. As thousands of macrophages die, they release their lysosomal contents, leading to fever, internal bleeding, septic shock, and rapid death.

In humans, more than 95% of naturally occurring anthrax is the cutaneous form. The incubation period for cutaneous anthrax is 1 to 15 days. Infection is initiated with the introduction of the spores through a break in the skin. After ingestion by macrophages at the site of entry, the spores germinate and give rise to the vegetative form, which multiplies extracellularly and forms a capsule and exotoxins. Skin infections initially resemble insect bites, then develop into a papular vesicle (figure 38.27*b*), and finally into an ulcer with a necrotic center called an **eschar** (figure 38.27*c*). The eschar dries and falls off in 1 to 2 weeks with little scarring. Without antibiotic treatment, mortality can be as high as 20%. Therapy is with ciprofloxacin, penicillin, or doxycycline. Treatment should continue for 7 to 10 days with the naturally acquired disease or for 60 days in the case of bioterrorism. With proper antibiotic treatment, mortality for cutaneous anthrax is very rare.

In inhalation anthrax, the spores (1 to 2 μm in diameter) are inhaled and lodge in the alveolar spaces where they are engulfed by alveolar macrophages. The spores survive phagocytosis and germinate within the endosome; the bacteria then spread to regional lymph nodes and eventually the bloodstream. Pulmonary anthrax results in massive pulmonary edema, hemorrhage, and respiratory arrest. Once the bacteria enter the bloodstream, they begin producing exotoxin. The medial lethal inhalation dose for humans has been estimated to be about 8,000 spores.

The classic clinical description of inhalation anthrax is that of a two-phase illness. In the initial phase, which follows an incubation period of 1 to 6 days, the disease appears as a nonspecific illness characterized by mild fever, malaise, nonproductive cough, and some chest pain. The second phase begins abruptly and involves a higher fever, acute dyspnea (shortness of breath), and cyanosis (oxygen deficiency). This stage progresses rapidly, with septic shock, associated hypothermia, and death occurring within 24 to 36 hours from respiratory failure. Treatment (the same antibiotics as for cutaneous anthrax) is successful only if begun before a critical concentration of toxin has accumulated. One reason this form of anthrax is so difficult to treat is that the symptoms appear after *B. anthracis* has already multiplied and started to produce large amounts of the tripartite exotoxin. Thus although antibiotics may kill the bacterium or suppress its growth, the exotoxin can still eventually kill the patient. Sixteen of the 18 cases of inhalation anthrax reported in the United States between 1900 and 1978 were fatal. Five of the 22 cases in 2001 were fatal.

The symptoms of gastrointestinal anthrax appear 2 to 5 days after the ingestion of undercooked meat containing spores and include nausea, vomiting, fever, and abdominal pain. The manifestations progress rapidly to severe, bloody diarrhea. The primary lesions are ulcerative enabling *B. anthracis* to become bloodborne. Mortality is greater than 50 percent.

In the past, the diagnosis of anthrax was made on the basis of clinical findings and the history of exposure to animal products. The significance of exposure history has changed with the 2001 delivery of weaponized anthrax spores through the mail. Presumptive identification in sentinel laboratories of the Laboratory Response Network (LRN) is based on the direct Gram stained smear of a skin lesion, cerebrospinal fluid, or blood that shows encapsulated, broad, gram-positive bacilli. Presumptive identification is also made on the basis of growth and biochemical characteristics of cultures: large, flat, nonhemolytic colonies; nonmotile; positive for catalase and positive for capsule production. Confirmatory diagnosis is performed by PCR and serological tests for toxins at a reference laboratory of the LRN.

Between 20,000 and 100,000 cases of anthrax are estimated to occur worldwide annually; in the United States, the annual incidence was 127 cases in the early part of the 20th century. However, it subsequently declined to less than 1 case per year—a rate maintained for 20 years. Until 2001, there had not been a case of inhalation anthrax in the United States for more than 20 years. Thus the 2001 occurrence of 22 cases of anthrax (including five deaths) has spotlighted the real concern about anthrax as a weapon of bioterrorism.

Vaccination of animals, primarily cattle, is an important control measure. However, people with a high occupational risk, such as those who handle infected animals or their products, including hides and wool, should be immunized with the cell-free vaccine obtainable from the CDC. United States military personnel also receive the vaccine.

**Brucellosis (Undulant Fever)

Brucellosis is a zoonotic disease caused by *Brucella* species; usually *B. abortus, B. melitensis, B. suis,* or *B. canis. Brucella* spp. are tiny, faintly staining, gram-negative coccobacilli. They are routinely grown on sheep blood agar, are urease positive, and positive for nitrate reduction. Sentinel labs should rule out *Oligella ureolytica* and *Haemophilus influenzae,* both tiny, gram-negative coccobacilli. Suspicion of *Brucella* species by laboratory personnel requires biosafety level-3 precautions as it readily aerosolizes. Notification of the Public Health LRN is required because it is considered a select agent. Class *Alphaproteobacteria* (section 22.1); Bioterrorism preparedness (section 36.9)

Brucella is commonly transmitted through consumption of contaminated animal products or abrasions of the skin from handling infected mammals (cattle, sheep, goats, pigs, and rarely from dogs). Humans are generally infected by (1) ingesting food or water that is contaminated with *Brucella,* (2) inhaling the organism, or (3) having the bacteria enter the body through skin wounds. The most common route is ingestion of contaminated milk products. Direct person-to-person spread of brucellosis is extremely rare. However, infants may be infected through their mother's breast milk; sexual transmission of brucellosis has also been reported. Although uncommon, transmission of brucellosis may also occur through transplantation of contaminated blood or tissue.

In the United States, *Brucella* infections (primarily *B. melitensis*) occur more frequently when individuals ingest unpasteurized milk or dairy products. Brucellosis also has occurred in laboratory workers—culturing the organisms concentrates them and increases the risk of their aerosolization. Naturally occurring cases in the United States are typically reported from California, Florida, Texas and Virginia. For the past 10 years, approximately 100 cases of brucellosis have been reported annually. Most of these cases have been in abattoir (slaughterhouse) workers, meat inspectors, animal handlers, veterinarians, and laboratorians. Areas of higher risk are those with no or limited animal control programs, including the Mediterranean Basin (Portugal, Spain, Southern France, Italy, Greece, Turkey, North Africa), South and Central America, Eastern Europe, Asia, Africa, the Caribbean, and the Middle East. Important for tourists is the potential infection through unpasteurized cheeses, sometimes called "village cheeses." A newer controversy has erupted in the northwestern United States over the transmission of *Brucella,* endemic in the wild bison and elk populations, to otherwise *Brucella*-free cattle.

Brucellosis, in the acute ($<$ 8 weeks from onset) form, presents as nonspecific, flu-like symptoms including fever, sweats, malaise, anorexia, headache, myalgia, and back pain. In the undulant (rising and falling) form ($<$1 year from onset), symptoms of brucellosis include undulant fevers, arthritis, and testicular inflammation in males. Neurologic symptoms may occur acutely in up to 5% of the cases. In the chronic form ($>$1 year from onset), brucellosis symptoms may include chronic fatigue syndrome, depression, and arthritis. Mortality is low, less than 2%. Treatment is usually with doxycycline and rifampin in combination for 6 weeks to prevent recurring infection. Sequellae of brucellosis are variable and include granulomatous hepatitis, peripheral arthritis, spondylitis, anemia, leukopenia, thrombocytopenia, meningitis, uveitis, optic neuritis, papilledema, and endocarditis.

Psittacosis (Ornithosis)

Psittacosis (ornithosis) is a worldwide infectious disease of birds that is transmissible to humans. It was first described in association with parrots and parakeets, both of which are psittacine birds. The disease is now recognized in many other birds—among them, pigeons, chickens, ducks, and turkeys—and the general term ornithosis [Latin *ornis,* bird] is used.

Ornithosis is caused by *Chlamydophilia (Chlamydia) psittaci.* Humans contract this disease either by handling infected birds or by inhaling dried bird excreta that contains viable *C. psittaci.* Ornithosis is recognized as an occupational hazard within the poultry industry, particularly to workers in turkey-processing plants. After entering the respiratory tract, the chlamydiae are transported to the cells of the liver and spleen. They multiply within these cells and then invade the lungs, where they cause inflammation, hemorrhaging, and pneumonia. Phylum *Chlamydiae* (section 21.5)

Laboratory diagnosis is either by isolation of *C. psittaci* from blood or sputum, or by serological studies. Treatment is with tetracycline. Because of antibiotic therapy, the mortality rate has dropped from 20 to 2%. Less than 100 cases of ornithosis are reported annually in the United States. Prevention and control has been by chemoprophylaxis (tetracycline) for pet birds and poultry, although this can lead to the development of antibiotic resistance and is discouraged.

**Tularemia

The gram-negative bacterium *Francisella tularensis* is widely found in animal reservoirs in the United States and causes the disease **tularemia** (from Tulare, a county in California where the disease was first described). It may be transmitted to humans by biting arthropods (ticks, deer flies, or mosquitoes), direct contact with infected tissue (rabbits), inhalation of aerosolized bacteria, or ingestion of contaminated food or water. However, tularemia is most often transmitted through contact with infected animals; it is called rabbit fever in the central United States because it is often a disease of hunters. After an incubation period of 2 to 10 days, a primary ulcerative lesion appears at the infection site, lymph nodes enlarge, and a high fever develops. Class *Gammaproteobacteria* (section 22.3)

Diagnosis is made by national reference laboratories using PCR or culture of the bacterium and fluorescent antibody and agglutination tests; treatment is with streptomycin, tetracycline, or aminoglycoside antibiotics. Prevention and control involve public education, protective clothing, and vector control. An attenuated live vaccine is available from the U.S. Army for high-risk laboratory workers. Fewer than 200 cases of tularemia are reported annually in the United States. Importantly, *F. tularensis* is a microorganism of concern as a biological threat agent. Because public health preparedness efforts in the United States have shifted toward a stronger defense against biological terrorism, public health and medical management protocols following a potential release of tularemia are now in place.

38.7 Dental Infections

Some microorganisms found in the oral cavity are discussed in section 30.3 and presented in figure 30.17. Of this large number, only a few bacteria can be considered true dental pathogens, or **odontopathogens.** These few odontopathogens are responsible for the most common bacterial diseases in humans: tooth decay and periodontal disease.

Dental Plaque

The human tooth has a natural defense mechanism against bacterial colonization that complements the protective role of saliva. The hard enamel surface selectively absorbs acidic glycoproteins (mucins) from saliva, forming a membranous layer called the **acquired enamel pellicle.** This pellicle, or organic covering, contains many sulfate (SO_4^{2-}) and carboxylate ($-COO^-$) groups that confer a net negative charge to the tooth surface. Because most bacteria also have a net negative charge, there is a natural repulsion between the tooth surface and bacteria in the oral cavity. Unfortunately, this natural defense mechanism breaks down when dental plaque formation occurs.

Dental plaque formation begins with the initial colonization of the pellicle by *Streptococcus gordonii, S. oralis,* and *S. mitis* **(figure 38.28)**. These bacteria selectively adhere to the pellicle by specific ionic, hydrophobic, and lectin-like interactions. Once the tooth surface is colonized, subsequent attachment of other bacteria results from a variety of specific coaggregation reactions **(figure 38.29)**. **Coaggregation** is the result of cell-to-cell recognition between genetically distinct bacteria. Many of these interactions are mediated by a lectin (a carbohydrate-binding protein) on one bacterium that interacts with a complementary carbohydrate on another bacterium. The most important species at this stage are *Actinomyces viscosus, A. naeslundii,* and *Streptococcus gordonii.* After these species colonize the pellicle, a microenvironment is created that allows *Streptococcus mutans* and *S. sobrinus* to become established on the tooth surface by attaching to these initial colonizers **(figure 38.30)**. Microbial growth in natural environments: Biofilms (section 6.6)

S. mutans and *S. sobrinus* produce extracellular enzymes (glucosyltransferases) that polymerize the glucose moiety of sucrose into a heterogeneous group of extracellular, water-soluble and water-insoluble glucan polymers and other polysaccharides. The fructose by-product can be used in fermentation. **Glucans** are branched-chain polysaccharides composed of glucose units; many glucans synthesized by oral streptococci have glucose monomers held together by α $(1 \rightarrow 6)$ or α $(1 \rightarrow 3)$ linkages. They act like a cement to bind bacterial cells together, forming a plaque ecosystem. (Dental plaque is one of the most dense collections of bacteria in the body; perhaps the source of the first human microorganisms to be seen under a microscope, by Anton van Leeuwenhoek, in the 17th century.) Once plaque becomes established, the surface of the tooth becomes anoxic. This leads to the growth of strict anaerobic bacteria (*Bacteroides melaninogenicus, B. oralis,* and *Veillonella alcalescens*), especially between opposing teeth and the dental-gingival crevices (figure 38.30).

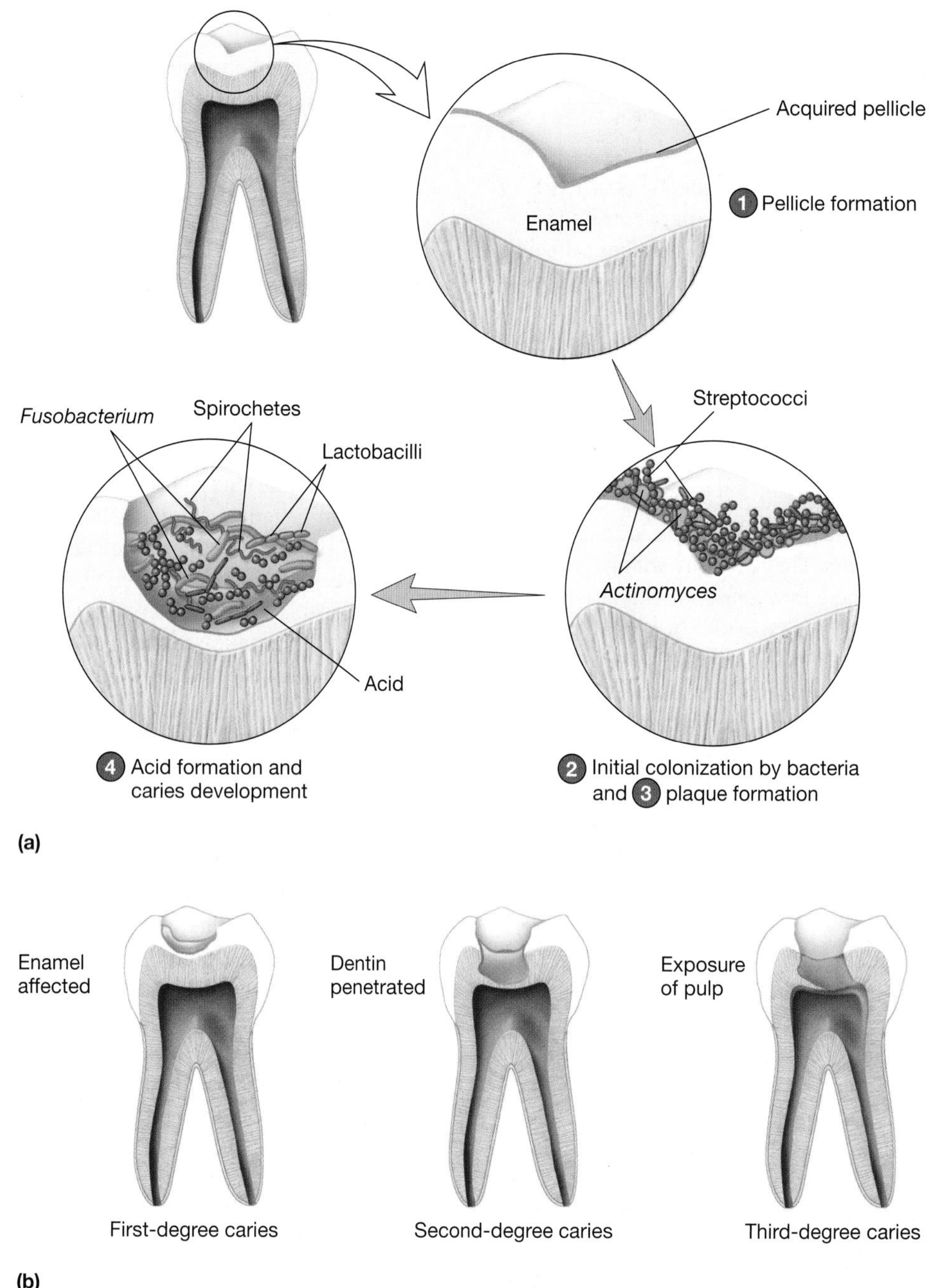

Figure 38.28 Stages in Plaque Development and Cariogenesis. **(a)** A microscopic view of pellicle and plaque formation, acidification, and destruction of tooth enamel. **(b)** Progress and degrees of cariogenesis.

After the microbial plaque ecosystem develops, bacteria produce lactic and possibly acetic and formic acids from sucrose and other sugars. Because plaque is not permeable to saliva, the acids are not diluted or neutralized, and they demineralize the enamel to produce a lesion on the tooth. It is this chemical lesion that initiates dental decay.

Dental Decay (Caries)

As fermentation acids move below the enamel surface, they dissociate and react with the hydroxyapatite of the enamel to form soluble calcium and phosphate ions. As the ions diffuse outward, some reprecipitate as calcium phosphate salts in the tooth's surface layer to create a histologically sound outer layer overlying a porous

Figure 38.29 The Formation of Dental Plaque on a Freshly Cleaned Tooth Surface. Diagrammatic representation of the proposed temporal relationship of bacterial accumulation and multigeneric coaggregation during the formation of dental plaque on the acquired enamel pellicle. Early tooth surface colonizers coaggregate with each other, and late colonizers coaggregate with each other. With a few exceptions, early colonizers do not recognize late colonizers. After the tooth surface is covered with the earliest colonizers, each newly added bacterium becomes a new surface for recognition by unattached bacteria.

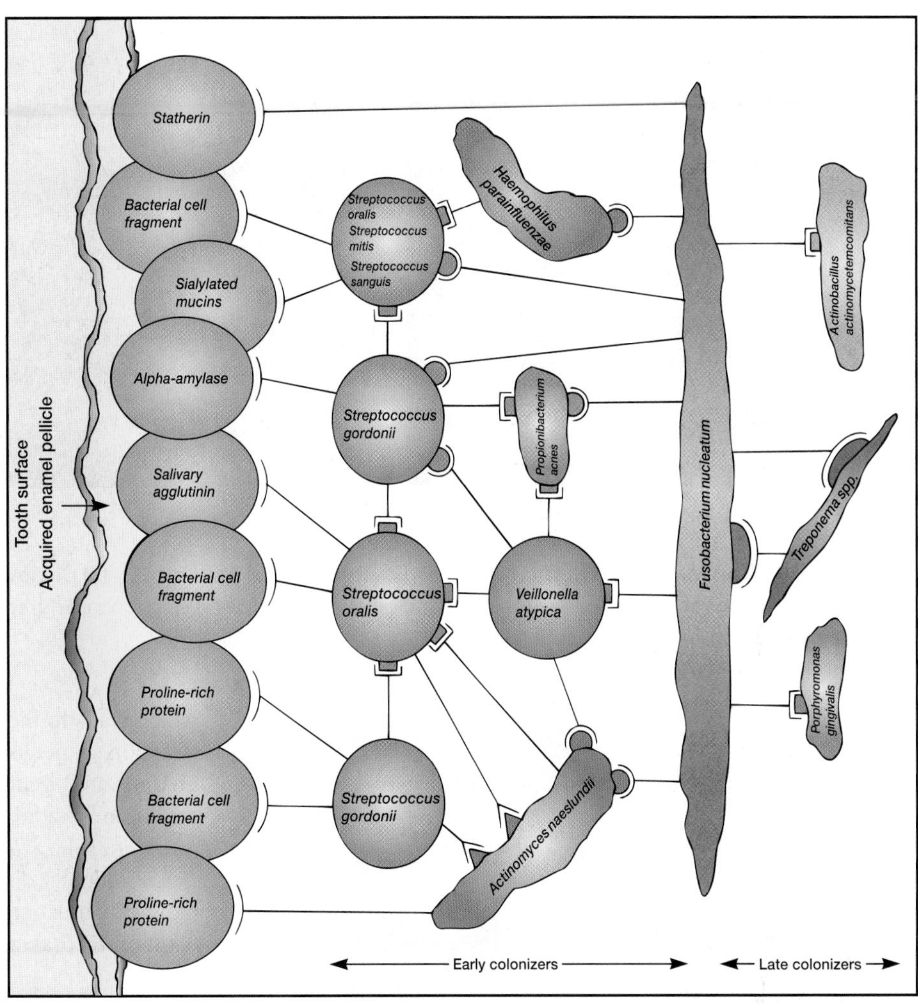

subsurface area. Between meals and snacks, the pH returns to neutrality and some calcium phosphate reenters the lesion and crystallizes. The result is a demineralization-remineralization cycle.

When an individual eats fermentable foods high in sucrose for prolonged periods, acid production overwhelms the repair process and demineralization is greater than remineralization. This leads to dental decay or **caries** [Latin, rottenness]. Once the hard enamel has been breached, bacteria can invade the dentin and pulp of the tooth and cause its death.

No drugs are available to prevent dental caries. The main strategies for prevention include minimal ingestion of sucrose; daily brushing, flossing, and rinsing with mouthwashes; and professional cleaning at least twice a year to remove plaque. The use of fluorides in toothpaste, drinking water, and mouthwashes, or fluoride and sealants applied professionally to the teeth, protects against lactic and acetic acids and reduces tooth decay.

Periodontal Disease

Periodontal disease refers to a diverse group of inflammatory diseases that affect the periodontium, and is the most common chronic infection in adults. The **periodontium** is the supporting structure of a tooth and includes the cementum, the periodontal membrane, the bones of the jaw, and the gingivae (gums). The gingiva is dense fibrous tissue and its overlying mucous membrane that surrounds the necks of the teeth. The gingiva helps to hold the teeth in place. Disease is initiated by the formation of **subgingival plaque,** the plaque that forms at the dentogingival margin and extends down into the gingival tissue. Colonization of the subgingival region is aided by the ability of *Porphyromonas gingivalis* to adhere to substrates such as adsorbed salivary molecules, matrix proteins, epithelial cells, and bacteria in biofilms on teeth and epithelial surfaces. Binding to these substrates is mediated by *P. gingivalis* fimbrillin, the structural subunit of the major fimbriae. *P. gingivalis* does not use sugars as an energy source, but requires hemin as a source of iron and peptides for energy and growth. The bacterium produces at least three hemagglutinins and five proteases to satisfy these requirements. It is the proteases that are responsible for the breakdown of the gingival tissue. A number of other bacterial species contribute to tissue damage. The result is an initial inflammatory reaction known as **periodontitis,** which is caused by the host's immune response to both

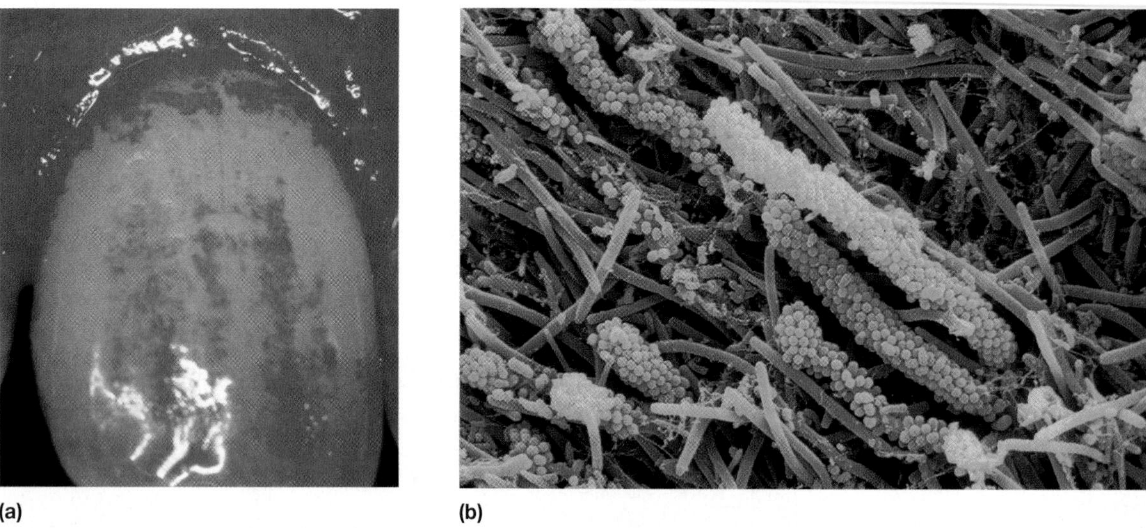

Figure 38.30 The Macroscopic and Microscopic Appearance of Plaque. **(a)** Disclosing tablets containing vegetable dye stain heavy plaque accumulations at the junction of the tooth and gingiva. **(b)** Scanning electron micrograph of plaque with long filamentous forms and "corn cobs" that are mixed bacterial aggregates.

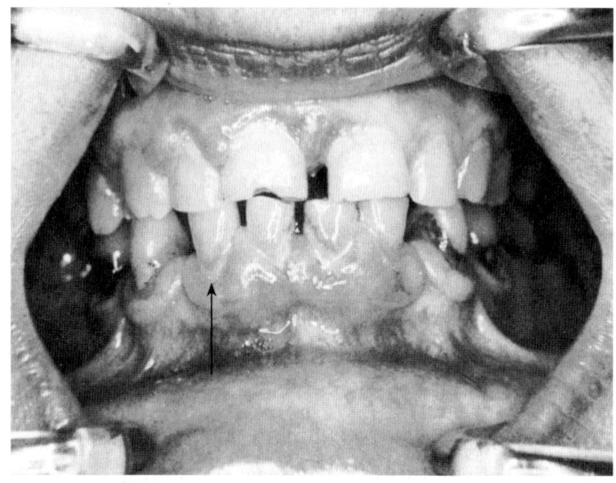

Figure 38.31 Periodontal Disease. Notice the plaque on the teeth (arrow), especially at the gingival (gum) margins, and the inflamed gingiva.

the plaque bacteria and the tissue destruction. This leads to swelling of the tissue and the formation of periodontal pockets. Bacteria colonize these pockets and cause more inflammation, which leads to the formation of a periodontal abscess; bone destruction, or **periodontosis;** inflammation of the gingiva, or **gingivitis;** and general tissue necrosis **(figure 38.31)**. If the condition is not treated, the tooth may fall out of its socket. Phylum *Bacteroidetes* (section 21.7)

Periodontal disease can be controlled by frequent plaque removal; by brushing, flossing, and rinsing with mouthwashes; and at times, by oral surgery of the gums and antibiotics.

1. Define sepsis and septic shock. How are microorganisms thought to cause shock?
2. How can humans acquire anthrax? Brucellosis?
3. Describe the symptoms of the disease as related to the infection process for anthrax and brucellosis.
4. How is ornithosis transmitted?
5. Describe the disease of tularemia. Why is tularemia considered a potential agent of bioterrorism?
6. Name some common odontopathogens that are responsible for dental caries, dental plaque, and periodontal disease. Be specific.
7. How does plaque formation occur? Dental decay?
8. How can caries and periodontal diseases be prevented?

Summary

Although only a small percentage of all bacteria are responsible for human illness, the suffering and death they cause are significant. Each year, millions of people are infected by pathogenic bacteria using the four major modes of transmission: airborne, arthropod-borne, direct contact, and food-borne and waterborne. As the fields of microbiology, immunology, pathology, pharmacology, and epidemiology have expanded current understanding of the disease process, the incidence of many human illnesses has decreased. Many bacterial infections, once leading causes of death, have successfully been brought under control in most developed countries. Alternatively, several are increasing in incidence throughout the world. The bacteria emphasized in this chapter and the diseases they cause are as follows:

38.1 Airborne Diseases

a. A number of infectious diseases are caused by bacteria as a result of their transmission through air. These include: chlamydial pneumonia (*Chlamydophilia pneumoniae*), diphtheria (*Corynebacterium diphtheriae*) (**figure 38.1**), Legionnaires' disease and Pontiac fever (*Legionella pneumophila*), meningitis (*Haemophilus influenzae* type b), *Neisseria meningitidis,* and *Streptococcus pneumoniae*, *M. avium-M. intracellulare* pneumonia and tuberculosis (*M. tuberculosis*) infection (**figure 38.3**), pertussis (*Bordetella pertussis*), mycoplasmal pneumonia (*Mycoplasma pneumoniae*), and streptococcal diseases (*Streptococcus* spp.) (**figures 38.5-38.7**).

38.2 Arthropod-Borne Diseases

a. Bacteria can also be transmitted to humans as a result of their interaction with arthopod vectors. Ehrlichiosis (*Ehrlichia chaffeensis*), epidemic (louse-borne) typhus (*Rickettsia prowazekii*), endemic (murine) typhus (*Rickettsia typhi*), Lyme disease (*Borrelia burgdorferi*) (**figure 38.8**), plague (*Yersinia pestis*) (**figure 38.9**), Q fever (*Coxiella burnetii*), and Rocky Mountain spotted fever (*Rickettsia rickettsii*) are transmitted to humans by ticks, lice, and fleas of animals.

38.3 Direct Contact Diseases

a. The direct contact of an uninfected human with sources of bacteria (including infected humans) can result in the transmission of bacteria and result in disease. Direct contact of skin, mucus membranes, open wounds or body cavities can lead to bacterial colonization and disease.

b. Examples of direct contact diseases caused by bacteria include: gas gangrene or clostridial myonecrosis (*Clostridium perfringens*) (**figure 38.11**); Group B streptococcal disease (*Streptococcus agalactiae*); inclusion conjunctivitis (*Chlamydia trachomatis*); leprosy (*Mycobacterium leprae*) (**figure 38.13**); peptic ulcer disease (*Helicobacter pylori*); staphylococcal diseases (*Staphylococcus aureus*) (**figures 38.16** and **38.17**); sexually transmitted diseases like bacterial vaginosis (*Gardnerella vaginalis*), chancroid (*Haemophilus ducreyi*), genitourinary mycoplasmal diseases (*Ureaplasma urealyticum, Mycoplasma hominis*), gonorrhea (*Neisseria gonorrhoeae*), lymphogranuloma venereum (*Chlamydia trachomatis*) (**figure 38.19**), nongonococcal urethritis (various microorganisms) and syphilis (*Treponema pallidum*) (**figure 38.20**); tetanus (*Clostridium tetani*), and trachoma (*Chlamydia trachomatis*) (**figure 38.22**).

38.4 Food-Borne and Waterborne Diseases

a. Food and water can serve as vehicles that transport bacteria to humans. Ingestion of contaminated food and water often results in infections of the gastrointestinal tract. Some common bacterial infectious diseases of the intestinal tract are botulism (*Clostridium botulinum*) (**figure 38.23**), gastroenteritis (*Campylobacter jejeuni* and other bacteria), cholera (*Vibrio cholerae*), listeriosis (*Listeria monocytogenes*), salmonellosis (*Salmonella* serovar Typhimurium) and typhoid fever (*Salmonella* serovar Typhi), shigellosis (*Shigella* spp.) staphylococcal food poisoning (*Staphylococcus aureus*), and traveler's diarrhea (*Escherichia coli*) (**figure 38.25**).

b. Some diseases are caused by the bacterial action on the cells of the intestine. Other diseases result from bacterial toxins.

38.5 Sepsis and Septic Shock

a. Gram-positive bacteria, fungi, and endotoxin containing gram-negative bacteria can initiate the pathogenic cascade of sepsis leading to septic shock (**figure 38.26**).

b. Gram-negative sepsis is most commonly caused by *E. coli,* followed by *Klebsiella* spp., *Enterobacter* spp., and *Pseudomonas aeruginosa.*

38.6 Zoonotic Diseases

a. Diseases of animals that are transmitted to humans are called zoonoses (sing. zoonsis). Anthrax (*Bacillus anthracis*) (**figure 38.27**), brucellosis (*Brucella* species), ornithosis (*Chlamydophilia psittaci*), and tularemia (*Francisella tularensis*) are a few examples of bacterial zoonotic diseases.

38.7 Dental Infections

a. Dental plaque formation begins on a tooth with the initial colonization of the acquired enamel pellicle by *Streptococcus gordonii, S. oralis,* and *S. mitis.* Other bacteria then become attached and form a plaque ecosystem (**figure 38.28**). The bacteria produce acids that cause a chemical lesion on the tooth and initiate dental decay or caries.

b. Periodontal disease is a group of diverse clinical entities that affect the periodontium. Disease is initiated by the formation of subgingival plaque, which leads to tissue inflammation known as periodontitis and to periodontal pockets. Bacteria that colonize these pockets can cause an abscess, periodontosis, gingivitis, and general tissue necrosis (**figures 38.29-38.31**).

Key Terms

acquired enamel pellicle 991
anthrax 987
aseptic meningitis syndrome 950
atypical pneumonia 955
bacille Calmette-Guerin (BCG) 955
bacterial (septic) meningitis 950
bacterial vaginosis 971
botulism 979
Bright's disease 958
brucellosis 990
bubo 962
bubonic plague 962
campylobacteriosis 979
caries 993
caseous lesion 954
cellulitis 957
chancre 976
chancroid 971
chlamydial pneumonia 948
cholera 983
choleragen 983
clostridial myonecrosis 965
clue cells 971
coaggregation 991
congenital syphilis 976
cutaneous anthrax 988
cutaneous diphtheria 949
dental plaque 991
diffusely adhering *E. coli* (DAEC) 987
diphtheria 948
DPT (diphtheria-pertussis-tetanus) vaccine 949
effacing lesions 986
ehrlichiosis 960
endemic (murine) typhus 961
enteroaggregative *E. coli* (EAggEC) 987
enterohemorrhagic *E. coli* (EHEC) 987
enteroinvasive *E. coli* (EIEC) 986
enteropathogenic *E. coli* (EPEC) 986
enterotoxigenic *E. coli* (ETEC) 986
enterotoxin 979
epidemic (louse-borne) typhus 960
erysipelas 957
eschar 989
exfoliative toxin (exfoliatin) 970
food-borne infection 979
food intoxication 979
food poisoning 979
gas gangrene 965
gastritis 967
gastroenteritis 979
gastrointestinal anthrax 988
genital ulcer disease 971
Ghon complex 954
gingivitis 994
glomerulonephritis 958
glucans 991
gonococci 974
gonorrhea 974
group A streptococcus (GAS) 956
group B streptococcus (GBS) 965
gummas 976
Hansen's disease 966
hemolytic uremic syndrome 987
impetigo 957
inclusion conjunctivitis 966
Legionnaires' disease (legionellosis) 949

lepromatous (progressive) leprosy 966
leprosy 966
listeriosis 984
Lyme disease (LD, Lyme borreliosis) 961
lymphogranuloma venereum (LGV) 975
meningitis 950
miliary tuberculosis 954
multidrug-resistant strains of tuberculosis (MDR-TB) 954
myositis 957
necrotizing fasciitis 957
nongonococcal urethritis (NGU) 976
odontopathogens 991
ophthalmia neonatorum (conjunctivitis of the newborn) 975
opportunistic infection 958
ornithosis 991
pannus 979
pelvic inflammatory disease (PID) 975
peptic ulcer disease 967
periodontal disease 993
periodontitis 993
periodontium 993
periodontosis 994
pertussis 955
pharyngitis 958
plague 962
pneumonic plague 962
Pontiac fever 950
psittacosis 991
pulmonary anthrax 988
Q fever 964
reactivation tuberculosis 954
rheumatic fever 958
Rocky Mountain spotted fever 964
salmonellosis 984
sepsis 987
septic shock 987
shigellosis 985
slime 968
staphylococcal food poisoning 985
staphylococcal scalded skin syndrome (SSSS) 969
streptococcal pharyngitis 958
streptococcal pneumonia 958
subgingival plaque 993
superantigen 969
tetanolysin 978
tetanospasmin 978
tetanus 978
tonsillitis 958
toxic shocklike syndrome (TSLS) 958
toxic shock syndrome (TSS) 969
trachoma 978
transovarian passage 964
traveler's diarrhea 986
tubercles 954
tuberculoid (neural) leprosy 966
tuberculosis (TB) 951
tuberculous cavity 954
tularemia 991
typhoid fever 984
vasculitis 960
venereal syphilis 976

Critical Thinking Questions

1. Why is tetanus a concern only when one has a deep puncture-type wound and not a surface cut or abrasion?
2. Think about our modern, Western lifestyles. Can you name and describe bacterial diseases that result from this life of relative luxury? Refer to *Infections of Leisure,* second edition, edited by David Schlossberg (1999), published by the American Society for Microbiology Press.
3. You have been assigned the task of eradicating gonorrhea in your community. Explain how you would accomplish this.
4. You are a park employee. How would you prevent visitors from acquiring arthropod-borne diseases?
5. Visit the World Heath Organization website at www.who.org and identify an infectious disease that is a problem in a developing country but not in the United States. List the reasons why the disease is not controlled in the developing country but is in North America. What policies and initiatives would you implement if you were in charge of the reducing the mortality and morbidity rate of this disease?

Learn More

Davis, D. H., and Elzer, P. H. (2002). *Brucella* vaccines in wildlife. *Vet. Microbiol.* 90:533–44.

Fux, C. A.; Costerton, J. W.; Stewart, P. S.; and Stoodley, P. 2005. Survival strategies of infectious biofilms. *Trends Microbiol.* 13:34–40.

Golden, M. R., and Manhart, L. F. 2005. Innovative approaches to the prevention and control of bacterial sexually transmitted infections. *Infect. Dis. Clin. North Am.*19:513–40.

Jenkinson, H. F., and Lamont, R. J. 2005. Oral microbial communities in sickness and in health. *Trends Microbiol.* 13:590–95.

Kirn, T. J.; Jude, B. A.; and Taylor, R. K. 2005. A colonization factor links *Vibrio cholerae* environmental survival and human infection. *Nature* 438:863–66.

Liu, Y. M.; Chi, C. Y.; Ho, M. W.; Chen, C. M.; Liao, W. C.; Ho, C. M.; Lin, P. C.; and Wang, J. H. 2005. Microbiology and factors affecting mortality in necrotizing fasciitis. *J. Microbiol. Immunol. Infect.* 38:430–5.

Marketon, M. M.; DePaolo, R. W.; DeBord, K. L.; Jabri, B.; and Schneewind, O. 2005. Plague bacteria target immune cells during infection. *Science.* 309:1739–76.

Moayeri, M., and Leppla, S. H. 2004. The roles of anthrax toxin in pathogenesis. *Curr. Opin. Microbiol.* 7:19–24.

Oyston, P. C. F.; Sjostedt, A.; and Titball, R. W. 2004. Tularaemia: Bioterrorism defence renews interest in *Francisella tularensis. Nature Rev. Microbiol.* 2:967–72.

Russell, D. G.; Purdy, G. E.; Owens, R. M.; Rohde, K. H.; and Yates, R. M. 2005. *Mycobacterim tuberculosis* and the four-minute phagosome. *ASM News* 10: 459–63.

Soriani, M.; Santi, I.; Taddei, A.; Rappuoli, R.; Grandi, G.; and Telford, J. L. 2006. Group B streptococcus crosses human epithelial cells by a paracellular route. *J. Infect. Dis.* 193:241–50.

Smith, R. P. 2006. Current diagnosis and treatment of lyme disease. *Compr. Ther.* 31:284–90.

Taylor, Z.; Nolan, C. M.; Blumberg, H. M.; American Thoracic Society; Centers for Disease Control and Prevention; and Infectious Diseases Society of America. 2005. Controlling tuberculosis in the United States. Recommendations from the American Thoracic Society, CDC, and the Infectious Diseases Society of America. *MMWR Recomm. Rep.* 4:1–81.

Tomaso, H.; Bartling, C.; Al Dahouk, S.; Hagen, R. M.; Scholz, H. C.; Beyer, W.; and Neubauer, H. 2006. Growth characteristics of *Bacillus anthracis* compared to other *Bacillus* spp. on the selective nutrient media Anthrax Blood Agar® and Cereus Ident Agar®. *Syst. Appl. Microbiol.* 29:24–28.

Please visit the Prescott website at www.mhhe.com/prescott7 for additional references.

39 Human Diseases Caused by Fungi and Protists

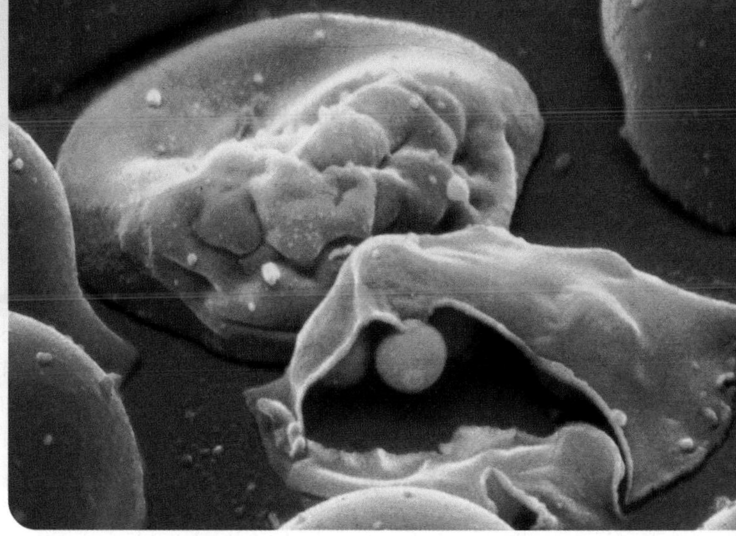

Malaria parasites (yellow) bursting out of red blood cells. Malaria is one of the worst scourges of humanity. Indeed, malaria has played an important part in the rise and fall of nations (see the chapter opening quote), and has killed untold millions the world over. Despite the combined efforts of 102 countries to eradicate malaria, it remains the most important disease in the world today in terms of lives lost and economic burden.

Preview

- Fungal diseases (mycoses) are usually divided into four groups according to the level of infected tissue and mode of entry into the host: (1) superficial, (2) cutaneous, (3) subcutaneous, and (4) systemic.
- About 20 different protists cause human diseases that afflict hundreds of millions of people throughout the world.
- While seemingly different, eucaryotic fungal and protist pathogens gain access to humans by transmission routes previously identified for bacteria and viruses: air, arthropods, direct contact, food, water, and the host itself.
- The systemic mycoses, which are typically transmitted through air, are the most serious of the fungal infections in the normal host because they can disseminate throughout the body. Examples include blastomycosis, coccidioidomycosis, cryptococcosis, and histoplasmosis.
- Fungal pathogens do not appear to be transmitted by arthropods; however, several notable protist pathogens are. These include *Leishmania, Plasmodium,* and *Trypanosoma.*
- A number of fungal and protist pathogens are transmitted by direct contact. Examples include the superficial mycoses, which occur mainly in the tropics and include black piedra, white piedra, and tinea versicolor; the cutaneous mycoses generally called ringworms, tineas, or dermatomycoses (occurring worldwide and representing the most common fungal diseases in humans); and the dermatophytes, which cause the subcutaneous mycoses (chromomycosis, maduromycosis, and sporotrichosis).
- Fungal and protist diseases can also be transmitted through food and water. Examples include amebiasis, cryptosporidiosis, and giardiasis.
- Opportunistic diseases typically arise from the endogenous microbial flora when the host can no longer control them. The opportunistic mycoses can create life-threatening situations in the compromised host. Examples of these diseases include aspergillosis, candidiasis, microsporidiosis and *Pneumocystis* pneumonia.

In this chapter we describe some of the fungi and protists that are pathogenic to humans and discuss the clinical manifestations, diagnosis, epidemiology, pathogenesis, and treatment of the diseases caused by them. The biology of these organisms is covered in chapters 25 and 26, respectively. This chapter follows the format of the previous two chapters in identifying their diseases by route of transmission.

39.1 Pathogenic Fungi and Protists

Fungi are eucaryotic saprophytes that are ubiquitous in nature. Although hundreds of thousands of fungal species are found in the environment, only about 50 produce disease in humans. **Medical mycology** is the discipline that deals with the fungi that cause human disease. These fungal diseases, known as **mycoses** [s., mycosis; Greek *mykes,* fungus], are typically divided into five groups according to the route of infection: superficial, cutaneous, subcutaneous, systemic, and opportunistic mycoses (**tables 39.1** and **39.2**). Superficial, cutaneous, and subcutaneous mycoses are direct contact infections of the skin, hair, and nails. Systemic mycoses are fungal infections that have disseminated to visceral tissues. Except for *Cryptococcus neoformans,* which has only a yeast form, the fungi that cause the systemic or deep mycoses are dimorphic—they exhibit a parasitic yeast-like phase (Y) and a saprophytic mold or mycelial phase (M). Most systemic mycoses are acquired by the inhalation of spores from soil in which the mold-phase of the fungus resides. If a susceptible person inhales enough spores, an infection begins as a lung lesion, becomes chronic, and spreads through the bloodstream to

Historians believe that malaria has probably had a greater impact on world history than any other infectious disease, influencing the outcome of wars, various population movements, and the development and decline of various civilizations.

—*Lynne S. Garcia*

Table 39.1 Examples of Some Medically Important Fungi

Group	Pathogen	Location	Disease
Superficial mycoses	*Piedraia hortae*	Scalp	Black piedra
	Trichosporon beigelii	Beard, mustache	White piedra
	Malassezia furfur	Trunk, neck, face, arms	Tinea versicolor
Cutaneous mycoses	*Trichophyton mentagrophytes, T. verrucosum, T. rubrum*	Beard hair	Tinea barbae
	Trichophyton, Microsporum canis	Scalp hair	Tinea capitis
	Trichophyton rubrum, T. mentagrophytes, Microsporum canis	Smooth or bare parts of the skin	Tinea corporis
	Epidermophyton floccosum, T. mentagrophytes, T. rubrum	Groin, buttocks	Tinea cruris (jock itch)
	T. rubrum, T. mentagrophytes, E. floccosum	Feet	Tinea pedis (athlete's foot)
	T. rubrum, T. mentagrophytes, E. floccosum	Nails	Tinea unguium (onychomycosis)
Subcutaneous mycoses	*Phialophora verrucosa, Fonsecaea pedrosoi*	Legs, feet	Chromoblastomycosis
	Madurella mycetomatis	Feet, other areas of body	Maduromycosis
	Sporothrix schenckii	Puncture wounds	Sporotrichosis
Systemic mycoses	*Blastomyces dermatitidis*	Lungs, skin	Blastomycosis
	Coccidioides immitis	Lungs, other parts of body	Coccidioidomycosis
	Cryptococcus neoformans	Lungs, skin, bones, viscera, central nervous system	Cryptococcosis
	Histoplasma capsulatum	Within phagocytes	Histoplasmosis
Opportunistic mycoses	*Aspergillus fumigatus, A. flavus*	Respiratory system	Aspergillosis
	Candida albicans	Skin or mucous membranes	Candidiasis
	Pneumocystis jiroveci	Lungs, sometimes brain	*Pneumocystis* pneumonia
	Encephalitozoon, Nosema, Vitta forma, Pleistophora, Enterocytozoon, Trachipleistophora, Microsporidium	Lungs, sometimes brain	Microsporidiosis

other organs (the target organ varies with the species). The *Fungi* (chapter 26)

Protozoa, single-celled eucaryotic chemoorganotrophs, have become adapted to practically every type of habitat on the face of the Earth, including the human body. Many protists are transmitted to humans by arthropod vectors or by food and water vehicles. However, some protozoan diseases are transmitted by direct contact. Although fewer than 20 genera of protists cause disease in humans (**tables 39.3** and **39.4**), their impact is formidable. For example, there are over 150 million cases of malaria in the world each year. In tropical Africa alone, malaria is responsible for the deaths of more than a million children under the age of 14 annually. It is estimated that there are at least 8 million cases of trypanosomiasis, 12 million cases of leishmaniasis, and over 500 million cases of amebiasis yearly. There is also an increasing problem with *Cryptosporidium* and *Cyclospora* contamination of food and water supplies (**table 39.5**). More of our population is elderly, and a growing number of persons are immunosuppressed due to HIV infection, organ transplantation, or cancer chemotherapy. These populations are at increased risk for protozoan infections. The protists (chapter 25)

While seemingly different, fungi and protists share a number of phenotypic features that serve them in their ability to cause infection: microscopic size, eucaryotic physiology, cell walls or wall-like structures, alternative stages for survival outside of the host, degradative enzymes, and others. Some fungi and protists also share transmission routes. We now discuss diseases of fungi and protists based on how they are acquired by the human host.

39.2 Airborne Diseases

Blastomycosis is the systemic mycosis caused by *Blastomyces dermatitidis,* a fungus that grows as a budding yeast in humans but as a mold on culture media and in the environment. It is found predominately in moist soil enriched with decomposing organic debris, as in the Mississippi and Ohio River basins. *B. dermatitidis* is endemic in parts of the south-central, southeastern and midwestern United States. Additionally, microfoci have been reported in Central and South America and in parts of Africa. The disease occurs in three clinical forms: cutaneous, pulmonary, and disseminated. The initial infection begins when blastospores are inhaled into the lungs. The fungus can then spread rapidly, especially to the skin, where cutaneous ulcers and abscess formation occur (**figure 39.1**). *B. dermatitidis* can be isolated from pus and biopsy sections. Diagnosis requires the demonstration of thick-walled, yeast-like cells, 8 to 15 μm in diameter. Complement-fixation, immunodiffusion, and skin tests (blastomycin) are also useful. Amphotericin B (Fungizone), itraconazole (Sporanox), or ketoconazole (Nizoral) are the drugs of choice for treatment. Surgery may be necessary for the drainage of large abscesses. Mortality is

Table 39.2 Examples of Some Human Fungal Diseases Recognized Since 1974

Year	Fungus	Disease
	Molds	
1974	*Phialophora parasitica*	Phaeohyphomycosis
1992	*Penicillium marneffei*	Disseminated infection
	Yeasts	
1985	*Enterocytozoon bieneusi*	Diarrhea, microsporidiosis
1989	*Candida lusitaniae*	Fungemia
1989	*Malassezia furfur*	Fungemia
1990	*Rhodotorula rubra*	Fungemia
1991	*Candida ciferrii*	Fungemia
1993	*Hansenula anomala*	Fungemia
1993	*Trichosporon beigelii*	Fungemia
1993	*Encephalitozoon cuniculi*	Disseminated microsporidiosis
1996	*Trachipleistophora hominis*	Disseminated microsporidiosis

Table 39.4 Examples of Human Protozoan Diseases Recognized Since 1976

Year	Protozoan	Disease
1976	*Cryptosporidium parvum*	Acute and chronic diarrhea, cryptosporidiosis
1986	*Cyclospora cayatanensis*	Persistent diarrhea
1991	*Babesia* spp.	Atypical babesiosis
1998	*Brachiola vesicularum*	Myositis

Table 39.3 Examples of Some Medically Important Protozoa

Morphological Group	Pathogen	Disease
Amoebae	*Entamoeba histolytica*	Amebiasis, amebic dysentery
	Acanthamoeba spp., *Naegleria fowleri*	Amebic meningoencephalitis
Apicomplexa	*Cryptosporidium parvum*	Cryptosporidiosis
	Cyclospora cayetanensis	Cyclosporidiosis
	Isospora belli	Isosporiasis
	Plasmodium falciparum, P. malariae, P. ovale, P. vivax	Malaria
	Toxoplasma gondii	Toxoplasmosis
Ciliates	*Balantidium coli*	Balantidiasis
Blood and tissue flagellates	*Leishmania tropica*	Cutaneous leishmaniasis
	L. braziliensis	Mucocutaneous leishmaniasis
	L. donovani	Kala-azar (visceral leishmaniasis)
	Trypanosoma cruzi	American trypanosomiasis
	T. brucei gambiense, T. brucei rhodesiense	African sleeping sickness
Digestive and genital organ flagellates	*Giardia intestinalis*	Giardiasis
	Trichomonas vaginalis	Trichomoniasis

Table 39.5 Water-Based Protozoan Pathogens That Can Be Maintained in the Environment Independent of Humans

Organism	Reservoir	Comments
Acanthamoeba	Sewage sludge disposal areas	Can cause granulomatous amebic encephalitis (GAE); keratitis, corneal ulcers
Cryptosporidium	Many species of domestic and wild animals	Causes acute enterocolitis; important with immunologically compromised individuals; cysts resistant to chemical disinfection; not antibiotic sensitive
Cyclospora cayetanensis	Waters—does not withstand drying; possibly other reservoirs	Causes long-lasting (43 days average) diarrheal illness; infection self-limiting in immunocompetent hosts; sensitive to prompt treatment with sufonamide and trimethoprim
Giardia intestinalis	Beavers, sheep, dogs, cats	Major cause of early spring diarrhea; important in cold mountain water
Naegleria fowleri	Warm water (hot tubs), swimming pools, lakes	Inhalation in nasal passages; central nervous system infection; causes primary amebic meningoencephalitis (PAM)

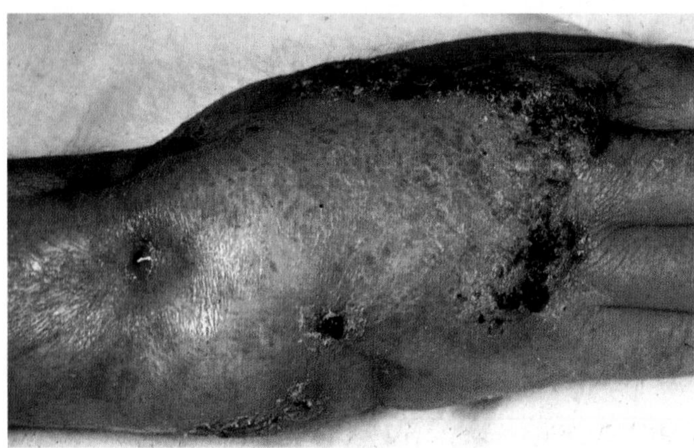

Figure 39.1 Systemic Mycosis. Blastomycosis of the forearm caused by *Blastomyces dermatitidis.*

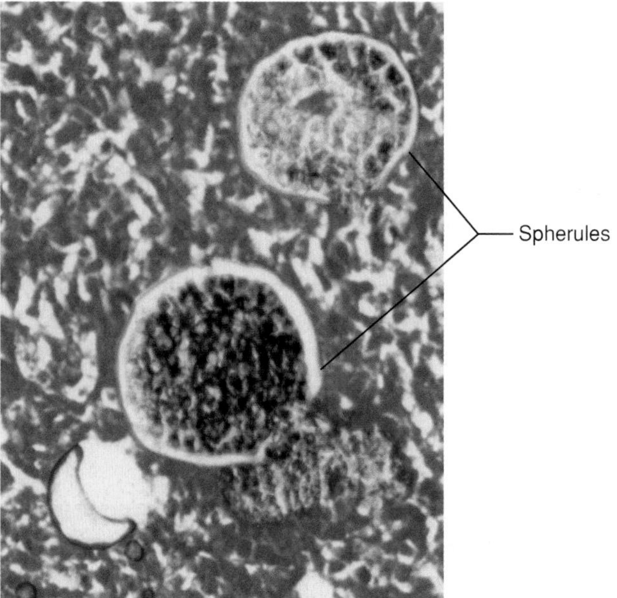

Figure 39.2 Systemic Mycosis: Coccidioidomycosis. *Coccidioides immitis* mature spherules filled with endospores within a tissue section; light micrograph (×400).

about 5%. There are no preventive or control measures. Antifungal drugs (section 34.7)

Coccidioidomycosis, also known as valley fever, San Joaquin fever, or desert rheumatism because of the geographical distribution of the fungus, is caused by *Coccidioides immitis. C. immitis* exists in the semi-arid, highly alkaline soils of the southwestern United States and parts of Mexico and South America. It has been estimated that in the United States about 100,000 people are infected annually, resulting in 50 to 100 deaths. Endemic areas have been defined by massive skin testing with the antigen coccidioidin, where 10 to 15% positive cases are reported. In the soil and on culture media, this fungus grows as a mold that forms arthroconidia at the tips of hyphae (*see figure 26.8*). Because arthroconidia are so abundant in these endemic areas, immunocompromised individuals can acquire the disease by inhalation as they simply move through the area. Wind turbulence and even construction of outdoor structures have been associated with increased exposure and infection. In humans, the fungus grows as a yeast-forming, thick-walled spherule filled with spores (**figure 39.2**). Most cases of coccidioidomycosis are asymptomatic or indistinguishable from ordinary upper respiratory infections. Almost all cases resolve in a few weeks, and a lasting immunity results. A few infections result in a progressive chronic pulmonary disease or disseminated infections of other tissues; the fungus can spread throughout the body, involving almost any organ or site.

Diagnosis is accomplished by identification of the large spherules (approximately 80 μm in diameter) in pus, sputum, and aspirates. Culturing clinical samples in the presence of penicillin and streptomycin on Sabouraud agar (used to isolate fungi) also is diagnostic. Newer methods of rapid confirmation include the testing of supernatants of liquid media cultures for antigens, serology, and skin testing. Miconazole (Lotrimin), itraconazole, ketoconazole, and amphotericin B are the drugs of choice for treatment. Prevention involves reducing exposure to dust (soil) in endemic areas. Antifungal drugs (section 34.7); Identification of microorganisms from specimens (section 35.2)

Cryptococcosis is a systemic mycosis caused by *Cryptococcus neoformans*. This fungus always grows as a large, budding yeast. In the environment, *C. neoformans* is a saprophyte with a worldwide distribution. Aged, dried pigeon droppings are an apparent source of infection. Cryptococcosis is found in approximately 15% of AIDS patients. The fungus enters the body by the respiratory tract, causing a minor pulmonary infection that is usually transitory. Some pulmonary infections spread to the skin, bones, viscera, and the central nervous system. Once the nervous system is involved, cryptococcal meningitis usually results. Diagnosis is accomplished by detection of the thick-walled, spherical yeast cells in pus, sputum, or exudate smears using India ink to define the organism (**figure 39.3**). The fungus can be easily cultured on Sabouraud dextrose agar. Identification of the fungus in body fluids is made by immunologic procedures. Treatment includes amphotericin B or itraconazole. There are no preventive or control measures.

Histoplasmosis is caused by *Histoplasma capsulatum* var. *capsulatum*, a facultative, parasitic fungus that grows intracellularly. It appears as a small, budding yeast in humans and on culture media at 37°C. At 25°C it grows as a mold, producing small microconidia (1 to 5 μm in diameter) that are borne singly at the tips of short conidiophores (*see figure 26.8*). Large macroconidia (8 to 16 μm in diameter) are also formed on conidiophores (**figure 39.4*a***). In humans, the yeastlike form grows within phagocytic cells (figure 39.4*b*). *H. capsulatum* var. *capsulatum* is found as the mycelial form in soils throughout the world and is localized in areas that have been contaminated with bird or bat excrement. The microconidia can become airborne when contaminated soil is disturbed. Infection ensues when the microconidia are inhaled. Histoplasmosis is not, however, transmitted from an infected person. Within the United States, histoplasmosis is endemic within the Mississippi, Kentucky, Tennessee, Ohio, and Rio Grande River basins. More than 80% of the people who reside in parts of these areas have antibodies against the fungus. It has been estimated that in endemic areas of the United States, about 500,000 individuals are infected annually: 50,000 to 200,000 become ill; 3,000 require hospitalization; and about 50 die. The total number of infected individuals may be over 40 million in the United States alone. Histoplasmosis is a common disease among poultry farmers, spelunkers (people who explore caves), and bat guano miners (bat guano is used as fertilizer).

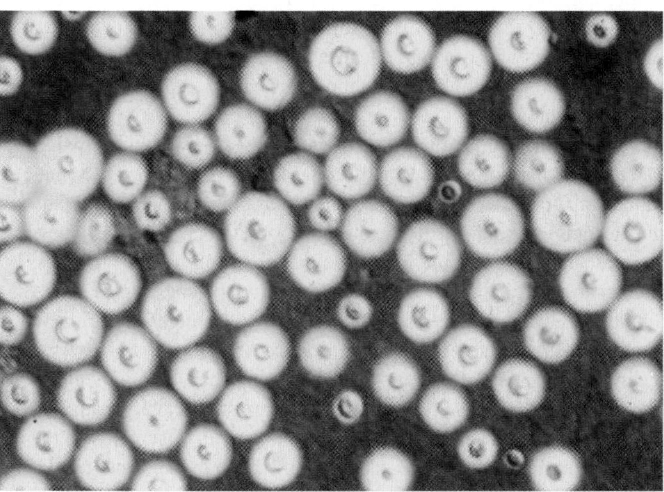

Figure 39.3 Systemic Mycosis: Cryptococcosis. India ink preparation showing *Cryptococcus neoformans*. Although these microorganisms are not budding, they can be differentiated from artifacts by their doubly refractile cell walls, distinctly outlined capsules surrounding all cells, and refractile inclusions in the cytoplasm; light micrograph (×150).

Humans acquire histoplasmosis from airborne microconidia that are produced under favorable environmental conditions. Microconidia are most prevalent where bird droppings—especially from starlings, crows, blackbirds, cowbirds, sea gulls, turkeys, and chickens—have accumulated. It is noteworthy that the birds themselves are not infected because of their high body temperature; their droppings simply provide the nutrients for this fungus. Only bats and humans demonstrate the disease and harbor the fungus.

Because histoplasmosis is a disease of the monocyte-macrophage system, many organs of the body can be infected (*see figure 31.3*). More than 95% of "histo" cases have either no symptoms or mild symptoms such as coughing, fever, and joint pain. Lesions may appear in the lungs and show calcification; the disease may resemble tuberculosis. Most infections resolve on their own. Only rarely does the disease disseminate. Laboratory diagnosis is accomplished by complement-fixation tests and isolation of the fungus from tissue specimens. Most individuals with this disease exhibit a hypersensitive state that can be demonstrated by the histoplasmin skin test. Currently the most effective treatment is with amphotericin B, ketoconazole, or itraconazole. Prevention and control involve wearing protective clothing and masks before entering or working in infested habitats. Soil decontamination with 3 to 5% formalin is effective where economically and physically feasible. Antifungal drugs (section 34.7)

39.3 Arthropod-Borne Diseases

Malaria

The most important human parasite among the protozoa is *Plasmodium*, the causative agent of **malaria** (**Disease 39.1**). It has been estimated that more than 300 million people are infected each year, and over 1 million die annually of malaria in Africa alone. About 1,000 cases are reported each year in the United States, divided between returning U.S. travelers and non-U.S.

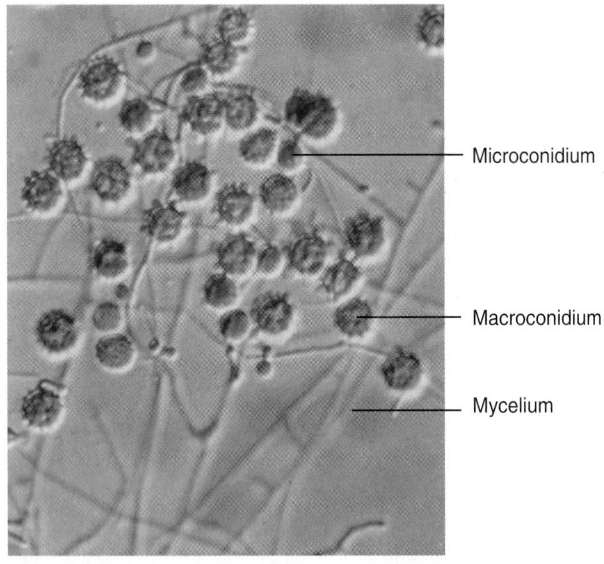

(a)

Figure 39.4 Morphology of *Histoplasma capsulatum* var. *capsulatum*. **(a)** Mycelia, microconidia, and chlamydospores as found in the soil. These are the infectious particles; light micrograph (×125). **(b)** Yeastlike cells in a macrophage. Budding *H. capsulatum* within a vacuole. Tubular structures, ts, are observed beneath the cell wall, cw; electron micrograph (×23,000).

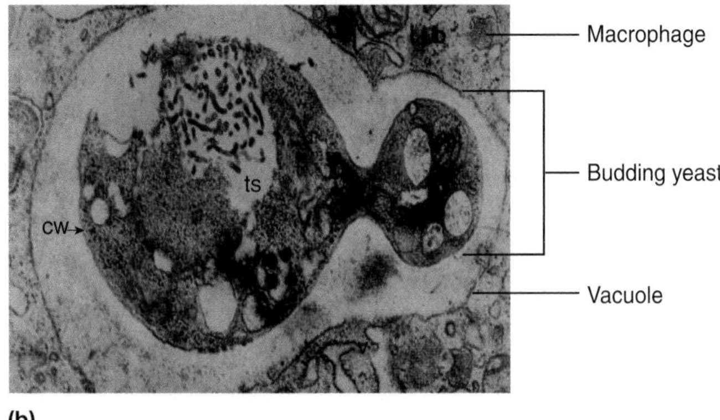

(b)

Disease

39.1 A Brief History of Malaria

No other single infectious disease has had the impact on humans that malaria has. The first references to its periodic fever and chills can be found in early Chaldean, Chinese, and Hindu writings. In the late 5th century B.C., Hippocrates described certain aspects of malaria. In the 4th century B.C., the Greeks noted an association between individuals exposed to swamp environments and the subsequent development of periodic fever and enlargement of the spleen (splenomegaly). In the 17th century the Italians named the disease *mal' aria* (bad air) because of its association with the ill-smelling vapors from the swamps near Rome. At about the same time, the bark of the quinaquina (cinchona) tree of South America was used to treat the intermittent fevers, although it was not until the mid-19th century that quinine was identified as the active alkaloid. The major epidemiological breakthrough came in 1880, when French army surgeon Charles Louis Alphonse Laveran observed gametocytes in fresh blood. Five years later the Italian histologist Camillo Golgi observed the multiplication of the asexual blood forms. In the late 1890s Patrick Manson postulated that malaria was transmitted by mosquitoes. Sir Ronald Ross, a British army surgeon in the Indian Medical Service, subsequently observed developing plasmodia in the intestine of mosquitoes, supporting Manson's theory. Using birds as experimental models, Ross definitively established the major features of the life cycle of *Plasmodium* and received the Nobel Prize in 1902.

Human malaria is known to have contributed to the fall of the ancient Greek and Roman empires. Troops in both the U.S. Civil War and the Spanish-American War were severely incapacitated by the disease. More than 25% of all hospital admissions during these wars were malaria patients. During World War II malaria epidemics severely threatened both the Japanese and Allied forces in the Pacific. The same can be said for the military conflicts in Korea and Vietnam.

In the 20th century efforts were directed toward understanding the biochemistry and physiology of malaria, controlling the mosquito vector, and developing antimalarial drugs. In the 1960s it was demonstrated that resistance to *P. falciparum* among West Africans was associated with the presence of hemoglobin-S (Hb-S) in their erythrocytes. Hb-S differs from normal hemoglobin-A by a single amino acid, valine, in each half of the Hb molecule. Consequently these erythrocytes—responsible for sickle cell disease—have a low binding capacity for oxygen. Because the malarial parasite has a very active aerobic metabolism, it cannot grow and reproduce within these erythrocytes.

In 1955 the World Health Organization began a worldwide malarial eradication program that finally collapsed by 1976. Among the major reasons for failure were the development of resistance to DDT by the mosquito vectors and the development of resistance to chloroquine by strains of *Plasmodium.* Scientists are exploring new approaches, such as the development of vaccines and more potent drugs. For example, in 1984 the gene encoding the sporozoite antigen was cloned, permitting the antigen to be mass-produced by genetic engineering techniques. In 2002, the complete DNA sequences of *P. falciparum* and *Anopheles gambiae* (the mosquito that most efficiently transmits this parasite to humans in Africa) were determined. Together with the human genome sequence, researchers now have in hand the genetic blueprints for the parasite, its vector, and its victim. This has made possible a holistic approach to understanding how the parasite interacts with the human host, leading to new antimalarial strategies including vaccine design. Overall, no greater achievement for molecular biology could be imagined than the control of malaria—a disease that has caused untold misery throughout the world since antiquity and remains one of the world's most serious infectious diseases.

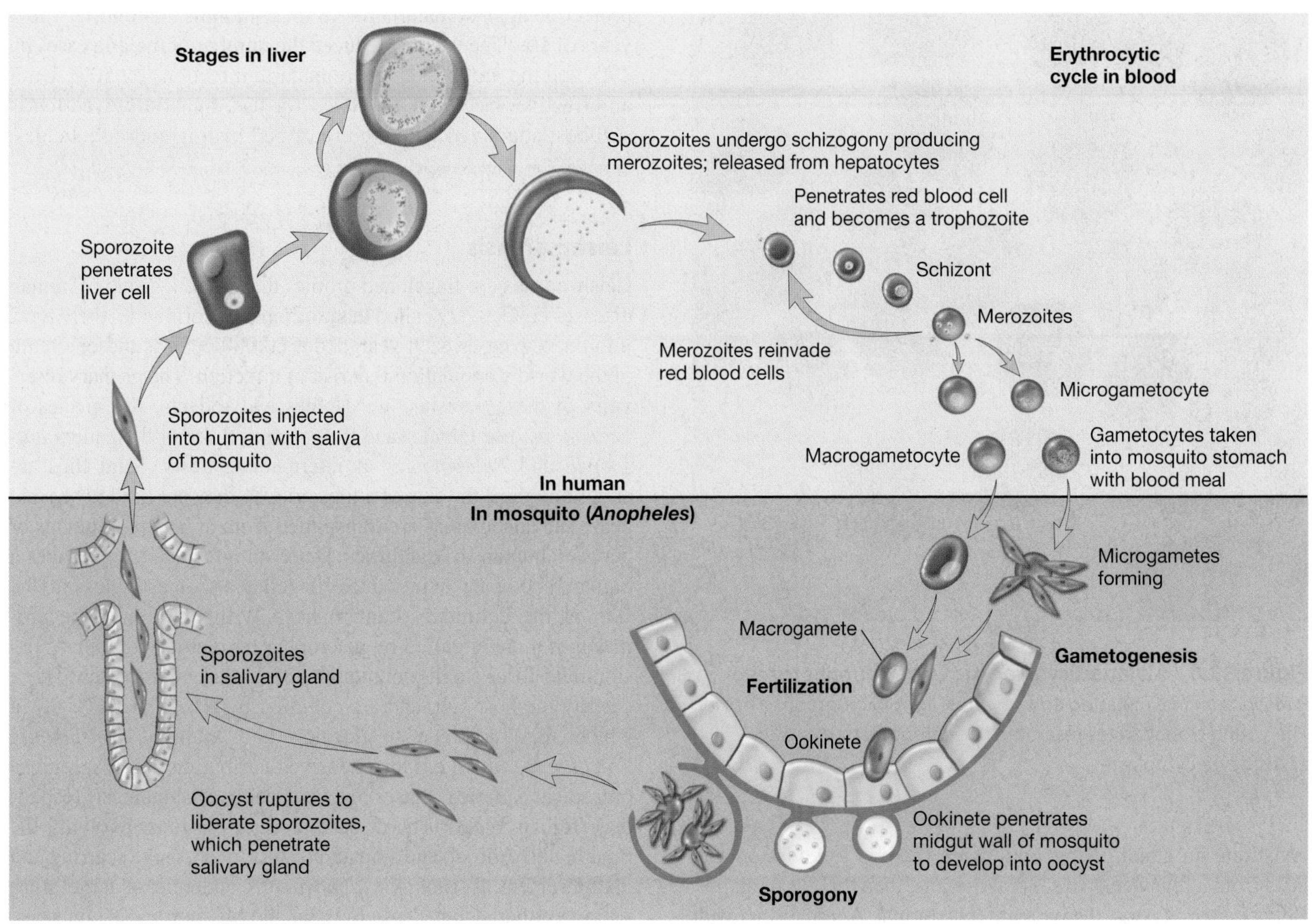

Figure 39.5 Malaria. Life cycle of *Plasmodium vivax.*

citizens. Human malaria is caused by four species of *Plasmodium: P. falciparum, P. malariae, P. vivax,* and *P. ovale*. The life cycle of *P. vivax* is shown in **figure 39.5.** The parasite first enters the bloodstream through the bite of an infected female *Anopheles* mosquito. As she feeds, the mosquito injects a small amount of saliva containing an anticoagulant along with small haploid sporozoites. The sporozoites in the bloodstream immediately enter hepatic cells of the liver. In the liver they undergo multiple asexual fission (schizogony) and produce merozoites. After being released from the liver cells, the merozoites attach to erythrocytes and penetrate these cells. Protists classification: *Alveolata* (section 25.6)

Once inside the erythrocyte, the *Plasmodium* begins to enlarge as a uninucleate cell termed a trophozoite. The trophozoite's nucleus then divides asexually to produce a schizont that has 6 to 24 nuclei. The schizont divides and produces mononucleated merozoites. Eventually the erythrocyte lyses, releasing the merozoites into the bloodstream to infect other erythrocytes. This erythrocytic stage is cyclic and repeats itself approximately every 48 to 72 hours or longer, depending on the species of *Plasmodium* involved. The sudden release of merozoites, toxins, and erythrocyte debris triggers an attack of the chills, fever, and sweats characteristic of malaria. Occasionally, merozoites differentiate into macrogametocytes and microgametocytes, which do not rupture the erythrocyte. When these are ingested by a mosquito, they develop into female and male gametes, respectively. In the mosquito's gut, the infected erythrocytes lyse and the gametes fuse to form a diploid zygote called the ookinete. The ookinete migrates to the mosquito's gut wall, penetrates, and forms an oocyst. In a process called sporogony, the oocyst undergoes meiosis and forms sporozoites, which migrate to the salivary glands of the mosquito. The cycle is now complete, and when the mosquito bites another human host, the cycle begins anew.

The pathological changes caused by malaria involve not only the erythrocytes but also the spleen and other visceral organs. Classic symptoms first develop with the synchronized release of merozoites and erythrocyte debris into the bloodstream, resulting in the malarial paroxysms—shaking chills, then burning fever followed by sweating. It may be that the fever and chills are caused partly by a malarial toxin that induces macrophages to release TNF-α and interleukin-1. Several of these paroxysms

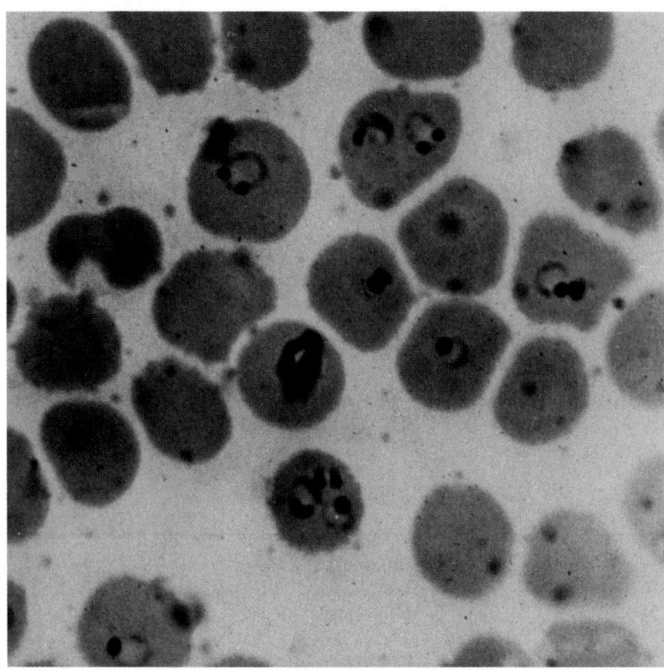

Figure 39.6 Malaria: Erythrocytic Cycle. Trophozoites of *P. falciparum* in circulating erythrocytes; light micrograph (×1,100). The young trophozoites resemble small rings resting in the erythrocyte cytoplasm.

constitute an attack. After one attack there is a remission that lasts from a few weeks to several months, then there is a relapse. Between paroxysms, the patient feels normal. Anemia can result from the loss of erythrocytes, and the spleen and liver often hypertrophy. Children and nonimmune individuals can die of cerebral malaria. Chemical mediators in nonspecific (innate) resistance: Cytokines (section 31.6)

Diagnosis of malaria is made by demonstrating the presence of parasites within Wright- or Giemsa-stained erythrocytes (**figure 39.6**). When blood smears are negative, serological testing can establish a diagnosis of malaria. Outside the United States, rapid diagnostic tests using species-specific antibodies are used for the diagnosis of malaria; these tests are not approved for use by the FDA. Specific recommendations for treatment are region-dependent. Treatment includes administration of chloroquine, amodiaquine, or mefloquine. These drugs suppress protozoan reproduction and are effective in eradicating erythrocytic asexual stages. Primaquine has proved satisfactory in eradicating the exoerythrocytic stages. However, because resistance to these drugs is occurring rapidly, more expensive drug combinations are now being used. One example is Fansidar, a combination of pyrimethamine and sulfadoxine. It is worth noting that individuals who are traveling to areas where malaria is endemic should receive chemoprophylactic treatment with chloroquine (**figure 39.7**). A credible vaccine against malaria was reported in 2005. Clinical trial data showed that Mosquirix provided partial protection against malaria for 18 to 21 months, in children 1 to 4 years of age. The vaccine reduced the number of malaria cases by 29% and the number of severe malaria infections by 50%. Until an effective vaccine is approved for use, malaria prevention is still best attempted with the use of bed netting and insecticides. Identification of microorganisms from specimens (section 35.2)

Leishmaniasis

Leishmanias are flagellated protists that cause a group of human diseases collectively called **leishmaniasis.** Worldwide, there are 2 million new cases each year, about 60,000 deaths, and one-tenth of the world's population is at risk of infection. The primary reservoirs of these parasites are canines and rodents. All species of *Leishmania* use female sand flies such as those of the genera *Lutzomyia* and *Phlebotomus* as intermediate hosts. Sand flies are about one-third the size of a mosquito so they are hard to see and hear. The leishmanias are transmitted from animals to humans or between humans by sand flies. When an infected sand fly takes a human blood meal, it introduces flagellated promastigotes into the skin of the definitive (human) host. Within the skin, the promastigotes are engulfed by macrophages, multiply by binary fission, and form small, nonmotile cells called amastigotes. These destroy the host cell, and are engulfed by other macrophages in which they continue to develop and multiply. *Leishmania braziliensis,* which has an extensive distribution in forest regions of tropical America, causes mucocutaneous leishmaniasis (espundia) (**figure 39.8*a***). The disease produces lesions involving the mouth, nose, throat, and skin and results in extensive scarring and disfigurement. *Leishmania donovani* is endemic in large areas within northern China, eastern India, the Mediterranean countries, the Sudan, and Latin America. It produces visceral leishmaniasis (kala-azar), which involves the monocyte-macrophage system and often results in intermittent fever and enlargement of the spleen and liver. Individuals who recover develop a permanent immunity. Protists classification: *Euglenozoa* (section 25.6)

Leishmania tropica and *L. mexicana* occur in the more arid regions of the Eastern Hemisphere and cause cutaneous leishmaniasis. *L. mexicana* is also found in the Yucatan Peninsula (Mexico) and has been reported as far north as Texas. In this disease a relatively small, red papule forms at the site of each insect bite—the inoculation site. These papules are frequently found on the face and ears. They eventually develop into crustated ulcers (figure 39.8*b*). Healing occurs with scarring and a permanent immunity.

Laboratory diagnosis of leishmaniasis is based on finding the parasite within infected macrophages in stained smears from lesions or infected organs. Culture and serological tests are also available for diagnosis of leishmaniasis. Treatment includes pentavalent antimicrobial compounds (Pentostam, Glucantime). Vector and reservoir control and aggressive epidemiological surveillance are the best options for prevention and containment of this disease. To date, there are no approved vaccines against leishmaniasis, although one experimental vaccine induces a 30%

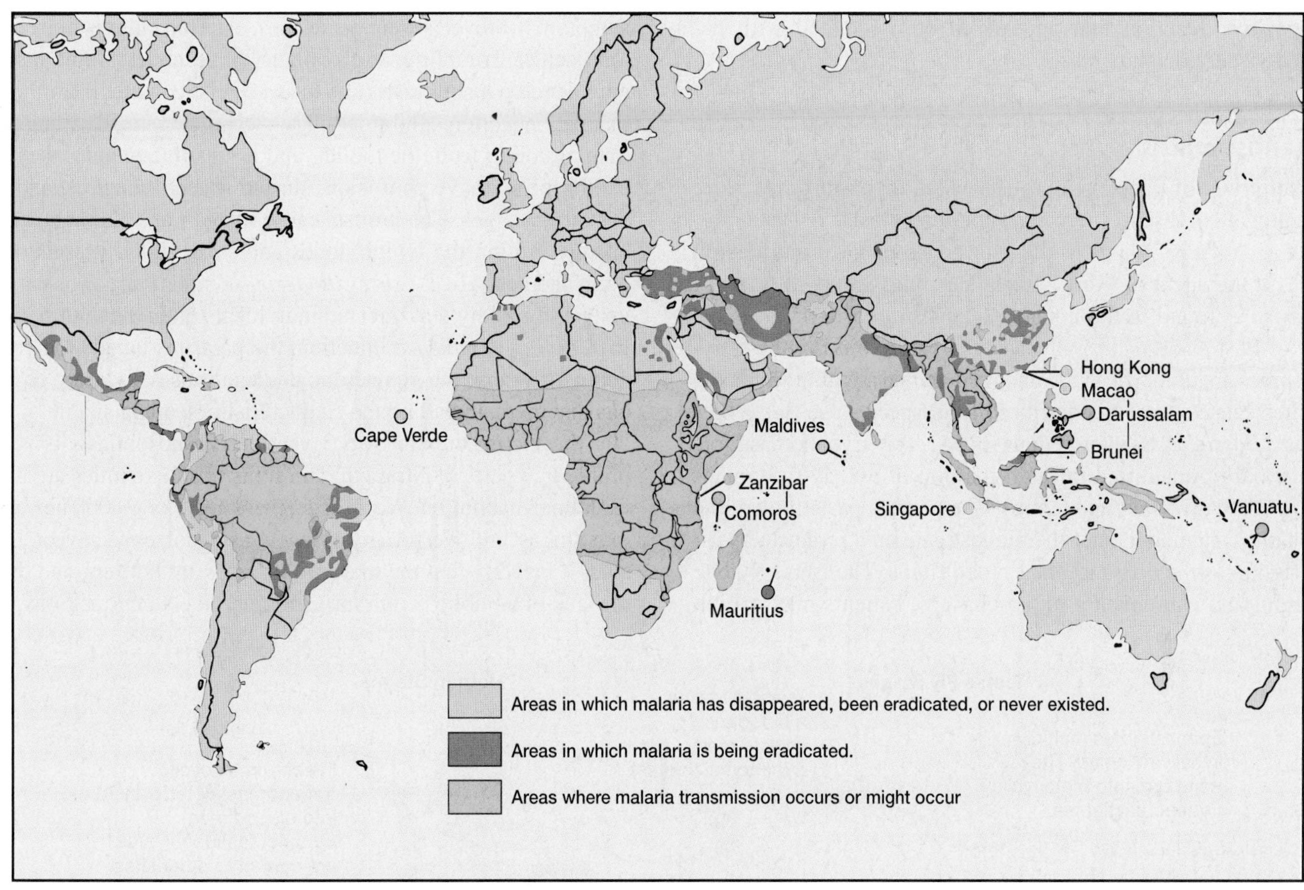

Figure 39.7 Geographic Distribution of Malaria. Notice that malaria is endemic around the equator. *Source: Data from the* World Health Statistics Quarterly, *41:69, 1988, World Health Organization, Switzerland.*

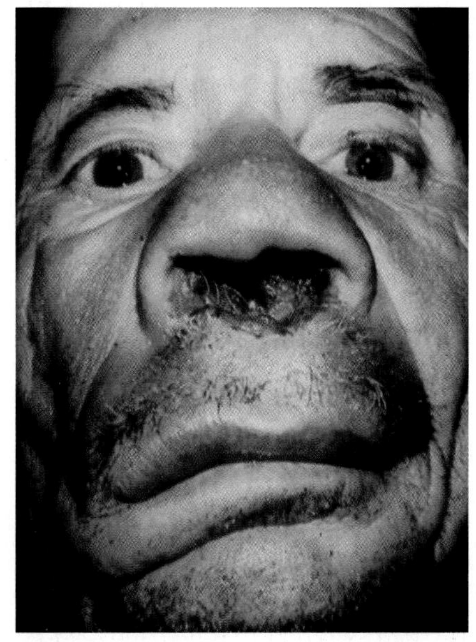

(a)

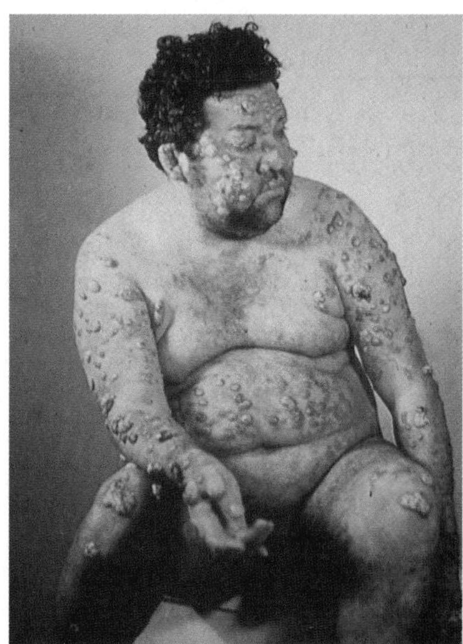

(b)

Figure 39.8 Leishmaniasis. **(a)** A person with mucocutaneous leishmaniasis, which has destroyed nasal septum and deformed the nose and lips. **(b)** A person with diffuse cutaneous leishmaniasis.

skin-positive reaction that appears to be associated with decreased visceral leishmaniasis.

Trypanosomiasis

Another group of flagellated protists called **trypanosomes** cause the aggregate of diseases termed **trypanosomiasis.** *Trypanosoma brucei gambiense,* found in the rainforests of west and central Africa, is the agent of West African sleeping sickness. *T. brucei rhodesiense,* found in the upland savannas of east Africa, is the agent of East African sleeping sickness. Reservoirs for these trypanosomes are domestic cattle and wild animals, within which the parasites cause severe malnutrition. Both species use tsetse flies (genus *Glossina*) as intermediate hosts. The trypanomastigote parasites are transmitted through the bite of the fly to humans (**figure 39.9,** *also see figure 25.6*). The protists pass through the lymphatic system and enter the bloodstream, and replicate by binary fission as they pass to other blood fluids. The tsetse fly bite is painful and can develop into a chancre. Patients may exhibit symptoms of fever, severe headaches, extreme fatigue, muscle and joint aches, irritability, and swollen lymph nodes. Some patients may develop a skin rash. The disease gets its name from the fact that patients often exhibit lethargy—characteristically lying prostrate, drooling from the mouth, and insensitive to pain; they also exhibit progressive confusion, slurred speech, seizures, and personality changes. The protists cause interstitial inflammation and necrosis within the lymph nodes and small blood vessels of the brain and heart. In *T. brucei rhodesiense* infection, the disease develops so rapidly that infected individuals often die within a year. In *T. brucei gambiense* infection, the parasites invade the central nervous system, where necrotic damage causes a variety of nervous disorders, including the characteristic sleeping sickness. Usually the victim dies in 2 to 3 years. Trypanosomiasis is such a problem in parts of Africa that millions of square miles are not fit for human habitation. Worldwide, there are over 40,000 new cases of both East and West African sleeping sicknesses each year. However, it is likely that the majority of cases are not reported due to the lack of a public health infrastructure in endemic regions. As a

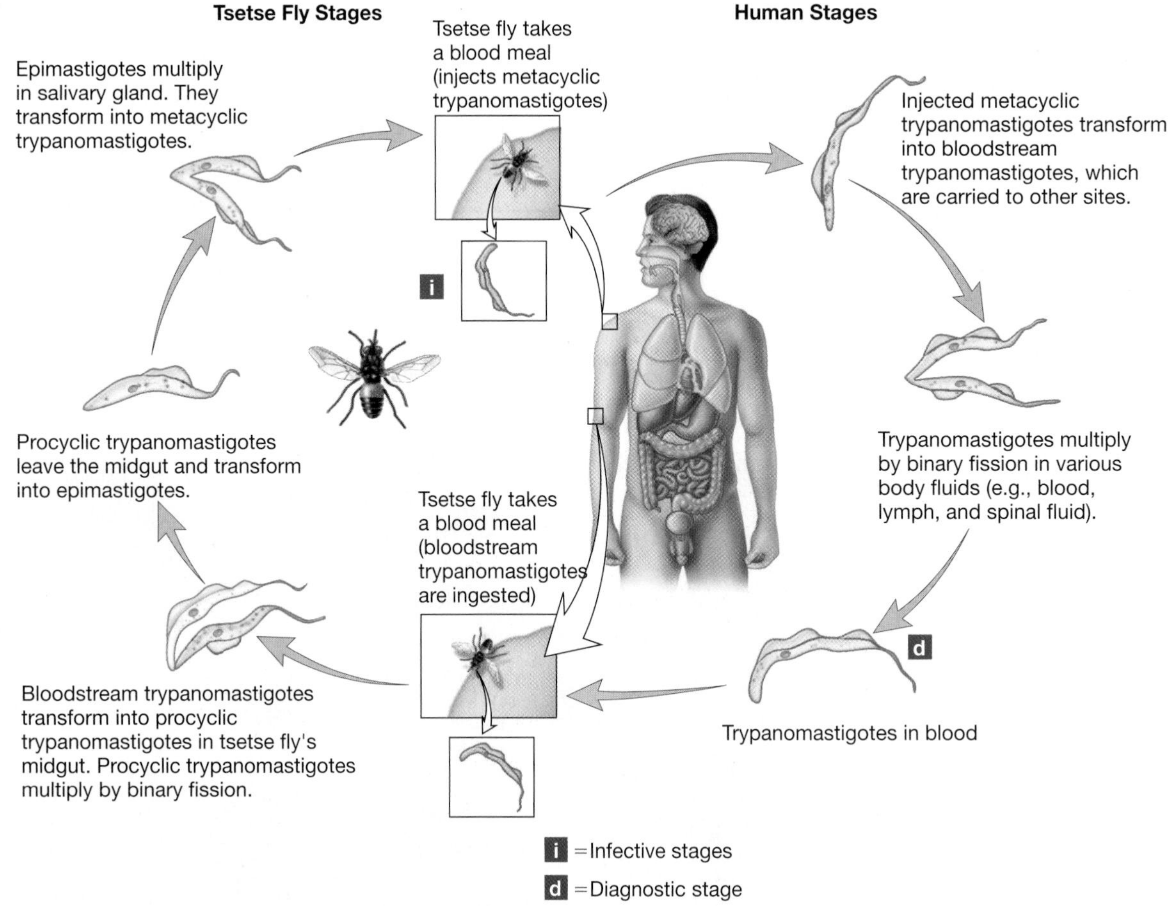
Tsetse Fly Stages
Human Stages
Tsetse fly takes a blood meal (injects metacyclic trypanomastigotes)
Epimastigotes multiply in salivary gland. They transform into metacyclic trypanomastigotes.
Injected metacyclic trypanomastigotes transform into bloodstream trypanomastigotes, which are carried to other sites.
i
Procyclic trypanomastigotes leave the midgut and transform into epimastigotes.
Trypanomastigotes multiply by binary fission in various body fluids (e.g., blood, lymph, and spinal fluid).
Tsetse fly takes a blood meal (bloodstream trypanomastigotes are ingested)
d
Bloodstream trypanomastigotes transform into procyclic trypanomastigotes in tsetse fly's midgut. Procyclic trypanomastigotes multiply by binary fission.
Trypanomastigotes in blood
i = Infective stages
d = Diagnostic stage

Figure 39.9 Life Cycle of *Trypanosoma brucei*. This trypanosome is transmitted to humans through the bite of the tsetse fly. The metacyclic trypanomastigotes enter the bloodstream where they disseminate to various tissue sites. They reproduce in the human and are ingested by tsetse flies as part of the blood meal. In the fly gut, they transform into procyclic trypanomastigotes and then into epimastigotes. Epimastogotes migrate to the salivary glands where they transform into metacyclic trypanomastigotes that can be passed on during feeding.

result, more than 100,000 new cases per year are likely. In the United States, only 21 cases of trypanosomiasis have been reported since 1967; all of these patients were travelers to Africa.

T. cruzi causes **American trypanosomiasis (Chagas' disease)**, which occurs in the tropics and subtropics of continental America. The parasite uses the triatomine (kissing) bug as a vector (**figure 39.10**). As the triatomine bug takes a blood meal, the parasites are discharged in the insect's feces. Some trypanosomes enter the bloodstream through the wound and invade the liver, spleen, lymph nodes, and central nervous system. Cell invasion stimulates the trypanosome's transformation into amastigotes, resulting in clinical manifestations of infection. The bloodstream trypanomastigotes do not replicate, however, until they enter a cell (becoming amastigotes) or are ingested by the arthropod vector (becoming epimastigotes). In some parts of Latin America, a high percentage of heart disease is due to parasitized cardiac cells. There are 16 to 18 million new cases each year and over 50,000 deaths.

Trypanosomiasis is diagnosed by finding motile parasites in fresh blood, spinal fluid, or skin biopsy and by serological testing. Treatment for African trypanosomiasis uses suramin and pentamidine for non-nervous system involvement and melarsoprol when the nervous system is involved. Currently there is no drug suitable for Chagas' disease, although nifurtimox (Lampit) and benznidazole have shown some value. Vaccines are not useful because the parasite is able to change its protein coat (antigenic shift) and evade the immunologic response. Protist classification: *Euglenozoa* (section 25.6)

1. Besides their route of transmission, how else are human fungal diseases categorized?
2. Why are fungal infections of the lungs potentially life-threatening?
3. Why is *Histoplasma capsulatum* found in bird feces but not within the birds themselves? What are the public health implication of this?
4. What flagellated protists invade the blood? What diseases do they cause?

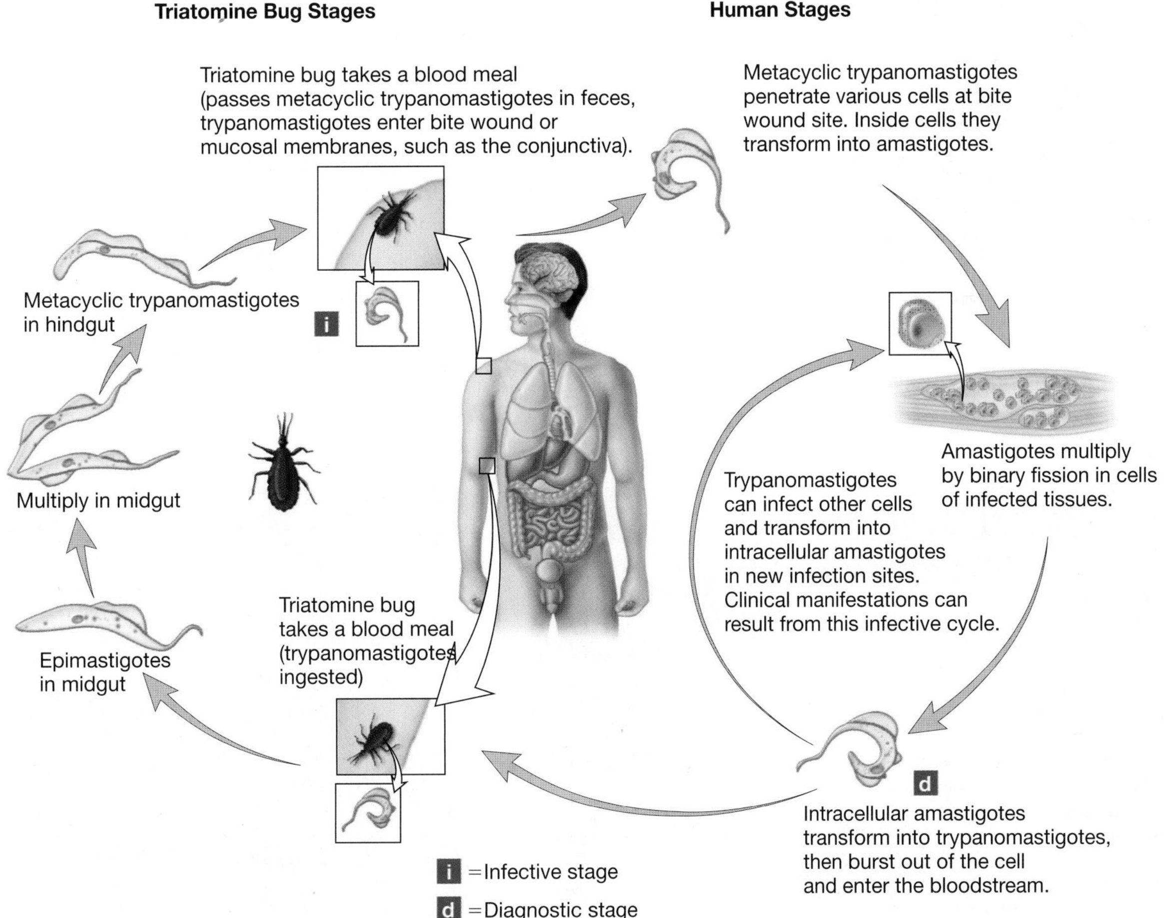
Triatomine Bug Stages
Human Stages
Triatomine bug takes a blood meal (passes metacyclic trypanomastigotes in feces, trypanomastigotes enter bite wound or mucosal membranes, such as the conjunctiva).
Metacyclic trypanomastigotes penetrate various cells at bite wound site. Inside cells they transform into amastigotes.
Metacyclic trypanomastigotes in hindgut
i
Amastigotes multiply by binary fission in cells of infected tissues.
Trypanomastigotes can infect other cells and transform into intracellular amastigotes in new infection sites. Clinical manifestations can result from this infective cycle.
Multiply in midgut
Triatomine bug takes a blood meal (trypanomastigotes ingested)
Epimastigotes in midgut
d
Intracellular amastigotes transform into trypanomastigotes, then burst out of the cell and enter the bloodstream.
i = Infective stage
d = Diagnostic stage

Figure 39.10 Life Cycle of *Trypanosoma cruzi*. This trypanosome is transmitted to humans from the triatomine bug. The metacyclic trypanomastigotes are shed in the bug feces, which enter the host through the bite wound or mucous membranes. The trypanomastigotes penetrate host cells and transform into amastigotes. After replication by binary fission, amastigotes transform into trypanomastigotes, which burst from their host cell to disseminate throughout the host. Blood-borne trypanomastigotes can be ingested by triatomine bugs as part of a blood meal, transform into epimastigotes in the bug's midgut, and transform into metacyclic trypanomastigotes in the hindgut.

39.4 Direct Contact Diseases

Superficial Mycoses

The superficial mycoses are extremely rare in the United States, and most occur in the tropics. The fungi responsible are limited to the outer surface of hair and skin and hence are called superficial. Infections of the hair shaft are collectively called **piedras** (Spanish for stone because they are associated with the hard nodules formed by mycelia on the hair shaft). For example, **black piedra** is caused by *Piedraia hortae* and forms hard, black nodules on the hairs of the scalp (**figure 39.11**). **White piedra** is caused by the yeast *Trichosporon beigelii* and forms light-colored nodules on the beard and mustache. Some superficial mycoses are called **tineas** [Latin for grub, larva, worm], the specific type is designated by a modifying term. Tineas are superficial fungal infections involving the outer layers of the skin, nails, and hair. **Tinea versicolor** is caused by the yeast *Malassezia furfur* and forms brownish-red scales on the skin of the trunk, neck, face, and arms. Treatment involves removal of the skin scales with a cleansing agent and removal of the infected hairs. Good personal hygiene prevents these infections.

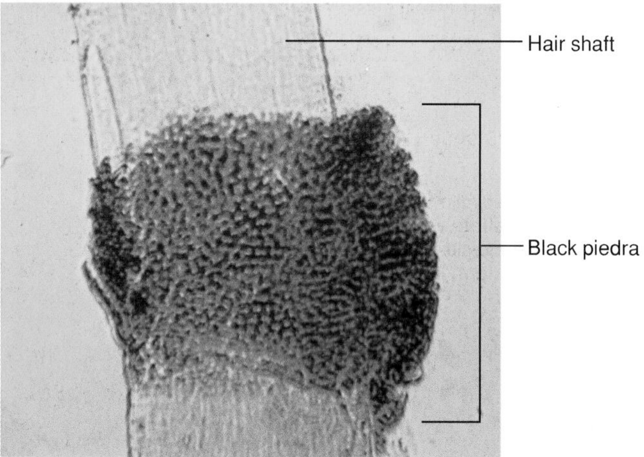

Figure 39.11 Superficial Mycosis: Black Piedra. Hair shaft infected with *Piedraia hortae;* light micrograph (×200).

Cutaneous Mycoses

Cutaneous mycoses—also called **dermatomycoses, ringworms,** or tineas—occur worldwide and represent the most common fungal diseases in humans. Three genera of cutaneous fungi, or **dermatophytes,** are involved in these mycoses: *Epidermophyton, Microsporum,* and *Trichophyton*. Diagnosis is by microscopic examination of biopsied areas of the skin cleared with 10% potassium hydroxide and by culture on Sabouraud dextrose agar. Treatment is with topical ointments such as miconazole (Monistat-Derm), tolnaftate (Tinactin), or clotrimazole (Lotrimin) for 2 to 4 weeks. Griseofulvin (Grifulvin V) and itraconazole (Sporanox) are the only oral antifungal agents currently approved by the FDA for treating dermatophytoses.

Tinea barbae [Latin *barba,* the beard] is an infection of the beard hair caused by *Trichophyton mentagrophytes* or *T. verrucosum.* It is predominantly a disease of men who live in rural areas and acquire the fungus from infected animals. **Tinea capitis** [Latin *capita,* the head] is an infection of the scalp hair (**figure 39.12*a***). It is characterized by loss of hair, inflammation, and scaling. Tinea capitis is primarily a childhood disease caused by *Trichophyton* or *Microsporum* species. Person-to-person transmission of the fungus occurs frequently when poor hygiene and overcrowded conditions exist. The fungus also occurs in domestic animals, who can transmit it to humans. A Wood's lamp (a UV light) can help with the diagnosis of tinea capitis because fungus-infected hair fluoresces when illuminated by UV radiation (figure 39.12*b*).

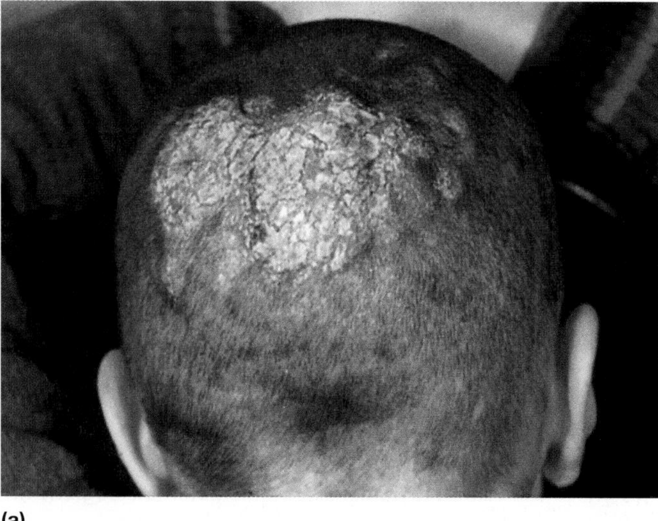

(a)

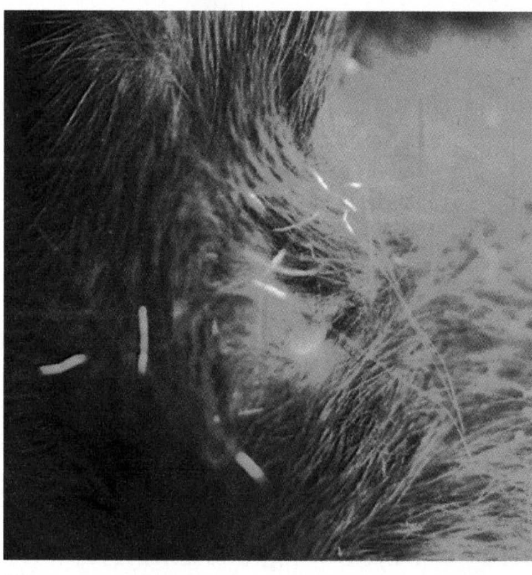

(b)

Figure 39.12 Cutaneous Mycosis: Tinea Capitis. **(a)** Ringworm of the head caused by *Microsporum audouinii.* **(b)** Close-up using a Wood's light (a UV lamp).

Tinea corporis [Latin *corpus,* the body] is a dermatophytic infection that can occur on any part of the skin (**figure 39.13**). The disease is characterized by circular, red, well-demarcated, scaly, vesiculopustular lesions accompanied by itching. Tinea corporis is caused by *Trichophyton rubrum, T. mentagrophytes,* or *Microsporum canis.* Transmission of the disease agent is by direct contact with infected animals and humans or by indirect contact through fomites (inanimate objects).

Tinea cruris [Latin *crura,* the leg] is a dermatophytic infection of the groin (**figure 39.14**). The pathogenesis and clinical manifestations are similar to those of tinea corporis. The responsible fungi are *Epidermophyton floccosum, T. mentagrophytes,* or *T. rubrum.* Factors predisposing one to recurrent disease are moisture, occlusion, and skin trauma. Wet bathing suits, athletic supporters (**jock itch**), tight-fitting slacks, panty hose, and obesity are frequently contributing factors.

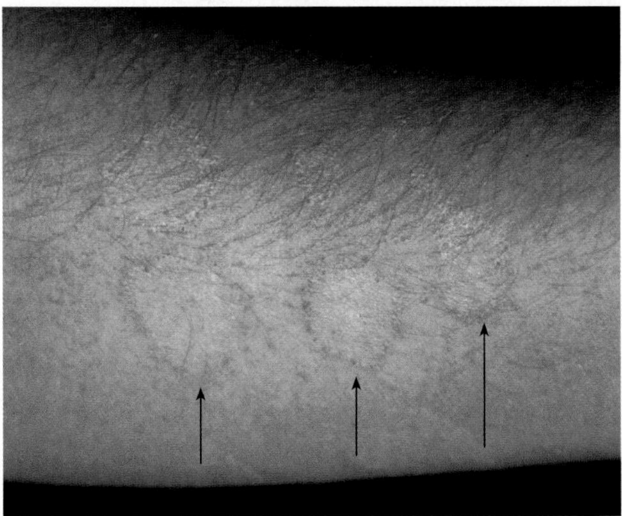

Figure 39.13 Cutaneous Mycosis: Tinea Corporis. Ringworm of the body—in this case, the forearm—caused by *Trichophyton mentagrophytes.* Notice the circular patches (arrows).

Tinea pedis [Latin *pes,* the foot], also known as **athlete's foot,** and **tinea manuum** [Latin *mannus,* the hand] are dermatophytic infections of the feet (**figure 39.15**) and hands, respectively. Clinical symptoms vary from a fine scale to a vesiculopustular eruption. Itching is frequently present. Warmth, humidity, trauma, and occlusion increase susceptibility to infection. Most infections are caused by *T. rubrum, T. mentagrophytes,* or *E. floccosum.* Tinea pedis and tinea manuum occur throughout the world, are most commonly found in adults, and increase in frequency with age.

Tinea unguium [Latin *unguis,* nail] is a dermatophytic infection of the nail bed (**figure 39.16**). In this disease the nail becomes discolored and then thickens. The nail plate rises and separates from the nail bed. *Trichophyton rubrum* or *T. mentagrophytes* are the causative fungi.

Subcutaneous Mycoses

The dermatophytes that cause subcutaneous mycoses are normal saprophytic inhabitants of soil and decaying vegetation. Because they are unable to penetrate the skin, they must be introduced into the subcutaneous tissue by a puncture wound. Most infections involve barefooted agricultural workers. Once in the subcutaneous tissue, the disease develops slowly—often over a period of years. During this time the fungi produce a nodule that eventually ulcerates and the organisms spread along lymphatic channels, producing more subcutaneous nodules. At times, such nodules drain to the skin surface. The administration of oral 5-fluorocytosine,

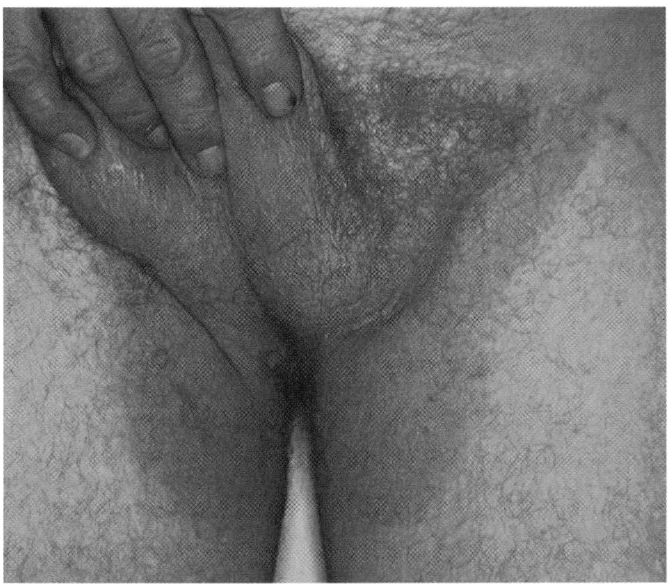

Figure 39.14 Cutaneous Mycosis: Tinea Cruris. Ringworm of the groin caused by *Epidermophyton floccosum.*

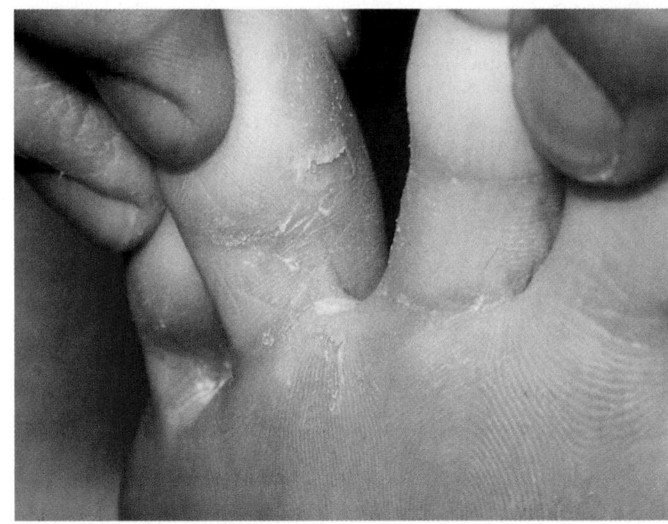

Figure 39.15 Cutaneous Mycosis: Tinea Pedis. Ringworm of the foot caused by *Trichophyton rubrum, T. mentagrophytes,* or *Epidermophyton floccosum.*

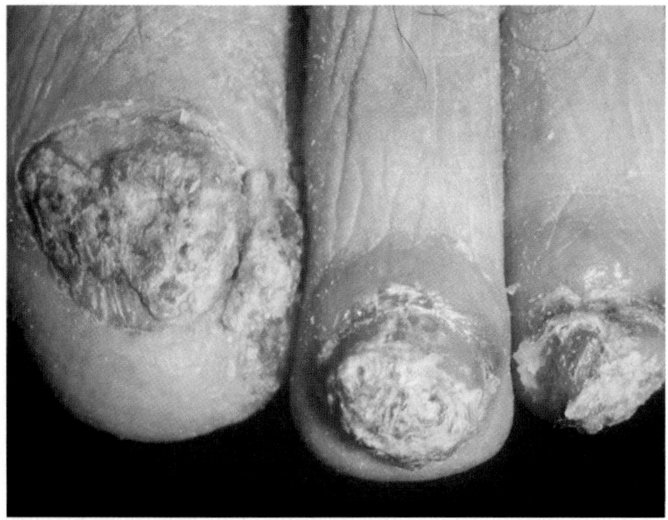

Figure 39.16 Cutaneous Mycosis: Tinea Unguium. Ringworm of the nails caused by *Trichophyton rubrum*.

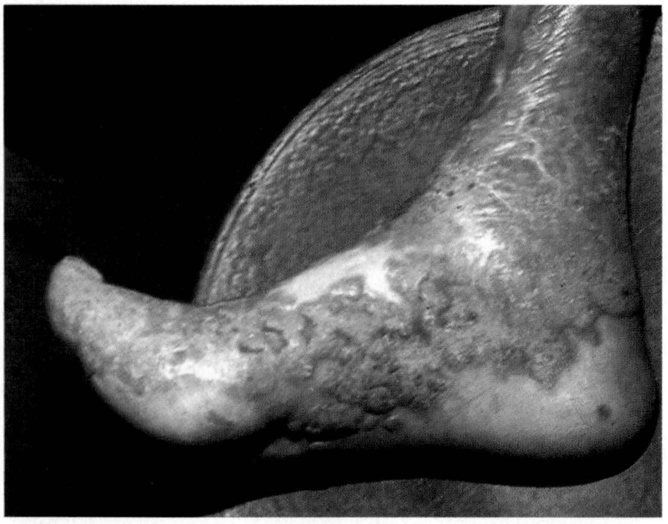

Figure 39.17 Subcutaneous Mycosis. Chromoblastomycosis of the foot caused by *Fonsecaea pedrosoi*.

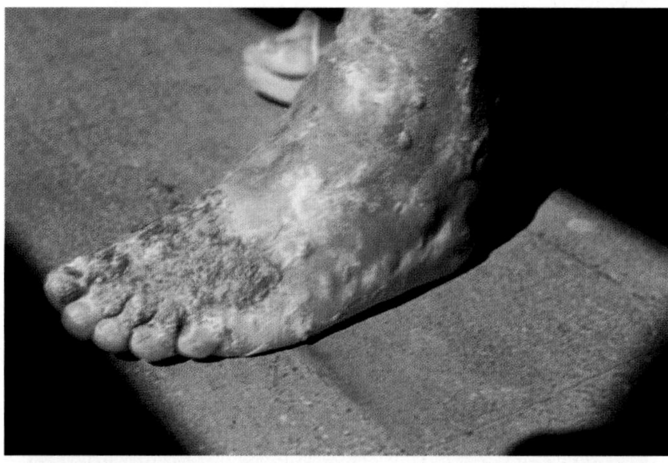

Figure 39.18 Subcutaneous Mycosis. Eumycotic mycetoma of the foot caused by *Madurella mycetomatis*.

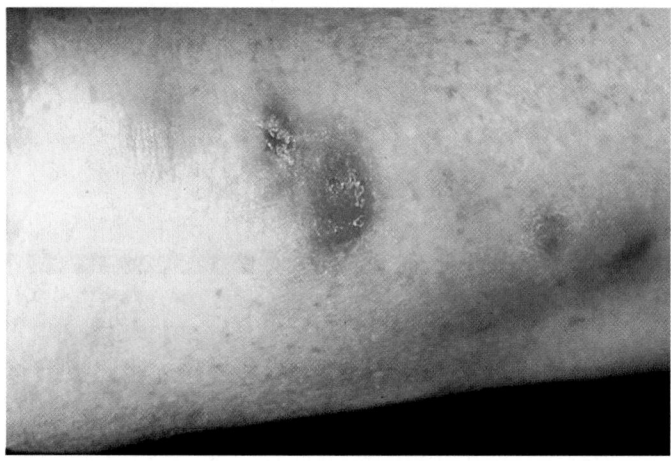

Figure 39.19 Subcutaneous Mycosis. Sporotrichosis of the arm caused by *Sporothrix schenckii*.

iodides, or amphotericin B, and surgical excision, are the usual treatments. Diagnosis is accomplished by culture of the infected tissue.

One type of subcutaneous mycosis is **chromoblastomycosis.** The nodules are dark brown. This disease is caused by the black molds *Phialophora verrucosa* and *Fonsecaea pedrosoi*. These fungi exist worldwide, especially in tropical and subtropical regions. Most infections involve the legs and feet (**figure 39.17**). Another subcutaneous mycosis is **maduromycosis,** caused by *Madurella mycetomatis,* which is distributed worldwide and is especially prevalent in the tropics. Because the fungus destroys subcutaneous tissue and produces serious deformities, the resulting infection is often called a **eumycotic mycetoma,** or fungal tumor (**figure 39.18**). One form of mycetoma, known as Madura foot, occurs through skin abrasions acquired while walking barefoot on contaminated soil.

Sporotrichosis is the subcutaneous mycosis caused by the dimorphic fungus *Sporothrix schenckii*. The disease occurs throughout the world and is the most common subcutaneous mycotic disease in the United States. The fungus can be found in the soil, on living plants, such as barberry shrubs and roses, or in plant debris, such as sphagnum moss, baled hay, and pine-bark mulch. Infection occurs by a puncture wound from a thorn or splinter contaminated with the fungus. The disease is an occupational hazard for florists, gardeners, and forestry workers. It is not spread from person to person. After an incubation period of 1 to 12 weeks, a small, red papule arises and begins to ulcerate (**figure 39.19**). New lesions appear along lymph channels and can remain localized or spread throughout the body, producing **extracutaneous sporotrichosis.**

Sporotrichosis is typically treated by ingestion of potassium iodide or itraconazole (Sporanox) until the lesions are healed, usually several weeks. Preventative measures include gloves and other protective clothing, as well as avoidance of contaminated landscaping materials, especially sphagnum moss.

Toxoplasmosis

Toxoplasmosis is a disease caused by the protist *Toxoplasma gondii*. This apicomplexan protist has been found in nearly all animals and most birds; cats are the definitive host and are required for completion of the sexual cycle (**figure 39.20**). Animals shed oocysts in the feces; the oocysts enter another host by way of the nose or mouth; and the parasites colonize the intestine. Toxoplasmosis also can be transmitted by the ingestion of raw or undercooked meat, congenital transfer, blood transfusion, or a tissue transplant. Originally toxoplasmosis gained public notice when it was discovered that in pregnant women the protist might also infect the fetus, causing serious congenital defects or death. Most cases of toxoplasmosis are asymptomatic. Adults usually complain of an "infectious mononucleosis-like" syndrome. In immunoincompetent or immunosuppressed individuals, it frequently results in fatal disseminated disease with a heavy cerebral involvement.

Acute toxoplasmosis is usually accompanied by lymph node swelling (lymphadenopathy) with reticular cell hyperplasia (enlargement). Pulmonary necrosis, myocarditis, and hepatitis caused by tissue necrosis are common. Retinitis (inflammation of the retina of the eye) is associated with necrosis due to the proliferation of the parasite within retinal cells. Toxoplasmosis is found worldwide, although most of those infected do not exhibit symptoms of the disease because the immune system usually prevents illness. However, *Toxoplasma* in the immunocompromised, such

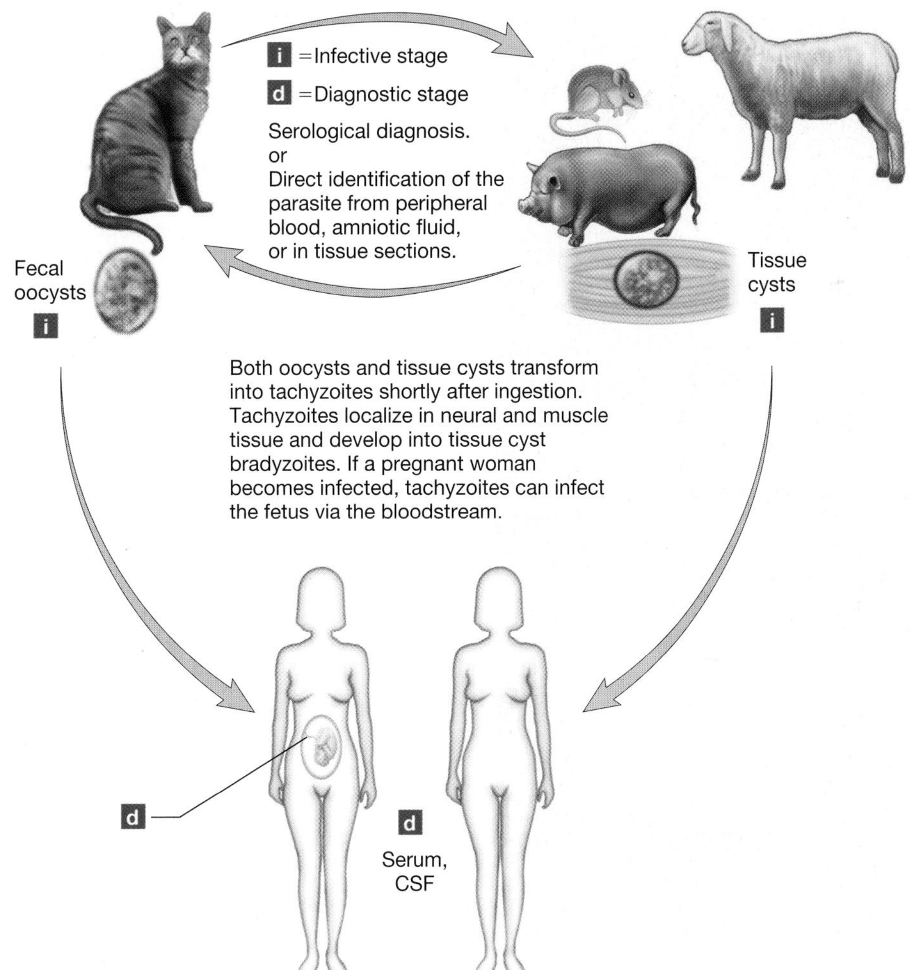

Figure 39.20 Life Cycle of *Toxoplasma gondii*. *Toxoplasma gondii* is a protozoan parasite of numerous mammals and birds. However, the cat, the definitive host, is required for completion of the sexual cycle. Oocysts are shed in the feces of infected animals where they may be ingested by another host. Ingested oocysts transform into tachyzoites, which migrate to various tissue sites via the bloodstream.

as AIDS or transplant patients, can produce a unique encephalitis with necrotizing lesions accompanied by inflammatory infiltrates. It continues to cause more than 3,000 congenital infections per year in the United States. Protist classification: *Alveolata* (section 25.6)

Laboratory diagnosis of toxoplasmosis is by serological tests. Epidemiologically, toxoplasmosis is ubiquitous in all higher animals. Treatment of toxoplasmosis is with a combination of pyrimethamine (Daraprim) and sulfadiazine. Prevention and control require minimizing exposure by the following: avoiding eating raw meat and eggs, washing hands after working in the soil, cleaning cat litterboxes daily, keeping household cats indoors if possible, and feeding them commercial food.

Trichomoniasis

Trichomoniasis is a sexually transmitted disease caused by the protozoan flagellate *Trichomonas vaginalis* (**figure 39.21**). It is one of the most common sexually transmitted diseases, with an estimated 7 million cases annually in the United States and 180 million cases annually worldwide. In response to the parasite, the body accumulates leukocytes at the site of the infection. In females this usually results in a profuse, purulent vaginal discharge that is yellowish to light cream in color and characterized by a disagreeable odor. The discharge is accompanied by itching. Males are generally asymptomatic because of the trichomonacidal action of prostatic secretions; however, at times a burning sensation occurs during urination. Diagnosis is made in females by microscopic examination of the discharge and identification of the protozoan. Infected males demonstrate protozoa in semen or urine. Treatment is by administration of metronidazole (Flagyl). Protist classification: *Parabasalia* (section 25.6)

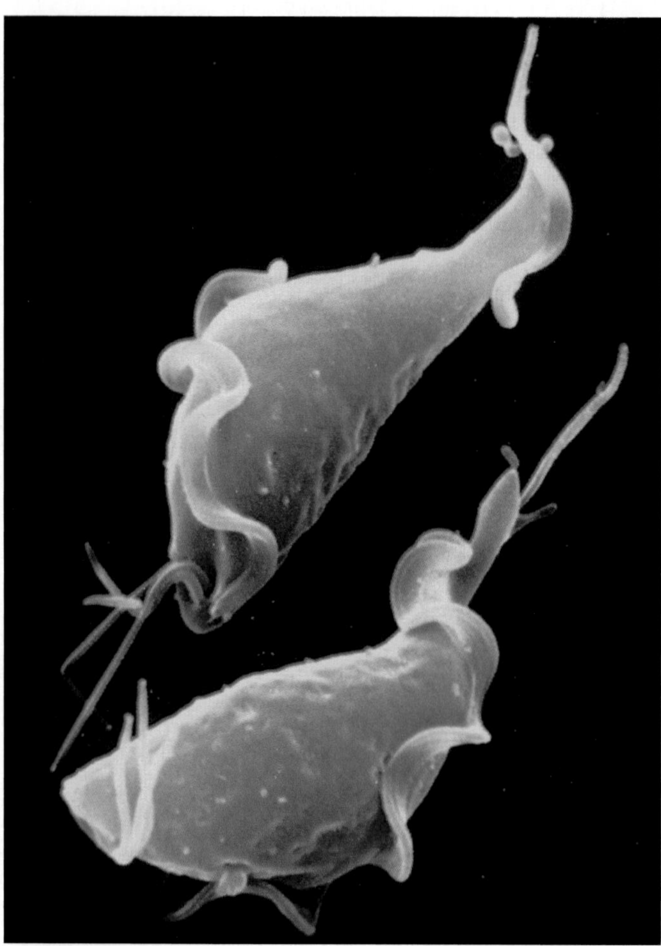

Figure 39.21 Trichomoniasis. *Trichomonas vaginalis,* showing the characteristic undulating membranes and flagella; scanning electron micrograph (×12,000).

1. Describe two piedras that infect humans?
2. Briefly describe the major tineas that occur in humans.
3. Describe the three types of subcutaneous mycoses that affect humans.
4. In what two ways does *Toxoplasma* affect human health?
5. How would you diagnose trichomoniasis in a female? In a male?

39.5 Food-Borne and Waterborne Diseases

Amebiasis

It is now accepted that two species of *Entamoeba* infect humans: the nonpathogenic *E. dispar* and the pathogenic *E. histolytica. E. histolytica* is responsible for **amebiasis (amebic dysentery).** This very common parasite is endemic in warm climates where adequate sanitation and effective personal hygiene are lacking. Within the United States about 3,000 to 5,000 cases are reported annually. However, it is a major cause of parasitic death worldwide; about 500 million people are infected and as many as 100,000 die of amebiasis each year. Protists classification: *Entamoebida* (section 25.6)

Infection occurs by ingestion of mature cysts from fecally contaminated water, food, or hands, or from fecal exposure during sexual contact. After excystation in the lower region of the small intestine, the metacyst divides rapidly to produce eight small trophozoites (**figure 39.22**). These trophozoites move to the large intestine where they can invade the host tissue, live as commensals in the lumen of the intestine, or undergo encystation. In many hosts, the trophozoites remain in the intestinal lumen, resulting in an asymptomatic carrier state with cysts shed with the feces.

If the infective trophozoites invade the intestinal tissues, they multiply rapidly and spread laterally, while feeding on erythrocytes, bacteria, and yeasts. The invading trophozoites destroy the epithelial lining of the large intestine by producing a cysteine protease. This protease is a virulence factor of *E. histolytica* and may play a role in intestinal invasion by degrading the extracellular matrix and circumventing the host immune response through cleavage of secretory immunoglobulin A (sIgA), IgG, and complement factors. These cysteine proteases are encoded by at least seven genes, several of which are found in *E. histolytica* but not *E. dispar.* Lesions (ulcers) are characterized by minute points of entry into the mucosa and extensive enlargement of the lesion after penetration into the submucosa. *E. histolytica* also may invade

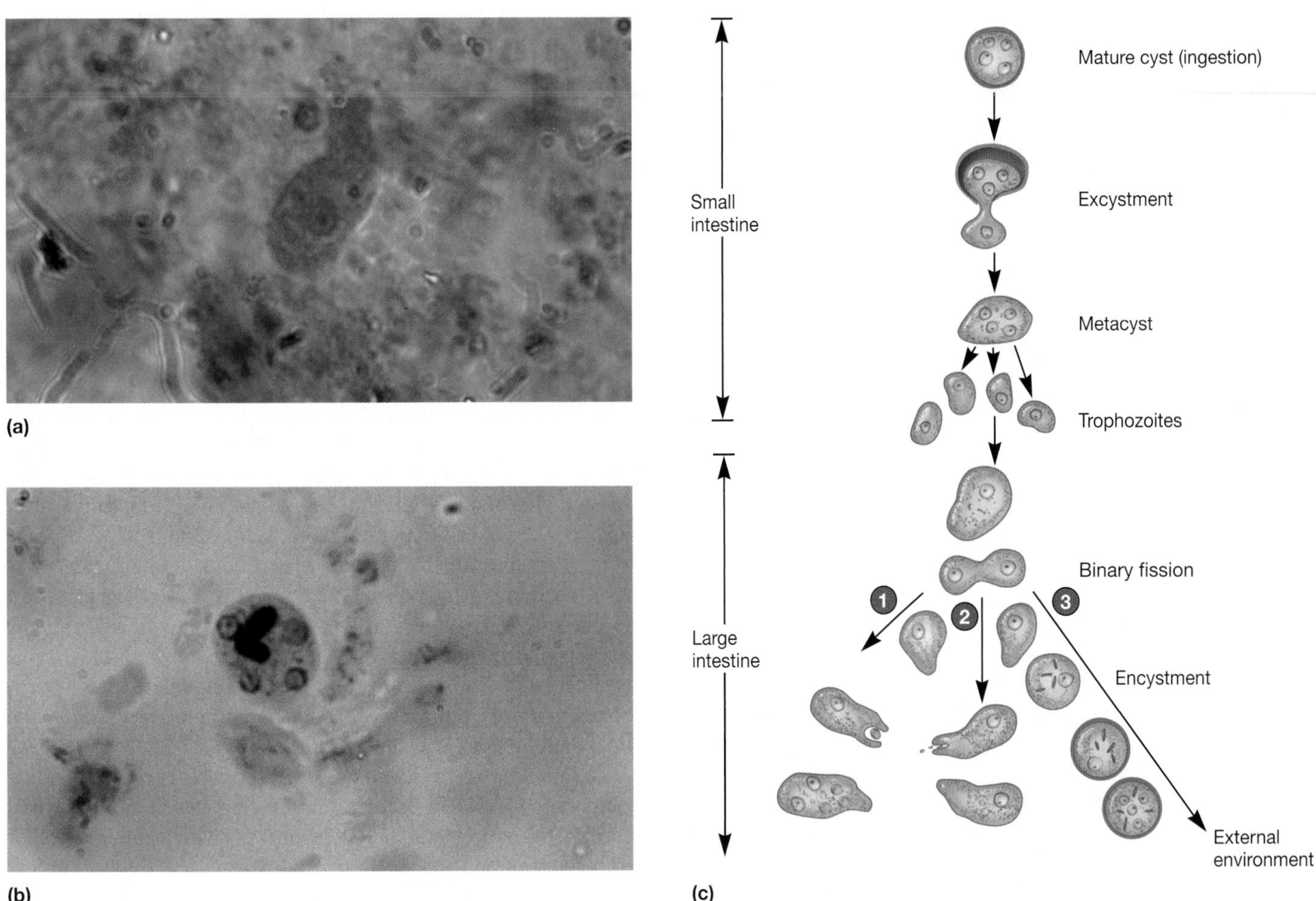

Figure 39.22 Amebiasis Caused by *Entamoeba histolytica*. **(a)** Light micrographs of a trophozoite (×1,000) and **(b)** a cyst (×1,000). **(c)** Life cycle. Infection occurs by the ingestion of a mature cyst. Excystment occurs in the lower region of the small intestine and the metacyst rapidly divides to give rise to eight small trophozoites (only four are shown). These enter the large intestine, undergo binary fission, and may (1) invade the host tissues, (2) live in the lumen of the large intestine without invasion, or (3) undergo encystment and pass out of the host in the feces.

and produce lesions in other tissues, especially the liver, to cause hepatic amebiasis. However, all extraintestinal amebic lesions are secondary to those established in the large intestine. The symptoms of amebiasis are highly variable, ranging from an asymptomatic infection to fulminating dysentery (exhaustive diarrhea accompanied by blood and mucus), appendicitis, and abscesses in the liver, lungs, or brain.

Laboratory diagnosis of amebiasis can be difficult and is based on finding trophozoites in fresh, warm stools and cysts in ordinary stools. Serological testing for *E. histolytica* is available but often unreliable. The therapy for amebiasis is complex and depends on the location of the infection within the host and the host's condition. Asymptomatic carriers who are passing cysts should always be treated with iodoquinol or paromomycin because they represent the most important reservoir of the protozoan in the population. In symptomatic intestinal amebiasis, metronidazole (Flagyl) or iodoquinol (Yodoxin) are the drugs of choice. Prevention and control of amebiasis is achieved by practicing good personal hygiene and avoiding water or food that might be contaminated with human feces. Viable cysts in water can be destroyed by hyperchlorination or iodination. Anitprotozoan drugs (section 34.9)

Amebic Meningoencephalitis

Free-living amoebae of the genera *Naegleria* and *Acanthamoeba* are facultative (opportunistic) parasites responsible for causing **primary amebic meningoencephalitis** in humans. They are among the most common protists found in freshwater and moist soil. In addition, several *Acanthamoeba* spp. are known to infect the eye, causing a chronically progressive, ulcerative *Acanthamoeba* **keratitis**—inflammation of the cornea—that may result in blindness. Wearers of soft contact lenses may be predisposed to this infection and should take care to prevent contamination of their lens-cleaning and soaking solutions. Diagnosis of these infections is by demonstration of amoebae in clinical specimens. Most freshwater amoebae are resistant to commonly used antimicrobial agents. These amoebae are reported in fewer than 100 human disease cases annually in the United States, although the incidence (especially of *Acanthamoeba* keratitis) is likely higher.

Cryptosporidiosis

The first case of human **cryptosporidiosis** was reported in 1976. The protist responsible was identified as *Cryptosporidium parvum,* a protozoan classified as an emerging pathogen by the CDC. In 1993 *C. parvum* contaminated the Milwaukee, Wisconsin water supply and caused severe diarrheal disease in about 400,000 individuals, the largest recognized outbreak of waterborne illness in U.S. history. *Cryptosporidium* ("hidden spore cysts") is found in about 90% of sewage samples, in 75% of river waters, and in 28% of drinking waters. Protist classification: *Alveolata* (section 25.6)

C. parvum is a common coccidial, apicomplexan protist found in the intestine of many birds and mammals. When these animals defecate, oocysts are shed into the environment. If a human ingests food or water that is contaminated with the oocysts, excystment occurs within the small intestine and sporozoites enter epithelial cells and develop into merozoites. Some of the merozoites subsequently undergo sexual reproduction to produce zygotes, and the zygotes differentiate into thick-walled oocysts. Oocyst release into the environment begins the life cycle again. A major problem for public health arises from the fact that the oocysts are only 4 to 6 μm in diameter—much too small to be easily removed by the sand filters used to purify drinking water. *Cryptosporidium* also is extremely resistant to disinfectants such as chlorine. The problem is made worse by the low infectious dose, around 10 to 100 oocysts, and the fact that the oocysts may remain viable for 2 to 6 months in a moist environment.

The incubation period for cryptosporidiosis ranges from 5 to 28 days. Diarrhea, which characteristically may be cholera-like, is the most common symptom. Other symptoms include abdominal pain, nausea, fever, and fatigue. The pathogen is routinely diagnosed by fecal concentration and acid-fast stained smears. (The thick-walled oocysts can be stained by the same method used to stain mycobacterial species.) No chemotherapy is available and patients are simply rehydrated. Although the disease usually is self-limiting in healthy individuals, patients with late-stage AIDS or who are immunocompromised in other ways may develop prolonged, severe, and life-threatening diarrhea.

Cyclospora

Cyclosporiasis is caused by the unicellular coccidian protist *Cyclospora cayetanensis* (previously known as the cyanobacterium-like or coccidia-like body). The disease is most common in tropical and subtropical environments although it has been reported in most countries. In Canada and the United States, cyclosporiasis has been responsible for affecting over 3,600 people in at least 11 food-borne outbreaks since 1990. *Cyclospora* was first identified in 1979. Infection is worldwide. A number of cyclosporiasis outbreaks have been linked to contaminated produce.

The disease presents with frequent, sometimes explosive, diarrhea. Often the patient exhibits loss of appetite, cramps, and bloating due to substantial gas production, nausea, vomiting, fever, fatigue, and substantial weight loss. Patients may report symptoms for days to weeks with decreasing frequency and then relapses. The protozoa infect the small intestine with a mean incubation period of approximately one week. Cyclosporan oocysts in freshly passed feces are not infective (**figure 39.23**). Thus direct oral-fecal transmission does not occur. Instead the oocysts must differentiate into sporozoites after days or weeks at temperatures between 22 and 32°C. Sporozoites can enter the food chain when oocyst-contaminated water is used to wash fruits and vegatables prior to their transport to market. Once ingested, the sporozoites are freed from the oocysts and invade intestinal epithelial cells where they replicate asexually. Sexual development is completed when sporozoites mature into new oocysts and are released into the intestinal lumen to be shed with the feces.

Laboratory identification of cyclosporiasis is by the identification of oocycts in feces. Identification may require several specimens over several days. Treatment is with a combination of trimethoprim and sulfamethoxazole (Bactrim or Septra), unless contraindicated, and fluids to restore water lost through diarrhea. Prevention of cyclosporiasis is by avoidance of contaminated food and water. No vaccine is available.

Giardiasis

Giardia intestinalis is a flagellated protist (discovered by van Leeuwenhoek in 1681 when he examined his own stools) (**figure 39.24*a***). It was initially named *Cercomonas intestinalis* by Lambl in 1859 and renamed *Giardia lamblia* by Stiles in 1915, in honor of Professor A. Giard of Paris and Dr. F. Lambl of Prague. However, *Giardia intestinalis* is considered to be the proper name for this protist. *G. intestinalis* causes the very common intestinal disease **giardiasis,** which is worldwide in distribution and affects children more seriously than adults. In the United States this protist is the most common cause of epidemic waterborne diarrheal disease (about 30,000 cases yearly). Approximately 7% of the population are asymptomatic carriers who shed cysts in their feces. *G. intestinalis* is endemic in child day-care centers in the United States, with estimates of 5 to 15% of diapered children being infected. Transmission occurs most frequently by cyst-contaminated water supplies. Epidemic outbreaks have been recorded in wilderness areas, suggesting that humans may be infected from pristine stream water with *Giardia* harbored by rodents, deer, cattle, or household pets. This im-

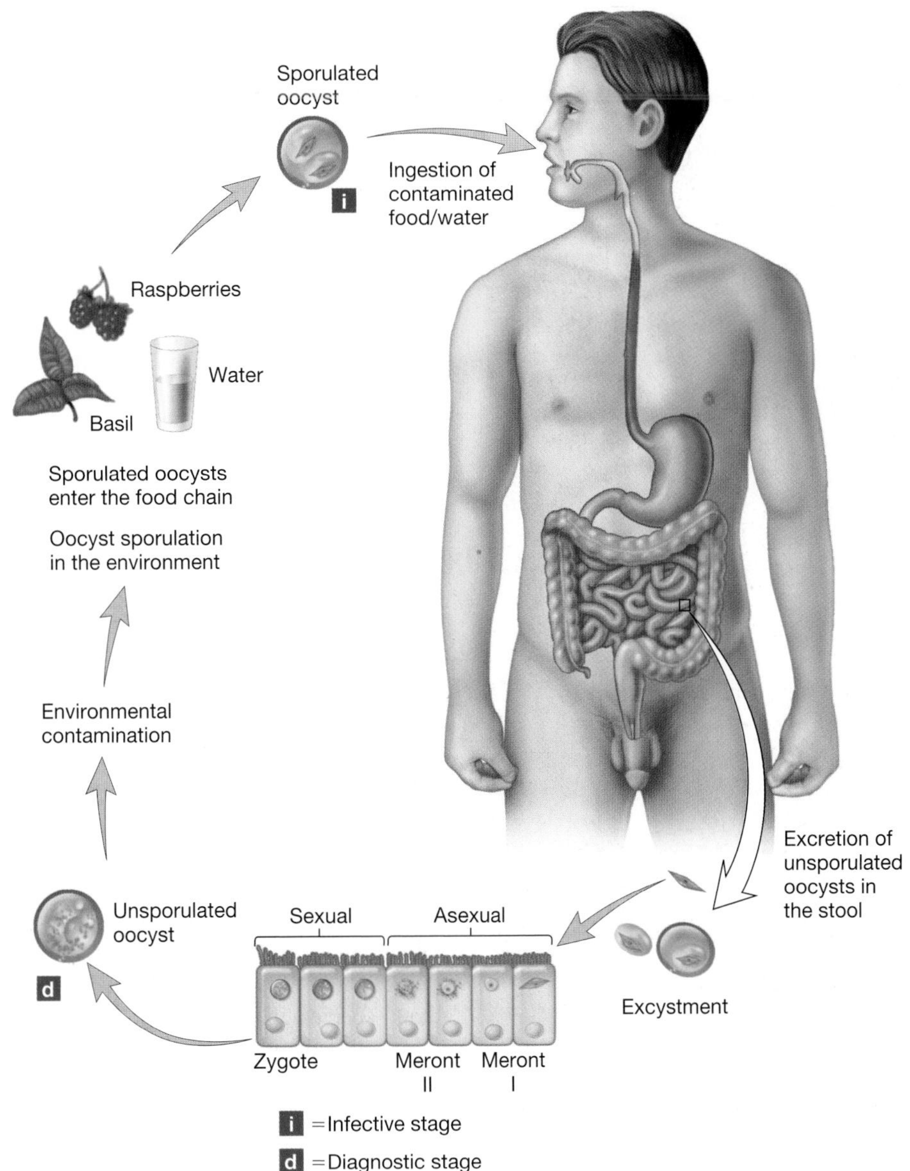

Figure 39.23 Life Cycle of *Cyclospora cayetanenis.* Unsporulated oocysts, shed in the feces of infected animals, sporulate and contaminate food and water. Once ingested, the sporulated oocysts excyst, penetrate host cells, and reproduce by binary fission (merogony). Sexual stages unite to form zygotes resulting in new oocysts (unsporulated) that shed in the feces. *Some of the elements in this figure were created based on an illustration by Ortega et al.* Cyclospora cayetanensis. *In: Advances in Parasitology: Opportunistic protozoa in humans. San Diego: Academic Press; 1998. p. 399–418.*

plies that human infections also can be a zoonosis. As many as 200 million humans may be infected worldwide. Protist classification: *Fornicata* (section 25.6)

Following ingestion, the cysts undergo excystment in the duodenum, forming trophozoites. The trophozoites inhabit the upper portions of the small intestine, where they attach to the intestinal mucosa by means of their sucking disks (figure 39.24*b*). The ability of the trophozoites to adhere to the intestinal epithelium accounts for the fact that they are rarely found in stools. It is thought that the trophozoites feed on mucous secretions and reproduce to form such a large population that they interfere with nutrient absorption by the intestinal epithelium. Giardiasis varies in severity, and asymptomatic carriers are common. The disease can be acute or chronic. Acute giardiasis is characterized by severe diarrhea, epigastric pain, cramps, voluminous flatulence ("passing gas"), and anorexia. Chronic giardiasis is characterized by intermittent diarrhea, with periodic appearance and remission of symptoms.

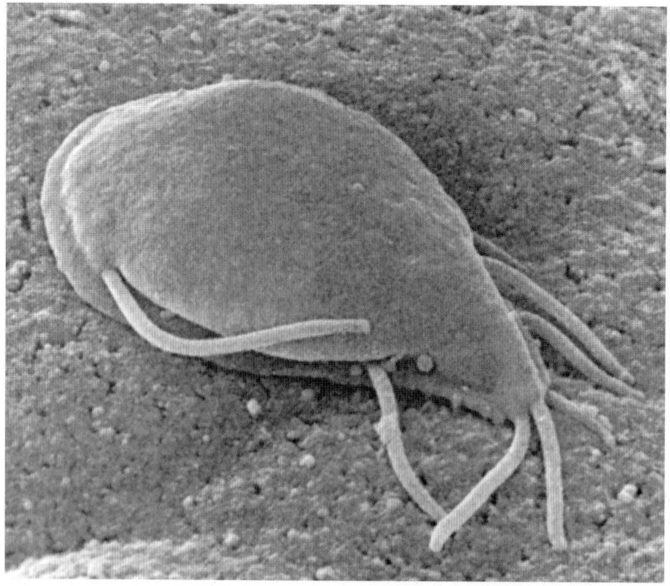

(a)

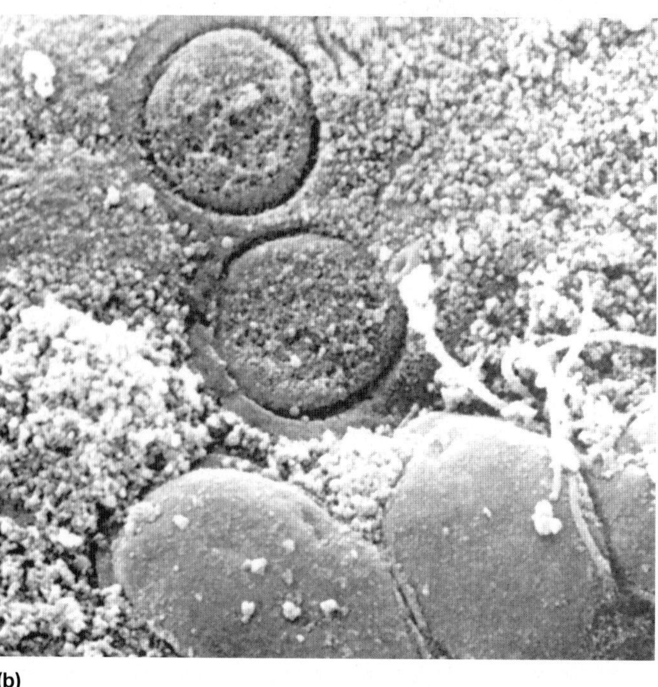

(b)

Figure 39.24 Giardiasis. **(a)** *Giardia intestinalis* adhering to the epithelium by its sucking disk; scanning electron micrograph. **(b)** Upon detachment from the epithelium, the protozoa often leave clear impressions on the microvillus surface (upper circles); scanning electron micrograph.

Laboratory diagnosis is based on the identification of trophozoites—only in the severest of diarrhea—or cysts in stools. A commercial ELISA test is also available for the detection of *G. intestinalis* antigen in stool specimens. Quinacrine hydrochloride (Atabrine) and metronidazole (Flagyl) are the drugs of choice for adults, and furazolidone is used for children because it is available in a liquid suspension. Prevention and control involve proper treatment of community water supplies, especially the use of slow sand filtration because the cysts are highly resistant to chlorine treatment. Water purification and sanitary analysis (section 41.1) Wastewater treatment (section 41.2)

39.6 Opportunistic Diseases

An **opportunistic microorganism** is generally harmless in its normal environment but becomes pathogenic in a compromised host. A **compromised host** is seriously debilitated and has a lowered resistance to infection. There are many causes of this condition: malnutrition, alcoholism, cancer, diabetes, leukemia, another infectious disease (e.g., HIV/AIDS), trauma from surgery or injury, an altered microbiota from the prolonged use of antibiotics (e.g., in vaginal candidiasis), and immunosuppression (e.g., by drugs, hormones, genetic deficiencies, cancer chemotherapy, and old age). Opportunistic mycoses may start as normal flora or ubiquitous environmental contaminants. The importance of opportunistic fungal pathogens is increasing because of the expansion of the immunocompromised patient population.

Aspergillosis

Of all the fungi that cause disease in human hosts, none are as widely distributed in nature as the *Aspergillus* species. *Aspergillus* is omnipresent—it is found wherever organic debris occurs, especially in soil, decomposing plant matter, household dust, building materials, some foods, and water. *Aspergillus fumigatus* is the usual cause of **aspergillosis.** *A. flavus* is the second most important species, particularly in invasive disease of immunosuppressed patients. Invasive disease typically results in pulmonary infection (with fever, chest pain, and cough) that disseminates to the brain, kidney, liver, bone, or skin. In immunocompetent patients, *Aspergillus* typically causes allergic sinusitis, allergic bronchitis, or a milder, localized bronchopulmonary infection. Characteristics of the fungal divisions: *Ascomycota* (section 26.6)

The major portal of entry for *Aspergillus* is the respiratory tract. Inhalation of conidiospores can lead to several types of pulmonary aspergillosis. One type is allergic aspergillosis. Infected individuals may develop an immediate allergic response and suffer asthma attacks when exposed to fungal antigens on the conidiospores. In bronchopulmonary aspergillosis, the major clinical manifestation of the allergic response is a bronchitis resulting from both type I and type III hypersensitivities. Although tissue invasion seldom occurs in bronchopulmonary aspergillosis, *Aspergillus* often can be cultured from the sputum. A most common manifestation of pulmonary involvement is the occurrence of colonizing aspergillosis, in which *Aspergillus* forms colonies within the lungs that develop into "fungus balls" called aspergillomas. These consist of a tangled mass of hyphae growing in a circumscribed area. From the pulmonary focus, the fungus may spread, producing disseminated aspergillosis in a variety of tissues and organs (**figure 39.25**). In patients whose resistance is severely compromised, invasive aspergillosis may occur and fill the lung with fungal hyphae. Immune disorders: Hypersensitivities (section 32.11)

Laboratory diagnosis of aspergillosis depends on identification, either by direct examination of pathological specimens or by isolation and characterization of the fungus. Successful therapy depends on treatment of the underlying disease so that host resistance increases. Treatment is with itraconazole. Antifungal drugs (section 34.7)

Candidiasis

Candidiasis is the mycosis caused by the dimorphic fungus *Candida albicans* (**figure 39.26*a***) or *C. glabrata*. In contrast to the other pathogenic fungi, *C. albicans* and *C. glabrata* are members of the normal microbiota within the gastrointestinal tract, respiratory tract, vaginal area, and mouth. In healthy individuals they do not produce disease because growth is suppressed by other microbiota and other host resistance mechanisms. However, if anything upsets the normal microbiota and immunecompetency, *Candida* may multiply rapidly and produce candidiasis. *Candida* species are important nosocomial pathogens. In some hospitals they may represent almost 10% of nosocomial bloodstream infections. Because *Candida* can be transmitted sexually, it is also listed by the CDC as a sexually transmitted disease.

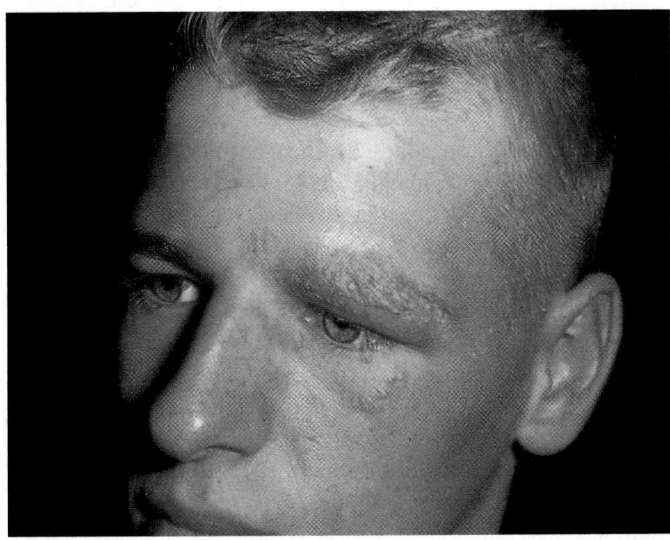

Figure 39.25 An Opportunistic Mycosis. Aspergillosis of the eye caused by *Aspergillus fumigatus*.

No other mycotic pathogen produces as diverse a spectrum of disease in humans as does *Candida* (**Disease 39.2**). Most infections involve the skin or mucous membranes. This occurs because *Candida* is a strict aerobe and finds such surfaces very suitable for growth. Cutaneous involvement usually occurs when the skin becomes overtly moist or damaged.

Oral candidiasis, or **thrush** (figure 39.26*b*), is a fairly common disease in newborns. It appears as many small, white flecks that cover the tongue and mouth. At birth, newborns do not have a normal microbiota in the oropharyngeal area. If the mother's vaginal area is heavily colonized with *Candida,* the upper respiratory tract of the newborn becomes colonized during passage through the birth canal. Thrush occurs because growth of *Candida* cannot be inhibited by the other microbiota. Once the newborn has developed his or her own normal oropharyngeal microbiota, thrush becomes uncommon. **Paronychia** and **onychomycosis** are associated with *Candida* infections of the subcutaneous tissues of the digits and nails, respectively (figure 39.26*c*). These infections usually result from continued immersion of the appendages in water.

Intertriginous candidiasis involves those areas of the body, usually opposed skin surfaces, that are warm and moist: axillae, groin, skin folds. **Napkin (diaper) candidiasis** is typically found in infants whose diapers are not changed frequently and therefore are not kept dry. **Candidal vaginitis** can result as a complication of diabetes, antibiotic therapy, oral contraceptives, pregnancy, or any other factor that compromises the female host. Normally the omnipresent lactobacilli can control *Candida* by the low pH they create. However, if their numbers are decreased by any of the

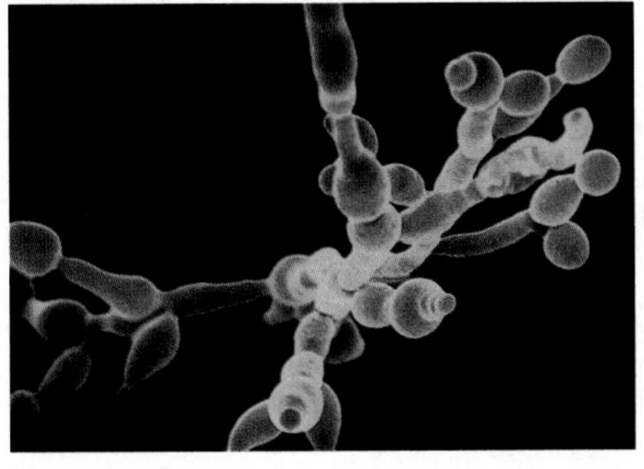

(a)

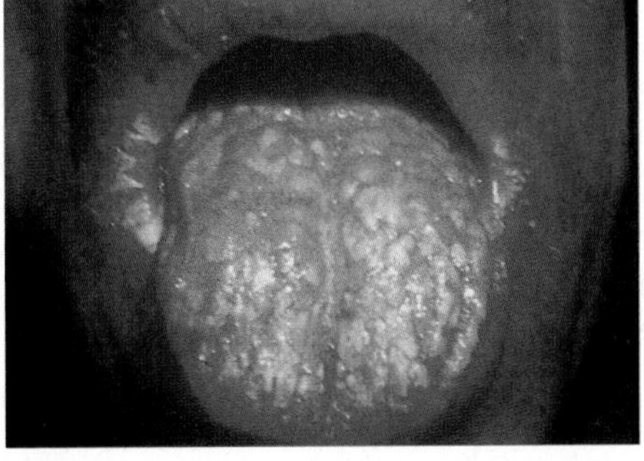

(b)

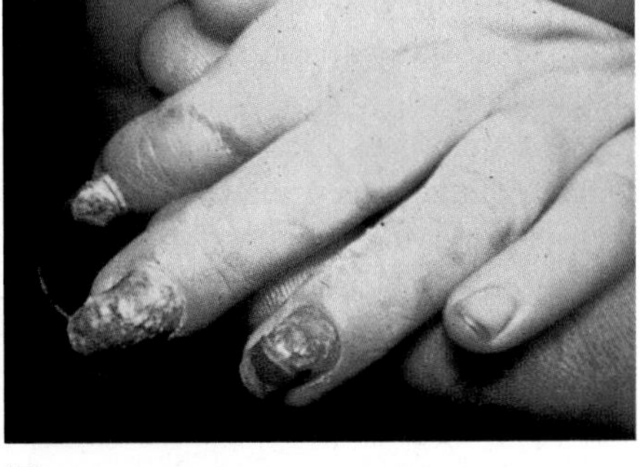

(c)

Figure 39.26 Opportunistic Mycoses Caused by *Candida albicans*. **(a)** Scanning electron micrograph of the yeast form (×10,000). Notice that some of the cells are reproducing by budding. **(b)** Thrush, or oral candidiasis, is characterized by the formation of white patches on the mucous membranes of the tongue and elsewhere in the oropharyngeal area. These patches form a pseudomembrane composed of spherical yeast cells, leukocytes, and cellular debris. **(c)** Paronychia and onychomycosis of the hands.

Disease

39.2 The Emergence of Candidiasis

Written descriptions of oral lesions that were probably thrush date back to the early 1800s. In 1839 Bernard Langenbeck in Germany described the organism he found in oral lesions of a patient as "Typhus-Leichen" (typhus bodies). By 1841 Emil Berg established the fungal etiology of thrush by infecting healthy babies with what he called "aphthous membrane material." In 1843 Charles Robin gave the organism its first name: *Oidium albicans.* Since then, more than 100 synonyms have been used for this fungus; of them all, *Candida albicans,* the name Roth Berkhout proposed in 1923, has persisted.

In 1861 Albert Zenker described the first well-documented case of systemic candidiasis. Historically, the most interesting period for candidiasis research coincided with the introduction of antibiotics. Since then there have been documented cases of this fungus involving all tissues and organs of the body, as well as an increase in the overall incidence of candidiasis. Some *Candida*-associated infections and diseases include AIDS arthritis, endophthalmitis, meningitis, myocarditis, myositis, and peritonitis. Besides the widespread use of antibiotics, other therapeutic and surgical procedures such as organ transplants and prostheses have been important in the expanding worldwide incidence of candidiasis.

aforementioned factors, *Candida* may proliferate, causing a curdlike, yellow-white discharge from the vaginal area. *Candida* can be transmitted to males during intercourse and lead to **balanitis;** thus it also can be considered a sexually transmitted disease. Balanitis is a *Candida* infection of the male glans penis and occurs primarily in uncircumcised males. The disease begins as vesicles on the penis that develop into patches and are accompanied by severe itching and burning.

Diagnosis of candidiasis is sometimes difficult because (1) this fungus is a frequent secondary invader in diseased hosts, (2) a mixed microbiota is most often found in the diseased tissue, and (3) no completely specific immunologic procedures for the identification of *Candida* currently exist. Mortality is almost 50% when *Candida* invade the blood or disseminate to visceral organs, as occasionally seen in immunocompromised patients.

There is no satisfactory treatment for candidiasis. Cutaneous lesions can be treated with topical agents such as sodium caprylate, sodium propionate, gentian violet, nystatin, miconazole, and trichomycin. Ketoconazole, amphotericin B, fluconazole, itraconazole, and flucytosine also can be used for systemic candidiasis.

Microsporidia

Microsporidia is a term used to describe obligate, intracellular fungi that belong to the phylum *Microspora*. Microsporidosis is an emerging infectious disease, found mostly in HIV patients. The taxonomy of these microogranisms is unsettled and many still consider them protists. More than 1,500 species have been catalogued, in 143 genera; at least 14 species are known to be human pathogens. Several domestic and feral animals appear to be reservoirs for several species that infect humans. Increasingly recognized as opportunistic infectious agents, microsporidia infect a wide range of vertebrate and invertebrate hosts. One unifying characteristic of the microsporidia is their production of a highly resistant spore, capable of surviving long periods of time in the environment. Spore morphology varies with species, but most resemble enteric bacteria. Microsporidial spores recovered from human infections are oval to rodlike, measuring 1 to 4 micrometers. Microsporidia also possess a unique organelle known as the **polar tubule,** which is coiled within the spore (**figure 39.27;** also see *figure 26.17*). Characteristics of the fungal divisions: *Microsporidia* (section 26.6)

Infection of a host cell results when the microsporidia extrudes its polar tubule from within the spore. Contact with a eucaryotic cell membrane allows the polar tubule to bore through the membrane. A sudden increase in spore calcium results in the injection of the sporoplasm (cytoplasm-like contents) through the polar tubule into the host cell. Inside the cell, the sporoplasm condenses and undergoes asexual multiplication. Multiplication of the sporoplasm is species-specific, occurring within the host cytoplasm or within a vacuole, and is usually completed through binary (merogony) or multiple (shizogony) fission. Once the sporoplasm has multiplied, it directs the development of new spores (sporogony) by first encapsulating meronts or shizonts (products of asexual multiplication) within a thick spore coating. New spores increase until the host cell membrane ruptures, releasing the progeny.

Microsporidia infections can result in a wide variety of patient symptoms. These include hepatitis, pneumonia, skin lesions, diarrhea, weight loss, and wasting syndrome. Diagnosis is based on clinical manifestations, and identification of microsporidia in gram-stained or giemsa-stained specimens (although the latter is somewhat difficult). Where possible by electron microscopy, identification can be made based on the characteristic polar tubule coiled within a spore. Molecular identification is also possible using PCR and primers to the small subunit ribosomal RNA (*see figure 19.10*). Treatment of microsporidiosis is not well defined. However, some successes have been reported with the use of albendazole, which inhibits tubulin and ATP synthesis in helminths, metronidazole, or thalidomide (a toxic, immunosupressant drug).

Pneumocystis Pneumonia

Pneumocystis is a fungus found in the lungs of a wide variety of mammals. Although it was previously classified as a protist, its rRNA, DNA sequences (from several genes) and biochemical

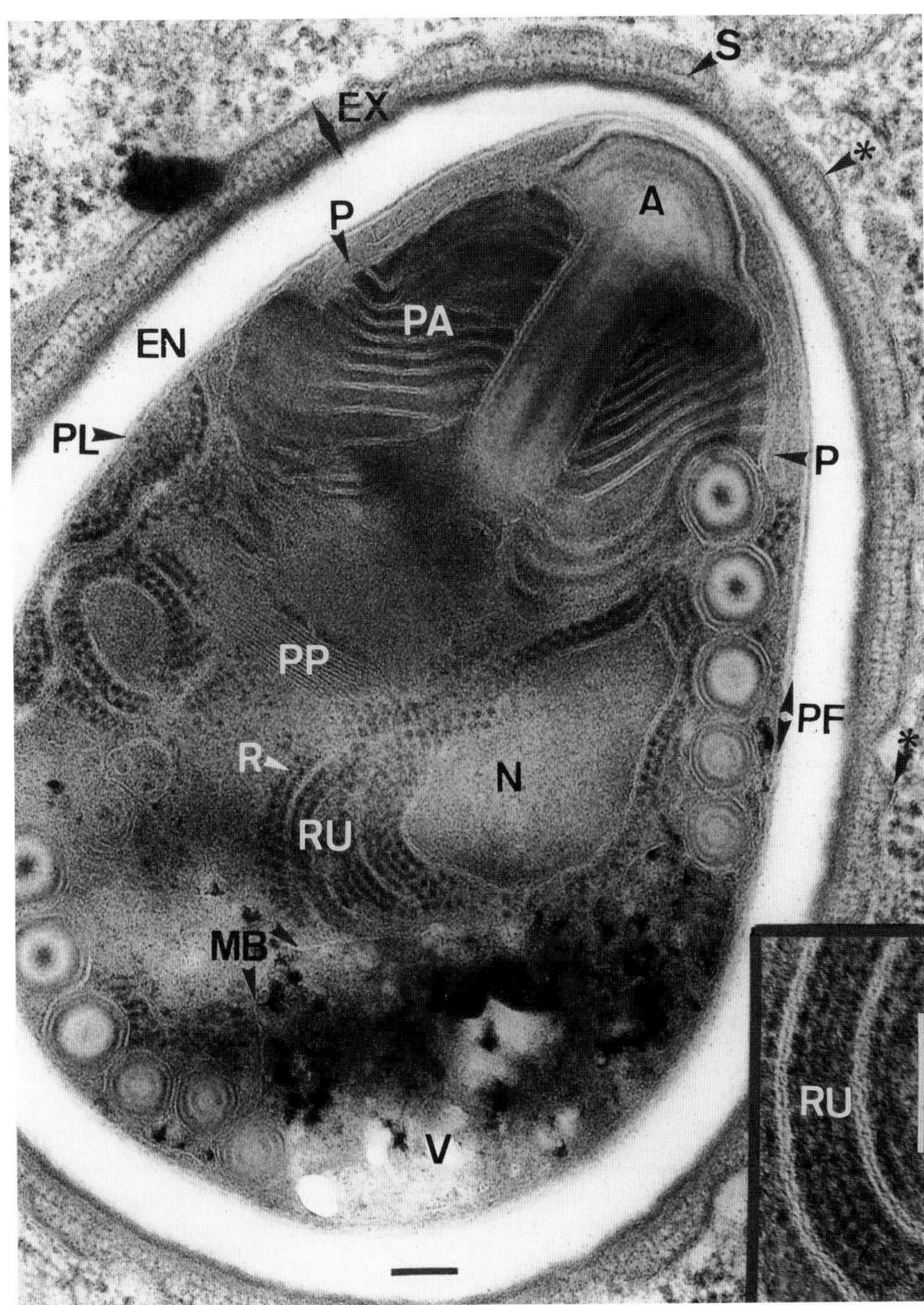

Figure 39.27 Microsporidia. Transmission electron micrograph of an *Episeptum inversum* spore. The infection apparatus has three components: PF, the polar filament, which is attached to A, the anchoring disc; and PA and PP the anterior and posterior parts of the polaroplast, a system of membrane-bound sacs. Two wall layers are seen: EX, an external electron-dense exospore layer, EN a wide endospore layer containing chitin, as well as MB, an internal plasma membrane. N: nucleus, R: ribosomes, RU rough endoplasmic reticulum (see inset); S: septum of the exospore coat; V, posterior vacuole, P, polar sac, and * outermost layer of exospore.

analyses have shown that *Pneumocystis* is more closely related to fungi than to protists. The life cycle of *Pneumocystis* is presented in **figure 39.28,** although some aspects of its development are not well known. The disease that this fungus causes has been called ***Pneumocystis* pneumonia** or ***Pneumocystis carinii* pneumonia (PCP).** Its name was changed to *Pneumocystis jiroveci* in honor of the Czech parasitologist, Otto Jirovec, who first described this pathogen in humans. In recent medical literature, the acronym PCP for the disease has been retained despite the loss of the old species name (*carinii*). PCP now stands for *Pneumocystis* pneumonia.

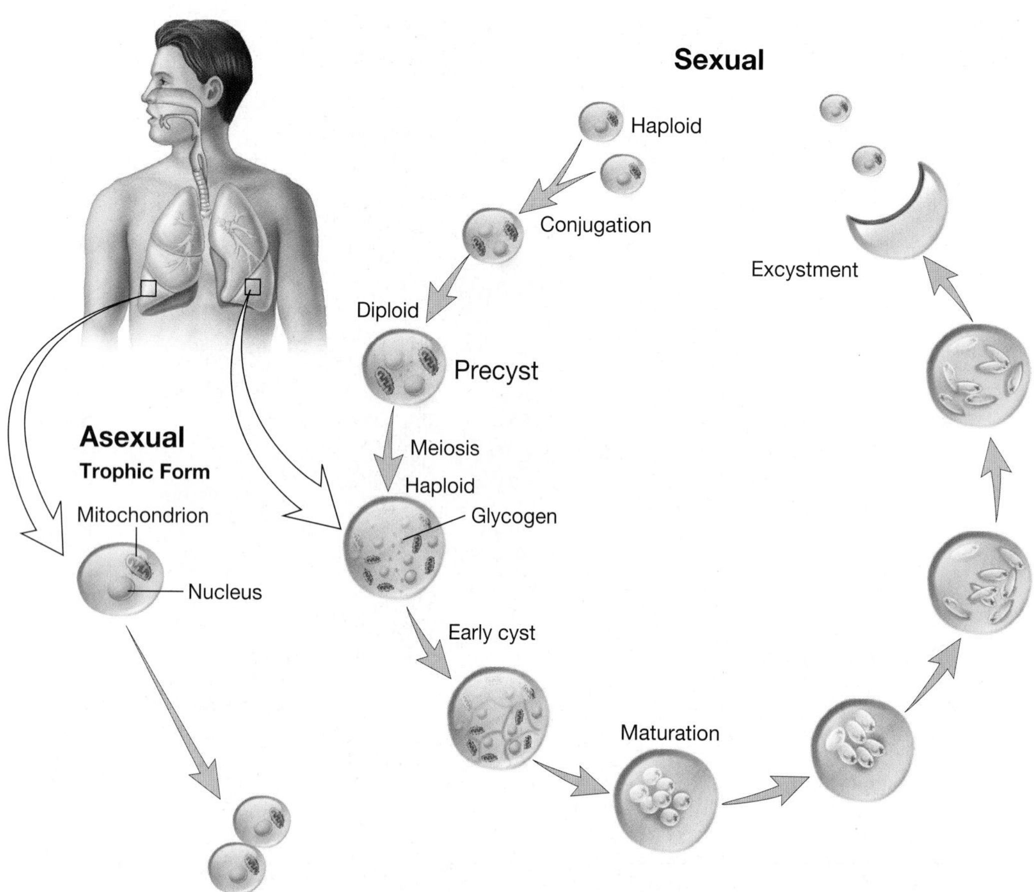

Figure 39.28 Life Cycle of *Pneumocystis jiroveci*. The *Pneumocystis* fungus exhibits both sexual and asexual reproductive stages. Sexual reproduction occurs when haploid cells undergo conjugation followed by meiosis. Asexual reproduction occurs by binary fission. Both forms of reproduction can occur within the human host.

Serological data indicate that most humans are exposed to *Pneumocystis* by age 3 or 4. However, PCP occurs almost exclusively in immunocompromised hosts. Extensive use of immunosuppressive drugs and irradiation for the treatment of cancers and following organ transplants accounts for the formidable prevalence rates for PCP. This pneumonia also occurs in premature, malnourished infants and in more than 80% of AIDS patients. Both the organism and the disease remain localized in the lungs—even in fatal cases. Within the lungs, *Pneumocystis* causes the alveoli to fill with a frothy exudate.

Laboratory diagnosis of *Pneumocystis* pneumonia can be made definitively only by microscopically demonstrating the presence of the microorganisms in infected lung material or by a PCR analysis. Treatment is by means of oxygen therapy and either a combination of trimethoprim and sulfamethoxazole (Bactrim, Septra), atovaquone (Mepron), or trimetrexate (Neutrexin). Prevention and control are through prophylaxis with drugs in susceptible persons.

1. How do infections caused by *Entamoeba histolytica* occur?
2. What is the most common cause of epidemic waterborne diarrheal disease?
3. Describe in detail the life cycle of the malarial parasite.
4. Why are some mycotic diseases of humans called opportunistic mycoses?
5. What parts of the human body can be affected by *Candida* infections?
6. Describe the infection process used by microsporidia.
7. When is *Pneumocystis* pneumonia likely to occur in humans?

Summary

39.1 Pathogenic Fungi and Protists

a. Human fungal diseases, or mycoses, can be divided into five groups according to the level and mode of entry into the host. These are the superficial, cutaneous, subcutaneous, systemic, and opportunistic mycoses (**tables 39.1** and **39.2**).

b. Protozoa are responsible for some of the most serious human diseases that affect hundreds of millions of people worldwide (**tables 39.3** and **39.4**) and can also be grouped by route of disease transmission.

39.2 Airborne Diseases

a. Most systemic mycoses that occur in humans are acquired by inhaling the spores from the soil where the free-living fungi are found. Four types can occur in humans: blastomycosis (**figure 39.1**), coccidioidomycosis (**figure 39.2**), cryptococcosis, (**figure 39.3**) and histoplasmosis (**figure 39.4**).

39.3 Arthropod-Borne Diseases

a. The most important human parasite among the sporozoa is *Plasmodium,* the causative agent of malaria (**figure 39.5**). Human malaria is caused by four species of *Plasmodium: P. falciparum, P. vivax, P. malariae,* and *P. ovale.*

b. The flagellated protozoa that are transmitted by arthropods and infect the blood and tissues of humans are called hemoflagellates. Two major groups occur: the leishmanias, which cause the diseases collectively termed leishmaniasis (**figure 39.8**), and the trypanosomes (**figures 39.9** and **39.10**), which cause trypanosomiasis.

39.4 Direct Contact Diseases

a. Superficial mycoses of the hair shaft are collectively called piedras. Two major types are black piedra (**figure 39.11**) and white piedra. Tinea versicolor is a third common superficial mycosis.

b. The cutaneous fungi that parasitize the hair, nails, and outer layer of the skin are called dermatophytes, and their infections are termed dermatophytoses, ringworms, or tineas. At least seven types can occur in humans: tinea barbae (ringworm of the beard), tinea capitis (ringworm of the scalp; **figure 39.12**), tinea corporis (ringworm of the body; **figure 39.13**), tinea cruris (ringworm of the groin; **figure 39.14**), tinea pedis (ringworm of the feet; **figure 39.15**), tinea manuum (ringworm of the hands), and tinea unguium (ringworm of the nails; **figure 39.16**).

c. The dermatophytes that cause the subcutaneous mycoses are normal saprophytic inhabitants of soil and decaying vegetation. Three types of subcutaneous mycoses can occur in humans: chromoblastomycosis (**figure 39.17**), maduromycosis (**figure 39.18**), and sporotrichosis (**figure 39.19**).

d. Toxoplasmosis is a disease caused by the protozoan *Toxoplasma gondii.* It is one of the major causes of death in AIDS patients (**figure 39.20**).

e. Trichomoniasis is a sexually transmitted disease caused by the protozoan flagellate *Trichomonas vaginalis* (**figure 39.21**).

39.5 Food-Borne and Waterborne Diseases

a. *Entamoeba histolytica* is the amoeboid protozoan responsible for amebiasis. This is a very common disease in warm climates throughout the world. It is acquired when cysts are ingested with contaminated food or water (**figure 39.22**).

b. Fresh water parasites like *Naegleria* and *Acanthamoeba* can cause primary amebic meningioencephalitis.

c. *Cryptosporidium parvum* is a common coccidial apicomplexan parasite that causes severe diarrheal disease. It is acquired from contaminated food or water.

d. *Cyclospora cayetanensis* is shed in the feces of a current host and can only infect the gastrointestinal tract of a new host after it has developed at 22–32°C for several days or weeks. It can be acquired from contaminated water used to rinse fruit or vegetables (**figure 39.23**).

e. *Giardia intestinalis* is a flagellated protozoan that causes the common intestinal disease giardiasis (**figure 39.24**). This disease is distributed throughout the world, and in the United States it is the most common cause of waterborne diarrheal disease.

39.6 Opportunistic Diseases

a. An opportunistic organism is one that is generally harmless in its normal environment but can become pathogenic in a compromised host. The most important opportunistic mycoses affecting humans include systemic aspergillosis (**figure 39.25**), candidiasis (**figure 39.26*b***), and *Pneumocystis* pneumonia (**figure 39.28**). An emerging opportunistic fungal disease, especially of the immunocompromised, is caused by the group of unique microbes known as the microsporidia. They live as a spore form and have a unique organelle used for infecting new host cells (**figure 39.27**).

Key Terms

amebiasis (amebic dysentery) 1012
American trypanosomiasis 1007
aspergillosis 1016
athlete's foot 1009
balanitis 1018
black piedra 1008
blastomycosis 999
candidal vaginitis 1017
candidiasis 1017
Chagas' disease 1007
chromoblastomycosis 1010
coccidioidomycosis 1000
compromised host 1016
cryptococcosis 1001
cryptosporidiosis 1014
cyclosporiasis 1014
dermatomycosis 1008
dermatophyte 1008
eumycotic mycetoma 1010
extracutaneous sporotrichosis 1010
giardiasis 1014
histoplasmosis 1001
intertriginous candidiasis 1017
jock itch 1009
keratitis 1013
leishmania 1004
leishmaniasis 1004
maduromycosis 1010
malaria 1001
medical mycology 997
microsporidia 1018
mycosis 997
napkin (diaper) candidiasis 1017
onychomycosis 1017
opportunistic microorganism 1016
oral candidiasis 1017
paronychia 1017
piedra 1008
Pneumocystis carinii pneumonia (PCP) 1019
Pneumocystis pneumonia 1019
polar tubule 1018
primary amebic meningoencephalitis 1013
ringworm 1008
sporotrichosis 1010
thrush 1017
tinea 1008
tinea barbae 1008
tinea capitis 1008
tinea corporis 1009
tinea cruris 1009
tinea manuum 1009
tinea pedis 1009
tinea unguium 1009
tinea versicolor 1008
toxoplasmosis 1011
trichomoniasis 1012
trypanosome 1006
trypanosomiasis 1006
white piedra 1008

Critical Thinking Questions

1. Compare and contrast treatment of diseases caused by fungi with those caused by viruses or bacteria.
2. What is one distinct feature of fungi that could be exploited for antibiotic therapy?
3. Why do you think most fungal diseases in humans are not contagious?
4. Trypanosomes are notorious for their ability to change their surface antigens frequently. Given the kinetics of a primary immune response (primary antibody production), how often would the surface antigen need to be changed to stay "ahead" of the antibody specificity? Why shouldn't it change the expression every time transcription occurs?

Learn More

Breman, J. G. 2001. The ears of the hippopotamus: Manifestations, determinants and estimates of the malaria burden. *Am. J. Tropical Med. Hygiene.* 64:1–11.

Casadevall, A.; Steenbergen, J. N.; and Nosanchuk, J. D. 2003. 'Ready made' virulence and 'dual use' virulence factors in pathogenic environmental fungi—the *Cryptococcus neoformis* paradigm. *Curr. Opin. Microbiol.* 6:332–37.

Gull, K. 2003. Host-parasite interactions and typanosome morphogenesis: A pocket full of goodies. *Curr. Opin. Microbiol.* 6:365–70.

Herwaldt, B. L. 2000. *Cyclospora cayetanensis:* A review, focusing on the outbreaks in the 1990s. *Clin. Infect. Dis.* 31:1040–57.

Herwaldt, B. L. 1999. Leishmaniasis. *Lancet* 354:1191–99.

Ho, A. Y.; Lopez, A. S.; Eberhart, M. G.; Finkel, B. S.; da Silva, A. J. 2002. Outbreak of cyclosporiasis associated with imported raspberries, Philadelphia, PA. *Emerging Infectious Diseases* 8:738–88.

McGovern, T. W.; Williams, W.; Fitzpatrick, J. E.; Cetron, M. S.; Hepburn, B. C.; Gentry, R. H. 1995. Cutaneous manifestations of African trypanosomiasis, *Arch Dermatol.* 131:1178–82.

Perfect, J. R. 2005. Weird fungi. *ASM News* 71:407–12.

Ravdin, J. I. 1995. Amebiasis. *Clin. Infect. Dis.* 20:1453–66.

Romani, L.; Bistoni, F.; and Puccetti, P. 2003. Adaptation of *Candida albicans* to the host environment: The role of morphogenesis in virulence and survival in mammalian hosts. *Curr. Opin. Microbiol.* 6:338–43.

Shah, S.; Filler, S.; Causer, L. M.; Rowe, A. K.; Bloland, P. B.; Barber, A. M., et al. 2002. Malaria surveillance-United States. *MMWR Surveillance Summary* 53:21–34.

Stringer, J. R.; Beard, C. B.; Miller, R. F.; and Wakefield, A. E. 2002. A new name (*Pneumocystis jiroveci*) for *Pneumocystis* from humans. *Emerg. Infect. Dis.* 8:891–96.

Woods, J. P. 2003. Knocking on the right door and making a comfortable home: *Histoplasma capsulatum* intracellular pathogenesis. *Curr. Opin. Microbiol.* 6: 327–31.

Please visit the Prescott website at www.mhhe.com/prescott7 for additional references.

40 Microbiology of Food

Large tanks used for wine production. Fermentations can be carried out in such open-air units in temperate regions. After completion of the fermentation, the fresh wine will be transferred to barrels for storage and aging.

Preview

- Food spoilage is a major problem in all societies. It can occur at any point in the course of food production, transport, storage, or preparation. Food-borne toxins are of increasing concern, especially with increases in international shipments and extended storage of food products before use. Growth of fungi can result in the synthesis of toxins such as aflatoxins, fumonisins, and ergot alkaloids. Algal-derived toxins can be transmitted to humans through freshwater and marine-derived food products.
- Foods can be preserved by physical, chemical, and biological processes. Refrigeration does not significantly reduce microbial populations but only retards spoilage. Pasteurization results in a pathogen-free product with a longer shelf life.
- Chemicals can be added to foods to control microbial growth. Such chemicals include sugar, salt, and many organic chemicals that affect specific groups of microorganisms. Microbial products, such as bacteriocins, can be added to foods to control spoilage organisms.
- Foods can transmit a wide range of diseases to humans. In a food infection, the food serves as a vehicle for the transfer of the pathogen to the consumer, in whom the pathogen grows and causes disease. With a food intoxication, the microorganisms grow in the food and produce toxins that can then affect the consumer.
- Foods that are consumed raw pose a risk of disease transmission if care is not taken in production, storage, and transport. Without adequate care, major disease outbreaks can occur because foods travel around the globe in extremely short times.
- Detection of food-borne pathogens is carried out using classic culture techniques, as well as immunological and molecular procedures.
- Dairy products, grains, meats, fruits, and vegetables can be fermented. Lactic acid bacteria are the principal microbes involved in milk fermentation; fungi are also used.
- Wines are produced by the direct fermentation of fruit juices or musts. For fermentation of cereals and grains, starches and proteins contained in these substrates must first be hydrolyzed to provide substrates for alcoholic fermentation.
- The making of bread, sauerkraut, sufu, pickles, and many other foods also involves the use of complex fermentation processes. When chopped plant materials are fermented, silages are created, which can be stored and used by animals.
- Microbial cells can be used as food sources and food amendments. These include mushrooms, cyanobacteria such as *Spirulina*, and yeasts. There is an increasing interest in probiotics—the use of microorganisms to change the microbial community in the intestine. Microbial colonization of surfaces in the intestine plays a critical role in these processes.

Foods, microorganisms, and humans have had a long and interesting association that developed long before recorded history. Foods are not only of nutritional value to those who consume them but often are ideal culture media for microbial growth. Microorganisms can be used to transform raw foods into gastronomic delights, including chocolate, cheeses, pickles, sausages, and soy sauce. Wines, beers, and other alcoholic products also are produced through microbial activity. On the other hand, microorganisms can degrade food quality and lead to spoilage. Importantly, foods also can serve as vehicles for disease transmission. The detection and control of pathogens and food spoilage microorganisms are important parts of food microbiology. During the entire sequence of food handling, from the producer to the final consumer, microorganisms can affect food quality and human health. In this chapter we consider the two opposing roles of microorganisms in food production and preservation.

Tell me what you eat, and I will tell you what you are.

—Brillat-Savarin

40.1 Microorganism Growth in Foods

Foods, because they are nutrient-rich, are excellent environments for the growth of microorganisms. Microbial growth is controlled by factors related to the food itself, called **intrinsic factors,** and also to the environment where the food is stored, described as **extrinsic factors,** as shown in **figure 40.1.**

The intrinsic or food-related factors include pH, moisture content, water activity or availability, oxidation-reduction potential, physical structure of the food, available nutrients, and the possible presence of natural antimicrobial agents. Extrinsic or environmental factors include temperature, relative humidity, gases (CO_2, O_2) present, and the types and numbers of microorganisms present in the food.

Intrinsic Factors

Food composition is a critical intrinsic factor that influences microbial growth. If a food consists primarily of carbohydrates, fungal, rather than bacterial, growth predominates and spoilage does not result in major odors. Thus foods such as breads, jams, and some fruits first show spoilage by fungal growth. In contrast, when foods contain large amounts of proteins and/or fats (for example, meat and butter), bacterial growth can produce a variety of foul odors. One only need think of rotting eggs. This anaerobic breakdown of proteins yields foul-smelling amine compounds and is called **putrefaction.** One major source of odor is the organic amine cadaverine (imagine the origin of that name). Degradation of fats ruins food as well. The production of short-chained fatty acids from fats renders butter rancid and foul smelling.

The pH of a food also is critical because a low pH favors the growth of yeasts and molds. In neutral or alkaline pH foods, such as meats, bacteria are more dominant in spoilage and putrefaction. Depending on the major substrate present in a food, different types of spoilage may occur (**table 40.1**). The influence of environmental factors on growth (section 6.5)

The presence and availability of water also affect the ability of microorganisms to colonize foods. Simply by drying a food, one can control or eliminate spoilage processes. Water, even if present, can be made less available by adding solutes such as sugar and salt. Water availability is measured in terms of water activity (a_w). This represents the ratio of relative humidity of the air over a test solution compared with that of distilled water. When large quantities of salt or sugar are added to food, most microorganisms are dehydrated by the hypertonic conditions and cannot grow (**table 40.2;** *see also table 6.4*). Even under these adverse conditions, osmophilic and xerophilic microorganisms may spoil food. **Osmophilic** [Greek *osmus,* impulse, and *philein,* to love] **microorganisms** grow best in or on media with a high osmotic concentration, whereas **xerophilic** [Greek *xerosis,* dry, and *philein,* to love] **microorganisms** prefer a low a_w environment and may not grow under high a_w conditions.

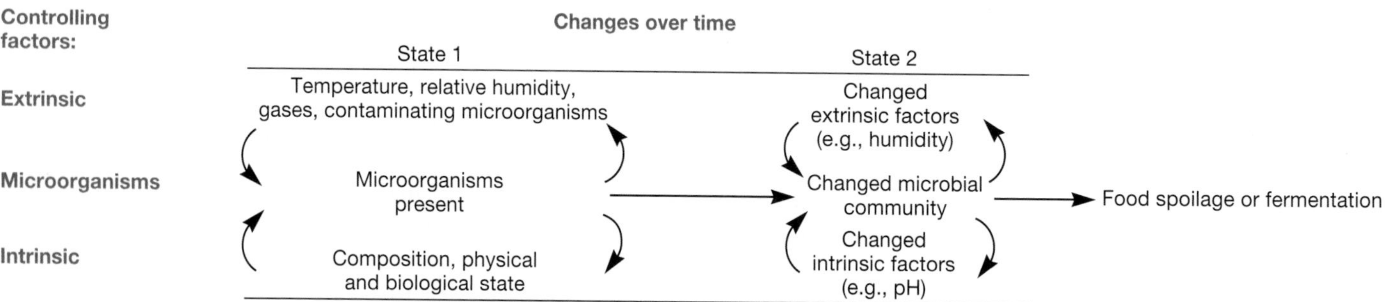

Figure 40.1 Intrinsic and Extrinsic Factors. A variety of intrinsic and extrinsic factors can influence microbial growth in foods. Time-related successional changes occur in the microbial community and the food.

Table 40.1 Differences in Spoilage Processes in Relation to Food Characteristics

Substrate	Food Example	Chemical Reactions or Processes[a]	Typical Products and Effects
Pectin	Fruits	Pectinolysis	Methanol, uronic acids (loss of fruit structure, soft rots)
Proteins	Meat	Proteolysis, deamination	Amino acids, peptides, amines, H_2S, ammonia, indole (bitterness, souring, bad odor, sliminess)
Carbohydrates	Starchy foods	Hydrolysis, fermentations	Organic acids, CO_2, mixed alcohols (souring, acidification)
Lipids	Butter	Hydrolysis, fatty acid degradation	Glycerol and mixed fatty acids (rancidity, bitterness)

[a]Other reactions also occur during the spoilage of these substrates.

Table 40.2 Approximate Minimum Water Activity Relationships of Microbial Groups and Specific Organisms Important in Food Spoilage

Organisms	a_w	Organisms	a_w
Groups		**Groups**	
Most spoilage bacteria	0.9	Halophilic bacteria	0.75
Most spoilage yeasts	0.88	Xerophilic molds	0.61
Most spoilage molds	0.80	Osmophilic yeasts	0.61
Specific Microorganisms		**Specific Microorganisms**	
Clostridium botulinum, type E	0.97	*Candida scottii*	0.92
Pseudomonas spp.	0.97	*Trichosporon pullulans*	0.91
Acinetobacter spp.	0.96	*Candida zeylanoides*	0.90
Escherichia coli	0.96	*Geotrichum candidum*	~0.90
Enterobacter aerogenes	0.95	*Trichothecium* spp.	~0.90
Bacillus subtilis	0.95	*Byssochlamys nivea*	~0.87
Clostridium botulinum, types A and B	0.94	*Staphylococcus aureus*	0.86
Candida utilis	0.94	*Alternaria citri*	0.84
Vibrio parahaemolyticus	0.94	*Pencillium patulum*	0.81
Botrytis cinerea	0.93	*Eurotium repens*	0.72
Rhizopus stolonifer	0.93	*Aspergillus conicus*	0.70
Mucor spinosus	0.93	*Aspergillus echinulatus*	0.64
		Zygosaccharomyces rouxii	0.62
		Xeromyces bisporus	0.51

Adapted from James M. Jay. 2000. *Modern Food Microbiology,* 6th edition. Reprinted by permission of Aspen Publishers, Inc. Gaithersburg, MD. Tables 3–5, p. 42.

The oxidation-reduction potential of a food also influences spoilage. When meat products, especially broths, are cooked, they often have lower oxidation-reduction potentials—that is, they present a reducing environment for microbial grow. These products with their readily available amino acids, peptides, and growth factors are ideal media for the growth of anaerobes, including *Clostridium* (*see table 38.6*).

The physical structure of a food also can affect the course and extent of spoilage. The grinding and mixing of foods such as sausage and hamburger not only increase the food surface area, but also distribute contaminating microorganisms throughout the food. This can result in rapid spoilage if such foods are stored improperly. Vegetables and fruits have outer skins (peels and rinds) that protect them from spoilage. Often spoilage microorganisms have specialized enzymes that help them weaken and penetrate protective peels and rinds, especially after the fruits and vegetables have been bruised.

Many foods contain natural antimicrobial substances, including complex chemical inhibitors and enzymes. Coumarins found in fruits and vegetables exhibit antimicrobial activity. Cow's milk and eggs also contain antimicrobial substances. Eggs are rich in the enzyme lysozyme that can lyse the cell walls of contaminating gram-positive bacteria (*see figure 31.17*).

Herbs and spices often possess significant antimicrobial substances; generally fungi are more sensitive than most bacteria. Sage and rosemary are two of the most antimicrobial spices. Aldehydic and phenolic compounds that inhibit microbial growth are found in cinnamon, mustard, and oregano. Other important inhibitors are garlic, which contains allicin, cloves, which have eugenol, and basil, which contains rosmarinic acid. However, spices also can sometimes contain pathogenic and spoilage organisms. Enteric bacteria, *B. cereus, Clostridium perfringens,* and *Salmonella* species have been detected in spices. Microorganisms can be eliminated or reduced by ethylene oxide sterilization. This treatment can result in *Salmonella*-free spices and herbs and a 90% reduction in the levels of general spoilage organisms. The use of chemical agents in control (section 7.5)

Unfermented green and black teas also have well-documented antimicrobial properties because of their polyphenol contents, which apparently are diminished when the teas are fermented. Such unfermented teas are active against bacteria, viruses, and fungi and may have anticancer properties. The term "fermentation" is a misnomer when applied to tea because it does not involve the metabolic processes of a specific microbe or group of microbes. Tea "fermentation" involves the drying and spreading of tea leaves in a cool, humid environment where plant enzymes oxidize compounds within the leaves. After one to five hours, the leaves are heated to inactivate the enzymes and remove remaining water. The longer the tea is allowed to "ferment," the more caffeine the end product contains.

Extrinsic Factors

Temperature and relative humidity are important extrinsic factors in determining whether a food will spoil. At higher relative humidities microbial growth is initiated more rapidly, even at lower temperatures (especially when refrigerators are not maintained in a defrosted state). When drier foods are placed in moist environments, moisture absorption can occur on the food surface, eventually allowing microbial growth.

The atmosphere in which food is stored also is important. This is especially true with shrink-packed foods because many plastic films allow oxygen diffusion, which results in increased growth of surface-associated microorganisms. Excess CO_2 can decrease the solution pH, inhibiting microbial growth. Storing meat in a high CO_2 atmosphere inhibits gram-negative bacteria, resulting in a population dominated by the lactobacilli.

The observation that food storage atmosphere is important has led to the development of **modified atmosphere packaging (MAP).** Modern shrink-wrap materials and vacuum technology make it possible to package foods with controlled atmospheres. These materials are largely impermeable to oxygen. This prolongs shelf-life by a factor of two to five times compared to the same product packaged in air. With a carbon dioxide content of 60% or greater in the atmosphere surrounding a food, spoilage fungi will not grow, even if low levels of oxygen are present. Recently, it has been found that high-oxygen MAP also may be effective. This is due to the formation of the superoxide (O_2^-) anion inside cells under these conditions. The superoxide anion is then transformed to highly toxic peroxide and oxygen, resulting in antimicrobial effects. Some products currently packaged using MAP technology include delicatessen meats and cheeses, pizza, grated cheese, some bakery items, and dried products such as coffee.

1. What are some intrinsic factors that influence food spoilage and how do they exert their effects?
2. What are the effects of food composition on spoilage processes?
3. Why might sausage and other ground meat products provide a better environment for the growth of food spoilage organisms than solid cuts of meats?
4. List some antimicrobial substances found in foods. What is the mechanism of action of lysozyme?
5. What primary extrinsic factors can determine whether food spoilage will occur?
6. What are the major gases involved in MAP? How are their concentrations varied to inhibit microbial growth?

40.2 Microbial Growth and Food Spoilage

Because foods are such excellent sources of nutrients, if the intrinsic and extrinsic conditions are appropriate, microorganisms grow rapidly and convert an attractive and appealing food into a sour, foul-smelling, or fungus-covered mass. Microbial growth in and on foods can lead to visible changes, including a variety of colors caused by spoilage organisms, which often have been associated with "miracles" and "witchcraft." One of the most famous is the report of "blood" on communion wafers and other bread, called the "Miracle of Bolsena," which occurred in 1263. The riddle was eventually solved by Bartolomeo Bizio in 1879, when he described the red-pigmented bacterium responsible for this phenomenon. He also named the bacterium *Serratia marcescens.*

Meat and dairy products, with their high nutritional value and the presence of easily metabolized carbohydrates, fats, and proteins provide ideal environments for microbial spoilage. Proteolysis and putrefaction are typical results of microbial spoilage of such high-protein materials. Unpasteurized milk undergoes a predictable four-step microbial succession during spoilage; acid production by *Lactococcus lactis* subsp. *lactis* is followed by additional acid production associated with the growth of more acid tolerant organisms such as *Lactobacillus.* At this point yeasts and molds become dominant and degrade the accumulated lactic acid, and the acidity gradually decreases. Eventually protein-digesting bacteria become active, resulting in a putrid odor and bitter flavor. The milk, originally opaque, eventually becomes clear (**figure 40.2**).

In comparison with meat and dairy products, most fruits and vegetables have a much lower protein and fat content and undergo a different kind of spoilage. Readily degradable carbohydrates favor spoilage by bacteria, especially bacteria that cause soft rots, such as *Erwinia carotovora,* which produces hydrolytic enzymes. The high oxidation-reduction potential and lack of reduced conditions permits aerobes and facultative anaerobes to contribute to the decomposition processes. Bacteria do not seem important in the initial spoilage of whole fruits; instead such

Figure 40.2 Spoilage of a Dairy Product. Fresh (left) and curdled (right) milk are shown. The curdled milk has undergone a natural four-step sequence of spoilage organism activity. The spoilage process has produced acidic conditions that have denatured and precipitated the milk casein to yield typical, separated curds and whey.

spoilage often is initiated by molds. These organisms have enzymes that contribute to the weakening and penetration of the protective outer skin.

Food spoilage problems occur with minimally processed, concentrated frozen citrus products. These are prepared with little or no heat treatment, and major spoilage can be caused by *Lactobacillus* and *Leuconostoc* spp., which produce diacetyl-butter flavors. *Saccharomyces* and *Candida* can also spoil juices. Concentrated juice has a decreased water activity (a_w = 0.8 to 0.83), and when kept frozen at about −10°C, juices can be stored for long periods. However, when concentrated juices are diluted with water that contains spoilage organisms, or if the juice is stored in improperly washed containers, problems can occur. Also, microorganisms in the frozen concentrated juices can begin the spoilage process after addition of water. Ready-to-serve (RTS) juices present other problems as the a_w values are sufficiently high to allow microbial growth. This is especially true with extended storage at refrigeration temperatures. Although pasteurization results in some flavor loss, most juices are now routinely pasteurized (see section 40.3).

Molds are a special problem for tomatoes. Even the slightest bruising of the tomato skin, exposing the interior, will result in rapid fungal growth. Frequently observed genera include *Alternaria, Cladosporium, Fusarium,* and *Stemphylium.* This growth affects the quality of tomato products, including tomato juices and ketchups.

Molds can rapidly grow on grains and corn when these products are stored in moist conditions. The moldy bread pictured in **figure 40.3*a*** shows extensive fungal hyphal development and sporulation. The green growth most likely is *Penicillium;* the black growth is characteristic of *Rhizopus stolonifer* (*see figure 26.10*). Contamination of grains by the ascomycete *Claviceps purpura* causes **ergotism,** a toxic condition. Hallucinogenic alkaloids produced by this fungus can lead to altered behavior, abortion, and death if infected grains are eaten. Ergotism is discussed in chapter 26.

Fungus-derived carcinogens include the aflatoxins and fumonisins. **Aflatoxins** are produced most commonly in moist grains and nut products. Aflatoxins were discovered in 1960, when 100,000 turkey poults died from eating fungus-infested peanut meal. *Aspergillus flavus* was found in the infected peanut meal, together with alcohol-extractable toxins termed aflatoxins. These flat-ringed planar compounds intercalate with the cells' nucleic acids and act as frameshift mutagens and carcinogens. This occurs primarily in the liver, where they are converted to unstable derivatives. At the present time, a total of 18 aflatoxins are known. The most important are shown in **figure 40.4.** Of these, aflatoxin B_1 is the most common and the most potent carcinogen.

Aflatoxins B_1 and B_2, after ingestion by lactating animals, are modified in the animal body to yield the aflatoxins M_1 and M_2. If cattle consume aflatoxin-contaminated feeds, these also can appear in milk and dairy products. The aflatoxins are potent liver or hepatocarcinogens, and have been linked to effects on immunocompetence, growth, and disease resistance in livestock and laboratory animals. The major aflatoxin types and their derivatives can be separated by chromatographic procedures and can be recognized under UV light by their characteristic fluorescence. Besides their importance in grains, they have also been observed in beer, cocoa, raisins, and soybean meal.

(a)

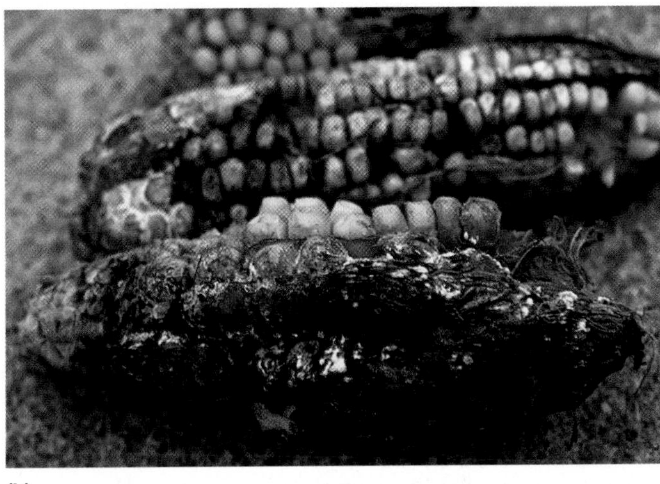

(b)

Figure 40.3 Food Spoilage. When foods are not stored properly, microorganisms can cause spoilage. Typical examples are fungal spoilage of **(a)** bread and **(b)** corn. Such spoilage of corn is called ear rot and can result in major economic losses.

Ultimately, the critical concern is the amounts of aflatoxins that are ingested. Diet appears to be related to aflatoxin exposure: the average aflatoxin intake in the typical European-style diet is 19 ng/day, whereas for some Asian diets it is estimated to be 103 ng/day. Aflatoxin sensitivity also can be influenced by prior disease exposure. Individuals who have had hepatitis B have a 30-fold higher risk of liver cancer upon exposure to aflatoxins than individuals who have not had this disease. This association illustrates an emerging link between inflammation and cancer. It has been observed that prevention of hepatitis B infections by vaccination and reduction of carrier populations will help to significantly control the potential effects of aflatoxins in foodstuffs.

The **fumonisins** are fungal contaminants of corn that were first isolated in 1988. These are produced by *Fusarium moniliforme*

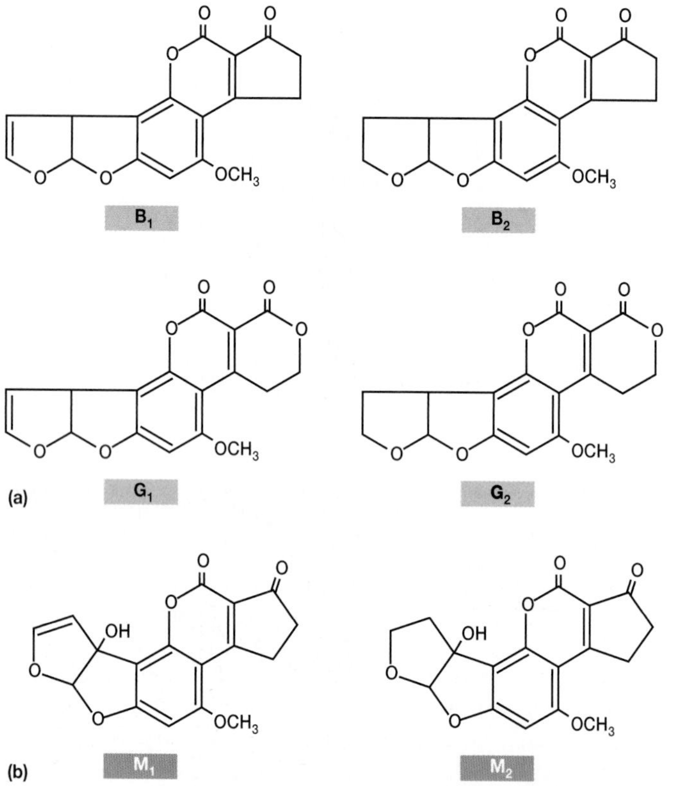

Figure 40.4 Aflatoxins. When *Aspergillus flavus* and related fungi grow on foods, carcinogenic aflatoxins can be formed. These have four basic structures. **(a)** The letter designations refer to the color of the compounds under ultraviolet light after extraction from the grain and separation by chromatography. The B_1 and B_2 compounds fluoresce with a blue color, and the G_1 and G_2 appear green. **(b)** The two type M aflatoxins are found in the milk of lactating animals that have ingested type B aflatoxins.

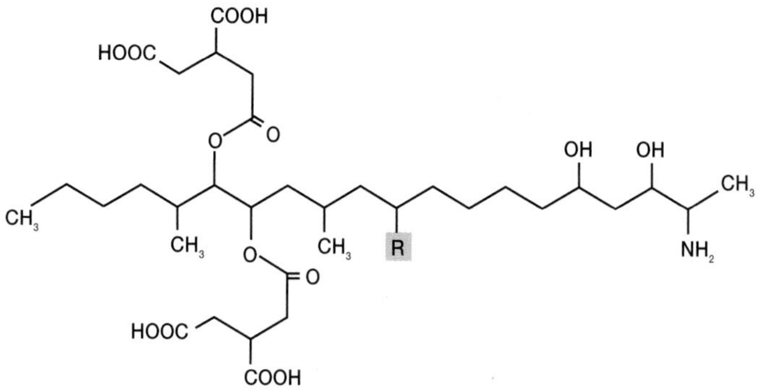

Figure 40.5 Fumonisin Structure. The basic structure of fumonisins FB1 and FB2 produced by *Fusarium moniliforme,* a fungal contaminant that can grow in improperly stored corn. A total of at least ten different fumonisins have been isolated. These are strongly polar compounds that cause diseases in domestic animals and also in humans. FB1, R = OH; FB2, R = H.

and cause leukoencephalomalacia in horses (also called "blind staggers"—it is fatal within 2 to 3 days), pulmonary edema in pigs, and esophageal cancer in humans. The fumonisins function by disrupting the synthesis and metabolism of sphingolipids, important compounds that influence a wide variety of cell functions.

There are at least ten different fumonisins; the basic structures of fumonisins FB1 and FB2 are shown in **figure 40.5.** Corn and corn-based feeds and foods, including cornmeal and corn grits, often are contaminated. The fumonisins inhibit ceramide synthase, a key enzyme for the proper use of fatty subtances in the cell. Thus it is extremely important to store corn and corn products under dry conditions where these fungi cannot develop.

Eucaryotic microorganisms can synthesize potent toxins other than aflatoxins and fumonisins. **Algal toxins** contaminate fish and thus affect the health of marine animals higher in the food chain; they also can contaminate shellfish and fin fish, which are later consumed by humans. Most toxins are produced by the protists dinoflagellates, but some diatoms also are toxic. Major human diseases that result from algal toxins in marine products include amnesic, diarrhetic, and paralytic shellfish poisoning (**table 40.3**). These complex toxins, most of which are temperature stable, are known to cause peripheral neurological system effects, often in less than one hour after ingestion.
Disease 25.1: Harmful algal blooms (HABs)

1. Describe in general how food spoilage occurs. What factors influence the nature of the spoilage organisms responsible?
2. Why do concentrated citrus juices present such interesting spoilage problems?
3. What fungal genus produces ergot alkaloids? What conditions are required for the synthesis of these substances?
4. Aflatoxins are produced by which fungal genus? How do they damage animals that eat the contaminated food?
5. What microbial genus produces fumonisins and why are these compounds of concern? If improperly stored, what are the major foods and feeds in which these chemicals might be found?
6. What is the usual route by which humans consume algal toxins? What are the major groups of protists that produce these complex substances?

40.3 Controlling Food Spoilage

With the beginning of agriculture and a decreasing dependence on hunting and gathering, the need to preserve surplus foods became essential to survival. The use of salt as a meat preservative and the production of cheeses and curdled milks was introduced in Near Eastern civilization as early as 3000 B.C. The production of wines and the preservation of fish and meat by smoking also were common by this time. Despite a long tradition of efforts to preserve food from spoilage, it was not until the nineteenth century that the microbial spoilage of food was studied systemati-

Table 40.3 Toxic Syndromes Associated with Marine Algal Toxins

Syndrome	Causative Organism(s)	Primary Vector	Toxin Type
Parasitic shellfish poisoning	*Alexaandrium* spp. *Gymnodinium* spp. *Pyrodinium* spp.	Shellfish	Saxitoxins
Neurotoxic shellfish poisoning	*Gymnodinium breve*	Shellfish	Brevitoxins
Ciguatera fish poisoning	*Gambierdiscus toxicus*	Reef fish	Ciguatoxins
Amnesic shellfish poisoning	*Pseudo-nitzchia* spp.	Shellfish	Domoic acid
Diarrhetic shellfish poisoning	*Dinophysis* spp. *Prorocentrum* spp.	Shellfish	Dinophysistoxins Okadaic acid
Estuary syndrome	*Pfiesteria piscicida*	Water	Unknown

Source: F. M. van Dolah, 2000. Marine algal toxins: Origins, health effects, and their increased occurrence. *Environ. Health Perspect.* 108(Suppl. 1):133–141. Table 1, p. 134.

Table 40.4 Basic Approaches to Food Preservation

Approach	Examples of Process
Removal of microorganisms	Avoidance of microbial contamination; physical filtration, centrifugation
Low temperature	Refrigeration, freezing
High temperature	Partial or complete heat inactivation of microorganisms (pasteurization and canning)
Reduced water availability	Water removal, as with lyophilization (freeze drying); use of spray dryers or heating drums; decreasing water availability by addition of solutes such as salt or sugar
Chemical-based preservation	Addition of specific inhibitory compounds (e.g., organic acids, nitrates, sulfur dioxide)
Radiation	Use of ionizing (gamma rays), nonionizing (UV), and electronic beam radiation
Microbial product–based inhibition	The addition of substances such as bacteriocins to foods to control food-borne pathogens

cally. Louis Pasteur established the modern era of food microbiology in 1857, when he showed that microorganisms cause milk spoilage. Pasteur's work in the 1860s proved that heat could be used to control spoilage organisms in wines and beers. The golden age of microbiology (section 1.4)

Foods can be preserved by a variety of methods (**table 40.4**). It is vital to eliminate or reduce the populations of spoilage and disease-causing microorganisms and to maintain the microbiological quality of a food with proper storage and packaging. Contamination often occurs after a package or can is opened and just before the food is served. This can provide an ideal opportunity for growth and transmission of pathogens, if care is not taken.

Removal of Microorganisms

Microorganisms can be removed from water, wine, beer, juices, soft drinks, and other liquids by filtration. This can keep bacterial populations low or eliminate them entirely. Removal of large particulates by prefiltration and centrifugation maximizes filter life and effectiveness. Several major brands of beer are filtered rather than pasteurized to better preserve the flavor and aroma of the original product.

Low Temperature

Refrigeration at 5°C retards microbial growth, although with extended storage, microorganisms eventually grow and produce spoilage. Slow microbial growth at temperatures below −10°C has been described, particularly with fruit juice concentrates, ice cream, and some fruits. Some microorganisms are very sensitive to cold and their numbers are reduced. Thus although refrigeration slows the metabolic activity of most microbes, it does not lead to significant decreases in overall microbial populations.

High Temperature

Controlling microbial populations in foods by means of high temperatures can significantly limit disease transmission and spoilage. Heating processes, first used by Nicholas Appert in 1809 (**Historical Highlights 40.1**), provide a safe means of preserving foods, particularly when carried out in commercial canning operations (**figure 40.6**). Canned food is heated in special containers called retorts at about 115°C for intervals ranging from 25 to over 100 minutes. The precise time and temperature depend on the nature of the food. Sometimes canning does not kill all microorganisms, but only those that will spoil the food (remaining bacteria are unable to grow due to acidity of the food, for example). After heat treatment the cans are cooled as rapidly as possible, usually with cold water. Quality control and processing effectiveness are sometimes compromised, however, in home

Historical Highlights

40.1 An Army Travels on Its Stomach

The movement and maintenance of large numbers of military personnel have always been limited by food supplies. The need to maintain large numbers of troops under hostile and inclement conditions led the French government in 1795 to offer a prize of 12,000 francs to the individual who could preserve foods for use under field conditions. Eventually the prize was awarded to Nicholas Appert, a candy maker, for his development of a heating process in which meats and other products could be preserved under sealed conditions.

Appert's work was based on the assumptions that heating and boiling control "ferments" and that sealing the food in bottles before heating it avoids the effects of air on spoilage. Despite Leeuwenhoek's earlier work, Appert did not have the concept of microorganisms to assist him in explaining the effectiveness of his process. His containers were large glass bottles, sealed with laminated corks and fish glue. With extreme care and attention to detail, he was able to heat these bottles in boiling water to provide food that could be stored for several years. Appert's work was an important foundation for the later studies of Louis Pasteur.

Figure 40.6 Food Preparation for Canning. Microbial control is important in the processing and preservation of many foods. Worker pouring peas into a large, clean vat during the preparation of vegetable soup. After preparation the soup is transferred to cans. Each can is heated for a short period, sealed, processed at temperatures around 110–121°C in a canning retort to destroy spoilage microorganisms, and finally cooled.

processing of foods, especially with less acidic (pH values greater than 4.6) products such as green beans or meats. The use of physical methods in control (section 7.4)

Despite efforts to eliminate spoilage microorganisms during canning, sometimes canned foods become spoiled. This may be due to spoilage before canning, underprocessing during canning, and leakage of contaminated water through can seams during cooling. Spoiled food can be altered in such characteristics as color, texture, odor, and taste. Organic acids, sulfides, and gases (particularly CO_2 and H_2S) may be produced. If spoilage microorganisms produce gas, both ends of the can will bulge outward. Sometimes the swollen ends can be moved by thumb pressure (soft swells); in other cases the gas pressure is so great that the ends cannot be dented by hand (hard swells). It should be noted that swelling is not always due to microbial spoilage. Acid in high-acid foods may react with the iron of the can to release hydrogen and generate a hydrogen swell. Hydrogen sulfide production by *Desulfotomaculum* can cause "sulfur stinkers."

Pasteurization involves heating food to a temperature that kills disease-causing microorganisms and substantially reduces the levels of spoilage organisms. In the processing of milk, beers, and fruit juices by conventional low-temperature holding (LTH) pasteurization, the liquid is maintained at 62.8°C for 30 minutes. Products can also be held at 71°C for 15 seconds, a high-temperature, short-time (HTST) process; milk can be treated at 141°C for 2 seconds for ultra-high-temperature (UHT) processing. Shorter-term processing results in improved flavor and extended product shelf life. Such heat treatment is based on a statistical probability that the number of remaining viable microorganisms will be below a certain level after a particular heating time at a specific temperature. This process is discussed in detail in section 7.4.

Water Availability

Dehydration, such as lyophilization to produce freeze-dried foods, is a common means of eliminating microbial growth. The modern process is simply an update of older procedures in which grains, meats, fish, and fruits were dried. The combination of free-water loss with an increase in solute concentration in the remaining water makes this type of preservation possible.

Chemical-Based Preservation

Various chemical agents can be used to preserve foods, and these substances are closely regulated by the U.S. Food and Drug Administration and are listed as being "generally recognized as safe" or **GRAS** (**table 40.5**). They include simple organic acids, sulfite, ethylene oxide as a gas sterilant, sodium nitrite, and ethyl formate. These chemical agents may damage the microbial plasma membrane or denature various cell proteins. Other compounds

Table 40.5 Major Groups of Chemicals Used in Food Preservation

Preservatives	Approximate Maximum Use Range	Organisms Affected	Foods
Propionic acid/propionates	0.32%	Molds	Bread, cakes, some cheeses, inhibitor of ropy bread dough
Sorbic acid/sorbates	0.2%	Molds	Hard cheeses, figs, syrups, salad dressings, jellies, cakes
Benzoic acid/benzoates	0.1%	Yeasts and molds	Margarine, pickle relishes, apple cider, soft drinks, tomato ketchup, salad dressings
Parabens[a]	0.1%	Yeasts and molds	Bakery products, soft drinks, pickles, salad dressings
SO_2/sulfites	200–300 ppm	Insects and microorganisms	Molasses, dried fruits, wine, lemon juice (not used in meats or other foods recognized as sources of thiamine)
Ethylene/propylene oxides	700 ppm	Yeasts, molds, vermin	Fumigant for spices, nuts
Sodium diacetate	0.32%	Molds	Bread
Dehydroacetic acid	65 ppm	Insects	Pesticide on strawberries, squash
Sodium nitrite	120 ppm	Clostridia	Meat-curing preparations
Caprylic acid	—	Molds	Cheese wraps
Ethyl formate	15–200 ppm	Yeasts and molds	Dried fruits, nuts

From James M. Jay. 2000. *Modern Food Microbiology,* 6th edition. Reprinted by permission of Aspen Publishing, Frederick, Md.

[a]Methyl-, propyl-, and heptyl-esters of *p*-hydroxybenzoic acid.

interfere with the functioning of nucleic acids, thus inhibiting cell reproduction.

The effectiveness of many of these chemical preservatives depends on the food pH. As an example, sodium propionate is most effective at lower pH values, where it is primarily undissociated and able to be taken up by lipids of microorganisms. Breads, with their low pH values, often contain sodium propionate as a preservative. Chemical preservatives are used with grain, dairy, vegetable, and fruit products. Sodium nitrite is an important chemical used to help preserve ham, sausage, bacon, and other cured meats by inhibiting the growth of *Clostridium botulinum* and the germination of its spores. This protects against botulism and reduces the rate of spoilage. Besides increasing meat safety, nitrite decomposes to nitric acid, which reacts with heme pigments to keep the meat red in color. Current concern about nitrite arises from the observation that it can react with amines to form carcinogenic nitrosamines.

Radiation

Radiation, both ionizing and nonionizing, has an interesting history in relation to food preservation. Ultraviolet radiation is used to control populations of microorganisms on the surfaces of laboratory and food-handling equipment, but it does not penetrate food. The major method used for radiation sterilization of food is gamma irradiation from a cobalt-60 source; however, cesium-137 is used in some facilities. Gamma radiation has excellent penetrating power, but must be used with moist foods because the radiation produces peroxides from water in the microbial cells, resulting in oxidation of sensitive cellular constituents. This process of **radappertization,** named after Nicholas Appert, can extend the shelf life of seafoods, fruits, and vegetables. To sterilize meat products, commonly 4.5 to 5.6 megarads are used.

Electron beams can also be used to irradiate foods. The electrons are generated electrically, so they can be turned on only when needed. Also, this approach does not generate radioactive waste. On the other hand, electron beams do not penetrate food items as deeply as does gamma radiation. It is important to note that regardless of the radiation source (gamma rays or electron beams), the food itself does not become radioactive.

Microbial Product-Based Inhibition

There is increasing interest in the use of **bacteriocins** for the preservation of foods. Bacteriocins are bactericidal proteins active against closely related bacteria, which bind to specific sites on the cell, and often affect cell membrane integrity and function. The only currently approved product is nisin. Nisin, produced by some strains of *Lactococcus lactis,* is a small hydrophobic protein. It is nontoxic to humans and affects mainly gram-positive bacteria, especially *Enterococcus faecalis.* Nisin can be used particularly in low-acid foods to improve inactivation of *Clostridium botulinum* during the canning process or to inhibit germination of any surviving spores.

Bacteriocins have a wide variety of names, depending on the organisms that produce them. They can function by several mechanisms. They may dissipate the proton motive force (PMF) of a susceptible bacterium. Some form pores in bacterial plasma membranes and promote the release of low-molecular-weight molecules. Bacteriocins may also inhibit protein or RNA synthesis. Bacteriocin addition to foods such as cheddar cheese can lead to a two- to threefold reduction in *Listeria monocytogenes* in 180-day-old cheeses. Chemical mediators in nonspecific (inate) resistance: Bacteriocins (section 31.6)

1. Describe the major approaches used in food preservation.
2. What types of chemicals can be used to preserve foods?
3. Nitrite is often used to improve the storage characteristics of prepared meats. What toxicological problems may result from the use of this chemical?
4. Under what conditions can ultraviolet light and gamma radiation be used to control microbial populations in foods and in food preparation? What is radappertization?
5. In principle, how do bacteriocins such as nisin function? What bacterial genus produces this important polypeptide?

40.4 Food-Borne Diseases

Food-borne illnesses impact the entire world. In the United States, based on recent information from the Centers for Disease Control and Prevention, annual incidences of food-related diseases involve 76 million cases, of which only 14 million can be attributed to known pathogens. Food-borne diseases result in 325,000 hospitalizations and at least 5,000 deaths per year. Since 1942, the number of recognized food-borne pathogens has increased over fivefold. Are these new microorganisms? In most cases, these pathogens are simply agents that we now can describe, based on an improved understanding of microbial diversity. Recent estimates indicate that Noroviruses, *Campylobacter jejuni* and *Salmonella* are the major causes of food-borne diseases. In addition, *Escherichia coli* O157:H7 and *Listeria* are important food-related pathogens.

Many diseases transmitted by foods, or food poisonings, are discussed in chapters 37 through 39, and only a few of the more important food-borne bacterial pathogens are mentioned here. There are two primary types of food-related diseases: food-borne infections and food intoxications. All of these food-borne diseases are associated with poor hygienic practices. Whether by water or food transmission, the fecal-oral route is maintained, with the food providing the vital link between hosts. Fomites, such as sink faucets, drinking cups, and cutting boards, also play a role in the maintenance of the fecal-oral route of contamination.

Food-Borne Infection

A **food-borne infection** involves the ingestion of the pathogen, followed by growth in the host, including tissue invasion and/or the release of toxins. The major diseases of this type are summarized in **table 40.6** (*see also table 38.6*).

Salmonellosis results from ingestion of a variety of *Salmonella* serovars, particularly Typhimurium and Enteritidis. Gastroenteritis is the disease of most concern in relation to foods such as meats, poultry, and eggs, and the onset of symptoms occurs after an incubation time as short as 8 hours. *Salmonella* infection can arise from contamination by workers in food-processing plants and restaurants, as well in canning processes (**Historical Highlights 40.2**).

Campylobacter jejuni is considered a leading cause of acute bacterial gastroenteritis in humans. This important pathogen is often transmitted by uncooked or poorly cooked poultry products. For example, transmission often occurs when kitchen utensils and containers are used for chicken preparation and then for salads. Contamination with as few as 10 viable *C. jejuni* cells can lead to the onset of diarrhea. *C. jejuni* also is transmitted by raw milk, and the organism has been found on various red meats. Thorough cooking of food prevents its transmission.

Listeriosis, caused by *Listeria monocytogenes,* was responsible for the largest meat recall in U.S. history—27.4 million pounds. In 2002, a seven-state listeriosis outbreak was linked to deli meats and hot dogs produced at a single meat-processing plant in Pennsylvania. Pregnant women, the young and old, and immunocompromised individuals are especially vulnerable to *L. monocytogenes* infections. In this outbreak, seven deaths, three stillbirths, and 46 illnesses were caused by consumption of contaminated meats. Food scientists matched the strain of *L. monocytogenes* found in the contaminated food products with samples obtained from floor drains in the Wampler packaging plant. This prompted the recall of 27.4 million pounds of meats that had been distributed over a five-month period to stores, restaurants, and school lunch programs. Following the outbreak, the plant closed for a month and the Wampler brand name was phased out. However, during this time, the company failed to test its meats to definitively show that its products (not just its drains) were contaminated with the offending *L. monocytogenes* strain. This episode prompted the U.S. Department of Agriculture (USDA) to step up its environmental testing program for *L. monocytogenes* so that it now tests plants that do not regularly submit data to the USDA. It also performs surprise inspections of those that do. The USDA advises people at risk of contracting listeriosis to avoid eating soft cheeses (e.g., Feta, Brie, Camembert), refrigerated smoked meats like lox, as well as deli meats and undercooked hot dogs. As a final note, the plant in Pennsylvania was closed in early 2006, having never recovered from the $100 million cost and the damage to its reputation caused by the 2002 outbreak.

Escherichia coli is an important food-borne disease organism. Enteropathogenic, enteroinvasive, and enterotoxigenic types can cause diarrhea (*see figure 38.25*). *E. coli* O157:H7 with its specific LPS O-antigen (O) and flagellar (H) antigen, is thought to have acquired enterohemorrhagic genes from *Shigella,* including the genes for shigalike toxins. This produced a new pathogenic strain, first discovered in 1982 and now known around the world. The pathogen is spread by the fecal-oral route, and an infectious dose appears to be only 500 bacteria. Enterohemorrhagic *E. coli* has been found in meat products such as hamburger and salami, in unpasteurized fruit drinks, on fruits and vegetables, and in untreated

Table 40.6 Major Food-Borne Infectious Diseases

Disease	Organism	Incubation Period and Characteristics	Major Foods Involved
Salmonellosis	*S. enterica* serovars Typhimurium and Enteritidis	8–48 hr Enterotoxin and cytotoxins	Meats, poultry, fish, eggs, dairy products
Arcobacter diarrhea	*Arcobacter butzleri*	Severe diarrhea, recurrent cramps	Meat products, especially poultry
Campylobacteriosis	*Campylobacter jejuni*	Usually 2–10 days Most toxins are heat-labile	Milk, pork, poultry products, water
Listeriosis	*L. monocytogenes*	Varying periods Related to meningitis and abortion; newborns and the elderly especially susceptible	Meat products, especially pork and milk
Escherichia coli diarrhea and colitis	*E. coli,* including serotype O157:H7	24–72 hr Enterotoxigenic positive and negative strains; hemorrhagic colitis	Undercooked ground beef, raw milk
Shigellosis	*Shigella sonnei, S. flexneri*	24–72 hr	Egg products, puddings
Yersiniosis	*Yersinia enterocolitica*	16–48 hr Some heat-stable toxins	Milk, meat products, tofu
Plesiomonas diarrhea	*Plesiomonas shigelloides*	1–2 hr	Uncooked mollusks
Vibrio parahaemolyticus gastroenteritis	*V. parahaemolyticus*	16–48 hr	Seafood, shellfish

Historical Highlights

40.2 Typhoid Fever and Canned Meat

Minor errors in canning have led to major typhoid outbreaks. In 1964 canned corned beef produced in South America was cooled, after sterilization, with nonchlorinated water; the vacuum created when the cans were cooled drew *S. enterica* serovar Typhi into some of the cans, which were not completely sealed. This contaminated product was later sliced in an Aberdeen, Scotland food store, and the meat slicer became a continuing contamination source; the result was a major epidemic that involved 400 people. The *S. enterica* serovar Typhi was a South American strain, and eventually the contamination was traced to the contaminated water used to cool the cans. This case emphasizes the importance of careful food processing and handling to control the spread of disease during food production and preparation.

well water. The newspapers are filled with reports of million-pound lots of beef being recalled due to *E. coli* contamination. Even if the contaminated beef does not reach consumers, this economic loss, due to poor hygiene, has many negative effects on cattle producers and meat processors.

Prevention of food contamination by *E. coli* O157:H7 is essential from the time of production until consumption. Hygiene must be monitored carefully in larger-volume slaughterhouses where contact of meat with fecal material can occur. Even fruits and vegetables should be handled with care because disease outbreaks have been caused by domestic and imported produce. Caution also is essential at the point of use. For example, avoidance of food contamination by hands and utensils is critical. Utensils used with raw foods should not contact cooked food; proper cleaning of cutting boards and utensils minimizes contamination.

Virus contamination is always a potential problem. This is based on transmission by waters or by lack of hygiene in food preparation and direct contamination by food processors and handlers. Similar situations occur with protozoan pathogens, discussed in chapter 39. Virus contamination has become a severe problem on many cruise ships, where Noroviruses have been involved in outbreaks, with person-to-person contact and possibly foods implicated in these ultimately avoidable occurrences.

An infectious agent of increasing worldwide concern with respect to food safety is a prion that causes new variant

Creutzfeldt-Jakob disease (vCJD). This is one of a group of progressively degenerative neuronal diseases termed transmissible spongiform encephalopathies (TSEs), and is associated with beef cattle. It is often called "mad cow disease." A major problem in controlling new vCJD is the lack of reliable detection methods. The major means of vCJD transmission between animals is the use of mammalian tissue in ruminant animal feeds; at the present time, there are significant problems in detecting such prohibited animal products in ruminant feeds. Prions (section 18.10); Prion diseases (section 37.6)

Foods that are transported and consumed in an uncooked state are an increasingly important source of food-borne infection. The problem becomes more serious because of rapid movement of people and products around the world. International trade in uncooked foods, aided by rapid air transport, provides many opportunities for disease transmission. Fresh foods such as sprouts, seafood, and raspberries pose significant hazards, which are discussed here.

Sprouts are a popular and attractive garnish to complement a variety of foods. Unfortunately, if sprouts are not germinated in pathogen-free waters and grown under sanitary conditions, major growth of pathogens can occur. Often sprouts are produced in areas of the world where there is poor control of water quality and sanitation. Contaminated alfalfa, beans, watercress, mungbean, mustard, and soybean sprouts can be major sources of typhoid and cholera.

Shellfish and finfish also present major concerns. Raw sewage can contaminate shellfish-growing areas; in addition, waterborne pathogens such as *Vibrio* are more prevalent in the water column during the warm months (e.g., in Chesapeake Bay on the mid-Atlantic coast of the United States). Viruses also can be a problem. Oysters are filter feeders that process several liters of water per day, leading to the potential concentration of at least 100 types of enteric viruses. Reverse transcriptase PCR can be used to detect RNA viruses in oysters based on the presence of their nucleic acids. However, the inability of molecular techniques to differentiate between infectious and noninfectious particles is a major problem. For example, UV treatment can inactivate many RNA viruses without eliminating the PCR signal, although the virions no longer replicate in a suitable tissue culture environment (**figure 40.7**). Heavy rainfall in shellfish areas can cause runoff of pathogens from adjacent septic systems and contaminate coastal waters. Often it is necessary to ban shellfish harvesting until the animals void pathogens from their digestive systems. Alternatively, shellfish from contaminated areas can be moved to clean waters to allow them to clean their digestive systems. The polymerase chain reaction (section 14.3)

Raspberries provide an important example of another major problem: the rapid air transport of raw agricultural products around the world. Major outbreaks of *Cyclospora cayetanensis* poisoning have been traced to raspberries imported from Central America into the United States and Canada. In the growth and harvesting process, the raspberries become contaminated, resulting in serious diarrhea in affected individuals. This organism has a complex life cycle, which is not fully understood at the present time. In comparison with *Giardia* and *Cryptosporidium,* which are infective immediately after being shed in feces, *Cyclospora* is not immediately infectious; sporulation or maturing requires 12 hours after release from the body. The mature infective cyst has two sporocysts (**figure 40.8**), an important criterion for confirming the presence of this protist on foods or in the environment. Protist classification: *Excavata* (section 25.6); Food-borne and waterborne diseases (section 39.5)

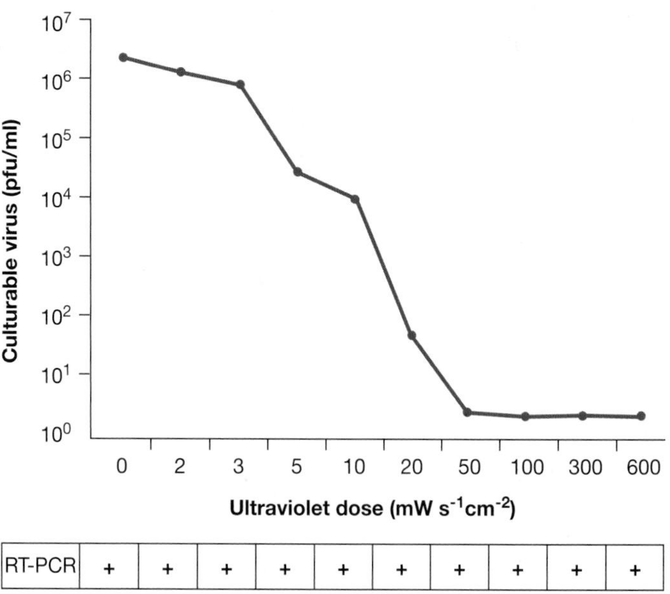

Figure 40.7 Cultural versus Molecular-Based Virus Detection. Comparison of plaque-forming ability and recovery of RNA from poliovirus type 2 by reverse transcriptase-PCR with varying UV doses. Even when plaque-forming ability is lost, nucleic acids are still detectable.

Food-Borne Intoxications

Microbial growth in food products also can result in a **food intoxication,** as summarized in table 38.6. Intoxication produces symptoms shortly after the food is consumed because growth of the disease-causing microorganism is not required. Toxins produced in the food can be associated with microbial cells or can be released from the cells.

Most *Staphylococcus aureus* strains cause a staphylococcal enteritis related to the synthesis of extracellular toxins. These are heat-resistant proteins, so heating does not usually render the food safe. The effects of the toxins are quickly felt, with disease symptoms occurring within 2 to 6 hours. The main reservoir of *S. aureus* is the human nasal cavity. Frequently *S. aureus* is transmitted to a person's hands and then is introduced into food during preparation. Growth and enterotoxin production usually occur when contaminated foods are held at room temperature for several hours.

Three gram-positive rods are known to cause food intoxications: *Clostridium botulinum, C. perfringens,* and *Bacillus cereus*

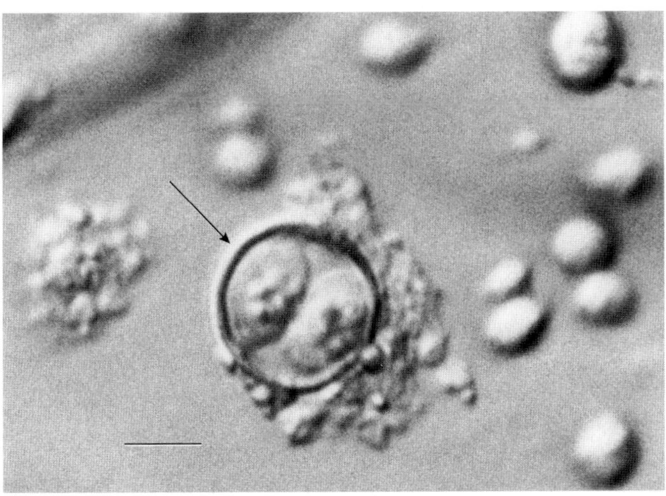

Figure 40.8 ***Cyclospora cayetanensis*, an Important Contaminant of Raw Foods.** *C. cayetanensis* can be recognized in waste waters and after recovery from contaminated foods due to the occurrence of an oocyst with two sporocysts. Bar = 5 μm.

(*see table 38.6*). *C. botulinum* poisoning is discussed in chapter 38, and *C. perfringens* intoxication is described here.

Clostridium perfringens food poisoning is one of the more widespread food intoxications. These microorganisms, which produce exotoxins, must grow to levels of approximately 10^6 bacteria per gram or higher in a food to cause disease. At least 10^8 bacteria must be ingested. They are common inhabitants of soil, water, food, spices, and the intestinal tract. Upon ingestion the cells sporulate in the intestine. The enterotoxin is a spore-specific protein and is produced during the sporulation process. Enterotoxin can be detected in the feces of affected individuals. *C. perfringens* food poisoning is common and occurs after meat products are heated, which results in O_2 depletion. If the foods are cooled slowly, growth of the microorganism can occur. At 45°C, enterotoxin can be detected 3 hours after growth is initiated. Onset of the symptoms—watery diarrhea, nausea, and abdominal cramps—usually occurs in about 8 to 16 hours.

Baked potatoes served in aluminum foil can provide a unique environment for pathogenic microorganisms. Potatoes, even after washing, are covered by *C. botulinum,* which naturally occurs in the soil. If the aluminum foil-covered potatoes are not heated sufficiently in the baking process, surviving clostridia can proliferate after removal of the potatoes from the oven and rapidly produce toxins.

Bacillus cereus also is of concern in starchy foods. It can cause two distinct types of illnesses depending on the type of toxin produced: an emetic illness characterized by nausea and vomiting with an incubation time of 1 to 6 hours, and a diarrheal type, with an incubation of 4 to 16 hours. The emetic type is often associated with boiled or fried rice, while the diarrheal type is associated with a wider range of foods.

1. What are the major food-borne diseases in the United States?
2. Discuss the major characteristic of a food-borne infection in terms of the time required between ingestion of the pathogen and the onset of the disease. Why does this occur?
3. What common food is often related to *Campylobacter*-caused gastroenteritis? What means can be used to control the occurrence of this disease from this source?
4. Give some of the sources of *E. coli* 0157:H7 that have been of concern in terms of disease transmission.
5. Why is new variant Creutzfeldt-Jakob disease (vCJD) of such concern?
6. What are some uncooked foods that have been implicated in food-borne disease transmission?
7. What are some of the major genera involved in food-borne intoxications?
8. How does a food-borne intoxication differ from a food-borne infection?

40.5 Detection of Food-Borne Pathogens

A major problem in maintaining food safety is the need to rapidly detect microorganisms in order to curb outbreaks that can affect large populations. This is especially important because of widescale distribution of perishable foods. Standard culture techniques may require days to weeks for positive identification of pathogens. Identification is often complicated by the low numbers of pathogens compared with the background microflora. Furthermore, the varied chemical and physical composition of foods can make isolation difficult. Fluorescent antibody, enzyme-linked immunoassays (ELISAs), and radioimmunoassay techniques have proven of value. These can be used to detect small amounts of pathogen-specific antigens. Clinical microbiology and immunology (chapter 35)

Molecular techniques also are increasingly used in identification. These methods are valuable for three purposes: (1) to detect the presence of a single, specific pathogen; (2) to detect viruses that cannot be grown conveniently; and (3) to identify slow-growing or nonculturable pathogens. Recombinant DNA technology (chapter 14)

Pathogens are frequently identified by detecting specific DNA or RNA base sequences with oligonucleotide probes. These usually are 14 to 40 bases in length and are specific for the pathogen of interest. They may be created by generating fragments with restriction endonucleases or through direct chemical synthesis. Probes are labeled by linking them to a variety of enzymatic, isotopic, chromogenic, or luminescent/fluorescent markers. A major advantage of their use is the speed with which specific microorganisms can be detected in a set of cultures, as shown in **figure 40.9.** In this example, a hydrophobic grid-membrane system has been used. The *Listeria monocytogenes* cultures are radioactive, indicating that they have bound the probe, while other *Listeria* species do not show probe binding.

Another example of the use of molecular techniques is provided by pathogenic *E. coli.* Currently *E. coli* O157:H7 is isolated

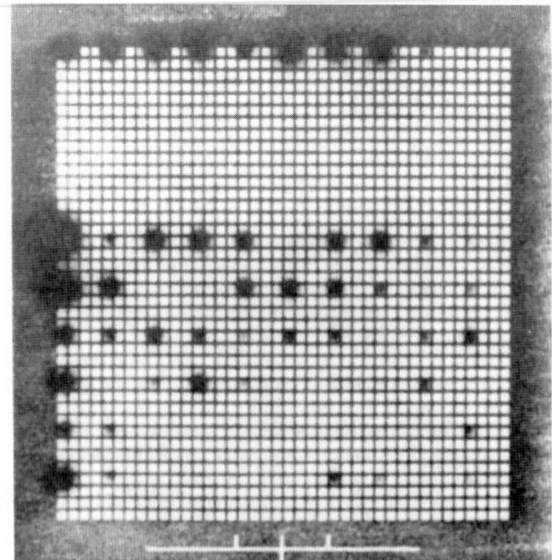

Figure 40.9 Molecular Probes and Food Microbiology. Autoradiogram of a radioactively-labeled *Listeria monocytogenes* probe against 100 *Listeria* cultures. Only the *Listeria monocytogenes* cultures show sequence homology and binding with the DNA probe, darkening the autoradiogram film. The other *Listeria* spp. do not react with the probe.

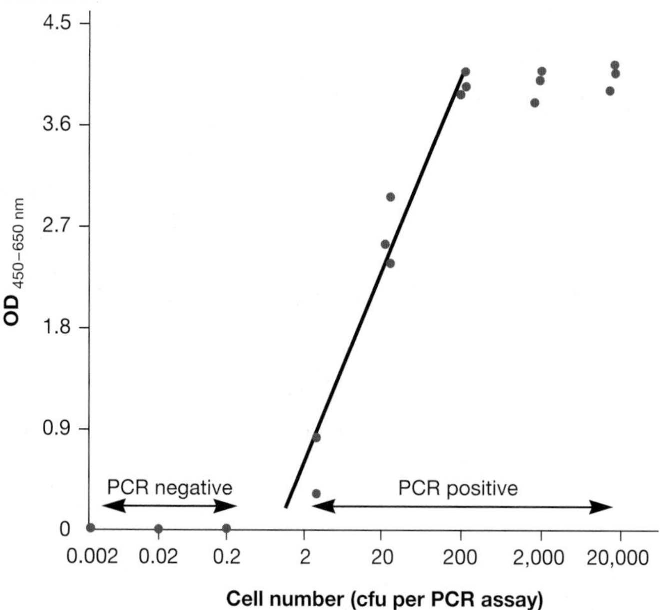

Figure 40.10 Polymerase Chain Reaction (PCR)-Based Pathogen Detection. Comparison of PCR sensitivity and growth for *Salmonella enterica* subsp. *enterica* serovar Agona detection. The Probalia PCR system can detect as few as two colony forming units (CFU) of the pathogen. OD = optical density.

and identified using selective culture media, rapid identification kits, rapid probe-based identification procedures, serotype-specific probes, and PCR techniques. These molecular techniques also enable the detection of a few target cells in large populations of background microorganisms. For example, by using the polymerase chain reaction, as few as 10 toxin-producing *E. coli* cells can be detected in a population of 100,000 cells isolated from soft-cheese samples.

As few as two colony forming units of *Salmonella* can be detected by PCR (**figure 40.10**). This makes it possible to confirm *Salmonella* presence within 24 hours, whereas 3 to 4 days is needed for presumptive identification with standard culture procedures. Confirmation of *Salmonella* presence would then require additional time. Frequently, to improve the sensitivity and increase the speed of this method, a pre-enrichment step is used before PCR. PCR is also used for rapid detection of other food-borne pathogens. For instance, the recovery of specific 159 and 1,223 base pair PCR products, which can be separated electrophoretically, has made it possible to detect *Campylobacter jejuni* and *Arcobacter butzleri* in the same sample within 8 hours. Gel electrophoresis (section 14.4); Techniques for determining microbial taxonomy and phylogeny: Molecular characteristics (section 19.4)

A major advance in the detection of food-borne pathogens is the use of standardized pathogen DNA patterns, or "food-borne pathogen fingerprinting." The Centers for Disease Control and Prevention in the United States has established a program, called **PulseNet,** in which pulsed-field gel electrophoresis (PFGE) is used under carefully controlled and duplicated conditions to determine the distinctive DNA pattern of each bacterial pathogen. With this uniform procedure, it is possible to link pathogens associated with disease outbreaks in different parts of the world to a specific food source. Data from around the world are being used in **FoodNet,** an active surveillance network, to follow nine major food-borne diseases. Using the FoodNet approach, it is possible to trace the course and cause of infection in days and not weeks. As an example, a *Shigella* outbreak in three different areas of North America was traced to Mexican parsley that had been tainted with polluted irrigation water. This program has resulted in more rapid establishment of epidemiological linkages and a decreased occurrence of many of these important food-borne diseases.

1. How is the polymerase chain reaction used in pathogen detection?
2. How are PulseNet and FoodNet used in the surveillance of food-borne diseases?

40.6 Microbiology of Fermented Foods

Over the last several thousand years, fermentation has been a major way of preserving food. Microbial growth, either of natural or inoculated populations, causes chemical and/or textural changes to form a product that can be stored for extended periods. The fermentation process also is used to create new, pleasing food flavors and odors (**Techniques & Application 40.3**).

The major fermentations used in food microbiology are the lactic, propionic, and ethanolic fermentations. These fermentations

Techniques & Applications

40.3 Chocolate: The Sweet Side of Fermentation

Chocolate could be characterized as the "world's favorite food," and yet few people realize that fermentation is an essential part of chocolate production. The Aztecs were the first to develop chocolate fermentation, serving a chocolate drink made from the seeds of the chocolate tree, *Theobroma cocao* [Greek *theos,* god and *broma,* food, or "food of the gods"]. Chocolate trees now grow in West Africa as well as South America.

The process of chocolate fermentation has changed very little over the past 500 years. Each tree produces large pods that each contain 30 to 40 seeds in a sticky pulp (see **Box Figure**). Ripe pods are harvested and slashed open to release the pulp and seeds. The sooner the fermentation begins, the better the product, so fermentation occurs on the farm where the trees are grown. The seeds and pulp are placed in "sweat boxes" or in heaps in the ground and covered, usually with banana leaves.

(a)

(b)

Cocoa Fermentation. (a) Cocoa pods growing on the cocoa tree. Each pod is 13 to 15 cm in length and contains 30 to 40 seeds in a sticky white pulp. **(b)** Seeds and pulp are fermented in boxes covered with banana leaves for 5 to 7 days and then dried in the sun, as shown here. Chocolate cannot be produced without fermentation.

Like most fermentations, this process involves a succession of microbes. First, a community of yeasts, including *Candida rugosa* and *Kluyveromyces marxianus,* hydrolyzes the pectin that covers the seeds and ferments the sugars to release ethyl alcohol and CO_2. As the temperature and the alcohol concentration increase, the yeasts are inhibited and lactic acid bacteria increase in number. The mixture is stirred to aerate the microbes and ensure an even temperature distribution. Lactic acid production drives the pH down; this encourages the growth of bacteria that produce acetic acid as a fermentation end product. Acetic acid is critical to the production of fine chocolate because it kills the sprout inside the seed and releases enzymes that cause further degradation of proteins and carbohydrates, contributing to the overall taste of the chocolate. In addition, acetate esters, derived from acetic acid, are important for the development of good flavor. Fermentation takes five to seven days. An experienced chocolate grower knows when the fermentation is complete—if it is stopped too soon the chocolate will be bitter and astringent. On the other hand, if fermentation lasts too long, microbes start growing on the seeds instead of in the pulp. "Off-tastes" arise when the gram-positive bacterium *Bacillus* and the filamentous fungi *Aspergillis, Penicillium,* and *Mucor* hydrolyze lipids in the seeds to release short-chain fatty acids. As the pH begins to rise, the bacteria of the genera *Pseudomonas, Enterobacter,* and *Escherichia* also contribute to bad tastes and odor.

After fermentation, the seeds, now called beans, are spread out to dry. Ideally this is done in the sun, although drying ovens are also used. The oven-drying method is considered inferior because the beans can acquire a smoky taste. The dried beans are brown and lack the pulp. They are bagged and sold to chocolate manufacturers, who first roast the beans to further reduce the bitter taste and kill most of the microbes (some *Bacillus* spores may remain). The beans are then ground and the nibs—the inner part of each bean—are removed. The nibs are crushed into a thick paste called a chocolate liquor, which contains cocoa solids and cocoa butter, but no alcohol. Cocoa solids are brown and have a rich flavor, and cocoa butter has a high fat content and is off-white in color. The two components are separated and the cocoa solids can be sold as cocoa for baking and hot chocolate, while the cocoa butter is used to make white chocolate or sold to cosmetics companies for use in lipsticks and lotions. However, the bulk of these two components will be used to make chocolate. The cocoa solids and butter are reunited in controlled ratios and sugar, vanilla, and other flavors are added. The better the fermentation, the less sugar needs to be added (and the more expensive the chocolate will be).

The final product, delicious chocolate, is a combination of over 300 different chemical compounds. This mixture is so complex that no one has yet been able to make synthetic chocolate that can compete with the natural fermented plant (note that artificial vanilla is readily available). Microbiologists and food scientists are studying the fermentation process to determine the role of each microbe. But like the chemists, they have had little luck in replicating the complex, imprecise fermentation that occurs on cocoa farms. In fact, the finest, most expensive chocolate starts as cocoa on farms where the details of fermentation have been handed down through generations. Chocolate production is truly an art as well as a science, while eating it is simply divine.

are carried out with a wide range of cultures, many of which have not been characterized. Fermentations (section 9.7)

Fermented Milks

Throughout the world, at least 400 different fermented milks are produced. These fermentations are carried out by mesophilic, thermophilic, and probiotic lactic acid bacteria, as well as by yeasts and molds as noted in **table 40.7.** Only major examples of these fermentation types will be discussed in this section.

Lactic Acid Bacteria

The majority of fermented milk products rely on **lactic acid bacteria (LAB).** The art of fermentation developed long before the science, and fermented milks were produced for thousands of years before Louis Pasteur discovered lactic acid fermentation. Pasteur's work enabled the development of pure LAB starter cultures and the industrialization of milk fermentation. LAB include species belonging to the genera *Lactobacillus, Lactococcus, Leuconostoc,* and *Streptococcus* (**figure 40.11**). These bacteria are low G + C gram-positives that tolerate acidic conditions, are nonsporing, and are aerotolerant with a strictly fermentative metabolism. Class *Bacilli:* Order *Lactobacillales* (section 23.5)

Table 40.7 Major Categories and Examples of Fermented Milk Products

Category	Typical Examples
I. Lactic fermentations	
Mesophilic	Buttermilk
	Cultured buttermilk
	Långofil
	Tëtmjolk
	Ymer
Thermophilic	Yogurt, laban, zabadi, labneh, skyr Bulgarian buttermilk
Probiotic	Biogarde, Bifighurt Acidophilus milk, yakult Cultura-AB
II. Yeast-lactic fermentations	Kefir, koumiss, acidophilus-yeast milk
III. Mold-lactic fermentations	Viili

Source: Table 3.1, p. 58. In B. A. Law, editor. 1997. *Microbiology and Biochemistry of Cheese and Fermented Milk,* 2nd ed. New York: Chapman and Hall.

Mesophilic

Mesophilic milk fermentations result from similar manufacturing techniques, in which acid produced through microbial activity causes protein denaturation. To carry out the process, one usually inoculates milk with the desired starter culture (**Techniques & Applications 40.4**); incubates it at optimum temperature (approximately 20 to 30°C), and then stops microbial growth by cooling. *Lactobacillus* spp. and *Lactococcus lactis* cultures are used for aroma and acid production. The organism *Lactococcus lactis* subsp. *diacetilactis* converts milk citrate to diacetyl, which gives a buttery flavor to the finished product. The use of these microorganisms with skim milk produces cultured buttermilk, and when cream is used, sour cream is the result.

Thermophilic

In addition to mesophilic milk fermentations, thermophilic fermentations can be carried out at temperatures around 45°C. An important example is yogurt production. Yogurt is one of the most popular fermented milk products in the United States and is produced commercially and at home with yogurt-making kits. In commercial production, nonfat or low-fat milk is pasteurized, cooled to 43°C or lower, and inoculated with a 1:1 ratio of *Streptococcus thermophilus* and *Lactobacillus delbrueckii* subspecies *bulgaricus (L. bulgaricus). S. thermophilus* grows more rapidly at first and renders the milk anaerobic and weakly acidic. *L. bulgaricus* then acidifies the milk even more. Acting together, the two species ferment almost all of the lactose to lactic acid and flavor the yogurt with diacetyl (*S. thermophilus*) and acetaldehyde (*L. bulgaricus*).

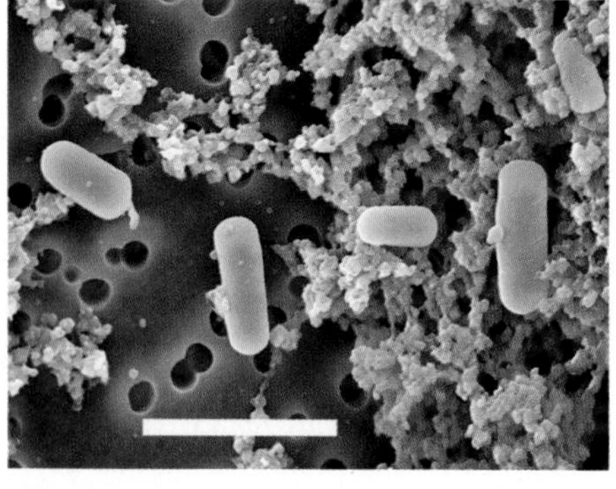
(a)

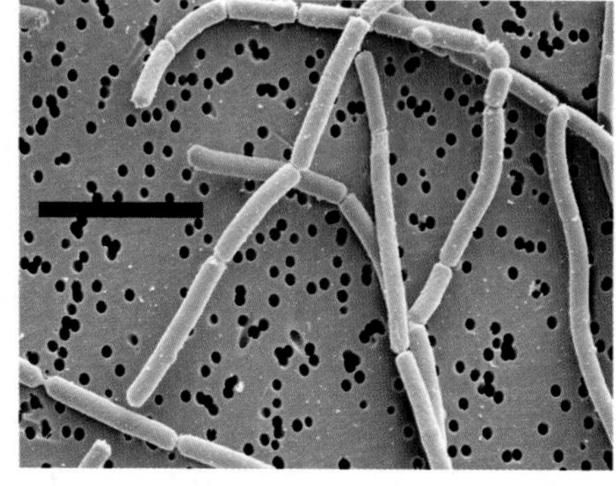
(b)

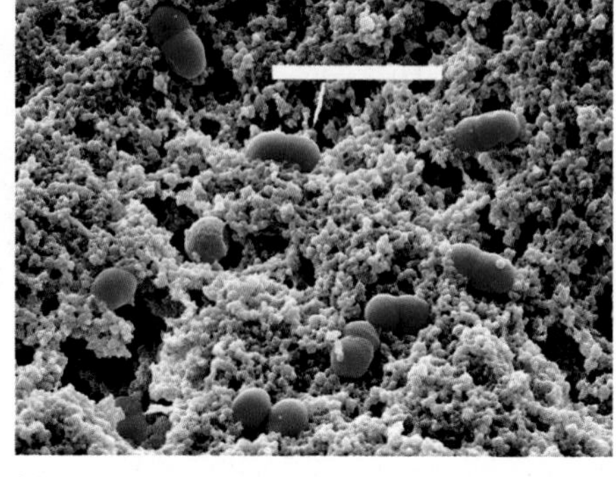
(c)

Figure 40.11 Lactic Acid Bacteria (LAB). Colored scanning electron micrographs of LAB used as starter cultures. **(a)** *Lactobacillus helveticus.* **(b)** *Lactobacillus delbrueckii* subspecies *bulgaricus.* **(c)** *Lactococcus lactis.* The bacteria are supported by filters, seen as holes in the background. Scale bar = 5μm

Techniques & Applications

40.4 Starter Cultures, Bacteriophage Infections, and Plasmids

Cultures of lactic acid bacteria, called **starter cultures,** are added to milk during the preparation of many dairy products. For example, *Streptococcus lactis* and *S. cremoris* are used in the production of cheese. One of the greatest problems for the dairy industry is the presence of bacteriophages that destroy these starter cultures. Lactic acid production by a heavily phage-infected starter culture can come to a halt within 30 minutes. The industry has tried to overcome this problem by practicing aseptic techniques in order to reduce phage contamination, and by selecting for phage-resistant bacterial cultures.

Most efforts at control have not been successful in the longer term. It has been found that the very aseptic techniques and phage-resistant pure cultures used to attempt to solve this problem actually were parts of the problem. The most stable and dependable cultures, called P starter cultures, contain the bacteriophages in a lysogenic state. When cultures are grown without phages or under aseptic conditions (L starters), they lose their phage resistance. The key to this riddle appears to be plasmids, which encode products that block phage adsorption. The loss of the plasmids in a subpopulation of the bacteria allows the phage carrier state to be established. New phages can develop by acquiring restriction enzymes (*see section 14.1*) from plasmids. These modified phages again become lytic, establishing a new equilibrium in the population.

Other control approaches are being tested. Antisense RNA is now used in an attempt to provide an agent against bacteriophage genes to help in the constant struggle between lactic acid bacteria and their phages.

Fruits or fruit flavors to be added are pasteurized separately and then combined with the yogurt. Freshly prepared yogurt contains about 10^9 bacteria per gram.

Probotics

The health benefits of fermented foods like yogurt have been touted for a great number of years. However, only recently have rigorous studies on the effects of certain bacteria that are either commensals or mutualists in the human intestine been explored. Techniques & Applications 30.3: Probiotics for human and animals

Microorganisms such as *Lactobacillus* and *Bifidobacterium* are being used in the rapidly developing area of **probiotics,** the addition of microorganisms to the diet in order to provide health benefits beyond basic nutritive value. The possible health benefits of the use of such microbial dietary adjuvants include immunomodulation, control of diarrhea, anticancer effects, and possible improvement of Crohn's disease (inflammatory bowel disease). These bacteria may also influence antigen presentation, uptake, and possible degradation. Probiotics have become a more attractive treatment option because the rate of antibiotic resistance among pathogens continues to climb. In addition, disease ecologists have come to recognize that intestinal microflora can be a contributing factor for certain conditions (e.g., Crohn's disease). Microbial interactions (section 30.1)

Acidophilus milk is produced by using *Lactobacillus acidophilus. L. acidophilus* may modify the microbial flora in the lower intestine, thus improving general health, and it often is used as a dietary adjunct, especially for lactose intolerant persons. Many microorganisms in fermented dairy products stabilize the bowel microflora, and some appear to have antimicrobial properties. The exact nature and extent of health benefits of consuming fermented milks may involve minimizing lactose intolerance, lowering serum cholesterol, and possibly exhibiting anticancer activity. Several lactobacilli have antitumor compounds in their cell walls. Such findings suggest that diets including lactic acid bacteria, especially *L. acidophilus,* may contribute to the prevention of colon cancer. Normal microbiota of the human body: Large intestine (section 30.3)

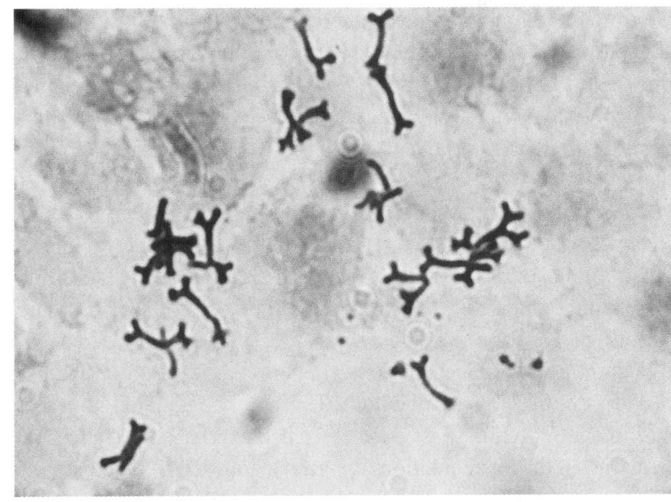

Figure 40.12 Bifidobacteria. Cultured milks are increasing in popularity. A light micrograph of *Bifidobacterium,* a microorganism suggested to provide many health benefits.

Another interesting group used in milk fermentations are the bifidobacteria. The genus *Bifidobacterium* contains irregular, nonsporing, gram-positive rods that may be club-shaped or forked at the end (**figure 40.12**). Bifidobacteria are nonmotile, anaerobic, and ferment lactose and other sugars to acetic and lactic acids. They are typical residents of the human intestinal tract and many beneficial properties are attributed to them. Bifidobacteria are thought to help maintain the normal intestinal balance, while improving lactose tolerance; to possess antitumorigenic activity; and to reduce serum cholesterol levels. In addition, some believe that they promote calcium absorption and the synthesis of

Figure 40.13 Examples of Bifid-Amended Dairy Products. These are produced in many countries.

B-complex vitamins. It has also been suggested that bifidobacteria reduce or prevent the excretion of rotaviruses, a cause of diarrhea among children. *Bifidobacterium*-amended fermented milk products, including yogurt, are now available in various parts of the world (**figure 40.13**).

Yeast-Lactic Fermentation

Yeast-lactic fermentations include **kefir,** a product with an ethanol concentration of up to 2%. This unique fermented milk originated in the Caucasus Mountains and it is produced east into Mongolia. Kefir products tend to be foamy and frothy, due to active carbon dioxide production. This process is based on the use of kefir "grains" as an inoculum. These are coagulated lumps of casein that contain yeasts, lactic acid bacteria, and acetic acid bacteria. In this fermentation, the grains are used to inoculate the fresh milk and then recovered at the end of the fermentation. Originally, kefir was produced in leather sacks hung by the front door during the day, and passersby were expected to push and knead the sack to mix and stimulate the fermentation. Fresh milk could be added occasionally to maintain activity.

Mold-Lactic Fermentation

Mold-lactic fermentation results in a unique Finnish fermented milk called viili. The milk is placed in a cup and inoculated with a mixture of the fungus *Geotrichium candidum* and lactic acid bacteria. The cream rises to the surface, and after incubation at 18 to 20°C for 24 hours, lactic acid reaches a concentration of 0.9%. The fungus forms a velvety layer across the top of the final product, which also can be made with a bottom fruit layer.

Cheese Production

Cheese is one of the oldest human foods and is thought to have been developed approximately 8,000 years ago. About 2,000 distinct varieties of cheese are produced throughout the world, representing approximately 20 general types (**table 40.8** and **figure 40.14**). Often cheeses are classified based on texture or hardness as soft cheeses (cottage, cream, Brie), semisoft cheeses (Muenster, Limburger, blue), hard cheeses (cheddar, Colby, Swiss), or very hard cheeses (Parmesan). All cheese results from a lactic acid fermentation of milk, which results in coagulation of milk proteins and formation of a curd. Rennin, an enzyme from calf stomachs, but now produced by genetically engineered microorganisms, can also be used to promote curd formation. After the curd is formed, it is heated and pressed to remove the watery part of the milk (called the whey), salted, and then usually ripened (**figure 40.15**). The cheese curd can be packaged for ripening with or without additional microorganisms.

Lactococcus lactis is used as a starter culture for a number of cheeses including Gouda (figure 40.14*a*) and cheddar (figure 40.15). Starter culture density is often over 10^9 colony-forming units (CFUs) per gram of cheese before ripening. However, the high salt, low pH, and the temperatures that characterize the cheese microenvironment reduce these numbers rather quickly. This enables other bacteria, sometimes called **nonstarter lactic acid bacteria (NSLAB)** to grow; their numbers can reach 10^7 to 10^9 CFUs/g after several months of aging. Thus both starter and nonstarter LAB contribute to the final taste, texture, odor, and appearance of the cheese.

In some cases, molds are used to further enhance the cheese. Obvious examples are Roquefort and blue cheese. For these cheeses, *Penicillium roqueforti* spores are added to the curds just before the final cheese processing. Sometimes the surface of an already formed cheese is inoculated at the start of ripening; for example, Camembert cheese is inoculated with spores of *Penicillium camemberti.* The final hardness of the cheese is partially a function of the length of ripening. Soft cheeses are ripened for only about 1 to 5 months, whereas hard cheeses need 3 to 12 months, and very hard cheeses like Parmesan require 12 to 16 months ripening.

The ripening process also is critical for Swiss cheese. Gas production by *Propionibacterium* contributes to final flavor development and hole or eye formation in this cheese. Some cheeses are soaked in brine to stimulate the development of specific fungi and bacteria; Limburger is one such cheese.

Meat and Fish

Besides the fermentation of dairy products, a variety of meat products, especially sausage, can be fermented: country-cured hams, summer sausage, salami, cervelat, Lebanon bologna, fish sauces (processed by halophilic *Bacillus* species), izushi, and katsuobushi. *Pediococcus cerevisiae* and *Lactobacillus plantarum* are most often involved in sausage fermentations. Izushi is based on the fermentation of fresh fish, rice, and vegetables by *Lactobacillus* spp.; katsuobushi results from the fermentation of tuna by *Aspergillus glaucus.* Both meat fermentations originated in Japan.

1. What are the major types of milk fermentations?
2. Briefly describe how buttermilk, sour cream, and yogurt are made.
3. What is unique about the morphology of *Bifidobacterium?* Why is it used in milk?

Table 40.8 Major Types of Cheese and Microorganisms Used in Their Production

Cheese (Country of Origin)	Contributing Microorganisms[a]: Earlier Stages of Production	Later Stages of Production
Soft, unripened		
Cottage	*Lactococcus lactis*	*Leuconostoc cremoris*
Cream	*L. cremoris, L. diacetylactis, Streptococcus thermophilus, L. delbrueckii* subspecies *bulgaricus*	
Mozzarella (Italy)	*S. thermophilus, L. bulgaricus*	
Soft, ripened		
Brie (France)	*Lactococcus lactis, L. cremoris*	*Penicillium camemberti, P. candidum, Brevibacterium linens*
Camembert (France)	*L. lactis, L. cremoris*	*Penicillium camemberti, B. linens*
Semisoft		
Blue, Roquefort (France)	*Lactococcus lactis, L. cremoris*	*P. roqueforti*
Brick, Muenster (United States)	*L. lactis, L. cremoris*	*B. linens*
Limburger (Belgium)	*L. lactis, L. cremoris*	*B. linens*
Hard, ripened		
Cheddar, Colby (Britain)	*Lactococcus lactis, L. cremoris*	*Lactobacillus casei, L. plantarum*
Swiss (Switzerland)	*L. lactis, L. helveticus, S. thermophilus*	*Propionibacterium shermanii, P. freudenreichii*
Very hard, ripened		
Parmesan (Italy)	*Lactococcus lactis, L. cremoris, S. thermophilus*	*Lactobacillus bulgaricus*

[a]*Lactococcus lactis* stands for *L. lactis* subsp. *lactis. Lactococcus cremoris* is *L. lactis* subsp. *cremoris*, and *Lactococcus diacetilactis* is *L. lactis* subsp. *diacetilactis.*

4. How and where are kefir and viili made?
5. What major steps are used to produce cheese? How is the cheese curd formed in this process? What is whey? How does Swiss cheese get its holes?
6. Which fungal genus is often used in cheese making? What cheeses are produced using this genus?
7. Give a microbial genus used in meat fermentations.

Production of Alcoholic Beverages

A variety of plants that contain adequate carbohydrates can be used to produce alcoholic beverages. When carbohydrates are available in readily fermentable form, the fermentation can be started immediately. For example, grapes are crushed to release the juice or **must,** which can be allowed to ferment without further delay. The must also can be treated by pasteurization or the use of sulfur dioxide, and then the desired microbial culture added.

In contrast, before cereals and other starchy materials can be used as substrates for the production of alcohol, their complex carbohydrates must be hydrolyzed. They are mixed with water and incubated in a process called **mashing.** The insoluble material is then removed to yield the **wort,** a clear liquid containing fermentable sugars and other simple molecules. Much of the art of beer and ale production involves the controlled hydrolysis of protein and carbohydrates to provide the desired body and flavor of the final product.

Wines and Champagnes

Wine production, or the science of **enology** [Greek *oinos,* wine, and *ology,* the science of], starts with the collection of grapes, continues with their crushing and the separation of the liquid (must) before fermentation, and concludes with a variety of storage and aging steps (**figure 40.16**). All grapes have white juices. To make a red wine from a red grape, the grape skins are allowed to remain in contact with the must before fermentation to release their skin-coloring components. Wines can be produced by using the natural grape skin microorganisms, but this natural mixture of bacteria and yeasts gives unpredictable fermentation results. To avoid such problems, fresh must is treated with a sulfur dioxide fumigant and a desired strain of the yeast *Saccharomyces cerevisiae* or *S. ellipsoideus* is added. After inoculation the juice is fermented for 3 to 5 days at temperatures between 20 and 28°C. Depending on the alcohol tolerance of the yeast strain (the alcohol eventually kills the yeast that produced it), the final product may contain 10 to 14% alcohol. Clearing and development of flavor occur during the aging process. The malolactic fermentation is an important part of wine production. Grape juice contains high levels of organic acids, including malic and tartaric acids. If the levels of these acids are not decreased during the fermentation process, the wine will be too acidic, and have poor stability and "mouth feel." This essential fermentation is carried out by the bacteria *Leuconostoc oenos, L. plantarum, L. hilgardii, L. brevis,* and *L. casei.* The activities of

(a)

(b)

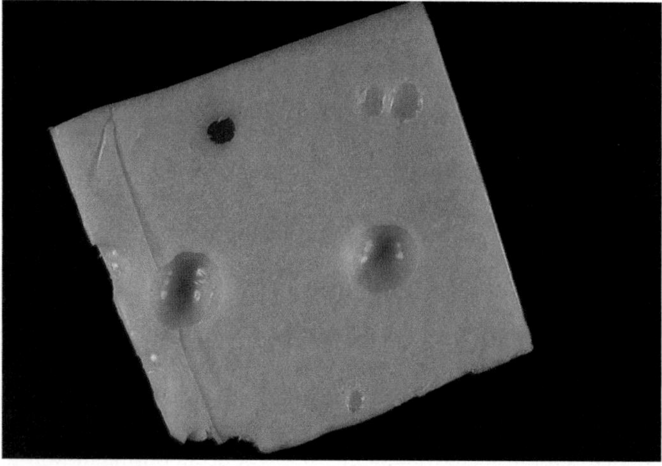

(c)

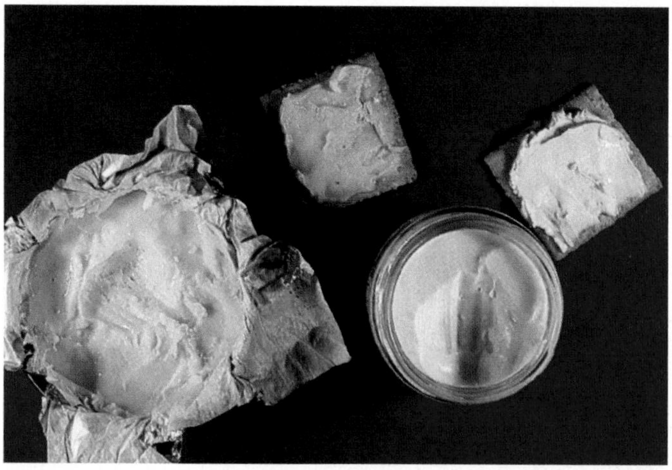

(d)

(e)

Figure 40.14 Cheese. A vast array of cheeses are produced around the world using microorganisms. **(a)** Gouda (top left) and cheddar cheese (lower right). Note the typical indentations on the surface of Gouda caused by the cheesecloth and red wax covering. **(b)** Roquefort cheese crumbled for use in salad dressing. The dark areas are the result of extensive *Penicillium* growth. **(c)** Swiss cheese, a hard, ripened cheese, contains holes formed by carbon dioxide from a *Propionibacterium* fermentation. **(d)** Brie (left) and Limburger (right) cheeses are soft, ripened cheeses. Ripening results from the surface growth of microorganisms like *Penicillium camemberti* (Brie) and *Brevibacterium linens* (Limburger). **(e)** Cottage cheese and cream cheese (spread on crackers) are soft, unripened cheeses. They are sold immediately after production, and the curd is consumed without further modification by microorganisms.

these microbes transform malic acid (a four-carbon tricarboxylic acid) to lactic acid (a three-carbon monocarboxylic acid) and carbon dioxide. This results in deacidification (pH increase), improvement of flavor stability, and in some cases the possible accumulation of bacteriocins in the wines. Characteristics of the fungal division: *Ascomycota* (section 26.6); Fermentations (section 9.7)

A critical part of wine making involves the choice of whether to produce a dry (no remaining free sugar) or a sweeter (varying amounts of free sugar) wine. This can be controlled by regulating the initial must sugar concentration. With higher levels of sugar, alcohol will accumulate and inhibit the fermentation before the sugar can be completely used, thus producing a sweeter wine.

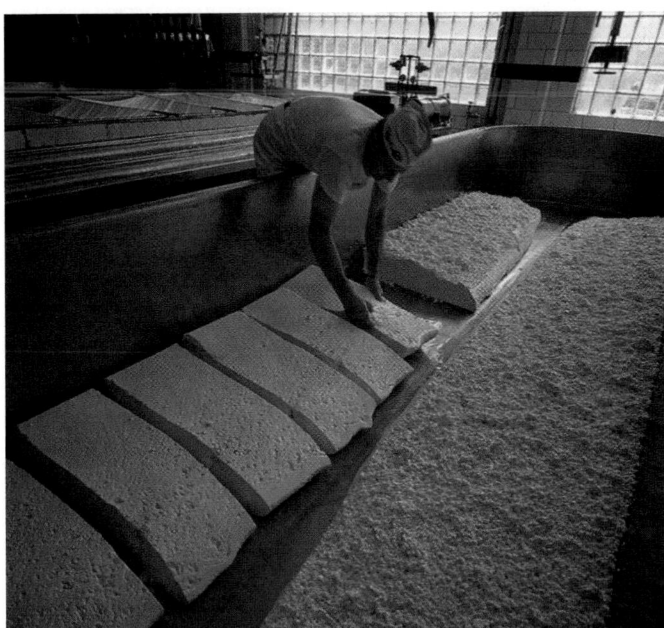

Figure 40.15 Cheddar Cheese Production. Cheddar, a village in England, has given its name to a cheese made in many parts of the world. "Cheddaring" is the process of turning and piling the curd to express whey and develop desired cheese texture.

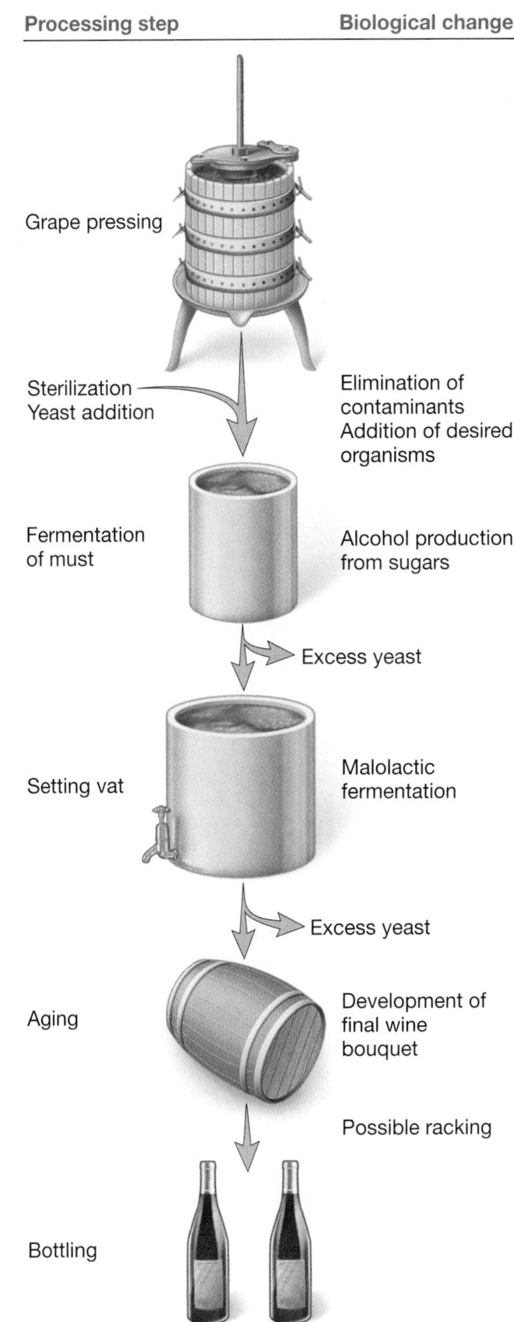

Figure 40.16 Wine Making. Once grapes are pressed, the sugars in the juice (the must) can be immediately fermented to produce wine. Must preparation, fermentation, and aging are critical steps.

During final fermentation in the aging process, flavoring compounds accumulate and influence the bouquet of the wine.

Microbial growth during the fermentation process produces sediments, which are removed during **racking.** Racking can be carried out at the time the fermented wine is transferred to bottles or casks for aging or even after the wine is placed in bottles.

Many processing variations can be used during wine production. The wine can be distilled to make a "burned wine" or brandy. *Acetobacter* and *Gluconobacter* can be allowed to oxidize the ethanol to acetic acid and form a **wine vinegar.** In the past an acetic acid generator was used to recirculate the wine over a bed of wood chips, where the desired microorganisms developed as a surface growth. Today the process is carried out in large aerobic submerged cultures under much more controlled conditions.

Natural champagnes are produced by continuing the fermentation in bottles to produce a naturally sparkling wine. Sediments that remain are collected in the necks of inverted champagne bottles after the bottles have been carefully turned. The necks of the bottles are then frozen and the corks removed to disgorge the accumulated sediments. The bottles are refilled with clear champagne from another disgorged bottle, and the product is ready for final packaging and labeling.

Beers and Ales

Beer and ale production uses cereal grains such as barley, wheat, and rice. The complex starches and proteins in these grains must be changed to a more readily usable mixture of simpler carbohydrates and amino acids. This process, shown in **figure 40.17,** involves germination of the barley grains and activation of their enzymes to produce a **malt.** The malt is then mixed with water and the desired grains, and the mixture is transferred to the mash tun or cask in order to hydrolyze the starch to usable carbohydrates. Once this process is completed, the **mash** is heated with hops (dried flowers of the female vine *Humulus lupulis*), which were originally added to the mash to inhibit spoilage microorganisms (**figure 40.18**). The hops also provide flavor and assist in clarification of the wort. In this heating step the hydrolytic enzymes are inactivated and the wort can be **pitched**—inoculated—with the desired yeast.

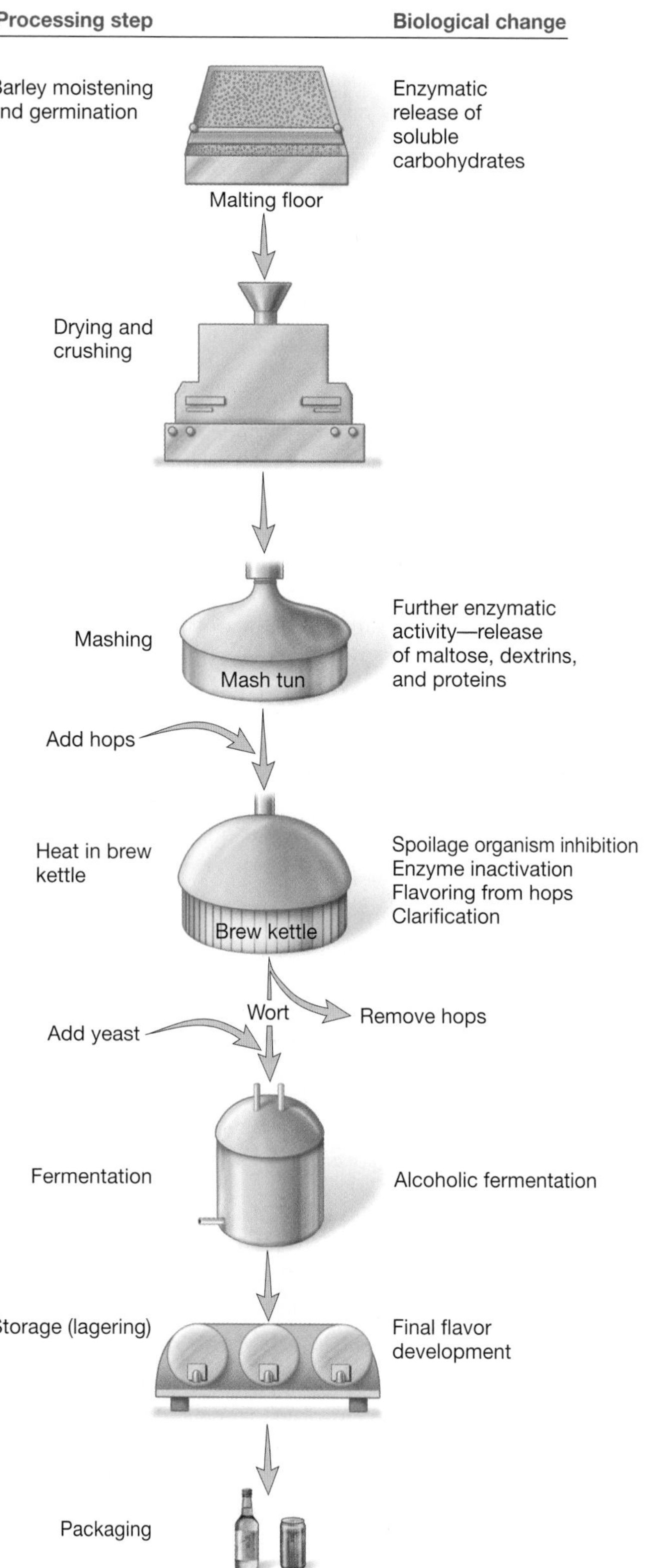

Figure 40.17 Producing Beer. To make beer, the complex carbohydrates in the grain must first be transformed into a fermentable substrate. Beer production thus requires the important steps of malting, and the use of hops and boiling for clarification, flavor development, and inactivation of malting enzymes, to produce the wort.

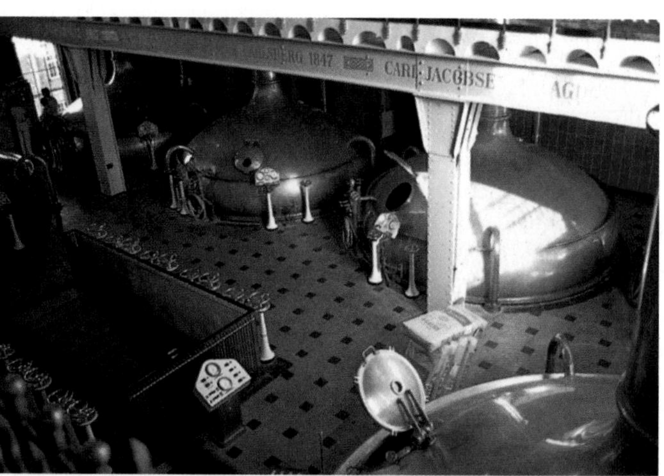

Figure 40.18 Brew Kettles Used for Preparation of Wort. In large-scale processes, copper brew kettles can be used for wort preparation, as shown here at the Carlsberg Brewery, Copenhagen, Denmark.

Most beers are fermented with **bottom yeasts,** related to *Saccharomyces pastorianus,* which settle at the bottom of the fermentation vat. The beer flavor also is influenced by the production of small amounts of glycerol and acetic acid. Bottom yeasts require 7 to 12 days of fermentation to produce beer with a pH of 4.1 to 4.2. With a top yeast, such as *Saccharomyces cerevisiae,* the pH is lowered to 3.8 to produce ales. Freshly fermented (green) beers are aged or **lagered,** and when they are bottled, CO_2 is usually added. Beer can be pasteurized at 140°F or higher or sterilized by passage through membrane filters to minimize flavor changes.

Distilled Spirits

Distilled spirits are produced by an extension of beer production processes. The fermented liquid is boiled, and the volatile components are condensed to yield a product with a higher alcohol content than beer. Rye and bourbon are examples of whiskeys. Rye whiskey must contain at least 51% rye grain, and bourbon must contain at least 51% corn. Scotch whiskey is made primarily of barley. Usually a **sour mash** is used; the mash is inoculated with a homolactic (lactic acid is the major fermentation product) bacterium such as *Lactobacillus delbrueckii* subspecies *bulgaricus* (figure 40.11*b*), which can lower the mash pH to around 3.8 in 6 to 10 hours. This limits the development of undesirable organisms. Vodka and grain alcohols are also produced by distillation. Gin is vodka to which resinous flavoring agents—often juniper berries—have been added to provide a unique aroma and flavor.

Production of Breads

Bread is one of the most ancient of human foods, and is produced with the help of microorganisms. The use of yeasts to leaven bread is carefully depicted in paintings from ancient Egypt. A bakery at the Giza Pyramid area, from the year 2575 B.C., has been excavated. It is estimated that 30,000 people a day were provided with bread from this bakery. Samples of bread from

2100 B.C. are on display in the British Museum. In breadmaking, yeast growth is carried out under aerobic conditions. This results in increased CO_2 production and minimum alcohol accumulation. The fermentation of bread involves several steps: alpha- and beta-amylases present in the moistened dough release maltose and sucrose from starch. Then a baker's strain of the yeast *Saccharomyces cerevisiae,* which produces maltase, invertase, and zymase enzymes, is added. The CO_2 produced by the yeast results in the light texture of many breads, and traces of fermentation products contribute to the final flavor. Usually bakers add sufficient yeast to allow the bread to rise within 2 hours—the longer the rising time, the more additional growth by contaminating bacteria and fungi can occur, making the product less desirable.

By using more complex assemblages of microorganisms, bakers can produce special breads such as sour doughs. The yeast *Saccharomyces exiguus,* together with a *Lactobacillus* species, produces the characteristic acidic flavor and aroma of such breads.

Bread products can be spoiled by *Bacillus* species that produce ropiness. If the dough is baked after these organisms have grown, stringy and ropy bread will result, leading to decreased consumer acceptance.

Other Fermented Foods

Many other plant products can be fermented, as summarized in **table 40.9.** These include sufu, which is produced by the fermentation of tofu, a chemically coagulated soybean milk product. To carry out the fermentation, the tofu curd is cut into small chunks and dipped into a solution of salt and citric acid. After the cubes are heated to pasteurize their surfaces, the fungi *Actinimucor elegans* and some *Mucor* species are added. When a white mycelium develops, the cubes, now called pehtze, are aged in salted rice wine. This product has achieved the status of a delicacy in many parts of the Western world. Another popular product is tempeh, a soybean mash fermented by *Rhizopus.*

Sauerkraut or sour cabbage is produced from wilted, shredded cabbage, as shown in **figure 40.19.** Usually the mixed microbial community of the cabbage is used. A concentration of 2.2 to 2.8% sodium chloride restricts the growth of gram-negative bacteria while favoring the development of the lactic acid bacteria. The primary microorganisms contributing to this product are *Leuconostoc mesenteroides* and *Lactobacillus plantarum.* A predictable microbial succession occurs in sauerkraut's development. The activities of the lactic acid-producing cocci usually cease when the acid content reaches 0.7 to 1.0%. At this point *Lactobacillus plantarum* and *Lactobacillus brevis* continue to function. The final acidity is generally 1.6 to 1.8, with lactic acid comprising 1.0 to 1.3% of the total acid in a satisfactory product.

Pickles are produced by placing cucumbers and such components as dill seeds in casks filled with a brine. The sodium chloride concentration begins at 5% and rises to about 16% in 6 to 9 weeks. The salt not only inhibits the growth of undesirable bacteria but also extracts water and water-soluble constituents from the cucumbers. These soluble carbohydrates are converted to lactic acid. The fermentation, which can require 10 to 12 days, involves the development of the gram-positive bacteria *L. mesenteroides, Enterococcus faecalis, Pediococcus cerevisiae, L. brevis,* and *L. plantarum. L. plantarum* plays the dominant role

Table 40.9 Fermented Foods Produced from Fruits, Vegetables, Beans, and Related Substrates

Foods	Raw Ingredients	Fermenting Microorganisms	Area
Coffee	Coffee beans	*Erwinia dissolvens, Saccharomyces* spp.	Brazil, Congo, Hawaii, India
Gari	Cassava	*Corynebacterium manihot, Geotrichum* spp.	West Africa
Kenkey	Corn	*Aspergillus* spp., *Penicillium* spp., lactobacilli, yeasts	Ghana, Nigeria
Kimchi	Cabbage and other vegetables	Lactic acid bacteria	Korea
Miso	Soybeans	*Aspergillus oryzae, Zygosaccharomyces rouxii*	Japan
Ogi	Corn	*Lactobacillus plantarum, Lactococcus lactis, Zygosaccharomyces rouxii*	Nigeria
Olives	Green olives	*Leuconostoc mesenteroides, Lactobacillus plantarum*	Worldwide
Ontjom	Peanut presscake	*Neurospora sitophila*	Indonesia
Peujeum	Cassava	Molds	Indonesia
Pickles	Cucumbers	*Pediococcus cerevisiae, L. plantarum, L. brevis*	Worldwide
Poi	Taro roots	Lactic acid bacteria	Hawaii
Sauerkraut	Cabbage	*L. mesenteroides, L. plantarum, L. brevis*	Worldwide
Soy sauce	Soybeans	*Aspergillus oryzae* or *A. soyae, Z. rouxii, Lactobacillus delbrueckii*	Japan
Sufu	Soybeans	*Actinimucor elegans, Mucor* spp.	China
Tao-si	Soybeans	*A. oryzae*	Philippines
Tempeh	Soybeans	*Rhizopus oligosporus, R. oryzae*	Indonesia, New Guinea, Surinam

Adapted from James M. Jay. 2000. *Modern Food Microbiology,* 6th edition. Reprinted by permission of Aspen Publishing, Frederick, Md.

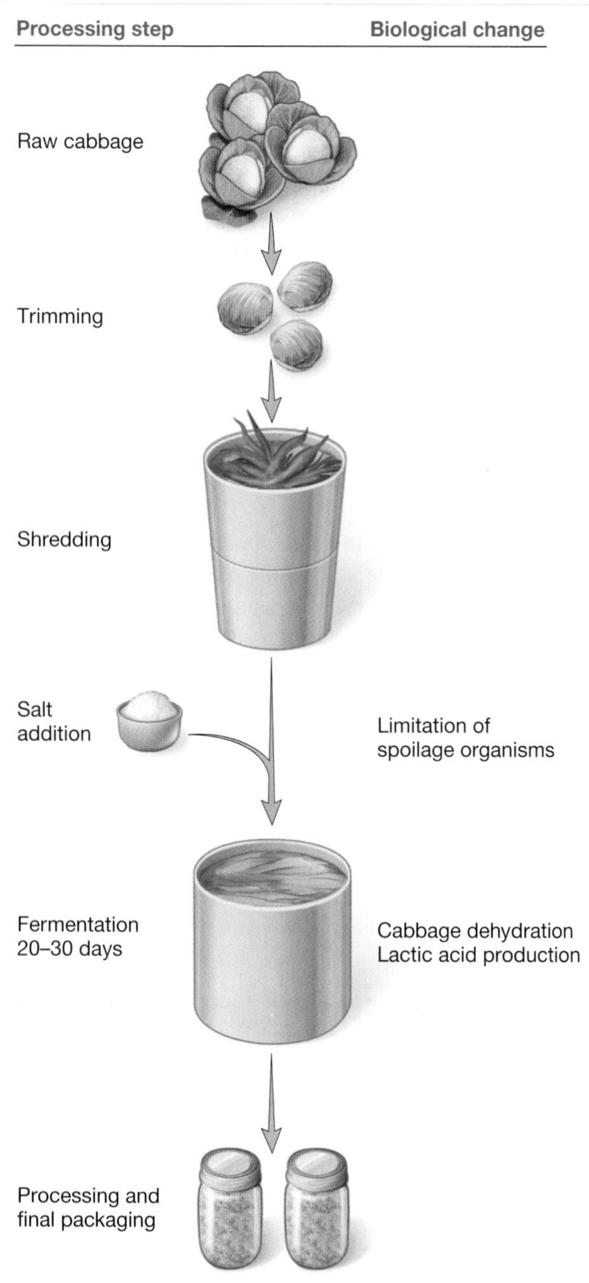

Figure 40.19 Sauerkraut. Sauerkraut production employs a lactic acid fermentation. The basic process involves fermentation of shredded cabbage in the presence of 2.25–2.5% by weight of salt to inhibit spoilage organisms.

in this fermentation process. Sometimes, to achieve more uniform pickle quality, natural microorganisms are first destroyed and the cucumbers are fermented using pure cultures of *P. cerevisiae* and *L. plantarum.*

Grass, chopped corn, and other fresh animal feeds, if stored under moist anoxic conditions, will undergo a lactic-type mixed fermentation that produces pleasant-smelling **silage.** Trenches or more traditional vertical steel or concrete silos are used to store the silage. The accumulation of organic acids in silage can cause rapid deterioration of these silos. Older wooden stave silos, if not properly maintained, allow the outer portions of the silage to become oxic, resulting in spoilage of a large portion of the plant material.

1. Describe and contrast the processes of wine and beer production. How are red wines produced when the juice of all grapes is white?
2. How do champagnes differ from wines?
3. Describe how distilled spirits like whiskey are produced.
4. How are bread, sauerkraut, and pickles produced? What microorganisms are most important in these fermentations?

40.7 Microorganisms as Foods and Food Amendments

Besides microorganisms' actions in fermentation as agents of physical and biological change, they themselves can be used as a food source. A variety of bacteria, yeasts, and other fungi have been used as animal and human food sources. Mushrooms (*Agaricus bisporus*) are one of the most important fungi used directly as a food source. Large caves provide optimal conditions for the production of this delicacy (**figure 40.20**). Microorganisms can be used directly as a food source or as a supplement to other foods and are then called single-cell protein. One of the more popular microbial food supplements is the cyanobacterium *Spirulina.* It is used as a food source in Africa and is now being sold in United States health food stores as a dried cake or powdered product.

An interesting application of probiotic microbes (primarily *Lactobacillus acidophilus*) is to decrease *E. coli* occurrence in beef cattle. The desired bacteria are sprayed on feed, and the cattle have been shown to have markedly lower (60% in some experiments) carriage of the toxic *E. coli* strain O157:H7. This can

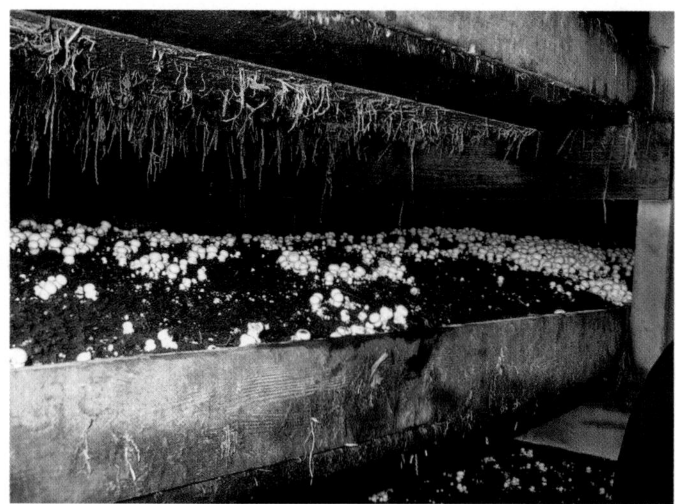

Figure 40.20 Mushroom Farming. Growing mushrooms requires careful preparation of the growth medium and control of environmental conditions. The mushroom bed is a carefully developed compost, which can be steam sterilized to improve mushroom growth.

make it easier to produce beef that will meet current standards for microbiological quality at the time of slaughter. Techniques & Applications 30.3: Probiotics for humans and animals

In addition, there is a greater appreciation of the role of oligosaccharide polymers or **prebiotics,** which are not processed until they enter the large intestine. The combination of prebiotics and probiotic microorganisms is described as a **synbiotic** system. This synbiotic combination can result in an increase in the levels of butyric and propionic acids, as well as an increase in *Bifidobacterium* in the human intestine. The butryrate, especially, may play a role in the possible beneficial effects of probiotics on intestinal processes.

Probiotics are used successfully with poultry. For instance, the USDA has designated a probiotic *Bacillus* strain for use with chickens as GRAS (p. 1030). Feeding chickens a strain of *Bacillus subtilis* (Calsporin) leads to increased body weight and feed conversion. There is also a reduction in coliforms and *Campylobacter* in the processed carcasses. It has been suggested that this probiotic decreases the need for antibiotics in poultry production and pathogen levels on farms. *Salmonella* can be controlled by spraying a patented blend of 29 bacteria, isolated from the chicken cecum, on day-old chickens. As they preen themselves, the chicks ingest the bacterial mixture, establishing a functional microbial community in the cecum and limiting *Salmonella* colonization of the gut in a process called **competitive exclusion.** In 1998, this product, called PREEMPT, was approved for use in the United States by the Food and Drug Administration.

1. What conditions are needed to have most efficient production of edible mushrooms? What is the scientific name of the most important fungus used for this purpose?
2. A cyanobacterium is widely used as a food supplement. What is this genus, and in what part of the world was it first used as a significant food source?
3. What are prebiotics, probiotics, and synbiotics?
4. What probiotic has recently been recognized as GRAS?

Summary

40.1 Microorganism Growth in Foods

a. Most foods, especially when raw, provide an excellent environment for microbial growth. This growth can lead to spoilage or preservation, depending on the microorganisms present and environmental conditions.

b. The course of microbial development in a food is influenced by the intrinsic characteristics of the food itself—pH, salt content, substrates present, water presence and availability—and extrinsic factors, including temperature, relative humidity, and atmospheric composition (**figure 40.1**).

c. Microorganisms can spoil meat, dairy products, fruits, vegetables, and canned goods in several ways. Spices, with their antimicrobial compounds, sometimes protect foods.

d. Modified atmosphere packaging (MAP) is used to control microbial growth in foods and to extend product shelf life. This process involves decreasing oxygen and increased carbon dioxide levels in the space between the food surface and the wrapping material.

40.2 Microbial Growth and Food Spoilage

a. Food spoilage is a major concern throughout the world. This can occur at any point in the food production process: growth, harvesting, transport, storage, or final preparation. Spoilage also can occur if foods are not stored properly.

b. Fungi, if they can grow in foods, especially cereals and grains, can produce important disease-causing chemicals, including the carcinogens aflatoxins (**figure 40.4**) and fumonisins (**figure 40.5**), and ergot alkaloids (mind-altering drugs).

c. Prior illness with hepatitis B can increase susceptibility to liver cancer from aflatoxins. Control of hepatitis B is suggested to be more critical than control of aflatoxins.

d. Algal toxins can be transmitted to humans by marine products. These can have severe amnesic, diarrhetic and neurotoxic effects.

40.3 Controlling Food Spoilage

a. Foods can be preserved in a variety of physical and chemical ways, including filtration, alteration of temperature (cooling, pasteurization, sterilization), drying, the addition of chemicals, radiation, and fermentation (**table 40.4**).

b. There is an increasing interest in using bacteriocins for food preservation. Nisin, a product of *Lactococcus lactis,* is the major substance approved for use in foods. Bacteriocins are especially important for control of *Listeria monocytogenes.*

40.4 Food-Borne Diseases

a. Foods can be contaminated by pathogens at any point in the food production, storage, or preparation processes. Pathogens such as *Salmonella, Campylobacter, Listeria,* and *E. coli* can be transmitted by the food to the susceptible consumer, where they grow and cause disease, or a food-borne infection (**table 40.6**). If the pathogen grows in the food before consumption and forms toxins that affect the food consumer without further microbial growth, the disease is a food-borne intoxication. Examples are intoxications caused by *Staphylococcus, Clostridium,* and *Bacillus.*

b. Noroviruses, *Campylobacter,* and *Salmonella* are thought to be the most important causes of food-borne illness. The cause of most food-borne illnesses is not known.

c. *E. coli* O157:H7 is an enterohemorrhagic bacterium that produces shigalike verocytotoxins, which especially affect the young. Proper food handling and thorough cooking are critical in control.

d. New variant Creutzfeldt-Jakob disease (vCJD) is of increasing worldwide concern as a food-borne infectious agent, which is related to the occurrence of "mad cow disease." The major means of vCJD transmission between animals is the use of mammalian tissue in ruminant animal feeds. There are difficulties in detecting such prohibited animal products in ruminant feeds.

e. Raw foods such as sprouts, seafood, and raspberries provide routes for disease transmission. Increases in the international shipment of fresh foods contribute to this problem. *Cyclospora* is a protozoan of concern, and the major source is contaminated waters (**figure 40.8**).

40.5 Detection of Food-Borne Pathogens

a. Detection of food-borne pathogens is a major part of food microbiology. The use of immunological and molecular techniques such as DNA and RNA hybridization, PCR, and pulsed field electrophoresis is making it possible to link disease occurrences to a common infection source (**figures 40.9** and **40.10**). PulseNet and FoodNet programs are being used to coordinate these control efforts.

40.6 Microbiology of Fermented Foods

a. Dairy products can be fermented to yield a wide variety of cultured milk products (**table 40.7**). These include mesophilic, therapeutic, thermophilic, lacto-ethanolic, and mold-lactic products.

b. Growth of lactic acid-forming bacteria (**figure 40.11**), often with the additional use of rennin, can coagulate milk solids. These solids can be processed to yield a wide variety of cheeses, including soft unripened, soft ripened, semisoft, hard, and very hard types (**table 40.8** and **figure 40.14**). Both bacteria and fungi are used in these cheese production processes.

c. Wines are produced from pressed grapes and can be dry or sweet, depending on the level of free sugar that remains at the end of the alcoholic fermentation (**figure 40.16**). Champagne is produced when the fermentation, resulting in CO_2 formation, is continued in the bottle.

d. Beer and ale are produced from cereals and grains. The starches in these substrates are hydrolyzed, in the processes of malting and mashing, to produce a fermentable wort. *Saccharomyces cerevisiae* is a major yeast used in the production of beer and ale (**figure 40.17**).

e. Many plant products can be fermented with bacteria, yeasts, and molds. Important products are breads, soy sauce, sufu, and tempeh (**table 40.9**). Sauerkraut and pickles are produced in a fermentation process in which natural populations of lactobacilli play a major role (**figure 40.19**).

40.7 Microorganisms as Foods and Food Amendments

a. Microorganisms themselves can serve as an important food source. Mushrooms (*Agaricus bisporus*) are one of the most important fungi used as a food source. *Spirulina,* a cyanobacterium, also is a popular food source sold in specialty stores.

b. Many microorganisms, including some of those used to ferment milks, can be used as food amendments or microbial dietary adjuvants. Microorganisms such as *Lactobacillus* and *Bifidobacterium,* termed probiotics, can be used with oligopolysaccharides, termed prebiotics, to yield synbiotics. Several types of probiotic microorganisms are being used successfully in poultry production.

Key Terms

aflatoxin 1027
algal toxin 1028
bacteriocin 1031
bottom yeast 1044
competitive exclusion 1047
enology 1041
ergotism 1027
extrinsic factor 1024
food-borne infection 1032
food intoxication 1034
FoodNet 1036
fumonisin 1027
GRAS 1030
intrinsic factor 1024
kefir 1040
lactic acid bacteria (LAB) 1038
lager 1044
malt 1043
mash 1043
mashing 1041
modified atmosphere packaging (MAP) 1026
must 1041
nonstarter lactic acid bacteria (NSLAB) 1040
osmophilic microorganism 1024
pasteurization 1030
pitching 1043
prebiotic 1047
probiotic 1039
PulseNet 1036
putrefaction 1024
racking 1043
radappertization 1031
silage 1046
sour mash 1044
starter culture 1039
synbiotic 1047
wine vinegar 1043
wort 1041
xerophilic microorganism 1024

Critical Thinking Questions

1. Fresh lemon slices are often served with raw or steamed seafood (oysters, crab, shrimp). From a food microbiology perspective, provide an explanation for their being served. Are there other examples in either the cooking or the serving of foods that not only enhance flavor, but might have an antimicrobial strategy? Consider the example of marinades.
2. You are going through a salad line in a cafeteria at the end of the day. Which types of foods would you tend to avoid, and why?
3. Why were aflatoxins not discovered before the 1960s? Do you think this was the first time they had grown in a food product to cause disease?
4. What advantage might the shigalike toxin give *E. coli* O157:H7? Can we expect to see other "new" pathogens appearing, and what should we do, if anything?
5. Keep a record of what you eat for a day or two. Determine if the food, beverages, and snacks you ate could have been produced (at any level) without the aid of microorganisms. Indicate at what level(s) microorganisms were deliberately used. Be sure to consider ingredients such as citric acid, which is produced at the industrial level by several species of fungi.
6. Colonization of a susceptible human is critical to food-borne disease microorganisms. How might it be possible to modify foods to decrease these attachment processes?

Learn More

Barrett, J. R. 2000. Mycotoxins: of molds and maladies. *Environ. Health Perspect.* 108:A20–A23.

Beale, B. 2002. Probiotics: Their tiny worlds are under scrutiny. *The Scientist* 16(15):20–22.

Broadbent, J. R., and Steele, J. L. 2005. Cheese flavor and the genomics of lactic acid bacteria. *ASM News* 71:121–28.

Burgess, C.; O'Connell-Motherway, M.; Sybesma, W.; Hugenholtz, J.; and van Sinderen, D. 2004. Riboflavin production in *Lactococcus lactis:* Potential for in situ production of vitamin-enriched foods. *Appl. Environment. Microbiol.* 70: 5769–77.

Farmworth. E. R. 2003. *Handbook of fermented functional foods.* Boca Raton. FL: CRC Press.

Fung, D. Y. C. 2000. Food spoilage and preservation. In *Encyclopedia of microbiology,* 2d ed., vol. 2, J. Leading, ed-in-chief, 412–20. San Diego: Academic Press.

Henry, S. H.; Bosch, F. X.; Troxell, T. C.; and Bolger, P. M. 2000. Reducing liver cancer—global control of aflatoxin. *Science* 286:2453–54.

Hui, Y. H.; Pierson, M. D.; and Gorham, J. R. 2001. *Foodborne disease handbook,* 2d ed., vol. 1., *Bacterial pathogens.* New York: Marcel Dekker.

Hui, Y. H.; Sattar, S. A.; Murrell, K. D.; Nip, W.-K.; and Stanfield, P. S. 2001. *Foodborne disease handbook,* 2d ed., vol. 2. *Viruses, parasites, pathogens and HACCP.* New York: Marcel Dekker.

Montville, T. J., and Matthews, K. 2004. *Food microbiology: An introduction.* Washington, D.C.: ASM Press.

Robinson, R. K.; Batt, C. A.; and Patel, P. D. 2000. *Encyclopedia of food microbiology.* San Diego; Academic Press.

Rodríguez-Lázaro, D.; Jofré, A.; Aymerich, T.; Hugas, M.; and Pla, M. 2004. Rapid quantitative analysis of *Listeria monocytogenes* in meat products by real-time PCR. *Appl. Environment. Microbiol.* 70:6299–301.

Please visit the Prescott website at www.mhhe.com/prescott7 for additional references.

41 Applied and Industrial Microbiology

Biodegradation often can be facilitated by changing environmental conditions. Polychlorinated biphenyls (PCBs) are widespread industrial contaminants that accumulate in anoxic river muds. Although reductive dechlorination occurs under these conditions, oxygen is required to complete the degradation process. In this experiment, muds are being aerated to allow the final biodegradation steps to occur.

Preview

- In this text, "applied microbiology" refers to the use of microorganisms in a variety of ecosystems to perform a specific process that helps meet a specific goal. Thus wastewater treatment and bioremediation are types of applied microbiology.
- The presence of clean freshwater is critical to all terrestrial organisms. Contaminated waters are a significant source of disease and worldwide death. Water purification systems must be designed and maintained to cleanse wastewater and ensure that human, agricultural, and industrial wastes do not foul our planet.
- Sewage treatment can be carried out using large vessels where mixing and aeration can be controlled (conventional treatment). Constructed wetlands, where aquatic plants and their associated microorganisms are used, now are finding widespread applications in the treatment of liquid wastes.
- Indicator organisms, which usually die off at slower rates than many disease-causing microorganisms, can be used to evaluate the microbiological quality of water.
- Groundwater is an important source of drinking water, especially in suburban and rural areas. In too many cases, this resource is being contaminated by disease-causing microorganisms and nutrients, especially from septic tanks.
- Microorganisms are used in industrial microbiology to create a wide variety of products and to assist in maintaining and improving the environment.
- Most work in industrial microbiology has traditionally been carried out using microorganisms isolated from nature or modified through mutations. Microorganisms with specific genetic characteristics are now more commonly genetically engineered to meet desired objectives.
- Protein evolution is used to generate variants with desired functions. Both in vivo and in vitro approaches are used.
- In controlled growth systems, different products are synthesized during growth and after growth is completed. Most antibiotics are produced after the completion of active growth.
- Antibiotics and other microbial products continue to contribute to animal and human welfare. Newer products include anticancer drugs. Combinatorial biology is making it possible to produce hybrid antibiotics with unique properties.
- The products of industrial microbiology also include bulk chemicals that are used as food supplements and acidifying agents. Other products are used as biosurfactants and emulsifiers in a wide variety of applications.
- The use of microbes to degrade toxic compounds in the environment is called bioremediation. Anaerobic degradation processes are important for the initial modification of many compounds, especially those with chlorine and other halogenated functions. Degradation can produce simpler or modified compounds that may not be less toxic than the original compound.
- Bacteria, fungi, and viruses are increasingly employed as biopesticides, thus reducing dependence on chemical pesticides.

In this final chapter, we use the term "applied microbiology," to refer to the use of microbes in their natural environment to perform processes useful to humankind. Such processes include wastewater treatment and bioremediation, both of which are discussed in this chapter. Like applied microbiology, industrial microbiology involves the use of microorganisms to achieve specific goals. Industrial microbiology, however, generally focuses on products such as pharmaceutical and medical compounds (e.g., antibiotics, hormones, transformed steroids), solvents, organic acids, chemical feedstocks, amino acids, and enzymes that have economic value. The microorganisms employed by industry have been isolated from nature, and in many cases, were modified using classic mutation-selection procedures. Genetic engineering has replaced this more traditional approach to developing microbial strains of industrial importance.

The microbe will have the last word.

—Louis Pasteur

In developed countries, the processes and products of applied and industrial microbiology are taken for granted, but this is not true globally. For instance, the provision of clean drinking water and the sanitary treatment of contaminated water is beyond reach for an alarmingly high number of people. According to the World Health Organization and UNICEF, over 1 billion people worldwide do not have access to safe, drinkable water and about 40% of the world's population lacks basic sanitation. We begin this chapter by presenting ways in which water can be purified so that it can be consumed without fear of disease transmission. We then describe several approaches to treating wastewater to keep our rivers, streams, lakes, and groundwater—which may be sources of drinking water—clean. The remainder of the chapter focuses on industrial microbiology. It is clear that many concepts presented throughout the text are integrated in these important subjects.

41.1 Water Purification and Sanitary Analysis

Many important human pathogens are maintained in association with living organisms other than humans, including many wild animals and birds. Some of these bacterial and protozoan pathogens can survive in water and infect humans. As examples, *Vibrio vulnificus, V. parahaemolyticus,* and *Legionella* are of continuing concern. When waters are used for recreation or are a source of seafood that is consumed uncooked, the possibility for disease transmission exists. In many countries, such waters are the source of drinking water. Food-borne and waterborne viral diseases (section 37.4); Food-borne and waterborne bacterial diseases (section 38.4); Food-borne and waterborne diseases of fungi and protists (section 39.5)

Water purification is a critical link in controlling disease transmission in waters. As shown in **figure 41.1,** water purification can involve a variety of steps, depending on the type of impurities in the raw water source. Usually municipal water supplies are purified by a process that consists of at least three or four steps. If the raw water contains a great deal of suspended material, it often is first routed to a **sedimentation basin** and held so that sand and other very large particles can settle out. The partially clarified water is then mixed with chemicals such as alum and lime and moved to a **settling basin** where more material precipitates out. This procedure is called **coagulation** or flocculation and removes microorganisms, organic matter, toxic contaminants, and suspended fine particles. After these steps the water is further purified by passing it through **rapid sand filters,** which physically trap fine particles and flocs. This removes up to 99% of the remaining bacteria. After filtration the water is treated with a disinfectant. This step usually involves chlorination, but ozonation is becoming increasingly popular. When chlorination is employed, the chlorine dose must be large enough to leave residual free chlorine at a concentration of 0.2 to 2.0 mg/liter. A concern is the creation of **dis-**

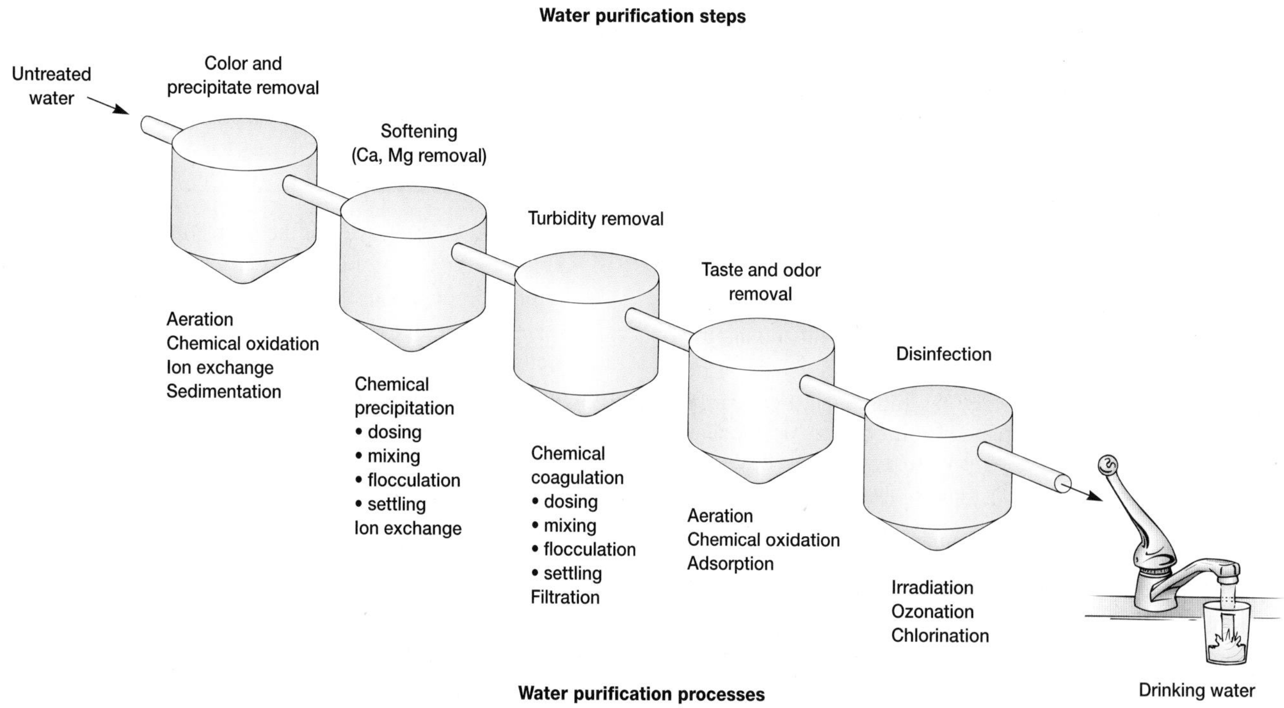

Figure 41.1 Water Purification. Several alternatives can be used for drinking water treatment depending on the initial water quality. A major concern is disinfection: chlorination can lead to the formation of *d*isinfection-*b*y*p*roducts (DBPs), including potentially carcinogenic *tri*h*a*lo*m*ethanes (THMs).

infection by-products (DBPs) such as **trihalomethanes (THMs)** that are formed when chlorine reacts with organic matter. Some of these compounds are carcinogens.

This purification process removes or inactivates disease-causing bacteria and indicator organisms (coliforms). Unfortunately, the use of coagulants, rapid filtration, and chemical disinfection often does not remove cysts of the protist *Giardia intestinalis, Cryptosporidium* oocysts, *Cyclospora,* and viruses. *Giardia,* a cause of human diarrhea, is now recognized as the most common identified waterborne pathogen in the United States. More consistent removal of *Giardia* cysts, which are about 7 to 10 by 8 to 12 μm in size, can be achieved with **slow sand filters.** This treatment involves the slow passage of water through a bed of sand in which a microbial layer covers the surface of each sand grain. Waterborne microorganisms are removed by adhesion to the gelatinous surface microbial layer (**Techniques & Applications 41.1**). Food-borne and waterborne diseases: Giardiasis (section 39.5)

In the last few years, *Cryptosporidium* has become a significant problem. This protozoan parasite is smaller than *Giardia* and is even more difficult to remove from water. A major source of *Giardia* and *Cryptosporidium* contamination of soils, plant materials, and waters is the Canada goose, protected as a migratory bird. Each goose produces about 0.68 kg of manure per day, and geese are documented carriers of these important pathogens. The goose population also is rapidly increasing. For example, in the Atlantic flyway, there are at least a million geese, and these are estimated to be increasing at 17% per year. They gather on golf courses, pastures, croplands, and vulnerable watersheds used for public water supplies, which of course adds to the problem.

Viruses in drinking water also must be inactivated or removed. Coagulation and filtration reduce virus levels about 90 to 99%. Further inactivation of viruses by chemical oxidants, high pH, and photooxidation may yield a reduction as great as 99.9%. None of these processes, however, is considered sufficient protection. New standards for virus inactivation are being developed. Bacteriophages, which can be easily grown and assayed, now are being used to monitor disinfection. If sufficient reductions in bacteriophage infectivity occur with a given disinfection process, it is assumed that viruses capable of infecting humans will also be reduced to satisfactory levels.

1. What steps are usually taken to purify drinking water?
2. Why is chlorination, although beneficial in terms of bacterial pathogen control, of environmental concern?
3. Which important waterborne pathogens are not controlled reliably by chlorination?

Sanitary Analysis of Waters

Monitoring and detection of indicator and disease-causing microorganisms are a major part of sanitary microbiology. Bacteria from the intestinal tract generally do not survive in the aquatic environment, are under physiological stress, and gradually lose their ability to form colonies on differential and selective media. Their die-out rate depends on the water temperature, the effects of sunlight, the populations of other bacteria present, and the chemical composition of the water. Procedures have been developed to attempt to "resuscitate" these stressed coliforms using selective and differential media.

A wide range of viral, bacterial, and protozoan diseases result from the contamination of water with human and other animal fecal wastes. Although many of these pathogens can be detected directly, environmental microbiologists have generally used **indicator organisms** as an index of possible water contamination by human pathogens. Researchers are still searching for the "ideal" indicator organism to use in sanitary microbiology. These are among the suggested criteria for such an indicator:

1. The indicator bacterium should be suitable for the analysis of all types of water: tap, river, ground, impounded, recreational, estuary, sea, and waste.
2. The indicator bacterium should be present whenever enteric pathogens are present.
3. The indicator bacterium should survive longer than the hardiest enteric pathogen.

Techniques & Applications

41.1 Waterborne Diseases, Water Supplies, and Slow Sand Filtration

Slow sand filtration, in which drinking water is passed through a sand filter that develops a layer of microorganisms on its surface, has had a long and interesting history. After London's severe cholera epidemic of 1849, Parliament, in an act of 1852, required that the entire water supply of London be passed through slow sand filters before use.

The value of this process was shown in 1892, when a major cholera epidemic occurred in Hamburg, Germany, and 10,000 lives were lost. The neighboring town, Altona, which used slow sand filtration, did not have a cholera epidemic. Slow sand filters were installed in many cities in the early 1900s, but the process fell into disfavor with the advent of rapid sand filters, chlorination, and the use of coagulants such as alum. Slow sand filtration, a time-tested process, is regaining favor because of its filtration effectiveness and lower maintenance costs. Slow sand filtration is particularly effective for the removal of *Giardia* cysts. For this reason slow sand filtration is used in many mountain communities where *Giardia* is a problem.

4. The indicator bacterium should not reproduce in the contaminated water and produce an inflated value.
5. The assay procedure for the indicator should have great specificity; in other words, other bacteria should not give positive results. In addition, the procedure should have high sensitivity and detect low levels of the indicator.
6. The testing method should be easy to perform.
7. The indicator should be harmless to humans.
8. The level of the indicator bacterium in contaminated water should have some direct relationship to the degree of fecal pollution.

Coliforms, including *Escherichia coli,* are members of the family *Enterobacteriaceae.* These bacteria make up approximately 10% of the intestinal microorganisms of humans and other animals and have found widespread use as indicator organisms. They lose viability in freshwater at slower rates than most of the major intestinal bacterial pathogens. When such "foreign" enteric indicator bacteria are not detectable in a specific volume (100 ml) of water, the water is considered **potable** [Latin *potabilis,* fit to drink], or suitable for human consumption. Class *Gammaproteobacteria:* Order *Enterobacteriales* (section 22.3)

The coliform group includes *E. coli, Enterobacter aerogenes,* and *Klebsiella pneumoniae.* Coliforms are defined as facultatively anaerobic, gram-negative, nonsporing, rod-shaped bacteria that ferment lactose with gas formation within 48 hours at 35°C. The original test for coliforms that was used to meet this definition involved the presumptive, confirmed, and completed tests, as shown in **figure 41.2.** The presumptive step is carried out by means of tubes inoculated with three different sample volumes to give an estimate of the **most probable number (MPN)** of coliforms in the water. The complete process, including the confirmed and completed tests, requires at least 4 days of incubations and transfers.

Unfortunately the coliforms include a wide range of bacteria whose primary source may not be the intestinal tract. To address this difficulty, tests have been developed that allow waters to be tested for the presence of **fecal coliforms.** These are coliforms derived from the intestine of warm-blooded animals, which can grow at the more restrictive temperature of 44.5°C.

To test for coliforms and fecal coliforms, and more effectively recover stressed coliforms, a variety of simpler and more specific tests have been developed. These include the membrane filtration technique, the **presence-absence (P-A) test** for coliforms and the related Colilert **defined substrate test** for detecting both coliforms and *E. coli.*

The **membrane filtration technique** has become a common and often preferred method of evaluating the microbiological characteristics of water. The water sample is passed through a membrane filter. The filter with its trapped bacteria is transferred to the surface of a solid medium or to an absorptive pad containing the desired liquid medium. Use of the proper medium enables the rapid detection of total coliforms, fecal coliforms, or fecal enterococci by the presence of their characteristic colonies. Samples can be placed on a less selective resuscitation medium, or incubated at a less stressful temperature, prior to growth under the final set of selective conditions. An example of a resuscitation step is the use of a 2 hour incubation on a pad soaked with lauryl sulfate broth, as is carried out in the LES Endo procedure. A resuscitation step often is needed with chlorinated samples, where the microorganisms are especially stressed. The advantages and disadvantages of the membrane filter technique are summarized in **table 41.1.** Membrane filters have been widely used with water that does not contain high levels of background organisms, sediment, or heavy metals.

More simplified tests for detecting coliforms and fecal coliforms are now available. The presence-absence test (P-A test) can be used for coliforms. This is a modification of the MPN procedure, in which a larger water sample (100 ml) is incubated in a single culture bottle with a triple-strength broth containing lactose broth, lauryl tryptose broth, and bromcresol purple indicator. The P-A test is based on the assumption that no coliforms should be present in 100 ml of drinking water. A positive test results in the production of acid (a yellow color) and constitutes a positive presumptive test requiring confirmation.

To test for both coliforms and *E. coli,* the related Colilert defined substrate test can be used. A water sample of 100 ml is added to a specialized medium containing *o*-nitrophenyl-β-D-galactopyranoside (ONPG) and 4-methylumbelliferyl-β-D-glucuronide (MUG) as the only nutrients. If coliforms are present, the medium will turn yellow within 24 hours at 35°C due to the hydrolysis of ONPG, which releases *o*-nitrophenol, as shown in **figure 41.3.** To check for *E. coli,* the medium is observed under long-wavelength UV light for fluorescence. When *E. coli* is present, MUG is modified to yield a fluorescent product. If the test is negative for the presence of coliforms, the water is considered acceptable for human consumption. The main change from previous standards is the requirement to have water free of coliforms and fecal coliforms. If coliforms are present, fecal coliforms or *E. coli* must be tested for.

Molecular techniques are now used routinely to detect coliforms in waters and other environments, including foods. 16S rRNA gene-targeted primers for coliforms enable the detection of one colony-forming unit (CFU) of *E. coli* per 100 ml of water, if a short enrichment step precedes the use of the PCR amplification. This allows the differentiation of nonpathogenic and enterotoxigenic strains, including the shigalike-toxin producing *E. coli* O157:H7. PCR (section 14.3)

In the United States a set of general guidelines for microbiological quality of drinking waters has been developed, including standards for coliforms, viruses, and *Giardia* (**table 41.2**). If unfiltered surface waters are being used, one coliform test must be run each day when the waters have higher turbidities.

Other indicator microorganisms include **fecal enterococci.** The fecal enterococci are increasingly being used as an indicator of fecal contamination in brackish and marine water. In salt water these bacteria die at a slower rate than the fecal coliforms, providing a more reliable indicator of possible recent pollution. Class *Bacilli:* Order *Lactobacillales* (section 23.5)

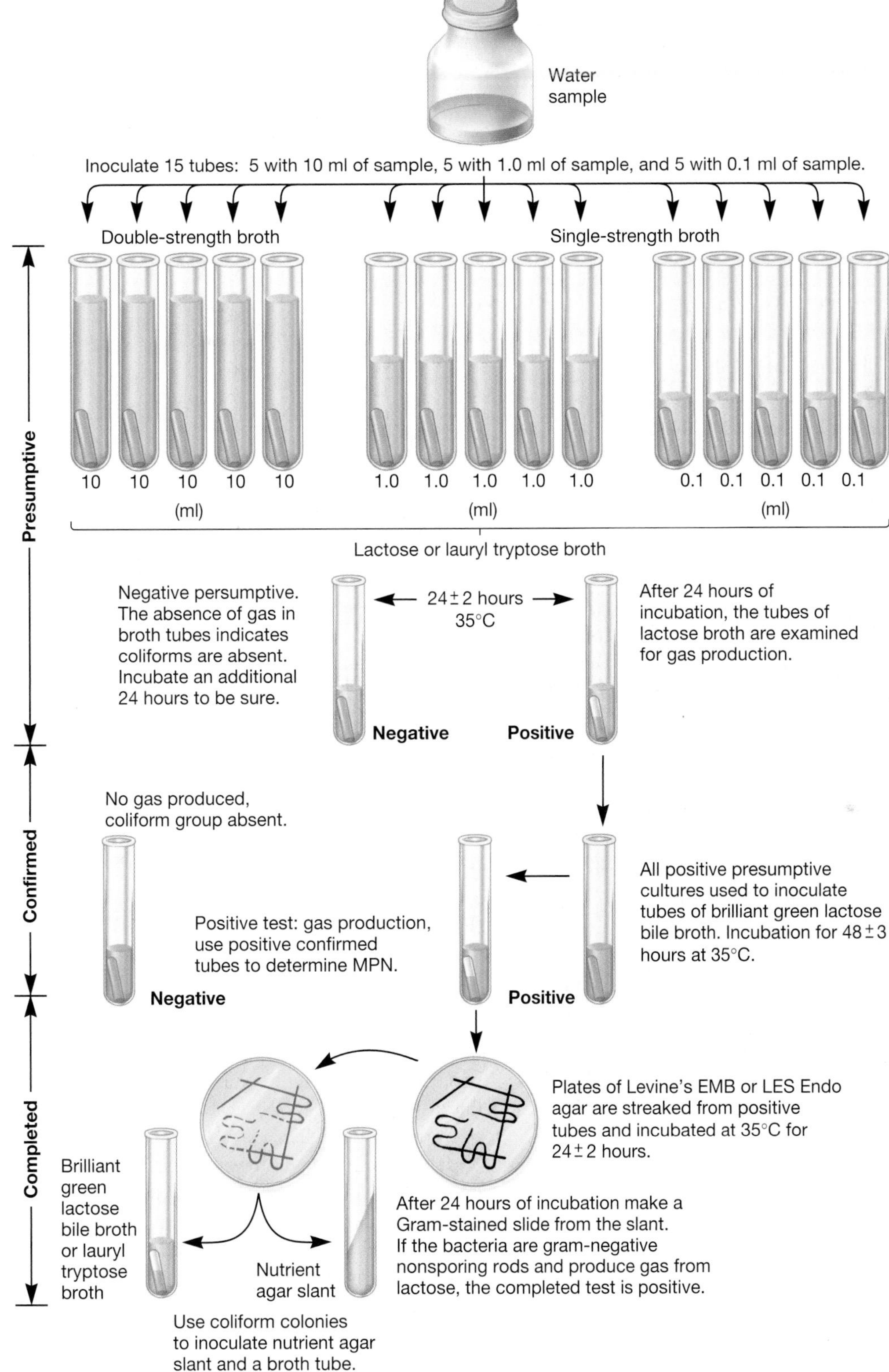
Water
sample
Inoculate 15 tubes: 5 with 10 ml of sample, 5 with 1.0 ml of sample, and 5 with 0.1 ml of sample.
Double-strength broth
Single-strength broth
10 10 10 10 10
(ml)
1.0 1.0 1.0 1.0 1.0
(ml)
0.1 0.1 0.1 0.1 0.1
(ml)
Lactose or lauryl tryptose broth
Presumptive
Negative persumptive. The absence of gas in broth tubes indicates coliforms are absent. Incubate an additional 24 hours to be sure.
24 ± 2 hours
35°C
After 24 hours of incubation, the tubes of lactose broth are examined for gas production.
Negative
Positive
Confirmed
No gas produced, coliform group absent.
Positive test: gas production, use positive confirmed tubes to determine MPN.
All positive presumptive cultures used to inoculate tubes of brilliant green lactose bile broth. Incubation for 48 ± 3 hours at 35°C.
Negative
Positive
Completed
Plates of Levine's EMB or LES Endo agar are streaked from positive tubes and incubated at 35°C for 24 ± 2 hours.
Brilliant green lactose bile broth or lauryl tryptose broth
Nutrient agar slant
After 24 hours of incubation make a Gram-stained slide from the slant. If the bacteria are gram-negative nonsporing rods and produce gas from lactose, the completed test is positive.
Use coliform colonies to inoculate nutrient agar slant and a broth tube.

Figure 41.2 The Multiple-Tube Fermentation Test. The multiple-tube fermentation technique has been used for many years for the sanitary analysis of water. Lactose broth tubes are inoculated with different water volumes in the presumptive test. Tubes that are positive for gas production are inoculated into brilliant green lactose bile broth in the confirmed test, and positive tubes are used to calculate the most probable number (MPN) value. The completed test is used to establish that coliform bacteria are present.

Table 41.1 Advantages and Disadvantages of the Membrane Filter Technique for Evaluation of the Microbial Quality of Water

Advantages
Good reproducibility
Single-step results often possible
Filters can be transferred between different media
Large volumes can be processed to increase assay sensitivity
Time savings are considerable
Ability to complete filtrations on site
Lower total cost in comparison with MPN procedure
Disadvantages
High-turbidity waters limit volumes sampled
High populations of background bacteria cause overgrowth
Metals and phenols can adsorb to filters and inhibit growth

Source: Data from L. S. Clesceri, et al., *Standard Methods for the Examination of Water and Wastewater,* 20th edition, pages 9–56, 1998. American Public Health Association, Washington, D.C.

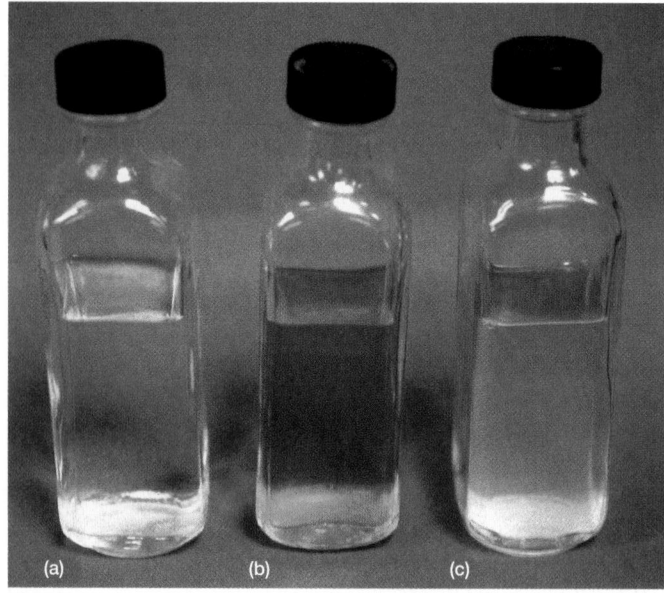

Figure 41.3 The Defined Substrate Test. This much simpler test is used to detect coliforms and fecal coliforms in single 100 ml water samples. The medium uses ONPG and MUG (see text) as defined substrates. **(a)** Uninoculated control. **(b)** Yellow color due to the presence of coliforms. **(c)** Fluorescent reaction due to the presence of fecal coliforms.

1. What is an indicator organism, and what properties should it have?
2. How is a coliform defined? How does this definition relate to presumptive, confirmed, and completed tests?
3. How does one differentiate between coliforms and fecal coliforms in the laboratory?

Table 41.2 Current Drinking Water Standards in the United States

Agent	Allowable Maximum Contaminant Level Goal (MCLG) or Maximum Contaminant Level (MCL)
Coliforms	MCLG = 0
	MCL = No more than 5% positive total coliform samples/month for water systems that collect > 40 samples/month. For water systems that collect < 40 routine samples/month, no more than 1 can be coliform positive. Every sample that has total coliforms must be analyzed for fecal coliforms. There cannot be any fecal coliforms or *E. coli.*
Cryptosporidium	MCLG = 0
Giardia intestinalis	MCLG = 0
Legionella	MCLG = 0
Viruses (enteric)	MCLG = 0

Source: Environmental Protection Agency, USA, July, 2002.

4. In what type of environment is it better to use fecal enterococci rather than fecal coliforms as an indicator organism? Why?
5. What are the advantages and disadvantages of membrane filters for microbiological examinations of water?
6. Why has the defined substrate test with ONPG and MUG been accepted as a test of drinking water quality?

41.2 Wastewater Treatment

Waters often contain high levels of organic matter from industrial and agricultural wastes (e.g., from food processing, petrochemical and chemical plants, and plywood plant resin wastes), and from human wastes. It is necessary to remove organic matter by the process of wastewater treatment. Depending on the effort given to this task, it may still produce waters containing nutrients and some microorganisms, which can be released to rivers and streams.

The process of wastewater treatment, when performed at a municipal level, must be monitored to ensure that waters released into the environment do not pose environmental and health risks. Our discussion of wastewater treatment must therefore begin with the means by which water quality is monitored. We then discuss large-scale wastewater treatment processes followed by home treatment systems.

Measuring Water Quality

Carbon removal during wastewater treatment can be measured several ways, including (1) as **total organic carbon (TOC),** (2) as chemically oxidizable carbon by the **chemical oxygen de-**

mand (COD) test, or (3) as biologically usable carbon by the **biochemical oxygen demand (BOD)** test. The TOC includes all carbon, whether or not it is usable by microorganisms. This is determined by oxidizing the organic matter in a sample at high temperature in an oxygen stream and measuring the resultant CO_2 by infrared or potentiometric techniques. The COD gives a similar measurement, except that lignin often will not react with the oxidizing chemical, such as permanganate, that is used in this procedure. The BOD test, in comparison, measures only the portion of the total carbon that can be oxidized by microorganisms in a 5-day period under standard conditions.

The biochemical oxygen demand is an indirect measure of organic matter in aquatic environments. It is the amount of dissolved O_2 needed for microbial oxidation of biodegradable organic matter. When O_2 consumption is measured, the O_2 itself must be present in excess and not limit oxidation of the nutrients (**table 41.3**). To achieve this, the waste sample is diluted to assure that at least 2 mg/liter of O_2 are used while at least 1 mg/liter of O_2 remains in the test bottle. Ammonia released during organic matter oxidation can also exert an O_2 demand in the BOD test, so nitrification or the **nitrogen oxygen demand (NOD)** is often inhibited by 2-chloro-6-(trichloromethyl) pyridine (nitrapyrin). In the normal BOD test, which is run for 5 days at 20°C on untreated samples, nitrification is not a major concern. However, when treated effluents are analyzed, NOD can be a problem.

In terms of speed, the TOC is fastest, but less informative in terms of biological processes. The COD is slower and involves the use of wet chemicals with higher waste chemical disposal costs. The TOC, COD, and BOD provide different but complementary information on the carbon in a water sample. It is critical to note that these measurements, concerned with carbon removal, do not directly address concerns for removal of minerals such as nitrate, phosphate, and sulfate from waters. These minerals have global impacts on cyanobacterial and algal growth in lakes, rivers, and the oceans by contributing to the process of eutrophication. The removal of dissolved organic matter and possibly inorganic nutrients, plus inactivation and removal of pathogens, are important parts of wastewater treatment. Marine and freshwater environments: Nutrient cycling (section 28.1)

Table 41.3 The Biochemical Oxygen Demand (BOD) Test: A System with Excess and Limiting Components

Components in Excess at the End of the Incubation Period
Nitrogen
Phosphorus
Iron
Trace elements
Microorganisms
Oxygen
Component Limiting at the End of the Incubation Period
Organic matter

1. What are TOC, COD, and BOD and how are these similar and different?
2. What factors can lead to a nitrogen oxygen demand (NOD) in water?
3. What components should limit the reactions in a BOD test, and what components should not limit reaction rates? Why?
4. What minerals can contribute to eutrophication?

Wastewater Treatment Processes

The aerobic self-purification sequence that occurs when organic matter is added to lakes and rivers can be carried out under controlled conditions in which natural processes are intensified. This often involves the use of large basins (conventional sewage treatment) where mixing and gas exchange are carefully controlled.

An aerial photograph of a modern sewage treatment plant is shown in **figure 41.4*a*. Wastewater treatment** involves a number of steps that are spatially segregated (figure 41.4*b*). The first three steps are called primary, secondary, and tertiary treatment (**table 41.4**). At the end of the process, the water is usually chlorinated (itself an emerging environmental and human health problem) before it is released.

Primary treatment physically removes 20 to 30% of the BOD that is present in particulate form. In this treatment, particulate material is removed by screening, precipitation of small particulates, and settling in basins or tanks. The resulting solid material is usually called **sludge.**

Secondary treatment promotes the biological transformation of dissolved organic matter to microbial biomass and carbon dioxide. About 90 to 95% of the BOD and many bacterial pathogens are removed by this process. Several approaches can be used in secondary treatment to biologically remove dissolved organic matter. All of these techniques involve similar microbial activities. Under oxic conditions, dissolved organic matter will be transformed into additional microbial biomass plus carbon dioxide. When microbial growth is completed, under ideal conditions the microorganisms will aggregate and form a settleable stable floc structure. Minerals in the water also may be tied up in microbial biomass. When microorganisms grow, flocs can form. As shown in **figure 41.5*a*,** a healthy settleable floc is compact. In contrast, poorly formed flocs have a network of filamentous microbes that will retard settling (figure 41.5*b*).

When these processes occur with lower O_2 levels or with a microbial community that is too young or too old, unsatisfactory floc formation and settling can occur. The result is a **bulking sludge,** caused by the massive development of filamentous bacteria such as *Sphaerotilus* and *Thiothrix,* together with many poorly characterized filamentous organisms. These important filamentous bacteria form flocs that do not settle well, and thus produce effluent quality problems. Class *Betaproteobacteria:* Order *Burkholderiales* (section 22.2), Class *Gammaproteobacteria:* Order *Thiotrichales* (section 22.3)

An aerobic **activated sludge** system (**figure 41.6*a***) involves a horizontal flow of materials with recycling of sludge—the active biomass that is formed when organic matter is oxidized and

(a)

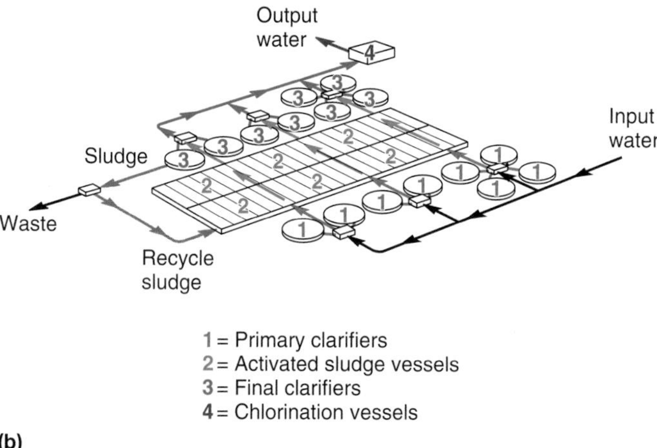

(b)

Figure 41.4 An Aerial View of a Modern Conventional Sewage Treatment Plant. Sewage treatment plants allow natural processes of self-purification that occur in rivers and lakes to be carried out under more intense, managed conditions in large concrete vessels. **(a)** A plant in New Jersey. **(b)** A diagram of flows in the plant.

Table 41.4 Major Steps in Primary, Secondary, and Tertiary Treatment of Wastes

Treatment Step	Processes
Primary	Removal of insoluble particulate materials by settling, screening, addition of alum and other coagulation agents, and other physical procedures
Secondary	Biological removal of dissolved organic matter
	Trickling filters
	Activated sludge
	Lagoons
	Extended aeration systems
	Anaerobic digesters
Tertiary	Biological removal of inorganic nutrients
	Chemical removal of inorganic nutrients
	Virus removal/inactivation
	Trace chemical removal

degraded by microorganisms. Activated sludge systems can be designed with variations in mixing. In addition, the ratio of organic matter added to the active microbial biomass can be varied. A low-rate system (low nutrient input per unit of microbial biomass), with slower growing microorganisms, will produce an effluent with low residual levels of dissolved organic matter. A high-rate system (high nutrient input per unit of microbial biomass), with faster growing microorganisms, will remove more dissolved organic carbon per unit time, but produce a poorer quality effluent.

Aerobic secondary treatment also can be carried out with a **trickling filter** (figure 41.6*b*). The waste effluent is passed over rocks or other solid materials upon which microbial biofilms have developed, and the microbial community degrades the organic waste. A sewage treatment plant can be operated to produce less sludge by employing the **extended aeration** process (figure 41.6*c*). Microorganisms grow on the dissolved organic matter, and the newly formed microbial biomass is eventually consumed to meet maintenance energy requirements. This requires extremely large aeration basins and long aeration times. In addition, with the biological self-utilization of the biomass, minerals originally present in the microorganisms are again released to the water.

All aerobic processes produce excess microbial biomass, or sewage sludge, which contains many recalcitrant organics. Often the sludges from aerobic sewage treatment, together with the materials settled out in primary treatment, are further treated by anaerobic digestion. Anaerobic digesters are large tanks designed to operate with continuous input of untreated sludge and removal of the final, stabilized sludge product. Methane is vented and often burned for heat and electricity production. This digestion process involves three steps: (1) the fermentation of the sludge components to form organic acids, including acetate; (2) production of the methanogenic substrates: acetate, CO_2, and hydrogen; and finally, (3) methanogenesis by the methane producers. These methanogenic processes, summarized in **table 41.5,** involve critical balances between electron acceptors and donors. To function most efficiently, the hydrogen concentration must be maintained at a low level. If hydrogen and organic acids accumulate, methane production can be inhibited, resulting in a stuck digester. Phylum *Euryarchaeota:* Methanogens (section 20.3)

Anaerobic digestion has many advantages. Most of the microbial biomass produced in aerobic growth is used for methane production in the anaerobic digester. Also, because the process of methanogenesis is energetically very inefficient, the microbes must consume about twice the nutrients to produce an equivalent biomass as that of aerobic systems. Consequently, less sludge is

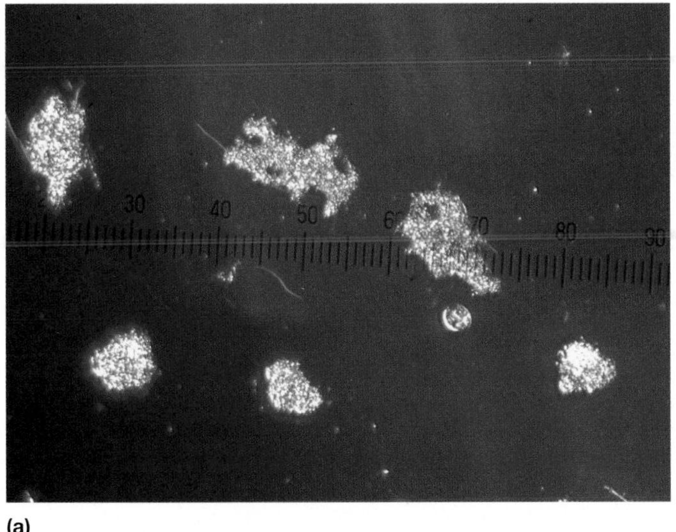
(a)

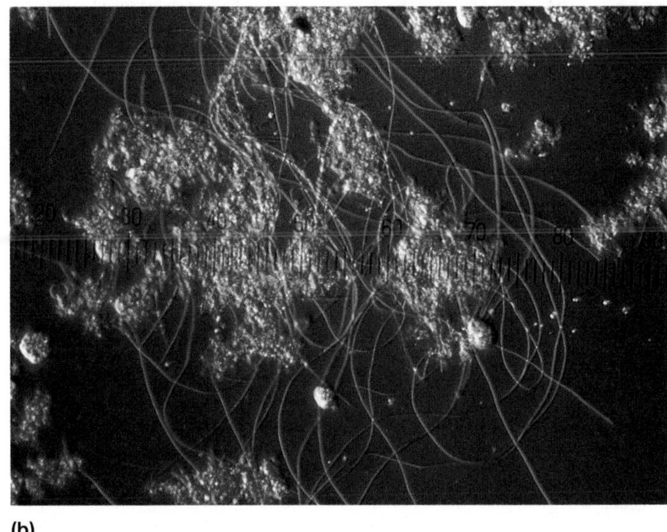
(b)

Figure 41.5 Proper Floc Formation in Activated Sludge. Microorganisms play a critical role in the functioning of activated sludge systems. **(a)** The operation is dependent on the formation of settleable flocs. **(b)** If the plant does not run properly, poorly settling flocs can form due to such causes as low aeration, sulfide, and acidic organic substrates. These flocs do not settle properly because of their open or porous structure. As a consequence, the organic material is released with the treated water and lowers the quality of the final effluent.

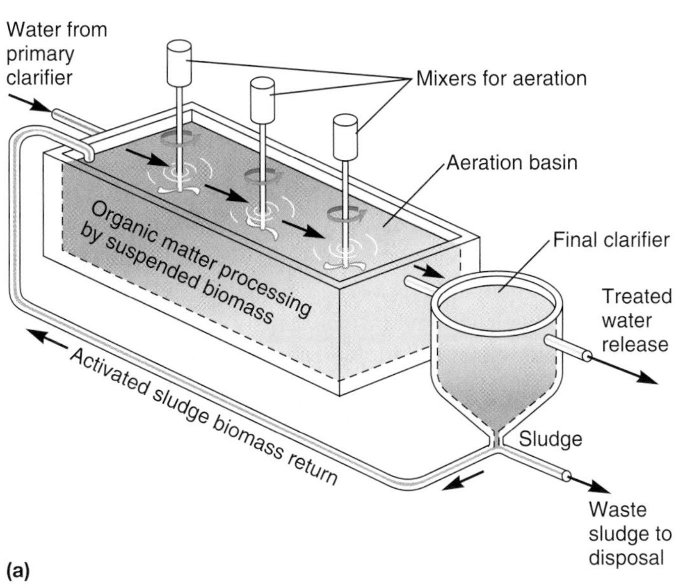

(a)

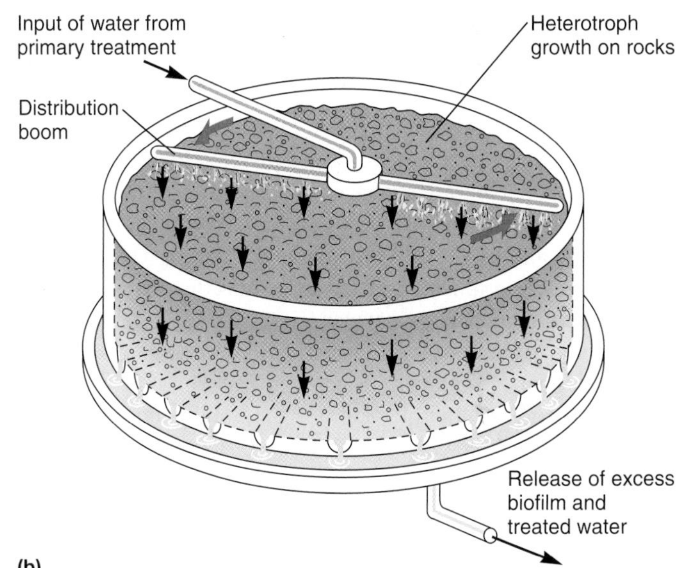

(b)

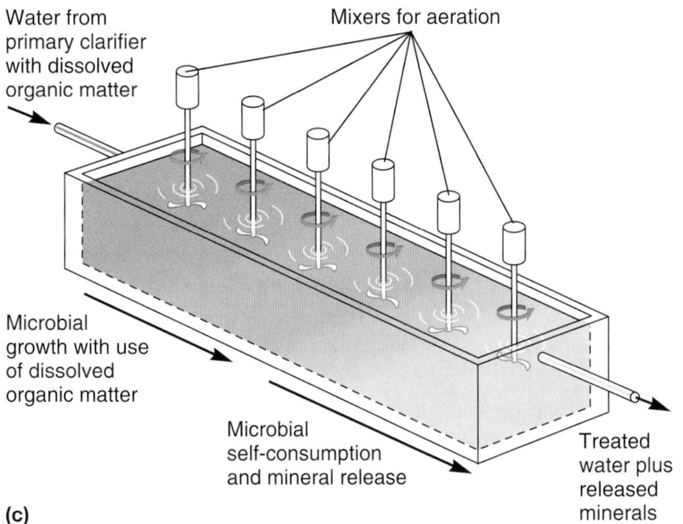

(c)

Figure 41.6 Aerobic Secondary Sewage Treatment. **(a)** Activated sludge with microbial biomass recycling. The biomass is maintained in a suspended state to maximize oxygen, nutrient, and waste transfer processes. **(b)** Trickling filter, where waste water flows over biofilms attached to rocks or other solid supports, resulting in transformation of dissolved organic matter to new biofilm biomass and carbon dioxide. Excess biomass and treated water flow to a final clarifier. **(c)** An extended aeration process, where aeration is continued beyond the point of microbial growth, allows the microbial biomass to self-consume due to microbial energy of maintenance requirements. (The extended length of the reactor allows this process of biomass self-consumption to occur.) Minerals originally incorporated in the microbial biomass are released to the water as the process occurs.

Table 41.5 Sequential Reactions in the Anaerobic Digester

Process Step	Substrates	Products	Major Microorganisms
Fermentation	Organic polymers	Butyrate, propionate, lactate, succinate, ethanol, acetate,[a] H_2,[a] CO_2[a]	*Clostridium* *Bacteroides* *Peptostreptococcus* *Peptococcus* *Eubacterium* *Lactobacillus*
Acetogenic reactions	Butyrate, propionate, lactate, succinate, ethanol	Acetate, H_2, CO_2	*Syntrophomonas* *Syntrophobacter*
Methanogenic reactions	Acetate	$CH_4 + CO_2$	*Acetobacterium* *Methanosarcina*
	H_2 and HCO_3^-	CH_4	*Methanothrix* *Methanobrevibacter* *Methanomicrobium* *Methanogenium* *Methanobacterium* *Methanococcus* *Methanospirillum*

[a]Methanogenic substrates produced in the initial fermentation step.

produced and it can be easily dried. Dried sludge removed from well-operated anaerobic systems can even be sold as organic garden fertilizer. In contrast, sludge can be dangerous if the system is not properly managed. Then heavy metals and other environmental contaminants can be concentrated in the sludge. There may be longer-term environmental and public health effects from disposal of this material on land or in water.

Tertiary treatment further purifies wastewaters. It is particularly important to remove nitrogen and phosphorus compounds that can promote eutrophication. Organic pollutants can be removed with activated carbon filters. Phosphate usually is precipitated as calcium or iron phosphate (for example, by the addition of lime). Excess nitrogen may be removed by "stripping," volatilization as NH_3 at high pHs. Ammonia itself can be chlorinated to form dichloramine, which is then converted to molecular nitrogen. In some cases, biological processes can be used to remove nitrogen and phosphorus. A widely used process for nitrogen removal is denitrification. Here nitrate, produced under aerobic conditions, is used as an electron acceptor under conditions of low oxygen with organic matter added as an energy source. Nitrate reduction yields nitrogen gas and nitrous oxide (N_2O) as the major products. Currently there is a great deal of interest in anaerobic nitrogen removal processes. These include the anammox process where ammonium ion (used as the electron donor), is reacted with nitrite (the electron acceptor) produced by partial nitrification. The anammox process can convert up to 80% of the beginning ammonium ion to N_2 gas. To remove phosphorus, oxic and anoxic conditions are used alternately in a series of treatments, and phosphorus accumulates in specially adapted microbial biomass as polyphosphate. Tertiary treatment is expensive and is usually not employed except where necessary to prevent obvious ecological disruption. Biogeochemical cycling: Nitrogen cycle (section 27.2)

Wetlands are a vital natural resource and a critical part of our environment, and increasingly efforts are being made to protect these fragile aquatic communities from pollution. A major means of wastewater treatment is the use of **constructed wetlands,** where the basic components of natural wetlands (soils, aquatic plants, waters) are used as a functional waste treatment system. Constructed wetlands now are increasingly employed in the treatment of liquid wastes and for bioremediation, which is discussed in section 41.6. This system uses floating, emergent, or submerged plants, as shown in **figure 41.7.** The aquatic plants provide nutrients in the root zone, which can support microbial growth. Especially with emergent plants, the root zone can be maintained in an anoxic state in which sulfide, synthesized by *Desulfovibrio* using root zone organic matter as an energy source, can trap metals. Constructed wetlands also are being used to treat acid mine drainage (AMD) in many parts of the world. Higher-strength industrial wastes also can be treated.

1. Explain how primary, secondary, and tertiary treatments are accomplished.
2. What is bulking sludge? Name several important microbial groups that contribute to this problem.
3. What are the steps of organic matter processing that occur in anaerobic digestion? Why is acetogenesis such an important step?
4. After anaerobic digestion is completed, why is sludge disposal still of concern?
5. Why might different aquatic plant types be used in constructed wetlands?

Home Treatment Systems

Groundwater, or water in gravel beds and fractured rocks below the surface soil, is a widely used but often unappreciated water resource. In the United States groundwater supplies at least 100 million people with drinking water, and in rural and suburban ar-

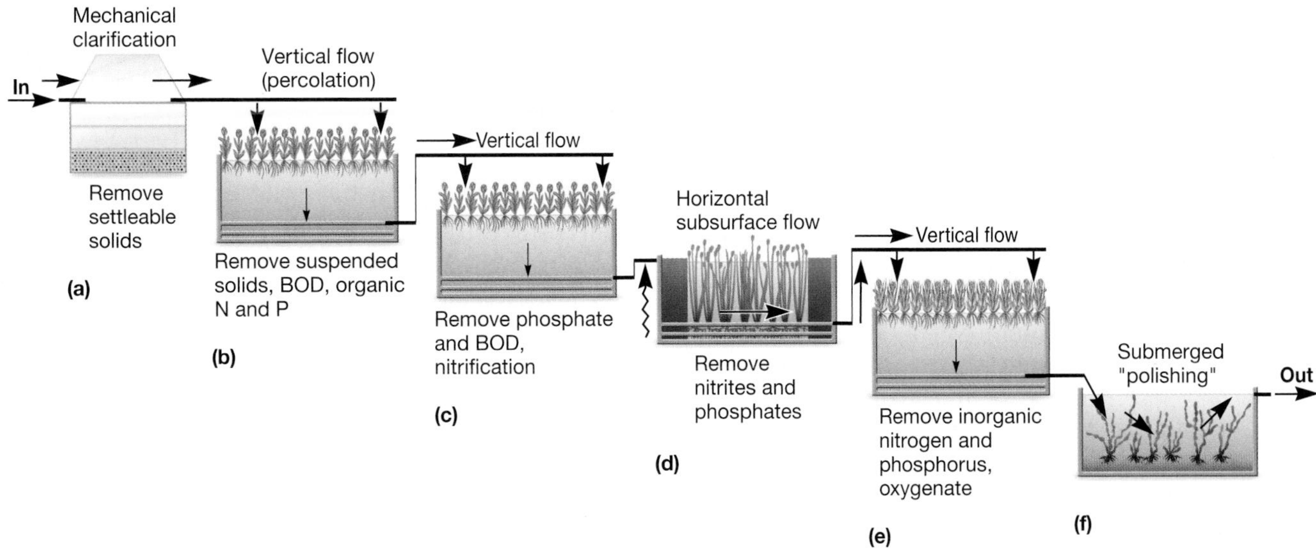

Figure 41.7 Constructed Wetland for Wastewater Treatment. Multistage constructed wetland systems can be used for organic matter and phosphate removal **(a)**. Free-floating macrophytes **(b,c,e)**, such as duckweed and water hyacinth, can be used for a variety of purposes. Emergent macrophytes **(d),** such as bulrush, allow surface flow as well as vertical and horizontal subsurface flow. Submerged vegetation **(f)**, such as waterweed, allows final "polishing" of the water. These wetlands also can be designed for nitrification and metal removal from waters.

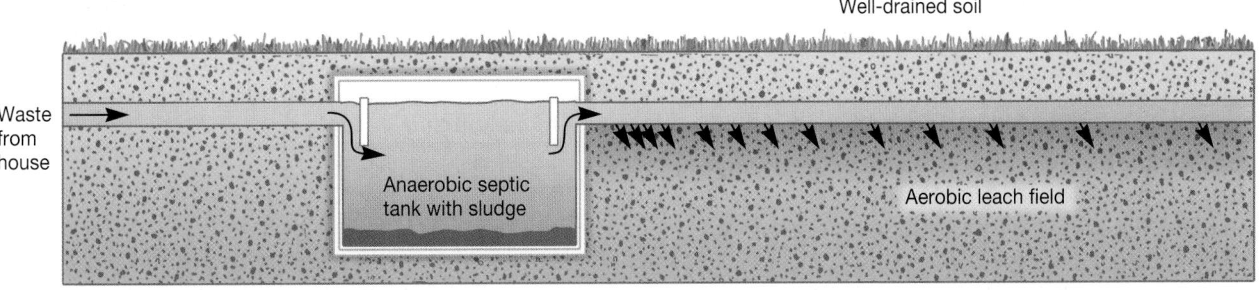

Figure 41.8 The Conventional Septic Tank Home Treatment System. This system combines an anaerobic waste liquefaction unit (the septic tank) with an aerobic leach field. Biological oxidation of the liquefied waste takes place in the leach field, unless the soil becomes flooded.

eas beyond municipal water distribution systems, 90 to 95% of all drinking water comes from this source.

The great dependence on this resource has not resulted in a corresponding understanding of microorganisms and microbiological processes that occur in the groundwater environment. Increasing attention is now being given to predicting the fate and effects of groundwater contamination on the chemical and microbiological quality of this resource. Pathogenic microorganisms and dissolved organic matter are removed from water during subsurface passage through adsorption and trapping by fine sandy materials, clays, and organic matter. Microorganisms associated with these materials—including predators such as protozoa—can use the trapped pathogens as food. This results in purified water with a lower microbial population.

This combination of adsorption-biological predation is used in home treatment systems (**figure 41.8**). Conventional **septic tank** systems include an anaerobic liquefaction and digestion step that occurs in the septic tank itself (the tank functions as a simple anaerobic digester). This is followed by organic matter adsorption and entrapment of microorganisms in an aerobic leach-field where biological oxidation occurs. A septic tank may not operate correctly for several reasons. If the retention time of the waste in the septic tank is too short, undigested solids move into the leach field, gradually plugging the system. If the leach field floods and becomes anoxic, biological oxidation does not occur, and effective treatment ceases. Other problems can occur, especially when a suitable soil is not present and the septic tank outflow from a conventional system drains too rapidly to the deeper subsurface.

Fractured rocks and coarse gravel materials provide little effective adsorption or filtration. This may result in the contamination of well water with pathogens and the transmission of disease. In addition, nitrogen and phosphorus from the waste can pollute the groundwater. This leads to nutrient enrichment of ponds, lakes, rivers, and estuaries as the subsurface water enters these environmentally sensitive water bodies.

Domestic and commercial on-site septic systems are now being designed with nitrogen and phosphorus removal steps. Nitrogen is usually removed by nitrification-denitrification processes with organic matter provided by sawdust or a similar material. Currently used systems function with essentially no maintenance for up to 8 years. For phosphorus removal, a reductive iron dissolution process can be used. With the need to control nitrogen and phosphorus releases from septic systems around the world, there will be increased emphasis on use of these and similar technologies in the future.

Subsurface zones also can become contaminated with pollutants from other sources. Land disposal of sewage sludges, illegal dumping of septic tank pumpage, improper toxic waste disposal, and runoff from agricultural operations all contribute to groundwater contamination with chemicals and microorganisms.

Many pollutants that reach the subsurface will persist and may affect the quality of groundwater for extended periods. Much research is being conducted to find ways to treat groundwater in place—**in situ treatment.** As will be further discussed, microorganisms and microbial processes are critical in many of these remediation efforts.

1. In rural areas, approximately what percentage of the water used for human consumption is groundwater?
2. What factors can limit microbial activity in subsurface environments? Consider the energetic and nutritional requirements of microorganisms in your answer.
3. How, in principle, are a conventional septic tank system and a leach-field system supposed to work? What alternatives are available to remove nitrogen and phosphorus from effluents? What factors can reduce the effectiveness of this system?

41.3 Microorganisms Used in Industrial Microbiology

The use of microorganisms in industrial microbiology follows a logical sequence. It is necessary first to identify or create a microorganism that carries out the desired process in the most efficient manner. This microorganism or its cloned genes are then used, either in a controlled environment such as a fermenter or in complex natural systems, such as in soils or waters, to achieve specific goals.

Thus the first task for an industrial microbiologist is to find a suitable microorganism, one that is genetically stable, easy to maintain and grow, and well suited for extraction or separation of desired products. A wide variety of alternative approaches are available, ranging from isolating microorganisms from the environment to using sophisticated molecular techniques to modify an existing microorganism. Here we present some of the commonly used approaches.

Finding Microorganisms in Nature

Until relatively recently, microbial cultures used in industrial microbiology were most often obtained from natural materials such as soil samples, waters, and spoiled bread and fruit. Cultures from all areas of the world continue to be examined to identify new strains with desirable characteristics. Interest in hunting for new microorganisms, or **bioprospecting,** continues today.

Less than 1% of the microbial species estimated to exist in most environments has been isolated or cultured (**table 41.6**). With increased interest in microbial diversity, microbial ecology, and especially in microorganisms from extreme environments (**Techniques & Applications 41.2**), microbiologists are exploring new ways to grow these previously uncultured microbes. For instance, single-cell gel microencapsulation allows nutrient diffusion and microbial communication, while minimizing overgrowth by more rapid-growing competitors.

Genetic Manipulation of Microorganisms

Genetic manipulations are used to produce microorganisms with new and desirable characteristics. The classical methods of genetic exchange coupled with recombinant DNA technology (*see chapters 13 and 14*) play a vital role in the development of cultures for industrial microbiology.

Mutagenesis

Once a promising microorganism is found, a variety of techniques can be used for its improvement, including chemical mutagens, ultraviolet light, and transposon mutagenesis. As an example, the first cultures of *Penicillium notatum,* which could be

Table 41.6 Estimates of the Percent "Cultured" Microorganisms in Various Environments

Environment	Estimated Percent Cultured
Seawater	0.001–0.100
Freshwater	0.25
Mesotrophic lake	0.1–1.0
Unpolluted estuarine waters	0.1–3.0
Activated sludge	1–15
Sediments	0.25
Soil	0.3

Source: D. A. Cowan. 2000. Microbial genomes—the untapped resource. *Tibtech* 18:14–16. Table 2, p. 15.

Techniques & Applications

41.2 The Potential of Thermophilic Archaea in Biotechnology

There is great interest in the characteristics of archaea isolated from the outflow mixing regions above deep hydrothermal vents that release water at 250 to 350°C. This is because these hardy organisms can grow at temperatures as high as 121°C. The problems in growing these microorganisms in a laboratory are formidable. For example, to grow some of them, it is necessary to use special culturing chambers and other specialized equipment to maintain water in the liquid state at these high temperatures.

Such microorganisms, termed hyperthermophiles, with optimum growth temperatures of 85°C or above, confront unique challenges in nutrient acquisition, metabolism, nucleic acid replication, and growth. Many are anaerobes that depend on elemental sulfur or ferric ion as electron acceptors. Enzyme stability is critical. Some DNA polymerases are inherently stable at 140°C, whereas many other enzymes are stabilized in vivo with unique thermoprotectants. When these enzymes are separated from their protectant, they lose their unique thermostability.

These enzymes may have important applications in methane production, metal leaching and recovery, and in immobilized enzyme systems. In addition, the possibility of selective stereochemical modification of compounds normally not in solution at lower temperatures may provide new routes for directed chemical syntheses. This is an exciting and expanding area of the modern biological sciences to which environmental microbiologists can make significant contributions.

grown only under stationary conditions, yielded low concentrations of penicillin. In 1943 a strain of *Penicillium chrysogenum* was isolated—strain NRRL 1951—which was further improved through mutation (**figure 41.9**). Today most penicillin is produced with *Penicillium chrysogenum,* grown in aerobic stirred fermenters, which gives 55-fold higher penicillin yields than the original static cultures. Mutations and their chemical basis (section 13.1); Transposable elements (section 13.5)

Short lengths of chemically synthesized DNA sequences can be inserted into recipient microorganisms by the process of **site-directed mutagenesis.** This can create small genetic alterations leading to a change of one or several amino acids in the target protein. Such minor amino acid changes have been found to lead, in many cases, to unexpected changes in protein characteristics, and have resulted in new products such as more environmentally resistant enzymes and enzymes that can catalyze desired reactions. These approaches are part of the field of **protein engineering.** Synthetic DNA (section 14.2)

Protoplast Fusion

Most yeasts and molds are asexual or of a single mating type, which decreases the chance of random mutations that could lead to strain degeneration. **Protoplast fusion** can be used in genetic studies with these microorganisms. Protoplasts—cells lacking a cell wall—are prepared by growing the cells in an isotonic solution while treating them with enzymes, including cellulase and beta-galacturonidase. The protoplasts are then regenerated using osmotic stabilizers such as sucrose. After regeneration of the cell wall, the new protoplasm fusion product can be used in further studies.

A major advantage of the protoplast fusion technique is that protoplasts of different microbial species can be fused, even if they are not closely linked taxonomically. For example, protoplasts of *Penicillium roquefortii* have been fused with those of *P. chrysogenum.* Even yeast protoplasts and erythrocytes can be fused.

1. Why is the recovery of previously uncultured microorganisms from the environment an important goal?
2. What is protoplast fusion and what types of microorganisms are used in this process?
3. What is the goal of protein engineering?

Transfer of Genetic Information between Different Organisms

The transfer and expression of genes between different organisms can give rise to novel metabolic processes and products. This is part of the rapidly developing field of **combinatorial biology** (**table 41.7**). An important early example of this approach was the creation of the "superbug," patented by A. M. Chakarabarty in 1974, which had an increased capability of hydrocarbon degradation. Similarly, the genes for antibiotic production can be transferred to a microorganism that produces another antibiotic, or even to a non-antibiotic-producing microorganism. Other examples are the expression, in *E. coli,* of the enzyme creatininase from *Pseudomonas putida* and the production of pediocin, a bacteriocin, in a yeast used in wine fermentation for the purpose of controlling bacterial contaminants. When functional genes from one organism are transcribed and translated in another, it is called **heterologous gene expression.**

Heterologous gene expression can improve production efficiency and minimize the purification steps required before the product is ready for use. For example, recombinant baculoviruses can be replicated in insect larvae to achieve rapid large-scale production of a desired virus or protein. Transgenic plants may be used to manufacture large quantities of a variety of metabolic products. A gene encoding a foot-and-mouth disease virus antigen has been incorporated into *E. coli,* enabling the expression of this genetic information and synthesis of the gene product for use in vaccine production (**figure 41.10**).

Genetic information transfer allows the production of specific proteins and peptides without contamination by other products

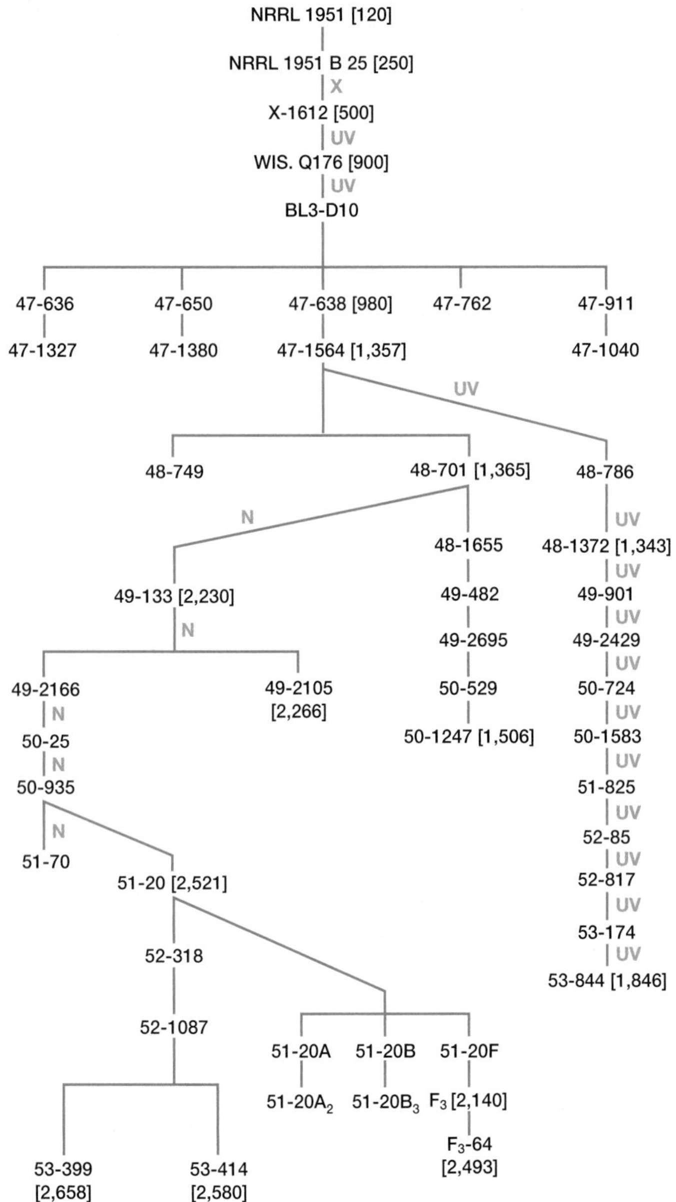

Figure 41.9 Mutation Makes It Possible to Increase Fermentation Yields. A "genealogy" of the mutation processes used to increase penicillin yields with *Penicillium chrysogenum* using X-ray treatment (X), UV treatment (UV), and mustard gas (N). By using these mutational processes, the yield was increased from 120 International Units (IU) to 2,580 IU, a 20-fold increase. Unmarked transfers were used for mutant growth and isolation. Yields in international units/ml in brackets.

that might be synthesized in the original organism. This approach can decrease the time and cost of recovering and purifying a product. Another major advantage of engineered protein production is that only biologically active stereoisomers are produced. This specificity is required to avoid the possible harmful side effects of inactive stereoisomers.

1. What is combinatorial biology and what is the basic approach used in this technique? What types of major products have been created using combinatorial biology?
2. Why might one want to insert a gene in a foreign cell and how is this done?
3. Why is it important to produce specific isomers of products for use in animal and human health?

Modification of Gene Expression

In addition to inserting new genes in organisms, it also is possible to modify gene regulation by modifying regulatory molecules or the DNA sites to which they bind. These approaches make it possible to overproduce a wide variety of products, as shown in **table 41.8.**

The modification of gene expression also can be used to intentionally alter metabolic pathways by inactivation or deregulation of specific genes, which is the field of **pathway architecture.** Understanding pathway architecture makes it possible to design a pathway that will be most efficient by avoiding slower or energetically more costly routes. This approach has been used to improve penicillin production by ***m*****etabolic *path*****way *e*****ngineering (MPE).**

An interesting development in modifying gene expression, which illustrates **metabolic control engineering,** is that of altering controls for the synthesis of lycopene, an important antioxidant thought to protect against some kinds of cancers and normally present at high levels in tomatoes. In this case, an engineered regulatory circuit was designed to control lycopene synthesis in response to the internal metabolic state of *E. coli.* Another recent development is the use of modified gene expression to produce variants of the antibiotic erythromycin. Blocking specific biochemical steps in pathways for the synthesis of an antibiotic precursor results in modified final products (**figure 41.11**). These altered products, which have slightly different structures, are tested for their possible antimicrobial effects. In addition this approach enables a better understanding of the structure-function relationships of antibiotics.

Protein Evolution

One of the newest approaches for creating novel metabolic capabilities in a given microorganism is protein evolution, which employs **forced evolution, adaptive mutations,** and **in vitro evolution** (**table 41.9**). Forced evolution and adaptive mutation involve the application of specific environmental stresses to "force" microorganisms to mutate and adapt, thus creating microorganisms with new biological capabilities. The mechanisms of these adaptive mutational processes include DNA rearrangements in which transposable elements and various types of recombination play critical roles.

In vitro evolution starts with purified nucleic acids rather than a whole organism. DNA templates (e.g., mutagenized versions of

Table 41.7 Combinatorial Biology in Biotechnology: The Expression of Genes in Other Organisms to Improve Processes and Products

Property or Product Transferred	Microorganism Used	Combinatorial Process
Ethanol production	*Escherichia coli*	Integration of pyruvate decarboxylase and alcohol dehydrogenase II from *Zymomonas mobilis.*
1,3-Propanediol production	*E. coli*	Introduction of genes from the *Klebsiella pneumoniae dha* region into *E. coli* makes possible anaerobic 1,3-propanediol production.
Cephalosporin precursor synthesis	*Penicillium chrysogenum*	Production of 7-ADC and 7-ADCA[a] precursors by incorporation of the expandase gene of *Cephalosoporin acremonium* into *Penicillium* by transformation.
Lactic acid production	*Saccharomyces cerevisiae*	A muscle bovine lactate dehydrogenase gene (LDH-A) expressed in *S. cerevisiae.*
Xylitol production	*S. cerevisiae*	95% xylitol conversion from xylose was obtained by transforming the *XYLI* gene of *Pichia stipitis* encoding a xylose reductase into *S. cerevisiae,* making this organism an efficient organism for the production of xylitol, which serves as a sweetener in the food industry.
Creatininase[b]	*E. coli*	Expression of the creatininase gene from *Pseudomonas putida* R565. Gene inserted in a plasmid vector.
Pediocin[c]	*S. cerevisiae*	Expression of bacteriocin from *Pediococcus acidilactici* in *S. cerevisiae* to inhibit wine contaminants.
Acetone and butanol production	*Clostridium acetobutylicum*	Introduction of a shuttle vector into *C. acetobutylicum* results in acetone and butanol formation.

[a]7-ACA = 7-aminocephalosporanic acid; 7-ADCA = 7-aminodecacetoxycephalosporonic acid.

[b]T.-Y. Tang; C.-J. Wen; and W.-H. Liu. 2000. Expression of the creatininase gene from *Pseudomonas putida* RS65 in *Escherichia coli. J. Ind. Microbiol. Biotechnol.* 24:2–6.

[c]H. Schoeman; M. A. Vivier; M. DuToit; L. M. Y. Dicks; and I. S. Pretorius. 1999. The development of bactericidal yeast strains by expressing the *Pediococcus acidilactici* pediocin gene (pedA) in *Saccharomyces cerevisiae. Yeast* 15:647–656.

Adapted from S. Ostergaard; L. Olsson; and J. Nielson. 2000. Metabolic engineering of *Saccharomyces cerevisiae.* Microbiol. Mol. Biol. Rev. 64(1):34–50.

genes whose product is of interest) are transcribed in vitro by a phage RNA polymerase into RNA molecules that are selected based on their capacity to perform a specific function. The enzyme reverse transcriptase is then used to copy the selected RNA molecules into cDNA, which can then be amplified by PCR. After a number of such cycles, a gene that might be of industrial importance will "evolve."

Laboratory-based protein evolution was introduced in the 1970s. Since that time, recombinant DNA approaches have not only provided more efficient means of generating mutations, but **high-throughput screening (HTS)** now enables the rapid selection of a single desirable mutant or molecule from tens of thousands of newly constructed strains, molecules, or compounds. HTS employs a combination of robotics and computer analysis to screen samples, usually present in 96-well microtitre plates for a specific trait. Not only is HTS used to sift through whole cells for a particular phenotype, it is also essential for the identification of new natural and synthetic compounds that have a desired activity. The combination of molecular biological approaches and HTS has propelled protein evolution to a new level of efficiency not previously envisioned.

Preservation of Microorganisms

Once a microorganism or virus has been selected or created to serve a specific purpose, it must be preserved in its original form for further use and study. Periodic transfers of cultures have been used in the past, but this can lead to mutations and phenotypic changes in microorganisms. To avoid these problems, a variety of culture preservation techniques may be used to maintain desired culture characteristics (**table 41.10**). **Lyophilization,** or freeze-drying, and storage in liquid nitrogen are frequently employed with microorganisms. Although lyophilization and liquid nitrogen storage are complicated and require expensive equipment, they allow microbial cultures to be stored for years without loss of viability or an accumulation of undesirable mutations.

1. What types of recombinant DNA techniques are being used to modify gene expression in microorganisms?
2. Define metabolic control engineering, metabolic pathway engineering, forced evolution, and adaptive mutations.
3. What is high-throughput screening and why has it become so important?
4. What approaches can be used for the preservation of microorganisms?

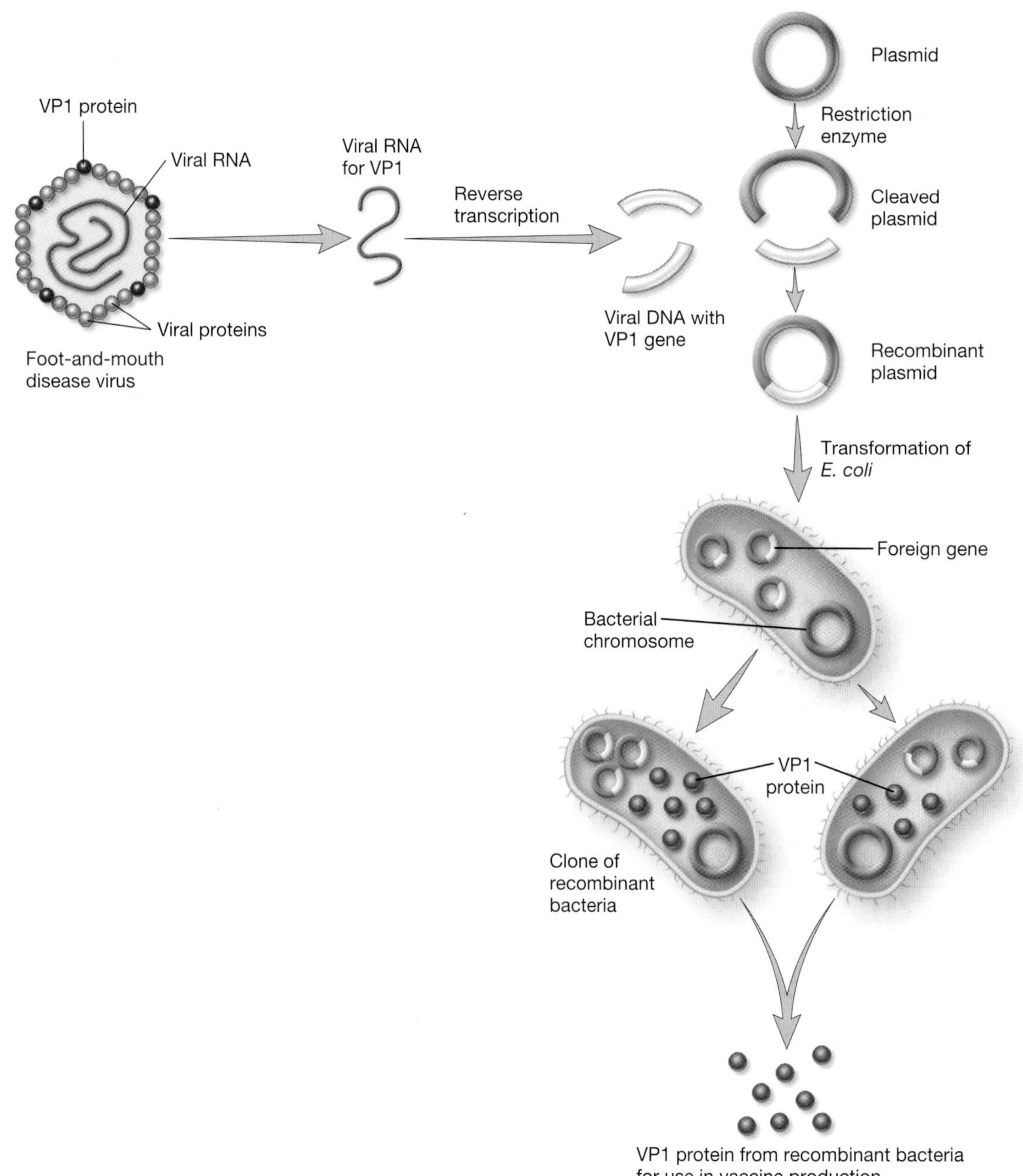

Figure 41.10 Recombinant Vaccine Production. Genes coding for desired products can be expressed in different organisms. By the use of recombinant DNA techniques, a foot-and-mouth disease vaccine is produced through cloning the vaccine genes into *E. coli.*

41.4 Microorganism Growth in Controlled Environments

For many industrial processes, microorganisms must be grown using specifically designed media under carefully controlled conditions, including temperature, aeration, and nutrient feeding. The development of appropriate culture media and the growth of microorganisms under industrial conditions are the subjects of this section.

Before proceeding, it is necessary to clarify terminology. The term **fermentation,** used in a physiological sense in earlier sections of the book, is employed in a much more general way in relation to industrial microbiology and biotechnology. As noted in **table 41.11,** the term can have several meanings. To industrial microbiologists, fermentation means the mass culture of microorganisms (or even plant and animal cells). Industrial fermentations requires the development of appropriate culture media and the transfer of small-scale technologies to a much larger scale.

Table 41.8 Examples of Recombinant DNA Systems Used to Modify Gene Expression

Product	Microorganism	Change
Actinorhodin	*Streptomyces coelicolor*	Modification of gene transcription
Cellulase	*Clostridium* genes in *Bacillus*	Increased secretion through chromosomal DNA amplification
Recombinant protein albumin	*Saccharomyces cerevisiae*	Fusion to a high-production protein
Heterologous protein	*Saccharomyces cerevisiae*	Use of the inducible strong hybrid promoter UAS_{gal}/CYCl
Enhanced growth rate[a]	*Aspergillus nidulans*	Overproduction of glyceraldehyde-3-phosphate dehydrogenase
Amino acids[b]	*Corynebacterium*	Isolation of biosynthetic genes that lead to enhanced enzyme activities or removal of feedback regulation

[a,b]S. Ostergaard; L. Olsson; and J. Nielson. 2000. Metabolic engineering of *Saccharomyces cerevisiae*. *Microbiol. Mol. Biol. Rev.* 64(1):34–50. Table 1, p. 35.

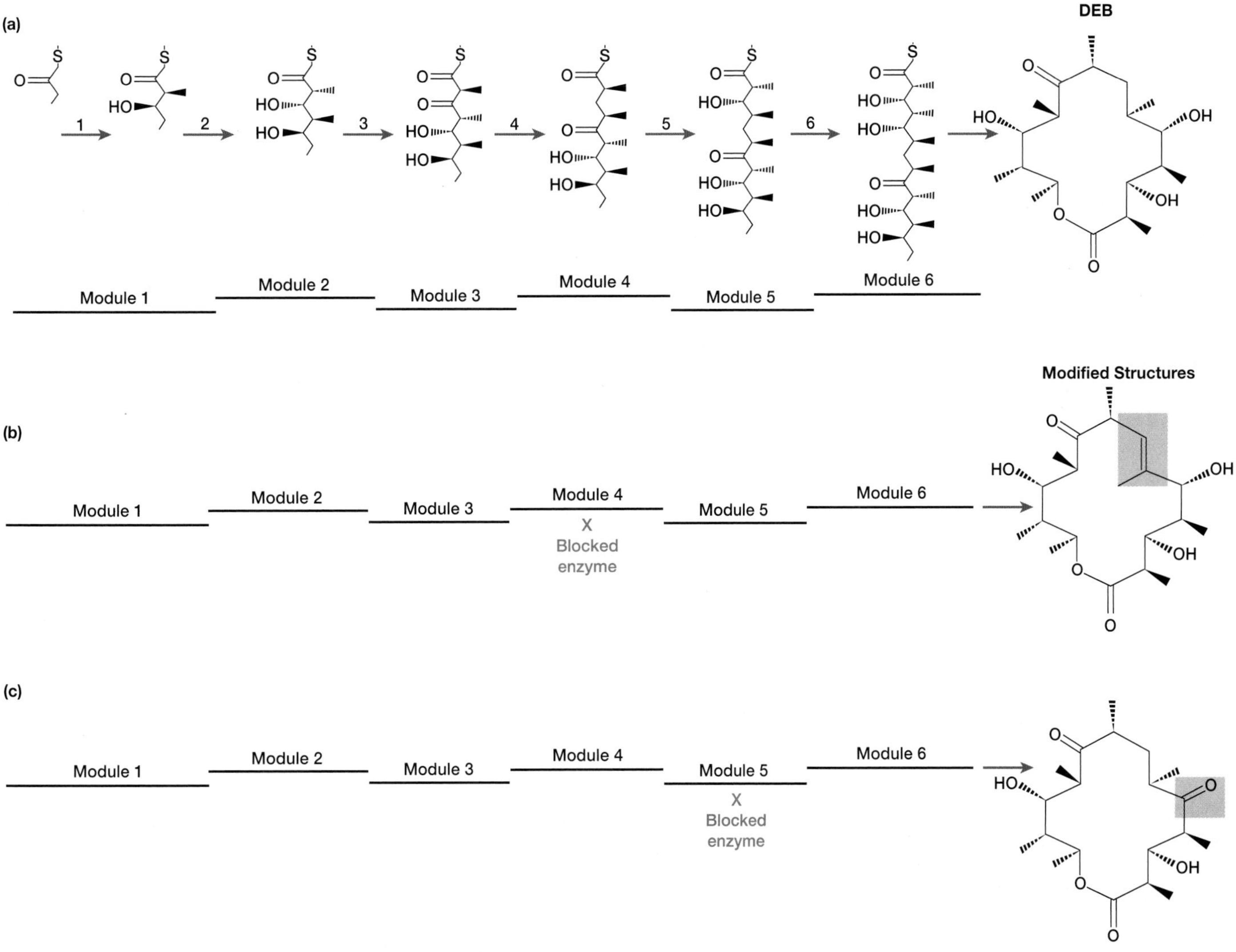

Figure 41.11 Metabolic Engineering to Create Modified Antibiotics. **(a)** Model for six elongation cycles (modules) in the normal synthesis of 6-deoxyerythonilide B (DEB), a precursor to the important antibiotic erythromycin. **(b)** Changes in structure that occur when the enoyl reductase enzyme of module 4 is blocked. **(c)** Changes in structure that occur when the keto reductase enzyme of module 5 is blocked. These changed structures (the highlighted areas) may lead to the synthesis of modified antibiotics with improved properties.

Table 41.9 Protein Evolution in Bacteria

Genetic Engineering Mechanisms	DNA Changes Mediated
Localized SOS mutagenesis	Base substitutions, frameshifts
Adapted frameshifting	−1 frameshifting
Tn*5*, Tn*9*, Tn*10* precise excision	Reciprocal recombination of flanking 8/9 bp repeats; restores original sequence
In vivo deletion, inversion, fusion, and duplication formation	Generally reciprocal recombination of short sequence repeats; occasionally nonhomologous
Type II topoisomerase recombination	Deletions and fusions by nonhomologous recombination, sometimes at short repeats
Site-specific recombination (type I topoisomerases)	Insertions, excisions/deletions, inversions by concerted or successive cleavage-ligation reactions at short sequence repeats; tolerates mismatches
Transposable elements (many species)	Insertions, transpositions, replicon fusions, adjacent deletions/excisions, adjacent inversions by ligation of 3′ OH transposon ends of 5′ PO_4 groups from staggered cuts at nonhomologous target sites
DNA uptake (transformation competence)	Uptake of single strand independent of sequence, or of double-stranded DNA carrying species identifier sequence

Adapted from J. A. Shapiro. 1999. Natural genetic engineering, adaptive mutation, and bacterial evolution. In *Microbial Ecology of Infectious Disease,* E. Rosenberg, editor, 259–75. Washington, D.C.: American Society for Microbiology. Derived from Table 2, pp. 263–64.

Table 41.10 Methods Used to Preserve Cultures of Interest for Industrial Microbiology and Biotechnology

Method	Comments
Periodic transfer	Variables of periodic transfer to new media include transfer frequency, medium used, and holding temperature; this can lead to increased mutation rates and production of variants
Mineral oil slant	A stock culture is grown on a slant and covered with sterilized mineral oil; the slant can be stored at refrigerator temperature
Minimal medium, distilled water, or water agar	Washed cultures are stored under refrigeration; these cultures can be viable for 3 to 5 months or longer
Freezing in growth media	Not reliable; can result in damage to microbial structures; with some microorganisms, however, this can be a useful means of culture maintenance
Drying	Cultures are dried on sterile soil (soil stocks), on sterile filter paper disks, or in gelatin drops; these can be stored in a desiccator at refrigeration temperature, or frozen to improve viability
Freeze-drying (lyophilization)	Water is removed by sublimation, in the presence of a cryoprotective agent; sealing in an ampule can lead to long-term viability, with 30 years having been reported
Ultrafreezing	Liquid nitrogen at −196°C is used, and cultures of fastidious microorganisms have been preserved for more than 15 years

Table 41.11 Fermentation: A Word with Many Meanings for the Microbiologist

1. Any process involving the mass culture of microorganisms, either aerobic or anaerobic
2. Any biological process that occurs in the absence of O_2
3. Food spoilage
4. The production of alcoholic beverages
5. Use of an organic substrate as the electron donor and acceptor
6. Use of an organic substrate as an electron donor, and of the same partially degraded organic substrate as an electron acceptor
7. Growth dependent on substrate-level phosphorylation

Medium Development

The medium used to grow a microorganism is critical because it can determine the level of microbial growth and product formation. In order to maximize competitiveness, lower-cost crude materials are used as sources of carbon, nitrogen, and phosphorus (**table 41.12**). Crude plant hydrolysates often are used as complex sources of carbon, nitrogen, and growth factors. By-products from the brewing industry frequently are employed because of their lower cost and greater availability. Other useful carbon sources include molasses and whey from cheese manufacture. Culture media (section 5.7)

Table 41.12 Major Components of Growth Media Used in Industrial Processes

Source	Raw Material
Carbon and energy	Molasses
	Whey
	Grains
	Agricultural wastes (corncobs)
Nitrogen	Corn-steep liquor
	Soybean meal
	Stick liquor (slaughterhouse products)
	Ammonia and ammonium salts
	Nitrates
	Distiller's solubles
Vitamins	Crude preparations of plant and animal products
Iron, trace salts	Crude inorganic chemicals
Buffers	Chalk or crude carbonates
	Fertilizer-grade phosphates
Antifoam agents	Higher alcohols
	Silicones
	Natural esters
	Lard and vegetable oils

The levels and balance of minerals (especially iron) and growth factors can be critical in medium formulation. For example, biotin and thiamine, by influencing biosynthetic reactions, control product accumulation in many fermentations. The medium also may be designed so that carbon, nitrogen, phosphorus, iron, or a specific growth factor will become limiting after a given time during the fermentation. In such cases the limitation often causes a shift from growth to production of desired metabolites.

Growth of Microorganisms in an Industrial Setting

Once a medium is developed, the physical environment for optimum microbial growth in the mass culture system must be defined. This often involves precise control of agitation, temperature, pH, and oxygenation. Phosphate buffers can be used to control pH while also providing a source of phosphorus. Oxygen limitations can be critical in aerobic growth processes.

The O_2 concentration and flux rate must be sufficiently high to have O_2 in excess within the cells. This is especially true when a dense microbial culture is growing. When filamentous fungi and actinomycetes are cultured, aeration can be even further limited by filamentous growth (**figure 41.12**). Such filamentous growth results in a viscous, plastic medium, known as a **non-Newtonian broth,** which offers even more resistance to stirring and aeration.

It is essential to assure that these physical factors are not limiting microbial growth. This is most critical during **scaleup,** where a successful procedure developed in a small shake flask is modified for use in a large fermenter. The microenvironment of the small culture must be maintained despite increases in the culture volume. If a successful transition is made from a process originally developed in a 250 ml Erlenmeyer flask to a 100,000 liter reactor, then the process of scaleup has been carried out successfully.

Microorganisms can be grown in culture tubes, shake flasks, and stirred fermenters or other mass culture systems. Stirred fermenters can range in size from 3 or 4 liters to 100,000 liters or larger, depending on production requirements. A typical industrial stirred fermentation unit is illustrated in **figure 41.13*b*.** Not only must the medium be sterilized but aeration, pH adjustment,

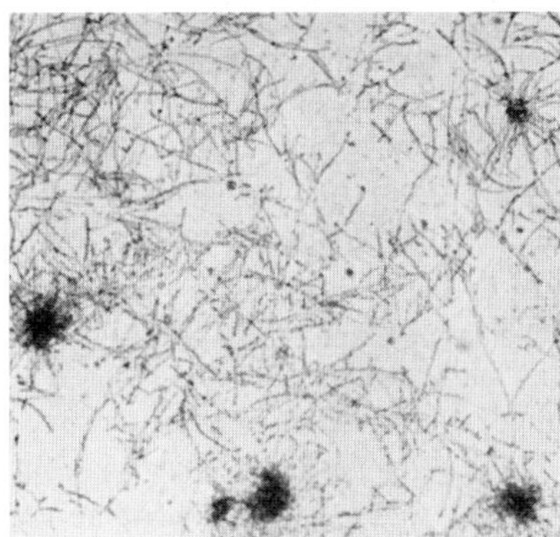

(a)

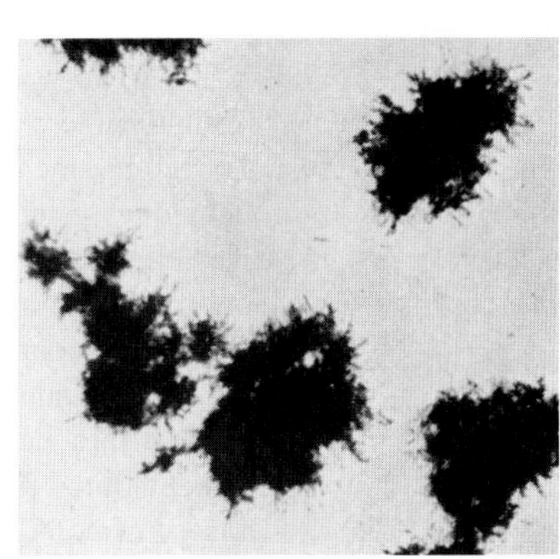

(b)

Figure 41.12 Filamentous Growth During Fermentation. Filamentous fungi and actinomycetes can change their growth form during the course of a fermentation. The development of pelleted growth by fungi has major effects on oxygen transfer and energy required to agitate the culture. **(a)** Initial culture. **(b)** After 18 hours growth.

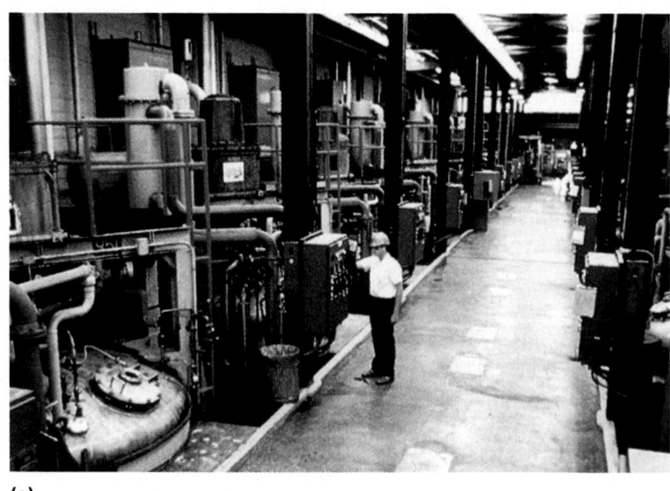

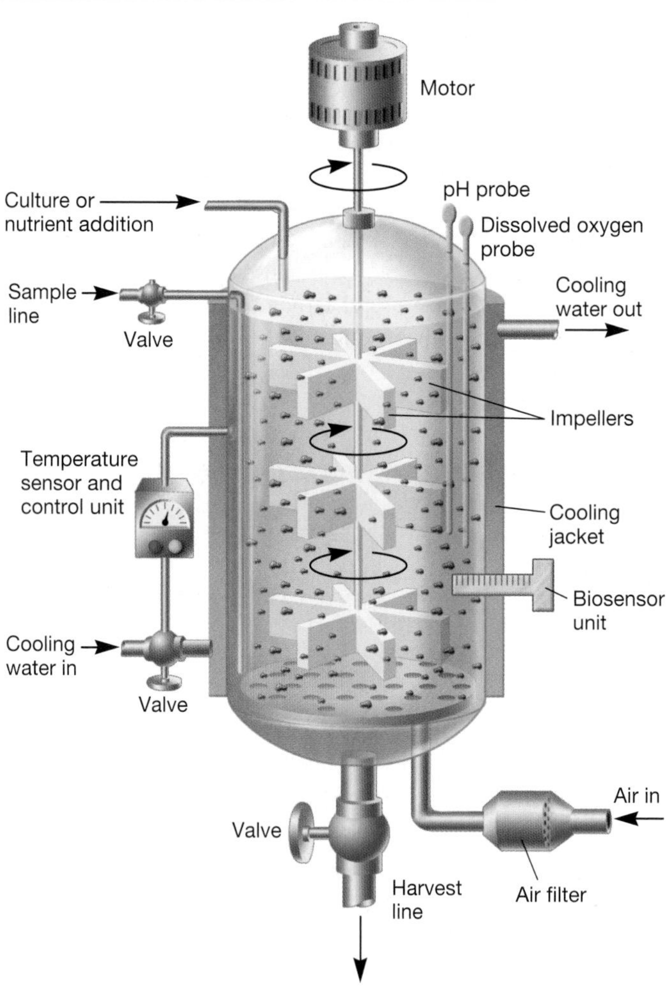

Figure 41.13 Industrial Stirred Fermenters. **(a)** Large fermenters used by a pharmaceutical company for the microbial production of antibiotics. **(b)** Details of a fermenter unit. This unit can be run under oxic or anoxic conditions, and nutrient additions, sampling, and fermentation monitoring can be carried out under aseptic conditions. Biosensors and infrared monitoring can provide real-time information on the course of the fermentation. Specific substrates, metabolic intermediates, and final products can be detected.

sampling, and process monitoring must be carried out under rigorously controlled conditions. When required, foam control agents must be added, especially with high-protein media. Computers are used to monitor outputs from probes that determine microbial biomass, levels of critical metabolic products, pH, input and exhaust gas composition, and other parameters. Environmental conditions can be changed or held constant over time, depending on the goals for the particular process.

Frequently a critical component in the medium, often the carbon source, is added continuously—**continuous feed**—so that the microorganism will not have excess substrate available at any given time. This is particularly important with glucose and other carbohydrates. If excess glucose is present at the beginning of a fermentation, it can be catabolized to yield ethanol, which is lost as a volatile product and reduces the final yield. This can occur even under oxic conditions.

Besides the traditional stirred aerobic or anaerobic fermenter, other approaches can be used to grow microorganisms. These alternatives, illustrated in **figure 41.14,** include lift-tube fermenters (figure 41.14*a*), which eliminate the need for stirrers that can be fouled by filamentous fungi. Also available is solid-state fermentation (figure 41.14*b*), in which a particulate substrate is kept moist to maintain a thin surface water film where microbes can grow and oxygen is available. In various types of fixed- (figure 41.14*c*) and fluidized-bed reactors (figure 41.14*d*), the microorganisms are associated with inert surfaces as biofilms and medium flows past the fixed or suspended particles.

Dialysis culture units also can be used (figure 41.14*e*). These units allow toxic waste metabolites or end products to diffuse away from the microbial culture and permit new substrates to diffuse through the membrane toward the culture. Continuous culture techniques using chemostats (figure 41.14*f*) can markedly improve cell outputs and rates of substrate use because microorganisms can be maintained in a continuous logarithmic phase. However, continuous maintenance of an organism in an active growth phase is undesirable in many industrial processes.

Microbial products often are classified as primary and secondary metabolites. As shown in **figure 41.15, primary metabolites** consist of compounds related to the synthesis of microbial

Figure 41.14 Alternate Methods for Mass Culture. In addition to stirred fermenters, other methods can be used to culture microorganisms in industrial processes. In many cases these alternate approaches will have lower operating costs and can provide specialized growth conditions needed for product synthesis.

(a) Lift-tube fermenter
Density difference of gas bubbles entrained in medium results in fluid circulation

Air in

(b) Solid-state fermentation
Growth of culture without presence of added free water

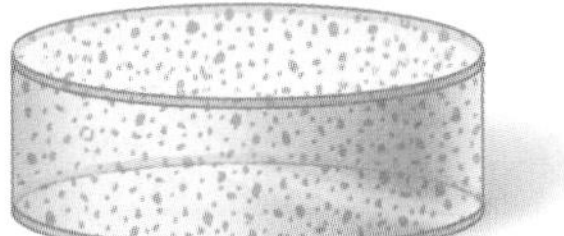

(c) Fixed-bed reactor
Microorganisms on surfaces of support material; flow can be up or down

Flow in
Fixed support material
Flow out

(d) Fluidized-bed reactor
Microorganisms on surfaces of particles suspended in liquid or gas stream–upward flow

Flow out
Suspended support particles
Flow in

(e) Dialysis culture unit
Waste products diffuse away from the culture. Substrate may diffuse through membrane to the culture

Membrane
Culture
Medium or buffer

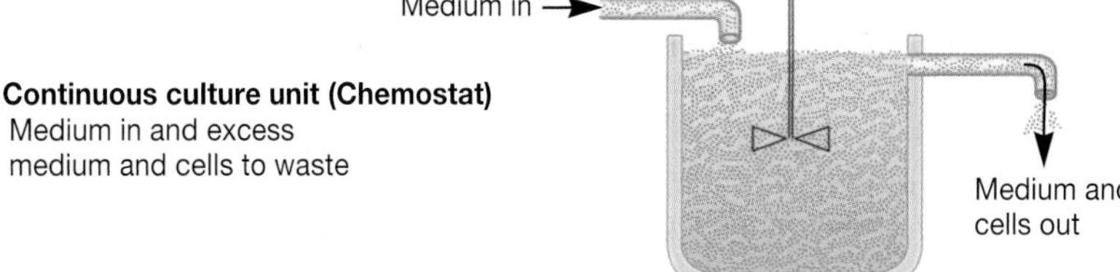

cells during balanced growth. They include amino acids, nucleotides, and fermentation end products such as ethanol and organic acids. In addition, industrially useful enzymes, either associated with the microbial cells or exoenzymes, often are synthesized by microorganisms during growth.

Secondary metabolites usually accumulate during the period of nutrient limitation or waste product accumulation that follows the active growth phase. These compounds have only a limited relationship to the synthesis of cell materials and normal growth. Most antibiotics and the mycotoxins fall into this category. The growth curve (section 6.2)

1. How is the cost of media reduced during industrial operations? Discuss the effect of changing balances in nutrients such as minerals, growth factors, and the sources of carbon, nitrogen, and phosphorus.
2. What are non-Newtonian broths and why are these important in fermentations?
3. Discuss scaleup and the objective of the scaleup process.
4. What parameters can be monitored in a modern, large-scale industrial fermentation?
5. Besides the aerated, stirred fermenter, what other alternatives are available for the mass culture of microorganisms in industrial processes? What is the principle by which a dialysis culture system functions?

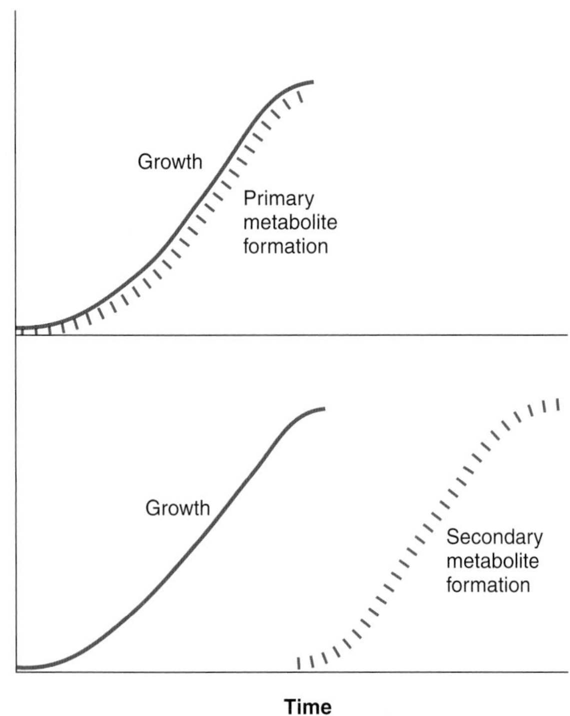

Figure 41.15 Primary and Secondary Metabolites. Depending on the particular organism, the desired product may be formed during or after growth. Primary metabolites are formed during the active growth phase, whereas secondary metabolites are formed after growth is completed.

Table 41.13 Major Microbial Products and Processes of Interest in Industrial Microbiology

Substances	Microorganisms
Industrial Products	
Ethanol (from glucose)	*Saccharomyces cerevisiae*
Ethanol (from lactose)	*Kluyveromyces fragilis*
Acetone and butanol	*Clostridium acetobutylicum*
2,3-butanediol	*Enterobacter, Serratia*
Enzymes	*Aspergillus, Bacillus, Mucor, Trichoderma*
Agricultural Products	
Gibberellins	*Gibberella fujikuroi*
Food Additives	
Amino acids (e.g., lysine)	*Corynebacterium glutamicum*
Organic acids (citric acid)	*Aspergillus niger*
Nucleotides	*Corynebacterium glutamicum*
Vitamins	*Ashbya, Eremothecium, Blakeslea*
Polysaccharides	*Xanthomonas*
Medical Products	
Antibiotics	*Penicillium, Streptomyces, Bacillus*
Alkaloids	*Claviceps purpurea*
Steroid transformations	*Rhizopus, Arthrobacter*
Insulin, human growth hormone, somatostatin, interferons	*Escherichia coli, Saccharomyces cerevisiae,* and others (recombinant DNA technology)
Biofuels	
Hydrogen	Photosynthetic microorganisms
Methane	*Methanobacterium*
Ethanol	*Zymomonas, Thermoanaerobacter*

41.5 Major Products of Industrial Microbiology

Industrial microbiology has provided products that have profoundly changed our lives and life spans. They include industrial and agricultural products, food additives, products for human and animal health, and biofuels (**table 41.13**). Antibiotics and other natural products used in medicine and health have made major contributions to the improved well-being of animal and human populations. Only major products in each category are discussed here.

Antibiotics

Many antibiotics are produced by microorganisms, predominantly by actinomycetes in the genus *Streptomyces* and by filamentous fungi (*see table 34.2*). We now discuss the synthesis of several of the most important antibiotics to illustrate the critical role of medium formulation and environmental control in the production of these important compounds. Antimicrobial chemotherapy (chapter 34)

Penicillin and Semisynthetic Penicillins

Although penicillin is less widely used because of antibiotic resistance, this drug produced by *Penicillium chrysogenum,* is an excellent example of a fermentation for which careful adjustment of the medium composition is used to achieve maximum yields. Provision of the slowly hydrolyzed disaccharide lactose, in combination with limited nitrogen availability, stimulates a greater accumulation of penicillin after growth has stopped (**figure 41.16**). The same result can be achieved by using a slow continuous feed of glucose. If a particular penicillin is needed, the specific precursor is added to the medium. For example, phenylacetic acid is added to maximize production of penicillin G, which has a benzyl side chain (*see figure 34.5*). The fermentation pH is maintained around neutrality by the addition of sterile alkali, which assures maximum stability of the newly synthesized penicillin. Once the fermentation is completed, normally in 6 to 7 days, the broth is separated from the fungal mycelium and processed by absorption, precipitation, and crystallization to yield the final product. This basic product can then be modified by chemical procedures to yield a variety of semisynthetic penicillins.

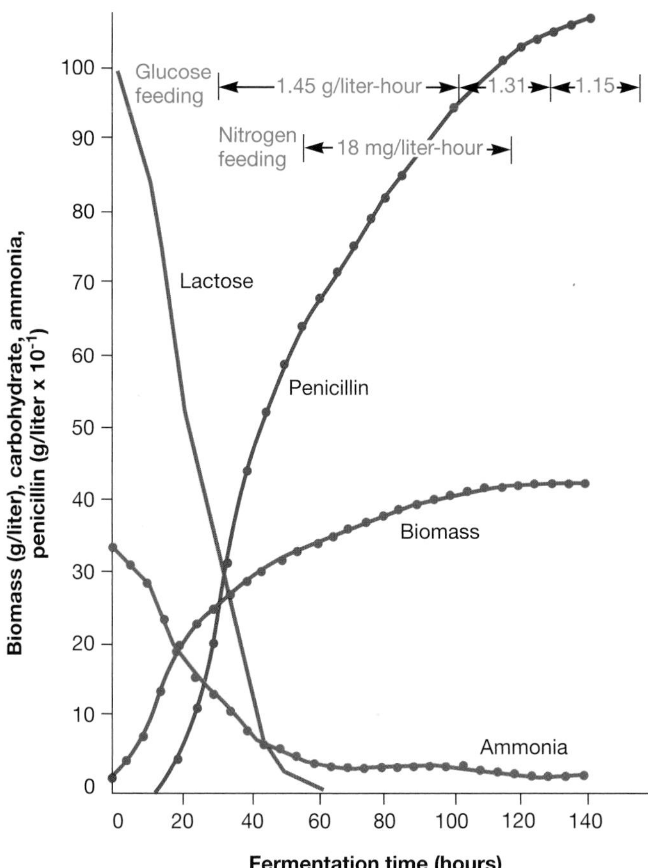

Figure 41.16 Penicillin Fermentation Involves Precise Control of Nutrients. The synthesis of penicillin begins when nitrogen from ammonia becomes limiting. After most of the lactose (a slowly catabolized disaccharide) has been degraded, glucose (a rapidly used monosaccharide) is added along with a low level of nitrogen. This stimulates maximum transformation of the carbon sources to penicillin. The scale factor is presented using the convention recommended by the ASM. That is, a number on the axis should be multiplied by 0.10 to obtain the true value.

Streptomycin

Streptomycin is a secondary metabolite produced by *Streptomyces griseus*. In this fermentation a soybean-based medium is used with glucose as a carbon source. The nitrogen source is thus in a combined form (soybean meal), which limits growth. After growth the antibiotic levels in the culture begin to increase (**figure 41.17**) under conditions of controlled nitrogen limitation.

Amino Acids

Amino acids such as lysine and glutamic acid are used in the food industry as nutritional supplements in bread products and as flavor-enhancing compounds such as monosodium glutamate (MSG).

Amino acid production is typically carried out by means of **regulatory mutants,** which have a reduced ability to limit synthesis of a specific amino acid or key intermediate. Normal microorganisms avoid overproduction of biochemical intermediates by the careful regulation of cellular metabolism. Production of glutamic acid and several other amino acids is now carried out using mutants of *Corynebacterium glutamicum* that lack, or have only a limited ability to process, the TCA cycle intermediate α-ketoglutarate (*see appendix II*) to succinyl-CoA as shown in **figure 41.18.** A controlled low biotin level and the addition of fatty acid derivatives results in increased membrane permeability and excretion of high concentrations of glutamic acid. The impaired bacteria use the glyoxylate pathway to meet their needs for essential biochemical intermediates, especially during the growth phase. After growth becomes limited because of changed nutrient availability, an almost complete molar conversion (or 81.7% weight conversion) of isocitrate to glutamate occurs. Synthesis of amino acids (section 10.5)

Lysine, an essential amino acid (i.e., required in the human diet) is used to supplement cereals and breads. It was originally produced in a two-step microbial process. This has been replaced by a single-step fermentation in which the bacterium *C. glutamicum,* blocked in the synthesis of homoserine, accumulates lysine. Over 44 g/liter can be produced in a 3-day fermentation.

Organic Acids

Organic acid production by microorganisms is important in industrial microbiology and illustrates the effects of trace metal levels and balances on organic acid synthesis and excretion. Citric, acetic, lactic, fumaric, and gluconic acids are major products (**table 41.14**). Until microbial processes were developed, the major source of citric acid was citrus fruit. Today most citric acid is produced by microorganisms; 70% is used in the food and beverage industry, 20% in pharmaceuticals, and the balance in other industrial applications.

The essence of citric acid fermentation involves limiting the amounts of trace metals such as manganese and iron to stop *Aspergillus niger* growth at a specific point in the fermentation. The medium often is treated with ion exchange resins to ensure low and controlled concentrations of available metals, and is carried out in aerobic stirred fermenters. Generally, high sugar concentrations (15 to 18%) are used, and copper has been found to counteract the inhibition of citric acid production by iron above 0.2 ppm. The success of this fermentation depends on the regulation and functioning of the glycolytic pathway and the tricarboxylic acid cycle. After the active growth phase, when the substrate level is high, citrate synthase activity increases and the activities of aconitase and isocitrate dehydrogenase decrease. This results in citric acid accumulation and excretion by the stressed microorganism.

In comparison, the production of gluconic acid involves a single microbial enzyme, glucose oxidase, found in *Aspergillus niger*. *A. niger* is grown under optimum conditions in a corn-steep liquor medium. Growth becomes limited by nitrogen, and the resting cells transform the remaining glucose to gluconic acid in a single-step reaction. Gluconic acid is used as a carrier for calcium and iron and as a component of detergents.

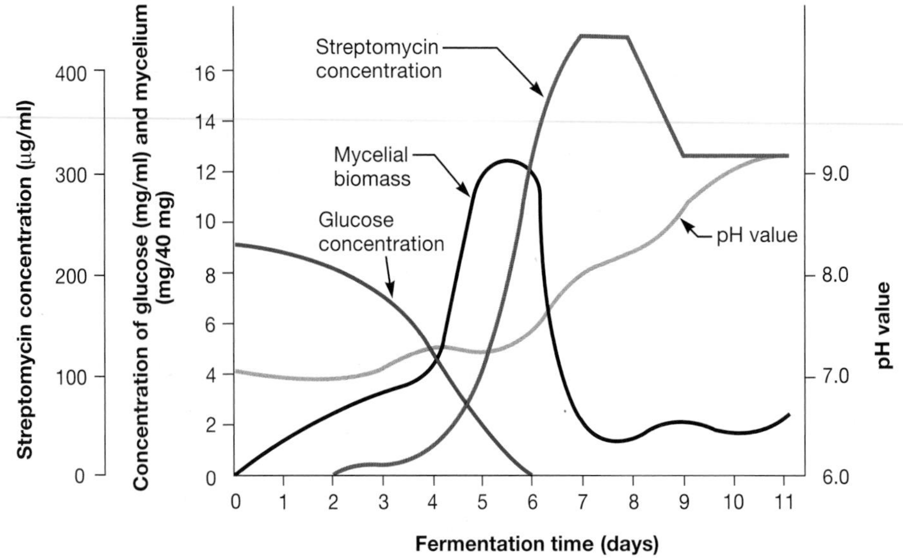

Figure 41.17 Streptomycin Production by *Streptomyces griseus*. Depletion of glucose leads to maximum antibiotic yields.

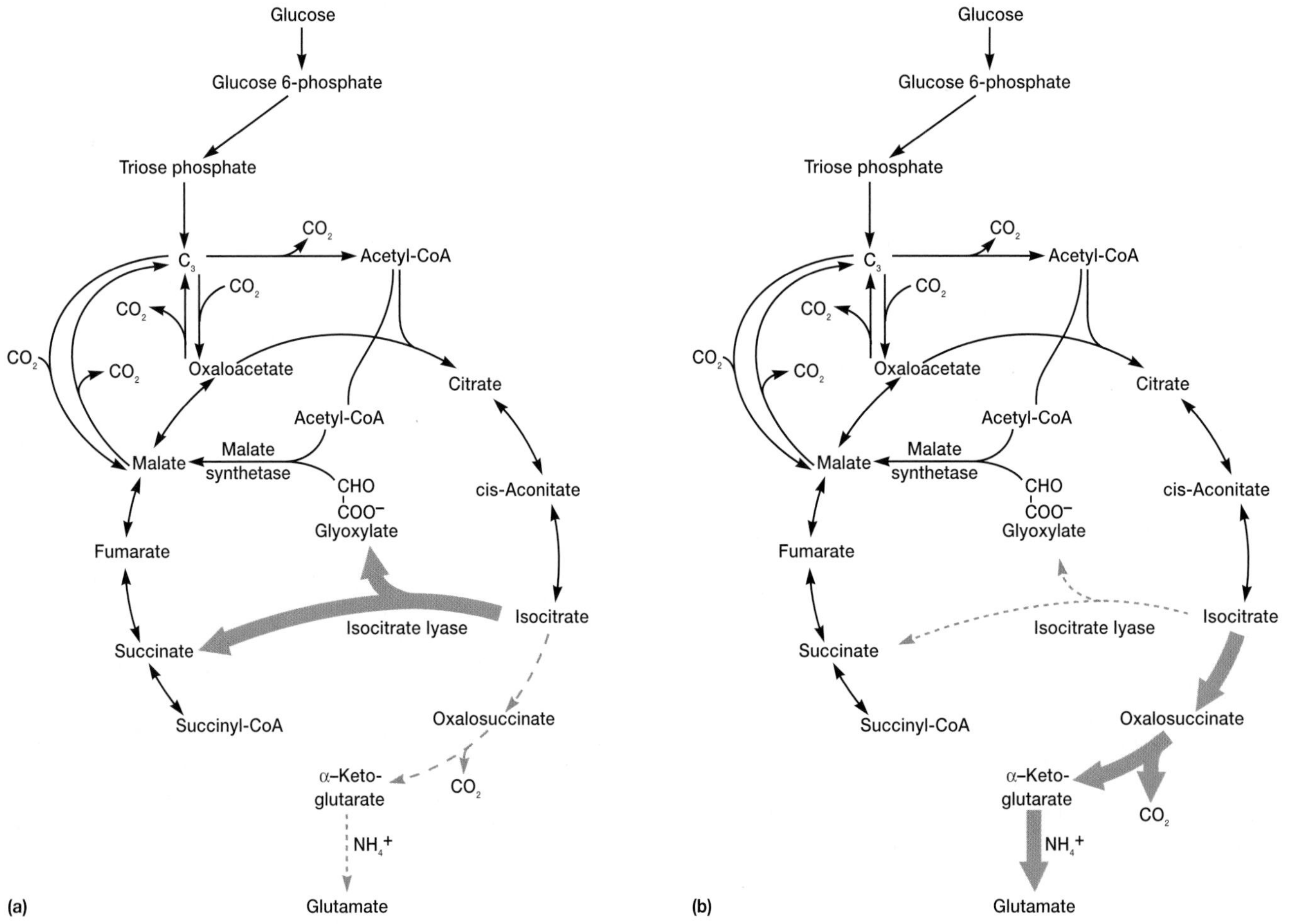

Glucose
Glucose 6-phosphate
Triose phosphate
C_3
CO_2
Acetyl-CoA
CO_2
CO_2
CO_2
CO_2
Oxaloacetate
Citrate
Acetyl-CoA
Malate
Malate synthetase
CHO
COO⁻
Glyoxylate
cis-Aconitate
Fumarate
Isocitrate
Isocitrate lyase
Succinate
Succinyl-CoA
Oxalosuccinate
α–Keto-glutarate
CO_2
NH_4^+
Glutamate
(a)

Glucose
Glucose 6-phosphate
Triose phosphate
C_3
CO_2
Acetyl-CoA
CO_2
CO_2
CO_2
CO_2
Oxaloacetate
Citrate
Acetyl-CoA
Malate
Malate synthetase
CHO
COO⁻
Glyoxylate
cis-Aconitate
Fumarate
Isocitrate
Isocitrate lyase
Succinate
Succinyl-CoA
Oxalosuccinate
α–Keto-glutarate
CO_2
NH_4^+
Glutamate
(b)

Figure 41.18 Glutamic Acid Production. The sequence of biosynthetic reactions leading from glucose to the accumulation of glutamate by *Corynebacterium glutamicum.* Major carbon flows are noted by blue arrows. **(a)** Growth using the glyoxylate bypass to provide critical intermediates in the TCA cycle. **(b)** After growth is completed, most of the substrate carbon is processed to glutamate (note shifted bold arrows). The dashed lines indicate reactions that are being used to a lesser extent.

Table 41.14 Major Organic Acids Produced by Microbial Processes

Product	Microorganism Used	Representative Uses	Fermentation Conditions
Acetic acid	*Acetobacter* with ethanol solutions	Wide variety of food uses	Single-step oxidation, with 15% solutions produced; 95–99% yields
Citric acid	*Aspergillus niger* in molasses-based medium	Pharmaceuticals, as a food additive	High carbohydrate concentrations and controlled limitation of trace metals; 60–80% yields
Fumaric acid	*Rhizopus nigricans* in sugar-based medium	Resin manufacture, tanning, and sizing	Strongly aerobic fermentation; carbon-nitrogen ratio is critical; zinc should be limited; 60% yields
Gluconic acid	*Aspergillus niger* in glucose-mineral salts medium	A carrier for calcium and sodium	Uses agitation or stirred fermenters; 95% yields
Itaconic acid	*Aspergillus terreus* in molasses-salts medium	Esters can be polymerized to make plastics	Highly aerobic medium, below pH 2.2; 85% yields
Kojic acid	*Aspergillus flavus-oryzae* in carbohydrate-inorganic N medium	The manufacture of fungicides and insecticides when complexed with metals	Iron must be carefully controlled to avoid reaction with kojic acid after fermentation
Lactic acid	Homofermentative *Lactobacillus delbrueckii*	As a carrier for calcium and as an acidifier	Purified medium used to facilitate extraction

Specialty Compounds for Use in Medicine and Health

In addition to the bulk products that have been produced over the last half century, such as antibiotics, amino acids, and organic acids, microorganisms are used for the production of nonantibiotic specialty compounds. These include hormones, antitumor and immunomodulatory agents, ionophores, and special compounds that influence bacteria, fungi, protists, insects, and plants (**table 41.15**). In all cases, it is necessary to produce and recover the products under carefully controlled conditions to assure that these medically important compounds reach the consumer in a stable, effective condition.

1. What critical limiting factors are used in the penicillin and streptomycin fermentations?
2. What are regulatory mutants and how were they used to increase the production of glutamic acid by *Corynebacterium?*
3. What is the principal limitation created to stimulate citric acid accumulation by *Aspergillus niger?*
4. Give some important specialty compounds that are produced by the use of microorganisms.

Biopolymers

Biopolymers are microbially produced polymers, primarily polysaccharides, used to modify the flow characteristics of liquids and to serve as gelling agents. These are employed in many areas of the pharmaceutical and food industries.

At least 75% of all polysaccharides are used as stabilizers, for the dispersion of particulates, as film-forming agents, or to promote water retention in various products. Polysaccharides help maintain the texture of many frozen foods, such as ice cream, that are subject to drastic temperature changes. These polysaccharides must maintain their properties under the pH conditions in the particular food and be compatible with other polysaccharides. They should not lose their physical characteristics if heated.

Biopolymers also include (1) dextrans, which are used as blood expanders and absorbents; (2) *Erwinia* polysaccharides used in paints; (3) polyesters, derived from *Pseudomonas oleovorans,* which are a feedstock for specialty plastics; (4) cellulose microfibrils, produced by an *Acetobacter* strain, that are used as a food thickener; (5) polysaccharides such as scleroglucan that are used by the oil industry as drilling mud additives; (6) xanthan polymers, which enhance oil recovery by improving water flooding and the displacement of oil. This use of xanthan gum, produced by *Xanthomonas campestris,* represents a large potential market for this microbial product.

Of special note are the cyclodextrins, which have a unique structure, as shown in **figure 41.19** for α-cyclodextrin. They are cyclic oligosaccharides whose sugars are joined by α-1,4 linkages. Cyclodextrins can be used for a wide variety of purposes because these cyclical molecules bind with substances and modify their physical properties. For example, cyclodextrins will increase the solubility of pharmaceuticals, reduce their bitterness, and mask chemical odors. Cyclodextrins also can be used as selective adsorbents to remove cholesterol from eggs and butter, to protect spices from oxidation, or as stationary phases in gas chromatography.

Biosurfactants

Biosurfactants are amphiphilic molecules that possess both hydrophobic and hydrophilic regions; thus they partition at the

Table 41.15 Nonantibiotic Specialty Compounds Produced by Microorganisms

Compound Type	Source	Specific Product	Process/Organism Affected
Polyethers	*Streptomyces cinnamonensis*	Monensin	Coccidiostat, rumenal growth promoter
	S. lasaliensis	Lasalocid	Coccidiostat, ruminal growth promoter
	S. albus	Salinomycin	Coccidiostat, ruminal growth promoter
Avermectins	*S. avermitilis*		Helminths and arthropods
Statins	*Aspergillus terreus*	Lovastatin	Cholesterol-lowering agent
	Penicillium citrinum + actinomycete[a]	Pravastatin	Cholesterol-lowering agent
Enzyme inhibitors	*S. clavaligerus*	Clavulanic acid	Penicillinase inhibitor
	Actinoplanes sp.	Acarbose	Intestinal glucosidase inhibitor (decreases hyperglycemia and triglyceride synthesis)
Bioherbicides	*S. hygroscopicus*	Bialaphos	
Immunosuppressants	*Tolypocladium inflatum*	Cyclosporin A	Organ transplants
	S. tsukabaensis	FK-506	Organ transplants
	S. hygroscopicus	Rapamycin	Organ transplants
Anabolic agents	*Gibberella zeae*	Zearalenone	Farm animal medication
Uterocontractants	*Claviceps purpurea*	Ergot alkaloids	Induction of labor
Antitumor agents	*S. peuceticus* subsp. *caesius*	Doxorubicin	Cancer treatment
	S. peuceticus	Daunorubicin	Cancer treatment
	S. caespitosus	Mitomycin	Cancer treatment
	S. verticillus	Bleomycin	Cancer treatment

[a]Compactin, produced by *Penicillium citrinum*, is changed to pravastatin by an actinomycete bioconversion.

Based on: A. L. Demain. 2000. Microbial biotechnology. *Tibtech* 18:26–31; A. L. Demain. 2000. Pharmaceutically active secondary metabolites of microorganisms. *App. Microbiol. Biotechnol.* 52:455–463; G. Lancini and A. L. Demain. 1999. Secondary metabolism in bacteria: Antibiotic pathways regulation and function. In *Biology of the Prokaryotes*, J. W. Lengeler, G. Drews, and H. G. Schlegel, editors, 627–51. New York: Thieme.

interface between fluids that differ in polarity, such as oil and water. For this reason, they are used for emulsification, increasing detergency, wetting and phase dispersion, as well as for solubilization. These properties are especially important in bioremediation, oil spill dispersion, and enhanced oil recovery (EOR). The most widely used microbially produced biosurfactants are glycolipids. These are carbohydrates that bear long-chain aliphatic acids or hydroxy aliphatic acids. They can be isolated as extracellular products from a variety of microorganisms including pseudomonads and yeasts.

Many biosurfactants also have antibacterial and antifungal activity. The amphipathic nature of these molecules promotes their ability to disrupt plasma membranes. Some biosurfactants even inactivate enveloped viruses. In addition, their capacity to prevent microbial adhesion to sites of invasion has generated interest in their use as potential protective agents. Although these and other medical applications of biosurfactants are still being investigated, these molecules hold promise for the future. They can be genetically engineered with relative ease and the continued rise in resistance to antimicrobial agents has placed a tremendous need for new products.

Bioconversion Processes

Bioconversions, also known as **microbial transformations** or **biotransformations,** are minor changes in molecules, such as the insertion of a hydroxyl or keto function or the saturation/desaturation of a complex cyclic structure, that are carried out by nongrowing microorganisms. The microorganisms thus act as **biocatalysts.** Bioconversions have many advantages over chemical procedures. A major advantage is stereochemical; the biologically active form of a product is made. Enzymes also carry out very specific reactions under mild conditions, and larger water-insoluble molecules can be transformed. Unicellular bacteria, actinomycetes, yeasts, and molds have been used in various bioconversions. The enzymes responsible for these conversions can be intracellular or extracellular. Cells can be produced in batch or continuous culture and then dried for direct use, or they can be prepared in more specific ways to carry out desired bioconversions.

A typical bioconversion is the hydroxylation of a steroid (**figure 41.20**). In this example, the water-insoluble steroid is dissolved in acetone and then added to the reaction system that contains the pregrown microbial cells. The course of the modification is monitored, and the final product is extracted from the medium and purified.

Biotransformations carried out by free enzymes or intact nongrowing cells do have limitations. Reactions that occur in the absence of active metabolism—without reducing power or ATP continuously available—are primarily exergonic reactions. If ATP or reductants are required, an energy source such as glucose must be supplied under carefully controlled nongrowth conditions.

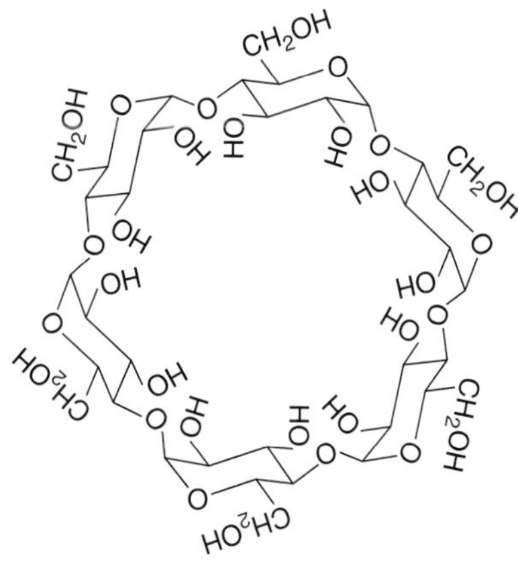

α-Cyclodextrin

Figure 41.19 **α-Cyclodextrin.** The structure of α-cyclodextrin, produced by *Thermoanaerobacter*. There also are β and γ forms, of similar structure but larger; these compounds have many applications in medicine and industry.

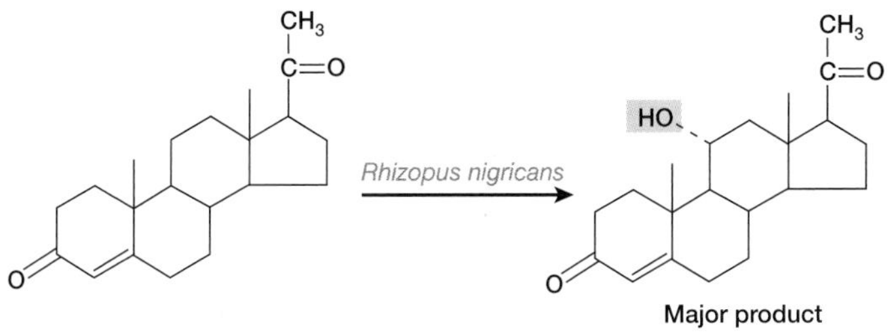

Rhizopus nigricans

Major product

Figure 41.20 **Biotransformation to Modify a Steroid.** Hydroxylation of progesterone in the 11α position by *Rhizopus nigricans*. The steroid is dissolved in acetone before addition to the pregrown fungal culture.

1. Discuss the major uses for biopolymers and biosurfactants.
2. What are cyclodextrins and why are they important additives?
3. What are bioconversions or biotransformations? Describe the changes in molecules that result from these processes.

41.6 Biodegradation and Bioremediation by Natural Communities

The metabolic activities of microbes can also be exploited in complex natural environments such as waters, soils, or high organic matter-containing composts where the physical and nutritional conditions for microbial growth cannot be completely controlled, and a largely unknown microbial community is present. Examples are (1) the use of microbial communities to carry out biodegradation, bioremediation, and environmental maintenance processes; and (2) the addition of microorganisms to soils or plants for the improvement of crop production. We discuss both of these applications in this section.

Biodegradation and Bioremediation Processes

Before discussing **biodegradation** processes carried out by natural microbial communities, it is important to consider definitions. Biodegradation has at least three outcomes (**figure 41.21**): (1) a minor change in an organic molecule leaving the main structure still intact, (2) fragmentation of a complex organic molecule in such a way that the fragments could be reassembled to yield the original structure, and (3) complete mineralization, which is the transformation of organic molecules to mineral forms. Biogeochemical cycling (section 27.2)

The removal of toxic industrial products in soils and in aquatic environments has become a daunting and necessary task. Compounds such as perchloroethylene (PCE), trichloroethylene (TCE), and polychlorinated biphenyls (PCBs) are common contaminants. These compounds adsorb onto organic matter in the environment, making decontamination using traditional approaches difficult or ineffective. The use of microbes to transform these contaminants to nontoxic degradation products is called **bioremediation.** In order to understand how bioremediation takes place at the level of an ecosystem, we first must consider the biochemistry of biodegradation.

Degradation of complex compounds requires several discrete stages, usually performed by different microbes. Initially contaminants are converted to less-toxic compounds that are more readily degraded. The first step for many contaminants, including organochloride pesticides, alkyl solvents, and aryl halides, is **reductive dehalogenation.** This is the removal of a halogen substituent (e.g., chlorine, bromine, fluorine) while at the same time adding electrons to the molecule. This can occur in two ways. In hydrogenolysis, the halogen substituent is replaced by a hydrogen atom. Alternatively, dihaloelimination removes two halogen substituents from adjacent carbons while inserting an additional bond between the carbons. Both processes require an electron donor. The dehalogenation of PCBs uses electrons derived from water; alternatively hydrogen can be the electron donor for the dehalogenation of different chlorinated compounds. Major genera that carry out this process include *Desulfitobacterium*, *Dehalospirillum*, and *Desulfomonile*.

Reductive dehalogenation usually occurs under anoxic conditions. In fact, humic acids, polymeric residues of lignin decomposition that accumulate in soils and waters, have been found to play a role in anaerobic biodegradation processes. They can serve as electron acceptors under what are called "humic-acid-reducing conditions." The use of humic acids as electron acceptors has been observed with the anaerobic dechlorination of vinyl chloride and dichloroethylene. Soils, plants, and nutrients (section 29.2)

(a) Minor change (dehalogenation)
Cl
Cl
O — CH2 — COOH + HOH
Cl⁻ + OH
OH
O — CH2 — COOH
(b) Fragmentation
Cl
Cl
O — CH2 — COOH + HOH
Cl
Cl
OH + HOCH2 — COOH
(c) Mineralization
Cl
Cl
O — CH2 — COOH
CO2 + 2Cl⁻ + HOH

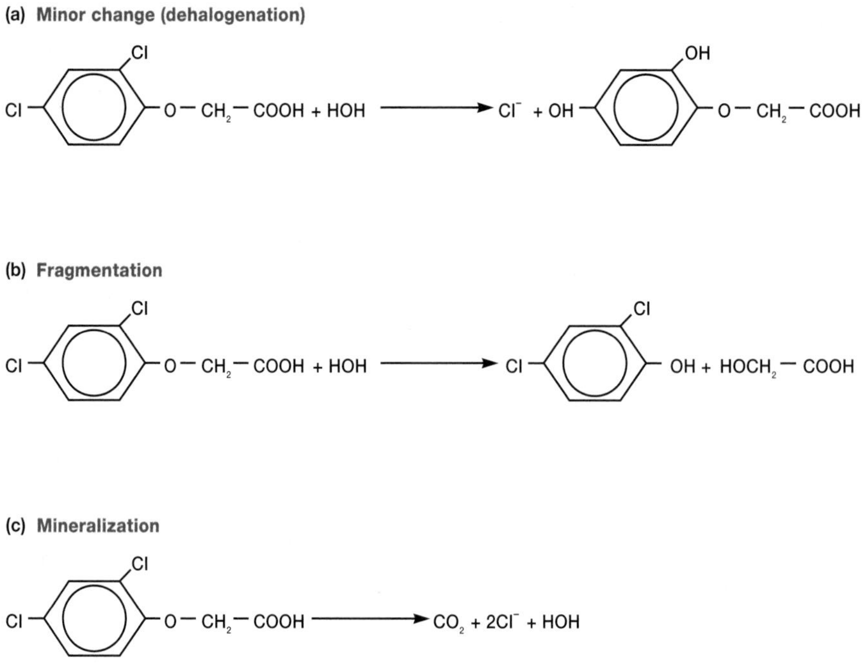

Figure 41.21 Biodegradation Has Several Meanings. Biodegradation is a term that can be used to describe three major types of changes in a molecule. **(a)** A minor change in the functional groups attached to an organic compound, such as the substitution of a hydroxyl group for a chlorine group. **(b)** An actual breaking of the organic compound into organic fragments in such a way that the original molecule could be reconstructed. **(c)** The complete degradation of an organic compound to minerals.

Once the anaerobic dehalogenation steps are completed, degradation of the main structure of many pesticides and other xenobiotics often proceeds more rapidly in the presence of O_2. Thus the degradation of halogenated toxic compounds generally requires the action of several microbial genera, sometimes referred to as a consortium.

Structure and stereochemistry are critical in predicting the fate of a specific chemical in nature. When a constituent is in the meta as opposed to the ortho position, the compound will be degraded at a much slower rate. The ***meta* effect** is shown in **figure 41.22.** This stereochemical difference is the reason that the common lawn herbicide 2,4-dichlorophenoxyacetic acid (2,4-D), with a chlorine in the ortho position, will be largely degraded in a single summer. In contrast, 2,4,5-trichlorophenoxyacetic acid, with a constituent in the meta position, will persist in the soils for several years, and thus is used for long-term brush control.

An important aspect of managing biodegradation is the recognition that many of the compounds that are added to environments are **chiral,** or possess asymmetry and handedness. Microorganisms often can degrade only one isomer of a substance; the other isomer will remain in the environment. At least 25% of herbicides are chiral. Thus it is critical to add the herbicide isomer that is effective and also degradable. Studies have shown that microbial communities in different environments will degrade different enantiomers. Changes in environmental conditions and nutrient supplies can alter the patterns of chiral form degradation.

Microbial communities change their characteristics in response to the addition of inorganic or organic substrates. If a particular compound, such as a herbicide, is added repeatedly to a microbial community, the community adapts and faster rates of degradation can occur—a process of acclimation (**figure 41.23**). The adaptive process often is so effective that enrichment culture-based approaches, established on the principles elucidated by Beijerinck, can be used to isolate organisms with a desired set of capabilities. For example, a microbial community can become so efficient at rapid herbicide degradation that herbicide effectiveness is diminished. To counteract this process, herbicides can

Chemical structure
Approximate time to degrade in soil
O — CH2 — COOH
Cl
Cl
2,4-D
3 months
(a)
O — CH2 — COOH
Cl
Cl
Cl
2,4,5-T
2–3 years
Blocked meta position
(b)

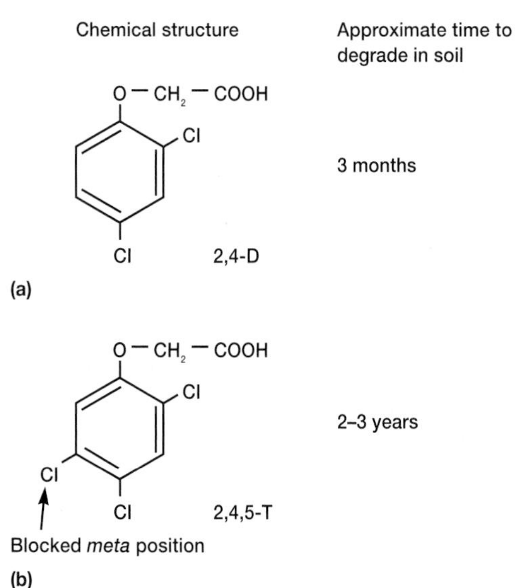

Figure 41.22 The *Meta* Effect and Biodegradation. Minor structural differences can have major effects on the biodegradability of chemicals. The *meta* effect is an important example. **(a)** Readily degradable 2,4-dichlorophenoxyacetic acid (2,4-D) with an exposed meta position on the ring degrades in several months; **(b)** recalcitrant 2,4,5-trichlorophenoxyacetic acid (2,4,5-T) with the blocked meta group, can persist for years.

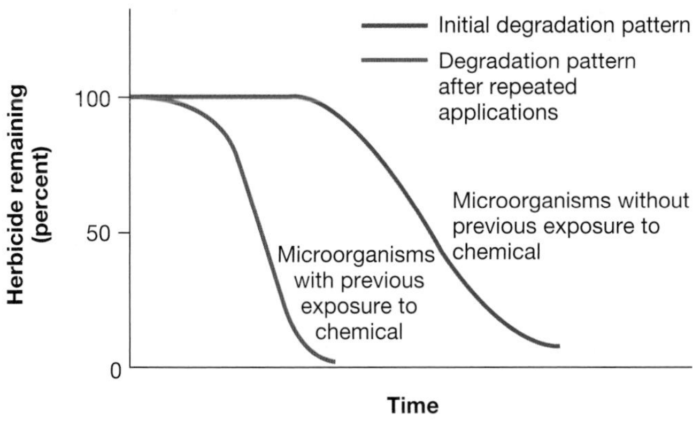

Figure 41.23 Repeated Exposure and Degradation Rate. Addition of an herbicide to a soil can result in changes in the degradative ability of the microbial community. Relative degradation rates for an herbicide after initial addition to a soil, and after repeated exposure to the same chemical.

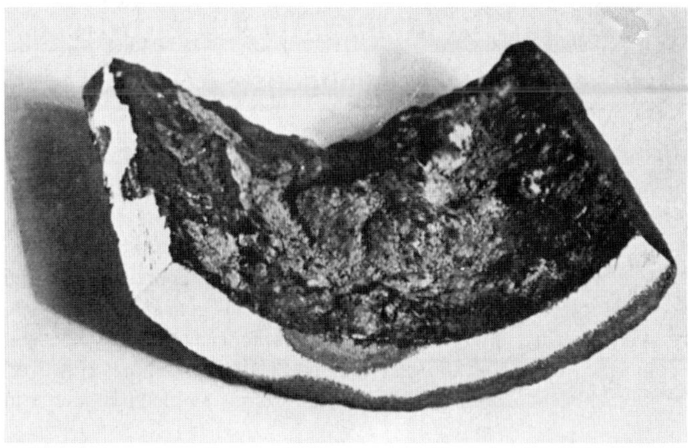

Figure 41.24 Microbial-Mediated Metal Corrosion. The microbiological corrosion of iron is a major problem. The graphitization of iron under a rust bleb on the pipe surface allows microorganisms, including *Desulfovibrio,* to corrode the inner surface.

be changed to alter the microbial community, thus preserving the effectiveness of the chemicals. The development of industrial microbiology and microbial ecology (section 1.5)

Degradation processes that occur in soils also can be used in large-scale degradation of hydrocarbons or wastes from agricultural operations. The waste material is incorporated into the soil or allowed to flow across the soil surface, where degradation occurs.

Unfortunately such degradation processes do not always reduce environmental problems. In fact, the partial degradation or modification of an organic compound may not lead to decreased toxicity. An example of this process is the microbial metabolism of 1,1,1-trichloro-2,2-*bis*-(p-chlorophenyl) ethane (DDT), a xenobiotic or foreign (chemically synthesized) organic compound. Degradation removes a chlorine function to give 1,1-dichloro-2,2-*bis*(*p*-chlorophenyl)ethylene (DDE), which is still of environmental concern. Another important example is the degradation of trichloroethylene (TCE), a widely used solvent. If this is degraded under anoxic conditions, the dangerous carcinogen vinyl chloride can be synthesized.

$$Cl_2 = CHCl \rightarrow ClHC = CH_2$$

Biodegradation also can lead to widespread damages and financial losses. Metal corrosion is a particularly important example. The microbially mediated corrosion of metals is particularly critical where iron pipes are used in waterlogged anoxic environments or in secondary petroleum recovery processes carried out at older oil fields. In these older fields water is pumped down a series of wells to force residual petroleum to a central collection point. If the water contains low levels of organic matter and sulfate, anaerobic microbial communities can develop in rust blebs or tubercles (**figure 41.24**), resulting in punctured iron pipe and loss of critical pumping pressure. Microorganisms that use elemental iron as an electron donor during the reduction of CO_2 in methanogenesis contribute to the corrosion of soft iron (**Microbial Diversity & Ecology 41.3**). Because of the wide range of interactions that occur between microorganisms and metals, the need to develop strategies to deal with microbial corrosion problems is critical.

1. Give alternative definitions for the term biodegradation.
2. What is reductive dehalogenation? Describe humic acids and the role they can play in anaerobic degradation processes.
3. Why is the "*meta* effect" important for understanding biodegradation?
4. Discuss chirality and its importance for understanding degradation effects in the environment.
5. What are some of the important microbial groups involved in iron corrosion?

Stimulating Biodegradation

Bioremediation usually involves stimulating the degradative activities of microorganisms already present in contaminated waters or soils. However, natural microbial communities may not be able to carry out biodegradation processes at a desired rate due to limiting physical or nutritional factors. For example, biodegradation often is limited by low oxygen levels. Nitrogen, phosphorus, and other needed nutrients also may be limiting. In these cases, it is necessary to determine the limiting factors, and supply the needed materials or modifiy the environment.

As shown in **figure 41.25,** monitoring and recovery wells are put into place so that the nutrient status and rates of biodegradation can be determined by periodic sampling. Often it is found that the addition of easily metabolized organic matter such as glucose increases biodegradation of recalcitrant compounds that are usually not used as carbon and energy sources by microorganisms. This process, termed **cometabolism,** can be carried out by simply adding easily catabolized organic matter such as glucose or cellulose and the compound to be degraded to a complex microbial community. Plants also may be used to provide the organic matter. Cometabolism is important in many different biodegradation systems.

Microbial Diversity & Ecology

41.3 Methanogens—A New Role for a Unique Microbial Group

The methanogens, an important group of archaea that produce methane, have had widespread impact on the Earth. In contrast to their essential role in producing reservoirs of natural gas, methanogens also contribute to the anaerobic corrosion of soft iron. The microbial group usually considered the major culprit in the anaerobic corrosion process is the bacterial genus *Desulfovibrio,* which can use sulfate as an electron acceptor and hydrogen produced in the corrosion process as an electron donor. However, methanogens can also use elemental iron as an electron source in their metabolism. It appears that corrosion may occur even without the presence of sulfate, which is required by *Desulfovibrio.* Rates of iron removal by the methanogens are around 79 mg/1,000 cm^2 of surface area in a 24-hour period. This may not seem a high rate, but in relation to the planned service life of metal structures in muds and subsurface soils—usually years and decades—such corrosion can become a major problem. Continuous efforts to improve protection of iron structures will be required in view of the diversity of iron-corroding microorganisms.

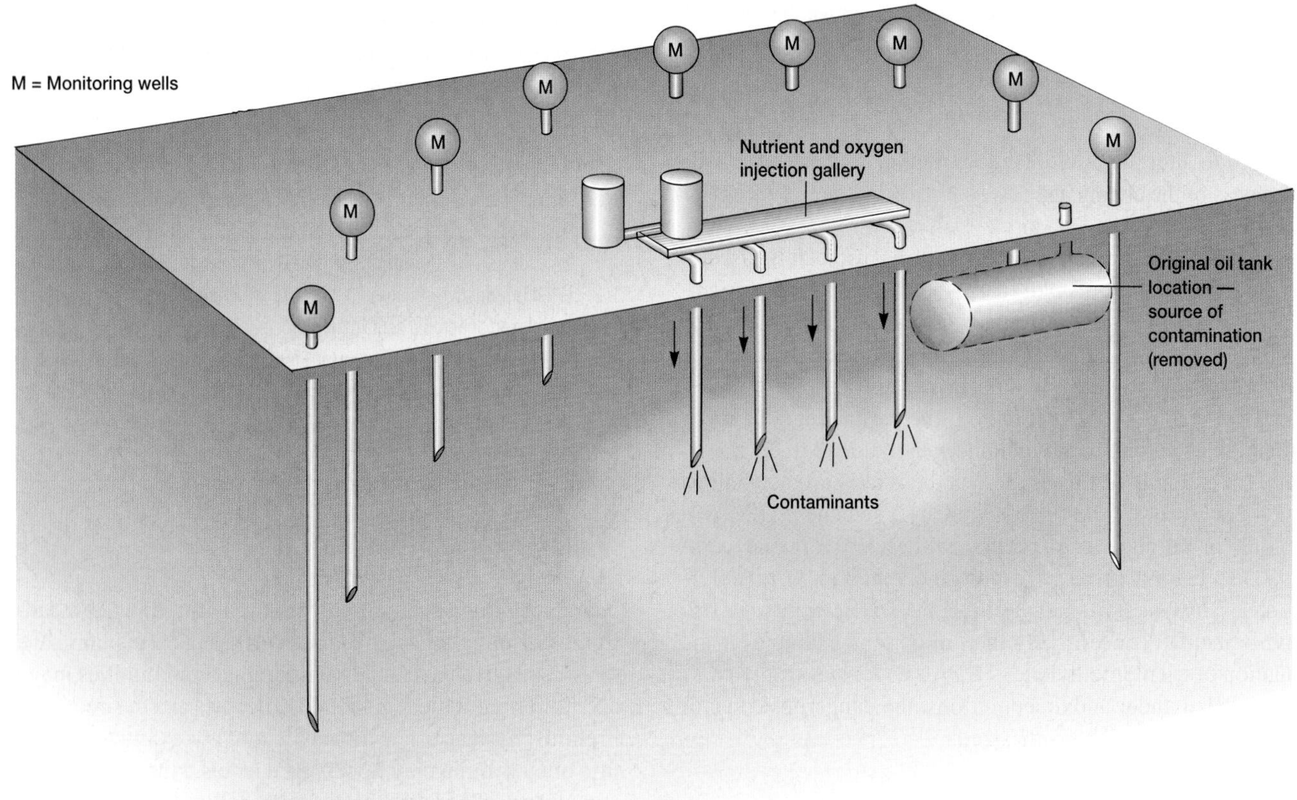

Figure 41.25 A Subsurface Engineered Bioremediation System. Monitoring and recovery wells are used to monitor the plume and its possible movement. Nutrients and oxygen (as air, peroxide, or other oxygen-releasing compounds) are added to the soil and groundwater to promote more efficient degradation of contaminants.

Stimulating Hydrocarbon Degradation in Waters and Soils

Experience with oil spills in marine environments illustrates these principles. In work with dispersed hydrocarbons in the ocean, contact between microorganisms, the hydrocarbon substrate, and other essential nutrients must be maintained. To achieve this, pellets containing nutrients and an oleophilic (hydrocarbon soluble) preparation are used. This technique accelerates the degradation of different crude oil slicks by 30 to 40%, in comparison with control oil slicks where the additional nutrients are not available.

A major challenge for this technology was the *Exxon Valdez* oil spill, which occurred in March 1989. This event resulted in the release of 11 million gallons (257,000 barrels or 38,800 metric tonnes) of oil into Prince William Sound, Alaska. Several different approaches were used to increase biodegradation. These included nutrient additions, chemical dispersants, biosurfactant additions, and the use of high-pressure steam. The use of a microbially produced glycolipid emulsifier proved very helpful.

These bioremediation approaches are also used in soils and sediments. For example, a unique two-stage process can be used to degrade PCBs in river sediments. First, partial dehalogenation of the PCBs occurs naturally under anoxic conditions. Then the muds are aerated to promote the complete degradation of the less chlorinated residues (chapter opening figure).

Phytoremediation

Phytoremediation, or the use of plants to stimulate the extraction, degradation, adsorption, stabilization or volatilization of contaminants is becoming an important part of biodegradation technology. A plant provides nutrients that allow cometabolism to occur in the plant root zone or rhizosphere (**figure 41.26**). Phytoremediation also includes plant contributions to degradation, immobilization, and volatilization processes, as noted in **table 41.16.** Transgenic plants may be employed in phytoremediation. Using cloning techniques with *Agrobacterium,* the *mer*A and *mer*B genes have been integrated into the mustard plant *Arabidopsis thaliana*, making it possible to transform extremely toxic organic mercury forms to elemental mercury, which is less of an environmental hazard. Transgenic tobacco plants have been constructed that express tetranitrate reductase, an enzyme from an explosive-degrading bacterium, thereby enabling the transgenic plants to degrade nitrate ester and nitro aromatic explosives. The genetically modified (GM) plants grow in solutions of explosives that unmodified plants cannot tolerate.

Techniques & Applications 14.2: Plant tumors and nature's genetic engineers

The power of microbial genetics was recently harnessed to improve phytoremediation of the toxin toluene. Researchers

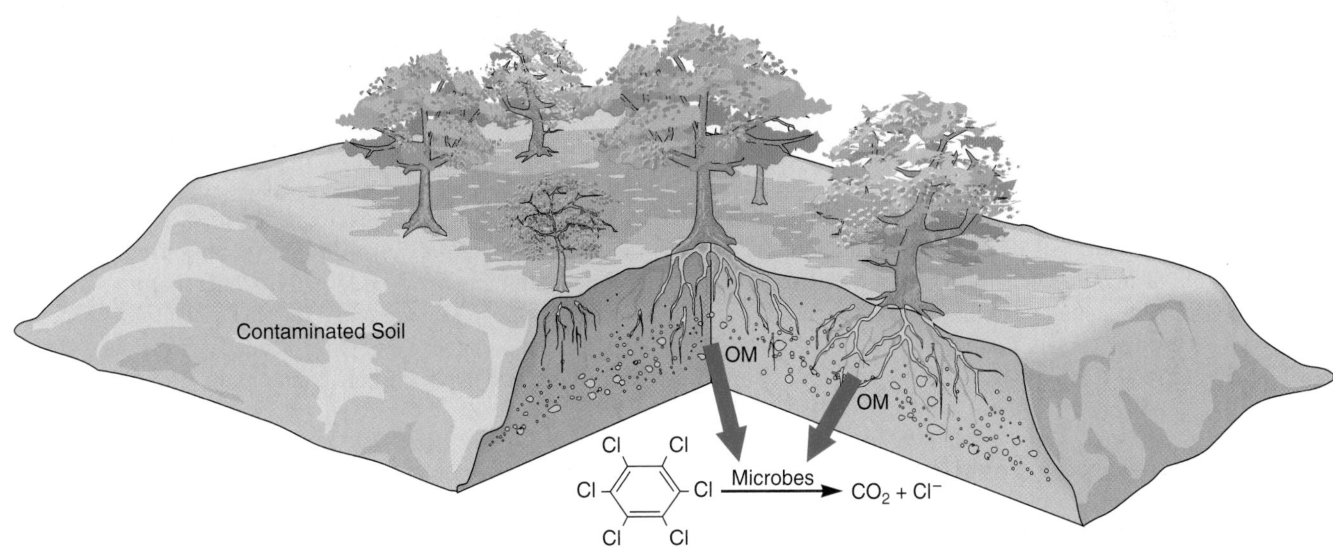

Figure 41.26 Phytoremediation. A conceptual view of a phytoremediation system, with a cut-away section of the root-soil zone. When organic matter (OM) is released from the plant roots, cometabolic processes can be carried out more efficiently by microbes, leading to enhanced degradation of contaminants. The mineralization of hexachlorobenzene is shown as an example.

Table 41.16 | Types of Phytoremediation

Process	Function
Phytoextraction	Use of pollutant-accumulating plants to remove metals or organics from soil by concentrating them in the harvestable plant parts
Phytodegradation	Use of plants and associated microorganisms to degrade organic pollutants
Rhizofiltration	Use of plant roots to absorb and adsorb pollutants, mainly metals, from water and aqueous waste streams
Phytostabilization	Use of plants to reduce the bioavailability of pollutants in the environment
Phytovolatilization	Use of plants to volatilize pollutants

Based on T. Macek; M. Mackova; and J. Kás. 2000. Exploitation of plants for the removal of organics in environmental remediation. *Biotechnol. Adv.* 18:23–34. P. 25.

noticed that microbes on the surface of the roots of yellow lupines only slowly degraded toluene. Scientists reasoned that if endophytic bacteria (i.e., bacteria growing within the plant roots) were genetically modified to efficiently degrade the toxin, their introduction into the plant would detoxify the surrounding soil. To accomplish this, conjugation was used to introduce a plasmid bearing the genes for toluene degradation into an endophytic strain of *Burkholderia cepacia*. Indeed, when the new toluene-degrading *B. capacia* bacteria colonized the yellow lupine, the host plant was protected from the toxic effects of toluene and reduced the amount of toluene that escaped into the atmosphere. Experiments such as these suggest that the field of phytoremediation will continue to provide important new approaches to decontaminating soils.

Metal Bioleaching

Bioleaching is the use of microorganisms, which produce acids from reduced sulfur compounds, to create acidic environments that solubilize desired metals for recovery. This approach is used to recover metals from ores and mining tailings with metal levels too low for smelting. Bioleaching carried out by natural populations of *Leptospirillum*-like species, *Thiobacillus thiooxidans,* and related thiobacilli, for example, allows recovery of up to 70% of the copper in low-grade ores. As shown in **figure 41.27,** this involves the biological oxidation of copper present in these ores to produce soluble copper sulfate.

It is apparent that nature will assist in bioremediation if given a chance. The role of microorganisms in biodegradation is now better appreciated. An excellent example is the xenobiotic metabolism of the versatile fungus *Phanerochaete chrysosporium* (**Microbial Diversity & Ecology 41.4**).

1. What is cometabolism and why is this important for degradation processes?
2. What factors must one consider when attempting to stimulate the microbial degradation of an oil spill in a marine environment?
3. Describe the major types of phytoremediation. What is the role of microorganisms in each of these processes?
4. How is bioleaching carried out and what microbial genera are involved?
5. What is unique about *Phanerochaete chrysosporium?* What does its name mean?

41.7 Bioaugmentation

Both in laboratory and field studies, attempts have been made to speed up existing microbiological processes by adding known active microorganisms to soils, waters, or other complex systems. This is called **bioaugmentation.** The microbes used in these experiments have been isolated from contaminated sites, taken from laboratory culture collections, or derived from uncharacterized enrichment cultures. For example, commercial culture preparations are available to facilitate silage formation and to improve septic tank performance.

The Impact of Protective Microhabitats

With the development of the "superbug" by A. M. Chakrabarty in 1974, there was initial excitement as it was hoped that such an improved microorganism might be able to degrade hydrocarbon pollutants very effectively. Chakrabarty's "superbug" was a laboratory pseudomonad that had been transformed with plasmids that encoded enzymes needed for efficient degradation of several hydrocarbon compounds. However, a critical

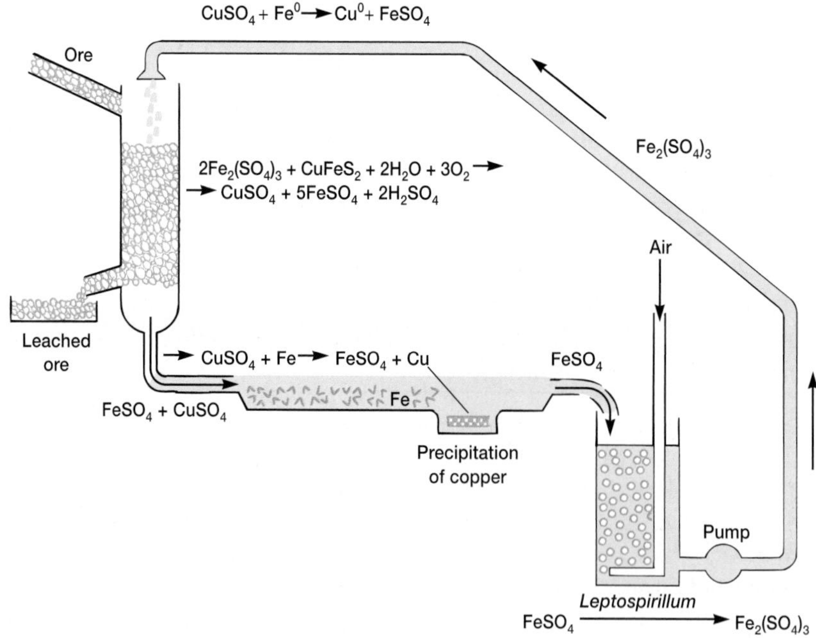

Figure 41.27 Copper Leaching from Low-Grade Ores. The chemistry and microbiology of copper ore leaching involve interesting complementary reactions. The microbial contribution is the oxidation of ferrous ion (Fe^{2+}) to ferric ion (Fe^{3+}). *Leptospirillum ferrooxidans* and related microorganisms are very active in this oxidation. The ferric ion then reacts chemically to solubilize the copper. The soluble copper is recovered by a chemical reaction with elemental iron, which results in an elemental copper precipitate.

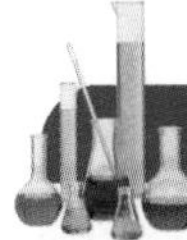

Microbial Diversity & Ecology

41.4 A Fungus with a Voracious Appetite

The basidiomycete *Phanerochaete chrysosporium* (the scientific name means "visible hair, golden spore") is a fungus with unusual degradative capabilities. This organism is termed a "white rot fungus" because of its ability to degrade lignin, a randomly linked phenylpropene-based polymeric component of wood. The cellulosic portion of wood is attacked to a lesser extent, resulting in the characteristic white color of the degraded wood. This organism also degrades a truly amazing range of xenobiotic compounds (nonbiological foreign chemicals) using both intracellular and extracellular enzymes.

As examples, the fungus degrades benzene, toluene, ethylbenzene, and xylenes (the so-called BTEX compounds), chlorinated compounds such as 2,4,5-trichloroethylene (TCE), and trichlorophenols. The latter are present as contaminants in wood preservatives and also are used as pesticides. In addition, other chlorinated benzenes can be degraded with or without toluenes being present. Even the insecticide Hydramethylnon is degraded.

How does this microorganism carry out such feats? Apparently most degradation of these xenobiotic compounds occurs after active growth, during the secondary metabolic lignin degradation phase. Degradation of some compounds involves important extracellular enzymes including lignin peroxidase, manganese-dependent peroxidase, and glyoxal oxidase. A critical enzyme is pyranose oxidase, which releases H_2O_2 for use by the manganese-dependent peroxidase enzyme. The H_2O_2 also is a precursor of the highly reactive hydroxyl radical, which participates in wood degradation. Apparently the pyranose oxidase enzyme is located in the interperiplasmic space of the fungal cell wall, where it can function either as a part of the fungus or be released and penetrate into the wood substrate. It appears that the nonspecific enzymatic system that releases these oxidizing products degrades many cyclic, aromatic, and chlorinated compounds related to lignins.

We can expect to continue hearing of many new advances regarding this organism. Potentially valuable applications being studied include growth in bioreactors where intracellular and extracellular enzymes can be maintained in the bioreactor while liquid wastes flow past the immobilized fungi.

point, which was not considered, was the actual location, or microhabitat, where the microbe had to survive and function. Engineered microorganisms were added to soils and waters with the expectation that rates of degradation would be stimulated as these microorganisms established themselves. Generally such additions led to short-term increases in rates of the desired activity, but typically after a few days the microbial community responses were similar in treated and control systems. After many unsuccessful attempts, it was found that the lack of effectiveness of such added cultures was due to at least three factors: (1) the attractiveness of laboratory-grown microorganisms as a food source for predators such as soil protozoa, (2) the inability of these added microorganisms to contact the compounds to be degraded, and (3) the failure of the added microorganisms to survive and compete with indigenous microorganisms. Such a modified microorganism may be less fit to compete and survive because of the additional energetic burden required to maintain the extra DNA.

Attempts have been made to make such laboratory-grown cultures more capable of survival in a natural environment by growing them in low-nutrient media or starving the microorganisms before adding them to an environment. These approaches select for mutant strains that can better survive under more natural conditions. However, this has not solved the problem. In recent years, there has been less interest in simply adding microorganisms to environments without considering the specific niche or microenvironment in which they are to survive and function. This has led to the field of **natural attenuation,** which emphasizes the use of natural microbial communities in the environmental management of pollutants.

Microorganism additions to natural environments can be more successful if the microorganism is added together with a microhabitat that gives the organism physical protection, as well as possibly supplying nutrients. This makes it possible for the microorganism to survive in spite of the intense competitive pressures that exist in the natural environment, including pressure from protozoan predators. Microhabitats may be either living or inert. Microbial interactions (section 30.1)

Specialized living microhabitats include the surface of a seed, a root, or a leaf. Here higher nutrient fluxes and rates of initial colonization by the added microorganisms, can protect the added microbe from the fierce competitive conditions in the natural environment. Examples include the use of *Bacillus thuringiensis* (p. 1083) and *Rhizobium*. In order to ensure that *Rhizobium* is in close association with the legume, seeds are coated with the microbe using an oil-organism mixture, or *Rhizobium* is placed in a band under the seed where the newly developing primary root will penetrate. Microorganism associations with vascular plants: The rhizobia (section 29.5)

Recently it has been found that microorganisms can be added to natural communities together with protective inert microhabitats. As an example, if microbes are added to a soil with microporous glass, the survival of added microorganisms can be markedly enhanced. Other microbes have been observed to create their own microhabitats. Microorganisms in the water column overlying PCB-contaminated sand-clay soils have been observed to create their own "clay hutches" by binding clays to their outer surfaces with exopolysaccharides. Thus the application of principles of microbial ecology can facilitate the successful management of microbial communities in nature.

1. What factors might limit the ability of microorganisms, after addition to a soil or water, to persist and carry out desired functions?
2. What types of microhabitats can be used with microorganisms when they are added to a complex natural environment?
3. Compare natural and inert microhabitats. Under what conditions might each be used?

41.8 Microbes As Products

So far we have discussed the use of microbial products like antibiotics and organic acids, or the use of microbial communities, to meet defined goals. However, single microbial species can be marketed as valuable products. Perhaps the most common example is the inoculation of legume seeds with rhizobia to ensure efficient nodulation and nitrogen fixation, as discussed previously. Here we introduce several other microbes and microbial structures that are of industrial and/or agricultural relevance.

Nanotechnology

Diatoms have aroused the interest of nanotechnologists. These photosynthetic protists produce intricate silica shells that differ according to species (**figure 41.28**). Nanotechnologists are interested in diatoms because they create precise structures at the micrometer scale. Three-dimensional structures in nanotechnology are currently built plane by plane and meticulous care must be taken to etch each individual structure to its final, exact shape. Diatoms, on the other hand, build directly in three dimensions and do so while growing exponentially, making them attractive for nanotechnology. There have been a number of ideas and approaches to harness these microbial "factories," but one technique is especially fascinating. Diatoms shells are incubated at 900°C in an atmosphere of magnesium for several hours. Amazingly, this results in an atom-for-atom substitution of silicon with magnesium without loss of 3D structure. Thus silicon oxide, which is of little use in nanotechnology, is converted to highly useful magnesium oxide. Protist classification: *Stramenopiles* (section 25.6)

Magnetotactic bacteria are also of interest to nanotechnologists. Magnetosomes are formed by certain bacteria through the accumulation of iron into magnetite. The sizes and shapes of the magnetosomes differ among species, but like diatom shells, they

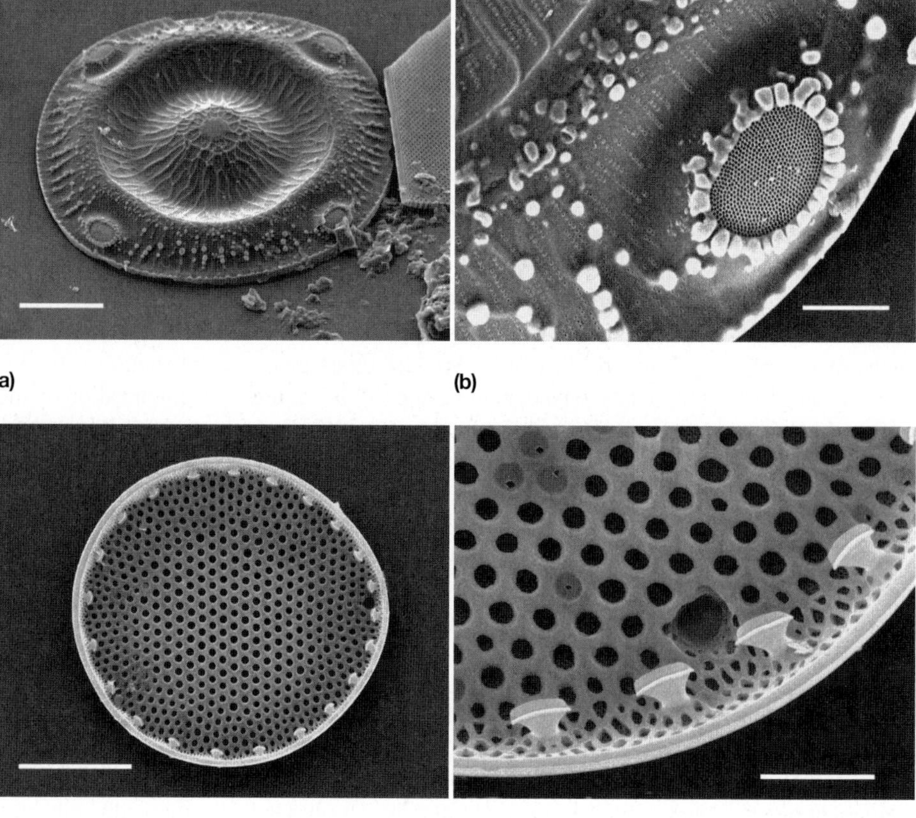

Figure 41.28 Marine Diatom Surface Features. **(a)** *Glyphodiscus stellatus*; scale bar is 20 μm. **(b)** *G. stellatus* close-up; scale bar is 5 μm. **(c)** *Roperia tesselata*; scale bar is 10 μm. **(d)** *R. tesselata* close-up; scale bar is 5 μm.

are perfectly formed despite the fact that they are only tens of nanometers in diameter. Although tiny, magnetic beads can be chemically synthesized, they are not as precisely formed and lack the membrane that surrounds magnetosomes. This membrane enables the attachment of useful biological molecules like enzymes and antibodies. Potential applications for magnetosomes currently under investigation include their use as a contrast medium to improve magnetic resonance tomography (MRI) and as biological probes to detect cancer at early stages. The cytoplasmic matrix: Inclusion bodies (section 3.3)

Biosensors

A rapidly developing area of biotechnology, is that of **biosensor** production. In this field of bioelectronics, living microorganisms (or their enzymes or organelles) are linked with electrodes, and biological reactions are converted into electrical currents (**figure 41.29**). Biosensors have been developed to measure specific components in beer, to monitor pollutants, to detect flavor compounds in food, and to study environmental processes such as changes in biofilm concentration gradients. It is possible to measure the concentration of substances from many different environments (**table 41.17**). Applications include the detection of glucose, acetic acid, glutamic acid, ethanol, and biochemical oxygen demand. Biosensors have been developed using immunochemical-based detection systems. These new biosensors will detect pathogens, herbicides, toxins, proteins, and DNA. Since the bioterrorism attacks of 2001, the U.S. government has stepped up funding for research and development of biosensors capable of detecting minute levels of potential airborne pathogens. Many of these biosensors are based on the use of a streptavidin-biotin recognition system (**Techniques & Applications 41.5**). Microbial tidbits 35.2: Biosensors

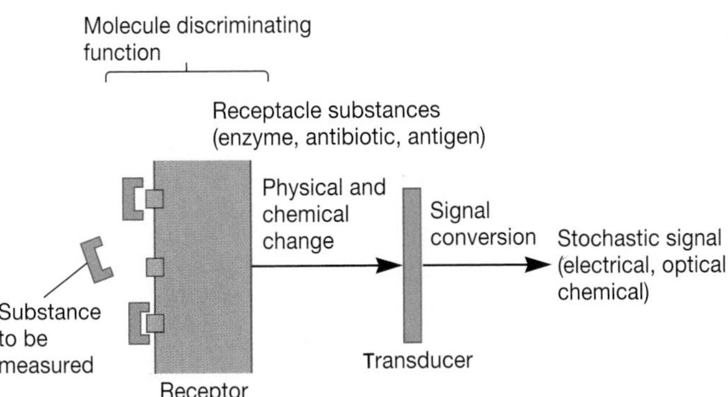

Figure 41.29 Biosensor Design. Biosensors are finding increasing applications in medicine, industrial microbiology, and environmental monitoring. In a biosensor a biomolecule or whole microorganism carries out a biological reaction, and the reaction products are used to produce an electrical signal.

Table 41.17	Biosensors: Potential Biomedical, Industrial, and Environmental Applications
Clinical diagnosis and biomedical monitoring	
Agricultural, horticultural, veterinary analysis	
Detection of pollution, and microbial contamination of water	
Fermentation analysis and control	
Monitoring of industrial gases and liquids	
Measurement of toxic gas in mining industries	
Direct biological measurement of flavors, essences, and pheromones	

1. What are biosensors and how do they detect substances?
2. What areas are biosensors being used in to assist in chemical and biological monitoring efforts?
3. Describe streptavidin-biotin systems and how they work. Why is this technique important?

Biopesticides

There has been a long-term interest in the use of bacteria, fungi, and viruses as **bioinsecticides** and **biopesticides** (**table 41.18**). These are defined as biological agents, such as bacteria, fungi, viruses, or their components, which can be used to kill a susceptible insect. In this section, we discuss some of the major uses of bacteria, fungi, and viruses to control populations of insects.

Bacteria

Bacterial agents include a variety of *Bacillus* species; however, *B. thuringiensis* is most widely used. This bacterium is only weakly toxic to insects as a vegetative cell, but during sporulation, it produces an intracellular protein toxin crystal, the parasporal body, that can act as a microbial insecticide for specific insect groups. Class *Bacilli*: Order *Bacillale*: (section 23.5)

The parasporal crystal, after exposure to alkaline conditions in the insect hindgut, fragments to release the protoxin. After this reacts with a protease enzyme, the active toxin is produced (**figure 41.30**). Six of the active toxin units integrate into the plasma membrane (figure 41.30*b,c*) to form a hexagonal-shaped pore through the midgut cell, as shown in figure 41.30*d*. This leads to the loss of osmotic balance and ATP, and finally to cell lysis.

B. thuringiensis can be grown in fermenters. The spores and crystals are released into the medium when cells lyse. The medium is then centrifuged and made up as a dust or wettable powder for application to plants. This insecticide, known as Bt, has been used on a worldwide basis for over 40 years. Unlike chemical insecticides, Bt does not accumulate in the soil or in nontarget animals. Rather it is readily lost from the environment by microbial and abiotic degradation.

41.5 Streptavidin-Biotin Binding and Biotechnology

Egg white contains many proteins and glycoproteins with unique properties. One of the most interesting, which binds tenaciously to biotin, was isolated in 1963. This glycoprotein, called avidin due to its "avid" binding of biotin, was suggested to play an important role: making egg white antimicrobial by "tying up" the biotin needed by many microorganisms. Avidin, which functions best under alkaline conditions, has the highest known binding affinity between a protein and a ligand. Several years later, scientists at Merck & Co., Inc. discovered a similar protein produced by the actinomycete *Streptomyces avidini,* which binds biotin at a neutral pH and which does not contain carbohydrates. These characteristics make streptavidin an ideal binding agent for biotin, and it has been used in an almost unlimited range of applications, as shown in the **Box figure.** The streptavidin protein is joined to a probe. When a sample is incubated with the biotinylated binder, the binder attaches to any available target molecules. The presence and location of target molecules can be determined by treating the sample with a streptavidin probe because the streptavidin binds to the biotin on the biotinylated binder, and the probe is then visualized. This detection system is employed in a wide variety of biotechnological applications, including use as a nonradioactive probe in hybridization studies and as a critical component in biosensors for a wide range of environmental monitoring and clinical applications.

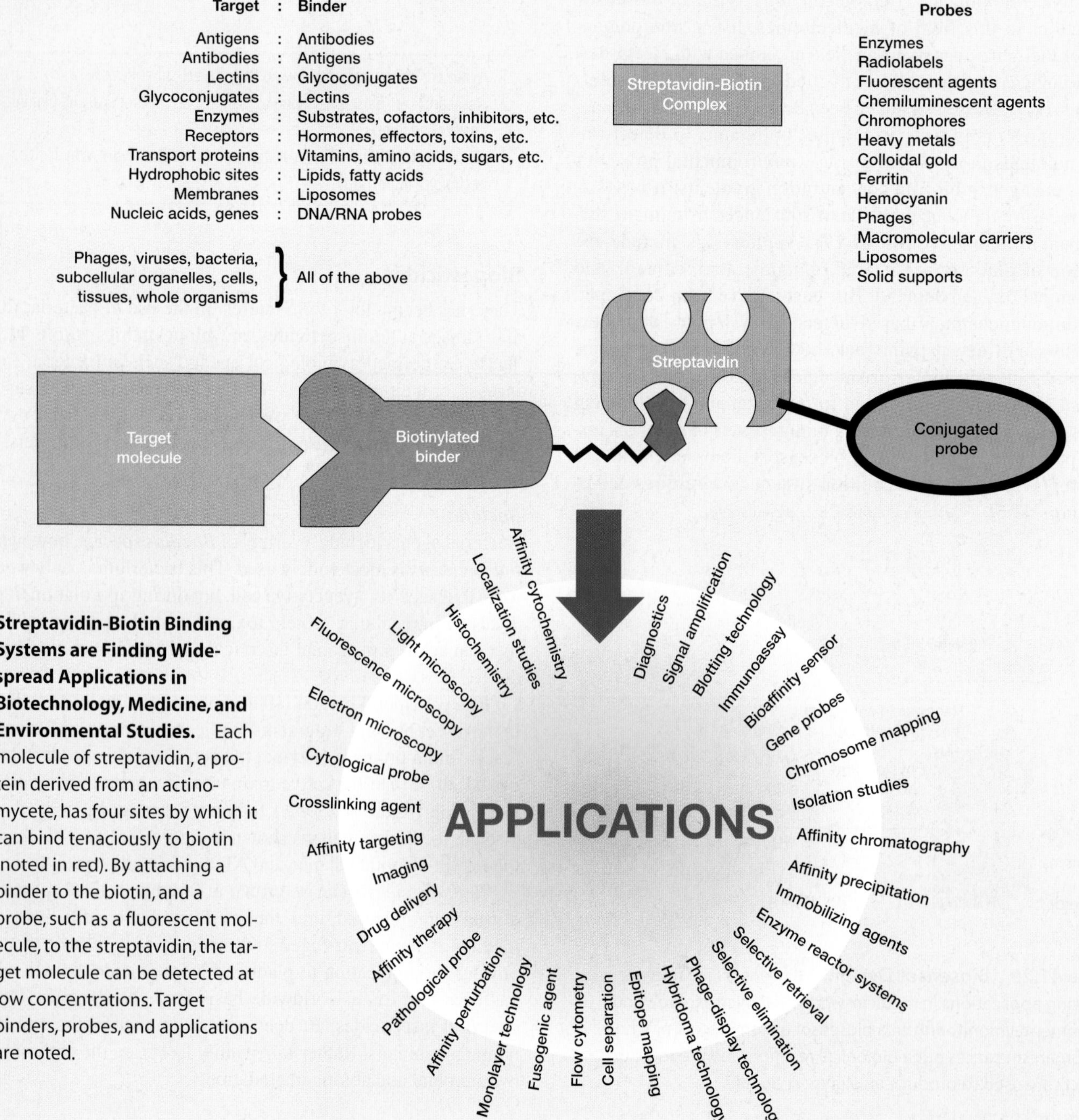

Streptavidin-Biotin Binding Systems are Finding Widespread Applications in Biotechnology, Medicine, and Environmental Studies. Each molecule of streptavidin, a protein derived from an actinomycete, has four sites by which it can bind tenaciously to biotin (noted in red). By attaching a binder to the biotin, and a probe, such as a fluorescent molecule, to the streptavidin, the target molecule can be detected at low concentrations. Target binders, probes, and applications are noted.

Table 41.18 The Use of Bacteria, Viruses, and Fungi As Bioinsecticides: An Older Technology with New Applications

Microbial Group	Major Organisms and Applications
Bacteria	*Bacillus thuringiensis* and *Bacillus popilliae* are the two major bacteria of interest. *Bacillus thuringiensis* is used on a wide variety of vegetable and field crops, fruits, shade trees, and ornamentals. *B. popilliae* is used primarily against Japanese beetle larvae. Both bacteria are considered harmless to humans. *Pseudomonas fluorescens,* which contains the toxin-producing gene from *B. thuringiensis,* is used on maize to suppress black cutworms.
Viruses	Three major virus groups that do not appear to replicate in warm-blooded animals are used: nuclear polyhedrosis virus (NPV), granulosis virus (GV), and cytoplasmic polyhedrosis virus (CPV). These viruses are more protected in the environment.
Fungi	Over 500 different fungi are associated with insects. Infection and disease occur primarily through the insect cuticle. Four major genera have been used. *Beauveria bassiana* and *Metarhizium anisopliae* are used for control of the Colorado potato beetle and the froghopper in sugarcane plantations, respectively. *Verticillium lecanii* and *Entomophthora* spp., have been associated with control of aphids in greenhouse and field environments. *Coelomyces,* a chytrid, also is used for control of mosquitoes.

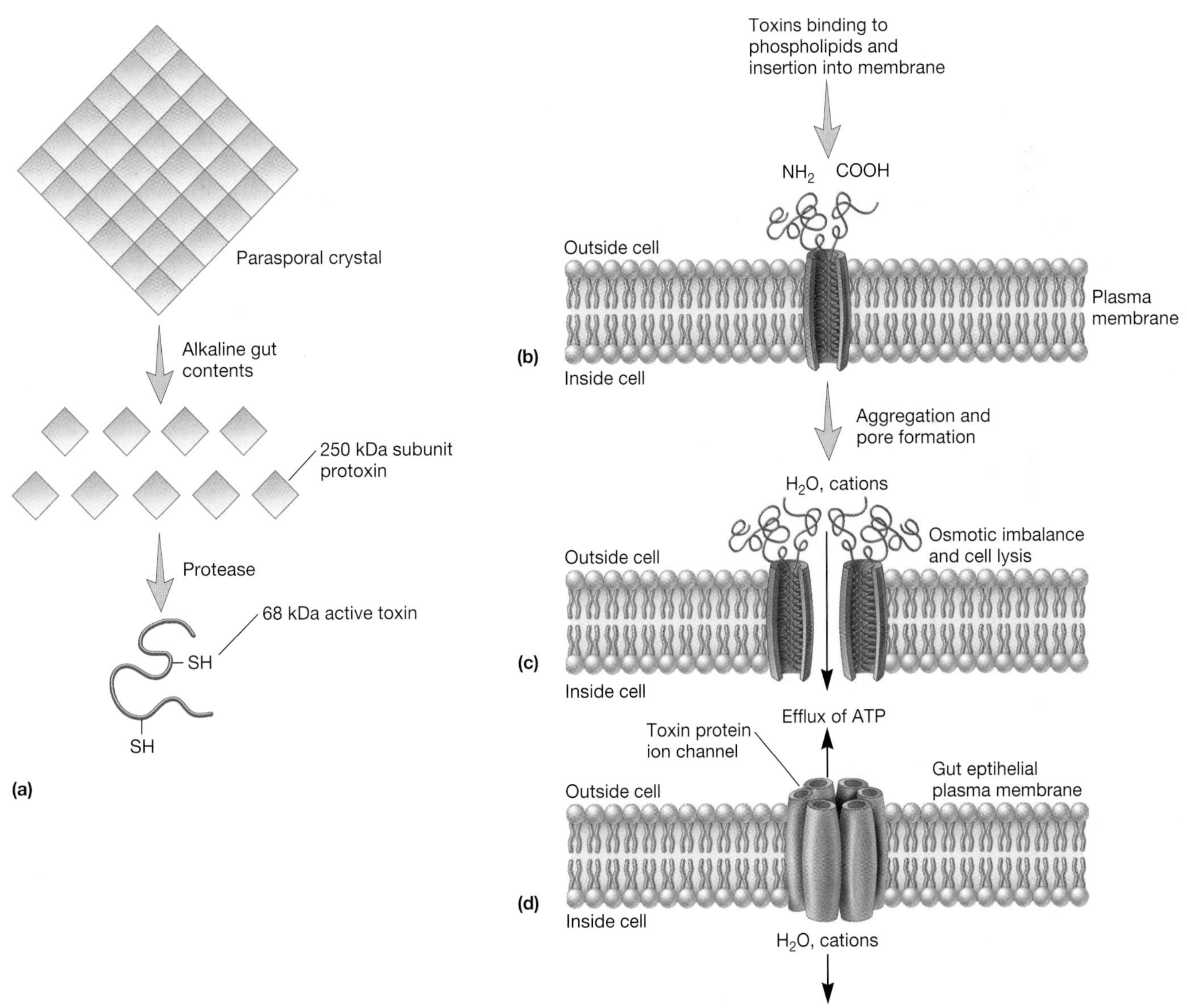

Figure 41.30 The Mode of Action of the *Bacillus thuringiensis* Toxin. **(a)** Release of the protoxin from the parasporal body and modification by proteases in the hindgut. **(b)** Insertion of the 68 kDa active toxin molecules into the membrane. **(c)** Aggregation and pore formation, showing a cross section of the pore. **(d)** Final hexagonal pore which causes an influx of water and cations as well as a loss of ATP, resulting in cell imbalance and lysis.

Unlike other genetically modified organisms (GMOs), transgenic plants expressing the *B. thuringiensis* toxin gene, *cry* (for *cry*stal), have generally been well accepted. In 1996, commercialized Bt-corn, Bt-potato, and Bt-cotton were introduced into the United States and soon farmers in other countries such as Australia, Canada, China, France, Indonesia, Mexico, Spain, and Ukraine followed. The widespread acceptance of these plants reflects the history of safe application of Bt as an insecticide without adverse environmental or health impacts. In addition, it is well understood that the Cry protein can only be activated in the target insect. Long-term studies have shown that Bt is nontoxic to mammals and is not an allergen in humans. One potential problem, the horizontal gene flow of the *cry* gene to weeds and other plants, has not been reliably demonstrated.

Viruses

Viruses that are pathogenic for specific insects include nuclear polyhedrosis viruses (NPVs), granulosis viruses (GVs), and cytoplasmic polyhedrosis viruses (CPVs). Currently over 125 types of NPVs are known, of which approximately 90% affect the *Lepidoptera*—butterflies and moths. Approximately 50 GVs are known, and they, too, primarily affect butterflies and moths. CPVs are the least host-specific viruses, affecting about 200 different types of insects. An important commercial viral pesticide is marketed under the trade name Elcar for control of the cotton bollworm *Heliothis zea*. Insect viruses (section 18.8)

Fungi

Fungi also can be used to control insect pests. Fungal bioinsecticides, as listed in table 41.18, are finding increasing use in agriculture. However, large-scale production of fungal insect pathogens is difficult and when introduced into natural ecosystems, they often do not persist, Nonetheless, the development of fungal biopesticides continues to progress.

1. What two important bacteria have been used as bioinsecticides?
2. Briefly describe how the *Bacillus thuringiensis* toxin kills insects.
3. What types of viruses are being used to attempt to control insects?
4. Which fungi presently are being used as biopesticides?

41.9 Impacts of Microbial Biotechnology

The use of microorganisms in industrial microbiology and biotechnology, as discussed in this chapter, does not take place in an ethical and ecological vacuum. Decisions to make a particular product, and also the methods used, can have long-term and often unexpected effects, as with the appearance of antibiotic-resistant pathogens around the world.

Microbiology is a critical part of the area of **industrial ecology,** concerned with tracking the flow of elements and compounds through the natural and social worlds, or the biosphere and the anthrosphere. Microbiology, especially as an applied discipline, should be considered within its supporting social world.

Microorganisms have been of immense benefit to humanity through their role in food production and processing, the use of their products to improve human and animal health, in agriculture, and for the maintenance and improvement of environmental quality. Other microorganisms, however, are important pathogens and agents of spoilage, and microbiologists have helped control or limit the activities of these harmful microorganisms. The discovery and use of beneficial microbial products, such as antibiotics, have contributed to a doubling of the human life span in the last century.

A microbiologist who works in any of these areas of biotechnology should consider the longer-term impacts of possible technical decisions. Our first challenge, as microbiologists, is to understand, as much as is possible, the potential impacts of new products and processes on the broader society as well as on microbiology. An essential part of this responsibility is to be able to communicate effectively with the various "societal stakeholders" about the immediate and longer-term potential impacts of microbial-based (and other) technologies.

1. Discuss possible ethical and ecological impacts of a particular product or process discussed in this chapter.Think in terms of the broadest possible impacts in your discussion of this problem.
2. Define industrial ecology.
3. What are the biosphere and anthrosphere? Why might you think the term anthrosphere was coined?

Summary

41.1 Water Purification and Sanitary Analysis

a. Water purification can involve the use of sedimentation, coagulation, chlorination, and rapid and slow sand filtration. Chlorination may lead to the formation of organic disinfection by-products, including trihalomethane (THM) compounds, which are potential carcinogens (**figure 41.1**).

b. *Cryptosporidium, Cyclospora,* viruses, and *Giardia* are of concern, as conventional water purification and chlorination will not always assure their removal and inactivation to acceptable limits.

c. Indicator organisms are used to indicate the presence of pathogenic microorganisms. Most probable number (MPN) and membrane filtration procedures are employed to estimate the number of indicator organisms present. Presence-absence (P-A) tests for coliforms and defined substrate tests for coliforms and *E. coli* allow 100 ml water volumes to be tested with minimum time and materials (**table 41.1; figures 41.2** and **41.3**).

d. Molecular techniques based on the polymerase chain reaction (PCR) can be used to detect waterborne pathogens such as Shigalike-toxin producing *E. coli* O157:H7, when a preenrichment step is used.

41.2 Wastewater Treatment

a. The biochemical oxygen demand (BOD) test is an indirect measure of organic matter that can be oxidized by the aerobic microbial community. In this assay, oxygen should never limit the rate of reaction. The chemical oxygen demand

(COD) and total organic carbon (TOC) tests provide information on carbon that is not biodegraded in the 5-day BOD test.

b. Conventional sewage treatment is a controlled intensification of natural self-purification processes, and it can involve primary, secondary, and tertiary treatment (**figure 41.4**).

c. Constructed wetlands involve the use of aquatic plants (floating, emergent, submerged) and their associated microorganisms for the treatment of liquid wastes (**figure 41.7**).

d. Home treatment systems operate on general self-purification principles. The conventional septic tank (**figure 41.8**) provides anaerobic liquefaction and digestion whereas the aerobic leach-field allows oxidation of the soluble effluent. These systems are now being designed to provide nitrogen and phosphorus removal, to lessen impacts of on-site sewage treatment systems on vulnerable marine and freshwaters.

e. Groundwater is an important resource that can be affected by pollutants from septic tanks and other sources. This vital water source must be protected and improved.

41.3 Microorganisms Used in Industrial Microbiology

a. Industrial microbiology has been used to manufacture such products as antibiotics, amino acids, and organic acids and has had many important positive effects on animal and human health. Most work in this area has been carried out using microorganisms isolated from nature or modified by the use of classic mutation techniques. Biotechnology involves the use of molecular techniques to modify and improve microorganisms.

b. Finding new microorganisms in nature for use in biotechnology is a continuing challenge. Only about 1% of the observable microbial community has been grown (**table 41.6**), but major advances in growing "uncultured" microbes are being made.

c. Selection and mutation continue to be important approaches for identifying new microorganisms. These well-established procedures are now being complemented by molecular techniques, including metabolic engineering and combinatorial biology. With combinatorial biology (**table 41.7**), it is possible to transfer genes from one organism to another organism, and to form new products.

d. Site-directed mutagenesis and protein engineering are used to modify gene expression. These approaches are leading to new and often different products with new properties (**figure 41.11**).

e. Protein evolution is of increasing interest. This involves exploiting microbial responses to stress in adaptive mutation and forced evolution, with the hope of identifying microorganisms with new properties. Alternatively, an in vitro approach can be used.

41.4 Microorganism Growth in Controlled Environments

a. Microorganisms can be grown in controlled environments of various types using fermenters and other culture systems. If defined constituents are used, growth parameters can be chosen and varied in the course of growing a microorganism. This approach is used particularly for the production of amino acids, organic acids, and antibiotics.

41.5 Major Products of Industrial Microbiology

a. A wide variety of compounds are produced in industrial microbiology that impact our lives in many ways (**table 41.13**). These include antibiotics, amino acids, organic acids, biopolymers such as the cyclodextrins, and biosurfactants (**figures 41.16–41.19**). Microorganisms also can be used as biocatalysts to carry out specific chemical reactions (**figure 41.20**).

b. Specialty nonantibiotic compounds are an important part of industrial microbiology and biotechnology. These include widely used antitumor agents (**table 41.15**).

41.6 Biodegradation and Bioremediation by Natural Communities

a. Microorganism growth in complex natural environments such as soils and waters is used to to carry out environmental management processes, including bioremediation, plant inoculation, and other related activities. In these cases, the microbes themselves are not final products.

b. Biodegradation is a critical part of natural systems mediated largely by microorganisms. This can involve minor changes in a molecule, fragmentation, or mineralization (**figure 41.21**).

c. Biodegradation can be influenced by many factors, including oxygen presence or absence, humic acids, and the presence of readily usable organic matter. Reductive dehalogenation proceeds best under anoxic conditions, and the presence of organic matter can facilitate modification of recalcitrant compounds in the process of cometabolism.

d. The structure of organic compounds influences degradation. If constituents are in specific locations on a molecule, as in the *meta* position (**figure 41.22**), or if varied structural isomers are present degradation can be affected.

e. Degradation management can be carried out in place, whether this be large marine oil spills, soils, or the subsurface (**figure 41.25**). Such large-scale efforts usually involve the use of natural microbial communities.

f. Degradation can lead to increased toxicity in many cases. If not managed carefully, widespread pollution can occur. Iron corrosion is a particular concern with methanogens and *Desulfovibrio* playing important roles in this process.

g. Plants can be used to stimulate biodegradation processes during phytoremediation. This can involve extraction, filtering, stabilization, and volatilization of pollutants (**figure 41.26** and **table 41.16**).

41.7 Bioaugmentation

a. Microorganisms can be added to environments that contain complex microbial communities with greater success if living or inert microhabitats are used. These can include living plant surfaces (seeds, roots, leaves) or inert materials such as microporous glass. *Rhizobium* is an important example of a microorganism added to a complex environment using a living microhabitat (the plant root).

41.8 Microbes As Products

a. Microorganisms are being used in a wide range of biotechnological applications such as nanotechnology and biosensors (**figures 41.28** and **41.29**).

b. Bacteria, viruses, and fungi can be used as bioinsecticides and biopesticides (**table 41.18**). *Bacillus thuringiensis* is an important biopesticide, and the BT gene has been incorporated into several important crop plants.

41.9 Impacts of Microbial Biotechnology

a. Industrial microbiology and biotechnology can have long-term and possibly unexpected positive and negative effects on the environment, and on animals and humans impacted by these technologies. Advances in biotechnology should be considered in a broad ecological and societal context, which is the focus of industrial ecology.

Key Terms

activated sludge 1055
adaptive mutation 1062
bioaugmentation 1080
biocatalyst 1074
biochemical oxygen demand (BOD) 1055
biodegradation 1075
bioinsecticides 1083
biopesticide 1083
biopolymer 1073
bioprospecting 1060
bioremediation 1075
biosensor 1083
biotransformation 1074
bulking sludge 1055
chemical oxygen demand (COD) 1054

Critical Thinking Questions

1. You wish to develop a constructed wetland for removal of metals from a stream at one site, and at another site, you wish to treat acid mine drainage. How might you approach each of these problems?
2. What alternatives, if any, can one use for protection against microbiological infection when swimming in polluted recreational water? Assume that you are part of a water rescue team.
3. What possible alternatives could be used to eliminate N and P releases from sewage treatment systems? What suggestions could you make that might lead to new technologies?
4. What further technological approaches will be required to culture microorganisms that have not yet been grown? Consider the roles of nutrient fluxes, communication molecules, and competition.
5. *Deinococcus radiodurans* is a species of bacteria that is highly resistant to radiation. Can you think of a biotechnological application? How would you test its utility?
6. Most commercial antibiotics are produced by actinomycetes, and only a few are synthesized by fungi and other bacteria. From physiological and environmental viewpoints, how might you attempt to explain this observation?
7. The terms biosphere and anthrosphere have been used, together with the term industrial ecology. How does microbial biotechnology relate to these concerns?
8. Discuss the risks of releasing genetically modified microbes or ones that are not natural to the particular environment. What would be your concerns? What precautions, if any, would you take?
9. Why, when a microorganism is removed from a natural environment and grown in the laboratory, will it usually not be able to effectively colonize its original environment if it is grown and added back? Consider the nature of growth media used in the laboratory in comparison to growth conditions in a soil or water when attempting to understand this fundamental problem in microbial ecology.
10. Why might *Bacillus thuringiensis* bioinsecticides be of interest in other areas of biotechnology? Consider the molecular aspects of their mode of action.

Learn More

Cameotra, S. S., and Makkar, R. S. 2004. Recent applications of biosurfactants as biological and immunological molecules. *Curr. Opin. Microbiol.* 7:262–66.

Compant, S.; Duffy, B.; Nowak, J.; Clément, C.; and Barka, E. A. 2005. Use of plant growth-promoting bacteria for biocontrol of plant diseases: Principles, action, and future prospects. *Appl. Env. Microbiol.* 71:4951–59.

de Maagd, R. A.; Bravo, A.; and Crickmore, N. 2001. How *Bacillus thuringiensis* has evolved specific toxins to colonize the insect world. *Trends Genet.* 17(4):193–99.

Drum, R. W., and Gordon, R. 2003. Star Trek replicators and diatom nanotechnology. *Trends Biotechnol.* 21:325–27.

Gross, R. A., and Kalra, B. 2002. Biodegradable polymers for the environment. *Science* 297:803–807.

Hurst, C. J.; Crawford, R. L.; Knudsen, G. R.; McInerney, M. J.; and Stetzenbach, L. D. 2002. *Manual of environmental microbiology,* 2nd ed. Washington, D.C.: ASM Press.

Rittmann, B. E., and McCarty, P. L. 2001. *Environmental biotechnology: Principles and applications.* New York: McGraw-Hill.

Selifonova, O.; Valle, F.; and Schellenberger, V. 2001. Rapid evolution of novel traits in microorganisms. *Appl. Environ. Microbiol.* 67:3645–49.

Shelton, A. M.; Zhao, J.-Z.; and Roush, R. T. 2002. Economic, ecological, food safety, and social consequences of Bt transgenic plants. *Annu. Rev. Entomol.* 47:845–81.

Teusink, B., and Smid, E. J. 2006. Modelling strategies for the industrial exploitation of lactic acid bacteria. *Nature Rev. Microbiol.* 4:46–56.

Wackett, L. P., and Hershberger, C. D. 2001. Biocatalysts and biodegradation: Microbial transformations of organic compounds. Washington, D.C.: ASM Press.

Yaun, L.; Kurek, I.; English, J.; and Keenan, R. 2005. Laboratory-directed protein evolution. *Microbiol. Molec. Biol. Rev.* 69:373–92.

Zhang, Y.-X.; Perry, K.; Vinci, V. A.; Powell, K.; Stemmer, W. P. C.; and del Cardayré, S. B. 2002. Genome shuffling leads to rapid phenotypic improvement in bacteria. *Nature* 415:644–46.

Please visit the Prescott website at www.mhhe.com/prescott7 for additional references.

Appendix I

A Review of the Chemistry of Biological Molecules

Appendix I provides a brief summary of the chemistry of organic molecules with particular emphasis on the molecules present in microbial cells. Only basic concepts and terminology are presented; introductory textbooks in biology and chemistry should be consulted for a more extensive treatment of these topics.

Atoms and Molecules

Matter is made of elements that are composed of atoms. An element contains only one kind of atom and cannot be broken down to simpler components by chemical reactions. An atom is the smallest unit characteristic of an element and can exist alone or in combination with other atoms. When atoms combine they form molecules. Molecules are the smallest particles of a substance. They have all the properties of the substance and are composed of two or more atoms.

Although atoms contain many subatomic particles, three directly influence their chemical behavior—protons, neutrons, and electrons. The atom's nucleus is located at its center and contains varying numbers of protons and neutrons (**figure AI.1**). Protons have a positive charge, and neutrons are uncharged. The mass of these particles and the atoms that they compose is given in terms of the atomic mass unit (AMU), which is 1/12 the mass of the most abundant carbon isotope. Often the term dalton (Da) is used to express the mass of molecules. It also is 1/12 the mass of an atom of ^{12}C or 1.661×10^{-24} grams. Both protons and neutrons have a mass of about one dalton. The atomic weight is the actual measured weight of an element and is almost identical to the mass number for the element, the total number of protons and neutrons in its nucleus. The mass number is indicated by a superscripted number preceding the element's symbol (e.g., ^{12}C, ^{16}O, and ^{14}N).

Negatively charged particles called electrons circle the atomic nucleus (figure AI.1). The number of electrons in a neutral atom equals the number of its protons and is given by the atomic number, the number of protons in an atomic nucleus. The atomic number is characteristic of a particular type of atom. For example, carbon has an atomic number of six, hydrogen's number is one, and oxygen's is eight (**table AI.1**).

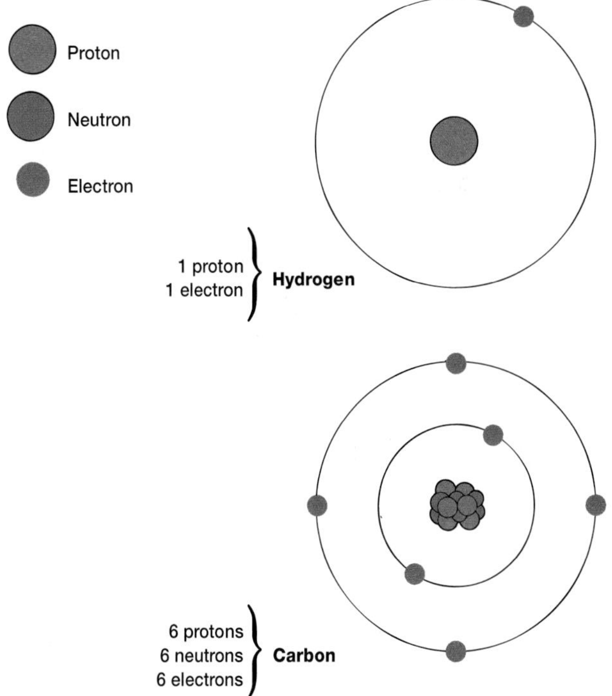

Figure AI.1 Diagrams of Hydrogen and Carbon Atoms. The electron orbitals are represented as concentric circles.

Table AI.1 Atoms Commonly Present in Organic Molecules

Atom	Symbol	Atomic Number	Atomic Weight	Number of Chemical Bonds
Hydrogen	H	1	1.01	1
Carbon	C	6	12.01	4
Nitrogen	N	7	14.01	3
Oxygen	O	8	16.00	2
Phosphorus	P	15	30.97	5
Sulfur	S	16	32.06	2

The electrons move constantly within a volume of space surrounding the nucleus, even though their precise location in this volume cannot be determined accurately. This volume of space in which an electron is located is called its orbital. Each orbital can contain two electrons. Orbitals are grouped into shells of different energy that surround the nucleus. The first shell is closest to the nucleus and has the lowest energy; it contains only one orbital. The second shell contains four orbitals, one circular and three shaped like dumbbells (**figure AI.2*a***). It can contain up to eight electrons. The third shell has even higher energy and holds more than eight electrons. Shells are filled beginning with the innermost and moving outward. For example, carbon has six electrons, two in its first shell and four in the second (figures AI.1 and AI.2*b*). The electrons in the outermost shell are the ones that participate in chemical reactions. The most stable condition is achieved when the outer shell is filled with electrons. Thus the number of bonds an element can form depends on the number of electrons required to fill the outer shell. Since carbon has four electrons in its outer shell and the shell is filled when it contains eight electrons, it can form four covalent bonds (table AI.1).

Chemical Bonds

Molecules are formed when two or more atoms associate through chemical bonding. Chemical bonds are attractive forces that hold together atoms, ions, or groups of atoms in a molecule or other substance. Many types of chemical bonds are present in organic molecules; three of the most important are covalent bonds, ionic bonds, and hydrogen bonds.

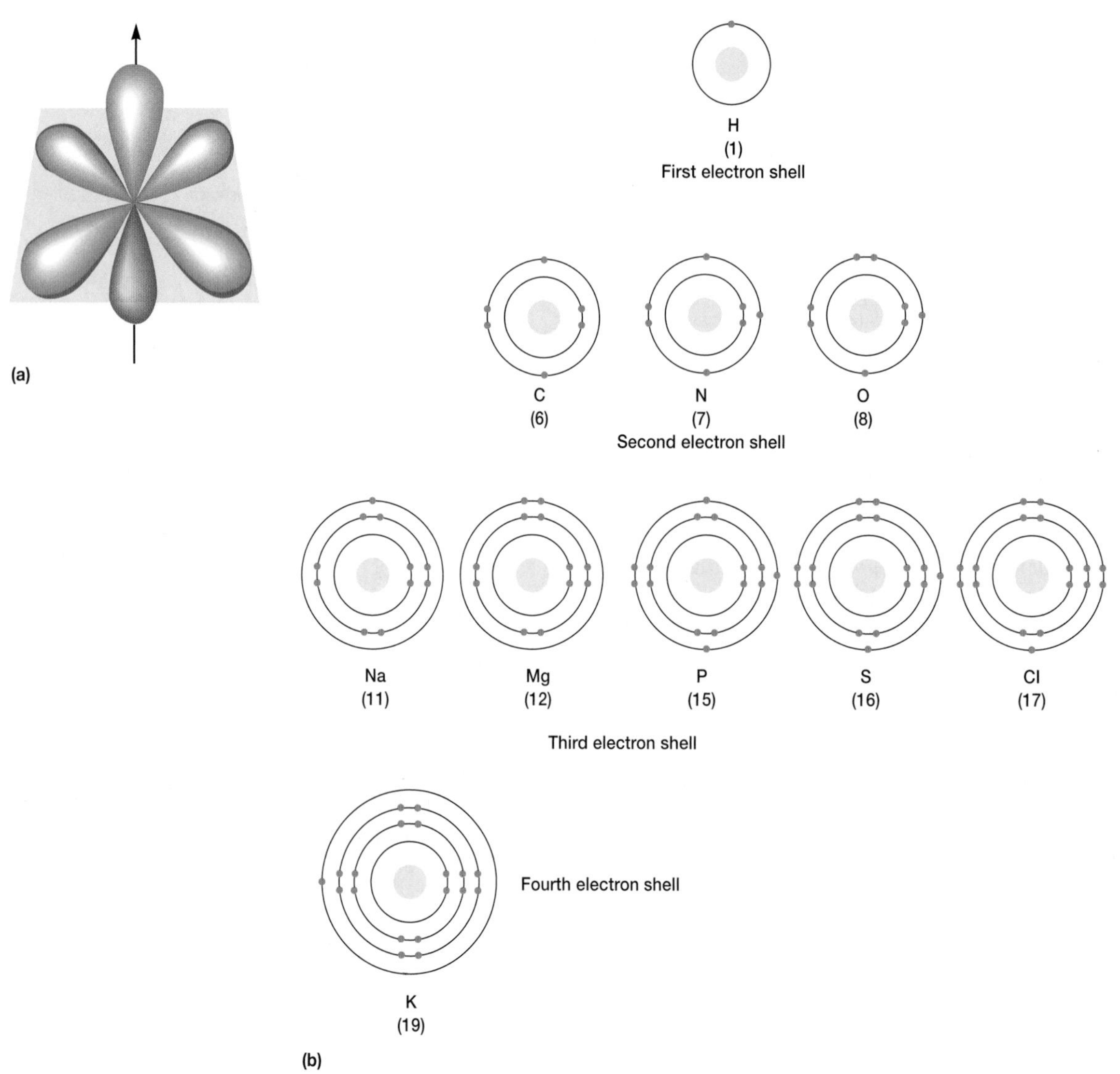

Figure AI.2 Electron Orbitals. **(a)** The three dumbbell-shaped orbitals of the second shell. The orbitals lie at right angles to each other. **(b)** The distribution of electrons in some common elements. Atomic numbers are given in parentheses.

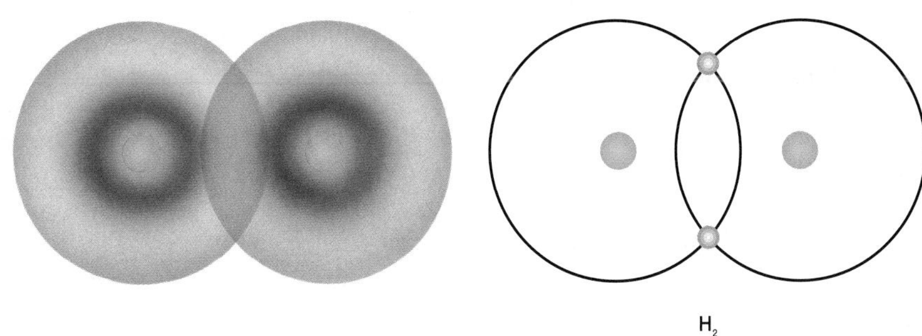

Figure AI.3 The Covalent Bond. A hydrogen molecule is formed when two hydrogen atoms share electrons.

$>C{=}O \cdots H{-}O{-}$

$\equiv N \cdots H{-}O{-}C({=}O){-}$

$>C{=}O \cdots H{-}N<$

$-O(H) \cdots H{-}N<$

$\equiv N \cdots H{-}N<$

Figure AI.4 Hydrogen Bonds. Representative examples of hydrogen bonds present in biological molecules.

(a) C_6H_{14} (Hexane)

(b) or C_6H_{12} (Cyclohexane)

(c) or C_6H_6 (Benzene)

Figure AI.5 Hydrocarbons. Examples of hydrocarbons that are **(a)** linear, **(b)** cyclic, and **(c)** aromatic.

In covalent bonds, atoms are joined together by sharing pairs of electrons (**figure AI.3**). If the electrons are equally shared between identical atoms (e.g., in a carbon-carbon bond), the covalent bond is strong and nonpolar. When two different atoms such as carbon and oxygen share electrons, the covalent bond formed is polar because the electrons are pulled toward the more electronegative atom, the atom that more strongly attracts electrons (the oxygen atom). A single pair of electrons is shared in a single bond; a double bond is formed when two pairs of electrons are shared.

Atoms often contain either more or fewer electrons than the number of protons in their nuclei. When this is the case, they carry a net negative or positive charge and are called ions. Cations carry positive charges and anions have a net negative charge. When a cation and an anion approach each other, they are attracted by their opposite charges. This ionic attraction that holds two groups together is called an ionic bond. Ionic bonds are much weaker than covalent bonds and are easily disrupted by a polar solvent such as water. For example, the Na^+ cation is strongly attracted to the Cl^- anion in a sodium chloride crystal, but sodium chloride dissociates into separate ions (ionizes) when dissolved in water. Ionic bonds are important in the structure and function of proteins and other biological molecules.

When a hydrogen atom is covalently bonded to a more electronegative atom such as oxygen or nitrogen, the electrons are unequally shared and the hydrogen atom carries a partial positive charge. It will be attracted to an electronegative atom such as oxygen or nitrogen, which carries an unshared pair of electrons; this attraction is called a hydrogen bond (**figure AI.4**). Although an individual hydrogen bond is weak, there are so many hydrogen bonds in proteins and nucleic acids that they play a major role in determining protein and nucleic acid structure.

Organic Molecules

Most molecules in cells are organic molecules, molecules that contain carbon. Since carbon has four electrons in its outer shell, it tends to form four covalent bonds in order to fill its outer shell with eight electrons. This property makes it possible to form chains and rings of carbon atoms that also can bond with hydrogen and other atoms (**figure AI.5**). Although adjacent carbons usually are connected by single bonds, they may be joined by double or triple bonds. Rings that have alternating single and double bonds, like the benzene ring, are called aromatic rings. The hydrocarbon chain or ring provides a chemically inactive skeleton to which more reactive groups of atoms may be attached. These reactive groups with specific properties are known as functional groups. They usually contain atoms of oxygen, nitrogen, phosphorus, or sulfur (**figure AI.6**) and are largely responsible for most characteristic chemical properties of organic molecules.

Organic molecules are often divided into classes based on the nature of their functional groups. Ketones have a carbonyl group within the carbon chain, whereas alcohols have a hydroxyl on the chain. Organic acids have a carboxyl group, and amines have an amino group (**figure AI.7**).

Organic molecules may have the same chemical composition and yet differ in their molecular structure and properties. Such molecules are called isomers. One important class of isomers is the stereoisomers. Stereoisomers have the same atoms arranged in the

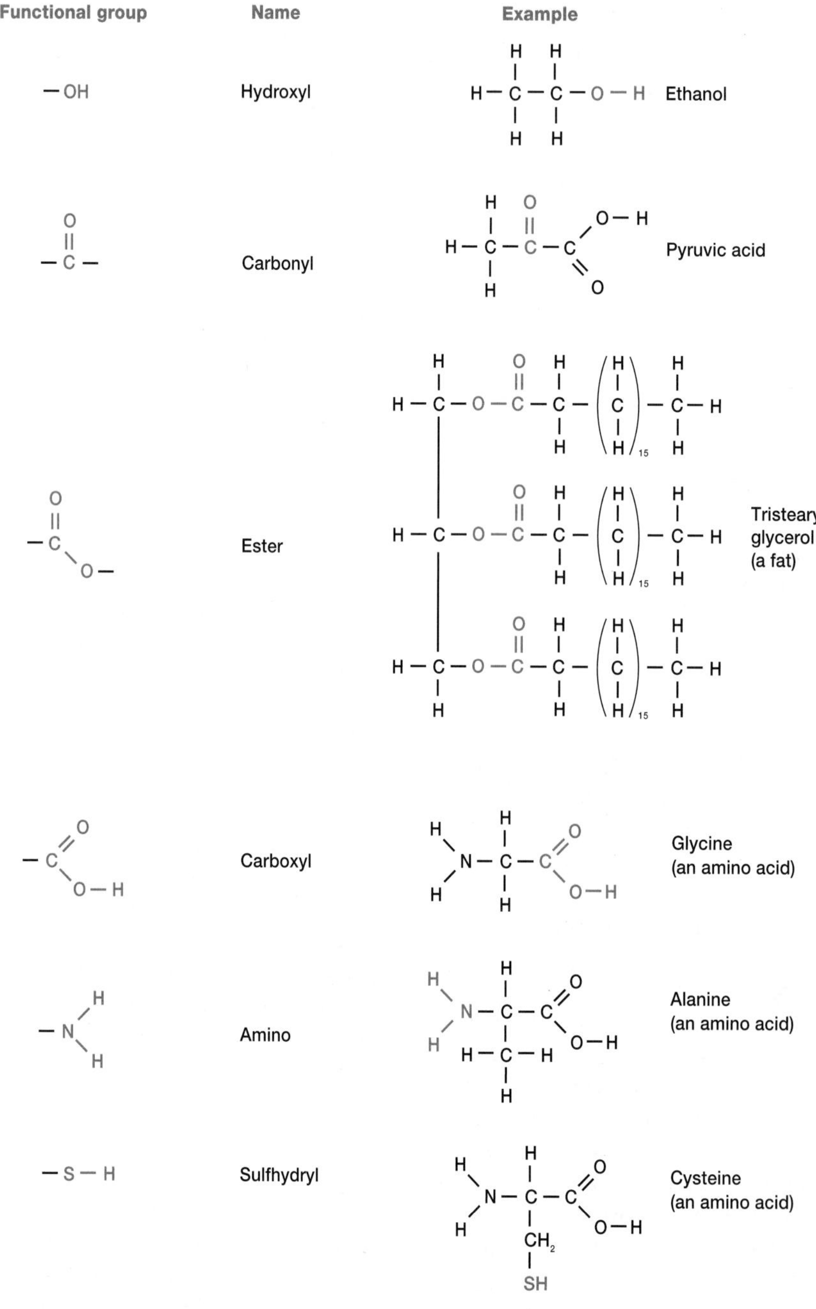

Figure AI.6 Functional Groups. Some common functional groups in organic molecules. The groups are shown in red.

Type of molecule	Example
Alcohol	CH_3-CH_2-OH
Aldehyde	$CH_3-C(=O)-H$
Amine	$CH_3-CH_2-NH_2$
Ester	$CH_3-C(=O)-O-CH_2-CH_3$
Ether	$CH_3-CH_2-O-CH_2-CH_3$
Ketone	$CH_3-C(=O)-CH_3$
Organic acid	$CH_3-C(=O)-OH$

Figure AI.7 Types of Organic Molecules. These are classified on the basis of their functional groups.

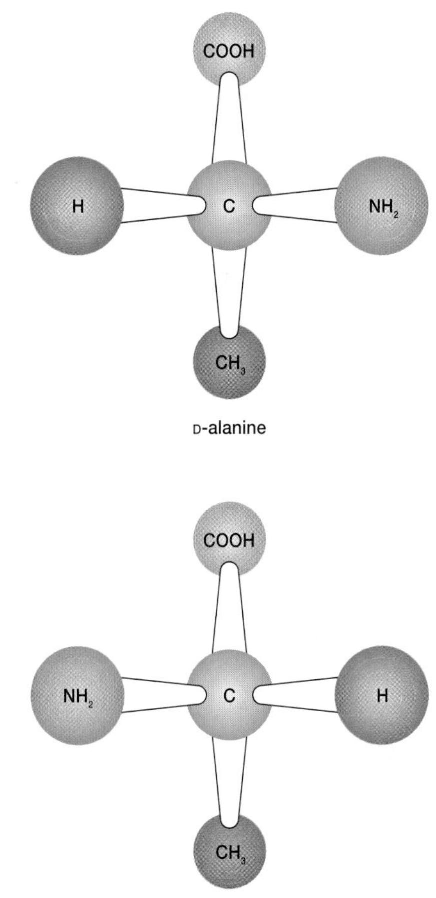

Figure AI.8 The Stereoisomers of Alanine. The α-carbon is in gray, L-alanine is the form usually present in proteins.

same nucleus-to-nucleus sequence but differ in the spatial arrangement of their atoms. For example, an amino acid such as alanine can form stereoisomers (**figure AI.8**). L-Alanine and other L-amino acids are the stereoisomer forms normally present in proteins.

Carbohydrates

Carbohydrates are aldehyde or ketone derivatives of polyhydroxy alcohols. The smallest and least complex carbohydrates are the simple sugars or monosaccharides. The most common sugars have five or six carbons (**figure AI.9**). A sugar in its ring form has two isomeric structures, the α and β forms, that differ in the orientation of the hydroxyl on the aldehyde or ketone carbon, which is called the anomeric or glycosidic carbon (**figure AI.10**). Microorganisms have many sugar derivatives in which a hydroxyl is replaced by an amino group or some other functional group (e.g., glucosamine).

Two monosaccharides can be joined by a bond between the anomeric carbon of one sugar and a hydroxyl or the anomeric carbon of the second (**figure AI.11**). The bond joining sugars is a glycosidic bond and may be either α or β depending on the orientation of the anomeric carbon. Two sugars linked in this way constitute a disaccharide. Some common disaccharides are maltose (two glucose molecules), lactose (glucose and galactose), and sucrose (glucose and fructose). If 10 or more sugars are linked together by glycosidic bonds, a polysaccharide is formed. For example, starch and glycogen are common polymers of glucose that are used as sources of carbon and energy (**figure AI.12**).

Lipids

All cells contain a heterogeneous mixture of organic molecules that are relatively insoluble in water but very soluble in nonpolar solvents such as chloroform, ether, and benzene. These molecules are called lipids. Lipids vary greatly in structure and include triacylglycerols, phospholipids, steroids, carotenoids, and many other types. Among other functions, they serve as membrane

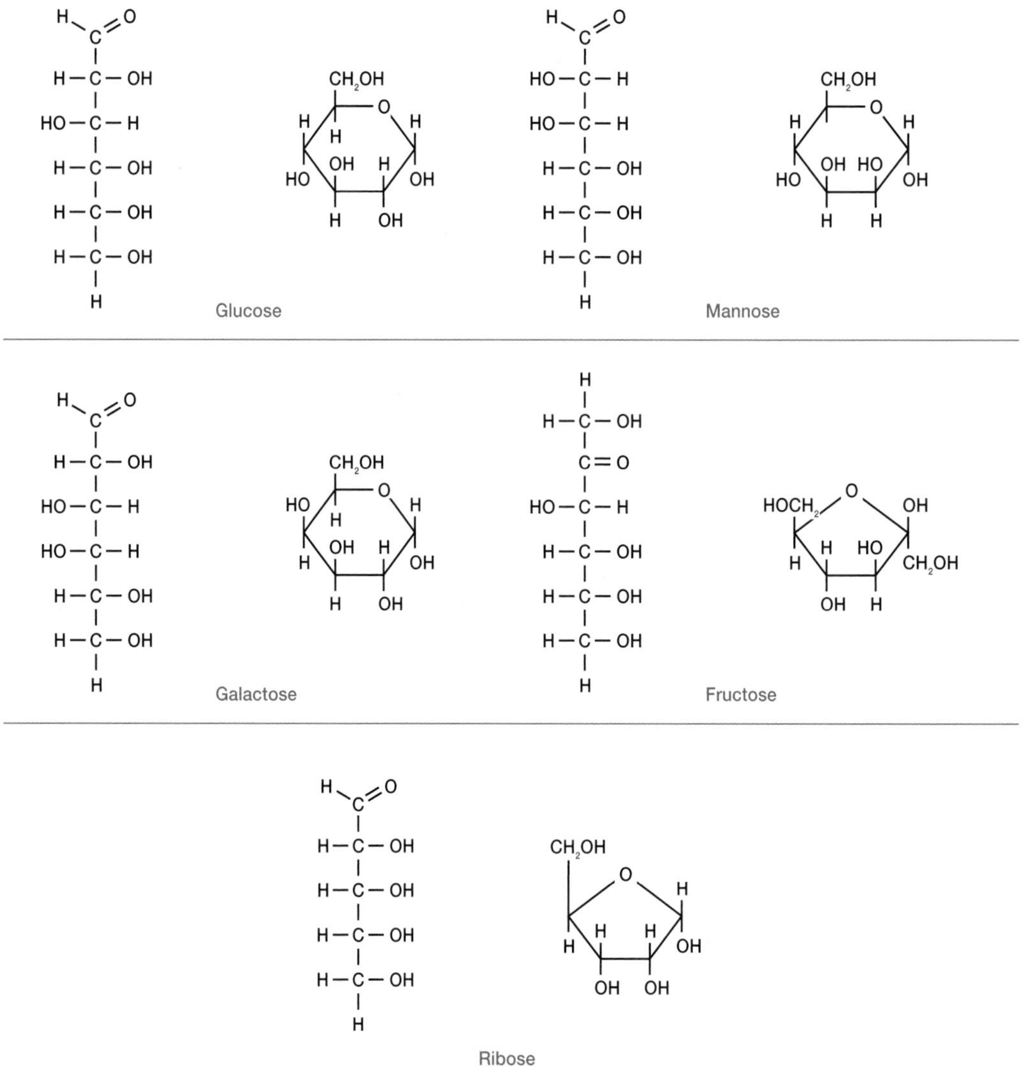

Figure AI.9 Common Monosaccharides. Structural formulas for both the open chains and the ring forms are provided.

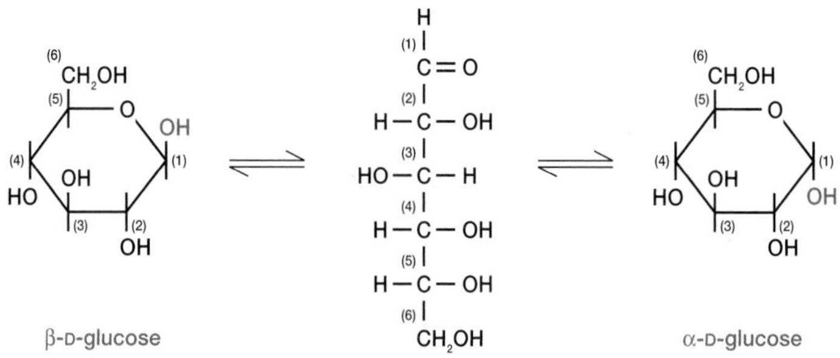

Figure AI.10 The Interconversion of Monosaccharide Structures. The open chain form of glucose and other sugars is in equilibrium with closed ring structures (depicted here with Haworth projections). Aldehyde sugars form cyclic hemiacetals, and keto sugars produce cyclic hemiketals. When the hydroxyl on carbon one of cyclic hemiacetals projects above the ring, the form is known as a β form. The α form has a hydroxyl that lies below the plane of the ring. The same convention is used in showing the α and β forms of hemiketals such as those formed by fructose.

components, storage forms for carbon and energy, precursors of other cell constituents, and protective barriers against water loss.

Most lipids contain fatty acids, which are monocarboxylic acids that often are straight chained but may be branched. Saturated fatty acids lack double bonds in their carbon chains, whereas unsaturated fatty acids have double bonds. The most common fatty acids are 16 or 18 carbons long.

Two good examples of common lipids are triacylglycerols and phospholipids. Triacylglycerols are composed of glycerol esterified to three fatty acids (**figure AI.13*a***). They are used to store carbon and energy. Phospholipids are lipids that contain at least one phosphate group and often have a nitrogenous constituent as well. Phosphatidylethanolamine is an important phospholipid frequently present in bacterial membranes (figure AI.13*b*). It is

Figure AI.11 Common Disaccharides. **(a)** The formation of maltose from two molecules of an α-glucose. The bond connecting the glucose extends between carbons one and four, and involves the α form of the anomeric carbon. Therefore, it is called an α (1 → 4) glycosidic bond. **(b)** Sucrose is composed of a glucose and a fructose joined to each other through their anomeric carbons, and αβ (1 → 2) bond. **(c)** The milk sugar lactose contains galactose and glucose joined by a β (1 → 4) glycosidic bond.

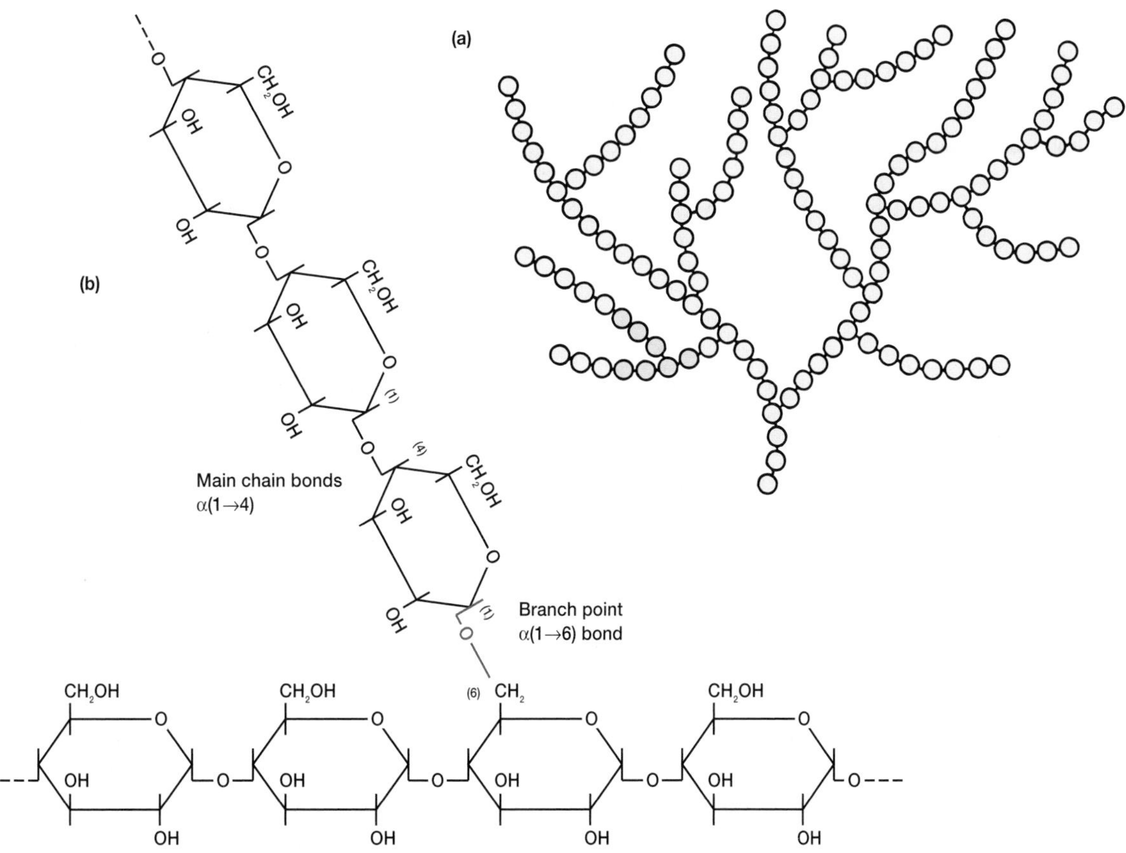

Figure AI.12 Glycogen and Starch Structure. **(a)** An overall view of the highly branched structure characteristic of glycogen and most starch. The circles represent glucose residues. **(b)** A close-up of a small part of the chain (shown in blue in part *a*) revealing a branch point with its α (1 → 6) glycosidic bond, which is colored blue.

(a)

$$
\begin{array}{l}
CH_2 - O - \overset{\displaystyle O}{\overset{\|}{C}} - R \\
| \\
CH - O - \overset{\displaystyle O}{\overset{\|}{C}} - R \\
| \\
CH_2 - O - \overset{\displaystyle O}{\overset{\|}{C}} - R
\end{array}
$$

(b)

$$
\begin{array}{l}
CH_2 - O - \overset{\displaystyle O}{\overset{\|}{C}} - R \\
| \\
CH - O - \overset{\displaystyle O}{\overset{\|}{C}} - R \\
| \\
CH_2 - O - \underset{\displaystyle O^-}{\underset{|}{\overset{\displaystyle O}{\overset{\|}{P}}}} - O - CH_2 - CH_2 - \overset{+}{N}H_3
\end{array}
$$

Figure AI.13 Examples of Common Lipids. (a) A triacylglycerol or neutral fat. **(b)** The phospholipid phosphatidylethanolamine. The R groups represent fatty acid side chains.

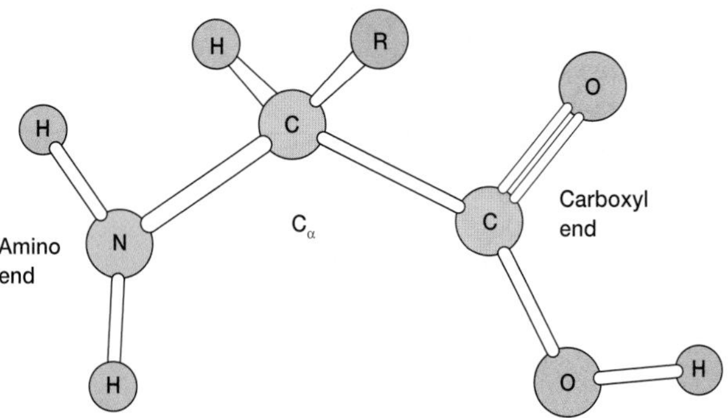

Figure AI.14 L-Amino Acid Structure. The uncharged form is shown.

composed of two fatty acids esterified to glycerol. The third glycerol hydroxyl is joined with a phosphate group, and ethanolamine is attached to the phosphate. The resulting lipid is very asymmetric with a hydrophobic nonpolar end contributed by the fatty acids and a polar, hydrophilic end. In cell membranes the hydrophobic end is buried in the interior of the membrane, while the polar-charged end is at the membrane surface and exposed to water.

Proteins

The basic building blocks of proteins are amino acids. An amino acid contains a carboxyl group and an amino group on its alpha carbon (**figure AI.14**). About 20 amino acids are normally found in proteins; they differ from each other with respect to their side chains (**figure AI.15**). In proteins, amino acids are linked together by peptide bonds between their carboxyls and α-amino groups to form linear polymers called peptides (**figure AI.16**). If a peptide contains more than 30 amino acids, it usually is called a polypeptide. Each protein is composed of one or more polypeptide chains and has a molecular weight greater than about 6,000 to 7,000.

Proteins have three or four levels of structural organization and complexity. The primary structure of a protein is the sequence of the amino acids in its polypeptide chain or chains. The structure of the polypeptide chain backbone is also considered part of the primary structure. Each different polypeptide has its own amino acid sequence that is a reflection of the nucleotide sequence in the gene that codes for its synthesis. The polypeptide chain can coil along one axis in space into various shapes like the α-helix (**figure AI.17**). This arrangement of the polypeptide in space around a single axis is called the secondary structure. Secondary structure is formed and stabilized by the interactions of amino acids that are fairly close to one another on the polypeptide chain. The polypeptide with its primary and secondary structure can be coiled or organized in space along three axes to form a more complex, three-dimensional shape (**figure AI.18**). This level of organization is the tertiary structure (**figure AI.19**). Amino acids more distant from one another on the polypeptide chain contribute to tertiary structure. Secondary and tertiary structures are examples of conformation, molecular shape that can be changed by bond rotation and without breaking covalent bonds. When a protein contains more than one polypeptide chain, each chain with its own primary, secondary, and tertiary structure associates with the other chains to form the final molecule. The way in which polypeptides associate with each other in space to form the final protein is called the protein's quaternary structure (**figure AI.20**).

The final conformation of a protein is ultimately determined by the amino acid sequence of its polypeptide chains. Under proper conditions a completely unfolded polypeptide will fold into its normal final shape without assistance.

Protein secondary, tertiary, and quaternary structure is largely determined and stabilized by many weak noncovalent forces such as hydrogen bonds and ionic bonds. Because of this, protein shape often is very flexible and easily changed. This flexibility is very important in protein function and in the regulation of enzyme activity. Because of their flexibility, however, proteins readily lose their proper shape and activity when exposed to harsh conditions. The only covalent bond commonly involved in the secondary and tertiary structure of proteins is the disulfide bond. The disulfide bond is formed when two cysteines are linked through their sulfhydryl groups. Disulfide bonds generally strengthen or stabilize protein structure but are not especially important in directly determining protein conformation.

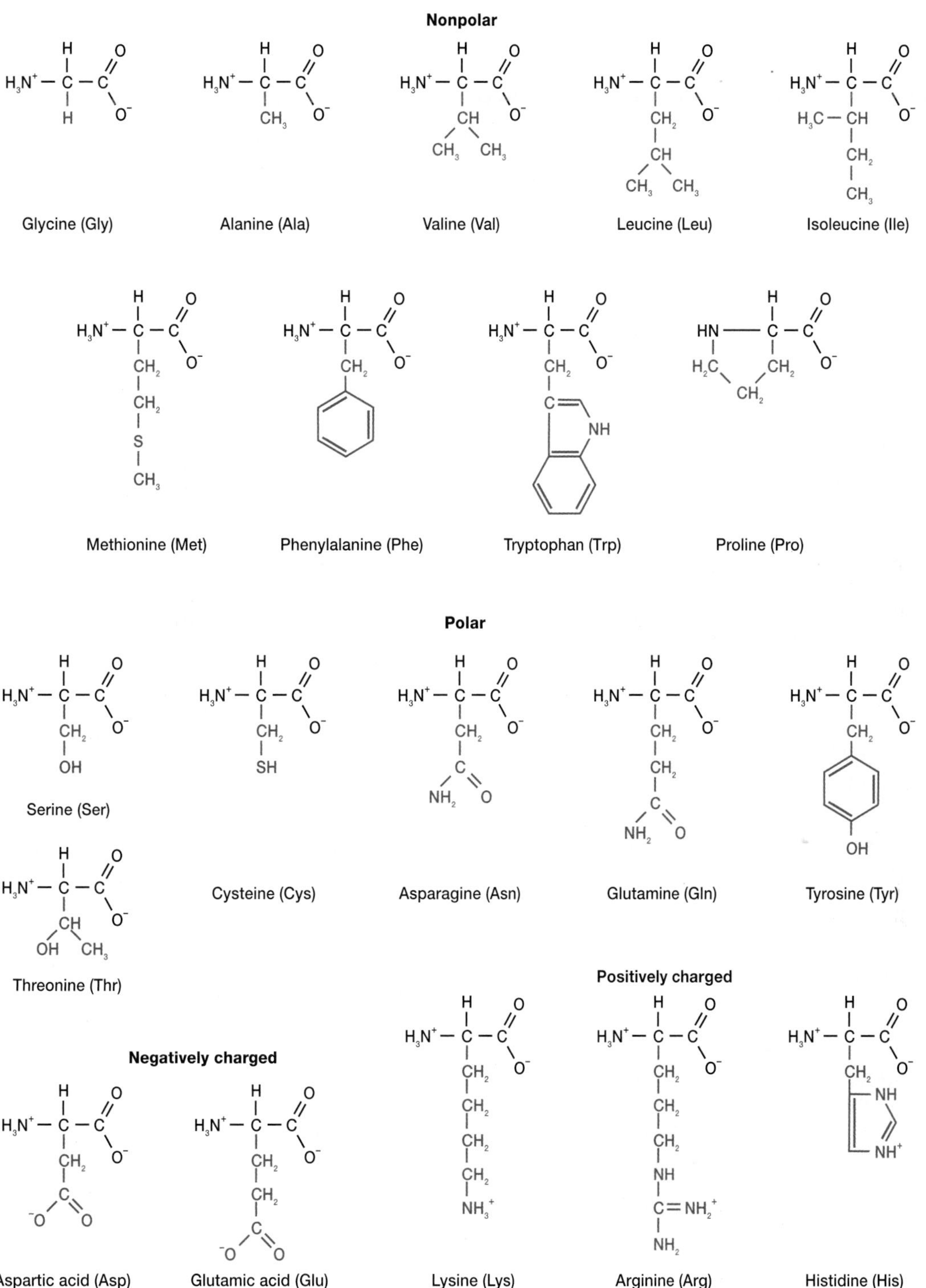

Figure AI.15 The Common Amino Acids. The structures of the α-amino acids normally found in proteins. Their side chains are shown in blue, and they are grouped together based on the nature of their side chains—nonpolar, polar, negatively charged (acid), or positively charged (basic). Proline is actually an imino acid rather than an amino acid.

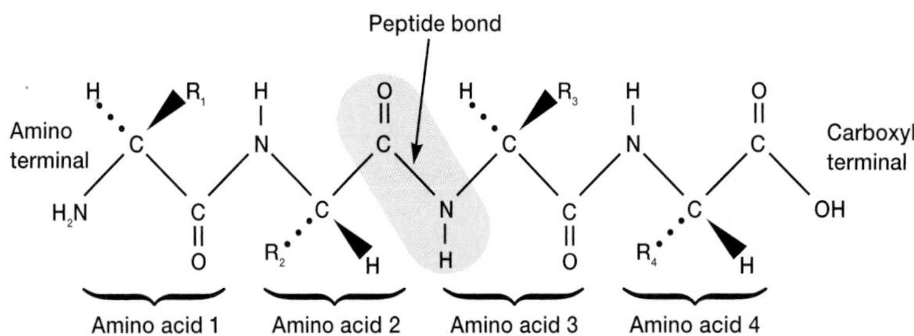

Figure AI.16 A Tetrapeptide Chain. The end of the chain with a free α-amino group is the amino or N terminal. The end with the free α-carboxyl is the carboxyl or C terminal. One peptide bond is shaded in blue.

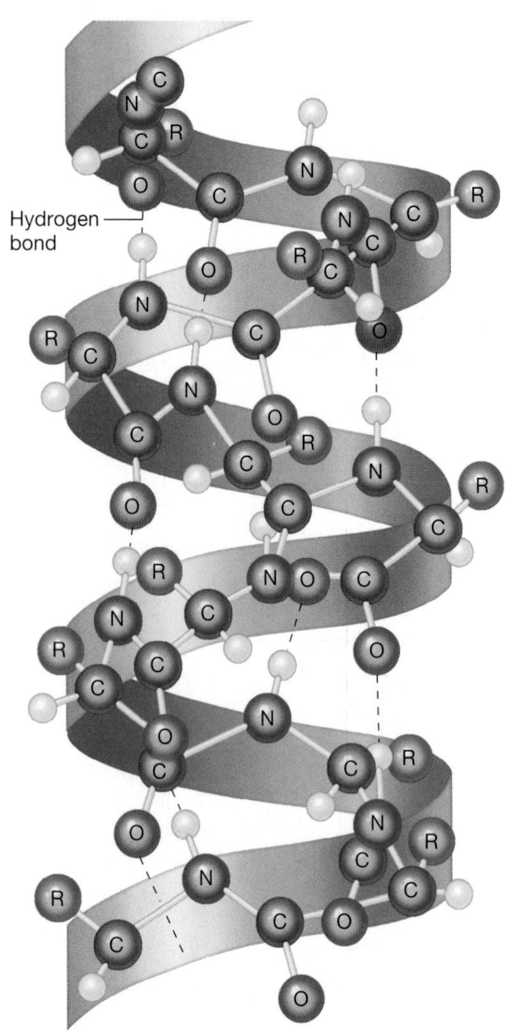

Figure AI.17 The α-Helix. A polypeptide twisted into one type of secondary structure, the α-helix. The helix is stabilized by hydrogen bonds joining peptide bonds that are separated by three amino acids.

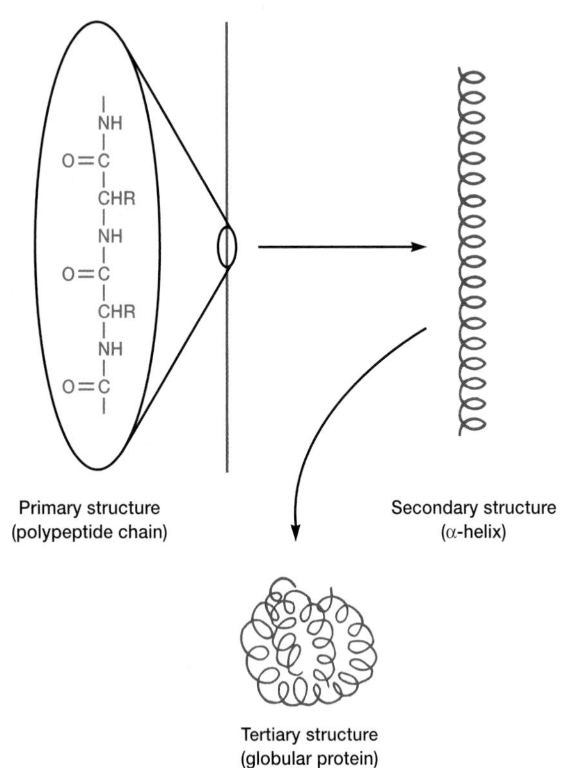

Figure AI.18 Secondary and Tertiary Protein Structures. The formation of secondary and tertiary protein structures by folding a polypeptide chain with its primary structure.

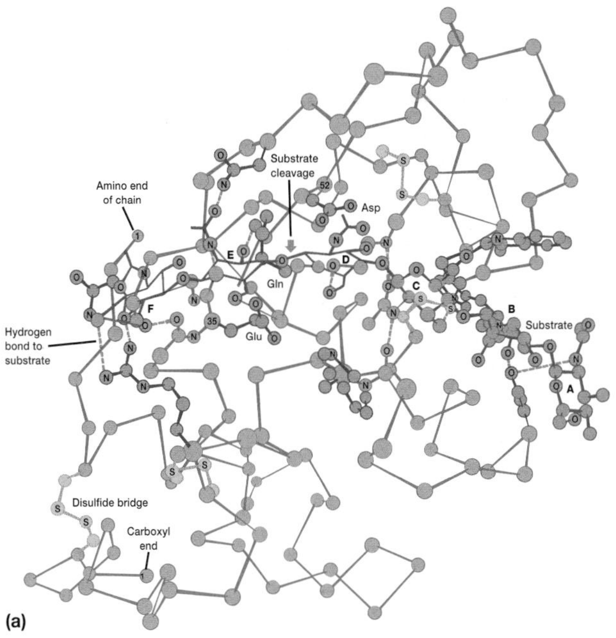

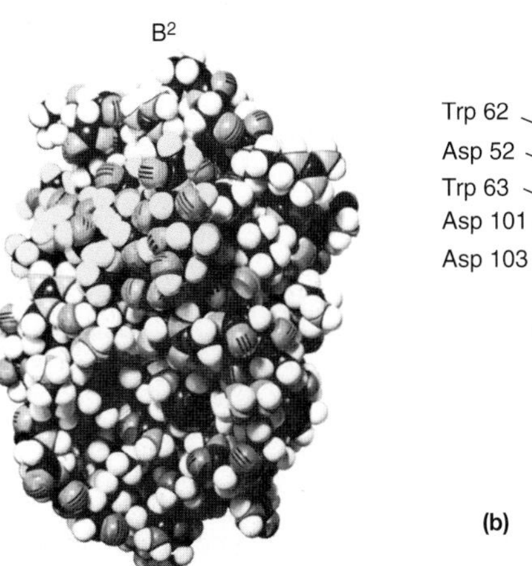

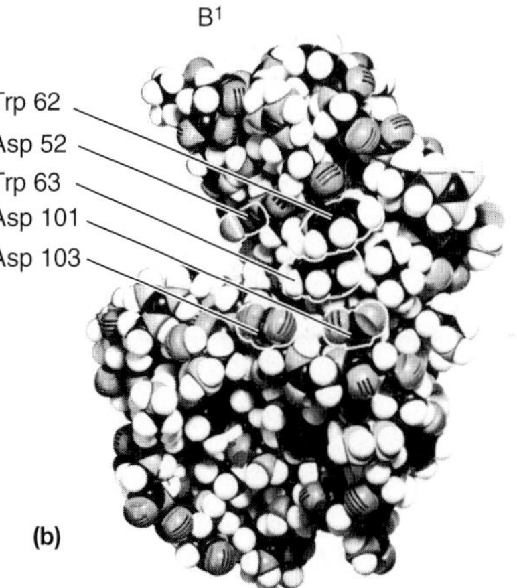

Figure AI.19 Lysozyme. The tertiary structure of the enzyme lysozyme. **(a)** A diagram of the protein's polypeptide backbone with the substrate hexasaccharide shown in blue. The point of substrate cleavage is indicated. **(b)** A space-filling model of lysozyme. The figure on the right shows the empty active site with some of its more important amino acids indicated. On the left the enzyme has bound its substrate (in pink).

Nucleic Acids

The nucleic acids, deoxyribonucleic acid (DNA) and ribonucleic acid (RNA), are polymers of deoxyribonucleosides and ribonucleosides joined by phosphate groups. The nucleosides in DNA contain the purines adenine and guanine, and the pyrimidine bases thymine and cytosine. In RNA the pyrimidine uracil is substituted for thymine. Because of their importance for genetics and molecular biology, the chemistry of nucleic acids is introduced earlier in the text. The structure and synthesis of purines and pyrimidines are discussed in chapter 10 (pp. 241–42). The structures of DNA and RNA are described in chapter 11 (pp. 252–53).

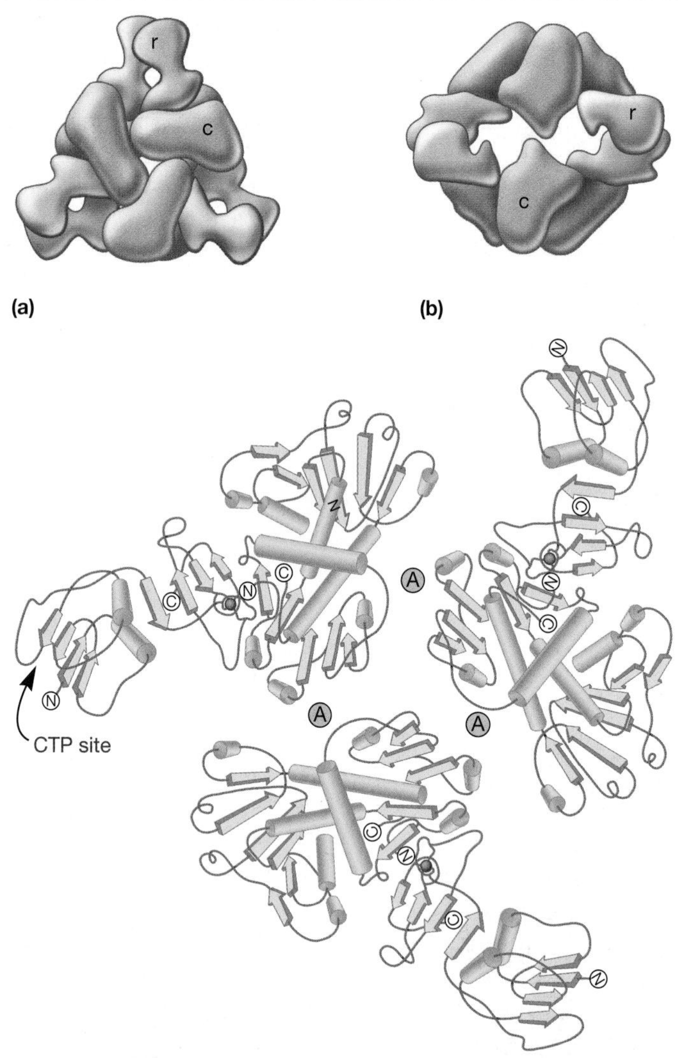

Figure AI.20 An Example of Quaternary Structure. The enzyme aspartate carbamoyltransferase from *Escherichia coli* has two types of subunits, catalytic and regulatory. The association between the two types of subunits is shown: **(a)** a top view, and **(b)** a side view of the enzyme. The catalytic (C) and regulator (r) subunits are shown in different colors. **(c)** The peptide chains shown when viewed from the top as in (a). The active sites of the enzyme are located at the positions indicated by A. (*See pp. 182–83 for more details.*) *(**a** and **b**) Adapted from Krause, et al., in* Proceedings of the National Academy of Sciences, *V. 82, 1985, as appeared in* Biochemistry, *3d edition by Lubert Stryer. Copyright © 1975, 1981, 1988. Reprinted with permission of W. H. Freeman and Company. (**c**) Adapted from Kantrowitz, et al., in* Trends in Biochemical Science, *V. 5, 1980, as appeared in* Biochemistry, *3d edition by Lubert Stryer. Copyright © 1975, 1981, 1988. Reprinted with permission of W. H. Freeman and Company.*

Appendix II
Common Metabolic Pathways

This appendix contains a few of the more important pathways discussed in the text, particularly those involved in carbohydrate catabolism. Enzyme names and final end products are given in color. Consult the text for a description of each pathway and its roles.

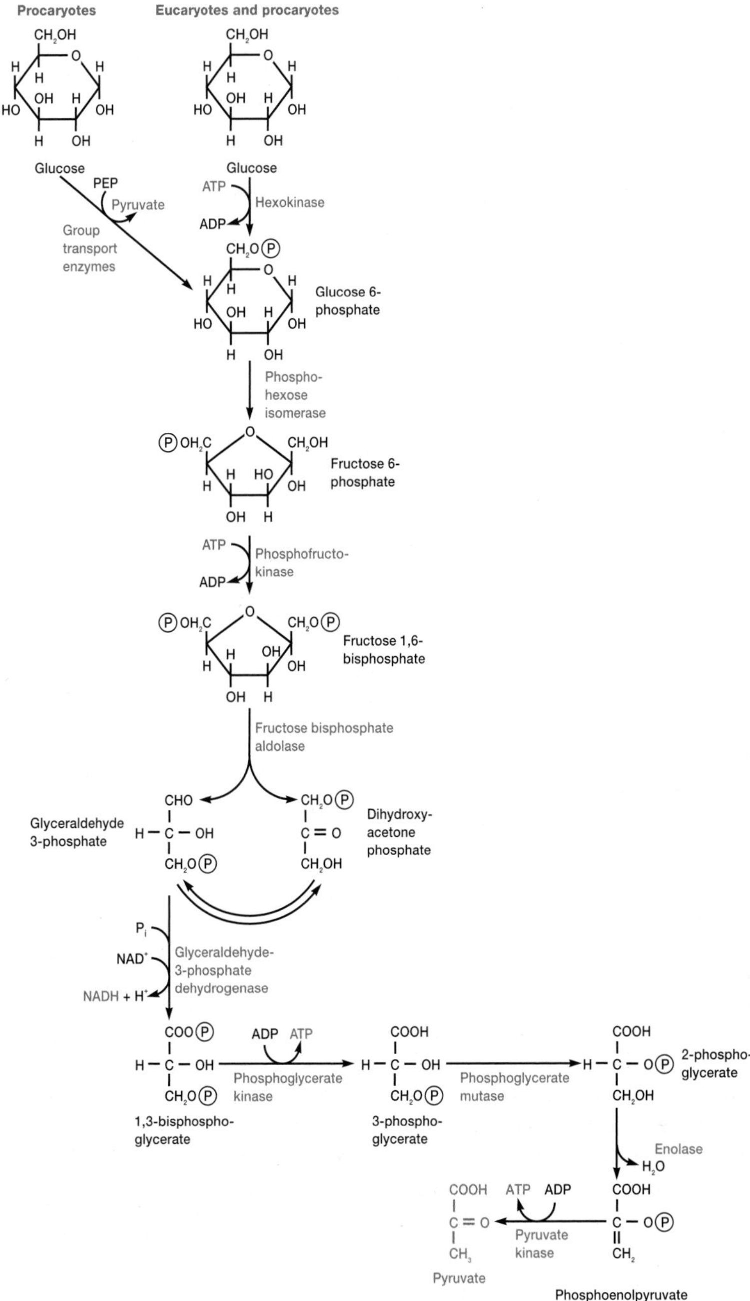

Figure AII.1 The Embden-Meyerhof pathway. This pathway converts glucose and other sugars to pyruvate and generates NADH and ATP. In some procaryotes glucose is phosphorylated to glucose 6-phosphate during group translocation transport across the plasma membrane.

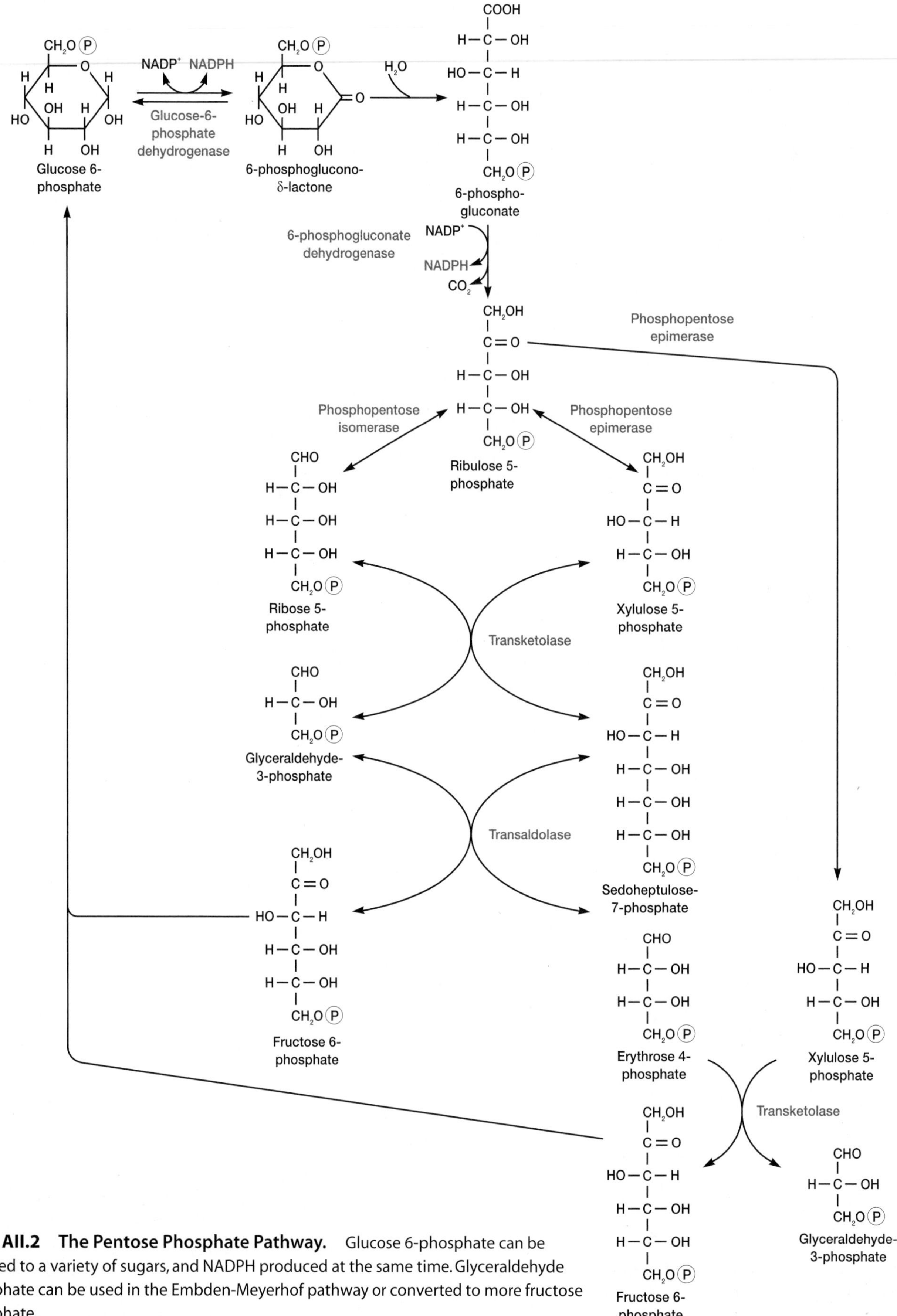

Figure AII.2 The Pentose Phosphate Pathway. Glucose 6-phosphate can be converted to a variety of sugars, and NADPH produced at the same time. Glyceraldehyde 3-phosphate can be used in the Embden-Meyerhof pathway or converted to more fructose 6-phosphate.

Glucose 6-phosphate

Glucose-6-phosphate dehydrogenase ($NADP^+$ → NADPH)

6-phosphoglucono-δ-lactone

Lactonase (H_2O)

6-phosphogluconate

6-phosphogluconate dehydrase (H_2O)

2-keto-3-deoxy-6-phosphogluconate

KDPG aldolase

Glyceraldehyde 3-phosphate

Pyruvate

Figure AII.3 The Entner-Doudoroff Pathway.

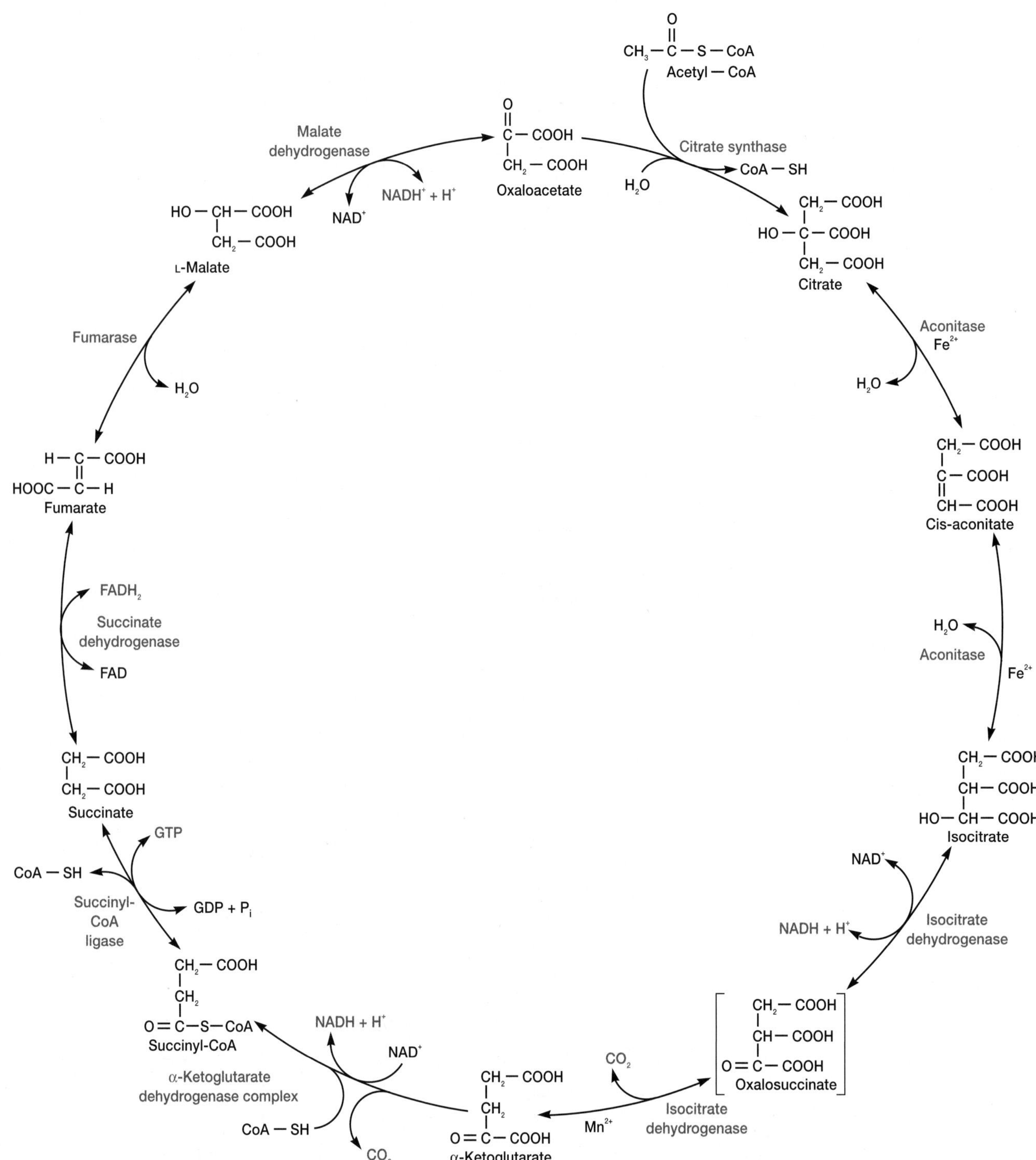

Figure AII.4 The Tricarboxylic Acid Cycle. Cis-aconitate and oxalosuccinate remain bound to aconitase and isocitrate dehydrogenase. Oxalosuccinate has been placed in brackets because it is so unstable.

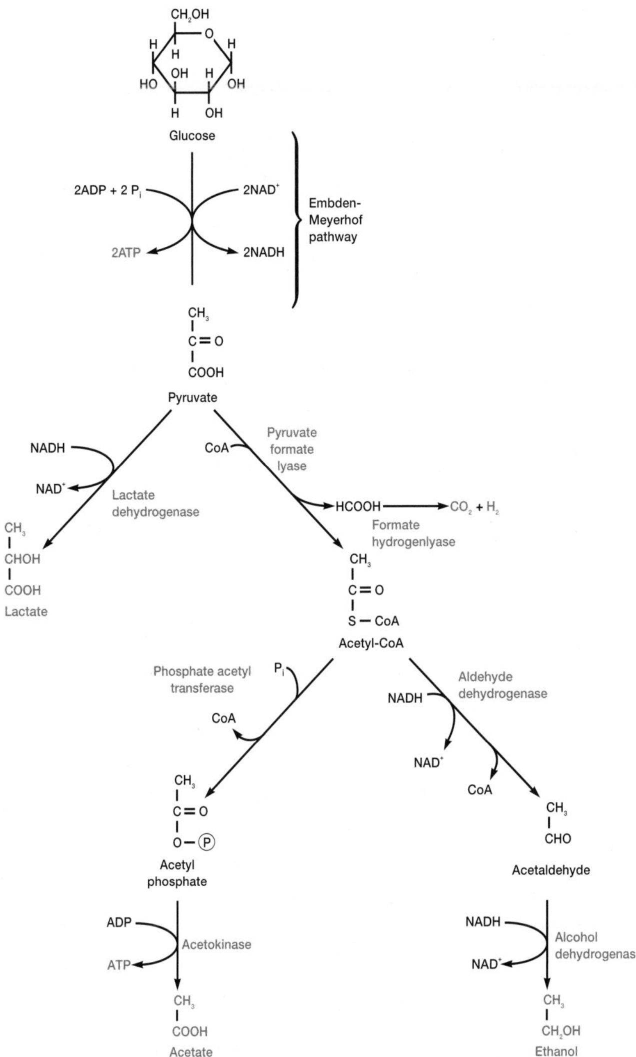

Figure AII.5 The Mixed Acid Fermentation Pathway. This pathway is characteristic of many members of the *Enterobacteriaceae* such as *E. coli*.

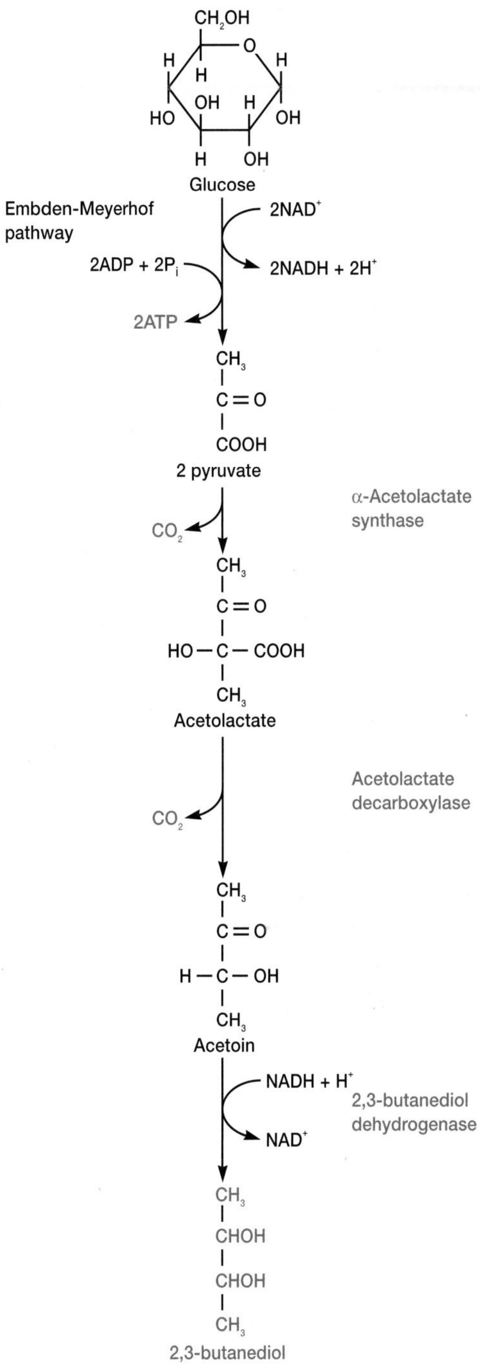

CH2OH
H
H
OH H
HO
OH
H OH
Glucose
Embden-Meyerhof pathway
2NAD+
2ADP + 2Pi
2NADH + 2H+
2ATP
CH3
C=O
COOH
2 pyruvate
α-Acetolactate synthase
CO2
CH3
C=O
HO—C—COOH
CH3
Acetolactate
Acetolactate decarboxylase
CO2
CH3
C=O
H—C—OH
CH3
Acetoin
NADH + H+
2,3-butanediol dehydrogenase
NAD+
CH3
CHOH
CHOH
CH3
2,3-butanediol

Figure AII.6 The Butanediol Fermentation Pathway. This pathway is characteristic of members of the *Enterobacteriaceae* such as *Enterobacter*. Other products may also be formed during butanediol fermentation.

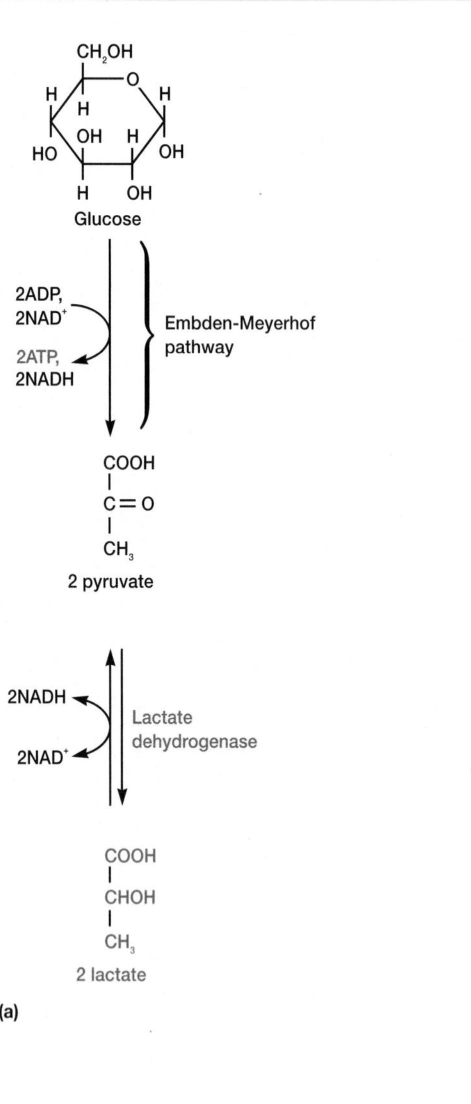

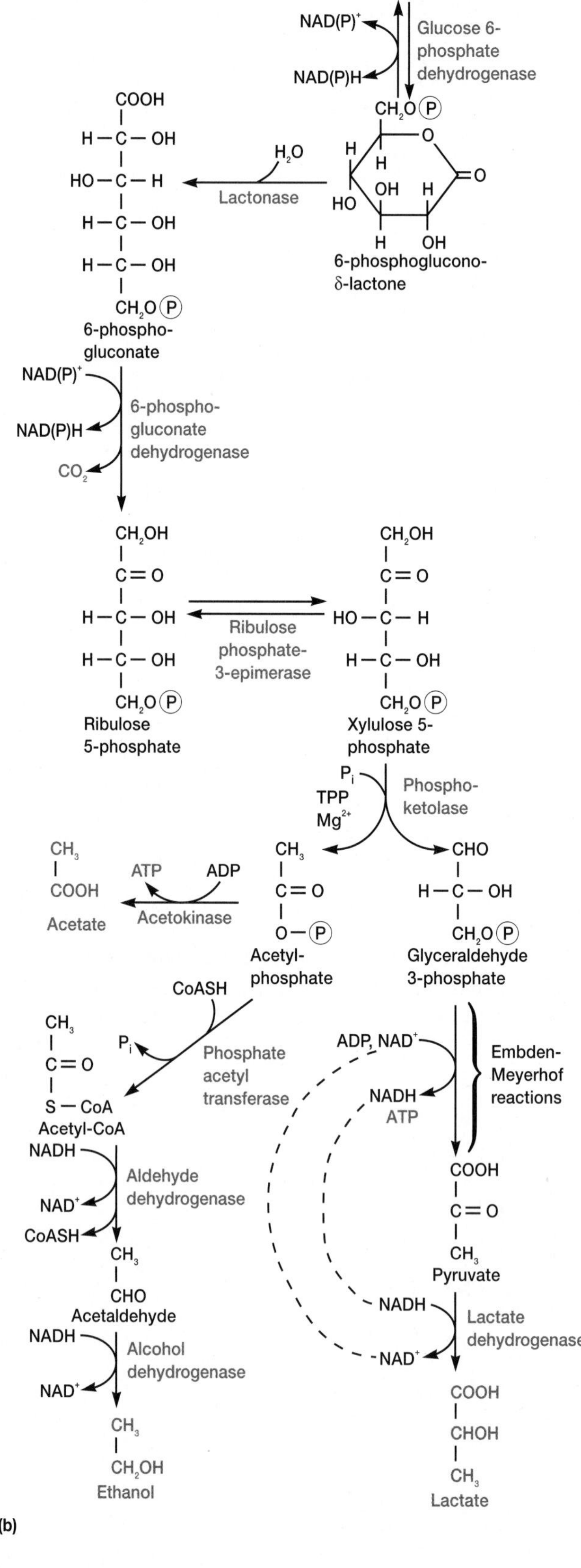

Figure AII.7 Lactic Acid Fermentations. **(a)** Homolactic fermentation pathway. **(b)** Heterolactic fermentation pathways.

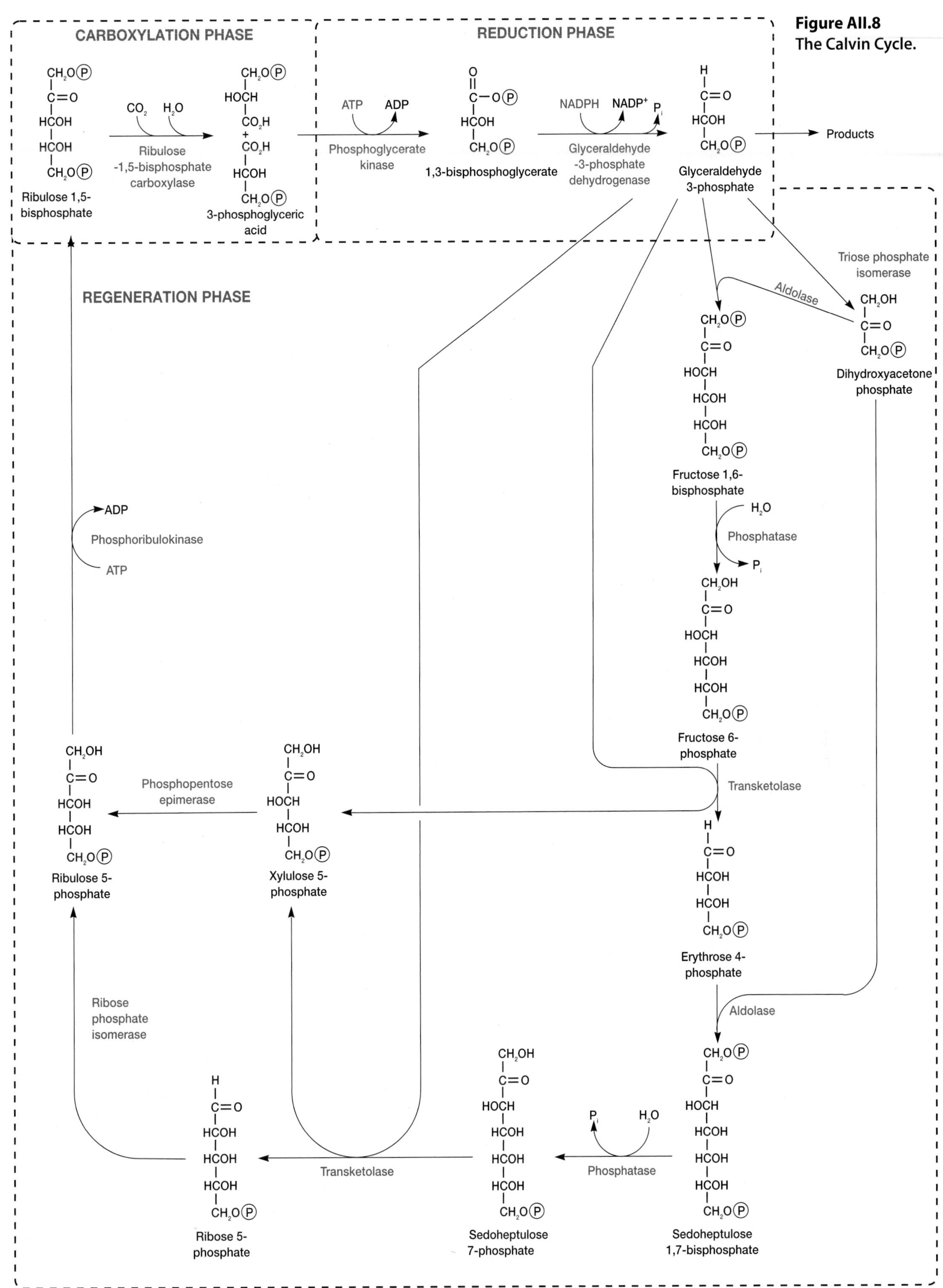

Figure AII.8
The Calvin Cycle.

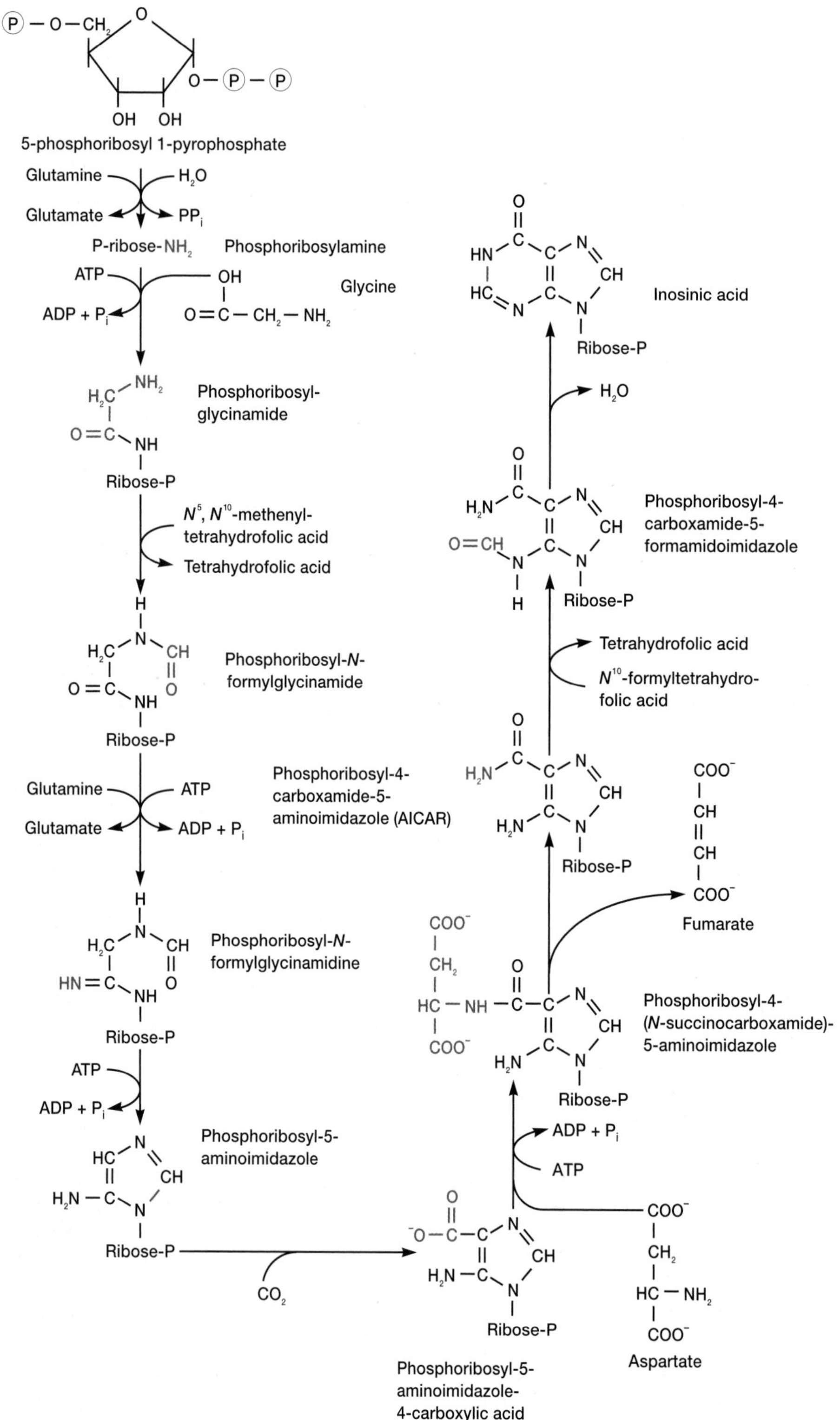

Figure AII.9 The Pathway for Purine Biosynthesis. Inosinic acid is the first purine end product. The purine skeleton is constructed while attached to a ribose phosphate.

Glossary

Pronunciation Guide

Many of the boldface terms in this glossary are followed by a phonetic spelling in parentheses. These pronunciation aids usually come from *Dorland's Illustrated Medical Dictionary.* The following rules are taken from this dictionary and will help in using its phonetic spelling system.

1. An unmarked vowel ending a syllable (an open syllable) is long; thus *ma* represents the pronunciation of *may; ne,* that of *knee; ri,* of *wry; so,* of *sew; too,* of *two;* and *vu,* of *view.*
2. An unmarked vowel in a syllable ending with a consonant (a closed syllable) is short; thus *kat* represents *cat; bed, bed; hit, hit; not, knot; foot, foot;* and *kusp, cusp.*
3. A long vowel in a closed syllable is indicated by a macron; thus *māt* stands for *mate; sēd,* for *seed; bīl,* for *bile; mōl,* for *mole; fūm,* for *fume;* and *fo͞ol,* for *fool.*
4. A short vowel that ends or itself constitutes a syllable is indicated by a breve; thus *ĕ-fekt′* for *effect, ĭ-mūn′* for *immune,* and *ŭ-klo͞od′* for *occlude.*

Primary (′) and secondary (″) accents are shown in polysyllabic words. Unstressed syllables are followed by hyphens.

Some common vowels are pronounced as indicated here.

ə	sofa	ĕ	met	ŏ	got
ā	mate	ī	bite	ū	fuel
ă	bat	ĭ	bit	ŭ	but
ē	beam	ō	home		

From Dorland's Illustrated Medical Dictionary. *Copyright © 1988 W. B. Saunders, Philadelphia, Pa. Reprinted by permission.*

A

AB toxins The structure and activity of many exotoxins based on the AB model. In this model, the B portion of the toxin is responsible for toxin binding to a cell but does not directly harm it. The A portion enters the cell and disrupts its function. (824)

ABC protein secretion pathway Transport systems that use ATP hydrolysis to drive translocation across the plasma membrane. When used for nutrient uptake, usually called ATP-binding cassette transport systems. (65)

accessory pigments Photosynthetic pigments such as carotenoids and phycobiliproteins that aid chlorophyll in trapping light energy. (217)

acellular slime mold Chemoorganotrophic protists with a distinctive life cycle that includes the streaming of protoplasm that moves in an amoeboid fashion. Cells within the multinucleate mass (called a plasmodium) lack cell walls. Also called *Myxogastria,* and were formerly considered fungi. (614)

acetyl-CoA pathway A biochemical pathway used by methanogens to fix CO_2. It is also used by acetogens to generate acetic acid. (506)

acetyl-coenzyme A (acetyl-CoA) A combination of acetic acid and coenzyme A that is energy rich; it is produced by many catabolic pathways and is the substrate for the tricarboxylic acid cycle, fatty acid biosynthesis, and other pathways. (198)

acid fast Refers to bacteria like the mycobacteria that cannot be easily decolorized with acid alcohol after being stained with dyes such as basic fuchsin. (26, 596)

acid-fast staining A staining procedure that differentiates between bacteria based on their ability to retain a dye when washed with an acid alcohol solution. (26)

acidic dyes Dyes that are anionic or have negatively charged groups such as carboxyls. (26)

acidophile (as′id-o-fīl″) A microorganism that has its growth optimum between about pH 0 and 5.5. (134)

acquired enamel pellicle A membranous layer on the tooth enamel surface formed by selectively adsorbing glycoproteins (mucins) from saliva. This pellicle confers a net negative charge to the tooth surface. (991)

acquired immune deficiency syndrome (AIDS) An infectious disease syndrome caused by the human immunodeficiency virus and is characterized by the loss of a normal immune response, followed by increased susceptibility to opportunistic infections and an increased risk of some cancers. (925)

acquired immune tolerance The ability to produce antibodies against nonself antigens while "tolerating" (not producing antibodies against) self-antigens. (802)

acquired immunity Refers to the type of specific (adaptive) immunity that develops after exposure to a suitable antigen or is produced after antibodies are transferred from one individual to another. (776)

actinobacteria (ak″tĭ-no-bak-tēr-e-ah) A group of gram-positive bacteria containing the actinomycetes and their high G + C relatives. (593)

actinomycete (ak″tĭ-no-mi′sēt) An aerobic, gram-positive bacterium that forms branching filaments (hyphae) and asexual spores. (589)

actinorhizae Associations between actinomycetes and plant roots. (704)

activated sludge Solid matter or sediment composed of actively growing microorganisms that participate in the aerobic portion of a biological sewage treatment process. The microbes readily use dissolved organic substrates and transform them into additional microbial cells and carbon dioxide. (1055)

activation energy The energy required to bring reacting molecules together to reach the transition state in a chemical reaction. (177)

active carrier An individual who has an overt clinical case of a disease and who can transmit the infection to others. (891)

active immunization The induction of active immunity by natural exposure to a pathogen or by vaccination. (778)

active site The part of an enzyme that binds the substrate to form an enzyme-substrate complex and catalyze the reaction. Also called the catalytic site. (177)

active transport The transport of solute molecules across a membrane against an electrochemical gradient; it requires a carrier protein and the input of energy. (107)

acute carrier *See* casual carrier.

acute infections Virus infections with a fairly rapid onset that last for a relatively short time. (461)

acute viral gastroenteritis An inflammation of the stomach and intestines, normally caused by Norwalk viruses, Noroviruses, adenoviruses, other caliciviruses, rotaviruses, and astroviruses. (939)

acyclovir (a-si′klo-vir) A synthetic purine nucleoside derivative with antiviral activity against herpes simplex virus. (856)

adaptive mutation Defined by the phenomenon observed in bacteria grown under a specific stress; such bacteria sometimes develop mutations that enable their survival at a higher rate than predicted by the natural mutation rate. (319, 1062)

adenine (ad′e-nēn) A purine derivative, 6-aminopurine, found in nucleosides, nucleotides, coenzymes, and nucleic acids. (241)

adenine arabinoside or **vidarabine** An antiviral agent used especially to treat keratitis and encephalitis caused by the herpes simplex virus. (855)

adenosine diphosphate (ADP) (ah-den′o-sēn) The nucleoside diphosphate usually formed upon the breakdown of ATP when it provides energy for work. (171)

adenosine 5′-triphosphate (ATP) The triphosphate of the nucleoside adenosine, which is a high energy molecule or has high phosphate group transfer potential and serves as the cell's major form of energy currency. (171)

adhesin (ad-he′zin) A molecular component on the surface of a microorganism that is involved in adhesion to a substratum or cell. Adhesion to a specific host tissue usually is a preliminary stage in pathogenesis, and adhesins are important virulence factors. (820)

adjuvant (aj′ə-vənt) Material added to an antigen to increase its immunogenicity. Common examples are alum, killed *Bordetella pertussis,* and an oil emulsion of the antigen, either alone (Freund's incomplete adjuvant) or with killed mycobacteria (Freund's complete adjuvant). (901)

adult T-cell leukemia A type of white blood cell cancer caused by the HTLV-1 virus. (935)

aerial mycelium The mat of hyphae formed by actinomycetes that grows above the substrate, imparting a fuzzy appearance to colonies. (589)

aerobe (a′er-ōb) An organism that grows in the presence of atmospheric oxygen. (139)

aerobic anoxygenic photosynthesis Photosynthetic process in which electron donors such as organic matter or sulfide, which do not result in oxygen evolution, are used under aerobic conditions. (650)

aerobic respiration (res″pĭ-ra′shun) A metabolic process in which molecules, often organic, are oxidized with oxygen as the final electron acceptor. (205)

aerotolerant anaerobes Microbes that grow equally well whether or not oxygen is present. (139)

aflatoxin (af″lah-tok′sin) A polyketide secondary fungal metabolite that can cause cancer. (1027)

agar (ahg′ar) A complex sulfated polysaccharide, usually from red algae, that is used as a solidifying agent in the preparation of culture media. (111)

agglutinates The visible aggregates or clumps formed by an agglutination reaction. (799, 876)

agglutination reaction (ah-gloo″tĭ-na′shun) The formation of an insoluble immune complex by the cross-linking of cells or particles. (799)

agglutinin (ah-gloo″tĭ-nin) The antibody responsible for an agglutination reaction. (799)

AIDS *See* acquired immune deficiency syndrome.

airborne transmission The type of infectious organism transmission in which the pathogen is truly suspended in the air and travels over a meter or more from the source to the host. (892)

akinetes Specialized, nonmotile, dormant, thick-walled resting cells formed by some cyanobacteria. (525)

alcoholic fermentation A fermentation process that produces ethanol and CO_2 from sugars. (208)

alga (al′gah) A common term for a series of unrelated groups of photosynthetic eucaryotic microorganisms lacking multicellular sex organs (except for the charophytes) and conducting vessels. Most are now considered protists. (605)

algicide (al′jĭ-sīd) An agent that kills algae. (151)

alkalophile A microorganism that grows best at pHs from about 8.5 to 11.5. (134)

allele An alternative form of a gene. (329)

allergen (al′er-jen) An antigen that induces an allergic response. (803)

allergic contact dermatitis An allergic reaction caused by haptens that combine with proteins in the skin to form the allergen that produces the immune response. (808)

allergy (al′er-je) *See* hypersensitivity.

allochthonous (ăl′ək-thə-nəs) Substances not native to a given environment. Nutrient influx into freshwater ecosystems (e.g., lakes and streams) is often of terrestrial, or allochthonous, origin. (682)

allograft (al′o-graft) A transplant between genetically different individuals of the same species. (810)

allosteric enzyme (al″o-ster′ik) An enzyme whose activity is altered by the noncovalent binding of a small effector or modulator molecule at a regulatory site separate from the catalytic site; effector binding causes a conformational change in the enzyme and its catalytic site, which leads to enzyme activation or inhibition. (181)

allotype Allelic variants of antigenic determinant(s) found on antibody chains of some, but not all, members of a species, which are inherited as simple Mendelian traits. (791)

alpha hemolysis A greenish zone of partial clearing around a bacterial colony growing on blood agar. (584, 828)

Alphaproteobacteria One of the five classes of proteobacteria, each with distinctive 16S rRNA sequences. This group contains most of the oligotrophic proteobacteria; some have unusual metabolic modes such as methylotrophy, chemolithotrophy, and nitrogen fixing ability. Many have distinctive morphological features. (540)

alternative complement pathway An antibody-independent pathway of complement activation that includes the C3–C9 components of the classical pathway and several other serum protein factors (e.g., factor B and properdin). (764)

alveolar macrophage A vigorously phagocytic macrophage located on the epithelial surface of the lung alveoli where it ingests inhaled particulate matter and microorganisms. (761)

amantadine (ah-man′tah-den) An antiviral agent used to prevent type A influenza infections. (855)

amebiasis (amebic dysentery) (am″e-bi′ah-sis) An infection with amoebae, often resulting in dysentery; usually it refers to an infection by *Entamoeba histolytica.* (1012)

amensalism (a-men′səl-iz-əm) A relationship in which the product of one organism has a negative effect on another organism. (732)

American trypanosomiasis *See* trypanosomiasis.

Ames test A test that uses a special *Salmonella* strain to test chemicals for mutagenicity and potential carcinogenicity. (325)

amino acid activation The initial stage of protein synthesis in which amino acids are attached to transfer RNA molecules. (276)

aminoacyl or **acceptor site (A site)** The ribosomal site that contains an aminoacyl-tRNA at the beginning of the elongation cycle during protein synthesis; the growing peptide chain is transferred to the aminoacyl-tRNA and lengthens by an amino acid. (284)

aminoglycoside antibiotics (am″ĭ-no-gli′ko-sīd) A group of antibiotics synthesized by *Streptomyces* and *Micromonospora,* which contain a cyclohexane ring and amino sugars; all aminoglycoside antibiotics bind to the small ribosomal subunit and inhibit protein synthesis. (845)

amoeboid movement Moving by means of cytoplasmic flow and the formation of pseudopodia (temporary cytoplasmic protrusions). (613)

amphibolic pathways (am″fe-bol′ik) Metabolic pathways that function both catabolically and anabolically. (194)

amphitrichous (am-fit′rĕ-kus) A cell with a single flagellum at each end. (67)

amphotericin B (am″fo-ter′i-sin) An antibiotic from a strain of *Streptomyces nodosus* that is used to treat systemic fungal infections; it also is used topically to treat candidiasis. (854)

anabolism (ah-nab′o-lizm″) The synthesis of complex molecules from simpler molecules with the input of energy. (168)

anaerobe (an-a′er-ōb) An organism that grows in the absence of free oxygen. (139)

anaerobic digestion (an″a-er-o′bik) The microbiological treatment of sewage wastes under anaerobic conditions to produce methane. (1056)

anaerobic respiration An energy-yielding process in which the electron transport chain acceptor is an inorganic molecule other than oxygen. (205)

anagenesis Changes in gene frequencies and distribution among species; the accumulation of small genetic changes within a population that introduces genetic variability but are not enough to result in either speciation or extinction. (477)

anammox reaction The coupled use of nitrite as an electron acceptor and ammonium ion as an electron donor under anaerobic conditions to yield nitrogen gas. (649)

anamnestic response (an″am-nes′tik) The recall, or the remembering, by the immune system of a prior response to a given antigen. (774)

anaphylaxis (an″ah-fĭ-lak′sis) An immediate (type I) hypersensitivity reaction following exposure of a sensitized individual to the appropriate antigen. Mediated by reagin antibodies, chiefly IgE. (803)

anaplasia The reversion of an animal cell to a more primitive, undifferentiated state. (460)

anaplerotic reactions (an′ah-plĕ-rot′ik) Reactions that replenish depleted tricarboxylic acid cycle intermediates. (239)

anergy (an′ər-je) A state of unresponsiveness to antigens. Absence of the ability to generate a sensitivity reaction to substances that are expected to be antigenic. (803)

anisogamy (ăn-ī-sŏg′ə-mē) In the sexual reproduction of certain protists, the union of gametes that are different in morphology or physiology. (609)

annotation The process of determining the location and potential function of specific genes and genetic elements in a genome sequence. (388)

anogenital condylomata (venereal warts) (kon″dĭ-lo″ mah-tah) Warts that are sexually transmitted and caused by types 6, 11, and 42 human papillomavirus. Usually occur around the cervix, vulva, perineum, anus, anal canal, urethra, or glans penis. (938)

anoxic (ə-nok′sik) Without oxygen present. (668)

anoxygenic photosynthesis Photosynthesis that does not oxidize water to produce oxygen; a form of photosynthesis characteristic of purple and green photosynthetic bacteria and heliobacteria. (218, 521)

antheridium (an″ther-id′e-um; pl., **antheridia**) A male gamete-producing organ, which may be unicellular or multicellular. (638)

anthrax (an′thraks) An infectious disease of warm-blooded animals, especially of cattle and sheep, caused by *Bacillus anthracis.* The disease can be transmitted to humans through contact with bacteria or spore-contaminated animal substances, such as hair, feces, or hides; and ingestion or inhalation of spores. Cutaneous anthrax is characterized by skin lesions that become necrotic. Inhalation anthrax is almost always fatal. (987)

antibiotic (an″tĭ-bi-ot′ik) A microbial product or its derivative that kills susceptible microorganisms or inhibits their growth. (164, 835)

antibody (immunoglobulin) (an′tĭ-bod″e) A glycoprotein made by plasma cells (mature B cells) in response to the introduction of an antigen. The antibody-binding region of an antibody molecule assumes a final configuration that is the three-dimensional mirror image of the antigen that stimulated its synthesis. Thus, the antibody can bind to the antigen with exact specificity. (744, 789)

antibody affinity The strength of binding between an antigen and an antibody. (774)

antibody-dependent cell-mediated cytotoxicity (ADCC) The killing of antibody-coated target cells by cells with Fc receptors that recognize the Fc region of the bound antibody. Most ADCC is mediated by NK cells that have the Fc receptor or CD16 on their surface. (748)

antibody-mediated immunity *See* humoral immunity.

antibody titer An approximation of the antibody concentration required to react with an antigen. (795)

anticodon triplet The base triplet on a tRNA that is complementary to the triplet codon on mRNA. (277)

antigen (an′tĭ-jen) A foreign (nonself) substance (such as a protein, nucleoprotein, polysaccharide, or sometimes a glycolipid) to which lymphocytes respond; also known as an immunogen because it induces the immune response. (744, 774)

antigen-binding fragment (Fab) "Fragment antigen binding." A monovalent antigen-binding fragment of an immunoglobulin molecule that consists of one light chain and part of one heavy chain, linked by interchain disulfide bonds. (790)

antigen processing The hydrolytic digestion of antigens to produce antigen fragments. Antigen fragments are often collected by the class I or class II MHC molecules and presented on the surface of a cell. Antigen processing can occur by proteasome action on antigens that entered a cell by means other than phagocytosis. This is known as endogenous antigen processing. Antigen processing can also occur during phagocytosis as antigens are degraded within the phagolysosome. This is known as exogenous antigen processing. (780)

antigenic determinant site (epitope) The molecular configuration of the variable region of an antibody molecule that interacts with the epitope of an antigen. (774)

antigenic drift A small change in the antigenic character of an organism that allows it to avoid attack by the immune system. (890, 916)

antigenic shift A major change in the antigenic character of an organism that makes it unrecognized by host immune mechanisms. (890, 916)

antigen-presenting cells Antigen-presenting cells (APCs) are cells that take in protein antigens, process them, and present antigen fragments to B cells and T cells in conjunction with class II MHC molecules so that the cells are activated. Macrophages, B cells, dendritic cells, and Langerhans cells may act as APCs. (780)

antimetabolite (an″tĭ-mĕ-tab′o-līt) A compound that blocks metabolic pathway function by competitively inhibiting a key enzyme's use of a metabolite because it closely resembles the normal enzyme substrate. (846)

antimicrobial agent An agent that kills microorganisms or inhibits their growth. (152)

antisense RNA A single-stranded RNA with a base sequence complementary to a segment of another RNA molecule that can specifically bind to the target RNA and alter its activity. (305)

antisepsis (an″tĭ-sep′sis) The prevention of infection or sepsis. (151)

antiseptic (an″tĭ-sep′tik) Chemical agents applied to tissue to prevent infection by killing or inhibiting pathogens. (151)

antiserum (an″tĭ-se′rum) Serum containing induced antibodies. (795)

antitoxin (an″tĭ-tok′sin) An antibody to a microbial toxin, usually a bacterial exotoxin, that combines specifically with the toxin, in vivo and in vitro, neutralizing the toxin. (799, 824)

apical complex (ap′ĭ-kal) A set of organelles characteristic of members of the protist subdivision Apicomplexa: polar rings, subpellicular microtubules, conoid, rhoptries, and micronemes. (619)

apicomplexan (a′pĭ-kom-plek′san) A protist that lacks special locomotor organelles but has an apical complex and a spore-forming stage. It is either an intra- or extracellular parasite of animals; a member of the subdivision Apicomplexa. (619)

apoenzyme (ap″o-en′zīm) The protein part of an enzyme that also has a nonprotein component. (176)

apoptosis (ap″o-to′sis) Programmed cell death. The fragmentation of a cell into membrane-bound particles that are eliminated by phagocytosis. Apoptosis is a physiological suicide mechanism that preserves homeostasis and occurs during normal tissue turnover. It causes cell death in pathological circumstances, such as exposure to low concentrations of xenobiotics and infections by HIV and various other viruses. (819)

aporepressor An inactive form of the repressor protein, which becomes the active repressor when the corepressor binds to it. (295)

appressorium A flattened region of hypha found in some plant-infecting fungi that aids in penetrating the host plant cell wall. Occurs in both pathogenic fungi and nonpathogenic mycorrhizal fungi. (640)

arbuscular mycorrhizal (AM) fungi The mycorrhizal fungi in a symbiotic fungus-root association that penetrate the outer layer of the root, grow intracellularly, and form characteristic much-branched hyphal structures called arbuscules. (698)

arbuscules Branched, treelike structures formed in cells of plant roots colonized by endotrophic mycorrhizal fungi. (699)

Archaea The domain that contains procaryotes with isoprenoid glycerol diether or diglycerol tetraether lipids in their membranes and archaeal rRNA (among many differences). (3, 503)

artificially acquired active immunity The type of immunity that results from immunizing an animal with a vaccine. The immunized animal now produces its own antibodies and activated lymphocytes. (778)

artificially acquired passive immunity The type of immunity that results from introducing either an animal antibodies that have been produced either in another animal or by in vitro methods. Immunity is only temporary. (778)

ascocarp (as′ko-karp) A multicellular structure in ascomycetes lined with specialized cells called asci in which nuclear fusion and meiosis produce ascospores. An ascocarp can be open or closed and may be referred to as a fruiting body. (638)

ascogenous hypha A specialized hypha that gives rise to one or more asci. (638)

ascogonium (as″ko-go′ne-um; pl., **ascogonia)** The receiving (female) organ in ascomycetous fungi which, after fertilization, gives rise to ascogenous hyphae and later to asci and ascospores. (638)

ascomycetes (as″ko-mi-se′tēz) A division of fungi that form ascospores. (637)

ascospore (as′ko-spor) A spore contained or produced in an ascus. (634)

ascus (as′kus) A specialized cell, characteristic of the ascomycetes, in which two haploid nuclei fuse to produce a zygote, which immediately divides by meiosis; at maturity an ascus will contain ascospores. (637)

aseptic meningitis syndrome *See* meningitis.

aspergillosis (as″per-jil-o′sis) A fungal disease caused by species of *Aspergillus.* (1016)

assimilatory reduction The reduction of an inorganic molecule to incorporate it into organic material. No energy is made available during this process. (649)

associative nitrogen fixation Nitrogen fixation by bacteria in the plant root zone (rhizosphere). (696)

athlete's foot *See* tinea pedis.

atomic force microscope A type of scanning probe microscope that images a surface by moving a sharp probe over the surface at a constant distance; a very small amount of force is exerted on the tip and probe movement is followed with a laser. (36)

atopic reaction A type I hypersensitivity response caused by environmental allergens. (803)

attenuated vaccine Live, nonpathogenic organisms used to activate adaptive immunity. The nonpathogenic organisms grow in the vaccinated individual without producing serious clinical disease while stimulating lymphocytes to produce antibody and activated T cells. (901)

attenuation (ah-ten″u-a′shun) 1. A mechanism for the regulation of transcription of some bacterial operons by aminoacyl-tRNAs. 2. A procedure that reduces or abolishes the virulence of a pathogen without altering its immunogenicity. (302)

attenuator A rho-independent termination site in the leader sequence that is involved in attenuation. (302)

atypical pneumonia An acute respiratory disease characterized by high fever and coughing. The pneumonia is atypical in that little fluid is found in the lungs. It is often caused by *Mycoplasma pneumoniae* and primarily affects children and young adults. (955)

autochthonous (ô-tŏk-thə-nəs) Substances (nutrients) that originate in a given environments. *See also* allochthonous. (682)

autoclave (aw′to-klāv) An apparatus for sterilizing objects by the use of steam under pressure. Its development tremendously stimulated the growth of microbiology. (153)

autogamy In the reproduction of certain protists, this form of self-fertilization involves the fusion of haploid nuclei or gametes derived from a single cell. (609)

autogenous infection (aw-toj′e-nus) An infection that results from a patient's own microbiota, regardless of whether the infecting organism became part of the patient's microbiota subsequent to admission to a clinical care facility. (909)

autoimmune disease (aw″to-ĭ-mūn′) A disease produced by the immune system attacking self-antigens. Autoimmune disease results from the activation of self-reactive T and B cells that damage tissues after stimulation by genetic or environmental triggers. (809)

autoimmunity (aw″to-ĭ-mun′ĭ-te) Autoimmunity is a condition characterized by the presence of serum autoantibodies and self-reactive lymphocytes. It may be benign or pathogenic. Autoimmunity is a normal consequence of aging; is readily inducible by infectious agents, organisms, or drugs; and is potentially reversible in that it disappears when the offending "agent" is removed or eradicated. (809)

autolysins (aw-tol′ĭ-sins) Enzymes that partially digest peptidoglycan in growing bacteria so that the peptidoglycan can be enlarged. (234)

autotroph (aw′to-trōf) An organism that uses CO_2 as its sole or principal source of carbon. (102)

auxotroph (awk′so-trōf) An organism with a mutation that causes it to lose the ability to synthesize an essential nutrient; because of the mutation the organism must obtain the nutrient or a precursor from its surroundings. (323)

avidity The combined strength of binding between an antigen and all the antibody-binding sites. (776)

axenic (a-zen′ik) Not contaminated by any foreign organisms; the term is used in reference to pure microbial cultures or to germfree animals. (734)

axial filament The organ of motility in spirochetes. It is made of axial fibrils or periplasmic flagella that extend from each end of the protoplasmic cylinder and overlap in the middle of the cell. The outer sheath lies outside the axial filament. (70, 532)

axopodium A thin, needlelike type of pseudopodium with a central core of microtubules. Found in the protists Radiolaria. (617)

B

bacille Calmette-Guerin (BCG) An attenuated form of *Mycobacterium tuberculosis* used in some countries as a vaccine for tuberculosis. (955)

bacillus (bah-sil′lus) A rod-shaped bacterium. (40)

bacteremia (bak″ter-e′me-ah) The presence of viable bacteria in the blood. (821)

Bacteria (bak-te′re-a) The domain that contains procaryotic cells with primarily diacyl glycerol diesters in their membranes and with bacterial rRNA. (2, 474)

bacterial artificial chromosome (BAC) A cloning vector constructed from the *E. coli* F-factor plasmid that is used to clone foreign DNA fragments. (370)

bacterial (septic) meningitis *See* meningitis. (950)

bacterial vaginosis (bak-te′re-əl vaj″ĭ-no′sis) Bacterial vaginosis is a sexually transmitted disease caused by *Gardnerella vaginalis, Mobiluncus* spp., *Mycoplasma hominis,* and various anaerobic bacteria. Although a mild disease, it is a risk factor for obstetric infections and pelvic inflammatory disease. (971)

bactericide (bak-tēr′ĭsid) An agent that kills bacteria. (151)

bacteriochlorophyll (bak-te″re-o-klo′ro-fil) A modified chlorophyll that serves as the primary light-trapping pigment in purple and green photosynthetic bacteria and heliobacteria. (218)

bacteriocin (bak-te′re-o-sin) A protein produced by a bacterial strain that kills other closely related bacteria. (53, 1031)

bacteriophage (bak-te′re-o-fāj″) A virus that uses bacteria as its host; often called a phage. (409, 427)

bacteriophage (phage) typing A technique in which strains of bacteria are identified based on their susceptibility to bacteriophages. (873)

bacteriorhodopsin A transmembranous protein to which retinal is bound; it functions as a light-driven proton pump performing photophosphorylation without chlorophyll or bacteriochlorophyll. Found in the purple membrane of halophilic archaea. (515)

bacteriostatic (bak-te″re-o-stat′ik) Inhibiting the growth and reproduction of bacteria. (151)

bacteroid (bak′tĕ-roid) A modified, often pleomorphic, bacterial cell within the root nodule cells of legumes; after transformation into a symbiosome it carries out nitrogen fixation. (701)

baeocytes Small, spherical, reproductive cells produced by pleurocapsalean cyanobacteria through multiple fission. (526)

balanced growth Microbial growth in which all cellular constituents are synthesized at constant rates relative to each other. (123)

balanitis (bal″ah-ni′tis) Inflammation of the glans penis usually associated with *Candida* fungi; a sexually transmitted disease. (1018)

barophilic (bar″o-fil′ik) or **barophile** Organisms that prefer or require high pressures for growth and reproduction. (141, 681)

barotolerant Organisms that can grow and reproduce at high pressures but do not require them. (141)

basal body The cylindrical structure at the base of procaryotic and eucaryotic flagella that attaches them to the cell. (67, 96)

base analogs Molecules that resemble normal DNA nucleotides and can substitute for them during DNA replication, leading to mutations. (319)

basic dyes Dyes that are cationic, or have positively charged groups, and bind to negatively charged cell structures. Usually sold as chloride salts. (26)

basidiocarp (bah-sid′e-o-karp″) The fruiting body of a basidiomycete that contains the basidia. (639)

basidiomycetes (bah-sid″e-o-mi-se′tēz) A division of fungi in which the spores are borne on club-shaped organs called basidia. (639)

basidiospore (bah-sid′e-ō-spōr) A spore borne on the outside of a basidium following karyogamy and meiosis. (634)

basidium (bah-sid′e-um; pl., **basidia**) A structure that bears on its surface a definite number of basidiospores (typically four) that are formed following karyogamy and meiosis. Basidia are found in the basidiomycetes and are usually clubshaped. (639)

basophil (ba′so-fil) A white blood cell in the granulocyte lineage. It is weakly phagocytic. Importantly, it synthesizes and stores vasoactive molecules (e.g., histamine) that are released in response to external triggers (*see* mast cell). (747)

batch culture A culture of microorganisms produced by inoculating a closed culture vessel containing a single batch of medium. (123)

B cell, also known as a **B lymphocyte** A type of lymphocyte derived from bone marrow stem cells that matures into an immunologically competent cell under the influence of the bursa of Fabricius in the chicken and bone marrow in nonavian species. Following interaction with antigen, it becomes a plasma cell, which synthesizes and secretes antibody molecules involved in humoral immunity. (748)

B-cell receptor (BCR) A transmembrane immunoglobulin complex on the surface of a B cell that binds an antigen and stimulates the B cell. It is composed of a membrane-bound immunoglobulin, usually IgD or a modified IgM, complexed with another membrane protein (the Ig-α/Ig-β heterodimer). (786)

benthic (ben′thic) Pertaining to the bottom of the sea or another body of water. (555)

beta hemolysis A zone of complete clearing around a bacterial colony growing on blood agar. The zone does not change significantly in color. (828)

β-lactam Referring to antibiotics containing a β-lactam ring. This includes the penicillins and cephalosporins. (844)

β-lactam ring The cyclic chemical structure composed of three carbon and one nitrogen elements. It has antibacterial activity, interfering with bacterial cell wall synthesis. (843)

β lactamase An enzyme that hydrolyzes the β-lactam ring rendering the antibiotic inactive. Sometimes called penicillinase. (843)

β-oxidation pathway The major pathway of fatty acid oxidation to produce NADH, $FADH_2$, and acetyl coenzyme A. (211)

Betaproteobacteria One of the five classes of proteobacteria, each with distinctive 16S rRNA sequences. Members of this subgroup are similar to the alpha-proteobacteria metabolically, but tend to use substances that diffuse from organic matter decomposition in anaerobic zones. (569)

binal symmetry The symmetry of some virus capsids (e.g., those of complex phages) that is a combination of icosahedral and helical symmetry. (412)

binary fission Asexual reproduction in which a cell or an organism separates into two identical daughter cells. (119, 543, 608)

binomial system The nomenclature system in which an organism is given two names; the first is the capitalized generic name, and the second is the uncapitalized specific epithet. (480)

bioaugmentation Addition of pregrown microbial cultures to an environment to perform a specific task. (1080)

biocatalysis The use of enzymes or whole microbes to perform chemical transformations on natural products. (1074)

biochemical oxygen demand (BOD) The amount of oxygen used by organisms in water under certain standard conditions; it provides an index of the amount of microbially oxidizable organic matter present. (1055)

biocrime The use of biological materials (organisms or their toxins) to subvert societal goals or laws. Lacing a restaurant salad bar with *Salmonella* to prevent citizens from voting is an example of a biocrime. (905)

biodegradation (bi″o-deg″rah-da′shun) The breakdown of a complex chemical through biological processes that can result in minor loss of functional groups, fragmentation into smaller constitutents, or complete breakdown to carbon dioxide and minerals. (1075)

biofilms Organized microbial communities consisting of layers of microbial cells associated with surfaces, often with complex structural and functional characteristics. Biofilms have physical/chemical gradients that influence microbial metabolic processes. They can form on inanimate devices (catheters, medical prosthetic devices) and also cause fouling (e.g., of ships' hulls, water pipes, cooling towers). (143, 653)

biogeochemical cycling The oxidation and reduction of substances carried out by living organisms and/or abiotic processes that results in the cycling of elements within and between different parts of the ecosystem (the soil, aquatic environment, and atmosphere). (644)

bioinformatics The highly interdisciplinary field that uses existing and develops new tools to manage and analyze large biological data sets including genome and protein sequences. Some areas of study include: sequence analysis, phylogenetic inference, genome database organization, pattern recognition and image analysis, and modeling macromolecular structures. (388)

bioinsecticide A pathogen that is used to kill or disable unwanted insect pests. Bacteria, fungi, or viruses are used, either directly or after manipulation, to control insect populations. (1083)

biologic transmission A type of vector-borne transmission in which a pathogen goes through some morphological or physiological change within the vector. (896)

bioluminescence (bi″o-loo″mĭ-nes′əns) The production of light by living cells, often through the oxidation of molecules by the enzyme luciferase. (557)

biomagnification The increase in concentration of a substance in higher-level consumer organisms. (652)

biopesticide The use of a microorganism or another biological agent to control a specific pest. (1083)

bioprospecting The collection, cataloging, and analysis of organisms including microorganisms and plants with the intent of finding a useful application and/or to document biodiversity. (1060)

bioremediation The use of biologically mediated processes to remove or degrade pollutants from specific environments. Bioremediation can be carried out by modification of the environment to accelerate biological processes, either with or without the addition of specific microorganisms. (1075)

biosensor A device for the detection of a particular substance (an analyte) that combines a biological receptor with a physicochemical detector. The receptor senses or captures the analyte. The receptor can be tissue, microorganism, organelle, cell receptor, enzyme, antibody, nucleic acid, etc. The detector reports the sensing or capture events and produces outputs that are physicochemical, optical, electrochemical, thermometric, piezoelectric, or magnetic in nature. (1083)

biosynthesis *See* anabolism.

biosynthetic-secretory pathway The process used by eucaryotic cells to synthesize proteins and lipids, followed by secretion or delivery to organelles or the plasma membrane. The pathway involves the endoplasmic reticulum, Golgi apparatus, and secretory vesicles. (86)

bioterrorism The intentional or threatened use of viruses, bacteria, fungi, or toxins from living organisms to produce death or disease in humans, animals, and plants. (905)

biotransformation or **microbial transformation** The use of living organisms to modify substances that are not normally used for growth. (1074)

black piedra (pe-a′drah) A fungal infection caused by *Piedraia hortae* that forms hard black nodules on the hairs of the scalp. (1008)

blastomycosis (blas″to-mi-ko′sis) A systemic fungal infection caused by *Blastomyces dermatitidis* and marked by suppurating tumors in the skin or by lesions in the lungs. (999)

B lymphocyte *See* B cell.

botulism (boch′oo-lizm) A form of food poisoning caused by a neurotoxin (botulin) produced by *Clostridium botulinum* serotypes A–G; sometimes found in improperly canned or preserved food. (979)

brevetoxins Toxins produced by certain bloom-forming dinoflagellates. These polycyclic compounds are lipid soluble and potent neurotoxins; associated with paralytic shellfish poisoning. (675)

bright-field microscope A microscope that illuminates the specimen directly with bright light and forms a dark image on a brighter background. (18)

Bright's disease *See* glomerulonephritis. (958)

broad-spectrum drugs Chemotherapeutic agents that are effective against many different kinds of pathogens. (837)

bronchial-associated lymphoid tissue (BALT) The type of defensive tissue found in the lungs. Part of the nonspecific (innate) immune system. (759)

bronchial asthma An example of an atopic allergy involving the lower respiratory tract. (804)

brucellosis (brōō′sə-lō′sĭs) A disease caused by the α-proteobacterium *Brucella.* The disease is characterized by cycles of (undulating) fever, sweating, weakness, and headache. It is transmitted to humans by direct contact with diseased animals or through ingestion of infected meat, milk, or cheese. It is also known as Bang's disease, Malta fever, Mediterranean fever, Rock fever, and undulant fever. (990)

bubo (bu′bo) A tender, inflamed, enlarged lymph node that results from a variety of infections. (962)

bubonic plague *See* plague.

budding A vegetative outgrowth of yeast and some bacteria as a means of asexual reproduction; the daughter cell is smaller than the parent. (543)

bulking sludge Sludges produced in sewage treatment that do not settle properly, usually due to the development of filamentous microorganisms. (1055)

bursa of Fabricius (bər′sə fə-bris′e-əs) Found in birds; the blind saclike structure located on the posterior wall of the cloaca; it performs a thymuslike function. A primary lymphoid organ where B-cell maturation occurs. Bone marrow is the equivalent in mammals. (750)

burst *See* rise period.

burst size The number of phages released by a host cell during the lytic life cycle. (430)

butanediol fermentation A type of fermentation most often found in the family *Enterobacteriaceae* in which 2,3-butanediol is a major product; acetoin is an intermediate in the pathway and may be detected by the Voges-Proskauer test. (209)

C

Calvin cycle The main pathway for the fixation (or reduction and incorporation) of CO_2 into organic material by photoautotrophs; it also is found in chemolithoautotrophs. (228)

campylobacteriosis A disease caused by species of the bacterium *Campylobacter,* primarily *Campylobacter jejuni,* characterized by an inflammatory, sometimes bloody, diarrhea and may present as a dysentery syndrome. (979)

cancer (kan′ser) A malignant tumor that expands locally by invasion of surrounding tissues and systemically by metastasis. (461)

candidal vaginitis Infection of the vagina caused by the fungus *Candida* sp. (1017)

candidiasis (kan″dĭ-di′ah-sis) An infection caused by *Candida* species of dimorphic fungi, commonly involving the skin. (1017)

capsid (kap′sid) The protein coat or shell that surrounds a virion's nucleic acid. (409)

capsomer (kap′so-mer) The ring-shaped morphological unit of which icosahedral capsids are constructed. (410)

capsule A layer of well-organized material, not easily washed off, lying outside the bacterial cell wall. (65)

carbonate equilibrium system The interchange among CO_2, HCO_3^-, and CO_3^{2-} that keeps oceans buffered between pH 7.6 to 8.2. (668)

carboxysomes Polyhedral inclusion bodies that contain the CO_2 fixation enzyme ribulose 1,5-bisphosphate carboxylase; found in cyanobacteria, nitrifying bacteria, and thiobacilli. (49, 229, 524)

caries (kar′e-ēz) Tooth decay. (993)

carotenoids (kah-rot′e-noids) Pigment molecules, usually yellowish in color, that are often used to aid chlorophyll in trapping light energy during photosynthesis. (217)

carrier An infected individual who is a potential source of infection for others and plays an important role in the epidemiology of a disease. (891)

caseous lesion (ka′se-us) A lesion resembling cheese or curd; cheesy. Most caseous lesions are caused by *Mycobacterium tuberculosis.* (954)

casual carrier An individual who harbors an infectious organism for only a short period. (892)

catabolism (kah-tab′o-lizm) That part of metabolism in which larger, more complex molecules are broken down into smaller, simpler molecules with the release of energy. (168)

catabolite repression (kah-tab′o-līt) Inhibition of the synthesis of several catabolic enzymes by a metabolite such as glucose. (308)

catalyst (kat′ah-list) A substance that accelerates a reaction without being permanently changed itself. (176)

catalytic site *See* active site.

catenanes (kăt′ə-nāns′) Circular, covalently closed nucleic acid molecules that are locked together like the links of a chain. (263)

cathelicidins Antimicrobial cationic peptides that are produced by a variety of cells (e.g., neutrophils, respiratory epithelial cells, and alveloar macrophages). They are linear, alpha-helical peptides that arise from precursor proteins having an N-terminal cathepsin L inhibitor domain and a C-terminus, which gives rise to the mature peptide of 12 to 80 amino acids. (736, 762)

catheter (kath′ĕ-ter) A tubular instrument for withdrawing fluids from a cavity of the body, especially one for introduction into the bladder through the urethra for the withdrawal of urine. (862)

caveola (ka-ve-o′lə) A small flask-shaped invagination of the plasma membrane formed during one type of pinocytosis. (86)

$CD4^+$ cell *See* T-helper cell.

$CD8^+$ cell *See* cytotoxic lymphyocyte.

CD95 Cell surface receptor that initiates apoptosis when ligand binds. *See also* Fas-FasL. (782)

cell cycle The sequence of events in a cell's growth-division cycle between the end of one division and the end of the next. In eucaryotic cells, it is composed of the G_1 period, the S period in which chromosomes are replicated, the G_2 period, and the M period (mitosis). (92, 119)

cell-mediated immunity The type of immunity that results from T cells coming into close contact with foreign cells or infected cells to destroy them; it can be transferred to a nonimmune individual by the transfer of cells. (814)

cellular slime molds Protists with a vegetative phase consisting of amoeboid cells that aggregate to form a multicellular pseudoplasmodium. They belong to the subdivision *Dictyostelia;* they were formerly considered fungi. (614)

cellulitis (sel″u-li′tis) A diffuse spreading infection of subcutaneous skin tissue caused by streptococci, staphylococci, or other organisms. The tissue is inflamed with edema, redness, pain, and interference with function. (957)

cellulose The major structural carbohydrate of plants cell walls; a linear ($\beta 1 \rightarrow 4$) glucan. (688)

cell wall The strong layer or structure that lies outside the plasma membrane; it supports and protects the membrane and gives the cell shape. (94)

central tolerance The process by which immune cells are rendered inactive. (803)

cephalosporin (sef″ah-lo-spōr′in) A group of β-lactam antibiotics derived from the fungus *Cephalosporium,* which share the 7-aminocephalosporanic acid nucleus. (844)

Chagas' disease *See* trypanosomiasis.

chancre (shang′ker) The primary lesion of syphilis occurring at the site of entry of the infection. (976)

chancroid (shang′kroid) A sexally transmitted disease caused by the γ-proteobacterium *Haemophilus ducreyi.* Chancroid is an important cofactor in the transmission of the AIDS virus. Also known as genital ulcer disease due to the painful circumscribed ulcers that form on the penis or entrance to the vagina. (971)

chaperone proteins Proteins that assist in the folding and stabilization of other proteins. Some are also involved in directing newly synthesized proteins to protein secretion systems, or to other locations in the cell. (284)

chemical oxygen demand (COD) The amount of chemical oxidation required to convert organic matter in water and wastewater to CO_2. (1054)

chemiosmotic hypothesis (kem″e-o-os-mot′ik) The hypothesis that a proton gradient and an electrochemical gradient are generated by electron transport and then used to drive ATP synthesis by oxidative phosphorylation. (202)

chemoheterotroph (ke″mo-het′er-o-trōf″) *See* chemoorganotrophic heterotrophs.

chemolithoautotroph A microorganism that oxidizes reduces inorganic compounds to derive both energy and electrons; CO_2 is the carbon source. Also called chemolithotrophic autotroph. (212)

chemolithoheterotroph A microorganism that uses reduced inorganic compounds to drive both energy and electrons; organic molecules are used as the carbon source. Also called mixotroph. (103)

chemoorganoheterotroph A microorganism that uses organic compounds as sources of energy, electrons, and carbon for biosynthesis. Also called chemoheterotroph and chemoorganotrophic heterotroph. (103)

chemoreceptors Special protein receptors in the plasma membrane or periplasmic space that bind chemicals and trigger the appropriate chemotaxic response. (71)

chemostat (ke′mo-stat) A continuous culture apparatus that feeds medium into the culture vessel at the same rate as medium containing microorganisms is removed; the medium in a chemostat contains one essential nutrient in a limiting quantity. (131)

chemotaxis (ke″mo-tak′sis) The pattern of microbial behavior in which the microorganism moves toward chemical attractants and/or away from repellents. (71)

chemotherapeutic agents (ke″mo-ther-ah-pu′tik) Compounds used in the treatment of disease that destroy pathogens or inhibit their growth at concentrations low enough to avoid doing undesirable damage to the host. (164, 835)

chemotrophs (ke′mo-trōfs) Organisms that obtain energy from the oxidation of chemical compounds. (103)

chickenpox (varicella) (chik′en-poks) A highly contagious skin disease, usually affecting 2- to 7-year-old children; it is caused by the varicella-zoster virus, which is acquired by droplet inhalation into the respiratory system. (914)

chiral (ki′rəl) Having handedness: consisting of one or another stereochemical form. (1076)

chitin (ki′tin) A tough, resistant, nitrogen-containing polysaccharide forming the walls of certain fungi, the exoskeleton of arthropods, and the epidermal cuticle of other surface structures of certain protists and animals. (631)

chlamydiae (klə-mid′e-e) Members of the genera *Chlamydia* and *Chlamydiophila:* gram-negative, coccoid cells that reproduce only within the cytoplasmic vesicles of host cells using a life cycle that alternates between elementary bodies and reticulate bodies. (531)

chlamydial pneumonia (klə-mid′e-əl noo-mo′ne-ə) A pneumonia caused by *Chlamydiophila pneumoniae.* Clinically, infections are mild and 50% of adults have antibodies to the chlamydiae. (948)

chloramphenicol (klo″ram-fen′ĭ-kol) A broad-spectrum antibiotic that is produced by *Streptomyces venzuelae* or synthetically; it binds to the large ribosomal subunit and inhibits the peptidyl transferase reaction. (846)

chlorophyll (klor′o-fil) The green photosynthetic pigment that consists of a large tetrapyrrole ring with a magnesium atom in the center. (216)

chloroplast (klo′ra-plast) A eucaryotic plastid that contains chlorophyll and is the site of photosynthesis. (90)

chlorosomes Elongated, intramembranous vesicles found in the green sulfur and nonsulfur bacteria that contain light-harvesting pigments. Sometimes called chlorobium vescicles. (523)

cholera (kol′er-ah) An acute infectious enteritis, endemic and epidemic in Asia, which periodically spreads to the Middle East, Africa, Southern Europe, and South America; caused by *Vibrio cholerae.* (983)

choleragen (kol′er-ah-gen) The cholera toxin; an extremely potent protein molecule made by strains of *Vibrio cholerae* in the small intestine after ingestion of feces-contaminated water or food. It acts on epithelial cells to cause hypersecretion of chloride and bicarbonate and an outpouring of large quantities of fluid from the mucosal surface. (983)

chromatic adaptation The capacity of cyanobacteria to alter the ratio of their light-harvesting or accessory pigments in response to changes in the spectral quality (wavelengths) of light. (525)

chromatin (kro′mah-tin) The DNA-containing portion of the eucaryotic nucleus; the DNA is almost always complexed with histones. It can be very condensed (heterochromatin) or more loosely organized and genetically active (euchromatin). (91)

chromoblastomycosis (kro″mo-blas″to-mi-ko′sis) A chronic fungal skin infection, producing wartlike nodules that may ulcerate. It is caused by the black molds *Phialophora verrucosa* or *Fonsecaea pedrosoi.* (1010)

chromogen (kro′me-jen) A colorless substrate that is acted on by an enzyme to produce a colored end product. (879)

chromophore group (kro″mo-fōr) A chemical group with double bonds that absorbs visible light and gives a dye its color. (26)

chromosomal nuclei Nuclei found in some protists in which the chromosomes remain condensed throughout the cell cycle. (609)

chromosomes (kro′mo-somz) The bodies that have most or all of the cell's DNA and contain most of its genetic information (mitochondria and chloroplasts also contain DNA and genes). (91)

chronic carrier An individual who harbors a pathogen for a long time. (892)

chytrids A term used to describe the *Chytridiomycota,* which are simple terrestrial and aquatic fungi that produce motile zoospores with single, posterior, whiplash flagella. (635)

cilia (sil′e-ah) Threadlike appendages extending from the surface of some protozoa that beat rhythmically to propel them; cilia are membrane-bound cylinders with a complex internal array of microtubules, usually in a 9 + 2 pattern. (95)

citric acid cycle *See* tricarboxylic acid (TCA) cycle.

class I MHC molecule *See* major histocompatibility complex.

class II MHC molecule *See* major histocompatibility complex.

class switching The change in immunoglobulin isotype (or class) secretion that results during B-cell and then plasma cell differentiation. (795)

classical complement pathway The antibody-dependent pathway of complement activation; it leads to the lysis of pathogens and stimulates phagocytosis and other host defenses. (766)

classification The arrangement of organisms into groups based on mutual similarity or evolutionary relatedness. (478)

clonal selection The process by which an antigen selects the best-fitting B-cell receptor, activating that B cell, resulting in the synthesis of antibody and clonal expansion. (798)

clone (klōn) A group of genetically identical cells or organisms derived by asexual reproduction from a single parent. (798)

clostridial myonecrosis (klo-strid′e-al mi″o-ne-kro′sis) Death of individual muscle cells caused by clostridia. Also called gas gangrene. (965)

clue cells Vaginal epithelial cells covered with bacteria. The name comes from the fact that identification of these cells offered a clue to the disease etiology. (971)

cluster of differentiation molecules (CDs) Functional cell surface proteins or receptors that can be measured in situ from peripheral blood, biopsy samples, or other body fluids. They can be used to identify leukocyte subpopulations. Some examples include interleukin-2 receptor (IL-2R), CD4, CD8, CD25, and intercellular adhesion molecule-1 (ICAM-1). (776)

coaggregation The collection of bacteria on a substrate such as a tooth surface because of cell-to-cell recognition of genetically distinct bacterial types. Many of these interactions appear to be mediated by a lectin on one bacterium that interacts with a complementary carbohydrate receptor on another bacterium. (991)

coagulase (ko-ag′u-las) An enzyme that induces blood clotting; it is characteristically produced by pathogenic staphylococci. (582)

coated vesicles The clathrin coated vesicles formed by receptor-mediated endocytosis. (87)

coccidioidomycosis (kok-sid″e-oi″do-mi-ko′sis) A fungal disease caused by *Coccidioides immitis* that exists in dry, highly alkaline soils. Also known as valley fever, San Joaquin fever, or desert rheumatism. (1000)

coccolithophore Photosynthetic protists belonging to the phylum Stramenopila. They are characterized by coccoliths—intricate cell walls made of calcite. (624)

coccus (kok′us, pl. **cocci,** kok′si) A roughly spherical bacterial cell. (39)

code degeneracy The genetic code is organized in such a way that often there is more than one codon for each amino acid. (275)

codon (ko′don) A sequence of three nucleotides in mRNA that directs the incorporation of an amino acid during protein synthesis or signals the stop of translation. (264)

coenocytic (se″no-sit′ik) Refers to a multinucleate cell or hypha formed by repeated nuclear divisions not accompanied by cell divisions. (119, 631)

coenzyme (ko-en′zīm) A loosley bound cofactor that often dissociates from the enzyme active site after product has been formed. (176)

cofactor The nonprotein component of an enzyme; it is required for catalytic activity. (176)

cold sore A lesion caused by the herpes simplex virus; usually occurs on the border of the lips or nares. Also known as a fever blister or herpes labialis. (931)

colicin (kol′ĭ-sin) A plasmid-encoded protein that is produced by enteric bacteria and binds to specific receptors on the cell envelope of sensitive target bacteria, where it may cause lysis or attack specific intracellular sites such as ribosomes. (763)

coliform (ko′lĭ-form) A gram-negative, nonsporing, facultative rod that ferments lactose with gas formation within 48 hours at 35°C. (1052)

colonization (kol″ə-nĭ-za′shən) The establishment of a site of microbial reproduction on an inanimate surface or organism without necessarily resulting in tissue invasion or damge. (820)

colony An assemblage of microorganisms growing on a solid surface such as the surface of an agar culture medium; the assemblage often is directly visible, but also may be seen only microscopically. (113)

colony forming units (CFU) The number of microorganisms that form colonies when cultured using spread plates or pour plates, an indication of the number of viable microorganisms in a sample. (130)

colony stimulating factor (CSF) A protein that stimulates the growth and development of specific cell populations (e.g., granulocyte-CSF stimulates granulocytes to be made from their precursor stem cells). (768)

colorless sulfur bacteria A diverse group of nonphotosynthetic proteobacteria that can oxidize reduced sulfur compounds such as hydrogen sulfide. Many are lithotrophs and derive energy from sulfur oxidation. Some are unicellular, whereas others are filamentous gliding bacteria. (550)

comedo (kom′ĕ-do; pl., **comedones**) A plug of dried sebum in an excretory duct of the skin. (737)

cometabolism The modification of a compound not used for growth by a microorganism, which occurs in the presence of another organic material that serves as a carbon and energy source. (1077)

commensal (kŏ-men′sal) Living on or within another organism without injuring or benefiting the other organism. (729)

commensalism (kŏ-men′sal-izm″) A type of symbiosis in which one individual gains from the association and the other is neither harmed nor benefited. (729)

common cold An acute, self-limiting, and highly contagious virus infection of the upper respiratory tract that produces inflammation, profuse discharge, and other symptoms. (932)

common-source epidemic An epidemic that is characterized by a sharp rise to a peak and then a rapid, but not as pronounced, decline in the number of individuals infected; it usually involves a single contaminated source from which individuals are infected. (889)

common vehicle transmission The transmission of a pathogen to a host by means of an inanimate medium or vehicle. (894)

communicable disease A disease associated with a pathogen that can be transmitted from one host to another. (888)

community An assemblage of different types of organisms or a mixture of different microbial populations. (643)

compatible solute A low-molecular-weight molecule used to protect cells against changes in solute concentrations (osmolarity) in their habitat; it can exist at high concentrations within the cell and still be compatible with metabolism and growth. (132)

competent A procaryotic cell that can take up free DNA fragments and incorporate them into its genome during transformation. (343)

competition An interaction between two organisms attempting to use the same resource (nutrients, space, etc.). (732)

competitive exclusion principle Two competing organisms overlap in resource use, which leads to the exclusion of one of the organisms. (732)

complementarity determining regions (CDRs) Hypervariable regions in an immunoglobulin protein that form the three-dimensional binding sites for epitope binding. (791)

complementary DNA (cDNA) A DNA copy of an RNA molecule (e.g., a DNA copy of an mRNA). (358)

complement system A group of plasma proteins that plays a major role in an animal's defensive immune response. (763)

complex medium Culture medium that contains some ingredients of unknown chemical composition. (111)

complex viruses Viruses with capsids having a complex symmetry that is neither icosahedral nor helical. (428)

composting The microbial processing of fresh organic matter under moist, aerobic conditions, resulting in the accumulation of a stable humified product, which is suitable for soil improvement and stimulation of plant growth. (1075)

compromised host A host with lowered resistance to infection and disease for any of several reasons. The host may be seriously debilitated (due to malnutrition, cancer, diabetes, leukemia, or another infectious disease), traumatized (from surgery or injury), immunosuppressed, or have an altered microbiota due to prolonged use of antibiotics. (740, 1016)

concatemer A long DNA molecule consisting of several genomes linked together in a row. (446)

conditional mutations Mutations with phenotypes that are expressed only under certain environmental conditions. (323)

confocal scanning laser microscope (CSLM) A light microscope in which monochromatic laser-derived light scans across the specimen at a specific level and illuminates one area at a time to form an image. Stray light from other parts of the specimen is blocked out to give an image with excellent contrast and resolution. (34)

congenital (neonatal) herpes An infection of a newborn caused by transmission of the herpesvirus during vaginal delivery. (934)

congenital rubella syndrome A wide array of congenital defects affecting the heart, eyes, and ears of a fetus during the first trimester of pregnancy, and caused by the rubella virus. (920)

congenital syphilis Syphilis that is acquired in utero from the mother. (976)

conidiospore (ko-nid′e-o-spōr) An asexual, thin-walled spore borne on hyphae and not contained within a sporangium; it may be produced singly or in chains. (633)

conidium (ko-nid′e-um; pl., **conidia**) *See* conidiospore.

conjugants (kon′joo-gants) Complementary mating types among protists that participate in a form sexual reproduction called conjugation. (620)

conjugation (kon″ju-ga′shun) 1. The form of gene transfer and recombination in procaryotes that requires direct cell-to-cell contact. 2. A complex form of sexual reproduction commonly employed by protists. (337, 620, 609)

conjugative plasmid A plasmid that carries the genes that enable its transfer to other bacteria during conjugation (e.g., F plasmid). (53, 334)

conjunctivitis of the newborn *See* ophthalmia neonatorum. (975)

consensus sequence A commonly occurring sequence of nucleotides within a genetic element such as the Pribnow box of bacterial promoters. (269)

consortium A physical association of two different organisms, usually beneficial to both organisms. (717)

constant region (C_L and C_H) The part of an antibody molecule that does not vary greatly in amino acid sequence among molecules of the same class, subclass, or type. (790)

constitutive mutant A strain that produces an inducible enzyme continually, regardless of need, because of a mutation in either the operator or regulator gene. (293)

constructed wetlands Intentional creation of marshland plant communities and their associated microorganisms for environmental restoration or to purify water by the removal of bacteria, organic matter, and chemicals as the water passes through the aquatic plant communities. (1058)

consumer An organism that feeds directly on living or dead animals, by ingestion or by phagocytosis. (652)

contact transmission Transmission of the pathogen by contact of the source or reservoir of the pathogen with the host. (892)

continuous culture system A culture system with constant environmental conditions maintained through continual provision of nutrients and removal of wastes. *See also* chemostat. (131)

contractile vacuole (vak′u-ōl) In protists and some animals, a clear fluid-filled cell vacuole that takes up water from within the cell and then contracts, releasing it to the outside through a pore in a cyclical manner. Contractile vacuoles function primarily in osmoregulation and excretion. (607)

convalescent carrier (kon″vah-les′ent) An individual who has recovered from an infectious disease but continues to harbor large numbers of the pathogen. (891)

cooperation A positive but not obligatory interaction between two different organisms. (726)

coral bleaching The loss of photosynthetic pigments by either physiological inhibition or expulsion of the coral photosynthetic endosymbiont, zooxanthellae, a dinoflagellate. Bleached corals may be temporarily or permanently damaged. (719)

corepressor (ko″re-pre′sor) A small molecule that inhibits the synthesis of a repressible enzyme. (295)

cortex The layer of the bacterial endospore that is particularly important in conferring heat resistance to the endospore. (73)

cosmid (koz′mid) A plasmid vector with lambda phage *cos* sites that can be packaged in a phage capsid; it is useful for cloning large DNA fragments. (370)

cristae (kris′te) Infoldings of the inner mitochondrial membrane. (89)

crossing-over A process in which segments of two adjacent DNA strands are exchanged; breaks occur in both strands, and the exposed ends of each strand join to those of the opposite segment on the other strand. (330)

crown gall disease A plant tumor, or gall, caused by the certain species of the α-proteobacterium *Agrobacterium,* most commonly *A. tumefaciens.* (546)

cryptin Antimicrobial peptides produced by Paneth cells in the intestines. (761)

cryptococcosis (krip″to-kok-o′sis) An infection caused by the basidiomycete, *Cryptococcus neoformans,* which may involve the skin, lungs, brain, or meninges. (1001)

cryptosporidiosis (krip″to-spo-rid″e-o′sis) Infection with protozoa of the genus *Cryptosporidium.* The most common symptoms are prolonged diarrhea, weight loss, fever, and abdominal pain. (1014)

crystallizable fragment (Fc) The stem of the Y portion of an antibody molecule. Cells such as macrophages bind to the Fc region; it also is involved in complement activation. (790)

cutaneous anthrax (ku-ta′ne-us an′thraks) A form of anthrax involving the skin. (988)

cutaneous diphtheria (ku-ta′ne-us dif-the′re-ah) A skin disease caused by *Corynebacterium diphtheriae* that infects wound or skin lesions, causing a slow-healing ulceration. (949)

cyanobacteria (si″ah-no-bak-te′re-ah) A large group of gram-negative bacteria that carry out oxygenic photosynthesis using a system like that present in photosynthetic eucaryotes. (524)

cyclic photophosphorylation (fo″to-fos″for-ĭ-la′shun) The formation of ATP when light energy is used to move electrons cyclically through an electron transport chain during photosynthesis; only photosystem I participates. (217)

cyclosporiasis A disease caused by infection with the protozoan parasite *Cyclospora cayetanensis.* Infection has been associated with ingestion of contaminated produce. *Cyclospora* oocysts contain two sporocysts, each containing two sporozoites. Unsporulated oocysts are passed in the stool, and sporulation occurs outside the host. (1014)

cyst (sist) A general term used for a specialized microbial cell enclosed in a wall. Cysts are formed by protists and a few bacteria. They may be dormant, resistant structures formed in response to adverse conditions or reproductive cysts that are a normal stage in the life cycle. (608)

cytochromes (si′to-krōms) Heme proteins that carry electrons, usually as members of electron transport chains. (174)

cytokine (si′to-kīn) A general term for proteins released by a cell in response to inducing stimuli, which are mediators that influence other cells. Produced by lymphocytes, monocytes, macrophages, and other cells. (748, 766)

cytokinesis Processes that apportion the cytoplasm and organelles, synthesize a septum, and divide a cell into two daughter cells during cell division. (121)

cytomegalovirus inclusion disease (si″to-meg″ah-lo-vi′rus) An infection caused by the cytomegalovirus and marked by nuclear inclusion bodies in enlarged infected cells. (933)

cytopathic effect (si″to-path′ik) The observable change that occurs in cells as a result of viral replication. Examples include ballooning, binding together, clustering, or even death of the cultured cells. (418, 470, 866)

cytoplasmic matrix (si″to-plaz′mik) The protoplasm of a cell that lies within the plasma membrane and outside any other organelles. In bacteria it is the substance between the cell membrane and the nucleoid. (48, 83)

cytoproct A specific site in certain protists (e.g., ciliates) where digested material is expelled. (608)

cytosine (si′to-sēn) A pyrimidine 2-oxy-4-aminopyrimidine found in nucleosides, nucleotides, and nucleic acids. (241)

cytoskeleton (si″to-skel′ĕ-ton) A network of microfilaments, microtubules, intermediate filaments, and other components in the cytoplasm of eucaryotic cells that helps give them shape. (83)

cytostome (si′to-stōm) A permanent site in a ciliate protist at which food is ingested. (608)

cytotoxic T lymphoryte (CTL) (si″to-tok′sik) A type of T cell that recognizes antigen in class I MHC molecules, destroying the cell on which the antigen is displayed. Also called $CD8^+$ cell. (782)

cytotoxin (si′to-tok′sin) A toxin or antibody that has a specific toxic action upon cells; cytotoxins are named according to the cell for which they are specific (e.g., nephrotoxin). (825)

D

Dane particle A 42 nm spherical particle that is one of three that are seen in hepatitis B virus infections. The Dane particle is the complete virion. (936)

dark-field microscopy Microscopy in which the specimen is brightly illuminated while the background is dark. (21)

dark reactivation The excision and replacement of thymine dimers in DNA that occurs in the absence of light. (215)

deamination (de-am″i-na′shun) The removal of amino groups from amino acids. (212)

decimal reduction time (*D* or *D* value) The time required to kill 90% of the microorganisms or spores in a sample at a specified temperature. (154)

decomposer An organism that breaks down complex materials into simpler ones, including the release of simple inorganic products. Often a decomposer such as an insect or earthworm physically reduces the size of substrate particles. (656)

defensin (de-fens′sin) Specific peptides produced by neutrophils that permeabilize the outer and inner membranes of certain microorganisms, thus killing them. (762)

defined medium Culture medium made with components of known composition. (111)

Deltaproteobacteria One of the five classes of proteobacteria. Chemoorganotrophic bacteria that usually are either predators on other bacteria or anaerobes that generate sulfide from sulfate and sulfite. (562)

denaturation (de-na″chur-a′shun) A change in protein shape that destroys its activity; the term is also sometimes applied to changes in nucleic-acid shape. (179)

denaturing gradient gel electrophoresis (DGGE) A technique by which DNA is rendered single-stranded (denatured) while undergoing electrophoresis so that DNA fragments of the same size can be separated according to nucleotide sequence rather than molecular weight. This is accomplished by preparing gels with a gradient of a chemical that denatures DNA. As fragments migrate from the negative to positive pole of the gel, they stop when they become single-stranded. (661)

dendritic cell (den-drit′ ik) An antigen-presenting cell that has long membrane extensions resembling the dendrites of neurons. These cells are found in the lymph nodes, spleen, and thymus (interdigitating dendritic cells); skin (Langerhans cells); and other tissues (interstitial dendritic cells). They express MHC class II and B7 costimulatory molecules and present antigens to T-helper cells. (747)

dendrogram A treelike diagram that is used to graphically summarize mutual similarities and relationships between organisms. (479)

denitrification (de-ni″trĭ-fĭ-ka′shən) The reduction of nitrate to gaseous products, primarily nitrogen gas, during anaerobic respiration. (205, 649)

dental plaque (plak) A thin film on the surface of teeth consisting of bacteria embedded in a matrix of bacterial polysaccharides, salivary glycoproteins, and other substances. (991)

deoxyribonucleic acid (DNA) (de-ok″se-ri″bo-nu-kle′ik) The nucleic acid that constitutes the genetic material of all cellular organisms. It is a polynucleotide composed of deoxyribonucleotides connected by phosphodiester bonds. (52, 252)

dermatomycosis (der′ma-to-mi-ko′sis) A fungal infection of the skin; the term is a general term that comprises the various forms of tinea, and it is sometimes used to specifically refer to athlete's foot (tinea pedis). (1008)

dermatophyte (der′mah-to-fit″) A fungus parasitic on the skin. (1008)

desensitization (de-sen″si-ti-za′shun) To make a sensitized or hypersensitive individual insensitive or nonreactive to a sensitizing agent (e.g., an allergen). (804)

desert crust A crust formed by microbial binding of sand grains in the surface zone of desert soil; primarily involves cyanobacteria. (695)

detergent (de-ter′jent) An organic molecule, other than a soap, that serves as a wetting agent and emulsifier; it is normally used as cleanser, but some may be used as antimicrobial agents. (163)

diatoms (di′ah-toms) Photosynthetic protists with siliceous cell walls called frustules. They constitute a substantial fraction of the phytoplankton. (621)

diauxic growth (di-awk′sik) A biphasic growth pattern or response in which a microorganism, when exposed to two nutrients, initially uses one of them for growth and then alters its metabolism to make use of the second. (308)

differential interference contrast (DIC) microscope A light microscope that employs two beams of plane polarized light. The beams are combined after passing through the specimen and their intereference is used to create the image. (23)

differential media (dif″er-en′shal) Culture media that distinguish between groups of microorganisms based on differences in their growth and metabolic products. (113)

differential staining procedures Staining procedures that divide bacteria into separate groups based on staining properties. (26)

diffusely adhering *E. coli* (DAEC) DAEC strains of *E. coli* adhere over the entire surface of epithelial cells and usually cause diarrheal disease in immunologically naive and malnourished children. (987)

dikaryotic stage (di-kar-e-ot′ik) In fungi, having pairs of nuclei within cells or compartments. Each cell contains two separate haploid nuclei, one from each parent. (634)

dilution susceptibility tests A method by which antibiotics are evaluated for their ability to inhibit bacterial growth in vitro. A standardized concentration of bacteria is added to serially diluted antibiotics and incubated. Tubes lacking additional bacterial growth suggest antibiotic concentrations that are bacteriocidal or bacteriostatic. (840)

dinoflagellate (di″no-flaj′e-lāt) A photosynthetic protist characterized by two flagella used in swimming in a spinning pattern. Many are bioluminescent and an important part of marine phytoplankton. (620)

diphtheria (dif-the′re-ah) An acute, highly contagious childhood disease that generally affects the membranes of the throat and less frequently the nose. It is caused by *Corynebacterium diphtheriae.* (948)

dipicolinic acid A substance present at high concentrations in the bacterial endospore. It is thought to contribute to the endospore's heat resistance. (575, 581)

diplococcus (dip″lo-kok′us) A pair of cocci. (39)

direct repair A type of DNA repair mechanism in which a damaged nitrogenous base is returned to its normal form (e.g., conversion of a thymine back to two normal thymine bases). (326)

disease (di-zez) A deviation or interruption of the normal structure or function of any part of the body that is manifested by a characteristic set of symptoms and signs. (885)

disease syndrome (sin′drŏm) A set of signs and symptoms that are characteristic of the disease. (888)

disinfectant (dis″in-fek′tant) An agent, usually chemical, that disinfects; normally, it is employed only with inanimate objects. (151)

disinfection (dis″in-fek′shun) The killing, inhibition, or removal of microorganisms that may cause disease. It usually refers to the treatment of inanimate objects with chemicals. (151)

disinfection by-products (DBPs) Chlorinated organic compounds such as trihalomethanes formed during chlorine use for water disinfection. Many are carcinogens. (1051)

dissimilatory nitrate reduction The process in which some bacteria use nitrate as the electron acceptor at the end of their electron transport chain to produce ATP. The nitrate is reduced to nitrite or nitrogen gas. (205)

dissimilatory reduction The use of a substance as an electron acceptor in energy generation. The acceptor (e.g., sulfate or nitrate) is reduced but not incorporated into organic matter during biosynthetic processes. (649)

dissolved organic matter (DOM) In aquatic and marine ecosystems, nutrients that are available in the soluble, or dissolved, state. (656)

DNA ligase An enzyme that joins two DNA fragments together through the formation of a new phosphodiester bond. (262, 367)

DNA microarrays Solid supports that have DNA attached in organized arrays and are used to evaluate gene expression. (389)

DNA polymerase (pol-im′er-ās) An enzyme that synthesizes new DNA using a parental nucleic acid strand (usually DNA) as a template. (259)

DNA vaccine A vaccine that contains DNA which encodes antigenic proteins. It is injected directly into the muscle; the DNA is taken up by the muscle cells and encoded protein antigens are synthesized. This produces both humoral and cell-mediated responses. (904)

domains (do-mān′) 1. Compact, self-folding, structurally independent regions of proteins (usually around 100–300 amino acids in length); large proteins may have two or more domains connected by less structured stretches of polypeptide. In the antibody molecule, they are the loops, along with about 25 amino acids on each side, that form compact, globular sections. 2. The primary taxonomic groups above the kingdom level; all living organisms may be placed in one of three domains. (288, 790)

double diffusion agar assay (Öuchterlony technique) An immunodiffusion reaction in which both antibody and antigen diffuse through agar to form stable immune complexes, which can be observed visually. (880)

doubling time *See* generation time.

DPT (diphtheria-pertussis-tetanus) vaccine A vaccine containing three antigens that is used to immunize people against diphtheria, pertussis or whooping cough, and tetanus. (949)

droplet nuclei Small particles (0 to 4 μm in diameter) that represent what is left from the evaporation of larger particles (10 μm or more in diameter) called droplets. (892)

***D* value** *See* decimal reduction time.

E

early mRNA Messenger RNA produced early in a virus infection that codes for proteins needed to take over the host cell and manufacture viral nucleic acids. (430)

Ebola hemorrhagic fever (a′bo-lə) An acute infection cause by the Ebola virus. The virus produces fever, bleeding, and shock in various degrees. Mortality is approximately 80%. (942)

eclipse period (e-klips′) The initial part of the latent period in which infected host bacteria do not contain any complete virions. (430)

ecosystem (ek″o-sis′tem) A self-regulating biological community and its associated physical and chemical environment. (643)

ectomycorrhizal Referring to a mutualistic association between fungi and plant roots in which the fungus surrounds the root tip with a sheath. (698)

ectoparasite (ek″to-par′ah-sīt) A parasite that lives on the surface of its host. (816)

ectoplasm In some protists, the cytoplasm directly under the cell membrane (plasmalemma) is divided into an outer gelatinous region, the ectoplasm, and an inner fluid region, the endoplasm. (607)

ectosymbiosis A type of symbiosis in which one organism remains outside of the other organism. (717)

effacing lesion (le′zhən) The type of lesion caused by enteropathogenic strains of *E. coli* (EPEC) when the bacteria destroy the brush border of intestinal epithelial cells. The term AE (attaching-effacing) *E. coli* is now used to designate true EPEC strains that are an important cause of diarrhea in children from developing countries and in traveler's diarrhea. (986)

ehrlichiosis (ar-lik″e-o′sis) A tick-borne (*Dermacentor andersoni, Amblyomma americanum*) rickettsial disease caused by *Ehrlichia chaffeensis.* Once inside leukocytes, a nonspecific illness develops that resembles Rocky Mountain spotted fever. (960)

electron acceptor A compound that accepts electrons in an oxidation-reduction reaction. Often called an oxidizing agent or oxidant. (172)

electron donor An electron donor in an oxidation-reduction reaction. Often called a reducing agent or reductant. (172)

electron transport chain A series of electron carriers that operate together to transfer electrons from donors such as NADH and $FADH_2$ to acceptors such as oxygen. Also called an electron transport system (ETS). (173, 200)

electrophoresis (e-lek″tro-fo-re′sis) A technique that separates substances through differences in their migration rate in an electrical field. (366, 393)

electroporation ((e-lek″tro-pə-ra′shən) The application of an electric field to create temporary pores in the plasma membrane in order to insert DNA into the cell and transform it. (371)

elementary body (EB) A small, dormant body that serves as the agent of transmission between host cells in the chlamydial life cycle. (531)

elongation cycle The cycle in protein synthesis that results in the addition of an amino acid to the growing end of a peptide chain. (283)

Embden-Meyerhof pathway (em′den mi′er-hof) A pathway that degrades glucose to pyruvate; the six-carbon stage converts glucose to fructose 1,6-bisphosphate, and the three-carbon stage produces ATP while changing glyceraldehyde 3-phosphate to pyruvate. (194)

embryonic stem cells Cells derived from an early embryo that are pluripotent; that is, they are capable of differentiating into any cell type. (376)

encystment (en-sis-′ta′shen) The formation of a cyst. (608)

endemic disease (en-dem′ik) A disease that is commonly or constantly present in a population, usually at a relatively steady low frequency. (886)

endemic (murine) typhus (mu′rin ti′fus) A form of typhus fever caused by the rickettsia *Rickettsia typhi* that occurs sporadically in individuals who come into contact with rats and their fleas. (961)

endergonic reaction (end″er-gon′ik) A reaction that does not spontaneously go to completion as written; the standard free energy change is positive, and the equilibrium constant is less than one. (170)

endocytosis (en″do-si-to′sis) The process in which a cell takes up solutes or particles by enclosing them in vesicles pinched off from its plasma membrane. It often occurs at regions of the plasma membrane are coated by proteins such as clathrin and caveolin. Endocytosis involving these proteins is called clathrin-dependent endocytosis and caveolae-dependent endocytosis, respectively. (86)

endogenote (en″do-je′nōt) The genome of a procaryotic cell that acts as a recipient during horizontal gene transfer; transferred DNA can integrate into the recipient's genome. (330)

endogenous infection (en-doj′ĕ-nus in-fek′shun) An infection by a member of an individual's own normal body microbiota. (769)

endogenous pyrogen (en-doj′ĕ-nus pi′ro-jen) A host-derived chemical mediator that acts on the hypothalamus, stimulating a rise in core body temperature (i.e., it stimulates the fever response). One example of an endogenous pyrogen is the white blood cell product interleukin-1. (769, 830)

endomycorrhizal Referring to a mutualistic association of fungi and plant roots in which the fungus penetrates into the root cells and arbuscules and vesicles are formed. (698)

endoparasite (en″do-par′ah-sīt) A parasite that lives inside the body of its host. (816)

endophyte (en′do-fīt) A microorganism living within a plant, but not necessarily parasitic on it. (696)

endoplasm *See* ectoplasm.

endoplasmic reticulum (ER) (en″do-plas′mik rĕ-tik′u-lum) A system of membranous tubules and flattened sacs (cisternae) in the cytoplasmic matrix of eucaryotic cells. Rough endoplasmic reticulum (RER) bears ribosomes on its surface; smooth endoplasmic reticulum (SER) lacks them. (84)

endosome (en′do-sōm) A membranous vesicle formed by endocytosis. (85)

endospore (en′do-spōr) An extremely heat- and chemical-resistant, dormant, thick-walled spore that develops within some gram-positive bacteria. (73)

endosymbiont (en″do-sim′be-ont) An organism that lives within the body of another organism in a symbiotic association. (717)

endosymbiosis (en″do-sim″bi-o′sis) A type of symbiosis in which one organism is found within another organism. (717)

endosymbiotic theory or **hypothesis** The theory that the eucaryotic organelles mitochondria and chloroplasts arose when bacteria established an endosymbiotic relationship with ancestral cells and then evolved into organelles. (476)

endotoxin (en″do-tox′sin) The lipid A component of gram-negative bacterial cell wall lipopolysaccharide (LPS) that is released from bacteria upon their death. Nanogram quantities can induce fever, activate complement and coagulation cascades, act as a mitogen to B cells, and stimulate cytokine release from a variety of cells. Systemic effects of endotoxin are referred to as endotoxic shock. (829)

endotoxin unit (E. U.) The endotoxin activity of 0.2 ng of Reference Endotoxin Standard, as defined by the U.S. Food and Drug Association. (830)

end product inhibition *See* feedback inhibition. (183)

energy The capacity to do work or cause particular changes. (169)

enhancer A site in the DNA to which a eucaryotic activator protein binds. (313)

enology The science of wine making. (1041)

enteric bacteria (enterobacteria) (en-ter′ik) Members of the family *Enterobacteriaceae* (gram-negative, peritrichous or nonmotile, facultatively anaerobic, straight rods with simple nutritional requirements); also used for bacteria that live in the intestinal tract. (558)

enteroaggregative *E. coli* (EAggEC) A toxin-producing strain of *E. coli* associated with persistent watery, bloody diarrhea and cramping in young children. EAggEC resemble ETEC strains in that the bacteria adhere to the intestinal mucosa and cause nonbloody diarrhea without invading or causing inflammation. However, EAggEC aggressively attack epithelial cells in culture, causing aggregation. (987)

enterohemorrhagic *E. coli* (EHEC) (en′tər-o-hem″ə-raj′ik) EHEC strains of *E. coli* (O157:H7) produce several cytotoxins that provoke fluid secretion in traveler's diarrhea. (987)

enteroinvasive *E. coli* (EIEC) (en′tər-o-in-va′siv) EIEC strains of *E. coli* cause traveler's diarrhea by penetrating and binding to the intestinal epithelial cells. EIEC may also produce a cytotoxin and enterotoxin. (986)

enteropathogenic *E. coli* (EPEC) (en′tər-o-path-o-jen′ik) EPEC strains of *E. coli* attach to the brush border of intestinal epithelial cells and cause a specific type of cell damage called effacing lesions that lead to traveler's diarrhea. (986)

enterotoxigenic *E. coli* (ETEC) (en′tər-o-tok″sĭ-jen′ik) ETEC strains of *E. coli* produce two plasmid-encoded enterotoxins (which are responsible for traveler's diarrhea) and are distinguished by their heat stability: heat-stable enterotoxin (ST) and heat-labile enterotoxin (LT). (986)

enterotoxin (en″ter-o-tok′sin) A toxin specifically affecting the cells of the intestinal mucosa, causing vomiting and diarrhea. (825, 979)

Entner-Doudoroff pathway A pathway that converts glucose to pyruvate and glyceraldehyde 3-phosphate by producing 6-phosphogluconate and then dehydrating it. (198)

entropy (en′tro-pe) A measure of the randomness or disorder of a system; a measure of that part of the total energy in a system that is unavailable for useful work. (169)

envelope (en′vĕ-lōp) 1. All the structures outside the plasma membrane in bacterial cells. 2. In virology it is an outer membranous layer that surrounds the nucleocapsid in some viruses. (412)

environmental genomics *See* metagenomics.

enzootic (en″zo-ot′ik) The moderate prevalence of a disease in a given animal population. (887)

enzyme (en′zīm) A protein catalyst with specificity for both the reaction catalyzed and its substrates. (176)

enzyme-linked immunosorbent assay (ELISA) A serological assay in which bound antigen or antibody is detected by another antibody that is conjugated to an enzyme. The enzyme converts a colorless substrate to a colored product reporting the antibody capture of the antigen. (877)

eosinophil (e″o-sin′o-fil) A polymorphonuclear leukocyte that has a two-lobed nucleus and cytoplasmic granules that stain yellow-red. A mobile phagocyte that is highly antiparasitic. (747)

epidemic (ep″ĭ-dem′ik) A disease that suddenly increases in occurrence above the normal level in a given population. (886)

epidemic (louse-borne) typhus (ep″ĭ-dem′ik tĭ′fus) A disease caused by *Rickettsia prowazekii* that is transmitted from person to person by the body louse. (960)

epidemiologist (ep″ĭ-de″me-ol′o-jist) A person who specializes in epidemiology. (886)

epidemiology (epi″-de″me-ol′o-je) The study of the factors determining and influencing the frequency and distribution of disease, injury, and other health-related events and their causes in defined human populations. (885)

epiphyte An organism that grows on the surface of plants. (696)

episome (ep′ĭ-sōm) A plasmid that can exist either independently of the host cell's chromosome or be integrated into it. (53, 334)

epitheca (ep″ĭ-the′kah) The larger of two halves of a diatom frustule (shell). (622)

epitope (ep′i-tōp) An area of the antigen molecule that stimulates the production of, and combines with, specific antibodies; also known as the antigenic determinant site. (774)

epizoonotic The word used to indicate a disease transferred from a human to an animal. (923)

epizootic (ep″ĭ-zo-ot′ik) A sudden outbreak of a disease in an animal population. (887)

epizootiology (ep″i-zo-ot″e-ol′o-je) The field of science that deals with factors determining the frequency and distribution of a disease within an animal population. (887)

Epsilonproteobacteria One of the five classes of proteobacteria, each with distinctive 16S rRNA sequences. Slender gram-negative rods, some of which are medically important (*Campylobacter* and *Helicobacter*). (567)

equilibrium (e″kwĭ-lib′re-um) The state of a system in which no net change is occurring and free energy is at a minimum; in a chemical reaction at equilibrium, the rates in the forward and reverse directions exactly balance each other out. (170)

equine encephalitis An inflammatory disease of the brain caused by a virus that is transmitted by a mosquito from an infected horse to a human. There is a vaccine for horses but to date there is no vaccine for humans. (922)

ergot (er′got) The dried sclerotium of *Claviceps purpurea*. Also, an ascomycete that parasitizes rye and other higher plants causing the disease called ergotism. (637)

ergotism (er′got-izm) The disease or toxic condition caused by eating grain infected with ergot; it is often accompanied by gangrene, psychotic delusions, nervous spasms, abortion, and convulsions in humans and in animals. (637, 1027)

erysipelas (er″ĭ-sip′ĕ-las) An acute inflammation of the dermal layer of the skin, occurring primarily in infants and persons over 30 years of age with a history of streptococcal sore throat. (957)

erythema infectiosum (er″ə-the′-mə) A disease in children caused by the parvovirus B19. This disease is common in children between 4 and 11 years of age and is sometimes called fifth disease, since it was the fifth of six erythematous rash diseases in children in an older classification. (935)

erythromycin (ĕ-rith″ro-mi′sin) An intermediate spectrum macrolide antibiotic produced by *Saccharopolyspora erythraea*. (846)

eschar (es′kar) A slough produced on the skin by a thermal burn, gangrene, or the anthrax bacillus. (989)

Eucarya The domain that contains organisms composed of eucaryotic cells with primarily glycerol fatty acyl diesters in their membranes and eucaryotic rRNA. (489)

eucaryotic cells (u″kar-e-ot′ik) Cells that have a membrane-delimited nucleus and differ in many other ways from procaryotic cells; protists, fungi, plants, and animals are all eucaryotic. (2, 96)

euglenids (u-gle′nids) A group of protists (super group *Excavata*) that includes chemoorganotrophs and photoautotrophs with chloroplasts containing chlorophyll *a* and *b*. They usually have a stigma and one or two flagella emerging from an anterior reservoir. (612)

Eumycetozoa Protists called cellular and acellular slime molds that were long thought to be fungi. *See also* acellular slime mold, cellular slime mold. (614)

eumycotic mycetoma (mi″se-to′mah) *See* maduromycosis. (1010)

eutrophic (u-trof′ik) A nutrient-enriched environment. (683)

eutrophication (u″tro-fĭ-ka′shun) The enrichment of an aquatic environment with nutrients. (684)

evolutionary distance A quantitative indication of the number of positions that differ between two aligned macromolecules, and presumably a measure of evolutionary similarity between molecules and organisms. (489)

exanthem subitum A term that means "sudden rash"; sometimes refers to "sixth disease." (934)

excision repair A type of DNA repair mechanism in which a section of a strand of damaged DNA is excised and replaced, using the complementary strand as a template. Two types are recognized: base repair and nucleotide excision repair. (326)

excystment (ek″sis-ta′shun) The escape of one or more cells or organisms from a cyst. (608)

exergonic reaction (ek″ser-gon′ik) A reaction that spontaneously goes to completion as written; the standard free energy change is negative, and the equilibrium constant is greater than one. (170)

exfoliative toxin (eks-fo′le-a″tiv) or **exfoliatin** (eks-fō″le-a′tin) An exotoxin produced by *Staphylococcus aureus* that causes the separation of epidermal layers and the loss of skin surface layers. It produces the symptoms of the scalded skin syndrome. (970)

exit site (E site) The location on a ribosome to which an empty (uncharged) tRNA moves from the P site before it finally leaves during protein synthesis. (284)

exoenzymes (ek″so-en′zīms) Enzymes that are secreted by cells. (58)

exogenote (eks″o-je′nōt) The piece of donor DNA that enters a procaryotic cell during horizontal gene transfer. (330)

exon (eks′on) The region in a split or interrupted gene that codes for RNA which ends up in the final product (e.g., mRNA). (273)

exotoxin (ek″so-tok′sin) A heat-labile, toxic protein produced by a bacterium as a result of its normal metabolism or because of the acquisition of a plasmid or prophage. It is usually released into the bacterium's surroundings. (824)

exponential phase (eks″po-nen′shul) The phase of the growth curve during which the microbial population is growing at a constant and maximum rate, dividing and doubling at regular intervals. (123)

expressed sequence tag (EST) A partial gene sequence unique to a gene that can be used to identify and position the gene during genomic analysis. (390)

expression vector A special cloning vector used to express a recombinant gene in host cells; the gene is transcribed and its protein synthesized. (372)

exteins Polypeptide sequences of precursor self-splicing proteins that are joined together during formation of the final, functional protein. They are separated from one another by intein sequences, which they flank. (288)

extracutaneous sporotrichosis (spo″ro-tri-ko′sis) An infection by the fungus *Sporothrix schenckii* that spreads throughout the body. (1010)

extreme barophilic bacteria Bacteria that require a high-pressure environment to function. (658)

extreme environment An environment in which physical factors such as temperature, pH, salinity, and pressure are outside of the normal range for growth of most microorganisms; these conditions allow unique organisms to survive and function. (658)

extremophiles Microorganisms that grow under harsh or extreme environmental conditions such as very high temperatures or low pHs. (132, 658)

extrinsic factor An environmental factor such as temperature that influences microbial growth in food. (1024)

F

facilitated diffusion Diffusion across the plasma membrane that is aided by a carrier protein. (106)

facultative anaerobes (fak′ul-ta″tiv an-a′er-ōbs) Microorganisms that do not require oxygen for growth, but do grow better in its presence. (139)

facultative psychrophile (fak′ul-ta″tiv si′kro-fīl) *See* psychrotroph. (138)

Fas-FasL pathway Fas is the nomenclature representing the CD95 receptor. FasL is the ligand to the CD95 receptor. (782)

***fas* gene** The gene that is active in target cells which are susceptible to killing by cells expressing the Fas ligand, a member of the TNF family of cytokines and cell surface molecules. (782)

fatty acid synthase (sin′thĕ-tās) The multienzyme complex that makes fatty acids; the product usually is palmitic acid. (242)

fecal coliform (fe′kal ko′lĭ-form) Coliforms whose normal habitat is the intestinal tract and that can grow at 44.5°C. They are used as indicators of fecal pollution of water. (1052)

fecal enterococci (fe′kal en″ter-o-kok′si) Enterococci found in the intestine of humans and other warm-blooded animals. (1052)

feedback inhibition A negative feedback mechanism in which a pathway end product inhibits the activity of an enzyme in the sequence leading to its formation; when the end product accumulates in excess, it inhibits its own synthesis. (183)

fermentation (fer″men-ta′shun) An energy-yielding process in which an organic molecule is oxidized without an exogenous electron acceptor. Usually pyruvate or a pyruvate derivative serves as the electron acceptor. (207, 1064)

fever A complex physiological response to disease mediated by pyrogenic cytokines and characterized by a rise in core body temperature and activation of the immune system. (769)

fever blister *See* cold sore. (931)

F factor The fertility factor, a plasmid that carries genes for bacterial conjugation and makes its *E. coli* host the gene donor during conjugation. (53, 336)

fifth disease A mild childhood illness presenting as flulike symptoms and a rash caused by the human parvovirus B19. (935)

filopodia Long, narrow pseudopodia found in certain amoeboid protists. (613)

fimbria (fim′bre-ah; pl., **fimbriae**) A fine, hairlike protein appendage on some gram-negative bacteria that helps attach them to surfaces. (66)

final host The host on/in which a parasite either attains sexual maturity or reproduces. (816)

first law of thermodynamics Energy can be neither created nor destroyed (even though it can be changed in form or redistributed). (169)

fixation (fik-sa′shun) The process in which the internal and external structures of cells and organisms are preserved and fixed in position. (25)

flagellin (flaj′ĕ-lin) The protein used to construct the filament of a bacterial flagellum. (67)

flagellum (flah-jel′um; pl., **flagella**) A thin, threadlike appendage on many procaryotic and eucaryotic cells that is responsible for their motility. (67, 95)

flat or **plane warts** Small, smooth, slightly raised warts. (938)

flavin adenine dinucleotide (FAD) (fla′vin ad′ĕ-nēn) An electron carrying cofactor often involved in energy production (for example, in the tricarboxylic acid cycle and the β-oxidation pathway). (173)

flow cytometry A tool for defining and enumerating cells using a capillary tube to control cell movement and a laser to detect cell size and morphology. Individual cells pass through the capillary tube and through the laser beam. Laser light detectors connected to a computer analyze the light patterns to identify cells. Fluorescent materials can also be used to label cells prior to evaluation. (881)

fluid mosaic model The currently accepted model of cell membranes in which the membrane is a lipid bilayer with integral proteins buried in the lipid, and peripheral proteins more loosely attached to the membrane surface. (46)

fluid-phase endocytosis A form of endocytosis in which a small portion of extracellular fluid is pinched off nonselectively and without concentration of the fluid contents. (86)

fluorescence in situ hybridization (FISH) A technique for identifying certain genes or organisms, in which specific DNA fragments are labeled with fluorescent dye and hybridized to the chromosomes of interest. (678)

fluorescence microscope A microscope that exposes a specimen to light of a specific wavelength and then forms an image from the fluorescent light produced. Usually the specimen is stained with a fluorescent dye or fluorochrome. (23)

fluorescent light (floo″o-res′ent) The light emitted by a substance when it is irradiated with light of a shorter wavelength. (23)

fomite (fo′mīt; pl., **fomites**) An object that is not in itself harmful but is able to harbor and transmit pathogenic organisms. Also called fomes. (820, 894)

food-borne infection Gastrointestinal illness caused by ingestion of microorganisms, followed by their growth within the host. Symptoms arise from tissue invasion and/or toxin production. (979, 1032)

food intoxication Food poisoning caused by microbial toxins produced in a food prior to consumption. The presence of living bacteria is not required. (979, 1034)

food poisoning A general term usually referring to a gastrointestinal disease caused by the ingestion of food contaminated by pathogens or their toxins. (979)

forced evolution *See* adaptive mutation.

F_1 particle Component of the ATPase on the inner mitochondrial membrane, which is the site of ATP synthesis by oxidative phosphorylation. (202)

F′ plasmid An F plasmid that carries some bacterial genes and transmits them to recipient cells when the F′ cell carries out conjugation. (339)

fragmentation (frag″men-ta′shun) A type of asexual reproduction among filamentous microbes in which hyphae break into two or more parts, each of which forms new hyphae. (120)

frameshift mutations Mutations arising from the loss or gain of a base or DNA segment, leading to a change in the codon reading frame and thus a change in the amino acids incorporated into protein. (323)

free energy change The total energy change in a system that is available to do useful work as the system goes from its initial state to its final state at constant temperature and pressure. (170)

French polio *See* Guillain-Barré syndrome. (918)

fruiting body A specialized structure that holds sexually or asexually produced spores; found in fungi and in some bacteria (e.g., the myxobacteria). (564)

frustule (frus′tūl) A silicified cell wall in the diatoms. (622)

fumonisin A family of toxins produced by mold belonging to the genus *Fusarium*. It primarily affects corn and it is known to be hepato- and nephrotoxic in animals. (1027)

fungicide (fun′jĭ-sīd) An agent that kills fungi. (151)

fungistatic (fun″jĭ-stat′ik) Inhibiting the growth and reproduction of fungi. (151)

fungus (fung′gus; pl., **fungi**) Achlorophyllous, heterotrophic, spore-bearing eucaryotes with absorptive nutrition; usually, they have a walled thallus. (3, 629)

***F* value** The time in minutes at a specific temperature (usually 250°F) needed to kill a population of cells or spores. (154)

G

gametangium gam-ĕ-tan′je-um; pl., **gametangia**) A structure that contains gametes or in which gametes are formed. (634)

Gammaproteobacteria One of the five classes of proteobacteria, each with distinctive 16S rRNA sequences. This is the largest subgroup and is very diverse physiologically; many important genera are facultatively anaerobic chemoorganotrophs. (551)

gamonts Gametic cells formed by protists when they undergo sexual reproduction. (609)

gas gangrene (gang′grēn) A type of gangrene that arises from dirty, lacerated wounds infected by anaerobic bacteria, especially species of *Clostridium*. As the bacteria grow, they release toxins and ferment carbohydrates to produce carbon dioxide and hydrogen gas. (965)

gastritis (gas-tri′tis) Inflammation of the stomach. (967)

gastroenteritis (gas″tro-en-ter-i′tis) An acute inflammation of the lining of the stomach and intestines, characterized by anorexia, nausea, diarrhea, abdominal pain, and weakness. It has various causes including food poisoning due to such organisms as *E. coli, S. aureus, Campylobacter* (camp lobacteriosis), and *Salmonella* species; consumption of irritating food or drink; or psychological factors such as anger, stress, and fear. Also called enterogastritis. (979)

gastrointestinal anthrax The intestinal disease form of anthrax characterized by nausea, loss of appetite, vomiting, fever, and followed by abdominal pain, vomiting of blood, and severe diarrhea. Intestinal anthrax is fatal in 25 to 60% of cases. (988)

gas vacuole A gas-filled vacuole found in cyanobacteria and some other aquatic bacteria that provides flotation. It is composed of gas vesicles, which are made of protein. (50)

gene (jēn) A DNA segment or sequence that codes for a polypeptide, rRNA, or tRNA. (251)

gene gun A device that uses high-pressure gas or another propellant to shoot a spray of DNA-coated microprojectiles into cells and transform them. Sometimes it is called a biolistic device. (371)

gene therapy The process by which human diseases are treated or potentially cured by the introduction of a gene(s) that encodes the gene product(s) that is either missing or mutated in the cells affected by the pathological condition. (376)

generalized transduction The transfer of any part of a procaryotic genome when the DNA fragment is packaged within a virus capsid by mistake. (345)

general secretion pathway (GSP) *See* Sec dependent pathway.

generation time The time required for a microbial population to double in number. (126)

genetic engineering The deliberate modification of an organism's genetic information by directly changing its nucleic acid genome. (357)

genital herpes (her′pēz) A sexually transmitted disease caused by the herpes simplex virus type 2. (933)

genital ulcer disease *See* chancroid. (971)

genome (je′nōm) The full set of genes present in a cell or virus; all the genetic material in an organism; a haploid set of genes in a cell. (247)

genome fusion hypothesis A hypothesis that seeks to explain the origin of the nucleus. It posits that certain archaeal and bacterial genes were combined to form a single eucaryotic genome. (475)

genomic fingerprinting A series of techniques based on restriction enzyme digestion patterns that enable the comparison of microbial species and strains and is thus useful in taxonomic identification. (478)

genomic library (je-nom′ik) The collection of clones that contains fragments which represent the complete genome of an organism. (370)

genomic reduction The decrease in genomic information that occurs over evolutionary time as an organism or organelle becomes increasingly dependent on another cell or a host organism. (732)

genomics The study of the molecular organization of genomes, their information content, and the gene products they encode. (383)

genotypic classification The use of genetic data to construct a classification scheme for the identification of an unknown species or the phylogeny of a group of microbes. (478)

genus (je′nəs) A well-defined group of one or more species that is clearly separate from other organisms. (481)

German measles *See* rubella.

germicide (jer′mĭ-sīd) An agent that kills pathogens and many nonpathogens but not necessarily bacterial endospores. (151)

germination (jer″mĭ-na′shun) The stage following spore activation in which the spore breaks its dormant state. Germination is followed by outgrowth. (75)

Ghon complex (gon) The initial focus of parenchymal infection in primary pulmonary tuberculosis. (954)

giardiasis (je″ar-di′ah-sis) A common intestinal disease caused by the parasitic protozoan *Giardia intestinalis.* (1014)

gingivitis (jin-jĭ-vi′tis) Inflammation of the gingival tissue. (994)

gingivostomatitis (jin″jĭ-vo-sto″mə-ti′tis) Inflammation of the gingiva and other oral mucous membranes. (931)

gliding motility A type of motility in which a microbial cell glides along a solid surface. (70, 525)

global regulatory systems Regulatory systems that simultaneously affect many genes and pathways. (307)

glomerulonephritis (glo-mer″u-lo-nĕ-fri′tis) An inflammatory disease of the renal glomeruli. (958)

glucans Polysaccharides composed of glucose units held together by glycosidic linkages. Some types of glucans have $\alpha(1\rightarrow3)$ and $\alpha(1\rightarrow6)$ linkages and bind bacterial cells together on teeth forming a plaque ecosystem. (991)

gluconeogenesis (gloo″ko-ne″o-jen′e-sis) The synthesis of glucose from noncarbohydrate precursors such as lactate and amino acids. (230)

glycocalyx (gli″ko-kal′iks) A network of polysaccharides extending from the surface of bacteria and other cells. (65)

glycogen (gli′ko-jen) A highly branched polysaccharide containing glucose, which is used to store carbon and energy. (49)

glycolysis (gli-kol′ĭ-sis) The conversion of glucose to pyruvic acid by use of the Embden-Meyerhof pathway, pentose phosphate pathway, or Entner-Douderoff pathway. (194)

glycolytic pathway (gli″ko-lit″ik) A pathway that converts glucose to pyruvic acid (e.g., Embden-Meyerhof pathway). (194)

glyoxylate cycle (gli-ok′sĭ-lat) A modified tricarboxylic acid cycle in which the decarboxylation reactions are bypassed by the enzymes isocitrate lyase and malate synthase; it is used to convert acetyl-CoA to succinate and other metabolites. (240)

gnotobiotic (no″to-bi-ot′ik) Animals that are germfree (microorganism free) or live in association with one or more known microorganisms. (734)

Golgi apparatus (gol′je) A membranous eucaryotic organelle composed of stacks of flattened sacs (cisternae), which is involved in packaging and modifying materials for secretion and many other processes. (85)

gonococci (gon′o-kok′si) Bacteria of the species *Neisseria gonorrhoeae*—the organism causing gonorrhea. (974)

gonorrhea (gon″o-re′ah) An acute infectious sexually transmitted disease of the mucous membranes of the genitourinary tract, eye, rectum, and throat. It is caused by *Neisseria gonorrhoeae.* (974)

Gram stain A differential staining procedure that divides bacteria into gram-positive and gram-negative groups based on their ability to retain crystal violet when decolorized with an organic solvent such as ethanol. (26)

grana (gra′nah) A stack of thylakoids in the chloroplast stroma. (90)

granulocyte A type of white blood cell that stores preformed molecules (enzymes and antimicrobial proteins) in vacuoles near the cell membrane. The appearance of these granulelike vacuoles in cells led to the term granulocyte. (746)

granuloma (gran″u-lo′mə) Term applied to nodular inflammatory lesions containing phagocytic cells. (757)

greenhouse gases Gases (e.g., CO_2, CH_4) released from the Earth's surface through chemical and biological processes that interact with the chemicals in the stratosphere to decrease the release of radiation from the Earth. This leads to global warming. (648)

green nonsulfur bacteria Anoxygenic photosynthetic bacteria that contain bacteriochlorophylls *a* and *c;* usually photoheterotrophic and display gliding motility. Include members of the phylum *Chloroflexi.* (523)

green sulfur bacteria Anoxygenic photosynthetic bacteria that contain bacteriochlorophylls *a,* plus *c, d* or *e;* photolithoautotrophic; use H_2, H_2S, or S as electron donor. Include members of the phylum *Chlorobi.* (523)

griseofulvin (gris″e-o-ful′vin) An antibiotic from *Penicillium griseofulvum* given orally to treat chronic dermatophytic infections of skin and nails. (854)

group A streptococcus (GAS) A gram-positive, coccus-shaped bacterium often found in the throat and on the skin of humans, having the A group of surface carbohydrate. Infections are relatively mild, but can cause other severe and even life-threatening diseases. *See* streptococcal pharyngitis. (956)

group B streptococcus (GBS) A gram-positive, coccus-shaped bacterium found occasionally on mucous membranes of humans, having the B group of surface carbohydrate. GBS is a common cause of pneumonia, meningitis, and sepsis. It can infect newborns in the first week of life ("early-onset") and cause life-threatening disease. A slightly less-serious "late-onset" form of disease that develops weeks to months after birth can also occur. (965)

group translocation A transport process in which a molecule is moved across a membrane by carrier proteins while being chemically altered at the same time. (109)

growth An increase in cellular constituents. (119)

growth factors Organic compounds that must be supplied in the diet for growth because they are essential cell components or precursors of such components and cannot be synthesized. (105)

guanine (gwan′in) A purine derivative, 2-amino-6-oxypurine, found in nucleosides, nucleotides, and nucleic acids. (241)

Guillain-Barré syndrome (ge-yan′bar-ra′) A relatively rare disease affecting the peripheral nervous system, especially the spinal nerves, but also the cranial nerves. The cause is unknown, but it most often occurs after an influenza infection or flu vaccination. Also called French Polio. (918)

gumma (gum′ah) A soft, gummy tumor occurring in tertiary syphilis. (976)

gut-associated lymphoid tissue (GALT) The defensive lymphoid tissue present in the intestines. *See* Peyer's patches. (759)

H

H-2 complex Term for the MHC in the mouse. (000)

halobacteria or **extreme halophiles** A group of archaea that have an absolute dependence on high NaCl concentrations for growth and will not survive at a concentration below about 1.5 M NaCl. (514)

halophile (hal′o-fīl) A microorganism that requires high levels of sodium chloride for growth. (133)

halotolerant The ability to withstand large changes in salt concentration. (673)

Hansen's disease *See* leprosy.

hantavirus pulmonary syndrome (HPS) The name given to an infectious lung disease caused by at least four different hantaviruses. (942)

hapten (hap′ten) A molecule not immunogenic by itself that, when coupled to a macromolecular carrier, can elicit antibodies directed against itself. (776)

harborage transmission The mode of transmission in which an infectious organism does not undergo morphological or physiological changes within the vector. (896)

harmful algal bloom (HAB) In an aquatic or marine ecosystem, the growth of a single population of phototroph, either a protist (e.g., diatom, dinoflagellate) or a cyanobacterium, that produces a toxin that is poisonous to other organisms, sometimes including humans. (676)

Harting net The area of nutrient exchange between ectomycorrhizal fungal hyphae and plant host cells. The fungal hyphae grow between plant root cells, enmeshing specific root cells by forming a hyphal network called the Harting net. (698)

hay fever Allergic rhinitis; a type of atopic allergy involving the upper respiratory tract. (803)

health (helth) A state of optimal physical, mental, and social well-being, and not merely the absence of disease and infirmity. (885)

healthy carrier An individual who harbors a pathogen, but is not ill. (892)

heat-shock proteins Proteins produced when cells are exposed to high temperatures or other stressful conditions. They protect the cells from damage and often aid in the proper folding of proteins. (287)

helical symmetry (hel′ĭ-kal) In virology this refers to a virus with a helical capsid surrounding its nucleic acid. (410)

helicases Enzymes that use ATP energy to unwind DNA ahead of the replication fork. (260)

hemadsorption (hem″ad-sorp′shun) The adherence of red blood cells to the surface of something, such as another cell or a virus. (866)

hemagglutination (hem″ah-gloo″tĭ-na′shun) The agglutination of red blood cells by antibodies or components of virus capsids. (422)

hemagglutinin (hem″ah-gloo′tĭ-nin) The antibody responsible for a hemagglutination reaction. (422)

hematopoesis The process by which blood cells develop into specific lineages from stem cells. Red and white blood cells and platelets develop from this process. (744)

hemolysin (he-mol′ĭ-sin) A substance that causes hemolysis (the lysis of red blood cells). At least some hemolysins are enzymes that destroy the phospholipids in erythrocyte plasma membranes. (828)

hemolysis (he-mol′ĭ-sis) The disruption of red blood cells and release of their hemoglobin. There are several types of hemolytic reactions when bacteria such as streptococci and staphylococci grow on blood agar. In α-hemolysis, a greenish zone of incomplete hemolysis forms around the colony. A clear zone of complete hemolysis without any obvious color change is formed during β-hemolysis. (828)

hemolytic uremic syndrome A kidney disease characterized by blood in the urine and often by kidney failure. It is caused by enterohemorrhagic strains of *Escherichia coli* O157:H7 that produce a Shiga-like toxin, which attacks the kidneys. (987)

hemorrhagic fever A fever usually caused by a specific virus that may lead to hemorrhage, shock, and sometimes death. (923)

hepatitis (hep″ah-ti′tis) Any infection that results in inflammation of the liver. Also refers to liver inflammation. (936)

hepatitis A (formerly infectious hepatitis) A type of hepatitis that is transmitted by fecal-oral contamination; it primarily affects children and young adults, especially in environments where there is poor sanitation and overcrowding. It is caused by the hepatitis A virus, a single-stranded RNA virus. (939)

hepatitis B (formerly serum hepatitis) This form of hepatitis is caused by a double-stranded DNA virus (HBV) formerly called the "Dane particle." The virus is transmitted by body fluids. (936)

hepatitis C A liver disease caused by a virus that is spread through infected blood, primarily in those who use illicit drugs and those who received blood transfusions prior to 1992. There is no vaccine. (937)

hepatitis D (formerly delta hepatitis) The liver diseases caused by the hepatitis D virus in those individuals already infected with the hepatitis B virus. (938)

hepatitis E (formerly enteric-transmitted NANB hepatitis) The liver disease caused by the hepatitis E virus. Usually, a subclinical, acute infection results; however, there is a high mortality in women in their last trimester of pregnancy. (940)

hepatitis G A liver disease caused a distant relative of hepatitis C virus. The virus appears to be transmitted through transfusions, though its role in acute and chronic hepatitis remains unclear. (938)

herd immunity The resistance of a population to infection and spread of an infectious agent due to the immunity of a high percentage of the population. (890)

herpes labialis *See* cold sore. (931)

herpetic keratitis (her-pet′ik ker″ah-ti′tis) An inflammation of the cornea and conjunctiva of the eye resulting from a herpes simplex virus infection. (932)

heterocysts Specialized cells of cyanobacteria that are the sites of nitrogen fixation. (525)

heteroduplex DNA A double-stranded stretch of DNA formed by two slightly different strands that are not completely complementary. (331)

heterokont flagella A pattern of flagellation found in the protist subdivision *Stramenopila,* featuring two flagella, one extending anteriorly and the other posteriorly. (621)

heterolactic fermenters (het″er-o-lak′tik) Microorganisms that ferment sugars to form lactate, and also other products such as ethanol and CO_2. (208)

heterologous gene expression The cloning, transcription, and translation of a gene that has been introduced (cloned) into an organism that normally does not possess the gene. (1061)

heterotroph (het′er-o-trōf″) An organism that uses reduced, preformed organic molecules as its principal carbon source. (102)

heterotrophic nitrification Nitrification carried out by chemoheterotrophic microorganisms. (000)

hexon or **hexamer** A virus capsomer composed of six protomers. (411, 426)

hexose monophosphate pathway (hek′sōs mon″o-fos′fāt) *See* pentose phosphate pathway. (196)

Hfr strain A bacterial strain that donates its genes with high frequency to a recipient cell during conjugation because the F factor is integrated into the bacterial chromosome. (339)

hierarchical cluster analysis The organization of microarray data such that induced and repressed genes are grouped separately. (402)

high-energy molecule A molecule whose hydrolysis under standard conditions makes available a large amount of free energy (the standard free energy change is more negative than about −7 kcal/mole); a high-energy molecule readily decomposes and transfers groups such as phosphate to acceptors. (171)

high-throughput screening (HTS) A system that combines liquid handling devices, robotics, computers, data processing, and a sensitive detection system to screen thousands of compounds for a single capability. It is often used by pharmaceutical companies to identify natural products that have potentially useful applications. (1063)

histatin An antimicrobial peptide composed of 24 to 38 amino acids, heavily enriched with histidine. Histatin enters the fungal cytoplasm where it targets mitochondria. (762)

histone (his′tōn) A small basic protein with large amounts of lysine and arginine that is associated with eucaryotic DNA in chromatin. Related proteins are observed in many archaeal species, where they form archaeal nucleosomes. (253)

histoplasmosis (his″to-plaz-mo′sis) A systemic fungal infection caused by *Histoplasma capsulatum* var *capsulatum*. (1001)

hives (hīvz) An eruption of the skin. (804)

HIV protease inhibitor A drug that prevents native HIV protease enzymes from cleaving HIV polyproteins into mature proteins required for HIV virion assembly. (856)

holdfast A structure produced by some bacteria (e.g., *Caulobacter*) that attaches them to a solid object. (544)

holoenzyme A complete enzyme consisting of the apoenzyme plus a cofactor. (176)

holozoic nutrition (hol″o-zo′ik) In this type of nutrition, nutrients (such as bacteria) are acquired by endocytosis and the subsequent formation of a food vacuole or phagosome. (606)

homolactic fermenters (ho″mo-lak′tik) Organisms that ferment sugars almost completely to lactic acid. (208)

homologous recombination Recombination involving two DNA molecules that are very similar in nucleotide sequence; it can be reciprocal or nonreciprocal. (331)

horizontal (lateral) gene transfer (HGT/LGT) The process in which genes are transferred from one mature, independent organism to another. In procaryotes, transformation, conjugation, and transduction are the mechanisms by which HGT can occur. (330, 391, 490)

hormogonia Small motile fragments produced by fragmentation of filamentous cyanobacteria; used for asexual reproduction and dispersal. (525)

host (hōst) The body of an organism that harbors another organism. It can be viewed as a microenvironment that shelters and supports the growth and multiplication of another organism. (743)

host-parasite relationship The symbiosis between a pathogen and its host. The term parasite is used to imply pathogenicity in this context. (816)

host restriction The degradation of foreign genetic material by nucleases after the genetic material enters a host cell. (330)

human herpesvirus 6 (HHV-6, type A and B) (hər′pēz) HHV-6 was initially called the human B-lymphotropic virus. It was later shown to have a marked tropism for CD4$^+$ T cells and was renamed HHV-6. HHV-6 causes exanthem subitum (roseola infantum or sixth disease) in infants and may be involved opportunistic infections in immunocompromised patients, hepatitis, lymphoproliferative diseases, synergistic interactions with HIV, lymphadenitis, and chronic fatigue syndrome. (934)

human immunodeficiency virus (HIV) A lentivirus of the family *Retroviridae* that is the cause of AIDS. (925)

human leukocyte antigen complex (HLA) The major histocompatibility protein antigens on the surface of cells of human tissues and organs is recognized by the immune system cells and therefore is important in the regulation of the immune response and graft rejection. This is the same as MHC class II. Also see major histocompatibility complex. (778)

human parvovirus B19 A small, single-stranded DNA virus that causes fifth disease in young children. (935)

humoral (antibody-mediated) immunity (hu′mor-al) The type of immunity that results from the presence of soluble antibodies in blood and lymph; also known as antibody-mediated immunity. (774)

hybridoma (hi″brī-do′mah) A fast-growing cell line produced by fusing a cancer cell (myeloma) to another cell, such as an antibody-producing cell. (864)

hydrogen hypothesis A thoery that considers the origin of the eucaryotes through the development of the hydrogenosome. It suggests the organelle arose as the result of an endosymbiotic anaerobic bacterium that produced CO_2 and H_2 as the products of fermentation. (476)

hydrogenosome An organelle found in some anaerobic protists that produce ATP by fermentation. (476)

hydrophilic (hi″dro-fil′ik) A polar substance that has a strong affinity for water (or is readily soluble in water). (45)

hydrophobic (hi″dro-fo′bik) A nonpolar substance lacking affinity for water (or which is not readily soluble in water). (45)

hyperendemic disease (hi″per-en-dem′ik) A disease that has a gradual increase in occurrence beyond the endemic level, but not at the epidemic level, in a given population; also may refer to a disease that is equally endemic in all age groups. (886)

hyperferremia Excessive iron in the blood. (769)

hypermutation A rapid production of multiple mutations in a gene or genes through the activation of special mutator genes. The process may be deliberately used to maximize the possibility of creating desirable mutants. (319)

hypersensitivity (hi″per-sen′si-tiv″i-te) A condition of increased immune sensitivity in which the body reacts to an antigen with an exaggerated immune response that usually harms the individual. Also termed an allergy. (803)

hyperthermophile (hi″per-ther′mo-fīl) A bacterium that has its growth optimum between 85°C and about 120°C. Hyperthermophiles usually do not grow well below 55°C. (139, 659)

hypha (hi′fah; pl., **hyphae**) The unit of structure of most fungi and some bacteria; a tubular filament. (589, 631)

hypoferremia (hi″po-fĕ-re′me-ah) Deficiency of iron in the blood. (769)

hypotheca (hi-po-theca) The smaller half of a diatom frustule. (622)

hypoxic (hi pok′sik) Having a low oxygen level. (668)

I

icosahedral In virology this term refers to a virus with an icosahedral capsid, which has the shape of a regular polyhedron having 20 equilateral triangular faces and 12 corners. (410)

identification (i-den″tĭ-fĭ-ka′shun) The process of determining that a particular isolate or organism belongs to a recognized taxon. (478)

idiotype (id′e-o-tīp′) A set of one or more unique epitopes in the variable region of an immunoglobulin that distinguishes it from immunoglobulins produced by different plasma cells. (791)

IgA Immunoglobulin A; the class of immunoglobulins that is present in dimeric form in many body secretions (e.g., saliva, tears, and bronchial and intestinal secretions) and protects body surfaces. IgA also is present in serum. (793)

IgD Immunoglobulin D; the class of immunoglobulins found on the surface of many B lymphocytes; thought to serve as an antigen receptor in the stimulation of antibody synthesis. (794)

IgE Immunoglobulin E; the immunoglobulin class that binds to mast cells and basophils, and is responsible for type I or anaphylactic hypersensitivity reactions such as hay fever and asthma. IgE is also involved in resistance to helminth parasites. (794)

IgG Immunoglobulin G; the predominant immunoglobulin class in serum. Has functions such as neutralizing toxins, opsonizing bacteria, activating complement, and crossing the placenta to protect the fetus and neonate. (792)

IgM Immunoglobulin M; the class of serum antibody first produced during an infection. It is a large, pentameric molecule that is active in agglutinating pathogens and activating complement. The monomeric form is present on the surface of some B lymphocytes. (792)

immobilization (im-mo″bil-i-za′shun) The incorporation of a simple, soluble substance into the body of an organism, making it unavailable for use by other organisms. (646)

immune complex (ĭ-mūn′kom′pleks) The product of an antigen-antibody reaction, which may also contain components of the complement system. (799)

immune surveillance (ĭ-mūn′sur-vāl′ans) The process by which cells of the immune system police the host for nonself antigens. (802)

immune system The defensive system in a host consisting of the nonspecific (innate) and specific (adaptive) immune responses. It is composed of widely distributed cells, tissues, and organs that recognize foreign substances and microorganisms and acts to neutralize or destroy them. (743)

immunity (ĭ-mu′nĭ-te) Refers to the overall general ability of a host to resist a particular disease; the condition of being immune. (802)

immunization The deliberate introduction of foreign materials into a host to stimulate an adaptive immune response. *See* vaccine. (901)

immunoblotting The electrophoretic transfer of proteins from polyacrylamide gels to nylon or polyvinyl difluoride (PVDF) filters to demonstrate the presence of specific proteins through reaction with labeled antibodies. (879)

immunodeficiency (im″u-no-dĕ-fish′en-se; pl., **immunodeficiencies**) The inability to produce a normal complement of antibodies or immunologically sensitized T cells in response to specific antigens. (811)

immunodiffusion A technique involving the diffusion of antigen and/or antibody within a semisolid gel to produce a precipitin reaction where they meet in proper proportions. Often both the antibody and antigen diffuse through the gel; sometimes an antigen diffuses through a gel containing antibody. (879)

immunoelectrophoresis (ĭ-mu″no-e-lek″tro-fore′sis; pl., **immunoelectrophoreses**) The electrophoretic separation of protein antigens followed by diffusion and precipitation in gels using antibodies against the separated proteins. (881)

immunofluorescence (im″u-no-floo″o-res′ens) A technique used to identify particular antigens microscopically in cells or tissues by the binding of a fluorescent antibody conjugate. (865)

immunoglobulin (Ig) (im″u-no-glob′u-lin) *See* antibody.

immunology (im″u-nol′o-je) The study of host defenses against invading foreign materials, including pathogenic microorganisms, transformed or cancerous cells, and tissue transplants from other sources. (743)

immunopathology (im″u-no-pə-thol′o-je) The study of diseases or conditions resulting from immune reactions. (817)

immunoprecipitation (im″u-no-pre-sip″i-ta′shun) A reaction involving soluble antigens reacting with antibodies to form a large aggregate that precipitates out of solution. (879)

immunotoxin (im′u-no-tok″sin) A monoclonal antibody that has been attached to a specific toxin or toxic agent (antibody + toxin = immunotoxin) and can kill specific target cells. (824)

impetigo (im″pə-ti′go) This superficial cutaneous disease, most commonly seen in children, is characterized by crusty lesions, usually located on the face; the lesions typically have vesicles surrounded by a red border. It is the most frequently diagnosed skin infection caused by *S. pyogenes* (impetigo can also be caused by *S. aureus*). (957)

inclusion bodies (1) Granules of organic or inorganic material in the cytoplasmic matrix of bacteria. (2) Clusters of viral proteins or virions within the nucleus or cytoplasm of virus-infected cells. (48, 461)

inclusion conjunctivitis (in-klu′zhun kon-junk″tĭ-vi′tis) An infectious disease that occurs worldwide. It is caused by *Chlamydia trachomatis* that infects the eye and causes inflammation and the occurrence of large inclusion bodies. (966)

incubation period The period after pathogen entry into a host and before signs and symptoms appear. (888)

incubatory carrier An individual who is incubating a pathogen but is not yet ill. (892)

index case The first disease case in an epidemic within a given population. (887)

indicator organism An organism whose presence indicates the condition of a substance or environment, for example, the potential presence of pathogens. Coliforms are used as indicators of fecal pollution. (1051)

inducer (in-dūs′er) A small molecule that stimulates the synthesis of an inducible enzyme. (294)

inducible enzyme An enzyme whose level rises in the presence of a small molecule that stimulates its synthesis or activity. (294)

industrial ecology The study of the ecology of industrial societies with a major focus on material cycling, energy flow, and the ecological impacts of such societies. (1086)

infantile paralysis (in′fan-til pah-ral′i-sis) *See* poliomyelitis.

infection (in-fek′shun) The invasion of a host by a microorganism with subsequent establishment and multiplication of the agent. An infection may or may not lead to overt disease. (816)

infection thread A tubular structure formed during the infection of a root by nitrogen-fixing bacteria. The bacteria enter the root by way of the infection thread and stimulate the formation of the root nodule. (701)

infectious disease Any change from a state of health in which part or all of the host's body cannot carry on its normal functions because of the presence of an infectious agent or its products. (816)

infectious disease cycle (chain of infection) The chain or cycle of events that describes how an infectious organism grows, reproduces, and is disseminated. (891)

infectious dose 50 (ID_{50}) Refers to the dose or number of organisms that will infect 50% of an experimental group of hosts within a specified time period. (817)

infectious mononucleosis (mono) (mon″o-nu″kle-o′sis) An acute, self-limited infectious disease of the lymphatic system caused by the Epstein-Barr virus and characterized by fever, sore throat, lymph node and spleen swelling, and the proliferation of monocytes and abnormal lymphocytes. (935)

infectivity (in″fek-tiv′i-te) Infectiousness; the state or quality of being infectious or communicable. (816)

inflammation (in″flah-ma′shun) A localized protective response to tissue injury or destruction. Acute inflammation is characterized by pain, heat, swelling, and redness in the injured area. (756)

influenza or flu (in″flu-en′zah) An acute viral infection of the respiratory tract, occurring in isolated cases, epidemics, and pandemics. Influenza is caused by three strains of influenza virus, labeled types A, B, and C, based on capsid antigens. (915)

Ingoldian fungi Aquatic hyphomycetes that often have a characteristic tetraradiate hyphal development form and which sporulate under water. Discovered by the British mycologist, C. T. Ingold. (672)

initial body *See* reticulate body (RB). (531)

innate or **natural immunity** *See* nonspecific resistance. (743)

insertion sequence (in-ser′shun se′kwens) A simple transposon that contains genes only for those enzymes, such as the transposase, that are required for transposition. (332)

***in silico* analysis** The study of physiology and/or genetics through the examination of nucleic acid and amino acid sequence. *See* bioinformatics. (338)

integration The incorporation of one DNA segment into a second DNA molecule to form a new hybrid DNA. Integration occurs during such processes as genetic recombination, episome incorporation into host DNA, and prophage insertion into the bacterial chromosome. (330)

integrins (in′tə-grin) A large family of α/β heterodimers. Integrins are cellular adhesion receptors that mediate cell-cell and cell-substratum interactions. Integrins usually recognize linear amino acid sequences on protein ligands. (756)

integron A genetic element with an attachment site for site-specific recombination and an integrase gene. It can capture genes and gene cassettes. (852)

inteins Internal intervening sequences of precursor self-splicing proteins that separate exteins and are removed during formation of the final protein. (288)

intercalating agents Molecules that can be inserted between the stacked bases of a DNA double helix, thereby distorting the DNA and inducing insertion and deletion (i. e., frameshift) mutations. (320)

interdigitating dendritic cell Special dendritic cells in the lymph nodes that function as potent antigen-presenting cells and develop from Langerhans cells. (744)

interferon (IFN) (in″tər-fēr′on) A glycoprotein that has nonspecific antiviral activity by stimulating cells to produce antiviral proteins, which inhibit the synthesis of viral RNA and proteins. Interferons also regulate the growth, differentiation, and/or function of a variety of immune system cells. Their production may be stimulated by virus infections, intracellular pathogens (chlamydiae and rickettsias), protozoan parasites, endotoxins, and other agents. (768)

interleukin (in″tər-loo′kin) A glycoprotein produced by macrophages and T cells that regulates growth and differentiation, particularly of lymphocytes. Interleukins promote cellular and humoral immune responses. (767)

intermediate filaments Small protein filaments, about 8 to 10 nm in diameter, in the cytoplasmic matrix of eucaryotic cells that are important in cell structure. (83)

intermediate host The host that serves as a temporary but essential environment for development of a parasite and completion of its life cycle. (816)

interspecies hydrogen transfer The linkage of hydrogen production from organic matter by anaerobic heterotrophic microorganisms to the use of hydrogen by other anaerobes in the reduction of carbon dioxide to methane. This avoids possible hydrogen toxicity. (729)

intertriginous candidiasis A skin infection caused by *Candida* species. Involves those areas of the body, usually opposed skin surfaces, that are warm and moist (axillae, groin, skin folds). (1017)

intoxication (in-tok″si-ka′shun) A disease that results from the entrance of a specific toxin into the body of a host. The toxin can induce the disease in the absence of the toxin-producing organism. (824)

intraepidermal lymphocytes T cells found in the epidermis of the skin that express the $\gamma\delta$ T-cell receptor. (759)

intranuclear inclusion body (in″trə-noo′kle-ər) A structure found within cells infected with the cytomegalovirus. (933)

intrinsic factors Food-related factors such as moisture, pH, and available nutrients that influence microbial growth. (1024)

intron (in′tron) A noncoding intervening sequence in a split or interrupted gene, which codes for RNA that is missing from the final RNA product. (273)

invasiveness (in-va′siv-nes) The ability of a microorganism to enter a host, grow and reproduce within the host, and spread throughout its body. (816)

ionizing radiation Radiation of very short wavelength and high energy that causes atoms to lose electrons or ionize. (141, 156)

isogamy The fusion of two morphologically and physiologically similar gametes during sexual reproduction in protists. (609)

isotype (i′so-tīp) A variant form of an immunoglobulin (e.g., an immunoglobulin class, subclass, or type) that occurs in every normal individual of a particular species. Usually the characteristic antigenic determinant is in the constant region of H and L chains. (791)

J

Jaccard coefficient (S_J) An association coefficient used in numerical taxonomy; it is the proportion of characters that match, excluding those that both organisms lack. (479)

J chain A polypeptide present in polymeric IgM and IgA that links the subunits together. (792)

jock itch *See* tinea cruris. (1009)

K

kallikrein An enzyme that acts on kininogen, releasing the active bradykinin protein. (757)

keratitis (ker″ah-ti′tis) Inflammation of the cornea of the eye. (1013)

kinetoplast (ki-ne′to-plast) A special structure in the mitochondrion of certain protists. It contains the mitochondrial DNA. (90)

kinetosome Intracellular microtubular structure that serves as the base of cilia in ciliated protists. Similar in structure to a centriole. Also referred to as a basal body. (608)

Kirby-Bauer method A disk diffusion test to determine the susceptibility of a microorganism to chemotherapeutic agents. (840)

Koch's postulates (koks pos′tu-lāts) A set of rules for proving that a microorganism causes a particular disease. (9)

Koplik's spots (kop′liks) Lesions of the oral cavity caused by the measles (rubeola) virus that are characterized by a bluish white speck in the center of each. (918)

Korarchaeota A proposed phylum in the *Archaea* domain. To date, it is based entirely on uncultured microbes that have been identified through 16S rRNA nucleotide sequences cloned directly from the environment. (511)

Korean hemorrhagic fever An acute infection caused by a virus that produces varying degrees of hemorrhage, shock, and sometimes death. (923)

Krebs cycle *See* tricarboxylic acid (TCA) cycle.

L

lactic acid fermentation (lak′tik) A fermentation that produces lactic acid as the sole or primary product. (208)

lactoferrin An iron-sequestering protein released from macrophages and neutrophils into plasma. (759)

lager Pertaining to the process of aging beers to allow flavor development. (1044)

lag phase A period following the introduction of microorganisms into fresh culture medium when there is no increase in cell numbers or mass during batch culture. (123)

Lancefield system (group) (lans′feld) One of the serologically distinguishable groups (as group A, group B) into which streptococci can be divided. (584, 876)

Langerhans cell Cell found in the skin that internalizes antigen and moves in the lymph to lymph nodes where it differentiates into a dendritic cell. (758)

Lassa fever An acute, contagious, viral disease of central western Africa. The disease is characterized by fever, inflammation, muscular pains, and difficulty swallowing. (942)

late mRNA Messenger RNA produced later in a virus infection, which codes for proteins needed in capsid construction and virus release. (431)

latent period (la′tent) The initial phase in the one-step growth experiment in which no phages are released. (429)

latent virus infections Virus infections in which the virus stops reproducing and remains dormant for a period before becoming active again. (461)

lateral gene transfer *See* horizontal gene transfer.

leader sequence A nontranslated sequence at the 5′ end of mRNA that lies between the operator and the initiation codon; it aids in the initiation and regulation of transcription. (265)

lectin complement pathway (lek′tin) An antibody-independent pathway of complement activation that is initiated by microbial lectins (proteins that bind carbohydrates) and includes the C3–C9 components of the classical pathway. (765)

leghemaglobin A heme-containing pigment produced in leguminous plants. It is similar in structure to vertebrate hemoglobin; however, it has a higher affinity for oxygen. It functions to protect nodule-forming, nitrogen-fixing bacteria from oxygen, which would poison their nitrogenase. (70)

legionellosis (le″jə-nel-o′sis) *See* Legionnaires' disease.

Legionnaires' disease (legionellosis) A pulmonary infection, caused by *Legionella pneumophila.* (949)

leishmanias (lēsh″ma′ne-ăs) Trypanosomal protists of the genus *Leishmania,* that cause the disease leishmaniasis. (1004)

leishmaniasis (lēsh″mah-ni′ah-sis) A group of human diseases caused by the protists called leishmanias. (612, 1004)

lepromatous (progressive) leprosy (lep-ro′mah-tus lep′ro-se) A relentless, progressive form of leprosy in which large numbers of *Mycobacterium leprae* develop in skin cells, killing the skin cells and resulting in the loss of features. Disfiguring nodules form all over the body. (966)

leprosy (lep′ro-se) or **Hansen's disease** A severe disfiguring skin disease caused by *Mycobacterium leprae.* (966)

lethal dose 50 (LD_{50}) Refers to the dose or number of organisms that will kill 50% of an experimental group of hosts within a specified time period. (423, 817)

leukemia (loo-ke′me-ah) A progressive, malignant disease of blood-forming organs, marked by distorted proliferation and development of leukocytes and their precursors in the blood and bone marrow. Certain leukemias are caused by viruses (HTLV-1, HTLV-2). (935)

leukocidin (loo″ko-si′din) A microbial toxin that can damage or kill leukocytes. (828)

leukocyte (loo′ko-sīt) Any colorless white blood cell. Can be classified into granular and agranular lymphocytes. (744)

lichen (li′ken) An organism composed of a fungus and either photosynthetic protists or cyanobacteria in a symbiotic association. (731)

lipid raft A microdomain in the plasma membrane that is enriched for particular lipids and proteins. (81)

lipopolysaccharide (LPSs) (lip″o-pol″e-sak′ah-rīd) A molecule containing both lipid and polysaccharide, which is important in the outer membrane of the gram-negative cell wall. (58)

listeriosis (lis-ter″e-o′sis) A sporadic disease of animals and humans, particularly those who are immunocompromised or pregnant, caused by the bacterium *Listeria monocytogenes.* (984)

lithotroph (lith′o-trōf) An organism that uses reduced inorganic compounds as its electron source. (103)

lobopodia Rounded pseudopodia found in some amoeboid protists. (613)

log phase *See* exponential phase.

lophotrichous (lo-fot′rĭ-kus) A cell with a cluster of flagella at one or both ends. (67)

Lyme disease (LD, Lyme borreliosis) (līm) A tick-borne disease caused by the spirochete *Borrella burgdorferi.* (961)

lymph node A small secondary lymphoid organ that contains lymphocytes, macrophages, and dendritic cells. It serves as a site for (1) filtration and removal of foreign antigens and (2) the activation and proliferation of lymphocytes. (750)

lymphocyte (lim′fo-sīt) A nonphagocytic, mononuclear leukocyte (white blood cell) that is an immunologically competent cell, or its precursor. Lymphocytes are present in the blood, lymph, and lymphoid tissues. *See* B cell and T cell. (748)

lymphocytic choriomeningitis (LCM) An enveloped, single-stranded RNA virus that causes a nonbacterial meningitis in mice and, occasionally, in humans. (942)

lymphogranuloma venereum (LGV) (lim″fo-gran″u-lo′mah) A sexually transmitted disease caused by *Chlamydia trachomatis* serotypes L_1–L_3, which affect the lymph organs in the genital area. (975)

lymphokine (lim′fo-kin) A biologically active glycoprotein (e.g., IL-1) secreted by activated lymphocytes, especially sensitized T cells. It acts as an intercellular mediator of the immune response and transmits growth, differentiation, and behavioral signals. (767)

lyophilization Freezing and dehydrating samples as a means of preservation. Many microorganisms and natural products can be lyophilized for long-term storage. Commonly referred to as freeze-drying. (1063)

lysis (li′sis) The rupture or physical disintegration of a cell. (61)

lysogenic (li-so-jen′ik) *See* lysogens.

lysogens (li′so-jens) Bacterial and archaeal cells that are carrying a provirus and can produce viruses under the proper conditions. (345, 438)

lysogeny (li-soj′e-ne) The state in which a viral genome remains within the bacterial or achaeal cell after infection and reproduces along with it rather than taking control of the host cell and destroying it. (345, 438)

lysosome (li′so-sōm) A spherical membranous eucaryotic organelle that contains hydrolytic enzymes and is responsible for the intracellular digestion of substances. (86)

lysozyme (li′sō-zīm) An enzyme that degrades peptidoglycan by hydrolyzing the $\beta(1 \rightarrow 4)$ bond that joins *N*-acetylmuramic acid and *N*-acetylglucosamine. (61, 759)

lytic cycle (lit′ik) A virus life cycle that results in the lysis of the host cell. (345, 428)

M

macroevolution Major evolutionary change leading to either speciation or extinction. (477)

macrolide antibiotic (mak′ro-līd) An antibiotic containing a macrolide ring, a large lactone ring with multiple keto and hydroxyl groups, linked to one or more sugars. (846)

macromolecule (mak″ro-mol′ĕ-kūl) A large molecule that is a polymer of smaller units joined together. (226)

macromolecule vaccine A vaccine made of specific, purified macromolecules derived from pathogenic microorganisms. (778)

macronucleus (mak″ro-nu′kle-us) The larger of the two nuclei in ciliate protists. It is normally polyploid and directs the routine activities of the cell. (609)

macrophage (mak′ro-fāj) The name for a large mononuclear phagocytic cell, present in blood, lymph, and other tissues. Macrophages are derived from monocytes. They phagocytose and destroy pathogens; some macrophages also activate B cells and T cells. (746)

maduromycosis (mah-du′ro-mi-ko′sis) A subcutaneous fungal infection caused by *Madurella mycetoma;* also termed an eumycotic mycetoma. (1010)

madurose The sugar derivative 3-O-methyl-D-galactose, which is characteristic of several actinomycete genera that are collectively called maduromycetes. (601)

magnetosomes Magnetite particles in magnetotactic bacteria that are tiny magnets and allow the bacteria to orient themselves in magnetic fields. (50)

maintenance energy The energy a cell requires simply to maintain itself or remain alive and functioning properly. It does not include the energy needed for either growth or reproduction. (177)

major histocompatibility complex (MHC) A chromosome locus encoding the histocompatibility antigens. Class I MHC molecules are cell surface glycoproteins present on all nucleated cells; class II MHC glycoproteins are on antigen-presenting cells. Class I MHC glycoproteins present endogenous antigens to $CD8^+$ T cells. Class II MHC glycoproteins present exogenous antigens to $CD4^+$ T cells. (778)

malaria (mah-la′re-ah) A serious infectious illness caused by the parasitic protozoan *Plasmodium.* Malaria is characterized by bouts of high chills and fever that occur at regular intervals. (1001)

malt (mawlt) Grain soaked in water to soften it, induce germination, and activate its enzymes. The malt is then used in brewing and distilling. (1043)

Marburg viral hemorrhagic fever An acute, infectious disease caused by the Marburg hemorrhagic fever virus. Symptoms include varying degrees of bleeding and shock. (942)

mash The soluble materials released from germinated grains and prepared as a microbial growth medium. (1043)

mashing The process in which cereals are mixed with water and incubated in order to degrade their complex carbohydrates (e.g., starch) to more readily usable forms such as simple sugars. (1041)

mast cell A white blood cell that produces vasoactive molecules (e.g., histamine) and stores them in vacuoles near the cell membrane where they are released upon cell stimulation by external triggers. Mast cells can bind IgE proteins by their Fc region; IgE capture (by the Fab region) of antigen acts as a trigger for the release of vasoactive molecules. (747)

mating type A strain of a eucaryotic organism that can mate sexually with another strain of the same species. Commonly refers to strains of fungal mating types (MAT) (e.g., MAT α and MAT **a** of *Saccharomyces cerevisiae.* (634)

M cell Specialized cell of the intestinal mucosa and other sites, such as the urogenital tract, that delivers the antigen from the apical face of the cell to lymphocytes clustered within the pocket in its basolateral face. (759)

mean growth rate constant (k) The rate of microbial population growth expressed in terms of the number of generations per unit time. (126)

measles (rubeola) (me′zelz) A highly contagious skin disease that is endemic throughout the world. It is caused by a morbilli virus in the family *Paramyxoviridae,* which enters the body through the respiratory tract or through the conjunctiva. (917)

medical mycology (mi-kol′o-je) The discipline that deals with the fungi that cause human disease. (997)

meiosis (mi-o′sis) The sexual process in which a diploid cell divides and forms two haploid cells. (94)

melting temperature (T_m) The temperature at which double-stranded DNA separates into individual strands; it is dependent on the G + C content of the DNA and is used to compare genetic material in microbial taxonomy. (483)

membrane attack complex (MAC) The complex complement components (C5b–C9) that create a pore in the plasma membrane of a target cell and leads to cell lysis. C9 probably forms most of the actual pore. (764)

membrane-disrupting exotoxin A type of exotoxin that lyses host cells by disrupting the integrity of the plasma membrane. (828)

membrane filter technique The use of a thin porous filter made from cellulose acetate or some other polymer to collect microorganisms from water, air, and food. (156, 1052)

memory cell An inactive lymphocyte clone derived from a sensitized B or T cell capable of an accentuated response to a subsequent antigen exposure. (786)

meningitis (men″in-ji′tis) A condition that refers to inflammation of the brain or spinal cord meninges (membranes). The disease can be divided into bacterial (septic) meningitis and aseptic meningitis syndrome (caused by nonbacterial sources). (950)

merozygote A partially diploid procaryotic cell produced by horizontal gene transfer; in most cases some of the extra genetic material is destroyed and some is incorporated into the recipient's chromosome by homologous recombination, restoring the haploid state. (330)

mesophile (mes′o-fīl) A microorganism with a growth optimum around 20 to 45°C, a minimum of 15 to 20°C, and a maximum about 45°C or lower. (138)

messenger RNA (mRNA) Single-stranded RNA synthesized from a nucleic acid template (DNA in cellular organisms, RNA in some viruses) during transcription; mRNA binds to ribosomes and directs the synthesis of protein. (251)

metabolic channeling (mĕt″ah-bol′ik) The localization of metabolites and enzymes in different parts of a cell. (180)

metabolic control engineering Modification of the controls for biosynthetic pathways without altering the pathways themselves in order to improve process efficiency. (1062)

metabolic pathway engineering (MPE) The use of molecular techniques to improve the efficiency of pathways that synthesize industrially important products. (1062)

metabolism (me-tab′o-lizm) The total of all chemical reactions in the cell; almost all are enzyme catalyzed. (167)

metachromatic granules (met″ah-kro-mat′ik) Granules of polyphosphate in the cytoplasm of some bacteria that appear a different color when stained with a blue basic dye. They are storage reservoirs for phosphate. Sometimes called volutin granules. (50)

metagenomics Also called environmental or community genomics, metagenomics is the study of genomes recovered from environmental samples without first isolating members of the microbial community and growing them in pure cultures. (402)

metastasis (mĕ-tas′tah-sis) The transfer of a disease like cancer from one organ to another not directly connected with it. (461)

methanogens (meth′ə-no-jens″) Strictly anaerobic archaea that derive energy by converting CO_2, H_2, formate, acetate, and other compounds to either methane or methane and CO_2. (510)

methanotrophy The ability to grow on methane as the sole carbon source; such a microorganism is called a methanotroph. (510)

methylotroph A bacterium that uses reduced one-carbon compounds such as methane and methanol as its sole source of carbon and energy. (544)

Michaelis constant (K_m) (mĭ-ka′lis) A kinetic constant for an enzyme reaction that equals the substrate concentration required for the enzyme to operate at half maximal velocity. (178)

microaerophile (mi″kro-a′er-o-fīl) A microorganism that requires low levels of oxygen for growth, around 2 to 10%, but is damaged by normal atmospheric oxygen levels. (139)

microbial ecology The study of microorganisms in their natural environments, with a major emphasis on physical conditions, processes, and interactions that occur on the scale of individual microbial cells. (643)

microbial loop The mineralization of organic matter synthesized by photosynthetic microorganisms through the activity of other microbes, such as bacteria and protozoa. This process "loops" minerals and carbon dioxide back for reuse by the primary producers and makes the organic matter unavailable to higher consumers. (656. 670)

microbial mat A firm structure of layered microorganisms with complementary physiological activities that can develop on surfaces in aquatic environments. (655)

microbial transformation (mi-kro′be-al) *See* bioconversion. (1074)

microbiology (mi″kro-bi-ol′o-je) The study of organisms that are usually too small to be seen with the naked eye. Special techniques are required to isolate and grow them. (1)

microbivory The use of microorganisms as a food source by organisms (e.g., protists) that can ingest or phagocytose them. (657)

microenvironment (mi″kro-en-vi′ron-ment) The immediate environment surrounding a microbial cell or other structure, such as a root. (653)

microevolution *See* anagenesis. (477)

microfilaments (mi″kro-fil′ah-ments) Protein filaments, about 4 to 7 nm in diameter, that are present in the cytoplasmic matrix of eucaryotic cells and play a role in cell structure and motion. (83)

micronucleus (mi″kro-nu′kle-us) The smaller of the two nuclei in ciliate protists. Micronuclei are diploid and involved only in genetic recombination and the regeneration of macronuclei. (609)

micronutrients Nutrients such as zinc, manganese, and copper that are required in very small quantities for growth and reproduction. Also called trace elements. (101)

microorganism (mi″kro-or′gan-izm) An organism that is too small to be seen clearly with the naked eye. (1)

microsporidia A primitive, fungal, obligate intracellular parasite of animals, primarily vertebrates. The infectious spore (0.5 to 2.0 μm) contains a coiled polar tubule that is used for injecting the spore contents into host cells where the sporoplasm undergoes mitotic division, producing more spores. Microsporidia are known to cause infections of the conjunctiva, cornea, and intestine. Chronic intractable diarrhea, fever, malaise, and weight loss symptoms are similar to those of cryptosporidiosis. (1018)

microtubules (mi″kro-tu′buls) Small cylinders, about 25 nm in diameter, made of tubulin proteins and present in the cytoplasmic matrix and flagella of eucaryotic cells; they are involved in cell structure and movement. (83)

miliary tuberculosis (mil′e-a-re) An acute form of tuberculosis in which small tubercles are formed in a number of organs of the body because *M. tuberculosis* is disseminated throughout the body by the bloodstream. Also known as reactivation tuberculosis. (954)

mineralization The conversion of organic nutrients into inorganic material during microbial growth and metabolism. (569, 646)

mineral soil Soil that contains less than 20% organic carbon. (688)

minimal inhibitory concentration (MIC) The lowest concentration of a drug that will prevent the growth of a particular microorganism. (840)

minimal lethal concentration (MLC) The lowest concentration of a drug that will kill a particular microorganism. (840)

minus, or **negative strand** The virus nucleic acid strand that is complementary in base sequence to the viral mRNA. (416)

mismatch repair A type of DNA repair in which a portion of a newly synthesized strand of DNA containing mismatched base pairs is removed and replaced, using the parental strand as a template. (326)

missense mutation A single base substitution in DNA that changes a codon for one amino acid into a codon for another. (320)

mitochondrion (mi″to-kon′dre-on) The eucaryotic organelle that is the site of electron transport, oxidative phosphorylation, and pathways such as the Krebs cycle; it provides most of a nonphotosynthetic cell's energy under aerobic conditions. It is constructed of an outer membrane and an inner membrane, which contains the electron transport chain. (88)

mitosis (mi-to′sis) A process that takes place in the nucleus of a eucaryotic cell and results in the formation of two new nuclei, each with the same number of chromosomes as the parent. (92)

mixed acid fermentation A type of fermentation carried out by members of the family *Enterobacteriaceae* in which ethanol and a complex mixture of organic acids are produced. (209)

mixotrophy A mode of metabolism in which oxidation of an inorganic substrate provides energy while an organic carbon is used, although this is sometimes supplemented by carbon fixation. Most commonly seen in certain protists. (103)

modified atmosphere packaging (MAP) Addition of gases such as nitrogen and carbon dioxide to packaged foods in order to inhibit the growth of spoilage organisms. (1026)

mold Any of a large group of fungi that cause mold or moldiness and that exist as multicellular filamentous colonies; also the deposit or growth caused by such fungi. Molds typically do not produce macroscopic fruiting bodies. (631)

molecular chaperones *See* chaperone proteins.

molecular chronometers Nucleic acid and protein sequences thought to gradually change over time in a random fashion and at a steady rate, and which might be used to determine phylogenetic relationships. (488)

monoclonal antibody (mAb) (mon″o-klōn′al) An antibody of a single type that is produced by a population of genetically identical plasma cells (a clone); a monoclonal antibody is typically produced from a cell culture derived from the fusion product of a cancer cell and an antibody-producing cell (a hybridoma). (799, 864)

monocyte (mon′o-sīt) A mononuclear phagocytic leukocyte that circulates briefly in the bloodstream before migrating to the tissues where it becomes a macrophage. (746)

monocyte-macrophage system The collection of fixed phagocytic cells (including macrophages, monocytes, and specialized endothelial cells) located in the liver, spleen, lymph nodes, and bone marrow. This system is an important component of the host's general nonspecific (innate) defense against pathogens. (746)

monokine (mon′o-kīn) A generic term for a cytokine produced by mononuclear phagocytes (macrophages or monocytes). (767)

monotrichous (mon-ot′rĭ-kus) Having a single flagellum. (67)

morbidity rate (mor-bid′i-te) Measures the number of individuals who become ill as a result of a particular disease within a susceptible population during a specific time period. (887)

mordant (mor′dant) A substance that helps fix dye on or in a cell. (26)

mortality rate (mor-tal′i-te) The ratio of the number of deaths from a given disease to the total number of cases of the disease. (887)

most probable number (MPN) The statistical estimation of the probable population in a liquid by diluting and determining end points for microbial growth. (1052)

mucociliary blanket The layer of cilia and mucus that lines certain portions of the respiratory system; it traps microorganisms up to 10 μm in diameter and then transports them by ciliary action away from the lungs. (761)

mucociliary escalator The mechanism by which respiratory ciliated cells move material and microorganisms, trapped in mucus, out of the pharynx, where it is spit out or swallowed. (761)

mucosal-associated lymphoid tissue (MALT) Organized and diffuse immune tissues found as part of the mucosal epithelium. It can be specialized to the gut (GALT), the bronchial system (BALT), or skin (SALT). (759)

multicloning site (MCS) A region of DNA on a cloning vector that has a number of restriction enzyme recognition sequences to facilitate the introduction, or cloning, of a gene. (367)

multi-drug-resistant strains of tuberculosis (MDR-TB) A multi-drug-resistant strain is defined as *Mycobacterium tuberculosis* resistant to isoniazid and rifampin, with or without resistance to other drugs. (954)

multilocus sequence typing (MLST) A method for genotypic classification of procaryotes within a single genus using nucleotide differences among five to seven housekeeping genes. (486)

mumps An acute generalized disease that occurs primarily in school-age children and is caused by a paramyxovirus that is transmitted in saliva and respiratory droplets. The principal manifestation is swelling of the parotid salivary glands. (919)

murein *See* peptidoglycan. (55)

must The juices of fruits, including grapes, that can be fermented for the production of alcohol. (1041)

mutagen (mu′tah-jen) A chemical or physical agent that causes mutations. (318)

mutation (mu-ta′shun) A permanent, heritable change in the genetic material. (317)

mutualism (mu′tu-al-izm″) A type of symbiosis in which both partners gain from the association and are unable to survive without it. The mutualist and the host are metabolically dependent on each other. (718)

mutualist (mu′tu-al-ist) An organism associated with another in an obligatory relationship that is beneficial to both. (718)

mycelium (mi-se′le-um) A mass of branching hyphae found in fungi and some bacteria. (40, 631)

mycobiont The fungal partner in a lichen. (731)

mycolic acids Complex 60 to 90 carbon fatty acids with a hydroxyl on the β-carbon and an aliphatic chain on the α-carbon; found in the cell walls of mycobacteria. (596)

mycologist (mi-kol′o-jist) A person specializing in mycology; a student of mycology. (629)

mycology (mi-kol′o-je) The science and study of fungi. (629)

mycoplasma (mi″ko-plaz′mah) Bacteria that are members of the class *Mollicutes* and order *Mycoplasmatales;* they lack cell walls and cannot synthesize peptidoglycan precursors; most require sterols for growth. (572)

Mycoplasma pneumoniae (mi″ko-plaz′mal nu-mo′ne-ah) A type of pneumonia caused by *Mycoplasma pneumoniae.* Spread involves airborne droplets and close contact. (896)

mycorrhizal fungi Fungi that form stable, mutualistic relationships on (ectomycorrhizal) or in (endomycorrhyizal) the root cells of vascular plants. The plants provide carbohydrate for the fungi, while the fungal hyphae extend into the soil and bring nutrients (e.g., nitrogen and phosphorus) to the plants, thereby enhancing plant nutrient uptake. (700)

mycorrhizosphere The region around ectomycorrhizal mantles and hyphae in which nutrients released from the fungus increase the microbial population and its activities. (700)

mycosis (mi-ko′sis; pl., **mycoses**) Any disease caused by a fungus. (629, 997)

mycotoxicology (mi-ko′tok″si-kol′o-je) The study of fungal toxins and their effects on various organisms. (629)

myeloma cell (mi″e-lo′mah) A tumor cell that is similar to the cell type found in bone marrow. Also, a malignant, neoplastic plasma cell that produces large quantities of antibodies and can be readily cultivated. (800)

myositis (mi″o-si′tis) Inflammation of a striated or voluntary muscle. (957)

myxobacteria A group of gram-negative, aerobic soil bacteria characterized by gliding motility, a complex life cycle with the production of fruiting bodies, and the formation of myxospores. (564)

Myxogastria *See* acellular slime molds.

myxospores (mik′so-spōrs) Special dormant spores formed by the myxobacteria. (565)

N

narrow-spectrum drugs Chemotherapeutic agents that are effective only against a limited variety of microorganisms. (837)

natural attenuation The decrease in the level of an enviromental contaminant that results from natural chemical, physical, and biological processes. (1081)

natural classification A classification system that arranges organisms into groups whose members share many characteristics and reflect as much as possible the biological nature of organisms. (478)

natural killer (NK) cell A type of white blood cell that has a lineage independent of the granulocyte, B-cell, and T-cell lineages. It is often called a non-B, non-T lymphocyte because it does not make antibody; nor does it make cytokines associated with T cells. NK cells are part of the innate immune system, exhibiting MHC-independent cytolytic activity against virus-infected and tumor cells. (748)

naturally acquired active immunity The type of active immunity that develops when an individual's immunologic system comes into contact with an appropriate antigenic stimulus during the course of normal activities; it usually arises as the result of recovering from an infection and lasts a long time. (776)

naturally acquired passive immunity The type of temporary immunity that involves the transfer of antibodies from one individual to another. (777)

necrotizing fasciitis (nek′ro-tīz″ing fas″e-i′tis) A disease that results from a severe invasive group A streptococcus infection. Necrotizing fasciitis is an infection of the subcutaneous soft tissues, particularly of fibrous tissue, and is most common on the extremities. It begins with skin reddening, swelling, pain, and cellulitis, and proceeds to skin breakdown and gangrene after 3 to 5 days. (957)

negative selection The process by which lymphocytes that recognize host (self) antigens undergo apoptosis or become anergic (inactive). (803)

negative staining A staining procedure in which a dye is used to make the background dark while the specimen is unstained. (26)

Negri bodies (na'gre) Masses of viruses or unassembled viral subunits found within the neurons of rabies-infected animals. (944)

neoplasia Abnormal cell growth and reproduction due to a loss of regulation of the cell cycle; produces a tumor in solid tissues. (461)

neurotoxin (nu″ro-tok′sin) A toxin that is poisonous to or destroys nerve tissue; especially the toxins secreted by *C. tetani, Corynebacterium diphtheriae,* and *Shigella dysenteriae.* (825)

neutrophil (noo′tro-fil) A mature white blood cell in the granulocyte lineage formed in bone marrow. It has a nucleus with three to five lobes and is very phagocytic. (747)

neutrophile (nu′ston″ik) Microorganisms that grow best at a neutral pH range between pH 5.5 and 8.0. (134)

niche The function of an organism in a complex system, including place of the organism, the resources used in a given location, and the time of use. (653)

nicotinamide adenine dinucleotide (NAD^+) (nik″o-tin′ah-mīd) An electron-carrying coenzyme; it is particularly important in catabolic processes and usually donates its electrons to the electron transport chain under aerobic conditions. (173)

nicotinamide adenine dinucleotide phosphate ($NADP^+$) An electron-carrying coenzyme that most often participates as an electron carrier in biosynthetic metabolism. (173)

Nipah virus A single-stranded RNA virus named for the Nipah village in Malaysia where it was first associated with disease. (943)

nitrification (ni″trĭ-fĭ-ka′shun) The oxidation of ammonia to nitrate. (213, 546, 648)

nitrifying bacteria (ni′trĭ-fi″ing) Chemolithotrophic, gram-negative bacteria that are members of several families within the phyllum *Proteobacteria* that oxidize ammonia to nitrite and nitrite to nitrate. (213, 546)

nitrogenase (ni′tro-jen-ās) The enzyme that catalyzes biological nitrogen fixation. (237)

nitrogen fixation The metabolic process in which atmospheric molecular nitrogen (N_2) is reduced to ammonia; carried out by cyanobacteria, *Rhizobium,* and other nitrogen-fixing procaryotes. (236, 648)

nitrogen oxygen demand (NOD) The demand for oxygen in sewage treatment, caused by nitrifying microorganisms. (1055)

nitrogen saturation point The point at which mineral nitrogen (e.g., NO_3^-, NH_4^+), when added to an ecosystem, can no longer be incorporated into organic matter through biological processes. (690)

nocardioforms Bacteria that resemble members of the genus *Nocardia;* they develop a substrate mycelium that readily breaks up into rods and coccoid elements (a quality sometimes called fugacity). (596)

nomenclature (no′men-kla″tūr) The branch of taxonomy concerned with the assignment of names to taxonomic groups in agreement with published rules. (478)

noncyclic photophosphorylation (fo″to-fos″for-i-la′shun) The process in which light energy is used to make ATP when electrons are moved from water to $NADP^+$ during photosynthesis; both photosystem I and photosystem II are involved. (218)

noncytopathic virus A virus that does not kill its host cell by viral release-induced lysis. (819)

nondiscrete microorganism A microorganism, best exemplified by a filamentous fungus, that does not have a defined and predictable cell structure or distinct edges and boundaries. The organism can be defined in terms of the cell structure and its cytoplasmic contents. (663)

nongonococcal urethritis (NGU) (u″rə-thri′tis) Any inflammation of the urethra not caused by *Neisseria gonorrhoeae.* (976)

nonsense codon A codon that does not code for an amino acid but is a signal to terminate protein synthesis. (275)

nonsense mutation A mutation that converts a sense codon to a nonsense or stop codon. (321)

nonspecific immune response (innate or **natural immunity)** *See* nonspecific resistance.

nonspecific resistance Refers to those general defense mechanisms that are inherited as part of the innate structure and function of each animal; also known as nonspecific, innate or natural immunity. (743)

normal microbiota (also **indigenous microbial population, microflora, microbial flora)** (mmi″kro-bi-o′tah) The microorganisms normally associated with a particular tissue or structure. (734)

nosocomial infection (nos″o-ko′me-al) An infection that develops within a hospital (or other type of clinical care facility) and is acquired during the stay of the patient. (908)

nuclear envelope (nu′kle-ar) The complex double-membrane structure forming the outer boundary of the eucaryotic nucleus. It is covered by pores through which substances enter and leave the nucleus. (91)

nucleic acid hybridization (nu-kle′ik) The process of forming a hybrid double-stranded DNA molecule using a heated mixture of single-stranded DNAs from two different sources; if the sequences are fairly complementary, stable hybrids will form. (483)

nucleocapsid (nu″kle-o-kap′sid) The nucleic acid and its surrounding protein coat or capsid; the basic unit of virion structure. (409)

nucleoid (nu′kle-oid) An irregularly shaped region in the procaryotic cell that contains its genetic material. (52)

nucleolus (nu-kle′o-lus) The organelle, located within the eucaryotic nucleus and not bounded by a membrane, that is the location of ribosomal RNA synthesis and the assembly of ribosomal subunits. (91)

nucleoside (nu′kle-o-sīd″) A combination of ribose or deoxyribose with a purine or pyrimidine base. (241)

nucleosome (nu′kle-o-sōm″) A complex of histones and DNA found in eucaryotic chromatin and some archaea; the DNA is wrapped around the surface of the beadlike histone complex. (253)

nucleotide (nu′kle-o-tīd) A combination of ribose or deoxyribose with phosphate and a purine or pyrimidine base; a nucleoside plus one or more phosphates. (241)

nucleus (nu′kle-us) The eucaryotic organelle enclosed by a double-membrane envelope that contains the cell's chromosomes. (91)

numerical aperture The property of a microscope lens that determines how much light can enter and how great a resolution the lens can provide. (19)

numerical taxonomy The grouping by numerical methods of taxonomic units into taxa based on their character states. (479)

nutrient (nu′tre-ent) A substance that supports growth and reproduction. (101)

nystatin (nis′tah-tin) A polyene antibiotic from *Streptomyces noursei* that is used in the treatment of *Candida* infections of the skin, vagina, and alimentary tract. (854)

O

O antigen A polysaccharide antigen extending from the outer membrane of some gram-negative bacterial cell walls; it is part of the lipopolysaccharide. (60)

obligate aerobes Organisms that grow only in the presence of oxygen. (139)

obligate anaerobes Microorganisms that cannot tolerate the presence of oxygen and die when exposed to it. (139)

odontopathogens Dental pathogens. (991)

Okazaki fragments Short stretches of polynucleotides produced during discontinuous DNA replication. (260)

oligonucleotide A short fragment of DNA or RNA, usually artificially synthesized, used in a number of molecular genetic techniques such as DNA sequencing, polymerase chain reaction, and Southern blotting. (361)

oligonucleotide signature sequence Short, conserved nucleotide sequences that are specific for a phylogenetically defined group of organisms. The signature sequences found in small subunit rRNA molecules are most commonly used. (485)

oligotrophic environment (ol″ĭ-go-trof′ik) An environment containing low levels of nutrients, particularly nutrients that support microbial growth. (142, 676)

oncogene (ong″ko-jēn) A gene whose activity is associated with the conversion of normal cells to cancer cells. (461)

oncovirus A virus known to be associated with the development of cancer. (463)

one-step growth experiment An experiment used to study the reproduction of lytic phages in which one round of phage reproduction occurs and ends with the lysis of the host bacterial population. (428)

onychomycosis (on″i-ko-mi-ko′sis) A fungal infection of the nail plate producing nails that are opaque, white, thickened, friable, and brittle. Also called ringworm of the nails and tinea unguium. Caused by *Trichophyton* and other fungi such as *C. albicans.* (1017)

oocyst (o′o-sist) Cyst formed around a zygote of malaria and related protozoa. (1014)

öomycetes (o″o-mi-se′tēz) A collective name for protists also known as the water molds. Formerly thought to be fungi. (622)

open reading frame (ORF) A sequence of DNA not interrupted by a stop codon and with an apparent promoter and ribosome binding site at the 5′ end and a terminator at the 3′ end. It is usually determined by nucleic acid sequencing studies. (388)

operator The segment of DNA to which the repressor protein binds; it controls the expression of the genes adjacent to it. (295)

operon (op′er-on) The sequence of bases in DNA that contains one or more structural genes together with the operator controlling their expression. (295)

ophthalmia neonatorum (of-thal′me-ah ne″o-nat-or-um) A gonorrheal eye infection in a newborn, which may lead to blindness. Also called conjunctivitis of the newborn. (975)

opportunistic microorganism or **pathogen** A microorganism that is usually free-living or a part of the host's normal microbiota, but which may become pathogenic under certain circumstances, such as when the immune system is compromised. (740, 816, 1016)

opsonization (op″so-ni-za′shun) The coating of foreign substances by antibody, complement proteins, or fibronectin to make the substances more readily recognized by phagocytic cells. (763, 792)

optical tweezer The use of a focused laser beam to drag and isolate a specific microorganism from a complex microbial mixture. (664)

oral candidiasis *See* thrush.

orchitis (or-ki′tis) Inflammation of the testes. (919)

organelle (or″gah-nel′) A structure within or on a cell that performs specific functions and is related to the cell in a way similar to that of an organ to the body. (79)

organotrophs Organisms that use reduced organic compounds as their electron source. (103)

origin of replication (*ori*) A site on a chromosome or plasmid where DNA replication is initiated. (120)

ornithosis *See* psittacosis.

ortholog A gene found in the genomes of two or more different organisms that share a common ancestry. The products of orthologous genes are presumed to have similar functions. (388)

osmophilic microorganisms (oz″mo-fil′ik) Microorganisms that grow best in or on media of high solute concentration. (1024)

osmotolerant Organisms that grow over a fairly wide range of water activity or solute concentration. (134)

osmotrophy A form of nutrition in which soluble nutrients are absorbed through the cytoplasmic membrane; found in procaryotes, fungi, and some protists. (632)

Ouchterlony technique *See* double diffusion agar assay.

outbreak The sudden, unexpected occurrence of a disease in a given population. (886)

outer membrane A special membrane located outside the peptidoglycan layer in the cell walls of gram-negative bacteria. (55)

ovular nuclei The morphology of nuclei found in some protists that is characterized by a large nucleus (up to 100 μm in diameter) with many peripheral nuclei. (609)

oxidation-reduction (redox) reactions Reactions involving electron transfers; the electron donor (reductant) gives electrons to an electron acceptor (oxidant). (172)

oxidative burst The generation of reactive oxygen species, primarily superoxide anion (O_2^-) and hydrogen peroxide (H_2O_2) by a plant or an animal, in response to challenge by a potential bacterial, fungal, or viral pathogen. *See* respiratory burst. (701)

oxidative phosphorylation (fos″for-ĭ-la′-shun) The synthesis of ATP from ADP using energy made available during electron transport initiated by the oxidation of a chemical energy source. (202)

oxygenic photosynthesis Photosynthesis that oxidizes water to form oxygen; the form of photosynthesis characteristic of plants, protists and cyanobacteria. (216, 520)

P

pacemaker enzyme The enzyme in a metabolic pathway that catalyzes the slowest or rate-limiting reaction; if its rate changes, the pathway's activity changes. (183)

pandemic (pan-dem′ik) An increase in the occurrence of a disease within a large and geographically widespread population (often refers to a worldwide epidemic). (887)

Paneth cell (pah′ net) A specialized epithelial cell of the intestine. Paneth cells secrete hydrolytic enzymes and antimicrobial proteins and peptides. (734, 761)

pannus (pan′us) A superficial vascularization of the cornea with infiltration of granulation tissue. (979)

paralog Two or more genes in the genome of a single organism that arose through duplication of a common ancestral gene. The products of paralogous genes generally have different, although frequently related, functions. (388)

parasite (par′ah-sīt) An organism that lives on or within another organism (the host) and benefits from the association while harming its host. Often the parasite obtains nutrients from the host. (816)

parasitism (par′ah-si″tizm) A type of symbiosis in which one organism benefits from the other and the host is usually harmed. (730)

parasporal body An intracellular, solid protein crystal made by the bacterium *Bacillus thuringiensis.* Upon ingestion by one of over 100 different insect species, the protein becomes extremely toxic. This toxin is the basis of the bacterial insecticide Bt. (580)

parenteral route (pah-ren′ter-al) A route of drug administration that is nonoral (e.g., by injection). (849)

parfocal (par-fo′kal) A microscope that retains proper focus when the objectives are changed. (18)

paronychia (par″o-nik′e-ah) Inflammation involving the folds of tissue surrounding the nail; usually caused by *Candida albicans.* (1017)

parsimony analysis A method for developing phylogenetic trees based on the estimation of the minimum number of nucleotide or amino acid sequence changes needed to give the sequences being compared. (489)

particulate organic matter (POM) Nutrients that are not dissolved or soluble, generally referring to freshwater and marine ecosystems. This includes microorganisms or their cellular debris after senscence or viral lysis. (671)

passive diffusion The process in which molecules move from a region of higher concentration to one of lower concentration as a result of random thermal agitation. (106)

passive immunization The induction of temporary immunity by the transfer of immune products, such as antibodies or sensitized T cells, from an immune vertebrate to a nonimmune one. (777)

Pasteur effect (pas-tur′) The decrease in the rate of sugar catabolism and change to aerobic respiration that occurs when microorganisms are switched from anaerobic to aerobic conditions. (207)

pasteurization (pas″ter-ĭ-za′-shun) The process of heating milk and other liquids to destroy microorganisms that can cause spoilage or disease. (153, 1030)

pathogen (path′o-jən) Any virus, bacterium, or other agent that causes disease. (734, 743, 816)

pathogen-associated molecular pattern (PAMP) Conserved molecular structures that occur in patterns on microbial surfaces. The structures and their patterns are unique to particular microorganisms and invariant among members of a given microbial group. (753)

pathogenicity (path″o-je-nis′ĭ-te) The condition or quality of being pathogenic, or the ability to cause disease. (734, 816)

pathogenicity island A 10 to 200 Kb segment of DNA in some pathogens that contains the genes responsible for virulence; often it codes for the type III secretion system that allows the pathogen to secrete virulence proteins and damage host cells. A pathogen may have more than one pathogenicity island. (822)

pathogenic potential The degree that a pathogen causes morbid signs and symptoms. (816)

pathway architecture The analysis, design, and modification of biochemical pathways to increase process efficiency. (1062)

pattern recognition receptor A receptor found on macrophages and other phagocytic cells that binds to pathogen-associated molecular patterns on microbial surfaces. (753)

ped A natural soil aggregate, formed partly through bacterial and fungal growth in the soil. (692)

pellicle (pel′ĭ-k′l) A relatively rigid layer of proteinaceous elements just beneath the plasma membrane in many protists. The plasma membrane is sometimes considered part of the pellicle. (94, 607)

pelvic inflammatory disease (PID) A severe infection of the female reproductive organs. The disease results when gonococci or chlamydiae infect the uterine tubes and surrounding tissue. (975)

penicillins (pen″ĭ-sil′ ins) A group of antibiotics containing a β-lactam ring, which are active against gram-positive bacteria. (61, 841)

penton or **pentamer** A capsomer composed of five protomers. (411)

pentose phosphate pathway (pen′tōs) The pathway that oxidizes glucose 6-phosphate to ribulose 5-phosphate and then converts it to a variety of three to seven carbon sugars; it forms several important products (NADPH for biosynthesis, pentoses, and other sugars) and also can be used to degrade glucose to CO_2. (196)

peplomer or spike (pep′lo-mer) A protein or protein complex that extends from the virus envelope and often is important in virion attachment to the host cell surface. (412)

peptic ulcer disease A gastritis caused by *Helicobacter pylori.* (967)

peptide interbridge (pep′tīd) A short peptide chain that connects the tetrapeptide chains in some peptidoglycans. (56)

peptidoglycan (pep″tĭ-do-gli′kan) A large polymer composed of long chains of alternating *N*-acetylglucosamine and *N*-acetylmuramic acid residues. The polysaccharide chains are linked to each other through connections between tetrapeptide chains attached to the *N*-acetylmuramic acids. It provides much of the strength and rigidity possessed by bacterial cell walls. (55)

peptidyl or **donor site (P site)** The site on the ribosome that contains the peptidyl-tRNA at the beginning of the elongation cycle during protein synthesis. (284)

peptidyl transferase The 23S rRNA ribozyme that catalyzes the transpeptidation reaction in protein synthesis; in this reaction, an amino acid is added to the growing peptide chain. (284)

peptones (pep′tōns) Water-soluble digests or hydrolysates of proteins that are used in the preparation of culture media. (111)

perforin pathway The cytotoxic pathway that uses perforin protein, which polymerizes to form membrane pores that help destroy cells during cell-mediated cytotoxicity. Perforin is produced by cytotoxic T cells and NK cells and stored in granules that are released when a target cell is contacted. (782)

period of infectivity Refers to the time during which the source of an infectious disease is infectious or is disseminating the pathogen. (891)

periodontal disease (per″e-o-don′tal) A disease located around the teeth or in the periodontium. (993)

periodontitis (per″e-o-don-ti′tis) An inflammation of the periodontium. (993)

periodontium (per″e-o-don′she-um) The tissue investing and supporting the teeth, including the cementum, periodontal ligament, alveolar bone, and gingiva. (993)

periodontosis (per″e-o-don-to′sis) A degenerative, noninflammatory condition of the periodontium, which is characterized by destruction of tissue. (994)

peripheral tolerance The inhibition of effector lymphocyte activity resulting in the lack of a specific response. (803)

periplasm (per′ĭ-plaz-əm) The substance that fills the periplasmic space. (55)

periplasmic flagella The flagella that lie under the outer sheath and extend from both ends of the spirochete cell to overlap in the middle and form the axial filament. Also called axial fibrils and endoflagella. (70, 532)

periplasmic space (per″i-plas′mik) or **periplasm** (per′ĭ-plazm) The space between the plasma membrane and the outer membrane in gram-negative bacteria, and between the plasma membrane and the cell wall in gram-positive bacteria. (55)

peristalsis The muscular contractions of the gut that propel digested foods and waste through the intestinal tract. (761)

peritrichous (pĕ-rit′rĭ-kus) A cell with flagella distributed over its surface. (67)

permease (per′me-ās) A membrane-bound carrier protein or a system of two or more proteins that transports a substance across the membrane. (106)

pertussis (pər-tus′is) An acute, highly contagious infection of the respiratory tract, most frequently affecting young children, usually caused by *Bordetella pertussis* or *B. parapertussis.* Consists of peculiar paroxysms of coughing, ending in a prolonged crowing or whooping respiration; hence the name whooping cough. (955)

petri dish (pe′tre) A shallow dish consisting of two round, overlapping halves that is used to grow microorganisms on solid culture medium; the top is larger than the bottom of the dish to prevent contamination of the culture. (117)

phage (fāj) *See* bacteriophage.

phagocytic vacuole (fag″o-sit′ik vak′u-ol) A membrane-delimited vacuole produced by cells carrying out phagocytosis. It is formed by the invagination of the plasma membrane and contains solid material. (608)

phagocytosis (fag″o-si-to′sis) The endocytotic process in which a cell encloses large particles in a membrane-delimited phagocytic vacuole or phagosome and engulfs them. (86, 752)

phagolysosome (fag″o-li′so-sōm) The vacuole that results from the fusion of a phagosome with a lysosome. (755)

phagosome A membrane-enclosed vacuole formed by the invagination of the cell membrane during endocytosis. (86)

phagovar (fag′o-var) A specific phage type. (873)

pharyngitis (far″in-ji′tis) Inflammation of the pharynx, often due to a *S. pyogenes* infection. (958)

phase-contrast microscope A microscope that converts slight differences in refractive index and cell density into easily observed differences in light intensity. (21)

phenetic system A classification system that groups organisms together based on the similarity of their observable characteristics. (478)

phenol coefficient test A test to measure the effectiveness of disinfectants by comparing their activity against test bacteria with that of phenol. (165)

phosphatase (fos′fah-tās″) An enzyme that catalyzes the hydrolytic removal of phosphate from molecules. (241)

phosphate group transfer potential A measure of the ability of a phosphorylated molecule such as ATP to transfer its phosphate to water and other acceptors. It is the negative of the $\Delta G^{o\prime}$ for the hydrolytic removal of phosphate. (171)

phospholipase An enzyme that hydrolyzes a specific ester bond in phospholipids. There are four types: A, B, C, and D. (828)

phosphorelay system A mechanism for regulating either transcription or enzyme activity that involves the transfer of a phosphate group from one molecule to another. Covalent addition of the phosphate group to enzymes or other proteins can either activate or inhibit their activity. Examples include the phosphotransferase system (PTS) of group translocation and two-component regulatory systems. (109, 300, 302)

photolithoautotroph An organism that uses light energy, an inorganic electron source (e.g., H_2O, H_2, H_2S), and CO_2 as its carbon source. Also called photolithotrophic autotroph. (103)

photoorganoheterotroph A microorganism that uses light energy, organic electron sources, and organic molecules as a carbon source. Also called photoorganotrophic heterotroph. (103)

photoreactivation (fo″to-re-ak″tĭ-va′shun) The process in which blue light is used by a photoreactivating enzyme to repair thymine dimers in DNA by splitting them apart. (326)

photosynthate Nutrient material that leaks from phototrophic organisms. Photosynthate contributes to the pool of dissolved organic matter that is available to microorganisms. (671)

photosynthesis (fo″to-sin′thĕ-sis) The trapping of light energy and its conversion to chemical energy, which is then used to reduce CO_2 and incorporate it into organic form. (214)

photosystem I The photosystem in eucaryotic cells and cyanobacteria that absorbs longer wavelength light, usually greater than about 680 nm, and transfers the energy to chlorophyll P700 during photosynthesis; it is involved in both cyclic photophosphorylation and noncyclic photophosphorylation. (217)

photosystem II The photosystem in eucaryotic cells and cyanobacteria that absorbs shorter wavelength light, usually less than 680 nm, and transfers the energy to chlorophyll P680 during photosynthesis; it participates in noncyclic photophosphorylation. (217)

phototaxis The ability of certain phototrophic bacteria to move, either by gliding or swimming motility, in response to a light source. In positive phototaxis, the microbe moves toward the light; in negative, it moves away. (525)

phototrophs Organisms that use light as their energy source. (103)

phycobiliproteins Photosynthetic pigments that are composed of proteins with attached tetrapyrroles; they are found in cyanobacteria. (217)

phycobilisomes Special particles on the membranes of cyanobacteria that contain photosynthetic pigments and electron transport chains. (524)

phycobiont (fi″ko-bi′ont) The photosynthetic protist or cyanobacterial partner in a lichen. (731)

phycocyanin (fi″ko-si′an-in) A blue phycobiliprotein pigment used to trap light energy during photosynthesis. (217)

phycoerythrin (fi″ko-er′i-thrin) A red photosynthetic phycobiliprotein pigment used to trap light energy. (217)

phycology (fi-kol′o-je) The study of algae. (605)

phyllosphere The surface of plant leaves. (696)

phylogenetic or **phyletic classification system** (fi″lo-jĕ-net′ik, fi-let′ik) A classification system based on evolutionary relationships rather than the general similarity of characteristics. (478)

phylogenetic tree A graph made of nodes and branches, much like a tree in shape, that shows phylogenetic relationships between groups of organisms and sometimes also indicates the evolutionary development of groups. (489)

phylotype A taxon that is characterized only by its nucleic acid sequence; generally discovered during metagenomic analysis. (402)

phytoplankton (fi″to-plank′ton) A community of floating photosynthetic organisms, largely composed of photosynthetic protists and cyanobacteria. (670)

phytoremediation The use of plants and their associated microorganisms to remove, contain, or degrade environmental contaminants. (1079)

picoplankton Planktonic microbes between 0.2 and 2.0 μm in size. This includes the cyanobacterial genera *Prochlorococcus* and *Synechococcus,* which together can account for over half the carbon fixation in some open ocean ecosystems. (670)

piedra (pe-a′drah) A fungal disease of the hair in which white or black nodules of fungi form on the shafts. (1008)

pitching Pertaining to inoculation of a nutrient medium with yeast, for example, in beer brewing. (1043)

plague ((plāg) An acute febrile, infectious disease, caused by the γ-proteobacterium *Yersinia pestis,* which has a high mortality rate; the two major types are bubonic plague and pneumonic plague. (962)

plankton (plank′ton) Free-floating, mostly microscopic microorganisms that can be found in almost all waters; a collective name. (606)

planktonic (adj.) *See* plankton.

plantar warts Viral infections of the epithelia comprising the sole of the foot. (938)

plaque (plak) 1. A clear area in a lawn of bacteria or a localized area of cell destruction in a layer of animal cells that results from the lysis of the bacteria by bacteriophages or the destruction of the animal cells by animal viruses. 2. The term also refers to dental plaque, a film of food debris, polysaccharides, and dead cells that cover the teeth. (418)

plasma cell A mature, differentiated B lymphocyte that synthesizes and secretes antibody; a plasma cell lives for only 5 to 7 days. (748, 786)

plasmalemma The plasma membrane in protists. (607)

plasma membrane The selectively permeable membrane surrounding the cell's cytoplasm; also called the cell membrane, plasmalemma, or cytoplasmic membrane. (42)

plasmid (plaz′mid) A double-stranded DNA molecule that can exist and replicate independently of the chromosome or may be integrated with it. A plasmid is stably inherited, but is not required for the host cell's growth and reproduction. (53, 334)

plasmid fingerprinting A technique used to identify microbial isolates as belonging to the same strain because they contain the same number of plasmids with the identical molecular weights and similar phenotypes. (875)

plasmodial (acellular) slime mold (plaz-mo′de-al) A member of the protist division *Amoebozoa (Myxogastria)* that exists as a thin, streaming, multinucleate mass of protoplasm, which creeps along in an amoeboid fashion. (614)

plasmodium (plaz-mo′de-um; pl., **plasmodia**) A stage in the life cycle of myxogastria protists; a multinucleate mass of protoplasm surrounded by a membrane. Also, a parasite of the genus *Plasmodium.* (614)

plasmolysis (plaz-mol′ĭ-sis) The process in which water osmotically leaves a cell, which causes the cytoplasm to shrivel up and pull the plasma membrane away from the cell wall. (61)

plastid (plas′tid) A cytoplasmic organelle of algae and higher plants that contains pigments such as chlorophyll, stores food reserves, and often carries out processes such as photosynthesis. (90)

pleomorphic (ple″o-mor′fik) Refers to bacteria that are variable in shape and lack a single, characteristic form. (41)

plus strand or **positive strand** The virus nucleicacid strand that is equivalent in base sequence to the viral mRNA. (416)

***Pneumocystis carinii* pneumonia (PCP);** (noo″mo-sis-tis) A type of pneumonia caused by the protist *Pneumocystis jiroveci.* (1019)

pneumonic plague *See* plague.

point mutation A mutation that affects only a single base pair in a specific location. (318)

polar flagellum A flagellum located at one end of an elongated cell. (67)

poliomyelitis (**polio** or **infantile paralysis**) (po″le-o-mi″e-li′tis) An acute, contagious viral disease that attacks the central nervous system, injuring or destroying the nerve cells that control the muscles and sometimes causing paralysis. (940)

poly-β-hydroxybutyrate (PHB) (hi-drok″se-bu′tĭ-rāt) A linear polymer of β-hydroxybutyrate used as a reserve of carbon and energy by many bacteria. (49)

polymerase chain reaction (PCR) An in vitro technique used to synthesize large quantities of specific nucleotide sequences from small amounts of DNA. It employs oligonucleotide primers complementary to specific sequences in the target gene and special heat-stable DNA polymerases. (362)

polymorphonuclear leukocyte (pol″e-mor″fo-noo′kle-ər) A leukocyte that has a variety of nuclear forms. (746)

polyphasic taxonomy An approach in which taxonomic schemes are developed using a wide range of phenotypic and genotypic information. (478)

polyribosome (pol″e-ri′bo-sōm) A complex of several ribosomes with a messenger RNA; each ribosome is translating the same message. (276)

Pontiac fever A bacterial disease caused by *Legionella pneumophila* that resembles an allergic disease more than an infection. First described from Pontiac, Michigan. *See* Legionnaires' disease. (950)

population An assemblage of organisms of the same type. (643)

porin proteins Proteins that form channels across the outer membrane of gram-negative bacterial cell walls. Small molecules are transported through these channels. (60)

postherpetic neuralgia The severe pain after a herpes infection. (915)

posttranscriptional modification The processing of the initial RNA transcript, pre-mRNA, to form mRNA. (272)

potable (po′tah-b′l) Refers to water suitable for drinking. (1052)

pour plate A petri dish of solid culture medium with isolated microbial colonies growing both on its surface and within the medium, which has been prepared by mixing microorganisms with cooled, still liquid medium and then allowing the medium to harden. (115)

precipitation (or **precipitin**) **reaction** (pre-sip″ĭ-ta′shun) The reaction of an antibody with a soluble antigen to form an insoluble precipitate. (799)

precipitin (pre-sip′ĭ-tin) The antibody responsible for a precipitation reaction. (799)

pre-mRNA In eucaryotes, the RNA transcript of DNA made by RNA polymerase II; it is processed to form mRNA. (272)

prevalence rate Refers to the total number of individuals infected at any one time in a given population regardless of when the disease began. (887)

Pribnow box A special base sequence in the promoter that is recognized by the RNA polymerase and is the site of initial polymerase binding. (269)

primary amebic meningoencephalitis An infection of the meninges of the brain by the free-living amoebae *Naegleria* or *Acanthamoeba.* (1013)

primary (frank) pathogen Any organism that causes a disease in the host by direct interaction with or infection of the host. (816)

primary metabolites Microbial metabolites produced during active growth of an organism. (1068)

primary producer Photoautotrophic and chemoautotrophic organisms that incorporate carbon dioxide into organic carbon and thus provide new biomass for the ecosystem. (656, 669)

primary production The incorporation of carbon dioxide into organic matter by photosynthetic organisms and chemoautotrophic bacteria. (656)

primary treatment The first step of sewage treatment, in which physical settling and screening are used to remove particulate materials. (1055)

primosome A complex of proteins that includes the enzyme primase, which is responsible for synthesizing the RNA primers needed for DNA replication. (260)

prion (pri′on) An infectious agent consisting only of protein; prions cause a variety of spongiform encephalopathics such as scrapie in sheep and goats. (468)

probe (prōb) A short, labeled nucleic acid segment complementary in base sequence to part of another nucleic acid, which is used to identify or isolate the particular nucleic acid from a mixture through its ability to bind specifically with the target nucleic acid. (358, 389)

probiotic A living organism that may provide health benefits beyond its nutritional value when ingested. (739, 1039)

procaryotic cells (pro″kar-e-ot′ik) Cells that lack a true, membrane-enclosed nucleus; bacteria are procaryotic and have their genetic material located in a nucleoid. (2, 97)

procaryotic species A collection of bacterial or archaeal strains that share many stable properties and differ significantly from other groups of strains. (480)

prodromal stage (pro-dro′məl) The period during the course of a disease in which there is the appearance of signs and symptoms, but they are not yet distinctive and characteristic enough to make an accurate diagnosis. (888)

promoter The region on DNA at the start of a gene that the RNA polymerase binds to before beginning transcription. (265)

propagated epidemic An epidemic that is characterized by a relatively slow and prolonged rise and then a gradual decline in the number of individuals infected. It usually results from the introduction of an infected individual into a susceptible population, and the pathogen is transmitted from person to person. (889)

prophage (pro′fāj) The latent form of a temperate phage that remains within the lysogen, usually integrated into the host chromosome. (345, 438)

prostheca (pros-the′kah) An extension of a bacterial cell, including the plasma membrane and cell wall, that is narrower than the mature cell. Found in the genus *Caulobacter,* an important model bacterium. (543)

prosthetic group (pros-thet′ik) A tightly bound cofactor that remains at the active site of an enzyme during its catalytic activity. (176)

protease (pro′te-ās) An enzyme that hydrolyzes proteins to their constituent amino acids. Also called a proteinase. (212)

proteasome A large, cylindrical protein complex that degrades ubiquitin-labeled proteins to peptides in an ATP-dependent process. (86)

protein engineering (pro′tēn) The rational design of proteins by constructing specific amino acid sequences through molecular techniques, with the objective of modifying protein characteristics. (1061)

protein modeling The process by which the amino acid sequence of a protein is analyzed using software that is designed to predict its three-dimensional structure. (394)

protein splicing The post-translational process in which part of a precursor polypeptide is removed before the mature protein folds into its final shape; it is carried out by self-splicing proteins that remove inteins and join the remaining exteins. (288)

Proteobacteria (pro″te-o-bak-tēr′e-ah) A large phylum of gram-negative bacteria, that 16S rRNA sequence comparisons show to be phylogenetically related; *Proteobacteria* contain the purple photosynthetic bacteria and their relatives and are composed of the α, β, γ, δ, and ϵ subgroups. (539)

proteome The complete collection of proteins that an organism produces. (393)

proteomics The study of the structure and function of cellular proteins. (393)

protist (pro′tist) Unicellular (and rarely acellular) eucaryotic organisms that lack cellular differentiation into tissues. Vegetative cell differentiation is limited to cells involved in sexual reproduction, alternate vegetative morphology, or resting states such as cysts. Protists vary in morphology and metabolism, including phototrophy, heterotrophy, and mixotrophy. Many phototrophic and mixotrophic forms (e.g., diatoms and dinoflagellates) are frequently referred to as algae. (3, 491)

protistology The study of protists. (605)

protomer An individual subunit of a viral capsid; a capsomer is made of protomers. (409)

proton motive force (PMF) The force arising from a gradient of protons and a membrane potential that is thought to power ATP synthesis and other processes. (202)

proto-oncogene The normal cellular form of a gene that when mutated or overexpressed results in or contributes to malignant transformation of a cell. (461)

protoplast (pro′to-plast) A bacterial or fungal cell with its cell wall completely removed. It is spherical in shape and osmotically sensitive. (48)

protoplast fusion The joining of cells that have had their walls weakened or completely removed. (1061)

prototroph (pro′to-trōf) A microorganism that requires the same nutrients as the majority of naturally occurring members of its species. (323)

protozoan or **protozoon** (pro″to-zo′an, pl. **protozoa**) A unicellular or acellular eucaryotic; chemoorganotrophic protist whose organelles have the functional role of organs and tissues in more complex forms. Protozoa vary greatly in size, morphology, nutrition, and life cycle. (3, 605)

protozoology (pro″to-zo-ol′o-je) The study of protozoa. (605)

proviral DNA Viral DNA that has been integrated into host cell DNA. In retroviruses it is the double-stranded DNA copy of the RNA genome. (457)

pseudopodium or **pseudopod** (soo″do-po′de-um) A nonpermanent cytoplasmic extension of the cell body by which amoeboid protists move and feed. (613)

psittacosis (ornithosis; sit″ah-ko′sis) A disease due to a strain of *Chlamydia psittaci,* first seen in parrots and later found in other birds and domestic fowl (in which it is called ornithosis). It is transmissible to humans. (991)

psychrophile (si′kro-fīl) A microorganism that grows well at 0°C and has an optimum growth temperature of 15°C or lower and a temperature maximum around 20°C. (137)

psychrotroph A microorganism that grows at 0°C, but has a growth optimum between 20 and 30°C, and a maximum of about 35°C. (138)

pulmonary anthrax (pul′mo-ner″e) A form of anthrax involving the lungs. Also known as woolsorter's disease. (988)

pulmonary syndrome hantavirus *See* hantavirus pulmonary syndrome. (923)

puncutated equilibria The observation based on the fossil record that evolution does not proceed at a slow and linear pace, but rather is periodically interrupted by rapid bursts of speciation and extinction driven by abrupt changes in environmental conditions. (477)

pure culture A population of cells that are identical because they arise from a single cell. (113)

purine (pu′rin) A basic, heterocyclic, nitrogen-containing molecule with two joined rings that occurs in nucleic acids and other cell constituents; most purines are oxy or amino derivatives of the purine skeleton. The most important purines are adenine and guanine. (241)

purple membrane An area of the plasma membrane of *Halobacterium* that contains bacteriorhodopsin and is active in photosynthetic light energy trapping. (515)

putrefaction (pu″trĕ-fak′shun) The microbial decomposition of organic matter, especially the anaerobic breakdown of proteins, with the production of foul-smelling compounds such as hydrogen sulfide and amines. (1024)

pyrenoid (pi′rĕ-noid) The differentiated region of the chloroplast that is a center of starch formation in some photosynthetic protists. (90, 608)

pyrimidine (pi-rim′i-dēn) A basic, heterocyclic, nitrogen-containing molecule with one ring that occurs in nucleic acids and other cell constituents; pyrimidines are oxy or amino derivatives of the pyrimidine skeleton. The most important pyrimidines are cytosine, thymine, and uracil. (241)

Q

Q fever An acute zoonotic disease caused by the rickettsia *Coxiella burnetii.* (964)

Quellung reaction The increase in visibility or the swelling of the capsule of a microorganism in the presence of antibodies against capsular antigens. (876)

quinolones A class of broad-spectrum antibiotics, derived from nalidixic acid, that bind to bacterial DNA gyrase, inhibiting DNA replication and transcription. This group of antibiotics is bacteriocidal. (847)

quorum sensing The process in which bacteria monitor their own population density by sensing the levels of signal molecules that are released by the microorganisms. When these signal molecules reach a threshold concentration, quorum-dependent genes are expressed. (144, 309)

R

rabies (ra′bēz) An acute infectious disease of the central nervous system, which affects all warm-blooded animals (including humans). It is caused by an ssRNA virus belonging to the genus *Lyssavirus* in the family *Rhabdoviridae.* (943)

racking The removal of sediments from wine bottles. (1043)

radappertization The use of gamma rays from a cobalt source for control of microorganisms in foods. (1031)

radioimmunoassay (RIA) (ra″de-o-im″u-no-as′a) A very sensitive assay technique that uses a purified radioisotope-labeled antigen or antibody to compete for antibody or antigen with unlabeled standard and samples to determine the concentration of a substance in the samples. (882)

rational drug design An approach to new drug development based a specific cellular macromolecule or target. (398)

reactivation tuberculosis *See* miliary tuberculosis. (954)

reactive nitrogen intermediate (RNI) Charged nitrogen radicals intermediate between various stable nitrogen molecules. (755)

reactive oxygen intermediate (ROI) Charged oxygen radicals intermediate between various stable oxygen molecules; also called reactive oxygen species (ROS). (755)

reading frame The way in which nucleotides in DNA and mRNA are grouped into codons or groups of three for reading the message contained in the nucleotide sequence. (264)

reagin (re′ah-jin) Antibody that mediates immediate hypersensitivity reactions. IgE is the major reagin in humans. (803)

real-time PCR A type of polymerase chain reaction (PCR) that quantitatively measures the amount of template in a sample as the amount of fluorescently labeled amplified product. (363)

receptor-mediated endocytosis A type of endocytosis that involves the specific binding of molecules to membrane receptors followed by the formation of coated vesicles. The substances being taken in are concentrated during the process. (86)

recombinant DNA technology The techniques used in carrying out genetic engineering; they involve the identification and isolation of a specific gene, the insertion of the gene into a vector such as a plasmid to form a recombinant molecule, and the production of large quantities of the gene and its products. (357)

recombinant-vector vaccine The type of vaccine that is produced by the introduction of one or more of a pathogen's genes into attenuated viruses or bacteria. The attenuated virus or bacterium serves as a vector, replicating within the vertebrate host and expressing the gene(s) of the pathogen. The pathogen's antigens induce an immune response. (904)

recombination (re″kom-bĭ-na′shun) The process in which a new recombinant chromosome is formed by combining genetic material from two organisms. (329)

recombinational repair A DNA repair process that repairs damaged DNA when there is no remaining template; a piece of DNA from a sister molecule is used. (329)

Redfield ratio The carbon-nitrogen-phosphorus ratio of marine microorganisms. This ratio is important for predicting limiting factors for microbial growth. (670)

red tides Red tides occur frequently in coastal areas and often are associated with population blooms of dinoflagellates. Dinoflagellate pigments are responsible for the red color of the water. Under these conditions, the dinoflagellates often produce saxitoxin, which can lead to paralytic shellfish poisoning. More commonly called harmful algal blooms (HABs). (621)

reducing power Molecules such as NADH and NADPH that temporarily store electrons. The stored electrons are used in anabolic reactions such as CO_2 fixation and the synthesis of monomers (e.g., amino acids). (168)

reductive dehalogenation The cleavage of carbon-halogen bonds by anaerobic bacteria that creates a strong electron-donating environment. (1075)

refraction (re-frak′shun) The deflection of a light ray from a straight path as it passes from one medium (e.g., glass) to another (e.g., air). (17)

refractive index (re-frak′tiv) The ratio of the velocity of light in the first of two media to that in the second as it passes from the first to the second. (17)

regulator T cell Regulator T cells control the development of effector T cells. Two types exist: T-helper cells ($CD4^+$ cells) and T-suppressor cells. There are three subsets of T-helper cells: T_H1, T_H2, and T_H0. T_H1 cells produce IL-2, IFN-γ, and TNF-β. They effect cell-mediated immunity and are responsible for delayed-type hypersensitivity reactions and macrophage activation. T_H2 cells produce IL-4, IL-5, IL-6, IL-10, IL-13. They are helpers for B-cell antibody responses and humoral immunity; they also support IgE responses and eosinophilia. T_H0 cells exhibit an unrestricted cytokine profile. (781)

regulatory mutants Mutant organisms that have lost the ability to limit synthesis of a product, which normally occurs by regulation of activity of an earlier step in the biosynthetic pathway. (1071)

regulon A collection of genes or operons that is controlled by a common regulatory protein. (307)

replica plating A technique for isolating mutants from a population by plating cells from each colony growing on a nonselective agar medium onto plates with selective media or environmental conditions, such as the lack of a nutrient or the presence of an antibiotic or a phage; the location of mutants on the original plate can be determined from growth patterns on the replica plates. (324)

replicase An RNA-dependent RNA polymerase used to replicate the genome of an RNA virus. (455)

replication (rep″lĭ-ka′shun) The process in which an exact copy of parental DNA (or viral RNA) is made with the parental molecule serving as a template. (251)

replication fork The Y-shaped structure where DNA is replicated. The arms of the Y contain template strand and a newly synthesized DNA copy. (256)

replicative form (RF) A double-stranded form of nucleic acid that is formed from a single-stranded virus genome and used to synthesize new copies of the genome. (436, 455)

replicon (rep'lĭ-kon) A unit of the genome that contains an origin for the initiation of replication and in which DNA is replicated. (257)

replisome (rep'lĭ-sōm) A protein complex or replication factory that copies the DNA double helix to form two daughter chromosomes. (120, 260)

repressible enzyme An enzyme whose level drops in the presence of a small molecule, usually an end product of its metabolic pathway. (294)

repressor protein (re-pres'or) A protein coded for by a regulator gene that can bind to the operator and inhibit transcription; it may be active by itself or only when the corepressor is bound to it. (295)

reproductive cloning Cloning with the intent of generating life. Human reproductive cloning is banned in most nations. (377)

reservoir (rez'er-vwar) A site, alternate host, or carrier that normally harbors pathogenic organisms and serves as a source from which other individuals can be infected. (818, 891)

reservoir host An organism other than a human that is infected with a pathogen that can also infect humans. (816)

resolution (rez″o-lu'shun) The ability of a microscope to separate or distinguish between small objects that are close together. (18)

respiration (res″ pĭ-ra' shən) An energy-yielding process in which the energy substrate is oxidized using an exogenous or externally derived electron acceptor. (192)

respiratory burst The respiratory burst occurs when an activated phagocytic cell increases its oxygen consumption to support the increased metabolic activity of phagocytosis. The burst generates highly toxic oxygen products such as singlet oxygen, superoxide radical, hydrogen peroxide, hydroxyl radical, and hypochlorite. (755)

respiratory syncytial virus (RSV; sin-sish'al) A member of the family *Paramyxoviridae* and genus *Pneumovirus;* it is a negative-sense ssRNA virus that causes respiratory infections in children. (919)

restriction enzymes Enzymes produced by host cells that cleave virus DNA at specific points and thus protect the cell from virus infection; they are used in carrying out genetic engineering. (357, 432)

reticulate body (RB) The cellular form in the chlamydial life cycle whose role is growth and reproduction within the host cell. (531)

reticulopodia Netlike pseudopodia found in certain amoeboid protists. (613)

retroviruses (re″tro-vi'rus-es) A group of viruses with RNA genomes that carry the enzyme reverse transcriptase and form a DNA copy of their genome during their reproductive cycle. (457)

reverse transcriptase (RT) An RNA-dependent DNA polymerase that uses a viral RNA genome as a template to synthesize a DNA copy; this is a reverse of the normal flow of genetic information, which proceeds from DNA to RNA. (358, 457)

reversible covalent modification A mechanism of enzyme regulation in which the enzyme's activity is either increased or decreased by the reversible covalent addition of a group such as phosphate or AMP to the protein. (183)

Reye's syndrome An acute, potentially fatal disease of childhood that is characterized by severe edema of the brain and increased intracranial pressure, vomiting, hypoglycemia, and liver dysfunction. The cause is unknown but is almost always associated with a previous viral infection (e.g., influenza or varicella-zoster virus infections). (918)

rheumatic fever (roo-mat'ik) An autoimmune disease characterized by inflammatory lesions involving the heart valves, joints, subcutaneous tissues, and central nervous system. The disease is associated with hemolytic streptococci in the body. It is called rheumatic fever because two common symptoms are fever and pain in the joints similar to that of rheumatism. (958)

rhizobia Any one of a number of alpha- and beta-proteobacteria that form symbiotic nitrogen-fixing nodules on the roots of leguminous plants. (701)

rhizomorph A macroscopic, densely packed thread consisting of individual cells formed by some fungi. A rhizomorph can remain dormant and/or serve as a means of fungal dissemination. (698)

rhizoplane The surface of a plant root. (696)

rhizosphere A region around the plant root where materials released from the root increase the microbial population and its activities. (696)

rho factor (ro) The protein that helps RNA polymerase dissociate from a rho-dependent terminator after it has stopped transcription. (270)

ribonucleic acid (RNA) (ri″bo-nu-kle'ik) A polynucleotide composed of ribonucleotides joined by phosphodiester bridges. (252)

ribosomal RNA (rRNA) The RNA present in ribosomes; ribosomes contain several sizes of single-stranded rRNA that contribute to ribosome structure and are also directly involved in the mechanism of protein synthesis. (269)

ribosome (ri'bo-sōm) The organelle where protein synthesis occurs; the message encoded in mRNA is translated here. (50)

riboswitch A site in the leader of an mRNA molecule that interacts with a metabolite or other small molecule that causes the leader to change its folding pattern. In some riboswitches, this change can attenuate transcription; in others, it affects translation—either positively or negatively. (304)

ribotyping Ribotyping is the use of *E. coli* rRNA to probe chromosomal DNA in Southern blots for typing bacterial strains. This method is based on the fact that rRNA genes are scattered throughout the chromosome of most bacteria and therefore polymorphic restriction endonuclease patterns result when chromosomes are digested and probed with rRNA. (874)

ribozyme An RNA molecule with catalytic activity. (472)

ribulose-1,5-bisphosphate carboxylase (ri'bu-lōs) The enzyme that catalyzes the incorporation of CO_2 in the Calvin cycle. (229)

Rift Valley fever An acute, viral disease of domestic animals and humans, transmitted by mosquitoes. The Rift Valley fever virus is a single-stranded RNA virus named for the trough stretching 4,000 miles from Jordan through eastern Africa to Mozambique. (922)

ringworm (ring'werm) The common name for a fungal infection of the skin, even though it is not caused by a worm and is not always ring-shaped in appearance. (1008)

rise period or **burst** The period during the one-step growth experiment when host cells lyse and release phage particles. (429)

RNA polymerase The enzyme that catalyzes the synthesis of mRNA under the direction of a DNA template. (269)

RNA world The theory that posits that the first self-replicating molecule was RNA and this led to the evolution of the first primitive cell. (472)

Rocky Mountain spotted fever A disease caused by *Rickettsia rickettsii.* (964)

rolling-circle replication A mode of DNA replication in which the replication fork moves around a circular DNA molecule, displacing a strand to give a 5' tail that is also copied to produce a new double-stranded DNA. (257)

root nodule Gall-like structures on roots that contain endosymbiotic nitrogen-fixing bacteria (e.g., *Rhizobium* or *Bradyrhizobium* is present in legume nodules). (701)

roseola infantum (ro-ze'o-lə) A skin eruption that produces a rose-colored rash in infants. Caused by the human herpesvirus 6. The disease is short-lived and characterized by a high fever of 3 to 4 days' duration. (934)

R plasmids or **R factors** Plasmids bearing one or more drug resistant genes. (53, 852)

rubella (German measles) A moderately contagious skin disease that occurs primarily in children 5 to 9 years of age that is caused by the rubella virus, which is acquired by droplet inhalation into the respiratory system; German measles. (920)

rubeola *See* measles.

rumen (roo-men) The expanded upper portion or first compartment of the stomach of ruminants. (724)

ruminant (roo'mĭ-nant) An herbivorous animal that has a stomach divided into four compartments and chews a cud consisting of regurgitated, partially digested food. (724)

run The straight line movement of a bacterium. (71)

S

salmonellosis (sal″mo-nel-o'sis) An infection with certain species of the genus *Salmonella,* usually caused by ingestion of food containing salmonellae or their products. Also known as *Salmonella* gastroenteritis or *Salmonella* food poisoning. (984)

sanitization (san″ĭ-ti-za′shun) Reduction of the microbial population on an inanimate object to levels judged safe by public health standards. (151)

saprophyte (sap′ro-fit) An organism that takes up nonliving organic nutrients in dissolved form and usually grows on decomposing organic matter. (632)

saprozoic nutrition (sap″ro-zo′ik) Having the type of nutrition in which organic nutrients are taken up in dissolved form; normally refers to animals or protists. (607)

SAR11 The most abundant microbe on Earth, this lineage of α-proteobacteria has been found in almost all marine ecosystems studied. The SAR11 isolate, *Pelagibacter ubique,* has been cultured and its genome sequenced and annotated. (678)

scaffolding proteins Special proteins that are used to aid procapsid construction during the assembly of a bacteriophage capsid and are removed after the completion of the procapsid. (433)

scanning electron microscope (SEM) An electron microscope that scans a beam of electrons over the surface of a specimen and forms an image of the surface from the electrons that are emitted by it. (30)

scanning probe microscope A microscope used to study surface features by moving a sharp probe over the object's surface (e.g., the scanning tunneling microscope). (35)

scarlatina (skahr″la-te′nah) *See* scarlet fever.

scarlet fever (scarlatina) (skar′let) A disease that results from infection with a strain of *Streptococcus pyogenes* that carries a lysogenic phage with the gene for erythrogenic (rash-inducing) toxin. The toxin causes shedding of the skin. This is a communicable disease spread by respiratory droplets. (956)

Sec-dependent pathway System that can transport proteins through the plasma membrane or insert them into the membrane. The bacterial translocon recognizes a signal peptide, and employs SecYEG, SecA, and chaperones such as SecB. (63)

secondary metabolites Products of metabolism that are synthesized after growth has been completed. Antibiotics are considered secondary metabolites. (589, 1069)

secondary treatment The biological degradation of dissolved organic matter in the process of sewage treatment; the organic material is either mineralized or changed to settleable solids. (1065)

second law of thermodynamics Physical and chemical processes proceed in such a way that the entropy of the universe (the system and its surroundings) increases to the maximum possible. (169)

secretory IgA (sIgA) The primary immunoglobulin of the secretory immune system. *See* IgA. (793)

secretory vacuole In protists and some animals, these organelles usually contain specific enzymes that perform various functions such as excystment. Their contents are released to the cell exterior during exocytosis. (86)

segmented genome A virus genome that is divided into several parts or fragments, each probably coding for a single polypeptide; segmented genomes are very common among the RNA viruses. (417)

selectable marker A gene whose wild-type or mutant phenotype can be determined by growth on specific media. (367)

selectins (sə-lek′tins) A family of cell adhesion molecules that are displayed on activated endothelial cells; examples include P-selectin and E-selectin. Selectins mediate leukocyte binding to the vascular endothelium. (756)

selective media Culture media that favor the growth of specific microorganisms; this may be accomplished by inhibiting the growth of undesired microorganisms. (112)

selective toxicity The ability of a chemotherapeutic agent to kill or inhibit a microbial pathogen while damaging the host as little as possible. (164, 837)

self-assembly The spontaneous formation of a complex structure from its component molecules without the aid of special enzymes or factors. (68, 227)

sensory rhodopsin Microbial rhodopsins that sense the spectral quality of light. Found in halophilic archaea (SRI and SRII) and cyanobacteria. *See* bacteriorhodopsin. (515)

sepsis (sep′sis) Systemic response to infection manifested by two or more of the following conditions: temperature $>$38 or $<$36°C; heart rate $>$90 beats per min; respiratory rate $>$20 breaths per min, or pCO_2 $<$32 mm Hg; leukocyte count $>$12,000 cells per ml^3 or $>$10% immature (band) forms. Sepsis also has been defined as the presence of pathogens or their toxins in blood and other tissues. (987)

septate (sep′tāt) Divided by a septum or cross wall; also with more or less regular occurring cross walls. (632)

septicemia (sep″tĭ-se′me-ah) A disease associated with the presence in the blood of pathogens or bacterial toxins. (567, 821)

septic shock (sep′tik) Sepsis associated with severe hypotension despite adequate fluid resuscitation, along with the presence of perfusion abnormalities that may include, but are not limited to, lactic acidosis, oliguria, or an acute alteration in mental status. Gram-positive bacteria, fungi, and endotoxin-containing gram-negative bacteria can initiate the pathogenic cascade of sepsis leading to septic shock. (987)

septic tank (sep′tik) A tank used to process domestic sewage. Solid material settles out and is partially degraded by anaerobic bacteria as sewage slowly flows through the tank. The outflow is further treated or dispersed in aerobic soil. (1059)

septum (sep′tum; pl., **septa**) A partition or crosswall that occurs between two cells in a procaryotic or fungal filament, or which partitions structures such as spores. A septum also divides a parent cell into two daughter cells during binary fission. (74, 120, 595, 633)

serial endosymbiotic theory (SET) A theory of eucaryotic origin that suggests that such cells arose by a series of discrete endosymbiotic steps, each endosymbiont giving rise to a different organelle. (477)

serology (se-rol′o-je) The branch of immunology that is concerned with in vitro reactions involving one or more serum constituents (e.g., antibodies and complement). (876)

serotyping A technique or serological procedure used to differentiate between strains (serovars or serotypes) of microorganisms that have differences in the antigenic composition of a structure or product. (876)

serum (se′rum; pl., **serums** or **sera**) The clear, fluid portion of blood lacking both blood cells and fibrinogen. It is the fluid remaining after coagulation of plasma, the noncellular liquid fraction of blood. (901)

serum resistance The type of resistance that occurs with bacteria such as *Neisseria gonorrhoeae* because the pathogen interferes with membrane attack complex formation during the complement cascade. (832)

settling basin A basin used during water purification to chemically precipitate out fine particles, microorganisms, and organic material by coagulation or flocculation. (1050)

severe acute respiratory syndrome (SARS) A disease caused by a single-stranded RNA coronavirus, characterized by fever, lower respiratory symptoms, and radiographic evidence of pneumonia. SARS has a mortality rate of approximately 10%. (920)

sex pilus (pi′lus) A thin protein appendage required for bacterial mating or conjugation. The cell with sex pili donates DNA to recipient cells. (67, 338)

sheath (shēth) A hollow tubelike structure surrounding a chain of cells and present in several genera of bacteria. (548)

shigellosis (shĭ′gəl-o′sis) The diarrheal disease that arises from an infection with *Shigella* spp. Often called bacillary dysentery. (985)

Shine-Dalgarno sequence A segment in the leader of procaryotic mRNA that binds to a special sequence on the 16S rRNA of the small ribosomal subunit. This helps properly orient the mRNAon the ribosome. (265)

shingles (herpes zoster) (shing′g′lz) A reactivated form of chickenpox caused by latent varicella-zoster virus. (915)

shuttle vector A DNA vector (e.g., plasmid, cosmid) that has two origins of replication, each recognized by a different microorganism. Thus the vector can replicate in both microbes. (367)

siderophore (sid′er-o-for″) A small molecule that complexes with ferric iron and supplies it to a cell by aiding in its transport across the plasma membrane. (109)

sigma factor A protein that helps bacterial RNA polymerase core enzyme recognize the promoter at the start of a gene. (269)

sign An objective change in a diseased body that can be directly observed (e.g., a fever or rash). (888)

signal peptide The special amino-terminal sequence on a peptide destined for transport that delays protein folding and is recognized in bacteria by the Sec-dependent pathway machinery. (63)

silage Fermented plant material with increased palatability and nutritional value for animals, which can be stored for extended periods. (1046)

silencer A site in the DNA to which a eucaryotic repressor protein binds. (313)

silent mutation A mutation that does not result in a change in the organism's proteins or phenotype even though the DNA base sequence has been changed. (320)

silicoflagellate Photosynthetic protists within the subdivision *Stramenopila* that have a complex internal skeleton made of silica. (000)

simple matching coefficient (S_{SM}) An association coefficient used in numerical taxonomy; the proportion of characters that match regardless of whether or not the attribute is present. (479)

single radial immunodiffusion (RID) assay An immunodiffusion technique that quantitates antigens by following their diffusion through a gel containing antibodies directed against the test antigens. (880)

site-specific recombination Recombination of nonhomologous genetic material with a chromosome at a specific site. (331)

sixth disease A viral disease of infants and young children caused by herpesvirus type 6. The disease presents with a sudden onset of high fever that lasts for days but suddenly subsides; a fine, red rash is often observed once the fever subsides. (934)

skin-associated lymphoid tissue (SALT) The lymphoid tissue in the skin that forms a first-line defense as a part of nonspecific (innate) immunity. (758)

S-layer A regularly structured layer composed of protein or glycoprotein that lies on the surface of many bacteria. It may protect the bacterium and help give it shape and rigidity. (66)

slime The viscous extracellular glycoproteins or glycolipids produced by staphylococci and *Pseudomonas aeruginosa* bacteria that allows them to adhere to smooth surfaces such as prosthetic medical devices and catheters. More generally, the term often refers to an easily removed, diffuse, unorganized layer of extracellular material that surrounds a procaryotic cell. (968)

slime layer A layer of diffuse, unorganized, easily removed material lying outside an archeal or bacterial cell wall. (65)

slime mold A common term for members of the protist divisions *Myxogastria* and *Dictyostelia.* (3)

slow sand filter A bed of sand through which water slowly flows; the gelatinous microbial layer on the sand grain surface removes waterborne microorganisms, particularly *Giardia,* by adhesion to the gel. This type of filter is used in some water purification plants. (1051)

slow virus disease A progressive, pathological process virus that remains clinically silent during a prolonged incubation period of months to years after which progressive clinical disease becomes apparent. (461)

sludge A general term for the precipitated solid matter produced during water and sewage treatment; solid particles composed of organic matter and microorganisms that are involved in aerobic sewage treatment (activated sludge). (1055)

small RNAs (sRNAs) Special small regulatory RNA molecules that do not function as messenger, ribosomal, or transfer RNAs. (273, 305)

smallpox (variola) Once a highly contagious, often fatal disease caused by a poxvirus. Its most noticeable symptom was the appearance of blisters and pustules on the skin. Vaccination has eradicated smallpox throughout the world. (920)

small subunit rRNA (SSU rRNA) The rRNA associated with the ribosomal small subunit: 16S rRNA in procaryotes and 18S rRNA in eucaryotes. Comparison of SSU rRNA nucleotide sequences (or that of the genes encoding these RNAs) is important for the taxonomic identification and phylogenetic analysis of microorganisms. (474)

snapping division A distinctive type of binary fission resulting in an angular or a palisade arrangement of cells, which is characteristic of the genera *Arthrobacter* and *Corynebacterium.* (594)

SOS response A complex, inducible process that allows bacterial cells with extensive DNA damage to survive, although often in a mutated form; it involves cessation of cell division, upregulation of severe DNA repair systems, and induction of translesion DNA synthesis. (327)

source The location or object from which a pathogen is immediately transmitted to the host, either directly or through an intermediate agent. (891)

Southern blotting technique The procedure used to isolate and identify DNA fragments from a complex mixture. The isolated, denatured fragments are transferred from an agarose gel to a nylon filter and identified by hybridization with probes. (358)

specialized transduction A transduction process in which only a specific set of bacterial or archaeal genes are carried to a recipient cell by a temperate virus; the cell's genes are acquired because of a mistake in the excision of a provirus during the lysogenic life cycle. (346)

species (spe′shēz) Species of higher organisms are groups of interbreeding or potentially interbreeding natural populations that are reproductively isolated. Procaryotic species are collections of strains that have many stable properties in common and differ significantly from other groups of strains. (480)

specific immune response (acquired, adaptive, or specific immunity) *See* acquired immunity. (744)

spheroplast (sfĕr′o-plast) A relatively spherical cell formed by the weakening or partial removal of the rigid cell wall component (e.g., by penicillin treatment of gram-negative bacteria). Spheroplasts are usually osmotically sensitive. (61)

spike *See* peplomer.

spirillum (spi-ril′um) A rigid, spiral-shaped bacterium. (40)

spirochete (spi′ro-kēt) A flexible, spiral-shaped bacterium with periplasmic flagella. (40)

spleen (splēn) A secondary lymphoid organ where old erythrocytes are destroyed and blood-borne antigens are trapped and presented to lymphocytes. (750)

spliceosome A complex of proteins that carries out RNA splicing. (274)

split or **interrupted gene** A structural gene with DNA sequences that code for the final RNA product (expressed sequences or exons) separated by regions coding for RNA absent from the mature RNA (intervening sequences or introns). (273)

spongiform encephalopathies Degenerative central nervous system diseases in which the brain has a spongy appearance; they appear due to prions. (469, 944)

sporadic disease (spo-rad′ik) A disease that occurs occasionally and at random intervals in a population. (886)

sporangiospore (spo-ran′je-o-spōr) A spore born within a sporangium. (590, 633)

sporangium (spo-ran′je-um; pl., **sporangia**) A saclike structure or cell, the contents of which are converted into an indefinite number of spores. It is borne on a special hypha called a sporangiophore. (73, 633)

spore (spōr) A differentiated, specialized form that can be used for dissemination, for survival of adverse conditions because of its heat and dessication resistance, and/or for reproduction. Spores are usually unicellular and may develop into vegetative organisms or gametes. They may be produced asexually or sexually and are of many types. (73, 572, 589, 632)

sporogenesis (spor′o-jen′ĕ-sis) *See* sporulation. (75)

sporotrichosis (spo″ro-tri-ko′sis) A subcutaneous fungal infection caused by the dimorphic fungus *Sporothrix schenckii.* (1010)

sporozoite The motile, infective stage of apicomplexan protists, including *Plasmodium,* the causative agent of malaria. (619)

sporulation (spor″u-la′shun) The process of spore formation. (75)

spread plate A petri dish of solid culture medium with isolated microbial colonies growing on its surface, which has been prepared by spreading a dilute microbial suspension evenly over the agar surface. (113)

sputum (spu′tum) The mucus secretion from the lungs, bronchi, and trachea that is ejected (expectorated) through the mouth. (862)

stalk (stawk) A nonliving bacterial appendage produced by the cell and extending from it. (543)

standard free energy change The free energy change of a reaction at 1 atmosphere pressure when all reactants and products are present in their standard states; usually the temperature is 25°C. (170)

standard reduction potential A measure of the tendency of a reductant to lose electrons in an oxidation-reduction (redox) reaction. The more negative the reduction potential of a compound, the better electron donor it is. (172)

staphylococcal food poisoning (staf″i-lo-kok′al) A type of food poisoning caused by ingestion of improperly stored or cooked food in which *Staphylococcus aureus* has grown. The bacteria produce exotoxins that accumulate in the food. (985)

staphylococcal scalded skin syndrome (SSSS) A disease caused by staphylococci that produce an exfoliative toxin. The skin becomes red (erythema) and sheets of epidermis may separate from the underlying tissue. (969)

starter culture An inoculum, consisting of a mixture of carefully selected microorganisms, used to start a commercial fermentation. (1039)

static Inhibiting or retarding the bacterial growth. (837)

stationary phase (sta′shun-er″e) The phase of microbial growth in a batch culture when population growth ceases and the growth curve levels off. (124)

stem-nodulating rhizobia Rhizobia that produce nitrogen-fixing structures above the soil surface on plant stems. These most often are observed in tropical plants and produced by *Azorhizobium.* (704)

sterilization (ster″ĭ-lĭ-za′shun) The process by which all living cells, viable spores, viruses, and viroids are either destroyed or removed from an object or habitat. (151)

stigma (stig′mah) A photosensitive region on the surface of certain protists that is used in phototaxis. (612)

strain A population of organisms that descends from a single organism or pure culture isolate. (480)

streak plate A petri dish of solid culture medium with isolated microbial colonies growing on its surface, which has been prepared by spreading a microbial mixture over the agar surface, using an inoculating loop. (113)

streptococcal pharyngitis (strep throat) Infection of the pharynx (throat) by the gram-positive bacterium *Streptococcus pyogenes. See also* group A streptococcus. (958)

streptococcal pneumonia An endogenous infection of the lungs caused by *Streptococcus pneumoniae* that occurs in predisposed individuals. (958)

streptolysin-O (SLO) (strep-tol′ĭ-sin) A specific hemolysin produced by *Streptococcus pyogenes* that is inactivated by oxygen (hence the "O" in its name). SLO causes beta-hemolysis of blood cells on agar plates incubated anaerobically. (828)

streptolysin-S (SLS) A product of *Streptococcus pyogenes* that is bound to the bacterial cell but may sometimes be released. SLS causes beta hemolysis on aerobically incubated blood-agar plates and can act as a leukocidin by killing white blood cells that phagocytose the bacterial cell to which it is bound. (828)

streptomycete Any high G + C gram-positive bacterium of the genera *Kitasatospora, Streptomyces,* and *Streptoverticillium.* The term is also often used to refer to other closely related families including *Streptosporangiaceae* and *Nocardiopsaceae.* (599)

streptomycin (strep′to-mi″sin) A bactericidal aminoglycoside antibiotic produced by *Streptomyces griseus.* (845)

strict anaerobes *See* obligate anaerobes.

stroma (stro′mah) The chloroplast matrix that is the location of the photosynthetic carbon dioxide fixation reactions. (90)

stromatolite (stro″mah-to′līt) Dome-like microbial mat communities consisting of filamentous photosynthetic bacteria and occluded sediments (often calcareous or siliceous). They usually have a laminar structure. Many are fossilized, but some modern forms occur. (473)

structural gene A gene that codes for the synthesis of a polypeptide or polynucleotide (i.e., rRNA, tRNA) with a nonregulatory function. (295)

subacute sclerosing panencephalitis Diffuse inflammation of the brain resulting from virus and prion infections. (918)

subgingival plaque (sub-jin′jĭ-val) The plaque that forms at the dentogingival margin and extends down into the gingival tissue. (993)

substrate-level phosphorylation The synthesis of ATP from ADP by phosphorylation coupled with the exergonic breakdown of a high-energy organic substrate molecule. (194)

substrate mycelium In the actinomycetes, hyphae that are on the surface and may penetrate into the solid medium on which the microbes are grown. (589)

subsurface biosphere The region below the plant root zone where microbial populations can grow and function. (711)

sulfate reduction The process of sulfate use as an oxidizing agent, which results in the accumulation of reduced forms of sulfur such as sulfide, or incorporation of sulfur into organic molecules, usually as sulfhydryl groups. (649)

sulfonamide (sul-fon′ah-mīd) A chemotherapeutic agent that has the SO_2-NH_2 group and is a derivative of sulfanilamide. (846)

superantigen Superantigens are toxic bacterial proteins that stimulate the immune system much more extensively than do normal antigens. They stimulate T cells to proliferate nonspecifically through simultaneous interaction with class II MHC proteins on antigen-presenting cells and variable regions on the β chain of the T-cell receptor complex. Examples include streptococcal scarlet fever toxins, staphylococcal toxic shock syndrome toxin-1, and streptococcal M protein. (785, 969)

superinfection (soo″per-in-fek′shun) A new bacterial or fungal infection of a patient that is resistant to the drug(s) being used for treatment. (438)

superoxide dismutase (SOD) (soo″per-ok′sīd dismu′tas) An enzyme that protects many microorganisms by catalyzing the destruction of the toxic superoxide radical. (140)

supportive media Culture media that are able to sustain the growth of many different kinds of microorganisms. (112)

suppressor mutation A mutation that overcomes the effect of another mutation and produces the normal phenotype. (320)

Svedberg unit (sfed′berg) The unit used in expressing the sedimentation coefficient; the greater a particle's Svedberg value, the faster it travels in a centrifuge. (50)

swab A wad of absorbent material usually wound around one end of a small stick and used for applying medication or for removing material from an area; also, a dacron-tipped polystyrene applicator. (862)

symbiosis (sim″bi-o′sis) The living together or close association of two dissimilar organisms, each of these organisms being known as a symbiont. (815)

symbiosome The final nitrogen-fixing form of rhizobia within root nodule cells. (701)

symptom (simp′təm) A change during a disease that a person subjectively experiences (e.g., pain, bodily discomfort, fatigue, or loss of appetite). Sometimes the term symptom is used more broadly to include any observed signs. (888)

syncytium A large, multinucleate cell formed by the fusion of numerous cells. (461)

syndrome *See* disease syndrome.

syngamy The fusion of haploid gametes. (609)

synthetic medium *See* defined medium.

syntrophism (sin′trōf-izəm) The association in which the growth of one organism either depends on, or is improved by, the provision of one or more growth factors or nutrients by a neighboring organism. Sometimes both organisms benefit. (726)

syphilis (sif′ĭ-lis) *See* venereal syphilis.

systematic epidemiology The field of epidemiology that focuses on the ecological and social factors that influence the development of emerging and reemerging infectious diseases. (898)

systematics (sis″te-mat′iks) The scientific study of organisms with the ultimate objective of characterizing and arranging them in an orderly manner; often considered synonymous with taxonomy. (478)

systemic lupus erythematosus (loo′pus er″ĭ-them-ah-to′sus) An autoimmune, inflammatory disease that may affect every tissue of the body. (811)

T

taxon (tak′son) A group into which related organisms are classified. (478)

taxonomy (tak-son′o-me) The science of biological classification; it consists of three parts: classification, nomenclature, and identification. (478)

TB skin test Tuberculin hypersensitivity test for a previous or current infection with *Mycobacterium tuberculosis.* (808)

T cell or **T lymphocyte** A type of lymphocyte derived from bone marrow stem cells that matures into an immunologically competent cell under the influence of the thymus. T cells are involved in a variety of cell-mediated immune reactions. (748)

T-cell antigen receptor (TCR) The receptor on the T cell surface consisting of two antigen-binding peptide chains; it is associated with a large number of other glycoproteins. Binding of antigen to the TCR, usually in association with MHC, activates the T cell. (781)

T-dependent antigen An antigen that effectively stimulates B-cell response only with the aid of T-helper cells that produce interleukin-2 and B-cell growth factor. (786)

teichoic acids (ti-ko′ik) Polymers of glycerol or ribitol joined by phosphates; they are found in the cell walls of gram-positive bacteria. (57)

temperate phages Bacteriophages that can infect bacteria and establish a lysogenic relationship rather than immediately lysing their hosts. (438)

template strand (tem′plat) A DNA or RNA strand that specifies the base sequence of a new complementary strand of DNA or RNA. (265)

terminator A sequence that marks the end of a gene and stops transcription. (266)

tertiary treatment (ter′she-er-e) The removal from sewage of inorganic nutrients, heavy metals, viruses, etc., by chemical and biological means after microbes have degraded dissolved organic material during secondary sewage treatment. (1058)

test A loose-fitting shell of an amoeba. (618)

tetanolysin (tet″ah-nol′ĭ-sin) A hemolysin that aids in tissue destruction and is produced by *Clostridium tetani.* (978)

tetanospasmin (tet″ah-no-spaz′min) The neurotoxic component of the tetanus toxin, which causes the muscle spasms of tetanus. Tetanospasmin production is controlled by a plasmid encoded gene. (978)

tetanus (tet′ah-nus) An often fatal disease caused by the anaerobic, spore-forming bacillus *Clostridium tetani,* and characterized by muscle spasms and convulsions. (978)

tetracyclines (tet″rah-si′klēns) A family of antibiotics with a common four-ring structure, which are isolated from the genus *Streptomyces* or produced semisynthetically; all are related to chlortetracycline or oxytetracycline. (845)

tetrapartite associations (tet″rah-par′tīt) A symbiotic association of the same plant with three different types of microorganisms. (707)

T_H 1 cell A $CD4^+$ T-helper cell that secretes interferon-gamma, interleukin-2, and tumor necrosis factor, influencing phagocytic cells. (782)

T_H 2 cell A $CD4^+$ T-helper cell that secretes interleukin (IL)-4, IL-5, IL-6, IL-9, IL-10, and IL-13, influencing growth and differentiation of lymphocytes. (782)

T_H 0 cell A $CD4^+$ T cell that secretes cytokines of T_H1 and T_H2 cells, implicating them as precursors to T_H1 and T_H2 cells. (782)

thallus (thal′us) A type of body that is devoid of root, stem, or leaf; characteristic of fungi. (631)

T-helper (T_H) cell A cell that is needed for T-cell-dependent antigens to be effectively presented to B cells. It also promotes cell-mediated immune responses. (781)

theory A set of principles and concepts that have survived rigorous testing and that provide a systematic account of some aspect of nature. (10)

therapeutic cloning Cloning genes for human gene therapy with the intent of treating human disease. (376)

therapeutic index The ratio between the toxic dose and the therapeutic dose of a drug, used as a measure of the drug's relative safety. (837)

thermal death time (TDT) The shortest period of time needed to kill all the organisms in a microbial population at a specified temperature and under defined conditions. (154)

thermoacidophiles A group of bacteria that grow best at acidic pHs and high temperatures; they are members of the *Archaea.* (508)

thermocycler The instrument in which the polymerase chain reaction is performed. (362)

thermophile (ther′mo-fīl) A microorganism that can grow at temperatures of 55°C or higher; the minimum is usually around 45°C. (138)

thrush Infection of the oral mucous membrane by the fungus *Candida albicans;* also known as oral candidiasis. (1017)

thylakoid (thi′lah-koid) A flattened sac in the chloroplast stroma that contains photosynthetic pigments and the photosynthetic electron transport chain; light energy is trapped and used to form ATP and NAD(P)H in the thylakoid membrane. (90)

thymine (thi′min) The pyrimidine 5-methyluracil that is found in nucleosides, nucleotides, and DNA. (241)

thymus (thi′məs) A primary lymphoid organ in the chest that is necessary in early life for the development of immunological functions. T-cell maturation takes place here. (749)

tick-borne encephalitis Inflammation of the central nervous system caused by tick-borne encephalitis virus, a member of the family Flaviviridae. (922)

T-independent antigen An antigen that triggers a B cell into immunoglobulin production without T-cell cooperation. (788)

tinea (tin′e-ah) A name applied to many different kinds of superficial fungal infections of the skin, nails, and hair, the specific type (depending on characteristic appearance, etiologic agent, and site) usually designated by a modifying term. (1008)

tinea capitis An infection of scalp hair by species of *Trichophyton* or *Microsporum.* (1008)

tinea corporis An infection of the smooth parts of the skin by either *Trichophyton rubrum, T. mentagrophytes,* or *Microsporum canis.* (1009)

tinea cruris An infection of the groin by either *Epidermophyton floccosum, Trichophyton mentagrophytes,* or *T. rubrum;* also known as jock itch. (1009)

tinea pedis A fungal infection of the foot by *Trichophyton rubrum, T. mentagrophytes,* or *E. floccosum;* also known as athlete's foot. (1009)

tinea unguium An infection of the nail bed by either *Trichophyton rubrum* or *T. mentagrophytes.* (1009)

Ti plasmid Tumor-inducing plasmid found in plant pathogenic species of the bacterium *Agrobacterium.* The genes for virulence (*vir* genes) and a region of DNA that is transferred to the infected plant (T DNA) reside on the Ti plasmid. The Ti plasmid has been modified to allow the construction of transgenic plants. (378, 706)

titer (ti′ter) Reciprocal of the highest dilution of an antiserum that gives a positive reaction in the test being used. (795)

T lymphocyte *See* T cell.

Toll-like receptor A type of pattern recognition receptor on phagocytes such as macrophages that triggers the proper response to different classes of pathogens. It signals the production of transcription factor NF κB, which stimulates formation of cytokines, chemokines, and other defense molecules. (753)

tonsillitis (ton′si-li′tis) Inflammation of the tonsils, especially the palatine tonsils often due to *S. pyogenes* infection. (958)

toxemia (tok-se′me-ah) The condition caused by toxins in the blood of the host. (824)

toxic shocklike syndrome (TSLS) A disease caused by an invasive group A streptococcus infection that is characterized by a rapid drop in blood pressure, failure of many organs, and a very high fever. It probably results from the release of one or more streptococcal pyrogenic exotoxins. (958)

toxic shock syndrome (TSS) (tok′sik) A staphylococcal disease that most commonly affects females who use ultra-absorbant tampons during menstruation. It is associated with the toxic shock syndrome toxin produced by strains of *Staphylococcus aureus.* (969)

toxigenicity (tok″sĭ-jĕ-nis′i-tē) The capacity of an organism to produce a toxin. (816)

toxin (tok′sin) A microbial product or component that injures another cell or organism. Often the term refers to a poisonous protein, but toxins may be lipids and other substances. (824)

toxin neutralization The inactivation of toxins by specific antibodies, called antitoxins, that react with them. (799)

toxoid (tok′soid) A bacterial exotoxin that has been modified so that it is no longer toxic but will still stimulate antitoxin formation when injected into a person or animal. (824, 901)

toxoplasmosis (tok″so-plaz-mo′sis) A disease of animals and humans caused by the parasitic protozoan, *Toxoplasma gondii.* (1011)

trachoma (trah-ko′mah) A chronic infectious disease of the conjunctiva and cornea, producing pain, inflammation and sometimes blindness. It is caused by *Chlamydia trachomatis* serotypes A–C. (978)

transamination (trans″am-i-na′shun) The removal of amino acid's amino group by transferring it to an α-keto acid acceptor. (212)

transcriptase (trans-krip′tās) An enzyme that catalyzes transcription; in viruses with RNA genomes, this enzyme is an RNA-dependent RNA polymerase that is used to make RNA copies of the RNA genomes. (438, 455)

transcription (trans-krip′shun) The process in which single-stranded RNA with a base sequence complementary to the template strand of DNA or RNA is synthesized. (251)

transcriptome All the messenger RNA that is transcribed from the genome of an organism under a given set of circumstances. (402)

transduction (trans-duk′shun) The transfer of genes between bacterial or archaeal cells by viruses. (345)

transfer host (trans′fer) A host that is not necessary for the completion of a parasite's life cycle, but is used as a vehicle for reaching a final host. (816)

transfer RNA (tRNA) A small RNA that binds an amino acid and delivers it to the ribosome for incorporation into a polypeptide chain during protein synthesis. (269)

transformation (trans″for-ma′shun) A mode of gene transfer in procaryotes in which a piece of free DNA is taken up by a cell and integrated into the its genome. (249, 342)

transgenic animal or **plant** An animal or plant that has gained new genetic information by the insertion of foreign DNA. It may be produced by such techniques as injecting DNA into animal eggs, electroporation of mammalian cells and plant cell protoplasts, or shooting DNA into plant cells with a gene gun. (371)

transient carrier *See* casual carrier.

transition mutations (tran-zish′un) Mutations that involve the substitution of a different purine base for the purine present at the site of the mutation or the substitution of a different pyrimidine for the normal pyrimidine. (319)

translation (trans-la′shun) Protein synthesis; the process by which the genetic message carried by mRNA directs the synthesis of polypeptides with the aid of ribosomes and other cell constituents. (251)

transmissible spongiform encephalopathies (TSE) A fatal, incurable, degenerative disease of the brain caused by prions and characterized by deteriorating mental and physical abilities. The disease is named for the altered mental state and spongelike appearance of the brain in infected individuals. (944)

transmission electron microscope (TEM) A microscope in which an image is formed by passing an electron beam through a specimen and focusing the scattered electrons with magnetic lenses. (29)

transovarian passage (trans″o-va′re-an) The passage of a microorganism such as a rickettsia from Generation to generation of hosts through tick eggs. No humans or other mammals are needed as reservoirs for continued propagation. (964)

transpeptidation 1. The reaction that forms the peptide cross-links during peptidoglycan synthesis. 2. The reaction that forms a peptide bond during the elongation cycle of protein synthesis. (233, 284)

transposable elements *See* transposon.

transposition (trans″po-zish′un) The movement of a piece of DNA around the chromosome. (331)

transposon (tranz-po′zon) A DNA element that carries the genes required for transposition and moves about the genome; if it contains genes other than those required for transposition, it may be called a composite transposon. Often the name is reserved only for transposable elements that also contain genes unrelated to transposition. (332)

transversion mutations (trans-ver′zhun) Mutations that result from the substitution of a purine base for the normal pyrimidine or a pyrimidine for the normal purine. (319)

traveler's diarrhea A type of diarrhea resulting from ingestion of viruses, bacteria, or protozoa normally absent from the traveler's environment. A major pathogen is enterotoxigenic *Escherichia coli.* (986)

tricarboxylic acid (TCA) cycle The cycle that oxidizes acetyl coenzyme A to CO_2 and generates NADH and $FADH_2$ for oxidation in the electron transport chain; the cycle also supplies carbon skeletons for biosynthesis. (198)

trichome (tri′kōm) A row or filament of microbial cells that are in close contact with one another over a large area. (537)

trichomoniasis (trik″o-mo-ni′ah-sis) A sexually transmitted disease caused by the parasitic protozoan *Trichomonas vaginalis.* (1012)

trickling filter A bed of rocks covered with a microbial film that aerobically degrades organic waste during secondary sewage treatment. (1056)

trimethoprim A synthetic antibiotic that inhibits production of folic acid by binding to dihydrofolate reductase. Trimethoprim has a wide spectrum of activity and is bacteriostatic. (847)

tripartite associations (tri-par′tīt) A symbiotic association of the same plant with two types of microorganisms. (707)

trophozoite (trof″ o-zo′īt) The active, motile feeding stage of a protozoan organism; in the malarial parasite, the stage of schizogony between the ring stage and the schizont. (608)

tropism (tro′piz-əm) (1) The movement of living organisms toward or away from a focus of heat, light, or other stimulus. (2) The selective infection of certain organisms or host tissues by a virus; results from the distribution of the specific receptor for a virus in different organisms or certain tissues of the host. (448, 819)

trypanosome (tri-pan′o-sōm) A protozoan of the genus *Trypanosoma.* Trypanosomes are parasitic flagellate protozoa that often live in the blood of humans and other vertebrates and are transmitted by insect bites. (1006)

trypanosomiasis (tri-pan″o-so-mi′ah-sis) An infection with trypanosomes that live in the blood and lymph of the infected host. (1006)

tubercle (too′ber-k′l) A small, rounded nodular lesion produced by *Mycobacterium tuberculosis.* (954)

tuberculoid (neural) leprosy (too-ber′ku-loid) A mild, nonprogressive form of leprosy that is associated with delayed-type hypersensitivity to antigens on the surface of *Mycobacterium leprae.* It is characterized by early nerve damage and regions of the skin that have lost sensation and are surrounded by a border of nodules. (966)

tuberculosis (TB) (too-ber″ku-lo′sis) An infectious disease of humans and other animals resulting from an infection by a species of *Mycobacterium* and characterized by the formation of tubercles and tissue necrosis, primarily as a result of host hypersensitivity and inflammation. Infection is usually by inhalation, and the disease commonly affects the lungs (pulmonary tuberculosis), although it may occur in any part of the body. (951)

tuberculous cavity (too-ber′ku-lus) An air-filled cavity that results from a tubercle lesion caused by *M. tuberculosis.* (954)

tularemia (too″lah-re′me-ah) A plaguelike disease of animals caused by the bacterium *Francisella tularensis* subsp. *tularensis* (Jellison type A), which may be transmitted to humans. (991)

tumble Random turning or tumbling movements made by bacteria when they stop moving in a straight line. (71)

tumor A growth of tissue resulting from abnormal new cell growth and reproduction (neoplasia). (461)

tumor necrosis factor (TNF) A cytokine produced by activated macrophages. Originally named for its cytotoxic effect on tumor cells, TNF has activities similar to those of interleukin-1, such as inducing inflammation, lipid metabolism, and coagulation. (767)

turbidostat A continuous culture system equipped with a photocell that adjusts the flow of medium through the culture vessel to maintain a constant cell density or turbidity. (132)

twiddle *See* tumble.

two-component phosphorelay system A signal transduction regulatory system that uses the transfer of phosphoryl groups to control gene transcription and protein activity. It has two major components: a sensor kinase and a response regulator. (300)

type I hypersensitivity A form of immediate hypersensitivity arising from the binding of antigen to IgE attached to mast cells, which then release anaphylaxis mediators such as histamine. Examples: hay fever, asthma, and food allergies. (803)

type II hypersensitivity A form of immediate hypersensitivity involving the binding of antibodies to antigens on cell surfaces followed by destruction of the target cells (e.g., through complement attack, phagocytosis, or agglutination). (805)

type III hypersensitivity A form of immediate hypersensitivity resulting from the exposure to excessive amounts of antigens to which antibodies bind. These antibody-antigen complexes activate complement and trigger an acute inflammatory response with subsequent tissue damage. (807)

type IV hypersensitivity A delayed hypersensitivity response (it appears 24 to 48 hours after antigen exposure). It results from the binding of antigen to activated T lymphocytes, which then release cytokines and trigger inflammation and macrophage attacks that damage tissue. Type IV hypersensitivity is seen in contact dermititis from poison ivy, leprosy, and tertiary syphilis. (807)

type I protein secretion pathway *See* ABC protein secretion pathway.

type II protein secretion pathway A system that transports proteins from the periplasm across the outer membrane of gram-negative bacteria. (65)

type III protein secretion pathway A system in gram-negative bacteria that secretes virulence factors and injects them into host cells. (65)

type strain The microbial strain that is the nomenclatural type or holder of the species name. A type strain will remain within that species should nomenclature changes occur. (480)

typhoid fever (ti-foid) A bacterial infection transmitted by contaminated food, water, milk, or shellfish. The causative organism is *Salmonella enterica* serovar Typhi, which is present in human feces. (984)

U

ultramicrobacteria Bacteria that can exist normally in a miniaturized form or which are capable of miniaturization under low-nutrient conditions. They may be 0.2 μm or smaller in diameter. (671)

ultraviolet (UV) radiation Radiation of fairly short wavelength, about 10 to 400 nm, and high energy. (142, 156)

universal phylogenetic tree A phylogenetic tree that considers the evolutionary relationships among organism from all three domains of life: *Bacteria, Archaea,* and *Eucarya.* (475)

uracil (u′rah-sil) The pyrimidine 2,4-dioxypyrimidine, which is found in nucleosides, nucleotides, and RNA. (241)

V

vaccine A preparation of either killed microorganisms; living, weakened (attenuated) microorganisms; or inactivated bacterial toxins (toxoids). It is administered to induce development of the immune response and protect the individual against a pathogen or a toxin. (901)

vaccinomics The application of genomics and bioinformatics to vaccine development. (901)

valence (va′lens) The number of antigenic determinant sites on the surface of an antigen or the number of antigen-binding sites possessed by an antibody molecule. (774)

vancomycin A glycopeptide antibiotic obtained from *Nocardia orientalis* effective only against gram-positive bacteria. Vancomycin is bactericidal because it binds to the D-alanine amino acids of peptidogylcan precursor units, inhibiting peptidoglycan synthesis and altering cell wall permeability. While the final cidal events are similar to those of penicillin, the mechanism of action is different. (845)

variable region (V_L and V_H) The region at the N-terminal end of immunoglobulin heavy and light chains whose amino acid sequence varies between antibodies of different specificity. Variable regions form the antigen binding site. (790)

vasculitis (vas″ku-li′tis) Inflammation of a blood vessel. (960)

vector (vek′tor) 1. In genetic engineering, another name for a cloning vector. A DNA molecule that can replicate (a replicon) and transports a piece of inserted foreign DNA, such as a gene, into a recipient cell. It may be a plasmid, phage, cosmid or artificial chromosome. 2. In epidemiology, it is a living organism, usually an arthropod or other animal, that transfers an infective agent between hosts. (358, 818, 892)

vector-borne transmission The transmission of an infectious pathogen between hosts by means of a vector. (896)

vehicle (ve′ĭ-k′l) An inanimate substance or medium that transmits a pathogen. (894)

venereal syphilis (ve-ne′re-al sif′ĭ-lis) A contagious, sexually transmitted disease caused by the spirochete *Treponema pallidum.* (976)

venereal warts *See* anogenital condylomata.

verrucae vulgaris (v′ĕ-roo′se vul-ga′ris; s. **verruca vulgaris**) The common wart; a raised, epidermal lesion with horny surface caused by an infection with a human papillomavirus. (938)

vesicular nucleus The most common nuclear morphology seen in protists, characterized by a nucleus 1 to 10 μm in diameter, spherical, with a distinct nucleolus and uncondensed chromosomes. (609)

viable but nonculturable (VBNC) microorganisms Microbes that have been determined to be living in a specific environment (either in nature or the laboratory) but are not actively growing and cannot be cultured under standard laboratory conditions. (125)

vibrio (vib′re-o) A rod-shaped bacterial cell that is curved to form a comma or an incomplete spiral. (40)

viral hemagglutination (vi′ral hem″ah-gloo″tĭ-na′shun) The clumping or agglutination of red blood cells caused by some viruses. (876)

viral hemorrhagic fevers (VHF) A group of illnesses caused by several distinct viruses, all of which cause symptoms of fever and bleeding in infected humans. (941)

viral neutralization An antibody-mediated process in which IgG, IgM, and IgA antibodies bind to some viruses during their extracellular phase and inactivate or neutralize them. (799)

viremia (vi-re′me-ə) The presence of viruses in the blood stream. (819)

viricide (vir′i-sīd) An agent that inactivates viruses so that they cannot reproduce within host cells. (151)

virion (vi′re-on) A complete virus particle that represents the extracellular phase of the virus life cycle; at the simplest, it consists of a protein capsid surrounding a single nucleic acid molecule. (409)

virioplankton Viruses that occur in waters; high levels are found in marine and freshwater environments. (679)

viroid (vi′roid) An infectious agent that is a single-stranded RNA not associated with any protein; the RNA does not code for any proteins and is not translated. (467)

virology (vi-rol′o-je) The branch of microbiology that is concerned with viruses and viral diseases. (407)

virulence (vir′u-lens) The degree or intensity of pathogenicity of an organism as indicated by case fatality rates and/or ability to invade host tissues and cause disease. (816)

virulence factor A bacterial product, usually a protein or carbohydrate, that contributes to virulence or pathogenicity. (816)

virulent viruses Viruses that lyse their host cells during the reproductive cycle. (345)

virus An infectious agent having a simple acellular organization with a protein coat and a nucleic acid genome, lacking independent metabolism, and reproducing only within living host cells. (3, 409)

virusoid An infectious agent that is composed only of single-stranded RNA that encodes some but not all proteins required for its replication; can be replicated and transmitted to a new host only if it infects a cell also infected by a virus that serves as a helper virus. (468)

vitamin An organic compound required by organisms in minute quantities for growth and reproduction because it cannot be synthesized by the organism; vitamins often serve as enzyme cofactors or parts of cofactors. (105)

volutin granules (vo-lu′tin) *See* metachromatic granules.

W

wart An epidermal tumor of viral origin. (938)

wastewater treatment The use of physical and biological processes to remove particulate and dissolved material from sewage and to control pathogens. (1055)

water activity (a_w) A quantitative measure of water availability in the habitat; the water activity of a solution is one-hundredth its relative humidity. (134)

water mold A common term for an öomycete. (3)

West Nile fever (encephalitis) A neurological viral disease that is spread from birds to humans by mosquitoes. It first appeared in the United States in 1999 and subsequently has been reported in almost every state. (924)

white blood cell (WBC) Blood cells having innate or acquired immune function. They are named for the white or buffy layer in which they are found when blood is centrifuged. (744)

white piedra A fungal infection caused by the yeast *Trichosporon beigelii* that forms light-colored nodules on the beard and mustache. (1008)

whole-cell vaccine A vaccine made from complete pathogens, which can be of four types: inactivated viruses; attenuated viruses; killed microorganisms; and live, attenuated microbes. (901)

whole-genome shotgun sequencing An approach to genome sequencing in which the complete genome is broken into random fragments, which are then individually sequenced. Finally the fragments are placed in the proper order using sophisticated computer programs. (384)

Widal test (ve-dahl′) A test involving agglutination of typhoid bacilli when they are mixed with serum containing typhoid antibodies from an individual having typhoid fever; used to detect the presence of *Salmonella typhi* and *S. paratyphi.* (876)

Winogradsky column A glass column with an anaerobic lower zone and an aerobic upper zone, which allows growth of microorganisms under conditions similar to those found in a nutrient-rich lake. (675)

wort The filtrate of malted grains used as the substrate for the production of beer and ale by fermentation. (1041)

X

xenograft (zen″o-graft) A tissue graft between animals of different species. (810)

xerophilic microorganisms (ze″ro-fil′ik) Microorganisms that grow best under low a_w conditions, and may not be able to grow at high a_w values. (1024)

Y

yeast (yēst) A unicellular, uninuclear fungus that reproduces either asexually by budding or fission, or sexually through spore formation. (631)

yeast artificial chromosome (YAC) Engineered DNA that contains all the elements required to propagate a chromosome in yeast and which is used to clone foreign DNA fragments in yeast cells. (370)

yellow fever An acute infectious disease caused by a flavivirus, which is transmitted to humans by mosquitoes. The liver is affected and the skin turns yellow in this disease. (924)

YM shift The change in shape by dimorphic fungi when they shift from the yeast (Y) form in the animal body to the mold or mycelial form (M) in the environment. (632)

Z

zidovudine A drug that inhibits nucleoside reverse transcriptase (also known as AZT, ZDV, or retrovir) used as an anti-HIV treatment. (856)

zoonosis (zo″o-no′sis; pl. *zoonoses*) A disease of animals that can be transmitted to humans. (892)

zooxanthella (zo″o-zan-thel′ah) A dinoflagellate found living symbiotically within cnidarians and other invertebrates. (620, 719)

Z ring A ring-shaped structure that forms on the cytoplasmic side of the plasma membrane during the bacterial cell cycle. Formation of the Z ring is the first step in formation of the septum, which will eventually divide the parent cell into two daughter cells. (122)

***z* value** The increase in temperature required to reduce the decimal reduction time to one-tenth of its initial value. (154)

zygomycetes (zi″go-mi-se′tez) A division of fungi that usually has a coenocytic mycelium with chitinous cell walls. Sexual reproduction normally involves the formation of zygospores. The group lacks motile spores. (635)

zygospore (zi′go-spōr) A thick-walled, sexual, resting spore characteristic of the zygomycetous fungi. (634)

zygote (zi′gōt) The diploid (2n) cell resulting from the fusion of male and female gametes. (614, 620)

Credits

Design Elements

Disease Box Icon: © Brand X Vol. 122/Getty Images; **Microbial Tidbits Box Icon:** Corbis RF; **Microbial Diversity and Ecology Box Icon:** © Vol. 29 PhotoDisc/Getty; **Techniques & Applications Box Icon:** © Vol. 4 PhotoDisc/Getty.

Photos

Chapter 1

Opener: © John D. Cunningham/Visuals Unlimited; **1.3a:** © Bettmann/Corbis; **1.3b(both):** © Kathy Park Talaro/Visuals Unlimited; **1.3c:** © Science VU/Visuals Unlimited; **1.4:** © John D. Cunningham/Visuals Unlimited; **1.6:** Corbis; **1.7:** American Society for Microbiology; **1.8:** North Wind Picture Archives; **1.9a:** Rita R. Colwell; **1.9b:** Dr. Robert G.E. Murray; **1.9c:** American Society for Microbiology Archives Collection; **1.9d:** Martha M. Howe; **1.9e:** Frederick C. Neidhardt; **1.9f:** Jean E. Brenchley.

Chapter 2

Opener: © George J. Wilder/Visuals Unlimited; **2.3:** Courtesy of Leica, Inc.; **2.4:** Courtesy of Nikon, Inc.; **2.8a:** © Charles Stratton/Visuals Unlimited; **2.8b:** © Robert Calentine/Visuals Unlimited; **2.8c:** ASM Microbelibrary.org. Photomicrograph by William Ghiorse; **2.8d:** © F. Widdel/Visuals Unlimited; **2.8e, 2.11:** © M. Abbey/Visuals Unlimited; **2.13a:** Courtesy of Molecular Probes, Eugene, OR; **2.13b:** © Richard L. Moore/Biological Photo Service; **2.13c:** © Evans Roberts; **2.14a:** © Kathy Park Talaro; **2.14b:** Harold J. Benson; **2.14c,d:** © Jack Bostrack/Visuals Unlimited; **2.14d:** © Manfred Kage/Peter Arnold, Inc.; **2.14f:** © A.M. Siegelman/Visuals Unlimited; **2.14g:** © David Frankhauser; **2.15b:** © Leon J. Le Beau/Biological Photo Service; **2.17a:** © George J. Wilder/Visuals Unlimited; **2.17b:** © Biology Media/Photo Researchers, Inc.; **2.17c:** © Harold Fisher; **2.18:** © William Ormerod/Visuals Unlimited; **2.20a,b:** © Fred Hossler/Visuals Unlimited; **2.22:** Courtesy of E. J. Laishley, University of Calgary; **2.24a:** © David M. Philips/Photo Researchers, Inc.; **2.24b:** © Paul W. Johnson/Biological Photo Service; **2.27:** © Driscoll, Yougquist & Baldeschwieler, Caltech/SPL/Photo Researchers, Inc.; **2.29a,b:** From Simon Scheuring (Scheuring S., Ringler P., Borgnia M., Stahlberg H., Müller D.J., Agre P., Engel A., "High resolution AFM topographs of the Escherichia coli water channel aquaporin," *Z. EMBO J.* 1999, 18:4981–4987).

Chapter 3

Opener: © E.C.S. Chan/Visuals Unlimited; **3.1a:** © Bruce Iverson; **3.1b:** Photo Researchers, Inc.; **3.1c:** © Arthur M. Siegelman/Visuals Unlimited; **3.1d:** © Thomas Tottleben/Tottleben Scientific Company; **3.1e:** Centers for Disease Control and Prevention; **3.2a–c:** © David M. Phillips/Visuals Unlimited; **3.2d:** Reprinted from *The Shorter Bergey's Manual of Determinative Bacteriology,* 8e, John G. Holt, Editor, 1977 © Bergey's Manual Trust. Published by Williams & Wilkins Baltimore, MD; **3.2e:** From Walther Stoeckenius: *Walsby's Square Bacterium: Fine Structures of an Orthogonal Procaryote;* **3.2f:** © Hans Hanert; p.43a,b: © Dr. Leon J. Le Beau; **3.8a:** American Society for Microbiology; **3.8b:** Reprinted *from The Shorter Bergey's Manual of Determinative Bacteriology,* 8e, John G. Holt, Editor, 1977 © Bergey's Manual Trust. Published by Williams & Wilkins Baltimore, MD; **3.12a,b:** Image courtesy of Rut Carballido-López and Jeff Errington; **3.13a:** © Ralph A. Slepecky/Visuals Unlimited; **3.13b:** National Research Council of Canada; **3.13c:** Reprinted from *The Shorter Bergey's Manual of Determinative Bacteriology,* 8e, John G. Holt, Editor, 1977 © Bergey's Manual Trust. Published by Williams & Wilkins Baltimore, MD; **3.14:** Courtesy of Daniel Branton, Harvard University; p. 61a: D. Balkwill and D. Maratea; p. 61b: Y. Gorby; p. 61c: Courtesy of Ralph Wolfe and A. Spormann, University of Illinois at Urbana-Champaign; **3.15:** Harry Noller, University of California, Santa Cruz; **3.16a:** © CNRI/SPL/Photo Researchers, Inc.; **3.16b:** © Dr. Gopal Murti/SPL/Photo Researchers, Inc.; **3.17(both):** © T.J. Beveridge/Biological Photo Service; **3.22:** Courtesy of M.R.J. Salton, NYU Medical Center; **3.27b:** From M. Kastowsky, T. Gutberlet, and H. Bradaczek, *Journal of Bacteriology,* 774:4798–4806, 1992; **3.28a,b:** Hiroshi Nikaido, *MMBR* 67(4):593–656 ASM/2003, Fig 2/p. 598; **3.30a,b:** From J.T. Staley, M.P. Bryant, N. Pfenning, and J.G. Holt (Eds.), *Bergey's Manual of Systematic Bacteriology,* Vol. 3. © 1989 Williams and Wilkins Co., Baltimore. Micrograph courtesy of D. Janekovic and W. Zillig; **3.34a,b:** © John D. Cunningham/Visuals Unlimited; **3.35:** © George Musil/Visuals Unlimited; **3.36:** Dr. Robert G.E. Murray; **3.37:** © Fred Hossler/Visuals Unlimited; **3.38a,b:** © E.C.S. Chan/Visuals Unlimited; **3.38c:** © George J. Wilder/Visuals Unlimited; **3.39c,d, 3.43, 3.44:** Courtesy of Dr. Julius Adler; **3.47:** American Society of Microbiology; **3.49(all):** Academic Press; **3.50:** American Society for Microbiology.

Chapter 4

Opener: © Arthur M. Siegelman/Visuals Unlimited; **4.1a:** © Eric Grave/Photo Researchers, Inc.; **4.1b:** © Carolina Biological Supply/Phototake; **4.1c:** © Arthur M. Siegelman/Visuals Unlimited; **4.1d:** © John D. Cunningham/Visuals Unlimited; **4.1e:** © Tom E. Adams/Visuals Unlimited; **4.1f:** © John D. Cunningham/Visuals Unlimited; **4.2:** © Richard Rodewald/Biological Photo Service; p. 84: Reprinted fig. 3a on page 98, L. Mahadevan & P. Matsudaira with permission from *Science,* Vol. 288: 94–98, April 7 © 2000 AAAS. Image courtesy of Lewis Tilney; **4.6:** © Manfred Schliwa/Visuals Unlimited; **4.7:** © B.F. King/Biological Photo Service; **4.8a:** © Henry C. Aldrich/Visuals Unlimited; **4.9b:** U.S. Department of Energy Genomics: GTL Program *http://www.ornl.gov/hgmis;* **4.11b:** Academic Press; **4.13a:** © Michael J. Dykstra/Visuals Unlimited; **4.13b:** Academic Press; **4.14a:** Prentice Hall, Upper Saddle River, New Jersey; **4.15:** Courtesy of Dr. Garry T. Cole, Univ. of Texas at Austin; **4.16:** © Don Fawcett/Visuals Unlimited; **4.18(both):** Dr. Jeremy Pickett-Heaps; **4.21:** National Research Council of Canada; **4.22:** © Karl Aufderheide/Visuals Unlimited; **4.23a:** © K.G. Murti/Visuals Unlimited; **4.24a:** © Ralph A. Slepecky/Visuals Unlimited; **4.24b:** © W.L. Dentler/Biological Photo Service.

Chapter 5

Opener: © Lauritz Jensen/Visuals Unlimited; **5.1a:** © John D. Cunningham/Visuals Unlimited; **5.1b:** From *ASM News* 53(2): cover, 187, American Society for Microbiology. Photo by H. Kaltwasser; **5.1c:** Shirley Sparling; **5.2a:** © Woods Hole Oceanographic Institution; **5.2b:** Image courtesy Mark Schneegurt; **5.9a,b–5.11b:** © Kathy Park Talaro; **5.13b:** Image courtesy Mark Schneegurt; **5.13c(both):** Dr. Eshel Ben-Jacob.

Chapter 6

Opener & 6.14a: Courtesy of Nagle Company; **6.14b:** © B. Otero/Visuals Unlimited; **6.14c:** Courtesy of Nagle Company; p. 138: © Science VU-D Foster, WHOI/Visuals Unlimited; **6.23:** Photo provided by ThermoForma of Marietta, Ohio; **6.26a,b:** Courtesy of Jeanne S. Poindexter, Long Island University; **6.27a:** Dr. Joachim Reitner; **6.27b:** From Ehrlich, et al. *ASM News* Vol. 70 #3 ASM 2004 Fig. 2 p. 129, image courtesy Garth D. Ehrlich, Ph.D; **6.30a:** Chris Frazee; **6.30b:** Margaret Jean McFall-Ngai.

Chapter 7

Opener: © Visuals Unlimited; **7.3a:** Courtesy of AMSCO Scientific, Apex, NC; **7.4:** © Raymond B. Otero/Visuals Unlimited; **7.6a:** Courtesy of Millipore Corporation; **7.7a:** Courtesy of Pall Ultrafine Filter Corporation; 7.7b: © Fred Hossler/Visuals Unlimited; **7.8a:** Photo provided by ThermoForma of Marietta, Ohio; **7.9:** © Tom Pantages; **7.13a:** Anderson Products, *www.anpro.com.*

Chapter 8

Opener: Reprinted by permission W.N. Lipscomb, Harvard University; **8.2:** © Artville CD; **8.17a,b:** John Wiley & Sons. Courtesy of Donald Voet; **8.26b,c:** Courtesy of David Eisenberg, UCLA; **8.28b:** Janine Maddock, University of Michigan.

Chapter 9

Opener & 9.32a: © The Nobel Foundation 1989.

Chapter 10

Opener & 10.21: From A.S. Moffat, "Nitrogenase Structure Revealed," *Science* 250: 1513, December 14, 1990. Photo by M.N. Georgiadis and D.C. Rees, Caltech.

Chapter 11

11.7: Illustrations, Irving Geis/Irving Geis Collection/Howard Hughes Medical Institute. Rights owned by Howard Hughes Medical Institute. Not to be used with out permission; **11.9a:** Reprinted with permission from *Nature* 389, 251, Macmillan Magazines Limited. Image courtesy Dr. Timothy J. Richmond; **11.11b:** From *Cold Spring Harbor Symposia of Quantitative Biology,* 28, p. 43 (1963); **11.13b:** Kriegstein, H.J. & Hogness, D.S. "Mechanism of DNA replication in Drosophila chromosomes: structure of replication forks and evidence for bidirectionality," *PNAS* 1974 Vol. 71. fig. 2 p. 137; **11.27a,b:** Reprinted with permission from Murakami, K.S.; Masuda, S.; and Darst, S.A. *Science* 296, May 17, 2002. © 2002 by the AAAS. Image courtesy Seth Darst; **11.27c:** Reprinted with permission from *Science* 292, June 8, 2001 Gnatt, A.L.; Cramer, P.; Jianhua, F.; Bushnell, D.A.; and Kornberg, R.D. © 2002 by the AAAs. Image courtesy Roger Kornberg and Dave Bushnell; **11.39a:** Steven McKnight and Oscar L Miller, Department of Biology, University of Virginia; **11.42:** From MA Rould, JJ Perona, DG Soll & T. Steitz; **11.45e,f:** Reprinted figures 2b,f,g from Yusupov, M. M.; Yusupov, G; Baucom, A; Lieberman, K; Earnest, T; Cate, J; and Noller, H. *Science* 292:883–896. May 4, 2001. © 2001 AAAS.

Chapter 12

Opener: Lewis et al., "Crystal Structure of the Lactose Operon Repressor and Its Complexes with DNA and Inducer." *Science* 271: 1247–54. Fig. 5A p. 250, 6a p. 1251, and 1j p. 153, the left illustration; 12.6b: Lewis et al. "Crystal Structure of the Lactose Operon Repressor and Its Complexes with DNA and Inducer." *Science,* 1996 March 1, Vol. 271 cover; **12.17b:** From S.C. Schultz, G.C. Shields and T.A. Steitz, "Crystal structure of a CAP–DNA Complex. The DNA is Bent by 90 degrees," *Science* 253: 1001–1007, Aug. 30, 1991. © 1991 by the AAAS.

Chapter 13

Opener: © Oliver Meckes/Photo Researchers, Inc.; **13.24:** Courtesy of Charles C. Brinton, Jr. and Judith Carnahan; **13.42b,c:** From *Molecular Biology of Bacterial Viruses* by Gunther S. Stent. © 1963 W.H. Freeman and Company Used with permission.

Chapter 14

Opener: © Argus Fotoarchiv/Peter Arnold, Inc.; **14.2:** Provided by Dr. A.K. Aggarwal (Mount Sinai School of Medicine) from Newman et al. *Science* 269, 656–663 (1995). © 1995 by the AAAS; **14.11b:** © Kathy Park Talaro; **14.13a,b:** © Huntington Potter and David Dressler/Time Life Pictures/Getty Images; **14.13c:** Reprinted with permission of Edvotek,Inc.; p. 375a–c: From: Sigal Ben-Yehuda and Richard Losick "Asymmetric Cell Division in B. subtilis Involves a Spiral-like Intermediate of the Cytokinetic Protein FtsZ; *Cell* 109:257–266, Cell Press 2002, p. 258 & 263 Figs 1 & 6.

Chapter 15

Opener: © Alfred Pasieka/Photo Researchers, Inc.; **15.8:** Image courtesy of Affymetrix, Inc. (Santa Clara, CA); **15.14:** Reprinted with permission from Fraser, C.M. et al. "The minimal gene complement of Mycoplasma genitalium," *Science* 270: 397–403. Fig. 1, page 398 © 1995 by the AAAS. Photo by The Institute for Genomic Research; **15.15:** Reprinted with permission from Fleishchman, R.E. et al. 1995. "Whole Genome random sequencing and assembly of Haemophilus influenzae Rd." *Science* 269:496–512. Fig. 1, page 507. © 1995 by the AAAS. Photo by the Institute for Genomic Research; **15.18:** Liu, Zhou, Omelchenko et al. "Transcritomedyamics," *Proceed Nat. Acad. Science* April 2003 vol. 100: 4191–96 Fig. 4193; **15.19a–d:** J. Hundelsman "Metagenomics," *MMBR* ASM *Press* Dec 2004 Vol. 68, fig 3 page 674.

Chapter 16

Opener: Courtesy of Robert C. Liddington and Stephen C. Harrison Harvard University; **16.3a:** © Dennis Kunkel/Phototake; **16.3c:** Courtesy of Gerald Stubbs and Keiichi Namba, Vanderbilt University; and Donald Caspar, Brandeis University; **16.4b:** © Dennis Kunkel/CNRI/Phototake; **16.5a:** Courtesy of Michael G. Rossmann, Purdue University; **16.5b:** © R. Feldman-Dan McCoy/Rainbow; **16.5c:** Courtesy of Harold Fisher, University of Rhode Island and Robley Williams, University of California at Berkeley; **16.5d:** © Science VU-NIH, R. Feldman/Visuals Unlimited; **16.7a,b:** Courtesy of Robert C. Liddington and Stephen C. Harrison, Harvard University; **16.9b:** © Harold Fisher; **16.10b:** © Chris Bjornberg/Photo Researchers, Inc.; **16.10c:** © Eye of Science/Photo Researchers, Inc.; **16.10d:** © Dr. Linda Stannard, UCT/Photo Researchers, Inc; **16.10e:** © R. Feldman-Dan McCoy/Rainbow; **16.11:** © K.G. Murti/Visuals Unlimited; **16.14a:** © Terry Hazen/ Visuals Unlimited; **16.14b:** © Dr. Jack Griffith; **16.15a–c:** M. Cooney; **16.16:** From S.E. Luria, *General Virology,* © 1978 John Wiley & Sons, Inc.; **16.17a:** © Runk/Schoenberger/Grant Heilman Photography, Inc.; **16.17b:** © Charles Marden Fitch; **16.20:** Courtesy of Janey S. Symington.

Chapter 17

Opener: © Oliver Meckes/Photo Researchers, Inc.; **Box 17.1:** Haring, et al. "Independent virus development outside a host." *Nature* Vol. 436 Nature Publishing 2005 Figure 1 p. 1101; **17.3(b1):** © Fred Hossler/Visuals Unlimited; **17.3(b2):** George Chapman, Georgetown University; **17.4f & 17.13:** © Lee D. Simon/Photo Researchers, Inc; **17.17:** © M. Wurtz/Photo Researchers, Inc.; **17.20a:** From F.D. Bushman, C. Shang, & M. Ptashne, "A Single Glutamic Acid Residue Plays a Key Rose in the Transcriptional Activation Function of Lambda Repressor." *Cell* 58: 1163–1171, September 22, 1989. Cell Press; **17.23a:** From A.K. Aggarwal, D.W. Rodgers, M. Drottar, M. Ptashne, and S.C. Harrison, "Recognition of a DNA Operator by the Repressor of Phage 434: A view at High Resolution." *Science* 242:889–907, Nov 11, 1988. © 1988 by the AAAS.

Chapter 18

Opener: Courtesy of Wayne Hendrickson, Columbia University; **18.10:** © K.G. Murti/Visuals Unlimited; **18.12a,b:** Center for Disease Control and Prevention; **18.15:** Courtesy of J.T. Finch, J.M. Kaper, USDA Agricultural Research Service; **18.17a,b:** Courtesy of Russell L. Steere, Advanced Biotechnologies, Inc.; **18.18:** © J.R. Adams/Visuals Unlimited.

Chapter 19

Opener: © Michael & Patricia Fogden/Minden Pictures; **19.1(all):** J. William Schopf. Reprinted with permission from *Science* 260 © 1993 Sept. 30, Fig 4a,f,g p. 643 AAAS; **19.2:** © Michael & Patricia Fogden/Minden Pictures; **19.4a,b:** Hardy et al., "Trichomonas hydrogenosomes contain the NADH dehydrogenase module of mitochondrial complex I" *Nature* 2004 Vol. 432 p. 618 figures 1a & 1c; **19.11:** Dr. Robert H. A. Coutts.

Chapter 20

Opener: Lansing Prescott; **20.2:** Prepared by Stephen D. Bell, MRC Cancer Cell Unit, UK; **20.5a,b:** Lansing Prescott; **20.6a,b:** Kazem Kashefi; **20.7a,b:** © Corale L. Brierley/Visuals Unlimited; **20.7c:** From J.T. Staley, M.P. Bryant, N. Pfenning and J.G. Holt (Eds.), *Bergey's Manual of Systematic Bacteriology,* Vol. 3. © 1989 Williams and Wilkins Co., Baltimore. Robinson, Dept. of Micro, U. of Cal., LA; **Box 20.1c:** Karl O. Stetter, M. Hohn; **20.9a:** © Friederich Widdell/Visuals Unlimited; **20.9b:** From J.T. Staley, M.P. Bryant, N. Pfenning and J.G. Holt (Eds.), *Bergey's Manual of Systematic Bacteriology,* Vol. © 1989 Williams and Wilkins Co., Baltimore; **20.9c:** © Henry C. Aldrich/Visuals Unlimited; p. 513: From *PNAS,* Vol. 99, pages 7663–7668, fig. 1, Orphan et al., May 2002. Image courtesy of Christopher H. House; **20.12a:** From J.T. Staley, M.P. Bryant, N. Pfenning and J.G. Holt, *Bergey's Manual of Systematic Bacteriology,* Vol. 3 © 1989 Williams and Wilkins Co., Baltimore, Prepared by G. Bentzen photographed by the Laboratory of Clinical Electron Microscopy, U. of Bergen; **20.12b:** Francisco Rodriguez-Valera; **20.14:** From J.T. Staley, M.P. Bryant, N. Pfenning and J.G. Holt, *Bergey's Manual of Systematic Bacteriology,* Vol. 3 © 1989 Williams and Wilkins Co., Baltimore.

Chapter 21

Opener: © Arthur M. Siegelman/Visuals Unlimited; **21.2:** R. Huber, H. Konig, K.O. Stetter; **21.3a:** From

J.T. Staley, M.P. Bryant, N. Pfenning and J.G. Holt, *Bergey's Manual of Systematic Bacteriology,* Vol. 3 © 1989 Williams and Wilkins Co., Baltimore; **21.3b:** From *Bergey's Manual of Systematic Bacteriology,* 2/e, vol. 1, Figure B4.4, part A page 400, Family I. Deinococcaceae, Battista, J.R., and Rainey, F.A. © Springer 2001. Image courtesy John R. Battista.; **21.5a:** From J.T. Staley, M.P. Bryant, N. Pfenning and J.G. Holt, *Bergey's Manual of Systematic Bacteriology,* Vol. 3 © 1989 Williams and Wilkins Co., Baltimore. Micrograph G. Cohen Bazire; **21.5b:** Reprinted from *The Shorter Bergey's Manual of Determinative Bacteriology,* 8e, John G. Holt, Editor, 1977 © Bergey's Manual Trust. Published by Williams & Wilkins; **21.6:** © Elizabeth Gentt/Visuals Unlimited; **21.7b:** From Carlsberg Research Communications 42:77–98, 1977 © Carlsberg Laboratories; **21.8a:** © T.E. Adams/Visuals Unlimited; **21.8b:** © Ron Dengler/Visuals Unlimited; **21.8c:** © M.I. Walker/Photo Researchers, Inc.; **21.8d:** © T.E. Adams/Visuals Unlimited; **21.9a:** © George J. Wilder/Visuals Unlimited; **21.9b:** Courtesy of Michael Richard, Colorado State Univ.; **21.9c:** P. Fay and N.J. Lang, *Proceedings of the Royal Society London.* B178: 185–192, 1971. Norma J. Lang, Univ. of California, Davis; **21.10a:** From J.T. Staley, M.P. Bryant, N. Pfenning and J.G. Holt (Eds.), *Bergey's Manual of Systematic Bacteriology,* Vol. © 1989 Williams and Wilkins Co., Baltimore. Micrograph courtesy of Ralph Lewin and L. Cheng; **21.10b:** Jean Whatley, *New Phytology* 79: 309–313, 1977; **21.11:** © John D. Cunningham/Visuals Unlimited; **21.12a,b:** Image courtesy John A. Fuerst and Richard I. Webb, from "Novel Compartmentalisation in Planctomycete Bacteria," *Microsc & Microanal,* 2004 10 (Suppl 2); **21.13:** © David M. Phillips/Visuals Unlimited; 21.14a: Reprinted from *The Shorter Bergey's Manual of Determinative Bacteriology,* 8e, John G. Holt, Editor, 1977 © Bergey's Manual Trust. Published by Williams & Wilkins Baltimore, MD; **21.14b:** © Arthur M. Siegelman/Visuals Unlimited; **21.14c:** Reprinted from *The Shorter Bergey's Manual of Determinative Bacteriology,* 8e, John G. Holt, Editor, 1977 © Bergey's Manual Trust. Published by Williams & Wilkins Baltimore, MD; **21.15(a2):** From S.C. Holt, *Microbiological Reviews* 42(1):117, 1978 American Society for Microbiology; **21.15c:** From M.P. Starr, et al. (Eds.), *The Prokaryotes,* Springer Verlag; **21.15d:** From S.C. Holt, *Microbiological Reviews* 42(1):122, 1978 American Society for Microbiology; **21.17a,b:** From S.C. Holt, *Microbiological Reviews* 42(1): 122, 1978 American Society for Microbiology; **21.18a–d:** From M.P. Starr et al. (Eds.), The Prokaryotes, Springer Verlag.

Chapter 22

Opener: © E.S. Anderson/Photo Researchers, Inc.; **22.3a:** © George J. Wilder/Visuals Unlimited; **22.3b,c:** Reprinted from *The Shorter Bergey's Manual of Determinative Bacteriology,* 8e, John G. Holt, Editor, 1977 © Bergey's Manual Trust. Published by Williams & Wilkins Baltimore, MD; **22.3d:** From M.P. Starr, et al. (Eds.), *The Prokaryotes,* Springer Verlag; **22.3e:** Reprinted from *The Shorter Bergey's Manual of Determinative Bacteriology,* 8e, John G. Holt, Editor, 1977 © Bergey's Manual Trust. Published by Williams & Wilkins Baltimore, MD; **22.4a,b:** From N.R. Krieg and J.G. Holt (Eds.), *Bergey's Manual of Systematic Bacteriology,* Vol. 1, 1984. Williams and Wilkins Co., Baltimore; **22.4c:** Courtesy of Dr. K.E. Hechemy; **22.4d:** From N.R. Krieg and J.G. Holt (Eds.), *Bergey's Manual of Systematic Bacteriology,* Vol. 1, 1984. Williams and Wilkins Co., Baltimore; **22.5:** From J.T. Staley, M.P. Bryant, N. Pfenning and J.G. Holt (Eds.), *Bergey's Manual of Systematic Bacteriology,* Vol. 3. © 1989 Williams and Wilkins Co., Baltimore; **22.7a:** © George J. Wilder/Visuals Unlimited; **22.7b,c:** Courtesy of Jeanne S. Poindexter, Long Island Univ.; **22.7d:** From J.T. Staley, M.P. Bryant, N. Pfenning and J.G. Holt (Eds.), *Bergey's Manual of Systematic Bacteriology,* Vol. 3. © 1989 Williams and Wilkins Co., Baltimore; **22.9a,b:** From N.R. Krieg and J.G. Holt (Eds.), *Bergey's Manual of Systematic Bacteriology,* Vol. 1, 1984. Williams and Wilkins Co., Baltimore; **22.10:** © John D. Cunningham/Visuals Unlimited; **22.11a,b:** © Woods Hole Oceanographic Institution; **22.11c:** S.W. Watson, Woods Hole Oceanographic Institution; **22.13a,b:** From M.P. Starr, et al. (Eds.), *The Prokaryotes,* Springer Verlag; **22.14a:** From van Veen, W.L., Mulder, E.G., Deinema, M.H., 1978. The Sphaerotilus Leptothrix Group of Bacteria. *Microbiological Reviews* 42: 329–356, fig 6, p. 334. American Society of Microbiology; **22.14b:** Mulder, E.G. & van Veen, W.L., "Investigations on the Sphaeerotilus-Leptothrix group." *Antonie von Leeuwenhoek Journal of Microbiology and Serology* 29: 121–153. Kluwer Publishers; **22.15a:** © Runk/Schoenberger/Grant Heilman Photography, Inc.; **22.15b:** © Thomas Tottleben/Tottleben Scientific Company; **22.16:** Reprinted by permission of Kluwer Academic Publishers from © J.G. Kuenen and H. Veldkamp/Martinus Nijhoff Publishers, 1972, *Antonie von Leeuwenhoek;* 22.18, **22.19a:** From M.P. Starr et al. (Eds.), *The Prokaryotes,* Springer-Verlag; **22.19b:** From J.T. Staley, M.P. Vryant, N. Pfenning and J.G. Holt (Eds.) *Bergey's Manual of Systematic Bacteriology,* Vol. 3 © 1986 Williams and Wilkins Co. Baltimore.; **22.20a:** From *ASM News* 53(2): cover, 187, American Society for Microbiology. Photo by H. Kaltwasser; **22.20b:** Shirley Sparling; **22.21:** Image courtesy Mark Schneegurt; **22.22b–d:** Original Micrographs courtesy of Ruth L. Harold and *Bacteriological Reviews;* 22.22e: Courtesy of Dr. Harkisan D. Raj; **22.23:** Courtesy of Michael Richard, Colorado State Univ.; **22.24a:** ASM Microbelibrary.org. Photo micrograph by William Ghiorse; **22.24b:** ASM Microbelibrary.org. Photo courtesy of Caroline Harwood, University of Iowa; **22.25:** © Christine Case/Visuals Unlimited; **22.26a:** © David M. Phillips/Visuals Unlimited; **22.26b, 22.27:** From N.R. Krieg and J.G. Holt (Eds.), *Bergey's Manual of Systematic Bacteriology,* Vol. 1, 1984. Williams and Wilkins Co., Baltimore; **22.28a:** © Kenneth Lucas, Steinhart Aquarium/Biological Photo Service; **22.28b,c:** Courtesy of James G. Morin, University of California–Los Angeles; **22.30a:** © Arthur M. Siegelman/Visuals Unlimited; **22.30b:** © E.S. Anderson/Photo Researchers, Inc.; **22.32a–c:** © F. Widdel/Visuals Unlimited; **22.33–22.34b:** Courtesy Dr. Jeffrey C. Burnham; **22.36b:** Jerry M. Kuner and Dale Kaiser, "Fruiting body morphogenesis in submerged cultures of Myxococeus xauthus," *J. Bacteriology,.* 151, 458–461, 1982; **22.35a–c:** From M.P. Starr et al. (Eds.), *The Prokaryotes,* Springer Verlag; **22.36(all):** Jerry M. Kuner and Dale Kaiser, "Fruiting body morphogenesis in submerged cultures of Myxococeus xauthus," *J. Bacteriology,* 151, 458–461, 1982; **22.37b,c:** © M. Dworkin-H. Reichenbach/Phototake; **22.37d:** © Patricia L. Grillione/Phototake; **22.38a,b:** Annette Summers Engel, Ph.D.

Chapter 23

Opener: © Arthur M. Siegleman/Visuals Unlimited; **23.3a:** © Michael G. Gabridge/Visuals Unlimited; **23.3b:** © David M. Phillips/Visuals Unlimited; **23.4:** © Michael G. Gabridge/Visuals Unlimited; **23.6a:** CNRI/Photo Researchers, Inc.; **23.6b:** © Dr. Tony Brain/Photo Researchers, Inc.; **23.7:** © Arthur M. Siegelman/Visuals Unlimited; **23.8:** © F. Widdel/Visuals Unlimited; **23.9a:** © Arthur M. Siegelman/Visuals Unlimited; **23.9b:** Courtesy of Molecular Probes, Inc.; **23.10a:** Courtesy of Dr. A.A. Yousten; **23.10b:** From H. de Barjac & J.F. Charles, "Une nouvelle toxine active sur les moustiques, presente dans des inclusions cristallines produites par Bacillus sphaericus." *C.R. Acad. Sci. Paris* ser. II: 296:905–910, 1983; **23.11a:** From M.P. Starr, et al. (Eds.), *The Prokaryotes,* Springer Verlag; **23.11b:** From S.T. Williams, M.E. Sharpe and J.G. Holt (Eds.), *Bergey's Manual of Systematic Bacteriology,* Vol. 4, © 1989 Williams and Wilkins Co., Baltimore; **23.12:** From J.G. Holt (Ed.), *The Shorter Bergey's Manual of Systematic Bacteriology,* Vol. 2. © 1986 Williams and Wilkins Co., Baltimore; 23.13a: © Bruce Iverson; **23.13b:** © Photo Researchers, Inc.; **23.14a,b:** © Arthur M. Siegelman/Visuals Unlimited; **23.14c:** © George J. Wilder/Visuals Unlimited; **23.15:** From M.P. Starr al. (Eds.), *The Prokaryotes,* Springer-Verlag; **23.17a:** © Thomas Tottleben/Tottleben Scientific Company; **23.17b:** © Photo Researchers, Inc.; **23.17c:** © M. Abbey/Visuals Unlimited; **23.18a–c:** © Fred E. Hossler/Visuals Unlimited.

Chapter 24

Opener: © Howard Berg/Visuals Unlimited; **24.3a–c:** From S.T. Williams, M.E. Sharpe and J.G. Holt (Eds.), *Bergey's Manual of Systematic Bacteriology,* Vol. 4, © 1989 Williams and Wilkins Co., Baltimore; **24.3d:** © Eli Lilly & Company. Used with Permission; **24.3e:** From S.T. Williams, M.E. Sharpe and J.G. Holt (Eds.), *Bergey's Manual of Systematic Bacteriology,* Vol. 4, © 1986 Williams and Wilkins Co., Baltimore; **24.6a:** © E.C.S. Chan/Visuals Unlimited; **24.6b:** © David M. Phillips/Visuals Unlimited; **24.7:** © Thomas Tottleben/Tottleben Scientific Company; **24.8a–d:** From J.G. Holt et al. (Eds.) *Bergey's Manual of Systematic Bacteriology,* Vol. 2, © 1986. Williams & Wilkins Baltimore; **24.9:** © Grant Heilman Photography; **24.10:** © John D. Cunningham/ Visuals Unlimited; **24.13b:** From Dr. Akio Seino, *Hakko to Kogyo (Fermentation and Industry)* 41 (3):3–4, 1983. Japan Bioindustry Association; **24.13d:** From S.T. Williams, M.E. Sharpe and J.G.

Holt (Eds.), *Bergey's Manual of Systematic Bacteriology,* Vol. 4, © 1989 Williams and Wilkins Co., Baltimore; **24.15a–c:** J.M. Willey; **24.16a–c:** From S.T. Williams, M.E. Sharpe and J.G. Holt (Eds.), *Bergey's Manual of Systematic Bacteriology,* Vol. 4, © 1989 Williams and Wilkins Co., Baltimore; **24.17a:** © Christine L. Case/Visuals Unlimited; **24.17b:** © Sherman Thompson/Visuals Unlimited; **24.18a–24.19a:** From S.T. Williams, M.E. Sharpe, and J.G. Holt (Eds.), *Bergey's Manual of Systematic Bacteriology,* Vol. 4, © 1989 Lippincott Williams and Wilkins Co., Baltimore; **24.19b:** © R. Howard Berg/Visuals Unlimited; **24.19c:** From S.T. Williams, M.E. Sharpe and J.G. Holt (Eds.), *Bergey's Manual of Systematic Bacteriology,* Vol. 4, © 1989 Williams and Wilkins Co., Baltimore; **24.20:** Staley, *Bergey's Manual Systematic Bacteriology,* Vol. 2, page 1418, figure 15.96a. Courtesy Prof. Bruno Biavati, Instituto Di Microbiologia.

Chapter 25

Opener: D.T. John et al. "Sucker-like structures on the pathogenic amoeba Naegleria Fowleri", *Applied Envir. Microbiol.* 47: 12–14 (image 3n). © 1984 American Society for Microbiology. Image courtesy of Thomas B. Cole; **25.4:** Courtesy S.W.B. Irwin, University of Ulster at Jordanstown, Northern Ireland; **25.6a:** © Manfred Kage/Peter Arnold, Inc.; **25.6b:** © Edward S. Ross; **25.8:** © John D. Cunningham/Visuals Unlimited; **25.9b:** © B. Beatty/Visuals Unlimited; **25.9c:** © Edward Degginger/Bruce Coleman, Inc.; **25.9d:** © Victor Duran/Visuals Unlimited; **25.9e:** © Sherman Thompson/Visuals Unlimited; **25.10b–e:** © Carolina Biological Supply/Phototake; **25.11:** © Phil A. Harrington/Peter Arnold, Inc.; **25.12b:** © Arthur M. Siegelman/Visuals Unlimited; **25.13:** © Manfred Kage/Peter Arnold, Inc.; **25.14:** © Richard Rowan/Photo Researchers, Inc.; **25.16b:** © David M. Phillips/Visuals Unlimited; **25.17a,b:** © Eric Grave/Photo Researchers, Inc.; **25.17c:** © Cabisco/Phototake; **25.19a:** © Dr. Anne Smith/SPL/Photo Researchers, Inc.; **25.19b:** © Jim Hinsch/ Photo Researchers, Inc.; **25.20:** © Natural History Museum, London.; **25.21a:** © M.I. Walker/Photo Researchers, Inc.; 25.21b: © John D. Cunningham/ Visuals Unlimited; 25.21c: © Manfred Kage/Peter Arnold, Inc.; 25.21d: © John D. Cunningham/Visuals Unlimited; 25.21e: © John D. Cunningham/Visuals Unlimited.

Chapter 26

Opener: © John D. Cunningham/Visuals Unlimited; **26.2:** © C. Gerald Van Dyke/Visuals Unlimited; **26.3a:** © Sherman Thompson/Visuals Unlimited; **26.3b:** © Richard Thom/Visuals Unlimited; **26.3c:** © William J. Werber/Visuals Unlimited; **26.5:** © John D. Cunningham/Visuals Unlimited; **26.6c:** Courtesy of Dr. Garry T. Cole, Univ. of Texas at Austin; **26.11a:** © John D. Cunningham/Visuals Unlimited; **26.11b:** © Robert Calentine/Visuals Unlimited; **26.11c:** © John D. Cunningham/Visuals Unlimited; **26.12a:** © J. Forsdyke, Gene Cox/SPL/Photo Researchers, Inc.; **26.13:** © David M. Phillips/Visuals Unlimited; **26.16:** J.K. Pataky, University of Illinois.

Chapter 27

Opener: Reprinted with permission from Edwards, K.J., Bond, P.L., Gihring, T.M., and Banfield, J.F. "An Archael Iron-oxidizing Extreme Acidophile Important in: Acid Mine Drainage," *Science* 287: 1796–2799. (10 March, 2000) Figure 3A, page 1798. © 2000 American Association for the Advancement of Science; Image courtesy of K.J. Edwards; **27.11:** P. Dirckx, MSU Center for Biofilm Engineering; **27.12:** Y. Cohen and E. Rosenberg, *Microbial Mats,* Fig 1a p. 4 1986. American Society for Microbiology; 27.13b: Jackie Parry; **27.13c:** © Eye of Science/Photo Researchers, Inc.; **27.14a:** © Pat Armstrong/Visuals Unlimited; **27.14b:** © Dan McCoy/Rainbow; **27.14c:** © John D. Cunningham/Visuals Unlimited; **27.15:** Reprinted with permission from Edwards, K.J., Bond, P.L., Tihring, T.,M., and Banfield, J.F. "An Archael Iron-oxidizing Extreme Acid Mine Drainage," *Science* 287: 1796–2799 (10 March, 2000) Fig 3A, page 179 © 2000 AAAS Image courtesy of K.E. Edwards; **27.16:** Courtesy of Molecular Probes, Inc.; **27.18:** Reprinted with permission from Nature 417:63 H. Huber et al. © 2002 *Nature;* Prof. Dr. K.O. Stetter, Dr. R. Rachel, Dr. H. Huber, University of Regensburg, Germany; **27.19a:** Frohlich, J., and H. Koening, 1999. "Rapid Isolation of Single Microbial Cells from mixed Natural and Laboratory Populations with Aid of a Micromanipulator," System. *Applied Microbiology* 2:249–257. Figure 4 page 253. Urban and Fisher Verlag. Photo courtesy Dr. Helmut Koenig.

Chapter 28

Opener: Reprinted with permission from Schulz; H.N., Brinkhoff, T., Ferdelman, T.G., Hernandez Marine, M., Teske, A., and Jorgensen, B.B. 1999. "Dense Populations of a Giant Sulfur Bacterium in Namibian Shelf Sediments," *Science* 284, 493–495, Fig 1. © 1999 American Association for the Advancement of Science. Image courtesy of Heide Schulz; 28.1: Used by permission per Dr. Roger Lukas, University of Hawaii Hawaii Ocean Time-series Study. National Science Foundation Grant OCE03-27513; 28.4: Reprinted with permission from Schulz; H.N., Brinkhoff, T., Ferdelman, T.G., Hernandez Marine, M., Teske, A., and Jorgensen, B.B. 1999. "Dense Populations of a Giant Sulfur Bacterium in Namibian Shelf Sediments," *Science* 284, 493-495, Fig 1. © 1999 American Association for the Advancement of Science. Image courtesy of Heide Schulz; **28.5a, b:** Reprinted from *FEMS Microbiol. Ecol.,* Vol. 28, 301–313, Fig's 1a,b,d; Jorgensen, B.B., and Gallardo, V.A., Thioploca sp.: "Filamentous Sulfur Bacteria with Nitrate Vacuoles", 1999, with permission from Elsevier Science. Photos courtesy of Bo B. Jorgensen; **28.6a:** Brec L. Clay; **28.9:** Burkholder Laboratory, North Carolina State University & Sea Grant National Media Relations; **28.11:** NASA; **28.13:** Photo by Susumu Honjo, Woods Hole Oceanographic Institution; **28.17b:** Reprinted with permission from Vincent et al., *Science* 286: 2094 (1999). © 2006 AAAS; **28.17c:** Reprinted with permission from Karl et al., Science 286: 2144–47 (10 Dec 1999). © 2006 AAAS.

Chapter 29

Opener: N.C. Schenck, *Methods & Principles of Micor-rhizal Research,* © 1992 American Phytopathological Society. Photo courtesy of Dr. Hugh Wilcox; **29.4:** Jo Handelsman; **29.6:** *Journal of Phycology* 32:774–782, Fig 1, p. 777, Garcia-Pichel, F. and Belnap, J. 1996. By Permission of the Journal of Phycology; **29.7:** © Sherman Thompson/Visuals Unlimited; **29.9:** N.C.Schenck, *Methods & Principles of Micorrhizal Research,* © 1992 American Phytopathological Society. Photo courtesy of Dr. Hugh Wilcox; **29.10:** © R.S. Hussey/Visuals Unlimited; **29.12:** Paola Bonfante/University of Turin; **29.13d:** Courtesy of Ray Tully, U.S. Department of Agriculture; **29.13f:** Courtesy of Dr. Ralph W.F. Hardy and the National Research Council of Canada; **29.13i,j & 29.14:** © John D. Cunningham/Visuals Unlimited; **29.15:** Dr. Bernard Dreyfus; **29.16:** Courtesy of Keith Clay, Indiana University-Bloomington; **29.18:** Courtesy of Dr. Sandor Sule, Plant Protection Institute, Hungary Academy of Sciences; **29.21:** © Michael & Patricia Fogden/Minden Pictures; 29.**25a,b:** From Andersson, M.A. et al., "Bacteria, Molds and Toxins in Water-damaged Building Materials," 63(2)387–393, Fig. 1, p. 388, *Applied and Environmental Microbiology,* © 2000 American Society for Microbiology. Image courtesy of Maria Andersson and Mirja.

Chapter 30

Opener: © Science VU/WHOI/Visuals Unlimited; p. 720a: © T. Wenseleers; p. 720b: World Heath Organization; 30.2a: © William J. Weber/Visuals Unlimited; **30.2b:** © M. Abbey/Visuals Unlimited; **30.3a:** © Stan Elms/Visuals Unlimited; **30.3b:** © Bob DeGoursey/Visuals Unlimited; **30.5a:** © WHOI/Visuals Unlimited; **30.9:** Craig Cary, University of Delaware; **30.10a,b:** © Woods Hole Oceanographic Institution, Woods Hole, MA; **30.11a:** Ott, J.A. Novak, R.F. Schiemer, U. Hentchel, M. Nebelsick, and M. Polz 1991. "Tackling the Sulfide Gradient; a Novel Strategy Involving Marine Nematodes and Chemoautotrophic Ectotsymbionts," *Marine Ecology* 12(3) 261–279, Figure 3, p. 266. Blackwell Wissenschafts-Verlag. Image courtesy of J. Ott and M. Polz; **30.11b:** Reprinted with permission of Blackwell Science, Inc, Fig 31.18b from Lengeler, J.W. et al., *Biology of Prokaryotes* 1999. Photo courtesy of J. Ott and M. Polz; **30.12b:** From Crane, Hecker, and Goluhev, "Heat Flow and Hydrothermal Vents in Lake Baikal, USSR," *Transactions of the American Geophysical Union (EOS)* 72(52) 585, Dec. 24, 1991. © by the American Geophysical Union; **30.14:** © John D. Cunningham/Visuals Unlimited; **30.15a:** © John Durham/SPL/Photo Researchers; 30.15c: From Currie et al., *Science* Vol 311: 81–83, Jan. 6 2006. Image courtesy Cameron Currie; **30.16b:** © Science Source/Photo Researchers, Inc.

Chapter 31

Opener: ©Jim Dowdalls/Photo Researchers, Inc.; **31.4:** Lennart Nilsson/Albert Bonniers Forlag AB; **31.5:** © David Scharf/Peter Arnold; **31.19b:** © Ellen

R. Dirksen/Visuals Unlimited; **31.23b:** © Reprinted from Wendell F. Rosse et al., "Immune Lysis of Normal Human and Parozysmal Nicturnal Hemoglogin-uria (PNH) Red Blood Cells," *Journal of Experimental Medicine,* 123:969, 1966. Rockefeller University Press.

Chapter 32

Opener: © Science Source/Photo Researchers, Inc.; **32.3(1–3):** © PhotoDisc RF/Getty; **32.3(4):** © Creatas/ PictureQuest; 32.5c,d: Courtesy of Dr. Paul Travers; **32.8b,c:** Courtesy of Dr. Gilla Kaplan, The Rockefeller University; **32.12b:** © R. Feldman-Dan McCoy/Rainbow; **32.27a:** © Stan Elms/Visuals Unlimited; **32.27b:** © Biophoto Associates/Photo Researchers; **32.32b:** © Kathy Park Talaro; **32.34:** © Kenneth Greer/Visuals Unlimited.

Chapter 33

Opener: © Alfred Pasieka/Peter Arnold; **33.3a:** © Stan Elms/Visuals Unlimited; **33.3b:** © Veronika Burmeister/Visuals Unlimited; **33.3c:** From M. Persi, J.C. Burham and J.L. Duhring, "Effects of Carbon Dioxide and pH on Adhesion of *Candida albicans* to Vaginal Epithelial Cells," *Infections and Immunity,* 30 (s): 82–90, Oct. 1985. American Society for Microbiology; **33.4d:** Prof. Guy R. Cornelis; **33.9a:** Daniel A. Portnoy, from 1989 *Journal of Cell Biology,* Rockefeller Press; 33.9b: Stevens, M.P. et al., *Molecular Microbiology* 56: 40–53 (Blackwell Publishing).

Chapter 34

Opener: © SPL/Photo Researchers; **34.1:** Christine L. Case, Skyline College; **34.2a:** Courtesy Becton-Dickinson Microbiology Systems; **34.2b:** © Lauritz Jensen/Visuals Unlimited; **34.4:** Etest® is a registered trademark of AB BIODISK and patented in all major markets.

Chapter 35

Opener & 35.1i: © Raymond B. Otero/Visuals Unlimited; **35.2b:** Courtesy Becton-Dickinson Microbiology Systems; **35.2c:** © Michael English/Custom Medical Stock Photo; **35.3:** Courtesy Becton-Dickinson Microbiology Systems; **35.4c:** © Syva Company. Reprinted by permission Syva Company, San Jose, CA; **35.4d:** Courtesy of Genetic System Corporation; **35.6a,b:** Analytab Products, A Division of Bio-Merieux, Inc.; **35.8d:** Courtesy Wadsworth Center, NYS Dept. of Public Health; **35.9a,b:** Centers for Disease Control and Prevention; **35.10:** © Raymond B. Otero/Visuals Unlimited; **35.15:** From Toad Tabaqchali, *Journal of Clinical Microbiology,* 1986, p. 380, ASM; **35.18d:** From N.R. Rose et al., *Manual of Clinical Laboratory Immunology,* 1992. American Society of Microbiology.

Chapter 36

Opener: Jim Gathany, Centers for Disease Control and Prevention; **36.9:** © Runk/Schoenberger/Grant Heilman Photography; p. 902: © Historic VU-NIH/Visuals Unlimited.

Chapter 37

Opener (left): Jim Gathany, Centers for Disease Control and Prevention; **Opener (right):** Cynthia Goldsmith, Centers for Disease Control and Prevention; **37.1b:** © John D. Cunningham/Visuals Unlimited; **37.2c:** ©Carroll H. Weiss/Camera M.D. Studios; **37.3a:** Armed Forces Institute of Pathology; **37.4:** © Biophoto Associates/Photo Researchers; **37.5:** © Carroll H. Weiss/Camera M.D. Studios; **37.6:** Armed Forces Institute of Pathology; p. 923: © Photo Researchers, Inc.; **37.10a:** © Carroll H. Weiss/Camera M.D. Studios; **37.10b:** © Science VU/Visuals Unlimited; **37.13:** © Carroll H. Weiss/Camera M.D. Studios; **37.14a,b:** From N.H. Olson et al., *Proceeding of the National Academy of Sciences,* 90:507. 1993. Photo courtesy of Dr. Michael Rossmann; **37.15:** Courtesy of Dan Wiedbrauk, Ph.D, Warde Medical Laboratory, Ann Arbor, Michigan; **37.16a:** © CDC/Science Source/Photo Researchers; **37.16b:** © Carroll H. Weiss/Camera M.D. Studios; **37.16c:** © Dr. P. Marazzi/Photo Researchers, Inc; **37.17:** © Dr. Brian Eyden/SPL/Photo Researchers; **37.18:** Barbara O'Connor; **37.19:** Courtesy of The National Institute of Health; **37.20a:** © Kenneth E. Greer/Visuals Unlimited; **37.20b:** © Carroll H. Weiss/Camera M.D. Studios; **37.20c,d:** © Kenneth E. Greer/Visuals Unlimited; **37.21:** © Eye of Science/Photo Researchers; p. 941: Corbis; **37.22a:** Science Photo/Custom Medical Stock.

Chapter 38

Opener: Courtesy of The Royal College of Surgeons Museum, Edinburg, Scotland; **38.1b–38.3b:** Centers for Disease Control and Prevention; **38.5:** © Carroll H. Weiss/Camera M.D. Studios; **38.6:** © 2005 The Regents of the University of California; **38.8a:** From *ASM News* 55(2) cover, 1986. American Society for Microbiology; **38.8b,c:** © CDC/Peter Arnold, Inc.; **38.9b,c:** Centers for Disease Control and Prevention; **38.10:** Department of Health & Human Services, courtesy of Dr. W. Burgdorfer; **38.11:** Courtesy of National Institute Slide Bank/The Welcome Center for Medical Sciences; **38.13:** © Science VU-WHO/ Visuals Unlimited; **38.14:** From V. Neman-Simha and F. Megrud, "In Vitro Model for Capylobacter pylori Adherence Properties," *Infection and Immunity,* 56(12):3329–3333, Dec. 1988. American Society for Microbiology; **38.15a:** From R. Baselga et al., "Staphylococcus Aureus: Implications in Colonization and Virulence," *Infection and Immunity,* 61(11) L4857–4862, 1993. © American Society for Microbiology; **38.15b:** Janice Carr, Centers for Disease Control and Prevention; **38.17a–f:** © Carroll H. Weiss/Camera M.D. Studios; **38.18:** © Arthur M. Siegelman/Visuals Unlimited; **38.19:** Armed Forces Institute of Pathology; **38.20a–c:** © Carroll H. Weiss/Camera M.D. Studios; **38.22:** Armed Forces Institute of Pathology; **38.24:** From Jacob S. Teppema, "In Vivo Adherence and Colonization of Vibrio cholerae Strains That Differ in Hemagglutinating Activity and Motility," *Journal of Infection and Immunity,* 55(9)2093–2102, Sept. 1987. Reprinted by permission of American Society for Microbiology; **38.27b:** CDC; **38.27c:** © Science VU/ Charles Stratton/Visuals Unlimited; **38.30a:** © R. Gottsegen/ Peter Arnold, Inc.; **38.30b:** © Stanley Flegler/Visuals Unlimited; **38.31:** © E.C.S. Chan/Visuals Unlimited.

Chapter 39

Opener: © Lennart Nilsson/Albert Bonniers Forlag AB; **39.1:** Reprinted by permission of Upjohn Co. from E.S. Beneke et al., 1984 *Human Mycosis in Microbiology;* **39.2:** © E.C.S. Chan/Visuals Unlimited; **39.3, 39.4a:** © Arthur M. Siegelman/Visuals Unlimited; **39.4b:** Armed Forces Institute of Pathology; **39.6:** Centers for Disease Control and Prevention; **39.8a,b:** Armed Forces Institute of Pathology; **39.11–39.12b:** © Everett S. Beneke/Visuals Unlimited; **39.13–39.16:** © Carroll H. Weiss/Camera M.D. Studios; **39.17:** © Everett S. Beneke/Visuals Unlimited; **39.18:** Reprinted by permission of Upjohn Co. from E.S. Beneke et al., 1984 *Human Mycosis in Microbiology;* **39.19:** © Everett S. Beneke/Visuals Unlimited; **39.21:** © David M. Phillips/Visuals Unlimited; **39.22a:** © Lauritz Jensen/Visuals Unlimited; **39.22b:** © Robert Calentine/Visuals Unlimited; **39.24a,b:** From M. Schaechter, G. Medoff, & D. Schlessinger (Eds.) *Mechanisms of Microbial Disease,* 1989. Williams and Wilkins; **39.25:** © Evertt S. Beneke/ Visuals Unlimited; **39.26a:** © David M. Phillips/ Visuals Unlimited; **39.26b,c:** © Evertt S. Beneke/ Visuals Unlimited; **39.27:** Archiv für Protistenkunde (1986, Vol. 131, 257–279), image courtesy J. I. Ronny Larsson.

Chapter 40

Opener: © Christiana Dittmann/Rainbow; **40.2:** Donald Klein; **40.3a:** © Tom E. Adams/Peter Arnold, Inc.; **40.3b:** © Martha Powell/Visuals Unlimited; **40.6:** © Photo by Mark Seliger, Courtesy of Campbell Soup Company; **40.8:** Reprinted from *Applied & Environment Microbiology* (64) 2248–2286, fig. 1, p. 2284, Sturbaum G., Ortega, Y.A., Gilman, R.H., Sterling, C.R., Caberea, L. and Klein, D.A., "Detection of Cyclospora Cayetanensis in Wastewater." © 1998 ASM. Image courtesy of Greg Sturbaum; **40.9:** From Peterkin, Idzigk, & Sharpe. "Screening DNA probes using the Hydrophobic Probe Grid-Membrane Filter, *Food Microbiology* 6:281-284, 1989. Academic Press, Inc. (London); p. 1037a,b: The Fairtrade Foundation; 40.11a–c: Jeffery Broadbent , Utah State University; **40.12:** © Elmer Koneman/Visuals Unlimited; **40.13:** Fr. D.B. Hughes and D.G. Hoover, *Food Technology,* April 1991, Fig 3, p. 79; **40.14a–e:** © John D. Cunningham/Visuals Unlimited; **40.15:** © Joe Munroe/Photo Researchers, Inc; **40.18:** © Vance Henry/Nelson Henry; **40.20:** © Stanley Flegler/Visuals Unlimited.

Chapter 41

Opener: Courtesy of General Electric Research and Development Center; **41.3:** Donald A. Klein; **41.4a:**

From D. Jenkins et al. *Manual of the Causes & Control of Activated Sludge Bulking & Forming,* 1986, US Environmental Protection Agency; **41.5a,b:** Cindy Wright-Jones, City of Ft. Collins, CO; **41.12a,b:** From B. Atkinson and Daoud; **41.13a:** Society for Industrial Microbiology; **41.24:** From J.R. Postgate, *The Sulphate-Reducing Bacteria,* Reprinted with the Permission of Cambridge University Press; **41.28a–d:** From Drum & Gordon, Trends Biotechnology, Vol. 21, No 8, 2003 pp. 235–328 Figure 2 p. 326, Elsevier.

Line Art/Tables/Illustrations

Chapter 1

1.1: From "A molecular view of microbial . . ." by Norman Pace from SCIENCE, Vol. 276, 1997, p. 735, figure 1. Copyright © 1997 AAAS. Reprinted by permission of AAAS; **1.2:** Steve Wagner.

Chapter 6

6.3: From Nelson, David L. and Michael M. Cox, PRINCIPLES OF BIOCHEMISTRY, 4/e. NY: W. H. Freeman; **6.4:** From "Dynamic Instability of a Bacterial Engine" by Jakob Moller–Jensen and Kenn Gerdes from SCIENCE, Vol. 306, November 5, 2004, p. 988, figure 1. © 2004 AAAS. Reprinted by permission of AAAS; **6.5:** From Weiss, David S., "Bacterial cell division and the septal ring" from MOLECULAR MICROBIOLOGY (2004) 54(3), 588–597; **6.8:** From Nystrom, Thomas, "Bacterial Senescence, Programmed Death, and Premeditated Sterility" from ASM News, Volume 71, Number 8, p. 363; **6.28:** From Bryers, James D. and Buddy D. Ratner, "Bioinspired Implant Materials Befuddle Bacterial" from ASM NEWS, Volume 70, Number 5, 2004.

Chapter 8

8.28a: From Bren, Anat and Michael Eisenbach, "How Signals Are Heard during Bacterial Chemotaxis: Protein-Protein Interactions in Sensory Signal Propagation" from JOURNAL OF BACTERIOLOGY, December 2000; **8.29:** From J. S. Parkinson, "Signal Amplification in Bacterial Chemotaxis through Receptor Teamwork" from ASM NEWS, Volume 70, Number 12, 2004.

Chapter 9

9.14: From Gao, Yi Qin, Wei Yang and Martin Karplus, "A Structure-Based Model for the Synthesis and Hydrolysis of ATP by F1-ATPase" from CELL, Vol. 123, pp. 195–205, October 2005.

Chapter 10

10.7: From Herter, Sylvia et al., "Autotrophic CO2 Fixation by Chloroflexus aurantiacus: Study of Glyoxylate Formation and Assimilation via the 3-Hydroxypropionate Cycle" from JOURNAL OF BACTERIOLOGY, July 2001, pp. 4305–4316.

Chapter 11

11.15: From BIOCHEMISTRY, 4/e by Lehninger, p. 958. New York: W. H. Freeman, 2005; **11.16:** From BIOCHEMISTRY, 4/e by Lehninger, p. 960. New York: W. H. Freeman, 2005; **11.18:** From BIOCHEMISTRY, 4/e by Lehninger, p. 962. New York: W. H. Freeman, 2005; **11.51:** From BIOCHEMISTRY, 4/e by Lehninger, p. 151. New York: W. H. Freeman, 2005.

Chapter 12

12.12: From "Bacterial gene regulation: from transcription attenuation to riboswitches and ribozymes" by Sabine Brantl in TRENDS IN MICROBIOLOGY, Vol. 12, No. 11, November 2004; **12.13:** From "RNA Sensors and Riboswitches: Self-Regulating Messages" by Eric C. Lai in CURRENT BIOLOGY, Vol. 13, R285–R291, April 1, 2003; **12.19:** From "Quorum Sensing in Gram-Negative Bacteria" by E. Peter Greenberg in ASM NEWS, July 1997; **12.20:** From "Interference with AI-2-mediated bacterial cell-cell communication" by Karina B. Xavier and Bonnie L. Bassler in NATURE, Vol. 437, September 2005; **12.21a:** From "Control of o factor activity during Bacillus subtilis sporulation" by Lee Kroos et al. in MOLECULAR MICROBIOLOGY, September 1998.

Chapter 13

13.29: From "F factor conjugation is a true type IV secretion system" by T.D. Lawley et al. in FEMS MICROBIOLOGY LETTERS 22 (2003).

Chapter 14

Table 14.1: From Strickberger, Monroe, W., GENETICS, 3rd Edition. © 1985, p. 354. Reprinted by permission of Prentice-Hall. Upper Saddle River, New Jersey.

Chapter 15

15.4: Figure 8–38a from MOLECULAR BIOLOGY OF THE CELL, 4/e by Alberts, Johnson, Lewis, Ruff et al., p. 506. © 2002. Reproduced by permission of Garland Science/Taylor & Francis; **15.5:** p. 15695 from "Membrane localization of MinD is mediated by a C-terminal motif that is conserved across eubacteria, archaea, and chloroplasts" by Tim H. Szeto et al. from PROCEEDINGS OF THE NATIONAL ACADEMY OF SCIENCE, November 2002. © 2002 National Academy of Sciences, U.S.A. Reprinted by permission; **15.6:** Source: The Ribosomoal Database Project; **15.10:** From "Another Extreme genome: how to live at pH0" by Maria Ciaramella et al. from TRENDS IN MICROBIOLOGY, Vol. 13, No. 2, February 2005, p. 49. Reprinted by permission of Elsevier; **15.16:** p. 2032 from "Comparative Genome Sequencing for Discovery of Novel Polymophisms in Bacillus anthracis" by Timothy D. Read et al. from SCIENCE, 14 June 2002 Vol. 296. © 2002 AAAS. Reprinted by permission; **15.20:** p. 70 From "Environmental Genome Shotgun Sequencing of the Sargasso Sea" by J. Craig Venter et al. in SCIENCE, April 2004, Vol. 304. © 2004 AAAS. Reprinted by permission.

Chapter 16

16.6: From MICROBIOLOGY, 3/e by Bernard D. Davis et al. © 1980 Harper & Row. Reprinted by permission of Lippincott Williams & Wilkins; **16.13:** The Structure of an Icosahedra Capsid from MICROBIOLOGY, Third Edition by Bernard D. Davis, et al. © 1980 by Harper & Row, Publishers, Inc. Reprinted by permission of HarperCollins Publishers, Inc.; **Table 16.1:** Modified from S. E. Luria, et al., GENERAL VIROLOGY, 3rd edition, 1983. John Wiley & Sons, Inc., New York, NY.

Chapter 17

17.1: "Major Bacteriophage Families and Genera" from Van Regenmortel, Fauquet, Bishop, et al., VIRUS TAXONOMY, 7th Report, 2000. Reprinted by permission of Elsevier; **17.6:** From PRINCIPLES OF VIROLOGY by Flint et al., p. 67, figure 3.1 ASM Press, 2004; **17.24:** From "Imbroglios of Viral Taxonomy: Genetic Exchange and Failings of Phenetic Approaches" by Jeffrey G. Lawrence et al. from JOURNAL OF BACTERIOLOGY, September 2002, p. 4896, figure 3. Reprinted by permission of American Society for Microbiology.

Chapter 18

Box 18.1: From "Adaptation of SARS Coronavirus to Humans" by Kathryn V. Holmes from SCIENCE, September 16, 2005, Vol. 309; **18.5:** From PRINCIPLES OF VIROLOGY, 2/e by Flint et al., figure 3.3c, p. 69. ASM Press, 2004; **18.6:** From PRINCIPLES OF VIROLOGY by S. J. Flint et al., ASM Press, 2000; **18.7:** From PRINCIPLES OF VIROLOGY by S. J. Flint et al., ASM Press, 2004; **18.8:** From PRINCIPLES OF VIROLOGY by S. J. Flint et al., ASM Press, 2004; **18.14:** From Van Regenmortel, Fauquet, Bishop, et al., VIRUS TAXONOMY, 7th Report, 2000. Reprinted by permission of Elsevier.

Chapter 19

Table 19.9: From R. H. Whittaker and L. Margulis, BIOSYSTEMS 10:3–18. © 1978 Elsevier Scientific Publishers. Reprinted by permission; **Table 19.10:** "Some Characteristic Differences between Gram-Negative and Gram-Positive Bacteria" from R.H. Whittaker and L. Margulis from BIOSYSTEMS 10:3–18, © 1978. Reprinted by permission of Elsevier; **19.3:** Figure 1, p. 735 from "A Molecular View of Microbial Diversity and the Biosphere" by Norman R. Pace from SCIENCE, 2 May 1997, Vol. 276. © 1997 AAAS. Reprinted by permission; **19.4:** From "The hydrogen hypothesis for the first eukaryote" by William Martin and Miklos Muller from NATURE, Vol. 392, March 5, 1998; **19.10:** Source: Data from C.P. Woese. MICROBIOLOGICAL REVIEWS, 51(2):221–227, 1987; **19.11:** From *www.msu.edu/~debruijn/dna1-4.htm;* **19.12:** From *www.msu.edu/*

~debruijn/dna1-4.htm; **19.15:** From Barkay and Smets, ASM NEWS, Vol. 71, 2005; **19.18:** Figure 16 from W. Ludwig & H-P Klenk in Boone, Castenholz and Garrity (Editors), BERGEY'S MANUAL OF SYSTEMIC BACTERIOLOGY, Volume 1, Second Edition, 2001, page 65. Reprinted by permission of Bergey's Manual Trust.

Chapter 20

Box 20.1: Figure a from Brochier et al. in THEORETICAL POPULATION BIOLOGY, Volume 61, Issue 4, pp. 409–422, © 2002. Reprinted with permission from Elsevier, Figure b from page 12987 from "The genome of Nanoarchaelim . . ." by Waters, John, Graham et al. in PROCEEDINGS OF THE NATIONAL ACADEMY OF SCIENCE, October 2003. © 2003 National Academy of Sciences, U.S.A. Reprinted by permission; **20.2:** From "Mechanism and regulation of transcription in archaea" by Stephen D. Bell and Stephen P. Jackson from CURRENT OPINION IN MICROBIOLOGY, 2001, Vol. 4, Figure 1, page 209; **20.4:** Figure A1.7 from H. Huber and K. O. Stettler in Boone, Castenholz and Garrity (Editors), BERGEY'S MANUAL OF SYSTEMATIC BACTERIOLOGY, Volume I, Second Edition, 2001, p. 180. Reprinted by permission of Bergey's Manual Trust; **20.8:** Figure 1, p. 633 from "Archaeal Phylogeny Based on Ribosomal Proteins" by Oriane Matte-Tailliez et al. in MOLECULAR BIOLOGY EVOLUTION, vol. 19(5):631–639, © 2002. Reprinted by permission of Oxford University Press.

Chapter 21

21.7(a): Illustration © Hartwell T. Crim, 1998; **21.12:** Figures 1, 2 and 3 from "Novel Compartmentalisation in Planctomycete Bacteria" by R.I. Webb et al. in MICROSCOPY AND MICROANALYSIS 10 (Suppl 2), 2004. Reprinted with the permission of Cambridge University Press; **22.2:** Figure 9 from W. Ludwig & H-P Klenk in Boone, Castenholz and Garrity (Editors), BERGEY'S MANUAL OF SYSTEMATIC BACTERIOLOGY, Volume 1, Second Edition, 2001, p. 62. Reprinted by permission of Bergey's Manual Trust.

Chapter 22

22.8: Figure 1, page 580 from "Cytokinesis Monitoring during Development: Rapid Pole-to-pole Shuttling of a Signaling Protein by Localized Kinase and Phasphatase in Caulobater" by Jean-Yves Matroule et al. in CELL, Vol. 118, September 3, 2004. Reprinted by permission of Elseiver; **22.12:** Figure 11 from W. Ludwig & H-P Klenk in Boone, Castenholz and Garrity (Editors), BERGEY'S MANUAL OF SYSTEMATIC BACTERIOLOGY, Volume 1, Second Edition, 2001, p. 63. Reprinted by permission of Bergey's Manual Trust; **Table 22.2:** From Brenner, D.J.; Krieg, N.R.; and Staley, J.T., Eds. 2005. Bergey's Manual of Systemic Bacteriology, 2nd ed. Vol. 2: The *Proteobacteria.* Garrity, G.M. Ed-in-Chief. New York: Reprinted by permission of Springer; **22.17:** Source: The Ribosomal Database Project; **22.31:** Figure 12 from W. Ludwig & H-P Klenk in Boone, Castenholz and Garrity (Editors), BERGEY'S MANUAL OF SYSTEMATIC BACTERIOLOGY, Volume 1, Second Edition, 2001, p. 63. Reprinted by permission of Bergey's Manual Trust; **22.36:** Source: *http://cmgm.stanford.edu/devbio/kaiserlab.*

Chapter 23

23.2: Source: The Ribosomal Database Project.

Chapter 24

24.4: E. Stackebrandt, F.A. Raineym, and N.L. Ward-Rainey. Proposal for a new hierarchic classification system, Actinobacteria, classis nov. Int. J. Syst. Bacteriol. 47(2):479–491, 1997, figure 3, p. 482; **24.5:** Source: The Ribosomal Database Project.

Chapter 25

25.1: From "A molecular view of microbial . . ." by Norman Pace from SCIENCE, Vol. 276, 1997, p. 735, figure 1. © 1997 AAAS. Reprinted by permission of AAAS.

Chapter 27

27.19(b): J. Frohlich and H. Konig, "Rapid isolation of single microbial cells from mixed natural and laboratory populations with the aid of a micromanipulator" in *System. Appl. Microbiol.* 2: 235, 1999. Reprinted by permission of Nature, via © Clearance Center.

Chapter 28

28.2: From "Carbonate Mysteries" by Henry Elderfield from SCIENCE, 31 May 2002, Vol. 296, p. 1617, figure 2. © 2002 AAAS. Reprinted by permission of AAAS; **28.8:** Source: *http://www.cwr.uwa.edu.au/cwr/outreach/envirowa/rivers/swan/change.html;* **28.12:** From "Stirring times in the Southern Ocean" by Sallie W. Chisholm from NATURE, Vol. 407, 12 October 2000, p. 685. Reprinted by permission of Nature, via © Clearance Center; Fig. 28.14: E. F. De Long, et al., "Visualization and enumeration of marine planktonic archaea and bacteria by using polyribonucleotide probes and fluorescent In Situ hybridization" in *Appl. Environ. Microbiol.* 65: 5560, 1999; **28.15:** From "Viruses in the sea" by Curtis A. Suttle from NATURE, Vol. 437, p. 358, figure 3; **28.16:** Figure from "Microbial Life Breathes Deep" by Edward F. DeLong from SCIENCE, 24 December 2004, Vol. 306, p. 2199. © 2004 AAAS. Reprinted by permission of AAAS.

Chapter 29

29.1: From PRINCIPLES AND APPLICATIONS OF SOIL MICROBIOLOGY by David M. Sylvia et al., Figure 1–2, p. 8; **29.11:** From "Nitrogen transfer int eh arbuscular mycorrhizal symbiosis" by Majula Govindarajulu et al. from NATURE, June 2005, Vol. 435, p. 819, figure 3; **29.20:** R. Conrad: "Soil microbial processes involved in production and consumption of atmospheric trace gasses. In *Adv. Micro. Ecol.* 14, 1995. Reprinted by permission of Springer-Verlag GmbH; Fig. 29.22: K.K. Lovely: Dissimilatory Fe (III) and MN (IV) reduction. In *Microbiol. Rev.* 55:269, 1991. Reprinted by permission of American Society for Microbiology; **29.23:** From J.M. Hunt, PETROLEUM GEOCHEMISTRY AND GEOLOGY, 2/e, © 1996. Reprinted by permission of W. H. Freeman and Company/Worth Publishers; **Table 29.1:** From E. W. Russell, SOIL CONDITIONS AND PLANT GROWTH, 10/e. © 1973 Longman Group Limited, Essex, United Kingdom. Reprinted by permission; **Table 29.3:** From Torsvik, V., Ovraes, L., and Thingstad, T.F. (2002) SCIENCE, Vol. 296: 1064–1066. © 2002 AAAS. Reprinted by permission of AAAS; **Table 29.5:** From J. W. Woldendorp, "The Rhizosphere as Part of the Plant-Soil System" in STRUCTURE AND FUNCTIONING OF PLAN POPULATIONS (Amsterdam, Holland: Proceedings, Royal Dutch Academy of Sciences, Natural Sciences Section: 2d Series, 1978) 70:243; **Table 29.7:** Data from Dr. D. Baker, MDS Panlabs and Dr. J. Dawson, University of Illinois. Personal communication; **Table 29.8:** From J. W. Lengler, G. Drews, H. G. Schlegel. 1999. BIOLOGY OF THE PROKARYOTES. Blackwell Science, Malden, Mass., table 34.4; **Table 29.9:** From R. Watling and D. B. Harper. 1998. MYCOLOGICAL RESEARCH 102(7): 769–87. Reprinted by permission of Elsevier.

Chapter 30

30.8: From W. W. Mohn and J. M. Tiedje, "Microbial Reductive Dehalogenation" in *Microbiological Reviews* 56(3), September 1992; **30.13:** R. Guerrero: "Predation as a prerequisite to organelle origin: Daptobecter as example. In *Symbiosis as a Source of Evolutionary Innovation: Specification and Morphogenesis*, L. Margulis and R. Fester, eds. Reprinted by permission of The MIT Press; **30.18:** From "Host-bacterial mutualism in the human intestine" by Fredrik Backhed et al. in SCIENCE, Vol. 307, 25 March 2005. ©2005 AAAS. Reprinted by permission of AAAS; **Table 30.1:** Adapted from L. Margulis and M. J. Chapman. 1998. Endosymbioses: Cyclical and permanent in evolution. TRENDS IN MICROBIOLOGY 6(9):342–46, tables 1, 2, and 3; **Table 30.2:** From E. G. Ruby, 1999. Ecology of a benign "infection": Colonization of the squid luminous organ by *Vibrio fischeri.* In MICROBIAL ECOLOGY AND INFECTIOUS DISEASE. E. Rosenberg, editor. American Society for Microbiology, Washington, DC, 217–31, table 1. Reprinted by permission.

Chapter 31

31.20: Figure of Schematic comparing b-defensin and cathelicidin DNA, messenger RNA and peptides by Robert Bals in JOURNAL OF RESPIRATORY RESEARCH, 1:141–150, 2000. Reprinted by permission of the author; **31.22:** Figure 2.19

from IMMUNOBIOLOGY, 5/e by Janeway, p. 57. © 2001. Reproduced by permission of Garland Science/ Taylor & Francis; **31.26:** From Nature Reviews Molecular Cell Biology 2: 627–633 (2001).

Chapter 32

32.15b: From Thomas J. Smith: Structure of a human rhinovirus-bivalently bound antibody complex. In *Proceedings of the National Academy of Science,* Vol. 90, pp. 7015–7018, August 1993. © 2003 National Academy of Sciences, U.S.A. Reprinted by permission; **32.21:** From *www.uccs.edu.*

Chapter 33

33.4a: From "YopT, a new Yersinia Yop effector protein, affects the cytoskeleton of host cells" by Maite Iriarte and Guy R. Cornelis from MOLECULAR MICROBIOLOGY, 1998, pp. 915–929; **33.4b:** From "Process of Protein Transport by the Type III Secretion System" by Partho Ghosh from MICROBIOLOGY AND MOLECULAR BIOLOGY REVIEWS, December 2004, pp. 771–795.

Chapter 34

34.17: From NAURE, Vol. 4, January 2006; **34.18:** From *www.bioteach.ubc.ca/Biodversity;* **34.19:** From *www.bioteach.ubc.ca/Biodversity;* **Box 34.2:** From EMERGING INFECTIOUS DISEASES, Volume 10, Number 3, March 2004.

Chapter 36

Table 36.3: Recommended Childhood and Adolescent Immunization Schedule, United States 2006. Department of Health and Human Services, Centers for Disease Control and Prevention: **Table 36.5:** Adapted from Goldsby, T. J. Kindt, and B. A. Osborne, KUBY IMMUNOLOGY, 2003. Reprinted by permission of W.H. Freeman and Company/Worth Publishers.

Chapter 37

37.7: UNAIDS.

Chapter 38

38.3a: From *www.hopkins-tb.org.* **38.27a:** Martin Enserink: *Science,* Vol. 294, Oct. 2001, pp. 490–491. © 2001 AAAS. Reprinted by permission of AAAS.

Chapter 39

39.7: Data from the *World Health Statistics Quarterly,* 41:69, 1988, World Health Organization, Switzerland.

Chapter 40

40.4a, b: M. A. Carlson, BIOSENSORS AND BIOELECTRONICS 14, 2000; **40.7:** Adapted from G. D. Lewis et al., "Influence of environemtnal factors on virus detection by RT-PCR and cell culture" in J. APPL. MICROBIOL. 88: 638. © 2000; **40.10:** Adapted from J. K. Wan, et al., "Probelia PCR system for rapid detection of Salmonella in milk powder and ricotta cheese in LET. APPL. MICROBIOL. 30:269. © 2000.

Chapter 41

41.9: Modified from Crueger and Crueger BIOTECHNOLOGY: A TEXTBOOK OF INDUSTRIAL MICROBIOLOGY, Second Edition. © Science Tech Publishers, Madison, WI, 1990; **41.11:** Adapted from S. S. D. Donadio, et al., "Recent Developments in the genetics of erythromycin formation" in INDUSTRIAL MICROORGANISMS: BASIC AND APPLIED MOLECULAR GENETICS, R. H. Baltz, et al., eds. ASM, Washington, DC 1993. Reprinted by permission of American Society for Microbiology; **41.18:** Modified from Crueger and Crueger BIOTECHNOLOGY: A TEXTBOOK OF INDUSTRIAL MICROBIOLOGY, Second Edition. © Science Tech Publishers, Madison, WI, 1990; **41.19:** Modified from S. Pedersen, L. Dijkhuizen, B. W. Dijkstra, B. F. Jensen, and S.T. Jorgensen, "A better enzyme for cyclodextrins" in CHEMTECH, 25: 19–25, Figure 1, p. 20.

Index

In this index, page numbers followed by a t designate tables; page numbers followed by an f refer to figures; page numbers set in boldface refer to major discussions.

A

C

D

F

G

H

I

J

M

O

P

W

Microorganism Pronunciation Guide

The pronunciation of each name is given in parentheses. The phonetic spelling system is explained at the beginning of the glossary (p. G-1)

Bacteria

Acetobacter (ah-se″to-bak′ter)
Acinetobacter (as″ĭ-net″o-bak′ter)
Actinomyces (ak″tĭ-no-mi′sēz)
Agrobacterium (ag″ro-bak-te′re-um)
Alcaligenes (al″kah-lij′ĕ-nēz)
Anabaena (ah-nab′ē-nah)
Arthrobacter (ar″thro-bak′ter)
Bacillus (bah-sil′lus)
Bacteroides (bak″tĕ-roi′dēz)
Bdellovibrio (del″o-vib′re-o)
Beggiatoa (bej″je-ah-to′ah)
Beijerinckia (bi″jer-ink′e-ah)
Bifidobacterium (bi″fid-o-bak-te′re-um)
Bordetella (bor″dĕ-tel′lah)
Borrelia (bŏ-rel′e ah)
Brucella (broo-sel′lah)
Campylobacter (kam″pĭ-lo-bak′ter)
Caulobacter (kaw″lo-bak′ter)
Chlamydia (klah-mid′e-ah)
Chlorobium (klo-ro′be-um)
Chromatium (kro-ma′te-um)
Citrobacter (sit″ro-bak′ter)
Clostridium (klo-strid′e-um)
Corynebacterium (ko-ri″ne-bak-te′re-um)
Coxiella (kok″se-el′lah)
Cytophaga (si-tof′ah-gah)
Desulfovibrio (de-sul″fo-vib′re-o)
Enterobacter (en″ter-o-bak′ter)
Erwinia (er-win′e-ah)
Escherichia (esh″er-i′ke-ah)
Flexibacter (flek″sĭ-bak′ter)
Francisella (fran-sĭ-sel′ah)
Frankia (frank′e-ah)
Gallionella (gal″le-o-nel′ah)
Haemophilus (he-mof′i-lus)
Halobacterium (hal″o-bak-te′re-um)
Hydrogenomonas (hi-dro″jĕ-no-mo′nas)
Hyphomicrobium (hi″fo-mi-kro′be-um)
Klebsiella (kleb″se-el′lah)
Lactobacillus (lak″to-bah-sil′lus)
Legionella (le″jun-el′ah)
Leptospira (lep″to-spi′rah)
Leptothrix (lep′to-thriks)
Leuconostoc (loo″ko-nos′tok)
Listeria (lis-te′re-ah)
Methanobacterium (meth″ah-no-bak-te′re-um)
Methylococcus (meth″il-o-kok′-us)
Methylomonas (meth″il-o-mo′nas)
Micrococcus (mi″kro-kok′us)
Mycobacterium (mi″ko-bak-te′re-um)
Mycoplasma (mi″ko-plaz′mah)
Neisseria (nīs-se′re-ah)
Nitrobacter (ni″tro-bak′ter)
Nitrosomonas (ni-tro″so-mo′nas)
Nocardia (no-kar′de-ah)
Pasteurella (pas″tĕ-rel′ah)
Photobacterium (fo″to-bak-te′re-um)
Propionibacterium (pro″pe-on″e-bak-te′re-um)
Proteus (pro′te-us)
Pseudomonas (soo″do-mo′nas)
Rhizobium (ri-zo′be-um)
Rhodopseudomonas (ro″do-soo″do-mo′nas)
Rhodospirillum (ro″do-spi-ril′um)
Rickettsia (rĭ-ket′se-ah)
Salmonella (sal″mo-nel′ah)
Sarcina (sar′sĭ-nah)
Serratia (sĕ-ra′she-ah)
Shigella (shi-gel′ah)
Sphaerotilus (sfe-ro′tĭ-lus)
Spirillum (spi-ril′um)
Spirochaeta (spi″ro-ke′tah)
Spiroplasma (spi″ro-plaz′mah)
Staphylococcus (staf″ĭ-lo-kok′us)
Streptococcus (strep″to-kok′us)
Streptomyces (strep″to-mi′sēz)
Sulfolobus (sul″fo-lo′bus)
Thermoactinomyces (ther″mo-ak″tĭ-no-mi′sēz)
Thermoplasma (ther″mo-plaz′mah)
Thiobacillus (thi″o-bah-sil′us)
Thiothrix (thi′o-thriks)
Treponema (trep″o-ne′mah)
Ureaplasma (u-re′ah-plaz″ma)
Veillonella (va″yon-el′ah)
Vibrio (vib′re-o)
Xanthomonas (zan″tho-mo′nas)
Yersinia (yer-sin′e-ah)
Zoogloea (zo″o-gle′ah)